W9-AOW-035

PROCEEDINGS OF THE

31st IEEE CONFERENCE ON DECISION AND CONTROL

DECEMBER 16-18, 1992
WESTIN LA PALOMA
TUCSON, ARIZONA, USA

IEEE
Control
Systems
Society

VOLUME 3 OF 4

92CH3229-2

THE 31ST IEEE CONFERENCE ON DECISION AND CONTROL

DECEMBER 16 – 18, 1992

SPONSORING ORGANIZATION:
IEEE Control Systems Society

COOPERATING ORGANIZATIONS:
Operations Research Society of America (ORSA)
Society for Industrial and Applied Mathematics (SIAM)

OPERATING COMMITTEE :

GENERAL CHAIRMAN	Tamer Başar	University of Illinois
PROGRAM CHAIRMAN	Sergio Verdú	Princeton University
PROGRAM VICE-CHAIRS	Robert R. Bitmead	Australian National University
	André L. Tits	University of Maryland
FINANCE	Peter B. Luh	University of Connecticut
PUBLICATIONS	Rodolfo Milito	AT&T Bell Labs
PUBLICITY	Tsu-Shuan Chang	University of California, Davis
LOCAL ARRANGEMENTS	Malur K. Sundareshan	University of Arizona
REGISTRATION	Sean P. Meyn	University of Illinois
EXHIBITS	Man-Feng Chang	General Motors Research Labs
WORKSHOPS	Ümit Özgüner	Ohio State University

PROGRAM COMMITTEE :

S. Verdú (Chair)
V. Anantharam
T. Banks
R. Bitmead (Vice-Chair)
M. Dahleh
B. Dickinson
D. Flamm
J. Freudenberg
M. Gevers
M. Ilić
P. Ioannou
A. Laub
G. Leugering (SIAM)
F. Lewis
J. Moore
G. Papavassilopoulos
V. Poor
P. Ramadge
G. Sasaki
E. Sontag
A. Tits (Vice-Chair)
C. White (ORSA)

FOREWORD

Welcome to this major annual meeting of the IEEE Control Systems Society – *31st IEEE Conference on Decision and Control* (CDC) – which continues to grow, as evidenced from the record number of submissions received this year. The technical program consists of 117 contributed and invited sessions, comprising over 900 papers by over 1400 authors. It was again prepared in cooperation with the Society for Industrial and Applied Mathematics (SIAM) and the Operations Research Society of America (ORSA), under the direction of the Program Chairman, Sergio Verdú. In addition to the contributed and invited sessions, the Conference features this year three plenary sessions, one on each day: On Wednesday morning, John N. Tsitsiklis will present a lecture with the title "*Complexity-Theoretic Aspects of Problems in Control Theory.*" On Thursday morning, Pravin Varaiya will present the second plenary, entitled "*Smart Cars on Smart Roads: Design and Evaluation*". On late Friday morning, Brian D. O. Anderson, the recipient of this year's Bode Prize, will deliver the Bode Lecture entitled, "*Controller Design: Getting from Theory to Practice.*"

This year, the CDC is preceded by five tutorial workshops, covering a broad spectrum of topics of current and emerging interest to the control systems community. They are all one-day workshops, distributed over two days – three on Monday, and two on Tuesday. Those planned for Monday are entitled, "*Intelligent Control with Applications*", "*Modeling and Scheduling in Semiconductor Fabrication*", and "*Identification and Robust Control Interplay*"; the two on Tuesday are "*Robust Control for State Space Systems*", and "*An Introduction to Control of Smart Structures*".

Several exhibits are planned for the entire duration of the Conference, which will include demonstrations of software for control system design and books from several publishers. The Awards Ceremonies, to be conducted during the Thursday's dinner banquet, will feature the presentation of several IEEE and CSS awards, and Fellow certificates.

As the number of submissions and proportionately the number of participants grew over the years, organizing the CDC has become a major undertaking, on both the administrative and technical sides. Our community is fortunate to have individuals who can devote their time and efforts freely toward successful completion of such tasks. Members of the Operating and Program Committees (whose names can be found on the facing page), the Editorial Board of the IEEE Transactions on Automatic Control, and hundreds of anonymous reviewers deserve our gratitude for many hours they have spent to ensure that we have a smooth running conference of top technical quality. Special recognitions go to André Tits, Vice Chair for Program, who did much more than his share of work, in the assembling of the Technical Program, and to Rodolfo Milito, Publications Chairman, who ably coped with the myriad of details required to produce the Advance and Final Programs, as well as this four-volume Proceedings. Special thanks also go to Mrs. Francie Bridges and Ms. Dixie Murphy, of the University of Illinois, for their efficient secretarial services during all phases of the organization of this CDC.

The Operating Committee has worked hard to create an environment for the participants to get the most benefit out of the Conference. We hope that these efforts were well worth it and that *you*, as a participant, will value this meeting as a memorable event, both scientifically and socially. Thank *you* for your participation in this year's CDC, and your contribution toward its success.

Tamer Başar
General Chairman

IEEE CONFERENCE ON DECISION AND CONTROL
PAST AND PRESENT

Following the tradition of last year's CDC, we are providing here a complete list of past conferences in this series, with titles, chairs and locations.[1] The CDC grew out of the former Symposium on Adaptive Processes, to become the premier conference in the field that it is today. At first it was associated with the Joint Automatic Control Conference (JACC – now the ACC) and then the National Electronics Conference (NEC). In the following listing GC denotes General Chair, PC stands for Program Chair, and SC is Symposium Chair. The proceedings of all past conferences may be perused at the IEEE Library, 345 47th Street, New York, NY 10017.

DISCRETE ADAPTIVE PROCESSES – SYMPOSIUM AND PANEL DISCUSSION (AIEE); part of 3rd JACC
GC: J. Sklansky
New York University, New York City, NY; 29 June 1962

SYMPOSIUM ON ADAPTIVE PROCESSES; part of NEC
GC: L. Kanal
McCormick Place, Chicago, IL; 28-29 October 1963

SYMPOSIUM ON ADAPTIVE PROCESSES; part of NEC
GC: F. J. Mullin
McCormick Place, Chicago, IL; 19-21 October 1964

SYMPOSIUM ON ADAPTIVE PROCESSES; part of NEC
GC: E. C. Jones, Jr., PC: G. Brown
McCormick Place, Chicago, IL; 25-27 October 1965

SYMPOSIUM ON ADAPTIVE PROCESSES; part of NEC
GC: F. N. Bailey, PC: J. C. Hancock
McCormick Place, Chicago, IL; 3-5 October 1966

SYMPOSIUM ON ADAPTIVE PROCESSES; part of NEC
GC: F. M. Waltz, PC: P. E. Mayes
International Amphitheater, Chicago, IL; 23-25 October 1967

IEEE SYMPOSIUM ON ADAPTIVE PROCESSES
PC: J. M. Mendel
UCLA, Los Angeles, CA; 16-18 December 1968

IEEE SYMPOSIUM ON ADAPTIVE PROCESSES
GC: J. B. Lewis, PC: G. J. McMurty
Pennsylvania State University, PA; 17-19 November 1969

1970 IEEE SYMPOSIUM ON ADAPTIVE PROCESSES (9th) DECISION AND CONTROL
PC: D. G. Lainiotis
University of Texas, Austin, TX; 7-9 December 1970

---

[1] This information was compiled last year by Panos Antsaklis with help from Jack Rugh and Len Shaw.

1971 IEEE CONFERENCE ON DECISION AND CONTROL
including the 10th SYMPOSIUM ON ADAPTIVE PROCESSES
GC: J. T. Tou, PC: S. K. Mitter, SC: J. M. Mendel
Americana Hotel, Miami Beach, FL; 15-17 December 1971

1972 IEEE CONFERENCE ON DECISION AND CONTROL
including the 11th SYMPOSIUM ON ADAPTIVE PROCESSES
GC: J. M. Mendel, PC: Y. C. Ho, SC: G. N. Saridis
Fontainebleau Motor Hotel, New Orleans, LA; 13-15 December 1972

1973 IEEE CONFERENCE ON DECISION AND CONTROL
including the 12th SYMPOSIUM ON ADAPTIVE PROCESSES
GC: J. S. Meditch, PC: D. G. Luenberger, SC: L. A. Gerhardt
Sheraton-Harbor Island Hotel, San Diego, CA; 5-7 December 1973

1974 IEEE CONFERENCE ON DECISION AND CONTROL
including the 13th SYMPOSIUM ON ADAPTIVE PROCESSES
GC: E. Axelband, PC: S. Kahne, SC: D. P. Lindorff
Del Webb's Towne House, Phoenix, AZ; 20-22 November 1974

1975 IEEE CONFERENCE ON DECISION AND CONTROL
including the 14th SYMPOSIUM ON ADAPTIVE PROCESSES
GC: J. B. Cruz, Jr., PC: J. B. Pearson, SC: G. Stein
Hyatt Regency Houston, Houston, TX; 10-12 December 1975

1976 IEEE CONFERENCE ON DECISION AND CONTROL
including the 15th SYMPOSIUM ON ADAPTIVE PROCESSES
GC: M. Athans, PC: E. R. Barnes, SC: T. Pavlidis
Sheraton-Sand Key Hotel, Clearwater, FL; 1-3 December 1976

1977 IEEE CONFERENCE ON DECISION AND CONTROL
including the 16th SYMPOSIUM ON ADAPTIVE PROCESSES
GC: K. S. Fu, PC: H. Sorenson, SC: T. Pavlidis
Fairmont Hotel, New Orleans, LA; 7-9 December 1977

1978 IEEE CONFERENCE ON DECISION AND CONTROL
including the 17th SYMPOSIUM ON ADAPTIVE PROCESSES
GC: Robert E. Larson, PC: Alan S. Willsky, SC: Jerry M. Mendel
Islandia Hyatt House Hotel, San Diego, CA; 10-12 January 1979

18th IEEE CONFERENCE ON DECISION AND CONTROL
including the SYMPOSIUM ON ADAPTIVE PROCESSES
GC: Stephan Kahne, PC: Alexander H. Levis, SC: Yaakov Bar-Shalom
Galt Ocean Mile Hotel, Ft. Lauderdale, FL; 12-14 December 1979

19th IEEE CONFERENCE ON DECISION AND CONTROL
including the SYMPOSIUM ON ADAPTIVE PROCESSES
GC: Pierre R. Belanger, PC: David L. Kleinman, SC: Richard V. Monopoli
The Regent Hotel, Albuquerque, NM; 10-12 December 1980

20th IEEE CONFERENCE ON DECISION AND CONTROL
including the SYMPOSIUM ON ADAPTIVE PROCESSES
GC: William R. Perkins, PC: Abraham H. Haddad, SC: K. S. Narendra
Vacation Village Hotel, San Diego, CA; 16-18 December 1981

21st IEEE CONFERENCE ON DECISION AND CONTROL
GC: Alexander H. Levis, PC: William S. Levine
Holiday Inn - International Drive, Orlando, FL; 8-10 December 1982

22nd IEEE CONFERENCE ON DECISION AND CONTROL
GC: James L. Melsa, PC: Steven I. Marcus
Marriott Hotel, San Antonio, TX; 14-16 December 1983

23rd IEEE CONFERENCE ON DECISION AND CONTROL
GC: Abraham H. Haddad, PC: Michael P. Polis
Las Vegas Hilton, Las Vegas, NV; 12-14 December 1984

24th IEEE CONFERENCE ON DECISION AND CONTROL
GC: Gene F. Franklin, PC: Anthony N. Michel
Bonaventure Hotel & Spa, Ft. Lauderdale, FL; 11-13 December 1985

25th IEEE CONFERENCE ON DECISION AND CONTROL
GC: Anthony Ephremides, co-GC: Spyros Tzafestas, PC: H. Vincent Poor
Atheneum Intercontinental Hotel, Athens, GREECE; 10-12 December 1986

26th IEEE CONFERENCE ON DECISION AND CONTROL
GC: William S. Levine, PC: John Baillieul
Westin Century-Plaza Hotel, Los Angeles, CA; 9-11 December 1987

27th IEEE CONFERENCE ON DECISION AND CONTROL
GC: Michael P. Polis, PC: William E. Schmitendorf
Hyatt Regency Austin on Town Lake, Austin, TX; 7-9 December 1988

28th IEEE CONFERENCE ON DECISION AND CONTROL
GC: Leonard Shaw, PC: Tamer Başar
Hyatt Regency Tampa Hotel, Tampa, FL; 13-15 December 1989

29th IEEE CONFERENCE ON DECISION AND CONTROL
GC: Charles J. Herget, PC: Raymond A. DeCarlo
Hilton Hawaiian Village, Honolulu, HI; 5-7 December 1990

30th IEEE CONFERENCE ON DECISION AND CONTROL
GC: Derek P. Atherton, PC: Panos J. Antsaklis
Metropole Hotel, Brighton, ENGLAND; 11-13 December 1991

31st IEEE CONFERENCE ON DECISION AND CONTROL
GC: Tamer Başar, PC: Sergio Verdú
Westin La Paloma, Tucson, AZ; 16-18 December 1992

THE 31st IEEE CONFERENCE ON DECISION AND CONTROL

PLENARY & BODE LECTURER BIOGRAPHIES

PLENARY 1 : John N. Tsitsiklis was born in Thessaloniki, Greece in 1958. He received the B.S. degree in Mathematics (1980), and the B.S. (1980), M.S. (1981) and Ph.D. (1984) degrees in Electrical Engineering, all from the Massachusetts Institute of Technology (*M.I.T.*), Cambridge, Massachusetts. During the academic year 1983-84, he was an acting assistant professor of Electrical Engineering at Stanford University, Stanford, California. Since 1984, he has been with the Department of Electrical Engineering and Computer Science at *M.I.T.*, where he is currently an associate professor. His research interests are in the areas of systems and control theory, parallel and distributed computation, and operations research. Dr. Tsitsiklis is a coauthor (with D. Bertsekas) of Parallel and Distributed Computation: Numerical Methods (1989). He has been a recipient of an IBM Faculty Development Award (1983), and NSF Presidential Young Investigator Award (1986), an Outstanding Paper Award by the IEEE Control Systems Society (for a paper coauthored with M. Athans), and of the Edgerton Faculty Achievement Award by M.I.T. (1989). He is an associate editor of Automatica and Applied Mathematics Letters, and has been an associate editor of the IEEE Transactions on Automatic Control.

PLENARY 2 : Pravin Varaiya is a professor of Electrical Engineering & Computer Sciences at the University of California, Berkeley. He received his Ph.D. in Electrical Engineering from the University of California at Berkeley. He is the author, with P. R. Kumar, of Stochastic Systems: Estimation, Identification, and Adaptive Control, Prentice-Hall, 1986 and editor, with A. Kurzhanski, of Discrete Event Systems: Models and Applications, Lecture Notes in Information Sciences, vol. 103, Springer, 1988. His area of research and teaching are in stochastic systems, communication networks, power systems and urban economics.

BODE LECTURE : Brian D.O. Anderson was born in Sydney, Australia, and received his undergraduate education at the University of Sydney, with majors in mathematics and electrical engineering. He subsequently obtained a Ph.D. degree in electrical engineering from Stanford University. Following completion of his education, he worked in industry in Silicon Valley and served as a faculty member in the department of electrical engineering at Stanford. In 1967, he joined the electrical engineering department at the University of Newcastle, Australia, where he remained until the end of 1981. At that time he became professor of Systems Engineering at the Australian National University. His research interests are in control engineering and signal processing, and he has co-authored a number of books in these fields, as well as research papers. He is a Fellow of the Royal Society, the Australian Academy of Science, Australian Academy of Technological Sciences and Engineering, and the Institute of Electrical and Electronic Engineers, and an Honorary Fellow of the Institute of Engineers, Australia. He holds a doctorate (honoris causa) from the Catholic University of Louvain, Belgium. He has served or is serving on government committees and councils including the Prime Minister's Science and Engineering Council, and is also a company board member. He is serving a term as President of the International Federation of Automatic Control from 1990 to 1993.

1992 IEEE CONFERENCE ON DECISION AND CONTROL

DAY	ROOM	1	2	3	4	5	6	7	8	9	10	11	12	13
	TIME	Canyon I	Finger Rock III	Canyon III	PrimRose	Lantana	Verbena	Indigo	Aster	Murphey I	Murphey II	Murphey III	Finger Rock I	Finger Rock II
	Wednesday 8:30 - 9:30 Plenary I Canyon II & IV J. Tsitsiklis "Complexity - Theoretic Aspects of Problems in Control Theory"													
WEDNESDAY	**WA** 10:00 to 12:30	Neural Networks in Control	Identification I	Nonlinear Systems Theory	Numerical Methods	Linear Multivariable Control	Linear Systems: Observers & Control	Robust Control I	Adaptive Control of Nonlinear Systems	Robotics: Flexible Manipulators & Path Planning	Robust Stabilization	Stabilizability of Distributed Parameter Systems	Scheduling of Flexible Manufacturing Systems	Transfer Function & Matrix Fraction Design
WEDNESDAY	**WM** 14:00 to 16:00	H_∞ Control: Discrete Time, Multirate, Sampled Data	Identification II	Aerospace Applications	Energy Systems	Robotics: Redundant & Constrained Robots	Applications to Space Station Design	Robustness Analysis with Parametric Uncertainty	Time Varying Parameters in Adaptive Control	Stability I	Nonlinear Regulation	Discrete Event Systems	Distributed Parameter Systems	Stabilization of Linear Systems
WEDNESDAY	**WP** 16:20 to 18:20	Neural Networks for Adaptive Control	Identification III	State Space H_∞ Control	Implicit Systems	Communication Networks	Model Approximation	Robust Control for Nonlinear Systems	Stochastic Theory-Adaptive Control	Linear Quadratic & Sampled Data Control	Nonlinear Control Applications	Control of Distributed Parameter Systems	Petri Nets for Discrete Event Systems	Filtering I
	Thursday 8:30 - 9:30 Plenary II Canyon II & IV P. Varaiya "Smart Cars on Smart Roads: Design and Evaluation"													
THURSDAY	**TA** 10:00 to 12:30	Adaptive Control & Applications	Boundary Control & Observation	Decentralized Control	Control Applications	Robotics: Robust & Adaptive Control	H_∞ Control	Computational Advances in Systems & Control	Optimal Control I	Robust Stability with Parametric Uncertainty	Nonholonomic Control	Game Theory & Optimization	Manufacturing Systems	Identification: Frequency Domain Techniques
THURSDAY	**TM** 14:00 to 16:00	Adaptive Control: New Approaches	Nonstandard Sampling & Hold	Control of Smart Structures Canyon II	Stochastic Control Techniques	Robotics: Force Control	Numerical Nethods for Robust Control	Sampled-Data Systems	Nonlinear control for Aerospace Systems	Robust Stability & Performance	Feedback Linearization	Nonlinear Stochastic Control	Discrete Event & Manufacturing Systems	Chaos in Control Systems
THURSDAY	**TP** 16:20 to 18:00	Robust Adaptive Control	Stochastic Control Theory	L1 & H_∞ Control	Discrete-Time Control	Estimation	Numerical Methods for H_∞ Control	H2 and H_∞ Estimation	Failure Detection Aerospace Systems	Discrete Event Systems	Variable Structure & Discrete Time Nonlinear Systems	Nonlinear Stochastic Control	Manufacturing Systems	Applications of Robust Control
	Thursday 19:30 - 22:30 Awards Banquet Canyon Ballroom													
FRIDAY	**FA** 8:30 to 11:00	Mathematical Aspects of Nonlinear Control Cottonwood	Intelligent Control	Industrial Control Applications Acacia	Robust Control of Constrained Systems	Robotics: Mobile & Teleoperated Robots	Neural Networks in Estimation & Control	Adaptive Control	Robust Stability with Mixed Perturbations	Large-Scale Systems & Model Reduction	Robust Control II	Voltage Stability in Power Systems	Hybrid Intelligent Control	Coupled & Multilink Distributed Parameter Systems
	Friday 11:30 - 12:30 Bode Prize Lecture Canyon Ballroom B. D. O. Anderson "Controller Design: Getting from Theory to Practice"													
FRIDAY	**FM** 14:00 to 16:00	Nonlinear Control I	Singular Perturbations	Linear Systems & Time-Varying Systems	Optimal Control of Distributed Parameter Systems	Robotics: Learning & Neural Control	Robust Performance	Stochastic Discrete Event Systems	Adaptive Control Methods	Passivity of Nonlinear Systems	Filtering II	Stochastic Control Algorithms	Timed Discrete Event Systems	Control of Fluid Flows
FRIDAY	**FP** 16:20 to 18:30	Nonlinear Control II	Stability II	Neural Network Applications	Distributed Parameter Systems	Adaptive & Variable Structure Control	Optimal Control II	Control System Design-A Factorization Approach	Adaptive Control Theory	Nonlinear Systems with I/O Derivatives	Estimation Algorithms and Time Series Analysis	Robust Control III	Discrete Event & Hybrid Systems	Tracking

TECHNICAL PROGRAM - WEDNESDAY

CODING

WA =	Wednesday morning
WM =	Wednesday midday
WP =	Wednesday afternoon
TA =	Thursday morning
TM =	Thursday midday
TP =	Thursday afternoon
FA =	Friday morning
FM =	Friday midday
FP =	Friday afternoon
(I) =	Paper invited by a session organizer
(SIAM) =	Paper contributed by the Society for Industrial and Applied Mathematics
(*) =	Paper not included in proceedings

WEDNESDAY MORNING
December 16, 1992

PLENARY SESSION I
8:30 - 9:30 Canyon II & IV

Chairs: T. Başar, *University of Illinois, U-C*
S. Verdú, *Princeton University*

Complexity-Theoretic Aspects of Problems in Control Theory
J.N. Tsitsiklis, *Massachusetts Institute of Technology*

WA-1 Room: Canyon I
NEURAL NETWORKS IN CONTROL

Organizers: P. Ioannou, M.A. Christodoulou
Chairs: P. Ioannou, *University of Southern California*
M.A. Christodoulou, *Tech. Univ. Crete*

WA-2 Room: Finger Rock III
IDENTIFICATION I

Chairs: A. Benveniste, *IRISA*
M. Gevers, *University of Louvain*

WA-3 Room: Canyon III
NONLINEAR SYSTEMS THEORY

Chairs: J.W. Grizzle, *University of Michigan*
V. Ramakrishna, *Princeton University*

WA-4 Room: PrimRose
NUMERICAL METHODS

Chairs: M. P. Misra, *Wright State University*
R. Patel, *Concordia University*

WA-5 Room: Lantana
LINEAR MULTIVARIABLE CONTROL

Chairs: D. Bensoussan, *University of Quebec*
P. Voulgaris, *University of Illinois, U-C*

WA-6 Room: Verbena
LINEAR SYSTEMS: OBSERVERS AND CONTROL

Chairs: V. Syrmos, *University of Hawaii*
M. Sznaier, *University of Central Florida*

WA-7 Room: Indigo
ROBUST CONTROL I

Chairs: J. Chen, *Georgia Institute of Technology*
M. Rotea, *Purdue University*

WA-8 Room: Aster
ADAPTIVE CONTROL OF NONLINEAR SYSTEMS

Chairs: J. Baillieul, *Boston University*
G. Kabuli, Integrated Systems Inc.

WA-9 Room: Murphey I
ROBOTICS: FLEXIBLE MANIPULATORS AND PATH PLANNING

Chairs: M.I. Jankovic, *Washington University*
G.N. Saridis, *Rensselaer Polytechnic Inst.*

WA-12 Room: Finger Rock I
SCHEDULING OF FLEXIBLE MANUFACTURING SYSTEMS

Organizers: S. Engell, P.B. Luh
Chairs: S. Engell, *Universität Dortmund*
P.B. Luh, *University of Connecticut*

WA-13 Room: Finger Rock II
TRANSFER FUNCTION AND MATRIX FRACTION DESIGN

Chairs: T. Georgiou, *University of Minnesota*
A. Stoorvogel, *Eindhoven Univ. of Tech.*

WM-1 Room: Canyon I
H∞ CONTROL: DISCRETE TIME, MULTIRATE, SAMPLED DATA

Chairs: B . Bamieh, *University of Illinois, U-C*
D.S. Flamm, *SRI International*

WM-2 Room: Finger Rock III
IDENTIFICATION II

Chairs: G. Gu, *Louisiana State University*
B. Wahlberg, *Linkoping University*

Chairperson:

WM-3 Room: Canyon III
AEROSPACE APPLICATIONS

Chairs: S .S . Banda, *Wright-Patterson AFB*
B. Morton, *Honeywell SRC*

WM-4 Room: PrimRose
ENERGY SYSTEMS

Chairs: E. Abed, *University of Maryland*, C. Park
A. Debs, *Georgia Inst. of Technology*

WM-8 Room: Aster
TIME VARYING PARAMETERS IN ADAPTIVE CONTROL

Organizers: V. Solo, J. Bentsman
Chairs: V. Solo, *Macquarie University*
J . Bentsman, *University of Illinois, U-C*

WM-9 Room: Murphey I
STABILITY I

Chairs: R.J . Ober, *University of Texas*
L. Saydy, *Ecole Polytechnique*

WM-13 Room: Finger Rock II
STABILIZATION OF LINEAR SYSTEMS

Chairs: N. Gündeş, *University of California, Davis*
P. Zagalak, *Czechoslovak Academy of Sciences*

WP-1 Room: Canyon I
NEURAL NETWORKS FOR ADAPTIVE CONTROL

Chairs: J . Sun, *Wayne State University*
K.S. Tsakalis, *Arizona State University*

WP-5 Room: Lantana
COMMUNICATION NETWORKS

Chairs: S. Stidham, *University of North Carolina*
J.E. Wieselthier, *Naval Research Lab.*

WP-6 Room: Verbena
MODEL APPROXIMATION

Organizers: A.C. Antoulas, S. Weiland,
Chairs: A.C. Antoulas, *Rice University*
S. Weiland, *Rice University*

WP-10 Room: Murphey II
NONLINEAR CONTROL APPLICATIONS

Chairs: M.D. Ilic, *MIT*
A. Teel, *Ecole des Mines*

WP-11 Room: Murphey III
CONTROL OF DISTRIBUTED PARAMETER SYSTEMS

Chairs: H.T. Banks, J.A. Burns,
Chairs: H.T. Banks, *North Carolina State University*
J.A. Burns, *Virginia Tech*

THURSDAY MORNING

December

PLENARY SESSION II

8:30 - 9:30 Canyon II & IV

Chairs: T. Başar, *Univ. of Illinois, Urbana-Champaign*
S. Verdú, *Princeton University*

Smart Cars on Smart Roads: Design and Evaluation
P. Varaiya, *University of California, Berkeley*

TA-1 Room: Canyon I

ADAPTIVE CONTROL AND APPLICATIONS

Chairs: A.M. Annaswamy, MIT
E. Ydstie, *University of Massachusetts*

TA-2 Room: Finger Rock III

BOUNDARY CONTROL & OBSERVATION

Chairs: E.B. Lee, *University of Minnesota*
E. Zuazua, *Univ. Complutense*

TA-3 Room: Canyon III
DECENTRALIZED CONTROL

Chairs: A. Datta, *Texas AM University*
W. Perkins, *University of Illinois*

TA-4 Room: PrimRose
CONTROL APPLICATIONS

Chairs: J. James, *Intermetrics, Inc.*
B. Paden, *University of California*

TA-7 Room: Indigo
COMPUTATIONAL ADVANCES IN SYSTEMS AND CONTROL

Organizers: W.S. Wong, P.S. Krishnaprasad
Chairs: W.S. Wong, *AT&T Bell Laboratories*
P.S. Krishnaprasad, *University of Maryland*

TA-8 Room: Aster
OPTIMAL CONTROL I

Chairs: F.M. Callier, *Fac. Univ. de Namur*
D.Q. Mayne, *University of California, Davis*

TA-11 Room: Murphey III
GAME THEORY & OPTIMIZATION

Chairs: A. Haurie, *University of Geneva*
G. Papavassilopoulos, *USC*

TA-12 Room: Finger Rock I
MANUFACTURING SYSTEMS

Chairs: T.S. Chang, *University of California*, Davis
F. Lewis, *University of Texas*

11:30 - 11:40
Online Parameter Optimization for a Multi-Product, Multi-Machine Manufacturing System 1676
J.S. Dhingra, G.L. Blankenship, *University of Maryland*

11:40 - 11:50
On Dynamic Control of Transportation Systems in a Manufacturing Environment 1678
M. Aicardi, A. Di Febbraro, R. Minciardi, R. Pesenti, *University of Genoa*

11:50 - 12:00
Optimal Policies for Two-Stage, Pull-Type Production/Inventory Systems 1682
M.B. Gursoy, H. Danhong, T. Altiok, *Rutgers University*

TA-13 Room: Finger Rock II
IDENTIFICATION: FREQUENCY DOMAIN TECHNIQUES

Chairs: J.S. Baras, *University of Maryland*
H.V. Poor, *Princeton University*

10:00 - 10:10
A New Algorithm for Optimal Estimation of Plant Parameters from Input-Output Data 1684
A.K. Shaw, *Wright State University*

10:10 - 10:30
General Formulation of a Prony Based Method for Simultaneous Identification of Transfer 1686
Functions and
D.A. Pierre, J.R. Smith, *Montana State University*, D.J. Trudnowski, *Battelle Pacific N.W. Labs*, J.W. Pierre, *University of Wyoming*

10:30 - 10:40
Identification in H Using Pick's Interpolation 1692
G. Gu, D. Xiong, K. Zhou, *Louisiana State University*

10:40 - 10:50
System Identification for H∞ Control 1694
N.P. Rubin, D.J.N. Limebeer, *Imperial College*

10:50 - 11:10
Parameter Identification of Large Spacecraft Systems Based on Frequency Characteristics 1696
D.R. Augestein, J.S. Baras, S.M. Fisher, *University of Maryland*

11:10 - 11:30
An Indirect Method for Transfer Function Estimation from Closed Loop Data 1702
P. Van den Hof, R.J.P. Schrama, O.H. Bosgra, *Delft University of Technology*

11:30 - 11:50
An Algorithm for State-Space Erequency Domain Identification without Windowing Distortions 1707
D.S. Bayard, *California Institute of Technology*

11:50 - 12:10
Reduced-Order Models for Continuous MIMO Systems via Rational Interpolations on a Minimal 1713
J . H . Lilly, *University of Louisville*

12:10 - 12:30
On Line H2, H∞ and Pointwise Uncertainty Bound Quantification in Identification of Restricted 1719
E.-W. Bai, S. Raman, *University of Iowa*

TM-1 Room: Canyon I
ADAPTIVE CONTROL: NEW APPROACHES

Chairs: R.R. Bitmead, *University of South Wales*
G.C. Goodwin, *University of Newcastle*

14:00 - 14:20
A Cyclic Switching Strategy for Parameter-Adaptive Control 1725
F. M. Pait, A. S. Morse, *Yale University*

14:20 - 14:40
Hysteresis Switching Adaptive Control of Linear Multivariable Systems 1731
S. R. Weller, G. C. Goodwin, *The University of Newcastle*

14:40 - 15:00
Remarks on Sufficient Information for Adaptive Nonlinear Regulation 1737
J.-B. Pomet, *Ecole Centrale de Nantes*

15:00 - 15:10
Adaptive Parameter Estimation for a Class of Distributed Parameter Systems with Persistence of Excitation 1742
M. Demetriou, I.G. Rosen, P. Ioannou, *University of Southern California*

15:10 - 15:30
Adaptive Pole Placement without Excitation Probing Signals for Discrete-Time Systems 1744
R. Lozano, X.-H. Zhao, *Université de Technologie de Compiègne*

15:30 - 15:50
Convergence of the Signed Output Error Adaptive Identifier 1750
J. Garnett, *Georgia Tech. Research Institute*, S. Dasgupta, *University of Iowa*, C. R. Johnson, *Cornell University*

15:50 - 16:10
Escape from Stable Equilibria in Blind Adaptive Equalizers 1756
M.R. Frater, *University College*, R.R. Bitmead, *Australian National University*, C.R. Johnson, *Cornell University*

TM-2 Room: Finger ROCK III
CONTROL SYSTEMS WITH NONSTANDARD SAMPLING AND HOLD MECHANISMS

Organizers: N. Schiavoni, M. Araki
Chairs: M . Araki, *University of Kyoto*
R. Scattolini, *Politecnico di Milano*

14:00 - 14:20
Application of Multilevel Multirate Sampled-Data Controllers to Simultaneous Pole Assignment Problem 1762
T. Hagiwara, M. Araki, H. Soma, *Kyoto University*

14:20 - 14:40
The Output Control Problem for Multirate Sampled-Data Systems 1768
P. Colaneri, R. Scattolini, N. Schiavoni, *Politecnico di Milano*

14:40 - 15:00
Robust Design of Reduced Order Multirate Compensators Using Constrained Optimization Techniques 1774
I.D. Apostolakis, D. Jordan, *The University of Connecticut*

15:00 - 15:20
An Introduction to Motion Planning under Multirate Digital Control 1780
S. Monaco, *Università di Roma "La Sapienza"*, D. Normand-Cyrot, *CNRS-ESE*

15:20 - 15:40
Optimal Tracking under Periodic Output Feedback Control 1786
G.L. Hsylop, *McDonnel Douglas*
H. Schättler, T.J. Tarn, *Washington University*

15:40 - 16:00
Control of Linear and Nonlinear Systems Using Generalized Sampled-Data Hold Functions ....*
C. Yang, *Michigan Technological University*, P.T. Kabamba, Y.S. Hung, *The University of Michigan*

TM-3 Room: Canyon II
CONTROL OF SMART STRUCTURES

Chairs: H.T. Banks, G.J. Knowles,
Chairs: H.T. Banks, *North Carolina State University*
G.J. Knowles, *Grumman Corporation*

14:00 - 14:20 (I)
Investigation of the Energy Tranfer and Consuumption of Adaptive Structures 1791
C.A. Rogers, C. Liang, F.P. Sun, *Virginia State University*

14:20 - 14:40 (I)
A Piezoelectric Actuator Model for Active Vibration and Noise Control in Thin Cylindrical Shells 1797
H. T. Banks, *North Carolina State University*, H.C. Lester, R.C. Smith, *NASA Langly Research Center*

14:40 - 15:00 (I)
Effects of Piezo-Actuator Thickness on the Active Vibration Control of a Beam ....*
S.J. Kim, J.D. Jones, *Purdue University*

15:00 - 15:20 (I)
Identification of Nonlinear Electromechanical Coupling in Smart Actuation: Electrostriction Effects ....*
S. Hanagud, J. Zhang, G. Kulkarm, *Georgia Institute of Technology*

15:20 - 15:40 (I)
Computational Methods for Identification in Structures with Piezoceramic Actuators and Sensors 1803
H.T. Banks, Y. Wang, *North Carolina State University*, D.J. Inman, *Virginia State University*, J.C. Slater, *State University of New York*

15:40 - 16:00 (I)
Computational Methods for Controlling a Timoshenko Beam ....*
J.A. Burns, M. Tadi, *VPI&SU*

16:00 - 16:20 (I)
Fault-Tolerant Vibration Control for Space Satellites Incorporating Smart Structures ....*
W. Van Nostrand, F. Austin, C. Huang, G. Knowles, *Grumman Corporate Research Center*

16:20 - 16:40 (I)
Combined Feedback and Adaptive Digital Control of Cylinder Vibration Using Smart Structures 1809
H.H. Cudney, J.K. Finefield, A. Sumali, *Virginia State University*

16:40 - 17:00 (I)
Control of Adaptive Optical Arrays .... *
M.J. Ealey, J. Wellman, *Litton-Itek Optical Systems*

17:00 - 17:20 (I)
Comparison of Vibration Control for Smart Antenna 1815
J.J. Dosch, D. Leo, *State University of New York*,
D. Inman, *Virginia State University*

17:20 - 17:40 (I)
Trunention Effect of Control Design Models 1821
J.P. Allen, J.P. Lauffer, *Sandia National Laboratories*

17:40 - 18:00 (I)
Experiments on Active Optical and Structural Control 1824
Z. Rahman, J. Spanos, J. Fanson, *California Institute of Technology*

TM-4 Room: PrimRose
STOCHASTIC CONTROL TECHNIQUES

Chairs: P.E. Caines, *McGill University*
S.R. Hall, *MIT*

14:00 - 14:20
Variable Structure Control of Discrete-Time Stochastic Systems 1830
F. Zheng, M. Cheng, W.-B. Gao, *Beijing University of Aeronautics and Astronautics*

14:20 - 14:40
Optimal Control with Actuator/Sensor Noise Strengths Related to Feedback and Kalman Gain 1836
A.K. Choudhury, M. Ruan, *Howard University*

14:40 - 15:00
Covariance Averaging in the Analysis of Uncertain Systems 1842
S.R. Hall, D.G. MacMartin, *MIT*,
D.S. Bernstein, *The University of Michigan*

15:00 - 15:20
Information Partitions, Deadlock, and Nonsequential Stochastic Control 1850
M.S. Andersland, *University of Iowa*, D. Teneketzis, *The University of Michigan*

15:20 - 15:40
Suboptimal Control for Nonlinear Stochastic Systems 1856
F.-Y. Wang, *University of Arizona*, G.N. Saridis, *RPI*

15:40 - 15:50
Chaos in Discrete Variable Structure System 1862
X. Yu, *University of Central Queensland*

TM-5 Room: Lantana
ROBOTICS: FORCE CONTROL

Chairs: D. Dawson, *Clemson University*
J.T.Y. Wen, *Rennsselaer Polytechnic Inst.*

14:00 - 14:10
Hierarchical Adaptive Control of Multiple Manipulators by Using contact Force Measurements 1864
W.-H. Zhu, Y.-G. Xi, Z.-J. Zhang, *Jiaotong University*

14:10 - 14:20
Robust Tracking Control of Rigid-Link Electrically Driven Robot 1866
J. Guldner, J.J. Carroll, D.M. Dawson, *Clemson University*, Z. Qu, *University of Central Florida*

14:20 - 14:40
A Stable Force/Position Controller for Robot Manipulators 1869
S. Chiaverini, B. Siciliano, L. Villani, *Università degli studi di Napoli "Federico II"*

14:40 - 15:00
Robust Motion and Force Control of Robot Manipulators in the Presence of Environmental Constraint Uncertainties 1875
B. Yao, S.P. Chan, D. Wang, *Nanyang Technological University*

15:00 - 15:20
Impedance Control for Dexterous Space Manipulators 1881
R. Colbaugh, K. Glass, *New Mexico State University*,
H. Seraji, *California Institute of Technology*

15:20 - 15:40
Comprehensive Modeling for Simultaneous Position and Force Control of Deformable Manipulators 1887
Y.-J. Lin, T.-S. Lee, *University of Akron*

15:40 - 15:50
Stability Characteristics of Force Controlled Manipulation in the Presence of Low Pass Filtering 1893
H.P. Qian, J. De Schutter, *Katholieke Universiteit Leuven*

TM-6 Room: Verbena
NUMERICAL METHODS FOR ROBUST CONTROL

Organizer: P. Van Dooren, *University of Illinois, U-C*
Chair: P. Van Dooren, *University of Illinois, U-C*

14:00 - 14:20 (1)
A Simplification of the Generalized Eigenspace Solution for Singular H∞ Problems 1897
K.C. Goh, M.G. Safonov, *University of Southern California*

14:20 - 14:40 (I)
On the Computation of the General Multiblock Problem 1903
I.J. Diaz-Bobillo, M.A. Dahleh, *Massachusetts Institute of Technology*

14:40 - 15:00 (I)
Computing Robustness of System Properties with Respect to Structured Real Matrix Perturbations 1909
M. Wicks, *GMI Engineering and Management Institute,*
R.A. DeCarlo, *Purdue University*

15:00 - 15:20 (I)
A Quadratically Convergent Local Algorithm on Minimizing Sums of the Largest Eigenvalues of a Symmetric Matrix 1915
M.K.H. Fan, B. Nekooie, *Georgia Institute of Technolgy*

15:20 - 15:40 (I)
On Computing the Eigenvalues of a Symplectic Pencil 1921
R. Patel, *Concordia University*

15:40 - 16:00 (I)
Numerical Solution of Large Scale Lyapunov Equations Using Krylov Subspace Methods 1927
I. Jaimoukha, E. Kasenally, D. Limebeer, *Imperial College*

16:00 - 16:20 (I)
Numerical Linear Algebra Techniques for large Scale Matrix Problems in Systems and Control 1933
P. Van Dooren, *University of Illinois*

TM-7 Room: Indigo
SAMPLED-DATA SYSTEMS

Chairs: C.E. de Souza, *University of Newcastle*
K. Sobel, *CCNY/CUNY*

14:00 - 14:10
A Unification of Periodic Time Varying and Multirate Controllers 1939
C. Zhang, *The University of Melbourne*

14:10 - 14:20
Optimal Input Matching in Sampled Data Control Systems 1941
K. Nordström, *Uppsala University*

14:20 - 14:30
Necessary and Sufficient Conditions for the Complete Reachability and Observability of Multirate Sampled-Data Systems 1944
S. Longhi, *Università di Anocona*

14:30 - 14:40
Soft Multirate Control 1946
P. Albertos, *University of Newcastle*

14:40 - 15:00
Limiting Properties of the Zeros of Sampled-Data Systems with Zero- and First-Order Holds 1949
T. Hagiwara, T. Yuasa, M. Araki, *Kyoto University*

15:00 - 15:20
Generalized Sample Hold Functions: Fact and Falacies 1955
A. Feuer, *Technion-Israel Institute of Technology*
G.C. Goodwin, *University of Newcastle*

15:20 - 15:40
Sampled-Data Observers with Generalized Holds for Unstable Plants 1961
W.M. Haddad, H.-H. Huang, *Florida Instit. of Tech.,*
D.S. Bernstein, *The University of Michigan*

15:40 - 15:50
On the Disturbance Rejection for Sampled-Data Discrete-time Linear Systems 1966
A.S. Buonocore, A. Tornambè, *Seconda Università di Roma "Tor Vergata"*

15:50 - 16:00
A Time Domain Approach to Performance Robustness of Sampled Data Systems Using the 1968
Delta Operator
J.E. Piou, K.M. Sobel, *The City College of New York*

TM-8 Room: Aster
APPLIED NONLINEAR CONTROL TECHNIQUES FOR AEROSPACE SYSTEMS

Organizers: C.Y. Huang, G.J. Knowles
Chair: C.Y. Huang, *Grumman Corporation*
G.J. Knowles, *Grumman Corporation*

14:00 - 14:20 (I)
Nonlinear Feedback Guidance Law for Aero-Assisted Orbit Transfer Maneuvers 1970
P. K .A. Menon, *Optimal Synthesis*

14:20 - 14:40 (1)
Higher Order Approximate Input-Output Linearization 1976
Z. Xu, *University of Southern California*, J. Hauser, *University of Colorado*, L.R. Hunt, *University of Texas*

14:40 - 15:00 (I)
Analysis of a Nonlinear Dynamic Inverse Control Strategy and Its Digital Implementation .... *
J. Reilly, W. Dayawansa, W.S. Levine, C. Huang, G. Knowles, *University of Maryland*

15:00 - 15:20 (I)
Stability of Dynamic Inversion Control Laws Applied to Nonlinear Aircraft Pitch-Axis Model .... *
B. Morton, D. Enns, *Honeywell System and Research Center*

15:20 - 15:40 (I)
Neural-Network-Based Nonlinear Inverse Flight Control 1982
C. Huang, J. Tylock, S. Engel, J. Whitson, J. Eilbert, *Grumman Corporation*

15:40 - 16:00 (I)
Dynamic Inversion/Structured Singular Value Control Laws for Aircraft .... *
G. Balas, W. Garrard, J. Reiner, *University of Minnesota*

TM-9 Room: Murphey I
ROBUST STABILITY & PERFORMANCE

Chairs: V. Balakrishnan, *Stanford University*
P.P. Khargonekar, *University of Michigan*

14:00 - 14:20
On Robust Stability with Structured Time-Invariant Perturbations 1987
B. Bamieh, *University of Illinois*, M. Dahleh, *University of California, Santa Barbara*

14:20 - 14:40
Uniform Stability and Performance in H_∞ 1991
A.P. Koshore, J.B. Pearson, *Rice University*

14:40 - 15:00
H_∞-Control of LFT Systems: An LMI Approach 1997
W.M. Lu, J.C. Doyle, *California Institute of Technology*

15:00 - 15:20
Does Robust Regulation Compromise H_2 Performance? 2002
J.L. Abedor, K.M. Nagpal, K. Poolla, *University of California, Berkeley*

15:20 - 15:30
Asymptotic Properties of the Maximum Entrophy Lyapunov Equation for Robust Stability and Performance analysis 2008
D.S. Bernstein, F. Tyan, *University of Michigan*, W.M. Haddad, *Florida Institute of Technology*, D.C. Hyland, *Government Aerospace Systems*

15:30 - 15:40
Existence and Uniqueness of Optimal Matrix Scalings 2010
V. Balakrishnan, S. Boyd, *Stanford University*

TM-10 Room: Murphey II
FEEDBACK LINEARIZATION

Chairs: O. Akhrif, *University of Quebec*
R. Murray, *California Inst. of Technology*

14:00 - 14:20
Stable Inversion of Nonlinear Nonminimum Phase Systems .... *
D. Chen, B. Paden, *University of California*

14:20 - 14:40
A Non-Exact Brunovsky Form and Dynamic Feedback Linearization 2012
J.-B. Pomet, C.H. Moog, E. Aranda, *Laboratoire d'Automatique de Nantes*

14:40 - 15:00
On the Exact Linearization of Second Order Bilinear Systems 2018
L. del Re, L. Guzzella, A. Astolfi, *Swiss Federal Institute of Technology*

15:00 - 15:10
Factorization Theory and Feedback Linearization of a Class of Nonlinear Systems 2023
A. Banos, M.A. Armada, *Instituto de Automática Industrial*, F.N. Bailey, *University of Minnesota*

15:10 - 15:30
Nonsingular Covering of Nonlinear Control Systems and Its Application to Feedback Linearization Along a Reference Trajectory 2027
S. Ishijima, A. Kojima, *Tokyo Metropolitan Institute of Technology*

15:30 - 15:40
Optimal Control of Feedback Linearizable Systems 2033
D.A. Schoenwald, *Oak Ridge National Laboratory*, Ü. Özgüner, *Ohio State University*

TP-4 Room: PrimRose
DISCRETE-TIME CONTROL

Chairs: Y. Peng, *University of Brussels*
R.E. Skelton, *Purdue University*

TP-5 Room: Lantana
ESTIMATION

Chairs: E.A. Jonckheere, *U.S.C.*
B. Pasik-Duncan, *University of Kansas*

TP-6 Room: Verbena
NUMERICAL METHODS FOR H∞ CONTROL

Chairs: D.J.N. Limebeer, *Imperial College*
A. Saberi, *Washington State University*

TP-10 Room: Murphey II
VARIABLE STRUCTURE AND DISCRETE TIME NONLINEAR SYSTEMS

Chairs: N. Olgac, *University of Connecticut*
R. Ortega, *University of Compiegne*

16:20 - 16:30
Sliding Modes Synthesis in Nonlinear Dynamic Systems with Application in Pulse Voltage Stabilizers 2366
Y.B . Shtessel, *University of South Carolina*

16:30 - 16:40
Sliding-Mode Observers Based on Equivalent Control Method 2368
S.V. Drakunov, *The Ohio State University*

16:40 - 17:00
Variable Structure Control Problems and the Theory of Singular Perturbations 2370
R. Johnson, P. Nistri, *Universita degli Studi di Firenze*

17:00 - 17:20
An Extended System Approach to Dynamical Pulse-Frequency-Modulation Control of Nonlinear Systems 2376
H. Sira-Ramérez, O. Llanes-Santiago, *Universidad de Los Andes*

17:20 - 17:40
Steady-State Errors in Discrete-Time Control Systems 2381
I.W. Sandberg, L.Y. Xu, *University of Texas*

17:40 - 17:50
Design of Discrete-Time Adaptive Sliding Mode Control 2387
G. Bartolini, A. Ferrara, *Universita di Genova*, V.I. Utkin, *Institute of Control Services*

17:50 - 18:10
Some Comments about Linearization under Sampling 2392
J.P. Barbot, D. Normand-Cyrot, N. Pantalos, *CNRS-ESE*, S. Monaco, *Università de Roma "La Sapienza"*

18:10 - 18:20
An Improved Sliding Mode Control 2398
N. Olgac, V.R. Iragavarapu, *University of Connecticut*

TP-11 Room: Murphey III
NONLINEAR STOCHASTIC CONTROL: NEW APPROACHES AND APPLICATIONS

Organizers: M.H.A. Davis, G. Burstein
Chairs: M.H.A. Davis, *Imperial College*
G. Burstein, *Imperial College*

16:20 - 16:40 (I)
On the Relation Between Partially Observed Stochastic Optimal Control and Deterministic Infinite Dimensional Optimal Control 2400
M.H.A. Davis, G. Burstein, *Imperial College of Science*

16:40 - 17:00 (I)
Recursive Parameter Estimation for Partially Observed Markov Chains .... *
R. Elliott, J. Moore, *The Australian National University*

17:00 - 17:20 (I)
Asymptotic Properties of Adaptive Control Laws for Partially Observed Controlled Diffusions .... *
N. Christopeit, *Universitat Bonn*

17:20 - 17:40 (I)
Nonsmooth Analysis on Stochastic Controls: A Survey 2406
X.Y. Zhou, *University of Toronto, Fudan University*

TP-12 Room: Finger Rock I
MANUFACTURING SYSTEMS

Organizers: S. Sethi, C. Temponi
Chairs: S. Sethi, *University of Toronto*
C. Temponi, *Old Dominion University*

16:20 - 16:40 (I)
Multilevel Hierarchical Controls in Dynamic Stochastic Marketing-Production Systems .... *
S. Sethi, *University of Toronto*, Q. Zhang, *University of Kentucky*

16:40 - 17:00 (1)
Nonnegative Systems for a Class of Production-Inventory Models 2412
C. Temponi, *Old Dominion University*, J. Sample, F. Lewis, *University of Texas*

17:00 - 17:20 (1)
A Numerical Approach to Solve Finite Horizon Optimal Control Problems Exhibiting the Turnpike Property 2417
P. Kopalle, *University of Arizona*, J. Assuncao, D.R. Lehmann, *Columbia University*

17:20 - 17:40 (I)
Stabilizing Scheduling Policies for a Single Machine System 2423
Y.H. Chen, *Georgia Institute of Technology*

TP-13 Room: Finger Rock II
APPLICATIONS OF ROBUST CONTROL

FRIDAY MORNING
December 18, 1992

FA-1 Room: Cottonwood
MATHEMATICAL ASPECTS OF NONLINEAR CONTROL

11:00 - 11:10
A New Methodology to Design Extended Kalman Filters. Application to Distillation Columns 2594
F. Viel, D. Bossanne, E. Busvelle, F. Deza, J.P. Gauthier, *Grand-Couronne Research Centre*

FA-4 Room: PrimRose
ROBUST CONTROL OF CONSTRAINED SYSTEMS

Organizers: M. Sznaier, A. Sideris
Chairs: M. Sznaier, *University of Central Florida*
A. Sideris, *University of California, Irvine*

8:30 - 8:50 (I)
Robust Control of a Compact Disc Player 2596
M. Steinbuch, G. Schootstra, *Philips Research Lab.*, O.H. Bosgra, *Delft University of Technology*

8:50 - 9:10 (I)
MIMO H∞ Control with Time Domain Constraints 2601
A. Sideris, *University of California*, H. Rotstein, *California Institute of Technology*

9:10 - 9:30 (I)
Robust Control of Systems with Input and State Constraints via LQG-Balancing 2607
E. Verriest, *Georgia Tech Lorraine*, G. Pajunen, G. Murali, *Florida Atlantic University*

9:30 - 9:50 (I)
New Absolute Stability Criteria for Robust Stability and Performance with Locally Slope Restricted Monotonic Nonlinearities 2611
D. Bernstein, *The University of Michigan*, W.M. Haddad, *Florida Institute of Technology*

9:50 - 10:10 (I)
Robust Control of Systems under Mixed Frequency/Time-Domain Constraints via Finite Dimensional Convex Optimization 2617
M. Sznaier, *University of Central Florida*, Z. Benzaid, *Embry Riddle Aeronautical University*

10:10 - 10:30 (I)
Controller Design with Actuator Constraints 2623
A.G. Tsirukis, M. Morari, *California Institute of Technology*

10:30 - 10:50 (I)
Minimum-Time Control for Uncertain Discrete-Time Linear Systems 2629
F. Blanchini, *Università degli Studi di Udine*

10:50 - 11:10 (I)
Feedforward Tracking for Uncertain Systems 2635
R.W. Benson, W.E. Schmitendorf, R.M. Dolphus, *University of California, Irvine*

11:10 - 11:30 (I)
Robust Linear Controllers under State and Control Constraints 2640
G. Bitsoris, E. Gravalou, *University of Patras*

FA-5 Room: Lantana
ROBOTICS: MOBILE AND TELEOPERATED ROBOTS

Chairs: P. Khosla, *Carnegie Mellon University*
K. Liu, *University of Texas*

8:30 - 8:50
Coordinating Locomotion and Manipulation of a Mobile Manipulator 2643
Y. Yamamoto, X. Yun, *University of Pennsylvania*

8:50 - 9:10
Stability and Transparency in Bilateral Teleoperation 2649
D. A. Lawrence, *University of Colorado*

9:10 - 9:30
Robustness Analysis of Nonlinear Biped Control Laws via Singular Perturbation Theory 2656
H. M. Hmam, D. A. Lawrence, *University of Colorado*

9:30 - 9:50
Real Time Classification of Teleoperation Data with a Neural Network 2662
P. Fiorini, *California Institute of Technology*, A. Giancaspro, G. Pasquariello, *Consiglio Nazionale delle Richerche*, S. Losito, *Agenzia Spaziale Italiana*

9:50 - 10:10
A Controller for the Dynamic Walk of a Biped Across Variable Terrain 2668
M. van de Panne, E. Fiume, Z. Vranesic, *University of Toronto*

10:10 - 10:30
Teleoperator Control System Design with Human in Control Loop and Telemonitoring Force Feedback 2674
H. S. Lee, *University of Southern California*
S. Lee, *University of Southern California, California Institute of Technology*

10:30 - 10:50
Modeling and Feedback Control of Nonholonomic Mobile Vehicles 2680
G. J. Pappas, K. J. Kyriakopoulos, *New York State Center for Advanced Technology* and *Rensselaer Polytechnic Institute*

10:50 - 10:10
Selfish and Coordinative Planning for Multiple Mobile Robots by Genetic Algorithm 2686
T. Shibata, T. Fukuda, K. Kosuge, F. Arai, *Nagoya University*

FA-6 Room: Verbena
NEURAL NETWORKS IN ESTIMATION & CONTROL

Chairs: C. Abdallah, *University of New Mexico*
K.S. Narendra, *Yale University*

FA-7 Room: Indigo
ADAPTIVE CONTROL

Chairs: E. Kamen, *Georgia Inst. of Technology*
I. Kanellakopoulos, *University of California*

FA-10 Room: Murphey II
ROBUST CONTROL II

Chairs: P.T. Kabamba, *University of Michigan*
R. Smith, *University of California, S-B*

FA-11 Room: Murphey III
VOLTAGE STABILITY IN POWER SYSTEMS NEW PERSPECTIVES

Organizers: M.A. Pai, D.J. Hill
Chairs: D.J. Hill, *University of Newcastle*
M .A. Pai, *University of Illinois*

FA-12 Room: Finger Rock I
HYBRID INTELLIGENT CONTROL

Organizers: R. Grossman, J.R. James
Chairs: R. Grossman, *University of Illinois*
J.R. James, *Intermetrics, Inc.*

FA-13 Room: Finger Rock II
MODELING, ANALYSIS AND CONTROL OF COUPLED AND MULTI-LINK DISTRIBUTED PARAMETER SYSTEMS

Organizers: J .E. Lagnese, G. Chen
Chairs: J.E. Lagnese, *Georgetown University*
G. Chen, *Texas A&M University*

9:50 - 10:10 (I)
Controllability of Networks of Elastic Strings and Beams ....*
G. Leugering, *Universitat Bayreuth*

10:10 - 10:30 (I)
On Hyperbolic Systems Associated with the Modelling and Control of Vibrating Networks 3003
J.E. Lagnese, *Georgetown University*, G. Leugering, *Universitaet Bayreuth,* E.J.P.G. Schmidt, *McGill University*

10:30 - 10:50 (I)
Boundary Controllability of Systems of Connected Strings and Vector Exponential Families 3009
S. Avdonin, Sergei A. Ivanov, *St. Petersburg State University*

BODE LECTURE
11:30 - 12:30 Canyon Ballroom

Chairs: A.H. Haddad, *Northwestern University*
M.K. Masten, *Texas Instruments*

Controller Design: Getting from Theory to Practice
B.D.O. Anderson, *Australian National University*

FM-1 Room: Canyon I
NONLINEAR CONTROL I

Chairs: J.H. Fu, *Wright State University*
R. Hunt, *University of Texas*

14:00 - 14:20
Two-Step High Performance Nonlinear Control Systems Design...3013
Y. Li, E. B. Lee, *University of Minnesota*

14:20 - 14:40
Liapunov Functions and Stability Criteria for Nonlinear Systems with Multiple Critical Eigenvalues 3019
J.-H. Fu, *Wright State University*

14:40 - 15:00
Nonlinear Controller to Reduce Resonance Effects During Despin of a Dual-Spin Spacecraft through Precession Phase Lock..3025
R. J. Kinsey, *The Aerospace Corporation,* D. L. Mingori, *University of California,* R. H. Rand, *Cornell University*

15:00 - 15:20
Global Robust Stabilization of Nonlinear Cascaded Systems 3031
J. Imura, T. Sugie, T. Yoshikawa, *Kyoto University*

15:20 - 15:30
A Possible Approach for Achieving Robust Tracking for Nonlinear Systems 3037
F. Delli Priscoli, *Universita di Roma "La Sapienza"*

15:30 - 15:40
Remarks on the Robustness of Disturbance Decoupling for Nonlinear Systems ....*
V. Ramakrishna, *University of California*

15:30 - 15:40
Asymptotic Output Tracking Through Singular Points for a Class of Uncertain SISO Nonlinear Systems...3040
Zvi Retchkiman, Jaime Alvarez, Rafael Castro, *C.I.N.V.E.S.T.A.Vdel I.P.N.*

15:40 - 15:50
On the Equilibrium Sets of Linear Systems with Saturating Feedback Control 3042
J. Alvarez, R. Suarez, J. Alvarez, *Universidad Autonoma Metropolitana - Iztapalapa*

FM-2 Room: Finger Rock III
SINGULAR PERTURBATIONS

Chairs: H.K. Khalil, *Michigan State University*
D. Taylor, *Georgia Inst. of Technology*

14:00 - 14:20
Suboptimal Control of Singularly Perturbed Systems and Periodic Optimization 3044
V. A. Gaitsgory, *Bar-Ilan University*

14:20 - 14:40
On the Commutativity of Decomposition and Discretization for Linear Singularly Perturbed Systems..3050
K. R. Shouse, D. G. Taylor, *Georgia Institute of Technology*

14:40 - 15:00
Near Optimum Regulators for Singularly Perturbed Systems with Random Parameters 3056
S. X. Shen, Q. Xia, M. Rao, Y. Ying, *University of Alberta,* Z. Gajic, *Rutgers University*

15:00 - 15:10
Estimates of the Singular Perturbation Parameter for Stability, Controllability, and Observability of Linear Systems 3062
S. M. Shahruz, *Berkeley Engineering Research Institute,* A. K. Packard, *University of California, Berkeley*

15:10 - 15:20
Dynamic Observer Design for Nonstandard Discrete-Time Singularly Perturbed Systems 3064
M.T. Beheshti, M.E. Sawan, *The Wichita State University*

FM-3 Room: Canyon III
LINEAR SYSTEMS AND TIME-VARYING SYSTEMS

FM-4 Room: PRIMROSE
OPTIMAL CONTROL OF DISTRIBUTED PARAMETER SYSTEMS

15:40 - 15:50 (SIAM)
The Minimal Time Function in Infinite Dimension .... *
O. Carja

15:50 - 16:00 (SIAM)
Properties of Relaxed Trajectories of Evolution Equations and Optimal Control 3126
X. Xiang, *Guizhou University*, N.U. Ahmed, *University of Ottowa*

FM-5 Room: Lantana
ROBOTICS: LEARNING AND NEURAL CONTROL

Chairs: R. Langari, *Texas AM University*
M. Sundareshan, *University of Arizona*

14:00 - 14:10
A Two-Time-Scale Neural Tracking Controller with Application to the Tracking Control of Rigid Arms 3128
W. Cheng, J.T. Wen, *RPI*

14:10 - 14:30
Optimal Robot Motions for Repetitive Tasks 3130
D. Simon, *TRW Space and Technology Group*

14:30 - 14:50
Learning Control for Impedance Controlled Manipulator 3135
Y. Maeda, *Fujitsu Laboratories LTD.*, H. Kano, *Tokyo Denki University*

14:50 - 15:10
Learning Control of Brushless DC Motors for Motion Control Applications 3141
L.Y. X. Ma, T. S. Low, T. H. Lee, Y. P. Ding, *National University of Singapore*

15:10 - 15:30
Learning Control for a Class of Nonlinear Systems and Its Application to a Direct Drive Manipulator 3147
H.-S. Ahn, K.-B. Kim, *Korea Institute of Science and Technology*, C.-H. Choi, *Seoul National University*

15:30 - 15:50
A Recurrent Neural Network-Based Adaptive Variable Structure Model Following Control of Multijointed Robotic Manipulators 3152
A. Karakasoğlu, M. K. Sundareshan, *University of Arizona*

FM-6 Room: Verbena
ROBUST² PERFORMANCE

Organizers: J.C. Doyle, M. Dahleh
Chairs: M. Dahleh, *University of California*
A. Vicino, *Universitá di L'Aquila*

14:00 - 14:20 (I)
Overview of Robust Stability and Performance Methods of Systems with Structured Mixed Perturbations 3158
M. Dahleh, *University of California*, J.C. Doyle, *Caltech*

14:20 - 14:40 (I)
Robust Analysis for Time Varying Systems 3163
J.S. Shamma, *University of Texas*

14:40 - 15:00 (I)
Robust Stability/Performance of Interconnected Interval Plants with Structured Norm Bounded Perturbations 3169
M. Dahleh, *University of California*, A. Tesi, *Università di Firenze*, A. Vicino, *Università di L'Aquila*

15:00 - 15:20 (I)
Mixed μ Problems and Branch and Bound Techniques 3175
M.P. Newlin, P.M. Young, *California Institute of Technology*

15:20 - 15:40 (I)
An Improved μ Lower Bounded via Adaptive Power Iteration 3181
J.E. Tierno, P.M. Young, *California Institute of Technology*

15:40 - 16:00 (I)
Mixed μ Upper Bound Computation Using LMI Optimization 3187
C. Beck, J. Doyle, *California Institute of Technology*

FM-7 Room: Indigo
STOCHASTIC DISCRETE EVENT SYSTEMS

Chairs: W.B. Gong, *University of Massachusetts*
S.G. Strickland, *University of Virginia*

14:00 - 14:10
Very Weak, P-Weak Majorization and Thei: Application 3193
P.D. Sparaggis, C.G. Cassandras, D. Towsely, *University of Massachusetts*

14:10 - 14:20
Alternate Representations of Stochastic Processes with Applications to Infinitesimal Perturbation Analysis 3195
S.G. Strickland, *University of Virginia*

FM-8 Room: Aster
ADAPTIVE CONTROL METHODS

Chairs: T. Duncan, *University of Kansas*
G. Tao, *University of Virginia*

FM-9 Room: Murphey I
RECENT RESULTS ON PASSIVITY OF NONLINEAR SYSTEMS AND IMPLICATIONS

Organizer: I.D. Landau
Chair: I.D. Landau, *Lab. d'Automatique de Grenoble*

15:00 - 15:20
Generalized Real-Time Temporal Logic Applied on Control of Nondeterministic Discrete Events Dynamic Systems 3369
B. I. Silva Jr., R. S. Mendes, *DCA/FEE/UNICAMP*

15:20 - 15:40
Communication Delays in Connections of Input-Output Discrete Event Processes 3374
S. Balemi, *Swiss Federal Institute of Technology (ETH)*

15:40 - 16:00
On the Existence of Petri Net Supervisors 3380
A. Giua, F. DiCesare, *Rensselaer Polytechnic Institute*

FM-13 Room: Finger Rock II
CONTROL OF FLUID FLOWS

Organizers: K. Ito, M.D. Gunzberger
Chairs: K. Ito, *University of Southern California*
M.D. Gunzberger, *VPI&SU*

14:00 - 14:20 (1)
Adaptive Control of a Functional Differential Equation Exhibiting Some Features of the Vortex Sheet 3386
S.B. Jiang, J.S. Gibson, *University of California,*
A.J. Pearlstein, *University of Illinois*
A. Glezer, *Georgia Institute of Technology*

14:20 - 14:40 (1)
Issues in Shape Optimization for the Navier-Stokes Equations 3390
M.D. Gunzberger, J.S. Peterson, *VPI&SU*

14:40 - 15:00 (1)
A Control Problem for Fluid Flow 3393
S. Kang, K. Ito, *University of Southern California*

15:00 - 15:20 (1)
Active Flow Control Relative to a Rotating Cylinder 3399
Y.-R. Ou, *Virginia Polytechnic Institute and State University*

15:20 - 15:40 (1)
Modeling and Optimal Design of a Chemical Vapor Deposition Reactor 3405
H.T. Tran, J.S. Scroggs, *North Carolina State University*

FP-1 Room: Canyon I
NONLINEAR CONTROL II

Chairs: F. Salam, *Michigan State University*
I.W. Sandberg, *University of Texas, Austin*

16:20 - 16:30 (SIAM)
Control of a Class of Nonlinear Systems with General Constraints 3410
M.E. Achhab, F.M. Callier, *Facultés de Namur*, V. Wertz, *Université Cathlique de Louvain*

16:30 - 16:50
Optimization Approach to Controlling Constrained Nonlinear Systems 3412
D. Rangel, T.-W. Wang, *University of Tennessee*

16:50 - 17:10
Fuzzy Systems as Nonlinear Dynamic System Identifiers Part II: Stability Analyis and Simulations 3418
L.-X. Wang, *University of California*

17:10 - 17:30
Semiglobal Stabilization of a Class of Nonlinear Systems using Output Feedback 3423
H. K. Khalil, *Michigan State University,* F. Esfandiari, *Detroit Engineering Center*

17:30 - 17:40
Output Feedback Controller for Systems Containing a Backlash 3429
D.H . Chyung, *University of Iowa*

17:40 - 17:50
Semi-Global Stabilization of Partially Linear Composite Systems via Output Feedback of the State of the Linear Part 3431
Z. Lin, A. Saberi, *Washington State University*

17:50 - 18:00
A note on observers for a bilinear system class 3434
R. Martínez, J. Alvarez, *Universidad Autónoma Metropolitana- Iztapalapa*

FP-2 Room: Finger Rock III
STABILITY II

Chairs: R.A. DeCarlo, *Purdue University*
A.N. Michel, *University of Notre Dame*

16:20 - 16:30 (SIAM)
Stability Analysis of Infinite Dimensional Discrete and Continuous Time Linear Systems 3436
R. Ober, Y. Wu, *University of Texas*

16:30 - 16:40
Asymptotic Stability of m-Switched Systems using Lyapunov Functions 3438
P. Peleties, R.A. DeCarlo, *Purdue University*

16:40 - 17:00
Asymptotic Stability of Discrete-Time Systems with Saturation Nonlinearities 3440
D. Liu, A. N. Michel, *University of Notre Dame*

17:00 - 17:20
On Stability of a Class of Discrete Nonlinear Systems 3446
X. Yang, Y. Stepanenko, *University of Victoria*

FP-3 Room: Canyon III
NEURAL NETWORK APPLICATIONS

Chairs: R.A. Milito, *AT&T Bell Labs.*
E.Y. Rodin, *Washington University*

FP-4 Room: PrimRose
DISTRIBUTED PARAMETER SYSTEMS

Chairs: J. Bentsman, *University of Illinois*
K. Morris, *University of Waterloo*

18:00 - 18:10 (SIAM)
Convergence of Controllers Designed Using Time-Domain Techniques 3521
K.A. Morris, *University of Waterloo*

FP-5 Room: Lantana
ADAPTIVE AND VARIABLE STRUCTURE CONTROL

Chairs: R.R. Mohler, *Oregon State University*
B. Shafai, *Northeastern University*

16:20 - 16:40
Nonlinear Multirate Adaptive Control of a Synchronous Motor 3523
G. Georgiou, A. Chelouah, D. Normand-Cyrot, *CNRS/ESE*, S. Monaco, *Università di Roma*

16:40 - 17:00
Identification of Coupled Markov Chain Model with Application 3529
R. A. Kennedy, S.-H. Chung, *Australian National University*

17:00 - 17:10
Adaptive Forced Balancing for Magnetic Bearing Control Systems 3535
S. Beale, B. Shafai, *Northeastern University*, P. LaRocca, E. Cusson, *The Charles Stark Draper Laboratory*

17:10 - 17:20
Requirements and Models for Adaptive Optics in Coherent Lidar Wind Measurements 3540
Robert Leland, *University of Alabama*

17:20 - 17:30
Experimental Implementation of MRAC for a DC Servo Motor 3542
A. Cerda, C. Abdallah, R. Jordan, *University of New Mexico*

17:30 - 17:50
Variable-Structure Control of Flexible AC Transmission Systems 3544
Y. Wang, W. Zhu, R. R. Mohler, R. Spée, *Oregon State University*

17:50 - 18:00
Applications of Variable Structure System Theory to Mechanical Systems 3550
P. Muraca, P. Pugliese, *Università della Calabria*

18:00 - 18:10
Discrete-Time Adaptive Control of Linear Plants with Uncertain Time Delay 3554
A. Ferrara, G. Bartolini, *University of Genova*

FP-6 Room: Verbena
OPTIMAL CONTROL II

Chairs: T. Runolfson, *Johns Hopkins University*
E.I. Verriest, *Georgia Inst. of Technology*

16:20 - 16:40
Robustness Studies of a Proximate Time-Optimal Controller 3559
L. Y. Pao, *The MITRE Corporation*, G. F. Franklin, *Stanford University*

16:40 - 17:00
New Global Regulation Performance Indices and Optimal Robust Eigenstructure 3565
D. Wang, H. Qiu, M. Rao, *University of Alberta*

17:00 - 17:20
Minimizing the Maximum Value of the Regulated Output to a Fixed Input 3571
D.E. Miller, *University of Waterloo*

17:20 - 17:30
Recursive Approach to the Optimal Control Problems for Singularly Perturbed Bilinear System 3577
X. Shen, Y. Ying, M. Rao, Q. Xia, *University of Alberta*

17:30 - 17:40
LQGPC - A Predictive Design as Tradeoff between LQG and GPC 3579
B. Taube, *Siemens AG*, B. Lampe, *Rostock University*

17:40 - 17:50
An Order-Reduction Approach for Singular Linear Quadratic Problem 3582
Q. Wang, C.-F. Lin, *American GNC Corporation*

17:50 - 18:00
An Alternative Approach to Discrete-Time Disturbance Attenuation 3584
K. Furuta, M. Wongsaisuwan, *Tokyo Institute of Technology*

18:00 - 18:10 (SIAM)
A Numerical Method for Solving Discrete Time Optimal Processes 3586
A.M. Tsirlin, V.A. Kazakov, *Program System Institute USSR Academy of Sciences*

18:10 - 18:20
A Two-Degree-of-Freedom Design of Optimal Servosystems 3588
Y. Fujisaki, M. Ikeda, *Kobe University*

FP-7 Room: Indigo
CONTROL SYSTEM DESIGN: A FACTORIZATION APPROACH

Organizers: M. Sebek, V. Kucera
Chairs: V. Kucera, *Czechoslovak Academy of Sci.*
M. Sebek, *Czechoslovak Academy of Sci.*

FP-8 Room: Aster
ADAPTIVE CONTROL THEORY

Chairs: M. Bodson, *Carnegie Mellon University*
I. Mareels, *Australian National University*

FP-9 Room: Murphey I
REPRESENTATIONS FOR NONLINEAR SYSTEMS THAT INVOLVE DERIVATIVES OF INPUTS AND OUTPUTS

Organizers: Y. Wang, E. Sontag
Chairs: Y. Wang, *Florida Atlantic University*
E. Sontag, *Rutgers University*

17:30 - 17:50 (I)
On Universal Observability 3669
S. Diop, *Rutgers University*

17:50 - 18:10 (I)
Nonlinear Input/Output Relations and Identifiability 3673
S.T. Glad, *Linköping University*

18:10 - 18:20 (I)
A Generalized State Space Decomposition of Nonlinear Systems 3676
G. Conte, A. M. Perdon, *Università di Ancona*, C. H. Moog, *Universitè di Nantes*, Y.F. Zheng, *East China Normal University*

FP-10 Room: Murphey II
ESTIMATION ALGORITHMS AND TIME SERIES ANALYSIS

Chairs: M.H.A. Davis, *Imperial College*
A. Lindquist, *Royal Inst.of Technology*

16:20 - 16:30
On the Nonlinear Dynamics of Fast Filtering Algorithms 3678
C.I. Byrnes, *Washington University*, A. Lindquist, Y. Zhou, *Royal Institute of Technology*

16:30 - 16:50
Efficient Square-Root Algorithms for PDA, IMM, and IMMPDA Filters 3680
V. Raghavan, K. R. Pattipati, Y. Bar-Shalom *University of Connecticut*

16:50 - 17:00
A Geometric Approach to the Reciprocal Realization Problem 3686
J.-Å. Sand, *Division of Optimization & Systems Theory*

17:00 - 17:10
A Stochastic Realization Approach to Order Determination in Time Series Modelling 3690
M.H.A. Davis, W-X. Zheng, *Imperial College*

17:10 - 17:30
Data Filtering, Reparametrization, and the Numerical Accuracy of Parameter Estimators 3692
G. Li, *Nanyang Technoligical University*, M. Gevers, *Louvain University*

17:30 - 17:50
A Novel High Resolution Parallel Spectral Estimation Method for Narrow-Band Signals 3698
W. Liu, R. Doraiswami, *University of New Brunswick*

17:50 - 18:00
Optimal and Robust Tracking of Regression Parameters 3703
B. Delyon, A. Juditsky, *Campus de Beaulieu, IRISA*
A. Nazin, *Institute for Control Sciences, Profsoyuznaya*

18:00 - 18:10
Combining Forecasts Using Recursive Equal Weighting and Linear Programming 3705
L. Wang, G. Libert, *Faculte Polytechnique de Mons*, B. Liu, *Tianjin University*

18:10 - 18:20
Non-Linear MAP Estimation using a Continuous State Viterbi Algorithm 3707
C.R. Champlin, *Motorola Strategic Electronics Division*, D.R. Morrell, *Arizona State University*

18:20 - 18:30
A Batch Least Squares Lattice Algorithm 3709
H. Aling, *Integrated Systems Inc.*

FP-11 Room: Murphey III
ROBUST CONTROL III

Chairs: J. Ackermann, D.L.R.
P. Colaneri, *Politecnico di Milano*

16:20 - 16:40
H_2 / H∞ Optimal Control Synthesis with Regional Pole Constraints 3711
Y. W. Wang, D. S. Bernstein, *The University of Michigan*

16:40 - 17:00
Mixed H_2 / H∞ Control for Continuous-Time Linear Systems 3717
J. C. Geromel, P. L. D. Peres, S. R. Souza, *LAC/DT/FEE- UNICAMP*

17:00 -17:10
On H∞ Control for Symmetric Systems 3723
L. Fortuna, G. Muscato and G. Nunnari, *Universitá di Catania*

17:10 - 17:20
A Time Domain Approach to the Robustness of Time Delay Systems 3726
W. Yu, K.M. Sobel, E.Y. Shapiro, *The City College of New York*

17:20 - 17:40
An Overview on the LQG/LTR Method using the Delta Operator 3728
M. Tadjine, M. M'Saad, L. Dugard, *Laboratoire d'Automatique de Grenoble*

17:40 - 18:00
Dynamic Output Feedback Simplex Control Design for Systems wtih Non-Matching Disturbances and Uncertainties 3734
B. M. Diong, J. V. Medanić, *University of Illinois*

18:00 - 18:10
Uncertain Linear Systems with Random Distrubances 3740
G. Yin, *Wayne State University*

18:40 - 18:50
On State Covariance Bounds for Linear Stochastic Uncertain Systems 3742
P. Bolzern, G.D. Nicolae, *Politecnico di Milano*, P. Colaneri, *Dipartimento di Elettronicae Informazione*

FP-12 Room: Finger Rock I
DISCRETE EVENT AND HYBRID SYSTEMS

Chairs: P.J. Antsaklis, *University of Notre Dame*
P. Varaiya, *University of California, Berkeley*

16:20 - 16:30
Suboptimal Control of a Class of Hybrid Systems 3744
V. Gaitsgory, *Bar-Ilan University*

16:30 - 16:40
Controllability Subsets of Live Rabin Automata 3746
J.G. Thistle, *University of California*

16:40 - 17:00
Modeling and Analysis of Hybrid Control Systems 3748
J A. Stiver, P. J. Antsaklis, *University of Notre Dame*

17:00 - 17:20
Properties of Behavioral Models for a Class of Hybrid Dynamical Systems 3752
L. E. Holloway, *University of Kentucky*, B. H. Krogh, *Carnegie Mellon University*

17:20 - 17:40
Automated Reasoning with Function Evaluations for COCOLOG 3758
S. Wang, P. E. Caines, *McGill University*

17:40 - 18:00
Recursive Computation of Limited Lookahead Supervisory Controls for Discrete Event Systems 3764
S.-L. Chung, S. Lafortune, *University of Michigan*, F. Lin, *Wayne State University*

18:00 - 18:20
Protocol Verification using Discrete-Event Systems 3770
K. Rudie, *University of Minnesota*, W. M. Wonham, *University of Toronto*

FP-13 Room: Finger Rock III
TRACKING

Chairs: P.K. Varshney, *Syracuse University*
P. Luh, *University of Connecticut*

16:20 - 16:40
A Stochastic Perturbation Analysis of Bias in the Extended Kalmar Filter as Applied to Bearings - Only Estimation 3778
M. J. Moorman, *Wright Laboratories*, T. E. Bullock, *University of Florida*

16:40 - 17:00
Tracking of a Reflective Target using Infrared Measurements and Laser Illumination 3784
P. S. Maybeck, T. D. Herrera, R. J. Evans, *Air Force Institute of Technology*

17:00 - 17:20
Interacting Multiple Bias Model Algorithm with Application to Tracking Maneuvering Targets 3790
W. D. Blair, G. A. Watson, *Dahlgren Division*

17:20 - 17:30
Integrated Probabilistic Data Association (IPDA) 3796
D. Musicki, R. J. Evans, S. Stankovic, *University of Melbourne*

17:30 - 17:40
On Multisensor Track Alignment 3799
E. J. Dela Cruz, A. T. Alouani, *Tennessee Technological University*, T. R. Rice and W. D.Blair, *Naval Surface Warfare Center*

17:40 - 17:50
A Tracking Algorithm for Maneuvering Targets 3801
T. C. Wang, P. K. Varshney, *Syracuse University*

17:50 - 18:00
Data Association and Tracking from Distributed Sensors using Hidden Markov Models and Evidential Reasoning 3803
F. Martinerie, P. Forster, *Thomson Sintra ASM*

18:00 - 18:20
Monocular 3-D Visual Tracking of a Moving Target by an Eye-in-Hand Robotic System 3805
N. Papanikolopoulos, *University of Minnesota*, B. Nelson, P. K. Khosla, *Carnegie Mellon University*

KEY: C-Chair, cC-coChair, O-Organizer, cO-coOrganizer, M-Moderator, cM-coModerator, Pa-Panelist, PS-Plenary Speaker, BL-Bode Lecture

Proceedings of the 31st Conference
on Decision and Control
Tucson, Arizona • December 1992

Asymptotic Stabilization of Three Dimensional Homogeneous Quadratic Systems

W. P. Dayawansa* C. F. Martin†, S. Samelson‡

Abstract

We give necessary and sufficient conditions for the asymptotic stabilizability of a generic class of single input, three dimensional, homogeneous quadratic systems.

Key Words: Asymptotic Stabilization, Nonlinear Systems, Homogeneous Systems, Quadratic Systems, Three Dimensional Systems.

1 Introduction

Asymptotic Stabilization of low dimensional nongeneric nonlinear systems is of much interest in nonlinear control theory since such systems occur naturally as the system evolving on a center manifold (see [Ca], [Ay] etc.). Within this class, those systems with nonvanishing quadratic part are generic, and therein lies our principle interest in the asymptotic stabilization problem for homogeneous quadratic systems.

It has been established that in a system of ordinary differential equations if the leading homogeneous part is asymptotically stable, then the overall system is locally asymptotically stable (see [Ha], and [He2] in the weighted homogeneous case). This motivates us to consider the use of homogeneous quadratic feedback in our problem. However, it is seen easily that no homogeneous quadratic system can be asymptotically stable. Hence we will broaden the class of feedback functions to include positively homogeneous functions, i.e. the defining homogeneity property holds only along positive rays.

There has been several notable studies of homogeneous and weighted homogeneous control sys-
tems (see [Ba],[Co],[Col],[DM2],[He2],[Ka2],[Sa] etc), and in particular on quadratic systems (see[Bo],[BT],[Ma],[Vi] etc). In a very recent paper[Vi] Vivalda gave necessary and sufficient conditions for the controllability of single input, three dimensional, homogeneous quadratic systems. In [BT] Bonnard and Tebkish devoted a section on the asymptotic stabilization of two and three dimensional homogeneous quadratic systems. Their results are complete in the two dimensional case, but restricted to a small class in the three dimensional case. In this paper we will give necessary and sufficient conditions for the asymptotic stabilizability of a generic class of single input homogeneous quadratic systems by using a positively homogeneous feedback function of degree two (Theorem 3.1).

Our analysis here is built upon some of the recent work on the stabilizability of low dimensional systems. Some topological conditions for stabilizability were derived by Brockett (see [Br]), and later extended by using a well known index theorem due to Krasnosel'skii and Zabreiko (see [KZ]) by Coron (see [Co]). In this paper we will use as our principle tools some necessary conditions (lemma 2.2) and some sufficient conditions (lemma 2.1) for the stabilizability of homogeneous systems derived in [Ka2] and [DM1].

*Department of Electrical Engineering and the Systems Research Center, University of Maryland, College Park, MD 20742 ; Supported in Part by NSF Grant #ECS 9096121 and the Engineering Research Center Program Grant # CD 8803012

†Department of Mathematics, Texas Tech University, Lubbock, TX 79409 ; Supported in Part by NSA Grant #MDA 904-90-8-4009

‡Army High Performance Computing Research Center, University of Minnesota, 1100 Washington Avenue South, Minneapolis, MN 55415 ; Supported in part by the US Army Contract: DAAL 03-89-C-0038

CH3229-2/92/0000-2463$1.00 © 1992 IEEE

2 Preliminaries

In this section we will describe three key lemmas related to the stabilization of homogeneous systems. The first one gives a sufficient condition, the second one gives a necessary condition for the stabilizability and the third one describes a class of deformations on the space of homogeneous systems, which will preserve the hypothesis of the first two lemmas. We will see in the next section that these lemmas are sufficient enough to determine the stabilizability of a generic class of homogeneous quadratic systems.

Let us consider an arbitrary single input C^1 homogeneous system in three state variables,

$$\begin{aligned} \dot{x} &= f(x,y), \\ \dot{y} &= u, \end{aligned} \qquad (2.1)$$

where, $x \in \Re^2, y \in \Re, u \in \Re$, and f is positively homogeneous of some degree p, i.e. $f(\lambda x, \lambda y) = (\lambda)^p f(x,y)$ for all $\lambda \geq 0$.

As has been pointed out by Coleman (see [Col],[Ka2],[DM1] etc.) it is convenient to first look at the restriction of f to the unit sphere S^2 in $\Re^3$. We will refer to $(0,0,1)$ and $(0,0,-1)$ as the poles of the unit sphere.

Much of the information regarding the stabilizability of a homogeneous system is contained in the topology of the subset of points in $\Re^3$ on which $f(x,y)$ is parallel to the $x-$ direction. This set is defined below.

For arbitrary $\delta \in \Re$ let $\hat{A}_\delta = \{(x,y) \in \Re^{n+1} | f(x,y) = \delta x\}$. Let $A_\delta = \hat{A}_\delta \cap S^2$. Let,

$$\begin{aligned} A_+ &= \cup_{\delta>0} A_\delta, \\ A_- &= \cup_{\delta<0} A_\delta, \\ A_{+0} &= A_+ \cup A_0 \\ A_\Re &= \cup_{\delta \in \Re} A_\delta \end{aligned} \qquad (2.2)$$

When we need to indicate that these sets are associated to a particular system Σ we will put Σ as a superscript. Let,

$$\begin{aligned} \Omega = \{\sigma : S^1 \to S^2 - \textit{poles} \mid & \ \sigma \text{ is } C^1 \text{ and} \\ & \sigma \text{ is transversal to the meridians}\} \end{aligned} \qquad (2.3)$$

Note that even if the system is homogeneous, i.e. $f(\lambda x, \lambda y) = \lambda^p f(x,y)$ for all $\lambda \in \Re$ (not just for $\lambda > 0$), one may have to use positively homogeneous feedback in order to stabilize the system. For example, no homogeneous quadratic system can be asymptotically stable, whereas there are asymptotically stable positively homogeneous quadratic systems. Therefore we will always consider *positively homogeneous* feedback functions in this paper.

Lemma 2.1 *If there is $\sigma \in \Omega$ such that $\sigma \cap A_{+0} = \emptyset$ and $\sigma \cap A_- \neq \emptyset$, then the system is asymptotically stabilizable by C^1 positively homogeneous feedback.*

We will skip the proof here. A complete proof of this lemma can be found in [Ka2] and [DM2].

Lemma 2.2 *Suppose that there exists a continuous curve $\mu : [0,1] \to S^2$ such that,*

(i) $\mu(0)$ = north pole and $\mu(1)$ = south pole,

(ii) $\mu \subset A_{+0}$.

Then the system does not admit a continuos positively homogeneous asymptotically stabilizing feedback function.

We will skip the proof here. A complete proof of this lemma can be found in [DM2].

Lemmas 2.1 and 2.2 imply that much information regarding the stabilizability of (2.1) is contained in $A_\Re$. We will now discuss one way to deform $A_\Re$ without loosing its information content.

Definition 2.1 A diffeomorphism $\phi : S^2 \to S^2$ is a *meridian preserving diffeomorphism* if it leaves all meridians invariant.

Lemma 2.3 *Let Σ_1 and Σ_2 be three dimensional homogeneous quadratic systems as in (2.1). Suppose that $\phi : S^2 \to S^2$ is a meridian preserving diffeomorphism such that $\phi(A_\mu^{\Sigma_1}) = A_\mu^{\Sigma_2}$ for $\mu = +, 0, -$ and $\Re$. Then Σ_1 satisfies the hypotheses of lemma 2.1 (respectively lemma 2.2) if and only if Σ_2 satisfies the hypotheses of lemma 2.1 (respectively lemma 2.2).*

Proof: Let $\sigma \in \Omega$ satisfies the hypotheses of lemma 2.1 for system Σ_1. Then $\phi \circ \sigma_1$ satisfies the hypotheses of lemma 2.1 for Σ_2. A similar argument works for lemma 2.2 as well. •

3 Asymptotic Stabilizability of Homogeneous Quadratic Systems

Here we consider the system,

$$d/dt(x,y) = F(x,y) + bu \quad (3.1)$$

where, $x \in \Re^2$, $y \in \Re$, $u \in \Re$, $b \in \Re^3$, F is a homogeneous quadratic vector field. We will assume that $b \neq 0$.

Without any loss of generality we will write (3.1) as,

$$\begin{aligned} \dot{x_1} &= f_1(x,y) = \sum_{j=1}^{3} a_{1,j}(x)y^{3-j} \\ \dot{x_2} &= f_2(x,y) = \sum_{j=1}^{3} a_{2,j}(x)y^{3-j} \\ \dot{y} &= u \end{aligned} \quad (3.2)$$

where $a_{i,1}$ are constants, $a_{i,2}$ are linear functions, and $a_{i,3}$ are homogeneous quadratic polynomials. We will denote (f_1, f_2) by f. We will impose certain genericity hypotheses (**H1** through **H5**) as we go along in order to avoid certain pathological cases. Analysis of the stabilizability of (3.2) requires somewhat finer arguments in these cases, and will be reported elsewhere in the near future.

The first of the genericity conditions is the following:

H1: $f(0,1) \neq 0$.

It is clear that we may apply a linear transformation in the x - space in order to bring forth the conditions,

$$\begin{aligned} a_{1,1} &= 1, \\ a_{2,1} &= 0, \end{aligned}$$

H2: $a_{2,2} \neq 0$.

Clearly, **H1** and **H2** are equivalent to the assumption that b, $[b,[b,F]]$ and $[b,[[b,[b,F]],F]]$ are linearly independent vectors.

Because of the assumption **H2**, we may redefine x_1 as $a_{2,2}(x)/(d/dx_1)a_{2,2}(x)$, and thus we may further assume that $a_{2,2} = x_1$. Now, we will redefine y as $[y + \{a_{2,3}(x_1,0) + (d/dx_2)|_{x_2=0}a_{2,3}(x_1,x_2)\}/x_1]$, and apply a suitable feedback transformation (to cancel out the extra terms in the $\dot{y}$ equation) to ensure that,

$$a_{2,3} = bx_2^2,$$

where b is a real constant.

H3: $b \neq 0$ and $a_{1,3}(x_1,0) \neq 0$.

Assumption **H3** ensures that A_0 (see (2.2) for the definition) does not meet the x_1 axis or the x_2 axis.

Now (3.2) has the structure,

$$\begin{aligned} \dot{x_1} &= y^2 + a_{1,2}(x)y + a_{1,3}(x) \\ \dot{x_2} &= x_1 y + bx_2^2 \\ \dot{y} &= u \end{aligned} \quad (3.3)$$

Since A_0 is given by the intersection of the two quadratics $f_1(x,y) = 0$ and $f_2(x,y) = 0$, the question of whether or not small neighborhoods of a point of A_0 contains points belonging to A_+ as well as A_- can be answered by analyzing the local geometry near points of intersection of the two quadratics, which is described by the multiplicity of the zeroes of the Sylvester resultant of f_1 and f_2. Here we confine ourselves to the simplest possible case by assuming that $f_1 = 0$ and $f_2 = 0$ intersect transversally at all real points of intersection (on $\Re P^1$). This is equivalent to the following:

H4: All real zeroes of the Sylvester resultant,

$$\lambda(x) := b^2x_2^4 - ba_{1,2}(x)x_1x_2^2 + a_{1,3}(x)x_1^2 \quad (3.4)$$

on $\Re P^1$ are simple.

Let us now examine the structure of $A_{\Re}$ (see (2.2) for the definition). It is the cubic curve,

$$x_2 f_1(x,y) - x_1 f_2(x,y) = 0 \quad (3.5)$$

Since no cubic terms in y appear in the above expression, it follows that on each meridian there are at most two points of $A_{\Re}$. By **H3** it follows that the branches of $A_{\Re}$ have the following asymptotic descriptions near the meridian passing through $(1,0,0)$.

$$\begin{aligned} \gamma_1 &\equiv \{(x_1, x_2, x_1^2/x_2)\},\ x_1 > 0 \\ \gamma_2 &\equiv \{(x_1, x_2, x_1^2/x_2)\},\ x_1 < 0 \\ \gamma_3 &\equiv \{(x_1, x_2, a_1, 3(1,0)x_2)\}, \end{aligned} \quad (3.6)$$

as $x_1 \to 0$. In particular, it follows that $A_\Re \neq \emptyset$. Note also that the antipodal map $(-id)$ maps A_λ onto $A_{-\lambda}$ for each $\lambda \in \Re$. Hence $(-id)(\gamma_i)$, $i = 1,2,3$ give the branches of $A_\Re$ near the meridian through $(-1,0,0)$.

Henceforth we will consider $A_\Re$ as a subset of the cylinder $S^1 \times \Re^* = \{(x,y) \in \Re^2 \times \Re^* \mid \|x\| = 1\}$, where $\Re^* := [-\infty,\infty]$. Horizontal coordinates of $S^1 \times \Re^*$ will be denoted by $\theta \in [-\pi,\pi]/\{-\pi,\pi\}$ and the vertical coordinates by $y \in \Re^*$. We remark that it follows from the expression for the cubic which describes $A_\Re$ that the only asymptotes to $A_\Re$ are the meridians passing through $(0,0)$ and $(\pi,0)$.

Let us consider (3.5) as a quadratic in y with coefficients which are functions of x. Branches of (3.5) meet exactly when the discriminant,

$$\begin{aligned}\mu(x) &= [x_2a_{1,2}(x) - x_1a_{2,2}(x)]^2 \\ &\quad -4x_2[a_{1,3}(x)x_2 - a_{2,3}(x)x_1] \qquad (3.7)\end{aligned}$$

has multiple zeroes.

H5: Real zeroes of $\mu(x)$ on $\Re P^1$ are simple.

This assumption guarantees that two real branches of (3.5) necessarily convert into complex conjugate branches after they meet each other.

Note that $\mu(x)$ cannot be negative definite since we have already seen in (3.6) that $A_\Re$ is nonempty. Also $\mu(x)$ is not identically equal to zero from **H5**. We will now consider the two remaining cases separately.

Let $pr_1 : S^1 \times S^1$, and $pr_2 : S^1 \times \Re^*$ be the projections onto the first and the second factor respectively.

We will first consider the case in which $\mu(x)$ is positive definite.

Lemma 3.1 *Suppose that $\mu(x)$ is positive definite and* **H1** *through* **H5** *are satisfied. Then, there exist continuous curves, $\sigma_i : [0,1] \to S^1 \times \Re^*$, $i = 1,2,3$ which are contained in $A_\Re$ and satisfy the following:*

(i) $\sigma_1(0) = (0,\infty)$ and $\sigma_1(1) = (\pi,\infty)$,

(ii) $\sigma_2(0) = (-\pi,-\infty)$ and $\sigma_2(1) = (0,\infty)$,

(iii) $\sigma_3(0) = (0,0)$ and $\sigma_3(1) = (2\pi,0)$,

(iv) $pr_1 \circ \sigma_i$, $i = 1,2,3$ are strictly monotone increasing,

(v) $pr_2 \circ \sigma_3([0,1]) \subset (-\infty,\infty)$,

(vi) $pr_2 \circ \sigma_i((0,1)) \subset (-\infty,\infty)$, $i = 1,2$.

Proof: Since the discriminant $\mu(x)$ of (3.5) is strictly positive on $S^1 \times \Re^*$, it follows that no two real branches of the algebraic curves (3.5) meet each other. We have concluded previously that for each $\theta \in [-pi,\pi]$ there can be at most two values of $y \in \Re^*$ such that (θ,y) lies on $A_\Re$, and unless $\theta \in \{0,\pi\}$, any such y in necessarily finite. Define σ_i as the branch which begins as γ_i for $i = 1,2,3$ (see 3.6). It is easily seen that γ_1 and $(-id)(\gamma_2)$ belong to the same branch which lies on σ_1, γ_2 and $(-id)(\gamma_1)$ belong to the same branch which lies on σ_2, and γ_3 and $(-id)(\gamma_3$ belong to a branch which lies on σ_3 and it encircles $S^1 \times \Re$ once. •

Lemma 3.2 *Suppose that $\mu(x)$ is positive definite and* **H1** *through* **H5** *are satisfied. Then (3.3) is asymptotically stabilizable by C^1 positively homogeneous feedback of degree two.*

Proof: By **H4** it follows that none of the three curves σ_i, $i = 1,2,3$ in lemma (3.1) belong to A_0. Since $(-id)(A_+) = A_-$ and vice versa, it follows that $A_- \neq \emptyset$. Let $(\theta_0,y_0) \in A_-$ and $y_0 \neq \infty$. Clearly it is possible to construct a meridian preserving diffeomorphism ϕ on $S^1 \times \Re^*$ which leaves (θ_0,y_0) invariant and such that the curve $y = y_0$ meets $\phi(A_\Re)$ only at (θ_o,y_0). Thus, it follows from lemmas 2.1 and 2.3 that the system (3.3) is asymptotically stabilizable by C^1, positively homogeneous feedback of degree two. •

We will now consider the case in which $\mu(x)$ is indefinite. Let us denote the two real zeroes of $\mu(cos(\theta),sin(\theta))$ closest to the origin as $\beta_1 < 0 < \beta_2$ (recall that **H3** guarantees that $\mu(1,0) \neq 0$). Similarly, let us denote the two nonzero real zeroes of $\lambda(cos(\theta),sin(\theta))$ closest to the origin as $\alpha_1 < 0 < \alpha_2$ ($\lambda(1,0) = 0$ always. We discard this zero at the origin in the definition of α_i, $i = 1,2$).

We have assumed that the real zeroes of $\mu(x)$ are simple (see **H5**). Thus it follows that the branches which contain γ_1 and γ_3 meet at some (θ_1,y_1) where $\theta_1 \in (0,\pi)$ and the branches which contain γ_2 and γ_3 meet at some (θ_2,y_2) where $\theta_2 \in (-\pi,0)$. Furthermore, $\theta_1 - \theta_2 < \pi$. Since (θ_1,y_1) and (θ_2,y_2) correspond to repeated solutions of (3.5) and since for each $\theta \in$

$[-\pi, \pi]$ there are at most two $y \in \Re^*$ such that $(cos(\theta), sin(\theta), y) \in A_\Re$, it follows that,

$$\theta_i = \beta_i, \; i = 1,2 \qquad (3.8)$$

Lemma 3.3 *Suppose that $\mu(x)$ is indefinite, and that* **H1** *through* **H5** *are satisfied. Let α_i, β_i, $i = 1,2$ be as above. Then (3.3) is asymptotically stabilizable by C^1 positively homogeneous feedback of degree two if and only if $\{\alpha 1, \alpha 2\} \cap [\beta_1, \beta 2] \neq \emptyset$.*

Proof: Let us define a continuous curve $\sigma :\rightarrow S^1 \times \Re^*$ which parametrizes the branch which contains γ_1, $i = 1,2,3$. The parametrization will be such that

$\sigma(0) = (0, \infty)$, $\sigma(1/2) = (0,0)$, $\sigma(1) = -\infty$.

Clearly points on σ near $(0, \infty)$ and $(0, -\infty)$ belong to A_+. Thus, if $\sigma \cap A_0 \neq \emptyset$, then the hypothesis of lemma 2.2 will be satisfied and the system (3.3) will not be asymptotically stabilizable by C^1 positively homogeneous feedback of degree two. But this is the case exactly when $\{\alpha_1, \alpha_2\} \cap [\beta_1, \beta_2] = \emptyset$.

Conversely, if $\{\alpha 1, \alpha 2\} \cap [\beta_1, \beta 2] \neq \emptyset$, then $A_0 \cap \sigma \neq \emptyset$. Moreover, at such points of intersection the curves $f_1(x) = 0$, and $f_2(x) = 0$ meet transversally, and hence on any neighborhood of such point there will be points which belong to A_+ and also points which belong to A_-. Also $((-id)\sigma) \cap A_- \neq \emptyset$. Let $(\theta_0, y_0) \in \sigma \cap A_-$, and $((\theta_1, y_1) \in (-id(\sigma) \cap A_-$. Since these are the only branches with points at infinity, it follows that it is possible to construct a meridian preserving diffeomorphism $\phi : S^1 \times \Re^* \rightarrow S^1 \times \Re^*$ which preserves (θ_0, y_0) and such that the curve $y = y_0$ meets $\phi(A_\Re)$ at (θ_0, y_0) and at $\phi((\theta_1, y_1)$. Thus by lemmas 2.1 and 2.3, the stabilizability of (3.3) follows. •

We have established the following:

Theorem 3.1 *Suppose that the genericity assumptions* **H1** *through* **H5** *are satisfied. Then the system (3.3) is asymptotically stabilizable by C^1 positively homogeneous feedback of degree two if and only if one of the following is satisfied.*

(i) The homogeneous quadratic polynomial $\mu(x)$ defined in **H4** *is positive definite.*

(ii) $\mu(x)$ is indefinite. Let us denote the two real zeroes of $\mu(cos(\theta), sin(\theta))$ closest to the origin as $\beta_1 < 0 < \beta_2$ and, the two nonzero real zeroes of $\lambda(cos(\theta), sin(\theta))$ closest to the origin as $\alpha_1 < 0 < \alpha_2$ respectively. Then $\{\alpha_1, \alpha_2\} \cap [\beta_1, \beta_2] \neq \emptyset$. •

4 Concluding Remarks

In this paper we have analyzed the asymptotic stabilization problem for a generic class of homogeneous quadratic systems in dimension three. The remaining cases can be analyzed by using similar, but somewhat more complex, arguments. We will report these details in a paper which is currently under preparation.

References

[AF] E. H. Abed and J.H. Fu, "Local feedback stabilization and bifurcation control,I, Hopf bifurcation" *Syst. and Cont. Lett.* **7**(1986) 11-17.

[ABS] A. Andreini, A. Bacciotti, G. Stefani, "Global Stabilizability of Homogeneous Vector Fields of Odd Degree," *Syst. Cont. Lett.*, **10**(1988), 251-256.

[Ay] D. Ayels, "Stabilization of a class of nonlinear systems by a smooth feedback," *Syst. Cont. Lett.*, **5**(1985), 181-191.

[AS] D. Ayels and M. Szafranski, "Comments on the stabilizability of the angular velocity of a rigid body," *Syst. and Cont. Lett.* **10**(1988), 35-39.

[Ba] J. Bailleul, "The geometry of homogeneous polynomial systems," *Nonlinear Analysis; Theory, Methods and Applications*, **17**(1991), 437-445.

[Bo] B. Bonnard, "Quadratic systems," *Mathematics of Control, Signals and Systems*, **4**(1991), 139-160.

[BT] B. Bonnard and H. Tebkish, "Quadratic control systems," *Proc. IEEE Conf. on Decision and Control*, Los Angeles, Dec 1987, 146-151.

[Br] R. Brockett, "Asymptotic stability and feedback stabilization" in *Differential Geometric Control Theory*, Birkhauser, Boston, 1983.

[Ca] J. Carr, *Applications of Center Manifold Theory*, Springer Verlag, New York, 1981.

[Col] C. Coleman, "Asymptotic Stability in 3-space," in *Contributions to the Theory of Nonlinear Oscillations*, Vol. **V**, eds. Cesari, LaSalle, Lefschetz, *Annals of Mathematics Studies*, Vol. **45**, Princeton University Press, Princeton, NJ, 1960.

[Co] J.-M. Coron, "A necessary condition for feedback stabilization," *Syst. Contr. Lett.*, **14**(1990), 227-232.

[CP] J.-M. Coron, L. Praly, "Adding an integrator for stabilization problem," *Syst. Contr. Lett.*, **17**(1991), 89-105.

[DM1] W. P. Dayawansa, "Asymptotic stabilization of low dimensional systems," in *Nonlinear Synthesis*, Progress in Systems and Control Theory, Vol. **9**, Birkhauser, Boston, 1991.

[DM2] W. P. Dayawansa and C. F. Martin,"Some sufficient conditions for the asymptotic stabilizability of three dimensional homogeneous systems," *Proc. IEEE Conf. on Decision and Control*, Tampa, Dec. 1989, 1366-1370.

[DM3] W.P. Dayawansa and C.F. Martin, "Asymptotic Stabilization of Two Dimensional Real Analytic Systems," *Systems and Control Letters*, **12**, (1989), 205-211.

[DMK] W. P. Dayawansa, C. Martin and G. Knowles, "Asymptotic stabilization of a class of smooth two dimensional systems," *SIAM J. on Control and Optimization*, **28**(1990), 1321-1349..

[Ha] W. Hahn, *Stability of Motion*, Springer Verlag, New York, 1967.

[He1] H. Hermes, "Asymptotic stabilization of planar nonlinear systems," *Syst. Contr. Lett.*, **17**(1991), 437-445.

[He2] H. Hermes, "Homogeneous coordinates and continuous asymptotically stabilizing control laws," in *Differential Equations, Stability and Control,* (ed. S. Elaydi), Lecture Notes in Applied Math 109, 249-260 Marcel Dekker Inc., 1991.

[Ka1] M. Kawski, "Stabilization of nonlinear systems in the plane," *Syst. Contr. Lett.*, **12**(1989), 169-175.

[Ka2] M. Kawski, "Homogeneous feedback laws in dimension three," *Proc. IEEE Conf. on Decision and Control* , Tampa, Dec 1989, 1370-1376.

[KZ] M. A. Krosnosel'skii and P. P. Zabreiko, *Geometric Methods of Nonlinear Analysis*, Springer Verlag, NY, 1984.

[Ma] L. Markus, "Quadratic differential equations and nonassociative algebras," in *Contributions to the Theory of Nonlinear Oscillations*, Vol. **V**, eds. Cesari, LaSalle, Lefschetz, *Annals of Mathematics Studies*, Vol. **45**, Princeton University Press, Princeton, NJ, 1960.

[Sa] N. Samardzija, "Stability properties of autonomous homogeneous polynomial differential systems," *J. Differential Eq.*, **48**(1983), 60-70.

[Vi] J. C. Vivalda, "On the controllability of quadratic systems in $\Re^3$," *Syst. Contr. Lett.*, **17**(1991), 123-130.

Proceedings of the 31st Conference
on Decision and Control
Tucson, Arizona • December 1992

Tracking Control of Nonlinear Systems with Disturbance Attenuation

R. Marino*, W. Respondek**, A.J. van der Schaft# and P. Tomei*

* Dept. of Electrical Eng., University of Roma, 'Tor Vergata',
Via della Ricerca Scientifica, 00133 Roma, Italy.
** Institute of Mathematics, Polish Academy of Sciences,
Sniadeckich 8, 00-950 Warsaw, Poland.
Dept. Applied Mathematics, University of Twente,
P.O. Box 217, 7500 AE Enschede, The Netherlands.

Abstract

Sufficient geometric conditions are given which lead to the explicit construction of a state feedback tracking control for single–input single–output nonlinear systems with bounded unmodelled disturbances entering nonlinearly. For any initial condition the output asymptotically tracks a bounded reference signal with bounded time derivatives with an arbitrary attenuation of the influence of the disturbance. The sufficient conditions are weaker than those presented in [6] and the technique of proof is also different.

1 Introduction

This paper provides sufficient conditions for the explicit construction of a state feedback control capable of forcing the output y to track a bounded reference signal $y_d(t)$ with an arbitrary attenuation of a bounded unmodelled time varying disturbance $\theta(t)$ for nonlinear systems

$$\begin{aligned} \dot{x} &= f(x) + g(x)u + q(x, \theta(t)), \quad x \in R^n, u \in R, \\ y &= h(x), \quad y \in R. \end{aligned} \tag{1}$$

In (1) $f : R^n \to R^n$, $g : R^n \to R^n$, $q : R^n \times \omega \to R^n$, $h : R^n \to R$ are smooth functions, $g(x) \neq 0$, $\forall x \in R^n$, x is the state, u is the input, $\theta : R^+ \to \Omega \in R^p$ is the disturbance, y is the output which is required to track a reference signal $y_d(t)$. This problem is also called almost disturbance decoupling, following the terminology introduced in [11] for linear systems: it is posed when the well understood disturbance decoupling (or rejection) problem, that is the design of a state feedback control which makes the output insensitive to unmodelled disturbances, is not solvable.

Necessary and sufficient geometric conditions for the solvability of the disturbance decoupling problem are obtained in [4] and [5], generalizing the results established in [12] and [1] for linear systems. Equivalent conditions based on the notion of characteristic indexes are given in [2] for linear systems and in [8] for nonlinear ones. When the disturbance decoupling problem is not solvable, it is natural to look for conditions which guarantee the attenuation of the influence of the disturbance on the output with any desired degree of accuracy. This problem, called almost disturbance decoupling, was posed and solved in [11] for linear systems in terms of necessary and sufficient geometric conditions. In particular, the problem turns out to be always solvable for single–input, single–output linear systems of type

$$\begin{aligned} \dot{x} &= Fx + gu + Q\theta \\ y &= hx \end{aligned} \tag{2}$$

where F and Q are $n \times n$ and $n \times p$ constant matrices, g and h^T are $n \times 1$ constant vectors. As pointed out in [11] the almost disturbance decoupling problem is related to high–gain feedback design since, when it is not (exactly) solvable, the higher the gains are the higher the disturbance attenuation results. In fact in [7] a parameterized state feedback control is explicitly obtained when the almost disturbance decoupling is solvable; for square and minimum–phase systems a parameterized output–feedback control is given in [10].

At the moment it is not known whether, as in the linear case, the almost disturbance decoupling problem is always solvable for single–input, single–output nonlinear systems (1). Sufficient conditions are obtained in [6] using differential geometric tools and singular perturbation techniques. The example (given in [6])

$$\begin{aligned} \dot{x}_1 &= x_2 + \theta_1(t) \\ \dot{x}_2 &= x_2^3\theta_2(t) + u(t) \\ y &= x_1 \end{aligned} \tag{3}$$

fails to satisfy the sufficient conditions in [6] and shows that the almost disturbance decoupling problem cannot be solved on the basis of linear approximations. The non–local nature of the problem is also pointed out by the example (also given in [6])

$$\begin{aligned} \dot{x}_1 &= \arctan x_2 + \theta(t) \\ \dot{x}_2 &= u(t) \\ y &= x_1 \end{aligned} \tag{4}$$

where disturbances $|\theta(t)| > \pi/2$ cannot be attenuated.

In this paper we provide sufficient conditions for disturbance attenuation which generalize those given

CH3229-2/92/0000-2469$1.00 © 1992 IEEE

in [6]: for instance, the example (3) satisfies the conditions given here. The control algorithm we develop generalizes the one given in [6] and coincides with it in special simpler cases even though the techniques of proof are entirely different. We do not use singular perturbation techniques; we determine special coordinates in which the control algorithm and a Lyapunov function are recursively built.

2 Main Result

Definition 2.1 *The control characteristic index of system (1) is defined as the integer* ρ *such that*

$$\begin{aligned} L_g L_f^i h(x) &= 0, \qquad 0 \le i \le \rho - 2, \forall x \in R^n, \\ L_g L_f^{\rho-1} h(x) &\neq 0, \qquad \forall x \in R^n. \end{aligned}$$

If $L_g L_f^i h(x) = 0, \forall i, \forall x \in R^n$, *then* $\rho = \infty$.

□

Definition 2.2 *The disturbance characteristic index* ν *of system (1) is defined as the integer such that*

$$\begin{aligned} L_q L_f^i h(x) &= 0, \qquad 0 \le i \le \nu - 2, \forall x \in R^n, \\ L_q L_f^{\nu-1} h(x) &\neq 0, \qquad \text{for some } \theta \in \Omega, \text{some } x \in R^n. \end{aligned}$$

□

As shown in [5] and [8] the exact disturbance decoupling problem is locally solvable if and only if $\nu > \rho$.
Assumption 1 We assume in the following that ρ is well defined and that $\nu \le \rho < \infty$, that is that the exact disturbance decoupling problem is not solvable.

Definition 2.3 *The tracking problem with disturbance attenuation is said to be solvable for system (1), if for any smooth bounded reference trajectory* $y_d(t)$, *with bounded time derivatives* $y_d^{(1)}, \ldots, y_d^{(\rho)}$, *for any bounded disturbance* $\theta(t) \in \Omega \subset R^p$, *and for any initial condition* $x(0)$ *a parameterized state feedback control law* $u = u(x, k, t)$ *exists such that* $\|x(t)\|$ *and the output error* $e(t) = y(t) - y_d(t)$ *are bounded* $\forall t \ge 0$, *and for every* $\epsilon > 0$ *and* $T > 0$ *there exists* $k(\epsilon, T)$ *such that*

$$\|e(t)\| \le \epsilon, \quad \forall t \ge T(\epsilon), \forall k > k(\epsilon, T).$$

□

Under Assumption 1 we can locally define a change of coordinates

$$\begin{aligned} z_1 &= h(x) \\ &\vdots \\ z_\rho &= L_f^{\rho-1} h(x) \\ z_{\rho+1} &= \phi_{\rho+1}(x) \\ &\vdots \\ z_n &= \phi_n(x) \end{aligned} \tag{5}$$

with $\phi_i(x)$, $\rho + 1 \le i \le n$, such that

$$< d\phi_i, g >= 0.$$

In new coordinates we have

$$\begin{aligned} \dot{z}_1 &= z_2 + L_q h(x) \\ &\vdots \\ \dot{z}_{\rho-1} &= z_\rho + L_q L_f^{\rho-2} h(x) \\ \dot{z}_\rho &= L_f^\rho h(x) + L_q L_f^{\rho-1} h(x) + L_g L_f^{\rho-1} h(x) u \\ \dot{z}_{\rho+j} &= L_f \phi_{\rho+j}(z) + L_q \phi_{\rho+j}(z, \theta(t)) \\ &\triangleq \beta_j(z, \theta(t)), \quad 1 \le j \le n - \rho \\ y &= z_1 \end{aligned} \tag{6}$$

in which $L_g L_f^{\rho-1} h(x) \neq 0$, $\forall x \in R^n$, according to Assumption 1 and Definition 2.1. Denoting $z_r = (z_{\rho+1}, \ldots, z_n)$ and $\beta = (\beta_1, \ldots, \beta_{n-\rho})$ the dynamics

$$\dot{z}_r = \beta(z_r, z_1(t), \ldots, z_\rho(t), \theta_1(t), \ldots, \theta_p(t)) \tag{7}$$

is called the tracking dynamics where $z_1(t), \ldots, z_\rho(t), \theta_1(t), \ldots, \theta_p(t)$ are the inputs. When $z_1 = \cdots = z_\rho = 0$ and $\theta = 0$ the tracking dynamics is the zero dynamics.
Assumption 2 The tracking dynamics (7) is bounded input bounded state.

Theorem 2.1 *Assume in addition to Assumptions 1 and 2 that the following conditions are satisfied for system (1):*

(i) there exist $\rho - \nu + 1$ *smooth functions* $\alpha_i(x)$, $\nu - 1, \le i \le \rho - 1$, *satisfying* $d\alpha_i \in span\{dh, dL_f h, \ldots, dL_f^i h\}$, *such that* $\forall x \in R^n$, $\forall \theta \in \Omega$,

$$|L_q L_f^i h| \le \alpha_i, \qquad \nu - 1 \le i \le \rho - 1;$$

(ii) the vector fields

$$\tilde{f} = f - \frac{1}{L_g L_f^{\rho-1} h} L_f^\rho h, \qquad \tilde{g} = \frac{1}{L_g L_f^{\rho-1} h} g$$

are complete.

Then, the problem of tracking with disturbance attenuation is solvable.

Proof. We consider the general case in which $\nu = 1$. By virtue of Assumption 1 and the additional condition (ii) the change of coordinates (5) is globally defined (see [6] and [3]) and system (1) can be globally transformed into (6). We introduce a new control variable v, defined as

$$v = L_g L_f^{\rho-1} h(x) u + L_f^\rho h(x) - y_d^{(\rho)} \tag{8}$$

which substituted in (6) gives

$$\begin{aligned} \dot{e}_1 &= e_2 + L_q h(x) \\ &\vdots \\ \dot{e}_{\rho-1} &= e_\rho + L_q L_f^{\rho-2} h(x) \\ \dot{e}_\rho &= L_q L_f^{\rho-1} h(x) + v \\ \dot{z}_r &= \beta(z_r, z_1(t), \ldots, z_\rho(t), \theta_1(t), \ldots, \theta_p(t)) \end{aligned} \tag{9}$$

where $e_i = z_i - y_d^{(i-1)}$, $1 \le i \le \rho$. Note that if $\nu > \rho$ the control (8) solves the exact disturbance decoupling problem. Define

$$e_2^* = -ke_1 - e_1\mu_1(t) \tag{10}$$

where μ_1 is a smooth function yet to be defined and $k > 0$. Consider the function

$$V_1 = \frac{1}{2}e_1^2 \tag{11}$$

The time derivative of V_1, with $e_2 = e_2^*$ in (9), is given by

$$\dot{V}_1 = -ke_1^2 - e_1^2\mu_1 + e_1 L_q h(x) \tag{12}$$

and, according to the inequality in (i), we have

$$\dot{V}_1 \le -ke_1^2 - e_1^2\mu_1 + |e_1|\alpha_0(z_1) \tag{13}$$

Since α_0 is a smooth function and y_d is bounded, we can write

$$\alpha_0(z_1) = \alpha_0(e_1 + y_d) = \alpha_0(y_d) + \bar{\alpha}_0(e_1, y_d)e_1 \tag{14}$$

in which

$$\bar{\alpha}_0(e_1, y_d) = \frac{\alpha_0(z_1) - \alpha_0(y_d)}{e_1}. \tag{15}$$

Hence, we choose μ_1 as a smooth function satisfying

$$\mu_1 \ge \bar{\alpha}_0(e_1, y_d). \tag{16}$$

From (13), we obtain

$$\dot{V}_1 \le -ke_1^2 + |e_1||\alpha_0(y_d)| \tag{17}$$

Therefore, if $\rho = 1$ the thesis is proved with

$$v = e_2^*. \tag{18}$$

In fact, we can write

$$\frac{\dot{V}_1}{V_1} \le -2k + \frac{|\alpha_0(y_d)|}{|e_1|}.$$

Recalling that y_d is bounded, $|\alpha_0(y_d)| < \gamma$, $\gamma > 0$; for every $|e_1| \ge \epsilon/2$, we obtain

$$\frac{\dot{V}_1}{V_1} \le -2k + 2\gamma/\epsilon,$$

which implies

$$V_1(t) \le V_1(0)e^{(-2k+2\gamma/\epsilon)t}.$$

For any $\epsilon > 0$, $T > 0$ there exists k which solves the problem.
If $\rho > 1$, we prove the following Claim.
Claim. Assume that for a given index i, $1 \le i \le \rho$, for the system

$$\begin{aligned} \dot{e}_1 &= e_2 + L_q h(x) \\ &\vdots \\ \dot{e}_i &= e_{i+1} + L_q L_f^{i-1} h(x) \end{aligned} \tag{19}$$

there exist i functions

$$\begin{aligned} &e_2^*(e_1, y_d), e_3^*(e_1, e_2, y_d, y_d^{(1)}), \ldots, \\ &e_{i+1}^*(e_1, \ldots, e_i, y_d, \ldots, y_d^{(i-1)}), \\ &e_j^*(0, \ldots, 0, y_d, \ldots, y_d^{(j-2)}) = 0, \quad 2 \le j \le i+1, \end{aligned} \tag{20}$$

such that in new coordinates ($M_j > 0$, $1 \le j \le i$)

$$\begin{aligned} \tilde{e}_1 &= e_1 \\ \tilde{e}_2 &= \frac{e_2 - e_2^*}{M_2} \\ &\vdots \\ \tilde{e}_i &= \frac{e_i - e_i^*}{M_i} \end{aligned} \tag{21}$$

the function

$$V_i = \frac{1}{2}\sum_{j=1}^{i} \tilde{e}_j^2 \tag{22}$$

has time derivative, with $e_{i+1} = e_{i+1}^*$ in (19), satisfying the inequality

$$\dot{V}_i \le -k \left\| \begin{bmatrix} \tilde{e}_1 \\ \vdots \\ \tilde{e}_i \end{bmatrix} \right\|^2 + \left\| \begin{bmatrix} \tilde{e}_1 \\ \vdots \\ \tilde{e}_i \end{bmatrix} \right\| \eta_i(y_d, \ldots, y_d^{(i-1)}, k) \tag{23}$$

with η_i a suitable smooth function such that

$$\lim_{k \to \infty} \frac{\eta_i(y_d, \ldots, y_d^{(i-1)}, k)}{k} = 0. \tag{24}$$

Then, for the system

$$\begin{aligned} \dot{e}_1 &= e_2 + L_q h(x) \\ &\vdots \\ \dot{e}_{i+1} &= e_{i+2} + L_q L_f^i h(x) \end{aligned} \tag{25}$$

there exists a function

$$\begin{aligned} &e_{i+2}^*(e_1, \ldots, e_{i+1}, y_d, \ldots, y_d^{(i)}), \\ &e_{i+2}^*(0, \ldots, 0, y_d, \ldots, y_d^{(i)}) = 0 \end{aligned} \tag{26}$$

such that in new coordinates ($M_{i+1} > 0$)

$$\tilde{e}_j, \quad 1 \le j \le i, \quad \tilde{e}_{i+1} = \frac{e_{i+1} - e_{i+1}^*}{M_{i+1}} \tag{27}$$

the function

$$V_{i+1} = \frac{1}{2}\sum_{j=1}^{i+1} \tilde{e}_j^2 \tag{28}$$

has time derivative, with $e_{i+2} = e_{i+2}^*$ in (25), satisfying the inequality

$$\begin{aligned} \dot{V}_{i+1} \le & -\frac{k}{2} \left\| \begin{bmatrix} \tilde{e}_1 \\ \vdots \\ \tilde{e}_{i+1} \end{bmatrix} \right\|^2 \\ & + \left\| \begin{bmatrix} \tilde{e}_1 \\ \vdots \\ \tilde{e}_{i+1} \end{bmatrix} \right\| \eta_{i+1}(y_d, \ldots, y_d^{(i)}, k) \end{aligned} \tag{29}$$

where η_{i+1} is a suitable smooth function such that

$$\lim_{k\to\infty} \frac{\eta_{i+1}(y_d,\ldots,y_d^{(i)},k)}{k} = 0. \tag{30}$$

Proof of the Claim. For convenience we adopt the following notations:

$$\begin{aligned} z_{(i)} &= [z_1,\ldots,z_i]^T, & z_{(i)}^d = [y_d,\ldots,y_d^{(i-1)}]^T \\ e_{(i)} &= [e_1,\ldots,e_i]^T, & \tilde{e}_{(i)} = [\tilde{e}_1,\ldots,\tilde{e}_i]^T \end{aligned}$$

From (21), (22) and (25) we obtain

$$\begin{aligned} \dot{V}_{i+1} &= -k\|\tilde{e}_{(i)}\|^2 + \|\tilde{e}_{(i)}\|\eta_i(z_{(i)}^d,k) + M_{i+1}\tilde{e}_i\tilde{e}_{i+1} \\ &\quad +\tilde{e}_{i+1}\dot{\tilde{e}}_{i+1} \end{aligned} \tag{31}$$

Since by (21) and (25),

$$\begin{aligned} \dot{\tilde{e}}_{i+1} &= \frac{1}{M_{i+1}}\frac{d}{dt}(e_{i+1}-e_{i+1}^*) \\ &= \frac{1}{M_{i+1}}\Bigg(e_{i+2}+L_qL_f^ih \\ &\quad -\sum_{j=1}^{i}\frac{\partial e_{i+1}^*}{\partial e_j}(e_{j+1}+L_qL_f^{j-1}h) \\ &\quad -\sum_{j=1}^{i}\frac{\partial e_{i+1}^*}{\partial y_d^{(j-1)}}y_d^{(j)}\Bigg) \end{aligned} \tag{32}$$

we define

$$\begin{aligned} e_{i+2}^* &= -M_{i+1}(k\tilde{e}_{i+1}+\tilde{e}_{i+1}\mu_{i+1}(t)+M_{i+1}\tilde{e}_i) \\ &\quad +\textstyle\sum_{j=1}^{i}\left(\frac{\partial e_{i+1}^*}{\partial e_j}e_{j+1}+\frac{\partial e_{i+1}^*}{\partial y_d^{(j-1)}}y_d^{(j)}\right) \end{aligned} \tag{33}$$

so that (32) with $e_{i+2}=e_{i+2}^*$ becomes

$$\begin{aligned} \dot{\tilde{e}}_{i+1} &= -k\tilde{e}_{i+1}+\mu_{i+1}\tilde{e}_{i+1}-\tilde{e}_iM_{i+1} \\ &\quad +\frac{1}{M_{i+1}}\left(L_qL_f^ih-\textstyle\sum_{j=1}^{i}\frac{\partial e_{i+1}^*}{\partial e_j}L_qL_f^{j-1}h\right) \end{aligned} \tag{34}$$

Substituting (34) into (31), we have (with $e_{i+2}=e_{i+2}^*$)

$$\begin{aligned} \dot{V}_{i+1} &= -k\|\tilde{e}_{(i)}\|^2+\|\tilde{e}_{(i)}\|\eta_i-k\tilde{e}_{i+1}^2-\mu_{i+1}\tilde{e}_{i+1}^2 \\ &\quad +\frac{\tilde{e}_{i+1}}{M_{i+1}}\left(L_qL_f^ih-\textstyle\sum_{j=1}^{i}\frac{\partial e_{i+1}^*}{\partial e_j}L_qL_f^{j-1}h\right) \end{aligned} \tag{35}$$

By assumption (ii), we have

$$|L_qL_f^jh| \le \alpha_j(z_{(j+1)}), \qquad 0\le j\le i, \tag{36}$$

Therefore, we can write

$$\begin{aligned} L_qL_f^ih &- \sum_{j=1}^{i}\frac{\partial e_{i+1}^*}{\partial e_j}L_qL_f^{j-1}h \le \alpha_i(z_{(i+1)}) \\ &+\sum_{j=1}^{i}\left|\frac{\partial e_{i+1}^*}{\partial e_j}\alpha_{j-1}(z_{(j)})\right| \\ &= \alpha_i(z_{(i+1)})-\alpha_i(z_{(i+1)}^d)+\alpha_i(z_{(i+1)}^d) \\ &+\sum_{j=1}^{i}\left|\frac{\partial e_{i+1}^*}{\partial e_j}\alpha_{j-1}(z_{(j)})-\left.\frac{\partial e_{i+1}^*}{\partial e_j}\right|_{e_{(i)}=0}\alpha_{j-1}(z_{(j)}^d)\right. \\ &+\left.\left.\frac{\partial e_{i+1}^*}{\partial e_j}\right|_{e_{(i)}=0}\alpha_{j-1}(z_{(j)}^d)\right| \end{aligned} \tag{37}$$

Since α_j, $0\le j\le i$, and e_{i+1}^* are smooth functions and $y_d^{(j)}(t)$, $0\le j\le i$, are bounded, there exist functions $\alpha_{i,j}(e_{(i+1)},z_{(i+1)}^d)$ (see [9], p. 39) such that

$$\alpha_i(z_{(i+1)})-\alpha_i(z_{(i+1)}^d)=\sum_{j=1}^{i+1}\alpha_{i,j}(e_{(i+1)},z_{(i+1)}^d)e_j \tag{38}$$

and functions $\beta_{j,l}(e_{(i)},z_{(i)}^d)$ such that

$$\begin{aligned} &\frac{\partial e_{i+1}^*}{\partial e_j}\alpha_{j-1}(z_{(j)})-\left.\frac{\partial e_{i+1}^*}{\partial e_j}\right|_{e_{(i)}=0}\alpha_{j-1}(z_{(j)}^d) \\ &= \sum_{l=1}^{i}\beta_{j,l}(e_{(i)},z_{(i)}^d)e_l, \quad 1\le j\le i. \end{aligned} \tag{39}$$

From (37)–(39), we obtain

$$\begin{aligned} L_qL_f^ih &- \sum_{j=1}^{i}\frac{\partial e_{i+1}^*}{\partial e_j}L_qL_f^{j-1}h \\ &\le \alpha_i+\sum_{j=1}^{i}\left|\left.\frac{\partial e_{i+1}^*}{\partial e_j}\right|_{e_{(i)}=0}\alpha_{j-1}\right| \\ &+\sum_{j=1}^{i+1}\alpha_{i,j}e_j+\sum_{j=1}^{i}\left|\sum_{l=1}^{i}\beta_{j,l}e_l\right| \end{aligned} \tag{40}$$

Since e_i are related to $\tilde{e}_i$ by the change of coordinates (21), (27) which preserves the origin, we can write for $1\le j\le i$

$$\begin{aligned} \sum_{j=1}^{i+1}\alpha_{i,j}(e_{(i+1)},z_{(i+1)}^d)e_j &= \sum_{j=1}^{i+1}\bar{\alpha}_{i,j}(e_{(i+1)},z_{(i+1)}^d)\tilde{e}_j \\ \sum_{l=1}^{i}\beta_{j,l}(e_{(i)},z_{(i)}^d)e_l &= \sum_{l=1}^{i}\bar{\beta}_{j,l}(e_{(i)},z_{(i)}^d)\tilde{e}_l. \end{aligned} \tag{41}$$

Therefore, defining

$$\begin{aligned} \Gamma_0(z_{(i+1)}^d) &= \alpha_i(z_{(i+1)}^d) \\ &\quad +\sum_{j=1}^{i}\left|\left.\frac{\partial e_{i+1}^*}{\partial e_j}\right|_{e_{(i)}=0}\alpha_{j-1}(z_{(j)}^d)\right| \\ \Gamma_1(e_{(i+1)},z_{(i+1)}^d) &= \frac{1}{M_{i+1}}\Bigg(\sum_{j=1}^{i+1}|\bar{\alpha}_{i,j}(e_{(i+1)},z_{(i+1)}^d)| \\ &\quad +\sum_{j=1}^{i}\sum_{l=1}^{i}|\bar{\beta}_{j,l}(e_{(i)},z_{(i)}^d)|\Bigg) \end{aligned} \tag{42}$$

from (35), we have

$$\dot{V}_{i+1} \le -k\|\tilde{e}_{(i)}\|^2 + \|\tilde{e}_{(i)}\|\eta_i - k\tilde{e}_{i+1}^2 - \mu_{i+1}\tilde{e}_{i+1}^2 + \frac{\Gamma_0}{M_{i+1}}|\tilde{e}_{i+1}| + |\tilde{e}_{i+1}|\Gamma_1\|\tilde{e}_{(i+1)}\| \tag{43}$$

Let M_{i+1} be defined as a constant such that

$$\left| \left. \frac{\partial e_{i+1}^*}{\partial e_j} \right|_{e_{(i)}=0} \right| \le M_{i+1}, \quad 1 \le j \le i, \tag{44}$$

which exists since y_d and its *derivatives* are bounded, so that

$$\lim_{k\to\infty} \frac{\Gamma_0(z_{(i+1)}^d)}{kM_{i+1}} = 0. \tag{45}$$

From (43), we obtain

$$\dot{V}_{i+1} \le -\frac{k}{2}\|\tilde{e}_{(i+1)}\|^2 + \|\tilde{e}_{(i)}\|\eta_i + \frac{\Gamma_0}{M_{i+1}}|\tilde{e}_{i+1}| - \begin{bmatrix} \tilde{e}_{(i+1)} \\ \tilde{e}_{i+1} \end{bmatrix}^T \begin{bmatrix} k/2 & -\Gamma_1/2 \\ -\Gamma_1/2 & \mu_{i+1} \end{bmatrix} \begin{bmatrix} \tilde{e}_{(i+1)} \\ \tilde{e}_{i+1} \end{bmatrix}$$

Therefore, choosing

$$\mu_{i+1} \ge \frac{\Gamma_1^2(e_{(i+1)}, z_{(i+1)}^d)}{2k} \tag{46}$$

the thesis is proved with

$$\eta_{i+1}(z_{(i+1)}^d, k) = \frac{\Gamma_0(z_{(i+1)}^d)}{M_{i+1}} + \eta_i(z_{(i)}^d, k). \tag{47}$$

□

Since we have shown that the hypotheses of the claim are true for $i = 1$, applying $(\rho - 1)$-times the claim we can construct a function

$$e_{\rho+1}^* = e_{\rho+1}^*(e_1, \ldots, e_\rho, y_d, \ldots, y_d^{(\rho-1)}) \tag{48}$$

which determines the final control v as

$$v = e_{\rho+1}^* \tag{49}$$

We also construct a change of coordinates

$$\tilde{e}_1 = e_1, \tilde{e}_2 = \frac{e_2 - e_2^*}{M_2}, \ldots, \tilde{e}_\rho = \frac{e_\rho - e_\rho^*}{M_\rho} \tag{50}$$

such that the function

$$V_\rho = \frac{1}{2}\sum_{j=1}^{\rho} \tilde{e}_j^2 \tag{51}$$

has time derivative satisfying the inequality (with k suitably redefined)

$$\dot{V}_\rho \le -k \left\| \begin{bmatrix} \tilde{e}_1 \\ \vdots \\ \tilde{e}_\rho \end{bmatrix} \right\|^2 + \left\| \begin{bmatrix} \tilde{e}_1 \\ \vdots \\ \tilde{e}_\rho \end{bmatrix} \right\| \eta_\rho(y_d, \ldots, y_d^{(\rho-1)}, k) \tag{52}$$

with

$$\lim_{k\to\infty} \frac{\eta_\rho(y_d, \ldots, y_d^{(\rho-1)}, k)}{k} = 0. \tag{53}$$

Therefore, $\tilde{e}_i$, $1 \le i \le \rho$, are bounded and consequently, since the reference signal $y_d(t)$ is bounded with its $\rho - 1$ time derivatives, $z_i(t)$, $1 \le i \le \rho$, are bounded. Since Assumption 2 holds $\|z_r(t)\|$ is also bounded. By virtue of (53), we have

$$|\eta_\rho| \le \gamma > 0,$$

so that for every $|e_1| \ge \epsilon/2$

$$\frac{\dot{V}_\rho}{V_\rho} \le -2k + 2\gamma/\epsilon,$$

which implies

$$e_1^2(t) \le V_\rho(t) \le V_\rho(0)e^{(-2k+2\gamma/\epsilon)t}.$$

For any $\epsilon > 0$, $T > 0$ there exists k which solves the problem. Therefore, the problem is solved by the feedback controller given by (8) and (49) with a suitable choice of k. □

Remark. Theorem 2.1 generalizes the main result in [6] in several ways. The disturbances are only allowed to enter linearly in [6] and are required to have bounded time derivatives while in this paper they may enter nonlinearly and no requirement is made on their time derivatives. While condition (ii) is common to both theorems the most important difference lies in condition (i) which considerably weakens the corresponding condition in two ways. The result in [6] requires for $\nu - 1 \le i \le \rho - 1$,

$$d(L_q L_f^i h) \in \operatorname{span}\{dh, \ldots, d(L_f^{\nu-1}h)\}, \tag{54}$$

while condition (i) in Theorem 2.1 only requires

$$d(L_q L_f^i h) \in \operatorname{span}\{dh, \ldots, d(L_f^i h)\}, \quad \nu-1 \le i \le \rho-1, \tag{55}$$

or even the weaker condition on some bounding functions

$$|L_q L_f^i h| \le \alpha_i, \quad \nu - 1, \le i \le \rho - 1,$$

with

$$d\alpha_i \in \operatorname{span}\{dh, \ldots, d(L_f^i h)\}.$$

For instance condition (55) applies to system (3) while the stronger condition (54) does not.

3 Example

Consider the system

$$\begin{aligned} \dot{x}_1 &= x_2 + \theta_1(t) \\ \dot{x}_2 &= x_2^3\theta_2(t) + u \\ y &= x_1 \end{aligned} \tag{56}$$

with $|\theta_1(t)| \le 1$ and $|\theta_2(t)| \le 1$, where $\theta(t) = [\theta_1(t), \theta_2(t)]^T$ is a disturbance signal. It is easily seen that system (31) satisfies the conditions of Theorem 2.1 with

$$\begin{aligned} \alpha_0 &= 1 \geq L_q h = \theta_1(t) \\ \alpha_1 &= x_2^3 \geq L_q L_f h = x_2^3 \theta_2(t). \end{aligned} \qquad (57)$$

Suppose that $y_d(t)$ is the desired output reference to be tracked. Define

$$e_1 = x_1 - y_d, \qquad e_2 = x_2 - \dot{y}_d. \qquad (58)$$

From (31) and (32), we have

$$\begin{aligned} \dot{e}_1 &= e_2 + \theta_1(t) \\ \dot{e}_2 &= x_2^3\theta_2(t) + u - \ddot{y}_d \triangleq x_2^3\theta_2(t) + v. \end{aligned}$$

Define as in (10)

$$e_2^* = -ke_1 - e_1\mu_1(t)$$

Since by (16)

$$\mu_1 \geq \bar{\alpha}_0(e_1, y_d) = 0$$

we choose $\mu_1 = 0$ so that

$$e_2^* = -ke_1.$$

Define as in (21) and (33)

$$\begin{aligned} \tilde{e}_2 &= \frac{e_2 - e_2^*}{M_2} \\ v &= -M_2(k\tilde{e}_2 + \tilde{e}_2\mu_2(t) + M_2 e_1) - ke_2 \end{aligned}$$

According to (44), we have

$$M_2 = k$$

and according to (42)

$$\begin{aligned} \Gamma_0 &= \dot{y}_d^3 + k \\ \Gamma_1 &= \frac{1}{k} 2k \left| \frac{x_2^3 - \dot{y}_d^3}{e_2} \right| = 2|e_2^2 + 3e_2\dot{y}_d + 3\dot{y}_d^2| \end{aligned}$$

which imply

$$\mu_2 = \frac{2}{k}(e_2^2 + 3e_2\dot{y}_d + 3\dot{y}_d^2)^2.$$

The final control u is given by

$$u = \ddot{y}_d - 2k^2(\tilde{e}_2 + e_1) - 2\tilde{e}_2(e_2^2 + 3e_2\dot{y}_d + 3\dot{y}_d^2)^2. \quad (59)$$

If we considered as in [6] the problem of stabilizing the linear approximation of system (56) we would obtain, using the same design technique, the following control law

$$u = -k(k\tilde{x}_2 + kx_1) - kx_2 = -2k(x_2 + kx_1)$$

which is very similar to the one obtained in [6]

$$u = -\frac{1}{\epsilon^2}(\epsilon x_2 + x_1) \qquad (60)$$

once we define $\epsilon = \dfrac{1}{k}$. In [6] it was shown that the control algorithm (60) does not guarantee almost disturbance decoupling for any bounded disturbance. Therefore, the nonlinear part of the control law (59) is crucial in order to obtain disturbance attenuation.

Acknowledgement

This work was supported in part by Ministero dell'Università e della Ricerca Scientifica e Tecnologica.

References

[1] G. Basile and G. Marro. Controlled and conditioned invariant subspaces in linear systems theory. *J. of Opt. Theory and Applications*, 3:306–315, 1969.

[2] S.P. Bhattacharyya. Disturbance rejection in linear systems. *Int. J. of Syst. Science*, 5:633–637, 1974.

[3] C.I. Byrnes and A. Isidori. Global feedback stabilization of nonlinear systems. In *IEEE 24th Conf. on Decision and Control*, pages 1031–1037, Ft. Lauderdale, 1985.

[4] R.M. Hirschorn. (A,B)–invariant distributions and disturbance decoupling of nonlinear systems. *SIAM J. Control Optimiz.*, 19:1–19, 1981.

[5] A. Isidori, A.J. Krener, C. Gori–Giorgi, and S. Monaco. Nonlinear decoupling via feedback: a differential geometric approach. *IEEE Trans. Automatic Control*, 26:331–345, 1981.

[6] R. Marino, W. Respondek, and A.J. van der Schaft. Almost disturbance decoupling for single–input single–output nonlinear systems. *IEEE Trans. Automatic Control*, 34:1013–1017, 1989.

[7] R. Marino, W. Respondek, and A.J. van der Schaft. A direct approach to almost disturbance and almost input-output decoupling. *Int. J. of Control*, 48:353–383, 1988.

[8] H. Nijmeijer and K. Tchon. *An Input-Output Characterization of Nonlinear Disturbance Decoupling.* Technical Report 502, Dept. Math. Twente Univ. Technol., Feb. 1985.

[9] H. Nijmeijer and A. van der Schaft. *Nonlinear Dynamical Control Systems.* Springer-Verlag, Berlin, 1990.

[10] A. Saberi. Output feedback control with almost disturbance decoupling property: a singular perturbation approach. *Int. J. of Control*, 45:1705–1722, 1987.

[11] J.C. Willems. Almost invariant subspaces: an approach to high-gain feedback design-Part I: Almost controlled invariant subspaces. *IEEE Trans. Automatic Control*, 26:235–252, 1982.

[12] W.M. Wonham and A.S. Morse. Decoupling and pole assignment in linear multivariable systems. *SIAM J. Control Optimiz.*, 8:1–18, 1970.

Proceedings of the 31st Conference
on Decision and Control
Tucson, Arizona • December 1992

Regularity and Minimum-Phase Properties in Nonlinear Asymptotic Tracking Loops

J. W. Grizzle[1], M. D. Di Benedetto[2], and F. Lamnabhi-Lagarrigue[3]

Abstract

If an analytic nonlinear plant is (1) dynamically input-output decouplable in a neighborhood of the origin, and is (2) minimum-phase, it is possible to achieve asymptotic tracking for an open set of output trajectories containing the origin in $C_m^N[0,\infty)$, the space of N-times continuously differentiable functions taking values in $\mathbb{R}^m$. When either of these sufficient conditions is not met, various authors have investigated approximate analytic solutions, discontinuous solutions and solutions for restricted sets of trajectories. We establish that, for locally controllable MIMO systems, under certain technical assumptions related to system inverses and zero-dynamics, conditions (1) and (2) are necessary for the existence of an analytic compensator which yields asymptotic tracking for an open set of output trajectories. Necessary conditions for input-output decouplability in a neighborhood of a given point are also discussed.

1. Introduction

Under the hypotheses that a given system possesses a vector relative degree (or that this can be achieved by dynamic compensation) and an asymptotically stable zero-dynamics, it is known how to construct internally stabilizing compensators such that the output of the system asymptotically approaches a desired trajectory, $y_d(t)$, for all $y_d(t)$ in the open neighborhood of the origin of $C_m^N[0,\infty)$ defined by $\sup_{t\geq 0}\{\|y_d(t)\|,\ldots,\|y_d^{(N)}(t)\|\} < \epsilon$, for $N \geq 0$ and $\epsilon > 0$ appropriately chosen. A possible approach consists of first input-output linearizing the system, rendering unobservable the system's zero-dynamics, and then applying standard linear theory to the resulting system [9]. It follows that one may not be able to find a solution to the tracking problem for two reasons: (1) the regularity conditions needed to construct an input-output linearizing compensator may not be met; (2) the system's zero-dynamics is not asymptotically stable.

If the trajectories to be followed are generated by a finite-dimensional exosystem [11, 13], or if discontinuous feedback is allowed [5], it is known that neither of these conditions, regularity nor asymptotically stable zero-dynamics, is necessary. In a similar manner, when one wishes to track a single output trajectory, singularities may not be an obstruction (see e.g., [10]). On the other hand, whenever the class of desired trajectories $y_d(t)$ is given by $\sup_{t\geq 0}\{\|y_d(t)\|,\ldots,\|y_d^{(N)}(t)\|\} < \epsilon$, a smooth compensator is sought, and the system falls into the situations (1) or (2) above, only approximate solutions to tracking are known; indeed, all current approaches are based upon approximating the system by one for which a vector relative degree can be achieved and which possesses an asymptotically stable zero-dynamics [6, 8, 18].

The goal of this paper is to establish to what extent such conditions are necessary for the existence of an analytic compensator yielding asymptotic tracking for an open set of output trajectories. In Section 2, the asymptotic and exact tracking problems are formulated and it is shown that any solution to the asymptotic tracking problem also provides a solution to the exact tracking problem from the origin. In Section 3, for the case of single-input single-output (SISO) systems, this relationship is exploited to

---

[1] Department of Electrical Engineering and Computer Science, The University of Michigan, Ann Arbor, MI 48109-2122, U.S.A.
Work supported in part by NSF Contract # ECS-88-96136, in part by the CNRS "poste-rouge" and in part by the University of Rome "La Sapienza".
[2] Dipartimento di Informatica e Sistemistica, Università di Roma "La Sapienza", Via Eudossiana, 18, 00184 Rome, Italy
Work supported by CNR.
[3] Laboratoire des Signaux et Systèmes, C.N.R.S., SUPELEC, 91192 Gif-sur-Yvette, Cedex, France
Work supported in part by European Contract # SC10433C(A)

CH3229-2/92/0000-2475$1.00 © 1992 IEEE

show that, whenever the plant is strongly accessible from the equilibrium, a well-defined relative degree is a necessary condition for achieving asymptotic tracking. For multi-input multi-output (MIMO) systems, under the stronger hypothesis that the linearization of the plant is controllable, it is necessary that a vector relative degree at zero can be achieved by dynamic compensation. For locally controllable MIMO systems, it is established that under certain "regularity conditions" associated with system inverses and zero-dynamics, the minimum-phase property is also a necessary condition for asymptotic tracking with internal stability. These results highlight the interest of approximate solutions to the asymptotic tracking problem when a plant has singularities in its input-output map or has a non-asymptotically stable zero-dynamics. In Section 4, it is emphasized that although necessary and sufficient conditions are known for when a vector relative degree can be achieved at a generic point of the state space, only sufficient conditions are known for achieving this at a given point. It is also shown that, in the particular case of a system having a generically invertible decoupling matrix, a vector relative degree can be achieved at a given point only if the decoupling matrix evaluated at that point is nonsingular (see also [20]).

2. A Relation Between Asymptotic and Exact Tracking

In this Section, the local asymptotic and exact tracking problems are formulated. It is proven that any compensator solving the former also provides a solution to the latter.

Consider the analytic plant P described by

$$P: \begin{array}{rcl} \dot{x} & = & f(x) + g(x)u = f(x) + \sum_{i=1}^{m} g_i(x)u_i \\ y & = & h(x) \, , \end{array} \tag{2.1}$$

where $x \in \mathbb{R}^n$, $u, y \in \mathbb{R}^m$, $f(0) = 0$ and the entries of f, the columns of g and the rows of h consist of analytic functions. An affine analytic dynamic compensator for P is a second system

$$C: \begin{array}{rcl} \dot{\xi} & = & c(x,\xi) + d(x,\xi)v \\ u & = & \gamma(x,\xi) + \delta(x,\xi)v \, , \end{array} \tag{2.2}$$

where $\xi \in \mathbb{R}^\nu$, $v \in \mathbb{R}^m$, $c(0,0) = 0$, $\gamma(0,0) = 0$, the entries of c, d, γ and δ consist of analytic functions. The closed-loop system (2.1)-(2.2) will often be denoted by $P \circ C$; it has input v and output y, sometimes denoted $y^{P \circ C}$. Let the class of desired output trajectories $y_d(t)$ be defined by $y_d \in C_m^N[0,\infty)$ and

$$\sup_{t \geq 0} \left\{ \|y_d(t)\|, \ldots, \|y_d^{(N)}(t)\| \right\} < \epsilon \tag{2.3}$$

for appropriate choices of $N \geq 0$ and $\epsilon > 0$. Let $Y_d(t) = [y_d'(t), \dot{y}_d'(t), \ldots, y_d^{(N)'}(t)]'$, where $'$ denotes transpose.

Definition 2.1 *a) The* ***local asymptotic tracking problem*** *for the plant P is to find an integer $N \geq 0$ and an $\epsilon > 0$ defining a class of output trajectories (2.3), an integer $q \geq 0$, an open neighborhood $\mathcal{O}_1$ of the origin in $\mathbb{R}^n \times \mathbb{R}^q \times \mathbb{R}^{(N+1)m}$ and two analytic functions $a : \mathcal{O}_1 \to \mathbb{R}^q$ and $\alpha : \mathcal{O}_1 \to \mathbb{R}^m$, with $a(0,0,0) = 0$, $\alpha(0,0,0) = 0$, defining the compensator Q*

$$Q: \begin{array}{rcl} \dot{z} & = & a(x,z,Y_d) \\ u & = & \alpha(x,z,Y_d) \, , \end{array} \tag{2.4}$$

such that the closed-loop $P \circ Q$ satisfies:

1) $\lim_{t\to\infty}(y^{P\circ Q}(t) - y_d(t)) = 0$, for all $(x_0, z_0) \in \mathcal{O}_2$, an open neighborhood of the origin in $\mathbb{R}^n \times \mathbb{R}^q$, and for all $y_d(t)$ satisfying (2.3);

2) the equilibrium (0,0) of the unforced system

$$\begin{array}{rcl} \dot{x} & = & f(x) + g(x)\alpha(x,z,0) \\ \dot{z} & = & a(x,z,0) \end{array} \tag{2.5}$$

is asymptotically stable.

b) The ***local exact tracking problem*** *for the plant P is to find an integer $N \geq 0$, $\epsilon > 0$ and $T > 0$ defining a class of output trajectories (2.3), an integer $q \geq 0$, an open neighborhood $\mathcal{O}_1$, of the origin in $\mathbb{R}^n \times \mathbb{R}^q \times \mathbb{R}^{(N+1)m}$, two analytic functions $a : \mathcal{O}_1 \to \mathbb{R}^q$, $\alpha : \mathcal{O}_1 \to \mathbb{R}^m$, with $a(0,0,0) = 0$, $\alpha(0,0,0) = 0$, defining the compensator (2.4), an open neighborhood $\mathcal{O}_3$ of the origin in $\mathbb{R}^{(N+1)m}$ and a map $\Phi : \mathcal{O}_3 \to \mathbb{R}^n \times \mathbb{R}^q$, with $\Phi(0) = 0$, providing the system's initialization, such that the closed-loop $P \circ Q$ satisfies:*

3) for each $y_d(t)$ in the class (2.3), whenever $P \circ Q$ is initialized at $(x_0, z_0) = \Phi(Y_d(0))$, then $y^{P\circ Q}(t) - y_d(t) = 0$ for $0 \leq t \leq T$.

c) If T can be taken equal to ∞ in (b), we refer to the ***long-term****, local exact tracking problem.*

d) The ***local exact tracking problem from the origin*** *is defined by (b) above with the following modifications:*

4) for each $y_d(t)$ in the class (2.3) satisfying $Y_d(0) = 0$, whenever $P \circ Q$

is initialized at $(x_0, z_0) = (0,0)$, *then* $y^{P\circ Q}(t) - y_d(t) = 0$ *for* $0 \le t \le T$. □

The following relation can be established between the asymptotic and exact tracking problems.

Theorem 2.2 *Consider the plant given by (2.1). Suppose that a compensator* Q *solves the local asymptotic tracking problem. Then, the same compensator* Q *solves the long-term local exact tracking problem from the origin. Moreover, the integer* N *describing the class of output trajectories can be taken to be the same, while the parameter* ϵ *is possibly reduced.*

Proof: See [7].

3. Regularity and minimum-phase properties of tracking Loops

In this Section, it is established that if a plant P has a controllable Jacobian linearization and the local exact tracking problem from the origin is solvable, then P can be dynamically input-output decoupled in a neighborhood of the origin. For the special case of SISO plants, the hypothesis on the controllability of the Jacobian linearization can be weakened to the strong accessibility of P from zero and it will be concluded that P necessarily has a well-defined relative degree at zero. It is also shown that, if P satisfies certain "regularity conditions" associated with the existence of a left-inverse at zero and has a well-defined zero-dynamics [12], the minimum-phase property is a necessary condition for achieving asymptotic tracking.

As far as the regularity properties are concerned, the following two theorems precisely state the results.

Theorem 3.1 *Consider a SISO plant* P *of the form (2.1) and suppose* P *is strongly accessible from zero. Then, if the local exact tracking problem from the origin is solvable,* P *has a well-defined relative degree at zero. In particular, this is the case if the local asymptotic tracking problem is solvable.*

Proof: See [7].

In the case of MIMO systems, one has to assume the stronger hypothesis that the plant has a controllable Jacobian linearization about the origin. This leads to the second important result of the Section.

Theorem 3.2 *Consider a MIMO plant* P *of the form (2.1) and suppose that* P *has a controllable Jacobian linearization about the origin. Then, if the local exact tracking problem from the origin is solvable,* P *is dynamically input-output decouplable in a neighborhood of the origin; more precisely, a vector relative degree at zero can be achieved for* P *by dynamic compensation. In particular, this is the case if the local asymptotic tracking problem is solvable.*

Proof: See [7].

To establish the connection between tracking and minimum-phase properties, we need to introduce the notion of "left-invertibility at zero." For this purpose, first note that any element y_d of $C_m^N[0,T]$ can be obtained as a solution of the system on $\mathbb{R}^{Nm}$

$$\begin{aligned} \dot{\xi}_1 &= \xi_2 \\ &\vdots \\ \dot{\xi}_N &= \xi_{N+1} \\ y_d &= \xi_1 , \end{aligned} \tag{3.1}$$

where ξ_{N+1} is continuous on $[0,T]$. Therefore, if a compensator Q of the form (2.4) solves the local exact tracking problem from the origin, the compensator

$$\overline{Q}: \begin{aligned} \dot{\xi}_1 &= \xi_2 \\ &\vdots \\ \dot{\xi}_N &= \xi_{N+1} \\ \dot{z} &= a(x,z,\xi_1,\dots,\xi_{N+1}) \\ u &= \alpha(x,z,\xi_1,\dots,\xi_{N+1}) \end{aligned} \tag{3.2}$$

is such that $y^{P\circ\overline{Q}}(t) = y_d(t), 0 \le t \le T$, whenever $P \circ \overline{Q}$ is initialized at 0, $\xi_{N+1}(0) = 0$ and $\sup_{0\le t\le T}\{\|\xi_1(t)\|,\dots,\|\xi_{N+1}(t)\|\} < \epsilon$. By appending integrators to ξ_{N+1} per $\dot{\xi}_{N+1} = v$ in (3.2), and letting Q^a denote the resulting compensator

$$Q^a: \begin{aligned} \dot{\xi}_1 &= \xi_2 \\ &\vdots \\ \dot{\xi}_N &= \xi_{N+1} \\ \dot{\xi}_{N+1} &= v \\ \dot{z} &= a(x,z,\xi_1,\dots,\xi_{N+1}) \\ u &= \alpha(x,z,\xi_1,\dots,\xi_{N+1}) , \end{aligned} \tag{3.3}$$

the system $P \circ Q^a$ is affine in v and exhibits an invertible linear input-output behavior from the origin.

Hence, the Jacobian linearization about the origin of $P \circ Q^a$ is invertible and this implies that the Jacobian linearization about the origin of P is invertible. It follows from [6] that the rank of P [4], ρ^*, equals m. In that case, it is known that, in a neighborhood of a generic point of the state space, the system possesses a left-inverse [14, 19]. To ensure that the origin is not an exceptional point, in addition to $\rho^* = m$ it is sufficient to suppose, for example, that the pair $(x_0 = 0, y \equiv 0)$ is locally strongly regular in the sense of [3]. In this case, it follows from [14, 19] that there exist, $\epsilon_2 > 0$ and an open neighborhood $\mathcal{O}$ of the origin of (2.1) such that if $u(t)$ and $v(t)$ are two inputs so that

(i) the associated state trajectories from the origin $x(t,0,u)$ and $x(t,0,v)$ remain in $\mathcal{O}$ for the interval of time $[0,T]$, where $T > 0$;

(ii) the associated output trajectories $y(t,0,u)$ and $y(t,0,v)$ coincide for $t \in [0,T]$; and

(iii) $\|y^{(j)}(t,0,u)\| < \epsilon_2,\ 0 \le j \le n-1,\ t \in [0,T]$;

then $u(t) = v(t)$, for $t \in [0,T]$. Whenever (i), (ii), and (iii) are satisfied, the system (2.1) is said to be **locally left-invertible at zero.**

The following result can then be established.

Theorem 3.3 *Consider a plant P of the form (2.1) and suppose*

(a) the Jacobian linearization of P about the origin is controllable;

(b) P possesses a zero-dynamics manifold Z^* [12];

(c) P is locally left-invertible at zero.

Then, if the local asymptotic tracking problem is solvable, P is minimum-phase. Moreover, if it is solvable with exponential stability (i.e., (2.5) is exponentially stable), then P is exponentially minimum-phase.

Proof: See [7].

A solution to the local asymptotic tracking problem is provided in [9] using a class of static state feedbacks. In the special case of SISO plants, their results can be stated as follows. Suppose

(H1) P has a well-defined relative degree at zero; and

(H2) P is minimum-phase.

Then, the class of feedback laws

$$u = \frac{1}{L_g L_f^{r-1} h(x)} \left[y_d^{(r)} - L_f^r h(x) + \sum_{i=0}^{r-1} \alpha_i \left(y_d^{(i)} - L_f^i h(x) \right) \right] ,$$

where r is the relative degree and α_i are real coefficients such that the polynomial $s^r + \alpha_{r-1}s^{r-1} + \ldots + \alpha_0$ is Hurwitz, solves the local asymptotic tracking problem. On the other hand, if one supposes that the Jacobian linearization of the plant P is controllable, then the results of Theorems 3.1 and 3.3 establish the necessity of the conditions (H1) and (H2) for the existence of a dynamic compensator of the form (2.4) solving the local asymptotic tracking problem. Indeed, from Theorem 3.1, it follows that (H1) must hold. By [12] and [14], one deduces that P therefore possesses a zero-dynamics as well as a well-defined full-order left-inverse at zero. Hence, by Theorem 3.3, (H2) also holds. This is formalized in the following Corollary.

Corollary 3.4 *Consider a SISO plant P of the form (2.1) and suppose that its Jacobian linearization about the origin is controllable. Then, the local asymptotic tracking problem is solvable if and only if (H1) and (H2) hold.* □

The main stumbling block for obtaining an extension of Corollary 3.4 to MIMO systems consists in the fact that the necessary conditions for the existence of a compensator yielding a vector relative degree in an open neighborhood of the origin are **unknown.** This point is discussed in the next Section.

4. Achieving vector relative degree at a point

It is known that, given a system of the from (2.1), a vector relative degree can be achieved at a generic point of the state space if, and only if, the system is right-invertible [2]. Only *sufficient* conditions are known for achieving a vector relative degree around a *given* point; these are related to the applicability of certain well-known algorithms (e.g. [2], [17]). The fact that none of these is necessary is illustrated by the following example.

Consider the system $\overline{\Sigma}$ described by

$$\begin{aligned} \dot{x}_1 &= x_2 + x_4 + x_2 u_1 \\ \dot{x}_2 &= x_3 \\ \dot{x}_3 &= x_5 + u_1 \\ \dot{x}_4 &= u_2 \\ \dot{x}_5 &= x_2 - x_5 , \end{aligned} \tag{4.4}$$

with outputs

$$\begin{aligned} \overline{y}_1 &= x_1 \\ \overline{y}_2 &= x_1 + x_2 . \end{aligned} \tag{4.5}$$

One computes

$$\begin{aligned}\dot{\bar{y}}_1 &= x_2 + x_4 + x_2 u_1 \\ \dot{\bar{y}}_2 &= x_2 + x_3 + x_4 + x_2 u_1\end{aligned},$$

so that the origin is **not** a regular point of any of the known systematic procedures for determining dynamic decoupling compensators [1, 17] or full-order left-inverses [14, 19]; the fact is that the matrix

$$\begin{bmatrix} x_2 & 0 \\ x_2 & 0 \end{bmatrix},$$

which in this case is the so-called decoupling matrix, does not have constant rank in an open neighborhood of the origin. In particular, the procedure of [9] for achieving asymptotic tracking is not immediately applicable. Nevertheless, we claim that (i) asymptotic tracking is achievable for this system, (ii) the system is dynamically input-output decouplable in a neighborhood of the origin; and (iii) the system has a well-defined (full or reduced order) left-inverse at zero.

Indeed, consider an associated system Σ consisting of the dynamics (4.4) with outputs

$$\begin{aligned} y_1 &= x_1 = h_1(x) \\ y_2 &= x_2 = h_2(x) \end{aligned}. \tag{4.6}$$

and observe that $\bar{y}_1 = y_1, \bar{y}_2 = y_1 + y_2$. The dynamic decoupling algorithm of [1] works in a neighborhood of the origin and provides the decoupling compensator

$$\begin{aligned} \dot{z} &= v_2 \\ u_1 &= z - x_5 \\ u_2 &= x_2(x_2 - x_5) + x_3(x_5 - z - 1) + v_1 \\ &\quad -x_2 v_2 , \end{aligned} \tag{4.7}$$

yielding the linear input-output behavior

$$\begin{aligned} y_1^{(2)} &= v_1 \\ y_2^{(3)} &= v_2 . \end{aligned}$$

In addition, the system's zero dynamics

$$\dot{\eta} = -\eta$$

is exponentially stable.

Let

$$\begin{bmatrix} \dot{x} \\ \dot{z} \end{bmatrix} = \tilde{f}(x,z) + \tilde{g}_1(x,z)v_1 + \tilde{g}_2(x,z)v_2$$

denote the closed-loop dynamics corresponding to (4.4) and (4.7). Then, according to [9], a compensator Q solving the asymptotic tracking problem is obtained by setting

$$\begin{aligned} v_1 &= y_{1d}^{(2)} + \textstyle\sum_{i=0}^{1} c_i \left(y_{1d}^{(i)} - L_{\tilde{f}}^i h_1\right) \\ v_2 &= y_{2d}^{(3)} + \textstyle\sum_{i=0}^{2} d_i \left(y_{2d}^{(i)} - L_{\tilde{f}}^i h_2\right) , \end{aligned} \tag{4.8}$$

with $s^2+c_1 s+c_0$ and $s^3+d_2 s^2+d_1 s+d_0$ Hurwitz.

To obtain from Q a tracking compensator for the original system $\bar{\Sigma}$, one makes the substitution

$$y_d = \begin{pmatrix} 1 & 0 \\ -1 & 1 \end{pmatrix} \bar{y}_d$$

in (4.8), where $\bar{y}_d$ is the desired trajectory for the output $\bar{y}$.

To obtain a dynamic input-output decoupling compensator for the original system $\bar{\Sigma}$ from (4.7), we set

$$\dot{v}_1 = \bar{v}_1 ,$$

so that

$$y^{(3)} = \begin{pmatrix} \bar{v}_1 \\ v_2 \end{pmatrix} = \begin{pmatrix} 1 & 0 \\ -1 & 1 \end{pmatrix} \bar{y}^{(3)} .$$

It follows that the compensator

$$\begin{aligned} \dot{z} &= w_2 - w_1 \\ \dot{v}_1 &= w_1 \\ u_1 &= z - x_5 \\ u_2 &= x_2(x_2 - x_5) + x_3(x_5 - z - 1) + v_1 \\ &\quad -x_2(w_2 - w_1) \end{aligned}$$

results in

$$\bar{y}^{(3)} = w ,$$

thus providing input-output decoupling.

To compute the full-order left-inverse for $\bar{\Sigma}$, we first compute it for Σ following the procedure described in [2] (the usual versions of the inversion algorithm [19] fail by loss of regularity), obtaining

$$\begin{aligned} u_1^* &= y_2^{(2)} - x_5 \\ u_2^* &= y_1^{(2)} - \dot{y}_2(1 + y_2^{(2)} - x_5) - y_2 y_2^{(3)} + (y_2)^2 \\ &\quad -y_2 x_5 \end{aligned} \tag{4.9}$$

For $\bar{\Sigma}$, the full-order left-inverse at zero is then obtained by setting $y_1^{(k)} = \bar{y}_1^{(k)}$ and $y_2^{(k)} = \bar{y}_2^{(k)} - \bar{y}_1^{(k)}$ in (4.9).

The procedure followed above for obtaining a dynamic decoupling compensator for $\bar{\Sigma}$ can be applied more generally and is actually quite systematic. Indeed, suppose that at some step k of the dynamic decoupling algorithm of [1] the decoupling matrix $D_k(\bar{x})$ ($\bar{x}$ denotes the extended state) does not have constant rank in a neighborhood of the origin, but that the dimension of the span over $\mathbb{R}$ of the rows of $D_k(\bar{x})$ is equal to the dimension of the span over the ring of analytic functions in $\bar{x}$ of the rows of $D_k(\bar{x})$. Let μ_k denote this dimension and suppose, furthermore, that μ_k is strictly less than m, the number of rows. Then, as in [15], there exists an invertible real matrix M_k such that $M_k D_k(\bar{x})$ has

$m - \mu_k$ rows of zeros. One can therefore differentiate the corresponding linear combinations of output derivatives until the input appears once again, and continue with the procedure of [1]. If, with this modification, the algorithm of [1] converges, then it can be shown that the original plant is dynamically input-output decouplable. On the other hand, if at some step k of the dynamic decoupling algorithm of [2] the decoupling matrix $D_k(\bar{x})$ is generically invertible, i.e., $\mu_k = m$, then a necessary and sufficient condition for achieving a vector relative degree at zero is that rank $D_k(0,0) = m$. More precisely, the following holds.

Theorem 4.1 ((see also [20])) *Consider a system P of the form (2.1) and suppose that its decoupling matrix $D(x)$ is generically invertible, i.e., $\det D(x) \neq 0$ for at least one point $x \in \mathbb{R}^n$. Then there exists a compensator C of the form (2.2) such that $P \circ C$ has a vector relative degree at zero if, and only if, rank $D(0) = m$.*

Proof: Suppose $\det D(p) \neq 0$ for some $p \in \mathbb{R}^n$. Let $R = \{r_1, \ldots, r_m\}$ be the list of relative degrees of P. Then,

Fact 1: there exists a one-dimensional affine compensator of the form (2.2) such that $P \circ C$ has a nonsingular decoupling matrix at $(p, 0)$ and the list of relative degrees of $P \circ C$ is $\{r_1 + 1, r_2, \ldots, r_m\}$

Proof of Fact 1: Letting $u = \beta(x)v$ where $\beta(x) := (\det D(x))D^{-1}(x)$ shows that it can be assumed w.l.o.g. that $D(x) = \begin{bmatrix} d_1(x) & & \\ & \ddots & \\ & & d_m(x) \end{bmatrix}$ with $d_i(p) \neq 0$ for $1 \leq i \leq m$. Define a dynamic compensator by

$$C: \begin{array}{l} \dot{u}_1 = w_1 \\ u_j = w_j \quad 2 \leq j \leq m \,. \end{array}$$

Then, for $P \circ C$, one computes

$$\begin{aligned} y_1^{(r_1)} &= L_f^{r_1} h_1(x) + d_1(x) u_1 \,, \\ y_1^{(r_1+1)} &= L_f \left[L_f^{r_1} h_1(x) + d_1(x) u_1 \right] + L_{g_1} \\ & \left[L_f^{r_1} h_1(x) + d_1(x) u_1 \right] u_1 + d_1(x) w_1 \\ & + \sum_{j=2}^{m} L_{g_j} \left[L_1^{f_1} h_1(x) + d_1(x) u_1 \right] w_j \,; \end{aligned}$$

for $2 \leq i \leq m$,

$$y_i^{(r_i)} = L_f^{r_i} h_1(x) + d_1(x) u_1 \,.$$

Thus, the decoupling matrix $\bar{D}$ of $P \circ C$ is,

$$\bar{D}(x, u_1) = \begin{bmatrix} d_1(x) & * & * & \cdots & * \\ 0 & d_2(x) & 0 & \cdots & 0 \\ \vdots & & \ddots & & \\ 0 & 0 & \cdots & & d_m(x) \end{bmatrix}$$

Therefore,

$$\det \bar{D}(p, 0) = \det D(p)$$

and the relative degrees of $P \circ C$ are $(r_1 + 1, r_2, \ldots, r_m)$. □

Fact 2: Given any $\bar{R} = \{\bar{r}_1, \ldots, \bar{r}_m\}$ such that $\bar{r}_i \geq r_i$, $i = 1, \ldots, m$, there exists a compensator C such that the decoupling matrix of $P \circ C$ is invertible at $(p, 0)$ and the list of relative degrees of $P \circ C$ is $\bar{R}$.

Proof of Fact 2: Consequence of Fact 1. □

From Fact 2, the relative degrees of $P \circ C$ can be assumed to be equal to $r_i + r$, for some $0 \leq r < \infty$. Then, if $D_{P \circ C}$ denotes the decoupling matrix of $P \circ C$ and D_C the decoupling matrix of the compensator C (from v to u), one has

$$D_{P \circ C} = D \cdot D_C$$

Indeed, Let $\bar{y} := y^{(R)} = L_f^R h(x) + D(x)u$. Then, by definition of r, for each $0 \leq k \leq r - 1$,

$$\bar{y}^{(k)} = \tilde{h}(x, u, \ldots, u^{(k-1)}) + D(x) u^{(k)}$$

and

$$\frac{\partial \bar{y}^{(k)}}{\partial v} = 0 \,.$$

Thus, since D is generically nonsingular,

$$\frac{\partial u^{(k)}}{\partial v} = 0$$

for each $0 \leq k \leq r - 1$. Therefore,

$$D_{P \circ C} := \frac{\partial}{\partial v} \bar{y}^{(r)} = D(x) \frac{\partial u^{(r)}}{\partial v} \,.$$

Since $D_{P \circ C}$ is generically nonsingular, it follows that $\frac{\partial u^{(k)}}{\partial v}$ is also, and thus

$$D_C = \frac{\partial u^{(k)}}{\partial v} \,.$$

Hence, rank $D_{P \circ C}(0) \leq$ rank $D(0)$, which implies the result. □

Remark 4.2: Theorem 4.1 is due to Santosusso in [20]; the only new thing here is a somewhat cleaner proof. Further necessary conditions for achieving a vector relative degree at a point were presented orally by Respondek at the IFAC Nonlinear Control Systems Design Symposium, June 1992, Bordeaux, France. A very interesting sufficient condition for achieving a vector relative degree at the origin has just been discovered by Martin [16]; it is weaker than the notion of strong regularity used in [6].

References

[1] J. Descusse and C.H. Moog, "Dynamic decoupling for right-invertible nonlinear systems," *System Control Lett.*, 8, 1987, pp. 345–349.

[2] M. D. Di Benedetto and J.W. Grizzle, "An analysis of regularity conditions in nonlinear synthesis," *Analysis and Optimization of Systems*, Lecture Notes in Control and Information Sciences, A. Bensoussan, J. L. Lions (Eds.), Vol. 144, 1990, pp. 843–850.

[3] M. D. Di Benedetto and J.W. Grizzle, "Intrinsic notions of regularity for local inversion, output nulling and dynamic extension of nonsquare systems," *Control-Theory and Advanced Technology*, Vol. 6, No. 3, 1990, pp. 357–381.

[4] M. Fliess, "A new approach to the noninteracting control problem in nonlinear systems theory," in *Proc. 23rd Allerton Conference*, University of Illinois, Monticello, IL, 1985, pp. 123–129.

[5] M. Fliess, P. Chantre, S. Abu El Ata, and A. Coic, "Discontinuous predictive control, inversion and singularities: Application to a heat exchanger," *Analysis and Optimization of Systems*, Lecture Notes in Control and Information Sciences, A. Bensoussan, J. L. Lions (Eds.), Vol. 144, Springer, Berlin, 1990 pp. 851–860.

[6] J. W. Grizzle and M. D. Di Benedetto, "Approximation by regular input-output maps," *IEEE Trans. on Automatic Control*, Vol. 37, No. 7, 1992, pp. 1052-1055.

[7] J. W. Grizzle, M. D. Di Benedetto, F. Lamnabhi-Lagarrigue "Necessary conditions for Asymptotic Tracking in Nonlinear Systems," college of Engineering, Control Group Reports, No. CGR-91-4, University of Michigan, September 1991.

[8] J. Hauser, S. Sastry, and P. Kokotović, "Nonlinear control via approximate input-output linearization: The ball and beam example," In *28th Conference on Decision and Control*, pp. 1987–1993, Tampa, FL, 1989.

[9] J. Hauser, S. Sastry, and G. Meyer, "Nonlinear controller design for flight control systems," In A. Isidori, editor, *Nonlinear Control Systems Design*, Pergamon Press, Oxford, 1990.

[10] R. M. Hirschorn and J. Davis, "Output tracking for nonlinear systems with singular points," *SIAM J. Control and Optimization*, Vol. 25, No. 3, 1987, pp. 547–557.

[11] J. Huang and W. J. Rugh, "On a nonlinear multivariable servomechanism problem," *Automatica*, Vol. 26, No. 6, 1990, pp. 963–972.

[12] A. Isidori, *Nonlinear control systems*, 2nd Edition, Communications and Control Engineering Series, Springer-Verlag, Berlin, Heidelberg, 1989.

[13] A. Isidori and C. I. Byrnes, "Output regulation of nonlinear systems," *IEEE Trans. Automatic Control*, Vol. AC-35, 1990, pp. 131–140.

[14] A. Isidori and C. H. Moog, "On the equivalent of the notion of transmission zeros," in Modelling and Adaptive Control, Proc. *International Institute of Applied Systems Analysis Conference*, Sopron, 1986, C. I. Byrnes and A. Kurszanski, eds., Lecture Notes in Control and Information Sciences, Vol. 105, Springer-Verlag, Berlin, New York, 1988.

[15] A. Isidori and A. Ruberti, "On the synthesis of linear input-output responses for nonlinear systems," *Systems and Control Lett.*, Vol. 4, 1984, pp. 17–22.

[16] P. Martin, "An intrinsic sufficient condition for regular decoupling," preprint, 1992.

[17] H. Nijmeijer and W. Respondek, "Dynamic input-output decoupling of nonlinear control systems," *IEEE Trans. Automatic Control*, Vol. AC-33, No. 11, 1988, pp. 1065–1070.

[18] J. J. Romano and S. N. Singh, "I-O map inversion, zero dynamics and flight control," *IEEE Trans. on Aerospace and Electronic Systems*, Vol. 26, No. 6, 1990, pp. 1022–1028.

[19] S. N. Singh, "A modified algorithm for invertibility in nonlinear systems," *IEEE Trans. Automatic Control*, Vol. AC-26, No. 2, 1981, pp. 595–598.

[20] G. L. Santosuosso, "Necessary conditions for the removal of singularities with dynamic state feedback," Proceedings *Nonlinear Control Systems Design Symposium 1992*, Bordeaux (France), June 1992, pp. 277-281.

Proceedings of the 31st Conference
on Decision and Control
Tucson, Arizona • December 1992

SEMI-GLOBAL STABILIZATION OF MINIMUM PHASE NONLINEAR SYSTEMS IN SPECIAL NORMAL FORM VIA LINEAR HIGH-AND-LOW-GAIN STATE FEEDBACK [1]

Zongli Lin & Ali Saberi
School of Electrical Engineering and Computer Science
Washington State University
Pullman, WA 99164-2752

ABSTRACT

We provide an alternative solution to the problem of semi-global stabilization of a class of minimum phase nonlinear systems which is considered in [17]. Our method yields a stabilizing *linear* state feedback law in contrast to a *nonlinear* state feedback law proposed in [17]. We eliminate the peaking phenomenon by inducing a specific time-scale structure in the linear part of the closed-loop system. This time-scale structure consists of a very slow and a very fast time-scale. The crucial component in our method is the relation between the slow and the fast time-scales. Our proposed linear state feedback control law has a *single* tunable gain parameter that allows for local asymptotic stability and regulation to the origin for any initial condition in some a priori given (arbitrarily large) bounded set.

1. INTRODUCTION

In this paper we consider the problem of semi-global feedback stabilization of partially linear composite systems of the form

$$\begin{aligned}\dot{\eta} &= f(\eta, \xi, u)\\ \dot{\xi} &= A\xi + Bu\\ y &= C\xi\end{aligned}$$

where $f(\eta, \xi, u)$ is a smooth (i.e., C^∞) function, A, B and C are constant matrices of appropriate dimensions and (A, B) is a controllable pair. As a class of nonlinear composite systems (e.g., [8], [13], [18]), the partially linear composite systems have become prominent because of recent results on partial feedback linearization, where $\dot{\eta} = f(\eta, 0, 0)$ is referred to as the "nonlinear zero dynamics" (e.g., [1], [2]). A main issue in the study of such composite systems is global or semi-global stabilization (e.g., [3], [4], [5], [6], [7], [9], [10], [11], [12], [14], [15], [16], [17]).

In this paper we continue the efforts of [3], [16], [17] on the study of the problem of semi-global stabilization of multi-input minimum phase nonlinear systems in the normal form:

$$\begin{aligned}\dot{\eta} &= f(\eta, \xi, u)\\ \dot{\xi}_1^i &= \xi_2^i\\ &\;\vdots\\ \dot{\xi}_{r_i}^i &= u_i\\ y_i &= \xi_1^i, \quad \text{for } i = 1, 2, \cdots, m\end{aligned} \tag{1.1}$$

$$\begin{aligned}\xi &= [(\xi^1)', (\xi^2)', \cdots, (\xi^m)']'\\ \xi^i &= [\xi_1^i, \xi_2^i, \cdots, \xi_{r_i}^i]', \text{ for } i = 1, 2, \cdots, m\end{aligned}$$

where the state $x = (\eta, \xi) \in \Re^n$ and f is smooth with $f(0,0,0) = 0$. By minimum phase it is meant that the equilibrium point $\eta = 0$ of the "nonlinear zero dynamics"

$$\dot{\eta} = f(\eta, 0, 0)$$

is globally asymptotically stable.

The works of [3] and [16] seek *linear* feedbacks which depend only on the linear state ξ and a tunable gain parameter that allows for local asymptotic stability and regulation to the origin for any initial condition in some a priori given (arbitrarily large) bounded set. Realizing that this is not always possible due to the peaking phenomenon, both of the works show that if the minimum phase nonlinear system (1.1) has the more specific form

$$\begin{aligned}\dot{\eta} &= f(\eta, y_1, \cdots, y_m)\\ \dot{\xi}_1^i &= \xi_2^i\\ &\;\vdots\\ \dot{\xi}_{r_i}^i &= u_i\\ y_i &= \xi_1^i, \quad \text{for } i = 1, 2, \cdots, m\end{aligned} \tag{1.2}$$

then such a family of linear feedbacks exists. This is due to possible elimination of the peaking phenomenon by excluding the dependency of f on the linear states ξ_j^i's other than the outputs y_i's.

A recent work, [17] has extended this result in a very interesting direction. In [17] the semi-global stabilization problem is solved for a more general class of systems than the system (1.2). Namely, the restriction on f is relaxed by allowing f to be dependent on any one state of each of the m chains of integrators. The peaking phenomenon in such systems is eliminated by stabilizing part of the linear system with a high-gain linear control and the remaining part of the linear system with a *small, bounded nonlinear control.* The crucial component of the result in [17] is the *arbitrarily small bounded control law* that stabilizes part of the linear system.

The goal of this paper is to show that semi-global stabilization of the same class of minimum phase nonlinear systems as considered in [17] can be done by *linear* state feedback laws which, of course, depend only on the linear states. Hence the small bounded nonlinear control laws as proposed in [17] are not necessary elements in semi-global stabilization of such systems. The fundamental issue

[1]This work is supported in part by Boeing Commercial Airplane Group.

CH3229-2/92/0000-2482$1.00 © 1992 IEEE

in designing such a linear state feedback law is to induce a specific time scale structure in the linear part of the closed-loop system. This time-scale structure consists of a very slow and a very fast time-scale, which are the result of a linear state feedback of the high and low gain nature. The slow and fast time scales must be chosen in a very careful fashion. In fact the relation between the slow and the fast time-scales is the most crucial part of our design, which enables us to eliminate peaking phenomenon. Furthermore, our proposed control law has a *single* tunable parameter which allows the regulation to the origin for any initial condition in some a priori given (arbitrarily large) bounded set to be achieved only by an appropriate choice of this parameter. This obviously represents a desirable feature from the practical point of view. We believe that our work points to a significant direction in the stabilization of nonlinear systems and might lead to further generalization of the class of minimum phase nonlinear systems which can be semi-globally stabilized.

2. PROBLEM STATEMENT

To clarify the problem at hand, we make the following definition:

Definition 2.1. *The system (1.1) is semi-globally stabilizable by state feedback if for any compact set of initial conditions X there exists a family of smooth state feedbacks*

$$u = \alpha(\eta, \xi) \tag{2.1}$$

such that the equilibrium $(0,0)$ *of the closed-loop system (1.1), and (2.1) is locally asymptotically stable and X is contained in the domain of attraction of* $(0,0)$.

The problem is to find a family of state feedback control laws which semi-globally stabilize the multi-input system in the following special normal form:

$$\begin{aligned} \dot{\eta} &= f(\eta, \xi^1_{j_1}, \cdots, \xi^m_{j_m}), \quad j_i \in \{1, 2, \cdots, r_i + 1\} \\ \dot{\xi}^i_1 &= \xi^i_2 \\ &\vdots \\ \dot{\xi}^i_{r_i} &= u_i \\ y_i &= \xi^i_1, \quad \text{for } i = 1, 2, \cdots, m \end{aligned} \tag{2.2}$$

where $\xi^i_{r_i+1} = u_i$. And this family of control laws should depend only on the linear state ξ. With respect to the outputs y_i's the system (2.2) is said to have vector relative degree $\{r_1, r_2, \cdots, r_m\}$. We define $r = \sum_{i=1}^m r_i$. We then have $\xi \in \Re^r$ and $\eta \in \Re^{n-r}$.

We also make the following standard assumption:

Assumption 1. The equilibrium point $\eta = 0$ of the nonlinear zero dynamics

$$\dot{\eta} = f(\eta, 0, \cdots, 0)$$

is globally asymptotically stable.

The distinguishing feature of the systems in the special normal form of (2.2) is that any one state of each of the m chains of integrators appears in the η dynamics. Systems of form (2.2) are more general than those of [3] in that the one state is not required to be the first state of the chain associated with y_i.

[17] gives a solution to this problem. This solution involves two phases, namely, i) for each of m chains of integrators, find a smooth arbitrarily small, bounded control $v_i(\xi^i_1, \cdots, \xi^i_{j_i-1})$ which globally stabilizes the first $j_i - 1$ integrators, and thus makes the system with the new output $\bar{y}_i = \xi^i_{j_i} - v_i(\xi^i_1, \cdots, \xi^i_{j_i-1})$ minimum phase on arbitrarily large set; ii) design a high-gain feedback control law to semi-globally stabilize the new minimum-phase system.

Our goal is to find an alternative solution to this problem which will involve *linear* feedbacks only. As we show later, our control laws will be combinations of linear high and low gain feedbacks.

3. SEMI-GLOBAL STABILIZATION — A LINEAR HIGH-AND-LOW-GAIN FEEDBACK APPROACH

We set out to give a family of state feedback control laws $u = F(\epsilon)\xi$ as follows:

For $i = 1$ to m,

$$u_i = -\sum_{k=1}^{j_i-1} \epsilon^{j_i-k} c^i_{j_i-k} \xi^i_{r_i-j_i+k+1} - \sum_{k=j_i}^{r_i} \frac{1}{\epsilon^{r_i-k+1}} d^i_{r_i-k+1} \tilde{\xi}^i_k \tag{3.1}$$

$$\tilde{\xi}^i_k = \xi^i_k + \sum_{l=1}^{j_i-1} \epsilon^{j_i-l} c^i_{j_i-l} \xi^i_{k-j_i+l}, \quad k = j_i, j_i+1, \cdots, r_i$$

where $\sum_{k=m}^n$ is taken to be zero when $m > n$, and where c^i_k's and d^i_k's are positive constants chosen arbitrarily subject to the constraint that the polynomials

$$p^i_s(s) := s^{j_i-1} + c^i_1 s^{j_i-2} + \cdots + c^i_{j_i-2} s + c^i_{j_i-1}$$

and

$$p^i_f(s) := s^{r_i-j_i+1} + d^i_1 s^{r_i-j_i} + \cdots + d^i_{r_i-j_i} s + d^i_{r_i-j_i+1}$$

are Hurwitz. The parameter $\epsilon > 0$ is to be specified later.

We then establish our main results in the following theorem.

Theorem 3.1. *Assume that the system (2.2) satisfies Assumption 1. Then the system (2.2) is semi-globally stabilized by the family of feedback control laws (3.1). That is, (3.1) locally asymptotically stabilizes (2.2) and for any compact set X of the state space (η, ξ) there exists an $\epsilon^* > 0$, such that for every ϵ, $0 < \epsilon \leq \epsilon^*$, X is contained in the domain of attraction of the closed-loop system.*

Before we prove this theorem, we need to establish the following two lemmas, the first of which is similar to Theorem 4.1 of [16]. Let us begin with a definition.

Definition 3.1. *Let $K = (K_1, K_2, \cdots, K_m)$ be an $m-$tuple of strictly positive real numbers, and let $\epsilon > 0$. We let $I(m, K, \epsilon)$ denote the set of all $\Re^m$-valued measurable functions $y(t) = (y_1(t), y_2(t), \cdots, y_m(t))$ on $[0, \infty)$ that satisfy the bounds*

$$|y_i(t)| \leq K_i(e^{-a_1 t/\epsilon} + \epsilon e^{-a_2 \epsilon t}), \quad \text{for } i = 1, 2, \cdots, m.$$

where a_1 and a_2 are positive constants independent of ϵ. Then a number ϵ is good for (R, K) if every solution $t \to x(t)$ of

$$\dot{x} = f(x, y)$$

corresponding to an input $y \in I(m, K, \epsilon)$ and an initial condition $x(0)$ such that $\|x(0)\| \leq R$, satisfies $x(t) \to 0$ as $t \to +\infty$. Moreover, the convergence is uniform with respect to $x(0) \in \{x : \|x\| \leq R\}$ and $y \in I(m, K, \epsilon)$.

We now have the the following lemmas,

Lemma 3.1. *Assume that the system $\dot{x} = f(x, 0)$ is globally asymptotically stable. Then for every $R > 0$, and every m-tuple $K = (K_1, K_2, \cdots, K_m)$ of strictly positive numbers, there exists an $\epsilon^* > 0$ such that every ϵ, $0 < \epsilon \leq \epsilon^*$, is good for (R, K).*

Proof : The proof is a slight modification of that of Lemma 4.1 in [16]. In particular, the two properties of the bounds on y_i's used in the proof are

1) Given any $\tau > 0$ and $\alpha > 0$, there exists an $\epsilon^* > 0$ such that for every ϵ, $0 < \epsilon \leq \epsilon^*$, $K_i(e^{-a_1 t/\epsilon} + \epsilon e^{-a_2 \epsilon t}) \leq \alpha$, $t \geq \tau$.
2) For any given ϵ and $\delta > 0$ there exists a $T > 0$ such that $K_i(e^{-a_1 t/\epsilon} + \epsilon e^{-a_2 \epsilon t}) \leq \delta$, $t \geq T$.

∎

Lemma 3.2. *For any $\epsilon > 0$, the state feedback control law (3.1) stabilizes the linear part of the system (2.2). Moreover, there exists an $\epsilon^* > 0$ such that for every ϵ, $0 < \epsilon \leq \epsilon^*$, the state $\xi^i_{j_i}$ is bounded by*

$$|\xi^i_{j_i}(t)| \leq K_0(e^{-a_1 t/\epsilon} + \epsilon e^{-a_2 \epsilon t})\|\xi(0)\|, \quad \text{for } i = 1, 2, \cdots, m. \tag{3.2}$$

Proof : Adopt the following change of variables and scaling, for $i = 1$ to m,

$$\bar{\xi}^i = [\xi^i_1, \xi^i_2, \cdots, \xi^i_{j_i-1}]', \quad \hat{\xi}^i = [\xi^i_{j_i}, \xi^i_{j_i+1}, \cdots, \xi^i_{r_i}]' \tag{3.3}$$

$$\bar{\xi}^i_k = \epsilon^{j_i-k-1}\xi^i_k, \quad k = 1, 2, \cdots, j_i - 1 \tag{3.4}$$

$$\hat{\xi}^i_k = \epsilon^{k-j_i}\tilde{\xi}^i = \epsilon^{k-j_i}(\xi^i_k + \sum_{l=1}^{j_i-1} \epsilon^{j_i-l} c^i_{j_i-l}\xi^i_{k-j_i+l}), \quad k = j_i, j_i+1, \cdots, r_i \tag{3.5}$$

Then the closed-loop system is written as,

$$\frac{1}{\epsilon}\dot{\bar{\xi}}^i_k = \bar{\xi}^i_{k+1}, \quad k = 1, 2, \cdots, j_i - 2 \tag{3.6}$$

$$\frac{1}{\epsilon}\dot{\bar{\xi}}^i_{j_i-1} = -c^i_{j_i-1}\bar{\xi}^i_1 - c^i_{j_i-2}\bar{\xi}^i_2 - \cdots - c^i_1\bar{\xi}^i_{j_i-1} + \frac{1}{\epsilon}\hat{\xi}^i_{j_i} \tag{3.7}$$

$$\epsilon\dot{\hat{\xi}}^i_k = \hat{\xi}^i_{k+1}, \quad k = j_i, j_i+1, \cdots, r_i - 1 \tag{3.8}$$

$$\epsilon\dot{\hat{\xi}}^i_{r_i} = -d^i_{r_i-j_i+1}\hat{\xi}^i_{j_i} - d^i_{r_i-j_i}\hat{\xi}^i_{j_i+1} - \cdots - d^i_1\hat{\xi}^i_{r_i} \tag{3.9}$$

This system is obviously stable for any $\epsilon > 0$ since it is a cascade of two stable subsystems (3.6)-(3.7) and (3.8)-(3.9).

To show (3.2), we put these two subsystems in different time scales, $\tau_s = \epsilon t$ and $\tau_f = t/\epsilon$ respectively. On these time scales, the closed-loop system (3.6)-(3.9) becomes,

$$\frac{d}{d\tau_s}\bar{\xi}^i_k = \bar{\xi}^i_{k+1}, \quad k = 1, 2, \cdots, j_i - 2 \tag{3.10}$$

$$\frac{d}{d\tau_s}\bar{\xi}^i_{j_i-1} = -c^i_{j_i-1}\bar{\xi}^i_1 - c^i_{j_i-2}\bar{\xi}^i_2 - \cdots - c^i_1\bar{\xi}^i_{j_i-1} + \frac{1}{\epsilon}\hat{\xi}^i_{j_i} \tag{3.11}$$

$$\frac{d}{d\tau_f}\hat{\xi}^i_k = \hat{\xi}^i_{k+1}, \quad k = j_i, j_i+1, \cdots, r_i - 1 \tag{3.12}$$

$$\frac{d}{d\tau_f}\hat{\xi}^i_{r_i} = -d^i_{r_i-j_i+1}\hat{\xi}^i_{j_i} - d^i_{r_i-j_i}\hat{\xi}^i_{j_i+1} - \cdots - d^i_1\hat{\xi}^i_{r_i} \tag{3.13}$$

Let

$$p^i_f(s) = (s - \mu^i_1)^{m^i_1}(s - \mu^i_2)^{m^i_2}\cdots(s - \mu^i_{\sigma^i})^{m^i_{\sigma^i}}$$

Then the transition matrix $\Phi^i(\tau_f)$ for the subsystem (3.12)-(3.13) is given by

$$\Phi^i(\tau_f) = \sum_{j=1}^{\sigma^i}\sum_{k=1}^{m^i_{\sigma^i}} P^i_{jk}\frac{\tau_f^{k-1}}{(k-1)!}e^{\mu^i_j\tau_f}$$

where P^i_{jk}'s are the residue matrices, independent of ϵ. Noting that

$$|\tau_f^{k-1}e^{\frac{1}{2}\mu^i_j\tau_f}| \leq \left(\frac{2k-2}{\mathrm{Re}\mu^i_j}\right)^{k-1}$$

we readily see that, for any $\epsilon > 0$,

$$\|\Phi^i(\tau_f)\| \leq \beta_1 e^{-a_1\tau_f} \tag{3.14}$$

where $a_1 = \min\{\frac{1}{2}|\mathrm{Re}\mu^i_j| \,:\, j = 1, 2, \cdots, \sigma^i, i = 1, 2, \cdots, m\}$ and β_1 is a positive constant independent of ϵ. In view of (3.5), we have, for $0 < \epsilon \leq 1$,

$$\|\hat{\xi}^i(0)\| \leq \beta_2\|\xi(0)\|, \text{ for some } \beta_2 > 0 \text{ independent of } \epsilon \tag{3.15}$$

It then follows from (3.14)-(3.15) that, for $0 < \epsilon \leq 1$,

$$\|\hat{\xi}^i(t)\| = \|\Phi^i(t/\epsilon)\hat{\xi}(0)\| \leq \beta_3 e^{-a_1 t/\epsilon}\|\xi(0)\|, \quad \text{for } i = 1, 2, \cdots, m. \tag{3.16}$$

where $\beta_3 = \beta_1\beta_2 > 0$ is independent of ϵ.

Now consider $\hat{\xi}^i_{j_i}$'s as the input to the subsystem (3.10)-(3.11) and let

$$p^i_s(s) = (s - \lambda^i_1)^{n^i_1}(s - \lambda^i_2)^{n^i_2}\cdots(s - \lambda^i_{\delta^i})^{n^i_{\delta^i}}$$

Then, it follows from the same arguments we used for the subsystem (3.12)-(3.13) that, for any $\epsilon > 0$, the transition matrix for the subsystem (3.10)-(3.11), $\Psi^i(\tau_s)$, satisfies

$$\|\Psi^i(\tau_s)\| \leq \beta_4 e^{-a_2\tau_s} \tag{3.17}$$

where $a_2 = \min\{\frac{1}{2}|\mathrm{Re}\lambda^i_j| \,:\, j = 1, 2, \cdots, \delta^i, i = 1, 2, \cdots, m\}$ and β_4 is a positive constant independent of ϵ. In view of (3.4), we have, for $0 < \epsilon \leq 1$,

$$\|\bar{\xi}(0)\| \leq \|\xi(0)\| \tag{3.18}$$

It then follows that, for $0 < \epsilon \leq 1$,

$$\begin{aligned}\|\bar{\xi}^i(t)\| &\leq \|\Psi^i(\epsilon t)\bar{\xi}(0)\| + \int_0^t \|\Psi(\epsilon(t-\rho))\|\|\hat{\xi}^i_{j_i}(\rho)|d\rho \\ &\leq \beta_4 e^{-a_2\epsilon t}\|\xi(0)\| + \beta_3\beta_4\int_0^t e^{-a_2\epsilon(t-\rho)}e^{-a_1\rho/\epsilon}d\rho\|\xi(0)\| \\ &\leq \beta_4 e^{-a_2\epsilon t}\|\xi(0)\| + \beta_3\beta_4 e^{-a_2\epsilon t}\int_0^t e^{-(a_1/\epsilon - a_2\epsilon)\rho}d\rho\|\xi(0)\|\end{aligned} \tag{3.19}$$

Let $\epsilon^* = min\{1, \sqrt{\frac{a_1}{2a_2}}\}$, (3.19) shows that, for $0 < \epsilon \leq \epsilon^*$,

$$
\begin{aligned}
\|\bar{\xi}^i(t)\| &\le \beta_4 e^{-a_2 \epsilon t}\|\xi(0)\| + \beta_3\beta_4 e^{-a_2 \epsilon t}\int_0^t e^{-\frac{a_1}{2\epsilon}\rho}d\rho\|\xi(0)\|, \\
&\le \beta_4 e^{-a_2 \epsilon t}\|\xi(0)\| + \frac{2\beta_3\beta_4\epsilon}{a_1}e^{-a_2\epsilon t}\|\xi(0)\| \\
&\le \beta_5 e^{-a_2\epsilon t}\|\xi(0)\|, \text{ for } i = 1, 2, \cdots, m \qquad (3.20)
\end{aligned}
$$

where $\beta_5 = \beta_4 + \frac{2\beta_3\beta_4\epsilon^*}{a_1} > 0$ is independent of ϵ. Now recall that

$$
\begin{cases} \xi^i_{j_i} = \hat{\xi}^i_{j_i} - \epsilon\sum_{l=1}^{j_i-1} c^i_{j_i-l}\bar{\xi}^i_l, & \text{if } j_i \le r_i \\ \xi^i_{j_i} = -\epsilon\sum_{l=1}^{r_i} c^i_{r_i-l+1}\bar{\xi}^i_l, & \text{if } j_i = r_i + 1 \end{cases}
$$

we immediately conclude from (3.16) and (3.20) that, for $0 < \epsilon \le \epsilon^*$,

$$
|\xi^i_{j_i}| \le \beta_3 e^{-a_1 t/\epsilon}\|\xi(0)\| + \epsilon\beta_5 e^{-a_2\epsilon t}\sum_{l=1}^{j_i-1} c^i_{j_i-l+1}\|\xi(0)\|, \text{ for } i = 1 \text{ to } m
$$

and the results follow. ∎

We are now ready to prove our main theorem.

Proof of Theorem 3.1: Let $X \subset \{x : \|x\| \le \gamma\} \subset \Re^n$, It then follows from Lemma 3.1 that there exists an ϵ_1^* such that every ϵ, $0 < \epsilon \le \epsilon_1^*$, is good for (γ, K), where $K := (K_1, K_2, \cdots, K_m)$ and $K_i = \gamma K_0$. Now Lemma 3.2 shows that there always exists an ϵ^*, $0 < \epsilon^* \le \epsilon_1^*$, such that for every ϵ, $0 < \epsilon \le \epsilon^*$, $\xi^i_{j_i} \in I(m, K, \epsilon)$, $i = 1, 2, \cdots, m$ for any initial condition belonging to the compact set X. This shows that any trajectory of the closed-loop system starting form X, $t \to (\eta(t), \xi(t))$, goes to zero as t approaches to infinity.

Finally, the proof of the local asymptotic stability is straightforward. See for example [3] and [18]. ∎

Remark 3.1. *The crucial component of our proof of Theorem 3.1 is to induce the time-scale structure $\{\epsilon t, t/\epsilon\}$ in the linear part of the closed-loop system. The time scale structure is chosen in such a way that the output y of the linear part of the closed-loop system satisfies the bound condition of Definition 3.1 and hence prevents the peaking phenomenon. This choice of time scale structure is not necessary unique. Also note that the poles of the linear part of the closed-loop system at the slow time scale ϵt and the fast time scale t/ϵ are the roots of $p^i_s(s) = 0$, $i = 1$ to m, and $p^i_f(s) = 0$, $i = 1$ to m, respectively. Hence there is no restriction on placing the poles of the closed-loop linear system at each time scale.*

4. EXAMPLES

To illustrate our stabilization scheme, we consider the three examples considered in [17].

Example 1. (Example 8.2 of [3])

$$
\begin{aligned}
\dot{\eta} &= -(1 - \eta\xi_2)\eta \\
\dot{\xi}_1 &= \xi_2 \\
\dot{\xi}_2 &= u
\end{aligned}
$$

Example 2. (Example 1.1 of [16])

$$
\begin{aligned}
\dot{\eta} &= -0.5(1 + \xi_2)\eta^3 \\
\dot{\xi}_1 &= \xi_2 \\
\dot{\xi}_2 &= u
\end{aligned}
$$

Example 3. (Example 4.1 of [17])

$$
\begin{aligned}
\dot{\eta} &= -\eta + \eta^2 u \\
\dot{\xi}_1 &= \xi_2 \\
\dot{\xi}_2 &= u
\end{aligned}
$$

Obviously all the three systems are in the special normal form we consider in this paper. The first two examples have essentially the same structure and are solved by the same family of feedback control laws,

$$
u = -\xi_1 - \frac{\epsilon^2 + 1}{\epsilon}\xi_2
$$

where the the poles of the fast and slow subsystems of the closed-loop linear system have been chosen to be $-\frac{1}{\epsilon}$ and $-\epsilon$ respectively.

In the third example, $j_1 = r_1 + 1$. For this case, no high-gain feedback is involved in the feedback laws and hence no fast subsystem is present in the closed-loop linear system. Choosing the poles of the linear slow subsystem to be $-\epsilon$ and -2ϵ, we obtain the following family of feedback control laws which semi-globally stabilizes the system,

$$
u = -2\epsilon^2\xi_1 - 3\epsilon\xi_2.
$$

The simulation results are shown in Figures 4.1-4.3.

5. CONCLUSIONS

We have solved the semi-global stabilization problem for a class of minimum phase nonlinear systems considered in [17] by linear state feedback control laws which depends only on the state of the linear part. We eliminated the peaking phenomenon by inducing a specific time-scale structure in the part of the closed-loop system. The proper choice of this time-scale structure was the key idea in our development.

References

[1] C.I. Byrnes and A. Isidori, "A frequency domain philosophy for nonlinear systems with applications to stabilization and adaptive control," *Proc. IEEE Conf. Decision Contr.*, Las Vegas, pp. 1569-1573, 1984.

[2] C.I. Byrnes and A. Isidori, "Local stabilization of minimum phase nonlinear systems," *Systems & Control Letters*, Vol. 10, pp. 9-17, 1988.

[3] C.I. Byrnes and A. Isidori, "Asymptotic stabilization of minimum phase nonlinear systems," *IEEE Trans. on Automatic Contr*, Vol. 36, pp. 1122-1137, 1991.

[4] C.I. Brynes, A. Isidori and J. Willems, "Passivity, feedback equivalence and the global stabilization of minimum phase nonlinear systems," *IEEE Trans. on Automatic Contr*, Vol. 36, pp. 1228-1239, 1991.

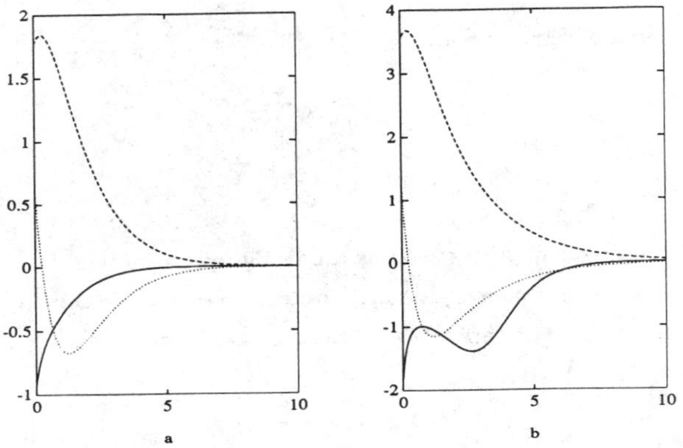

Figure 4.1: Example 1. The solid line is nonlinear state η, the dashed line is linear state ξ_1 and the dotted line is the linear state ξ_2. a) $\epsilon = 0.9$; b) $\epsilon = 0.5$

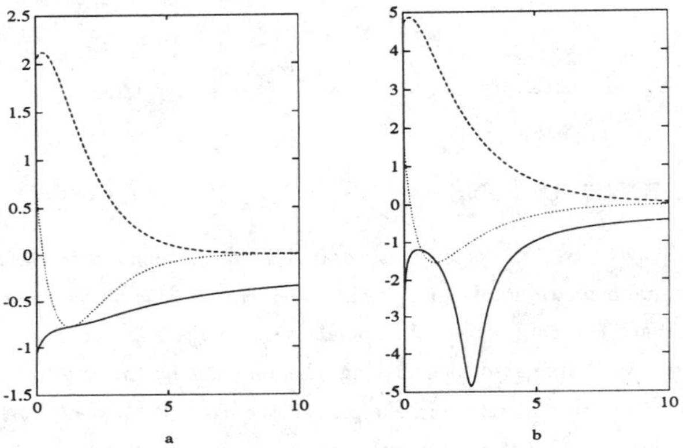

Figure 4.2: Example 2. The solid line is nonlinear state η, the dashed line is linear state ξ_1 and the dotted line is the linear state ξ_2. a) $\epsilon = 0.9$; b) $\epsilon = 0.5$

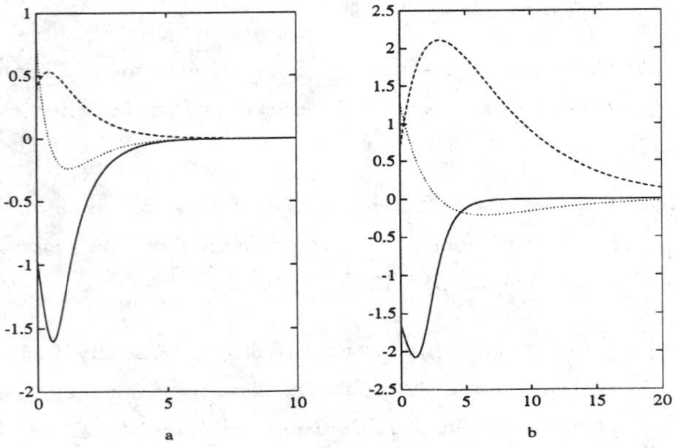

Figure 4.3: Example 3. The solid line is nonlinear state η, the dashed line is linear state ξ_1 and the dotted line is the linear state ξ_2. a) $\epsilon = 0.9$; b) $\epsilon = 0.2$

[5] H.K. Khalil and A. Saberi, "Adaptive stabilization of a class of nonlinear systems using high-gain feedback," *IEEE Trans. on Automatic Control*, Vol. 32, pp. 1031-1035, 1987.

[6] P.V. Kokotovic and H.J. Sussmann, "A positive real condition for global stabilization of nonlinear systems," *Systems & Control Letters*, Vol. 13, pp.125-133, 1989.

[7] R. Marino, "Feedback stabilization of single-input nonlinear systems," *Systems & Control Letters*, Vol. 10, pp 201-206, 1988.

[8] A.N. Michel and R.K. Miller, *Qualitative Analysis of Large Scale Dynamic Systems*, Mathematics in Engineering, Vol. 134, Academic Press, New York, 1977.

[9] L. Praly, B. d'Andrea Novel and J.M. Coron, "Lyapunov design of stabilizing controllers,' *Proc. 28th IEEE Conf. Decision Contr.*, pp. 1047-1052, 1989.

[10] A. Saberi, P.V. Kokotovic and H.J. Sussmann, "Global stabilization of partially linear composite systems," *SIAM J. Control and Optimization*, Vol. 28, No. 6, pp. 1491-1503, 1990.

[11] A. Saberi and Z. Lin, "Adaptive high-gain stabilization of 'minimum-phase' nonlinear systems," *Control-Theory and Advanced Technology*, Vol. 6, pp. 595-607, 1990.

[12] S. Sastry and A. Isidori, "Adaptive control of linearizable systems," *IEEE Trans. on Automatic Control*, Vol. 34, pp. 1123-1131, 1989.

[13] D.D. Siljak, *Large-Scale Dynamic Systems, Stability and Structure*, North-Holland, Amsterdam, 1978.

[14] E.D. Sontag, "Smooth stabilization implies coprime factorization," *IEEE Trans. Automatic Control*, April 1989.

[15] E.D. Sontag, "Remarks on stabilization and input-to-state stability," *Proc. 28th IEEE Conf. Decision Contr.*, pp. 1376-1378, 1989.

[16] H.J. Sussmann and P.V. Kokotovic, "The peaking phenomenon and the global stabilization of nonlinear systems"' *IEEE Transactions on Auto. Contr.*, Vol. 36, No. 4, pp 424-440, 1991.

[17] A.R. Teel, "Semi-global stabilization of minimum phase nonlinear systems in special normal forms," *Memorandum No. UCB/ERL M91/101, Electronics Research Laboratory, University of California, Berkeley*, 1991.

[18] M. Vidyasagar, "Decomposition techniques for large-scale systems with non-additive interactions: stability and stabilizability," *IEEE Transactions on Auto. Contr.*, Vol. 25, pp 773-779, 1980.

Proceedings of the 31st Conference
on Decision and Control
Tucson, Arizona • December 1992

Nonlinear Steady-State Tracking and Disturbance Rejection with Application to Power Systems

Jianliang Wang, Rujing Zhou
School of Electrical and Electronic Engineering
Nanyang Technological University
Singapore 2263

Abstract

Multi-input, multi-output nonlinear systems with unmeasurable disturbances are considered. Assuming that both the reference signal and the disturbance take constant values or are slowly time-varying, then nonlinear static feed-back is designed so that in steady-state, the output is equal to the corresponding reference signal, regardless what the disturbance is. This method is applied to an infinite bus, single machine power generator system.

1. Introduction

We consider multi-input, multi-output nonlinear system with unmeasurable disturbances. We assume that the reference signal that the system output is supposed to follow and the disturbance both are constant or are slowly time-varying. We will design a static nonlinear feedback such that, in steady-state, the output takes the value of the corresponding reference signal, regardless what the disturbance is. That is, in steady-state, the output tracks the reference signal and rejects the disturbance which is unknown. Huang and Rugh [3] studied this problem using dynamic nonlinear feedback. In this paper, we show that under certain conditions, this problem can be solved using only static nonlinear feedback. Finally, this method is applied to control an infinite bus, single machine power generator system, such that the closed-loop system poles are all of a certain distance to the left of the imaginary axis of the complex plane and the output (terminal voltage) is equal to the constant reference singal, regardless what the input mechanical power is.

2. Controller Design

We consider nonlinear systems with disturbance of the following form

$$\begin{aligned} \dot{x}(t) &= f(x(t),u(t),v(t)) \\ y(t) &= h(x(t),u(t),v(t)) \end{aligned} \tag{1}$$

where $x \in R^n$ is the state, $u \in R^m$ is the input, $y \in R^m$ is the output and $v \in R^q$ is the disturbance, and $m+q<n$.

Proposition 1: [7] Suppose that $f(\cdot,\cdot)$ and $h(\cdot,\cdot)$ are both twice continuously differentiable and

$$rank\left[\frac{\partial f}{\partial x}(0,0,0) \quad \frac{\partial f}{\partial u}(0,0,0) \quad \frac{\partial f}{\partial v}(0,0,0)\right] = n \tag{2}$$

Then there exists an open neighborhood $\Gamma \subset R^{m+q}$ containing the origin and continuously differentiable functions $U(\cdot)$, $V(\cdot)$, $X(\cdot)$, $Y(\cdot)$ such that for all $\alpha \in \Gamma$

$$\begin{aligned} f(X(\alpha),U(\alpha),V(\alpha)) &= 0 \\ h(X(\alpha),U(\alpha),V(\alpha)) &= Y(\alpha) \end{aligned} \tag{3}$$

That is, the nonlinear system has a family of constant operating points

$$\mathscr{E} = \left\{ [U(\alpha), V(\alpha), X(\alpha), Y(\alpha)],\ \alpha \in \Gamma \;\middle|\; \begin{array}{l} f(X(\alpha),U(\alpha),V(\alpha)) = 0 \\ h(X(\alpha),U(\alpha),V(\alpha)) = Y(\alpha) \end{array} \right\} \tag{4}$$

Proof: Consider the mapping

$$\begin{bmatrix} f(x,u,v) \\ h(x,u,v)-y \end{bmatrix}$$

Under the assumption (2), the Jacobian of this map has full row rank $m+n$.

$$rank \begin{bmatrix} \frac{\partial f}{\partial x}(0,\ 0,\ 0) & \frac{\partial f}{\partial u}(0,\ 0,\ 0) & \frac{\partial f}{\partial v}(0,\ 0,\ 0) & 0 \\ \frac{\partial h}{\partial x}(0,\ 0,\ 0) & \frac{\partial h}{\partial u}(0,\ 0,\ 0) & \frac{\partial f}{\partial v}(0,\ 0,\ 0) & -I_m \end{bmatrix} = m+n \tag{5}$$

Let $\alpha \in R^{m+q}$ consists of components of u, v, x and y, which correspond to the $m+q$ linearly dependent columns in (5). Hence by the inverse function theorem [5], we can solve (3) for the remaining $m+n$ variables (components) that correspond to the $m+n$ linearly independent columns in (5). That is, there exist functions $U(\cdot)$, $V(\cdot)$, $X(\cdot)$ and $Y(\cdot)$, and an open neighborhood Γ of $0 \in R^{m+n}$ such that (3) hold for all $\alpha \in \Gamma$.□

The linearization family of the nonlinear system (1) about the constant operating point family (4) is given by

$$\begin{aligned} \frac{dx_\delta}{dt}(t) &= A(\alpha)x_\delta(t) + B(\alpha)u_\delta(t) + E(\alpha)v_\delta(t) \\ y_\delta(t) &= C(\alpha)x_\delta(t) + D(\alpha)u_\delta(t) + F(\alpha)v_\delta(t) \end{aligned} \tag{6}$$

where

$$\begin{aligned} &x_\delta(t) = x(t) - X(\alpha), \quad y_\delta(t) = y(t) - Y(\alpha) \\ &u_\delta(t) = u(t) - U(\alpha), \quad v_\delta(t) = v(t) - V(\alpha) \end{aligned} \tag{7}$$

$$\begin{aligned} &A(\alpha) = \frac{\partial f}{\partial x}(X(\alpha),U(\alpha),V(\alpha)), \quad B(\alpha) = \frac{\partial f}{\partial u}(X(\alpha),U(\alpha),V(\alpha)) \\ &E(\alpha) = \frac{\partial f}{\partial v}(X(\alpha),U(\alpha),V(\alpha)), \quad C(\alpha) = \frac{\partial h}{\partial x}(X(\alpha),U(\alpha),V(\alpha)) \\ &D(\alpha) = \frac{\partial h}{\partial u}(X(\alpha),U(\alpha),V(\alpha)), \quad F(\alpha) = \frac{\partial h}{\partial v}(X(\alpha),U(\alpha),V(\alpha)) \end{aligned} \tag{8}$$

CH3229-2/92/0000-2487$1.00 © 1992 IEEE

We consider nonlinear control laws of the following form

$$u(t)=k(x(t),\ w(t)) \tag{9}$$

where $w(t)\in R^m$ is the external input with constant operating point family $W(\alpha)$. Its linearization family [8] is given by

$$u(t)-U(\alpha)=K(\alpha)[x(t)-X(\alpha)]+M(\alpha)[w(t)-W(\alpha)] \tag{10}$$

where

$$U(\alpha)=k(X(\alpha),W(\alpha))$$

$$K(\alpha)=\frac{\partial k}{\partial x}(X(\alpha),W(\alpha)) \tag{11}$$

$$M(\alpha)=\frac{\partial k}{\partial w}(X(\alpha),W(\alpha))$$

So, for the linearization of the closed-loop nonlinear system about any constant operating point in $\mathscr{E}$, the transfer function from $w_\delta(t)=w(t)-W(\alpha)$ to $y_\delta(t)$ is

$$H^1_\alpha(s)=[C(\alpha)+D(\alpha)K(\alpha)][sI_n-A(\alpha)-B(\alpha)K(\alpha)]^{-1}B(\alpha)M(\alpha)+D(\alpha)M(\alpha) \tag{12}$$

where I_n is the identity matrix of dimension n; and the transfer function from $v_\delta(t)$ to $y_\delta(t)$ is

$$H^2_\alpha(s)=[C(\alpha)+D(\alpha)K(\alpha)][sI_n-A(\alpha)-B(\alpha)K(\alpha)]^{-1}E(\alpha)+F(\alpha) \tag{13}$$

To achieve steady-state tracking and disturbance rejection, we require that, for all $\alpha\in\Gamma$,

$$H^1_\alpha(s)|_{s=0}=I_m\ ,\qquad H^2_\alpha(s)|_{s=0}=0_{m\times q} \tag{14}$$

Following the method of Preuss [4], we have that if

$$[M^{-1}-D\quad -F]\ [B\quad E]^\dagger\ [B\quad E]=[M^{-1}-D\quad -F] \tag{15}$$

then (14) can be satisfied by selecting

$$\begin{aligned} K(\alpha)&=-M(\alpha)[L(\alpha)A(\alpha)+C(\alpha)]\\ L(\alpha)&=[M^{-1}(\alpha)-D(\alpha)\quad -F(\alpha)]\ [B(\alpha)\quad E(\alpha)]^\dagger\\ &\quad+P(\alpha)\left[I_n-[B(\alpha)\quad E(\alpha)][B(\alpha)\quad E(\alpha)]^\dagger\right] \end{aligned} \tag{16}$$

where † denotes the Moore-Penrose pseudo-inverse of a matrix [1], $P(\alpha)$ is an arbitrary $m\times n$ matrix function of $\alpha\in\Gamma$, and $M(\alpha)$ is the (nonsingular) input gain in the feedback law (10). So we can select $P(\alpha)$ and $M(\alpha)$ such that $A(\alpha)+B(\alpha)K(\alpha)$ is stable. Substituting $L(\alpha)$ into $K(\alpha)$, we can write $K(\alpha)$ as

$$K(\alpha)=K_o(\alpha)+M(\alpha)P(\alpha)\tilde{C}(\alpha) \tag{17}$$

where

$$\begin{aligned} K_o(\alpha)&=[M(\alpha)D(\alpha)-I_m\quad M(\alpha)F(\alpha)]\ [B(\alpha)\quad E(\alpha)]^\dagger A(\alpha)\\ &\quad-M(\alpha)C(\alpha)\\ \tilde{C}(\alpha)&=(I_n-[B(\alpha)\quad E(\alpha)]\ [B(\alpha)\quad E(\alpha)]^\dagger)A(\alpha) \end{aligned} \tag{18}$$

and the parameterized closed-loop A matrix is

$$A_c(\alpha)=A(\alpha)+B(\alpha)K(\alpha)=\tilde{A}(\alpha)+\tilde{B}(\alpha)P(\alpha)\tilde{C}(\alpha) \tag{19}$$

where

$$\tilde{A}(\alpha)=A(\alpha)+B(\alpha)K_o(\alpha)\ ,\qquad \tilde{B}(\alpha)=B(\alpha)M(\alpha) \tag{20}$$

So, $M(\alpha)$ can be use to stabilize $\tilde{A}(\alpha)$ first; then $P(\alpha)$ is utilized to stabilize $A_c(\alpha)$.

Finally, we are to calculate a nonlinear control law of the form (9) such that (11) is satisfied with the specified $K(\alpha)$ and $M(\alpha)$. That is, we show that (10) is a feedback linearization family [6], and calculate a nonlinear feedback to "realize" it.

Proposition 2: Assume that, for all $\alpha\in\Gamma$, (15) is satisfied and

$$rank\left[\left(\frac{\partial x}{\partial\alpha}(\alpha)\right)^T\ \left(\frac{\partial u}{\partial\alpha}(\alpha)\right)^T\right]^T=m+q \tag{21}$$

(i.e., of full column rank). Then the parameterized linear feedback (10) with $K(\alpha)$ and $M(\alpha)$ specified as in (16) and with $Y(\alpha)=W(\alpha)$ is a linearization family, and a specific nonlinear feedback that satisfies (11) is given by

$$\begin{aligned} k(x(t),\ y(t),\ w(t))=U(\alpha)&+K(\alpha)[x(t)-X(\alpha)]\\ &+M(\alpha)[w(t)-Y(\alpha)]|_{\alpha=S^{-1}(s(t))} \end{aligned} \tag{22}$$

where $S^{-1}(.)$ is the inverse function of $S(\alpha)$; and $S(\alpha)$ and $s(t)$ consist of components of $X(\alpha)$, $Y(\alpha)$ and $x(t)$, $y(t)$, respectively.

Proof: Under the assumption (21), the necessary and sufficient condition for (10) to be a feedback linearization family is [6]

$$\frac{\partial U}{\partial\alpha}(\alpha)=K(\alpha)\frac{\partial X}{\partial\alpha}(\alpha)+M(\alpha)\frac{\partial W}{\partial\alpha}(\alpha)\ ,\qquad \text{for all } \alpha\in\Gamma \tag{23}$$

Since (6) is a linearization family, we have [7]

$$\begin{aligned} A(\alpha)\frac{\partial X}{\partial\alpha}(\alpha)+B(\alpha)\frac{\partial U}{\partial\alpha}(\alpha)+E(\alpha)\frac{\partial V}{\partial\alpha}(\alpha)&=\mathbf{0}\\ C(\alpha)\frac{\partial X}{\partial\alpha}(\alpha)+D(\alpha)\frac{\partial U}{\partial\alpha}(\alpha)+F(\alpha)\frac{\partial V}{\partial\alpha}(\alpha)&=\frac{\partial Y}{\partial\alpha}(\alpha) \end{aligned} \tag{24}$$

Substituting (23) into (24) and solving for $[\partial X/\partial\alpha](\alpha)$ gives

$$\begin{aligned} \frac{\partial X}{\partial\alpha}(\alpha)&=-[A(\alpha)+B(\alpha)K(\alpha)]^{-1}\\ &\quad\times\left[B(\alpha)M(\alpha)\frac{\partial W}{\partial\alpha}(\alpha)+E(\alpha)\frac{\partial V}{\partial\alpha}(\alpha)\right] \end{aligned} \tag{25}$$

Then, substitute the above result back into (23) and use the definitions in (12) and (13), we have

$$H^1_\alpha(0)\frac{\partial W}{\partial\alpha}(\alpha)+H^2_\alpha(0)\frac{\partial V}{\partial\alpha}(\alpha)=\frac{\partial Y}{\partial\alpha}(\alpha)\ ,\qquad \alpha\in\Gamma \tag{26}$$

By assumption (15), there exist $K(\alpha)$ and $L(\alpha)$ (hence $M(\alpha)$) so that (14) is satisfied. Therefore, we have $[\partial W/\partial\alpha](\alpha)=\partial Y/\partial\alpha(\alpha)$, and we can choose $W(\alpha)=Y(\alpha)$, $\alpha\in\Gamma$. So (10) is a feedback linearization family with $W(\alpha)=Y(\alpha)$ and $K(\alpha)$ and $M(\alpha)$ as in (16).

Furthermore, if the assumption (21) holds, we can construct a static nonlinear feedback as follows. Since

$$\begin{bmatrix}\frac{\partial X}{\partial\alpha}(\alpha)\\ \frac{\partial U}{\partial\alpha}(\alpha)\end{bmatrix}=\begin{bmatrix}I_n & 0_{n\times m}\\ K(\alpha) & M(\alpha)\end{bmatrix}\begin{bmatrix}\frac{\partial X}{\partial\alpha}(\alpha)\\ \frac{\partial W}{\partial\alpha}(\alpha)\end{bmatrix}$$

and $M(\alpha)$ is nonsingular for $\alpha\in\Gamma$, assumption (21) implies that

$$\left[\left(\frac{\partial X}{\partial\alpha}(\alpha)\right)^T\ \left(\frac{\partial W}{\partial\alpha}(\alpha)\right)^T\right]^T$$

is of full column rank $m+q$. Using the relation $W(\alpha)=Y(\alpha)$, we have

$$rank\left[\left(\frac{\partial X}{\partial\alpha}(\alpha)\right)^T\ \left(\frac{\partial Y}{\partial\alpha}(\alpha)\right)^T\right]^T=m+q\ ,\qquad \alpha\in\Gamma \tag{27}$$

Thus, we can form a locally invertible function $S(\alpha):\Gamma\to R^{m+q}$, by selecting $m+q$ functions from among the components of $X(\alpha)$ and $Y(\alpha)$ corresponding to linearly indpendent rows in (27), and denote its inverse function by $S^{-1}(\cdot)$. Let $s(t):R\to R^{m+q}$ be the function obtained by selecting the same components from $x(t)$, $y(t)$ and $w(t)$. Then it is easy to check that the particular nonlinear feedback control law (22) does satisfying (11).□

So $s(t)$ is used as the scheduling variable to schedule the gain of the feedback. Unlike the traditional gain scheduling, this nonlinear feedback provides an automatic gain scheduling by itself. Note that if any of the entries in $Y(\alpha)$ are selected to form $S(\alpha)$, then these entries in $y(t)$ are included in the nonlinear feedback control (22), which becomes a mixture of state and output feedback. We summarize the above development into the following theorem.

Theorem 1: Suppose that condition (21) and the continuity assumption of Proposition 1 both hold; and suppose further that (15) is satisfied. Then we can construct a nonlinear control law (22) such that the linearization family of the closed-loop nonlinear system about the constant operating point family (4) is such that the two transfer functions in (12) and (13) is such that

$$H^1_\alpha(s)|_{s=0} = I_m , \qquad H^2_\alpha(s)|_{s=0} = 0_{m\times q} \tag{14}$$

Remark: If (15) is not satisfied, then (16) will no longer be a solution to (14). Instead, (16) will be the best approximate solution such that the Eucledean norm $\|H^1_\alpha(s)|_{s=0}-I_m\|$ and $\|H^2_\alpha(s)|_{s=0}\|$ are both minimized. [1, p.104]

Now we discuss the stability and convergence of the closed-loop nonlinear systems. Let $\alpha^o(t)$: $R\to\Gamma$ and correspondingly, $x^o(t)=X(\alpha^o(t))$, $u^o(t)=U(\alpha^o(t))$, $v^o(t)=V(\alpha^o(t))$ and $y^o(t)=Y(\alpha^o(t))$. Obviously we have

$$\begin{aligned} f(x^o(t),u^o(t),v^o(t)) &= 0\\ h(x^o(t),u^o(t),v^o(t)) &= y^o(t)\\ k(x^o(t),y^o(t)) &= u^o(t)\end{aligned}$$

Then, a direct application of the stability and convergence theorem in [9] gives the following theorem on the stability and convergence of the closed-loop nonlinear system.

Theorem 2: Suppose that $f(x,u,v)$, $h(x,u,v)$ and $U(\alpha)$, $V(\alpha)$, $X(\alpha)$ and $Y(\alpha)$ are all twice continuously differentiable. Suppose further that for every operating point in (4), all eigenvalues of $A_c(\alpha)=A(\alpha)+B(\alpha)K(\alpha)$ have real parts no greater than $-\lambda$ for some $\lambda>0$. Then the closed nonlinear system with nonlinear control law (22) will be such that, for any given $\rho>0$, $T>0$, there exist $\delta_1(\rho)>0$, $\delta_2(\rho,T)>0$ such that if

$$\|x(t_0)-x^o(t_0)\|<\delta_1$$
$$\frac{1}{T}\int_t^{t+T}\left\|\begin{matrix}\dot{x}^o(\sigma)\\ \dot{v}^o(\sigma)\end{matrix}\right\|d\sigma<\delta_2, \qquad t>t_0$$

then

$$\|x(t)-x^o(t)\|<\rho, \qquad t>t_0$$
$$\|y(t)-y^o(t)\|<\rho, \qquad t>t_0$$

where $\|\cdot\|$ denotes vector norm. Furthermore, if for some $t_1>t_0$, $w(t)=W$, $v(t)=V$ for all $t>t_1$, then we have

$$\lim_{t\to\infty} y(t) = W$$

3 Application to a Power Generator

We consider an infinite bus, single machine power system which is described by the following differential equations [2]:

$$\begin{aligned}\dot{\delta}(t) &= \omega(t)\\ \dot{\omega}(t) &= -\frac{D}{H}\omega(t)-\frac{\omega_0}{H}[P_e(t)-P_m(t)]\\ \dot{E}_q'(t) &= \frac{k_f u_f(t)-E_q(t)}{T_{d0}}\end{aligned}$$

where $\delta(t)$ is the power angle of the generator; $\omega(t)$ is the relative angular speed of the generator; $E_q'(t)$ is the transient EMF in the quadrature axis of the generator; $P_m(t)$ is the mechanical input power; $P_e(t)$ is the active electrical power delivered by the generator, given by

$$P_e(t) = \frac{V_s E_q(t)}{x_{ds}}\sin\delta(t)$$

and E_q is the EMF in the quadrature axis, given by

$$E_q(t) = \frac{x_{ds}}{x'_{ds}}E'_q(t)-\frac{x_d-x'_d}{x'_{ds}}V_s\cos\delta(t)$$

The output of the system is taken as the generator terminal voltage

$$V_t(t)=\frac{1}{x_{ds}}\sqrt{x_s^2E_q^2(t)+V_s^2x_d^2+2x_sx_dx_{ds}P_e(t)\cot\delta(t)}$$

So the state of system is $x(t)=[\delta(t),\ \omega(t),\ E'_q(t)]^T$; the input (control) is $u(t)=u_f(t)$; and the output is $y(t)=V_t(t)$. The output (termainal voltage) V_t should always be 1, reguardless what P_m is. So we can take P_m as the disturbance $v(t)=P_m(t)$, and design a nonlinear control law such that for different constant values of (or slowly time-varying) P_m, output $y(t)$ is always 1 in steady-state.

Let $\alpha=[u_f\ \ P_m]\triangleq[\alpha_1\ \ \alpha_2]$. As $0\le P_m\le 1$, we can take the parameter set Γ as $R\times(0,1)$. The constant operating point family is given by

$$X(\alpha) = \begin{bmatrix}\arcsin\left(\dfrac{x_{ds}\alpha_2}{V_sk_c\alpha_1}\right)\\ 0\\ \dfrac{k_cx'_{ds}\alpha_1}{x_{ds}}+\dfrac{(x_d-x'_d)root_1}{k_cx_{ds}\alpha_1}\end{bmatrix}$$

$$U(\alpha) = \alpha_1 , \qquad V(\alpha) = \alpha_2 , \qquad Y(\alpha) = \frac{root_2}{x_{ds}}$$

where

$$root_1 = \sqrt{V_s^2k_c^2\alpha_1^2-x_{ds}^2\alpha_2^2} , \qquad root_2 = \sqrt{x_s^2k_c^2\alpha_1^2+V_s^2x_d^2+2x_sx_d\,root_1}$$

The linearization family is given by

$$A(\alpha)=\begin{bmatrix}0 & 1 & 0\\ -\dfrac{\omega_0 root_1}{x_{ds}H}-\dfrac{\omega_0(x_d-x'_d)x_{ds}\alpha_2^2}{Hx'_{ds}k_c^2\alpha_1^2} & -\dfrac{H}{D} & \dfrac{\omega_0x_{ds}\alpha_2}{Hx'_{ds}k_c\alpha_1}\\ -\dfrac{(x_d-x'_d)x_{ds}\alpha_2}{x'_{ds}T_{d0}k_c\alpha_1} & 0 & -\dfrac{x_{ds}}{x'_{ds}T_{d0}}\end{bmatrix}$$

$$C^T(\alpha)=\begin{bmatrix}\dfrac{x_s\alpha_2}{x'_{ds}k_c^2\alpha_1^2root_2}[(x_d-x'_d)(x_sk_c^2\alpha_1^2+x_droot_1)-x_dx'_{ds}k_c^2\alpha_1^2]\\ 0\\ \dfrac{x_s(x_sk_c^2\alpha_1^2+x_droot_1)}{x'_{ds}k_c\alpha_1root_2}\end{bmatrix}$$

$$B(\alpha)=\begin{bmatrix}0\\0\\\frac{k_c}{T_{d0}}\end{bmatrix},\quad E(\alpha)=\begin{bmatrix}0\\\frac{\omega_0}{H}\\0\end{bmatrix},\quad D(\alpha)=[0],\quad F(\alpha)=[0]$$

Since

$$[B\ \ E]^{\dagger}=\begin{bmatrix}0 & 0\\ 0 & \frac{\omega_0}{H}\\ \frac{k_c}{T_{d0}} & 0\end{bmatrix}^{\dagger}=\begin{bmatrix}0 & 0 & \frac{k_c}{T_{d0}}\\ 0 & \frac{H}{\omega_o} & 0\end{bmatrix}$$

(15) is satisfied. Denote

$$Q(\alpha) = \frac{H x_{ds} b_3(\alpha)}{\omega_0 k_c^2 \alpha_1 (x_d V_s^2 + x_s root_1)}$$

and let

$$M(\alpha) = \frac{x_{ds}' x_{ds} H\, T_{d0}\, root_2\, b_3(\alpha)}{x_s \omega_0 k_c^2 \alpha_1 (x_d V_s^2 + x_s\, root_1)} = \frac{x_{ds}' T_{d0}}{x_s} root_2\, Q(\alpha)$$

$$P(\alpha) = \begin{bmatrix}\dfrac{x_s \alpha_1\, root_1 + \left(\dfrac{x_{ds}(x_d - x_d')\alpha_2^2}{x_{ds}' k_c^2 \alpha_1^2} - \dfrac{H\, b_2(\alpha)}{\omega_0}\right) x_s x_{ds} \alpha_1}{x_{ds}^2 \alpha_2\, Q(\alpha)} \\ P_{12}(\alpha) \\ P_{13}(\alpha)\end{bmatrix}^T$$

Then we have $K(\alpha)=[K_{11}(\alpha)\ \ K_{12}(\alpha)\ \ K_{13}(\alpha)]$ with

$$K_{11}(\alpha) = \frac{x_{ds}(x_d - x'_d)\alpha_2}{x'_{ds} k_c^2 \alpha_1} - \left[x_s(x_d - x'_d) - x_d x'_{ds} + \frac{x_d(x_d - x'_d)\, root_1}{k_c^2 \alpha_1^2}\right] \times T_{d0}\alpha_2 Q(\alpha)$$

$$K_{12}(\alpha) = -\frac{T_{d0} x'_{ds} \alpha_1\, root_1}{x_{ds}^2 \alpha_2} + \frac{T_{d0} H x'_{ds} b_2(\alpha)\alpha_1}{x_{ds}\omega_0 \alpha_2} - \frac{T_{d0}(x_d - x'_d)\alpha_2}{k_c^2 \alpha_1}$$

$$K_{13}(\alpha) = \frac{x_{ds}}{k_c x'_{ds}} - \frac{T_{d0}(x_s k_c^2 \alpha_1^2 + x_d\, root_1)}{k_c \alpha_1} Q(\alpha)$$

and the above closed-loop characteristic polynomial is

$$\det[sI - [A(\alpha) + B(\alpha)K(\alpha)]] = s^3 + a(\alpha)\, b_3(\alpha) s^2 + b_2(\alpha) s + b_3(\alpha)$$

where

$$a(\alpha) = \frac{H x_{ds}(x_s k_c^2 \alpha_1^2 + x_d\, root_1)}{\omega_0 k_c^2 \alpha_1^2 (x_d V_s^2 + x_s\, root_1)}$$

Using Routh-Hurwitz criterior [8], it is easy to check that the closed-loop characteristic polynomial is stable for all $\alpha\in\Gamma$ if $b_2>1/a(\alpha)$, and $b_3>0$. A further analysis shows that in order to achieve a relative stability of r [8], i.e. all of poles are of a distance at least r to the left of the imaginary axis of the complex plane, we must have

$$b_3 > \frac{3r}{a(\alpha)}$$

$$\underline{b}_2 \triangleq \frac{b_3}{a(\alpha)b_3 - 2r} + 2r[a(\alpha)b_3 - 2r] < b_2 < \frac{b_3}{r} + [a(\alpha)b_3 - r]r \triangleq \overline{b}_2$$

Choose

$$b_3 = \frac{\sigma r}{a(\alpha)}, \quad \sigma>3$$

Then

$$\underline{b}_2 = \frac{\sigma}{\sigma-2}\frac{1}{a(\alpha)} + 2(\sigma-2)r^2, \qquad \overline{b}_2 = \frac{\sigma}{a(\alpha)} + (\sigma-1)r^2$$

Since we must have $\overline{b}_2(\alpha) > \underline{b}_2(\alpha)$ for all $\alpha\in\Gamma$, therefore, $r<\sqrt{\sigma/[(\sigma-2)a(\alpha)]}$, for all $\alpha\in\Gamma$. So there is a limit on how much relative stability we can achieve. Let the relative stability be

$$r(\alpha)=\mu(\alpha)\sqrt{\sigma/[(\sigma-2)a(\sigma)]}, \quad \mu(\alpha)=4/[3(\alpha_1+1)]<1$$

Taking $\sigma=3.01$ and plotting $r(\alpha)$ against α_1 for $0<\alpha_2<1$ shows that a relative stability of 4 can be achieved except for α_1 very close to 0. But it is easy to check that for α_1 very close to 0, we still have $b_2(\alpha)>1/a(\alpha)$, hence the closed system is still stable. By choosing

$$b_2(\alpha) = \frac{1}{2}[\underline{b}_2(\alpha) + \overline{b}_2(\alpha)]$$

we have the following parameterized linear feedback gains $K(\alpha)$ and $M(\alpha)$

$$K_{11}(\alpha) = \frac{(x_d - x_d')\alpha_2}{k_c^2 \alpha_1}\left[\frac{x_{ds}}{x_{ds}'} - T_{d0}\sigma r(\alpha)\right] + \frac{x_d x_{ds}' T_{d0}\sigma r(\alpha)\alpha_2\alpha_1}{x_s k_c^2 \alpha_1^2 + x_d\, root_1}$$

$$K_{12}(\alpha) = \frac{T_{d0} x_{ds}' \alpha_1}{x_{ds}\alpha_2}\left(\frac{H\, b_2(\alpha)}{\omega_0} - \frac{root_1}{x_{ds}}\right) - \frac{T_{d0}(x_d - x_d')\alpha_2}{k_c^2 \alpha_1}$$

$$K_{13}(\alpha) = \frac{x_{ds}}{k_c x_{ds}'} - \frac{T_{d0}\ \sigma r(\alpha)}{k_c}$$

$$M(\alpha) = \frac{T_{d0} x_{ds}'\ root_2\ \alpha_1\ \sigma r(\alpha)}{x_s(x_d k_c^2 \alpha_1^2 + x_d\, root_1)}$$

Next, since the Jacobian of $[X_1(\alpha), Y(\alpha)]^T$ is nonsingular, we can solve $x_1=X_1(\alpha)$ and $y=Y(\alpha)$ for α_1 and α_2 to get

$$\alpha_1 = \frac{\sqrt{x_{ds}^2 V_t^2 - x_d^2 V_s^2 \sin^2 x_1} - x_d V_s \cos x_1}{k_c x_s} \qquad (28)$$

$$\alpha_2 = \frac{V_s k_c \alpha_1}{x_{ds}}\sin x_1$$

The final nonlinear feedback control law is given by (27). Note that with α specified as above, $W(\alpha)$ evaluates to $y(t)=V_t(t)$ and $X_1(\alpha)$ evaluates to x_1. So this is a mixture of state and output feedback. The following parameters are used for simulation, with .

$\omega_0=314.159$, $D=5.0$, $H=8.0$, $T_{d0}=6.9$, $k_c=1.0$,
$x_d=1.863$, $x'_d=0.257$, $x_{ds}=2.4753$, $x'_{ds}=0.8693$,
$V_s=1.0$, $x_s=0.6123$, $\sigma=3.01$

The system is simulated with initial condition $\alpha(0)=1.23$ *rad* (70.6 deg), $\omega(0)=0$ and $E'_q(0)=0.86$. The mechanical power input (disturbance) P_m is taken as

$$P_m(t)=\begin{cases}1.0, & 0<t<5 \text{ sec;}\\ 0.5, & 5<t<10 \text{ sec;}\\ 0.2, & 10<t<15 \text{ sec.}\end{cases}$$

and the reference input $W(t)$ is fixed at 1. The response of $y(t)$ is given in Figure 1. We can see that the output voltage $V_t(t)$ (*i.e.*, $y(t)$) converges rapidly to the steady-state value of 1, regardless the value of the $P_m(t)$. The response of $\delta(t)$ is given in Figure 2.

Because of the complexity of the system and the resulting nonlinear control law, Taylor series expansion can be used to obtain approximate solutions. Rewrite the nonlinear control law (27) as

$$k(x(t),\ w(t)) = K_d(\alpha) + K_{12}(\alpha)x_2(t) + K_{13}(\alpha)x_3(t) + M(\alpha)[w(t) - y(t)]|_\alpha$$

where $|_\alpha$ denotes evaluation of α according to (28), and $K_d(\alpha) = \alpha_1 - K_{13}(\alpha)X_3(\alpha)$. Taking the cubic order Taylor expansion of $K_d(\alpha)$, $K_{12}(\alpha)$, $K_{13}(\alpha)$ and $M(\alpha)$ (with α specified as in (28)) about the $[x_1, y]$=[1.2324, 1], we obtain an approximate nonlinear control law. The responses of the closed-loop system using this approximate nonlinear controller are also given in Figures 1 and 2, which are very close to that using the full nonlinear control law.

4 Conclusion

The problem of steady-state tracking and disturbance rejection for nonlinear systems is studied using static feedback only. It is shown that under certain conditions, in steady state, the system output tracks the constant (or slowly time-varying) reference input while rejecting the constant (or slowly time-varying) disturbance. It is also shown that the tracking error can be made arbitrarily small if the reference input and the disturbance is sufficiently slowly time-varying. The method is applied to the control of an infinite bus, single machine power generator system.

5 References

[1] Ben-Israel, A., T.N.E. Greville, *Generalized Inverses: Theory and Applications*, John Wiley & Sons: New York, 1974.

[2] Bergan, A.R., *Power System Analysis*, Prentice-Hall: New Jersey, 1986.

[3] Huang, J., W.J. Rugh, "On a nonlinear servomechanism problem," *Automatica*, Vol. 26, No. 6, pp. 963 - 972, 1990.

[4] Preuss, H.-P., "Perfect steady-state tracking and disturbance rejection by constant state feedback," *Int. J. of Control*, Vol. 35, pp. 75 -94, 1982.

[5] Rudin, W., *Principles of Mathematical Analysis*, McGraw-Hill: New York, third edition, 1976.

[6] Wang, J., W.J. Rugh, "Feedback linearization families for nonlinear systems," *IEEE Transactions on Automatic Control*, Vol. AC-32, pp. 953 - 940, 1987.

[7] Wang, J., W.J. Rugh, "Parameterized linear systems and linearization families for nonlinear systems," *IEEE Trans. on Circuits and Systems*, Vol. CAS-34, pp. 650 - 657, 1987.

[8] Hostetter, G.H., C.J. savant, Jr., R.T. Stefani, *Design of Feedback Control Systems*, 2nd edition, Holt, Rinehart and Winston: Orlando, Florida, 1989.

[9] Lawrence, D.A., W.J. Rugh, "On a stability theorem for nonlinear systems with slowly varying inputs," *IEEE Trans. on Automatic Control*, Vol. AC-35, No. , pp. 860 - 864, 1990.

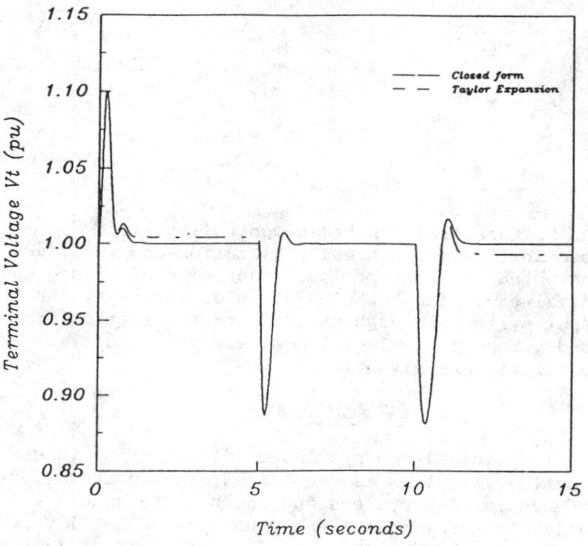

Figure 1: Output voltage response $V_t(t)$ for power system example.

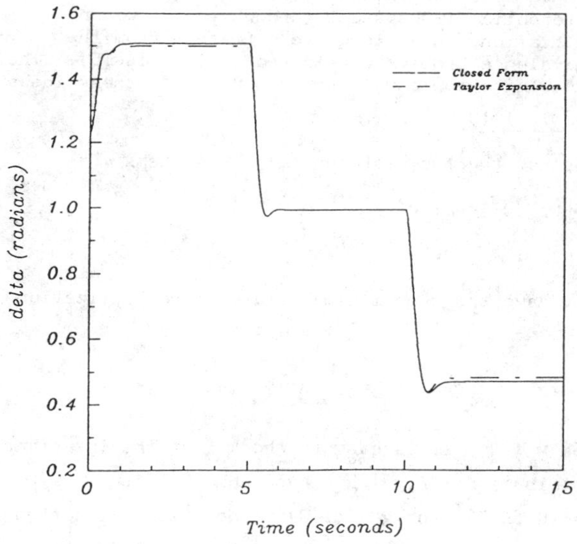

Figure 2: Response of $x_1(t)$ for power system example.

Proceedings of the 31st Conference on Decision and Control
Tucson, Arizona • December 1992

EQUIVALENCE OF INPUT-OUTPUT SYSTEMS

Xiaoyun Y. Lu and David. J. Bell

Department of Mathematics
UMIST
Manchester, M60 1QD, U.K.

ABSTRACT

Differential algebraic controllability, observability and minimal realization are considered upon which new concepts of (generic) equivalence and minimal equivalence of two given differential input-output systems through state space realization are based. Some criteria for equivalence and minimal equivalence are given.

1. INTRODUCTION

The equivalence problem of differential input-output(I-O) systems is considered. The relation between differential I-O systems and their state space realizations has been largely unexplored until recently [1-14]. By the realization of a differential I-O system we understand the realization of a set of I-O mappings which satisfy the I-O system. This is quite different from the realization of one I-O mapping. The realization of I-O systems is more formal and direct.

We consider in this paper the equivalence of two differential I-O systems through their common realization, which is quite different from the equivalence problem considered in [4,8,14,15]. The following example will help to motivate these ideas.

Example 1 [3]

Consider the two differential I-O systems

$$\Gamma_1 : (y^{(2)})^3 = 27\, u^3\, \dot{y}^2 \qquad (1)$$

$$\Gamma_2 : \quad u^2 y^{(4)} = u\, y^{(2)} u^{(2)} + 3u\dot{u}y^{(3)} - 3\,\dot{u}^2 y^{(2)} + 6u^5 \qquad (2)$$

They have the same minimal state space realization

$$\dot{x}_1 = u$$
$$\dot{x}_2 = x_1^3$$
$$y = x_2$$

when $x_1 \neq 0$. It is easy to check that the invariant distribution $\langle f,g \mid g \rangle$ has rank 2 ,with $f = (0\ ,x_1^3)^T$, and $g = (1,0)^T$. So this realization is differentially algebraically controllable. It is also differentially algebraically observable because (x_1,x_2) is a set of algebraically independent elements over $K\langle u\rangle$. So this is a differential algebraic minimal realization (cf. Section II below). Of course, it is also a minimal realization in the sense of [16,17].

From Example 1 a general problem may be posed: without realizing them, how can we check that two differential I-O systems produce the same input-state-output(I-S-O) realization (Σ,x), or the same minimal realization, for all $x \in X$, where X is a proper state manifold? At first sight, it may be thought that this will be the case if and only if one system can be obtained from the other by differentiation and elimination. But this is not true. For, consider the following two examples:

Example 2

$$\text{Given } \Gamma: \quad \begin{aligned} &\dot{y}_1 - 2u_2^2 + u_1 y_2 = 0 && (3)\\ &y_2^{(2)} - y_1^2 + u_1 u_2 = 0 && (4)\end{aligned}$$

If we only differentiate (3) or (4) respectively, we get

$$\Gamma_1 : \quad \begin{aligned} &y_1^{(2)} - 4u_2\dot{u}_2 + \dot{u}_1 y_2 + u_1\dot{y}_2 = 0\\ &y_2^{(2)} - y_1^2 + u_1 u_2 = 0\end{aligned}$$

$$\Gamma_2 : \quad \begin{aligned} &\dot{y}_1 - 2u_2^2 + u_1 y_2 = 0\\ &y_2^{(3)} - 2y_1\dot{y}_1 + \dot{u}_1 u_2 + u_1\dot{u}_2 = 0\end{aligned}$$

Clearly, Γ_1 and Γ_2 have the same minimal realization as Γ, but one can neither get Γ_1 from Γ_2, nor Γ_2 from Γ_1 simply by differentiation and elimination. In general, the situation is even more complicated.

Example 3

In Example 1, we can check that Γ_2 can be obtained from Γ_1 by differentiation and elimination. Clearly, $d^3/dt^3(\Gamma_1)$ has the same minimal realization as Γ_1, (and as Γ_2), but one can check that $d^3/dt^3(\Gamma_1)$ surprisingly cannot be obtained from Γ_2 under differentiation and elimination.

In this paper we use differential algebra [18] to consider differential algebraic systems [19,20,21]. Observability and controllability are considered further [22]. New concepts of (generic) equivalence and minimal equivalence of differential I-O systems (through state space representation) are defined. Then some criteria for equivalence and minimal equivalence are given.

II. CONTROLLABILITY, OBSERVABILITY AND MINIMALITY

The concept of controllability used in this paper is from [23,24]. Observability, minimality, and general notation are those used in [22].

Observability

We consider observability of nonlinear algebraic differential control systems from the realization point of view.i.e. we start from the given prime differential I-O system

$$\Gamma: \phi_i(y,u) = 0\ , \quad \phi_i \in K\{u,y\},\ i=1,2,\dots,p. \qquad (5)$$

$K = \mathbf{R}\langle t\rangle$ is the differential field generated by t over $\mathbf{R}$, and $K\{u,y\}$ is the differential algebra generated by differential indeterminates $y = (y_1,\dots,y_p)$, and $u = (u_1,\dots,u_m)$ over K. We assume that u is the set of generically independent inputs. This is equivalent to the fact that u can be chosen as the set of differential parametric indeterminates for the system Γ. Without loss of generality, let Γ be a prime system, i.e., Γ generates a prime differential ideal in $K\{u,y\}$. So we can obtain a differential field extension of K by Γ,

$$K\langle\Gamma\rangle = K\langle y_o,u\rangle = Q(\ K\{u,y\}/\ \{\Gamma\}\ ),$$

where Q(S) is the localization of differential integral domain S [25]. In this framework, a set of states is a set of generators of a (pure) algebraic field $K\langle y_o,u\rangle$

CH3229-2/92/0000-2492$1.00 © 1992 IEEE

(simply dropping differentiation d/dt in $K\langle y_o,u\rangle$) over $K\langle u\rangle$. In some sense $K\langle y_o,u\rangle$ is the generalization of the concept *observation space* defined in [11]. Actually, $K\langle y_o,u\rangle$ is just the jet space of inputs and outputs [26] because we can always assume that $K\langle x,u\rangle = K\langle y_o,u\rangle$. In general, state space equations may be in an implicit form and/or contain derivatives of u. This is unavoidable when considering the inverse system of a given I-S-O system.

Definition 4 Suppose an I-O system Γ is a prime system and (Σ,x) is its state space realization with state variables $x = (x_1,\dots,x_n)$, $x_i \in K\langle y_o,u\rangle$. Then Σ is said to be observable if x is an algebraic transcendence basis of $K\langle y_o,u\rangle$ over $K\langle u\rangle$.

Remark 5 Observability only depends on the choice of state variables. This choice determines both the relations among state variables and those between states and outputs. This definition keeps the sense of observability that different states produce different input-output mappings [22].

Lemma 6 Given a prime system Γ as in (5). Let

$$\mathbf{A}: A_1,\dots,A_p$$

be its characteristic set with A_i containing only $y_1,\dots,y_i$ with y_i of order n_i. Then the set $x = (x_1,\dots,x_n)$, with $n_1+\dots+n_p = n$, is a set of algebraically independent elements of $K\langle y_o,u\rangle = K\langle\Gamma\rangle$ over $K\langle u\rangle$, introduced canonically as follows:

$$x_1 = y_{o1},\ x_2 = \dot{y}_{o1},\dots,\ x_{n_1} = y_{o1}^{(n_1-1)},$$

$$x_{n_1+1} = y_{o2},\ x_{n_1+2} = \dot{y}_{o2},\dots,\ x_{n_1+n_2} = y_{o2}^{(n_2-1)}$$

$$\dots\dots\dots\dots \qquad (6)$$

$$x_{n-n_p+1} = y_{op},\dots\dots,x_n = y_{op}^{(n_p-1)}.$$

Proof. By [18] **A** has the form

$$A_1(y_1,u),\ A_2(y_1,y_2,u),\dots,A_p(y_1,y_2,\dots,y_p,u)$$

and $A_i \in [\Gamma]$ which is the prime ideal determined by Γ. If we let $A_1 = 0$, then $y_{o1}^{(n_1)}$ is implicitly determined by $x_1,\dots,x_n$ over $K\langle u\rangle$. If we let $A_2 = 0$, then $y_{o2}^{(n_2)}$ is implicitly determined by $x_1,x_2,\dots,x_{n_1+n_2}$. Similarly, we see that all $y_{oi}^{(k_i)}$ with $k_i = 1,\dots,n_i$; $i = 1,\dots,p$, are determined by $x = (x_1,\dots,x_n)$ over $K\langle u\rangle$. If we differentiate A_i, we see that all derivatives of y_{oi} with higher order than n_i are also determined by x, $i = 1,\dots,p$. We have proved that

$$\mathrm{Al.Trd}^{o}(K\langle y_o,u\rangle : K\langle u\rangle) = n.$$

So x is a set of algebraically independent elements of $K\langle y_o,u\rangle$ over $K\langle u\rangle$. Q.E.D.

As for the justification of using a characteristic set of Γ to replace Γ itself in the realization process, this will be discussed in Section III below.

Controllability and Minimality

As in the case of observability, if we consider the state space as a jet space of inputs and outputs, then we have the definition of controllability based on a given prime differential I-O system [21,23,24].

Definition 7 Given a prime system (5), it is said to be controllable if K is differentially algebraically closed in $K\langle\Gamma\rangle = K\langle y_o,u\rangle$.

Remark 8 Intuitively, Definition 7 means that there is no element $\xi \in K\langle y_o,u\rangle$ but $\xi \notin K$ such that ξ satisfies a differential polynomial in one variable.

If we start from an I-S-O system, eg.

$$\dot{x} = f(x,u)$$
$$y = h(x,u)$$

we can still adopt Definition 7 for controllability by simply taking the state equation

$$\dot{x} = f(x,u)$$

and constructing outputs as $\bar{y} = x$.

So an alternative to Definition 7 is one based upon an I-S system as follows.

Definition 9 Let the state space system

$$\Sigma:\ \dot{x} = f(x,u)$$

be prime in $K\{x,u\}$. It is controllable if K is differentially algebraically closed in $K\langle x_o,u\rangle = Q(K\{x,u\}/\{\Sigma\})$, where $\{\Sigma\}$ is the prime differential ideal determined by Σ.

Indeed, the work of [27] discussed the relation between Definition 9 and the definition of controllability in a differential geometric setting [17] for an affine I-S system (or an I-O system which has an affine realization)

$$\dot{x} = f(x) + \sum_{i=1}^{m} g_i(x)\,u_i\ ,$$

where $f(x) = (f_1(x),\dots,f_n(x))^T$, $g_i(x) = (g_{i1}(x),\dots,g_{in}(x))^T$. It is proved in [27] that Definition 9 is equivalent to the fact that the distribution

$$\Delta = \langle f,g_1,\dots,g_n \mid g_1,\dots,g_n\rangle$$

has full rank n.

The condition that Δ has full rank is stronger than the controllability rank condition [17] where the distribution $\hat{\Delta} = \langle f,g_1,\dots,g_n \mid f,g_1,\dots,g_n\rangle$ is adopted.

Remark 10 Suppose Γ is an arbitrarily given prime I-O system. If we choose the state $x = (x_1,\dots,x_n)$ as an algebraic transcendence basis of $K\langle\Gamma\rangle$, then whether the system (Σ,x) is controllable or not in the sense of Definition 9 depends on the structure of the I-O system. As discussed in [22], to assume controllability (and hence minimality) we sometimes need to shrink to a maximum subfield F of $K\langle y_o,u\rangle$

$$K \subset F \subseteq K\langle y_o,u\rangle$$

such that K is differentially algebraically closed in F. There is a possibility that some output y_{oi} may not appear in F. This is the case in geometric control theory.

The following result is a condition which excludes this possibility.

Proposition 11 Given a prime differential I-O system Γ as in (5), then in the realizaton of Γ, all outputs are included in F (minimal realization), if and only if Γ has a characteristic set of the form

$A_1(y_1,u)$

$A_2(y_1,y_2,u)$

$\ldots\ldots\ldots\ldots\ldots$

$A_p(y_1,\ldots,y_p,u)$

with the ranking

$$y_1<y_2<\ldots<y_p<\dot{y}_1<\ldots<\dot{y}_p<\ldots<u_1<\ldots<u_m<\dot{u}_1<\ldots<\dot{u}_m<\ldots \quad (7)$$

under any exchange of subscripts of $y_1,\ldots,y_p$.

Proof. Necessary condition: By [18], for some fixed order of subscripts on the y_i's and the ranking (7), Γ has a characteristic set of the form

$B(y_1)$

$B(y_1,y_2)$

$\ldots\ldots\ldots$

$B(y_1,y_2,\ldots,y_r)$

$B(y_1,\ldots,y_{r+1},u)$

$\ldots\ldots\ldots\ldots\ldots\ldots$

$B(y_1,\ldots,y_p,u)$.

If $r \neq 0$, then the system $B(y_1)=0$,..., $B(y_1,\ldots,y_r)=0$ defines an autonomous system, and clearly y_{o1} is an autonomous element. So any subfield of $K<y_o,u>$ containing $y_{o1},\ldots,y_{or}$, corresponds to an noncontrollable realization.

Sufficient condition: Assume $y_{jo} \notin F$ for some j. Then y_{jo} must be differential algebraic over K. Therefore, there exists a non-trivial differential polynomial $\beta(.)$ in one variable over K such that $\beta(.)$ is of minimal order and minimal degree and $\beta(y_{jo}) = 0$.

So if we choose $j = 1$ and reorder the subscripts of y_i's, we obtain a characteristic set of the form $D(y_1)$,.... This yields a contradiction. Q.E.D.

Remark 12 If a prime I-O system satisfies the condition of Proposition 11, then each output y_{oi} is affected by input u. This is not the controllability in Definiton 7.

In what follows, all prime systems satisfy the condition of Proposition 11.

Proposition 13 Consider a prime system Γ (5) with canonical state variables (6) and suppose that the state space representation (Σ,x) is controllable. Then, for an arbitrarily chosen state $\bar{x} = (\bar{x}_1,\ldots,\bar{x}_n)$ as an algebraic transcendence basis of $K<y_o,u>$, the realization is still controllable.

Proof. When X is an algebraic transcendence basis and (Σ,x) is controllable, then (Σ,x) is a minimal realization. This is also the case when x is canonically introduced as in Lemma 6. Any two minimal realizations of Γ differ by an algebraic transformation over $K<u>$ [22]. So (Σ,x) and $(\bar{\Sigma},\bar{x})$ must have the same property, controllable or noncontrollable, since an algebraic transformation over $K<u>$ transforms an autonomous elememt of $K<y_o,u>$ into an autonomous element. Q.E.D.

Example 14 The given system

$$\Gamma:\ uy^{(2)} - \dot{u}\dot{y} - u^3 = 0 \quad \text{(it is prime when } u \neq 0\text{)},$$

has a faithful realization (cf. Section III below)

$$\dot{x}_1 = x_2u + d_1u \qquad (d_1 \text{ is a constant})$$

$$\dot{x}_2 = u$$

$$y = x_1 .$$

It is easy to see that $\omega = \dot{x}_1 - x_2\dot{x}_2 - d_1\dot{x}_2$ is an autonomous element. Indeed $\dot{\omega} = 0$. So this system is not controllable.

If we introduce canonical state variables as in (6), we have a faithful realization

$$\dot{z}_1 = z_2$$

$$\dot{z}_2 = (\dot{u}/u)\ z_2 + u^2$$

$$y = z_1 .$$

Clearly, we have $z_1 = x_1$

$$z_2 = x_2u + d_1u.$$

Substituting in ω yields another autonomous element

$$\bar{\omega} = \dot{z}_1 - (\ (z_2-d_1u)/u - 1\ )\ \frac{d}{dt}(\ (z_2-d_1u)/u\ )\ ,$$

with $d/dt(\ \bar{\omega}\ ) = 0$.

From this example, we see that differential algebraic controllability has the advantage that it is adaptable to a system which contains derivatives of u, even in an implicit form. This cannot be achieved in differential geometric control theory.

III. FAITHFUL REALIZATION

A new concept was proposed in [22], namely the faithful realization of a given differential I-O system. We now discuss further the critical role this concept plays in the realization of differential I-O systems.

Definition 15 Given a differential I-O system Γ as in (5), let (Σ,x) be its state space representation. Let Ω be the differential algebraic manifold determined by Γ, and $\tilde{\Omega} = \bigcup_{x\in X}\Omega_x$, where Ω_x is the set of all input-output pairs given by the I-O mapping corresponding to x. If $\Omega = \tilde{\Omega}$ for a proper state manifold X, then we call (Σ,x) a *faithful realization* of Γ.

Remark 16 When Γ is prime, this is equivalent to saying that $K(\ x,\hat{u}\ ) = K<\Gamma>$.

In particular, if we choose $x = (x_1,\ldots,x_n) \subset K<\Gamma>$ as a set of generators, then $K(\ x,\hat{u})$ is a faithful realization, where $K(x,\hat{u})$ is the algebraic field generated by x and $\hat{u} = (\dot{u},u^{(2)},\ldots)$.

Remark 17 Proper choice of the state manifold gives much flexibility. In this sense, the *faithfulness* is a generic concept. In general, Ω and the total I-O manifold determined by (Σ,x) may differ by a lower dimensional algebraic set under the Zariski topology.

Lemma 18 Let Γ be a prime system as in (5) and **A** is a characteristic set of $[\Gamma]$ under the ranking (7). Then a faithful realization of **A** is a faithful realization of Γ. The reverse is also true.

Proof. (cf. [22])

If **A** is prime, then $[\mathbf{A}] = [\Gamma]$.

Now suppose **A** is not prime. Let S denote all the separants of **A** under different rankings (because the exchange of subscripts of y_i's changes the ranking system). Then S has a finite number of elements, say s_i, $i = 1,\ldots,r$. We know that Γ and the system **A** with $s_i \neq 0$ define the same differential algebraic manifold. So we have

$$\{ A \} = [\Gamma] \cap \Sigma_2 \cap \cdots \cap \Sigma_\tau$$

as an irredundant decomposition of the perfect differential ideal {**A**} into prime components [18]. $[\Gamma]$ contains no separant of **A** and each Σ_i ($i>1$) contains at least one separant of **A**. Because Σ_i contains **A** as well, it follows that $K\langle\Sigma_i\rangle$ ($i>1$) has strictly lower algebraic dimension than $K\langle\Gamma\rangle$ over $K\langle u\rangle$. If we introduce canonical state variables x as in (6), then $s_i \neq 0$, ($i = 1,\ldots,r$) just define the proper state manifold X. Clearly, with such a state x, the realization is faithful. Q.E.D.

Remark 19 This lemma also justifies the work in [9] in which the author tries to use characteristic sets to give the corresponding I-O system of an I-S-O system (elimination) and an I-S-O system of an I-O system (realization).i.e. the relation between the I-S-O system and the corresponding I-O system must be generic.

IV. EQUIVALENCE OF DIFFERENTIAL I-O SYSTEMS

Definitions

In our opinion, there are two kinds of equivalence problems arising from differential I-O systems. One can be called "exact equivalence". This equivalence problem compares two differential algebraic manifolds without resorting to the realization. In this case, two differential I-O systems Γ_1, Γ_2 are said to be equivalent if they produce the same I-O behaviour [4,8,14,15]. Algebraically, Γ_1 and Γ_2 are said to be equivalent in this sense if and only if they determine the same perfect differential ideal, i.e. $\{\Gamma_1\} = \{\Gamma_2\}$ in $K\{y,u\}$.

Equivalence problems considered below may be called "generic equivalence", through the state space realization.

Definition 20 Two differential I-O systems Γ_1 and Γ_2 are said to be *equivalent* if they have the same faithful realization (irrespective of initial conditions).

Motivated by Example 1 we have

Definition 21 Two differential I-O systems Γ_1 and Γ_2 are said to be *minimal equivalent* if they have the same minimal realization (irrespective of initial conditions).

Criteria for Equivalence

Proposition 22 Given two SISO systems Γ_1: ϕ ,and Γ_2: ψ with $\phi,\psi \in K\{y,u\}$. If ϕ and ψ are algebraically irreducible, then ϕ and ψ are equivalent if and only if $\phi = c.\psi$, $c \in K$.

Proof. By a result of [18,p.13] we have the prime decompositions

$$\{\phi\} = \Sigma_1 \cap \cdots \cap \Sigma_r$$

$$\{\psi\} = \tilde{\Sigma}_1 \cap \cdots \cap \tilde{\Sigma}_s ,$$

where each Σ_i ($i>1$) contains at least one separant of ϕ, and each $\tilde{\Sigma}_j$ ($j>1$) contains at least one separant of ψ. So the faithful realization of ϕ and ψ are determined by the general solutions Σ_1 and $\tilde{\Sigma}_1$ respectively. So ϕ and ψ have the same faithful realization if and only if $K\langle\Sigma_1\rangle \cong K\langle\tilde{\Sigma}_1\rangle$. i.e. $\Sigma_1 = \tilde{\Sigma}_1$. Clearly, ϕ ,$\psi \in \Sigma_1$ and ϕ and ψ have the same order because the highest order of ϕ and ψ are determined by the system order of Σ_1 and Σ_2 and they are the same. So ϕ is divisible by ψ and ψ is divisible by ϕ. So ϕ and ψ only differ by a factor in K. Q.E.D.

So in the SISO case, the "generic equivalence" of two algebraically irreducible systems is the same as the "exact equivalence". This is also true for prime systems in the MIMO case.

Corollary 23 Two prime systems Γ_1 and Γ_2 as in (5) are equivalent if and only if $[\Gamma_1] = [\Gamma_2]$. Alternatively, one system can be obtained from the other by differentiation and elimination.

Proof. Direcly from Remark 16. Q.E.D.

For two general I-O systems to be equivalent, the situation is more complicated and will be discussed in a future report.

Criteria for Minimal Equivalence

In this section, we restrict ourselves to prime systems. By the discussion in Section II, two prime systems Γ_1 and Γ_2 have the same minimal realization if and only if there exists a subfield F with

$$K \subset F \subseteq K\langle\Gamma_1\rangle ,$$
$$K \subset F \subseteq K\langle\Gamma_2\rangle ,$$

such that (i) F is maximum in both $K\langle\Gamma_1\rangle$ and $K\langle\Gamma_2\rangle$, and
(ii) K is differentially algebraically closed in F.

Because F is a field, then according to [25], there exists a prime differential system, say $\Gamma_o \subset K\{y,u\}$, such that $F = K\langle\Gamma_o\rangle$. Actually, we can take Γ_o as a set of generators of Ker τ, where τ is the specialization (epimorphism over K) $\tau : K\{y,u\} \longrightarrow F$. Because F is a field, $[\Gamma_o] = \text{Ker } \tau$ is a prime ideal in $K\{y,u\}$.

Definition 24 For a given prime I-O system $\Gamma \subset K\{y,u\}$, the prime I-O system Γ_o, corresponding to the minimal realization of Γ, is called the *minimal I-O system* with respect to Γ.

Lemma 25 For a given I-O system Γ, its corresponding minimal I-O system Γ_o is unique. Furthermore, $[\Gamma] \subset [\Gamma_o]$, and Γ and Γ_o have the same differential dimension and codimension.

Proof. Γ_o is unique up to equivalence because the subfield F, which corresponds to the minimal realization is unique.

As for the second statement, $Q(K\{y,u\}/[\Gamma_o])$ can be a subfield of $Q(K\{y,u\}/[\Gamma])$ only if $[\Gamma] \subset [\Gamma_o]$.

The last statement follows directly from the discussion in Proposition 11 and the assumption following. Q.E.D.

Corollary 26 Two prime I-O systems Γ_1 and Γ_2 have the same minimal realization (or are minimal equivalent) if and only if Γ_1 and Γ_2 have the same minimal I-O system.

Proof. Directly from Definition 24 and Lemma 25. Q.E.D.

Proposition 27 For a given prime system Γ, if there is no other prime system $\Gamma_o \subset K\{y,u\}$, such that $[\Gamma] \subsetneq [\Gamma_o]$, then Γ is a minimal I-O system, i.e. its faithful realization is a minimal realization.

Proof. For such a system Γ, consider

$$K \subset K\langle u\rangle \subset F \subset K\langle\Gamma\rangle$$

such that F is a maximum field in $K\langle\Gamma\rangle$ and K is differentially algebraically closed in F. Then $F \neq K\langle u\rangle$ by Proposition 11. Then $[\Gamma] \subset [\Gamma_o]$ from Lemma 25.

Now consider the morphism

$$\vartheta : K\{y,u\}/[\Gamma] \longrightarrow K\{y,u\}/[\Gamma_o] \text{ defined by}$$

$$\omega + [\Gamma] \longrightarrow \omega + [\Gamma_o], \quad \forall\, \omega \in K\{y,u\}.$$

It is easy to check that ϑ is a differential epimorphism.

$F \neq K\langle\Gamma\rangle \Rightarrow K\{y,u\}/[\Gamma] \cong K\{y,u\} \Rightarrow [\Gamma] \neq [\Gamma_o]$ and so contradicts the original assumption.

Proposition 28 Let ϕ and ψ be two SISO differential I-O systems. If ϕ and ψ have the same minimal realization, then there exist non-zero elements $\alpha_i, \beta_j \in K\{y,u\}$ such that

$$\sum_{i=0}^{M} \alpha_i \frac{d^i}{dt^i}(\phi) = \sum_{j=0}^{N} \beta_j \frac{d^j}{dt^j}(\psi) \qquad (8)$$

Proof. For $\phi \in K\{y,u\}$, we denote by $\partial(\phi)$ the order of ϕ with respect to y. Let A be the minimal I-O system corresponding to the common minimal realization of ϕ and ψ. Without loss of generality, we suppose $\partial(\phi) = m \geq \partial(\psi) = q$.

If $\partial(\psi) = \partial(A)$, then ψ is divisible by A [18,p.31] and so $A = c.\psi$ ($c \in K$) because ψ is algebraically irreducible. So we suppose $\partial(\psi) > \partial(A)$. By [18,p.64] we have unique expressions

$$S^{t_1}\phi = \sum_{\mu=1}^{q_1} c_\mu A^{i_{o\mu}} A_1^{i_{1\mu}} \cdots A_{m-n}^{i_{m-n,\mu}} \qquad (9)$$

$$S^{t_2}\phi = \sum_{\nu=1}^{q_2} c_\nu A^{k_{o\tau}} A_1^{k_{1\tau}} \cdots A_{q-n}^{k_{q-n,\tau}} \qquad (10)$$

where $A_i = d^i/dt^i(A)$, S is the separant of A, and all the powers are non-negative integers. Furthermore, $\partial(c_\tau), \partial(c_\mu) \leq \partial(A)$ and no term in the righthand side of (9) and (10) is free of A and its derivatives, and no coefficient c_μ or c_ν is divisible by A.

We rank the terms appearing in (9) according to their power sets in the following way:

$$(i_{o\mu}, i_{1\mu}, \dots, i_{m-n,\mu}) \qquad (11)$$

$$(i_{o\sigma}, i_{1\sigma}, \dots, i_{m-n,\sigma}) \qquad (12)$$

If e is the smallest number such that

$$i_{e\mu} = i_{e\sigma},\ i_{e+1,\mu} = i_{e+1,\sigma}, \dots, i_{m-n,\mu} = i_{m-n,\sigma},$$

but $i_{e-1,\mu} < i_{e-1,\sigma}$, then we say the term with power set (11) is lower than the term with power set (12). By assumption, no two terms in (9) have the same power set.

The highest term in (9) is called the leading term. $\psi \leq \phi$ if the leading term of ϕ is not lower than the leading term of ψ. The term "strictly lower" has the obvious meaning.

Now we begin an elimination process which leads to (8). By assumption, $\psi \leq \phi$.

(i) Step 1: Consider (9) as a polynomial of A_{m-n}. Clearly, we can choose non-zero $\phi_{11}, \phi_{12} \in K\{y,u\}$ with lowest possible order in y such that the remainder

$$R_1 = \phi_{11}(S^{t_1}\phi) - \phi_{12}\left(\frac{d^{m-q}}{dt^{m-q}}(S^{t_2}\psi)\right)^{i_{m-n,\mu}}$$

with $R_1 < \phi$ strictly holding, and R_1 has a similar expression as (9) with no term free of A and its derivatives because $R_1 \in [A]$.

If $\psi \leq R_1$, then we repeat this process by replacing ϕ with R_1 until we arrive at R_o^1 such that $R_o^1 < \psi$ strictly holds. We note that $R_o^1 \in [A]$, and R_o^1 is a linear combination of ϕ, ψ and their derivatives.

(ii) Step 2: If $\partial(A) < \partial(R_1^o)$, we use ψ and R_o^1 to take the place of ϕ and ψ in Step 1, and repeat Step 1. Then we arrive at R_o^2, with $R_o^2 \in [A]$ and is a linear combination of ϕ and ψ and their derivatives, and $R_o^2 < R_o^1$ strictly holds.

If $\partial(A) < \partial(R_o^2)$, then repeat this process by replacing ψ and R_o^1 with R_o^1 and R_o^2 respectively.

(iii) Repeat Step 2 if necessary until we finally arrive at R_o^* with $R_o^* < (1,0,\dots,0)$ strictly holding.

We have $R_o^* \in [A]$ and it is a linear combination of ϕ and ψ and their derivatives. In a similar expression of R_o^* in the form of (9), no term is free of A and its derivatives. This is possible only if $R_o^* \equiv 0$.

So we get (8) by removing all the terms containing ψ and its derivatives to the righthand side. Q.E.D.

V. CONCLUSION

Equivalence problems for two given differential I-O systems(Γ_1 and Γ_2)through their state space representation are considered in a differential algebraic framework. These problems are of two different kinds.

Γ_1 and Γ_2 are equivalent if they produce the same faithful realization. This kind of equivalence is actually an exact equivalence. Two prime systems Γ_1 and Γ_2 are equivalent if and only if they generate the same differential prime ideal.

Γ_1 and Γ_2 are said to be minimal equivalent if they produce the same differential algebraic minimal realization. This minimal realization corresponds to a unique prime I-O system. In the SISO case, if prime systems Γ_1 and Γ_2 are minimal equivalent, then $[\Gamma_1]$ and $[\Gamma_2]$ have a non-trivial intersection.

For MIMO systems, the necessary and sufficient condition for two prime I-O systems to be minimal equivalent will be discussed in a future report.

REFERENCES

[1] A.J. Van der Schaft, "A realization procedure for systems of nonlinear higher order differential equations", in Proceedings of the 10th IFAC World Congress, Munich, 1987, pp.97-102.

[2] A.J. Van der Schaft, "On realization of nonlinear systems described by higher order differential equations", Math.Systems Theory, Vol.19, pp.239-275, 1987.

[3] P.E. Crouch and F. Lamnabhi-Lagarrigue, "State space realizations of nonlinear systems defined by input-output differential equations", Lecture Notes in Control and Information Science, Vol.111, pp.138-149, 1988.

[4] A.J. Van der Schaft, "Transformations and representations of nonlinear systems" in Proceedings of the Conference 'Perspectives in Control Theory' (B. Jakubczyk et al., Eds.) Sielpia Poland 1988, pp.297-314.

[5] E.D. Sontag, "Bilinear realization is equivalent to existence of a singular affine differential i/o equation", Systems and Control Letters, Vol.11, pp.181-187, 1988.

[6] E.D. Sontag and Y.Wang, "Input-output equations and realizability" presented at the MTNS Symposium in Amsterdam, the Netherlands, June 19-23, 1989.

[7] A.J. Van der Schaft, "Representing a nonlinear state space system as a set of higher-order differential equations in the inputs and outputs", Systems and Control Letters, Vol. 12, pp. 151-160, 1989.

[8] A.J. Van der Schaft, "Transformations of nonlinear systems under external equivalence", in LNCIS 122, Springer Verlag, 1989, pp.33-43.

[9] S.T. Glad, "Nonlinear state space and input-output descriptions using differential polynomials", in LNCIS 122, Springer Verlag, 1989, pp. 182-189.

[10] S.T. Glad, "Differential algebraic modelling of nonlinear systems", in Report No. LiTH-ISY-1-1013, Linkoping University, 1989.

[11] Y. Wang and E.D. Sontag, "On two definitions of observation spaces", Systems and Control Letters, Vol. 13, pp. 279-289, 1989.

[12] A.J. Van der Schaft, "Structural properties of realizations of external differential systems", in *Proceedings of the Conference*, Nonlinear Control Systems Design (A. Isidori, Ed.), IFAC Symposia Series, Pergamon Press, Vol. 2, 1990, pp. 77-82.

[13] P.E. Crouch and F. Lamnabhi - Lagarrigue, "Realizations of input-output differential equations", presented at the MTNS Symposium in Japan, June, 1991.

[14] S. Diop, "Elimination in control theory", *Math. Control Signals Systems,* Vol.4, pp. 17-32, 1991.

[15] J.C.Willems, "Models for dynamics", *Dynamics Reported*, (U. Kirchgraber and H.O. Walther ed), John Wiley & Son Ltd., and B.G. Teubner, pp. 171-269, 1989.

[16] R. Hermann and A.J. Krener, "Nonlinear controllability and observability", IEEE Trans. Aut. Contr. Vol. 22, pp. 728-740, 1977.

[17] A. Isidori, *Nonlinear Control Systems*, Berlin: Springer Verlag, 1989.

[18] J.F. Ritt, *Differential Algebra*, New York: Dover, 1966.

[19] M. Fliess, "Nonlinear control theory and differential algebra", presented at the Conference on Modelling Adaptive Control, Sopron, Hungary, 1986.

[20] M. Fliess, "A differential algebraic approach to some problems of nonlinear control theory", Presented at the 26th IEEE Conference on Decision and Control, Los Angeles, U.S.A., 1987.

[21] M. Fliess, "State-variable representation revisited, application to some control problems", in *Proceedings of the Conference* Perspectives in Control Theory, Sielpia, Poland, 1988, pp. 26-38.

[22] X.Y. Lu and D.J.Bell, "Realization theory for differential algebraic input-output systems", Theoretical and Applied Mechanics Report TAM 8, UMIST, Manchester, England, 1991.

[23] J.F. Pommaret, *Lie Pseudogroups and Mechanics.* New York: Gordon and Breach, 1987, ch 7, pp. 507-527.

[24] M. Fliess, "Controllability revisited", to appear in a book dedicated to R.E. Kalman (A.C.Antoules Ed.), Springer Verlag, 1992.

[25] J.F. Pommaret, *Differential Galois Theory*, New York: Gordon and Breach, 1983, chs 1,3, pp. 203-257 and 447-557.

[26] D.J. Saunders, *The Geometry of Jet Bundles*, Cambridge University Press, 1989.

[27] A. Haddak, "Differential algebra and controllability", in *Proceedings of the Conference* Nonlinear Control Systems Design, IFAC Symposium, Capri, Italy, 1990, pp.13-16.

Proceedings of the 31st Conference
on Decision and Control
Tucson, Arizona • December 1992

OUTPUT REGULATION FOR NON LINEAR UNCERTAIN SYSTEMS

G. Bartolini*, P. Pydynowski*, T. Zolezzi+

*Department of Communication, Computer and System Sciences, Via Opera Pia 11A, 16145 Genova - ITALY. Fax +39 10 3532948, Tel +39 10 3532706, E-mail giob@dist.unige.it.

+Department of Mathematics, Via L. B. Alberti 4, 16145 Genova - ITALY. Tel +39 10 3538741, E-mail zolezzi@igecuniv.bitnet

Abstract

A control strategy proposed by the authors, in orders to reduce the effect of the chattering phenomenon in the control of uncertain dynamic system is extended to the case of non complete availability of system state. The proposed procedure is described, for the sake of simplicity by considering the regulation problem (the extension to the tracking problem is straightforward) of uncertain second order non linear systems.

Introduction

In previous works [1],[2] the case of non linear uncertain system with complete availability of the state have been dealt with by the authors using an approach inspired to the variable structure control theory. In [3] it is shown that with ideal discontinuous control law on suitable manifold in the state space it is possible to solve, with simple finite gain devices, problems whose classical solution would require continuous feedback control law with infinite loop gain.

The discontinuity has been considered a draw back of this approach and various attempts to smooth it have been presented in the literature [4] [5], with the obvious results that the control objectives are met only approximately and more complex analysis is required to describe the approximate behaviours.

The idea for the elimination of discontinuities, in the control law, presented by the authors consist in embedding the original system in a system of higher order in which the time derivative of the control explicitly appears. This signal is chosen to be discontinuous on a suitable manifold of the augmented state space following the a standard V.S.C. approach for the resulting system. Asymptotic stability or model tracking can be ensured for the original system which results to be controlled by a continuous signal.

The proposed procedure requires the introduction of a first order estimator and the asymptotic reaching of the sliding manifold can be guarantied by solving a differential inequality, which allows uncertainties in the plant equations. Additional features of this approach consist in avoiding oscillation of finite frequency and ambiguities in the description of the system's trajectories during sliding motion, which can appear if unmodelled dynamics are taken into account, as stressed in [6].

In this paper the approach is extended to the case of unavailable state by using a procedure very similar to that used to avoid chattering phenomenon.

For the sake of simplicity only the case of regulation problem of a second order uncertain non linear system with unavailable output derivative is considered. The estimation of the output derivative is carried out by a discontinuous estimator of a new kind giving rise to an error equation whose structure is identical to that analyzed in the previous works [1] [2]. In the literature similar results have been obtained only for perfectly known systems [7].

In the next sections a brief description of the previous work is presented while the last two sections contain a presentation of the estimation procedure and of complete dynamic output feedback scheme.

Problem Statement

We consider a second order nonlinear plant as a starting point to deal with more general case, as follows:

$$\frac{d^2y}{dt^2} = g(y,\dot{y},u) \tag{1}$$

and its canonical realization:

$$\dot{x}_1 = x_2 \tag{2a}$$

$$\dot{x}_2 = g(x_1,x_2,u) \tag{2b}$$

$y = x_1$ where $u \in U \subset R$.

Our approach consists in avoiding discontinuities on the control law without introduction "intentional" non idealities (dead zone, saturation, etc.) by using the following procedure. We perform the time derivative of equation (2b):

$x_3 = g(x_1,x_2,u)$

CH3229-2/92/0000-2498$1.00 © 1992 IEEE

$$\dot{x}_3 = g_x\dot{x} + g_u\dot{u} \qquad (2c)$$

If $\dot{x}_3$ is perfectly known the augmented system (2) could be dealt with standard V.S.C. method (in which the time derivatives $\dot{u}$ plays the role of the control law). That is the control derivatives $\dot{u}$ is chosen to be discontinuous on the manifold:

$$s(x_1,x_2,u) = - x_3 - c_2x_2 - c_1x_1 = 0 \qquad (3)$$

with amplitude depending on the possible residual uncertainties on $g_x(\cdot)$ and $g_u(\cdot)$.

The actual control signal u, the time integral of a function with discontinuities of the firs kind, results to be a continuous time function, thereby the chattering phenomenon would disappear.

Unfortunately the case in which x_3 is known is of no interest for our purpose because the feedback linearization approach [7] would suffices. Therefore we have considered the case in which is x_3 is unavailable, that is the case in which the plant equation is uncertain.

The available state case

Since $x_3=g(x_1,x_2,u)$ is unknown we consider an estimator $\hat{s}$ of the quantity s defined in (3):

$$\hat{s} = - z - c_2x_2 - c_1x_1 \qquad (4)$$

in which z is an estimation of x_3 provided by the first order estimator:

$$\begin{aligned}\dot{z} &= - c_2z + p\hat{s} - c_1x_2 = \\ &= - c_2(z+p) - (c_1+pc_2)x_2 - pc_1x_1 \qquad (5)\end{aligned}$$

We define $E=x_3-z$ as the estimation error, it results:

$$E = \hat{s} - s \qquad (6)$$

The control objective consists in achieving simultaneously the three conditions:

$$\lim_{t\to+\infty} \hat{s} = 0 \qquad (7a)$$

$$\lim_{t\to+\infty} E = 0 \qquad (7b)$$

$$|s| < \Delta \qquad \forall\, t \geq t_0 \qquad (7c)$$

where $\Delta>0$ is any fixed positive constant, $t=t_0$ is the first time instant when $\hat{s}$ crosses zero.

To this end let us consider the differential equations describing the evolution of $\hat{s}$ and E. With simple calculations we obtain:

$$\dot{\hat{s}} = - c_2E - p\hat{s} \qquad (8)$$

$$\dot{E} = - c_2E - p\hat{s} + [\frac{dg}{dt} + c_2g + c_1x_2] \qquad (9)$$

In a more compact form we have:

$$\begin{vmatrix} \dot{\hat{s}} \\ \dot{E} \end{vmatrix} = - \begin{vmatrix} p & c_2 \\ p & c_2 \end{vmatrix} \begin{vmatrix} \hat{s} \\ E \end{vmatrix} + \begin{vmatrix} 0 \\ 1 \end{vmatrix} \mu$$

where $\mu=\mu(x_1,x_2,\dot{u}) \triangleq$

$$=g_{x_1}x_2+g_{x_2}g+g_u\dot{u}+c_2g+c_1x_2 \qquad (10)$$

This equation is nonlinear since the input signal $\mu(t)$ depends in a nonlinear way on the state variables $\hat{s}$ and E.

As it results from the previous work, conditions (7) are fulfilled if, starting from the first time instance $(t=t_0)$ of the zero crossing of $\hat{s}$, the following differential inequality is satisfied:

$$\frac{\dot{\mu}}{\mu} < (c_2 + p) - \lambda \qquad (11)$$

The sign of m changes at any time instants at which $\hat{s}=0$. We require that:

$$\mu > \frac{(c_2+p)^2}{2c_2+p}\Delta \qquad (12a)$$

when commuting from $\hat{s}\leq 0$ to $\hat{s}>0$ and:

$$\mu < - \frac{(c_2+p)^2}{2c_2+p}\Delta \qquad (12b)$$

when commuting from $\hat{s}\geq 0$ to $\hat{s}<0$.

Here Δ, λ are any fixed positive constants $(\lambda<c_2+p<2p)$, while $\dot{\mu}$ is defined everywhere except in a countable set of time instants. In these instants $\dot{\mu}$ has a discontinuity of the first kind.

The existence of the infinite sequence of such time intervals is proved by analyzing the behaviour of the system (2). The constants Δ and c_2 are arbitraly chosen while the parameter p must be chosen in order to find a feasible signal $\dot{\mu}$ which satisfies inequality (11) in accordance with known bound of system uncertainties.

In the previous analysis the complete availability of the state (x_1,x_2) is assumed. In the next section we shall deal with the problem of estimating the component x_2, assumed unavailable, by means of a finite gain estimation.

A finite gain observer for uncertain system

Contrarily to what happens with the previous case where the unknown quantity x_3 is a function of the state and is bounded by known function of the same argument, in this case the unknown quantity is x_2, a component of the state. We shall use an analogous procedure to asymptotically estimate the unavailable state variable and to use its estimate in the previous control scheme. In order to obtain an estimate of x_2 let us introduce a second order dynamic system:

$$\dot{z}_1 = z_2 \tag{13a}$$

$$\dot{z}_2 = - cz_2 + w \tag{13b}$$

Consider the dynamics of the following error signals:

$$e_1 = x_1 - z_1 \tag{14a}$$

$$e_2 = x_2 - z_2 \tag{14b}$$

where e_2 is unavailable.

By using a approach similar to the previous one, we define a first order estimator:

$$\dot{z}_3 = - cz_3 - q\hat{\sigma} = - (c + q)z_3 - qce_1 \tag{15}$$

where $\sigma = e_2 + ce_1$, $\hat{\sigma} = z_3 + ce_1$ are respectively the ideal sliding manifold and the approximated one.

By defining $E_1 = \hat{\sigma} - \sigma$, we have:

$$E_1 + x_2 = z_3 + z_2 \tag{16}$$

thus, if

$$\lim_{t\to+\infty} E_1 = 0 \quad \text{then} \quad z_3 + z_2 \to x_2.$$

Analogously to the previous case, the equations describing the evolution of $\hat{\sigma}$ and E_1 are:

$$\dot{\hat{\sigma}} = - cE_1 - q\hat{\sigma} \tag{17}$$

$$\dot{E}_1 = - cE_1 - q\hat{\sigma} + \mu_1 \tag{18}$$

where $\mu_1 = \mu_1(w,x_1,x_2,u) = w - (g + cx_2)$ (19)

These equations are satisfied if the differential inequality:

$$\frac{\dot{\mu}_1}{\mu_1} < (c + q) - \lambda_1 \tag{20}$$

holds, requiring as before that;

$$\mu_1 > \frac{(c+q)^2}{2c+q} \Delta_1 \tag{21a}$$

from $\hat{\sigma} \leq 0$ to $\hat{\sigma} > 0$, and

$$\mu_1 < - \frac{(c+q)^2}{2c+q} \Delta_1 \tag{21b}$$

from $\hat{\sigma} \geq 0$ to $\hat{\sigma} < 0$,

where Δ_1, λ_1 are any fixed positive constants ($\lambda_1 < c+q < 2q$), while $\dot{\mu}_1$ is defined except in a countable set of time instants. In these instants $\dot{\mu}_1$ has a discontinuity of the first kind.

The a priori choice of the parameter q, sufficient to find a control law w causing inequality (20) to be fulfilled, requires some stronger assumption about the priori boundedness of the phase trajectories. For example the assumption that the phase trajectory are confined to a suitable open subset of R^2 could often be considered realistic (for example the case of manipulators with saturation level in the velocity components).

The dynamic output feedback

The connection between the schemes of the two previous section must be performed taking into account that, now, x_2 is not available. As a consequence of this fact $\hat{s}$ defined in (4) results to be a non measurable quantity.

In this case we choose:

$$\dot{z}' = - c_2 z' + p\,\hat{\sigma}' - c_1(z_2+z_3) \tag{22}$$

$$\hat{s}' = - z' - c_2(z_2+z_3) - c_1 x_1 \tag{23}$$

where is now available.

Taking into account equation (16):

$$\dot{\hat{s}}' = - c_2\dot{z}' - c_2(\dot{E}_1 + \dot{x}_2) - c_1 x_2$$

by simple calculation we obtain:

$$\dot{\hat{s}}' = - c_2 E - p\hat{s}' - c_2\dot{E}_1 + c_1 E_1 \tag{24}$$

$$\dot{E} = - c_2 E - p\hat{s}' + \mu \tag{25}$$

where

$$\mu = \mu(x_1,x_2,u,\dot{u}) \triangleq \frac{dg}{dt} + c_2 g + c_1(z_2+z_3) \tag{26}$$

Note that the system (24), (25) is different from the previous ones (8), (9) and (17), (18) due to the term $- c_2\dot{E}_1 + c_1 E_1$ appearing in (24).which is discontinuous on the.manifold $\hat{\sigma}=0$. Since a formal treatment of the equivalence between equations (24), (25) and (8), (9) is beyond the scope of this paper we shall give only a qualitative description of such a treatment.

Equation (24) is discontinuous therefore its solution must be related to the Filippov's solution concept or to the Utkin equivalence control method. Further since the discontinuous term appears affinely the approximability theory [6], [8] also applies.

REFERENCES

[1] Bartolini G., "Chattering phenomena in discontinuous control systems", Int. Journ. Systems SCI, vol. 20, n. 12, 1989.

[2] Bartolini G., Pydynowski P., "Asymptotic linearization of uncertain non linear system by means of continuous control", Int. Journ. of Robust and Non Linear Control, to appear.

[3] Young K.K.D., Kokotovic P.V. and Utkin V.I., "A singular perturbation analysis of high gain feedback systems", IEEE Trans. Automat. Contr., n. 22, pp . 931-939, 1979.

[4] A.S.I. Zinober, O.M.E. El-Ghezawi, and S. A. Billings, "Multivariable variable-structure adaptive model-following control systems", IEEE Proc., Jan. 1982.

[5] M. Corless and G. Leitmann, "Continuous state feedback guaranteeing uniform ultimate boundedness for uncertain dynamic systems", IEEE Trans. Automat. Contr., vol. AC-26, 1981.

[6] Utkin V.I., "Sliding Modes in Control and Optimization", Springer Verlag, in press.

[7] Isidori A., Byrnes C.I., "Output regulation of non linear systems", IEEE Trans. Automat. Contr., vol. AC-35, 1990.

[8] Bartolini B., Zolezzi T., "Control of nonlinear variable structure systems", Journ. of Math. Analysis and Applications, vol. 118, n. 1. Aug. 1986.

Proceedings of the 31st Conference
on Decision and Control
Tucson, Arizona • December 1992

Controlling Chaotic Continuous-Time Systems via Feedback

Xiaoning Dong and **Guanrong Chen**

Department of Electrical Engineering

University of Houston, Houston, TX 77204

Abstract. *In this paper, we study how to design a conventional nonlinear feedback controller to drive a chaotic trajectory of the well-known (continuous-time) Duffing system to any of its inherent periodic orbits. The more convenient, yet more difficult, linear feedback controllers design for the same purpose is left to the authors' forthcoming paper [3], which is a continuous counterpart of our earlier paper [1], where we studied the conventional linear feedback controllers design for chaotic discrete-time systems.*

1. Introduction

In the past few years, there has been some increasing interest in the control of chaotic (nonlinear) systems, in both physics and engineering communities [1-5,8-10,12-13]. One of the motivations for such research is, as pointed out in the present authors' earlier paper [1], that the capability of controlling the chaotic brain wave "may be the chief property that makes the brain different from an artificial-intelligence machine" [7].

We have developed in [1] some new ideas and techniques for the control of chaotic discrete-time systems using the conventional engineering feedback control methods, where the target position is an equilibrium point of the chaotic system. In this paper, we further extend this successful approach to continuous-time chaotic systems. We study how the conventional canonical feedback control techniques can control the chaotic trajectory of a continuous-time nonlinear system to converge to both its equilibrium points and, much more significantly, its multi-periodic orbits. More specifically, we describe in detail our investigation of the well-known chaotic Duffing equation, with emphasis on the control of its chaotic trajectory to one of its multi-periodic orbits. The motivation for this research is two-fold: First, the controllers design for driving a chaotic trajectory to a periodic orbit is more difficult and hence more interesting from a theoretical point of view, and second, it has more significant physical meaning in the sense that, for example, when a periodic oscillation of a machine is expected to be recovered from its chaotic situations during the motion and then to be maintained thereafter. We discuss nonlinear feedback controllers design in this paper, and leave the linear feedback design, a more useful yet more difficult one, to our forthcoming paper [3].

In 1918, Duffing [6] introduced a nonlinear oscillator, with a cubic stiffness term, to describe the hardening spring effect observed in many mechanical problems. Since then, this equation has become one of the most popular models, like the well-known van der Pol equation, in the studies of nonlinear oscillations, bifurcations and chaos. Duffing's equation has been modified, by Moon and Holmes [11] for example, in different manners afterwards. In this paper, to be more general we consider a modified Duffing equation of the form

$$\ddot{x} + p\dot{x} + p_1 x + x^3 = q\cos(\omega t)\,. \qquad (1)$$

where $p(>0)$, p_1, q and ω are real parameters.

In Section 2 below, we first review some observations on the dynamic behavior of Duffing's equation. Our interest is then to study how to design a conventional nonlinear feedback controller which can drive a chaotic trajectory of the system back to an inherent periodic orbit. In Section 3, the design of nonlinear feedback controllers for driving chaotic trajectories of Duffing's equation to its periodic orbits are discussed, leaving the linear controllers design, which is more convenient in applications but more difficult to achieve, to our another paper [3].

2. Dynamics of Duffing's Equation

Consider Duffing's equation (1). By introducing $y = \dot{x}$, this equation can be rewritten as

$$\begin{cases} \dot{x} = y \\ \dot{y} = -p_1 x - x^3 - py + q\cos(\omega t)\,. \end{cases} \qquad (2)$$

Some typical periodic and chaotic solutions of Eq.(2), when displayed in the x-y phase plane, are shown in Fig.1, where $p = 0.4$, $p_1 = -1.1$, $\omega = 1.8$, and

(a) $q = 0.620$ (period 1) (b) $q = 1.498$ (period 2)
(c) $q = 1.800$ (chaotic) (d) $q = 2.100$ (chaotic)
(e) $q = 2.300$ (period 1) (f) $q = 7.000$ (period 1)

As shown in the figures, with parameters $p = 0.4$, $p_1 = -1.1$, $q = 2.1$, and $\omega = 1.8$, the Duffing system has a chaotic response.

3. Control of the Chaotic Duffing Equation

Our interest is in studying how to control this kind of chaotic trajectories (e.g., those of Fig.1(d)) when it appears, to one of

CH3229-2/92/0000-2502$1.00 © 1992 IEEE

the inherent periodic orbits of the system (e.g., that of Fig.1(a) or Fig.1(b)) by designing a conventional feedback controller. In Eq.(2), let $(\bar{x},\bar{y}) = (\bar{x}(t),\bar{y}(t))$ be one of its periodic orbits that we are targeting. We want to be able to control the system, so that for any given $\varepsilon > 0$ there exists a $T_\varepsilon > 0$ such that

$$|x(t) - \bar{x}(t)| \le \varepsilon \quad \text{and} \quad |y(t) - \bar{y}(t)| \le \varepsilon \quad \text{for all } t \ge T_\varepsilon . \qquad (3)$$

For this purpose, we consider the conventional nonlinear feedback controller of the form $u = h(x)$, where $h(\cdot)$ is a nonlinear function in general, which is added to the second equation of the original system (2). Then, we obtain the following "controlled Duffing equation":

$$\begin{cases} \dot{x} = y \\ \dot{y} = -p_1 x - x^3 - py + q\cos(\omega t) + h(x) . \end{cases} \qquad (4)$$

Observe that the periodic orbit $(\bar{x},\bar{y})$ is itself a solution of the original equation. Subtracting (2), with (x,y) being replaced by $(\bar{x},\bar{y})$ therein, from (4), and denoting $X = x - \bar{x}$ and $Y = y - \bar{y}$, we arrive at

$$\begin{cases} \dot{X} = Y - pX \\ \dot{Y} = -p_1 X - (x^3 - \bar{x}^3) + h(x) . \end{cases} \qquad (5)$$

Our design of the nonlinear feedback controller $h(x)$ will be based on Eq.(5), see Fig.2 for the closed-loop configuration of the feedback control system.

In the following, we describe our feedback controllers:

(1) $h(x) = -K(x-\bar{x}) + x^3 - \bar{x}^3$ with the constant feedback gain K to be determined.

Using this nonlinear feedback control, the controlled Duffing equation (5) reduces to a linear system in the form

$$\begin{cases} \dot{X} = Y - pX \\ \dot{Y} = -(K + p_1)X . \end{cases} \qquad (6)$$

Since its Jacobian is $J = \begin{bmatrix} -p & 1 \\ -(K+p_1) & 0 \end{bmatrix}$, which has the eigenvalues

$$\lambda_{1,2} = \frac{-p \pm \sqrt{p^2 - 4(K+p_1)}}{2} ,$$

by choosing the feedback gain $K > -p_1$, we can ensure the asymptotical stability of the controlled system (6), so that $X \to 0$ and $Y \to 0$ as $t \to \infty$, or the goal (3) is achieved.

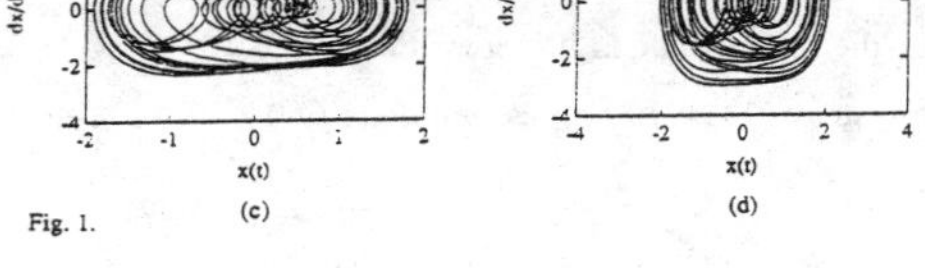

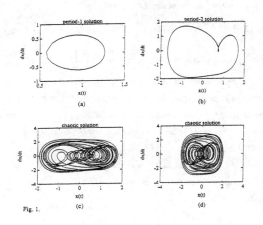

Fig. 1.

(2) $h(x) = -K(x-\bar{x}) + 3\bar{x}^2(x-\bar{x}) + 3\bar{x}(x-\bar{x})^2$ with the constant feedback gain K to be determined.

Using this nonlinear feedback control, the controlled Duffing equation (5) becomes

$$\begin{cases} \dot{X} = Y - pX \\ \dot{Y} = -(K + p_1)X - X^3 . \end{cases} \qquad (7)$$

Consider the Lyapunov function

$$U(X,Y) = \frac{K+p_1}{2}X^2 + \frac{1}{4}X^4 + \frac{1}{2}Y^2 , \qquad (8)$$

we can verify that

$$\begin{aligned} \dot{U} &= [(K+p_1)X + X^3]\dot{X} + Y\dot{Y} \\ &= [(K+p_1)X + X^3][Y - pX] + Y[-(K+p_1)X - X^3] \\ &= (K+p_1)XY - p(K+p_1)X^2 + X^3Y - pX^4 \\ &\quad - (K+p_1)XY - X^3Y \\ &= -p(K+p_1)X^2 - pX^4 \le 0 , \end{aligned}$$

where equality holds if and only if $X \equiv 0$. This means that the controlled Duffing equation (7) is asymptotically stable, so that $X \to 0$ and $Y \to 0$ as $t \to \infty$, or the goal (3) is achieved.

References

[1] G.Chen and X.Dong: On feedback control of chaotic dynamic systems. Int. J. of Bifur. Chaos, 2 (1992) 407-411.

[2] G.Chen and X.Dong: Controlling Chua's circuits (invited paper), J. of Circ. Sys. Comput., 1993, to appear.

[3] G.Chen and X.Dong: On feedback control of chaotic continuous-time systems, submitted.

[4] X.Dong and G.Chen: Control of discrete-time chaotic systems, Proceedings of the 1992 Amer. Contr. Conf., Chicago, IL, June 24-26, 1992, 2234-2235.

[5] W.L.Ditto, S.N.Rauseo, and M.L.Spano: Experimental control of chaos, Phys. Rev. Lett., 65 (1990) 3211-3214.

[6] G.Duffing: *Erzwungene Schwingungen bei Veränderlicher Eigenfrequenz.* F.Vieweg u.Sohn, Braunschweig, 1918.

[7] W.J.Freeman: The physiology of perception, Scientific American, Feb. (1991) 78-85.

[8] A.W.Hübler: Ph.D. Dissertation, Department of Physics, Technical University of Münich, Germany, Nov., 1987.

[9] E.R.Hunt: Stabilizing high-period orbits in a chaotic system, Phys. Rev. Lett., 67 (1991) 1953-1955.

[10] E.A.Jackson: On the control of complex dynamic systems, Physica D, 50 (1991), 341-366.

[11] F.C.Moon and P.J.Holmes: Magnetoelastic strange attractor. J. Sound Vib., 65 (1979) 275-301.

[12] E.Ott, C.Grebogi, and J.A.Yorke: Controlling chaos, Phys. Rev. Lett., 64 (1990) 1196-1199.

[13] T.L.Vincent and J.Yu: Control of a chaotic system, Dynamics and Control, 1 (1991) 35-52.

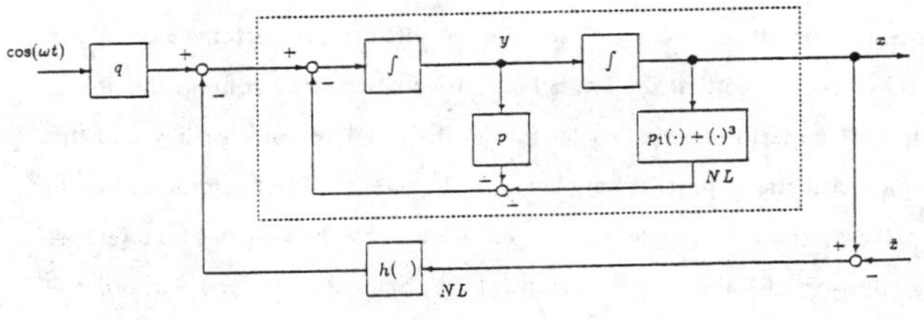

Fig 2.

Proceedings of the 31st Conference
on Decision and Control
Tucson, Arizona • December 1992

ROBUST HIGHER ORDER REPETITIVE LEARNING CONTROL ALGORITHM FOR TRACKING CONTROL OF DELAYED REPETITIVE SYSTEMS

Chen, Yangquan Sun, Mingxuan Huang, Baojian Dou, Huifang
(Xi' an Institute of Technology, Box. 209, 710032 Xi' an, P. R. China)

ABSTRACT

Iterative learning control (ILC) concept has been received increasing attention as an alternative for controlling uncertain dynamic systems in a simple manner. It is recursive online control method that relies on less calculation and requires less a priori knowledge about the system dynamics. In this paper, a robust higher order PID—type ILC algorithm is presented for tracking control of delayed nonlinear time—varying MIMO repetitive systems. A convergency proof is given in more general case. When initial state bias exist, a repetitive ILC scheme (i. e., forward learning and backward learning) is proposed to make the algorithm more robust with respect to this bias. Simulational results indicate that the proposed method converges more faster than previous methods. The effect of system delay (or even multi—delays and time varying delays) on the ILC convergency is very small. Examples are provided to show the proposed method is with wider effectiveness and with great applicable potentials.

Keywords: Iterative algorithm; tracking control; delayed systems; repetitive systems; learning control method.

I. INTRODUCTION

Recently, iterative learning control (ILC) concept has been received increasing attention as an alternative for controlling uncertain repetitive dynamic systems in a simple manner [1—5]. The ILC concept stands for the repeatability of operating a given objective system and the possibility of improving the command input on the basis of previous actual operation data. It is recursive online control method that relies on less calculation and requires less a priori knowledge about the system dynamics. From the internal mechanism of ILC, if the system is causal and repetitive, the measureable system output $y_i(t)$ (the subscript i refers to i —th iterative learning) can track onto the desired trajectory $y_d(t)$ within the given ε —bound through iterative learning process where the control u$_i$t) is improved by using previous operation information such as previous control (trial), error, error derivative anderror integral etc. The idea is very attractive in robotic applications, such as arc—welding, spray painting and appearance inspection.

In this paper, an N—order PID—type ILC algorithm is presented. Itmeans that to get $(i+1)$—th control $u_{i+1}(t)$, previous N iterations control information $\{u_i(t), u_{i+1}(t), \ldots, u_{i-N+1}(t)\}$, error information $\{e_i(t), e_{i-1}(t), \ldots, e_{i-N+1}(t)\}$ and its derivative and integral are to be used. Here, $e_i(t) = y_d(t) - y_i(t)$. In [1], the 1—st order D—type ILC algorithm is pro—posed and applied to time invariant robotic systems. It is further developed in [2] where 1—st order PD, PI and PID type ILC algorithm are proposed. Both in [1] and in [2], only the proof about convergency of $y_i(t)$derivative is given. [3] is the first to use higher order P—type ILC algorithm for tracking control of time—varying systems. It was demonstratedthat higher order ILC scheme did yield better performance of convergency. But it is claimed that the system output equation must contain the lineart-erm of control. This term plays a crucial role in the proof of convergency and this restriction is not realistic in practical application. In[1—3], the assumption that at every ILC process, the initial state and output are accurately known and $x_i(0) = x_d(0)$ *and* $y_i(0) = y_d(0)$ are very important. [4] proposed a robust 1—st order D—type ILC algorithm. It is proved that when there exist state disturbances and initial state bias, through ILC the asymptotic error are bounded; when the bounds on the state disturbances and bounds on initial state bias tend to zero, the error bounds also tend to zero. 1--st order PD— and DI—type ILC are also discussed. The so—called "robustness of ILC algorithm" means that when state disturbances and initial state inaccuracy (or error, bias etc.) exist in ILC process, the $e_i(t)$ is asymptotically bounded and the control error and state error are also with asymptotic bounds. ILC concept is proved to be widely effective. The class of system considered in [4] is fairly general. With respect to LTI systems, [5] conducted that ILC is not essential.

A robust higher order PID—type ILC algorithm is presented for tracking control of delayed nonlinear time—varying MIMO repetitive systems. A convergency proof is given in more general case. When initial state bias exist, a repetitive ILC scheme (i. e., forward learning and backward learning) is proposed to make the algorithm more robust with respect to this bias. Simulational results indicate that the proposed method converges more faster than previous methods. The effect of system delay (or even multi—delays or time varying delays) on the ILC convergency is very small.

Examples are provided to show the proposed method is with wider effectiveness and with great applicable potentials.

II. ROSBEST HIGHER ORDER ILC ALGORITHM

Consider nonlinear delayed repetitive systems as follows

$$\begin{cases} \dot{x}_i(t) = f(x_i(t), x_i(t-t_{d1}), t) + B(x_i(t), x_i(t-t_{d2}), t) \cdot u_i(t) + w_i(t) \\ y_i(t) = g(x_i(t), t) + v_i(t) \end{cases} \quad (1)$$

where, $t \in [0, T]$, i denotes the i —th repetition of system (1). $x_i(t) \in R^n, u_i(t) \in R^m, w_i(t) \in R^n$, $v_i(t) \in R^m$. Suppose when $t \in [-t_d, 0]$, $x(t) = \psi(t)$, where $t_d = \max\{t_{d_1}, t_{d_2}\}$

Here, two assumptions are made for the discussions in the sequel.

Assumption 1 (A1): functions f, g, g_x, g_t, and B satisfy a Lipschitz continuity condition and functions g_x *and* B are bounded on $[0, T]$.

Assumption 2 (A2): disturbances $w_i(t)$, $v_i(t)$ *and* $\dot{v}_i(t)$ are with bounds on $[0, T]$.

A1—A2 are reasonable and can be satisfied in most real systems.

The proposed robust higher order ILC algorithm is

$$u_{i+1}(t) = (1-\gamma)\sum_{k=1}^{N} P_k u_{i-k+1}(t) + \gamma u_0(t) + \sum_{k=1}^{N}\{Q_k e_{i-k+1}(t) + R_k \dot{e}_{i-k+1}(t) + S_k \cdot \int_0^t e_{i-k+1}(\tau) d\tau \quad (2)$$

where, N is the order of ILC algorithm, P_k, Q_k, R_k, $S_k$are constant learning parametric matrices of proper dimensions. $0 \leqslant \gamma < 1$. The introduced γ is to restrain the larger frustration of control function at the beginniing of this ILC algorithm. In application, γ can be selected to be a monototic sharp—decreasing—to—zero function w. r. t. ILC times i.

We give out the main result in Theorem i.

Theorem 1: For system (1) satisfying assumptions (A1)—(A2), given attainable $y_d(t)$, if conditions

CH3229-2/92/0000-2504$1.00 © 1992 IEEE

(C1): $\sum_{k=1}^{N} P_k = I_r$

(C2): there exist ρ_k satisfying

$$\| (1-\gamma)P_k - R_k g_x B \| \leqslant \rho_k, \quad \forall (x,t) \in R^n \times [0,T]$$

and $\sum_{k=1}^{N} \rho_k = \rho < 1$

are satisfied and suppose the initial state bias bounded, when $i \to \infty$, the $\| u_d(t) - u_i(t) \|$, $\| x_d(t) - x_i(t) \|$ *and* $\| y_d(t) - y_i(t) \|$ obtained from ILC algorithm (2) are with bounds asymptotically; when the bounds on the state disturbances, bounds on initial state bias and bounds on output noise tend to zero, all above bounds tend to zero.

All the norm appeared in Theorem 1 defined as follows

for vector $f = (f^{(1)}, \ldots, f^{(r)})$, the vector norm $\| f \| = \max_{1 \leqslant i \leqslant r} | f^{(i)} |$;

and for matrix $G = (g^{(ij)})_{r \times r}$, the matrix norm $\| G \| = \max_{1 \leqslant i \leqslant r} \{ \sum_{j=1}^{r} |g^{(ij)}| \}$

Proof: For brevity of the proof, in the following, denote $(*)_d$ for the desired trajectory of function vector $*$, $(*)_l$ for the trajectory of $(i-k+1)$ —th ILC process, $(*)_x$ for partial derivative w. r. t. x and $(*)_t$ for p. d. w. r. t. time t, if not mentioned.

From ILC algorithm (2) and condition (C1) we have

$$\begin{aligned} u_d - u_{i+1} = & \sum_{k=1}^{N} [(1-\gamma)P_k - R_k g_{xl} B_l](u_d - u_l) + \gamma \cdot (u_d - u_0) \\ & + \sum_{k=1}^{N} \{ Q_k(g_d - g_l - v_l) + S_k \cdot \int_0^t (g_d - g_l - v_l) d\tau \\ & + R_k \cdot [(g_{xd} - g_{xl}) \cdot (f_d + B_d u_d) + g_{xl}(f_d - f_l) \\ & + g_{xl}(B_d - B_l) u_d + g_{td} - g_{tl} - g_{xl} w_l - \dot{v}_l] \} \end{aligned} \quad (3)$$

Estimating the norm about eq. (3) yields

$$\begin{aligned} & \| u_d - u_{i+1} \| \\ & \leqslant \sum_{k=1}^{N} \| (1-\gamma)P_k - R_k g_{xl} B_l \| \cdot \| u_d - u_l \| + \gamma \cdot \| u_d - u_0 \| \\ & + \sum_{k=1}^{N} \{ \| Q_k \| (\| g_d - g_l \| + \| v_l \|) + \| S_k \| \cdot \int_0^t \| g_d - g_l - v_l \| d\tau \\ & + \| R_k \| [\| g_{xd} - g_{xl} \| \cdot \| f_d + B_d u_d \| + \| g_{xl} \| \cdot \| f_d - f_l \| \\ & + \| g_{xl} \| \| B_d - B_l \| \cdot \| u_d \| + \| g_{td} - g_{tl} \| \\ & + \| g_{xl} \| \cdot \| w_l \| + \| \dot{v}_l \|] \} \end{aligned} \quad (4)$$

Based on assumptions (A1)—(A2) we can denote

$$b_d \triangleq \sup_{t \in [0,T]} \| f_d + B_d u_d \|, \quad b_{ud} \triangleq \sup_{t \in [0,T]} \| u_d \|, \quad b_v \triangleq \sup_{t \in [0,T]} \| v \|$$

$$b_{\dot{v}} \triangleq \sup_{t \in [0,T]} \| \dot{v} \|, \quad b_w \triangleq \sup_{t \in [0,T]} \| w \|, \quad b_{gx} \triangleq \sup_{t \in [0,T]} \| g_x \|$$

and

$$b_{1k} \triangleq \| Q_k \| k_g + \| R_k \| (b_d k_{gx} + b_{gx} k_f + b_{gx} b_{ud} + k_{gt})$$

$$b_{2k} \triangleq \| R_k \| b_{gx} k_f, \quad b_{3k} \triangleq \| R_k \| b_{gx} b_{ud}, \quad b_{4k} \triangleq \| S_k \| k_g$$

$$b_{rdk} \triangleq \| Q_k \| b_v + \| R_k \| (b_{gx} b_w + b_{\dot{v}}) + \| S_k \| \cdot T \cdot b_v$$

where k_* is the corresponding Lipschitz constant of function $(*)$. From condition (C2) and by above notations, we obtain

$$\begin{aligned} & \| u_d(t) - u_{i+1}(t) \| \\ & \leqslant \sum_{k=1}^{N} \rho_k \| u_d(t) - u_l(t) \| + \gamma \cdot \| u_d(t) - u_0(t) \| \\ & + \sum_{k=1}^{N} (b_{1k} \cdot \| x_d(t) - x_l(t) \| + b_{2k} \| x_d(t - t_{d1}) - x_l(t - t_{d_1}) \| \\ & + b_{3k} \| x_d(t - t_{d_2}) - x_l(t - t_{d_2}) \| + b_{4k} \cdot \int_0^t \| x_d(\tau) - x_l(\tau) \| d\tau + b_{rdk} \end{aligned} \quad (5)$$

Applying Bellman—Gronwall Lemma [3], we have

$$\begin{aligned} & \| x_d(t) - x_i(t) \| \\ & = \| x_d(0) - x_i(0) + \int_0^t [f_d + B_d u_d - (f_i + B_i u_i + w_i)] d\tau \| \\ & \leqslant \| x_d(0) - x_i(0) \| e^{b_5 t} + \int_0^t e^{b_5 (t-\tau)} (b_B \| u_d(\tau) - u_i(\tau) \| + b_w) d\tau \end{aligned} \quad (6)$$

where, $b_5 \triangleq 2 \cdot k_f + 2b_{ud} \cdot k_B$, $b_B = \sup_{t \in [0,T]} \| B \|$

By substituting (6) into (5), referring to eq. (2) we get

$$\begin{aligned} \| \delta u_{i+1} \| \leqslant & \sum_{k=1}^{N} \rho_k \cdot \| \delta u_l \| + \gamma \cdot \| \delta u_0 \| + \sum_{k=1}^{N} \{ b_k [\| \delta x(0) \| \cdot e^{b_5 t} \\ & + \int_0^t e^{b_5 (t-\tau)} \cdot (b_B \cdot \| \delta u_l \| + b_w) d\tau] + b_{rdk} \} \end{aligned} \quad (7)$$

where $\delta(*)_{i+1} = (*)_d - (*)_{i+1}$, $b_k \triangleq b_{1k} + b_{2k} + b_{3k} + b_{4k} \cdot T$

Define λ —norm of a function $h(t)$ as $\| h(t) \|_\lambda \triangleq \sup_{t \in [0,T]} e^{-\lambda t} \cdot \| h(t) \|$ taking multiplication of $e^{-\lambda t}$ to each side of eq. (7) yields

$$\begin{aligned} \| \delta u_{i+1} \|_\lambda \leqslant & \sum_{k=1}^{N} \rho_k \cdot \| \delta u_l \|_\lambda + \gamma \cdot \| \delta u_0 \|_\lambda \\ & + \sum_{k=1}^{N} \{ b_k \| \delta x(0) \| e^{(b_5 - \lambda)t} + b_{rdk} e^{-\lambda t} \\ & + b_k b_w \int_0^t e^{-\lambda \tau} \cdot e^{(b_5 - \lambda)(t-\tau)} d\tau \\ & + b_k b_B \int_0^t \| \delta u_l \|_\lambda e^{(b_5 - \lambda)(t-\tau)} d\tau \end{aligned} \quad (8)$$

Note that the integration terms in eq. (8) are strict monototic increasing w. r. t. time t and the λ—norm of constant c is also equal to c. We thus get

$$\| \delta u_{i+1} \|_\lambda \leqslant \sum_{k=1}^{N} \bar{\rho}_k \cdot \| \delta u_l \|_l + \varepsilon$$

where $\bar{\rho}_k \triangleq \rho_k + \frac{b_k b_B}{\lambda - b_5} (1 - e^{(b_5 - \lambda)T})$

$$\varepsilon \triangleq \gamma \| \delta u_0 \|_\lambda + \sum_{k=1}^{N} \left\{ b_k \cdot \| \delta x_{(0)} \| + b_{rdk} + \frac{b_k b_w}{\lambda - b_5} (1 - e^{(b_5 - \lambda)T}) \right\} \quad (9)$$

From condition (C2) we know $\sum_{k=1}^{N} \rho_k < 1$, then it is obvious that there exist a λ ($\lambda > b_5$) to hold $\sum_{k=1}^{N} \bar{\rho}_k < 1$. So, based on the Lamma 1 in Appendix we have

$$\lim_{i \to \infty} \sup_{t \in [0,T]} \| \delta u_i(t) \|_\lambda \leqslant \frac{1}{1-\rho} \cdot \varepsilon \quad (10)$$

Also, from eq. (6) and eq. (1) can we know $\| \delta x_i \|_\lambda$, $\| \delta y_i \|_\lambda$ are with bounds on [0,T]. It is necessary to point out that from the expression of ε, when state disturbance bounds b_w, output noise bounds b_v and $b_{\dot{v}}$, initial state bias $\| \delta x(0) \|$ and λ tend to zero, the asymptotic errorbounds $\| u_d(t) - u_i(t) \|$, $\| x_d(t) - x_i(t) \|$, $\| y_d(t) - y_i(t) \|$ also tend to zero.

Remark 1: The state disturbances, output noise, initial state bias and selection of λ do not affect the proposed ILC algorithm' s convergency. But they affect the con verged error bounds directly.

Remark 2: The proposed ILC algorithm itself can not reject the distur bances, noise and bias. But when once they did not appear any more in the coming ILC process es, the algorithm can recovery from them.

Remark 3: As higher order ILC algorithm supplies more flexibility inchoosing learn ing parametric matrices, more faster ILC convergency can beexpected.

Remark 4: Higher order PID—type ILC algorithm can learn the past experiences conprehensively.

Remark 5: The ILC idea is widely effective. The system class is not restricted to eq. (1).

Remark 6: The theory about convergency speed of ILC is still an openproblem.

Remark 7: For a specified system, the selection of learning parametricmatrices must be considered specifically by taking Theorem 1 as a baseline.

III. A AREPETITIVE ILC SCHEME

In Section—II we know the proposed ILC algorithm itself can not reject the state disturbances, output noise and initial state bias (see Remark 1—2). In this Section, a repetitive ILC (RILC) scheme is proposeed to eliminate the effect of initial state bias iteratively and repetitively.

We know, any way, state disturbances and output noises can be attenuated by some means such as filtering, compensation and etc.. To ILC, the effect of state disturbances and output noises is not so vital as toother intelligent control methods. But for repetitive systems, to apply the ILC efficiently, the initial state in every ILC process should be the sameand known, which is emergently assumed at the beginning of every existingliteratures on ILC. The initial state bias affects the ILC result significantly.

For a given $y_d(0)$, there exist $x_d(0)$ if the inverse output functiong^{-1} exists. If g^{-1} does not exist, $x_d(0)$ can not be determined; if g^{-1} is not in explicit form, the obtained $x_d(0)$ are with numerical errors. In real cases, $x_i(0)$ would be subjected to disturbance such as poor accuracy of repetitive positioning. It is necessary to combine a scheme within ILC to track the initial state $x_i(0)$ converging to $x_d(0)$. In the following, we give out an experienced RILC scheme (subscript F, B denote forward andbackward ILC process respectively).

(1). With $x(0) \neq x_d(0)$ and the beginning information of forward ILC (i. e. preceding past information is originally given or set to zero.), perform forward ILC of the first round RILC. Forward ILC parametric metrices areproperly selected to make the ILC converged trajectories satisfy the fol—lowing condition

$$|x(0) - x_d(0)| > |x_F^{(1)}(T) - x_d(T)| \tag{11}$$

(2). Taking $x_F^{(1)}(T)$ as initial point and with the beginning information of backward ILC (i. e. preceding past information is originally given or set to zero.), perform backward ILC of the first round RILC. The backward ILC parametric matrices are properly selected to let the converged ILC trajectories satisfy the condition as follows

$$|x_{(1)F}(T) - x_d(T)| > |x_B^{(1)} - x_d(0)| \tag{12}$$

(3). Let $k=2$. In every round of RILC, the forward and backward ILC parameters are unchanged.

(4). With initial state $x_B^{(k-1)}(0)$, carry out forward ILC of k —th round of RILC, taking the corresponding information of ($k-1$)—th round RILC as the beginning information of current round of ILC iteration processes. When ILC converges to given precision, we get $x_F^{(k)}(T)$

(5). With initial state $x_F^{(k)}(T)$, carry out backward ILC of k —th round of RILC, taking the corresponding information of ($k-1$) —th round RILC as the beginning information of current round of ILC iteration processes. When ILC-converges to given precision, we get $x_B^{(k)}(0)$

(6). Let $k=k+1$, goto step (4) until in forward ILC process, given trackingerror accuracy is reached.

The aforementioned RILC scheme is with practical background. The forward and backward learning parameters may be different and they must beselected to satisfy conditions (11)—(12), which are easy to meet in ourexperiments. As shown in simulation Section, the inaccurate initial statetends to real value (here we call it desired value $x_d(0)$) rapidly through RILC scheme. In each round of RILC, by the proposed robust higher order ILC algorithm, only a few iterations of the forward and backward learning control are needed. Again, the mathematic description of RILC and its convergency proof are interesting open problems.

IV. SIMULATION STUDIES

In this Section, two examples are give to demonstate the effectiveness of the proposed robust higher ILC algorithm eq. (2) of this paper. In example 1, the same linear time varying plant as that of Reference [1] and [3] is considered. The con vergency and performances of recovering from state and output disturbances of ILC algorithms are compared. When initial state bias exists, the result of tracking de sired initial stateby a proposed RILC scheme in Section III are also given. In exam ple 2,a nonlinear delayed system, which is not described by eq. (1),is considered.

Example 1. ([1],[3]) The control mission is, for a given system

$$\begin{bmatrix} \dot{x}_1(t) \\ \dot{x}_2(t) \end{bmatrix} = \begin{bmatrix} 0 & 1 \\ -(2+5t) & -(3+2t) \end{bmatrix} \begin{bmatrix} x_1(t) \\ x_2(t) \end{bmatrix} + \begin{bmatrix} 0 \\ 1 \end{bmatrix} u(t)$$

$$[x_1(0), x_2(0)]^T = [0,0]^T \quad ; \quad t \in [0,1]$$

$$y(t) = [0,1] \cdot [x_1(t), x_2(t)]^T \tag{4.1}$$

and desired output trajectory to be tracked

$$y_d(t) = 12t^2(1-t) \quad , \quad t \in [0,1] \tag{4.2}$$

to find the optimal control $u(t)$ satisfying

$$e_b \stackrel{\Delta}{=} \sup_{t\in[0,T]} \|y_d(t) - y_i(t)\| \leq \varepsilon^* \tag{4.3}$$

where, ε^* is given error bound.

<1>. Convergency comparison of ILC algorithms

As in Reference [3], let $\varepsilon^* = 0.06$, the system is state disturbancefree and output noise free and with no initial bias. Let $u_0(t) = 0$ and all quantities before 0 —th ILC be 0. Three ILC algorithms are compared. They are of [1], [3] and of this paper. See Table—1.1 for the information about ILC algorithms such as the or der of ILC, learning parameters and type ofILC. We have built up a general pur pose ILC simulation software in IBM/PCMS—FORTRAN4.0 with good automatic graphics interaction. The comparabilityof ILC algorithms are carefully considered. Table—1.1 shows the convergency comparisons where RMS(e) denotes root mean square errors and e_b is defined in eq. (4.3). We can see from Table—1.1 that the algorithm of this paper converges much more rapidly. [1] is D—type ILC and this paper is PID—type. They all converge faster than [3] which purely uses P informa tion. It is understandable that for faster ILC convergency, the D information isnec essary. When PID—type ILC employed, it is clear that the claim in [3]that the system output equation must contain the linear term of controlis not neccessory.

<2>. Comparisons of recovery from disturbances

Suppose the ILC algorithms converge to satisfactory accuracy whena disturbance d (t)

$$d(t) = [1 + sign(t-0.5)] * (t-0.5)/2. \tag{4.4}$$

is introduced. d(t) is added in the coming ILC process and then dismissedin all forthcoming ILC processes. Under the same settings of <1>, we compare the ILC property of recovery from disturbances. Three ways of disturbance adding are con sidered, i. e., adding in state equations, adding inoutput equation and simultaneous ly adding. Via simulation runs, we get Table—1.2.1 — 1.2.3 which show that the recovery from state disturbance is much more faster than from output distur bance. It is interesting to note that this paper recovers faster than [1] but [3] is the fastest. To explainthis phenomenon mathematically is beyond the scope of this paper.

<3> Repetitive ILC result when initial state bias exists.

Under the same conditions in <1>, here let $x_2(0)=2.0$ which is faraway from real initial state 0. Set forward learning parameters the samewith that in Table—1.1 and backward learning parameters the same withforward learning pa rameters. By using a RILC scheme in Section III, 4rounds of RILC is enough for converging to given tracking accuracy. SeeTable—1. 3 and Fig. 1 for details. We can see that the RILC is very robustto initial state bias, which could not achieved by other existing ILC algorithms. In Fig. 1, the plotted output trajectories are converged onesin each round of ILC (forward and backward), e. g., 2B denotes converged output trajectory of backward ILC in 2nd round of RILC, which needs 3 iterations (see line 4 and col. 2, "B. 3" in Table1. 3)

Example 2. Consider the following nonlinear delayed system

$$\dot{x}(t) = -5[1 + x(t-\tau_1)]^{x(t-\tau_2)} \cdot tanh(u) \ , t \in [0,1]$$
$$y(t) = x(t) \qquad (4.5)$$
$$x(t) = 0 \ , t \leqslant 0$$

and the desired output trajectory is in triangular waveform

$$y_d(t) = \begin{cases} t \ , & t \in [0, 0.2] \\ -t + 0.4, & t \in (0.2, 0.4] \\ 0, & t \in (0.2, 0.4] \\ t - 0.6, & t \in (0.2, 0.4] \\ -t + 1, & t \in (0.2, 0.4] \end{cases} \qquad (4.6)$$

<1> With no delays

First, suppose there are no disturbances, noises and bias, let ε^* 0. 005. Through several trials, we choose the ILC parameters as follows $N = 2$, PID—type ILC algorithm. $p_1 = 0.98$, $p_2 = 0.02$; $q_1 = 0.005$, $q_2 = 0.001$; $r_1 = -0.145$, $r_2 = -0.005$; $s_1 = -0.8$, $s_2 = -0.1$.

All beginning information of ILC is set to 0. Simulation shows just 2 iterations are required to reach the given accuracy, see Table—2. 1.

In iteration no. 3, a disturbance $d(t)$ is added to output equation,

$$d(t) = [1 + sign(t - 0.5)]/5 \qquad (4.7)$$

and then removed the coming iterations, i. e., in iteration 4,5,…… After 8 iterations, the ILC algorithm recovers from disturbance and converges to desired trajectory within given ε^* bound. See Fig. 2 for the ILC recovery processes.

Now we consider the RILC process. Suppose initial state x(0) isbiased to 0. 2, let forward learning parameters the same as above and backward learning parameters be the same with forward ones. Simulation show 3 rounds of RILC are enough to converge to given accuracy. The RILC processes are described in Fig. 3. See Fig. 1 for a reference.

It is necessary to indicate that system (4. 5) can not be concludedby eq. (1). From above simulations we can say that ILC concept and a proposed RILC scheme are widerly effective.

<2> With delays

Not lossing generality, for convenience of discussion, assume $\tau_1 = \tau_1 = \tau_2 = \tau$. Consider 4 cases, i. e. $\tau = 0, 0.1, 0.6, 1.25$ seperatively. Under the same settings of <1> in this example, carrying out simulation runs with disturbance, noise and bias free, we get comparative results in Table—2. 1. We can see that the ILC algorithm is still effective to delayed system. Furthermore, the system state delay does not affect the ILC convergency significantly. This can be deduced from the proof of Theorem 1. In general, even if the state delay is time—varying or in multi—delay form, the proposed ILC algorithm is very robust to them. Further simulation runs show, in these cases, recovery from disturbance and RILC are robust, too.

V. CONCLUSIONS

Through the discussions and great deal of simulations in this paper, we can conclude in the following:

(1) the proposed robust higher order PID—type ILC algorithm supplies the possibility of faster ILC convergency as the flexibility for choosing learning parameters is extended.

(2) a proposed repetitive ILC scheme is a very attrative way for on—linee liminating the effect of initial state bias by incorporating a real initial state tracking process into ILC algorithm.

(3) ILC and corresponding RILC is widerly effective with respect to wider class of repetitive systems.

(4) the system state delays do not significantly affect ILC convergency and the effectiveness of RILC process.

Further researches include:

(1) theories of ILC convergency speed.

(2) mathematical analysis of ILC disturbance recovering speed and RILCmechanism.

(3) selecting ILC parameters varying w. r. t. time and times of iteration.

(4) based on decision manner of human beings, for a class of systems, toset up new types of ILC laws based on PID information. In previous literatures and in this paper, the control updating law is in the linear form of PID information. Nonlinear PID information based higher order control up dating laws can be developed.

References

[1]. Arimoto, S., Kawamura, S., and Miyazaki, F., Bettering Operation of Robots by Learning. J. of Robotic Systems. Vol. 1 Summer, 1984. pp. 123—140

[2]. Arimoto, S., Miyazaki, and F. Kawamura, S., Motion Control of Robotic Manipulator Based on Motor Program Learning, 2nd IFAC Symposium on Robot Control. 1988 (SYROCO' 88), 169—176

[3]. Bien, Z. and Huh, K. M., Higher — order iterative learning control algorithm, IEE Proc. —D. 1989, 136(3), pp. 105—112

[4]. G. Heinzinger, D. Fenwick, B. Paden, and F. Miyazaki, Robust Learning Control, Proc. of the 28th IEEE CDC, Dec. 1989. Tempa Florida, pp. 436—440.

[5]. K. L. Moore, Iterative Learning for Trajectory Control, Proceedings of the 28th IEEE CDC, Tempa, Florida, Dec. 1989. pp. 860—865.

Appendix

Lemma 1: Suppose a real positive series $\{a_n\}_1^\infty$ satisfies

$$a_n \leqslant \rho_1 a_{n-1} + \rho_2 a_{n-2} + \cdots + \rho_N a_{n-N} + \varepsilon, \ (n = N+1, N+2, \cdots)$$

where, $\rho_i \geqslant 0$, $(i = 1, 2, \cdots, N)$, $\varepsilon \geqslant 0$ and $\rho = \sum_{i=1}^{N} \rho_i < 1$

then the following holds

$$\overline{\lim_{n \to \infty}} a_n \leqslant \frac{\varepsilon}{1-\rho}$$

Proof: Denote $a_{n1} = max\{a_{n-1}, a_{n-2}, \cdots, a_{n-N}\}$, we have

$$a_n \leqslant \rho_1 a_{N-1} + \rho_2 a_{n-2} + \cdots\cdots + \rho_N a_{n-N} + \varepsilon \leqslant \rho a_{n1} + \varepsilon$$

Similarily, denoting $a_{n2} = max\{a_{n_1-1}, + \cdots\cdots a_{n_1-N}$ yields

$$a_{n1} \leqslant \rho a_{n_2} + \varepsilon$$

So we get

$$a_n \leqslant \rho^2 a_{n_2} + \rho\varepsilon + \varepsilon$$

In general, we can get

$$a_n \leqslant \rho^m a_{n_m} + \rho^{m-1} \cdot \varepsilon + \rho^{m-2} \cdot \varepsilon + \cdots\cdots \rho\varepsilon + \varepsilon$$

If an m is selected to satisfy $n_m \leqslant N$, we know $[\frac{n}{N}] - 1 \leqslant m \leqslant n - N$

Let $M = max\{a_1, \cdots, a_N\}$, we obtain

$$a_n \leqslant \rho^m \cdot M + \frac{1-\rho^m}{1-\rho} \cdot \varepsilon \leqslant M\rho^{\frac{n}{N}-2} + \frac{1-\rho^{n-N}}{1-\rho} \cdot \varepsilon$$

So,

$$\overline{\lim_{n\to\infty}} a_n \leqslant \frac{\varepsilon}{1-\rho}$$

which completes the proof of Lemma 1.

Table 1. 1 Convergency comparison with disturbances free.

ILC time	Arimoto' s D—type N=1; p=1 q=1		N=2; p1=1. 1, p2=—0. 1 q1=1. 4, q2=—0. 15		higher order PID—type algorithm of this paper N=2; p1=1. 05, p2=—0. 05 s1=9; q1=1. 1, q2=0. 01 s2=1; r1=0. 9, r2=0. 1	
	RMS(e)	e_b	RMS(e)	e_b	RMS(e)	e_b
0	1. 160	1. 777	1. 160	1. 777	1. 160	1. 777
1	1. 065	1. 727	. 9258	1. 409	. 3509	. 5668
2	. 8497	1. 700	. 7396	1. 117	. 03388	. 05186
3	. 5751	1. 452	. 5903	. 8832		
4	. 3359	. 9837	. 4700	. 6962		
5	. 1722	. 5620	. 3749	. 5473		
6	. 07864	. 2792	. 2985	. 4298		
7	. 03236	. 1231	. 2376	. 3382		
8	. 01211	. 04880	. 1892	. 2680		
9			. 1507	. 2154		
10			. 1202	. 1761		
11			. 09614	. 1544		
12			. 07719	1352		
13			. 06235	. 1180		
14			. 05082	. 1030		
15			. 04195	. 09011		
16			. 03519	. 07913		
17			. 03012	. 06992		
18			. 02634	. 06223		
19			. 02355	. 05583		

Table 1. 2. 1 Comparisons of recoveries from disturbance (Arimoto' s)

ILC times	state disturbance		output disturbance		combined disturbance	
	RMS(e)	e_b	RMS(e)	e_b	RMS(e)	e_b
8	. 01211	. 04880	. 01211	. 04880	. 01211	. 04880
9	. 00858	. 03719	. 2047	. 4824	. 2056	. 4628
10			. 09178	. 1580	. 09116	. 1584
11			. 06343	. 1509	. 06441	. 1473
12			. 03590	. 1073	. 03695	. 1090
13			. 01678	. 05862	. 01737	. 06042
14					. 00686	. 02683

Table 1. 2. 2 Comparisons of recoveries from disturbance (this paper)

ILC times	state disturbance		output disturbance		combined disturbance	
	RMS(e)	e_b	RMS(e)	e_b	RMS(e)	e_b
2	. 03388	. 05186	. 03388	. 05186	. 03388	. 05186
3	. 04256	. 06545	. 2183	. 4699	. 2196	. 4503
4	. 01097	. 02043	. 1312	. 3284	. 1318	. 3123
5			. 06794	. 1728	. 06827	. 1667
6			. 01356	. 04154	. 01415	. 04246

Table 1. 2. 3 Comparisons of recoveries from disturbance (Bien' s)

ILC times	state disturbance		output disturbance		combined disturbance	
	RMS(e)	e_b	RMS(e)	e_b	RMS(e)	e_b
19	.02355	.05583	.02355	.05583	.02335	.05583
20	.01935	.04504	.2144	.5505	.2150	.5309
21			.03132	.05263	.03198	.05450

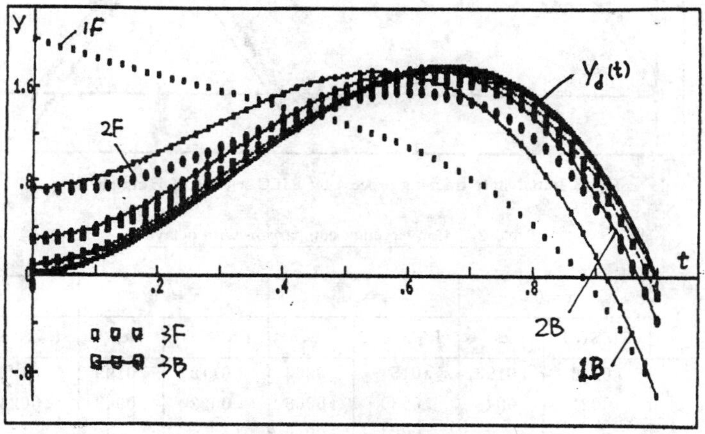

Fig. 1 Initial state tracking by proposed RILC scheme of Example 1.

Table 1. 3 Repetitive ILC process of Example 1. $[x_2(0)=2.0,\ x_{2d}(0)=0.0]$

Repetitive ILC round number	Total forward or backward ILC times	RMS(e)	e_b	$x_2(0)$	$x_2(T)$
1	F. 2	1.094	2.000	2.000	—.9936
1	B. 3	.5689	.9936	.7279	—.9936
2	F. 3	.3912	.7279	.7279	—.3725
2	B. 3	.2254	.3725	.2975	—.3725
3	F. 4	.1623	.2975	.2975	—.1671
3	B. 5	.09132	.1671	.08548	—.1671
4	F. 5	.04848	.08548	.08548	—.05912
4	B. 5	.03177	.05912	.02329	—.05912

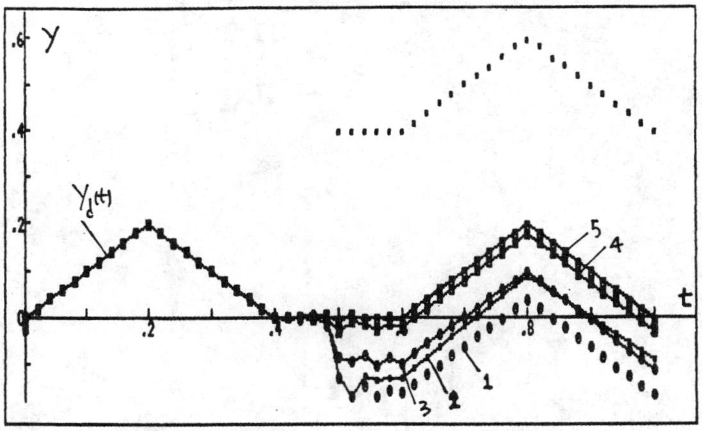

Fig. 2 Recovery processes from output disturbance of Example 2.

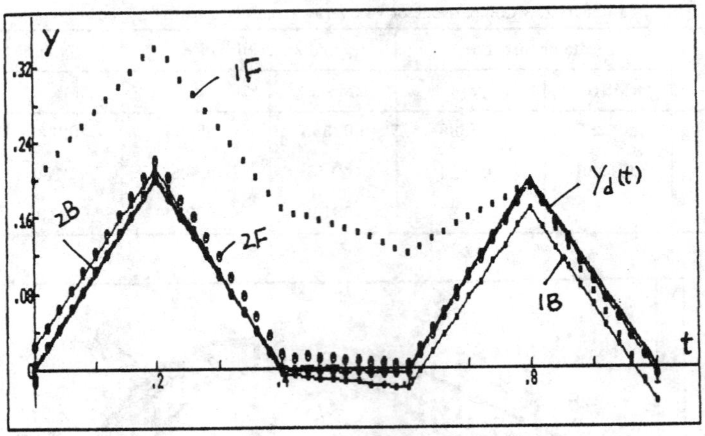

Fig. 3 Initial state tracking process by RILC scheme of Example 2.

Table 2. 1 Convergency comparison with delays

ILC time	τ=0		τ=. 1		τ=. 6		τ=1. 25	
	RMS(e)	e_b	RMS(e)	e_b	RMS(e)	e_b	RMS(e)	e_b
1	. 0116	. 0196	. 0154	. 0263	. 0118	. 0199	. 0120	. 0196
2	. 0021	. 0048	. 0030	. 0068	. 0022	. 0049	. 0023	. 0049
3			. 0004	. 0010				

Proceedings of the 31st Conference
on Decision and Control
Tucson, Arizona • December 1992

STABLE ADAPTIVE FUZZY CONTROL OF NONLINEAR SYSTEMS

Li-Xin Wang

Department of Electrical Engineering and Computer Science
University of California at Berkeley
Berkeley, CA 94720

Abstract

A fuzzy adaptive controller is synthesized from a collection of fuzzy IF-THEN rules. The parameters of the membership functions characterizing the linguistic terms in the fuzzy IF-THEN rules change according to some adaptive law for the purpose of controlling a plant to track a reference trajectory. These fuzzy IF-THEN rules are either collected from experienced human operators or generated automatically during the adaptation procedure. In this paper, a direct adaptive fuzzy controller is designed for a general higher-order nonlinear continuous system through the following three steps: first, define some fuzzy sets whose membership functions cover the state space; then, use fuzzy IF-THEN rules from human experts and some arbitrary rules to construct an initial adaptive fuzzy controller in which some parameters are free to change; finally, develop an adaptive law to adjust the free parameters based on a Lyapunov synthesis approach. It is proven that: 1) the closed-loop system using this adaptive fuzzy controller is globally stable in the sense that all signals involved are bounded, and 2) the tracking error converges to zero asymptotically if the minimum approximation error of the fuzzy controller to an optimal controller is squared integrable along the state trajectory. Finally, we apply this direct adaptive fuzzy controller to control an unstable system.

1 Introduction

Fuzzy logic controllers [3] are generally considered applicable to plants that are mathematically poorly-understood and where experienced human operators are available for providing qualitative "rules of thumb." Although achieving many practical successes, fuzzy control has not been viewed as a rigorous science due to a lack of formal synthesis techniques which guarantee the very basic requirements of global stability and acceptable performance. Although there has been some research on the stability analysis of fuzzy control systems, it has been assumed that the mathematical model of the plant is known, which contradicts the very basic premise of fuzzy control systems, i.e., to control mathematically poorly-understood processes. In fact, if the mathematical model of a plant is known, then conventional linear and nonlinear control methods should be given higher priority. Fuzzy control should be useful in the situations where: 1) there is no acceptable mathematical model for the plant, and 2) there are experienced human operators who can satisfactorily control the plant and can provide qualitative control rules in terms of vague and fuzzy sentences. There are many practical situations where both 1) and 2) are true.

The adaptive fuzzy controller developed in this paper is suitable for these situations. The basic architecture of our adaptive fuzzy controller is a standard fuzzy logic controller [3] used in most fuzzy control systems, plus a supervisory control which fires only when the state hits some (large) boundaries. The initial adaptive fuzzy controller is constructed from the fuzzy IF-THEN rules provided by human experts and some arbitrary rules; an adaptive law is then used to update the parameters of the adaptive fuzzy controller during the adaptation procedure. If the fuzzy IF-THEN rules from human experts provide good control strategies, then the adaptation procedure will converge very quickly; on the other hand, if there are no linguistic rules from human experts, then our adaptive fuzzy controller becomes a regular nonlinear adaptive controller, similar to the radial basis function adaptive controller [5,6], neural network adaptive controller [4-6], etc.

2 Some General Discussions on Fuzzy Control

Fuzzy control is by far the most successful application of fuzzy sets and systems theory to practical problems. The present interest in fuzzy theory is largely due to the successful applications of the fuzzy logic controllers to a variety of consumer products and industrial systems. Why fuzzy control has been so successful ?

2.1 Why Fuzzy Control ?

Theoretical reasons for why fuzzy control:

- As a general rule, a good engineering approach should be capable of making effective use of all the available information. If the mathematical model of a system is too hard to obtain (this is true for many practical systems), then the most important information comes from two sources: 1) sensors which provide numerical measurements of key variables, and 2) human experts who provide linguistic descriptions about the system and control instructions. Fuzzy controllers, by design, provide a systematic and efficient framework to incorporate linguistic fuzzy information from human experts. Conventional controllers, however, cannot incorporate the linguistic fuzzy information into their designs. If in some situations the most important information comes from human experts, then fuzzy control is the best choice.
- Fuzzy control is a model-free approach, i.e., it does not require a mathematical model of the system under control. Control engineers are now facing more and more complex systems, and the mathematical models of these systems are more and more difficult to obtain. Thus, model-free approaches become more and more important in control engineering. Conventional control also has some model-free approaches, e.g., nonlinear adaptive control and PID control. Fuzzy control provides yet another model-free approach.
- Fuzzy control provides nonlinear controllers, which are well-justified due to the Universal Approximation Theorem in [9,10], i.e., these fuzzy logic controllers are general enough to perform any nonlinear control actions. Therefore, by carefully choosing the parameters of the fuzzy controllers, it is always possible to design a fuzzy controller that is just suitable for the nonlinear system under control.

Practical reasons for why fuzzy control:

- Easy to understand. Because fuzzy control emulates human control strategy, its principle is easy to understand for non-control specialists. During the last two decades, conventional control theory has been using more and more advanced mathematical tools. This is needed in order to solve difficult problems in a rigorous fashion; however, this also results in fewer and fewer practical engineers who can understand the theory. Therefore, practical engineers who are in the frond line of designing consumer products tend to use the approaches which are simple and easy to understand. Fuzzy control is just such an approach.
- Simple to implement. Fuzzy logic systems, which are heart of fuzzy control, admit a high degree of parallel implementation. Many fuzzy VLSI chips have been developed which make the implementation of fuzzy controllers simple and fast.
- Cheap to develop. From a practical point of view, the developing cost is one of the most important criterion for a successful product. Because fuzzy control is easy to understand, the time to learn the approach is short, i.e., the "software cost" is low. Also, because fuzzy control is simple to implement, the "hardware cost" is also low. Additionally, there are software tools available for designing fuzzy controllers. Thus, fuzzy control is an approach that has a high performance/cost ratio.

2.2 Why Adaptive Fuzzy Control ?

Fuzzy controllers are supposed to work in situations where there is a

CH3229-2/92/0000-2511$1.00 © 1992 IEEE

large uncertainty or unknown variation in plant parameters and structures. Generally, the basic objective of adaptive control is to maintain consistent performance of a system in the presence of these uncertainties. Therefore, advanced fuzzy control should be adaptive.

What is adaptive fuzzy control ? Roughly speaking, if a controller is constructed from adaptive fuzzy systems (an adaptive fuzzy system is a fuzzy logic system equipped with a training (adaptation) algorithm), it is called an adaptive fuzzy controller. An adaptive fuzzy controller can be a single adaptive fuzzy system, or be constructed from several adaptive fuzzy systems.

How an adaptive fuzzy controller compares with a conventional adaptive controller ? The most important advantage of adaptive fuzzy control over conventional adaptive control is that: adaptive fuzzy controllers are capable of incorporating linguistic fuzzy information from human operators, whereas conventional adaptive controllers cannot. This is especially important for the systems with a high degree of uncertainty, e.g., chemical process, aircraft, etc., because although these systems are difficult to control from a control theoretical point of view, they are often successfully controlled by human operators. How can human operators successfully control such a complex system without a mathematical model in their minds? If we ask the human operators what are their control strategies, they may just tell us a few control rules in fuzzy terms and some linguistic descriptions about the behavior of the system under various conditions which are, of course, also in fuzzy terms. Although these fuzzy control rules and descriptions are not precise and may not be sufficient for constructing a successful controller, they provide very important information about how to control the system and how the system behaves. Adaptive fuzzy control provides a tool for making use of the fuzzy information in a systematic and efficient manner.

3 Description of Fuzzy Logic Systems

Figure 1 shows the basic configuration of the fuzzy logic systems considered in this paper. The fuzzy logic system performs a mapping from $U \subset R^n$ to R. We assume that $U = U_1 \times \cdots \times U_n$, where $U_i \subset R$, $i = 1, 2, ..., n$. We now present a detailed description of each of the four blocks in the fuzzy logic system of Fig. 1.

The *fuzzy rule base* consists of a collection of fuzzy IF-THEN rules:

$$R^{(l)}: \quad IF \ x_1 \ is \ F_1^l \ and \ \cdots \ and \ x_n \ is \ F_n^l; \quad THEN \ y \ is \ G^l, \tag{1}$$

where $\underline{x} = (x_1, ..., x_n)^T \in U$ and $y \in R$ are the input and output of the fuzzy logic system, respectively, F_i^l and G^l are labels of fuzzy sets in U_i and R, respectively, and $l = 1, 2, ..., M$. Each fuzzy IF-THEN rule of (1) defines a fuzzy implication [3] $F_1^l \times \cdots \times F_n^l \rightarrow G^l$ which is a fuzzy set defined in the product space $U \times R$. Based on generalizations of implications in multivalue logic, many fuzzy implication rules were proposed in the fuzzy logic literature. Here we quote four commonly used fuzzy implication rules [3]:

- Min-operation rule of fuzzy implication:

$$\mu_{F_1^l\times\cdots\times F_n^l \rightarrow G^l}(\underline{x}, y) = min[\mu_{F_1^l\times\cdots\times F_n^l}(\underline{x}), \mu_{G^l}(y)]; \tag{2}$$

- Product-operation rule of fuzzy implication:

$$\mu_{F_1^l\times\cdots\times F_n^l \rightarrow G^l}(\underline{x}, y) = \mu_{F_1^l\times\cdots\times F_n^l}(\underline{x})\mu_{G^l}(y); \tag{3}$$

- Arithmetic rule of fuzzy implication:

$$\mu_{F_1^l\times\cdots\times F_n^l \rightarrow G^l}(\underline{x}, y) = min[1, 1 - \mu_{F_1^l\times\cdots\times F_n^l}(\underline{x}) + \mu_{G^l}(y)]; \tag{4}$$

- Maximum rule of fuzzy implication:

$$\mu_{F_1^l\times\cdots\times F_n^l \rightarrow G^l}(\underline{x}, y) = max[min(\mu_{F_1^l\times\cdots\times F_n^l}(\underline{x}), \mu_{G^l}(y)), 1 - \mu_{F_1^l\times\cdots\times F_n^l}(\underline{x})]; \tag{5}$$

where $\mu_{F_1^l\times\cdots\times F_n^l}(\underline{x})$ is defined by

$$\mu_{F_1^l\times\cdots\times F_n^l}(\underline{x}) = \mu_{F_1^l}(x_1) \star \cdots \star \mu_{F_n^l}(x_n), \tag{6}$$

where "$\star$" denotes the t-norm [3] which corresponds to the conjunction "and" in (1): the most commonly used operations for the t-norm are:

$$u \star v = \begin{cases} min(u, v) & fuzzy \ intersection \\ uv & algebric \ product \\ max(0, u + v - 1) & bounded \ product \end{cases} \tag{7}$$

The *fuzzy inference engine* performs a mapping from fuzzy sets in U to fuzzy sets in R, based upon the fuzzy IF-THEN rules in the fuzy rule base and the compositional rule of inference [12]. Let A_x be an arbitrary fuzzy set in U: then, each $R^{(l)}$ of (1) determines a fuzzy set, $A_x \circ R^{(l)}$, in R based on the following sup-star composional rule of inference

$$\mu_{A_x \circ R^{(l)}}(y) = sup_{\underline{x}\in U}[\mu_{A_x}(\underline{x}) \star \mu_{F_1^l\times\cdots\times F_n^l \rightarrow G^l}(\underline{x}, y)]. \tag{8}$$

where $\star$ is the t-norm (7), and $\mu_{F_1^l\times\cdots\times F_n^l \rightarrow G^l}(\underline{x}, y)$ is determined by the fuzzy implication rules of (2)-(5). The final fuzzy set $A_x \circ (R^{(1)}, ..., R^{(M)})$ determined by all the M rules in the fuzzy rule base is obtained by combining $\mu_{A_x \circ R^{(l)}}(y)$ of (8) for $l = 1, 2, ..., M$ using fuzzy disjunction:

$$\mu_{A_x \circ (R^{(1)}, ..., R^{(M)})}(y) = \mu_{A_x \circ R^{(1)}}(y) \dot{+} \cdots \dot{+} \mu_{A_x \circ R^{(M)}}(y), \tag{9}$$

where $\dot{+}$ denotes the t-conorm [3]: the most commonly used operations for $\dot{+}$ are

$$u \dot{+} v = \begin{cases} max(u, v) & fuzzy \ union \\ u + v - uv & algebraic \ sum \\ min(1, u + v) & bounded \ sum \end{cases} \tag{10}$$

If we use the product-operation (3) and choose $\star$ in (6) and (8) to be algebraic product, then the inference is called *product infernece*. Using product inference, (8) becomes

$$\mu_{A_x \circ R^{(l)}}(y) = sup_{\underline{x}\in U}[\mu_{A_x}(\underline{x})\mu_{F_1^l}(x_1) \cdots \mu_{F_n^l}(x_n)\mu_{G^l}(y)]. \tag{11}$$

The *fuzzifier* maps a crisp point $\underline{x} = (x_1, ..., x_n)^T \in U$ into a fuzzy set A_x in U. There are (at least) two possible choices of this mapping:

(i) A_x is a fuzzy singleton with support $\underline{x}$, i.e., $\mu_{A_x}(\underline{x}') = 1$ for $\underline{x}' = \underline{x}$ and $\mu_{A_x}(\underline{x}') = 0$ for all other $\underline{x}' \in U$ with $\underline{x}' \neq \underline{x}$; and,

(ii) $\mu_{A_x}(\underline{x}) = 1$ and $\mu_{A_x}(\underline{x}')$ decreases from one as $\underline{x}'$ moves away from $\underline{x}$, e.g., $\mu_{A_x}(\underline{x}') = exp[-\frac{(\underline{x}'-\underline{x})^T(\underline{x}'-\underline{x})}{\sigma^2}]$, where σ^2 is a parameter characterizing the shape of $\mu_{A_x}(\underline{x}')$.

The *defuzzifier* maps fuzzy sets in R to a crisp point in R. There are (at least) three possible choices of this mapping:

(i) *maximum defuzzifier*, defined as

$$y = argsup_{y'\in R}(\mu_{A_x \circ (R^{(1)}, ..., R^{(M)})}(y')), \tag{12}$$

where $\mu_{A_x \circ (R^{(1)}, ..., R^{(M)})}(y')$ is given by (9);

(ii) *centroid defuzzifier* (which is the most commonly used defuzzifier in the literature), defined as

$$y = \frac{\sum_{l=1}^{M} \bar{y}^l(\mu_{A_x \circ R^{(l)}}(\bar{y}^l))}{\sum_{l=1}^{M}(\mu_{A_x \circ R^{(l)}}(\bar{y}^l))}, \tag{13}$$

where $\bar{y}^l$ is the point in R at which $\mu_{G^{(l)}}(y)$ achieves its maximum value, and $\mu_{A_x \circ R^{(l)}}(y)$ is given by (8); and,

(iii) *modified centroid defuzzifier*, defined as

$$y = \frac{\sum_{l=1}^{M} \bar{y}^l(\mu_{A_x \circ R^{(l)}}(\bar{y}^l)/\sigma^l)}{\sum_{l=1}^{M}(\mu_{A_x \circ R^{(l)}}(\bar{y}^l)/\sigma^l)}, \tag{14}$$

where σ^l is a parameter characterizing the shape of $\mu_{G^l}(y)$ such that the narrower the shape of $\mu_{G^l}(y)$, the smaller is σ^l: for example, if $\mu_{G^l}(y) = exp[-(\frac{y-\bar{y}^l}{\sigma^l})^2]$, then σ^l is such a parameter.

From the above we see that the fuzzy logic systems of Fig. 1 comprise a very rich class of static systems mapping from $U \subset R^n$ to R, because within each block there are many different choices, and many combinations of these choices can result in useful subclasses of fuzzy logic systems. We now consider one subclass of fuzzy logic systems which will be used as building blocks of our adaptive fuzzy controller.

Lemma 1: The fuzzy logic systems with *centroid defuzzifier* (13), *product-inference* (11), and *singleton fuzzifier* are of the following form:

$$y(\underline{x}) = \frac{\sum_{l=1}^{M} \bar{y}^l(\prod_{i=1}^{n} \mu_{F_i^l}(x_i))}{\sum_{l=1}^{M}(\prod_{i=1}^{n} \mu_{F_i^l}(x_i))}, \tag{15}$$

where $\bar{y}^l$ is the point at which μ_{G^l} achieves its maximum value, and we assume that $\mu_{G^l}(\bar{y}^l) = 1$.

Proof: Using the centroid defuzzifier (13), we have

$$y(\underline{x}) = \frac{\sum_{l=1}^{M} \bar{y}^l(\mu_{A_x \circ R^{(l)}}(\bar{y}^l))}{\sum_{l=1}^{M}(\mu_{A_x \circ R^{(l)}}(\bar{y}^l))}. \tag{16}$$

where $\mu_{A_x \circ R^{(l)}}(\bar{y}^l)$ is given by the product inference (11). If we use the singleton fuzzifier, we have $\mu_{A_x}(\underline{x}') = 1$ for $\underline{x}' = \underline{x}$ ($\underline{x}$ is the input crisp point to the fuzzy logic system) and $\mu_{A_x}(\underline{x}') = 0$ for all other $\underline{x}' \in U$; therefore, the "sup" in (11) is achieved at $\underline{x}' = \underline{x}$, and (11) can be simplified to

$$\mu_{A_x \circ R^{(l)}}(\bar{y}^l) = \prod_{i=1}^{n} \mu_{F_i^l}(x_i) \tag{17}$$

(we assume that $\mu_{G^l}(\bar{y}^l) = 1$). Substituting (17) into (16), we obtain (15). Q.E.D.

If we fix the $\mu_{F_i^l}(x_i)$'s and view the $\bar{y}^l$'s as adjustable parameters, then (15) can be written as

$$y(\underline{x}) = \underline{\theta}^T \underline{\xi}(\underline{x}), \tag{18}$$

where $\underline{\theta} = (\bar{y}^1, ..., \bar{y}^M)^T$ is a parameter vector, and $\underline{\xi}(\underline{x}) = (\xi^1(\underline{x}), ..., \xi^M(\underline{x}))^T$ is a regressive vector with the regressor $\xi^l(\underline{x})$ (which is called fuzzy basis function in [10]) defined as

$$\xi^l(\underline{x}) = \frac{\prod_{i=1}^{n} \mu_{F_i^l}(x_i)}{\sum_{l=1}^{M}(\prod_{i=1}^{n} \mu_{F_i^l}(x_i))}. \tag{19}$$

4 Basic Ideas of Constructing Stable Direct Adaptive fuzzy Controllers

Consider the nth-order nonlinear systems of the form

$$x^{(n)} = f(x, \dot{x}, ..., x^{(n-1)}) + bu, \qquad y = x, \tag{20}$$

where f is an unknown continuous function, b is a positive unknown constant, and $u \in R$ and $y \in R$ are the input and output of the system, respectively. We assume that the state vector $\underline{x} = (x_1, x_2, ..., x_n)^T = (x, \dot{x}, ..., x^{(n-1)})^T \in R^n$ is available for measurement. The control objective is to force y to follow a given bounded reference signal $y_m(t)$, under the constraint that all signals involved must be bounded. More specifically, we have

Control Objectives: Determine a feedback control $u = u(\underline{x}|\underline{\theta})$ (based on fuzzy logic systems) and an adaptive law for adjusting the parameter vector $\underline{\theta}$ such that:

(i) the closed-loop system must be globally stable in the sense that all variables, $\underline{x}(t)$, $\underline{\theta}(t)$ and $u(\underline{x}|\underline{\theta})$, must be uniformly bounded, i.e., $|\underline{x}(t)| \le M_x < \infty$, $|\underline{\theta}(t)| \le M_\theta < \infty$ and $|u(\underline{x}|\underline{\theta})| \le M_u < \infty$ for all $t \ge 0$, where M_x, M_θ and M_u are design parameters specified by the designer; and,

(ii) the tracking error, $e \equiv y_m - y$, should be as small as possible under the constraints in (i).

We now show the basic ideas of how to construct a direct adaptive fuzzy controllers to achieve these control objectives. To begin, let $\underline{e} = (e, \dot{e}, ..., e^{(n-1)})^T$ and $\underline{k} = (k_n, ..., k_1)^T \in R^n$ be such that all roots of the polynomial $h(s) = s^n + k_1 s^{n-1} + \cdots + k_n$ are in the open left-half plane. If the function f and constant b are known, then the control law

$$u^* = \frac{1}{b}[-f(\underline{x}) + y_m^{(n)} + \underline{k}^T \underline{e}] \tag{21}$$

applied to (20) results in

$$e^{(n)} + k_1 e^{(n-1)} + \cdots + k_n e = 0 \tag{22}$$

which implies that $lim_{t\to\infty} e(t) = 0$ — a main objective of control. Since f and b are unknown, the optimal control u^* of (21) cannot be implemented. Our purpose is to design a fuzzy logic system to approximate this optimal control.

Suppose that the control u is the summation of a fuzzy control $u_c(\underline{x}|\underline{\theta})$ and a supervisory control $u_s(\underline{x})$:

$$u = u_c(\underline{x}|\underline{\theta}) + u_s(\underline{x}), \tag{23}$$

where $u_c(\underline{x}|\underline{\theta})$ is a fuzzy logic system in the form of (15) (or equivalently (18)), and $u_s(\underline{x})$ will be determined later in this section (for the reasons why u_s is called a supervisory control, see the discussions at a later point in this section). Substituting (23) into (20), we have

$$x^{(n)} = f(\underline{x}) + b[u_c(\underline{x}|\underline{\theta}) + u_s(\underline{x})]. \tag{24}$$

Now adding and substructing bu^* to (24) and after some straightforward manipulation we obtain the error equation governing the closed-loop system:

$$e^{(n)} = -\underline{k}^T \underline{e} + b[u^* - u_c(\underline{x}|\underline{\theta}) - u_s(\underline{x})], \tag{25}$$

or equivalently,

$$\dot{\underline{e}} = \Lambda_c \underline{e} + \underline{b}_c[u^* - u_c(\underline{x}|\underline{\theta}) - u_s(\underline{x})], \tag{26}$$

where

$$\Lambda_c = \begin{bmatrix} 0 & 1 & 0 & 0 & \cdots & 0 & 0 \\ 0 & 0 & 1 & 0 & \cdots & 0 & 0 \\ \cdots & \cdots & \cdots & \cdots & \cdots & \cdots & \cdots \\ 0 & 0 & 0 & 0 & \cdots & 0 & 1 \\ -k_n & -k_{n-1} & \cdots & \cdots & \cdots & \cdots & -k_1 \end{bmatrix}, \underline{b}_c = \begin{bmatrix} 0 \\ \cdots \\ 0 \\ b \end{bmatrix}. \tag{27}$$

Define $V_e = \frac{1}{2}\underline{e}^T P \underline{e}$, where P is a symmetric positive definite matrix satisfying the Lyapunov equation

$$\Lambda_c^T P + P\Lambda_c = -Q, \tag{28}$$

where $Q > 0$. Using (28) and the error equation (26), we have

$$\begin{aligned} \dot{V}_e &= -\frac{1}{2}\underline{e}^T Q \underline{e} + \underline{e}^T P \underline{b}_c[u^* - u_c(\underline{x}|\underline{\theta}) - u_s(\underline{x})] \\ &\le -\frac{1}{2}\underline{e}^T Q \underline{e} + |\underline{e}^T P \underline{b}_c|(|u^*| + |u_c|) - \underline{e}^T P \underline{b}_c u_s. \end{aligned} \tag{29}$$

Our task now is to design u_s such that $\dot{V}_e \le 0$. In order to do so, we need the following assumption:

Assumption 1: We can determine a function $f^U(\underline{x})$ and a constant b_L such that $|f(\underline{x})| \le f^U(\underline{x})$ and $0 < b_L \le b$.

We construct the supervisory control $u_s(\underline{x})$ as follow:

$$u_s(\underline{x}) = I_1^* sgn(\underline{e}^T P \underline{b}_c)[|u_c| + \frac{1}{b_L}(f^U + |y_m^{(n)}| + |\underline{k}^T \underline{e}|)], \tag{30}$$

where $I_1^* = 1$ if $V_e > \bar{V}$ ($\bar{V}$ is a constant specified by the designer, see discussions at a later point), and $I_1^* = 0$ if $V_e \le \bar{V}$. Because $b > 0$, $sgn(\underline{e}^T P \underline{b}_c)$ can be determined; also, all other terms in (30) can be determined, thus the supervisory control u_s of (30) can be implemented. Substituting (30) and (21) into (29) and considering the $I_1^* = 1$ case, we have

$$\begin{aligned} \dot{V}_e \le & -\frac{1}{2}\underline{e}^T Q \underline{e} + |\underline{e}^T P \underline{b}_c|[\frac{1}{b}(|f| + |y_m^{(n)}| + |\underline{k}^T \underline{e}|) + |u_c| - |u_c| \\ & - \frac{1}{b_L}(f^U + |y_m^{(n)}| + |\underline{k}^T \underline{e}|)] \le -\frac{1}{2}\underline{e}^T Q \underline{e} \le 0. \end{aligned} \tag{31}$$

Therefore, using the supervisory control u_s of (30), we always have $V_e \le \bar{V}$. Because $P > 0$, the boundedness of V_e implies the boundedness of $\underline{e}$ which in turn implies the boundedness of $\underline{x}$.

From (30) we see that the u_s is nonzero only when the error function V_e is greater than the positive constant $\bar{V}$. That is, if the closed-loop system with the pure fuzzy controller u_c is well-behaved in the sense that the error is not big (i.e., $V_e \le \bar{V}$), then the supervisory control u_s is zero; on the other hand, if the system tends to be unstable (i.e., $V_e > \bar{V}$), then the supervisory control u_s begins operating to force $V_e \le \bar{V}$. In this way, the control u_s is like a *supervisor*; this is why we call the u_s a *supervisory control*.

Next, we replace the $u_c(\underline{x}|\underline{\theta})$ by the fuzzy logic system (18) and develop an adaptive law to adjust the parameter vector $\underline{\theta}$. Define the optimal parameter vector:

$$\underline{\theta}^* \equiv argmin_{|\underline{\theta}|\le M_\theta}[sup_{|\underline{x}|\le M_x}|u_c(\underline{x}|\underline{\theta}) - u^*|]. \quad (32)$$

and the "minimum approximation error":

$$w \equiv u_c(\underline{x}|\underline{\theta}^*) - u^*. \quad (33)$$

The error equation (26) can be rewritten as

$$\begin{aligned}\dot{\underline{e}} &= \Lambda_c\underline{e} + \underline{b}_c[u_c(\underline{x}|\underline{\theta}^*) - u_c(\underline{x}|\underline{\theta})] - \underline{b}_c u_s(\underline{x}) - \underline{b}_c w \\ &= \Lambda_c\underline{e} + \underline{b}_c\underline{\phi}^T\underline{\xi}(\underline{x}) - \underline{b}_c u_s - \underline{b}_c w,\end{aligned} \quad (34)$$

where $\underline{\phi} \equiv \underline{\theta}^* - \underline{\theta}$ and $\underline{\xi}(\underline{x})$ is the fuzzy basis function (19). Define the Lyapunov function candidate

$$V = \frac{1}{2}\underline{e}^T P\underline{e} + \frac{b}{2\gamma}\underline{\phi}^T\underline{\phi}. \quad (35)$$

where γ is a positive constant. Using (34) and (28). we have

$$\dot{V} = -\frac{1}{2}\underline{e}^T Q\underline{e} + \underline{e}^T P\underline{b}_c(\underline{\phi}^T\underline{\xi}(\underline{x}) - u_s - w) + \frac{b}{\gamma}\underline{\phi}^T\dot{\underline{\phi}}. \quad (36)$$

Let $\underline{p}_n$ be the last column of P. then from (27) we have

$$\underline{e}^T P\underline{b}_c = \underline{e}^T\underline{p}_n b. \quad (37)$$

Substituting (37) into (36). we have

$$\dot{V} = -\frac{1}{2}\underline{e}^T Q\underline{e} + \frac{b}{\gamma}\underline{\phi}^T[\gamma\underline{e}^T\underline{p}_n\underline{\xi}(\underline{x}) + \dot{\underline{\phi}}] - \underline{e}^T P\underline{b}_c u_s - \underline{e}^T P\underline{b}_c w. \quad (38)$$

If we choose the adaptive law:

$$\dot{\underline{\theta}} = \gamma\underline{e}^T\underline{p}_n\underline{\xi}(\underline{x}). \quad (39)$$

then (38) becomes

$$\dot{V} \le -\frac{1}{2}\underline{e}^T Q\underline{e} - \underline{e}^T P\underline{b}_c w. \quad (40)$$

where we use the facts $\underline{e}^T P\underline{b}_c u_s \ge 0$ (observe (30)) and $\dot{\underline{\phi}} = -\dot{\underline{\theta}}$. This is the best we can achieve. In order to guarantee $|\underline{\theta}| \le M_\theta$. we will use a projection algorithm to modify the basic adaptive law (39); the details are given in Section 5.

In summary of the above. the overall scheme of our direct adaptive fuzzy controller is shown in Fig. 2.

Finally. we specify the linguistic information which can be directly incorporated into the direct adaptive fuzzy controller.

Assumption 2: There are L fuzzy control rules from human experts in the following form:

$$R_c^{(r)}: IF\ x_1\ is\ A_1^r\ and\ \cdots\ and\ x_n\ is\ A_n^r.\ THEN\ u\ is\ C^r, \quad (41)$$

where A_i^r and C^r are labels of fuzzy sets in R. and $r = 1.2.\ldots.L$.

5 Design and Stability Analysis of the Direct Adaptive Fuzzy Controller

Design of Direct Adaptive Fuzzy Controller:

Step 1: Off-line Preprocessing

- Specify the $k_1.\ldots.k_n$ such that all roots of $s^n + k_1 s^{n-1} + \cdots + k_n = 0$ are in the open left-half plane. Specify a positive definite $n \times n$ matrix Q. Solve the Lyapunov equation (28), e.g., using the method of [11]. to obtain a symmetric $P > 0$. Specify the design parameters M_θ. M_x and M_u based on practical constraints.

Step 2: Initial Controller Construction

- Define m_i fuzzy sets $F_i^{l_i}$ whose membership functions $\mu_{F_i^{l_i}}$ uniformly cover U_i. where $l_i = 1.2.\ldots.m_i$ and $i = 1.2.\ldots.n$. The $U = U_1 \times \cdot \times U_n$ is usually chosen to be $\{\underline{x} \in R^n : |\underline{x}| \le M_x\}$. We require that the $F_i^{l_i}$'s include the A_i^r's in (41).
- Construct the fuzzy rule base for the fuzzy logic system $u_c(\underline{x}|\underline{\theta})$ which consists of $m_1 \times m_2 \times \cdots \times m_n$ rules whose IF parts comprise all the possible combinations of the $F_i^{l_i}$'s for $i = 1.2.\ldots.n$. Specifically. the fuzzy rule base of $u_c(\underline{x}|\underline{\theta})$ consists of rules:

$$\begin{aligned}R_c^{(l_1,\ldots,l_n)}: &\quad IF\ x_1\ is\ F_1^{l_1}\ and\ \cdots\ and\ x_n\ is\ F_n^{l_n}, \\ &\quad THEN\ u_c\ is\ G^{(l_1,\ldots,l_n)},\end{aligned} \quad (42)$$

where $l_i = 1.2.\ldots.m_i$. $i = 1.2.\ldots.n$. and $G^{(l_1,\ldots,l_n)}$ are fuzzy sets in R which are specified as follows: if the IF part of (42) agrees with the IF part of (41). set $G^{(l_1,\ldots,l_n)}$ equal to the corresponding C^r in (41): otherwise. set $G^{(l_1,\ldots,l_n)}$ arbitrarily with the constraint that the centers of $G^{(l_1,\ldots,l_n)}$ (which correspond to the $\bar{y}^l$ parameters) are inside the constraint sets $\{\underline{\theta} : |\underline{\theta}| \le M_\theta\}$. Therefore. *the initial adaptive fuzzy controller is constructed from the fuzzy control rules (41).*

- Construct the fuzzy basis functions

$$\xi^{(l_1,\ldots,l_n)}(\underline{x}) = \frac{\prod_{i=1}^n \mu_{F_i^{l_i}}(x_i)}{\sum_{l_1=1}^{m_1}\cdots\sum_{l_n=1}^{m_n}(\prod_{i=1}^n \mu_{F_i^{l_i}}(x_i))}. \quad (43)$$

and collect them into a $\prod_{i=1}^n m_i$-dimensional vector $\underline{\xi}(\underline{x})$ in a natural ordering for $l_1 = 1.2.\ldots.m_1.\ldots.l_n = 1.2.\ldots.m_n$. Collect the points at which $\mu_{G^{(l_1,\ldots,l_n)}}$'s achieve their maximum values. in the same ordering as $\underline{\xi}(\underline{x})$, into a vector $\underline{\theta}$. The $u_c(\underline{x}|\underline{\theta})$ is constructed as

$$u_c(\underline{x}|\underline{\theta}) = \underline{\theta}^T\underline{\xi}(\underline{x}). \quad (44)$$

Step 3: On-line Adaptation

- Apply the feedback control (23) to the plant (20), where u_c is given by (44). and u_s is given by (30).
- Use the following adaptive law to adjust the parameter vector $\underline{\theta}$:

$$\dot{\underline{\theta}} = \begin{cases}\gamma\underline{e}^T\underline{p}_n\underline{\xi}(\underline{x})\ if\ (|\underline{\theta}| < M_\theta)\ or\ (|\underline{\theta}| = M_\theta\ and\ \underline{e}^T\underline{p}_n\underline{\theta}^T\underline{\xi}(\underline{x}) \le 0) \\ P\{\gamma\underline{e}^T\underline{p}_n\underline{\xi}(\underline{x})\}\ if\ (|\underline{\theta}| = M_\theta\ and\ \underline{e}^T\underline{p}_n\underline{\theta}^T\underline{\xi}(\underline{x}) > 0),\end{cases} \quad (45)$$

where the projection operator $P\{*\}$ is defined as [1]:

$$P\{\gamma\underline{e}^T\underline{p}_n\underline{\xi}(\underline{x})\} = \gamma\underline{e}^T\underline{p}_n\underline{\xi}(\underline{x}) - \gamma\underline{e}^T\underline{p}_n\frac{\underline{\theta}\underline{\theta}^T\underline{\xi}(\underline{x})}{|\underline{\theta}|^2}. \quad (46)$$

The following theorem shows the properties of this direct adaptive fuzzy controller.

Theorem 1: Consider the nonlinear plant (20) with the control (23). where u_c is given by (44). and u_s is given by (30). Let the parameter vector $\underline{\theta}$ be adjusted by the adaptive law (45). and let Assumptions 1 and 2 be true. Then. the overall control scheme (shown in Fig.2) guarantees the following properties:

(i) $|\underline{\theta}(t)| \le M_\theta$.

$$|\underline{x}(t)| \le |\underline{y}_m| + (\frac{2V}{\lambda_{min}})^{1/2}. \quad (47)$$

and

$$|u(t)| \le 2M_\theta + \frac{1}{b_L}[f^U + |y_m^{(n)}| + |\underline{k}|(\frac{2V}{\lambda_{min}})^{1/2}]. \quad (48)$$

for all $t \ge 0$. where λ_{min} is the minimum eigenvalue of P. and $\underline{y}_m = (y_m, \dot{y}_m, \ldots, y_m^{(n-1)})^T$.

(ii)

$$\int_0^t |\underline{e}(\tau)|^2 d\tau \le a + c\int_0^t |w(\tau)|^2 d\tau \quad (49)$$

for all $t \ge 0$. where a and c are constants. and w is minimum approximation error defined by (33).

(iii) If w is squared integrable. i.e.. $\int_0^\infty |w(t)|^2 dt < \infty$. then $lim_{t\to\infty} |\underline{e}(t)| = 0$.

Proof of this theorem is given in the Appendix.

Remark 1: For many practical control problems. the state $\underline{x}$ and control u are required to be constrained within certain regions. For given constraints M_x and M_u, we can specify the design parameters $\underline{k}$, M_θ, Q and V, based on (47) and (48), such that the state $\underline{x}$ and control u are within the constraint sets.

Remark 2: From Step 2 we see that the fuzzy control rules (41) are incorporated into the adaptive fuzzy controller by constructing the initial controller based on them. If the linguistic rules (41) provide good control strategy, then the initial u_c should be close to the optimal control u^*: as a result, we can hope that the closed-loop system behaves approximately like (22). If no linguistic information is available, our adaptive fuzzy controller is still a well-performing nonlinear adaptive controller, in the sense of having the properties (i)-(iii) of Theorem 1. In summary, *good linguistic information can help us to construct a good initial controller so that we can have a fast adaptation*; we will show an example in Section 6 to illustrate this point.

Remark 3: From (iii) of Theorem 1 we see that in order for the tracking error $e(t)$ converge to zero, we require that the "minimum approximation error w" defined by (33) is small (in the sense of squared integrable). Based on the Universal Approximation Theorem of [9 ,10], we can say that if we use sufficient number of rules to construct u_c, the w should be small.

6 Simulations

Example 1: In this example, we use our direct adaptive fuzzy controller to regulate the plant

$$\dot{x}(t) = \frac{1 - e^{-x(t)}}{1 + e^{-x(t)}} + u(t) \qquad (50)$$

to the origin, i.e., $y_m \equiv 0$. It is clear that the plant (50) is unstable if without control, because if $u(t) \equiv 0$, then $\dot{x} = \frac{1-e^{-x}}{1+e^{-x}} > 0$ for $x > 0$, and $\dot{x} = \frac{1-e^{-x}}{1+e^{-x}} < 0$ for $x < 0$. We choose $\gamma = 1$, $M_x = 3$, $M_\theta = 3$, $b_L = 0.5 < 1 = b$, and $f^U = 1 \geq \frac{1-e^{-x(t)}}{1+e^{-x(t)}}$. In Step 1, we define 6 fuzzy sets over the interval $[-3, 3]$, with labels $N3$, $N2$, $N1$, $P1$, $P2$ and $P3$, and membership functions $\mu_{N3}(x) = 1/(1 + exp(5(x + 2)))$, $\mu_{N2}(x) = exp(-(x + 1.5)^2)$, $\mu_{N1}(x) = exp(-(x + 0.5)^2)$, $\mu_{P1}(x) = exp(-(x - 0.5)^2)$, $\mu_{P2}(x) = exp(-(x-1.5)^2)$ and $\mu_{P3}(x) = 1/(1+exp(-5(x-2)))$, which are shown in Fig. 3. In Step 2, we consider two cases: 1) there are no fuzzy control rules, and the initial $\theta_i(0)$'s are chosen randomly in the interval $[-2, 2]$; and, 2) there are two fuzzy control rules:

$$R_1: \; IF \; x \; is \; N2, \; THEN \; u \; is \; PB, \qquad (51)$$

$$R_2: \; IF \; x \; is \; P2, \; THEN \; u \; is \; NB, \qquad (52)$$

where $\mu_{PB}(u) = exp(-(u-2)^2)$, and $\mu_{NB}(u) = exp(-(u+2)^2)$. These two rules are obtained by considering the fact that our problem is to control $x(t)$ to zero: therefore, if x is negative, then the control $u(x)$ should be "positive big" (PB) so that it may happen that $\dot{x} > 0$ (see (50)): on the other hand, if x is positive, then the control $u(x)$ should be "negative big" (NB) so that it may happen that $\dot{x} < 0$. We used the MATLAB command "ode23" to simulate the overall control system. We chose the initial state $x(0) = 1$. Figures 4 and 5 show the $x(t)$ for the cases without and with the linguistic control rules (51) and (52), respectively. We see from Figs. 4 and 5 that: 1) our direct adaptive fuzzy controller could regulate the plant to the origin without using the fuzzy control rules, and 2) by incorporating the fuzzy control rules, the speed of convergence became much faster. We also simulated for other initial conditions, and the results were very similar: we do not show them in order to clear the figures and make the comparison easier.

In the simulations shown in Figs. 4 and 5, the state $x(t)$ did not hit the boundary $|x| = 3$, therefore the supervisory control u_s never fired. Now we change $M_x = 1.5$ and keep all other parameters to be the same as in the simulation shown in Fig. 4 (i.e., without using the fuzzy control rules). The simulation result for this case is shown in Fig. 6. We see that the supervisory control u_s did force the state to be inside the constraint set $|x| \leq 1.5$.

It is interesting to observe how this bounded control was achieved. From Fig. 6 we see that as soon as the state hit the boundary, the supervisory control u_s began operating to force the state back to the constraint set. Since as soon as the state was inside the constraint set, the supervisory control stopped operating, which resulted in the state hit the boundary again. The continuation of this kind of back-and-forth operation resulted in the "holding" of the state around the boundary, as shown in Fig. 6 in the intervals $t \in [0.5, 2.5]$ and $t \in [5.5, 7]$ (approximately). That is, the supervisory control could prevent the system from being unstable, but could not regulate the state to the origin. The encouraging observation is that during these "holding periods", the adaptive law adjusted the parameters of the fuzzy controller u_c, and, finally, the fuzzy control "recovered" and finished the control task — regulating the state to the origin.

We may view the fuzzy controller u_c as a lower-level operator and the supervisory control u_s as a higher-level supervisor. If the operator could perform successful control, the supervisor just observes and does not take any action. If somehow the operator could not control the system well, the supervisor takes actions to prevent catastrophic consequences. During this period, the operator learns to correct his/her mistakes and regains the control. After back to normal, the supervisor becomes an observer again. From Fig. 6 we see that our adaptive fuzzy control scheme operated in exactly the way as described above.

7 Conclusions

In this paper, we developed a direct adaptive fuzzy controller which: 1) does not require an accurate mathematical model of the system under control, 2) is capable of incorporating fuzzy control rules directly into the controllers, and 3) guarantees the global stability of the resulting closed-loop system in the sense that all signals involved are uniformly bounded. We provide the specific formula of the bounds so that controller designers can determine the bounds based on their requirements. We used the direct adaptive fuzzy controller to regulate an unstable system to the origin: the simulation results show that: 1) the direct adaptive fuzzy controller could perform successful control without using any fuzzy control rules, and 2) after incorporating some fuzzy control rules into the controllers, the adaptation speed became much faster. We also showed explicitly how the supervisory control forced the state to be within the constraint set and how the adaptive fuzzy controller learned to regain control.

8 Acknowledgements

The author would like to thank Prof. Jerry M. Mendel and Prof. Lotfi A. Zadeh for their encouragement. This work was supported by the Rockwell International Science Center.

9 REFERENCES

[1] Goodwin, G. C. and D. Q. Mayne, "A parameter estimation perspective of continuous time model reference adaptive control," *Automatica*, Vol. 23, pp. 57-70, 1987.

[2] Isidori, A., *Nonlinear Control Systems*, Springer-Verlag, Berlin, 1989.

[3] Lee, C. C., "Fuzzy logic in control systems: fuzzy logic controller, part I and II," *IEEE Trans. on Syst., Man, and Cybern.*, Vol.SMC-20, No.2, pp.404-435, 1990.

[4] Narendra, K. S. and K. Parthasarathy, "Identification and control of dynamical systems using neural networks," *IEEE Trans. on Neural Networks*, Vol.1, No.1, pp.4-27, 1990.

[5] Polycarpou, M. M. and P. A. Ioannou, "Identification and control nonlinear systems using neural network models: design and stability analysis," USC EE-Report 91-09-01, 1991.

[6] Sanner, R. M. and J. E. Slotine, "Gaussian networks for direct adaptive control," *Proc. American Control Conf.*, pp. 2153-2159, 1991.

[7] Sastry, S. and M. Bodson, *Adaptive Control: Stability, Convergence, and Robustness*, Englewood Cliffs, NJ, Prentice-Hall, 1989.

[8] Slotine, J. E. and W. Li, *Applied Nonlinear Control*, Prentice-Hall, Inc., NJ, 1991.

[9] Wang, L. X., "Fuzzy systems are universal approximators," *Proc. IEEE International Conf. on Fuzzy Systems*, San Diego, pp. 1163-1170, 1992.

[10] Wang, L. X. and J. M. Mendel, "Fuzzy basis functions, universal approximation, and orthogonal least squares learning," *IEEE Trans. on Neural Networks*, to appear, Sept., 1992.

[11] Wang, L. X. and J. M. Mendel, "Three-dimensional structured networks for matrix equation solving," *IEEE Trans. on Computers*, Vol. 40, No. 12, pp. 1337-1346, 1991.

[12] Zadeh, L. A., "Outline of a new approach to the analysis of complex systems and decision processes," *IEEE Trans. on Systems, Man, and Cybern.*, Vol. SMC-3, No. 1, pp. 28-44, 1973.

A Appendix: Proof of Theorem 1

Proof of Theorem 1: (i) To prove $|\underline{\theta}| \leq M_\theta$. let $V_\theta = \frac{1}{2}\underline{\theta}^T\underline{\theta}$. If the first line of (45) is true. we have either $|\underline{\theta}| < M_\theta$ or $\dot{V}_\theta = \gamma\underline{\epsilon}^T\underline{p}_n\underline{\theta}^T\underline{\xi}(\underline{x}) \leq 0$ when $|\underline{\theta}| = M_\theta$. i.e.. we always have $|\underline{\theta}| \leq M_\theta$: if the second line of (45) is true. we have $|\underline{\theta}| = M_\theta$ and $\dot{V}_\theta = \gamma\underline{\epsilon}^T\underline{p}_n\underline{\theta}^T\underline{\xi}(\underline{x}) - \gamma\underline{\epsilon}^T\underline{p}_n\frac{|\underline{\theta}|^2\underline{\theta}^T\underline{\xi}(\underline{x})}{|\underline{\theta}|^2} = 0$, i.e.. $|\underline{\theta}_f| \leq M_f$. Therefore. we always have $|\underline{\theta}(t)| \leq M_\theta$. $\forall t \geq 0$.

In Section 4 we proved that $V_\epsilon \leq \bar{V}$: therefore. $\frac{1}{2}\lambda_{min}|\underline{e}|^2 \leq \frac{1}{2}\underline{\epsilon}^T P\underline{\epsilon} \leq \bar{V}$. i.e.. $|\underline{e}| \leq (\frac{2\bar{V}}{\lambda_{min}})^{1/2}$. Since $\underline{\epsilon} = \underline{y}_m - \underline{x}$. we have $|\underline{x}| \leq |\underline{y}_m| + |\underline{e}| \leq |\underline{y}_m| + (\frac{2\bar{V}}{\lambda_{min}})^{1/2}$. which is (47). Finally, we prove (48). Since $u_c(\underline{x}|\underline{\theta})$ is a weighted average of the elements of $\underline{\theta}$, we have $|u_c(\underline{x}|\underline{\theta})| \leq |\underline{\theta}| \leq M_\theta$. Therefore. from (23) and (30) we have (48).

(ii) From (38). (45) and (46). we have

$$\dot{V} = -\frac{1}{2}\underline{\epsilon}^T Q\underline{\epsilon} - \underline{\epsilon}^T P\underline{b}_c u_s - \underline{\epsilon}^T P\underline{b}_c w + I_1\underline{\epsilon}^T\underline{p}_n b\frac{\underline{\phi}^T\underline{\theta}\underline{\theta}^T\underline{\xi}(\underline{x})}{|\underline{\theta}|^2}, \quad \text{(A.1)}$$

where $I_1 = 0(1)$ if the first (second) line of (45) is true. Now we show that the last term of (A.1) are nonpositive. If $I_1 = 0$, the conclusion is trivial. Let $I_1 = 1$. which means that $|\underline{\theta}| = M_\theta$ and $\underline{\epsilon}^T\underline{p}_n\underline{\theta}^T\underline{\xi}(\underline{x}) > 0$, we have $\underline{\phi}^T\underline{\theta} = (\underline{\theta}^* - \underline{\theta})^T\underline{\theta} = \frac{1}{2}[|\underline{\theta}^*|^2 - |\underline{\theta}|^2 - |\underline{\theta} - \underline{\theta}^*|^2] \leq 0$. since $|\underline{\theta}| = M_\theta \geq |\underline{\theta}^*|$. Therefore. the last term of (A.1) is nonpositive, and we have

$$\dot{V} \leq -\frac{1}{2}\underline{\epsilon}^T Q\underline{\epsilon} - \underline{\epsilon}^T P\underline{b}_c u_s - \underline{\epsilon}^T P\underline{b}_c w. \quad \text{(A.2)}$$

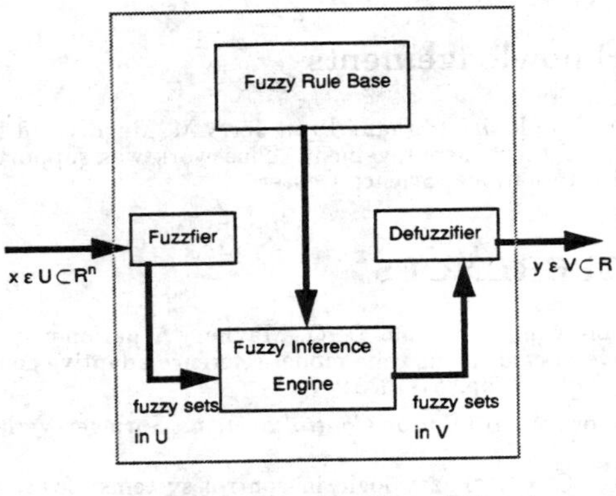

Figure 1: Basic configuration of fuzzy logic systems.

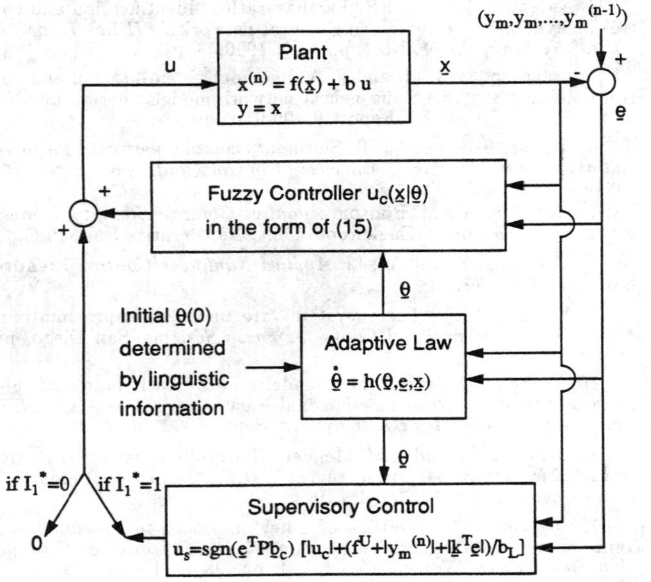

Figure 2: The overall scheme of direct adaptive fuzzy control.

From (30) we have $\underline{\epsilon}^T P\underline{b}_c u_s \geq 0$: therefore. (A.2) can be further simplified to

$$\begin{aligned} \dot{V} &\leq -\frac{1}{2}\underline{\epsilon}^T Q\underline{\epsilon} - \underline{\epsilon}^T P\underline{b}_c w \\ &\leq -\frac{\lambda_{Qmin} - 1}{2}|\underline{\epsilon}|^2 - \frac{1}{2}[|\underline{\epsilon}|^2 + 2\underline{\epsilon}^T P\underline{b}_c w + |P\underline{b}_c w|^2] + \frac{1}{2}|P\underline{b}_c w|^2 \\ &\leq -\frac{\lambda_{Qmin} - 1}{2}|\underline{\epsilon}|^2 + \frac{1}{2}|P\underline{b}_c w|^2. \end{aligned} \quad \text{(A.3)}$$

where λ_{Qmin} is the minimum eigenvalue of Q. Integrating both sides of (A.3) and assuming that $\lambda_{Qmin} > 1$ (since Q is determined by the designer. we can choose such a Q). we have

$$\int_0^t |\underline{\epsilon}(\tau)|^2 d\tau \leq \frac{2}{\lambda_{Qmin} - 1}[|V(0)| + |V(t)|] + \frac{1}{\lambda_{Qmin} - 1}|P\underline{b}_c|^2 \int_0^t |w(\tau)|^2 d\tau. \quad \text{(A.4)}$$

Define $a = \frac{2}{\lambda_{Qmin}-1}[|V(0)| + sup_{t\geq 0}|V(t)|]$ and $c = \frac{1}{\lambda_{Qmin}-1}|P\underline{b}_c|^2$. (A.4) becomes (49) (note that $sup_{t\geq 0}|V(t)|$ is finite because $\underline{\epsilon}$ and $\underline{\phi}$ are all bounded).

(iii) If $w \in L_2$. then from (49) we have $\underline{\epsilon} \in L_2$. Because we have proven that all the variables in the right hand side of (34) are bounded. we have $\dot{\underline{\epsilon}} \in L_\infty$. Using the Barbalat's Lemma [7] (if $\underline{\epsilon} \in L_2 \cap L_\infty$ and $\dot{\underline{\epsilon}} \in L_\infty$. then $lim_{t\to\infty}|\underline{\epsilon}(t)| = 0$). we have $lim_{t\to\infty}|\underline{\epsilon}(t)| = 0$. Q.E.D.

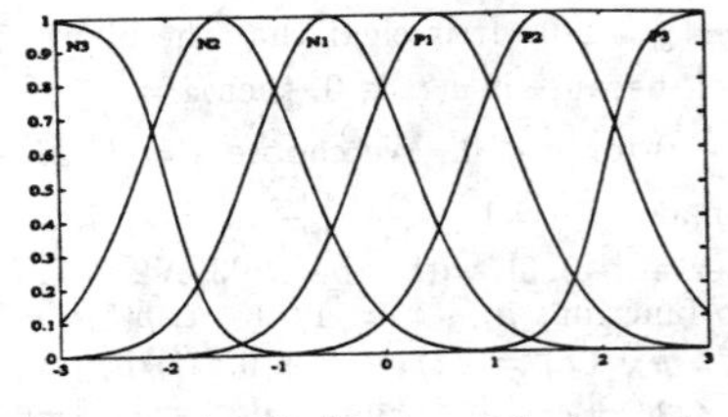

Figure 3: Fuzzy membership functions defined over the state space

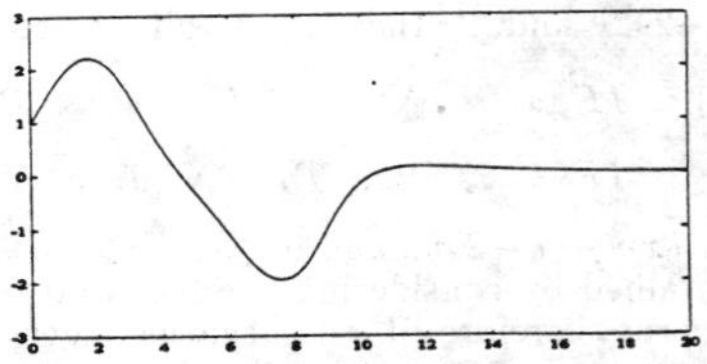

Figure 4: Closed-loop system state $x(t)$ using the direct adaptive fuzzy controller for the plant (50) without incorporating any fuzzy control rules.

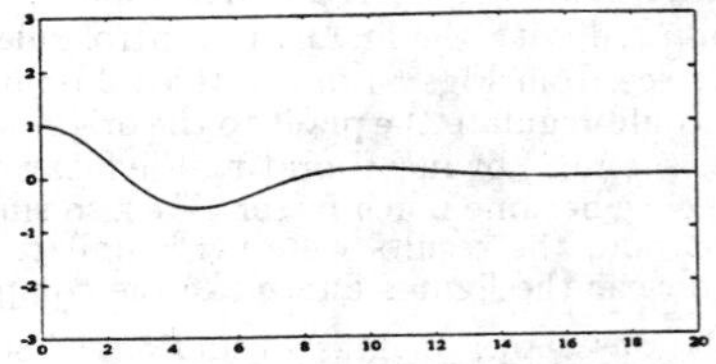

Figure 5: Closed-loop system state $x(t)$ using the direct adaptive fuzzy controller for the plant (50) after incorporating the fuzzy control rules (51) and (52).

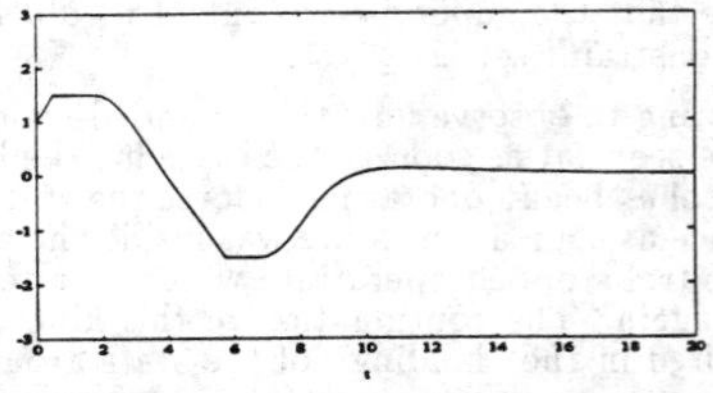

Figure 6: Closed-loop system state $x(t)$ for the same situation as Fig. 4 except $M_x = 1.5$.

Proceedings of the 31st Conference
on Decision and Control
Tucson, Arizona • December 1992

Adaptive Fuzzy Dominant–Pole Placement Control

Kyriakos Kyriakides **Anthony Tzes**
Department of Mechanical Engineering
Polytechnic University
333 Jay Street
Brooklyn, NY 11201

Abstract

The design of an adaptive fuzzy dominant-pole placement controller is addressed in this paper. Under a priori known noise specifications, a convex region in the z-domain containing the transfer function parameters can be estimated through application of the Set Membership Recursive Least Squares identification algorithm. The degree of uncertainty on the estimated parameters is reflected through the area of this region. Fuzzy membership functions are assigned to cells partitioning a universe of discourse, which includes the identified region. The knowledge data base consists of feedback gains required to place the closed loop poles at predefined locations, and a rule based controller infers the control input variable weighting each gain with the integral of the identified region under the membership functions. The aforementioned controller is demonstrated in simulation studies on an example system.

I. Introduction

Modern control theory evolved under the assumption of a crisp, deterministic and precise in character goal function. However, despite the development of advanced mathematical tools, control theory fails to cope with systems which lack either a quantitative input–output description, or, a well defined control objective. If the system asserts to be ambiguous, then the modeling language has to be adapted to model appropriately its inherent characteristics.

Fuzzy set theory provides a strict mathematical framework in which conceptual phenomena can be precisely and rigorously studied. Such a framework provides a natural way of dealing with problems in which the source of imprecision is the absence of sharply defined criteria, rather than the presence of random variables. Fuzzy logic systems can be used to assist in the storage and manipulation (management) of the vague and uncertain portion of knowledge.

Fuzzy Logic Controllers (FLCs) [1, 2], operating in an uncertain fuzzy environment, manifest the following attributes : 1) separate domain-specific knowledge (Knowledge Base) and problem solving methodology (Inference Engine), 2) expandable and modifiable knowledge base (representing both declarative and procedural knowledge), 3) simple and user-transparent control strategy, 4) ability to reason under uncertainty conditions and insufficient information, and 5) capability of probabilistic reasoning. The principal components of a FLC consist of: a fuzzification interface, a knowledge base, the decision making logic, and the defuzzification interface.

This material is based upon work supported by the NSF under Grant No 91-12362

In this paper, a self–tuning fuzzy controller for a second–order system is considered. Self–tuning fuzzy control schemes attempt to decrease the effects of plant uncertainty by identifying the system on–line, thereby yielding a closed loop with reduced sensitivity and improved performance over non-adaptive algorithms. Such schemes therefore consist of two components–the controller and the estimator.

The estimator must provide a reliable system estimate in the presence of unmodeled dynamics and unmeasurable disturbances. Set membership (SM) identification technique [3, 4, 5, 6] was used to provide a feasible set of system parameters which is consistent with the measurement data and the inherent model structure. Based on an assumed knowledge on the noise energy constraints in the identified system, the estimate can be restricted within a subset of the parameter space. Specifically, the unknown parameter vector can be contained within a hyper–ellipsoid whose volume reflects the uncertainty around the nominal model.

Fuzzy controllers inquire the uncertainty information provided by the identifier to infer the appropriate control action. In fuzzy logic, the associated universe of discourse encloses the estimated hyper–ellipsoid. Fuzzy membership functions are assigned to layers partitioning this area. The knowledge data base consists of feedback gains required to place the closed loop poles at predefined locations, and a rule based controller infers the output variable from the weighted integral of this area on the membership functions.

II. Problem Statement

In this paper we address the issue of feedback control on a linear plant with unknown parameters. In general, the class of linear systems whose input-output behavior is dominated by a pair of complex poles, can be approximated with a second order differential (or difference) equation. Considering any element of this class, our objective is the synthesis of a closed-loop system with certain dynamic characteristics in order, for example, to follow a given reference signal. The design approach consists of two separate parts. The first part is involved with the identification of the unknown second order plant. At this point, interference of measurement noise into the identification process, should be taken under consideration.

CH3229-2/92/0000-2517$1.00 © 1992 IEEE

The second part is the design of the controller, which, naturally, dependents on that of the identifier. The filter design is based on the SM estimation, while the controller is a hybrid unit whose components are a fuzzy-rulebase and a cell pole-placement technique.

A. *The Identification Issue*

Since the system dynamics are unknown, the design of a proper identification algorithm is the central topic of the problem. The system under consideration has the following mathematical description

$$\begin{aligned} y(n) &= -a_1 y(n-1) - a_2 y(n-2) + b_1 u(n-1) \\ d(n) &= y(n) + \tilde{v}(n) \end{aligned}$$

where $\tilde{v}(n)$ is the random sequence of measurement noise. Without loss of generality, we assume knowledge of the noise level. In order to proceed with a filter design we combine the previous equations in one as below

$$\begin{aligned} d(n) = & - a_1 d(n-1) - a_2 d(n-2) \\ & + a_1 \tilde{v}(n-1) + a_2 \tilde{v}(n-2) \\ & + b_1 u(n-1). \end{aligned} \qquad (1)$$

The noise terms in the RHS of (1) can be incorporated in one noise component, by rewriting (1) in the following form

$$d(n) = -a_1 d(n-1) - a_2 d(n-2) + b_1 u(n-1) + v(n) \qquad (2)$$

where $v(n)$ is the filtered version of the measurement noise. We point out that the mathematical content of (2) is the evolution of the measured output due to desired and undesired inputs. An output error filter design [7] can be easily applied on (2). Assuming the following filter equation

$$y_0(n) = -\hat{a}_1 y_0(n-1) - \hat{a}_2 y_0(n-2) + b_1 u(n-1)$$

we can engage a RLS scheme in order to minimize the estimation error $e(n) = d(n) - y_0(n)$. The choice of a conventional identification method does not take into consideration the noise interference. In other words minimization of the output error does not imply convergence of the filter parameters to their true values. Only in the cases of low level measurement noise, i.e. $d(n) \approx y(n)$, one can claim that the estimated parameters approach their correct values.

B. *The controller issue*

As we mentioned before, the design of the controller depends on the choice of the identification scheme. In our case, where relatively high noise interference is compensated with a set membership identification scheme, the output of the estimator filter is the current information about the area of the parameter space which includes the true parameter vector. In order to take advantage of this information we engage a type of cell pole-placement control law. The general idea is to assign certain feedback gains to several cells of the unit disk, in order to cover a sufficiently large region. At this point a very rough estimation of the location of the open loop poles is desirable.

Selection of overlapping cells defines a fuzzy-partition of the unit disk. This kind of partition requires a fuzzy-logic decision making, for a given point of the parameter space to be a member of any one of the participating cells. The choice of overlapping cells serves merely to avoid any discontinuities in the rate of change of the feedback gains along any path on the area covered by the cells. We point out that the choice of a cell control strategy [8] associates a small number of pre-defined feedback gains with a large area of the z-domain.

Having these notions in mind we define a building block for the construction of the fuzzy-rulebase. This element consists of three overlapping cells. Every cell infers the feedback gain calculated for an open-loop parameter at its midpoint. A descriptive presentation of the proposed adaptive-fuzzy control scheme is shown in Figure 1.

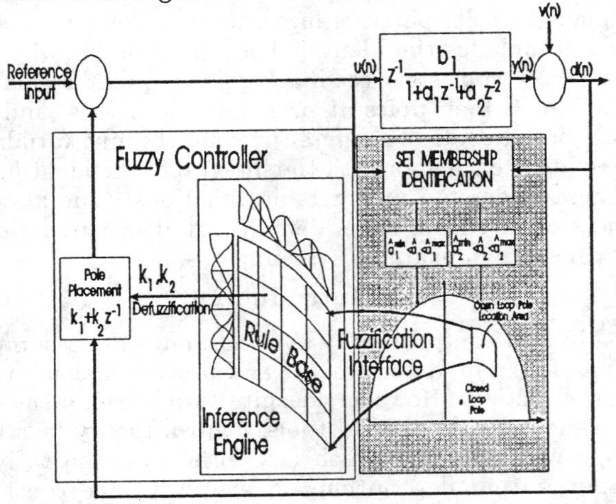

Figure 1: Adaptive Fuzzy Pole Placement Control

III. Transfer Function Identification Method

The proposed adaptive scheme involves an identification method, which appears to be superior from the conventional RLS estimation algorithms [9]. Historically, the SM algorithm evolved from a state estimator [5, 10], to a recursive filter [3]. The approach has been developed for the l-th order ARMA model

$$\begin{aligned} d(n) = & - a_1 d(n-1) - \cdots - a_l d(n-l) \\ & + b_0 u(n) + \cdots + b_m u(n-m) \\ & + v(n) \end{aligned} \qquad (3)$$

with the fundamental assumption of a priori knowledge of the noise level. According to the usual notation in recursive schemes, we denote the vector of the unknown parameters as,

$$\theta^T = [a_1 \cdots a_l \quad b_0 \quad b_1 \cdots b_m].$$

Similarly, the vector of the measured output is

$$\bar{d}_n^T = [d(1) \cdots d(n)]$$

and the regression vector is

$$\bar{x}_n^T = [-d(n-1)\cdots-d(n-l) \quad u(n)\cdots u(n-m)].$$

Now, equation (3) takes the following matrix form,

$$\bar{d}_n = X_n\theta + \bar{v}$$

where ${X_n}^T = [\bar{x}_1\cdots\bar{x}_n]$ and $\bar{v}^T = [v(1)\cdots v(n)]$. In most applications the energy content of the noise can be estimated. A convenient measure of this energy is the norm of the vector $\bar{v}$,

$$\bar{v}^T\bar{v} = (\bar{d}_n^T - \theta^T {X_n}^T)(\bar{d}_n - X_n\theta^T). \tag{4}$$

Defining ${P_n}^{-1} \equiv {X_n}^T X_n$ and $\theta_c \equiv P_n {X_n}^T \bar{d}_n$ we rewrite (4) as

$$\bar{v}^T\bar{v} = (\theta-\theta_c)^T {P_n}^{-1}(\theta-\theta_c) - {\theta_c}^T {P_n}^{-1}\theta_c + \bar{d}_n^T\bar{d}_n. \tag{5}$$

If the noise sequence does not exceed a certain level, $1/{\gamma^2}_n$, at the particular time n, then we can right

$$v^2(n) \leq \frac{1}{{\gamma^2}_n}.$$

Summing over the first n steps, we have

$$\bar{v}^T\bar{v} \leq F(n), \quad F(n) \equiv \sum_{i=1}^{n} \frac{1}{{\gamma^2}_i} \tag{6}$$

and in particular, if γ_i is constant[1] we can write,

$$\bar{v}^T\bar{v} \leq \frac{n}{{\gamma^2}_n}.$$

Combining (6) with (5) we have the following inequality

$$(\theta-\theta_c)^T {P_n}^{-1}(\theta-\theta_c) \leq G(n),$$
$$G(n) \equiv F(n) + {\theta_c}^T {P_n}^{-1}\theta_c - \bar{d}_n^T\bar{d}_n. \tag{7}$$

The terms of inequality (7) can be expressed in a recursive form (note that P_n and $\theta_c(n)$ have the usual RLS representation),

$$P_n = P_{n-1} - \frac{P_{n-1}\bar{x}_n\bar{x}_n^T P_{n-1}}{1+\bar{x}_n^T P_{n-1}\bar{x}_n},$$

$$\theta_c(n) = \theta_c(n-1) + P_{n-1}[d(n-1) - {\theta_c}^T(n-1)\bar{x}_n]\bar{x}_n$$

and

$$G(n) = G(n-1) + [F(n) - F(n-1)] - \frac{[d(n-1) - {\theta_c}^T(n-1)\bar{x}_n]^2}{1-\bar{x}_n^T P_{n-1}\bar{x}_n}.$$

The recursion of $G(n)$ enables the SM algorithm to control the convergence of θ_c to its true value. Under certain conditions one can investigate the behavior of the proposed recursion. For the purpose of this paper we have chosen a criterion in order to decouple the RLS iteration from the recursion of $G(n)$. According to this criterion we reject new data whenever $G(n) \leq 0$, *or* $\lambda_{min} \leq 0$, where λ_{min} is the minimum eigenvalue of the matrix $\frac{1}{G(n)}{P_n}^{-1}$. For example, if $G(n) > 0$ and $\lambda_{min} \leq 0$, then the hyper-ellipsoid of (7) has a negative principal length. Therefore the current data are not consistent with the previous step, which located the true value of the parameter vector as an interior point of the hyper-ellipsoid. We emphasize that the above criterion allows the algorithm to follow a possible change in the system dynamics detected by the RLS recursion. On the other hand the criterion protects the estimation from a biased convergence of the RLS due to correlated input–output signals (e.g. in the case of feedback). Other criteria that could be used to discard the corrupted data points utilize an optimum forgetting factor [4].

Once the bounding ellipsoid is well defined we can find the rectangular area of the parameter space containing the true vector. For the design objectives of this paper the ellipsoid is two dimensional. In this case the extreme values of the correct parameters can be easily found from the following quadratic equation,

$$(\theta-\theta_c)^T \frac{{P_n}^{-1}}{G(n)}(\theta-\theta_c) = 1.$$

For a system of higher order the calculation of the parameter bounds [10] becomes more difficult, yet the solution of the problem is feasible. In general the output of the SM filter consists of these extreme values, and for the second order case has the following form,

$$a_{i,max} = \hat{a}_i + p_i, \quad a_{i,min} = \hat{a}_i - p_i; \quad i = 1,2 \tag{8}$$

where $p_i > 0$ is the projection of the corresponding principal axis on the axis of the parameter space. We point out that $\hat{a}_i$ may not be the current RLS estimate, the criterion of data rejection will update it or not.

IV. The Design of a Fuzzy Pole Placement Control Law

The central idea of a fuzzy controller design is the creation of the decision rulebase. The inference engine induced by that rulebase is not governed by the conventional (single-valued) logic. Instead, the decision making is based on fuzzy (multi-valued) logic. The necessity of this reasoning is due to a lack of ability in identifying (with certain possibility) the membership of an input-element to the rulebase. In the literature, a large class of controllers, which are based on fuzzy-logic, possesses an expert inference rule. This approach has some disadvantages, among them the fact that the controller is very specialized and depends on the ideal operator, whose performance tries to imitate in an automated manner. A smaller class of fuzzy controllers is designed similarly to the well known PD and PID schemes [11, 12, 13].

[1] Constant γ_i corresponds to a noise sequence with a uniform first order probability density function.

These techniques are often associated with an expertise. This is done either directly, by simply assigning an ideal control action to a pair of error/change-of-error measurements, or indirectly by adjusting the gains of the individual control parts so that certain specifications are satisfied. In all of these cases, the tuning of the controller is based on the input–output system behavior.

The control law suggested in this paper relies on a pole placement technique. In this sense, no expertise is required and the area of application contents all linear plants with a dominant second order behavior. The advantage of this control policy is the direct modification of the system dynamics to any desirable dynamic behavior. In addition the pole placement technique is easily implemented through a cell control strategy [14, 8]. This combined approach, which essentially corresponds to gain scheduling, assigns to each cell the feedback gains sufficient to locate the closed-loop poles in a neighborhood of the design point. Consider the system of (2) with the following feedback

$$u(n-1) = k_1 d(n-1) + k_2 d(n-2) + r(n-1).$$

Assuming that $a_{1,des}$ and $a_{2,des}$ are the desired closed-loop parameters, the feedback gains are given as follows,

$$k_i = \frac{a_i - a_{i,des}}{b_1}; \quad i = 1, 2.$$

Since the open-loop parameters are unknown, the controller is dependent on the identification scheme in order to infer a pair of (k_1, k_2). In a typical RLS estimation algorithm the feedback gains are functions of the current iteration.

A. *Fuzzy Partition of the Parameter Space*

In order to understand the need for a fuzzy partition of the parameter space the designer should think in a linguistic manner. Since there is uncertainty about the proximity of the estimated parameter vector to its true value, a linguistic-type description of this distance can be engaged. Suppose that we select several points of the parameter space for which the feedback gains place the closed loop poles in their specific location. As we have shown before, for parameter vectors "away" from these points the closed loop eigenvalues will deviate from their desirable placement. Here, we introduce the fuzzy-set description of the term "away" (or "close"), in order to deal with a quantity whose exact value is unknown.

The grade of membership of the estimated parameter vector in a particular fuzzy-set decreases (increases) as the vector moves "away" ("close") from (to) the point of full grade. In this sense we program the controller to weight the gain assigned to the identified parameter with the value of the membership function, associated with the corresponding fuzzy-set. This weighting process continues until all of the fuzzy-sets, which share the particular element, are encountered. As an example, we consider the fuzzy-reasoning for the membership of the parameters $\hat{a}_i$, $i = 1, 2$, identified by a RLS scheme. Assuming that the parameter is related to three fuzzy-sets, $j = 1, \ldots, 3$, the rulebase takes the following form,

- if $\hat{a}_i$ is $A_{i,j}$, then the gain is $k_{i,j}$.

Since a_1 and a_2 are independent, we infer the k's involving only one of the a's. The next step is to weight the inferred gains with the value of the corresponding membership function. The weighted average of these gains is the defuzzified output of the rulebase.

B. *Control Defuzzification*

The rulebase which operates with fuzzy-logic infers several outcomes for a single event. Therefore, there is a need for a defuzzification interface between the controller and the actuator. As we mentioned before the final output of the inference engine is in a form of a weighted sum over the set of inferred gains. Denoting by F_i the weight associated with the fuzzy-set, i, we write the usual defuzzification average as below,

$$u_f(n) = \frac{\sum_{i=1}^{s} \tilde{u}_i F_i}{\sum_{i=1}^{s} F_i}, \tag{9}$$

where s is the number of fuzzy sets in the basic building block, and the control action $\tilde{u}_i$ is in feedback form,

$$\tilde{u}_i(n) = k_{i,1} d(n) + k_{i,2} d(n-1). \tag{10}$$

Combining the linear equations (9) and (10) we get the inferred control in a compact form,

$$u_f(n) = k_1 d(n) + k_2 d(n-1)$$

where the k's are expressed as weighted averages

$$k_j = \frac{\sum_{i=1}^{s} k_{i,j} f_i}{\sum_{i=1}^{s} f_i}; \qquad j = 1, 2.$$

C. *Selection of the Membership Functions*

The choice of the membership functions is important for the adequate behavior of the controller along any path of the fuzzy-partition. The usage of certain types of membership functions is standard in the literature of fuzzy-logic applications. In this section our main concern is the shape of these functions. Linear segments are in most cases a convenient choice. Because of the feedback structure of the controller, discontinuities in the first derivative of the defuzzified gain are likely to cause impulsive responses in the output, similar to the case of a non-fuzzy cell control action. Parabolic segments [15], have several advantages drawn from the fuzzy-set theory. Among their advantages is the smooth behavior of the first derivative. In addition these functions have already been incorporated in applications of feedback control based on fuzzy-logic [13]. Given the shape of the membership curves, an appropriate choice of the curvature parameters is desirable. At this point the designer should encounter the sensitivity of the closed loop root-locus in variations of the filter parameters. The building block of the fuzzy partition consists of three fuzzy-sets, as shown in Figure 2.

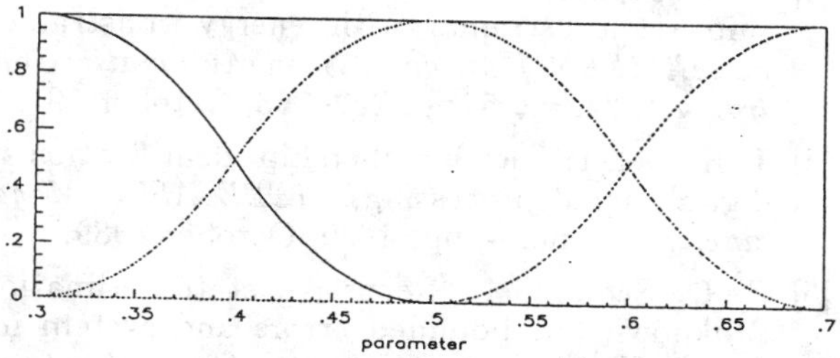

Figure 2: Parabolic Membership Functions

The general form of the parabolic membership functions, denoted by S, Π and Z, is the following

$$S(t;a,b,c) = \begin{cases} 0 & t < a \\ 2(\frac{t-a}{c-a})^2 & a < t < b \\ 1 - 2(\frac{t-c}{a-c})^2 & b < t < c \\ 1 & c < t \end{cases}$$

$$\Pi(t;b,c) = \begin{cases} S(t;c-b,c-\frac{b}{2},c) & t < c \\ 1 - S(t;c,c+\frac{b}{2},c+b) & c < t \end{cases}$$

$$Z(t;a,b,c) = 1 - S(t;a,b,c)$$

D. Interface between the SM Filter and the Fuzzy-Partition

In the previous section we observed that the SM identification algorithm estimates a rectangular area on the parameter space, which includes the true parameter vector. This information should be incorporated by the controller. Hence, there is a need for an interface between the distributed output of the estimator and the functions of the fuzzy-set partition of the parameter space. The proposed interface is based on the evaluation of the integral under the corresponding membership curve with the values of $a_{i,min}$ and $a_{i,max}$ from (8) as the integration limits. The following scheme gives an idea for the scanning of the partitioned space using $a_{i,min}$ as the reference point, $j = 1, \ldots, 3$,

- if $a_{i,min}$ is $A_{i,j}$, evaluate $\int_{a_{i,min}}^{a_{i,max}} f_{i,j}(t)dt$.

In order to take a measure for the grade of membership of the interval $(a_{i,min}, a_{i,max})$ to a certain fuzzy-set, we normalize the integrals evaluated before with the length of the above interval. In other words, $i = 1, 2$,

$$F_j = \max_i \left[\int_{a_{i,min}}^{a_{i,max}} \frac{f_{i,j}(t)}{(a_{i,max} - a_{i,min})} dt \right]$$

is the weighting factor of the corresponding gain in the defuzzification process. One can observe that as $a_{i,max} - a_{i,min}$ approaches zero, (i.e. as the identified rectangle converges to a point), the weighting factor tends to the value of the membership function at that particular point.

V. Simulation Results

A second order single-input single-output (SISO) plant transfer function is chosen to show the performance of the proposed fuzzy controller,

$$H(z) = \frac{b_1 z^{-1}}{1 + a_1 z^{-1} + a_2 z^{-2}}.$$

In order to demonstrate the effectiveness of the SM identification algorithm we choose a system with a moderate oscillatory open-loop response. In particular $a_1 = -1.4$ and $a_2 = b_1 = 0.55$. The reference input signal, $r(n)$, is of an alternating-step type, with unit amplitude and period 60 steps. The output measurements are corrupted with uniform white noise corresponding to a 20db SNR.

The fuzzy-rulebase consists of 9 fuzzy-cells (i.e. 3 three fuzzy-sets in each parameter direction). A trial and error procedure enabled as to choose appropriate lengths for the building block and its fuzzy-components. Larger areas on the unit disk can be covered through proper augmentation of neighboring building blocks. As a trade-off, the number of rules and gains needed to be stored in the data base of the controller should increase considerably. Here we use one building block, which includes the true value of the system parameters. We can always enclose the correct parameter vector within the fuzzy-rulebase, by performing the initial steps of the SM identification off-line. In this example the control action starts at the beginning of the 10th step. We emphasize that the true parameter vector can lay anywhere inside the building block. This rulebase choice needs 3 k_1-type and 3 k_2-type feedback gains. These gains form the data base of the controller.

Results from an open-loop identification of (2) are shown in Figure 3. The closed-loop system is sim-

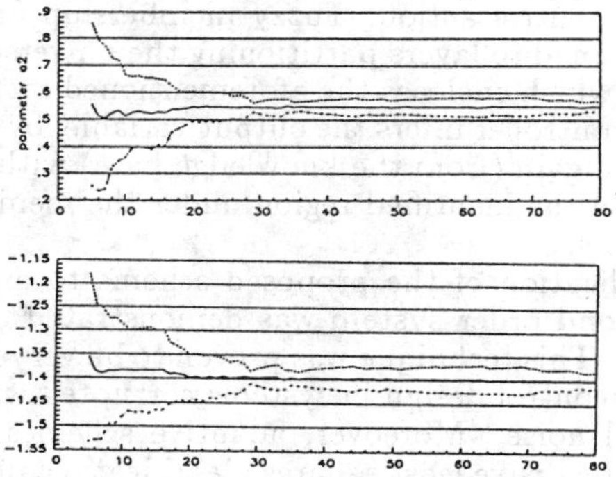

Figure 3: Open Loop SM and RLS Identification

ulated with the settings of the fuzzy controller discussed before. From Figure 4 we observe that the RLS filtering in the on-line identification scheme converges to a biased solution. In contrast, the SM filtering has the ability to exclude any inconsistent data points from updating the principal lengths of the bounding ellipsoid.

The closed loop responses are depicted in Figure 5, where the response with the adaptive fuzzy

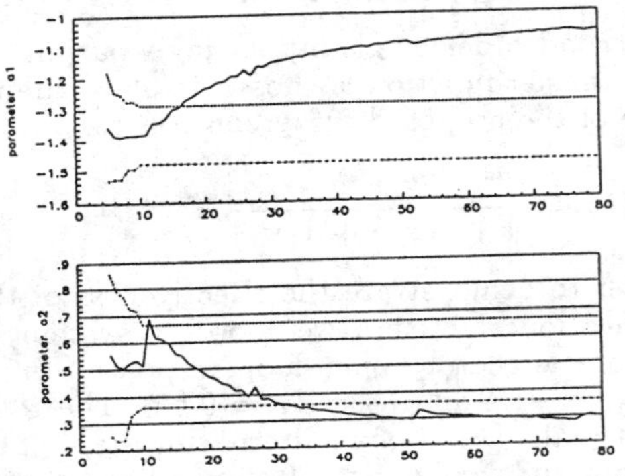

Figure 4: Closed Loop SM and RLS Identification

controller (solid line) matches closely that of an ideal pole placement controller (noiseless output).

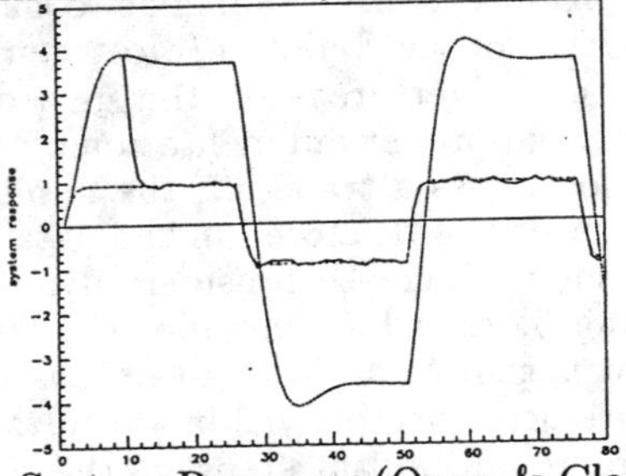

Figure 5: System Response (Open & Closed Loop)

VI. Conclusion

An adaptive fuzzy controller for dominant–pole placement has been presented. The proposed scheme consists of an estimator and a fuzzy controller. Set membership identification technique was used to provide a feasible set of the transfer function system parameters, consistent with the measurement data. The fuzzy controller utilizes this information to infer the control action. Fuzzy membership functions are assigned to layers partitioning the universe of discourse which encloses the aforementioned set. A rule based controller infers the output variable, by weighting each gain (from the knowledge base) with the integral of the identified region under the membership functions.

Application of the proposed scheme to an example second order system was demonstrated and discussed. This technique was proven to be very promising for control design in systems corrupted with substantial noise. Moreover, adaptive schemes relying on the recursive least squares identification algorithm suffer from the persistent excitation problem; in the advocated SM algorithm, corrupted data, either from noise contamination or due to a non–exciting input, can be discarded.

VII. References

[1] C. C. Lee, "Fuzzy Logic in Control Systems: Fuzzy Logic Controller –Part I," *IEEE Transactions on Systems, Man, and Cybernetics*, vol. 20, no. 2, pp. 404–418, March 1990.

[2] C. C. Lee, "Fuzzy Logic in Control Systems: Fuzzy Logic Controller –Part II," *IEEE Transactions on Systems, Man, and Cybernetics*, vol. 20, no. 2, pp. 419–435, March 1990.

[3] E. Fogel, "System identification via membership set constraints with energy constrained noise," *IEEE Transactions on Automatic Control*, vol. 24, no. 5, pp. 752–758, October 1979.

[4] J. R. Deller, "Set membership identification in digital signal processing," *IEEE ASSP Magazine*, vol. 6, no. 4, pp. 4–20, October 1989.

[5] F. C. Schweppe, "Recursive state estimation :Unknown but bounded errors and system inputs," *IEEE Transactions on Automatic Control*, vol. 13, no. 1, pp. 22–28, February 1968.

[6] M. Milanese and G. Belforte, "Estimation theory and uncertainty intervals evaluation in presence of unknown but bounded errors: Linear families of models and estimators," *IEEE Transactions on Automatic Control*, vol. 27, no. 2, pp. 408–414, April 1982.

[7] J. J. Shynk, "Adaptive IIR Filtering," *IEEE ASSP Magazine*, vol. 6, no. 2, pp. 4–21, April 1989.

[8] E. H. Mandami, "Application of fuzzy algotithms for simple dynamic plant," *Proc. IEE*, vol. 121, no. 12, pp. 1585–1588, 1974.

[9] G. C. Goodwin and K. S. Sin, *Adaptive Filtering Prediction and Control.* Engelwood Cliffs: Prentice-Hall, 1984.

[10] H. S. Witsenhausen, "Sets of possible states of linear systems given perturbed observations," *IEEE Transactions on Automatic Control*, vol. 13, no. 5, pp. 556–558, October 1968.

[11] K. L. Tang and R. J. Mulholland, "Comparing fuzzy logic with classical controller designs," *IEEE Transactions on Systems, Man, and Cybernetics*, vol. 17, no. 6, pp. 1085–1087, 1987.

[12] Y. Dote, "Stability analysis of variable-structured PI controller by fuzzy logic for servo system," in *Proceedings of the IEEE Conference on Decision and Control*, pp. 1217–1218, Brighton, England, December 1991.

[13] H. Kang and G. Vachtsevanos, "Model reference fuzzy control," in *Proceedings of the IEEE Conference on Decision and Control*, pp. 751–756, Tampa, FL, December 1989.

[14] B. H. Wang and G. Vachtsevanos, "Fuzzy logic control: A systematic design methodology," in *Proceedings of the IEEE Conference on Decision and Control*, pp. 1219–1220, Brighton, England, December 1991.

[15] L. A. Zadeh, "Fuzzy sets as a basis for a theory of possibility," *Fuzzy Sets Systems*, vol. 1, no. 1, pp. 3–28, 1978.

Proceedings of the 31st Conference
on Decision and Control
Tucson, Arizona • December 1992

FUZZY LEARNING CONTROL FOR ANTI-SKID BRAKING SYSTEMS*

Jeffery R. Layne†, Kevin M. Passino‡, and Stephen Yurkovich

Department of Electrical Engineering
The Ohio State University
2015 Neil Avenue
Columbus, OH 43210

Abstract

Although anti-skid braking systems (ABS) are designed to optimize braking effectiveness while maintaining steerability, their performance often degrades for harsh road conditions (e.g., icy/snowy roads). In this paper we introduce the idea of using the fuzzy model reference learning control (FMRLC) technique[1] for maintaining adequate performance even under such adverse road conditions. This controller utilizes a learning mechanism which observes the plant outputs and adjusts the rules in a direct fuzzy controller so that the overall system behaves like a "reference model" which characterizes the desired behavior. The performance of the FMRLC-based ABS is demonstrated by simulation for various road conditions (wet asphalt, icy) and "split road conditions" (the condition where, e.g., emergency braking occurs and the road switches from wet to icy or vice versa).

I. Introduction

Anti-skid Braking Systems (ABS) present a challenging control problem since there can be significant brake/automotive system parameter variations (e.g., due to brake pad coefficient of friction changes or road slope variations) and environmental influences (e.g., due to adverse road conditions). While conventional control approaches[2, 3, 4] and even direct fuzzy/knowledge based approaches[5, 6, 7, 8] have been successfully implemented, their performance will still degrade when adverse road conditions are encountered. The basic reason for this performance degradation is that the control algorithms have limited ability to *learn* how to compensate for the wide variety of road conditions that exist. In this paper we will investigate the role that *learning controllers* can take in enabling ABS to compensate for adverse road conditions.

A "learning system" possesses the capability to improve its performance over time by interaction with its environment. A learning control system is designed so that its "learning controller" has the ability to improve the performance of the closed loop system by generating command inputs to the plant and utilizing feedback information from the plant. The learning mechanism in the fuzzy model reference learning control (FMRLC) system that we design for the ABS will monitor the performance of a fuzzy controller and tune it to adapt to adverse road conditions as they are encountered. This FMRLC was first introduced in [1] and it grew from ideas in linguistic self-organizing control (SOC)[9] and conventional model reference adaptive control (MRAC)[10]. In fact, it has provided significant improvements over the SOC approach for enhanced performance feedback and knowledge base modification and has compared favorably to the MRAC for a ship steering application[1, 11].

In this paper we illustrate that the FMRLC provides an effective solution to the problem of compensating for certain adverse road conditions. We begin by describing the ABS under consideration. Next, we illustrate the FMRLC performance on dry asphalt, wet asphalt, and an icy surface. Finally, we study FMRLC performance for split road conditions. In particular, we study braking effectiveness when there is a transition (in both directions) between icy and wet road surfaces.

In Section II we will overview the FMRLC technique. In Section III we describe the ABS problem while in Section IV we provide simulation results that give an initial assessment of the performance of the FMRLC for ABS. Section V provides some concluding remarks.

II. Fuzzy Model Reference Learning Control

The FMRLC, which is shown in Figure 1, utilizes a learning mechanism that (i) observes data from a fuzzy control system (i.e., $\underline{y}_r(kT)$ and $\underline{y}(kT)$), (ii) characterizes its current performance, and (iii) automatically synthesizes and/or adjusts the fuzzy controller so that some pre-specified performance objectives are met. These performance objectives are characterized via the *reference model* shown in Figure 1. In an analogous manner to conventional MRAC where conventional controllers are adjusted, the learning mechanism seeks to adjust the fuzzy controller so that the closed-loop system (the map from $\underline{y}_r(kT)$ to $\underline{y}(kT)$) acts like a pre-specified reference model (the map from $\underline{y}_r(kT)$ to $\underline{y}_m(kT)$). Next we describe each component of the FMRLC in more detail.

A. *The Fuzzy Controller*

The process in Figure 1 is assumed to have r inputs denoted by the r - dimensional vector $\underline{u}(kT) = [u_1(kT) \ldots u_r(kT)]^t$ (T is the sample period) and s outputs denoted by the s - dimensional vector $\underline{y}(kT) = [y_1(kT) \ldots y_s(kT)]^t$. Most often the inputs to the fuzzy controller are generated via some function of the plant output $\underline{y}(kT)$ and reference input $\underline{y}_r(kT)$. Figure 1 shows a special case of such a map that was found useful in many applications. The inputs to the fuzzy controller are the error $\underline{e}(kT) = [e_1(kT) \ldots e_s(kT)]^t$ and change in error $\underline{c}(kT) = [c_1(kT) \ldots c_s(kT)]^t$ defined as

$$\underline{e}(kT) = \underline{y}_r(kT) - \underline{y}(kT), \quad (1)$$

$$\underline{c}(kT) = \frac{\underline{e}(kT) - \underline{e}(kT-T)}{T}, \quad (2)$$

respectively, where $\underline{y}_r(kT) = [y_{r_1}(kT) \ldots y_{r_s}(kT)]^t$ denotes the desired process output.

In fuzzy control theory, the range of values for a given controller input or output is often called the "universe of discourse" [12]. Often, for greater flexibility in fuzzy controller implementation, the universes of discourse for each process input are "normalized" to the interval $[-1, +1]$ by means of constant scaling factors. For our fuzzy controller design, the gains $\underline{g}_e$, $\underline{g}_c$, and $\underline{g}_u$ were employed to normalize the universe of discourse for the error $\underline{e}(kT)$, change in error $\underline{c}(kT)$, and controller output $\underline{u}(kT)$, respectively (e.g., $\underline{g}_e = [g_{e_1}, \ldots, g_{e_s}]^t$ so that $g_{e_i}e_i(kT)$ is an input to the fuzzy controller).

We utilize r multiple-input single-ouput (MISO) fuzzy controllers, one for each process input u_i (equivalent to using one MIMO controller). The knowledge base for the fuzzy controller associated with the nth process input is generated from **IF-THEN** control rules of the form:

*This work was partially supported by a grant from the Ohio Transportation Research Endowment Program.

†J. Layne gratefully acknowledges the support of the U.S. Air Force Palace Knight Program. J. Layne has recently joined Wright Laboratories, WL/AAAS-3, Wright Patterson AFB, Ohio 45433-6543

‡K. Passino was supported in part by an Engineering Foundation Research Initiation Grant. Please address all correspondence to K.Passino (email: passino@eagle.eng.ohio-state.edu).

CH3229-2/92/0000-2523$1.00 © 1992 IEEE

$$\begin{aligned}&\textbf{If } \tilde{e}_1 \text{ is } \tilde{E}_1^j \textbf{ and } \dots \textbf{ and } \tilde{e}_s \text{ is } \tilde{E}_s^k \textbf{ and}\\ &\quad \tilde{c}_1 \text{ is } \tilde{C}_1^l \textbf{ and } \dots \textbf{ and } \tilde{c}_s \text{ is } \tilde{C}_s^m\\ &\quad\quad \textbf{Then } \tilde{u}_n \text{ is } \tilde{U}_n^{j,\dots,k,l,\dots,m},\end{aligned}$$

where $\tilde{e}_a$ and $\tilde{c}_a$ denote the *linguistic variables* associated with controller inputs e_a and c_a, respectively, $\tilde{u}_n$ denotes the linguistic variable associated with the controller output u_n, $\tilde{E}_a^b$ and $\tilde{C}_a^b$ denote the bth *linguistic value* associate associated with $\tilde{e}_a$ and $\tilde{c}_a$, respectively, and $\tilde{U}_n^{j,\dots,k,l,\dots,m}$ denotes the *consequent linguistic value* associated with $\tilde{u}_n$. The above control rule may be quantified by utilizing fuzzy set theory to obtain a fuzzy implication of the form:

$$\textbf{If } E_1^j \textbf{ and } \dots \textbf{ and } E_s^k \textbf{ and } C_1^l \textbf{ and } \dots \textbf{ and } C_s^m \textbf{ Then } U_n^{j,\dots,k,l,\dots,m},$$

where E_a^b, C_a^b, and $U_n^{j,\dots,k,l,\dots,m}$ denote the fuzzy sets that quantify the *linguistic statements* "$\tilde{e}_a$ is $\tilde{E}_a^b$", "$\tilde{c}_s$ is $\tilde{C}_s^m$", and "$\tilde{u}_n$ is $\tilde{U}_n^{j,\dots,k,l,\dots,m}$", respectively. This fuzzy implication can be represented by a fuzzy relation

$$R_n^{j,\dots,k,l,\dots,m} = (E_1^j \times \dots \times E_s^k) \times (C_1^l \times \dots \times C_s^m) \times U_n^{j,\dots,k,l,\dots,m} \quad (3)$$

A set of such rules forms the "rule-base" which characterizes how to control a dynamical system. We use triangular membership functions for the input and output (normalized) universes of discourse, Zadeh's compositional rule of inference, and the standard center-of-gravity (COG) defuzzification technique[12].

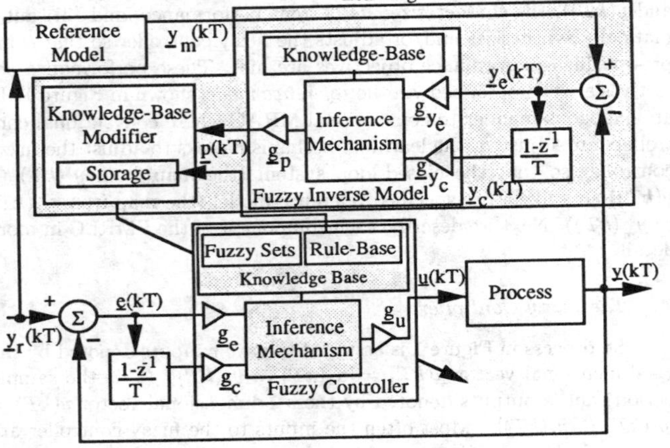

Figure 1: Architecture for the FMRLC

B. The Reference Model

The reference model provides a capability for quantifying the desired performance of the process. In general, the reference model may be any type of dynamical system (linear or non-linear, time-invariant or time-varying, discrete or continuous time, etc.). The performance of the overall system is computed with respect to the reference model by generating an error signal $\underline{y}_e(kT) = [y_{e_1} \dots y_{e_s}]^t$ where

$$\underline{y}_e(kT) = \underline{y}_m(kT) - \underline{y}(kT). \quad (4)$$

Given that the reference model characterizes design criteria such as rise time and overshoot and the input to the reference model is the reference input $\underline{y}_r(kT)$, the desired performance of the controlled process is met if the learning mechanism forces $\underline{y}_e(kT)$ to remain very small for all time; hence, the error $\underline{y}_e(kT)$ provides a characterization of the extent to which the desired performance is met at time $t = kT$. If the performance is met ($\underline{y}_e(kT) \approx 0$) then the learning mechanism will not make significant modifications to the fuzzy controller. On the other hand if $\underline{y}_e(kT)$ is big, the desired performance is not achieved and the learning mechanism must adjust the fuzzy controller. Next we describe the operation of the learning mechanism.

C. The Learning Mechanism

As previously mentioned, the learning mechanism performs the function of modifying the knowledge base of a direct fuzzy controller so that the closed loop system behaves like the reference model. These knowledge base modifications are made by observing data from the controlled process, the reference model, and the fuzzy controller. The learning mechanism consists of two parts: a fuzzy inverse model and a knowledge base modifier. The fuzzy inverse model performs the function of mapping $\underline{y}_e(kT)$ (representing the deviation from the desired behavior), to the relative changes in to process inputs $\underline{p} = [p_1 \dots p_r]^t$ that are necessary to force $\underline{y}_e(kT)$ to zero. The knowledge base modifier performs the function of modifying the fuzzy controller's knowledge base to affect the needed changes in the process inputs. More details of this process are discussed next.

The *fuzzy inverse model* was developed in [1] by investigating methods to alleviate the problems with using the inverse process model in the linguistic SOC framework of Procyk and Mamdani [9]. Procyk and Mamdani's use of the inverse process model depended on the use of an explicit mathematical model of the process and ultimately restrictive assumptions about the underlying physical process (which cause significant difficulties in applying their approach). Using the fact that most often a control engineer will know how to roughly characterize the inverse model of the plant, the authors in [1] introduce the idea of using a fuzzy system to map $\underline{y}_e(kT)$ and possibly functions of $\underline{y}_e(kT)$ (or process operating conditions), to the necessary changes in the process inputs $\underline{p}(kT)$. This map is called the "fuzzy inverse model" since information about the plant inverse dynamics is used in its specification. Note that similar to the fuzzy controller, the fuzzy inverse model shown in Figure 1 contains normalizing scaling factors, namely $\underline{g}_{y_e}$, $\underline{g}_{y_c}$, and $\underline{g}_p$, for each universe of discourse. Selection of the normalizing gains can be very critical with respect to the overall performance of the system and a gain selection procedure is given in [11].

Given that $g_{y_{e_i}} y_{e_i}$ and $g_{y_{c_i}} y_{c_i}$ are inputs to the fuzzy inverse model, the knowledge base for the fuzzy inverse model associated with the nth process input is generated from fuzzy implications of the form:

$$\textbf{If } Y_{e_1}^j \text{ and } \dots \text{ and } Y_{e_s}^k \text{ and } Y_{c_1}^l \text{ and } \dots \text{ and } Y_{c_s}^m \textbf{ Then } P_n^{j,\dots,k,l,\dots,m},$$

where $Y_{e_a}^b$ and $Y_{c_a}^b$ denote the bth fuzzy set for the error y_{e_a} and change in error y_{c_a}, respectively, associated with the ath process output and $P_n^{j,\dots,k,l,\dots,m}$ denotes the consequent fuzzy set for this rule describing the necessary change in the nth process input. As with the fuzzy controller we utilize triangular membership functions for both the input and output universes of discourse, Zadeh's compositional rule of inference, and COG defuzzification.

The *knowledge base modifier* for the FMRLC also grew from research performed on the linguistic SOC [9, 1]. In the linguistic SOC framework, knowledge base modification was performed on the overall fuzzy relation ($R_n = \bigcup_{j,\dots,k,l,\dots,m} R_n^{j,\dots,k,l,\dots,m}$) used to implement the fuzzy controller. However, this method of knowledge base modification can be computationally complex due to the fact that R_n is generally a very large array. In [1] the authors presented a new knowledge base modification algorithm which increases computational efficiency by modifying the membership functions of consequent fuzzy sets $U_n^{j,\dots,k,l,\dots,m}$ rather than the fuzzy relation array R_n.

Given the information about the necessary changes in the input as expressed by the vector $\underline{p}(kT)$, the knowledge base modifier changes the knowledge base of the fuzzy controller so that the previously applied control action will be modified by the amount $\underline{p}(kT)$. Therefore, consider the previously computed control action, which contributed to the present good/bad system performance. Note that $\underline{e}(kT-T)$ and $\underline{c}(kT-T)$ would have been the process error and change in error, respectively, at that time. Likewise, $\underline{u}(kT-T)$ would have been the controller output at that time. The controller output which would have been desired is expressed by

$$\underline{\bar{u}}(kT-T) = \underline{u}(kT-T) + \underline{p}(kT). \quad (5)$$

By modifying the fuzzy controller's knowledge base we may force the fuzzy controller to produce this desired output given similar controller inputs.

Assume that only symmetric membership functions are defined for the fuzzy controller's output so that $c_n^{j,\ldots,k,l,\ldots,m}$ denotes the center value of the membership function associated with the fuzzy set $U_n^{j,\ldots,k,l,\ldots,m}$. Knowledge base modification is performed by shifting centers of the membership functions of the fuzzy sets $U_n^{j,\ldots,k,l,\ldots,m}$ which are associated with the fuzzy implications that contributed to the previous control action $\underline{u}(kT-T)$. This modification involves shifting these membership functions by an amount specified by $\underline{p}(kT) = [p_1(kT) \ \ldots \ p_r(kT)]^t$ so that

$$c_n^{j,\ldots,k,l,\ldots,m}(kT) = c_n^{j,\ldots,k,l,\ldots,m}(kT-T) + p_n(kT). \tag{6}$$

The degree of contribution for a particular fuzzy implication whose fuzzy relation is denoted $R_n^{j,\ldots,k,l,\ldots,m}$ is determined by its "activation level", defined

$$\begin{aligned}\delta_n^{j,\ldots,k,l,\ldots,m}(t) = \min\{&\mu_{E_1^j}(e_1(t)), \ldots, \mu_{E_s^k}(e_s(t)),\\ &\mu_{C_1^l}(c_1(t)), \ldots, \mu_{C_s^m}(c_s(t))\},\end{aligned} \tag{7}$$

where μ_A denotes the membership function of the fuzzy set A. Only those fuzzy implications $R_n^{j,\ldots,k,l,\ldots,m}(kT-T)$ whose activation level $\delta_n^{j,\ldots,k,l,\ldots,m}(kT-T) > 0$ are modified. All others remain unchanged (this allows for local learning and hence memory).

III. Anti-skid Braking Systems

The objective of the FMRLC-based ABS system is to regulate wheel slip to maximize the coefficient of friction between the tire and road for any given road surface. In general, the coefficient of friction μ during a braking operation can be described as a function of slip, λ, which for a braking operation is defined as

$$\lambda(t) = \frac{\frac{V_v(t)}{R_w} - \omega_w(t)}{\frac{V_v(t)}{R_w}}, \tag{8}$$

where $\omega_w(t)$ is the angular velocity of the wheel, $V_v(t)$ is the velocity of the vehicle, and R_w is the radius of the tire. Since the term $\frac{V_v(t)}{R_w}$ is the angular velocity of the vehicle with respect to the tire angular velocity, we will sometimes denote this quantity by $\omega_v(t)$. The braking coefficient of friction as a function of slip $\mu(\lambda)$ was measured in [13, 14]. The results of these experiments were approximated for dry asphalt, wet asphalt, and ice as shown in Figure 2 (we use this data in all our simulations in Section IV). As one would expect, the braking coefficient of friction is greatest for dry asphalt, slightly reduced for wet asphalt, and greatly reduced for ice. From Equation (8) and Figure 2 observe

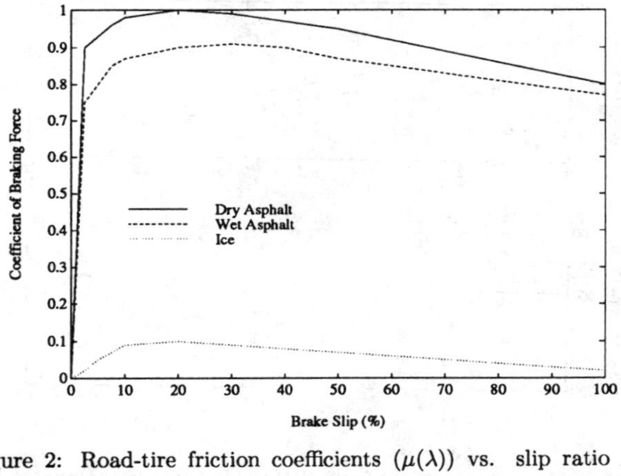

Figure 2: Road-tire friction coefficients ($\mu(\lambda)$) vs. slip ratio λ for various road surfaces.

that 0% slip represents the free rolling wheel condition ($\omega_w = \omega_v$ and $\mu(\lambda) = 0$) while 100% corresponds to a wheel that is locked up ($\omega_w = 0$). From Figure 2 we see that for the three road conditions shown, $\mu(\lambda)$ is maximized for $\lambda \approx 20\%$; hence we will always seek to regulate slip to 20% to maximize the coefficient of friction between the tire and the road. During normal vehicle operation, the road conditions are constantly changing. Since the road surface directly affects the braking characteristic, a controller design which compensates for all possible types of road conditions is difficult (especially for split road conditions).

A simplified model for a vehicle, a single wheel, and its braking system was employed for this research. The process model contains both linear vehicle dynamics and one-wheel rotational dynamics where wind resistance effects and all the vertical dynamics associated with the suspension system are assumed negligible. The differential equation which describes the motion of the wheels can be determined by summing the rotational torques which are applied to the wheel; hence,

$$\dot{\omega}_w(t) = \frac{1}{J_w}[-T_b(t) - \omega_w(t)B_w + T_t(t)], \tag{9}$$

where J_w is the rotational inertia of the wheel, B_w is the viscous friction of the wheel, $T_b(t)$ is the braking torque, and $T_t(t)$ is the torque generated due to slip between the wheel and the road surface. In general, $T_t(t)$ is a function of the force $F_t(t)$ exerted between the wheel and the road surface, or $T_t(t) = R_w F_t(t)$, where R_w is the radius of the wheel. The vehicle dynamics are determined by summing the total forces applied to the vehicle during a normal braking operation to obtain

$$\dot{V}_v(t) = \frac{-1}{M_v}[F_t(t) + B_v V_v(t) + F_\theta(\theta)], \tag{10}$$

where M_v is the mass of the vehicle, B_v is the vehicle viscous friction, g is the gravitational acceleration constant, $F_\theta(\theta)$ is the force applied to the car which results from a vertical gradient in the road so that $F_\theta(\theta) = M_v g \sin(\theta(t))$ where θ is the angle of inclination of the road. The force $F_t(t)$ is generally expressed as a function of the coefficient of friction and the normal force on the wheel, or $F_t(t) = \mu(\lambda)N_v(\theta)$ where $N_v(\theta)$ is the normal force applied to the tire. For this model we assume that the vehicle has 4 wheels and the weight of the vehicle is evenly distributed among these wheels. As a result, the normal force $N_v(\theta)$ may be expressed by $N_v(\theta) = \frac{M_v g}{4}\cos(\theta(t))$.

The braking system parameters used in this study are vehicle mass $M_v = 4 \times 342\ kg$, viscous friction associated with the linear motion of the vehicle $B_v = 6\ N\ s$, rotational inertia of the wheel $J_w = 1.13\ N\ m\ s^2$, rolling radius of the wheel $R_w = 0.33\ m$, viscous friction associated with the motion of the wheel $B_w = 4\ N\ s$, and $g = 9.8\ m\ s^2$[4].

Since slip is the controlled parameter of the braking system, we desire to measure this quantity. However, currently it is difficult and/or impossible to measure slip directly so it is necessary to estimate it. We will assume that sensors for measuring vehicle acceleration and wheel speed are available for estimating slip as is done in [7, 5]. Equation (8) may be rewritten to obtain

$$\omega_w(t) = (1 - \lambda(t))\ \omega_v(t). \tag{11}$$

Taking the time derivative of Equation (11) yields

$$\dot{\omega}_w(t) = (1 - \lambda(t))\ \dot{\omega}_v(t) - \dot{\lambda}(t)\ \omega_v(t), \tag{12}$$

where $\dot{\omega}_w(t)$ is related to the vehicle linear acceleration $a_v(t)$ by

$$\dot{\omega}_w(t) = \frac{\dot{V}_v(t)}{R_w} = \frac{a_v(t)}{R_w}. \tag{13}$$

Substituting Equation (13) and the fact $\omega_v = \frac{V_v}{R_w}$ into Equation (12), we obtain

$$\dot{\omega}_w(t) = (1 - \lambda(t))\ \frac{a_v(t)}{R_w} - \dot{\lambda}(t)\ \frac{V_v(t)}{R_w}. \tag{14}$$

Thus, by rearranging Equation (12) we can solve for the wheel slip derivative $\dot{\lambda}(t)$ which yields

$$\dot{\lambda}(t) = \left(\frac{1-\lambda(t)}{V_v(t)}\right) a_v(t) - \left(\frac{R_w}{V_v(t)}\right) \dot{\omega}_w(t) \tag{15}$$

and a general approach for estimating slip (we used simple Euler integration for the implementation of this technique in our simulations). Above we have illustrated one possible method for approximating slip. Through some investigations we have found that the FMRLC also works well for other slip estimation methods similar to those described in [15, 2, 3].

IV. Fuzzy Model Reference Learning Control for ABS

A. *FMRLC Design*

In this Section we describe the FMRLC that we will use for the ABS. For the FMRLC-based ABS we use $e(kT) = \lambda_r(kT) - \lambda(kT)$ where $\lambda_r(kT) = 20\%$ ($T = 1\text{ms}$) and $c(kT)$ is defined in Equation (2). We utilize a direct fuzzy controller that has 11 fuzzy sets with membership functions uniformly distributed on each (normalized) input universe of discourse. All membership functions used in our FMRLC are triangular shaped with a base-width of 0.4 (except when it is appropriate to use trapezoidal shapes for the outermost regions of the universes of discourse). The triangular membership functions for the fuzzy controller output (normalized) universe of discourse are initially set to be centered at zero indicating that the fuzzy controller initially does not know how to specify the control input (this is what the FMRLC will learn how to do). The "normalizing" controller gains for the error, change in error, and the controller output are chosen to be $g_e = 1$, $g_c = \frac{1}{1000}$, and $g_u = 2200$, respectively (see [11] for an explanation of this choice).

The reference model for this process was chosen to be

$$\frac{d\,\lambda_m(t)}{dt} = -10.0\ \lambda_m(t) + 10.0\ \lambda_r(t). \tag{16}$$

The inputs to the fuzzy inverse model include the error and change in error between the reference model and the wheel slip expressed as

$$\lambda_e(kT) = \lambda_m(kT) - \lambda(kT), \tag{17}$$

$$\lambda_c(kT) = \frac{\lambda_e(kT) - \lambda_e(kT-T)}{T}, \tag{18}$$

respectively. For these inputs, 11 fuzzy sets are defined with triangular shaped membership functions which are evenly distributed on the appropriate universes of discourse. The "normalizing" controller gains associated with $\lambda_e(kT)$, $\lambda_c(kT)$, and $p_i(kT)$ are chosen to be $g_{\lambda_e} = 1$, $g_{\lambda_c} = \frac{1}{1000}$, and $g_{p_i} = 2200$, respectively (see [11] for an explanation of this choice). In a typical braking system, an increase in the braking torque $T_b(kT)$, will generally result in an increase in the wheel slip. This implies that the incremental relationship between the process inputs and outputs is monotonically increasing. Consequently, the knowledge base array shown below in Table 1 was employed for the fuzzy inverse model. In Table 1, Λ_e^j is the jth fuzzy set associated with the error signal λ_e and Λ_c^k is the kth fuzzy set associated with the change in error signal λ_c. For convenience, rather than listing the indices i for $P_i^{j,k}$ in the body of the table, we list the center values of triangular membership functions corresponding to the fuzzy inverse model output fuzzy sets $P_i^{j,k}$.

B. *Performance for Various Road Conditions*

The FMRLC described above was simulated for the automotive ABS system. The results of this simulation for wet asphalt and for an icy surface are shown below in Figures 3 and 4, respectively. For these simulation results, only one brake was applied. The braking action was initiated when the vehicle was moving 25 $\frac{meters}{sec}$ (appox. 56 $\frac{miles}{hour}$) on a level surface ($\theta = 0$) and we desire to regulate slip to 20%. (Due to the fact that the wheel and vehicle velocity are nearly zero at low speeds, the magnitude of slip tends to goes to infinity as the vehicle speed approaches zero. This often causes determination of the slip to become very sensitive at slow speeds and as a result slip is very difficult to control at slow speeds. Therefore, as is standard in the literature, we only performed the simulation until the vehicle is slowed to approximately 5 $\frac{meters}{sec}$.)

Table 1: Knowledge base array table employed in the fuzzy inverse model for the ABS system.

	$P_i^{j,k}$	Λ_c^k −5	−4	−3	−2	−1	+0	+1	+2	+3	+4	+5
Λ_e^j	−5	−1	−1	−1	−1	−1	−1	−.8	−.6	−.4	−.2	0
	−4	−1	−1	−1	−1	−1	−.8	−.6	−.4	−.2	0	.2
	−3	−1	−1	−1	−1	−.8	−.6	−.4	−.2	0	.2	.4
	−2	−1	−1	−1	−.8	−.6	−.4	−.2	0	.2	.4	.6
	−1	−1	−1	−.8	−.6	−.4	−.2	0	.2	.4	.6	.8
	0	−1	−.8	−.6	−.4	−.2	0	.2	.4	.6	.8	1
	+1	−.8	−.6	−.4	−.2	0	.2	.4	.6	.8	1	1
	+2	−.6	−.4	−.2	0	.2	.4	.6	.8	1	1	1
	+3	−.4	−.2	0	.2	.4	.6	.8	1	1	1	1
	+4	−.2	0	.2	.4	.6	.8	1	1	1	1	1
	+5	0	.2	.4	.6	.8	1	1	1	1	1	1

Note that for the wet asphalt case, the reference model output and the braking system slip value tracked almost perfectly. As a result, the system does not exhibit the limit cycle effect which many ABS systems are designed for. Also, the braking torque for this case was very smooth. The controller seems to have found the appropriate level of braking torque which needs to be applied to the wheels to maintain

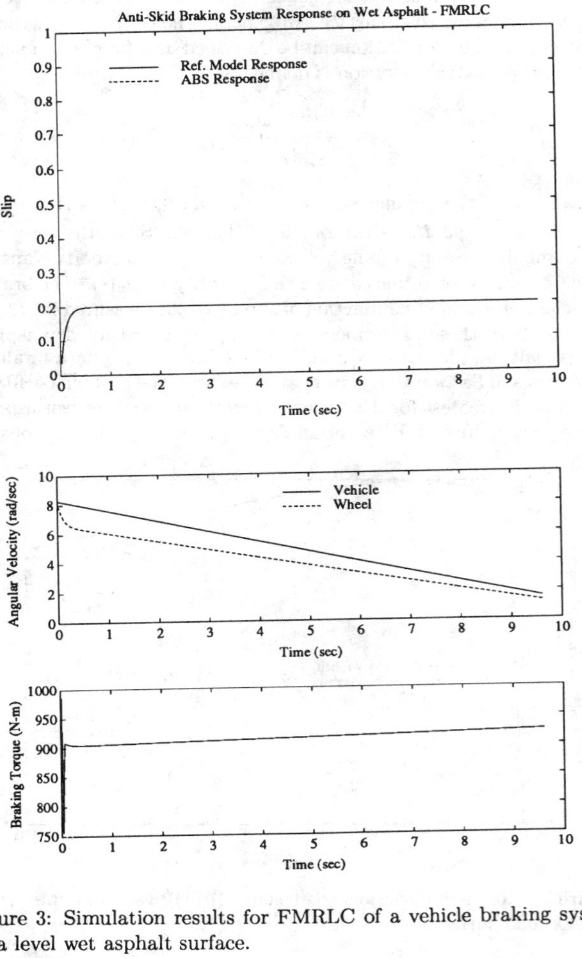

Figure 3: Simulation results for FMRLC of a vehicle braking system on a level wet asphalt surface.

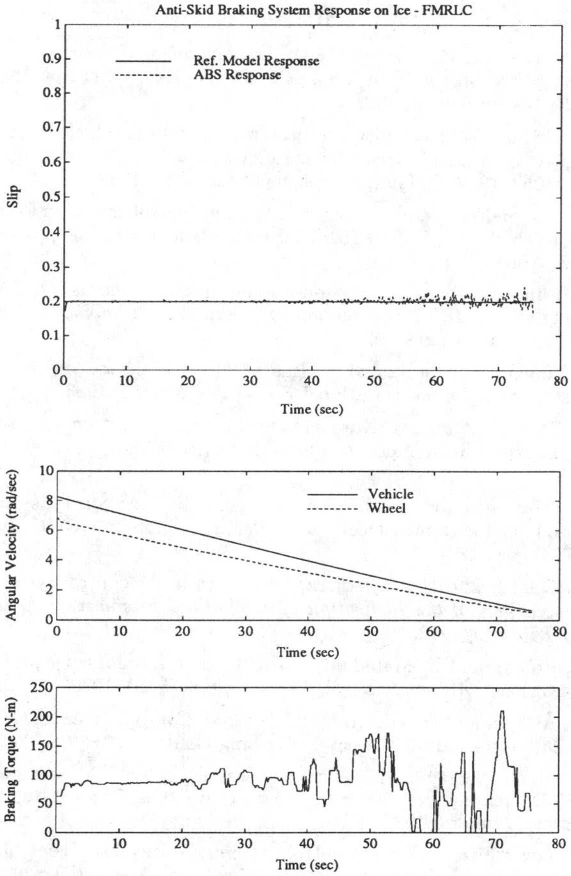

Figure 4: Simulation results for FMRLC of a vehicle braking system on a level icy surface.

a slip of 20%.

Although the simulation results for the icy surface shown in Figure 4 are likely to be considered acceptable by most control engineers, they are not as good as the results obtained for wet asphalt road condition. In general, it is very difficult to control slip on an icy surface due to the fact that a very small braking torque is likely to cause lock-up. This becomes even more difficult at slower speeds since determination of the wheel slip becomes very sensitive. This fact is clearly illustrated by the simulation results shown in Figure 4 where "good" control of slip was achieved at the faster vehicle velocity and degrades slightly as the vehicle slows.

Table 2 below is provided to illustrate the potential of the ABS system described above by comparing the stopping distance which resulted for the FMRLC algorithm with the case where the wheel is locked-up. Note that a substantial decrease in the stopping distance is obtained on all road surfaces which were considered in Figure 2 (the plots for dry asphalt were omitted in the interest of space as they were similar in shape to the wet asphalt case).

Table 2: Stopping distance for a single wheel ABS system implemented using FMRLC Vs. a single wheel lock-up braking action.

	Stopping Distance (meters)	
Road Surface	**FMRLC**	**Lock-up**
Dry Asphalt	130.2611	153.4057
Wet Asphalt	147.5513	159.5036
Ice	990.0790	2714.800

C. Split Road Conditions

The next set of simulations was performed to illustrate the effectiveness of the FMRLC algorithm for split road conditions. Here we consider two very likely real world scenarios. The first involves the situation where the brakes are applied on wet asphalt and during the braking action the road surface suddenly becomes icy. Notice that during the initial braking action, the wet asphalt would allow for a relatively large braking torque without lock-up occurring. However, when the vehicle reaches the icy road condition braking torque must be quickly reduced to prevent lockup. This large system variation presents a very demanding controller modification on the FMRLC algorithm. However, the simulation results for this scenario shown below in Figure 5 illustrate that the FMRLC algorithm is capable of dealing with such drastic process variations.

The second case involves the reverse of the situation described above. In this case, the brakes are applied on an icy surface and during the braking action the road surface suddenly becomes wet asphalt. This situation would require the FMRLC to reconfigure itself to increase the torque when the vehicle reaches the wet asphalt. Figure 6 below illustrates the simulation result for this scenario. Once again the FMRLC was successful.

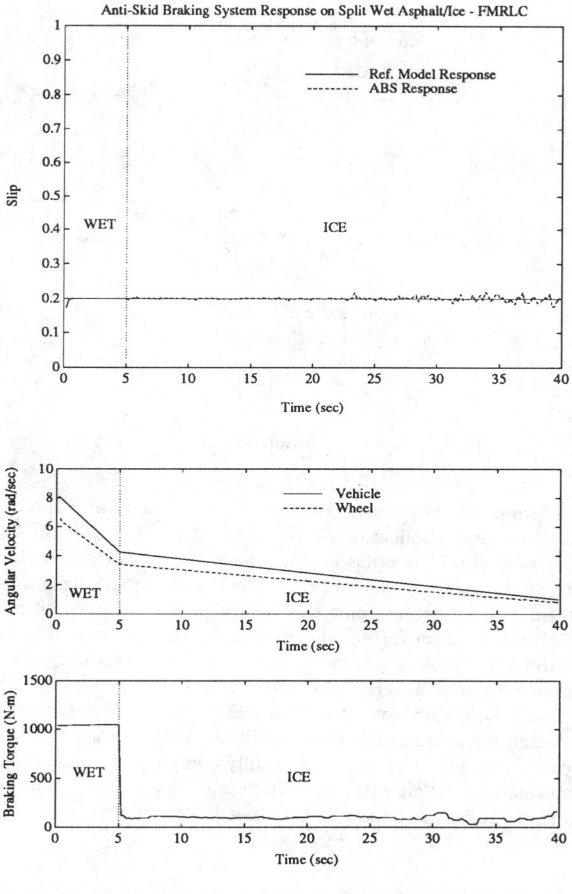

Figure 5: Simulation results for FMRLC of a vehicle braking system on a level split wet asphalt/icy surface.

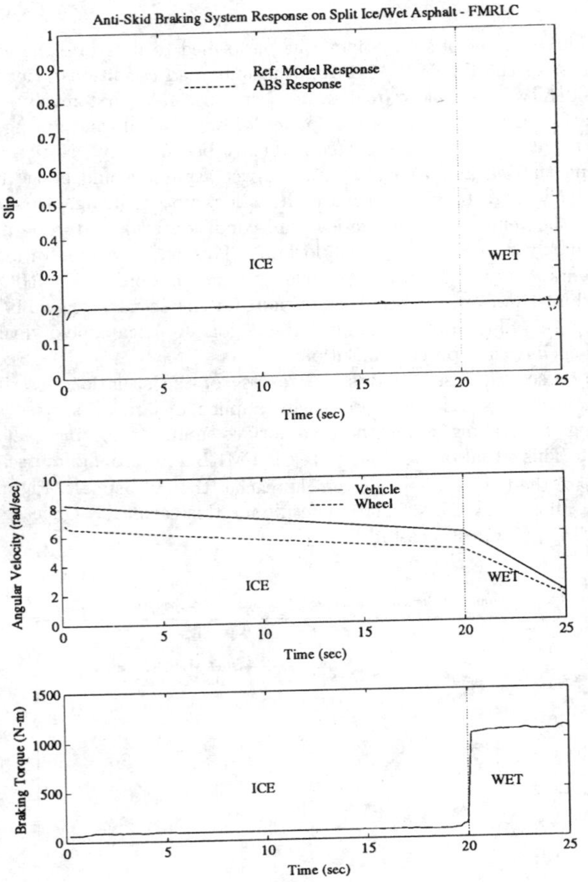

Figure 6: Simulation results for FMRLC of a vehicle braking system on a level split icy/wet asphalt surface.

V. Conclusions

The principal objective of this paper was to illustrate the design methodology and application of the new FMRLC algorithm for an automotive anti-skid braking system which is subjected to harsh road conditions. While the behavior of a conventional braking system varies significantly for different road and operating conditions, the results obtained in this paper (although somewhat preliminary) indicate that the FMRLC provides a promising approach to maintaining effective braking even under adverse road conditions (e.g., icy and split road conditions). Directions for future research include: (i) testing the FMRLC design for other harsh road conditions (e.g., snowy roads, split road conditions involving snow), (ii) fully comparing the approach to conventional control algorithms for ABS, and (iii) studying implementation characteristics of the FMRLC-based ABS system.

VI. References

[1] J. Layne and K. Passino, "Fuzzy model reference learning control," *Proceedings of the 1st IEEE Conference on Control Applications*, September 1992.

[2] H. Tan and M. Tomizuka, "A discrete-time robust vehicle traction controller design," *1988 American Controls Conference*, vol. 3, pp. 1053–1058, Pittsburgh, Pennsylvania, June 1989.

[3] H. Tan and M. Tomizuka, "Discrete-time controller design for robust vehicle traction," *IEEE Control Systems Magazine*, pp. 107–113, April 1990.

[4] R. Fling and R. Fenton, "A describing-function approach to antiskid design," *IEEE Transactions on Vehicular Technology*, vol. 30, pp. 134–144, August 1981.

[5] S. Yoneda, Y. Naitoh, and H. Kigoshi, "Rear brake lock-up control system of Mitsubishi Starion," *SAE paper 830482*, 1983.

[6] T. Tabo, N. Ohka, H. Kuraoka, and M. Ohba, "Automotive antiskid system using modern control theory," *IECON*, pp. 390–395, 1985.

[7] H. Takahashi and Y. Ishikawa, "Anti-skid braking control system based on fuzzy inference." *U.S. Patent*, Number*4842342*, June 1989.

[8] R. Guntur and H. Ouwerkerk, "Adaptive brake control system," *Proceedings of the Institution of Mechanical Engineers*, vol. 186, pp. 855–880, 1972.

[9] T. Procyk and E. Mamdani, "A linguistic self-organizing process controller," *Automatica*, vol. 15, no. 1, pp. 15–30, 1979.

[10] K. Åström and B. Wittenmark, *Adaptive Control*. Reading, Massachusetts: Addison-Wesley Publishing Company, 1989.

[11] J. Layne, "Fuzzy model reference learning control," Master's thesis, Department of Electrical Engineering, The Ohio State University, March 3 1992.

[12] C. Lee, "Fuzzy logic in control systems: Fuzzy logic controller-part I," *IEEE Trans. on Systems, Man. and Cybernetics*, vol. 20, pp. 404–418, March/April 1990.

[13] J. Harned, L. Johnston, and G. Scharpf, "Measurement of tire brake force characteristics as related to wheel slip (antilock) control system design," *SAE paper 690214*, 1986.

[14] S. Rhee, "Friction coefficient of automotive friction materials – its sensitivity to load speed and temperature," *SAE paper 740415*, 1974.

[15] E. Leaphart, "A DSP based hybrid simulator for evaluating antilock brake system control designs," Master's thesis, The Ohio State University, 1991.

Proceedings of the 31st Conference
on Decision and Control
Tucson, Arizona • December 1992

Object Oriented Structuring of Real Time Systems

M. Fabian, B. Lennartson

Chalmers University of Technology
S-412 96 Gothenburg, Sweden
e-mail: fabian@contrl.chalmers.se

ABSTRACT

We present an approach for structuring large real-time control systems, building on object-oriented techniques, c.f. Adiga (1989). The strategy separates the control of the individual physical resources from the synchronization of the system as a whole. This is achieved by building software models of the resources - internal objects. Local control of the individual resources is encapsulated within these objects, only a general well-defined interface is outwardly visible. The software models communicate through high-level messages, specially designed for expressing the synchronizing aspects of the system. An application-specific entity - a controller - administers the message handling, object synchronization and operator-interface.

INTRODUCTION

The current state of the art in implementing and structuring control of production systems, e.g. batch processes or discrete event manufacturing systems suffers from the combinatorial complexity that arises as the number of I/O-signals increase. The low-level mechanisms available, logical functions, GRAFCET, etc., make no provision for describing the synchronization aspects of the control of the fabrication process, on a practical enough abstraction-level. Rather, this must be intermingled with the local control of the individual units.

Separation of the systems synchronizing necessity from the individual physical resources control strategy, facilitates the implementation of real-time control systems. By building software models that encapsulate specific functionality, and interacts through messages, the coupling of the resources from each other can be relaxed. This aids in the structuring of the overall system, as the software models can be raised to equal levels of 'intelligence', whereby the synchronizing entity only needs to consider high level messages, distinct from the low level functions that are needed to implement this 'intelligence'.

PHYSICAL AND SOFTWARE OBJECTS

A real-time system can be viewed as consisting of two parts, see fig 1 - a physical system dissectable into physical subsystems, and a software system, supervising, controlling and synchronizing the physical system, making sure it behaves according to some given specification. This monitoring demands, apart from the information flow between the systems in the guise of the control and sensor signals, also some mapping between the control system and the physical system.

Distinct physical subsystems exhibit similar general behavior on certain abstraction levels, although they may crave vastly different control structures on lower levels to achieve this. This general behavior can be abstracted out into a template to which all objects with this behavior adhere - a class. These templates can be implemented and tested once, and then reused in different applications. Instantiations of these classes constitute software models - internal objects - of the physical subsystems - the external objects. Supervisory control over the external objects are routed through the internal ones, so every physical subsystem must map onto (at least) one internal object.

Though linked to each other by information and/or production flows, the physical subsystems are distinct individual units, not necessarily aware of each others presence. This linkage is a manifestation of the synchronization aspects of the system-control, and is only explicit within the control system. Letting this coupling be expressed by high-level messages exchanged between the internal models, relaxes this linkage and facilitates structuring of the control software. The programmer only has to consider these high-level messages, unnecessitating any concerns of low-level implementational specifics, when devising the control software.

THE OBJECT MODEL

The internal objects must present a class-general interface outwards, an abstract representation of the

CH3229-2/92/0000-2529$1.00 © 1992 IEEE

corresponding resource. This representation is effected as a state-automata, describing the aspects of the resource's dynamic behavior necessary for the synchronization with the other internal objects. Thus the internal object must encapsulate specific control actions that implement the outwardly presented functionality. Similar physical devices, for instance, from different vendors need different control structures to achieve alike results.

These two distinct aspects can be conceptualized by splitting the internal object into two parts - a general and a specific part, see fig 2. The general part participates in the message-exchanging between the internal objects, as well as exchanging messages with the specific part. Also, the general part gives some feedback, preferably graphical, as to what is going on in the real world. This would include showing the state of the external object as depicted by the incorporated state-machine. The visual feedback would also include a representation of the process going on within the resource. This representation will have to be rather abstract, since similar resources can accomplish the same end-results in many different ways, all with different appearances. This is a price we have to pay for generality. The benefit is that the general part is only class-specific, inheritable and specializable for new applications.

The specific part implements the promised functionality, by interacting with the general part and the external object. If the external object is of sufficient sophisticacy the interaction could consist of only relaying the message received from the general part to the external object. However, most often it would be the case that the external object is not so 'intelligent', so the specific part would have to transform the message from the general part into a set of control actions understood by the external object. Thus, the specific part is the portion of the internal object that would have to be adapted for each physical resource, according to vendor, model, etc., while the general part would be the same for all similar resources.

THE CONTROLLER

Each physical resource is a self-contained unit which must be synchronized by some entity for useful work to be accomplished. This entity has to know all the units within the system, and must therefore be taylor-made for the particular application. This controller, though, does not, and need not know how each object does what it has to do, nor even exactly what they do. It only has to synchronize the objects so that a special overall worksequence is carried out. To the controller the internal objects are black boxes, receiving and responding to messages. This distribution of the control of the individual work routines simplifies the implementation of the controller as well as of the whole system.

In fact, recent research has shown that by distributing the controller it can itself be generalized, driven by adaptable recipes governing the work process. This distribution is accomplished by creating production-units, each routing its own way through the workprocess, shifting from a plant-related to a product-related view of fabrication process control, see Fabian (1992).

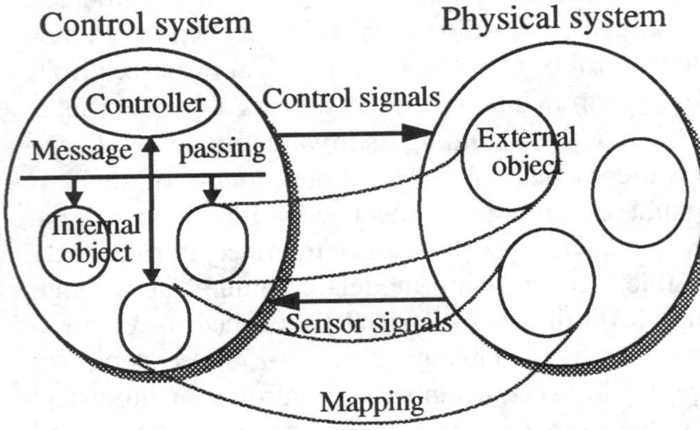

Figure 1. Physical and software objects. Mapping between the physical and software systems.

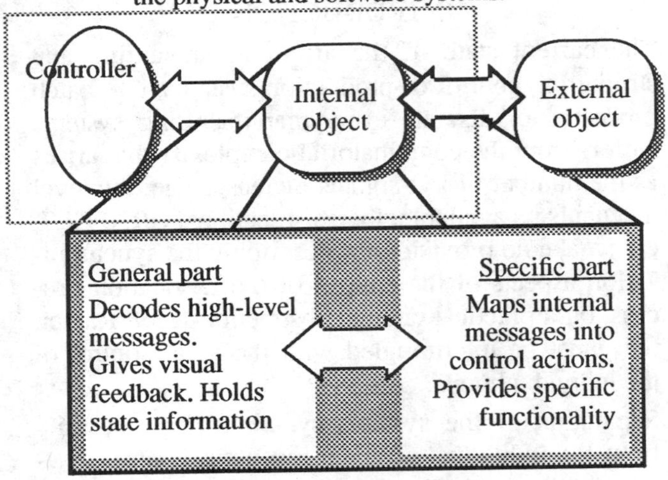

Figure 2. Structure of the internal objects.

REFERENCES

Adiga, S. (1989). *Software Modeling of Manufacturing Systems - a Case for an Object Oriented Approach*, Annals of Operatiions Research, 17, 1-4, 363-378

Fabian, M., B. Lennartson, (1992). *Distributed Objects for Real Time Control Systems*, submitted to 12th IFAC World Congress, 1993, Sydney, Australia

Proceedings of the 31st Conference
on Decision and Control
Tucson, Arizona • December 1992

Point to point polynomial control of linear systems by learning

Pasquale Lucibello
Dipartimento di Elettronica, Informatica e Sistemistica
Università degli Studi della Calabria — 87036 Arcavacata di Rende, Cosenza, Italy

Abstract — State transfer for linear systems by means of learning control is considered. An algorithm, which uses only the positional error at the end of a trial, is introduced. The computed control law is polynomial in time.

I. INTRODUCTION

In this paper we consider the problem of finding, by means of a learning algorithm, a control which brings, in a given time interval, the state of a linear system from one point to another in the state space. This problem differs from the one usually considered in the learning literature [1-14], which deals with exact output reproduction over a given time interval. Consistently, the algorithm herein introduced employes only the positional error at the end of a trial to update the control for the next one.

After a generic formulation of this problem, we consider the specific case of controls which are polynomial in time. Finally, for illustration, we apply the developed procedure to the repositioning control of a flexible arm.

II. POINT TO POINT CONTROL

Consider a linear control system

$$\dot{x}(t) = A\,x(t) + B\,u(t) + p(t), \quad x \in R^n, \quad u \in R^m, \qquad (2.1)$$

where x is the state, u the piece-wise continuous control and p a piece-wise continuous disturbance.

Given an arbitrary pair of points $x_1, x_2 \in R^n$, a control which steers the state from x_1 to x_2 in a given time interval [0, T] exists if and only if the system is controllable.

More than one control may realize the desired state transfer with no constraints on state trajectory and control. In particular, by means of the constant of variation formula, one of such controls is computed as

$$u(t) = B'\,e^{A'(T-t)}\,W^{-1}\left(x_f - e^{AT}x_i - \int_0^T e^{A(T-\tau)}p(\tau)\,d\tau\right),\ 0 \le t \le T,$$

where the map

$$W = \int_0^T e^{A(T-\tau)}B\,B'\,e^{A'(T-\tau)}\,d\tau$$

is invertible under the hypothesis that the system is controllable. (The prime denotes transposition.)

A control polynomial in time is computed as follows. Consider the system in its controllability canonical form

$$\frac{d^n z}{dt^n} = -\sum_{i=0}^{n-1} a_i \frac{d^i z}{dt^i} + u, \qquad (2.2)$$

where the a_i are the coefficients of the characteristic polynomial of A and u is scalar. Let

$$w_i = \frac{d^i z}{dt^i},\ i = 0, ..., n-1,$$

and denote by w° the initial state and by w_f the final state. Let w_{fd} the deviation at time T due to the disturbance and define

$$w^f = w_f - w_{fd}.$$

Let

$$\pi(t) = \sum_{k=0}^{2n-1} \gamma_k\, t^k \in R$$

be the unique polynomial in time such that

$$\left[\frac{d^i\pi}{dt^i}\right]_{t=0} = w_i^0,\ i = 0, ..., n-1,$$

$$\left[\frac{d^i\pi}{dt^i}\right]_{t=T} = w_i^f,\ i = 0, ..., n-1.$$

By substituting π(t) in 2.2, a polynomial control of degree less than or equal to (2n - 1) is computed which, owing to linearity, steers the state from w° to w_f in the face of the disturbance. By means of the Lemma of Heymann [15], this procedure can be extended to the multi-input case.

The linear and invertible relationship which links the coefficients γ_i with w° and w^f is given by

$$\begin{pmatrix} w^o \\ w^f \end{pmatrix} = \begin{pmatrix} M_{11} & 0 \\ M_{21} & M_{22} \end{pmatrix} \begin{pmatrix} \beta_1 \\ \beta_2 \end{pmatrix},$$

where

$$M_{11}(i, j) = 0 \quad \text{for} \quad i \ne j,$$
$$M_{11}(i, i) = i \quad \text{for} \quad i, j = 1, ..., n,$$
$$M_{21}(i, j) = 0 \quad \text{for} \quad i < j,$$
$$M_{21}(j, i) = \frac{(j-1)!}{(j-i)!}T^{(j-1)} \quad \text{for} \quad j \ge i, \quad i, j = 1, ..., n,$$
$$M_{22}(j, i) = \frac{(j+n-1)!}{(j+n-i)!}T^{(j+n-1)} \quad i, j = 1, ..., n,$$

and

$$\beta_1 = \begin{pmatrix} \gamma_0 \\ \dots \\ \gamma_{n-1} \end{pmatrix}, \quad \beta_2 = \begin{pmatrix} \gamma_n \\ \dots \\ \gamma_{2n-1} \end{pmatrix}$$

The vector of the coefficients of the polynomial control, ordered from lower to higher powers, is given by

$$\theta = G\,\gamma,$$

with

$$G(k, k+i) = \frac{(k+i-1)!}{(k-1)!}a_i,$$

i = 0, ..., n; k = 1, ..., 2 n - i + 1. Moreover since γ is uniquely determined by w° and w^f, we have that there exist linear maps L_1 and L_2 such that

$$\theta = L_1\,w^o + L_2\,w^f = L_1\,w^o + L_2\,(w_f - w_{fd}). \qquad (2.3)$$

Notice that, because of controllability is preserved under arbitrary, even if sufficiently small in norm, linear plant perturbations, the solution maps computed continuously depend on plant parameters in a neighborhood of their nominal values.

Denote by N_1 and N_2 the linear maps defined by 2.3 in non canonical coordinates. Consider 2.1 initialized at x_1 and driven by a polynomial control law of vector of coefficients θ

$$x(T) = e^{AT}x_1 + \int_0^T e^{A(T-t)}B\sum_{i=0}^{2n-1}\theta(k)\,t^k\,dt + \int_0^T e^{A(T-t)}p(t)\,dt = H\,x_1 + P\,\theta + x_{fd}$$

For $\theta = N_1\,x_1 + N_2\,(x_2 - x_{fd})$, we have $x(T) = x_2$ and

CH3229-2/92/0000-2531$1.00 © 1992 IEEE

$$x_2 = H x_1 + P \{N_1 x_1 + N_2 (x_2 + x_{fd})\} + x_{fd}.$$

Hence, $P N_1 = - H$ and $P N_2 = I$, that is N_2 is a left inverse of P.

III. THE LEARNING ALGORITHM

Consider 2.1 initialized at x_1 and driven by a polynomial control law of vector of coefficients $\theta(k)$, during the k-th trial. The state reached at the end of the k-th trial is given by

$$x_f(k) = H x_1 + P \theta(k) + x_{fd}.$$

Update the control by means of

$$\theta(k + 1) = \theta(k) - F (x_f(k) - x_2), \quad (3.1)$$

with x_2 the state to be reached at time T. We have

$$x_f(k + 1) = H x_1 + P \theta(k) - P F (x_f(k) - x_2) + x_{fd} = x_f(k) - P F (x_f(k) - x_2).$$

Hence for the error

$$e(k) = x_f(k) - x_2$$

we have

$$e(k + 1) = (I - P F) e(k).$$

If the system were exactly known, by setting $F = N_2 (I - D)$, $|D| < 1$, the error would converge to zero, being N_2 a left inverse of P. In general, since N_2 continuously depends on plant parameters, the condition

$$\| I - P N_2 (I - D) \| < 1$$

is satisfied for arbitrary linear plant perturbations, which are sufficiently small in norm. Hence, the feedback learning law defined by 3.1, with $F = N_2 (I - D)$, $|D| < 1$, is robust with respect to linear plant perturbations.

IV. A CASE STUDY

The flexible arm depicted in Fig.1 consists of two rigid rods: the first one is connected to the basement through a revolute and actuated joint; the second one is connected to the first through another revolute joint where a spring-damper mechanism is located. The plane of motion is horizontal.

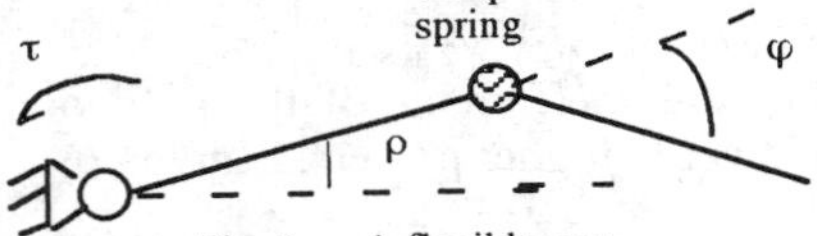

Fig. 1 — A flexible arm

We denote by ρ and φ the first and second joint rotations, respectively, and by τ the control torque. For small φ and small angular velocities, the linearized Lagrange equations of motion are given by

$$\begin{pmatrix} i_t & i_2 + l_1 S \\ i_2 + l_1 S & i_2 \end{pmatrix} \begin{pmatrix} \ddot{\rho} \\ \ddot{\varphi} \end{pmatrix} = \begin{pmatrix} \tau - d_1 \dot{\rho} \\ - d_2 \dot{\varphi} - k \varphi \end{pmatrix},$$

where l_1 and l_2 are the lengths of the rods, d_1 and d_2 are the joint viscous damping coefficients, k is the spring stiffness, S is the static moment of the second rod, i_1 and i_2 the rods moments of inertia and

$$i_t = i_1 + i_2 + 2 l_1 S + m_2 l_1^2,$$

with m_2 the second rod mass.

Collect the state of the arm system in the vector

$$x = (\rho, \varphi, \dot{\rho}, \dot{\varphi})'.$$

It is required to move the arm from $x_1 = (0, 0, 0, 0)'$ to $x_2 = (\pi/2, 0, 0, 0)$, in two seconds. It is also required that the torque is zero at time zero as well as t = 2 sec, in order to start from and arrive at a true equilibrium point.

To comply with this last specification, we dynamically extend the plant to include the torque in the state space, that is we set

$$\dot{\tau} = u,$$

with u the new control. In the extended state space, the starting point is then $(x_1, 0)$ and the arrival point $(x_2, 0)$.

The following numerical values are used: $l_1 = 5$ m, $l_2 = 7$ m, $m_2 = 1$ kg, $i_1 = 1$ kg m^2, k = 40 N m/rad, $d_1 = 0.1$ N m sec/rad, $d_2 = 0.05$ N m sec/rad. Assuming a uniform distributed mass for the second rod, we also have:

$$i_2 = \frac{1}{3} m_2 l_2^2, \quad S = \frac{1}{2} m_2 l_2.$$

The robustness of the learning algorithm with respect to perturbations of the second rod mass has been then investigated. The map D of the learning feedback law has been set equal to r I, r ∈ (-1, 1). In Fig.2, the norm

$$\| I - P N_2 (I - D) \| = \max_k |\lambda_k|,$$

with λ_k the eigenvalues of the map between brackets, has been plotted versus the ratio of the estimated mass with the true one. We notice that for values of r close to zero, we have a faster rate of convergence in a neighborhood of the nominal point one. However, the algorithm is less robust, in the sense that its interval of convergence is smaller. This points out a conflict between the rate of convergence and the interval of robustness.

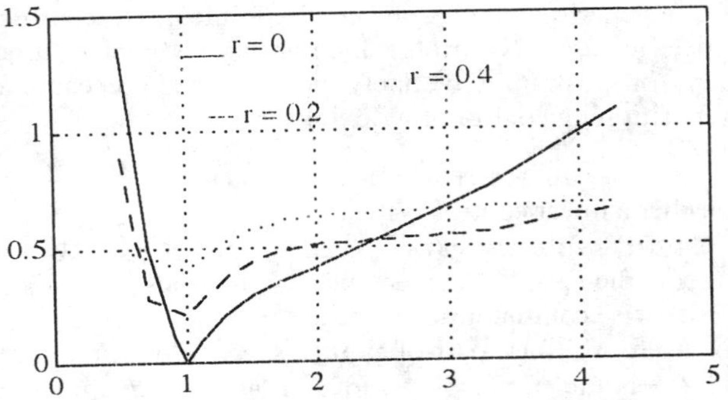

Fig. 2 — Algorithm robustness w.r.t. mass variations

REFERENCES

[1] S. Arimoto, S. Kawamura, F. Miyazaki, Bettering operations of robots by learning, *J. Robotic Systems*, vol. 1 (1984), pp. 123-140

[2] J.J. Craig, Adaptive control of manipulators through repeated trials, *Proc. Amer. Control Conf.*, 1984

[3] G. Casalino, G. Bartolino, A learning procedure for the control of movements of robotic manipulators, *Proc. IASTED Symp. on Robotics and Autom.*, Amsterdam, The Netherlands, June 1984

[4] S. Arimoto, S. Kawamura, F. Miyazaki, S. Tamaki, Learning control theory for dynamical systems, *Proc. 24th IEEE Conf. on Decision and Control*, 1985

[5] J. Craig, *Adaptive control of mechanical manipulators*, Addison-Wesley, NY (1988)

[6] S. Kawamura, F. Miyazaki, S. Arimoto, Application of learning method for dynamic control of robot manipulator, *Proc. 24th IEEE Conf. on Decision and Control*, 1985

[7] S. Arimoto, S. Kawamura, F. Miyazaki, Realization of robot motion based on the learning method, *IEEE Systems Man Cybernetics*, vol. 18 (1988), pp. 126-134

[8] M. Togai, O. Yamano, Analysis and design of an optimal learning control scheme for industrial robots: a discrete system approach, *Proc. 24th IEEE Conf. on Decision and Control*, 1985

[9] T. Mita, E. Kato, Iterative control and its application to motion control of robot arm - A direct approach to servo-problems, *Proc. 24th IEEE Conf. on Decision and Control*, 1985

[10] L. Hideg, R. Judd, Frequency domain analysis of learning systems, *Proc. 27th IEEE Conf. on Decision and Control*, 1988

[11] J. Hauser, Learning control for a class of nonlinear systems, *Proc. 26th IEEE Conf. on Decision and Control*, 1987

[12] L. Hideg, R. Judd, R. Van Til, Stability analysis for learning systems, *Proc. American Control Conf.*, 1990

[13] K.L. Moore, M. Dahleh, S.P. Bhattacharyya, Iterative learning for trajectory control, *Proc. 28th IEEE Conf. on Decision and Control*, 1989

[14] P. Lucibello, Inversion of linear square systems by learning, *Proc. 30th IEEE Conf. on Decision and Control*, 1991

[15] W.M. Wonham, *Linear multivariable control. A geometric approach*, Springer-Verlag, 2nd edition, 1979

Proceedings of the 31st Conference
on Decision and Control
Tucson, Arizona • December 1992

Output regulation of flexible mechanisms by learning

Pasquale Lucibello
Dipartimento di Elettronica, Informatica e Sistemistica
Università della Calabria — 87036 Arcavacata di Rende, Cosenza, Italy

Abstract – Output regulation of flexible mechanisms by means of learning algorithms is investigated. The spillover effect, which may adversely affect closed loop performance, is dealt with. An example is worked out.

I. INTRODUCTION

In the contest of repetitive tasks operation of robots, some researchers have introduced a novel control scheme known as learning. Control algorithms improving task performance of robots from one trial to the next one have been independently presented in [1, 2, 3]. The idea is that of using tracking position errors as well as velocity and acceleration errors of a trial to update the control of the next trial. Exact state trajectory tracking is expected as the number of trials tends to infinity. Properties of this novel control scheme have been further investigated in [4, 5, 6, 7, 8, 9, 10].

Researchers have also recognized that this type of control can be generalized to control systems different from the robotic ones. Strictly positive linear systems have been considered in [4] and some classes of nonlinear systems with Lipschitz dynamics in [4, 11]. In [12], a Lyapunov stability analysis of learning algorithms for single input linear systems is presented. In [13] convergence conditions of learning algorithms for linear time invariant systems are investigated. In [14] algorithms for the inversion of square linear systems by learning are presented and in [15] point to point control of linear systems is addressed.

In this paper the problem of output regulation of mechanisms composed by flexible bodies connected through lower kinematic pairs is considered. The motion that such mechanisms undergo is the sum of a rigid motion plus a small elastic deviation. The state of the system has to be steered from one equilibrium point to another in a given time interval, so that at the end of the interval not only the system is in equilibrium but in addition a given function of the mechanism position (the output) takes on a prescribed value. This problem is different from the ones previously addressed in the literature, not only because it deals with flexible mechanisms, but also because it cares only for the output and the equilibrium at a given time instant. The error used to improve the system performance from one trial to the next one is then the output error and the state derivative at the end of the trial and not the trajectory error during the complete trial.

The spillover effect, which may adversely affect the performance of the learning algorithm, is also dealt with. It is shown that high frequency dynamics can be neglected with minor detriment of system performance. An example is finally worked out to illustrate the learning control and the spillover effect.

II. LEARNING CONTROL

II.1 Let R^{2p} be the state space of the mechanism and h: $R^p \to R^n$ the output, function of the mechanism position. Let U be the Banach space of piece-wise continuous functions on the time interval [0, T] with values in R^n, the space of the external forces acting on the mechanism, equipped with the sup norm on [0, T].

Given the initial state, a control $u \in U$ transfers the state of the mechanism and its derivative to uniquely determined points at time T. Let L: $U \to R^n \times R^p$ denote the map which assigns to a control u the output and the state derivative at time T and assume, without loss of generality, that the output at time T must be zero. The control problem to be solved consists in finding a zero of

$$L(u) = 0. \qquad (2.1)$$

Under the hypothesis that (2.1) admits at least one solution, one may apply one of the existing technique for nonlinear equations in functional spaces. In particular if L is Fréchet differentiable, Newton-Raphson methods can be applied. This hypothesis is satisfied if the output function is differentiable and the linearized equations of motion around a system trajectory exist and admit unique solutions. Assume that L is differentiable and consider a point $u_0 \in U$, one has

$$L(u) = L(u_0) + DL(u_0)\,(u - u_0) + o(|u - u_0|),\ u, u_0 \in B(\varepsilon),$$
$$B(\varepsilon) = \{ u \mid |u - u_0| < \varepsilon\},$$

with ε sufficiently small and DL the Fréchet derivative of L. By neglecting second order terms, one may think to solve for $Du = u - u_0$ the equation

$$L(u) = L(u_0) + DL(u_0)\, Du = 0.$$

Assume that there exists a linear map Q: $R^n \times R^p \to U$ such that the map $DL(u_0)Q$ is invertible and $Q\,[DL(u_0)Q]^{-1}$ is a right inverse of $DL(u_0)$. (Under the hypothesis that the linearized system is controllable and some other minor hypothesis, the existence of one of such maps is shown in the Appendix.) Set $Du = Q\, v_1$ and solve for v_1

$$v_1 = -\,[DL(u_0)\,Q]^{-1}\, L(u_0).$$

By applying the control $u = u_0 + Q\, v_1$, one expects that $L(u_0 + Q\, v_1)$ is closer to zero than $L(u_0)$. By iterating this procedure, one expects that the sequences

plant sequence:

$$y_i = L(u_0 + Q\, v_i), \qquad (2.2a)$$

control updating sequence:

$$v_{i+1} = v_i - [DL(u_0)\,Q]^{-1}\, y_i\,,\ v_0 = 0, \qquad (2.2b)$$

are convergent. In this case, one says that 2.2 define a learning algorithm.

CH3229-2/92/0000-2533$1.00 © 1992 IEEE

Classically, the contraction mapping principle can be used to establish conditions under which this scheme is convergent. The following theorem is adapted from [16].

Theorem 1 Suppose that

(i) the Fréchet derivative of L: $U \to R^n \times R^p$ exists at u_0;

(ii) there exists Q: $R^n \times R^p \to U$ such that $[DL(u_0)\, Q]$ is invertible;

(iii) for each $u = u_0 + Q\, v$, $v \in \{ v \mid |v| < \delta\}$, the Fréchet derivative of L: $U \to R^n \times R^p$ exists;

(iv) $\alpha = \sup | I - [DL(u_0)\, Q]^{-1}\, DL(u_0 + Q\, v)\, Q | < 1$, $v \in \{ v \mid |v| < \delta\}$, with I the identity map and δ and α linked by 2.3.

Then the equation

$$L(u_0 + Q\, v) = 0$$

has a solution $v^* \in \{ v \mid |v| < \delta\}$ to which the sequence 2.2 converges.

Proof.

From 2.2 one gets

$$v_{i+1} = v_i - [DL(u_0)\, Q]^{-1}\, L(u_0 + Q\, v_i) = S(v_i),\ v_0 = 0.$$

It is now proved that the map S is a contraction mapping on $\{ v \mid |v| < \delta\}$. The Fréchet derivative of S is given by

$$DS = I - [DL(u_0)\, Q]^{-1}\, DL(u_0 + Q\, v)\, Q.$$

For any two point v', v" $\in \{ v \mid |v| < \delta\}$, one has

$$|S(v') - S(v")| \le \sup |DS(v' + s\, (v" - v')|\ |v" - v'|,\ 0 \le s \le 1,$$

by the mean value theorem and hypothesis (iii). But, by hypothesis (iv)

$$|S(v') - S(v")| \le \alpha\ |v" - v'|,\ a < 1,$$

and S is a contraction on $\{ v \mid |v| < \delta\}$ has claimed. Moreover, since S is a contraction, one has

$$|v^* - v_n| \le \frac{\alpha^n}{1-\alpha} |v_n - v_0| = \frac{\alpha^n}{1-\alpha} |v_n|.$$

Hence the domain of the solution is bounded by

$$|v^*| \le \frac{1}{1-\alpha} |v_1| = \frac{1}{1-\alpha} \left| [DL(u_0)\, Q]^{-1}\, L(u_0) \right| = \delta. \qquad (2.3)$$

◊

Remark 1 This scheme implies the uniform convergence and hence convergence is guaranteed for sufficiently small nonpersistent perturbations [17]. In particular, an approximate right inverse of the Fréchet derivative can be used. ◊

Remark 2 Different schemes can be derived by using different maps Q and different Newton-Raphson schemes. To each map Q corresponds a different control. ◊

Remark 3 Notice that the dynamics of the learning algorithm are discrete over the countable set of the trials and this suggests to refer to the map $Q\, [DL(u_0)\, Q]^{-1}$ as the learning feedback. ◊

II.2 The design of feedback laws are classically done on the basis of a more or less accurate dynamic model of the system. The reader may refer to [18] for one of the early and quite comprehensive paper on modelling of flexible mechanism composed by flexible bodies connected by lower kinematic pairs and modelling is then not addressed in the sequel. On the contrary, the effectiveness of a controller designed on the basis of a finite dimensional model neglecting high frequency dynamics is still a controversial question. The main concern is the possible instability of the controlled system due to neglected dynamics. This problem, in the case of output regulation of linear mechanical systems as $t \to \infty$, has been studied in [19] and named spillover. In the following this issue is investigated by means of a singular perturbation approach.

The flexible mechanisms herein dealt with undergo a motion which may be named as quasi-rigid, that is a motion which is the sum of a rigid motion plus a small elastic deviation. The elastic displacement field is expressed as the sum of a finite number of independent components (elastic degrees of freedom). Different rigid motions can be selected for the same mechanism motion [20], obtaining different motion representations which are linked by coordinates transformations. Accurate modelling may require a large number of elastic degrees of freedom, which may not be practical to take into account when designing the controller (*). When closing the feedback loop, however, neglected dynamics may become unstable. This phenomena is the one which may adversely affect the performance of high gain force control schemes designed on the basis of rigid mechanism models [22]. Rigid models or models with a low number of elastic degrees of freedom relies upon the idea that high frequency vibrations have fast stable dynamics and small displacement amplitude. This fact can be investigated by means of singularly perturbed models [23].

Let $q \in R^n$ be the vector of variables defining the rigid motion and $x \in R^s$ the vector of variables defining the elastic motion. According to the hypothesis of small elastic displacements, the dependency of the material points velocities on the elastic displacements is neglected and the kinetic energy is taken equal to

$$T = 0.5\, b_{ij}(q)\, \dot{q}_i\, \dot{q}_j + 0.5\, m_{hk}(q)\, \dot{x}_h\, \dot{x}_k + s_{ik}(q)\, \dot{q}_i\, \dot{x}_k,$$

where the tensors b_{ij} and m_{hk} are symmetric and positive definite for each value of q and the Einstein summation convention is from now on adopted. For each q, T is positive definite. The potential energy is taken equal to

$$U = g(q) + \eta_k(q)\, x_k + 0.5\, k_{hk}\, x_h\, x_k,$$

where k_{hk} is the symmetric and positive definite stiffness tensor. The other terms take into account the gravitational potential energy. It is further assumed that the virtual work of the forces delivered by the actuators is given by

$$\delta L_a = \tau_i\, \delta q_i,$$

where $\tau \in R^n$ is the vector of actuator forces. This hypothesis entails a particular selection of the rigid motion.

(*) In case of system with distributed mass and elasticity, it is known that the greater the number of independent components used to describe the elastic displacement field the smaller the difference between the motion of the infinite dimensional model and the finite dimensional one [21].

The virtual work of the internal damping forces is taken equal to

$$\delta L_d = c_{hk}\,\dot{x}_h\,\delta x_k,$$

where c_{hk} is the symmetric and positive definite damping tensor. The dynamic model is given by the Lagrange equations of motion

$$\begin{aligned}
& b_{ij}\,\ddot{q}_j + s_{ik}\,\ddot{x}_k + (b_{ij,s} - 0.5\,b_{sj,i})\,\dot{q}_s\,\dot{q}_j + (s_{ik,s} - s_{sk,i})\,\dot{q}_s\,\dot{x}_k - \\
& 0.5\,m_{hk,i}\,\dot{x}_h\,\dot{x}_k + g_{,i} + \eta_{k,i}\,x_k = \tau_i, \\
& m_{hk}\,\ddot{x}_k + s_{ik}\,\ddot{q}_i + m_{hk,s}\,\dot{q}_s\,\dot{x}_k + s_{ik,s}\,\dot{q}_s\,\dot{q}_i + \\
& c_{hk}\,\dot{x}_k + k_{hk}\,x_k + \eta_k = 0,
\end{aligned}$$

where the comma denote derivative with respect to a component of q and the functional dependency on q has been omitted for notational simplicity.

Consider now the mechanical system obtained by locking the rigid motion and neglecting the gravitational forces

$$m_{hk}\,\ddot{x}_k + c_{hk}\,\dot{x}_k + k_{hk}\,x_k = 0.$$

Assume that the damping tensor is proportional to the stiffness tensor and denote by ϑ_{ij} the q-dependent tensor of the real natural modes of this vibrating system. Make the change of coordinates

$$x_j = \vartheta_{jh}\,p_h,$$

and multiply by ϑ_{jh}. The system is diagonalized as

$$\ddot{p}_h + 2\,\zeta_h\,\omega_h\,\dot{p}_h + \omega_h^2\,p_h = 0, \quad h = 1, \ldots, s,$$

if proper normalized eigenvectors are used. Here, ω_h are the ordered natural circular frequencies and ζ_h the damping factors. Note that these quantities are dependent on q, i.e. on the position where the rigid motion has been frozen.

Assume that the tensor ϑ is two times differentiable with respect to q and consider the same change of coordinates when the rigid motion is not frozen, one has

$$\begin{aligned}
\dot{x}_j &= \vartheta_{jh}\,\dot{p}_h + \vartheta_{jh,i}\,\dot{q}_i\,p_h, \\
\ddot{x}_j &= \vartheta_{jh}\,\ddot{p}_h + \vartheta_{jh,i}\,\ddot{q}_i\,p_h + 2\,\vartheta_{jh,i}\,\dot{q}_i\,\dot{p}_h + \vartheta_{jh,is}\,\dot{q}_i\,\dot{q}_s\,p_h,
\end{aligned}$$

Assume that the modes greater or equal to $s^* < s$ are substantially stiff with respect to the first ones, that is let $\omega_h =: \omega_h/\sqrt{\varepsilon}$, $h \geq s^*$, with ε small, and $\zeta_h =: \zeta_h/\sqrt{\varepsilon}$, $h \geq s^*$, to maintain the same level of modal damping. Next let $\varepsilon\,\phi_h = p_h$, $h \geq s^*$, and singularly perturb [22] the system by letting $\varepsilon \rightarrow 0$. The system splits in a slow system and in a fast system. The slow system is given by

$$\begin{aligned}
& b_{ij}\,\ddot{q}_j + s_{ik}\,\ddot{\xi}_k + (b_{ij,s} - 0.5\,b_{sj,i})\,\dot{q}_s\,\dot{q}_j + (s_{ik,s} - s_{sk,i})\,\dot{q}_s\,\dot{\xi}_k - \\
& 0.5\,m_{hk,i}\,\dot{\xi}_h\,\dot{\xi}_k + g_{,i} + \eta_{k,i}\,\xi_k = \tau_i, \\
& \ddot{p}_r + \vartheta_{hr}\,\{s_{ih}\,\ddot{q}_i + m_{hk,s}\,\dot{q}_s\,\dot{\xi}_k + s_{ik,s}\,\dot{q}_s\,\dot{q}_i + c_{hk}\,\dot{\xi}_k + \\
& m_{hk}\,(2\,\vartheta_{kw,s}\,\dot{q}_s\,\dot{p}_w + \vartheta_{kw,is}\,\dot{q}_i\,\dot{q}_s\,p_w + \vartheta_{kw,i}\,\ddot{q}_i\,p_w)\} + \\
& \omega_r^2\,p_r + \eta_k\,\vartheta_{kr} = 0, \\
& \vartheta_{hz}\,\{s_{ih}\,\ddot{q}_i + m_{hk,s}\,\dot{q}_s\,\dot{\xi}_k + s_{ik,s}\,\dot{q}_s\,\dot{q}_i + \\
& m_{hk}\,(2\,\vartheta_{kw,s}\,\dot{q}_s\,\dot{p}_w + \vartheta_{kw,is}\,\dot{q}_i\,\dot{q}_s\,p_w + \vartheta_{kw,i}\,\ddot{q}_i\,p_w)\} + \\
& 2\,\zeta_z\,\omega_z\,\dot{\phi}_z + \omega_z^2\,\phi_z + \eta_k\,\vartheta_{kr} = 0,
\end{aligned}$$

where

$$\begin{aligned}
& \xi_j = \vartheta_{jw}\,p_w, \quad \dot{\xi}_j = \vartheta_{jw}\,\dot{p}_w + \vartheta_{jw,i}\,\dot{q}_i\,p_w, \\
& \ddot{\xi}_j = \vartheta_{jw}\,\ddot{p}_w + \vartheta_{jw,i}\,\ddot{q}_i\,p_w + 2\,\vartheta_{jw,i}\,\dot{q}_i\,\dot{p}_w + \vartheta_{jw,is}\,\dot{q}_i\,\dot{q}_s\,p_w,
\end{aligned}$$

and the indexes r, w and z are limited to

$$r, w = 1, \ldots, m^*\text{-}1, \qquad z = m^*, \ldots, m.$$

The fast system is given by

$$\chi_k + 2\,\zeta_k\,\omega_k\,\chi_k = 0,$$

where χ_k is the fast transient of ϕ_k, $k = m^*, \ldots, m$. Since the fast system is globally exponentially stable, Tikhonov's Theorem [23] applies and the solutions to the slow system approximate the unperturbed system trajectories within an ε-approximation.

From this analysis one sees that if the commanded torques are not function of the higher modes of oscillation, no instability of these modes occurs. This result is consistent with the one reported in [19].

This rule can be improved according to the following considerations. In case of regulation for $t \rightarrow \infty$, since the slow transient is exponentially stable, stability is preserved as long as the disturbances, arising from a functional dependency of the control on ϕ, are sufficiently small. Similarly, in case of learning control, the discrete dynamics of ϕ are uniformly convergent if no functional dependency of the updated law on $\phi(T)$ is allowed. Indeed, in this case, the dynamics of ϕ are linear and their eigenvalues are zero. This implies that stability is preserved [17] whenever the disturbance due to a functional dependency on $\phi(T)$ is sufficiently small.

In conclusion, feedback gains from high frequency neglected modes must be small; condition that has to be satisfied by proper modelling, sensors location and control loop implementation, without adversely affecting closed loop behavior.

It is also worth to notice that the perturbation induced by the neglected dynamics is nonpersistent. For, suppose that at the end of the time interval the desired output is exactly reproduced and the measured velocities and accelerations are zero. In this case the control is not updated and at the next trial the same output and measured velocities and accelerations at time T are obtained. This means that the set of points characterized by exact output reproduction and zero measured velocities and accelerations is an invariant set with respect to the learning dynamics. However, it should not be overlooked the fact that the physical system does not reach an equilibrium point, since the state derivative of the neglected dynamics does not necessarily vanish, being in general of order ε.

II.3 The material so far presented allows the development of a learning scheme for the control of a flexible mechanism, provided that the conditions of Theorem 1 or similar conditions are satisfied.

Theorem 2

Suppose that the output function is continuously differentiable and $\partial h/\partial q$ is invertible. Under this hypothesis there exists $\tau^0(t)$, $t \in [0, T]$, which accomplishes output regulation at time T of the mechanism with infinitely stiff bodies. Let $\pi(t)$ be the state trajctory corresponding to $\tau^0(t)$.

Assume that the dynamics of the linearized system continuously depend on the state in a neighborhood of $\pi(t)$.

Suppose that the system linearized at $\pi(t)$ is controllable. Then, for sufficiently small elastic displacements, the iterative scheme 2.2, with starting point $\tau^0(t)$, is convergent.

Proof

Since the linearized system continuously depend on the state in a neighborhood of $\pi(t)$ and is controllable, the first three hypothesis of Theorem 1 hold.

Moreover, because of DL(u) is continuous at u_0, with reference to Theorem 1, there exists a constant γ such that $\alpha \leq \gamma\, \delta$.

Let Q: $R^n \times R^p \rightarrow U$ be a linear operator such that $[DL(u_0)\, Q]$ is invertible and consider the term

$$v_1 = - [DL(u_0)\, Q]^{-1} L(u_0).$$

This is arbitrarily small for arbitrarily small elastic displacements. But from 2.3 the lowest value of δ for small $|v_1|$ is $\delta \approx |v_1|$. Hence, for sufficiently small elastic displacements, $\alpha < 1$ and the iterative scheme 2.2 is convergent. ◊

Under the hypothesis of Theorem 2, a way of estimating the Fréchet derivative without integrating the equations motion is the following. Let $\theta(t)$, $t \in [0, T]$, be a vector of joint trajectories, two times differentiable, which accomplish output regulation at time T of the mechanism with infinitely stiff bodies. The corresponding torque is given by

$$\tau_i^0 = b_{ij}\, \ddot{\theta}_j + (b_{ij,s} - 0.5\, b_{sj,i})\, \dot{\theta}_s\, \dot{\theta}_j + g_{,i},$$

where the q-dependent tensors are evaluated at $\theta(t)$. Make the change of coordinates $q(t) = \theta(t) + \delta q(t)$ and, consistently with the hypothesis of small elastic motion, neglect second order terms in δq, x and their derivatives to rewrite the equations of motion as

$$b_{ij}\, \delta\ddot{q}_j + s_{ik}\, \ddot{x}_k + (b_{ij,s} - 0.5\, b_{sj,i})\, \delta\dot{q}_s\, \dot{\theta}_j + (s_{ik,s} - s_{sk,i})\, \dot{\theta}_s\, \dot{x}_k +$$
$$b_{ij,k}\, \ddot{\theta}_j\, \delta q_k + s_{ik}\, \ddot{x}_k + (b_{ij,sk} - 0.5\, b_{sj,ik})\, \dot{\theta}_s\, \dot{\theta}_j\, \delta q_k +$$
$$(b_{ij,s} - 0.5\, b_{sj,i})\, \dot{\theta}_s\, \delta\dot{q}_j + g_{,ik}\, \delta q_k + \eta_{k,i}\, x_k = \tau_i - \tau_i^0 = \delta\tau_i$$
$$m_{hk}\, \ddot{x}_k + s_{ih}\, \delta\ddot{q}_i + s_{ih,k}\, \ddot{\theta}_i\, \delta q_k + m_{hk,s}\, \dot{\theta}_s\, \dot{x}_k +$$
$$s_{ih,sk}\, \dot{\theta}_s\, \dot{\theta}_i\, \delta q_k + s_{ih,s}\, (\delta\dot{q}_s\, \dot{\theta}_i + \dot{\theta}_s\, \delta\dot{q}_i) + c_{hk}\, \dot{x}_k + k_{hk}\, x_k +$$
$$\eta_{h,k}\, \delta q_k = - \eta_k - s_{ih,s}\, \dot{\theta}_s\, \dot{\theta}_i - s_{ih}\, \ddot{\theta}_i,$$

where the q-dependent tensors are evaluated at $\theta(t)$. Using this linearized system an estimate of $DL(\tau^0)$ is obtained.

III. AN EXAMPLE

To illustrate the learning control and the spillover effect, an example is presented. Consider the planar mechanism depicted in Fig.1: a one link flexible arm. It consists of two equal rigid rods with revolute joints. A damped spring of stiffness k and damping coefficient c is placed in the joint connecting the two rods. The relative clockwise rotation (x) of the rods is assumed small. A control torque (τ) is applied at the joint connecting the mechanism to the basement. The clockwise rotation of this joint is denoted by q. The constant gravity field is parallel to the plane of motion. The point of minimum gravitational potential energy is q = 0, x = 0.

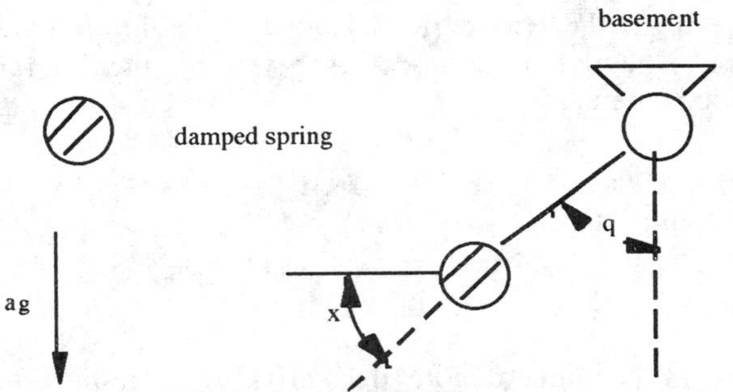

Fig.1 - A one link flexible arm

Let $J = (m\, l^2)/3$ denote the moment of inertia of a rod with respect to one of its end and $S = m\, l/2$ its static moment, where m is the total distributed mass of the rod and l its length. For small elastic rotations, the equations of motion are given by

$$J_t\, \ddot{q} + (J + l\, S)\, \ddot{x} + m\, l\, a_g\, (2 \sin(q) + 0.5\, x \cos(q)) = \tau,$$
$$(J + l\, S)\, \ddot{q} + J\, \ddot{x} + c\, \dot{x} + k\, x + 0.5\, m\, l \sin(q) = 0,$$

where $J_t = (2\, J + m\, l^2 + 2\, l\, S)$ and a_g is the gravity acceleration. As output the angular end point position is taken

$$z = q + 0.5\, x.$$

Consider the design of a learning regulator by neglecting the flexibility, that is on the basis of the equation

$$J_t\, \ddot{q} + 2\, m\, l\, a_g \sin(q) = \tau,$$

Let

$$\tau = 2\, m\, l\, a_g \sin(q) + J_t\, w$$
$$\dot{w} = v, \quad w(0) = 0,$$

one has

$$\frac{d^3 q}{dt^3} = v, \quad \ddot{q}(0) = 0.$$

Compute v(t), $t \in [0, T]$,0 so that

$$z(T) = z°, \quad \dot{z}(T) = 0, \quad \ddot{z}(T) = 0,$$

with z° the output to be reached at time T. These conditions would imply the equilibrium if the mechanism was rigid. Suppose that at the end of the first trial

$$z_0(T) \neq z°, \quad \dot{z}_0(T) \neq 0, \quad \ddot{z}_0(T) \neq 0,$$

where the subscript denotes the trial number, and update the control by inverting the linear system

$$\frac{d^3\delta q}{dt^3} = \delta v, \quad \delta \ddot{q}(0) = 0, \quad \delta\dot{q}(0) = 0, \quad \delta q(0) = 0,$$

so that at time T

$$\delta q(T) = - z_0(T) + z°, \quad \delta\dot{q}(T) = - \dot{z}_0(T), \quad \delta \ddot{q}(T) = - \ddot{z}_0(T).$$

By iterating this procedure, a learning algorithm of the type introduced in section II is obtained.

More than one control steers the above linear system. One of this is given by

$$\delta v(t) = 6\, a + 24\, b\, t + 60\, e\, t^2,$$

to which corresponds the trajectory

$$\delta q(t) = a\, t^3 + b\, t^4 + e\, t^5.$$

At the k-th trial the coefficients a, b and e are updated by solving

$$a\, T^3 + b\, T^4 + e\, T^5 = - z_k(T) + z°,$$
$$3\, a\, T^2 + 4\, b\, T^3 + 5\, e\, T^4 = - \dot{z}_k(T),$$
$$6\, a\, T + 12\, b\, T^2 + 20\, e\, T^3 = - \ddot{z}_k(T).$$

This learning control scheme has been numerically simulated with $m = 2$ kg, $l = 0.5$ m, $a_g = 10$ m/sec^2. The elastic constant has been varied from 20 N/rad to 1 N/rad and the damping coefficient set equal to $0.02 \sqrt{(k\ J)}$. The mechanism was required to move from the rest position $q = 0$, $x = 0$ at time zero, to the end point position $z = \pi/2$, with zero end point velocity and acceleration, at time $T = 1$ sec. Beside from the neglected dynamics, no other perturbations have been considered. However, the same qualitative results hold for any sufficiently small nonpersistent perturbations.

All numerical simulations reported have been performed with the MATLAB® routine "ode45", which uses a Runge-Kutta-Fehlberg integration method. The relative tolerance of accuracy was set equal to 10^{-7}.

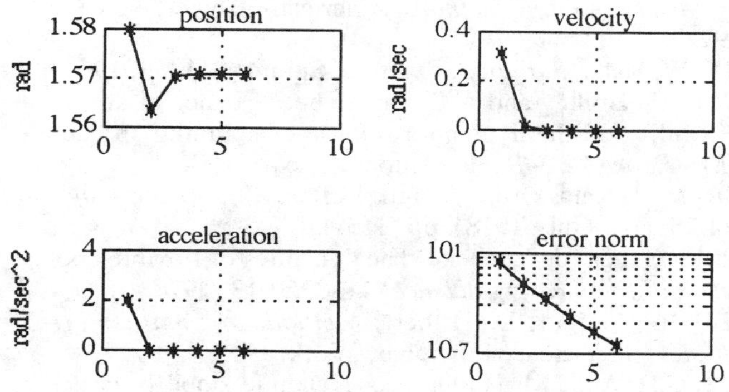

Fig. 2 - k = 20. End point angular position, velocity, acceleration and error norm versus trials

In Fig.2, the end point angular position, velocity and acceleration, with k = 20, are reported.

Convergence is quite fast in spite of the neglected dynamics. The end point desired position is practically achieved in three iterations, with zero velocity and zero acceleration. The semi-logarithmic graph of the euclidean norm

$$\sqrt{[z(T) - \pi/2]^2 + [\dot{z}(T)]^2 + [\ddot{z}(T)]^2}$$

versus the trial numbers evidences the rate of convergence. The amplitude of the elastic rotation during the movement was less than 0.25 rad, consistently with the hypothesis of small elastic rotation.

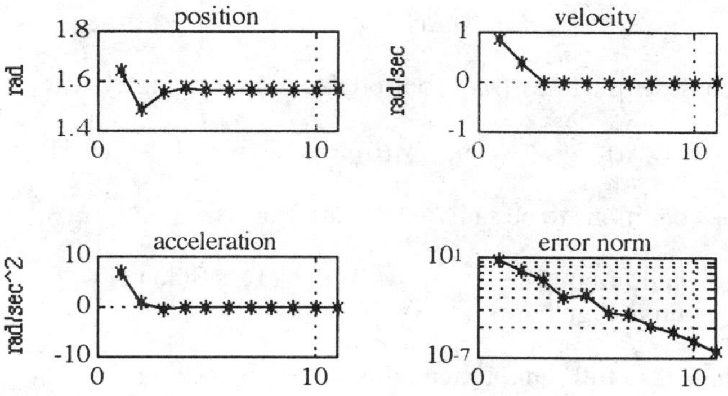

Fig. 3 - k = 5. End point angular position, velocity, acceleration and error norm versus trials

A slower rate of convergence was achieved with k = 5, as shown in Fig. 3. In this case, the maximum amplitude of the elastic rotation during the movement was about 1 rad, which is somewhat large to be considered small.

With k = 1 convergence was lost, value for which however the elastic rotation during the movement cannot be considered small. For greater repositioning times, smaller elastic rotations are expected and hence convergence with smaller stiffness values.

Even if convergence is achieved with k as low as 5 N/rad, due to the neglected dynamics, the mechanism is not truly at rest at time T = 1 sec. However, the larger the stiffness the smaller the norm of the state derivative of the mechanism as shown in Fig.4.

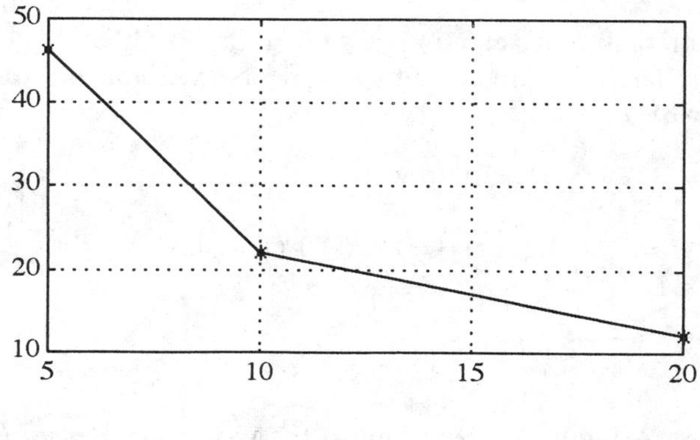

Fig.4 - Norm of arm state derivative at time T versus stiffness (N/rad)

IV. SUMMARY

The problem of regulating the output of a flexible mechanism by means of learning algorithms has been discussed.

A general methodology has been presented and its applicability in the face of neglected high frequency dynamics investigated.

The learning algorithm introduced is of the gradient type and its convergence has been at least proven in the case of sufficiently small elastic displacements.

The performance of the proposed learning algorithm has been illustrated by means of an academic example.

APPENDIX

Consider the linearized, controllable, time varying system

$$\dot{w} = A(t)\, w + B(t)\, u, \quad w(0) = w^{\circ}, \quad w \in R^{p}, \quad u \in R^{n}$$

The conditions to be satisfied at time T are

$$\begin{aligned} &\text{equilibrium:} && A(T)\, w(T) + B(T)\, u(T) = 0, \\ &\text{output regulation:} && C\, w(T) = y \in R^{n}, \end{aligned}$$

where C is full rank. Extend this system by setting

$$\dot{u} = v, \quad u(0) = u^{\circ}.$$

Let

$$z = \begin{pmatrix} x \\ u \end{pmatrix}$$

be the extended state and denote by B_e the new input matrix and by $\Phi(t, s)$ the state transition matrix of the extended system. One has

$$z(T) = \Phi(t, 0)\, z^{\circ} + \int_0^T \Phi(t, s)\, B_e(s)\, v(s)\, ds.$$

Assume that the square matrix

$$C_e = \begin{pmatrix} C & 0 \\ A(T) & B(T) \end{pmatrix}$$

is full rank and set $v(s) = B_e'(s)\, \Phi'(T, s)\, C_e'\, \chi$. Since controllability is preserved by integrators extension, one can solve for χ

$$\chi = W^{-1}\, (y - C_e\, F(t, 0)\, w^{\circ}),$$

$$W = C_e \int_0^T \Phi(T, s)\, B_e(s)\, B_e'(s)\, \Phi'(T, s)\, ds\, C_e'.$$

REFERENCES

[1] S. Arimoto, S. Kawamura, F. Miyazaki, Bettering operations of robots by learning, *J. Robotic Systems*, vol. 1 (1984), pp. 123-140

[2] J.J. Craig, Adaptive control of manipulators through repeated trials, *Proc. Amer. Control Conf.*, 1984

[3] G. Casalino, G. Bartolino, A learning procedure for the control of movements of robotic manipulators, *Proc. IASTED Symp. on Robotics and Autom.*, Amsterdam, The Netherlands, June 1984

[4] S. Arimoto, S. Kawamura, F. Miyazaki, S. Tamaki, Learning control theory for dynamical systems, *Proc. 24th IEEE Conf. on Decision and Control*, 1985

[5] J. Craig, *Adaptive control of mechanical manipulators*, Addison-Wesley, NY (1988)

[6] S. Kawamura, F. Miyazaki, S. Arimoto, Application of learning method for dynamic control of robot manipulator, *Proc. 24th IEEE Conf. on Decision and Control*, 1985

[7] S. Arimoto, S. Kawamura, F. Miyazaki, Realization of robot motion based on the learning method, *IEEE Systems Man Cybernetics*, vol. 18 (1988), pp. 126-134

[8] M. Togai, O. Yamano, Analysis and design of an optimal learning control scheme for industrial robots: a discrete system approach, *Proc. 24th IEEE Conf. on Decision and Control*, 1985

[9] T. Mita, E. Kato, Iterative control and its application to motion control of robot arm - A direct approach to servo-problems, *Proc. 24th IEEE Conf. on Decision and Control*, 1985

[10] L. Hideg, R. Judd, Frequency domain analysis of learning systems, *Proc. 27th IEEE Conf. on Decision and Control*, 1988

[11] J. Hauser, Learning control for a class of nonlinear systems, *Proc. 26th IEEE Conf. on Decision and Control*, 1987

[12] L. Hideg, R. Judd, R. Van Til, Stability analysis for learning systems, *Proc. American Control Conf.*, 1990

[13] K.L. Moore, M. Dahleh, S.P. Bhattacharya, Iterative learning for trajectory control, *Proc. 28th IEEE Conf. on Decision and Control*, 1989

[14] P. Lucibello, Inversion of linear square systems by learning, *Proc. 30th IEEE Conf. on Decision and Control*, 1991

[15] P. Lucibello, Point to point control of linear systems by learning, *Univ. Calabria, Dip. Sistemi, Internal Report*, 1991

[16] S.M. Roberts, J.S. Shipman, *Two-point boundary value problems: shooting methods*, American Elsevier, New York, 1972

[17] W. Hahn, *Stability of motion*, Spinger-Verlag, 1967

[18] C.S. Bodley and A.C. Park, The influence of structural flexibility on the dynamic response of spinning spacecraft, *AIAA Paper 72-348*, San Antonio, Texas, 1972

[19] M. J. Balas, Active control of flexible systems, *JOTA*, vol.25, no.3 (July 1978), pp. 415-436

[20] B. Fraeus de Veubeke, The dynamics of flexible bodies, *Int. J. of Engineering Science*, 14, 895-913, 1976

[21] R. Courant, D. Hilbert, *Methods of mathematical physics*, Interscience Publ., New York, 1953

[22] C.H. An, J.M. Hollerbach, Dynamic stability issues in force control of manipulators, *Proc. IEEE Conf. on Robotics and Automation*, 1987

[23] P. Kokotovic, H.K. Khalil, J. O'Reilly, *Singular perturbation method in control: analysis and design*, Academic Press, 1986

Proceedings of the 31st Conference
on Decision and Control
Tucson, Arizona • December 1992

A Hybrid Control Architecture with Fuzzy Interface for Intelligent Control

F. Lin, J. Sun and L. Y. Wang

Department of Electrical and Computer Engineering
Wayne State University, Detroit, MI 48202

Abstract

We propose a new control architecture for hybrid systems with both continuous variable and discrete-event components. The architecture includes a conventional controller, a supervisor and a fuzzy interpreter. The conventional controller is used to control the continuous variable part of the system, while the task of the supervisor is to control the discrete-event part of the system and to make decisions on which control algorithm to be used by the conventional controller. The fuzzy interpreter links the conventional controller to the supervisor. Our approach is suitable for systems which operate at very different operating conditions and need to have proper decision making mechanisms.

1 Introduction

This paper aims at the development of a paradigm of intelligent control systems that can operate under a wide range of operating conditions and involve man-machine interactions. Several approaches to intelligent control have been proposed, for example, in [1,2 and the reference therein]. Our approach can be used to control hybrid systems with different types of uncertainties and design specifications. The operation of such systems involves the processing of two different types of signals: continuous variable signals (conventional signals for short) and discrete-event signals (event signals for short). The conventional signals are in the form of numerical values and represent, say, currents, voltages, displacements, velocities, etc.. The event signals, on the other hand, are in linguistic or logical form and can represent occurrences of events such as failures in components, connection or disconnection of communication channels, implementation of control algorithms, etc..

Control of a complex hybrid system involves both conventional continuous variable control and discrete-event control. The control objectives and methods are very different for conventional control and discrete-event control. In conventional control, the objective is to stablize or to optimize the system subject to uncertainties that are usually topological (i.e. measured by norms or metrics). The objective of discrete-event control is to ensure the system's optimality and/or reliability under ever changing operating conditions and/or failures in components. Conventional control has traditionally been studied using transfer functions and state space representations. Many approaches have been proposed and studied intensively. A wide range of tools for conventional controller design are available[5,6,10,12]. Systematic studies of discrete-event systems, on the other hand, started only about ten years ago. During the past ten years, several approaches has been proposed including supervisory control [4,8,9,11]. In supervisory control, a discrete-event system to be controlled is modeled by an automaton. The behavior of the system is usually described by the language generated by the automaton. Each sequence in the language represents a possible trajectory of the system. A supervisor is then designed, whose task is to enable or disable some controllable events based on a record of occurrences of some observable events, so that the closed loop system achieves a given control objective. In supervisory control, the control objective is also described by languages. This framework captures many aspects of discrete-event systems and will be used in this paper.

The main control architecture proposed in this paper for tackling hybrid systems consists of a supervisor and a conventional controller. Since they deal with different aspects of the system and process information and signals using totally different methods, an interface has to be deviced to link them. How to interpret a conventional signal is a problem to be resolved by the interface. We will use a fuzzy interpreter to serve as the interface. The fuzzy interpreter takes continuous signals as its input and produces appropriate event signals to the supervisor. For example, a sudden change in the output of the system may imply the failure of a component. Therefore this

CH3229-2/92/0000-2539$1.00 © 1992 IEEE

conventional signal may infer an event signal such as "machine A is broken" or " channel B is disconnected". Form the supervisor's point of view, this conventional signal is fuzzy because several possible interpretations may be drawn from it. Therefore, the fuzzy logic is a natural tool to design and implement the interpreter.

In Section 2, we will propose a control architecture that includes a supervisor, a conventional controller and a fuzzy interpreter. In Section 3, we will describe the mathematical model of each part and interconnections between them. Since the design of the conventional controller is well understood, we will concentrate our discussions on the designs of the supervisors and the fuzzy interpreter. So in Section 4, we will discuss the issues related to the design of the supervisor. In Section 5, we will describe the fuzzy interpreter and various ways of defuzzification. Finally, in Section 6, we will discuss issues related to our approach.

2 Control Architecture

In this paper, a new architecture for the control of hybrid systems will be developed which consists of conventional controllers, supervisors, and interpreters. A typical example of the architecture is shown in Figure 1.

Including both conventional and discrete-event control in the architecture is an obvious consequence of the hybrid nature of most practical systems to be controlled. The conventional controller is designed by traditional methods to control the continuous variable part of the system. The supervisor acts like a decision maker. It monitors the structural changes of the system caused by component failures and dramatically changing operating condition, and controls the discrete-event part of the system and commands the reconfiguration of the conventional controller. Consequently, the output of the supervisor has two functions: one is to control the discrete-event part of the system and the other is to direct the configuration mechanism of the conventional controller.

The necessity of introducing interpreters in the architecture stems from the fact that a supervisor and a conventional controller in a hybrid control system deal with different aspects of the system, and process information and signals using totally different methods. Hence an interface must be devised to link them. The interpreter takes conventional signals as its input and produces appropriate event signals to the supervisor. It also translates control commands from the supervisor into suitable control actions for the conventional control subsystems, including algorithm selection, structure reconfiguration, operating point transitions, etc. For example, a sudden change in the measurement of a conventional signal may imply the failure of a component. In other words, this conventional signal may imply an event such as "machine A is broken" or " channel B is disconnected." We will use a fuzzy logic approach to model the interpreter. The rationale for this is as follows. First, from the supervisor's point of view, event information contained in the conventional signal is fuzzy, because several possible interpretations may be drawn from observations of the conventional signal. Second, conventional models of observers and estimators depend on system descriptions using differential or difference equations. As a result, they are not readily extendible to observers in discrete-event systems, let alone to hybrid systems. By contrast, both conventional and discrete-event observers can be unified in a fuzzy logic framework. Finally, a fuzzy logic framework can potentially integrate human intelligence into the observer design. Therefore, fuzzy logic is a natural tool for modeling and implementing the interpreter.

The conventional and discrete-event parts of a hybrid system are tightly coupled. For instance, the essential notion of states in the discrete-event part depends heavily on the conventional part. Each state in the discrete-event part represents a distinct operating point of the conventional part. The transition among operating points of the system can be caused by sources of different origins. For example, an automobile engine often operates under very different weather conditions and speeds. Variations in temperature or engine speed change the operating points of the engine, and in turn its dynamics and characterization. Although the engine operating points may drift continuously, to reduce the complexity of control design we may divide the set of all possible operating points into a finite number of subsets. A typical operating point is then selected in each subset, producing a state in the discrete-event part. Similar situations can develop due to failures of components, such as a flat tire or a failure in a fuel injector. Based on the information on states and events from the interpreter, the supervisor can then take appropriate control actions.

3 A Mathematical Model for Hybrid Systems

The class of hybrid systems we will consider is described by the following hybrid system models. In the rest of this paper, conventional control systems and discrete-event systems will be abbreviated by CCS and DES, respectively.

CCS is modeled by an internal state-space model

$$\dot{x}(t) = A(s_t)x(t) + B(s_t)u(t)$$
$$y(t) = C(s_t)x(t),$$

where $u \in R^m, y \in R^k, x \in R^n = X$ are input, output and state, respectively. We assume that the system starts at $t = t_0$ with initial condition $x(t_0) = x_0$. s_t in the model is the trajectory of discrete events in the time interval $[0, t]$, and represents the dependence of the conventional part of the hybrid system on the discrete-event part. For instance, a breakdown of a communication channel is a discrete-event which changes parameters in A, B, and C.

We use formal languages to describe a DES G. Let Σ be a nonempty finite alphabet, which represents events. We denote by Σ^* the set of all finite strings of elements of Σ, including the empty string ϵ. A string represents a possible trajectory of the system. Any subset of Σ^* is called a language over Σ. The set of all possible free trajectories of the system G, defined as the uncontrolled process language $L(G)$, describes the behavior of the uncontrolled DES.

The dynamics of DES is governed by state transition function $\delta : \Sigma \times Q \times X \times R \to Q$, where Q is the set of DES states with initial state q_0, R is the time interval $[0, \infty)$. For a given history of DES states $q_t = \{q(t_i) \in Q, i = 1, \cdots, n;\ \ t_i \in [0, t]\}$ and CCS states $x_t = \{x(\tau) \in X, \tau \in [0, t]\}$, and current event $\sigma \in \Sigma$, δ determines the next DES state of the system. The explicit inclusion of $X \times R$ in δ indicates the effect of CCS on DES in a hybrid system. The state history $\omega_t = (q_t, x_t)$ induces a trajectory of events of DES. Such induction is characterized by an event mapping

$$\mu : Q \times X \times R \to \Sigma^*.$$

Let $\Omega_t := Q \times X \times [0, t]$. Then, for $\omega_t \in \Omega_t$, $s_t = \mu(\omega_t)$ is an event trajectory in $[0, t]$.

4 Supervisor

For a system with event set Σ and uncontrolled process language $L(G)$, the action of a supervisor may be characterized by a mapping

$$\gamma : L(G) \to 2^{\Sigma}$$

That is, a supervisor assigns an enabled event set to each string $s \in L(G)$. A supervisor can be realized by a recognizer and a feedback map from states to the power set of Σ ([9,11]). In the sequel, for simplicity, we do not distinguish between a supervisor and the characterizing mapping γ.

Since some events are disabled, the language generated by the closed-loop system $L(G, \gamma)$ is defined as follows:

$$\epsilon \in L(G, \gamma)$$
$$(\forall s \in L(G, \gamma)) s\sigma \in L(G, \gamma) \Leftrightarrow \sigma \in \gamma$$

In other words, a string $s\sigma$ is a possible trajectory in $L(G, \gamma)$ if and only if s is a possible trajectory and σ is enabled by γ.

Our objective is to design a supervisor that guarantees the stability of the hybrid system (assuming that conventional controllers are properly designed) and achieves optimality if possible. To formally specify this objective, we denote by L_1 the set of event trajectories under which the CCS is controllable and observable; and L_2 the set of event trajectories under which the CCS is stabilizable and detectable. Formally,

$$L_1 = \{s_t \in L(G) : (A(s_t), B(s_t)) \text{ is controllable and } (A(s_t), C(s_t)) \text{ is observable }\}$$

$$L_2 = \{s_t \in L(G) : (A(s_t), B(s_t)) \text{ is stabilizable and } (A(s_t), C(s_t)) \text{ is detectable }\}.$$

Realization of L_2 by supervisory control will guarantee stabilizability of the CCS by output feedback, while that of L_1 will provide design freedom in the CCS to achieve performance. We call L_1 and L_2 the *desirable language* and the *tolerable language* respectively. The control objectives of the supervisor therefore are to ensure that the controlled system

(1) Never goes beyond the tolerable language L_2;

(2) Achieves as much as possible of the desirable language L_1 under (1);

(3) Tolerates as little as possible under (2).

To achieve these objectives is a nontrivial task because not all events are controllable: some transitions from one state to another may not be preventable (e.g., a failure may not be preventable). Thus we partition the event set Σ into controllable event set Σ_c and uncontrollable event set Σ_{uc}: $\Sigma = \Sigma_c \dot{\cup} \Sigma_{uc}$ and introduce controllability [11] of a language. Intuitively, a language is said to be controllable if uncontrollable events satisfy the following condition: After a "good" trajectory, no uncontrollable event that can possibly occur will lead to a "bad" trajectory. Formally, a language K is controllable (with respect to $L(G)$ and Σ_{uc}) if

$$(\forall s \in K)(\forall \sigma \in \Sigma_{uc}) s\sigma \in L(G) \Rightarrow s\sigma \in K$$

It has been shown that a supervisor γ exists such that $L(G, \gamma) = K$ if and only if K is controllable.

If K is not controllable, then we can find the supremal controllable sublanguage of K, which is the largest subset of K satisfying the controllability condition, or we can find the infimal controllable superlanguage of K, which is the smallest controllable superset of K satisfying the controllability condition. The supremal controllable sublanguage of K is denoted by $K^{\uparrow}$ and the infimal controllable superlanguage of K is denoted by $K^{\downarrow}$. As shown in [8], the above control objectives can be achieved by a supervisor γ satisfying

$$L(G, \gamma) = (B_1 \cap B_2^{\uparrow})^{\downarrow}.$$

5 Fuzzy Interpreter

The discrete sensors for supervisors are installed to detect the states and occurrence of events for the supervisor. In many practical applications, however, the exact event or state, such as failures of components or the operating point of systems, are rather difficult and/or expensive to detect directly on-line. Instead, they are often inferred from some conventional signals such as the error, the derivative of the error, the required control, etc. Here, we discuss a fuzzy interpreter which interprets the discrete state and occurrence of events from conventional signals.

Like most fuzzy controllers, the fuzzy interpreter consists of three parts: fuzzification, fuzzy associate memory (FAM or fuzzy interpretation rules) and the defuzzification scheme, as shown in Figure 2. As we will show later, depending on the situation, the defuzzification can be performed before or after the supervisor, or combined with the supervisor. For each numerical value which is sampled from the conventional signal $y(t)$, the fuzzifier translates it into fuzzy sets $Y_1, Y_2, \cdots, Y_s$ and their fit values $m_1, m_2, \cdots, m_s$ respectively, where Y_i corresponds to linguistic terms like ZERO, POSITIVE, NEGATIVE, etc.. The FAM is a bank of interpretation rules, which are in the IF-THEN form as:

$$\text{If } y_1 = Y_{1i}, y_2 = Y_{2j}, \cdots, \text{ then } q = q_k, \sigma = \sigma_l$$

where $q_k \in Q$ is a discrete state in the supervisor and $\sigma_l \in \Sigma' = \Sigma \cup \{\epsilon\}$ is an event or the empty sequence ϵ to denote that no event has occurred. For example, if the continuous inputs to the fuzzy interpreter are error e and its derivative $\dot{e}$, and their fuzzy sets and fuzzy set values are specified in Figure 3, then the rule "If $e = Z, \dot{e} = N$ then $q = q_1, \sigma = \sigma_3$ " corresponding to the linguistic statement: "if error is small (or zero) and its derivative is negative, then the state is q_1 and the event σ_3 has occurred". The output of the FAMs are fuzzy sets which define the possible states and events occurred together with a confidence value, that is

$$q = (q_1/\alpha_1, q_2/\alpha_2, \cdots, q_n/\alpha_n)$$

$$\sigma = (\sigma_1/\beta_1, \sigma_2/\beta_2, \cdots, \sigma_m/\beta_m)$$

The confidence values α_i, β_j are calculated using fuzzy operations. A typical scheme in calculating α_i, β_j is the min-max scheme, see [4,7] for details. In general α_i, β_j are nonzero, which indicates that more than one event and state can be possibly interpreted from the signal. It should be noted that the states q_i and events σ_j are nonfuzzy to the supervisor, but they are treated as fuzzy sets in the interpreter.

Since the supervisor can only accept discrete states and events, the defuzzification scheme in Figure 2 is different from the conventional defuzzifier used in fuzzy controllers which generally results in a crisp output value. We use the following defuzzification schemes:

- Maximum confidence defuzzification: we encode the fuzzy sets q, σ according to the confidence of q (or σ) in the fuzzy sets q_i (or σ_j). The set corresponding to the maximum confidence value is used as the output value of the defuzzifier and is sent to the supervisor, i.e.,

  $$q_d = q_m, \quad m = arg\{max_i(\alpha_i)\}$$
  $$\sigma_d = \sigma_n, \quad n = arg\{max_j(\beta_j)\}$$

  We can set a threshold for the confidence value in the defuzzifier. If the maximum confidence is less than the threshold value, the defuzzifier will send a signal to the supervisor indicating that the interpretation is not very reliable and request more data to be sampled.

- Most-likelihood centroid defuzzifier: Since all the states in Q and events in Σ, which are defined in the supervisor, are considered as fuzzy set in the interpreter, we can define the "universe of discourse" $\bar{Q}, \bar{\Sigma}$ for Q and Σ. An example is shown in Figure 4. It should be noted that the quantification of the fuzzy numbers in Figure 4 is rather fictitious, it only facilitates the defuzzification process. As far to the questions such as what shape the membership function should take, how much overlap there should be between the neighbors, whether q_1 should be next to q_2 instead of q_3, etc., the answers will depend on the problems at hand. Here, for simplicity, we assume that all membership functions assume the triangular symmetric form. Then according to the conventional centroid defuzzification scheme, we calculate:

$$\bar{q} = \frac{\sum \alpha_i \bar{q}_i}{\sum \alpha_i}, \quad \bar{\sigma} = \frac{\sum \beta_j \bar{\sigma}_j}{\sum \beta_j}$$

where $\bar{q}_i$ is the value of $\bar{Q}$ at which the fuzzy number $\bar{q}$ assumes full member of q_i. (Since we are using the triangular membership, $\bar{q}$ is the full member of q_i only at $\bar{q} = \bar{q}_i$). We then interprete the discrete state or event from the centroid value $\bar{q}, \bar{\sigma}$ by associating it with the closest fuzzy set q_i, i.e.,

$$q_d = q_m, \quad m = arg\{min_i \|\bar{q} - q_i\|\}$$
$$\sigma_d = \sigma_n, \quad n = arg\{min_j \|\bar{\sigma} - \sigma_j\|\}$$

for some fuzzy norm $\|\cdot\|$. When q_i, σ_j all have triangular symmetric membership functions, we can just use the central values $\bar{q}_i, \bar{\sigma}_j$ and the outputs of the defuzzifier become:

$$q_d = q_m, \quad m = arg\{min_i |\bar{q} - \bar{q}_i|\}$$
$$\sigma_d = \sigma_n, \quad n = arg\{min_j |\bar{\sigma} - \bar{\sigma}_j|\}$$

If $\bar{q}, \bar{\sigma}$ is in equal distance to two different states or events, then a signal is sent to the supervisor indicating that the interpretation is impossible and more sample data is needed. For example, if in Figure 4 the output of the FAM is given by

$$q = (q_1/0, q_2/0.2, q_3/0.4, q_4/0.5, q_5/0.2)$$

then according to the conventional centroid defuzzification scheme,

$$\bar{q} = \frac{0\times(-2)+0.2\times(-1)+0.4\times0+0.5\times1+0.2\times2}{0+0.2+0.4+0.5+0.2} = 0.46$$

Therefore, the output of the defuzzifier to the supervisor is

$$q_d = q_3$$

If the maximum confidence is used, then

$$q_d = q_4$$

- Supervised defuzzification. The defuzzification can be integrated with the supervisor such that the supervisor can take action based on the fuzzy input from the FAMs. Three possible supervised defuzzifications are proposed as follows:

  – Disjunction defuzzification: Consider a special case where the supervisor $\gamma : Q \to 2^{\Sigma}$ is a state feedback control. The for the disjunctive defuzzification we take $\gamma(q)$ to be the union of $\gamma(q_i)$, where q_i is a possible state with confidence greater than a specified threshold α_o.

  $$\gamma(q) = \bigcup_{\alpha_i > \alpha_o} \gamma(q_i)$$

  – Conjunction defuzzification: We take $\gamma(q)$ to be the intersection of $\gamma(q_i)$, where q_i is a possible state with confidence greater than a specified threshold α_o.

  $$\gamma(q) = \bigcap_{\alpha_i > \alpha_o} \gamma(q_i)$$

  – Linear interpolation: If $\gamma(q_i)$ is to employ a specified conventional control algorithm, then the supervisor can, instead of using a single control algorithm, combine these control algorithms which would be employed under the states q_i, where q_i is a possible state with confidence greater than a specified threshold α_o.

  $$f(\gamma(q)) = \frac{\sum_{\alpha_i > \alpha_o} \alpha_i f(\gamma(q_i))}{\sum_{\alpha_i > \alpha_o} \alpha_i}$$

  where $f(\gamma(q_i))$ is the output of the conventional controller using the algorithm specified by $\gamma(q_i)$.

6 Conclusion

This paper exploits a hybrid control architecture consisting of a supervisor, a conventional controller, and a fuzzy interpreter. The reason for using a supervisor is to introduce decision making into the control loop and hence bring "intelligence" into the system. This architecture will allow the system to handle uncertainties. In practical control problems, one encounters structured and unstructured uncertainties, which change system dynamics significantly during the course of operation. Classical control methodologies are based on the assumption that the system structure is fixed from the outset. This assumption is, however, violated frequently in practical control systems. Many systems are not single-structured. Such multi-structured systems operate in different operating points. Each operating point corresponds to a state in DES. A convention control algorithm is designed for each operating point. Switching from one operating point to another is done by the supervisor. Various information regarding the operation of the system is abstracted as events and reported to the supervisor by the fuzzy interpreter. Based on the information on the history of the system, the supervisor will make appropriate decisions for the next actions. In particular, it will decide which control algorithm to use.

7 Acknowledgement

This research is supported in part by the National Science Foundation under grants ECS-9008947, ECS-9110984 and ECS-9209001.

8 References

[1] P.J. Antsklis, K.M. Passino and S.J. Wang, 1991. An introduction to autonomous control systems, *Control Systems Magazine* Vol. 11, No. 4, pp. 5-13.

[2] K.J. Astrom, 1988. Toward intelligent control, Keynote Speech, Proc. 1988 ACC Conference.

[3] J.J. Buckley, 1991. Fuzzy I/O controllers, *Fuzzy Sets and Systems.*

[4] S.-L. Chung, S. Lafortune and F. Lin, 1991. Limited lookahead policies in supervisory control of discrete event systems. Report No. CGR-70, The University of Michigan.

[5] G. F. Franklin, J. D. Powell and B. Emani-Naeini, 1986. *Feedback Control of Dynamic Systems*, Prentice Hall.

[6] P. V. Kokotovic, 1991. "The Joy of Feedback: Nonlinear and Adaptive", 1991 Bode Prize Lecture, Control Systems Magazine, pp7–17.

[7] B. Kosko, 1992. *Neural Networks and Fuzzy Systems*, Prentice Hall.

[8] S. Lafortune and F. Lin, 1991. On tolerable and desirable behaviors in supervisory control of discrete event systems. *Discrete Event Dynamic Systems: Theory and Applications, 1(1)*, pp. 61-92.

[9] F. Lin and W. M. Wonham, 1990, 'Decentralized control and coordination of discrete-event systems with partial observation', *IEEE Transactions on Automatic Control, 35(12)*, pp. 1330-1337.

[10] K. Ogata, 1990. *Modern Control Engineering*, second edition, Prentice Hall.

[11] R. J. Ramadge and W. M. Wonham, 1987. Supervisory control of a class of discrete event processes. *SIAM J. Control and Optimization, 25(1)*, pp. 206-230.

[12] G. Zames, 1988. Feedback organization, learning and complexity in H^∞, Plenary address, *1988 ACC Conference.*

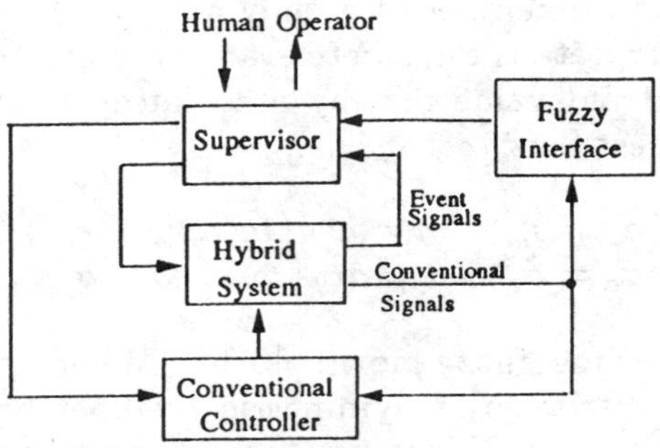

Figure 1. Control Architecture

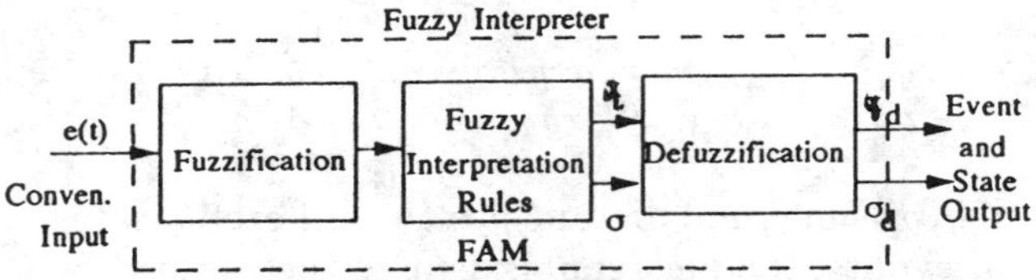

Figure 2. Fuzzy Interface for the Supervisor

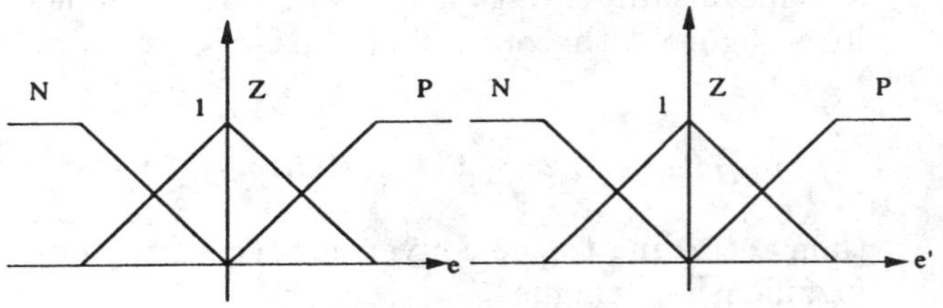

Figure 3. Fuzzy Numbers for the Conventional Input

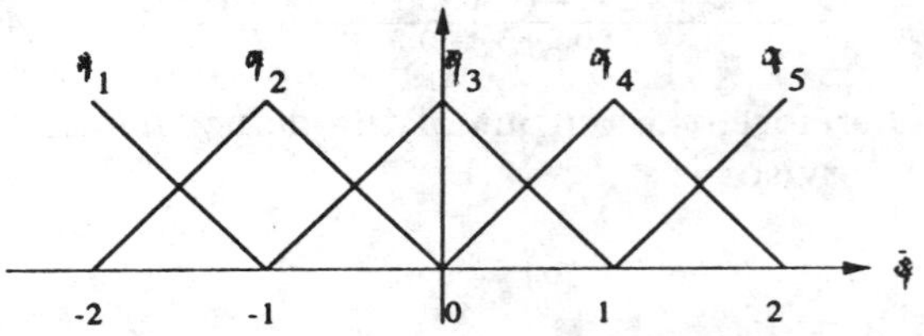

Figure 4. Fuzzy Numbers for the State (Event) Output

Proceedings of the 31st Conference
on Decision and Control
Tucson, Arizona • December 1992

ITERATIVE LEARNING CONTROL - CONVERGENCE USING HIGH GAIN FEEDBACK

David H.Owens,
Centre for Systems and Control Engineering, University of Exeter,
North Park Road, Exeter EX4 4QF, England

Abstract

Convergence theorems for iterative learning control systems to provide a well defined convergence criterion parameterized by a single gain parameter. The convergence is in the weak topology of $L_2^m(0,T)$ with T finite and applies to both finite dimensional systems and a class of infinite- dimensional systems.

Key words - learning control, iterative learning control

Introduction

One aspect of learning that has grown out of the area of robotics is that of iterative learning control (ILC) [1] where the basic problem for the control algorithm is to learn inputs that generate required outputs from a dynamical system by repeated trials and updating of control inputs from trial to trial. More formally, given a dynamical system

$$\dot{x}(t) = Ax(t) + Bu(t) \quad \in R^n, \qquad y(t) = Cx(t) \quad \in R^m \qquad (1)$$

and a desired output signal $r(t), 0 \le t \le T$, construct a causal sequence of experiments that generates a sequence of input signals $(u_k(t))_{k\ge0}$ and outputs $(y_k(t))_{k\ge0}$ with the properties that

(a) The input $u_{k+1}(t)$ is generated from only known trial data and the previous trial input $u_k(t)$.

(b) The experiment has the convergence property

$$\lim_{k\to\infty} (r(\cdot) - y_k(\cdot)) = 0 \qquad (2)$$

with respect to the chosen topology of a given linear space of output signals.

The update law will be defined in terms of the error sequence $(e_k(\cdot))_{k\ge0}$ ($e_k = r - y_k$ for all k). In contrast to other papers in the area this paper develops the update law

$$u_{k+1}(t) = u_k(t) + (Ke_{k+1})(t), \qquad 0 \le t \le T \qquad (3)$$

where K is a causal "learning" operator feeding back the **current** trial error. The physical meaning of this form of learning law is that current trial feedback obtained during the trial by normal feedback mechanisms is used for updating the control input.

Assuming zero initial conditions on each trial and the stability of L, the error propogates from trial to trial according to the discrete evolution law in $L_2^m(0,T)$,

$$e_{k+1} = (I - L)e_k, \qquad k \ge 0 \qquad (4)$$

where

$$L = (I + GK)^{-1}GK \qquad (5)$$

is the feedback operator on each trial and $G(s) = C(sI - A)^{-1}B$ is the operator (transfer function) describing systems i/o dynamics on each trial. Padieu and Su [1] relate the convergence of the error (via sufficient conditions) to the frequency domain properties of the operators in a manner similar to the theory of repetitive dynamical systems (Owens et al [2], [3]). Neither approaches do enables the analysis of the above class of learning rules as they demand that the operator $I - L$ has H_∞ norm or spectral radius strictly less than unity at almost all frequencies. This paper introduces a new theoretical approach for a class of linear systems in state space form that does not require this condition. The approach has an alternative operator-theoretical form related to an extended notion of positive reality which is also outlined in the paper.

The core of the paper is convergence results for the algorithm described. Convergence in the norm topology of the Hilbert space $L_2(0,T)$ seems to be difficult to prove but convergence in the weak topology is proven.

Basic Theoretical Results

The convergence of the algorithm is intuitively dependent on the systems dynamics. The following result identifies requirements for relative degree one, minimum-phase systems in terms of "sufficiently high gain" control loops.

Theorem 1: (L_2 - Weak Convergence of ILC)

Suppose that the system $S(A,B,C)$ is minimum-phase with CB square and nonsingular with spectrum in the open right-half complex plane. Suppose also that the iterative learning law is

$$u_{k+1}(t) = u_k(t) + Ke_{k+1}(t), \qquad 0 \le t \le T < \infty \qquad (6)$$

where K is a scalar, positive gain. Then the resultant ILC system has the property that ,for sufficiently high "gain" K, the error sequence satisfies the inequalities

$$\sum_{k=0}^{\infty} ||v_k||^2_{L_2^m(0,T)} < \infty \qquad (slow \quad change) \qquad (7)$$

$$||e_{k+1}||^2_{L_2^m(0,T)} \le ||e_k||^2_{L_2^m(0,T)} \qquad (Monotonicity) \qquad (8)$$

where $v_{k+1} = Le_k = -(e_{k+1} - e_k)$ for $k \ge 0$ hence converges in norm to zero as $k \to \infty$ in $L_2 \cap L_\infty$.

Moreover, the error sequence $(e_k)_{k\ge0}$ is uniformly bounded in L_2 and the output sequence converges to the desired signal r in the **weak** topology on $L_2^m(0,T)$.

Proof: For simplicity, the proof is provided for the case of $m = 1$ only.

$$||e_{k+1}||^2 = ||(I - L)e_k||^2 = ||e_k||^2 - 2 < v_{k+1}, e_k > + ||v_{k+1}||^2 \qquad (9)$$

The map $e_k \to v_{k+1}$ has the state space model

$$\dot{z}_k(t) = (A - BKC)z_k(t) + BKe_k(t), \quad z_k(0) = 0 \qquad (10)$$

$$v_{k+1} = Cz_k(t) \qquad (11)$$

The work of Owens et al (Ilchmann [4]) gives the inequality,

CH3229-2/92/0000-2545$1.00 © 1992 IEEE

$$\frac{1}{2}v_{k+1}^2(t) \leq (M-CBK)||v_{k+1}||^2_{L_2(0,t)} + CBK < v_{k+1}, e_k >_{L_2^m(0,t)} \quad (12)$$

for some $M \geq 0$ independent of T, K. Hence for $(2M - CBK) > 0$

$$< v_{k+1}, e_k >_{L_2^m(0,T)} \geq (2CBK)^{-1}v_{k+1}^2(T) + (1 - \frac{M}{CBK})||v_{k+1}||^2_{L_2(0,T)} \quad (13)$$

Using this in equation (10) gives

$$||e_{k+1}||^2 \leq ||e_k||^2 - (1 - \frac{2M}{CBK})||v_{k+1}||^2 - \frac{1}{CBK}v^2(T) \quad (14)$$

which proves the required inequalities by an inductive argument. The proof that $v_k \to 0$ in $L_2 \cap L_\infty$ now follows from (8) and (13).

It remains to prove the weak convergence of e_k to the demand r in $L_2^m(0,T)$. Evaluate the behaviour of

$$< f, \dot{v}_{k+1} > = - < \dot{f}, v_{k+1} > + [f(t)v_{k+1}(t)]_0^T \quad (15)$$

valid for any $f \in L_2(0,T) \cap L_\infty(0,T)$ with $\dot{f} \in L_2(0,T)$. By choosing f in the dense subspace of finite linear combinations of the natural Fourier basis in $L_2(0,T)$, it follows that $< f, \dot{v}_{k+1} > \to 0$ on a dense subset of $L_2(0,T)$ and hence that $(\dot{v}_{k+1})_{k\geq 0}$ converges **weakly** to zero in $L_2(0,T)$. Using the representation due to Logemann and Owens [5]

$$\dot{v}_{k+1} = (A_{11} - CBK)v_{k+1} + Hv_{k+1} + CBKe_k \quad (16)$$

with H a bounded, causal linear operator on $L_2(0,\infty)$ and hence $L_2(0,T)$. Consequently $e_k \to 0$ weakly and the result is proved.

Remark: The proof relied on the inequality (13) and the representation (16) due to Logemann and Owens [5]. Equation (13) is essentially a positive real condition which, together with (16) enables the application of the results to a class of infinite dimensional systems. The details are omitted for brevity.

The result provides quite useful information on convergence in terms of the monotonicity of $||e_k||$ and the slow variation condition. The weak convergence of the result is disappointing but simulation experiments indicate that it is likely to be good enough for practical applications.

The result relates to a different control scheme and covers the multivariable case but in other respects is related to that of Padieu who uses the operator $L = GK$ rather than the loop feedback $L = (I + GK)^{-1}GK$ used in this paper. Hence Padieu cannot utilize high gain feedback to aid in convergence. A comparison is however possible through the notion of positive real (p.r.) systems. More precisely, with $m = 1$, and G p.r. $I - L = (I + GK)^{-1}$ is p.r. with gain < 1 at all but a finite number of isolated frequencies and the point at infinity which is precisely the condition required by Padieu for $I - GK$. It is achieved easily and naturally with the algorithm proposed in this paper.

The connection with positive real conditions can also be obtained from the state space point of view. To underpin this observation, suppose there exists a gain K^* such that L is p.r.. Then (Anderson [6] and Owens et al [7]) there exists a Liapunov matrix P satisfying $PB = C^T$, a Liapunov function $V = \frac{1}{2}z_k^T Pz_k$ and a constant $\mu > 0$ such that

$$\tfrac{1}{2}(CB)^{-1}v_{k+1}^2(t) + (K - K^*)||v_{k+1}||^2_{L_2(0,t)} \leq V(t) + \mu \textstyle\int_0^t V(s)ds$$

$$+ (K - K^*)||v_{k+1}||^2_{L_2(0,t)} \leq \quad K < v_{k+1}, e_k > \quad (17)$$

which is just the inequality (14) with $2M$ replaced by K^* if $K \geq K^*$. The proof remains valid in this case and, if G is p.r., $K^* \to 0$ and the algorithm converges for all positive gains $K \geq K^*$. Note also that this analysis indicates that $(z_k(t))_{k\geq 0} \to 0$ in $L_2(0,T) \cap L_2(0,T)$ indicating behaviours of the state compatible with the output behaviours.

Theorem 2 (Convergence of the input sequence)

Suppose $r = Gu_\infty$ with $u_\infty \in L_2^m(0,T)$. Then, under the conditions of theorem 1, **if the input sequence $(u_k)_{k\geq 0}$ is bounded in $L_2(0,T)$**, it converges weakly to u_∞ in $L_2^m(0,T)$.

Proof: We have, for any $f \in L_2^m(0,T)$, and with G^* denoting the adjoint operator of G,

$$< f, G(u_\infty - u_k) > = < G^*f, (u_\infty - u_k) > \to 0 \quad (18)$$

as $k \to \infty$. As the orthogonal complement of the range of G^* is the null space of G (and hence $\{0\}$), the range of G^* is dense. The result now follows trivially.

The boundedness assumption on the input sequence relates, intuitively, to the avoidance of impulsive solutions.

Conclusions

The paper has provided a convergence theory for iterative learning control based on the use of high gain current trial feedback for the special case of relative degree one, MIMO minimum-phase systems. The results are related to those of Padieu and Su via the notion of positive real systems. In particular positive real systems are easily arranged to have convergent learning by simple proportional learning rules of arbitrary positive gain.

Acknowledgements

To the University of Exeter Research Fund for a short term Research Assistantship for Herr D.Neuffer (Stuttgart). Future work with Dr.E.Rogers of the University of Southampton will be supported by the SERC under contract number GR/H/48286. Dr.R.Chapman (Mathematics, Exeter University) advised on the mathematical background to the proof of theorem 2.

References

[1] Padieu and Su:"An H_∞ approach to learning control", Int. J. Adapt. Control and Sig. Proc., 1990(4), pp 465-474.

[2] J.B.Edwards and D.H.Owens:"Analysis and control of multipass processes", Research Studies Press, Taunton, England, 1982.

[3] E.Rogers and D.H.Owens:"Stability analysis for linear repetitive processes", Springer-verlag, Berlin, April 1992.

[4] A.Ilchmann:"Non-identifier-based control of dynamical systems - a survey", IMA J. Math. Control and Inf., 1991(8), pp 321-366.

[5] H.Logemann and D.H.Owens:"Input-output theory of high-gain adaptive stabilization of infinite dimensional systems", Int. J. Adapt. Control and Sig. Proc., 1988(2),pp 193-216.

[6] B.D.O.Anderson:"A systems theoretical criterion for positive real systems", J.SIAM Control, 5(2), 1967, pp 171-182.

[7] D.H.Owens, D.Pratzel-Wolters and A.Ilchmann:"Positive real structure and high gain adaptive stabilization", IMA J. Math. Control and Inf., 1987(4), pp 167-181.

Proceedings of the 31st Conference
on Decision and Control
Tucson, Arizona • December 1992

A Systematic Design Method for Fuzzy Logic Control With Application to Automotive Idle Speed Control

George Vachtsevanos, Shehu S. Farinwata and Hoon Kang
School of Electrical Engineering
Georgia Institute of Technology
Atlanta, Georgia 30332-0250

Key Words: Automobile Engine, Idle Speed Control, Fuzzy Control, Systematic Design

We investigate a systematic design procedure of an automatic rule generation for dynamic systems such as a nonlinear engine dynamic model. By 'automatic rule generation' we mean clustering or collection of such meaningful transitional relations from one conditional subspace to another. Elements that result from such transitions are anticipated according to the applied action of each rule. Data required for this transitional set of rules can be collected via (*i*) a priori information such as experimental results, (*ii*) numerical simulation runs based on dynamic models, and (*iii*) heuristics. However, the main advantage of the automatic rule generation scheme is that reliability can be potentially increased even in the presence of large-grained uncertainty in the investigated system. Specifications - accuracy and precision - of the system tolerances can be arbitrarily adjusted and are a function of the resolution of the design parameters. Fuzzy logic based feedback control is a suitable goal of our automatic rule generation scheme. Membership functions stored in a fuzzy logic controller can be easily modified and updated without tedious re-evaluation of different dynamic models.

Engine Model

Consider a two-state dynamic model given by

$$\dot{e}_1 = f_1(e_1, e_2, \delta, \theta, T_d), \qquad \dot{e}_2 = f_2(e_1, e_2, \delta, \theta, T_d) \tag{1}$$

where e_1, e_2 are the states (or the errors); f_1, f_2 are the nonlinear mappings; and δ, θ are the control inputs. T_d is a scalar disturbance term. For the two-stage engine model we are considering, we have explicitly [1-2]:

$$\dot{P} = k_P(\dot{m}_{ai} - \dot{m}_{ao})$$

$$\dot{N} = k_N(T_i - T_l) \tag{2}$$

$$\dot{m}_{ai} = (1+0.907\theta + 0.0998\theta^2)\, g(\,)$$

$$\dot{m}_{ao} = -0.0005968N - 0.1336P + 0.0005341NP + 0.00000175NP^2$$

$$T_i = -39.22 + \frac{325024}{120N}\dot{m}_{ao} - 0.0112\delta^2 + 0.000675\delta N(2\pi/60) + 0.635\delta + 0.0216N(2\pi/60) - 0.000102N^2(2\pi/60)^2$$

$$T_L = (N/263.17)^2 + T_d$$

$$g(P) = \begin{cases} 1 & P < 50.66 \\ 0.0197(101.325P - P^2)^{\frac{1}{2}} & P \geq 50.66 \end{cases} \tag{3}$$

where P is the manifold pressure in KPa and N is engine speed in RPM. These equations are highly nonlinear. The dynamic equations can be expressed in the general form

$$\dot{x} = f(x(t), u(t), t) \tag{4}$$

We define the state vector $x \in \mathbb{R}^2$, and the control vector $u \in \mathbb{R}^2$, where

$$x = \begin{pmatrix} P \\ N \end{pmatrix}, \quad u = \begin{pmatrix} \theta \\ \delta \end{pmatrix} \tag{5}$$

- $\theta \in [5, 35]$ degrees, is the throttle angle (idling limits),
- $\delta \in [10,45]$ degrees, is the spark advance in degrees,
- $T_d \in [0,61]$ Nm, is the applied accessory load torque.

The nominal (equilibrium) operating point is typically determined on the basis of operating conditions, minimum energy dissipation, etc. (in the test engine considered it is set at N = 750 rpm at a corresponding equilibrium pressure, P).

Automatic Rule Generation

From the admissible controls, we select finite representative constant values of δ's and θ's, and we denote them as δ_i and θ_j. These crisp numbers will be fuzzified later unless the performance of the controller with crisp inputs is satisfactory. Moreover, we choose finite representative points in the state space to anticipate the trajectories from one subspace to another. The countless trajectories are called a 'manifold', and one set of data is collected by setting d_i and q_j to be constant. Such data are repeatedly collected as they are used next in obtaining a rule base for feedback regulation of x_1 and x_2. We denote by E_{1m} and E_{2m} the *m*-th and *n*-th intervals of x_1 and x_2, respectively. Linguistically, we define

$$L_{mn} = E_{1m} \times E_{2n}$$

and then L_{mn} is a finite region in the state space.

By applying fixed controls δ_i and θ_j, a set of transitional relations is obtained; for example as

$$(\delta_1, \theta_1) : L_{12} \rightarrow L_{34}$$
$$(\delta_1, \theta_1) : L_{23} \rightarrow L_{33}$$
$$\ldots$$
$$(\delta_1, \theta_1) : L_{55} \rightarrow L_{73}$$

We continue to change (δ_i, θ_j) and exhaustively gather a finite number of transitional relations. When we store the information above, we determine and attach, for a transition, such performance quantities as the required transition time, expended energy and squared error. The general form of the performance index that we have employed is given below.

$$J = \sum_{i,j}^{i+1,J+1} [a(eTe) + b(uTu) + \tau]t, \quad \text{for all } e \in L_{mn} \tag{6}$$

CH3229-2/92/0000-2547$1.00 © 1992 IEEE

where u is the control input applied while in L_{mn} in order to transition to the next cell-group. On changing (δ_i, θ_j), we may end up in the same L_{mn} as the source and also for the next transition. We call this an 'invariant manifold'. To avoid an infinite transit time while in such a cell-group, a time limit is set after which system evolution in the cell-group space resumes. It is therefore clear that the target cell-group L_{mn} (the specified goal) must be an invariant manifold for some (δ_i, θ_j)-pair. This is also necessary for convergence and asymptotic stability. This is equivalent to the 'reachability condition' in modern control theory. The target is denoted by L*. The vector fields of the nonlinear dynamic system are utilized and use is made again of fuzzy optimality concepts to obtain the optimal fuzzy controls for each cell-group. However, due to the approximate nature of the model chosen, the phase portrait assignment algorithm leads to sub-optimal conditions for the fuzzy clusters.

Fuzzy Controller Structure

The quantities of interest are the pressure and speed errors and their time derivatives. The derivatives are not used at the moment but will be at a later time. However, we do have provisions for them in the program. The three perturbations, namely, spark advance, throttle angle and accessory load define the premises and consequents. So, we are proposing a five-premise and two-consequent structure. The general form of the rules is given below:

If (e_N is F_1) & (e_p is F_2) & ($\dot{e}_N$ is F_3) & ($\dot{e}_p$ is F_4) & (ΔT_d is F_5) THEN ($\Delta\delta$ is G_1) & ($\Delta\theta$ is G_2).

where F_i and G_i are the consequent and premise linguistic terms respectively. We denote these as follows:

Premise:	Consequent:
$x_1 = F_1(e_N)$	$y_1 = G_1(\Delta\delta)$
$x_2 = F_2(e_p)$	$y_2 = G_2(\Delta\theta)$
$x_3 = F_3(\Delta T_d)$	
$x_4 = F_4(\dot{e}_N)$	
$x_5 = F_5(\dot{e}_p)$	

We denote by P_i for i=1 to 5, the premise matrices, and C_1, C_2 the consequent matrices. As stated at the beginning, we are not looking at the derivatives at the moment, so x_4 and x_5 drop out. Then

$$P_i = \begin{bmatrix} x_i^{1T} \\ \vdots \\ x_i^{kT} \end{bmatrix} \quad for\ i = 1,...,3$$

$$C_j = \begin{bmatrix} y_j^{1T} \\ \vdots \\ y_j^{kT} \end{bmatrix} \quad for\ j = 1,...,2 \tag{7}$$

where k is the number of rules.

If we congregate the 3-premise fuzzy vectors into one fuzzy set, it becomes a 3-dimensional premise hypercube of the fuzzy inputs given by

$$x^k[1,...,3] = x_1^k[1] \otimes x_2^k[2] \otimes x_3^k[3] \tag{8}$$

Similarly, the 2-dimensional consequent fuzzy hypercube is formed as

$$y_k[1,2] = y_1^k[1] \otimes y_2^k[2] \tag{9}$$

Degree of Fulfillment, d_{if} for the ith Premise

$$d_f^i = P_i \circ \tilde{x}_i, \quad for\ i = 1,...,3 \tag{10}$$

The overall degree of fulfillment d_f is then formed as:

$$d_f = d_f^1 \otimes ... \otimes d_f^3 \tag{11}$$

The jth fuzzy output vector $\tilde{y}_j$ is obtained using the max-min operation with respect to the consequent matrix C_i.

$$so,\ \tilde{y}_j = C_j \circ d_f, \quad j = 1,2 \tag{12}$$

The above fuzzy output vector is then converted to the corresponding crisp data using any of the defuzzification strategies such as the center of gravity method.

We have employed this method to effect the idle speed control of the engine model described herein, using 56 rules. Preliminary results showed desirable properties namely, stability, and robustness with respect to small accessory load perturbations. Detailed results will be published at a later time.

References

1. A.W. Olbrot and B.K. Powell, "Robust Design and Analysis of Third and Fourth Order Time Delay Systems with Application to Automotive Idle Speed Control," *1989 ACC*, vol. 2, pp. 1029-39, 1989.

2. B.K. Powell and J. Cook, "Non-linear Low Frequency Phenomenological Engine Modeling and Analysis," *1987 ACC*, pp. 332-340, June 1987.

3. P.R. Crossley and J.A. Cook, "A Nonlinear Engine Model for Drivetrain System Development," *IEE Control Conference*, Scotland, UK, March, 1991.

4. C.C. Lee, "Fuzzy Logic in Control Systems: Fuzzy Logic Controller-Part I," (and Part (II)), *IEEE Transaction on Systems, Man, and Cybernetics*, vol. 20, no. 2, pp. 404-435, March/April, 1990.

5. Y.Y. Chen and T.C. Tsao, "A Description of the Dynamical Behavior of Fuzzy Systems," *IEEE Transactions on Systems, Man, and Cybernetics*, vol. 19, No. 4, pp. 745-755, July/August 1989.

6. H. Kang and G.J. Vachtsevanos, "Nonlinear Fuzzy Control Based on the Vector Fields of the Phase Portrait Assignment Algorithm," *Proc. of the ACC '90*, pp. 1479-1484, San Diego, 1990.

7. H. Kang and G.J. Vachtsevanos, "Fuzzy Hypercubes: Linguistic Learning/Reasoning Systems for Intelligent Control and Identification," *Proc. of the 30th IEEE CDC*, pp. 1200-1205, Brighton, England, December, 1991.

8. S.M. Smith and D.J. Comer, "Automated Calibration of a Fuzzy Logic Controller Using a Cell State Space Algorithm," *IEEE Control Systems Magazine*, Vol. 2, No. 5, pp. 18-28, August 1991.

9. M. Abate and N. Dosio, "Use of Fuzzy Logic for Engine Idle Speed Control," *SAI Trans. 900594*, Vol. 99, pp. 107-114, 1990.

Proceedings of the 31st Conference
on Decision and Control
Tucson, Arizona • December 1992

FUZZY OPTIMIZATION METHOD FOR STEADY-STATE CONTROL OF COMPLEX SYSTEMS

Assann SIDAOUI, Zdenek BINDER, René PERRET
Laboratoire d'Automatique de Grenoble
CNRS-UA 228. ENSIEG-INPG
B.P. 46, 38402 Saint Martin d'Hères

ABSTRACT

Considering the steady-state optimization problem pertaining to complex systems, this paper proposes application of fuzzy techniques to optimise systems which cannot usually be described in precise mathematical terms. Based on Bellman and Zadeh's fuzzy model of decisions [1], a fuzzy optimization methodology is proposed and illustrated by a numerical example.

I. INTRODUCTION

A common method of keeping an industrial process at an optimum operating point is to compute optimum values of controller set points using a mathematical description. As is often the case, valid and accurate models of the processes to be controlled are not available. Adaptive techniques are employed to update the model used in the optimization problem [2]. Furthermore, the algorithms proposed are too complex and their implementaton difficult [3].

In recent years, several control studies have appeared in the literature, see [4] for reviews, which attempt to overcome the problem of mathematical models deficiencies by using qualitative knowledge of the processes in the design of a control algorithm. All these studies are based on the use of Zadeh's fuzzy sets theory [5].

Our interest concerns application of fuzzy techniques to optimize systems which are "too complex to admit of precise mathematical analysis". Fuzzy models of representation are used. Such models are limited in accuracy and complexity, but are rigourous qualitative descriptions of the essential features of the process behaviour.

Based on Bellman and Zadeh's approach, a new optimization methodology is proposed in this paper. After formulating the optimization problem, we develop the proposed methodology. A numerical example is finally presented to illustrate our approach.

II. THE STEADY-STATE OPTIMIZATION PROBLEM

The considered system under steady-state control is represented in figure 1 where $c_i \in C_i$, $u_i \in U_i$ and $y_i \in Y_i$ are respectively steady-state control, interconnection input and interconnection output vectors, i=1,...,N and N is the number of subsystems.Each subsystem is described by the input-output mapping

$$y_i = F_i(c_i, u_i) \quad (1)$$

Interconnexions between subsystems are described by:

$$u_i = \sum_{j=1}^{N} H_{ij} y_j \qquad H_{ij} \in \{0,1\} \qquad i = 1,\cdots,N \quad (2)$$

H is the interconnexions matrix.

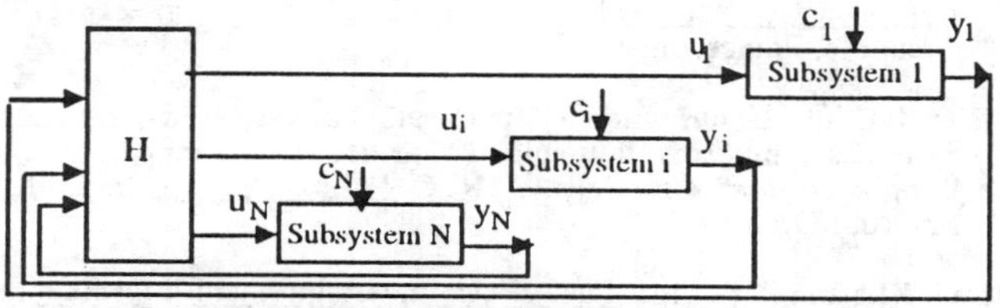

Figure 1. Global system representation.

In each subsystem, constraints can be present, i.e.

$$(c_i, u_i, y_i) \in CUY_i = \{(c_i, u_i, y_i) : g_i (c_i, u_i, y_i) \le 0 \} \quad (3)$$

where g_i is a constraint function vector , i = 1,..., N.

Finally, with each subsystem a performance function Q_i is associated, and the overall performance function (to be minimised) is

$$Q(c,u,y) = \sum_{i=1}^{N} Q_i(c_i,u_i,y_i) \quad (4)$$

the joint description of the optimizing control problem is:

$$\begin{cases} \underset{c,u,y}{\text{Min}} \ Q(c,u,y) \\ \quad \text{s.t.} \quad y = F(c,u) \\ \qquad\quad u = Hy \\ \qquad\quad g(c,u,y) \le 0 \end{cases} \quad (5)$$

III. THE FUZZY APPROACH

III.1 Fuzzy system modeling

A fuzzy model is taken to mean a finite set of Rules $R^{(j)}$ (j = 1,...,n) which together form an algorithm R for determining outputs of the process from its inputs according to some control actions.

$$R^{(j)} : \text{ if } (c,u) \in \tilde{C}^j \times \tilde{U}^j \text{ then } y \in \tilde{Y}^j \quad (6)$$

$$R = R^{(1)} \cup \ldots \cup R^{(n)} \quad (7)$$

where $\tilde{C}^j$, $\tilde{U}^j$ and $\tilde{Y}^j$ are fuzzy sets defined respectively on the universes of discourses C, U and Y.

(6) and (7) are equivalent to:

$$\mu_{R^{(j)}}(c,u,y) = \mu_{\tilde{C}^j}(c) \wedge \mu_{\tilde{U}^j}(u) \wedge \mu_{\tilde{Y}^j}(y) \quad (8)$$

$$\mu_R(c,u,y) = \bigvee_{1 \le j \le n} \mu_{R^{(j)}}(c,u,y) \quad (9)$$

where $\wedge,\vee$ stand for sup (max) and inf (min) operators respectively. μ_R, μ_A represent membership functions.

To calculate the fuzzy output $\tilde{Y}$, given fuzzy input $\tilde{U}$ and fuzzy control $\tilde{C}$, the compositional rule of inference [6] is used:

$$\tilde{Y} = (\tilde{C},\tilde{U}) \circ R \quad (10)$$

where $\circ$ is the max - min composition of fuzzy relations.

We can rewrite (10) in the following form:

$$\mu_{\tilde{Y}}(y) = \bigvee_{(c,u) \in C \times U} [\, \mu_{\tilde{C}}(c) \wedge \mu_{\tilde{U}}(u) \wedge \mu_R(c,u,y) \,] \quad (11)$$

It is easy to verify that (10) is also equivalent to:

$$\tilde{Y} = \bigcup_{1 \le j \le n} \tilde{Y}^j \quad \text{where } \tilde{Y}^j = (\tilde{C},\tilde{U}) \circ R^{(j)} \quad (12)$$

III.2 Bellman and Zadeh's fuzzy model of decisions

X being the set of possible actions and Y the set of possible effects or outcomes, both constraints and goals are treated as fuzzy sets. The fuzzy constraints are defined on X and the fuzzy goals on Y. This fuzziness allows the decision maker to frame the goals and constraints in vague, linguistic terms which may more accurately reflect the state of knowledge concerning these.

A function f can then be defined as a mapping from the set of actions X to the set of outcomes Y. $f : X \rightarrow Y$ such that a fuzzy goal defined on set Y induces a corresponding fuzzy goal G' on set X thus

$$\mu_{G'}(u) = \mu_G(F(u))$$

A fuzzy decision D may then be defined as the choice that satisfies both the fuzzy goals and constraints:

$$D = G' \cap C$$

which can easily be extented for any number of goals and constraints.

Once a fuzzy decision has been arrived at, it may be necessary to choose the "best" single crisp alternative from this fuzzy set. This may be accomplished in a straightforward manner by choosing the alternative $x \in X$ that attaints the maximum membership grade in D.

CH3229-2/92/0000-2549$1.00 © 1992 IEEE

IV. THE FUZZY OPTIMIZATION METHODOLOGY

We propose to use fuzzy techniques for developing a new optimization method applicable to fuzzy systems. The main idea of our approach consists in transforming (5) into a satisfaction problem of fuzzy goals and constraints, and characterizing the fuzzy control that satisfies "at best" both goals and constraints.

Taking into account the fuzzy model R, (5) is replaced by:

$$\begin{cases} \underset{c,u,y}{\text{Min}} \quad Q(c,u,y) \\ \quad \text{s.t.} \quad y = (c,u) \circ R \\ \qquad\quad u = Hy \\ \qquad\quad g(c,u,y) \le 0 \end{cases} \tag{13}$$

The global system is composed of N subsystems. Every subsystem is modelled by a fuzzy model R_i (i=1,...,N):

$$R = R_1 \times \ldots \times R_N \tag{14}$$

where $$R_i = R_i^{(1)} \cup \ldots \cup R_i^{(n_i)} \tag{15}$$

and n_i = number of rules of model R_i .

Performance functions Q_i and local constraints g_i (i=1,...N) are are supposed not fuzzy. However, fuzzification can be introduced without throwing the validity of the proposed methodology back into question. Interconnections constraints are also not fuzzy. To take them into account, it is essential to develop "separable" structures of computing adapted to the resolution of large-scale problems.Let us consider the following sets:

$$\begin{aligned} Cg &= \{(c,u,y) \,/\, u = Hy,\ g(c,u,y) \le 0\} \\ Cm &= \{(c,u,y) \,/\, y = (c,u) \circ R\} \end{aligned} \tag{16}$$

Only Cm is fuzzy. Its membership function is:

$$\begin{aligned} \mu_{Cm}(c,u,y) = \mu_R(c,u,y) &= \bigwedge_{1\le i\le N} \left(\mu_{R_i}(c_i,u_i,y_i)\right) \\ &= \bigwedge_{1\le i\le N} \bigvee_{1\le j\le n_i} \left(\mu_{R_i^{(j)}}(c_i,u_i,y_i)\right) \end{aligned} \tag{17}$$

Function Q is the sum of the local performance functions. We have to suppose that Q admits a finite minimum on the feasible crisp set Cg:

$$Q^* = \underset{(c,u,y) \in Cg}{\text{Min}} Q(c,u,y) \qquad Q^* = (Q^*_1,\ldots,Q^*_N) \tag{18}$$

We define the fuzzy objectives G_i : "Q_i close to Q_i*" by:

$$\mu_{G_i}(c_i,u_i,y_i) = \phi\left(Q_i(c_i,u_i,y_i) - Q^*_i\right) \tag{19}$$

where $\phi(Q) = \exp(-Q/2d^2) \qquad d \neq 0$

The global objective G is satisfied if the all objectives G_i are too:

$$\mu_G(c,u,y) = \bigwedge_{1\le i\le N} \left(\mu_{G_i}(c_i,u_i,y_i)\right) \tag{20}$$

Therefore, solutions of our fuzzy optimization problem are :

$$(c^*,u^*,y^*) = \text{Arg.} \underset{(c,u,y)\in Cg}{\text{Max}} \left(\mu_G(c,u,y) \wedge \mu_{Cm}(c,u,y)\right) \tag{21}$$

where μ_G is defined by (20) and μ_{Cm} by (17).

It is easy to see that :

$$\mu_G(x) \wedge \mu_{Cm}(x) = \bigwedge_{1\le i\le N} \left(\mu_{G_i}(x_i) \wedge \mu_{R_i}(x_i)\right) \tag{22}$$

where x = (c,u,y) and $x_i = (c_i,u_i,y_i)$.

(22) is a constrained optimization problem. Resolution of this problem by using ordinary mathematical programming is difficult because of the special form of the function to be optimised which is obtained by several compositions of min and max operators. One way proposed to overcome this difficulty [7] consist in replacing a problem like:

$$\underset{x \in C}{\text{Max}} \quad \mu_A(x) \tag{23}$$

where μ_A is a membership function and C a crisp set of constraints by:

$$\underset{\substack{x\in C \\ \mu_A(x) \ge \lambda}}{\text{Max}} \lambda \tag{24}$$

Obviously, $0 \le \lambda \le 1$. This transformation leads both to a method of solving the optimization problem and of expression of 'the grade of membership' of the solution (i.e. λ).

Remark:
$$\begin{aligned} \mu_{G_i}(x_i) \wedge \mu_{R_i}(x_i) &= \mu_{G_i}(x_i) \wedge \left(\bigvee_{1\le j\le n_i} \mu_{R_i^{(j)}}(x_i)\right) \\ &= \bigvee_{1\le j\le n_i} \left(\mu_{G_i}(x_i) \wedge \mu_{R_i^{(j)}}(x_i)\right) \end{aligned}$$

V. NUMERICAL EXAMPLE

To illustrate et make understand the mechanism of our approach, we have used it to resolve the following simple problem [4]

$$\begin{cases} \underset{c,y}{\text{Min}} \quad \left(2(c-2)^2 + (y-1)^2\right) \\ \quad \text{s.t.} \quad y = c \circ R \\ \qquad\quad 0 \le c \le 10 \end{cases} \tag{25}$$

where: $R = R^{(1)} \cup R^{(2)}$ (see figure 2).

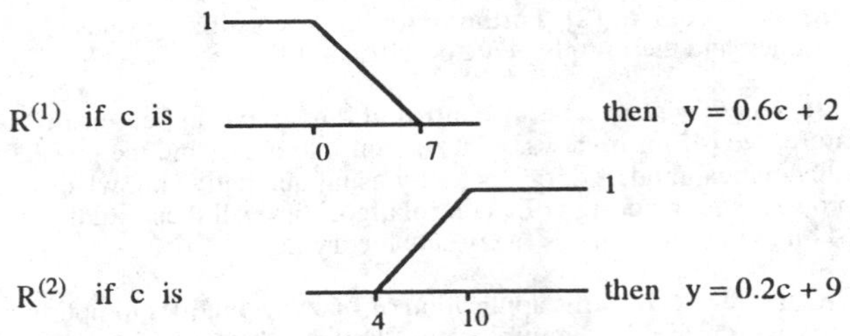

Figure 2. The fuzzy model used in the numerical example.

The solution of this problem depends on the subjective definition of the fuzzy objectives G_i (see eq. (19)) taking d = 4, we find [4]:

$c^* \approx 1.1 \qquad y^* = 0.6c^*+2 \qquad \mu^* = (7-c^*)/7$

VI CONCLUSION

A new methodology for resolving optimising control problems of complex or ill-defined systems has been presented in this paper, based on fuzzy systems modeling and Bellman and Zadeh's fuzzy model of decisions. Extension of this approach to large-scale problems requires first elaboration of new optimization procedures adapted to fuzzy logic formalism, the major obstacle being the "non regular" form of the functions to be optimised. The rapid development in fuzzy programming methods would provide efficient tools for optimization of large-scale fuzzy systems.

References

[1] R. Bellman and L.A. Zadeh, "Decision making in a fuzzy environment", Management Science, 17, pp. 144-164.1970.

[2] P.D. Roberts and J. Lin, "Potential for hierarchical optimising control in the process industry", Proceedings of the first European Control Conference (ECC 91), vol. 1, pp. 213-217, Grenoble, France, July 1991.

[3] A. Sidaoui, "Contribution à l'optimisation hiérarchisée des grands systèmes complexes. Poursuite d'objectifs et prise en compte de l'imprécision des modèles", INPG Ph.D. Thesis, Grenoble, France, March 1992.

[4] G. Klir and A.F. Tina, Fuzzy sets, uncertainty, and information, Prentice-Hall International Editions.1988.

[5] L. Zadeh, "Fuzzy sets", Inform. Control, vol. 8, pp 338-353, 1965

[6] L. Zadeh, "Outline of a new approach to the analysis of complex systems and decision processes", IEEE Trans. on Sys., Man, and Cybernetics, vol. SMC-3, n°1, pp. 28-44, 1973.

[7] H.J. Zimmermann, Fuzzy Set Theory and its Applications, Kluwer-Nijhoff, Boston. 1985.

Proceedings of the 31st Conference
on Decision and Control
Tucson, Arizona • December 1992

FA3 - 8:30

PARALLEL DECOUPLING OF A HIGH-PURITY DISTILLATION COLUMN

Ole B. Gjøsæter and Bjarne A. Foss

The University of Trondheim
The Norwegian Institute of Technology
Division of Engineering Cybernetics
N-7034 Trondheim, Norway
Email: obg@itk.unit.no / baf@itk.unit.no

Abstract

In this paper we present a design technique that can be applied to process control systems with both high and low RGA entries. We have called this technique parallel decoupling. The controller is designed on the basis of a nominal process model, $G(s)$. We assume that the real process, $G_P(s)$, contains a bandwidth constraint due to time delays or right-half plane zeros. For simplicity of exposition this will be illustrated by norm-bounded perturbations of the nominal process model.

A high-purity distillation column with large RGA entries at all frequencies is used as a design example, and μ-analysis is used to evaluate the quality of the design compared to other methods.

Nomenclature

$G(s)$	nominal linear process
$G_P(s)$	process with uncertainty
$C(s)$	controller
$C_f(s)$	reference prefilter
$\mathbf{d}$	disturbance
$\mathbf{e}$	control error
$\mathbf{r}$	reference input
$\mathbf{u}$	control input
$\mathbf{y}$	output
$\hat{\mathbf{y}}$	weighted output
σ	singular value
γ	condition number
D_1, D_2	diagonal matrices
NP	nominal performance
RS	robust stability
RP	robust performance

1 Introduction

There are two factors which, in particular, complicate the control of multivariable systems. These are the combination of strong crosscouplings and time delays. High-purity distillation columns exhibit these characteristics and are hence, difficult to control. The crosscouplings cannot be fully compensated by a decoupler, for instance. This is because model uncertainty combined with strong crosscoupling and non-minimum phase effects limits the obtainable bandwidth of the control loop.

In this paper we will present a lower-order controller structure, and apply it to an ill-conditioned plant. This structure will be analysed both in the frequency domain through μ-analysis, see e.g. Morari and Zafiriou (1989) [4], and in the time domain through simulations. The structure is also compared to existing solutions.

μ-analysis is a useful tool for analysing robust performance in the the presence of model uncertainty. However, μ-optimal (or close to optimal) controllers tend to be of rather high order and are thus difficult to apply in practice.

Most advanced control methods do not explicitly take uncertainty into account. Therefore, they have a tendency to fail when confronted with difficult control problems. This paper will present a simple design technique that takes uncertainty into account.

The rest of this paper is structured as follows. The characteristics of ill-conditioned plants are briefly discussed. Then, a parallel decoupler controller structure is derived and analysed. Next, the control problem and existing solutions are presented. This includes a simulation study. Finally, some conclusions are presented.

2 Ill-conditioned Plants

In the following we will limit ourselves to linear 2×2-systems of the form:

$$\mathbf{y}(s) = G(s)\mathbf{u}(s) \tag{1}$$

We assume that the entries of the nominal process transfer function, $G(s)$, are minimum phase.

The RGA (relative gain array) matrix, $\Lambda(s)$, is given by:

$$\Lambda(s) = \begin{pmatrix} \lambda_{11} & 1-\lambda_{11} \\ 1-\lambda_{11} & \lambda_{11} \end{pmatrix} \tag{2}$$

The RGA matrix can be computed using the formula

$$\Lambda(s) = G(s) \times (G^{-1}(s))^T \tag{3}$$

where $\times$ denotes element by element multiplication (Schur product).

CH3229-2/92/0000-2551$1.00 © 1992 IEEE

The RGA matrix, $\Lambda(s)$, has the following properties, see e.g. Grosdidier et al. (1985) [1]:

1) It is independent of diagonal scaling. Matematically $\Lambda(D_1GD_2) = \Lambda(G)$.

2) All row and column sums equal one.

3) Any permutation of rows or columns in G results in the same permutations in the RGA.

4) Relative perturbations in elements in G, and in its inverse are related by $d(G^{-1})_{ji}/(G^{-1})_{ji} = \lambda_{ij}dg_{ij}/g_{ij}$.

The system is chosen such that the λ_{11} is positive. This is no constraint as in general this can be done by interchanging the control inputs.

$$\lambda_{11} = \left(\frac{1}{1-\frac{g_{12}g_{21}}{g_{11}g_{22}}}\right) \tag{4}$$

Equation (4) shows that the elements of the RGA matrix becomes large if $g_{12}g_{21} \approx g_{11}g_{22}$. In this case the condition number of G, $\gamma(G)$, is also large. This implies that the plant is ill-conditioned. The plant is fundamentally difficult to control since time delays or right-half plane zeros, which are always present, limit the gain. Therefore, even though the RGA entries of the plant model usually decline as a function of frequency, the control system has to operate in a frequency range with corresponding high value RGA entries. This implies that the performance will be limited whatever controller is used. More specifically, a decoupler cannot be used since its RGA value is prohibitively high.

A typical feature of plants with high RGA entries at all frequencies is that the standard design techniques seem to fail, see Maciejowski (1989) [3]. Therefore, there is no simple design technique with high performance. Powerful analysis techniques like H_∞ exist, and they are very useful to define the achievable performance for the problem.

3 Parallel Control of Plants with High RGA Entries

The class of models in which the process is included is given by:

$$G_P(j\omega) = G(j\omega)(I + w_I(j\omega)\Delta_I), \bar{\sigma}(\Delta_I) < 1, \quad \forall\omega \geq 0 \tag{5}$$

$$\Delta_I = \begin{pmatrix} \Delta_1 & 0 \\ 0 & \Delta_2 \end{pmatrix} \tag{6}$$

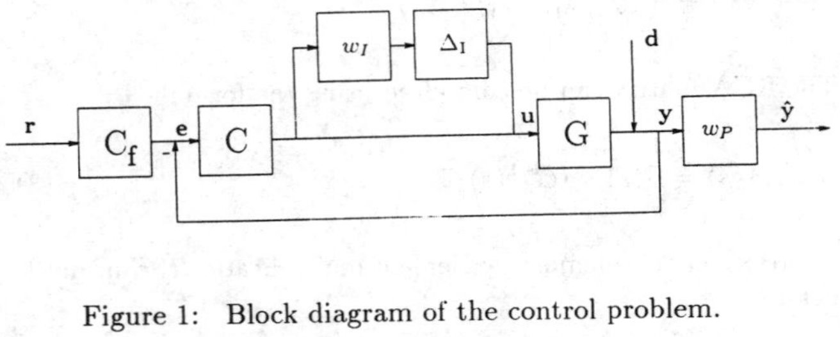

Figure 1: Block diagram of the control problem.

This model and uncertainty is shown in Fig. 1, w_I being a scalar function. We assume that we have diagonal input uncertainty since Δ_I is a diagonal matrix. w_P is the performance weight.

Definitions

Some important terms from robust control need to be specified.

Nominal performance (NS): the system fulfills the performance requirements without uncertainty.

Robust stability (RS): the system is stable for all defined uncertainty.

Robust performance (RP): the system fulfills the performance requirements for all defined uncertainty.

Review of Inverse-based Control

The following inverse-based controller has been studied by Skogestad et al. (1988) [5]:

$$C(s) = \frac{k}{s}G^{-1}(s) \tag{7}$$

The error matrix is given by:

$$G(s)w_I(s)\Delta_I G^{-1}(s) = \frac{w_I(s)}{det(G(s))}$$

$$\begin{pmatrix} g_{11}g_{22}\Delta_1 - g_{12}g_{21}\Delta_2 & g_{11}g_{12}(\Delta_2 - \Delta_1) \\ g_{21}g_{22}(\Delta_1 - \Delta_2) & g_{22}g_{11}\Delta_2 - g_{12}g_{21}\Delta_1 \end{pmatrix} \tag{8}$$

As we see the error matrix is related to the uncertainty, and the system will therefore be very sensitive to input uncertainty. The important factor, however, is that the high RGA entries implies that $det(G) = g_{11}g_{22} - g_{12}g_{21}$ becomes a small number. Hence, the the error matrix becomes large.

We choose the following input uncertainty: $\Delta_1 = -\Delta_2 = \Delta$, this is the worst case uncertainty combination. Also we get the following error term:

$$G(s)w_I(s)\Delta_I G^{-1}(s) = \frac{w_I(s)}{g_{11}g_{22} - g_{12}g_{21}}$$

$$\begin{pmatrix} \Delta(g_{11}g_{22} + g_{12}g_{21}) & -2\Delta g_{11}g_{12} \\ \Delta g_{21}g_{22} & -\Delta(g_{11}g_{22} + g_{12}g_{21}) \end{pmatrix} \tag{9}$$

From (9) we get large entries of the error matrix compared to one, which are the entries in the nominal case ($G{*}G^{-1} = I$). This shows that the decoupler is very sensitive to input uncertainty when $G(s)$ has large RGA entries.

Characteristics of Process Control Systems

We based the proceeding algorithm on the following general observations of process control systems:

- The uncertainty is frequency dependent. The uncertainty is less at lower frequencies than at higher frequencies.

- At high frequencies uncertainty will dominate, such that decoupling is difficult. Hence, diagonal control will be close to optimal.

In addition, if the system has large RGA entries we have:

- At low frequency a limited directionality compensation (decoupling) may contribute to improving performance.

- It is necessary that the controller preserves ill-conditioning of the nominal open loop transfer function, GC. This implies that GC gets high RGA entries, and $C(s)$ gets low RGA entries.

Parallel Decoupling

These observations can be formulated in the following parallel controller structure:

$$C(s) = \frac{k_{dec}}{s(T_f s+1)^n det(G(s))} \begin{pmatrix} k_x * g_{22} & -g_{12} \\ -g_{21} & k_y * g_{11} \end{pmatrix} + \begin{pmatrix} k_1 * \frac{T_{d1}s+1}{0.1T_{d1}s+1} & 0 \\ 0 & k_2 * \frac{T_{d2}s+1}{0.1T_{d2}s+1} \end{pmatrix} \quad k_x, k_y \geq 1.0 \quad (10)$$

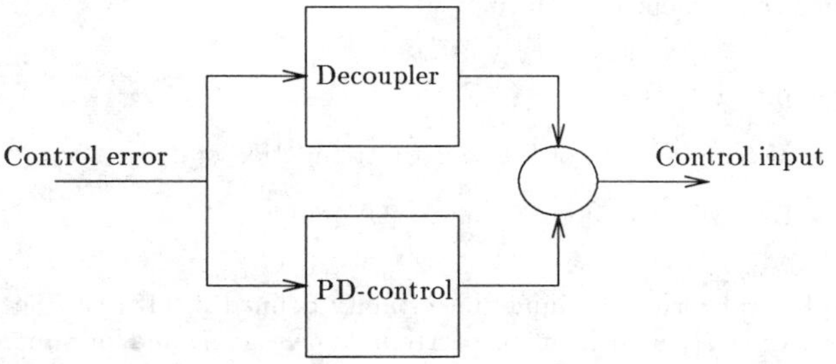

Figure 2: Block diagram of the parallel decoupler structure.

As seen from Fig. 2 the controller consists of two blocks. A modified decoupler, and a PD-controller in parallel. We recognize that the parallel decoupler is model-based only at the lower frequencies. The original decoupler, (7), is modified in two ways. First, the diagonal elements are multiplied by suitable constants, such that the diagonal dominance of the decoupler is increased. In this way the nominal open loop transfer function, GC, remains ill-conditioned after compensation. This will reduce the the sensitivity with respect to input uncertainty at the expense of nominal performance. (When applied to a plant with high RGA entries a controller with low RGA entries will not be especially sensitive to input uncertainty, but the performance will nominally not be so good). Second, the modified decoupler is low-pass filtered. The motivation for the low-pass filter is the fact that at higher frequencies the uncertainty will even dominate the modified decoupler as seen from (8) and (9). This is necessary to maintain robust performance.

In essence the decoupler performs the control at low frequency, and the PD-controller is performing the control at high frequency.

The RGA of the Parallel Decoupler

The need to increase the diagonal dominance of the decoupler is directly related to the RGA of the process as a function of frequency. If the RGA entries of the plant fall off as a function of frequency, this will reduce the need for change.

The RGA element for the modified decoupler, $\lambda_{11,dec}$, is given by:

$$\lambda_{11,dec} = \frac{k_x k_y}{k_x k_y - (1 - \frac{1}{\lambda_{11}})} \quad (11)$$

We see that appropriate choices of k_x and k_y reduce the RGA of the controller such that the sensitivity is reduced with respect to input uncertainty. The decoupler defined in (7) tries to remove all the ill-conditioning of the process which is not possible. This implies that $k_x = k_y = 1.0$ in (11). We see that the RGA element for the decoupler in this case is equal to the RGA element of the process, $G(s)$.

Characteristics of the Parallel Decoupler

The main characteristics of the controller are:

- Nominal performance is sacrificed in order to improve robust performance.

- The filter constant, T_f, is chosen in accordance with the bandwidth of the closed loop system. The order of the low-pass filter, n, is chosen such that the controller becomes realizable.

- The increase in the diagonal elements of the decoupler is chosen in accordance with the RGA number of the process.

An essential characteristic of the controller is its parallel structure. Most inverse-based controllers have a series or cascaded structure as shown in Fig. 3, where the diagonal PID-controller is used to obtain integral action. The parallel structure gives increased flexibility since the two controller parts interact differently than in the series structure. Hence, it is possible to dedicate the two different parts of the control to performing different tasks in the frequency plane as is done in the proposed controller.

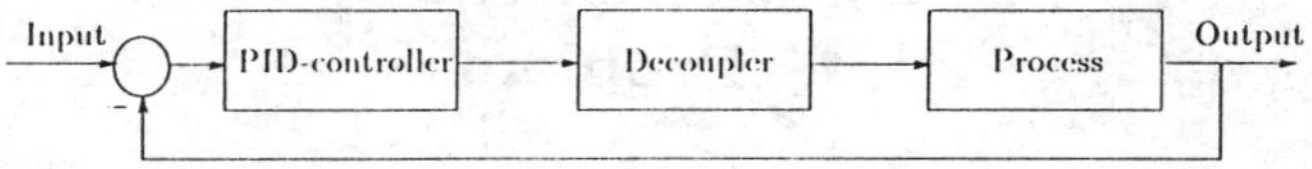

Figure 3: Conventional series decoupler structure.

The RGA element of the parallel decoupler, $\lambda_{11,C}(s)$, becomes $\lambda_{11,C}(0) = \lambda_{11,dec}(0)$ at steady state, and $|\lambda_{11,C}(s)| \to 1.0$ at higher frequencies. This implies that the parallel decoupler becomes non-interactive at higher frequencies. In this way we see that the PD-controller changes the RGA element as a function of frequency of the decoupler in Fig. 2. In Fig. 3 we observe that the decoupler is multiplied by a diagonal PID-controller.

This PID-controller cannot change the RGA element of the decoupler since the RGA matrix is independent of the diagonal scaling, property 1) of the RGA matrix. This is a fundamental difference between the two controller structures.

In summary, we can observe that one part, the decoupler, utilizes the available model information at low frequency, and the other part, the PD-controller, is active at high frequency.

Analysis of the Robustness with Respect to Diagonal Input Uncertainty

The following open loop transfer function is obtained from the decoupling part of the parallel decoupler:

$$GC_{dec} = \frac{k_{dec}}{s(T_f s+1)^n} \begin{pmatrix} 1+(k_x-1)\lambda_{11} & (k_y-1)\frac{g_{12}}{g_{22}}\lambda_{11} \\ (k_x-1)\frac{g_{21}}{g_{11}}\lambda_{11} & 1+(k_y-1)\lambda_{11} \end{pmatrix} \tag{12}$$

Equation (12) shows that the diagonal dominance of the nominal transfer function is decreased compared to the conventional inverse-based decoupler.

We define

$$C^*_{dec} = \frac{1}{det(G(s))} \begin{pmatrix} k_x * g_{22} & -g_{12} \\ -g_{21} & k_y * g_{11} \end{pmatrix} \tag{13}$$

and get

$$GC^*_{dec} = \begin{pmatrix} 1+(k_x-1)\lambda_{11} & (k_y-1)\frac{g_{12}}{g_{22}}\lambda_{11} \\ (k_x-1)\frac{g_{21}}{g_{11}}\lambda_{11} & 1+(k_y-1)\lambda_{11} \end{pmatrix} \tag{14}$$

The error term matrix then becomes :

$$Gw_I(s)\Delta_I C^*_{dec} = \frac{w_I(s)}{det(G(s))} \begin{pmatrix} g_{11}g_{22}\Delta_1 k_x - g_{12}g_{21}\Delta_2 & g_{11}g_{12}(\Delta_2 k_y - \Delta_1) \\ g_{21}g_{22}(\Delta_1 k_x - \Delta_2) & g_{22}g_{11}\Delta_2 k_y - g_{21}g_{12}\Delta_1 \end{pmatrix} \tag{15}$$

When we compare the entries of (14) and (15) we see that the uncertainty will be much less dominant here compared to the conventional decoupler, cf. the discussion of (9).

4 Presentation of the Control Problem

A linear plant is called ill-conditioned when the condition number of the transfer function matrix of the system is large. This is a property that is only found in multivariable systems. It means that the gain of the plant is highly dependent on the input direction. This implies that the plant has lowgain and highgain directions. These directions may be found by singular value decomposition of the model.

In order to get adequate gain in the lowgain direction, high control gain should be applied. However, as shown by Skogestad et al. (1988) [5] uncertainty may change the input to the highgain direction. This may have intolerable consequences on the control performance.

A high-purity distillation column is an example of a plant with these characteristics. This paper uses the following model of a high-purity distillation column. The model is presented in Skogestad et al. (1988) [5].

$$G(s) = \frac{1}{75s+1} \begin{pmatrix} 0.878 & -0.864 \\ 1.082 & -1.096 \end{pmatrix} \tag{16}$$

The condition number, $\gamma(G)$, of the plant (the ratio of the largest to the smallest singular value) is:

$$\gamma(G) = \frac{\overline{\sigma}(G)}{\underline{\sigma}(G)} = 141.7 \tag{17}$$

The following input uncertainty weight, $w_I(s)$, is applied

$$w_I(s) = 0.2\frac{5s+1}{0.5s+1} \tag{18}$$

This means that we have 20% input uncertainty at low frequency and 200% uncertainty at high frequency.

The performance weight, $w_P(s)$, is chosen as:

$$w_P(s) = 0.5\frac{10s+1}{10s} \tag{19}$$

This is equivalent to requiring:

- Integral action
- Maximum peak of the sensitivity function of 2
- Bandwidth of approximately 0.05 rad/min.

In physical terms the input uncertainty defined in (18) can be approximately equivalent to partly unknown gains and/or time delays. In subsequent simulations we will select our process from the following model class:

$$G_P(s) = G(s) \begin{pmatrix} l_1 e^{-\theta_1 s} & 0 \\ 0 & l_2 e^{-\theta_2 s} \end{pmatrix} \tag{20}$$

where $0.8 \le l_1 \le 1.2$, $0.8 \le l_2 \le 1.2$, $0 \le \theta_1 \le 1.0$, $0 \le \theta_2 \le 1.0$.

5 Analysis and Simulation

Skogestad et al. (1988) [5] discusses solutions to the above defined control problem. It is shown how a decoupler cannot be used since it has an unacceptable robust performance.

When applying a PI-controller, the gain has to be significantly reduced due to the bandwidth limitation introduced by a delay in the input. If the gain is reduced too much, then performance becomes unacceptably low. It should be noted, however, that the PI-controller has considerably better robust performance than a decoupler.

The μ-optimal controller has the best performance. This is, however, a higher-order controller and therefore mostly of theoretical interest. This is because it is favourable that the tuning parameters have physical interpretation in practical applications. Nevertheless, it is useful as a measure to define the achievable controller performance for this problem.

We propose the following controller for the previously defined control problem, cf. (10):

$$C(s) = k_{dec} * \frac{75s+1}{s(T_f s+1)} \begin{pmatrix} k_x * 39.942 & -31.487 \\ 39.432 & -k_y * 31.997 \end{pmatrix} + k * \frac{T_d s+1}{0.1 * T_d s+1} \begin{pmatrix} 1 & 0 \\ 0 & -1 \end{pmatrix} \quad (21)$$

The following tuning parameters are chosen:

$k_{dec} = 0.044$
$k = 44$
$T_d = 0.47$
$k_x = 2.0$
$k_y = 2.0$
$T_f = 7.0$

Robustness Analysis of the Parallel Decoupler

With the above chosen tuning parameters we get from (14):

$$GC^*_{dec} = \begin{pmatrix} 36.1 & 27.7 \\ 43.3 & 36.1 \end{pmatrix} \quad (22)$$

The error matix becomes

$$Gw_I(s)\Delta_I C^*_{dec} =$$

$$0.2\frac{5s+1}{0.5s+1} \begin{pmatrix} 35.1 * k_x + 34.1 & -27.7 * (k_y + 1.0) \\ 43.2 * (k_x + 1.0) & -35.1 * k_y - 34.1 \end{pmatrix} \quad (23)$$

if we choose $\Delta_1 = -\Delta_2 = 1.0$ in (15) which is the worst case uncertainty combination for this problem. We get the following error matrix at steady state:

$$Gw_I(0)\Delta_I C^*_{dec} = \begin{pmatrix} 20.9 & -16.6 \\ 25.9 & -20.9 \end{pmatrix} \quad (24)$$

We see that the uncertainty influences this decoupler much less than the original conventional decoupler. At steady state (22) and (24) give us a relative change of about 60% in the entries of the open loop transfer matrix. This is much less than for the conventional decoupler.

The RGA of the Parallel Decoupler

We also observe from (23) that at high frequencies, the error terms will dominate even for this modified decoupler. Therefore, the low-pass filter is necessary. In Fig. 4 we see $|\lambda_{11,C}|$ for the parallel decoupler with and without the low-pass filter $\frac{1}{T_f s+1}$, see (21). Also the low-pass filter contributes to making the controller diagonal in the frequency range where uncertainty dominates. $|\lambda_{11,C}| = 1.0$ implies that the controller is diagonal.

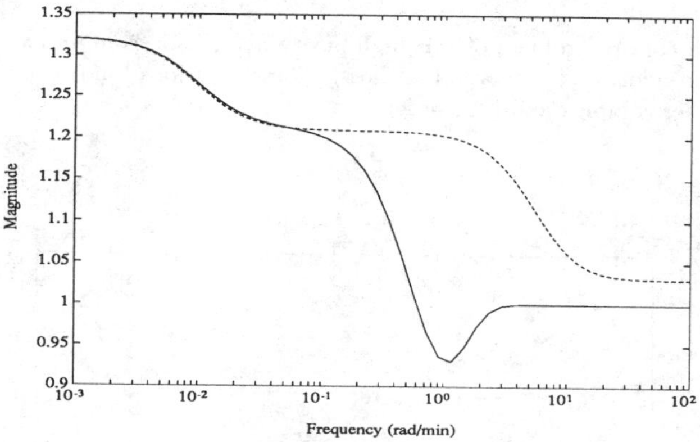

Figure 4: $|\lambda_{11,C}|$ for the parallel decoupler with low-pass filter (solid), and without low-pass filter (broken).

The PD-controller will be active at higher frequencies. In this frequency range it is not possible to correct for directionality because the uncertainty will dominate totally. The PD-controller does not have high gain in any particular direction, and therefore will not be especially sensitive to input uncertainty in any particular direction.

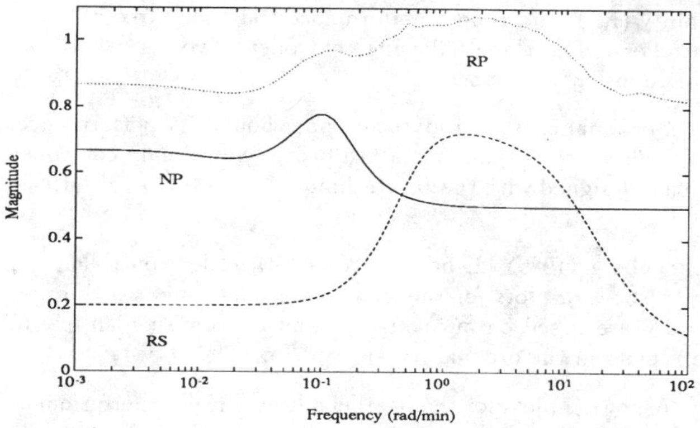

Figure 5: μ-plots for the "μ-optimal" controller.

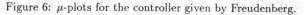

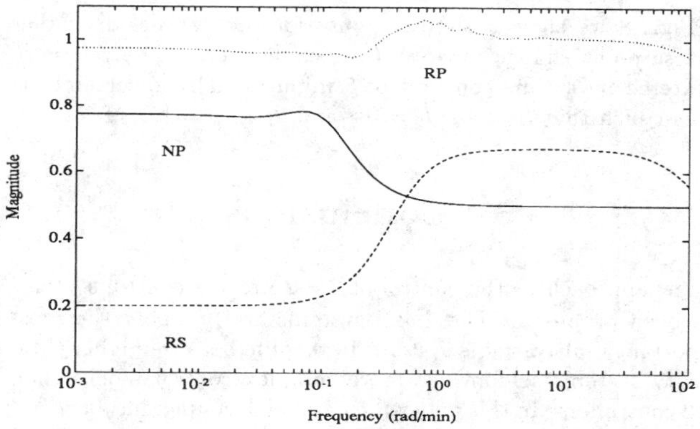

Figure 6: μ-plots for the controller given by Freudenberg.

Comparison between the Parallel Decoupler and other Methods

The control problem of this high-purity distillation column has been addressed by several authors. These solutions tend to be rather complex and high order.

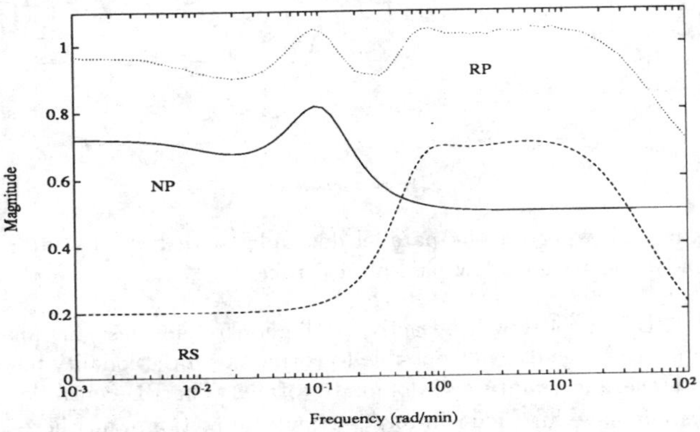

Figure 7: μ-plots for the parallel decoupler.

μ-plots in the figures show nominal performance (NP), robust stability (RS), and robust performance (RP) for three different controllers. The three following controllers give essentially the same robust performance.

The μ-optimal controller gives an upper bound for performance. In Fig. 5 we see μ-plots for a 6th order "μ-optimal" controller which is designed with the H_∞-technique, Skogestad et al. (1988) [5].

Freudenberg (1989) [2] has derived a 5th order controller. In Fig. 6 shows μ-plots for this controller. The μ-plots in Figures 5 and 6 are based on our software, and are almost identical to the μ-plots in the original papers.

Fig. 7 shows μ-plots for the parallel decoupler. The performance of this controller is very close to optimal despite the relatively simple parametrization. More important is the fact that the tuning parameters in the parallel decoupler have a physical interpretation. This is of decisive importance in practical applications.

In Fig. 8 we show a time response for the parallel decoupler to a setpoint change in y_1 at time equals zero. The reference is filtered by a time constant of 5 minutes. The uncertainty is chosen such that $l_1 = 1.2$, $l_2 = 0.8$ and $\theta_1 = \theta_2 = 1.0$, cf. (20).

6 Conclusions

A new approach to the control of 2×2 process control systems has been proposed. The 2×2 systems are probably the most important multivariable systems in real life because higher complexity systems seldom use more complex control blocks than 2×2 controllers. In this paper the parallel decoupler has been applied to an ill-conditioned, high-purity distillation column with large RGA entries at all frequencies. The controller utilizes the fact that the nominal open loop transfer function, GC, needs to remain ill-conditioned after compensation.

The μ-optimal controller has better performance, but this is a higher-order controller and therefore mostly of theoretical interest.

The main advantage of the parallel decoupler is its simple parametrization. In fact, there is no efficient simple design technique for plants with high RGA entries. Preliminary results show that the parallel decoupler structure gives relatively good performance to process control systems with low RGA entries as well.

An important feature of the controller is its parallel structure. This is essential in order to dedicate the two different control parts to two different control tasks in the frequency plane.

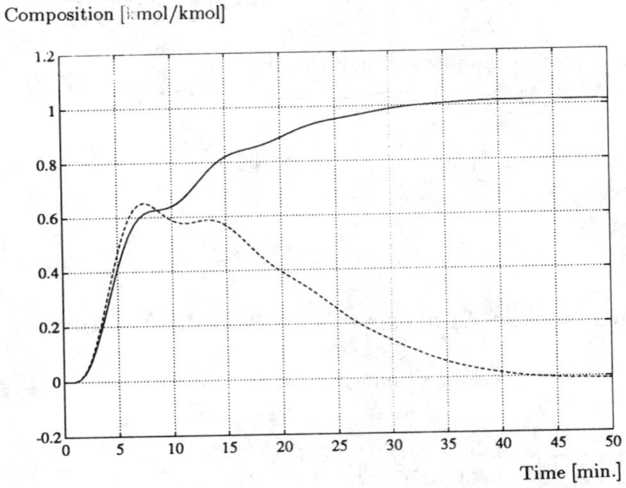

Figure 8: y_1 (solid) and y_2 (broken) for the parallel decoupler.

Acknowledgements

The authors would like to thank Sigurd Skogestad, The Norwegian Institute of Technology, and Tørris Digernes, Aker Elektro a.s. for their stimulating discussions.

References

[1] Grosdidier, P., Morari, M. and Holt, B. R., (1985) *Closed-Loop Properties from Steady-State Gain Information.* Ind. Eng. Chem. Fundam., **24**, 221-235

[2] Freudenberg, J. S., (1989) *Ill-conditioned Plant Analysis and Design.* Int. J. Control, **49**, p. 873-903

[3] Maciejowski, J. M. (1989) *Multivariable Feedback Design.* Addison-Wesley.

[4] Morari, M. and Zafiriou, E., (1989) *Robust Process Control.* Prentice-Hall.

[5] Skogestad, S., Morari, M. and Doyle, J. C., (1988) *Robust Control of Ill-conditioned Plants: High Purity Distillation.* IEEE Transactions on Automatic Control, **33**, p. 1092-1105 (Also see correction to μ-optimal controller in **34**, 672)

Proceedings of the 31st Conference
on Decision and Control
Tucson, Arizona • December 1992

SUPERVISION OF A STEEL STRIP RINSING PROCESS

B. Sohlberg

University College of Falun Borlänge
P.O.Box 100 44
781 10 Borlänge, Sweden
E-mail: bso@hfb.se

ABSTRACT

This paper deals with condition supervision of a steelstrip rinsing process. The rinsing process is a dynamic nonlinear process. Modelling and identification of the process is based on apriori knowledge about the process and measured data from the process, known as grey-box identification.

Some parts of the process wear out and are changed after manual inspection. In the model, the worn parts are modelled explicitly and estimated from measured data from the process. Data are collected before and after the worn parts are exchanged. The estimation is first made by off line identification by optimizing the likelihood function. Secondly the estimation is made by using an Extended Kalman Filter. The result of the estimation is used to give a basis for a decision of which worn parts are to be exchanged.

Key-words: Grey box identification, Extended Kalman Filter, Supervision, Rinsing Process.

1. INTRODUCTION

Many industrial processes have parts which wear out while the process is running. These parts are usually manually inspected and an operator has to decide whether a part should be exchanged or not. Normally a decision is taken based on experience or practice. If the worn parts are changed too often it gives rise to extra costs for spare parts and if the parts are changed too seldom, the process will not behave proporly.

In this paper we use a model of the rinsing process based on apriori knowledge about the process and unknown parts of the process are identified using measured data from the process.

Wear out parts of the process are modelled explicitly and have to be estimated. Estimation is done with measured data collected before and after exchange of worn out parts. To investigate the invariance of the unknown parts, estimation is first done off line by optimizing the likelihood funktion. Secondly the estimates is made by the Extended Kalman Filter technique.

The result of the estimation is used to compute a measure of the condition of the wear out parts. By comparing the condition it seems possible to decide which parts are in best condition and which parts are in worst condition. The compound influence of the worn out parts is discussed.

2. DESCRIPTION OF THE RINSING PROCESS

The rinsing process is a part of the pickling line at Swedish Steel Corporation. The pickling line consists of tanks of hydrochloric acid and rinsing tanks. The steel strip continuously passes first the acid tanks and directly afterwards the strip is cleaned by passing the rinse tanks. Depending on several production factors the amount of acid transferred from the acid tanks into the rinse tanks varies, this means that the purity of the rinse water, and indirectly the purity of the strip, depends on the actual production.

The rinsing process consists of five rinse zones. The acid is transferred from the last acid tank into rinse zone 1 via the strip. In the zone the strip is rinsed by a circulated flow of rinse water. This means that the liquid retained on the strip has a lower concentration of acid at the end of the rinse zone. The squeezer rolls between the rinse zones are there to reduce the amount of liquid transferred. Squeezer rolls are iron rolls with surface of rubber. The strip is continuously rinsed when passing the five rinse zones, fig. 1. Clean water is fed into rinse tank 5. The equivalent amount of rinse water flows from tank 5 to tank 4 and so on. A flow of clean water is also fed into tank 2.

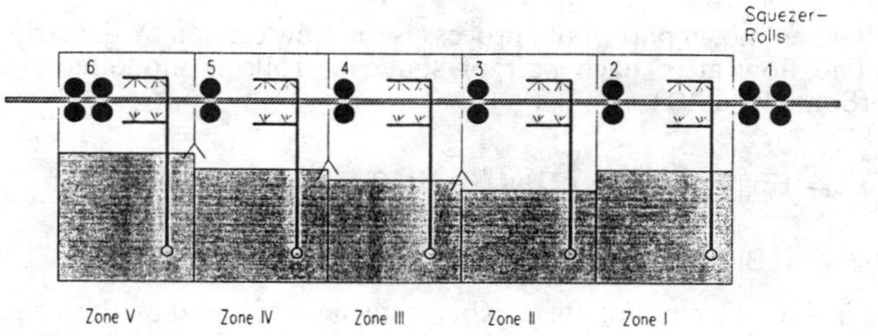

Fig. 1. Rinsing process.

3. GREY BOX MODEL

A mathematical model of a physical process can conventionally be made in one of two different ways, (Bohlin 1991):

- The process can be completely described by mathematical equations as differential equations, algebraic equations, logical relations and so on. This type of model is described as "white box model".
- The process can be seen as completely unknown. It is not necessary to use a model structure that reflects the physical property of the process. During the identification, a model selected from a group of standard models is used to describe the process. The parameters of the model are adapted to input-output data from an experiment with the process.

The amount of apriori knowledge varies from process to process. Between the "white box" model and the "black box" model there is a grey zone, where it is possible to talk about a "grey box" model, (Bohlin 1991).

CH3229-2/92/0000-2557$1.00 © 1992 IEEE

For the rinsing process each rinse zone is considered as a subsystem, these subsystems are coupled by different flows, fig. 2, (Stein 1988).

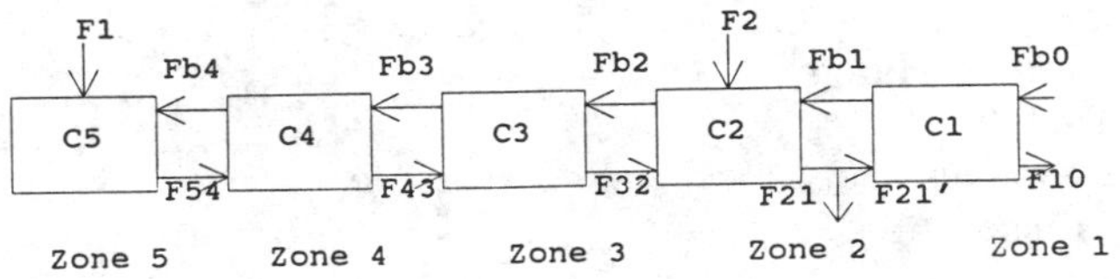

Fig. 2: Diagram of five-step rinse process.

In fig. 2 there are different types of flows. F54 .. F21 represent the main flow of rinse water between the tanks, F1 and F2 are the controllable flows of clean water into tank 5 and tank 2. Fb0 .. Fb4 are flows created by the strip through the rinsing process.

The concentration of hydrochloric acid C1 .. C5 in tank 1 .. tank 5 is measured by conductivity transducers. At low levels of concentration the relation between acid and conductivity is linear, (Norton 1970).

The amount of hydrochloric acid in the tanks is modelled by the principle of mass balance, (Kushner 1976). The concentration in each tank will be a state if the mixing in the tanks is perfect. Inputs to the process consist of the three production factors, strip velocity, strip width, strip thickness and the two flows of clean water. Furthermore, the states and the outputs are influenced by noise, which is modelled as discrete time white noise.

The unknown part of the process is the flow carried by the strip. The flow after each pair of squeezer rolls is modelled by, (Sohlberg 1991):

$$Fb_i = Kb_i * Bv * Bb + Kt_i * Bv * (Bt - Btoff_i) * \alpha \qquad (3.1)$$

$\alpha = 1$ if $Bt > Btoff_i$ else $\alpha = 0$

i is the rolls number, fig. 1, Kb_i is thickness of liquid on the strip, Kt_i is thickness of liquid on the edge of the strip, $Btoff_i$ is an offset parameter. Bv is strip velocity, Bb is strip width and Bt is strip thickness. In equation (3.1) Kb_i, Kt_i and $Btoff_i$ are unknown parameters and need to be estimated by data measured from the process.

4. OFF LINE ESTIMATION

4.1 Measured data from the process

The purpose of collecting input-output data from the process is to achieve data which can be used to estimate the unknown parameters in equation (3.1). The interest is focused on in what way changes of squeezer rolls influence the flow made by the strip. Furthermore it is of interest whether the function of the squeezer rolls changes during a longer period.

Three measurement sequences are collected. Sequence #1 is collected before the change of squeezer rolls and consists of 570 samples, sequence #2 is collected after the change of squeezer rolls and consists of 800 samples, sequence #3 is collected directly after sequence #2 and consists of 700 samples. Squeezer rolls 1,2,4 and 6 were changed and squeezer rolls 3 and 5 were not changed between sequence #1 and #2.

During the collection of data, the process was controlled by the ordinare controller. The flow of clean water fed into tank 5 is controlled by a feed-forward controller, proportional to the product of strip velocity and strip width. The flow of clean water fed into tank 2 is controlled by a on-off controller. If the conductivity in tank 2 is higher than an upper limit or the liquid level in tank 2 is lower than a lower limit, 1 m3/h of clean water is fed into tank 2.

4.2 Choice of process model

Estimation of the unknown parameters in equation (3.1) is made by using a software named Cypros/Kalmax, (Camo 1987). It is based on the maximum likelihood identification method.

Equation (3.1) gives one unknown parameter to estimate for the flow made by the top- and bottomside of the strip and two parameters to estimate the flow beside the strip. Totally, this gives too many parameters to estimate when the estimation is to be made on line. The number of unknown parameters describing the flow beside the strip is reduced by setting one of the unknown parameters at a constant value. Since there are two parameters there are two possibilities. The model is chosen which gives largest value of the likelihood function, (Bohlin 1991).

The flow made by the strip after the double pair of squeezer rolls 6 is very small and can be neglected. The flow after the double pair of squeezer rolls 1 is also very small, but due to high concentration in the liquid film on the strip it cannot be neglected. Adaptation to measured data gives the result, that no connection between this flow and strip thickness.

The identification yields also that the estimate of unknown parameters describing the flow after squeezer rolls 5 are not reproducable. So this flow is modelled only by the flow on the top- and bottomside of the strip, which gives the largest value of the likelihood function compared to the case when the flow is modelled only by a flow besides the strip. The model gives a total of 8 parameters to estimate.

4.3 Results

The result of the estimation of the 8 unknown parameters with the three measurement sequences is presented in fig. 3. Between sequence #1 and #2 there was a planned stop of the process and squeezer rolls 1, 2, 4, and 6 were exchanged. Squezeer rolls 3 and 5 were not exchanged at the planned stop. Some comments can be made for the different pairs of squeezer rolls:

- **Pair of squeezer rolls 1**

These rolls were changed during the planned stop. From fig. 3 it is seen that the new rolls are better than the old rolls. Kb1 is smaller for the new rolls which means that the liquid film is thinner after the change of rolls

- **Pair of squeezer rolls 2**

These rolls were changed during the planned stop. From fig. 3 it is seen that Btoff2 is increased for the new rolls. This means that less flow goes besides the strip with the new rolls. Comparing the estimate of Btoff2 for measurement sequences #2 and #3, the rolls had hardened. There is no change in film thickness on the strip.

- **Pair of squeezer rolls 3**

These rolls were not changed during the planned stop. From fig. 3 it can be seen that Btoff3 has decreased between measurement sequences #1 and #2. This means that the flow beside the strip has increased somewhat due to the fact that wears of the rolls have increased between measurements.

• **Pairs of squeezer rolls 4**

These rolls were changed during the planned stop. From fig. 3 it can be seen that Btoff4 is less and Kb4 is larger for the new rolls. This means that the new rolls are worse than the old rolls.

• **Pair of squeezer rolls 5**

These rolls were not changed during the planned stop. Kb5 has increased much between the three measurement sequences. This means that the rolls have worn out when the process is running.

5. ON LINE ESTIMATION

5.1 Extend Kalman Filter

The unknown parameters of the model need to be estimated on line because they are going to be used for deciding whether a pair of squeezer rolls are good or bad. The Extended Kalman Filter is an algoritm used in the field of nonlinear filtering, (Jazwinski 1970). Here, the filter is used to estimate both the state of the process and the unknown parameters.

Introduce a vector $\theta(k)$, consist of the unknown parameters to be estimated:

$$\theta = [\mathrm{Kb1, Kb2, Btoff2, Kb3, Btoff3, Kb4, Btoff4, Kb5}]^T \quad (5.1)$$

For simplicity in notation, the sampleindex k has been dropped in equation (5.1).

The vector $\theta(k)$ is modelled as:

$$\theta(k+1) = \theta(k) + v_\theta(k) \quad (5.2)$$

where $\theta(k)$ is a 8-dim vector, with an initial state which is gaussian with mean value $\theta(0)$ and covariance matrix $P_\theta(0)$, and $v_\theta(k)$ is discrete time white noise with zero mean and covariance matrix R_θ.

5.2 Results

An Extended Kalman Filter has been simulated with the same data sequences that were used in off line estimation. R_1 and R_2 are the same as in chapter 4. R_θ and P(0) are chosen by trial and error. The result of the simulation with measurement sequence #1 is presented in fig. 4a.

From fig. 4a it can be seen that the estimated parameters converge to the estimated values achieved during off line estimation.

The measure sequences #2 and #3 are brought together in a long sequence to give the opportunity to study the wear out effect of the squeezer rolls. A simulation shows that the estimate of the parameters follows the changes, which were detected during off line estimation when comparing the results between sequence #2 and #3, fig.4b.

6. BASIS FOR A DECISION TO CHANGE SQUEEZER ROLLS

6.1 Efficiency of squeezer rolls

During the planned stop squeezer rolls 2 and 4 were exchanged; rolls 1 and 6 are always exchanged at the planned stop. From the discusion in previous chapters it is pointed out that squeezer rolls 2 were in bad condition and squeezer rolls 4 were in good condition. Therefore it is not simpel matter to decide by manual inspection which squeezer rolls are to be exchanged.

The parameters estimated on line are used in this chapter to compute the flow carried by the strip after the squeezer rolls. This flow is nonlinearly dependent on the productionparameters strip velocity, strip width and strip thickness, equation (3.1). Therefore the flow carried with the strip is related to a nominal stationary input and a nominal stationary state, which is representative for the process.

At the nominal stationary state, the reduction o f concentration between two tanks is formulated as:

$$\eta_i = \frac{C_{i-1}}{C_i} \qquad i = 2, .., 5 \qquad (6.1)$$

where C_i is the concentration in tank i, figure 2.

By using the principle of mass balance the reduction at the nominal stationary state is achieved by:

$$\eta_i = 1 + \frac{F1+F2}{F_{i-1}} \qquad i = 2 \qquad (6.2a)$$

$$\eta_i = 1 + \frac{F1}{F_{i-1}} \qquad i = 3, .., 5 \qquad (6.2b)$$

where F1 and F2 are input flows and Fb_i is the flow done with the strip.

The reduction of concentration between the tanks is equal to the reduction of conductivity, because the conductivity is proportional to the concentration. Then the reduction of conductivity from tank 1 to tank 5 can be formulated as:

$$\mathrm{Led1} = \eta_2 * \eta_3 * \eta_4 * \eta_5 * \mathrm{Led5} \qquad (6.3)$$

where Led1 and Led 5 represent the conductivity in tank 1 and tank 5.

A reasonable value of the conductivity in tank 5 is 1 mS/m to achive a proper result of the rinsing process. Take the natural logarithm of equation (6.3) and let Led5 be 1, then equation (6.3) can be formulated as:

$$\mathrm{Led1} = \ln(\eta_2) + \ln(\eta_3) + \ln(\eta_4) + \ln(\eta_5) \qquad (6.4)$$

Equation (6.4) gives the condition to achieve a conductivity of 1 mS/m in tank 5 at the nominal stationary state. Notice that the condition can be achived in many ways. This means that a bad pair of squeezer rolls can be compensated by an other efficient pair of squeezer rolls.

6.2 Results

The reduction is computed based on the estimated parameters achieved from an Extended Kalaman Filter. The nominal inputs are chosen so the influence of all inputs are considered. This means that the strip thickness needs to exceed a lower limit, equation (3.1), to have an influence on the flow done by the strip. The nominal inputs are then chosen as the mean values of the inputs under the condition that the strip thickness exceeds 3.0 mm.

The natural logarithm of the reduction of concentration or conductivity is computed from equation (3.1) and equation (6.2). The estimated parameters achieved at the end of sequence #1, #2 and #3 are used in equation (3.1). The result of the reduction between the tanks is presented in table 1.

Squeezer rolls	Sequence #1	Sequence #2	Sequence #3
2	1.71	2.99	2.99
3	2.24	1.84	1.80
4	2.56	2.18	2.02
5	1.73	1.45	1.24
Sum	8.24	8.46	8.05

Table 1: Logaritm of the reduction between the tanks done by the squeezer rolls.

From table 1 it can be seen that after sequence #1 the reduction made by pair of squeezer rolls 2 is smallest. This means that these rolls were in worst condition compared to the other rolls. It can also be seen that the pair of squeezer rolls 4 were in best condition. Notice that both pairs of squeezer rolls were changed after sequenze #1.

After exchange of squeezer rolls 2, you can see that the efficiencies of the these rolls has increased. Notice that after change of squezeer rolls 4, the efficiency of these rolls has decreased. Comparison of the sum of reduction between sequence #2 and #3 shows that the efficiency of squeezer rolls has decreased, this means that the rolls are worn out.

The mean value of the conductivity in tank 1 is about 3300 mS/m. Based on this value of the conductivity and the result of table 1, it is possible to compute the conductivity in the other tanks. This gives a conductivity profile of the process, fig. 5.

From fig. 5 it can be seen can that the conductivity in tank 2 has decreased after the change of squeezer rolls 2. The improvement is partly lost in the later stage of the process because the efficiency of squeezer rolls 3 and 5 has decreased.

7. CONCLUSIONS

First, estimation of unknown parameters describing the flow done by the strip is made by off line estimation. The result shows that it is possible to estimate the effect of the wear out of the squeezer rolls. It also shows that it is a difficult to decide which squeezer rolls should be excanged only by manual inspection of the rolls, because both worn out rolls and rolls not worn out are exchanged.

Secondly, the estimation of the unknown parameters is made by an Extended Kalman Filter. Simulation shows that the filter estimates the parameters as well as off line estimation.

Finally, a basis for decision of which squezer rolls should be exchanged is presented. This gives information about which pair of squeezer rolls is in best and which is in worst condition. It also gives the compound effect of all pairs of squeezer rolls to achieve a proper result when the process is running.

ACKNOWLEDGMENT

The author would like to thank Professor T. Bohlin, Royal Institute of technology, Sweden, for his many valuable comments and helpful discussions. The research descriebed in this paper was supported by the Swedish Technology Development Board, the University College of Falun Borlänge and Swedish Steel Corporation.

REFERENCES

Bohlin T. (1991). *Interactive System Identification; Prospects and Pitfalls*. Springer-Verlag, Berlin.

Camo A/S (1987). *Manual for Cypros/Kalmax*, Tronheim, Norway

Jawinski, A. H. (1970). *Stochastic processes and filtering theory*, Acadenic Press, New York

Kushner J. B. (1976) *Water and waste control for the plating shop*. Gardner Publications Inc USA.

Norton H. (1970) *Sensor and analyzer handbook*. Prentice Hall.

Sohlberg B. (1991). *Computer aided modelling of a rinsing process*. IMACS Symposium MCTS Casablanca -1991.

Stein B. (1988). *Recurative rinsing a mathematical approach*, Metal Finishing January 1988.

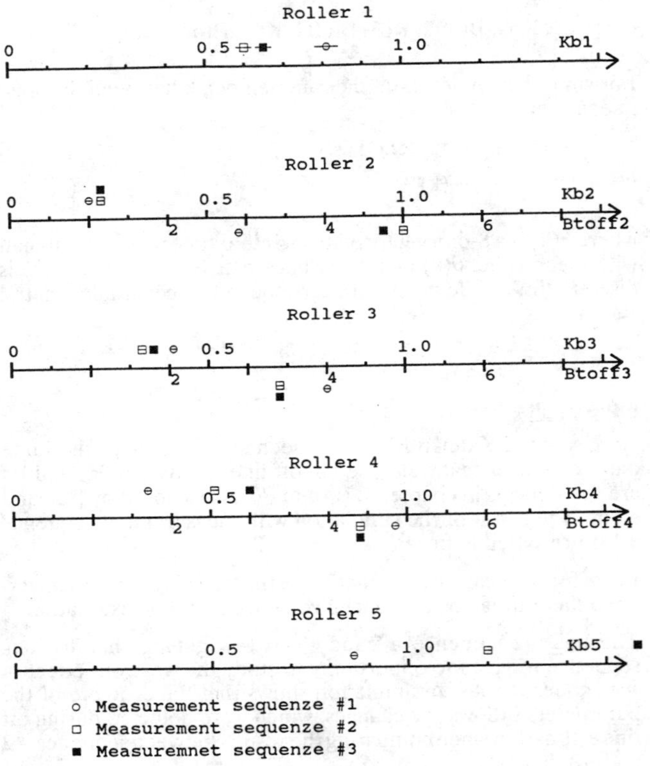

Fig. 3. Result from off line estimation.

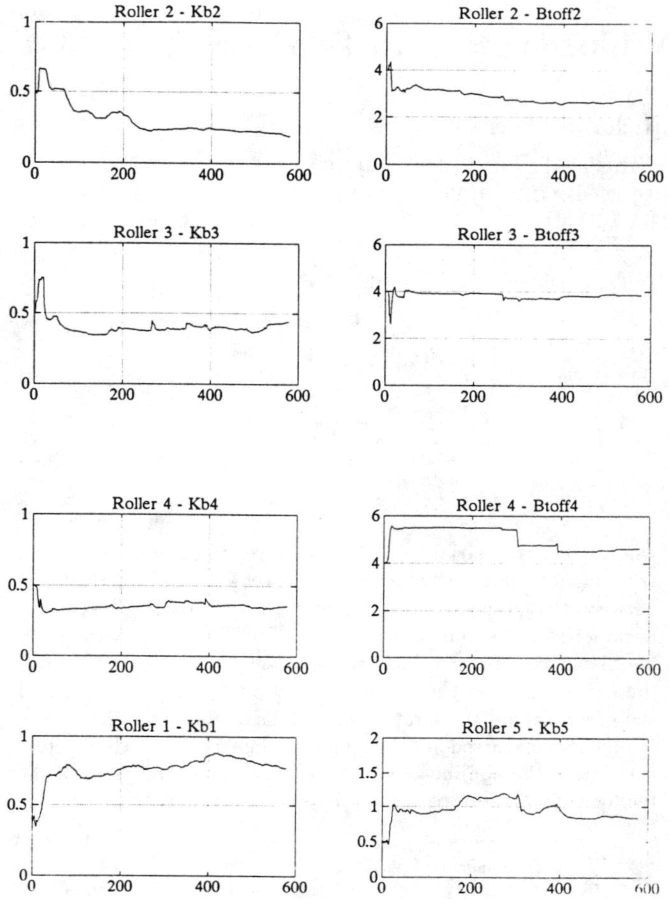

Fig. 4a. Results from on line estimation, measurement sequence #1

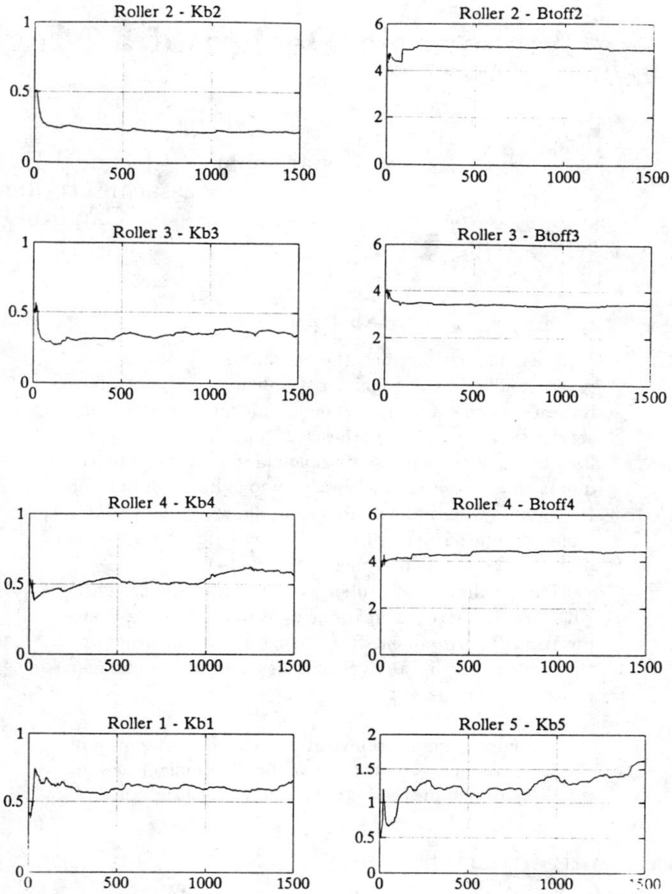

Fig. 4b. Results from on line estimation, measurement sequence #2 and #3.

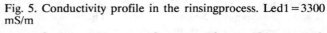

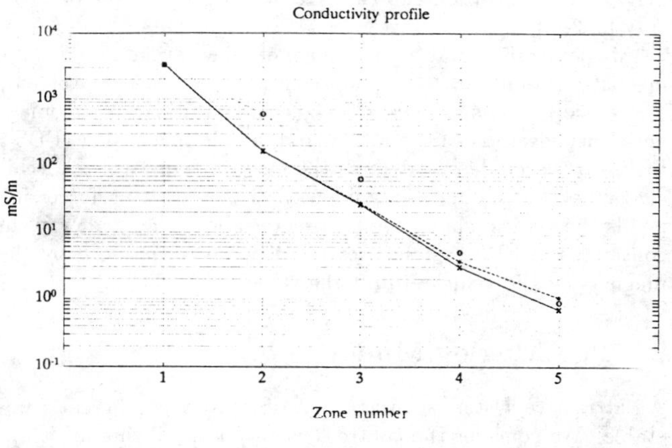

Fig. 5. Conductivity profile in the rinsingprocess. Led1=3300 mS/m

.... Sequence #1 ______ Sequence #2 -------- Sequence #3

FA3 - 9:10

Proceedings of the 31st Conference
on Decision and Control
Tucson, Arizona • December 1992

Automatic Design of a Maglev Controller in State Space

Feng Zhao[1] Richard Thornton
Department of Electrical Engineering and Computer Science
Massachusetts Institute of Technology
Cambridge, MA 02139

Abstract

This paper describes the automatic synthesis of a global nonlinear controller for stabilizing a magnetic levitation system—a simplified model for the German Transrapid system. The synthesized control system stabilizes the maglev vehicle with much larger initial displacements than those allowed in a previous linear design for the same system. The simulation shows that our nonlinear design outperforms the linear design by a factor of 20 on the nominal model of the system.

The nonlinear controller is automatically designed with the **Control Engineer's Workbench**, a computational environment integrating a suite of programs that automatically analyze and design global nonlinear control systems.

Keywords: Magnetic levitation; control system synthesis; state-space methods; artificial intelligence; numeric/symbolic processing; computer-aided design.

1 Introduction

Magnetically levitated trains provide a high speed, very low friction alternative to conventional trains with steel wheels on steel rails. Several experimental maglev systems in Germany and Japan have demonstrated that this mode of transportation can profitably compete with air travel. More importantly, maglev transportation can ease traffic congestion and save energy [2, 3, 5].

Maglev transportation uses magnetic levitation and electromagnetic propulsion to provide contactless vehicle movement. There are two basic types of magnetic levitation: electromagnetic suspension (EMS) and electrodynamic suspension (EDS). In EMS, the guideway attracts the electromagnets of the vehicle that wraps around the guideway. The attracting force suspends the vehicle about one centimeter above the guideway. In contrast, the EDS system uses repulsive force, induced by the magnets on the vehicle, to lift the vehicle.

2 The Maglev Model

An attractive system such as the EMS system is inherently unstable. We consider the control design for stabilizing an EMS-mode train traveling on a guideway—a simplified model for the German Transrapid experimental system. The Transrapid system is schematically shown in Figure 1. It uses attractive magnetic forces to counterbalance gravitational forces.

The state equations for the magnetically levitated vehicle and the guideway are described by

$$\begin{cases} \dfrac{dx}{dt} = \dfrac{z(V_i - Rx)}{L_0 z_0} + \dfrac{xy}{z} \\ \dfrac{dy}{dt} = g - \dfrac{L_0 z_0 x^2}{2mz^2} \\ \dfrac{dz}{dt} = y \end{cases} \qquad (1)$$

where the state variables x, y, and z represent coil current in the magnet, vertical velocity of the vehicle, and vertical gap between the guideway and the vehicle, respectively. The control parameter is the coil input voltage V_i. The other parameters are the mass of the vehicle m, the coil resistance R, the coil inductance L_0 and the vertical gap z_0 at the equilibrium, and the gravitational acceleration g. Details of the derivation of the model are discussed in [4]. The nonlinearities of the system come from the nonlinear inductance due to the geometry of the magnet and the inverse square magnetic force law.

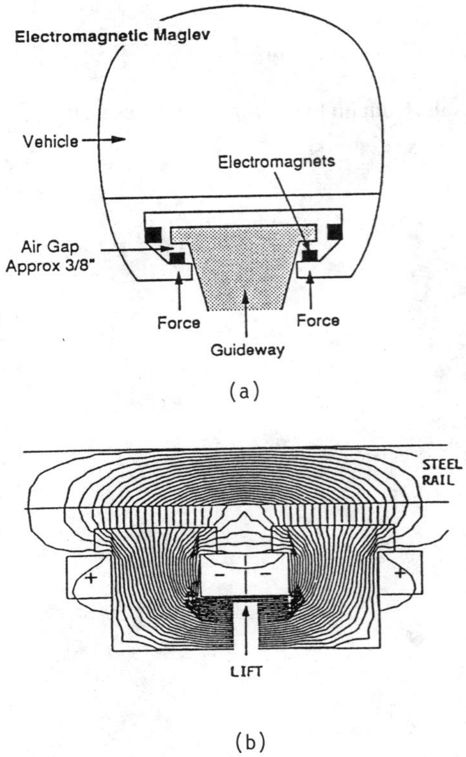

Figure 1: EMS maglev system for high-speed ground transportation, representing a simplified drawing of the German Transrapid design. (a) Electromagnetic suspension (from [MTAC Report 1989]); (b) Detail of a suspension magnet, superimposed on the field distribution (from [Eastham 1989]).

[1]Zhao's current address is: Department of Computer and Information Science, Ohio State University, 2036 Neil Avenue, Columbus, OH 43210. Tel. 614-292-1553.

CH3229-2/92/0000-2562$1.00 © 1992 IEEE

The system has one equilibrium state at which the magnetic force exactly counterbalances the force due to gravity and the vehicle has no vertical velocity and acceleration. However, the equilibrium is a saddle node which is not stable. The control objective, therefore, is to stabilize the vehicle traveling down the guideway and to maintain a constant distance between the vehicle and the guideway despite any roughness in the guideway. The available control input is the coil input voltage V_i in the model (1). We further assume that V_i is produced by a buck converter capable of delivering any voltage from 0 to 300 volts.

A linear control design for the maglev system described in [4] uses the pole-placement method. The system is first linearized around the equilibrium. The linearized system has unstable poles, i.e., the poles in the right-half of s-plane. A linear feedback is introduced to move the poles to the desired locations in the left-half of the s-plane. Such a control design can bring the system back to the equilibrium with an initial displacement of up to 0.2mm from the equilibrium. The linear controller saturates at the beginning for larger initial displacements. This is because the linearized model no longer approximates the original system well in regions far away from the equilibrium. A global, nonlinear control law such as a bang-bang control that respects the nonlinearity of the system must therefore precede the linear feedback control. However, the real challenge for the nonlinear design is to determine the global control law specifying, for instance, the switching points.

3 The Control Engineer's Workbench

Our global nonlinear control law for the maglev system is designed with computational tools. We have constructed a computational environment, **the Control Engineer's Workbench**, integrating a suite of programs that automatically analyze and design high-performance, global controllers for a large class of nonlinear systems [8]. These programs combine powerful techniques from numerical and symbolic computations with novel representation and reasoning mechanisms of artificial intelligence. The two major components in the Workbench—the Phase Space Navigator and MAPS—work together to visualize and model the state-space geometry and topology of a given system. They reason about and manipulate the state-space geometry and topology and search for optimal control paths connecting initial state and the desired state for the system. The Workbench represents the result of design and analysis in a symbolic form manipulable by other programs, and produces a high-level summary meaningful to professional engineers. It also presents the result in a graphical form.

The Workbench employs a state-space design method that synthesizes control systems in state space or phase space. The design method computationally exploits dynamical systems' nonlinearities in terms of state-space geometries and topology. It uses the novel technique of **flow pipes** to group infinite numbers of distinct trajectories into a manageable discrete set that becomes the basis for establishing reference trajectories, and navigates the system along the planned reference trajectories. The state-space design approach requires powerful computational tools that are able to identify, extract, represent, and manipulate qualitative features of state space, and is embodied in programs comprised in the Workbench. References [6, 7] detail the analysis and synthesis algorithms used in the programs MAPS and Phase Space Navigator of the Control Engineer's Workbench.

Given a model of a physical system and a control objective, the Workbench analyzes the system and designs a control law achieving the control objective. A user typically interacts with the Workbench in the following way.

The user first tells the Workbench about the system: he inputs a system model in terms of a differential equation, parameter values, and bounds on state variables for analysis in the form of a state-space region. The user also tells the Workbench about the requirements on the control design: he specifies the desired state for the system to settle in, the initial states of the system, the allowable control parameter values, and the constraints on the control responses.

The user then asks the Workbench to analyze the system within the parameter ranges of the model. The Workbench visualizes the totality of the behaviors of the system over the parameter ranges; it represents the qualitative aspects of the system in a data structure and reports to the user a high-level, symbolic summary of the system behaviors and, if necessary, a graphic visualization of the state-space qualitative features.

Next, the user instructs the Workbench to synthesize a control law for the system, subject to the specified design requirements. The Workbench searches for the global control paths that connect the initial states of the system and the desired state, using the qualitative description about the system. More specifically, the search is conducted in a collection of discrete flow pipes representing trajectory flows in state space. After the global control paths are established, the Workbench determines the controllable region of the system and the switching surfaces where control parameters should change values. A synthesized control reference trajectory consists of a sequence of trajectory segments, each of which is under interval-constant control.

4 State-Space Control Trajectory Design

We describe a nonlinear control design — a switching-mode control — in state space for the maglev system with large initial displacements from the equilibrium. We will show that this controller can be automatically designed with the Control Engineer's Workbench. The nonlinear controller brings the system to the vicinity of the equilibrium and then switches to the linear controller.

For the purpose of demonstration, we assume that the vehicle is displaced from the equilibrium in the direction further away from the guideway. We will concentrate on the global design of the control reference trajectories and assume that a linear feedback design is available as soon as the system enters the capture region of the linear controller.

4.1 Modeling state-space geometry

The global control law is designed by analyzing and modeling the state-space geometry of the system. The Workbench explores the state space of the system and characterizes the state space with stability regions and trajectory flow pipes. It composes the state spaces for different control parameter values and uses flow pipes to synthesize a composite state space.

The state variables x, y, and z in the model are scaled by 1, 10^3, and 2×10^4, respectively. The parameters of model are assumed to be: $L_0 = 0.1h$, $z_0 = 0.01m$, $R = 1\Omega$, $m = 10000kg$, and $g = 9.8m/sec^2$, typical of a large vehicle lift magnet. Assume the power supply delivers 140 volts, *i.e.*, $V_i = 140$, at the equilibrium. The Workbench explores the state space in a region bounded by the box $\{(x, y, z)|x \in [0, 400], y \in [-300, 350], z \in [0, 600]\}$ and finds the following equilibrium point:

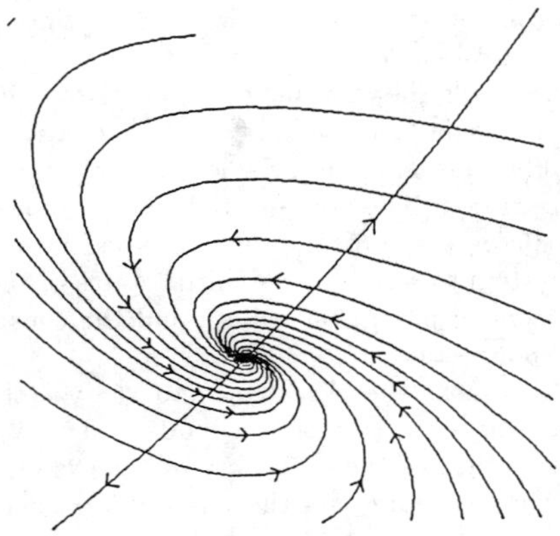

Figure 2: Stable and unstable manifolds of the saddle for V_i = 140 (yz-projection).

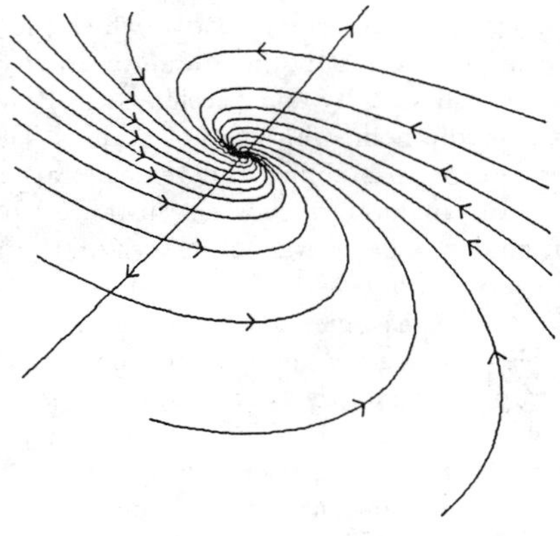

Figure 3: Stable and unstable manifolds of the saddle for V_i = 300 (yz-projection).

```
saddle:        #(140. 0. 200.)
eigenvalues:   -17.004+22.963i
               -17.004-22.963i
               24.007
eigenvectors:  #(.23604 .97174 0)
               #(.51331 -.55588 .65384)
               #(.30157 .73255 .61027)
```

With the information about the stable eigenvectors of the saddle, the Workbench computes the stable manifold of the saddle, a two-dimensional surface. The Workbench generates a set of trajectories evenly populating the stable manifold to approximate the surface. The trajectories are obtained by backward integrations from initial points in a small neighborhood of the saddle. This neighborhood lies within the plane spanned by the stable eigenvectors of the saddle.

Figure 2 shows the trajectories on the stable and unstable manifolds of the saddle in the yz-projection of the state space. The stable manifold is two-dimensional and the unstable one is one-dimensional. The stable manifold separates the state space into two halves: trajectories in the upper-half approach $z \rightarrow \infty$ along one of the unstable trajectories, corresponding to the case in which the vehicle falls off the rail; and trajectories in the lower-half approach $z = 0$ plane along the other unstable trajectory, corresponding to the case in which the train collides with the rail.

4.2 Synthesizing a global stabilization law

For an initial displacement above or below the equilibrium, the uncontrolled system will follow either a trajectory traveling upwards with increasing z and leaving the bounding box or a trajectory traveling downwards and hitting the $z = 0$ plane. To stabilize the system at the equilibrium, it is necessary to synthesize a new vector field on both sides of the stable manifold so that trajectories travel towards the stable manifold of the saddle in the new vector field. We consider only the top-half here.

The Workbench first explores the state space of the model for different values of V_i and concludes that the larger the V_i is, the further away the stable manifold is from the $z = 0$ plane. For $V_i = v > 140$, the region sandwiched by the stable manifold of $V_i = 140$ and that of $V_i = v$ has the desired property—the vector field of $V_i = v$ in this region is pointed towards the stable manifold of $V_i = 140$. When $v = 300$, the region is maximized.

Similarly, the Workbench finds that the model with $V_i = 300$ has a saddle node at (300., 0., 428.57). The stable manifold of the saddle has a similar structure as that of $V_i = 140$ case, but with larger z coordinate. The yz-projection of the stable manifold and unstable trajectories for $V_i = 300$ is shown in Figure 3.

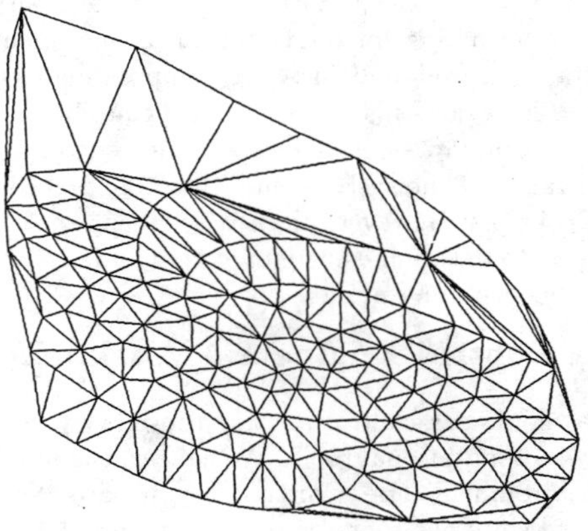

Figure 4: The sandwiched region in yz-projection.

Figure 4 shows the yz-projection of the sandwiched region discussed above. The region is bounded by three pieces of triangulated surfaces, the top boundary shown in Figure 5(a), the side one in Figure 5(b), and the bottom one in Figure 5(c), respectively. The top boundary represents the stable manifold of the saddle for $V_i = 300$ and the bottom one for $V_i = 140$. The trajectories of $V_i = 300$ flow into the region from the side boundary in Figure 5(b) and leave the region at the bottom boundary in Figure 5(c). There is no flow across the top boundary in Figure 5(a).

The Workbench determines that the above region consists of two trajectory flows: the one for $V_i = 300$ that flows into the region on the side boundary and pierces through the bottom boundary, and the one for $V_i = 140$ on the bottom boundary that approaches the desired equilibrium in the limit. With the flow-pipe characterization of the state-space trajectory flows, the Workbench searches for control trajectories in this set of flow pipes and finds a sequence of flow pipes that lead to the desired goal: the composite of the trajectory flow for $V_i = 300$ and the flow for $V_i = 140$, glued together at the stable manifold of $V_i = 140$ represented by the bottom boundary of Figure 5(c). As a result, all the trajectories of $V_i = 300$ within the region can be brought to the equilibrium by switching to $V_i = 140$ as soon as the trajectories hit the bottom boundary. We call this region the controllable region for the system, as shown in Figure 4.

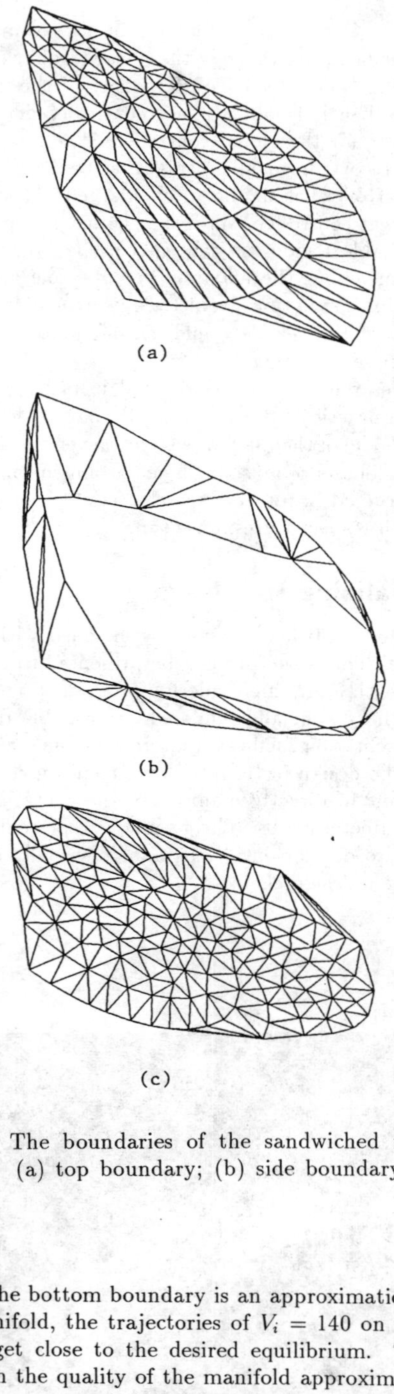

Figure 5: The boundaries of the sandwiched region in yz-projection: (a) top boundary; (b) side boundary; (c) bottom boundary.

Since the bottom boundary is an approximation to the true stable manifold, the trajectories of $V_i = 140$ on the boundary can only get close to the desired equilibrium. The closeness depends on the quality of the manifold approximation. As the trajectories enter a small neighborhood of the equilibrium, we use a linear feedback controller, such as the one discussed in [4], to stabilize the system at the equilibrium. Figure 6 shows the synthesized control reference trajectories originating at different initial displacements from the equilibrium: 1mm, 4mm, 4.5mm, and 5mm. The controller is able to recover from the first three initial points that are within the region. The last point is outside the region and thus uncontrollable; the current in the magnet can not build up fast enough to keep up with the ever-increasing airgap.

For example, the control law for the initial displacement of 1mm is specified as:

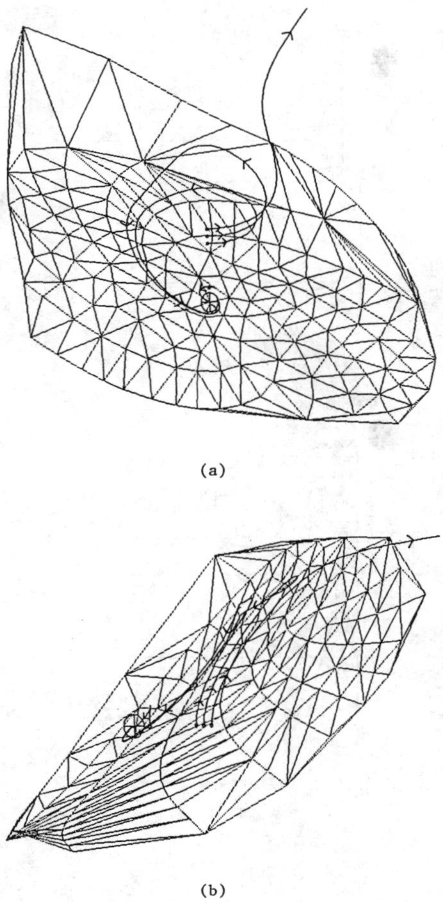

Figure 6: The synthesized control reference trajectories originating from four different initial states, together with the controllable region: (a) yz-projection; (b) zx-projection.

```
((time 0.) (switching-state #(140. 0. 220.))
 (control 300.))
((time .0134) (switching-state #(163. 4.44 221.))
 (control 140.))
((time .127) (switching-state #(139. .0789 202.))
 (control *local-control*))
```

This control law consists of a sequence of tuples, each of which specifies the time instance for each switching, the switching state, and the corresponding control value during the following time interval.

4.3 Evaluating the control design

The synthesized global control law is a switching-mode one that changes the control parameter at the switching surface—the stable manifold of $V_i = 140$. It is able to bring trajectories originating from any states within the controllable region to a local neighborhood of the saddle.

The responses of the controller with respect to the four different initial displacements are shown in Figure 7. The vertical axis of each graph represents state variables x, y, and z as in the maglev model (1), one for each curve, and the horizontal axis represents the time. For all the controllable initial displacements, the controller is able to bring the system back to the

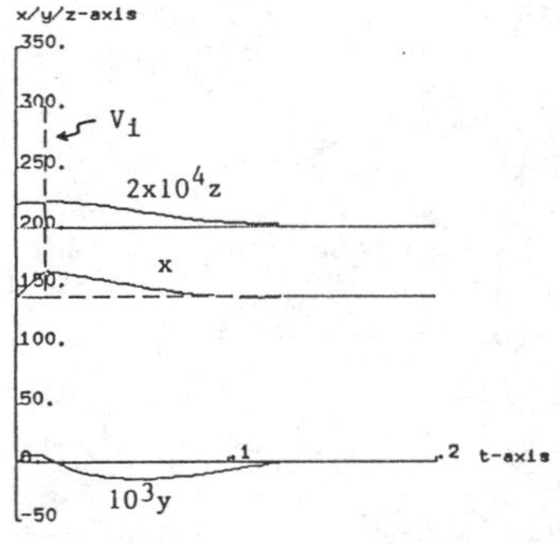

(a) dz=1mm

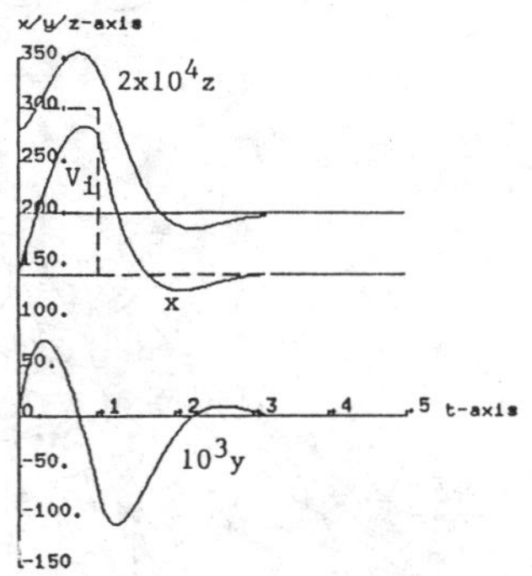

(b) dz=4mm

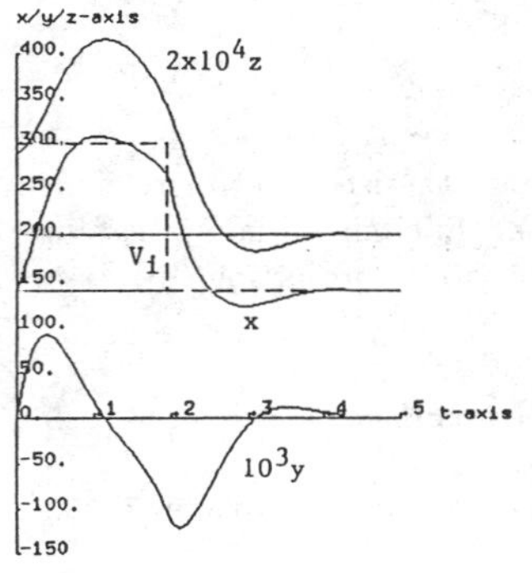

(c) dz=4.5mm

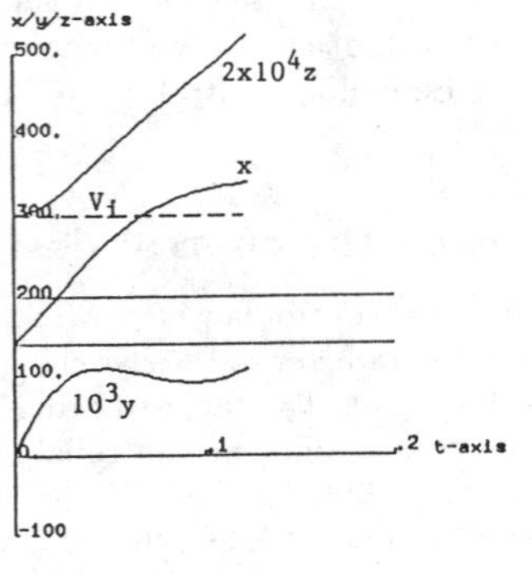

(d) dz=5mm

Figure 7: Simulation of the nonlinear control design for different initial displacements.

equilibrium with errors less than 0.2mm in displacement—a distance within the capture range of the linear feedback controller.

By exploring the state-space geometries of the maglev system, the Workbench is able to automatically determine the switching points for the global controller. The linear feedback controller can recover from only displacements of less than 0.2mm. The global controller has significantly enlarged the operating region of the linear controller. With the geometric representation of the controllable region in state space, the Workbench precisely determined that the maximum recoverable displacement is 4.55mm. Our simulation geometrically explained the observation in [4] that the vehicle would fall off the rails with a displacement of 5mm or larger.

Many issues remain to be addressed in order to make the control design practical. Since our control law is designed with the nominal model of the maglev system, the effect of uncertainties in the model and of noise in the environment on the design needs to be studied in future research. The design can also be optimized with respect to response time.

4.4 Visualizing the design

The synthesized controllable region is modeled with a polyhedral structure. This structure can be presented to engineers in a visual way with a graphic rendering program. Figure 8 shows a picture of the graphically rendered controllable region. The graphical presentation facilitates interactive, incremental modifications to the design in terms of the geometry. For example, one might want to directly manipulate the controllable region by tuning parameter knobs. The geometric representation helps the designers to develop intuitions and to visually explore the effects of certain design choices.

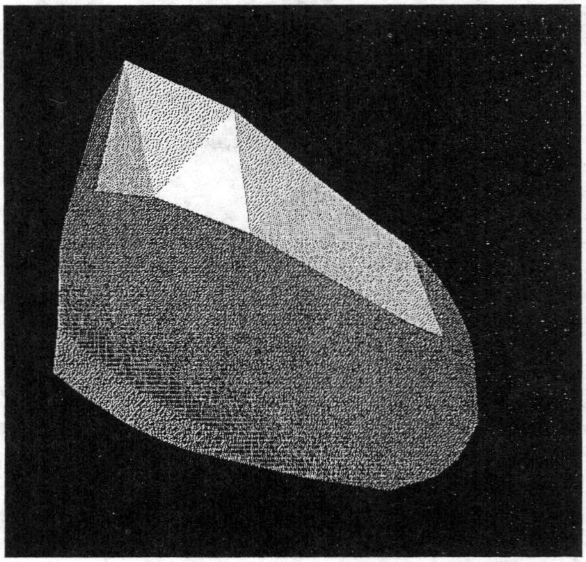

Figure 8: The controllable region for the maglev control design, rendered with the light source straight on. Hidden faces and lines are removed.

4.5 Implementation of the controller

The control design described above has been computationally simulated only. How will this design be implemented on the real system and used in real-time?

The control law specifying the switching surfaces in state space can be compiled into a table. The control execution will be a table lookup and a geometric inequality test. At each step of the execution, the state of the system is sensed and checked against the switching surface. If the state is on the switching surface, the corresponding control value for the next time interval is read from the table and applied to the physical system. The implementation does not have to be very different from that for a dynamic programming one.

5 Summary

We have described the automatic design of a high-quality controller for stabilizing an EMS-mode maglev vehicle traveling above a guideway with the Control Engineer's Workbench. The synthesized control reference trajectory consists of a sequence of trajectory segments, connected at intermediate points where the control voltage switches. At run-time, the controller will cause the system to track the reference trajectory and reactively correct local deviation from the desired trajectory. We have illustrated the state-space method for designing the global switching points of the nonlinear controller. The simulation showed that our design allows the maglev train to operate with much larger disturbances in the airgap than the classical linear feedback design does. This work demonstrates that the difficult control synthesis task can be automated, using computer programs that actively exploit knowledge of nonlinear dynamics and state space and combine powerful numerical and symbolic computations with artificial intelligence techniques.

6 Acknowledgment

The authors would like to thank Hal Abelson, Gerry Sussman, George Verghese, and Jeff Lang for discussions. J. Choi implemented the graphic rendering program. This paper describes research done at the Artificial Intelligence Laboratory of the Massachusetts Institute of Technology, supported in part by the Advanced Research Projects Agency of the Department of Defense under Office of Naval Research contract N00014-89-J-3202, and in part by the National Science Foundation grant MIP-9001651.

References

[1] H. Abelson, M. Eisenberg, M. Halfant, J. Katzenelson, E. Sacks, G.J. Sussman, J. Wisdom, and K. Yip, "Intelligence in Scientific Computing." *CACM*, 32(5), May 1989.

[2] T. R. Eastham, "Maglev: A Realistic Option for the Nineties." In *Magnetic Levitation Technology for Advanced Transit Systems,* A collection of papers presented at SAE Conference and Exposition on Future Transportation Technology, Society of Automotive Engineers, 1989.

[3] The Maglev Technology Advisory Committee, "Benefits of Magnetically Levitated High-Speed Transportation for the United States." Volume 1, Executive Report for the United States Senate Committee on Environment and Public Works. Published by Grumman Corp., Bethpage, NY, June 1989.

[4] R. Thornton, *Electronic Circuits.* MIT course notes, 1991.

[5] R. Thornton, "Why U.S. Needs a Maglev System." *Technology Review*, April 1991.

[6] F. Zhao, "Extracting and Representing Qualitative Behaviors of Complex Systems in Phase Spaces." *Proc. of the 12th International Joint Conference on Artificial Intelligence*, Morgan Kaufmann, 1991.

[7] F. Zhao, "Phase Space Navigator: Towards Automating Control Synthesis in Phase Spaces for Nonlinear Control Systems", *Proc. of the 3rd IFAC International Workshop on Artificial Intelligence in Real Time Control*, Pergaman Press, 1991.

[8] F. Zhao, *Automatic Analysis and Synthesis of Controllers for Dynamical Systems Based on Phase Space Knowledge.* PhD Thesis, Technical Report AI-TR-1385, M.I.T. Artificial Intelligence Lab, 1992.

Proceedings of the 31st Conference
on Decision and Control
Tucson, Arizona • December 1992

Control Strategy for Temperature Tracking in Rapid Thermal Processing of Semiconductor Wafers

POOGYEON PARK, CHARLES D. SCHAPER AND THOMAS KAILATH
Department of Electrical Engineering Stanford University
Stanford, CA 94305-4055

Abstract

A multivariable control strategy is developed and applied to a three-zone lamp, three-point sensor rapid thermal processing (RTP) system. The strategy is based on a physics-based nonlinear model of wafer heating. A feedforward mechanism is used to predict temperature transients and a feedback mechanism is used to correct for errors in the prediction and reduce spatial temperature nonuniformities. Gain-scheduling is employed to compensate for nonlinear energy transfer phenomena.

Experimental results are presented that show a controlled ramp-and-hold from 20°C to 900°C at a rate of 45°C/second with less than 15°C nonuniformity during the ramp and less than 1°C nonuniformity of the time-averaged temperatures during the hold as measured by three thermocouples placed at the center, 1-inch radius and 1.75-inch radius on a four-inch diameter wafer. The results are obtained in a one atmosphere environment with 1 ℓpm of N_2 gas flow. The power to each of three lamp zones is manipulated by the automatic control strategy to achieve the results.

RTP design issues are also discussed in terms of control authority or the ability to generate a wide range of energy flux profiles to achieve temperature uniformity for different processing conditions. Improvements to an experimental RTP system are presented that improve control authority.

1 Introduction

Rapid thermal processing (RTP) systems are currently being developed for single-wafer manufacturing of integrated circuits. The process performs thermal related fabrication steps such as annealing, formation of thin dielectric films and chemical vapor deposition. The advantages associated with RTP have been well documented [1, 2].

In order to achieve slip-free and uniform processing, it is necessary to maintain near uniform temperature distribution over the wafer during both steady-state and transient (fast ramping of wafer temperature) situations. Furthermore, this uniform distribution must be achieved for a range of operating conditions including different pressures, gasses, and processing temperatures. In this manner, an RTP system is available for flexible manufacturing applications that can adapt and optimize to changes in processing specifications.

Recent innovations in the design of RTP systems have provided the ability to achieve temperature uniformity over a range of processing conditions [3]. The basic requirement is the ability to vary the spatial energy flux distribution radiating to the wafer as the necessities for wafer temperature uniformity change as a function of operating conditions. To achieve this requirement, one approach is the use of multiple concentric circular rings of lamps that can be manipulated independently, (for example, see [4]). An example of this design is the Stanford RTM (Rapid Thermal Multiprocessor), of which a cross-sectional schematic of the three-zone lamp heating system is shown in Figure 1.

In this paper, an automatic multivariable controller is described and applied to a three-zone lamp, three-point sensor RTP systems. It is believed that this multivariable control strategy is the first to be successfully applied to a multi-zone lamp, multi-point sensor RTP system. Previous approaches employed scalar control systems using only a single power supply and sensor. This scalar approach has been shown to be inadequate to achieve the necessary objectives to make RTP feasible for a semiconductor manufacturing environment.

The control strategy is based on a first-principles model of the energy transport mechanisms in RTP systems. The model is developed specifically for the purpose of control. It retains the nonlinear effects of wafer heating while achieving a parsimonious structure. The coefficients of the model can be obtained from a small number of identification experiments.

The controller employs a feedforward mechanism since the desired trajectory is known *a priori* and a model is available. Errors in the feedforward trajectory due to modeling error and disturbances are compensated by feedback control. The feedback control is developed using the nonlinear model and employs gain-scheduling since the time-constants and gains of the system change by a factor of ten over the desired processing range.

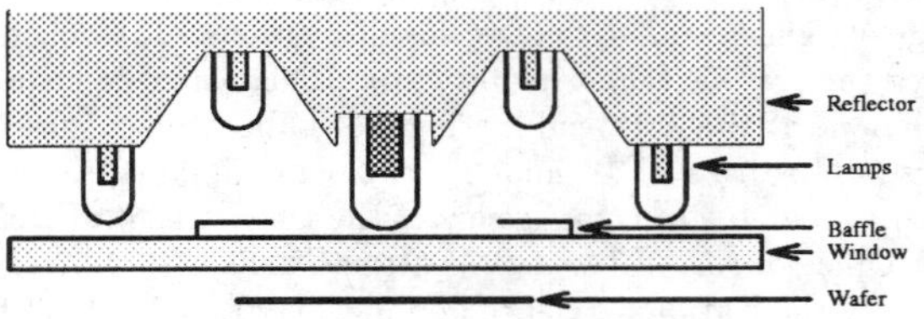

Figure 1: Cross-section of lamp

2 Model Development

Modeling of the heat transfer characteristics of RTP systems is useful for several applications including:

□ Lamp and chamber design — A model can assist in evaluating alternative designs before they are built. In this case, a detailed model is needed to determine the energy flux requirements to achieve wafer temperature uniformity at steady-state and transient situations under a range of operating conditions. This model may take the chamber and reflector geometry into account, in addition to the three mechanisms of wafer heating, radiation, convection and conduction energy transfer. ∎

CH3229-2/92/0000-2568$1.00 © 1992 IEEE

□ Design of automatic temperature controllers — Models can be used to synthesize advanced process control systems. It is necessary that these models take the nonlinear behavior of wafer temperature heating into account. (This nonlinear behavior is described below.) A model structure that is more parsimonious than the one used for design can be used for control law synthesis and real-time prediction. The model needs to be accurate over a wide range of operating conditions (pressures, temperatures, flow rates) in order to achieve precision control (+/−5°C) although the controller can be designed to be robust to a certain magnitude of modeling error. ■

□ Temperature measurement — Models can be used for advanced signal processing of sensor readings for control and diagnostic applications. In this application, the model can be used to predict wafer temperature at some locations on the wafer where a sensor is not present. Furthermore, the signal processing method can be used to filter noise in the sensor signal and reduce biased temperature readings (for example, a pyrometer with an incorrect estimate of emissivity can be improved using model-based signal processing techniques [5].) ■

There are two possible approaches to develop a model for control. In the first approach, the RTP system is viewed as a black-box. The physical laws governing the energy transfer in an RTP system are not employed by this approach. Instead, the model is developed by fitting experimental data to an assumed correlation that relates the input signals to the output signals . A black-box model of the RTP system would then relate the signals sent to the lamps to the sensors that record temperature. These black-box models are usually linear in structure. Because of the substantial (input/output) nonlinearities of wafer heating due to radiative heat transfer, a series of linear models is needed. A control strategy can then use this series of linear models to adapt to the nonlinearities.

In this paper, the second approach to modeling for control is pursued. In this approach, the physical laws that describe wafer heating are employed. For an RTP system, a physics-based model of the system behavior can be derived and is shown to be nonlinear in structure. With this approach, the underlying nonlinear structure of wafer heating is captured by a single global model. Experimental data can then be fit to the various unknown parameters of this physics-based structure. This single fundamental nonlinear model is used instead of a multitude of linearized black-box models.

The development of a physical model of an RTP system is complex. In Figure 2, the relationship between the systems inputs and the system outputs is presented. This relationship essentially consists of three blocks. In the first block, a voltage signal (0 — 5 Volts) is sent to the lamps. In the Stanford RTM, three concentric rings of lamps can be manipulated independently. In this case, three voltage signals are sent to the system. This voltage signal is then converted into radiative power by heating a tungsten filament lamp. There are modeling dynamics associated with the transfer of electrical energy to radiative energy. In the second block, the radiative power from the lamps is absorbed by the semiconductor wafer. In addition, convective, conductive and additional radiative heat transfer mechanisms that take place in the wafer. The energy flux to the wafer is influenced by the chamber, process gasses, pressure, and lamps. In the third block, a sensor measures the temperature of the wafer. This measurement may be based on a radiance signal to a pyrometer or an electrical signal from a thermocouple.

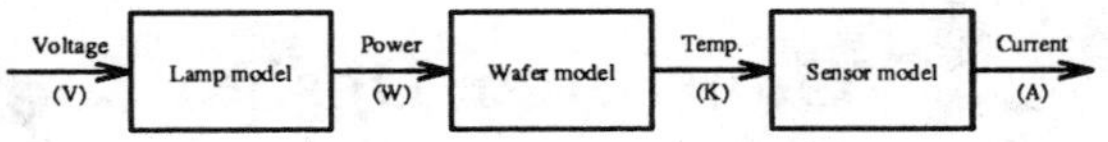

Figure 2: Physical model diagram

The model developed in this section was first developed in [6, 7].

2.1 Lamp Dynamics

The dynamics of the tungsten-halogen bulbs are modeled. The radiative power from the lamps is given by

$$P = \epsilon_f \sigma A_f T_f^4 \qquad (1)$$

where A_f is the surface area of the filament and T_f is an m-dimensional vector of filament temperatures. Assuming that convective and conductive losses are negligible, an energy balance on the filament yields

$$m_f C_{p,f} \dot{T}_f = -\epsilon_f \sigma A_f T_f^4 + L \qquad (2)$$

where L is the power to the lamp. Thus, we have

$$\tau_l \dot{P} = -P + L \qquad (3)$$

where the local time constant of the lamp is given by

$$\tau_l = \frac{\rho_f D C_{p,f}}{16 \epsilon_f \sigma T_f^3} \qquad (4)$$

where ρ_f is the density of the filament and D is the diameter of the filament.

2.2 Wafer Dynamics

State that the equation is derived in [7]. The resulting state description of the process is expressed as

$$\dot{T} = -\tilde{\mathbf{A}}_{rad} T^4 - \mathbf{A}_{conv}(T - T_a) + \mathbf{B}P \qquad (5)$$

2.3 Sensor Dynamics

In the RTP system, accurate measurement of temperature is a critical issue. In this subsection, we consider two standard sensors, the pyrometer and the thermocouple.

It is important to consider the dyanamic models of these temperature sensors. For most noninvasive sensors, such as a pyrometer, the response time for a detected change in wafer surface temperature is fast (less than 100 milliseconds). In this case the measured temperature for a pyrometer can be modeled as

$$T_m(t) = T(t) + n(t) \qquad (6)$$

where $n(t)$ is random noise. In some pyrometric equipment, analog filters are present to smooth out the effects of noise. Typically, a low pass filter is used. The measured temperature is given by

$$\tau_p \dot{T}_m(t) = -T_m(t) + T(t) + n(t) \qquad (7)$$

where τ_p is the time constant of the pyrometric equipment.

To test control strategies or to calibrate noninvasive temperature sensors, thermocouples are often used. These thermocouples are bonded to the wafer. They represent a bulk measurement rather than a surface measurement. The heat transfer effects for a thermocouple can be represented as a series of first-order differential equations which can be modeled by a time-delay equation,

$$T_m(t) = T(t - \theta_s) + n(t) \qquad (8)$$

where θ_s is the time-delay associated with the thermocouple. Typically the time delay ranges from 200 to 500 milliseconds. This amount is substantial when used to evaluate control systems for trajectories ranging from 50°C/second to 150°C/sec. It is well-known that time-delay substantially reduces the achievable levels of performance for process control systems.

3 Controller Design

The main goal of an automatic temperature control system for RTP is to achieve spatial temperature uniformity while tracking prespecified trajectories.

In the RTP system, there are many difficulties in controlling wafer temperature to the required precision of $+/-5°C$. These difficulties include:

□ Limitation of lamp power. The safety of the lamp bulbs requires that the ramping rate be limited below a specified value which is a function of the magnitude of the current sent to the bulbs. The range of each lamp power therefore corresponds to zero and an upper bound. ■

□ Modeling error of the gain matrix. The system engages the digital-analog converter to transduce the desired power command into a current value that is sent to the lamp bulbs. Becuase the relationship between input current and output lamp power is not well known, modeling error is introduced to the process gain. ■

□ Thermal memory effects of the reactor. It acts as a heat container with slow dynamics. The result is a time-varying change in the operating points. This feature prevents the use of well-defined look-up tables that correlate powers to steady-state temperature. In addition, a model needs to be linearized as a function of operating points that are time-varying. ■

□ Nonlinear heating properties of the wafer. The linearized model at certain operating points has coefficients which is a highly nonlinear function of the temperature. The time-constants and gain vary by an order of magnitude over the processing range. ■

□ Limitations of sensor equipment. The sensors have measurement delay, dynamics and noise that are a function of wafer temperature. ■

□ Unmeasured disturbances from the chamber window, chamber walls, gas flows and so on. ■

Among these difficulties, the limiting factor to control design is the overall time delay in the system. Through experiments, our system has about 0.6 second time delay. Considering that the system uses 10 Hz sampling and a ramping-rate of 45°C/second is desired, it is difficult to get good uniform performance for overall temperature range control by using a classic predictive controller. For example, if we use a classic predictive controller, a 6 step prediction of states is needed. However, the operating points are variant and therefore a linearized model produces error on the estimated 6 states. It results in large nonuniformity during tracking.

3.1 Control startegy

The objective of the controller is to track a desired temperature trajectory with wafer temperature uniformity. A typical desired temperature trajectory is shown in Figure 3. This trajectory is repeated for each wafer. The gain and (local) time-constant associated with wafer heating change by a factor of more than ten over the trajectory as shown in [7].

The difficulties in achieving good uniformity during tracking the desired trajectory require a controller divided into two blocks, feedforward and feedback part. The former takes charge of global control(tracking) and the latter, local control(maintaining uniformity, rejecting disturbances and reducing the tracking error). There are many candidates for the feedback part of controller such as LQ regulator, PID regulator, and so on. In our case, an integrator-type controller is used due to its simplicity. The fixed temperature trajectory and the similar initial conditions make power trajectories similar to each other, from run to run. Therefore, the feedforward part of the controller is based on the experimental power table which has been already generated by the previous experiments. A block diagram of the strategy is shown in Figure 4. Gain-scheduling is employed to adjust the feedback control parameters to compensate for the nonlinearities. A physics-based nonlinear model [7] of rapid thermal processing is used to design a control system.

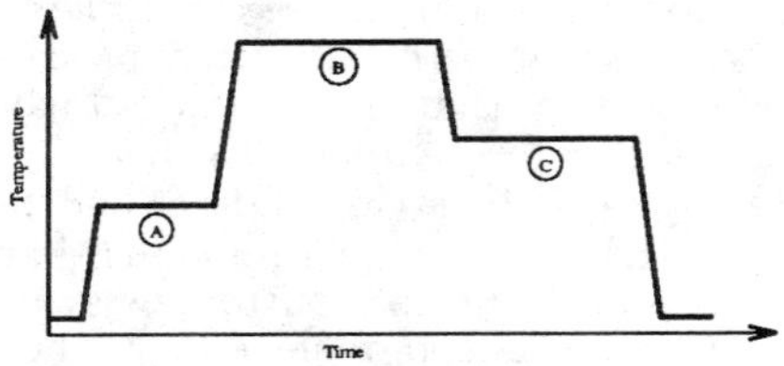

Figure 3: Typical temperature trajectory

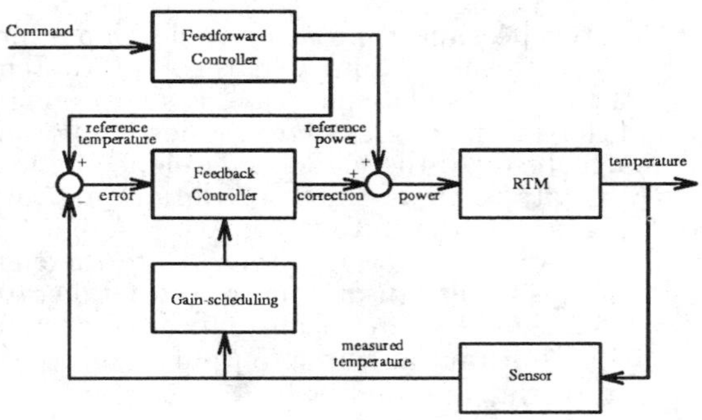

Figure 4: Control strategy diagram

3.2 Feedforward Controller Design

In this subsection, the feedforward controller is designed. This controller which is used to achieve the global tracking control consists of two parts, a steady-state power table and a compensating power for ramping-up.

The equations describing the heat transfer characteristics of the RTP system can be expressed as follows,

$$\dot{T} = \mathcal{F}(T) + K(T)P \tag{9}$$

where T is an $N \times 1$ vector and denotes each temperature sensor, $\mathcal{F}(T)$ is a $N \times 1$ vector and each entry is a function describing radiative, convective and conductive heat transfer, $K(T)$ is an $N \times N$ matrix which can be identified from the system gain, and P is an $N \times 1$ vector of lamp powers. For convenience, the notation denoting time is dropped keeping in mind, for example, that T and P are functions of time.

At steady-state conditions, Eqn 9 is given by,

$$0 = \mathcal{F}(T_s) + K(T_s)P_s \tag{10}$$

where T_s and P_s denote $N \times 1$ vectors of steady-state temperature and pressure, respectively. Subtracting Eqn 10 from Eqn 9, we have

$$\dot{T} = \mathcal{F}(T) - \mathcal{F}(T_s) + K(T)P - K(T_s)P_s \tag{11}$$

We will soon show that P_s as a function of T_s can be determined from experiments. We denote this function as $P_s|_{T_s}$ and rewrite Eqn 11

$$\dot{T} = \mathcal{F}(T) - \mathcal{F}(T_s) + K(T)P - K(T_s)\ P_s|_{T_s} \quad (12)$$

We now consider the reference temperature trajectory denoted by T_r for each sensor. Since P_s is known as a function of T_s, P_s is also known as a function of T_r. Eqn 12 is rewritten as

$$\dot{T}_r = K(T_r)(P - P_s|_{T_r}) \quad (13)$$

Solving Eqn 13 for the desired power P, we get

$$P = P_s|_{T_r} + K(T_r)^{-1}\dot{T}_r \quad (14)$$

This feedforward relation consists of two terms. The first term accounts for the steady-state power to hold the wafer at a given temperature. The second term provides the power to heat the wafer along the desired temperature ramp.

3.3 Feedback Controller Design

In this subsection, a feedback controller is designed for use in a feedforward/feedback control strategy RTP systems. The feedback controller is used to achieve the correction for modeling errors involved in the design of the feedforward controller and the local control such as maintaining uniformity, rejecting disturbances and reducing the tracking error. In addition, the feedback controller can be used in tracking mode as a set-point controller without the use of the feedforward action. This mode can be employed to generate $P_s|_{T_s}$, for example.

A nonlinear physics-based model of RTP systems is used and is developed in Chapter 2. The form of this nonlinear model linearized at temperature vector T is represented in multivariable form with Laplace notation as

$$T(s) = G(s)P(s) \quad (15)$$

where

$$G(s) = e^{-\theta s}G_1(s)KG_2(s)$$

$$G_1(s) = diag[\tfrac{1}{\tau_i s+1}] \ \ and\ \ G_2(s) = diag[\tfrac{1}{\lambda_i s+1}]$$

and where θ is the time delay due to the lamps and sensors, K is the temperature dependent gain matrix, τ_i is the temperature dependent (local) time constant of the wafer at sensor position $i = 1, ..., N$ and λ_i is the power dependent (local) time constant of the lamp. The gain and time constants of the wafer are dependent on emissivity, heat capacity and temperature at which the linearization is performed.

The shape of an open-loop gain matrix determines the performance of a closed-loop system such as the rejection of disturbance and measurement noise and the robustness on the perturbation of system matrices. The equation describing error due to disturbances and noise is given by

$$e(s) = (I + G(s)G_c(s))^{-1}(G(s)d(s) + m(s)) \quad (16)$$

where $e(s)$ denotes an $N \times 1$ vector of errors due to disturbances and measurement noise, $G_c(s)$ an $N \times N$ feedback controller transfer function, $d(s)$ an $N \times 1$ vector of arbitrary disturbances and $m(s)$ an $N \times 1$ vector of measurement noise. The open-loop gain matrix denotes $G_o(s) = G(s)G_c(s)$. It is desirable to make the open-loop gain matrix a great magnitude at high frequencies and a small magnitude at low frequencies since the disturbances are composed up of low frequency components like chamber window heating loss, gas flow, and so on and the measurement noises contain high frequcueny components like electric surges.

The following procedure is used to design the feedback controller. The system matrix $G(s)$ is factored as

$$G(s) = G_+(s)G_-(s) \quad (17)$$

where $G_+(s)$ contains any time delays and right-half plane zeros. for this problem. We can then employ following controller

$$G_c(s) = G_-^{-1}(s)F(s) \quad (18)$$

where $F(s)$ is an $N \times N$ matrix of filters that are designed so that the open-loop gain matrix, now given by

$$G_o(s) = e^{-\theta s}F(s), \quad (19)$$

has a great magnitude at high frequencies and a small magnitude at low frequencies and the controller is realizable. For our application, the following simple filter is selected,

$$F(s) = \frac{1}{\tau_c s}I \quad (20)$$

where τ_c is the (only) tuning parameter. This approach is similar to the internal model control (IMC) and minimal prototype approaches in process control. It can be easily seen that with this choice of filter, $G_c(s)$ is comprised of controllers with a PID structure. This structure will guarantee zero offset regulation despite disturbances, measurement noises and even modeling error.

Because the controller is based on a nonlinear physically model, the temperature dependency of the parameters (gains and time constants) of $G_c(s)$ is known. Consequently, the controller parameters can be continuously scheduled as a function of temperature. The mechanism works by measuring temperature and then updating $G_c(s)$ according to an analytic relationship. This gain-scheduling provides improved control. For example, the controller involves the inverse of the gain matrix. The gains decrease by a factor of ten over the temperature range from 400°C to 900°C. Thus, the feedback controller gain will automatically increase as the temperature of the wafer follows the desired trajectory.

As noted above, this feedback control methodology can be used to generate the steady-state temperature T_s versus power P_s relation that is used for the feedforward controller described in the previous subsection. In this case, the tuning parameter τ_c corresponds to the closed-loop (local) time constant of an exponential response to any set point change.

4 Experimental Results

Original rapid thermal processing systems used a single power supply for wafer heating. This power supply was connected to either a single arc lamp or to a bank of linear lamps. It was found that slip-free processing was difficult to obtain because of severe temperature nonuniformities at high temperatures. It was shown in [3, 4, 8, 9] that a multivariable lamp actuation (or power supply) system could achieve slip-free processing over a range of operating conditions. In a multivariable system, the energy flux to the semiconductor wafer could be varied dynamically to account for the inherent nonlinearities of radiative heating.

A cross-sectional schematic of the multiple-actuation Stanford RTM is shown in Figure 1. The outer ring of lamps consists of twenty-four one-kilowatt bulbs. The intermediate ring of lamps consists of twelve one-kilowatt bulbs. The central lamp is a two-kilowatt bulb. Three power supplies are available to control the power to each lamp ring. The reflector head is water-cooled and the surface is gold-plated. The window is made of quartz. The wafer is four inches in diameter and supported by pins.

An annular gold-plated stainless steel baffle was added to improve the controllability (or the ability to generate a wider variety of energy flux profiles to the wafer). Prior to the addition of the baffle, studies indicated that the intermediate and outer lamp rings were largely coupled. The energy differential between the intermediate and outer zones had little influence on the wafer temperature profile. For the case where no baffle is present, Figure 5 (**a**) presents the percentage power to each of the three zones that achieves equal temperature readings on three thermocouples bonded along a common wafer radius (at radii positions 0, 1, and 1.75 inches). As can be seen, it was impossible to achieve equal temperature readings at all three locations at temperatures below 650°C since the intermediate zone would saturate to zero. This saturation in effect was a loss in controllability and degrees of freedom. The three lamp heating system essentially operated as a two lamp heating system.

However, the baffle provided a twofold effect of attenuating direct radiation from 1) the intermediate lamps to the outer portion of the wafer and 2) the outer lamps to the intermediate portion of the wafer. Consequently, the baffle allowed the outer lamps to provide a higher degree of wafer edge heating while minimizing the contribution of energy to the interior locations of the wafer. Because of the increased surface area at the wafer edge, less energy flux is required than the rest of the wafer during heating. In Figure 5 (**b**), the improvement in the range of achievable uniformity is demonstrated for the case with the baffle present. The range of achievable uniformity is increased which can be seen by comparison to Figure 5 (**a**). Identification studies also indicated an improvement to the weakest direction of control (as measured by the minimum singular value of the gain matrix) by a factor of five [10].

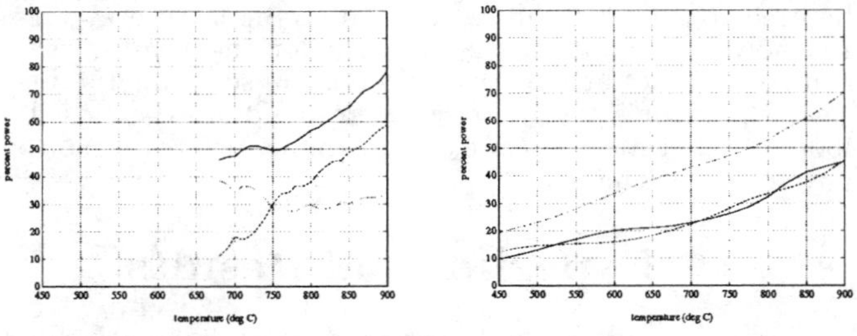

Figure 5: Lamp Powers w/(**a**) and w/o(**b**) a baffle

The multivariable, gain-scheduled, feedforward/feedback control strategy was applied to the Stanford Rapid Thermal Multiprocessor (RTM). The schematic of the lamp heating system is shown in Figure 1. Three thermocouples were bonded to a four-inch diameter wafer at radii locations of 0, 1, and 1.75 inches.

The objective of the control experiment was the following: *Starting from room temperature, ramp to 900 °C at a rate of 45 °C/second and then hold at 900 °C for five minutes. Conduct the experiment at one atmosphere pressure with 100 sccm N_2 flow rate.* This specification corresponds to a thermal oxidation step where oxygen is used instead of nitrogen. Oxygen was not used because it would damage the thermocouple wafer. Thermocouples were used since accurate noninvasive sensors were not yet available. A 45°C/second ramp was selected because earlier studies indicated that it was the maximum achievable ramping rate with temperature uniformity for the present control authority of the RTM. The differential energy flux provided by three lamp zones was not sufficient to achieve a higher ramping rate and still retain temperature uniformity.

The following procedure was used to achieve the control objective:

□ The gain-scheduled, integral action, feedback controller was used to obtain the relation of steady-state uniform temperature (T_s) to steady-state powers (P_s) over the temperature range from 20°C to 950°C at 50°C increments. It was important to generate this relation by:

(1) First, the feedback controller was used alone to exponentially rise from room temperature to the desired temperature. The feedback controller was designed so that the time to settle at the desired temperature was equal to the total temperature change divided by the desired ramp rate.

(2)Next, the wafer was returned returned to room temperature.

This two step-procedure was then repeated until the range of temperatures at the specified increment was covered. This approach was important since the feedforward powers necessary to achieve a desired temperature were representative of the conditions under which the wafer was to be processed. That is, the system (*i.e.* quartz window, walls) was in a transient slow-heating condition and was not allowed to come to thermal equilibrium. The difference in necessary power to hold wafer temperature uniform was substantially dependent on the system temperatures. ■

□ The relation of T_s to P_s and the physics-based model was then used to obtain the feedforward trajectory. The feedback controller was wrapped around this feedforward trajectory with a specified closed-loop time constant specified as 1.5 seconds. ■

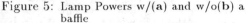

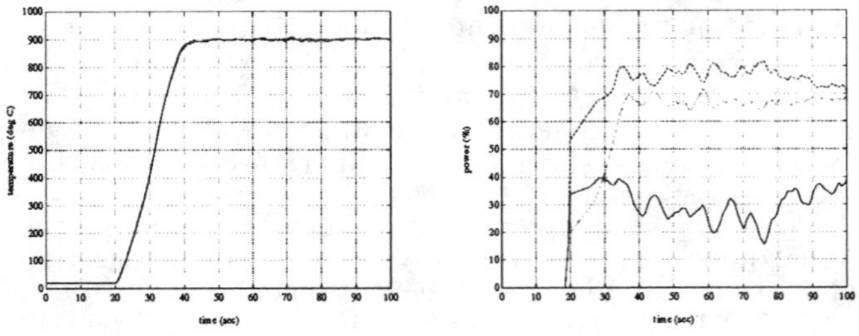

Figure 6: Trajectory and Power over 100 sec.

The controlled ramp for the first 100 seconds is examined in Figure 6 (**a**) where the three thermocouple measurements are shown for the ramp and hold. The corresponding powers are shown in Figure 6 (**b**). The sensitivity of the controller to the sensor noise can be seen in this result. This sensitivity can easily be reduced by adding a filter to the sensor. It is interesting to note the time delay of the system by comparing the powers to the temperature response as the ramp started. Approximately a two second total delay existed in the beginning of the response. Of this delay, 1.5 seconds is due to a power surge protection scheme on the lamps which subsides after the percentage power to the lamps surpasses fifteen. A 0.5 second delay remains and is due to the thermocouples and lamps. This delay is very substantial when one considers a desired ramp rate of 45°C/second.

The nonuniformity of the controller throughout the entire ramp and hold trajectory is studied. In Figure 7 (**a**), the temperature corresponding to the three thermocouple sensors is plotted for the five minute trajectory. In Figure 7 (**b**) the maximum temperature difference as measured by subtracting the minimum of the three thermocouple readings from the maximum is plotted as a function of time. The maximum nonuniformity during the ramp is approximately 15°C. The mean temperature corresponding to this error is approximately 350°C. This mean temperature is low and the nonuniformity will not effect processing nor damage the wafer. The substantial sensor noise of the thermocouples can also be seen when the mean temperature is 900°C.

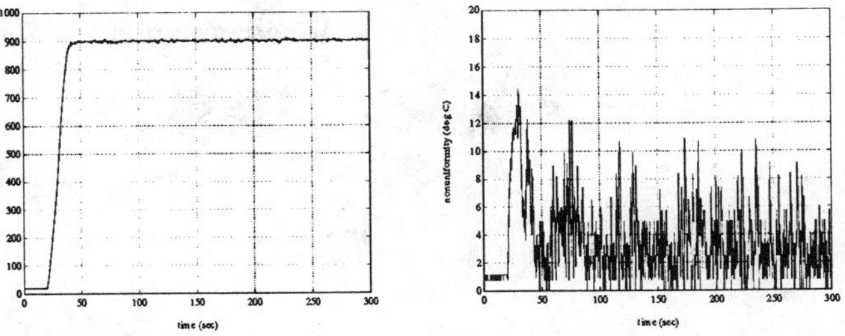

Figure 7: Trajectory and uniformity over 5 min.

The capability of the controller to hold the wafer temperature at a desired processing temperature despite slow heating modes is examined. As seen in Figure 7 (**a**), the multivariable controller was able to hold the temperature about the desired value of 900°C. Although the sensors were quite noisy in the atmospheric environment and had resolution limited to approximately 0.5°C, the average temperature during the hold portion of the ramp for the three sensors was, 900.9°C, 900.7°C and 900.8°C. This result is important since the average temperature is nearly uniform and will result in uniform processing. The reason for the temperatures being higher that the targeted 900°C is explained by examining the slow heating modes of the chamber walls and quartz window. The temperature of the quartz window and chamber base of the RTM are plotted in Figure 8 (**a**) and (**b**), respectively. This slow heating of the components of the RTM act as slow disturbances to the wafer. Because the gain of the controller was reduced to compensate for time delay, a slightly higher average temperature than the desired 900°C was achieved. However, without a feedback control with integral action, the wafer temperature would have drifted substantially to greater than 950°C.

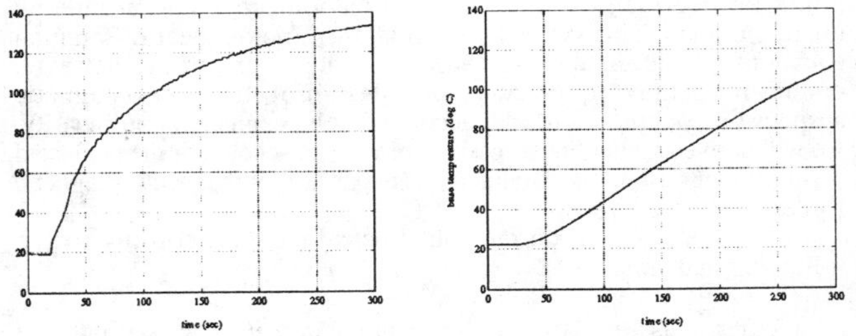

Figure 8: Temp. of Quartz window and Base

In summary, the controller met the objective and performed satisfactory in the presence of numerous challenges including substantial time delays, saturating actuators, sensor noise, system nonlinearities, and slow disturbances. Extensions to the control strategy include an adaptation feature to fine-tune the nonlinear physics-based model using data from a cycle. The controllers are then fine-tuned from the adapted model.

5 Conclusions

A physics-based nonlinear model of wafer heating in RTP systems has been explained. A multivariable control strategy for tracking the temperature along a desired trajectory with uniformity has been described. The advantage of this strategy is that it can be used for the nonlinear RTP system where processing occurs over a wide range. Even with the extreme constraints of the RTP system, a controlled ramp was achieved from 20°C to 900°C at a rate of 45°C/second with less than 15°C peak-to-peak nonuniformity during the ramp. It is believed that this multivariable control strategy is the first to be successfully implemented in a multi-zone lamp, multi-point sensor RTP system.

6 Acknowledgements

We thank Len Booth and Paul Gyugyi for help with the experimental work. Young Man Cho performed model identification. We also thank Steve Norman and Professors Boyd, Franklin and Saraswat for helpful discussions. This work was supported by the Advanced Research Projects Agency of the Department of Defense and was monitored by the Air Force Office of Scientific Research.

REFERENCES

[1] M. Moslehi, Single-wafer optical processing of semiconductors: Thin insulator growth for integrated electronic device applications, *Appl. Phys. A*, **46**:255–273, 1988.

[2] K. Saraswat, Center for Research on Manufacturing Science and Technology for VLSI, Annual report, Center for Integrated Systems, Stanford, CA, 1989.

[3] P. Apte and K. Saraswat, Rapid thermal processing uniformity using multivariable control of a circularly symmetric three zone lamp, submitted, 1991.

[4] S. Norman, C. Schaper, and S. Boyd, Improvement of temperature uniformity in rapid thermal processing systems using multivariable control, In *Mater. Res. Soc. Proc.: Rapid Thermal and Integrated Processing*. Materials Research Society, April 1991.

[5] Y. Cho, C. Schaper, and T. Kailath, *In-Situ* temperature estimation in rapid thermal processing systems using extended Kalman Filtering, In *Mater. Res. Soc. Proc.: Rapid Thermal and Integrated Processing*. Materials Research Society, April 1991.

[6] C. Schaper, Y. Cho, P. Park., S. Norman, P. Gyugyi, G. Hoffmann, S. Balemi, S. Boyd, G. Franklin, T. Kailath, and K. Saraswat, Modeling and control of a rapid thermal processor, In *SPIE Rapid Thermal and Integrated Processing*, 1991.

[7] C. Schaper, Y. Cho, and T. Kailath, Low-order modeling and dynamic characterization of rapid thermal processing, *Applied Physics A*, **54**:317–326, 1992.

[8] S. Norman, RTP control system analysis using convex optimization, Technical report, Stanford Univ. Information Systems Lab., 1992.

[9] S. Norman, Optimization of transient temperature uniformity in RTP systems, *IEEE Trans. Electron Devices*, **39**:205–207, 1992.

[10] Y. Cho and T. Kailath, Model identification in rapid thermal processing systems, Submitted to IEEE Trans. Semiconductor Manufacturing, 1991.

Proceedings of the 31st Conference
on Decision and Control
Tucson, Arizona • December 1992

μ-Synthesis of an Electromagnetic Suspension System

Masayuki Fujita†, Toru Namerikawa‡, Fumio Matsumura‡, and Kenko Uchida§

† School of Information Science, Japan Advanced Institute of Science and Technology, Hokuriku, Tatsunokuchi, Ishikawa 923-12, Japan

‡ Department of Electrical and Computer Engineering, Kanazawa University, 2-40-20 Kodatsuno, Kanazawa 920, Japan

§ Department of Electrical Engineering, Waseda University, 3-4-1 Okubo, Shinjuku, Tokyo 169, Japan

Abstract

This paper deals with μ -synthesis of an electromagnetic suspension system. Firstly, the issue of modeling a real physical electromagnetic suspension system is discussed. We derive a nominal model as well as a set of models in which the real system is assumed to reside. Different model structures and possible model parameter values are fully employed to determine unstructured additive plant perturbation, which directly yield the uncertainty weighting. Secondly, based on the set of plant models, we setup the robust performance control objectives. Thirdly, using μ -Analysis and Synthesis Toolbox, we make use of the *D-K* iteration approach for the controller design. Finally, implementing the controller with a digital signal processor μ PD77230, a lot of experiments are carried out in order to evaluate the robust performance of this design.

1. Introduction

The progress of feedback control theory [1], as well as the concepts of LFTs, LMIs, and μ [2], now develops rigorous mathematical techniques to cope with uncertainty. Hence in this situation, experimental research of μ -theory is concerned extremely. Until now, advanced control theories are not applied to magnetic suspension technology very much. In this paper, we will evaluate a μ -controller experimentally on an electrical magnetic suspension system. We derive a nominal model as well as a set of models in which the real system is assumed to reside. The method to model the plant as belonging to a family or set plays a key role for systematic robust control design. For the design, we first setup the robust performance control objectives as a structured singular value test. Next, we make use of the so-called $D-K$ iteration approach, using the μ -Analysis and Synthesis Toolbox [1]. Finally, a lot of experiments are carried out in order to evaluate that the closed-loop system achieves robust performance against various external disturbances.

2. Electromagnetic Suspension System

Consider an electromagnetic suspension system shown in Figure 1. An electromagnet is located at the upper part of the experimental system. Utilizing this electromagnetic force, we will suspend the iron ball stably. Note that this electromagnetic suspension system is essentially unstable. Hence feedback control is indispensable. As a gap sensor, a standard induction probe of eddy-current type is placed at the bottom.

A digital signal processor (DSP)-based real-time controller is implemented with NEC μ PD77230, which can execute one instruction in 150 ns with 32-bit floating point arithmetic. The data acquisition board MSP-77230 consists of a 12-bit A/D converter and a 12-bit D/A converter with the maximum conversion speed of 10.5 μ s and 1.5 μ s, respectively.

The applications of this electromagnetic suspension technology include magnetically levitated vehicles, magnetic bearings, and so on.

2.1 Model Structure

This section will employ four different model structures for the electromagnetic suspension system depicted in Figure 1. Among them, the simplest one will be used for control law calculations as a nominal model; while the others will be used to determine uncertainty. All of the models are finite-dimensional, linear, and time-invariant of the following state space form:

$$\dot{\boldsymbol{x}} = A\boldsymbol{x} + Bu, \quad y = C\boldsymbol{x}. \tag{1}$$
$$\boldsymbol{x} = [x \ \ \dot{x} \ \ i]^T, \ u = e, \ y = x.$$

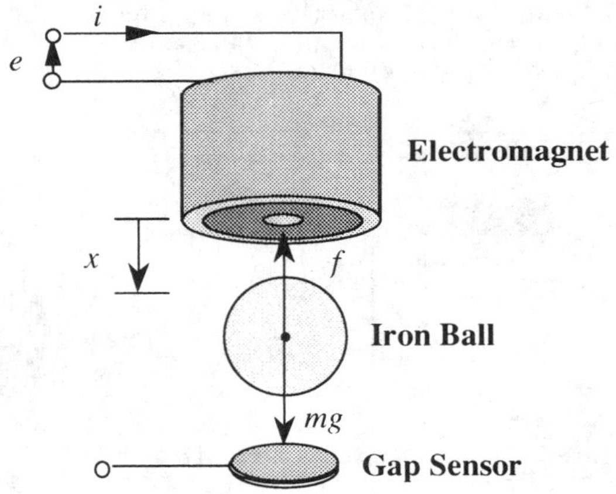

Figure 1 Electromagnetic Suspension System

- f : electromagnetic force
- m : mass of the iron ball
- X : steady-state gap between the magnet and the iron ball
- x : displacement from the steady-state gap
- L : inductance of the electromagnet
- R : resistance of the electromagnet
- E : steady-state voltage of the electromagnet
- e : displacement from the steady-state voltage
- I : steady-state current of the electromagnet
- i : displacement from the steady-state current
- k, x_0, L_0: constants determined by experiments

Applying the laws of physics, we first obtain ideal mathematical models for the real electromagnetic suspension system.

However, due to the idealizing assumptions that we make, two types of mathematical models can be derived hereafter, which are composed of nonlinear differential equations. We define them Type(A), and Type(B), respectively. It is needless to say, even if a model is written down by nonlinear differential equations, that there are always discrepancies between the ideal mathematical model and the real system.

Since the behavior of the electromagnetic force is highly nonlinear, we then employ the linearization procedure around an operating point. This yields linear models from nonlinear differential equations. The linear approximation of the attraction force is, however, crucial for achieving robust control design of the electromagnetic suspension system. In order to account for the neglected nonlinearity, we will derive another linear model which takes account of the neglected second-order term in the Taylor series expansion of the electromagnetic force.

Thus, we will derive four linear models according to the following manners:

(A1) we assume that L = constant; and we approximate the nonlinearity of the electromagnetic force up to the first-order term in the Taylor series expansion.

(A2) we assume that L = constant; and we approximate the nonlinearity of the electromagnetic force up to the second-order term in the Taylor series expansion.

CH3229-2/92/0000-2574$1.00 © 1992 IEEE

(B1) we assume that $L = L(x)$; and we approximate the nonlinearity of the electromagnetic force up to the first-order term in the Taylor series expansion.

(B2) we assume that $L = L(x)$; and we approximate the nonlinearity of the electromagnetic force up to the second-order term in the Taylor series expansion.

It is noted that Type (A) denotes L=constant; while Type (B) denotes $L = L(x)$: the inductance is a function of the gap.

(A1) Assume L=constant. Then, the electromagnetic suspension system can be written by the following nonlinear differential equations [4, 5]

$$m\frac{d^2x}{dt^2}=mg-f,\quad f=k\left(\frac{i}{x+x_0}\right)^2,\quad Ri+L\frac{di}{dt}=e. \tag{2}$$

From (2), we can obtain the linear model (3) by approximating the nonlinearity of the electromagnetic force up to the first-order term in the Taylor series expansion

$$A=\begin{bmatrix}0 & 1 & 0\\ \frac{2kI^2}{m(X+x_0)^3} & 0 & -\frac{2kI}{m(X+x_0)^2}\\ 0 & 0 & -\frac{R}{L}\end{bmatrix},\quad B=\begin{bmatrix}0\\0\\ \frac{1}{L}\end{bmatrix},\quad C=\begin{bmatrix}1\\0\\0\end{bmatrix}^T \tag{3}$$

(A2) From (2), we can further obtain another linear model (4) by approximating the nonlinearity of the electromagnetic force up to the second-order term in the Taylor series expansion

$$A=\begin{bmatrix}0 & 1 & 0\\ \frac{2kI^2}{m(X+x_0)^3}\Delta y & 0 & -\frac{2kI}{m(X+x_0)^2}\Delta y\\ 0 & 0 & -\frac{R}{L}\end{bmatrix},\quad B=\begin{bmatrix}0\\0\\ \frac{1}{L}\end{bmatrix},\quad C=\begin{bmatrix}1\\0\\0\end{bmatrix}^T$$

$$\Delta y=I-\frac{3}{2}\Delta x+\frac{I}{2}\Delta i,\quad \Delta x=\frac{x}{X+x_0},\quad \Delta i=\frac{i}{I} \tag{4}$$

In this way, we deal with state variables due to the second-order term as fixed numbers and include them in the matrix A

(B1) On the other hand, assume $L = L(x)$: i.e., the inductance L of the electromagnet is a function of the gap x. In this case, the electromagnetic suspension system can be written by the following nonlinear differential equations [4, 5]

$$m\frac{d^2x}{dt^2}=mg-f,\quad f=k\left(\frac{i}{x+x_0}\right)^2 \tag{5a}$$

$$e=Ri+\frac{d}{dt}(Li),\quad L=\frac{2k}{x_0+x}+L_0 \tag{5b}$$

From (5), the linear model (6) will be derived by approximating the nonlinearity of the electromagnetic force up to the first-order term in the Taylor series expansion

$$A=\begin{bmatrix}0 & 1 & 0\\ \frac{2kI^2}{m(X+x_0)^3} & 0 & -\frac{2kI}{m(X+x_0)^2}\\ 0 & \frac{2kI}{(X+x_0)\{2k+L_0(X+x_0)\}} & -\frac{R(X+x_0)}{2k+L_0(X+x_0)}\end{bmatrix}$$

$$B=\begin{bmatrix}0\\0\\ \frac{X+x_0}{2k+L_0(X+x_0)}\end{bmatrix},\quad C=\begin{bmatrix}1\\0\\0\end{bmatrix}^T. \tag{6}$$

(B2) Moreover, the linear model (7) can be derived by approximating the nonlinearity of the electromagnetic force in (5) up to the second-order term in the Taylor series expansion

$$A=\begin{bmatrix}0 & 1 & 0\\ \frac{2kI^2}{m(X+x_0)^3}\Delta y & 0 & -\frac{2kI}{m(X+x_0)^2}\Delta y\\ 0 & \frac{2kI(1-2\Delta x+\Delta i)}{(X+x_0)\{2k(1-\Delta x)+L_0(X+x_0)\}} & -\frac{R(X+x_0)}{2k(1-\Delta x)+L_0(X+x_0)}\end{bmatrix}$$

$$B=\begin{bmatrix}0\\0\\ \frac{X+x_0}{L_0(X+x_0)+2k(1-\Delta x)}\end{bmatrix},\quad C=\begin{bmatrix}1\\0\\0\end{bmatrix}^T. \tag{7}$$

In the same way as (A2), we treat state variables due to the second-order term as fixed numbers and include them in the matrix A

Thus, we obtain four linear model structures: (A1), (A2), (B1), and (B2) for the electromagnetic suspension system.

2.2 Model Parameters

The model parameters are determined by measurements and/or experiments. There the value of each parameter is by no means deterministic. In order to account for unpredictable perturbations in the model parameters, we will set the nominal value as well as the possible max./min. value of each parameter in every linear model.

To obtain the possible max./min. value of each parameter, consider the steady-state gap X=5.0 mm (nominal). Now let us perturb it with X=4.5 mm and X=5.5 mm (perturbed ±0.5 mm). And, for these cases, we measure the values of the parameters. The results of measurements are shown in Table 1. We also obtained the value of each parameter if X=4.0 mm and X=6.0 mm (perturbed ± 1.0 mm), but perturbations were too big.

2.3 Nominal Model

Here we setup a nominal model of the electromagnetic suspension system. Four model structures and possible values of model parameters have been obtained. With these preliminaries, we will form the nominal model using the simplest (A1) model structure and the nominal model parameter (X=5.0[mm] case). This model will play a key role in controller calculations. Its state space form is then of the following form

$$A_{nom}=\begin{bmatrix}0 & 1 & 0\\ 4481 & 0 & -18.43\\ 0 & 0 & -45.69\end{bmatrix},\quad B_{nom}=\begin{bmatrix}0\\0\\1.969\end{bmatrix},\quad C_{nom}=\begin{bmatrix}1\\0\\0\end{bmatrix}^T. \tag{8}$$

And the corresponding transfer function is

$$G_{nom}=\frac{-36.27}{(s+66.94)(s-66.94)(s+45.69)}. \tag{9}$$

Table 1 Parameters

Parameter	Max. Value	Nominal Value	Min. Value
m [kg]	- - -	1.75	- - -
X [m]	5.5×10^{-3}	5.0×10^{-3}	4.5×10^{-3}
I [A]	1.18	1.06	0.93
x [m]	5.0×10^{-4}	0	-5.0×10^{-4}
i [A]	1.17×10^{-1}	0	-1.26×10^{-1}
L [H]	5.57×10^{-1}	5.08×10^{-1}	4.65×10^{-1}
R [Ω]	$2.37\times10^{+1}$	$2.32\times10^{+1}$	$2.27\times10^{+1}$
k [Nm^2/A^2]	3.35×10^{-4}	2.90×10^{-4}	2.53×10^{-4}
x_0 [m]	-3.32×10^{-4}	-6.41×10^{-4}	-9.42×10^{-4}
L_0 [H]	3.96×10^{-1}	3.75×10^{-1}	3.54×10^{-1}

2.4 Model Perturbations

There is no need to say that discrepancies exist between the nominal model and the real plant depicted in Figure 1. These uncertainties may be due to the parameter identification errors, and the neglected nonlinearities in the electromagnet. In order to account for such inaccuracies, we should consider not only a nominal model but also a set of plant models in which the real system is assumed to reside. Model perturbations are defined in this section.

Considering only unstructured uncertainties, we take no account of structured uncertainties in this case. We choose not to model the uncertainties in detail, but rather to get all unstructured uncertainties together into 1-full block uncertainty.

In order to estimate the quantities of additive model perturbations, we employ differences of gain between the nominal transfer function and the perturbed transfer function with only one parameter changed and the others fixed. Plural parameters are not assumed to change.

For example, model 1a denotes a perturbed model with the (A1) model structure and with the nominal model parameters in Table 1 except k=k_max (max. value of k). In such a way, the following 24 perturbed models have been employed.

model 1a	: (A1)	k_max	model 1b	: (A1)	k_min
model 2a	: (A1)	x_0_max	model 2b	: (A1)	x_0_min
model 3a	: (A1)	R_max	model 3b	: (A1)	R_min
model 4a	: (A1)	L_max	model 4b	: (A1)	L_min
model 5a	: (A2)	x_max	model 5b	: (A2)	x_min
model 6a	: (A2)	i_max	model 6b	: (A2)	i_min
model 7a	: (B1)	k_max	model 7b	: (B1)	k_min
model 8a	: (B1)	L_0_max	model 8b	: (B1)	L_0_min
model 9a	: (B1)	x_0_max	model 9b	: (B1)	x_0_min
model 10a	: (B1)	R_max	model 10b	: (B1)	R_min
model 11a	: (B2)	x_max	model 11b	: (B2)	x_min
model 12a	: (B2)	i_max	model 12b	: (B2)	i_min

With these notations, we can define the corresponding perturbed transfer functions $G_{ij\_perturb}$ $(1 \le i \le 12,\ j = a,b)$ in an obvious way. For example, $G_{1a\_perturb}$ is a transfer function of the model 1a . Now let

$$\Delta_{ij} := G_{ij\_perturb} - G_{nom} \quad (1 \le i \le 12,\ j = a,b). \tag{10}$$

Each magnitude of these additive perturbations $|\Delta_{ij}(j\omega)|$ are plotted by dotted lines in Figure 2.

2.5 Set of Plant Models

Now let us consider the set of plant models for the electromagnetic suspension system in Figure 1. In this paper, we assume the following form (see Figure 3)

$$\boldsymbol{G} := \left\{ G_{nom} + \Delta_{add} W_{add} : \|\Delta_{add}\|_\infty \le 1 \right\} \tag{11}$$

in which the real plant is assumed to reside. Here the transfer function $\Delta_{add} W_{add}$ represents the potential differences between the nominal model G_{nom} and the actual behavior of the real plant. All of the uncertainty is captured in the normalized, unknown transfer function Δ_{add}. As shown, the unstructured model uncertainty is represented as the additive plant perturbations.

From dotted lines in Figure 2, it is natural to choose the uncertainty weighting W_{add} as follows

$$W_{add} = \frac{1.4 \times 10^{-5}(1+s/8)(1+s/170)(1+s/420)}{(1+s/30)(1+s/35)(1+s/38)} \tag{12}$$

Here it should be noted that the magnitude of the uncertainty weighting W_{add} covers all the model perturbations shown in Figure 2. Although the model perturbations caused by neglected nonlinearities and model parameter errors are explicitly dealt with above, we will assume that this uncertainty weighting W_{add} also covers the other possible perturbations such as the unmodeled dynamics, the errors associated with digital controller implementation, and so on. These perturbations generally have a large gain in the high frequency range, hence the uncertainty weighting W_{add} is chosen to have relatively large gain in that range.

3. Design

3.1 Control Objectives

Consider the feedback structure shown in Figure 4. The dashed box represents the transfer function of the real electromagnetic suspension system, which is unstable in nature. Hence, our principal control objective is in its stabilization. In fact, we would like to design a stabilizing controller K not only for the nominal model G_{nom} but for all the possible plant models $G \in \boldsymbol{G}$. This robust stability requirement is equivalent to

$$\left\| W_{add\_r} K (I + G_{nom} K)^{-1} W_{add\_l} \right\|_\infty < 1. \tag{13}$$

It is noted in Figure 4 that we factor the uncertainty weighting as $W_{add} = W_{add\_l} W_{add\_r}$ where $W_{add\_l} = 1.0 \times 10^{-5}$.

As shown, the electromagnetic suspension system is subject to the various disturbance forces. Hence the performance of this feedback system can be evaluated using the sensitivity function: $S := (I + GK)^{-1}$. In order to reject the disturbances at low frequency band, the performance weighting W_{perf} is now chosen as (see Figure 5)

$$W_{perf} = \frac{200.0}{1 + s/0.1} \tag{14}$$

where we factor $W_{perf} = W_{perf\_l} W_{perf\_r}$ with $W_{perf\_l} = 1.0 \times 10^{-5}$. In practical situation, we would like to achieve this performance requirement for all the possible plant models $G \in \boldsymbol{G}$. This yields

$$\left\| W_{perf\_r} (I + GK)^{-1} W_{perf\_l} \right\|_\infty < 1,\ \forall G \in \boldsymbol{G} \tag{15}$$

Now, the control objective is to find a controller K such that the closed-loop system remains internally stable for every $G \in \boldsymbol{G}$, and in addition the weighted sensitivity function satisfies the performance (15) for all $G \in \boldsymbol{G}$. This is the robust performance objective.

The above control objectives exactly fit in the μ-synthesis framework by introducing a fictitious uncertainty block Δ_{perf}. The appended uncertainty block Δ_{perf} is used to incorporate the robust performance calculation. Rearranging the feedback structure in Figure 4, we can build the interconnection structure shown in Figure 6. The open-loop interconnection P in Figure 6 is often referred to as the generalized plant.

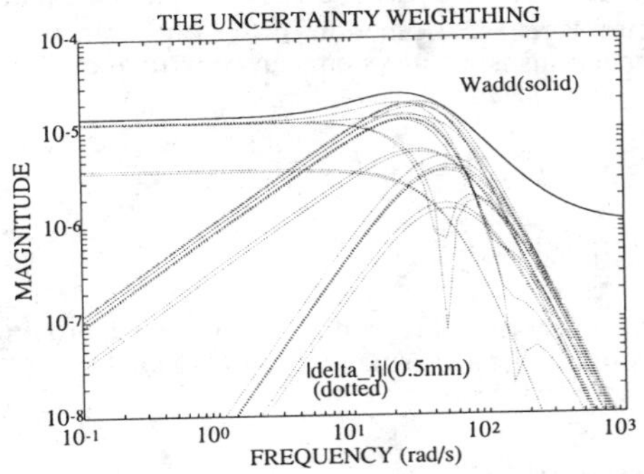

Figure 2 Model Perturbations and the Uncertainty Weighting

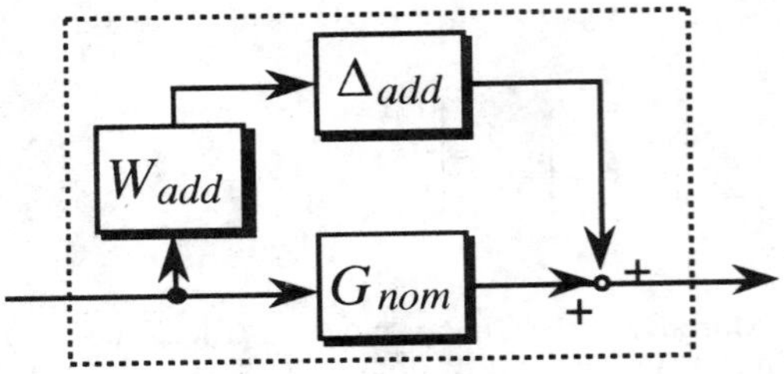

Figure 3 Set of Plant Models

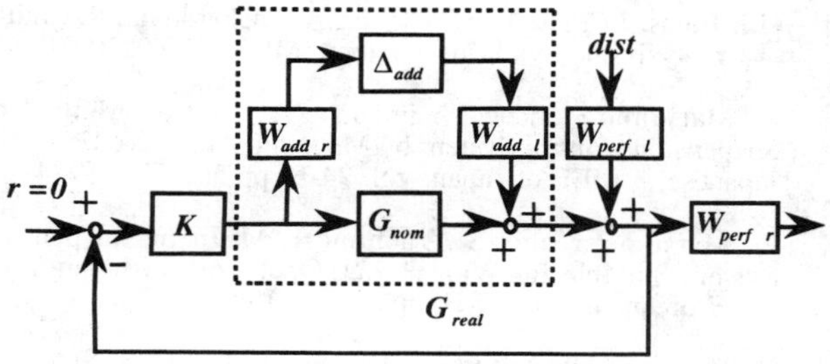

Figure 4 Feedback Structure

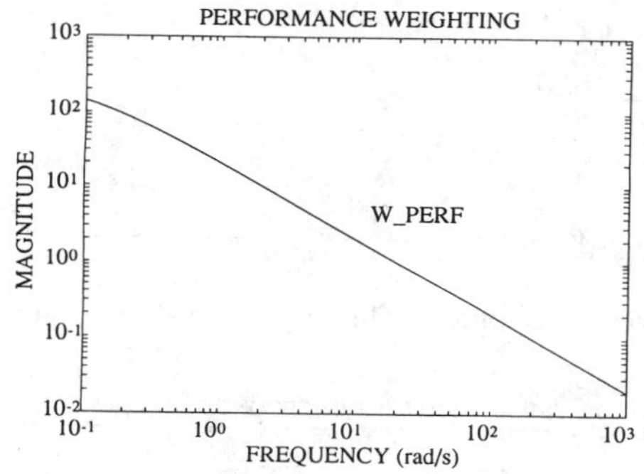

Figure 5 Performance Weighting

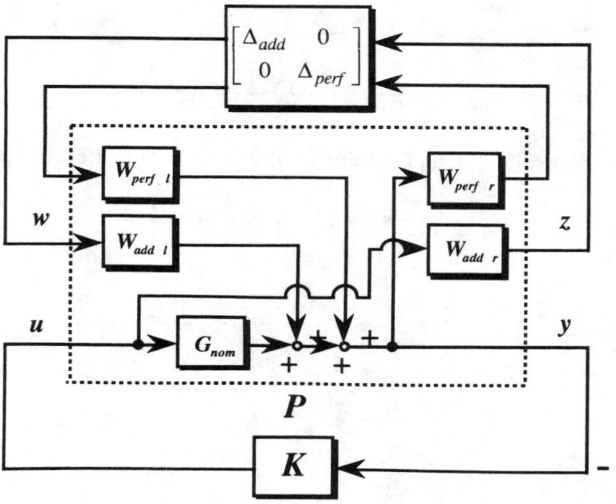

Figure 6 Interconnection Structure

3.2 Structured Singular Value μ

First, define a block structure $\underline{\Delta}$ as

$$\underline{\Delta}:=\left\{\begin{bmatrix}\Delta_{add} & 0\\ 0 & \Delta_{perf}\end{bmatrix}:\Delta_{add}\in C,\Delta_{perf}\in C\right\} \tag{16}$$

Next, with P partitioned as

$$P=\begin{bmatrix}P_{11} & P_{12}\\ P_{21} & P_{22}\end{bmatrix} \tag{17}$$

in an obvious way, let $F_l(P,K)$ denote a linear fractional transformation on P by K, where

$$F_l(P,K):=P_{11}+P_{12}K(I-P_{22}K)^{-1}P_{21} \tag{18}$$

(see Figure 6).

Then, robust performance condition is equivalent to the following structured singular value test

$$\sup_{\omega\in R}\mu_{\Delta}(F_l(P,K)(j\omega))<1 \tag{19}$$

Recall that the structured singular value $\mu_{\Delta}(M)$ is defined as

$$\mu_{\Delta}(M):=\frac{1}{\min\{\overline{\sigma}(\Delta):\Delta\in\underline{\Delta},\det(I-M\Delta)=0\}} \tag{20}$$

for a matrix $M\in C^{2\times 2}$ (in this case).

It is not known how to obtain a controller K achieving the structured singular value test (19) directly. Hence, our approach taken here is the so-called $D-K$ iteration. Using the known upper bound for μ, we can attempt to find a stabilizing controller K and a scaling matrix D such that

$$\|DF_l(P,K)D^{-1}\|_{\infty} \tag{21}$$

in minimized.

3.3 μ-Synthesis

The $D-K$ iteration involves a sequence of minimizations over either K or D while holding the other fixed, until a satisfactory controller is constructed. First, for $D=I$ fixed, the controller K is synthesized using the well-known state space H_{∞} optimization method. Let $P_1=P$ denote the given open-loop interconnection structure in Figure 6, and $F_l(P,K)$ be the closed-loop transfer function from the disturbances w to the errors z. Then, solving the following H_{∞} control problem

$$\|F_l(P_1,K)\|_{\infty}<\gamma_1,\quad \gamma_1=1.3 \tag{22}$$

yields the central controller K_1 below

$$K_1=\frac{-5.22\times10^{8}(s+12.46)(s+30.0)(s+35.0)}{(s+0.10)(s+31.6-j5.12)(s+31.6+j5.12)(s+39.77)}\times\frac{(s+38.0)(s+45.69)(s+66.94)}{(s+315.2-j329.6)(s+315.2+j329.6)(s+734.7)} \tag{23}$$

Thus, the first step of the $D-K$ iteration amounts to the standard H_{∞} (sub)optimal control design. Here we try to assess robust performance of this closed-loop system using μ-analysis associated with the block structure (16). The maximum singular value and μ of the closed-loop transfer function $F_l(P_1,K_1)$ are plotted in Figure 7. It is noteworthy to point out that the peak value of the μ plot is not less than 1. This reveals that the closed-loop system with the H_{∞} mixed sensitivity controller K_1 does not achieve robust performance.

Next, the above calculations of μ produce a scaling matrix at each frequency so as to minimize (21). These data will be fit with a stable, minimum-phase, real-rational function. The resulting scaling matrix D will be absorbed into the interconnection structure with multiplication and inverse. In this design, we try to fit the curve using a 1st order transfer function.

Now, let P_2 denote the new open-loop interconnection structure absorbing the scaling matrix D. This time, from the following H_{∞} control problem

$$\|F_l(P_2,K)\|_{\infty}<\gamma_2,\quad \gamma_2=1.0 \tag{24}$$

we can calculate the controller K_2 as follows

$$K_2 = \frac{-8.01\times10^9(s+10.54)(s+15.75)}{(s+0.10)(s+19.59-j5.32)(s+19.59+j5.32)} \times \frac{(s+30.0)(s+35.0)(s+38.0)}{(s+38.48-j2.70)(s+38.48+j2.70)(s+176.6)} \times \frac{(s+45.69)(s+66.94)(s+169.6)}{(s+420.1-j272.8)(s+420.1+j272.8)(s+8180)} \quad (25)$$

The maximum singular value and μ of this closed-loop transfer function are plotted in Figure 8. Since the value of μ is less than 1 in Figure 8, robust performance is now achieved for the closed-loop system with the μ controller K_2. Further, in order to confirm the accomplishment of nominal performance, robust stability and robust performance with the controller K_2, the magnitudes of the frequency responses for the following three transfer functions $W_{perf}S$, $W_{add}K_2S$, and $|W_{perf}S|+|W_{add}K_2S|$ (where $S=(I+G_{nom}K_2)^{-1}$) are plotted in Figure 9. As in Figure 9, the magnitude plots of all these transfer functions are less than 1, we can also find that nominal performance, robust stability and robust performance are achieved.

4 . Experimental Results

The designed continuous-time controllers K_1 and K_2 are discretized via the well known Tustin transform at the sampling rate of $45\mu s$ and $60\mu s$, respectively.

We evaluate robust performance as well as robust stability of the closed-loop system with the response against several external disturbances. These experimental results are shown in Figure 10 through Figure 25. There the disturbances are added to the experimental system as an applied voltage in the electromagnet. It is noted that there are 8-types of disturbances. The magnitude of each disturbance in fact amounts to 8.58 N, 17.15 N, 25.73 N, and 34.30 N, respectively, for both upward and downward directions. Because the steady-state force in the electromagnet is 17.15 N in this experiments, so these disturbances are big enough to evaluate the robustness of this design.

These experimental results show that the closed-loop systems with both the controllers K_1 and K_2 remain stable against all these disturbances. Hence the robust stability requirement would be satisfied. We can also find good transient responses against relatively small disturbances for both the controllers K_1 and K_2. Hence the nominal performance requirement would be satisfied, too.

However, the responses with K_1 deteriorate very much against relatively large disturbances (see Figure 16 and Figure 24). While the responses with K_2 maintain good transient responses against all these disturbances (see Figure 17 and Figure 25). Therefore, these experimental results show that the closed-loop system with the μ controller K_2 certainly achieves robust performance.

5 . Conclusions

In this paper, we experimentally evaluate a controller designed by μ-synthesis with an electromagnetic suspension system. We have obtained a nominal model as well as a set of models in which the real system is assumed to reside. Firstly, different model structures were derived based on the several idealizing assumptions for the real system. Secondly, for every model, the nominal value as well as the possible max./min. value of each parameter was determined by measurements. Thirdly, a nominal model was naturally chosen. Then, model perturbations were defined to account for uncertainties such as neglected nonlinearities and model parameter errors. Fourthly, we defined a family of plant models where unstructured additive perturbation was employed. The method to model the plant as belonging to a family or set plays a key role for systematic robust control design. Fifthly, we setup robust performance objective as a structured singular value test. Next, for the design, the $D-K$ iteration approach was employed. Finally, a lot of experimental results showed that the closed-loop system with the μ-controller achieves not only nominal performance and robust stability, but in addition robust performance. For real practical applications of μ, we need further experimental evaluations [6].

References

[1] J.C. Doyle, B.A. Francis, and A.R. Tannenbaum, Feedback Control Theory, Macmillan Publishing Company, 1992.

[2] J.C. Doyle, A. Packard, and K. Zhou., "Review of LFTs, LMIs, and μ," Proc. 30th IEEE Conference on Decision and Control, Brighton, England, 1991.

[3] G.J. Balas, J.C. Doyle, K. Glover, A. Packard, and R. Smith, μ-Analysis and Synthesis Toolbox, MUSYN Inc., 1991.

[4] F. Matsumura and S. Yamada ,"A Control Method of Suspension Control System by Magnetic Attractive Force (in Japanese)," J.IEE of Japan, vol. 94-B, pp.567-574, 1974.

[5] F. Matsumura and S. Tachimori,"Magnetic Suspension System Suitable for Wide Range Operation (in Japanese)," J.IEE of Japan, vol. 99-B, pp. 25-32, 1978.

[6] M. Fujita, F. Matsumura, and T. Namerikawa," μ-Synthesis of a Flexible Beam Magnetic Suspension System," Proc. Third International Symposium on Magnetic Bearings, Virginia, USA, pp. 495-504, 1992.

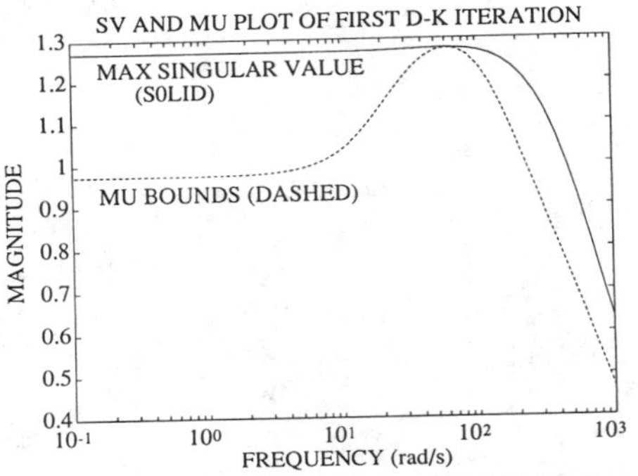

Figure 7 $\bar{\sigma}$ and μ plot of the first D-K iteration

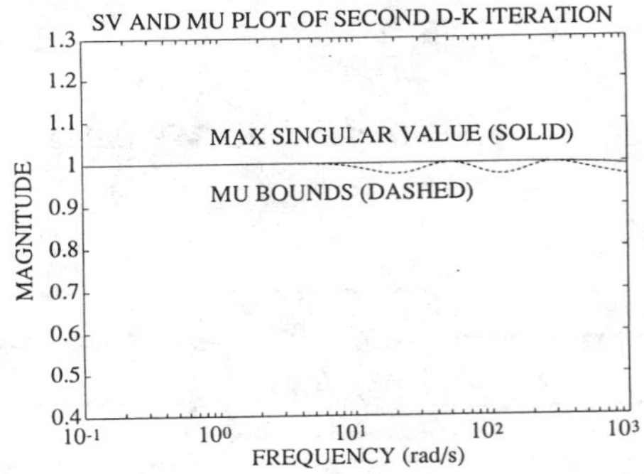

Figure 8 $\bar{\sigma}$ and μ plot of the second D-K iteration

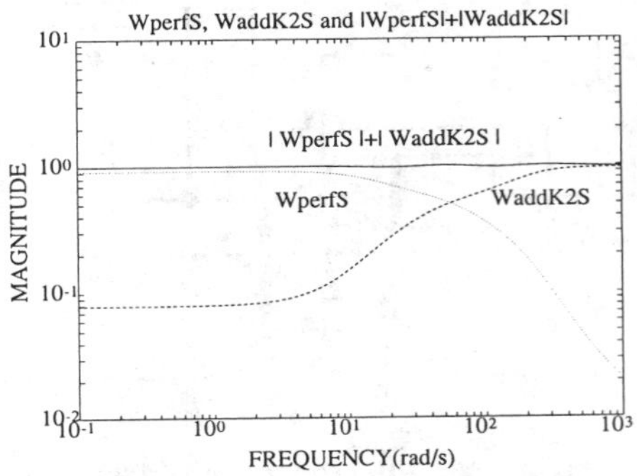

Figure 9 $W_{perf}S$, $W_{add}K_2S$, and $|W_{perf}S|+|W_{add}K_2S|$

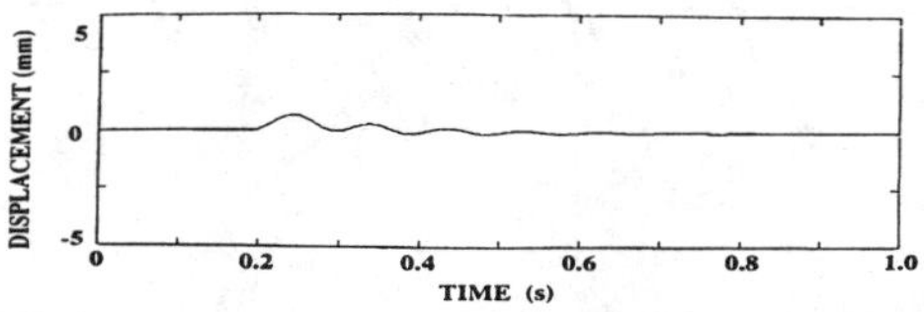

Figure 10 Response to step disturbance with K1 (-8.58N)

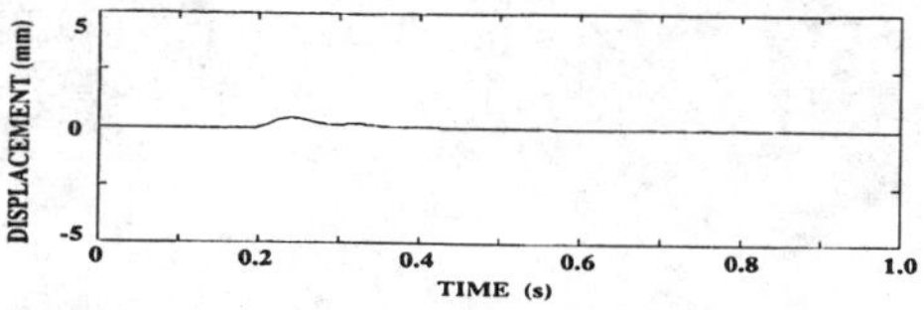

Figure 11 Response to step disturbance with K2 (-8.58N)

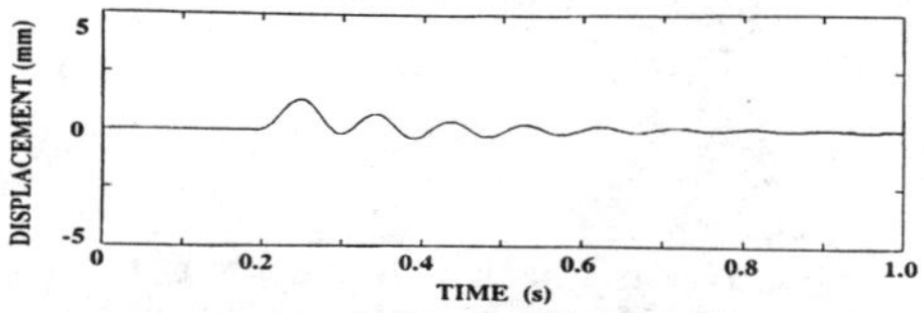

Figure 12 Response to step disturbance with K1 (-17.15N)

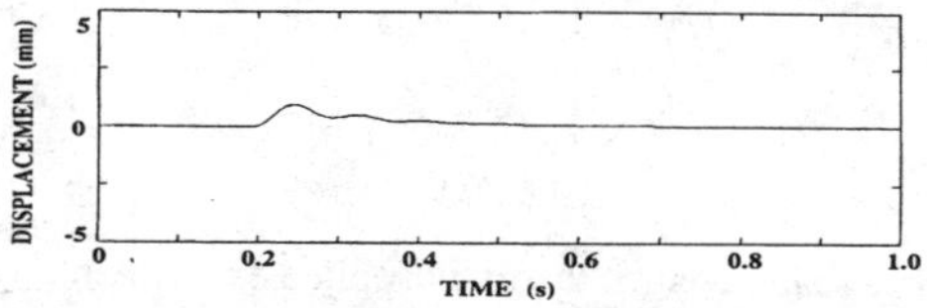

Figure 13 Response to step disturbance with K2 (-17.15N)

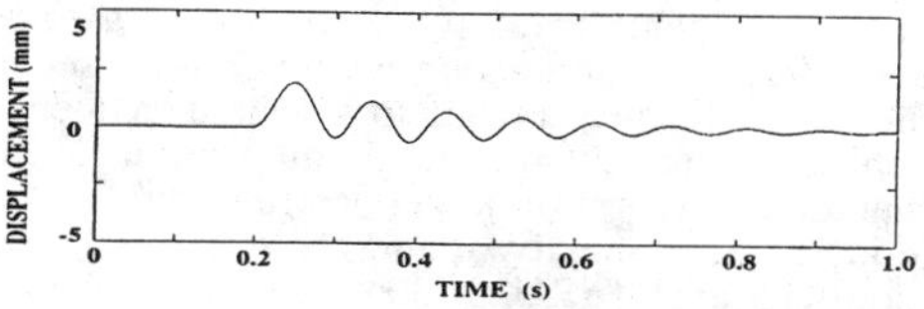

Figure 14 Response to step disturbance with K1 (-25.73N)

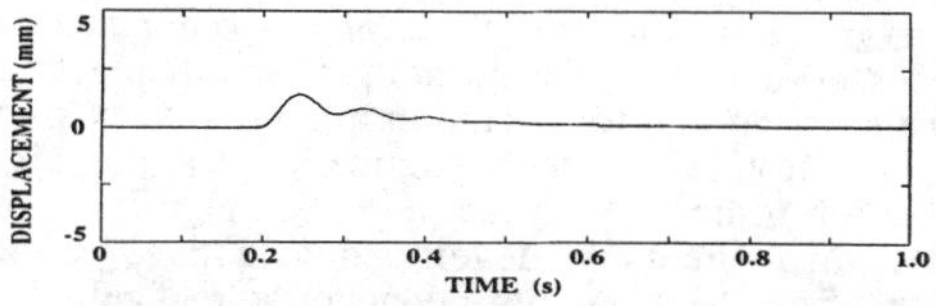

Figure 15 Response to step disturbance with K2 (-25.73N)

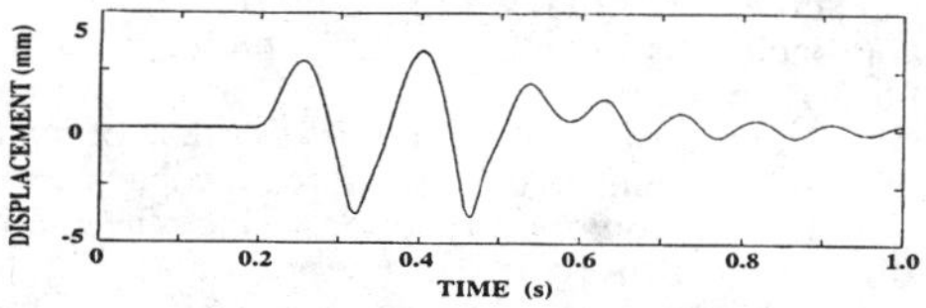

Figure 16 Response to step disturbance with K1 (-34.30N)

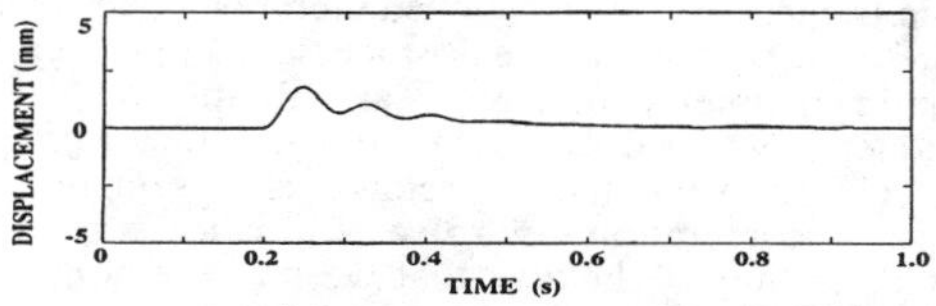

Figure 17 Response to step disturbance with K2 (-34.30N)

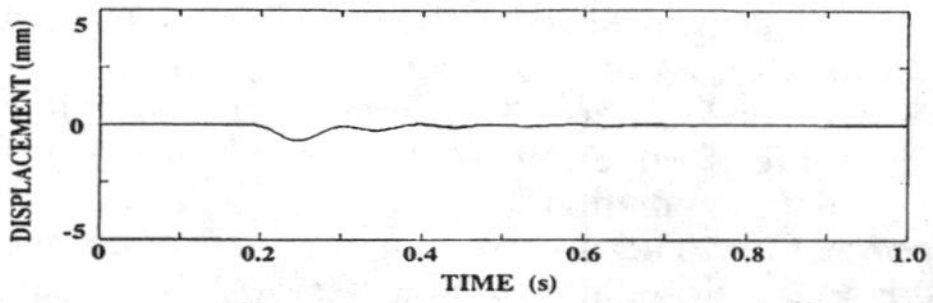

Figure 18 Response to step disturbance with K1 (+8.58N)

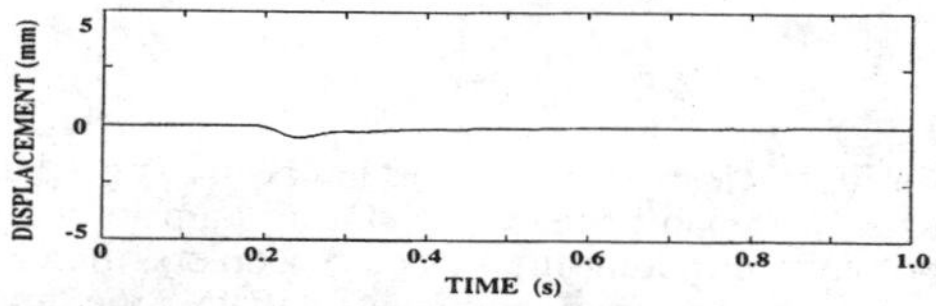

Figure 19 Response to step disturbance with K2 (+8.58N)

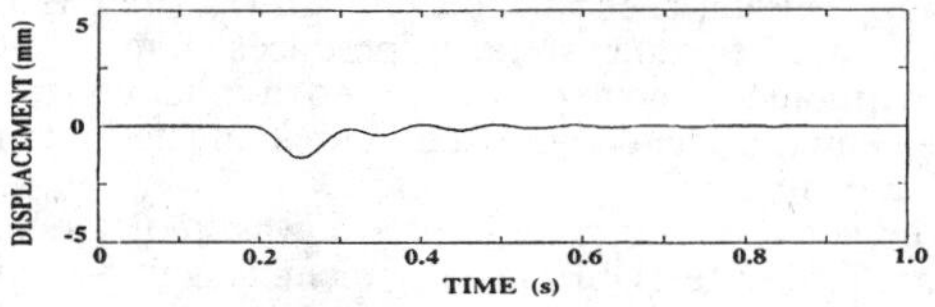

Figure 20 Response to step disturbance with K1 (+17.15N)

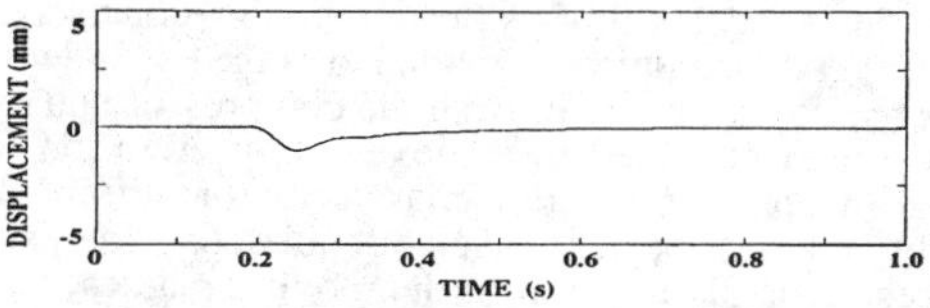

Figure 21 Response to step disturbance with K2 (+17.15N)

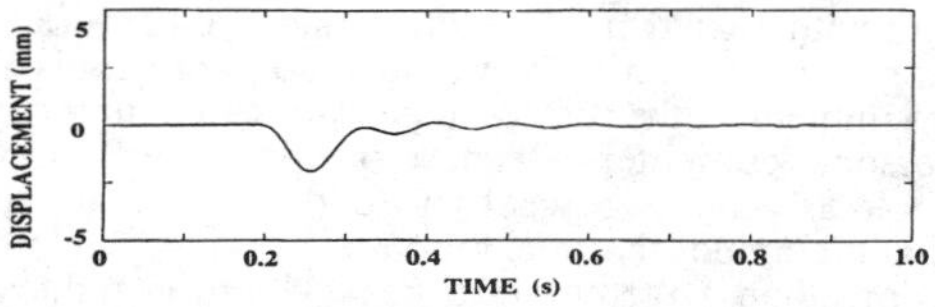

Figure 22 Response to step disturbance with K1 (+25.73N)

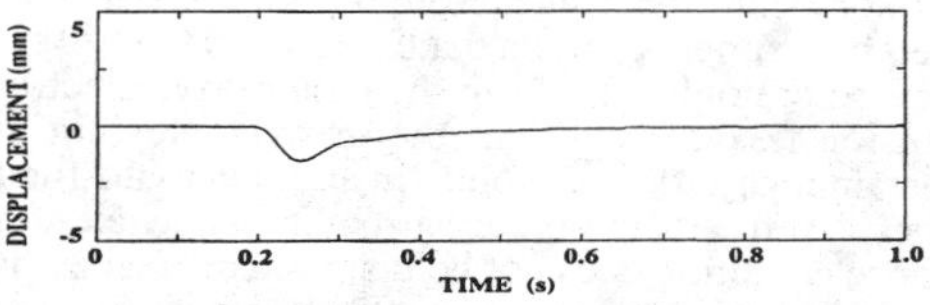

Figure 23 Response to step disturbance with K2 (+25.73N)

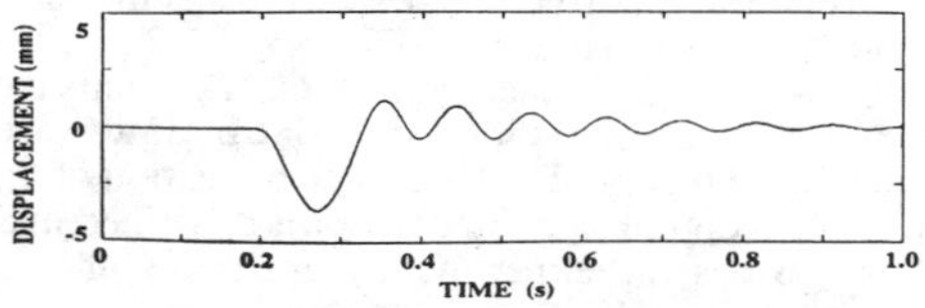

Figure 24 Response to step disturbance with K1 (+34.30N)

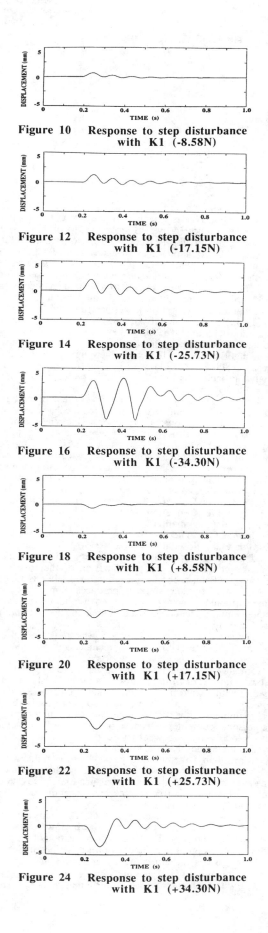

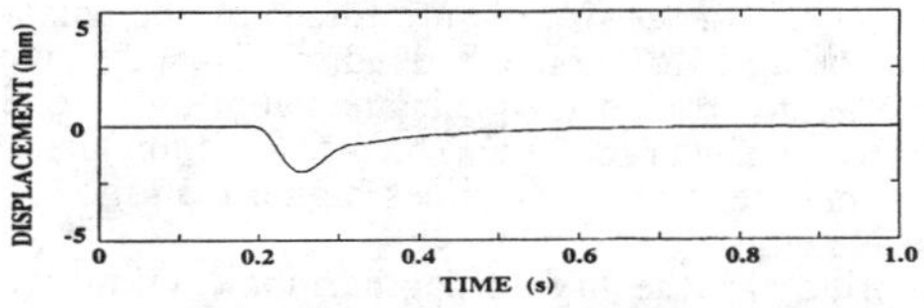

Figure 25 Response to step disturbance with K2 (+34.30N)

Proceedings of the 31st Conference
on Decision and Control
Tucson, Arizona • December 1992

DEMONSTRATION OF ACTIVE VIBRATION CONTROL OF THE HUGHES CRYOCOOLER TESTBED

Yeong Wei Andy Wu

Hughes Aircraft Company
Electro-Optical & Data Systems Group
P. O. Box 902
El Segundo, CA 90245

Abstract

Narrow-band control theory is applied experimentally to the Hughes Stirling-cycle cryocooler expander. The cryocooler expander designed and built by Hughes consists of a displacer and a balancer, which are mechanically placed in a back-to-back configuration and commanded to move in opposite directions in order to cancel the vibration forces produced by the motions of the displacer and the balancer. The existing control design employing the position loop servos failed to provide adequate attenuation for the high-order harmonic vibration forces. A vibration control algorithm based on narrow-band control theory is developed to actively control the high-order vibrations. In this control concept, a load cell measuring the high-order vibration forces is used as a feedback element and a servo compensator is designed to produce an infinite open loop gain at each harmonic frequency. The developed vibration control algorithm is implemented with a digital signal processor, and integrated and tested on the cryocooler expander. Peak vibration at the second harmonic frequency was reduced by more than a factor of 25 in the test. The effect of sampled-data parameters such as sampling rate and quantization on the vibration control performance was evaluated. After extensive testing we believe that we have demonstrated the technical feasibility of applying the proposed active vibration control to meet the next generation cryocooler vibration specfications.

1.0 Introduction

A Stirling-cycle cryogenic cooler has been designed and built by Hughes under company funded IR&D program to support the development of near-term long-life space cryocoolers. In addition to achieve the required thermodynamic effeciency, the low froce vibration has been identified to be one of the key technical challenges for these coolers.

The designed Stirling-cycle cryocooler consists of three basic modules: (1) the compressor module, (2) the expander module, and (3) the servo control module. Both the compressor and expander modules have been designed to achieve a low vibration level by incorporating an opposed reaction mass/actuator within the same housing to obtain a near perfect balance in all active forces, such as pressure forces, dynamic reaction of the moving masses, diaghragm spring forces, motor current lead spring forces, dissipative forces due to eddy currents in the motor, dissipative regenerator pressure drop forces, and motor electromechanical forces. The imbalance forces are then controlled by a simple position matching servo control system.

Because of nonlinearities in the motor drive electronics, the piston suspension flexures, and gas thermodynamics, the resulting vibrations contain high-order harmonics in addition to the fundamental drive frequency. At these high-order harmonic frequencies, the above dynamic balance condition does not hold any more. Hence one has to rely on the position loop servos to attenuate these high-order harmonic vibrations.

This force imbalance becomes much worse in the expander module due to the absence of pressure forces on the balancer. Our recent test data on the IR&D cooler had indicated that the pure position loop servos failed to provide adequate attenuations for the high-order harmonics in the expander module. The measured imbalance force in the expander module is about 40 times higher than the requirement level of 0.01 Newton.

In principle, the high-order harmonic vibrations can be further reduced by increasing the position loop servo bandwidth. However, the motor drive dynamics and internal structural resonances of the suspension system will prevent us from implementing higher bandwidth servos.

An innovative approach to controlling the cryocooler vibration has been reported by JPL recently [1]. In this concept, a reaction mass actuator (Ling Shaker) in a back-to-back configuration is used to cancel the vibration forces produced by the BAe cryocooler. The cancellation is accomplished by incorporating a force sensor output in the feedback in conjunction with a bank of second-order undamped filters. Each filter is tuned to a desired frequency so that the closed loop gain is zero at the corresponding harmonic frequency. SatCon Technology Corporation in cooperation with JPL staff had designed, coded these narrow-band control algorithms for implementation with a TI TMS320C30 processor, and successfully demonstrated the performance improvement of these algorithms over the existing algorithm on the JPL cryocooler vibration test facility. In reference 2, an active vibration control using frequency shaped cost functional approach was developed, and applied to a helicopter vibration control system.

In this paper, we extend the JPL's idea of using force sensors as feedback elements to design a force feedback controller to perform the vibration suppression function. Since the use of closed position loop servos can relax some manufacturing tolerance requirements, such as suspension system stiffness and motor parameters in the expander module design, we derive the force feedback control law in the presence of closed position loop servos. As shown in the following section, the resultant control law consists of three major blocks: a notch filter; an inverse filter, and a bank of second-order undamped filters. The notch filter is used to isolate the fundamental drive frequency from the high-order harmonic frequencies so that the position loop servos will not interact with the vibration loop servo. The inverse filter is used to compensate the phase loss due to the closed position loop servos. If the bandwidth of the closed position loop servo is much less than the controlled harmonic frequencies, the transfer function of the inverse filter becomes unity. Each second-order undamped filter is tuned to a specific harmonic frequency so that it produces an infinite open loop gain at the harmonic frequency. Three second-order undamped filters were implemented in the demonstration to attenuate up to the fourth harmonic force vibration.

The active vibration control law was implemented using a digital signal processor (DSP) development system provided by the Star Semiconductor. The development system has an easy-to-use high-performance DSP consisting of four independent general signal processors with 1Kx24-bit program memory, a shared central memory unit with 1Kx24-bit data RAM, and two (2) serial I/O ports. The development system also provides software tools that allow designers to implement the servo compensation in minutes through the use of a personal computer. Becuase of this powerful development tools, it took us only two weeks to code and debug the control algorithm, and conduct the integration test on the IR&D cryocooler testbed. More than a factor of 25 peak vibration reduction at the second harmonic frequency was obtained during the test. In overall, we have demonstrated that the peak imbalance force of less than 0.04 Newton at all harmonic frequencies can be achieved using the developed vibration control algorithm.

Following this introduction, Section II presents the expander dynamics and derives the vibration control law based on the narrow-band control theory. Detailed implementation of the vibration control law using a digital signal processor is described in Section III. Section IV describes the experimental apparatus and test results. Performance limitations of the current hardware/software development system along with the recommendations for future studies are stated in the last section.

CH3229-2/92/0000-2580$1.00 © 1992 IEEE

2.0 Derivation of Vibration Control Law

The complicated dynamics of a cryocooler system requires multiple degrees of freedom (DOF) to completely describe vibrations in all axes. However, since we are interested in the vibrational energy along the axial axis, we will only consider a single degree of freedom motion in the following analysis.

Figure 2 shows all the essential blocks of a single DOF dynamic model of a Stirling-cycle cooler. The model describes the connections between compressor module (with two pistons), expander module (with displacer and balancer), and thermodynamic coupling. As shown in the figure, each moving mass is subject to a set of forces as defined by Newton's law. The pressure forces produced by the thermodynamic effect are applied to all the moving mass except the balancer. As a result, there is an intrinsic force imbalance in the expander module. More detailed description of the cryocooler dynamics can be found in reference 3. In the following, we will focus on the expander module dynamics.

During the cryogenic operation, both the displacer and the balancer are driven at a single frequency (35Hz) with an adjustable amplitude and phase. Position sensing devices (LVDTs) are used and fed back to the command summing junction, where they are compared with the reference signals. The position errors are both compensated by the compensation networks. The compensated outputs are then used to command the current loops for driving the linear motors. The amplitude and phase corrections are applied to the balancer position command so that the residual vibration forces acting on the expander housing structure are minimized. However, because of nonlinearities in the motor drive electronics, the piston suspension fluxures, and the gas thermodynamics, the resulting vibration forces contain high-order harmonics in addition to the fundamental drive frequency (35Hz). It is the purpose of this research work to develop a control scheme that provides an effective means to suppress these high-order harmonic vibration forces.

When driven by a single frequency sinusoid, the net motion of either the displacer or the balancer is produced only at the fundamental drive frequency and its harmonics. Because of the essentially time-invariant nature of the cryocooler, the amplitude of these harmonics remain nearly stationary. Therefore, the net position can be accurately modelled as a summation of the fundamental drive frequency and its harmonics:

$$x(t) = \sum_{i=1}^{n} a_i \sin(2\pi i f_D t) = x_f(t) + \tilde{x}_D(t) \qquad (1)$$

with $a_1 = 1$; $x_f(t) = \sin(2\pi f_D t)$; $\tilde{x}_D(t) = \sum_{i=2}^{n} a_i \sin(2\pi i f_D t)$

In order to simplify the mathematical expression, we will assume that there is only one harmonic frequency in the following analysis. Neverthless, the analysis stated below can be easily extended to the general case.

Figure 3 shows the simplified servo block diagram that describes the dynamic responses of the displacer and the balancer motions subject to a commanded reference signal. The motor current drive dynamics is neglected here since its frequency response is far beyond our interest. As shown in the figure, the net force acting on the expander module housing is given by:

$$\Delta F = M_d \ddot{x}_d + M_b \ddot{x}_b$$

or in frequency domian:

$$\Delta F(s) = M_d s^2 x_d(s) + M_b s^2 x_b(s) \qquad (2)$$

where $x_d(s)$, $x_b(s)$ can be obtained as follows:

$$x_d = \frac{G_d P_d}{1 + G_d P_d} x_c + \frac{a_d}{1 + G_d P_d} x_D \qquad (3)$$

$$x_b = \frac{-\alpha G_b P_b}{1 + G_b P_b} x_c + \frac{a_b}{1 + G_b P_b} x_D \qquad (4)$$

It is noted that the dependent variable "s" is omitted in the above equations. If we assume that both the displacer and the balancer have an identical closed loop response, then by substituting equations (3) and (4) into equation (2) one obtains:

$$\Delta F = (M_d - \alpha M_b) s^2 H_c x_c + (a_d M_d + a_b M_b) \frac{s^2 x_D}{1 + GP} \qquad (5)$$

where $$H_c = \frac{G_d P_d}{1 + G_d P_d} = \frac{G_b P_b}{1 + G_b P_b} = \frac{GP}{1 + GP} \qquad (6)$$

From equation (5), it is clear that if the parameter α is set to be equal to M_d / M_b, then the net force, ΔF will only be generated by the disturbance force F_D, or $(a_d M_d + a_b M_b) s^2 x_D$. In the compressor module, $M_d = M_b$ and $a_d = - a_b$ due to virtually identical (opposed) pressure forces and identical moving masses, the compressor module has a well balancing condition even at harmonic frequencies. On the other hand, since the balancer is not exposed to the pressure force, the absolute value of a_d is not equal to the absolute value of a_b. As a result, the only vibration force suppression in this case comes from the wide-band error rejection characteristics of a closed position loop servo:

$$\Delta F = \frac{F_D}{1 + GP} \qquad (7)$$

In order to provide an adequate vibration force attenuation, the position loop servo needs to have a higher control bandwidth, which may not be realizable since the control bandwidth is limited by the motor drive dynamics and the structural stiffness of the suspension system.

Suppose we add one more current command, I_{b2} to the balancer motor drive as indicated in Figure 4, then the net force acting on the expander module housing becomes:

$$\Delta F = \frac{F_D}{1 + GP} + \frac{M_b s^2 P}{1 + GP} I_{b2} \qquad (8)$$

After introducing this additional current command, our control objective here is then to find a feedback control law as a function of load cell output, or $I_{b2} = -G_f(s)\, \Delta F$ such that the net force is driven to zero at all the disturbance frequencies. Based on the classical feedback control theory, in order to accomplish this control objective, the vibration force compensation $G_f(s)$ should be chosen such that the resultant open loop transfer function (from the disturbance force input F_D to the measured force output ΔF) produces an infinite gain at the disturbance frequencies [4]. Having this theorem in mind, let $G_f(s)$ be expressed as:

$$G_f(s) = H_I(s)\, G_D(s) \qquad (9)$$

with $$H_I(s) = \frac{1 + GP}{M_b s^2 P} \qquad (10)$$

then equation (8) becomes:

$$(1 + G_D(s))\, \Delta F = \frac{F_D}{1 + GP}$$

or $$\Delta F = \left(\frac{1}{1 + G_D(s)}\right)\left(\frac{F_D}{1 + GP}\right) \qquad (11)$$

It can be shown that if

$$G_D(s) = \frac{K_D s}{s^2 + \omega_D^2} \qquad (12)$$

then $$\Delta F = \left(\frac{s^2 + \omega_D^2}{s^2 + K_D s + \omega_D^2} \right) \left(\frac{F_D}{1 + G P} \right) \tag{13}$$

and thus the net force will be driven to zero at the frequency f_D for any positive constant K_D. Equations (9), (10), and (12) constitude the basic vibration control law that produces a zero net force at a single discrete frequency. For a general system with multiple discrete frequencies, the transfer function $G_D(s)$ becomes:

$$G_D(s) = \sum_{i=1}^{n} \frac{K_{Di}\, s}{s^2 + \omega_{Di}^2} \tag{14}$$

As long as there exists a set of constant parameters K_{Di} so that the closed loop system, equation (11) is stable, then the net force will be zero at all discrete frequencies.

3.0 Implementation of Vibration Control Law

In order to illustrate this simple vibration control concept and demonstrate its performance improvement over the existing algorithms, we implemented the developed vibration control law, equations (9), (10), and (14) with a digital signal processor (DSP) development system, and integrated and tested the algorithm on the Hughes IR&D cryocooler.

3.1 Servo Characterization

The purpose of the inverse filter $H_I(s)$, given by equation (10) is to compensate the dynamics produced by the closed position loop servos. To design this inverse filter we first measured the two transfer functions: G(s) and G(s)P(s)/1+G(s)P(s), through the use of HP spectrum analyzer. Figure 5 shows the magnitude and phase plots of these two transfer functions. We then synthesized the transfer functions using the following poles and zeros generated by the HP spectrum analyzer which best fitted the measured magnitude and phase:

$$G(s) = \frac{1 + \frac{s}{2\pi(36)}}{1 + \frac{s}{2\pi(213)}} \tag{15}$$

$$\frac{G(s)P(s)}{1+G(s)P(s)} = \frac{1 + \frac{s}{2\pi(117)}}{\left(1 + \frac{s}{2\pi(38+107j)}\right)\left(1 + \frac{s}{2\pi(38-107j)}\right)\left(1 + \frac{s}{2\pi(262+24j)}\right)\left(1 + \frac{s}{2\pi(262-24j)}\right)} \tag{16}$$

Finally, the inverse filter is obtained by dividing Equation (15) by Equation (16). After adding two integrators in the denominator, $H_I(s)$ is given by:

$$H_I(s) = \frac{\left(1+\frac{s}{2\pi(36)}\right)\left(1+\frac{s}{2\pi(38+107j)}\right)\left(1+\frac{s}{2\pi(38-107j)}\right)\left(1+\frac{s}{2\pi(262+24j)}\right)\left(1+\frac{s}{2\pi(262-24j)}\right)}{s^2\left(1+\frac{s}{2\pi(117)}\right)\left(1+\frac{s}{2\pi(213)}\right)} \tag{17}$$

3.2 Digital Signal Processor Development System

A DSP development system provided by the Star Semiconductor was used to implement the above vibration control algorithm. The development system has an easy-to-use high-performance DSP which consists of four independent general signal processors with 1Kx24-bit program memory, a shared memory unit with 1Kx24-bit data RAM, and two serial I/O ports. The processor running at a 20MHz clock has a 24-bit fixed-point architecture which handles all signal scaling automatically. Figure 6 shows the processor functional diagram. As shown in the diagram, the system has an Access Port to provide the link between the interface module, such as a PC-based computer, and the processor. This feature allows the designer to dynamically modify system parameter values such as gains in a real time fashion. A built-in, software-directed signal probe is also provided to allow the designer to select internal signals to a dedicated output port for display or measurement.

The development system also provides powerful software tools that enables the direct transformation of block diagrams to production-ready signal-processing systems in minutes. As shown in Figure 7, four basic blocks: gain, first-order filter, second-order filter, and summation blocks were used in generating the transfer function, Equation (9). A 35-Hz notch filter was added to equation (9) in serial for the purpose of isolating the fundamental drive frequency from the vibration harmonic frequencies. Three second-order lightly-damped filters with Q = 1000 were implemented to provide a narrow-band attenuation at harmonic frequencies of 70Hz, 105Hz, and 140Hz. The inverse filter was modified by (i) approximating the pure integrator with a first-order filter having a near origin pole, and (ii) adding a high-pass filter in front of the approximated integrator to avoid a possible signal situation caused by a DC offset. Figure 8 shows the magnitude and phase plots of the transfer function as defined in Figure 7 with a sampling rate of 19.5KHz. From the phase plot, it indicates that the DSP has a fixed time delay of approximately 400 microseconds. This time delay is primarily caused by a scheduling software in handling the input and output data.

4.0 Experimental Results of Vibration Control Algorithm

The testbed as shown in Figure 9 consists of a cryocooler system (compressor/expander modules), an electronics rack, a HP personal computer, the Star DSP, two HP spectrum analyzers (one monitors the load cell output and the other monitors the output from the DSP), and all necessary power supplies. One of the two load cells mounted on the expander module housing was connected to the Star DSP, and the analog output signal from the Star DSP was connected to the test summing amplifier that drived the balancer linear motor (see Figure 10). These two connections represent the required vibration control loop. Before activating the vibration control loop, the amplitude and the phase of the reference signal were adjusted so that a near perfect balance condition at the fundamental drive frequency was obtained. Figure 11 shows the power spectrum density of the measured imbalance force from one of the load cells when the expander operated at 75% of the maximum stroke and the above balance condition was reached. To ensure a smooth transition when closing the vibration control loop, a set of reduced gains (0.2, 0.08, 0.1 as indicated in Figure 7) was used in the three lightly-damped filters before the full gains (1.0, 0.32, 1.0) were applied. Table 1 summarizes the performance results. The results indicate that the peak vibration at the second harmonic frequency (70Hz) was reduced by more than a factor of 25, and both the third and the fourth harmonics were driven to a force level of less than 0.01 Newton RMS by using the proposed vibration control algorithm. As indicated in Figure 12, the imbalance force vibrations were dominanted by the sixth harmonics (210Hz) after closing the vibration control loop. Because of the inherent data latency (400 microseconds) within the DSP, we did not attempt to control the sixth harmonics. However, the data latency can be minimized with a more efficient scheduling software. Hence, we believe that the total imbalance vibration forces of less than 0.01 Newton RMS can be achieved to meet the next generation cryocooler vibration specifications with the developed vibration control algorithm. We also investigated the effect of sampling rate and quantization on the vibration performance. Table 2 summaries the results for the sampling rate effect and Table 3 summaries the results for the A/D, D/A quantization effect. As one expects, the vibration performance at the fourth harmonic (140Hz) is more sensitive to the sampling rate as compared to the second harmonic (70Hz). As indicated in the tables, the system became unstable when the sampling rate was below 2.44 KHz, and with a higher sampling rate, a lower A/D, D/A bit resolution can be used to perform the vibration control.

Although a hybrid configuration was implemented in this test for the purpose of a quick demonstration, the command generation, the position control loop servos should be integrated with the vibration control loop servos, and a full digital implementation with a digital signal processor should be realized in the future. This implementation approach will allow the designers to adaptively adjust the filter parameters to compensate the possible fundamental drive frequency variation.

5.0 Conclusions

Narrow-band control theory has been applied to develop a vibration control algorithm for controlling the high-order harmonic vibrations of a cryocooler system. The developed vibration control algorithm has been implemented with a digital signal processor development system, and integrated and tested on the Hughes IR&D cryocooler testbed. A peak vibration reduction of more than 25 at the second harmonic frequency was obtained by using the implemented vibration control algorithm. Based on the test data we believe that we have demonstrated the technical feasibility of appling the proposed active vibration control to meet the next generation cryocooler vibration specifications. Applications of modern control design techniques, such as LQG/LTR, H_∞, and ℓ_1 control design techniques [5,6,7], to the active suppression of the Stirling-cycle cryocooler vibration will be investigated and evaluated in the future.

6.0 References

[1]: B. G. Johnson and M. S. Gaffney,"Demonstration of Active Vibration Control on the JPL Cryocooler Vibration Testbed," Final Report, SatCon Technology Corporation, Cambridge, MA. August 1991. Also in Proceedings of 1992 ACC, pp 1630-1631.

[2]: J. V. R. Prasad, A. J. Calise, and E. V. Byrns, Jr., "Active Vibration Control Using Fixed Order Dynamic Compensation with Frequency Shaped Cost Functions," IEEE Control System Magazine, Vol. II, April 1991, pp 71-78.

[3]: " 65 Degree K Standard Spacecraft Cryocooler Program Final Report," Hughes Aircraft Company, El Segundo, CA 90245, May 1990.

[4]: L. A. Sievers and A. H. von Flotow, "Comparison and Extensions of Control Methods for Narrowband Disturbance Rejection," Proceeding of the ASME Winter Annual Meeting, NCA-Vol. 8, American Society of Mechanical Engineers, New York, pp 11-22.

[5]: M. Athans, "A Tutorial on the LQG/LTR Method," Proc. American Control Conference, Seattle, WA, June 1986, pp 1289-1296.

[6]: J. C. Doyle, K. Glover, P. P. Khargonekar, and B. A. Francis, "State-Space Solutions to Standard H_2 and H_∞ Control Problems", IEEE Trans. on Automatic Control Vol. 34, No. 8 August 1989, pp 831-847.

[7]: M. A. Hahleh and B. Pearson, Jr., "ℓ_1-Optimal Compensators for Continuous-Time Systems," IEEE Trans. on Automatic Control, Vol. AC-32, No. 10, October 1987, pp 889-895.

Table 1 Summary of Vibration Control Performance

Harmonic Frequency	Measured Vibration Force (RMS)	
	Without vibration control loop	With vibration control loop
70Hz	0.362 N	0.0143 N
105 Hz	0.129 N	0.008 N
140 Hz	0.097 N	0.01 N
175 Hz	0.08 N	0.04 N
210 Hz	0.075 N	0.043 N

Table 2 Vibration Performance as a Function of Sampling Rate

Sampling Rate	Residual Froce Vibration (mN RMS)	
	70 Hz	140 Hz
19.53 KHz	14.30	10.55
9.76 KHz	15.24	20.63
4.88 KHz	15.15	35.30
2.44 KHz	unstable	unstable

Table 3 Vibration Performance as a Function of A/D and D/A Resolution

Sampling Rate	A/D	D/A	Residual Force Vibration (mN RMS)	
			70 Hz	140 Hz
19.53 KHz	16-bit	16-bit	14.30	10.55
19.53 KHz	10-bit	16-bit	14.41	10.73
19.53 KHz	10-bit	12-bit	14.34	10.59
19.53 KHz	8-bit	16-bit	14.64	10.40
19.53 KHz	8-bit	12-bit	14.38	10.81
4.88 KHz	16-bit	16-bit	15.15	35.30
4.88 KHz	10-bit	16-bit	15.82	35.40
4.88 KHz	10-bit	14-bit	15.63	34.80
4.88 KHz	10-bit	12-bit	15.21	34.89
4.88 KHz	8-bit	16-bit	unstable	unstable

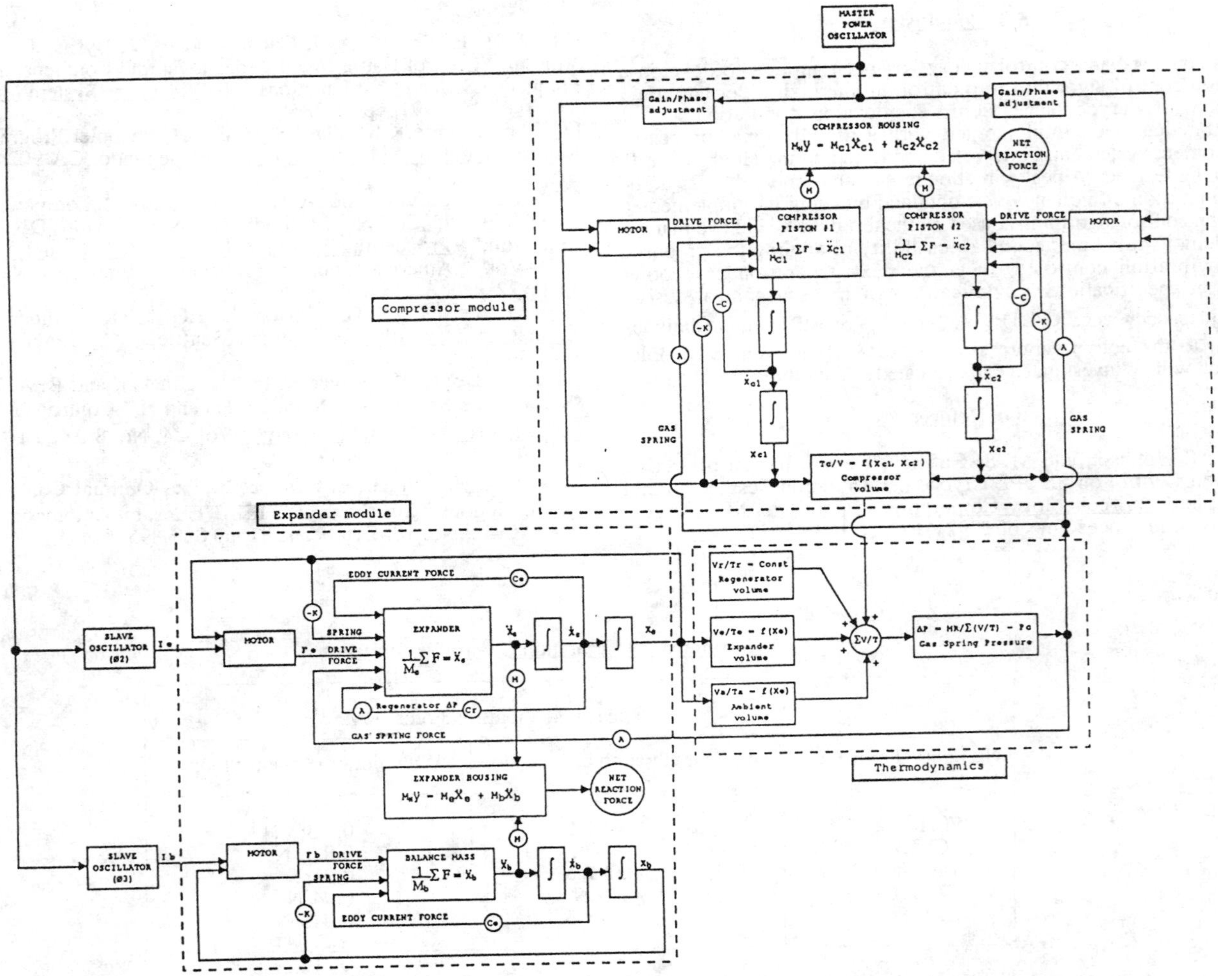

Figure 2 Block Diagram of Compressor and Expander Modules with Coupling Thermodynamics

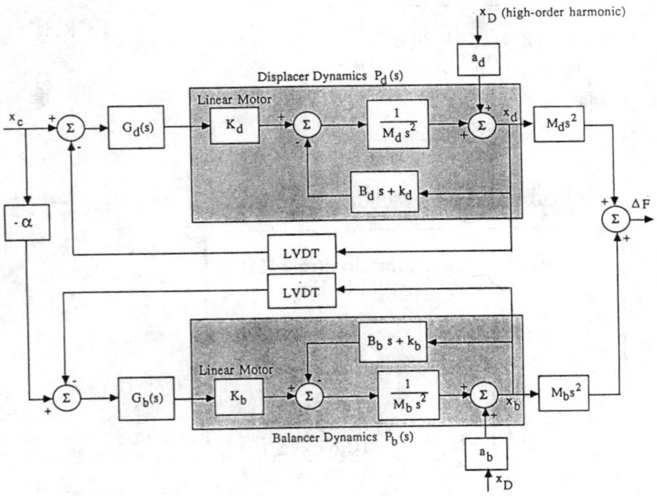

Figure 3 Servo Block Diagram of Expander Module

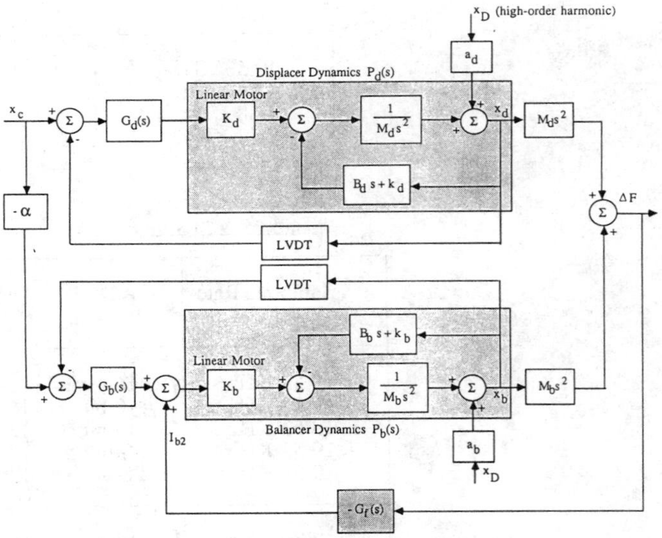

Figure 4 Servo Block Diagram of Expander Module with a Force Sensor Feedback

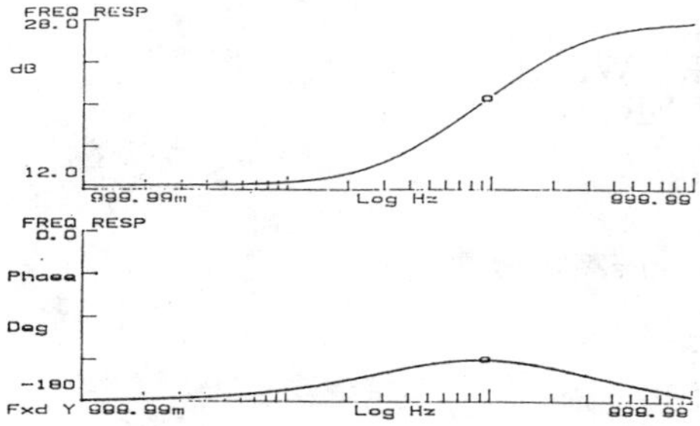

Figure 5(a) Magnitude and Phase Plots of G(s)

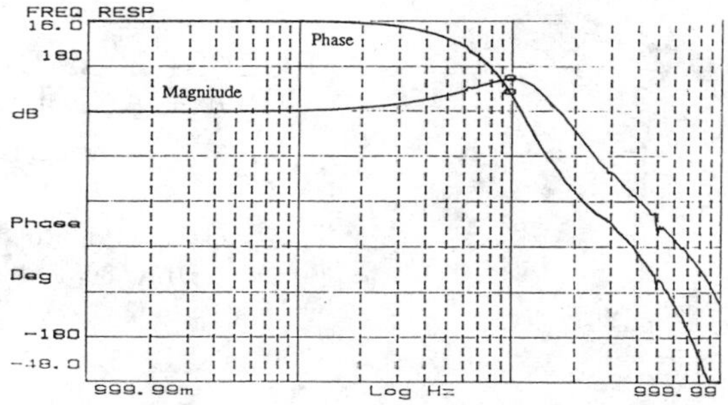

Figure 5(b) Magnitude and Phase Plots of G(s)P(s)/1+G(s)P(s)

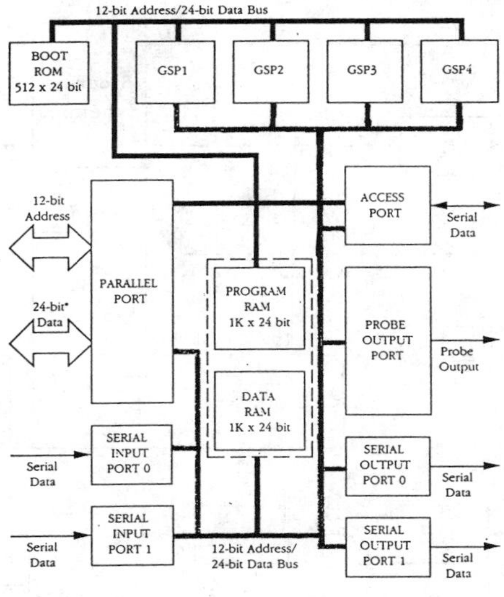

* 8/16/24 bit selectable

Figure 6 Star DSP Functional Diagram

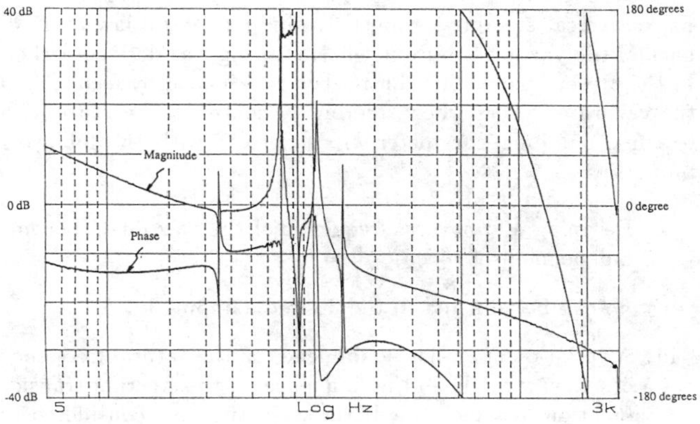

Figure 8 Frequency Response of Vibration Control Compensation Network

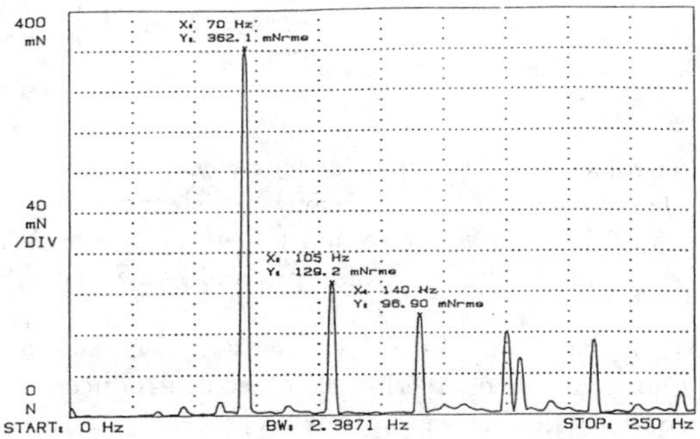

Figure 11 Vibration Force Spectra of Cryocooler Expander - without Vibration Control Loop

Figure 9 Hughes IR&D Cryocooler Active Vibration Control Testbed

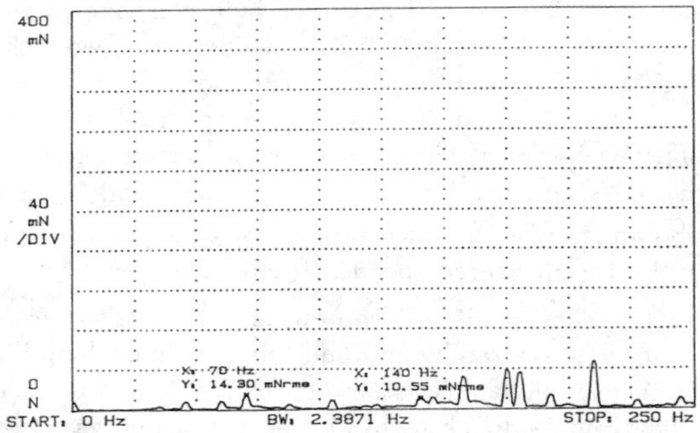

Figure 12 Vibration Force Spectra of Cryocooler Expander - with Vibration Control Loop Closed

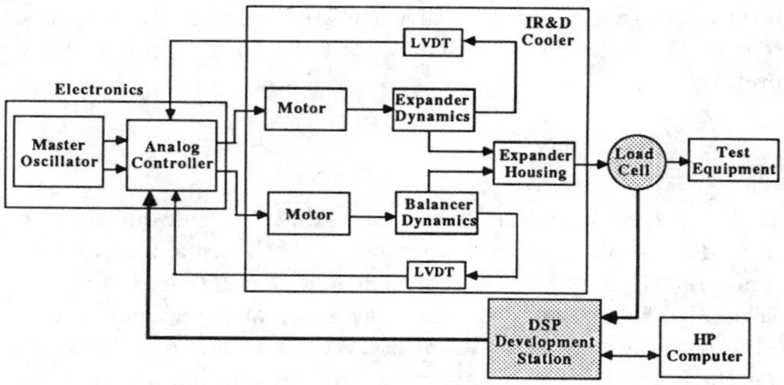

Figure 10 Integration with Hughes IR&D Cryocooler Testbed

Proceedings of the 31st Conference
on Decision and Control
Tucson, Arizona • December 1992

ROBUST YAW DAMPING OF CARS WITH FRONT AND REAR WHEEL STEERING

Juergen Ackermann

Institute for Robotics and System Dynamics,
DLR, 8031 Oberpfaffenhofen, Germany

ABSTRACT

For active car steering a robust decoupling control law by feedback of the yaw rate to front wheel steering was derived earlier. In the present paper this control law is extended by feedback of the yaw rate to rear wheel steering. A controller structure with one free damping parameter k_D is derived with the following properties:

i) Damping and natural frequency of the yaw mode become independent of speed,

ii) k_D can be adjusted to the desired damping level,

iii) A variation of k_D has no influence on the natural frequency of the yaw mode and no influence on the steering transfer function by which the driver keeps the car - considered as a mass point - on his planned path.

I. INTRODUCTION

Recently several car manufacturers (Nissan, Honda, Mazda, Mitsubishi, Toyota, Daihatsu, BMW) have introduced additional rear-wheel steering for automobiles. Thus a second actuator is becoming available for car steering dynamics. How do we control engineers use it?

Fig. 1 shows three examples of feedforward and feedback control system structures. The block "steering dynamics" has the inputs δ_f and δ_r, the steering angle of the front and rear wheels. Outputs of interest are the sideslip angle β, the lateral acceleration a_f at the front axle and the yaw rate r. δ_S is the command signal from the steering wheel, and d is a yaw disturbance, resulting for example from side wind or from braking with different road-tire friction at the left and right wheels (μ-split braking).

The most common control system structure is that of Fig. 1.a.

The front wheel steering is unchanged, i.e. $\delta_f = \delta_S$. The prefilter F_r for generating the rear wheel steering angle δ_r is usually scheduled by the measured car velocity v. For example in [1] a prefilter

$$F_r(s,v) = K(v)\frac{1+T_D(v)s}{1+T_1(v)s} \tag{1}$$

is determined from the requirement that the sideslip angle β should be zero (i.e. neutral steer instead of the usual understeer). The prefilter may also be replaced by a gain $K(v,r,\ldots)$ that is scheduled by measured velocity, yaw rate, lateral acceleration at different longitudinal positions etc. At low speeds K is negative, i.e. the front and rear wheels are steered in opposite directions for better maneuverability. For higher speeds the wheels are steered in the same direction, i.e. K is positive [2], [3].

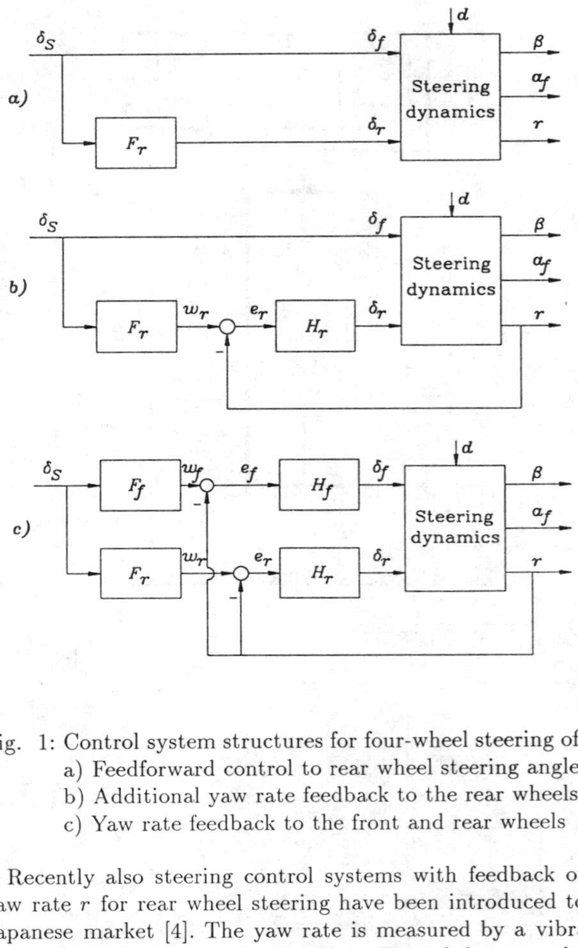

Fig. 1: Control system structures for four-wheel steering of cars
a) Feedforward control to rear wheel steering angle δ_r
b) Additional yaw rate feedback to the rear wheels
c) Yaw rate feedback to the front and rear wheels

Recently also steering control systems with feedback of the yaw rate r for rear wheel steering have been introduced to the Japanese market [4]. The yaw rate is measured by a vibration gyro. The additional feedback path in Fig. 1.b has two obvious advantages: a) The rear wheels are steered not only by the driver but also automatically to reduce the influence of disturbances d, e.g. the reaction time of the driver before he compensates the influence of a sudden side wind is avoided. b) Feedback via the compensator $H_r(s)$ allows pole shifting, for example the yaw damping can be increased. The pole shifting problem must be solved robustly, however, since there are uncertain parameters like speed, mass and friction between tire and road surface in steering dynamics. For the uncertain speed also gain scheduling of $H_r(s,v)$ by the measured speed is feasible.

Additional design freedom becomes available, if also the front wheels are actively steered by an underlying feedback loop like in Fig. 1.c. In fact in [5], [6] a car steering theorem for robust decoupling by yaw rate feedback to front wheel steering was proven. It is based on some modelling assumptions for the steering dynamics and reads:

CH3229-2/92/0000-2586$1.00 © 1992 IEEE

The feedback compensator

$$H_f(s) = \frac{1}{s} \tag{2}$$

makes the yaw mode unobservable from the lateral acceleration a_f of the front axle.

The control law (2) can be implemented by using an electric or hydraulic steering actuator without position feedback. Thus no analog or digital compensator or steer-by-wire implementation is needed. It was shown in [5] that the decoupling effect is independent of vehicle parameters or operating conditions like speed, mass and friction between tire and road surface, i.e. decoupling is robust. There is no need to tune the decoupling control law (2) to the specific vehicle or to schedule it by the operating conditions. In fact any modification of (2) would destroy the decoupling property.

There is however a price that we have to pay for the use of the control law (2): It changes the yaw damping and there is no way to increase the yaw damping by front wheel steering unless we sacrifice decoupling. But we can use rear wheel steering to achieve any desired damping. In the present paper a controller structure $H_r(s, v)$ is given that yields speed-independent yaw damping in combination with the decoupling control law (2).

In section II the model and the decoupling control law are reviewed and notations and model assumptions are introduced, section III describes the main result for robust yaw damping, prefilters for sideslip ange $\beta \equiv 0$ are given in section IV, and conclusions are contained in section V.

II. STEERING DYNAMICS AND ROBUST DECOUPLING

The essential features of car steering dynamics in a horizontal plane are described by the "single-track model" (or "two wheel model") by Riekert and Schunck [7], for derivations see for example [8], [9]. It is obtained by lumping the two front wheels into one wheel in the center line of the car, the same is done with the two rear wheels, see Fig. 2.

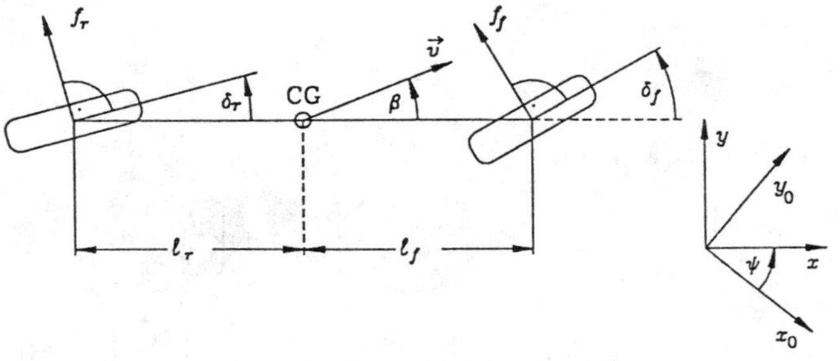

Fig. 2: Single–track model for car steering

In Fig. 2 $\vec{v}$ is the velocity vector at the center of gravity (CG), it has the absolute value $v = |\vec{v}|$. We assume $v > 0$ because the vehicle is not controllable for $v = 0$. The angle β between center line and velocity vector is called "sideslip angle". The distance between CG and front axle (resp. rear axle) is ℓ_f (resp. ℓ_r), and the sum

$$\ell = \ell_r + \ell_f \tag{3}$$

is the wheelbase.

In the horizontal plane of Fig. 2 an inertially fixed coordinate system (x_0, y_0) is shown together with a vehicle fixed coordinate system (x, y) that is rotated by a "yaw angle" ψ. In the dynamic equations the "yaw rate" $r = \dot{\psi}$ will appear. For small steering and sideslip angles and constant velocity the model can be linearized. Here we make an additional assumption: Let the longitudinal mass distribution be equivalent to concentrated masses at the front and rear axles. Then the total mass m is related to the moment of inertia J with respect to a perpendicular axis through CG by

$$J = m\ell_r\ell_f. \tag{4}$$

The linearized steering dynamics is described by the following state space model [5], [8], [9]

$$\begin{bmatrix} \dot{\beta} \\ \dot{r} \end{bmatrix} = \begin{bmatrix} a_{11} & a_{12} \\ a_{21} & a_{22} \end{bmatrix} \begin{bmatrix} \beta \\ r \end{bmatrix} + \begin{bmatrix} b_{11} & b_{12} \\ b_{21} & b_{22} \end{bmatrix} \begin{bmatrix} \delta_f \\ \delta_r \end{bmatrix} \tag{5}$$

where

$$\begin{aligned} a_{11} &= -(c_r + c_f)/mv \\ a_{12} &= -1 + (c_r\ell_r - c_f\ell_f)/mv^2 \\ a_{21} &= (c_r\ell_r - c_f\ell_f)/m\ell_r\ell_f \\ a_{22} &= -(c_r\ell_r^2 + c_f\ell_f^2)/mv\ell_r\ell_f \\ b_{11} &= c_f/mv \\ b_{12} &= c_r/mv \\ b_{21} &= c_f/m\ell_r \\ b_{22} &= -c_r/m\ell_f \end{aligned}$$

The parameters c_r and c_f are the rear and front "cornering stiffnesses". They vary in particular with the friction between tire and road surface. The characteristic polynomial of the dynamics matrix in (5) may be written as

$$\begin{aligned} p_0(s) &= \omega_0^2 + 2D_0\omega_0 s + s^2 \\ \omega_0^2 &= \frac{c_r c_f \ell^2 + mv^2(c_r\ell_r - c_f\ell_f)}{m^2v^2\ell_r\ell_f} \\ D_0 &= \frac{\ell(c_r\ell_r + c_f\ell_f)}{2\sqrt{\ell_r\ell_f[c_r c_f\ell^2 + mv^2(c_r\ell_r - c_f\ell_f)]}} \end{aligned} \tag{6}$$

Both the yaw damping D_0 and the natural frequency ω_0 decrease with increasing speed v. For high-speed vehicles sufficient damping may be achieved constructively by a long wheelbase ℓ.

The effect of the decoupling control law (2) can be shown by the following steps:

i) Augment the state space model (5) by

$$\dot{\delta}_f = e_f \tag{7}$$

ii) Transform the state space model to the new state vector $[a_f, r, \delta_f]^T$, where

$$\begin{aligned} a_f &= v(r + \dot{\beta}) + \ell_f\dot{r} \\ &= v(r + a_{11}\beta + a_{12}r + b_{11}\delta_f + b_{12}\delta_r) \\ &\quad + \ell_f(a_{21}\beta + a_{22}r + b_{21}\delta_f + b_{22}\delta_r) \\ &= d(-\beta - \ell_f r/v + \delta_f)\ , \quad d = \ell c_f/m\ell_r \end{aligned} \tag{8}$$

a_f is the lateral acceleration of the front axle.

iii) Close the feedback loop by

$$e_f = w_f - r \tag{9}$$

see Fig. 1.c.

As a result of steps i), ii), iii), the following model is obtained.

$$\begin{bmatrix} \dot{a}_f \\ - \\ \dot{r} \\ \dot{\delta}_f \end{bmatrix} = \begin{bmatrix} d_{11} & | & 0^* & 0^* \\ - & - & - & - \\ d_{21} & | & d_{22} & d_{23} \\ 0 & | & -1 & 0 \end{bmatrix} \begin{bmatrix} a_f \\ - \\ r \\ \delta_f \end{bmatrix} + \begin{bmatrix} d & 0^* \\ - & - \\ 0 & b_{22} \\ 1 & 0 \end{bmatrix} \begin{bmatrix} w_f \\ \delta_r \end{bmatrix} \tag{10}$$

$$a_f = \begin{bmatrix} 1 & | & 0^* & 0^* \end{bmatrix} \begin{bmatrix} a_f \\ r \\ \delta_f \end{bmatrix}$$

$$\begin{aligned} d_{11} &= -\ell c_f / m v \ell_r \\ d_{21} &= -(c_r \ell_r - c_f \ell_f)/c_f \ell_f \ell \\ d_{22} &= -c_r \ell / m v \ell_f \\ d_{23} &= c_r / m \ell_f \end{aligned}$$

The zeros marked with asterisks indicate the canonical form introduced by Kalman [10] and Gilbert [11] for the separation of observable and unobservable (or controllable and uncontrollable) subsystems. Thus we can read off (10) that

- the states r and δ_f are unobservable from a_f and
- a_f is not controllable from δ_r.

Thus the steering dynamics has been split into two subsystems

(a) the lateral motion of the front axle described by

$$\dot{a}_f = d_{11} a_f + (\ell c_f / m \ell_r) w_f \tag{11}$$

(b) the yaw motion described by

$$\begin{bmatrix} \dot{r} \\ \dot{\delta}_f \end{bmatrix} = \begin{bmatrix} d_{22} & d_{23} \\ -1 & 0 \end{bmatrix} \begin{bmatrix} r \\ \delta_f \end{bmatrix} + \begin{bmatrix} d_{21} \\ 0 \end{bmatrix} a_f + \begin{bmatrix} 0 & b_{22} \\ 1 & 0 \end{bmatrix} \begin{bmatrix} w_f \\ \delta_r \end{bmatrix} \tag{12}$$

The driver has to control only subsystem (a). He keeps the car – considered as a mass point at the front axle – on top of his planned path by generation of a lateral acceleration via the transfer function

$$a_f(s) = G_f(s) w_f(s) = G_f(s) F_f(s) \delta_S(s) \tag{13}$$

$$G_f(s) = \frac{d}{s - d_{11}} = \frac{v}{1 + Ts}, \quad T = \frac{m v \ell_r}{\ell c_f}$$

The decoupled yaw motion (b) has the characteristic polynomial

$$p_I(s) = \omega_I^2 + 2 D_I \omega_I s + s^2 \tag{14}$$

$$\omega_I^2 = \frac{c_r}{m \ell_f}, \quad D_I = \frac{\ell}{2v} \sqrt{\frac{c_r}{m \ell_f}}$$

It is interesting to compare the yaw damping D_0 for the uncontrolled vehicle and D_I for the decoupled vehicle. Setting both dampings equal and solving for the limiting speed yields

$$v_\ell^2 = \frac{c_r^2 \ell_r \ell^2}{m \ell_f (c_f \ell_f + 3 c_r \ell_r)} \tag{15}$$

For $v < v_\ell$ the decoupled vehicle has better damping, for $v > v_\ell$ the decoupled vehicle has lower damping. For typical vehicle data v_ℓ is in the domain of operation. Thus the safety-critical problem of decreasing $D_0(v)$ is getting worse by decoupling feedback (2).

The natural frequency is now independent of the speed. A comparison with (6) gives the same natural frequency $\omega_I = \omega_0$ at a speed

$$v_\omega = \ell \sqrt{\frac{c_r}{m \ell_f}} \tag{16}$$

This speed is about twice as high as v_ℓ of (15) but still in the domain of operation for typical vehicle data. For $v < v_\omega$ the uncontrolled vehicle has a higher natural frequency, for $v > v_\omega$ it has a lower natural frequency than the vehicle with decoupling control.

III. ROBUST YAW DAMPING BY REAR WHEEL STEERING

The main idea is formulated as the following car steering theorem for robust yaw damping by yaw rate feedback to rear wheel steering [12].

The controller

$$\delta_r = (\ell/v - k_D)(w_r - r) \tag{17}$$

yields velocity independent yaw eigenvalues.

Remark:
Note that this controller is not generic like (2). It must be tuned to the specific vehicle by substituting the wheelbase ℓ and it must be scheduled by the measured velocity v. w_r is the reference input for the rear wheel steering feedback system, see Fig. 1.c. The controller parameter k_D can be set to adjust the desired damping level.

Proof:
Substituting (17) into (12) yields

$$\begin{bmatrix} \dot{r} \\ \dot{\delta}_f \end{bmatrix} = \begin{bmatrix} d_{22} - (\ell/v - k_D) b_{22} & d_{23} \\ -1 & 0 \end{bmatrix} \begin{bmatrix} r \\ \delta_f \end{bmatrix} + \begin{bmatrix} d_{21} \\ 0 \end{bmatrix} a_f + \begin{bmatrix} 0 & b_{22}(\ell/v - k_D) \\ 1 & 0 \end{bmatrix} \begin{bmatrix} w_f \\ w_r \end{bmatrix}$$

and with (5) and (10)

$$\begin{bmatrix} \dot{r} \\ \dot{\delta}_f \end{bmatrix} = \begin{bmatrix} -k_D c_r / m \ell_f & c_r / m \ell_f \\ -1 & 0 \end{bmatrix} \begin{bmatrix} r \\ \delta_f \end{bmatrix} + \begin{bmatrix} d_{21} \\ 0 \end{bmatrix} a_f + \begin{bmatrix} 0 & b_{22}(\ell/v - k_D) \\ 1 & 0 \end{bmatrix} \begin{bmatrix} w_f \\ w_r \end{bmatrix} \tag{18}$$

The characteristic polynomial is now

$$\begin{aligned} p_{II}(s) &= \omega_{II}^2 + 2 D_{II} \omega_{II} s + s^2 \\ \omega_{II}^2 &= \frac{c_r}{m \ell_f} \\ D_{II} &= \frac{k_D}{2} \sqrt{\frac{c_r}{m \ell_f}} \end{aligned} \tag{19}$$

$\omega_{II} and D_{II}$ do not depend on v, q.e.d.

Note that the natural frequency has not been changed by the control law (17), i.e. $\omega_{II} = \omega_I$.

Example:
The worst damping D_{II}^- is obtained for the minimum cornering stiffness c_r^- (i.e. the value for icy road) and the maximum mass m^+ (i.e. the mass of the fully loaded vehicle). Assume that we want to have damping $1/\sqrt{2}$ for this most critical case, then

$$\frac{1}{\sqrt{2}} = \frac{k_D}{2}\sqrt{\frac{c_r^-}{m^+\ell_f}}$$

$$k_D = \sqrt{\frac{2m^+\ell_f}{c_r^-}} \tag{20}$$

IV. PREFILTER

Similar as in [1] the required prefilter ratio for sideslip angle $\beta \equiv 0$ can be calculated as follows.

$$\frac{F_r}{F_f} = K(v) \cdot \frac{1 + T_D(v)s}{(1 + T_1(v)s)(1 + T_2(v)s)} \tag{21}$$

with

$$K(v) = -1 + \frac{\ell_r - mv^2\ell_f/\ell c_r}{\ell - k_D v}$$

$$T_D(v) = \frac{m\ell_f\ell_r}{\left(k_D - \frac{\ell_f}{v}\right)\ell c_r - mv\ell_f}$$

$$T_1(v) = \frac{mv\ell_r}{\ell c_f}$$

$$T_2(v) = \frac{\ell_f}{v}$$

At the speed $v = \ell/k_D$ the rear feedback loop gain goes to zero and the prefilter gain goes to infinity. In order to avoid this effect, the feedback gain $\ell/v - k_D$ must be implemented in the rear feedback path, see Fig. 3.

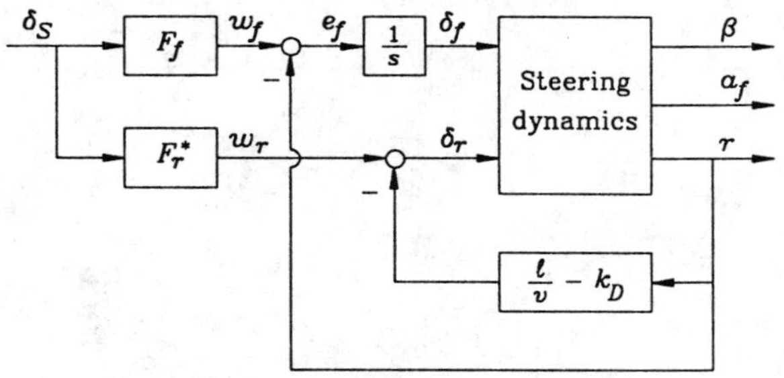

Fig. 3: Control system structure for robust decoupling and robust yaw stabilization of 4WS cars

The modified prefilter ratio F_r^*/F_f then has a gain

$$\begin{aligned} K^*(v) &= K(v)(\ell - k_D v) \\ &= -\ell_f + k_D v - mv^2\ell_f/\ell c_r \end{aligned}$$

There are different opinions among automotive engineers about the requirement $\beta \equiv 0$, but similar calculations can be done for other requirements.

The front prefilter F_f may be a scheduled gain $F_f = 1/v$ to compensate for the gain v of the transfer function $G_f(s)$ in (13). Then the driver commands the lateral acceleration a_f at the front axle by the steering wheel input. For avoiding steer-by-wire, $F_f = 1$ is the choice, in this case the steering wheel angle is a reference input for the yaw rate. Eventually it can only be determined in road tests, what prefilter the human driver likes best.

V. CONCLUSIONS

The car steering control system of Fig. 3 with some simplifying assumptions for the car steering model has useful properties that are summarized as follows

i) the yaw mode has damping and natural frequency independent of the car velocity v. The desired damping level can be set by the controller parameter k_D.

ii) The yaw mode has no influence on the lateral acceleration a_f of the front axle; a_f is controlled by the driver via a stable first order low pass filter. He only has to keep a mass point at the front axle on top of his planned path by a_f. He does not have to care about the automatically controlled yaw mode.

For the derivation of the above result some simplifying assumptions have been made. Further robustness studies are needed in order to analyze the effects of non-ideally satisfied assumptions.

Significant safety advantages of the new car steering control system can be expected for two reasons:

- In the conventional car the transfer function from the steering wheel to the lateral acceleration at the front axle has two poles and two zeros which vary widely with the uncertain parameters. The driver has to learn the reactions of his car at different velocities, load and road conditions. In contrast the transfer function (13) has become simple and robust.

- Yaw motions induced by side-wind or μ-split braking are automatically controlled by yaw rate feedback to rear wheel steering. The driver does not have to care about yaw motions, thus his reaction time to the above disturbances does not matter. The yaw eigenvalues are speed-invariant and nicely damped by the choice of k_D. Due to the integral action in the active front-wheel steering a gust of side wind does not change the stationary heading of the car, if the driver just keeps the steering wheel straight.

References

[1] E. Donges, R. Aufhammer, P. Fehrer and T. Seidenfuß, Funktion und Sicherheitskonzept der Aktiven Hinterachskinematik von BMW, Automobiltechnische Zeitschrift, No. 10, pp. 580-587, 1990.

[2] B. Los, A. Matheis, J.E. Nametz und R.E. Smith, Konstruktion und Entwicklung eines Fahrzeugs mit mikroprozessor–geregelter Allradlenkung, VDI Bericht Nr. 650, Duesseldorf 1987, S. 239–257.

[3] S. Sano, Y. Furukawa, S. Shiraishi, Four wheel steering system with rear wheel steer angle controlled as a function of steering wheel angle, Society of Automotive Engineers (SAE), paper 860625, International Congress, Detroit, 24.-28. Feb. 1986.

[4] H. Inoue, H. Harada and Y. Yokoya, Allradlenksystem im Toyota Soarer, Conference "Allradlenkung bei Personenwagen", Haus der Technik, Essen, Dec. 3-4, 1991.

[5] J. Ackermann, Robust car steering by yaw rate control, Proc. of the 29th Conference on Decision and Control (CDC), Honolulu, Hawaii, Dec. 1990, pp. 2033-2034.

[6] J. Ackermann, Verfahren zum Lenken von Strassenfahrzeugen mit Vorder- und Hinterradlenkung, Patentanmeldung 6.9.1990, Nr. P 40 28 320.8-21 Deutsches Patentamt.

[7] P. Riekert und T.E. Schunck, Zur Fahrmechanik des gummibereiften Kraftfahrzeugs, Ing. Archiv 1940.

[8] M. Mitschke, Dynamik der Kraftfahrzeuge, Vol. C. Berlin: Springer, 1990

[9] A. Zomotor, Fahrwerktechnik: Fahrverhalten, Vogel–Verlag, Wuerzburg, 1987.

[10] R.E. Kalman, Mathematical description of linear dynamical systems, J. SIAM on Control, 1963, S. 152–192.

[11] E.G. Gilbert, Controllability and observability in multivariable control systems, J. SIAM on Control, 1963, S. 128–151.

[12] J. Ackermann, Ein Regelungsverfahren zur Gierdämpfung von Strassenfahrzeugen mit Allradlenkung Patentanmeldung 3.3.1992, Nr. P 4206654.9, Deutsches Patentamt.

Proceedings of the 31st Conference
on Decision and Control
Tucson, Arizona • December 1992

Position Control of a Plastic Injection Moulding Machine via Feedback Linearization

B. Bona, L. Giacomello, C. Greco, A. Malandra
Dipartimento di Automatica e Informatica, Politecnico di Torino
Corso Duca degli Abruzzi, 24 - 10129 Torino, Italy

Abstract

This paper presents the application of the nonlinear feedback linearization method to the position control of the movable platen in a plastic injection moulding machine. The dynamic model of the platen movement during the injection cycle presents many different nonlinearities. Some are due to the hydraulic subsystem while other depends on the movable platen kinematic chain. Both were linearized according to the theory, and the satisfactory results of many simulated test show the applicability of these concepts in this important industrial process.

1 Introduction

In the last years important theoretical results have been obtained in the field of nonlinear control using the feedback linearization technique [1], [2], which has obtained a widespread success in robotic applications [3], [4]. Therefore it is certainly interesting to study the behaviour and applicability of these techniques to other industrial processes.

Along this line, the paper presents an application to the position control of the movable platen of an industrial plastic injection moulding machine (PIMM). The main components of a PIMM are: a) the barrell of the injection unit, b) the injection unit, c) the mould, and d) the movable platen.

Position and velocity of the platen play a considerable role in the injection cycle and are affected by the variability of the mould mass, by the nonlinearities of the kinematic chain and of the hydraulic system, and by the variability of the friction acting on the piston.

The hydraulic command system is composed by a proportional flow control servovalve and a proportional pressure servovalve; each one is controlled by the current flowing into the command coil. The combined action of these two valves on the hydraulic fluid allows to command the piston of an hydraulic power cylinder. The movable platen, supporting half of the mould, is connected to the piston through a complex kinematic chain, with a highly nonlinear input/output characteristics: the platen position and its velocity are the two variables to be controlled.

Other nonlinearities due to saturations, dead zones, quadratic characteristics of the hydraulic orifices, are considered, while the variability of some physical parameters, both in time and across different machines of the same type will not be considered here, as a research study, aimed to apply a model reference adaptive control to a low order linearized version of the system, is still under way.

Although the system has two inputs and two outputs, the design complexity has been reduced avoiding the control of the moving platen velocity; in this way only the current into the command coil of the flow control valve has been considered, while the current commanding the pressure valve is kept fixed to a known predefined setting value.

2 The Model

Technical and economical reasons have suggested to use for this preliminary study a simulated model of the PIMM; therefore a seventh order dynamical model has been described using SIMNON. This model was obtained from a more detailed model of order 16, and reducing it, considering both the physical knowledge of the process and a complete set of identification experiments, conducted on real and simulated data.

The resulting model, shown in Fig. 1, consists of six blocks. Block 1: proportional pressure valve, with input i_p constant; Block 2: proportional flow valve, with input $i_q(t)$; Block 3: first hydraulic cylinder nozzle, Block 4: second hydraulic cylinder nozzle, Block 5: intrinsic feedback between Blocks 3, 4 and Block 1, Block 6: hydraulic piston and kinematic chain. Each one of them includes some static and/or dynamic nonlinearities. The nonlinear system is represented by the following state equations in which v_i are auxiliary variables, c_i are constants and d_i are suitable parameters:

Block 1

$$\begin{aligned}
\dot{x}_1 &= \frac{d_1 v_1 - x_1}{d_3} \\
v_1 &= d_4 - v_2 - v_3 \frac{\mathrm{d}q}{\mathrm{d}\Delta p} \\
v_4 &= x_1 - i_p d_5 \\
v_3 &= \begin{cases} 0 & v_4 < 0 \\ y & 0 < v_4 < c_1 \\ c_1 & v_4 > c_1 \end{cases}
\end{aligned}$$

Block 2

$$\begin{aligned}
\dot{x}_2 &= \frac{d_2 v_5 - d_6 x_2 - x_3}{d_7} \\
\dot{x}_3 &= x_2 \\
v_5 &= \begin{cases} i_q + c_2 & i_q < -c_2 \\ 0 & -c_2 < i_q < c_2 \\ i_q - c_2 & i_q > c_2 \end{cases}
\end{aligned}$$

Blocks 3 and 4

$$\begin{aligned}
\dot{x}_4 &= \frac{d_8 x_6 - d_{10} d_{16} \sqrt{\frac{2\,|x_4|}{\rho}}\operatorname{sign}(x_4)}{d_{14}} - \\
&\quad \frac{d_{12} d_{16} y_{sw} \sqrt{\frac{2\,|v_6|}{\rho}}\operatorname{sign}(v_6)}{d_{14}} \\
\dot{x}_5 &= \frac{-d_9 x_6 - d_{11} d_{16} \sqrt{\frac{2\,|x_5|}{\rho}}\operatorname{sign}(x_5)}{d_{15}} + \\
&\quad \frac{b_1 d_{16} y_{sw} \sqrt{\frac{2\,|v_7|}{\rho}}\operatorname{sign}(v_7)}{d_{15}} \\
v_6 &= \begin{cases} x_1 - x_4 & y_{sw} < 0 \\ x_4 & y_{sw} > 0 \end{cases} \\
v_7 &= \begin{cases} x_5 & y_{sw} < 0 \\ x_1 - x_5 & y_{sw} > 0 \end{cases}
\end{aligned}$$

Block 6

$$\begin{aligned}
\dot{x}_6 &= \frac{d_9 x_5 - d_8 x_4 - [d_{17} + d_{18} v_8^2(x_7)] x_6}{d_{19} + v_8^2(x_7) d_{20}} - \\
&\quad \frac{v_8(x_7) v_9(x_7) d_{20} x_6^2}{d_{19} + v_8^2(x_7) d_{20}} \\
\dot{x}_7 &= x_6
\end{aligned}$$

CH3229-2/92/0000-2591$1.00 © 1992 IEEE

Block 5

$$\text{if} \quad y_{sw} < 0 \quad \text{then} \quad v_2 = b_1 c y_{sw} \sqrt{\frac{2\,|x_1 - x_5|}{\rho}} \operatorname{sign}(x_1 - x_5)$$

$$\text{if} \quad y_{sw} > 0 \quad \text{then} \quad v_2 = -d_{12} c y_{sw} \sqrt{\frac{2\,|x_1 - x_4|}{\rho}} \operatorname{sign}(x_1 - x_4)$$

where $x_1 \dots x_7$ are the seven states considered, $u = i_q$ is the input and $y = x_7$ is the output.

When the current input i_q has a periodic trapezoidal velocity profile, with period $T = 2\,\mathrm{s}$, and i_p is kept constant, the control should fulfill the following specifications: 1) steady-state error < 1 mm, over a 322 mm stroke, 2) settling time around 0.5 s, 3) no overshoot.

3 Linearized control

A preliminary analysis showed that an overall straightforward feedback linearization of the entire system was impossible for analytical reasons; a step-by-step approach was therefore preferred.

3.1 Kinematic chain control

The movable platen kinematic chain (Block 6) was considered first, with the aim of linearizing its dynamic behaviour. The actual platen position vs piston position is shown in Fig. 2: the equivalent platen mass, seen by the piston, can vary at each stroke in a ratio of 1 to 100. The input to this block is the force u_{us} applied to the piston, while the output is the piston position x_7 (or, equivalently, the platen position, related to x_7 by the nonlinear function in Fig. 2).

The feedback linearization of this block was straightforward with the input being the following analytical control law:

$$u_{us} = v_{10} \left[\frac{1}{v_{10}} (v_{11} x_6 + v_{12} x_6^2) + \ddot{x}_{7,d} - \beta_1 (x_6 - x_{6,d}) - \beta_2 (x_7 - x_{7,d}) \right]$$

where v_{10}, v_{11} and v_{12} are suitable nonlinear functions of x_2.

The results of the subsequent pole placement control of the linearized equation are presented in Fig. 4; these results show a limited but nonzero steady-state position error, which in any case is inside the requested specifications.

3.2 Hydraulic cylinder control

The next step was to proceed with the linearization and control of Blocks 1-3-4-5, in order to obtain good tracking capabilities between the switching input signal y_{sw} and the output u_{us}. In this case it was necessary to consider two different sub-cases, i.e. for $y_{sw} > 0$ and for $y_{sw} < 0$, which corresponds to two different state equations, arising from the nonlinear dynamics of the hydraulic command circuit, when it opens or closes the platen. Moreover it was necessary to obtain the espression for the desired value of the signal u_{us}, called $u_{us,d}$, and its time derivative $\dot{u}_{us,d}$. The analytical control law is:

$$y_{sw} = \frac{d_{8,14}(d_8 x_6 - d_{10} w_1) + d_{9,15}(d_9 x_6 + d_{11} w_2) + \dot{u}_{us,d} - \gamma_0 (u_{us} - u_{us,d})}{d_{8,14} d_{12} w_2 + d_{9,15} b_1 w_4}$$

where the w_i are variables that switch with the signal y_{sw} and dependent on the states variables x_1, x_4 and x_5. The plot of $u_{us,d}$ is shown in Fig. 4c.

The analysis of the error dynamics of the linearized system showed an internal stable behaviour. A typical result of the system response between y_{sw} and x_7 is shown in Fig. 5, which presents a steady state error less than 1 mm.

3.3 Global control

The last step of the analysis was devoted to the inclusion of Block 2 into the overall structure. This block, representing the flow regulation valve, is described by a second order system with two static nonlinearities (a dead zone and a saturation effect). Without entering here into further details, different approaches to signal conditioning were carried out to overcome these nonlinearities; experiments were also conducted introducing a state feedback law on the valve, which is certainly a complication in terms of practical measurability of the two internal states of the valve. The blosk diagram is shown in Fig. 3.

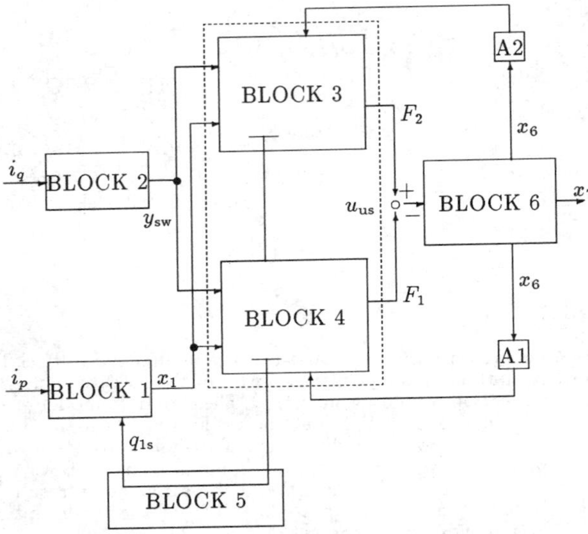

Figure 1: Position command of the PIMM.

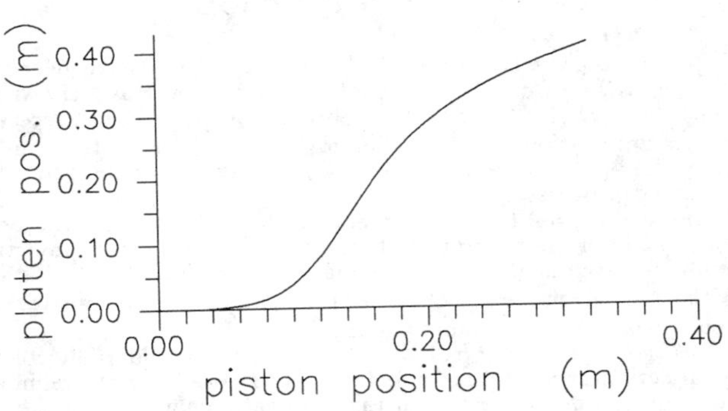

Figure 2: The kinematic transmission nonlinear function.

4 Results and further developments

In conclusions, the best results obtained so far are shown in Figg. 4-6. At present, further analysis and simulation should be carried out in order to verify if a dynamic observer of the valve states, as well as of other umeasurable states, is of some interest, or simpler approaches can be used without unacceptable deterioration of the control performances.

Another further step will consist in the implementation of this architecture to the real PIMM, in order to tune and validate the entire design, and to analyze the computational constraints and other practical aspects that were omitted in this preliminary study.

References

[1] A. Isidori. *Nonlinear Control Systems: An Introduction*, Springer-Verlag, Berlin, 2-nd edition, 1989.

[2] H. Nijmeijer, and A.J. van der Schaft. *Nonlinear Dynamical Control Systems*, Springer-Verlag, New York, 1990.

[3] J.J.E. Slotine, and W. Li. *Applied Nonlinear Control*, Prentice-Hall International, Englewood Cliffs, 1991.

[4] M.W. Spong, and M. Vidyasagar. *Robot Dynamics and Control*, John Wiley & Sons, New York, 1989.

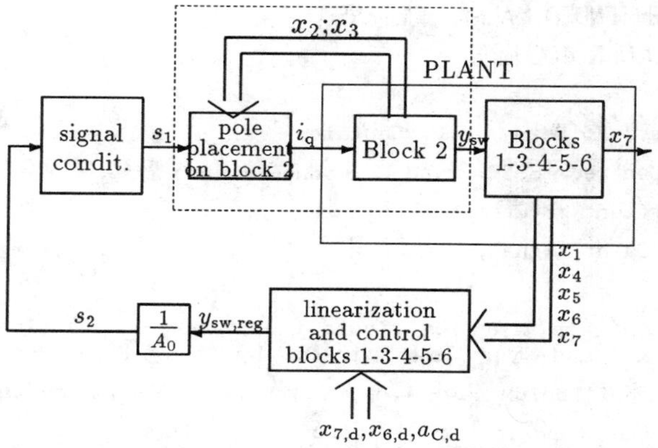

Figure 3: Global control with pole placement on Block 2.

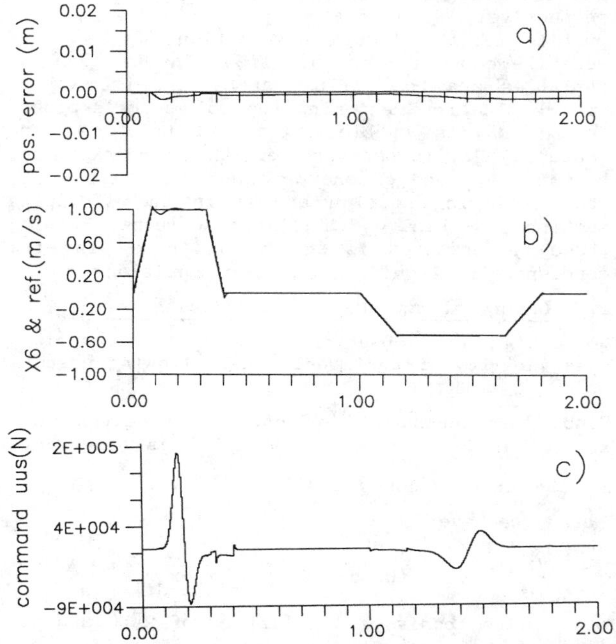

Figure 4: Results after feedback linearization and pole placement control of Block 6.

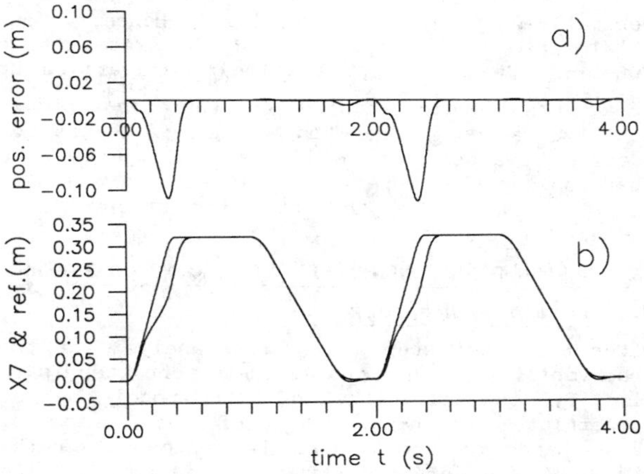

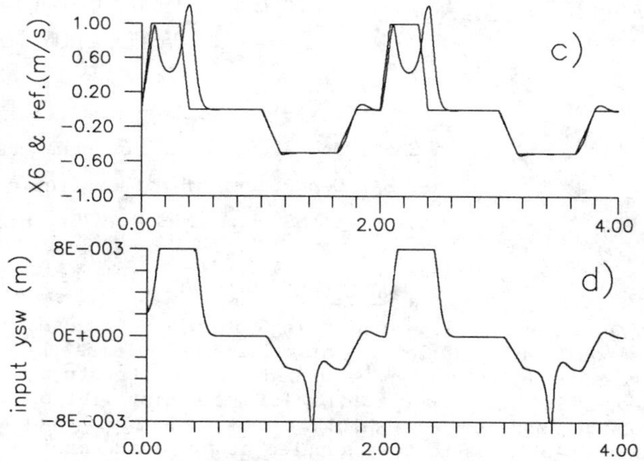

Figure 5: System response after feedback linearization and pole placement control of Blocks 1-3-4-5.

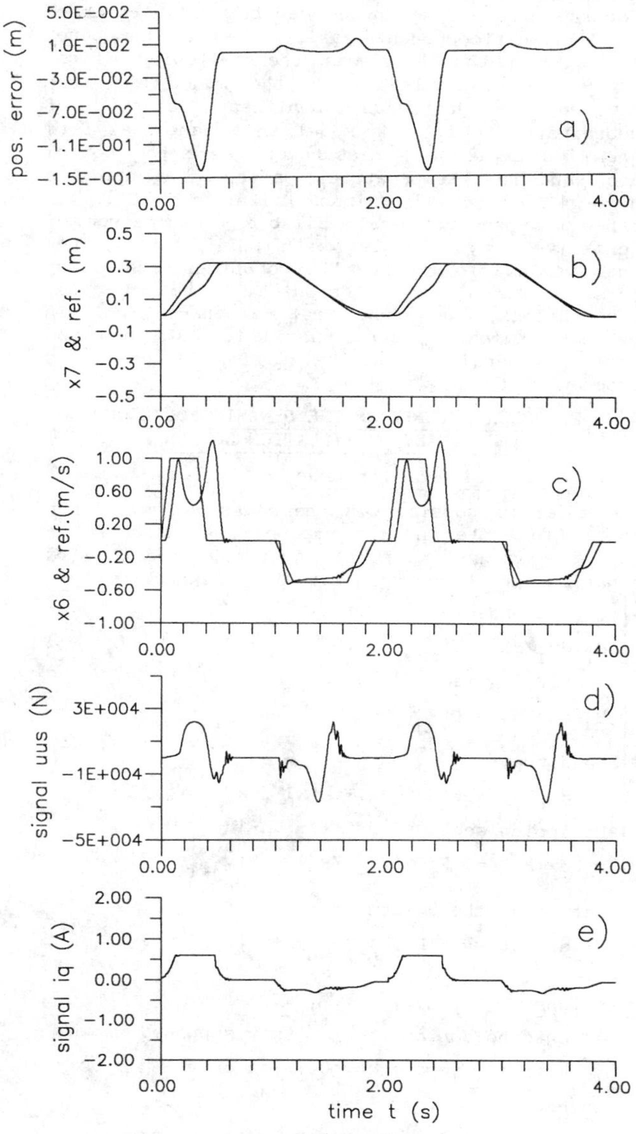

Figure 6: The response of the globally controlled system.

Proceedings of the 31st Conference
on Decision and Control
Tucson, Arizona • December 1992

A NEW METHODOLOGY TO DESIGN EXTENDED KALMAN FILTERS APPLICATION TO DISTILLATION COLUMNS

F. Viel[*+], D. Bossanne[*], E. Busvelle[*], F. Deza[*], J.P. Gauthier[*+]

* Shell Research, Grand-Couronne Research Centre, 76530 Grand-Couronne, France

+ INSA, Department of Mathematics and Department of Chemical Engineering, AMS, URA CNRS D 1378, 76131 Mont Saint Aignan, France

ABSTRACT

An exponentially converging observer for the composition profile of distillation columns is presented. Its design is based on the structural properties of observability of the distillation columns models. We exhibit a new canonical form that enables us to take supplementary measurements (temperature of sensitive trays) into account.

1. INTRODUCTION

In this work, we apply a new nonlinear estimation technique developed in (1), (2), and (3) to distillation columns. All the reasoning is done for the binary case but it can be easily extended to the multicomponent case. In a previous paper (4), the methodology was the following: first study the observability of the model and then perform a judicious nonlinear change of coordinates to put the model in a canonical form when the available measurements are the bottom and condenser temperatures; after this apply a high-gain extended Kalman filter (HG-EKF). If other measurements are available (temperature of sensitive trays) it is worth while taking them into account to improve the robustness. By doing this we lose the canonical form of (5). Nevertheless, we show that another high-gain extended Kalman filter applied to this new and more general canonical form converges exponentially.

2. OBSERVABILITY OF THE DYNAMIC MODEL OF THE BINARY DISTILLATION COLUMN

2.1. *Presentation of the model*

The model we consider is the classical model for the reflux rate and the vapour rate, that is to say the constant molal overflow model. Its equations are based upon material balances:

Total condenser:

$$H_1.dx_1/dt = V(y_2 - x_1)$$

Rectifying section: $j=2,\dots,f-1$,

$$H_j.dx_j/dt = L(x_{j-1} - x_j) + V(y_{j+1} - y_j)$$

Feed tray:

$$H_f.dx_f/dt = F(ZF - x_f) + L(x_{f-1} - x_f) + V(y_{f+1} - y_f)$$

Stripping section: $j=f+1,\dots,n-1$,

$$H_j.dx_j/dt = (F+L)(x_{j-1} - x_j) + V(y_{j+1} - y_j)$$

Bottom of the column:

$$H_n.dx_n/dt = (F+L)(x_{n-1} - x_n) + V(x_n - y_n)$$

Notations:

H_j: liquid holdup on the jth tray supposed known

x_j, y_j: liquid, vapour composition on the jth tray

f: number of the feed tray

F, L, V: feed, reflux and vapour rate (measured)

ZF: feed composition (measured)

On each tray the liquid and vapour compositions, x_j and y_j, are linked by the liquid-vapor equilibrium law, i.e., $y_j = k(x_j)$. We assume that the function k is monotonic, i.e., we do not consider azeotropic distillation. The condenser and the bottom are assigned to tray 1 and tray n respectively. The state of the model is the profile of the liquid composition of the more volatile component on each tray. The two control variables are the reflux rate L and the vapour rate V. The feed entering the column is supposed to be at its bubble point. In practice, one measures the temperature of the condenser and bottom trays and of one or "sensitive" trays, in the rectifying section and/or in the stripping section. For binary distillation columns, knowing these temperatures is equivalent to knowing the corresponding liquid or vapour compositions.

2.2. *Observability properties of the model*

To design our observers, we first study the observability of the model. The following results hold:

Since the condenser and bottom temperatures are known, that is to say, x_1 and x_n, x is observable as soon as $V \neq 0$ and $F+L \neq 0$.

Indeed, we have:

$$y_2 = H_1/V.dx_1/dt + x_1$$

$$y_{j+1} = y_j + (H_j\, dx_j/dt + L\,(x_j - x_{j-1}))/V$$

Thus, on the basis of the history of x_1, L and V, i.e. their derivatives, one can reconstruct formally the substate $(x_2,\dots,x_f)$ as soon as $V \neq 0$. Symmetrically, one can express the substate $(x_f,\dots,x_{n-1})$ as a function of x_n, F, L, V and their derivatives, providing that $F+L \neq 0$. Hence, we can obtain an explicit expression of the state x as a function of x_1, x_n, F, L, V and their derivatives up to a certain order. Thus, we see that for strictly positive values for F, L and V, that is to say for practical values, the model is observable. Note that knowing the temperatures on "sensitive" trays is of no use in establishing the observability property of our model.

3. DESIGN OF EXPONENTIALLY CONVERGING OBSERVERS

3.1. *Design methodology*

After the first step above (i.e. analysis of the observability of the model), the second step is to simplify the structure of the model by an appropriate nonlinear change of coordinates ϕ. This change of coordinates is deduced from the study of the observability of the model. The reasoning for the derivation of this nonlinear change of coordinates $w = \phi(x)$ is the following: consider the first output x_1, the top product composition. When we differentiate it with respect

CH3229-2/92/0000-2594$1.00 © 1992 IEEE

to time, we get $dx_1/dt = V(y_2-x_1)/H_1$. Thus, the new information about the state is in $y_2 = k(x_2)$. Denoting w_1 for x_1 and w_2 for y_2, one has:

$$dw_1/dt = V/H_1\, w_2 + (-V/H_1)w_1.$$

Differentiating w_2, the new information obtained about the state is in the term y_3. Iterating this procedure up to the feed tray, one both verifies the observability of the substate $(x_2,\ldots,x_f)$ and derives the corresponding nonlinear expressions for ϕ. Performing the same reasoning for the second output x_n gives us the remaining expressions for ϕ. Thus, ϕ is defined by:

$w_1 = x_1$,

$w_i = k(x_i)$, $(i=2,\ldots,f-1$; rectifying section)

$w_j = x_j$, $(j=f,\ldots,n$; stripping section)

We suppose here that without the loss of generality, we have two supplementary measurements that are equivalent to the vapour composition of the sensitive tray in the rectifying section w_r and the liquid composition of the sensitive tray in the stripping section w_s. In these new coordinates the model is rewritten as:

$$\Sigma \begin{cases} dw/dt = A(w,V,F+L).w + g(w,F,L,V,ZF) = f(w,u) \\ y \;=\; (w_1, w_r, w_s, w_n) = Cw \end{cases}$$

$$A(w,V,F+L) = \mathrm{diag}(V.A_1, V.A_2, (F+L).A_3, (F+L).A_4)$$

$(A_1)_{1,j} = \delta_{2,j}\, w_1/H_1$

$(A_1)_{i,j} = \delta_{i+1,j}\, k'(k^{-1}(w_i))/H_i$ $\quad (r-1)\times(r-1)$

$(A_2)_{i,j} = \delta_{i+1,j}\, k'(k^{-1}(w_i))/H_i$ $\quad (f-r)\times(f-r)$

$(A_3)_{i,j} = \delta_{i,j+1}\, / H_i$ $\quad (s-f+1)\times(s-f+1)$

$(A_4)_{i,j} = \delta_{i,j+1}\, / H_i$ $\quad (n-s)\times(n-s)$

$g = (g_1, g_2, g_3, g_4)$

$$g_1 = \begin{bmatrix} g_1^1\,(w_1,V) \\ g_1^2\,(w_1,w_2,L,V) \\ \vdots \\ g_1^{r-1}(w_1,..,w_r,L,V) \end{bmatrix}, \quad g_2 = \begin{bmatrix} g_2^r\,(w_1,..,w_r,L,V) \\ \vdots \\ g_2^{f-2}(w_1,..,w_{f-2},L,V) \\ g_2^{f-1}(w,L,V) \end{bmatrix},$$

$$g_3 = \begin{bmatrix} g_3^f\,(w,F,L,V,ZF) \\ g_3^{f+1}(w_{f+1},..,w_n,F,L,V) \\ \vdots \\ g_3^{s-2}(w_{s-2},..,w_n,F,L,V) \\ g_3^{s-1}(w_{s-1},..,w_n,F,L,V) \end{bmatrix}, g_4 = \begin{bmatrix} g_4^s(w_{s-1},..,w_n,F,L,V) \\ g_4^{s+1}(w_{s+1},..,w_n,F,L,V) \\ \vdots \\ g_4^{n-1}(w_{n-1},w_n,F,L,V) \\ g_4^n\,(w_n,F,L,V) \end{bmatrix}$$

The third step of our methodology is to apply a high-gain (with a large fictitious state noise) extended Kalman filter to the model obtained after this nonlinear change of coordinates.

Since the form obtained, Σ, is not the canonical form as described in (2) or (5), we have to find an appropriate new linear change of coordinates Δ that ensures the convergence of the high-gain extended Kalman filter.

Set $\Theta > 0$, $\quad \Delta_1$ for $\mathrm{diag}(\Theta^{f-3},\ldots,\Theta^{f-r-1})$,

Δ_2 for $\mathrm{diag}(\Theta^{f-r-1},\ldots,\Theta,1)$,

Δ_3 for $\mathrm{diag}(1,\Theta,\ldots,\Theta^{s-f})$,

Δ_4 for $\mathrm{diag}(\Theta^{s-f},\ldots,\Theta^{n-f-1})$,

Δ for $\mathrm{diag}(\Delta_1,\Delta_2,\Delta_3,\Delta_4)$.

Set Q an nxn symmetric positive definite matrix and $R = \mathrm{diag}(r_1,r_2,r_3,r_4)$ with $r_1,r_2,r_3,r_4 > 0$.

Denote $\Theta\, \Delta^{-1}Q\Delta^{-1}$ by Q_Θ and

$\mathrm{diag}(r_1/\Theta^{2(f-2)-1}, r_2/\Theta^{2(f-r)-1}, r_3/\Theta^{2(s-f+1)-1}, r_4/\Theta^{2(n-f)-1})$ by R_Θ.

Assume that:

H1: the state w is bounded.

H2: functions under consideration are globally Lipschitz with respect to w uniformly with respect to the input u.

Theorem

As soon as Θ is large enough and under assumptions H1, H2, the following "high gain" extended Kalman filter:

$$\text{HG-EKF} \begin{cases} d\hat{w}/dt = f(\hat{w},u) - PC'R_\Theta^{-1}(C\hat{w} - y) \\ dP/dt = Q_\Theta + Pf^*(\hat{w},u)' + f^*(\hat{w},u)P - PC'R_\Theta^{-1}CP \end{cases}$$

is an exponential observer for the state of Σ.

(The proof can be easily deduced from (3).)

3.2. *Robustness property*

One can show easily that under bounded perturbations of the assumed model, the error of this observer is bounded by a bound proportional to the bound on the perturbations. But this bound is also proportional to $\sup(\Theta^{f-3},\Theta^{n-f-1})$ and it is smaller than that obtained in (4) (without any supplementary measurements).

4. CONCLUSION

Our approach enables us to build a new high-gain extended Kalman filter on a more general canonical form because we introduce supplementary measurements in the design of this observer. We can summarize the properties of this observer:

its exponential convergence,

its improved robustness to bounded modelling errors.

References

(1) Gauthier, J.P., Hammouri, H. & Othman, S., A simple observer for nonlinear systems - application to bioreactors. *IEEE Trans. Aut. Control*, **37** , 1992.

(2) Deza, F., 1991, Contribution to the synthesis of exponential observers. *PhD Thesis, INSA Rouen, France.*

(3) Deza, F., Gauthier, J.P., Busvelle, E. & Rakotopara D. High gain estimation for nonlinear systems. *Systems and Control Letters*, **18**, 1992.

(4) Deza, F., Busvelle, E, Gauthier, J.P., Exponentially converging observers for distillation columns and internal stability of the dynamic output feedback, *to appear in Chemical Engineering Science, September 1992.*

(5) Gauthier, J.P. & Bornard, G., 1981, Observability for any u(t) of a class of nonlinear systems. *IEEE Trans. Auto. Control*, **26**, 922-926.

Proceedings of the 31st Conference
on Decision and Control
Tucson, Arizona • December 1992

Robust Control of a Compact Disc Player

Maarten Steinbuch, Gerrit Schootstra
Philips Research Laboratories
P.O.Box 80.000, 5600 JA Eindhoven, The Netherlands

Okko H. Bosgra
Mech. Eng. Systems and Control Group
Delft University of Technology
Mekelweg 2, 2628 CD Delft, The Netherlands.

Abstract

This paper considers the design of robust controllers for a Compact Disc mechanism. Using μ-synthesis a controller has been designed for good track-following. The design problem involves time-domain constraints on signals and robustness requirements for norm-bounded structured plant uncertainty. It is shown that by using weighting functions in the μ-framework this problem can be solved, but on the cost of many design iterations. The μ-controller has been implemented in an experimental set-up, with a digitally controlled Compact Disc player.

1 Introduction

A Compact Disc player is an optical decoding device that reproduces high-quality Audio from a digitally coded signal recorded as a spiral track on a reflective disc [1]. Recently, other optical data systems (CD-ROM, Optical Data Drives) and combined Audio/Video (CD-Interactive) applications have emerged. An increasing amount of all CD based applications is for portable use. Power consumption and shock sensitivity play an significant role in the design of the mechanical, optical and electronic parts of portable CD systems. In this paper we will concentrate on the possible improvements of the track-following behavior of a Compact Disc player, using robust control design techniques.

2 Compact Disc Mechanism

In Fig.1 a schematic view of a Compact Disc mechanism is shown.

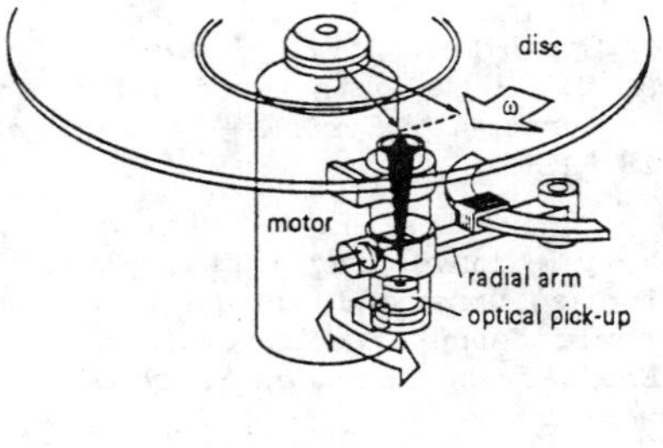

Fig. 1: Schematic view of a rotating arm Compact Disc mechanism

The mechanism is composed of a turn-table dc-motor for the rotation of the Compact Disc, and a balanced radial arm for track following. The radial actuator is a permanent-magnet/coil system. An optical element is mounted at the end of the radial arm. A diode generates a laser beam that passes through a series of optical lenses to give a spot on the disc surface. An objective lens, suspended by two parallel leaf springs, can move in a vertical direction to give a focusing action. Four photodiodes provide position-error information. Both focus and radial positioning are realized with active control loops. Since the dynamic interaction between both loops is relatively low, they can be analyzed independently. In this paper we will concentrate on the radial servo system.

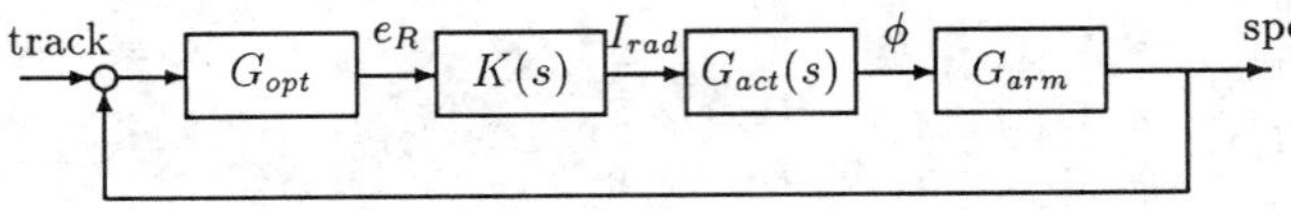

Fig. 2: Configuration of the radial control loop

In Fig.2 a block-diagram of the radial control loop is shown. The difference between the track position and the spot position is detected by the optical system; it generates a radial error (e_R) signal via a gain G_{opt}. A controller $K(s)$ feeds the system with the current I_{rad}. This in turn generates a torque resulting in an angular acceleration. The transfer function from the current I_{rad} to the angular displacement ϕ of the arm is called $G_{act}(s)$. A (non-linear) gain G_{arm} relates the angular displacement with the spot movement in the radial direction. Only the control-error signal e_R is available for measurement. Neither the true spot position, which can be interpreted as the system output, nor the track position are available as signals. In current systems $K(s)$ is a PID controller [2]. The radial servo system has a design bandwidth of 500 Hz, a compromise value in which several conflicting factors are taken into account: (i) accommodation of mechanical shocks acting on the player, (ii) achievement of the required disturbance attenuation at the rotational frequency (4-8 Hz) of the disc, necessary to cope with significant disc eccentricity ($\leq 300\mu m$), (iii) playability of discs containing faults, (iv) audible noise generated by the actuator and (v) power consumption.

However, for extreme disturbance situations occurring during portable use it is necessary to investigate whether improvements of the servo behavior are possible. In this paper we will use μ-

CH3229-2/92/0000-2596$1.00 © 1992 IEEE

synthesis for the design of a radial controller, and we will show experimental results using a DSP (Digital Signal Processor) implementation of the controller.

3 Modeling

3.1 Dynamic model of the radial actuator system

A measured frequency response of the radial system ($G(s) = G_{opt}G_{arm}G_{act}(s)$) is given in Fig.3.

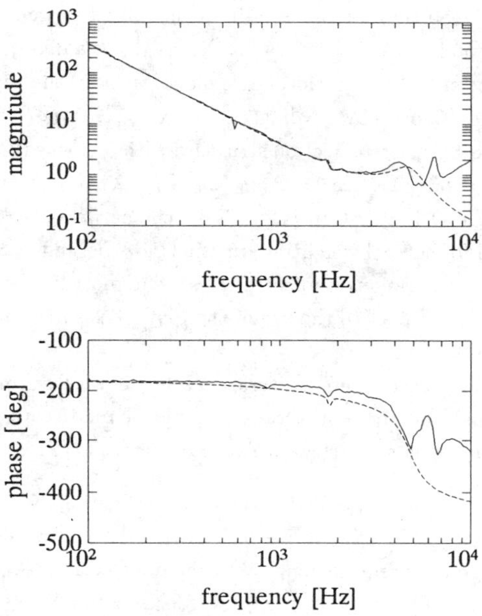

Fig. 3: Measured frequency response of the radial actuator system (—), and of the identified model with delay (- -)

At low frequencies the actuator transfer from current input I_{rad} to position error output e_R is a critically stable system with a phase lag of 180^0 (rigid body mode). At higher frequencies (1.7, 4 and 6 kHz) the measurement shows parasitic dynamics. Experimental modal analyses and finite element calculations have revealed that these phenomena are due to mechanical resonances of the radial arm and mounting plate (flexible bending and torsional modes).

Using frequency domain based system identification a 6th order model has been found. Since the controller is going to be implemented digitally, a delay corresponding to a sample rate of 40 kHz is added to the model. The frequency response of the model is also shown in Fig.3.

3.2 Uncertainty modeling

An important issue in the design of the mechanics, optics and electronics is the impact of production tolerances, since Compact Disc mechanisms are cheap mass-produced consumer electronic devices. To achieve cost price reduction, we want the system to be robust for the resulting variations of the characteristics of the system.

The most important system variations we want to account for are: (i) gain variations, and (ii) uncertainties in the frequencies of the parasitic resonances, especially those at 1.7 and 4 kHz.

The uncertainty modeling of gain variations can be done by a multiplicative parametric uncertainty at the input (or output): $\dot{x} = Ax + B(I + \Delta_{gain})u = Ax + Bu + Bw$, $w = \Delta_{gain}z$, $z = u$, with a scalar uncertainty block Δ_{gain}. The modeling of the variations of the frequencies of the parasitic resonances is not as straightforward. One possibility is the use of additive or multiplicative frequency response embedding; however, this may result in high order shaping functions. For this reason we would like to model these uncertainties as parametric perturbations [5]. Therefore, we transform the identified model into a modal state-space form. Consider a part of the A matrix corresponding to one resonance:

$$A_{sec} = \begin{bmatrix} \lambda & 0 \\ 0 & \bar{\lambda} \end{bmatrix} \tag{1}$$

with λ the complex eigenvalue and $\bar{\lambda}$ its complex conjugate. By using the similarity transformation

$$T = \begin{bmatrix} 1 & 1 \\ \lambda & \bar{\lambda} \end{bmatrix} \tag{2}$$

we obtain a real-valued A matrix with the standard canonical form for second order systems:

$$\tilde{A}_{sec} = \begin{bmatrix} 0 & 1 \\ -\omega_0^2 & -2\beta\omega_0 \end{bmatrix} = \begin{bmatrix} 0 & 1 \\ -c/m & -d/m \end{bmatrix} \tag{3}$$

with ω_0 the undamped resonance frequency, β the relative damping and with c the spring stiffness, d the damping and m the mass associated with the resonance. By perturbing entry (2,1) or (2,2) we can separately change the stiffness and the damping. Here we only consider a varying frequency, and as uncertainty model we then obtain:

$$\dot{x} = \tilde{A}_{sec}x + \begin{bmatrix} 0 \\ -\omega_0^2 s_{freq} \end{bmatrix} w \tag{4}$$

$$z = \begin{bmatrix} 1 & 0 \end{bmatrix} x \tag{5}$$

$$w = \Delta_{freq} z \tag{6}$$

Notice that this uncertainty modeling procedure results in a scalar (multiplicative) block Δ_{freq}. Typical value of the scaling parameter s_{freq} is 0.1, i.e. 10% frequency shift.

For the CD player the final uncertainty structure has 3 scalar blocks: Δ_{par}=diag(Δ_{gain}, $\Delta_{1.7kHz}$, Δ_{4kHz}).

Besides these parametric uncertainties, the identified design model of the radial servo system has an error with respect to the measurement. We will model this as an additive uncertainty, bounded by a high pass transfer function.

4 Performance Specification

Two major sources of disturbances are track-position irregularities and external mechanical shocks. The track position specifications are based on standardization of CDs: the allowable radial deviation is 300 μm (eccentricity) and the allowable radial acceleration is 0.4 m/s^2. The external mechanical shocks are not specified. Measurements show that during portable use disturbance signals occur ranging from 5 Hz up to frequencies of 150 Hz, modeled as half sine-wave shocks with a duration of 100 ms during running or jogging and 3 ms when putting the player on a table.

The maximum allowable track-following error is $0.1\mu m$, and the actuator current I_{rad} should have a high roll-off above the bandwidth to prevent occurrence of audible noise generated by the actuator.

The control problem can therefore be formulated as designing a controller $K(s)$ such that for the l_∞-bounded disturbance signals, the error is l_∞-bounded, the input to the actuator has a high roll-off and the performance is robust for the assumed plant uncertainty.

A method that can handle time-domain constraints on signals and induced-norm bounds for structured uncertainties simultaneously, other than using nonlinear optimization, does not exist. In this paper we will investigate how the μ-synthesis framework can be used to solve this control problem.

5 μ-Synthesis

5.1 Choice of standard plant

For clarity of exposition, let us assume that the radial system can be described by a double integrator. For low frequent reduction we want to add two integrating filters, and we need one lead for stabilization. For high frequency robustness we want to have a first order roll-off in the controller. These loop-shaping arguments result in a 4th order controller. Which standard plant can give the same result? Assume that we solve the H_∞ mixed sensitivity problem

$$\left\| \begin{bmatrix} W_1 S \\ W_2 T \end{bmatrix} \right\|_\infty < 1 \qquad (7)$$

with $T = GKS$. In order to have two integrating actions in the controller the weighting function W_1 has to have four states. (Note that the system itself gives already a +2 slope in the sensitivity function). For a first order roll-off in the controller W_2 needs three orders by the same argument. The resulting H_∞ controller will have nine states! Reducing this to four states by using model reduction techniques is not straightforward.

Another commonly used design problem is

$$\left\| \begin{bmatrix} W_1 S \\ W_2 KS \end{bmatrix} \right\|_\infty < 1 \qquad (8)$$

For this standard plant W_1 still needs four states, but W_2 only needs one; the resulting controller will have seven states. A further reduction can be obtained by weighting SG instead of S, hence making use of the slope -2 in the system. However, this can not be combined with weighting of KS, without also weighting S and KSG. The following general problem results:

$$\left\| \begin{bmatrix} W_1 SG & W_1 S \\ W_2 KSG & W_2 KS \end{bmatrix} \right\|_\infty < 1 \qquad (9)$$

To overcome the problem of also weighting S and KSG we associate with this standard plant a structured performance block: diag(Δ_{SG}, Δ_{KS}). Using the μ-methodology we are now able to generate input/output scalings so that we effectively weight only SG and KS. Notice that the use of a structured performance block is only beneficial if the associated D-scaling has a low order (< 2).

5.2 Application to the CD player

For the Compact Disc radial servo system the performance requirements as stated in the previous section are used to make choices for W_1 and W_2. The resulting weighting W_1 is low-pass with high gains up to 150 Hz. For the high-frequent roll-off as well as for the additive (modeling) uncertainty we design W_2 as a high-pass filter. Besides the structured performance block, we have the three parametric uncertainties Δ_{par}=diag(Δ_{gain}, $\Delta_{1.7kHz}$, Δ_{4kHz}), resulting in an overall μ-interconnection structure with 5 blocks (2 complex, 3 real): Δ=diag(Δ_{SG}, Δ_{KS}, Δ_{gain}, $\Delta_{1.7kHz}$, Δ_{4kHz})

Using the D-K iteration [4, 6, 7, 8] scheme, μ-controllers were calculated for several weighting functions for SG and KS, including the three parametric uncertainties. The most important conclusions with respect to the use of the μ-synthesis method for this problem are:

- Convergence of the D-K iteration scheme was fast (most cases 2 steps), but the first iteration should be made with care; one possibility is to start with $\Delta_{par} = 0$ or $\Delta_{par} = I$.
- The structured performance block, weighting SG and KS, was very successful to prevent high order controllers due to performance requirements. The related D-scalings could often be approximated with a zeroth order function.
- The final μ-controllers did have high orders due to the dynamic D-scalings associated with the parametric uncertainties Δ_{par}. However, using balanced model-reduction low order controllers (6 states) could be found without degrading the performance too much.
- The mapping of the time-domain constraints into performance specifications using weighting functions did take many design iterations. The procedure followed was to choose weighting functions, calculate the μ-controller and disturb the closed-loop system with realistic track signals. By checking the error-signal e_R it was analyzed which frequency part of the signal caused the largest amplitude. At this frequency the weighting of SG was adjusted and the procedure was started over again. The real problem was to find out a trade-off between disturbance attenuation and robustness requirements.
- The frequency-range of interest (1 Hz to 10 kHz) in this application did cause numerical problems. For this reason, internal balancing and time-scaling have been applied to the standard plant.

One of the results is shown in Fig.4, and is compared to a 1 kHz bandwidth PID controller. Clearly, the μ-controller has a better

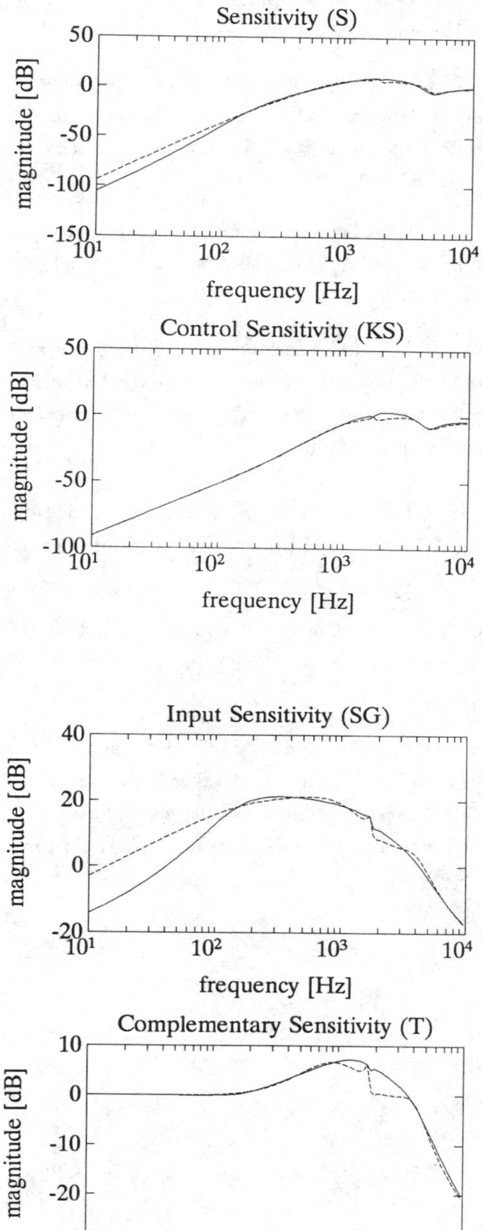

Fig. 4: Closed-loop transfer functions of the radial system, with the μ-controller (—) and with a PID (- -)

low-frequent disturbance rejection compared to the PID solution. The controller transfer functions are given in Fig.5. Especially, the resonance at 1.7 kHz is effectively compensated by the μ-controller.

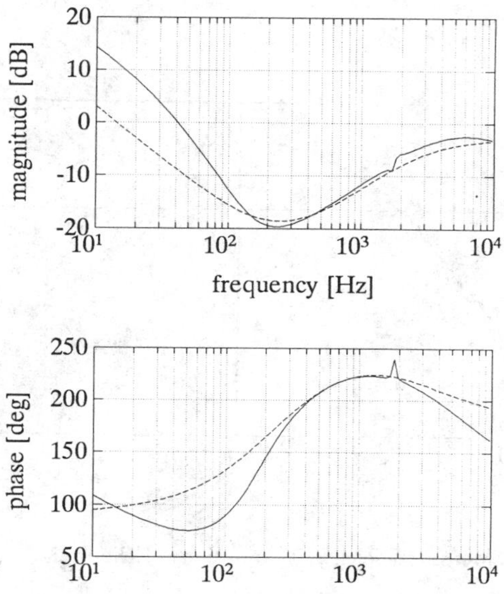

Fig. 5: Transfer functions of the μ-controller (—) and of the PID controller (- -)

Finally, in Fig. 6 the σ, μ plot for the standard plant with μ-controller is shown. The difference between σ and μ is large for all frequencies. The most critical frequency range in this problem is between 1 and 4 kHz. The real μ calculation [7] is not much lower than the complex μ case.

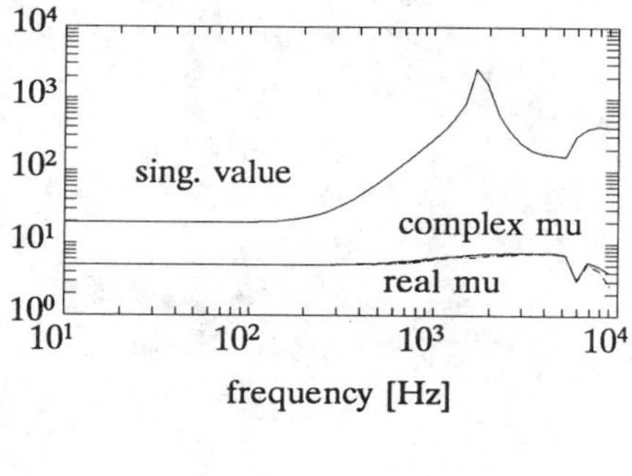

Fig. 6: Maximum singular value and μ for the CD μ-control design

6 Implementation Results

Using a fast DSP environment it is possible to control the CD player using digital servo loops. The designed μ-controller was discretized and implemented at a sampling rate of 40 kHz. Measured frequency responses of the system are given in Fig.7.

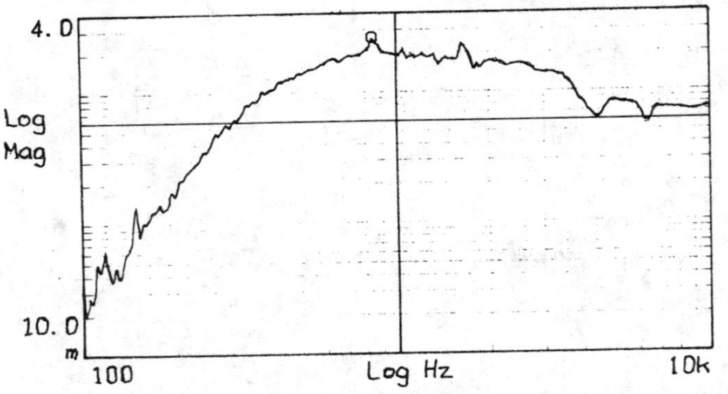

Fig. 7: Measured sensitivity functions of the radial system

7 Conclusions

In this paper μ-synthesis is applied to the radial servo system of a Compact Disc player. The design problem involves time-domain constraints on signals and robustness requirements for norm-bounded structured plant uncertainty. It is shown that weighting functions in the μ-framework can be used to achieve this goal, but on the cost of many design iterations. The μ-controller has been implemented in an experimental set-up, with a digitally controlled Compact Disc player, showing satisfactory performance.

References

[1] Bouwhuis, G. et al., *Principles of optical disc systems*, 1985, Adam Hilger Ltd.

[2] Draijer, W., M. Steinbuch and O.H. Bosgra, "Adaptive control of the radial servo system of a Compact Disc player", *IFAC Automatica*, 28, 1992, pp.455-462.

[3] Doyle, J.C. and K. Glover, "State space formulae for all stabilizing controllers that satisfy an H_∞-norm bound and relations to risk sensitivity", *Systems and Control Letters*, 11, 1988, pp. 167-172.

[4] Doyle, J.C., "Advances in multivariable control", *Lecture Notes of the ONR/Honeywell Workshop* 1984, Honeywell, Minnesota.

[5] Steinbuch, M., J.C. Terlouw and O.H. Bosgra, "Robustness analysis for real and complex perturbations applied to an electro-mechanical system", *Proc. 1991 American Control Conference*, pp. 556-561.

[6] Packard, A.K., G.J. Balas, K. Glover and J.C. Doyle, "Theory and Application of Robust Multivariable Control", *Short Course, Delft*, June, 1990.

[7] Balas, G.J., J.C. Doyle, K. Glover, A.K. Packard and R. Smith, *"μ-Analysis and Synthesis Toolbox"*, MUSYN Inc., Minneapolis, 1991.

[8] Doyle, J.C., K. Lenz and A. Packard "Design examples using μ-synthesis: Space Shuttle lateral axis FCS during re-entry", *NATO ASI Series, vol.F34: Modeling, Robustness and sensitivity reduction in Control Systems*, (Ed. R.Curtain), 1987.

Proceedings of the 31st Conference
on Decision and Control
Tucson, Arizona • December 1992

MIMO H_∞-Control with Time Domain Constraints

Hèctor Rotstein
Dept. of Electrical Engineering
California Institute of Technology
Pasadena, CA 91125

Athanasios Sideris
Dept. of Mechanical & Aerospace Engineering
University of California, Irvine
Irvine, CA 92717

Abstract

Standard $\mathcal{H}_\infty$ optimization cannot handle specifications or constraints on the time response of a closed-loop system exactly. In this paper, the problem of $\mathcal{H}_\infty$ optimization subject to time domain constraints over a finite horizon is considered. More specifically, given $\gamma > 0$, and a set of fixed inputs $\{w^i\}$, it is required to find a controller such that a closed-loop transfer matrix has an $\mathcal{H}_\infty$-norm less than γ, and the time responses y^i to the signals w^i belong to some prespecified sets Ω^i. First, the one-block constrained $\mathcal{H}_\infty$ optimal control problem is reduced to a finite dimensional, convex minimization problem and a standard $\mathcal{H}_\infty$ optimization problem. Then, the general four-block $\mathcal{H}_\infty$ optimal control problem is solved by reduction to the one-block case. The objective function is constructed via state space methods and some properties of $\mathcal{H}_\infty$ optimal constrained controllers are given. It is shown how satisfaction of the constraints over a finite horizon can imply good behavior overall. An efficient computational procedure based on the ellipsoid algorithm is also discussed.

1 Introduction

In spite of its success and popularity, $\mathcal{H}_\infty$ has some drawbacks, that were quickly recognized by the researchers in the area. One such drawback is that all specifications must be translated into the frequency domain by the use of weighting functions that attempt to capture the essence of each specification; in the absence of a formal procedure, this step may turn out to be quite hard and time consuming. This is especially true in the case of time-domain specifications given, for example, in terms of envelopes for step responses, hard bounds on actuator signals, etc. Moreover, the designer cannot assess by using standard $\mathcal{H}_\infty$ machinery, whether a given constrained problem has a feasible solution and thus achieve various trade-offs between specifications and establish the limits of control system performance.

The main purpose of this paper is to develop an approach by which time domain constraints can be explicitly (i.e. without first translating them to the frequency domain) incorporated into the $\mathcal{H}_\infty$ design framework. Previous works on the subject include [3, 6, 1, 4, 8]. In particular, in [8] asolution to a constrained, single-input single-output, discrete-time control problem is pursued. Time-domain constraints are imposed over a finite horizon and an $\mathcal{H}_\infty$ optimization problem subject to such additional time-domain constraints is formulated. This problem is then reduced to a finite-dimensional convex minimization problem and a standard $\mathcal{H}_\infty$ optimization problem. The overall solution is found by combining the solutions of the two sub-problems. This approach has several advantages over the other approaches mentioned above. First, the frequency domain objective is treated *exactly*, i.e., without resorting to any kind of frequency discretization. Secondly, the optimization problem that has to be solved in order to get a solution is *finite dimensional* and not semi-infinite as in [6]. Thirdly, the use of analytical tools allows the investigation of some of the characteristics of the optimal solution. Finally, the optimization problem can be constructed by simple and numerically stable state-space methods, and both the objective function and its gradient can be reliably computed.

This paper extends the results of [8] in the multivariable case. We first derive a decomposition of the constrained discrete-time $\mathcal{H}_\infty$ optimization problem to a finite-dimensional, convex minimization problem and a standard $\mathcal{H}_\infty$ optimization in the so-called one-block case. We then treat the general, four-block, constrained $\mathcal{H}_\infty$ optimal control problem by reducing it to the one-block case via an allpass embedding argument [5, 7]. Considering constraints only over a finite horizon may seem to be reasonable, since the asymptotic stability of the closed loop system implies a decay in the time domain responses. However in [8], it is shown that constrained $\mathcal{H}_\infty$ optimal solutions may exhibit undesirable time-domain behavior after the horizon, regardless of how long the horizon might be taken. This behavior can be avoided by considering certain sub-optimal solutions and this remains true in the multivariable case as well.

The paper is organized as follows. In Section 2, a general framework for $\mathcal{H}_\infty$-Optimization subject to time domain specifications is considered. In Section 3, the unconstrained problem is reviewed. In particular, it is shown how to reduce the four-block problem to the one-block case. In Section 4, the constrained one-block case is treated by closely following the SISO case considered in [8]. In Section 5 the general constrained problem is studied; the constrained four-block problem is reduced to a constrained one-block problem by using the material in Section 3, and then this problem is solved by using the results of Section 4. Section 6 contains our conclusions.

Notation

L_∞ denotes the Lebesgue space of vector valued complex functions which are essentially bounded on the unit circle. By $\mathcal{H}_\infty(\mathcal{H}_\infty^\sim)$ we denote the set of stable (antistable) functions $G(z) \in \mathcal{L}_\infty$, i.e. $G(z)$ with all its poles inside (outside) the unit disk. Therefore, z^{-1} represents the unit delay operator. A real-rational transfer function may be written as

$$G(z) = G_p(z) + G_i(z) \qquad (1-1)$$

where $G_p(z)$ is proper and $G_i(z)$ is a polynomial in z. Let $\mathcal{RH}_\infty$ denote the set of stable, proper, real rational transfer functions with a state space realization, i.e., $G(z) \in \mathcal{RH}_\infty$ may be written as

$$G(z) = D + C(zI - A)^{-1}B = \left(\begin{array}{c|c} A & B \\ \hline C & D \end{array}\right) \qquad (1-2)$$

For example, if

$$A_f^n = \begin{pmatrix} 0 & I_l & 0 & \cdots & 0 \\ 0 & 0 & I_l & \cdots & 0 \\ \vdots & \vdots & \vdots & \cdots & \vdots \\ 0 & 0 & \cdots & 0 & I_l \\ 0 & 0 & 0 & 0 & 0 \end{pmatrix} \in \mathcal{R}^{nl\times nl} \qquad (1-3)$$

and $E_1^n \doteq [I_l\ 0 \cdots 0]^t$, $E_n^n \doteq [0 \cdots 0\ I_l]^t$, then the function $z^{-n}I_l$ has a realization:

$$z^{-n}I_l = \left(\begin{array}{c|c} A_f & E_n^n \\ \hline E_1^{nt} & 0 \end{array}\right) \qquad (1-4)$$

and if $G(z)$ is an FIR, then

$$G(z) = \sum_{i=0}^{n-1} G_i z^{-i} = \left(\begin{array}{c|c} A_f^{n-1} & E_{n-1}^{n-1} \\ \hline [G_{n-1}\cdots G_1] & G_0 \end{array}\right) \qquad (1-5)$$

A_f^n and E_i^n will be denoted A_f and E_i respectively whenever the block-dimension is clear from the context. If $G \in \mathcal{RL}_\infty$, then $G = G_s + G_u$, where $G_s \in \mathcal{RH}_\infty$ is strictly proper and $G_u \in \mathcal{RH}_\infty^\sim$. Moreover, the operation $[\cdot]_-$ denotes taking the stable part of a system in $\mathcal{RL}_\infty$, i.e., $[G]_- = G_s$. If $G \in \mathcal{RH}_\infty$, G can be expanded as:

$$G(z) = D + \sum_{i=0}^{\infty} CA^iBz^{-(i+1)} \qquad (1-6)$$

CH3229-2/92/0000-2601$1.00 © 1992 IEEE

For notational convenience, we will sometimes write $G_0 = D$ and $G_i = CA^{i-1}B,\ i = 1, 2, \cdots$, and define $\mathbf{G_n} = \left[\begin{array}{ccc} G_0 & \cdots & G_{n-1}\end{array}\right]$. It will be assumed in the following that all transfer matrices are in $\mathcal{RL}_\infty$. Also define

$$G^\sim(z) \doteq G(1/z)^t \qquad (1-7)$$

then if $det(A) \neq 0$, and $G(z)$ is given by 1-2,

$$G^\sim(z) = \left(\begin{array}{c|c} A^{-t} & A^{-t}C^t \\ \hline -B^{-t}A^{-t} & D^t - B^tA^{-t}C^t \end{array}\right) \qquad (1-8)$$

Note that if $det(A) = 0$, then $G^\sim(z)$ will have a polynomial part.

2 Problem formulation

In this section, the problem of minimizing the infinity norm of a closed-loop transfer function subject to time domain constraints over a finite horizon is formulated. Consider a transfer matrix P with input vector $[w_t^t\ w_f^t\ u^t]^t$ and output vector $[z_t^t\ z_f^t\ y^t]^t$, i.e. it holds

$$\begin{aligned} z_t &= P_{11}w_t + P_{12}w_f + P_{13}u \\ z_f &= P_{21}w_t + P_{22}w_f + P_{23}u \\ y &= P_{31}w_t + P_{32}w_f + P_{33}u \end{aligned}$$

Here P is obtained from the nominal plant and describes the effect of the vectors of exogenous signals $[w_t^t\ w_f^t]^t$, and control action u on the vector of outputs $[z_t^t\ z_f^t]^t$ and measurements y. In this model, the signals w_f and z_f will be associated with the $\mathcal{H}_\infty$ optimization and the signals w_t and z_t with the time-domain constraints. Now suppose that the loop is closed with a controller K so that $u = Ky$. Then, consider the closed-loop transfer matrices

$$\begin{aligned} T_{z_tw_t} &= P_{11} + P_{13}K(I - P_{33}K)^{-1}P_{31} \\ T_{z_fw_f} &= P_{22} + P_{23}K(I - P_{33}K)^{-1}P_{32} \end{aligned}$$

that by using the discrete-time version of the Youla Parametrization Lemma can be parametrized as

$$\begin{aligned} T_{z_tw_t} &= T_{11} - T_{13}QT_{31} \\ T_{z_fw_f} &= T_{22} - T_{23}QT_{32} \end{aligned}$$

where the T_{ij}'s are constructed from the data of the problem and $Q \in \mathcal{RH}_\infty$. In the previous parametrization we can assume that T_{23} and T_{32} are inner and co-inner transfer matrices respectively [2].

Time Domain Specifications Suppose that for some fixed input vector of signals w_t, one wants to constrain the j-th output of the system so that the first N samples satisfy some specified bounds. For instance, w_t can be the vector $w_t = [0\ 0\ \cdots\ w_t^i\ \cdots 0]^t$ where w_t^i is a step and one wants the j-th output to remain between some given bounds. In fact, by linearity, the general case can be recovered using superposition. Let $T_{z_tw_t}^{ji} = t_{11}^{ji} - t_{13}^{j\cdot}Qt_{31}^{\cdot i}$, where lower letters are used to stress that the functions are just scalars and $t_{13}^{j\cdot}$, $t_{31}^{\cdot i}$ denote row and column vectors with entries t_{13}^{ji} and t_{31}^{ji} respectively. To simplify the notation, let $a \doteq t_{11}^{ji}$, $b \doteq t_{13}^{j\cdot}$, $c \doteq t_{31}^{\cdot i}$, with $a = \sum_{s=0}^\infty a_s z^{-s}$, $b = \sum_{s=0}^\infty b_s z^{-s}$, $c = \sum_{s=0}^\infty c_s z^{-s}$ and $Q = \sum_{s=0}^\infty Q_s z^{-s}$. Then:

$$a - bQc = \sum_{k=0}^{\infty}(a_k - \sum_{r=0}^{k}\sum_{s=0}^{r} b_{r-s}Q_sc_{k-r})z^{-k}$$

In particular $(a - bQc)_k = a_k - \sum_{r=0}^k \sum_{s=0}^r b_{r-s}Q_sc_{k-r}$ implying that convex (linear) constraints on the input-output Markov parameters translate into convex (linear) constraints on the Q_s.

Let $\mathbf{Q_n} \doteq [Q_0 \cdots Q_{n-1}]$ denote the matrix formed by the first n terms of the expansion of $Q(z)$. Then, 2-1 implies that the time domain constraints considered in this section translate easily into constraints on the matrix $\mathbf{Q_n}$. The set where $\mathbf{Q_n}$ must lie in order to satisfy the time domain constraints is called Ω^n.

Frequency Domain Specifications Consider now the transfer function matrix between w_f and z_f. Then, it is well known that one can write:

$$\|T_{w_fz_f}\|_\infty = \left\| \begin{array}{cc} G_{11} & G_{12} \\ G_{21} & G_{22} - Q^\sim \end{array} \right\|_\infty \qquad (2-1)$$

where $G = \left[\begin{array}{cc} G_{11} & G_{12} \\ G_{21} & G_{22}\end{array}\right]$ is a stable transfer matrix depending on P. Recall that when all the G_{ij} are non zero, 2-1 is called a *four-block problem* [2] whereas if $G_{11} = G_{12} = G_{21} = 0$, then 2-1 is called an one-block problem. Using this nomenclature, the case studied in [8] corresponds to a scalar one-block problem.

We now formulate the time-domain constrained $\mathcal{H}_\infty$-Control problem (TDC$\mathcal{H}_\infty$) as

Find $Q(z) \in \mathcal{H}_\infty$ *so that* $\|T_{w_fz_f}\|_\infty \leq \gamma$ *and* $\mathbf{Q_n} \in \Omega^n$

If one is interested in the optimal case, i.e., determining the minimum γ for which (TDC$\mathcal{H}_\infty$) has a solution, then one can follow the usual unconstrained procedure, by which γ is iteratively adjusted to reach the optimal value as closely as desired. In what follows, we assume without loss of generality that $\gamma = 1$.

3 The Unconstrained $\mathcal{H}_\infty$-control Problem

In this section, the solution to the unconstrained $\mathcal{H}_\infty$-control problem is reviewed. Since solutions for the one-block $\mathcal{H}_\infty$ optimization are easier to compute, this special problem is treated first. A solution to the four-block $\mathcal{H}_\infty$ problem is then found by reducing it to a special one-block case.

3.1 One-block $\mathcal{H}_\infty$-optimal Control

Consider first the frequency domain specification $\|T_{w_fz_f}\|_\infty \leq \gamma$ for the one-block problem (i.e., the case $G = G_{22}$). Let $G = \left(\begin{array}{c|c} A & B \\ \hline C & D\end{array}\right)$ $\in \mathcal{RH}_\infty$, and let X and Y denote the controllability and observability grammians of G respectively. Then X and Y are symmetric and positive semidefinite, and solve the Lyapunov equations:

$$\begin{aligned} X &= AXA^t + BB^t \\ Y &= A^tYA + C^tC \end{aligned}$$

If $\|G\|_H$ denotes the Hankel-norm of G, then $\|G\|_H = \sqrt{\rho(XY)}$. For later reference, let $x \doteq X^{1/2}$ and $y \doteq Y^{1/2}$, with $(\cdot)^{1/2}$ denoting the positive square root.

Theorem 1 *With the notation above:*

a) There exist $Q \in \mathcal{RH}_\infty$ *such that* $\|G - Q^\sim\|_\infty \leq 1$ *if and only if* $\|G\|_H \leq 1$.

b) Assume that $\|G\|_H \leq 1$*, and let*

$$\begin{aligned} D_Q &= D^t - E_D^t \\ A_Q &= A - BC_Q \\ B_Q &= AXC^t + BE_D^t \end{aligned} \qquad (3-2)$$

with C_Q *solving*

$$C_Q(I - XY) = E_D^tC + B^tYA \qquad (3-3)$$

and E_D *such that the matrix*

$$D_e = \left[\begin{array}{cc} yAx & yB \\ Cx & E_D \end{array}\right] \qquad (3-4)$$

is a contraction (i.e., $\overline{\sigma}(D_e) \leq 1$*). Then* $Q \doteq \left(\begin{array}{c|c} A_Q & B_Q \\ \hline C_Q & D_Q\end{array}\right) \in$ $\mathcal{RH}_\infty$*, and is such that* $\|G - Q^\sim\|_\infty \leq 1$.

Proof: Part a of the theorem is standard; see [7] for a proof of part b.

□

3.2 Four-block $\mathcal{H}_\infty$-optimal Control

The solution to the four-block problem given here consists of two steps [5, 7]. The first step is to reduce the problem to a special one-block case, by embedding G into a larger transfer matrix G_a. The second is to compute a solution to the four-block problem, by applying Theorem 1 to G_a and choosing a particular solution. The embedding is obtained as follows. Let

$$G = \begin{bmatrix} G_{11} & G_{12} \\ G_{21} & G_{22} \end{bmatrix} = \left(\begin{array}{c|cc} & m_1 & m_2 \\ A & B_1 & B_2 \\ \hline C_1 & D_{11} & D_{12} \\ C_2 & D_{21} & 0 \end{array} \right) \begin{array}{c} \\ \\ p_1 \\ p_2 \end{array} \qquad (3-5)$$

Note that $D_{22} = 0$, and it is well known that this can be assumed without loss of generality. Assume in what follows that

$$\|G_{11}\ G_{12}\|_\infty < 1 \quad \text{and} \quad \left\| \begin{matrix} G_{11} \\ G_{21} \end{matrix} \right\|_\infty < 1 \qquad (3-6)$$

Define:

$$\begin{aligned} A_X &\doteq A + (B_1 D_{11}^t + B_2 D_{12}^t)(I - D_{11}D_{11}^t - D_{12}D_{12}^t)^{-1} C_1 \\ Q_X &\doteq -C_1^t (I - D_{11}D_{11}^t - D_{12}D_{12}^t)^{-1} C_1 \\ R_X &\doteq [B_1\ B_2] \left(\begin{bmatrix} I & 0 \\ 0 & I \end{bmatrix} - \begin{bmatrix} D_{11}^t \\ D_{12}^t \end{bmatrix} [D_{11}\ D_{12}] \right)^{-1} \begin{bmatrix} B_1^t \\ B_2^t \end{bmatrix} \end{aligned}$$

and

$$\begin{aligned} A_Y &\doteq A + B_1(I - D_{11}^t D_{11} - D_{21}^t D_{21})^{-1}(D_{11}^t C_1 + D_{21}^t C_2) \\ Q_Y &\doteq -B_1(I - D_{11}^t D_{11} - D_{21}^t D_{21})^{-1} B_1^t \\ R_Y &\doteq [C_1^t\ C_2^t] \left(\begin{bmatrix} I & 0 \\ 0 & I \end{bmatrix} - \begin{bmatrix} D_{11} \\ D_{21} \end{bmatrix} [D_{11}^t\ D_{21}^t] \right)^{-1} \begin{bmatrix} C_1 \\ C_2 \end{bmatrix} \end{aligned}$$

With these definitions, consider the two Discrete Time Algebraic Riccati equations:

$$\begin{aligned} X &= A_X X A_X^t - A_X X Q_X (I + X Q_X)^{-1} X A_X^t + R_X \qquad (3\text{-}7) \\ Y &= A_Y^t Y A_Y - A_Y^t Y Q_Y (I + Y Q_Y)^{-1} Y A_Y + R_Y \qquad (3\text{-}8) \end{aligned}$$

associated with the spectral factorization problems

$$\begin{aligned} G_{13}G_{13}^\sim &= I - G_{11}G_{11}^\sim - G_{12}G_{12}^\sim \qquad (3\text{-}9) \\ G_{31}^\sim G_{31} &= I - G_{11}^\sim G_{11} - G_{12}^\sim G_{12} \qquad (3\text{-}10) \end{aligned}$$

respectively. Finally, define

$$\begin{aligned} D_{13} &\doteq (I - D_{11}D_{11}^t - D_{12}D_{12}^t - C_1 X C_1^t)^{1/2} \qquad (3\text{-}11) \\ B_3 &\doteq -(AXC_1^t + B_1 D_{11}^t + B_2 D_{12}^t) D_{13}^{-t} \qquad (3\text{-}12) \\ D_{31} &\doteq (I - D_{11}^t D_{11} - D_{21}^t D_{21} - B_1^t Y B_1)^{1/2} \qquad (3\text{-}13) \\ C_3 &\doteq -D_{31}^{-t}(B_1^t Y A + D_{11}^t C_1 + D_{21}^t C_2) \qquad (3\text{-}14) \end{aligned}$$

By Assumption 3-6, Equations 3-7 and 3-8 have a unique symmetric positive definite solution such that $A - C_1 D_{13}^{-1} B_3$ and $A - C_3 D_{31}^{-1} B_1$ are stable; then $G_{13} \doteq \left(\begin{array}{c|c} A & B_3 \\ \hline C_1 & D_{13} \end{array} \right)$ and $G_{31} \doteq \left(\begin{array}{c|c} A & B_1 \\ \hline C_3 & D_{31} \end{array} \right)$ are the canonical spectral factors of the spectral factorization problems 3-9 and 3-10 respectively.

Theorem 2 *[7] Assume that* $\|G_{11}\ G_{12}\|_\infty < 1$, $\left\| \begin{matrix} G_{11} \\ G_{21} \end{matrix} \right\|_\infty < 1$ *and let* X *and* Y *be the symmetric positive definite solutions to 3-7 and 3-8 respectively. Then there exist* $Q \in \mathcal{RH}_\infty$ *such that* $\left\| \begin{matrix} G_{11} & G_{12} \\ G_{21} & G_{22} - Q^\sim \end{matrix} \right\|_\infty \leq 1$ *if and only if* $\rho(XY) < 1$.

□

The additional assumptions in Theorem 2 are introduced in order to rule out a special kind of optimality that requires a more sophisticated treatment, although the theory actually extends to such a case as well.

Consider now the augmented system

$$G_a = \begin{bmatrix} G_{11} & G_{12} & G_{13} \\ G_{21} & G_{22} & G_{23} \\ G_{31} & G_{32} & G_{33} \end{bmatrix} \doteq \left(\begin{array}{c|ccc} A & B_1 & B_2 & B_3 \\ \hline C_1 & D_{11} & D_{12} & D_{13} \\ C_2 & D_{21} & 0 & 0 \\ C_3 & D_{31} & 0 & 0 \end{array} \right) \qquad (3-15)$$

Theorem 3 *[7] For each solution* $Q \in \mathcal{RH}_\infty$ *of* $\left\| G - \begin{bmatrix} 0 & 0 \\ 0 & Q^\sim \end{bmatrix} \right\|_\infty \leq 1$ *there exists a solution* $Q_a \in \mathcal{RH}_\infty$ *that solves the one-block* $\mathcal{H}_\infty$ *problem* $\|G_a - Q_a^\sim\| = 1$, *and* $Q = \begin{bmatrix} 0 & I & 0 \end{bmatrix} Q_a \begin{bmatrix} 0 \\ I \\ 0 \end{bmatrix}$. *More specifically, one such solution can be constructed by computing* $Q_a = \left(\begin{array}{c|c} A_Q & B_Q \\ \hline C_Q & D_Q \end{array} \right)$ *using the formulas in Theorem 1, with* B_a, C_a *and* D_a *replacing* B, C *and* D, *and taking*

$$E_D = \begin{bmatrix} D_{11} & D_{12} & D_{13} \\ D_{21} & E_{22} & E_{23} \\ D_{31} & E_{32} & E_{33} \end{bmatrix} \qquad (3-16)$$

for some E_{22}, E_{23}, E_{32} *and* E_{33} *that make the matrix*

$$D_a = \begin{bmatrix} yAx & yB_1 & yB_2 & yB_3 \\ C_1 x & D_{11} & D_{12} & D_{13} \\ C_2 x & D_{21} & E_{22} & E_{23} \\ C_3 x & D_{31} & E_{32} & E_{33} \end{bmatrix} \qquad (3-17)$$

a contraction; then $Q = [0\ I\ 0] Q_a \begin{bmatrix} 0 \\ I \\ 0 \end{bmatrix}$ *gives a solution to the four-block problem*

Proof: It is shown in [5, 7], that to any $Q \in \mathcal{RH}_\infty$ that solves the four-block problem, there corresponds a solution $Q_a = \begin{bmatrix} 0 & 0 & 0 \\ 0 & Q_a & * \\ 0 & * & * \end{bmatrix}$ that solves the specially constructed one block problem $\|G_a - Q_a^\sim\|_\infty \leq 1$, and the converse obviously holds by taking Q as the compression stated above. It is then shown in [5, 7] how to pick those solutions to the one-block problem that have the first block row and column zero, by taking E_D in the form of the theorem, with E_{11}, E_{12}, E_{21} and E_{22} stable transfer functions that make D_e a contraction; in particular, taking these matrices as constants gives the solution in the theorem. For more details, the reader is referred to [7].

□

In solving the four-block *constrained* $\mathcal{H}_\infty$ problem, we follow an approach analogous to the one reviewed in this section. In Section 4, a solution to the one-block *constrained* problem is presented, extending the SISO results in [8]. In Section 5, the four-block *constrained* problem is reduced to the one-block case using the approach outlined in Section 3.2.

4 The One-Block Constrained $\mathcal{H}_\infty$-control Problem

In this section, the solution to the time domain constrained $\mathcal{H}_\infty$ one-block control problem is studied. For notational simplicity, it is assumed that G is square with l inputs and l outputs. The results here extend those derived for a SISO problem in [8], and are required for solving the four-block problem.

Let $G = \left(\begin{array}{c|c} A & B \\ \hline C & D \end{array} \right) \in \mathcal{RH}_\infty$, and suppose that some $\mathbf{Q}_\mathbf{n}^* = [Q_0\ Q_1\ \cdots\ Q_{n-1}]$ of appropriate dimensions is given. Consider the system defined as

$$G(z, \mathbf{Q}_n) \doteq z^{-n}[G(z) - \sum_{i=0}^{n-1} Q_i^t z^i] \qquad (4-1)$$

and let $\varphi(\mathbf{Q_n}) \doteq \|G(z; \mathbf{Q_n})\|_H$. Consider the following realization for 4-1:

$$G(\cdot; \mathbf{Q_n}) = \left(\begin{array}{cc|c} A & BE_1^t & 0 \\ 0 & A_f & E_n \\ \hline C & -\hat{\mathbf{Q}}_\mathbf{n} & 0 \end{array} \right) \doteq \left(\begin{array}{c|c} A_e & B_e \\ \hline C_e & D_e \end{array} \right) \qquad (4-2)$$

where $\hat{\mathbf{Q}}_\mathbf{n} = [Q_0^t\ Q_1^t\ \cdots\ Q_{n-1}^t]$, and let X_e and Y_e denote the controllability and observability grammians of 4-2. Then,

Theorem 4 *With the previous notation,*

a) There exists $Q(z) \in \mathcal{RH}_\infty$ *such that* $\mathbf{Q_n} = \mathbf{Q_n^*}$ *and* $\|G(z) - Q(z)^\sim\|_\infty \le 1$ *if and only if* $\varphi(\mathbf{Q_n}) = \|G(z, \mathbf{Q_n^*})\|_H \le 1$.

b) There exists stabilizing controller solving ($TDC\mathcal{H}_\infty$), *with* $G = G_{22}$ *and* $\gamma = 1$, *if and only if* $\min_{\mathbf{Q_n} \in \Omega^n} \varphi(\mathbf{Q_n}) \le 1$

c) $\varphi(\mathbf{Q_n}) = \overline{\sigma}(W_1(\mathbf{Q_n}))$ *where*

$$W_1(\mathbf{Q_n}) = \begin{bmatrix} yA^n x & CA^{n-1}x & \cdots & CAx & Cx \\ yA^{n-1}B & CA^{n-2}B & \cdots & CB & -Q_0^t \\ yA^{n-2}B & CA^{n-3}B & \cdots & -Q_0^t & -Q_1^t \\ \vdots & \vdots & \ddots & \vdots & \vdots \\ yAB & CB & \cdots & -Q_{n-3}^t & -Q_{n-2}^t \\ yB & -Q_0^t & \cdots & -Q_{n-2}^t & -Q_{n-1}^t \end{bmatrix} \quad (4-3)$$

□

By the previous discussion, one has first to find $\mathbf{Q_n^*}$ that satisfies $\varphi(\mathbf{Q_n^*}) \le 1$ —this is done by solving a convex minimization problem in $\mathbf{Q_n}$, and then to find a solution to the "standard" one-block $\mathcal{H}_\infty$ minimization problem $\|G(\cdot, \mathbf{Q_n^*}) - Q_t^\sim\|_\infty \le 1$ —this can be done, for example, by using Theorem 1. Then a solution to problem $TDC\mathcal{H}_\infty$ can be obtained as $Q(z) = \sum_{i=0}^{n-1} Q_i z^{-i} + z^{-n} Q_t(z)$. This procedure has some disadvantages. First, if G has degree r, then $G(z; \mathbf{Q_n})$ has degree $ln + r$ and therefore one needs to manipulate large matrices in order to compute $Q_t(z)$. This makes the computation time-consuming and potentially sensitive to numerical errors. Moreover, one would expect that the ln poles at 0, which are fictitiously introduced to formulate the problem, should cancel. This was established in [8] for a SISO problem, but the proof there does not extend to the multivariable case. Moreover, the proof does not provide a constructive way for cancelling such modes. The following theorem gives a realization for $Q(z) = \sum_{i=0}^{n-1} Q_i z^{-i} + z^{-n} Q_t(z)$ of order $r + nl$ that circumvents all such difficulties.

Theorem 5 *Assume that* $\mathbf{Q_n} = [Q_0 \ \cdots \ Q_{n-1}]$ *is such that* $\rho = \|G(z; \mathbf{Q_n})\|_H \le 1$. *Let* $x_e = \begin{pmatrix} x & 0 \\ 0 & I \end{pmatrix}$, $y_e = Y_e^{1/2}$, *and*

$$A_Q = A_e - B_e C_Q$$

$$B_Q = \begin{pmatrix} AXC^t - BQ_0 \\ -Q_1 \\ \vdots \\ -Q_{n-1} \\ D_Q \end{pmatrix}$$

where C_Q *satisfies*

$$C_Q(I - X_e Y_e) = -D_Q C_e + B_e^t Y_e A_e$$

and $D_Q \doteq -E_D^t$, *where* E_D *is such that the matrix*

$$D_a = \begin{bmatrix} yA_e x & yB_e \\ C_e x & E_D \end{bmatrix} \quad (4-4)$$

is a contraction. Then

$$Q \doteq \left(\begin{array}{c|c} A_Q & B_Q \\ \hline [0 \ \ -E_1^t] & Q_0 \end{array} \right) \quad (4-5)$$

is such that $Q \in \mathcal{RH}_\infty$, $\|G(z) - Q(z)^\sim\|_\infty \le 1$ *and the first* n *terms in the expansion of* $Q(z)$ *are* $Q_0, Q_1, \cdots, Q_{n-1}$.

□

In standard $\mathcal{H}_\infty$ optimization, the multiplicity of the largest singular value associated with the Hankel norm is generically one. On the other hand, in the constrained case the multiplicity can be much larger. This follows because the process of minimizing the largest singular value of W_1 tends to cluster the singular values together, and hence to enlarge the multiplicity; intuitively, as the constraints become less binding, one expects the multiplicity to rise. In [8] it was shown that in the limiting case in which the unconstrained solution satisfies the time domain constraints, the multiplicity is increased by n. A similar property can be established for the general one-block case.

5 The Four-block Constrained $\mathcal{H}_\infty$-Control Problem

In this section the four-block constrained $\mathcal{H}_\infty$-control problem is considered. The main idea is first to reduce the four-block problem to the one-block case by using the procedure of Section 3, and then apply the results for the constrained one-block case derived in Section 4.

5.1 Problem Transformation

Consider first the frequency domain specification $\|T_{w_f z_f}\|_\infty \le \gamma$ (recall that $\gamma = 1$). Let G have a state space realization

$$G = \left(\begin{array}{c|cc} & m_1 & m_2 \\ A & B_1 & B_2 \\ \hline C_1 & D_{11} & D_{12} \\ C_2 & D_{21} & 0 \end{array} \right) \begin{array}{c} \\ \\ p_1 \\ p_2 \end{array} \quad (5-1)$$

Assume, as in Section 3, that $\max\left\{ \|G_{11} \ G_{12}\|_\infty, \left\| \begin{matrix} G_{11} \\ G_{21} \end{matrix} \right\|_\infty \right\} < 1$. Note that this assumption holds if the unconstrained optimal norm is strictly less than one; since one expects that the constrained problem will usually achieve a minimum norm strictly larger than the unconstrained one, the condition will usually hold with no difficulty (unless an unconstrained solution satisfies the time domain specifications). In order to reduce the problem to the one-block case, consider the discrete-time algebraic Riccati equations 3-7 and 3-8 with positive definite solutions X and Y respectively. From Theorem 2, there exist a $Q \in \mathcal{RH}_\infty$ such that $\left\| \begin{matrix} G_{11} & G_{12} \\ G_{21} & G_{22} - Q^\sim \end{matrix} \right\|_\infty \le 1$ if and only if $\rho(XY) \le 1$. Let B_3, C_3, D_{13} and D_{31} be defined by 3-12, 3-14, 3-11 and 3-13. By the previous assumption, these matrices are well defined. Let G_a be as in 3-15, and consider the one-block problem associated with G_a. By Theorem 3 there exists a solution to the four-block problem if and only if there exists a solution to this one-block problem of the form $\begin{bmatrix} 0 & 0 & 0 \\ 0 & * & * \\ 0 & * & * \end{bmatrix}$. Let then Q, Q_{23}, Q_{32}, $Q_{33} \in \mathcal{RH}_\infty$ be such that

$$\left\| \begin{matrix} G_{11} & G_{12} & G_{13} \\ G_{21} & G_{22} - Q^\sim & G_{23} - Q_{32}^\sim \\ G_{31} & G_{32} - Q_{23}^\sim & G_{33} - Q_{33}^\sim \end{matrix} \right\|_\infty \le 1. \quad (5-2)$$

Consider $Q_t(z) \doteq \sum_{i=0}^\infty Q_{i+n} z^{-i}$, $Q_{23t}(z) \doteq \sum_{i=0}^\infty Q_{23,i+n} z^{-i}$, and define: $G_{12}^n \doteq G_{12}(z) z^{-n}$, $G_{22}(z; \mathbf{Q_n}) \doteq [G_{22}(z) - \sum_{i=0}^{n-1} Q_i^t z^i] z^{-n}$ and $G_{32}(z; \mathbf{Q_{23n}}) \doteq [G_{32}(z) - \sum_{i=0}^{n-1} Q_{23,i}^t z^i] z^{-n}$. Then Theorem 4 can be used to establish the following result.

Theorem 6 *Suppose that* Q_i, $i = 1, 2, \cdots, n-1$ *are given. Then there exist* Q_i, $i = n+1, n+2, \cdots$ *such that* $Q(z) \doteq \sum_{i=0}^\infty Q_i z^{-i}$ *satisfies*

$$\left\| \begin{matrix} G_{11} & G_{12} \\ G_{21} & G_{22} - Q(z)^\sim \end{matrix} \right\|_\infty \le 1$$

if and only if there exist $\mathbf{Q_{23n}} = [Q_{23,0}, \cdots Q_{23,n-1}]$ *such that* $\varphi(\mathbf{Q_n}; \mathbf{Q_{23n}}) \le 1$, *where*

$$\varphi(\mathbf{Q_n}; \mathbf{Q_{23n}}) = \|G_a(\mathbf{Q_n}; \mathbf{Q_{23n}})\|_H \quad (5\text{-}3)$$

$$= \left\| \begin{matrix} G_{11} & G_{12}^n & G_{13} \\ G_{21} & G_{22}(\cdot; \mathbf{Q_n}) & G_{23} \\ G_{31} & G_{32}(\cdot; \mathbf{Q_{32n}}) & G_{33} \end{matrix} \right\|_H \quad (5\text{-}4)$$

□

5.2 State-Space Computation of the Objective Function

In this section, an expression for $\varphi(\cdot)$ in terms of the state-space realization of G_a and the matrices Q_i, $Q_{23,i}$ is given, following the formula derived in the one-block case. Although the number of variables in the problem has been enlarged by the inclusion of the $Q_{23,i}$'s, we will show that condition 5-2 implies that these variables are uniquely determined by the Q_i's.

Consider the following realization for $G_a(\cdot; \mathbf{Q_n}, \mathbf{Q_{23n}})$:

$$G_a(\cdot; \mathbf{Q_n}, \mathbf{Q_{23n}}) = \left(\begin{array}{cc|ccc} A & B_2E_1^t & B_1 & 0 & B_3 \\ 0 & A_f & 0 & E_n & 0 \\ \hline C_1 & D_{12}E_1^t & D_{11} & 0 & D_{13} \\ C_2 & -\hat{\mathbf{Q}}_\mathbf{n} & D_{21} & 0 & 0 \\ C_3 & -\hat{\mathbf{Q}}_\mathbf{23n} & D_{31} & 0 & 0 \end{array}\right) = \left(\begin{array}{c|c} A_e & B_e \\ \hline C_e & D_e \end{array}\right) \quad (5-5)$$

where

$$\hat{\mathbf{Q}}_\mathbf{n} \doteq [Q_0^t \; Q_1^t \; \cdots \; Q_{n-1}^t]$$
$$\hat{\mathbf{Q}}_\mathbf{23n} \doteq [Q_{23,0}^t \; Q_{23,1}^t \; \cdots \; Q_{23,n-1}^t]$$

The realization for G_a in 5-5 is similar to the one in 4-2. Let X_e and Y_e denote the controllability and observability grammians of G_a. The following expression for $\varphi(\mathbf{Q_n}, \mathbf{Q_{23n}})$ can be obtained.

Lemma 1 *Let W_1 be defined as:*

$$W_1(\mathbf{Q_n}, \mathbf{Q_{23n}}) =$$
$$\begin{bmatrix} yA^nx & yA^{n-1}B_2 & \cdots & yAB_2 & yB_2 \\ C_1A^{n-1}x & C_1A^{n-2}B_2 & \cdots & C_1B_2 & D_{12} \\ C_1A^{n-2}x & C_1A^{n-3}B_2 & \cdots & D_{12} & 0 \\ \vdots & \vdots & & \vdots & \vdots \\ C_1Ax & C_1B_2 & \cdots & 0 & 0 \\ C_1x & D_{12} & \cdots & \cdots & 0 \\ C_2A^{n-1}x & C_2A^{n-2}B_2 & \cdots & C_2B_2 & -Q_0^t \\ C_2A^{n-2}x & C_2A^{n-3}B_2 & \cdots & -Q_0^t & -Q_1^t \\ \vdots & \vdots & \vdots & & \vdots \\ C_2Ax & C_2B_2 & \cdots & \cdots & -Q_{n-2}^t \\ C_2x & -Q_0^t & \cdots & \cdots & -Q_{n-1}^t \\ C_3A^{n-1}x & C_3A^{n-2}B_2 & \cdots & C_3B_2 & -Q_{23,0}^t \\ C_3A^{n-2}x & C_3A^{n-3}B_2 & \cdots & -Q_{23,0}^t & -Q_{23,1}^t \\ \vdots & \vdots & & \vdots & \vdots \\ C_3Ax & C_3B_2 & \cdots & \cdots & -Q_{23,n-1}^t \\ C_3x & -Q_{23,0}^t & \cdots & \cdots & -Q_{23,n-1}^t \end{bmatrix}$$

Then

$$\varphi(\mathbf{Q_n}, \mathbf{Q_{23n}}) = \overline{\sigma}[W_1(\mathbf{Q_n}, \mathbf{Q_{23n}})]^2 \quad (5-6)$$

□

The expressions for the objective function obtained so far, depend on the original variables of the problem and on the $Q_{23,i}$'s that were introduced in the transformation of the problem discussed in the previous section. However, the structure of the problem implies that once the Q_i's are fixed, the $Q_{23,i}$'s are completely determined. Indeed,

Lemma 2 *The auxiliary variables Q_{23}^i, $i = 0, \cdots, n-1$ satisfy*

$$Q_{23,i} = [B_2^tYA_{13}^iB_1 + D_{12}^tC_{13}A_{13}^iB_1 - Q_0D_{21} - \sum_{j=0}^{i-1} Q_jC_{23}A_{13}^{i-1-j}B_1]D_{31}^{-1}$$

where $A_{13} \doteq A - B_1D_{31}^{-1}C_3$, $C_{13} \doteq C_1 - D_{11}D_{31}^{-1}C_3$ and $C_{23} = C_2 - D_{21}D_{31}^{-1}C_3$.

□

The following theorem gives an alternative objective function, that depends only on the original variables of the problem.

Theorem 7 *Let*

$$W_2(\mathbf{Q_n}) =$$
$$\left[\begin{array}{c|cccc|cccc} yA^nx & yA^{n-1}B_1 & \cdots yAB_1 & & yB_1 & yA^{n-1}B_2 & \cdots & yAB_2 & yB_2 \\ \hline C_1A^{n-1}x & C_1A^{n-3}B_1 & \cdots C_1B_1 & & D_{11} & C_1A^{n-2}B_2 & \cdots & C_1B_2 & D_{12} \\ C_1A^{n-2}x & C_1A^{n-3}B_1 & \cdots & D_{11} & 0 & C_1A^{n-3}B_2 & \cdots & D_{12} & 0 \\ \vdots & \vdots & \vdots & \ddots & \vdots & \vdots & \ddots & \vdots & \vdots \\ C_1x & D_{11} & 0 & \cdots & 0 & D_{12} & 0 & \cdots & 0 \\ \hline C_2A^{n-1}x & C_2A^{n-2}B_1 & \cdots C_2B_1 & & D_{21} & C_2A^{n-2}B_2 & \cdots & C_2B_2 & -Q_0^t \\ C_2A^{n-2}x & C_2A^{n-3}B_1 & \cdots & D_{21} & 0 & C_2A^{n-3}B_2 & \cdots & -Q_0^t & -Q_1^t \\ \vdots & \vdots & \vdots & \ddots & \vdots & \vdots & \ddots & \vdots & \vdots \\ C_2x & D_{21} & 0 & \cdots & 0 & -Q_0^t & -Q_2^t & \cdots & -Q_{(n-1)^t} \end{array}\right]$$

Then $\rho(W_0) \leq 1$ if and only if $\overline{\sigma}(W_2) \leq 1$.

□

With a slight abuse of notation, let $W_0(\mathbf{Q_n})$ and $W_1(\mathbf{Q_n})$ denote $W_0(\mathbf{Q_n}, \mathbf{Q_{23n}}(\mathbf{Q_n}))$ and $W_1(\mathbf{Q_n}, \mathbf{Q_{23n}}(\mathbf{Q_n}))$ respectively. Theorem 8 summarizes the results as follows.

Theorem 8 *Consider the problems*

$$\mu_1 = \min_{\mathbf{Q_n} \in \Omega^n} \overline{\sigma}[W_1(\mathbf{Q_n})]$$
$$\mu_2 = \min_{\mathbf{Q_n} \in \Omega^n} \overline{\sigma}[W_2(\mathbf{Q_n})]$$

and assume that Ω^n is a convex set. Then:

a) There exist a solution to Problem TDC$\mathcal{H}_\infty$ if and only if either one of the μ_i's is less than or equal to one.

b) The minimization problems defined by the μ_i's and Ω^N are convex.

5.3 State Space Formula for a Solution

Theorem 8 shows how to check whether the four-block constrained $\mathcal{H}_\infty$-control problem has a solution (i.e., μ_i has to be less than or equal to one), and how to calculate the first n terms Q_0, $Q_1, \cdots, Q_{n-1}$ of the expansion of such a solution. Having the optimal Q_i's, and the corresponding $Q_{23,i}$'s computed from Lemma 2, we can apply Theorem 3 to compute a solution Q_{at} that solves the one-block problem $\|G_a(\cdot; \mathbf{Q_n}, \mathbf{Q_{23n}}) - Q_{at}^{\sim}\|_\infty \leq 1$. Then, by taking the compression $Q_t = \begin{bmatrix} 0 & I & 0 \end{bmatrix} Q_{at} \begin{bmatrix} 0 \\ I \\ 0 \end{bmatrix}$, we have that $Q(z) = \sum_{i=0}^{n-1} Q_iz^{-i} + z^{-n}Q_t(z)$ solves the four-block TDC$\mathcal{H}_\infty$ problem. Theorem 5 can be extended to obtain a realization for $Q(z)$ without the fictitious poles at zero introduced in the formulation of the problem. To see this, consider again the realization 5-5 for G_a, and suppose that a solution to the one-block problem is computed by using Theorem 1. From Theorem 3, and in order to get a solution to the four-block problem, take

$$E_D = \begin{bmatrix} D_{11} & 0 & D_{13} \\ D_{21} & E_{22} & E_{23} \\ D_{31} & E_{32} & E_{33} \end{bmatrix}$$

for some E_{ij}, $i, j =$ 2, 3 that make

$$D_a = \begin{bmatrix} y_eA_ex_e & y_e\begin{bmatrix} B_1 \\ 0 \end{bmatrix} & y_e\begin{bmatrix} 0 \\ E_n \end{bmatrix} & y_e\begin{bmatrix} B_3 \\ 0 \end{bmatrix} \\ \begin{bmatrix} C_1X & D_{12}E_1^t \end{bmatrix} & D_{11} & 0 & D_{13} \\ \begin{bmatrix} C_2X & -\hat{\mathbf{Q}}_\mathbf{n} \end{bmatrix} & D_{21} & E_{22} & E_{23} \\ \begin{bmatrix} C_3X & -\hat{\mathbf{Q}}_\mathbf{23n} \end{bmatrix} & D_{31} & E_{32} & E_{33} \end{bmatrix}$$

a contraction. Here $x_e = \begin{bmatrix} x & 0 \\ 0 & I \end{bmatrix}$ and $y_e = Y_e^{1/2}$. Let Q_{at} be such that $\|G_a - Q_{at}^{\sim}\|_\infty \leq 1$ and let

$$Q_{at} = \left(\begin{array}{c|c} A_Q & B_Q \\ \hline C_Q & D_Q \end{array} \right) = \left(\begin{array}{c|ccc} A_Q & B_{Q_1} & B_{Q_2} & B_{Q_3} \\ \hline C_{Q_1} & 0 & 0 & 0 \\ C_{Q_2} & 0 & -E_{22}^t & -E_{32}^t \\ C_{Q_3} & 0 & -E_{23}^t & -E_{33}^t \end{array} \right) \qquad (5-7)$$

with A_Q, B_Q and C_Q constructed as in Theorem 3. We can then show that $Q_t = \begin{bmatrix} 0 & I & 0 \end{bmatrix} Q_{at} \begin{bmatrix} 0 \\ I \\ 0 \end{bmatrix} = \left[\begin{array}{c|c} A_Q & B_{Q_2} \\ \hline [C_{Q_{21}} \;\; C_{Q_{22}}] & -E_{22}^t \end{array} \right]$ and $Q(z) = \sum_{i=0}^{\infty} Q_i z^{-i} + z^{-n} Q_t(z)$. The following theorem gives expessions for A_Q, B_{Q_2} and a state space realization for $Q(z)$.

Theorem 9 *Suppose* Q_i, $i = 0, 1, \cdots, n-1$ *are such that* $\overline{\sigma}[W_2(\mathbf{Q_n})] \leq 1$, *and let*

$$A_Q = \begin{bmatrix} A - B_3 C_{Q_{31}} & B_2 E_1^t - B_3 C_{Q_{32}} \\ -E_n C_{Q_{21}} & A_f - E_n C_{Q_{22}} \end{bmatrix} \qquad (5-8)$$

and $B_{Q_2} =$

$$\begin{bmatrix} (A - B_3 D_{13}^{-1} C_1) X C_2^t - (B_2 + B_3 D_{13}^{-1} D_{12}) Q_0 + (B_1 + B_3 D_{13}^{-1} D_{11} D_{21}^t) \\ -Q_1 \\ -Q_2 \\ \vdots \\ -Q_{n-1} \\ E_{22}^t \end{bmatrix}$$

Then,

$$Q = \left[\begin{array}{c|c} A_Q & B_{Q_2} \\ \hline -\begin{bmatrix} 0 & E_1^t \end{bmatrix} & Q_0 \end{array} \right] \qquad (5-9)$$

is such that

$$\left\| \begin{bmatrix} G_{11} & G_{12} \\ G_{21} & G_{22} - Q^{\sim} \end{bmatrix} \right\|_\infty \leq 1$$

Moreover, $Q(z) \in \mathcal{RH}_\infty$ *and* $Q(z) = \sum_{i=0}^{\infty} Q_i z^{-i} + z^{-n} Q_t(z)$, *where* $Q_t(z) \in \mathcal{RH}_\infty$.

□

6 Conclusions

In this paper, the problem of $\mathcal{H}_\infty$ optimization subject to time domain constraints is considered. More specifically, given $\gamma > 0$, and a set of fixed inputs $\{w^i\}_{i=1}^{t}$, the problem of designing a controller such that a closed-loop transfer matrix achieves an $\mathcal{H}_\infty$ norm less than or equal to γ, and the time response y^i to the signal w^i satisfies given constraints over a finite horizon is studied. Although the time-domain constraints considered are imposed over a finite horizon, it is shown that one can easily obtain solutions with good behavior overall.

The $\mathcal{H}_\infty$ optimization component of this problem is transformed to the so-called general distance problem by employing the Youla parametrization lemma. First, the one-block case subject to the time domain constraints is considered and is reduced to *i)* a finite-dimensional, convex, non-differentiable, minimization problem, from which the first n coefficients of the expansion of the optimal value of the Youla parameter $Q(z)$ are computed, and *ii)* a standard $\mathcal{H}_\infty$ optimization problem, from which the "tail" of the optimal $Q(z)$ is found. Theorems 4 and 5 give state space procedures for the evaluation of the objective function of the minimization problem, and the computation of a solution respectively.

Next, the general four-block, constrained $\mathcal{H}_\infty$ optimization problem is reduced to an equivalent constrained one-block problem by using the corresponding reduction for the unconstrained case [5, 7] and which is reviewed in Section 3. Then, the previous decomposition is again established (see Theorem 6). Lemma 1, and Theorem 7 give state space formulas for the evaluation of two alternative objective functions for the finite-dimensional minimization, and Theorem 9 gives state space formulas for the construction of a constrained optimal solution.

The controllers resulting from this procedure are rational and usually of high order. However, we found that the high order is due to spurious modes that can be easily removed without any loss of performance. Thus the effective complexity of the constrained optimal controllers is comparable to that of unconstrained $\mathcal{H}_\infty$ optimal controllers.

We believe that a key advantage offered by our approach is the directness with which time-domain specifications are incorporated in the $\mathcal{H}_\infty$ optimization framework. Thus, such specifications can be transparently traded-off against frequency domain ones and precise limits of performance can be established. This translates to significant savings in effort on the control designer's part, as tedious design iterations due to the approximate formulation of time domain specifications in the frequency domain are circumvented.

References

[1] S. Boyd, V. Balakrishnan, C. Barratt, N. Khraishi, X. Li, D. Meyer and S. Norman, "A New CAD Method and Associated Architectures for Linear Controllers", *IEEE Transactions on Automatic Control*, vol. 33, No.3, March 1988.

[2] J. Doyle, *Lecture Notes in Advances in Multivariable Control*, ONR/Honeywell Workshop, Minneapolis, 1984.

[3] C. Gustafson and C. Desoer (1983). "Controller Design for Linear Multivariable Feedback Systems with Stable Plants, Using Optimization with Inequality Constraints". *International Journal of Control*, vol. 37, no. 5, pp. 881-907.

[4] J. Helton and A. Sideris, "Frequency Response Algorithms for $\mathcal{H}_\infty$ Optimization with Time Domain Constraints", *IEEE Transactions on Automatic Control*, vol. 34, No.4, April 1989.

[5] D. Kavranoglu and A. Sideris, "A Simple Solution to $\mathcal{H}_\infty$ Optimization Problems", *Proceedings of American Control Conference*, Pittsburgh Pennsylvania, 1989.

[6] E. Polak and S. Salcudean, "On the Design of Linear Multivariable Feedback Systems Via Constrained Nondifferentiable Optimization in $\mathcal{H}_\infty$ Spaces", *IEEE Transactions on Automatic Control*, Vol.34, No.3, March 1989.

[7] H. Rotstein and A. Sideris (1992). "Discrete-time $\mathcal{H}_\infty$-Control: The One-Block Problem". *Proceedings of the Conference on Decision and Control*, Tuscon, AZ.

[8] A. Sideris and H. Rotstein (1992). "Single Input-Single Output $\mathcal{H}_\infty$-Control with Time Domain Constraints". to appear in*Automatica*, also in *Proceedings of the Conference on Decision and Control*, Honolulu, HI.

Proceedings of the 31st Conference
on Decision and Control
Tucson, Arizona • December 1992

ROBUST CONTROL OF SYSTEMS WITH INPUT AND STATE CONSTRAINTS VIA LQG-BALANCING

Erik I. Verriest
Georgia Tech Lorraine
Technopole Metz-2000
Metz, France 57070

Grazyna A. Pajunen
Department of Electrical Engineering
Florida Atlantic University
Boca Raton, Florida 33431

Gopinathan Murali
Department of Electrical Engineering
Florida Atlantic University
Boca Raton, Florida 33431

ABSTRACT

A new approach to robust control of systems with input and state constraints is suggested. First, a robustification is obtained by embedding the nominal model in a whole class of extended systems. This embedding is suggested as an "inverse" to the model reduction technique based on balanced truncation (project of dynamics). The LQG-balancing is favored because of its "naturalness" with respect to the given constraints. Moreover, open loop balancing is not appropriate because of its restriction to stable systems. In the final step, a saturating controller is designed for the class of non-nominal systems based on positive invariance [4,5].

1. INTRODUCTION

The problem of robust stabilization and null-controllability of open-loop unstable time multi-input systems with constraints on the states and the inputs is addressed in this paper. An alternative to the framework proposed in [1] is suggested. The focus is on the analytically simpler case of quadratic constraints on the states and inputs. The feasibility of using these results for problems with linear constraints, by using a suitable embedding of the linear constraints in the quadratic constraints, was demonstrated in an earlier work [4].

We start from a nominal unstable model, which is only an approximation of the true but otherwise unknown model. The uncertainty is assumed to step uniquely from unmodeled (but linear) dynamics. We will show that a previously derived sufficient condition guaranteeing for the nominal open-loop system for the existence of a linear controller such that the closed loop is positive invariant, extends to this uncertain case. The main idea is an inversion of the model reduction technique via balanced truncation (project of dynamics) [9]. For stable open-loop systems, these ideas are developed in Section 2.

As however in this problem the nominal open-loop system is unstable, balancing in the usual sense cannot be performed as the gramians are nonconverging integrals. Alternative formulations exist, e.g., using finite interval gramians (the Sliding Interval Balancing in [2]), but their use may be rather adhoc. Another idea is provided by the so-called LQG-balancing, for which the open loop does not need to be stable. By the nature of the problem, quadratic costs need to be introduced on the inputs and states as well as covariance matrices for the plant and measurement noises. As these are not given in the original problem set-up, these are arbitrary design parameters. However, we argue that because of the constraints on states and inputs itself, these four design matrices should be directly related to the weight matrices on the quadratic input and state constraints. LQG-balancing is developed in Section 3. Section 4 shows how to choose the covariances and costs related to the constraints.

In Section 5, the full robust control problem is attached, and the criterion for positive invariance derived. Then the optimal control law is computed and the admissible set of initial conditions is given such that the input and state constraints are satisfied. Finally, in Section 5, the domain of feasible initial conditions is enlarged, if possible, by allowing a saturating control. The saturation is with respect to the given quadratic constraints, hence not component-wise. These ideas are illustrated in the simulation.

2. EXTENSION VIA BALANCED REALIZATION

Given a stable minimal system (A, B, C), it is well known that a state space transformation T exists, so that in the new coordinates the realization (TAT^{-1}, TB, CT^{-1}) is balanced, i.e., its reachability and observability gramians are equal and diagonal, say $\wedge$. The reachability and observability gramians for a system (A, B, C) are respectively

$$R = \int_0^\infty e^{At}BB'e^{A't}\,dt \qquad O = \int_0^\infty e^{At}CC'e^{A't}\,dt \; .$$

Without loss of generality, it will be assumed that the diagonal elements of the canonical gramian $\wedge$ satisfy $\lambda_1 \geq \lambda_2 \geq \ldots \geq \lambda_n$. The model reduction technique based on the so-called *Projection of Dynamics* consists of truncating the A, B, and C matrices of the balanced realization consistent with the dominant part of the canonical gramian [3]. Stochastic as well as deterministic motivations can be given for this methodology. In [2] a *real dimension*, i.e., a real number representing the "effective order" of the system was suggested, based on the relative values of the canonical elements λ_i. For instance, if m elements are equal, and the remaining $n-m$ are of order ε, then the real dimension in the limit for $\varepsilon \to 0$ is m. Based on this notion a system of order n would be truncated to order n_r, where n_r is the integer above the real dimension ρ of the system. Conversely, given a system of order n, it may itself be an approximation, via balanced model reduction, of a system of higher order N, whose real dimension is just below N. For continuous time systems, this means that the given balanced realization (A, B, C) should be extended to a system

$$A_e = \begin{bmatrix} A & A_{12} \\ A_{21} & A_2 \end{bmatrix} \quad B_e = \begin{bmatrix} B \\ B_2 \end{bmatrix} \quad C_e = [C \;\; C_2]$$

The subsystem (A_2, B_2, C_2) is itself balanced with canonical gramian $\wedge_2$, which is chosen to satisfy the real-dimension criterion above, and for simplicity can be taken to be proportional to the identity matrix. It should be noted that the canonical gramian does not uniquely specify the balanced realization [6]. In the SISO case there are 2n free parameters, whereas the canonical gramian specifies only n degrees of freedom. A signature matrix (n signs) and for instance the b-vector complete the specification of the realization. The coupling terms A_{12} and A_{21} of the extended system then follow from the Lyapunov equations. Thus, given the extended canonical gramian, there exists a whole class of systems whose truncation leads to the nominal system. If the extended system were driven by white noise, then the state covariance is exactly (see [2,11]).

$$\begin{vmatrix} \wedge & 0 \\ 0 & \wedge_2 \end{vmatrix}$$

so that in the partitioned dynamical equations

$$\dot{x} = Ax + A_{12}x_2 + Bu$$

and

$$y = Cx + C_2x_2 \; ,$$

the extended terms $A_{12}x_2$ and C_2x_2 act as uncorrelated (with x) perturbations of the nominal system. It is suggested to keep the signal to noise ratios above some threshold:

$$\| C_2 \wedge_2 C_{12}' \| \leq \varepsilon \| C \wedge C' \|$$

and

$$\| A_{12} A_2 A_{12}' \| \leq \varepsilon \| A \wedge A' \| \; .$$

This leads then to a continuum of systems of order N in the neighborhood of the nominal nth order system. By construction each system in this set is a potential model incorporating the "unmodeled dynamics", up to some tolerance level (determined by ε and $\wedge_2$).

CH3229-2/92/0000-2607$1.00 © 1992 IEEE

3. LQG BALANCING

Balancing in its usual form is not possible for unstable systems since the gramian matrices do not converge. To remedy this one can . look at finite balancing [2], although this seems rather ad hoc, and not well motivated for the application at hand. However, instead of balancing the open loop gramians, once can also balance the Riccatians (the solutions to a control and a filtering Riccati equation rather than the Lyapunov equations of the open loop). This is described in [10,7]. Again, nice interpretations leading to the motivation for closed loop model reduction can be given. In this paper, it is the inverse, i.e., the model extension, that is needed. In [8,9] the conditions for feasibility of such a reduction were derived. In particular, it was discovered that the canonical Riccatian (the common solution to both Riccati-equations of the LQG-balanced system) does not provide the complete information. In fact, this is also true in the open-loop balancing, as shown in [12] and by Kabamba [6]. In the inverse problem of model extension these conditions provide bounds on the set of extended systems in a neighborhood about the nominal system in the same way as described for the open loop balancing.

4. EXTENSION VIA LQG-BALANCING

Let us first introduce the problem. Given an unstable nominal system

$$\dot{x}=Ax+Bu$$

It is desired to control this system with a control satisfying

$$u'Qu\leq 1$$

to a domain satisfying the state constraint

$$x'Px\leq 1$$

where P and Q are positive definite. Let v = Su and z = Rx, where R and S are respectively the square roots of P and Q and z can be interpreted as some fictitious output while v is a new input. Both constraints are now normalized.

In order to use the LQG balanced extension of the previous section, quadratic cost weight matrices and noise covariances need to be introduced first for the nominal problem. We suggest, again motivated by the analysis in [8,9], to use exactly the weight matrices of the constraints, i.e., specifying the performance index

$$\int_0^\infty [x'Px+u'Qu]\,dt,$$

assuming an equal importance of satisfying the state and the input constraint. Moreover, this mapping of an LQG-problem seems natural in view of the application to constrained control. As for the noise characteristics, the input uncertainty being dominated by max u'Qu = 1, we can consider the input term as white noise which should then have covariance Q^{-1}/m, where m is the input dimension. Indeed the expected value of the term u'Qu is then 1. If we consider the "output" z, it has covariance I if the constraint is satisfied. A perturbation of the constraint can be described as an equivalent "observation noise", which thus has its covariance proportional to P^{-1}. The control and filtering Riccati equations for the resulting LQG-problem are respectively

$$A'X+XA+C'PC-XBQ^{-1}B'X=0$$

and

$$AY+YA'+BQ^{-1}B'-YC'PCY=0$$

Not only has the saturated control problem been recast into a LQG problem, but the LQG form displays a remarkable symmetry: The only parameters in the problem are here A, $BQ^{-1}B'$ and C'PC. (In fact C = I).

Now the constraint control problem can be robustified by the LQG-balancing technique explained above. The LQG problem is first transformed to balanced form, for which the matrices X = Y = Ω are diagonal and the resulting balanced matrices A, $BQ^{-1}B'$ and C'PC are extended to A_e, $(BQ^{-1}B')_e = B_eQ^{-1}B'_e$ and $(C'PC)_e = C'_eP_eC_e$, from which a set of extended systems (A_e, B_e, C_e) follows.

5. ROBUST LINEAR CONTROL

Now all the bases are laid down to robustify our approach in [4,5]. Take the extended problem (in balanced coordinates) with state vector partitioned as $\chi = [x', x_2']'$

$$\dot{\chi}=A_eX+B_eu$$

but with control restricted to

$$u=K_1x$$

as the unmodeled dynamics is indeed not accessible (the state components of x_2). As explained in [5], the problem simplifies if we take the canonical form for the saturation problem. For the nominal problem, it transformed to a closed loop dynamic matrix N − MD, where D is some normalized gain, and M is diagonal, padded with zeros. A sufficient condition for the existence of a linear controller making the state constraint set positively invariant was easily characterized. (The derivation was actually done for the discrete case, but the ideas are similar for the continuous case where positive invariance for a gain K is guaranteed if the matrix (A − BK)'P + P(A − BK) is negative definite). The restricted form of the control that is necessary here leads to a restriction of the admissible D-matrices in the canonical problem. Robustification lies then in the fact that a gain D is found which makes all possible extensions in the neighborhood of the nominal system positively invariant.

6. ROBUST SATURATED CONTROL

This extends the ideas of [4,5] directly to the robust case by enlarging the domain of feasible initial conditions, by letting the linear control derived in the previous section be quadratically saturated, i.e., one considers the non linear control

$$u=\frac{Kx}{max(1,\|Kx\|_Q)}$$

All details of this method are shown in a technical report [13] and will be submitted for publication.

7. EXAMPLE

Consider the nominal first order model

$$\dot{x}=\frac{3}{4}x+u$$

with the constraints |x| ≤ 1 and |u| ≤ 1.

Step 1. Obtain the LQG Model

The system remain the same except for the additive noise. Its performance index is set to

$$\int_0^\infty (x^2+u^2)\,dt$$

and the covariances Σ_w and Σ_v, respectively, of the plant noise w and the observation noise v, are both 1. Note that in the nominal model y = x The solution of the control and the filter ARE's are respectively x = 2 and y = 2.

Step 2: LQG Balancing

Since x = y = 2 = Ω, the nominal model is already balanced.

Step 3: Balanced Extension

All the second order models into which the nominal system can be embedded up to some tolerance (as determined by the canonical Riccatian) are obtained, For a fixed tolerance let the canonical Riccatian thus be

$$\Omega_e = \begin{bmatrix} 2 & 0 \\ 0 & \varepsilon \end{bmatrix}$$

Using the constraints of balanced parametrization the family of extended systems is given by

$$A_e = \begin{bmatrix} \frac{3}{4} & \frac{2\varepsilon^2 c - \varepsilon(\gamma + 4b) + 2p}{\varepsilon^2 - 4} \\ \frac{2\varepsilon^2 b - \varepsilon(p + 4c) + 2\gamma}{\varepsilon^2 - 4} & \frac{\varepsilon^2(b^2 + c^2) - (p^2 + \gamma^2 + n^2 + \delta^2)}{4\varepsilon} \end{bmatrix}$$

$$B_e = \begin{bmatrix} 1 \\ b \end{bmatrix} \qquad C_e = [1\ ,\ c]$$

$$P_e = \begin{bmatrix} 1 & p \\ p & p^2 + n^2 \end{bmatrix} \quad (\Sigma_w)_e = \begin{bmatrix} 1 & \gamma \\ \gamma & \gamma^2 + \delta^2 \end{bmatrix}$$

Q_e and $(\Sigma_v)_e$ of course remain respectively Q and Σ_v as the number of inputs and outputs remain unchanged.

The "free" parameters are (recall that we keep ε fixed), b, c, γ p and one of n or δ_2 since the latter are bound by the equation

$$p^2 + n^2 + \varepsilon^2 c^2 = \gamma^2 + \delta^2 + \varepsilon^2 b^2$$

These parameters are not entirely free, i.e., arbitrary on $\mathbb{R}^5$ since the extended model must be such that its reduction to a first order model using the LQG balanced truncation is feasible. These conditions were derived in [9, equations (28) and (29)] which in our present notation are respectively.

$$\Omega_1 \left[(\Sigma_{w,e})_{11} + \Omega_1^2 (B_e' Q^{-1} B_e)_{11} \right] >> \Omega_2 \left[(\Sigma_{w,e})_{22} + \Omega_2^2 (B_e' Q^{-1} B_e')\right]$$

$$\Omega_1 \left[(P_e)_{11} + \Omega_1^2 (c_e \Sigma_v^{-1} c_e')_{11} >> \Omega_2 \right] \left[(P_e)_{22} + \Omega_2^2 (c_e \Sigma_v^{-1} c_e')_{22} \right]$$

In the present example, that is

$$10 >> \varepsilon(\gamma^2 + \delta^2 + \varepsilon^2 b^2)$$

$$10 >> \varepsilon(P^2 + n^2 + \varepsilon^2 c^2)$$

Step 4: Mapping to an Extended Constraint Control Problem

Since, also the number of inputs is the same as for the nominal system (here 1), the symmetry explained in Section 4 has been lost since P_e and $(\Sigma_w)_e$ have full rank in general.

If, however, we impose the symmetry constraints

$$p^2 = \gamma^2$$

$$b^2 = c^2$$

Then we have approximately (noting that $\varepsilon << 2$) as dictated

$$A_e = \begin{bmatrix} \frac{3}{4} & \varepsilon b - \frac{p}{2} \\ \varepsilon b \sigma_b - \sigma_p \frac{p}{2} & \frac{\varepsilon^2 b^2 - (p^2 + n^2)}{2\varepsilon} \end{bmatrix} \quad B_e = \begin{bmatrix} 1 \\ b \end{bmatrix}$$

$C_e = [1 \quad \sigma_b b\ ]$ and $n^2 = \delta^2$, with σ_b and $\sigma_p \in \{-1, 1\}$

The extended system constraints are

$$X_e' \begin{bmatrix} 1 & p \\ p & p^2 + n^2 \end{bmatrix} X_e \leq 1$$

$$|u| \leq 1$$

As discussed, the feasibility conditions require

$$10 >> \varepsilon p^2 + \varepsilon^3 b^2$$

Letting this be "$\varepsilon/10$" times smaller, then we find the parametrization of all systems in the extended family by

$$(p, b) \varepsilon M \subseteq \mathbb{R}^2$$

$$\sigma_b, \sigma_p \in \{-1,\ +1\}$$

where M is the interior of the ellipse

$$p^2 + \varepsilon^2 b^2 = 1$$

Step 5: Constraint Control Design

Note that in the present set up x_2 pertains to the unmodeled dynamics. Hence, using state estimator will not work since the exact dynamics are unknown.

Hence the control should only work with the observed part, i.e.,

$$u = Kx$$

where K is such that for all θ in the parameter set M, (the ellipse in the example) we have that

$$\left[A_e(\theta) + B_e(\theta) K C_e(\theta) \right]' p_e(\theta) + p_e(\theta) \left[A_e(\theta) + B_e(\theta) K C_e(\theta) \right]$$

is negative definite. If such K (independent of θ!!) can be found, it means the linear controller makes the sets $\{ x \mid x'p(\theta)\, x = \alpha \}$ positive invariant for all systems in the neighborhood of the nominal one.

Then check $u'Qu = K^2 x^2$ which must be smaller than 1. This implies $|x| < 1/|k|$ restricts further the feasible domain. So, now enlarge this by taking

$$u = \begin{Bmatrix} 1 & \text{if } |Kx| > 1 \\ Kx & \text{if } |Kx| \leq 1 \end{Bmatrix}$$

in this example.

REFERENCES

[1] M. Sznaier and A. Sideris, "Suboptimal Norm Based Robust Control of Constrained Systems with an H_∞ Cost", Proceedings of 30th IEEE Conference on Decision and Control, 1991, pp. 2280-2286.

[2] E. I. Veriest and T. Kailath, "On Generalized Balanced Realizations", IEEE Transactions Automatic Control, 1983, pp. 833-844.

[3] J. M. Maciejowski, Multivariable Feedback Design, Addison Wesley, 1989.

[4] E. I. Verriest and G. A. Pajunen, "Saturating Control of Unstable Open Loop Systems", Proceedings of the 30th IEEE Conference on Decision and Control, 1991, pp. 2872-2877.

[5] G. A. Pajunen and E. I. Verriest, "Canonical Form of Control Systems with Saturated Actuator", Proceedings of the 209th Annual Allerton Conference on Communication, Control and Computing, 1991, pp. 420-429.

[6] P. T. Kabamba, "Balanced Forms: Conanicity and Parametrization", IEEE Transactions Automatic Control, AC-30, no. 11, pp. 1106-1109, November 1985.

[7] E. A. Jonckheere and L. M. Silverman, "A New Set of Invariants for Linear Systems: Applications to Reduced Order Compensator Design", IEEE Transactions Automatic Control,, AC-28, pp. 953-964, 1983.

[8] E. I. Verriest, "Model Reduction via Balancing and Connections with Other Methods", in Modeling and Applications of Stochastic Processes, U. B. Desai (ed.), Kluwer Academic Publishers, pp. 123-154, 1986.

[9] E. I. Verriest, "Reduced Order LQG Design, Conditions for Feasibility", Proceedings of the 1986 IEEE Conference on Decision and Control, 1986, pp. 1765-1769.

[10] E. I. Verriest, "Low Sensitivity Design and Optimal Order Reduction for the LQG-Problem", Proceedings of the 1981 Midwest Symposium on Stochastic Systems, 1981, pp. 365-369.

[11] E. I. Verriest, "Uncertainty Equivalent Reduced Order Models for Discrete Systems", Proceedings of the 1985 IEEE Conference on Decision and Control, 1985, pp. 1238-1239.

[12] E. I. Verriest, "The Structure of Multivariable Balanced Realizations", Proceedings of the 1983 IEEE International Symposium on Circuits and Systems, 1983, pp. 110-113.

[13] E. I. Verriest, G. A. Pajunen and M. Gopinathan, "Robust Control of Systems with Input and State Constraints via LQG-Balancing", Internal Report, Georgia Institute of Technology, School of Electrical Engineering, EIV-9-92.

Proceedings of the 31st Conference
on Decision and Control
Tucson, Arizona • December 1992

New Absolute Stability Criteria for Robust Stability and Performance With Locally Slope-Restricted Monotonic Nonlinearities

Wassim M. Haddad
Department of Mechanical and
Aerospace Engineering
Florida Institute of Technology
Melbourne, FL 32901

Dennis S. Bernstein
Department of Aerospace Engineering
The University of Michigan
Ann Arbor, MI 48109-2140

1. Introduction

For a robustness result to be useful for either analysis or synthesis it must be both flexible in addressing a large class of uncertainty structures and restrictive in excluding uncertainties that are not physically meaningful [BH1]. For example, robustness tests based upon small gain theory are often conservative since they admit nonlinear time-varying uncertainty [HB2]. Such tests admit time-varying uncertainty due to the fact that they are based upon a single fixed Lyapunov function. One approach that excludes time variation from the allowable uncertainty is based upon Lyapunov functions that depend upon the uncertainty. For example, the classical Popov criterion [P] is based upon the nonlinearity-dependent Lur'e-Postnikov Lyapunov function

$$V(x,\phi) = x^{\mathrm{T}}Px + \int_0^y \phi(\sigma)\mathrm{d}\sigma, \tag{1.1}$$

where $y = Cx$ and $\phi(\cdot)$ is a time-invariant, sector-bounded memoryless nonlinearity. For the case of linear uncertainty $\phi(y) = Fy$, the Lyapunov function (1.1) becomes

$$V(x,F) = x^{\mathrm{T}}Px + Fx^{\mathrm{T}}C^{\mathrm{T}}Cx, \tag{1.2}$$

which, as was pointed out in [HB3], is a *parameter-dependent Lyapunov function.*

Although a nonlinearity-dependent Lyapunov function as in (1.1) restricts the time variation of the nonlinearity, there remains an enormous gap between an arbitrary uncertain nonlinearity $\phi(y)$ and a linear uncertainty $\phi(y) = Fy$. If the uncertainly is known to be linear, it is thus desirable to modify the Lyapunov function to *exclude* uncertain nonlinearities in much the same way that the Lur'e-Postnikov Lyapunov function exludes time-varying uncertainties.

Our goal is thus to exclude uncertain nonlinearities in order to reduce conservatism with respect to constant linear parametric uncertainty. In the classical absolute stability literature a variety of results are available for accomplishing this task. For example, refinements of the Popov criterion for monotonic and odd monotonic nonlinearities are developed in [BW], [NC], [NN1], [NN2], [NT1], [NT2], [TS], [TSR], [ZF]. These results are extensively discussed in the important book [NT2]. Unfortunately, after the publication of [NT2], activity along this line of research apparently ceased. The purpose of the present paper is to present *new* results in the spirit of [NT2].

The novel feature of the present paper is to develop new robust stability conditions based upon the Lyapunov function

$$V(x,\phi) = x^{\mathrm{T}}Px + \phi^{\mathrm{T}}(y)\phi(y), \tag{1.3}$$

where $y = Cx$. This Lyapunov function is clearly positive definite, excludes time-variation of $\phi(\cdot)$ just as the Lur'e-Postnikov Lyapunov function does, and, unlike the Lur'e-Postnikov Lyapunov function, requires that $\phi(\cdot)$ be differentiable. However, the novel feature of (3) is that the expression for $\dot{V}(x)$ involves the *local* slope $\phi'(y)$. Thus, it turns out that satisfying the condition $\dot{V}(x) \leq 0$ requires a restriction on the sign and magnitude of $\phi'(y)$. These restrictions therefore exclude large classes of nonlinearities from consideration. Table 1 shows several Lyapunov functions along with their admissible nonlinearities. For the case of linear uncertainty $\phi(y) = Fy$, (1.3) yields a new parameter-dependent Lyapunov function of the form

$$V(x,F) = x^{\mathrm{T}}Px + x^{\mathrm{T}}C^{\mathrm{T}}F^{\mathrm{T}}FCx. \tag{1.4}$$

Since the Lyapunov function (1.3) or, equivalently, (1.4) for linear uncertainty, provides the framework for addressing *locally* slope-restricted monotonic nonlinearities, it potentially provides a less conservative test for constant real parameter uncertainty. Specifically, in this case the nonlinear set is a much better approximation to the linear uncertainty set, and thus provides a framework for reducing conservatism, as numerically demonstrated in [SW].

The contents and scope of this paper are as follows. In Section 2 we establish definitions and mathematical preliminaries. In Section 3 we present absolute stability criteria for locally slope-restricted monotonic nonlinearities. Unlike multivariable extensions of the Popov criterion [MA1], [MA2] and absolute stability criteria for monotonic nonlinearities [NN2], [HHHB], which only consider decoupled nonlinearities, we allow a fully coupled nonlinearity structure. Furthermore, our results are fundamentally different from earlier results on absolute stability criteria for monotonic and odd monotonic nonlinearities [NC], [NN1], [TSR], [NT2], [ZF] which involve suitable stability multipliers realizable by driving-point impedances of passive electrical networks which exhibit interlacing pole-zero patterns on the negative real axis. Specifically, our stability multipliers involve the parahermitian conjugate of the linear part of the system. In addition, unlike the results of [NC], [NN1], [TSR] we do not require pole-zero cancellation between the stability multiplier and the plant transfer function which is often quite restrictive+. In Section 4 we combine (1.1) and (1.3) to construct a new Lur'e-Postnikov Lyapunov function for an extended Popov criterion with locally slope-restricted monotonic nonlinearities. In Sections 5-7 we consider the robust stability and performance analysis problems for real parametric uncertainty. Specifically, using a parameter-dependent Lyapunov function framework for linear uncertainty we make explicit connections between absolute stability theory and robust stability and H_2 performance.

$\mathbb{R}, \mathbb{R}^{r\times s}, \mathbb{R}^r$	real numbers, $r \times s$ real matrices, $\mathbb{R}^{r\times 1}$
$\mathbb{C}, \mathbb{C}^{r\times s}, \mathbb{C}^r$	complex numbers, $r \times s$ complex matrices, $\mathbb{C}^{r\times 1}$
$\mathbb{E}, \mathrm{tr}, 0_{r\times s}$	expectation, trace, $r \times s$ zero matrix
$\bar{\lambda}$	complex conjugate of $\lambda \in \mathbb{C}$
$I_r, (\ )^{\mathrm{T}}, (\ )^*$	$r \times r$ identity, transpose, complex conjugate transpose
$(\ )^{-\mathrm{T}}, (\ )^{-*}$	inverse transpose, inverse complex conjugate transpose
$\rho(\ ), \sigma_{\max}(\ )$	spectral radius, largest singular value
$\mathbb{S}^r, \mathbb{N}^r, \mathbb{P}^r$	$r \times r$ symmetric, nonnegative-definite, positive-definite matrices
$Z_1 \leq Z_2, Z_1 < Z_2$	$Z_2 - Z_1 \in \mathbb{N}^r, Z_2 - Z_1 \in \mathbb{P}^r, Z_1, Z_2 \in \mathbb{S}^r$
$\lVert Z\rVert_{\mathrm{F}}$	$[\mathrm{tr}\ ZZ^*]^{1/2}$ (Frobenius matrix norm)
$\lVert G(s)\rVert_2$	$[(1/2\pi)\int_{-\infty}^{\infty} \lVert G(j\omega)\rVert_{\mathrm{F}}^2 \mathrm{d}\omega]^{1/2}$
$\lVert G(s)\rVert_\infty$	$\sup_{\omega\in\mathbb{R}} \sigma_{\max}[G(j\omega)]$

+ An alternative approach that does not preclude the use of pole-zero cancellation but, however, allows for greater flexibility in the stability multiplier is the augmentation technique discussed in [NT2] and [HHHB]. This approach, however, results in increased system dimensionality.

This research was supported in part by the National Science Foundation under Research Initiation Grant ECS-9109558 and the Air Force Office of Scientific Research under Grant F49620-92-J-0127.

CH3229-2/92/0000-2611$1.00 © 1992 IEEE

2. Mathematical Preliminaries

In this section we establish definitions and several key lemmas. In this paper a *real-rational matrix function* is a matrix whose elements are rational functions with real coefficients. Furthermore, a *transfer function* is a real-rational matrix function each of whose elements is *proper*, i.e., finite at $s = \infty$. A *strictly proper transfer function* is a transfer function that is zero at infinity. Finally, an *asymptotically stable transfer function* is a transfer function each of whose poles is in the open left half plane. The space of asymptotically stable transfer functions is denoted by RH_∞, i.e., the real-rational subset of H_∞. Let

$$G(s) \sim \left[\begin{array}{c|c} A & B \\ \hline C & D \end{array}\right]$$

denote a state space realization of a transfer function $G(s)$, that is, $G(s) = C(sI - A)^{-1}B + D$. The notation "$\overset{\min}{\sim}$" is used to denote a minimal realization. In addition, the parahermitian conjugate $G^{\sim}(s)$ of $G(s)$ has the realization

$$G^{\sim}(s) \sim \left[\begin{array}{c|c} -A^{\mathrm{T}} & C^{\mathrm{T}} \\ \hline -B^{\mathrm{T}} & D^{\mathrm{T}} \end{array}\right].$$

A square transfer function $G(s)$ is called *positive real* ([AV], p. 216) if 1) all poles of $G(s)$ are in the closed left half plane, and 2) $G(s)+G^*(s)$ is nonnegative definite for all $\mathrm{Re}[s] > 0$. A square transfer function $G(s)$ is called *strictly positive real* [W], [LJ] if 1) $G(s)$ is asymptotically stable, and 2) $G(j\omega) + G^*(j\omega)$ is positive definite for all real ω. Finally, a square transfer function $G(s)$ is *strongly positive real* if it is strictly positive real and $D + D^{\mathrm{T}} > 0$, where $D \triangleq G(\infty)$. Recall that a minimal realization of a positive real transfer function is stable in the sense of Lyapunov, while a minimal realization of a strictly positive real transfer function is asymptotically stable.

For notational convenience in the paper, $G(s)$ will denote an $\ell \times m$ transfer function with input $u \in \mathbb{R}^m$, output $y \in \mathbb{R}^\ell$, and internal state $x \in \mathbb{R}^n$. Next we state the well-known positive real lemma used to characterize positive realness in the state-space setting [A].

Lemma 2.1 (Positive Real Lemma). $G(s) \overset{\min}{\sim} \left[\begin{array}{c|c} A & B \\ \hline C & D \end{array}\right]$ is positive real if and only if there exist matrices P, L, and W with P positive-definite such that

$$0 = A^{\mathrm{T}}P + PA + L^{\mathrm{T}}L, \tag{2.1}$$
$$0 = PB - C^{\mathrm{T}} + L^{\mathrm{T}}W, \tag{2.2}$$
$$0 = D + D^{\mathrm{T}} - W^{\mathrm{T}}W. \tag{2.3}$$

Lemma 2.2 (Strict Positive Real Lemma). Let $G(s) \overset{\min}{\sim} \left[\begin{array}{c|c} A & B \\ \hline C & D \end{array}\right]$. Then the following statements are equivalent:

i) $G(s)$ is strictly positive real;

ii) there exist matrices P, L, and W with P positive definite such that (2.1)-(2.2) are satisfied, the pair (A, L) is observable, and rank $\hat{G}(j\omega) = m$, $\omega \in \mathbb{R}$, where $\hat{G}(s) \overset{\min}{\sim} \left[\begin{array}{c|c} A & B \\ \hline L & W \end{array}\right]$.

Proof. See [V, p. 474]. □

Next, suppose that $D + D^{\mathrm{T}} > 0$. Then, since $W^{\mathrm{T}}W = D + D^{\mathrm{T}}$, it follows that $W^{\mathrm{T}}W$ is nonsingular and (2.2) implies

$$W^{\mathrm{T}}L = C - B^{\mathrm{T}}P. \tag{2.4}$$

Using (2.4) and noting that $W(W^{\mathrm{T}}W)^{-1}W^{\mathrm{T}}$ is an orthogonal projection so that $I \geq W(W^{\mathrm{T}}W)^{-1}W^{\mathrm{T}}$ and hence $L^{\mathrm{T}}L \geq L^{\mathrm{T}}W(W^{\mathrm{T}}W)^{-1}W^{\mathrm{T}}L$ it follows from (2.1) that

$$0 \geq A^{\mathrm{T}}P + PA + (C - B^{\mathrm{T}}P)^{\mathrm{T}}(W^{\mathrm{T}}W)^{-1}(C - B^{\mathrm{T}}P)$$

or, since $(W^{\mathrm{T}}W)^{-1} = (D + D^{\mathrm{T}})^{-1}$,

$$0 \geq A^{\mathrm{T}}P + PA + (C - B^{\mathrm{T}}P)^{\mathrm{T}}(D + D^{\mathrm{T}})^{-1}(C - B^{\mathrm{T}}P).$$

Lemma 2.3. Let $G(s) \overset{\min}{\sim} \left[\begin{array}{c|c} A & B \\ \hline C & D \end{array}\right]$. Then the following statements are equivalent:

i) A is asymptotically stable and $G(s)$ is strongly positive real;

ii) $D + D^{\mathrm{T}} > 0$ and there exist positive-definite matrices P and R such that

$$0 = A^{\mathrm{T}}P + PA + (C - B^{\mathrm{T}}P)^{\mathrm{T}}(D + D^{\mathrm{T}})^{-1}(C - B^{\mathrm{T}}P) + R. \tag{2.5}$$

Lemma 2.4. Let $M, N \in \mathbb{C}^{m\times m}$ be such that $M + M^* \geq 0$ and $N + N^* > 0$. Then $\det(I_m + MN) \neq 0$.

3. Absolute Stability for Locally Slope-Restricted Monotonic Nonlinearities

In this section we introduce the absolute stability problem. Specifically, given $G(s) \overset{\min}{\sim} \left[\begin{array}{c|c} A & B \\ \hline C & D \end{array}\right]$, the problem is to derive conditions involving *only* the transfer function $G(s)$ such that for a certain class of nonlinearities ϕ: $\mathbb{R}^m \to \mathbb{R}^m$ the negative feedback interconnection of $G(s)$ and ϕ is globally asymptotically stable. The class of nonlinearities Φ we consider involves multivariable sector-bounded locally slope-restricted time-invariant memoryless nonlinearities. Let $M \in \mathbb{R}^{m\times m}$ be a given positive-definite matrix, $N, \mu \in \mathbb{R}^{m\times m}$ be given nonnegative-definite matrices, and define

$$\Phi \triangleq \{\phi\colon \mathbb{R}^m \to \mathbb{R}^m\colon \ \phi^{\mathrm{T}}(y)[M^{-1}\phi(y) - y] \leq 0, \quad \phi(\cdot) \text{ is differentiable,}$$
$$N\phi'(y) = \phi'^{\mathrm{T}}(y)N, \text{and } 0 \leq N\phi'(y) \leq \mu, y \in \mathbb{R}^m\},$$

where $\phi'(y) \in \mathbb{R}^{m\times m}$ denotes the Jacobian of ϕ. Note that in the special case in which $N = \mathrm{diag}(N_1, \ldots, N_m)$, $M = \mathrm{diag}(M_1, M_2, \ldots, M_m)$, and $\phi(y) = [\hat{\phi}_1(y_1)\ \hat{\phi}_2(y_2)\ \cdots\ \hat{\phi}_m(y_m)]^{\mathrm{T}}$ is decoupled, it follows that each component $\hat{\phi}_i(y_i)$ of ϕ satisfies

$$0 \leq \hat{\phi}_i(y_i)y_i \leq M_i y_i^2, \quad 0 \leq N_i\hat{\phi}_i'(y_i) \leq \mu_i, \quad y_i \in \mathbb{R}, \quad i = 1, \ldots, m.$$

Unlike the multivariable extensions of the Popov criterion [D], [G], [MA1], [MA2] which assume decoupled nonlinearities, the set Φ allows fully coupled nonlinearities.

Remark 3.1. Note that in the case in which $\phi'(y)$ is symmetric the condition $N\phi'(y) = \phi'^{\mathrm{T}}(y)N$, for all $\phi \in \Phi$ can always be satisfied by choosing $N = I_m$. Alternatively, if $\phi(y) = [\hat{\phi}_1(y_1)\ \hat{\phi}_2(y_2)\cdots\hat{\phi}_m(y_m)]^{\mathrm{T}}$ then N can be an arbitrary nonnegative-definite diagonal matrix.

Theorem 3.1. Let $G(s) \overset{\min}{\sim} \left[\begin{array}{c|c} A & B \\ \hline C & 0 \end{array}\right]$. If $M^{-1} - \frac{1}{2}\mu + G(s) - \frac{1}{2}[sG(s)]^{\sim}\mu[sG(s)]$ is strictly positive real, then there exist matrices P, L, and W with P positive definite satisfying

$$0 = A^{\mathrm{T}}P + PA + A^{\mathrm{T}}C^{\mathrm{T}}\mu CA + L^{\mathrm{T}}L, \tag{3.1}$$
$$0 = B^{\mathrm{T}}P - C + B^{\mathrm{T}}C^{\mathrm{T}}\mu CA + W^{\mathrm{T}}L, \tag{3.2}$$
$$0 = 2M^{-1} - B^{\mathrm{T}}C^{\mathrm{T}}\mu CB - \mu - W^{\mathrm{T}}W. \tag{3.3}$$

Furthermore, if there exist matrices P, L, and W with P positive definite satisfying (3.1)-(3.3), then, for all $\phi \in \Phi$, the negative feedback interconnection of $G(s)$ and ϕ is globally asymptotically stable with a Lyapunov function given by

$$V(x) = x^{\mathrm{T}}Px + \phi^{\mathrm{T}}(y)N\phi(y). \tag{3.4}$$

Remark 3.2. If $M^{-1} - \frac{1}{2}\mu + G(s) - \frac{1}{2}[sG(s)]^{\sim}\mu[sG(s)]$ is strongly positive real, then it follows from Lemma 2.3 that $W^{\mathrm{T}}W > 0$, and hence (3.1)-(3.3) are equivalent to the single Riccati equation

$$0 = A^{\mathrm{T}}P + PA + A^{\mathrm{T}}C^{\mathrm{T}}\mu CA + (C - B^{\mathrm{T}}C^{\mathrm{T}}\mu CA - B^{\mathrm{T}}P)^{\mathrm{T}}$$
$$\cdot [2M^{-1} - B^{\mathrm{T}}C^{\mathrm{T}}\mu CB - \mu]^{-1}(C - B^{\mathrm{T}}C^{\mathrm{T}}\mu CA - B^{\mathrm{T}}P) + R \tag{3.12}$$

where $R > 0$. This equation will be used for robust controller synthesis in Section 7.

Remark 3.3. In the SISO case the frequency domain test given by Theorem 3.1 has a geometric interpretation involving a circle test with a frequency dependent radius and center in the Nyquist plane. To see this, let $G = x + jy$ and note that Theorem 3.1 requires $2M^{-1} - \mu + 2\mathrm{Re}[G(j\omega)] - \omega^2\mu|G(j\omega)|^2 > 0$, $\omega \in \mathbb{R}$. For $\omega^2\mu \neq 0$, this condition can be written as

$$x^2 + y^2 < \frac{2}{\omega^2\mu}x + \frac{2}{\omega^2\mu M} - \frac{1}{\omega^2} \tag{3.13}$$

or, equivalently,

$$\left[x - \frac{1}{\omega^2\mu}\right]^2 + y^2 < \frac{1}{\omega^2}\left[\frac{2 - \mu M}{\mu M} + \frac{1}{\omega^2\mu^2}\right], \tag{3.14}$$

which corresponds to a circle with radius $\sqrt{\frac{1}{\omega^2}(\frac{1}{\omega^2}\mu + \frac{2-\mu M}{\mu M})}$ and center $(\frac{1}{\omega^2}\mu, 0)$ for each $\omega \in \mathbb{R}$.

4. Lur'e-Postnikov Lyapunov Functions for an Extended Popov Criterion With Slope-Restricted Monotonic Nonlinearities

In this section we use a modified Lur'e-Postnikov Lyapunov function to generalize the multivariable Popov criterion. As in Section 3 we consider sector-bounded locally slope-restricted time-invariant nonlinearities with the additional restriction that the nonlinearities be decoupled. Let $M \in \mathbb{R}^{m\times m}$ be a given positive-definite matrix, $N_i, \mu_i, i = 1, \ldots, m$, be nonnegative scalars, and define

$$\Phi_{\mathrm{d}} \triangleq \Big\{\phi\colon \mathbb{R}^m \to \mathbb{R}^m\colon \phi^{\mathrm{T}}(y)[M^{-1}\phi(y) - y] \le 0, \phi(y) = [\phi_1(y_1)\ \phi_2(y_2) \cdots \phi$$

$$\phi(\cdot) \text{ is differentiable, and } 0 \le N_i\phi_i'(y_i) \le \mu_i, \quad i = 1, \ldots, m, y \in \mathbb{R}'$$

In the special case $M = \operatorname{diag}(M_1, M_2, \ldots, M_m), M_i > 0, i = 1, \ldots, m$, it follows that each component $\phi_i(y_i)$ of ϕ satisfies

$$0 \le \phi_i(y_i)y_i \le M_i y_i^2, \quad 0 \le N_i\phi_i'(y_i) \le \mu_i, \quad y_i \in \mathbb{R}, \quad i = 1, \ldots, m.$$

In this case $\phi'(y) = \operatorname{diag}(\phi_1'(y_1), \phi_2'(y_2), \ldots, \phi_m'(y_m))$ and is nonnegative definite. For convenience in stating the next result we define $\mu \triangleq \operatorname{diag}(\mu_1, \mu_2, \ldots, \mu_m)$ and $N \triangleq \operatorname{diag}(N_1, N_2, \ldots, N_m)$.

Theorem 4.1. Let $G(s) \overset{\min}{\sim} \left[\begin{array}{c|c} A & B \\ \hline C & 0 \end{array}\right]$. If there exists a nonnegative-definite diagonal matrix $\hat{N}$ such that $M^{-1} - \frac{1}{2}\mu + (I + \hat{N}s)G(s) - \frac{1}{2}[sG(s)]^{\sim}\mu[sG(s)]$ is strictly positive real, then there exist matrices P, L, and W with P positive definite satisfying

$$0 = A^{\mathrm{T}}P + PA + A^{\mathrm{T}}C^{\mathrm{T}}\mu CA + L^{\mathrm{T}}L, \tag{4.1}$$
$$0 = B^{\mathrm{T}}P - (C + \hat{N}CA - B^{\mathrm{T}}C^{\mathrm{T}}\mu CA)^{\mathrm{T}} + W^{\mathrm{T}}L, \tag{4.2}$$
$$0 = (M^{-1} + \hat{N}CB) + (M^{-1} + \hat{N}CB)^{\mathrm{T}} - B^{\mathrm{T}}C^{\mathrm{T}}\mu CB - \mu - W^{\mathrm{T}} \tag{4.3}$$

Furthermore, if there exist matrices P, L, and W with P positive definite satisfying (4.1)-(4.3), then, for all $\phi \in \Phi_{\mathrm{d}}$, the negative feedback interconnection of $G(s)$ and ϕ is globally asymptotically stable with a Lyapunov function given by

$$V(x) = x^{\mathrm{T}}Px + 2\sum_{i=1}^{m}\int_0^{y_i}\phi_i(\sigma)\hat{N}_i \mathrm{d}\sigma + \phi^{\mathrm{T}}(y)N\phi(y). \tag{4.4}$$

Remark 4.1. Setting $\mu = 0$ in Theorem 4.1 yields the condition that $M^{-1} + (I + \hat{N}s)G(s)$ is strictly positive real, which is the multivariable version of the Popov criterion [HB2], [HB3], [MA1], [MA2].

Remark 4.2. If $M^{-1} - \frac{1}{2}\mu + (I + \hat{N}s)G(s) - \frac{1}{2}[sG(s)]^{\sim}\mu[sG(s)]$ is strongly positive real, then it follows from Lemma 2.3 that $W^{\mathrm{T}}W > 0$, and (4.1)-(4.3) are equivalent to the single Riccati equation

$$\begin{aligned} 0 = A^{\mathrm{T}}P + PA + A^{\mathrm{T}}C^{\mathrm{T}}\mu CA + (C + \hat{N}CA - B^{\mathrm{T}}C^{\mathrm{T}}\mu CA - B^{\mathrm{T}}P)^{\mathrm{T}} \\ \cdot[(M^{-1} + \hat{N}CB) + (M^{-1} + \hat{N}CB)^{\mathrm{T}} - B^{\mathrm{T}}C^{\mathrm{T}}\mu CB - \mu]^{-1} \\ \cdot (C + \hat{N}CA - B^{\mathrm{T}}C^{\mathrm{T}}\mu CA - B^{\mathrm{T}}P) + R, \end{aligned} \tag{4.11}$$

where $R > 0$.

Remark 4.3. For the SISO case, the frequency domain test of Theorem 4.1 has a geometric interpretation involving a circle test with a frequency dependent radius in the Nyquist plane reminiscent to the classical off-axis circle criterion [CN]. To see this, let $G = x + jy$ and note that Theorem 4.1 requires $2M^{-1} - \mu + 2\mathrm{Re}[1 + j\omega\hat{N}]G(j\omega) - \omega^2\mu|G(j\omega)|^2 > 0$, $\omega \in \mathbb{R}$. It now follows that the above condition for $\omega^2\mu \ne 0$ implies

$$x^2 + y^2 - \frac{2}{\omega^2\mu}x + \frac{2\hat{N}}{\omega\mu}y < \frac{2}{\omega^2\mu M} - \frac{1}{\omega^2} \tag{4.12}$$

or, equivalently,

$$\left[x - \frac{1}{\omega^2\mu}\right]^2 + \left[y + \frac{\hat{N}}{\omega\mu}\right]^2 < \frac{1}{\omega^2}\left[\frac{1 + \omega^2\hat{N}^2}{\omega^2\mu^2} + \frac{2 - \mu M}{\mu M}\right], \tag{4.13}$$

which corresponds to a circle with radius $\sqrt{\frac{1}{\omega^2}(\frac{1+\omega^2\hat{N}^2}{\omega^2\mu^2} + \frac{2-\mu M}{\mu M})}$ and center $(\frac{1}{\omega^2\mu}, \frac{-\hat{N}}{\omega\mu})$ for each $\omega \in \mathbb{R}$.

5. Robust Stability and Performance Analysis for Real Parametric Uncertainty

As discussed in the Introduction, in order to address the constant real parameter uncertainty problem it is crucial to restrict the allowable time variations of the uncertainty. In the next two sections we develop a unified framework for robust stability and performance that specifically captures this effect and draws connections to the new absolute stability criteria for locally slope restricted monotonic nonlinearities developed in the earlier sections. First, however, we present the robust stability and performance problems.

Let $\mathcal{U} \subset \mathbb{R}^{n\times n}$ denote a set of perturbations ΔA of a given nominal dynamics matrix $A \in \mathbb{R}^{n\times n}$. Within the context of robustness analysis, it is assumed that A is asymptotically stable and $0 \in \mathcal{U}$. We begin by considering the question of whether or not $A + \Delta A$ is asymptotically stable for all $\Delta A \in \mathcal{U}$.

Robust Stability Problem. Determine whether the linear system

$$\dot{x}(t) = (A + \Delta A)x(t), \quad t \in [0, \infty), \tag{5.1}$$

is asymptotically stable for all $\Delta A \in \mathcal{U}$.

To consider the problem of robust performance, we introduce an external disturbance model involving white noise signals as in standard LQG (H_2) theory. The robust performance problem concerns the worst-case H_2 norm, that is, the worst-case (over $\mathcal{U}$) of the expected value of a quadratic form involving outputs $z(t) = Ex(t)$, where $E \in \mathbb{R}^{q\times n}$, when the system is subjected to a standard white noise disturbance $w(t) \in \mathbb{R}^d$ with weighting $D \in \mathbb{R}^{n\times d}$.

Robust Performance Problem. For the disturbed linear system

$$\dot{x}(t) = (A + \Delta A)x(t) + Dw(t), \quad t \in [0, \infty), \tag{5.2}$$
$$z(t) = Ex(t), \tag{5.3}$$

where $w(\cdot)$ is a zero-mean d-dimensional white noise signal with intensity I_d, determine a performance bound β satisfying

$$J(\mathcal{U}) \triangleq \sup_{\Delta A\in\mathcal{U}} \limsup_{t\to\infty} \mathbb{E}\{\|z(t)\|_2^2\} \le \beta. \tag{5.4}$$

As shown in Section 8, (5.2) and (5.3) may denote a control system in closed-loop configuration subjected to external white noise disturbances and for which $z(t)$ denotes the state and control regulation error.

Of course, since D and E may be rank deficient, there may be cases in which a finite performance bound β satisfying (5.4) exists, whereas (5.1) is not asymptotically stable over $\mathcal{U}$. In practice, however, robust performance is mainly of interest when (5.1) is robustly stable. In this case the performance $J(\mathcal{U})$ involves the steady-state second moment of the state. The following result is immediate. For convenience define the $n \times n$ nonnegative-definite matrices

$$R \triangleq E^{\mathrm{T}}E, \quad V \triangleq DD^{\mathrm{T}}.$$

Lemma 5.1. Suppose $A + \Delta A$ is asymptotically stable for all $\Delta A \in \mathcal{U}$. Then

$$J(\mathcal{U}) = \sup_{\Delta A\in\mathcal{U}} \operatorname{tr} Q_{\Delta A}R, \tag{5.5}$$

where the $n \times n$ matrix $Q_{\Delta A} \triangleq \lim_{t\to\infty} \mathbb{E}[x(t)x^{\mathrm{T}}(t)]$ is given by

$$Q_{\Delta A} = \int_0^\infty e^{(A+\Delta A)t}Ve^{(A+\Delta A)^{\mathrm{T}}t}\mathrm{d}t, \tag{5.6}$$

which is the unique, nonnegative-definite solution to

$$0 = (A + \Delta A)Q_{\Delta A} + Q_{\Delta A}(A + \Delta A)^{\mathrm{T}} + V. \tag{5.7}$$

In order to draw connections with traditional Lyapunov function theory, we express the H_2 performance measure in terms of a dual variable $P_{\Delta A}$ for which the roles of $A + \Delta A$ and $(A + \Delta A)^{\mathrm{T}}$ are interchanged.

Proposition 5.1. Suppose $A + \Delta A$ is asymptotically stable for all $\Delta A \in \mathcal{U}$. Then $J(\mathcal{U}) = \sup_{\Delta A\in\mathcal{U}} \operatorname{tr} P_{\Delta A}V$, (5.8)

where $P_{\Delta A} \in \mathbb{R}^{n\times n}$ is the unique, nonnegative-definite solution to

$$0 = (A + \Delta A)^{\mathrm{T}}P_{\Delta A} + P_{\Delta A}(A + \Delta A) + R. \tag{5.9}$$

6. Robust Stability and Performance Via Parameter-Dependent Lyapunov Functions

The key step in obtaining robust stability and performance is to bound the uncertain terms $\Delta A^{\mathrm{T}} P_{\Delta A} + P_{\Delta A}\Delta A$ in the Lyapunov equation (5.9) by means of a parameter-dependent bounding function $\Omega(P, \Delta A)$ which guarantees robust stability by means of a family of Lyapunov functions. As shown in [HB3], this framework corresponds to the construction of a parameter-dependent Lyapunov function that guarantees robust stability. As discussed in [HB3], a key feature of this approach is the fact that it constrains the class of allowable time-varying uncertainties thus reducing conservatism in the presence of constant real parameter uncertainty. The following result is fundamental and forms the basis for all later developments.

Theorem 6.1. Let Ω_0: $\mathbb{N}^n \to \mathbb{S}^n$ and P_0: $\mathcal{U} \to \mathbb{S}^n$ be such that

$$\Delta A^{\mathrm{T}} P + P\Delta A \le \Omega_0(P) - [(A+\Delta A)^{\mathrm{T}} P_0(\Delta A) + P_0(\Delta A)(A+\Delta A)], \quad \Delta A \in \mathcal{U}, \quad P \in \mathbb{N}^n, \tag{6.1}$$

and suppose there exists $P \in \mathbb{N}^n$ satisfying

$$0 = A^{\mathrm{T}} P + PA + \Omega_0(P) + R \tag{6.2}$$

and such that $P + P_0(\Delta A)$ is nonnegative definite for all $\Delta A \in \mathcal{U}$. Then

$$(A + \Delta A,\ E) \text{ is detectable}, \quad \Delta A \in \mathcal{U}, \tag{6.3}$$

if and only if

$$A + \Delta A \text{ is asymptotically stable}, \quad \Delta A \in \mathcal{U}. \tag{6.4}$$

In this case,

$$P_{\Delta A} \le P + P_0(\Delta A), \quad \Delta A \in \mathcal{U}, \tag{6.5}$$

where $P_{\Delta A}$ is given by (5.9). Therefore,

$$J(\mathcal{U}) \le \operatorname{tr} PV + \sup_{\Delta A \in \mathcal{U}} \operatorname{tr} P_0(\Delta A) V. \tag{6.6}$$

If, in addition, there exists $\bar{P}_0 \in \mathbb{S}^n$ such that

$$P_0(\Delta A) \le \bar{P}_0, \quad \Delta A \in \mathcal{U}, \tag{6.7}$$

then

$$J(\mathcal{U}) \le \beta, \tag{6.8}$$

where

$$\beta \triangleq \operatorname{tr}[(P + \bar{P}_0)V]. \tag{6.9}$$

Proof. See [HB3]. □

Note that with $\Omega(P, \Delta A) \triangleq \Omega_0(P) - [(A + \Delta A)^{\mathrm{T}} P_0(\Delta A) + P_o(\Delta A)(A + \Delta A)]$ condition (6.1) can be written as

$$\Delta A^{\mathrm{T}} P + P\Delta A \le \Omega(P, \Delta A), \quad \Delta A \in \mathcal{U}, \quad P \in \mathbb{N}^n, \tag{6.1$'$}$$

where $\Omega(P, \Delta A)$ is a function of the uncertain parameters ΔA. For convenience we shall say that $\Omega(\cdot,\cdot)$ is a parameter-dependent bounding function or, to be consistent with [BH1], a parameter-dependent Ω-bound. One can recover the standard guaranteed cost bound or parameter-independent Ω-bound as developed in [BH1] by setting $P_0(\Delta A) \equiv 0$ so that $\Omega(P, \Delta A) \equiv \Omega_0(P)$ and therefore $\Delta A^{\mathrm{T}} P + P\Delta A \le \Omega_0(P)$ for all $\Delta A \in \mathcal{U}$. Finally, since we do not assume that $P_0(0) = 0$, it follows that $\Omega_0(P)$ need not be nonnegative definite. If, however, $P_0(0) = 0$, then $\Omega_0(P) \ge 0$ for all nonnegative-definite P. To apply Theorem 6.1, we first specify a function $\Omega_0(\cdot)$ and an uncertainty set $\mathcal{U}$ such that (6.1)$'$ holds. If the existence of a nonnegative-definite solution P to (6.2) can be determined analytically or numerically and the detectability condition (6.3) is satisfied, then robust stability is guaranteed and the performance bound (6.9) can be computed.

Finally, we establish connections between Theorem 6.1 and Lyapunov function theory. Specifically, we show that a parameter-dependent Ω-bound establishing robust stability is equivalent to the existence of a parameter-dependent Lyapunov function which also establishes robust stability. To show this, assume there exists a positive-definite solution to (6.2), let P_0: $\mathcal{U} \to \mathbb{N}^n$, and define the parameter-dependent Lyapunov function

$$V(x) \triangleq x^{\mathrm{T}}(P + P_0(\Delta A))x. \tag{6.15}$$

Note that since P is positive definite and $P_0(\Delta A)$ is nonnegative definite, $V(x)$ is positive definite. Thus, the corresponding Lyapunov derivative is given by

$$\begin{aligned}\dot{V}(x) &= x^{\mathrm{T}}[(A+\Delta A)^{\mathrm{T}}(P + P_0(\Delta A)) + (P + P_0(\Delta A))(A + \Delta A)]x \\ &= x^{\mathrm{T}}[A^{\mathrm{T}}P + PA + \Delta A^{\mathrm{T}} P + P\Delta A + A^{\mathrm{T}} P_0(\Delta A) + P_0(\Delta A)A \\ &\qquad + \Delta A^{\mathrm{T}} P_0(\Delta A) + P_0(\Delta A)\Delta A]x\end{aligned} \tag{6.16}$$

or, equivalently, using (6.2)

$$\begin{aligned}\dot{V}(x) = -x^{\mathrm{T}}[\Omega_0(P) - \{\Delta A^{\mathrm{T}} P + P\Delta A + A^{\mathrm{T}} P_0(\Delta A) + P_0(\Delta A)A \\ + \Delta A^{\mathrm{T}} P_0(\Delta A) + P_0(\Delta A)\Delta A\} + R]x.\end{aligned} \tag{6.17}$$

Thus, using (6.1) it follows that $\dot{V}(x) \le 0$ so that $A + \Delta A$ is stable in the sense of Lyapunov. To show asymptotic stability using La Salle's Theorem [L] we need to show that $\dot{V}(x) = 0$ implies $x = 0$. Note that $\dot{V}(x) = 0$ implies $Rx = 0$, or, equivalently, $Ex = 0$. Thus, with $\dot{x}(t) = (A + \Delta A)x(t)$, $Ex = 0$ and the detectability assumption (6.3), it follows from the PBH test that $x = 0$. Hence asymptotic stability is established.

7. Construction of Parameter-Dependent Lyapunov Functions: A Unification Between Absolute Stability Criteria and Robust Stability and Performance

Having established the theoretical basis for our approach, we now assign explicit structure to the set $\mathcal{U}$ and the parameter-dependent bounding function $\Omega(\cdot,\cdot)$. Specifically, the uncertainty set $\mathcal{U}$ is defined by

$$\mathcal{U} \triangleq \{\Delta A \in \mathbb{R}^{n\times n} : \ \Delta A = B_0 F C_0, \ \ F \in \mathcal{F}\}, \tag{7.1}$$

where $\mathcal{F}$ satisfies

$$\mathcal{F} \subseteq \hat{\mathcal{F}} \triangleq \{F \in \mathbb{R}^{m_0\times m_0} : \ \ F^{\mathrm{T}}(M^{-1} + M^{-\mathrm{T}})F \le F + F^{\mathrm{T}}\}, \tag{7.2}$$

and where $B_0 \in \mathbb{R}^{n\times m_0}$, $C_0 \in \mathbb{R}^{m_0\times n}$ are fixed matrices denoting the structure of the uncertainty, M is a given $m_0 \times m_0$ invertible matrix, and $F \in \mathbb{R}^{m_0\times m_0}$ is an uncertain matrix. Note that we do not exclude the possibility that $\mathcal{F}$ is equal to $\hat{\mathcal{F}}$. However, for flexibility, $\mathcal{F}$ may be a specified proper subset of the right hand side of (7.2).

Next, we digress slightly to provide simplified characterizations of the set $\mathcal{F}$. Define the subset $\hat{\mathcal{F}}_0$ of $\hat{\mathcal{F}}$ by

$$\hat{\mathcal{F}}_0 = \{F \in \hat{\mathcal{F}} : \ \ \det(I - M^{-1}F) \ne 0\}.$$

Proposition 7.1. The set $\hat{\mathcal{F}}_0$ is equivalently characterized by

$$\hat{\mathcal{F}}_0 = \{F \in \mathbb{R}^{m_0\times m_0} : \ F = (I + \hat{F}M^{-1})^{-1}\hat{F}, \text{ where } \hat{F} \in \mathbb{R}^{m_0\times m_0}, \ \hat{F} + \hat{F}^{\mathrm{T}} \ge 0, \text{ and } \det(I + \hat{F}M^{-1}) \ne 0\}.$$

In the special case that M is positive definite, it follows from Lemma 2.4 that the condition $\det(I + \hat{F}M^{-1}) \ne 0$ in the definition of $\hat{\mathcal{F}}_0$ is automatically satisfied. Furthermore, in the case where M is positive definite we have the following norm bound inequality on F.

Lemma 7.1. Let $F \in \mathcal{F}$ and assume that $M \in \mathbb{P}^{m_0}$. Then

$$\sigma_{\max}(F) \le \sigma_{\max}(M). \tag{7.3}$$

Lemma 7.2. Let $F \in \mathbb{S}^{m_0}$ and $M \in \mathbb{P}^{m_0}$. Then there exists an invertible matrix $S \in \mathbb{R}^{m_0\times m_0}$ such that SMS^{T} and SFS^{T} are both diagonal.

Lemma 7.3. Let $F \in \mathbb{S}^{m_0}$ and $M \in \mathbb{P}^{m_0}$. Then $FM^{-1}F \le F$ if and only if $0 \le F \le M$.

Thus, in the case in which F is symmetric and M is positive definite, the set $\hat{\mathcal{F}}$ defined by (7.2) becomes

$$\hat{\mathcal{F}}_S \triangleq \{F \in \mathbb{S}^{m_0} : \ \ 0 \le F \le M\}$$

Note that if F in $\mathcal{F}$ is constrained to have the diagonal structure $\operatorname{diag}[F_1, F_2, \ldots, F_{m_0}]$, then $0 \le F_i \le M_i$, $i = 1, \ldots, m_0$, where $M = \operatorname{diag}[M_1, \ldots, M_{m_0}]$. More generally, F may have repeated elements and/or blocks on the diagonal of the form $\operatorname{diag}[F_1, F_1, F_1, F_2, \ldots, F_{m_0}]$. Finally, in accordance with our assumption in Section 5 that $0 \in \mathcal{U}$, we shall assume that $0 \in \mathcal{F}$. Also for convenience we assume that $M \in \mathcal{F}$.

For the structure of $\mathcal{U}$ satisfying (7.1), the parameter dependent bound $\Omega(\cdot,\cdot)$ satisfying (6.12) can now be given a concrete form. However, since the elements ΔA in $\mathcal{U}$ are parameterized by the elements F in $\mathcal{F}$, for convenience in the following results we shall write $P_0(F)$ in place of $P_0(\Delta A)$. Finally, we introduce a key definition that will be used in subsequent developments.

Definition 7.1. Let $M, N \in \mathbb{R}^{m_o \times m_o}$. Then $\mathcal{F}$ and N are *compatible* if $F^{\mathrm{T}}N$ is symmetric for all $F \in \mathcal{F}$. Furthermore, $\mathcal{F}$ and N are *strongly compatible* if, in addition, $F^{\mathrm{T}}N$ is nonnegative-definite for all $F \in \mathcal{F}$.

Finally, for the remainder of this paper we assume for simplicity that M is positive definite. Note that in this case it follows from Lemma 7.1 that there exist μ, $\hat{\mu}$, and $\bar{\mu} \in \mathbb{S}^{m_o}$ such that $F^{\mathrm{T}}N \leq \mu$, $F^{\mathrm{T}}\hat{N} \leq \hat{\mu}$, and $F^{\mathrm{T}}NF \leq \bar{\mu}$ for all $F \in \mathcal{F}$.

Proposition 7.2. Let $\hat{N} \in \mathbb{R}^{m_o \times m_o}$ and $N \in \mathbb{N}^{m_o}$ be such that $\mathcal{F}$ and $\hat{N}$ are compatible, $\mathcal{F}$ and N are strongly compatible, and

$$[M^{-1} - \hat{N}C_0B_0 + (M^{-1} - \hat{N}C_0B_0)^{\mathrm{T}} - B_0^{\mathrm{T}}C_0^{\mathrm{T}}\mu C_0B_0 - \mu] > 0. \quad (7.7)$$

Then the functions

$$\begin{aligned}\Omega_0(P) = {} & A^{\mathrm{T}}C_0^{\mathrm{T}}\mu C_0A + (C_0 + \hat{N}C_0A + B_0^{\mathrm{T}}C_0^{\mathrm{T}}\mu C_0A + B_0^{\mathrm{T}}P)^{\mathrm{T}} \\ & \cdot[(M^{-1} - \hat{N}C_0B_0) + (M^{-1} - \hat{N}C_0B_0)^{\mathrm{T}} - B_0^{\mathrm{T}}C_0^{\mathrm{T}}\mu C_0B_0 - \mu]^{-1} \\ & \cdot(C_0 + \hat{N}C_0A + B_0^{\mathrm{T}}C_0^{\mathrm{T}}\mu C_0A + B_0^{\mathrm{T}}P),\end{aligned}$$

$$P_0(F) = C_0^{\mathrm{T}}F^{\mathrm{T}}\hat{N}C_0 + C_0^{\mathrm{T}}F^{\mathrm{T}}NFC_0, \quad (7.9)$$

satisfy (6.1) with $\mathcal{U}$ given by (7.1).

Next, using Theorem 6.1 and Proposition 7.2 we have the following immediate result.

Theorem 7.1. Let $\hat{N} \in \mathbb{R}^{m_o \times m_o}$ and $N \in \mathbb{N}^{m_o}$ be such that $\mathcal{F}$ and $\hat{N}$ are strongly compatible and $\mathcal{F}$ and N are strongly compatible and such that (7.7) is satisfied. Furthermore, suppose there exists a nonnegative-definite matrix P satisfying

$$\begin{aligned}0 = {} & A^{\mathrm{T}}P + PA + A^{\mathrm{T}}C_0^{\mathrm{T}}\mu C_0A + (C_0 + \hat{N}C_0A + B_0^{\mathrm{T}}C_0^{\mathrm{T}}\mu C_0A + B_0^{\mathrm{T}}P)^{\mathrm{T}} \\ & \cdot[(M^{-1} - \hat{N}C_0B_0) + (M^{-1} - \hat{N}C_0B_0)^{\mathrm{T}} - B_0^{\mathrm{T}}C_0^{\mathrm{T}}\mu C_0B_0 - \mu]. \\ & \cdot(C_0 + \hat{N}C_0A + B_0^{\mathrm{T}}C_0^{\mathrm{T}}\mu C_0A + B_0^{\mathrm{T}}P) + R.\end{aligned}$$

Then

$$(A + \Delta A,\ E) \text{ is detectable}, \quad \Delta A \in \mathcal{U}, \quad (7.12)$$

if and only if

$$A + \Delta A \text{ is asymptotically stable}, \quad \Delta A \in \mathcal{U}. \quad (7.13)$$

In this case,

$$J(\mathcal{U}) \leq \operatorname{tr}[P + C_0^{\mathrm{T}}\hat{\mu}C_0 + C_0^{\mathrm{T}}\bar{\mu}C_0)V]. \quad (7.14)$$

Proof. The result is a direct specialization of Theorem 6.1 and Proposition 7.2. We only note that $P_0(\Delta A)$ now has the form $P_0(F) = C_0^{\mathrm{T}}F^{\mathrm{T}}\hat{N}C_0 + C_0^{\mathrm{T}}F^{\mathrm{T}}NFC_0$. Since by assumption $F^{\mathrm{T}}\hat{N} \geq 0$ for all $F \in \mathcal{F}$ it follows that $P + P_0(F)$ is nonnegative definite for all $F \in \mathcal{F}$ as required by Theorem 6.1. □

Next, we establish connections between the parameter-dependent bounding function formed from (7.8) and the absolute stability criteria for slope restricted monotonic nonlinearities developed in Sections 3 and 4. Furthermore, by exploiting results from positivity theory it is possible to guarantee the existence of a positive-definite solution to (7.11). Specifically, using Lemma 2.3 we have the following sufficient condition for the existence of a solution to (7.11).

Theorem 7.2. Let $\hat{G}(s) = M^{-1} - \frac{1}{2}\mu + (I + \hat{N}s)G(s) - \frac{1}{2}[sG(s)]^{\sim}\mu[sG(s)]$, where $G(s) \overset{\min}{\sim} \left[\begin{array}{c|c} A & -B_0 \\ \hline C_0 & 0 \end{array}\right]$. If A is asymptotically stable and $\hat{G}(s)$ is strongly positive real, then there exists an $n \times n$ matrix $P > 0$ satisfying (7.11). Conversely, if there exists $P > 0$ satisfying (7.11) for all $R > 0$, then A is asymptotically stable and $\hat{G}(s)$ is strongly positive real.

8. Robust Controller Synthesis Via Parameter-Dependent Lyapunov Functions

Robust Stability and Performance Problem. Given the nth-order stabilizable plant with constant real-valued plant parameter variations

$$\dot{x}(t) = (A + \Delta A)x(t) + Bu(t) + Dw(t), \quad t \in [0, \infty), \quad (8.1)$$

$$y(t) = Cx(t), \quad (8.2)$$

where $u(t) \in \mathbb{R}^m$, $w(t) \in \mathbb{R}^d$, and $y(t) \in \mathbb{R}^\ell$, determine a static output feedback control law

$$u(t) = Ky(t) \quad (8.3)$$

that satisfies the following design criteria:

i) the closed-loop system (8.1), (8.2) is asymptotically stable for all $\Delta A \in \mathcal{U}$, that is, $A + BKC + \Delta A$ is asymptotically stable for all $\Delta A \in \mathcal{U}$; and

ii) the performance functional

$$J(K) \triangleq \sup_{\Delta A \in \mathcal{U}} \limsup_{t\to\infty} \frac{1}{t}\mathbb{E}\left\{\int_0^t [x^{\mathrm{T}}(s)R_1x(s) + u^{\mathrm{T}}(s)R_2u(s)]\mathrm{d}s\right\} \quad (8.4)$$

is minimized.

Remark 8.1. The cost functional (8.4) is identical to the standard LQR theory with the exception of the supremum for evaluating worst-case quadratic performance over $\mathcal{U}$.

For each variation $\Delta A \in \mathcal{U}$, the closed-loop system (8.1)-(8.3) can be written as

$$\dot{x}(t) = (\tilde{A} + \Delta A)x(t) + Dw(t), \quad t \in [0, \infty), \quad (8.5)$$

where

$$\tilde{A} \triangleq A + BKC, \quad (8.6)$$

and where the white noise disturbance has intensity $V = DD^{\mathrm{T}}$. Finally, note that if $\tilde{A} + \Delta A$ is asymptotically stable for all $\Delta A \in \mathcal{U}$ for a given K, then (8.4) can be written as

$$J(K) = \sup_{\Delta A \in \mathcal{U}} \limsup_{t\to\infty} \mathbb{E}[x^{\mathrm{T}}(t)\tilde{R}x(t)] = \sup_{\Delta A \in \mathcal{U}} \operatorname{tr} Q_{\Delta A}\tilde{R} = \sup_{\Delta A \in \mathcal{U}} \operatorname{tr} P_{\Delta A}V, \quad (8.7)$$

where

$$\tilde{R} \triangleq R_1 + K^{\mathrm{T}}R_2K \quad (8.8)$$

and $P_{\Delta A}$ satisfies (5.9) with A replaced by $\tilde{A}$ and R replaced by $\tilde{R}$.

To apply Theorem 7.1 to controller synthesis we consider the performance bound (6.9) in place of the actual worst-case performance $J(K)$ as in Theorem 7.1 with A, R replaced by $\tilde{A}$ and $\tilde{R}$ to address the closed-loop control problem. This leads to the following optimization problem.

Optimization Problem. Determine $K \in \mathbb{R}^{m\times\ell}$ that minimizes

$$\mathcal{J}(K) \triangleq \operatorname{tr}[(P + C_0^{\mathrm{T}}\hat{\mu}c_0 + C_0^{\mathrm{T}}\bar{\mu}C_0)V] \quad (8.9)$$

subject to

$$\begin{aligned}0 = {} & \tilde{A}^{\mathrm{T}}P + P\tilde{A} + \tilde{A}^{\mathrm{T}}C_0^{\mathrm{T}}\mu C_0\tilde{A} + (C_0 + \hat{N}C_0\tilde{A} + B_0^{\mathrm{T}}C_0^{\mathrm{T}}\mu C_0\tilde{A} + B_0^{\mathrm{T}}P)^{\mathrm{T}} \\ & \cdot[(M^{-1} - \hat{N}C_0B_0) + (M^{-1} - \hat{N}C_0B_0)^{\mathrm{T}} - B_0^{\mathrm{T}}C_0^{\mathrm{T}}\mu C_0B_0 - \mu] \\ & \cdot(C_0 + \hat{N}C_0\tilde{A} + B_0^{\mathrm{T}}C_0^{\mathrm{T}}\mu C_0\tilde{A} + B_0^{\mathrm{T}}P) + \tilde{R}\end{aligned} \quad (8.10)$$

Proposition 8.1. If $P \in \mathbb{N}^n$ and $K \in \mathbb{R}^{m\times\ell}$ satisfy (8.10) and the detectability condition (6.3) holds, then $\hat{A} + \Delta A$ is asymptotically stable for all $\Delta A \in \mathcal{U}$, and

$$J(K) \leq \mathcal{J}(K). \quad (8.11)$$

Next, we present sufficient conditions for robust stability and performance for the static output feedback case. For arbitrary $P, Q \in \mathbb{R}^{n\times n}$ define the notation

$$\hat{C}_0 \triangleq C_0 + \hat{N}C_0A + B_0^{\mathrm{T}}C_0^{\mathrm{T}}\mu C_0A,$$

$$R_0 \triangleq (M^{-1} - \hat{N}C_0B_0) + (M^{-1} - \hat{N}C_0B_0)^{\mathrm{T}} - B_0^{\mathrm{T}}C_0^{\mathrm{T}}\mu C_0B_0 - \mu,$$

$$A_P \triangleq A + B_0R_0^{-1}\hat{C}_0$$

$$R_{2a} \triangleq R_2 + B^{\mathrm{T}}C_0^{\mathrm{T}}\mu C_0B + (\hat{N}C_0B + B_0^{\mathrm{T}}C_0^{\mathrm{T}}\mu C_0B)^{\mathrm{T}}R_0^{-1}(\hat{N}C_0B + B_0^{\mathrm{T}}C_0^{\mathrm{T}}\mu C_0B_0),$$

$$\begin{aligned}P_a \triangleq {} & B^{\mathrm{T}}P + B^{\mathrm{T}}C_0^{\mathrm{T}}\hat{N}^{\mathrm{T}}R_0^{-1}B_0^{\mathrm{T}}P + B^{\mathrm{T}}\mu C_0\hat{C}_0B_0R_0^{-1}B_0^{\mathrm{T}}P + B^{\mathrm{T}}C_0^{\mathrm{T}}\mu C_0A \\ & + (\hat{N}C_0B + B_0^{\mathrm{T}}C_0^{\mathrm{T}}\mu C_0B)^{\mathrm{T}}R_0^{-1}\hat{C}_0\end{aligned}$$

$$\nu \triangleq QC^{\mathrm{T}}(CQC^{\mathrm{T}})^{-1}C, \quad \nu_\perp = I_n - \nu,$$

when the indicated inverses exist.

Theorem 8.1. Assume $R_0 > 0$ and assume $\mathcal{F}$ and $\hat{N}$ are strongly compatible and $\mathcal{F}$ and N are strongly compatible. Furthermore, suppose there exist $n \times n$ nonnegative-definite matrices P, Q such that $CQC^{\mathrm{T}} > 0$ and

$$=A_P^{\mathrm{T}}P + PA_P + R_1 + \hat{C}_0^{\mathrm{T}}R_0^{-1}\hat{C}_0 + PB_0R_0^{-1}B_0^{\mathrm{T}}P - P_a^{\mathrm{T}}R_{2a}^{-1}P_a + \nu_\perp^{\mathrm{T}}P_a^{\mathrm{T}}R_{2a}^{-1}P_a\nu_\perp$$

$$=[A_P - (I + B_0R_0^{-1}NC_0 + B_0R_0^{-1}B_0^{\mathrm{T}}C_0^{\mathrm{T}}\mu C_0)BR_{2a}^{-1}P_a\nu]Q + Q[A_P - (I + B_0R_0^{-1}NC_0 + B_0R_0^{-1}B_0^{\mathrm{T}}C_0^{\mathrm{T}}\mu C_0)BR_{2a}^{-1}P_a\nu]^{\mathrm{T}} +$$

and let K be given by

$$K = -R_{2a}^{-1}P_aQC^{\mathrm{T}}(CQC^{\mathrm{T}})^{-1}. \tag{8.14}$$

Then $(\tilde{A} + \Delta A, \tilde{R})$ is detectable for all $\Delta A \in \mathcal{U}$ if and only if $\tilde{A} + \Delta A$ is asymptotically stable for all $\Delta A \in \mathcal{U}$. In this case the closed-loop system performance (8.7) satisfies the bound

$$J(K) \le \operatorname{tr}[(P + C_0^{\mathrm{T}}\hat{\mu}C_0 + C_0^{\mathrm{T}}\bar{\mu}C_0)V]. \tag{8.15}$$

Theorem 8.1 provides constructive sufficient conditions that yield static output feedback controllers for robust stability and performance. These conditions comprise a system of one modified algebraic Riccati equation and one modified Lyapunov equation in variables P and Q, respectively. When solving (8.12) and (8.13) numerically, the matrices M, μ, and $\hat{N}$ and the structure matrices B_0 and C_0 appearing in the design equations can be adjusted to examine tradeoffs between performance and robustness. To further reduce conservatism, one can view the multiplier matrices $\hat{N}$ and N as free parameters and optimize the H_2 performance bound $\mathcal{J}$ with respect to $\hat{N}$ and N. In particular, setting $\partial\mathcal{L}/\partial\hat{N} = 0$ and $\partial\mathcal{L}/\partial N = 0$ (or, equevelently, $\partial\mathcal{J}/\partial\hat{N} = 0$ and $\partial\mathcal{J}/\partial N = 0$) yields

$$0 = \frac{1}{2}MC_0VC_0^{\mathrm{T}} + R_0^{-1}(C_0 + \hat{N}C_0\tilde{A} + B_0^{\mathrm{T}}C_0^{\mathrm{T}}\mu C_0\tilde{A} + B_0^{\mathrm{T}}P)Q\tilde{A}^{\mathrm{T}}C_0^{\mathrm{T}} + R_0^{-1}(C_0 + \hat{N}C_0\tilde{a} + B_0^{\mathrm{T}}C_0^{\mathrm{T}}\mu C_0\tilde{A} + B_0^{\mathrm{T}}P)Q(C_0 + \hat{N}C_0\tilde{A} + B_0^{\mathrm{T}}C_0^{\mathrm{T}}\mu C_0\tilde{A} + B_0^{\mathrm{T}}P)^{\mathrm{T}}R_0^{-1}B_0^{\mathrm{T}}C_0^{\mathrm{T}},$$

$$\begin{aligned}0 = {} & MC_0VC_0^{\mathrm{T}} + MC_0\tilde{A}Q\tilde{A}^{\mathrm{T}}C_0^{\mathrm{T}} \\ & + 2MC_0B_0R_0^{-1}(C_0 + \hat{N}C_0\tilde{A} + B_0^{\mathrm{T}}C_0^{\mathrm{T}}MNC_0\tilde{A} + B_0^{\mathrm{T}}P)^{\mathrm{T}}Q\tilde{A}^{\mathrm{T}}C_0^{\mathrm{T}} \\ & C_0B_0R_0^{-1}(C_0 + \hat{N}C_0\tilde{A} + B_0^{\mathrm{T}}C_0^{\mathrm{T}}MNC_0\tilde{A} + B_0^{\mathrm{T}}P)Q(C_0 + \hat{N}C_0\tilde{A} + B_0^{\mathrm{T}}C_0^{\mathrm{T}}MNC_0\tilde{A} \\ & + B_0^{\mathrm{T}}P)^{\mathrm{T}}R_0^{-1}B_0^{\mathrm{T}}C_0^{\mathrm{T}} \\ & + MR_0^{-1}(C_0 + \hat{N}C_0\tilde{A} + B_0^{\mathrm{T}}C_0^{\mathrm{T}}MNC_0\tilde{A} + B_0^{\mathrm{T}}P)Q(C_0 + \hat{N}C_0\tilde{A} + B_0^{\mathrm{T}}C_0^{\mathrm{T}}MNC_0A + B_0^{\mathrm{T}}P)^{\mathrm{T}}R_0^{-1},\end{aligned}$$

where Q satisfies (8.18). Now, the basic approach is to employ a numerial algorithm to design the optimal controller and the multiplers $\hat{N}$ and N simultaneously, thus avoiding the need to iterate between controller design and optimal multiplier evaluation.

[A] B. D. O. Anderson, A System Theory Criterion for Positive Real Matrices, *SIAM J. Contr. Optim,.* **5** (1967), 171-182.

[AV] B. D. O. Anderson and S. Vongpanitlerd, *Network Analysis and Synthesis: A Modern Systems Theory Approach*, Prentice-Hall, Englewood Cliffs, NJ, 1973.

[BH1] D. S. Bernstein and W. M. Haddad, Robust Stability and Performance Analysis for State Space Systems Via Quadratic Lyapunov Bounds, *SIAM J. Matrix Anal. Appl.*, **11** (1990), 239-271.

[BH2] D. S. Bernstein and W. M. Haddad, Is There More to Robust Control Theory Than Small Gain?, *Proc. Amer. Contr. Conf.*, Chicago, IL, June 1992, 83-84.

[BHH] D. S. Bernstein, W. M. Haddad, and D. C. Hyland, Small Gain Versus Positive Real Modeling of Real Parameter Uncertainty, *AIAA J. Guid. Contr. Dyn.*, **15** (1992), 538-540.

[BW] R. W. Brockett and J. L. Willems, Frequency Domain Stability Criteria I and II, *IEEE Trans. Autom. Contr.*, **AC-10** (1965), 255-261, 407-413.

[D] C. A. Desoer, A Generalization of the Popov Criterion, *IEEE Trans. Autom. Contr.*, **10** (1965), 182-185.

[DV] C. A. Desoer, and M. Vidyasagar, *Feedback Systems: Input-Output Properties*, Academic Press, New York, 1975.

[G] Lj. T. Grujić, Solutions for the Lurie-Postnikov and Aizerman Problems, *Int. J. Systems Sci.*, **9** (1978), 1359-1372.

[HB1] W. M. Haddad, and D. S. Bernstein, Robust Stabilization with Positive Real Uncertainty: Beyond the Small Gain Theorem, *Sys. Contr. Lett.*, **17** (1991), 191-208.

[HB2] W. M. Haddad and D. S. Bernstein, Explicit Construction of Quadratic Lyapunov Functions for the Small Gain, Positivity, Circle, and Popov Theorems and Their Application to Robust Stability, *Control of Uncertain Dynamic Systems*, S. P. Bhattacharyya and L. H. Keel, Eds., CRC Press, Boca Raton, FL. (1991), 149-173.

[HB3] W. M. Haddad and D. S. Bernstein, Parameter-Dependent Lyapunov functions, Constant Real Parameter Uncertainty, and the Popov Criterion in Robust Analysis and Synthesis, *Proc. IEEE Conf. Dec. Contr.*, Brighton, U.K., December (1991), 2274-2279, 2632-2633.

[HCB] W. M. Haddad, E. G. Collins, Jr., and D. S. Bernstein, Robust Stability Analysis Using the Small Gain, Circle, Positivity, and Popov Theorems: A Comparative Study, submitted to *IEEE Trans. Autom. Contr.*

[HHHB] W. M. Haddad, J. P. How, S.R. Hall, and D. S. Bernstein, Extensions of Mixed-μ Bounds to Monotonic and Odd Monotonic Nonlinearities Using Absolute Stability Theory, submitted for publication.

[LJ] R. Lozano-Leal and S. Joshi, Strictly Positive Real Transfer Functions Revisited, *IEEE Trans. Autom. Contr.*, **35** (1990), 1243-1245.

[MA1] A. I. Mees and D. P. Atherton, The Popov Criterion for Multiple-Loop Feedback Systems, *IEEE Trans. Autom. Contr.*, **25** (1980), 924-928.

[NC] K. S. Narendra and Y.-S. Cho, Stability of Feedback systems Containing a Single Odd Monotonic Nonlinearity, *IEEE Trans. Autom. Contr.*, **AC-12** (1967), 448-450.

[MA2] J. B. Moore and B. D. O. Anderson, A Generalization of the Popov Criterion, *J. Franklin Inst.*, **285** (1968), 488-492.

[NN1] K. S. Narendra and C. P. Neuman, Stability of a Class of Differential Equations with a Single Monotone Nonlinearity, *SIAM J. Control*, 4 (1966), 295-308.

[NN2] K. S. Narendra and C. P. Neuman, Stability of Continuous Time Dynamical Systems with m-Feedback Nonlinearities, *AIAA Journal*, **5** (1967), 2021-2027.

[NT1] K. S. Narendra and J. H. Taylor, Lyapunov Functions for Nonlinear Time-Varying Systems, *Information and Control*, **12** (1968), 378-393.

[NT2] K. S. Narendra and J. H. Taylor, *Frequency Domain Criteria for Absolute Stability*, Academic Press, New York, 1973.

[P] V. M. Popov, Absolute Stability of Nonlinear Systems of Automatic Control, *Auto. and Remote Control*, **22** (1962), 857-875.

[SW] M. G. Safonov and G. Wyetizmer, Computer-Aided Stability Analysis Renders Popov Criterion Obsolete, *IEEE Trans. Autom. Contr.*, **AC-32** (1987), 1128-1131.

[TS] M. A. L. Thathachar and M. D. Srinath, Some Aspects of the Luré Problem, *IEEE Trans. Autom. Contr.*, **AC-12** (1967), 451-453.

[TSR] M. A. L. Thathachar, M. D. Srinath, and H. K. Ramaprige, On a Modified Luré Problem, *IEEE Trans. Autom. Contr.*, **AC-12** (1967), 731-740.

[V] M. Vidyasagar, *Nonlinear Systems Analysis*, Prentice-Hall, Englewood Cliffs, NJ, 1993.

[ZF] G. Zames, and P. L. Falb, Stability Conditions for Systems with Monotone and Slope-Restricted Nonlinearities, *SIAM J. Contr. Optim.*, **4** (1968), 89-108.

Proceedings of the 31st Conference
on Decision and Control
Tucson, Arizona • December 1992

Robust Control of Systems under Mixed Time/Frequency Domain Constraints via Convex Optimization

Mario Sznaier
Electrical Engineering, University of Central Florida, Orlando, Fl, 32816–2450
email: msznaier@frodo.engr.ucf.edu
Zoubir Benzaid
Department of Mathematics, Embry Riddle Aeronautical University, Daytona Beach, Fl 32114

Abstract

A successful controller design paradigm must take into account both model uncertainty and design specifications. Model uncertainty can be addressed using the $\mathcal{H}_\infty$ or l_1 robust control theory. However, these frameworks cannot accommodate the realistic case where the design specifications include both time and frequency domain constraints. In this paper we propose an approach that takes explicitly into account both mixed time/frequency domain constraints and model uncertainty. This is achieved by minimizing a set-induced operator norm, subject to additional frequency domain performance specifications such as bounds on the $\mathcal{H}_2$ or $\mathcal{H}_\infty$ norm of relevant transfer functions. We show that this formulation results in a convex optimization problem that can be exactly solved. Thus, the conservatism inherent in some previous approaches is eliminated.

1. Introduction

A large number of control problems require designing a controller capable of achieving acceptable performance under system uncertainty and design specifications, usually including both time and frequency domain constraints. However, despite its practical importance, this problem remains to a large extent still open, even in the simpler case where the system under consideration is linear.

The problem of controlling linear systems under time domain constraints has been solved only in the rather idealized case where the dynamics are completely known (see for instance [1–2] and references therein). Clearly such an assumption can be too restrictive, resulting in controllers that are seldom suitable for real–world applications.

During the last decade a large research effort has led to procedures for designing robust controllers capable of achieving desirable properties under various classes of model uncertainty. In particular, a powerful framework has been developed, addressing the issues of robust stability and robust performance in the presence of norm–bound perturbations by minimizing an $\mathcal{H}_\infty$ bound [3]. The $\mathcal{H}_\infty$ framework, combined with μ–analysis [4] (in order to exploit the structure of the uncertainty) has been successfully applied to a number of hard practical control problems (see for instance [5]). However, in spite of this success, it is clear that plain $\mathcal{H}_\infty$ control can only address a subset of the common performance requirements since, being a frequency domain method, it can not address time domain specifications. Recently some progress has been made in this direction [6–7]. However, most of the proposed methods rely on a number of approximations, and this may preclude finding a solution if the design specifications are tight. In [8–9] time–domain constraints over a finite horizon are incorporated into an $\mathcal{H}_\infty$ optimal control problem which is then exactly solved. However, at this stage constraints over an infinite horizon can be handled only indirectly.

A different approach to robust control has been pursued in [10–11], where robustness and disturbance rejection are approached using the l_1 optimal control theory introduced by Vidyasagar [10] and developed by Pearson and coworkers [11]. These methods are attractive since they allow for an explicit solution to the robust performance problem. However, they cannot accommodate some common classes of frequency domain specifications (such as $\mathcal{H}_2$ or $\mathcal{H}_\infty$ bounds).

M. Sznaier was supported in part by NSF under grant ECS–9211169 and by Florida Space Grant Consortium

We recently proposed [12–13] to approach time–domain constrained systems using an operator norm–theoretic approach. In this framework, robustness against model uncertainty and satisfaction of time–domain constraints are achieved by minimizing a set induced operator norm, subject to additional frequency domain constraints. In this paper we generalize the framework of [12–13] by eliminating some of the approximations used there and by considering the more general case of output feedback controllers. The main result of the paper shows that for the case of $\mathcal{H}_2$ or $\mathcal{H}_\infty$ constraints, the resulting optimization problem can be cast into a finite–dimensional convex optimization form. This approach eliminates most of the conservatism inherent in previously proposed methods.

2. Preliminaries

2.1 Notation

By l_1 we denote the space of real sequences $\{q_i\}$, equipped with the norm $\|q\|_1 = \sum_{k=0}^{\infty} |q_k| < \infty$. Given a sequence $q \in l_1$ we will denote its Z–transform by $Q(z)$. $\mathcal{L}_\infty$ denotes the Lebesgue space of complex valued transfer matrices which are essentially bounded on the unit circle with norm $\|T(z)\|_\infty \stackrel{\Delta}{=} \sup_{|z|=1} \sigma_{max}(T(z))$. $\mathcal{H}_\infty$ ($\mathcal{H}_\infty^-$) denotes the set of stable (antistable) complex matrices $G(z) \in \mathcal{L}_\infty$, i.e. analytic in $|z| \geq 1$ ($|z| \leq 1$). $\mathcal{H}_2$ denotes the space of complex matrices square integrable in the unit circle and analytic in $|z| > 1$, equipped with the norm:

$$\|G\|_2^2 = \frac{1}{2\pi} \oint_{|z|=1} Trace\{G(z)'G(z)\} z dz$$

where $'$ indicates transpose conjugate. The prefix $\mathcal{R}$ denotes real rational transfer matrices. Throughout the paper we will use packed notation to represent state–space realizations, i.e.

$$G(z) = C(zI - A)^{-1}B + D \stackrel{\Delta}{=} \left(\begin{array}{c|c} A & B \\ \hline C & D \end{array}\right)$$

Given two transfer matrices $T = \begin{pmatrix} T_{11} & T_{12} \\ T_{21} & T_{22} \end{pmatrix}$ and Q with appropriate dimensions, the lower *linear fractional transformation* is defined as:

$$\mathcal{F}_l(T,Q) \stackrel{\Delta}{=} T_{11} + T_{12}Q(I - T_{22}Q)^{-1}T_{21}$$

For a transfer matrix $G(z)$, $G^{\sim} \stackrel{\Delta}{=} G'(\frac{1}{z})$. Finally, $\underline{x}$ indicates that x is a vector quantity.

2.2 Definitions and Preliminary Results

• **Def. 1:** Consider the linear, discrete time, autonomous system modeled by the difference equation:

$$\underline{x}_k = \phi_k \underline{x}_o, \; k = 0, 1 \ldots \; \phi_0 = I_n \qquad (S^a)$$

subject to the constraint $\underline{x} \in \mathcal{G} \subset R^n$, where $\mathcal{G}$ is a compact, convex, balanced set containing the origin. The system (S^a) is *Constrained Stable* if for *any* point $\tilde{\underline{x}} \in \mathcal{G}$, the trajectory $\underline{x}_k(\tilde{\underline{x}})$ originating at $\tilde{\underline{x}}$ remains in $\mathcal{G}$ for all k.

CH3229-2/92/0000-2617$1.00 © 1992 IEEE

• **Def. 2:** Consider the family of linear discrete time systems modeled by the difference equation:

$$\underline{x}_k = \phi_{k\Delta}\underline{x}_o,\ k = 0,1 \qquad (S^a_\Delta)$$

where $\phi_{k\Delta}$ belongs to some family $\mathcal{P} \subset R^{n*n}$ described by the parameter Δ which takes values in a set $\mathcal{D}$. The system (S^a) is *Robustly Constrained Stable* with respect to the family $\mathcal{P}$ if every element of $\mathcal{P}$ is constrained–stable.

• **Def. 3:** The *Minkowsky Functional* p of a convex, balanced, set $\mathcal{G}$ containing the origin in its interior is defined by

$$p(\underline{x}) = \inf_{r>0}\left\{r: \frac{\underline{x}}{r} \in \mathcal{G}\right\} \qquad (1)$$

A well known result in functional analysis (see for instance [14]) establishes that p defines a seminorm in R^n. Furthermore, when $\mathcal{G}$ is compact, this seminorm becomes a norm. In the sequel we will denote this norm as $\|\underline{x}\|_\mathcal{G} \stackrel{\Delta}{=} p(\underline{x})$ and we will use its properties to establish some of the key results in the paper. Of particular importance are the facts that induced norms are submultiplicative and that all finite dimensional matrix norms are equivalent [15]. The $\|.\|_\mathcal{G}$ also provides a connection with Lyapunov theory. As we show in the next lemma, (S^a) is constrained–stable iff ϕ is a contraction in this norm.

• **Lemma 1:** Consider the system (S^a). Let $\phi \stackrel{\Delta}{=} \{\phi_k\}$ and denote by $\|.\|_\mathcal{G}$ the operator norm induced in R^{n*n} by $\mathcal{G}$, i.e. $\|A\|_\mathcal{G} = \sup_{\|\underline{x}\|_\mathcal{G}=1} \frac{\|A\underline{x}\|_\mathcal{G}}{\|\underline{x}\|_\mathcal{G}}$. Finally, let $\|\phi\|_\mathcal{G} \stackrel{\Delta}{=} \sup_k \|\phi_k\|_\mathcal{G}$. Then the system (S^a) is constrained stable *iff* $\|\phi\|_\mathcal{G} \leq 1$

Proof: The proof follows immediately from Def. 1 by noting that:

$$\begin{aligned}\|\phi\|_\mathcal{G} \leq 1 &\iff \|\phi_k\|_\mathcal{G} \leq 1\ \forall\ k\\ &\iff \|\phi_k\bar{\underline{x}}\|_\mathcal{G} \leq 1\ \forall\ k,\ \|\bar{\underline{x}}\|_\mathcal{G} \leq 1\\ &\iff \|\underline{x}_k(\bar{\underline{x}})\|_\mathcal{G} \leq 1 \forall \bar{\underline{x}} \in \mathcal{G} \iff \underline{x}_k(\bar{\underline{x}}) \in \mathcal{G} \forall\ k\end{aligned} \qquad (2)$$

where $\underline{x}_k(\bar{\underline{x}})$ denotes the trajectory that originates in $\bar{\underline{x}}$ ⋄.

Remark 1: For the special case of systems subject to additive parametric uncertainty:

$$\underline{x}_{k+1} = (A+\Delta)\underline{x}_k \qquad (3)$$

where Δ is constant, it can be easily shown that the condition $\|\phi\|_\mathcal{G} \leq 1$ reduces to $\|A+\Delta\|_\mathcal{G} \leq 1$, the necessary and sufficient condition derived in [16].

Remark 2: The operator norm defined in Lemma 1 can be extended to $\mathcal{RH}^{n*n}_\infty$ as follows: Let $\Phi(z) \in \mathcal{RH}^{n*n}_\infty$ and let $\{\phi_k\} = Z^{-1}\{\Phi(z)\}$. Then we can define:

$$\|\Phi(z)\|_\mathcal{G} \stackrel{\Delta}{=} \sup_k \|\phi_k\|_\mathcal{G} \qquad (4)$$

Note that since $\Phi(z) \in \mathcal{RH}^{n*n}_\infty$, $\{\phi_i\} \in l^{n*n}_1$. Since $\|\phi_k\|_1$ is uniformly bounded with respect to k, it follows from the equivalence of all finite–dimensional matrix norms [15] that $\|\phi_k\|_\mathcal{G}$ is also uniformly bounded, hence $\|\Phi\|_\mathcal{G}$ is finite.

2.3 The Uncertainty Model

In this paper we will consider systems subject to *unstructured* multiplicative dynamic model uncertainty. Thus, if we denote by $\Phi^o(z)$ the z–transform of (S^a), then the family of systems under consideration will be modeled as:

$$\begin{aligned}\mathcal{P}_\delta &= \{\Phi(z): \Phi(z) = \Phi^o(z)(I_n + \Delta),\ \Delta \in \mathcal{D}_\delta\}\\ \mathcal{D}_\delta &= \{\Delta \in \mathcal{RH}^{n*n}_\infty: \|\Delta\|_{\mathcal{G},1} \leq \delta\}\end{aligned} \qquad (5)$$

where

$$\|\Delta\|_{\mathcal{G},1} \stackrel{\Delta}{=} \sum_{i=0}^{\infty} \|\Delta_i\|_\mathcal{G} \qquad (6)$$

Note that since $\Delta \in \mathcal{RH}^{n*n}_\infty$, $\{\Delta_i\} \in l^{n*n}_1$. Hence $\|\Delta\|_{\mathcal{G},1}$ is finite.

In section 2.5 we will show that the uncertainty description (5) includes as a special case the additive parametric uncertainty (3).

2.4 The Mixed Performance Robust Control Problem:

Consider the LTI system represented by the following state–space realization:

$$\begin{aligned}\underline{x}_{k+1} &= A\underline{x}_k + B_1\underline{\omega}_k + B_2\underline{u}_k\\ \underline{\zeta}_k &= C_1\underline{x}_k + D_{12}\underline{u}_k\\ \underline{y}_k &= C_2\underline{x}_k + D_{21}\underline{\omega}_k\end{aligned} \qquad (S)$$

subject to the constraint:

$$\underline{x}_k \in \mathcal{G} \subset R^n$$

where the pairs (A, B_2) and (A, C_2) are stabilizable and detectable respectively, D_{12} has full column rank, D_{21} has full row rank, $\underline{x} \in R^n$ represents the state; $\underline{\zeta} \in R^q$ represents the variables subject to performance specifications; $\underline{y} \in R^p$ represents the outputs available to the controller, $\underline{u} \in R^m$ represents the control input; and where $\underline{\omega} \in R^r$ contains other external inputs of interest such as disturbances or commands. Then, the basic problem that we address in this paper is the following: Given the nominal system (S) subject to model uncertainty of the form (5), with additional frequency–domain performance specifications of the form:

$$\|W(z)T_{\zeta w}(z)\|_* \leq 1 \qquad (P)$$

where $*$ indicates either $\mathcal{H}_2$ or $\mathcal{H}_\infty$, and $W(z)$ is a suitable weighting function, find a *dynamic output–feedback controller:*

$$\begin{aligned}\hat{\underline{x}}_{k+1} &= F\hat{\underline{x}}_k + G\underline{y}_k\\ \underline{u}_k &= H\hat{\underline{x}}_k + E\underline{y}_k\end{aligned} \qquad \hat{\underline{x}}_o = 0 \qquad (C)$$

such that the resulting closed–loop system is robustly constrained stable (i.e. for all members of the family (5) $\underline{x}_k \in \mathcal{G}$ for any initial condition $\underline{x}_o \in \mathcal{G}$) and satisfies the performance specifications (P)

2.5 Constrained Stability Analysis

In this section we consider constrained stability in the presence of model uncertainty and we introduce a constrained robustness measure. We begin by deriving a bound on the $\|.\|_\mathcal{G}$ of the dynamics Φ for all the elements of the family $\mathcal{P}_\delta$ and showing that this bound is tight.

• **Theorem 1:** Consider the family of systems $\mathcal{P}_\delta$ (5). Then:

$$\|\Phi\|_\mathcal{G} \leq \|\Phi^o\|_\mathcal{G}(1+\delta) \qquad (7)$$

and there exist at least one $\hat{\Phi} \in \mathcal{P}_\delta$ such that (7) is an equality.

Proof: Let $\Psi(z) = \Phi(z)\Delta(z)$. Then:

$$\begin{aligned}\|\Psi_k\|_\mathcal{G} &= \|\sum_{i=0}^{k}\phi_i\Delta_{k-i}\|_\mathcal{G} \leq \sum_{i=0}^{k}\|\phi_i\|_\mathcal{G}\|\Delta_{k-i}\|_\mathcal{G}\\ &\leq (\sup_k\|\phi_k\|_\mathcal{G})\sum_{i=0}^{\infty}\|\Delta_i\|_\mathcal{G} = \|\Phi\|_\mathcal{G}\|\Delta\|_{\mathcal{G},1}\end{aligned} \qquad (8)$$

From (8) it follows that:

$$\|\Phi\|_\mathcal{G} = \|\Phi^o(I+\Delta)\|_\mathcal{G} \leq \|\Phi^o\|_\mathcal{G}(1+\|\Delta\|_{\mathcal{G},1})$$

Finally, let $\hat{\Delta} \stackrel{\Delta}{=} \delta I_n$. Then $\hat{\Phi} = \Phi^o(I_n + \hat{\Delta}) \in \mathcal{P}$ and $\|\hat{\Phi}\|_\mathcal{G} = \|\Phi^o\|_\mathcal{G}(1+\delta)$ ⋄.

Remark 3: The uncertainty description (5) includes as a special case systems subject to additive parametric uncertainty (3) in the sense that if there exist $\tilde{\Delta}$, $\|\tilde{\Delta}\|_\mathcal{G} \leq \delta$ such that $\|A+\tilde{\Delta}\|_\mathcal{G} = 1$ then $\Delta_o = \frac{\tilde{\Delta}}{z} \in \mathcal{D}_\delta$ and it can be easily shown that $\|\Phi(z)\|_\mathcal{G} = \|\Phi^o(z)(I+\Delta_o(z))\|_\mathcal{G} = 1$.

Corollary: The family of systems described by (5) is constraint stable *iff*

$$\delta \leq \frac{1-\|\Phi^o\|_\mathcal{G}}{\|\Phi^o\|_\mathcal{G}}$$

This result can be used to define a quantitative measure of robustness in terms of the "size" of the smallest destabilizing perturbation as follows:

• **Def. 4:** Consider the system (S^a). The *constrained stability measure*, ϱ is defined as:

$$\varrho \stackrel{\Delta}{=} 1 - \|\Phi\|_{\mathcal{G}}$$

From Theorem 1 it follows that the family of systems described by (5) is constraint stable iff $\delta \leq \frac{\varrho}{1-\varrho}$. Thus a larger value of ϱ indicates a system capable of accommodating larger model uncertainty.

2.6 Effect of Disturbances

In the previous section we defined a measure of stability in terms of the smallest destabilizing model perturbation and we showed that in order to maximize this measure, $\|\Phi\|_{\mathcal{G}}$ should be minimized. This analysis neglected the effect of the disturbances $\underline{\omega}$ on the states $\underline{x}$. In this section we consider this effect and we show that it can be minimized by minimizing Φ. It follows then that achieving min $\|\Phi\|_{\mathcal{G}}$ is desirable both in terms of maximizing robustness against model uncertainty *and* minimizing the effects of perturbations.

• **Lemma 2:** Let $\underline{x}(z) = T_{x\hat{\omega}}(z)\hat{\underline{\omega}}(z)$, where $\hat{\underline{\omega}} \stackrel{\Delta}{=} B_1\underline{\omega}$ and $\underline{\omega} \in l_1$. Then

$$\|\underline{x}(z)\|_{\mathcal{G}} \stackrel{\Delta}{=} \sup_k \|\underline{x}_k\|_{\mathcal{G}} \leq \|T_{x\hat{\omega}}\|_{\mathcal{G}} \|\hat{\underline{\omega}}\|_{1,\mathcal{G}}$$

where:

$$\|\hat{\underline{\omega}}\|_{1,\mathcal{G}} \stackrel{\Delta}{=} \sum_{k=1}^{\infty} \|\hat{\underline{\omega}}_k\|_{\mathcal{G}}$$

Proof: The proof is similar to the proof of Theorem 1 $\diamond$.

• **Lemma 3:** Consider the system:

$$\underline{x}_{k+1} = A\underline{x}_k + \hat{\underline{\omega}}_k \qquad (S_w)$$

with initial condition $\underline{x}_o$. Then $\|T_{x\hat{\omega}}\|_{\mathcal{G}} = \|T_{xx_o}\|_{\mathcal{G}} \stackrel{\Delta}{=} \|\Phi\|_{\mathcal{G}}$.

Proof: The system (S_w) is equivalent to:

$$\begin{aligned} \tilde{\underline{x}}_{k+1} &= A\tilde{\underline{x}}_k + \hat{\underline{\omega}}_k + A\delta_{k,o}\underline{x}_o \\ \underline{x}_k &= \tilde{\underline{x}}_k + \delta_{ko}\underline{x}_o \end{aligned}$$

with initial condition $\tilde{\underline{x}}_o = 0$. Taking z transforms yields:

$$\underline{x}(z) = (I - \frac{A}{z})^{-1}\underline{x}_o + (zI - A)^{-1}\hat{\underline{\omega}} \qquad (9)$$

Hence we have $T_{x\hat{\omega}}(z) = \frac{1}{z}\Phi(z)$ and, by taking inverse z transforms, $(T_{x\hat{\omega}})_k = \phi_{k-1}$. It follows that $\|T_{x\hat{\omega}}\|_{\mathcal{G}} = \sup_k \|(T_{x\hat{\omega}})_k\|_{\mathcal{G}} = \|\Phi\|_{\mathcal{G}}$ $\diamond$.

Remark 4: From Lemmas 2 and 3 it follows that by minimizing $\|\Phi\|_{\mathcal{G}}$ (and hence maximizing the constrained robustness measure), we are also minimizing the effects of the disturbances $\underline{\omega}$ upon the states.

3. Robust Constrained Control Synthesis

From sections 2.5 and 2.6 it follows that a robust controller guaranteeing satisfaction of the state constraints in the presence of model uncertainty can be obtained by maximizing the constrained stability measure. In [16] we showed that for the simpler case of static full–state feedback and uncertainty limited to a conic set this approach yields well–behaved optimization problems. However, in most cases maximizing the robustness measure does not necessarily guarantee a design that meets desirable specifications. Moreover, good performance and good robustness properties are usually conflicting design objectives which must be traded–off. Hence, a better design can be achieved by selecting a set of specifications and then using the extra degrees of freedom that may be available in the problem to maximize the robustness measure over the set of all controllers that satisfy the given specifications for the nominal plant. Thus, the design problem will have the general form of a non–differentiable constrained minimization problem. In this section we show that: i) with an appropriate parametrization of all the achievable closed–loop maps, this optimization problem is convex and ii) For the $\mathcal{H}_2$ and $\mathcal{H}_\infty$ cases the structure of the problem can be used to cast it into a finite dimensional convex optimization.

3.1 Problem Transformation

The system (S) is equivalent to:

$$\begin{aligned} \tilde{\underline{x}}_{k+1} &= A\tilde{\underline{x}}_k + B_1\underline{\omega}_k + B_2\underline{u}_k + A\delta_{k,o}\underline{x}_o \\ \underline{x}_k &= \tilde{\underline{x}}_k + \delta_{k,o}\underline{x}_o \\ \hat{z}_k &= C_1\underline{x}_k + C_1\delta_{k,o}\underline{x}_o + D_{12}\underline{u}_k \\ \underline{y}_k &= C_2\underline{x}_k + C_2\delta_{k,o}\underline{x}_o + D_{21}\underline{\omega}_k \end{aligned} \qquad (S_o)$$

with initial condition $\tilde{\underline{x}}_o = 0$. It can be easily shown (see for instance [17]) that if the pairs (A, B_2) and (A, C_2) are stabilizable and detectable respectively, then the set of all internally stabilizing controllers and all achievable closed–loop transfer functions can be parametrized in terms of a free parameter $\tilde{Q} \in \mathcal{RH}_\infty$ respectively as:

$$\begin{aligned} K &= \mathcal{F}_l(J, \tilde{Q}) \\ T_{\underline{x}x_o}(z) \stackrel{\Delta}{=} \Phi(z) &= \mathcal{F}_l(T_x, \tilde{Q}) = T_{11}^x + T_{12}^x\tilde{Q}T_{21}^x \\ T_{\zeta\underline{\omega}}(z) &= \mathcal{F}_l(T_f, \tilde{Q}) = T_{11}^f + T_{12}^f\tilde{Q}T_{21}^f \end{aligned}$$

where J, T_x and T_f have the following state–space realizations:

$$J = \left(\begin{array}{c|cc} A + B_2F + LC_2 & -L & B_2 \\ \hline F & 0 & I \\ -C_2 & I & 0 \end{array}\right)$$

$$T_x = \left(\begin{array}{cc|cc} A_F & -B_2F & A & B_2 \\ 0 & A_L & A_L & 0 \\ \hline I & 0 & I & 0 \\ 0 & C_2 & C_2 & 0 \end{array}\right) \qquad (10)$$

$$T_f = \left(\begin{array}{cc|cc} A_F & -B_2F & B_1 & B_2 \\ 0 & A_L & B_1 + LD_{21} & 0 \\ \hline C_1 + D_{12}F & -D_{12}F & 0 & D_{12} \\ 0 & C_2 & D_{21} & 0 \end{array}\right)$$

$$\begin{aligned} A_F &= A + B_2F \\ A_L &= A + LC_2 \end{aligned}$$

where F and L are such that A_F and A_L are stable. A suitable choice for F and L is provided in the next lemma.

• **Lemma 4:** Let

$$\begin{aligned} F &\stackrel{\Delta}{=} -(R + B_2'XB_2)^{-1}(D_{12}'C_1 + B_2'XA) \\ L &\stackrel{\Delta}{=} -(D_{21}B_1' + C_2YA')'(S + C_2YC_2')^{-1} \end{aligned} \qquad (11)$$

where $R \stackrel{\Delta}{=} D_{12}'D_{12}$, $S \stackrel{\Delta}{=} D_{21}D_{21}'$ and X and Y are the unique positive solutions to the following Riccati equations:

$$\begin{aligned} &-(B_2'XA + D_{12}'C_1)'(D_{12}'D_{12} + B_2'XB_2)^{-1}(B_2'XA + D_{12}'C_1) \\ &+ A'XA - X + C_1'C_1 = 0 \end{aligned} \qquad (12)$$

$$\begin{aligned} &-(C_2YA' + D_{21}B_1')'(D_{21}D_{21}' + C_2YC_2')^{-1}(C_2YA' + D_{21}B_1') \\ &AYA' - Y + B_1B_1' = 0 \end{aligned} \qquad (13)$$

Then F stabilizes the pair (A, B_2), L' stabilizes the pair (A', C_2') and $T_{\zeta w} = G_cB_1 - NFG_f + UR_B^{\frac{1}{2}}\tilde{Q}R_L^{\frac{1}{2}}V$, where:

$$
\begin{aligned}
G_c &= \left(\begin{array}{c|c} A_F & I \\ \hline C_1 + D_{12}F & 0 \end{array}\right) \\
G_f &= \left(\begin{array}{c|c} A_L & B_1 + LD_{21} \\ \hline I & 0 \end{array}\right) \\
R_B &= D_{12}' D_{12} + B_2' X B_2 \\
R_L &= D_{21} D_{21}' + C_2 Y C_2' \\
U &= N R_B^{\frac{-1}{2}} \\
V &= R_L^{\frac{-1}{2}} M \\
N &= \left(\begin{array}{c|c} A_F & B_2 \\ \hline C_1 + D_{12}F & D_{12} \end{array}\right) \\
M &= \left(\begin{array}{c|c} A_L & B_1 + LD_{21} \\ \hline C_2 & D_{21} \end{array}\right)
\end{aligned} \tag{14}
$$

Moreover, the following properties hold: i) U and V are *inner* and *co-inner* respectively, i.e. $U^{\sim}U = I, VV^{\sim} = I$; ii) G_cB_1 is orthogonal to N and G_f is orthogonal to M.

Proof: The proof follows from standard state–space manipulations and is omitted for space reasons.

3.2 Robust Constrained Stability Optimization

In this section we show that, in the absence of performance constraints, maximum constrained robustness is achieved by *constant* state feedback. Thus in this case the optimally robust controller can be found by using the simple design procedure proposed in [16]. We will use this result to show that solving the mixed performance robust control problem requires considering only a finite number of inequalities.

• **Lemma 5:** Assume that there exists F such that $\|A_F\|_{\mathcal{G}} < 1$. Then:

$$\min_{K(z)\ \mathrm{stab}} \|\Phi(z)\|_{\mathcal{G}} = \min_F \|A + B_2F\|_{\mathcal{G}} \tag{15}$$

Proof: From (10) we have that:

$$\Phi(z) = T_{11}^x + T_{12}^x \tilde{Q} T_{21}^x \tag{16}$$

Assume, by eliminating redundant outputs if necessary, that C_2 has full row rank and define:

$$Q \triangleq \left(\tilde{Q}C_2 - F\right)\left(I - \frac{A_L}{z}\right)^{-1} \tag{17}$$

Since A_L is stable (17) defines a *bijection* over $\mathcal{RH}_\infty$. In terms of Q, $\Phi(z)$ is given by:

$$\Phi(z) = \left\{I_n + (zI_n - A_F)^{-1}(A_F + B_2Q)\right\} \tag{18}$$

Let Q_i denote the coefficients of the impulse response of the transfer matrix Q. From (18) we have that:

$$
\begin{aligned}
\phi_0 &= I_n \\
\phi_k &= A_F^k + \sum_{i=0}^{k-1} A_F^{k-1-i} B_2 Q_i
\end{aligned} \tag{19}
$$

Let $\mu = \max_k \|\phi_k\|_{\mathcal{G}} = \|\Phi(z)\|_{\mathcal{G}}$. Then from (19) we have:

$$\mu \geq \|\phi_1\|_{\mathcal{G}} = \|A + B_2(F + Q_0)\|_{\mathcal{G}} \tag{20}$$

Hence

$$\min_{Q \in \mathcal{RH}_\infty} \mu \geq \min_{\hat{F}} \|A + B_2\hat{F}\|_{\mathcal{G}} \triangleq \mu^* \tag{21}$$

The proof is completed by noting that for $F = \hat{F}$ and $Q(z) = 0$ we have $\mu = \|\phi_1\|_{\mathcal{G}} = \mu^*$ since $\|.\|_{\mathcal{G}}$ is submultiplicative, $\|A_F\|_{\mathcal{G}} \leq 1$, and $\phi_k = A_F^k$ ⋄.

• **Lemma 6:** If $Q(z)$ is such that its impulse response satisfies $\|q_i\|_2 \leq C_q\delta^i$, with $\delta < 1$, then there exists N, independent of Q, such that:

$$\|\Phi(z)\|_{\mathcal{G}} = \sup_{1 \leq k \leq N} \|\phi_k\|_{\mathcal{G}}$$

Proof: Let ρ denote the spectral radius of A_F. Since A_F is stable, $|\rho| < 1$ and it can be easily shown that there exist $C_a, 1 > \lambda > \delta$ such that $\|A_F^k\|_2 \leq C_a\lambda^k$. Hence

$$
\begin{aligned}
\|\phi_k\|_2 &\leq \|A_F^k\|_2 + \sum_{i=0}^{k-1} \|A_F^{k-1-i}\|_2 \|B_2\|_2 \|Q_i\|_2 \\
&\leq C_a\lambda^k + \frac{C_aC_q\|B_2\|_2}{\lambda - \delta}\lambda^k \triangleq M\lambda^k
\end{aligned} \tag{22}
$$

Let μ_u denote the minimum (over all $Q \in \mathcal{RH}_\infty$) of $\|\Phi\|_{\mathcal{G}}$, obtained by solving the optimization problem (15). Clearly $\mu_u \leq \|\Phi\|_{\mathcal{G}}$. From the equivalence of all finite dimensional matrix norms [15] it follows that there exist c such that $\|.\|_{\mathcal{G}} \leq c\|.\|_2$. Hence, by selecting N such that:

$$cM\lambda^N < \mu_u \tag{23}$$

we have that $\|\phi_k\|_{\mathcal{G}} \leq \mu_u \leq \|\Phi\|_{\mathcal{G}}\ \forall\ k \geq N$ and therefore $\max_k \|\phi_k\|_{\mathcal{G}}$ is achieved for some $k < N$ ⋄.

In the sequel we consider the following special cases: i) The perturbation $\underline{\omega}$ is a bounded power spectral signal and the performance variable $\underline{\zeta}$ is a bounded power signal (or alternatively $\underline{\omega} \in l_2$ and $\underline{\zeta} \in l_\infty$), hence the appropriate induced norm is $\|T_{\zeta w}\|_2$; and ii) $\underline{\omega}$ and $\underline{\zeta} \in l_2$, resulting in $\|T_{\zeta w}\|_\infty$. The dual interpretation of the disturbances as l_1 signals, for the purpose of constrained stability, and as bounded spectral power or bounded power, for the purpose of performance, is similar to the approach used by Bernstein and Haddad [18] to address the mixed $\mathcal{H}_2/\mathcal{H}_\infty$ problem.

3.3 $\mathcal{H}_\infty$ Performance Criterion

In this case the mixed performance control problem can be stated as:

$$\min_{Q \in \mathcal{RH}_\infty} \|T_{11}^x + T_{12}^x \tilde{Q} T_{21}^x\|_{\mathcal{G}} \tag{$\mathcal{H}_\infty$}$$

subject to:

$$\|T_{11}^f + T_{12}^f \tilde{Q} T_{21}^f\|_\infty \leq \gamma$$

Since U is inner and V is co–inner, we can find $U_\perp, V_\perp$ such that $(U\ U_\perp)$ and $(V\ V_\perp)'$ are unitary. Since (pre)post–multiplication by a unitary matrix preserves the ∞ norm we have that:

$$
\begin{aligned}
\|T_{\zeta w}\|_\infty &= \| \begin{pmatrix} U \\ U_\perp \end{pmatrix}^{\sim} T_{\zeta w} \begin{pmatrix} V & V_\perp \end{pmatrix}^{\sim} \|_\infty \\
&= \| \begin{pmatrix} U^{\sim}(G_cB_1 - NFG_f)V^{\sim} + R_B^{\frac{1}{2}}\tilde{Q}R_L^{\frac{1}{2}} & U^{\sim}(G_cB_1 - NFG_f)V_\perp^{\sim} \\ U_\perp^{\sim}(G_cB_1 - NFG_f)V^{\sim} & U_\perp^{\sim}(G_cB_1 - NFG_f)V_\perp^{\sim} \end{pmatrix} \|
\end{aligned} \tag{24}
$$

Note that, in general, we have a 4-block general distance problem. In this paper, for simplicity, we will limit ourselves to the special case where the system is square and right invertible. In this case U, V are unitary and (24) reduces to:

$$\|T_{\zeta w}\|_\infty = \|U^{\sim}(G_cB_1 - NFG_f)V^{\sim} + R_B^{\frac{1}{2}}\tilde{Q}R_L^{\frac{1}{2}}\|_\infty \triangleq \|R + Q_B\|_\infty \tag{25}$$

where $R \triangleq U^{\sim}(G_cB_1 - NFG_f)V^{\sim}$ and $Q_B \triangleq R_B^{\frac{1}{2}}\tilde{Q}R_L^{\frac{1}{2}}$.

Problem ($\mathcal{H}_\infty$) is a convex optimization problem in $\mathcal{RH}_\infty$. However, even though this space is complete, it is not compact. Therefore a minimizing solution to ($\mathcal{H}_\infty$) may not exists. Motivated by this difficulty we introduce the additional constraint that all the poles of the closed–loop system must lay in the closed δ–disk, where $\delta < 1$ is given. Thus, the original problem is modified to:

$$\min_{\tilde{Q} \in \mathcal{RH}_\delta} \|\Phi(z)\|_{\mathcal{G}} = \|T_{11}^x + T_{12}^x \tilde{Q} T_{21}^x\|_{\mathcal{G}} \tag{$\mathcal{H}_\infty^\delta$}$$

subject to:

$$\Phi(z) \in \mathcal{RH}_\delta$$

$$\|T_{11}^f + T_{12}^f \tilde{Q} T_{21}^f\|_\delta \leq \gamma$$

where $\mathcal{RH}_\delta = \{Q(z) \in \mathcal{RH}_\infty : Q(z) \text{ analytic in } |z| \geq \delta\}$ and $\|G(z)\|_\delta = \sup_{|z|=\delta} \sigma_{max}(G(z)))$.

• **Theorem 2:** Let $Q_F = \sum_{i=0}^{N-1} q_i z^{-i}$ be given. Then, the condition that there exist $Q_R \in \mathcal{RH}_\delta$ such that $\|R+Q\|_\delta \le \gamma$, where $Q = Q_F + z^{-N} Q_R$ and R has all its poles outside the disk $|z| \le \delta$, is equivalent to a *convex* constraint of the form $\|\mathcal{Q}\|_2 \le \gamma$ where:

$$\begin{aligned}
\mathcal{Q} &= W^{\frac{1}{2}} \begin{pmatrix} I & 0 \\ 0 & \mathcal{H}' \end{pmatrix} L_c^{\frac{1}{2}} \\
Lc &= \begin{pmatrix} L_{11}^C & L_{12}^C \\ L_{12}^{C\prime} & L_{22}^C \end{pmatrix} \\
L_{11}^C &= L_o^C \\
L_{12}^C &= -\left((A_R')^{N-1} C_R' \quad (A_R')^{N-2} C_R' \dots \quad C_R' \right) \\
L_{22}^C &= I_N \\
W'^{\frac{1}{2}} W^{\frac{1}{2}} &= \begin{pmatrix} L_o^0 & \mathcal{A} \\ \mathcal{A}' & I \end{pmatrix} \\
\mathcal{A} &= \left(A_R^{-N} B_R \quad A_R^{-(N-1)} B_R \dots A_R^{-1} B_R \right) \\
\mathcal{H} &= \begin{pmatrix} H_N & H_{N-1} & \dots & \dots & H_1 \\ & H_N & H_{N-1} & \dots & H_2 \\ & & \ddots & & \\ & & & H_N & H_{N-1} \\ & & & & H_N \end{pmatrix} \\
H_i &= q_{N-i} + B_R' (A_R')^{N-1-i} C_R' \qquad 1 \le i \le N-1 \\
H_N &= q_0 + D_R \\
R &= \left(\begin{array}{c|c} \delta A_R & \delta^{\frac{1}{2}} B_R \\ \hline \delta^{\frac{1}{2}} C_R & D_R \end{array} \right)
\end{aligned} \tag{26}$$

and where L_o^0 and L_o^C are the solutions to the following Lyapunov equations:

$$\begin{aligned}
A_R L_o^0 A_R' - L_o^0 &= B_R B_R' \\
A_R' L_o^C A_R - L_o^C &= (A_R')^N C_R' C_R (A_R)^N
\end{aligned} \tag{27}$$

Proof: Consider first the case where $\delta = 1$. Let $G \stackrel{\Delta}{=} R + Q_F$. The proof follows by noting that, given Q_F, there exist $Q_R \in \mathcal{RH}_\infty$ such that $\|T_{\zeta w}\|_\infty \le \gamma$ *iff* the corresponding *unconstrained* 1 block Nehari approximation problem has a solution, i.e. if:

$$\begin{aligned}
\min_{q_R \in \mathcal{RH}_\infty} \|G + z^{-N} q_R\|_\infty &= \min_{q_R \in \mathcal{RH}_\infty} \|z^N G + q_R\|_\infty \\
&= \min_{q_R \in RH_\infty^-} \|z^{-N} G^\sim + q_R\|_\infty \\
&= \Gamma_H (z^{-N} G^\sim) \le \gamma
\end{aligned} \tag{28}$$

where Γ_H indicates the maximum Hankel singular value and where we used the facts that z^N is an inner function and that the best stable approximation to a given function coincides with the best antistable approximation to its conjugate. In order to compute Γ_H we need to compute the observability L_o and controllability L_c grammians of the stable part $\mathcal{G}$ of $z^{-N} G^\sim$. In [13] we showed, through some lengthy computations, that these grammians can be computed explicitly. Furthermore, L_c is independent of Q_F and L_o is given by:

$$\begin{aligned}
L_o &= \begin{pmatrix} L_o^0 & \mathcal{A}\mathcal{H}' \\ \mathcal{H}\mathcal{A}' & \mathcal{H}\mathcal{H}' \end{pmatrix} \\
&= \begin{pmatrix} I & 0 \\ 0 & \mathcal{H} \end{pmatrix} \begin{pmatrix} L_o^0 & \mathcal{A} \\ \mathcal{A}' & I \end{pmatrix} \begin{pmatrix} I & 0 \\ 0 & \mathcal{H}' \end{pmatrix}
\end{aligned}$$

Hence:

$$\begin{aligned}
L_c^{\frac{1}{2}} L_o L_c^{\frac{1}{2}} &= \mathcal{Q}' \mathcal{Q} \\
\mathcal{Q} &\stackrel{\Delta}{=} W^{\frac{1}{2}} \begin{pmatrix} I & 0 \\ 0 & \mathcal{H}' \end{pmatrix} L_c^{\frac{1}{2}}
\end{aligned} \tag{29}$$

From Nehari Theorem it follows that:

$$\begin{aligned}
\|T_{\zeta w}\|_\infty \le \gamma &\iff \rho^{\frac{1}{2}} \left(L_c^{\frac{1}{2}} L_o L_c^{\frac{1}{2}} \right) \le \gamma \\
&\iff \|\mathcal{Q}\|_2 \le \gamma
\end{aligned} \tag{30}$$

where ρ indicates the spectral radius. The proof is completed by noting that the case $\delta < 1$ follows by using the transformation $z = \delta \hat{z}$ ⋄.

By combining the results of Lemma 6 and Theorem 2, it follows that for $\mathcal{H}_\infty$ constraints, the mixed performance robust control problem can be solved by solving a finite–dimensional convex optimization problem and an unconstrained Nehari approximation problem. This result is summarized in the following theorem:

• **Theorem 3:** $Q^o(z) = Q_F^o(z) + z^{-N} Q_R^o(z)$ solves problem $(\mathcal{H}_\infty)$ **iff** $Q_F^o(z) = \sum_{i=0}^{N-1} Q_i z^{-i}$ solves the finite–dimensional convex optimization problem:

$$\min_{Q_i} \left\{ \max_{1 \le k \le N_\delta} \|\phi_k\|_{\mathcal{G}} \right\}$$

subject to:

$$\|\mathcal{Q}\|_2 \le \gamma$$

and Q_R^o solves the unconstrained Nehari approximation problem:

$$\min_{Q_R \in \mathcal{RH}_\delta} \|z^{-N} G^\sim + Q_R\|_\delta$$

where $G = R + Q_F^o$ and N is selected according to Lemma 6.

Proof: Since Q satisfies the constraint $\|Q + R\|_\delta \le \gamma$ then $\|Q\|_\delta \le \|R\|_\delta + \gamma \stackrel{\Delta}{=} C_q$. Since $Q(z)$ is analytical inside the closed disk $|z| \le \delta$ we have that:

$$Q_k = \frac{1}{2\pi j} \oint_\tau Q(z) z^{k-1} dz \tag{31}$$

where τ is the circle with radius δ. From (31) it follows that:

$$\|Q_k\|_2 \le \|Q\|_\delta \delta^k \le C_q \delta^k \tag{32}$$

The proof follows now from Theorem 2 and Lemma 6⋄.

3.4 $\mathcal{H}_2$ Performance Criterion

Let q_i denote the coefficients of the impulse response of $R_B^{\frac{1}{2}} \tilde{Q} R_L^{\frac{1}{2}}$. Then, from Lemma 4 it follows that, given $\gamma \ge \|G_c B_1 - NFG_f\|_2$, all stabilizing controllers yielding $\|T_{\zeta w}\|_2 \le \gamma$ can be parametrized in terms of $\tilde{Q}$, where q_i satisfy the following constraint:

$$\sum_{i=0}^{i=\infty} \|R_B^{\frac{1}{2}} q_i R_L^{\frac{1}{2}}\|_F^2 \le \gamma^2 - \|G_c B_1 - NFG_f\|_2^2 \tag{32}$$

where $\|.\|_F$ denotes the Frobenius norm. This results follows immediately from (14) and the orthogonality of $G_c B_1, U$, G_f, V by noting that $\|Q(z)\|_2^2 = \sum_{i=0}^{\infty} \|Q_i\|_F^2$.

As in the $\mathcal{H}_\infty$ case, to avoid the difficulties due to the non–compactness of $\mathcal{RH}_\infty$, the original problem is modified to:

$$\min_{\tilde{Q} \in \mathcal{RH}_\delta} \|\Phi(z)\|_{\mathcal{G}} = \|T_{11}^x + T_{12}^x \tilde{Q} T_{21}^x\|_{\mathcal{G}} \tag{$\mathcal{H}_2^\delta$}$$

subject to:

$$\begin{aligned}
\sum_{i=0}^{i=\infty} \|R_B^{\frac{1}{2}} q_i R_L^{\frac{1}{2}}\|_F^2 &\le \gamma^2 - \|G_c B_1 - NFG_f\|_2^2 \\
\|q_i\|_2 &\le C_q \delta^i
\end{aligned} \tag{33}$$

Remark 5: The additional constraint puts an upper bound on $\|Q\|_\infty$, since it forces $\|Q\|_\infty \le \frac{C_q}{1-\delta}$. This additional constraint improves the robustness by bounding $\|T_{\zeta w}\|_\infty$.

• **Theorem 4:** Let $Q(z) = \sum_{k=1}^{\infty} Q_k z^{-k}$. Then the solution to the problem $(\mathcal{H}_2^\delta)$ has $Q_k = 0$ for $k > N$.

Proof: The proof follows from Lemma 6 by noting that ϕ_k depends only on Q_i, $i \le k$. Since optimization of $\|\Phi\|_{\mathcal{G}}$ requires considering only ϕ_k, $k \le N$, terms in the impulse response of $Q(z)$ corresponding to $k > N$ can only increase the $\mathcal{H}_2$ cost while not affecting $\|\Phi\|_{\mathcal{G}}$.

4. Conclusions

Most realistic control problems involve both some type of time–domain constraints and certain degree of model uncertainty. Model uncertainty can be successfully addressed through frequency–domain constraints combined with $\mathcal{H}_\infty$ techniques. However the standard $\mathcal{H}_\infty$ formalism cannot handle time–domain constraints. Alternatively, the recently developed l_1 robust control theory can be used to deal with model uncertainty. Although this framework can incorporate time domain constraints, it cannot handle frequency domain specifications.

In this paper we propose to approach model uncertainty and time–domain constraints using an operator norm induced by the constraints to assess the stability properties of a family of systems. The proposed controller design method results in a convex optimization problem, where additional frequency domain constraints can be imposed. We showed that when these additional constraints have the form of an $\mathcal{H}_2$ or $\mathcal{H}_\infty$ bound, the resulting problem can be transformed into a finite dimensional optimization and solved exactly. Thus, the conservatism inherent in some previous approaches is eliminated. Although here we considered only the simpler case of a one–block problem, we anticipate that the results will extend naturally to the 4–block case.

Perhaps the most severe limitation of the proposed method is that may result in very large order controllers (roughly $2N$) necessitating some type of model reduction. Preliminary results suggest that substantial order reduction can be accomplished without performance degradation. Research is currently under way addressing this issue and pursuing the extension of the formalism to structured uncertainty.

References

[1]. M. Sznaier and M. J. Damborg, "Heuristically Enhanced Feedback Control of Constrained Discrete Time Linear Systems," *Automatica*, Vol 26, 3, pp 521–532, 1990.

[2]. M. Vassilaki, J. C. Hennet and G. Bitsoris, "Feedback Control of Discrete–Time Systems under State and Control Constraints", *Int. J. Control*, 47, pp. 1727–1735, 1988.

[3]. G. Zames, "Feedback and Optimal Sensitivity: Model Reference Transformations, Multiplicative Seminorms and Approximate Inverses", *IEEE Trans. Autom. Contr.*, Vol 26, 2, pp. 301–320, 1981.

[4]. J. Doyle, "Analysis of Feedback Systems with Structured Uncertainties," *IEE Proc.*, Vol 129, Pt. D, 6, pp. 252–250, 1982.

[5]. S. Skogestad, M. Morari and J. Doyle, "Robust Control of Ill–Conditioned Plants: High–Purity Distillation," *IEEE Trans. Autom. Contr.*, 33, 12, pp. 1092–1105, 1988.

[6]. E. Polak and S. Salcudean, "On the Design of Linear Multivariable Feedback Systems Via Constrained Nondifferentiable Optimization in H_∞ Spaces," *IEEE Trans. Automat. Contr.*, Vol 34, 3, pp 268–276, 1989.

[7]. S. Boyd and C. Barratt, "Linear Controller Design: Limits of Performance," Prentice Hall Information and Systems Sciences Series, Englewood Cliffs, New Jersey, 1991.

[8]. J. W. Helton and A. Sideris, "Frequency Response Algorithms for H_∞ Optimization with Time Domain Constraints," *IEEE Trans. Autom. Contr.*, Vol 34, 4, pp. 427–434, 1989.

[9]. A. Sideris and H. Rotstein, "H_∞ Optimization with Time Domain Constraints over a Finite Horizon," *Proc. of the 29th IEEE CDC*, Hawaii, Dec 5-7 1990, pp. 1802–1807.

[10]. M. Vidyasagar, "Optimal Rejection of Persistent Bounded Disturbances," *IEEE Trans. Autom. Contr.*, 31, pp. 527–535, 1986.

[11]. M. Khammash and J. B. Pearson, "Performance Robustness of Discrete–Time Systems with Structured Uncertainty," *IEEE Trans. Autom. Contr.*, 36, pp. 398–412, 1991.

[12]. M. Sznaier and A. Sideris, "Norm Based Robust Dynamic Feedback Control of Constrained Systems, " *Proc. of the First IFAC Symposium on Design Methods for Control Systems,* Zurich, Switzerland, September 4–6, 1991, pp. 258–263.

[13]. M. Sznaier and A. Sideris, "Suboptimal Norm Based Robust Control of Constrained Systems with an H_∞ Cost," *Proc. of the 30th IEEE CDC,* Brighton, England, December 11–13 1991, pp. 2280–2286.

[14]. J. B. Conway, "A Course in Functional Analysis," *Vol 96 in Graduate Texts in Mathematics,* Springer–Verlag, New–York, 1990.

[15]. R. A. Horn and C. R. Johnson, *Matrix Analysis,* Cambridge University Press, 1985.

[16]. M. Sznaier, *"Norm Based Robust Control of State–Constrained Discrete–Time Linear Systems,"* IEEE Transactions Autom. Contr., 37, 7, pp. 1057–1062, 1992.

[17]. J. Doyle, "Lecture Notes in Advances in Multivariable Control," *ON-R/Honeywell Workshop,* Minneapolis, MN., 1984.

[18]. D. S. Bernstein and W. M. Haddad, "LQG Control with an H_∞ Performance Bound: A Riccati Equation Approach," *IEEE Trans. Autom. Contr.,* 34, pp. 293–305, 1989.

Proceedings of the 31st Conference
on Decision and Control
Tucson, Arizona • December 1992

Controller Design with Actuator Constraints

Athanasios G. Tsirukis
and
Manfred Morari

Chemical Engineering 210-41
California Institute of Technology
Pasadena, CA 91125

Abstract

The Infinite-Horizon Model Predictive Controller (IH-MPC) can globally stabilize discrete linear systems in the presence of actuator constraints, as long as their eigenvalues lie in the closed unit disk. These conditions are proven to be necessary and sufficient, i.e. if the system cannot be stabilized by IH-MPC, then there exists no other stabilizing control law. Finally, the form of Infinite-Horizon Optimal Controller (IH-OC), which is a limiting case of IH-MPC, coupled with an asymptotic observer is shown to stabilize certain classes of linear systems with output feedback, in the presence of asymptotically vanishing disturbances.

1. Introduction

Recently, the problem of control design in the presence of pointwise in time actuator constraints is receiving increased attention. It was shown by Sontag and Sussmann, [1], that continuous linear systems with saturated controls can be smoothly stabilized under certain natural controllability and eigenvalue conditions. Based on their exitence result, Teel, [2], proposed a control design that stabilizes single-input multiple integrators of any dimension. His design method was extended by Yang et al., [3], to stabilize linear systems that satisfy the conditions underlined in [1]. In parallel, optimization-based designs, as for example Model Predictive Control (MPC), [4], also known as moving horizon control and receding horizon control, have attracted significant attention, mainly due to their inherent ability to satisfy actuator constraints. The availability of cheap, powerful computation has significantly contributed to the acceptance of MPC by the process industry, [4]. Recently, Muske and Rawlins, [5], proved that the Infinite-Horizon Model Predictive Controller guarantees closed-loop stability for stable discrete linear systems with feasible input and state constraints. Michalska and Mayne, [6], proposed a generalization of the Infinite-Horizon Optimal Control design that stabilizes nonlinear systems. Another optimization-based design for linear systems was proposed by Yang and Polak, [7], which was guaranteed to be stable under the presence of disturbances and plant uncertainty. In the present paper we show that the controllability and eigenvalue conditions of [1] carry over as necessary and sufficient conditions to the case of discrete linear systems controlled by IH-MPC. In addition, some limiting results on the stability of these systems in the presence of asymptotically vanishing disturbances (CICS) and output feedback are reported. Due to space limitations, the proofs of the main theorems will only be outlined.

2. Problem Definition

In this paper we consider discrete, linear systems of the type:

$$x_{k+1} = Ax_k + Bu_k \; ; \quad x_0 = x^* \tag{1}$$

where $x_k \in \Re^n$, $u_k \in \Re^m$, $A \in \Re^{n\times n}$ and $B \in \Re^{n\times m}$. We assume that x_k is measured, unless otherwise specified. In addition, linear constraints are imposed on the manipulated variables:

$$|u_{kj}| \leq \epsilon, \quad j = 1, 2, \ldots, m \tag{2}$$

$$D_0 u_k + D_1 u_{k-1} + \cdots + D_r u_{k-r} \leq d \tag{3}$$

where $\epsilon > 0$, $d \in \Re^p$ and $D_i \in \Re^{p\times m}$. In the remainder of the present section we will define the Model Predictive Controller and its Infinite-Horizon variant.

Definition 1. Consider a dynamic system of the form (1). Then, at time k , MPC is based on the solution of the following optimization problem:

$$\min_{z,v} \quad \sum_{j=0}^{H} z_j^T R z_j + v_j^T S v_j \tag{4}$$

s.t.

$$z_0 = x_k \tag{5}$$

$$z_{j+1} = Az_j + Bv_j \tag{6}$$

$$||v_j||_\infty \leq \epsilon \tag{7}$$

$$D_0 v_j + D_1 v_{j-1} + \cdots + D_r v_{j-r} \leq d \tag{8}$$

$$j = 0, 1, \ldots, H$$

CH3229-2/92/0000-2623$1.00 © 1992 IEEE

where R and S are symmetric positive definite matrices. Of the $H+1$ control moves computed at the kth time-instance, only the first one is implemented: $u_k = v_0$. At the next time instance, when x_{k+1} becomes available, problem (4)-(8) is solved with the new initial condition to provide u_{k+1}.

A drawback of MPC is that it does not necessarily lead to closed-loop stable systems. This deficiency is removed by introducing the Infinite-Horizon MPC, which is based on the minimization of the following infinite horizon quadratic objective function:

$$\sum_{j=0}^{\infty} z_j^T R z_j + v_j^T S v_j \qquad (9)$$

To avoid an infinite-dimensional quadratic program, we set the input to 0 for time-instances $j \geq N$:

$$v_j = 0, \; j \geq N \qquad (10)$$

If A contains unstable modes, then we partition A as follows:

$$A = VJV^{-1} = [V_u \; V_s] \begin{bmatrix} J_u & 0 \\ 0 & J_s \end{bmatrix} \begin{bmatrix} \tilde{V}_u \\ \tilde{V}_s \end{bmatrix}$$

where J_s contains the eigenvalues of A that lie in the open unit disk, $(|\lambda_i| < 1)$, and J_u contains the remaining eigenvalues of A. If A contains unstable modes and if there exists a sequence $\{v_j, j = 0, 1, \ldots, N-1\}$ such that (9) is bounded, then the following equation must be satisfied:

$$\tilde{V}_u x_N = 0 \qquad (11)$$

otherwise (9) would become unbounded. Clearly, the addition of equations (11) as constraints to the optimization formulation does not alter the location of the optimum. In summary, the IH-MPC formulation, $P(N, x_k)$, is comprised of the objective function (9) and constraints (5)-(8), (10) and (11). $P(N,x)$ contains two parameters: the initial state x and the number of iterations N. The particular case of IH-MPC with $N = \infty$ is called the Infinite-Horizon Optimal Controller. We denote by $\phi_N(x)$ the objective function at the optimum of $P(N,x)$ and by $\varphi(x)$ the objective function at the optimum of the Infinite Horizon Optimal Control formulation, $P(\infty, x)$.

3. Stability Results

In the present section we summarize the main results on the stability of IH-MPC applied to constrained linear systems of the type (1)-(3), in the form of three theorems. Central to the development of IH-MPC is the feasibility of $P(N,x)$ with respect to constraints (11). Theorem 1 presents some necessary and sufficient conditions on the eigenvalues of A, under which $P(N,x)$ is guaranteed to be feasible. These results naturally lead to the stability of IH-MPC, presented in Theorem 2. Finally, some additional implications of Theorem 1 are outlined by Theorem 3 and Corollaries 1 and 2.

Theorem 1. Consider a discrete linear dynamic system of the form (1), where (A, B) is controllable. Denote by $\lambda_i \in \mathcal{C}$ the eigenvalues of A. Then,

1. If λ_i lie in the closed unit disk, i.e. $|\lambda_i| \leq 1$, then $\forall x^* \in \Re^n$ and $\forall \epsilon > 0$, $\exists N \in \mathcal{Z}$ finite, and a sequence of controls $\{u_k; k = 0, 1, 2, \ldots, N-1\}$ with $|u_{kj}| \leq \epsilon$, $j = 1, 2, \ldots, m$, such that if $x_0 = x^*$ then $x_N = 0$.
2. If at least one λ_j lies outside the closed unit disk, i.e. $|\lambda_j| > 1$, then $\forall \epsilon > 0$, $\exists x^*(\epsilon) \in \Re^n$ such that $\forall N \in \mathcal{Z}$, $N > 0$, and every sequence of controls $\{u_k; k = 0, 1, 2, \ldots, N-1\}$ with $|u_{kj}| \leq \epsilon$, $j = 1, 2, \ldots, m$, if $x_0 = x^*$ then $x_N \neq 0$.

Proof. Notice that constraints (3) have not been included in the statement of Theorem 1, because they can easily be transformed to sufficient conditions on the magnitude of the control variables, of the form (2).

1) The proof is centered around the case where A is a full Jordan block with $\lambda = 1$. All the other cases are direct consequences. The kth iterate can be written as follows:

$$\begin{aligned} x_k &= Ax_{k-1} + Bu_{k-1} \\ &= A^k x_0 + A^{k-1}Bu_0 + \cdots + ABu_{k-2} + Bu_{k-1} \\ &= A^k[x_0 + A^{-1}Bu_0 + \cdots + A^{-k}Bu_{k-1}] \end{aligned}$$

Notice that A^{-1} exists, by assumption. Define the following set:

$$\begin{aligned} \mathcal{X}_k \doteq \{x \in \Re^n : \; & x = A^{-1}Bu_0 + \cdots + A^{-k}Bu_{k-1}, \\ & |u_{tj}| \leq \epsilon, \; t = 0, 1, \ldots, k-1, \\ & j = 1, 2, \ldots, m\} \end{aligned}$$

In essence, $\mathcal{X}_k$ is the set of all points in $\Re^n$ that are created by all the possible combinations of values of the bounded manipulated variables in k iterations. We need to prove that $\forall x \in \Re^n$, $\exists$ $k \in \mathcal{Z}$, such that $\mathcal{X}_k$ contains x. If this happens, then for every bounded initial condition x_0, $\exists N^*$ such that $-x_0 \in \mathcal{X}_{N^*}$. Hence, $\exists$ a control sequence, $\{u_0, u_1, \ldots, u_{N^*-1}\}$, $|u_{ij}| \leq \epsilon$, such that:

$$\begin{aligned} -x_0 &= A^{-1}Bu_0 + \cdots + A^{-N^*}Bu_{N^*-1} \quad \Rightarrow \\ x_{N^*} &= A^{N^*}[x_0 + A^{-1}Bu_0 + \cdots + A^{-N^*}Bu_{N^*-1}] \\ &= A^{N^*}[x_0 - x_0] \\ &= 0 \end{aligned}$$

We will study the structure of $\mathcal{X}_k$. Notice that $\mathcal{X}_k = \mathcal{A}_k(\mathcal{U}_k)$, where $\mathcal{A}_k \doteq [A^{-1}B \;\; A^{-2}B \;\; \cdots \;\; A^{-k}B]$ and $\mathcal{W}_k = \{w \in \Re^{km} : |w_i| \leq \epsilon, \; i = 1, \ldots, km\}$. Since (A, B) is controllable, rank$(\mathcal{A}_n) = n$ and $\mathcal{X}_k$ is a convex body in $\Re^n$, $k \geq n$. Therefore, $\exists\, \delta > 0$ such that the closed n-dimensional cube:

$$c(\delta) = \{x \in \Re^n : ||x||_\infty \leq \delta\} \qquad (12)$$

is a subset of $\mathcal{X}_n$. Suppose that $k = ln$, where $l \in \mathcal{Z}$ and $l \geq 1$. Then,

$$\begin{aligned} \mathcal{X}_{ln} = \{x \in \Re^n : x = \; & A^{-1}Bu_0 + \cdots + A^{-n}Bu_{n-1} + \\ & A^{-n-1}Bu_n + \cdots + A^{-2n}Bu_{2n-1} + \\ & \cdots + \\ & A^{-(l-1)n-1}Bu_{(l-1)n} + \cdots + A^{-ln}Bu_{ln-1}, \\ & |u_{ij}| \leq \epsilon\} \\ = \{x \in \Re^n : x = \; & y_0 + A^{-n}y_1 + \cdots + A^{-(l-1)n}y_{l-1}, \\ & y_i \in \mathcal{X}_n\} \\ \supseteq \{x \in \Re^n : x = \; & y_0 + A^{-n}y_1 + \cdots + A^{-(l-1)n}y_{l-1}, \\ & ||y_i|| \leq \delta\} \doteq \mathcal{X}'_{ln} \end{aligned}$$

In order to prove that $\mathcal{X}'_{ln}$ (consequently $\mathcal{X}_{ln}$) can grow to contain any x, we identify inside $\mathcal{X}'_{ln}$ a convex polyhedron, $\mathcal{X}''_{ln}$, whose vertices are the intercepts of $\mathcal{X}'_{ln}$ with the coordinate axes. The intercept of $\mathcal{X}'_{ln}$ with the ith coordinate axis is given by the objective function value at the optimum of the following linear program:

$$\max_{y,z} \quad y_i \tag{13}$$
$$s.t.$$
$$y = \sum_{t=0}^{l-1} A^{-tn} z_t$$
$$y_j = 0, \quad j = 1, \ldots, n, \quad j \neq i$$
$$|z_{tj}| \leq \delta, \quad t = 0, \ldots, l-1, \quad j = 1, \ldots, n$$

Let α_{li} denote the objective function value at the optimum of (13). Given the special structure of A, we were able to identify a feasible solution of (13), in which the coordinates of the z_t vectors are 1 and 0. Then, we proved that the horizon length, $N(p)$, necessary to achieve $\alpha_{li} = p$, where p is any integer, $i = 1, 2, \ldots, n$, is upper-bounded as follows:

$$N(p) \leq O(n^{2^{n-1}-1} p^{2^{n-1}}) \tag{14}$$

Hence, if the initial condition of system (1) is $x_0 = [0, 0, \cdots, p]^T$ then $P(N, x_0)$ is feasible. Notice that (14) is always finite, as long as p is bounded. Since any p can be achieved in a finite number of iterations, we can easily show that $\exists$ a finite l, such that X''_{ln} contains any bounded $x \in \Re^n$. The same is true for $\mathcal{X}_{ln}$.

2) A direct consequence of this analysis is that if at least one eigenvalue is outside the closed unit disk, $\mathcal{X}_k$ cannot grow to include certain $x \in \Re^n$, $\forall k \in \mathcal{Z}$. □

The essence of Theorem 1 is that for systems with eigenvalues on the closed unit disk and for every bounded initial condition x_0, there exists an integer $N(x_0)$ such that the problem $P(N', x_0)$ is feasible, for $N' \geq N$. This is not true if at least one of the eigenvalues of A lies outside the closed unit disk. Based on Theorem 1 we will prove that IH-MPC can globally asymptotically stabilize discrete, constrained linear systems with eigenvalues on the closed unit disk. Before proceeding with the statement of Theorem 2, we present a number of properties for $\phi_N(x)$ and $\varphi(x)$, upon which the proof of Theorem 2 will be based.

3.1 Properties of $\phi_N(x)$

1. $\phi_N(x)$ is defined $\forall x \in \mathcal{X}_N$
2. $\phi_N(x) \geq 0$.
3. $\phi_N(x) = 0 \Leftrightarrow x = 0$.
4. $\forall N \geq N(x)$, $\phi_N(x) \geq \phi_{N+1}(x)$.
5. $\phi_N(x)$ is continuously differentiable, [8]
6. $\phi_N(x)$ is convex in $\mathcal{X}_N$, [8]
7. If $\{z_j,\ j = 1, \ldots, N-1\}$ is the sequence of states generated by the solution of $P(N, x)$, then:
   - If $P(N, x)$ is well defined, then so are $P(N, z_j)$, $j = 1, \ldots, N-1$.
   - $\phi_N(z_1) \leq \phi_N(x) - x^T R x - v_0^T S v_0$. In general:
   $$\phi_N(z_i) \leq \phi_N(x) - \sum_{j=0}^{i-1} z_j^T R z_j - \sum_{j=0}^{i-1} v_j^T S v_j$$
8. If A is similar to the following structure:
   $$\begin{bmatrix} A_1 & 0 \\ 0 & A_2 \end{bmatrix} \tag{15}$$
   where A_1 is unitary and A_2 is stable, then $\forall x_0 \in \Re^n$ bounded, $\exists\ 0 < N^*(x_0) < \infty$, such that for $k \geq N^*(x_0)$:
   $$\phi_k(x_0) \leq \beta_0 ||x_0||^3 + \beta_1 ||x_0||^2 + \beta_2 ||x_0|| + \beta_3 \tag{16}$$
   where $\beta_i \geq 0$.

   Proof. The proofs of properties 1-7 are easy. The proof of property 8 will be outlined. It is easy to show that:
   $$||x_j||^2 \leq ||x_0||^2 + 2jm\bar{\sigma}(B)\epsilon||x_0|| + j^2 m^2 \bar{\sigma}(B)^2 \epsilon^2 \tag{17}$$
   where $j = 0, 1, \ldots, N$. From the definition of $\phi_N(x_0)$:
   $$\phi_N(x_0) \ \leq \ \bar{\sigma}(R) \sum_{j=0}^{N} ||x_j||^2 + \bar{\sigma}(S) m \epsilon N \tag{18}$$
   If A has the special structure (15), then the number of iterations necessary to bring the system to 0, $N(x_0)$, grows linearly with $||x_0||$. Combining this statement with (17) and (18) we anticipate that the dependence of $\phi_N(x_0)$ on $||x_0||$ is third-order. □
9. $\{\phi_k(x),\ k \geq N\}$ is a positive sequence of decreasing functions, lower-bounded by 0:
   $$\phi_N(x) \geq \phi_{N+1}(x) \geq \cdots \geq 0$$
   Therefore, it has a limit, i.e. a function, $\varphi |\ \Re^n \rightarrow \Re$, which is identical to the solution of the Infinite Horizon Optimal Control formulation.

Properties 1-6, 8 are easily proved for $\varphi(x)$, whereas property 7 is a strict equality

$$\varphi(z_i) = \varphi(x) - \sum_{j=0}^{i-1} z_j^t R z_j - \sum_{j=0}^{i-1} v_j^t R v_j$$

otherwise, the optimality of sequence $\{z_i,\ i = 0, 1, \ldots, \infty\}$ would be violated. In addition,

$$\lim_{||x|| \rightarrow \infty} \varphi(x) = \infty \tag{19}$$

due to the convexity of $\varphi(x)$. From properties 2, 3, 5 and equation (19) we conclude that $\varphi(x)$ is a positive definite function for all x in $\Re^n$. A consequence of property 8 for $\varphi(x)$ is that:

$$\lim_{||x|| \rightarrow \infty} \frac{\varphi(x)}{||x||^3} \leq \beta_0 \tag{20}$$

By a direct application of l'Hôpital's rule to (20), we conclude that $\exists M > 0$, such that for $k \geq N^*(x_0)$ and $||x_0|| > M$

$$||\nabla \varphi(x_0)|| \leq \gamma ||x_0||^2 \tag{21}$$

where $\gamma \geq 0$.

Theorem 2. Consider a discrete linear dynamic system of the type (1) with eigenvalues in the closed unit disk. Suppose that the control variables are subject to a set of constraints of the form (2) and (3). Then, for every bounded initial condition x_0, there exists $N^*(x_0)$ such that IH-MPC with horizon $N \geq N^*$ is asymptotically stabilizing.

Proof. Easy, based on property 7, which proves that $\phi_N(x)$ is a Lyapunov function of the closed-loop system. □.

Theorem 2 is a generalization of the results presented in [5], proving that IH-MPC is globally stabilizing for systems with (possibly) multiple eigenvalues on the unit circle. From a different viewpoint, Theorem 2 confirms the results of [1] for discrete linear systems stabilized by the IH-MPC. Theorem 1 carries some additional implications. We prove next that if IH-MPC cannot stabilize system (1) from $x_0 = x^*$, then there exists no feasible control law that can stabilize the system. This is a corollary of the following theorem.

Theorem 3. Consider a discrete linear dynamic system of the form (1). Then, a necessary and sufficient condition for the existence of a control law that satisfies constraints (2)-(3) and stabilizes system (1) from x_0, is the existence of a bounded control sequence $\{v_k; k = 0, 1, \ldots, N-1\}$, $N \in \mathcal{Z}, N < \infty$, which can drive system (1) from x_0 to $x_N = 0$.

Proof. *(a)* If $\exists$ a bounded, feasible sequence $\{v_k; k = 0, 2, \ldots, N-1\}$, $N \in \mathcal{Z}, N < \infty$, such that $x_N = 0$, then, clearly, the sequence

$$\begin{aligned} \{u_k|\ u_k &= v_k,\ k \leq N-1 \ \textit{and} \\ u_k &= 0,\ k \geq N\} \end{aligned}$$

satisfies constraints (2), (3) and asymptotically stabilizes system (1):

$$\lim_{k\to\infty} ||x_k|| = 0$$

(b) Suppose that $\exists$ a control sequence $\{u_k; k = 0, 1, \ldots, \infty\}$, which satisfies constraints (2), (3) and asymptotically stabilizes system (1):

$$\lim_{k\to\infty} ||x_k|| = 0 \tag{22}$$

From (22) we know that $\exists M \in \mathcal{Z}$, such that $||x_M|| \leq \delta$ where δ is defined in (12). Clearly, since $x_M \in \mathcal{X}_n$, $\exists$ a sequence $\{w_k, k = 0, 1, \ldots, n-1\}$ that satisfies constraints (2), (3), such that:

$$\begin{aligned} -x_M &= A^{-1}Bw_0 + \cdots + A^{-n+1}Bw_{n-1} \quad \Rightarrow \\ x_{M+n} &= A^{n-1}[x_M + A^{-1}Bw_0 + \cdots + A^{-n+1}Bw_{n-1}] \\ &= A^{n-1}[x_M - x_M] \\ &= 0 \end{aligned}$$

Then, the following sequence:

$$\begin{aligned} \{v_k|\ v_k &= u_k,\ k \leq M-1 \ \textit{and} \\ v_k &= w_{k-M},\ M \leq k \leq M+n-1\} \end{aligned}$$

is clearly feasible and produces $x_{M+n} = 0$. □

The following propositions are direct consequences of Theorems 1 and 2.

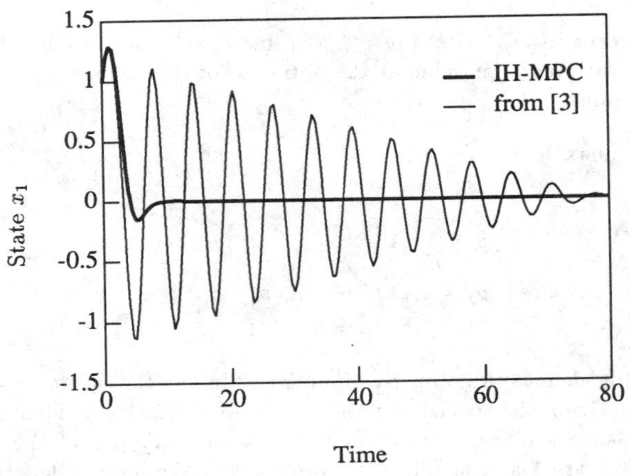

Figure 1: Time-evolution of x_1

Corollary 1. If system (1) subject to constraints (2) and (3) cannot be asymptotically stabilized from $x_0 = x^*$ by IH-MPC, then there exists no other feasible control law that can asymptotically stabilize (1).

Corollary 2. A necessary and sufficient condition for the existence of a feasible, globally asymptotically stabilizing controller for system (1) subject to constraints (2) and (3) is that $|\lambda_i(A)| \leq 1$.

3.2 Example

In the present section we compare the performance of the IH-MPC applied to the stabilization of an oscillator with multiplicity two, with the performance of the control algorithm presented in [3] for the same system. The continuous system is: $\dot{x} = Ax + bu$, where

$$A = \begin{bmatrix} 0 & 1 & 0 & 0 \\ -1 & 0 & 1 & 0 \\ 0 & 0 & 0 & 1 \\ 0 & 0 & -1 & 0 \end{bmatrix}, \quad b = \begin{bmatrix} 0 \\ 0 \\ 0 \\ 1 \end{bmatrix} \tag{23}$$

and u satisfying the constraint $|u| \leq \epsilon = 1$. In order to apply the IH-MPC controller, the system was discretized with a sampling time of 0.1. Using the notation of Definition 1, the selected objective function weights are: $R = I$ and $S = 10$. The two different control laws were applied to the stabilization of the system from the initial condition: $x_0 = [1, 0.5, 0.5, 1]^T$. For this initial condition, $N(x_0) = 50$. Figure 1 depicts the time-evolution of state x_1 for the two closed-loop systems. The behavior of the other three states is similar. The corresponding control actions are shown in Figure 2. We immediately notice that even though both control laws are able to stabilize the system, the difference in performance is striking. The qualitative features of the control actions are similar, at least in the beginning. By being more aggressive initially, the IH-MPC is able to stabilize the system almost an order of magnitude sooner than the controller that was proposed in [3]. We should keep in mind however that the last control law was designed to ensure stability but not performance. The general control algorithm that is proposed in [3]

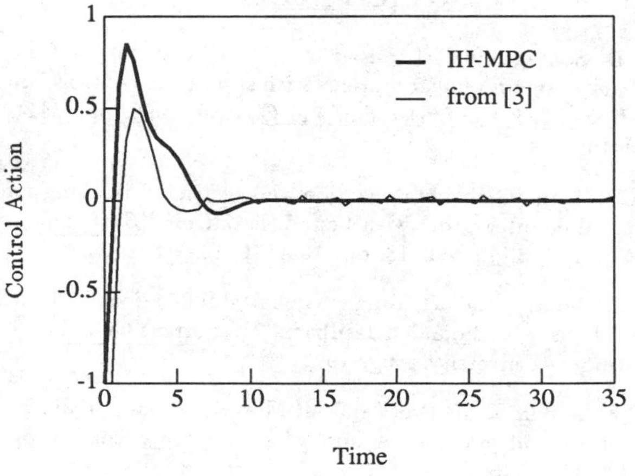

Figure 2: Time-evolution of control action

contains a number of design parameters which can be adjusted by the user, in order to achieve better performance. The performance shown in Figures 1, 2 is insensitive to small variations in R, S, N and ϵ.

4. Disturbance Rejection

In the present section we will examine the stability properties of system (1) controlled by the Infinite Horizon Optimal Controller, when an unknown, asymptotically vanishing disturbance is injected in the system. We were able to prove that IH-OC asymptotically stabilizes systems of the form (1), when A has a special structure.

Theorem 4. Consider a discrete linear dynamic system of the form:

$$x_{k+1} = Ax_k + Bu_k + e_k \qquad (24)$$

where A is similar to structure (15), (A, B) is controllable and e_k is an unknown disturbance, such that $\lim_{k\to\infty} e_k = 0$. Suppose that the control variables are subject to a set of constraints of the form (2) and (3). Then, the Infinite Horizon Optimal Controller is globally asymptotically stabilizing.

Proof. If we assume that at the kth time-instance the states of the system are exactly known, then:

$$\begin{aligned} x_k &= z_0 \\ x_{k+1} &= Ax_k + Bu_k + e_k = z_1 + e_k \end{aligned} \qquad (25)$$

where z are the estimates of the future states x, computed by the optimization problem $P(\infty, x_k)$. We will provide an estimate of $\varphi(x_{k+1})$ based on $\varphi(x_k)$. Applying the mean-value theorem to $\varphi(x_{k+1})$, we get:

$$\varphi(x_{k+1}) \le \varphi(x_k) - \underline{\sigma}(R)||x_k||^2 + ||\nabla_x\varphi(c)||\ ||e_k|| \qquad (26)$$

where $c = \theta x_{k+1} + (1-\theta)z_1$, $0 \le \theta \le 1$. Based on (26) we will prove that the closed-loop system is asymptotically stable. From the definition of c:

$$||c|| \le \bar{\sigma}(A)||x_k|| + \bar{\sigma}(B) + \theta||e_k|| \qquad (27)$$

A substitution of (21) and (27) in (26) yields:

$$\begin{aligned} \varphi(x_{k+1}) \le\ & \varphi(x_k) - \underline{\sigma}(R)||x_k||^2 + \\ & \gamma_0' e_k^0||x_k||^2 + \gamma_1' e_k^1||x_k|| + \gamma_2' e_k^2 \end{aligned} \qquad (28)$$

where $\gamma_i' \ge 0$ and $e_k^i \ge 0$, $\lim_{k\to\infty} e_k^i = 0$. Notice that the negative term in the RHS of (27) grows as $O(||x_k||^2)$. The three last terms in the RHS of (28) are products of terms that grow at most as $O(||x_k||^2)$ and sequences $||e_k^i||$, which converge to 0. Clearly, after a certain time-instance, the negative term prevails and $\varphi(x)$ can serve as a Lyapunov function of the closed-loop system. □

If we use the terminology of Sontag and Yang in [3], then Theorem 4 proves the "convergent input implies convergent state" (CICS) property of discrete linear systems that have the special structure (15) and which are regulated by the Infinite-Horizon Optimal Controller.

5. State Estimation

A simple extension of the reasoning presented in the proof of Theorem 4 can be used to prove that IH-OC coupled with an asymptotic observer is able to stabilize systems of the form (15), when the initial state x_0 is unknown.

Theorem 5. Consider a discrete linear dynamic system of the form

$$\begin{aligned} x_{k+1} &= Ax_k + Bu_k + e_k \\ y_k &= Cx_k \end{aligned}$$

where A is similar to (15), $y_k \in \Re^m$ and $C \in \Re^{m\times n}$. Suppose that (A, B) is controllable, (C, A) is observable and e_k is an unknown disturbance with $\lim_{k\to\infty} e_k = 0$. Finally, assume that the control variables are subject to a set of constraints of the form (2) and (3). If the initial state of the system, x_0, is unknown, then the Infinite-Horizon Optimal Controller (φ) with an asymptotic observer asymptotically stabilize the system.

Proof. Since (C, A) is observable, we can construct an asymptotic observer, which provides asymptotically convergent estimates of the states, $\hat{x}_k$, by feeding back y_k information. If we call the estimation error g_k, then the observer has the following behavior:

$$\begin{aligned} x_k &= \hat{x}_k + g_k \\ g_{k+1} &= Dg_k + e_k \end{aligned} \qquad (29)$$

The observer matrix D is designed to be stable. Hence,

$$\lim_{k\to\infty} g_k = 0$$

The estimate $\hat{x}_k$ is fed as the initial condition to the IH-OC formulation, which calculates the control moves. Therefore,

$$\begin{aligned} x_{k+1} &= Ax_k + Bu_k \\ &= A\hat{x}_k + Bu_k + Ag_k \\ &= \hat{x}_{k+1} + h_k \end{aligned}$$

where $h_k = Ag_k$ and $\lim_{k\to\infty} h_k = 0$. In conclusion,

$$\begin{aligned} x_k &= z_0 + g_k \\ x_{k+1} &= z_1 + h_k \end{aligned} \qquad (30)$$

Notice that (30) is a generalization of (25). Because φ is differentiable, we can apply the mean-value theorem to get:

$$\varphi(x_{k+1}) = \varphi(z_1) + \nabla\varphi(c_1)h_k \tag{31}$$

where $c_1 = \theta_1 x_{k+1} + (1-\theta_1)z_1$, $0 < \theta_1 < 1$. Substituting (30) in (31), and reapplying the mean-value theorem, we get:

$$\begin{aligned}
\varphi(x_k+1) &= \varphi(z_0) - z_0^t R z_0 - u_0^t S u_0 + \nabla\varphi(c_1)h_k \\
&= \varphi(x_k) - z_0^t R z_0 - u_0^t S u_0 \\
&\quad + \nabla\varphi(c_1)h_k + \nabla\varphi(c_0)g_k \\
&\leq \varphi(x_k) - (x_k - g_k)^t R(x_k - g_k) + \\
&\quad ||\nabla\varphi(c_1)||\,||h_k|| + ||\nabla\varphi(c_0)||\,||g_k|| \\
&\leq \varphi(x_k) - \underline{\sigma}(R)||x_k||^2 + 2\bar{\sigma}(R)||x_k||\,||g_k|| + \\
&\quad \bar{\sigma}(R)||g_k||^2 + ||\nabla\varphi(c_1)||\,||h_k|| \\
&\quad + ||\nabla\varphi(c_0)||\,||g_k||
\end{aligned} \tag{32}$$

where $c_0 = \theta_0 x_k + (1-\theta_0)z_0$, $0 \leq \theta_0 \leq 1$. We immediately notice the similarity of (32) and (26). From the definitions of c_1 and c_0:

$$||c_1|| \leq \bar{\sigma}(A)||x_k|| + \bar{\sigma}(B) + \theta_1||h_k|| \tag{33}$$

$$||c_0|| \leq ||x_k|| + \theta_0||g_k|| \tag{34}$$

If we use (33) and (34) and apply in (32) the same reasoning as in Theorem 4, then we prove that the closed-loop system is asymptotically stable. □

6. Discussion

We must underline that the stability proofs in Theorems 4 and 5 cover only the limiting form of Infinite-Horizon Optimal Controller, which is not, in general, easily implementable. Part of our future research effort will be devoted to generalizing Theorems 4 and 5 for the IH-MPC, i.e. to creating finite horizon strategies for which the disturbance rejection property and the stability with state-estimation will hold.

The reader should also notice that the disturbance rejection results, presented in Theorem 4, cover discrete linear systems that have the special structure (15). The inability to extend the result to more general systems, eg. Jordan blocks with unit eigenvalue, stems from the fact that, for the latter systems, the number of iterations, N, necessary to ensure that $x_N = 0$, cannot be upper-bounded by a linear function of $||x_0||$, as in the special case covered by Theorem 4. The available upper bound behaves as a higher-order polynomial of $||x_0||$, given by (14). Clearly, for more general systems property 8 cannot be asserted, and the reasoning of Theorem 4 is not applicable. In the future, we will try to understand whether this limitation can be attributed to the reasoning procedure, or is inherent to the design of Infinite Horizon Model Predictive Controller.

The third point we would like to address concerns the close relation between the disturbance rejection proof in Theorem 4 and the proof of stability with state-estimation in Theorem 5.

References

[1] E.D. Sontag and H.J. Sussmann, "Nonlinear output feedback design for linear systems with saturating controls," in Proc. IEEE Conf. Decision and Control, 1990, pp. 3414-3416.

[2] A.R. Teel, "Global stabilization and restricted tracking for multiple integrators with bounded controls," Syst. and Contr. Letters, vol. 18, pp. 165-171, 1992.

[3] E.D. Sontag and Y. Yang, "Global stabilization of linear systems with bounded feedback," Report SYCON-91-09, Rutgers University, Aug. 1991.

[4] C.E. Garcia, D.M. Prett and M. Morari, "Model predictive control and practice– A survey," Automatica, vol. 25, pp. 335-348, 1989.

[5] J. B. Rawlings and K. R. Muske, "The stability of constrained receding horizon control," IEEE Trans. Automat. Contr., Submitted, 1992.

[6] D.Q. Mayne and H. Michalska, "An implementable receding horizon controller for the stabilization of nonlinear systems," in Proc. IEEE Conf. Decision and Control, 1990, pp. 3396-3397.

[7] T.H. Yang and E. Polak, "Moving horizon control of linear systems with input saturation and plant uncertainty," Memorandum No. UCB/ERL M91/84, University of California, Berkeley, Sep. 1991.

[8] V.F. Dem'yanov and V.N. Malozemov, Introduction to Minimax, New York: Dover Publications, Inc., 1974 ch. VI, pp. 187-188.

Proceedings of the 31st Conference
on Decision and Control
Tucson, Arizona • December 1992

Minimum-Time Control for Uncertain Discrete-Time Linear Systems.(*)

Franco Blanchini

(1) Dipartimento di Matematica ed Informatica Università degli Studi di Udine,
Via Zanon 6, 33100, Udine, Italy.

Abstract - This paper deals with linear discrete-time uncertain systems with control and state constraints. We consider the problem of reaching and ultimately bounding the state in an assigned target set in minimum time. A solution based on the construction of the domain of attraction to the target set in terms of controllability sets is given. A linear programming algorithm for the construction of such a domain is proposed for the case in which linear constraints are considered.

1 Introduction

One of the most crucial problems arising in control engineering is that the mathematical model of the plant to be controlled has often uncertain parameters. Moreover, the system is in many cases affected by disturbances which cannot be reasonably modelled by fixed input functions. These are reasons why the control design is often performed with incomplete informations. One of the most attractive approaches to cope with this situation, is to assume that the uncertain data can take any value in an assigned bounding set and to design the controller assuring that given specifications are satisfied for all possible uncertainty variations in this set. For a complete exposition of the literature on this subject, the reader is referred to [11].
An other problem of practical relevance is that control and state variables are subject to constraints [2]-[14] which have to be considered in the design stage. The problem of the constrained control in the presence of uncertainties is a subject that has been already considered in literature. In [4][5] it has been considered the problem of confining the system state in an assigned polyhedral set via linear state feedback control. In [6] it has been proposed a method to design a nonlinear control assuring convergence of the system state to an assigned target set.
In this paper, we consider the problem of designing a state-feedback control with the goal of reaching and bounding the state in a target set under the condition that the state and the control do not violate their constraints.
We search for a time-optimal control in the sense that it minimizes the time value for which the inclusion condition into the target set is guaranteed since then on.

(*) This work has been supported by the C.N.R. contract 92.02887.CT07

To pursue this objective, we work in a game theoretic framework similar to that proposed in [3] and [7], where the theory of the min-max controllers has been developed. As a result, we extend the techniques presented in [8] and [9] to solve the minimum-time problem for discrete-time linear systems without uncertainties.

2 Problem statement

In the sequel we call a C-set a convex and compact set containing the origin as an interior point.
The model we deal with is a discrete-time uncertain system of the form

$$x(t+1)=A(w(t))\,x(t) + B(w(t))\,u(t) + E\,v(t) \qquad (1)$$

where $x(t)\in\mathcal{R}^n$, $u(t)\in\mathcal{R}^q$, $v(t)\in\mathcal{R}^p$, $w(t)\in\mathcal{R}^m$, are respectively the system state, the control input, the disturbance input and the parameter uncertainty vector. The functions v(t) and w(t) are supposed to be unknown but subject to the constraints

$$v(t)\in\mathcal{V}\subset\mathcal{R}^p,\ \ w(t)\in\mathcal{W}\subset\mathcal{R}^m, \qquad t\geq 0, \qquad (2)$$

where, $\mathcal{V}$ and $\mathcal{W}$ are assigned compact sets. The control u(t) and the state x(t) are also assumed to be constrained in assigned compact C-sets

$$u(t)\in\mathcal{U} \subset \mathcal{R}^q, \qquad t\geq 0, \qquad (3a)$$
$$x(t)\in\mathcal{X} \subset \mathcal{R}^n, \qquad t\geq 0. \qquad (3b)$$

We assume that the entries of A(w) and B(w) are continuous functions of w and that the matrix B(w) has full column rank for some $w\in\mathcal{W}$.
In the case that the system is not affected by uncertainties, the minimum-time problem has the following well-known formulation: given an initial state $x(0)\in\mathcal{X}$ find a control function $u(t)\in\mathcal{U}$ which minimizes T(x(0)) such that x(T(x(0)))=0 with the condition that the constraints (3a) (3b) are not violated. This problem is formulated in an open-loop sense that is the control function is determined from the knowledge of the initial state. For practical reasons, it is important to find a feedback solution say a function $u=\phi(x)$ such that T(x(0)) is minimized for all possible initial states (for more

CH3229-2/92/0000-2629$1.00 © 1992 IEEE

details the reader is referred to specialized literature, see for instance [10]). If no uncertainties are present, the feedback solution is optimal if and only if the control function u(t) obtained as $u(t)=\phi(x(t))$ is optimal in the open-loop sense for every $x(0)\in\mathscr{X}$.

If the system model is uncertain the problem is more complicated. The first reason is that, the requirement that the zero state is exactly reached is not reasonable. Indeed, unless for trivial cases, it is not possible to find a control assuring that the state reaches the origin since the functions w(t) and v(t) are unknown. Moreover, even if $x(T)=0$ is fulfilled for some T, in view of the disturbance v, such a condition does not hold in general for $t>T$. What we can reasonably do, is to require that the state reaches a given target set in minimum time and that it remains confined in it for all future times.

An other fundamental difference, is that the open-loop and the feedback formulations are not equivalent if the system is uncertain. Indeed, the possibility of driving the state in the target set via a feedback control from a given initial state, does not implies that we can achieve the same goal using an open-loop control.

Here, we formulate the problem in a feedback sense. To this aim we introduce the following definitions.

Definition 1. The state x_0 is controllable to the compact set $\mathscr{G}\subset\mathscr{X}$ if the there exists a control $u=\phi(x)$ and $T\geq 0$ such that the solution x(t) of (1) with initial condition $x(0)=x_0$ is such that

$$u(t)\in\mathscr{U},\ x(t)\in\mathscr{X},\ t\geq 0, \tag{4}$$
$$x(T)\in\mathscr{G}, \tag{5}$$

for all $w(t)\in\mathscr{W}$ and $v(t)\in\mathscr{V}$, $t=0,1,..,T-1$.

Definition 2. The state x_0 is ultimately boundable in the compact set $\mathscr{G}\subset\mathscr{X}$ if there exists a control $u=\phi(x)$ and $T\geq 0$ such that the solution x(t) of (1) with initial condition $x(0)=x_0$ is such that

$$u(t)\in\mathscr{U},\ x(t)\in\mathscr{X},\ t\geq 0, \tag{6}$$
$$x(t)\in\mathscr{G},\ t\geq T, \tag{7}$$

for all $w(t)\in\mathscr{W}$ and $v(t)\in\mathscr{V}$, $t\geq 0$.

The difference between the two definition lies on the fact that the first one requires that the state is contained in $\mathscr{G}$ at the time T, while the second one requires that the state is constrained in $\mathscr{G}$ for $t\geq T$.

Given a control ϕ the following indices are defined

$$J_1(x_0) = \min\{T \text{ for which (4) (5) are fulfilled for all w(t) and v(t) as in (2) }\} \tag{8a}$$

$$J_2(x_0) = \min\{T \text{ for which (6) (7) are fulfilled for all w(t) and v(t) as in (2) }\} \tag{8b}$$

If a strategy ϕ does not assure that the conditions (4)(5) (respectively (6)(7)) are fulfilled for some finite T, then we assume $J_1(x_0)=\infty$ (respectively $J_2(x_0)=\infty$). We name $J_1(x_0)$ reaching time and $J_2(x_0)$ capture time respectively. Of course we have $J_1(x_0)\leq J_2(x_0)$.

We remark that these indices refer to the worst case. In other words, let $T=J_1(x_0)$ (or $T=J_1(x_0)$), then the condition $x(t)\in\mathscr{G}$ is guaranteed for $t=T$ (or $t\geq T$) for every $v(t)\in\mathscr{V}$ and $w(t)\in\mathscr{W}$ while if $T<J_1(x_0)$ (or $T<J_1(x_0)$) there exists two sequences $v(t)\in\mathscr{V}$ and $w(t)\in\mathscr{W}$ such that the condition $x(t)\in\mathscr{G}$ is not satisfied for $t=T$ (or for some $t\geq T$) (but it may be satisfied for other sequences).

We are able now formulate the following problems.

Problem 1. (Minimum-Time Controllability Problem) Find a feedback control strategy $u=\phi(x)$ which minimizes $J_1(x_0)$ for all $x_0\in\mathscr{X}$.

Problem 2. (Minimum-Time Ultimate Boundedness Problem) Find a feedback control strategy $u=\phi(x)$ which minimizes $J_1(x_0)$ for all $x_0\in\mathscr{X}$.

These problems are clearly of a game-theoretic nature. There is a player P1 (the controller) who has the goal of reaching (or confining) the state in the target set in minimum time and a player P2, who has the goal of delaying as much as possible this occurrence.

In section 3 we analyze both problems from a theoretical point of view and we show under which condition they are equivalent. We use this condition to reduce Problem 2 to Problem 1. In section 4, we propose an algorithm for the solution in the case in which the sets $\mathscr{V},\mathscr{W},\mathscr{X},\mathscr{U}$, are defined by linear constraints.

3 Reachability of the target set

Henceforth we assume that the target set $\mathscr{G}$ is a C-set. To solve our problems we introduce the following functions which associate to x the optimal value of Problem 1 and Problem 2 if $x(0)=x$, say the minimum reaching time and the minimum capture time that can be obtained via feedback control starting from x

$$\varphi_1(x) = \min\{T:\ \exists\ \phi:\ \mathscr{R}^n\rightarrow\mathscr{R}^q:\ \text{if } x(0)=x \text{ then } \phi(x(t))\in\mathscr{U},\ x(t)\in\mathscr{X},\ t\geq 0,\ x(T)\in\mathscr{G}\ \forall\ v(t)\in\mathscr{V} \text{ and } w(t)\in\mathscr{W}\ \} \tag{9a}$$

$$\varphi_2(x) = \min\{T:\ \exists\ \phi:\ \mathscr{R}^n\rightarrow\mathscr{R}^q:\ \text{if } x(0)=x \text{ then } \phi(x(t))\in\mathscr{U},\ x(t)\in\mathscr{X},\ t\geq 0,\ x(k)\in\mathscr{G},\ k\geq T\ \forall\ v(t)\in\mathscr{V} \text{ and } w(t)\in\mathscr{W}\} \tag{9b}$$

Define the sets

$$\mathscr{R}_k = \{x\in\mathscr{X}:\ \varphi_1(x) = k\} \tag{10}$$

so that a control function ϕ is a solution for Problem 1 if and only if it fulfills the following property

$$A(w)\,x + B(w)\,\phi(x) + E\,v \in \mathscr{R}_h,\quad h<k,\ \phi(x)\in\mathscr{U},\quad \text{for all } x\in\mathscr{R}_k \text{ and all } w\in\mathscr{W},\ v\in\mathscr{V}. \tag{11}$$

We see that the set of all the feedback control functions ϕ that solve Problem 1 is characterized by (11) and then the determination of the sets $\mathcal{R}_k$ in principle solves the problem. We can characterize the sets $\mathcal{R}_k$ as follows. Given the set $\mathcal{S}$, define the controllability set to $\mathcal{S}$ as

$$\mathcal{Q}(\mathcal{S}) = \{x: \ \exists\, u(x) \in \mathcal{U} \text{ such that } A(w)\,x + B(w)\,\phi(x) + E\,v(t) \in \mathcal{S} \text{ for all } w \in \mathcal{W}, v \in \mathcal{V}\}$$

Define the sets $\mathcal{C}_k$ in the following recursive way

$$\mathcal{C}_0 = \mathcal{G},\ \mathcal{C}_{k+1} = \mathcal{Q}(\mathcal{C}_k) \cap \mathcal{X} \qquad (12)$$

Each set $\mathcal{C}_k$ (the controllability sets in k steps) has the property to be the set of all states x(t) for which a control $u(x) \in \mathcal{U}$ can be chosen such that $x(t+1) \in \mathcal{C}_{k-1}$ for all $w(t) \in \mathcal{W}$, $v(t) \in \mathcal{V}$. This means that x can be controlled to $\mathcal{G}$ in k steps if and only if $x \in \mathcal{C}_k$. This means that

$$\varphi_1(x) = \min\{\, k: \ x \in \mathcal{C}_k \,\}.$$

The condition of reaching the target set in T steps may be not of practical interest because what we desired in practice is to ultimately bounding the state in some neighborood of the origin for all times. So we introduce the following definition which is important to establish a connection between Problem 1 and Problem 2.

Definition 3. A set $\mathcal{S} \subset \mathcal{X}$ is said to be positively invariant (PI) if $\mathcal{S} \subset \mathcal{Q}(\mathcal{S})$.

If $\mathcal{S}$ is PI, there exists a control $u=\phi(x)$ such that if $x(0) \in \mathcal{S}$, then the conditions $x(t) \in \mathcal{S}$, $u(t) \in \mathcal{U}$, $t \geq 0$, hold for all w(t) and v(t) as in (2). The following theorem holds.

Theorem 1. The following statements are equivalent

i) $\mathcal{G}$ is a PI set
ii) $\varphi_1(x)=\varphi_2(x)$ for all $x \in \mathcal{X}$.
iii) $\mathcal{C}_k \subset \mathcal{C}_{k+1}$ for all k.

Proof. The implication (i) $\Rightarrow$ (ii), is motivated by the fact that to bound the state in T steps in a PI set, is equivalent to reach such a set in T steps since the permanence of the state in it can be guaranteed by definition. Conversely, if ii) holds for each $x \in \mathcal{X}$, then for each $x \in \mathcal{G}$ we have that $\varphi_2(x)=\varphi_1(x)=0$ and so there exist a control such that if $x(0) \in \mathcal{G}$ then $x(t) \in \mathcal{G}$. We prove now that (iii) is equivalent to (i). Suppose that (i) holds, and $x(0) \in \mathcal{C}_k$. This is equivalent to say that we can reach the set $\mathcal{G}$ in k steps. But since $\mathcal{G}$ is PI, we can also assure that $x(k+1) \in \mathcal{G}$ and then $x(0) \in \mathcal{C}_{k+1}$. Conversely, if (iii) holds, in particular we have $\mathcal{G} = \mathcal{C}_0 \subset \mathcal{C}_1$, say $\mathcal{C}_0 \subset \mathcal{Q}(\mathcal{C}_0)$ and this implies that $\mathcal{G}$ is PI. □

Assume that $\mathcal{G}$ is positively invariant and that a control ϕ^* is known assuring that $x(t) \in \mathcal{G}$, for each $x(0) \in \mathcal{G}$. Under this condition, Problem 1 and Problem 2 for such a target set are *equivalent* in the sense that if we know a solution for one of them, then we know a solution for the other having the same cost for all initial state. Indeed, if ϕ_2 solves Problem 1, then applying the control $\phi(x) = \phi_2(x)$ for $x \notin \mathcal{G}$ and $\phi(x) = \phi^*(x)$ for $x \in \mathcal{G}$, we ultimate bound the state in $\mathcal{G}$ in $\varphi_1(x(0))$ steps for all $x(0) \in \mathcal{X}$. From (ii) we have that ϕ solves Problem 2. On the other hand, if ϕ solves Problem 2, then the condition $x(t) \in \mathcal{G}$, is assured for $t \geq \varphi_2(x(0))$. In particular, in view of property (ii) we have that for $t=\varphi_1(x(0))=\varphi_2(x(0))$ $x(t) \in \mathcal{G}$ and then ϕ is an optimal solution for Problem 1. Conversely, if $\mathcal{G}$ is not PI, (ii) does not hold, and the two problems are not equivalent.

Let us consider now Problem 1 for a generic target set $\mathcal{G}$ not necessarily PI.

We say that the PI set $\mathcal{P}_m \subset \mathcal{G}$ is *maximal* in $\mathcal{G}$ if every other PI set $\mathcal{S} \subset \mathcal{G}$ is such that $\mathcal{P} \subset \mathcal{S}_m$.

Theorem 2. The state x_0 is ultimately boundable in the target set $\mathcal{G} \subset \mathcal{X}$ if and only if x_0 is controllable to the set $\mathcal{P} \subset \mathcal{G}$ where $\mathcal{P}$ is a PI set contained in $\mathcal{G}$.

Proof. Assume that with $u=\phi_1(x)$ we reach the PI set $\mathcal{P}$ from x_0 in T steps. Since $\mathcal{P}$ is PI, there exists $u=\phi_2(x)$ such that $x(t) \in \mathcal{P} \Rightarrow x(t+1) \in \mathcal{P}$. So the control x

$$\phi(x) = \begin{cases} \phi_1(x), & \text{if } x \notin \mathcal{P} \\ \phi_2(x), & \text{if } x \in \mathcal{P} \end{cases}$$

ultimately bounds x_0 in $\mathcal{G}$. Conversely, according to [6], a control can assure condition (7), if and only if $x(T) \in \mathcal{P}_m$ where $\mathcal{P}_m \subset \mathcal{G}$ is the maximal PI set in $\mathcal{G}$. □

As an immediate consequence of Theorem 1 we have the following Proposition.

Corollary 1. To solve Problem 2 it is necessary and sufficient to solve Problem 1 for the target set $\mathcal{P}_m$.

To find an expression for $\mathcal{P}_m$, define the controllability sets $\mathcal{G}_k$ as in (12) replacing $\mathcal{X}$ by $\mathcal{G}$

$$\mathcal{G}_0 = \mathcal{G},\ \mathcal{G}_{k+1} = \mathcal{Q}(\mathcal{G}_k) \cap \mathcal{G}. \qquad (13a)$$

then the maximal PI set contained in $\mathcal{G}$ is [6]

$$\mathcal{P}_m = \bigcap_{i=0}^{\infty} \mathcal{G}_i \qquad (13b)$$

We remark that $\mathcal{P}_m$ may be an empty set and, in this case, $\varphi_2(x_0)=\infty$ for all $x_0 \in \mathcal{X}$. Conversely, if $\mathcal{P}_m$ is nonempty, for all $x_0 \in \mathcal{P}_m$ we have $\varphi_2(x_0)=0$. So the following question arises: which are the initial conditions from which we can ultimately bound the state in $\mathcal{G}$ or, equivalently, from which we can reach $\mathcal{P}_m$?

Definition 4. We say that the set $\mathcal{A}(\mathcal{G}) \subset \mathcal{X}$ is the domain of attraction to $\mathcal{G}$ for system (1) if every initial state $x_0 \in \mathcal{A}(\mathcal{G})$ can be ultimately bounded in $\mathcal{G}$, that is there exists a control strategy ϕ such that the corresponding $J_2(x_0)$ is finite.

Clearly, we have that $\mathcal{P}_m \subset \mathcal{A}(\mathcal{G})$ and $\mathcal{A}(\mathcal{G}) = \mathcal{A}(\mathcal{P}_m)$.
Since Problem 2 reduces to that of reaching $\mathcal{P}_m$ in minimum time, we consider now Problem 1 under the assumption that the target set $\mathcal{G}$ is PI .
So let $\mathcal{G}$ be a PI set. From the conditions (ii) and (iii) of Theorem 1 we have that the sets $\mathcal{R}_k$ are given by

$$\mathcal{R}_k = \{x \in \mathcal{X}: \varphi_1(x) = \varphi_2(x) = k\} = \mathcal{C}_{k+1} \backslash \mathcal{C}_k ,$$

where the expression $\backslash \mathcal{C}_k$ means the intersection with the complement of $\mathcal{C}_k$.
The sets $\mathcal{C}_k$ play a fundamental role since they define the function $\varphi_2(x)$ which characterizes the domain of attraction. Indeed the condition $x(0) \in \mathcal{A}(\mathcal{G})$, is equivalent by construction to $x(0) \in \mathcal{C}_k$ for some k. So, if $\mathcal{G}$ is PI, $\varphi_2(x(0)) = \varphi_1(x(0)) = \min\{k: x(0) \in \mathcal{C}_k\} < \infty$. Moreover, in view of (11), the sets $\mathcal{C}_k$ characterize the set of all feedback strategies which solve the problem of steering the state in $\mathcal{G}$ in minimum time. We summarize all the considerations above in the following theorem.

Theorem 3. Given a target set $\mathcal{G}$ the domain of attraction $\mathcal{A}(\mathcal{G})$ is nonempty if and only if the maximal PI set $\mathcal{P}_m$ contained in $\mathcal{G}$ is nonempty. The set $\mathcal{A}(\mathcal{G})$ is

$$\mathcal{A}(\mathcal{G}) = \{x: \varphi_2(x) < \infty\} = \bigcup_{i=0}^{\infty} \mathcal{R}_i$$

where $\mathcal{R}_k = \mathcal{C}_k \backslash \mathcal{C}_{k-1}$ and $\mathcal{C}_k$ are defined as in (12) with the condition $\mathcal{C}_0 = \mathcal{P}_m$. In this case, there exists a feedback control strategy that solves Problem 2 for each state in $\mathcal{A}(\mathcal{G})$. The set of all controls which solve the problem is given by all the functions ϕ which fulfill condition (11) for all $x \in \mathcal{A}(\mathcal{G})$. □

In the next section, we cope with the problem of the construction of the domain of attraction in the case in which the control inputs, the state and the uncertainties are linearly constrained.

4 Costruction of the domain of attraction via linear programming

Henceforth, we assume that the sets $\mathcal{X}, \mathcal{U}, \mathcal{V}$, are polyhedral C-sets. Moreover we assume that A(w) and B(w) are matrix polyhedra, say they have the form

$$A(w) = \sum_{k=1}^{m} w_k A_k, \quad B(w) = \sum_{k=1}^{m} w_k B_k,$$

$$w_k \geq 0, \qquad \sum_{k=1}^{m} w_k = 1,$$

where the matrices A_k and B_k are assigned.
Assume now that a polyhedral C-set $\mathcal{S}$ is given as

$$\mathcal{S} = \{x: F_i x \leq g_i, \; i=1, .., s, \text{ or synthetically } F x \leq g\},$$

and define the set

$$\mathcal{S}^* = \{x: F_i x \leq g_i - \delta_i = g_i^*, \; i=1,.., s, \text{ or } F x \leq g - \delta = g^*\} \quad (14)$$

where

$$\delta_i = \max_{v \in \mathcal{V}} F_i E v$$

According to [6], the controllability set $\mathcal{Q}(\mathcal{S})$ to $\mathcal{S}$ is the projection of the set

$$\mathcal{Z} = \{[x,u]: u \in \mathcal{U} \text{ and } F A_k x + F B_k u \leq g^*, k=1,.., m\} \subset \mathcal{R}^n \times \mathcal{R}^q$$

on the subspace related to the first n components say

$$\mathcal{Q}(\mathcal{S}) = \{x: \exists u: [x,u] \in \mathcal{Z}\} \subset \mathcal{R}^n .$$

Since $\mathcal{Z}$ is a closed polyhedral set, its projection $\mathcal{Q}(\mathcal{S})$ is also a closed polyhedron. An algorithm based on the Furier-Motzkin elimination method to achieve the projection of a polyhedral set assigned in terms of its delimiting planes, has been proposed in [9].
So, if a target polyhedral C-set $\mathcal{G}$ is given, the sets $\mathcal{C}_k$ as defined in (12) (which are the results of projections and intersections of polyhedral sets) are also polyhedra and they are compact since they are included in $\mathcal{X}$. By computing these sets we solve Problem 1 so we can use linear programming to synthesize the control.
Let now consider Problem 2. If $\mathcal{P}_m$ is determined, we have simply to solve Problem 1 taking $\mathcal{P}_m$ as target set. If such a set is polyhedral, the procedure described above for the determination of the sets $\mathcal{C}_k$ (and then of the sets $\mathcal{R}_k$) does apply. This condition is true if (but not only if) the condition $\mathcal{G}_i = \mathcal{G}_{i+1}$ is satisfied for some i, which is equivalent to $\mathcal{G}_i = \mathcal{P}_m$. Unfortunately the set $\mathcal{P}_m$ is not in general polyhedral but it is only convex, even if in the largest part of the considered examples a polyhedral $\mathcal{P}_m$ has been found. However, the sequence of sets $\mathcal{G}_i$ in (13a) converge to their intersection $\mathcal{P}_m$ and then we can approximate $\mathcal{P}_m$ by one of the sets $\mathcal{G}_i$ which is polyhedral. It is clear that from a practical point of view it makes no difference since $\mathcal{P}_m$ may be approximated by a polyhedral set with an arbitrary precision. So, in practice, to compute $\mathcal{P}_m$ by expression (12) we introduce a tolerance, and we perform the computation of the sets $\mathcal{G}_i$ until the condition $\mathcal{G}_i = \mathcal{G}_{i+1}$ is satisfied with the assigned tolerance. Moreover, in [6] it is shown how,

introducing a certain contractivity factor, a "near-maximal" polyhedral PI set contained in $\mathcal{G}$ can be found. In this case, we can obtain a suboptimal solution by reaching a PI set which is not maximal contained in $\mathcal{G}$.
In view of these considerations, we study Problem 1 for a given a polyhedral PI target set $\mathcal{G}$. This problem can be solved by the computation of the sets $\mathcal{R}_i$ which, in view of the positive invariance of $\mathcal{G}$, are $\mathcal{R}_i = \mathcal{C}_i \backslash \mathcal{C}_{i-1}$ where the sets $\mathcal{C}_i$ are convex polyhedra. The sets $\mathcal{R}_i$ form a covering for the domain of attraction $\mathcal{A}(\mathcal{G})$ to $\mathcal{G}$ and the sets $\mathcal{C}_i$ converge to $\mathcal{A}(\mathcal{G})$ so, we can stop the procedure for the construction of the sets $\mathcal{A}(\mathcal{G})$ when the condition $\mathcal{C}_{i+1} = \mathcal{C}_i$ is fulfilled with a certain tolerance. Assume now that the polyhedra $\mathcal{C}_i$ are known

$$\mathcal{C}_i = \{ x: \Psi^{(k)} x \leq \rho^{(k)} \}$$

The control is derived by implementing the following procedure which has to be solved on-line.

Procedure 1. (On-line control computation)
Step 1. For each $x \in \mathcal{A}(\mathcal{G})$ compute $\varphi_2(x)$, say find the minimum $k \geq 0$ such that $x \in \mathcal{R}_k$.
Step 2. Compute u such that $A(w)\, x + B(w)\, u + E\, v \in \mathcal{C}_{k-1}$ if $k>0$ and $A(w)\, x + B(w)\, u + E\, v \in \mathcal{R}_0$ if $k=0$.

With the convention $[\Psi^{(-1)}, \rho^{(-1)}] = [\Psi^{(0)}, \rho^{(0)}]$ both conditions of Step 2 may be written as

$$u \in \mathcal{L}_k(x) = \{u: \Psi^{(k-1)} [A_k\, x + B_k\, u] \leq \rho^{(k-1)} - \delta^{(k-1)},\ k=1,..,m\} \qquad (15)$$

where $\delta^{(k-1)}$ is a vector defined as in (14). These conditions do not define a single control value u(x) but they define a set of possible control vectors we denote by $\mathcal{L}_k(x)$. The feedback control laws that solve the problem are all the functions ϕ such that

$$\phi(x) \in \mathcal{L}_k(x) \quad \text{for all } x \in \mathcal{R}_k . \qquad (16)$$

Any control function with this property assures that if $x(t) \in \mathcal{R}_k$ then $x(t+1) \in \mathcal{C}_{k-1}$ and then $x(t+1) \in \mathcal{R}_h$ with $0 \leq h < k$, for all $v \in \mathcal{V}$ and $w \in \mathcal{W}$.
Among all the function which fulfill (16) we can choose those which minimize $\| u \|$ to derive a minimum effort control. If the norm is suitably chosen, there is an unique solution u for each given x. For instance, if we choose $\| . \|$ as the Euclidean norm, then the function ϕ is unique and it is known as minimal selection [1].
The feedback control is obtained by computing u on-line by solving the linear programming problem of selecting a vector in the set $\mathcal{L}_k(x)$. Such a linear programming problem has the entries of u as unknown and the inequalities (15) as constraints. If the control space has a low dimension (as it is often the case), the solution on-line of this problem does not require an high computing power. In particular, for single-input systems, this problem consists on finding a point in the intersection of a $s(k-1) \times m + 2$ half-intervals where with $s(k-1)$ is the number of constraints which define $\mathcal{C}_{k-1}$. One important feature of the technique suggested here is that it may be used jointly with previous methods for the computation on-line of the control such as those in [12] [13].

5 Numerical example

Let us consider the water reservoir system of Fig. 1.

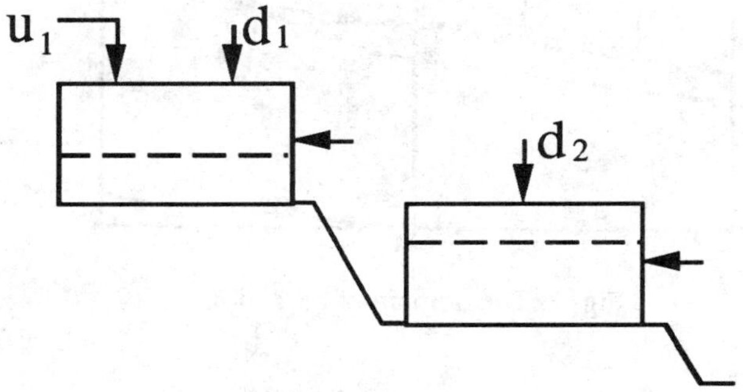

Fig. 1. Two reservoir system

Such a system is described by the following equations:

$$x_1(t+1) = x_1(t) - w_1 (x_1(t) - x_2(t)) + d_1(t) + u(t)$$
$$x_2(t+1) = x_2(t) + w_1 (x_1(t) - x_2(t)) + w_2\ x_2(t) + d_1(t)$$

where $x_1(t)$ and $x_2(t)$ are the water level of the upper and the lower reservoir respectively, u(t) is the water flow control input, while $d_1(t)$ and $d_2(t)$ are disturbance flows. We assume that the following constraints are given

$$1 \leq w_1(t) \leq 2 , \quad 1 \leq w_2(t) \leq 2,$$
$$-0.1 \leq d_1(t) \leq 0.1 , \quad -0.1 \leq d_2(t) \leq 0.1$$
$$-3 \leq x_1(t) \leq 3 , \quad -3 \leq x_2(t) \leq 3 , \quad -1 \leq u(t) \leq 1.$$

and we consider the target set

$$\mathcal{G} = \{ x_1 , x_2 : -1 \leq x_1 \leq 1, \ -1 \leq x_2 \leq 1 \ \}.$$

The maximal PI set in $\mathcal{G}$ and the domain of attraction are both polyhedral sets and they are given by

$$\mathcal{P}_m = \{ x_1 , x_2 : -1 \leq x_1 \leq 1 , -1 \leq x_2 \leq 1 , -1 \leq -0.111\ x_1 + x_2 \leq 1 \}$$

$$\mathcal{A}(\mathcal{G}) = \mathcal{C}_{10} = \{ x_1 , x_2 : -3 \leq x_1 \leq 3 , -3 \leq x_2 \leq 3 , -3 \leq -0.111\ x_1 + x_2 \leq 3, -3 \leq -0.222\ x_1 + x_2 \leq 3.333 \}$$

The set $\mathcal{A}(\mathcal{G})$ of attraction to $\mathcal{P}_m$ and some of the isocost sets $\mathcal{R}_k$ are represented in Fig. 2.

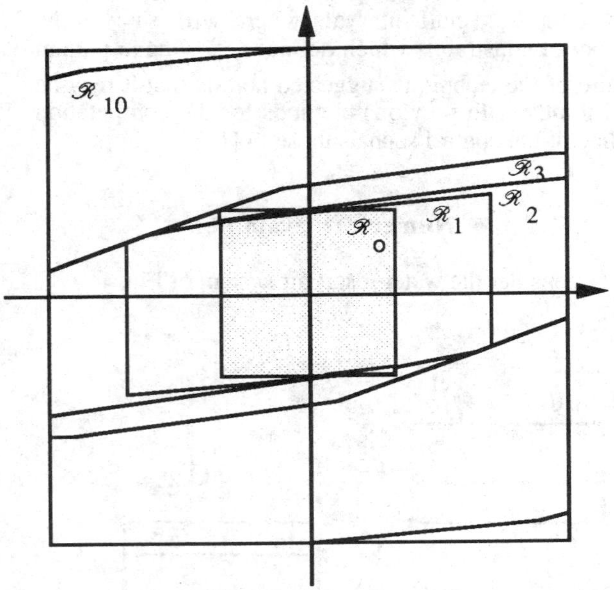

Fig. 2. The domain of attraction

6. Conclusions

In this paper we considered the problem of reaching and keeping the state of an uncertain linear discrete-time system in an assigned target set without violating some control and state constraints. The main contribute of the paper is to extend some techniques to cope with control and state constraints presented in previous references in order to guarantee a robust performance to the control.
The proposed control is time-optimal in the sense it minimize the worst-case reaching time.
The suggested technique is not conservative in the sense that, if the formulated problem has a solution, then it gives the class of all controls that solve the problem.
The control computation is based on an off-line part in which the controllability regions to the target set are determined and on an on-line part in which the data computed off-line are used to produce the control.
If linearly-constrained systems and polyhedral bounding sets are considered, the controllability regions are polyhedral sets and their computation can be performed via linear programming.
A limit of the procedure is that there is no in general an upper bound for the number of planes defining these sets. These sets are involved on-line for the control computation which requires the solution of a linear programming problem, having a number of constrains proportional to the number of their delimiting planes.
On the other hand, the control components which are the variables of this linear programming problem, are in many real cases not numerous and, in all these cases, the problem can be easily solved by commonly available computing power machines.

References.

[1] P. Aubin and A. Cellina, "Differential Inclusions", Springer-Verlag: Berlin, 1984.

[2] A. Benzaouia and C. Burgat, "Regulator Problem for Linear Discrete-Time Systems with non-symmetrical Constrained Control," International Journal of Control, vol. 48, 2441-2451, 1988.

[3] D. P. Bertsekas and I. B. Rhodes, "On the min-max Reachability of Target Set and Target Tubes", Automatica, Vol. 7, pp. 233-247, 1971.

[4] F. Blanchini, "Feedback Control for Linear Systems with State and Control Bounds in the Presence of Disturbance," IEEE Transaction on Automatic Control, Vol. 35, 1131-1135, 1990.

[5] F. Blanchini, "Constrained Control for Uncertain Systems, " Journal of Optimization Theory and Application, Vol. 71, no. 3. pp.465-484, December 1991.

[6] F. Blanchini "Ultimate Boundedness Control for Discrete-Time Uncertain System via Set Induced Lyapunov Functions", 30° IEEE Conference on Decision and Control, Brighton, U.K., pp. 1755-1760, December 11-13, 1991.

[7] Glover J.D. and Schweppe F.C. "Control of Linear Dynamic Systems with Set Constrained Disturbances", IEEE Transactions on Automatic Control, Vol. AC-16, pp. 411-423, 1971.

[8] P. O. Gutman and M. Cwikel, "Admissible sets and Feedback Control for Discrete-Time Linear Dynamical Systems with Bounded Control and States", IEEE Transactions on Automatic Control, Vol. AC-31, pp. 373-376, 1986.

[9] S.S. Keerthy and E. G. Gilbert, "Computation of Minimum-time Feedback Control Laws for Discrete-Time Systems with state and Control Constraints", IEEE Transactions on Automatic Control, Vol. AC-32, No. 5, pp. 432-435, 1987.

[10] G. Leitmann, "The calculus of Variations and Optimal Control" Plenum Press Inc., Angelo Miele Editor, 1986.

[11] D. D. Siljak, "Parameter Space Methods for Robust Control Design: A Guided Tour" IEEE Transactions on Automatic Control, Vol. 34, no. 7, pp. 674-688, July 1989.

[12] M. Sznaier and M. J. Damborg, Heuristically Enhanced Feedback Control of Constrained Discrete-Time Systems, Automatica, Vol. 26, No. 3, pp. 521-532, 1990.

[13] M. Sznaier and M. Damborg, An Analog "Neural Net" Based Suboptimal Controller for Constrained Discrete-Time Linear Systems" Automatica, Vol. 28, No.1, 139-144, 1992.

[14] M. Vassilaky, J. C. Hennet and G. Bitsoris, "Feedback Control of Linear Discrete Time Systems under State and Control Constraints," International Journal of Control, Vol. 47, 1727-1735, 1988.

Proceedings of the 31st Conference
on Decision and Control
Tucson, Arizona • December 1992

Feedforward Tracking for Uncertain Systems *

R. W. Benson, W. E. Schmitendorf, and R. M. Dolphus
Department of Mechanical and Aerospace Engineering
University of California – Irvine
Irvine, CA 92717
714-856-8451

Abstract

In this paper we consider the tracking problem for uncertain systems with time-invariant uncertainty. We propose a feedforward tracking approach for both full state feedback and measurement feedback problems. The design is reduced to a standard H_∞ problem.

1 Introduction

This paper provides a tracking result for uncertain systems with time-invariant uncertainty. The tracking control law is based on the feedforward approach due to O'Brien and Broussard [1]. A similar robust control problem was considered in Hopp and Schmitendorf [2] for the full state feedback case with time-varying uncertainty. The disadvantage of that approach, however, is that the desired ball of convergence cannot be specified *a priori*. For matched systems, this problem was overcome in Dolphus and Schmitendorf [3]. An augmented system approach to the full state feedback case was also considered in Schmitendorf and Barmish [4].

The results of this paper differ from the ones mentioned above in that the desired ball of convergence is pre-specified regardless of whether the system is matched or not, and the results apply to the full state and output feedback problems. A special case of this paper was presented in [5] where feedforward control was assumed to have no effect on the uncertain terms (i.e., $w \equiv 0$, where w will be defined later).

2 Preliminaries

In this section results are presented for a feedforward tracking control law for the following uncertain system

$$\begin{aligned} \dot{x}(t) &= (A + \Delta A(r))x(t) + Bu(t) \qquad (1) \\ z(t) &= C_1 x(t) \\ y(t) &= C_2 x(t) \end{aligned}$$

where $x \in R^n$ is the state, $u \in R^m$ is the control input, $z \in R^{p_1}$ is the controlled output, and $y \in R^{p_2}$ is the measured output. The nominal matrices A, B, C_1, and C_2 are known. The vector $r \in \mathcal{R} \subset R^{n_u}$ is the vector of parameter uncertainties, and $\mathcal{R}$ is a prespecified compact set. The uncertain matrix $\Delta A(r)$ is assumed to satisfy

$$\Delta A(r) = DF(r)E \qquad (2)$$

where $D \in R^{n \times q}$ and $E \in R^{r \times n}$ are known matrices and $F(r) \in R^{q \times r}$ satisfies

$$F^T(r)F(r) \le \bar{r}^2 I \qquad \forall r \in \mathcal{R} \qquad (3)$$

where $\bar{r}$ is known.

The design goal is to find a robust feedforward tracking control law for (1) which will drive the tracking error to a prespecified neighborhood of the origin in finite time. In particular, given a scalar $N > 0$, we wish to find a feedback control law such that given a constant reference input z^* satisfying $\|z^*\|_2 < N$, then for $T > 0$ sufficiently large and $\epsilon > 0$ prespecified, we have $\max_i |z_i(t) - z_i^*| < \epsilon$ for all $t > T$ and all $r \in \mathcal{R}$.

Following ideas from [1], the *ideal plant response* for the nominal system is determined. We assume that z^* is given and satisfies $\|z^*\|_2 < N$. Then the following equations are solved for x^* and u^*,

$$\begin{aligned} 0 &= Ax^* + Bu^* \qquad (4) \\ z^* &= C_1 x^*. \end{aligned}$$

or,

$$\begin{bmatrix} 0 \\ z^* \end{bmatrix} = X \begin{bmatrix} x^* \\ u^* \end{bmatrix}, \qquad (5)$$

*This research was supported by a grant from the California Space Institute

CH3229-2/92/0000-2635$1.00 © 1992 IEEE

where X is given by

$$X \triangleq \begin{bmatrix} A & B \\ C_1 & 0 \end{bmatrix}.$$

We assume that A, B, and C_1 are full rank. If $p_1 > m$ then, in general, a solution does not exist. So we assume that $p_1 \leq m$. If $p_1 = m$, then the unique solution may be obtained by a standard matrix inversion of X. If $p_1 < m$ then there is no unique solution. The solution which minimizes $||x^*||_2^2 + ||u^*||_2^2$ may be obtained by solving the following equation,

$$\begin{aligned} \begin{bmatrix} x^* \\ u^* \end{bmatrix} &= X^T\{XX^T\}^{-1}\begin{bmatrix} 0 \\ z^* \end{bmatrix} \qquad (6) \\ &\triangleq \begin{bmatrix} \Omega_{11} & \Omega_{12} \\ \Omega_{21} & \Omega_{22} \end{bmatrix}\begin{bmatrix} 0 \\ z^* \end{bmatrix}. \end{aligned}$$

We assume throughout that a solution to (4) exists.

For notational convenience the following quantities are defined,

$$\begin{aligned} \hat{x} &\triangleq x - x^* \qquad (7) \\ v &\triangleq u - u^* \\ \hat{z} &\triangleq z - z^* \\ \hat{y} &\triangleq y - y^*, \end{aligned}$$

where $y^* = C_2 x^*$.

Recalling that $\Delta A(r) = DF(r)E$, we define $w \triangleq F(r)Ex^* \in R^q$. Based on the quantities defined in (7), the system can be written as

$$\begin{aligned} \dot{\hat{x}} &= (A + \Delta A(r))x + Bu - Ax^* - Bu^* \qquad (8) \\ &= (A + \Delta A(r))\hat{x} + Bv + Dw \\ \hat{z} &= C_1 x - C_1 x^* = C_1 \hat{x} \\ \hat{y} &= C_2 x - C_2 x^* = C_2 \hat{x}. \end{aligned}$$

Now the tracking problem becomes the problem of determining a feedback control law $v = K(s)\hat{y}$ which stabilizes the system (8) and attenuates the effect of the disturbance w, which results from the uncertainty. Then, $u = v + u^*$ will produce the desired tracking result. We will denote the closed loop transfer function from the disturbance w to the tracking output $\hat{z}$ for a given $r \in \mathcal{R}$ by $F_{\hat{z}w}(s,r)$.

3 Results on Norms and Norm Bounds

In this section we discuss various norms for signals and systems and their relationships. We first define the signal norms which are used in this work.

We define the normed vector space L_p^n where $1 \leq p < \infty$ as the set of real-valued functions $f : R_+ \rightarrow R^n$ which satisfy $||f(\cdot)||_p < \infty$, where

$$||f(\cdot)||_p \triangleq \left(\int_0^\infty \sum_{i=1}^n |f_i(t)|^p dt\right)^{\frac{1}{p}}. \qquad (9)$$

We also define the normed vector space L_∞^n as the set of real-valued functions $f : R_+ \rightarrow R^n$ which satisfy $||f(\cdot)||_\infty < \infty$, where

$$||f(\cdot)||_\infty \triangleq \max_i \sup_{t \geq 0} |f_i(t)|. \qquad (10)$$

We will make use of two norms for linear time-invariant (LTI) systems. Let u be the system input ($u : R_+ \rightarrow R^m$) and g be the impulse response matrix ($g : R_+ \rightarrow R^{p\times m}$). Assume that for each $i = 1, \cdots, p$ and $j = 1, \cdots, m$ that the ij^{th} element of g, denoted g_{ij}, is in L_1^1 defined above. The linear operator $\bar{G}$ is defined by

$$\bar{G} : u \rightarrow \bar{G}u \triangleq \int_0^t g(t-\tau)u(\tau)d\tau. \qquad (11)$$

This is simply the output $y(t)$ due to the input $u(t)$ with zero initial conditions.

The Laplace transform of $g(t)$ will be denoted by $G(s)$. Since $g(t)$ is the impulse response of the system, it follows that $G(s)$ is the transfer function for the system. The transfer function norms are defined by,

$$||g(\cdot)||_1 \triangleq \max_i \int_0^\infty \sum_{j=1}^m |g_{ij}(\tau)|d\tau \qquad (12)$$

and

$$||G(\cdot)||_\infty \triangleq \sup_{\omega \in R} \bar{\sigma}(G(j\omega)) \qquad (13)$$

where $\bar{\sigma}(A)$ denotes the largest singular value of the matrix A.

We now discuss the concept of an induced norm. The induced norm of an operator $H : x \rightarrow Hx$ on a normed vector space with norm $||\cdot||_p$ (where $||\cdot||_p$ is defined above) is defined by

$$||H||_{p_i} \triangleq \sup_{||x||_p = 1} ||Hx||_p. \qquad (14)$$

For a linear system given by $\bar{G}$, the following two facts will be useful.

Fact 1 $||\bar{G}||_{2_i} = ||G(\cdot)||_\infty$

Fact 2 $||\bar{G}||_{\infty_i} = ||g(\cdot)||_1$

Remark: The definitions and facts given here are taken primarily from Desoer and Vidyasagar [6], where Facts 1 and 2 are given as an exercise for the special case where $m = p$.

The next result is a generalization of Theorem 1 of [7], due to Boyd and Doyle. Although that result is for discrete time systems, the same result holds for continuous time systems.

Lemma 1 *If $G(s) \in C^{p\times m}$ is a strictly proper, rational, BIBO stable transfer function matrix, and $g(t) \in R^{p\times m}$ is the corresponding impulse response matrix, then $||g(\cdot)||_1 \leq 2mn||G(\cdot)||_\infty$, where n is the McMillan degree of G.*

Proof: Note that

$$||g(\cdot)||_1 = \max_i \sum_{j=1}^m ||g_{ij}(\cdot)||_1 \leq \max_i \sum_{j=1}^m 2n_{ij}||G_{ij}(\cdot)||_\infty$$

by Theorem 1 of [7], and where n_{ij} is the McMillan degree of G_{ij}, and G_{ij} is the Laplace transform of g_{ij}. From Fact 3, p. 74 of Callier and Desoer [8], it follows that $n_{ij} \leq n$. Hence,

$$||g(\cdot)||_1 \leq \max_i \sum_{j=1}^m 2n||G_{ij}(\cdot)||_\infty.$$

We now claim that $||G_{ij}(\cdot)||_\infty \leq ||G(\cdot)||_\infty$. The claim follows from the fact that for any complex matrix A, $\max_{i,j}|a_{ij}| \leq \bar{\sigma}(A)$, where a_{ij} is the ij^{th} element of A. Hence,

$$\begin{aligned} ||g(\cdot)||_1 &\leq \sum_{j=1}^m 2n||G(\cdot)||_\infty \\ &= 2nm||G(\cdot)||_\infty. \end{aligned}$$

□

The McMillan degree of a transfer function $G(s)$ is the degree of the pole polynomial of the Smith-McMillan form of $G(s)$, and is the order of a minimal realization of $G(s)$ [9].

Remark: This lemma provides a useful bound on the transfer function one-norm. This result is significant because one is often interested in bounding the induced infinity norm of a transfer function (i.e. the L_1 norm) since it provides a quantitative measure of the worst case deviation of the output from the origin resulting from a bounded input. Mathematically, this is expressed by $||z||_\infty \leq ||g(\cdot)||_1||w||_\infty$. However, techniques for designing controllers that bound the L_1 norm of a transfer function are not as clean or developed as are techniques for bounding the H_∞ norm of a transfer function (see [10, 11]). Lemma 1 shows that by bounding the H_∞ norm of the transfer function, the L_1 norm is also being bounded.

4 Main Result

Assume that a control law of the following form is given,

$$\begin{aligned} \dot{\theta} &= A_c\theta + B_c\hat{y} \\ v &= C_c\theta + D_c\hat{y}. \end{aligned} \qquad (15)$$

Then the closed loop system would be given by

$$\begin{bmatrix} \dot{\hat{x}} \\ \dot{\theta} \end{bmatrix} = \begin{bmatrix} A+\Delta A(r)+BD_cC_2 & BC_c \\ B_cC_2 & A_c \end{bmatrix}\begin{bmatrix} \hat{x} \\ \theta \end{bmatrix} + \begin{bmatrix} D \\ 0 \end{bmatrix} w$$

$$\hat{y} = \begin{bmatrix} C_1 & 0 \end{bmatrix}\begin{bmatrix} \hat{x} \\ \theta \end{bmatrix}. \qquad (16)$$

We will denote the system, input, and output matrices in (16) by $A_{cl}(r)$, D_{cl}, and C_{cl}, respectively.

We now give the following definition for robustly stabilizing a system with an H_∞ -norm bound.

Definition 1 *Let the constant $\gamma > 0$ be given. Then the uncertain system in (8)* is robustly stabilizable with an H_∞ -norm bound γ *if there exists a fixed LTI proper feedback law (15) such that for any $r \in \mathcal{R}$*

1. *$A_{cl}(r)$ is a stable matrix;*
2. *$\gamma > ||C_{cl}(sI - A_{cl}(r))^{-1}D_{cl}||_\infty \triangleq ||F_{\hat{z}w}(s,r)||_\infty$.*

Now we provide the main result of the paper.

Theorem 1 *Let $\epsilon > 0$ be given. If the uncertain system (8) is robustly stabilizable with an H_∞ -norm bound $\gamma > 0$ via a feedback control law (15), where*

$$\gamma < \frac{\epsilon}{2nq\bar{r}N||\Omega_{12}||_2||E||_2}, \qquad (17)$$

then with this control, $\max_i |\hat{z}_i(t)| < \epsilon$ for all $t > 0$ and $r \in \mathcal{R}$, provided $\hat{x}(0) = 0$.

Proof: Suppose there exists a robustly stabilizing control law such that $||F_{\hat{z}w}(s,r)||_\infty < \gamma$ for all $r \in \mathcal{R}$. Let $f_{\hat{z}w}(t,r)$ denote the inverse Laplace transform of $F_{\hat{z}w}(s,r)$. Then, by Lemma 1 of Section 3, we have

$$\begin{aligned} ||f_{\hat{z}w}(\cdot,r)||_1 &\leq 2nq||F_{\hat{z}w}(\cdot,r)||_\infty \leq 2nq\gamma \\ &\leq \frac{2nq\epsilon}{2nq\bar{r}N||\Omega_{12}||_2||E||_2} \\ &= \frac{\epsilon}{\bar{r}N||\Omega_{12}||_2||E||_2}. \end{aligned}$$

From Fact 2 of Section 3, we have

$$||\hat{z}(\cdot)||_\infty \leq ||f_{\hat{z}w}(\cdot)||_1||w(\cdot)||_\infty.$$

But recall that w is time invariant. Hence, $||w(\cdot)||_\infty = \max_i |w_i|$. Observing that for any vector $x \in R^n$ that $||x||_\infty \leq ||x||_2$, it follows that

$$||\hat{z}(\cdot)||_\infty \leq \frac{\epsilon}{\bar{r}N||\Omega_{12}||_2||E||_2}||FEx^*||_2,$$

but, for any $r \in \mathcal{R}$,

$$\begin{aligned} ||F(r)Ex^*||_2 &\leq ||F(r)||_2||Ex^*||_2 \leq \bar{r}||Ex^*||_2 \\ &\leq \bar{r}||E||_2||x^*||_2 \leq \bar{r}N||E||_2||\Omega_{12}||_2. \end{aligned}$$

Thus $\max_i |\hat{z}_i(\cdot)| \leq \epsilon$.

□

Remark: If the state initial condition $\hat{x}$ is not zero, then given any $r \in \mathcal{R}$ and $\hat{\epsilon} > 0$ there exists a $T > 0$ such that for all $t > T$ then $||\hat{z}(t)||_\infty < \epsilon + \hat{\epsilon}$. This is true since the general solution is

$$\hat{z}(t) = C_{cl}e^{A_{cl}t}\hat{x}(0) + \int_0^t C_{cl}e^{A_{cl}(t-\tau)}D_{cl}w d\tau \triangleq \hat{z}_1 + \hat{z}_2.$$

The first term, $\hat{z}_1$, decays to zero because of robust stability, and the second term, $\hat{z}_2$, is equivalent to the response with zero initial conditions, which based on the above theorem satisfies $\max_i |\hat{z}_{2_i}| < \epsilon$ for all $t > 0$.

Now we provide a technique for robustly stabilizing an uncertain system with an H_∞ -norm bound of γ. This result is due to Xie, Fu, and DeSouza [12]. First consider the following system,

$$\begin{aligned} \dot{\hat{x}}(t) &= A\hat{x}(t) + \left[\sqrt{\alpha}\bar{r}D \;\; \tfrac{1}{\gamma}D\right]\tilde{w}(t) + Bv(t) \quad (18) \\ \tilde{z}(t) &= \begin{bmatrix} \frac{1}{\sqrt{\alpha}}E \\ C_1 \end{bmatrix}\hat{x}(t) \\ \hat{y}(t) &= C_2\hat{x}(t) \end{aligned}$$

where $\tilde{z}$ and $\tilde{w}$ are the modified controlled outputs and disturbance inputs to account for the uncertainty, and $\alpha > 0$ is a design parameter. The following theorem provides the desired result.

Theorem 2 *[12] Let the constant $\gamma > 0$ and the controller (15) be given. Then the uncertain system (8) is robustly stabilizable with an H_∞ -norm bound γ via (15) if and only if there exists a constant $\alpha > 0$ such that the closed loop system corresponding to (18) with the controller in (15) is stable with unitary H_∞ -norm bound.*

One technique for stabilizing (18) with a unitary H_∞ -norm bound is due to [13] with the scalings of [14]. In the output feedback case, if there exist positive definite symmetric matrices P_c and P_o which satisfy

$$P_cA + A^TP_c - P_c(\frac{1}{\mu_c}BB^T - (\alpha\bar{r}^2 + \frac{1}{\gamma^2})DD^T)P_c + \frac{1}{\alpha}E^TE + C_1^TC_1 = 0, \quad (19)$$

$$P_oA + A^TP_o - (\frac{1}{\mu_o}C_2^TC_2 - (\frac{1}{\alpha}E^TE + C_1^TC_1)) + (\alpha\bar{r}^2 + \frac{1}{\gamma^2})P_oDD^TP_o = 0, \quad (20)$$

and

$$P_o > P_c \quad (21)$$

for some $\mu_c > 0$ and $\mu_o > 0$, then the following control law stabilizes (18) with unitary H_∞ -norm bound,

$$\begin{aligned} \dot{\theta}(t) &= (A + BK + (\alpha\bar{r}^2 + \frac{1}{\gamma^2})DD^TP_c + ZLC_2)\theta(t) \\ &\quad -ZL\hat{y}(t) \\ v(t) &= K\theta(t), \quad (22) \end{aligned}$$

where

$$\begin{aligned} K &= -\frac{1}{\mu_c}B_2^TP_c \\ L &= -\frac{1}{\mu_o}P_o^{-1}C_2^T \\ Z &= (I - P_o^{-1}P_c)^{-1}. \end{aligned}$$

Conversely, if (18) is stabilizable with unitary H_∞ -norm bound, then there exist $\mu_c > 0$ and $\mu_o > 0$ such that (19 - 20) have solutions $P_c > 0$ and $P_o > 0$ such that (21) will be satisfied.

Example: Consider the following system from [2],

$$\begin{aligned} \dot{x} &= \begin{bmatrix} -1+r & 1 \\ 2 & 1 \end{bmatrix}x + \begin{bmatrix} 0 \\ 1 \end{bmatrix}u \quad (23) \\ z &= \begin{bmatrix} 1 & 0 \end{bmatrix}x \\ y &= \begin{bmatrix} 1 & 0 \end{bmatrix}x \end{aligned}$$

where $|r| \leq 0.1$. Let $D = \begin{bmatrix} 1 & 0 \end{bmatrix}^T$ and $E = \begin{bmatrix} 1 & 0 \end{bmatrix}$. Assume that z^* such that $|z^*| \leq 1$ (i.e., $N = 1$) is given. Solving (4) for x^* and u^* yields,

$$x^* = \begin{bmatrix} z^* \\ z^* \end{bmatrix}, \quad u^* = -3z^*.$$

We choose a prespecified accuracy of $\epsilon = 0.2$. From (17), noting that $||\Omega_{12}|| = \sqrt{2}$, it is required that $\gamma < 0.5/sqrt(2)$. The Riccati equations in (19) and (20) are solved with $\mu_c = \mu_o = 1 \times 10^{-4}$ and $\alpha = 0.1$, yielding,

$$\begin{aligned} P_c &= \begin{bmatrix} 0.98710 & 0.041215 \\ 0.041215 & 3.2003 \times 10^{-3} \end{bmatrix} \\ P_o &= \begin{bmatrix} 35.46 & -0.12542 \\ -0.12542 & 0.062491 \end{bmatrix}. \end{aligned}$$

The resulting control law is then formed from (22).

References

[1] O'Brien, M. J. and Broussard, J. R., "Feedforward Control to Track the Output of a Forced Model," *Proceedings of the IEEE Conference on Decision and Control*, 1978, pp.1149-1155.

[2] Hopp, T.H. and Schmitendorf, W.E., "Design of a Linear Controller for Robust Tracking and Model Following," *J. of Dynamic Systems, Measurement, and Control*, Vol. 112, pp. 552-558, 1990.

[3] Dolphus, R.M. and Schmitendorf, W.E., "A Non-Iterative Riccati Approach to Robust Control Design," *Proceedings of the American Control Conference*, 1990, pp.916 - 918.

[4] Schmitendorf, W.E. and Barmish, B.R., "Robust Asymptotic Tracking for Linear Systems with Unknown Parameters," *Automatica*, Vol. 22, No. 33, 1986, pp.355-360.

[5] Schmitendorf, W.E., Dolphus, R.M., Benson, R.W., "Quadratic Stabilization and Tracking: Applications to the Benchmark Problem," *Proceedings of the American Control Conference*, 1992, pp. 2055-2058.

[6] Desoer, C.A. and Vidyasagar, M., *Feedback Systems: Input-Output Properties*, Academic Press, New York, 1975.

[7] Boyd, S. and Doyle, J., "Comparison of Peak and RMS Gains for Discrete-Time Systems," *Systems and Control Letters*, Vol. 9, 1987, pp. 1-6.

[8] Callier, F.M. and Desoer C.A., *Multivariable Feedback Systems*, Springer-Verlag, New York, 1982.

[9] Maciejowski, J.M., *Multivariable Feedback Design*, Addison-Wesley, Wokingham, England, 1989.

[10] Vidyasagar, M., "Optimal Rejection of Persistant Bounded Disturbances," *IEEE Trans. Automat. Contr.*, Vol. AC-31, pp.527-534, 1986.

[11] Dahleh, M.A. and Pearson, J.B., "L^1 -Optimal Compensators for Continuous-Time Systems," *IEEE Trans. Automat. Contr.*, Vol. AC-32, pp.889-895, 1987.

[12] Xie, L., Fu, M., DeSouza, C.A., "H_∞ Control and Quadratic Stabilization of Systems with Parameter Uncertainty via Output Feedback," *IEEE Trans. Automat. Contr.*, Vol. AC-37, pp.1253 -1256,1992.

[13] Doyle, J., Glover K., Khargonekar, P.P., and Francis, B.A., "State-Space Solutions to the Standard H_2 and H_∞ Control Problems," *IEEE Trans. Automat. Contr.*, Vol. AC-34, pp.831-847, 1989.

[14] Safonov, M.G., Limebeer, D.J.N., Chiang, R.Y., "Simplifying the H_∞ Theory Via Loop-Shifting, Matrix-Pencil and Descriptor Concepts," *Int. J. Control*, Vol. 50, No. 6, pp. 2467-2488, 1989.

Proceedings of the 31st Conference
on Decision and Control
Tucson, Arizona • December 1992

Robust Linear Controller Under State and Control Constraints

George BITSORIS and Eliana GRAVALOU

Control Systems Laboratory
Electrical Engineering Department
University of Patras
26500 Patras-GREECE

Abstract

In this paper the Robust Linear Constrained Regulation Problem is investigated. An optimization method is established for the determination of linear state feedback control laws which transfer asymptotically to the origin any initial state belonging to a prespecified convex set, while linear constraints on the control vector are satisfied for any perturbed system.

1.Introduction

The problem of designing state feedback controls to stabilize linear systems with parameter uncertainties has been widely studied over the last decade. In recent years, the design of constrained controllers has also been the object of an intensive research effort. However, little work has been done for the development of efficient design methods of linear robust constrained regulators. The robust regulation under linear states constraints has been studied by Sznaier and Sideris [13]. A linear programming approach for the design of linear states feedback controls for systems with state and control constraints in the presence of additive disturbances have been studied by Blanchini [6,7].

In the present paper, the Linear Constrained Regulation Problem in the presence of parameter uncertainties is investigated. The problem consists in the determination of linear state feedback control laws which transfer asymptotically to the equilibrium all initial states belonging to a prespecified convex set, while linear constraints on the control vector are respected for any perturbed system.

The paper is organized as follows: In section 2, the formulation of the Linear Robust Constrained Regulation Problem (LRCRP) is introduced. In the next section, a linear programming approach to the design of linear constrained regulators is presented. Finally in section 4, the robustness of controlled systems is investigated. Then, an algorithm for the determination of robust constrained controllers is suggested.

2. Problem Statement

Throughout the paper upper case letters denote real matrices, lower case letters denote column vectors or scalars, $R^n(C^n)$ denotes the real (complex) n-space and $R^{nxm}(C^{nxm})$ denotes the set of all nxm real (complex) matrices. For two real vectors $x=[x_1\ x_2\ \ldots\ x_3]^T$ and $y=[y_1\ y_2\ \ldots\ y_3]^T$, $x\leq y$ $(x<y)$ is equivalent to $x_i\leq y_i$ $(x_i<y_i)$, $i=1,2,\ldots n$. For a matrix $H\in R^{mxm}$, $H^+=(h^+_{ij})$ with $h^+_{ij}=\max(h_{ij},0)$ and $H^-=(h^-_{ij})$ with $h^-_{ij}=-\min(h_{ij},0)$. Finally, $\lambda(A)$ denotes the eigenvalues of matrix A.

We consider linear discrete-time systems described by the state equations

$$x(k+1) = Ax(k) + Bu(k) \tag{S}$$

where $x\in R^n$, $u\in R^m$, $A\in R^{nxn}$, $B\in R^{nxm}$ and $k\in T$, T being the time set $T=\{0,1,2,\ldots\}$.

It is known that in most of the practical applications, the state and control vectors must satisfy physical constraints. So, we shall assume that the control vector is constrained to satisfy linear inequalities of the form

$$-d_2 \leq u \leq d_1 \tag{1}$$

where d_1, d_2 are real vectors with positive components.

Since global constrained stabilization by linear state feedback is not possible, a set of initial states is also given. The initial states of interest belong to a convex subset of R^n defined by the linear inequalities

$$-w_2 \leq Gx_0 \leq w_1 \tag{2}$$

where $G\in R^{qxn}$ with rank(G)=n and w_1, w_2 are real vectors with positive components.

As it is shown , the matrix A characrerizing the dynamics of the system is not always perfectly known due to either nonlinearities or unpredicted variations. So, we assume that the matrix A can have any value in the interval (A_l,A_u) where A_l and A_u denote lower and upper bounds of matrix A respectively.

The problem to be solved is the determination of a linear state feedback control law u=Fx such that for any matrix A in the interval (A_l,A_u) all initial states of the resulting closed-loop system

$$x(k+1) = (A+BF)x(k) \tag{3}$$

which satisfy inequalities (2) are transferred asymptotically to the origin while the control vector respects constraints (1). Due to the linearity of the system and the control and states constraints, this problem is termed the Linear Robust Constrained Regulation Problem (LRCRP).

The determination of sufficient conditions on the bounds of the admissible control effort and of those of the interval of variation of matrix A guaranteeing the existence of a solution to the LCRP is a related problem. Finally in the case where these conditions are not satisfied the following problems are posed:

a)Given a set of initial states and the bounds A_l and A_u of the variation of matrix A, estimate the minimal bounds of the admissible control effort so that a solution to the LRCRP exists. It is the problem of the determination of the characteristics of the linear robust constrained controller.

b)Given a set of initial states and the bounds of the admissible control effort, determine the maximal interval of variation (A_l,A_u) of matrix A, so that a solution to the LRCRP exists. It is the problem of robustness analysis of linear constrained controllers.

3. Design of linear constrained regulators

The design of linear constrained regulators for discrete time systems has been studied for the first time by Gutman and Hagander [10] using the theory of ellipsoidal positively invariant sets. The recent development of the theory of polyhedral positively invariant sets [2,3] however, is more adapted to this problem due to the linearity of both state and control constraints and of the control law. In this context, Bitsoris and Vassilaki established a design method based on nonlinear programming [4]. More complicated approaches are those based on eigenstructure assignment [5,12]. Other approaches for some classes of linear systems have been proposed in [1,8,11]. However, the simplest design method is that consisting in solving a set of linear inequalities [9,15]. This can be achieved by defining and solving a related linear programming problem. This method will be adopted for the design of robust regulators.

Let $R(G,w_1,w_2)$ denote the set of initial states defined by the linear inequalities (2):

$$R(G,w_1,w_2) = \{x\in R^n: -w_2 \leq Gx \leq w_1\}$$

According to this notation the set of states where a linear control law u=Fx respects the constraints (1) is denoted by $R(F,d_1,d_2)$.

Proposition 1: [15]

The control law u=Fx is a solution to the LCRP for system (S), if $R(G,w_1,w_2)$ is a positively invariant set for the resulting closed-loop system, $|\lambda_i(A+BF)|<1$ $i=1,2,\ldots,n$, and $-d_2\leq Fx^{(i)}\leq d_1$ $i=1,2,\ldots,q$, where $x^{(i)}$ denote the vertices of the polyhedral set $R(G,w_1,w_2)$.

Proposition 2: [3]

The set $R(G,w_1,w_2)$ is positively invariant with respect to the closed-loop system (3) if and only if there exists a matrix $H\in R^{nxn}$ such that

$$GA + GBF = (H^+ - H^-)G \tag{4}$$

CH3229-2/92/0000-2640$1.00 © 1992 IEEE

$$\begin{bmatrix} H^+ & H^- \\ H^- & H^+ \end{bmatrix} \begin{bmatrix} w_1 \\ w_2 \end{bmatrix} \le \begin{bmatrix} w_1 \\ w_2 \end{bmatrix} \tag{5}$$

Proposition 3: [3]

If inequality (5) is satisfied strictly, then the equilibrium x=0 of the closed-loop system (3) is asymptotically stable.

According to the above propositions, a constrained stabilizing control law u=Fx for system S can be obtained by determining a matrix F satisfying the relations

$$GA + GBF = (H^+ - H^-)G$$

$$\begin{bmatrix} H^+ & H^- \\ H^- & H^+ \end{bmatrix} \begin{bmatrix} w_1 \\ w_2 \end{bmatrix} \le \varepsilon \begin{bmatrix} w_1 \\ w_2 \end{bmatrix}$$

$$-d_2 \le Fx^{(i)} \le d_1 \qquad i=1,2,\dots,q,$$

$$0 \le \varepsilon < 1$$

where H^+, H^- and ε are undetermined parameters [15].

4.Robustness of Linear Constrained Regulators

Let $A_0 = \dfrac{A_l + A_u}{2}$

Then any matrix in the interval (A_l,A_u) can be written as

$$A = A_0 + \Delta A$$

where ΔA satisfy the inequalities

$$-A_\delta \le \Delta A \le A_\delta \text{ with } A_\delta = \frac{A_u - A_l}{2}$$

A constrained stabilizing control law $u=F_0x$ for the nominal system

$$x(k+1) = A_0 x(k) + Bu(k)$$

can be obtained by determining a matrix F_0 satisfying the relations

$$GA_0 + GBF_0 = (H_0^+ - H_0^-)G \tag{6}$$

$$\begin{bmatrix} H_0^+ & H_0^- \\ H_0^- & H_0^+ \end{bmatrix} \begin{bmatrix} w_1 \\ w_2 \end{bmatrix} \le \varepsilon_0 \begin{bmatrix} w_1 \\ w_2 \end{bmatrix} \tag{7}$$

$$-d_2 \le F_0 x^{(i)} \le d_1 \qquad i=1,2,\dots,q, \tag{8}$$

$$0 \le \varepsilon_0 < 1 \tag{9}$$

where H_0^+, H_0^- and ε_0 are undetermined parameters.

The problem now is the determination of the additional conditions which guarantee that the above control law is a solution to the LCRP for any matrix $A_0+\Delta A$, ΔA being in the interval $(-A_\delta,A_\delta)$

Proposition 4:

Let ε_0 be the value of the parameter resulting by solving relations (6)-(9). Let also ε_Δ be the parameter satisfying the inequality

$$\begin{bmatrix} \Delta H^+ & \Delta H^- \\ \Delta H^- & \Delta H^+ \end{bmatrix} \begin{bmatrix} w_1 \\ w_2 \end{bmatrix} \le \varepsilon_\Delta \begin{bmatrix} w_1 \\ w_2 \end{bmatrix} \tag{10}$$

for all pairs $(\Delta A, \Delta H)$ such that

$$G(\Delta A) = (\Delta H^+ - \Delta H^-)G \tag{11}$$

and

$$-A_\delta \le \Delta A \le A_\delta$$

If $\varepsilon_0 + \varepsilon_\Delta \le 1$ then the control law $u = F_0 x$ is a solution to the LRCRP.

Proof:

According to Proposition 2, the polyhedral set $R(G,w_1,w_2)$ is positively invariant with respect to the closed-loop system

$$x(k+1) = (A + BF_0)x(k) \tag{12}$$

for a matrix $A \in (A_l, A_u)$ if and only if there exists a matrix H such that

$$GA + GBF_0 = (H^+ - H^-)G \tag{13}$$

$$\begin{bmatrix} H^+ & H^- \\ H^- & H^+ \end{bmatrix} \begin{bmatrix} w_1 \\ w_2 \end{bmatrix} \le \begin{bmatrix} w_1 \\ w_2 \end{bmatrix} \tag{14}$$

Now, by combining (6)-(7) with (10)-(11) we get

$$G(A_0 + \Delta A) + GBF_0 = (H_0^+ - H_0^- + \Delta H^+ - \Delta H^-)G \tag{15}$$

$$\begin{bmatrix} H_0^+ + \Delta H^+ & H_0^- + \Delta H^- \\ H_0^- + \Delta H^- & H_0^+ + \Delta H^+ \end{bmatrix} \begin{bmatrix} w_1 \\ w_2 \end{bmatrix} \le (\varepsilon_0 + \varepsilon_\Delta) \begin{bmatrix} w_1 \\ w_2 \end{bmatrix} \tag{16}$$

Moreover, $(H_0^+ - H_0^- + \Delta H^+ - \Delta H^-)^+ \le H_0^+ + \Delta H^+$ and $(H_0^+ - H_0^- + \Delta H^+ - \Delta H^-)^- \le H_0^- + \Delta H^-$.

Therefore (16) becomes

$$\begin{bmatrix} (H_0^+ - H_0^- + \Delta H^+ - \Delta H^-)^+ & (H_0^+ - H_0^- + \Delta H^+ - \Delta H^-)^- \\ (H_0^+ - H_0^- + \Delta H^+ - \Delta H^-)^- & (H_0^+ - H_0^- + \Delta H^+ - \Delta H^-)^+ \end{bmatrix} \begin{bmatrix} w_1 \\ w_2 \end{bmatrix} \le (\varepsilon_0 + \varepsilon_\Delta) \begin{bmatrix} w_1 \\ w_2 \end{bmatrix} \tag{17}$$

Setting $A=A_0+\Delta A$ and $H=H_0+\Delta H$ and combining relations (15),(17) with (13),(14) we conclude that if $\varepsilon_0+\varepsilon_\Delta \le 1$ then the polyhedron $R(G,w_1,w_2)$ is positively invariant with respect to the closed-loop system (12) for any matrix A in the interval (A_l,A_u). Furthermore, since inequality (8) is independent of matrix A and is satisfied by matrix F_0, we conclude that all the conditions of Proposition 1 are satisfied for any matrix $A \in (A_l,A_u)$. Therefore the control law $u=F_0x$ is a solution to the LRCRP. □

By combining this result with those of section 3 we can establish the following design algorithm of Linear Robust Constrained Regulators.

Step 1

Solve the linear programming problem

$$\min_{\Delta H^+, \Delta H^-, \varepsilon_\Delta} \{\varepsilon_\Delta\}$$

under constraints

$$\begin{bmatrix} \Delta H^+ & \Delta H^- \\ \Delta H^- & \Delta H^+ \end{bmatrix} \begin{bmatrix} w_1 \\ w_2 \end{bmatrix} \le \varepsilon_\Delta \begin{bmatrix} w_1 \\ w_2 \end{bmatrix}$$

$$G(\Delta A) = (\Delta H^+ - \Delta H^-)G$$

and

$$-A_\delta \le \Delta A \le A_\delta$$

Step 2

Solve the linear inequalities

$$GA_0 + GBF_0 = (H_0^+ - H_0^-)G \tag{18}$$

$$\begin{bmatrix} H_0^+ & H_0^- \\ H_0^- & H_0^+ \end{bmatrix} \begin{bmatrix} w_1 \\ w_2 \end{bmatrix} \le (1 - \varepsilon_\Delta^*) \begin{bmatrix} w_1 \\ w_2 \end{bmatrix} \tag{19}$$

$$-d_2 \le F_0 x^{(i)} \le d_1 \qquad i=1,2,\dots,q, \tag{20}$$

where ε_Δ^* is the optimal value of ε_Δ determined in Step 1.

It is clear that Step 2 is quite analogous to the design algorithm of linear constrained regulators for the nominal system. The only difference is the factor $1-\varepsilon_\Delta^*$ in the right side member of inequality (19). Due to this factor the eigenvalues of the resulting closed-loop system are in the disk $|\lambda_j(A+BF)| \le 1-\varepsilon_\Delta^*$ [15]. So, the first step of the algorithm is the determination of the parameter ε_Δ^* which defines the admissible domain where the eigenvalues of the nominal system must be located for the requirements of the constrained control problem to be satisfied by any system whose matrix A is in the interval (A_l,A_u). In order to enlarge the domain of admissible eigenvalues $\lambda(A+BF)$ it is necessary to minimize the parameter ε_Δ.

Example 1.

Consider the second order system

$$x(k+1) = Ax(k) + Bu(k)$$

where

$$\begin{bmatrix} 0.70 & 0.48 \\ -0.42 & 1.00 \end{bmatrix} \le A \le \begin{bmatrix} 0.90 & 0.52 \\ -0.38 & 1.40 \end{bmatrix}, \quad B = \begin{bmatrix} 0 \\ 1 \end{bmatrix} \qquad (21)$$

and the initial state set $R(G,w_1,w_2)$ with

$$G = \begin{bmatrix} 1.4 & 0.6 \\ 0.4 & 0.9 \end{bmatrix} \quad \text{and} \quad w_1 = w_2 = \begin{bmatrix} 2.0 \\ 1.3 \end{bmatrix}$$

The control constraint is $-7 \le Fx(k) \le 7$.
Then

$$A_\delta = \begin{bmatrix} 0.1 & 0.02 \\ 0.02 & 0.2 \end{bmatrix} \quad A_0 = \begin{bmatrix} 0.8 & 0.5 \\ -0.4 & 1.2 \end{bmatrix}$$

and the solution of the linear programming problem stated in step 1 results in the optimal $\varepsilon_\Delta^*=0.4843$. With this value of ε_Δ we have to determine a solution to inequalities (18)-(20). Such a solution is

$$F_0 = [-0.4752 \quad -1.645]$$

With the control law $u_0=F_0x$ all initial states belonging to the set $R(G,w_1,w_2)$ are transferred asymptotically to the origin for any matrix A satisfying (21).

If it is not possible to assign eigenvalues in the disk $|\lambda_i(A+BF)| \le 1-\varepsilon_\Delta^*$ it is necessary to enlarge the domain of admissible controls. In this case the problem is the minimization of the maximal control effort required for the transfer to the equilibrium of any initial state belonging to the set $R(G,w_1,w_2)$. If the control effort is characterized by the sum $u_1^2+u_2^2+...+u_m^2$, step 2 must be replaced by the following nonlinear optimization problem:

$$\min_{H_0^+,H_0^-,d_{i1},d_{i2},\ i=1,...,m} \left\{ \sum_{i=1}^{m} (d_{i1}^2+d_{i2}^2) \right\}$$

under constraints

$$GA_0 + GBF_0 = (H_0^+ - H_0^-)G$$

$$\begin{bmatrix} H_0^+ & H_0^- \\ H_0^- & H_0^+ \end{bmatrix} \begin{bmatrix} w_1 \\ w_2 \end{bmatrix} \le (1-\varepsilon_\Delta^*) \begin{bmatrix} w_1 \\ w_2 \end{bmatrix}$$

$$-d_2 \le F_0x^{(i)} \le d_1 \qquad i=1,2,...,q,$$

The problem of robustness analysis can be solved by a similar approach. Let us assume that we are given a nominal system

$$x(k+1) = A_0x(k) + Bu(k),$$

a set of initial states defined by (1) and the control constraints (2). Consider also the matrix of variation ΔA_0. The problem of robustness analysis can be formulated as follows: Determine the maximal set of uncertainties

$$\Delta(\delta) = \{\Delta A \in R^{nxn}: -\delta\,\Delta A_0 \le \Delta A \le \delta\,\Delta A_0\}$$

$\delta>0$, so that there exists a control law $u=F_0x$ such that all initial states satisfying (2) are transferred asymptotically to the origin while the control constraints (1) are respected for any system with $A=A_0+\Delta A$ and $\Delta A \in \Delta(\delta)$. This problem can be solved by applying the following algorithm composed of two steps:

Step 1:

Solve the linear programming problem

$$\min_{H_0^+,H_0^-,\varepsilon} \{\varepsilon\}$$

under constraints

$$GA_0 + GBF_0 = (H_0^+ - H_0^-)G$$

$$\begin{bmatrix} H_0^+ & H_0^- \\ H_0^- & H_0^+ \end{bmatrix} \begin{bmatrix} w_1 \\ w_2 \end{bmatrix} \le \varepsilon \begin{bmatrix} w_1 \\ w_2 \end{bmatrix}$$

$$-d_2 \le F_0x^{(i)} \le d_1 \qquad i=1,2,...,q,$$

Step 2:

Let ε^* be the optimal value of ε obtained in step 1. Then solve the following linear programming problem to determine the maximal value of the parameter δ:

$$\max_{\Delta H^+,\Delta H^-,\delta} \{\delta\}$$

under constraints

$$G(\Delta A) = (\Delta H^+ - \Delta H^-)G$$

$$-\delta\,\Delta A_0 \le \Delta A \le \delta\,\Delta A_0$$

and

$$\begin{bmatrix} \Delta H^+ & \Delta H^- \\ \Delta H^- & \Delta H^+ \end{bmatrix} \begin{bmatrix} w_1 \\ w_2 \end{bmatrix} \le (1-\varepsilon^*) \begin{bmatrix} w_1 \\ w_2 \end{bmatrix}$$

5. Conclusions

In this paper, the Linear Robust Constrained Regulation Problem was investigated. It was shown that the theory of polyhedral positively invariant sets can provide simple solutions to both the analysis and synthesis of linear constrained regulators for systems with parameter uncertainties. For both problems, a solution can be obtained by applying simple linear programming algorithms.

REFERENCES

[1] Benzaouia,A. and C.Burgat, "Regulator problem for linear discrete-time systems with nonsymmetrical constrained control", Int.J.Contr., vol.48, pp.2441-2451, 1988.

[2] Bitsoris,G., Positively invariant polyhedral sets of discrete-time linear systems, Int.J.Contr., vol.47, pp.1713-1726, 1988.

[3] Bitsoris,G., "On the positive invariance of polyhedral sets for discrete-time systems", Syst.Contr.Lett., vol.11, pp.243-248, 1988.

[4] Bitsoris,G. and M.Vassilaki, "An optimization approach to the Linear Constrained Regulation Problem", Int.J.Syst.Science, vol.22, No 10, pp.1953-1960, 1991.

[5] Bitsoris,G. and M.Vassilaki, "The Linear Constrained Regulation for discrete-time systems", Preprints of 11th IFAC World Congress, Tallin, Estonia, vol.2, pp.287-292, 1990.

[6] Blanchini,F., "Feedback control for linear time-invariant systems with state and-control bounds in the presence of disturbances", IEEE Trans.Aut.Contr., vol 35, No 11, pp.1231-1234, 1990.

[7] Blanchini,F., "Ultimate boundedness control for uncertain discrete-time systems via set-induced Lyapunov functions", Proc. 30th IEEE CDC, Brighton-England, vol.2, pp.1755-1760, 1991.

[8] Chegancas,J. and C.Burgat, "Regulateur P-invariant avec contraintes sur les commandes", Actes du congres Automatique 1985 d'AFCET, pp.193-203, 1986.

[9] Gravalou,E. and G.Bitsoris, "Constrained Regulation of linear systems: An algorithm", Proc. 30th IEEE CDC, Brighton-England, vol.2, pp.1744-1747, 1991.

[10] Gutman, P.O. anf P.Hagander, "A new design of constrained controllers for linear systems",IEEE Trans.Aut.Contr., vol AC-30, No 1, pp.22-33, 1985.

[11] Hennet,J.C.,and J.P.Beziat, "Invariant regulators for a class of constrained linear systems", Preprints of 11th IFAC World Congress, Tallin, Estonia, vol.2, pp.299-303, 1990.

[12] Hennet,J.C.,and E.B.Castelan, "Invariance and stability by state feedback for constrained linear systems", Proc. ECC 1991, Grenoble, vol.1, pp.367-373, 1991.

[13] Sznaier M., and A.Sideris, "Suboptimal norm based robust control of constrained systems with an H cost", Proc 30th IEEE CDC, Brighton-England, vol.3, pp.2280-2286, 1991.

[14] Sznaier M., and M.J.Damborg, "Heuristically enhanced feedback control of constrained discrete-time linear systems", Automatica, vol.26, pp.521-532, 1990.

[15] Vassilaki,M., J.C.Hennet and G.Bitsoris, "Feedback control of linear discrete-time systems under state and control constraints", Int.J.Contr., vol.47, pp.1727-1735,1988.

Proceedings of the 31st Conference
on Decision and Control
Tucson, Arizona • December 1992

COORDINATING LOCOMOTION AND MANIPULATION OF A MOBILE MANIPULATOR

Yoshio Yamamoto and Xiaoping Yun
General Robotics and Active Sensory Perception (GRASP) Laboratory
University of Pennsylvania
3401 Walnut Street, Room 301C
Philadelphia, PA 19104-6228

ABSTRACT

A mobile manipulator in this study is a manipulator mounted on a mobile platform. Assuming the end point of the manipulator is guided, *e.g.*, by a human operator to follow an arbitrary trajectory, it is desirable that the mobile platform is able to move as to position the manipulator in certain preferred configurations. Since the motion of the manipulator is unknown *a priori*, the platform has to use the measured joint position information of the manipulator for motion planning. This paper presents a planning and control algorithm for the platform so that the manipulator is always positioned at the preferred configurations measured by its manipulability. Simulation results are presented to illustrate the efficacy of the algorithm. Also the algorithm is implemented and verified on a real mobile manipulator system. The use of the resulting algorithm in a number of applications is also discussed.

1 Introduction

When a human writes across a board, he positions his arm in a comfortable writing configuration by moving his body rather than reaching out his arm. Also when humans transport a large and/or heavy object coorperatively, they tend to prefer certain configurations depending on various factors, *e.g.*, the shape and the weight of the object, the transportation velocity, the number of people involved in a task, and so on. Therefore when a mobile manipulator performs a manipulation task, it is desirable to bring the manipulator into certain preferred configurations by appropriately planning the motion of the mobile platform. If the trajectory of the manipulator end point in a fixed coordinate system (world coordinate system) is known *a priori*, then the motion of the mobile platform can be planned accordingly. However, if the motion of the manipulator end point is unknown *a priori*, *e.g.*, driven by a visual sensor or guided by a human operator, then the path planning has to be made locally and in real time rather than globally and off-line. This paper presents a planning and control algorithm for the platform in the latter case, which takes the measured joint displacement of the manipulator as the input for motion planning, and controls the platform in order to bring the manipulator into a preferred operating region. While this region can be selected based on any meaningful criterion, the manipulability measure [1] is utilized in this study. By using this algorithm, the mobile platform will be able to "understand the intention of its manipulator and respond accordingly."

This control algorithm has a number of immidiate applications. First, a human operator can easily move around the mobile manipulator by "dragging" the end point of the manipulator while the manipulator is in the free mode (compensating the gravity only). Second, if the manipulator is force-controlled, the mobile manipulator will be able to push against and follow an external moving surface. Third, when two mobile manipulators transport a large object with one being the master and the other being slave, this algorithm can be used to control the slave mobile manipulator to support the object and follow the motion of the master, resulting in a cooperative control algorithm for two mobile manipulators.

Although there has been a vast amount of research effort on mobile platforms (commonly referred to as mobile robots) in the literature, the study on mobile manipulators is very limited. Joshi and Desrochers considered a two link manipulator on a moving platform subject to random disturbances in its orientation [2]. Wien studied dynamic coupling between a planar vehicle and a one-link manipulator on the vehicle [3]. Dubowsky, Gu, and Deck derived the dynamic equations of a fully spatial mobile manipulator with link flexibility [4]. Recently, Hootsmans proposed a mobile manipulator control algorithm (the Mobile Manipulator Jacobian Transpose Algorithm) for a dynamically coupled mobile manipulator [5]. He showed that with the algorithm the manipulator could successfully compensate a trajectory error caused by vehicle's passive suspension with the help of limited sensory information from mobile vehicle.

What makes the coordination problem of locomotion and manipulation a difficult one is twofold. First, a manipulator and a mobile platform, in general, have different dynamic characteristics, namely, a mobile platform has slower dynamic response than a manipulator. Second, a wheeled mobile platform is subject to nonholonomic constraints while a manipulator is usually unconstrained. These two issues must be taken into consideration in developing a planning and control algorithm.

Dynamic modeling of mechanical systems with nonholonomic constraints is richly documented by work ranging from Neimark and Fufaev's comprehensive book [6] to more recent developments (see for example, [7]). However, the literature on control properties of such systems is sparse [8]. The interest in control of nonholonomic systems has been stimulated by the recent research in robotics. The dynamics of a wheeled mobile robot is nonholonomic [9], and so is a multi-arm system manipulating an object through the whole arm manipulation [10].

Bloch and McClamroch [8] first demonstrated that a nonholonomic system cannot be feedback stabilized to a single equilibrium point by a smooth feedback. In a follow-up paper [11], they showed that the system is small-time locally controllable. Campion *et al* [12] showed that the system is controllable regardless of the structure of nonholonomic constraints. Motion planning of mobile robots has been an active topic in robotics in the past several years [13, 14, 9, 15, 16]. Nevertheless, much less is known about the dynamic control of mobile robots with nonholonomic constraints and the developments in this area are very recent [17, 18, 19].

In this paper, we first present the theoretic formulation of a general nonholonomic system. Next we apply the formulation to the specific mobile platform used for the experiments in order to derive the dynamic equations. Then we describe the path planning algorithm and show the simulation and experimental results followed by concluding remark.

2 Nonholonomic Systems

2.1 Dynamic Equations of Motion

Consider a mechanical system with n generalized coordinates q subject to m bilateral constraints whose equations of motion are described by

$$M(q)\ddot{q} + V(q,\dot{q}) = E(q)\tau - A^T(q)\lambda \tag{1}$$

where $M(q)$ is the $n \times n$ inertia matrix, $V(q,\dot{q})$ is the vector of position and velocity dependent forces, $E(q)$ is the $n \times r$ input transformation matrix[1], τ is the r-dimensional input vector, $A(q)$ is the $m \times n$ Jacobian matrix, and λ is the vector of constraint forces. The m constraint equations of the mechanical system can be written in the form

$$C(q,\dot{q}) = 0 \tag{2}$$

If a constraint equation is in the form $C_i(q) = 0$, or can be integrated into this form, it is a holonomic constraint. Otherwise it is a kinematic (not geometric) constraint and is termed nonholonomic.

We assume that we have k holonomic and $m-k$ nonholonomic independent constraints, all of which can be written in the form of

$$A(q)\dot{q} = 0 \tag{3}$$

[1] $E(q)$ is an identity matrix in most cases. However, if the generalized coordinates are chosen to be some variables other than the joint variables, or if there are passive joints without actuators, it is not an identity matrix.

CH3229-2/92/0000-2643$1.00 © 1992 IEEE

Let $s_1(q), \cdots, s_{n-m}(q)$ be a set of smooth and linearly independent vector fields in the null space of $A(q)$, *i.e.*,

$$A(q)s_i(q) = 0 \qquad i = 1, \ldots, n-m.$$

Let $S(q)$ be the full rank matrix made up of these vectors

$$S(q) = [s_1(q) \quad \cdots \quad s_{n-m}(q)] \tag{4}$$

and let Δ be the distribution spanned by these vector fields

$$\Delta = span\{s_1(q), \quad \cdots, \quad s_{n-m}(q)\}$$

It follows that $\dot{q} \in \Delta$. Δ may or may not be involutive. For that reason, we let Δ^* be the smallest involutive distribution containing Δ. It is clear that $dim(\Delta) \leq dim(\Delta^*) = k$. There are three possible cases (as observed by Campion, *et al.* in [12]). First, if $k = m$, that is, all the constraints are holonomic, then Δ is involutive itself. Second, if $k = 0$, that is, all the constraints are nonholonomic, then Δ^* spans the entire space. Finally, if $0 < k < m$, the k constraints are integrable and k components of the generalized coordinates may be eliminated from the motion equations. In this case, $dim(\Delta^*) = n - k$.

2.2 State Space Representation

We now consider the mechanical system given by (1) and (3). Since the constrained velocity is always in the null space of $A(q)$, it is possible to define $n - m$ velocities $\nu(t) = [\nu_1 \quad \nu_2 \cdots \nu_{n-m}]$ such that

$$\dot{q} = S(q)\nu(t) \tag{5}$$

These velocities need not be integrable.

Differentiating Equation (5), substituting the expression for $\ddot{q}$ into (1), and premultiplying by S^T, we have

$$S^T(MS\dot{\nu}(t) + M\dot{S}\nu(t) + V) = S^T E\tau \tag{6}$$

by noting that

$$A(q)\dot{q} = 0. \tag{7}$$

Using the state space variable $x = [q^T \quad \nu^T]^T$, we have

$$\dot{x} = \begin{bmatrix} S\nu \\ f_2 \end{bmatrix} + \begin{bmatrix} 0 \\ (S^TMS)^{-1}S^TE \end{bmatrix}\tau \tag{8}$$

where $f_2 = (S^TMS)^{-1}(-S^TM\dot{S}\nu - S^TV)$. Assuming that the number of actuator inputs is greater or equal to the number of the degrees of freedom of the mechanical system ($r \geq n - m$), and $(S^TMS)^{-1}S^TE$ has rank $n - m$, we may apply the following nonlinear feedback

$$\tau = ((S^TMS)^{-1}S^TE)^+(u - f_2) \tag{9}$$

where the superscript + denotes the generalized matrix inverse. The state equation simplifies to the form

$$\dot{x} = f(x) + g(x)u \tag{10}$$

where $f(x) = \begin{bmatrix} S(q)\nu \\ 0 \end{bmatrix}, \quad g(x) = \begin{bmatrix} 0 \\ I \end{bmatrix}$.

2.3 Control Properties

The following two properties of the system (10) have been established in [12, 8, 11] for the special case in which *all* constraints are nonholonomic.

Theorem 1 *The nonholonomic system (10) is controllable.*

Theorem 2 *The equilibrium point $x = 0$ of the nonholonomic system (10) can be made Lagrange stable, but can not be made asymptotically stable by a smooth state feedback.*

In the rest of this section, we discuss the more general case in which Equation (3) consists of both holonomic and nonholonomic constraints.

Theorem 3 *System (10) is not input-state linearizable by a state feedback if at least one of the constraints is nonholonomic.*

Proof: The system has to satisfy two conditions for input-state linealization: the strong accessibility condition and the involutivity condition [20]. It is shown below that the involutivity condition is not satisfied.

Define a sequence of distributions

$$D_j = span\{L_f^i g \mid i = 0, 1, \ldots, j-1\}, \quad j = 1, 2, \ldots$$

Then the involutivity condition requires that the distributions D_1, D_2, ..., D_{2n-m} are all involutive. Note that the dimension of the state variable is $2n - m$. $D_1 = span\{g\}$ is involutive since g is constant. Next we compute

$$L_f g = [f, g] = \frac{\partial g}{\partial x} f - \frac{\partial f}{\partial x} g = -\begin{bmatrix} S(q) \\ 0 \end{bmatrix}$$

Since the distribution Δ spanned by the columns of $S(q)$ is not involutive, the distribution $D_2 = span\{g, L_f g\}$ is not involutive. Therefore, the system is not input-state linearizable. □

Although a system with nonholonomic constraints is not input-state linearizable, it is input-output linearizable if a proper set of output equations are chosen. Consider the position control of the system, *i.e.*, the output equations are functions of position state variable q only. Since the number of the degrees of freedom of the system is instantaneously $n - m$, we may have at most $n - m$ independent position outputs equations.

$$y = h(q) = [h_1(q) \quad \cdots \quad h_{n-m}(q)] \tag{11}$$

The necessary and sufficient condition for input-output linearization is that the decoupling matrix has full rank [20]. With the output equation (11), the decoupling matrix $\Phi(x)$ for the system is the $(n - m) \times (n - m)$ matrix

$$\Phi(q) = J_h(q)S(q) \tag{12}$$

where $J_h = \frac{\partial h}{\partial q}$ is the $(n-m) \times n$ Jacobian matrix. $\Phi(x)$ is nonsingular if the rows of J_h are independent of the rows of $A(q)$.

To characterize the zero dynamics and achieve input-output linearization, we introduce a new state space variable z defined as follows

$$z = T(x) = \begin{bmatrix} z_1 \\ z_2 \\ z_3 \end{bmatrix} = \begin{bmatrix} h(q) \\ L_f h(q) \\ \bar{h}(q) \end{bmatrix} = \begin{bmatrix} h(q) \\ \Phi(q)\nu \\ \bar{h}(q) \end{bmatrix} \tag{13}$$

where $\bar{h}(q)$ is an m-dimensional function such that $[J_h^T \quad J_{\bar{h}}^T]$ has full rank. It is easy to verify that $T(x)$ is indeed a diffeomorphism and thus a valid state space transformation. The system under the new state variable z is characterized by

$$\dot{z}_1 = \frac{\partial h}{\partial q}\dot{q} = z_2 \tag{14}$$

$$\dot{z}_2 = \dot{\Phi}(q)\nu + \Phi(q)u \tag{15}$$

$$\dot{z}_3 = J_{\bar{h}}S\nu = J_{\bar{h}}S(J_hS)^{-1}z_2 \tag{16}$$

Utilizing the following state feedback

$$u = \Phi^{-1}(q)(v - \dot{\Phi}(q)\nu) \tag{17}$$

we achieve input-output linearization as well as input-output decoupling by noting the observable part of the system

$$\dot{z}_1 = z_2 \qquad \dot{z}_2 = v \qquad y = z_1$$

The unobservable zero dynamics of the system is (obtained by substituting $z_1 = 0$ and $z_2 = 0$)

$$\dot{z}_3 = 0 \tag{18}$$

which is clearly Lagrange stable but not asymptotically stable.

3 Mobile Platform

3.1 Constraint Equations

In this subsection, we derive the constraint equations for a LABMATE[2] mobile platform. The platform has two driving wheels (the center ones) and four passive supporting wheels (the corner ones). The two driving wheels are independently driven by two DC motors, respectively. The following notations will be used in the derivation of the constraint equations and dynamic equations (see Figure 1).

[2]LABMATE is a trademark of Transitions Research Corporation.

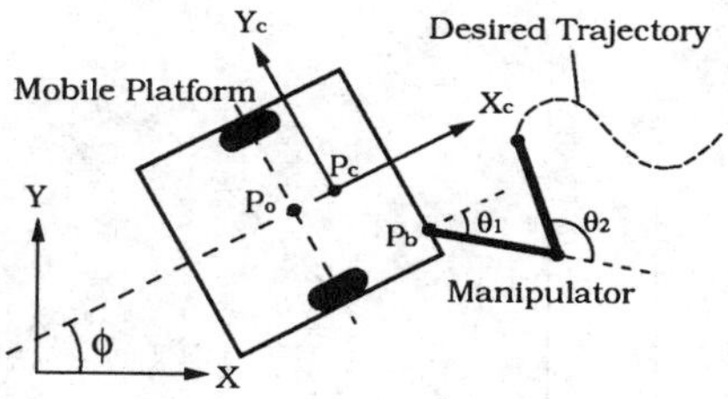

Figure 1: Schematic of the mobile manipulator.

P_o: the intersection of the axis of symmetry with the driving wheel axis;
P_c: the center of mass of the platform;
P_b: the location of the manipulator on the platform;
P_r: the reference point to be followed by the mobile platform;
d: the distance from P_o to P_c;
b: the distance between the driving wheels and the axis of symmetry;
r: the radius of each driving wheel;
m_c: the mass of the platform without the driving wheels and the rotors of the DC motors;
m_w: the mass of each driving wheel plus the rotor of its motor;
I_c: the moment of inertia of the platform without the driving wheels and the rotors of the motors about a vertical axis through P_o;
I_w: the moment of inertia of each wheel and the motor rotor about the wheel axis;
I_m: the moment of inertia of each wheel and the motor rotor about a wheel diameter.

There are three constraints. The first one is that the platform must move in the direction of the axis of symmetry, *i.e.*,

$$\dot{y}_c \cos\phi - \dot{x}_c \sin\phi - d\dot{\phi} = 0 \tag{19}$$

where (x_c, y_c) is the coordinates of the center of mass P_c in the world coordinate system, and the ϕ is the heading angle of the platform measured from the X-axis of the world coordinates. The other two constraints are the rolling constraints, *i.e.*, the driving wheels do not slip.

$$\dot{x}_c \cos\phi + \dot{y}_c \sin\phi + b\dot{\phi} = r\dot{\theta}_r \tag{20}$$

$$\dot{x}_c \cos\phi + \dot{y}_c \sin\phi - b\dot{\phi} = r\dot{\theta}_l \tag{21}$$

where θ_r and θ_l are the angular displacement of the right and left wheels, respectively.

Let $q = (x_c, y_c, \phi, \theta_r, \theta_l)$, the three constraints can be written in the form of

$$A(q)\dot{q} = 0$$

where

$$A(q) = \begin{bmatrix} -\sin\phi & \cos\phi & -d & 0 & 0 \\ -\cos\phi & -\sin\phi & -b & r & 0 \\ -\cos\phi & -\sin\phi & b & 0 & r \end{bmatrix} \tag{22}$$

It is straightforward to verify that the following matrix

$$S(q) = [s_1(q),\ s_2(q)] = \begin{bmatrix} c(b\cos\phi - d\sin\phi) & c(b\cos\phi + d\sin\phi) \\ c(b\sin\phi + d\cos\phi) & c(b\sin\phi - d\cos\phi) \\ c & -c \\ 1 & 0 \\ 0 & 1 \end{bmatrix}$$

satisfies $A(q)S(q) = 0$, where the constant $c = \frac{r}{2b}$. Computing the Lie bracket of $s_1(q)$ and $s_2(q)$ we obtain

$$s_3(q) = [s_1(q),\ s_2(q)] = \begin{bmatrix} -rc\sin\phi \\ rc\cos\phi \\ 0 \\ 0 \\ 0 \end{bmatrix}$$

which is not in the distribution Δ spanned by $s_1(q)$ and $s_2(q)$. Therefore, at least one of the constraints is nonholonomic. We continue to compute the Lie bracket of $s_1(q)$ and $s_3(q)$

$$s_4(q) = [s_1(q),\ s_3(q)] = \begin{bmatrix} -rc^2\cos\phi \\ -rc^2\sin\phi \\ 0 \\ 0 \\ 0 \end{bmatrix}$$

which is linearly independent of $s_1(q)$, $s_2(q)$, and $s_3(q)$. However, the distribution spanned by $s_1(q)$, $s_2(q)$, $s_3(q)$ and $s_4(q)$ is involutive. Therefore, we have

$$\Delta^* = span\{s_1(q),\ s_2(q),\ s_3(q),\ s_4(q)\} \tag{23}$$

It follows that, among the three constraints, two of them are nonholonomic and the third one is holonomic. To obtain the holonomic constraint, we subtract Equation (21) from Equation (20).

$$2b\dot{\phi} = r(\dot{\theta}_r - \dot{\theta}_l) \tag{24}$$

Integrating the above equation and properly choosing the initial condition of θ_r and θ_l, we have

$$\phi = c(\theta_r - \theta_l) \tag{25}$$

which is clearly a holonomic constraint equation.

3.2 Dynamic Equations

We now derive the dynamic equation for the mobile platform. The Lagrange equations of motion of the platform with the Lagrange multipliers λ_1, λ_2, and λ_3 are given by

$$m\ddot{x}_c - m_c d(\ddot{\phi}\sin\phi + \dot{\phi}^2\cos\phi) - \lambda_1\sin\phi - (\lambda_2 + \lambda_3)\cos\phi = 0 \tag{26}$$

$$m\ddot{y}_c + m_c d(\ddot{\phi}\cos\phi - \dot{\phi}^2\sin\phi) + \lambda_1\cos\phi - (\lambda_2 + \lambda_3)\sin\phi = 0 \tag{27}$$

$$-m_c d(\ddot{x}_c\sin\phi - \ddot{y}_c\cos\phi) + I\ddot{\phi} - d\lambda_1 + b(\lambda_3 - \lambda_2) = 0 \tag{28}$$

$$I_w\ddot{\theta}_r + \lambda_2 r = \tau_r \tag{29}$$

$$I_w\ddot{\theta}_l + \lambda_3 r = \tau_l \tag{30}$$

where

$$\begin{aligned} m &= m_c + 2m_w \\ I &= I_c + 2m_w(d^2 + b^2) + 2I_m \end{aligned}$$

and τ_r and τ_l are the torques acting on the wheel axis generated by the right and left motors respectively. These five equations of motion can easily be written in the form of Equation (1). The matrix $A(q)$ has been defined in Equation (22). The matrices $M(q)$, $V(q,\dot{q})$, and $E(q)$ are given by

$$M(q) = \begin{bmatrix} m & 0 & -m_c d\sin\phi & 0 & 0 \\ 0 & m & m_c d\cos\phi & 0 & 0 \\ -m_c d\sin\phi & m_c d\cos\phi & I & 0 & 0 \\ 0 & 0 & 0 & I_w & 0 \\ 0 & 0 & 0 & 0 & I_w \end{bmatrix}$$

$$V(q,\dot{q}) = \begin{bmatrix} -m_c d\dot{\phi}^2\cos\phi \\ -m_c d\dot{\phi}^2\sin\phi \\ 0 \\ 0 \\ 0 \end{bmatrix}, \quad E(q) = \begin{bmatrix} 0 & 0 \\ 0 & 0 \\ 0 & 0 \\ 1 & 0 \\ 0 & 1 \end{bmatrix}$$

In this case, owing to the choice of $S(q)$ matrix, we have

$$\nu = \begin{bmatrix} \nu_1 \\ \nu_2 \end{bmatrix} = \begin{bmatrix} \dot{\theta}_r \\ \dot{\theta}_l \end{bmatrix}$$

The state variable is then

$$x = [x_c \ \ y_c \ \ \phi \ \ \theta_r \ \ \theta_l \ \ \dot{\theta}_r \ \ \dot{\theta}_l]$$

Using this state variable, the dynamics of the mobile platform can be represented in the state space form, Equation (8).

3.3 Output Equations

While the state equation of a dynamic system is uniquely, modulo its representation, determined by its dynamic characteristics, the output equation is chosen in such a way that the tasks to be performed by the dynamic system can be *conveniently specified* and that the controller design can be *easily accomplished.* For example, if a 6-DOF robot manipulator is to perform pick-and-place or trajectory tracking tasks, the six-dimensional joint position vector or the 6-dimensional Cartesian position and orientation vector is normally chosen as the output equation. In this section, we present the output equation for the mobile platform and discuss its properties.

It is convenient to define a platform coordinate frame X_c-Y_c at the center of mass of at the mobile platform, with X_c in the forward direction of the platform. We may choose an arbitrary point P_r with respect to the platform coordinate frame X_c-Y_c as a reference point.

The mobile platform is to be controlled so that the reference point follows a desired trajectory. Let the reference point be denoted by (x_r^c, y_r^c) in the platform frame X_c-Y_c. Then the world coordinates (x_r, y_r) of the reference point are given by

$$x_r = x_c + x_r^c \cos\phi - y_r^c \sin\phi \quad (31)$$
$$y_r = y_c + x_r^c \sin\phi + y_r^c \cos\phi \quad (32)$$

The selection of the reference point for the purpose of coordinating locomotion and manipulation is discussed in the following section. Having chosen the reference point, x_r^c and y_r^c are constant. By taking the coordinates of the reference point to be the output equation

$$y = h(q) = [x_r \;\; y_r]^T \quad (33)$$

we have a trajectory tracking problem studied in [17, 18]. The corresponding decoupling matrix for this output is

$$\Phi(q) = J_h(q)S(q) = \begin{bmatrix} \Phi_{11} & \Phi_{12} \\ \Phi_{21} & \Phi_{22} \end{bmatrix} \quad (34)$$

where

$$\Phi_{11} = c((b - y_r^c)\cos\phi - (d + x_r^c)\sin\phi) \quad (35)$$
$$\Phi_{12} = c((b + y_r^c)\cos\phi + (d + x_r^c)\sin\phi) \quad (36)$$
$$\Phi_{21} = c((b - y_r^c)\sin\phi + (d + x_r^c)\cos\phi) \quad (37)$$
$$\Phi_{21} = c((b + y_r^c)\sin\phi - (d + x_r^c)\cos\phi) \quad (38)$$

Since the determinant of the decoupling matrix is $\det(\Phi(q)) = -\frac{r^2(d+x_r^c)}{2b}$, it is singular if and only if $x_r^c = -d$, that is, the point P_r is located on the wheel axis. Therefore, trajectory tracking of a point on the wheel axis including P_o is not possible as pointed out in [18]. This is clearly due to the presence of nonholonomic constraints. Choosing x_r^c not equal to $-d$, we may decouple and linearize the system.

Since $S^T E = I_{2\times2}$, the nonlinear feedback, Equations (9) and (17), in this case is simplified to

$$\tau = (S^T M S)u + S^T M \dot{S}\nu + S^T V \quad (39)$$

and

$$u = \Phi^{-1}(q)(v - \dot{\Phi}(q)\nu) \quad (40)$$

The linearized and decoupled subsystems are

$$\ddot{y}_1 = v_1 \quad (41)$$
$$\ddot{y}_2 = v_2 \quad (42)$$

4 Motion Planning

For simplicity, a two link planar manipulator is considered in this discussion. Let θ_1 and θ_2 be the joint angles and L_1 and L_2 be the link length of the manipulator. Also let the coordinates of the manipulator base with respect to the platform frame X_c-Y_c be denoted by (x_b^c, y_b^c). We let the reference point to the end point of the manipulator at a preferred configuration. We choose the configuration that maximizes the manipulability measure of the manipulator. If we specify the position of the end point as the desired trajectory for the reference point, the mobile platform will move in such a way that the manipulator is brought into the preferred configuration. The manipulability measure is defined as [1]

$$w = \sqrt{\det(J(\theta)J^T(\theta))} \quad (43)$$

where θ and $J(\theta)$ denote the joint vector and Jacobian matrix of the manipulator. If we consider non-redundant manipulators, the equation (42) reduces to

$$w = \mid \det J(\theta) \mid \quad (44)$$

For the two-link manipulator shown in Figure 1, the manipulability measure w is

$$w = \mid \det J \mid = L_1 L_2 \mid \sin\theta_2 \mid \quad (45)$$

Note that the manipulability measure is maximized for $\theta_2 = \pm 90°$ and arbitrary θ_1. We choose $\theta_2 = +90°$ and $\theta_1 = -45°$ to be the preferred configuration, denoting them by θ_{1r} and θ_{2r}. Then the coordinates of the reference point with respect to the platform frame X_c-Y_c is given by

$$x_r^c = x_b^c + L_1\cos\theta_{1r} + L_2\cos(\theta_{1r} + \theta_{2r}) \quad (46)$$
$$y_r^c = y_b^c + L_1\sin\theta_{1r} + L_2\sin(\theta_{1r} + \theta_{2r}) \quad (47)$$

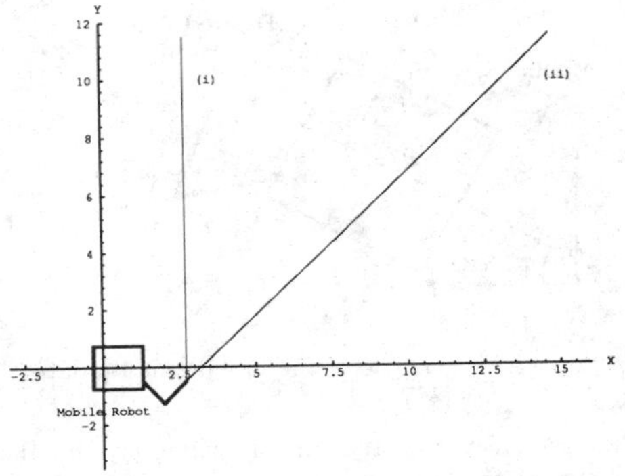

Figure 2: Two desired trajectories for simulations.

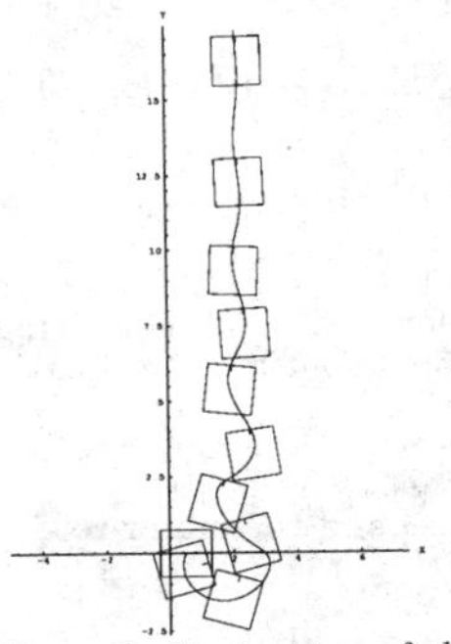

Figure 3: Trajectory of the point P_o for experiment (i)

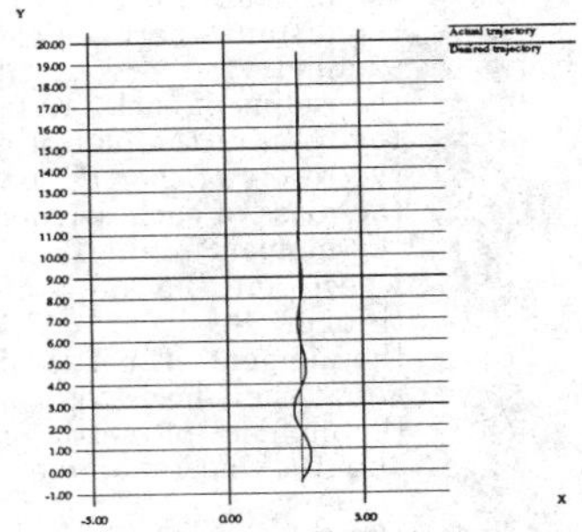

Figure 4: Desired and actual trajectories of the reference point for experiment (i)

We emphasize that x_r^c and y_r^c are constant and will be used in the representation of the output equation (33). The manipulator is regarded as a passive device whose dynamics is neglected. It is assumed that a human operator drags the end effector of the manipulator. The position of the end effector is given as the desired trajectory for the reference point P_r. The manipulator will be kept in the preferred configuration provided that the reference point is able to follow the desired trajectory. Any tracking error of the reference point will leave the manipulator out of the preferred configuation, resulting in a drop in manipulability measure. To count for measurement and communication delay, the current position of the end effector is made available to the mobile platform a fixed number of sampling periods later (five periods in the simulation). Further, before given to mobile platform as the desired trajectory, the position data of the end effector is approximated by piecewise polynomial functions generated in real time by *singular value decomposition*. This approximation is to eliminate high frequency (noise) components and to allow differentiation of discrete data in order to obtain desired velocity for the reference point.

5 Simulation

The mobile platform is initially directed toward positive X-axis at rest and the initial configuration of the manipulator is $\theta_1 = -45°$ and $\theta_2 = 90°$. Two different paths used for the simulation are shown in Figure 2. The velocity along the paths is constant.

1. A straight line perpendicular to the X-axis or the initial forward direction of the mobile platform,

2. A forward slanting line by 45 degree from X-axis.

The sampling rate is 0.01 sec. The linear state feedback gains for the two subsystems (41) and (42) are chosen so that the overall system has a natural frequency $\omega_n = 2.0$ and a damping ratio $\zeta = 1.2$. The higher damping ratio is to simulate the slow response of the mobile platform. For each simulation, we plot the trajectory of P_o, the trajectory of the reference point P_r, the manipulability measure, the joint angles of the manipulator, the heading angle of the platform, and the velocity of the P_o.

1. Figure 3 shows the trajectory of point P_o, in which a box[3] and a notch on one side represent the mobile platform and its forward direction, respectively. Note that the desired trajectory is

[3] These boxes are not equally distributed in time.

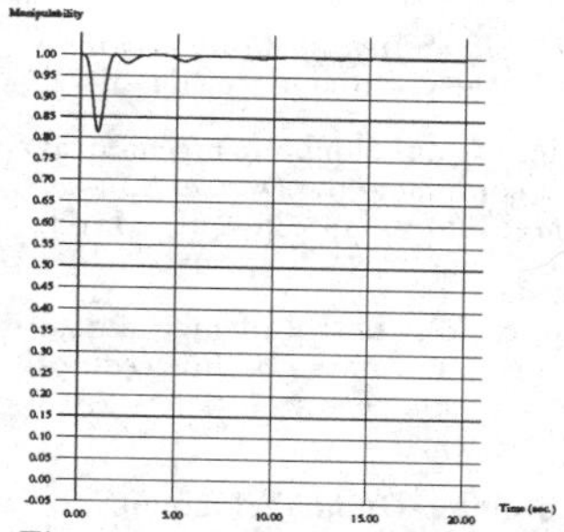

Figure 5: Manipulability measure for experiment (i)

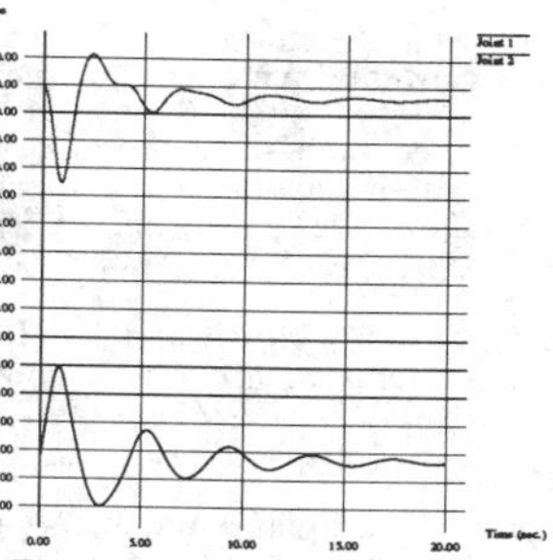

Figure 6: Joint angles for experiment (i)

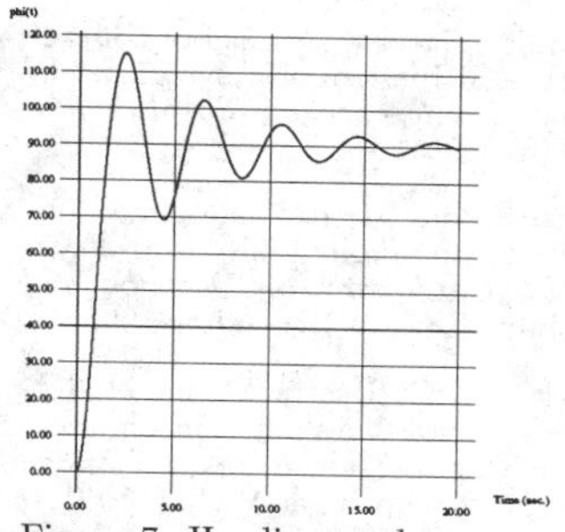

Figure 7: Heading angle for experiment (i)

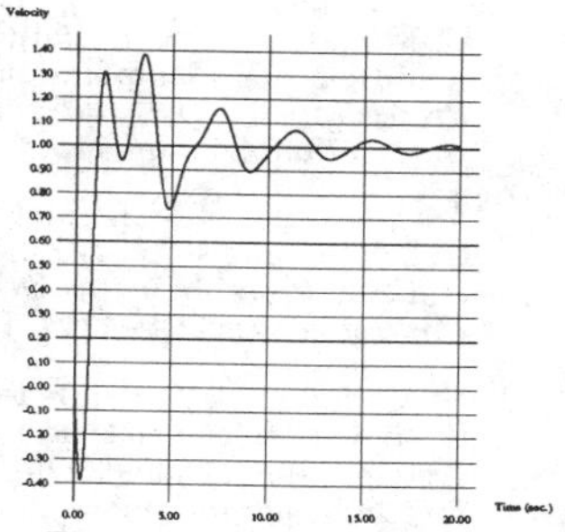

Figure 8: Velocity of the point P_o for experiment (i)

given for the reference point P_r. P_o has no desired trajectory. Figure 4 shows the desired and actual trajectories of the reference point P_r. The manipulability measure, the joint angles, the heading angle, and the velocity of point P_o are shown in Figure 5, 6, 7, and 8, respectively. Figure 5 shows a little degradation of manipulability measure corresponding to the early maneuver by the mobile platform. Figure 7 shows that the heading angle rapidly increases and exceeds 90° at the beginning, and evenually settles at 90°. The negtive value in Figure 8 indicates that the mobile platform moved backwards for a short period of time at the very beginning in order to achieve the needed heading angle. Note that the motion of the platform, or exactly the trajectory of point P_o is not planned. Therefore, the exhibited backward motion is not explicitly planned and is a consequence of the control algorithm.

2. The results for the slanting trajectory are shown in Figure 9 through 12. Figure 10 shows that the reference point follows the desired trajectory successfully. From Figure 11, the degradation of manipulability measure is smaller than that of the previous case as expected. Figure 12 indicates that that no backward motion occurs.

6 Experiment

The algorithm stated above is implemented with an experimental mobile manipulator. The system consists of a PUMA 250 6-DOF manipulator and a LABMATE platform (Figure 13). For simplicity only the first three joints of the manipulator are taken into account, *i.e.*, no wrist joints are considered. The sampling rates of PUMA 250 and

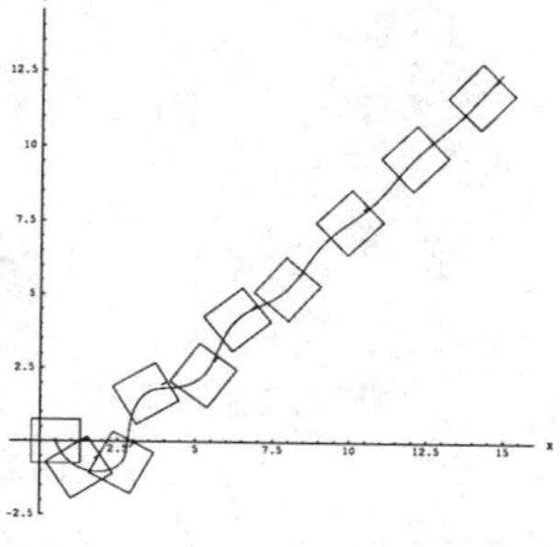

Figure 9: Trajectory of the point P_o for experiment (ii)

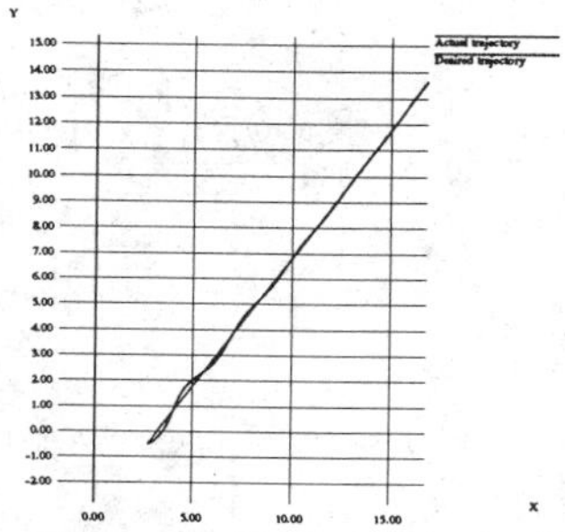

Figure 10: Desired and actual trajectries of the reference point for (ii)

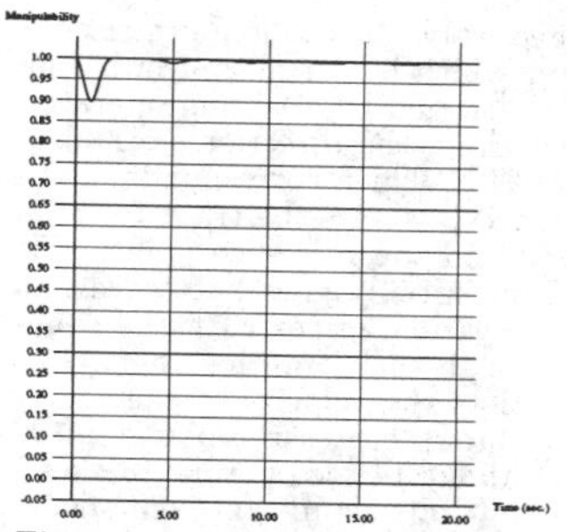

Figure 11: Manipulability measure for experiment (ii)

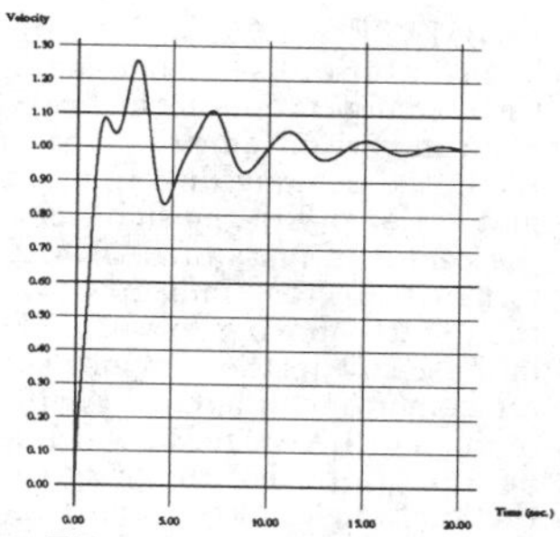

Figure 12: Velocity of the point P_o for experiment (ii)

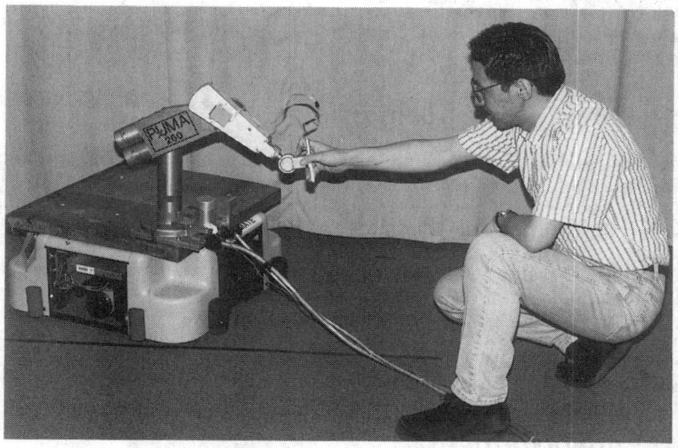

Figure 13: Mobile manipulator used in the experiments.

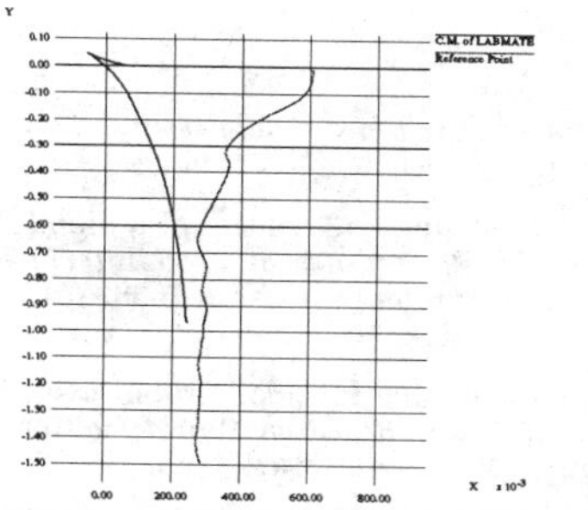

Figure 14: Trajectories of the P_o of LABMATE and the reference point

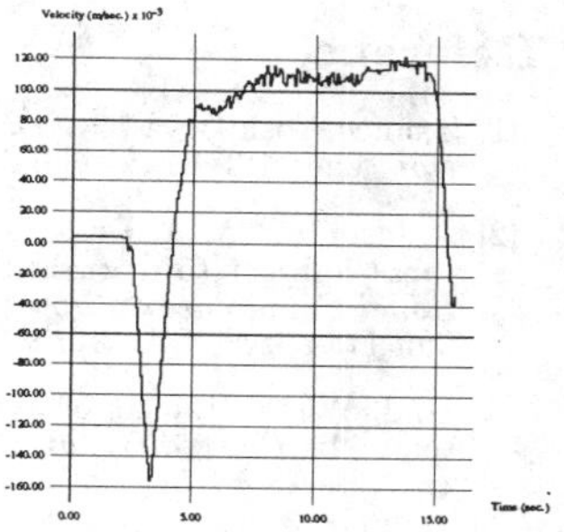

Figure 15: Velocity of the point P_o of LABMATE

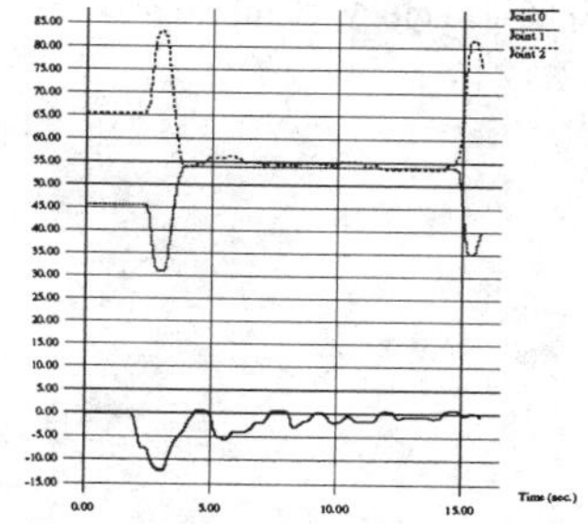

Figure 16: Joint angles of PUMA 250

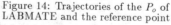

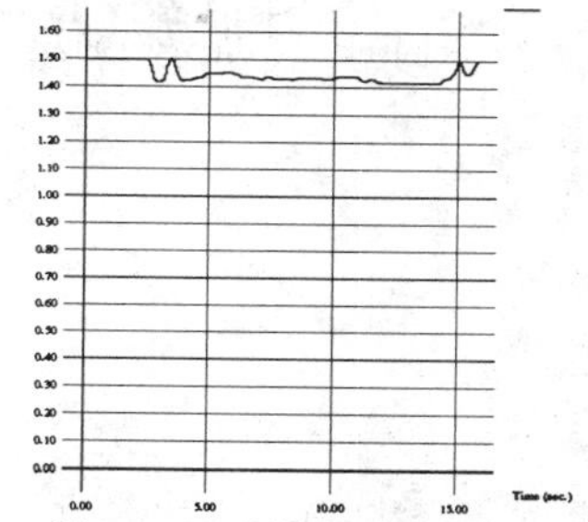

Figure 15: Manipulability measure

LABMATE are 250 and 16 Hz, respectively. In the experiment the end effector of the mobile manipulator which is at rest and in an optimal configuration in the beginning is dragged by a human operator. For comparison purpose it is dragged along the direction normal to the initial heading direction of LABMATE, which corresponds to the first trajectory in the simulations. Figure 14 shows the trajectories of the center of mass of LABMATE (P_o) and the reference point. The former trajectory indicates the platform initially goes backward and then starts moving forward. This observation agrees with the simulation result in the previous section though their transient behaviors are somewhat different. Figure 15 depicts the velocity of the center of mass of LABMATE, which also exhibits the presense of the initial backup. Note that dragging ceases at about 14 sec. Figure 16 shows the joint angles of the manipulator. The joints significantly change in the early stage, and then remain almost constant after the platform reaches an approximately constant velocity. Manipulability measure is shown in Figure 17. The manipulability slightly drops in the beginning and is maintained at the same level while the platform is in motion. It then comes back to a nearly optimal configuration after dragging stops. The slight degradation during motion is mainly due to the communication delay.

7 Concluding Remarks

We presented a planning and control algorithm for coordinating motion of a mobile manipulator. The design criterion was to control the mobile platform so that the manipulator is maintained at a configuration which maximizes the manipulability measure. We verified the effectiveness of our method by simulations on two representative trajectories. The algorithm was implemented with an actual mobile manipulator and tested on one of the trajectories for comparison purpose. For future work, we will investigate the integration of the proposed method and force control. An alternative path planning approach will be explored as mentioned in the previous section such that the maneuverability of mobile platform is taken into consideration as well.

Acknowlegement

This work is in part supported by NSF Grants CISE/CDA-90-2253, CISE/CDA 88-22719, and MSS-9157156, NATO Grant CRG 911041, Navy Grant N0014-88-K-0630, AFOSR Grants 88-0244 and 88-0296, Army/DAAL 03-89-C-0031PRI, and the University of Pennsylvania Research Foundation.

References

[1] Tsuneo Yoshikawa. *Foundations of Robtics: Analysis and Control.* The MIT Press, Cambridge, Massachusetts, 1990.

[2] J. Joshi and A. A. Desrochers. Modeling and control of a mobile robot subject to disturbances. In *Proceedings of 1986 International Conference on Robotics and Automation*, pages 1508–1513, San Francisco, CA, April 1986.

[3] G. J. Wiens. Effects of dynamic coupling in mobile robotic systems. In *Proceedings of SME Robotics Research World Conference*, pages 43–57, Gaithersburg, Maryland, May 1989.

[4] S. Dubowsky, P-Y. Gu, and J. F. Deck. The dynamic analysis of flexibility in mobile robotic manipulator systems. In *Proceedings of the Eighth World Congress on the Theory of Machines and Mechanisms*, page , Prague, Czechoslovakia, August 1991.

[5] N. A. M. Hootsmans. *The Motion Control of Manipulators on Mobile Vehicles.* PhD thesis, Department of Mechanical Engineering, Massachusetts Institute of Technology, Cambridge, Massachusetts, January 1992.

[6] Ju I. Neimark and N. A. Fufaev. *Dynamics of Nonholonomic Systems.* American Mathematical Society, Providence, RI, 1972.

[7] S. K. Saha and J. Angeles. Dynamics of nonholonomic mechanical systems using a natural orthoganal complement. *Transactions of the ASME, Journal of Applied Mechanics*, 58:238–243, March 1991.

[8] Anthony Bloch and N. H. McClamroch. Control of mechanical systems with classical nonholonomic constraints. In *Proceedings of 28th IEEE Conference on Decision and Control*, pages 201–205, Tampa, Florida, December 1989.

[9] J. Barraquand and Jean-Claude Latombe. On nonholonomic mobile robots and optimal maneuvering. In *Proceedings of Fourth IEEE International Symposium on Intelligent Control*, Albany, NY, September 1989.

[10] V. Kumar, X. Yun, E. Paljug, and N. Sarkar. Control of contact conditions for manipulation with multiple robotic systems. In *Proceedings of 1991 International Conference on Robotics and Automation*, Sacramento, CA, April 1991.

[11] Anthony Bloch, N. H. McClamroch, and M. Reyhanoglu. Controllability and stability properties of a nonholonomic control system. In *Proceedings of 29th IEEE Conference on Decision and Control*, pages 1312–1314, Honolulu, Hawaii, December 1990.

[12] G. Campion, B. d'Andrea-Novel, and G. Bastin. Controllability and state feedback stabilization of non holonomic mechanical systems. In C. Canudas de Wit, editor, *Lecture Notes in Control and Information Science*, pages 106–24, Springer-Verlag, 1991.

[13] J. P. Laumond. Finding collision-free smooth trajectories for a non-holonomic mobile robot. In *10th International Joint Conference on Artificial Intelligence*, pages 1120–1123, Milano, Italy, 1987.

[14] Z. Li and J. F. Canny. *Robot Motion Planning with Nonholonomic Constraints.* Technical Report Memo UCB/ERL M89/13, Electronics Research Laboratory, University of California, Berkeley, CA, February 1989.

[15] Jean-Claude Latombe. *Robot Motion Planning.* Kluwer Academic Publishers, Boston, MA, 1991.

[16] G. Lafferriere and H. Sussmann. Motion planning for controllable systems without drift. In *Proceedings of 1991 International Conference on Robotics and Automation*, pages 1148–1153, Sacramento, CA, April 1991.

[17] B. d'Andrea-Novel, G. Bastin, and G. Campion. Modelling and control of non holonomic wheeled mobile robots. In *Proceedings of 1991 International Conference on Robotics and Automation*, pages 1130–1135, Sacramento, CA, April 1991.

[18] C. Samson and K. Ait-Abderrahim. Feedback control of a nonholonomic wheeled cart in cartesian space. In *Proceedings of 1991 International Conference on Robotics and Automation*, pages 1136–1141, Sacramento, CA, April 1991.

[19] C. Canudas de Wit and R. Roskam. Path following of a 2-DOF wheeled mobile robot under path and input torque constraints. In *Proceedings of 1991 International Conference on Robotics and Automation*, pages 1142–1147, Sacramento, CA, April 1991.

[20] H. Nijmeijer and A. J. van der Schaft. *Nonlinear Dynamic Control Systems.* Springer-Verlag, New York, 1990.

Proceedings of the 31st Conference
on Decision and Control
Tucson, Arizona • December 1992

Stability and Transparency in Bilateral Teleoperation

Dale A. Lawrence

Dept. of Aerospace Engineering Sciences
University of Colorado
Boulder, CO 80309

Abstract

Many applications of telerobots are characterized by significant communication delays between operator commands and resulting robot actions at a remote site. A high degree of telepresence is also desired to enable operators to safely conduct teleoperation tasks. This paper provides tools for analyzing teleoperation system performance and stability when communication delays are present. A general multivariable system architecture is utilized which includes all four types of data transmission between master and slave: force and position in both directions. It is shown that a proper use of all four channels is of critical importance in achieving high performance telepresence in the sense of accurate transmission of task impedances to the operator. It is also shown that transparency and robust stability (passivity) are conflicting design goals in teleoperation systems. Achieved transparency and stability properties of two common architectures, as well as a recent "passivated" approach and a new "transparency optimized" architecture are quantitatively compared on one degree of freedom examples.

Introduction

Teleoperation has the potential to play a significant role in future remote and hazardous operations, such as space construction, satellite servicing, underwater platforms and vehicles, mining, etc. In addition, teleoperation can allow human involvement at scales much larger and smaller than is possible directly, e.g. in handling large loads or in micromanipulation.

However, current teleoperation systems are not capable of replacing direct human manipulation in most tasks. Experience with current systems reveals a surprising performance gap [1] between direct manipulation, and operating the same task via a telerobot, where task completion times are typically two orders of magnitude larger. Much harder to quantify is the degree of telepresence, or the "feel" of the remote site available to the operator through the teleoperator device. Even the best current systems have a characteristically "mushy" feel, and extensive amounts of training are necessary to operate them safely and efficiently. When communication delays are present, performance further degrades to the point where, at some large delay, direct telemanipulation becomes impossible. Thus, the desire to utilize the manipulation capability of a human operator must be balanced against the limits of technology in providing telepresence performance. To determine the role of teleoperation in practical applications, the limits of performance in the presence of delay must be more fully understood. This is the subject of the present paper.

The limits of telepresence performance in bilateral teleoperation are explored using the concrete notion of transparency (Section 1). When applied to the general teleoperation architecture in Section 2, this notion of transparency leads to specific guidelines for improving performance (Section 3). Since stability often limits performance, a general analysis approach based on passivity ideas is discussed in Section 4. These stability results extend those of [6] to allow the destabilizing effects of communication delay to be reduced in any teleoperation architecture. Section 5 examines the stability/performance trade off for four teleoperator architectures.

1. Transparency Performance

Ideally, the teleoperation system should be completely transparent, so the operator feels he is directly interacting with the remote task [4,17]. When in contact with the task, the slave positions V_e and forces F_e are related by the impedance Z_e of the task (slave environment).

$$F_e = Z_e(V_e) \tag{1}$$

The operator's force on the master F_h and the master's motion V_h should have the same relationship, i.e. for the same forces $F_e = F_h$ we want the same motions $V_e = V_h$. This requires that the impedance Z_t transmitted to or "felt" by the operator satisfies the transparency condition

$$Z_t = Z_e \, . \tag{2}$$

Figure 1 is a general model of a teleoperation system. The task impedance Z_e is transmitted to the operator through the teleoperation system T, which is a two port relating slave forces and positions to master forces and positions. In this paper, we concentrate on linearized behavior. This allows the two-port to be characterized in the frequency domain using Laplace transforms using the general "hybrid" matrix formulation [4,5,17]

$$\begin{bmatrix} F_h(s) \\ V_h(s) \end{bmatrix} = \begin{bmatrix} H_{11}(s) & H_{12}(s) \\ H_{21}(s) & H_{22}(s) \end{bmatrix} \begin{bmatrix} V_e \\ -F_e \end{bmatrix} \tag{3}$$

CH3229-2/92/0000-2649$1.00 © 1992 IEEE

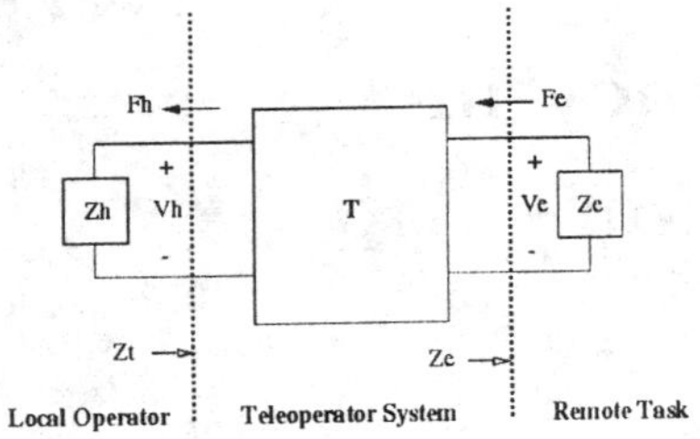

Figure 1: General two-port model of a bilateral teleoperation system.

Solving for F_h and V_h and using (1) we obtain

$$F_h = \underbrace{(H_{11} - H_{12}Z_e)(H_{21} - H_{22}Z_e)^{-1}}_{Z_t} V_h \tag{4}$$

We will examine how practical limitations affect the H parameters, hence the ability of the teleoperator system to transmit the task impedance to the operator.

2. A General Architecture

A bilateral teleoperation system consists of the master and slave mechanical systems, often with some degree of "self control", i.e. control loops closed separately around master and slave. For example, in the position-position architecture, master and slave each have position control loops to enable tracking of position commands. Other control loops are constructed by establishing communication links between master and slave. In general, both positions and forces can be communicated bilaterally, as well as various filtered versions of positions and forces. Figure 2 shows a block diagram of the entire teleoperation system, including master, slave, bilateral communication, operator, and task (environment) dynamics. Common teleoperation systems captured by the general structure of Figure 2 are the position-position [2], position-force [3] architectures. Others include [4], which has a generalized position-position form, [5,6] which discuss modified position-force strategies, and [13] which proposes a force-force architecture. Other structures utilizing some subset of the four communication paths C_1, C_2, C_3, C_4 have been suggested, but have not seen wide use. The general architecture [17] is also equivalent to Figure 2.

Although arguments have been made for preferring one architecture over another, e.g. using notions of operator information capacity [7], quantitative performance comparisons have been difficult to obtain. In this paper, the general architecture of Figure 2 is used to analyze and quantitatively compare various teleoperation schemes in terms of transparency performance and stability.

Figure 2 and (3) yield

$$H_{11}=(Z_m+C_m)(C_1+C_3Z_m+C_3C_m)^{-1}(Z_s+C_s-C_3C_4)+C_4 \tag{5}$$

$$H_{12}=-(Z_m+C_m)(C_1+C_3Z_m+C_3C_m)^{-1}(I-C_3C_2)-C_2 \tag{6}$$

$$H_{21}=(C_1+C_3Z_m+C_3C_m)^{-1}(Z_s+C_s-C_3C_4) \tag{7}$$

$$H_{22}=-(C_1+C_3Z_m+C_3C_m)^{-1}(I-C_3C_2) \tag{8}$$

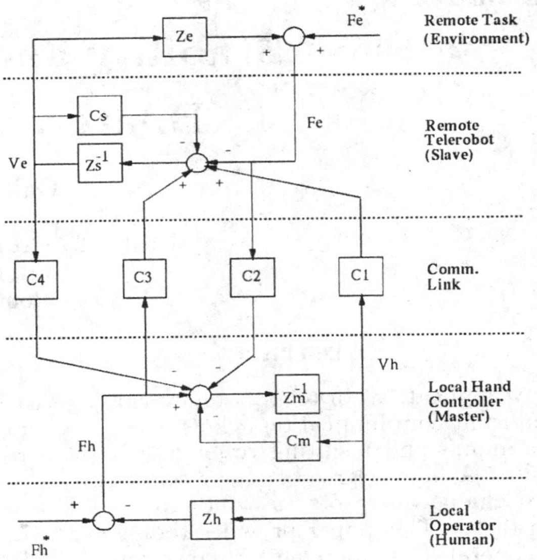

Figure 2: General bilateral teleoperator system block diagram, showing all subsystem dynamic blocks.

Using (5)-(8) in (4), a general expression for the transparency can be obtained. This expression can be used to quantitatively compare the transparency performance of competing teleoperation architectures (see Section 5). It can also be used as a design tool to improve or optimize transparency, as shown below.

3. Optimizing for Transparency

The key to achieving the high levels of transparency is the removal of the H_{11} and H_{22} terms from (4). This, in turn, requires the use of *all four* information channels in Figure 2 (position and force channels in both directions). This architecture does not correspond to existing two-channel topologies [7]; it is truly a "four-channel bilateral architecture". Such a four-channel architecture was independently proposed by Yokokohji (see, e.g. [17]). In particular, we need

$$C_3C_2 = I \tag{9}$$

to remove Z_e from the denominator of (4), and

$$C_4 = -(Z_m + C_m) \quad ; \quad C_1 = (Z_s + C_s)\,, \tag{10}$$

so the H_{11} term in the numerator of (4) will vanish, making Z_t a linear function of Z_e:

$$Z_t = -H_{12}Z_eH_{21}^{-1} = C_2Z_e\,. \tag{11}$$

Within the constraint $C_3C_2 = I$, non-unity impedance scale factors can be obtained. For example, in remote operation of heavy machinery or construction robots, a small C_2 is desired. In micro-fabrication or microsurgery, a C_2 larger than unity may be required to scale impedances up to human levels of sensitivity.

Observe that the control laws (10) for C_1 and C_4 require acceleration measurements, since Z_m and Z_s contain s terms due to the inertial part of the impedance. Good low frequency transparency can be obtained using simpler control laws which only require position and velocity measurements to implement C_1 and C_4:

$$C_1 = \hat{Z}_s + C_s \quad C_4 = -(\hat{Z}_m + C_m) \qquad (12)$$

where the "hat" impedances only contain terms in the form As^i, $i \leq 0$. At higher frequencies, the value of Z_t becomes increasingly inaccurate relative to Z_e. Although higher bandwidths of accurate transparency imply a larger degree of telepresence, quantitative measures of desired bandwidths have yet to be determined. Eventually, stability becomes a limiting factor in achievable bandwidths. This is the subject of the next section.

4. Stability Analysis via Passivity

The basic feedback structure of the teleoperator system of Figure 2 can be seen more clearly by defining the variables

$$F_{ce} = C_3F_h + C_1V_h - F_e^* = C_3F_h^* - F_e^* + (C_1 - C_3Z_h)V_h \qquad (13)$$

$$F_{ch} = -C_2F_e - C_4V_e + F_h^* = -C_2F_e^* + F_h^* - (C_4 + C_2Z_e)V_e \qquad (14)$$

and rearranging the block diagram as in Figure 3. A precise necessary and sufficient condition for stability is extremely difficult to obtain, since all paths in Figure 3 are multivariable. However, a sufficient condition for stability based on passivity arguments is well-suited to this application, since the subsystems $(Z_s + Z_e + C_s)^{-1}$ and $(Z_m + Z_e + C_m)^{-1}$ are often passive, and the design of C_1, C_2, C_3, and C_4 for good transparency usually implies strong coupling from master to slave (large loop gains in Figure 3). In contrast, the sufficient conditions for stability resulting from small gain arguments are not likely to be satisfied. Since this approach would result in extremely conservative design criteria (poor transparency), it is not considered here.

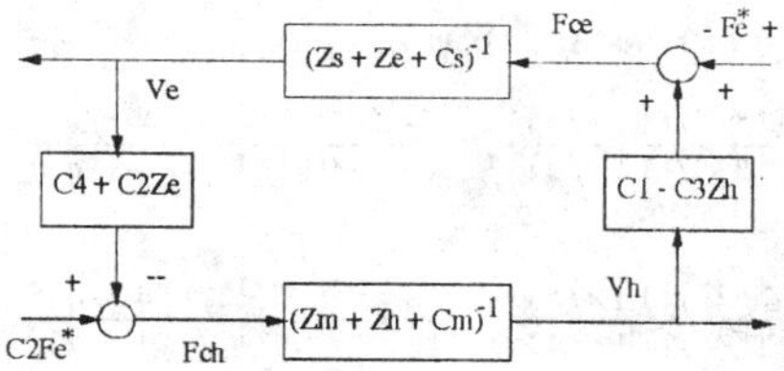

Figure 3: Single loop representation of the general teleoperator system.

Unfortunately, an application of the "passivity theorem" [15] to the system in Figure 3 as in [17] fails when the communication link contains delays. This is easy to see in one degree of freedom case, where C_1 through C_4 are scalars multiplied by the delay operator $e^{-s\tau}$, where τ is the one-way delay between master and slave. The loop gain in Figure 3 contains a factor of $e^{-2s\tau}$, and cannot have a positive real part over all frequency. However, by suitable reformulation of the information transmitted, it is possible to retain passivity in the presence of communication delays. The development below was motivated by [6] which treats a position-force teleoperator architecture. The approach given here is more general, since it can be implemented on any teleoperator system in the form of Figure 2. Also, the development is more elementary, and does not rely on electric circuit equivalents of teleoperator dynamics.

If $\mathcal{T}$ is a vector of through variables (e.g forces) and $\mathcal{A}$ is the corresponding vector of across variables (e.g. velocities), then the system is passive [6,15] if

$$\int_0^r \mathcal{T}^T \mathcal{A}\, dt \geq 0\,, \quad \forall r > 0 \qquad (15)$$

The key to designing a passive communication link in the presence of delay is the correct association of signals with through and across variables. This can be accomplished via the schematic representation of the teleoperator system as shown in Figure 4. In this representation, the inputs to the communication link are the across source $\mathcal{A}_1$ at the master, and the through source $\mathcal{T}_2$ at the slave. These sources are dependent on the velocities of master and slave, respectively. Outputs are the through variable $\mathcal{T}_1$ at the master and the across variable $\mathcal{A}_2$ at the slave. The force sources F_{ch} and F_{ce} depend on these communication link outputs.

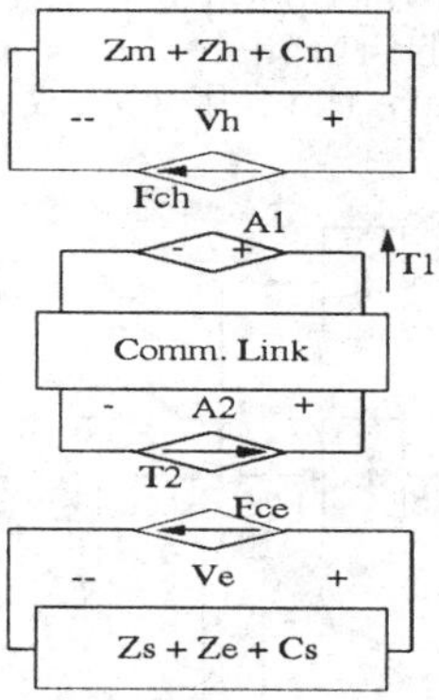

Figure 4: Schematic arrangement of the general teleoperator system, showing through and across variables in the communication link.

There are several possible relationships between communication link signals and teleoperator forces and velocities which preserve the teleoperator structure of Figure 2. One which leads to an implementable set of control laws is obtained as follows:

$$\mathcal{A}_1 = (C_1 - C_3Z_h)V_h \qquad (16)$$

$$F_{ch} = -\mathcal{T}_1 + F_h^* - C_2F_e^* \qquad (17)$$

$$F_{ce} = \mathcal{A}_2 + F_e^* + C_3F_h^* \qquad (18)$$

$$\mathcal{T}_2 = (C_4 + C_2Z_e)V_e \qquad (19)$$

Note that when there is no delay in the communication link, we will want $\mathcal{T}_1 = \mathcal{T}_2$ and $\mathcal{A}_2 = \mathcal{A}_1$ to recover the original system of Figure 3. Since the operator which maps communication link inputs to outputs will not be passive in the presence of delay, we reformulate the system so that the communication link maps across variables to through variables. With $\mathcal{T}^T = [\mathcal{T}_1\ ,\ -\mathcal{T}_2]$ and $\mathcal{A}^T = [\mathcal{A}_1\ ,\ \mathcal{A}_2]$, we find that (ignoring the external inputs F_e^* and F_h^* for the moment)

$$\mathcal{A} = R(s)(-\mathcal{T}) = \begin{bmatrix} P1(s) & 0 \\ 0 & P4(s) \end{bmatrix} \tag{20}$$

where $P1(s) = (C_1 - C_3 Z_h)(Z_m + Z_h + C_m)^{-1}$ and $P4(s) = (Z_s + Z_e + C_s)(C_4 + C_2 Z_e)^{-1}$. Then the teleoperator part of the system is passive if $R(s)$ is a positive real transfer function. (A result of applying the Parseval theorem to the integral (15) [15]). Similarly, by defining the communication link transfer function $L(s)$ by

$$\mathcal{T} = L(s)\mathcal{A} \tag{21}$$

then this part of the system is passive if $L(s)$ is positive real. The overall teleoperator system can then be transformed into the through versus across form of Figure 5, and we have the following result.

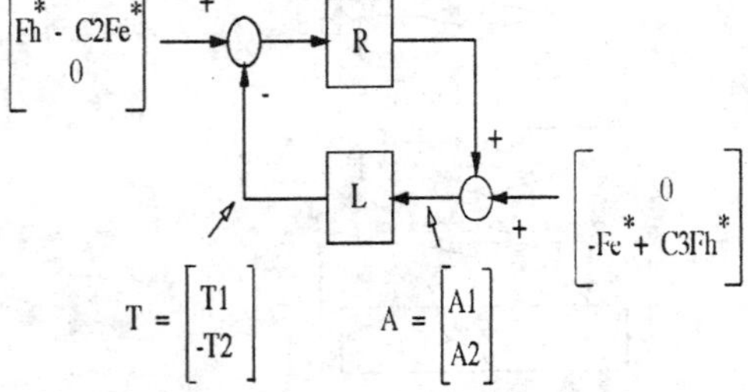

Figure 5: Transformed feedback system which provides a passive transfer function for the communication link L.

THEOREM 1 With bounded inputs F_h^* and F_e^*, all signals in the overall teleoperator system, including the communication link, are bounded provided the following conditions hold.

a) $R(s)$ is a positive real transfer function.

b) $L(s)$ is a stable, strictly positive real transfer function.

c) The transfer functions $(C_4 + C_2 Z_e)(Z_s + Z_e + C_s)^{-1}$, $(Z_s + Z_e + C_s)^{-1}$, $(C_1 - C_3 Z_h)(Z_m + Z_e + C_m)^{-1}$, and $(Z_m + Z_e + C_m)^{-1}$ are stable.

Proof: Since $R(s)$ may be non-proper, the conditions of the passivity theorem as stated in [15] do not hold. However, since all system signals are uniquely defined for any external inputs (the teleoperator system is an interconnection of causal linear systems), the manipulations of input-output inner products in the proof of the passivity theorem carry through. Thus, $\mathcal{T}_1$, $\mathcal{A}_1$, $\mathcal{T}_2$, and $\mathcal{A}_2$ are in L^2 when $F_h^* - C_2 F_e^*$ and $F_e^* + C_3 F_h^*$ are in L^2. Linearity, time-invariance, and condition c) imply that all system signals are bounded when inputs are bounded. $\Delta\Delta\Delta$

Remark: The *strict* passivity (strictly positive real) and finite gain condition on at least one subsystem in Theorem 1 is essential: Consider the case where $R = L = 1/s$. Both subsystems are passive, but neither is strictly passive nor do they have finite gain. The feedback connection has imaginary axis poles, and is not bounded-input bounded-output stable.

Hence, when R is positive real, all that is necessary for guaranteed stability in the presence of delay is that L is stable and strictly positive real. An example of such a passive delay operator is an electrical transmission line, which is the basis of the idea in [6]. Here we add a "loss factor" σ to achieve stability as well as strict passivity, which are required to claim bounded-input bounded-output stability as discussed in the remark following Theorem 1. Also, a low-pass effect in the communication channel is desired for robust high frequency behavior. Such an operator L can be indirectly described via the so-called scattering operator S, which satisfies

$$(\mathcal{A} - \mathcal{T}) = S(s)(\mathcal{A} + \mathcal{T}) \tag{22}$$

where

$$S(s) = \begin{bmatrix} 0 & \frac{e^{-(s+\sigma)\tau}}{1+s/a} \\ \frac{e^{-(s+\sigma)\tau}}{1+s/a} & \end{bmatrix}, \quad \sigma,\ a > 0 \tag{23}$$

The resulting expression for $L(s)$ is

$$\begin{bmatrix} \frac{1+e^{-2(s+\sigma)\tau}/(1+s/a)^2}{1-e^{-2(s+\sigma)\tau}/(1+s/a)^2} & \frac{-2e^{-(s+\sigma)\tau}/(1+s/a)}{1-e^{-2(s+\sigma)\tau}/(1+s/a)^2} \\ \frac{-2e^{-(s+\sigma)\tau}/(1+s/a)}{1-e^{-2(s+\sigma)\tau}/(1+s/a)^2} & \frac{1+e^{-2(s+\sigma)\tau}/(1+s/a)^2}{1-e^{-2(s+\sigma)\tau}/(1+s/a)^2} \end{bmatrix} \tag{24}$$

which has poles to the left of $\min(a(e^{-\epsilon\tau} - 1), -\sigma + \epsilon)$ for any $\epsilon > 0$, hence is stable. It can also be verified that $L(j\omega) + L^T(-j\omega)$ is uniformly positive definite over all ω, as required in Theorem 1. Since the algebra is tedious, we can appeal to the loop transformation theorem [15,p. 216] which shows that L is strictly passive if S has an induced Euclidean norm less than one. This norm is given by the largest singular value of $S(j\omega)$ over all frequency, or equivalently by $\sup_\omega |e^{-(j\omega+\sigma)\tau}/(1 + j\omega/a)| \leq e^{-\sigma\tau} < 1$. Thus Theorem 1 is satisfied by the form (24) for the communication link.

To implement the strictly passive communication link, rearrange (22) to obtain the transfer function $K(s)$.

$$\begin{bmatrix} \mathcal{T}_1 \\ -\mathcal{A}_2 \end{bmatrix} = K(s) \begin{bmatrix} \mathcal{T}_2 \\ \mathcal{A}_1 \end{bmatrix} \tag{25}$$

Using the source relationships (16)-(19), and noting that the definitions (13)-(14) imply

$$F_{ch} = (Z_m + Z_h + C_m)V_h \tag{26}$$

$$F_{ce} = (Z_s + Z_e + C_s)V_e\ , \tag{27}$$

we can write the teleoperator system equations in the form

$$(Z_m + Z_h)V_h = -C_m' V_h - (C_4' + C_2' Z_e)e^{-s\tau} V_e \tag{28}$$

$$(Z_s + Z_e)V_e = -C_s'V_e + (C_1' - C_3'Z_h)e^{-s\tau}V_h \quad (29)$$

where the new control laws (the "primed" C's) are related to the original control laws as follows. Define the filters

$$H_1(s) = \frac{2e^{-\sigma\tau}/(1+s/a)}{1+e^{-2(s+\sigma)\tau}/(1+s/a)^2}$$

$$H_2(s) = \frac{1-e^{-2(s+\sigma)\tau}/(1+s/a)^2}{1+e^{-2(s+\sigma)\tau}/(1+s/a)^2}$$

Then

$$C_i' = H_1(s)C_i\ , \quad i = 1,2,3,4 \quad (30)$$

$$C_s' = C_s + H_2(s)(C_4 + C_2Z_e) \quad (31)$$

$$C_m' = C_m + H_2(s)(C_1 - C_3Z_h) \quad (32)$$

Thus, the original control laws C_m and C_s on the master and slave are augmented with terms that depend on the delay in the communication channel ($H_2(s)$), and terms which depend on the original (undelayed) control laws between master and slave (C_1 through C_4). The new (delayed) control laws which link master and slave are simply the undelayed ones, altered by the addition of the filter $H_1(s)$.

Note that $H_2(s)C_2Z_eV_e$ is available as a filtering on the measured environmental force: $H_2(s)C_2F_e$, and (34) can be implemented by filtering the measured hand force F_h. Hence the impedances Z_t and Z_e are not required.

The filters $H_1(s)$ and $H_2(s)$ can be implemented via analog delay lines, or by transmitting additional signals over the communication channel and back. In a digital implementation, the required delays are realized via software buffering.

Although Theorem 1 provides only sufficient conditions for stability, it is useful because it applies to multivariable systems. Recent research on "almost passive" systems [12] also promises to extend the application of these ideas. Even in cases where Theorem 1 cannot be used, the idea of using a passive communication link has merit, so at least the L in Figure 5 does not "generate energy". Although the addition of such an L can indeed destroy stability, contrary to the implication in [6] (unless R is also passive), this approach would seem a very benign way to accommodate communication delay. Including such an L provided improved stability in the simulation examples provided in Section 5, hence was included in every case for ease of comparison.

5. Transparency vs. Passivity

Several common teleoperation architectures are quantitatively compared in this section, using the impedance Z_t reflected to the operator and passivity of $R(s)$ from Figure 5. The position-position [2], position-force [3], "passivated" position-force [6], and the new "transparency optimized" architectures will be examined. This comparison uses one-degree-of-freedom models, designated by lower case versions of the blocks in Figure 2. The "primed" control laws (30)-(32), are used to guarantee a passive communication channel in each of the following cases, with

$$\tau = 0.05\ sec. \quad \sigma = 0.02 \quad a = 20\ rad/sec \quad (33)$$

Transparency performance is measured by fixing z_e at the nominal value,

$$z_e = 100/s\ lb.sec/in \quad (34)$$

and examining the frequency domain behavior of the achieved z_t. Mechanical parameters corresponding to a typical slave manipulator and a relatively light, small master were chosen

$$z_m = 0.1s\ lb.sec/in \quad z_s = 0.50s\ lb.sec/in \quad (35)$$

together with the human impedance

$$z_h = 0.1s + 1.0 + 1.0/s \quad lb.sec/in \quad (36)$$

which corresponds to an operator with the master firmly in hand, but with very little arm tension.

Relative stability of each architecture is evaluated using the passivity approach of Section 4. When $R(s)$ is non-passive, stability is not guaranteed, and must be determined by other means. In these scalar cases, the Nyquist criterion was used to verify the stability of all simulated systems (see [16] for more details).

5.1 Position-Position Architecture

The control parameters were chosed to provide good position control on the slave, and moderate position control on the master:

$$c_1 = c_s = 10 + 50/s\ lb.sec/in \quad c_2 = c_3 = 0$$

$$c_4 = -c_m = -2 - 10/s\ lb.sec/in$$

The resulting impedance transmitted to the operator shown in Figure 6. Ideally, the ratio z_t/z_e should have a magnitude of unity (0 dB) and a phase of 0 rad.

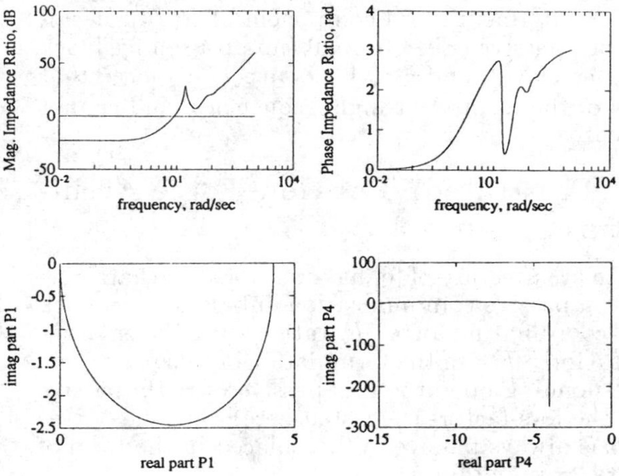

Figure 6: Impedance ratio z_t/z_e and passivity functions $P1$ and $P4$ in a typical position-position architecture.

It is clear that the transmitted stiffness is very different from the stiffness of z_e. The plots of the two components of the $R(s)$ matrix from (20) show that passivity conditions for stability according to Theorem 1 are not satisfied (real parts are not always positive). Since C_4 is "negative" in the expression (20) for $P4$, , a non-passive R is a characteristic of the position-position architecture.

5.2 Position–Force Architecture

Using the same nominal system parameters as the position-position case, except that

$$c_m = 0.1\ lb.sec/in \quad ; \quad c_4 = 0 \quad ; \quad c_2 = 1$$

(corresponding to a typical control on the master in the position-force case), the frequency response of z_t/z_e is shown in Figure 7.

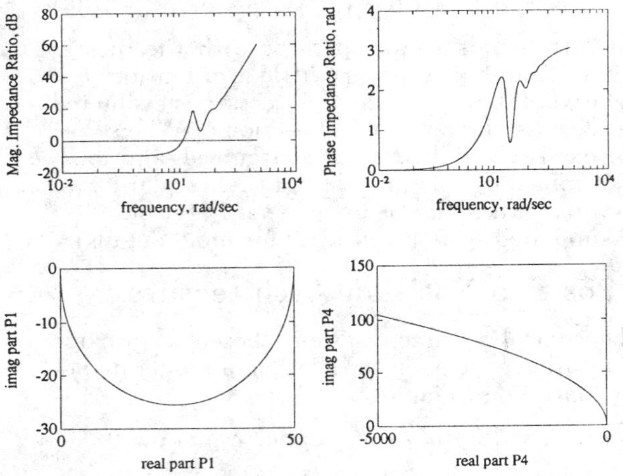

Figure 7: Impedance ratio z_t/z_e in a typical position-force teleoperator architecture.

We see that the low frequency error between task impedance and transmitted impedance can be quite large. Note that the $P4$ component of $R(s)$ is decidedly not passive. Since $c_4 = 0$ (no position feedback from the slave), and $c_2 z_e$ looks like $1/s$, we see that $P4$ is of the form s^2 at high frequencies, and cannot be passive.

5.3 "Passivated" Position–Force Architecture

The architecture of [6] has two essential characteristics: a passive communication link, and a feedback of a "coordinating force" F_s rather than the environmental force F_e. In the formulation of this paper, the corresponding operator L is passive (strictly passive with the loss-factors suggested here), and the operator R is always passive. When placed in the form of Figure 2, we obtain

$$c_1 = c_s = 10 + 50/s \quad ; \quad c_2 = \alpha_f + 1 = \text{constant}$$

$$c_3 = c_m = 0 \quad ; \quad c_4 = 0.5s$$

The presence of the s term in c_4 due to z_s allows $P4$ from (20) to look like a (positive) constant at high frequencies. This approach results in guaranteed stability for the teleoperator system, but poor transparency as shown in Figure 8. This simulation used the same values for system parameters as in the earlier cases, and $\alpha_f = 100$. This architecture has similar transparency performance as the position-position case, but much better passivity properties.

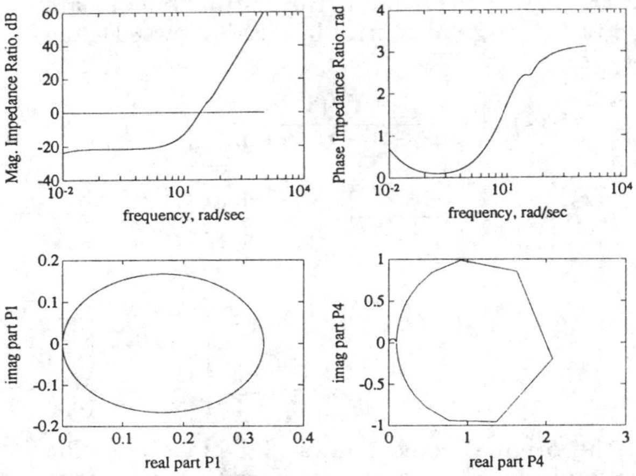

Figure 8: Impedance ratio z_t/z_e in the "passivated" position-force teleoperator architecture.

5.4 Transparency–Optimized Architecture

The control laws (9)-(10) which provide perfect transparency are simplified as suggested in (12) to obviate acceleration measurements. Using the values of z_m, z_c, c_m and c_s as in the position-position architecture, the performance of the "transparency optimized" architecture is shown in Figure 9. This

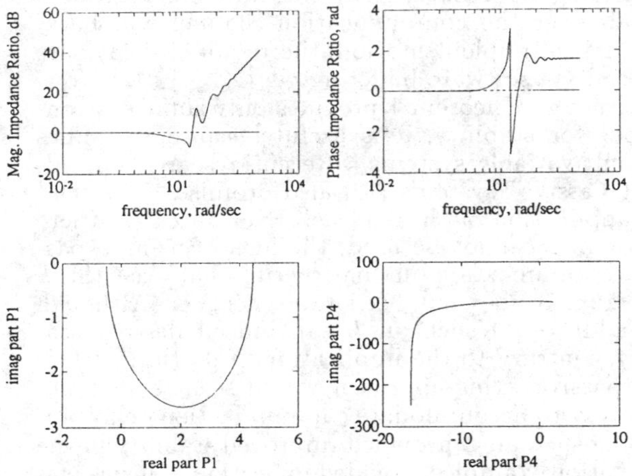

Figure 9: Transparency Ratio of the "transparency optimized" teleoperator architecture.

architecture provides extremely accurate impedance transmission at low frequencies, indicating that this system would faithfully reproduce task stiffness to the operator. As in the other architectures, the transparency begins to degrade at about 10 rad/sec. Note that perfect low frequency transparency has been obtained with physically reasonable control laws using this general architecture of Figure 2, unlike [4] which requires infinite gains to achieve the same performance. Even though the system is stable, the plots of the $P1$ and $P4$ components of the $R(s)$ matrix reveal that the system is not passive. Optimal transparency requires that c_4 in $P4$ is "negative" according to (12), preventing this architecture from ever satisfying the conditions of Theorem 1. Thus, passivity (and a certain degree of stability) has been traded for an increase in transparency performance.

Conclusion

Analysis tools were provided in this paper for examining transparency and stability of a general teleoperation architecture. These tools provide guidance in selecting control laws to optimize transparency, and to mitigate the effects of communication delays between master and slave.

Several bilateral teleoperation architectures were quantitatively compared in terms of transparency and passivity. The common position-position and position-force architectures provide poor transparency, even at low frequencies, and poor stability properties (neither one can provide a passive teleoperator block R for use in Theorem 1). A "passivated" version of the position-force scheme [6] provides significant improvement in passivity, hence stability, but does not improve transparency. An example of a "transparency optimized" design was given which required the use of all four possible information channels. This architecture provides accurate low frequency transparency, but cannot provide a passive operator R. Hence passivity (stability) and transparency are conflicting objectives in teleoperator system design. Unfortunately, the question of what levels of transparency are necessary for efficient task execution has seen little quantitative study, hence the stability/transparency trade-off remains an important area for further research.

The simplified scalar (one degree of freedom) examples presented in this paper only provide guidance in the design of realistic (multivariable) teleoperation systems; the detailed specification of desired levels of transparency and stability remains a difficult problem. Recent approaches using "robust" multivariable notions, such as small gain theory [10], passivity [11], and combinations [12,14] provide some tools, but the resulting designs can often be overly conservative (preventing high performance) or not very robust (can allow arbitrarily small stability margins). Shaping recent developments in multivariable control theory to suit the teleoperation problem has a potentially large payoff. Results in this area may significantly extend the capabilities of remote manipulation systems to the point where they can be effective replacements for humans in many manipulation tasks.

References

[1] J. Vertut and P. Coiffet, *Robot Technology, Teleoperations and Robotics*, Vol 3a, Prentice-Hall, 1984.

[2] R. C. Goertz, *et al*, "The ANL Model 3 Master-Slave Manipulator Design and Use in a Cave", *Proc. 9th Conf. Hot Lab. Equip.*, 1961, pp. 121.

[3] C. R. Flatau, "SM 229, a New Compact Servo Master-Slave Manipulator", *Proc. 25th Remote Sys. Tech. Div. Conf.*, 1977, pp. 169.

[4] G. J. Raju, G. C. Verghese, and T. B. Sheridan, "Design Issues in 2-Port Network Models of Bilateral Remote Teleoperation", *Proc. IEEE Intl. Conf. Rob. Auto.*, 1989, pp. 1317–1321.

[5] B. Hannaford, "Stability and Performance Tradeoffs in Bi-Lateral Telemanipulation", *Proc. IEEE Intl. Conf. Rob. Auto.*, 1989, pp. 1764–1767.

[6] R. J. Anderson and M. W. Spong, "Bilateral Control of Teleoperators with Time Delay", *IEEE Trans. Auto. Cont.*, Vol 34, No. 5, May, 1989, pp. 494–501.

[7] T. L. Brooks, "Telerobotic Response Requirements", *Proc. IEEE Conf. Systems, Man, Cyber.*, 1990.

[8] N. Hogan, "Controlling Impedance at the Man/Machine Interface", *Proc. IEEE Intl. Conf. Rob. Auto.*, 1989, pp. 1626–1631.

[9] H. Kazerooni, "Human/Robot Interaction via the Transfer of Power and Information Signals", *Proc. IEEE Intl. Conf. Rob. Auto.*, 1989, pp. 1632.

[10] H. Kazerooni, "Fundamentals of Robust Compliant Motion of Manipulators", *IEEE Trans. Rob. Auto.*, Vol. RA-2, no. 2, 1986.

[11] J. E. Colgate and N. Hogan, "Robust Control of Dynamically Interacting Systems", *Int. J. Control*, Vol. 48, No. 1, pp. 65.

[12] J. D. Chapel and R. Su, "Attaining Impedance Control Objectives Using H^∞ Design Methods", *Proc. IEEE Intl. Conf. Rob. Auto.*, 1991, pp. 1482.

[13] P. M. Bobgan and H. Kazerooni, "Achievable Dynamic Performance in Telerobotic Systems", *Proc. IEEE Intl. Conf. Rob. Auto.*, 1991, pp. 2040.

[14] J. E. Colgate, "Power and Impedance Scaling in Bilateral Manipulation", *Proc. IEEE Intl. Conf. Rob. Auto.*, 1991, pp. 2292.

[15] C. A. Desoer and M. Vidyasagar, *Feedback Systems: Input Output Properties*, Academic Press, New York, 1975.

[16] D. A. Lawrence, "Designing Teleoperator Architectures for Transparency" *Proc. IEEE Conf. Rob. Auto.*, Nice, 1992, pp. 1406–1411.

[17] Y. Yokokohji and T. Yoshikawa, "Bilateral Control of Master-Slave Manipulators for Ideal Kinesthetic Coupling" *Proc. IEEE Conf. Rob. Auto.*, Nice, 1992, pp. 849–858.

Proceedings of the 31st Conference
on Decision and Control
Tucson, Arizona • December 1992

Robustness Analysis of Nonlinear Biped Control Laws Via Singular Perturbation Theory

Hatem M. Hmam Dale A. Lawrence

Dept of Aerospace Engineering Sciences
University of Colorado
Boulder, Colorado 80309, USA

Abstract

This paper addresses the gait and stability analysis of a complex biped model using nonlinear techniques. The overall dynamics of the system are treated analytically using singular perturbation theory and phase plane techniques. The "unperturbed" dynamical system corresponds to a "reduced" model of the biped vehicle where only a few components play a major role in determining the gross behavior of the biped dynamics. Since the dynamics of the reduced model are relatively easy to understand analytically, the approach followed in this research is to design control laws for the "reduced" biped via nonlinear limit cycles, and treat more complicated system models as perturbations to this reduced model. The reduced control law is shown to be robust to small perturbation. However, simulations indicate that the robustness to dynamical perturbation can be quite large.

Introduction

Studies in biped locomotion may be classified into two broad categories. While some researchers have attempted a full modeling of natural neuromuscular motor control systems [1,2], others chose to understand and design artificial control systems that can walk and run [3-8]. To the extent that motion trajectories obtained from both controlled systems (the one controlled with muscle systems and the one controlled by artificial joint torques) are the same, the muscle system and the artificial actuators must solve similar control problems. Our work belongs to the second category of research and the strategy used is to construct feedback control laws using the properties of limit cycles [6,10] to generate robust, "natural" locomotion. By natural locomotion, it is meant that the legged vehicle moves relying fundamentally on its intrinsic mechanics. This idea corresponds to the philosophy of passive dynamics in bipedal walking espoused by McGeer[11] on the role of elasticity in animal locomotion. This paper differs in that a feedback control law is added which provides a large region of stability around the passive trajectory. Moreover, the overall behavior, e.g. step height, gait speed , gait type, etc., are controlled by manipulating a few control limit cycle parameters (refer to [7,8]). This facilitates the incorporation of "intelligent" capabilities at higher hierarchical levels, which need only manipulate control parameters as often as the overall vehicle objectives change. The concept of using limit cycles has been used before by Katoh and Mori [6], but their usage did not fit them within the framework of the biped mechanics. In other words, their control method forces the biped system to artificially evolve within the constraints of the devised oscillators and not as a result of its natural interaction with gravity. Bay and Hemami [10] used a set of coupled Van der Pol oscillators to model the central pattern generator (CPG) of living organisms for particular motor tasks such as walking and jumping maneuvers. These oscillators are capable of generating stable limit cycles. By running numerous computer simulations, they fine tuned the oscillator parameters until a close match occurred between the oscillator output signals and joint kinematics data obtained from physiological experiments.

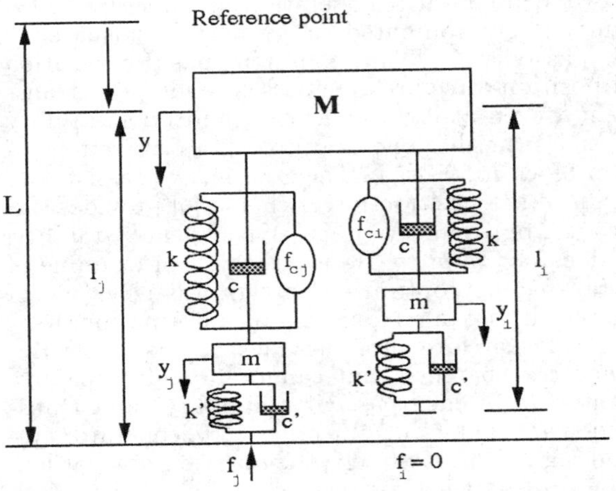

Figure 1: Perturbed model of a biped system, including foot mass and compliance.

In this paper, we focus on the analytic study of a fairly complex biped motion in the frontal

CH3229-2/92/0000-2656$1.00 © 1992 IEEE

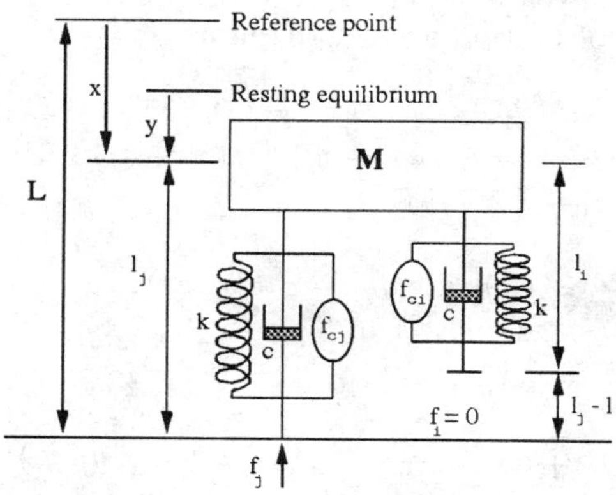

Figure 2: Dynamical model of the biped vehicle

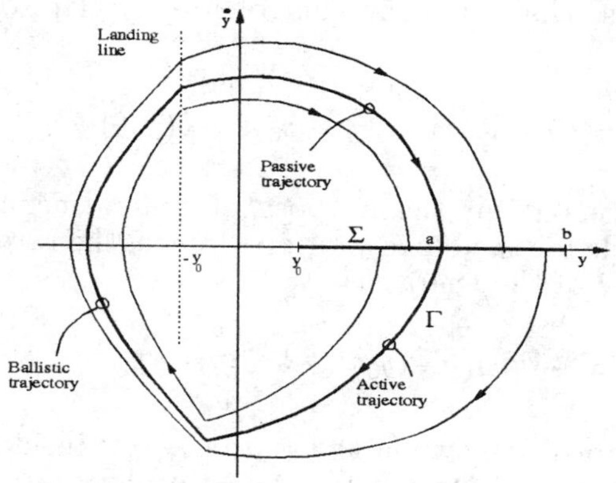

Figure 3: Biped motion exhibiting a stable limit cycle (thick solid line).

plane (see Figure 1) under nonlinear actuation laws. The essential difficulty of the analytic study is the presence of several different types of dynamics in a single locomotion cycle, depending on whether the biped system is airborne or standing on one leg. Although the system state evolves in a sixth-dimensional phase space, the foot components of the model have a small effect on the dynamics of the overall system. This paper provides a rigorous analysis of this high-dimensional nonlinear dynamics via singular perturbation theory [11,12]. In essence, it is shown that the gross behavior of the system is portrayed by a simpler reduced second order model as illustrated in Figure 2. The idea of dynamical reduction has been investigated before by Furushu and Masubuchi [9]. By closing the system loop of the linearized motion equations with relatively large gain PD joint controllers, they showed that the whole biped system behaved like an inverted pendulum. Section 1 provides a list of the motion equations in different phases of a running gait. Section 2 outlines the gait stability analysis for the reduced system [7,8]. Section 3 outlines a rigorous proof of the gait stability of the perturbed system which includes the foot dynamics. Finally, the simulations and some concluding remarks are listed in the simulation and conclusion sections.

1 Equations of Motion

We analyze the biped motion of a running gait, which includes two types of dynamics: body supported by one leg (single stance phase), and freely flying motion (ballistic phase). For the biped model of Figure 1, assuming the leg indexed by j is supporting the body, the equations of motion of the biped system are given by

$$M\ddot{y} + 2c\dot{y} + 2ky = k(y_j + y_i) + c(\dot{y}_j + \dot{y}_i) + f_{c_j} + f_{c_i}, \quad (1)$$

$$m\ddot{y}_j + (c+c')\dot{y}_j + (k+k')y_j = ky + c\dot{y} - f_{c_j} \quad (2)$$

The motion of the swinging leg is governed by

$$m\ddot{y}_i + c\dot{y}_i + ky_i = ky + c\dot{y} - f_{c_i} + (m + \frac{M}{2})g \quad (3)$$

The force of contact of the stance leg (indexed by j) is

$$f_j = c'\dot{y}_j + k'y_j + (m + \frac{M}{2})g \quad (4)$$

The ballistic phase is characterized by the following equations of motion:

$$M\ddot{y} + 2c\dot{y} + 2ky = k(y_j + y_i) + c(\dot{y}_j + \dot{y}_i) + f_{c_j} + f_{c_i}, \quad (5)$$

$$m\ddot{y}_j + c\dot{y}_j + ky_j = ky + c\dot{y} - f_{c_j} + (m + \frac{M}{2})g \quad (6)$$

$$m\ddot{y}_i + c\dot{y}_i + ky_i = ky + c\dot{y} - f_{c_i} + (m + \frac{M}{2})g \quad (7)$$

The next section outlines the control approach of the reduced system of Figure 2.

2 Stability Analysis of the Reduced System

In this section, only an informal outline of the stability control analysis for the running motion of the reduced model is provided (refer to the model of Figure 2). For more details, the reader may consult [7,8]. Roughly speaking, in the course of the running gait, three different dynamics occur in a single locomotion cycle. These dynamics are the passive ($f_{c_j} = 0$), the active, and the ballistic dynamics (conservative mode).

From a dynamical viewpoint, the passive mode dynamics is dissipative in nature. Hence, if E denotes the total energy $\frac{1}{2}M\dot{y}^2 + \frac{1}{2}k(y - y_0)^2$, then $\dot{E} = -c\dot{y}^2 < 0$. During the active mode dynamics, the nonlinear control force (refer to [7,8])

$$f_{c_j} = c\dot{y} + \vartheta\left(r - a(y - y_0)^2 - b(\dot{y})^2\right)\dot{y}$$

is constructed for the purpose of generating a limit cycle for the biped motion. Hence, the rate of change of the energy is

$$\dot{E} = -\vartheta\left(a(y - y_0)^2 + b(\dot{y})^2 - r\right)\dot{y}^2$$

which is positive[1] when the state $(y, \dot{y})$ is inside the limit cycle. During the ballistic phase, the body dynamics are conservative and no energy change occurs.

Thus, given the topological properties of these three dynamics, an initial state outside the limit cycle tends to spiral towards the limit cycle. However, an initial state inside the limit cycle tends to propagate outwards away from the equilibrium point $(y_0, 0)$. Hence, using the Poincaré-Bendixon theorem [13,14], it follows there exists a piecewise limit cycle Γ that constitutes the locomotion cycle of the running gait as illustrated in Figure 3. For more details about the reduced dynamical analysis, refer to [7,8]. The following section shows rigorously that the perturbed system motion under the control of the reduced system actuators is stable as long as the foot dynamics are of small significance compared with the rest of the biped system.

3 Stability of the Perturbed System

3.1 Perturbation Formulation of the Biped Dynamics

It is assumed in Figure 1 that the feet have small mass and are equipped with large stiff dampers (m is small, k' and c' are large). We also assume that the feet are of negligible vertical length compared with the length of the legs. Expressing the equations of motion of the complete system (body, legs, and feet) in a dimensionless form by putting $\tau = \omega t$,

$$\omega = \sqrt{\frac{k}{M}}, \quad u = \sqrt{k^2 + (c\omega)^2}, \quad v_1 = \frac{m\omega^2}{u},$$

$$v_2 = \frac{u}{\sqrt{k'^2 + (c'\omega)^2}}, \quad p = \frac{(c + c')\omega}{\sqrt{k'^2 + (c'\omega)^2}} = O(1),$$

$$q = \frac{(k + k')}{\sqrt{k'^2 + (c'\omega)^2}} = O(1), \quad s = \frac{c}{M\omega}.$$

[1] The energy rate of change is negative outside the limit cycle.

Hence, in the case of the single stance phase, the equations of motion are rewritten as

$$\begin{aligned} \frac{dy}{d\tau} &= z, \quad \frac{dy_j}{d\tau} = z_j, \quad \frac{dy_i}{d\tau} = z_i, \\ \frac{dz}{d\tau} &= -2sz - 2y + (y_j + y_i) + s(z_j + z_i) \\ &\quad + \tfrac{1}{M\omega^2}(f_{c_j} + f_{c_i}), \end{aligned} \tag{8}$$

$$\begin{aligned} v_1 v_2 \frac{dz_j}{d\tau} &= -\left(pz_j + qy_j - v_2 h(y, z)\right), \\ v_1 \frac{dz_i}{d\tau} &= -\tfrac{c\omega}{u} z_i - \tfrac{k}{u} y_i + \tfrac{k}{u} y + \tfrac{c\omega}{u} z - \tfrac{f_{c_i}}{u} \\ &\quad + \tfrac{1}{u}\left(\tfrac{u}{\omega^2} v_1 + \tfrac{M}{2}\right) g, \end{aligned} \tag{9}$$

where $h(y, z) = \frac{ky + c\omega z - f_{c_j}}{u}$, $X = (y, z, y_j, y_i)^T$ is the slow state vector and $Z = (z_j, z_i)^T$ the fast state vector. The associated unperturbed dynamical system is obtained by setting $\Upsilon = (v_1, v_2)^T$ to zero. Hence,

$$\begin{aligned} \frac{d\bar{y}}{d\tau} &= \bar{z}, \quad \frac{d\bar{y}_j}{d\tau} = \frac{-q\bar{y}_j}{p}, \\ \frac{d\bar{z}}{d\tau} &= -s\bar{z} - \bar{y} + (1 - s\tfrac{q}{p})\bar{y}_j \\ &\quad + \frac{\bar{f}_{c_j}}{M\omega^2} + \frac{g}{2\omega^2}, \\ \frac{d\bar{y}_i}{d\tau} &= -\frac{k}{c\omega}\bar{y}_i + \frac{k}{c\omega}\bar{y} + \bar{z} - \frac{\bar{f}_{c_i}}{c\omega} + \frac{M}{2c\omega}g. \end{aligned} \tag{10}$$

$$\begin{aligned} \bar{z}_j &= \frac{-q\bar{y}_j}{p}, \\ \bar{z}_i &= -\frac{k}{c\omega}\bar{y}_i + \frac{k}{c\omega}\bar{y} + \bar{z} - \frac{\bar{f}_{c_i}}{c\omega} - \frac{M}{2c\omega}g, \end{aligned} \tag{11}$$

where $\bar{X} = (\bar{y}, \bar{z}, \bar{y}_j, \bar{y}_i)^T$, $\bar{Z} = (\bar{z}_j, \bar{z}_i)^T$ are the unperturbed state vectors.

Similar equations may be derived for the ballistic phase. However this time, the foot viscosity and elasticity play no role and therefore v_2 is omitted from the dimensionless motion equations. The jacobian matrices with respect to the fast vector Z have real negative eigenvalues for both single stance and ballistic dynamics. Thus, for an initial condition [11,12] $Z(0)$ inside the basin of attraction, the dynamics of Z is asymptotically stable.

3.2 Stability Analysis of the Perturbed System

The natural setup for the perturbed dynamics is a sixth-dimensional phase space. The *poincaré* analysis is restricted on a two-dimensional phase plane representing the projection of the body dynamics on the plane (y, z). Let Σ be the *poincaré* section $[y_0, b)$ as in Figure 3. Let $P = (y(0), z(0))$ be any arbitrary point on Σ in the projected phase plane (y, z), and let $Q = \phi(P)$ its first mapped return on the *poincaré* section using the dynamics of the reduced system. Let $\psi_\Upsilon(P)$ be the first mapped return of P on the *poincaré* section obtained by projecting the perturbed dynamics of a single locomotion cycle on the phase plane (y, z). The next proposition shows that the perturbed

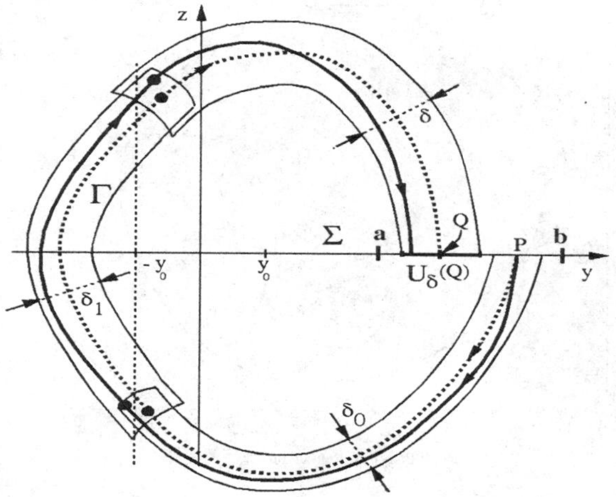

Figure 4: First return maps. The dashed line is associated to the reduced map. The solid line corresponds to the perturbed map.

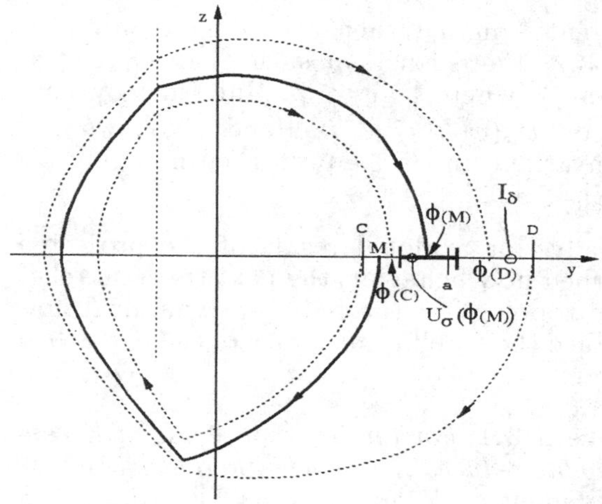

Figure 5: Figure illustrating the invariance of I_δ under the poincare map ψ_Υ for sufficiently small Υ.

poincaré mapping ψ_Υ is continuous with respect to perturbation parameters as they approach zero.

Proposition 3.1 *For any open neighborhood $U_\delta(Q) = (y_Q - \delta, y_Q + \delta)$, there exists $\rho > 0$ small enough such that for any $\Upsilon = (v_1, v_2)^T$, with $\|\Upsilon\| < \rho$, $\psi_\Upsilon(P) \in U_\delta(Q)$.*

Proof: (Outline) Informally, this proposition says that for any small open interval $U_\delta(Q)$ centered at the point Q, the perturbed mapping of the point P through the series of dynamics of one locomotion cycle can be made to fall inside the interval $U_\delta(Q)$ for small enough perturbation vector $\Upsilon = (v_1, v_2)^T$ (refer to Figure 4). A detailed proof of this proposition is lengthy and is omitted here. However, the proof can be provided by the authors. Roughly speaking, given an open tube C_Γ^δ around the unperturbed trajectory Γ and of radius δ, we show that a perturbed trajectory can "fit" inside the tube provided that the perturbation vector Υ is small enough. Hence, we need to show that the perturbed active mode, the ballistic, and the passive mode trajectories can be fit inside the tube. Moreover, the events where a switching between phases occur are of crucial importance for the rigorous proof. □

The result of the last proposition 3.1 can be relaxed to hold on a small open neighborhood of P. In other words, there exists an open interval I_P centered at P such that $\psi_\Upsilon(M) \in U_\delta(Q)$ for all $M \in I_P$. The next proposition shows that one Υ (small enough) is sufficient to approximate the dynamics of the perturbed model by those of the unperturbed model on a whole closed interval of initial states.

Proposition 3.2 *For any closed subset I in (y_0, b), there exists ρ, small enough such that $\psi_\Upsilon(M) \in U_\delta(\phi(M))$ for any point $M \in I$ and $\|\Upsilon\| < \rho$.*

Proof: The interval I is closed and bounded, hence it is compact. It follows that among the infinite family $\{I_M\}$ that cover I, a finite number of these open intervals $\{I_{M_k}\}$ $k = 1 \ldots n$ can cover I. Let $\rho = min\{\rho_k\}$, $k = 1 \ldots n$. Thus for all $\|\Upsilon\| < \rho$, we have necessarily $\psi_\Upsilon(M) \in U_\delta(\phi(M))$ for any point $M \in I$. □

We did not address till this point the topological aspects of the perturbed system dynamics. However, given the attraction behavior of the flow of the reduced body dynamics to the limit cycle Γ, it follows necessarily that the body perturbed dynamics will acquire this attraction behavior. This is true, since proposition 3.2 shows that the projected perturbed trajectory approaches uniformly its reduced counterpart as Υ approaches zero. The next proposition about the invariance[2] of the flow propagation follows due to the flow properties of the reduced dynamics.

Proposition 3.3 *For any interval I_δ in (y_0, b) of radius δ centered at a^3, there exists $\rho > 0$, small enough such that $\psi_\Upsilon(I_\delta) \subset I_\delta$ for any $\|\Upsilon\| < \rho$.*

Proof: Let C, D the two boundary points (see Figure 5) of the interval I_δ. Consider mapping once both states $X_C = (y_C, 0)$, $X_D = (y_D, 0)$, associated to the interval boundaries using the reduced dynamics ϕ. Let $d_I = \min\{|y_{\phi(C)} -$

[2] A set S is invariant under ψ_Υ if $\psi_\Upsilon(S) \subset S$.
[3] $a = \Gamma \cap \Sigma$.

$y_C|, |y_{\phi(D)} - y_D|\}$ and put $\sigma = d_I/2$. If M is any point from I_δ, then according to proposition 3.2, there exists ρ such that $\psi_\Upsilon(M) \in U_\sigma(\phi(M))$, where $\|\Upsilon\| < \rho$. But then by construction, $U_\sigma(\phi(M))^4 \subset I_\delta$. Hence it follows that I_δ is invariant under the perturbed mapping ψ_Υ for $\|\Upsilon\| < \rho$. □

Finally, for a global result of the projected perturbed flow behavior, the next theorem relies on the topology of the reduced dynamical flow to deduce the stability property of the perturbed system.

Theorem 3.1 *For any open tube C_Γ^δ in R^2 enclosing the reduced limit cycle, there exists $\rho > 0$, small enough such that for any $\|\Upsilon\| < \rho$, the trajectory of the propagating vector $(y, z)^T$ of the perturbed dynamics is eventually located inside C_Γ^δ.*

Proof: Consider the previously contructed parametrized *poincaré* map $\psi_\Upsilon(.) : \Sigma \to \Sigma$. For a starting state at P on Σ, it was shown in proposition 3.1 that $\psi_\Upsilon(P)$ is continuous[5] at $\Upsilon = 0$. Under the dynamics of the reduced model, the evolving point $(\bar{y}, \bar{z})$ asymptotically approaches the reduced limit cycle Γ. Thus, the propagating partial state $(\bar{y}, \bar{z})$, starting from the point P, eventually crosses the interval $C_\Gamma^\delta \cap \Sigma$ after a finite time t_n corresponding to n locomotion cycles. The intercept point corresponds to the iterated map $\phi^n(P)$. For that number of iterations n, we construct the mapping $\psi_{(.)}^n(P) : \Sigma \to \Sigma$. Obviously, when $\Upsilon = 0$, it follows $\psi_{(0)}^n(P) = \phi^n(P)$. On the other hand, the nth composition of $\psi_{(\Upsilon)}(P)$ is continuous at $\Upsilon = 0$. Therefore, necessarily there exists Υ_1 such that $\psi_{\Upsilon_1}^n(P) \in C_\Gamma^\delta \cap \Sigma$. It is still not clear that the interval $C_\Gamma^\delta \cap \Sigma$ is invariant under ψ_{Υ_1}. But thanks to proposition 3.3, there exists $\Upsilon_2 > 0$ such that $C_\Gamma^\delta \cap \Sigma$ is invariant under ψ_{Υ_2}. The choice of $\rho = \min\{\Upsilon_1, \Upsilon_2\}$ guarantees $\psi_\Upsilon^r(P) \in C_\Gamma^\delta \cap \Sigma$ for $r \geq n$, and $\|\Upsilon\| < \rho$. □

As a result of this theorem, we conclude that the perturbed biped motion is indeed stable. This property is in fact passed over from the reduced system dynamics which is highly stable due to the presence of the active limit cycle.

4 Computer Simulations

Computer simulations of the performance of the actual biped system have been written in MATLAB. The current simulations were performed on a legged system with $g = 10N/Kg$, $l_0 = 0.7m$, $M = 2Kg$, $L = 1m$, $k = 100N/m$, $c = 4Ns/m$, $a = 2000$, $b = 40$, $r = 200$, and $\vartheta = 0.2$.

The performance of the reduced system has been tested (see [7,8]) on two different gaits namely walking and running vertically in the frontal plane. Moreover, it has been shown for each gait that the motion occurs on a piecewise stable limit cycle. The perturbed biped system was equipped with feet of mass $m = 0.2Kg$ each, two stiff springs $K' = 2000N/m$, and two dampers $c' = 100Ns/m$. Figure 6 illustrates the perturbed as well as the reduced dynamics portrayed by the body on a level ground. As shown in the figure, even though the ratio of the foot to body mass is 10% (20% when consider-

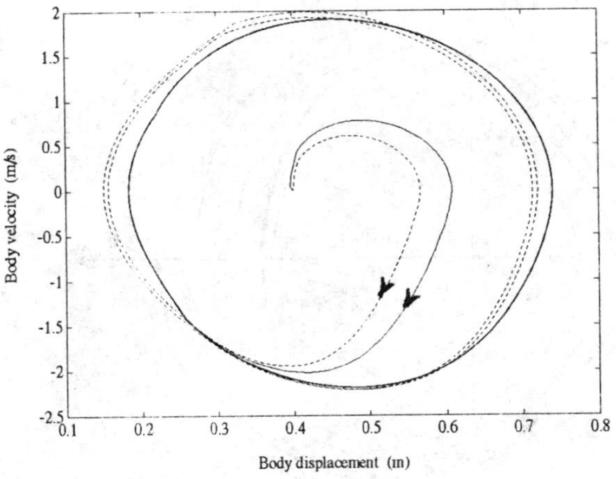

Figure 6: Phase portrait of the biped with foot dynamics running in the frontal plane. The dashed orbit is associated to the reduced system dynamics.

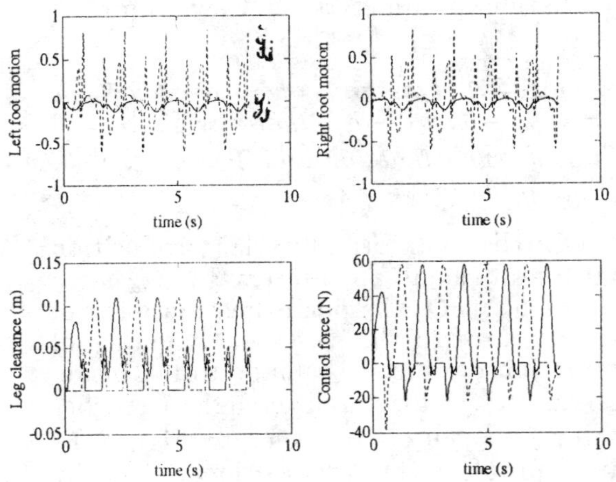

Figure 7: (a) Foot kinematics. (b) Control force and leg clearance history.

[4] It should remembered that the reduced mapping ϕ maps points from Σ closer to the fixed point a.

[5] $\psi = \phi$ when $\upsilon = 0$.

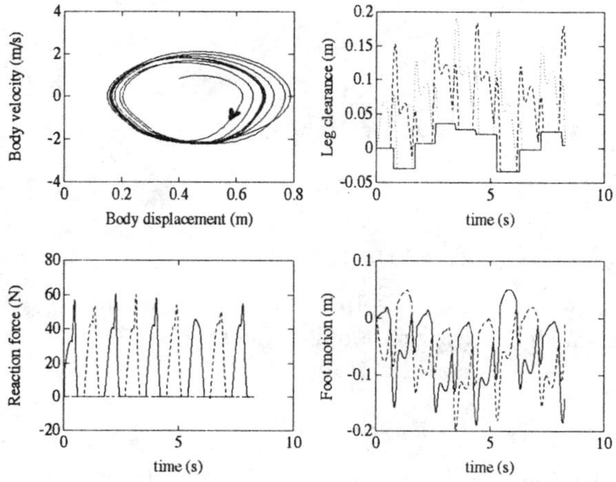

Figure 8: Performance of the perturbed system on a vibrating ground.

ing both feet), the foot dynamics did not significantly alter the limit cycle behavior (running). However, because of the weight of the feet, the body experienced a slight loss in altitude during the ballistic phase. The reaction forces of the perturbed and reduced biped systems are alike because the foot accelerations are too small to have an impact on the ground reaction. The foot dynamics, the leg clearance and the control forces employed to stabilize the biped motion are sketched in Figure 7. Notice that at some larger foot mass, the running gait degenerates to a walking gait.

Finally, the full perturbed model is tested (simulation) on a rough (vibrating) ground. The foot mass used is $m = 0.1Kg$. As illustrated in Figure 8 (top right corner), the biped system was able to sustain the running motion. These computer simulations show that despite the dynamical alterations introduced by the added foot systems and despite the roughness of the environment, the nonlinear controller was able to autonomously stabilize the biped motion without any prior knowledge about the environment.

Conclusion

In a previous paper [7], nonlinear actuators that produce robust biped motion were constructed for the reduced system of Figure 2. In this paper, both phase plane techniques and singular perturbation theory were used to pursue the stability analysis of the perturbed biped system of Figure 1. It was proven using singular perturbation theory that the reduced stable motion and other dynamical properties are "passed over" to the perturbed system provided the added foot dynamics are of small significance compared with the overall system dynamics. But computer simulation showed that the perturbation parameters can be fairly large without the desired biped motion being significantly altered.

References

[1] J. M. Winters and S. L-W Woo, *Multiple Muscle Systems: Biomechanics and Movement Organization*, Springer-Verlag, New York, 1990.

[2] F. E. Zajac, "Muscle and tendon: properties, models, scaling, and application to biomechanics and motor control", *CRC Critical Reviews in Biomedical Engineering*, Vol. 17, 1989, pp. 359-411.

[3] H. Hemami, R. Tomovic, and A. Z. Ceranowicz, "Finite state control of planar bipeds with application to walking and sitting", *J. of Bioengineering*, Vol. 2, pp. 477-494, 1978.

[4] M. H. Raibert, *Legged robots that balance*, MIT Press, Cambridge, 1986.

[5] T. McGeer, "Passive bipedal running", *Proceedings Royal Society of London*, Vol B.240, 1990, pp. 107-134.

[6] R. Katoh and M. Mori, "Control method of biped locomotion giving asymptotic stability of trajectory", *Automatica* Vol. 20, 1984, pp. 405-414.

[7] H. M. Hmam and D. A. Lawrence, "Biped Control Via Nonlinear Dynamics", 29th *Allerton Conference on Communication, Control, and Computing*, October 1991.

[8] H.M. Hmam, *Biped Control Via Nonlinear Dynamics*, University of Cincinnati, June 1992.

[9] J. Furusho, and M. Masibuchi, "A theoretically motivated reduced order model for the control of dynamic biped locomotion", *J. of Dynamic Systems, Measurement, and Control*, Vol. 109, June 1987, pp. 155-163.

[10] J. S. Bay, and H. Hemami, "Modeling of a neural pattern generator with coupled nonlinear oscillators", *J. of Biomedical Engineering*, Vol. 34, No. 4, April 1987, pp. 299-306.

[11] P. V. Kokotovic, H. K. Khalil, and J. O'Reilly, *Singular Perturbation Methods in Control: Analysis and Design*, Academic Press, London, 1986.

[12] A. B. Vasil'eva, "Asymptotic behavior of solutions to certain problems involving nonlinear differential equations containing a small parameter multiplying the highest derivatives", *Russian Mathematical surveys*, Vol. 18, 1963, pp. 13-81.

[13] D. W. Jordan and P. Smith, *Nonlinear Ordinary Differential Equations*, Clarendron Press, Oxford, 1987.

[14] M. W. Hirsh and S. Smale, *Differential equations, Dynamical Systems and Linear Algebra*, Academic Press, New York, 1974.

Proceedings of the 31st Conference
on Decision and Control
Tucson, Arizona • December 1992

Real-Time Classification of Teleoperation Data with a Neural Network

Paolo Fiorini

Jet Propulsion Laboratory
California Institute of Technology
Pasadena, CA, USA

Antonio Giancaspro

Istituto per le Tecnologie Informatiche Spaziali
Consiglio Nazionale delle Ricerche
Matera, Italy

Sergio Losito

Centro di Geodesia Spaziale
Agenzia Spaziale Italiana
Matera, Italy

Guido Pasquariello

Istituto per la Elaborazione dei Segnali e delle Immagini
Consiglio Nazionale delle Ricerche
Bari, Italy

Abstract

In recent years, research in telemanipulation has been primarily concerned with problems related to the control of teleoperators and to their Man-Machine interface. With the possible need of extensive teleoperation for the deployment and maintenance of space systems the attention must focus on techniques to improve operators performance during telemanipulation. This paper describes the development of a monitoring program that can be used in the future to evaluate operator performance and provide symbolic feedback about task progress.

A classifier has been designed to recognize teleoperation task phases, independently of variations due to differences in working conditions and in phases features. We have used Neural Networks to recognize task phases by using force data. Two network architectures have been tested in simulation on real teleoperation data and the one with the best performance has been implemented on a teleoperation system. During tests on actual telemanipulation tasks, the classifier had a lower recognition percentage than the simulated tests, but it showed an unexpected generalization capability. It was in fact able to correctly segment tasks whose phase sequence was significantly different from those in the training data.

1 Introduction

Control Systems owe their popularity to the capability of compensating for output errors by using feedback signals. Unfortunately, this approach does not fully apply to teleoperation systems where output errors depend on the operator's perception and understanding of the teleoperated task. In teleoperation, several types of feedback are used. Feedback signals generated by actuator sensors are used for low level control, but are of limited interest to the operator. *Visual feedback* is usually available to the operator and provides much information about the status of the remote task. *Kinesthetic force feedback* let the operator feel some of the forces and torques applied to the remote manipulator. Experiments [6] have shown that this type of feedback can increase operator's performance during telemanipulation. Unfortunately, none of the above feedback types provides abstract information about task progress or guides towards its successful completion. They all describe instantaneous situations and their interpretation in terms of global task progress is left to the operator's judgment.

It can be hypothesized that a symbolic feedback describing the task progress and the current level of performance can provide additional help to the operator, eliminating some of the human dependent errors. This type of feedback would mimic the case of supervised teleoperation training, where trainees are guided through a teleoperated task by an experienced operator who provides conceptual feedback and practical advice in the form of warnings and direct commands. Similarly, an automatic supervisory system would monitor the global evolution of the task and provide real-time feedback about the current operator's performance as compared to a stored model of the task execution. The first step towards achieving these capabilities is the addition to a teleoperator system of a real-time task segmentation program that can consistently recognize the various phases of a task and label them correctly, independently of differences in the working conditions and in their characteristics.

This paper presents the results of simulations and tests of teleoperation task segmentations done with two *Artificial Neural Network* architectures. The input to the networks consisted of the force measurements. The two architectures have been tested off-line using a simulated data stream and the one with the best recognition rate has been interfaced to the Advanced Teleoperation (ATOP) System at the Jet Propulsion Laboratory (JPL). The first architecture is based on the concept of turning temporal sequences into spatial patterns and the second one extends this model by adding connections

CH3229-2/92/0000-2662$1.00 © 1992 IEEE

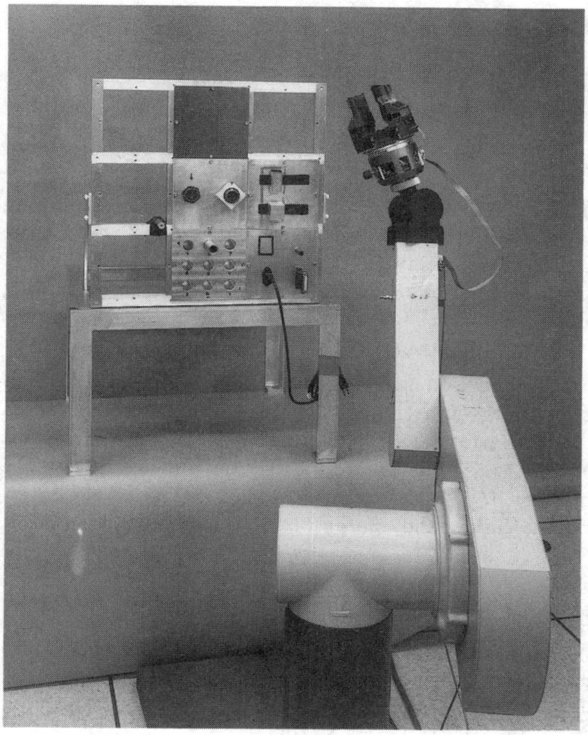

Figure 1: The Puma Manipulator with the Task Board

from the output nodes to the input layer as proposed in [8] using *partially recurrent networks*. These networks have mainly feed-forward connections, but some feedback connections are allowed, whose weights can be either trainable or fixed [7]. Generally, these connections are made to a suitable number of *context units*, which are organized as complete layers or as parts of other layers [2], [9] and that are used to remember past states. In the area of segmentation of teleoperation tasks, one successful approach consisted of modeling the underlying mental process guiding the task with a Markov Process whose states are not directly observable [5].

The paper is organized as follows. The rest of the introduction gives a short description of the teleoperator system used in the tests and of the task characteristics. The computations embedded into the network are presented in section 2, while section 3 describes the architectures. Section 4 present the results of the off-line simulations and of the real-time tests. The conclusions and our future research and development directions conclude the paper.

1.1 System and Task Descriptions

Data used for the off-line simulations were collected during experiments with the *Enhanced Six-Axis Breadboard* (ESAB) teleoperation system [1]. This system consisted of an upgraded version of the JPL-Stanford Force Reflecting Hand Controller, a Puma 560 manipulator and the Model A JPL Smart Hand. The Smart Hand was a single-degree-of-freedom gripping mechanism mounted in series with a six-axis Maltese Cross force/torque sensor [3]. The Smart Hand and the Puma robot were controlled with two JPL Universal Motor Controllers (UMC), which performed all coordinate transformations, master and slave motor control and operator interface functions. The *ESAB* system was configured with a local site and a remote site. The local site consisted of the control room equipped with the Force Reflecting Hand Controller and three monitors displaying views of the remote site, since no direct view of the task area was allowed. The remote site consisted of the Puma manipulator equipped with the Smart Hand. The experiments consisted of a series of idealized operations performed on a modular task board (figure 1) which had provisions for *peg-in-hole* insertion and extraction with different levels of difficulty, four types of *electrical connector* insertions and extractions and a *velcro* attachment. During the experiments, forces and torques were measured by the sensor at 100 Hz rate and stored in 6-dimensional arrays for off-line analysis. The real-time tests were conducted on the *Advanced Teleoperation* (ATOP) system which is a dual-arm version of the previous system with improved sensing and man-machine interface. For compatibility with the previous experiments, the tests used the ATOP in a single-arm mode. The force sensing was performed by the Model C Smart Hand, with a sampling rate of about 60 Hz.

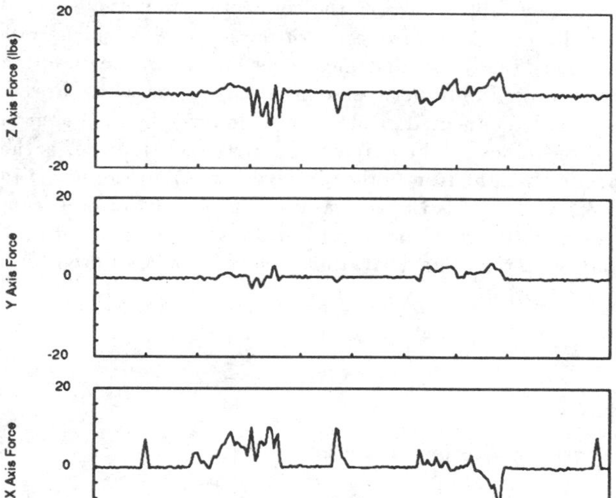

Figure 2: Force Profiles in a Peg-in-Hole Task

The task used during simulations and experiments was the *peg-in-hole*. It consisted of a precise sequence of steps that the operator had to follow: 1) move to a marked location on the task board, 2) tap the peg on the task board, 3) move to the hole opening, 4) insert and release the peg, 5) move to the back of the peg, 6) tap the peg end, 7) move near the peg, 8) regrasp and extract the peg, 9) move back to the mark, 10) tap the mark, 11) move to a final position.

This sequence of steps is reflected in the structure of the data collected during the experiment, as shown in figure 2 representing the force readings of the peg-in-hole task. In particular the force along the approach direction of the manipulator, the *X-axis* in the plots, clearly shows the phases' sequence.

2 Neural Computations

Artificial *Neural Networks* (NN) are systems consisting of a large number of interconnected units, often called *neurons* for the similarity of these systems to biological ones. Their implementation can be either with hardware circuits or with software routines on general purpose computers. The system described in this paper belongs to this last group. The network is usually equipped with a set of input units to feed the data into the network and with a set of output units, from which the result of the *NN* computation can be read. Hereafter, $(\xi_1^\mu, \ldots, \xi_l^\mu)$ will denote a generic input pattern to a *NN* with l input terminals, while the corresponding n outputs will be represented by $(\hat{\zeta}_1^\mu, \ldots, \hat{\zeta}_n^\mu)$. In its most common form, a *NN* is able to store P input-output pattern associations, of the type:

$$(\xi_1^\mu, \ldots, \xi_l^\mu) \rightarrow (\hat{\zeta}_1^\mu, \ldots, \hat{\zeta}_n^\mu), \quad \mu = 1, \ldots, P$$

by finding suitable values of the connection weights.

The basic architectures are *feed-forward NN* in which units are arranged in layers: all connections have the same direction and are allowed only between contiguous layers which are, in general, fully connected. Both connection weights and neuron activation values can be written in matrix notation: w_{ij}^m is the connection weight from unit j in layer $m-1$ to the unit i in layer m, while V_i^m is the activation value of unit i in layer m. The value m normally ranges from 0, the input layer, to M, the output layer. The activation value V_i^m is computed by

$$V_i^m = g(net_i^m)$$

where

$$net_i^m = \sum_j w_{ij}^m V_j^{m-1}$$

is the *net input* to unit i, and

$$g(x) = \frac{1}{(1 + exp(-2\beta x))}$$

is the *sigmoid activation function* and the parameter β is a *gain factor*.

Given an input $(\xi_1^\mu, \ldots, \xi_l^\mu)$, the corresponding net outputs, $(\hat{\zeta}_1^\mu, \ldots, \hat{\zeta}_n^\mu)$ can be computed by feed-forwarding the input values through the network. This calculation can be carried out with the following steps:

1. $V_k^0 = \xi_k^\mu$, for all k in the *input* layer;
2. for each layer $m = 1, \ldots, M$:
$$V_i^m = g(net_i^m) = g(\sum_j w_{ij}^m V_j^{m-1})$$
3. $\hat{\zeta}_k^\mu = V_i^M$, for all k in the *output* layer;

A generic *NN* can be trained with a supervised learning algorithm in which the net outputs can be compared to the known correct answers. In this way, it is necessary to give a prescription for updating the weights w_{ij}^m given a training data set

$$\{\xi_1^\mu, \ldots, \xi_l^\mu, \zeta_1^\mu, \ldots, \zeta_n^\mu\}_{\mu=1,\ldots,P}$$

A typical supervised learning algorithm is based on minimizing an energy function E, by using the gradient descent method. In particular, if $(\zeta_1^\mu, \ldots, \zeta_n^\mu)$ is the desired output pattern, $(\hat{\zeta}_1^\mu, \ldots, \hat{\zeta}_n^\mu)$ is the actual *NN* output in reply to the input pattern $(\xi_1^\mu, \ldots, \xi_l^\mu)$ and $\mathbf{w}$ is the matrix of the interconnection weights, the *energy function* can be written as:

$$E[\mathbf{w}] = \frac{1}{2}\sum_\mu \sum_i (\zeta_i^\mu - \hat{\zeta}_i^\mu)^2$$

if the weights are updated after all patterns have been presented to the *NN* inputs (*batch learning*), or

$$E[\mathbf{w}] = \frac{1}{2}\sum_i (\zeta_i^\mu - \hat{\zeta}_i^\mu)^2$$

if the weights are updated after each pattern is presented to the *NN* inputs (*incremental learning*).

In both cases, the gradient descent algorithm consists of changing each w_{ij}^m by an amount Δw_{ij}^m proportional to the gradient of $E[\mathbf{w}]$, so as to slide downhill on the surface defined by the energy function:

$$\Delta w_{ij}^m = -\eta \frac{\partial E}{\partial w_{ij}^m}$$

An efficient algorithm to perform this computation is the *Back Propagation* method [10]. For each training pattern

$$(\xi_1^\mu, \ldots, \xi_l^\mu, \zeta_1^\mu, \ldots, \zeta_n^\mu)$$

with net output

$$(\hat{\zeta}_1^\mu, \ldots, \hat{\zeta}_n^\mu)$$

we compute the so called *delta error* on the unit i of the output layer M by the following formula:

$$\delta_i^M = \begin{cases} g'(net_i^M)[\zeta_i^M - V_i^M] & \text{if } |\zeta_i^M - V_i^M| \geq \varepsilon_l \\ 0 & \text{otherwise} \end{cases}$$

where ε_l is the admissible *error on learning* and $g' = \frac{\partial g}{\partial x}$.

The delta errors on the hidden layer units $m = M - 1, \ldots, 2$, are then computed recursively as:

$$\delta_i^{m-1} = g'(net_i^{m-1}) \sum_j w_{ji}^m \delta_j^m$$

Weights are updated as:

$$w_{ij}^m(t) = w_{ij}^m(t-1) + \Delta w_{ij}^m(t)$$

where

$$\Delta w_{ij}^m(t) = \eta \delta_i^m V_j^{m-1}$$

A *momentum term* [7] is also used to give each connection w_{ij}^m some momentum so that it tends to change in the *average downhill* direction, avoiding sudden oscillations. This scheme is implemented by adding to the weights change Δw_{ij}^m a contribution from the previous time step, that is:

$$\Delta w_{ij}^m(t) = \eta \delta_i^m V_j^{m-1} + \alpha \Delta w_{ij}^m(t-1)$$

The *momentum parameter* α must be between 0 and 1.

3 Learning Time Sequences

The problem of *learning time sequences* has been the subject of extensive research in the area of neural network. One of the applications studied is *sequences recognition*: a particular output pattern has to be produced when a specific input sequence is observed. This is also appropriate for *telemanipulation tasks segmentation*, where the output has to indicate task phases corresponding to different sensor observations. The following subsections describe the network architectures used to associate sensor signals to the corresponding task phases. In both cases, the supervised learning algorithm is carried out using the state sequences computed by the Viterbi decoder described in [5] and the input data set is reduced from the original 6-dimensional arrays to the single *X-axis* force signal.

3.1 Time-Delay Neural Networks

According to this formulation, the telemanipulation task segmentation can be performed by associating a window of finite length l of signal sensors, $x(t-(l-1)\Delta t), .., x(t)$, to the state classified at time t, $s(t)$.

We have used a two-layer feed-forward *NN*. This network has l inputs, one for each signal of the window, formally:

$$\xi_i^\mu = x(t-(l-i)\Delta t), \qquad i = 1, .., l$$

The output layer has as many units as necessary according to the selected representation for $s(t)$. For n states, the best choice is to have n output units: the state k is codified by keeping active the output unit k and quiescent the others. Formally, the code of state $s(t)$ is represented as a vector $(\sigma_1(t), .., \sigma_n(t))$ so that for $s(t) = k$, it will be:

$$\zeta_j^\mu = \sigma_j(t) = \begin{cases} 1 & \text{if } j = k \\ 0 & \text{otherwise} \end{cases} \qquad j = 1, \ldots, n$$

A hidden layer of $2m+1$ units is always used. A *shift register* is connected to the input layer. When a new sample of the sensor signal is fed to the network, all other samples are shifted to the right and the less recent sample is shifted out. This architecture is sometimes called **time-delay neural network** (see figure 3).

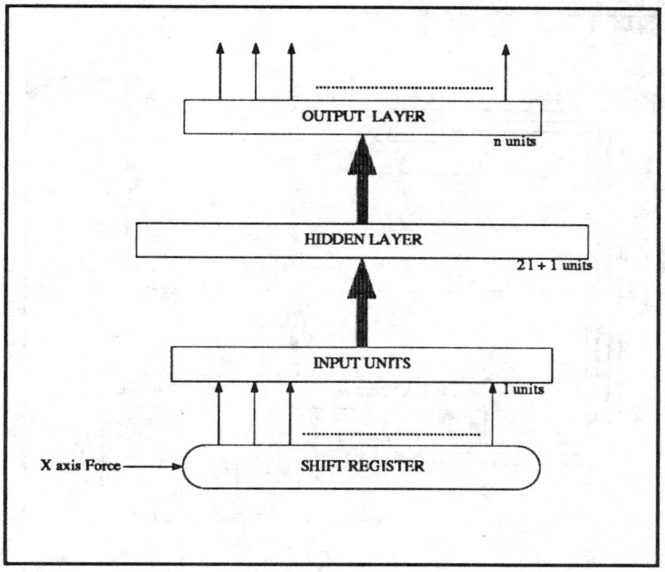

Figure 3: Time-Delay Neural Network

This network learns the following associations:

$$(x(t-(l-1)\Delta t), .., x(t)) \rightarrow (\sigma_1(t), .., \sigma_n(t))$$

where $(\sigma_1(t), \ldots, \sigma_n(t))$ is the state computed by the Viterbi algorithm in correspondence of the x-force sensor signal $x(t)$. The window length (l value) is a critical factor in order to obtain good results.

During tests, the interpretation of a net output

$$(\hat{\sigma}_1(t), \ldots, \hat{\sigma}_n(t))$$

proceeds in the following way. First, for all $i = 1, \ldots, n$:

$$\hat{\sigma}_i(t) := \begin{cases} 1 & \text{if } \hat{\sigma}_i(t) \geq (1-\varepsilon_t) \\ 0 & \text{otherwise} \end{cases}$$

where ε_t is the admissible *error on tests*. Each vector

$$(\hat{\sigma}_1(t), \ldots, \hat{\sigma}_n(t))$$

having exactly one element $\hat{\sigma}_k(t) = 1$ is classified as state k. Special classes account for vectors having none or many elements equal to 1.

3.2 Partially Recurrent Networks

The temporal evolution of the task can be better represented by an extension of the previous time-delay architecture, called *Partially Recurrent NN*. In this case, a number of units equal to the number of output units has been added to the input layer in order to form a set of context units. These units represents the task phase already executed – *the previous state*. A

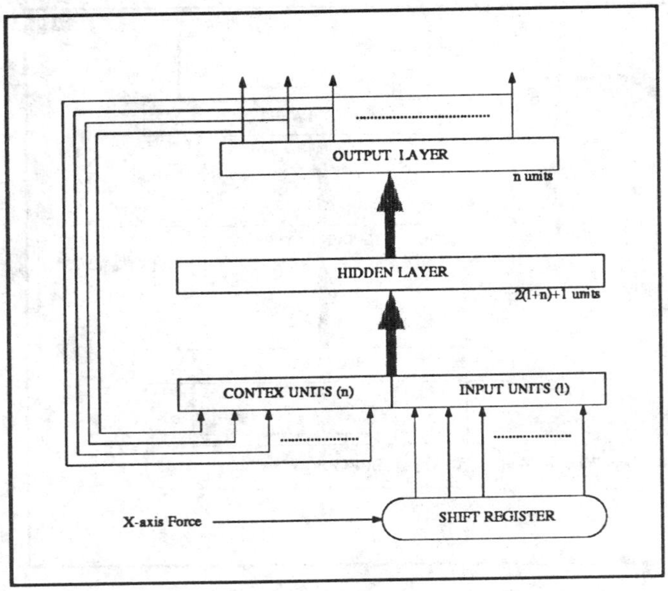

Figure 4: Partially Recurrent Network with Fixed Connections

set of fixed weight connections have been established among the output and context layer units (see figure 4).

Training has been carried out to associate the previous state and the window of sensor signals to the current state; formally:

$$\sigma_1(t-\Delta t), .., \sigma_n(t-\Delta t), x(t-(l-1)\Delta t), .., x(t) \rightarrow \sigma_1(t), .., \sigma_n(t)$$

During tests, the network is used in a slightly different way. In fact, except for the initial state $(\sigma_1(t_0), .., \sigma_n(t_0))$, the input to the context units representing the previous state, $(\sigma_1(t_k - \Delta t), .., \sigma_n(t_k - \Delta t))$, is computed from the preceding net output, $(\hat{\sigma}_1(t_{k-1}), .., \hat{\sigma}_n(t_{k-1}))$, through the feedback connections.

This computation is performed by evaluating

$$\nu = \max_{1 \leq i \leq n} \hat{\sigma}_i(t_{k-1})$$

and by setting:

$$\sigma_i(t_k - \Delta t) = \begin{cases} 1 & \text{if } i = \nu \\ 0 & \text{otherwise} \end{cases}$$

for all $i = 1, .., n$. During tests, the net outputs are classified as in the case of time-delay architecture.

4 Simulation Results

All simulations were carried out on a VAX 3900 minicomputer by using a custom Back Propagation algorithm. The

gain factor	β	1/2
learning rate	η	1
momentum parameter	α	0.9
error on learning	ε_l	0.2
error on test	ε_t	0.5
sampling rate	Δt	0.1
input window length	l	10 (15)
input layer	$l, l+n$	10 (15), 21 (26)
hidden layer	$2(l+n)+1$	43, 53
output layer	n	11

Table 1: Network Parameters

values of the different parameters involved in the computation are shown in table 1 and the results of the simulations are discussed in [4].

The Time-Delay architecture needed a high number of iteration to learn the associations and after 2000 epochs of training, the correct recognition was still not greater than 64.1% for $m = 10$ and 78.0% for $m = 15$. The Partially Recurrent architecture gave good results both in training and in tests. The percentage of correct recognition in training approached 100% in a few number of epochs (about 200) both for $m = 10$ and $m = 15$.

5 Results of the Real-Time Tests

To verify the simulation results and test the feasibility of this approach in a real system, the *partially recurrent NN* architecture has been interfaced to the telemanipulator of the JPL-ATOP. The teleoperator configuration is significantly different from the one used to collect the training data [1] and therefore these tests also verified the robustness of the approach to variations in the supporting hardware.

Figure 5 shows as a dotted line the output of the real-time segmentation performed by the network during an actual peg-in-hole task. The solid line represents the correspondence between task's phases and corresponding data. The time response of the network is evident in the lag between homologous transitions between the solid and the dotted lines and it depends primarily on the computer used to collect data and perform the network calculations. Several experiments have been carried out and the results have been quite encouraging with a percentage of correct segmentations approximately equal to 65%. This value means that within the Peg-in-Hole task, consisting of eleven phases, the network was misclassifying, on the average, about 3.85 task's phases. Figure 5 refers to the Peg-in-Hole phases' sequence described in Section 1: the values on the plot y-axis are the phases' indices. Approximately at time 29 sec. the network misclassified the end of the *extract* phase (index n. 8) that should have occurred at time 34 sec. Since the network was trained on a single experiment it was very sensitive to small changes in the duration and/or amplitude of the force signal. Furthermore the temporal dependency was not as strong as it should have been

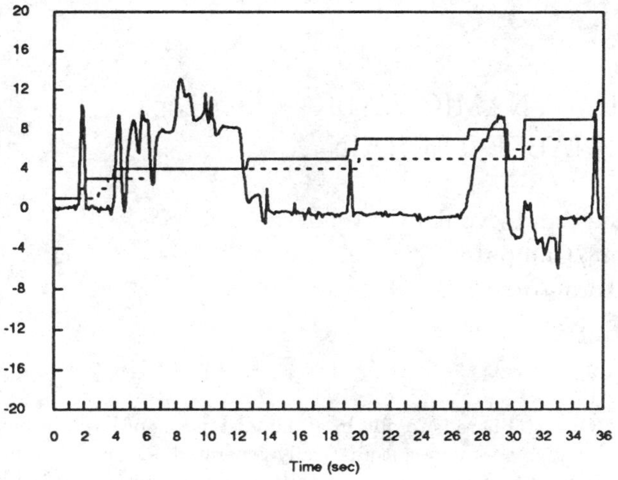

Figure 5: Segmentation in the real-time experiment

and the network wrongly assigned data to a phase that occurred earlier in the sequence, as shown in figure 5. On the other hand, the network showed an unexpected capability to recover after misclassifying some phase and it was also able to follow tasks whose phase sequence was very different from the training task. During peg extraction, for example, if the operator decided to regrasp the peg, the network was sometimes able to make the transition from *extraction* to *insertion* and again to *extraction*.

6 Conclusions

In this paper, a neural network approach to the real-time segmentation of telemanipulation tasks have been presented as a first step towards the definition and implementation of on-line monitoring of teleoperation. A partially recurrent network with fixed feedback connections has performed well in simulation and the results of the real-time experiments have been encouraging: more extensive training and architectural upgrades are needed to make the network more robust. These aspects are part of our current research in the area of network architectures. The use of this type of information for generating feedback messages to the operator requires the definition of on-line quality measures and of symbolic messages.

7 Acknowledgment

The simulations have been carried out at the Center for Space Geodesy of the Italian Space Agency (*ASI*) in Matera, Italy. The experiments have been carried out at the Jet Propulsion Laboratory (*JPL*), California Institute of Technology, under contract with the National Aeronautic and Space Administration. The authors would like to thank Dr. Paul Schenker and Dr. Antal Bejczy of *JPL* for their support during the course of the experiments.

References

[1] A.K. Bejczy, Z. Szakaly, and W.K. Kim. A laboratory breadboard system for dual arm teleoperation. In *Third Annual Workshop on Space Operations, Automation and Robotics*, number NASA Conf. Publication 3509, JSC, Houston, TX, July 25-27 1989.

[2] J.L. Elman. Finding structure in time. *Cognitive Science*, (14), 1990.

[3] P. Fiorini. A versatile hand for manipulators. *IEEE Control Systems Magazine*, 8(5), October 1988.

[4] P. Fiorini et al. Neural networks for the off-line segmentation of teleoperation tasks. In *International Symposium on Intelligent Control*, Glasgow, UK, August, 11-13 1992.

[5] B. Hannaford and P. Lee. Hidden markov model analysis of force-torque information in telemanipulation. *International Journal of Robotics Research*, 10(5), October 1991.

[6] B. Hannaford, L. Wood, D. McAffee, and H. Zak. Performance evaluation of a six-axis generalized force-reflecting teleoperator. *IEEE Transaction on Systems, Man and Cybernetics*, 21(3), May/June 1991.

[7] J. Hertz, A. Krogh, and R.G. Palmer. *Introduction to the Theory of Neural Computation.* Addison-Wesley, Redwood City, CA, 1991.

[8] M.I. Jordan. *Serial Order: A Parallel, Distribuited Processing Approach.* Erlbaum, Hillside, 1989.

[9] D. T. Pham and X. Liu. Dynamic system modelling using partially recurrent neural networks. *Journal of System Engineering*, (2):90–97, 1992.

[10] D.E. Rumelhart, G.E. Hinton, and R.J. Williams. Learning internal representation by error propagation. In D.E. Rumelhart and J.L. McClelland, editors, *Parallel Distribuited Processing*, volume 1. MIT Press, Cambridge, MA, 1986.

Proceedings of the 31st Conference
on Decision and Control
Tucson, Arizona • December 1992

A CONTROLLER FOR THE DYNAMIC WALK OF A BIPED ACROSS VARIABLE TERRAIN

Michiel van de Panne, Eugene Fiume*, and Zvonko G. Vranesic
Dept. of Electrical Engineering/Computer Science*
University of Toronto
Toronto, Ontario, M5S 1A4, Canada

Abstract

A set of independant controllers is used to control the dynamic walk of a biped. Our planar model has 11 links and 13 degrees of freedom. The most important controller is that used to control the stance leg. A state-space control table is calculated for this purpose. This lookup table contains an optimal control solution over a predefined region of state-space and is calculated using dynamic programming. A series of control tables are generated corresponding to various slopes of terrain. Linear interpolation between control tables is used to generate the desired control for any arbitrary slope. The control scheme is shown to be capable of generating the internal torques necessary to make the biped model walk across smoothly varying terrain as well as climbing up stairs.

1 Introduction

The synthesis of bipedal locomotion has received attention from a fair number of researchers in the past. The proposed synthesis methods have usually relied on reference trajectories obtained from human gaits or use an inverted pendulum model to perform the necessary control analysis and synthesis.

An example of the use of a reference trajectory for the control of a biped may be found in [3]. Video recordings of angles and angular velocities are recorded for a human walk and used as reference inputs for the control system.

Much work has been done on the control of biped systems using a series of inverted pendulum motions. In these systems, the length of the stance leg (ie the weight-bearing leg) is assumed to be held constant, hence the inverted pendulum. A stable gait is achieved by carefully controlling the swing-leg to achieve the desired stride length. If the biped is gaining too much speed, a larger stride than normal is taken. Similarly, a smaller stride is taken to gain speed. Examples using this principle of control can be found in [4] and [6]. It is also possible to create an entirely passive walk, such that walking occurs as a natural mode of a mechanical device placed on a slight incline[5].

Optimal control approaches have also been used. These are probably closest in nature to the ideas we shall present. In [2] a quadratic performance criteria is used to calculate a trajectory that is optimal with respect to this criteria. The method produces an optimal trajectory, not an optimal controller, and thus it is presumably necessary to recalculate the solution when the system is somehow perturbed. The solution optimizes the trajectory over three separate stages and uses optimal solutions for the stages to approximate a reasonable final solution, which is possibly suboptimal.

Our goal here is to obtain an understanding of how a robust walking gait may be automatically generated, so we choose not to rely on any reference trajectories. We are also interested in producing a controller general enough to be able to walk a biped model across varying terrain. An inverted pendulum model is not well suited for modelling ascending and descending motions of a biped because the leg length varies greatly throughout such motions, and the motions are not symmetrical about the point of support. As such, we choose to avoid restricting ourselves to an inverted pendulum for our analysis and synthesis. Lastly, we are interested in calculating a usable controller that can react appropriately to external disturbances. This precludes the use of costly optimization methods that determine only a single optimal trajectory.

2 The Biped Model

The biped model we shall use is shown in Figure 1. The various parameters for the links are given in Table 1. The origin of the local coordinate system used for defining the location of the center of mass (given as x and y in the table) is located at the point of attachment of the link to its parent link. For example, the origin of the coordinate system for links 9 and 11 is located at the elbow joint. The model is restricted to motion in the sagittal plane. All the joints are thus pin joints. Each of the 11 links has a degree of freedom in its angular position and the trunk has an additional two degrees of freedom in its location, yielding a total of 13 degrees of freedom. The state of the system can thus be represented by a unique vector of 26 numbers, as each degree of freedom requires both a position and a velocity specification.

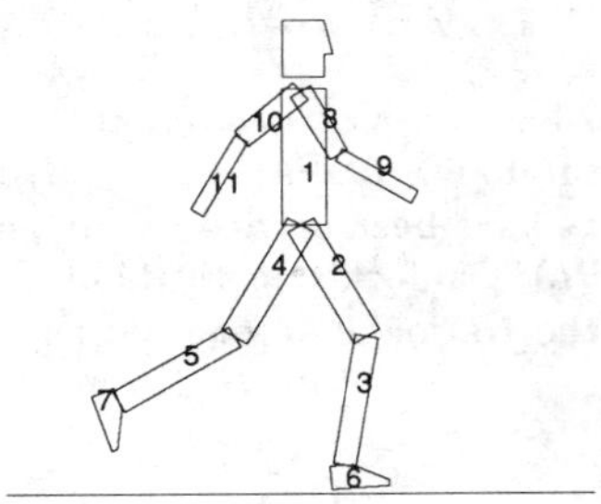

Figure 1: The biped model

link	name	length (m)	mass (kg)	I (kgm^2/s)	x (m)	y (m)
1	trunk	0.50	63	1.48	0	0.24
2	upper leg	0.45	5	0.09	0	-0.225
3	lower leg	0.45	4	0.07	0	-0.225
4	upper leg	0.45	5	0.09	0	-0.225
5	lower leg	0.45	4	0.07	0	-0.225
6	foot	–	1	0.01	0.04	-0.03
7	foot	–	1	0.01	0.04	-0.03
8	upper arm	0.26	2	0.02	0	-0.13
9	lower arm	0.30	1	0.01	0	-0.15
10	upper arm	0.26	2	0.02	0	-0.13
11	lower arm	0.30	1	0.01	0	-0.15

Table 1: Parameters for the biped model

The foot has the dimensions shown in Figure 2a. The toes are not explicitly included in our model. Their function is assumed to be one of keeping the ball of the foot in stable contact with the ground as the heel lifts off. Our explicit modelling of the floor contact forces ensures this stable contact, and therefore we have chosen not to include the toes as extra links. When the heel of the foot lifts off the ground, the foot is free to rotate about its front point, representative of the ball of the foot.

The biped figure is able to exert internal torques at each pin joint. Gravity and the forces exerted by the floor are the only external forces involved. The forces exerted by the floor on the model are calculated using a spring-and-damper model. Figure 2b illustrates the principle.

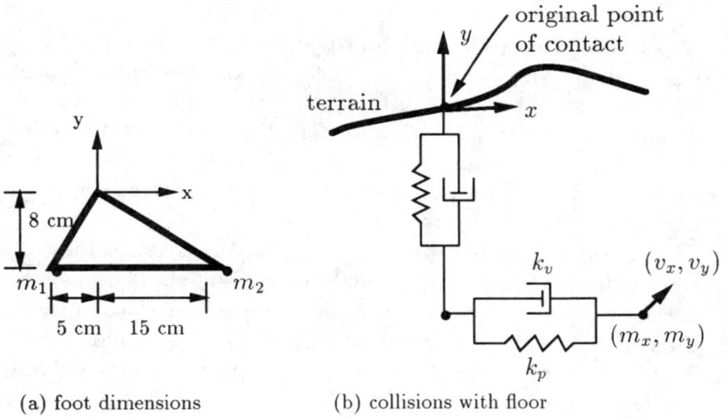

Figure 2: Modeling of the foot

External forces exerted by the floor are calculated for two points on each foot, indicated by points m_1 and m_2 in Figure 2a. The positions and velocities of these *monitor* points are used to calculate the external forces applied at these points as follows:

$$F_x = -(m_x - p_x)k_p - v_x k_v$$

$$F_y = -(m_y - p_y)k_p - v_y k_v$$

Typical spring and damper constants are $k_p = 10^5 N/m$ and $k_v = 10^3 Ns/m$. This creates a suitable stiff floor that functions effectively when used in a simulator with an adaptive time step. The advantage of modeling the biped as a free body in the air and explicitly calculated ground forces is that the same formulation of the equations of motion can be used throughout the entire walking motion. There is no need to calculate the magnitude of any impulsive forces that occur upon impact of the feet with the ground. These are effectively approximated by the spring-and-damper model with a suitably small time step.

The equations of motion for our biped are too complex to derive by hand. In addition, we wish to have an efficient implementation to allow for experimenting with the walking controller in a reasonably interactive fashion. We have written a *dynamics compiler*[7] that produces a linear set of equations of the form $Ax = b$ given a physical description of the articulated figure as input. A and b are dependent on the internal torques, external forces, the physical properties of the links, and the state of the links. x represents the vector of unknown accelerations. The accelerations are numerically integrated to determine the new velocity and position of the links.

The output of the dynamics compiler gives the symbolic value of each of the elements of the matrix A and the vector b. Values of common subexpressions are precalculated to avoid unnecessary duplicate calculations. In our implementation, we use the recursive Newton-Euler formulation[1], and the values of A and b are output as lines of 'C' code so that the equations of motion can be compiled. LU decomposition is used to solve the linear system of equations for the accelerations. An adaptive time step allows for maximum simulation speed while retaining numerical stability during collisions with the floor. The time step is allowed to vary between 0.00025 and 0.02 seconds.

3 Control Decomposition

A simplified block diagram of the biped system and its controller is shown in Figure 3. The controller uses the current state of the biped as well as information from the environment, namely the step height, to calculate the vector of 10 inputs to the biped system. The control vector is calculated at the same rate as the simulation of the biped system.

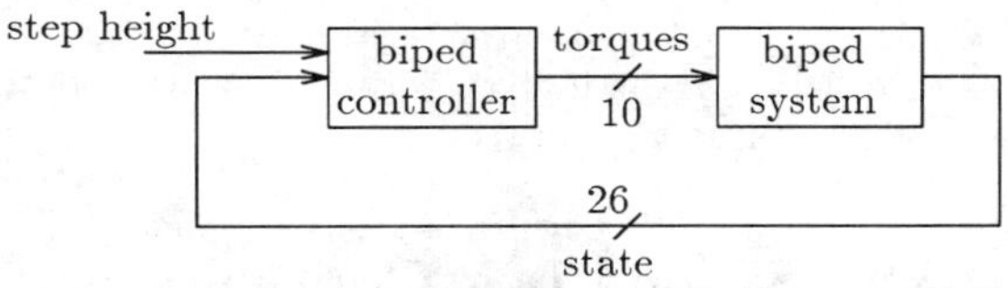

Figure 3: Simplified block diagram

Some form of decomposition is necessary to make the creation of a suitable controller a tractable proposition. Figure 4 shows the decomposition used. We shall begin with some general comments on the decomposition, and then proceed with more detailed descriptions of the specific subcontrollers involved.

We shall use *modular* to describe controllers that operate independently. A set of such controllers determine their output only as a function of their own states, which are non-intersecting substates of the entire system state. It can be seen from Figure 4 that the body controller is modular. We shall also define the term *hierarchical* to describe controllers that use common information to determine their output. In Figure 4 the controllers for the stance leg, swing leg, and arms all make use of a quantity we call the *phase*[1] of the motion and thus qualify as being hierarchical. The stance and swing leg also share information on the height of the next step to be taken. If the stride length were to be made variable, they would also share this information. We shall restrict ourselves to a fixed stride length here, however. The three 'hierarchical' controllers are also almost modular because they only make use of their own substate in addition to the phase.

[1]The phase is a number between 0 and 1 describing the progress of a stride

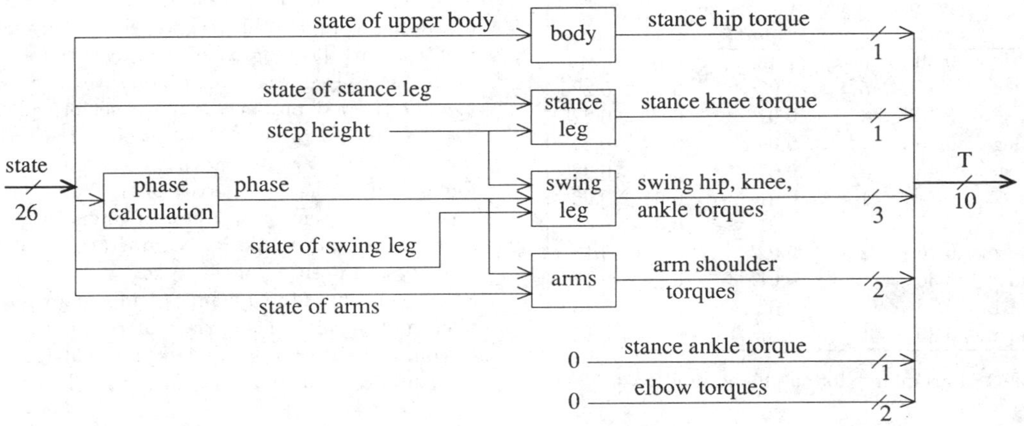

Figure 4: Control decomposition

In designing such a modular control system, we require that the interference of controllers with each other is not significant enough to prevent the control system from working as a whole. We are not able to prove the stability of our controller decomposition in a formal sense, but we will show it functions well in practice.

The most important subcontroller in our case is the stance-leg controller. In previous work on the dynamic walk of bipeds, the use of inverted pendulum models usually results in a straight stance leg being the desired position. The swing leg is used to control the stride length and hence ultimately provides the necessary control over the gait. We choose to fix the stride length and use the knee of the stance leg to provide the necessary gait control. While this has some disadvantages in creating natural-looking gaits for level terrain, it allows for robust control of gaits involving motion over variable terrain.

4 Stance Leg Control

The stance leg controller uses information from the portion of the state representing the state of the stance leg. In effect, the stance controller sees only a very much simplified model of the biped. This is shown in Figure 5. Mass M represents the combined masses of the trunk, arms, and swing leg. Masses m_2 and m_3 are as defined in Table 1. We shall choose cartesian coordinates to represent the state of the model of Figure 5. The state of the system is given by (p_x, p_y, v_x, v_y) and is calculated using the substate of the stance leg in the biped.

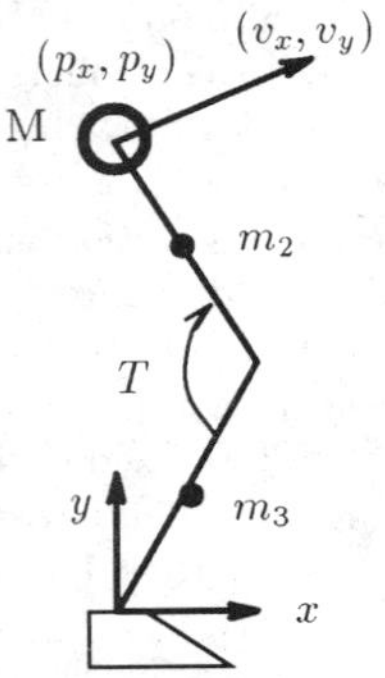

Figure 5: Model used for stance leg control

The state space of all possible states of the simplified model can be represented as a 4-dimensional space. We shall define a region in this space that contains the states of interest and call it our controller *domain*. Figure 6 shows the x and y dimensions of a 4-dimensional controller domain. The bounds placed in the x-dimension will serve to define the stride length as well as the point at which the stride and swing legs are exchanged. This exchange happens as soon as the hip exits the right-hand side of the controller domain (ie: $p_x > x_2$. The stride length is given by $x_2 - x_1$. y_1 and y_2 bound the hip-heights of interest. Similarly, the velocities (v_x and v_y) of interest must also be bounded.

A *destination* state, shown as point D in Figure 6, must also be specified. An optimization problem to be solved may now be informally stated as follows: from any state within the controller domain, move the system state from the current state to the destination state in an optimal way. The optimization function we use is $f_{opt} = \int_0^t | T | \, dt$, where T is the input we can provide, namely the torque on the stance knee.

For solving the optimization problem, the edges of the controller domain at x_1 and x_2 should be connected. The topology of the state space is cylindrical in the x-dimension. This captures the fact that walking is a periodic motion and thus allows optimization over the entire period of the motion while only fixing one point on the path, namely the destination state.

Because the system is non-linear, the state space is discretized and dynamic programming is used to perform the optimization.

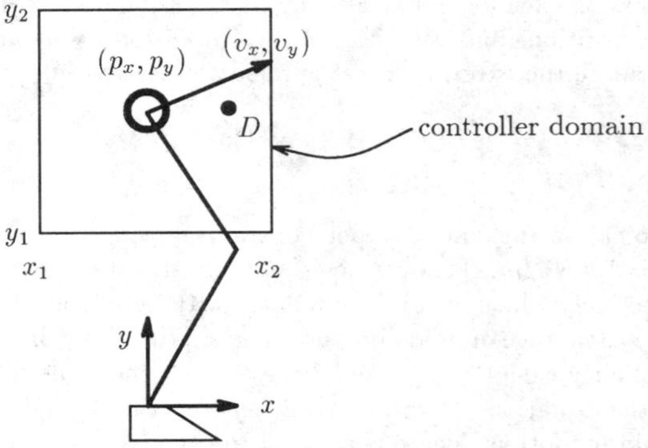

Figure 6: Defining the controller

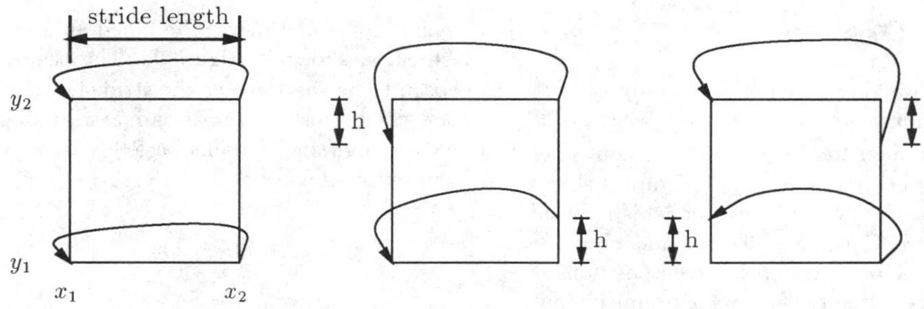

Figure 7: Controllers for various slopes

There is insufficient space here to describe this process in detail and the reader is referred to [8] for details. The optimization is a time-consuming process because of the 4-dimensional space. Because of this, the optimal solution for all states in the controller domain is calculated once and then stored in a table. This control (lookup) table then provides fast access to optimal control values when the controller is being used.

The interesting feature of this method is that it can be readily used to generate control values for ascending or descending strides. In this case, the topology of the state space becomes cylindrical with an offset. Figure 7 illustrates this principle. Consider a walk on ascending terrain. At the end of a stride, the foot of what will be the new stance leg is a step height h above the foot of the old stance leg. Thus, when the exchange of swing and stance legs occurs, the body will be lower by the step height h. This leads to a cylindrical topology with an offset, as shown in Figure 7. In this way, generating controllers for ascending and descending is no different than generating a controller for a level walk.

The control tables for the stance leg are typically discretized into 10 steps along each dimension, thus making for 10^4 entries. A series of 7 control tables are calculated for slopes ranging from -0.3625 to $+0.3625$. When using the control tables, linear interpolation is used within control tables to retrieve appropriate control values, and also between control tables to calculate suitable controls for arbitrary slopes. The stance-leg control tables are the key to capabilities of the variable terrain biped walker.

5 Other Subcontrollers

The function of the swing-leg controller is to bring the swing-leg foot down one stride length past the current stance-leg foot while maintaining suitable clearance over the ground. The step height of the ground one stride length ahead is obtained from the environment and used in helping plan the trajectory. No specific trajectory is planned for the swing leg. Instead, two points are specified as desired points for the swing-leg ankle to pass through. These are shown in Figure 8.

Point p_1 is a point in midstride with suitable ground clearance. Point p_2 is the desired location of the ankle at landing. Point p_1 should be reached halfway through the given stride ($phase = 0.5$), and point p_2 should be reached just before the end of the stride ($phase = 1.0$). Using the current phase, a desired velocity of the swing ankle can be determined such that the ankle will reach the desired point at the right time. Inverse dynamics of the swing leg is used to obtain the torques necessary

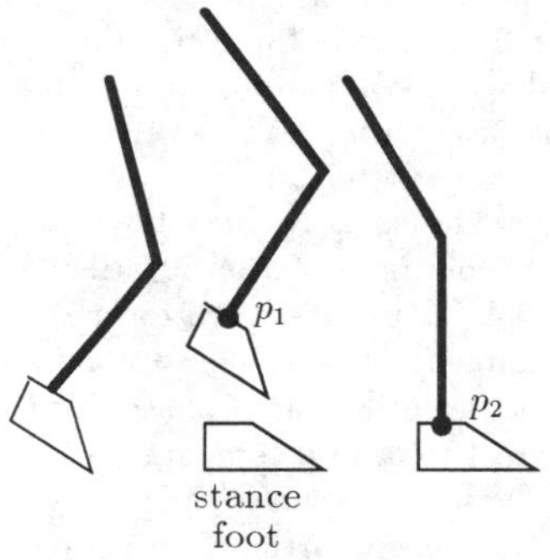

Figure 8: Swing leg control

to achieve the desired velocity of the swing-leg ankle. Note that the inverse dynamics of the swing leg assumes that the motion of the swing leg is independent of that of the rest of the biped. This is not completely true in reality, of course, but it proves to be a reasonable assumption, as our simulation results will verify. The swing-leg ankle is prepared for the slope of the coming step by using a simple proportional-derivative (PD) controller to generate the appropriate ankle torque.

The shoulders of the arms are driven by simple PD controllers to swing in phase with their 'opposite' leg. Because our model is restricted to the sagittal plane, the arm motion does not serve the same purpose as in real walking, namely to alleviate twisting motions of the body.

The trunk is maintained vertical by a PD controller using the hip of the stance leg to exert the necessary torque. Movement of the swing leg has a small but visible reactionary effect on the motion of the trunk, which is stabilized using the PD controller. The small oscillations of the trunk appear quite natural, however, when animations of the biped motion are viewed.

6 Results

The biped is capable of walking over both smoothly varying terrain and up stairs of various pitches. The stride length is at present fixed to 25 cm, although we expect that the controller will work equally well for other stride lengths. Figure 9 shows the results of five walks. For clarity, the biped model is reduced to a stick figure, the arm motion is not shown, and only the motion of one leg is shown. The time step between positions is 0.083 seconds for all five walks. Figure 9f shows one time instant of the simulation shown in Figure 9e with a more representative model of the biped. The calculation of the control, simulation, and display takes 3 seconds of real time for every second of simulated time on a Silicon Graphics IRIS workstation. This speed allows the system to be used in an interactive fashion. The user can draw the terrain to be traversed and then immediately view a walk across the terrain.

As mentioned previously, a stance-leg control table is generated for each of seven different slopes. Figures 9a–c show three of these slopes. In order to be able to cross variable terrain such as that shown in Figure 9d, we use linear interpolation between the seven control tables in order to be able to calculate stance-leg control for arbitrary slopes. The stride length is constant across the variable terrain. The speed is slightly lower for ascending and descending than for level terrain. This occurs as a result of the chosen destination state and because of the optimization function that was chosen.

Although the biped model is physically realistic in terms of its dimensions and dynamics, there remain some artifacts that appear unnatural when compared with a human gait. Some of these are now discussed in a qualitative fashion. When walking steeply uphill on smooth terrain such as in Figure 9a, people tend to use their ankles so as to place their weight over the ball of the stance foot for most of the duration of the stride. In contrast, our model has passive (zero torque) ankles during the stance phase, meaning the heel is likely to remain in contact with the ground during the entire stance phase. The trunk also seems to remain more erect than in a comparable human making such a steep ascent. These two artifacts are likely related, however, because placing the weight further forward over the ball of the foot requires the trunk to lean forward to keep the centre of mass over the support point.

Walks across level terrain occur with the knees more bent than in the equivalent human gait. The origin of this artifact is unknown as of yet. For descending walks, the gait resembles that of a person descending on a slippery surface. This occurs as a result of the way the stance-leg control table is calculated. It is assumed that the velocity of mass M (ie: the system momentum) in Figure 5 remains unchanged when the exchange of swing and stance legs occurs. When the state passes through the right-hand side of the controller domain in Figure 6, it reappears on the left side unchanged except for the new value of p_x. This does not always hold true for humans descending a hill because the instant at which the swing leg hits the ground to become the new stance leg is a convenient point to instantly dissipate some of the forward and downward momentum. This is not possible when the terrain is slippery and hence our model closely represents this case.

The controller drives the biped up a set of stairs in a natural fashion, as shown in Figure 9e. The step width of the stairs was chosen to be the same as the stride length of our controller. The good results for this terrain are attributable to the fact that the ankle in humans remains largely passive when climbing stairs, as in our controller.

(a)

(b)

(c)

(d)

(e)

(f)

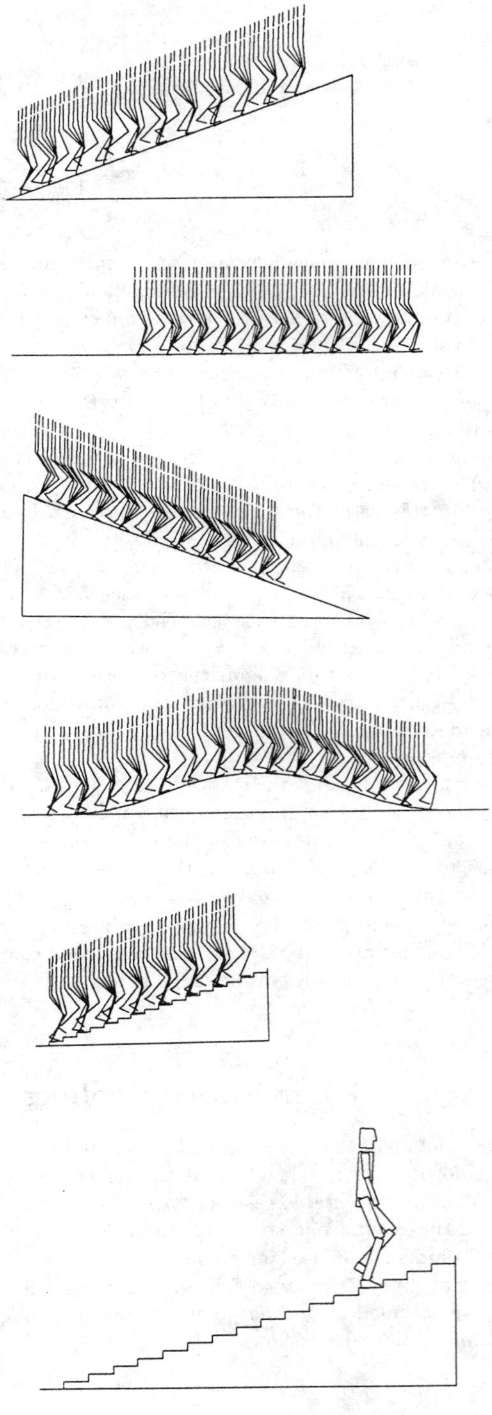

Figure 9: Results

7 Conclusions

A controller capable of driving a fully dynamic simulation of an 11-link biped across variable terrain has been presented. The control decomposition makes several assumptions about the independence of the motion of various links of the biped. These assumptions are put to the test when the biped motion is simulated as a whole. The results show that the control decomposition used is a useful one.

The stance-leg controller is the most important subcontroller in giving our dynamic biped model the ability to walk across variable terrain. The method used for generating the stance-leg control table is able to deal with ascending and descending gaits in the same fashion as level gaits. The control tables calculated for the stance leg are a useful technique for using the results of a family of optimal control solutions in real time. Dynamic programming is a useful technique for calculating the control tables. In effect, the control problem is converted into a path-planning problem throught state space.

Additional research needs to be performed into determining the changes necessary to make the resulting motion more natural. A study of the stability of the model across a range of terrain slopes, stride lengths, and optimization functions is necessary. A more quantitative comparison with human bipedal gaits would be very interesting to perform.

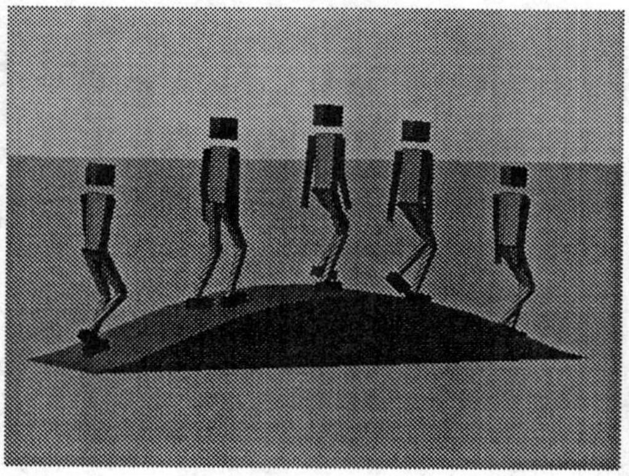

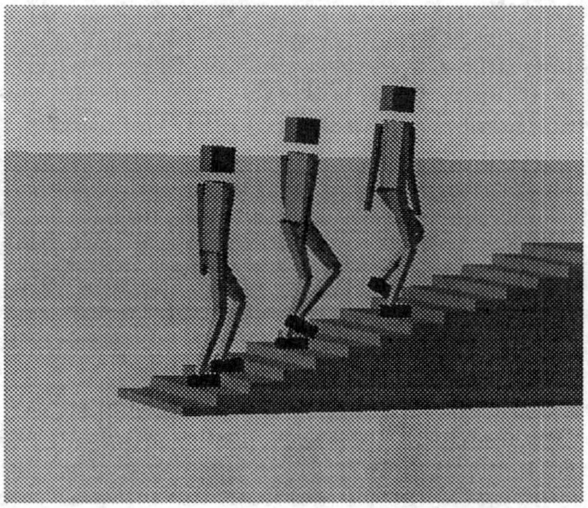

References

[1] H. Asada and J. J. Slotine. *Robot Analysis and Control.* John Wiley and Sons, 1986.

[2] C. Chow and D. Jacobson. Studies of human locomotion via optimal programming. *Mathematical Biosciences*, (10):239–306, 1971.

[3] H. Hemami and R. Farnsworth. Postural and gait stability of a planar five link biped by simulation. *IEEE Transactions on Automatic Control*, pages 452–458, June 1977.

[4] S. Kitamura, Y. Kurematsu, and M. Iwata. Motion generation of a biped locomotive robot using an inverted pendulum model and neural networks. *Proceedings of the IEEE Conference on Decision and Control*, pages 3308–3312, 1990.

[5] Tad McGeer. Passive walking with knees. *Proceedings of IEEE Internation Conference on Robotics and Automation*, pages 1640–1645, 1990.

[6] H. Miura and I. Shimoyama. Dynamic walk of a biped. *The International Journal of Robotics Research*, 3(2):60–74, 1984.

[7] M. van de Panne. Motion synthesis for simulation-based animation. Master's thesis, University of Toronto, 1989.

[8] Michiel van de Panne, Eugene Fiume, and Zvonko Vranesic. Reusable motion synthesis using state-space controllers. *Proceedings of SIGGRAPH '90 (Aug. 1990). ACM Computer Graphics*, 24(4):225–234, 1990.

Proceedings of the 31st Conference
on Decision and Control
Tucson, Arizona • December 1992

Teleoperator Control System Design with Human in Control Loop and Telemonitoring Force Feedback

Sukhan Lee[1,2] and Hahk Sung Lee[2] *

Jet Propulsion Laboratory[1]
California Institute of Technology
Pasadena, CA 91109

Dept. of EE-Systems and Computer Science[2]
University of Southern California
Los Angeles, CA 90089-0781

Abstract

This paper presents a new design methodology of advanced teleoperator control system using the paradigm of *telemonitoring*. Human dynamics reacting to visual and kinesthetic force feedback is modeled and incorporated into control loop for designing and evaluating teleoperator control systems. The monitoring force feedback is proposed for achieving a new kinesthetic coupling between human operator and manipulators under significant communication time delay and shared control.

Comparative study shows: 1) teleoperator control systems designed without human dynamics in the control loop become invalid when a human is added to the control loop, 2) teleoperator control systems aimed at achieving *telepresence* or *passive* system suffer from either poor performance or instability under communication time delay.

Based on the comparative study, we propose an optimal structure of teleoperator system for resolving considerable communication time delay.

1 Introduction

An advanced teleoperator control system must be designed with the consideration of "*human dynamics in the control loop*", with the designed control system supporting "*a proper coordination between the operator and the manipulator*" to achieve control objectives under shared control and communication time delays. In doing so, the control system will be able to reduce operator's control effort or burden by fully utilizing the capabilities of the dexterous manipulator and/or autonomous system.

Previous teleoperator control systems designed based on *telepresence* [1, 2] or *passivity* [3] may not provide a concerted man-machine coordination in achieving the above control objectives under shared control and time delay. This is because: 1) most control systems have been designed without the consideration of human dynamics reacting to visual and kinesthetic force feedback in teleoperation. The human dynamic behavior is crucial in determining overall control performance and system stability, and 2) under the autonomous compliance/force control performed by the remote manipulator in the form of shared control, the previous kinesthetic feedback may bring a conflict control action between man and manipulators.

As opposed to other control systems, the teleoperator control system has to deal with the inherent time delay existing in communication between the control and remote stations: Time delay of up to hundreds of milliseconds is common for many of the earth-bound or space-borne remote teleoperators. While, time delay of up to several seconds (on the order of 2 to 8 seconds) exists for the earth-to-orbit telemanipulation of manipulator/vehicle systems. The time delay existing in kinesthetic and visual coupling between the human operator and master arm located in the control station and the slave arm located in the remote station may seriously degrade the fidelity of the teleoperator and may incur control instability. It thus becomes essential that we design the system to support a high performance of teleoperation under time delay constraints.

The existing methods for dealing with time delay can be categorized into the following:

- The construction of a "*real-time teleoperator control system*" which has delay-tolerant control law based on two-port network model [3]
- The use of a "*teleprogramming facility*" with the concept of time and position clutches [4], the elegant predictive graphical display [5], or knowledge-based sensor programming [6]

The first approach aims at maintaining system stability under time delay in real time by controlling the force feedback signal of the control station, as well as the position or velocity command signals of the remote station. Whereas, the latter approach aims at establishing an indirect control link between the control and remote stations through the use of a simulated task environment (graphical display) at the control station. The first approach does not need to have a simulation environment, but suffers from performance degradation, especially when dealing with larger time delays [7, 8]. The latter approach may bring about a high quality of task performance, provided that the modeled simulation environment is accurate or that the remote manipulator is capable of carrying out

*The author would like to thank for the financial support granted by HYOSUNG INDUSTRIES CO., LTD. in KOREA.

sensor-based local autonomy for automatically correcting possible errors in teleprogramming.

This paper focuses on designing a teleoperator control system supporting high quality of teleoperator under the time delay of up to several seconds. Especially, we will show that 1) A controller designed without human in the control loop may be in flaw or become invalid when human is in the control loop. This is because without human in the control loop, the dynamics of human reacting to time delayed visuo/kinesthetic stimuli, which is crucial in determining overall control behavior, is totally neglected. 2) Under the compliance and force control implemented at the remote manipulator in the form of shared control, the conventional kinesthetic coupling between man and machine based on telepresence may cause an adverse effect on performance. This is because the human reaction to the kinesthetic stimulus of environmental force without understanding the role of the automatic control shared by the manipulator may amplify control errors.

2 Human in Control Loop

Although there have been many publications [9, 10] on human dynamic behavior in teleoperation, it is clear that the human dynamic models which account for the aforementioned distinctive features of teleoperation do not truly represent the human dynamics involved in telemanipulation [11]. This is because under teleoperation, the operator internally generates the most appropriate trajectory for the task in response not only to positional but also to force feedback signals. The existence of visual and force feedback in bilaterally controlled teleoperation prompts the need to model the human dynamics reacting to both visual and kinesthetic (force) stimuli.

We presented a simple teleoperation rule of human operator in terms of human visual and kinesthetic force tracking mechanisms [8] by combining the experimental results previously obtained for visual as well as force tracking. As shown in Fig. 1, the human dynamic behavior involved in visual tracking in teleoperation is modeled as a PD controller followed by pure time delay component of human reaction time, by which the position discrepancy is transformed to the human visual tracking force. The human dynamics involved in force tracking is modeled similarly.

We conjecture that the fusion of the two sensory modalities takes place by fusing the two sources of intentional forces generated for visual and force trackings as follows:

$$\boldsymbol{f}_h = \alpha \boldsymbol{f}_{hv} + (1-\alpha)\boldsymbol{f}_{hf} \quad (1)$$

where α, $\alpha \leq 1$, represents the fusion coefficient, and may be a function of both task conditions and operator's experience; $\boldsymbol{f}_{hv}, \boldsymbol{f}_{hf}$ represent respectively human intentional forces reacting to visual and kinesthetic stimuli.

The combined force (1) will now work as a driving force for the human arm interacting with the environment. Thus, the combination of two tracking models represents the dynamics of human arm reacting to both visual and force feedbacks in teleoperation:

$$\alpha \boldsymbol{f}_{hv} \quad + \quad (1-\alpha)\boldsymbol{f}_{hf} + \boldsymbol{f}_{eh}$$

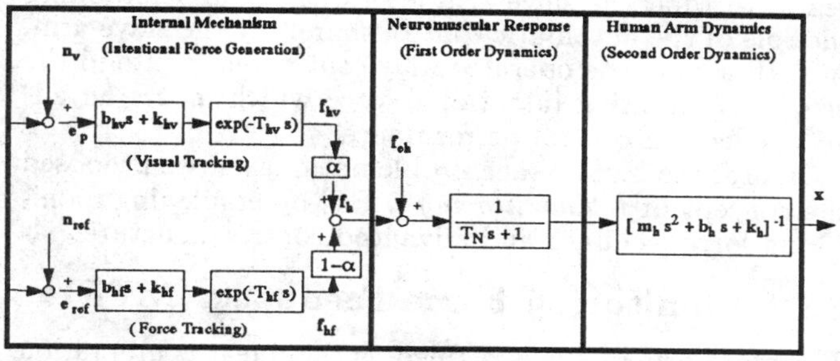

Figure 1: Human Arm Dynamic Model in Teleoperation

$$= (T_n s + 1)(\boldsymbol{m}_h s^2 + \boldsymbol{b}_h s + \boldsymbol{k}_h)\boldsymbol{x}. \quad (2)$$

where $\boldsymbol{f}_{eh}$ represents the interaction force applied to the human arm; the term, $(T_n s+1)$, represents the first order response lag in the neuromuscular system; $\boldsymbol{m}_h, \boldsymbol{b}_h$ and $\boldsymbol{k}_h$ represent the human arm physical mass, damping and spring coefficient matrices, respectively.

The human arm dynamic model (2) indicates that human controls his/her arm in such a way as to compromise the motion discrepancy and the contact force discrepancy. Fig. 1 shows the conceptual block diagram of the proposed human arm dynamic model. The proposed model not only allows us to design a teleoperator control system with explicit human dynamics in the control loop under visual and kinesthetic coupling between man and machine, but also enables us to incorporate human control errors (due to physical fatigue, lack of experience, and misjudgment, etc.) in the design and evaluation of a control system.

For simplicity, we further simplify the human dynamics given by (2) as follows [8]:

$$\boldsymbol{f}_h + \boldsymbol{f}_{eh} = \boldsymbol{m}_h \ddot{\boldsymbol{x}}. \quad (3)$$

with

$$\boldsymbol{f}_{hv} = \boldsymbol{b}_{hv}\dot{\boldsymbol{e}}_p + \boldsymbol{k}_{hv}\boldsymbol{e}_p, \quad (4)$$

$$\boldsymbol{f}_{hf} = \boldsymbol{b}_{hf}\dot{\boldsymbol{e}}_{ref} + \boldsymbol{k}_{hf}\boldsymbol{e}_{ref}. \quad (5)$$

Note that we put the same weight on both visual and kinesthetic (force) stimuli fusion coefficient.

3 Telemonitoring

Conventional telepresence is to give operator the best feeling on what is going on manipulator site, so that the operator can generate high fidelity of command. Under the time delay, however, the original good of telepresence becomes obscured due to modification of sensed signal for stability. The other important fact we considered is that under shared control, there are two sources (human and machine) of decision making which should be properly coordinated. This is a problematic example of shared control with an autonomous slave arm performing compliance and force control, where the conventional kinesthetic force feedback (the direct feedback of sensed force to the operator) may confuse the operator in terms of evaluating his/her position command. This is because the operator reacts to what the slave arm feels without understanding the role of the automatic control shared by the slave arm. In such a case, the operator may continuously extend the error in generating intentional force which incurs an adverse effect on control performance.

In order to handle such problematic cases, we proposed the concept of *telemonitoring* [7, 12] by employing monitoring force feedback and advanced control structure.

3.1 Monitoring Force Feedback (MFF)

The monitoring force is defined as the linear combination of two forces generated based on the position error and the force error. The position error vector, $\boldsymbol{e}_s$, is defined as the discrepancy between the position command generated by the C-station, $\boldsymbol{x}_{ds}$, and the current slave arm position , $\boldsymbol{x}_s$, (i.e., $\boldsymbol{e}_s = \boldsymbol{x}_{ds} - \boldsymbol{x}_s$). Then, the force, $\boldsymbol{f}_{rp}$, is obtained by

$$\boldsymbol{f}_{rp} = \boldsymbol{G}_{rp}\boldsymbol{e}_s, \tag{6}$$

where $\boldsymbol{G}_{rp}$ is a transfer function matrix of the monitoring force compensator which is design parameter.

The force error vector, $\boldsymbol{e}_f$, is defined as the discrepancy between the desired contact force, $\boldsymbol{f}_d$, and the feedback force, $\boldsymbol{f}_r$, which is related to the sensed interacting force of the slave arm end-effector, $\boldsymbol{f}_{es}$. For the ease and comfortable teleoperation, the feedback force, $\boldsymbol{f}_r$, is transformed as follows:

$$\boldsymbol{f}_r = -\boldsymbol{k}_{cs}\boldsymbol{f}_{es}, \tag{7}$$

where $\boldsymbol{k}_{cs}$ represents a positive definite diagonal matrix for the force transformation.

Thus, the force error vector, $\boldsymbol{e}_f$, is obtained by

$$\boldsymbol{e}_f = \boldsymbol{f}_d - \boldsymbol{f}_r = \boldsymbol{f}_d + \boldsymbol{k}_{cs}\boldsymbol{f}_{es}. \tag{8}$$

Finally, two forces are combined into the monitoring force, $\boldsymbol{f}_{ref}$, by the following linear combination:

$$\boldsymbol{f}_{ref} = \boldsymbol{e}_f - \boldsymbol{f}_{rp}. \tag{9}$$

For the kinesthetic coupling between the C-station and the R-station, the monitoring force will be incorporated into the master arm dynamics.

It should be noted that the monitoring force is interpreted as a reaction force due to either a contact force error or a position error of the slave arm, and coupled to the operator through the master arm dynamics. As opposed to the conventional kinesthetic coupling based on telepresence, the operator can monitor slave arm regarding interaction force error, as well as position tracking error of the slave arm end-effector.

3.2 Advanced Control Structure

In the proposed system, the master in control station and slave arm in remote station are designed in such a way that manipulators become active partners of the human operator in supporting perception, decision-making and cooperative task execution.

1. Control (C) Station

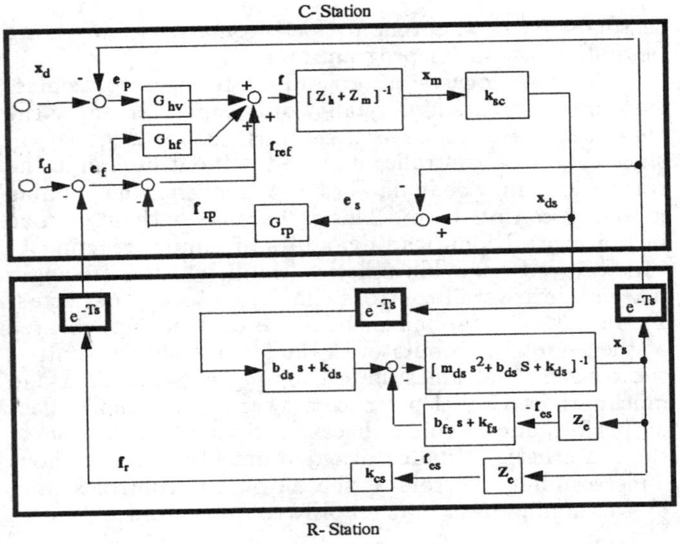

Figure 2: Structure of the Proposed Teleoperator System

The operator has limited capabilities in providing stiffness and damping effect for stabilizing the system under large disturbance force. Furthermore, for system robustness, the motion of the master arm should be generated through the proper dynamic coordination between the C-station and R-station instead of solely relying upon the operator. For such coordination, the master arm dynamics needs to maintain a desirable relation between the sensory information provided by R-station and the intentional force applied by the human operator. This prompts the generalized impedance (GI) of master arm to be defined as follows:

$$\boldsymbol{f}_{em} + \boldsymbol{f}_{ref} = \boldsymbol{m}_{dm}\ddot{\boldsymbol{x}}_m + \boldsymbol{b}_{dm}\dot{\boldsymbol{x}}_m, \tag{10}$$

where $\boldsymbol{m}_{dm}$ and $\boldsymbol{b}_{dm}$ are master arm parameters of the GI specified as positive definite diagonal matrices; and $\boldsymbol{f}_{ref}$ represents the monitoring force vector given by (9).

Under the dynamic interaction with the master arm, the operator feels $\boldsymbol{f}_{ref}$ as a reaction force through the master arm dynamics. Hence, $\boldsymbol{f}_{hf}$ given by (5), is rewritten as

$$\boldsymbol{f}_{hf} = \boldsymbol{b}_{hf}\dot{\boldsymbol{f}}_{ref} + \boldsymbol{k}_{hf}\boldsymbol{f}_{ref}. \tag{11}$$

Finally, position command of the slave arm, $\boldsymbol{x}_{ds}$, is obtained by

$$\boldsymbol{x}_{ds} = \boldsymbol{k}_{sc}\boldsymbol{x}_m, \tag{12}$$

where $\boldsymbol{k}_{sc}$ represents a positive definite diagonal matrix for the position transformation.

By combining above dynamics, the dynamics of the C-station is represented by

$$\boldsymbol{f}_h + \boldsymbol{f}_{ref} = (\boldsymbol{m}_h + \boldsymbol{m}_{dm})\ddot{\boldsymbol{x}}_m + \boldsymbol{b}_{dm}\dot{\boldsymbol{x}}_m \tag{13}$$

2. Remote (R) Station

Although the human has the strength in understanding the task mechanism, recognizing objects and planning

Table 1: Parameters of Proposed Teleoperator System

parameter group	parameter values
human operator	$m_h = 0.15, b_{hv} = 0.5, k_{hv} = 7.0$ $b_{hf} = 0.5, k_{hf} = 1.0$
master arm	$m_{am} = 1.0, u_{bm} = 0.5\dot{x}_m$ $m_{dm} = 0.05, b_{dm} = 4.0, u_{bm} = 0.5\dot{x}_m$
slave arm	$m_{as} = 3.0, u_{bs} = 0.0$ $m_{ds} = 1.0, b_{ds} = 8.0, k_{ds} = 10.0$ $b_{fs} = 0.75, k_{fs} = 0.0$
scaling factor	$k_{sc} = 1.0, k_{cs} = 0.2$

trajectories under global constraints, he/she may produce control errors in generating the intentional force due to various factors such as fatigue, weakness in precise estimation of position and so on. Thus, one should consider the motion command given by the C-station as a *contaminated* command. Under such situation, the slave arm needs to be able to cope with contaminated commands, especially when a dynamic interaction between slave arm and environment occurs. This requires that the slave arm should follow a given command precisely to complete a given task, and also be able to make a compromise between the command and the interaction force applied to the slave arm within the framework of autonomous system.

To achieve such a partially autonomous system, let us define the GI of slave arm as follows:

$$\begin{aligned} \boldsymbol{b}_{fs}\dot{\boldsymbol{f}}_{es} + \boldsymbol{k}_{fs}\boldsymbol{f}_{es} &= \boldsymbol{m}_{ds}\ddot{\boldsymbol{x}}_s + \boldsymbol{b}_{ds}(\dot{\boldsymbol{x}}_s - \dot{\boldsymbol{x}}_{ds}) \\ &\quad + \boldsymbol{k}_{ds}(\boldsymbol{x}_s - \boldsymbol{x}_{ds}), \qquad (14) \end{aligned}$$

where $\boldsymbol{b}_{fs}, \boldsymbol{k}_{fs}, \boldsymbol{m}_{ds}, \boldsymbol{b}_{ds}$ and $\boldsymbol{k}_{ds}$ are the slave arm parameters of the GI; $\boldsymbol{x}_s$, $\dot{\boldsymbol{x}}_s$ represent respectively the actual position and velocity vectors; and $\boldsymbol{f}_{es}$, $\dot{\boldsymbol{f}}_{es}$ represent the interaction force vector sensed and its derivative, respectively.

3.3 Stability Analysis

The overall structure of the proposed teleoperator system for a single degree of freedom can be represented by Fig. 2. For stability analysis, let us define various transfer functions, $G_d(s) = x_{ds}/f = k_{sc}/D_m$, $G_s(s) = x_s/x_{ds} = N_s/D_s$, and $G_{rp}(s) = f_{rp}/e_s = N_{rp}/D_{rp}$. By block diagram manipulations, the slave arm position, x_s, can be represented in terms of the system inputs, x_d and f_d. When a time delay exists in the system, the loop transfer function becomes

$$\begin{aligned} \hat{L} = \frac{e^{-2\tau s} \cdot k_{sc}}{D_{rp} D_s D_m} \cdot [G_{hv} D_{rp} N_s + (G_{hf} + 1) \\ \{D_{rp} N_s k_{cs} Z_e + N_{rp}(D_s e^{2\tau s} - N_s)\}] \qquad (15) \end{aligned}$$

In order to guarantee system stability under the communication time delay of τ seconds, the Nyquist diagram for $\hat{L}$ should not encircle the point $(-1, 0)$ on the complex plane through all frequency range.

In general, the time delay incurs an additional phase lag to the loop transfer function due to $e^{-2\tau s}$ which reduces the phase margin of the system. As shown in (15), however, the numerator of monitoring force compensator, N_{rp}, introduces an additional phase leading effect. The term $e^{2\tau s}$ compensates, to some extent, for phase lagging effect so that the proposed system is robust to communication time delay. As an example, we showed in [8] that the proposed MFF significantly enhances system stability and that MFF is applicable to stabilizing previous teleoperator control systems. Furthermore, for fixed parameters of master and slave arms, we have freedom in shaping the loop transfer function by using G_{rp}, k_{cs} and k_{sc} as shown in (15) which is crucial to the trade-off between robustness and performance.

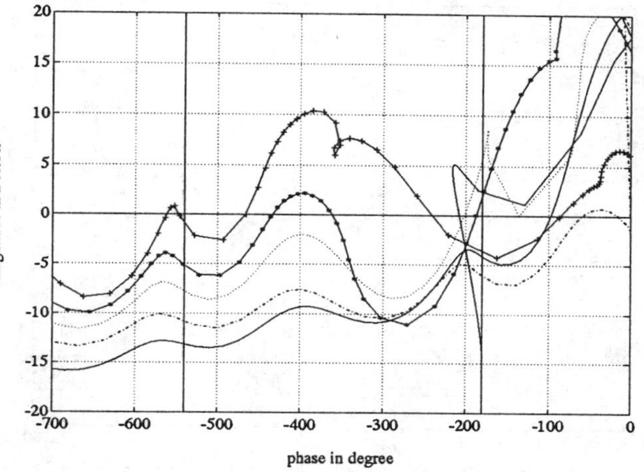

Figure 3: The Nichols diagram for a passive teleoperator system under various time delays of 0.7 (solid), 1.2 (dotted), 1.4 (dashdot), 1.6 (*) and 2.0 (+) *sec*, respectively

4 Comparative Study

In [3], a teleoperator system was constructed based on passivity concept and its control law was derived based on characteristics of the lossless transmission line in order to maintain the communication channel as a passive system. It was proven that both master and slave arms are asymptotically stable in velocity sense regardless of communication time delay. It was also concluded that the closed-loop of the proposed system is stable for any passive environment and passive human operator dynamics. In teleoperation, however, the human operator generates the intentional force based on the sensory information such as a visual display on monitor, which incurs additional time delay in the closed-loop system. We will show that although the closed-loop teleoperator system including master and slave arms, communication channel, environment, and human operator is passive, there still exist several other factors which may incur instability.

For simplicity, we use one dimensional joy-stick as the master arm and a mobile vehicle as the slave arm. The designed parameters of the proposed teleoperator system are shown in Table 1 and closed-loop dynamics of master and slave arms of passive teleoperator are shown as follows:

master arm: $\ddot{x}_m + 4.0\dot{x}_m = f_h - f_{md}$

slave arm: $3.0\ddot{x}_s + 4.0\dot{x}_s = f_s + (1 + 2.25)f_{es}$ with

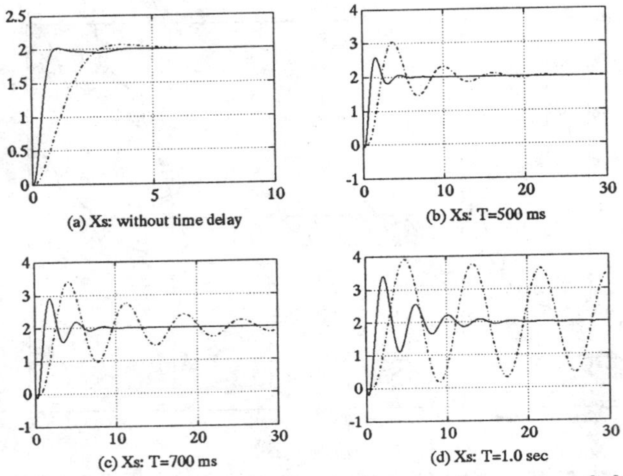

Figure 4: The position (x_s) tracking performance of the proposed (solid) and passive teleoperator system (dashed)

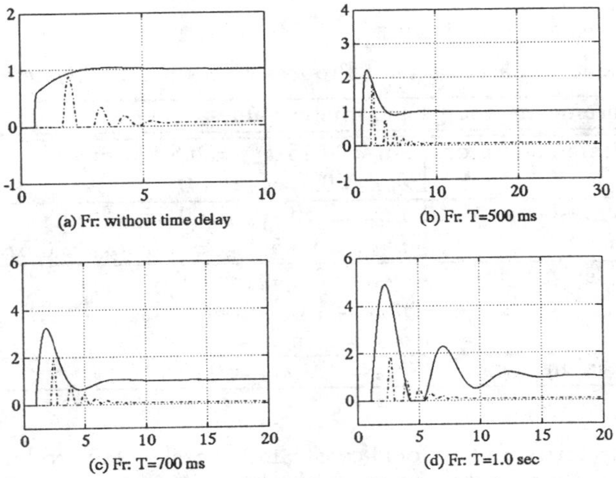

Figure 5: The force tracking performance of the proposed (solid) and passive teleoperator (dashed)

$$f_s = 20.0(\dot{x}_{ds} - \dot{x}_s) + 30.0(x_{ds} - x_s)$$

communication law:

$$f_{md}(t) = f_s(t-\tau) + \dot{x}_m(t) - \dot{x}_{ds}(t-\tau)$$
$$\dot{x}_{ds}(t) = \dot{x}_m(t-\tau) + f_{md}(t-\tau) - f_s(t)$$

Note that parameters for passive teleoperator system are designed based on [3], and parameters for human operator are the same as the ones in Table 1.

Fig. 3 shows the Nichols diagram for the passive teleoperator system for free space navigation with $10^{-1} \leq \omega \leq 10^{1}$. As shown in Fig. 3, this system cannot guarantee stability around 1.2 and 1.6 *sec* of time delay. This result illustrates that the passive teleoperator system designed without consideration of human dynamics incurs instability when the human operator is in the control loop, especially there exists time delay on visual display channel.

Fig. 4 shows the position tracking performance of the proposed and the passive teleoperator systems in free space navigation, and Fig. 5 shows the force tracking performance at contact case. Compared to the proposed control system, in both cases, the passive teleoperator system suffers from relatively severe oscillation. This is because the passive teleoperator system where the communication channel is modeled as a lossless transmission line without impedance matching, incurs wave reflection. Such wave reflection contaminates the C-station's command generation. To avoid this problem requires an explicit impedance matching between the control structures and transmission line, which violates the system passivity, especially, in multiful degree of freedom case.

5 Guideline For Optimal Design

As shown in comparative study, the passive teleoperator system also cannot guarantee system stability for arbitrary time delay involved in the visual display, the constraint of the passivity may not necessarily be imposed on controller design. In addition, the passive teleoperator system suffers from poor performance, especially with large time delay since there is little freedom to choose optimal control parameters due to excess constraints to be satisfied for stability. While the controllers designed with predictive model to cope with time delay may offer optimal performance, however, it is subject to instability or failure depending on the accuracy of predictive model.

As an preliminary effort of combining features of real-time teleoperator system and teleprogramming methodology, we propose an optimal design procedure for the teleoperator control system under time delay based on Smith's principle [13] .

The Smith's principle defined an ideal response of the system under time delay of T as a time-shifted version of $y(t)$ (i.e., $y(t-T)$) where $y(t)$ is the system response for the delay-free environment. This idea, which leads directly to a design method, has a strong advantage that the designed system based on this principle preserves stability property under time delay case. As opposed to teleprogramming approach [4, 5], this method can handle contingencies or uncertainties involved in system modeling in real time.

By applying Smith's method to Fig. 2, we obtained the modified structure shown in Fig. 6 where G_{so} and Z_{eo} represent respectively simulated models of the slave arm, G_s, and the environment, Z_e.

The simulated slave arm position, x_{so}, will be fed back to the C-station through the loop-1 as shown in Fig. 6. If there exist modeling errors either in slave arm or environment, the error between actual slave arm position, x_s, and simulated slave arm position, x_{so}, will be fed back to the C-station through the loop-2.

Under modeling error, the loop transfer function, $\hat{L}$, is represented by

$$\begin{aligned} G_{d1}G_{so}[G_{hv} &+ (G_{hf}+1)(k_{cs}Z_{eo} - G_{rp})] \\ &+ G_{hv}G_{d1}(G_{so} - G_s) \cdot e^{-2\tau s} \quad (16) \end{aligned}$$

with $G_{d1} = G_d/[1 + (G_{hf}+1)G_{rp}G_d]$.

The first term in (16) is determined by parameters of C-station and simulated environment, and the second term by C-station and modeling error in slave arm which affects system stability under modeling error. It thus requires

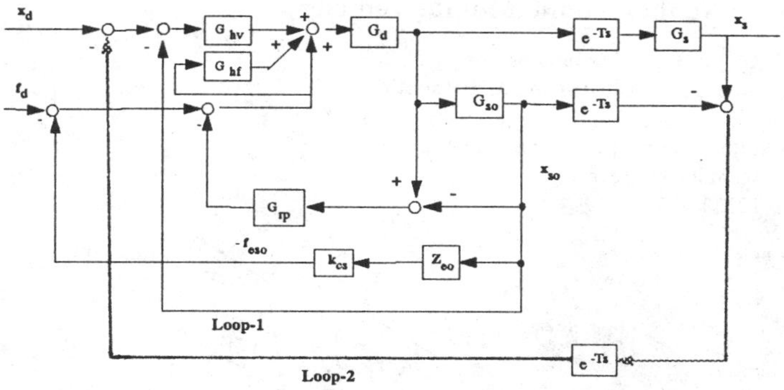

Figure 6: Modified Structure of the Proposed Teleoperator

trade-off between performance and robustness in designing teleoperator control system. We propose the following design principle to cope with robustness/performance trade-off under time delay: *Optimize the parameters involved in position/force coordination to maximize the performance index, while maintaining the stability up to the known maximum bound on modeling error and time delay.*

The original system stability and performance determined by the first term of (16) (i.e., designed without consideration of system uncertainties), is affected by the second term. For a given maximum bounds on modeling error and T, however, a suboptimal control performance can be achieved by tuning the optimal parameters. One possible design procedure can be described as follows:

First, design optimal parameters satisfying a desired specification. Second, for given bounds on modeling error and T, check system stability by plotting Nyquist diagrams for loop transfer function given by in (16). For the last step, if the designed parameter violates system stability, then go to the first step, and redesign optimal parameters by relaxing some of the desired specification requirements.

6 Conclusion

A new design concept of telemonitoring has been proposed for an advanced teleoperator control system.

The human dynamics reaction to visual and force feedback is modeled and incorporated into the controller design and evaluation. This allows us to incorporate various human control errors in simulation. The dynamic characteristics of the master and the slave arms are actively modified in such a way as to systematically implement the desirable dynamic characteristics within the framework of generalized impedance control. The monitoring force feedback is redefined here as a combination of position and force discrepancies, and yields a proper coordination between two stations in achieving control objectives under shared control. The design procedure for achieving an optimal structure of teleoperator control system under time delay and modeling error is proposed.

References

[1] Y. Yokokohji and T. Yoshikawa. Bilateral control of master-slave manipulators for ideal kinesthetic coupling: Formulation and experiment. In *Proceedings of the IEEE International Conference on Robotics and Automation*, Vol. 1, pp. 849-858, May 1992.

[2] B. Hannaford. Stability and performance tradeoffs in bi-lateral telemanipulation. In *Proceedings of the IEEE International Conference on Robotics and Automation*, pp. 1764-1767 1989.

[3] R. J. Anderson and M. W. Spong. Asymptotic stability for force reflecting teleoperators with time delay. In *Proceedings of the IEEE International Conference on Robotics and Automation*, pp. 1618-1625 1989.

[4] L. Conway, R. A. Volz, and M. W. Walker. Teleautonomous system: Projecting and coordinating intelligent action at a distance. In *The IEEE Transactions on Robotics and Automation*, Vol. 6, No. 2, pp. 146-158, Apr. 1990.

[5] R. Paul, T. Lindsay, and C. Sayers. Time delay insensitive teleoperation. In *IEEE/RSJ International Conference on Intelligent Robots and Systems*, pp. 247-254, Jul. 1992.

[6] G. Hirzinger, J. Heindl, and K. Landzettel. Predictive and knowledge-based telerobotic control concepts. In *Proceedings of the IEEE International Conference on Robotics and Automation*, Vol.3, pp. 1768-1777 1989.

[7] Sukhan Lee and Hahk Sung Lee. An advanced teleoperator control system: Design and evaluation. In *Proceedings of the IEEE International Conference on Robotics and Automation*, Vol. 1, pp. 859-864, May 1992.

[8] Sukhan Lee and Hahk Sung Lee. Advanced teleoperator control system design with human in control loop. In *the 1992 ASME Winter Annual Meeting*, Issues in the development of Kinesthetic Displays for Teleoperation and Virtual Environments, to be published in Nov. 1992.

[9] T. B. Sheridan and W. R. Ferrell. *Man-Machine Systems: Information, Control, and Decision Models of Human Performance.* The MIT Press, pp. 187-297 1974.

[10] D. T. McRuer. Human dynamics in man-machine systems. In *Automatica*, Vol. 16, pp. 237-253 1980.

[11] T. L. Brooks. Telerobot response requirements: A position paper on control response requirements for the FTS telerobot. Technical report, STX Corporation, No. STX/ROB/90-03, Mar. 1990.

[12] Sukhan Lee and Hahk Sung Lee. Modeling, design and evaluation for advanced teleoperator control systems. In *IEEE/RSJ International Conference on Intelligent Robots and Systems*, Vol. 2, pp. 881-888, Jul. 1992.

[13] O. J. M. Smith. *Feed-back Control System.* McGraw-Hill, 1958.

FA5 - 10:30

Proceedings of the 31st Conference
on Decision and Control
Tucson, Arizona • December 1992

Modeling and Feedback Control of Nonholonomic Mobile Vehicles

George J. Pappas and Kostas J. Kyriakopoulos
N.Y. State Center for Advanced Technology in Automation and Robotics
and
Electrical, Computer and Systems Engineering Department
Rensselaer Polytechnic Institute
Troy, NY 12180-3590

Abstract

A dynamic model for mobile vehicles moving on nonplanar surfaces under nonholonomic constraints is derived. Dynamic constraints are considered. The Invariant Manifold Technique is introduced for steering a nonholonomic mobile vehicle to an invariant manifold which is relatively locally-locally controllable. A closed loop control strategy is formulated and applied to the planar, kinematic case. Simulation results follow.

1 Introduction

In general, mobile vehicles are systems that satisfy nonholonomic constraints, ie. the variations of their generalized coordinates are constrained by nonintegrable equations. The growing use of wheeled mobility in numerous applications such as transportation vehicles, ferrying mobile robots and surveilance robots motivates the need for strategies for automatic parking, docking and path tracking and correction in high speeds in both structured and unstructured environments.

Dynamic models of wheeled mobile systems have been derived in [1], however their derivation is restricted to planar motion and does not include dynamic constraints. The control of nonholonomic systems has been considered in [1, 2, 3, 4, 5] and others, while [6, 7] have specified their control algorithms to nonholonomically constrained mobile robots.

In this paper, the general case of a mobile vehicle, moving on nonplanar terrains, under nonholonomic and dynamic constraints is considered. Conditions to avoid rolling without slipping and skidding are presented. Then, the **Invariant Manifold Technique** is presented. According to this control strategy the nonholonomic systems are driven to the final configuration by moving along intersecting manifolds. On those manifolds, the system has to be relatively locally-locally controllable. Presently, the control scheme has been derived only for kinematic control of mobile vehicles.

In section 2, the dynamic model of a nonholonomic mobile vehicle is derived while in section 3 the Invariant Manifold Technique is introduced and necessary conditions for its application are derived. Section 4 describes a closed loop formulation of the proposed control strategy that is then used to drive a planar mobile vehicle to a final state. Simulation results that include tracking of reference trajectories that are either holonomic or nonholonomic are presented.

2 Modeling

A mobile vehicle is a kinematic mechanism that is composed of a body and rolling wheels. The kinematics and dynamics of a mobile vehicle can be modeled based on the assumption that the wheels are ideally rolling. Conditions to achieve rolling without slipping and skidding have been presented in [8].

2.1 Kinematic Modeling

The development of Alexander and Maddocks [9] is based on the assumption of planar motion. In order for a vehicle to have ideal rolling, any two wheels i, j of the vehicle must satisfy the **rolling compatibility condition**:

$$\omega(x_i - x_j) = r(\dot{\vartheta}_i - \dot{\vartheta}_j) \quad (1)$$

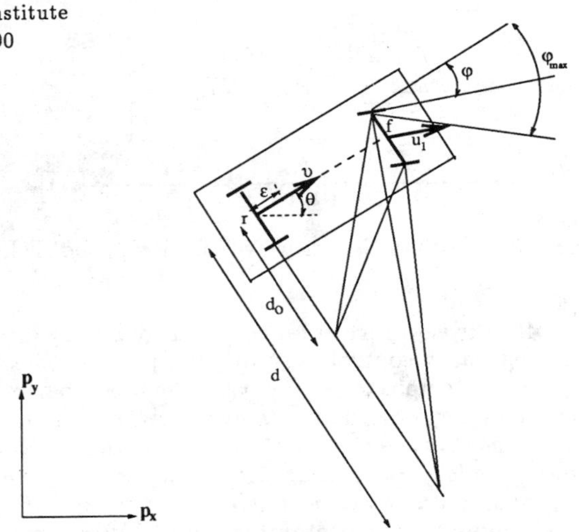

Figure 1: Planar wheel driven mobile vehicle

where x_k is the position of wheel k with respect to some world coordinate frame, $\dot{\vartheta}_k$ is its respective rotational velocity and ω is the rotational velocity of the vehicle, and r is the radius of the wheels.

In [2, 4, 10] the following set of equations has been used as model of the planar, wheel driven vehicle shown in Figure 1:

$$\dot{p_x} = \cos\theta\cos\phi \cdot u_1 \qquad \dot{\phi} = u_2 \quad (2)$$
$$\dot{p_y} = \sin\theta\cos\phi \cdot u_1 \quad \dot{\theta} = \tfrac{1}{l}\sin\phi \cdot u_1 \quad (3)$$

where p_x, p_y are the cartesian coordinates of point R of the rear wheel axis of the vehicle, θ is its orientation. Notice that ϕ is the steering angle of a "virtual" wheel (see Fig. 1) located at point F of the front wheel axis. This wheel is "created" by the requirement that in order to have ideal rolling without skidding all the wheels axis' should intersect at the same point. The control inputs are u_1, the translational velocity of point F and u_2 the steering velocity of the "virtual" wheel. If, for the time, we ignore the dynamics of the steering and set

$$v = u_1 \cdot \cos\phi \qquad \omega = \dot{\theta} = \frac{1}{l}\sin\phi \cdot u_1 \quad (4)$$

Then the following simplified (unicycle) model is obtained

$$\begin{aligned} \dot{p_x} &= v\cos\theta \\ \dot{p_y} &= v\sin\theta \end{aligned} \qquad \dot{\theta} = \omega$$

where p_x, p_y are the cartesian coordinates of point R of the rear wheel axis of the vehicle, θ is the orientation of the vehicle, v is the translational velocity of R and ω the angular velocity of the vehicle. Although the dynamics of steering are ignored at the kinematic level they are considered in the dynamic level as presented in the next section.

Equations (5) are a special case of the more general model

$$\begin{aligned} \dot{p_x} &= v\cos\theta - \varepsilon\omega\sin\theta \\ \dot{p_y} &= v\sin\theta + \varepsilon\omega\cos\theta \end{aligned} \qquad \dot{\theta} = \omega$$

which describes the motion of any point E on the vehicle that is ε distance on the y axis away from the rear wheel axis coordinate frame.

CH3229-2/92/0000-2680$1.00 © 1992 IEEE

2.2 Dynamic Modeling

The simplifying assumptions made to obtain the dynamic model of the vehicle are:

- No **slipping** : rolling compatibility conditions satisfied.
- No **skidding** [1]
- The rotational kinetic energy of the rotating wheels is not considered.
- **Frictionless motion** (w.r.t its kinematic mechanism).
- The dynamics of the suspension and traction system were not considered.

The Lagrangian, which is equal to the kinetic energy, of the moving vehicle is:

$$K = \frac{1}{2} \cdot I \cdot \omega^2 + \frac{1}{2} \cdot m \cdot v^2 \tag{5}$$

where m is its mass, I is the moment of inertia of the vehicle around the vertical axis passing through point R of the rear wheel axis. The rotational velocity of the vehicle is

$$\omega = vf \tag{6}$$

where f is the curvature of the trajectory of the system. Therefore the Lagrangian becomes

$$K = \frac{1}{2} \cdot I \cdot f^2 \cdot v^2 + \frac{1}{2} \cdot m \cdot v^2 \tag{7}$$

From Eq. (5) we have $v^2 = \dot{p_x}^2 + \dot{p_y}^2$ and the Lagrangian becomes

$$L = \frac{1}{2} \cdot (m + I \cdot f^2) \cdot (\dot{p_x}^2 + \dot{p_y}^2)$$

The Euler-Lagrange equations, assuming no nonconservative forces, such as friction, are:

$$\begin{aligned} \frac{d}{dt}\frac{\partial L}{\partial \dot{p_x}} - \frac{\partial L}{\partial p_x} &= u_x + R_x \\ \frac{d}{dt}\frac{\partial L}{\partial \dot{p_y}} - \frac{\partial L}{\partial p_y} &= u_y + R_y \end{aligned} \tag{8}$$

subject to the following nonholonomic constraint

$$\begin{bmatrix} \sin\theta & -\cos\theta \end{bmatrix} \begin{bmatrix} \dot{p_x} \\ \dot{p_y} \end{bmatrix} = A(\theta) \begin{bmatrix} \dot{p_x} \\ \dot{p_y} \end{bmatrix} = 0 \tag{9}$$

which is easily derived from Eq. (5) or (5). The generalized forces u_x, u_y are applied in the $\vec{p_x}, \vec{p_y}$ directions respectively (from Figure 1). R_x, R_y represent the reaction forces of the constraints in their respective directions and are equal to [12] :

$$R_x = \sum_{i=1}^{r} \lambda_i a_{i1} \qquad R_y = \sum_{i=1}^{r} \lambda_i a_{i2} \tag{10}$$

where r is the number of independent nonholonomic constraints, a_{ij} is the i,j element of the constraint matrix $A(\theta)$ and λ_i is the Lagrange multiplier associated with independent constraint i. In this case there is only one independent constraint and therefore only one Lagrange multiplier is associated. Eliminating the Lagrange multiplier results in the following equation:

$$\left[\frac{d}{dt}\frac{\partial L}{\partial \dot{p_x}} - \frac{\partial L}{\partial p_x}\right]\cos\theta + \left[\frac{d}{dt}\frac{\partial L}{\partial \dot{p_y}} - \frac{\partial L}{\partial p_y}\right]\sin\theta = u_2 \tag{11}$$

[1] An extended treatment of dynamic constraints can be found in [11]

where $u_2 = u_x \cos\theta + u_y \sin\theta$ and is simply the projection of generalized forces u_x, u_y on the direction of the translational velocity. From (8)

$$\begin{aligned} \frac{d}{dt}\frac{\partial L}{\partial \dot{p_x}} - \frac{\partial L}{\partial p_x} &= (m + If^2)\ddot{p_x} + (If\dot{f})\dot{p_x} \\ \frac{d}{dt}\frac{\partial L}{\partial \dot{p_y}} - \frac{\partial L}{\partial p_y} &= (m + If^2)\ddot{p_y} + (If\dot{f})\dot{p_y} \end{aligned} \tag{12}$$

Combining (11),(12) and algebraic manipulation result in

$$\dot{v} = \left[u_2 - If\dot{f}v\right]/(m + If^2)$$

A state space model describing the dynamic behavior of a vehicle, moving on a plane, can now be obtained. The configuration variables are: p_x, p_y, the cartesian coordinates of point R of the vehicle with respect to the world reference frame, θ, the orientation of the vehicle,f the curvature of the trajectory and v, the translational velocity of vehicle. The control inputs to the system are $u_1 = \dot{f}$ and u_2. If Eq. (5) are used , the state equations are

$$\begin{aligned} \dot{p_x}(t) &= v(t)\cos\theta(t) \\ \dot{p_y}(t) &= v(t)\sin\theta(t) \\ \dot{\theta}(t) &= f(t)v(t) \\ \dot{f}(t) &= u_1(t) \\ \dot{v}(t) &= [u_2(t) - If(t)v(t)u_1(t)]/(m + If^2) \end{aligned} \tag{13}$$

or in the more general case (using Eq. (5)),

$$\begin{aligned} \dot{x}(t) &= v(t)\cos\theta(t) - \varepsilon f(t)v(t)\sin\theta \\ \dot{y}(t) &= v(t)\sin\theta(t) + \varepsilon f(t)v(t)\cos\theta \\ \dot{\theta}(t) &= f(t)v(t) \\ \dot{f}(t) &= u_1(t) \\ \dot{v}(t) &= [u_2(t) - If(t)v(t)u_1(t)]/(m + If^2) \end{aligned} \tag{14}$$

Notice that the first three equations describe the kinematic behavior of the vehicle (see model (5)) while the last two describe a simplified version of the dynamic behavior of the vehicle. From now on the first three equations will be referered to as the **kinematic subsystem** while the last two as the **dynamic subsystem**.

3 Invariant Manifold Technique

3.1 The Basic Idea

The model given by equations (5) is typical of nonholonomic systems with no drift term

$$\dot{x} = F(x)u \tag{15}$$

where $x \in \Re^n$ is the state vector and $u \in \Re^m$ is the control input. A fundamental research issue is the stabilization of those systems. The basic idea for closed loop control of the specific nonholonomic system (5) is presented in a way that suggests its possible application to a wider class of nonholonomic systems.

Nonholonomic systems, although globally controllable, do not possess the property of local-local controllability. The main idea of this present work is the decomposition of the state space into intersecting manifolds on which the system is relatively locally-locally controllable. Steering to a final desired state x_f is achieved by moving on those manifolds (**Invariant Manifolds**) through well known feedback control techniques. The concept of local-local controllability that is used is defined in [2]. Relative local-local controllability needs to be defined :

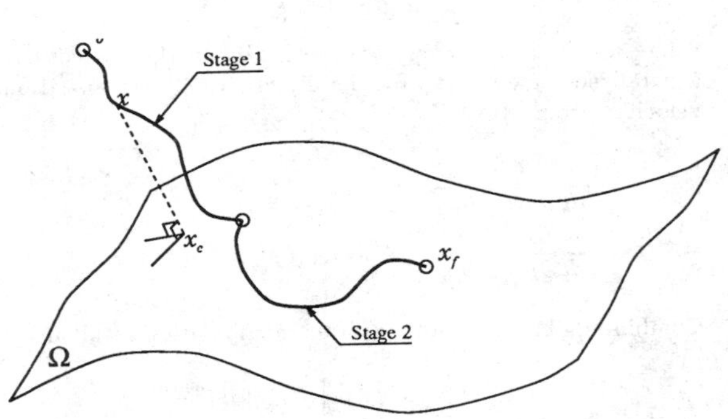

Figure 2: Stages 1 and 2 of the motion.

Definition 1 *Let Ω be a manifold defined by*

$$\Omega \triangleq \{x \in \Re^n | \Phi(x) = 0,\ \Phi(\cdot) \in \Re^k\} \tag{16}$$

where the components $\Phi^i(x)$ $i = 1, ..., k$ of vector $\Phi(x)$ are assumed to be mutually independent. System (15) is said to locally-locally controllable relative to Ω at x if and only if $\forall \varepsilon > 0$ $\exists \delta > 0$ such that all points in the δ-neighborhood of x relative to Ω can be linked to x by a trajectory of (15) that does not leave the ε-neighborhood.

Consider a manifold Ω relative to which system (15) is locally-locally controllable (Figure 2). Assume that the final desired state x_f is on Ω (i.e $x_f \in \Omega$). The Invariant Manifold Technique consists of the following two stages

Stage 1 : Steering of the nonholonomic system from an initial state x_0 onto Ω.

Stage 2 : Steering of the system on Ω to the desired final state x_f.

Motion during the individual stages, 1 and 2, can be controlled with smooth feedback. The switching from Stage 1 to 2 is in accordance with Brockett's idea of nonexistence of smooth feedback law [13]. The requirement for local-local controllability on Ω is essential for the existence of smooth feedback law to lead the system to the final configuration.

If the vector set of equations defining Ω is of dimension k, ($\Phi(\cdot) \in \Re^k$) then Ω is of dimension $(n-k)$, or $\dim(\Omega) = n - k$. In Stage 2,

$$\Phi(x(t)) = 0\ \forall t \Rightarrow \dot{\Phi}(x) = (\nabla\Phi)^T F(x) \cdot u = 0 \tag{17}$$

Since $(\nabla\Phi)^T F(x) \in \Re^{k\times m}$, existence of at least one solution for u requires that $k \leq m$. A necessary condition for relative local-local controllability on Ω (assuming that $\Phi^i(x)$ $i = 1, ..., k$ are mutually independent) is

$$\dim(\Omega) = n - k \leq m. \tag{18}$$

Equations (17,18) suggest that a necessary condition for applying the invariant manifold technique is that

$$m \geq \frac{n}{2}. \tag{19}$$

This condition is independent of k, the dimension of vector $\Phi(\cdot)$ defining the manifold.

While in Stage 1, the system is moving in space $\Re^n - \Omega$. Since $\dim(\Re^n - \Omega) = n - (n-k) = k$ then a necessary condition to move from x_0 onto Ω using smooth feedback control $u \in \Re^m$ is $k \leq m$. In the case that this condition is not satisfied then a number r of Invariant manifolds $\Omega_i = \{x \in \Re^n | \Phi_i(x) = 0,\ \Phi_i(\cdot) \in \Re^{k_i}\}$ $i = 1, ..., r$, each with $\dim(\Omega_i) = n - k_i$, will be needed. Thus, the more general necessary condition

$$n - \sum_{i=1}^{r}(n - k_i) \leq m \tag{20}$$

is obtained. Notice that (20) includes (3.1). Further investigation of the domain of solution of the Invariant Manifold Technique is necessary, and sufficient conditions must be derived.

3.2 Feedback Control for Stage 1

In order to move the system onto Ω, a Lyapunov function candidate, (representing the notion of distance) $V(x)$ from $x \in \Re^n$ to Ω is defined

$$V(x) \triangleq \min_{x_m \in \Omega} \|x - x_m\|_P^2 \tag{21}$$

where P is some symmetric, positive definite matrix. If

$$x_c \triangleq \arg \min_{x_m \in \Omega} \|x - x_m\|_P^2 \tag{22}$$

(see fig. 2) then $V(x) = \|x - x_c\|_P^2$ Notice that from its definition $x_c = x_c(x)$. Define

$$R(x) \triangleq x - x_c \quad \in \Re^n \tag{23}$$

From the definition of Ω, equation (23) and assuming that $\Phi(x)$ is of full rank, we obtain

$$R(x) = x - x_c = \sum_{i=1}^{k} \alpha_i \nabla\Phi(x_c) \tag{24}$$

The following system of $n + k$ equations

$$\begin{aligned} x - x_c &= \textstyle\sum_{i=1}^{k} \alpha_i \nabla\Phi(x_c) \\ \Phi(x_c) &= 0 \end{aligned} \tag{25}$$

can be solved and the $n + k$ unknowns $x_c, \alpha_1, .., \alpha_k$ can be obtained.

The time derivative of $V(x)$ is

$$\begin{aligned} \dot{V}(x) &= 2(x - x_c)^T \cdot P \cdot (\dot{x} - \dot{x}_c) = 2R^T(x) \cdot P \cdot \dot{R}(x) \\ &= 2R^T(x) \cdot P \cdot \nabla R(x) \cdot F(x) u \end{aligned} \tag{26}$$

If the control law $u = -K \cdot R(x)$ (where $K \in \Re^{m\times n}$) is applied then $\dot{V}(x)$ becomes

$$\dot{V}(x) = -2R^T(x) \cdot P \cdot \nabla R(x) \cdot F(x) \cdot K \cdot R(x) \tag{27}$$

Thus, a sufficient condition for global asymptotic convergence from any $x \in \Re^n$ to Ω is

$$P \cdot \nabla R(x) \cdot F(x) \cdot K > 0 \tag{28}$$

for appropriately selected matrix K and $\forall x \in \{\Re^n - \Omega\}$.

Upon completion of Stage 1, system (15) is arbitrarily close to Ω. The Lipschitzian continuity of the system during Stage 2, as a result of the employed smooth feedback control, guarantees that it can be driven close to the final state x_f.

3.3 Application of the Invariant Manifold Technique

Reconsider model (5) that describes the motion of a vehicle. Let $x = [p_x\ p_y\ \theta]^T$ be the state vector. The desired final state is the origin ($x_f = [0\ 0\ 0\ ]^T$). Since the system is of third order and two controllers are available, necessary condition (19) is satisfied. The Invariant Manifold, Ω, is constructed by setting $\omega = 0$ and integrating equations (5) by making sure that $x_f \in \Omega$. The result is

$$\Omega \triangleq \{x \in \Re^3 | \phi_1(x) = 0\ \phi_2(x) = 0\} \tag{29}$$

where

$$\phi_1(x) = \theta = 0 \tag{30}$$

$$\phi_2(x) = p_y = 0 \tag{31}$$

$$\tag{32}$$

Equations (25) give $x_c = [p_{xc}\ p_{yc}\ \theta_c]^T$, α_1 and α_2 in closed form

$$p_{xc} = p_x \quad (33)$$

$$p_{yc} = 0 \quad (34)$$

$$\theta_c = 0 \quad (35)$$

$$\alpha_1 = \theta \quad (36)$$

$$\alpha_2 = -p_y \quad (37)$$

and $R(x)$ is now given by

$$R(x) = x - x_c = [0\ p_y\ \theta]^T \quad (38)$$

The distance of any state x from Ω is

$$V(x) = R^T(x) \cdot R(x) \quad (39)$$

The time derivative of distance $V(x)$ is

$$\dot{V}(x) = R^T(x)\nabla R(x)F(x)u = R^T(x)\begin{bmatrix} 0 & 0 \\ \sin\theta & \varepsilon\cos\theta \\ 0 & 1 \end{bmatrix} u \quad (40)$$

Applying $u = -K \cdot R(x)$ results in

$$\dot{V}(x) = -2R^T(x)\begin{bmatrix} 0 & 0 \\ \sin\theta & \varepsilon\cos\theta \\ 0 & 1 \end{bmatrix}\begin{bmatrix} k_{11} & k_{12} & k_{13} \\ k_{21} & k_{22} & k_{23} \end{bmatrix} R(x) \quad (41)$$

or $\dot{V}(x) = -2 \cdot R^T(x) \cdot D(x) \cdot R(x)$ where $D(x)$ is

$$D(x) = \begin{bmatrix} 0 & 0 & 0 \\ \varepsilon k_{11}\cos\theta + k_{21}\sin\theta & \varepsilon k_{22}\cos\theta + k_{12}\sin\theta & \varepsilon k_{23}\cos\theta + k_{13}\sin\theta \\ k_{21} & k_{22} & k_{23} \end{bmatrix} \quad (42)$$

Since $R(x) = [0\ p_y\ \theta]^T$, convergence can be tested by simply considering the definiteness of the following submatrix, $Q(x)$,

$$Q(x) = \begin{bmatrix} \varepsilon k_{22}\cos\theta + k_{12}\sin\theta & \varepsilon k_{23}\cos\theta + k_{13}\sin\theta \\ k_{22} & k_{23} \end{bmatrix} \quad (43)$$

the symmetric part, $Q_s(x)$ of which is

$$Q_s(x) = \begin{bmatrix} \varepsilon k_{22}\cos\theta + k_{12}\sin\theta & \frac{1}{2}(\varepsilon k_{23}\cos\theta + k_{13}\sin\theta + k_{22}) \\ \frac{1}{2}(\varepsilon k_{23}\cos\theta + k_{13}\sin\theta + k_{22}) & k_{23} \end{bmatrix} \quad (44)$$

By considering the principal minors we see that sufficient conditions for $Q_s(x) \geq 0$ are

$$\varepsilon k_{22}\cos\theta + k_{12}\sin\theta \geq 0 \quad (45)$$

$$-\frac{k_{22}^2}{4} + \frac{\varepsilon k_{22}k_{23}\cos\theta}{2} - \frac{\varepsilon^2 k_{23}^2\cos^2\theta}{4} - \frac{k_{13}k_{22}\sin\theta}{2} + k_{12}k_{23}\sin\theta - \frac{k_{13}^2\sin^2\theta}{4} - \frac{\varepsilon k_{13}k_{23}\sin 2\theta}{4} \geq 0 \quad (46)$$

To satisfy (45) we choose

$$k_{22} = l_{22}\cos\theta \quad (47)$$

$$k_{12} = l_{12}\sin\theta \quad (48)$$

where l_{12}, l_{22} are positive constants. It is fairly easy to show that if

$$k_{13} = 0 \quad (49)$$

$$k_{23} = \frac{l_{22}}{\varepsilon} \quad (50)$$

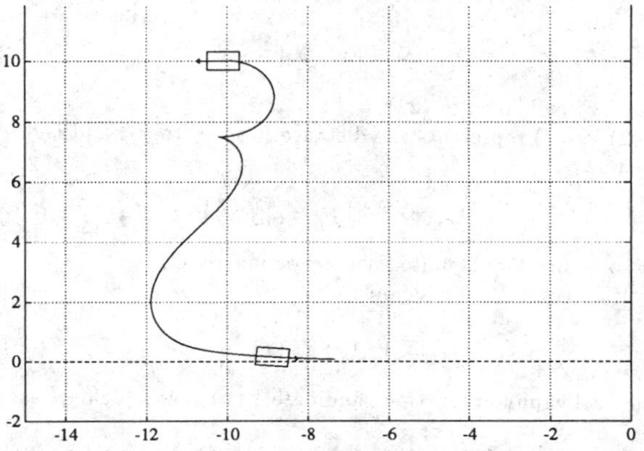

Figure 3: Stage 1: Convergence to the Invariant Manifold

then 46 is satisfied and becomes equality when $\sin\theta = 0$. Therefore $Q_s(x)$ (and Q) is a positive semidefinite matrix and the system under such a control will be driven to the largest invariant manifold S where $\dot{V} = 0$. But

$$\sin\theta = 0 \Rightarrow \theta = 0 \quad or \quad \theta = \pm\pi \quad (51)$$

Therefore $\omega \to 0$ and

$$\underline{\text{if } \theta \to 0:}\ \omega \to 0 \ \Rightarrow\ -k_{22}p_y - k_{23}\theta \to 0 \Rightarrow\ \Rightarrow\ k_{22}p_y \to 0 \Rightarrow p_y \to 0 \quad (52)$$

$$\underline{\text{if } \theta \to \pm\pi:}\ \omega \to 0 \ \Rightarrow\ -k_{22}p_y - k_{23}\theta \to 0 \overset{(47,50)}{\Rightarrow}\ \Rightarrow\ l_{22}p_y - \frac{l_{22}}{\varepsilon}(\pm\pi) \to 0 \Rightarrow p_y \to \pm\frac{\pi}{\varepsilon} \quad (53)$$

If $p_y(0) = y_0 + \varepsilon \cdot \cos\theta_0$ is the position of the point initially where y_0, θ_0 is the initial configuration of the robot then

$$V_i = \frac{1}{2}y_0^2 + \frac{1}{2}\varepsilon^2\sin^2\theta_0 + \varepsilon y_0\sin\theta_0 + \frac{1}{2}\theta_0^2 \quad (54)$$

is the initial ($t = 0$) Lyapunov function value. The value of the lyapunov function at those undesirable configurations is

$$V_e = \frac{(\pm\pi)^2}{\varepsilon^2} + (\pm\pi)^2 = \frac{\pi^2}{\varepsilon^2} + \pi^2 \quad (55)$$

From the above equations becomes obvious the fact that it is easy to choose ε so that $V_e > V_i$.

Since $R(x) = [0\ p_y\ \theta]^T$, constants k_{11} and k_{21} do not contribute in the control law and can be conveniently set to zero. The resulting control law is

$$\begin{aligned} v &= -l_{12} \cdot \sin\theta \cdot p_y \\ \omega &= -l_{22} \cdot \cos\theta \cdot p_y - \tfrac{l_{22}}{\varepsilon} \cdot \theta \end{aligned} \quad (56)$$

The convergence to the invariant manifold is shown in figure 3.

4 Closed Loop Control Strategy

During Stage 1 of the Invariant Manifold Technique, system (15) is driven to Ω. Although Stage 1 guarantees global asymptotic stability to the Ω, the requirement to decrease the distance from x_f is not guaranteed. In order to achieve that, two metrics are defined:

1) $V_{xc}(x)$ represents the distance from a point $x \in \Re^n$ to Ω and was defined in section 3 as

$$V_{xc}(x) \triangleq \min_{x_m \in \Omega} \|x - x_m\|_P^2 \quad (57)$$

or

$$V_{xc}(x) = \|x - x_c\|_P^2 \tag{58}$$

where x_c is defined in (22).

2) $V_{cf}(x)$ represents the distance from x_c to x_f, and is defined as

$$V_{cf}(x) \triangleq \|x_f - x_c\|_Q^2 \tag{59}$$

for some positive definite, symmetric matrix Q.

If a vector $\mathcal{D}(x)$ is defined as

$$\mathcal{D}(x) \triangleq [V_{xc}(x)\ V_{cf}(x)]^T \tag{60}$$

Then a Lyapunov function candidate $V(x)$ can be defined as

$$V(x) = \|\mathcal{D}(x)\|_\infty^w \triangleq \max\{w_1 V_{xc}(x), w_2 V_{cf}(x)\} \tag{61}$$

where $w = [w_1\ w_2]^T$ appropriate weighting constant vector. Two cases are obviously possible.

Case 1 : $w_1 \cdot V_{xc} > w_2 \cdot V_{cf}$:

In the previous section it was shown that if a control input of the form (56) is applied then $\dot{V}_{xc} \leq 0$

Case 2 : $w_1 \cdot V_{xc} < w_2 \cdot V_{cf}$:

Since $x_f = 0$ $V_{cf}(x) = x_c^T \cdot Q \cdot x_c$ where x_c is defined in equation (37) and Q is the following symmetric matrix

$$Q = \begin{bmatrix} f_3 & 0 & 0 \\ 0 & 1 & 0 \\ 0 & 0 & 1 \end{bmatrix} \tag{62}$$

and $f_3 > 0$ is strictly greater than zero in order to guarantee positive definiteness of Q. Thus

$$\dot{V}_{cf}(x) = 2x_c^T \cdot Q \cdot \dot{x}_c \tag{63}$$

or after simple calculations

$$\dot{V}_{cf}(x) = f_3 p_x(v\cos\theta - \omega\varepsilon\sin\theta) \tag{64}$$

If a control law

$$\begin{aligned} v &= -K_v \cdot p_x \cdot sign(\cos\theta) \\ \omega &= -K_\omega \cdot p_x \cdot sign(\sin\theta) \end{aligned} \tag{65}$$

is employed (K_v, K_ω), then

$$\dot{V}_{cf}(x) = -f_3 \cdot p_x^2 \cdot (K_v|\cos\theta| + K_\omega\varepsilon|\sin\theta|) \leq 0 \tag{66}$$

LaSalle's invariance principle guarantees the fact that the system is globally asymptotically convergent to the manifold $\Omega_x = \{x \in \Re^3 \ni \dot{V}_{cf}(x) = 0\} = \{x \in \Re^3 \ni p_x = 0\}$.

Thus, the switching strategy is

$$u(x) = \begin{cases} \text{given by eq. (56) if} & V(x) = V_{xc}(x) \\ \text{given by eq. (65) if} & V(x) = V_{cf}(x) \end{cases} \tag{67}$$

Obviously, $x(t) \to \Omega \cap \Omega_x = x_f$.

4.1 Implementation Issues

The proposed switching strategy tries to lead the system to the invariant manifold Ω but at the same time to move it as close as possible to the final state x_f. The presented application is to demonstrate the principle. The choice of Ω is fundamental in determining the "shape" of the converging trajectory to x_f.

Figure 4 depicts the fact that the particular choice of Ω is satisfactory for the case of leading a vehicle to the final state. Furthermore, if the proposed closed loop strategy is used for tracking a completely nonholonomic reference trajectory the response is plotted in Figure 5 where the desired and actual

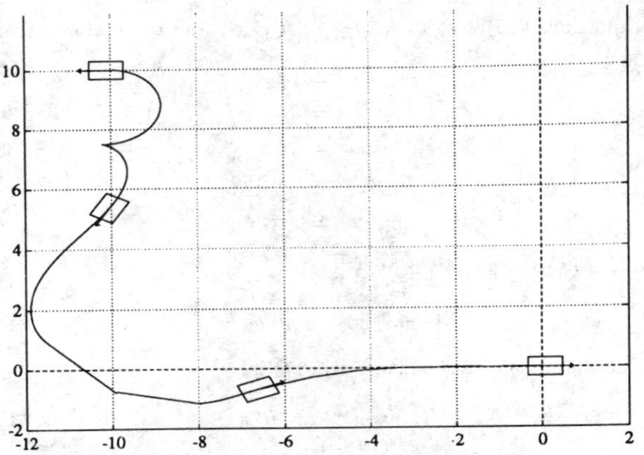

Figure 4: Convergence to $x_f = [0\ 0\ 0]^T$.

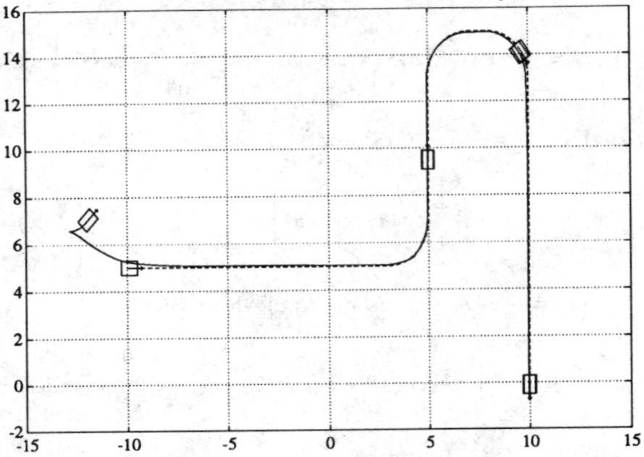

Figure 5: Tracking of completely nonholonomic trajectory

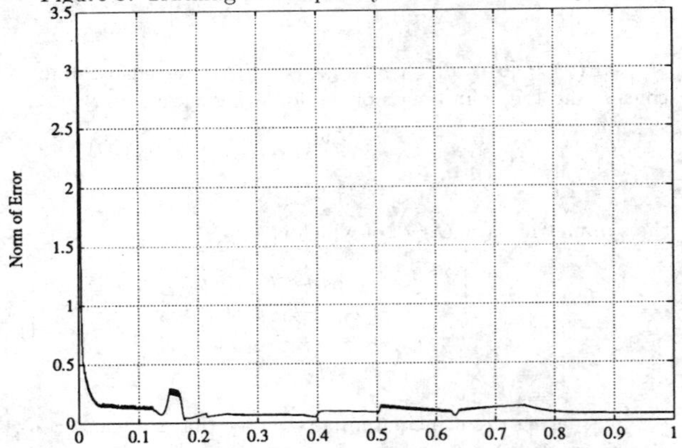

Figure 6: Error $\|(p_x - x_r)^2 + (p_y - y_r)^2 + (\theta - \theta_r)^2\|_2$ plot.

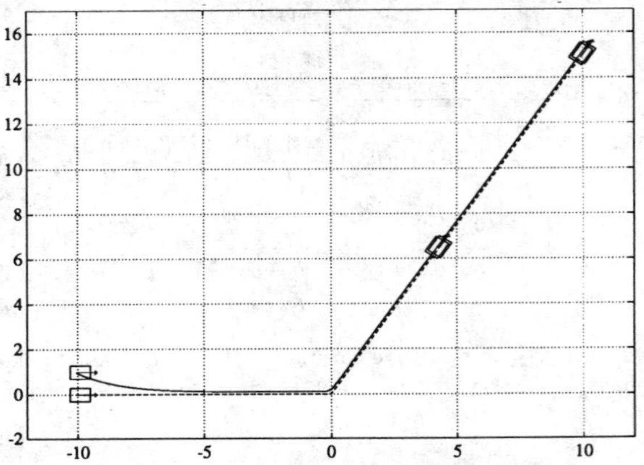

Figure 7: Tracking of holonomic trajectory

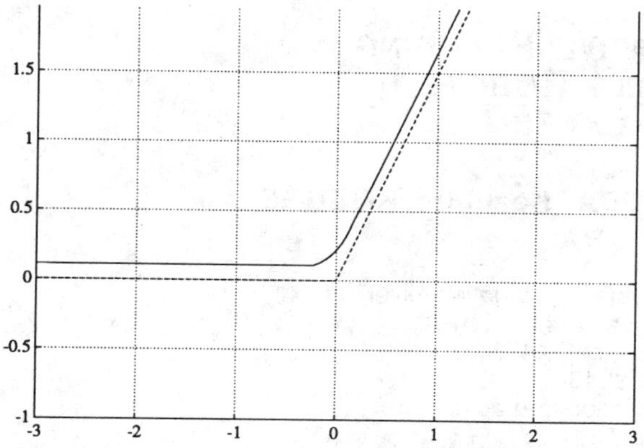

Figure 8: Magnification of the previous figure around 0

trajectories are shown, while a Euclidean norm error profile along the trajectory is shown in Figure 6.

One of the significant advantages of this control strategy is that it could be used even in the case when the reference trajectory is holonomic. This can greatly simplify the planning stage of a given motion task but on the other hand may require a more appropriate choice of Ω. Preliminary simulation results are shown in Figure 7 where system (5) is asked to violate its nonholonomic constraint during a corner turn. Notice in figure 8 the smoothness of the turn and the error because of the fact that the controller is "P" type.

5 Conclusion - Issues for further research

A complete dynamic model of mobile vehicles moving on non-planar surfaces under nonholonomic and dynamic constraints has been presented. Kinematic control of such systems was examined and the Invariant Manifold Technique was introduced. A closed loop control strategy that is based on the Invariant Manifold Technique was formulated and applied to the control of nonholonomic vehicles. Tracking control of such vehicles was demonstrated with both holonomic and nonholonomic reference trajectories.

The proposed control strategy for nonholonomic vehicles must be extended to the dynamic case. In order to do so, investigation of continuity properties of the translational and rotational velocities resulting from the control strategy (67) is needed as well as incorporation of control and state constraints. The obtained controllers from the chosen model representation and Lyapunov functions are of the "P" form thus give slagish behavior and steady state errors. "PD" or "PID" type of controllers have to be obtained. The sensing problem of detection of violation of nonholonomic constraints (slipping, skidding) must also be considered.

Sufficient conditions for the application of the Invariant Manifold Technique must be derived and the possibility of relaxing constraint (19) should be investigated. The Invariant Manifold Technique can be extended and applied to other classes of nonholonomic systems such as vehicles with trailers, systems that are clearly of high complexity and more difficult to control. Other issues of relevant interest are the use of optimization and obstacle avoidance techniques.

Acknowledgement

This work has been conducted in the New York State Center for Advanced Technology (CAT) in Automation and Robotics at Rensselaer Polytechnic Institute. The CAT is funded by a grant from the New York State Science and Technology Foundation.

References

[1] B. d'Andrea-Novel G. Campion and G. Bastin. Modelling and state feedback control of nonholonomic mechanical systems. In *Proceedings of the 1991 IEEE Conference on Decision and Control*, December 1991.

[2] L. Gurvits C. Fernandes and Z.X. Li. Foundations of nonholonomic motion planning.

[3] M.Reyhanoglu A. M. Bloch and N. H. McClamroch. Control and stabilization of nonholonomic caplygin dynamic systems. In *Proceedings of the 1991 IEEE Conference on Decision and Control*, December 1991.

[4] R. M. Murray and S. S. Sastry. Steering nonholonomic systems using sinusoids. In *Proceedings of the 1990 IEEE Conference on Decision and Control*, December 1990.

[5] G. Lafferriere and H. Sussmann. Motion planning for controllable systems without drift. In *Proceedings of the 1991 IEEE International Conference on Robotics and Automation*, April 1991.

[6] C. Canudas de Wit and O. J. Sordalen. Exponential stabilization of mobile robots with nonholonomic constraints. In *Proceedings of the 1991 IEEE Conference on Decision and Control*, December 1991.

[7] C. Samson and K. Ait-Abderrahim. Feedback control of a nonholonomic wheeled cart in cartesian space. In *Proceedings of the 1991 IEEE International Conference on Robotics and Automation*, April 1991.

[8] K. J. Kyriakopoulos and G. N. Saridis. An integrated collision prediction and avoidance scheme for mobile robots in nonstationary environments. *To appear in Automatika.*

[9] J.C. Alexander and J.H. Maddocks. On the kinematics of wheeled mobile robots. *The International Journal of Robotics Research*, October 1989.

[10] A. W. Divelbiss and J. Wen. A new optimization method with perturbation refinement for nonholonomic motion planning. In *Proceedings of the 1992 American Control Conference*, June 1992.

[11] George J. Pappas and K. J. Kyriakopoulos. Modeling and feedback control of nonholonomic mobile vehicles. In *Proceedings of the 1992 Conference in Decision and Control*, December 1992.

[12] Ju. I. Neimark and F.A. Fufaev. *Dynamics of Nonholonomic Systems*, volume 33 of *Translations of Mathematical Monographs*. A.M.S, 1972.

[13] R. Brockett. *Control Theory and Singular Riemannian Geometry*. New Directions in Applied Mathematics. Springer, 1981.

[14] K. J. Kyriakopoulos and G. N. Saridis. Optimal motion planning for collision avoidance of mobile robots in nonstationary environments. In *Proceedings of the 1992 American Control Conference*, June 1992.

Proceedings of the 31st Conference
on Decision and Control
Tucson, Arizona • December 1992

SELFISH AND COORDINATIVE PLANNING FOR MULTIPLE MOBILE ROBOTS BY GENETIC ALGORITHM

Takanori SHIBATA, Toshio FUKUDA, Kazuhiro KOSUGE, and Fumihito ARAI

Dept. of Mechano-Informatics and Systems, Nagoya University
1 Furo-cho, Chikusa-ku, Nagoya, 464-01, JAPAN
Phone +81-52-781-5111 ext. 4478
Fax +81-52-781-9243
e-mail: d43131a@nucc.cc.nagoya-u.ac.jp

Key Words: Multiple Robots, Genetic Algorithm, Planning, Coordination

ABSTRACT This paper presents a new strategy for coordination of multiple autonomous robots by using Genetic Algorithms (GAs). When a mobile robot moves from a point to a goal point, it is necessary to plan an optimal or feasible path for itself avoiding obstructions in its way and minimizing a cost such as time, energy, and distance. This planning is referred to as the selfish-planning. When many robots move in a same space, it is necessary to select a most reasonable path so as to avoid collisions with other robots and to minimize the cost. This planning is referred to as the coordinative-planning. The GAs are search algorithms based on the mechanics of natural selection and natural genetics. The GAs are applied to both the selfish-planning and the coordinative-planning of multiple mobile robots.

1. INTRODUCTION

Autonomous robots which work without human operators are required in robotic fields. In order to achieve tasks, autonomous robots have to be intelligent and should decide their own action. When the autonomous robot decides its action, it is necessary to plan optimally depending on their tasks and coordinating among other robots. In the case of a mobile robot, it is necessary to plan a collision-free path minimizing a cost such as time and distance. In the case of multiple mobile robots, it is necessary for each robot to determine a most reasonable path coordinating with other robots. This paper presents a strategy for coordination in path-planning of multiple mobile robots.

When an autonomous robot moves from a point to a target point in its given environment, it is necessary to plan an optimal or feasible path avoiding obstructions in its way and minimizing a cost such as time or distance. Therefore, the major task for path planning of one mobile robot is to search a collision-free path. Many works on this topic have been carried out for the path planning of one robot among stationary obstacles [1-8]. Habib and Asama proposed MAKLINK graphs which is based on a free-link's concept to construct available free space within robot's environment [8] in terms of free convex area. The graph is less complex for searching a collision-free path than other methods, because the numbers of nodes and links in the graph are less than that in other graphs [8]. However, as number of nodes increases in the graph, the search space becomes wide and it is more difficult to search an optimal path using graph search techniques. These approaches are not so effective in planning with respect to computing time.

Moreover, when many robots move in a same space, it is necessary for each robot to select a most reasonable path so as to avoid collisions with other robots and to minimize the total cost of all robots. If each robot selects its optimal path selfishly, collisions or dead-lock states may occur. Some researches use a rule that a robot waits for other robots as coordination to avoid collisions or dead-locks [9-11]. However, if many robots pass a same road or point frequently, the waiting time becomes large amount. Therefore, when multiple mobile robots move in a same area, selection of an optimal path for each robot is not always effective. In order to solve the problem, this paper proposes a new approach for path planning of multiple mobile robots using Genetic Algorithms (GA) [13].

GAs are search algorithms based on the mechanics of natural selection and natural genetics. In this paper, the GAs are applied to the path-planning of each robot at first, in which order-based and coded strings are used, and in which a scalar objective function is used to compute *fitness* of the solution represented by the string [14]. Each robot explores some kinds of paths *selfishly* which are optimal and feasible for itself. This planning is referred to as the ***selfish-planning***. Moreover, each robot selects the best path of all which it explored before. This path is selected by *coordinating* among robots so as to avoid collisions with other robots and dead-lock using a rule, and to minimize the total cost. This planning is referred to as the ***coordinative-planning***. For this selection, the GA is also used. Simulation results show effectiveness of the proposed approaches.

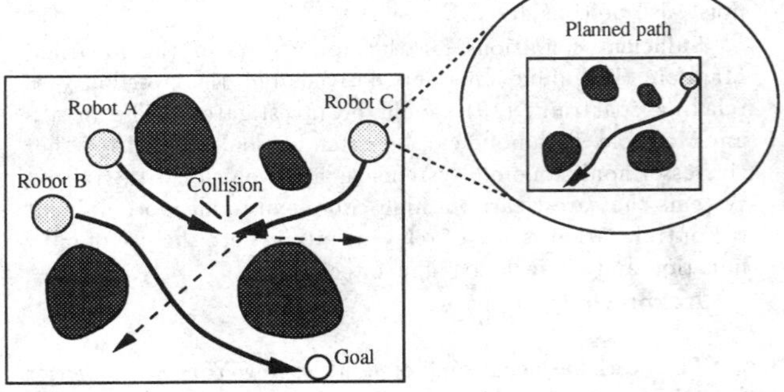

Fig. 1 Collision between robots according to *selfish-planning*

2. CONCEPT OF SELFISHNESS AND COORDINATION IN MULTIPLE ROBOTS

When an autonomous robot works at its tasks, it is necessary to have knowledge and to decide its action by itself. For intelligent control of robots, a hierarchical intelligent control has been proposed. At some tasks, a single autonomous robot can not achieve the task by itself because of its functional limitation. In this case, it is necessary to work cooperatively together with other robots. When many robots work together at some tasks, configuration of the tasks can be classified into two types. One is a centralized task at which one master robot exists and other robots obey the master. The another is a decentralized task at which each robot work cooperatively. In the centralized task, the master robot is required to have many functions and if it breaks down, the task cannot be achieved. On the other hand, in the decentralized task, each robot has each function and since plural robots work cooperatively for the one task, it is easy to deal with various tasks. Moreover, if one robot breaks down, other robots can help or replace the broken robot. Therefore, the decentralized task is superior to the centralized task. However, it is difficult to achieve the decentralized task, because interaction among robots influences sometimes unfavorably when public resources are limited. If there is no interaction, each robot had better work optimally for its purpose, so that the total task should be achieved optimally. That is, each robot should work selfishly. Unless, coordination among the robots is important as well as selfishness. In this paper, both selfishness and coordination are dealt with for an optimal plan at the decentralized task. For illustration, this paper deals with path-planning for multiple mobile robots.

When there is a single mobile robot in a space, it may plan a path only for itself. However, if multiple mobile robots plan paths for themselves respectively when they move in a same space, they may collide with each other. Therefore, it is necessary for them to coordinate in planning. Most previous research discussed in the case of one robot for a collision-free path against stationary obstacles. That is, they dealt with *selfish-planning*. Certainly, there are some research about multiple robots. However, they were very complex and time-consuming planning by using some rules. This thesis proposes new approaches for both *selfish-planning* and *coordinative-planning* using GAs, respectively. Each robot makes its own plans selfishly at first. Then, every robot selects most reasonable ways as the result of coordination by hierarchically taking account of the results of each robot's selfish-planning.

3. ENVIRONMENT REPRESENTATION FOR ROBOTS

The major task for path-planning for single mobile robot is to search a collision-free path. Habib and Asama proposed the MAKLINK graphs which is based on a free-link concept to construct available free space within robot's environment [8] in terms of free convex area. The graph is less complex for searching a collision-free path than other methods, because the numbers of nodes and links in the graph are less than that in other graph. Therefore, the graph is used to model the known environment of the robot.

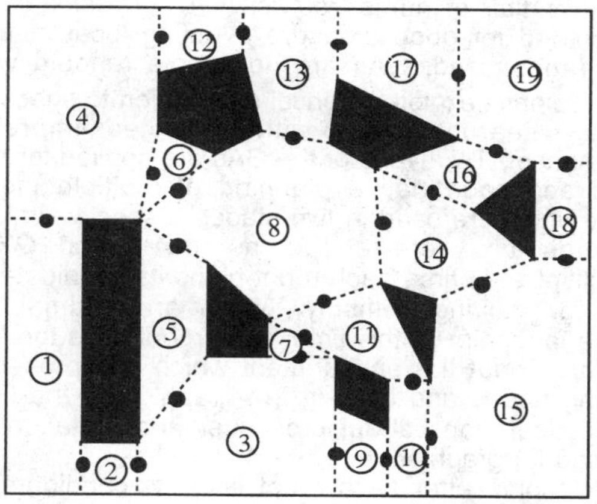

Fig. 2 Environment with obstacles and free-convex areas

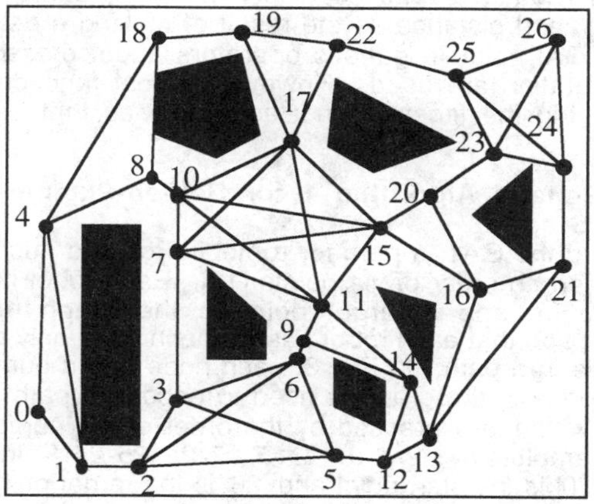

Fig. 3 A graph for representation of the robot's environment

A simple model of robot's environment is used, in which all the obstacles in the environment are represented by a set of vertices to describe each polygonal obstacle. The constructed model represents the knowledge of the robot about its environment in terms of an external map. This external map is then given to each robot in the same space. An internal representation of the world is built through the map manager as an internal map which is suitable to be used for free space structuring and preparing the required information for path-planning. Figure 2 shows an example of robots' environment with some stationary obstacles, in which free-convex areas with free-links are decided and each free-convex area is numbered. Figure 3 shows the graph in which each node is numbered and length of each link is used to calculate a fitness in GAs.

4. GENETIC ALGORITHMS FOR SELFISH-PLANNING AND COORDINATIVE-PLANNING

GAs are search algorithms based on the mechanics of natural selection and natural genetics [13]. They combine survival of the fittest among string structures with a structured yet randomized information exchange to form a search algorithm with some of the innovative flair of human search. An occasional new part is tried for good measure avoiding local minima. While randomized, GAs are no simple random walk. They efficiently exploit historical information to speculate on new search points with expected improved performance. Two types of the GAs are applied for both selfish and coordinative planning of multiple mobile robots, which are used in two stages respectively. The plannings are referred to as ***GA1*** and ***GA2***, respectively. At first, each robot plans its candidates of paths for moving selfishly, which are optimal and feasible for itself. At this time, each robot uses the GA1 with data about the environment which is represented by using nodes and links in the graph. For the GA1, genetic operations about crossover and mutation are proposed for graph search.

Second, after each robot finds its candidates of the collision-free path with obstacles, the multiple robots coordinate selecting each path by using the GA2 so as to avoid collisions among themselves and dead-lock, and to reduce a total cost such as time, energy for moving, and distance as the result of making a detour. In the GA2, simple genetic operators about crossover and mutation are used. However, a cost function for calculating the fitness is not as simple as that in the GA1.

4-1. Genetic Algorithm 1 for Selfish-Planning in First Stage

In the GA1, a path for a mobile robot is encoded based on an order of via points [13]. Each robot has a start point and a target point in the graph under assumption that each robot passes each point only once or none in a path. In Fig. 3, each node has a number. The nodes in the graph is used to encode a path as a string which is expressed by the order of the numbers, for example, *'0-1-2-6-9-11-15-17-22-25-26.'*. In this string, '0' is the start point and '26' is the target point for a mobile robot. These points are selected randomly at first, while adjacent numbers must be connected with a link in the graph. Since the order-based strings are used [14], specialized operations of crossover and mutation are proposed.

In the crossover, a string which has a good fitness is randomly selected as a parent, and a number in the string is also randomly selected at first. Then, one more string which also has a good fitness is selected as a second parent at random. If the second parent has the same number selected in the first parent, each string is exchanged the part of strings after that number. If not, other string is selected as the second parent, and same operation is done. Followings are examples:

Parent 1: 0-1-2-6-9-11-**15**-17-22-25-26
Parent 2: 0-4-8-10-**15**-20-23-26 (1)

Each parents have the point '15' and exchange the part of each string,

Child 1: 0-1-2-6-9-11-**15**-20-23-26
Child 2: 0-4-8-10-**15**-17-22-25-26 (2)

After the crossover, we check the children whether each string has same numbers. If so, we cut out a part of the string between the same numbers as follows:

Child 3: 0-1-2-6-9-**11**-15-10-7-**11**-17-22-25-26 (3)

after cut out operation

Child 3': 0-1-2-6-9-**11**-17-22-25-26 (4)

In the mutation, a position in each string is selected at random based on a probability of mutation which is low. Then, the number of nodes is selected randomly for following positions which are connected sequentially.

Distance of a path indicated by a string is used for *fitness* of each string. Therefore, as the distance decreases, the fitness increases. To calculate the fitness, the length of the link between numbers in the string is summed as followings:

String: 0-1-2-6-9-11-15-17-22-25-26
Fitness in GA1 (distance):

$$F_1(k) = d_{0-1} + d_{1-2} + d_{2-6} + \cdots + d_{25-26}, \quad (5)$$

where $F_1(k)$ is the fitness value of the *k*th string and d_{i-j} is distance between points/node *i* and *j*, which is equal to a length of link.

Eventually, each mobile robot plans optimal and feasible paths *selfishly* by using the GA1. This paper assumes that each mobile robot plans some paths in its environment using GA1.

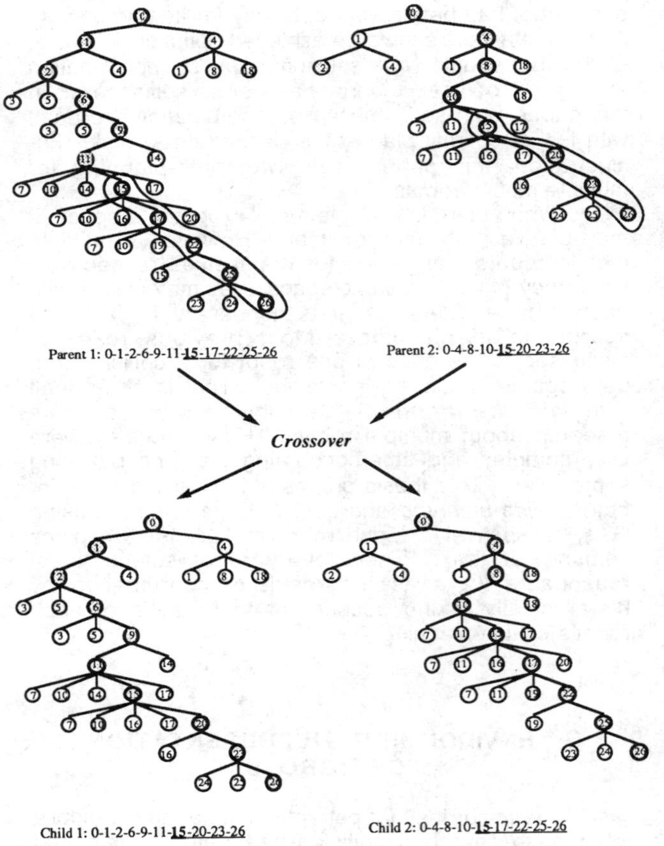

Fig. 4 Crossover operation for the order based coding in GA1

4-3. Genetic Algorithm 2 for Coordinative-Planning

GA2 is used for coordinative path-planning of multiple mobile robots after selfish planning of each robot. Simple encoding is used for a string. Each code in a string is integer whose position means each robot's number, and which itself means a selected path for the robot. This paper assumes that there are ten robots in a space and each robot has several candidates of path which are decided in the first stage selfishly by using the GA1. Therefore, the string consists of integer-codes for the ten robots, and each code has several kinds of integer for the several candidates.

Operations of crossover and mutation are simple. For the crossover, binary templates are used for selection of exchange positions. For example, in the case that there are ten robots and each robot has four candidates of path, the crossover is operated as follows:

Parent 1: 4-2-1-3-3-1-1-1-4-2
Parent 2: 1-4-2-2-4-4-4-2-2-1 (6)

Template: 0-0-1-1-1-0-1-0-1-0 (7)

Child 1: 4-2-2-2-4-1-4-1-2-2
Child 2: 1-4-1-3-3-4-1-2-4-1 (8)

For the mutation, each code in each string is changed at random based on a low probability of mutation.

The calculation for the fitness of each string is not as simple as that in the GA1. The free-convex areas are used to avoid collisions among robots on moving and dead-lock as a result of coordination. Only one robot can enter in a free-convex area and other robots which would like to enter the same free-convex area have to wait at the point on the edge of the area. Therefore, the fitness takes account of not only distance but also time for waiting. The robot may make a detour to avoid much waiting time. The robot selects not the first candidate of the path but the second or other candidate of the path. However, if the robot does not select the first candidate of the path, the distance becomes longer and the amount of the cost increases. The fitness can also take account of energy for acceleration or slowing down. The cost is calculated as the fitness. The cost is an extended distance as the result of making a detour, waiting time, and energy for acceleration and deceleration. This paper deals with both extended distance comparing with the path length of first candidate and waiting time as scalar value for calculation of the fitness. The cost as the fitness in the GA2 is calculateed as follows:

Fitness in GA2 (distance and other *cost*):

$$F_{2(k)} = Cost_A + Cost_B + \cdots + Cost_K + \cdots + Cost_N \qquad (9)$$

where $Cost_K$ is the cost of the robot K such as extended distance as the result of making a detour, waiting time, and energy for acceleration and deceleration. We deal with both extended distance comparing with the path length of first candidate and waiting time as scalar value for calculation of the fitness.

In the GA2, the rule for fitness is very simple, but very effective and efficient for coordinative path-planning of multiple mobile robots.

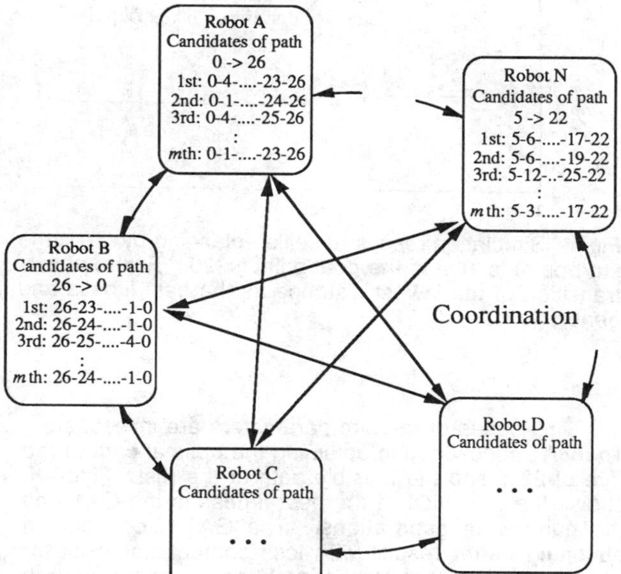

Fig. 5 Coordination among multiple robots by using GA2 Each robot has some candidates of paths by selfish-planning.

5. Simulation Results

5-1. Selfish Path-planning

5-1-1. Random Search

The proposed GA1 is compared with random search method for graph search at first. In random search, a string is generated randomly following the graph and evaluated the fitness as well as that in the GA1. The string which has better fitness is selected. The number of nodes in the graph is 27 and number of combination as the size of search space is 27! if there is no restriction of the graph. Therefore, it is difficult to find out an optimal path. In the simulation, the start point is '0' and the goal point is '26.' In this case, the size of search space is about 620,000. By 50,000 iteration, the random search method could not find out the optimal path in the 50 cases of random parameters though succeeded in obtaining the feasible path.

5-1-2. GA1 for Selfish-planning

The GA1 is used for planning of collision-free path of single mobile robot as the selfish-path-planning in a space with stationary obstacles. Each robot which had different start point and target point from other robots succeeded in producing optimal and feasible paths for itself. Search space of the graph was very large. In the simulation, the start point is '0' and the goal point is '26' and the size of search space is about 620,000. In the GA1, 50 strings are used for a population, and calculated 100 generations with 80% probability of the crossover and 20% probability of the mutation for each string.

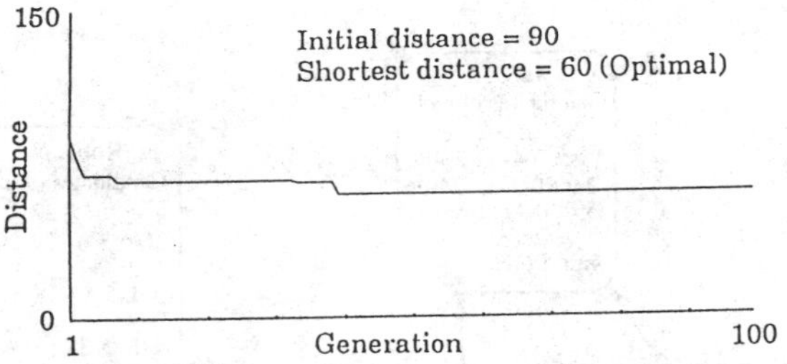

Fig. 6 Simulation results of selfish-planning by GA1: the start point is '0' and the goal point is '26.' Figure shows transition of the lowest distance as the best fitness and generations.

50 cases of random parameters are investigated. The GA1 succeeded in obtaining the optimal path at the rate of 22% and the feasible path at the rest. Figure 6 shows the transition of the best fitness in the GA1 and the number of generations. The GA1 succeeded in obtaining better results with less computation cost for searching than random search as a graph search technique. As the numbers of nodes increase, the search space becomes much more complex. However, the GA1 can search optimal and feasible paths easily.

5-2. Coordinative Path-planning by GA2

After path-planning of each robot by using the GA1, every robot plan coordinately to avoid colliding with each other on moving and dead-lock states by using the GA2. In simulation, the numbers of robots were ten, and each robot had planed its own several candidates of paths for itself by using the GA1. In the GA2, each string had integer codes according to the number of robots and each code indicated the selected path for each robot. Fitness was calculated while taking account of collision and dead-lock with the simple rule. The cost was waiting time at the edge of the free convex area to wait for passing of other robot in the area, and the extended distance is the result of making a detour. Figure 7 shows paths in the case that each robot selects the optimal path selfishly. Table 1 shows the optimal paths. The cost defined is very large because every robot pass frequently at the center part of the map and have to wait other robots passing the free-convex area following the rule to avoid collisions and dead-lock.

In the GA2, 50 strings were used for a population, and calculated 100 generations with 50% probability of the crossover and 10% probability of the mutation for each code. The cases where each robot has 2, 3, 4, or 5 candidates for path planed selfishly by using the GA1 were investigated. For each case, 50 cases of random parameters were investigated, and each sting were set to select 1st paths for all robots at the first generation. Figure 8 shows those in the case that each robot has five candidates. Table 2 shows the planned paths when each robot had five candidates. The averages of the best fitness (lowest cost) of each case were normalized by the fitness/cost of the case that each robot selects the first candidate. Table 3 shows the result. In the case of 5 candidates, each robot did not select the 5th candidate of the path. Therefore, 4 candidates were enough to plan coordinately in this situation. The size of search space of 5 candidates was 5^{10} in the case that each robot of 10 robots had 5 candidates.

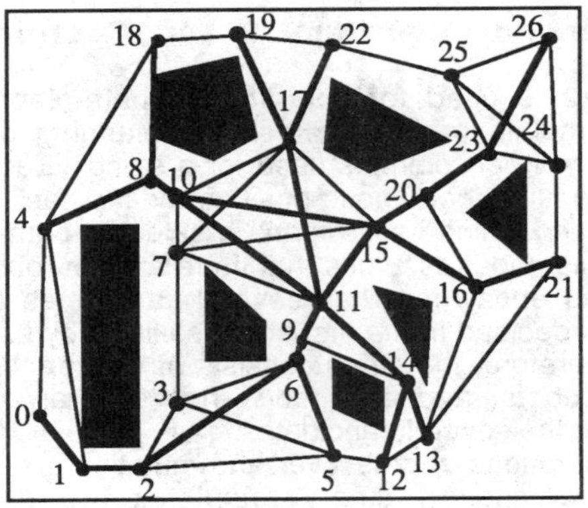

Fig. 7 Paths: each robot selects the optimal path selfishly

Table 1 Optimal path for each robot

No. of Robot	Start Point	Goal Point	Optimal Path
1	0	26	0-1-2-6-9-11-15-20-23-26
2	26	0	26-23-20-15-11-9-6-2-1-0
3	22	12	22-17-11-14-12
4	12	22	12-14-11-17-22
5	5	19	5-6-9-11-17-19
6	19	5	19-17-11-9-6-5
7	4	21	4-8-10-15-16-21
8	21	4	21-16-15-10-8-4
9	18	13	18-8-10-11-14-13
10	13	18	13-14-11-10-8-18

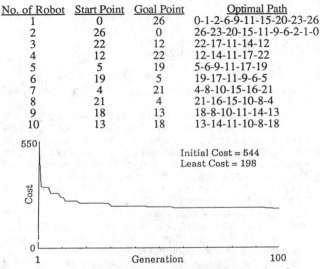

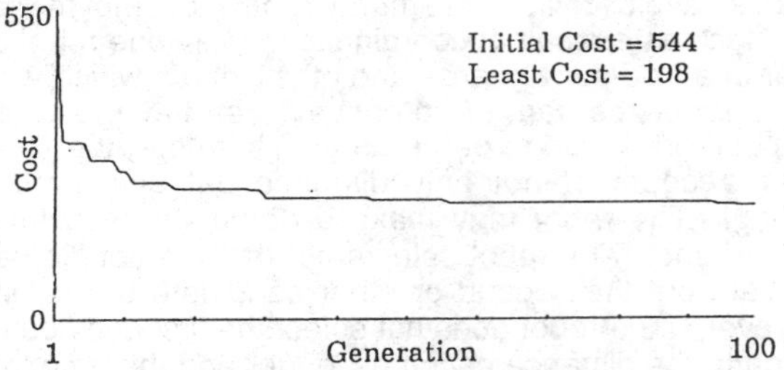

Fig. 8 Simulation results of coordinative-planning by using GA2 in the case that each robot has five candidates of paths. Figure shows transition of the lowest total cost as the best fitness and generations.

Table 2 Planned paths:
Each robot has five candidates of path.

No. of Robot	No. of path	Path
1	2	0-1-2-5-12-13-16-21-24-26
2	4	26-25-22-19-18-8-4-0
3	1	22-17-11-14-12
4	2	12-13-21-24-25-22
5	1	5-6-9-11-17-19
6	1	19-17-11-9-6-5
7	1	4-8-10-15-16-21
8	2	21-24-25-22-19-18-4
9	4	18-4-0-1-2-3-5-12-13
10	2	13-14-9-6-3-7-10-8-18

Table 3 Comparison of Results by GA2

	1 cand.	2 cand.	3 cand.	4 cand.	5 cand.
Rate	1	0.432	0.386	0.360	0.361

6. Conclusions

This paper proposed a new strategy for path planning of multiple mobile robots. The strategy consisted of coordinative planning of every robot combined with selfish planning of each robot. The Genetic Algorithms are applied to both the selfish-planning and the coordinative-planning and altered to suit for both plannings. By the approaches, multiple mobile robots can obtain good plans avoiding collisions with each other and dead-lock state and decreasing total cost for moving.

References

[1] N. J. Nilson, A Mobile Automation; an Application of Artificial Intelligence Techniques, Proc. of IJCAI, pp. 509-520 (1969)

[2] T. Lozano-Perez and M. A. Wesley, An algorithm for Planning Collision Free Paths among Obstacles, Communication ACM, Vol. 22, pp. 560- 570 (1979)

[3] R. A. Brooks, Solving the Find Path Problem by Good Representation of Free Space, IEEE Trans. on Sys., Man, and Cyb., Vol. 13, No. 3, pp. 190-197 (1983)

[4] L. Fu and D. Liu, An Efficient Algorithm for Finding a Collision Free Path among Polyhedral Obstacles, Journal of Robotic Systems, Vol. 7 (1), pp. 129-137 (1990)

[5] H. Noborio, Several path-planning algorithms of a mobile robot for an uncertain work-space and their evaluation, Proc. of the IEEE Int'l Workshop on Intelligent Motion Control, Vol. 1, pp. 289-294 (1990)

[6] S. Lee and J. Park, Cellular Robotic Collision-Free Path Planning, IEEE Fifth Int'l Conf. on Advanced Robotics, pp. 539-544 (1991)

[7] D. K. Cho and M. J. Chung, Intelligent Motion Control Strategy for a Mobile Robot in the presence of Moving Obstacles, Proc. of IEEE/RSJ Int'l Workshop on Intelligent Robotics and Systems IROS '91, Vol. 2, pp. 541-546 (1991)

[8] M. K. Habib and H. Asama, Efficient Method to Generate Collision Free Paths for Autonomous Mobile Robot Based on New Free Space Structuring Approach, Proc. of IEEE/RSJ Int'l Workshop on Intelligent Robotics and Systems IROS '91, Vol. 2, pp. 563-567 (1991)

[9] J. Wang and G. Beni, Distributed Computing Problems in Cellular Robotic Systems, IEEE Int'l Workshop on Intelligent Robots and Systems (IROS '90), pp. 819-826 (1990)

[10] J. Wang, Fully Distributed Traffic Control Strategies for Many-AGV Systems (IROS '91), pp. 1191-1204 (1991)

[11] S. Yuta, and S. Premvuti, Consideration on Cooperation of Multiple Autonomous Mobile Robots - Introduction to Modest Cooperation - , Proc. 1991 IEEE International Conference on Advanced Robotics (ICAR'91), pp. 545-549, (1991)

[12] A. H. Bond and L. Gasser (Ed.), readings in Distributed Artificial Intelligence, Morgan Kaufmann, (1988)

[13] D. E. Goldberg, Genetic Algorithms in Search, Optimization, and Machine Learning, Addison Welsey (1989)

[14] L. Davis (ed.), Handbook of Genetic Algorithms, Van Nostrand Reinhold (1991)

[15] T. Shibata, T. Fukuda et al., New Strategy for Hierarchical Intelligent Control of Robotic manipulator: - Hybrid Neuromorphic and Symbolic Control -, Proc. of IJCNN '91-Singapore, Vol. 1, pp. 107-112 (1991)

[16] T. Fukuda, T. Ueyama, and F. Arai, Control Strategy for a Network of Cellular Robots - Determination of a Master Cell for Cellular Robotic Network based on a Potential Energy -, Proc. 1991 IEEE International Conference on Robotics and Automation, pp. 1616-1621, (1991)

Proceedings of the 31st Conference
on Decision and Control
Tucson, Arizona • December 1992

AUTOMATIC TUNING FOR A TELEOPERATED ARM CONTROLLER*

Reid L. Kress and John F. Jansen
Robotics & Process Systems Division
Oak Ridge National Laboratory
P.O. Box 2008, Building 7601
Oak Ridge, Tennessee 37831-6304 USA

Abstract

This paper addresses the problem of determining an optimal set of gains for a controller for a teleoperated arm. Specifically, an automatic tuning technique was applied and investigated for tuning an independent-joint proportional-derivative controller for a teleoperated manipulator. The Hooke and Jeeves method is used in conjunction with a one-dimensional search routine in the tuning algorithm. The algorithm was used to optimize gains for a two-link teleoperator simulation and the results of several optimizations were used to determine the best form for an input trajectory and cost function. The desired joint angle trajectory is taken from low-pass filtered step inputs with randomly generated magnitudes, which vary at a predetermined interval. Both positive and negative angles are generated, but they are constrained to lie within the manipulator work space. It was determined that the cost function should be based on tracking error, peak position error over the entire desired path, overshoot, actuator torque bounds, and gain limits. The optimized gains obtained from the simulation were applied to an actual teleoperator and some improvement was seen.

1. Introduction

The problem of tuning a controller is as old as control system design itself. Even for simple designs such as proportional-derivative (PD) controllers for teleoperators, the gains must be set to obtain some desired performance. Heuristic methods are available for tuning controllers; however, most involve a considerable investment of the designer's time, especially if a large number of parameters result from complicated control algorithms. When a controller is tuned for a teleoperated manipulator, the problem is compounded because of the complex nonlinearities present in real systems. Inertia can vary several hundred percent during a single motion, and inertia differences between joints are even greater. Viscous friction, coulomb friction, gear ratios, actuator limits, and link stiffness can all vary from joint to joint. These variations require that each joint of a teleoperator be tuned independently. For arms with several joints (seven joints is not uncommon, because of interest in redundant positioning), tuning time can become excessive. A tuning technique that could determine the gain settings automatically with an intelligent search technique is highly desirable. A recent effort in automatic tuning is the work of Chen[1,2], where the method of Hooke and Jeeves was used to tune a Puma 560 robot arm. Chen[2] employed a cost function using tracking error, peak error, the first and second derivatives of the tracking error, and final error. The method worked well although it occasionally "got stuck" without finding an optimal set of gains. This paper extends the work of Chen by suggesting the specific form of the desired trajectory and the cost function for use with teleoperated manipulators.

2. Tuning Algorithm and Simulations

2.1 Algorithm

The problem of automatically tuning a control algorithm for a teleoperator is cast in the form of minimizing a cost function, which depends upon the performance of the arm. For example, a simple cost function could be to minimize the difference between the actual and desired trajectory for each joint over a restricted set of gain settings (stability must be maintained). Solving this type of problem requires the use of techniques from the field of nonlinear programming. The method used in this study is the Hooke and Jeeves method, an unconstrained optimization technique that does a multidimensional search without taking a derivative of the cost function[3]. A flowchart showing the major steps of the algorithm is presented in Fig. 1a. A 1-D search in the direction of each of the parameters to be optimized is done, followed by another 1-D search in the direction of the maximum rate of change. A flowchart for the 1-D search routine is shown in Fig. 1b. The parameter optimization terminates when the gains change a limited amount or after a fixed number of iterations of the minimization searches.

2.2 Simulations

The first tests of the tuning algorithm were done on a two-degree-of-freedom (2-DOF) model of a teleoperated manipulator. These tests were designed to determine the needs of the tuning algorithm. The model contained two revolute joints; the first rotates about a horizontal axis (called the pitch joint) and the second rotates about a vertical axis (called the yaw joint). Teleoperators are highly nonlinear systems; therefore, several of the dominate nonlinearities of typical manipulators were included in the simulation (e.g. inertia variation as a

* Managed by Martin Marietta Energy Systems, Inc., for the U.S. Department of Energy under contract DE-AC05-84OR21400. Research sponsored by the National Aeronautics and Space Administration, Langley Research Center.

CH3229-2/92/0000-2692$1.00 © 1992 IEEE

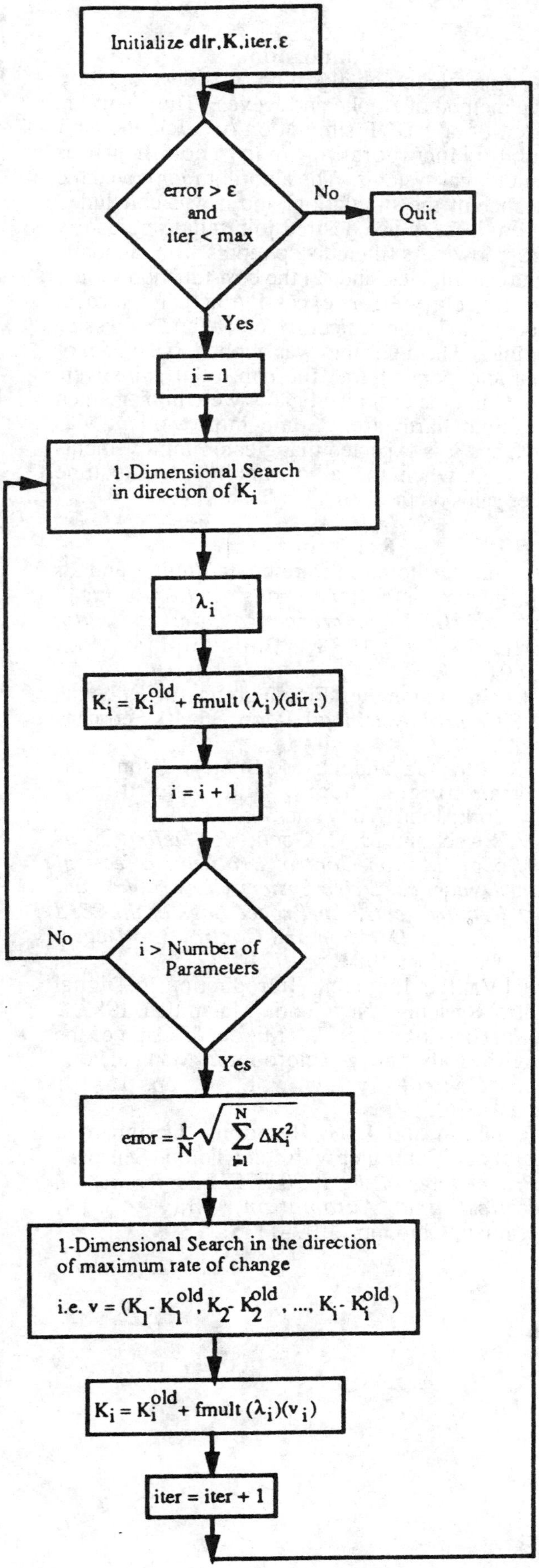

Fig. 1a. Hooke and Jeeves Algorithm

Legend

a,b	= bounds on λ in 1-D search
conv	= convergence criterion in 1-D search
dir	= direction vector
error	= change in parameter values
fmult	= scale factor
iter	= iteration counter
K	= parameter vector with N parameters
max	= maximum number of iterations
PI	= performance index (cost function)
ε	= purturbation magnitude in 1-D search
λ	= magnitude of motion in **dir**

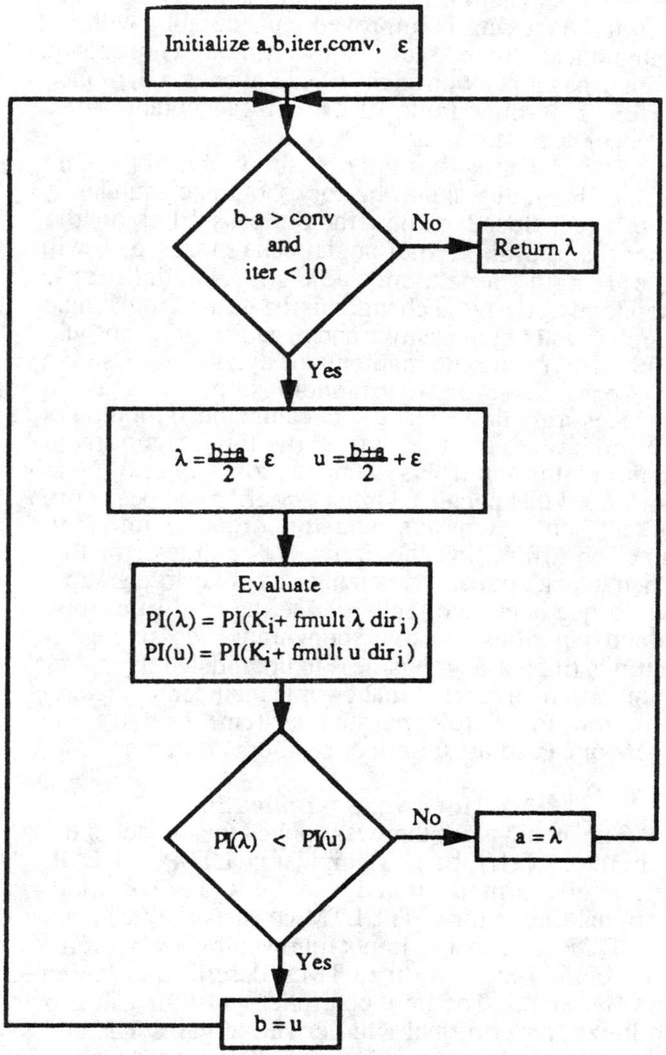

Fig. 1b. 1-D Search Algorithm

function of joint position, Coulomb friction, and actuator saturation)[4]. Gravity effects were assumed to be suitably compensated, and Coriolis effects were assumed to be negligible. A simple PD controller was modeled for each joint as shown in Eq. (1):

$$T_i = KP_i\left(\theta_{Desired\ i} - \theta_i\right) + KV_i\left(\dot{\theta}_{Desired\ i} - \dot{\theta}_i\right) \quad (1)$$

where i is either 1 or 2, T_i is the joint torque command, $\theta_{Desired\ i}$ is the desired joint trajectory, θ_i is the actual joint trajectory, and KP_i and KV_i are the joint position and velocity gains, respectively.

A typical result from a simulation of the tuning algorithm is shown in Figs. 2 and 3. Figure 2 shows how the gains vary for each iteration. An iteration includes the 1-D search in each of the gain directions as well as the 1-D search in the direction of maximum rate of change of the cost function. The gains in Fig. 2 were bounded; however, this optimization terminated because the magnitude of the change in all of the gains was less than a predetermined threshold. Figure 3 shows the trajectory of joint 2 using a stable set of gains without automatic gain tuning and after automatic tuning. Tracking is improved considerably with the automatically tuned set of gains. Although not shown, joint 1 has a peak error of 0.1 rad even for the tuned case. This magnitude of error is acceptable for a teleoperated manipulator.

The 2-DOF simulations show the following results. First, the input trajectory should be a randomly generated closed path with low-pass filter on the changes in desired joint angle. The closed path will avoid gain dependence on initial conditions. Randomly generated changes in the desired joint angle amplitudes in both positive and negative directions will arbitrarily excite the manipulator dynamics. Using a low-pass filter on the random steps[5] will avoid unnecessarily large changes in commanded input. For optimization of the gains of the teleoperated manipulator control system, a cost function was developed that penalized trajectory error and peak error (like Chen), excessive actuator torques (similar in function to limiting the derivatives of the error like Chen in that both are penalizing excessive acceleration, but torque is more directly tied to the hardware), plus added penalties for overshoot and excessive gain settings (increases robustness to unmodeled dynamics at high frequencies). Final error is included in the cost function for a teleoperated system, because the operator can adjust for minor positional differences.

3. Hardware Application

The tuning algorithm was applied to a model of the Laboratory Telerobotic Manipulator (LTM), a 7-DOF telerobotic arm designed and built at Oak Ridge National Laboratory. The LTM consists of three pairs of 2-DOF (pitch/yaw) joints followed by a wrist roll.[6] The control system for the LTM is described in Jansen and Herndon.[7] The joint controller had four gains to optimize: proportional and derivative gains KP and KV, an inner-loop damping gain K_{Vi}, and a torque feedback gain K_T. The results from tuning the 2-DOF model simulating the LTM's control system were implemented on the LTM hardware. Some improvement over the set of nonoptimized gains were observed. Greater improvements are expected in the future when the algorithm is applied directly to the actual teleoperator hardware instead of simply to a model.

4. Conclusions

An automatic tuning algorithm was developed by using the method of Hooke and Jeeves. The algorithm was tested on a 2-DOF simulation of a teleoperated manipulator incorporating many nonlinearities important to real systems. The algorithm improved the performance of the simulation, and it was concluded that (1) it is best to use desired joint angle trajectory taken from low-pass filtered step inputs with randomly generated magnitudes, and (2) the cost function should penalize trajectory error, excessive actuator torque, overshoot, maximum trajectory error, and excessive gain settings. The algorithm was applied to a model of an actual teleoperated arm, the controller gains were optimized, and the optimized gains were implemented on the actual hardware. Some improvement was observed, and it is expected that greater improvements will be seen when the algorithm is used to tune controller gains on the actual hardware.

References

[1] Y. Chen, "Automatic Parameter Tuning and Its Implementation for Robot Arms," in *Proceeding of the 1988 IEEE Conference Robotics and Auto,* April 24–29, 1988, Philadelphia, Pa., pp. 691–698.

[2] Y. Chen, "Parameter Fine-Tuning for Robots," *IEEE Control System* vol. 9, pp. 35–40, February 1989.

[3] M. S. Bazaraa and C. M. Shetty, "Nonlinear Programming, Theory and Algorithms," New York: John Wiley and Sons, 1979.

[4] L. M. Sweet and M. C. Good, "*Re-Definition of the Robot Motion Control Problem: Effects of Plant Dynamics, Drive System Constraints, and User Requirements,*" in *Proceedings of the 23rd Conference on Decision and Control*, Las Vegas, Nev., December 1984, pp. 724–732.

[5] H. F. VanLandingham, "Introduction to Digital Control Systems," New York: Macmillan, 1985.

[6] W. W. Hankins and R. W. Mixon, "Design of the Langley Laboratory Telerobotic Manipulator," *IEEE Control System.* vol. 10, pp. 16–19, August 1990.

[7] J. F. Jansen and J. N. Herndon, "Design of a Telerobotic Controller with Joint Torque Sensors," in *Proceedings of the 1990 IEEE Conference on Robotics and Automation,* May 13-18, Cincinnati, Ohio, pp. 1109–1115.

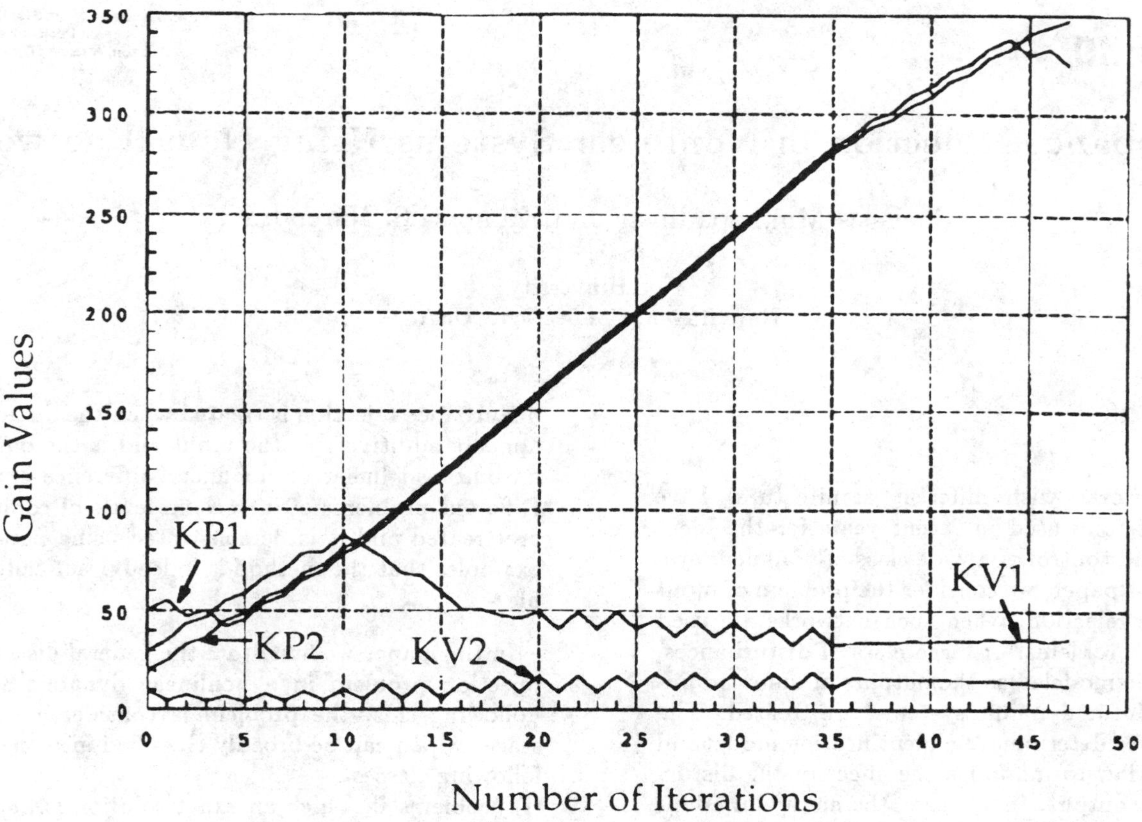

Fig. 2. Gain optimization for 2-DOF simulation

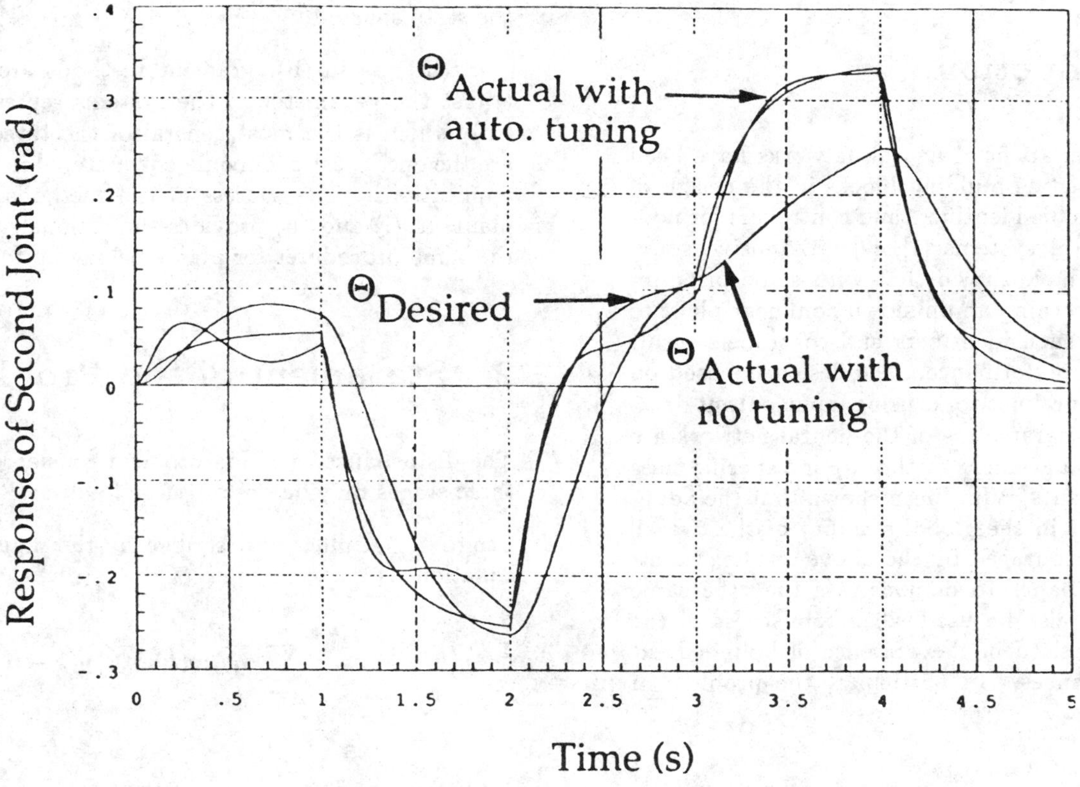

Fig. 3. Trajectories for joint 2 simulation

Proceedings of the 31st Conference
on Decision and Control
Tucson, Arizona • December 1992

Disturbance Rejection in Nonlinear Systems Using Neural networks

Snehasis Mukhopadhyay and Kumpati S. Narendra

Yale University
New Haven, Connecticut, USA

Abstract

Neural networks with different architectures have been successfully used in recent years for the identification and control of a wide class of nonlinear systems. In this paper, we consider the problem of input disturbance rejection, when such networks are used in practical problems. A large class of disturbances, which can be modeled as the outputs of unforced linear or nonlinear dynamic systems, are treated. The objective is to determine the identification model and the control law to minimize the effect of the disturbance at the output. In all cases, the method used involves expansion of the state space of the disturbance-free plant in an attempt to eliminate the effect of the disturbance. Several stages of increasing complexity of the problem are discussed in detail. Two simulation studies based on the results discussed are included towards the end of the paper.

1 Introduction

In recent years, artificial neural networks have been successfully used as building blocks in the design of practically feasible identifiers and controllers for nonlinear dynamical systems [1], [2]. Extensive simulation studies have shown that with some prior information concerning an unknown nonlinear plant to be controlled, such identifiers and controllers result in satisfactory performance. The design is based on training the neural networks using input-output data. Generally, the parameters of the neural network are adjusted using a gradient method. For a specific class of nonlinear plants, it has been shown that the adaptive laws result in the global stability of the overall system [3]. Encouraged by the above results, we attempt in this paper to demonstrate that the same procedure can also be used when robustness of the controller is desired in the presence of bounded external disturbances. In particular, the problem of disturbance rejection is treated when the disturbance appears additively at the input and is the output of an unforced linear or nonlinear difference equation. Different problems of increasing levels of complexity are treated and it is demonstrated using non-trivial examples that the method is indeed practically feasible.

In this paper we first state the general disturbance rejection problem in a nonlinear dynamic system. Following this, the problem is considered in three stages which can be broadly classified into one of the following classes :
(i) problems in which an exact solution is known to exist and stable adaptive laws can be determined,
(ii) problems in which a solution exists but stable adaptive laws cannot be designed to find the solution, necessitating the use of gradient methods, and
(iii) problems in which further assumptions concerning the plant have to be made to assure the existence of a solution to the control problem in some region of the state space.

In (iii), as in (ii), gradient methods are used to adjust the parameters. The systems represented in (iii), which is the most general of the three classes, are the ones most commonly encountered in practical applications. The success of the methods used for plants in (i) and (ii) provides the rationale for using the same procedures for plants belonging to (iii).

2 Statement of the Problem

The disturbance rejection problem considered in the three stages may be described as follows:

Stage 1 : A plant is described by the input-output equation

$$y(k+1) = \sum_{i=0}^{N} \alpha_i f_i[y(k), \ldots, y(k-n+1)] \quad (1)$$

CH3229-2/92/0000-2696$1.00 © 1992 IEEE

$$+ \sum_{j=0}^{m-1} \beta_j[u(k-j)+v(k-j)]$$

$$v(k+1) = \sum_{i=0}^{p-1} \gamma_i v(k-i)$$

where f_i are known functions, $u(k)$ and $y(k)$ are the input and output respectively at time k and $v(k)$ is an additive disturbance at time k. It is assumed that $v(k)$ is the solution of a linear homogeneous difference equation and that the coefficients α_i, β_j and γ_i are unknown. This problem may be considered as a direct extension of that encountered in linear adaptive control. The results of this problem can be extended to the case where the plant is described by

$$\begin{aligned} y(k+1) &= f[y(k),\ldots,u(k-n+1)] \\ &+ \sum_{j=0}^{m-1} \beta_j[u(k-j)+v(k-j)] \end{aligned} \quad (2)$$

where f is an unknown function.

Stage 2 : In stage 2 we consider the case where

$$\begin{aligned} y(k+1) &= f[y(k),\ldots,y(k-n+1)] \\ &+ \beta_0[u(k)+v(k)] \\ v(k+1) &= g[v(k),\ldots,v(k-p+1)] \end{aligned} \quad (3)$$

Stage 3 : Finally in stage 3 we have the general case where the system as well as the disturbance dynamics are nonlinear.

$$\begin{aligned} &y(k+1) = f[y(k),\ldots,y(k-n+1),[u(k)+v(k)] \\ &\quad ,\ldots,[u(k-m+1)+v(k-m+1)]] \\ &v(k+1) = g[v(k),\ldots,v(k-p+1)] \end{aligned} \quad (4)$$

The representation given in Stage 1 is amenable to theoretical analysis. The functions $f_i[.]$ $(i = 1,\ldots,N)$ are assumed to be globally Lipschitz and known. Only the coefficients α_i $(i = 1,\ldots,N)$ and β_j $(j = 0,\ldots,(m-1))$ are unknown. The corresponding disturbance-free case was considered in [3] in detail. In this case, adaptive laws can be designed to ensure global stability of the overall system for the disturbance-free case. It can be extended to the case when a disturbance is present and is of the form described by equation (1).

When the function $f[.]$ is unknown and has to be estimated, either radial basis functions networks (RBFN) or multilayer neural networks (MNN) can be used for this purpose in the identification model. When an RBFN is used, $f[.]$ is approximated as

$$\hat{f}[y(k),\ldots,y(k-n+1)] = \sum_{i=1}^{N} c_i R_i[y(k),\ldots,y(k-n+1)] \quad (5)$$

where $R_i[.]$ are the basis functions and c_i are adjustable parameters (weights) of the network. Since the adjustable parameters contribute linearly to the output, the same technique used in Stage 1 can be used for this problem also.

A special case where the disturbance can be described by a nonlinear model is considered in Stage 2. In this case, the output $y(k+1)$ explicitly depends only on $u(k)$ and not on its past values. Also, $u(k)$ is linear in the r.h.s. of equation (3). An exact solution can be demonstrated to exist for this case. In an adaptive context, multilayer neural networks can be used to identify and control the plant in the presence of such disturbances.

Finally, in Stage 3, the general problem of disturbance rejection is considered where both plant and disturbance models are nonlinear. The disturbance is still assumed to be additive. The existence of an exact solution cannot be demonstrated in this stage. Simplifying assumptions have to be made in order to find an approximate but satisfactory solution.

3 Stage 1

In stage 1, the disturbance-free plant is represented by the difference equation

$$y(k+1) = \sum_{i=1}^{N} \alpha_i f_i[Y(k)] + D(z)u(k) \quad (6)$$

where $Y(k) = [y(k), y(k-1),\ldots,y(k-n+1)]^T$ and $D(z) = \beta_0 + \beta_1 z^{-1} + \ldots + \beta_{m-1} z^{-(m-1)}$, z^{-j} denoting a delay of j. In [3], it has been shown that for the plant represented by equation (6), under suitable assumptions, globally bounded adaptive laws can be designed. These assumptions are : (i) The integers n and m are known. (ii) $f_i[.], (i = 1,2,\ldots,N)$ is a known globally Lipschitz function, (iii) the unknown parameters are $\alpha_i (i = 1,\ldots,N)$ and $\beta_j (j = 1,\ldots,(m-1))$. (Assumption (iii) implies that β_0 is assumed to be known and nonzero. This assumption is made to simplify the ensuing discussions. The modifications needed when β_0 is unknown can be found in [3]), and (iv) The polynomial $D(z)$

is strictly stable. (Throughout this paper, a polynomial $P(z) = p_0 + p_1 z^{-1} + \ldots + p_{l-1} z^{-(l-1)}$ is defined to be stable if the roots of $z^{l-1}P(z)$ lie on or within the unit circle in the complex plane, with the roots on the unit circle being simple. The polynomial $P(z)$ is defined to be strictly stable if all the roots of $z^{l-1}P(z)$ lie strictly within the unit circle.)

The approach used and the corresponding adaptive laws are outlined below. The unknown parameters $\alpha_i (i = 1, \ldots, N)$ and $\beta_j (j = 1, \ldots, m-1)$ in equation (6) are estimated as $\hat{\alpha}_i(k)$ and $\hat{\beta}_j(k)$ using an identification model described by the equation

$$\hat{y}(k+1) = \sum_{i=1}^{N} \hat{\alpha}_i(k) f_i[Y(k)] + \sum_{j=1}^{m-1} \hat{\beta}_j(k) u(k-j) + \beta_0 u(k) \tag{7}$$

Defining $\theta = [\alpha_1, \ldots, \alpha_N, \beta_1, \ldots, \beta_{n-1}]^T$, $\hat{\theta}(k)$ as the estimate of θ, and $\omega(k)$ as the vector $\omega(k) = [f_1[Y(k)], \ldots, f_N[Y(k)], u(k-1), \ldots, u(k-m+1)]^T$, the plant and the identification model equations can be written as

Plant : $y(k+1) = \theta^T \omega(k) + \beta_0 u(k)$

Ident. Model : $\hat{y}(k+1) = \hat{\theta}^T(k)\omega(k) + \beta_0 u(k)$.

Defining $e(k) = \hat{y}(k) - y(k)$, the adaptive laws for adjustment of the parameter vector $\hat{\theta}$ is given by

$$\hat{\theta}(k+1) = \hat{\theta}(k) - \eta(k) \frac{e(k+1)\omega(k)}{1 + \|\omega(k)\|^2} \tag{8}$$

where $\eta(k)$ is chosen so that $0 < \eta(k) < 2$.

The control input u(k) is computed as

$$u(k) = \frac{1}{\beta_0}\{y^*(k+1) - \sum_{i=1}^{N} \hat{\alpha}_i(k) f_i[Y(k)] - \sum_{j=1}^{m-1} \hat{\beta}_j(k) u(k-j)\}$$

In [3], it is shown that the adaptive algorithm described above results in global stability and the asymptotic convergence of the output error to zero.

The Disturbance Rejection Problem :

When an external disturbance is present, the overall system is described by the equations

$$\begin{aligned} y(k+1) &= \textstyle\sum_{i=1}^{N} \alpha_i f_i[Y(k)] + D(z)[u(k) + v(k)] \\ v(k+1) &= R(z)v(k) \end{aligned} \tag{9}$$

where $R(z) = \gamma_0 + \gamma_1 z^{-1} + \ldots + \gamma_{p-1} z^{-(p-1)}$. This implies that the disturbance is generated as the output of an unforced linear difference equation. It is assumed that the disturbance is bounded, implying that the polynomial $R_1(z) = 1 - z^{-1}R(z)$ is stable.

Noting from equation (9) that

$$D(z)v(k) = y(k+1) - \sum_{i=1}^{N} \alpha_i f_i[Y(k)] - D(z)u(k), \forall k,$$

the output of the system given by equation (9) can also be written as

$$\begin{aligned} y(k+1) &= \sum_{i=1}^{\bar{N}} \bar{\alpha}_i \bar{f}_i[\bar{Y}(k)] + R(z)y(k) + \bar{D}(z)u(k) \\ &= \sum_{i=1}^{\bar{\bar{N}}} \bar{\bar{\alpha}}_i \bar{\bar{f}}_i[\bar{Y}(k)] + \bar{D}(z)u(k) \end{aligned} \tag{10}$$

where $\bar{Y}(k) = [y(k), y(k-1), \ldots, y(k-p-n+1)]^T$, $\bar{N} = (p+1)N$, $\bar{\bar{N}} = (p+1)N + p$. The functions $\bar{\bar{f}}_i[.]$ are, again, known globally Lipschitz functions. $\bar{D}(z) = D(z)R_1(z) = D(z)[1 - z^{-1}R(z)] = [\bar{\beta}_0 + \bar{\beta}_1 z^{-1} + \ldots + \bar{\beta}_{m+p-1} z^{-(m+p-1)}]$. $\bar{D}(z)$, being the product of a stable and a strictly stable polynomials, is stable. $\bar{\beta}_0$ is equal to β_0 which was assumed to be known. Hence, the same procedure as described earlier for the disturbance-free case, can be employed to adaptively control the plant in the presence of disturbances to ensure that the tracking error tends to zero.

Comment: When the plant and the disturbance are given by

$$\begin{aligned} y(k+1) &= f[Y(k)] + D(z)[u(k) + v(k)] \\ v(k+1) &= R(z)v(k) \end{aligned} \tag{11}$$

the composite system can be expressed as

$$y(k+1) = \bar{f}[\bar{Y}(k)] + \bar{D}(z)u(k) \tag{12}$$

where

$$\bar{Y}(k) = [y(k), y(k-1), \ldots, y(k-p-n+1)]^T$$

$$\bar{f}[\bar{Y}(k)] = f[Y(k)] + \sum_{i=1}^{p} \gamma_{i-1}(y(k-i+1) - f[Y(k-i)])$$

$$\bar{D}(z) = D(z)R_1(z) = D(z)[1 - z^{-1}R(z)] = \sum_{i=0}^{m+p-1} \bar{\beta}_i z^{-i}$$

When the nonlinear function $f[.]$ is not known, as mentioned earlier, either an RBFN or an MNN can be used to estimate it. Both these classes of networks have been shown to approximate any continuous function over a compact domain arbitrarily closely. If an RBFN is used to approximate the function $\bar{f}[.]$, the output of the network with fixed centers and widths can be expressed as

$$N[\bar{Y}(k)] = \sum_{i=1}^{N} \alpha_i R_i[\bar{Y}(k)] \qquad (13)$$

where N is the number of basis functions used, $R_i[.](i = 1, \ldots, N)$ represent the basis functions and $\alpha_i, (i = 1, \ldots, N)$ are the adjustable weights of the RBFN. If the desired function can be exactly represented by an RBFN, the adaptive control problem using RBFN reduces to the problem discussed in Stage 1. If this is not the case, it is possible to set up an RBFN to approximate $\bar{f}[.]$ in a compact domain. The resultant approximation error after the RBFN has converged, can be treated as a small bounded disturbance and as shown in [3], using a dead-zone in the adaptive laws, stability can be assured. However, the use of RBFN becomes prohibitive in problems of higher order and complexity. This is particularly so in a disturbance rejection problem because the order of the overall system (*i.e.* the sum of the orders of the plant and the disturbance generating system) may be high even for low order plants. This, in turn, will require the number of basis functions to increase exponentially. Adaptive placement of radial basis functions to deal with the exponential complexity, is still an open research problem. In such high dimensional disturbance rejection problems, multilayer neural networks can be used. Gradient methods (back propagation) will be used for the adjustment of the parameters of a multilayer network in all cases [1].

4 Stage 2

In stage 1, the disturbance dynamics was assumed to be linear. This was essential to show the existence of a solution. In this section, we consider a special case of a more complex problem where the disturbance is modeled as the output of a stable unforced nonlinear system. Well known nonlinear difference equations *e.g* Van der Pol equation and chaotic time series equations (*e.g.* Mackey-Glass equation), are examples of such nonlinear models of disturbance. The special case that is considered here, corresponds to a plant whose disturbance-free dynamical representation depends explicitly only on $u(k)$ and not on its past values. Further, such dependence is assumed to be linear. In this particular case, a solution can be demonstrated to exist for the problem of complete disturbance rejection, even though the disturbance dynamics is complex and nonlinear.

Let the plant and disturbance generating system be described by Plant : $y(k+1) = f[Y(k)] + u(k) + v(k)$ Distrbance: $v(k+1) = g[v(k), \ldots, v(k=p+1)]$. Eliminating the disturbance, the following equivalent description of the system is obtained

$$y(k+1) = \bar{f}[\bar{Y}(k), \bar{U}(k-1)] + u(k) \qquad (14)$$

where $\bar{Y}(k) = [y(k), \ldots, y(k-n-p+1)]^T$ and $\bar{U}(k-1) = [u(k-1), \ldots, u(k-p)]^T$. Hence, a procedure similar to that used in stage 1 can be adopted in this stage also.

5 Stage 3

If the plant with a disturbance $v(k)$ is represented by

$$y(k+1) = f[Y(k), U(k) + V(k)]$$

and the disturbance dynamics is given by

$$v(k+1) = g[\bar{V}(k)]$$

where $Y(k) = [y(k), \ldots, y(k-n+1)]^T$, $U(k) = [u(k), \ldots, u(k-n+1)]^T$, $V(k) = [v(k), \ldots, v(k-n+1)]^T$ and $\bar{V}(k) = [v(k), \ldots, v(k-p+1)]^T$, it is not known a priori whether the overall system can be represented as

$$y(k+1) = \bar{f}[\bar{Y}(k), \bar{U}(k)] \qquad (15)$$

where $\bar{Y}(k) = [y(k), \ldots, y(k-s)]^T$ and $\bar{U}(k) = [u(k), \ldots, u(k-s)]^T$, for a finite positive integer s. Even assuming the existence of such a value for s, the lack of knowledge about the exact value of s precludes the use of an identification model. However, in a practical situation, identification models with increasing values of s may be trained and tested for performance, to find a large enough s which results in a desired performance.

6 Simulation Results

In [4], detailed simulation studies are reported for the different stages considered. In this paper, we include one such experiment where the problem belongs to stage 2. In addition, we discuss the results of a simulation experiment where the disturbance is generated by a linear dynamical system excited by white noise. Although a detailed analysis has not been carried out for this case, the preliminary results definitely show the efficacy of the suggested methods even for such stochastic disturbances.

Example 1: A second order plant in the presence of an external disturbance is given by

$$y(k+1) = f[y(k), y(k-1)] + u(k) + v(k)$$

where $f[y(k), y(k-1)] = \frac{\cos\{y(k)\}}{1+\sin^2\{y(k-1)\}} + \sin\{y(k-1)\}\exp\{-y^2(k)\} + \frac{y(k)}{1+y^2(k)}$

The disturbance is assumed to be generated as the output of a second-order nonlinear system, which is described the following equations

$$x_{v1}(k+1) = x_{v1}(k) + 0.2x_{v2}(k)$$

$$x_{v2}(k+1) = -0.2x_{v1}(k) + x_{v2}(k) - 0.1(x_{v1}^2(k) - 1)x_{v2}(k)$$

$$v(k) = x_{v1}(k)$$

The nonlinear difference equation is a discrete-time van der Pol equation. The identification model of the plant in the presence of the disturbance is given by $\hat{y}(k+1) = NN[y(k), .., y(k-q-1), u(k-1), .., u(k-q)] + u(k)$ where $NN(.)$ is realized by an MNN $\in \mathcal{N}^3_{(2q+2):40:20:1}$. For complete rejection of the disturbance, q should be greater than or equal to 2.

The desired output is assumed to be the output of the reference model

$$y_m(k+1) = 0.6y_m(k) + r(k)$$

with the reference input $r(k) = \sin\frac{2\pi k}{10} + \sin\frac{2\pi k}{25}$. The control input is computed as $u(k) = y_m(k+1) - NN[y(k), \ldots, y(k-q-1), u(k-1), \ldots, u(k-q)]$

The identification was carried out off-line in the presence of the disturbance with uniformly random input in the range $[-1, 1]$. The weights of the MNN were adjusted using static back-propagation for 500,000 time units using a step size of 0.05. The results of the tracking problem for $q = 0$ (no disturbance rejection) and $q = 2$ (full disturbance rejection) are shown in Figs. 1 and 2 respectively. It is evident that $q = 2$ results in much better response than that obtained with $q = 0$.

Example 2: The plant in this case is given by

$$y(k+1) = \frac{y(k)}{1+y^2(k)} + u(k) + v(k)$$

The disturbance is generated by the following linear dynamic system forced by a white noise $u_d \in [-0.1, 0.1]$ $d_1(k+1) = d_2(k)$
$d_2(k+1) = -0.81d_1(k) - 1.8d_2(k) + u_d(k)$
$v(k) = d_1(k)$
Since the poles of this system (0.9 and 0.9) are very near the unit circle, the disturbance $v(k)$ is a colored noise taking values in the interval $[-8, -12]$. The identification model in this case is described by $\hat{y}(k+1) = N[y(k), y(k-1), y(k-2)] + u(k) + \hat{\theta}_1 u(k-1) + \hat{\theta}_2(k-2)$. The parameters of the MNN $N(.) \in \mathcal{N}^3_{3,20,10,1}$ as well as the estimates $\hat{\theta}_1$ and $\hat{\theta}_2$ were adjusted off-line for 500,000 steps using gradient methods. In the control phase, the control input was computed as $u(k) = y^*(k+1) - N[y(k), y(k-1), y(k-2)] - \hat{\theta}_1 u(k-1) - \hat{\theta}_2 u(k-2)$ where $y^*(k+1)$ is the desired output at instant $(k+1)$.

The results without and with disturbance rejection are shown in Figs. 3 and 4 respectively. The performance improves dramatically with the modifications suggested for the rejection of the disturbance.

References

[1] K. S. Narendra and K. Parthasarathy, "Identification and control of dynamical systems using neural networks", *IEEE Trans. Neural Networks*, vol. 1, pp. 4–27, Mar. 1990.

[2] K. S. Narendra and S. Mukhopadhyay, "Intelligent Control Using Neural Networks", *Proceedings of the American Control Conference*, pp. 1069–1073, 1991.

[3] K. S. Narendra and K. Parthasarathy, "Stable Adaptive Control of a Class of Discrete-Time Nonlinear Systems Using Radial Basis Neural Networks", Center for Systems Science, Yale University, CT, Technical Report No. 9103, 1991.

[4] S Mukhopadhyay and K. S. Narendra. "Disturbance Rejection in Nonlinear Systems Using Neural Networks", Tech. Report No. 9114, Center for Systems Science, Yale University. Also to appear in *IEEE Trans. on Neural Networks*.

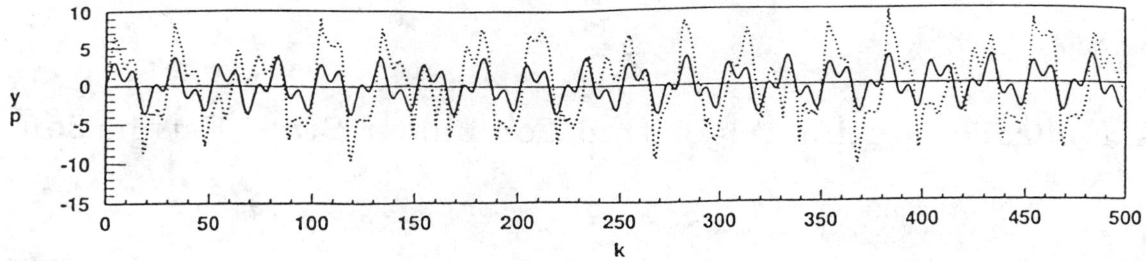

Figure 1: Example 1. Simulation results with $q = 0$

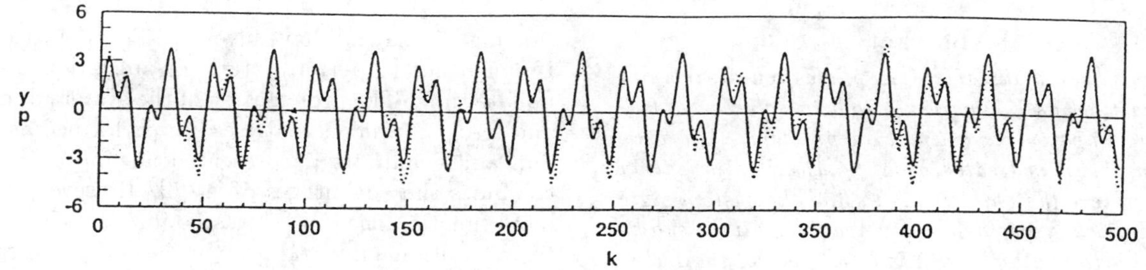

Figure 2: Example 1. Simulation results with $q = 2$

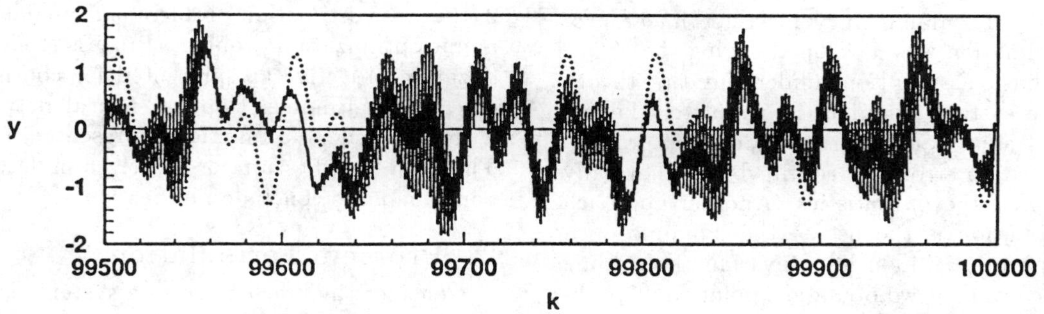

Figure 3: Example 2. Simulation results without any disturbance rejection

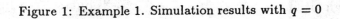

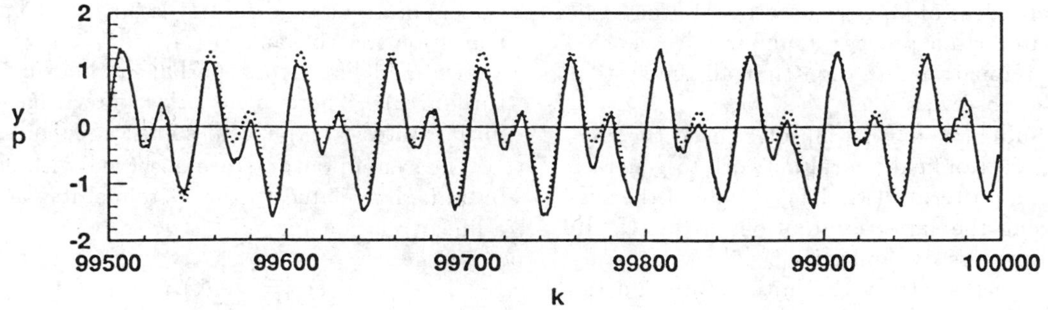

Figure 4: Example 2. Simulation results with disturbance rejection

Proceedings of the 31st Conference
on Decision and Control
Tucson, Arizona • December 1992

A Neurocomputing Algorithm For Linear State Estimation

A.T. Alouani

Electrical Engineering Department
Tennessee Technological University
Cookeville, TN 38505

Q. Sun

Manufacturing Center
Tennessee Technological University
Cookeville, TN 38505

Abstract

In this paper a linearized Hopfield neural network (HNN) is used as a computing tool to solve a continuous time linear state estimation problem. The estimation problem is treated as a dynamic optimization problem, where the objective is to find the system state that optimizes a performance measure. It is shown that by apprtopriate choice of the weights of the HNN, the optimal state can be obtained as the scaled output of a HNN.

1 Introduction

The theory of Kalman-Bucy filter for linear process state estimation was developed during the 1960s, [1]–[2]. The basic assumptions underlying this theory were: (1) the system dynamics is completely known and modelled as a Gauss-Markov stochastic process; (2) the observation errors are assumed to be normally distributed with known statistics. Another approach to state estimation is the use of the dynamic optimization concept, Bellman [3]. Invariant imbedding was used to convert a two boundary point value problem into a Cauchy problem to derive the filter. No knowledge about the observation error was assumed, however, complete knowledge of process dynamics was required and the process was assumed to be deterministic. The main ideas of [3] were used by Abutaleb [4], where linear uncertain process dynamics was considered with an implicit assumption that all the system states must be observed.

Recently, Sudharasanan [5]-[6] introduced the concept of using a a Hopfield neural network to generate the maximum a posteriori (MAP) estimate of the system state, using the same assumptions as in [1]. In addition, the process dynamics in [5]-[6] is a near deterministic linear one. Only the initial state and the observation error are stochastic and must be normally distributed for a quadratic objective function to exist. The process dynamics must be discrete in time for the derivation to be valid (for otherwise, the state transition matrix may not have inverse). This approach uses Hopfield net to perform static optimization at every update time. The propagation of the system dynamics and error covariance matrices are performed as in [1]. Sun and coauthors [7] presented an approach which combines the advantages of [4]-[6]. However, the sigmoid function used in [7] was nonlinear. This has led to an additional integral term in the energy function of the HNN. Therefore this energy function is quadratic only when this integral term can be neglected. This paper uses a linear sigmoid function to eliminate this error. This paper is organized as follow. In section 2, the state estimation problem is formulated as a dynamic optimization problem. Process model and optimization objective are presented for continuous-time case. A continuous Hopfield neural network model with linear sigmoid function is presented in Section 3. The HNN based estimator is derived in Section 4. The paper paper is concluded in Section 5.

2 Problem Formulation

Consider the linear uncertain system:

$$\dot{X}(t) = A(t)X(t) + W(t), \tag{1}$$

where $X(t)$ is an $n \times 1$ vector which represents the system state at time t, $A(t)$ is an $n \times n$ known state transition matrix, and $W(t)$, an $n \times 1$ vector of unknown modeling errors (perhaps unknown system disturbances). There is no prior knowledge about the initial states $X(t_0)$ except some vague guesses.

The system outputs are observed with errors. The outputs are related to the system states according to a linear relationship:

$$Y(t) = C(t)X(t) + V(t), \tag{2}$$

where $Y(t)$ is an $m \times 1$ vector of observed system outputs, $C(t)$ is an $m \times n$ known observation matrix, $V(t)$ is an $m \times 1$ vector of observation errors.

CH3229-2/92/0000-2702$1.00 © 1992 IEEE

The system model described by (1) and (2) is general enough to cover a wide range of practical situations. For example, to use linear system theory to a nonlinear system, the nonlinear system is usually linearized into a quasi-linear model. The modeling errors can be included in $W(t)$. A linear system with uncertain parameters or input disturbances can be collectively put into $W(t)$ as well.

This model is similar, in form, to the model used by Kalman and Bucy in deriving Kalman-Bucy filter [2]. The essential difference is the relaxation of the assumptions about the knowledge of the disturbance term $W(t)$, the observation error term $V(t)$, and the initial state $X(t_0)$.

The ultimate objective is to find the state estimates of the system, such that the integral of a quadratic function of both modeling error and observation error are minimized over the time period of consideration, subject to the dynamic constraint posed by (1) and (2). The optimization measure is:

$$J = \frac{1}{2}\int_{t_0}^{t_f} [(Y(t) - C(t)X(t))^T Q (Y(t) - C(t)X(t)) + W(t)^T R W(t)]dt, \quad (3)$$

where $[t_0, t_f]$, $t_0 < t_f$, is the time period of interest; Q and R are $m \times m$ and $n \times n$ symmetric positive definite matrices, respectively. They are the estimator design parameters. Since the integrand is a nonnegative function of the estimate of the system states, the integral is nonnegative for any X. This optimization problem is well-defined.

3 Hopfield Net Model

A general continuous Hopfield network [8] has a dynamics of

$$\frac{du_i}{dt} = -\frac{u_i}{\tau_i} + \sum_{j=1}^{n} T_{ij} v_j + I_i, \quad i = 1, \ldots, n \quad (4)$$

where n is the number of neurons; u_i is the state of the ith neuron; $\tau_i(> 0)$ is the time constant of the ith neuron; T_{ij} is the post-synaptic strength from jth neuron to ith neuron; v_i is the output of ith neuron, I_i is the bias at the ith neuron. The outputs of the neurons are related to the states of the neurons through sigmoid functions

$$v_i = g_i(u_i), \quad i = 1 \ldots n. \quad (5)$$

In [6], the following sigmoid functions were used,

$$g_i(s) = g(s) = tanh(\alpha s) = \frac{e^{\alpha s} - e^{-\alpha s}}{e^{\alpha s} + e^{-\alpha s}}, \quad \forall i, \quad (6)$$

where α is a positive constant.

Using the above sigmoid, the network has an energy function,

$$E = -\frac{1}{2}\sum_{i=1}^{n}\sum_{j=1}^{n} T_{ij} v_i v_j - \sum_{i=1}^{n} I_i v_i + \sum_{i=1}^{n} \frac{1}{\tau_i}\int_0^{v_i} g_i^{-1}(v)dv. \quad (7)$$

As can be seen from (8), the energy function is not quadratic because of the presence of the integral term. As discussed in [11], there are means to reduce the size of this term. However, this may lead to implementation problems. To get around this problem, the following linear sigmoid function [6] is used

$$g_i(u_i) = \beta u_i \text{ if } |u_i| \leq A, \quad (8)$$
$$g_i(u_i) = \beta A \text{ if } |u_i| \geq A, \quad (9)$$
$$g_i(u_i) = -\beta A \text{ if } |u_i| \leq -A, \quad (10)$$

The energy function associated with the above sigmoid is

$$E = -\frac{1}{2}\sum_{i=1}^{n}\sum_{j=1}^{n}(T_{ij} - D_n) v_i v_j - \sum_{i=1}^{n} I_i v_i, \quad (11)$$

where D_n is a nxn diagonal matrix with elements equal to α_i, given by

$$\alpha_i = \frac{1}{\beta \tau_i}, \quad (12)$$

4 A Hopfield-Net Based Estimator

The philosophy of this estimator is as follows. First, treat the estimation problem as a linear quadratic servo-mechanism problem [8]–[9]. Then find the synaptic weights and input vector of the HNN so that its output corresponds to the system state estimate.

The Hamiltonian associated with (3) is

$$H(X, W, \lambda) = \frac{1}{2}[(Y - CX)^T Q (Y - CX) + W^T R W] + \lambda^T (AX + W). \quad (13)$$

The first order necessary conditions for optimization are:

$$\frac{\partial H}{\partial W} = R\hat{W} + \hat{\lambda} = 0, \forall t; \quad (14)$$
$$\frac{\partial H}{\partial X} = -C^T Q (Y - CX) + A^T \hat{\lambda} = -\dot{\hat{\lambda}}, \forall t; \quad (15)$$
$$\frac{\partial H}{\partial \lambda} = \dot{\hat{X}} = A\hat{X} + \hat{W}, \forall t. \quad (16)$$

with boudary conditions,

$$\lambda^T(t_0)\zeta_x(t_0) - \lambda^T(t_f)\zeta_x(t_f) = 0, \quad (17)$$

where ζ_x denotes the first order variation of X from $\hat{X}$ at time.

The boundary conditions may be as follows.

$$\lambda(t_0) = 0 \quad (18)$$
$$\zeta_x(t_f) = 0. \quad (19)$$

The reason for (18) is that the initial state is not known. Eq.(19) can be justified by the fact that if the optimal solution is to be found, then at time t_f, this solution is equal to $X(t_f)$.

Using (14)-(16),

$$\hat{W} = -R^{-1}\hat{\lambda}, \quad \forall t; \quad (20)$$
$$\dot{\hat{\lambda}} = C^TQ(Y - C\hat{X}) - A^T\hat{\lambda}, \quad \forall t; \quad (21)$$
$$\dot{\hat{X}} = A\hat{X} + \hat{W}, \quad \forall t. \quad (22)$$

Let

$$\hat{\lambda} = P\hat{X} - \xi. \quad (23)$$

Where P, an $n \times n$ symmetric matrix of real function of time, and ξ, an $n \times 1$ vector of real function of time, are to be determined later. Following [9], it can be shown that P and ξ satisfy

$$\dot{P} = -PA + PR^{-1}P - C^TQC - A^TP \quad (24)$$
$$\dot{\xi} = (PR^{-1} - A^T)\xi - C^TQY \quad (25)$$

If$\hat{W}$ is substituted back into (3), one gets

$$J = \frac{1}{2}\int_{t_0}^{t_f} K[X(t)]dt, \quad (26)$$

where $K[.]$ is the utility function kernel, given by

$$\begin{aligned} K(\hat{X}) &= \frac{1}{2}[(Y - C\hat{X})^TQ(Y - C\hat{X}) + \hat{W}^TR\hat{W}] \\ &= \frac{1}{2}(Y - C\hat{X})^TQ(Y - C\hat{X}) + \\ &\quad \frac{1}{2}[-R^{-1}(P\hat{X} - \xi)]^TR[-R^{-1}(P\hat{X} - \xi)] \\ &= \frac{1}{2}\hat{X}^T(C^TQC + PR^{-1}P)\hat{X} - \\ &\quad (Y^TQC + \xi^TR^{-1}P)\hat{X} + \\ &\quad \frac{1}{2}Y^TQY + \frac{1}{2}\xi^TR^{-1}\xi \end{aligned} \quad (27)$$

This kernel is quadratic in $\hat{X}$and is positive for all $t \in [t_0, t_f]$. It is obvious that, for all $f(t) > 0, \quad \forall t \in [t_0, t_f]$,

$$\int_{t_0}^{t_f} f_1(t)dt \leq \int_{t_0}^{t_f} f_2(t)dt,$$
$$\text{if } f_1(t) \leq f_2(t), \quad \forall t \in [t_0, t_f]. \quad (28)$$

Therefore, to minimize the utility function, it is sufficient to minimize the kernel with respect to $\hat{X}$ at all time $t \in [t_0, t_f]$. Now the problem has been converted into a static optimization problem.

Noting the similarity between the utility function kernel $K[.]$ and the HNN energy function (11), one can use a HNN to produce the solution to our optimization problem. This is done by adjusting the synaptic weights and bias vector according to:

$$T = -[(C^TQC + PR^{-1}P - D_n)] \quad (29)$$
$$I = [(Y^TQC + \xi^TR^{-1}P)]. \quad (30)$$

The following algorithm summarizes how to use HNN to solve for the estimation problem.

Given the system matricies A and C, and the design matricies Q and R,

1. use (22) to solve for P

Given the measuremnt at time i,

2. use (23) to solve for ξ
3. use (28) to adapt T
4. use (29) to compute I

The initialization of the Hopfield net is arbitrary. A possible choice would be zero initial values for u_i, $i = 1, \ldots, n$, and so are v_i, $i = 1, \ldots, n$, because of the large slopes for g_i's at $u_i = 0, i = 1, \ldots, n$.

The estimator structure is shown in Figure 1.

5 Summary and Conclusions

This paper combines the use of dynamic optimization and Hopfield neural network to solve an estimation problem for continuous time linear systems. It was found that the network parameters can be automatically adapted on-line, given the system model and the estimator design parameters Q and R. The work reported in this paper is in progress to analyse the performance of this estimator.

Several issues related to this work remain to be answered. These include

1. How to adjust the estimator design Q and R to get the required performance performance of the estimator.

2. How does the estimator perform relative to existing linear state estimators.

3. What are the computational requirements of this estimator.

References

[1] Kalman, R. E., "A new approach to linear filtering and prediction problems," *Trans. ASME, J. Basic Eng.*, Series 82D, pp. 35–45, March, 1960.

[2] Kalman, R. E. and Bucy, R. S., "New results in linear filtering and prediction theory," *Journal of Basic Eng.*, pp. 95–108, March 1961.

[3] Bellman, R. E., et al., "Invariant imbedding and nonlinear filtering theory," *J. Astronaut Sci.*, Vol. 13, No. 3, pp. 110–115, May/June, 1966.

[4] Abutaleb, A. S., "New results in Sridhar filtering theory," *J. Franklin Inst.*, Vol. 322, pp. 229–240, Oct. 1986.

[5] Sudharsanan, S. I. and Sundareshan, M. K., "Neural Network computational algorithms for least squares estimation problems," *Proceedings of the International Joint Conference on Neural Networks,* Washington D. C., June, 1989.

[6] Sudharsanan, S. I. and Sundareshan, M. K., "Maximium a posteriori state estimation: A neural processing algorithm," *Proc. 28th IEEE Conference on Decision and Control,* Tampa, FL, December, 1989.

[7] Sun, Q., A.T. Alouani, T.R. Rice, and J. Gray, "Linear state estimation: A neurocomputing approach," *Proc. 1992 American Control Conference,* Chicago, IL, June 1992.

[8] Hopfield, J. J., "Neurons with graded response have collective computational properties like those of two-state neurons," *Proc. of the Natl. Academy of Sciences,* 81, pp. 3088–3092, 1984.

[9] Singh, M. G. and Titli, A., *Systems: Decomposition, Optimization and Control,* Pergamon Press, Oxford, England, 1978.

[10] Sage, A. P., *Optimum Systems Control,* Prentice-Hall, Inc., Englewood Cliffs, N.J., pp. 281, 1968.

[11] Copeland, B. R., "Global minima within the Hopfield hypercube," *Proc. of Intl. Joint Conf. on Neural Networks,* Washington DC, January 15–19, 1990.

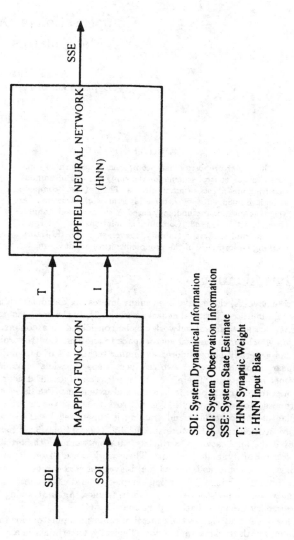

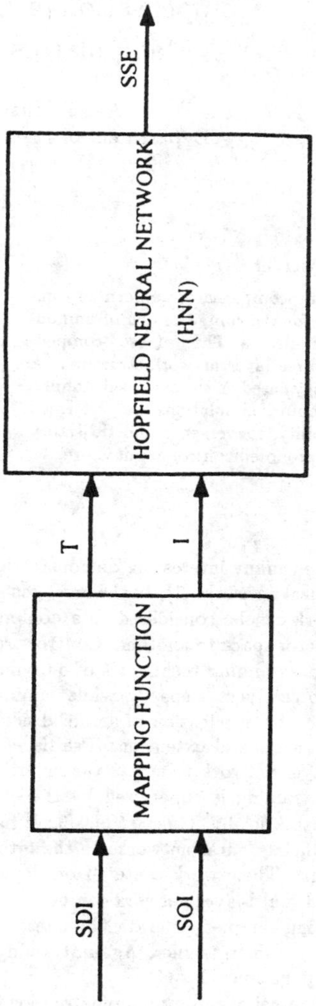

SDI: System Dynamical Information
SOI: System Observation Information
SSE: System State Estimate
T: HNN Synaptic Weight
I: HNN Input Bias

Figure 1. HHN Based State Estimator

FA6 - 9:10

Proceedings of the 31st Conference
on Decision and Control
Tucson, Arizona • December 1992

Compensation of Unstructured Uncertainty in Manipulators using Neural Networks

Aaron Kuan and Behnam Bavarian
Department of Electrical and Computer Engineering
University of California,Irvine
Irvine, California 92717

Abstract

This paper proposes a neurocompensator augmented computed torque control scheme for the compensation of unmodeled frictional effects in manipulators. The proposed compensator is implemented by a three layer network structure. A weight adaptation methodology based on the extended Kalman filter algorithm is used. Computer simulations are performed to verify and study the stability, convergence and trajectory tracking performance of the proposed control architecture.

1 Introduction

In recent years, there have been many interesting demonstrations of the capabilities of artificial neural networks [1]. In the most general sense, an artificial neural network can be considered as a nonlinear function which maps one real vector space to another. Control problems can generally be considered as finding techniques of deriving a nonlinear mapping between two real vector spaces, while minimizing some cost criteria. Therefore, the application of artificial neural networks to control problems is a natural extension. Psaltis *et al.* [2] proposed to train a muilti-layer network to act as the inverse of the plant to be controlled and training is supervised by the *error backpropagation* rule [1]. More specifically, Guez *et al.* [3, 4] have proposed the application of artificial neural networks to the intelligent control of dynamic systems. Their work centered on demonstrating that a continuous valued multi-layer network can be trained to control a dynamic system, using samples generated by linear and nonlinear control laws as well as a human trainer. Again, training is supervised by the error backpropagation rule.

This paper investigates the control of robot manipulators for the continuous path tracking application. Typically, uncertainty in a manipulator dynamic model can be divided into two categories, *structured* and *unstructured uncertainty. Structured uncertainty* is defined as the case of a correct structural model with inaccurate parameter values. In general, the parameter set consists of parameters such as link masses, link lengths, link center of gravity locations and moments of inertia of each link with respect to various link axes. *Unstructured uncertainty* is defined to represent effects not known to the designer, or those that are too complex to model. Unmodeled effects such as friction, motor backlash and external disturbance are typical entries to this category. This paper investigates the effects of frictional forces on manipulator control, and proposes a compensator implemented by artificial neural networks, i.e. a neurocompensator, to compensate for these effects. An extended Kalman filter based algorithm is employed for neural network training.

We begin our discussion by introducing the formulation of the manipulator dynamics and the computed torque control technique in Section 2. The proposed neurocompensator is also presented in the same section. Detail derivation of the extended Kalman filter based training methodologies and stability proof can be found in [5]. The neurocompensator is applied to the computed torque controller as an augmentation to the feedback loop of the control architecture. Performance of the proposed control architecture is studied using the computer model of a simplified 3-link 3-DOF manipulator. Simulation results are presented and discussed in Section 3. Section 4 summarizes the contribution of this paper and outlines future extension of this research.

2 Problem Formulation

A robot manipulator is typically modeled as a serial chain of n rigid bodies. In general, one end of the chain is fixed to some reference surface while the other end is free, thus forming an open kinematic chain of moving rigid bodies. The vector equation of motion of such a device can be written in the form

$$M(\mathbf{q})\ddot{\mathbf{q}} + C(\mathbf{q},\dot{\mathbf{q}})\dot{\mathbf{q}} + G(\mathbf{q}) + F(\dot{\mathbf{q}}) = \mathbf{T}. \quad (1)$$

This vector equation represents the dynamics of an interconnected chain of ideal rigid bodies, where $\mathbf{T}$ is the $n\times 1$ vector of joint actuator torques and $\mathbf{q}$ is the $n\times 1$ vector of generalized joint positions. $M(\mathbf{q})$ is an $n\times n$ matrix, usually referred to as the manipulator mass matrix containing the kinetic energy functions of the manipulator. $C(\mathbf{q},\dot{\mathbf{q}})$ represents torques arising from centrifugal and *Coriolis* forces. $G(\mathbf{q})$ represents torques due to gravitational effects when the manipulator is moving in its work space. $F(\dot{\mathbf{q}})$ represents local frictional effects which are in general unstructured and difficult to model. In this paper we will assume that the frictional forces in the manipulator are largely viscous, and

$$F(\dot{\mathbf{q}}) = F_m(\dot{\mathbf{q}}) + \mathbf{T}_{uf}, \quad (2)$$

where F_m is a diagonal matrix of viscous friction coefficients and $\mathbf{T}_{uf}$ is a vector of unstructured friction effects.

The nonlinear dynamic model of a manipulator, Equation 1, is used to compute the control torque inputs [6]

$$\mathbf{T} = \hat{M}(\mathbf{q})\ddot{\mathbf{q}}^* + \hat{C}(\mathbf{q},\dot{\mathbf{q}})\dot{\mathbf{q}} + \hat{G}(\mathbf{q}) + \hat{F}(\dot{\mathbf{q}}), \quad (3)$$

where the quantities with "$\wedge$" are estimates of the true parameters, and

$$\ddot{\mathbf{q}}^* = \ddot{\mathbf{q}}_d + K_v\dot{\mathbf{E}} + K_p\mathbf{E}. \quad (4)$$

In Equation 4, the servo error $\mathbf{E}$ is defined as

$$\mathbf{E} = \mathbf{q}_d - \mathbf{q}, \quad (5)$$

where $\mathbf{q}_d$ is the desired joint position vector, and K_v and K_p are $n\times n$ constant, diagonal gain matrices with $k_{vj} > 0$ and $k_{pj} > 0$. Equation 3 is referred to as the computed torque method of manipulator control. Any desired trajectory of the manipulator end effector is assumed known and can be expressed as time functions of joint positions, velocities, and accelerations, $\mathbf{q}_d(t)$, $\dot{\mathbf{q}}_d(t)$ and $\ddot{\mathbf{q}}_d(t)$, respectively.

The control law of Equation 3 is selected because in the ideal situation of perfect knowledge of parameter values, the closed-loop error dynamics of the joints are decoupled in the form

$$\ddot{\mathbf{E}} + K_v\dot{\mathbf{E}} + K_p\mathbf{E} = 0, \quad (6)$$

which is derived by substituting Equations 3 and 4 into Equation 1. Therefore, in this ideal situation, the matrices K_v and K_p may be selected to place closed-loop poles of each joint, to achieve desired manipulator performance. This desired performance is contingent on the assumption that the dynamic model and estimated parameters used to generate the control law must match those of the actual system, if the benefits of linearizing and decoupling effects of the computed torque servo are to be realized. However, the frictional forces

CH3229-2/92/0000-2706$1.00 © 1992 IEEE

are in general unstructured and difficult to model. In this paper, we will use an artificial neural network to "learn" and compensate for the effects of the unmodeled frictional forces. In other words, the basic computed torque control law consisting of the first three known terms on the right hand side of Equation 3 will be augmented by a neural network approximation of the unknown term $F(\dot{\mathbf{q}})$.

The network model selected to implement the proposed neuro-compensator is a three layer feedforward network structure. This structure has an input layer, a single hidden layer of nonlinear processing units and an output layer of nonlinear processing units. Therefore, there is one variable weight matrix and one bias vector for the hidden layer and output layer processing units, respectively. The hyperbolic tangent function has been chosen as the nonlinearity in the processing units. Therefore, the input-output relationship of the network is

$$\mathbf{y} = \tanh(W(\tanh(V\mathbf{x} + \mathbf{b}) + \mathbf{d})), \tag{7}$$

where $\mathbf{y}$ and $\mathbf{x}$ are the input and output vectors, respectively, $\mathbf{V}$ and $\mathbf{W}$ are the hidden and output layer weight matrices, respectively, and $\mathbf{b}$ and $\mathbf{d}$ are the hidden and output layer processing units bias vectors, respectively. The weights and biases of the neural network are nonlinearly related to the output, hence, the updating of weights and biases can be viewed as a nonlinear estimation problem. We propose to use the extended Kalman filter algorithm [7], a modification of the linear Kalman filter algorithm to solve this problem.

The algorithm is presented here without proof, details of which can be found in [5]. If we lump all the weights and biases of the neural network into a single weight vector $\mathbf{a}$ and consider it a "state "vector to be estimated, then, the following equations describe the "state" and the "observation dynamic" of the neural network system

$$\mathbf{a}(k+1) = \mathbf{a}(k) + \mathbf{N}_s, \tag{8}$$

$$\mathbf{g}(k) = \mathbf{f_k}(\mathbf{a}(k), \mathbf{x}(k)) + \mathbf{N}_m, \tag{9}$$

in which $\mathbf{a}(k)$ denotes the weight vector and $\mathbf{g}(k)$ is the desired output of the network at time instant k. $\mathbf{f_k}$, whose arguments are the weight vector $\mathbf{a}$ and the input vector $\mathbf{x}$ may be time varying, but in our case is equal to the right hand side of Equation 7. It is assumed that the network output and "state" are corrupted by Gaussian white noise $\mathbf{N}_m$ with covariance matrix $R(k)$, and $\mathbf{N}_s$ with covariance matrix $Q(k)$, respectively. At the beginning of each training cycle, the input $\mathbf{x}$ is set to one of the input in the training set of input-output pairs, and the desired output $\mathbf{g}$ is set to the corresponding output $\mathbf{y}$. During each training cycle, the Kalman filter system is iterated one or more number of times while the values of $\mathbf{x}$ and $\mathbf{g}$ are held fixed. The governing iterative algorithm is described in the following equations:

$$\hat{\mathbf{a}}(k+1) = \hat{\mathbf{a}}(k) + K(k)[\mathbf{g}(k) - \mathbf{f}_k(\hat{\mathbf{a}}(k), \mathbf{x}(k))], \tag{10}$$

$$K(k) = P(k)H^T(k)[R(k) + H(k)P(k)H^T(k)]^{-1}, \tag{11}$$

$$P(k+1) = P(k) - K(k)H(k)P(k). \tag{12}$$

In the above equations, $\hat{\mathbf{a}}(k)$ denotes the estimates of $\mathbf{a}(k)$, K is the Kalman gain, P is the error covariance matrix related to $\mathbf{a}(k)$, and H is the Jacobian of $\mathbf{f}_k$ of Equation 9, with respect to the weight vector $\mathbf{a}$ evaluated at the current estimate $\hat{\mathbf{a}}(k)$. It should be noted that the form of the weight estimate update rule in Equation 10 is similar to many other algorithms such as the "delta rule". However, the Kalman filter algorithm is essentially a second order algorithm in the sense that it makes use of the covariance information, which is second order statistics. Also, the algorithm is gobal implying that a change in the current estimate of a certain weight, potentially affects the changes in all other weight estimates. The global coupling is a result of the off diagonal entries of the covariance matrix, which affects the computation of the Kalman gain. This usually makes the algorithn converge more rapidly than first order algorithms, such as the gradient decent backpropagation algorithm.

3 Simulations

Simulations are performed to study in detail the extended Kalman filter algorithm of neural network training and the neurocompensator augmented computed torque control scheme. A three link, three degree of freedom simplified manipulator, shown in Figure 1, is used in the simulations. The manipulator is modeled as three rigid links with point masses at locations indicated by center of gravity sysbols in Figure 1. The dynamic model is in the general form of Equation 1. The manipulator mass and length parameters are assumed to be known accurately, while the frictional term F is unmodeled. In the simulation, frictional torques in the following form is included in the manipulator dynamics

$$F(\dot{\mathbf{q}}) = V(\dot{\mathbf{q}}) + C\mathsf{sign}(\dot{\mathbf{q}}), \tag{13}$$

where V and C are diagonal matrices of coefficients of viscous and coulomb friction, respectively. In our simulations, the torques associated with frictional effects in joint 2 and 3 of the manipulator are to be captured by the neural network, while joint 1 is assumed to be free of frictional effects.

The network training architecture is shown in Figure 2. Two neural networks are used to capture the torques associated with frictional effects in joint 2 and 3. Each network has one input unit and one output unit, and a hidden layer of twenty units resulting in a total of 61 elements in the weight set. The desired training trajectory of joint position is in the form of

$$q_{id}(t) = K_i e^{-\frac{1}{\alpha}(t-\mu_i)^2},$$

for $i = 2, 3$.

The computed torque control law is generated using *a priori* knowledge of the manipulator model as decribed in Section 2, i.e.

$$\mathbf{T} = \hat{M}(\mathbf{q})\ddot{\mathbf{q}}^* + \hat{C}(\mathbf{q}, \dot{\mathbf{q}})\dot{\mathbf{q}} + \hat{G}(\mathbf{q}). \tag{14}$$

Notice that Equation 14 is identical to Equation 3 except for the term $\hat{F}$, since it is unmodeled. As the manipulator is driven by $\mathbf{T}$ of Equation 14 to executed the desired trajectories, its position vector $\mathbf{q}$ and velocity vector $\dot{\mathbf{q}}$ are recorded at selected time instants, and used to generate torque history $\hat{\mathbf{T}}$ via the known dynamic model. The difference between the control torque $\mathbf{T}$ and the torque generated via "inverse dynamic" $\hat{\mathbf{T}}$, which is identified as $\mathbf{dT}$, is also recored at the same time instants. $\mathbf{dT}$ is the frictional torque vector to be captured by the neural networks. The sets of elements of the vector $\dot{\mathbf{q}}$ and the corresponding sets of elements of the frictional torque vector $\mathbf{dT}$, form the input-output mappings to be learned by the neural networks. That is, the sets of elements of $\dot{\mathbf{q}}$ are used as input x to the network and the sets of elements of $\mathbf{dT}$ as desired output g as describe in Equations 8 and 9.

The top halves of Figures 3 and 4 show the tracking performance of the basic model based computed torque control law without knowledge the frictional effect terms. Trajectory tracking is achieved to some extend with considerable tracking errors. Figure 5 shows the frictional torque set dT_2 of joint 2 generated during training and the corresponding velocity set $\dot{q}_2$. Figure 7 shows results of the training in plots of the actual unmodeled frictional torque of joint 2 and its neural network approximation. Figure 6 shows the frictional torque set dT_3 of joint 3 generated during training and the corresponding velocity set $\dot{q}_3$. Figure 8 shows results of the training in plots of the actual unmodeled frictional torque of joint 3 and its neural network approximation. Figure 9 shows the training errors of joint 2 (top half) and joint 3 (botom half) at each training instant.

After the completion of training, the neural networks can be considered approximator of of $F(\dot{\mathbf{q}})$ in Equation 13 which we will call $\hat{F}(\dot{\mathbf{q}})$. An augmentation to the computed torque control law can be generated by adding $\hat{F}(\dot{\mathbf{q}})$ to the right hand side of Equation 14. With the neurocompensator in place, the augmented control law is used to drive the manipulator so that the same trajectory is followed.

The bottom halves of Figures 3 and 4 show significant improvement in the tracking performance after addition of the approximated frictional terms to the control law.

4 Conclusions

We have presented a second order estimation rule based training algori'hm for three layer artificial neural networks, for compensation of unmodeled frictional torques in a nonlinear robot manipulator system. Results from the simulations show that the training algorithm derived via the extended Kalman eilter is stable and convergence is also verified. The simulations also verified the stability of the computed torque control law augmented by the neurocompensator approximating unmodeled frictional effects. Similar results were obtained in an earlier paper [8] where a one link one degree of freedom simplified manipulator model was used.

In another paper [9], we proposed a similar control scheme, where the neurocompensator is implemented by an *Adaline* network trained by a *Lyapunov* method based algorithm. The control architecture and training algorithm were also verified by simulation using a three link, three degrees of freedom universal manipulator. In both schemes we propose to implement the model based computed torque control with fixed "dedicated" hardware, which accepts joint variables as inputs and generates control torques as outputs. When accurate *a priori* knowledge of model parameters are available, they can be used in the dynamic model and need not be updated. Therefore, the computed torque controller can be realized by "fixed" hardware. From the derivation of the neurocompensator in Section 2, it is clear that the inputs to the neural network are manipulator joint velocities. Therefore, the only recurssive component in our proposed architecture is the computer model of the three layer neural networks. Any future changes in frictional effects can re-approximated by the neural networks, which in turn generate updated compensation torques without having to "tamper" with the fixed hardware.

In general, parameters related to link moments of inertia are also difficult to model accurately. Therefore, we are also studying the learning ability of the proposed neurocompensator when these parameters are to be learned.

References

[1] Rumelhart, D., Hinton G. E. and Williams, R. J., "Learning Internal Representations by Error Propagation", in D. E. Rumelhart and J. L. McCleland, Eds., *Parallel Distributed Processing: Explorations in the Microstructure of Cognition*, Vol 1: Foundations. MIT Press 1986.

[2] Psaltis, D., Sideris, A. and Yamamura, A., "Neural Controllers", *Proc. IEEE 1st Int. Conf. Neural Networks*, San Diego, Jun 1987.

[3] Guez, A., Eilbert, J. and Kam, M., "Neuromorphic Architecture for Adaptive Robot Control: a preliminary analysis", *Proc. IEEE 1st Int. Conf. Neural Networks*, Vol. 1, pp 567, Jun 1987.

[4] Guez, A. and Selinsky J., "A Trainable Neuromorphic Controller", *Journal Rob. Sys.*, 5(4), pp 363-388, 1988.

[5] Wabgaonkar, H. M., "Synthesis of Neural Network Based Associative Memories", *PHD Dissertation.*, Univercity of Calif., Irvine, 1991.

[6] Criag J. J., *Adaptive Control of Mechanical Manipulators*, Addison-Wesley 1988.

[7] Singhal, S. and Wu, L., "Training Multi-layer Perceptrons with the Extended Kalman Algorithm", in *Advances in Neural Information Processing Systems I, Touretzky, D.S. (ed.) pp.133-140*, Morgan Kauffmann, 1989.

[8] Kuan A. and Bavarian B., "Compensation of Unmodeled Friction in Manipulators using Neural Networks", submitted to the *International Joint Conference on Neural Networks*; Jun 1992.

[9] Kuan A. and Bavarian B., "A Neural Network Adaptive Compensator for Variable Load Compensation in Robot Manipulators", Proceedings of the *Singapore International Conference on Intelligent Control and Instrumentation*; Singapore, Feb 17-21, 1992.

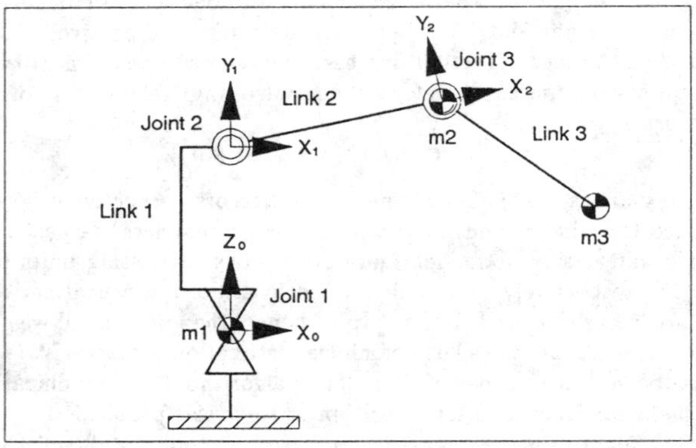

Figure 1: A Simplified Manipulator Model.

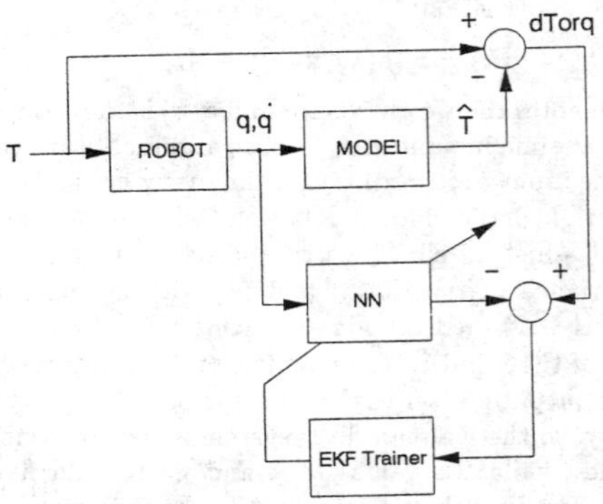

Figure 2: Neural Network Training Architecture.

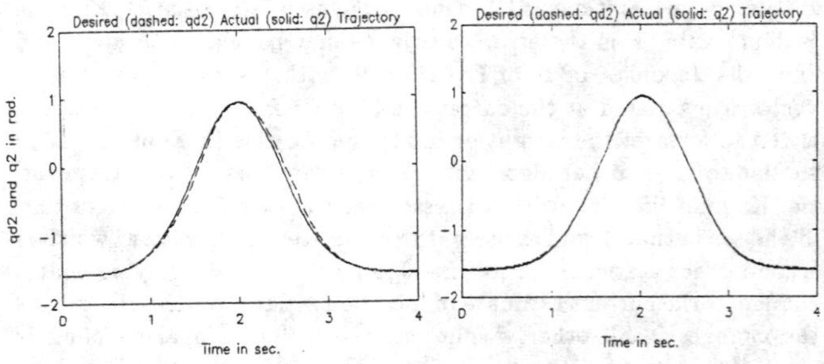

Figure 3: Desired and Actual Joint 2 Position Trajectories.

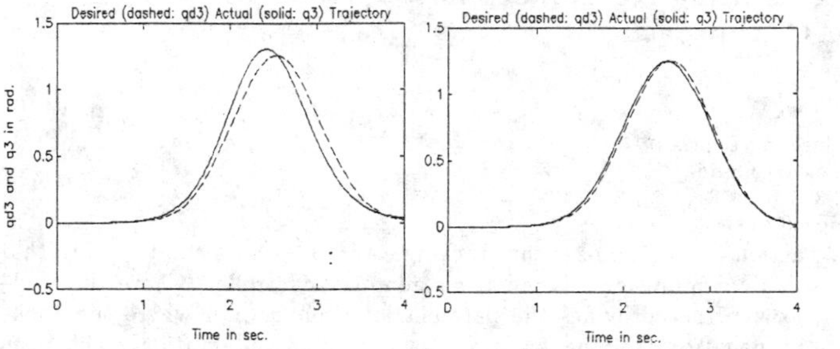

Figure 4: Desired and Actual Joint 3 Position Trajectories.

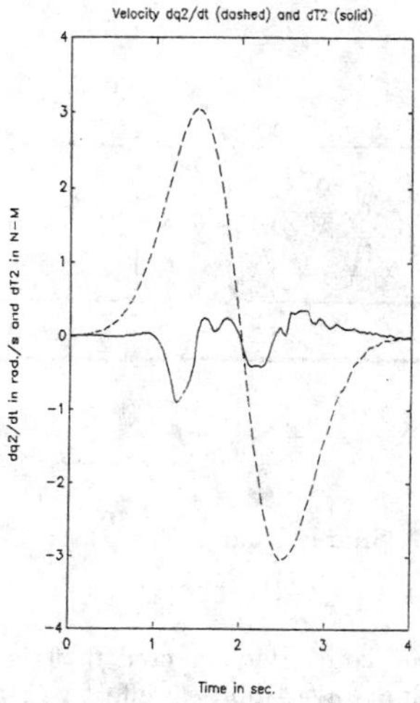

Figure 5: Velocity and Frictional Torque Time History, Joint 2.

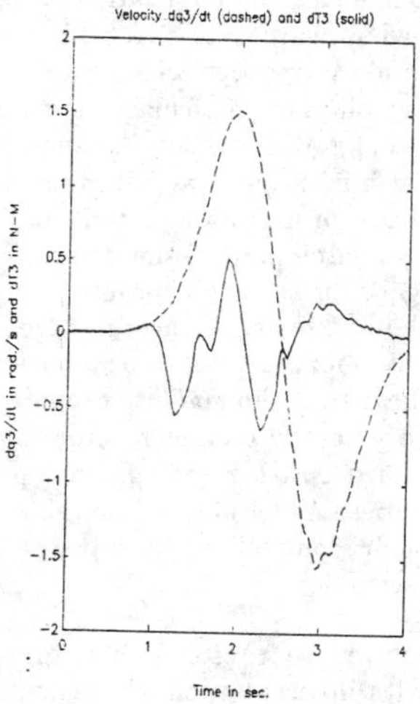

Figure 6: Velocity and Frictional Torque Time History, Joint 3.

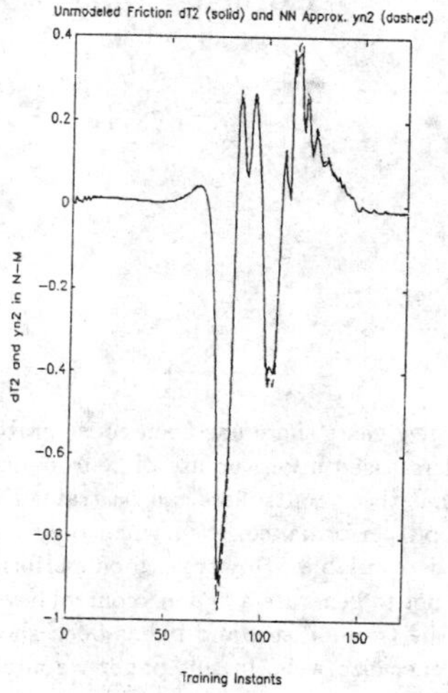

Figure 7: Training Results, Joint 2.

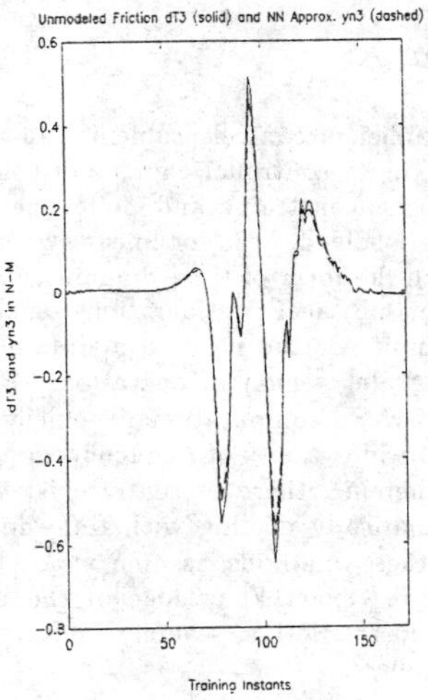

Figure 8: Training Results, Joint 3.

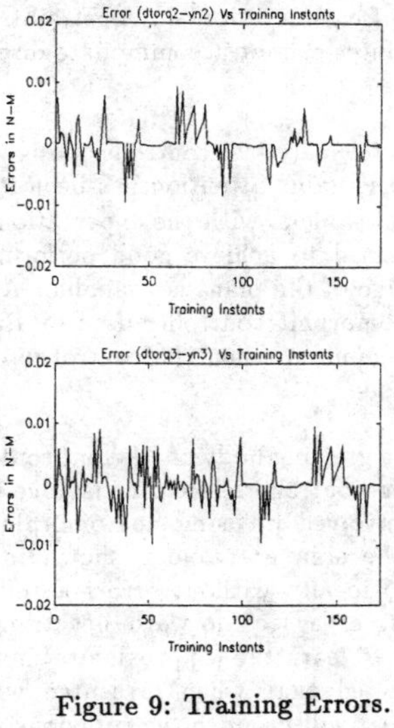

Figure 9: Training Errors.

Proceedings of the 31st Conference
on Decision and Control
Tucson, Arizona • December 1992

Robust Control of Dynamic Systems using Neuromorphic Controllers: A CMAC Approach

Mario Sznaier

Electrical Engineering Dept.,
University of Central Florida,
Orlando, Fl 32816–2450
msznaier@frodo.engr.ucf.edu

Abstract

During the last few years there has been considerable interest in the use of trainable controllers based upon the use of neuron–like elements, with the expectation being that these controllers can be trained, with relatively little effort, to achieve good performance, even when only minimal knowledge of the plant dynamics is available. However, good performance hinges on the ability of the neural net to generate a "good" control law even when the input does not belong to the training set, and it has been shown that neural–nets do not necessarily generalize well. In this paper we address this problem by proposing a feedback controller based upon the use of a CMAC neural net. We show that the proposed controller has good generalization properties. Moreover, by proper choice of the training set the resulting closed–loop system is guaranteed to be robustly stable with respect to model uncertainty.

1. Introduction

A substantial number of control problems can be summarized as the problem of designing a controller capable of achieving acceptable performance under design constraints and model uncertainty. However, the problem is far from solved, even for linear systems. In most cases engineering goals take the form of time–domain constraints reflecting both performance and physical considerations, such as the presence of "hard" actuator limits or the need to maintain the states of the plant confined to a "safe" region of operation. However, although there currently exist several computationally–efficient design methods capable of handling a wide variety of frequency–domain specifications ([1] and references therein), there presently exist very few methods that allow for systematically dealing with time–domain constraints. Moreover, most of these methods assume exact knowledge of the dynamics involved (i.e. exact knowledge of the model). Such an assumption can be too restrictive, severely limiting the applicability of the resulting controllers.

On the other hand, during the last decade a large research effort led to procedures for designing robust controllers, capable of achieving stability under various classes of plant perturbations while, at the same time, satisfying frequency–domain constraints. However, most of these design procedures cannot accommodate directly time domain constraints [1].

As an alternative to analytical controller design methods, during the last few years considerable attention has been focused on the use of neural–net based controllers, with the expectation being that these controllers can be trained to achieve good performance, even when only minimal knowledge of the plant is available. As an example, we can mention the neuromorphic controller used by Barto et. al. [2] to control an inverted pendulum when the control force is restricted to have bounded magnitude.

Although there is a growing body of different control configurations using neural–net based controllers (see for instance [3] and references therein), the issues involved in using neuromorphic controllers can be illustrated with the simple topology shown in figure 1. Here the goal is to follow, ideally without error, a reference trajectory. The controller consist of a feed–forward net trained, using back–propagation methods, to learn the (approximate) inverse of the plant. Therefore, it is expected that when presented with the reference trajectory, the controller will produce the (previously learned) control action that minimizes the tracking error. It is important to note that this configuration resembles an adaptive controller, where the feed–forward net provides the parametric structure and where the back–propagation training functions as a gradient adaptation mechanism [4].

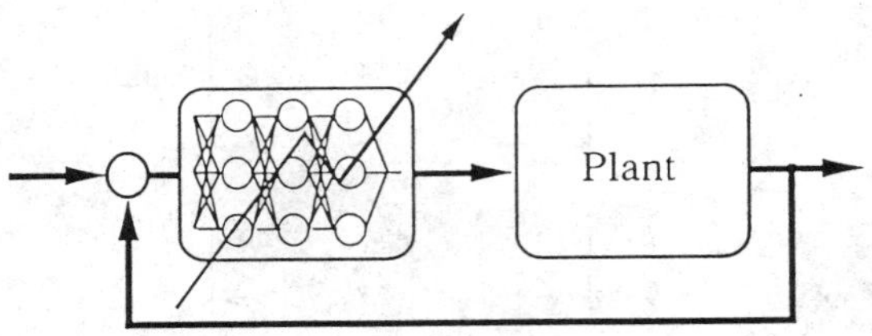

Figure 1. A Simple Neural–Net Based Controller

The success of the controller depicted in figure 1 hinges on the following issues: 1)plant invertibility 2)effective learning and 3)ability of the neural net to generalize, i.e. to provide an appropriate output even when the input is not a member of the training set. Plant invertibility has been exhaustively studied in control theory and therefore poses no new problems. However, the other two issues remain largely ignored in neural net applications to controls. Moreover, the combination of feed–forward nets and back–propagation training exhibits some undesirable properties which may adversely affect the performance of the controller, in particular the stability of the closed–loop system. It is well known (see for instance [5]) that the error surfaces can have local minima and multitude of areas with shallow slopes. Hence the back–propagation algorithm is not guaranteed to converge to a global minimum of the error. Furthermore, even when it does, convergence may take a prohibitively large amount of time, due to the shallow regions. Concerning generalization, it has been shown [6] that feed–forward nets *do not* necessarily generalize well. Therefore, it follows that the *stability* properties of the resulting closed–loop system are generally unknown. Since most critical control applications require "hard", rather than factual proof of closed-loop stability, these difficulties are a major stumbling block preventing the use of neuromorphic controllers, in spite of their potential to outperform classical controllers.

In this paper we propose to solve these problems by using a neural net (CMAC) with inherent good generalization properties and by incorporating a–priori knowledge of the plant dynamics into the design and training processes. We show that by using this knowledge, the resulting neuromorphic controller is capable of robustly stabilizing a family of plants. Furthermore, we give bounds on the mismatch (in the sense of a norm) between the nominal plant (used for the initial training of the network) and the actual plant such that stability of the closed loop system is guaranteed.

The paper is organized as follows: In section II we introduce some required concepts and we present a formal definition to our problem. In section III we briefly describe the CMAC network. Section IV contains

Supported in part by a grant from Florida Space Grant Consortium

CH3229-2/92/0000-2710$1.00 © 1992 IEEE

the bulk of the theoretical results. The main result of the section shows that by incorporating information about the plant dynamics into the design and training processes, the resulting controller is guaranteed to stabilize a family of systems. In section V we present an example of application. Finally, in section VI we summarize our results and indicate directions for future research.

2. Problem Formulation and Preliminary Results

2.1 Statement of the Problem

Consider the family of discrete–time systems represented by the following state-space realization:

$$\underline{x}_{k+1} = A(q)\underline{x}_k + B(q)\underline{u}_k \qquad (P)$$

where $\underline{x}$ represents the state, $\underline{u}$ represents the control input, _ indicates a vector quantity, $q \in \mathcal{Q}$ compact represents uncertainty, and where the dynamics satisfy the following conditions:

$$A(q) = A_o + \Delta_A(q),\ B(q) = B_o + \Delta_B(q) \qquad (U)$$

Finally, assume that the uncertainties $\Delta_A(q)$ and $\Delta_B(q)$ are norm–bounded by:[‡]

$$\begin{aligned} \max_{q\in\mathcal{Q}} \|\Delta_A(q)\| &\leq \Delta_1, \\ \max_{q\in\mathcal{Q}} \|\Delta_B(q)\underline{u}\| &\leq \Delta_2(\underline{u}), \end{aligned} \qquad (1)$$

Then, the basic control problem that we address in this paper is the following:

• **Robust Constrained Control Synthesis Problem:** Given the family (P) find a *feedback* controller such that, for all $q \in \mathcal{Q}$, the resulting closed–loop system satisfies the following specifications:

i) The states remain confined to a region $\mathcal{G} \subset R^n$, where $\mathcal{G}$ is a compact, convex balanced set (i.e such that $\underline{x} \in \mathcal{G} \Rightarrow \lambda\underline{x} \in \mathcal{G}$ for $|\lambda| \leq 1$) containing the origin in its interior.

ii) The control effort is constrained by $\underline{u}_k \in \Omega \subset R^m$, where Ω is a compact, convex set containing the origin in its interior.

iii) Given an open, convex, target set O containing the origin in its interior, the system is driven to O, for any initial condition $\underline{x}_o \in \mathcal{G}$. This performance specification is closely related to the concept of practical stability [8].

Remark 1: Note that due to the existence of state and control constraints, this control problem does not admit, in general, a closed–form solution. Hence it is specially well suited for a training–based approach, requiring only knowledge of the appropriate control action for a finite given sets of inputs.

2.2 Definitions and Preliminary Results

In this subsection we introduce the definitions required to analyze the properties of the closed–loop system obtained when using a CMAC–based controller. These ideas, illustrated in figure 2, formalize the concept of "quantization" of state–space.

• **Def. 1:** Consider a closed set $\mathcal{G} \subseteq R^n$. A family $\mathcal{C}$ of closed sets C_i is called a *closed cover* of $\mathcal{G}$ if $\mathcal{G} \subseteq \bigcup_i C_i$

‡ Since all finite dimensional norms are equivalent [7], it is unnecessary to specify the actual norm. We will use this freedom in selecting the norm in section IV.

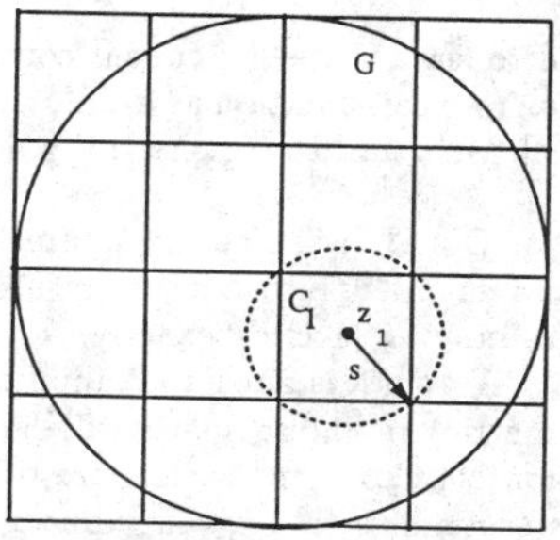

Figure 2: A Closed Cover Formed by Square Boxes of Size s

• **Def. 2:** Consider a closed set $\mathcal{G} \subseteq R^n$ and a closed cover $\mathcal{S} = \{S_i\}$. A quantization χ of $\mathcal{G}$ is a set $\chi = \{z_i\}$ containing exactly one element from each set S_i.

• **Def. 3:** Given a quantization χ of a set $\mathcal{G}$, the *size* of the quantization with respect to some norm $\mathcal{N}$ defined in $\mathcal{G}$ is defined as:

$$s = \min_i \{r : C_i \subseteq B(\underline{z}_i, r)\ \forall i\}$$

where $B(\underline{z}_i, r)$ indicates the $\mathcal{N}$-norm ball centered at $\underline{z}_i$ and with radius r.

Consider now the case where the sets of the family $\mathcal{C}$ that defines a quantization χ have pairwise disjoint interiors (i.e. $\text{int}(C_i) \cap \text{int}(C_j) = \emptyset, i \neq j$). In this case, $\mathcal{C}$ induces an equivalence relation in $\mathcal{G}$ as follows:

• **Def. 4:** Consider a closed cover $\mathcal{C}$ of $\mathcal{G}$ with pairwise disjoint interiors, and two points $\underline{x}_1, \underline{x}_2 \in \mathcal{G}$. $\underline{x}_1$ and $\underline{x}_2$ are *equivalent modulo* $\mathcal{C}$ if $\exists\ i$ such that $\underline{x}_1$ and $\underline{x}_2 \in \text{int}(C_i)$. To complete the partition of $\mathcal{G}$ into equivalence classes, we assign the points that are in $C_i \cap C_j$ (i.e. in the common boundary) *arbitrarily* to either one of the classes. Two points equivalent modulo $\mathcal{C}$ will be denoted as $\underline{x}_1 \equiv \underline{x}_2$.

• **Def. 5:** Consider a quantization $\chi = \{\underline{z}_i\}$ of a given set $\mathcal{G}$. It follows from Definitions 2 and 4 that for *any* point $\underline{x} \in \mathcal{G}$ there exists an element $\underline{z} \in \chi$ such that $\underline{z} \equiv \underline{x}$. We will define the operator that assigns $\underline{x} \rightarrow \underline{z}$ as the *quantization* operator and we will denote it as: $\underline{z} = \chi(\underline{x})$.

Finally, we show that the set $\mathcal{G}$ induces a norm in R^n. This norm will be used to design a CMAC–based controller, guaranteed to stabilize the family (P).

• **Def. 6:** [9] The *Minkowsky Functional* (or gauge) p of a convex set $\mathcal{G}$ containing the origin in its interior is defined by

$$p(\underline{x}) = \inf_{r>0} \left\{ r : \frac{\underline{x}}{r} \in \mathcal{G} \right\} \qquad (2)$$

A well known result in functional analysis (see for instance [9]) establishes that p defines a seminorm in R^n. Furthermore, when $\mathcal{G}$ is compact, this seminorm becomes a norm. In the sequel, we will denote this norm as $\|\underline{x}\|_{\mathcal{G}} \triangleq p(\underline{x})$

3. The CMAC Neural Net

3.1 Description of CMAC

In this section we provide a brief description of the Cerebellar Model Articulation Controller (CMAC) neural net. The reader is referred to [10–11] for more details. Originally introduced by Albus [10] for learning to control a robotic arm, the CMAC network has often been overlooked by the Neural Net community, mainly because it was considered impractical. However, in the last few years it has become the focus of growing interest, prompted by the disadvantages of back–propagation mentioned earlier. In particular CMAC has been successfully used to learn state-dependent control actions. Among the recent applications we can mention as examples the work of Miller and coworkers [12–13], Ersu and coworkers [14], and Moody [15].

A diagram of CMAC is shown in figure 3. The input space X is discretized and mapped into a "conceptual" memory M, in such a way that each input $\underline{x}$ excites exactly A^* association cells in M. The mapping $S : X \rightarrow M$ is such that inputs that are close (in the sense of some metric) in input space will have their corresponding sets of association cells overlap, with more overlap for closer inputs. The output corresponding to a given input $\underline{x}$ is obtained by adding the contents of the A^* association cells excited. The number A^* can be thought as the ratio of generalization width to quantization width. A larger A^* provides for better generalization at the price of larger memory requirements or reduced resolution for individual input patterns. Finally, in real implementations, to reduce the memory requirement the "conceptual memory" M is mapped, (using hashing) into a physical memory M'.

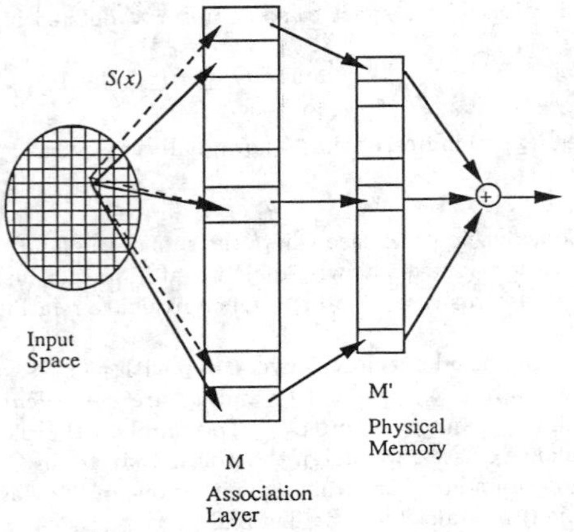

Figure 3. Diagram of a CMAC Neural Net

In order to complete the description of CMAC, a training rule must be provided. Consider, for simplicity the case where the output is a scalar u. Let $\underline{x}$ be the input, u_d the desired output and u^k the output of the net after the kth iteration. Then, the simplest (one-shot error correction) CMAC rule evenly distributes the error among the contents of the A^* association cells excited by the input pattern, i.e.:

$$w_i^{k+1} = w_i^k + \frac{u_d - u^k}{A^*}, \; i \in S(\underline{x}) \tag{3}$$

where $S(\underline{x})$ indicates the set of association cells excited by the input $\underline{x}$ and w_i is the content of the ith cell. It has recently been shown [16] that, in the absence of collisions in the hashing map, this training scheme amounts to solving a system of simultaneous equations using the Gauss-Seidel iterative procedure. Therefore, the training is guaranteed to converge, provided that some mild structural restrictions are observed.

It has been argued that the CMAC structure has the potential to provide for good generalization properties through the overlapping of the sets of association cells. For a new input $\underline{x}$ which is close to the learned inputs $\underline{x}_1, \underline{x}_2, ...\underline{x}_k$ the association cells $S(\underline{x})$ will have some ovelap with the sets $S(\underline{x}_1), S(\underline{x}_2)...S(\underline{x}_k)$ and therefore a natural interpolation will occur. Let $u_1, ...u_k$ denote the corresponding learned outputs. The overlapping of the sets guarantees that the output $u = \text{CMAC}(\underline{x}) = \sum_{i \in S(\underline{x})} w_i$ will be close in some sense to the learned outputs u_i. However, it *does not necessarily* imply that u will belong to the convex hull of the points u_i as illustrated by the simple example shown in figure 4. There we have a situation where the set $S(\underline{x}) \subset S(\underline{x}_1) \cup S(\underline{x}_2)$. However, $u(\underline{x}) = 2$ which *does not* belong to the segment $\overline{u(\underline{x}_1)u(\underline{x}_2)}$. One can easily envision a situation where the generated control action (which has opposite direction to those corresponding to the closest training points ($\underline{x}_1$ and $\underline{x}_2$)) can move the system in the wrong direction. It follows that, unless provisions are made during training to ensure that a situation similar to this cannot arise, the generalization properties of CMAC are not enough to guarantee stability of the closed-loop system.

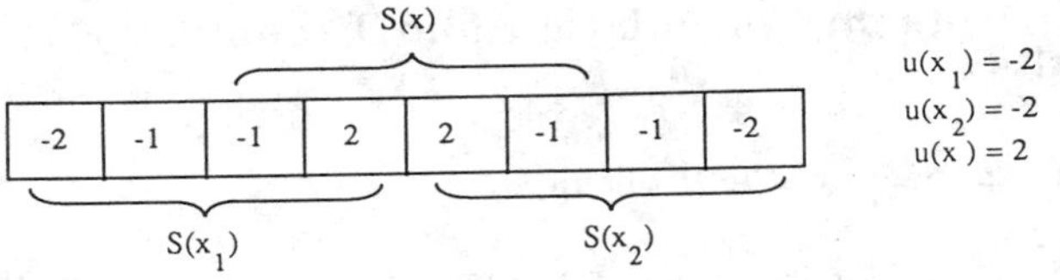

Figure 4. Example Illustrating Potential Problems when Using a CMAC

In this paper, to avoid this problem we will make the following assumption: There exist a set $\mathcal{B}$ of inputs such that i) every association cell in M is excited by at least one input $\underline{b} \in \mathcal{B}$ ii) for $\underline{b}_i, \underline{b}_j \in \mathcal{B}$, $S(\underline{b}_i) \bigcap S(\underline{b}_j) = \emptyset$, i.e, the set of association cells corresponding to the elements of $\mathcal{B}$ are mutually disjoint. The set $\mathcal{B}$ will be called a *basis* for CMAC.

When the training set is limited to the set $\mathcal{B}$, then the learning process (3) becomes trivial, converging in one iteration. It may be argued then that in this case the whole CMAC approach becomes trivial, since the storage capacity of the network appears to be under-utilized. However, note that two of the key features that make CMAC attractive, namely robustness to unit failure (obtained by distributing the information among several cells) and speed of computation remain intact. Furthermore, as we show in next section, by limiting the training set to $\mathcal{B}$, CMAC is guaranteed to produce an appropriate output for each possible point of the input space, thus making unnecessary the use of training patterns outside $\mathcal{B}$.

3.2 Designing a CMAC for Control Systems Applications

From the definitions of section 2.2 and the description above, it follows that a CMAC architecture is conceptually equivalent to considering a quantization χ of the input space and mapping all the elements of an equivalence class to the same set of association cells. Hence, a CMAC design can proceed as follows:

1) Select a closed cover $\mathcal{C}$ for $\mathcal{G}$. For simplicity, in the sequel we consider the case where $\mathcal{C}$ is formed by n-dimensional hypercubes, with sides of size δx parallel to the coordinate axes. Thus, each equivalence class $C(\underline{x}^-, \underline{x}^+)$ is formed by all the vectors $\underline{x}$ such that their components satisfy: $x_i \in [x_i^-, x_i^+)$, $i = 1, \ldots n$, where $x_i^+ - x_i^- = \delta x$.

2) Form a quantization by selecting one element from each equivalence class. We will select as representative of the class $C(\underline{x}^-, \underline{x}^+)$ the point $\underline{z} \triangleq \frac{1}{2}(x_1^- + x_1^+, x_2^- + x_2^+, \ldots, \underline{x}_n^- + \underline{x}_n^+)$. Thus, given an input vector $\underline{x}$, the corresponding $\underline{z} = \chi(\underline{x})$ can be easily found by discretizing each coordinate x_i of $\underline{x}$ with resolution δx.

The operation of the resulting CMAC can be described by the composition of the quantization and association maps, i.e:

$$\begin{aligned} A^* &= S \circ \chi(\underline{x}) \\ \underline{u} &\triangleq \text{CMAC}(\underline{x}) = \sum_{A^*} w_i \end{aligned} \tag{4}$$

To complete the description, we need to determine the number of association cells excited by each input pattern and the mapping S from input-space to association-cells space. We will choose A^* according to the following formula: $A^* \triangleq c_q^n$, were the integer c_q is a design parameter. Each input pattern will be mapped to the c_q^n cells starting at its equivalence class and extending c_q units in each direction, as shown in figure 5. Finally, we will choose as basis $\mathcal{B}$ a subset of $\{\underline{z}_i\}$ with coordinates spaced $c_q\delta x$. Hence, each group of A^* association cells corresponds to intervals of dimension δx along each coordinate axis, and contains c_q^n equivalence classes. It is easily seen that this choice of sets, along with rule (4) guarantees that for each point in input space (after discretizing) the output is given by the linear combination of the outputs of the corresponding basis points, i.e. if

$$\underline{z} = \chi(\underline{x}) = \sum_{i=1}^{2^n} \lambda_i \underline{b}_i;\ \lambda_i \geq 0; \sum \lambda_i = 1$$

then:

$$\text{CMAC}(\underline{x}) = \sum \lambda_i \text{CMAC}(\underline{b}_i) \tag{5}$$

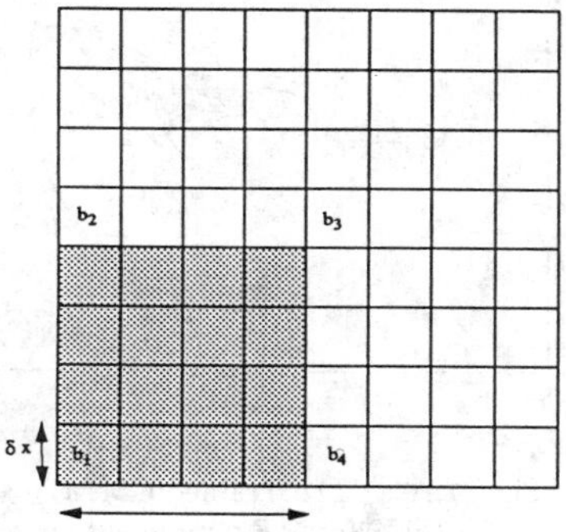

Figure 5. Determining the State Space Regions Corresponding to Each Set of Association Cells

4. Theoretical Results

In this section we present the basic theoretical results on solving the Robust Constrained Control Problem using CMAC. The main result of this section shows that by a proper choice of the parameters δx and c_q, the design procedure of section 3.2 leads to a controller guaranteed to satisfy the requirements specified in section 2.1. The proof proceeds along the following steps: i) obtain a lower bound on the amount that the norm of the present state of the system can be decreased in one step ii) use this bound to show that for each element of the basis there exist a control law that decreases the norm of the present state. iii) show that by proper choice of the basis this is true for any state. It follows then that the control law generated by CMAC is guaranteed to stabilize the system. These ideas are formalized in Lemmas 1, 2 and 3 (proved in the Appendix) and in Theorem 1.

We will address first the case were no collisions occur during hashing and then we will indicate how to modify our results to take the effects of hashing into account.

• **Lemma 1:** Consider a target set $O \subset \mathcal{G}$ and let O_1 be the subset formed by the equivalence classes entirely contained in O, i.e $O_1 \triangleq \cup C_i,\ C_i \subseteq O$ (see figure 6). Let:

$$\Lambda = \min_{\underline{x} \in \mathcal{G} - O_1} \left\{ \lambda > 0 : (\frac{1}{\lambda}\underline{x}) \in \partial\mathcal{G} \right\} \tag{6}$$

where $\partial\mathcal{G}$ denotes the boundary of the set $\mathcal{G}$. Then:

$$\min_{\underline{x}\in\mathcal{G}-O_1} \left\{ \|\underline{x}\|_\mathcal{G} - \min_{\underline{u}\in\Omega} \{\|A_o\underline{x} + B\underline{u}\|_\mathcal{G} + \Delta_1\|\underline{x}\|_\mathcal{G}\} \right\} >$$
$$\Lambda \min_{\underline{y}\in\partial\mathcal{G}} \left\{ 1 - \min_{\underline{u}\in\Omega} \{\|A_o\underline{x} + B_o\underline{u}\|_\mathcal{G} + \Delta_2(\underline{u}) + \Delta_1\} \right\} \tag{7}$$

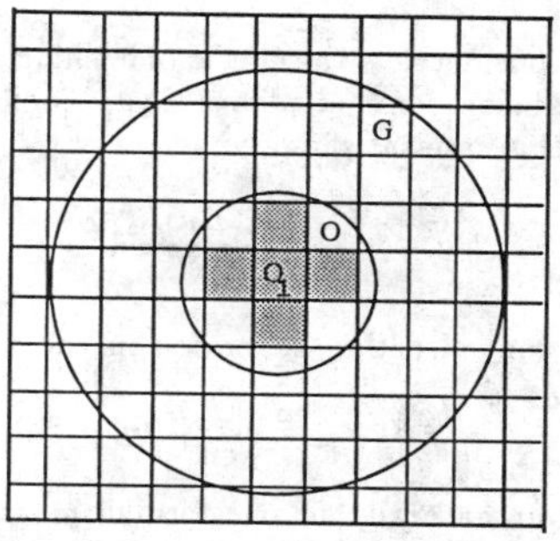

Figure 6: The Sets $\mathcal{G}$, O and O_1 (shadowed) of Lemma 1

• **Lemma 2:** Assume that:

$$1 - \max_{\|\underline{x}\|_\mathcal{G}=1} \left\{ \min_{\underline{u}\in\Omega} \{\|A_o\underline{x} + B_o\underline{u}\|_\mathcal{G} + \Delta_2(\underline{u}) + \Delta_1\} \right\} = \delta > 0 \tag{8}$$

Then, for each element $\underline{b}_i \in \mathcal{B}$, $\underline{b}_i \in \mathcal{G} - O_1$ there exist an admissible control law $\underline{u}_i$ such that $\|A(q)\underline{b}_i + B(q)\underline{u}\|_\mathcal{G} < \|\underline{b}_i\|_\mathcal{G}$ for all $q \in \mathcal{Q}$.

• **Lemma 3:** Assume that condition (8) holds. Then, δx and c_q can be selected such that, for any input $\underline{x}_o \in \mathcal{G} - O$ the control action $\underline{u}_o$ generated by CMAC is admissible and such that $\|A(q)\underline{x}_o + B(q)\underline{u}_o\|_\mathcal{G} < \|\underline{x}_o\|_\mathcal{G}$

• **Theorem 1:** Assume that (8) holds and define the vector $\underline{\delta x} \triangleq (\delta x, \delta x, \ldots \delta x)$. If the CMAC design parameters $\underline{\delta x}$ and c_q are selected such that:

$$\|\underline{\delta x}\|_\mathcal{G} < \frac{\Lambda\delta}{c_q - \frac{1}{2} + \frac{\|A\|_\mathcal{G}}{2}} \tag{9}$$

and the resulting CMAC is trained according to (3), then the resulting controller solves the robust constrained control problem.

Proof: Let $\underline{x}_o$ be an arbitrary initial condition in $\mathcal{G}$. If $\underline{x}_o \in O$ the theorem is trivial, so consider the case where $\underline{x}_o \notin O$. Then, from Lemma 3 it follows that, as long as $\underline{x}_k \notin O$, the sequence $\mathcal{U} = \{\underline{u}_o, \underline{u}_1 \ldots\}$ of control actions generated by CMAC is admissible and such that:

$$\begin{aligned} \|\underline{x}_1\|_\mathcal{G} &< \|\underline{x}_0\|_\mathcal{G} - \mu \\ \|\underline{x}_2\|_\mathcal{G} &< \|\underline{x}_1\|_\mathcal{G} - \mu \\ &\vdots \\ \|\underline{x}_m\|_\mathcal{G} &< \|\underline{x}_{m-1}\|_\mathcal{G} - \mu \end{aligned} \tag{10}$$

where $\mu = \Lambda\delta - [c_q - 1 + \frac{1}{2}(1 + \|A\|_\mathcal{G})]\ \|\underline{\delta x}\|_\mathcal{G} > 0$. It follows then that there exists n_o such that $\underline{x}_{n_o} \in O$. Furthermore, since $\underline{x}_o \in \mathcal{G}$ then $\|\underline{x}_o\|_\mathcal{G} \leq 1$. Hence $\|\underline{x}_i\|_\mathcal{G} < 1$ which implies that $\underline{x}_i \in \mathcal{G}$ for all i. Therefore all the requirements specified in section 2.1 are satisfied $\diamond$.

•**Corollary 1:** The size δx of the quantization introduced in Theorem 1 is inversely proportional to Λ. Hence, as the size of the target set gets smaller, the number of cells increases, while the size of the state–space region that they cover decreases. However, note that the target set O is achieved through a sequence of *intermediate* sets $O_i, i = 1, 2 \ldots, n$ with $O_1 \equiv int(\mathcal{G})$ and $O_n \equiv O$. Since Λ in (6) can be thought of as a lower bound of the ratio of the norm of the next state of the system to the norm of the present state, it follows that to guarantee the practical stability of the closed–loop system, it suffices to choose:

$$\Lambda = \max_i \Lambda_i;\ \Lambda_i = \min_{\underline{x}\in\overline{O}_i - O_{i+1}} \left\{ \lambda > 0 : (\frac{1}{\lambda}\underline{x}) \in \partial\overline{O}_i \right\} \tag{11}$$

where $\overline{O}$ denotes the closure of O.

•**Corollary 2:** Consider now the effects of hashing. Assume that there is at most one collision per equivalence class. Then, the control action $\underline{u}_c$ generated in this case satisfies:

$$\|\underline{u}_c - \underline{u}_o\|_{\mathcal{G}} \leq \frac{2(Bu)_{\max}}{c_q^n} \stackrel{\Delta}{=} \delta u$$

where $\underline{u}_o$ is the control action generated in the absence of collisions and where:

$$(Bu)_{\max} \stackrel{\Delta}{=} \max_{\underline{u}\in\Omega} \|B\underline{u}\|_{\mathcal{G}}$$

Hence collisions fit naturally in our formalism as another source of uncertainty. It follows that in this case stability can be guaranteed by selecting:

$$\|\underline{\delta x}\|_{\mathcal{G}} \leq \frac{\Lambda\delta - \delta u}{c_q - \frac{1}{2} + \frac{\|A\|_{\mathcal{G}}}{2}}$$

and c_q large enough so $\delta u < \Lambda\delta$.

5. A Simple Example

Consider the problem of bringing the angular velocity of a spinning space station with a single axis of symmetry from an initial condition $\underline{x}_o$, $\|\underline{x}_o\|_2 = R_x$ to a final state such that $\|\underline{x}_f\|_2 \leq R_f$. This situation can model the case where a CMAC–based controller is used to bring a system to some region (for instance a region where the constraints are not binding) where some relatively easy to design controller can take over. The nominal system can be represented by [17]:

$$A_o = \begin{pmatrix} \cos T & \sin T \\ -\sin T & \cos T \end{pmatrix} \; B_o = \begin{pmatrix} \sin T & (1-\cos T) \\ (\cos T - 1) & \sin T \end{pmatrix}$$

$$\mathcal{G} = \{\underline{x} \in R^2 : \|\underline{x}\|_2 \leq R_x\},\; \Omega = \{\underline{u} \in R^2 : \|\underline{u}\|_2 \leq 1\} \quad (12)$$

where T is the sampling interval. Note that in this case $\|\underline{x}\|_{\mathcal{G}} = \frac{\|\underline{x}\|_2}{R_x}$. Let $\|\Delta A\|_2$ and $\|\Delta B\|_2\|\underline{u}\|_2$ be the uncertainty bounds introduced in (1). Then:

$$\begin{aligned} \|A_o\underline{x}_k + B_o\underline{u}\|_2 + \|\Delta B\|_2\|\underline{u}\|_2 &= (\|\underline{x}_k\|_2^2 + 2\underline{x}^T A_o^T B_o \underline{u} + \underline{u}^T B_o^T B_o \underline{u})^{\frac{1}{2}} \\ &+ \|\Delta B\|_2\|\underline{u}\|_2 = (\|\underline{x}_k\|_2^2 + 2\underline{x}^T B_o^T \underline{u} + \alpha^2\|\underline{u}\|_2^2)^{\frac{1}{2}} + \|\Delta B\|_2\|\underline{u}\|_2 \end{aligned} \quad (13)$$

where $\alpha^2 \stackrel{\Delta}{=} 2(1-\cos T)$. It can be easily shown that, as long as $\|\underline{x}\|_2 > \alpha$ and $\|\Delta B\|_2 < \alpha$, the minimum over $\underline{u} \in \Omega$ of (13) is achieved by selecting: $\underline{u}_o = \frac{-B_o\underline{x}_k}{\|B_o\underline{x}_k\|_2}$. For this value $\underline{u}_o$ we have:

$$\|A_o\underline{x}_k + B_o\underline{u}_o\|_2 + \|\Delta B\|_2\|\underline{u}_o\|_2 = \|\underline{x}_k\|_2 - \alpha + \|\Delta B\|_2 \quad (14)$$

From (14) it follows that:

$$\begin{aligned} \delta &= 1 - \max_{\|\underline{x}\|_{\mathcal{G}}=1} \left\{ \min_{\underline{u}\in\Omega} \{\|A_o\underline{x} + B_o\underline{u}\|_{\mathcal{G}} + \|\Delta B\|_2\|\underline{u}_o\|_2 + \|\Delta A\|_2\} \right\} \\ &= \frac{\alpha - \|\Delta B\|_2 - \|\Delta A\|_2}{R_x} \end{aligned} \quad (15)$$

Since in this case $\|.\|_{\mathcal{G}}$ is simply the euclidian norm scaled by R_x it follows that $\|A\|_{\mathcal{G}} = 1$. Hence, from (9) we have that:

$$\|\underline{\delta x}\|_{\mathcal{G}} = \frac{\|\underline{\delta x}\|_2}{R_x} < \frac{\Lambda\delta}{c_q} = \frac{\Lambda(\alpha - \Delta)}{c_q R_x} \quad (16)$$

where $\Delta \stackrel{\Delta}{=} \|\Delta A\|_2 + \|\Delta B\|_2$. Hence:

$$\delta x \leq \frac{\Lambda(\alpha - \Delta)}{\sqrt{2}c_q} \quad (17)$$

Since the norm of the present state of the system can be decreased at each stage by Λ it follows that δx should be selected such that:

$$\delta x \leq R_x - \Lambda R_x = R_x(1 - \Lambda) \quad (18)$$

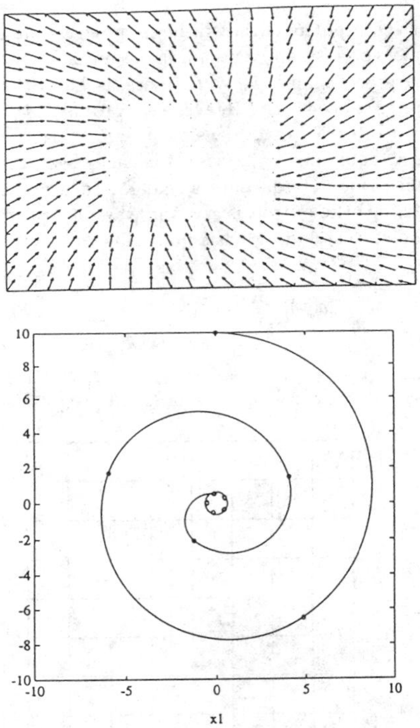

Figure 7: a) The Trained CMAC
b) Trajectory for the Simple Example with $\underline{x}_o = (0,\ 10)$

to guarantee that the present and next state of the system are in different equivalence classes. Hence, the region $\|\underline{x}\|_2 \leq \alpha$ can be reached by designing a CMAC such that:

$$\frac{\Lambda(\alpha - \Delta)}{\sqrt{2}c_q} \leq R_x(1 - \Lambda) \iff \Lambda \leq \frac{1}{1 + \frac{\alpha - \Delta}{R_x\sqrt{2}c_q}} \quad (19)$$

In our case selecting $T = 2.5$ sec., $c_q = 3$ and $R_x = 10$ yields $\alpha = 1.898$ $\Lambda = 0.9615$ and $\delta x = 0.3848$

Figure 7 a) shows the contents of the association cells for the trained CMAC (i.e. the two dimensional control vector), the untrained square in the center corresponding to the region O_1. Figure 7 b) shows a sample trajectory driving the system from the boundary of the admissible region to the target set.

6. Conclusions

During the last few years, there has been considerable interest in the use of trainable controllers based upon the use of neuron like elements. These controllers can be trained, for instance by presenting several instances of *"desirable"* input–output pairs, to achieve good performance, even in the face of poor or minimal modeling. However, the use of neuromorphic controllers has been hampered by the facts that good performance hinges on the ability of the neural–net to generalize the input–output mapping to inputs that are not part of the training set. Through examples [6], it has been shown that neural–nets *do not* necessarily generalize well. Therefore, it follows that the *stability* properties of the closed–loop system are unknown. Moreover, it is conceivable that poor generalization capabilities may result in limit cycles or even in destabilizing control laws. In this paper we address these problems by proposing a controller based upon a neural–net (CMAC) with good generalization properties. Using the similarity of CMAC with quantization of state–space, we develop an analytical framework to investigate the properties of the resulting closed–loop system. These theoretical results are used to show that by incorporating partial a–priori information about the plant in the design process, the resulting controller is guaranteed to stabilize a family of plants. Perhaps the most valuable contribution of this paper results from the qualitative aspects of equation (9), that identify the factors that affect any controller based upon the quantization of state–space (independently of the specific implementation). Most notably, through the norm of the operator that appears in (9), it is possible to formalize the idea of *"poor"* modeling and to design a *"robust"* controller capable of accommodating modeling errors and disturbances.

There are several questions that remain open. The proposed design process is guaranteed to yield a stabilizing controller for all possible members of the family (P). However, since it is based upon a "worst-case" approach, it may achieve so at the expense of performance. Since one of the main reasons for using neural–net based controllers is their ability to yield good performance with imperfect models, the proposed off–line training may be combined with an on–line training (such as the one proposed in [18]) with the goal of improving performance. This research direction is currently being pursued.

Finally, the results of Theorem 1 that guarantee stability can be overly restrictive in some cases, since they result from a "worst–case" type analysis. A relaxed version of these conditions will be highly desirable.

7. References

[1]. M. Sznaier, "Set Induced Norm Based Robust Control Techniques," to appear in *Advances in Control and Dynamic Systems,* C. T. Leondes Editor, Academic Press, 1992.

[2]. A. G. Barto, R. S. Sutton, and C. W. Anderson, "Neuronlike adaptive elements that can solve difficult learning control problems," *IEEE Trans. Syst. Man. Cybern.,* Vol SMC–13, pp. 834–846, 1983.

[3]. T. Miller III, R. S. Sutton and P. J. Werbos editors, *Neural Networks for Control,* MIT Press, Cambridge, Mass., 1990.

[4]. D. Psaltis, A. Sideris and A. Yamamura, "A Multilayered Neural Network Controller," *IEEE Control Systems Magazine,* Vol 8, 3, pp 17–21, April 1988.

[5]. R. Hecht–Nielsen, *Neurocomputing,* Addison–Wesley, 1991.

[6]. Sideris, A., "Controlling Dynamical Systems by Neural Networks," *28th IEEE Conf. on Decision and Control,* Tampa, Florida, 1989.

[7]. R. A. Horn and C. R. Johnson, *Matrix Analysis,* Cambridge University Press, 1985.

[8]. B. R. Barmish, M. Corless and G. Leitmann, "A New Class of Stabilizing Controllers for Uncertain Dynamical Systems," *SIAM Journal Contr. Opt.,* Vol 21, 2, pp 246–255, 1983.

[9]. J. B. Conway, *A Course in Functional Analysis,* Vol 96 in Graduate Texts in Mathematics, Springer–Verlag, New–York, 1990.

[10]. J. S. Albus, "A New Approach to Manipulator Control: The Cerebellar Model Articulation Controller (CMAC)," *Trans. ASME J. Dynamic Syst. Meas. Contr.,* Vol 97, pp. 220–227, 1975.

[11]. J. S. Albus, "Data Storage in the Cerebellar Model Articulation Controller (CMAC)," *Trans. ASME J. Dynamic Syst. Meas. Contr.,* Vol 97, pp. 228–233, 1975.

[12]. W. T. Miller III, "Real–Time Application of Neural Networks for Sensor Based Control of Robots with Vision," *IEEE Trans. Syst. Man. Cybern.,* Vol SMC–19, pp. 825–831, 1989.

[13]. W. T. Miller III, F. H. Glantz and L. G. Kraft, "CMAC: An Associative Neural Network Alternative to Backpropagation," *Proc IEEE,* Vol 78, pp. 1561–1567, 1990.

[14]. E. Ersu and J. Militzer, "Real–Time Implementation of an Associative Memory–Based Learning Control Scheme for Non–Linear Multivariable Processes," *Proc. 1st Measurements and Control Symposium on Applications of Multivariable Systems Techniques,* pp. 109–119, 1984.

[15]. J. Moody, "Fast Learning in Multi–Resolution Hierarchies," *Advances in Neural Information Processing Systems 1,* D. S. Touretzky Editor, Morgan Kauffman Publishers, 1989.

[16]. Y. F. Wong and A. Sideris, "Learning Convergence in the Cerebellar Model Articulation Controller," to appear in *IEEE Trans. on Neural Networks,* 1992.

[17]. M. Sznaier and M. Damborg, "Heuristically Enhanced Feedback Control of Constrained Discrete Time Linear Systems," *Automatica,* Vol 26, No 3, pp. 521–532, 1990.

[18]. C. S. Lin and H. Kim, "CMAC–Based Adaptive Critic Self–Learning Control," *IEEE Trans. on Neural Networks,* Vol 2, 5, pp. 530–533, 1991.

Appendix. Proofs of Lemmas 1, 2 and 3

Proof of Lemma 1: Given any $\underline{x} \in \mathcal{G} - O_1$ it can be expressed as $\lambda_o \underline{y}_o$ with $\underline{y}_o \in \partial\mathcal{G}$ and $0 < \lambda_o \leq 1$. Then:

$$\begin{aligned}
\|\underline{x}\|_{\mathcal{G}} - \min_{\underline{u}\in\Omega} &\left\{\|A_o\underline{x} + B\underline{u}\|_{\mathcal{G}} + \Delta_1\|\underline{x}\|_{\mathcal{G}}\right\} \\
&= \|\lambda_o\underline{y}_o\|_{\mathcal{G}} - \min_{\underline{u}\in\Omega}\left\{\|A_o\lambda_o\underline{y}_o + B\underline{u}\|_{\mathcal{G}} + \Delta_1\|\lambda_o\underline{y}_o\|_{\mathcal{G}}\right\} \\
&\geq \|\lambda_o\underline{y}_o\|_{\mathcal{G}} - \min_{\underline{u}\in\Omega}\left\{\|\lambda_o A_o\underline{y}_o + B(\lambda_o\underline{u})\|_{\mathcal{G}} + \lambda_o\Delta_1\|\underline{y}_o\|_{\mathcal{G}}\right\} \\
&= \lambda_o\left\{\|\underline{y}_o\|_{\mathcal{G}} - \min_{\underline{u}\in\Omega}\left\{\|A_o\underline{y}_o + B\underline{u}\|_{\mathcal{G}}\right\} + \Delta_1\|\underline{y}_o\|_{\mathcal{G}}\right\}\right\} \\
&\geq \min_{\substack{\underline{y}\in\partial\mathcal{G} \\ \lambda\underline{y}\in\mathcal{G}-O_1}} \lambda\left\{\|\underline{y}\|_{\mathcal{G}} - \min_{\underline{u}\in\Omega}\left\{\|A_o\underline{y} + B_o\underline{u}\|_{\mathcal{G}} + \Delta_2(\underline{u}) + \Delta_1\|\underline{y}\|_{\mathcal{G}}\right\}\right\} \\
&\geq \Lambda \min_{\underline{y}\in\partial\mathcal{G}}\left\{1 - \min_{\underline{u}\in\Omega}\left\{\|A_o\underline{y} + B_o\underline{u}\|_{\mathcal{G}} + \Delta_2(\underline{u}) + \Delta_1\right\}\right\}
\end{aligned} \tag{A1}$$

since $\|\underline{y}\|_{\mathcal{G}} = 1$, $\underline{y} \in \partial\mathcal{G}$ and $0 \leq \lambda_o \leq 1$ ◇

Proof of Lemma 2: From the hypothesis and Lemma 1 it follows that:

$$\begin{aligned}
\max_{\underline{u}\in\Omega}\{\|\underline{b}_i\|_{\mathcal{G}} - \|A\underline{b}_i + B\underline{u}\|_{\mathcal{G}}\} &= \|\underline{b}_i\|_{\mathcal{G}} - \min_{\underline{u}\in\Omega}\{\|A\underline{b}_i + B\underline{u}\|_{\mathcal{G}}\} \\
\geq \|\underline{b}_i\|_{\mathcal{G}} - \min_{\underline{u}\in\Omega}&\{\|A_o\underline{b}_i + B_o\underline{u}_i\|_{\mathcal{G}} + \Delta_2(\underline{u}) + \Delta_1\|\underline{b}_i\|_{\mathcal{G}}\} \\
\geq \Lambda\left\{1 - \max_{\underline{x}\in\partial\mathcal{G}}\left\{\min_{\underline{u}\in\Omega}\|A_o\underline{x} + B_o\underline{u}\|_{\mathcal{G}} + \Delta_2(\underline{u}) + \Delta_1\right\}\right\} &= \Lambda\delta > 0
\end{aligned} \tag{A2}$$

Proof of Lemma 3: From the definition of quantization it follows that there exists $\underline{z}_o \in \chi$ such that $\underline{x}_o \equiv \underline{z}_o$. Write $\underline{z}_o$ as $\sum_{i=1}^{2^n}\lambda_i\underline{b}_i$ where $0 \leq \lambda_i \leq 1$, $\sum\lambda_i = 1$ and where $\underline{b}_i$ are the vertices of the smallest hypercube containing $\underline{z}_o$. It can be easily shown, for instance by induction on n (the dimension of the space) that if $\lambda_i \neq 0$ then $\|\underline{z}_o - \underline{b}_i\|_\infty \leq (c_q - 1)\delta x$. Hence, for these λ_i $\|\underline{z}_o - \underline{b}_i\|_{\mathcal{G}} \leq (c_q - 1)\|\underline{\delta x}\|_{\mathcal{G}}$, where $\underline{\delta x} \stackrel{\Delta}{=} (\delta x, \delta x, \ldots \delta x)$. Denote by $\underline{u}(\underline{x})$ the control action generated by CMAC Then:

$$\underline{u}(\underline{x}_o) = \underline{u}(\underline{z}_o) = \sum_{i=1}^{2^n}\lambda_i\underline{u}(\underline{b}_i) \in \Omega$$

since Ω is convex. Consider now:

$$\begin{aligned}
\|A\underline{z}_o + B\underline{u}\|_{\mathcal{G}} = \|A\sum\lambda_i\underline{b}_i + B\sum\lambda_i\underline{u}(\underline{b}_i)\|_{\mathcal{G}} &\leq \sum\lambda_i\|A\underline{b}_i + B\underline{u}(\underline{b}_i)\|_{\mathcal{G}} \\
&\leq \sum\lambda_i\left(\|\underline{b}_i\|_{\mathcal{G}} - \Lambda\delta\right) \\
&\leq \sum\lambda_i\left(\|\underline{z}_o\|_{\mathcal{G}} + (c_q - 1)\|\underline{\delta x}\|_{\mathcal{G}} - \Lambda\delta\right) \\
&= \|\underline{z}_o\|_{\mathcal{G}} + (c_q - 1)\|\underline{\delta x}\|_{\mathcal{G}} - \Lambda\delta
\end{aligned} \tag{A3}$$

Hence:

$$\begin{aligned}
\|\underline{x}_o\|_{\mathcal{G}} - \|A\underline{x}_o + B\underline{u}_o\|_{\mathcal{G}} &\geq \|\underline{z}_o\|_{\mathcal{G}} - \frac{1}{2}\|\underline{\delta x}\|_{\mathcal{G}} - \|A\underline{z}_o + B\underline{u}_o\|_{\mathcal{G}} \\
&- \frac{1}{2}\|A\|_{\mathcal{G}}\|\underline{\delta x}\|_{\mathcal{G}} \geq \Lambda\delta - \left[c_q - 1 + \frac{1}{2}(1 + \|A\|_{\mathcal{G}})\right]\|\underline{\delta x}\|_{\mathcal{G}}
\end{aligned} \tag{A4}$$

where $\|A\|_{\mathcal{G}} = \sup_{\|\underline{x}\|_{\mathcal{G}}=1}\|A\underline{x}\|_{\mathcal{G}}$. It follows that, if:

$$\|\underline{\delta x}\|_{\mathcal{G}} < \frac{\Lambda\delta}{c_q - \frac{1}{2} + \frac{\|A\|_{\mathcal{G}}}{2}} \tag{A5}$$

then $\|A\underline{x}_o + B\underline{u}\|_{\mathcal{G}} < \|\underline{x}_o\|_{\mathcal{G}}$ ◇.

FA6 - 9:50

Proceedings of the 31st Conference
on Decision and Control
Tucson, Arizona • December 1992

USING SELF-ORGANIZING ARTIFICIAL NEURAL NETWORKS FOR SOLVING UNCERTAIN DYNAMIC NONLINEAR SYSTEM IDENTIFICATION AND FUNCTION MODELING PROBLEMS

Jeffrey J. Garside, Timothy L. Ruchti, and Ronald H. Brown

(414)288-1609, (414)288-1609, (414)288-3501

Marquette University

1515 West Wisconsin Avenue

Milwaukee, WI 53233

KEYWORDS: **Artificial Neural Network, Associative Memory, Control System Identification**

ABSTRACT: Associative memory artificial neural networks have been shown to be a powerful modeling and identification tool with successful results in various control applications. Related to many of these structures are Kohonen topology-preserving self-organizing neural networks (KNNs) and numerical relaxation techniques to solve ordinary differential equations. Because of this relationship, these KNN maps can be used to approximate or represent an m-dimensional mapping where n≤m. This paper describes novel implementations of this KNN structure as it is applied to nonlinear functions, control system identifications, and switched reluctance motor torque modelings. Specifically, this paper examines novel training paradigms including a procedure for initializing and resetting neuron weights, incorporating *a priori* knowledge into a KNN, and preferentially training specific areas of a KNN. Several functions are modeled as examples of the implementation and properties of this technique. Also, the KNN is used to model a nonlinear mapping embedded in a series-parallel control identifier. Finally, a 2-dimensional KNN is used to successfully estimate the torque in a switched reluctance motor.

I. Introduction:

Although considerable study has been devoted to the development of techniques suitable for controlling and identifying uncertain dynamic linear systems [1], only limited methods are presently available for certain dynamic systems with a high degree of nonlinearity. Recently backpropagation artificial neural networks have had a high degree of success in this area. However, some restrictions still remain that prohibit final application of these techniques. These obstacles include computationally intensive algorithms, slow rates of convergence, restrictions on the type of nonlinearities that can be approximated, and a lack of an established technique for selecting network architecture [2]. An alternative strategy is to approximate a given dynamic nonlinear system using self-organizing topology-preserving KNNs [2,3] that are related to the relaxation techniques in [9]. With these maps and enhanced methods for training and initializing them, estimations of a general class of dynamic nonlinear systems and functions can be achieved.

This paper presents a means of using KNNs as function modelers and system identifiers. In Section 2, training techniques, including initialization and resetting of neuron weights, and the use of *a priori* knowledge and preferential training are presented. These methods along with learning rules are used to configure the KNNs into conditions and states that allow them to represent a desired function or identify a system. In Section 3 a function with jump discontinuities is estimated by these KNNs. In Section 4 a series-parallel identifier, as described by Narendra [4] is used in conjunction with the KNN to model an unknown nonlinearity inherent in the control structure. In Section 5, the torque of a switched reluctance motor is modeled with a 2-dimensional KNN. Finally, in Section 6, results, conclusions, and areas for further study are discussed.

Notation will be as follows: a one-dimensional KNN map will be called $G(i,w)$ where i indicates the neuron number and w corresponds to the x, y, and/or z values of the neuron weights. A two-dimensional KNN map will be called $G(i,j,w)$ where i and j are the row index and column index of the neuron, respectively, and w is again the x, y, and/or z values of the neuron weights.

Also, the RMS error of the KNN estimation of the system or function evaluated at the neuron weight values will be referred to as simply the RMS error. The RMS change in the neuron weights, that is the RMS difference in the weights at some training iterations t and t-N, will be referred to as the RMS neuron change.

CH3229-2/92/0000-2716$1.00 © 1992 IEEE

II. Training and Initialization Techniques:

A. Initialization and Resetting of Neuron Weights:

Assume that all points in a region of the *xz*-plane are of equal interest, $x \in [a\ b]$ and $z \in [c\ d]$. Then it is desirable to have an (N+1)×(N+1) grid of neurons approximate that region. The map of neurons, *G(i,j,w)*, is created in a rectangular grid-like pattern. Each neuron has its weights initialized to:

$$\begin{aligned} G(i,j,x) &= i(b-a)/N + a \\ G(i,j,y) &= 0.0 \\ G(i,j,z) &= j(d-c)/N + c \end{aligned} \tag{1}$$

where these equations hold for all i and j. This general form can be used for one dimensional mapping of a system or function of the form *y=f(x)* by simply eliminating the j subscript and the z weight in (1); it is this type of network that will be used in the subsequent sections.

It is important to realize that this grid pattern can be maintained after an arbitrary amount of training. The neuron weights can be drawn back to their desired uniform values by use of a Hebbian-type learning rule as in:

$$\begin{aligned} G(i,j,x) &= \alpha x^{\text{desired}} + (1-\alpha)G(i,j,x^{\text{old}}) \\ G(i,j,z) &= \alpha z^{\text{desired}} + (1-\alpha)G(i,j,z^{\text{old}}) \end{aligned} \tag{2}$$

where these equations again hold for all i and j. Alternatively, these values could simply be re-initialized according to (1).

It is not necessary to lock these neurons in uniform grid-like structures. If it is desirable to have more neurons in specific subregions, then simply initialize them as such and periodically draw the neurons back to the desired values as in (2) or re-initialize the x and z weights according to (1). As an example of the initialization of a non-uniform initialization, the two-dimensional KNN in Figure 1 has been initialized according to:

$$\begin{aligned} G(i,j,x) &= \sqrt{i} \quad for\ i=1 \ldots 10, \\ G(i,j,y) &= \sqrt{j} \quad for\ j=1 \ldots 10. \end{aligned} \tag{3}$$

This type of an implementation would be useful if it is known *a priori* that the function the KNN is modeling changes more rapidly in the neighborhood of (3,3) than in the neighborhood of (1,1) and would require more neurons in that region. Some of these nonuniform initialization techniques and relaxation methods can be found in [9].

B. *A Priori* Knowledge in a Neural Network:

It is desirable to have a neural network approximate a function as closely as possible. Since often exact values of the function at specific points are known, this information should be exploited during training of a neural network. For example, if it is known that a function, *y=y(x)*, is of the form:

$$y(x) = \sum_{k=1}^{N} b_k \sin(2\pi k x) \tag{4}$$

then it is also known that the curve passes through the points of the form (±i,0), for any integer i.

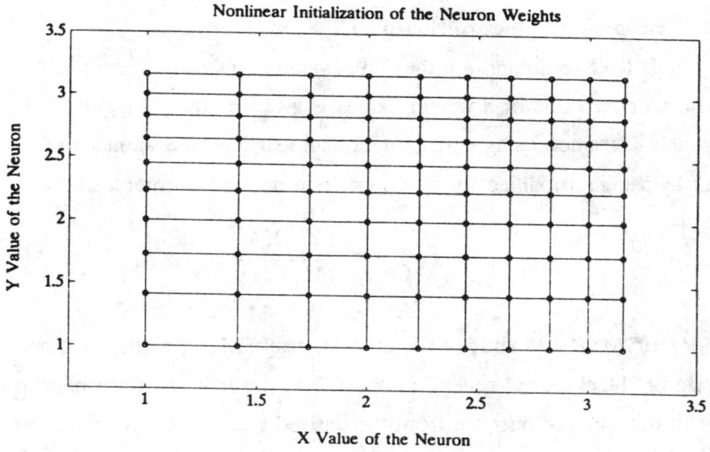

Figure 1 An example of nonuniformly initializing the KNN weights.

This information can be used to bias the network toward specific weight values or even set the weights of neurons. This *a priori* knowledge gives the network an advantage as it trains. To implement this *a priori* information in a neural network, periodically set the known neuron weights to the exact theoretical values or train specific neuron weights toward the known values. This forces proper convergence of the neuron weights to the desired, known values and is done in addition to the normal training techniques.

C. Preferential Training of Neurons in Specific Regions:

In many of the examples and results of training Kohonen networks there is a boundary effect visible that causes a slight contraction of the map at its edges [2,3]. The density of weight vectors, or neurons, is distorted at the borders. To reduce this effect, typically the size of the network and hence the number of neurons was increased or a conscience mechanism was implemented [6]. However, this is not necessary. By preferentially selecting the edges of the region during training, that is, using points on the region boundary as the random training sample for the network, these neurons are drawn to the edges and develop properly, without this undesirable contraction effect.

Preferential training of the edges of the region is used to draw these edge neurons to their desired values. If initialization and resetting of neuron weights is in effect, preferential training will aid in training the edge neurons to appropriate values. This aids in modeling the boundaries of the region where extreme values often occur.

It is important to avoid training these boundary areas too often or the neurons will be trained toward the boundaries and a pronounced contraction of the map toward its boundary will occur there; too little training and contraction away from the boundary persists. Results indicate that for preferential training a good estimation of the probability density that should be assigned to an edge during training is:

$$P_e \approx 2 \times N_e \,/\, N_t \tag{5}$$

where P_e is the probability density of a given edge, N_e is the desired number of neurons on that edge, and N_t is the total number of neurons in the map. P_e may have to be increased slightly if contraction at boundaries persist, or it may have to be decreased slightly if the neurons inside the edge neurons are also drawn to the boundary.

D. Training *Y* Values of Neurons:

The KNN estimation of the actual value of a system or function at the specific x and/or z weight values is given by the y weight. This weight is trained using either a Hebbian learning rule, similar to (2), or by using a modified form of a gradient descent learning rule as in:

$$G(i,j,y) = \alpha \frac{(Y_p - \hat{Y}_p)}{(i^2+j^2+1)} \tag{6}$$

for $i=i^*-M...i^*+M$ and $j=j^*-M...j^*+M$, where M is the neighborhood size or "block party" region, i^* and j^* are the indices of the neuron with the smallest distance from the desired value, and Y_p and $\hat{Y}_p$ are system outputs that are not necessarily equal to the y values of the neuron weights. There are any number of training rules that may be implemented; these are just one that has been implemented for demonstration in this paper. Other types of rules can be found in [3].

Therefore, now that the initialization of an KNN and the training of the neuron weights have been discussed, an implementation of these strategies is now in order.

III. Nonlinear Function Modeling Example:

As an example of these training paradigms, a nonlinearity given by (7) will be estimated.

$$\begin{aligned} &1.\ y = -0.3x^{0.3} \\ &2.\ y = 1.125 - 0.3x - 0.015\sin(\pi x) \\ &3.\ y = 0.3 - 0.15\sin(\pi x) \\ &4.\ y = -0.35\left(1 + e^{-3(x-7.5)}\sin(1.5\pi x + 0.5\pi)\right) \end{aligned} \tag{7}$$

where the first equation is for $0.0 \le x < 2.5$, the second for $2.5 \le x < 5.0$, the third $5.0 \le x < 7.5$, and the fourth $7.5 \le x \le 10.0$. Notice that there are jump discontinuities in the function value at 2.5, 5.0, and 7.5. Recall that a backpropagation neural network has only been shown to be able to model a function with continuous values [7].

The KNN $G(i,w)$ is initialized to the following values:

$$G(i,x) = 10i/N \quad \text{and} \quad G(i,y) = 0.0 \tag{8}$$

for $i=0$ to N and N=200. The KNN is now trained with a random x value from [0 10] and a y value that corresponds to the function in equation (7) evaluated at x.

Shown in Figure 2 is the KNN after 500 training iterations. Those neurons still located at points of the form $(x,0)$ have not yet trained. In Figure 3, the KNN is shown after 100,000 training iterations. The network has converged to an RMS error of 1E-5. The network has achieved an excellent estimation of the function, even in the neighborhood of the discontinuities. In Figure 4 the RMS error in the neuron weights is shown as a function of training iteration.

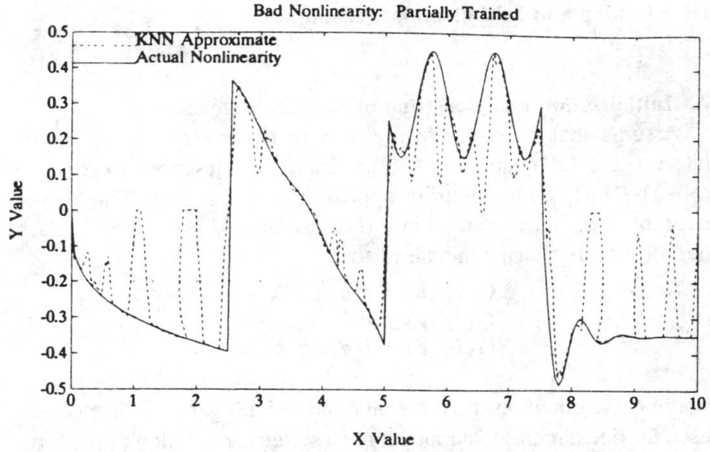

Figure 2 The KNN estimate of the bad nonlinearity after 500 training iterations.

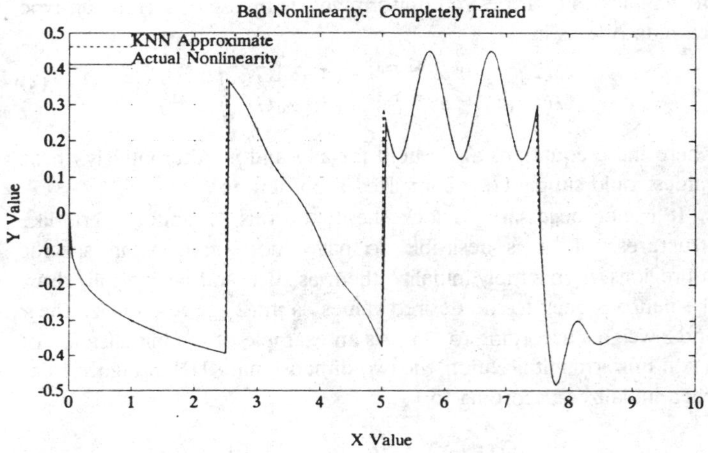

Figure 3 The KNN estimation of the bad nonlinearity after training is complete.

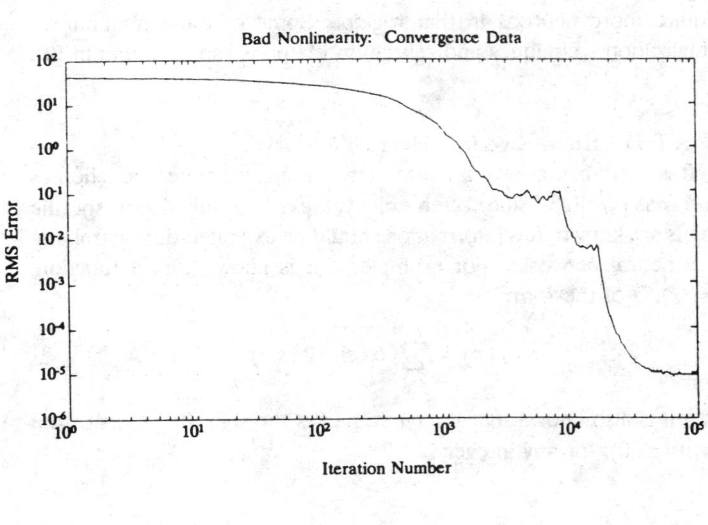

Figure 4 The RMS error vs. training iteration.

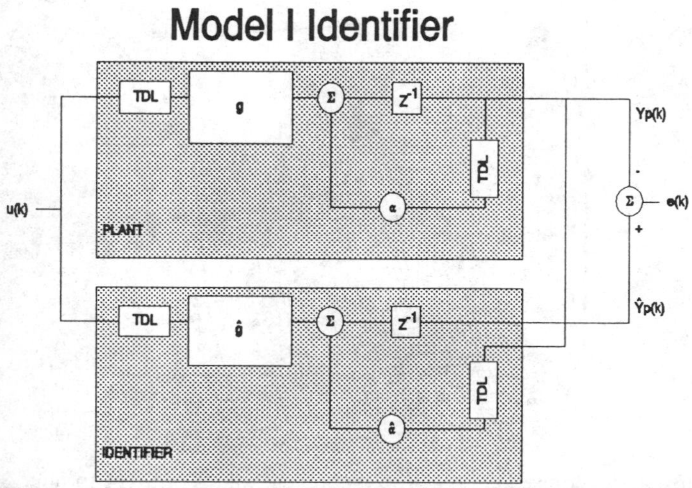

Figure 5 The series-parallel identification structure of Example 1.

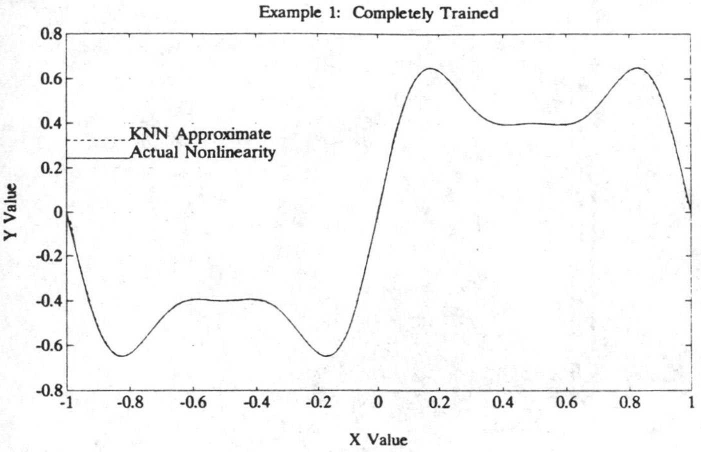

Figure 6 The KNN estimation of nonlinearity after training is complete.

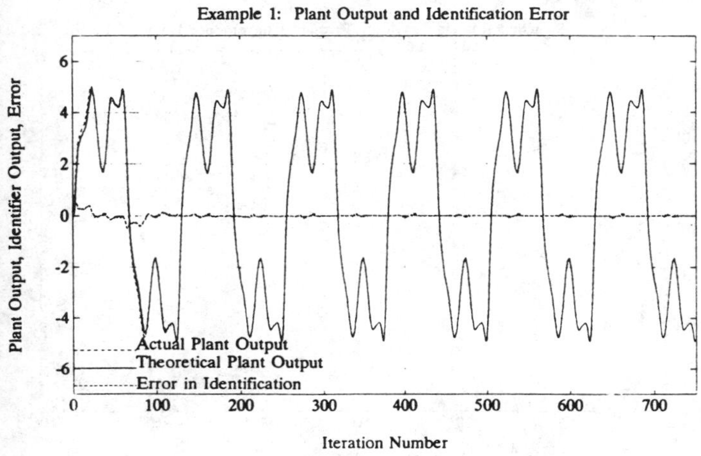

Figure 7 The output of the plant, the output of the identifier, and the error in identification.

IV. Implementation into a System Identification Model:

As a practical example of this type of an KNN structure, the KNN has been embedded into the identification system shown in Figure 5, taken from [4]. Notice that the KNN is used only to approximate the plant nonlinearity, G, and that the actual Y_P is fed back internally in the plant and fed back from the plant to the identifier. The output error, $e(k)$, calculated from the difference between Y_P and $\hat{Y}_P$, is used to train the KNN even though this error is not directly related to the KNN's output.

The plant is governed by the difference equation:

$$\begin{aligned} y_p(k+1) &= 0.3y_p(k) + 0.6y_p(k-1) + f[u(k)], \\ f(u) &= 0.6\sin(\pi u) + 0.3\sin(3\pi u) + 0.1\sin(5\pi u). \end{aligned} \quad (9)$$

By examination of the system poles (both of which are inside the unit circle of the Z plane) it can be seen that the unforced linear system is asymptotically stable and hence it is guaranteed that a bounded input will result in a bounded output, or the system is said to possess BIBO stability. Consequently, a sinusoidal input will yield a bounded output. The input, $u(k)$, is given by:

$$u(k) = \sin(2\pi k/250), \quad k = 0,\ 1,\ 2, \cdots. \quad (10)$$

Therefore, by applying a sinusoidal input in the form of equation (10) to the system, the output of the system will not become unbounded.

To identify the plant, a series-parallel model is used. The model is given by:

$$\hat{y}_p(k+1) = 0.3y_p(k) + 0.6y_p(k-1) + N[u(k)] \quad (11)$$

where $N[u(k)]$ is the KNN estimate of $f(u(k))$, and $Y_P(k)$ and $Y_P(k\text{-}1)$ are outputs from the plant.

As can be seen from Figure 6 and Figure 7, this KNN implementation has amazingly good results. The final identified nonlinearity is exemplary, as can be seen in Figure 6. The tracking of the system is almost immediate as can be seen in Figure 7 which simultaneously shows the output of the plant and the identifier. The difference in these two values quickly becomes small.

The KNN implemented has 76 neurons. If the number of neurons is selected too large, the nature of $u(k)$ will leave some of the neurons trained improperly. That is, since $u(k)$ only takes on a small, finite set of values (125 in this implementation) the input range of x is not completely spanned, not all neurons will train properly. If an alternative input, $u_o(k)$, completely spans the range of x, any size KNN could be used as long as the learning rates and neighborhood blockparty sizes are modified properly.

V. A Two-Dimensional Modeling of the Torque of a Switched Reluctance Motor:

The estimation of the torque generated by a SRM is of great value in control applications. As a practical application of the KNN with *a priori* knowledge, preferential training, and initialization and resetting of neuron weights, the theoretical SRM torque is modeled.

For the purpose of realistic study, a detailed model developed by Taylor [5] was utilized to generate torque versus current and position data. The model considers all significant nonlinearities. It is given by:

$$T(i,\theta) = \sum_{j=1}^{M}\sum_{k=1}^{N} C(j,k)\sin(j\theta)\left(k\frac{i}{4} + e^{-(k\frac{i}{4})} - 1\right) \qquad (12)$$

where θ is rotor position in electrical radians and i is the current in amperes. The theoretical torque surface is represented in Figure 8.

Assigning x to position, θ, and z to current, i, the xz-plane in the range of $x \in [0\ \pi]$, $z \in [0\ 4]$ represents the uniform probability density function from which training examples are selected, and the y value of the neuron weights represents the torque at the corresponding x and z values. The neural network is initialized to be a grid of uniformly spaced neurons on the xz-plane encompassing the designated x and z ranges. A neural network, G(j,k,w), of (N+1)×(N+1) neurons is initialized according to:

$$\begin{aligned} G(j,k,x) &= \pi j/N, \\ G(j,k,y) &= 0.0, \\ G(j,k,z) &= 4k/N, \end{aligned} \qquad (13)$$

where these equations hold for integers j and k in the range [0 40]. In this application N=40. Periodically, the x, y, and z values of the neuron weights on the $x=\theta=0.0$, $x=\theta=\pi$, and $z=i=0.0$ curves are locked back on to their initial, theoretically correct values, as are the x and z values of all other neurons. The edge representing maximum current or $z=4.0$, is preferentially trained during about 5% of the training iterations.

Uniformly distributed random values in the region are selected, the corresponding theoretical torque calculated, and the neuron with the minimum euclidean distance from its weights to the random training sample and its neighborhood have their weights updated.

Training continues until the RMS and maximum error values have fallen to an acceptable level. The error is shown in Figure 10. When the resetting of the neuron weights is discontinued, the neurons initially exhibit exponential convergence onto the theoretical torque surface until the RMS error reaches a minimum value, which is similar to what happened in other RMS error plots. The final weight values are not necessarily in the perfect grid pattern that they were locked onto before this time, but the neurons will remain in some region surrounding their desired x and z values and the map continues to exhibit the grid-like structure, as shown in Figure 9.

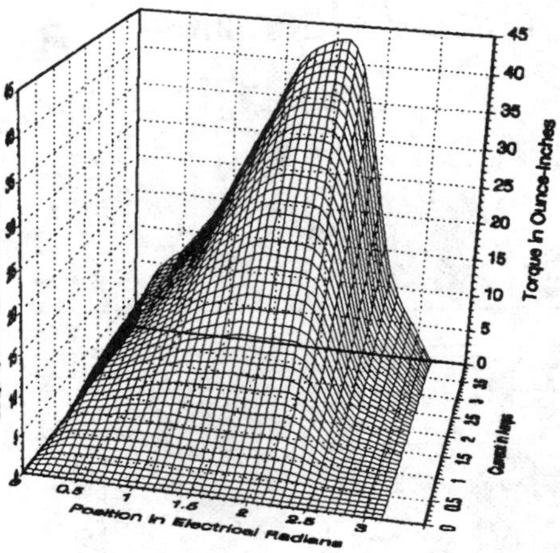

Figure 8 A representation of the theoretical values of SRM torque according to (12).

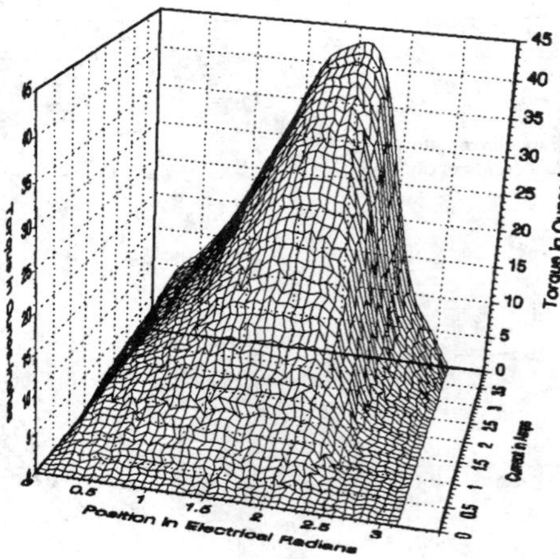

Figure 9 The KNN estimation of SRM torque after training is complete.

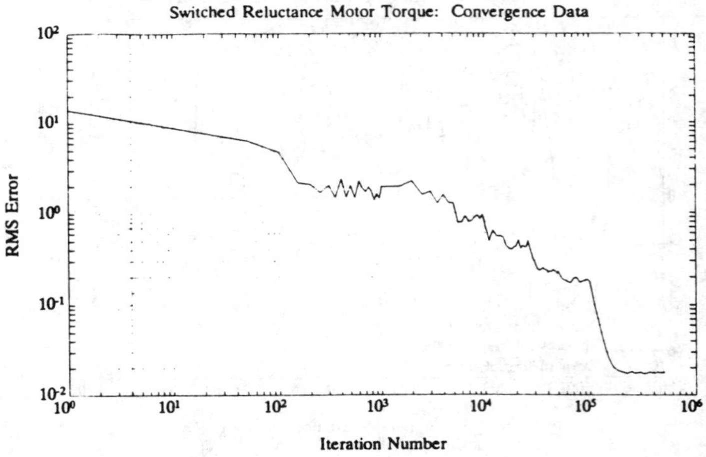

Figure 10 The RMS error vs. training iteration number for the SRM torque estimate.

VI. Conclusion:

The remarkable results in the previous sections demonstrate this new method for training KNNs is a powerful tool that can be applied to nonlinear control and system identification problems. In the static torque identification example, it was experimentally shown that the torque at any given current and position can be estimated by using the weights of the neurons and interpolation techniques. Since the quality of control is intimately linked with the system model [8], it can be expected that this topological representation will be an extremely valuable contribution to the field of adaptive control.

It is important to realize that this KNN did not have to be trained with theoretical values; these values could have been actual values measurements at points representing a random sampling over the position and current ranges. For practical purposes and with the objective of presenting RMS error data, theoretical values were chosen.

The new training techniques presented in this paper have not yet realized their full potential. Our research has not yet been able to ascertain how ill conditioned a function must be before the neuron map will be unable to estimate it. These findings indicate that functions containing step discontinuities can be modeled using this method, a nonlinearity that feedforward neural networks presently have not been proven able to model [7].

It is critical to know the region boundaries for application of the preferential training technique. The input region must also be known to implement initialization and resetting of neuron weights for the proper selection of a grid pattern, and for the initialization and modification of the neuron weights.

As closer approximations are desired, as in most KNN applications, more neurons become necessary. A typical KNN may require more neurons than a corresponding feedforward neural network. However, the training of an KNN takes significantly less time than training a corresponding backpropagation neural network because of the computational overhead of the backpropagation training algorithms used in feedforward neural networks. The example in Section 4 took less than 10 minutes to complete 25,000 training iterations and the application in Section 6 took less than 20 hours to complete 600,000 training iterations, both written in Turbo C and running on a 386-25 MHz PC with a math coprocessor.

The potential for a great deal of research in this area exists. For example, this technique could be used for off line system identification, but is it also applicable to on line identification problems? What is the most advantageous scheme for applying initialization and resetting of neuron weights? How is the contraction of the neuron map at region boundaries best avoided? How should the learning rate and the neighborhood size be chosen and when and how should these values be altered? There are many areas of research in this field that have not been fully investigated.

VII. References:

[1] K. J. Astrom and P. Eykhoff, "System identification - a survey," Automatica, vol. 7, pp. 123-162, 1971.

[2] R. Hecht-Nielsen, Neurocomputing. Reading, MA: Addison-Wesley, 1990.

[3] T. Kohonen, Self-organization and associative memory, Springer Series in Information Sciences, vol. 8, Third Edition, Springer-Verlag, Berlin, 1989.

[4] K. S. Narendra and K. Parthasarathy, "Adaptive identification of dynamical systems using neural networks," Proceedings: 28th Conference on Decision and Control, Tampa, FL, pp. 1737-1738, Dec., 1989.

[5] D. G. Taylor, Feedback control of uncertain nonlinear systems with applications to electric machinery and robotics manipulators, Ph.D. Thesis, Department of Electrical Engineering, University of Illinois at Urbana-Champaign, IL, 1988.

[6] D. Desieno, "Adding a conscience to competitive learning," Proc. Int. Conf. on Neural Networks, **I**, pp. 117-124, IEEE Press, New York, July 1988.

[7] A. N. Kolmogorov, "On the representation of continuous functions of many variables by superposition of continuous functions of one variable and addition" [in Russian], Dokl, Akad., Nauk USSR, **114**, pp. 953-956, 1957.

[8] P. J. Werbos, "Neural networks for control and system identification," Proc. of the 28th Conf. CDC., Tampa, FL, WA10, pp. 260-265, December 1989.

[9] W. H. Press, Numerical recipes (FORTRAN), Cambridge University press, Cambridge, MA, 1989.

[10] J. J. Garside, R. H. Brown, T. L. Ruchti, and X. Feng, "Nonlinear estimation of torque in switched reluctance motors using grid locking and preferential training techniques on self-organizing neural networks," IJCNN '92, (Baltimore, MD), June 1992, pp. II 811-816.

Proceedings of the 31st Conference
on Decision and Control
Tucson, Arizona • December 1992

ANALYSIS AND SYNTHESIS OF A CLASS OF DISCRETE-TIME NEURAL NETWORKS WITH NONLINEAR INTERCONNECTIONS*

J. Si and A. N. Michel

College of Engineering
University of Notre Dame
Notre Dame, IN 46556

ABSTRACT

In the present paper a qualitative analysis and a synthesis procedure for a class of discrete-time, synchronous neural networks with nonlinear interconnections (of polynomial form) are presented. Networks generated by the present synthesis procedure have in general greater memory storage capacity than corresponding networks with linear interconnecting structure.

The qualitative results which constitute the basis of the synthesis procedure include a simple stability test for the equilibrium points of the class of networks considered. Conditions under which such networks are globally stable have not been identified at this time.

In order to simplify the presentation, we confine ourselves in the present paper to second order polynomial network interconnections; however, the qualitative results and the synthesis procedure developed herein apply to networks with higher order polynomial interconnections as well.

I. INTRODUCTION

Significant progress has been made during the past few years in neural networks research for associative memory. Most of this research addresses the stability, storage capacity and the design of neural networks. It has been shown in [2] that the capacity of a discrete Hopfield associative memory of length n, when presented with a number of m independent random fundamental memories, is about $n/(4\ln^n)$. The authors of [1,2] show that the capacity of first order interconnected neural networks (i.e., neural networks with linear interconnections) is of order O(n). It is also shown in [3] that the storage capacity of a neural network of order n with energy function of degree d is $O(n^{d-1})$. The design methods proposed in [7] will generate an identity interconnection matrix when the prototype vectors are linearly independent and/or when the rank of the prototype vectors is n. In this case the network becomes useless. As reported in [6], even if this extreme case is avoided, the attractivity of the stored states decreases with an increasing number of stored patterns. This problem can be alleviated to a large extent by making use of "spurious" states [6]. Another possible solution consists in "gardening" the state space [6] in order to increase the extent of the basins of attraction of the desired memories. In any case, no dramatic improvements are realized by these attempts, since the storage capacity of feedback neural networks with linear interconnections is O(n).

In feedback neural networks, the above difficulties can be circumvented at the expense of increased complexity, by introducing high order (i.e., nonlinear) interactions between neurons. Alternatively, in feed-through networks one might also use multi-layer, first order neural networks. The well known backward error propagation algorithm has allowed great increase in the application of multi-layer, first order networks. However, the performance of such networks in terms of learning speed still prevents their application to some practical problems. Existing works which extend the perceptron learning rule and the back propagation algorithm in the design of high order feed-through networks include the results given in [4,5].

In the present paper, we consider a class of synchronous, discrete-time neural networks described by a system of first order ordinary difference equations with nonlinear interconnections (also called high order interconnections). The interconnections are characterized by polynomials. The results of the present paper are applicable to high order polynomials. However, to simplify matters, we confine ourselves in the present paper to neural networks with interconnections described by second order polynomial nonlinearities.

As in the case of neural networks with linear interconnections, the outer product method utilized in [5] does not necessarily guarantee that desired memories will be stored as equilibria in high order neural networks. As in the case of neural networks with linear interconnecting structure, the pseudo-inverse method utilized in [6] does guarantee that desired memories will be stored as equilibria in high order neural networks; however, these equilibria need not necessarily be asymptotically stable. The synthesis procedure developed in the present paper guarantees to store specified memories as asymptotically stable equilibrium points.

CH3229-2/92/0000-2722$1.00 © 1992 IEEE

II. NEURAL NETWORK MODELS

We consider a class of synchronous discrete-time neural networks endowed with nonlinear interconnections described by a system of first order ordinary difference equations, given by

$$x_i(k+1) = \mathrm{sat}\,(F_i(k)) \qquad \text{(H)}$$

with

$$F_i(k)=\sum_{j=1}^{n} C_{ij}x_j(k)+\sum_{j=1}^{n-1}\sum_{l=j+1}^{n} D_{ijl}x_j(k)x_l(k)+I_i \qquad (2.1)$$

for $i=1,\dots,n$, and $k=0,1,2,\dots$, where $x=(x_1,\dots,x_n)^T\in R^n$, $C_{ij}\in R$, $D_{ijl}\in R$, and $I_i\in R$. The sat function is employed to represent neurons and is given by

$$\mathrm{sat}(\sigma)=\begin{cases} 1 & \text{when } \sigma>1 \\ \sigma & \text{when } -1\le\sigma\le 1 \\ -1 & \text{when } \sigma<-1 \end{cases}.$$

The weight C_{ij} is used to transmit information from neuron j to neuron i while the weight D_{ijl} transmits information from the neurons j and l to neuron i. The term I_i denotes an external bias.

III. STABILITY RESULTS

This section presents the asymptotic stability results of the system considered herein.

3.1. Preliminaries

In characterizing the notion of solution of system (H), we will find it useful to employ the notation established in the following.

Let

$$\Delta=\{\xi=(\xi_1,\dots,\xi_n)^T : \xi_i = \pm 1 \text{ or } 0,\ 1\le i\le n\}.$$

For each m, $0\le m\le n$, let

$\Delta_m=\{\xi=(\xi_1,\dots,\xi_n)^T\in\Delta:\ \xi_{\mu(i)}=0,\ 1\le i\le m$ and $\xi_{\mu(i)}=\pm1,\ m<i\le n$, for some $\mu\in \mathrm{sym}(n)\}$, where sym(n) is the symmetric group of order n.

Also for a given $\xi\in\Delta$, let

$U(\xi)=\{x=(x_1,\dots,x_n)^T\in R^n:\ |x_i|<1$ if $\xi_i=0$, and $x_i=\xi_i$ if $\xi_i\neq 0\}$.

Let

$$B^n = \{x\in R^n:\ x_i=\pm1,\ i=1,\dots,n\}$$

and let

$$D^n = \{x\in R^n:\ |x_i|\le 1,\ i=1,\dots,n\}.$$

Definition 3.1 1) Consider $\xi\in\Delta_m$, $0<m\le n$, with $\mu\in \mathrm{sym}(n)$, such that $\xi_{\mu(i)}=0$, $1\le i\le m$, and $\xi_{\mu(i)}=1$ or -1, $m<i\le n$. Also, consider the system defined by

$$x_{\mu(i)}(k+1)=\sum_{j=1}^{n} C_{\mu(i)\mu(j)}x_{\mu(j)}(k)+\sum_{j=1}^{n-1}\sum_{l=j+1}^{n} D_{\mu(i)\mu(j)\mu(l)}x_{\mu(j)}(k)x_{\mu(l)}(k)+I_{\mu(i)} \qquad (H_\xi)$$

for $i=1,\dots,m$.

In (H_ξ), we let $x_I=(x_{\mu(1)},\dots,x_{\mu(m)})^T$ and $x_{II}=(x_{\mu(m+1)},\dots,x_{\mu(n)})^T$, where $-1<x_{\mu(i)}<1$, for $i=1,\dots,m$, and $x_{\mu(i)}=\xi_{\mu(i)}$, $i=m+1,\dots,n$. System (H_ξ) is said to be a *reduced system (of order m)* of system (H) over the region $U(\xi)$.

2) For $\xi\in\Delta_m$, a function $\varphi: \{0,1,\dots,k_0\}\to U(\xi)$, is said to be a *(local) solution* of system (H) if the vector function $\varphi_I=(\varphi_{\mu(1)},\dots,\varphi_{\mu(m)})^T$ is a solution of the reduced system (H_ξ), i.e.,

$$\varphi_{\mu(i)}(k+1)=\sum_{j=1}^{n} C_{\mu(i)\mu(j)}\varphi_{\mu(j)}(k) + \sum_{j=1}^{n-1}\sum_{l=j+1}^{n} D_{\mu(i)\mu(j)\mu(l)}\varphi_{\mu(j)}(k)\varphi_{\mu(l)}(k)+I_{\mu(i)} \qquad (3.1)$$

for $i=1,\dots,m$, and $k\in\{0,1,2,\dots,k_0\}$, and

$$\min\{[\sum_{j=1}^{n} C_{\mu(i)\mu(j)}\varphi_{\mu(j)}(k)+\sum_{j=1}^{n-1}\sum_{l=j+1}^{n} D_{\mu(i)\mu(j)\mu(l)}\varphi_{\mu(j)}(k)\varphi_{\mu(l)}(k)+I_{\mu(i)}]\times\xi_{\mu(i)}\}\ge 1 \qquad (3.2)$$

for $i=m+1,\dots,n$, and $k\in\{0,1,2,\dots,k_0\}$.

In particular, if $\xi\in\Delta_m$, $m<n$, the solution φ is said to be in a *saturated mode.* ∎

Let Z^+ denote the non-negative integers.

Definition 3.2 For $x_0\in D^n$, a function $\varphi=\varphi(.,x_0)$: $Z^+\to D^n$ is said to be a *solution of (H) starting at x_0* if

a) $\varphi(0,x_0)=x_0$, and

b) there are countably many sets $E_i=\{T_i,\dots,T_i+k_i\}$, $T_i\in Z^+$, $k_i\in Z^+$, $T_0=0$, such that $E_i\cap E_j=\emptyset$ when $i\neq j$, and $\cup E_i=Z^+$, and $\varphi(\cdot)$ restricted to each E_i is a (local) solution as defined in Definition 3.1. ∎

Definition 3.3 a) A vector $x_e=(x_{1e},\dots,x_{ne})^T\in D^n$ is said to be an *equilibrium* of system (H) if the function $\varphi=\varphi(.,x_e)$: $Z^+\to D^n$ defined by $\varphi(k,x_e)\equiv x_e$ is a solution of (H). In particular, if $|x_{\mu(i)e}|<1$, we call $x_{\mu(i)e}$ a *non-saturated component* of x_e and if $|x_{\mu(i)e}|=1$, we call $x_{\mu(i)e}$ a *saturated component* of x_e.

b) Let x_e be an equilibrium point of system (H).

i) x_e is said to be *stable* if for every $\varepsilon>0$, there is a $\delta=\delta(\varepsilon)>0$ such that $|\varphi(k,x)-x_e|<\varepsilon$ for all $k\in Z^+$ whenever $|x-x_e|<\delta$.

ii) x_e is said to be *asymptotically stable* if it is stable and if there is an $\eta>0$ such that $\lim_{k\to\infty}|\varphi(k,x)-x_e|=0$ whenever $|x-x_e|<\eta$.

iii) x_e is said to be *unstable* if it is not stable. ∎

3.2. Transformation of an Equilibrium

In the present section we address the stability properties of equilibrium points of system (H) in the sense of Lyapunov. In this type of analysis, it is convenient to assume that a given equilibrium under investigation, say x_e, is located at the origin. If this is not the case, then we can always transform the system (H) to an equivalent system (P), such that $p_e=0$ is an equilibrium of (P) and p_e corresponds to x_e under this transformation. In the present subsection we provide the details of such a transformation in terms of system (H_ξ) (rather than system (H)) to obtain a corresponding system (P_ξ) (rather than system (P)).

In order to accomplish the transformation described above for the reduced system (H_ξ), we first substitute $x_{\mu(j)}(k) = \pm 1$ into (H_ξ), for $m<j\leq n$. We obtain,

$$x_{\mu(i)}(k+1)=\sum_{j=1}^{m} C'_{\mu(i)\mu(j)}x_{\mu(j)}(k)+ \sum_{j=1}^{m-1}\sum_{l=j+1}^{m} D'_{\mu(i)\mu(j)\mu(l)}x_{\mu(j)}(k)x_{\mu(l)}(k)+I'_{\mu(i)} \quad (3.3)$$

for $i=1,\ldots,m$, where $C'_{\mu(i)\mu(j)}$, $D'_{\mu(i)\mu(j)\mu(l)}$ and $I'_{\mu(i)}$ are determined in an obvious way.

Given $\xi\in\Delta_m$, it is easy to see that if $x^T=(x_I{}^T, x_{II}{}^T)^T$ is an equilibrium of (H) then $x_I\in R^m$ is an equilibrium of (H_ξ), i.e.,

$$f_{\mu(i)}(x_I)\overset{\Delta}{=} \sum_{j=1}^{m} C'_{\mu(i)\mu(j)}x_{\mu(j)}+ \sum_{j=1}^{m-1}\sum_{l=j+1}^{m} D'_{\mu(i)\mu(j)\mu(l)}x_{\mu(j)}x_{\mu(l)}+I'_{\mu(i)}-x_{\mu(i)}=0 \quad (3.4)$$

$i=1,\ldots,m.$

For $\xi\in\Delta_m$, $m\geq 1$, let

$$p_{\mu(i)}(k)=x_{\mu(i)}(k)-x^*_{\mu(i)},\ i=1,\ldots,m, \quad (T1)$$

and

$$p_{\mu(i)}(k)=x^*_{\mu(i)}=\xi_{\mu(i)},\ i=m+1,\ldots,n, \quad (T2)$$

where $x^*\in D^n$, and x^* is an equilibrium point of (H). Then for $i=1,\ldots,m$,

$$p_{\mu(i)}(k+1)+x^*_{\mu(i)}=\sum_{j=1}^{m} C'_{\mu(i)\mu(j)}\{p_{\mu(j)}(k)+x^*_{\mu(j)}\}+ \sum_{j=1}^{m-1}\sum_{l=j+1}^{m} D'_{\mu(i)\mu(j)\mu(l)}\{p_{\mu(j)}(k)+x^*_{\mu(j)}\}\{p_{\mu(l)}(k)+ x^*_{\mu(l)}\}+I'_{\mu(i)}$$

and for $i=m+1,\ldots,n$,

$$\min\{[\sum_{j=1}^{n} C_{\mu(i)\mu(j)}x^*_{\mu(j)}+ \sum_{j=1}^{n-1}\sum_{l=j+1}^{n} D_{\mu(i)\mu(j)\mu(l)}x^*_{\mu(j)}x^*_{\mu(l)}+I_{\mu(i)}]\times\xi_{\mu(i)}\}\geq 1. \quad (3.5)$$

Since x^* is an equilibrium point of (H), $x^*_I=(x^*_{\mu(1)},\ldots,x^*_{\mu(m)})^T$ must satisfy equation (H_ξ). We thus have,

$$p_{\mu(i)}(k+1) = \sum_{j=1}^{m} C'_{\mu(i)\mu(j)}\, p_{\mu(j)}(k) + \sum_{j=1}^{m-1}\sum_{l=j+1}^{m} D'_{\mu(i)\mu(j)\mu(l)}p_{\mu(j)}(k)p_{\mu(l)}(k)+ \sum_{j=1}^{m-1}\sum_{l=j+1}^{m} D'_{\mu(i)\mu(j)\mu(l)}\, p_{\mu(j)}(k)x^*_{\mu(l)}+ \sum_{j=1}^{m-1}\sum_{l=j+1}^{m} D'_{\mu(i)\mu(j)\mu(l)}\, p_{\mu(l)}(k)x^*_{\mu(j)}.$$

Let

$$A_{\mu(i)\mu(j)}=\sum_{l=j+1}^{m} D'_{\mu(i)\mu(j)\mu(l)}x^*_{\mu(l)}, \quad (3.6)$$

for $i=1,\ldots,m$, $j=1,\ldots,m-1$,

$$A_{\mu(i)\mu(j)}=0, \quad (3.7)$$

for $i=1,\ldots,m$, $j=m$,

$$B_{\mu(i)\mu(j)}=\sum_{k=1}^{j-1} D'_{\mu(i)\mu(k)\mu(j)}x^*_{\mu(k)}, \quad (3.8)$$

for $i=1,\ldots,m$, $j=2,\ldots,m$,

$$B_{\mu(i)\mu(j)} = 0, \quad (3.9)$$

for $i=1,\ldots,m$, $j=1$.

Then,

$$p_{\mu(i)}(k+1) = \sum_{j=1}^{m}[A_{\mu(i)\mu(j)}+B_{\mu(i)\mu(j)}+ C'_{\mu(i)\mu(j)}]p_{\mu(j)}(k)$$

for $i=1,\ldots,m$, which has an equilibrium $p_I{}^*=(p^*_{\mu(1)},\ldots,p^*_{\mu(m)})^T=0$ corresponding to the equilibrium $x_I{}^*$ for the reduced system (H_ξ).

We will use the transformed system (P_ξ) and (3.5) to study the stability of an equilibrium point x^* of (H).

Remark 3.1 When analyzing the stability properties of a given equilibrium point x^* of (H), we will assume that this equilibrium point is isolated. By the Inverse Function Theorem, this is true if in a neighborhood of x^* the matrix

$$J_\xi(x_I) = \begin{bmatrix} \frac{\partial f_{\mu(1)}(x_I)}{\partial x_{\mu(1)}} & \cdots & \frac{\partial f_{\mu(1)}(x_I)}{\partial x_{\mu(m)}} \\ \cdots & \cdots & \cdots \\ \frac{\partial f_{\mu(m)}(x_I)}{\partial x_{\mu(1)}} & \cdots & \frac{\partial f_{\mu(m)}(x_I)}{\partial x_{\mu(m)}} \end{bmatrix}$$

is nonsingular, where for a given $\xi\in\Delta_m$, $m\geq 1$, the functions $f_{\mu(i)}$ are defined in (3.4). ∎

3.3. Asymptotic Stability of an Equilibrium

We are now in a position to establish the following stability result for system (H).

Theorem 3.1 For $\xi\in\Delta_m$, let $x^*\in U(\xi)$ be an equilibrium point of system (H) and let

$$r_\xi= \min_{m+1\leq i\leq n}\{ \sum_{j=1}^{n} C_{\mu(i)\mu(j)}x^*_{\mu(j)}+ \sum_{j=1}^{n-1}\sum_{l=j+1}^{n} D_{\mu(i)\mu(j)\mu(l)}x^*_{\mu(j)}x^*_{\mu(l)}+I_{\mu(i)}\}\times\xi_{\mu(i)}. \quad (3.10)$$

Let

$$A=[\ A_{\mu(i)\mu(j)}\ ],\ B=[\ B_{\mu(i)\mu(j)}\ ],\ C'=[\ C'_{\mu(i)\mu(j)}\ ] \quad (3.11)$$

where the elements of $A \in R^{m\times m}$, $B \in R^{m\times m}$ and $C' \in R^{m\times m}$ are determined in (P_ξ).

1) for $0<m\leq n$, $\xi \in \Delta_m$, if the eigenvalues of the m×m matrix (A+B+C') are all within the unit circle, then x* is an asymptotically stable equilibrium point of system (H);

2) if m=0 and $\xi \in \Delta_0 = B^n$, then $x^*=\xi$ is an asymptotically stable equilibrium point of system (H).

Because of space limitation, the proof is given elsewhere.

Theorem 3.1 suggests a means of testing the asymptotic stability of an equilibrium point of system (H). We summarize a procedure of determining the asymptotic stability of any of the equilibrium points of system (H).

Stability Test 3.1

Given $\xi \in \Delta_m$.

Case 1, m≥1.

1) Substitute $x_{\mu(i)}=\xi_{\mu(i)}$, $m+1\leq i\leq n$, into (H_ξ) to obtain (3.3), and thus, the coefficients $C'_{\mu(i)\mu(j)}$, $D'_{\mu(i)\mu(j)\mu(l)}$, and $I'_{\mu(i)}$, for $i, j, l \in \{1,...,m\}$.

2) Solve equation (3.4) to determine $x^*_{\mu(i)}$, i=1,...,m, which is an equilibrium of the reduced system (of order m) of system (H).

3) Substitute $x^*=(x^*_{\mu(1)},..., x^*_{\mu(m)}, \xi_{\mu(m+1)},..., \xi_{\mu(n)})^T$ into equation (3.5). If $r_\xi<1$, x* is not an equilibrium point of (H). If $r_\xi>1$, continue.

4) Perform the transformations (T1) and (T2) on (3.3).

5) Calculate the coefficients in equations (3.6)-(3.9) to obtain (P_ξ).

6) If all eigenvalues of (A+B+C') are within the unit circle, x* is asymptotically stable. Otherwise not.

Case 2, m=0.

7) If $r_\xi>1$, then $x^*=\xi$ is an asymptotically stable equilibrium point.

IV. SYNTHESIS PROCEDURE

In the present section we develop a synthesis procedure for system (H) which is in the spirit of some of the work in [7].

We will find it convenient to express system (H) in the form

$$x(k+1)=\text{sat}(Fy(k)+I), \qquad (H_\tau)$$

where $x=(x_1,...,x_n)^T \in D^n$, $y \in D^{n+s}$, $F \in R^{n\times(n+s)}$, $I=(I_1,...,I_n)^T \in R^n$ and $\text{sat}(\theta)=(\text{sat}(\theta_1),..., \text{sat}(\theta_n))^T$ when $\theta \in R^n$. The first n components of y(k) are identical to the components of x(k) while the remaining s components of y(k) are determined by the nonlinear interconnections. As mentioned earlier in section 1, we will confine ourselves in the present paper to second order polynomial nonlinearities. For this case, the last s components of the vector y(k) are taken from the set of $\frac{n^2-n}{2}$ terms given by

$$x_1x_2, x_1x_3, x_1x_4, ..., x_1x_n,$$
$$x_2x_3, x_2x_4, ..., x_2x_n,$$
$$......$$
$$x_{n-1}x_n.$$

The first n columns of the matrix F are identical to the columns of the matrix $C=[C_{ij}]$ while the remaining columns are determined by the coefficients D_{ijl} given in (H).

We seek to determine a set of coefficients F_{ij}, i=1,...,n, j=1,...,n+s, and a set of external bias terms I_i, i=1,...,n for system (H_τ) such that 1) a specified set of vectors $\alpha^1,...,\alpha^m$ belonging to B^n are reliably stored in (H_τ) as asymptotically stable equilibrium points, where m>n is allowed; and 2) the total number of spurious asymptotically stable equilibrium points (i.e., undesired asymptotically stable equilibria of system (H_τ)) is as small as possible. To accomplish this we invoke the results of the preceding section. Specifically, if for $x^q = \alpha^q \in B^n$, q=1,...,m, it is true that

$$\tau_1 x_i^q=\sum_{j=1}^n F_{ij}\alpha_j^q+\sum_{j=1}^{n-1}\sum_{l=j+1}^n F_{ik}\alpha_j^q\alpha_l^q+I_i, \qquad (4.1)$$

for k=n+1,...,n+s, i=1,...,n, $\tau_1>1$, then by Theorem 3.1, part 2, $r_\xi=\min_{1\leq i\leq n}(\tau_1\alpha_i^q)\times(\alpha_i^q)>1$, i.e., α^q, q=1,...,m will be asymptotically stable equilibria for system (H_τ).

Suppose that we are given m vectors in B^n, say, $\alpha^1,...,\alpha^m$. In the following we establish a synthesis procedure which guarantees that the vectors $\alpha^1,...,\alpha^m$ are stored as asymptotically stable equilibrium points of system (H_τ). Let β^q be augmented vectors of α^q, q=1,...,m, where the first n components of β^q are identical to the components of α^q, and the remaining s components of β^q are chosen from the set $\alpha_1^q\alpha_2^q ,..., \alpha_1^q\alpha_n^q, \alpha_2^q\alpha_3^q,..., \alpha_{n-1}^q\alpha_n^q$. For example, $\beta^q=[\alpha_1^q, \alpha_2^q,..., \alpha_n^q, \alpha_1^q\alpha_2^q,..., \alpha_1^q\alpha_n^q, \alpha_2^q\alpha_3^q,..., \alpha_{n-1}^q\alpha_n^q]^T$.

We will require that $\text{rank}\{\beta^1,...,\beta^m\}<n+s$. Let

$$L=\text{Span}(\beta^1-\beta^m ,..., \beta^{m-1}-\beta^m)$$

and

$$L_a=\text{Aspan}(\beta^1 ,..., \beta^m).$$

Then L is the linear subspace of R^{n+s} generated by the (m-1) vectors $\beta^1-\beta^m,\ldots,\beta^{m-1}-\beta^m$, L_a is the affine subspace of R^{n+s} generated by the vectors $\beta^1,\ldots,\beta^m$, and $L_a=L+\beta^m$. Assume that k=dim(L), $\{u^1,\ldots,u^k\}$ is an orthonormal basis of L, and $\{u^{k+1},\ldots,u^{n+s}\}$ is an orthonormal basis of $L^{\perp}$. Let

$$F_a^+=[F_{aij}^+]_{(n+s)\times(n+s)}=\Sigma_{i=1}^{k}u^i u^{i\,T}$$

$$F_a^-=[F_{aij}^-]_{(n+s)\times(n+s)}=\Sigma_{i=k+1}^{n+s}u^i u^{i\,T}$$

and let

$$F_{ij}^+=F_{aij}^+\ ,\ i=1,\ldots,n,\ j=1,\ldots,n+s$$

$$F_{ij}^-=F_{aij}^-\ ,\ i=1,\ldots,n,\ j=1,\ldots,n+s.$$

The F and I matrices for system (H_τ) are now determined by

$$F=\tau_1F^+-\tau_2F^- \tag{4.2}$$

and

$$I=\tau_1\alpha^m-F\beta^m \tag{4.3}$$

where $\tau_1>1$ and $\tau_2>-1$ are parameters. For this class of synthesized systems (H_τ), with F and I determined by the preceding equations (4.2) and (4.3), we have the following results.

Theorem 4.1 For $\tau_1>1$ and $\tau_2>-1$, if $\beta\in B^{n+s}\cap L_a$, then α, which is determined by the first n components of β, is an asymptotically stable equilibrium point of the synthesized system

$$x(k+1)=\text{sat}(Fy(k)+I)\quad k=0,1,2,\ldots \qquad (H_\tau)$$

In particular, this is true for α^q, q=1,...,m, given in equation (4.1). ∎

Because of space limitation, the proof is given elsewhere.

We conclude the present section by summarizing the synthesis procedure developed above.

Synthesis procedure

Suppose we are given m vectors $\alpha^1,\ldots,\alpha^m$ in B^n which are to be stored as asymptotically stable equilibrium points in an n dimensional neural network (H_τ). We proceed as follows.

1) Calculate the augmented vectors $\beta^1,\beta^2,\ldots,\beta^m$, where the first n components of β^q are identical to the components of α^q, and the remaining s components of β^q are chosen from the set $\alpha_1^q\alpha_2^q,\ldots,\alpha_1^q\alpha_n^q,\ \alpha_2^q\alpha_3^q,\ldots,\ \alpha_{n-1}^q\alpha_n^q$. For example, *one possible* set of vectors β^q, q=1,...,m, is given by

$$\beta^q=[\alpha_1^q,\ \alpha_2^q,\ldots,\ \alpha_n^q,\ \alpha_1^q\alpha_2^q,\ldots,\ \alpha_1^q\alpha_n^q,\ \alpha_2^q\alpha_3^q,\ldots,\ \alpha_{n-1}^q\alpha_n^q]^T,\ q=1,\ldots,m.$$

Let

$$\gamma^q=\beta^q-\beta^m,\quad q=1,\ldots,m-1.$$

2) Perform a Gram-Schmidt orthogonalization on the vectors $\gamma^1,\ldots,\gamma^{m-1}$, $e^1,\ldots,e^{n+s}$, where $\{e^1,\ldots,e^{n+s}\}$ is the natural basis of R^{n+s}. The resulting first (n+s) orthogonal vectors, $u^1,u^2,\ldots,u^k,\ldots,u^{n+s}$, form a basis of R^{n+s}, where k=largest number of linearly independent vectors in the set $\{\gamma^1,\gamma^2,\ldots,\gamma^{m-1}\}$, and $u^1,\ldots,u^k$ is an orthonormal basis of Span $(\gamma^1,\ldots,\gamma^{m-1})$.

3) The coefficients F_{ij} and I_i are determined by:

$$F_{ij}=\tau_1\Sigma_{r=1}^{k}u_i^ru_j^r-\tau_2\Sigma_{r=k+1}^{n+s}u_i^ru_j^r,$$

where τ_1 and τ_2 are two real numbers with $\tau_1>1$, and τ_2 is a large positive number, and

$$I_i=\tau_1\alpha_i^m-\Sigma_{j=1}^{n+s}F_{ij}\beta_j^m.$$

Then $\alpha^1,\ldots,\alpha^m$ will be stored as asymptotically stable equilibrium points in system

$$x(k+1)=\text{sat}(Fy(k)+I) \qquad (H_\tau)$$

V. AN EXAMPLE

We now present a specific example to demonstrate the applicability of the results presented herein. We employ a class of discrete-time neural networks with nonlinear interconnections to implement an encoder.

The encoder problem has been widely used as a benchmarking problem in the neural networks community for testing backpropagation and other methods. The general encoding problem involves finding an efficient set of hidden units to encode a large number of input/output patterns. The number of hidden units is intentionally made small to obtain efficient encoding. An extensive and frequently cited simulation study is given in [8], where In particular, a 10-5-10 encoder network is considered for different learning parameters.

In the present paper, we use a class of discrete-time neural networks with nonlinear interconnections to implement a 10 dimensional encoder. The proposed neural network, which implements the encoder, has 10 distinct patterns as desired memories. Each of the 10 patterns has only one of the 10 units turned on (set to 1), the orther bits are turned off (set to -1). In vector form, these 10 patterns are,

$$\alpha^i=[-1,\ldots,-1,\ 1,\ -1,\ldots,-1]^T,$$

where only the ith element of the vector has a value of 1.

From analysis as well as simulation results, it is known that in general the 10 patterns can not be stored reliably as asymptotically stable equilibria in a 10 dimensional neural network with linear interconnecting structure. With only some of the nonlinear terms utilized, all the patterns were stored as asymptotically stable equilibria. In particular, we attempted to use as small of a number of interconnetions as possible. From our previous analysis, for a 10 dimensional network, with 10 desired patterns to be stored, only 1 high order term is required for the reliable storage of the 10 patterns.

Simulations verified that with 1 randomly selected high order term, the 10 patterns are indeed stored as asymptotically stable equilibrium points. Furthermore, to demonstrate the reliablity of the network, we attempted to recognize the 10 key patterns from noisy input patterns. The simulation was carried out as follows. The interconnection matrix F is computed from the synthesis procedure of Section 4, with external bias vector I=0, and $\tau_1=1.1$, $\tau_2=0$. We made 3 differrent sets of simulation runs. In each case we randomly selected a single second order interconnection term. The noisy input pattern was generated by adding zero mean Gaussian noise with a given specific standard deviation to the key pattern. For each of the given standard deviations, SD=0.2, SD=0.4, SD=0.6, SD=0.8, 100 randomly selected inputs for the 10 key patterns were used as initial conditions for the discrete-time system (H_τ). The results were then averaged and tabulated as shown.

Table 1

rate in percentage \ noise type	Zero Mean Gaussian Noise			
	SD=0.2	SD=0.4	SD=0.6	SD=0.8
converge to desired memories	100	79.33	41.33	24
converge to spurious states	0	20.67	58.67	75.67
not convergent [1]	0	0	0	0.33

[1] means the system does not converge within 100 time iterations.

It is worthwhile to note that although the encoder problem has been used as a benchmarking problem in the neural network community for testing backpropagation and similar networks, as the author of [8] notes, there is no single schedule that works for problems of various types and of different sizes, and there is no theory that allows one to choose a good schedule for a problem before one attempts to run it. Our experience suggests that, an encoder can usually be implemented by the class of neural networks with nonlinear interconnections presented herein.

VI. CONCLUSION

In the present paper we first addressed the qualitative properties of a class of synchronous, discrete-time neural networks with second order polynomial interconnections (Section 3). The results of this analysis served as the basis of a synthesis procedure for these networks (Section 4). We demonstrated the applicability of the above results by means of a specific example.

Networks of the type considered herein have in general greater memory storage capacity than corresponding networks with linear interconnecting structure.

To simplify our presentation, we confined ourselves to networks with second order polynomial interconnections; however, the results and methodology advanced herein apply also to neural networks with higher order polynomial interconnections.

REFERENCES

[1] Y. Abu-Mostafa, J. Jacques, "Information Capacity of the Hopfield Model", *IEEE Trans. on Information Theory*, Vol. IT-31, No.4, July 1985, pp.461-464.

[2] R. McEliece, E. Posner, E. Rodemich, S. Venkatesh, "The Capacity of the Hopfield Associative Memory", *IEEE Trans. on Information Theory*, Vol. IT-33, No.4, July, pp.461-482.

[3] P. Baldi, S. Venkatesh, "Number of Stable Points for Spin-Glass and Neural Networks of Higher Orders", *Physical Review Letters*, Vol.58, No.9, March 2, 1987, pp.913-916.

[4] T. Maxwell, C. Giles, Y. Lee, H. Chen, "Nonlinear Dynamics of Artificial Neural Systems", *AIP Conference Proceeding*, 1986, pp.299-304.

[5] C. Giles, T. Maxwell, "Learning, Invariance, and Generalization in High-Order Neural Networks", *Applied Optics*, Vol.26, No.23, Dec. 1987, pp.4972-4978.

[6] I. Guyon, L. Personnaz, J. Nadal, G. Dreyfus, "High Order Neural Networks for Efficient Associative Memory Design", in D. Z. Anderson (Ed.), *Neural Information Processing Systems, AIP Conference Proceeding*, Denver, Co., 1987, pp.233-241.

[7] A. N. Michel, J. Si, G. Yen, "Analysis and Synthesis of a class of Discrete-Time Neural Networks Described on Hypercubes", *IEEE Trans. on Neural Networks,* Vol. 2, No. 1, Jan. 1991, pp. 32-46.

[8] S. E. Fahlman, "Faster-Learning Variations on Back-Propagation: An Empirical Study", *Proceedings of the 1988 Connectionist Models Summer School*, D. Touretzky, G. Hinton, and T. Sejnowski (Eds.), Morgan Kaufmann Publishers, San Mateo, CA., 1988.

[9] R. K. Miller, A. N. Michel, *Ordinary Differential Equations,* Academic Press, New York, 1972.

* Supported in part by NSF under grant ECS 91-07728.

FA6 - 10:30

Proceedings of the 31st Conference
on Decision and Control
Tucson, Arizona • December 1992

ESTIMATION OF ARTIFICIAL NEURAL NETWORK PARAMETERS FOR NONLINEAR SYSTEM IDENTIFICATION

Timothy L. Ruchti, Ronald H. Brown, and Jeffrey J. Garside
(414/288-1609), (414/288-3501), (414/288-1609)

Department of Electrical and Computer Engineering
Marquette University
1515 West Wisconsin Avenue
Milwaukee, WI 53233

Abstract **- The structure of feedforward artificial neural networks (ANNs) has been shown to be suitable for representing arbitrary piecewise-continuous nonlinear mappings that are part of dynamic systems. However, the utility of this structure for nonlinear system identification is intimately linked with the ability to determine its parameters on the basis experimental observations. In this paper, a unified framework for representing ANN training algorithms is developed by considering weight selection as a parameter estimation problem. Three existing ANN training strategies are reviewed within this framework including gradient-descent backpropagation, the extended Kalman algorithm, and the recursive least squares method. A strikingly different approach to error backpropagation is presented resulting in the development of a novel method of backward signal propagation and target state generation for embedded layers. The proposed technique is suitable for implementation with a linear Kalman based update algorithm and is applied with a time-varying method of covariance modification for the elimination of transients associated with initial conditions. Results from a nonlinear identification experiment demonstrate an increased rate of convergence in comparison with backpropagation. In addition, the new algorithm displayed similar rates of parameter convergence and a decreased computational overhead compared to the extended Kalman algorithm.**

I. INTRODUCTION

System identification is concerned with the representation of a given system through the construction and parameterization of a model on the basis of experimental observation [1]. Although extensive effort has been devoted to the development of powerful techniques for the analysis and identification of uncertain linear systems, analogous methods are yet to be developed for systems that exhibit significant nonlinearity [1,11]. Artificial neural networks (ANNs) are an emerging technology that contribute a general structure for representing nonlinear mappings to the field of system identification. As a result of a parallel organization and an inherent nonlinearity, feedforward ANNs have the ability to model any piecewise-continuous nonlinear mapping to an arbitrary degree of accuracy with properly selected parameters [2,7]. Therefore, it has been established that ANNs are a suitable structure for representing a general unknown continuous nonlinearity in a dynamic system [12,18].

The method of parameterizing an ANN on the basis of observable information is not straightforward and currently restricts the utilization of this technology. Although the popular method of backpropagation has attained success in a variety of applications [4,5,8,12], it suffers from the inherent limitations of gradient search techniques, namely, slow rates of convergence, noise sensitivity, and the inability to distinguish between local and global minimal points [7]. Therefore, new methods must be developed which supply ANNs with the ability to learn and track time-varying nonlinearities before the full potential of this structure can be exploited.

This paper addresses the problem of ANN weight selection by contributing a unified parameter estimation based framework for representing ANN training algorithms. Under this framework, a new algorithm is developed which is suitable for application to dynamic nonlinear system identification. Generalizing from the work of [14, 17], this method partitions an ANN into linear sub-components allowing the application of linear estimation techniques. The salient feature of this design is an innovative method of backpropagating target states to embedded layers facilitating the utilization of a linear Kalman filter to estimate ANN parameters. In addition, a unique method of compensating for transients associated with initial conditions is realized. The new technique is compared to the traditional method of gradient-descent backpropagation [15], and the recently proposed extended Kalman [3,16] and recursive least-squares algorithms [14,17] in a nonlinear system identification problem.

II. FEEDFORWARD ARTIFICIAL NEURAL NETWORKS

An ANN is a computational structure, consisting of simple interconnected processing elements called neurons, that performs a nonlinear mapping from an input vector space $U \subset \mathbb{R}^m$ to an output vector space $Y \subset \mathbb{R}^n$. In a multilayer feedforward ANN the neurons, illustrated in Fig. 3, are organized into cascaded layers containing no feedback or lateral connections. A block diagram representation of such a network is shown in Fig. 1 consisting of a total of N layers, including an input layer to which the input vector $u^1(k) \in U$ is presented, a number of hidden layers, and an output layer generating $y^N(k) \in Y$. The index k refers to the observation or alternately the discrete-time index of a dynamic system.

A given layer, L, as represented in Fig. 2, consists of n^L neurons and a set of weights or synapses $\{ w^L_{ij} : i=1 \text{ to } n^L, j=1 \text{ to } n^{L-1} \}$ organized into a matrix $W^L(k)$. Each weight w^L_{ij} is an adjustable parameter which propagates and scales the output of the jth neuron in the previous layer to the ith neuron in the Lth layer. At time k the propagation of an input, $u^L(k)$, through a layer is defined by a linear transformation and a nonlinear mapping. The linear transformation is a consequence of the synaptic connections between adjacent layers and is given by

$$x^L(k) = W^L(k) u^L(k) = W^L(k) y^{L-1}(k) \tag{2.1}$$

where $u^L(k) \in \mathbb{R}^{n^{L-1}}$ is the input to the layer, $x^L(k) \in \mathbb{R}^{n^L}$ defines the activation level or state of the layer and $W^L(k)$ is the n^L by n^{L-1} weight matrix. The nonlinear vector function, $\mathbf{f}: \mathbb{R}^{n^L} \rightarrow \mathbb{R}^{n^L}$, is any well-defined nonlinear mapping which performs the transformation

$$y^L(k) = \mathrm{f}^L(x^L(k)) = \mathrm{f}^L(W^L(k) u^L(k)) \tag{2.2}$$

where $y^L(k) \in \mathbb{R}^{n^L}$ is the output of the layer. Although there are many effective activation functions, this paper employs the conventional sigmoidal nonlinearity which maps $[-\infty \ \infty] \rightarrow [-1 \ 1]$ according to

$$\mathrm{f}_i(x_i) = \frac{1 - e^{-x_i T}}{1 + e^{-x_i T}} \qquad \forall x_i \in x \,, \ \forall f_i \in f(\cdot) \tag{2.3}$$

where T is a scalar representing the temperature or excitability of a particular neuron.

CH3229-2/92/0000-2728$1.00 © 1992 IEEE

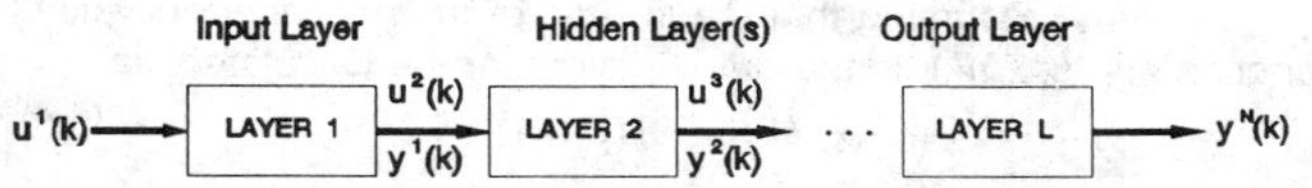

Figure 1. *A block diagram representation of a multilayer feedforward artificial neural network.*

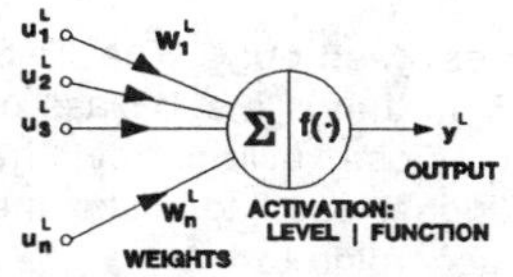

Figure 2. *A single neuron.*

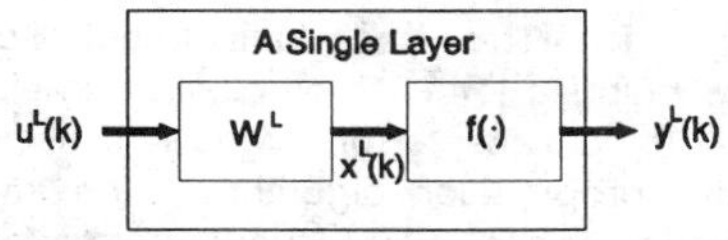

Figure 3. *A single layer of a feedforward artificial neural network.*

III. TRAINING: A PARAMETER ESTIMATION APPROACH

In a system identification application, a feedforward ANN is trained to represent an unknown nonlinear mapping by adjusting its weights according to an ongoing stream of observations $\{u_d^L(k), y_d^L(k): k = 1,...,N\}$. Although this training mechanism is generally referred to as supervised learning, it is equivalent to the process of parameter estimation. This approach follows from the definition of weights as adjustable parameters selected according to a predefined error criterion function.

At each time instant, when new observations are made available, it is desirable to incorporate the additional information into the current parameter estimate without recalculating the influence of prior observations sample by sample. An efficient update algorithm that performs this task for a particular weight matrix is specified by the following recursive equation

$$W^L(k) = W^L(k-1) + \alpha e^L(k) M(k)^T \tag{3.1}$$

where $M(k) \in \mathbb{R}^{n^{L-1} \times 1}$ is the algorithm gain matrix, α is a scalar constant modifying the step size, and $e^L(k)$ is the prediction error given by

$$e^L(k) = y_d^L(k) - y^L(k) = y_d^L(k) - f^L(W^L(k) u^L(k)) \tag{3.2}$$

Typically, $M^L(k)$ is related to prior observations and/or the gradient of $y^L(k)$ with respect to $W^L(k)$. While (3.1) is actually a subclass of a more general recursive parameter estimation algorithm [11], it is sufficient to serve as a framework for the ANN training algorithms discussed subsequently.

A. Gradient-Descent Backpropagation

Gradient based algorithms make an adjustment of the parameters of a given weight matrix, $W^L(k)$, in the direction of the negative gradient of the *kth* sum square prediction error given by

$$J(k) = \frac{1}{2} e^{N^T}(k) e^N(k) \tag{3.3}$$

where $e^N(k)$ is the prediction error of the network output. Taking the partial of (3.3) with respect to a given weight, w_{ij}^L, leads to the recursive update equation

$$w_{ij}^L(k) = w_{ij}^L(k-1) + \alpha e^T(k) \frac{\partial y(k)}{\partial w_{ij}^L(k)} \tag{3.4}$$

where α is the step size which is in the same form as the general algorithm in (3.1).

The method of backpropagation is an efficient method for calculating the partial derivatives of (3.4) [15]. The algorithm is developed by first defining $\delta^L(k)$ according to

$$\delta_i^L(k) = -\frac{\partial J(k)}{\partial x_i^L(k)} = \frac{\partial J(k)}{\partial o_i^L(k)} \frac{df_i^L(x_i^L(k))}{dx_i^L(k)} \tag{3.5}$$

Then, in the output layer

$$\delta_i^L(k) = e(k)^T \frac{df^L(x_i^L(k))}{dx_i^L(k)} \tag{3.6}$$

while for any other layer:

$$\delta_i^L(k) = \frac{df(x_i^L(k))}{dx_i^L(k)} W_i^{L+1}(k) \delta^{L+1}(k) \tag{3.7}$$

where W_i^{L+1} is the *i*th row of the weight matrix in the *(L+1)*th layer. Hence, the partial derivatives of (3.4) are implicitly calculated in (3.6) and (3.7) as part of a recursive error propagation algorithm. Substituting these into (3.4) produces the recursive parameter update law

$$w_{ij}^L(k+1) = w_{ij}^L(k) + \alpha \delta_i^L(k) u_j^L(k) \tag{3.8}$$

Although this technique is powerful, it suffers from the typical limitations of gradient methods, namely, inefficient step size selection causing slow rates of convergence, the inability to distinguish global from local minimal points and a sensitivity to noise.

B. The Extended Kalman Update Algorithm

As shown above, gradient-descent estimation techniques adjust the parameters on the basis of the current error independent of prior observations. An alternate approach is to concatenate all the network parameters into a single vector $\theta(k)$ and to define an operator, NET(·), which performs the complete mapping function of a multilayer feedforward ANN on the basis of this parameter vector and the input, $u(k)$, according to

$$y = \mathrm{NET}(\theta, u) = f^N(W^N f^{N-1}(W^{N-1} \cdots f^1(W^1 u) \cdots)) \tag{3.9}$$

where k has been omitted for notational simplicity. The parameter estimation problem is then defined as the determination of $\theta(p)$ which minimizes the sum square prediction error of all prior observations embodied in the criterion function

$$J(p) = \sum_{k=1}^{p} \left(y_d(k) - y(k)\right)^T \left(y_d(k) - y(k)\right) \tag{3.10}$$

where $y_d(k)$ is the observed output vector and p denotes the number of training samples. Assume that the observation is of the form:

$$y_d(k) = \mathrm{NET}(\theta^*, u(k)) + v(k)$$

where $\theta^*(k)$ is the desired parameter vector minimizing (3.10), $v(k)$ is the measurement noise with covariance $R(k)$, and that the nonlinearities of (3.9) are sufficiently smooth. Then the suboptimal extended Kalman filter can be applied directly to the estimation problem through the equations:

$$\begin{aligned}
\hat{\theta}(k) &= \hat{\theta}(k-1) + K(k)\left(y_d(k) - y(k)\right) \\
K(k) &= \frac{P(k-1)H^T(k)}{\lambda(k)R(k) + H^T(k)P(k-1)H(k)} \\
P(k) &= \frac{1}{\lambda(k)}\left(I - K(k)H^T(k)\right)P(k-1) \\
H(k) &= \frac{\partial\, \mathrm{NET}(\theta(k), u(k))}{\partial \theta(k)} \Big|_{\theta(k) = \hat{\theta}(k)}
\end{aligned} \tag{3.12}$$

where $\hat{\theta}(k)$ is the estimated parameter vector, $P(k)$ is the approximate error covariance matrix, $H(k)$ is the Hessian matrix containing the partial derivative of the network output with respect to each individual weight, and $\lambda(k)$ is a time-varying forgetting factor.

This technique has been found to converge quickly in comparison with gradient-descent techniques, is insensitive to uncorrelated noise, and has a better success rate than does the gradient algorithm [3,16]. However, the computational overhead is significantly increased since the covariance matrix is the same dimension as the number of adjustable weights in the network. Furthermore, the weight updating requires a centralized computing facility eliminating the advantage of the inherent parallelism of artificial neural networks.

IV. THE LINEAR KALMAN FILTER

The objective of this section is to develop a parameter estimation technique for feedforward ANNs which maintains the computational simplicity of gradient techniques but yet achieves the high rates of convergence characteristic of the extended Kalman algorithm. The basic approach follows from a generalization of the recursive least-squares (RLS) technique [14, 17] which prescribes a layer by layer application of the RLS estimator. The design methodology presented here consists of three steps: the design of a linear estimation algorithm, the determination of suitable desired or target input and output states for embedded layers, and the formation of techniques to eliminate transients resulting from the initial conditions.

A. Kalman Based Parameter Estimation

The problem of determining the weights of a particular layer of a feedforward ANN can be transformed into a parameter estimation problem by viewing the network as a cascade of separate segments called layers. As noted by [14, 17], the selection of the weights in each layer can be further transformed into a *linear* parameter estimation problem by partitioning individual layers into linear and nonlinear sub-components defined by (2.1) and (2.2), respectively, and solving each linear estimation problem individually.

The determination of suitable input-output target states for embedded linear sub-components, denoted $u_d^L(k)$ and $y_d^L(k)$, is not a trivial task. However, for the moment assume that appropriate target states are available. Additionally define $W_d^L(k)$ as the desired linear transformation of the Lth layer such that the measurement or observation of $x_d^L(k)$ given the input $u_d^L(k)$ is found through

$$x_d^L(k) = W_d^L(k)u_d^L(k) + v(k) \tag{4.1}$$

where $v(k)$ is the measurement noise with covariance $R(k)$ and zero mean. Then the optimal solution for the estimate $W^L(k)$ which minimizes the error criterion function

$$J^L(p) = \frac{1}{2}\sum_{k=1}^{p}\left(x_d^L(k) - W^L(p)u_d^L(k)\right)^T\left(x_d^L(k) - W^L(p)u_d^L(k)\right)$$

is the linear Kalman filter given by

$$\begin{aligned} W(k) &= W(k-1) + \alpha\left(x_d(k) - x(k)\right)K^T(k) \\ K(k) &= \frac{P(k-1)u_d(k)}{\lambda(k)R(k) + u_d^T(k)P(k-1)u_d(k)} \\ P(k) &= \frac{1}{\lambda(k)}\left(I - K(k)u_d^T(k)\right)P(k-1) \end{aligned} \tag{4.3}$$

where $P(k)$ is the covariance matrix, $K(k)$ is the Kalman gain, $\lambda(k)$ is a time-varying forgetting factor, α is the step size, and the layer index, L, has been omitted for notational simplicity. This defines a linear recursive estimation algorithm with same form as (3.1) for determining the non-stationary weights of an ANN layer by layer. Since the estimation problem is partitioned into subsections,an algorithm results that is computationally less burdensome than the extended Kalman method. Furthermore, since this algorithm assumes that the estimated matrix is non-stationary, it is expected that it will enable the tracking of a time-varying system.

B. Target State Selection

It follows from the discussion above that the application of the Kalman filter to the estimation of weight parameters is straight forward if the appropriate target states are available. Unfortunately, this is not the case. In the input layer, the applied vector input $u_d(k)$ is clearly available. However, the desired input state into each subsequent layer is unavailable since $u_d(k)$ is propagated through untrained or partially trained layers.

Similarly, assuming that the inverse of the nonlinear activation function exists, $x_d^L(k)$, in the output layer, can be calculated via

$$x_d^N(k) = f^{-1}(y_d(k)) \tag{4.4}$$

All other target outputs are not prescribed and alternate strategies must be developed.

1. The Recursive Least Squares Method: The RLS techniques [14, 17], employing an algorithm that is a sub-class of (4.3), defines error signals similar to those utilized by the backpropagation algorithm. The error signal used to make the update in the output layer is determined according to

$$e^N(k) = f^{-1}(y_d^N(k)) - x^N(k) \tag{4.5}$$

where N is the number of layers in the ANN, $\mathbf{f}(\cdot)$ is the sigmoidal nonlinearity defined in (2.3), $y_d(k)$ is the desired output, and $x^N(k)$ is the activation level of the output layer. All other layers are updated according to the backpropagated error, $\delta^L(k)$, defined by (3.7).

2. Error Signal Backpropagation: This paper presents a fundamentally different approach by defining a pseudo or suboptimal target input-output pair to each linear subsection of a given layer as $(\bar{u}^L(k), \bar{x}^L(k))$. Then the pseudo target states for the input to the first layer and the output of the output or Nth layer are defined by

$$\bar{u}^1(k) = u_d(k), \qquad \bar{x}^N(k) = f^{-1}\left(y_d(k)\right) \tag{4.6}$$

In addition, the suboptimal input target vector to each linear element is defined according to the signal resulting from the forward propagation of $u_d(k)$:

$$\bar{u}^L(k) = u^L(k) \qquad \text{for } L = 2, 3, \cdots, N \tag{4.7}$$

The selection of $\bar{x}^L(k)$ is not as straight forward and is the innovation presented in this section. Assume that $\bar{x}^{L+1}(k)$ has been found and it is desired to determine an optimal selection for $\bar{x}^L(k)$ considering the weights of the subsequent layer and the current activation level of the Lth layer. The first step is to determine $\bar{y}^L(k)$ such that $\bar{x}^L(k)$ can be calculated through the inversion of $\mathbf{f}^L(\cdot)$ described by (4.4). With this in mind the following criteria for selecting $\bar{y}^L(k)$ are proposed:

1. Select $\bar{y}^L(k)$ such that the inverse
$$\bar{x}(k) = f^{-1}(\bar{y}^L(k)) \tag{4.8}$$
exists.

2. The selection of $\bar{y}^L(k)$ must be made so as to closely approximate $\bar{x}_{L+1}(k)$ after forward propagation through the current matrix estimate, $W^{L+1}(k)$:
$$\bar{x}^{L+1} \approx W^{L+1}(k)\bar{y}^L(k) \tag{4.9}$$

3. In general there may be a family of solutions or there may be no solution which satisfies (4.8) and (4.9) . In either case, select a pseudo vector $\bar{y}^L(k)$ that is close to $y^L(k)$.

These criteria are embodied in the following cost function:

$$J^L(\bar{y}^L) = \frac{1}{2}(y^L - \bar{y}^L)^T(y^L - \bar{y}^T) + \frac{\beta}{2}(x^{L+1} - W^{L+1}\bar{y}^L)^T(x^{L+1} - W^{L+1}\bar{y}^L) \tag{4.10}$$

where $\beta > 0$ is chosen such that criterion (1) is satisfied and the index k is neglected for notational simplicity. The optimal selection for $\bar{y}^L(k)$ according to the above criteria is defined as that vector which minimizes equation (4.10). Therefore, the partial derivative of (4.10) with respect to $\bar{y}^L(k)$ is set to zero:

$$\frac{\partial J^L}{\partial \bar{y}^L} = -(y^L - \bar{y}^L) - \beta W^{L+1^T}(\bar{x}^{L+1} - W^{L+1}\bar{y}^L) = 0 \qquad (4.11)$$

Solving this for $\bar{y}^L(k)$ yields

$$\bar{y}^L = \left[I + \beta W^{L+1^T} W^{L+1}\right]^{-1}\left[y^L + \beta W^{L+1^T}\bar{x}^{L+1}\right] \qquad (4.12)$$

By proper choice of β a solution can always be found that satisfies (4.8). Intuitively it can be seen that large values of β with cause (4.12) to approximate the pseudo inverse of (4.9) while small values of β will make (4.12) approximately equal to $y^L(k)$.

While the matrix inversion of (4.12) is a drawback of this approach there are several inherent properties that can be exploited. First, the inverse in (4.12) is guaranteed to exist and the matrix is positive definite symmetrical. Therefore, numerical methods can be applied which will reduce the number of computations significantly. Secondly, if $W^L(k)$ is n^L by n^{L-1} and $n^L < n^{L-1}$, the matrix-inversion lemma can be applied to (4.12) to produce

$$\bar{y}^L = \left[I + \beta W^{L+1^T}(I + \beta W^{L+1}W^{L+1^T})^{-1}W^{L+1}\right]\left[y^L + \beta W^{L+1^T}\bar{x}^{L+1}\right]$$

which involves a n^L by n^L matrix inversion rather than a n^{L-1} by n^{L-1} inversion. In an ANN with a single output, this results in a scalar division.

C. Covariance Modification

When the network is randomly initialized, original training pairs rapidly become obsolete as the adjacent layers adapt their weights according to their perceived observations. Hence, the $P(k)$ matrix for a particular layer, containing the sum total of all prior observations rapidly becomes outdated. Since the Kalman filter minimizes the prediction error of all prior observations, parameter estimates become biased towards initial conditions and transient pseudo observations as defined in the previous section. The solution to this dilemma is the modification of the covariance matrix in each layer through the forgetting factor $\lambda(k)$. The forgetting factor, $\lambda(k)$, is used to prioritize more recent observations by scaling $P(k)$ in the following manner

$$P^{-1}(k) = \lambda(k)P^{-1}(k-1) + u(k)u^T(k) \qquad (4.14)$$

where $\lambda(k) \epsilon (0,1]$. A smaller value for $\lambda(k)$ gives better responsiveness to changes but a larger variance in parameter estimates. However, when the input is not persistently exciting or when $\lambda(k)$ is kept small for many samples, $P(k)$ will grow exponentially which can lead to the temporary instability of the estimator. This phenomena is avoided by determining $\lambda(k)$ as follows

$$\lambda(k) = 1 - \gamma \frac{e^2(k)}{1 + u^T(k)P(k-1)u(k)} \qquad (4.15)$$

where γ is a constant which is inversely proportional to the expected covariance of the prediction error, $e(k)$ [5]. As a result, $\lambda(k)$ remains close to one when P(k) is already large and the estimator is sensitive to parameter variations. A lower bound, λ_o, on $\lambda(k)$ is used to avoid the forgetting factor from becoming excessively small (or even negative), resulting in large fluctuations of the estimates in spite of small prediction errors. To ensure the avoidance of estimator wind-up an upper limit is placed on the trace of the $P(k)$ matrix.

V. RESULTS

In this section simulation results from a nonlinear plant identification problem are presented in which a feedforward ANN was utilized to approximate a highly nonlinear mapping. The unknown nonlinear function, chosen due to its demanding characteristics, is

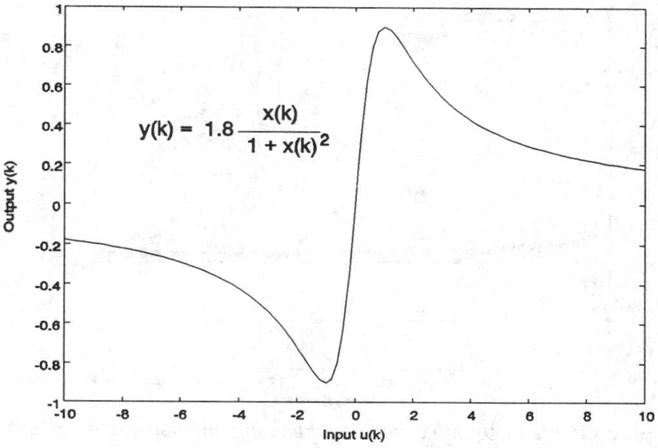

Figure 4. *Exemplar nonlinear mapping used in the simulation examples.*

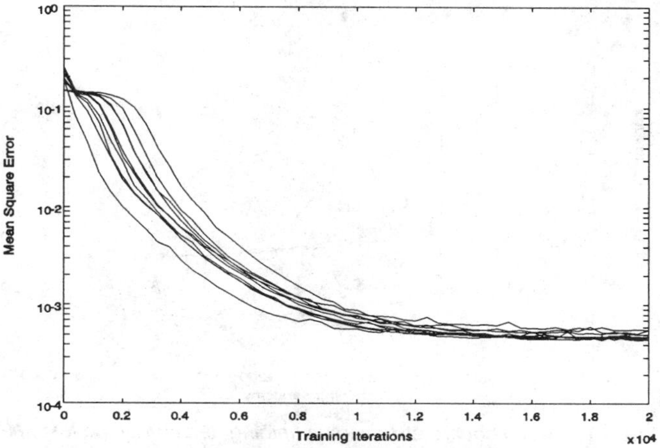

Figure 5. *Mean square error versus training iteration of the test set during training via backpropagation.*

shown in Fig. 4 and is given by

$$y_d(k) = 1.8\frac{u_d(k)}{1 + u_d^2(k)} \qquad (4.16)$$

where $y_d(k) \in [-0.9\ 0.9]$ when $u_d(k) \in [-10\ 10]$. All four of the previously discussed parameter estimation techniques were employed to train an ANN, with an architecture of 1-8-4-1 to approximate (5.1). To demonstrate the average or normal error convergence, ten experiments, starting from different random weight initializations as described below, were performed for each of the estimation techniques. The training set was a different sequence of random points on the function for each experiment generated with $u_d(k) \in [-10\ 10]$ uniformly distributed. The test set, used to evaluate the performance of each ANN during training through a mean square error (MSE) measurement, consisted of a constant set of 134 points evenly spaced on the range of the input, $u_d(k)$.

A. Gradient Descent Backpropagation

The gradient descent algorithm described in Section III.A was applied with a constant step size, α, of 0.25 in each layer. Initially the weights were set uniformly in the interval [-1,1] and the MSE versus training iteration for each of the ten experiments is shown in Figure 4. Convergence in each experiment to a MSE near 5.0E-4 occurred after approximately about 20,000 training iterations which is comparable with results presented in the literature [12]. As will be shown by the subsequent experiments, lower error rates are possible

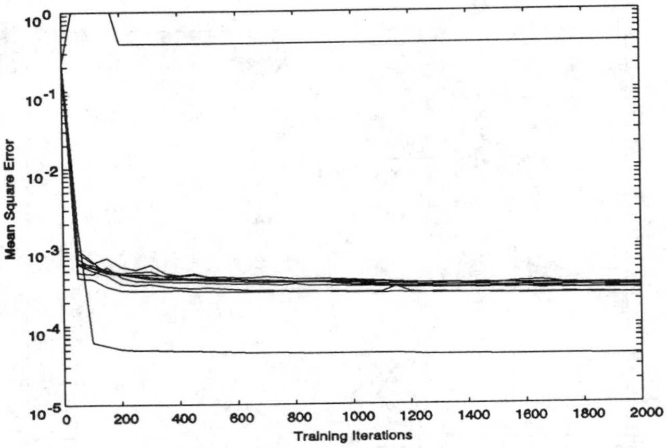

Figure 6. *Mean square error versus training iteration of the test set during training via the extended Kalman algorithm.*

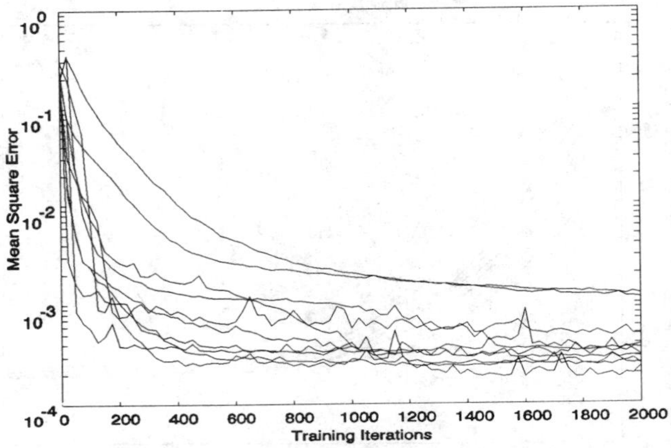

Figure 7. *Mean square error versus training iteration of the test set during training via the recursive least squares algorithm.*

for the same network architecture, hence this convergence is to a local minimal point.

B. Extended Kalman Algorithm

The extended Kalman parameter estimation technique, introduced in Section III.B, was implemented with a step size of 0.5 and a forgetting factor, described by (4.15), with γ=0.5. Initially the weights were uniformly distributed in the interval [-0.1, 0.1] and the covariance matrix was set to $1000 \cdot \mathbf{I}$. The convergence plots of the ten experiments are shown in Fig. 6. The results reveal an extremely fast convergence, requiring 100 times fewer training iterations than the gradient-descent backpropagation algorithm. However, there are two singular experiments represented in the figure. First, one experiment seemed to diverge to a final MSE of 0.4 which is greater than that of the initial conditions. The second curve converged to a MSE an order of magnitude lower than any of the others. The reason for the divergence is the result of estimator buildup which can be corrected by covariance monitoring. The experiment which achieved the low MSE demonstrates that all other ANNs converged to a local minimum.

C. Recursive Least Squares

The recursive least squares method was employed as described in Section IV.B.1 and by [14, 17] with the learning rates $\alpha\mathbf{1}$ = 10, $\alpha\mathbf{2}$ = 5, and $\alpha\mathbf{3}$ = 1 corresponding to the first, second and third layers and a global forgetting factor set to 0.99. Initially, the covariance matrices were set to $100 \cdot \mathbf{I}$ and the weights were uniformly distributed in the interval [-1 1]. The results of the ten experiments, shown in Fig. 7, demonstrates a marked improvement over the gradient method although convergence is not as fast as the extended Kalman method.

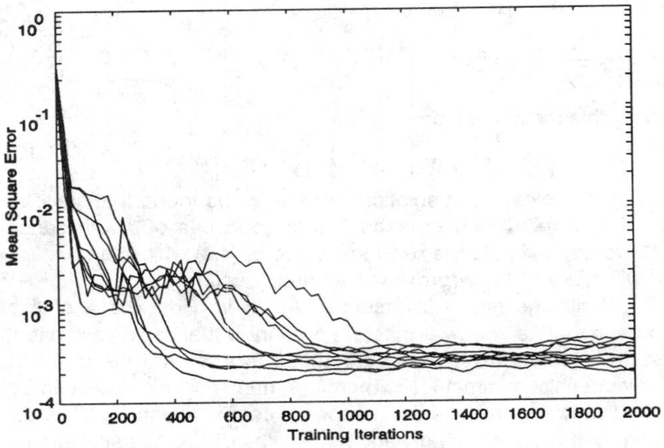

Figure 8. *Mean square error versus training iteration of the test set during training via the linear Kalman algorithm.*

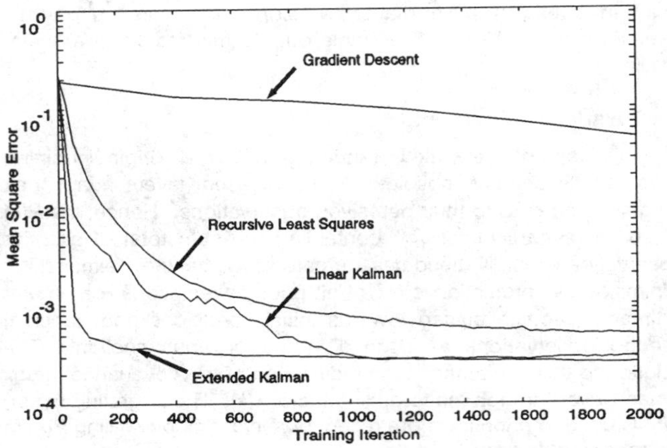

Figure 9. *Comparison of the mean square error versus training iteration of the four parameter estimation algorithms.*

D. Linear Kalman with Target State Backpropagation

The linear Kalman technique with error signal backpropagation, developed in Section IV, was implemented with a step size of 0.25 for each of the three layers and a measurement error covariance, $R(k)$, of 1. The constant, β, of equation (4.12), was set to 2 and γ of 18 was initially set to 1000 and later decreased to 1 after the initial transients subsided. The covariance matrices were initialized to $100 \cdot \mathbf{I}$ and the weights were initially set as follows: in layer one all weights were set to 10, in layer 2 the weights were uniformly distributed between [-1, 1], and in layer three the weights were uniformly distributed between [-0.1, 0.1]. The results of the linear Kalman curves are shown in Fig. 8. demonstrating an extremely fast rate of convergence, similar to that of the extended Kalman. The use of the error dependent forgetting factor was instrumental in driving the MSE low during the first few hundred training iterations until the bias caused by the initial conditions was eliminated. Different initial weight matrices, particularly in the first layer caused some experiments, not shown in this paper, to remain in local minima well above the final MSE of the experiments illustrated in Fig. 8. While these results are extremely encouraging, it is clear that further research is necessary before conclusive statements can be made concerning the new method of signal backpropagation employed by the linear Kalman technique. However, the example shown in this paper indicates that this method is extremely promising.

E. Comparison

Fig. 9 shows a comparison of the average MSE versus training iteration of the four methods over the first 2000 training iterations. In the case of the extended Kalman curve, the singular divergent experiment was excluded since it appeared to be inconsistent with other results. Clearly the extended Kalman method is the fastest followed by the linear Kalman, although the RLS algorithm is quite comparable. The computational overhead of the extended Kalman algorithm is undesirable which makes the linear Kalman or the RLS methods more feasible.

VI. CONCLUSIONS

The improved rates of convergence demonstrated by the three recent ANN training algorithms described in this paper testify to the suitability of multilayer feedforward ANNs in nonlinear system identification and control. The new technique, described as a linear Kalman algorithm employing target state backpropagation, was shown in the nonlinear identification example to be comparable in speed with the extended Kalman technique although it is computationally simpler because it partitions the network into linear sub-components.

The novel approach to signal backpropagation presented in this paper was shown in the example to be effective for parameterizing the weights of embedded layers. This approach, while quite different from the traditional method of error backpropagation is an evolving technique for which more research is necessary before conclusive statements can be made. However, due to the nature of the backpropagated signal it is expected that this technique will be more suitable for tracking time-varying systems. Furthermore, the characteristic positive-definite element of (4.12) seems to lend to a Lyapunov based stability analysis.

The treatment of ANN training algorithms as a parameter estimation problem will produce further development of ANN training algorithms due to the abundance of past research on parameter estimation. Furthermore, many of the theorems relating the convergence and consistency of parameter estimates in the linear case may be generalized in the future to the partitioned linear case presented in this paper.

VII. REFERENCES

[1] K.J. Astrom and P. Eykhoff, "System Identification - A Survey." Automatica, vol. 7, pp. 123-162, 1971.

[2] E. K. Blum and L. K. Li, "Approximation Theory and Feedforward Networks," Neural Networks Sci., vol. 4, pp. 511-515, 1991.

[3] S. Chen, C. F. Cowan, S. A. Billings, and P. M. Grant, "Parallel Recursive Prediction Error Algorithm for Training Layered Neural Networks." Int. J. Control, vol. 51, no. 8, pp. 1215-1228, 1990.

[4] Fu-Chuang Chen and H. K. Khalil, "Adaptive Control of Nonlinear Systems Using Neural Networks," Proceedings: 29th Conference on Decision and Control, Honolulu, Hawaii, pp. 1707-1712, Dec., 1990.

[5] S. R. Chu, R. Shoureshi, and M. Tenorio, "Neural Networks for System Identification," IEEE Control Systems Magazine, pp. 31-35, April, 1990.

[6] T. R. Fortesque, L. S. Kershenbaum, and B. E. Ydstie, "Implementation of Self-tuning Regulators with Variable Forgetting Factors," Automatica, vol. 17, no. 6, pp. 831-835, 1981.

[7] R. Hecht-Nielsen, Neurocomputing. Reading, MA: Addison-Wesley, 1990.

[8] J. J. Helferty, S. Biswas, and M. Maund, "Experiments in Adaptive Control Using Artificial Neural Networks," Proceedings: IEEE International Conf. on Sys. Eng., Pittsburgh, PA, pp. 339-342, Aug., 1990.

[9] R. P. Lippmann, "An Introduction to Computing with Neural Networks," IEEE ASSP Magazine, pp. 4-22, April, 1987.

[10] L. Ljung and S. Gunnarsson, "Adaption and Tracking in System Identification - A Survey." Automatica, vol. 26, pp. 7-21, 1990.

[11] L. Lung, System Identification Theory For the User. Englewood Cliffs, New Jersey: Prentice Hall, 1987.

[12] K. S. Narendra and K. Parthasarathy, "Identification and Control of Dynamic Systems Using Neural Networks," IEEE Trans. Neural Networks, vol. 1, no. 1, pp. 4-26, March 1990.

[13] K. S. Narendra and K. Parthasarathy, "Adaptive Identification of Dynamical Systems using Neural Networks," Proceedings: 28th Conference on Decision and Control, Tampa, FL, pp. 1737-1738, Dec., 1989.

[14] F. Palmieri and S. A. Shah, "A New Algorithm For Training Multilayer Perceptron." IEEE Conf. Sys. Eng., pp. 427-428, 1989.

[15] Rumelhart, D. E., Hinton, G. E., and R. J. Williams, "Learning internal representations by error propagation," Parallel Distributed Processing: Explorations in the Microstructure of Cognition, vol. 1, MA: MIT Press, pp. 318-362.

[16] S. Singhal and L. Wu, "Training Feed-Forward Networks with the Extended Kalman Algorithm." Proc. ICASSP, pp. 1187-1190, 1989.

[17] R. S. Scalero, "A Fast New Algorithm for Training Feed-Forward Neural Networks, Ph.D. Dissertation, Florida Institute of Technology, 1989.

[18] P. J. Werbos, "Neural Networks for Control and System Identification," Proc. of the 28th Conf. CDC., Tampa, FL, WA10, pp. 260-265, December, 1989.

FA6 - 10:50

Proceedings of the 31st Conference
on Decision and Control
Tucson, Arizona • December 1992

Learning Control of an Inverted Pendulum Using Neural Networks

S. Kawaji, T. Maeda, and N. Matsunaga

Department of Electrical Engineering and Computer Science
Faculty of Engineering, Kumamoto University
Kurokami 2–39–1, Kumamoto 860, JAPAN

Abstract

When neural network is used for the control, there exist several problems from the view point of control engineering. In this paper, it is pointed out that the utilization of some knowledge about the properties of the controlled object is necessary and effective. As an example, a hierarchical control system of an inverted pendulum using a neural network is proposed. From the structural properties of the object, the controller consists of pendulum part driven by neural network, and cart part generating the virtual signal to control the cart. The simulation and experimental results show that our method is very effective.

1. Introduction

In recent years, control methods using neural networks (NN)[1],[2] have been attempted in order to control a given system successfully through trials. In the methods, it is usually assumed that no knowledge of the controlled object is available. However, if the controlled object is thus treated as a black box, the correlation between the controlled object and their controller is not obvious. This implies that there may arise several problems from a viewpoint of control enginnering[3]. Especially, the generation of teaching signal for learning can not be easily realized, and the learning can not be evaluated.

In this paper, several problems in the control using NN are discussed, and it is pointed out that the utilization of some knowledge about the properties of the controlled object is necessary and effective. As an example, a hierarchical control system of an inverted pendulum using NN is proposed. First, NN is discussed from the viewpoint of general design equation, and several problems in the control using NN are explored. Secondly, the structural properties of the inverted pendulum are revealed, and a control system using NN based on its properties is constructed, which consists of pendulum part driven by neural network, and cart part generating the virtual signal to control the cart. Finally, the effectiveness of the proposed control system is confirmed by the simulation and experimental results.

2. Learning System using NN and a Few Problems

Let's discuss the control using NN from the viewpoint of general design equation[3]. In general, in designing a control system for plants, the following design equation holds[4].

$$\{M : \text{control objective}\} = \{C : \text{controller}\} \times \{P : \text{plant}\}$$

For example, the design equations for open-loop and closed-loop systems are described as, respectivly,

$$M = \mathrm{ser}\,(P,\ C) \tag{1}$$

$$M = \mathrm{inv}\,[\,\mathrm{par}\,[\,I,\ \mathrm{inv}\,\{\,\mathrm{ser}\,(P,\ C)\}]] \tag{2}$$

where ser $(a,\ b)$, and par $(a,\ b)$ are series, pararell connection of a and b, respectively. inv (#) is the inverse of #. Solving (1) and (2), we get

$$C = \mathrm{ser}\,[\,\mathrm{inv}\,(P),\ M] \tag{3}$$

$$C = \mathrm{ser}\,[\,\mathrm{inv}\,(P),\ \mathrm{inv}\,[\,\mathrm{par}\,\{-I,\ \mathrm{inv}\,(M)\}]] \tag{4}$$

It is remarked that (3) and (4) contain the following properties.

- Any expressions of M, P, C are allowed, if the calculations in the equations are available and their results exist physically.
- The structure of controller should be influenced by desired objectives and charactaristics of the plant.

In the control of NN, these properties should be also valid. Now we consider a learning system using multi-layered NN with the back propagation algorithm as shown in Fig.1. The objective of this system is to obtain a control law which stabilizes the system. The controller using NN, which we call "neuro-controller", works as a nonlinear feedback controller.

CH3229-2/92/0000-2734$1.00 © 1992 IEEE

NN is learned by the teaching signal which is appropriately generated. The learning algorithm can be described in programming language style.

```
{C}_0 = { preknowledge }
{do}{
        ΔC_n = BP[{ evaluation }_n]
        {C}_{n+1} = {C}_n + ΔC_n
} while ( until satisfaction in objective )
```

From this description, we found that in control system with neuro-controller the characteristics of the controlled object is implicitly evaluated by trials. Thus, in this control system, the generation of the teaching signal is very important, because the dynamical properties of the controlled object should be reflected in the teaching signal.

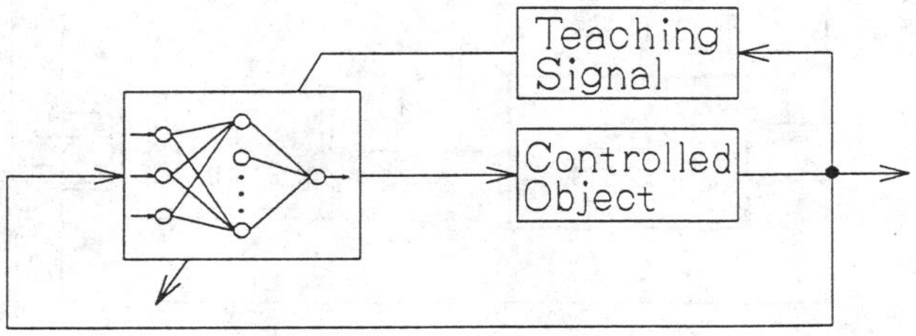

Fig.1 Learning System using NN

However, the controlled object is usually treated as a black box, so the following problems arise :

(1) The design method of the teaching signal can not be easily established.

(2) As the controlled object become more complicated, many neurons will be required.

Therefore, it is necessary to construct the controller based on both the control objective and the properties of controlled object. Fortunately, some knowledge about the controlled object can be usually obtained in real control, so we can utilize sufficiently such a priori knowledge in control using NN. As an example, we consider a control system of an inverted pendulum.[5]~[7]

3. Inverted Pendulum

The inverted pendulum is used as a good example to verify a control theory, which is illustrated in Fig.2. By using the Lagrange's method, the equations of motion is obtained as follows.

$$\begin{cases} (M+m)\ddot{r} + m\ell\cos\theta\ddot{\theta} = -F\dot{r} + m\ell\dot{\theta}^2\sin\theta + f \\ m\ell\cos\theta\ddot{r} + (J+m\ell^2)\ddot{\theta} = -C\dot{\theta} + mg\ell\sin\theta \end{cases} \tag{5}$$

where

M : Mass of the cart
m : Mass of the pendulum
F : Friction constant between the cart and the monorail
C : Friction constant at the pivot
J : Moment of inertia with respect to the center of gravity of the pendulum
ℓ : Length between the pivot and the center of gravity
g : Acceleration of gravity

The control objective is to keep the pendulum in its upright position and the cart in the specified position of the monorail, simultaneously. In other words, the objective is to keep four state variables $\{r, \dot{r}, \theta, \dot{\theta}\}$ at the origin. Because of the complexity of dynamical interaction, control methods based on an exact model of the inverted pendulum, such as linear feedback, may be desirable. However, these methods require the physical parameters of the system. If the models and their parameters can not be obtained, "learning methods " are expected to be useful.

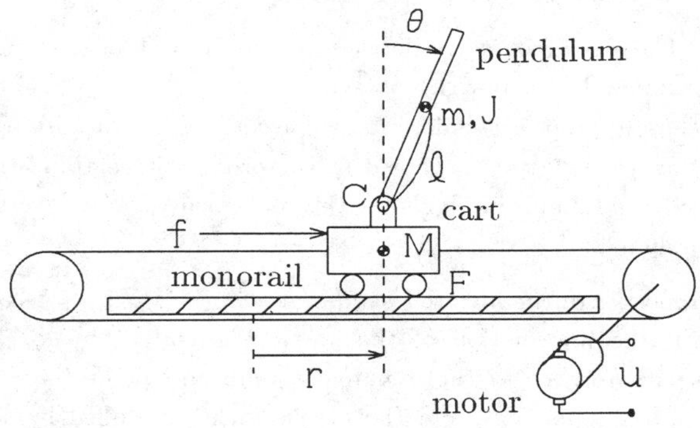

Fig.2 Inverted pendulum system

4. Hierarchical Control System Using NN

The performance of learning system is highly dependent on representation of its input. In this paper, three-layered network is used with an input layer, a hidden layer and an output layer. If the number of the neurons of the input layer is equal to that of the state variables, the construction of the total system become very complex (e.g. the learning scheme in [6]).

In order to reduce the number of the neurons and to simplify the design method of the teaching signal, any information delt with the NN should be reduced. Therefore we will construct a hierarchical control system, where information of the pendulum is used in the neuro-controller, and one of the cart is treated in upper level. First, the pendulum will be stabilized, so the neuro-controller which stabilize only the pendulum is constructed in the lower level. Secondly, our original objective is to stabilize the pendulum and the cart simultaneously, so the following knowledge about the inverted pendulum is utilized in upper level.

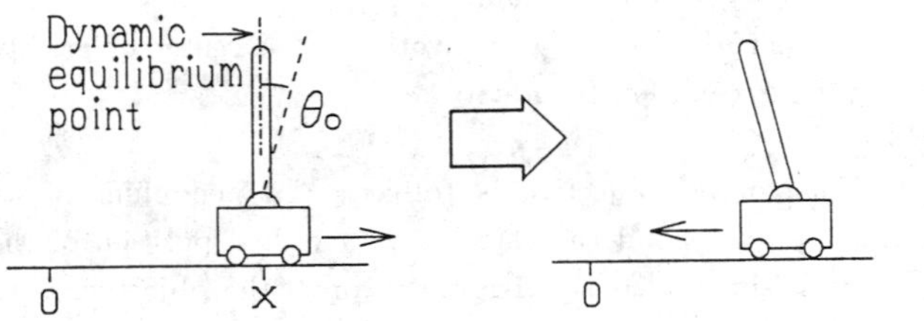

Fig.3 Movement of the inverted pendulum

From this property, in the case that the cart is in the right (left) side to the destination (i.e. the origin of the cart), following movement can be utilized.

(i) By adding positive θ_o to θ, the cart moves to the right direction. So the pendulum falls down in the left direction.

(ii) The cart moves to the left direction to keep the pendulum in its upright position.

If θ_o is appropriately determined according to the state of the cart, the cart is converged to the origin with stabilization of the pendulum. Fig.3 shows the movement of the inverted pendulum.

<Properties of the inverted pendulum>

If the information of the angle θ is shifted by θ_o with respect to the dynamic equilibrium point (i.e. upright position) and the pendulum is controlled by using the shifted information, then the cart moves to the direction of θ_o.

This implies that we should construct a regulator for the variable $(\theta+\theta_o, \dot{\theta})$ to control the cart. Because the pendulum is controlled so that $\theta+\theta_o \to 0$, i.e. $\theta \to -\theta_o$, θ_o is regarded as equiliblium point for the pendulum system. So we call θ_o the virtual equilibrium point. If θ_o is appropriately determined in upper level, our objective will be finally accomplished. As a result, the hierarchical control system is constructed as shown in Fig.4. The design method of the controller is described in following subsection.

4-1. Generation of Virtual Equilibrium Point

In the following, $f(t)$ is some function f with respect to time t, and $f(n) = f(nT)$ with sampling interval T.

From the consideration in previous section, the virtual equiliblium point is a mapping from the position r and velocity $\dot{r}$ to the new variable θ_o. Here θ_o is generated in linear form as

$$\theta_o(n) = a\{r(n) + b\dot{r}(n)\} \tag{6}$$

at time nT, where $\{a, b\}$ are the parameter which we must decide properly.

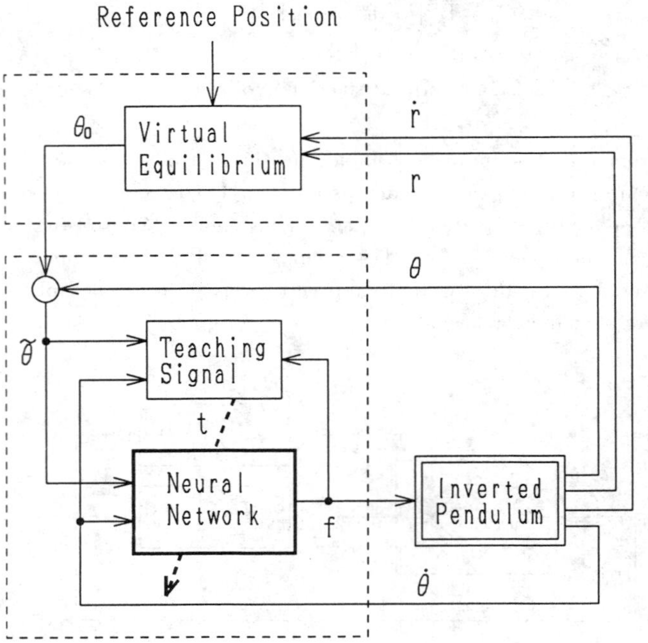

Fig.4 Hierarchical control system of an inverted pendulum

4-2. Neuro-controller and Teaching Signal for Learning

NN is constructed by the input layer of 2 neurons, the hidden layer of 4 neurons, and the output layer of 1 neuron. The output of the neuro-controller is the force which will be applied to the inverted pendulum. The activation function, which characterize the input-output relation on the each neuron, of the hidden layer is sigmoid function and of the output layer is linear function.

Learning procedure in our control system is realized based on backpropagation algorithm [4] with teaching signal generated by evaluation of the state of the inverted pendulum. Accroding to the present state, the desirable state at one sampling interval after is defined, which we call "state-reference". The teaching signal is decided by next two steps.

< Step 1: at time nT >

Define $\tilde{\theta}(n)$ as

$$\tilde{\theta}(n) = \theta_o(n) + \theta(n) \tag{7}$$

The state-reference at one smpling interval after, which conforms following differential equation, is decided according to $\{\tilde{\theta}(n), \dot{\theta}(n)\}$.

$$\ddot{\theta} + 2\zeta\omega_n\dot{\theta} + \omega_n^2\theta = 0 \tag{8}$$

That is, the solution of the eq.(4), $\{\theta_{ref}(n+1), \dot{\theta}_{ref}(n+1)\}$, is the state-reference with the initial value $\{\tilde{\theta}(n), \dot{\theta}(n)\}$.

< Step 2: at time $(n+1)T$ >

The output of the neuro-controller with respect to the input $\{\ \tilde{\theta}(n),\ \dot{\theta}(n)\ \}$ is represented as

$$f(n) = F\{\tilde{\theta}(n), \dot{\theta}(n)\} \tag{9}$$

where F is a nonlinear function representing the input-output relation of the NN.

The state $\{\tilde{\theta}(n+1), \dot{\theta}(n+1)\}$ is generated as a result that $f(n)$ is applied to the inverted pendulum. So these values will not be equal to the state-reference $\{\theta_{ref}(n+1), \dot{\theta}_{ref}(n+1)\}$ and the following errors

$$\begin{cases} e(n+1) = \theta_{ref}(n+1) - \tilde{\theta}(n+1) \\ \dot{e}(n+1) = \dot{\theta}_{ref}(n+1) - \dot{\theta}(n+1) \end{cases} \tag{10}$$

exist. Therefore, in order to decrease these errors $f(n)$ is modified by Δf given as

$$\Delta f = -K_1 e(n+1) - K_2 \dot{e}(n+1) \tag{11}$$

where K_1, K_2 are parameters related to a learning rate. Consequently, teaching signal $t(n)$ is calculated as

$$t(n) = f(n) + \Delta f \tag{12}$$

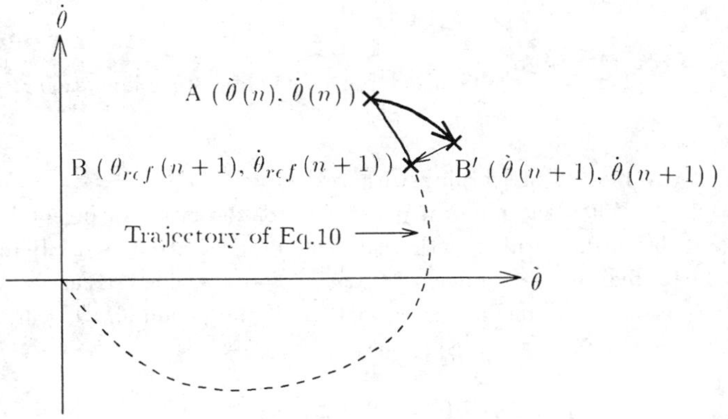

Fig.5 The phase plane of the reference error

The proposed hierarchical control system has the following advantages.

(a) Since the pendulum subsystem and the cart subsystem are separately treated by utilizing the properties of the inverted pendulum, many neurons are not required.

(b) The speed of convergence of the cart can be controlled by adjusting the parameters $\{a, b\}$.

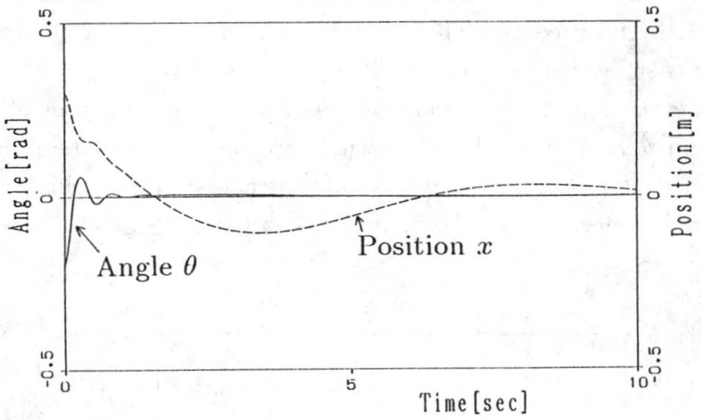

Fig.6 Result at 15th trial

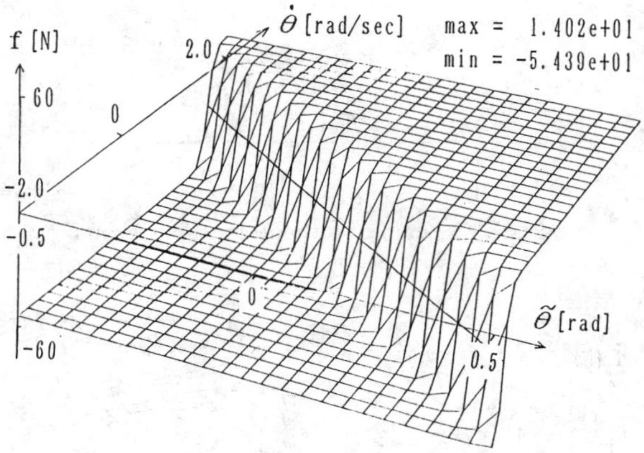

Fig.7 Output of the controller after 15th trial

5. Simulation and Experimental Results

5-1. Simulation Results

In this section, simulation results are shown. Recall that our control objective is to keep the pendulum in its upright position and the cart in the specified position of the mono-rail simultaneously, in other words, to avoid "failure" which is defined by $|\theta| > 1.0$ [rad] or $|x| > 1.0$ [m]. If the controller receives a failure signal within 10 [sec] or the inverted pendulum is stabilized during 10 [sec], the pendulum is reset to its initial condition, and next learning trial begins. The parameters are chosen as

virtual equilibrium	$a = 0.05$, $b = 1.0$
teaching signal	$\zeta = 1.0$, $\omega_n = 30.0$
	$K_1 = 20.0$, $K_2 = 0.3$
sampling interval	$T = 10$[msec]

for the inverted pendulum with $M = 4.4$, $m = 0.1$, $\ell = 0.115$, $F = 19.2$, $C = 0.0005$, $J = 0.0021$. Initial condition is $\{r, \dot{r}, \theta, \dot{\theta}\} = \{0.3, 0.0, -0.2, 0.0\}$, and the initial weights of NN are settled by small random values at beginning of trials.

Simulation results are given as follows. The stabilization of the inverted pendulum was failed at 1st ~ 4th trials, but was succeeded at 5th trial. It is observed that as many trials are attempted, the response is improved. The responses of r and θ at 15th trial are shown in Fig.6. The output of the controller after 15th trial has a tendency such as inclination toward negative as shown in Fig.7.

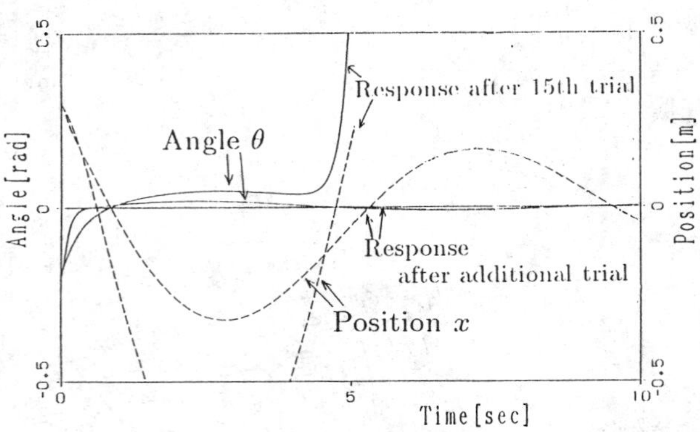

Fig.8 Response without teaching signal

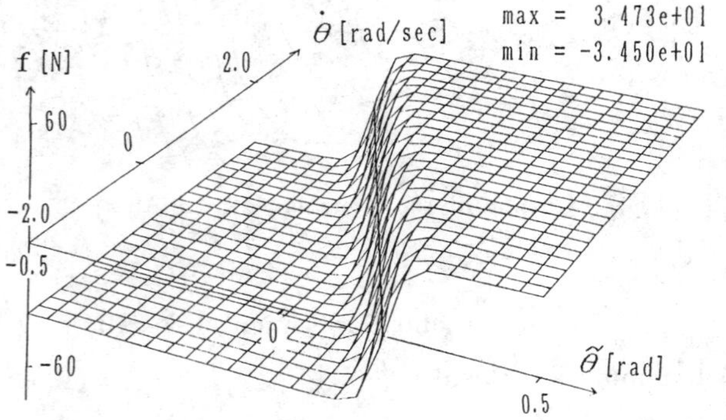

Fig.9 Output of the controller after additional trials

Now we have following question to confirm that NN has obtained control law in itself just as expected. " If NN has obtained the control law as expected, it requires the teaching signal no longer. Can it stabilize without teaching signal ? " So we omitted the teaching signal after 15th trial. Then the neuro-controller failed in stabilizing the pendulum as shown in Fig.8. This implies that the control law is not sufficiently obtained. It is supposed that this failure occured because of the fixed initial condition of the inverted pendulum through trials. Therefore, additional trials at $\{r, \dot{r}, \theta, \dot{\theta}\} = \{-0.3, 0.0, 0.2, 0.0\}$ were attemptted. After additional 10 trials, the neuro-controller succeded in stabilizing without teaching signal as shown in Fig.8. Notice that the output of the controller has no inclination of its output as shown in Fig.9.

<Remark> Note that $\{a, b\}$ are free parameters after learning. Therefore, as $\{a, b\}$ are chosen bigger after learning, the transient response of the cart is remarkably improved.

5-2. Experimental Results

We attempt the learning control to the real inverted pendulum. The state variables r and θ are measured with potentiometers and $\dot{r}$ and $\dot{\theta}$ are calculated by the difference of r and θ, respectively. Real-time control is realized by using Multi-task Real-time Control Operating System (MRCOS)[8], which is developed on MS-DOS with PC-9801 (NEC) by the authors. First, results are shown by using the controller after trials in simulation. Next, results of learning are shown with the real system.

(i) Results using the controller learned by simulation

When the physical parameters of the system can be obtained, the method using the controller learned by simulation is useful. The inverted pendulum was well stabilized by using the controller learned by simulation as shown in Fig.10.

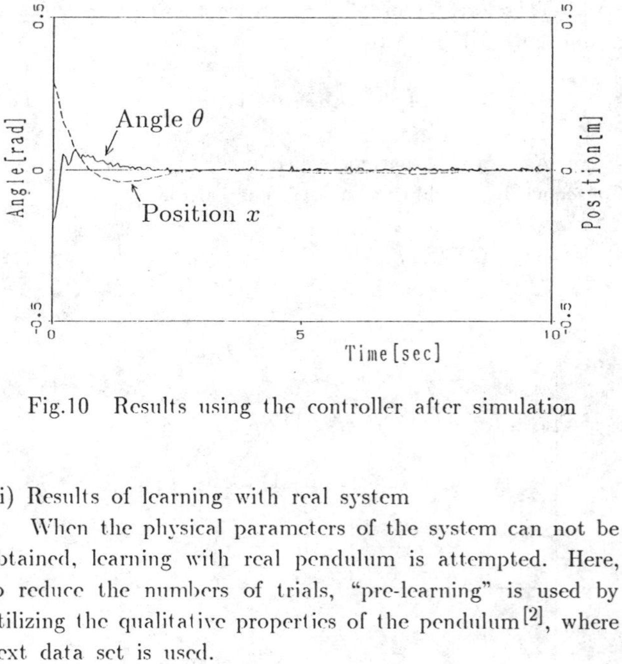

Fig.10 Results using the controller after simulation

(ii) Results of learning with real system

When the physical parameters of the system can not be obtained, learning with real pendulum is attempted. Here, to reduce the numbers of trials, "pre-learning" is used by utilizing the qualitative properties of the pendulum[2], where next data set is used.

$$f = \begin{cases} \pm fc & (\theta = \pm 0.5) \\ \pm 0.6 fc & (\theta = \pm 0.4) \\ \pm 0.2 fc & (\theta = \pm 0.2) \\ 0.0 & (\theta = 0.0) \end{cases} \qquad (13)$$

fc : some constant value

This data set implies the idea: if the angle of the pendulum is positive (negative), positive (negative) force should be applied to the cart.

The initial condition is similar to one in simulation. The control parameters are

$a = 0.05,\ b = 1.0,\ \zeta = 1.0,\ \omega_n = 30.0$

$K_1 = 10.0,\ K_2 = 0.1$

The controller succeeded in stabilizing the inverted pendulum at 45 trial without teaching signal as in Fig.11, where a is chosen 2.0. These results show the proposed learning method is effective for the real pendulum.

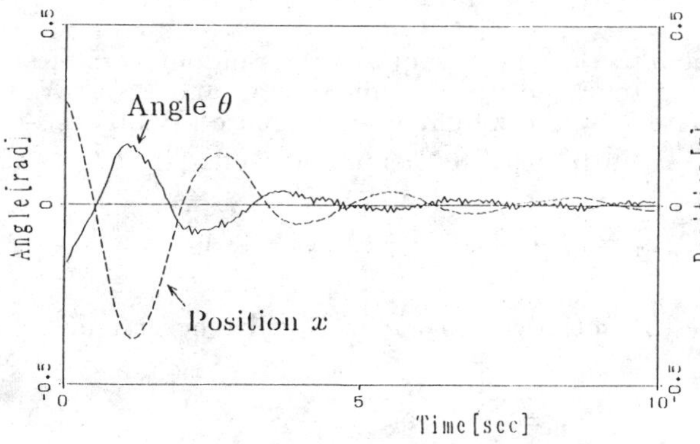

Fig.11 Results of learning with real system

6. Conclusion

In this paper, several problems in the control using NN have been pointed out, and a method of utilization of some knowledge about the properties of the controlled object has been indicated. As an example, we have proposed a hierarchical control system of the inverted pendulum using NN. From the structural properties of the system, the controller consists of pendulum part driven by neural network, and cart part generating the virtual signal. As a result, the neuro-controller can be constructed easily and be learned rapidly. The effectiveness of our control system has been confirmed by the simulation and experimental results.

References

[1] W.T.Miller Ed. : Neural Networks for Control, The MIT Press, 1989

[2] P.J.Antsaklis, Ed. : Special Issue on Neural Networks in Control Sytems, *IEEE Control System Magazine*, Vol.10, No.3, pp.3–87, 1990

[3] S.Kawaji et al. : On Learning Control using Neural Networks, *Proc. of KACC'91*, Seoul, pp.1318–1353, 1991

[4] T.Kitamori : Alternative Approaches for Fuzzyness, *J. of Japan Soc. for Fuzzy Theory and Systems*, Vol.3, No.1, pp.27–34, 1991

[5] C.W.Anderson : Learning to Control an Inverted Pendulum Using Neural Networks, *IEEE Control System Magazine*, Vol.9, No.3, pp.31–37, 1989

[6] N.Ikeda et al. : Learning Control for Stabilization of an Inverted Pendulum Using a Multi-Layered Neural Networks, *Trans. of the Institute of Systems, Control and Information Engineers (in Japan)*, Vol.3, No.12, pp.405–413, 1990

[7] T.Ishida et al. : Learning Control of an Inverted Pendulum Using a Neural Network, *Proc. of IECON'91*, Kobe, pp.1401–1404, 1991

[8] T.Shiotsuki, S.Kawaji : Design and Implementation of VSS Controller on Personal Computers, *Proc. of KACC'88*, Seoul, pp.848 851, 1988

FA7 - 8:30

Proceedings of the 31st Conference
on Decision and Control
Tucson, Arizona • December 1992

WEIGHTED ESTIMATION AND TRACKING FOR ARMAX MODELS

Bernard Bercu

Laboratoire de Statistiques, Bat. 425 Mathématiques, Université de Paris-Sud,
91 405 Orsay Cedex, France.

ABSTRACT

For a complex multivariate **ARMAX** model, we study the weighted least squares algorithm which improves the usual least squares algorithm by the choice of suitable ponderations. Concerning adaptive tracking problems, we ensure both strong consistency of the estimator and control optimality. We also precise almost sure rates of convergence.

I. WEIGHTED ESTIMATION

Let $(\Omega, \mathcal{A}, P)$ be a probability space with a filtration $\mathbf{F} = (\mathcal{F}_n)_{n\geq 0}$. We consider the complex multivariate **ARMAX** model of order (p, q, r)

$$A(R)Y_n = B(R)U_n + C(R)\varepsilon_n, \qquad (1)$$

where Y, U and ε are the output (d_1), the input (d_2) and the driven noise (d_1) respectively. Set for the shift-back operator R

$$A(R) = I_{d_1} - A_1 R - \cdots - A_p R^p, \qquad (2)$$
$$B(R) = B_1 R + \cdots + B_q R^q, \qquad (3)$$
$$C(R) = I_{d_1} + C_1 R + \cdots + C_r R^r, \qquad (4)$$

where A_i, B_j, C_k are unknown matrices. Assume that the control $U = (U_n)$ and the noise $\varepsilon = (\varepsilon_n)$ are adapted to $\mathbf{F}$ and that ε is a martingale difference sequence such that

$$\sup_{n\geq 0} E\left[\, \| \varepsilon_{n+1} \|^2 | \, \mathcal{F}_n \,\right] < \sigma^2 \quad \text{a.s.} \qquad (5)$$

where σ^2 is deterministic. We make use of

$${}^t\Psi_n = \left({}^tY_n^p,\ {}^tU_n^q,\ {}^t\varepsilon_n^r \right), \qquad (6)$$

where we write ${}^tY_n^p = ({}^tY_n, \ldots, {}^tY_{n-p+1})$ and idem for U_n^q and ε_n^r. Let $\hat\theta_n$ be an estimator of θ with

$${}^*\theta = (A_1, \ldots, A_p, B_1, \ldots, B_q, C_1, \ldots, C_r). \qquad (7)$$

The noise ε is predicted by the *a posteriori* error

$$\hat\varepsilon_n = Y_n - {}^*\hat\theta_n \Phi_{n-1}, \qquad (8)$$
$${}^t\Phi_n = \left({}^tY_n^p,\ {}^tU_n^q,\ {}^t\hat\varepsilon_n^r \right). \qquad (9)$$

Let $a = (a_n)$ be a sequence of random variables adapted to $\mathbf{F}$, positive, nonincreasing and < 1. We estimate θ by the weighted least square (**WLS**) estimator $\hat\theta_n$ introduced by Bercu and Duflo [2]

$$\hat\theta_{n+1} = \hat\theta_n + a_n S_n^{-1} \Phi_n^* \left(Y_{n+1} - {}^*\hat\theta_n \Phi_n \right) \qquad (10)$$

where the initial value $\hat\theta_0$ is arbitrarily chosen and

$$S_n = \sum_{k=0}^{n} a_k \Phi_k {}^*\Phi_k + S \qquad (11)$$

with any positive definite hermitian and deterministic matrix S. We also make use of

$$f_n = a_n {}^*\Phi_n S_n^{-1} \Phi_n, \quad s_n = \sum_{k=0}^{n} \| \Phi_k \|^2 + s \qquad (12)$$

$$\pi_n = {}^*\theta \Psi_n - {}^*\hat\theta_n \Phi_n \qquad (13)$$

where $s = Tr(S)$. The sequence a is said to be admissible if Δ is integrable where

$$\Delta_n = \sum_{k=0}^{n} a_k f_k, \quad \Delta = \lim_{n\to+\infty} \Delta_n. \qquad (14)$$

For exemple, we can choose $a_n = s_n^{-\gamma}$, $a_n = (\log s_n)^{-1-\gamma}$ with $\gamma > 0$.

II. STRONG CONSISTENCY

We use the traditional assumption of passivity:

(A_1) $C^{-1} - \frac{1}{2} I_{d_1}$ is strictly positive real.

Theorem 1. For the **WLS** algorithm, assume that (A_1) is satisfied and a is admissible. Then, we have

$$\| S_n^{\frac{1}{2}} (\hat\theta_{n+1} - \theta) \|^2 = O(\Delta_n) \quad \text{a.s.} \qquad (15)$$

$$\sum_{k=0}^{n} \lambda_{min} S_{k-1} \| \hat\theta_{k+1} - \hat\theta_k \|^2 = O(\Delta_n) \quad \text{a.s.} \qquad (16)$$

CH3229-2/92/0000-2740$1.00 © 1992 IEEE

$$\sum_{k=0}^{n} a_k \parallel \Phi_k - \Psi_k \parallel^2 = O(\Delta_n) \quad \text{a.s.} \tag{17}$$

$$\sum_{k=0}^{n} a_k (1 - f_k) \parallel \pi_k \parallel^2 = O(\Delta_n) \quad \text{a.s.} \tag{18}$$

Corollary 1. The **WLS** estimator given by (10) is strongly consistent on $\{\lambda_{min} S_n \longrightarrow +\infty\}$ and we have

$$\parallel \hat{\theta}_{n+1} - \theta \parallel^2 = O(\{\lambda_{min} S_n\}^{-1}) \quad \text{a.s.} \tag{19}$$

Theorem 2. For the **WLS** algorithm, assume that (A_1) is satisfied and a is admissible. Then, we have

$$E\left[\sup_{n \geq 0} \parallel S_n^{\frac{1}{2}}(\hat{\theta}_{n+1} - \theta) \parallel^2\right] < +\infty \tag{20}$$

$$E\left[\sum_{n=0}^{\infty} \lambda_{min} S_{n-1} \parallel \hat{\theta}_{n+1} - \hat{\theta}_n \parallel^2\right] < +\infty \tag{21}$$

$$E\left[\sum_{n=0}^{\infty} a_n \parallel \Phi_n - \Psi_n \parallel^2\right] < +\infty \tag{22}$$

$$E\left[\sum_{n=0}^{\infty} a_n (1 - f_n) \parallel \pi_n \parallel^2\right] < +\infty. \tag{23}$$

Corollary 2. The prediction errors sequence given by (13) satisfies

$$E\left[\sum_{n=0}^{\infty} (a_n^{-1} + \parallel \Phi_n \parallel^2)^{-1} \parallel \pi_n \parallel^2\right] < +\infty. \tag{24}$$

More particularly, if $s_n^{-1} \leq \rho a_n$ with $\rho > 0$, then

$$E\left[\sum_{n=0}^{\infty} \frac{\parallel \pi_n \parallel^2}{s_n}\right] < +\infty. \tag{25}$$

Finally, we also have

$$E\left[\sum_{n=0}^{\infty} (a_n^{-1} + \frac{\parallel \Phi_n \parallel^2}{\lambda_{min} S_{n-1}})^{-1} \parallel \pi_n \parallel^2\right] < +\infty. \tag{26}$$

III. ADAPTIVE TRACKING

The goal of adaptive tracking is to inforce the output $Y = (Y_n)$ to follow a given reference trajectory $y = (y_n)$. We use the traditional adaptive tracking control (**ATC**) such that

$$y_{n+1} = {}^*\hat{\theta}_n \Phi_n. \tag{27}$$

It is well known that the **ATC** is almost surely defined if the $\mathcal{F}_n$ conditional distribution of ε_{n+1} is absolutely continuous with respect to the Lebesgue measure. In all the sequel, we also assume that the driven noise ε has constant conditional covariance matrix Γ and satisfies the strong law of large numbers. We finally need the usual assumption of causality:

(A_2) $d_2 \leq d_1$ and the matrix B_1 is of full rank d_2. Moreover, if B_+ denotes the left inverse of B_1 and if $D(R) = B_+ R^{-1} B(R)$ for the shift-back operator R, then D is causal.

Theorem 3. For the **WLS** algorithm, assume that (A_1) and (A_2) are satisfied. For the tracking trajectory y, suppose that y_{n+1} is $\mathcal{F}_n$ measurable and

$$\sum_{k=1}^{n} \parallel y_k \parallel^2 = O(n) \quad \text{a.s.} \tag{28}$$

If a is admissible with $s_n^{-1} = O(a_n)$, then the **ATC** is optimal.

We finally give a last theorem when the tracking trajectory y is strongly exciting. If ε has finite conditional moment of order > 2, we can also obtain a similar result using the same approach than the one of Bercu and Duflo [2] or of Guo and Chen [4]. We use the following assumption of irreducibility:

(A_3) The matrix $B_+ B_q$ is regular and the matricial polynomials $B_+ A$, $B_+ C$ and D are left coprime.

Theorem 4. For the **WLS** algorithm, assume that (A_1) to (A_3) are satisfied and that Γ is regular. For the tracking trajectory y, suppose that y_{n+1} is $\mathcal{F}_{n-p-s}$ and $\mathcal{F}_{n-r-s}$ measurable with $\parallel y_n \parallel^2 = o(n)$ and that (28) is satisfied. Moreover, suppose that y is strongly exciting with order $p+s+1$ i.e.

$$\liminf \lambda_{min}\left(\frac{1}{n} \sum_{k=p+s}^{n} y_k^{p+s+1} \, {}^* y_k^{p+s+1}\right) > 0 \quad \text{a.s.} \tag{29}$$

If a is admissible with $s_n^{-1} = O(a_n^2)$, then the **ATC** is optimal with the convergence rate $(na_n)^{-1}$. Finally, the **WLS** estimator $\hat{\theta}_n$ is strongly consistent and

$$\parallel \hat{\theta}_{n+1} - \theta \parallel^2 = O(\frac{1}{na_n}) \quad \text{a.s.} \tag{30}$$

Remark. If $a_n = (\log s_n)^{-1-\gamma}$ with $\gamma > 0$, we find the same convergence rate for the strong consistency and the optimality that is $n^{-1}(\log n)^{1+\gamma}$.

REFERENCES

[1] B. Bercu Estimation pondérée et poursuite pour les modèles ARMAX, note au CRAS, Série 1, 314 (1992), pp. 403-406.

[2] B. Bercu et M. Duflo, Moindres carrés pondérés et poursuite, Ann. Inst. H. Poincaré, 28 (1992)

[3] M. Duflo, Méthodes récursives aléatoires, Masson, Paris, 1990.

[4] L. Guo, H. F. Chen, The Aström-Wittenmark self-tuning regulator revisited and ELS-based adaptive trackers, IEEE Trans. Automat. Control, 36 (1991), pp. 802-812.

Proceedings of the 31st Conference
on Decision and Control
Tucson, Arizona • December 1992

Discrete-time Bayesian adaptive control problems with complete observations

Omar Zane

Department of Mathematics
University of Kansas
66045 - Lawrence, Kansas

Abstract : The problem studied is a nonlinear Bayesian stochastic adaptive control problem with finite horizon, discrete-time and complete observations. Using an algorithm based on a suitable approximation we can find ε-optimal controls for the original problem; the results obtained from the application of the algorithm and a comparison with an exact solution are given.

1. Introduction

We shall consider a Bayesian finite horizon, discrete-time stochastic adaptive control problem with state equation $(n = 0,1,....,N-1)$

(1) $$x_{n+1} = f_1(x_n,u_n) + \theta f_2(x_n,u_n) + \sigma(x_n,u_n)\, w_{n+1}$$

and cost function

(2) $$J(u^{N-1}) = E\left\{ \sum_{n=1}^{N} g(x_n,u_{n-1}) \right\}$$

where θ is an unknown parameter described by a random variable with known prior density and $u^{N-1} = (u_0,...,u_{N-1})$.

The problem is to find a sequence of admissible controls $\hat{u}^{N-1} = (\hat{u}_0, ...\hat{u}_{N-1})$ in such a way that for any other sequence of admissible controls $u^{N-1} = (u_0, ...u_{N-1})$ we have

(3) $$J(\hat{u}^{N-1}) \leq J(u^{N-1})$$

In [4] the authors, using a variational method, give the solution for the particular case in which the prior density of θ is normal with mean 1 and variance 1, $x_0 = 0$, the state equation is $(n = 0,1,.......,N-1)$

(4) $$x_{n+1} = x_n + u_n + \theta + w_{n+1}$$

and the cost function is

(5) $$J(u^{N-1}) = E\left\{ \sum_{n=1}^{N} x_n^2 + u_{n-1}^2 \right\}$$

The main peculiarities of this case are that $f_1(x,u)$ is linear in x and u, $f_2(x,u)$ is constant and g is quadratic; any attempt to solve the problem (1)-(2) in cases that are just a little less peculiar raises complications even if the functions f_1, f_2, σ, g and the initial density function of θ are relatively simple; an example is the case in which the prior density of θ is normal with mean 0 and variance 1; the state equation is $(n = 0,1,...,N-1)$:

(6) $$x_{n+1} = 1 - \theta\, u_n + w_{n+1}$$

and the cost function is

(7) $$J(u^{N-1}) = E\left\{ \sum_{n=1}^{N} x_n^2 \right\}$$

treated in [4] where an optimal control could not be determined (a problem essentially equivalent to (4)-(5) is discussed in [2] with the same conclusion).

In order to overcome the difficulties and solve some of these problems we define a sequence of approximating problems that we are able to solve in such a way that we can get an algorithm for constructing, for every $\varepsilon > 0$, a sequence of controls that are ε-optimal for the original problem, namely, admissible controls $\hat{u}_\varepsilon^{N-1}$ such that

(8) $$J(\hat{u}_\varepsilon^{N-1}) \leq \inf J(u^{N-1}) + \varepsilon$$

where the infimum is taken over all possible sequences of admissible controls.

2. Assumptions

In problem (1)-(2) we assume that the prior density function of θ is normal with known mean $\bar{\theta}_0$ and variance $\bar{R}_0^2$; x_0 is a random variable that admits a moment generating function with known prior density function; the admissible controls u_n are functions of the available information $I_n = \{\bar{\theta}_0, \bar{R}_0, x_0, u_0, x_1, ...u_{n-1}, x_n\}$, that take values in a compact set U and are measurable with respect to the σ-algebra $\mathcal{F}^n$ generated by $x^n := \{x_0, ...x_{n-1}, x_n\}$.

Moreover we make the following assumptions on the functions f_1, f_2, σ, g: there exist continuous functions $A_1(\cdot)$, $A_2(\cdot)$, $B(\cdot)$, $D(\cdot)$, $G(\cdot)$ and a constant C such that

A.1 : for every u, $|f_1(x,u)| \leq A_1(u)|x| + A_2(u)$

A.2 : for every u, $|f_2(x,u)| \leq B(u)$ and $\lim_{|x|\to\infty} f_2(x,u)$ exists

A.3 : for every u, $0 < C \leq \sigma(x,u) \leq D(u)$ and $\lim_{|x|\to\infty} \sigma(x,u)$ exists

A.4 : f_1, f_2, σ are continuous in u and Lipschitz continuous in x uniformly with respect to u

A.5 : g is continuous in x and u, bounded from below and such that for every x and every u we have $g(x,u) \le G(u)\, e^{|x|}$

Remark : Note that both the problem given by equations (4)-(5) and the problem given by equations (6)-(7) satisfy these Assumptions.

3. Approximation

The approach to the approximation step is based on a result from [1]; it says that if we have an approximating problem with cost function $J_{(k)}$ and a δ-optimal control for it then this solution is ε-optimal control for J:

Proposition 1 ([1]) Given $\varepsilon > 0$, if for all admissible controls u^{N-1} there is $\delta,\ 0 < \delta < \frac{\varepsilon}{2}$ such that

$$\left| J(u^{N-1}) - J_{(k)}(u^{N-1}) \right| \le \frac{\varepsilon}{2} - \delta \tag{9}$$

and $(u^{k,\delta})^{N-1}$ is such that

$$J_{(k)}\left((u^{k,\delta})^{N-1}\right) \le \inf_{u^{N-1}} J_{(k)}(u^{N-1}) + \delta \tag{10}$$

then

$$J\left((u^{k,\delta})^{N-1}\right) \le \inf_{u^{N-1}} J(u^{N-1}) + \varepsilon \qquad \# \tag{11}$$

We can now see that under the Assumptions made in Section 2 the condition (9) is satisfied; the following result combines Theorem 3.1 and Theorem 3.2 in [6] :

Theorem 1 ([6]) For the problem described above under Assumptions A.1-A.5 there exists a sequence of approximating problems with cost functions $\left\{ J_{(k)}(u^{N-1}) \right\}_{k \in N}$ such that for every admissible control

$$\left| J(u^{N-1}) - J_{(k)}(u^{N-1}) \right| \le N\, G\, Z_k \tag{11}$$

where Z_k is such that $\lim_{k \to \infty} Z_k = 0$. #

4. Algorithm and Computation

From Section 3 we have that if we can determine a δ-optimal control for $J_{(k)}$ then using that control we can determine an ε-optimal control for J.

An algorithm for finding a δ-optimal control for $J_{(k)}$ and constructing an ε-optimal control for J is given in [6]. In order to evaluate the performance of this algorithm we have applied it to problem (4)-(5) and compared the results that we have obtained with the results that we get using the optimal solution given in [4].

This is done in two steps: first we find the optimal value for the cost function using the optimal controls in [4] , call it J*, and find the optimal value for the cost function for an approximated problem, $\hat{J}^*$; then using a simulation we compare the value of the cost function that we get applying the optimal control in [4], call it J^*_{sim}, with the value of the cost function that we get applying the controls that we determine using the algorithm given in [6], $\hat{J}^*_{sim}$, (generating the same values for the noise and the unknown parameter in the two cases).

As we expect from Proposition 1 and Theorem 1, the computational results show that as we increase the number of elements in the partition that is used for the approximation of the identity function, the values of $\hat{J}^*$ and $\hat{J}^*_{sim}$ get closer and closer to the values J^* and J^*_{sim} respectively (see [7] for details on the construction of the approximating problems and the application of the algorithm).

The following values are an example of the computational results for the case when N is equal to 3, with 75 elements in the partition and 100 control values. The values are

$$J^* = 12.051 \tag{12a}$$

$$\hat{J}^* = 12.052 \tag{12b}$$

$$J^*_{sim} = 12.049 \tag{12c}$$

$$\hat{J}^*_{sim} = 12.061 \tag{12d}$$

References

[1] G.B. Di Masi and W.J. Runggaldier. An Approach to Discrete-Time Stochastic Control Problems under Partial Observation, *SIAM J. Contr. Optimiz.*, **25** (1987), 38-48.

[2] P.R. Kumar and P.Varaiya. *Stochastic Systems: Estimation, Identitfication and Adaptive Control*, Englewood Cliffs, NJ: Prentice-Hall, 1986.

[3] B. Pasik-Duncan. *On Adaptive Control,* Warsaw: Central School of Planning and Statistics Publishers, 1986.

[4] R. Rishel and L. Harris. An Algorithm for a Solution of a Stochastic Adaptive Linear Quadratic Optimal Control Problem, *IEEE Trans. Automat. Contr.*, **AC-31** (1986), 1165-1170.

[5] W.J. Runggaldier and O. Zane. Approximations for Discrete-Time Adaptive Control: Construction of ε-Optimal Controls, *Math. Control Signals Systems*, **4** (1991), 269-291.

[6] O. Zane. An Algorithm for the Construction of ε-Optimal Controls for Discrete-Time Bayesian Adaptive Control Problems with Complete Observations, to appear on *J. Math. Systems, Estimation and Control* .

[7] O. Zane. Computational Results for Discrete-Time Bayesian Adaptive Control Problems with Complete Observations, in preparation.

Proceedings of the 31st Conference
on Decision and Control
Tucson, Arizona • December 1992

A METHODOLOGY FOR THE ADAPTIVE CONTROL OF MARKOV CHAINS UNDER PARTIAL STATE INFORMATION

Emmanuel Fernández-Gaucherand †, Aristotle Arapostathis ‡.
and Steven I. Marcus §

SUMMARY

We consider a stochastic adaptive control problem where complete state information *is not available* to the controller. The system is modelled as a *finite stochastic automaton* (FSA) [PAZ], [DOB]. These models are a slight generalization of the more common partially observable controlled Markov chain models as presented in, e.g. [BE], [KV]. A controlled FSA is described by the quintuplet $\langle \mathbf{X}, \mathbf{Y}, \mathbf{U}, \{P(y \mid u) : (y,u) \in \mathbf{Y} \times \mathbf{U}\}, c\rangle$; here $\mathbf{X} = \{1,2,\ldots,N_{\mathbf{X}}\}$ is the finite set of internal states, $\mathbf{Y} = \{1,2,\ldots,N_{\mathbf{Y}}\}$ is the set of observations (or messages), $\mathbf{U} = \{1,2,\ldots,N_{\mathbf{U}}\}$ is the set of decisions (or controls), and $c(\cdot,\cdot)$ is the one-stage cost function. For each pair $(y,u) \in \mathbf{Y} \times \mathbf{U}$, we have that $P(y \mid u) := [p_{i,j}(y \mid u)]$ is a $N_{\mathbf{X}} \times N_{\mathbf{X}}$ matrix, such that

$$p_{i,j}(y \mid u) \geq 0, \qquad \sum_{y=1}^{N_{\mathbf{Y}}} \sum_{j=1}^{N_{\mathbf{X}}} p_{i,j}(y \mid u) = 1, \quad \forall i \in \mathbf{X}, u \in \mathbf{U}.$$

If at time t the automaton is in state $X_t = i$ and decision $U_t = u$ is made, then by the beginning of the next decision time the automaton would have evolved to state $X_{t+1} = j$, and output a message $Y_{t+1} = y$ with probability $p_{i,j}(y \mid u)$. The cost incurred in this process is $c(i,j)$. We refer to [ABFGM], [DOB], [PAZ] for more details.

At decision time t, the information available to the decision-maker is

$$I_t := \{p_0, U_0, Y_1, U_1, \ldots, U_{t-1}, Y_t\} = (I_{t-1}, U_{t-1}, Y_t),$$

where $p_0 \in S_{N_{\mathbf{X}}} := \{p \in \mathbb{R}^{N_{\mathbf{X}}} \mid p^{(i)} \geq 0, \sum_i p^{(i)} = 1\}$ is the initial state distribution. It is well known that the partially observable optimal control problem for a FSA, under several optimality criteria, can be transformed into an *equivalent completely observable problem*, in terms of an *information state process* [ABFGM], [BE], [KV], as follows. Given $p_0 \in S_{N_{\mathbf{X}}}$, compute recursively

$$p_{t+1} = T(Y_{t+1}, p_t, U_t), \qquad t \in \mathbb{N},$$

where,

$$T(y,p,u) := \frac{P(y \mid u)p}{\mathbf{1}'P(y \mid u)p};$$

here we have $\mathbf{1} = (1,1,\ldots,1)'$. The process $\{p_t\}$ is a controlled Markov chain and equals the conditional distribution of the internal state X_t given I_t [ABFGM], [BE]. The "new" state is then taken as the process $\{p_t\}$.

A (stationary separated) policy π is a rule for making decisions, based on $\{p_t\}$, i.e. $\pi : S_{N_{\mathbf{X}}} \to \mathbf{U}$, and $U_t = \pi(p_t)$. The stochastic optimal control problem of interest to us is that of finding a policy π^*, optimal with respect to the long-run expected average cost (AC) performance criterion, which for a given policy π and initial state distribution $p_0 \in S_{N_{\mathbf{X}}}$ is given as

$$J(p_0;\pi) := \lim_{N\to\infty} \frac{1}{N} \mathbf{E}^{\pi}_{p_0}\left\{ \sum_{t=0}^{N-1} \overline{c}(p_t, U_t) \right\},$$

where $\overline{c}(p,u) := p'\big(c(1,u),\ldots,c(N_{\mathbf{X}},u)\big)'$. The infinite horizon optimal control problem for FSA under an AC criterion has been studied by the authors and others [ABFGM]. Since the state space $S_{N_{\mathbf{X}}}$ is a general (Borel) space, then this problem can be thought of as falling into the realm of completely observable *controlled Markov processes* (CMP), with general (Borel) state space, c.f. [ABFGM]. However, the problem with partial information has a very rich structure which is not fully utilized by following the latter approach [ABFGM], [BE], [FAM1], [FG]. Furthermore, many of the assumptions used in the literature on the AC control problem for general state space CMP require some form of strong ergodicity for the controlled process $\{p_t\}$, *under all stationary control policies.* This is not satisfied for many applications of much interest, c.f. [FAM1]. Hence the idea of viewing the (equivalent) FSA problem as a general state space completely observable CMP very often is not advantageous at all, for many purposes. This is especially true for the case of *parametric adaptive control* of FSA. In this situation, the model depends on some unknown parameter θ_0, which we denote as $P_{\theta_0}(y \mid u)$; the parameter takes values in some (Borel) parameter space Θ. Hence, the *true* conditional probability depends on this parameter, i.e.:

$$p_{t+1} = T(Y_{t+1}, p_t, U_t; \theta_0).$$

Therefore, since the true value of the parameter is unknown to the controller, $\{p_t\}$ *cannot be computed* and thus the equivalent problem is *not completely observable* anymore.

Although a very interesting problem with much potential for applications [BDO], [BTE], [FAM2], [WAK] there is very little available in the literature concerning the adaptive control of FSA. Recently, the above adaptive control problem has been studied by the authors [FAM2], [FG]. In [FAM2], a complete analysis for a particular case study has been reported; the methodology used has been generalized in [FG] as follows: we adopt an "enforced certainty equivalence" approach which involves recursively computing estimates $\{\hat{\theta}_t\}$ of the unknown parameter, and using at each decision time the latest available estimate to compute

$$\hat{p}_{t+1} = T(Y_{t+1}, \hat{p}_t, U_t; \hat{\theta}_t), \quad \hat{p}_0 = p_0. \tag{1}$$

We assume that the solution to the stochastic optimal control problem is known, for each $\theta \in \Theta$, which is expressed as follows; see [ABFGM], [BE], [KV].

† Systems and Industrial Engineering Department, The University of Arizona, Tucson, Arizona 85721 (emmanuel@sie.arizona.edu).

‡ Department of Electrical and Computer Engineering, The University of Texas at Austin, Austin, Texas 78712-1084 (ari@emx.utexas.edu).

§ Systems Research Center & Electrical Engineering Department, The University of Maryland, College Park, Maryland 20742 (marcus@src.umd.edu).

CH3229-2/92/0000-2750$1.00 © 1992 IEEE

Assumption A.1: For each θ, there is a bounded solution $(\rho_\theta^*, h_\theta)$, with $\rho_\theta^* \in \mathbb{R}$, to the corresponding *average cost optimality equation* (ACOE)

$$\rho_\theta^* + h_\theta(p) = \min_{u \in \mathbf{U}} \left\{ \overline{c}(p,u) + \sum_{y \in \mathbf{Y}} \mathbf{1}' P(y \mid u) p \, h_\theta(T(y,p,u)) \right\}.$$

Under the above assumption, there exists a set of optimal policies $\mathcal{OP} = \{\pi^*(\cdot\,;\theta)\}_{\theta \in \Theta}$, see [ABFGM]. The certainty equivalent adaptive policy is given as follows.

- **Adaptive Policy:** Given a sequence of estimates $\{\hat{\theta}_t\}_{t=0}^{\infty}$ of θ_0, compute the control action at each time t by

$$U_t = \pi^*(\hat{p}_t; \hat{\theta}_t),$$

where $\hat{p}_t$ is computed recursively using (1).

The above adaptive policy will be denoted by π^a. Under a set of assumptions, it was shown in [FG] that the adaptive policy π^a is *self-optimizing* with respect to the AC criterion, i.e. it achieves the same asymptotic average performance as the optimal policy $\pi^*(\cdot\,;\theta_0) \in \mathcal{OP}$ corresponding to the true parameter. The other assumptions used in [FG] are the following.

Assumption A.2: The parameter set Θ is compact; $(\rho_\theta^*, h_\theta)$ are continuous and bounded, both in p and in θ.

Assumption A.3: $P_\theta(y \mid u)$ is continuous in θ, for each $(y,u) \in \mathbf{Y} \times \mathbf{U}$.

Assumption A.5: We have that $\hat{p}_t \underset{t\to\infty}{\longrightarrow} p_t$, in probability, for all p_0.

We can now prove our main result.

Theorem: Under Assumptions **A.1-A.5**, π^a is self-optimizing with respect to the AC criterion.

Proof: Let $\Phi_\theta(\cdot\,,\cdot)$ denote Mandl's discrepancy function, corresponding to the parameter value $\theta \in \Theta$, i.e. for $p \in S_{N_\mathbf{X}}$ and $u \in \mathbf{U}$ (see [ABFGM], [FAM1])

$$\Phi_\theta(p,u) := \overline{c}(p,u) + \sum_{y \in \mathbf{Y}} \mathbf{1}' P_\theta(y \mid u) p \, h_\theta(T(y,p,u;\theta)) - \rho_\theta^* - h_\theta(p).$$

Then by the assumptions made, $\Phi_\theta(p,u)$ is continuous in both $p \in S_{N_\mathbf{X}}$ and $\theta \in \Theta$. Furthermore, since Θ is compact, then $\Phi_\theta(p,u)$ is uniformly continuous and bounded in $(p,\theta) \in S_{N_\mathbf{X}} \times \Theta$, and thus $\Phi_{\hat{\theta}_t}(\hat{p}_t, u)$ is uniformly integrable, for each $u \in \mathbf{U}$. Therefore, for each $u \in \mathbf{U}$, we have

$$\mathbb{E}_{p_0}^{\pi^a} \left\{ \left| \Phi_{\hat{\theta}_t}(\hat{p}_t, u) - \Phi_{\theta_0}(p_t, u) \right| \right\} \underset{t\to\infty}{\longrightarrow} 0,$$

and since $\mathbf{U}$ is finite, then

$$\mathbb{E}_{p_0}^{\pi^a} \left\{ \Phi_{\theta_0}(p_t, \pi(\hat{p}_t; \hat{\theta}_t)) \right\} \underset{t\to\infty}{\longrightarrow} 0, \qquad (2)$$

where we used the fact that $\Phi_{\hat{\theta}_t}(\hat{p}_t, \pi(\hat{p}_t; \hat{\theta}_t)) = 0$, since $\pi(\cdot\,;\theta) \in \mathcal{OP}$ minimizes the corresponding ACOE, for the parameter value $\theta \in \Theta$. The result then follows from (2); see [ABFGM, Theorem 6.3]. $\square$

Let us briefly examine the assumptions used in deriving the result above. Verifiable conditions on the model specifications exist in the literature that imply Assumption **A.1** holds [ABFGM], [FAM1]. Assumption **A.3** is easy to verify, and holds trivially if the parameterization of the model is taken in terms of the entire matrices $P_\theta(y \mid u)$. Assumption **A.4** depends on the parameter estimation scheme used, and is very problem-specific [FAM2]. The continuity required in Assumption **A.2** depends to a large extent on the continuity required in Assumption **A.3**, and on some ergodic properties of the model [FAM2]. Finally, it is clear that even if Assumptions **A.1-A.4** hold, it may be the case that Assumption **A.5** does not. Under continuity with respect to the parameterization, and in the presence of converging parameter estimates, this last assumption will hold if there is some type of, e.g. *regenerative* behavior for the processes $\{p_t\}$ and $\{\hat{p}_t\}$, such that at some times both processes are reset to the same value. This type of behavior occurs naturally in some inventory, queueing and machine replacement problems [BE], [ABFGM], [FAM2].

ACKNOWLEDGEMENTS: This work was supported in part by the Texas Advanced Technology Program under Grants No. 003658-093 and No. 003658-186, in part by the Air Force Office of Scientific Research under Grants AFOSR-91-0033, F49620-92-J-0045, F49620-92-J-0083, and in part by the National Science Foundation under Grants CDR-8803012 and INT-9201430.

REFERENCES

[ABFGM] A. Arapostathis, V. Borkar, E. Fernández-Gaucherand, M.K. Ghosh and S.I. Marcus, Discrete-Time Controlled Markov Processes with Average Cost Criterion: A Survey, to appear in *SIAM Journal on Control & Optimization*).

[BE] D.P. Bertsekas, *Dynamic Programming: Deterministic and Stochastic Models,* Prentice-Hall, Englewood Cliffs, 1987.

[BDO] J.S. Baras and A.J. Dorsey, Stochastic Control of Two Partially Observed Competing Queues, *IEEE Transactions on Automatic Control,* **AC-26** (1981) 1106-1117.

[BTE] F.J. Beutler and D. Teneketzis, Routing in Queueing Networks Under Imperfect Information: Stochastic Dominance and Thresholds, *Stochastics & Stochastics Reports,* **26** (1989) 81–100.

[DOB] E.-E. Doberkat, *Stochastic Automata: Stability, Nondeterminism, and Prediction,* Springer-Verlag, Berlin, 1981.

[FAM1] E. Fernández-Gaucherand, A. Arapostathis, and S.I. Marcus, On the Average Cost Optimality Equation and the Structure of Optimal Policies for Partially Observable Markov Decision Processes, *Annals of Operations Research* **29** (1991) 439–470.

[FAM2] E. Fernández-Gaucherand, A. Arapostathis and S.I. Marcus, Analysis of an Adaptive Control Scheme for a Partially Observed Controlled Markov Chain, to appear in *IEEE Transactions in Automatic Control.*

[FG] E. Fernández-Gaucherand, *Controlled Markov Processes on the Infinite Planning Horizon: Optimal & Adaptive Control,* Ph.D. Dissertation, The University of Texas at Austin, August 1991.

[KV] P.R. Kumar and P. Varaiya, *Stochastic Systems: Estimation, Identification and Adaptive Control,* Prentice-Hall, Englewood Cliffs, 1986.

[PAZ] A. Paz, *Introduction to Probabilistic Automata,* Academic Press, New York, 1971.

[WAK] K. Wakuta, Optimal Control of an M/G/1 Queue with Imperfectly Observed Queue Length when the Input Source is Finite, *Journal of Applied Probability,* **28** (1991) 210-220.

Proceedings of the 31st Conference
on Decision and Control
Tucson, Arizona • December 1992

Adaptive Control of I.I.D. Processes and Markov Chains on a Compact Control Set

Rajeev Agrawal

Department of Electrical and Computer Engineering
University of Wisconsin–Madison
Madison, WI 53706–1691
agrawal@engr.wisc.edu

Abstract

In this paper we consider the multi-armed bandit problem and the adaptive control of Markov chains with continuous arms that are chosen from a compact subset of $\mathbb{R}^d$. We devise a learning scheme based on a kernel estimator. Using this learning scheme, we construct a class of certainty equivalence control with forcing schemes and derive asymptotic upper bounds on their learning loss.

1. Introduction

We consider stochastic adaptive control problems with the *learning loss* criterion. This criterion was first used in the context of the multi-armed bandit problem ([1], [2], [3], [4]), and then for the more general class of stochastic adaptive control problems modeled by adaptive control of Markov chains ([5], [6], [7]). All of these papers deal with only finite control sets (number of *arms*). In this paper we consider both the multi-armed bandit problem and the adaptive control of Markov chains with compact control sets. This problem is much more difficult because the builtin learning task is now infinite dimensional whereas previously it was only finite dimensional. We construct various classes of adaptive control schemes and derive upper bounds on their learning loss. To the best of our knowledge these are the best rates available to date. Moreover, the rates obtained by us are still stronger than the $o(n)$ required for optimality with respect to the average-cost-per-unit-time criterion.

The approach taken in this paper is similar to that of [6]. In particular, we construct a class of certainty equivalence control with forcing schemes for the multi-armed bandit problem in Section 2 and for the adaptive control of Markov chains in Section 3.

2. The Multi-armed Bandit Problem

Consider a (memory-less) discrete-time stochastic system modeled by a controlled i.i.d. process, i.e.,

$$\mathsf{P}\{X_n \in B | U_1, X_1, \ldots, U_{n-1}, X_{n-1}, U_n\} = \mathsf{P}\{X_n \in B | U_n\} \tag{2.1}$$

where $\{U_n, X_n\}_{n=1}^{\infty}$ is the chronological sequence of controls and states. The states X_n take values in some arbitrary set $\mathcal{X}$, and the controls U_n are chosen from a compact set $\mathcal{U} \subset \mathbb{R}$ (say $\mathcal{U} = [0,1]$). There is a one-step reward, $r(U_n, X_n)$, associated with each pair $(U_n, X_n), n \geq 1$, where $r : \mathcal{U} \times \mathcal{X} \to \mathbb{R}$. Let $\mathsf{E}[r(U_n, X_n)|U_n] = m(U_n)$, where $m(u)$ is finite for each $u \in \mathcal{U}$. Then we can write

$$r(U_n, X_n) = m(U_n) + w(U_n), \tag{2.2}$$

where by (2.1) the sequence $\{w(U_n)\}_{n=1}^{\infty}$ is independent and zero-mean given $\{U_n\}_{n=1}^{\infty}$. For simplicity we will assume that $\{w(U_n) =: w_n\}_{n=1}^{\infty}$ is i.i.d., as well as independent of $\{U_n\}_{n=1}^{\infty}$. The problem is to design an adaptive control scheme $\gamma = \{\gamma_n\}_{n=1}^{\infty}$, i.e., $U_n = \gamma_n(U_1, X_1, \ldots, U_{n-1}, X_{n-1})$, so as to "maximize" the total reward

$$J_n := \sum_{i=1}^{n} r(U_i, X_i) = \sum_{i=1}^{n} m(U_i) + \sum_{i=1}^{n} w_n \ , \tag{2.3}$$

as $n \to \infty$. Clearly if $m : \mathcal{U} \to \mathbb{R}$ were known with a point of maximum at $u^* \in \mathcal{U}$, then the scheme $\gamma_n \equiv u^*$ would be optimal in the sense that $J_n \overset{\text{st}}{\leq} nm(u^*) + \sum_{i=1}^{n} w_n$. In the absence of the knowledge of m it is desirable to approach this performance as closely as possible. For this purpose define the *learning loss*

$$L_n := \sum_{i=1}^{n} m(u^*) - m(U_i) \ . \tag{2.4}$$

We also define the *ϵ-learning loss*

$$L_n^{\epsilon} := \sum_{i=1}^{n} I\{m(u^*) - m(U_i) > \epsilon\} \ . \tag{2.5}$$

The objective is to design adaptive control schemes for which the learning loss (or ϵ-learning loss) increases slowly. Before we proceed with the design of the adaptive control schemes, we will address just the learning aspect of the problem.

2.1 The Learning Scheme

The ultimate objective of the learning part of the problem is to estimate the point(s) of maximum, u^*, of the function m. The only assumption we make on the function m is that it be Lipschitz continuous. Since we do not make any unimodality assumptions we will need a "global search" strategy. The approach we take to this problem is to first obtain an "uniformly good" estimate of the function m, and the use the point(s) of maximum of the estimate of m as an estimate of the point(s) of maximum of m. The problem of estimating the function $m : \mathcal{U} \to \mathbb{R}$ on the basis of "noisy" measurements $\{Y_n = m(U_n) + w_n\}_{n=1}^{\infty}$ taken at a sequence of points $\{U_n\}_{n=1}^{\infty} \in \mathcal{U}$ is the standard nonparametric regression problem in statistics (see [8], and references therein).

Here we will use a fixed (nearly) equispaced design. In particular the points $\{u_n\}_{n=1}^{\infty}$ are chosen as follows:

$$u_n = 0.n_1 n_2 n_3 \ldots \tag{2.6}$$

where

$$n - 1 = \ldots n_3 n_2 n_1 \tag{2.7}$$

is the binary representation of $n-1$. We use a window estimator which is a special case of the more general class of Nadaraya-Watson kernel estimators. Thus,

$$\hat{m}_n(u) = \frac{\sum_{i=1}^{n} K_{h_n}(u - u_i) Y_i}{\sum_{i=1}^{n} K_{h_n}(u - u_i)}, \tag{2.8}$$

where

$$K_{h_n}(u) = h_n^{-1} K(u/h_n), \tag{2.9}$$

$$K(u) = I\{|u| \leq 1/2\} \tag{2.10}$$

is the window kernel, and $\{h_n\}$ is a sequence of bandwidths to be specified later. Let

CH3229-2/92/0000-2752$1.00 © 1992 IEEE

$$d_\infty(\hat{m}, m) = \sup_{u \in \mathcal{U}} |\hat{m}(u) - m(u)| \quad (2.11)$$

be the uniform metric. It is well known that the above estimator is uniformly consistent (under mild conditions on the distribution of w_n), i.e., $d_\infty(\hat{m}_n, m) \to 0$ w.p.1 for a wide range of bandwidth sequences. We also provide the following (w.p.1) and (L_p) rates of convergence which strengthen the usual (in P) rates found in the literature.

Theorem 2.1 *Assume that m is Lipschitz continuous with Lipschitz constant L, and that $w_1, w_2, \dots$ is an i.i.d. sequence with mean zero, variance σ^2, and a finite moment generating function in some neighborhood of 0. Then, for the above window estimator with the bandwidth sequence $h_n = h(\log n/n)^{1/3}$, $h > 0$, we have*

$$\overline{\lim_{n\to\infty}}\, d_\infty(\hat{m}_n, m)/(\log n/n)^{1/3} \le c(L,\sigma,h) \quad \text{a.s. and in } L_p, p \ge 1 \quad (2.12)$$

where $c(L,\sigma,h) = (L/2)h + (4\sigma/\sqrt{3})\sqrt{h}$. Also, for the above window estimator with any bandwidth sequence h_n such that $h_n \to 0$ and $\log h_n/nh_n \to 0$ as $n \to \infty$, we have

$$\overline{\lim_{n\to\infty}} \frac{1}{nh_n} \log P(d_\infty(\hat{m}_n, m) > \epsilon) \le -2\Gamma(\epsilon) \quad (2.13)$$

for all $\epsilon > 0$, for some $\Gamma(\epsilon) > 0$.

Proof. See [9].

2.2 The Adaptive Control Scheme

In this section we construct a class of certainty equivalence control with forcing type adaptive control schemes based on the learning schemes constructed in the previous section. Let $\{b_i\}_{i=1}^\infty$ be a positive integer valued sequence to be specified later. Define the related sequence $\{a_i\}_{i=1}^\infty$ as follows:

$$a_i := 1 + \sum_{k=1}^{i-1}(b_k + 1) = \sum_{k=1}^{i-1} b_k + i, \quad i \ge 1. \quad (2.14)$$

At times a_i, $i \ge 1$ use (force) the ith control u_i from the fixed design sequence of the previous section. Let $\hat{u^*}_i = \text{argmin}_{u\in U}\, \hat{m}_i(u)$ be the estimate for u^* based on the observations made at times a_k, $1 \le k \le i$. Use the control $\hat{u^*}_i$ from time $a_i + 1$ to the time $a_{i+1} - 1$, i.e., for b_i times. Thus,

$$U_{a_i} = u_i, \quad U_n = \hat{u^*}_i \text{ for } a_i + 1 \le n \le a_{i+1} - 1, \quad i \ge 1. \quad (2.15)$$

The following theorem provide upper bounds on the learning loss associated with the class of schemes constructed above.

Theorem 2.2 *Assume that the conditions of Theorem 2.1 hold. Then, for the certainty equivalence control with forcing type scheme constructed above, with the learning scheme of Section 2.1 with the bandwidth sequence $h_i = h(\log i/i)^{1/3}$, and with the sequence $b_i = \lfloor (i/\log i)^{1/3} \rfloor$, we have*

$$L_n = o(n^{\frac{3}{4}+\delta}) \text{ for all } \delta > 0, \text{ a.s. and in } L_p, p \ge 1, \quad (2.16)$$

Also, for the same scheme with any bandwidth sequence h_i such that $h_i \to 0$ and $\log h_i/ih_i \to 0$ as $i \to \infty$, and with the sequence $b_i = \lfloor e^{i(h_i)^2} \rfloor$, we have

$$\mathsf{E}[L_n^\epsilon] = O((h_n)^{-2} \log n), \quad (2.17)$$

for all $\epsilon > 0$.

Proof. See [9].

Note that so far we have considered $\mathcal{U} = [0,1]$. We can generalize these results to the d-dimensional case. However, the strongest rates that we obtain are for the a.s. case. The rates obtained for L_p are somewhat weaker (See [10] for details).

3. Adaptive Control of Markov Chains

In this section we extend the results obtained for the multi-armed bandit problem to the more general setting of controlled Markov chains. Thus now

$$\mathsf{P}\{X_{n+1} \in B | X_0, U_0, \dots, X_n, U_n\} = \mathsf{P}\{X_{n+1} \in B | X_n, U_n\}, \quad (3.1)$$

where $\{X_n, U_n\}_{n=0}^\infty$ is the chronological sequence of controls and states. The states X_n take values in a finite set $\mathcal{X}$, and the controls U_n are chosen from a compact set $\mathcal{U} \subset \mathbb{R}$ (say $\mathcal{U} = [0,1]$). There is a one-step reward, $r(X_n, U_n)$, associated with each pair $(X_n, U_n), n \ge 0$, where $r : \mathcal{X} \times \mathcal{U} \to \mathbb{R}$. We assume that the under any stationary control law, the resulting Markov chain is ergodic, and that the reward and transition probabilities are Lipschitz continuous functions of the control.

By viewing control laws as *arms*, and by using each control law over complete recurrence intervals, we transform this problem into a higher-dimensional multi-armed bandit problem. In particular, we construct two classes of certainty equivalence control with forcing schemes. For the first class of schemes we obtain the following upper bound on th learning loss: $L_n = o(n^\alpha)$ w.p.1 and in $L_p, p \ge 1$, for some $0 < \alpha < 1$, where α depends only on $|\mathcal{X}|$. For the second class of schemes we obtain upper bound on the ϵ-learning loss: $\mathsf{E}[L_n^\epsilon] = o((\log n)^{1+\delta})$ for any $\delta > 0$.

Remark. See [10] for details.

Acknowledgements

This research was supported by NSF Grant No. ECS-8919818.

References

[1] T. L. Lai and H. Robbins, "Asymptotically efficient adaptive allocation rules," *Adv. Appl. Math.*, vol. 6, pp. 4–22, 1985.

[2] V. Anantharam, P. Varaiya, and J. Walrand, "Asymptotically efficient allocation rules for the mutiarmed bandit problem with multiple plays; Part I: IID rewards," *IEEE Trans. Automat. Contr.*, vol. 32, no. 11, pp. 968–975, 1987.

[3] V. Anantharam, P. Varaiya, and J. Walrand, "Asymptotically efficient allocation rules for the mutiarmed bandit problem with multiple plays; Part II: Markovian rewards," *IEEE Trans. Automat. Contr.*, vol. 32, no. 11, pp. 975–982, 1987.

[4] R. Agrawal, M. Hegde, and D. Teneketzis, "Asymptotically efficient adaptive allocation rules for the multi-armed bandit problem with switching cost," *IEEE Trans. Automat. Contr.*, vol. AC-33, pp. 899–906, Oct. 1988.

[5] R. Agrawal, D. Teneketzis, and V. Anantharam, "Asymptotically efficient adaptive allocation schemes for controlled Markov chains: Finite parameter space," *IEEE Trans. Automat. Contr.*, vol. AC-34, pp. 1249–1259, Dec. 1989.

[6] R. Agrawal and D. Teneketzis, "Certainty equivalence control with forcing: Revisited," *Syst. Contr. Lett.*, vol. 13, pp. 405–412, Dec. 1989.

[7] R. Agrawal, "Minimizing the learning loss in adaptive control of Markov chains under the weak accessibility condition," *J. Appl. Prob.*, vol. 28, pp. 779–790, Dec. 1991.

[8] W. Härdle, *Applied Nonparametric Regression*. Cambridge University Press, 1990.

[9] R. Agrawal, "The multi-armed bandit problem with continuous arms drawn from the real line." Submitted for Publication, 1992.

[10] R. Agrawal, "Adaptive control of i.i.d. processes and Markov chains with a multidimensional control set." Submitted for Publication, 1992.

Proceedings of the 31st Conference
on Decision and Control
Tucson, Arizona • December 1992

PERFORMANCE OF STOCHASTIC ADAPTIVE CONTROL IN THE PRESENCE OF LARGE HIGH-FREQUENCY UNMODELLED DYNAMICS

Miloje Radenkovic
Department of Electrical Engineering
University of Colorado at Denver
Denver, CO 80204
U.S.A.

Anthony N. Michel
Department of Electrical Engineering
University of Notre Dame
Notre Dame, IN 46556
U.S.A.

1 Introduction

Robust stochastic adaptive control in the presence of unmodelled dynamics and external disturbances is considered. it is assumed that multiplicative and additive system perturbations as well as external disturbances are complex, unstructured and large at high frequencies. Global stability is established without requiring the minimum phase assumption and persistency exciting conditions. The self-stabilization mechanism of the proposed adaptive control algorithm is characterized analytically.

2 Robust Stochastic Adaptive Control

2.1. Problem Statement

Let us consider the following stochastic discrete time SISO system with unmodelled dynamics

$$\begin{aligned} A_\theta(q^{-1})y(t) &= B_\theta(q^{-1})[1+\Delta_1(q^{-1})W_1(q^{-1})]u(t-1) \\ &+ A_\theta(q^{-1})\Delta_2(q^{-1})W_2(q^{-1})u(t-1) \\ &+ [1+q^{-1}\Delta_3(q^{-1})W_3(q^{-1})]\omega(t) \end{aligned} \quad (3.1)$$

where $\{y(t)\}, \{u(t)\}$ and $\{\omega(t)\}$ are output, input and stochastic disturbance sequences, respectively, while q^{-1} represents the unit delay operator. Polynomials $A_\theta(q^{-1})$ and $B_\theta(q^{-1})$ are given by

$$\begin{aligned} A_\theta(q^{-1}) &= 1 + a_1 q^{-1} + \cdots + a_{n_A} q^{-n_A} \\ B_\theta(q^{-1}) &= b_0 + b_1 q^{-1} + \cdots + b_{n_B} q^{-n_B}, (b_0 \neq 0). \end{aligned} \quad (3.2)$$

In the notation of the polynomials $A_\theta(q^{-1})$ and $B_\theta(q^{-1})$ we use the subscript θ to emphasize that the corresponding part of the system model (3.1) is structured parametrically, where by θ we denote the parameter vector

$$\theta^T = [a_1, \ldots, a_{n_A};\ b_0, b_1, \cdots, b_{n_B}]. \quad (3.3)$$

In (3.1) the $\Delta_i(q^{-1})W_i(q^{-1})$, $i = 1,2$ are multiplicative and additive perturbations. Regarding transfer functions $\Delta_i(z)W_i(z)$, $i = 1,2,3$, we assume:

(S_1) transfer functions $\Delta_i(z)W_i(z)$, $i = 1,2,3$, are causal and stable and the transfer functions $W_i(z)$, $i = 1,2,3$, are known, while

$$\| \Delta_i(z) \|_{H^\infty} \leq 1,\ i = 1,2,3.$$

Let $\omega(t)$ be a stochastic process defined on the underlying probability space $\{\Omega, \mathcal{F}, P\}$ and introduce the following assumptions:

(S_2) If $\mathcal{F}_t$ is the σ-algebra generated by $\{\omega(1), \ldots, \omega(t)\}$, then for $t \geq 1$

$$\begin{aligned} E\{\omega(t+1)|\mathcal{F}_t\} &= 0 & (a.s.) \\ E\left\{\omega(t+1)^2|\mathcal{F}_t\right\} &= \sigma_\omega^2 & (a.s.) \\ \sup_t E\left\{|\omega(t+1)|^{2+\eta}|\mathcal{F}_t\right\} &\leq k_\omega < \infty,\ \eta > 0. & (a.s.) \end{aligned}$$

The structure of Eq. (3.1) is designed to capture the following frequency dependent nature of the *a priory* knowledge about the system. In the system description we consider frequency interval $0 \leq f \leq \pi$. The same holds for the $-\pi \leq f \leq 0$.

(i) The system model is well known for frequencies f where $0 < f \leq f_1$, implying that $G_\theta(z) = B_\theta(z)/A_\theta(z), (z = e^{if}, i^2 = -1)$ is independent of θ. For $0 < f \leq f_1$, $\| W_i(z) \|_{H^\infty} \leq \gamma$, $i = 1,2,3$, where $\gamma > 0$ is a sufficiently small number compared with $|G_\theta(z)|$.

(ii) In the frequency range $f_1 < f \leq f_2, G_\theta(z)$ is highly dependent on θ, and $\| W_i(z) \|_{H^\infty} \leq \gamma$, $i = 1,2,3$. The dynamics in this frequency range may include parts of the known structure but unknown parameters such as resonant modes whose damping coefficients are unknown, and resonant frequencies can vary between f_1 and f_2.

(iii) In the range $f > f_2, |\Delta_i(e^{if})W_i(e^{if})|$, $i = 1,2,3$ are large and the system is dominated by those dynamics which are *unknown in both structure and complexity.*

Thus we will assume that there exists an unknown parameter set Θ^* so that for every $\theta \in \Theta^*$, the transfer function $B_\theta(z)/A_\theta(z)$ matches well the dynamics of the system at low frequencies $f \leq f_2$. From the above system model description it is also obvious that at frequencies $f > f_2$, unstructured and complex external stochastic disturbances are possible.

Classical non-adaptive controllers can produce a large loop gain and achieve satisfactory performances at frequencies $f \leq f_1$, but must reduce the loop gain neglecting performances in order to preserve stability at frequencies $f > f_1$. Thus the bandwidth of the closed-loop system is defined by the intermediate frequency dynamics whose structure is known.

Since the adaptive controller permanently estimates the parameter vector θ, it could be used to extend the bandwidth of the system to f_2, thus extending the usefulness of the system model up to frequency f_2. Such an adaptive controller must still provide enough gain attenuation at fequencies above f_2 in order to preserve stability in the face of the unstructured uncertainties. Obviously we need a *frequency selective* adaptive controller with capabilities of

(i) achieving large loop gains at low frequencies $f \leq f_2$,

(ii) adapting to unknown and possibly time-varying parameters in order to maintain stability and performances at intermediate frequencies $f_1 < f \leq f_2$, and

(iii) providing attenuation to preserve stability at high frequencies, $f > f_2$.

Such a frequency selective controller which will behave as an adaptive controller for some frequency range and as a nearly fixed controller in another frequency range will be proposed in this section.

Since the transfer functions $W_i(z)$, $i = 1,2,3$, are known, we are able to design a low-pass finite impulse response filter (FIR), $\tilde{W}(z)$, so that for $f \leq f_2, |\tilde{W}(e^{if})| \cong 1$ and for $f > f_2, |\tilde{W}(e^{if})| \leq \frac{\gamma}{T_0}$ where

$$T_0 = \max_{1 \leq i \leq 2} \{\| W_i(z) \|_{H^\infty}\}, \quad (3.4)$$

or, if we want good disturbance rejection at high frequencies,

$$T_0 = \max_{1 \leq i \leq 3} \{\| W_i(z) \|_{H^\infty}\} \quad (3.5)$$

and γ is the upper bound on $\| W_i(z) \|_{H^\infty}$, $i = 1,2,3$ at low frequencies $f \leq f_2$. The idea of this design procedure is that at low frequencies $f \leq f_2$

$$\left|\tilde{W}(e^{if})\frac{B_\theta(e^{if})}{A_\theta(e^{if})}\right| \cong \left|\frac{B_\theta(e^{if})}{A_\theta(e^{if})}\right|, \quad (3.6)$$

and for frequencies $f > f_2$

$$\left|\tilde{W}(e^{if})W_i(e^{if})\right| \leq \gamma,\ i = 1,2,3. \quad (3.7)$$

For frequencies close to f_2, conditions (3.6) and (3.7) may be difficult to achieve, because we need a filter with a short transient band. Actually this is a matter of designer skills and the specific system that has to be controlled. In our considerations we will assume that (3.6) and (3.7) are satisfied. We will also assume that $\tilde{W}(z)$ is a

CH3229-2/92/0000-2754$1.00 © 1992 IEEE

stable polynomial. Specifically, if we design a filter with amplitude characteristics described by Eqs. (3.6) and (3.7) and some of its zeros appear to be unstable, we will use a filter where the unstable zeros are replaced by their reciprocal values. These two filters have different phase characteristics but the same amplitude characteristics and that is what we are interested in. Let the low-pass filter $\tilde{W}(z)$ have the following form

$$\tilde{W}(z^{-1}) = w_0 + w_1 z^{-1} + \cdots + w_{nw} z^{-nw}. \tag{3.8}$$

Our objective is to design an adaptive controller as a function of initial conditions and measurements to stabilize the system and for a given reference signal $y^*(t)$, minimize the functional criterion

$$J = \lim_{N\to\infty} \frac{1}{N}\sum_{t=1}^{N}\left\{P(q^{-1})[(y_f(t+1) - y_f^*(t+1))] + Q(q^{-1})u(t)\right\}^2 \tag{3.9}$$

where the polynomials $P(q^{-1})$ and $Q(q^{-1})$ are chosen by the designer, while

$$y_f(t) = \tilde{W}(q^{-1})y(t),\ y_f^*(t) = \tilde{W}(q^{-1})y^*(t) \tag{3.10}$$

where $\tilde{W}(q^{-1})$ is the low-pass FIR filter defined by Eq. (3.8).

Using prior information related to the transfer function $B_\theta(z)/A_\theta(z)$, for $f \leq f_2$, we can choose polynomials $P(q^{-1})$ and $Q(q^{-1})$ so that the following assumption is satisfied:

(S_3) $D(z^{-1}) = A_\theta(z^{-1})Q(z^{-1}) + B_\theta(z^{-1})\tilde{W}(z^{-1})P(z^{-1})$ has zeros outside the unit disc, and $b_0 w_0 + q_0 \neq 0$.

We shall assume that reference signal $y^*(t)$ satisfies:

(S_4) $\{y^*(t)\}$ is a bounded deterministic sequence defined for $t \geq 0$, i.e., there exists a number m_1 such that $|y^*(t)| \leq m_1$ for all $t \geq 1$.

Note that system (3.1) can be written in the form

$$A_\theta(q^{-1})y_f(t+1) = B_\theta(q^{-1})\tilde{W}(q^{-1})u(t) + \omega(t+1) + \gamma(t) \tag{3.11}$$

where $\gamma(t)$ is given by

$$\begin{aligned}\gamma(t) &= [B_\theta(q^{-1})\Delta_1(q^{-1})W_1(q^{-1}) + A_\theta(q^{-1})\Delta_2(q^{-1})W_2(q^{-1})]\\ &\quad \tilde{W}(q^{-1})u(t) + [\tilde{W}(q^{-1}) - 1]\omega(t+1)\\ &+ \Delta_3(q^{-1})W_3(q^{-1})\tilde{W}(q^{-1})\omega(t).\end{aligned} \tag{3.12}$$

Equation (3.11) can be written in the form

$$\begin{aligned} z(t) &= \theta_0^T\phi(t) + Q(q^{-1})u(t) + q[P(q^{-1}) - 1]y_f(t)\\ &- P(q^{-1})y_f^*(t+1) + \gamma(t)\end{aligned} \tag{3.13}$$

where

$$\theta_0^T = [a_1, \cdots, a_{n_A};\ b_0, b_1, \cdots, b_{n_B}], \tag{3.14}$$

$$\phi(t)^T = [-y_f(t), \cdots, -y_f(t-n_A+1);\ u_f(t), \cdots, u_f(t-n_B)], \tag{3.15}$$

$$u_f(t) = \tilde{W}(q^{-1})u(t), \tag{3.16}$$

$$z(t) = e(t+1) - \omega(t+1) \tag{3.17}$$

and

$$e(t+1) = P(q^{-1})[y_f(t+1) - y_f^*(t+1)] + Q(q^{-1})u(t). \tag{3.18}$$

2.2. Robust Adaptive Control

Using prior information about transfer function $B_\theta(z)/A_\theta(z)$ we are able to determine the compact convex set Θ^0 which contains all θ from the set Θ^*. Recall that Θ^* is the set containing the unknown θ for which transfer function $B_\theta(z)/A_\theta(z)$ matches the system dynamics well at low frequencies $f \leq f_2$. Thus $\Theta^* \subset \Theta^0$, and the following assumption holds:

(S_5) The compact convex set Θ^0 which contains the true parameters θ_0, the sign of $r_0 = b_0 w_0 + q_0$, and a lower bound, $r_{0,\min}$, on the magnitude of $|r_0|$, are known. Without loss of generality we assume that $r_0 > 0$ and $r_{0,\min} > 0$.

For the estimation of θ_0 we propose the following stochastic gradient-type algorithm

$$\begin{aligned}\hat{\theta}(t+1) &= \mathcal{P}\left\{\hat{\theta}(t) + \frac{\bar{a}}{\bar{r}(t)}\phi(t)[P(q^{-1})(y_f(t+1) - y_f^*(t+1))\right.\\ &+ \left.Q(q^{-1})u(t)]\right\},\ 0 < \bar{a} < 1\end{aligned} \tag{3.19}$$

where $\mathcal{P}\{\cdot\}$ projects orthogonally onto Θ^0, so that $\mathcal{P}\{\theta\} \in \Theta^0$ for all $\theta \in R^{n_A+n_B+1}$, and there exists a finite constant d_0 so that $\|\hat{\theta}(t) - \theta_0\|^2 \leq d_0$ and $r_0(t) \geq r_{0,\min} > 0$ for all $t > 0$. The algorithm gain sequence $\bar{r}(t)$ is given by

$$\bar{r}(t) = \max\left\{2\max_{1\leq\tau\leq t}\|\phi(\tau)\|^2, r(t)^{1-\varepsilon} + t^{1-\varepsilon}\right\},\quad 0 < \varepsilon < \frac{1}{2} \tag{3.20}$$

where

$$r(t) = r(t-1) + \|\phi(t)\|^2,\quad r(0) > 1. \tag{3.21}$$

Since θ_0 is unknown, as adaptive control law we use the "certainty equivalence" controller,

$$\hat{\theta}(t)^T\phi(t) + Q(q^{-1})u(t) + q[P(q^{-1}) - 1]y_f(t) = P(q^{-1})y_f^*(t+1). \tag{3.22}$$

Since by assumption (S_1) $\Delta_i(z)W_i(z)$, $i = 1,2,3$ are stable transfer functions and by assumption (S_3) $D(z)$ is stable we can define the following H^∞-norms:

$$\begin{aligned} C_A &= \left\|\frac{A_\theta(z)}{D(z)}\right\|_{H^\infty},\ C_P = \left\|\frac{P(z)}{D(z)}\right\|_{H^\infty},\\ C_{APW} &= \left\|\frac{A(z)P(z)\tilde{W}(z)}{D(z)}\right\|_{H^\infty},\ C_{AP} = \left\|\frac{A_\theta - P(z)}{D(z)}\right\|_{H^\infty},\\ \gamma_1 &= \|B_\theta(z)W_1(z)\tilde{W}(z)\|_{H^\infty},\ \gamma_2 = \|A_\theta(z)W_2(z)\tilde{W}(z)\|_{H^\infty},\\ C_W &= \|\tilde{W}(z) - 1\|_{H^\infty},\ \gamma_3 = \|W_3(z)\tilde{W}(z)\|_{H^\infty}.\end{aligned} \tag{3.23}$$

Regarding the intensity of unmodelled dynamics, we introduce the following assumption:

(S_6) $C_P(\gamma_1 + \gamma_2) < 1$, $\rho_1 = 1 - \frac{\bar{a}}{2} - C_\gamma > 0$, $C_\gamma = (\gamma_1 + \gamma_2)C_A/[1 - (\gamma_1 + \gamma_2)C_P]$ The above assumption actually implies two requirements:

1) At low frequencies $f \leq f_2, W_i(e^{if})$, $i = 1,2$, should satisfy the condition $|W_i(e^{if})| \leq \gamma$, where $\gamma > 0$ is a small number.

2) At high frequencies $f > f_2$, performances of the designed low pass filter $\tilde{W}(z^{-1})$ should provide that $|W_i(e^{if})\tilde{W}(e^{if})| \leq \gamma$.

The global stability result is given in the following theorem.

Theorem 3.1: Let the assumptions (S_1) – (S_8) hold. Then

1) $$\lim_{N\to\infty}\sup_N \frac{1}{N}\sum_{t=1}^{N} z(t)^2 \leq (1 + \bar{C}_{41})\frac{16}{\rho_1^2}\Sigma_\nu \quad \text{(a.s.)} \tag{3.24}$$

where $0 < \bar{C}_{41} < \infty, \rho_1$ is defined by assumption (S_6), and

$$\begin{aligned}\Sigma_\nu &= \frac{1}{(1 - (\gamma_1 + \gamma_2)C_P)^2}\{(\gamma_1 + \gamma_2)C_{APW}m_1\\ &+ [\gamma_3 + C_W + (\gamma_1 + \gamma_2)C_{AP}]\sigma_\omega\}^2.\end{aligned} \tag{3.25}$$

2)

$$\begin{aligned}&\lim_{N\to\infty}\sup_N \frac{1}{N}\sum_{t=1}^{N}\left\{P(q^{-1})[y_f(t+1) - y_f^*(t+1)] + Q(q^{-1})u(t)\right\}^2\\ &\leq \sigma_\omega^2 + (1 + \bar{C}_{41})\frac{16}{\rho_1^2}\Sigma_\nu^2\ (a.s.).\end{aligned} \tag{3.26}$$

3) $$\lim_{N\to\infty}\sup_N \frac{1}{N}\sum_{t=1}^{N}\|\phi(t)\|^2 \leq \bar{C}_{42} < \infty \quad \text{(a.s.)}. \tag{3.27}$$

Proof: Proof of this theoorem is omitted due to space limitation and can be found in [1].

References

[1] M. S. Radenkovic and A. N. Michel, "Robust Stochastic Adaptive Control in the Presence of Unmodelled Dynamics with Large Amplitudes at High Frequencies", *Technical Report*, UCD, pp. 91-03, 1991.

Proceedings of the 31st Conference
on Decision and Control
Tucson, Arizona • December 1992

MODEL REFERENCE ADAPTIVE CONTROL WITH UNKNOWN HIGH FREQUENCY GAIN

Rogelio Lozano and Rubén G. Moctezuma

U.T.C. - HEUDIASYC
URA CNRS n° 817
B.P. 649 - 60206 Compiègne
FRANCE

Abstract: This paper presents an indirect model reference adaptive control for minimum phase linear systems of arbitrary order with unknown high frequency gain. It is proved that the (modified) estimate of the high frequency gain has a uniform positive lower bound. The problem has been solved by using the Least Squares covariance matrix properties to define an appropriated modification of the parameters estimates.

1. INTRODUCTION.

Globally stable model reference adaptive controllers were first proposed in the late 1970's [3], [4], [15], [11], [2]. These results are applicable to any minimum phase linear system, nevertheless we require the system's order , relative degree and high frequency gain sign to be known.

We will focus our atention in the approaches that can be used to relax the assumption on the high frequency gain which has never been completely justified in the general case. The problem was motivated by Morse (1983) and a first solution was given by [16] for a first order plant. Later an interesting solution was proposed in [9]. The Nussbaum's gain technique was also extended to the arbitrary order case in [14].

An alternative way to solve the problem was presented in [8] based on a particular modification of the parameters estimates and hysteresis switching to avoid division by zero. The technique follows the work in [7] for discrete time systems, however the problem of existence of solutions that arises naturally in continuous-time systems due to the discontinuities introduced in the control scheme has not been adressed. In [6] a solution is presented for adaptive stabilization of a (nonlinear) first order system without *a priori* information on the plan parameters. The issue of existence of solutions was properly solved.

In [13] an interesting solution for the problem is presented based on hysteresis switching in the commutation between a set of possible adaptive controllers. In fact, their approach relaxes not only the assumption on the high frequency gain but also the one on the relative degree. However, boundedness of all the estimates is guaranted only if the reference input converges to zero.

This paper presents an indirect model reference adaptive control scheme for minimum phase linear systems of arbitrary order with unknown high frequency gain. The results could be interpreted as an extension of those presented in [6] for higher order systems but interestingly enough the algorithm proposed here requires neither signals normalization nor the definition of an augmented error. Global convergence is established in the sense that the plant output asymptotically tracks the reference model output using a bounded input. . It is also shown that the high frequency gain (modified) estimated has an uniformly non-zero lower bound without requiring any *a priori* knowledge on the system's high frequency gain.

Relaxation of the assumption on the high frequency gain sign is a problem in its own right. For clarity of presentation we have prefered to deal with systems in the ideal situation, i.e. without noise. It has been proved in [5] that the technique employed here can also be used to cope with bounded noise.

The outline of the paper is has follows: Section 2 introduces the system parameter estimation and its useful properties, Section 3 present the indirect model reference adaptive control and the procedure to avoid singularities in the control law, Section 4 is devoted to the convergence analysis of the closed loop system.

2. SYSTEM PARAMETER ESTIMATION

Consider the following single input-single output LTI plant with unknown parameters

$$A^* y = B^* u \tag{1}$$

with

$$A^* = D^n + a_1^* D^{n-1} + \ldots + a_n^*$$
$$B^* = b_0^* D^m + b_1^* D^{m-1} + \ldots + b_m^*$$

where $n \geq m$ and $D = \frac{d}{dt}$. We make the following assumptions

1. n and m are known.
2. $B^*(D)$ is a Hurwitz polynomial.
3. $b_0^* \neq 0$.

Since the derivatives of u and y are not supposed to be available we introduce a filter F to be able to estimate the parameters of the system (1). Let us define the filtered signals y_f and u_f as follows

$$F y_f = y \tag{2.a}$$
$$F u_f = u \tag{2.b}$$

with $F = D^n + f_1 D^{n-1} + \ldots + f_n$ Hurwitz. In the rest of the paper we will use the notation $u_f^{(i)}$ for $D^i u_f$. Equations (2) and (3) can also be described by the following state space realizations

$$\dot{x}_1 = A_F x_1 + B_F u \tag{3}$$
$$u_f = C_F^T x_1$$

$$\dot{x}_2 = A_F x_2 + B_F y \tag{4}$$
$$y_f = C_F^T x_2$$

Define the state

$$x_3 = A^* x_2 - B^* x_1 \tag{5}$$

Then from (3) and (4) and using (1) and (5) we obtain

$$\dot{x}_3 = A^* \dot{x}_2 - B^* \dot{x}_1 = A_F x_3 \tag{6}$$

Also from (3) and (4) we can show that

$$A^* y_f - B^* u_f = C_F^T x_3 = C_F^T e^{A_F t} x_3(0) \tag{7}$$

The initial conditions $x_3(0)$ in (5) depends on $x_1^{(i)}(0)$, $x_2^{(i)}(0)$ for i=1,...,n. From (3) and (4) we see that $x_1^{(i)}(0)$, $x_2^{(i)}(0)$ depend on

CH3229-2/92/0000-2756$1.00 © 1992 IEEE

$u^{(i-1)}(0)$ and $y^{(i-1)}(0)$ which are unknown. Then $x_3(0)$ is unknown and we denote it as $x_3(0)=x_0^*$.

Expression (7) can also be written as

$$y_f^{(n)} = \theta^{*T}\phi \tag{8}$$

where we have defined the unknown parameter vector

$$\theta^{*T}=[b_0^*, \ldots, b_m^*, a_1^*, \ldots, a_n^*, x_0^{*T}] \tag{9}$$

and the regressor

$$\phi^T = [u_f^{(m)}, \ldots, u_f, -y_f^{(n-1)}, \ldots, -y_f, C_F^T e^{A_F t}] \tag{10}$$

Eventhough the term $C_F^T e^{A_F t}$ converges exponentially to zero we include it in the regressor ϕ to obtain expression (8) without any extra vanishing term. The parameter vector θ^* can now be estimated using the following standard Least Squares estimation algorithm

$$e = y_f^{(n)} - \theta^T\phi \tag{11}$$

$$\dot{\theta} = P\phi e \tag{12}$$

$$\dot{P} = -P\phi\phi^T P \tag{13}$$

where e is the prediction error, P the inverse of the covariance matrix and

$$\theta^T=[b_0, \ldots, b_m, a_1, \ldots, a_n, x_0^T] \tag{14}$$

is the estimate of θ^*. The parametric error is defined as

$$\tilde{\theta} = \theta - \theta^* \tag{15}$$

The following lemma establishes some important properties of the Least Squares algorithm

Lemma 1 : The parameter estimation algorithm defined in (9) - (11) is such that along the solutions of (1) the following properties hold

1. $0 < P \le P(0)$, $P\phi \in L_2$ and P converges.
2. $P^{-1}\tilde{\theta}=P^{-1}\tilde{\theta}(0)$.
(16)
3. θ converges.
4. $e \in L_2$.

5. If $\phi, \dot{\phi} \in L_\infty$ then e and $P\phi$ converge to zero.

Proof : See [6] and [17].

♦♦♦

3. ADAPTIVE CONTROL SCHEME

In this section we present a Model Reference Adaptive Control algorithm based on the estimated plant model obtained through a particular estimate modification

Parameter modification

Let us now define the following parameter modification

$$\bar{\theta} = \theta + P\beta \tag{17}$$

with

$$\bar{\theta}^T = [\bar{b}_0, \ldots, \bar{b}_m, \bar{a}_1, \ldots, \bar{a}_n, \bar{x}_0^T] \tag{18}$$

Introducing (17) into (11) we get

$$e = y_f^{(n)} - (\bar{\theta} - P\beta)^T\phi \tag{19}$$

which can be written as

$$Ay_f = Bu_f + w_2 \tag{20}$$

where

$$A = D^n + \bar{a}_1 D^{n-1} + \ldots + \bar{a}_n$$

$$B = \bar{b}_0 D^m + \bar{b}_1 D^{m-1} + \ldots + \bar{b}_m$$

and

$$w_2 = e - \beta^T P\phi + C_F^T e^{A_F t}\bar{x}_0 \tag{21}$$

The term β in (17) is defined as follows (see also figure 1) where

$$p^T = [1\ 0\ \ldots\ 0]P \tag{22}$$

is the first row of the P matrix and ε is an arbitrary constant such that $0 < \varepsilon << 1$. Such a modification has an hysteresis which is introduced to prevent the discontinuities in β from occuring an infinite number of times or infinitely often. The value of the hysteresis width is $\varepsilon\,|b| + \varepsilon\,\|p\|$. See [6] for further details.

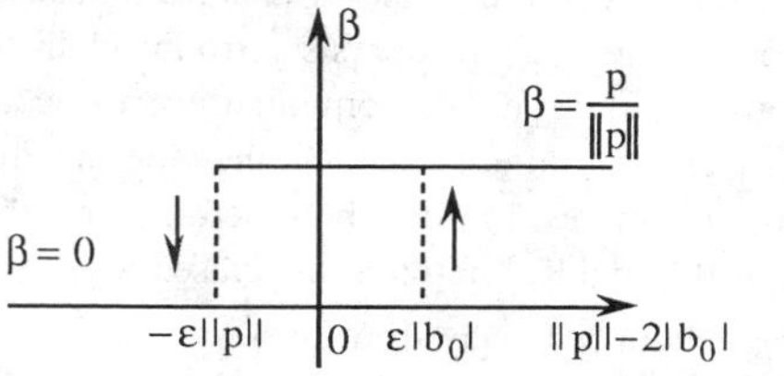

Figure 1. Parameter estimates modification procedure.

Some properties of the estimates modification procedure defined previously are given in the next lemma.

Lemma 2 : The estimates correction procedure defined in (17) and (22) is such that along the solutions of (1) the following results hold

1. $|b_0| + \|p\| \ge b'$ (23)

with

$$b' = \frac{|b_0^*|}{\max[1, P^{-1}\tilde{\theta}(0)}\tag{24}$$

2. $|\bar{b}_0| \ge \frac{1-\varepsilon}{3+\varepsilon}$. (25)

3. β and $\bar{\theta}$ converge and the discontinuities in β occur a finite number of times.

Proof : See Lozano and Brogliato (1992).

♦♦♦

Controller Design

The controller for the plant (1) is defined as

$$SBu_f = -Ry_f + r^M \tag{26}$$

with

$$S = D^{n-m} + s_1 D^{n-m-1} + \ldots + s_{n-m}$$

$$R = r_0 D^{n-1} + r_1 D^{n-2} + \ldots + r_{n-1}$$

$$B = \bar{b}_0 D^m + \bar{b}_1 D^{m-1} + \ldots + \bar{b}_m$$

where $Cy^M = r^M$ defines a reference model and R,S satisfies the following identity

$$A(\lambda)S(\lambda) + R(\lambda) = C(\lambda)F(\lambda) \tag{27}$$

with $C=D^{n-m} + c_1 D^{n-m-1} + \ldots + c_{n-m}$ a Hurwitz polynomial whose zeros represent the desired closed loop poles . It is noteworthy to point out that in (28) we have written explicitly the polynomials in the λ operator instead of the D operator to remark that this is an algebraic relation of the coefficients on the involved polynomials where the time varying nature of A plays no role.

Equation (27) can be written as

$$u_f^{(n)} = -\frac{1}{\bar{b}_0}(g_1 u_f^{(n-1)} + \ldots + g_n u_f + r_0 y_f^{(n-1)} + \ldots + r_{n-1} y_f - r^M) \tag{28}$$

where $SBu_f = g_0 u_f^{(n)} + g_1 u_f^{(n-1)} + \ldots + g_n u_f$ and $\bar{b}_0 = g_0$. Then the controller (29) provide us with $u_f^{(n)}$ which together with (2.b) allow the computation of the control input u.

In view of Lemma 2 we can also conclude that the controller (29) is free of singularities in the sense that the estimate $\bar{b}_0$ is bounded away from zero.

Existence of Solutions.

The existence of solutions is proved by using arguments similar to those in Lozano, Brogliato (1992). Briefly a solution is obtained by concatenating the solutions defined between the time instants at which commutations occur. The hysteresis introduced in the estimates modification procedure avoids commutations to occur an infinite number of times or infinitely often. On the other hand boundedness of the plant estimates implies boundedness of the controller parameters in B,S and R. Therefore the closed loop system is linear and since $\bar{b}_0$ is bounded away from zero we conclude that the plant input and output can not grow faster than exponentially. We can then state the following Lemma

Lemma 3 : The plant input and output do not grow faster than exponentially.

♦♦♦

4. CONVERGENCE ANALYSIS.

To proceed with the convergence analysis the closed loop system (7), (20) and (27) is expressed in a state space representation as

$$\dot{x} = A_{cl}x + v_2 + Br^M \tag{29}$$
$$y_f = C_{cl}^T x$$

with

$$A_{cl} = \left[\begin{array}{ccccccccc}
-\bar{a}_1 & \cdots & -\bar{a}_n & 0 & \cdots & 1 & 0 & \cdots & 0 \\
1 & & \vdots & \vdots & \vdots & \vdots & \vdots & & \vdots \\
\vdots & & 1 & \vdots & \vdots & \vdots & \vdots & & \vdots \\
\hline
-r_0 & \cdots & -r_{n-1} & -s_1 & \cdots & -s_{n-m} & 0 & \cdots & 0 \\
\vdots & & \vdots & 1 & & \vdots & \vdots & & \vdots \\
\vdots & & \vdots & \vdots & & 1 & \vdots & & \vdots \\
\hline
\frac{a_1^* - \bar{a}_1}{b_0^*} & \cdots & \frac{a_n^* - \bar{a}_n}{b_0^*} & 0 & \cdots & \frac{1}{b_0^*} & -\frac{b_1^*}{b_0^*} & \cdots & -\frac{b_m^*}{b_0^*} \\
\vdots & & \vdots & \vdots & & \vdots & \vdots & 1 & \vdots \\
\vdots & & \vdots & \vdots & & \vdots & \vdots & & 1\ 0
\end{array}\right] \tag{30}$$

$$B = \mathrm{diag}[1\ 0 \ldots 0\ \frac{1}{b_0^*}\ 0 \ldots 0] \tag{31}$$

$$x^T = \left[y_f^{(n-1)}, \ldots, y_f \,|\, (Bu_f)^{(n-m-1)}, \ldots, Bu_f \,|\, u_f^{(m-1)}, \ldots, u_f\right] \tag{32}$$

$$v_2^T = [w_2, 0, \ldots, 0, \frac{w_2 - w_0}{b_0^*}, 0, \ldots, 0] \tag{33}$$

where $w_0 = C_F^T e^{A_F t} x_3(0)$ and C_{cl}^T is obvious. Since $\bar{\theta}$ converges we conclude that $A_{cl} \rightarrow A_s$ as $t \rightarrow \infty$, or equivalently

$$A_{cl} = A_s + \Delta \tag{34}$$

with $\Delta \rightarrow 0$ as $t \rightarrow \infty$ and A_s a constant matrix.Using (32) the closed loop system (29) can be rewritten as

$$\dot{x} = (A_s + \Delta)x + v_2 + Br^M \tag{35}$$
$$y_f = C_{cl}^T x$$

where $v_2 \in L_2$ (See (21) and Lemmas 1, 2). Note that if the coefficients of A and B in (20) were constant, the control law in (26) would be such that 2n - m closed loop poles would be located exactly at the values given by the zeros of CF in (27). Therefore 2n - m eigenvalues of the A_{cl} matrix in (29) would be equal to the zeros of CF. As we will see later, the remaining m eigenvalues of A_{cl} correspond to the zeros of B^*. These arguments lead us to conclude that, eventhough A_{cl} is time varying, its eigenvalues are equal to the zeros of CFB^* at every fixed time. This does not allow us to conclude directly the stability of system (31) but insures us that A_s in (35) is a stable matrix. Since $\Delta \rightarrow 0$ and $v_2 \in L_2$ we can expect x to converge to a certain bounded signal depending only on r^M and A_s. Note that A_s depends only on the final values of A and B in (20). Searching for the function describing x in the limit we were lead to use the artifice of defining an unreal system described as follows.

Let us define the following system for which we will define a controller

$$A^* y_a = B^* u_a \tag{36}$$

where A^* and B^* are the same polynomials as in (1) but y_a and u_a denote the new system output and input respectively. Define the filtered signals y_{fa} and u_{fa} as follows

$$Fy_{fa} = y_a \tag{37}$$
$$Fu_{fa} = u_a \tag{38}$$

with F as in (2). Using a similar procedure to obtain (7) we have from (36), (37) and (38)

$$A^* y_{fa} = B^* u_{fa} + w_{0a} \tag{39}$$

with $w_{0a} = C_F^T e^{A_F t} x_{3a}(0)$. Let us now define an estimated model of (36) in the limit and assume that y_a and u_a are such that

$$A_\infty y_a = B_\infty u_a \tag{40}$$

where A_∞ and B_∞ denote the final values of A and B. Combining the above equation with (37) and (38) we obtain an equation similar to (40)

$$A_\infty y_{fa} = B_\infty u_{fa} + w_{2a} \tag{41}$$

where w_{2a} is an exponentially decaying term. The controller for (41) can be defined as

$$S_\infty B_\infty u_{fa} = -R_\infty y_{fa} + r^M \tag{42}$$

with S_∞ and A_∞ satisfying

$$A_\infty S_\infty + R_\infty = CF \tag{43}$$

Then, using (41) to (43) we have

$$\begin{aligned} CFy_{fa} &= (A_\infty S_\infty + R_\infty) y_{fa} \\ &= S_\infty(B_\infty u_{fa} + w_{2a}) + R_\infty y_{fa} \\ &= r^M + S_\infty w_{2a} \end{aligned} \tag{44}$$

Since CF is Hurwitz and w_{2a} is an exponentially decayig term we conclude that y_{fa} is bounded. Premultiplying (44) by A^* and using (39) we get

$$\begin{aligned} CFA^* y_{fa} &= CF(B^* u_{fa} + w_{0a}) \\ &= A^* r^M + A^* S_\infty w_{2a} \end{aligned}$$

from which we finally obtain

$$CFB^* u_{fa} = A^* r_d^M + A^* S_\infty w_{2a} - CFw_{0a} \tag{45}$$

and u_{fa} is also bounded. From the above and assumption 2 we conclude that the closed loop system (39), (41) and (42) is stable and its poles are located at the zeros of CFB^*. A state space representation for this closed loop system can be obtained as was done to obtain (29)

$$\dot{x}_a = A_s x_a + v_{2a} + Br^M \tag{46}$$
$$y_{fa} = C_{cl}^T x_a$$

with A_s defined in (34), B and C_{cl} as in (29) and

$$v_{2a}^T = [w_{2a}, 0, \ldots, 0, \frac{w_{2a} - w_{0a}}{b_0^*}, 0, \ldots, 0] \tag{47}$$

an exponentially decaying term. In view of (46) the system state x_a defined as

$$x_a^T = \left[y_{fa}^{(n-1)},\ldots,y_{fa} \vdots (Bu_{fa})^{(n-m-1)},\ldots,Bu_{fa} \vdots u_{fa}^{(m-1)},\ldots,u_{fa}\right] \quad (48)$$

is bounded. Let us now define the state error

$$x_e = x - x_a \quad (49)$$

Introducing (35) and (46) into the above we get

$$\dot{x}_e = (A_s + \Delta)x_e + \alpha_0 + \alpha_2 \quad (51)$$

where

$$\alpha_0 = \Delta x_a, \qquad \alpha_2 = v_2 - v_{2a} \quad (52)$$

It is clear that $\alpha_0 \to 0$ and $\alpha_2 \in L_2$ since $\Delta \to 0$, x_a is bounded and $v_2, v_{2a} \in L_2$.

The following lemma establishes that the state x of the time variant system (35) approaches asymptotically to the state x_a of the time invariant system (47), i.e. $x_e \to 0$ as $t \to \infty$.

Lemma 4 : Consider the system in (50) with $\Delta \to 0$, $\alpha_0 \to 0$ and $\alpha_2 \in L_2$, then $x_e \to 0$ as $t \to \infty$.

Proof : See appendix A.

♦♦♦

Finally we can state the main result of this paper.

Theorem : Consider the plant (1) in closed loop with the adaptive control defined in sections 2 and 3, then

1. All the states of the adaptive system are bounded functions of time.
2. $y - y^M \to 0$ as $t \to \infty$.

Proof : See appendix B.

♦♦♦

5. CONCLUSIONS.

This paper has presented a globally convergent indirect model reference adaptive control for minimum phase systems of arbitrary order with arbitrary high frequency gain.

It was proved that the high frequency gain (modified) estimated has a uniform positive lower bound without requiring any information on the plant parameters. The proposed algorithm requires neither signals normalization nor the definition of an augmented error.

Current research is focused in the extension of the proposed technique to systems with unknown relative degree and multivariable systems.

REFERENCES.

[1]Desoer, C.A., M. Vidyasagar (1975). Feedback Systems: Input Output Properties. *Academic Press, New York.*

[2]Egardt B. (1980). Stability analysis of discrete-time adaptive control schemes. *IEEE Trans. Aut. Control*, AC-25, 710-717.

[3]Feuer, A. and A.S.Morse. (1978). Adaptive control of single input-single output linear systems. *IEEE Trans. Aut. Control*, AC-23, 557-569.

[4]Goodwin G.C., P.J. Ramadge and P.E. Caines (1980). Discrete-time multivariable adaptive control. *IEEE Trans. Aut. Control*, AC-25, 449-456.

[5]Lozano R. (1989). Robust adaptive regulation without persistence excitation. *IEEE Trans. Aut. Control*, AC-34, 1260-1267.

[6]Lozano, R. and B.Brogliato (1992). Adaptive control of a simple nonliear system without *a priori* information on the parameters. *IEEE Trans. Aut. Control,* AC-37, 30-37.

[7]Lozano R., J. Collado. Adaptive control for systems with bounded disturbances. *IEEE Trans. on Automatic Control*, Vol. 34, 225-228.

[8]Lozano R., J. Collado, S. Mondié (1990). Model reference adaptive control without a priori knowledge of the high frequency gain. *IEEE Trans. on Automatic Control,* AC-35, 71-78.

[9]Mårtensson, B. (1985). The order of any stabilizing regulator is sufficient *a priori* information for adaptive stabilization. *Systems and Control Letters* 6(2), pp. 85-91.

[10]Middleton, R.H. and P.V.Kokotovic. (1991). Boundeness properties of simple indirect adaptive control systems. *Proceeding of the 1991 American Control Conference, Boston..*

[11]Morse, A.S. (1980). Global stability of parameter-adaptive control systems. *IEEE Trans. Aut. Control*, AC-25, 433-439.

[12]Morse, A.S. (1983). Recent problems in parameter adaptive control, in : *I.D. Landau, Ed. Outils et Modèles Mathématiques pour l'Automatique, l'Analyse de Systèmes et le Traitement du Signal, Editions du CNRS*, Vol. 3, pp. 733-740.

[13]Morse A.S., D.Q. Mayne and G. C. Goodwin (1991). Application of hysteresis switching algorithm in parameter adaptive control. *Proc. of the 30th IEEE-CDC, Brighton.*

[14]Mudgett, R.D. and A.S.Morse. (1985). Adaptive stabilization of linear systems with unknown high frequency gains. *IEEE Trans. Aut. Control*, AC-30, 549-554.

[15]Narendra, K.S., Y.H. Lin. and L.S.Valavani. (1980). Stable adaptive controller design, Part II: Proof of stability. *IEEE Trans. Aut. Control*, AC-25, 440-448.

[16]Nussbaum, R.D. (1983). Some remarks on a conjecture in parameter adaptive control. *Systems and Control Letters* 3(5), pp. 243-246.

[17]Sastry, S. and M. Bodson (1989). Adaptive control, stability, convergence and boundedness. Prentice Hall.

[18]Thomopoulos S.C.A. and Y.N.M. Papadakis (1990). On the existence of solutions in adaptive control. *Proc. of the 29th IEEE-CDC, Honolulu.*

APPENDIX A

Proof of lemma 4.

Let us decompose x_e in (50) as follows

$$x_e = z_0 + z_2 \quad (A.1)$$

with

$$\dot{z}_0 = (A_s + \Delta)z_0 + \alpha_0 \quad (A.2)$$

$$\dot{z}_2 = (A_s + \Delta)z_2 + \alpha_2 \quad (A.3)$$

Since A_s is a stable matrix, $\exists\, P_L, Q > 0$ such that

$$A_s^T P_L + P_L A_s = -Q \quad (A.4)$$

Define the candidate Lyapunov function for each of the systems (A.2) and(A.3)

$$V_i = z_i^T P_L z_i \qquad i = 0,2 \quad (A.5)$$

From (A.2) and (A.5) we have

$$\dot{V}_i = [\, z_i^T (A_s^T + \Delta^T) + \alpha_i^T] P_L z_i + z_i^T P_L [\, (A_s + \Delta)\, z_i + \alpha_i]$$
$$\leq - z_i^T Q\, z_i + 2\| z_i \|^2 \|P_L \Delta\| + 2\| z_i \| \|P_L \alpha_i\| \quad (A.6)$$

Note that

$$z_i^T Q\, z_i \geq \lambda_1 \| z_i \|^2 \quad (A.7)$$

with $\lambda_1 = \lambda_{min}(Q)$ the smallest eigenvalue of Q, and

$$\lambda_2 \|z_i\|^2 \le z_i^T P_L z_i = V_i$$
$$\le \lambda_3 \|z_i\|^2 \le \frac{\lambda_3}{\lambda_1} z_i^T Q z_i \quad \text{(using (A.7)} \qquad (A.8)$$

where $\lambda_2 = \lambda_{min}(P_L)$ and $\lambda_3 = \lambda_{max}(P_L)$ are the smallest and largest eigenvalue of P_L respectively. Introducing (A.7) and (A.8) into (A.6) we obtain

$$\dot{V}_i \le -\frac{\lambda_1}{\lambda_3} V_i + 2\frac{V_i}{\lambda_2} \|P_L \Delta\| + 2\left[\frac{V_i}{\lambda_2}\right]^{1/2} \|P_L \alpha_i\| \qquad (A.9)$$

Since $\alpha_0 \in L_\infty$ and $\alpha_2 \in L_2$ then z_0 and z_2 in (A.2) and (A.3) do not have finite scape time. Then, since $\Delta \to 0$, there exists a finite time t_n such that $\|\Delta\|$ is small enough that the following inequality holds

$$\frac{2\|P_L \Delta\|}{\lambda_2} \le \frac{\lambda_1}{2\lambda_3} \qquad t > t_n \qquad (A.10)$$

Introducing (A.10) in (A.9) it follows that

$$\dot{V}_i \le -\frac{\lambda_1}{2\lambda_3} V_i + 2\left[\frac{V_i}{\lambda_2}\right]^{1/2} \|P_L \alpha_i\| \quad t > t_n \qquad (A.11)$$

For system (A.2), i.e. for i = 0 we obtain

$$\dot{V}_0 \le -\frac{\lambda_1}{2\lambda_3} V_0^{1/2}(V_0^{1/2} - \delta) \qquad (A.12)$$

with

$$\delta = \frac{4\|P_L \alpha_0\| \lambda_3}{\lambda_1 \lambda_2^{1/2}} \qquad (A.13)$$

For $V_0^{1/2} > \delta$, $\dot{V}_0 < 0$ and thus in the limit V_0 is bounded by δ. Since δ in (A.13) converges to zero so does V_0. Consequently z_0 converges to zero too.
In order to study the stability of (A.3), i.e. for i = 2 we use the inequality $2ab \le a^2 + b^2 \ \forall$ a,b to decompose the last term in the RHS of (A.11) as follows

$$2V_2^{1/2}\left[\frac{\lambda_1}{4\lambda_3}\right]^{1/2}\left[\frac{\lambda_1}{4\lambda_3}\right]^{-1/2} \frac{\|P_L \alpha_2\|}{\lambda_2^{1/2}}$$
$$\le \frac{\lambda_1}{4\lambda_3} V_2 + \frac{4\lambda_3}{\lambda_1 \lambda_2} \|P_L \alpha_2\|^2 \qquad (A.14)$$

Introducing (A.14) in (A.11) for i = 2 we obtain

$$\dot{V}_2 \le -\frac{\lambda_1}{4\lambda_3} V_2 + \frac{4\lambda_3}{\lambda_1 \lambda_2} \|P_L \alpha_2\|^2 \qquad (A.15)$$

Since $\alpha_2 \in L_2$, integrating both sides of the above equation it follows that $V_2 \in L_1$. From (A.15) we also get $\dot{V}_2 \in L_1$. Therefore we conclude that $V_2 \to 0$, $z_2 \to 0$ and x_e in (A.1) converges to zero.

♦♦♦

APPENDIX B.

Proof of the theorem.

1). In view of Lemma 4, x is bounded and converges to x_a, i.e. (see (32) and (48))

$$\lim_{t\to\infty} ;\ )(y_f^{(i)} - y_{fa}^{(i)}) \to 0 \qquad \text{for } i = 0,\dots,n-1$$

$$\lim_{t\to\infty} (u_f^{(j)} - u_{fa}^{(j)}) \to 0 \qquad \text{for } j = 0,\dots,m-1$$

(B.1)

$$\lim_{t\to\infty} [(Bu_f)^{(k)} - (Bu_{fa})^{(k)}] \to 0 \qquad \text{for } k = 0,\dots,n-m-1$$

Noting that

$$Bu_f = \bar{b}_0 u_f^{(m)} + \dots + \bar{b}_m u_f \qquad (B.2)$$

is bounded (see (B.1)) and $\bar{b}_0$ is different from zero (see Lemma 2) it follows that also $u_f^{(m)}$ is bounded. This implies boundedness of ϕ (see 10) and $y_f^{(n)}$(see (8)). Derivating (B.2) we obtain

$$(Bu_f)^{(1)} = \dot{\bar{b}}_0 u_f^{(m)} + \bar{b}_0 u_f^{(m+1)} + \dots + \dot{\bar{b}}_m u_f + \bar{b}_m u_f^{(1)} \qquad (B.3)$$

In order to prove that $u_f^{(m+1)}$ is bounded in the above equation we require the first derivatives of the modified parameter estimates to be bounded. Recall that the discontinuities in β occur only a finite number of times and that after a finite time β is either equal to zero or equal to $p/\|p\|$ (see Lemma 2). In order to be able to properly define the derivative of the modified estimates and since we know that there exists no finite escape time (see Lemma 3), we will start our analysis, without loss of generality, after the discontinuities in β have already disappeared.
Derivating (17) we have

$$\dot{\bar{\theta}} = \dot{\theta} + \dot{P}\beta + P\dot{\beta} \qquad (B.4)$$

which can be written as (see (12), (13) and (A.6))

$$\dot{\bar{\theta}} = -P\phi\phi^T\tilde{\theta} - P\phi\phi^T P\beta + P\dot{\beta} \qquad (B.5)$$

From fig. 1 β has two different values in the limit, i.e. $\beta = 0$ and $\beta = p/\|p\|$. For $\beta = 0$ we have $\dot{\beta} = 0$. For $\beta = p/\|p\|$, $\dot{\beta}$ is

$$\dot{\beta} = \frac{\dot{p}}{\|p\|} - \frac{pp^T\dot{p}}{\|p\|^3} \qquad (B.6)$$

From (13) and the above and since p is bounded

$$\|\dot{\beta}\| \le k_1 \|[1\ 0\dots0]\ P\phi\phi^T P\| \qquad (B.7)$$

where k_1 is a positive constant. Therefore $\dot{\beta}$ is bounded. From the above and since P, ϕ, $\tilde{\theta}$ and β are bounded we conclude that $\dot{\bar{\theta}}$ in (B.5) is also bounded. Therefore $u_f^{(m+1)}$ in (B.3) is also bounded since $\bar{\theta}$, $Bu_f \in L_\infty$ and $u_f^{(j)}$ is bounded for j = 0,...,m.

Boundedness of $u_f^{(m+1)}$ and $y_f^{(n)}$ implies boundedness of $\dot{\phi}$ (see (10)). Using property 5 in Lemma 1 it follows that e and $P\phi$ converge to zero. Therefore x_e in (50) (see also (21) and (33)) converges to zero which in turn implies that (B.1) holds for i=n, j=m and k=n-m. Therefore y (see (2.a)) is bounded.
In order to prove that u is bounded we will prove first that $u_f^{(j)} \in L_\infty$ for j = m+2,...,n. Derivating (B.3) we obtain

$$(Bu_f)^{(2)} = \bar{b}_0^{(2)} u_f^{(m)} + 2\bar{b}_0^{(1)} u_f^{(m+1)} + \bar{b}_0 u_f^{(m+2)} + \dots$$
$$\dots + \bar{b}_m^{(2)} u_f + 2\bar{b}_m^{(1)} u_f^{(1)} + \bar{b}_m u_f^{(2)} \qquad (B.8)$$

Since $(Bu_f)^{(2)}$, $\bar{\theta}$, $\dot{\bar{\theta}}$ and $u_f^{(j)}$ for j = 0,...,m+1 are all bounded we must only show that the second derivatives of the modified parameter estimates are bounded to conclude boundedness of $u_f^{(m+2)}$.

Derivating (B.5) we have that $\bar{\theta}^{(2)}$ can be expressed as

$$\bar{\theta}^{(2)} = f_\theta(\tilde{\theta}, P, \phi, \dot{\phi}, \beta, \dot{\beta}, \ddot{\beta}) \qquad (B.9)$$

where f_θ is a continuous function of its arguments. The derivative of $\dot{\beta}$ in (B.6) can be written as

$$\overline{\beta}^{(2)} = f_\beta(P, \phi, \dot{\phi}) \tag{B.10}$$

where f_β is also a continuous function of its arguments. From the above and (B.6) we conclude that (B.9) can be expressed as

$$\overline{\theta}^{(2)} = f_\theta(\tilde{\theta}, P, \phi, \dot{\phi}) \tag{B.11}$$

Then boundedness of $\tilde{\theta}$, P, ϕ and $\dot{\phi}$ implies $\overline{\theta}^{(2)} \in L_\infty$. Therefore $u^{(m+2)}$ in (B.8) is also bounded. Since $\dot{\beta} \in L_\infty$ we conclude also that $y_f^{(n+1)}$ in (8) is bounded which in turn implies that $\phi^{(2)} \in L_\infty$. Repeating this procedure for j=m+3,…,n and noting that the (j-m)th derivative of (B.11) can be written as

$$\overline{\theta}^{(j-m)} = f_\theta(\tilde{\theta}, P, \phi,\ldots, \phi^{(j-m-1)}) \tag{B.12}$$

we obtain $u_f^{(j)} \in L_\infty$ for j = 0,…,n. From the above and (2.b) we finally conclude that u is bounded.

2). From (37),(44) and since $Cy^M = r^M$ we have

$$C(y_a - y^M) = S_\infty w_{2a} \tag{B.3}$$

from which we conclude $y_a - y^M \rightarrow 0$ since w_{2a} is an exponentially decaying term. We have proved above that $y_f^{(i)} \rightarrow y_{fa}^{(i)}$ for i= 1,…,n , then (see (2) and (37))

$$y - y_a = F(y_f - y_{fa}) \rightarrow 0 \tag{B.4}$$

We finally conclude that $y - y^M \rightarrow 0$.

♦♦♦

Proceedings of the 31st Conference
on Decision and Control
Tucson, Arizona • December 1992

ROBUSTNESS OF A SIMPLE INDIRECT CONTINUOUS TIME ADAPTIVE CONTROLLER IN THE PRESENCE OF BOUNDED DISTURBANCES

Changyun Wen

School of Electrical and Electronic Engineering
Nanyang Technological University
Nanyang Ave. 2263, SINGAPORE

Abstract

This paper reports a new result on the robustness of an indirect continuous adaptive controller just using estimator projection. It is shown that for arbitrarily bounded disturbances, global stability is ensured without any assumption on signals. Moreover, the robust adaptive controller can still retain the properties that earlier unmodified conventional adaptive controllers have when there is no disturbance. The apriori knowledge for the implementation of the simple adaptive controller is the (large) ranges of unknown parameters in the nominal system model.

1 Introduction

Stability of adaptive control for a linear time invariant system with unknown parameters was established at the end of seventies or in the early of eighties [1] - [4]. It was realized that, when bounded external disturbances are present, the unmodified conventional adaptive control algorithms used in the stability proof could result in the parameter estimation error growing in an unbounded solution [5] and [4]. This is not surprising since the adaptive system is nonlinear. So far, although some efforts have been made to develop robust adaptive control algorithms for systems with both bounded and unbounded modelling errors [5] - [14], robust continuous time adaptive control in the presence of bounded external disturbances still requires either priori information regarding the bound on the disturbance for the implementation of the algorithm ([4] and [5]) in order to guarantee boundedness of signals or persistent excitation of reference signals in order for the system to have an unique equilibrium point at the origin when external disturbances disappear [12]. For the adaptive algorithm in [4] and [5], boundedness was only shown for a restricted class of disturbances the amplitude of which satisfies the bound for a deadzone built into the adaptive controller. Furthermore, most schemes concerning robustness against bounded disturbances are in the class of direct model reference type [12].

Generally speaking, indirect adaptive control has more flexibility of choosing parameter estimation algorithms and control laws. However, the analysis is more complicated. In this paper, we will examine the robustness properties of a simple indirect continuous time adaptive controller. The only modification on the basis of earlier conventional adaptive algorithms is the use of a projection operation to constrain parameter estimates in a convex compact region. It is shown that for arbitrarily bounded disturbances, global stability of the adaptive system in the sense of bounded input bounded state stable is guaranteed. Moreover, in contrast with some other robust algorithms as in [4], [5] and [12], the above robust adaptive controller can still retain the properties of earlier unmodified conventional adaptive controllers [4] without any restriction on the signals in the closed loop such as persistence of excitation, when no disturbance appears. The apriori knowledge for the implementation of the simple adaptive controller is the (large) ranges of unknown parameters in the nominal system model.

*This work was supported by NTU under the Applied Research Project Grant RP 23/92

2 Plant Model

Suppose the class of systems considered can be mathematically described by

$$y(t) = \frac{B(D)}{A(D)}u(t) + d(t) \tag{1}$$

where $y(t)$ and $u(t)$ are system output, input, $d(t)$ denotes external disturbances, $A(D)$ and $B(D)$ are polynomials of differential operator D which satisfies, i.e.

$$\begin{aligned} A(D) &= D^n + a_{n-1}D^{n-1} + \ldots + a_0 \\ B(D) &= b_m D^m + b_{m-1}D^{m-1} + \ldots + b_0, \end{aligned}$$

and $m < n$.

Without loss of generality, we take $m = n - 1$. If $m < n - 1$, then $b_{m+1} = \ldots = b_{n-1} = 0$ in this case.

For the plant given in (1) we make the following assumptions concerning our priori knowledge on the plant parameters.

Assumption 2.1
The coefficients of $A(D)$ and $B(D)$ are inside a known (large) compact convex region $\mathcal{C}$ which has the property that polynomials $\hat{A}(D)$ and $\hat{B}(D)$ induced by an arbitrary (nonzero) parameter vector in the space $\mathcal{C}$ are uniformly coprime.

Comments 2.1:

- The coprimeness of $\hat{A}(D)$ and $\hat{B}(D)$ is only required when pole assignment control law synthesis is used. It can be removed when some other control strategies [4] are employed;
- The size of $\mathcal{C}$ is not necessary small.

For disturbance $d(t)$, we have the following boundedness assumption.

Assumption 2.2:
Disturbance $d(t)$ is bounded.

CH3229-2/92/0000-2762$1.00 © 1992 IEEE

Comments 2.2:
The bounds for $d(t)$ are not required for the implementation of the adaptive controller proposed later. Thus the adaptive controller can withstand arbitrarily bounded external disturbances for a globally stable system. This is contrast with some other adaptive algorithms (see [4] and [5]) which need a bound of the external disturbances to design an adaptive controller and thus global stability were only shown for a restricted class of external disturbances with magnitudes satisfying the bounds built in the deadzones of the adaptive controllers

We now further process model (1) by transforming it to the following form

$$Ay = Bu + Ad \tag{2}$$

A filter $\frac{1}{F(D)}$ is introduced to avoid differentiation of the input and output signals used for feedback as well as disturbances which may be discontinuous and/or include noise. The following filtered variables are defined:

$$\begin{aligned} y_f(t) &= \frac{1}{F(D)}y(t) \\ u_f(t) &= \frac{1}{F(D)}u(t) \\ d_f(t) &= \frac{A(D)}{F(D)}d(t) \end{aligned}$$

where $F(D)$ is a monic Hurwitz polynomial with $\partial_F = \partial_A = n$. Then Equation (2) becomes

$$Ay_f(t) = Bu_f(t) + d_f(t) \tag{3}$$

From the stability of $\frac{A}{F}$, we can show that for an arbitrary bounded disturbance, there exists a constant d such that

$$|d_f(t)| \leq d \tag{4}$$

Note that constant d is not required to be known, but will be used as an auxiliary element in the stability analysis given in later sections.

Now from (3), we have

$$\begin{aligned} y &= Fy_f \\ &= (F-A)y_f + Bu_f + d_f \\ &= \phi^T\theta_* + d_f \end{aligned} \tag{5}$$

where

$$\begin{aligned} \phi^T(t) &= [D^{n-1}y_f(t), ..., y_f, D^{n-1}u_f(t), ...u_f(t)] \\ \theta_*^T &= [f_{n-1} - a_{n-1}, ..., f_0 - a_0, b_{n-1}, ..., b_0] \end{aligned}$$

and $f_i, i = 0, ..., n-1$ are the coefficients of $F(D)$.

The control problem is now formulated as the design of an adaptive controller for plant (1) subjected to Assumption 2.1 and 2.2 so that all signals in the closed loop system are bounded for arbitrarily bounded trajectory $y^*(t)$ and bounded disturbances and output $y(t)$ follows the given trajectory $y^*(t)$ as close as possible. In addition, when disturbances disappear the adaptive controller should retain the properties of earlier unmodified conventional adaptive controllers without any additional requirement.

3 Adaptive Control Scheme

In this section, an indirect adaptive control scheme is proposed. The adaptive controller consists of two modules: an parameter estimator and a linear controller designed based on *Certainty Equivalence Principle* [4]. We now present these two modules separately.

3.1 Parameter Estimator

The following estimation algorithm in [15] and [4] called normalized gradient algorithm with projection is introduced to the estimator.

$$\dot{\hat{\theta}}(t) = \mathcal{P}\{\frac{\phi(t)e(t)}{1+\phi^T(t)\phi(t)}\} \tag{6}$$

where $e(t)$ is the prediction error defined as

$$e(t) = y(t) - \phi^T(t)\hat{\theta}(t), \tag{7}$$

$\mathcal{P}\{.\}$ denotes a projection operation. It was shown in [15] that such an operation can ensure the estimated parameter vector $\hat{\theta}(t) \in \mathcal{C}$ for all t if $\hat{\theta}(0) \in \mathcal{C}$ (for more details on the projection operation, see [15]).

About the estimator in (6) and (7), we have

Lemma 3.1
Suppose M_0 is a positive constant s.t. $d/M_0 \leq \delta$. The estimator (6) and (7), applied to systems given in (1), has the following properties:

1. If $\|\phi(t)\| > M_0$, for all $t > t_0$ and $\|\phi(t_0)\| = M_0$ then

   (a)

   $$\begin{aligned} \bar{e}(t) &= \frac{e(t)}{(1+\phi^T(t)\phi(t))^{1/2}} \\ &\leq k_\theta + \delta \qquad \text{for } t \geq t_0 \end{aligned} \tag{8}$$

   and

   (b)

   $$\int_{t_0}^{t} \bar{e}^2(\tau)d\tau \leq k + \alpha_1(t-t_0) \tag{9}$$

   where

   $$k = \frac{1}{2}k_\theta^2 \tag{10}$$

   $$\alpha_1 = (k_\theta + \delta)\delta \tag{11}$$

2.

   $$\|\dot{\hat{\theta}}(t)\| \leq |\bar{e}(t)| \tag{12}$$

Proof:

1. (a) From (5) and (7), we get

   $$e(t) = -\tilde{\theta}^T(t)\phi(t) + d_f(t) \tag{13}$$

   where

   $$\tilde{\theta} = \hat{\theta} - \theta_*$$

   Then

   $$|e(t)| \leq k_\theta\|\phi(t)\| + d$$

   Thus

   $$\begin{aligned} |\bar{e}(t)| &\leq \frac{k_\theta\|\phi(t)\| + d}{(1+\phi^T(t)\phi(t))^{1/2}} \\ &\leq k_\theta + \delta \\ &\qquad \text{for } t \geq t_0 \end{aligned}$$

   (b) We consider the function $v(t) = \frac{1}{2}\tilde{\theta}^T(t)\tilde{\theta}(t)$

   Then (6) and (13) yield

   $$\begin{aligned} \dot{v}(t) &\leq -\bar{e}^2(t) + \frac{|d_f(t)||e(t)|}{1+\phi^T(t)\phi(t)} \\ &\leq -\bar{e}^2 + \frac{d}{(1+\phi^T(t)\phi(t))^{1/2}}\bar{e}(t) \end{aligned}$$

   where equality holds save for those times when the projection operator $\mathcal{P}$ is invoked [15]. Under the assumption of the lemma, we have

$$\int_{t_0}^{t} \tilde{e}^2(\tau)d\tau \le -\int_{t_0}^{t} \dot{v}d\tau + \alpha_1 \int_{t_0}^{t} d\tau$$
$$\le k + \alpha_1(t - t_0)$$

2.

$$\|\dot{\hat{\theta}}(t)\| \le \frac{\|\phi(t)\|\,|e(t)|}{1 + \|\phi(t)\|^2}$$
$$\le |\tilde{e}(t)|$$

Again the first inequality is an equality save for when the projection is invoked.

□

Comments 3.1:

1. α_1 in (11) can be made small by a sufficiently large number M_0. M_0 is used here for the purpose of stability analysis only. It is not a design parameter.

2. The least square version is more commonly used in practical algorithms [4]. Similar properties for this estimator can be derived by defining a different Lyapunov type function $v(t)$ as in [15], but the analysis is more tedious.

3. When disturbance disappears, $d_f = 0$ and so $d = 0$. In this case, $\delta = 0$, then properties given in Lemma 3.1 are exactly the same as those of unmodified conventional estimators [4]. As for indirect adaptive control, an adaptive controller can be decomposed into 'modules' of parameter estimation and control law synthesis. Our modification is only on the estimator part and thus our adaptive controller will retain the results of unmodified conventional adaptive controllers without any additional requirement in the absence of disturbances.

3.2 Control Law Synthesis

One of the advantages of indirect adaptive control is that one can have flexibility of choosing control laws. For the module of controller synthesis in the adaptive controller, there are many possible choices of schemes such as model reference control (for minimum phase plants), classic three term control law, pole assignment, linear quadratic optimal control and so on [4]. Here we use a pole assignment strategy for analysis. The control $u(t)$ is given by

$$\hat{L}(D)u(t) = \hat{P}(D)(y^*(t) - y(t)) \quad (14)$$

or

$$\hat{L}(D)u_f(t) = \hat{P}(D)(y_f^*(t) - y_f(t)) \quad (15)$$

where $y^*(t)$ is the setpoint and $y_f^* = \frac{1}{F}y^*$, $\hat{L}$ and $\hat{P}$ are polynomials in D of the form

$$\hat{L}(D) = D^{n_l} + l_{n_l-1}D^{n_l-1} + ... + l_0$$
$$\hat{P}(D) = p_{n_p}D^{n_p} + p_{n_p-1}D^{n_p-1} + ... + p_0$$

and determined from the following Diophantine Equation.

$$\hat{A}(t)\hat{L}(t) + \hat{B}(t)\hat{P}(t) = A^* \quad (16)$$

where A^* is a monic polynomial in D of degree $2n$ having its zeros to be the required closed loop poles and chosen by users. The degrees of $\hat{L}$ and $\hat{P}$ are, respectively, $n_l = n$ and $n_p = n - 1$. Clearly (14) yields a strictly proper control law and can be implemented as

$$u(t) = (F - \hat{L})u_f - \hat{P}(y_f - y_f^*) \quad (17)$$

4 Stability Analysis

In this section, we will analyze the adaptive system presented in the last two sections. The equation describing the closed loop system can be obtained by combining (15) with (7).

$$D\phi(t) = \hat{A}_c\phi(t) + b_1e(t) + b_2r(t) \quad (18)$$

where

$$b_1^T = [1, 0, ..., 0]$$
$$b_1^T = [0, ...0, 1, ..., 0]$$

$$r(t) = \frac{\hat{P}}{F}y^*(t) \quad (19)$$

$$\hat{A}_c = \begin{bmatrix} -\hat{a}_{n-1} & \cdots & -\hat{a}_1 & -\hat{a}_0 & \hat{b}_{n-1} & \cdots & \hat{b}_1 & \hat{b}_0 \\ 1 & \cdots & 0 & 0 & 0 & \cdots & 0 & 0 \\ \vdots & \ddots & \vdots & \vdots & \vdots & \vdots & \vdots & \vdots \\ 0 & \cdots & 1 & 0 & 0 & \cdots & 0 & 0 \\ -\hat{p}_{n-1} & \cdots\cdots & -\hat{p}_1 & -\hat{p}_0 & -\hat{l}_{n-1} & \cdots & -\hat{l}_1 & -\hat{l}_0 \\ 0 & \cdots & 0 & 0 & 1 & \cdots & 0 & 0 \\ \vdots & \vdots & \vdots & \vdots & \vdots & \ddots & \vdots & \vdots \\ 0 & \cdots & 0 & 0 & 0 & \cdots & 1 & 0 \end{bmatrix} \quad (20)$$

Since $\hat{P}(t)$ is bounded, then $r(t)| \le c_p|y^*(t)|$ where c_p is a constant.

From Lemma 3.1, we can obtain

Lemma 4.1 The matrix $\bar{A}_c(t)$ defined in (20) satisfies

1. $\bar{A}_c(t)$ is bounded $\forall t$.

2. If $\|\phi(\tau)\| > M_0, \tau > t_0$ and $\|\phi(t_0)\| = M_0$, then
$$\int_{t_0}^{t} \|\bar{A}_c(\tau)\|^2 d\tau \le \bar{k}(k + \alpha_1(t - t_0)) \quad \text{for } t \ge t_0$$
where $\bar{k}$ is a constant.

3. The eigenvalues of $\bar{A}_c(t)$ are the zeros of $A^* \forall t$.

Proof:

1. This follows from Assumption 2.1.

2. From Assumption 2.1 and Lemma 3.1, we can obtain the result.

3. This is easy to verify from (20).

□

From Lemma 6.2 in [11] and Lemma 4.1 above, we can show that $\exists c > 0, \sigma > 0$ such that the transition matrix of the homogeneous part of (18), denoted $\Phi(t, \tau)$, satisfies

$$\|\Phi(t, \tau)\| \le ce^{-\sigma(t-\tau)} \quad \text{for } t \ge \tau \ge t_0 \quad (21)$$

if $\|\phi(\tau)\| > M_0, \tau > t_0$, $\|\phi(t_0)\| = M_0$ and $\alpha_1 \le \bar{\alpha}_1^*$ where $\bar{\alpha}_1^*$ is a sufficiently small number. From (11), the bound $\bar{\alpha}_1{}^*$ is equivalent to a bound $\bar{\delta}^*$ where $\bar{\delta}^*$ is just a sufficiently small number.

Now we are in the position to present our stability result.

Theorem 4.1 Consider the adaptive system consisting of plant (1), estimator (6) and (7) and controller (14) to (16). Under Assumption 2.1 and 2.2, then the system is globally stable in the sense that $y(t)$ and $u(t)$ (and , hence all states) are bounded $\forall t$ for all finite initial states, any bounded y^* and arbitrarily bounded external disturbances.

Proof:
Notice that for any bounded initial conditions $\phi(0)$, set points y^* and disturbances $d(t)$, there always exists a number M_0 such that $\|\phi(0)\| \leq M_0, \|r(t)\|_\infty \leq M_0$ and $\frac{d}{M_0} \leq \delta$ for a sufficiently small δ, where $r(t)$ is given by (19).

In order to apply Lemma 3.1 and the exponential stability property of $\bar{A}_c(t)$ in the closed loop equation (18), and also inspired by the techniques used in [13] and [14], we choose an intermediate sufficiently large constant M_0 satisfying the above constraint and divide the time interval $\Re_+$ into two subsequences

$$\begin{aligned} \Re_1 &:= \{t \in \Re_+ | \|\phi(t)\| > M_0\} \\ \Re_2 &:= \{t \in \Re_+ | \|\phi(t)\| \leq M_0\} \end{aligned}$$

Clearly, the result is proved if we can show that $\|\phi(t)\|$ is bounded for $t \in \Re_1$. To do this, we choose a t_0 so that $\|\phi(t_0)\| = M_0$ and $\tau \in \Re_1$ for $\tau > t_0$. Also constrain the initial time (say 0) to be in $\Re_2$.

Now we outline the proof of the boundedness of $\phi(t)$ for $t \in \Re_1$ as follows.

The general solution of (18) is

$$\phi(t) = \Phi(t,t_0)\phi(t_0) + \int_{t_0}^{t} \Phi(t,\tau)B_1(e(\tau)+r(\tau))d\tau \quad (22)$$

Using (21) and inequality

$$|e(\tau)| \leq (1+\|\phi(\tau)\|)|\bar{e}(\tau)|$$

we have

$$\begin{aligned} \|\phi(t)\| \leq\ & ce^{-\sigma(t-t_0)}M_0 + \int_{t_0}^{t} ce^{-\sigma(t-\tau)}[|\bar{e}(\tau)|\|\phi(\tau)\| \\ & +|\bar{e}(\tau)| + M_0]d\tau \end{aligned} \quad (23)$$

In the following, all $c_i, i = 1, 2, \ldots$ denote constants without further clarification.

Squaring both sides of (23), applying the Schwartz inequality and noting that $e^{-\sigma(t-t_0)} \leq 1$ for $t \geq t_0$, we get

$$\begin{aligned} \|\phi(t)\|^2 \leq\ & c_1e^{-\sigma(t-t_0)}M_0^2 + c_2\int_{t_0}^{t} e^{-\sigma(t-\tau)}[|\bar{e}(\tau)|^2\|\phi(\tau)\|^2 \\ & +|\bar{e}(\tau)|^2 + M_0^2]d\tau \end{aligned} \quad (24)$$

Multiplying both sides of (24) by $e^{\sigma t}$ gives

$$e^{\sigma t}\|\phi(t)\|^2 \leq s^2(t) + c_2\int_{t_0}^{t} e^{\sigma\tau}\|\phi(\tau)\|^2|\bar{e}(t)|^2 d\tau \quad (25)$$

where

$$s^2(t) = c_1e^{\sigma t_0}M_0^2 + c_2\int_{t_0}^{t} e^{\sigma\tau}(|\bar{e}(\tau)|^2 + M_0^2) \quad (26)$$

Then the Grownwall Lemma [16] can be applied to (25) to yield

$$\begin{aligned} \|\phi(t)\|^2 &\leq e^{-\sigma t}s^2(t) + c_2\int_{t_0}^{t} e^{-\sigma t}\bar{e}^2(\tau)s^2(\tau)e^{\int_\tau^t c_2\bar{e}^2(\tau_1)d\tau_1}d\tau \\ &\leq e^{-\sigma t}s^2(t) + c_2\int_{t_0}^{t} e^{-\sigma\tau}\bar{e}^2(\tau)s^2(\tau)c_3e^{-\sigma_c(t-\tau)}d\tau \quad (27) \\ & \quad \text{using (9)} \end{aligned}$$

where $\sigma_c = \sigma - c_2\alpha_1$. Then from (11), it is clear that there exist a constant $\bar{\bar{\delta}}^*$ which are small enough to guarantee that $\sigma - c_4\alpha_1 > 0$ for $\delta \leq \bar{\bar{\delta}}^*$. Let $\delta^* = \min\{\bar{\delta}^*, \bar{\bar{\delta}}^*\}$.

Now consider the terms $e^{-\sigma t}s^2(t)$ and $e^{-\sigma\tau}\bar{e}^2(\tau)s^2(\tau)$ appearing in (27).

$$\begin{aligned} e^{-\sigma t}s^2(t) &\leq c_1e^{-\sigma(t-t_0)}M_0^2 + c_2\int_{t_0}^{t} e^{-\sigma(t-\tau}((k_\theta+\delta^*)^2 + M_0^2) \\ &\leq c_4M_0^2 + c_5 \qquad \text{for } \delta \leq \delta^* \quad (28) \\ e^{-\sigma\tau}s^2(\tau)\bar{e}^2(\tau) &\leq (c_4M_0^2 + c_5)(k_\theta+\delta^*)^2 \\ &\leq c_6(c_4M_0^2 + c_5) \quad (29) \end{aligned}$$

Substituting (28) and (29) into (27), we get

$$\begin{aligned} \|\phi(t)\|^2 &\leq (c_4M_0^2 + c_5) + c_7\int_{t_0}^{t}(c_4M_0^2 + c_5)e^{-\sigma_c(t-\tau)}d\tau \\ &\leq c_8M_0^2 + c_9 \end{aligned} \quad (30)$$

Thus $\phi(t)$ is bounded by M_0 in $\Re_1$ if $\delta \leq \delta^*$. Note that M_0 only depends on initial values of $\phi(0)$, bounds of y^* and d_f (The dependence of M_0 on d_f is to make $\delta \leq \delta^*$ satisfied for a sufficiently small constant δ^*). As all of those are bounnded, so is M_0 and $\phi(t)$. □

Comments 4.1

1. For a given system, there always exists a sufficiently large constant M_0 such that $\|\phi(0)\| \leq M_0, \|r(t)\|_\infty \leq M_0$ and $\|d_f\|_\infty \leq \delta^* M_0$ for any bounded initial condition, set point and disturbance $d(t)$, where $\delta^* = \min\{\bar{\delta}^*, \bar{\bar{\delta}}^*\}, \bar{\delta}^*$ and $\bar{\bar{\delta}}^*$ are sufficiently small numbers to ensure (21) and (27) satisfied.
2. As stated earlier, we do not need to know M_0 while being aware of its existence and role as an auxiliary variable in proving our result. The established bounds for $\phi(t)$ depends on M_0 which essentially implies that $\phi(t)$ depends on the initial conditions, bounds on the setpoints and disturbances.

5 Conclusion

In this paper, we studied a simple indirect continuous time adaptive control algorithm including of a gradient estimator, subject to parameter projection as the only modification in the estimator. Pole assignment control law is used to design the controller. The only apriori information required for the implementation of the algorithm is the range that each unknown parameter of the plant lies in, which is quite reasonable.

It has been shown that for arbitrary bounded disturbances, the above adaptive controller can globally stabilise the unknown plant without any requirement on the signals in the closed loop such as persistence of excitation. Also when external disturbances disappear, the adaptive controller will give the properties of those unmodified conventional adaptive controllers without any additional assumption.

Research is still in progress to study the robustness of the simple adaptive controller mentioned above against unbounded modelling errors such as higher order unmodelled dynamics. It was proved that the discrete time version of the adaptive controller does have the robustness properties of counteracting the effects of unbounded modelling errors in [14]. However due to the induction techniques used in [14], we are unable to show the robustness of the adaptive controller in the presence of unbounded modelling errors for continuous time systems though we believe that there is such a result.

6 References

[1] G.C. Goodwin, P.J. Ramadge and P.E. Caines, "Discrete time multivariable adaptive control", IEEE Transactions on Auto. Control, vol 25, pp449-456, 1980.

[2] A.S. Morse, "Global stability of parameter adaptive systems", IEEE Transactions on Auto. Control, vol 25, pp433-439, 1980.

[3] K. S. Narendra, Y.H. Lin and L.S. Valavani, "Stable adaptive controller design - part II: proof of stability", IEEE Transactions on Auto. Control, vol 25, pp440-448, 1980.

[4] G.C. Goodwin and K.S. Sin, Adaptive Filtering Prediction and Control, Prentice-Hall, 1984.

[5] B. Egardt, Stability of Adaptive Controllers, Springer-Verlag, New York, 1979.

[6] P.A. Ioannou and K.S. Tsakalis, "A robust direct adaptive controller", IEEE Trans. Auto. Control, vol. 31, pp1033 - 1043, 1986.

[7] G. Kreisselmeier and B.D.O. Anderson, "Robust model reference adaptive control", IEEE Trans. Auto. Control, vol. 31, pp127 - 133, 1986.

[8] L. Praly, "Robustness of indirect adaptive control based on pole-placement design". in Proc. of IFAC Workshop on Adaptive Control, San Francisco, 1983, (preliminary version).

[9] K.J. Astrom, "Adaptive feedback control", Proc. of IEEE, vol75, pp185-217, 1987.

[10] R. Ortega and T. Yu, "Theoretical results on robustness of direct adaptive controllers: a survey", in Proc. of 10th World Congress on Automatic Control, vol. 10, pp1-15, Munich, West Germany, 1987.

[11] R.H. Middleton, G.C. Goodwin, D.J. Hill and D.Q. Mayne, "Design issues in adaptive control", IEEE Trans. Auto. Control, vol. 33, pp50-58, 1988.

[12] K.S. Narendra and A.M. Annaswamy, Stable Adaptive Systems, Prentice-Hall, 1989.

[13] C. Wen and D.J. Hill, "Robustness of adaptive control without deadzones, data normalization or persistence of excitation", Automatica, vol.25, pp943-947, 1989.

[14] C. Wen and D.J. Hill, "Global boundedness of discrete-time adaptive control just using estimator projection", Automatica, November, 1992.

[15] G.C. Goodwin and D.Q. Mayne, "A parameter estimation perspective of continuous time model reference adaptive control", Automatica, September 1986.

[16] C.A. Desoer and M. Vidyasagar, Feedback Systems: Input-Output Properties, Academic Press, New York, 1975.

Proceedings of the 31st Conference
on Decision and Control
Tucson, Arizona • December 1992

Continuous Input-Output Robust Tracking Control of SISO Continuous-Time Systems with Fast Time-Varying Parameters: A Model Reference Approach

Zhihua Qu
Department of Electrical Engineering
University of Central Florida
Orlando, FL 32826, U.S.A.

J.F.Dorsey
School of Electrical Engineering
Georgia Institute of Technology
Atlanta, GA 30332, U.S.A.

Darren M. Dawson
Department of Electrical and Computer Engineering
Clemson University
Clemson, SC 29634, U.S.A.

Abstract

In this paper the output tracking problem of time-varying systems is investigated. A system under consideration contains time-varying parameters which may change arbitrarily fast and is subjected to nonlinear disturbances. A robust controller requiring only input-output measurement is proposed. The control guarantees exponential stability or stability of uniform ultimate boundedness which can be made through choosing a design constant to be arbitrarily close to asymptotic stability. The resulting control is continuous, uniformly bounded, and designed by a recursive mapping procedure. The only information of unknown systems required by the robust control approach are the bounding function of nonlinear disturbances and the bounds on the parameters and their derivatives, which represent the distinct features of the proposed result. That is, time-varying parameters are not restricted to be either slow time-varying or of known structure.

1 Introduction

Among different analysis and control design techniques, Lyapunov's second method has its dominant role since it can be used for general nonlinear and time-varying systems. If the plant under consideration has fast time-varying parameters and is subject to unknown disturbance, a robust control designed using Lyapunov's direct method will be the natural choice. Robust control design characterized by pioneer work [2,4] is usually carried out in terms of state space model and often requires certain structural conditions on uncertainties, the matching conditions. Although robust control approach can be widely applied by its nature, it has two major limitations. First, the matching conditions [2,4] are more or less required, even though recent progress has been reported in [10,12,14,15,16] and the references therein. Second, most robust control schemes requires full-state feedback except for some special cases shown in [3] and the references therein, and state tracking may not necessarily imply output tracking since the output matrix may contain unknown parameters. As an example, asymptotic output tracking results reported in [18,19] require state feedback, are applicable only for linear time-invariant systems, and impose the restrictions that the disturbance is constant and that the reference signal is generated from unit step function by finite integration or differentiation.

The approach to overcome these limitations is to develop robust control design procedure based on input-output dynamic description rather than state space model. Such an idea was explored in [11] in which a general procedure for designing robust control, called model reference robust control, was introduced for basically time-invariant systems. The design procedure integrates model reference control scheme into robust control methodology. The intention of this paper to extend the idea to time-varying systems. The extension is shown to solve the long-standing problem, output tracking problem of fast time-varying systems under disturbances.

The idea of model reference control (MRC) has been widely studied in the literature of linear control theory, especially adaptive control of linear time-invariant systems [8,9,17,20]. In the past decade, extensive research effort has been made in order to extend model reference adaptive control (MRAC) to time-varying systems. It is only recently that some success has been achieved using MRAC approach. The most recent and important result was given by [24]. It was shown in [24] that linear time-varying systems can be stabilized under MRAC if the parameters are either slow time-varying or are unknown constant multiples of known fast-changing time functions. Although this result provides a MRC structure which guarantees exact model matching for general linear time-varying systems with known parameters, it (and its later result [25]) fails to remove the major assumption of parameters being slow time-varying. Also, it has been well documented that MRAC have performance degradation and even instability under bounded, additive disturbances. Furthermore, if there is any nonlinearity, stability results from MRAC method will be local since certainty equivalence principle is used. All these facts illustrate the need of finding a better alternative for controlling fast time-varying systems.

There has been some work on combining model reference control with other control design methods, such as variable structure control (VSC). It has been shown in [1] that a discontinuous VSC can guarantee asymptotic stability when the relative degree of the plant is one. However, VSC may present intensive chattering and can not guarantee existence of classical solution.

In this paper we consider analysis and design of I/O robust control for time-varying systems. We first investigate how to combine MRC of time-varying systems with robust control design. Although the MRC structure in [24] achieves exact model matching, it will be shown in this paper that the structure in [24] is not adequate for robust control design. This leads us to introduce a new MRC structure which will be integrated into a recursive mapping procedure to generate a robust controller. The controller is shown to have major advantages over MRAC and VSC schemes. The advantages include continuity of control law, conceptual simplicity, easier implementation, guaranteed robust stability and performance under bounded nonlinear disturbances, allowing arbitrarily fast time-varying parameters, straightforward stability proof. These properties indicate that the proposed robust control gives as first time a complete solution to output tracking of fast time-varying systems.

This paper is organized as follows. Section two contains the formulation of output tracking problem. Perfect tracking of linear time-varying under perfect knowledge is discussed in section three, in which the existing MRC structure is briefly reviewed and a new MRC structure is then proposed. In section four, examples are first discussed to show intuitively why robust control works better for time-varying systems than any other technique. In section five, input-output robust controller and its design procedure is developed.

2 Problem Formulation

To simplify our discussion, we shall restrict our attention in this paper to single-input single-output systems whose possible nonlinearities depend on the system output and which dynamics is described by the following input-output differential equation:

$$A_p(s,t)[y(t)] = b_{p0}(t)B_p(s,t)[u(t)+d(y,t)] \triangleq k_p(t)B_p(s,t)[u(t)+d(y,t)], \tag{1}$$

where $A_p(s,t)$ and $B_p(s,t)$ are linear time-varying (LTV), monic polynomial differential operators (PDOs) given by

$$A_p(s,t) = s^n + a_{p1}(t)s^{n-1} + \cdots + a_{pn}(t), \qquad B_p(s,t) = s^m + b_{p1}(t)s^{m-1} + \cdots + b_{pm}(t),$$

the symbol s denotes the differential operator d/dt, $m \le n-1$, and $d(y,t)$ denotes any bounded nonlinear time-varying uncertainties or disturbances. The rules and properties of PDOs are given in [21].

With respect to the model of the TV plant, we make the following assumptions:

A.1 The indices n and m are constant and exactly known, and the LTV PDOs $A_p(s,t)$ and $B_p(s,t)$ are right coprime [24]. Moreover, the sign of $k_p(t) \triangleq b_{p0}(t)$ is fixed and assumed without loss of any generality to be positive.

A.2 The parameters of the system, $a_{pi}(t)$ and $b_{pj}(t)$, $1 \le i \le n$ and $0 \le j \le m$, are unknown but continuously differentiable up to $(\max\{4n-2m-3, 5n-2m-5\})$-th order. The parameters may drift arbitrarily fast, but the values of the parameters and their derivatives of necessarily high orders belong to known compact sets in $\mathbf{R}$. For mathematical convenience, compact sets can be enlarged to be polygon, that is, there are known constants $\underline{a}_{pi}$, $\overline{a}_{pi}$, $\underline{b}_{pj}$, $\overline{b}_{pj}$, $\underline{\xi}_{aik}$, $\overline{\xi}_{aik}$, $\underline{\xi}_{bjk}$, and $\overline{\xi}_{bjk}$, such that $\forall 1 \le i \le n$, $0 \le j \le m$, $1 \le k \le \max\{3n-m-3, 4n-m-5\}$, and $t \ge 0$,

$$\underline{a}_{pi} \le a_{pi}(t) \le \overline{a}_{pi}, \qquad \underline{b}_{pj} \le b_{pj}(t) \le \overline{b}_{pj},$$

$$\underline{\xi}_{aik} \le \frac{d^k a_{pi}(t)}{dt^k} \le \overline{\xi}_{aik}, \qquad \underline{\xi}_{bjk} \le \frac{d^k b_{pj}(t)}{dt^k} \le \overline{\xi}_{bjk}.$$

where $\underline{b}_{p0} > 0$. The upper and lower bounds do not have to be constant.

A.3 The disturbance $d(y,t)$ is continuous and bounded in size by a known, well-defined, and $(n-m-1)$-times differentiable function $\rho(y,t)$:

$$\|d(y,t)\| \le \rho(y,t) \;\; \forall (y,t),$$

where $\|\cdot\|$ denotes standard Euclidean norm [22].

A.4 The PDO $B_p(s,t)$ is exponentially stable with rate no larger than $-\gamma_B$ for some $\gamma_B > 0$ [24].

Remark: In an upcoming paper [13], model reference robust control is developed for linear time-invariant system whose order is not known, which relative order is upper bounded, which parameters are not limited inside known compact sets, and which has significant unmodelled dynamics. The treatment therein can also be used for fast time-varying systems, that is, Assumption A.2 can be eliminated, Assumptions A.1 and A.4 can be relaxed. However, we choose to proceed our discussion without making such a complication in order to reveal basic MRRC design for time-varying systems. It should be noted that Assumption A.4 is necessary for tracking under any reference signal.

The control objective is to find a continuous feedback controller to guarantee output tracking under nonlinear uncertainties, while the controller requires only input and output measurements. The performance for output tracking is at least uniformly ultimately bounded stability, which can be made arbitrarily close to asymptotic stability, or exponential convergence, if achievable. Because the system under consideration is fast time-varying and contains nonlinear uncertainties, the prevailing approach is robust control methodology. In [11], a robust controller has been designed for time-invariant and some simple time-varying systems. We intend to extend the previous discussions of robust control design in [11] to general time-varying systems in the form of (1).

In order to specify the transient performance of output tracking, the model following control (MRC) formulation is integrated into robust control design. That is, robust control is to make the output of system (1) track the output of a reference model under any given uniformly bounded reference signal $r(t)$, in the presence of significant uncertainty $d(y,t)$ and under fast drifting of system parameters. The reference model is chosen to be a time-invariant linear system described by the transfer function

$$\frac{Y_m(s)}{R(s)} = b_{m0}\frac{s^{m_m} + b_{m1}s^{m_m-1} + \cdots + b_{mm_m}}{s^{n_m} + a_{m1}s^{n_m-1} + \cdots + a_{mn_m}} \triangleq k_m\frac{B_m(s)}{A_m(s)} \triangleq G_m(s), \tag{2}$$

where $r(t)$ represents a continuous and uniformly bounded reference input signal. It is assumed that both $A_m(s)$ and $B_m(s)$ be coprime and Hurwitz polynomials [5]. It is well known from linear system theory that perfect tracking is feasible only if $n_m - m_m \ge n - m$. Assuming $n_m - m_m = n - m$ does not lose any generality since, if otherwise, $G_m(s)$ in (2) can always be obtained by multiplying reference signal by a proper and stable transfer function. In the

This work is supported in part by U.S. National Science Foundation under grants MSS-9110034 and IRI-9111258.

CH3229-2/92/0000-2767$1.00 © 1992 IEEE

following discussion, it is assumed for simplicity that $n_m = n - m$ and $m_m = 0$, that is, $B_m(s) \equiv 1$.

In the next section, we begin robust control design with model matching under perfect knowledge. The existing results on model reference control of known, linear time-varying systems will first be reviewed, that is, the non-standard structure proposed in [24]. Then, a modification on the non-standard controller structure in [24] will be made in order to introduce a new controller structure. The new controller structure will be used in section five to give a complete solution to robust control of time-varying systems.

3 MRC of LTV Systems with Perfect Knowledge

In this section, it is assumed that there be no uncertainty, i.e., $d(y,t) = 0\ \forall(y,t)$. It is also assumed that perfect knowledge of the LTV plant be available. It is obvious from intuition that MRC of unknown systems is not solvable unless the MRC problem is first solved under perfect knowledge.

Under perfect knowledge, the standard MRC structure in [8,9,17,20] is shown to be sufficient for linear time-invariant systems. It was then realized ([23] and the references therein) that the standard MRC structure is not adequate for LTV systems. Recently, a non-standard MRC structure was proposed in [24], as shown in Figure 1, which can be used to achieve exact matching between the closed I/O description and the desired transfer function. The difference between the standard MRC structure and the MRC structure in [24] is that the auxiliary signals are defined differently. As in Figure 1, the two auxiliary signals w_1 and w_2 are defined by

$$\dot{w}_1 = A_0 w_1 + \theta_1(t) u \quad (3)$$
$$\dot{w}_2 = A_0 w_2 + \theta_2(t) y \quad (4)$$

where $w_1, w_2 \in \mathbf{R}^{(n-1)}$ are auxiliary state vectors, $A_0 \in \mathbf{R}^{(n-1)\times(n-1)}$ is a constant stable matrix with

$$\det[sI - A_0] \triangleq P(s).$$

The output matrix $B_0 \in \mathbf{R}^{(n-1)}$ is a constant vector such that the pair (A_0, B_0) is observable. The MRC law is given by

$$u(t) = k(t)r(t) + \theta_0(t)y + B_0^T w_1 + B_0^T w_2, \quad (5)$$

where $k(t), \theta_0(t) \in \mathbf{R}$, together with $\theta_1(t), \theta_2(t) \in \mathbf{R}^{(n-1)}$, are time-varying parameters in the controller.

It has been shown in [24] that the problem of exact I/O operator matching can be achieved under the MRC system in Figure 1. This result can be summarized by the following lemma.

Lemma 1: [24] *Consider the system represented by (1) whose parameters are known and satisfy Assumptions A.1, A.2, and A.4. Then, there exists bounded parameters in control (5), $\theta_0^*(t), \theta_1^*(t), \theta_2^*(t), k^*(t)$, so that the closed loop LTV plant in Figure 1 with control (5) is internally stable and its I/O operator $r(t) \to y$ is the same as the transfer function $G_m(s)$ of the reference model (2). Moreover, the control parameters are the solutions of the following PDO equations:*

$$N_1(s,t) = P(s) - N_3(s,t)B_p(s,t) \quad (6)$$
$$N_3(s,t)k_p^{-1}(t)A_p(s,t) - N_2(s,t) = P(s)k(t)A_m(s)/k_m, \quad (7)$$

where adj(R) denotes the adjoint matrix of an invertible matrix R, and

$$N_1(s,t) = B_0^T \mathrm{adj}(sI - A_0)\theta_1(t), \qquad N_2(s,t) = B_0^T \mathrm{adj}(sI - A_0)\theta_2(t) + P(s)\theta_0(t).$$

The step-by-step procedure of solving control parameters is the following: Let $k^*(t) = k_m/k_p(t)$; solve for monic PDOs $N_2(s,t)$ and $N_3(s,t)$ from equation (7); determine monic PDO $N_1(s,t)$ from equation (6); find control parameters from the definitions of $N_1(s,t)$ and $N_2(s,t)$. It is easy to see that control parameters contain the derivatives of the plant parameters of up to at most $(\max\{2n-2, 3n-4\})$-th order.

Although the structure in Figure 1 gives a complete solution to MRC problem of LTV systems under perfect knowledge, it will be shown section five that the structure is not adequate for designing robust control if the relative degree between $A(s,t)$ and $B(s,t)$ is greater than one. Such a difficulty is caused by the fact that PDOs are generally not commutative. Inspired by the MRC structure in Figure 1 which is improved for LTV systems from the standard MRC structure, we propose the new MRC structure shown in Figure 2 in which the auxiliary signals are redefined to be

$$\dot{w}_1 = A_0 w_1 + \alpha(s)\theta_1(t)\frac{1}{\alpha(s)}u \quad (8)$$
$$\dot{w}_2 = A_0 w_2 + \alpha(s)\theta_2(t)\frac{1}{\alpha(s)}y \quad (9)$$

where $\alpha(s)$ is any monic Hurwitz polynomial of order p, $w_1, w_2 \in \mathbf{R}^{(n-1+p)}$ are modified auxiliary state vectors, $A_0 \in \mathbf{R}^{(n-1+p)\times(n-1+p)}$ with $\det[sI - A_0] = P(s)$, and $B_0 \in \mathbf{R}^{(n-1+p)}$ is chosen such that (A_0, B_0) is observable. Similar to the discussions in [24], it can be shown that the MRC structure in Figure 2 also gives a solution to exact matching of I/O description between the closed loop plant and the reference system. Such a result is summarized by the following lemma whose proof is conceptually the same as that of Lemma 1 and therefore omitted.

Lemma 2: *Consider the system represented by (1) whose parameters are known and satisfy Assumptions A.1, A.2, and A.4. Let $\alpha(s)$ be an arbitrary but given Hurwitz polynomial of s. The closed system in Figure 2 is the plant under the control:*

$$u(t) = \alpha(s)k(t)\frac{1}{\alpha(s)}r(t) + \alpha(s)\theta_0(t)\frac{1}{\alpha(s)}y + B_0^T w_1 + B_0^T w_2, \quad (10)$$

where $w_1(t)$ and $w_2(t)$ are given in (8) and (9), and $k(t), \theta_0(t) \in \mathbf{R}$, $\theta_1(t), \theta_2(t) \in \mathbf{R}^{(n-1+p)}$, are time-varying parameters in the controller. Then, there exists bounded parameters in the control $\theta_0^*(t), \theta_1^*(t), \theta_2^*(t), k^*(t)$, so that the closed loop LTV plant under control (10) is internally stable and its I/O operator $r(t) \to y$ is the same as the transfer function $G_m(s)$ of the reference model (2). Moreover, the parameters in the controller are the solutions of the following PDO equations:

$$N_1(s,t) = P(s)\alpha(s) - N_3(s,t)B_p(s,t)\alpha(s) \quad (11)$$
$$N_3(s,t)k_p^{-1}(t)A_p(s,t)\alpha(s) - N_2(s,t)\alpha(s) = P(s)\alpha(s)k(t)A_m(s)/k_m, \quad (12)$$

where

$$N_1(s,t) = B_0^T \mathrm{adj}(sI - A_0)\alpha(s)\theta_1(t), \qquad N_2(s,t) = B_0^T \mathrm{adj}(sI - A_0)\alpha(s)\theta_2(t) + P(s)\alpha(s)\theta_0(t).$$

Controller parameters in Figure 2 can be determined by the same procedure as that following Lemma 1. It is easy to see that control parameters contain the derivatives of the plant parameters of up to at most $(\max\{2n+2p-2, 3n+2p-4\})$-th order, where p is the order of $\alpha(s)$ in s. Later, the MRC structure in Figure 2 will be shown to be better for robust control design than that in [24].

As in the LTI case [8,9,17,20], the discussions in this paper is based on the I/O operator (PDO or transfer function) method which inherently assumes zero initial conditions for all internal states of the system. Fortunately, this treatment does not imply that zero initial conditions are required since, for a globally internally stable system in which nonlinearities can be isolated as an additive term at the input, non-zero initial conditions only contribute to the system output (or the solution of the state) an additive term which decays to zero exponentially. As in the TV case [24], cancellation of stable PDOs is validated by the following lemma proven therein.

Lemma 3: [24] *Let $L(s,t)$ be an exponential stable PDO with rate $-\beta_L < 0$. Consider the following two systems with I/O pairs (y_1, u) and (y_2, u) satisfying*

$$L(s,t)y_1 = L(s,t)u, \quad y_2 = L(s,t)x_2, \ L(s,t)x_2 = u,$$

where x_2 is an intermediate variable. Then, the differences $y_2 - u$ and $y_1 - u$ are terms exponentially decaying to zero with rate at least $-\beta_L$.

These observations imply that robust stability result to be developed the following sections hold for any initial conditions.

For notational briefness in the subsequent discussions, let

$$\theta(t) = [\ k(t)\ \ \theta_0(t)\ \ \theta_1^T(t)\ \ \theta_2^T(t)\ ]^T, \quad w(t) = [\ r(t)\ \ y(t)\ \ w_1^T(t)\ \ w_2^T(t)\ ]^T. \quad (13)$$

where $\theta \in \mathbf{R}^{2n}$ is the vector containing all the perfect control parameters, and $w \in \mathbf{R}^{2n}$ the vector of all the signals including reference, auxiliary and output signals. Therefore, the control laws (5) and (10) can be written in a compact form as

$$u(t) = U_1(\theta(t), w(t)), \quad (14)$$

and

$$u(t) = U_2(\theta(t), w(t)), \quad (15)$$

respectively. Suppose that the solution of controller parameters with perfect knowledge of the plant is denoted by θ^* (from either Lemma 1 or Lemma 2). Then, it follows from the above discussions that the control $u(t) = U(\theta^*(t), w^*(t))$ ($U(\theta^*(t), w^*(t))$ denotes either $U_1(\theta^*(t), w^*(t))$ or $U_2(\theta^*(t), w^*(t))$) guarantees $y \to y_m$ as $t \to \infty$ for any $r(t)$ under the condition that $d(y,t) = 0$. The question now is what robust control should be used if the plant is unknown and if $d(y,t) \neq 0$. The answer to this question is the subject of section five. In section five, we shall first investigate robust control design based on the new MRC structure in Figure 2. After developing a recursive procedure for designing I/O robust control, we shall show why the MRC structure in Figure 1 is not sufficient for robust control.

Before proceeding methematical development of MRRC for time-varying systems, we choose to devote the next section to intuitive explanations through two simple examples on what are the forms of robust control laws to be used and why these robust controllers work.

4 Why Robust Control Works?

Let us introduce an *important* notation for subsequent analysis. Let $|p|$ or $\|p\|$ denote the magnitude of p, depending whether p is a scalar or vector. Let $|||p|||$ represent an upper (known) bound on the magnitude of p whenever p is unknown.

To illustrate the basic idea of how input-output robust control works, consider the following two examples. They are second-order systems with relative degree one and two, respectively. Again, for simplicity, we assume that $d(y,t)$ be zero in this section.

Example 1: A second-order LTV system:

$$\ddot{y} + a_1(t)\dot{y} + a_2(t)y = \dot{u} + b_1(t)u,$$

where $a_1(t)$, $a_2(t)$, $b_1(t) > 0$ are time-varying parameters. The plant output y is required to tracking the output of the reference model

$$\dot{y}_m + 2y_m = r,$$

where r is any given bounded and continuous reference input. Note that the reference model is strictly proper [9].

As will be shown later, $\alpha(s) = 1$ should be chosen for this system. It follows from (10), (11), and (12) that the following control is a perfect nominal tracking control (i.e., $\lim_{t\to\infty}(y_m - y) = 0$)

$$\dot{w}_1 = -w_1 + \theta_1^* u$$
$$\dot{w}_2 = -w_2 + \theta_2^* y$$
$$u^*(t) = r(t) + \theta_0^* y + w_1 + w_2,$$

where θ_i^* are given by

$$\theta_1^*(t) = 1 - b_1(t), \quad \theta_0^*(t) = a_1(t) - 3 \quad \theta_2^*(t) = a_2(t) - 2 - \theta_0^*(t) - \dot{\theta}_0^*(t).$$

Since perfect knowlege of system parameters is not available, we can not let $u = u^*(t)$. However, since $u^*(t)$ is the perfect control, we can rewrite

$$u = (u - \theta_0^* y - w_1 - w_2) + \theta_0^* y + w_1 + w_2,$$

which implies that

$$\dot{y} + 2y = u + r(t) - u^*(t).$$

Define the error signal $e(t) = y_m(t) - y(t)$ yields

$$\dot{e} + 2e = -u + u^*(t). \quad (16)$$

It is noted that $u^*(t)$ can be viewed as a bounded uncertainty since θ_i^* can be easily bounded using the bounds on the system parameters. The uncertainty $u^*(t)$ satisfies the matching conditions [2]. So, robust control can be designed to compensator for $u^*(t)$.

As will be shown later, the proposed MRRC for this example has the following form:

$$u(t) = \frac{e|||u^*(t)|||}{|e| \cdot |||u^*(t)||| + \epsilon e^{-\beta t}}|||u^*(t)|||, \quad (17)$$

As will be shown next section, the above control is continuous and can compensate for any continuous uncertainty $u^*(t)$ which size is bounded by given $|||u^*(t)|||$. The resulting performance is exponential stability.

To illustrate how robust control compensates for time-varying functions, we choose to simulate here system (16) under (17) (simulation of MRRC for this example will be done in full in the next section). Figure 3 shows the simulation using SIMNON© in which the control parameters are $\epsilon = 1$, $\beta = 0.1$, and the bounding function is

$$|||u^*(t)||| = 1 + 2|y| + \int_{t_0}^{t} e^{-(t-\tau)}[|y(\tau)| + |u(\tau)|]d\tau.$$

The "uncertainty" is chosen to be

$$u^*(t) = 0.5\sin t + y\cos 2t + \int_{t_0}^{t} e^{-(t-\tau)}[y(\tau)\cos(3\tau) + \sin(3\tau)u(\tau)]d\tau.$$

We have the following intuitive explanation on why the above robust control guarantees exponential asymptotic stability. It is noted that $|u| \le |||u^*(t)|||$ which is the maximum size of uncertainty. The ratio

$$\frac{e|||u^*(t)|||}{|e|\cdot|||u^*(t)||| + \epsilon e^{-\beta t}}$$

could converge to any value between -1 and 1. Since the uncertainty is assumed to be continuous, the robust control has the capability of implicitly learning smooth time-varying functions in the sense that the above ration approach in the limit the time-varying ration $u^*(t)/|||u^*(t)|||$. □

Example 2: Consider LTV system:

$$\ddot{y} + a_1(t)\dot{y} + a_2(t)y = u,$$

where $a_1(t)$ and $a_2(t)$ are time-varying parameters. The plant output y is required to tracking the output of the reference model

$$\ddot{y}_m + 3\dot{y}_m + 2y_m = r.$$

It will be shown later that polynomial $\alpha(s)$ should be of first order in order to find a well-defined I/O robust control for this example. However, to reveal basic ideas, we shall consider here $\alpha(s) = 1$. By (10), (11), and (12), the following control ensures perfect nominal tracking control, i.e., $\lim_{t\to\infty}(y_m - y) = 0$.

$$\begin{aligned}\dot{w}_1 &= -w_1 + \theta_1^* u\\ \dot{w}_2 &= -w_2 + \theta_2^* y\\ u^*(t) &= r(t) + \theta_0^* y + w_1 + w_2,\end{aligned}$$

where θ_i^* are given by

$$\theta_1^*(t) = a_1(t) - 3, \quad \theta_0^*(t) = -5 + a_1(t) + \dot{a}_1(t) + a_2(t) - \theta_1^*(t)a_1(t),$$

$$\theta_2^*(t) = a_2(t) + \dot{a}_2(t) - \theta_1^*(t)a_2(t) - \theta_0^*(t) - \dot{\theta}_0^*(t) - 2.$$

Due to the lack of perfect knowlege, we can not let $u = u^*(t)$. However, since $u^*(t)$ is the perfect control, the system can be rewritten as

$$\ddot{y} + 3\dot{y} + 2y = u + r(t) - u^*(t),$$

or

$$\ddot{e} + 3\dot{e} + 2e = -u + u^*(t). \tag{18}$$

It is noted that $u^*(t)$ can again be viewed as a bounded uncertainty.

Unlike Example 1, robust control can not be designed directly. This is because, although the uncertainty is matched [2], the reference model as well as (18) is not strictly proper. That is, a robust control designed using standard robust control theory will require measurement not only y but also $\dot{y}$, which violates the objective of I/O robust control. To get around this problem, let

$$\overline{u} = \frac{1}{s+1.5}u$$

which yields

$$\begin{aligned}\ddot{e} + 3\dot{e} + 2e &= -\dot{\overline{u}} - 1.5\overline{u} + u^*(t)\\ \dot{\overline{u}} &= -1.5\overline{u} + u.\end{aligned}$$

The above system can be viewed as two cascaded, strictly proper subsystems. Since the subsystems are strictly proper, output robust control can be designed for each one of them, and cascaded connection can be used to generate the overall control $u(t)$ in a similar fashion as those in [10,15,16]. The treat-off is that the uncertainty $u^*(t)$ is now not matched but satisfies the generalized matching conditions [16].

As shown in [10,15,16], the loss of matching conditions for uncertainty implies that asymptotic stability or expoenential stability can not be achieved in general. An intuitive explanaton is the following. The intermediate control variable $\overline{u}$ could be designed in a similar form as that in (17). As explained in Example 1, a robust control in the form of (17) can compensate any continuous $u^*(t)$. There is no limitation on the uncertainty $u^*(t)$ except its size bounding function $|||u^*(t)|||$. However, inside the given size bounding function, the uncertainty may vary continuous but arbitrarily fast, this implies that a robust control in the form of (17) has the capability to change arbitrarily fast as well in order to match up with the change in the uncertainty. Therefore, the time derivative of $\overline{u}$ can not be bounded apriori. Thus, to achieve exponential stability for plant of high relative degree, the control u may be excessively large since $u = \dot{\overline{u}} + 1.5\overline{u}$. Because of this reason, exponential stability for plant of relative degree greater than one will not pursued in this paper. The control objective for these systems is to make the tracking error arbitrarily small to achieve any given accuracy requirement. □

With the basic ideas of robust control illustrated by the above examples, we can proceed mathematical development of MRRC design.

5 I/O Robust Tracking Control of TV Systems

When the parameters of the plant are unknown, we propose to design a robust control which guarantees stability and performance even under nonlinear uncertainty $d(y,t)$. We shall investigate robust control design based on the new MRC structure in Figure 2, from which insufficiency of the MRC structure in Figure 1 for robust control design becomes obvious.

Note that any control $u(t)$ can be rewritten as

$$u(t) = U_2(\theta^*(t), w^*(t)) + [u(t) - U_2(\theta^*(t), w^*(t))], \tag{19}$$

where $U_2(\theta^*, w^*(t))$ is given by (15) and obtained from Lemma 2 (and Figure 2) if perfect knowledge of the plant is available. Due to the existence of uncertainty $d(y,t)$, the total input to the plant is $u(t) + d(y,t)$. Thus, the auxiliary signals have to be modified to be

$$\dot{w}_1^* = A_0 w_1^* + \alpha(s)\theta_1^*(t)\frac{1}{\alpha(s)}[u + d(y,t)] \tag{20}$$

$$\dot{w}_2^* = A_0 w_2^* + \alpha(s)\theta_2^*(t)\frac{1}{\alpha(s)}y \tag{21}$$

where A_0 and B_0 are the same as those in (8). Note that w_1^* depends on uncertainty, therefore can not be calculated, and should be bounded.

Since the vectors of ideal parameters and signals $\theta^*(t)$ and $w^*(t)$ satisfies Lemma 2, the plant output under control (19) must be

$$y(t) = G_m(s)\left\{r(t) + \left[\alpha(s)k^*(t)\frac{1}{\alpha(s)}\right]^{-1}[u(t) - U_2(\theta^*(t), w^*(t)) + d(y,t)]\right\}, \tag{22}$$

where $k^*(t) = k_m/k_p(t)$, and $\alpha(s)$ is chosen to be monic polynomial of $(n-m-1)$-th order. Let us define the output tracking error $e(t)$ to be

$$e(t) = y_m(t) - y(t),$$

where $y_m(t) = G_m(s)r(t)$. Note that

$$\left[\alpha(s)k^*(t)\frac{1}{\alpha(s)}\right]\cdot\left[\alpha(s)\frac{1}{k^*(t)}\frac{1}{\alpha(s)}\right] = 1.$$

It then follows from Lemma 3 that the dynamics of the output tracking error $e(t)$ can be rewritten to be

$$\begin{aligned}e(t) &= G_m(s)\left[\alpha(s)\frac{1}{k^*(t)}\frac{1}{\alpha(s)}\right][-u(t) + U_2(\theta^*(t), w^*(t)) - d(y,t)]\\ &= \overline{G}_m(s)\frac{1}{k^*(t)}\left[-\frac{1}{\alpha(s)}u(t) + \overline{d}(y,u,t)\right] &(23)\\ &= \overline{G}_m(s)\frac{1}{k^*(t)}\left[-\overline{u}(t) + \overline{d}(y,u,t)\right], &(24)\end{aligned}$$

where

$$\overline{r}(t) = \frac{1}{\alpha(s)}r(t), \quad \overline{u}(t) = \frac{1}{\alpha(s)}u(t), \quad \overline{y}(t) = \frac{1}{\alpha(s)}y(t),$$

$$\overline{d}(y,u,t) = \frac{1}{\alpha(s)}[U_2(\theta^*(t), w^*(t)) - d(y,t)].$$

The transfer function $\overline{G}_m(s)$ in (24) is defined by

$$\overline{G}_m(s) = G_m(s)\alpha(s) \triangleq k_m\frac{\overline{B}_m(s)}{A_m(s)}, \tag{25}$$

where $\alpha(s)$ is chosen from now on to be a monic and Hurwitz polynomial in s of degree $n-m-1$, and $G_m(s)$ is the reference model. In order to design a robust controller requiring only input and output data, it is necessary to have $\overline{G}_m(s)$ be a strictly positive real (SPR) transfer function. The conditions for a transfer function with relative degree one to be SPR can be found on P.64 of [9]. If $n-m-1=0$, it is assumed that $G_m(s)$ be SPR. If $G_m(s)$ has a relative degree larger than one, $\alpha(s)$ is chosen such that $\overline{G}_m$ and is SPR. It is well known that, if $\overline{G}_m(s)$ is a stable, minimum phase but not SPR, a SPR transfer function can be generated by filtering the reference input.

The term $\overline{d}(y,u,t)$ in (24) stands for the total uncertainty in the closed loop system, and can be bounded as follows: assuming $w_1^*(0) = w_2^*(0) = r(0) = 0$ with loss of any generality,

$$\begin{aligned}|\overline{d}(y,u,t)| &= \left|k^*(t)\frac{1}{\alpha(s)}r(t) + \theta_0^*(t)\frac{1}{\alpha(s)}y + \frac{1}{\alpha(s)}B_0^T w_1^*(t) + \frac{1}{\alpha(s)}B_0^T w_2^*(t) - \frac{1}{\alpha(s)}d(y,t)\right|\\ &\le |||k^*(t)|||\cdot|\overline{r}(t)| + |||\theta_0^*(t)|||\cdot|\overline{y}(t)| + \int_{t_0}^{t} h_o(t-\tau)|||\theta_1^*(\tau)|||\cdot|\overline{u}(\tau)|d\tau\\ &\quad + \int_{t_0}^{t} h_o(t-\tau)|||\theta_2^*(\tau)|||\cdot|\overline{y}(\tau)|d\tau\\ &\quad + \int_{t_0}^{t}[h_\alpha(t-\tau) + h_1(t-\tau)|||\theta_1^*(\tau)|||]\cdot\rho(y(\tau),\tau)d\tau\\ &\triangleq |||\overline{d}(y,u,t)|||,\end{aligned}$$

where $h_\alpha(t), h_o(t) \ge 0$ are stable impulse responses defined as follows. $h_\alpha(t)$ has almost the same expression as the inverse Laplace transform of $1/\alpha(s)$ except that all trigonometric functions are removed and that some (or all) its coefficients are made positive in order to ensure $h_\alpha(t) \ge 0$. $h_o(t)$ is defined in the same way as $h_\alpha(s)$ but is based on $\|B_0^T e^{A_0 t}\|$. $h_1(t) \ge 0$ is the convolution of $h_\alpha(t)$ and $h_o(t)$.

As will be shown later, if $n-m-1 > 0$, the bounding function $|||\overline{d}(y,u,t)|||$ is required to be differentiable. In this case, we can choose

$$\begin{aligned}|||\overline{d}(y,u,t)||| &= 1 + 0.5|||k^*(t)|||^2\cdot|\overline{r}(t)|^2 + 0.5|||\theta_0^*(\tau)|||^2\cdot\overline{y}^2(t)\\ &\quad + \int_{t_0}^{t} h_o(t-\tau)|||\theta_1^*(\tau)|||\cdot|\overline{u}(\tau)|d\tau + \int_{t_0}^{t} h_o(t-\tau)|||\theta_2^*(\tau)|||\cdot|\overline{y}(\tau)|d\tau\\ &\quad + \int_{t_0}^{t}[h_\alpha(t-\tau) + h_1(t-\tau)|||\theta_1^*(\tau)|||]\cdot\rho(y(\tau),\tau)d\tau.\end{aligned}$$

By definition, $h_\alpha(t)$, $h_o(t)$, and $h_1(t)$ are differentiable. It is worth emphasizing here that the time derivative of $|||\overline{d}(y,u,t)|||$ does not contain $u(t)$ but rather the filtered version of u. This can be seen by noting

$$\frac{d}{dt}\int_{t_0}^{t} h_o(t-\tau)|||\theta_1^*(\tau)|||\cdot|\overline{u}(\tau)|d\tau = \int_{t_0}^{t} h_2(t-\tau)|||\theta_1^*(\tau)|||\cdot|\overline{u}(\tau)|d\tau, \tag{26}$$

where $h_2(t)$ is the inverse Laplace transform of the product of s and the Laplace transform of $h_o(t)$.

Based on the error dynamics (24), we can proceed our robust control design. The following analysis is an extension of the results in [11] to fast TV systems. We shall discuss separately two disjoint and complement cases: $n-m-1=0$ and $n-m-1>0$. If $n-m-1=0$, the PDOs of the plant $A(s,t)$ and $B(s,t)$ have relative degree one. In this case, $\alpha(s) = 1$, and a simple design of robust control is given by the following theorem.

Theorem 1: *Consider the system represented by (1) which satisfies Assumptions A.1 to A.4. Then, the plant output tracking error $e(t)$ converges to zero exponentially under the robust control*

$$\overline{u}(t) = \frac{\mu(e,y,u,t)}{2\underline{b}_{p0}(|\mu(e,y,u,t)| + \epsilon e^{-\beta t})}g(y,u,t), \tag{27}$$

where $\beta, \epsilon > 0$ are constants, $\overline{b}_{p0}$ and $\underline{b}_{p0}$ are upper and lower bounds of $b_{p0}(t)$ (i.e., $k_p(t)$) as defined in Assumption A.3, and

$$g(y,u,t) = 2\bar{b}_{p0}\|\|\overline{d}(y,u,t)\|\|, \qquad \mu(e,y,u,t) = e(t)g(y,u,t).$$

Furthermore, the control $\overline{u}(t)$ is uniformly bounded and uniformly continuous.
Proof: Suppose that the triple $\{A, B, C\}$ is a minimal realization of the transfer function $\overline{G}_m(s)$, i.e., the realization

$$\begin{aligned} \dot{x}_e &= Ax_e + B\left[-\overline{u}(t) + \overline{d}(y,u,t)\right]/k^*(t) \\ e(t) &= Cx_e, \end{aligned} \tag{28}$$

is controllable and observable. Since $\overline{G}_m(s)$ is SPR, it follows from the Kalman-Yakubovich lemma in [7] that, given a symmetric positive definite (s.p.d.) matrix Q, there exists a s.p.d. matrix P such that

$$A^TP + PA = -Q, \qquad PB = C^T.$$

Choose V to be the Lyapunov function defined as

$$V_e = k_m x_e^T P x_e.$$

Taking its time derivative along the trajectories of the system yields

$$\begin{aligned} \dot{V}_e &= k_m x_e^T(PA + A^TP)x_e + 2x_e^T PB\frac{k_m}{k^*(t)}\left[\overline{d}(y,u,t) - \overline{u}(t)\right] \\ &= -k_m x_e^T Q x_e + 2e(t)k_p(t)\left[\overline{d}(y,u,t) - \overline{u}(t)\right] \\ &\le -k_m x_e^T Q x_e + 2|e(t)|\bar{b}_{p0}\left|\overline{d}(y,u,t)\right| - 2k_p(t)e(t)u(t) \\ &\le -k_m x_e^T Q x_e + [|e(t)|g(y,u,t) - 2k_p(t)e(t)u(t)] \\ &= -k_m x_e^T Q x_e + |\mu(e,y,u,t)| - \frac{k_p(t)}{\underline{b}_{p0}}\frac{|\mu(e,y,u,t)|^2}{|\mu(e,y,u,t)| + \epsilon e^{-\beta t}} \\ &\le -k_m x_e^T Q x_e + |\mu(e,y,u,t)| - \frac{|\mu(e,y,u,t)|^2}{|\mu(e,y,u,t)| + \epsilon e^{-\beta t}} \\ &= -k_m x_e^T Q x_e + \frac{|\mu(e,y,u,t)|}{|\mu(e,y,u,t)| + \epsilon e^{-\beta t}}\epsilon e^{-\beta t} \\ &\le -\frac{\lambda_{min}(Q)}{\lambda_{max}(P)}V_e + \epsilon e^{-\beta t}. \end{aligned}$$

Now, let us define

$$s_e(t) = \dot{V}_e + \lambda_e V_e - \epsilon e^{-\beta t},$$

where $\lambda_e \triangleq \lambda_{min}(Q)/\lambda_{max}(P)$. It follows from the above Lyapunov analysis that $s_e(t) \le 0$. Solving the above first-order differential equation yields

$$\begin{aligned} V_e(t) &= e^{-\lambda_e(t-t_0)}V_e(t_0) + \int_{t_0}^t e^{-\lambda_e(t-\tau)}(s_e(t) + \epsilon e^{-\beta\tau})d\tau \\ &\le e^{-\lambda_e(t-t_0)}V_e(t_0) + \epsilon\int_{t_0}^t e^{-\lambda_e(t-\tau)}e^{-\beta\tau}d\tau \\ &= e^{-\lambda_e(t-t_0)}V_e(t_0) + \begin{cases} \frac{\epsilon e^{-\beta t_0}}{\lambda_e-\beta}(e^{-\beta(t-t_0)} - e^{-\lambda_e(t-t_0)}) & \text{if } \lambda_e \ne \beta \\ \epsilon(t-t_0)e^{-\beta t} & \text{if } \lambda_e = \beta \end{cases}, \end{aligned}$$

Therefore, V_e converges to zero exponentially, and so does the state x_e. Consequently, the output tracking error $e(t)$ converges to zero exponentially.

To show $\overline{u}$ is uniformly bounded, note that

$$|\overline{u}| \le \frac{\bar{b}_{p0}}{\underline{b}_{p0}}\|\|\overline{d}(y,u,t)\|\|.$$

Recall that e converges to zero, that $h_1(t)$, $h_o(t)$, and $h_\alpha(t)$ are corresponding to stable transfer functios, that the parameters θ_i^* are bounded, and that $r(t)$ is uniformly bounded. It then follows from the expression of $\|\|\overline{d}(y,u,t)\|\|$ that

$$|\overline{u}| \le C + \frac{\bar{b}_{p0}}{\underline{b}_{p0}}\int_{t_0}^t h_o(t-\tau)\|\|\theta_1^*(\tau)\|\| \cdot |\overline{u}(\tau)|d\tau$$

for some constant C. Since $h_o(t)$ has finite integration over infinite horizon, it follows from Gronwall-Bellman inequality [7] that $|\overline{u}|$ is uniformly bounded.

It is worth proving that, in the limit as $t \to \infty$, control (27) **does not** become a variable-structure (switching) type control. This can be seen from the following argument. Since the control (27) is continuous at every finite instant of time, there exists a classical solution $x_e(t)$ for system (28). It follows that x_e is uniformly continuous since x_e converges to zero. It then follows from Barbalat Lemma in [20] that $\dot{x}_e$ converges to zero as time approaches infinity. Taking the limit on both side of (28) yields

$$\lim_{t\to\infty}[\overline{u}(t) - \overline{d}(y,u,t)] = 0,$$

which implies that the control is continuous at $t = \infty$ as long as the uncertainties are continuous. It is worth recalling that the uncertainties are assumed to be continuous. This result reflects ability and adequacy of robust control (27). This also shows that the control (27) is superior to min-max control [4] and saturation control [2] and is not a simple mixture of them. □

The control $\overline{u}(t)$ in (27) has several nice properties. First, it is uniformly continuous, uniformly bounded, and ensures existence of a classical solution to the system. The control $\overline{u}(t)$ also guarantees robust stability and performance in the presence of additive nonlinear uncertainties. Moreover, the control can handle fast TV systems in which parameter drifting can be arbitrarily large but bounded.

Example 1:(Cont.) To verify the theoretical analysis, we make the following choices for simulation purpose:

$$b_1(t) = 2 - \sin(t), \quad a_1(t) = \cos(3t), \quad a_2(t) = \sin(3t), \quad d(y,t) = \sin(t) + y\cos(t) - 0.5y^2,$$

and zero initial conditions. The reference input is chosen to be $r(t) = \sin(2t)$. The robust control is implemented using

$$\rho(y,t) = 1+y^2, \quad \epsilon = \beta = 1.0, \quad \|\|\overline{d}\|\| = 2+y^2+2|y| + \int_{t_0}^t e^{-(t-\tau)}[|u(\tau)|+|y(\tau)|+2+2y^2(\tau)]d\tau.$$

The simulation results are shown in Figure 4 in which the plots are consistent with theoretical results. □

It follows from Theorem 1 and equations (27) and (24) that the control guaranteeing exponentially stability under nonlinear uncertainties is

$$u(t) = \alpha(s)\overline{u}(t). \tag{29}$$

Although equation (29) appears to give a general form of a robust tracking controller, the resulting control (29) requires time derivatives of $\overline{u}(t)$ when $n - m - 1 > 0$, which in turn requires measurement of output derivatives of orders up to $n - m - 1$. So, in the case of high relative degree, robust controller should be redesigned.

Let $l = n - m - 1 > 0$. In order not to require measurements of output derivatives, we need modify the control problem posed in Theorem 1. Let us first introduce the new state variables z_i as the state in the controllable canonical realization of the transfer function $1/\alpha(s)$. That is, if $\alpha(s) = s^l + \alpha_1 s^{l-1} + \cdots + \alpha_l$, we define

$$z_1 = \overline{u}, \qquad \dot{z}_i = z_{i+1}, \qquad \forall i = 1, \cdots, l-1.$$

The ultimate robust control (29) can then be represented in terms of the state space equation as

$$\begin{aligned} \dot{z}_1 &= z_2, \\ \dot{z}_i &= z_{i+1}, \quad \forall i = 2, \cdots, l-1, \\ \dot{z}_l &= -\alpha_1 z_l - \cdots - \alpha_l z_1 + u - \theta^T w(t) \\ &= -\alpha_1 z_l - \cdots - \alpha_l z_1 + u(t), \\ \overline{u} &= z_1. \end{aligned} \tag{30}$$

Using the new state variables z_i, the state space realization (28) can be rewritten as

$$\begin{aligned} \dot{x}_e &= Ax_e + B\left[-z_1 + \overline{d}(y,u,t)\right]/k^*(t) \\ e(t) &= C_1 x_e. \end{aligned} \tag{31}$$

Furthermore, it follows from definitions of z_i and (24) that

$$\begin{aligned} \dot{e} &= s\overline{G}_m(s)\frac{1}{k^*(t)}\left[-\overline{u}(t) + \overline{d}(y,u,t)\right], \\ \dot{z}_i &= \frac{s^i}{\alpha(s)}u(t), \qquad \forall i = 1, \cdots, l-1, \end{aligned} \tag{32}$$

where $\overline{d}(y,u,t)$ and its derivatives can be bounded in a similar fashion as proceeded in (24).

With state-space equations (30), (31) and (32) in hand, we can proceed with input-output robust design for TV systems with high relative degree. The design procedure is a straightforward application of the Lyapunov's second method, and the robust control is generated by the following recursive, nonlinear mappings:

$$\begin{aligned} v_1 &= \frac{\mu_1(e,y,u,t)|\mu_1(e,y,u,t)|}{2\underline{b}_{p0}\left(\mu_1^2(e,y,u,t) + \epsilon_1^2\right)}g_1(y,u,t), \\ v_2 &= v_1 - z_1 + \frac{\mu_2(e,y,u,t)|\mu_2(e,y,u,t)|}{2\left(\mu_2^2(e,y,u,t) + \epsilon_2^2\right)}g_2(y,u,t), \\ v_i &= v_{i-2} - z_{i-2} + v_{i-1} - z_{i-1} + \frac{\mu_i(e,y,u,t)|\mu_i(e,y,u,t)|}{2\left(\mu_i^2(e,y,u,t) + \epsilon_i^2\right)}g_i(y,u,t), \\ v_{l+1} &= \alpha_1 z_l + \cdots + \alpha_l z_1 + v_{l-1} - z_{l-1} + v_l - z_l \\ &\quad + \frac{\mu_{l+1}(e,y,u,t)|\mu_{l+1}(e,y,u,t)|}{2\left(\mu_{l+1}^2(e,y,u,t) + \epsilon_{l+1}^2\right)}g_{l+1}(y,u,t), \end{aligned} \tag{33}$$

where $i = 3, \cdots, l$, $\epsilon_j > 0$, $j = 1, \cdots, l+1$, are constants, and

$$g_1(y,u,t) = 2\bar{b}_{p0}\|\|\overline{d}(y,u,t)\|\|, \qquad \mu_1(e,y,u,t) = e(t)g_1(y,u,t),$$

$$g_2(y,u,t) = \bar{b}_{p0}^2 + |e|^2 + 2\|\|\dot{v}_1\|\|, \qquad \mu_2(e,y,u,t) = (v_1 - z_1)g_2(y,u,t),$$

$$g_i(y,u,t) = 2\|\|\dot{v}_{i-1}\|\|, \qquad \mu_i(e,y,u,t) = (v_{i-1} - z_{i-1})g_i(y,u,t), \qquad i = 3, \cdots l,$$

$$g_{l+1}(y,u,t) = 2\|\|\dot{v}_l\|\|, \qquad \mu_{l+1}(e,y,u,t) = (v_l - z_l)g_{l+1}(y,u,t).$$

It should be noted that every step in the recursive mapping basically involves finding a bounding function of $|\dot{v}_i|$ for obtaining v_{i+1}. The term $|\dot{v}_i|$ can be bounded by first developing bounds for the first-order partial derivatives of v_i with respect to its variables, and then by determining the bounds for the first-order time derivatives of its variables using the relations in (32). In summary, $\|\|\dot{v}_i\|\|$ can be found in a similar way as that of finding $\|\|\overline{d}(y,u,t)\|\|$, and differentiability can be guaranteed by properly choosing the bounding functions. Refer to [11] for more details on this subject.

We are now in a position to state the main result of this paper.

Theorem 2: *Consider the system represented by (1) which satisfies Assumptions A.1 to A.4. Then, under the robust control $u(t)$ given by*

$$u(t) = v_{l+1}, \tag{34}$$

where v_{l+1} is defined by the outcome of the mapping procedure (33), the plant output tracking error $e(t)$ is globally and uniformly ultimately bounded. More specifically, the magnitude of the output tracking error $e(t)$ is bounded from above by an exponentially decaying time function, and, as time approaches infinity, becomes no larger than a design parameter ϵ. Furthermore, the robust control $u(t)$ and all internal variables of the system are all uniformly continuous and globally, uniformly bounded.

Proof: Choose V to be the Lyapunov function defined as

$$V = V_e + \sum_{i=1}^{l}(z_i - v_i)^2 = k_m x_e^T P x_e + \sum_{i=1}^{l}(z_i - v_i)^2,$$

where V_e is the same as that defined in the proof of theorem 1. Taking its time derivative along the trajectories of the system given by (30) and (31) yields

$$\begin{aligned} \dot{V} &= k_m x_e^T(PA + A^TP)x_e + 2x_e^T PB\frac{k_m}{k^*(t)}\left[-z_1 + \overline{d}(y,u,t)\right] + \sum_{i=1}^{l}2(z_i - v_i)(\dot{z}_i - \dot{v}_i) \\ &= -k_m x_e^T Q x_e + 2e(t)k_p(t)\left[-v_1 + \overline{d}(y,u,t)\right] - 2e(t)k_p(t)[z_1 - v_1] \\ &\quad + \sum_{i=1}^{l-1}2(z_i - v_i)(z_{i+1} - \dot{v}_i) + 2(z_l - v_l)(u(t) - \alpha_1 z_l - \cdots - \alpha_l z_1 - \dot{v}_l) \\ &= -k_m x_e^T Q x_e + 2e(t)k_p(t)\left[-v_1 + \overline{d}(y,u,t)\right] + 2(z_1 - v_1)(v_2 - k_p(t)e(t) - \dot{v}_1) \\ &\quad + \sum_{i=2}^{l-1}2(z_i - v_i)(v_{i+1} + z_{i-1} - v_{i-1} - \dot{v}_i) \\ &\quad + 2(z_l - v_l)(u(t) - \alpha_1 z_l - \cdots - \alpha_l z_1 + z_{l-1} - v_{l-1} - \dot{v}_l). \end{aligned}$$

It follows that

$$
\begin{aligned}
2e(t)k_p(t)\left[-v_1+\overline{d}(y,u,t)\right] &\le |\mu_1(e,y,u,t)| - \frac{|\mu_1(e,y,u,t)|^3}{|\mu_1(e,y,u,t)|^2+\epsilon_1^2} \\
&= \frac{|\mu_1(e,y,u,t)|\epsilon_1^2}{|\mu_1(e,y,u,t)|^2+\epsilon_1^2} \\
&\le \epsilon_1,
\end{aligned}
$$

since, if $|\mu_1(e,y,u,t)| \ge \epsilon_1$,

$$\frac{|\mu_1(e,y,u,t)|\epsilon_1}{|\mu_1(e,y,u,t)|^2+\epsilon_1^2} \le \frac{|\mu_1(e,y,u,t)|\epsilon_1}{|\mu_1(e,y,u,t)|^2} = \frac{\epsilon_1}{|\mu_1(e,y,u,t)|} \le 1,$$

and if $|\mu_1(e,y,u,t)| \le \epsilon_1$,

$$\frac{|\mu_1(e,y,u,t)|\epsilon_1}{|\mu_1(e,y,u,t)|^2+\epsilon_1^2} \le \frac{|\mu_1(e,y,u,t)|\epsilon_1}{\epsilon_1^2} = \frac{|\mu_1(e,y,u,t)|}{\epsilon_1} \le 1.$$

By mimicking the above derivation, we can show that

$$2(z_1-v_1)\left(v_2-k_p(t)e(t)-\dot{v}_1\right) \le -2(z_1-v_1)^2+\epsilon_2,$$

$$2(z_i-v_i)(v_{i+1}+z_{i-1}-v_{i-1}-\dot{v}_i) \le -2(z_i-v_i)^2+\epsilon_{i+1}, \quad i\in\{2,\cdots,l-1\},$$

$$2(z_l-v_l)(u(t)-\alpha_l z_l-\cdots-\alpha_l z_1+z_{l-1}-v_{l-1}-\dot{v}_l) \le -2(z_l-v_l)^2+\epsilon_{l+1}.$$

Therefore, we have

$$\dot{V} \le -k_m x_e^T Q x_e - 2\sum_{i=1}^{l}(z_i-v_i)^2 + \sum_{j=1}^{l+1}\epsilon_j \le \lambda V + \lambda\underline{\lambda}c\epsilon^2,$$

where C is the output matrix defined in (28),

$$\lambda = \frac{\min\{k_m\lambda_{min}(Q),2\}}{\max\{k_m\lambda_{max}(P),1\}}, \quad \underline{\lambda} = \max\{k_m\lambda_{min}(P),1\}, \quad c = CC^T, \quad \epsilon = \sqrt{\frac{1}{c\lambda\underline{\lambda}}\sum_{i=1}^{l+1}\epsilon_i}.$$

Now, let

$$s(t) = \dot{V} + \lambda V - \lambda\underline{\lambda}c\epsilon^2.$$

It follows that $s(t)\le 0$. Solving the above first-order differential equation yields

$$
\begin{aligned}
V(t) &= e^{-\lambda(t-t_0)}V(t_0) + \int_{t_0}^{t} e^{-\lambda(t-\tau)}(s(t)+\lambda\underline{\lambda}c\epsilon^2)d\tau \\
&\le e^{-\lambda(t-t_0)}V(t_0) + \underline{\lambda}c\epsilon^2\int_{t_0}^{t} e^{-\lambda(t-\tau)}\lambda d\tau \\
&= e^{-\lambda(t-t_0)}V(t_0) + \underline{\lambda}c\epsilon^2(1-e^{-\lambda(t-t_0)}) \\
&\to \underline{\lambda}c\epsilon^2, \quad \text{as } t\to\infty.
\end{aligned}
$$

Therefore, V is uniformly ultimately bounded. Therefore, all state variables including the output tracking error $e(t)$ are globally and uniformly ultimately bounded. The convergence of $V(t)$ is obviously exponentially, and so is the convergence of $e(t)$. Moreover, it is easy to show that

$$\lim_{t\to\infty}\sup_{\tau\ge t}\|x_e\|^2 \le c\epsilon^2 \quad \Longrightarrow \quad \lim_{t\to\infty}\sup_{\tau\ge t}\|e\| \le \epsilon.$$

It follows from (34) and (33) that

$$|v_1| \le g_1(e,y,u,t).$$

By first noting the fact that $v_1-\overline{u}$ is uniformly bounded and then using the same argument as that in Theorem 1, one can show that the intermediate (or fictitious) control variables $v_i(t)$, $i=1,\cdots,l$, and consequently the actual control $u(t)=v_{l+1}$ are all globally and uniformly bounded. □

The control (34) has the same properties as $\overline{u}(t)$ in Theorem 1. It is obvious that Theorem 1 is a special case of Theorem 2. Unlike the exponential stability result in Theorem 1, the stability result in Theorem 2 is uniform ultimate boundedness about a ball of radius ϵ centered at origin in the state space. The constant ϵ is a design parameter that can be arbitrarily selected. The uniform ultimate bounded stability result approaches asymptotic stability as $\epsilon\to 0$. In general, it is not possible to achieve exponential stability for system (1) with relative degree greater than one. For conditions under which exponential stability can be achieved, refer to the discussions in [10] for general nonlinear uncertain systems.

The idea of designing robust control using a nonlinear mapping such as (33) was originally proposed in [10]. It is worth mentioning that designing control through a mapping procedure was also reported in adaptive control theory, for example, the backstepping procedure reported in [6] and the references therein. Although both mapping procedures are conceptually similar, the mapping (33) is mathematically much more general, since it can be used to handle unknown nonlinear functional of the state and time, and include the backstepping procedure as a special case.

Before concluding the discussions in this section, let us look at why the MRC structure proposed by Tsakalis and Ioannou and shown in Figure 1 is not directly applicable to robust control design. Note that any control $u(t)$ can be rewritten as

$$u(t) = U_1(\theta^*(t),w^*(t)) + [u(t) - U_1(\theta^*(t),w^*(t))], \tag{35}$$

where $U_1(\theta^*,w^*(t))$ is given by (14) and obtained from Lemma 1 (and Figure 1) if perfect knowledge of the plant is available. Since the vectors of ideal parameters and signals $\theta^*(t)$ and $w^*(t)$ satisfies Lemma 1, the plant output under control (35) must be

$$y(t) = G_m(s)\left\{r(t) + \frac{1}{k^*(t)}[u(t)-U_1(\theta^*(t),w^*(t))+d(y,t)]\right\}, \tag{36}$$

It then follows from definition of output tracking error that

$$
\begin{aligned}
e(t) &= G_m(s)\frac{1}{k^*(t)}[-u(t)+U_1(\theta^*(t),w^*(t))-d(y,t)] \\
&= \overline{G}_m(s)\left[-\frac{1}{\alpha(s)}\frac{1}{k^*(t)}u(t)+\overline{\overline{d}}(y,u,t)\right],
\end{aligned}
\tag{37}
$$

where

$$\overline{\overline{d}}(y,u,t) = \frac{1}{\alpha(s)}\frac{1}{k^*(t)}[U_1(\theta^*(t),w^*(t))-d(y,t)].$$

As proceeded before, we can design an input-output robust controller if and only if the following two conditions hold: (1) The transfer function $\overline{G}_m(s)$ is SPR; (2) The variable substitution $\overline{\overline{u}} = k'(t)\frac{1}{\alpha(s)}\frac{1}{k^*(t)}u(t)$, for some $k'(t)$ bounded away from zero, can be used to generate a recursive mapping for the ultimate control variable $u(t)$. The first condition is easy to satisfied. It is easy to verify that a nonlinear mapping similar to (33) exists if and only if $\overline{\overline{u}}$ can be calculated by properly choosing known or unknown $k'(t)$. Since the two PDOs $\frac{1}{\alpha(s)}$ and $\frac{1}{k^*(t)}$ are not commute, the second condition is satisfied if and only if one of the following conditions holds:

(i) $k^*(t)$ is time-invariant, which implies that $k_p(t)=k_p$ is a constant.

(ii) $n-m=1$ in which case $\alpha(s)=1$.

(iii) $k_p(t)=k_p c(t)$ where $c(t)$ is a known time function.

In other words, the MRC structure in Figure 1 fails, after combined with robust control design, to be applicable to the general case that the system is of high relative degree and has unknown time-varying gain $k_p(t)$. As a further evidence, it follows from (26) that, if the MRC structure in Figure 1 is used, $\overline{u}$ in (26) has to be replaced by u. This makes the recursive mapping not well defined since the bouding functions of uncertainties will depend explicitly on the control input or even its derivatives.

Conclusion

In this paper we consider input-output robust control problem of fast time-varying systems with significant nonlinear uncertainties. Inspired by the MRC structure for known LTV systems proposed by Tsakalis and Ioannou, we first propose a new MRC structure which is a minor modification of the existing one. We then integrate the new MRC structure into the recursive robust control design methodology in [10,11] to generate a robust controller. The resulting control has several significant features: the control is continuous, uniformly bounded, requires only input-output measurement, guarantees existence of a classical solution for the system, ensures exponential stability or stability of uniform ultimate boundedness which can be made arbitrarily close to asymptotic stability. The most fundamental difference between this result and the existing ones (most of them using adaptive control approach) is that system parameters are **not** required by the proposed robust control scheme to be either slow time-varying or fast time-varying with known structure [24]. Another feature of MRRC, which is important both theoretically and practically, is that the designed control is always uniformly continuous.

Comparing with the results in model reference adaptive control, the robust controller is much simpler and remains effective even under significant nonlinear uncertainty. The implementation of robust controllers requires neither adaptation laws nor explicit calculation of auxiliary signals. The proof in this paper is merely a straightforward application of Lyapunov argument.

References

[1] C.-J. Chien and L.-C. Fu, "A new approach to model reference control of a class of aybitrarily fast time-varying unknown plants," *Automatica*, Vol.28, No.2, pp.437-440, 1992.

[2] M.J.Corless and G.Leitmann "Continuous State Feedback Guaranteeing Uniform Ultimate Boundedness for Uncertain Dynamic Systems," *IEEE Trans. Automat. Contr.*, Vol.26, No.5, pp.1139-1144, May, 1981.

[3] D.M.Dawson, Z.Qu, and J.Carroll, "On the state observation and output feedback problems for nonlinear uncertain dynamical systems," *Systems & Control Letters*, Vol.18, No.2, pp.217-222, 1992.

[4] S.Gutman, "Uncertain Dynamical Systems — A Lyapunov Min-Max Approach," *IEEE Trans. Automat. Contr.*, Vol.24, No.3, pp.437-443, June, 1979.

[5] T.Kailath, *Linear Systems*, Englewood Cliffs, Prentice Hall, Inc., 1980.

[6] I.Kanellakpoulos, P.V.Kokotovic, and A.S.Morse, "Systematic design of adaptive controllers for feedback linearizable systems," *IEEE Trans. Automat. Contr.*, Vol.36, No.11, pp.1241-1253, November, 1991.

[7] H.K.Khalil, *Nonlinear Systems*, New York, Macmillan Publishing Company, 1991.

[8] Y.D.Landau, *Adaptive Control - The Model Reference Approach*, Marcel Dekker, 1979.

[9] K.S.Narendra and A.M.Annaswamy, *Stable Adaptive Systems*, Prentice-Hall, 1989.

[10] Z.Qu and D.M.Dawson, "Lyapunov direct design of robust tracking control for classes of cascaded nonlinear uncertain systems without matching conditions," *The 30th IEEE Conference on Decision and Control*, pp.2521-2526, Brighton, U.K. December 1991.

[11] Z.Qu, D.M.Dawson, and J.F.Dorsey, "Model reference robust control of a class of SISO systems," submitted to *IEEE Transactions on Automatic Control*, 2/5/92, revised 9/15/92. The preliminary version of the paper was presented at *1992 American Control Conference*, Chicago, IL., pp. 1182-1186, June, 1992.

[12] Z.Qu and J.F.Dorsey, "Robust control of generalized dynamic systems without matching conditions," *Transactions of ASME, Journal of Dynamic Systems, Measurement, And Control*, Vol.113, No.4, pp.582-589, December, 1991.

[13] Z.Qu, "Model reference robust control of SISO systems with significant unmodelled dynamics," submitted to *1993 American Control Conference*, 8/30/92.

[14] Z.Qu, "Global stabilization of nonlinear systems with a class of unmatched uncertainties," *Systems & Control Letter*, Vol.18, No.3, pp.301-307, May, 1992.

[15] Z.Qu and D.M.Dawson, "Robust control design of a class of cascaded nonlinear uncertain systems," *Control of Systems with Inexact Dynamic Models, 1991 ASME Winter Annual Meeting*, pp.63-71, Atlanta GA. December 1991.

[16] Z.Qu, "Robust control of nonlinear uncertain systems under generalized matching conditions," submitted to *Automatica*, 4/21/91, revised 5/18/92.

[17] S. Sastry and M. Bodson, *ADAPTIVE CONTROL: Stability, Convergence, and Robustness*, Prentice-Hall, Inc. 1989.

[18] W.E.Schmitendorf and B.R. Barmish, "Robust asymptotic tracking for linear systems with unknown parameters," *Automatica*, Vol. 22, pp.355-360, 1986.

[19] W.E.Schmitendorf and B.R. Barmish, "Guaranteed asymptotic output stability for systems with constant disturbance," *J. of Dynamic Systems, Measurement, and Control*, Vol. 109, pp.186-189, 1987.

[20] J.J.Slotine, *Applied Nonlinear Control*, Englewood Cliffs, NY, Prentice-Hall, Inc. 1991.

[21] A.V.Solodov, *Linear Automatic Control Systems with Varying Parameters*, New York: American Elsevier, 1966.

[22] G.W.Stewart, *Introduction to Matrix Computations*, New York: Academic 1973.

[23] K.S.Tsakalis and P.A.Ioannou, "Adaptive control of linear time-varying plants," *Automatica*, Vol.23, pp.459-468, July, 1987.

[24] K.S.Tsakalis and P.A.Ioannou, "Adaptive control of linear time-varying plants: A new model reference controller structure," *IEEE Transactions on Automatic Control*, Vol.34,

pp.1038-1046, October, 1989.

[25] K.S.Tsakalis and P.A.Ioannou, "A new adaptive control scheme for time-varying plants," *IEEE Transactions on Automatic Control*, Vol.35, No.6, pp.697-705, 1989.

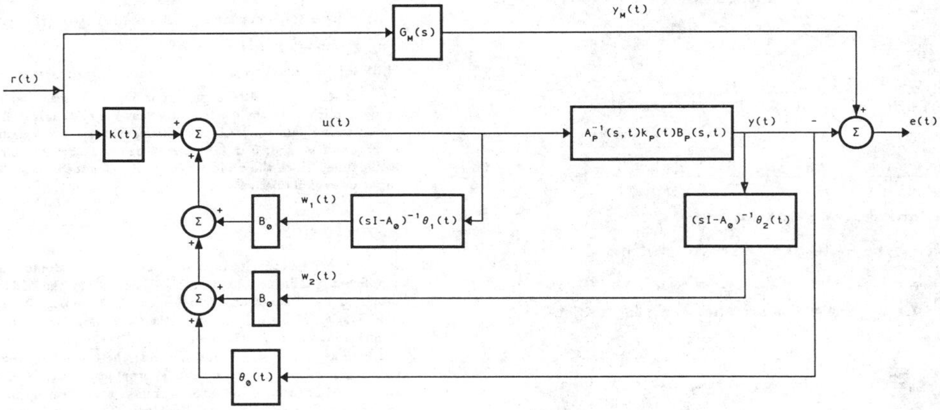

Figure 1: The MRC structure for known LTV systems proposed by Tsakalis and Ioannou

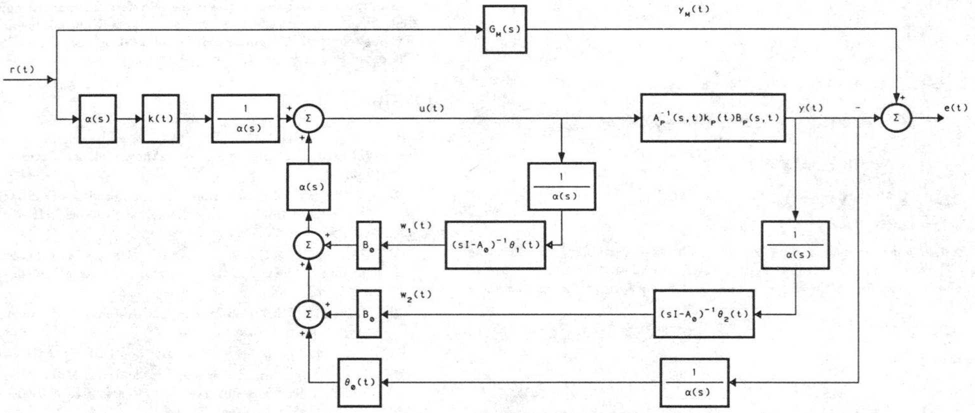

Figure 2: A new MRC structure for known LTV systems for time-varying systems

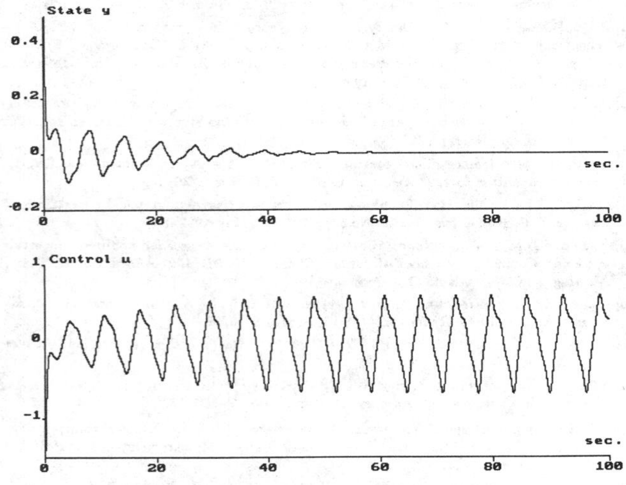

Figure 3: Illustration of robust control performance

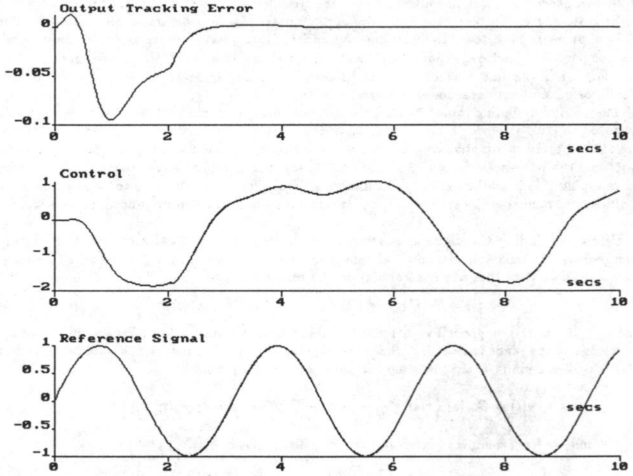

Figure 4: Simulation of Example 1

Proceedings of the 31st Conference
on Decision and Control
Tucson, Arizona • December 1992

ADAPTIVE CONTROL OF SYSTEMS OF THE FORM $\dot{x} = \theta_1^{*T} f(x) + \theta_2^{*T} g(x)\, u$ WITH REDUCED KNOWLEDGE OF THE PLANT PARAMETERS

B.BROGLIATO* and R.LOZANO#

*LAG ; URA CNRS 228
BP 46 Domaine universitaire
38402 St Martin d'Hères - FRANCE

#UT Compiègne; Heudiasyc URA CNRS 817
Centre de Recherches de Royallieu, BP 649
60206 Compiègne - FRANCE

Abstract
This paper presents an adaptive control strategy for a class of first order nonlinear systems of the form $\dot{x} = \theta_1^{*T} f(x) + \theta_2^{*T} g(x)\, u$. It is assumed that the system remains controllable for all values of x, but the sign of $\theta_2^{*T} g(x)$ is unknown. Non globally Lipschitz nonlinearities are tolerated in the system through g(x).

1) INTRODUCTION

Adaptive control techniques applied to nonlinear systems have recently emerged in the literature see e.g. [3]-[4]. Concerning relaxation of the *a priori* knowledge of the sign of the term premultiplying the control input (for nonlinear systems linear in the input), in [5] the author presented the first scheme applied to general nonlinear first order systems, based on Nussbaum's gain. Recently the ideas in [6] have been clarified and applied to adaptive control of a simple nonlinear system of order one [1].
In this paper, we present an adaptive control scheme for a class of nonlinear first order systems of the form $\dot{x} = \theta_1^{*T} f(x) + \theta_2^{*T} g(x)\, u$. The work that follows can be seen as an extension of our previous work on the subject [1].

2) PROBLEM FORMULATION

Consider the following class of nonlinear systems

$$\dot{x} = \theta_1^{*T} f(x) + \theta_2^{*T} g(x)\, u \tag{1}$$

where $x, u \in \mathbf{R}$, θ_1^* and θ_2^* are constant vectors of dimension n-m and m respectively.
First let us rewrite (1) as

$$\dot{x} = \theta_1^{*T} f(x) + \theta_2^{*T} g_n(x)\, \bar{u}$$

where

$$g_n(x) = \frac{g(x)}{|g(x)|} \quad , \quad \bar{u} = |g(x)|\, u$$

As long as $g(x) \neq 0$, both systems are strictly equivalent one to each other but, as we shall see later, the second form is more suitable for our purpose. We now make the following assumptions on the system

A.1
$$|f(x)| \leq k|x| + k \qquad \text{for some } k>0$$

A.2 g(x) is a smooth vector function

A.3
$$|g(x)| \geq \varepsilon \qquad \left|\theta_2^{*T} g_n(x)\right| \geq \varepsilon$$

Following the ideas in [2] we define a domain $D^0(\varepsilon)$ as

$$D^0(\varepsilon) = \left\{ v \in \mathbf{R}^m : \left| v^T g_n(x) \right| \geq \varepsilon ,\, x \in \mathbf{R} \right\}$$

Note that by assumption A.3 θ_2^* belongs to $D^0(\varepsilon)$. Let us now introduce a second domain denoted $D(\varepsilon_1, \varepsilon_2, z)$:

$$D(\varepsilon_1, \varepsilon_2, z) = \left\{ v \in \mathbf{R}^m : \frac{|z^T v|}{|v|} \geq 1 - \varepsilon_1,\ |v| \geq \varepsilon_2 \right\}$$

with $|z| = 1$, $\varepsilon_2 > 0$, $0 \leq \varepsilon_1 < 1$

CH3229-2/92/0000-2773$1.00 © 1992 IEEE

which allows us to introduce our last assumption on the system

A.4 A vector $z \in \mathbf{R}^m$, constants $\varepsilon_2 > 0$, $\varepsilon_1 \in [0, 1)$, $\varepsilon'_1 \in [0, 1)$ with $\varepsilon_1 - \varepsilon'_1 = h_1 > 0$ are known such that

$$D(\varepsilon_1, \varepsilon_2, z) \subset D(\varepsilon'_1, \varepsilon_2, z) \subseteq D(\varepsilon)$$

$$\text{and } \overset{*}{\theta_2} \in D(\varepsilon_1, \varepsilon_2, z)$$

The desired state trajectory is given by the linear model

$$\dot{x}_m = -a\, x_m + r$$

where $a > 0$ and r is a bounded reference input. With $e = x - x_m$, then from (1) we obtain

$$\dot{e} = -a\, e + \theta^{*T} \phi + a\, x - r$$

where

$$\theta^{*T} = \left[\, \theta_1^{*T}, \theta_2^{*T} \right] \text{ and } \phi^T = \left[\, f^T(x), g_n^T(x)\, \bar{u} \,\right]$$

We can also write

$$e = h * \left(\theta^{*T} \phi + a\, x - r \right)$$

where h*w denotes the convolution between h and w, and h is the inverse Laplace transform of $H(s) = (s + a)^{-1}$. If the control input verifies

$$\bar{u} = \frac{-\theta_1^{*T} f(x) - a\, x + r}{\theta_2^{*T} g_n(x)}$$

Then e will converge asymptotically towards zero. Since the parameters in θ^* are unknown, and in order to get a control input that is still implementable, the adaptive control law will be given as follows

$$\bar{u} = \frac{-\bar{\theta}_1^{T} f(x) - a\, x + r}{\bar{\theta}_2^{T} g_n(x)}$$

The modification in the standard estimated parameters θ is given by

$$\bar{\theta} = \theta + c = \theta + \begin{bmatrix} c_1 \\ c_2 \end{bmatrix}$$

where $c_1 \in \mathbf{R}^{n-m}$, $c_2 \in \mathbf{R}^m$ and θ is the standard Least-squares (LS) estimates vector. c is a correction vector that we shall define in the next section.

The aim of the modification is to insure that the modified estimates $\bar{\theta}_2$ belong to $D(\varepsilon''_1, \varepsilon_3, z)$ with $\varepsilon_3 > 0$, $\varepsilon'_1 < \varepsilon''_1 < \varepsilon_1$ (i.e. $D(\varepsilon'_1, \varepsilon_2, z) \supset D(\varepsilon''_1, \varepsilon_2, z) \supset D(\varepsilon'_1, \varepsilon_2, z)$.

In other words, c should be designed such that the following inequalities are satisfied

$$C_{r1}: \frac{\left| z^T \bar{\theta}_2 \right|}{\left| \bar{\theta}_2 \right|} \geq 1 - \varepsilon''_1 \quad \text{and} \quad C_{r2}: \left| \bar{\theta}_2 \right| \geq \varepsilon_3$$

error equation

We get (see e.g. [1] for details)

$$e_a = \tilde{\theta}^T \xi \qquad \xi = h * (\phi)$$

3) PARAMETERS ESTIMATION AND CORRECTION PROCEDURE

The following normalized Least-Squares estimation algorithm will be used to update θ

$$\dot{\theta} = -\frac{P\,\xi}{m^2} e_a \quad \dot{P} = -\frac{P\,\xi\,\xi^T P}{m^2} \;, \quad P(0) > 0$$

$$m = \max_{0 \leq \tau \leq t} \left(1 + | x(\tau) | \right) \tag{2}$$

We are now able to write the full set of ordinary differential equations that describe our system

$$\dot{x} = \theta_1^{*T} f(x) + \theta_2^{*T} g_n(x) \left[\frac{-\bar{\theta}_1^{T} f(x) - a\, x + r}{\bar{\theta}_2^{T} g_n(x)} \right]$$

$$\dot{x}_m = -a\,x_m + r \ , \quad x_m(0) = x(0)$$

$$\dot{z}_1 = -a\,z_1 + c^T \phi \ , \quad z_1(0) = 0 \qquad (3)$$

$$\dot{z}_2 = -a\,z_2 - \dot{\theta}^T \xi \ , \quad z_2(0) = 0$$

$$\dot{\xi} = -a\,\xi + \phi \ , \quad \xi(0) = 0$$

and (2)

Parameters estimates modification

Let us first define the matrices P_1 and P_2 as

$$P_1 = E_1 P \quad \text{and} \quad P_2 = E_2 P$$

where $E_1 = [\, I_{n-m} , 0_{(n-m)xm}] \in \mathbf{R}^{(n-m)x\,n}$

$$E_2 = [\, 0_{mx(n-m)} , I_m \,] \in \mathbf{R}^{mx\,n}$$

Let us now assume that rank(P_2) = $r \leq m$ and, without loss of generality, that the first r columns of P_2 are independent. We thus define a new full rank matrix $\overline{P}_2$ as follows

$$\overline{P}_2 = P_2\,\overline{E}_2 = E_2\,P\,\overline{E}_2$$

where

$$\overline{E}_2^T = [\, I_r , 0_{rx(n-r)} \,] \in \mathbf{R}^{rxn}$$

The choice of the modification

Let us recall a property of the estimation algorithm in (2).

lemma 1

Assume that the solution of the system (3) exists in a domain $D_1 = [0, t_1] \times B$, where B is a ball centered at the origin in the full-state space of (3). Then as long as the solution exists we have

$$P^{-1}\,\tilde{\theta} = P^{-1}(0)\,\tilde{\theta}(0) = -\beta^* \qquad (4)$$

and this result holds independently of the size of B.

<u>proof</u> please see [1], lemma 2.

∇

Premultiplying both sides of (4) by E_2 we obtain

$$\theta_2^* = \theta_2 + \overline{P}_2\,\overline{\beta}^*$$

where $\overline{\beta}^* = \left(\overline{\beta}_1^*, \ldots , \overline{\beta}_r^*\right)^T$ is calculated from β^* as follows

$$\overline{\beta}_i^* = \beta_i^* + \sum_{k=r+1}^{n} \beta_k^*\,\alpha_k^i \quad , 1 \leq i \leq r$$

the α_k^i 's , k = r + 1, ..., n and i = 1, ... , r are the components of the last n - r columns of P_2 in the base composed by its first independent columns. Now we use the fact that $\theta_2^* \in D(\varepsilon_1, \varepsilon_2, z)$. Thus

$$C_{r1} \quad \frac{\left|z^T \theta_2^*\right|}{\left|\theta_2^*\right|} = \frac{\left|z_\perp^T \theta_{2\perp}^* + z_p^T \overline{P}_2 \left(\overline{\beta}^* + \overline{P}_2^+ \theta_2\right)\right|}{\left(\left\|\theta_{2\perp}^*\right\|^2 + \left\|\overline{P}_2 \left(\overline{\beta}^* + \overline{P}_2^+ \theta_2\right)\right\|^2\right)^{\frac{1}{2}}} \geq 1 - \varepsilon_1$$

$$C_{r2} \quad \left\|\theta_{2\perp}^*\right\|^2 + \left\|\overline{P}_2 \left(\overline{\beta}^* + \overline{P}_2^+ \theta_2\right)\right\|^2 \geq \varepsilon_2^2 \qquad (5)$$

Let us choose c as

$$c_2 = \overline{P}_2\,\overline{\beta} = P_2\,\beta \quad \text{with } \beta^T = \left[\,\overline{\beta}_1, \ldots, \overline{\beta}_r\, 0 \ldots 0\,\right] \qquad (6)$$

$$c_1 = P_1\,\beta$$

where k is a scalar (possibly time-varying) to be designed later. Introducing (6) into C_{r1} we obtain

$$\frac{\left|z^T \overline{\theta}_2\right|}{\left\|\overline{\theta}_2\right\|} = \frac{\left|z_\perp^T \theta_{2\perp}\right| + \left|\left\|z_p\right|^2 k\right|}{\left(\left\|\theta_{2\perp}\right\|^2 + \left|z_p\right\|^2 k^2\right)^{\frac{1}{2}}} = \frac{\left|z_\perp^T \theta_{2\perp}\right| + \left\|z_p\right\| k'}{\left(\left\|\theta_{2\perp}\right|^2 + k'^2\right)^{\frac{1}{2}}}$$

with $k' = k \left\|z_p\right\|$

In view of (5) it follows that there exists a

constant k'

$$\frac{\left|z_\perp^T \theta_{2\perp}\right| + \left|z_p\right| k'}{\left(\left|\theta_{2\perp}\right|^2 + k'^2\right)^{\frac{1}{2}}} \geq 1 - \varepsilon_1 \qquad (7)$$

which can also be written as

$$A k'^2 + B k' + C \geq 0 \qquad (8)$$

Let us denote r_1 and r_2 the roots (real or complex) of the polynomial in the RHS of (8). Consider now the following set of values for k'

$$k'_0 = \varepsilon_2 \quad k'_1 = 0 \quad k'_2 = \max\left[\, r_1, \varepsilon_2 \,\right]$$

$$k'_3 = \max\left[-\frac{C}{B}, \varepsilon_2 \right] \qquad k'_4 = \max\left[\, r_1, r_2 \,\right]$$

$$k'_5 = \max\left[\, r_1, r_2, \varepsilon_2 \,\right] \qquad (9a)$$

and the corresponding set of values for $\bar{\beta}$

$$\bar{\beta}_i = -\bar{P}_2^+ \theta_2 - \bar{P}_2^+ z \frac{k'_i}{|z_p|} \qquad (9b)$$

$$\bar{\beta}_{i+6} = -\bar{P}_2^+ \theta_2 + \bar{P}_2^+ z \frac{k'_i}{|z_p|}, \quad 0 \leq i \leq 5$$

The next lemma states one of the main results of the paper

Lemma 2

Consider the set of possible modifications in (9). Then as long as a solution of the system (3) exists and is unique, there always exists at least one bounded $\bar{\beta}_i$ in (9) such that conditions C_{r1} and C_{r2} are satisfied, i.e.

$$C_{r1i} = \frac{\left| z^T\left(\theta_2 + \bar{P}_2 \bar{\beta}_i\right)\right|}{\left|\theta_2 + \bar{P}_2 \bar{\beta}_i\right|} \geq 1 - \varepsilon_1 - \frac{h_1}{2}$$

$$C_{r2i} = \left|\theta_2 + \bar{P}_2 \bar{\beta}_i\right| \geq \varepsilon_3 > 0$$

where ε_3 is defined in the next procedure.

∇

We now propose an algorithm for the choice of $\bar{\beta}$. The first part is a procedure that enables us to obtain a suitable modification vector $\bar{\beta}_s$ at each time instant; the second part is an algorithm used to switch from the current value of $\bar{\beta}$ to the value $\bar{\beta}_s$ chosen by the procedure running in parallel.

step 1
$k' = k'_1 = 0$

If $C_{r11} \geq 1 - \varepsilon_1 - h_1/2$ and $C_{r21} \geq (1 - \varepsilon_1)\varepsilon_2 - h_2$, where h_1 is defined in assumption A.4 and h_2 is such that $0 < h_2 < (1 - \varepsilon_1)\varepsilon_2$, then $\bar{\beta}_s = \bar{\beta}_1$.

step 2
$k' = k'_0 = \varepsilon_2$

If $C_{r10} \geq 1 - \varepsilon_1$ then $\bar{\beta}_s = \bar{\beta}_0$

If $C_{r16} \geq 1 - \varepsilon_1$ then $\bar{\beta}_s = \bar{\beta}_6$

step 3
$k' = k'_2$ or $k' = k'_3$

Then $\bar{\beta} \in \left\{ \bar{\beta}_2, \bar{\beta}_3, \bar{\beta}_8, \bar{\beta}_9 \right\} = S_3$

$\bar{\beta}_s = \bar{\beta}_i$ such that $C_{r1i} \geq C_{r1j}$ for all $\bar{\beta}_j \in S_3$, $j \neq i$

step 4
$k' = k'_4$ or $k' = k'_5$

Then $\bar{\beta} \in \left\{ \bar{\beta}_4, \bar{\beta}_5, \bar{\beta}_{10}, \bar{\beta}_{11} \right\} = S_4$

$\bar{\beta}_s = \bar{\beta}_i$ such that $C_{r1i} \geq C_{r1j}$ for all $\bar{\beta}_j \in S_4$, $j \neq$

comments
i) The first step aims at avoiding any division by zero in $\bar{\beta}$.

ii) One sees that the criterion at step 1 is different from the ones at steps 2, 3 or 4. This is due to the fact that we have to choose $k' = 0$ when $|z_p|$ is close to zero in order to avoid any division by zero. Conversely, we must guarantee that if C_{r11} or C_{r21} do not verify the criterion, then $|z_p|$ is clearly strictly positive.

∇

We now describe the switching algorithm that allows us to guarantee that at each time instant the current modification is suitable, while avoiding infinite switching during finite time interval.

Let us denote the current modification as $\bar{\beta}_i$. Then

$$\bar{\beta} = \bar{\beta}_j \text{ if } \bar{\beta}_j = \bar{\beta}_s \text{ and } \{C_{r1i} = 1 - \varepsilon'_1$$
$$\text{or } C_{r2i} = (1 - \varepsilon_1)\varepsilon_2 - h_2' \text{or } |\bar{\beta}_i| \geq |\bar{\beta}_s| + \delta)$$

where

$$0 < h_2 < h'_2 < (1 - \varepsilon_1)\varepsilon_2 \quad \text{and } \delta > 0 \text{ arbitrary}$$

and bounded.

comments

i) It may happen that the current value of $\bar{\beta}$ either becomes unbounded or goes to zero while still satisfying C_{r1} (i.e. the modified parameters estimates vector has the right direction, but it becomes too small in norm or it diverges). This is the reason why we have included informations about $|\bar{\theta}_2|$ and $|\bar{\beta}_i|$ in the switching algorithm.

4) EXISTENCE AND UNIQUENESS OF SOLUTIONS

Lemma 3

The closed-loop control system differential equations (3) have a solution over the interval $[0, T]$ with $T \in \mathbf{R}^+$. Moreover the state $\bar{x}$ of the closed-loop system (3) does not grow faster than exponentially.

proof Please see [1] lemma 5.

5) CONVERGENCE ANALYSIS

Lemma 4

1) $\frac{e_a}{m} \in L_2$, $\frac{P\xi}{m} \in L_2$

2) $\frac{\phi}{m}$ and $\frac{\xi}{m}$ are bounded

3) P and θ converge

4) $\bar{\theta}_2$ converges and the discontinuities in $\bar{\beta}$ occur a finite number of times.

∇

We can now establish the main result of the paper

Lemma 5

1) $\frac{e}{m} \in L_2$

2) e asymptotically converges towards zero, and all the signals in the closed-loop system are bounded.

6) FEASIBILITY AND EXAMPLES

The whole analysis relies on exact knowledge of the rank of the matrix P_2. However, it may happen that P_2 is not "well-conditioned". It may even be difficult to find the right rank of P_2. In the following we outline a method that enables us to overcome that difficulty. First recall that we have

$$\theta_2^* = \theta_2 + P_2\beta^* \tag{10}$$

and assuming that the first r columns of P_2 are independent - the other n-r ones belong to the space spanned by $P_{21} \ldots P_{2r}$ - (10) can be rewritten as

$$\theta_2^* = \theta_2 + \bar{P}_2\bar{\beta}^* \tag{11}$$

Assume now that the following situation occurs: $\{P_{21} \ldots P_{2(r-1)}\}$ clearly form a set of (r-1) independent vectors. $P_{2(r+1)} \ldots P_{2n}$

clearly belong to the space spanned by $\{P_{21} \ldots P_{2(r-1)}\}$. But it is difficult to establish whether P_{2r} is independent of $P_{21} \ldots P_{2(r-1)}$ or not. It can be shown that we can write for some γ^*

$$P_2\beta^* = P_2^{r-1}\gamma^* + \varepsilon_r \quad , \quad \gamma^{*T} = (\gamma_1^* \ldots \gamma_{r-1}^*)$$

ε_r is proportional to the part of P_{2r} which is orthogonal to the image of P_2^{r-1}, which by assumption is close to zero. Now still using the fact that θ_2^* belongs to $D(\varepsilon_1, \varepsilon_2, z)$ it follows that provided ε_r is small enough, there exists a bounded vector γ^* such that the modified parameter vector

$$\overline{\theta}_2 = \theta_2 + P_2^{r-1}\gamma^*$$

belongs to a domain $D(\varepsilon''_1, \varepsilon_2, z)$ contained in $D(\varepsilon'_1, \varepsilon_2, z)$. Moreover the matrix P_2^{r-1} is "well-conditioned" in the sense that its rank is clearly known - we have taken into account only the columns which are clearly independent -.Therefore the whole analysis can be done using the "reduced"matrix P_2^{r-1} instead of the "ideal" matrix $P_2^r = \overline{P}_2$. Then, instead of considering a set of eleven possible modifications in (9), we can augment that set with modifications computed with P_2^{r-1}, i.e. we obtain a set of 22 possible values for $\overline{\beta}$. The four-steps procedure described in section 3 has to be started with the values computed with P_2^{r-1}, and then continued with the values computed with $P_2^r = \overline{P}_2$.

The above procedure clearly applies to the case studied in [1], using the just above remark on feasibility of the scheme. However the scheme presented here results in a different scheme from the one presented in [1].

Consider the system in (1) with

$$g^T(x) = [1, x^2] \ , \ g_n^T(x) = \left[\frac{1}{(1+x^2)^{1/2}}, \frac{x^2}{(1+x^2)^{1/2}}\right]$$

The domains $D^0(\varepsilon)$ and $D(\varepsilon_1, \varepsilon_2, z)$ are depicted below.

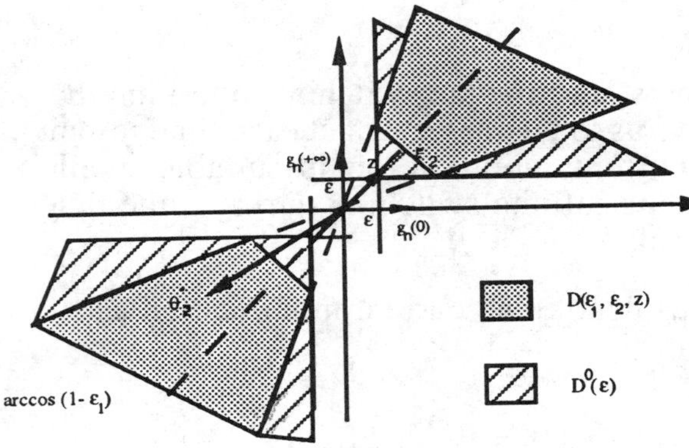

bibliography

[1] Lozano R., Brogliato B. *"Adaptive control of a simple nonlinear system with reduced a priori knowledge of the plant parameters"* IEEE Trans. Aut. Contr., January 1992.

[2] Lozano R., Brogliato B. *"Adaptive control of flexible joint manipulators"* IEEE Trans. Aut. Contr., February 1992.

[3] Kanellakopoulos I., Kokotovic P.V., Morse S. *"Systematic design of adaptive controllers for feedback linearizable nonlinear systems"* IEEE Trans. Aut. Contr., November 1991.

[4] Sastry S.S., Isidori A. *"Adaptive control of linearizable systems"* IEEE Trans. Aut. Contr., November 1989.

[5] Martensson B. *"Remarks on adaptive stabilization of first order nonlinear systems"* Syst. Contr. Lett., 14, 1990.

[6]Lozano R., Collado J., Mondie S. *"Model reference adaptive control without a priori knowledge of the high frequency gain"* IEEE Trans. Aut. Contr., January 1990.

Proceedings of the 31st Conference
on Decision and Control
Tucson, Arizona • December 1992

ADAPTIVELY DISTURBANCE ATTENUATION DESIGN OF ROBUSTLY STABLE CONTROL SYSTEM

Pria Mulyono, Hiromitsu Ohmori and Akira Sano
Department of Electrical Engineering, Keio University,
3-14-1 Hiyoshi, Kohoku-ku, Yokohama 223, Japan

Abstract

This paper proposes a direct adaptive control scheme of robustly stable control systems. The adaptive controller feeds back a disturbance estimate via a filter, which includes free parameters of the robust stabilizing controller. The proposed scheme extends fixed robust stabilizing controllers, in that the free parameters can be adjusted adaptively. Since disturbance is estimated in on-line manner, the disturbance can be attenuated without using *a prior* knowledge of disturbance or its generator. The effectiveness of the proposed control system design is confirmed by numerical simulations.

I. Introduction

The novel class of adaptive controller is proposed by Tay et al. [1],[2]. The adaptive controllers, one of them is called adaptive disturbance estimate feedback (DEF), are the blending techniques of on-line adaptive control with off-line designed controller so that the performance of the latter is enhanced.

In control system design, we should be encountered with uncertainty in the model of the plant or perturbed plant. The model uncertainty can sometimes make the designed control system unstable. Thus, it is necessary to design control system for the plant with uncertainty, such that the control system be stable. This is well-known as the robust stability problem discussed in [3].

Zames et al.[5] proposed the minimization of the sensitivity function that is measured in a weighted H_∞ norm. The selection of the weight is a specific problem and therefore it is difficult to provide a definitive set of rules for building and modifying weights. For certain minimization problems, the method for a weighting function selection is proposed in [7]. Unfortunately, the selection need *a prior* knowledge of the disturbance or at least its estimate. On the other hand, the disturbance may change during the operation of the system. Therefore, it is necessary that the control system design can adaptively attenuate the disturbance and always maintain the robust stability. This is the aim of this paper.

Since the DEF controller in [1] can not insure the robust stability of control systems with model uncertainty, this paper proposes a new type of DEF controller by which the robust stability of the control system is always maintained. It is known any robust stabilizing controller can be parametrized in free parameter S. Therefore, by adaptively adjusting the free parameters in S, the disturbance can be attenuated adaptively with keeping the control system robustly stable.

II. Problem Statement

Consider a plant with additive plant uncertainty in the presence of small bounded disturbance as:

$$y(k) = G(z)u(k) + d(k); \quad G(z) = G_0(z) + \Delta(z) \quad (1)$$

where u is input, y is output, and d is disturbance. $G(z), G_0(z)$, and $\Delta(z)$ are additive perturbed plant, nominal plant and additive plant uncertainty, respectively. Suppose that

(A1) Nominal plant $G_0(z)$ is known, no poles and zeros on unit circle, proper, pulse transfer function.

(A2) Additive plant uncertainty $\Delta(z)$ is stable, which satisfies $\mathcal{D} \equiv \{\Delta(z) : |\Delta(e^{j\omega})| < |r_a(e^{j\omega})|, \ ^\forall\omega\}$, where $r_a(z)$ is an upper bound of the uncertainty, outer and known.

(A3) $\{G_0(z), r_a(z)\}$ is robust stabilizable.

(A4) $d(k)$ is unknown uniformly bounded disturbance.

On the above assumptions, feedback system of Fig. 1(a) will be made robustly stable, and secondly the control system that can attenuate unknown disturbance by adaptively adjusting the free parameters in S of the designed robust stabilizing controller, will be designed. Here, $r(k)$ is a reference input.

III. Direct Adaptive Control Scheme

Let $\|X\|_\infty = \sup_{\omega\in\mathcal{R}}[\lambda_{max}(X^*(e^{j\omega})X(e^{j\omega}))]^{1/2}$, $\mathcal{RH}_\infty$ is a set of proper asymptotically stable rational functions, $\mathcal{BH}_\infty$ is a set of $\mathcal{H}_\infty$ functions satisfying $\|X\|_\infty < 1$. Any robust stabilizing controller C can be parametrized in a free parameter $S(z) \in \mathcal{BH}_\infty$ [8], as described in Fig. 1(a), where

$$K(z) = \begin{bmatrix} K_{11}(z) & K_{12}(z) \\ K_{21}(z) & K_{22}(z) \end{bmatrix} \quad (2)$$

includes informations of the nominal plant G_0 and the upper bound function r_a.

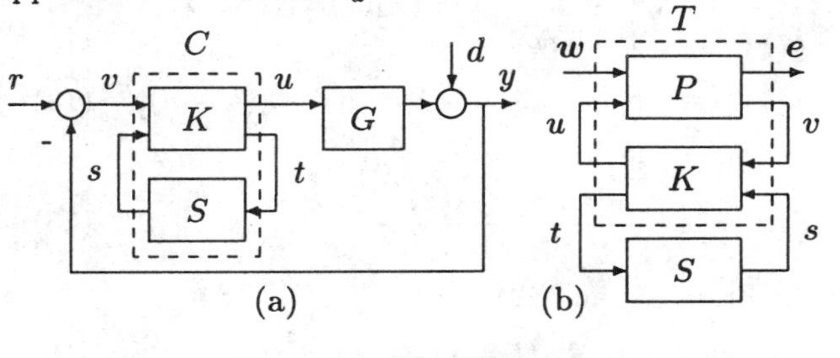

Fig. 1 Feedback system

Let consider a new direct adaptive scheme for adjusting free parameter $S(z)$ adaptively such that unknown disturbance d can be attenuated.

Let denote augmented plant as P. Then, Fig. 1(a) can be described as in Fig. 1(b). Here, $\boldsymbol{w}(k) = [r(k)\ d(k)]^T$, and $\boldsymbol{e}(k)$ is vector should be minimized. Then,

$$\begin{bmatrix} e \\ v \end{bmatrix} = P(z)\begin{bmatrix} w \\ u \end{bmatrix} = \begin{bmatrix} P_{11}(z) & P_{12}(z) \\ P_{21}(z) & P_{22}(z) \end{bmatrix}\begin{bmatrix} w \\ u \end{bmatrix} \quad (3)$$

$$\begin{bmatrix} u \\ t \end{bmatrix} = K(z)\begin{bmatrix} v \\ s \end{bmatrix} = \begin{bmatrix} K_{11}(z) & K_{12}(z) \\ K_{21}(z) & K_{22}(z) \end{bmatrix}\begin{bmatrix} v \\ s \end{bmatrix} \quad (4)$$

$$s = S(z)t. \quad (5)$$

Eliminating v, u of (3) and (4)

$$\begin{bmatrix} e \\ t \end{bmatrix} = T(z)\begin{bmatrix} w \\ s \end{bmatrix} = \begin{bmatrix} T_{11}(z) & T_{12}(z) \\ T_{21}(z) & T_{22}(z) \end{bmatrix}\begin{bmatrix} w \\ s \end{bmatrix} \quad (6)$$

is obtained, where

$$T_{11}(z) = P_{11}(z) + \frac{P_{12}(z)K_{11}(z)P_{21}(z)}{1 - P_{22}(z)K_{11}(z)}$$

$$T_{12}(z) = \frac{P_{12}(z)K_{21}(z)}{1 - P_{22}(z)K_{11}(z)}, \quad T_{21}(z) = \frac{K_{21}(z)P_{21}(z)}{1 - P_{22}(z)K_{11}(z)}$$

$$T_{22}(z) = \frac{K_{21}(z)P_{22}(z)K_{12}(z)}{1 - P_{22}(z)K_{11}(z)} + K_{22}(z)$$

Note that $s = S(z)t$. Then, transfer function from $\boldsymbol{w}$ to $\boldsymbol{e}$, let $T_s(z)$, is

$$\boldsymbol{e} = T_s(z)\boldsymbol{w}, \quad T_s(z) = T_{11}(z) + \frac{T_{12}(z)S(z)T_{21}(z)}{1 - T_{22}(z)S(z)} \quad (7)$$

Finally, subtituting $T_{11}(z), \cdots, T_{22}(z)$ to (7), and simple manipulation, we have the following lemma.

Lemma 1: $\boldsymbol{e}$ can be described as follows (let omit (z)).

$$\boldsymbol{e} = \left(P_{11} + \frac{P_{12}K_{11}P_{21}}{1 - P_{22}K_{11}}\right)\boldsymbol{w} + \frac{P_{12}K_{12}}{1 - P_{22}K_{11}}St \quad (8)$$

(Proof is ommitted).

The following theorem is basic of the propose adaptive rule.

Theorem 1: Let free parameter S be a FIR filter described by

$$S(z) = \theta_0 + \theta_1 z^{-1} + \cdots + \theta_n z^{-n} \quad (9)$$

then

$$e(k) = \zeta(k) + \varphi^T(k)\boldsymbol{\theta} \quad (10)$$

is obtained, where

$$\begin{aligned}
\zeta(k) &\equiv \left(P_{11}(z) + \frac{P_{12}(z)K_{11}(z)P_{21}(z)}{1 - P_{22}(z)K_{11}(z)}\right) w(k) && (11)\\
\boldsymbol{\xi}(k) &\equiv F(z)t(k) && (12)\\
\varphi(k) &\equiv [\boldsymbol{\xi}(k),\ \boldsymbol{\xi}(k-1),\ \cdots,\ \boldsymbol{\xi}(k-n)]^T && (13)\\
\boldsymbol{\theta} &\equiv [\theta_0,\ \theta_1,\ \cdots,\ \theta_n]^T && (14)\\
F(z) &\equiv \frac{P_{12}(z)K_{12}(z)}{1 - P_{22}(z)K_{11}(z)} && (15)
\end{aligned}$$

Proof : From (5) and (9) ,

$$s(k) = \theta_0 t(k) + \theta_1 t(k-1) + \ldots + \theta_n t(k-n) \quad (16)$$

is obtained. From Lemma 1 and (11)–(16)

$$\begin{aligned}
e(k) &= \zeta(k) + F(z)S(z)t(k)\\
&= \zeta(k) + (F(z)\theta_0 t(k) + \ldots + F(z)\theta_n t(k-n))\\
&= \zeta(k) + [\boldsymbol{\xi}(k)\ \ \boldsymbol{\xi}(k-1)\ \ \ldots\ \ \boldsymbol{\xi}(k-n)]\boldsymbol{\theta}\\
&= \zeta(k) + \varphi^T(k)\boldsymbol{\theta}
\end{aligned}$$

follows. □

Remark 1: For (15), $-G_0$ is used instead of P_{22}.

Remark 2: Since free parameter $S(z)$ is FIR function, $S(z)$ is always stable if its coeficients do not diverge.

To generate optimal $S(z)$ recursively, the following recursive least squares estimation algorithm is applied.i

$$\begin{aligned}
\hat{\boldsymbol{\theta}}(k) &= \hat{\boldsymbol{\theta}}(k-1) + \Gamma(k)\varphi(k)\epsilon(k), && (17)\\
\epsilon(k) &= \hat{\zeta}(k) - \varphi(k)^T\hat{\boldsymbol{\theta}}(k-1) && (18)\\
\hat{\zeta}(k) &= e(k) - F(z)s(k) && (19)\\
\Gamma(k) &= \Gamma(k-1) - \Gamma(k-1)\varphi(k)\\
&\quad \times[I + \varphi^T(k)\Gamma(k-1)\varphi(k)]^{-1}\varphi^T(k)\Gamma(k-1) && (20)
\end{aligned}$$

Note that $\Gamma(0) > 0$ is symmetric matrix. Infinity norm of S with adjustable parameters $\bar{\boldsymbol{\theta}}(k)$ generated by the above least squares algorithm, is not always less than 1. To insure the restriction on S always holds, the generated parameters must be transformed appropriately, by the following theorem.

Theorem 2: For free parameters of (9), sufficiency condition of the restriction on $S \in \mathcal{BH}_\infty$, i.e. $\|S\|_\infty < 1$ is

$$\sum_{i=0}^{n} \theta_i^2 < \frac{1}{\sqrt{n+1}} \quad . \quad (21)$$

Proof : Let rewrite (9) as follows.

$$S = \boldsymbol{\theta}^T\boldsymbol{\ell}(z), \quad \boldsymbol{\ell}(z) \equiv [1\ \ z^{-1}\ \ \cdots\ \ z^{-n}]^T \quad (22)$$

Then

$$\|\boldsymbol{\ell}\|_\infty = \sqrt{n+1}, \ \|\boldsymbol{\theta}\|_\infty = \sqrt{\sum_{i=0}^{n}\theta_i^2}, \ \|S\|_\infty \le \|\boldsymbol{\theta}^T\|_\infty\|\boldsymbol{\ell}\|_\infty.$$

Therefore if (21) holds, the restriction on S is maintained. □

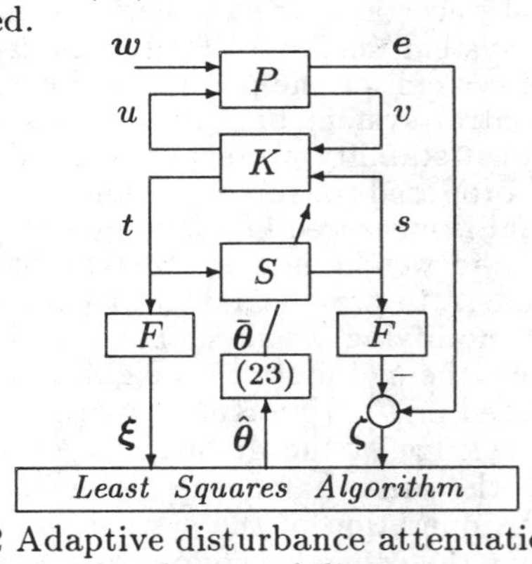

Fig. 2 Adaptive disturbance attenuation system with robust stability structure

To guarantee the restriction on S it is necessary adjusting magnitude of the adjustable parameter vector. Actually, the following algorithm is included to the recursive least squares algorithm.

$$\bar{\boldsymbol{\theta}}(k) = \begin{cases} \dfrac{\gamma\hat{\boldsymbol{\theta}}(k)}{\sqrt{(n+1)\sum_{i=0}^{n}\hat{\theta}_i^2(k)}}, & \|\hat{\boldsymbol{\theta}}\|_\infty \ge \dfrac{1}{\sqrt{n+1}} \\ \hat{\boldsymbol{\theta}}(k), & \|\hat{\boldsymbol{\theta}}\|_\infty < \dfrac{1}{\sqrt{n+1}} \end{cases} \quad (23)$$

Here, $\gamma < 1$. Fig. 2 describes the proposed structure.

Remark 3: The transformation just changes magnitude of vector $\hat{\boldsymbol{\theta}}$. Its direction is always preserved.

Remark 4: Adaptive rule in (17)-(20) describe minimization at infinite time on $\mathcal{L}_2$ norm performance in frequency domain.

IV. Numerical Simulation

Consider a nominal plant $G_0(z) = 1/(z-2)$, an upper bound function $r_a(z) = (10z-8)/(14.5z+8.5)$ and an actual plant $G = \{(z+0.5874)(z+0.0005)\}/\{z(z-2.0138)(z-0.0498)\}$.

Purpose of this simulation is robust stabilizing the actual plant and minimizing the following performance function.

$$I(k) = \frac{1}{k}\sum_{i}^{k}\{\rho u(k)^2 + (r(k) - y(k))^2\}, \quad \rho = 0.01 \quad (24)$$

Based on (24), the augmented plant P is constructed and the proposed control system is simulated. Reference input $r(k)$ and disturbance $d(k)$ both are step function and its magnitude are 8 and 3, respectively. Figure 3 describes a simulation result when the adaptive structure is not used ,i.e. $S(z) = 0$. Figure 4 is result when the proposed system is used. In this case, it is confirmed that when adaptive parameter $\bar{\theta}(k)$ converged as in Fig. 5, gain of the controller $C(z)$ in lower frequency is high. Therefore, output $y(k)$ converged to reference input $r(k)$.

V. Conclusions

In this paper, the direct adaptive controller of the robust stable control system is presented. In the system, an adaptive structure is blended with robust stable controller. The adaptive structure estimates the disturbance input and adjusts the parameters of free parameter of the robust stabilizing controller such that the disturbance can be attenuated. The key feature of the proposed design method is that the robust stable control system can attenuate unknown disturbance in an on-line manner without *a prior* knowledge of the disturbance.

References

[1] T.T. Tay and J.B. Moore, "Enhancement of Fixed Controller via Adaptive-Q Disturbance Estimate Feedback," *Automatica*, vol. 27, No. 1, pp. 39-53, 1991.

[2] T.T. Tay and J.B. Moore, "Performance Enhancement of Two-degree-of-freedom Controller via Adaptive techniques," *Int. J. Adaptive Control and Signal Processing*, vol. 4, pp. 69-84, 1990.

[3] H. Kimura, "Robust Stability for a Class of Transfer Functions," *IEEE Trans. Auto. Contr.*, AC-29, No. 9, pp. 788-793, 1984.

[4] G. Zames, "Feedback and Optimal Sensitivity: Model Reference Transformations, Multiplicative seminorms, and Approximate Inverses," *IEEE Trans. Auto. Contr.*, AC-26, No. 2, pp. 301-320, 1981.

[5] B.A. Francis and G. Zames, "Feedback, Minimax Sensitivity, and Optimal Robustness," *IEEE Trans. Auto. Contr.*, AC-28, No. 5, pp. 585-601, 1983.

[6] B.A Francis, "A Course on H_∞," *Lecture Notes in Control on Information Sciences*, Springer-Verlag, Boston, 1987.

[7] I. Postlethwaite, M.C Tsai and D.W. Gu, "Weighting Function selection in H_∞ Design," *IFAC World Congress*, Tallinn, 1990.

[8] R. Kondo, S. Hara and T. Itou, "Characterization of Discrete Time of H_∞ Controller via Bilinear Transformation," *Proceeding of the 29th Conference on Decision and Control*, Honolulu, Hawai, December 1990.

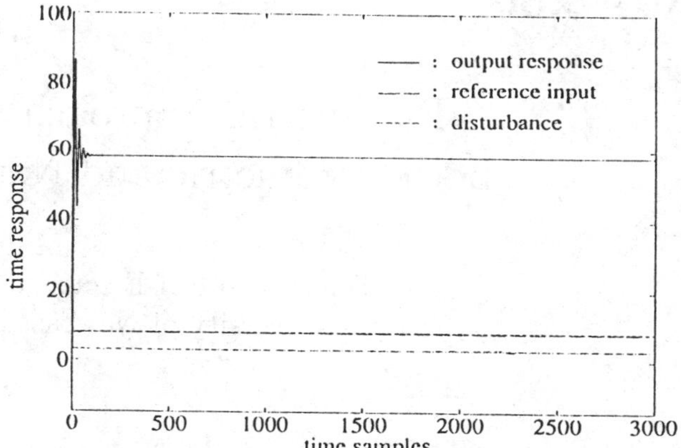

Fig. 3 Output response of robustly stable system (non-adaptive case)

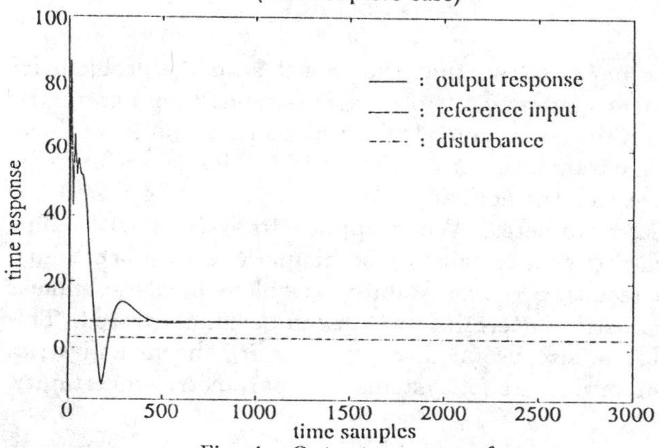

Fig. 4 Output response of adaptive disturbance attenuation scheme

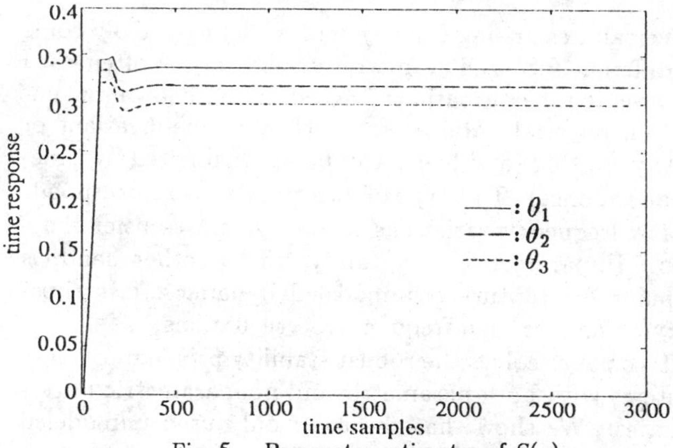

Fig. 5 Parameter estimates of $S(z)$

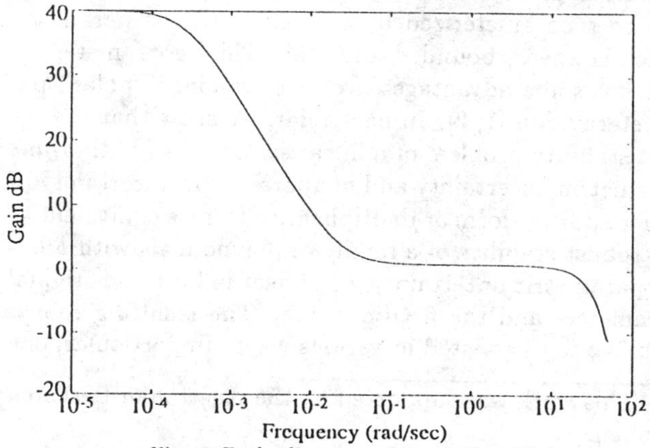

Fig. 6 Bode diagram of controller $C(z)$

Proceedings of the 31st Conference
on Decision and Control
Tucson, Arizona • December 1992

Parametric Approach to Robust Stability with both Parametric and Nonparametric Uncertainties *

Minyue Fu
Department of Electrical and Computer Engineering
University of Newcastle, N.S.W. 2308 Australia

Abstract

This paper shows that the robust stability problem for a linear system with both parametric and nonparametric uncertainties is equivalent to a robust stability problem with parametric uncertainty only. This is achieved by converting the nonparametric uncertainty into a fictitious linear parameter. When applied to systems with affine parametric uncertainty and nonparametric uncertainty, the resulting robust stability problem involves bilinear parametric uncertainty which can be simply tested. This result is also useful in computing H_∞ norm and strict positive realness for systems with parametric uncertainty.

1 Introduction

Uncertainties arising from system modeling usually come in different forms. Parametric uncertainty is often used to describe perturbations caused by unknown or uncertain physical parameters which are time-invariant or slowly drifting (and hence can be approximated by time-invariant ones). This type of uncertainty also corresponds to low-frequency variations in the system transfer function. Unparametric uncertainty, on the other hand, is popular for capturing unmodeled dynamics, fast time-varying and/or high-frequency perturbations.

This paper solves the robust stability problem of linear systems with both parametric and nonparametric uncertainties. We show that the commonly used unmodeled dynamics, either additive or multiplicative uncertainties, can be reparameterized by a single fictitious parameter which is linear, bounded and real. This reparameterization has some advantages over the standard "polar" parameterization [1, 2]. In particular, we show that the robust stability problem of a linear system with both *affine* parametric uncertainty and nonparametric uncertainty in either additive form or multiplicative form is equivalent to the robust stability of a family of polynomials with *bilinear* parametric uncertainty, i.e., linear in both the original parameters and the fictitious one. The resulting robust stability can be tested in various ways. In particular, one can use the Finite Zero Exclusion Principle in [3] which has the advantage of avoiding "frequency sweeping."

The development of our results is through an equivalence among the robust stability with nonparametric uncertainty, H_∞ performance, and strict positive realness (SPRness). For this reason, our results are applicable to the problems of testing the H_∞ performance and SPR property of transfer functions with parametric uncertainty. In other words, these problems are shown to be equivalent to the robust stability problem of a family of polynomials with parametric uncertainty with a fictitious (linear) parameter.

*This work was supported by the Australian Research Council.

2 Problem Formulation and Notation

Consider a single-input-single-output unity feedback system with the open loop transfer function in s given by $G(s,q)+\Delta(s)$, where $G(s,q)\in\mathcal{G}$ is the parametric part of the transfer function which depends an parameter vector q which belongs to a bounding set $Q\in\mathbf{R}^m$, $\Delta(s)\in\mathcal{D}$ is the nonparametric uncertainty in the additive form. We restrict ourselves to additive perturbations although multiplicative ones can be treated similarly. The following assumptions are required throughout the paper:

A1 $G(s,q)$ is nth-order, strictly proper and real-valued for all $q\in Q$, i.e.,

$$G(s,q)=\frac{n(s,q)}{d(s,q)}=\frac{\sum_{i=0}^{n-1}b_i(q)s^i}{\sum_{i=0}^{n}a_i(q)s^i} \quad (1)$$

where $a_i(q)$ and $b_i(q)$ are real-valued functions with $a_n(q)\neq 0$ for all $q\in Q$.

A2 $a_i(q)$ and $b_j(q)$ are continuous in q over Q, $0\le i\le n, 0\le j\le n-1$.

A3 $G(s,q)$ has no poles on the $j\omega$ axis for all $q\in Q$.

A4 $G(s,q)$ and $G(s,q)+\Delta(s)$ have the same number of poles in the closed right half plane, for all $\Delta(s)\in\mathcal{D}$.

A5 There exists some strictly proper, stable and minimum-phase, real-valued transfer function $\gamma(s)=\frac{\alpha(s)}{\beta(s)}$ such that

$$\mathcal{D}=\{\Delta(s)\text{ satisfying A4}: |\Delta(j\omega)|\le\gamma(j\omega)\}\,. \quad (2)$$

CH3229-2/92/0000-2782$1.00 © 1992 IEEE

The robust stability problem to be solved in this paper is to determine whether the closed-loop stability holds for all for all $q \in Q$ and $\Delta(s) \in \mathcal{D}$.

Notation: The convex set of two elements x and y is denoted by conv$[x, y]$, i.e., conv$[x, y] = \{\lambda x + (1-\lambda)y : \lambda \in [0, 1]\}$. Similarly, the convex family of two polynomials $a(s)$ and $b(s)$ will be denoted by conv$[a(s), b(s)]$.

3 Main Results

Several related results are to be developed. Theorem 1 shows that a rational function is SPR if and only if the convex family of two associated polynomials is robustly stable. This relationship simplifies a result by Rantzer [3] which needs two parameters to do the conversion. Theorem 2 generalizes Theorem 1 to testing the H_∞ norm of a rational function by using an equivalence between H_∞ performance and SPRness. Theorem 3, an application of Theorem 2, gives a parametric approach to the robust stability problem of linear systems with both parametric and nonparametric uncertainties.

Theorem 1. *Given two n-th order real-valued polynomials $a(s)$ and $b(s)$ with the same sign for their leading coefficients, $a(s)/b(s)$ is SPR if and only if* conv$[a(s), jb(s)]$ *is robustly stable (strictly Hurwitz).*

Proof. From the definition of SPRness, we know that $a(s)/b(s)$ is SPR if and only if (i) both $a(s)$ and $b(s)$ are stable, and (ii) $\mathrm{Re}[a(j\omega)/b(j\omega)] > 0$, $\forall -\infty < \omega < \infty$. Because the leading coefficients of $a(s)$ and $b(s)$ have the same sign, i.e., $a(j\infty)/b(j\infty) > 0$, the condition (ii) above can be replaced by

$$\mathrm{Re}[a(j\omega)/b(j\omega)] \neq 0, \ \forall -\infty < \omega < \infty ,$$

which is actually equivalent to

$$\frac{a(j\omega)}{b(j\omega)} + j\frac{t}{1-t} \neq 0, \ \forall t \in [0, 1) . \quad (3)$$

Using the nonzeroness of $a(j\omega)$ and $b(j\omega)$ (from the stability of $a(s)$ and $b(s)$), the inequality (3) holds if and only if

$$p(j\omega, t) = (1-t)a(j\omega) + jtb(j\omega) \neq 0 , \ \forall t \in [0, 1] . \quad (4)$$

Note that the leading coefficient of $p(s, t)$ is nonvanishing for all $t \in [0, 1]$. Finally, due to the zero exclusion principle (see [4], for example), the condition (i) and that in (4) are further equivalent to the robust stability of conv$[a(s), b(s)]$. $\nabla\nabla\nabla$

The following lemma which establishes the relationship between SPRness and H_∞ performance is well-known; see [5], for example.

Lemma 2. *Given a strictly proper transfer function $H(s) = n(s)/d(s)$ where $n(s)$ and $d(s)$ are real-valued polynomials, $||H(s)||_\infty < 1$ if and only if the transfer function $(d(s) - n(s))/(d(s) + n(s))$ is SPR.*

Following the lemma above and Theorem 1, we have the next result which relates the H_∞ performance of a transfer function to robust stability of a family of polynomials. The proof is omitted due to its obviousness.

Theorem 3. *Given a strictly proper transfer function $H(s) = n(s)/d(s)$ where $n(s)$ and $d(s)$ are real-valued polynomials, $||H(s)||_\infty < 1$ if and only if* conv$[d(s) - n(s), j(d(s) + n(s))]$ *is robustly stable.*

Now we return to the robust stability problem formulated in Section 2. The following result is a natural combination of the small gain theorem (see, [5], for example) and Theorem 2 above.

Theorem 4. *Consider the unity feedback uncertain system described in Section 2 which satisfies A1-A5. Then, the closed-loop system is robustly stable if and only if the following family of polynomials is robustly stable:*

$$\mathcal{H}_a = \{h_a(s, q, \lambda) : q \in Q, \lambda \in [0, 1]\} \quad (5)$$

where

$$\begin{aligned} h_a(s, q, \lambda) &= (1-\lambda)[(\beta - \alpha)d(s, q) + \beta n(s, q)] \\ &+ j\lambda[(\beta + \alpha)d(s, q) + \beta n(s, q)] . \end{aligned} \quad (6)$$

Remark 1. Note that the families of polynomials $\mathcal{H}_a$ and $\mathcal{H}_m$ will involve multilinear parametric uncertainty if the open-loop transfer function $G(s, q)$ is multilinear in q, or bilinear uncertainty if $G(s, q)$ is affine in q. Effecient numerical algorithms are available for testing the robust stability with these classes of uncertainty. In particular, we draw attention to those based on the Finite Zero Principle in [3] which require testing certain zero exclusion properties at a finite number of frequencies. Since the fictitious parameter λ introduced above is a linear parameter, robust stability test can be carried out efficiently.

REFERENCES

[1] B. R. Barmish and P. P. Khargonekar, "Robust stability of feedback control systems with uncertain parameters and unmodelled dynamics," *Math. Contr. Signals Syst.*, vol. 3, pp. 197–210, 1990.

[2] H. Chapellat, M. Dahleh, and S. P. Bhattacharyya, "Robust stability under structured and unstructed perturbations," *IEEE Tran. Auto. Contr.*, vol. 35, no. 10, pp. 1100–1108, 1990.

[3] A. Rantzer, "A finite zero exclusion principle," in *Proc. MTNS*, (Amsterdam, Netherlands), pp. 239–245, 1989.

[4] B. R. Barmish, "A generalization of Kharitonov's four polynomial concept for robust stability problems with linearly dependent coefficient perturbations," *IEEE Trans. Auto. Contr.*, vol. 34, pp. 157–165, Feb. 1989.

[5] C. S. Desoer and M. Vidyasagar, *Feedback Systems-Input Output Properties.* New York, NY: Academic Press, 1975.

Proceedings of the 31st Conference
on Decision and Control
Tucson, Arizona • December 1992

New Vertex Results on H_∞ Performance of Interval Plants and Interval Feedback Systems

C. V. Hollot

Department of Electrical and
Computer Engineering
University of Massachusetts
Amherst, Massachusetts 01003

R. Tempo

CENS-CNR
Politecnico di Torino
Corso Duca degli Abruzzi 24
10129 Torino (Italy)

Abstract

The worst case H_∞ norm of an interval plant weighted by $\frac{p(s)}{s^\ell(s+\alpha)}$ where ℓ is an integer, α is a real number and $p(s)$ is an arbitrary polynomial, is achieved at one of the Kharitonov plants. Similar results hold for weighted sensitivity and complementary sensitivity functions of interval feedback systems. In this case, the admissible weights are of the form $\frac{s+\beta}{s^\ell(s+\alpha)}$.

I. Introduction

Consider a proper transfer function with uncertain coefficients modelled by a so-called *interval plant* $P(s,q,r)$ described by

$$P(s,q,r) = \frac{N(s,q)}{D(s,r)} = \frac{q_0 + q_1 s + q_2 s^2 + \ldots + q_m s^m}{r_0 + r_1 s + r_2 s^2 + \ldots + s^n} \qquad (1)$$

where $q \in Q$, $r \in R$ and

$$Q = \left\{ q : q_i^- \le q_i \le q_i^+ : i = 0, 1, \ldots, m \right\}$$

and

$$R = \left\{ r : r_i^- \le r_i \le r_i^+ : i = 0, 1, \ldots, n-1 \right\}$$

are rectangles of parameters. In [2] and [3], it is shown that the H_∞ norm of stable interval plants enjoy the following property:

$$\max_{q\in Q, r\in R} \|P(s,q,r)\|_\infty = \max_{i,k=1,2,3,4} \|P_{ik}(s)\|_\infty \qquad (2)$$

where

$$P_{ik}(s) = \frac{N_i(s)}{D_k(s)}, \quad i,k = 1,2,3,4$$

are the *Kharitonov plants* and $N_i(s)$ and $D_k(s)$ for $i = 1,2,3,4$ are the *Kharitonov polynomials*; see [1]. In [4] it is proven that stable interval feedback systems enjoy

$$\max_{q\in Q, r\in R} \|S(s,q,r)\|_\infty = \max_{i,k=1,2,3,4} \|S_{ik}(s)\|_\infty ;$$
$$\max_{q\in Q, r\in R} \|T(s,q,r)\|_\infty = \max_{i,k=1,2,3,4} \|T_{ik}(s)\|_\infty \qquad (3)$$

where

$$S(s,q,r) = \frac{D(s,r)}{N(s,q) + D(s,r)}$$

is the *sensitivity function,*

$$T(s,q,r) = 1 - S(s,q,r)$$

is the *complementary sensitivity function* and where $S_{ik}(s)$ and $T_{ik}(s)$ are their Kharitonov counterparts. Recognizing the utility of (2) and (3) in establishing levels of H_∞ performance for uncertain transfer functions and feedback systems, this paper aims to generalize these results to *weighted* H_∞ norms. That is, we seek classes of weights $W_P(s)$, $W_S(s)$ and $W_T(s)$ for which the proper transfer functions $W_P(s)P(s,q,r)$, $W_S(s)S(s,q,r)$ and $W_T(s)T(s,q,r)$ satisfy the relations

$$\max_{q\in Q, r\in R} \|W_P(s)P(s,q,r)\|_\infty = \max_{i,k=1,2,3,4} \|W_P(s)P_{ik}(s)\|_\infty , \qquad (4)$$
$$\max_{q\in Q, r\in R} \|W_S(s)S(s,q,r)\|_\infty = \max_{i,k=1,2,3,4} \|W_S(s)S_{ik}(s)\|_\infty \qquad (5)$$

and

$$\max_{q\in Q, r\in R} \|W_T(s)T(s,q,r)\|_\infty = \max_{i,k=1,2,3,4} \|W_T(s)T_{ik}(s)\|_\infty . \qquad (6)$$

The usefulness of (2) comes when one evaluates the H_∞ performance of interval plants which arise, for example, in establishing the stability of a feedback loop containing both parametric uncertainty (as modelled by the interval plant) and a non-parametric uncertainty. In [4], extensions of (2) to include frequency-weighted interval plants $W_P(s)P(s,q,r)$ were studied. In addition, in [4] we stated that

$$\max_{q\in Q, r\in R} \left\| \frac{p(s)}{s^\ell(s+\alpha)} P(s,q,r) \right\|_\infty = \max_{i,k=1,2,3,4} \left\| \frac{p(s)}{s^\ell(s+\alpha)} P_{ik}(s) \right\|_\infty . \qquad (7)$$

In continuing this line of research, we now prove that

$$\max_{q\in Q, r\in R} \left\| \frac{s+\beta}{s^\ell(s+\alpha)} S(s,q,r) \right\|_\infty = \max_{i,k=1,2,3,4} \left\| \frac{s+\beta}{s^\ell(s+\alpha)} S_{ik}(s) \right\|_\infty ;$$
$$\max_{q\in Q, r\in R} \left\| \frac{s+\beta}{s^\ell(s+\alpha)} T(s,q,r) \right\|_\infty = \max_{i,k=1,2,3,4} \left\| \frac{s+\beta}{s^\ell(s+\alpha)} T_{ik}(s) \right\|_\infty \qquad (8)$$

We remark that the extremality conditions proved in this paper require the understanding of the stability properties of uncertain polynomials. Evenmore, some of the latest results in this area; e.g., the theory of stable "convex directions" in [7], are not directly applicable here. Indeed, the key idea in our proofs is to show the equivalence (from the stability point of view) between some uncertain complex coefficient polynomials and real coefficient polynomials. To this end, we use *boundary interpolation theory* to show this equivalence. These arguments form the core of our main results (see Lemma 1 and 2 as serving as the technical novelty of this paper).

The paper is organized as follows. In Section II, we study the extremality properties of H_∞ norm of weighted interval plants. These extreme point results are proved in Sections III. In Section IV, these results are extended to show extremality of the H_∞ norm of weighted sensitivity and complementary sensitivity.

II. Extremality Properties of the H_∞ Norm of Weighted Interval Plants

In this section, we present our first result which shows that extremality properties occur whenever the weighting function is of the form

$$\frac{p(s)}{s^\ell(s+\alpha)}.$$

CH3229-2/92/0000-2784$1.00 © 1992 IEEE

Theorem 1: (See Section III for proof) *Let $p(s)$ be an arbitrary polynomial, α a real nonnegative number and ℓ an integer. If*

$$\frac{p(s)}{s^{\ell}(s+\alpha)}P(s,q,r)$$

is proper and strictly stable for all $q \in Q$ and $r \in R$, then

$$\max_{q\in Q, r\in R}\left\|\frac{p(s)}{s^{\ell}(s+\alpha)}P(s,q,r)\right\|_{\infty} = \max_{i,k=1,2,3,4}\left\|\frac{p(s)}{s^{\ell}(s+\alpha)}P_{i,k}(s)\right\|_{\infty}. \quad (9)$$

□

Remarks: The specialist in extremal properties for polynomials and rational functions may be tempted to dismiss the vertex result in Theorem 1 on the grounds that the multiplier $s+\alpha$ is a *real convex direction*; e.g., see [5] and [7]. However, this notion is not applicable here – the theorem's result is non-trivial since the H_{∞} problem induces a special *complex* polynomial stability problem. To see this, consider a proper and strictly stable transfer function $P(s,q,r)$. It can be easily shown (see; e.g., [3]) that the parametric H_{∞} problem

$$\max_{q\in Q, r\in R}\left\|\frac{N(s,q)}{(s+\alpha)(D(s,r)}\right\|_{\infty} < 1 \quad (10)$$

is equivalent to the Hurwitzness of the complex polynomial

$$e^{j\phi}N(s,q)+(s+\alpha)D(s,r) \quad (11)$$

for all $q \in Q$, $r \in R$ and $\phi \in [0, 2\pi]$. Thus, the problem considered in Theorem 1 involves complex polynomials and not real ones. Additionally, to obtain the extremality result described in this theorem, the special structure of the complex polynomial in (11) must be exploited. To illustrate this fact more fully, the example below shows that the term $s+1$ does not give rise to extremality properties for *arbitrary* complex polynomials.

Example: (see [8]) Consider a family of complex polynomials of the form

$$p(s,\lambda) = (4.3176 - j0.8398) - (1.0111 + j15.1285)s -$$

$$(1.2272 + j6.3118)s^2 + \lambda(s+1) \quad (12)$$

where $\lambda \in [0,1]$. The vertex polynomials $p(s,0)$ and $p(s,1)$ are Hurwitz, but the interior polynomial $p(s,0.5)$ is *not* Hurwitz. Hence, the term $(s+1)$ does not generically support vertex results when complex polynomials are considered. However, this fact does not rule out vertex possibilities for (11) since the complex portion of (12) does not have the structure $e^{j\phi}N(s,q)+(s+\alpha)D(s,r)$ present in (11).

III. Proof of Theorem 1

To prove Theorem 1 we need a key lemma.

Lemma 1: *Let $N(s)$ and $D(s)$ be fixed polynomials, $q(s)$ a polynomial of degree less than the degree of $D(s)$ having only even or odd coefficients and $\alpha > 0$. If*

$$\frac{N(s)}{(s+\alpha)(D(s)\pm q(s))}$$

is proper and strictly stable, then

$$\max_{\lambda\in[-1,1]}\left\|\frac{N(s)}{(s+\alpha)(D(s)+\lambda q(s))}\right\|_{\infty} = \max_{\lambda=\pm 1}\left\|\frac{N(s)}{(s+\alpha)(D(s)+\lambda q(s))}\right\|_{\infty}. \quad (13)$$

Proof of Lemma 1: For notational simplicity, let

$$G(s,\lambda) \doteq \frac{N(s)}{(s+\alpha)(D(s)+\lambda q(s))}. \quad (14)$$

Proceeding by contradiction, assume that (13) does not hold. Then,

$$\max_{\lambda\in[-1,1]}\|G(s,\lambda)\|_{\infty} > \max_{\lambda=\pm1}\|G(s,\lambda)\|_{\infty}.$$

Scaling both sides by $\max_{\lambda\in[-1,1]}\|G(s,\lambda)\|_{\infty}$ and absorbing this factor in $N(s)$ we obtain

$$\max_{\lambda\in[-1,1]}\|G(s,\lambda)\|_{\infty} = 1$$

while

$$\max_{\lambda=\pm1}\|G(s,\lambda)\|_{\infty} < 1.$$

Without loss of generality, redefine λ so that the maximum of $\|G(s,\lambda)\|_{\infty}$ is achieved at $\lambda = 0$. This gives

$$\max_{\lambda\in[\alpha,\beta]}\|G(s,\lambda)\|_{\infty} = \|G(s,0)\|_{\infty} = 1 \quad (15)$$

and

$$\max_{\lambda=\alpha,\beta}\|G(s,\lambda)\|_{\infty} < 1 \quad (16)$$

for some $\alpha < 0 < \beta$.

Equation (15) implies that the Nyquist plot of $G(s,\lambda)$ is contained in the unit disk for all $\lambda \in [\alpha,\beta]$ with the plot $G(s,0)$ actually touching the disk boundary. Let ω^* be a frequency for which this occurs; i.e.,

$$|G(j\omega^*,0)| = 1.$$

There may be frequencies different from ω^* at which the Nyquist plot of $G(s,0)$ touches the unit circle. In this case, we list them as $\omega_i, i = 1,2,\ldots,N$. Since $G(s,0)$ is strictly stable, $|G(j\omega^*,0)| = 1$ and $|G(j\omega,0)| \le 1$ for all $\omega \ge 0$, then, for

$$\phi^* = \pi - \angle G(j\omega^*,0),$$

the solutions of the equation

$$1 + e^{j\phi^*}G(s,0) = 0$$

are all contained in $Re\ s < 0$ except for a finite number of $j\omega$-axis roots which include $s = j\omega^*$. Now, define

$$f(s,\lambda) \doteq 1 + e^{j\phi^*}G(s,\lambda). \quad (17)$$

Equations (15) and (16) imply that the λ root locus segment associated with $\{s : f(s,\lambda) = 0 \text{ for some } \lambda \in [\alpha,\beta]\}$ is wholly contained in $Re\ s \le 0$. Moreover, this root locus segment touches the imaginary axis at only a *finite* number of points. Indeed, if this were not the case, then this root locus segment would "flatten out" against the imaginary axis. More precisely, the root locus would be contained in the imaginary axis for all λ in some subinterval of $[\alpha,\beta]$. However, since none of the root locus lies in $Re\ s > 0$, such behavior violates the analyticity of the root locus branch functions and is thus ruled out. Consequently, for sufficiently small ϵ, $f(s,\lambda)$ has only zeroes in $Re\ s < 0$ for all $\lambda \in [-\epsilon,\epsilon]\backslash 0$. The zeroes of $f(s,0)$ are also strictly stable except for $j\omega$-axis roots which include $s = j\omega^*$.

Let $s(\lambda)$ denote the root function of $f(s,\lambda)$ for which $s(0) = j\omega^*$. From the previous discussion, $s(\lambda)$ satisfies

$$Re[s'(0)] = 0 \quad (18)$$

and

$$Re[s''(0)] < 0. \quad (19)$$

Define the function

$$\tilde{f}(s,\lambda) \doteq 1 + \frac{N(s)p_1(s)}{\gamma_1(s+\alpha)(D(s)p_2(s)+\lambda q(s)a(s))} \quad (20)$$

where $\gamma_1 \in (0,1)$ and the polynomials $p_1(s)$ and $p_2(s)$ are to be determined and where $a(s)$ is an antistable polynomial satisfying the condition

$$a(j\omega^*) = p_2(j\omega^*). \tag{21}$$

If $p_2(s)$ is Hurwitz and

$$\left.\frac{p_1(s)}{p_2(s)}\right|_{s=j\omega^*} = \gamma_1 e^{j\phi^*}$$

then $\tilde{f}(j\omega^*,0) = f(j\omega^*,0) = 0$ and we can take $\tilde{s}(\lambda)$ to be the root function of $\tilde{f}(s,\lambda)$ for which $\tilde{s}(0) = j\omega^*$. We now have the following claim.

Claim: *Given γ_1, $\gamma_2 \in (0,1)$ and polynomial $a(s)$ satisfying (21), there exist polynomials $p_1(s)$ and $p_2(s)$ with $p_2(s)$ and $a(s)$ Hurwitz such that*

$$\left\|\frac{p_1(s)}{p_2(s)}\right\|_\infty \leq 1, \tag{22}$$

$$\left.\frac{p_1(s)}{p_2(s)}\right|_{s=j\omega^*} = \gamma_1 e^{j\phi^*}, \tag{23}$$

$$\left.\frac{p_1(s)}{p_2(s)}\right|_{s=j\omega_i} = \gamma_2 e^{j\phi^*} \tag{24}$$

for $i = 1, 2, \ldots, N$,

$$\tilde{s}'(0) = s'(0) \tag{25}$$

and

$$\tilde{s}''(0) = s''(0). \tag{26}$$

Proof of Claim: With $a(s)$ chosen as in (21), we construct a polynomial $p_1(s)$ and a Hurwitz polynomial $p_2(s)$ such that (22)-(26) hold. To this end, we evaluate

$$\begin{aligned} s'(\lambda) &= -\frac{\partial f}{\partial \lambda}\left(\frac{\partial f}{\partial s}\right)^{-1}; \\ \tilde{s}'(\lambda) &= -\frac{\partial \tilde{f}}{\partial \lambda}\left(\frac{\partial f}{\partial \tilde{s}}\right)^{-1} \end{aligned} \tag{27}$$

and

$$\begin{aligned} s''(\lambda) &= -\frac{\partial^2 f}{\partial \lambda^2}\left(\frac{\partial f}{\partial s}\right)^{-1} + \frac{\partial f}{\partial \lambda}\frac{\partial^2 f}{\partial s^2}\left(\frac{\partial f}{\partial s}\right)^{-2}; \\ \tilde{s}''(\lambda) &= -\frac{\partial^2 \tilde{f}}{\partial \lambda^2}\left(\frac{\partial \tilde{f}}{\partial \tilde{s}}\right)^{-1} + \frac{\partial \tilde{f}}{\partial \lambda}\frac{\partial^2 \tilde{f}}{\partial \tilde{s}^2}\left(\frac{\partial \tilde{f}}{\partial \tilde{s}}\right)^{-2} \end{aligned} \tag{28}$$

where the root functions s and $\tilde{s}$ are implicitely defined by $f(s,\lambda) = 0$ and $\tilde{f}(s,\lambda) = 0$. From (27)-(28) we see that (25) and (26) hold if we require

$$\begin{aligned} \left.\frac{\partial f}{\partial \lambda}\right|_{\substack{\lambda=0\\ s=j\omega^*}} &= \left.\frac{\partial \tilde{f}}{\partial \lambda}\right|_{\substack{\lambda=0\\ s=j\omega^*}}; \\ \left.\frac{\partial^2 f}{\partial \lambda^2}\right|_{\substack{\lambda=0\\ s=j\omega^*}} &= \left.\frac{\partial^2 \tilde{f}}{\partial \lambda^2}\right|_{\substack{\lambda=0\\ s=j\omega^*}}; \\ \left.\frac{\partial f}{\partial s}\right|_{\substack{\lambda=0\\ s=j\omega^*}} &= \left.\frac{\partial \tilde{f}}{\partial \tilde{s}}\right|_{\substack{\lambda=0\\ s=j\omega^*}}; \\ \left.\frac{\partial^2 f}{\partial s^2}\right|_{\substack{\lambda=0\\ s=j\omega^*}} &= \left.\frac{\partial^2 \tilde{f}}{\partial \tilde{s}^2}\right|_{\substack{\lambda=0\\ s=j\omega^*}}. \end{aligned} \tag{29}$$

We thus seek $p_1(s)$ and a Hurwitz polynomial $p_2(s)$ such that (22)-(24) and (29) hold. For $f(s,\lambda)$ in (17) we compute

$$\begin{aligned} \left.\frac{\partial f}{\partial \lambda}\right|_{\substack{\lambda=0\\ s=j\omega^*}} &= -e^{j\phi^*}\left.\frac{q(s)N(s)}{D^2(s)(s+\alpha)}\right|_{s=j\omega^*}; \\ \left.\frac{\partial^2 f}{\partial \lambda^2}\right|_{\substack{\lambda=0\\ s=j\omega^*}} &= 2e^{j\phi^*}\left.\frac{q^2(s)N(s)}{D^3(s)(s+\alpha)}\right|_{s=j\omega^*}; \\ \left.\frac{\partial f}{\partial s}\right|_{\substack{\lambda=0\\ s=j\omega^*}} &= e^{j\phi^*}\left.\left(\frac{N(s)}{D(s)(s+\alpha)}\right)'\right|_{s=j\omega^*}; \\ \left.\frac{\partial^2 f}{\partial s^2}\right|_{\substack{\lambda=0\\ s=j\omega^*}} &= e^{j\phi^*}\left.\left(\frac{N(s)}{D(s)(s+\alpha)}\right)''\right|_{s=j\omega^*} \end{aligned} \tag{30}$$

while for $\tilde{f}(s,\lambda)$ in (20) we obtain

$$\begin{aligned} \left.\frac{\partial \tilde{f}}{\partial \lambda}\right|_{\substack{\lambda=0\\ s=j\omega^*}} &= -\left.\frac{p_1(s)}{\gamma_1 p_2(s)}\frac{q(s)N(s)}{D^2(s)(s+\alpha)}\right|_{s=j\omega^*}; \\ \left.\frac{\partial^2 \tilde{f}}{\partial \lambda^2}\right|_{\substack{\lambda=0\\ s=j\omega^*}} &= 2\left.\frac{p_1(s)}{\gamma_1 p_2(s)}\frac{q^2(s)N(s)}{D^3(s)(s+\alpha)}\right|_{s=j\omega^*}; \\ \left.\frac{\partial \tilde{f}}{\partial \tilde{s}}\right|_{\substack{\lambda=0\\ s=j\omega^*}} &= \left.\left(\frac{p_1(s)}{\gamma_1 p_2(s)}\frac{N(s)}{D(s)(s+\alpha)}\right)'\right|_{s=j\omega^*}; \\ \left.\frac{\partial^2 \tilde{f}}{\partial \tilde{s}^2}\right|_{\substack{\lambda=0\\ s=j\omega^*}} &= \left.\left(\frac{p_1(s)}{\gamma_1 p_2(s)}\frac{N(s)}{D(s)(s+\alpha)}\right)''\right|_{s=j\omega^*}. \end{aligned} \tag{31}$$

Using (30) and (31), a straightforward calculation shows that $p_1(s)$ and $p_2(s)$ satisfy (29) if they meet the following interpolation conditions:

$$\begin{aligned} \left.\frac{p_1(s)}{p_2(s)}\right|_{s=j\omega^*} &= \gamma_1 e^{j\phi^*}; \\ \left.\left(\frac{p_1(s)}{p_2(s)}\right)'\right|_{s=j\omega^*} &= 0; \\ \left.\left(\frac{p_1(s)}{p_2(s)}\right)''\right|_{s=j\omega^*} &= 0. \end{aligned} \tag{32}$$

Notice that (32) includes (23). Since γ_1 and γ_2 are less than one, it follows from boundary interpolation results, see for example Theorem 1.5 and Remark 1.7 in [9], that there exists a stable rational function $\frac{p_1(s)}{p_2(s)}$ satisfying (22), (24) and (32). This proves the claim.

Return to the proof of Lemma 1 and fix $\gamma_2 \in (0,1)$. For $\gamma_1 \in (0,1)$ sufficiently close to 1 and, for antistable $a(s)$ satisfying (21), it follows from the claim that there exist polynomials p_1, $p_2(s)$ and $a(s)$ such that the Nyquist plot of

$$\tilde{G}(s) \doteq \frac{N(s)p_1(s)}{\gamma_1(s+\alpha)D(s)p_2(s)} \tag{33}$$

is contained in the unit disk, touching $-1+j0$ only when $\omega = \omega^*$. Specifically, (23) implies that

$$\tilde{G}(j\omega^*) = e^{j\phi^*}G(j\omega^*,0) = -1$$

while (24) and γ_1 sufficiently close to 1 gives

$$|\tilde{G}(j\omega_i)| = \frac{\gamma_2}{\gamma_1}|G(j\omega_i)| < 1$$

for $i = 1, 2, \ldots, N$. Finally, for all other frequencies, (22), (29) and $\gamma_1 \approx 1$ guarantee that

$$|\tilde{G}(j\omega)| < 1.$$

Now, as a consequence of $\tilde{G}(s)$ being real, strictly stable and having Nyquist plot touching the unit circle at only at $-1+j0$, the rational function $\tilde{f}(s,0)$ in (20) has only zeroes in $Re\ s < 0$ except for a pair of imaginary roots at $s = j\omega^*$. Moreover, from (25) and (26), these two imaginary roots of $\tilde{f}(s,\lambda)$ migrate into the open left half plane when λ infinitesimally changes from zero. Consequently, there exists an $\tilde{\epsilon} > 0$ such that $\tilde{f}(s,\lambda)$ has zeroes only in the open left half plane for all $\lambda \in [-\tilde{\epsilon}, \tilde{\epsilon}]\backslash 0$. From this and (20) it follows that the uncertain polynomial

$$\gamma_1(s+\alpha)D(s)p_2(s) + N(s)p_1(s) + \lambda q(s)(s+\alpha)a(s)$$

has a zero at $s = j\omega^*$ when $\lambda = 0$ but is stable for all $\lambda \in [-\tilde{\epsilon}, \tilde{\epsilon}]\backslash 0$. However, this contradicts the fact that $q(s)(s+\alpha)a(s)$ is a convex direction (see [7]) and hence proves Lemma 1. □

Proof of Theorem 1: For fixed $r \in R$ and $\omega \in \mathbf{R}$, it is well-known that

$$\max_{q\in Q}\left|\frac{p(j\omega)N(j\omega,q)}{(j\omega)^\ell(j\omega+\alpha)D(j\omega,r)}\right| = \max_{i=1,2,3,4}\left|\frac{p(j\omega)N_i(j\omega)}{(j\omega)^\ell(j\omega+\alpha)D(j\omega,r)}\right|.$$

Thus, to prove the theorem it suffices to show that the relation

$$\max_{r\in R}\left\|\frac{p(s)N_i(s)}{s^\ell(s+\alpha)D(s,r)}\right\|_\infty = \max_{k=1,2,3,4}\left\|\frac{p(s)N_i(s)}{s^\ell(s+\alpha)D_k(s)}\right\|_\infty \quad (34)$$

holds for fixed $i = 1,2,3,4$.

The remainder of the proof is a straightforward application of Lemma 1. Clearly, relation (34) holds if and only if for each fixed $\gamma > 0$,

$$\max_{k=1,2,3,4}\left\|\frac{p(s)N_i(s)}{s^\ell(s+\alpha)D_k(s)}\right\|_\infty \le \gamma \quad (35)$$

implies

$$\max_{r\in R}\left\|\frac{p(s)N_i(s)}{s^\ell(s+\alpha)D(s,r)}\right\|_\infty \le \gamma. \quad (36)$$

We now show that (35) implies (36). For fixed $\gamma > 0$, (36) holds if and only if the complex coefficient polynomial

$$\gamma s^\ell(s+\alpha)D(s,r) + e^{j\phi}p(s)N_i(s) \quad (37)$$

is Hurwitz for all $r \in R$ and all $\phi \in [0, 2\pi]$. Recall that the Kharitonov polynomials $D_k(s)$, $k = 1,2,3,4$ satisfy the relation

$$\{D_k(s), k=1,2,3,4\} = \{D_0(s) \pm d_e(s) \pm d_o(s)\} \quad (38)$$

where $D_0(s)$ is a fixed polynomial and $d_e(s), d_o(s)$ are polynomials having only even or odd powers, respectively. For fixed $\phi^* \in [0, 2\pi]$, it follows from the Box Theorem [11] that (37) is Hurwitz for all $r \in R$ if and only if the polynomials

$$\gamma s^\ell(s+\alpha)[D_0(s) + \lambda_e d_e(s)] + e^{j\phi^*}p(s)N_i(s) \quad (39)$$

and

$$\gamma s^\ell(s+\alpha)[D_0(s) + \lambda_o d_o(s)] + e^{j\phi^*}p(s)N_i(s) \quad (40)$$

are Hurwitz for all $\lambda_e \in [-1,1]$ and $\lambda_o \in [-1,1]$. Since $\phi^* \in [0,2\pi]$ is arbitrary in (39) and (40), equation (36) holds if and only if

$$\max_{\lambda_e\in[-1,1]}\left\|\frac{p(s)N_i(s)}{s^\ell(s+\alpha)(D_0(s)+\lambda_e d_e(s))}\right\|_\infty \le \gamma \quad (41)$$

and

$$\max_{\lambda_o\in[-1,1]}\left\|\frac{p(s)N_i(s)}{s^\ell(s+\alpha)(D_0(s)+\lambda_o d_o(s))}\right\|_\infty \le \gamma. \quad (42)$$

Now, apply Lemma 1 to both (41) and (42). With (38), this implies that (36) holds if and only if

$$\left\|\frac{p(s)N_i(s)}{s^\ell(s+\alpha)D_k(s)}\right\|_\infty \le \gamma \quad (43)$$

for all $k = 1,2,3,4$. This shows that (35) implies (36) and completes the proof. □

IV. Extremality Properties of the H_∞ Norm of Weighted Sensitivity and Complementary Sensitivity Functions

As a companion to Theorem 1, we now show the extremality property enjoyed by the sensitivity and complementary sensitivity functions of an interval feedback system.

Theorem 2: *Let α, β and K be real nonnegative numbers and ℓ an integer. If*

$$K\frac{s+\beta}{s^\ell(s+\alpha)}S(s,q,r) \quad \text{and} \quad K\frac{s+\beta}{s^\ell(s+\alpha)}T(s,q,r)$$

are proper and strictly stable for all $q \in Q$ and $r \in R$, then

$$\max_{q\in Q, r\in R}\left\|K\frac{s+\beta}{s^\ell(s+\alpha)}S(s,q,r)\right\|_\infty = \max_{i,k=1,2,3,4}\left\|K\frac{s+\beta}{s^\ell(s+\alpha)}S_{i,k}(s)\right\|_\infty \quad (44)$$

and

$$\max_{q\in Q, r\in R}\left\|K\frac{s+\beta}{s^\ell(s+\alpha)}T(s,q,r)\right\|_\infty = \max_{i,k=1,2,3,4}\left\|K\frac{s+\beta}{s^\ell(s+\alpha)}T_{i,k}(s)\right\|_\infty. \quad (45)$$

We will prove only (44). To this end, we need a key technical lemma.

Lemma 2: *Let $N(s)$ and $D(s)$ be fixed polynomials, $q(s)$ a polynomial of degree less than the degree of $D(s)$ and α, β real nonnegative numbers. If*

$$K\frac{s+\beta}{s+\alpha}\frac{D(s)\pm q(s)}{N(s)+D(s)\pm q(s)}$$

is proper and strictly stable, then

$$\max_{\lambda\in[-1,1]}\left\|K\frac{s+\beta}{s+\alpha}\frac{D(s)+\lambda q(s)}{N(s)+D(s)+\lambda q(s)}\right\|_\infty =$$

$$\max_{\lambda=\pm 1}\left\|K\frac{s+\beta}{s+\alpha}\frac{D(s)+\lambda q(s)}{N(s)+D(s)+\lambda q(s)}\right\|_\infty. \quad (46)$$

Proof of Lemma 2: For notational simplicity, let

$$G(s,\lambda) = K\frac{s+\beta}{s+\alpha}\frac{D(s)+\lambda q(s)}{N(s)+D(s)+\lambda q(s)}. \quad (47)$$

Proceeding by contradiction, assume that (46) does not hold. Then,

$$\max_{\lambda\in[-1,1]}\|G(s,\lambda)\|_\infty > \max_{\lambda=\pm1}\|G(s,\lambda)\|_\infty.$$

Scaling both sides by $\max_{\lambda\in[-1,1]}\|G(s,\lambda)\|_\infty$ and absorbing this factor in $N(s)$ we obtain

$$\max_{\lambda\in[-1,1]}\|G(s,\lambda)\|_\infty = 1$$

while

$$\max_{\lambda=\pm1}\|G(s,\lambda)\|_\infty < 1.$$

Without loss of generality, redefine λ so that the maximum of $\|G(s,\lambda)\|_\infty$ is achieved at $\lambda = 0$. This gives

$$\max_{\lambda\in[\alpha,\beta]}\|G(s,\lambda)\|_\infty = \|G(s,0)\|_\infty = 1 \quad (48)$$

and

$$\max_{\lambda=\alpha,\beta}\|G(s,\lambda)\|_\infty < 1 \quad (49)$$

for some $\alpha < 0 < \beta$.

Equation (48) implies that the Nyquist plot of $G(s,\lambda)$ is contained in the unit disk for all $\lambda \in [\alpha,\beta]$ with the plot $G(s,0)$ actually touching the disk boundary. Let ω^* be a frequency for which this occurs; i.e.,

$$|G(j\omega^*,0)| = 1.$$

There may be frequencies different from ω^* at which the Nyquist plot of $G(s,0)$ touches the unit circle. In this case, we list them as $\omega_i, i = 1,2,\ldots,N$. Since $G(s,0)$ is strictly stable, $|G(j\omega^*,0)| = 1$ and $|G(j\omega,0)| \le 1$ for all $\omega \ge 0$, then, for

$$\phi^* = \pi - \angle G(j\omega^*,0),$$

the solutions of the equation

$$1 + e^{j\phi^*}G(s,0) = 0$$

are all contained in $Re\ s < 0$ except for a finite number of $j\omega$-axis roots which includes $s = j\omega^*$. Now, define

$$f(s,\lambda) \doteq 1 + e^{j\phi^*}G(s,\lambda). \tag{50}$$

Equations (48) and (49) imply that the λ root locus segment associated with $\{s : f(s,\lambda) = 0 \text{ for some } \lambda \in [\alpha,\beta]\}$ is wholly contained in $Re\ s \le 0$. Moreover, this root locus segment touches the imaginary axis at only a *finite* number of points. Indeed, if this were not the case, then this root locus segment would "flatten out" against the imaginary axis. More precisely, the root locus would be contained in the imaginary axis for all λ in some subinterval of $[\alpha,\beta]$. However, since none of the root locus lies in $Re\ s > 0$, such behavior violates the analyticity of the root locus branch functions and is thus ruled out. Consequently, for sufficiently small ϵ, $f(s,\lambda)$ has only zeroes in $Re\ s < 0$ for all $\lambda \in [-\epsilon,\epsilon]\backslash 0$. The zeroes of $f(s,0)$ are also strictly stable except for $j\omega$-axis roots which include $s = j\omega^*$.

Let $s(\lambda)$ denote the root function of $f(s,\lambda)$ for which $s(0) = j\omega^*$. From the previous discussion, $s(\lambda)$ satisfies

$$Re[s'(0)] = 0 \tag{51}$$

and

$$Re[s''(0)] < 0. \tag{52}$$

Define the function $\tilde{f}(s,\lambda)$

$$\tilde{f}(s,\lambda) \doteq 1 + K\frac{s+\beta}{s+\alpha}\frac{D(s)p_1(s) + \lambda q(s)a_1(s)}{\gamma_1[N(s)+D(s)]p_2(s) + \lambda q(s)a_2(s)} \tag{53}$$

where $\gamma_1(s) \in (0,1)$, $p_1(s)$ and $p_2(s)$ are polynomials to be determined and $a_1(s), a_2(s)$ are polynomials satisfying the conditions

$$\begin{aligned} a_1(j\omega^*) &= p_1(j\omega^*); \\ a_2(j\omega^*) &= p_2(j\omega^*). \end{aligned} \tag{54}$$

Claim: *Given $\gamma_1,\ \gamma_2 \in (0,1)$ and polynomials $a_1(s)$ and $a_2(s)$ satisfying (54), there exist polynomials $p_1(s)$ and $p_2(s)$ with $p_2(s)$ Hurwitz such that*

$$\left\|\frac{p_1(s)}{p_2(s)}\right\|_\infty \le 1, \tag{55}$$

$$\left.\frac{p_1(s)}{p_2(s)}\right|_{s=j\omega^*} = \gamma_1 e^{j\phi^*}, \tag{56}$$

$$\left.\frac{p_1(s)}{p_2(s)}\right|_{s=j\omega_i} = \gamma_2 e^{j\phi^*} \tag{57}$$

for $i = 1,2,\ldots,N$,

$$\tilde{s}'(0) = s'(0) \tag{58}$$

and

$$\tilde{s}''(0) = s''(0). \tag{59}$$

Proof of Claim: With $a_1(s)$ and $a_2(s)$ chosen as in (54) we construct a polynomial $p_1(s)$ and a Hurwitz polynomial $p_2(s)$ such that (55)-(59) hold. To this end, we evaluate

$$\begin{aligned} s'(\lambda) &= -\frac{\partial f}{\partial\lambda}\left(\frac{\partial f}{\partial s}\right)^{-1}; \\ \tilde{s}'(\lambda) &= -\frac{\partial \tilde{f}}{\partial\lambda}\left(\frac{\partial f}{\partial \tilde{s}}\right)^{-1} \end{aligned} \tag{60}$$

and

$$\begin{aligned} s''(\lambda) &= -\frac{\partial^2 f}{\partial\lambda^2}\left(\frac{\partial f}{\partial s}\right)^{-1} + \frac{\partial f}{\partial\lambda}\frac{\partial^2 f}{\partial s^2}\left(\frac{\partial f}{\partial s}\right)^{-2}; \\ \tilde{s}''(\lambda) &= -\frac{\partial^2 \tilde{f}}{\partial\lambda^2}\left(\frac{\partial \tilde{f}}{\partial \tilde{s}}\right)^{-1} + \frac{\partial \tilde{f}}{\partial\lambda}\frac{\partial^2 \tilde{f}}{\partial \tilde{s}^2}\left(\frac{\partial \tilde{f}}{\partial \tilde{s}}\right)^{-2} \end{aligned} \tag{61}$$

where the root functions s and $\tilde{s}$ are implicitely defined by $f(s,\lambda) = 0$ and $\tilde{f}(s,\lambda) = 0$. From (60)-(61) we see that (58) and (59) hold if we require

$$\begin{aligned} \left.\frac{\partial f}{\partial\lambda}\right|_{\substack{\lambda=0\\ s=j\omega^*}} &= \left.\frac{\partial \tilde{f}}{\partial\lambda}\right|_{\substack{\lambda=0\\ s=j\omega^*}}; \\ \left.\frac{\partial^2 f}{\partial\lambda^2}\right|_{\substack{\lambda=0\\ s=j\omega^*}} &= \left.\frac{\partial^2 \tilde{f}}{\partial\lambda^2}\right|_{\substack{\lambda=0\\ s=j\omega^*}}; \\ \left.\frac{\partial f}{\partial s}\right|_{\substack{\lambda=0\\ s=j\omega^*}} &= \left.\frac{\partial \tilde{f}}{\partial \tilde{s}}\right|_{\substack{\lambda=0\\ s=j\omega^*}}; \\ \left.\frac{\partial^2 f}{\partial s^2}\right|_{\substack{\lambda=0\\ s=j\omega^*}} &= \left.\frac{\partial^2 \tilde{f}}{\partial \tilde{s}^2}\right|_{\substack{\lambda=0\\ s=j\omega^*}}. \end{aligned} \tag{62}$$

We thus seek $p_1(s)$ and a Hurwitz polynomial $p_2(s)$ such that (55)-(57) and (62) hold. For $f(s,\lambda)$ in (50) we compute

$$\begin{aligned} \left.\frac{\partial f}{\partial\lambda}\right|_{\substack{\lambda=0\\ s=j\omega^*}} &= \left. e^{j\phi^*}\frac{s+\beta}{s+\alpha}\frac{q(s)N(s)}{(N(s)+D(s))^2}\right|_{s=j\omega^*}; \\ \left.\frac{\partial^2 f}{\partial\lambda^2}\right|_{\substack{\lambda=0\\ s=j\omega^*}} &= \left. -2e^{j\phi^*}\frac{s+\beta}{s+\alpha}\frac{q^2(s)N(s)}{(N(s)+D(s))^3}\right|_{s=j\omega^*}; \\ \left.\frac{\partial f}{\partial s}\right|_{\substack{\lambda=0\\ s=j\omega^*}} &= \left. e^{j\phi^*}\left(\frac{s+\beta}{s+\alpha}\frac{D(s)}{N(s)+D(s)}\right)'\right|_{s=j\omega^*}; \\ \left.\frac{\partial^2 f}{\partial s^2}\right|_{\substack{\lambda=0\\ s=j\omega^*}} &= \left. e^{j\phi^*}\left(\frac{s+\beta}{s+\alpha}\frac{D(s)}{N(s)+D(s)}\right)''\right|_{s=j\omega^*}. \end{aligned} \tag{63}$$

Similarly, we obtain

$$\begin{aligned} \left.\frac{\partial \tilde{f}}{\partial\lambda}\right|_{\substack{\lambda=0\\ s=j\omega^*}} &= \left. \frac{p_1(s)}{\gamma_1 p_2(s)}\frac{s+\beta}{s+\alpha}\frac{q(s)N(s)}{(N(s)+D(s))^2}\right|_{s=j\omega^*}; \\ \left.\frac{\partial^2 \tilde{f}}{\partial\lambda^2}\right|_{\substack{\lambda=0\\ s=j\omega^*}} &= \left. -2\frac{p_1(s)}{\gamma_1 p_2(s)}\frac{s+\beta}{s+\alpha}\frac{q^2(s)N(s)}{(N(s)+D(s))^3}\right|_{s=j\omega^*}; \\ \left.\frac{\partial \tilde{f}}{\partial \tilde{s}}\right|_{\substack{\lambda=0\\ s=j\omega^*}} &= \left. \frac{p_1(s)}{\gamma_1 p_2(s)}\left(\frac{s+\beta}{s+\alpha}\frac{D(s)}{N(s)+D(s)}\right)'\right|_{s=j\omega^*}; \\ \left.\frac{\partial^2 \tilde{f}}{\partial \tilde{s}^2}\right|_{\substack{\lambda=0\\ s=j\omega^*}} &= \left. \frac{p_1(s)}{\gamma_1 p_2(s)}\left(\frac{s+\beta}{s+\alpha}\frac{D(s)}{N(s)+D(s)}\right)'\right|_{s=j\omega^*}. \end{aligned} \tag{64}$$

Using (63) and (64), a straightforward calculation shows that $p_1(s)$ and $p_2(s)$ satisfy (62) if they meet the following interpolation conditions:

$$\begin{aligned} \left.\frac{p_1(s)}{p_2(s)}\right|_{s=j\omega^*} &= \gamma_1 e^{j\phi^*}; \\ \left.\left(\frac{p_1(s)}{p_2(s)}\right)'\right|_{s=j\omega^*} &= 0; \\ \left.\left(\frac{p_1(s)}{p_2(s)}\right)''\right|_{s=j\omega^*} &= 0. \end{aligned} \tag{65}$$

Notice that (65) includes (56). Since γ_1 and γ_2 are less than one, it follows from boundary interpolation results, see for example Theorem 1.5 and Remark 1.7 in [9], that there exists a stable rational function $\frac{p_1(s)}{p_2(s)}$ satisfying (55), (57) and (65). This proves the claim.

Return to the proof of Lemma 1 and fix $\gamma_2 \in (0,1)$. For $\gamma_1 \in (0,1)$ sufficiently close to 1 and $a_1(s)$, $a_2(s)$ satisfying (54), it follows from the claim that there exist polynomials p_1 and $p_2(s)$, $p_2(s)$ Hurwitz, such that the Nyquist plot of

$$\tilde{G}(s) \doteq \frac{N(s)p_1(s)}{\gamma_1(s+\alpha)D(s)p_2(s)} \tag{66}$$

is contained in the unit disk, touching $-1+j0$ only when $\omega = \omega^*$. Specifically, (56) implies that

$$\tilde{G}(j\omega^*) = e^{j\phi^*}G(j\omega^*, 0) = -1$$

while (57) and γ_1 sufficiently close to 1 gives

$$|\tilde{G}(j\omega_i)| = \frac{\gamma_2}{\gamma_1}|G(j\omega_i)| < 1$$

for $i = 1, 2, \ldots, N$. Finally, for all other frequencies, (55), (62) and $\gamma_1 \approx 1$ guarantee that

$$|\tilde{G}(j\omega)| < 1.$$

Now, as a consequence of $\tilde{G}(s)$ being real, strictly stable and having Nyquist plot touching the unit circle at only at $-1+j0$, the rational function $\tilde{f}(s,0)$ in (53) has only zeroes in $Re\ s < 0$ except for a pair of imaginary roots at $s = j\omega^*$. Moreover, from (58) and (59), these two imaginary roots of $\tilde{f}(s,\lambda)$ migrate into the open left half plane when λ infinitesimally changes from zero. Consequently, there exists an $\tilde{\epsilon} > 0$ such that $\tilde{f}(s,\lambda)$ has zeroes only in the open left half plane for all $\lambda \in [-\tilde{\epsilon}, \tilde{\epsilon}]\backslash 0$. From this and (53), it follows that the uncertain polynomial

$$D(s)\gamma_1[(s+\alpha)p_2(s) + K(s+\beta)p_1(s)] + N(s)\gamma_1(s+\alpha)p_2(s) + \lambda q(s)[K(s+\beta)a_1(s) + (s+\alpha)a_2(s)]. \tag{67}$$

has a zero at $s = j\omega^*$ when $\lambda = 0$ but is stable for all $\lambda \in [-\tilde{\epsilon}, \tilde{\epsilon}]\backslash 0$. Presently, $a_1(s)$ and $a_2(s)$ are chosen to satisfy (54). In addition they can be selected so that

$$\tilde{p}(s) \doteq K(s+\beta)a_1(s) + (s+\alpha)a_2(s) \tag{68}$$

is a convex direction. To see this, let c_0, c_1, d_0 and d_1 be real numbers and

$$\begin{aligned} a_1(s) &= s^2 + d_1 s + d_0; \\ a_2(s) &= s^2 + c_1 s + c_0. \end{aligned} \tag{69}$$

Now, to satisfy condition (54) we take

$$\begin{aligned} c_1 + d_1 &= 0; \\ c_0 - d_0 + \omega^{*2}(d_2 - c_2) &= 0; \\ c_1(1-K) + \beta + K\alpha &= 0; \\ \beta c_0 + K(\alpha d_0 - 1) &= 1. \end{aligned} \tag{70}$$

Thus, $\tilde{p}(s)$ in (68) becomes

$$\tilde{p}(s) = s^3(1+K) + s(c_0 + \beta c_1 + K(d_0 - \alpha c_1)] + (1+K).$$

It can be verified that this polynomial is a convex direction polynomial for some K, α, β; e.g., see [10].

Return to (67) using these newly computed values of $a_1(s)$ and $a_2(s)$. The same conclusion remains but $q(s)(s+\alpha)a(s)$ is a convex direction (see [7]) and hence proves Lemma 2. □

The proof of Theorem 2 follows from Lemma 2 in exactly the same way that the proof of Theorem 1 leans on Lemma 1. □

Acknowledgements

This work was partially supported by CENS-CNR of Italy. C. V. Hollot conducted research while visiting the Laboratory of Automatic Control at the University of Louvain, Louvain-La-Neuve, Belgium and the Institute for Automatic Control, Swiss Federal Institute of Technology, ETH-Zentrum, 8092, Zurich, Switzerland. He is grateful for their support.

References

[1] V. L. Kharitonov, "Asymptotic Stability of an Equilibrium Position of a Family of Systems of Linear Differential Equations," *Differentsial'nye Uravneniya*, Vol. 14, pp. 2086-2088, 1978.

[2] T. Mori and S. Barnett, "On Stability Tests for Some Classes of Dynamical Systems with Perturbed Coefficients," IMA Journal of Mathematical Control and Information, Vol. 5, pp. 117-123, 1988.

[3] H. Chapellat, M. Dahleh and S. P. Bhattacharyya, "Robust Stability Under Structured and Unstructured Perturbations," *IEEE Transactions on Automatic Control*, Vol. AC-35, pp. 1100-1108, 1990.

[4] C. V. Hollot and R. Tempo, "H_∞ Performance of Interval Plants and Interval Feedback Systems," *Proceedings of the International Workshop on Robustness of Systems with Parameter Uncertainties*, Ascona (Switzerland) April, 1992.

[5] C. V. Hollot and F. Yang, "Robust Stabilization of Interval Plants Using Lead or Lag Compensators," *Systems and Control Letters*, Vol. 14, no. 1, pp. 9-12, 1990.

[6] A. C. Bartlett, C. V. Hollot and L. Huang, "Root Locations for an Entire Polytope of Polynomials: It Suffices to Check the Edges," *Mathematics of Controls, Signals and Systems*, Vol. 1, no. 1, pp. 61-71.

[7] A. Rantzer, "Stability Conditions for a Polytope of Polynomials," *IEEE Transactions on Automatic Control*, Vol. 37, pp. 79-89, 1992.

[8] B. R. Barmish, R. Tempo, C. V. Hollot and H. I. Kang, "An Extreme Point Result for Robust Stability of a Diamond of Polynomials," to appear in *IEEE Transactions on Automatic Control*, 1992.

[9] P. P. Khargonekar and A. Tannenbaum, "Non-Euclidean Metrics and the Robust Stabilization of Systems with Parameter Uncertainties," *IEEE Transactions on Automatic Control*, Vol. AC-30, pp. 1005-1013, 1985.

[10] B. R. Barmish and H. I. Kang, "Extreme Point Results for Robust Stability of Interval Plants: Beyond First Order Compensators," to appear in *Automatica*; for a preliminary version, see the *Proceedings of IFAC Symposium on Design Methods for Control Systems*,

[11] H. Chapellat and S. P. Bhattacharyya, "A Generalization of Kharitonov's Theorem: Robust Stability of Interval Plants," *IEEE Transactions on Automatic Control*, Vol. AC-34, pp. 306-311, 1989.

Proceedings of the 31st Conference
on Decision and Control
Tucson, Arizona • December 1992

Shifted Popov Criterion and Stability Analysis of Fuzzy Control Systems

Eiko Furutani†, Masami Saeki‡ and Mituhiko Araki†

† Department of Electrical Engineering, Kyoto University
Yoshida-Honmachi, Sakyo-ku, Kyoto 606-01, Japan
‡ Faculty of Engineering, Hiroshima University
1-4-1 Kagamiyama, Higashi-Hiroshima 724, Japan

Abstract

In this paper, we derive a new stability criterion for the nonlinear feedback systems whose nonlinearity consists of the main term and the deviation term, where the main term is time-invariant and depends only upon the output of the linear part, but the deviation term is time-varying and/or depends on variables other than the output. To show a practical significance of the new criterion, we analyse a certain type of fuzzy control systems.

1 Introduction

As representative stability conditions of nonlinear feedback systems, we can give the "circle criterion" and the "Popov criterion"[1]. The former can be applied to the case of time-varying nonlinearity but its requirement is strict, while the latter can be applied only to the case of time-invariant nonlinearity but its requirement is less strict. The purpose of the present paper is to fill in the gap between the two criteria. Namely, we deal with the case in which the nonlinearity consists of the main term and the deviation term, where the main term is time-invariant and depends only upon the output y of the linear part, but the deviation term is time-varying and/or depends on variables other than y. For this class of nonlinear feedback systems, we derive a new stability criterion, which is named as "shifted Popov criterion". The new criterion converges to the ordinary Popov criterion in a graphical sense when the deviation term converges 0, and coincides with the circle criterion when the parameter θ in the criterion is set 0. Its requirement becomes considerably less conservative than that of the circle criterion if the deviation term is small. To show a practical significance of the new criterion, we analyse a certain type of fuzzy control systems.

2 Shifted Popov Criterion

In this section, we give our main result together with its graphical implication, from which the name "shifted Popov criterion" originates.

2.1 Main Theorem

Consider a nonlinear feedback system consisting of the linear part

$$\dot{x} = Ax - bu, \quad y = cx \tag{1}$$

and the nonlinear part

$$u = \varphi(y, x, t). \tag{2}$$

As for the linear part, the input u and the output y are both scalars, the state x is an n-vector, and (A, b, c) is a minimum realization of the transfer function

$$G(s) = c(sI - A)^{-1}b. \tag{3}$$

As for the nonlinear part, it is assumed that

$$\varphi(y, x, t) = \varphi_P(y) + \varphi_D(x, t); \tag{4}$$

where $\varphi_P(y)$ is a time-invariant function satisfying the sector condition

$$0 \leq \varphi_P(y)y \leq ky^2, \ \varphi_P(0) = 0 \tag{5}$$

and $\varphi_D(x, t)$ is a time-varying function of x satisfying

$$|\varphi_D(x, t)| \leq \varepsilon|y|. \tag{6}$$

Here, k is positive and ε is nonnegative. The assumption about the nonlinear part implies that the nonlinearity $\varphi(y, x, t)$ consists of the main term $\varphi_P(y)$ which is of the Popov type, and the deviation term $\varphi_D(x, t)$ which may be time-varying and dependent on all the state variables of the system but is bounded by $|y|$. The structure of the system is illustrated in Fig. 1, where the hatched area of the φ–y graph means that the value of φ may deviate from $\varphi_P(y)$ within this range.

By assumptions (5) and (6), the origin $x = 0$ is an equilibrium point of the system (1),(2). In the following, we say the system is stable (asymptotically stable) when $x = 0$ is stable (asymptotically stable). Concerning stability of the system (1),(2), we have the next theorem.

Theorem 1 Assume that A is Hurwitz. When A has multiple eigenvalues, assume that the pair $(A, c + \theta cA)$ remains observable for the following value of θ. Then,

CH3229-2/92/0000-2790$1.00 © 1992 IEEE

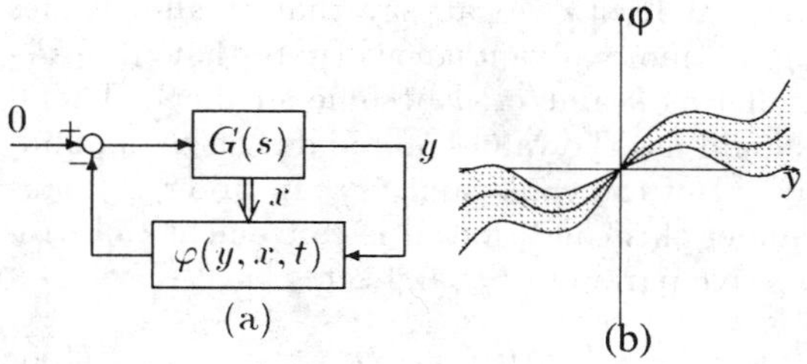

Figure 1: The structure of the system

the system (1),(2) is stable in the sense of Lyapunov if there are positive constants κ, μ and a constant θ such that

$$\kappa\mu \geq \varepsilon^2 \tag{7}$$

$$\frac{1}{k} + \mathrm{Re}\,\{(1+j\omega\theta)G(j\omega)\} - \frac{\kappa}{2}|(1+j\omega\theta)G(j\omega)|^2 - \mu(\frac{\kappa}{k}+\frac{1}{2})|G(j\omega)|^2 \geq 0 \quad \forall\omega \geq 0 \tag{8}$$

$$1 - \frac{k}{2}\{\kappa\theta^2(cb)^2 - 2\theta cb\} > 0. \tag{9}$$

In addition, the system is asymptotically stable in the large, if the inequality obtained from eq. (8) by replacing $j\omega$ with $j\omega - \delta$ holds true.

2.2 Graphical Meaning

We can interpret that eq. (7) restricts the choice of κ and μ, eq. (9) restricts the choice of θ, and eq. (8) restricts the behavior of the frequency response of the linear part. If ε is very small (i.e. if the deviation term of the nonlinear part is very small), κ and μ can be chosen very small. Then, the third and fourth terms of eq. (8) and the first term in the brace { } of eq. (9) become very small, and eqs. (8) and (9) converge to

$$\frac{1}{k} + \mathrm{Re}\,\{(1+j\omega\theta)G(j\omega)\} \geq 0 \tag{10}$$

$$1 + k\theta cb > 0 \tag{11}$$

respectively. Eqs. (10) and (11) are nothing but the inequalities required by the Popov criterion for the case $\varphi(y,x,t) = \varphi_P(y)$. Thus, we have seen that the requirements of Theorem 1 converge to those of the Popov criterion when the deviation term of the nonlinear part becomes very small.

To clarify the graphical meaning of eqs. (8) and (9), let the sum of the third and the fourth term of eq. (8) be denoted by

$$\gamma(\omega) = \frac{\kappa}{2}|(1+j\omega\theta)G(j\omega)|^2 + \mu(\frac{\kappa}{k}+\frac{1}{2})|G(j\omega)|^2 \tag{12}$$

and put

$$X(\omega) = \mathrm{Re}\,[G(j\omega)] - \gamma(\omega) \tag{13}$$

$$Y(\omega) = \omega\,\mathrm{Im}\,[G(j\omega)]. \tag{14}$$

Then, we can rewrite eq. (8) as

$$\frac{1}{k} + X(\omega) + \theta Y(\omega) \geq 0 \quad \forall\omega \geq 0. \tag{15}$$

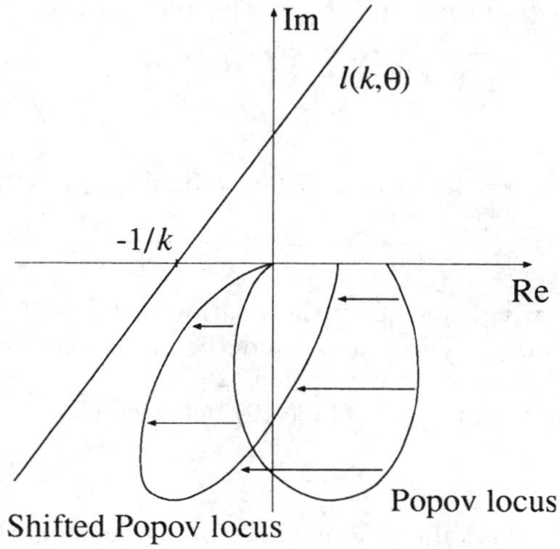

Figure 2: Popov locus and shifted Popov locus

Consider the locus of the point z on the complex plane which is determined by

$$\mathrm{Re}\,[z] = X(\omega),\ \mathrm{Im}\,[z] = Y(\omega). \tag{16}$$

We refer to this locus as **shifted Popov locus**, because it is obtained by shifting each point of the Popov locus for the distance of $\gamma(\omega)$ to the left (see Fig. 2). At the same time, consider the straight line on the same complex plane which passes the point $-1/k$ on the real axis and whose slope is $1/\theta$ (when $\theta = 0$, consider the line parallel to the imaginary axis). We denote this straight line by $l(k,\theta)$. Then, the inequality (15) is equivalent to the condition: "the shifted Popov locus be in the closed right side of $l(k,\theta)$ (see Fig. 2)." In addition, from the fact that the lefthand side of eq. (9) is nothing but the limit of the lefthand side of eq. (8) for $\omega \to \infty$, we can conclude that eq. (9) is equivalent to the condition: "the limiting point of the shifted Popov locus for $\omega \to \infty$ be strictly to the right of $l(k,\theta)$."

Since the two requirements of Theorem 1 can be expressed in terms of the relation between the shifted Popov locus and the straight line $l(k,\theta)$, we refer to Theorem 1 as the **shifted Popov criterion**.

2.3 Relation to the Circle Criterion

The circle criterion reads as follows[1].

Theorem 2 Concerning the system (1),(2), assume that $\varphi(y,x,t)$ satisfies the sector condition

$$k_1 y^2 \leq y \cdot \varphi(y,x,t) \leq k_2 y^2 \tag{17}$$

Then, the system is stable in the sense of Lyapunov if

$$\mathrm{Re}\,\frac{G(j\omega)}{1+k_1 G(j\omega)} + \frac{1}{k_2 - k_1} \geq 0 \ \forall\omega \geq 0 \tag{18}$$

holds true.

When the assumptions about the nonlinear part given in **2.1** (i.e. eqs. (4),(5),(6)) are satisfied, eq. (17) holds true for the following values of k_1 and k_2

$$k_1 = -\varepsilon,\ k_2 = k + \varepsilon \tag{19}$$

So, for our problem, eq. (18) of Theorem 2 turns out to be

$$(X_C(\omega)-c)^2+Y_C(\omega)^2 \le r^2 \tag{20}$$

where

$$X_C(\omega)=\mathrm{Re}\,G(j\omega),\ Y_C(\omega)=\mathrm{Im}\,G(j\omega) \tag{21}$$

$$c=\frac{1}{2}\left(\frac{1}{\varepsilon}-\frac{1}{k+\varepsilon}\right),\ r=\frac{1}{2}\left(\frac{1}{\varepsilon}+\frac{1}{k+\varepsilon}\right) \tag{22}$$

In the following, we clarify the relation of the shifted Popov criterion (Theorem 1) and the circle criterion (Theorem 2) using the expression (20).

Since the lefthand side of eq. (9) becomes 1 for

$$\theta = 0 \tag{23}$$

, we can use this value of θ in Theorem 1. Then, eq. (8) becomes

$$\frac{1}{k}+X_C-\left(\frac{\kappa\mu}{k}+\frac{\kappa+\mu}{2}\right)(X_C^2+Y_C^2)\ge 0\ \ \forall\omega>0. \tag{24}$$

Since $X_C^2+Y_C^2 \ge 0$, the above inequality becomes more plausible to be satisfied as

$$f(\kappa,\mu)=\left(\frac{\kappa\mu}{k}+\frac{\kappa+\mu}{2}\right) \tag{25}$$

is made smaller. Under the constraint (7), the minimum of $f(\kappa,\mu)$ is

$$\min_{(7)} f(\kappa,\mu)=\frac{\varepsilon^2}{k}+\varepsilon \tag{26}$$

and is attained for the values of κ and μ:

$$\kappa=\mu=\varepsilon. \tag{27}$$

For this choice of κ and μ, eq. (24) becomes

$$\frac{1}{k}+X_C-\left(\frac{\varepsilon^2}{k}+\varepsilon\right)(X_C^2+Y_C^2)\ge 0 \tag{28}$$

, which can be rewritten as

$$\left\{X_C+\frac{1}{2}\left(-\frac{1}{\varepsilon}+\frac{1}{k+\varepsilon}\right)\right\}^2+Y_C^2\le\left\{\frac{1}{2}\left(-\frac{1}{\varepsilon}-\frac{1}{k+\varepsilon}\right)\right\}^2. \tag{29}$$

The above is exactly same with eq. (20) (note eq. (22)). Thus, we have shown that the shifted Popov criterion (Theorem 1) coincides with the circle criterion (Theorem 2) if we set κ,μ,θ as given by eqs. (23) and (27).

2.4 Remarks on Application

Here, we consider how to analyse stability of a given system by applying Theorem 1. Unfortunately, we have no simple condition to assure existence of the values of κ,μ and θ which satisfy the three requirements (7),(8) and (9) of Theorem 1. Therefore, in application, we must search appropriate values of κ,μ and θ by a trial-and-error method. But the problem can be simplified a little more as explained in the following.

As shown in the previous subsection, the best choice of κ and μ for $\theta=0$ is given by eq. (27). For general values of θ, we cannot easily find the best choice exactly, but, at least, we can say that smaller values of κ and μ are more advantageous (note that $\gamma(\omega)$ becomes smaller as κ and/or μ become smaller). Therefore, we should use the values of κ and μ on the boundary ($\kappa\mu=\varepsilon^2$) of the constraint (7). In order to make a search under this condition, it is convenient to introduce a positive parameter ψ and set

$$\kappa=\varepsilon\psi,\ \mu=\varepsilon/\psi. \tag{30}$$

Note that $\psi=1$ gives the values of eq. (27).

If we use eq. (30), what we have to do is to search appropriate values of the two parameters θ and ψ for which eqs. (8) and (9) are satisfied. In order to make this search, we made a computer program. Examples of stability analysis using this program will be given in the presentation.

3 Application to Fuzzy Control Systems

The fuzzy logic was first used for the control of a dynamical plant by Mamdani[2] in mid-70's. Approximately after 10 years, its application called explosive attention in Japan and many trials were reported[5]. Especially, in order to show the advantage of the "fuzzy control" visually, the stabilization problem of an inverted pendulum on a moving cart was used, and it was exhibited (and believed) that the fuzzy controller can keep various pendulums standing upright on a running cart[6] and so, attains high robustness. But, later, it was proved[7] that the pendulum would roll down after the cart has run for a certain distance in the actual situation where the viscous-friction cannot be removed.[1] Partially because of this result and partially from experience, engineers started to realize the danger of using fuzzy controllers only based on intuitive reasoning, and arguments have been continued about advantages and disadvantages of the fuzzy controller.

If we restrict problems to the feedback control of dynamical plants, the main possible advantages of the fuzzy logic are

(a) To enable the non-specialist to understand the role of the feedback controller and to adjust it based on their understanding, and

(b) To derive a continuous quantity out of several signals of the on/off type.

The first advantage seems to be appreciated especially by operators and managers, and fuzzy controllers are actually being used in various practical plants. In order to guarantee the safety of such control systems, plant

[1]This distance depends on the coefficients of the viscous-friction and saturation level of the motor of the cart. For some experimental equipment, it is about 40 meters. The reason why the fuzzy controller was believed to be successful in the early experiments of the inverted pendulum was that the length of the rail on which the cart could move was a few meters and, for that length, the controller could keep the pendulum nearly upright. The paper[7] also clarified that insertion of an integral action can stabilize the system, and, in recent experiments, fuzzy controllers with integral action seem to be used frequently.

engineers had often to assure stability of the system beforehand by huge amount of simulation, and now they wish to have a theoretical tool for stability analysis. Kitamura[8] already made a contribution to this end by using the circle criterion. Our Theorem 1 can be applied in the same way and give a sharper result.

3.1 Fuzzy Controller

First, we give a brief explanation about the fuzzy controller which we deal with in our stability analysis. This controller was used by Mamdani[2] and its mathematical model was derived by Jang and Araki[7]. For details, refer to those references.

The fuzzy controller which we are going to study has two scalar inputs v_1 and v_2 and a scalar output u. The output u is determined by applying the fuzzy logic shown in Table 1. Each square of the table corresponds to a fuzzy rule. For instance, the entry "$R1 : NM$" in the left-upper square means:

Rule 1: If v_1 is NS and v_2 is NS, set u equal to NM.

where the signs NM, NS, ZR, PS, and PM mean fuzzy values of the variables which read "negative medium", "negative small", "zero", "positive small", and "positive medium", respectively.

Table 1: Rules of the fuzzy controller

$v_2 \setminus v_1$	NS	ZR	PS
NS	$R1 : NM$	$R4 : NS$	$R7 : ZR$
ZR	$R2 : NS$	$R5 : ZR$	$R8 : PS$
PS	$R3 : ZR$	$R6 : PS$	$R9 : PM$

Because of the way how u is determined from v_1 and v_2, the above fuzzy controller is called of the "mini-max and center-of-gravity" type. It was derived by Jang and Araki that the input-output characteristic of this fuzzy controller is given by the nonlinear functions of Table 2, where each function describes the characteristic in each part of the upper half of the input space shown in Fig. 3. The characteristic in the lower half part can be easily obtained from the fact that it is point-symmetric with respect to the origin. In Table 2, the result is given in terms of the normalized variables $v_1' = v_1/\alpha_1, v_2' = v_2/\alpha_2$ and the output u is given by

$$u = \beta f(v_1/\alpha_1, v_2/\alpha_2) \quad (31)$$

where α_1 and α_2 are the width of the membership functions for the inputs v_1 and v_2, and β is that for the output u.

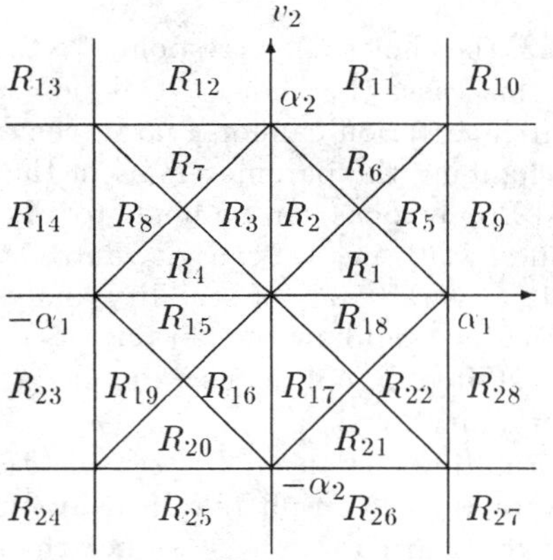

Figure 3: Division of the input space

Table 2: Nonlinear function which characterizes the fuzzy controller $u = \beta f(v_1', v_2')$

Region	Function $f(v_1', v_2')$ $v_1' = v_1/\alpha_1, v_2' = v_2/\alpha_2$
R_1	$\dfrac{-v_1'^2 + 3v_1' - v_2'^2 + 5v_2'}{2(-v_1'^2 + v_1' + v_2' + 1)}$
R_2	$\dfrac{-v_1'^2 + 5v_1' - v_2'^2 + 3v_2'}{2(-v_2'^2 + v_1' + v_2' + 1)}$
R_3	$\dfrac{v_1'^2 + 3v_1' - v_2'^2 + 3v_2'}{2(-v_2'^2 - v_1' + v_2' + 1)}$
R_4	$\dfrac{v_1'^2 + 3v_1' - v_2'^2 + 3v_2'}{2(-v_1'^2 - v_1' + v_2' + 1)}$
R_5	$\dfrac{v_1'^2 - v_1' - 3v_2'^2 + 5v_2' + 2}{2(-v_2'^2 - v_1' + v_2' + 2)}$
R_6	$\dfrac{-3v_1'^2 + 5v_1' + v_2'^2 - v_2' + 2}{2(-v_1'^2 + v_1' - v_2' + 2)}$
R_7	$\dfrac{-v_1'^2 + v_1' + v_2'^2 + v_2'}{2(-v_1'^2 - v_1' - v_2' + 2)}$
R_8	$\dfrac{-v_1'^2 + v_1' + v_2'^2 + v_2'}{2(-v_2'^2 + v_1' + v_2' + 2)}$
R_9	$\dfrac{-3v_2'^2 + 5v_2' + 2}{2(-v_2'^2 + v_2' + 1)}$
R_{10}	2
R_{11}	$\dfrac{-3v_1'^2 + 5v_1' + 2}{2(-v_1'^2 + v_1' + 1)}$
R_{12}	$\dfrac{-v_1'^2 + v_1' + 2}{2(-v_1'^2 - v_1' + 1)}$
R_{13}	0
R_{14}	$\dfrac{v_2'^2 + v_2' - 2}{2(-v_2'^2 + v_2' + 1)}$

The fuzzy controller described above is realized as a special-purpose electronic circuits[9, FZ-3000,OMRON] and also as a software in a μ-processor based controller[10, FRUITAX, Fuji Electric].

3.2 Transformation of Fuzzy Control Systems

Consider the fuzzy control system given in Fig. 4(a), where the fuzzy controller is as stated in the previous subsection. The linear plant can be described by a state equation

$$\dot{x} = Ax + bu, \; v_1 = c_1 x, \; v_2 = c_2 x \quad (32)$$

where

$$G_1(s) = c_1(sI - A)^{-1}b, \; G_2(s) = c_2(sI - A)^{-1}b. \quad (33)$$

If we introduce new variables:

$$\begin{aligned} \sigma_1 &= \alpha_2 v_1 + \alpha_1 v_2 = (\alpha_2 c_1 + \alpha_1 c_2)x \\ \sigma_2 &= \alpha_2 v_1 - \alpha_1 v_2 = (\alpha_2 c_1 - \alpha_1 c_2)x \end{aligned} \quad (34)$$

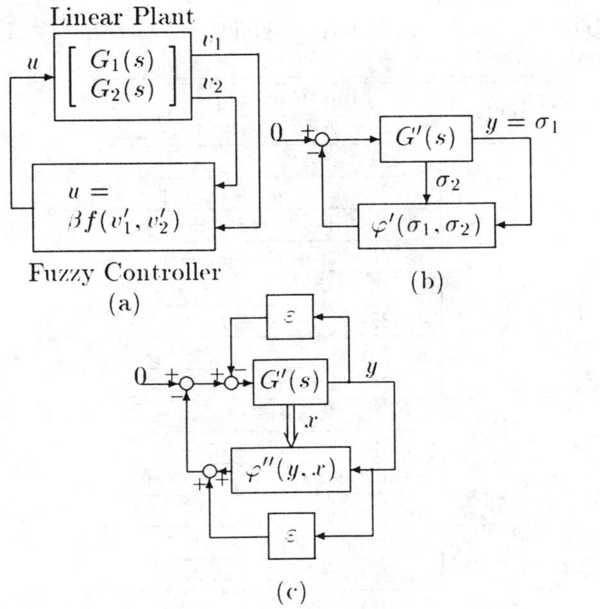

Figure 4: Fuzzy control system

the control system can be expressed in the form of Fig. 4(b) where

$$G'(s) = (\alpha_2 c_1 + \alpha_1 c_2)(sI - A)^{-1} b$$
$$= \alpha_2 G_1(s) + \alpha_1 G_2(s)$$
$$\varphi'(\sigma_1, \sigma_2) = \beta f\left(\frac{\sigma_1 + \sigma_2}{2\alpha_1\alpha_2}, \frac{\sigma_1 - \sigma_2}{2\alpha_1\alpha_2}\right).$$

If we restrict our consideration to the case where the variables v_1 and v_2 remain in the regions $R_1 \sim R_6$ and $R_{15} \sim R_{20}$ of Fig. 3, the value of $\varphi'(\sigma_1, \sigma_2)$ is mainly determined by σ_1 and fluctuates according as σ_2 changes. Namely, $\varphi'(\sigma_1, \sigma_2)$ can be depicted on the σ_1–u plane as in Fig. 5, where the value of $u = \varphi'(\sigma_1, \sigma_2)$ changes in the hatched region according to the change of σ_2. To be more exact

$$\begin{array}{c} 0 < \varphi'(\sigma_1, \sigma_2)\,\sigma_1 \le k'\sigma_1^2 \\ k' = 2.0\beta/\alpha_1\alpha_2 \end{array} \quad (35)$$

holds true and the amount of the fluctuation $\varphi_D{}'$ caused by σ_2 can be bounded as

$$\begin{array}{c} \|\varphi_D{}'\| \le \varepsilon|\sigma_1|, \\ \varepsilon = 0.7143\beta/\alpha_1\alpha_2. \end{array} \quad (36)$$

If we note that $\sigma_1 = y$ and σ_2 is a function of x, we can conclude that $\varphi'(\sigma_1, \sigma_2)$ can be expressed as

$$\varphi'(\sigma_1, \sigma_2) = \varphi_P{}''(y) + \varphi_D{}''(x) \quad (37)$$

where

$$\begin{array}{c} \varepsilon y^2 < y\varphi_P{}''(y) \le (k' - \varepsilon)y^2 \\ |\varphi_D{}''(x)| \le \varepsilon|y|. \end{array} \quad (38)$$

Finally, by applying the feedback-feedforward transformation shown in Fig. 4(c), we can transform the fuzzy control system into the form of Fig. 1 where

$$\begin{array}{c} G(s) = \dfrac{G'(s)}{1 + \varepsilon G'(s)} \\ k = k' - 2\varepsilon = 0.5715\beta/\alpha_1\alpha_2. \end{array} \quad (39)$$

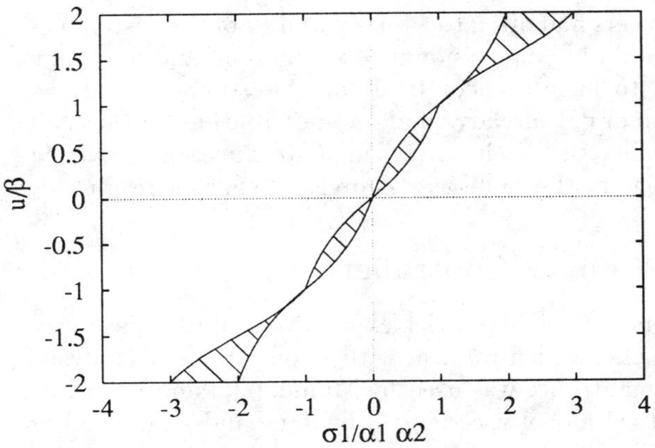

Figure 5: Region of the change of u

3.3 Stability Analysis and Remarks

If we apply Theorem 1 to the transformed system (Fig. 4(c)), we can analyse the stability of the fuzzy control system. At the first sight, the procedure of the transformation seems complicated. But, if we follow the equations, we can find that the shifted Popov locus can be immediately obtained from the parameters $\alpha_1, \alpha_2, \beta$ of the fuzzy controller and the transfer functions $G_1(s), G_2(s)$ of the linear plant. For instance, consider the case

$$G_1(s) = \frac{0.5(s+3)}{(s+1)(s^2+s+1)}$$
$$G_2(s) = \frac{0.5}{s^2+s+1}$$
$$\alpha_1 = 1, \alpha_2 = 1, \beta = 0.7.$$

Then,

$$G(s) = \frac{s+2}{s^3 + 2s^2 + 2.5s + 2}$$
$$\varepsilon = 0.5$$
$$k = 0.4$$

Fig. 6 shows the shifted Popov locus for this case for the two choices of θ, κ and μ:

① $\theta = 0, \kappa = 0.5, \mu = 0.5$

② $\theta = 0.5, \kappa = 0.5, \mu = 0.5$

As stated in 2.3, the choice ① corresponds to the circle criterion. In this case, the choice ① is not successful, i.e. the circle criterion cannot guarantee stability. However, by changing the parameters as in the choice ②, the shifted Popov locus can be made totally to the right of the line $l(k, \theta)$, i.e. we can guarantee stability by Theorem 1. Several trials of stability analysis will enable one to understand how the parameters $\alpha_1, \alpha_2, \beta$ effects on the shifted Popov locus (and, as a result, feasibility of stability).

The transformation (34) from (v_1, v_2) to (σ_1, σ_2) is due to Kitamura[8]. The result obtained by our Theorem 1 is always sharper (at least same) than Kitamura's, which uses the circle criterion as the basic tool.

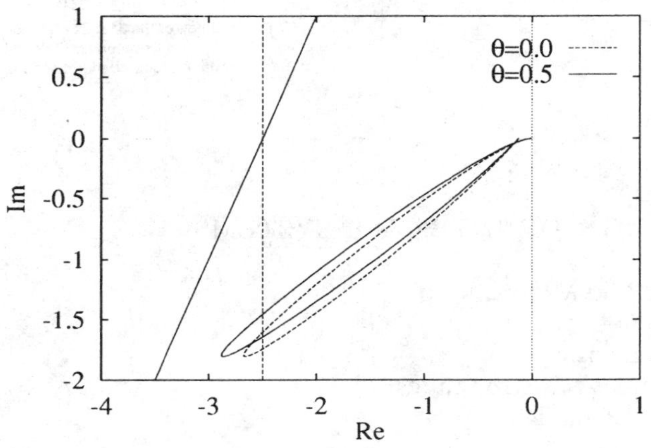

Figure 6: Stability analysis of fuzzy control system

There are reported other sorts of fuzzy controllers[11] in actual applications, which differ in the form of the membership functions and the way of fuzzy inference. However, most of them have the same basic property; i.e. by the variable transformation of the form eq. (34), the function $\varphi'(\sigma_1, \sigma_2)$ which describes the input-output relation of the fuzzy controller, turns to have the property as given by the inequalities (35) and (36) (at least in a certain neighborhood of the origin which has significant extension). So, our method of stability analysis can be extended to those cases. Here, it should be noted that, for stability analysis, it is not necessary to obtain $\varphi'(\sigma_1, \sigma_2)$ (i.e. $f(v_1/\alpha_1, v_2/\alpha_2)$) in the closed form as given in Table 2, but it is enough to know k' and ε. The values of k' and ε can be easily estimated based on numerical calculation.

The method is basically extendible to the fuzzy controller which has many (≥ 3) inputs. For instance, if the fuzzy controller has three inputs v_1, v_2 and v_3, the transformation of the form

$$\begin{aligned} \sigma_1 &= w_{11}v_1 + w_{12}v_2 + w_{13}v_3 \\ \sigma_2 &= \begin{bmatrix} w_{21}v_1 + w_{22}v_2 + w_{23}v_3 \\ w_{31}v_1 + w_{32}v_2 + w_{33}v_3 \end{bmatrix} \end{aligned} \quad (40)$$

will work to bring the problem into our framework. However, to find appropriate values of the coefficient w_{ij} would not be easy.

4 Conclusion

In this paper, we derived the "shifted Popov criterion", which fills in the gap between the Popov and circle criteria. It is useful in giving a sharp stability result especially when the nonlinearity mainly consists of the time-invariant function of the output of the linear part but is slightly deviated by a time-dependent and/or state-dependent function.

The proof of the theorem uses a lemma obtained in the composite-system analysis[3], and also a special form of the positive-real lemma[4]. Extension of the criterion to multivariable case is reported in a separate paper [12].

The result is generally useful for analysing nonlinear control systems. A typical class of nonlinear systems which need the help of the reported criterion is fuzzy control systems.

References

[1] K. S. Narendra, J. H. Taylor: *Frequency Domain Criteria for Absolute Stability*, Academic Press (1973)

[2] E. H. Mamdani: Application of fuzzy algorithms for control of simple dynamic plant, *Proceedings of IEE*, Vol. 121, No. 12, 1585/1588 (1974)

[3] M. Saeki, M. Araki, and B. Kondo: Local stability of composite systems –frequency-domain condition and estimate of the domain of attraction, *IEEE Trans. Automat. Contr.*, Vol. AC-25, No. 5, 936/940 (1980)

[4] A. Hattori and K. Kobayashi: The matrix forms of Yakubovich-Kalman-Lefschetz theorems derived from the Popov theory of positive systems, *IEEE Trans. Automat. Contr.*, Vol. AC-25, No. 1, 102/104 (1980)

[5] Special issue on fuzzy control, *Journal of the Society of Instrument and Control Engineers*, Vol. 28, No. 11 (1989)

[6] T. Yamakawa: Stabilization of an inverted pendulum by a high-speed fuzzy logic controller hardware system, *Fuzzy Sets and Systems 32*, 161/180 (1989)

[7] S. D. Jang and M. Araki: Mathematical analysis of fuzzy control systems and on possibility of industrial applications (in Japanese), *Transactions of the Society of Instrument and Control Engineers*, Vol. 26, No. 11, 1267/1274 (1990)

[8] S. Kitamura and T. Kurozumi: Extended circle criterion, and stability analysis of fuzzy control systems, *Fuzzy Engineering toward Human Friendly Systems*, Vol. 2, 634/643 (1991)

[9] Compact fuzzycom FZ-3000/3010, *Journal of the Society of Instrument and Control Engineers*, Vol. 28, No. 11, 1023/1024 (1989)

[10] Special issue on fuzzy technology, *Fuji Electric Journal*, Vol.63 (1990)

[11] H. Ogai et al.: Robust process control by fuzzy inference,*Preprints of the 34th Japan Joint Automatic Control Conference*, 7/10 (1991)

[12] M. Saeki and M. Araki: Robust stability of large-scale systems which contains time-invariant Lur'e type uncertainties in subsystems and norm bounded uncertainties in interconnections, to appear

Proceedings of the 31st Conference
on Decision and Control
Tucson, Arizona • December 1992

ROBUST STRICT POSITIVE REAL DESIGN OF INTERVAL SYSTEMS

Yingmin Jia, Weibing Gao and Mian Cheng

The 7th Research Division
Beijing University of Aeronautics & Astronautics
Beijing 100083, P.R.China

ABSTRACT

A new strict positive real criterion of rational functions, which is computational tractability, is proposed. By applying the criterion to a class of interval denominator systems, the related extreme point results for robust strict positive real stabilization with first order controller are derived, which shows that the controller design procedure for the interval systems can be greatly simplified.

1. INTRODUCTION

It is well known that the strict positive real concept plays an important role in robust control, adaptive control and network theory[1,2]. For example, a nominal strict positive real transfer function may tolerate large passive uncertainties without the loss of stability. Hence it is important to test the strict positive realness of a given transfer function. In system design theory, the more important work is how to design a strict positive real transfer function.

In this paper, robust strict positive real stabilization of interval systems is studied, i.e. considering how to design a single controller which both robustly stabilize an interval system under unity feedback and makes the closed–loop system transfer function be robustly strictly positive real. In fact, main results of this paper are further aspects of authors' work[3] where more general uncertainty descriptions are considered. Now, our attention is focused on interval systems and we want to know whether the simultaneous strict positive real stabilization of the extreme point systems of an interval system implies the robust strict positive real stabilization of the interval system. Obviously, we are going to develop a robust strict positive stabilization version of Kharitonov Theorem as its robust strict positive real analysis version given in [4].

Compared with [1] and [2], this paper is of the following characterizations: (1) In the former, given a real polynomial set Ω, consider when there exists a real polynomial, or more general, a real rational function $b(s)$ such that $a(s)/b(s)$ is strictly positive real for all $a(s)\in\Omega$. there, a "Tandem" design method in frequency domain is used while in present paper a feedback approach is adopted. (2) In the latter, a feedback design in state space framework is proceeded, but no uncertainty is involved in system models. Hence it is not a robust design.

This paper is organized as follows. In Section 2, some necessary preliminaries are introduced. In Section 3, a motivated result about a family of disc polynomials is improved and a new sufficient condition for strict positive realness of rational functions is also given. Main conclusions are obtained in Section 4. there, extreme point results for robust strict positive stabilization of a class of interval denominator systems with first order controllers are proved. The computational tractability of related conditions is analyzed in Section 5, which shows that the results of the paper are of important application value. Finally, the conclusions are given in Section 6.

2. PRELIMINARIES

In this Section, some basic facts about real interval polynomials will be introduced. Let $\delta(s)$ denote some real polynomial ($\delta^{even}(s)$ and $\delta^{odd}(s)$ denote its even part and odd part respectively), thus an interval polynomial may be written as

$$F(s)=\{\delta(s)=\delta_0+\delta_1 s+\delta_2 s^2+\ldots+\delta_n s^n \mid \delta_i \in[x_i,y_i],\ i=0,1,2,\ldots,n\} \quad (1)$$

Four Kharitonov extreme point polynomials relating to (1) are respectively

$$\begin{aligned} K^1 &= K^{even}_{min}(s)+K^{odd}_{min}(s),\\ K^2 &= K^{even}_{min}(s)+K^{odd}_{max}(s),\\ K^3 &= K^{even}_{max}(s)+K^{odd}_{min}(s),\\ K^4 &= K^{even}_{max}(s)+K^{odd}_{max}(s), \end{aligned}$$

here

$$\begin{aligned} K^{even}_{min}(s) &= x_0+y_2 s^2+x_4 s^4+y_6 s^6+x_8 s^8+\ldots\\ K^{even}_{max}(s) &= y_0+x_2 s^2+y_4 s^4+x_6 s^6+y_8 s^8+\ldots\\ K^{odd}_{min}(s) &= x_1 s+y_3 s^3+x_5 s^5+y_7 s^7+x_9 s^9+\ldots\\ K^{odd}_{max}(s) &= y_1 s+x_3 s^3+y_5 s^5+x_7 s^7+y_9 s^9+\ldots \end{aligned}$$

Write $\delta^e(\omega)=\delta^{even}(j\,\omega)$, $\delta^o(\omega)=\delta^{odd}(j\,\omega)/j\,\omega$,

CH3229-2/92/0000-2796$1.00 © 1992 IEEE

then the subscripts "min" and "max" in the above formulas represent facts that for any $\delta(s)\in F(s)$ and $\omega\in[0,\infty)$, have

$$K^{even}_{min}(j\ \omega) \leqslant \delta^{e}(\omega) \leqslant K^{even}_{max}(j\ \omega) \quad (2)$$

$$K^{odd}_{min}(j\ \omega)/j\ \omega \leqslant \delta^{o}(\omega) \leqslant K^{odd}_{max}(j\ \omega)/j\ \omega \quad (3)$$

The notable Kharitonov Theorem is that interval polynomial (1) is stable iff its four Kharitonov extreme point polynomials are stable.

Consider interval rational function

$$g(s) = F_a(s)/F_b(s) \quad (4)$$

$F_a(s)$, $F_b(s)$ are all interval polynomials as in (1).

Let K^i_a, K^i_b, i=1,2,3,4, be respectively the Kharitonov extreme point polynomials of $F_a(s)$, and $F_b(s)$, then sixteen Kharitonov extreme point rational functions with respect to interval rational function (4) may be denoted as follows

$$K^{ij}_g = K^i_a / K^j_b, \quad i, j = 1,2,3,4. \quad (5)$$

If $F_b(s)$, is an interval polynomial and $F_a(s)$, is a certain polynomial, (4) is known as an interval denominator rational function. Correspondingly, there are only four Kharitonov extreme point functions

$$F_a(s)/K^i_b, \quad i = 1,2,3,4 \quad (6)$$

Similarly, the definition of interval numerator rational function and its Kharitonov extreme point functions can be given and omitted here since in this paper only interval denominator case is studied. The existed results with only denominator uncertainties see [5] and references therein.

Another important concept used in this paper is the strict positive realness of rational functions. Though the standard definition of strict positive real rational functions is stated in many literature, for reference convenience it is introduced as follows.

Definition 1[6]. A real rational function is said to be strictly positive real if the following conditions are satisfied.

1. $p(s)$ has no poles in the closed right–half plane Re(s)≥0.
2. $Re(p(j\ \omega))>0 \quad \forall\ \omega\in R$.

Lemma 1[4]. A proper stable real rational function $p(s)=n(s)/d(s)$ is strictly positive real iff the following conditions hold.

1. $Re(p(0))>0$.
2. $n(s)$ is stable.
3. $d(s)+j\ \alpha n(s)$ is stable for all $\alpha\in R$.

Remark 1. Condition 3 in Lemma 1 shows that the stability of infinite complex polynomials must be tested in order to confirm the strict positive realness of a rational function. It is difficult to apply it since there are no stability criteria of complex polynomials which may be conveniently used, particularly, the α–value is unbounded. In other words, Lemma 1 is only a theoretical result.

The following result turns the stability test of infinite complex polynomials into a norm test.

Lemma 2 [7]. Suppose $p(s)=n(s)/d(s)$ is a proper stable rational function (real or complex) with degree n, then $\|p(s)\|_\infty < 1$ iff

1. $|a_n|<|b_n|$, here, a_n, b_n are the leading coefficients of numerator and denominator of p(s) respectively.
2. $d(s)+e^{j\theta}n(s), \quad \theta\in[0,2\pi)$, is stable.

Now we conclude this Section with a stable polynomial design result.

Lemma 3[8]. Assume f(s) is a positive coefficient stable polynomial with degree n, $\bar{f}(s)$ is a real polynomial with degree $\bar{n}$. If

1. $\bar{n}\leqslant n+1$.
2. when $\bar{n}=n+1$, the leading coefficient of $\bar{f}(s)$ is positive

then there exists $\varepsilon^* > 0$ such that $f_\varepsilon(s) = f(s) + \varepsilon\bar{f}(s)$ is also a positive coefficient stable polynomial for all $\varepsilon\in(0, \varepsilon^*)$.

3. AN ENLIGHTEN RESULT AND IMPROVEMENT

Since the main idea in this paper stems from [7], a more accurate description of its main result about the stability of a family of disc polynomials first given, then a sufficient condition of strict positive real test is obtained which will be useful for examining the extreme point results of robust strict positive real stabilization of a family of interval denominator systems.

Consider a family of disc polynomials denoted by

$$\delta(s) = \delta_0 + \delta_1 s + ... + \delta_n s^n, \quad \delta_i \in D_i, \ i = 0,1,2,...,n \quad (7)$$

here, D_i represents a disc centered at β_i with radius r_i. Obviously, (7) includes three special polynomials

1. $\beta(s) = \beta_0 + \beta_1 s + \ldots + \beta_{n-1} s^{n-1} + + \beta_n s^n$
2. $r_1(s) = r_0 - j\, r_1 s - r_2 s^2 + j\, r_3 s^3 + r_4 s^4 + \ldots$
3. $r_2(s) = r_0 + j\, r_1 s - r_2 s^2 - j\, r_3 s^3 + r_4 s^4 + \ldots$

Usually, polynomial 1 is known to be center polynomial.

Now assume $0 \notin D_n$, we can get the following Lemma.

Lemma 4[7]. A family of disc polynomials as in (7) are stable iff

1. $\beta(s)$ is stable.
2. $\|r_1(s) / \beta(s)\|_\infty < 1$ and $\|r_2(s) / \beta(s)\|_\infty < 1$.

Let ∂D_i denote the circle of D_i, consider a subset of (7)

$$\beta(s), \quad \delta(s) = \delta_0 + \delta_1 s + \ldots + \delta_n s^n, \qquad \delta_i \in \partial D_i, \quad i = 0,1,2,\ldots,n \tag{8}$$

Hereafter we call (8) the family of circle polynomials with center.

Theorem 1. (7) is stable iff (8) is stable

Proof: Similar to [7], we first prove that (8) is stable if and only if

1′. $\beta(s)$ is stable.

2′. $\|\sum_{i=0}^{n} r_i e^{j\theta_i} s^i / \beta(s)\|_\infty < 1, \quad \theta_i \in [0, 2\pi)$.

Obviously the sufficiency of 1′ and 2′ may directly be obtained from Rouch'e's Theorem. Since $\beta(s)$ is in (8), 1′ is necessary. Note that $0 \notin D_n$, then $|r_n / \beta_n| < 1$. Assume 2′ is not true, there must exist $\bar{\theta}_i \in [0, 2\pi)$, $i = 0,1,2,\ldots n$, such that

$$\|\sum_{i=0}^{n} r_i e^{j\bar{\theta}_i} s^i / \beta(s)\|_\infty \geqslant 1$$

Let $d(s) = \beta(s)$ and $n(s) = \sum_{i=0}^{n} r_i e^{j\bar{\theta}_i} s^i$.

From Lemma 2, there exists $\theta^* \in [0, 2\pi)$ making

$$\beta(s) + e^{j\theta^*} \sum_{i=0}^{n} r_i e^{j\bar{\theta}_i} s^i$$

be unstable which just locates at the circle. This contradicts the stability of (8) and the necessity of 2′ is proved. The remains of the proof is to prove the equivalence between conditions 1′–2′ and conditions 1–2 in Lemma 4. In fact, this may be easily obtained from the following formulas.

$$|\sum_{i=0}^{n} r_i e^{j\theta_i} (j\,\omega)^i| \leqslant \sum_{i=0}^{n} r_i (\omega)^i, \qquad \omega \geqslant 0$$

$$|\sum_{i=0}^{n} r_i e^{j\theta_i} (j\,\omega)^i| \leqslant \sum_{i=0}^{n} (-1)^i r_i (\omega)^i, \quad \omega < 0$$

and note that the upper bounds in the right sides of the above formulas may be reached at $\theta_i = -i\pi/2$ and $\theta_i = i\pi/2$ respectively. Thus the proof is completed.□

Though Theorem 1 more accurately describes the stability conditions of a family of disc polynomials than Lemma 4, it does not remove the constraint that all coefficients in (8) must vary independently. Generally, this independence will affect its applications. Observe Lemma 2, we have, every coefficient of condition 2 varies dependently in the corresponding circles with the same angle positions, which motivates us to change the related strict positive conclusions (see Lemma 1) represented by a family of stable polynomials into that denoted by module or norm.

Theorem 2(criterion). Given a proper stable real rational function $p(s) = n(s) / d(s)$ and assume $d(s) + n(s)$ is stable. then $p(s)$ is strictly positive real if

1. $Re(p(0)) > 0$.
2. $n(s)$ is stable.
3. $|(d(j\,\omega) - n(j\,\omega)) / (d(j\,\omega) + n(j\,\omega))| < 1$, $\omega \in R$.

Further, if $deg(d(s)) = deg(n(s))$, then condition 3 may be replaced by

3′ $\|(d(s) - n(s)) / (d(s) + n(s))\|_\infty < 1$

and conditions 1, 2, 3′ are also necessary.

Proof: From Lemma 1, we only need to prove that condition 3 implies $d(s) + j\,\alpha n(s)$ is stable for all $\alpha \in R$. First we point out an interesting fact in the classic control theory that the Nyquist plot of any transfer function of the form $k / (Ts + 1)$ in whole frequency range is a circle centered at $(k/2, 0j)$ with radius $k/2$. Its equation may be denoted by

$$(k/2) + (k/2)e^{j\theta}, \quad \theta \in [0, 2\pi).$$

If we consider $\omega \in R$, then this equation becomes

$$(k/2)+(k/2)e^{j\theta},\ \theta\in(-\pi,\pi).$$

Now condition 3 means

$$(d(j\omega)+n(j\omega))+e^{j\theta}(d(j\omega)-n(j\omega))\neq 0,\quad \theta\in[0,2\pi),\ \omega\in R \tag{9}$$

In particular, when $\theta\in(-\pi,\pi)$, (9) corresponds a family of fixed degree polynomials

$$(d(s)+n(s))+e^{j\theta}(d(s)-n(s)),\quad \theta\in(-\pi,\pi), \tag{10}$$

Note that when $\theta=0\in(-\pi,\pi)$, (10) becomes 2d(s) which is stable. In terms of Zero Exclusion Con– dition, (10) is stable. Obviously the stability of (10) is equivalent to that of

$$n(s)+(\frac{1}{2}+\frac{1}{2}e^{j\theta})(d(s)-n(s)),\ \theta\in(-\pi,\pi)$$

It follows from the aforementioned fact that $n(s)+(d(s)-n(s))/(1+j\alpha)$, $\alpha\in R$, is stable, i.e., $d(s)+j\alpha n(s)$, $\alpha\in R$, is stable. Similarly, the necessity and sufficiency of conditions 1, 2, 3′ can be proved.□

Remark 2. Compared with Lemma 1, the condition that $d(s)+n(s)$ is stable is added to Theorem 2 for the well–posedness of condition 3 and 3′. As we will see in the following, this condition can be satisfied by means of the design procedure in section 5.

4. EXTREME POINT RESULTS OF INTERVAL SYSTEMS

For convenience, write plant $p(s)=n_p(s)/d_p(s)$, controller $c(s)=n_c(s)/d_c(s)$ and the closed loop transfer function under unity feedback

$$G(s)=\frac{n_p(s)n_c(s)}{d_p(s)d_c(s)+n_p(s)n_c(s)}$$

Definition 2. The plant $p(s)$ is known as strictly positive real stabilizable if there exists a controller c(s) such that

1. $d_p(s)d_c(s)+n_p(s)n_c(s)$ is stable.
2. $G(s)$ is strictly positive real.

if $p(s)$ is an interval rational function as in (4) and satisfies conditions 1–2, then $p(s)$ is known as robustly strictly positive real stabilizable.

In terms of Definition 2, the following Theorem 3 is a direct result of Theorem 2.

Theorem 3. The plant $p(s)$ is strictly positive real stabilizable if there exists a controller c(s) such that

1. $Re(G(0))>0$ (11)
2. $n_p(s)n_c(s)$, $d_p(s)d_c(s)+n_p(s)n_c(s)$, $d_p(s)d_c(s)+2n_p(s)n_c(s)$ (12)

 are all stable.

3. $$\left|\frac{d_p(j\omega)d_c(j\omega)}{d_p(j\omega)d_c(j\omega)+2n_p(j\omega)n_c(j\omega)}\right|<1,\quad \omega\in R, \tag{13}$$

Hereafter we call (11)–(13) design requirements for strict positive real stabilization.

Lemma 5[9]. Write $F(x_1,x_2,\ldots,x_n)$
$=\{f(x_1,x_2,\ldots,x_n)\mid f(x_1,x_2,\ldots,x_n)$
$=\sum c_{i_1i_2\ldots i_q}x_{i_1}x_{i_2}\ldots x_{i_q}$,
$i_1\neq i_2\neq\ldots\neq i_q$,
$c_{i_1i_2\ldots i_q}$ is real, $1\leqslant q\leqslant n\}$.

If $a_i\leqslant x_i\leqslant b_i$, $i=1,2,\ldots,n$, and a_i, b_i are real, then $F(x_1,x_2,\ldots,x_n)$ attains an extremum at an extreme point of $(x_1,x_2,\ldots,x_n)$.

Lemma 6. Given an interval denominator system with relative degree 1, then a first order controller of the form $k(s+z)/(s+p)$ robustly stabilize the system iff it simultaneously stabilizes four Kharitonov extreme point systems of the system.

Remark 3. Since the relative degree of the interval system is 1, this shows that the system is strict proper and the leading coefficient interval of denominator polynomial does not contain 0. Hence Lemma 6 may be easily proved by means of the approach employed in Theorem 1 of [10] and omitted here.

Now we are in a position to state our extreme point results.

Theorem 4. An interval denominator system with relative degree 1 satisfies conditions (11)–(13) of Theorem 3 with first order controller of the form $k(s+z)/(s+p)$ iff its four extreme point systems satisfy them with the same controller.

Proof: From Lemma 6, the equivalence of conditions (11)–(13) between the interval system and its extreme systems is obvious. We only need to prove the sufficiency of condition (13). Assume $p(s)=n_p(s)/d_p(s)$ be arbitrary element of an interval denominator system, then

$$\left|\frac{d_p(j\,\omega)d_c(j\,\omega)}{d_p(j\,\omega)d_c(j\,\omega)+2n_p(j\,\omega)n_c(j\,\omega)}\right|<1,\quad \omega\in R$$

iff

$$|d_p(j\,\omega)d_c(j\,\omega)|<|d_p(j\,\omega)d_c(j\,\omega)+2n_p(j\,\omega)n_c(j\,\omega)|,\quad \omega\in R \tag{14}$$

Let $d_p(j\,\omega)=d_p^R+j\,d_p^I$, $d_c(j\,\omega)=d_c^R+j\,d_c^I$,

$$n_p(j\,\omega)=n_p^R+j\,n_p^I,\quad n_c(j\,\omega)=n_c^R+j\,n_c^I$$

then (14) is equivalent to the following (15)

$$(d_p^R d_c^R-d_p^I d_c^I)(n_p^R n_c^R-n_p^I n_c^I)+(d_p^R d_c^I+d_p^I d_c^R)(n_p^R n_c^I+n_p^I n_c^R)+|n_p(j\,\omega)n_c(j\,\omega)|^2>0,\quad \omega\geqslant 0 \tag{15}$$

Now fixing ω, it follows from (2)–(3) that d_p^R, d_p^I take value in some interval. In terms of Lemma 5 the minimum of the sum of first two items in the left side of (15) reaches at some extreme point system. Thus the extreme point systems of the interval system satisfy (15) at ω implies $p(s)$ also satisfies (15) at the same ω. Note that ω is arbitrary, the proof is completed.□

Corollary 1. Given an interval denominator system with relative degree 0. If it at least has a real zero, then the system satisfies conditions (11)–(13) with first order controller of the form $k/(s+p)$ iff its extreme point systems satisfy them with the same controller.

Theorem 4 and its corollary tell us that in order to complete robust strict positive real stabilization of an interval denominator system with first order controller it is enough to design a first order controller which makes the extreme point systems of the interval system satisfy conditions (11)–(13). Obviously this is a desirable extreme point results.

5. FEASIBILITY OF DESIGN

For the applications of extreme point results we will analyze the computational tractability of conditions (11)–(13). The following two assumptions are needed.

Assumption 1. plant and controller are all minimum phase, i.e. $n_p(s)$, $n_c(s)$ are all stable.

Assumption 2. The relative degree of the plant $p(s)$ is 0 or 1 so that the properness of the controller is guaranteed.

In fact from Lemma 1 and Definition 2, the above assumptions are just the necessary conditions for robust strict positive real stabilization. Therefore these assumptions are reasonable.

Theorem 5. Assume $p(s)$ satisfies assumptions 1–2, then there exists a proper controller $c(s)$ which completes strict positive real stabilization of the system.

In order to give the proof of Theorem 5, we first give the following proposition.

Proposition 1. Suppose $f_1(s)$ is a stable polynomial and $f_2(s)$ is also a real polynomial with the same leading coefficient sign as $f_1(s)$ and $deg(f_2)\leqslant deg(f_1)+1$, then there exists $\varepsilon^*>0$ such that for all $\varepsilon\in(0,\ \varepsilon_1)$, $f_1(s)+\varepsilon f_2(s)$ is stable and

$$\left|\frac{\varepsilon f_2(j\,\omega)}{f_1(j\,\omega)+\varepsilon f_2(j\,\omega)}\right|<1,\quad \omega\in R \tag{16}$$

Proof: Write $f_1(j\,\omega)=f_1^R+jf_1^I$, $f_2(j\,\omega)=f_2^R+jf_2^I$. Obviously (16) is equivalent to (17)

$$|f_1(j\,\omega)|^2+2\varepsilon(f_1^R f_2^R+f_1^I f_2^I)>0,\quad \omega\in R, \tag{17}$$

Since $deg(f_2)\leqslant deg(f_1)+1$, the degrees of ω in the left side two items of (17) either are the same or the first item is higher than the second item. Thus there exists $\varepsilon_1>0$ such that when $\varepsilon\in(0,\ \varepsilon_1)$ the leading coefficient sign about ω in the left side of (17) is the same as that of the first item. Hence some $\omega_m>0$ may be found which is independent of the choice of ε in $(0,\ \varepsilon_1)$ such that when $|\omega|>\omega_m$, (17) holds. Since $f_1(s)$ is stable, we have $min\{|f_1(j\,\omega)|:\ \omega\in R\}=A>0$. Therefore, ε_2 may be chosen such that

$$max\left\{|2\varepsilon(f_1^R f_2^R+f_1^I f_2^I)|:\ |\omega|\leqslant\omega_m\right\}<A$$

for $\varepsilon\in(0,\ \varepsilon_2)$, i.e. (17) is true. Note that if the conditions of Lemma 3 are satisfied we can also select $\varepsilon_3>0$ such that $f_1(s)+\varepsilon f_2(s)$ is stable for any $\varepsilon\in(0,\ \varepsilon_3)$. Thus, let $\varepsilon=min\{\varepsilon_1,\varepsilon_2,\varepsilon_3)$ which satisfies the demand.□

Proof of Theorem 5. Arbitrarily choose $n_c(s)$ which

is a stable polynomial with $deg(n_c(s)) \geqslant deg(d_p(s)) - deg(n_p(s)) - 1$. Then $d_c(s) = d_m s^m + ... + d_1 s + d_0$ may be designed as follows.

Step 1. Let $f_1(s) = 2n_p(s)n_c(s)$ or $n_p(s)n_c(s)$, $f_2(s) = d_p(s)$. From Lemma 3 and Proposition 1 we may design $\varepsilon_0 > 0$ satisfies

$$n_p(s)n_c(s) + \varepsilon_0 d_p(s), \quad 2n_p(s)n_c(s) + \varepsilon_0 d_p(s)$$

are stable and

$$\left|\frac{\varepsilon_0 d_p(j\,\omega)}{2n_p(j\,\omega)n_c(j\,\omega) + \varepsilon_0 d_p(j\,\omega)}\right| < 1, \quad \omega \in R.$$

Step 2. Let $f_1(s)$ is the same as in Step 1 and $f_2(s) = d_p(s)(s + \varepsilon_0)$. Similar to step 1, $\varepsilon_1 > 0$ may be designed such that

$$n_p(s)n_c(s) + \varepsilon_1 d_p(s)(s + \varepsilon_0),$$
$$2n_p(s)n_c(s) + \varepsilon_1 d_p(s)(s + \varepsilon_0)$$

are stable and for $\omega \in R$

$$\left|\frac{\varepsilon_1 d_p(j\,\omega)(j\,\omega + \varepsilon_0)}{2n_p(j\,\omega)n_c(j\,\omega) + \varepsilon_1 d_p(j\,\omega)(j\,\omega + \varepsilon_0)}\right| < 1.$$

Thus ε_2 may be designed by the same approach.

Step 3. In order to guarantee the stability of the controllers, first $\bar{\varepsilon}_3 > 0$ should be designed such that

$$\bar{\varepsilon}_3 s_3 + \varepsilon_2(s^2 + (\varepsilon_1(s + \varepsilon_0))$$

is stable, then let $f_1(s) = 2n_p(s)n_c(s)$,

$$f_2(s) = d_p(s)(\bar{\varepsilon}_3 s_3 + \varepsilon_2(s^2 + (\varepsilon_1(s + \varepsilon_0)))$$

Thus $\varepsilon_3 > 0$ may be obtained.

Step 4. Repeat the above steps until

$$m = deg(n_p(s)n_c(s)) - deg(d_p(s)) + 1.$$

we have $d_p(s) = \varepsilon_m(\bar{\varepsilon}_m s^m + \varepsilon_{m-1}(\bar{\varepsilon}_{m-1} s^{m-1}$

$$+ ... + \varepsilon_2(s^2 + (\varepsilon_1(s + \varepsilon_0)...)$$

and $d_i = \prod_{j=i}^{m} \varepsilon_j, \quad i = 0,1,2;$

$$d_i = \bar{\varepsilon}_i \prod_{j=i}^{m} \varepsilon_j, \quad i = 3,4,...m.$$

6. CONCLUSIONS

In the paper, a sufficient condition for strict positive real stabilization is given and the extreme point results of a class of interval denominator systems with respect to this condition are also obtained. In particular, the feasibility analysis of design requirements shows the results of this paper is of important application value. Further research should aim at the robust strict positive real stabilization of more general interval systems with first order or higher order controllers.

ACKNOWLEDGMENT

This work was supported by the National Natural Science Foundation of China.

REFERENCES

[1] Anderson, B.D.O., Dasgupta, S., Khargonekar, P., Kraus, F.J. and M. Mansour, "Robust Strict Positive Realness: Characterization and Construction", IEEE Trans. Circuits and SystemsVol.37, 1990, 869–876.

[2] Abdallah, C., Dorato, P. and Karni, S., "SPR Design Using Feedback", *Proc. ACC. 1991,* 1–6.

[3] Jia, Y.M., Gao, W.B. and Cheng, M., "Robust Strict Positive Real Stabilization Criteria for SISO Uncertain Systems", to appear on Systems and Control Letters, Vol.19, No.2, 1992.

[4] Chapellat, H., Dahleh, M. and Bhattacharyya, S.P., "On Robust Nonlinear Stability of Interval Control Systems", IEEE Trans. Autmat. Control, Vol.36, 1991, 59–67.

[5] Amillo, J. and Mata, F.A., "Robust Stabilization of Systems with Multiple Real Pole Uncertainties", IEEE Trans. Autmat. Control, Vol.36, 1991, 749–752.

[6] Gao, W.B. An Introduction to Nonlinear Control Systems, Science Press, Beijing, 1988.

[7] Chapellat, H., Dahleh, M. and Bhattacharyya, S.P.,"Robust Stability of a Family of Disc Polynomials", Int. J. Control, Vol.51, No.6, 1990, 1353–1362.

[8] Wei, K.H. and Yedavalli, R.K., "Robust Stabilizability for Linear System with both Parameter Variation and Unstructured Uncertainty", IEEE Trans. Autmat. Control, Vol.34, 1989, 149–156.

[9] Jiang, C.L., "Robust Stability in Linear State Space Models", Int. J. Control, Vol.48, No.2, 1988, 813–816.

[10] Barmish, B.R., Hollot, C.V., Kraus, F.J. and Tempo, R., "Extreme Point Results for Robust Stabilization of Interval Plants with First Order Compensators, IEEE Trans. Autmat. Control, Vol.37, 1992, 707–714.

FA8 - 9:40

Proceedings of the 31st Conference
on Decision and Control
Tucson, Arizona • December 1992

On Strict Passivity and its Application to Interpolation and $\mathcal{H}_\infty$ Control

Gjerrit Meinsma

Department of Applied Mathematics
University of Twente
P.O. Box 217, 7500 AE Enschede
The Netherlands

Michael Green

Department of Systems Engineering
Research School of Physical Sciences
Australian National University
G.P.O. Box 4, A.C.T. 2601, Australia

Abstract. We introduce the $\mathcal{L}_{2-}$-system and derive necessary and sufficient conditions for these systems to be strictly passive. Strictly passive $\mathcal{L}_{2-}$-systems are characterized as having a representation in terms of a co-J-lossless matrix. A state space proof is developed and provides a Riccati equation characterization of a strictly passive $\mathcal{L}_{2-}$-system, as well as a formula for the co-J-lossless matrix representation. Applications to Nevanlinna Pick interpolation and an $\mathcal{H}_\infty$ filtering problem are considered.

1 Introduction

The theory of positive (or negative) subspaces of an indefinite inner product space provides powerful results in interpolation theory and related areas such as $\mathcal{H}_\infty$ control (see [2], [3], [6]). Yet these ideas have remained the preserve of a handful of researchers, perhaps due to the abstractness of the problem formulation. Such abstractness, although necessary if the most general theorems are to be derived, is unnecessary in the more concrete world of rational systems.

In this paper, we consider rational systems described by ordinary linear differential equations. We use the Willems behavioral systems approach (see [17]) and represent systems by the AR description

$$R(\frac{d}{dt})w(t) = 0$$

in which R is a polynomial matrix and the vector w is the external signal. In other words, the external signal w is 'orthogonal' to the differential operator $R(d/dt)$.

In the applications we have in mind, closed loops are required to be stable. This leads us to focus our attention on the *unstable* 'behavior' of the system. For example, right half plane zeros determine the possible closed loops in the optimal sensitivity problem; right half plane poles do the same in optimal robust stabilization problems; and both can be formulated as optimal interpolation problems with the interpolation points in the right half plane ([4], [10], [12], [18]). An $\mathcal{L}_{2-}$-system is the idea we introduce to focus attention on this unstable 'behavior'. Roughly speaking, the behavior of an $\mathcal{L}_{2-}$-system is an unstable orthogonal complement of a differential operator $R(d/dt)$. We will argue that this provides a very convenient language for all sorts of $\mathcal{H}_\infty$ optimization problems. (See Section 4).

To illustrate our approach, consider the following constant matrix optimization problem. Given A and B, find Q such that $H = A + QB$ and $\|H\| < 1$, where $\|\cdot\|$ is the Euclidean induced norm. This problem boils down to showing that certain matrices have a specific inertia. If they have not, then no solution exists; if they have, then all solutions can be generated. This is most easily seen by using orthogonal complements.

If B is square and nonsingular, the problem always has a solution (set $Q = -AB^{-1}$), because there is no part of the space that cannot be affected by Q. When there are parts of the space Q cannot affect, these subspaces provide necessary conditions for the existence of a solution. Let $B_\perp$ be an orthogonal complement of B, so $BB_\perp = 0$. Then, for any Q, $HB_\perp = AB_\perp$. If $\|H\| < 1$, then

$$0 < B_\perp^*(I - H^*H)B_\perp = B_\perp^*(I - A^*A)B_\perp.$$

The right hand side is independent of Q, so the inequality $B_\perp^*(I - A^*A)B_\perp > 0$ is a necessary condition for the problem to have a solution. An equivalent statement is that

$$P^*JP > 0\,; \quad P = \begin{pmatrix} I \\ -A \end{pmatrix} B_\perp\,; \quad J = \begin{pmatrix} I & 0 \\ 0 & -I \end{pmatrix}.$$

To see that $P^*JP > 0$ is also sufficient for a solution to exist, note that

$$GP = 0\,; \quad G = \begin{pmatrix} B & 0 \\ A & I \end{pmatrix}.$$

Using inertia arguments, it may be shown that $P^*JP > 0$ and $GP = 0$ imply that $GJG^* = W\hat{J}W^*$ for some nonsingular matrix W and a signature matrix $\hat{J}$ which has the same number of negative eigenvalues as J. Define $(\,H_1 \quad Q_2\,) = (\,U \quad I\,)W^{-1}G$ in which $\|U\| < 1$ (for example, $U = 0$). Then $H_1H_1^* - Q_2Q_2^* = (\,U \quad I\,)\hat{J}(\,U \quad I\,)^* = UU^* - I < 0$. Therefore Q_2 is nonsingular and $\|Q_2^{-1}H_1\| < 1$. Define Q by $(\,Q \quad I\,) = Q_2^{-1}(\,U \quad I\,)W^{-1}$. Then $H = A + QB = Q_2^{-1}H_1$, so $\|H\| < 1$. (It is easily seen that all solutions are generated as U varies over the space if contractive matrices.)

This approach has a positive subspace interpretation: consider the subspace

$$\mathcal{B} = P\mathbb{C}^n$$

or, equivalently,

$$\mathcal{B} = \{\, w \mid Gw = 0 \,\}.$$

Then $\mathcal{B}$ is a positive subspace, where the indefinite inner product is $[x,y] = y^*Jx$, since there is an $\epsilon > 0$ such that $[w,w] = w^*Jw \geq \epsilon w^*w$ for all $w \in \mathcal{B}$. Note that $\mathcal{B}$ has representation in terms of a matrix M satisfying $MJM^* = \hat{J}$:

$$M\mathcal{B} = 0.$$

(Take $M = W^{-1}G$.)

Consider now the situation where G and Q are stable, rational matrices. The problem is now more difficult, since inverting stable matrices may introduce unstable poles. An other way of saying the same thing is that the essential properties of G can not always be recovered from an orthogonal complement P of G. Kimura *et al.* [11] have been particularly active in developing approaches to the solution of $\mathcal{H}_\infty$ control problems based on what they call J-lossless conjugation. Their J-lossless conjugation approach and our approach share the same fundamental idea and goal (using the simple orthogonal complement ideas which work for constant matrices in a more general setting), but differ in implementation. The crucial difference is the introduction of the $\mathcal{L}_{2-}$-system.

Throughout this paper we use the following notation:

$\mathbb{C}_-$, $\mathbb{C}_+$, $\mathbb{C}_0$	Open left half complex plane, open right half complex plane, imaginary axis.
A^*	Complex conjugate transpose of A.
$[z]$	The number of components of a vector signal z.
$\|z\|$	Euclidean norm of a vector z ($\|z\|^2 = z^*z$).

CH3229-2/92/0000-2802$1.00 © 1992 IEEE

D^{-L}, D^{-R}, $D_\perp$	Left inverse, right inverse and orthogonal complement of a constant matrix D: $D^{-L}D = I$, $DD^{-R} = I$, $D_\perp$ is a maximal full rank matrix such that $DD_\perp = 0$ or $D_\perp D = 0$, depending on whether D has more columns than rows or vice versa.
$H^\sim$, H^*	$H^\sim(s) = H(-\bar{s})^*$ and $H^*(s) = (H(s))^*$.
$\|H\|_\infty$	Supremum of the largest singular value of $H(s)$ over all $s \in \mathbb{C}_0$.
$\mathcal{L}_2(a,b)$, $\mathcal{L}_{2-}, \mathcal{L}_{2+}$	$\{w : (a,b) \to \mathbb{C}^m \mid \int_a^b w^*(t)w(t)dt < \infty\}$, $\mathcal{L}_{2-} = \mathcal{L}_2(-\infty,0)$, $\mathcal{L}_{2+} = \mathcal{L}_2(0,\infty)$.
$\mathcal{H}_2, \mathcal{H}_2^\perp$	The set of functions f analytic in $\mathbb{C}_+$ such that $\sup_{\sigma>0}\int_{-\infty}^{\infty}\|f(\sigma+j\omega)\|_2^2 d\omega < \infty$, the set of functions f such that $f^\sim \in \mathcal{H}_2$.
$\mathcal{L}_2$	$\mathcal{L}_{2-} \oplus \mathcal{L}_{2+}$, or $\mathcal{H}_2^\perp \oplus \mathcal{H}_2$.
$\mathcal{H}_\infty^{m\times p}$, $\mathcal{GH}_\infty^m$	The set of $m\times p$ matrix valued functions that are analytic and bounded in $\mathbb{C}_+$, the set of invertible elements in $\mathcal{H}_\infty^{m\times m}$ (we frequently omit explicit mention of dimensions).
π_-, π_+	Orthogonal projection from $\mathcal{L}_2$ to $\mathcal{H}_2^\perp$ (or $\mathcal{L}_{2-}$), $\pi_+ = 1 - \pi_-$.
$\hat{w}$	Left, right or two-sided Laplace transform of a time signal w.

A rational matrix G is *stable* if $G \in \mathcal{H}_\infty$; G is *antistable* if $G^\sim \in \mathcal{H}_\infty$. A rational matrix H is *contractive* if $\|H\|_\infty < 1$. If A is the constant 'A-matrix' in a state space description, then A *stable* (*antistable*) means that all eigenvalues of A lie in $\mathbb{C}_-$ ($\mathbb{C}_+$). $J_{p,q}$ is a diagonal signature matrix of the form

$$J_{p,q} = \begin{pmatrix} I_p & 0 \\ 0 & -I_q \end{pmatrix}. \tag{1}$$

An $(r+q)\times(p+q)$ rational matrix M is *co-$J_{p,q}$-lossless* if $MJ_{p,q}M^\sim = J_{r,q}$, and $MJ_{p,q}M^* \leq J_{r,q}$ in $\mathbb{C}_+$. Note that $J_{p,q}$ and $J_{r,q}$ are not the same, but they do have the same number of negative eigenvalues.

The notation $\|\cdot\|_2$ denotes the well known 2-norm. For the most part we will be dealing with the signal space $\mathcal{L}_{2-}$, so we will have $\|w\|_2^2 := \int_{-\infty}^0 w(t)^*w(t)\, dt$.

2 $\mathcal{L}_{2-}$-Systems

A dynamical system $\Sigma = (\mathbf{T}, W, \mathcal{B})$ is a triple, where $\mathbf{T}$ is the time axis, W the signal space, and $\mathcal{B} \subset W^{\mathbf{T}}$ the behavior, that is, the set of all possible trajectories that may occur. The systems we will look at are systems $\Sigma_- = (\mathbb{R}_-, \mathbb{R}^p, \mathcal{B}_-)$, with

$$\mathcal{B}_- = \{\, w \mid R(d/dt)w(t) = 0 \,\} \cap \mathcal{L}_{2-}, \tag{2}$$

in which R is a polynomial matrix. The frequency domain counterpart is a system $\hat{\Sigma}_- = (\mathbb{C}_-, \mathbb{C}^p, \hat{\mathcal{B}}_-)$ with

$$\hat{\mathcal{B}}_- = \{\, \hat{w} \mid G\hat{w} \in \mathcal{H}_2 \,\} \cap \mathcal{H}_2^\perp, \tag{3}$$

in which G is a stable rational matrix. It can be shown that for every polynomial matrix R there exists a stable G, and vice versa, such $\mathcal{B}_-$ and $\hat{\mathcal{B}}_-$ are isomorphic under the Laplace transform. The polynomial matrix R in (2) is said to define an *AR representation* of the $\mathcal{L}_{2-}$-system. We will drop the 'hats' as long as no confusion can arise, and we switch from time domain to frequency domain whenever it is convenient. We say that stable G in (3) defines an AR representation of the $\mathcal{L}_{2-}$-system.

Definition 2.1 A dynamical system $\Sigma_- = (\mathbf{T}, W, \mathcal{B}_-)$ is an *$\mathcal{L}_{2-}$-system* if $\mathbf{T} = \mathbb{R}_-$, $W = \mathbb{R}^p$, and $\mathcal{B}_- = \mathcal{B} \cap \mathcal{L}_{2-}$, for some $\mathcal{B} \subset W^{\mathbf{T}}$. Behaviors (2) are abbreviated to $\mathcal{B}_R$.

Dynamical systems $\hat{\Sigma}_- = (\mathbb{C}_-, \mathbb{C}^p, \hat{\mathcal{B}} \cap \mathcal{H}_2^\perp)$ are also referred to as $\mathcal{L}_{2-}$-systems. Behaviors (3) are abbreviated to $\mathcal{B}_G$. ∎

Note: the definition allows for more general $\mathcal{L}_{2-}$-systems than can be described by (2). In this paper, however, only $\mathcal{L}_{2-}$-systems described by (2) are considered. Signals w in a behavior are sometimes referred to as *external* signals, so as to distinguish them from other signals we may like to introduce.

Example 2.2

1. Suppose $y = Hu$ for some stable transfer matrix H. Let $w := \binom{u}{y}$ be the external signal. Then $G = (-H \;\; I)$ defines an AR representation of an $\mathcal{L}_{2-}$-system, with

   $$\mathcal{B}_G = \begin{pmatrix} I \\ \pi_- H \end{pmatrix} \mathcal{H}_2^\perp,$$

   where π_- is the orthogonal projection onto $\mathcal{H}_2^\perp$.

2. Suppose $G(s) = C(sI - (I - C))^{-1} - I$ is stable. Then $G\hat{w} \in \mathcal{H}_2$ for $\hat{w} \in \mathcal{H}_2^\perp$, iff $w(t) = Ce^{tI}x_0$, $x_0 \in \mathbb{C}^n$. So $\mathcal{B}_G = C(sI - I)^{-1}\mathbb{C}^n \subset \mathcal{H}_2^\perp$ has finite dimension. ∎

The advantages of using AR representations instead of the usual transfer matrices are: (1) AR representations are defined without an input/output partitioning of the external signal; and (2) AR representations can represent systems that can not be represented by a transfer matrix from inputs to outputs. The second example in Example 2.2 can not be written as $y = Hu$, with external signal $w = \binom{u}{y}$. Transfer matrices mapping inputs to outputs give rise to infinite dimensional behaviors (the input u can be chosen from an infinite dimensional space, usually). With AR representations behaviors in general consist of an infinite dimensional part and a finite dimensional part.

Lemma 2.3 ([14]) *Suppose G and $\bar{G}$ are two stable matrices that have full row rank on $\mathbb{C}_0 \cup \infty$. Then $\mathcal{B}_G = \mathcal{B}_{\bar{G}}$ if and only if $G = W\bar{G}$ for some $W \in \mathcal{GH}_\infty$. Furthermore, there exists a stable E such that $\mathcal{B}_G = \mathcal{B}_E$ with E having all its zeros in $\mathbb{C}_+$.* ∎

Definition 2.4 An $\mathcal{L}_{2-}$-system $\Sigma_- = (\mathbb{R}_-, \mathbb{R}^p, \mathcal{B}_-)$ is *strictly passive (SP) with respect to partitioning $w = \binom{u}{y}$* if there exists an $\epsilon > 0$ such that every w in $\mathcal{B}_-$ satisfies

$$\int_{-\infty}^{T} w^*(t)J_{[u],[y]}w(t)\; dt \;\geq\; \epsilon \int_{-\infty}^{T} w^*(t)w(t)\; dt, \tag{4}$$

for all time $T \in \mathbb{R}_-$. Inequality (4) is referred to as the strict passivity inequality (SP inequality). ∎

The systems we consider are time invariant, so we may restrict our attention to the case $T = 0$.

Example 2.5 Suppose $y = Hu$, $H \in \mathcal{H}_\infty$. Let $G = (H \;\; -I)$ and consider the $\mathcal{L}_{2-}$-system Σ_- with behavior $\mathcal{B}_- = \mathcal{B}_G$. Then Σ_- is SP with respect to the partitioning $w = \binom{u}{y}$ iff $\|H\|_\infty < 1$. ∎

The main result is formulated next.

Theorem 2.6 *Let G be a stable matrix that has full row rank on $\mathbb{C}_0 \cup \infty$. The $\mathcal{L}_{2-}$-system $\Sigma_- = (\mathbb{R}_-, \mathbb{R}^p, \mathcal{B}_G)$ is strictly passive with respect to the partitioning $w = \binom{u}{y}$ if and only if $\mathcal{B}_G = \mathcal{B}_M$ for some $M \in \mathcal{H}_\infty$ that is co-$J_{[u],[y]}$-lossless.* ∎

Proof 2.7 That SP implies the existence of a co-J-lossless matrix that defines the system is proven in the next section. Sufficiency is easy, see [14]. ∎

A co-J-lossless M can be constructed from a given G, if it exists:

Corollary 2.8 ([14]) *Suppose G is stable and has full row rank m on $\mathbb{C}_0 \cup \infty$. Then $\Sigma_- = (\mathbb{R}_-, \mathbb{R}^p, \mathcal{B}_G)$ is SP with respect to the partitioning $w = \binom{u}{y}$ if and only if $GJ_{[u],[y]}G^\sim = WJ_{m-[y],[y]}W^\sim$ has a solution $W \in \mathcal{GH}_\infty$ and $W^{-1}G$ is co-$J_{[u],[y]}$-lossless.* ∎

3 State Space Analysis of Strictly Passive $\mathcal{L}_{2-}$-Systems

In this section we determine necessary and sufficient conditions under which an $\mathcal{L}_{2-}$-system is strictly passive, expressed in terms of state space descriptions of the system. To begin, we consider the state space descriptions with which we work.

3.1 ONR's and DVR's

This subsection contains some basic results on two different state space representations of a system. The external signal will be denoted by w. Other signals, like the state, are there to enable a parameterization of all w's. In this subsection it is *not* assumed that w is in $\mathcal{L}_{2-}$; we will introduce this additional constraint in the next subsection.

Definition 3.1

1. An *output nulling state space system* is a system whose external signal w satisfies

$$\begin{aligned} \dot{x} &= Ax + Bw \\ 0 &= Cx + Dw. \end{aligned} \tag{5}$$

   No restrictions are imposed on the 'initial' state.

   The quadruple $\{A, B, C, D\}$ defines an *output nulling representation* (ONR) of the system.

2. A *driving variable state space system* is a system whose external signal w satisfies

$$\begin{aligned} \dot{x} &= \bar{A}x + \bar{B}v \\ w &= \bar{C}x + \bar{D}v. \end{aligned} \tag{6}$$

   No assumptions are imposed on the signal v or on the 'initial' state.

   The quadruple $\{\bar{A}, \bar{B}, \bar{C}, \bar{D}\}$ defines a *driving variable representation* (DVR) of the system.

3. The *zeros* of a quadruple $\{A, B, C, D\}$ defining an ONR or DVR of a system is the set of values s for which

$$\begin{pmatrix} A - sI & B \\ C & D \end{pmatrix} \tag{7}$$

   drops below normal rank. ∎

Example 3.2 Consider the system described by

$$\begin{aligned} \dot{x} &= \hat{A}x + \hat{B}u \\ y &= \hat{C}x + \hat{D}u. \end{aligned}$$

This is a DVR if the external signal is y and the 'initial' state is arbitrary. If the external signal is $w = \binom{u}{y}$, we have DVR

$$\begin{aligned} \dot{x} &= \hat{A}x + \hat{B}v \\ w &= \begin{pmatrix} 0 \\ \hat{C} \end{pmatrix} x + \begin{pmatrix} I \\ \hat{D} \end{pmatrix} v \end{aligned}$$

and ONR

$$\begin{aligned} \dot{x} &= \hat{A}x + (\hat{B} \;\; 0)\, w \\ 0 &= \hat{C}x + (\hat{D} \;\; -I)\, w. \end{aligned}$$ ∎

The 'output' $Cx + Dw$ in the ONR (5) is zero, so applying output injection amounts to doing nothing; DVR's are unaffected by state feedback, since this is just a redefinition of the driving variable.

An ONR or DVR is considered minimal if the associated matrix

$$\begin{pmatrix} A & B \\ C & D \end{pmatrix}$$

has minimal dimension (amongst all possible such representations). The following lemma describes the minimality conditions.

Lemma 3.3

1. *Suppose $\{A, B, C, D\}$ defines an ONR of a system Σ.*
   (a) *The quadruple $\{A + HC, B + HD, TC, TD\}$ defines an ONR of the same system Σ, for any H and any nonsingular T. (We call this transformation a regular output injection.)*
   (b) *There exists a quadruple $\{\hat{A}, \hat{B}, \hat{C}, \hat{D}\}$ defining an ONR of the system Σ such that $(\hat{C}, \hat{A})$ is observable and $\hat{D}$ has full row rank. (We call this a minimal ONR.)*
2. *Suppose $\{\bar{A}, \bar{B}, \bar{C}, \bar{D}\}$ defines a DVR of a system Σ.*
   (a) *The quadruple $\{\bar{A} + \bar{B}F, \bar{B}R, \bar{C} + \bar{D}F, \bar{D}R\}$ defines a DVR of the same system Σ, for any F and any nonsingular R. (We call this transformation a regular state feedback.)*
   (b) *There exists a quadruple $\{\hat{A}, \hat{B}, \hat{C}, \hat{D}\}$ defining a DVR of the system Σ which is strongly observable (i.e., $(\hat{C} + \hat{D}F, \hat{A} + \hat{B}F)$ observable for every F) and such that $\hat{D}$ has full column rank. (We call this a minimal DVR.)* ∎

Example 3.4 Controllability plays no role in the minimality of an ONR or DVR. Consider for example the system whose signals are of the form

$$w(t) = Ce^{At}x_0; \qquad x_0 \in \mathbb{R}^n.$$

This system has ONR quadruple $\{A, 0, -C, I\}$ and DVR quadruple $\{A, , C, \}$ (the 'driving variable' has null dimension). If (C, A) is observable, both the ONR and the DVR are minimal. ∎

The following lemma describes the relationship between minimal ONR's and DVR's of a system Σ:

Lemma 3.5

1. *Let $\{A, B, C, D\}$ define a minimal ONR of a system Σ. Then $\{\bar{A}, \bar{B}, \bar{C}, \bar{D}\}$ given by*

$$\begin{pmatrix} \bar{A} & \bar{B} \\ \bar{C} & \bar{D} \end{pmatrix} = \begin{pmatrix} A & B \\ 0 & I \end{pmatrix} \begin{pmatrix} I & 0 \\ -D^{-R}C & D_{\perp} \end{pmatrix} \tag{8}$$

   defines a minimal DVR of Σ. Furthermore, λ is a zero of $\{A, B, C, D\}$ if and only if it is an uncontrollable mode of $(\bar{A}, \bar{B})$.

2. *Let $\{\bar{A}, \bar{B}, \bar{C}, \bar{D}\}$ define a minimal DVR of a system Σ. Then $\{A, B, C, D\}$ given by*

$$\begin{pmatrix} A & B \\ C & D \end{pmatrix} = \begin{pmatrix} I & -\bar{B}\bar{D}^{-L} \\ 0 & -\bar{D}_{\perp} \end{pmatrix} \begin{pmatrix} \bar{A} & 0 \\ \bar{C} & -I \end{pmatrix} \tag{9}$$

   defines a minimal ONR of Σ. Furthermore, λ is a zero of $\{A, B, C, D\}$ if and only if it is an uncontrollable mode of $(\bar{A}, \bar{B})$. ∎

3.2 ONR's and DVR's of $\mathcal{L}_{2-}$-systems

Consider a DVR of a system Σ. For the external signal w to be in $\mathcal{L}_{2-}$, the stable part of the dynamics must not appear in w; all $\mathcal{L}_{2-}$-systems may be represented by DVR's which have antistable dynamics; and all $\mathcal{L}_{2-}$-systems may be represented by ONR's which have zeros only in $\mathbb{C}_+$:

Lemma 3.6

1. *Let w be the external signal of a system Σ.*

   (a) *$w \in \mathcal{L}_{2-}$ implies that $x \in \mathcal{L}_{2-}$ and $\lim_{t\to-\infty} x(t) = 0$, in which x is the state variable in any minimal ONR of Σ.*

   (b) *$w \in \mathcal{L}_{2-}$ implies that $v \in \mathcal{L}_{2-}$, $x \in \mathcal{L}_{2-}$ and that $\lim_{t\to-\infty} x(t) = 0$, in which x is the state variable and v is the driving variable in any minimal DVR of Σ.*

2. *Let Σ_- be an $\mathcal{L}_{2-}$-system.*

   (a) *Σ_- can be defined by a minimal ONR $\{A, B, C, D\}$ that has all it's zeros in $\mathbb{C}_+$. We call such an ONR an $\mathcal{L}_{2-}$-minimal ONR.*

   (b) *Σ_- can be defined by a minimal DVR $\{\bar{A}, \bar{B}, \bar{C}, \bar{D}\}$ such that $(\bar{A}, \bar{B})$ is antistabilizable. We call such a DVR an $\mathcal{L}_{2-}$-minimal DVR. (Note that by choice of an anti-stabilizing regular state feedback there exists a minimal DVR with antistable '$\bar{A}$-matrix'.)* ■

Proof 3.7

1. (a) Introduce an antistabilizing output injection H. Since $\dot{x} = (A + HC)x + (B + HD)w$ in which $A + HC$ is antistable and $w \in \mathcal{L}_{2-}$, we have $x \in \mathcal{L}_{2-}$. Also, $\dot{x} = Ax + Bw$ implies $\dot{x} \in \mathcal{L}_{2-}$. Thus $\lim_{t\to-\infty} x(t) = 0$.

   (b) Consider minimal DVR quadruple $\{\bar{A}, \bar{B}, \bar{C}, \bar{D}\}$ and let E be a nonsingular matrix such that $E\bar{D} = \binom{I}{0}$. So $Ew = E\bar{C}x + \binom{I}{0}v$. Apply regular state feedback $v = -E_1\bar{C}x + \hat{v}$, where E_1 is the upper row block of E. Now we have $E_1 w = \hat{v}$, so $\hat{v}$ is in $\mathcal{L}_{2-}$. Strong observability implies that $(E_2\bar{C}, \bar{A})$ is observable (E_2 is the lower block row of E), so there exists an H such that $\bar{A} + HE_2\bar{C}$ is antistable. Rewrite the dynamics as $\dot{x} = (\bar{A} + HE_2\bar{C})x + \hat{v} - HE_2 w$ and proceed as in Item (a).

2. Note that Items (a) and (b) are equivalent, by Lemma 3.5. We prove Item (b), the DVR result.

   Consider any minimal DVR $\{\hat{A}, \hat{B}, \hat{C}, \hat{D}\}$ defining Σ_-. By Item 1b, the state x and the driving variable v are both in $\mathcal{L}_{2-}$ and $x(-\infty) = 0$. The state corresponding to any uncontrollable mode $\lambda \in \mathbb{C}_- \cup \mathbb{C}_0$ of $(\hat{A}, \hat{B})$ is therefore identically zero. Removing all such modes from the DVR leaves the desired antistabilizable DVR. ■

3.3 Strictly passive $\mathcal{L}_{2-}$-systems

Consider an $\mathcal{L}_{2-}$-system Σ_- with external signal w partitioned as

$$w = \begin{pmatrix} u \\ y \end{pmatrix}. \tag{10}$$

Define the two signature matrices: $J = J_{[u],[y]}$, $\hat{J} = J_{m-[y],[y]}$.

Theorem 3.8 *Let $\{\bar{A}, \bar{B}, \bar{C}, \bar{D}\}$ define an $\mathcal{L}_{2-}$-minimal DVR of an $\mathcal{L}_{2-}$-system Σ_-.*

1. *Then Σ_- is SP with respect to the partitioning (10) if and only if*

   (a) *$\bar{D}^*J\bar{D} > 0$ and*

   (b) *The exists X such that*

   $$\begin{aligned}&\bar{A}^*X + X\bar{A} - \bar{C}^*J\bar{C} \\ &\quad + [X\bar{B} - \bar{C}^*J\bar{D}](\bar{D}^*J\bar{D})^{-1}[-\bar{D}^*J\bar{C} + \bar{B}^*X] = 0 \end{aligned} \tag{11}$$

   such that $\bar{A} + \bar{B}(\bar{D}^*J\bar{D})^{-1}[-\bar{D}^*J\bar{C} + \bar{B}^*X]$ is antistable and

   (c) *$X > 0$.*

2. *Furthermore, given an antistabilizing solution X to (11) and a nonsingular solution W to $W^*W = \bar{D}^*J\bar{D}$, define $\{A_2, B_2, C_2, D_2\}$ as*

$$\begin{pmatrix} A_2 & B_2 \\ C_2 & D_2 \end{pmatrix} = \begin{pmatrix} \bar{A} & \bar{B} \\ \bar{C} & \bar{D} \end{pmatrix} \begin{pmatrix} I & 0 \\ (\bar{D}^*J\bar{D})^{-1}[-\bar{D}^*J\bar{C} + \bar{B}^*X] & W^{-1} \end{pmatrix}. \tag{12}$$

Let D_M be a constant matrix such that $D_M\bar{D} = 0$ and $D_M J D_M^* = \hat{J}$. Then $M(s) := D_M - D_M C_2 X^{-1}(sI + A_2^*)^{-1}C_2^*J$ is a stable, rational matrix that defines an AR representation of Σ_-, and M is co-J-lossless iff $X > 0$. ■

Proof 3.9

(Item 1,(a)) Suppose Σ_- is SP. Consider driving variable $v(t) = \sqrt{\delta(t)}v_0$, with $x(t) = 0$ for $t < 0$. Then

$$v_0^*D^*JDv_0 = \int_{-\infty}^{0} w^*(t)Jw(t)\,dt \geq \epsilon \int_{-\infty}^{0} w^*(t)w(t)\,dt = \epsilon v_0^*D^*Dv_0.$$

Since D has full column rank, this implies $D^*JD > 0$. (This argument can be made precise by considering $\mathcal{L}_{2-}$ approximations to $\delta(t)$—we omit these tedious and unenlightening details).

To simplify the algebra for Items 2 and 3, consider the following regular state feedback. Let

$$\begin{pmatrix} A_1 & B_1 \\ C_1 & D_1 \end{pmatrix} = \begin{pmatrix} \bar{A} & \bar{B} \\ \bar{C} & \bar{D} \end{pmatrix} \begin{pmatrix} I & 0 \\ -(\bar{D}^*J\bar{D})^{-1}\bar{D}^*J\bar{C} & W^{-1} \end{pmatrix}. \tag{13}$$

in which W is a nonsingular solution to $W^*W = \bar{D}^*J\bar{D}$. The quadruple $\{A_1, B_1, C_1, D_1\}$ defines a DVR of the same system as the two DVR's differ by a regular state feedback. It suffices therefore to consider the DVR

$$\begin{aligned} \dot{x} &= A_1 x + B_1 v_1 \\ w &= C_1 x + D_1 v_1. \end{aligned}$$

It is easily checked that $D_1^*JD_1 = I$ and $D_1^*JC_1 = 0$.

(Item 1,(b)) Suppose the system Σ_- is SP. Let H be the Hamiltonian matrix

$$H = \begin{pmatrix} A_1 & B_1B_1^* \\ C_1^*JC_1 & -A_1^* \end{pmatrix}. \tag{14}$$

It may be shown that H has no eigenvalues on the imaginary axis. Since $(-A_1, B_1)$ is stabilizable, $B_1B_1^* \geq 0$ and H in (14) has no imaginary axis eigenvalue, it follows from standard Hamiltonian matrix results (see, for instance, [6], Chapter 7 or [5], Lemma 2) that there exists a Y such that

$$Y(-A_1) + (-A_1^*)Y + YB_1B_1^*Y - C_1^*JC_1 = 0$$

and $-A_1 + B_1B_1^*Y$ is stable. Setting $X = -Y$, we have

$$XA_1 + A_1^*X + XB_1B_1^*X - C_1^*JC_1 = 0 \tag{15}$$

and $A_1 + B_1B_1^*X$ is antistable. A straightforward calculation shows that X also satisfies (11) and $\bar{A} + \bar{B}(\bar{D}^*J\bar{D})^{-1}[-\bar{D}^*J\bar{C} + \bar{B}^*X]$ is antistable.

(Item 1,(c)) Suppose Σ_- is SP. Let X be as in Item (b) and consider the DVR $\{A_1, B_1, C_1, D_1\}$. The signals w, x and v_1 are in $\mathcal{L}_{2-}$ and $x(-\infty) = 0$. Completing the square gives

$$\int_{-\infty}^{0} w(t)^*Jw(t)\;dt \;=\; x(0)^*Xx(0) + \|v_1 - B_1^*Xx\|_2^2. \tag{16}$$

Now consider the driving variable $v_1 = B_1^*Xx$, which is in $\mathcal{L}_{2-}$ since X is antistabilizing. Let w be the external signal resulting from v_1, with 'initial' state $x(0)$. Then

$$\begin{aligned} x(0)^*Xx(0) &= \int_{-\infty}^{0} w(t)^*Jw(t)\,dt \quad \text{by (16)} \\ &\geq \epsilon \int_{-\infty}^{0} w(t)^*w(t)\,dt = \epsilon\, x(0)^*Qx(0) \end{aligned}$$

[illegible] $Q(A_1+B_1B_1^*X)+(A_1+B_1B_1^*X)^*Q = (C_1+D_1B_1^*X)^*(C_1+D_1B_1^*X)$. Note that $(C_1+D_1B_1^*X, A_1+B_1B_1^*X)$ is observable (by strong observability), so $Q>0$. Hence $X \geq \epsilon Q > 0$.

Conversely, suppose $X>0$. Write

$$\begin{aligned} \dot{x} &= (A_1+B_1B_1^*X)x + B_1(v_1 - v_{\mathrm{opt}}) \\ w &= (C_1+D_1B_1^*X)x + D(v_1 - v_{\mathrm{opt}}) \end{aligned}$$

in which $v_{\mathrm{opt}} = B_1^*Xx$. Since $A_1 + B_1B_1^*X$ is antistable, there exist constants $\beta \geq 0$ and $\gamma > 0$ such that

$$\|w\|_2^2 \leq \gamma\|v_1 - v_{\mathrm{opt}}\|_2^2 + \beta\|x(0)\|^2\,; \qquad \gamma > \frac{\beta}{\lambda_{\min}(X)}.$$

Hence

$$\begin{aligned} \int_{-\infty}^{0} w(t)^*Jw(t)\ dt &= x(0)^*Xx(0) + \|v_1 - v_{\mathrm{opt}}\|_2^2 \\ &\geq (\lambda_{\min}(X) - \frac{\beta}{\gamma})\|x(0)\|^2 + \frac{1}{\gamma}\|w\|_2^2 \geq \frac{1}{\gamma}\|w\|_2^2 \end{aligned}$$

which proves the system is SP.

(Item 2) Let $A_M = -A_2^*$, $B_M = C_2^*J$ and $C_M = -D_MC_2X^{-1}$, so $M(s) = C_M(sI - A_M)^{-1}B_M + D_M$. The DVR $\{A_2, B_2, C_2, D_2\}$ is normalized in the sense that $A_2^*X + XA_2 - C_2^*JC_2 = 0$, $B_2^*X - D_2^*JC_2 = 0$, $D_2^*JD_2 = I$. This can be used to show that $A_MX + XA_M^* + B_MJB_M^* = 0$, $B_MJD_M^* + XC_M^* = 0$, $D_MJD_M^* = \hat{J}$. It follows from [8], Theorem 5.3 that M is co-J-lossless iff $X>0$.

Remains to show that M defines an AR representation of Σ_-. We first show that $\{A_M, B_M, C_M, D_M\}$ defines an ONR of Σ_-. It follows from Lemma 3.5, Item 2 that

$$\begin{pmatrix} A_2 - B_2D_2^{-L}C_2 & B_2D_2^{-L} \\ -D_{2\perp}C_2 & D_{2\perp} \end{pmatrix} = \begin{pmatrix} A_2 - B_2D_2^*JC_2 & B_2D_2^*J \\ -D_MC_2 & D_M \end{pmatrix}$$

defines an ONR of Σ_-. Here we used that $D_{2\perp} := D_M$ is an orthogonal complement of D_2 and that $D_2^{-L} := D_2^*J$ is a left inverse of D_2. It is easily checked that this ONR transforms under output injection $H = C_2^*D_M^*\hat{J}$ into

$$\begin{pmatrix} A_2 - X^{-1}C_2^*JC_2 & X^{-1}C_2^*J \\ -D_MC_2 & D_M \end{pmatrix} = \begin{pmatrix} X^{-1}A_MX & X^{-1}B_M \\ C_MX & D_M \end{pmatrix}.$$

A state transformation gives the desired ONR.

In order to see the connection with AR representations, write $C_Mx + D_Mw$ as

$$\begin{aligned} y(t) &:= C_Mx(t) + D_Mw(t) \\ &= C_Me^{A_M(t+T)}x(-T) + \int_{-T}^{t} C_Me^{A_M(t-\tau)}B_Mw(\tau)\ d\tau + Dw(t). \end{aligned} \tag{17}$$

Letting T go to infinity and taking Laplace transforms shows that $\hat{y}(s) = M(s)\hat{w}(s)$. Due to the Paley-Wiener Theorem we have $y(t)=0$ for $t<0$ iff $\hat{y} = M\hat{w}$ is in $\mathcal{H}_2$. So $w \in \mathcal{L}_{2-}$ satisfies the ONR equations iff $\hat{w} \in \mathcal{B}_M$. ∎

Theorem 3.8 can be easily translated into a corresponding ONR result using Lemma 3.5. It is frequently the case that we would like to determine whether or not a given ONR defines an SP $\mathcal{L}_{2-}$-system without having to find an $\mathcal{L}_{2-}$-minimal representation. This is quite easily done:

Corollary 3.10 *Let $\{A,B,C,D\}$ define an ONR of an $\mathcal{L}_{2-}$-system Σ_-. Suppose D has full row rank, (C,A) is detectable and that*

$$\begin{pmatrix} A - sI & B \\ C & D \end{pmatrix}$$

has full row rank for all $s \in \mathbb{C}_0 \cup \infty$.

1. *Then Σ_- is SP if and only if*
   (a) *There is a nonsingular matrix W such that $DJD^* = W\hat{J}W^*$ and*
   (b) *There exists Q such that*
   $$\begin{aligned} &AQ + QA^* + BJB^* \\ &- [QC^* + BJD^*](DJD^*)^{-1}[CQ + DJB^*] = 0 \end{aligned} \tag{18}$$
   with $A - [QC^* + BJD^*](DJD^*)^{-1}C$ stable and
   (c) *$Q \geq 0$.*
2. *Furthermore, given a stabilizing solution Q to (18) and a nonsingular solution W to $W\hat{J}W^* = DJD^*$, define $\{A_M, B_M, C_M, D_M\}$ as*
   $$\begin{pmatrix} A_M & B_M \\ C_M & D_M \end{pmatrix} = \begin{pmatrix} I & -[BJD^* + QC^*](DJD^*)^{-1} \\ 0 & W^{-1} \end{pmatrix}\begin{pmatrix} A & B \\ C & D \end{pmatrix}.$$
   Then $M(s) := C_M(sI - A_M)^{-1}B_M + D_M$ is a stable rational matrix that defines an AR representation of Σ_-, and M is co-J-lossless iff $Q \geq 0$. ∎

4 Examples

Example 4.1 (Nevanlinna Pick Interpolation.) The aim of this example is to show a connection between SP $\mathcal{L}_{2-}$-systems and Nevanlinna Pick Interpolation problems (NPIP). We show that a certain NPIP has a solution iff a corresponding $\mathcal{L}_{2-}$-system is SP.

Consider the problem of finding $H \in \mathcal{H}_\infty$ such that $\|H\|_\infty < 1$ and such that a set of interpolation conditions is satisfied: $H(\zeta_i)a_i = b_i$ for $i \in \{1, \cdots, n\}$. We assume that all ζ_i lie in $\mathbb{C}_+$ and that $\zeta_i \neq \zeta_j$ if $i \neq j$. The claim is that this NPIP has a solution iff the $\mathcal{L}_{2-}$-system $\Sigma_- = \{\mathbb{C}_-,\ \mathbb{C}^p,\ \mathcal{B}_-\}$ with the n dimensional behavior

$$\mathcal{B}_- = \begin{pmatrix} a_1 & \cdots & a_n \\ b_1 & \cdots & b_n \end{pmatrix}\left(sI - \begin{pmatrix} \zeta_1 & & \\ & \ddots & \\ & & \zeta_n \end{pmatrix}\right)^{-1}\mathbb{C}^n \tag{19}$$

is SP with respect to partitioning $w = \binom{u}{y}$, with $[u] = [a_1]$, and $[y] = [b_1]$.

Suppose NPIP has a stable contractive solution H. The interpolation conditions on H imply that $(H\ \ -I)w \in \mathcal{H}_2$, for every $w \in \mathcal{B}_-$. In other words, $\mathcal{B}_-$ is a subset of the behavior of a system $\bar{\Sigma}_-$ with behavior $\mathcal{B}_{(H\ -I)}$. As H is stable and contractive, the system $\bar{\Sigma}_-$ is SP. As a result the SP inequality also holds on the subset $\mathcal{B}_-$. This proves that solvability of the NPIP implies SP of Σ_-.

Now suppose that Σ_- is SP. Then $\mathcal{B}_- = \mathcal{B}_M$ for some stable co-J-lossless M. Define $(\,H_1\ \ -H_2\,) := (\,U\ \ I\,)\,M$ where H_2 is square and H_1 and U have the same size as the solution H we are trying to find. Take U stable and contractive. Then $H_1H_1^* - H_2H_2^* \leq UU^* - I < 0$ in the closed right-half plane, by co-J-losslessness of M. It follows that H_2 is in $\mathcal{GH}_\infty$, and that $H = H_2^{-1}H_1$ is stable and contractive. For $w \in \mathcal{B}_-$, $Mw \in \mathcal{H}_2$, so $(\,H\ \ -I\,)\,w = H_2^{-1}(\,U\ \ I\,)(Mw) \in \mathcal{H}_2$ if $w \in \mathcal{B}_-$ as H_2^{-1} and U are stable. This holds in particular for

$$w = \begin{pmatrix} a_j \\ b_j \end{pmatrix}\frac{1}{s - \zeta_j} \in \mathcal{B}_-.$$

Thus $(H(s)a_j - b_j)/(s - \zeta_j)$ is in $\mathcal{H}_2$, which implies $H(s)a_j - b_j$ is zero at $s = \zeta_j$. In other words, H is stable contractive and satisfies the interpolation conditions as well. That is, H is a solution to the NPIP.

Note that there is a freedom in the construction of H. Every stable contractive U will give rise to a solution H to the NPIP. In fact, it may be shown that *all* solutions H to the NPIP are generated this way (see e.g. [1], [12]).

Using the state space results of the previous section, it follows (see (11)) that this NPIP is solvable iff the solution X to the Lyapunov equation

$$\begin{pmatrix} \zeta_1^* & & \\ & \ddots & \\ & & \zeta_n^* \end{pmatrix} X + X \begin{pmatrix} \zeta_1 & & \\ & \ddots & \\ & & \zeta_n \end{pmatrix} = \begin{pmatrix} a_1^* & b_1^* \\ \vdots & \vdots \\ a_n^* & b_n^* \end{pmatrix} J \begin{pmatrix} a_1 \cdots a_n \\ b_1 \cdots b_n \end{pmatrix}$$

is positive definite. This is a standard result and X is known as the Pick matrix. The co-J-lossless matrix M so that $\mathcal{B}_- = \mathcal{B}_M$ may be chosen to be

$$M(s) = \begin{pmatrix} a_1 \cdots a_n \\ b_1 \cdots b_n \end{pmatrix} X^{-1} \left(sI + \begin{pmatrix} \zeta_1^* & & \\ & \ddots & \\ & & \zeta_n^* \end{pmatrix} \right)^{-1} \begin{pmatrix} a_1^* & -b_1^* \\ \vdots & \vdots \\ a_n^* & -b_n^* \end{pmatrix} - I.$$

The general NPIP where interpolation with multiplicities are allowed may be handled using behaviors $C(sI-A)^{-1}\mathbb{C}^n$, with A a matrix in Jordan form. ■

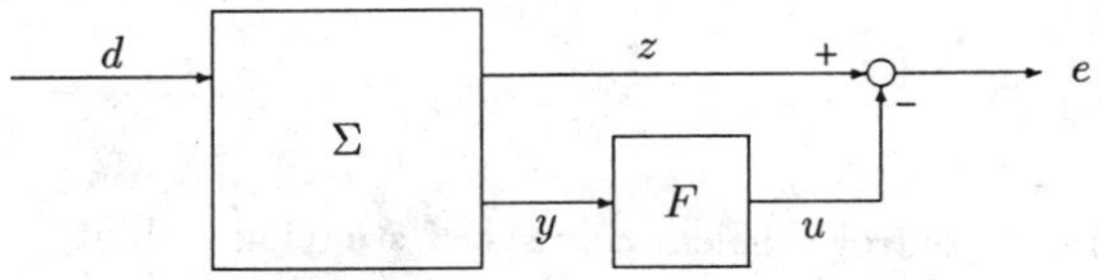

Figure 1: The $\mathcal{H}_\infty$ filtering configuration.

Example 4.2 ($\mathcal{H}_\infty$ filtering. Necessity results.) The $\mathcal{H}_\infty$ filtering problem is to find filters F such that the closed-loop system map is stable and contractive. (We estimate z with error $\|e\|_2 < \|d\|_2$.) We allow causal, homogeneous filters F. By homogeneous we mean that F maps the zero signal into the zero signal. The given system Σ is assumed to be of the usual type:

$$\Sigma : \begin{cases} \dot{x} &= Ax + Bd \\ z &= C_1 x \\ y &= C_2 x + Dd. \end{cases} \tag{20}$$

We assume without loss of generality that $D\begin{pmatrix} B^* & D^* \end{pmatrix} = \begin{pmatrix} 0 & I \end{pmatrix}$ (see [5], [13]).

The idea is to show that there exists a subset of the closed-loop behavior that does not depend on the filter. If the $\mathcal{H}_\infty$ filtering problem has a solution, i.e., if there is a filter such that the closed-loop system is SP, then certainly the SP inequality must hold on this filter independent subset. This provides a necessary condition for the $\mathcal{H}_\infty$ filtering problem to have a solution.

The $\mathcal{L}_{2-}$ behavior that is important here is the set of external signals (d,e) in $\mathcal{L}_{2-}$ that do not 'activate' the output $y(t)$ for $t<0$, so that there is nothing to filter for $t<0$. Then $u(t)=0,\ t<0$ and $e=z$, by causality and homogeneity of F. Consider therefore the behavior

$$\mathcal{B}_- = \{\ (d,e)\ |\ \begin{matrix} \dot{x} &=& Ax &+& Bd & \\ 0 &=& C_1 x & & & -Ie\ , \\ 0 &=& C_2 x &+& Dd & \end{matrix} \quad (d,e) \in \mathcal{L}_{2-}\ \}.$$

The behavior is of the ONR type. In order to be able to apply Corollary 3.10, we have to assume

1. $(\begin{pmatrix} C_1 \\ C_2 \end{pmatrix}, A)$ is detectable;

2. $\begin{pmatrix} A - sI & B & 0 \\ C_1 & 0 & -I \\ C_2 & D & 0 \end{pmatrix}$ has full row rank for all $s \in \mathbb{C}_0 \cup \infty$.

We assumed in the first place that $D(B^*\ D^*) = (0\ I)$ and this implies that the second assumption is equivalent to (A,B) not having uncontrollable modes on the imaginary axis.

It follows from Corollary 3.10 that under these assumptions the system with behavior $\mathcal{B}_-$ is SP iff

$$AQ + QA^* + Q(C_1^*C_1 - C_2^*C_2)Q + BB^* = 0 \tag{21}$$

has a stabilizing nonnegative definite solution Q.

Conversely, given a stabilizing nonnegative definite solution Q to (21) a filter that solves the problem can be constructed (see [5], [13]):

$$\begin{aligned} \dot{\hat{x}} &= A\hat{x} + QC_2^*(y - C_2\hat{x}) \\ u &= C_1\hat{x}. \end{aligned}$$

■

References

1. Ball J., I. Gohberg, and L. Rodman, *Interpolation of Rational Matrix Functions*, Operator Theory, **34**, Birkhäuser Verlag, Basel, 1990.

2. Ball J., and J. W. Helton, 'A Beurling-Lax theorem for the Lie group $U(m,n)$ which contains most classical interpolation theory,' *J. Operator Theory.*, **9**, 107-142, 1983.

3. Ball J., and J. W. Helton, 'Shift invariant subspaces, passivity, reproducing kernels and $\mathcal{H}^\infty$-optimization', Operator Theory: Adv. Appl., **35**, Birkhauser Verlag, Basel, 1988.

4. Doyle J. C., B. Francis and A. Tannenbaum, *Feedback Control Theory*, Macmillan Publishing Co., New York, 1991.

5. Doyle J. C., K. Glover, P. P. Khargonekar and B. A. Francis, 'State space solutions to standard $\mathcal{H}_2$ and $\mathcal{H}_\infty$ control problems', *IEEE Trans. Aut. Control,* **AC-34**, 831-847, 1989.

6. Francis B. A., *A Course in $\mathcal{H}_\infty$ Control Theory*, Springer Lecture Notes in Control and Information Sciences, **88**, Springer Verlag, Heidelberg, etc., 1987.

7. Green M., K. Glover, D. Limebeer and J. Doyle, 'A J-spectral factorization approach to $\mathcal{H}_\infty$ control', *SIAM J. Control and Opt.*, **28**, 1350-1371, 1990.

8. Green M., '$\mathcal{H}_\infty$ controller synthesis by J-lossless coprime factorization', *SIAM J. Control and Opt.*, **30**, 522-547, 1992.

9. Khargonekar P. P., 'State-space $\mathcal{H}_\infty$ control theory and the LQG problem,' In: A.C. Antoulas, *Mathematical System Theory - The influence of R.E. Kalman*, Springer Verlag, Berlin etc., 1991.

10. Kimura H., 'Robust stabilization for a class of transfer functions,' *IEEE Trans. Aut. Control,* **AC-29**, 788-793, 1984.

11. Kimura H., Y. Lu and R. Kawatani, 'On the structure of $\mathcal{H}^\infty$ control systems and related extensions,' *IEEE Trans. Aut. Control,* **AC-36**, 653-667, 1991.

12. Limebeer D. and B. Anderson, 'An interpolation theory approach to $\mathcal{H}^\infty$ controller degree bounds,' Linear Algebra and its Applications, **96**, 347-386, 1988.

13. Limebeer D. and U. Shaked, 'Minimax terminal state estimation and $\mathcal{H}_\infty$ filtering', submitted to *IEEE Trans. Aut. Control.*

14. Meinsma G., 'Strict passivity in the frequency domain,' *Second IFAC Workshop on System Structure and Control*, Prague, 1992.

15. Weiland S., *Theory of Approximation and Disturbance Attenuation for Linear Systems,* Doctoral dissertation, University of Groningen, the Netherlands, 1991.

16. Willems J. C., 'Least-squares stationary optimal control and the Riccati equation,' *IEEE Trans. Aut. Control,* **AC-16**, 621-634, 1971.

17. Willems J. C., 'Paradigms and puzzles in the theory of dynamical systems', *IEEE Trans. Aut. Control,* **AC-36**, 259-294, 1991.

18. Zames G., 'Feedback and optimal sensitivity: model reference transformations, multiplicative seminorms, and approximate inverses,' *IEEE Trans. Aut. Control,* **AC-26**, 301-320, 1981.

Proceedings of the 31st Conference
on Decision and Control
Tucson, Arizona • December 1992

ON THE ROBUST POPOV CRITERION FOR INTERVAL LUR'E SYSTEMS

Mohammed Dahleh*, Alberto Tesi** and Antonio Vicino***

* *Department of Mechanical Engineering, University of California, Santa Barbara - CA 93106 - USA*
** *Dipartimento di Sistemi e Informatica, Università di Firenze, Via S. Marta 3 - 50139 Firenze - Italy*
*** *Dipartimento di Ingegneria Elettrica, Università di L'Aquila, 67040 Poggio di Roio - L'Aquila - Italy*

ABSTRACT

In this paper a robust version of the classical Popov criterion is given for robust absolute stability of Lur'e systems including parametric uncertainty. The parametric dependence is modeled by an *interval plant* description. The generalized criterion is obtained in terms of a vertex result for the Popov strict positive realness condition.

I. INTRODUCTION

The extensive literature on robust stability of linear systems with parametric uncertainties inspired by the Theorem of Kharitonov, has stimulated research in the same direction for nonlinear systems. First contributions on robustness issues in the nonlinear field can be found in [1], while more recent contributions are in [2,3]. In all these papers, the important class of Lur'e control systems is considered. More precisely, the linear plant of the feedback system is assumed to be subject to parametric perturbations and *robust Absolute Stability* (AS) against parameter variations is studied by means of the classical circle or Popov criterion. Both criteria allow one to reduce the robust AS problem to a robust *Strict Positive Realness* (SPR) problem. This fact has stimulated work on robust SPR for parametric perturbations in recent years. Papers [4,5,6] are among the first references in the literature addressing the robust SPR problem for families of uncertain transfer functions.

The aim of this paper is to use a frequency domain approach ([7]) to prove a robust version of the Popov criterion for absolute stability of *interval-controller* Lur'e systems. This result provides sufficient conditions for absolute stability of the interval-controller Lur'e system in terms of the Popov condition applied to a finite number of one-dimensional segments of linear plants. For the important case of *interval* Lur'e systems, the paper establishes a sufficient condition for absolute stability requiring the application of the Popov condition to a finite number of linear plants independent of the order of the linear system.

II. NOTATION AND PRELIMINARIES

We recall that the classical Lur'e control problem assumes a stable linear system $L(s) = C(s)G(s)$ in the feedforward path of a feedback control scheme and a memoryless continuous nonlinear function $\phi(y)$ in the feedback path, subject to the so called "sector condition", i.e. such that $\phi \in \Phi_k$ where

$$\Phi_k = \{\phi : \phi(0) = 0 \ \text{ and } \ 0 \le y\phi(y) \le ky^2, \quad 0 < k < \infty\} .$$

The Lur'e control problem consists of studying stability in the large of the closed loop system for any $\phi \in \Phi_k$. If the system is stable in the large for all $\phi \in \Phi_k$, the closed loop system is said to be absolutely stable in the sector $[0, k]$.

It is well known that one of the most widely used *sufficient conditions* for absolute stability is provided by the fundamental Popov's criterion.

Popov Criterion: If there exists a real $\theta \in \mathbf{R}$ such that the following inequality holds

$$L_k(j\omega) = k^{-1} + \mathbf{Re}[(1 + j\omega\theta)C(j\omega)G(j\omega)] > 0, \ \forall\omega \ge 0 \quad (1)$$

then the system is absolutely stable. ■

The maximal k such that the Popov criterion is satisfied is called the *Popov sector* of the Lur'e system.

The Popov criterion relates AS to SPR of a suitable transfer function. In this context, the robust AS problem in the presence of coefficient perturbations in the linear plant reduces to the robust SPR problem. We will consider *interval-controller* Lur'e systems, i.e. Lur'e systems where the linear path is composed of the cascade of a linear controller and an *interval plant.*

We denote by $\mathcal{G}_I$ a proper interval plant, i.e.

$$\mathcal{G}_I = \{G(s) : G(s) = \frac{N(s)}{D(s)}, \ N(s) \in \mathcal{N}_I, \ D(s) \in \mathcal{D}_I\}$$

where $\mathcal{N}_I$ and $\mathcal{D}_I$ are interval polynomials, i.e.

$$\mathcal{N}_I = \{N(s) : N(s) = \sum_{i=0}^{m} b_i s^i, \ b_i \in [b_i^-, b_i^+], \ i = 0, \ldots, m\}$$

$$\mathcal{D}_I = \{D(s) : D(s) = s^n + \sum_{i=0}^{n-1} a_i s^i, \ a_i \in [a_i^-, a_i^+], \ i = 0, \ldots, n-1\} .$$

The transfer function $C(s) = N_c(s)/D_c(s)$ is the controller and the interval-controller family of transfer functions is given by

$$\mathcal{L} = \{L(s) : L(s) = C(s)G(s), \ G(s) \in \mathcal{G}_I\} .$$

For a given interval polynomial, we define two subsets playing a key role in robustness analysis. For a moment, we make specific reference to the numerator family, but similar definitions hold for the denominator family. The first set $\mathcal{N}_K = \{N_i(s), \ i = 1, \ldots, 4\}$ contains the four *Kharitonov polynomials*, while the second set

CH3229-2/92/0000-2808$1.00 © 1992 IEEE

$\mathcal{N}_{KS} = \{S_i^N,\ i = 1, \ldots, 4\}$, contains the four *Kharitonov segments* of polynomials.

Strictly related to the sets $\mathcal{N}_K, \mathcal{N}_{KS}, \mathcal{D}_K, \mathcal{D}_{KS}$ of the numerator and denominator families are the following subsets of transfer functions of $\mathcal{G}_I$

$$\mathcal{G}_K = \{G(s) : G(s) = \frac{N(s)}{D(s)},\ N(s) \in \mathcal{N}_K,\ D(s) \in \mathcal{D}_K\},$$

$$\mathcal{G}_{KS} = \{G(s) : G(s) = \tfrac{N(s)}{D(s)},\quad N(s) \in \mathcal{N}_K,\ D(s) \in \mathcal{D}_{KS} \text{ or } N(s) \in \mathcal{N}_{KS},\ D(s) \in \mathcal{D}_K\}.$$

The set $\mathcal{G}_K$ is called *Kharitonov (vertex) set* of $\mathcal{G}_I$ and $\mathcal{G}_{KS}$ is denoted as the *Kharitonov segment set* of $\mathcal{G}_I$. The former set is made of 16 plants, while the latter includes 32 segments of plants. Accordingly, we define the following two extremal sets of transfer functions

$$\mathcal{L}_{KS} = \{L(s) : L(s) = C(s)G(s),\ G(s) \in \mathcal{G}_{KS}\}$$
$$\mathcal{L}_K = \{L(s) : L(s) = C(s)G(s),\ G(s) \in \mathcal{G}_K\}\ .$$

Given a class of nonlinearities Φ_k and a family of stable transfer functions $\mathcal{L}$, the natural way of checking if the interval-controller Lur'e system is absolutely stable, is to apply at each $L(s) \in \mathcal{L}$ the Popov criterion (1). However, this calls for analyzing the nonlinear relation between the uncertain coefficients of the transfer functions and the parameter θ. In order to avoid this difficulty, we consider the following robust version of the Popov Criterion.

Robust Popov Criterion: Given a family of stable transfer functions $\mathcal{L}$, if there exists a real $\theta \in \mathbf{R}$ such that for each $L(s) \in \mathcal{L}$ the following inequality holds

$$L_k(j\omega) = k^{-1} + \mathbf{Re}[(1 + j\omega\theta)C(j\omega)G(j\omega)] > 0,\ \forall\omega \geq 0,$$

then the interval-controller Lur'e system is absolutely stable. ∎

We point out that since the above criterion requires the existence of a θ holding for all the transfer functions of the family, it may be conservative in assessing robust AS with respect to the case where θ is allowed to be dependent on $G(s)$. Nevertheless, the robust Popov criterion as stated above improves over the robust circle criterion, the only alternative criterion presently available in the literature ([2]).

III. MAIN RESULT

Proofs of the results presented in this section are omitted and they can be found elsewhere [8].

Theorem 1. A Lur'e control system is absolutely stable in the sector $[0, k]$ for all $L(s) \in \mathcal{L}$ if there exists a real θ verifying the robust Popov condition for $L(s) \in \mathcal{L}_{KS}$. ∎

In what follows, we discuss in more detail the case when $C(s)$ is a fixed order polynomial, i.e. $C(s) \equiv N_c(s)$. For this case, the following theorem which is the main contribution of the paper, provides a vertex result for the robust Popov problem.

Theorem 2. A Lur'e control system is absolutely stable in the sector $[0, k]$ for all $L(s) \in \mathcal{L}$ with $C(s) = N_c(s)$ if there exists a real θ verifying the robust Popov condition for $L(s) \in \mathcal{L}_K$. ∎

Remark 1. The importance of the above result lies in the fact that it reduces the problem of checking absolute stability of an interval family of Lur'e systems (via the Popov criterion) to a finite number of Popov tests, independently of the order of the uncertain linear part of the feedback system. This means, for example, that by using well-known graphical techniques, we need to draw only Popov plots of the 16 Kharitonov transfer functions of the family $\mathcal{L}_K$ to check that the robust Popov condition holds for the whole family $\mathcal{L}$.

Remark 2. Denote by k_l the maximal value of k such that there exists $\theta = \theta_l$ for which the Popov condition is satisfied for $L(s) \in \mathcal{L}_K$. In addition, let k_u be defined as $k_u = \min_{i=1,\ldots,16} k_i$, where k_i is the Popov sector corresponding to each of the Kharitonov plants of the family. Let $k_{\mathcal{G}}$ be the Popov sector of the interval Lur'e system defined as $k_{\mathcal{G}} = \min_{G \in \mathcal{G}_I} k_G$, where k_G is the Popov sector relative to the plant $G(s)$. It is easy to check that

$$k_l \leq k_{\mathcal{G}} \leq k_u\ . \tag{2}$$

We notice that the inequalities in (2) are sharp if the Popov line corresponding to k_l is tangent to only one out of the 16 Popov plots of the family $\mathcal{L}_K$.

IV. CONCLUSION

In this paper the recently developed techniques of analysis of interval systems are used to study the problem of robust stability in the presence of parametric and sector bounded nonlinear perturbations. A generalization of the classical Popov criterion is given, and the computation associated with this criterion is discussed.

ACKNOWLEDGEMENTS

This work was partially supported by funds of Ministero della Università e della Ricerca Scientifica e Tecnologica and CNR Special Project "Algoritmi ed Architetture per l' Identificazione e il Controllo Robusto e Adattivo".

REFERENCES

[1] Šiljak, D. D., *Nonlinear Systems: The Parameter Analysis and Design*. Wiley, New York, 1969.

[2] Chapellat, H., M. Dahleh, and S. P. Bhattacharyya, "On robust nonlinear stability of interval control systems", *IEEE Trans. on Automat. Contr.*, vol. AC-36, pp. 59-67, 1991.

[3] Tesi, A., and A. Vicino, "Robust absolute stability of Lur'e control systems in parameter space", *Automatica*, vol. 27, pp. 147-151, 1991.

[4] Dasgupta, S., "A Kharitonov like theorem for systems under nonlinear passive feedback", *Proc. 26th IEEE Conf. Decision and Contr.*, Los Angeles (CA), pp. 2062-2063, 1987.

[5] Bose, N. K., and J. F. Delansky, " Boundary implications for interval positive rational functions" *IEEE Trans. on Circuit and Systems*, vol. CAS-36, pp. 454-458, 1989.

[6] Šiljak, D. D.," Polytopes of nonnegative polynomials", *Proc. 1989 American Contr. Conf.*, Pittsburgh (USA), June 1989.

[7] Vicino, A., and A. Tesi, "Robust strict positive realness: new results for interval plant plus controller families", *Proc. 30th IEEE Conf. Decision and Contr.*, Brighton (UK), pp. 421-426, 1991 and to appear in *IEEE Trans. on Automat. Contr.*.

[8] Dahleh, M., A. Tesi and A. Vicino, "On the robust Popov criterion for interval Lur'e systems", Tech. Rep. DSI/RT-24/91, Università di Firenze, 1991 and to appear in *IEEE Trans. on Automat. Contr.*.

Proceedings of the 31st Conference
on Decision and Control
Tucson, Arizona • December 1992

Stability robustness bounds for linear uncertain systems —— a frequency domain approach

J. S. Luo and A. Johnson

Kramers Laboratory, Delft University of Technology
Prins Bernhardlaan 6, 2628 BW Delft, The Netherlands

Abstract

The systems to be discussed are described by state-space models, but with structured deterministic uncertainties. First we present a new sufficient condition which improves on the sufficient condition given by Juang et al. (1987a) for stability robustness of linear continuous-time systems. Then we extend the result to linear discrete-time systems by presenting a new sufficient condition which improves on the sufficient condition given by Juang et al. (1987b). Illustrative examples are given.

Notation

$\mathbb{R}^{n\times n}$: Space of real n×n matrix
$\mathbb{C}^{n\times n}$: Space of complex n×n matrix
A[≤]B: '≤' is applied element by element to two matrices
$\lambda(A)$: The eigenvalues of matrix A
$\rho(A)$: The spectral radius of matrix A
$\|A\|$: The spectral norm of matrix A
A^+: A matrix obtained by replacing the entries of A with their absolute values
I: n×n identity matrix

1. Introduction

The problem of stability robustness of a nominally stable linear time-invariant system which is subject to linear parametric uncertainties has been investigated in many papers. Recently, based on the fact that the eigenvalues of a matrix depend continuously on the elements of the matrix (Lancaster, 1969), Juang et al (1987a, 1987b) have presented sufficient conditions for stability robustness of linear systems described by a state-space model with structured and unstructured time-invariant uncertainties. The sufficient conditions given by Juang et al. (1987a, 1987b) are easy to understand and apply. However, their sufficient conditions for linear systems with structured uncertainties are given without considering that the bounded elements in an uncertain matrix may be linearly dependent functions of some uncertain parameters, which is the situation in most practical problems. In this paper we present a new sufficient condition which improves on the sufficient condition given by Juang et al. (1987a). We use a more general system model, including the dependent structure of uncertain parameters, and employ the similarity transformation to reduce conservatism. Then, we extend the result to linear discrete-time systems and propose a new sufficient condition which improves on the sufficient condition given by Juang et al. (1987b). The improvements are illustrated by examples. The rest of this papers is organized as follows: In section 2 we characterize the problem. Then, in sections 3 and 4, the new stability sufficient conditions are given for linear continuous-time and discrete-time systems with structured time-invariant uncertainties, respectively. Finally, we complete this paper with a brief conclusion in the last section.

2 Description of the problem

Consider a linear continuous-time or discrete-time system described by:

$$\dot{x}(t)=Ax(t) \tag{2,1}$$

or

$$x(k+1)=Gx(k) \tag{2,1'}$$

where $x\in\mathbb{R}^n$ is the state vector, and $A\in\mathbb{R}^{n\times n}$ or $G\in\mathbb{R}^{n\times n}$ is the time-invariant stable system matrix. It is assumed that the nominal stable system (2,1) or (2,1') is subject to linear parametric uncertainties in the entries of A or G described by ΔA or ΔG, respectively. Further we assume that ΔA or ΔG is a linear function of some uncertain parameters, i.e. ΔA or ΔG can be represented as

$$\Delta A=\sum_{i=1}^{m}\varepsilon_i\alpha_i A_i \tag{2.2}$$

or

$$\Delta G=\sum_{i=1}^{m}\varepsilon_i\gamma_i G_i \tag{2,2'}$$

where A_i or G_i are constant matrices determined by the structure of the uncertainties, ε_i are time-invariant uncertain parameters which are assumed to lie in the interval around zero, i.e.

$$\varepsilon_i\in[-\varepsilon,\ \varepsilon] \qquad i=1,\ 2,\dots,\ m. \tag{2,3}$$

and α_i or γ_i is a set of non-negative weights. The choices of α_i in (2,2) or γ_i in (2,2') allow the possibility of using different weights for the different parameters as may be required in any particular application.

For clarity we define $E_i\overset{\Delta}{=}\alpha_i A_i$ for continuous-time systems or $E_i\overset{\Delta}{=}\gamma_i G_i$ for discrete-time systems, respectively. Thus, the linear continuous-time or discrete-time systems with structured uncertainty are given by

$$\dot{x}(t)=(A+\sum_{i=1}^{m}\varepsilon_i E_i)x(t) \tag{2,4}$$

or

$$x(k+1)=(G+\sum_{i=1}^{m}\varepsilon_i E_i)x(k) \tag{2,4'}$$

Now we state the problem as follows:
To determine what the upper bounds of ε is for maintaining the asymptotic stability of the closed-loop system (2,4) or (2,4') for all ε_i(i=1, 2,..., m) described by (2,3)

3. Stability robustness of linear continuous-time systems

Lemma 3.1 (Ortega, 1972)

For every A, $B\in\mathbb{C}^{n\times n}$, if $A^+[\le]B^+$, then $\rho(A)\le\rho(B^+)$ (3,1)

Lemma 3.2 Assuming that the system is described by (2,4), where A is a strict Hurwitz matrix, then $A+\sum_{i=1}^{m}\varepsilon_i E_i$ remains a strict Hurwitz matrix for all ε_i (i=1,..., m) described by (2,3), if

$$\det(j\omega I-A-\sum_{i=1}^{m}\varepsilon_i E_i)\neq 0 \qquad \forall\varepsilon_i\in[-\varepsilon,\ \varepsilon]\quad i=1,\dots,\ m \tag{3,2}$$

for all

$$\omega\in\{\omega:\ 0\le\omega<\|A\|+\|A^{-1}\|^{-1}\} \tag{3,3}$$

Proof All the eigenvalues of A matrix lie in the left half of the complex plane, and the eigenvalues of $A+\sum_{i=1}^{m}\varepsilon_i E_i$ depend continuously on ε_i (i=1,..., m). This implies that if any eigenvalue of $A+\sum_{i=1}^{m}\varepsilon_i E_i$ lies in the right half of the complex plane, the root locus of $A+\sum_{i=1}^{m}\varepsilon_i E_i$ must intersect the imaginary axis. So there must at least exist

CH3229-2/92/0000-2810$1.00 © 1992 IEEE

one set of ε_i (i=1,..., m), which we denote ε_i^*, which makes $j\omega^*$ $(-\infty<\omega^*<+\infty)$ one of the eigenvalues of $A+\sum_{i=1}^{m}\varepsilon_i E_i$, i.e.

$$\det(j\omega^* I-A-\sum_{i=1}^{m}\varepsilon_i^* E_i)=0 \qquad (3,4)$$

Therefore, if condition (3,2) is satisfied for $-\infty<\omega<+\infty$, we know that $A+\sum_{i=1}^{m}\varepsilon_i E_i$ remains a strict Hurwitz matrix for all $\varepsilon_i\in[-\varepsilon, \varepsilon]$, (i=1,..., m).

However, the root locus of $A+\sum_{i=1}^{m}\varepsilon_i E_i$ is symmetric to the real axis in the complex plane, so it is sufficient for ω in (3,2) to be non-negative.

Now we prove that it is only required for ω in (3,2) to satisfy the inequality in (3,3). The sufficient condition given by Matin (1987) and Juang et al. (1987a) for the stability of linear systems with time-invariant unstructured uncertainty is:

$$\|E\|<\|(j\omega-A)^{-1}\|^{-1} \qquad \forall\omega\geq 0 \qquad (3,5)$$

Taking $E=\sum_{i=1}^{m}\varepsilon_i E_i$ and $\omega=0$ in (3,5), we get

$$\|\sum_{i=1}^{m}\varepsilon_i E_i\|<\|A^{-1}\|^{-1} \qquad (3,6)$$

Meanwhile, suppose $\omega^*>0$ in (3,4). We obtain

$$\det[I-(j\omega^*)^{-1}A-(j\omega^*)^{-1}\sum_{i=1}^{m}\varepsilon_i^* E_i]=0 \qquad (3,7)$$

which implies

$$\lambda_j[(j\omega^*)^{-1}A+(j\omega^*)^{-1}\sum_{i=1}^{m}\varepsilon_i^* E_i]=1 \qquad j\in(1,\ldots, m) \qquad (3,8)$$

which givess

$$\omega^*=\lambda_j[(j)^{-1}A+(j)^{-1}\sum_{i=1}^{m}\varepsilon_i^* E_i]$$

$$\leq\|(j)^{-1}A+(j)^{-1}\sum_{i=1}^{m}\varepsilon_i^* E_i\|$$

$$\leq\|A\|+\|\sum_{i=1}^{m}\varepsilon_i^* E_i\|$$

from (3,6) $\leq\|A\|+\|A^{-1}\|^{-1}$ □

Theorem 3.1 Suppose

(1) The system is described by (2,4), where A is a strict Hurwitz matrix

(2) M may be any similarity transformation matrix

Then, the system (2,4) is asymptotically stable for all ε_i described by (2,3), if

$$\varepsilon<\frac{1}{\rho\{\sum_{i=1}^{m}[M^{-1}(j\omega I-A)^{-1}E_i M]^+\}} \qquad 0\leq\omega<\|A\|+\|A^{-1}\|^{-1} \qquad (3,9)$$

Proof Because A is a strict Hurwitz matrix, $\lambda_i(j\omega I-A)\neq 0$ (i=1,..., n) for all ω. So $(j\omega I-A)^{-1}$ always exists.

If (3,9) is satisfied, then

$$\rho\{\sum_{i=1}^{m}\varepsilon[M^{-1}(j\omega I-A)^{-1}E_i M]^+\}<1$$

From Lemma 3.1

$$\Rightarrow \rho[\sum_{i=1}^{m}M^{-1}(j\omega I-A)^{-1}\varepsilon_i E_i M]<1$$

$$\Leftrightarrow \rho[\sum_{i=1}^{m}(j\omega I-A)^{-1}\varepsilon_i E_i]<1$$

$$\Rightarrow \lambda_k[\sum_{i=1}^{m}(j\omega I-A)^{-1}\varepsilon_i E_i]\neq 1 \qquad k=1,\ldots, m$$

$$\Leftrightarrow \lambda_k[I-\sum_{i=1}^{m}(j\omega I-A)^{-1}\varepsilon_i E_i]\neq 0 \qquad k=1,\ldots, m$$

$$\Leftrightarrow \det[I-\sum_{i=1}^{m}(j\omega I-A)^{-1}\varepsilon_i E_i]\neq 0$$

$$\Leftrightarrow \det[(j\omega I-A)^{-1}]\det[(j\omega I-A)-\sum_{i=1}^{m}\varepsilon_i E_i]\neq 0$$

where $\det[(j\omega I-A)^{-1}]\neq 0$ for all ω

$$\Leftrightarrow \det[j\omega I-A-\sum_{i=1}^{m}\varepsilon_i E_i]\neq 0 \qquad \forall\ 0\leq\omega<\|A\|+\|A^{-1}\|^{-1}$$

It follows from Lemma 3.2 that $A+\sum_{i=1}^{m}\varepsilon_i E_i$ remains a strict Hurwitz matrix for all ε_i (i=1,..., m) described by (2,3). The system (2,4) is asymptotically stable.

Remark 3.1 Taking M=I and $\sum_{i=1}^{m}E_i^+=E$, and sweeping the whole line $\omega\geq 0$, our sufficient condition (3,9) reduces to Juang et al's (1987a) sufficient condition.

Example 3.1 This example is used to show the improvements of our sufficient condition on Juang et al's (1987a) sufficient condition.

The system is described by (2,4), where

$$A=\begin{bmatrix} -4.5 & -1 & 0.5 \\ -1 & -3 & -1 \\ 2.5 & 2 & -1.5 \end{bmatrix}$$

with eigenvalues $\lambda(A)=-4, -2.5\pm 0.8660j$;

$$E_1=\begin{bmatrix} -0.3333 & -0.6667 & -0.3333 \\ 0.3333 & 0.6667 & 0.3333 \\ 0.6667 & 1.3333 & 0.6667 \end{bmatrix}$$

and

$$E_2=\begin{bmatrix} -0.6667 & -0.3333 & -0.6667 \\ 0.3333 & 0.3333 & 0.6667 \\ 1.3333 & 0.6667 & 1.3333 \end{bmatrix}$$

(1) The sufficient condition of Juang et al. (1987a)

Taking $E=E_1^++E_2^+=\begin{bmatrix} 1 & 1 & 1 \\ 1 & 1 & 1 \\ 2 & 2 & 2 \end{bmatrix}$, from Juang et al's (1987a) sufficient condition

$$\varepsilon<\frac{1}{\rho\{[(j\omega I-A)^{-1}]^+E\}} \qquad \forall\omega\geq 0$$

we get the stability robustness bound $\varepsilon<0.4179$

(2) Our sufficient condition (3,9)

Taking M=I, from our sufficient condition (3,9)

$$\varepsilon<\frac{1}{\rho\{\sum_{i=1}^{m}[M^{-1}(j\omega I-A)^{-1}E_i M]^+\}} \qquad 0\leq\omega<\|A\|+\|A^{-1}\|^{-1}=7.7432$$

we get the stability robustness bound $\varepsilon<0.7773$

To reduce conservatism we choose

$$M=\begin{bmatrix} -0.1667 & 0.25 & -0.5 \\ 0.6667 & 0 & 0 \\ -0.1667 & 0.25 & 0.5 \end{bmatrix}$$

Then, from our sufficient condition (3,9), we get the stability robustness bound $\varepsilon<1.4$ which is much less conservative than the result of Juang et al·. Moreover, because $\mathrm{MaxRe}\lambda(A+\varepsilon E_1+\varepsilon E_2)<0$ is a necessary condition of stability, from $\mathrm{MaxRe}\lambda(A+1.4E_1+1.4E_2)=0$ we know that we have attained the necessary and sufficient condition of system stability.

4. Stability robustness of linear discrete-time systems

Lemma 4.1 Assuming that the system is described by (2,4'), where G is a strict Schur matrix, then $G+\sum_{i=1}^{m}\varepsilon_i E_i$ remains a strict Schur matrix for all ε_i (i=1,..., m) described by (2,3), if

$$\det(e^{j\omega}I-G-\sum_{i=1}^{m}\varepsilon_i E_i)\neq 0 \qquad \forall\varepsilon_i\in(-\varepsilon,\varepsilon)\ \ i=1,\ldots, m \quad \forall 0\leq\omega\leq\pi \qquad (4,1)$$

Proof All the eigenvalues of the G matrix lie in unit circle of complex plane, and the eigenvalues of $G+\sum_{i=1}^{m}\varepsilon_i E_i$ depend continuously on ε_i (i=1,..., m). This implies that if any eigenvalue of $G+\sum_{i=1}^{m}\varepsilon_i E_i$ lies outside of the unit circle of the complex plane, the root locus of $G+\sum_{i=1}^{m}\varepsilon_i E_i$

must intersect the boundary of the unit circle and there must at least exist one set of ε_i (i=1,..., m), which we denote ε_i^*, which makes $e^{j\omega^*}$ ($0\le\omega^*\le2\pi$) one of eigenvalues of $G+\sum_{i=1}^{m}\varepsilon_iE_i$, i.e.

$$\det(e^{j\omega^*}I-G-\sum_{i=1}^{m}\varepsilon_i^*E_i)=0 \qquad (4,2)$$

Since the root locus of $G+\sum_{i=1}^{m}\varepsilon_iE_i$ is symmetric to the real axis in the complex plane, it is sufficient for us to assume $0\le\omega\le\pi$.

Therefore, if condition (4,1) is satisfied, we know that all the eigenvalues of $G+\sum_{i=1}^{m}\varepsilon_iE_i$ remain in the unit circle of the complex plane.

Theorem 4.1 Suppose

(1) The system is described by (2,4'), where G is a strict Schur matrix

(2) M may be any similarity transformation matrix

Then, the system (2,4') is asymptotically stable for all ε_i described by (2,3), if

$$\varepsilon<\frac{1}{\rho\{\sum_{i=1}^{m}[M^{-1}(e^{j\omega}I-G)^{-1}E_iM]^+\}} \qquad \forall\omega\in[0,\pi] \qquad (4,3)$$

Proof Because G is a strict Schur matrix, $\lambda_i(e^{j\omega}-G)\neq0$ (i=1,..., n) for all $\omega\in[0,\pi]$. So $(e^{j\omega}I-G)^{-1}$ always exists.

If (4,3) is satisfied, then

$$\rho\{\sum_{i=1}^{m}\varepsilon[M^{-1}(e^{j\omega}-G)^{-1}E_iM]^+\}<1$$

From Lemma 3.1

$$\Rightarrow \rho[\sum_{i=1}^{m}M^{-1}(e^{j\omega}I-G)^{-1}\varepsilon_iE_iM]<1$$

$$\Leftrightarrow \rho[\sum_{i=1}^{m}(e^{j\omega}I-G)^{-1}\varepsilon_iE_i]<1$$

$$\Rightarrow \lambda_k[\sum_{i=1}^{m}(e^{j\omega}I-G)^{-1}\varepsilon_iE_i]\neq1 \qquad k=1,\dots, m$$

$$\Leftrightarrow \lambda_k[I-\sum_{i=1}^{m}(e^{j\omega}I-G)^{-1}\varepsilon_iE_i]\neq0 \qquad k=1,\dots, m$$

$$\Leftrightarrow \det[I-\sum_{i=1}^{m}(e^{j\omega}I-G)^{-1}\varepsilon_iE_i]\neq0$$

$$\Leftrightarrow \det[(e^{j\omega}I-G)^{-1}]\det[(e^{j\omega}I-G)-\sum_{i=1}^{m}\varepsilon_iE_i]\neq0$$

where $\det[(e^{j\omega}I-G)^{-1}]\neq0$

$$\Leftrightarrow \det[e^{j\omega}I-G-\sum_{i=1}^{m}\varepsilon_iE_i]\neq0 \qquad \forall\omega\in[0,\pi]$$

It follows from Lemma 4.1 that $G+\sum_{i=1}^{m}\varepsilon_iE_i$ remains a strict Schur matrix for all ε_i (i=1,..., m) described by (2,3). The system (2,4') is asymptotically stable.

Remark 4.1 Taking M=I and $\sum_{i=1}^{m}E_i^+=E$, our sufficient condition (4,3) reduces to that of Juang et al's (1987b).

Example 4.1 This example is used to show the improvements of our sufficient condition (4,3) on Juang et al's (1987b) sufficient condition.

The system is described by (2,4'), where

$$G=\begin{bmatrix}0.95 & 0.15 & 0.85\\ 0.225 & 0.375 & 0.175\\ -0.575 & -0.125 & -0.425\end{bmatrix}.$$

with the eigenvalues $\lambda(G)=0.4,\ 0.25\pm0.0866j$; and

$$E_1=\begin{bmatrix}0 & 0 & 0\\ -0.75 & -0.25 & -1.25\\ 0.75 & 0.25 & 1.25\end{bmatrix} \qquad E_2=\begin{bmatrix}0 & 0 & 0\\ -1 & 0 & -1\\ 1 & 0 & 1\end{bmatrix}$$

(1) The result of the sufficient condition of Juang et al. (1987b). Taking

$$E=E_1^++E_2^+=\begin{bmatrix}0 & 0 & 0\\ 1.75 & 0.25 & 2.25\\ 1.75 & 0.25 & 2.25\end{bmatrix},$$

from Juang et al's (1987b) sufficient condition

$$\varepsilon<\frac{1}{\rho\{[(e^{j\omega}I-G)^{-1}]^+E\}} \qquad \forall\omega\in[0,\pi]$$

we get the stability robustness bound

$$\varepsilon<0.1668$$

(2) Our sufficient condition (4,3)

Taking M=I, from our sufficient condition (4,3)

$$\varepsilon<\frac{1}{\rho\{\sum_{i=1}^{m}[M^{-1}(e^{j\omega}I-A)^{-1}E_iM]^+\}} \qquad \forall\omega\in[0,\pi]$$

we get the stability robustness bound

$$\varepsilon<0.3$$

To reduce conservatism we choose

$$M=\begin{bmatrix}-1 & 1.75 & -1\\ 2 & -0.5 & -2\\ 1 & -0.75 & 1\end{bmatrix}$$

Then from our sufficient condition (4,3) we get the stability robustness bound

$$\varepsilon<0.38$$

Because $\lambda_{max}^+(G+\varepsilon E_1+\varepsilon E_2)<1$ is a necessary condition of stability, from $\lambda_{max}^+(G+0.38E_1+0.38E_2)=1$ we know that we have attained the necessary and sufficient condition of system stability.

5. Conclusion

In this paper we present new sufficient conditions for stability robustness of linear systems with time-invariant parametric uncertainties. Juang et al.'s (1987a,1987b) sufficient conditions are included as special cases of the proposed sufficient conditions. Illustrative examples are given to show that it is possible to attain the necessary and sufficient condition of the system stability by searching for the best similarity transformation matrix. However, how to find the best M matrix to get the largest stability robustness bound is still an important open problem.

References

[1] Juang, T.T., Kuo T.S. and Hsu, C.F. (1987a). New approach to time-domain analysis for stability robustness of dynamic systems. Int. J. Systems Sci., Vol. 18, pp1363-376.

[2] Juang, T.T., Kuo T.S. and Hsu, C.F. (1987b). Stability robustness analysis of digital control systems in state-space models. Int. J. Control, Vol. 46, pp1547-1556.

[3] Lancaster, P. (1969). Theory of matrices. Academic Press, New York.

[4] Matin, J.M. (1987). State-space measures for stability robustness. IEEE Trans. on Automatic Control, AC-32, pp509-512.

[5] Ortega, J.M. (1972). Numerical analysis. Academic Press, New York.

Proceedings of the 31st Conference
on Decision and Control
Tucson, Arizona • December 1992

Extensions of Mixed-μ Bounds to Monotonic and Odd Monotonic Nonlinearities Using Absolute Stability Theory: Part I

Wassim M. Haddad
Dept. of Mechanical and
Aerospace Engineering
Florida Institute of Technology
Melbourne, FL 32901

Jonathan P. How
Space Eng. Research Center
Dept. of Aero. and Astro.
Mass. Inst. of Technology
Cambridge, MA 02139

Steven R. Hall
Space Eng. Research Center
Dept. of Aero. and Astro.
Mass. Inst. of Technology
Cambridge, MA 02139

Dennis S. Bernstein
Dept. of Aerospace Engineering
The University of Michigan
Ann Arbor, MI 48109-2140

Abstract

In this paper we make explicit connections between classical absolute stability theory and modern mixed-μ analysis and synthesis. Specifically, using the parameter-dependent Lyapunov function of Haddad and Bernstein and the frequency dependent off-axis circle interpretation of How and Hall, we extend previous work on absolute stability theory for monotonic and odd monotonic nonlinearities to provide tight approximations for constant real parametric uncertainty. An immediate application of this framework is the generalization and re-interpretation of mixed-μ analysis and synthesis in terms of Lyapunov functions and Riccati equations. This observation is exploited to provide robust, reduced-order controller synthesis while avoiding the standard $D, N-K$ iteration.

1 Introduction

Many of the great landmarks of control theory are associated with the theory of absolute stability. The Aizerman conjecture and Lur'e problem as well as the circle and Popov criteria are extensively developed in the classical monographs by Aizerman and Gantmacher [1], Lefschetz [2], and Popov [3]. A more modern treatment is given in Safonov [4], while an excellent textbook treatment is presented in Vidyasagar [5]. The influence of absolute stability on the development of modern robust control is clearly evident from such works as Zames [6]. However, despite continued development of the theory as summarized in the important book by Narendra and Taylor [7], absolute stability has had lmited direct influence on the development of robust control theory. Since absolute stability theory concerns the stability of a system for classes of nonlinearities, which can readily be interpreted as an uncertainty model, Siljak [8] and Haddad and Bernstein [9], it is surprising that modern robust control did not take greater advantage of this wealth of knowledge. There appear to be (at least) three reasons for this state of affairs, namely, $\mathcal{H}_\infty$ theory, state space Lyapunov function theory, and linear uncertainty.

The development of $\mathcal{H}_\infty$ or bounded real theory as a key component of robust control theory focuses on small gain arguments for robustness guarantees. Although such conditions can be recast for sector-bounded uncertainty, such connections were rarely made [10]. Furthermore, the extensive development of state space Lyapunov function theory as in Leitmann [11], Khargonekar *et al.* [12], and Packard and Doyle [13], was seemingly remote from absolute stability theory, which involves frequency domain conditions, with an emphasis on graphical techniques. Finally, much of modern robust control theory is concerned with linear uncertainty, as distinct from the class of sector-bounded nonlinearities addressed by absolute stability theory.

Several recent developments now allow one to discern the relationship between the classical theory of absolute stability and the modern theory of robust control. First, the state space formulation of $\mathcal{H}_\infty$ or bounded real theory, as developed in Anderson and Vongpanitlerd [14], Petersen [15], and Doyle *et al.* [16], provides a better understanding of the time domain foundations of absolute stability theory. Next is the realization that absolute stability results such as the Popov criterion, when specialized to the linear uncertainty problem, are based on parameter-dependent Lyapunov functions, see Haddad and Bernstein [9, 17]. And, finally, is the development of upper bounds for mixed-μ theory due to Fan, Tits, and Doyle [18], that have recently been interpreted by How and Hall [19] in terms of frequency dependent off-axis circles, and thus connected directly to absolute stability theory.

*Research was funded in part by NSF Grant ECS-9109558, NASA Grant NAGW-2014, NASA SERC Grant NAGW-1335, and AFOSR Grant F49620-92-J-0127.

The purpose of the present paper is to make significant progress in understanding the relationship between classical absolute stability theory and modern μ-analysis and sythesis. Our goal is to generalize the results of How and Hall [19] by demonstrating that the frequency domain multipliers for the various classes of nonlinearities correspond to specific selections of the D, N-scales that arise in the mixed-μ problem. A key aspect of our development is the construction of parameter-dependent Lyapunov functions that support the mixed-μ results.

The principal limitation of the norm-based $\mathcal{H}_\infty$ theory resides in the fact that uncertainty phase information is discarded, so that constant real parametric plant uncertainty is captured as a nonparametric frequency-dependent uncertainty. In the time domain, nonparametric uncertainty is manifested as uncertain real parameters that may be time varying, and these can destabilize a system even when the parameter variations are confined to a region in which constant variations are nondestabilizing. Consequently, time-varying models of constant real parametric uncertainties are unnecessarily conservative. Thus, to address the constant real parameter uncertainty problem, it is crucial to restrict the allowable time-variation of the uncertainty. One approach is to construct refined Lyapunov functions that explicitly contain the uncertain parameters, an idea proposed in Ref. [9, 17] and developed in this paper using storage functions [19,20]. The form of the family of parameter-dependent Lyapunov functions $V(x, \Delta A) = x^T P(\Delta A)x$ is critical since the presence of ΔA restricts the allowable time-variation of the uncertain parameters, and thus exploits phase information. This approach is used in Ref. [9,17] to generalize the nonlinearity-dependent Lur'e-Postnikov Lyapunov function of the classical Popov criterion to a parameter-dependent Lyapunov function for constant real parameter uncertainty. Potentially less conservative tests for constant real parameter uncertainty can be obtained from similar generalizations of the Lyapunov functions for the slope restricted monotonic and odd monotonic nonlinearities. In this case, the linear uncertainty set will be a tight subset of the nonlinear set, and thus provide a framework to significantly reduce conservatism, as demonstrated by Safonov and Wyetizner [21].

In this paper, we extend the previous work on sector-bounded nonlinearities, restricting to the class of differentiable, slope bounded monotonic and odd monotonic memoryless nonlinearities. This class serves as a much better approximation to the case of constant real parametric uncertainty. We show in Section 5 that the choice of certain D, N-scales corresponds to the absolute stability criteria for the monotonic and odd monotonic nonlinearities due to Narendra, Cho, Neuman, Zames, and Thathachar in Refs. [22–26]. A direct benefit of these constructions is the new machinery for mixed-μ analysis and synthesis in terms of parameter-dependent Lyapunov functions and Riccati equations for full- and reduced-order compensator synthesis. Related optimality conditions arising from the chosen class of D, N-scales also play a role in the controller synthesis procedure. The overall framework thus provides an alternative approach to μ-synthesis, while avoiding the standard $D, N-K$ iteration.

2 Mathematical Preliminaries

In this section we establish definitions and notation. Let $\mathcal{R}$ and $\mathcal{C}$ denote the real and complex numbers, let $(\cdot)^{\mathrm{T}}$ and $(\cdot)^*$ denote transpose and complex conjugate transpose. Furthermore, we write $\|\cdot\|_2$ for the Euclidean norm, $\|\cdot\|_{\mathrm{F}}$ for the Frobenius matrix norm, $\sigma_{\max}(\cdot)$ for the maximum singular value, $\rho(\cdot)$ for the spectral radius, tr for the trace operator, and $M \geq 0$ ($M > 0$) to denote the fact that the Hermitian matrix M is nonnegative (positive) definite. In this paper a *real-rational matrix function* is a matrix whose elements are rational functions with real coefficients. Furthermore, a *transfer function* is a

CH3229-2/92/0000-2813$1.00 © 1992 IEEE

real-rational matrix function each of whose elements is *proper*, i.e., finite at $s = \infty$. A *strictly proper transfer function* is a transfer function that is zero at infinity. Finally, an *asymptotically stable transfer function* is a transfer function each of whose poles is in the open left half plane. The space of asymptotically stable transfer functions is denoted by $\mathcal{RH}_\infty$, i.e., the real-rational subset of $\mathcal{H}_\infty$. Let

$$G(s) \sim \left[\begin{array}{c|c} A & B \\ \hline C & D \end{array}\right] \tag{1}$$

denote a state space realization of a transfer function $G(s)$, that is, $G(s) = C(sI - A)^{-1}B + D$. The notation "$\overset{\min}{\sim}$" is used to denote a minimal realization. The $\mathcal{H}_2$ and $\mathcal{H}_\infty$ norms for a stable transfer function $G(s)$ are defined as

$$\|G(s)\|_2^2 \triangleq \tfrac{1}{2\pi}\textstyle\int_{-\infty}^{\infty} \|G(\jmath\omega)\|_F^2\, d\omega, \; \|G(s)\|_\infty \triangleq \sup_{\omega\in\mathcal{R}} \sigma_{\max}\left[G(\jmath\omega)\right].$$

A square transfer function $G(s)$ is *positive real* (Ref. [14], p. 216) if 1) all poles of $G(s)$ are in the closed left half plane, and 2) $G(s) + G^*(s)$ is nonnegative definite for $\mathrm{Re}[s] > 0$. A square transfer function $G(s)$ is *strictly positive real* [27,28] if 1) $G(s)$ is asymptotically stable, and 2) $G(\jmath\omega) + G^*(\jmath\omega)$ is positive definite for all real ω. Finally, a square transfer function $G(s)$ is *strongly positive real* if it is strictly positive real and $D + D^T > 0$, where $D \triangleq G(\infty)$. Recall that a minimal realization of a positive real transfer function is stable in the sense of Lyapunov [29], while a strictly positive real transfer function is asymptotically stable [27].

For notational convenience in the paper, G will denote an $\ell \times m$ transfer function with input $u \in \mathcal{R}^m$, output $y \in \mathcal{R}^\ell$, and internal state $x \in \mathcal{R}^n$. We will omit all matrix dimensions throughout, and assume that all quantities have compatible dimensions.

3 Supply rates, storage functions, and stability

Several definitions are necessary to develop the appropriate tools for the analysis framework. Consider a dynamical system $\mathcal{G}$ of the form

$$\dot{x}(t) = Ax(t) + Bu(t), \tag{2}$$
$$y(t) = g(x(t)), \tag{3}$$

where $u(t) \in \mathcal{R}^m$, $y(t) \in \mathcal{R}^l$, and $x(t) \in \mathcal{R}^n$. In the special case that the output function is linear, $g(x) = Cx$, then $\mathcal{G} = G(s)$ is an LTI system with realization of the form in Eq. 1 with $D = 0$.

For the dynamical system $\mathcal{G}$ of Eqs. 2 and 3, a function $r : \mathcal{R}^l \times \mathcal{R}^m \to \mathcal{R}$, is called a *supply rate* if it is locally integrable, so that $\int_{t_1}^{t_2} |r(y(\xi), u(\xi))|\, d\xi < \infty$ for all $t_1, t_2 > 0$, and if $\oint r(y(\xi), u(\xi))\, d\xi \geq 0$ for every path that takes the dynamical system from some initial state to the same final state. A more general form for the supply rate presented by Pinzoni and Willems [30] will be used in this paper. Under the new definition, the supply rate can be a function of the signals (u, y) and, if they exist, their time derivatives.

Definition 1. (Willems [20]). *A system $\mathcal{G}$ of the form in Eqs. 2 and 3, with states $x \in \mathcal{R}^n$ is said to be* dissipative with respect to a supply rate $r(\cdot,\cdot)$ *if there exists a nonnegative-definite function $V_s : \mathcal{R}^n \to \mathcal{R}$, called a* storage function, *that satisfies the dissipation inequality*

$$V_s(x(t_2)) \leq V_s(x(t_1)) + \int_{t_1}^{t_2} r(y(\xi), u(\xi))\, d\xi, \tag{4}$$

for all t_1, t_2 and for all $x(\cdot), y(\cdot)$, and $u(\cdot)$ satisfying Eqs. 2 and 3.

If $V_s(x)$ is a differentiable function, then an equivalent statement of dissipativeness of the system $\mathcal{G}$ with respect to the supply rate r is that

$$\dot{V}_s(x(t)) \leq r(y(t), u(t)), \qquad t \geq 0, \tag{5}$$

where $\dot{V}$ denotes the total derivative of $V(x)$ along the state trajectory $x(t)$ [20]. For a *strongly dissipative system*, Eq. 5 is replaced by the condition $\dot{V}_s(x(t)) < r(y(t), u(t))$ with a similar modification to Eq. 4. For the particular example of a mechanical system with force inputs and velocity outputs we can associate the storage function with the stored or available "energy" in the system, and the supply rate with the net flow of "energy" into the system. However, the concepts of the supply rates and storage functions also apply to more general systems for which this energy interpretation is no longer valid.

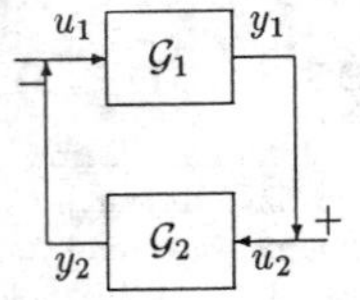

Fig. 1: Interconnected systems.

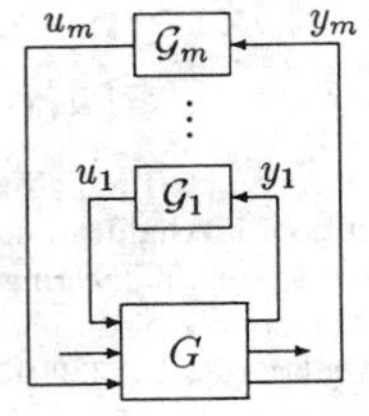

Fig. 2: Multiple uncertainties.

A variety of supply rates have been considered by Willems Refs. [20] and Hill and Moylan Refs. [31, 32]. An appropriate supply rate for testing the passivity of a system $y = G(s)u$ is $r(y, u) = u^T y$. This choice can be motivated by the example of a system with a force input and velocity output, so that the product $u^T y$ is a measure of power. For bounded gain tests, the supply rate is $r(y, u) = u^T u - \gamma^2 y^T y$. A motivation for this choice follows from the identity

$$\int_0^T r(y, u)\, dt = \int_0^T u^T u\, dt - \gamma^2 \int_0^T y^T y\, dt \tag{6}$$

which consists of two parts, the first associated with the energy at the system input, and the second with the weighted energy at the system output. If the integral on the left hand side of Eq. 6 is positive, indicating that the weighted output energy is less than the input energy at any time T, then the system is gain bounded since $\|G\|_\infty < \gamma^{-1}$.

As will now be shown, storage functions and supply rates provide a means for developing Lyapunov functions for determining stability of coupled feedback systems. In particular, if, for each subsystem, there exists a storage function that is dissipative with respect to appropriate supply rates, then these functions can be combined to form a Lyapunov function for the interconnected system. A more precise statement of this result for two systems interconnected as in Fig. 1 is provided by the following lemma.

Lemma 1. *Consider two dynamical systems $\mathcal{G}_1$ and $\mathcal{G}_2$ with state space representation as in Eqs. 2 and 3, and input-output pairs (u_1, y_1) and (u_2, y_2) respectively. Assume that the two systems are* neutrally connected *as illustrated in Fig. 1, so that $u_1 = -y_2$ and $u_2 = y_1$. Furthermore, associated with these systems are states x_1, x_2, supply rates $r_1(y_1, u_1)$, $r_2(y_2, u_2)$ and storage functions $V_{s1}(x_1)$, $V_{s2}(x_2)$ respectively. Suppose that at least one of $V_{s1}(x_1)$ and $V_{s2}(x_2)$ is positive definite, and that the supply rates satisfy $r_1(y_1, u_1) + r_2(y_2, u_2) = 0$, for all $u_1 = -y_2$ and $u_2 = y_1$. Then the solution $(x_1, x_2) = 0$ of the feedback interconnection of $\mathcal{G}_1$ and $\mathcal{G}_2$ is Lyapunov stable with Lyapunov function $V = V_{s1} + V_{s2}$.*

Proof. Since $V = V_{s1} + V_{s2}$ is at least the sum of a positive definite and a nonnegative-definite function, it is positive definite. Furthermore, $\dot{V}(x_1, x_2) = \dot{V}_{s1}(x_1) + \dot{V}_{s2}(x_2) \leq r_1(y_1, u_1) + r_2(y_2, u_2) = 0$. From the positive definiteness of V and the negative semidefiniteness of $\dot{V}$, it follows that V is a Lyapunov function that guarantees the Lyapunov stability of the combined system at $(x_1, x_2) = 0$. □

Next, we extend the results of Lemma 1 to the case of importance for this paper, where a single LTI system $G(s)$ is independently interconnected to m systems, as illustrated in Fig. 2. The systems in this example are special cases of the ones in Lemma 1 since one system is LTI and the other is block diagonal.

Corollary 1. *Consider an LTI system $G(s)$ with inputs u_i, outputs y_i, $i = 1, \ldots, m$ and states x. Introduce the dynamical systems $\mathcal{G}_i$*

$$\dot{x}_i(t) = A_i x_i(t) + B_i u_i(t), \tag{7}$$
$$y_i(t) = g_i(x_i(t)), \tag{8}$$

with supply rates $r_i(u_i(t), y_i(t))$ and storage functions $V_{si}(x_i(t))$. Define an overall supply rate $R(y_1(t), \ldots, y_m(t), u_1(t), \ldots, u_m(t)) = \sum_{i=1}^m \gamma_i r_i(u_i(t), y_i(t))$, *with $\gamma_i > 0$. If there exists a positive definite storage function $V_G(x)$ for the system $G(s)$ which is dissipative with respect to the negative of the overall supply rate, then the interconnected system is Lyapunov stable.*

Proof. The proof follows directly from Lemma 1. □

The results of Corollary 1 convert the problem of determining the stability of interconnected systems to that of determining supply rates and storage functions, and then testing for dissipativeness of the independent systems with respect to the supply rates. For the problem addressed here, where insights into determining the supply rates are available from the characteristics of the nonlinearities, this greatly simplifies the construction of the Lyapunov functions. Corollary 1 also allows us to incorporate both complex and real uncertainties by mixing the supply rates for the different dynamical systems. Note that, if at least one of the systems has a positive definite storage function that is strongly dissipative with respect to the corresponding supply rate, then the Lyapunov function is positive definite and $\dot{V}$ is negative definite, and then the combined system is asymptotically stable.

In the next section, we develop a framework in which different classes of nonlinear functions are incorporated into the stability criteria. The absolute stability criteria that are developed are then related to the robust stability and performance problem with real parametric uncertainty.

4 Stability robustness for refined nonlinearities

While graphical tests have been developed for SISO feedback systems with a single nonlinearity in the absolute stability literature, the goal in this paper is to employ a consistent framework from which one can develop both state space and frequency domain robust stability criteria for multivariable systems. The approach will use the concepts of supply rates, storage functions, and the dissipation inequality defined in Section 3. How and Hall [19] developed this framework from the original work of Willems [20], and applied it to multivariable nonlinearities with general sector constraints, extending the results of Popov. Haddad and Bernstein [9,17] have investigated the multivariable Popov criterion for robust stability and $\mathcal{H}_2$ performance for both linear uncertainties and time-invariant nonlinearities. The following analysis extends these results to encompass both monotonic and odd monotonic restrictions on the nonlinear functions.

The basis of the stability analysis tests is illustrated in Fig. 3, where $G(s)$ is an LTI system with realization $G(s) \sim \left[\begin{array}{c|c} A & B_0 \\ \hline C_0 & 0 \end{array}\right]$ and $f(\cdot)$ is a nonlinear function. As discussed in the Introduction, this nonlinearity is used to model the uncertainty in the system $G(s)$. The transfer function $W(s)$ is an appropriate frequency domain *stability multiplier* [6] which is selected based on the known properties of the memoryless nonlinearity $f(\cdot)$, such as gain or slope bounds, and its purpose is to modify the region of instability for the system, as discussed by Zames [6]. The system from $\hat{y}$ to u through $W^{-1}(s)$ and $f(\cdot)$ can be written as a dynamical system of the form in Eqs. 2 and 3. Furthermore, for the independently coupled case discussed previously, $W(s)$ will be a diagonal matrix, and $f(\cdot)$ is a component decoupled nonlinearity.

The process for determining the *absolute stability* of a system, stability for an entire class of nonlinear functions, can be broken down into several steps. For the particular case of interest, with m nonlinear functions independently interconnected to the LTI system $G(s)$, it follows from Corollary 1, that the storage functions and supply rates can be developed separately. For each input-output pair (u_i, y_i) of the system $G(s)$, a storage function V_{si} based on the states of the dynamics of the multiplier $W_i(s)$ must be shown to be dissipative with respect to the supply rate $r_i(u_i, y_i)$. Then, as in Corollary 1, the condition that the storage function for the linear system $G(s)$ be dissipative with respect to a supply rate that is the negative sum of the supply rates for the nonlinear systems leads to a condition for stability of the interconnected system. This criteria for stability can be interpreted as requiring a positive definite solution of an algebraic Riccati equation. Furthermore, combining the storage functions for the linear and nonlinear systems provides a parameter-dependent Lyapunov function (see Ref. [17]) for the interconnected system.

Since we consider constant, linear, real parameter uncertainties in the system $G(s)$, the class of allowable nonlinearities must be restricted to provide good approximations of these errors. While the Popov multiplier corresponds to sector bounded time-invariant nonlinearities, several authors have discussed the appropriate multipliers for monotonic and odd monotonic restrictions on the nonlinear functions. Narendra and Taylor [7], Brockett and Willems [33], Narendra and

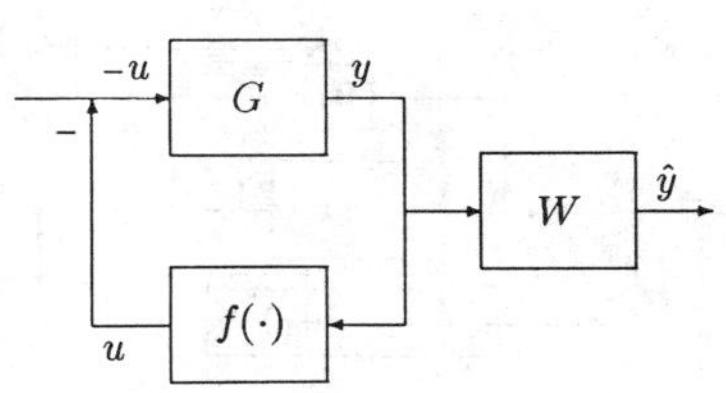

Fig. 3: Framework for stability tests with linear systems coupled to nonlinearities.

Cho [23], Zames and Falb [26], and Thathachar and Srinath [24,34], have developed suitable stability multipliers $W(s)$ for monotonic and odd monotonic nonlinear functions. These are given by the functions in the sets $\mathcal{W}_{RL}$ and $\mathcal{W}_{RC}$, which exhibit an interlacing pole-zero pattern on the negative real axis. The two are distinguished by which is closest to the origin, a pole ($\mathcal{W}_{RC}$) or a zero ($\mathcal{W}_{RL}$) [35]. The standard form of the multiplier for each $i = 1, \ldots, m$ is

$$W_i(s) = \alpha_{i0} + \beta_{i0}s + \sum_{j=1}^{m_{i1}} \alpha_{ij}\left(1 - \frac{\alpha_{ij}}{\beta_{ij}(s+\eta_{ij})}\right) + \sum_{j=m_{i1}+1}^{m_{i2}} \alpha_{ij}\left(1 + \frac{\alpha_{ij}}{\beta_{ij}(s+\eta_{ij})}\right), \quad (9)$$

where the coefficients α_{ij}, β_{ij}, and η_{ij} are nonnegative and satisfy $\eta_{ij}\beta_{ij} - \alpha_{ij} \geq 0$. To consider just monotonic nonlinearities, in Eq. 9, take $m_{i2} = m_{i1}$. For odd monotonic nonlinearities, it is also possible to include multipliers with terms that explicitly contain complex poles and zeroes. While the extra freedom associated with this extension will be discussed later, with the three components in Eq. 9, one can develop very general forms of the multiplier $W_i(s)$.

As discussed by How and Hall [19] and Haddad and Bernstein [9], the multiplier phase plays a crucial role in determining the conservativeness of the analysis test. The first two terms of Eq. 9 correspond to the standard Popov multiplier, and the phase angle of this multiplier increases monotonically from 0° and 90°. The first sum in Eq. 9 is a partial fraction expansion of a driving point impedance of a resistor-inductor (RL) network. While the phase for this class also lies between 0° to 90°, it is not a monotonically increasing function of frequency. The last summation in Eq. 9 is of the form of a driving point impedance of a resistor-capacitor (RC) network, with a pole closest to the origin, and phase between 0° and −90°.

As illustrated in Fig. 3, proving stability of the coupled system requires handling signals of the form $W(s)y$. While obtaining filtered outputs of this form is simple for the Popov multiplier, it is quite complicated for the multipliers in Eq. 9. In particular, with these extended multipliers it is necessary to augment the multiplier dynamics to the original system so that the filtered outputs to be defined later can be obtained directly from the augmented state vector. The resulting augmented matrix A_a then contains the poles of both the system $G(s)$ and the multipliers $W_i(s)$, $i = 1, \ldots, m$.

While much of the absolute stability theory has been developed for infinite sector or slope restrictions on the nonlinearity, the shifting approach discussed by Rekasius and Gibson [36] and Desoer and Vidyasagar [37] can be used to handle finite bounds. Define M_1, $M_2 \in \mathcal{R}^{m\times m}$ as diagonal matrices whose nonzero elements represent the upper and lower sector bounds for each input-output loop. The transformations illustrated in Fig. 4 convert the general slope restrictions (M_1, M_2) to a one-sided condition $(0, M_2 - M_1)$, and then finally to an infinite one $(0, \infty)$. For now we consider only the bounds $(0, M_2)$, and a later remark will consider the more general case. The following section outlines the process for shifting these sectors and augmenting the multiplier dynamics. Sections 4.2 and 4.3 then present the tests for stability for the monotonic and odd monotonic nonlinearities.

4.1 Multiplier Augmentation

We begin with a discussion of the transformations illustrated in Fig 4. In the following, take $M_1 = 0$ and $M_2 = M = \text{diag}(M_{11}, \ldots, M_{mm})$, and consider differentiable monotonic and odd monotonic nonlinear functions that satisfy the constraint that $\frac{df_i(\sigma)}{d\sigma} < M_{ii}$ for all values of σ. Note that this implies that $f_i(\sigma)$ satisfies the sector constraint $0 \leq \sigma f_i(\sigma) < M_{ii}\sigma^2$. From the figure, with $M_1 = 0$, observe

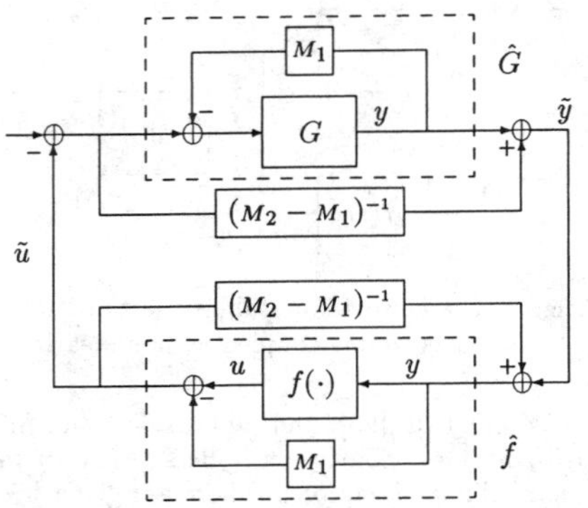

Fig. 4: System transformations and the definitions of $\hat{G}$, $\hat{f}(\cdot)$, $\tilde{u}$, and $\tilde{y}$. If the lower slope bound $M_1 = 0$, then $\hat{f}(\cdot) = f(\cdot)$, $\hat{G}(s) = G(s)$, and $\tilde{u} = u$.

that $f(y) = \tilde{f}(\tilde{y})$, $\tilde{f}(0) = 0$, and

$$\tilde{y} = y - M^{-1}\tilde{f}(\tilde{y}). \tag{10}$$

Desoer and Vidyasagar [37] present an excellent discussion of the existence and uniqueness of the solution of this equation. While these issues dominate the discussion for arbitrary nonlinearities, for sector bounded monotonic nonlinearities these properties are automatically satisfied [34]. Furthermore, for each nonlinearity, with $y_i \neq 0$

$$\frac{\tilde{f}_i(\tilde{y}_i)}{\tilde{y}_i} = \frac{f_i(y_i)/y_i}{1 - M_{ii}^{-1} f_i(y_i)/y_i}, \tag{11}$$

so that if $f_i(\cdot)$ is sector bounded by M_{ii}, then the equivalent condition for the shifted nonlinearity is $\tilde{y}_i \tilde{f}_i(\tilde{y}_i) \geq 0$. Also, by the chain rule,

$$\frac{\mathrm{d}\tilde{f}_i(\tilde{y}_i)}{\mathrm{d}\tilde{y}_i} = \frac{\mathrm{d}f_i(y_i)/\mathrm{d}y_i}{1 - M_{ii}^{-1}\mathrm{d}f_i(y_i)/\mathrm{d}y_i}, \tag{12}$$

so that if $f_i(\cdot)$ is differentiable and satisfies the slope restrictions $0 \leq \frac{\mathrm{d}f_i(\sigma)}{\mathrm{d}\sigma} < M_{ii}$, then $\tilde{f}_i(\cdot)$ is differentiable and satisfies $0 \leq \frac{\mathrm{d}\tilde{f}_i(\sigma)}{\mathrm{d}\sigma}$, and is thus monotonic. The same process holds for slope restricted odd monotonic nonlinearities.

The corresponding changes to the LTI system are also illustrated in Fig. 4. In particular, for $M_1 = 0$, the shifted system is given by

$$\tilde{G}(s) = G(s) + M^{-1}, \tag{13}$$

with inputs $-u$, and outputs $\tilde{y}$. Each transformed nonlinearity $\tilde{f}_i(\cdot)$ is restricted to lie in the first and third quadrants, so

$$\sigma \tilde{f}_i(\sigma) \geq 0, \qquad \sigma \in \mathcal{R}. \tag{14}$$

Furthermore, since the transformed nonlinearities are monotonic, they satisfy

$$0 \leq (\tilde{f}_i(\sigma_1) - \tilde{f}_i(\sigma_2))(\sigma_1 - \sigma_2), \ \sigma_1, \sigma_2 \in \mathcal{R}. \tag{15}$$

Having discussed the transformations, we can now proceed with the multiplier augmentation. As illustrated in Fig. 3 (with $G(s)$, $f(\cdot)$ replaced by $\tilde{G}(s)$, $\tilde{f}(\cdot)$), the stability tests are performed on the systems formed by combining the multiplier with the system $\tilde{G}(s)$ and its inverse with the nonlinearity $\tilde{f}(\cdot)$. The supply rate for these systems will then be a function of the new output (equivalent to $\hat{y}$ in Fig. 3), which is obtained by applying the appropriate multiplier to each element of the output of $\tilde{G}(s)$. Observing the form of the multiplier in Eq. 9, it can be seen that in the expression $W_i(s)\tilde{y}_i$, we obtain terms of the form

$$z_{ij} = \frac{\alpha_{ij}\tilde{y}_i}{\beta_{ij}(s + \eta_{ij})}. \tag{16}$$

Hence, the corresponding supply rate will involve real signals z_{ij} obtained by passing the system output $\tilde{y}_i$ through a parallel bank of decoupled low pass filters with time constants $1/\eta_{ij}$ and positive gains $\alpha_{ij}/\beta_{ij}\eta_{ij}$. For the system as in Eqs. 2 and 3, formed from $W^{-1}(s)$ and $\tilde{f}(\cdot)$, with z_{ij} as the system states, the dynamics of each term in the multiplier can be augmented to the system by rewriting Eq. 16 as

$$\dot{z}_{ij} + \eta_{ij} z_{ij} = \frac{\alpha_{ij}}{\beta_{ij}}\tilde{y}_i = \frac{\alpha_{ij}}{\beta_{ij}}\left(C_0 x - M^{-1}u\right)_i, \tag{17}$$

where $(\cdot)_i$ denotes to the ith row of $(\cdot)$. Writing the states in a vector $z_i^T \triangleq [z_{i1}, \ldots, z_{im_{i2}}]$, the dynamics associated with each multiplier $W_i(s)$ can be written as

$$\dot{z}_i = \begin{bmatrix} \hat{C}_i & A_i \end{bmatrix} \begin{bmatrix} x \\ z_i \end{bmatrix} - \hat{M}_i u, \tag{18}$$

where $A_i \triangleq \mathrm{diag}(-\eta_{ij})$, $j = 1, \ldots, m_{i2}$, and

$$\hat{C}_i \triangleq \begin{bmatrix} \frac{\alpha_{i1}}{\beta_{i1}} \\ \frac{\alpha_{i2}}{\beta_{i2}} \\ \vdots \\ \frac{\alpha_{im_{i2}}}{\beta_{im_{i2}}} \end{bmatrix} (C_0)_i, \quad \hat{M}_i \triangleq \begin{bmatrix} \frac{\alpha_{i1}}{\beta_{i1}} \\ \frac{\alpha_{i2}}{\beta_{i2}} \\ \vdots \\ \frac{\alpha_{im_{i2}}}{\beta_{im_{i2}}} \end{bmatrix} (M^{-1})_i, \tag{19}$$

where $(C_0)_i$ and $(M^{-1})_i$ denote the ith rows of the respective matrices.

With m input-output pairs to the system $G(s)$, we augment the multiplier dynamics to the shifted system $\tilde{G}(s)$ to obtain a state space representation of $\tilde{G}_a(s)$ given by

$$\dot{x}_a = A_a x_a - B_a u, \ \tilde{y} = C_a x_a - M^{-1}u, \tag{20}$$

where, $x_a \in \mathcal{R}^{n_a}$, $n_a \triangleq n + \sum_{i=1}^{m} m_{i2}$, A_a, B_a, and C_a are defined as

$$x_a = \begin{bmatrix} x \\ z_1 \\ z_2 \\ \vdots \\ z_m \end{bmatrix}, \ A_a = \begin{bmatrix} A & 0 & 0 & \cdots & 0 \\ \hat{C}_1 & A_1 & 0 & & 0 \\ \hat{C}_2 & 0 & A_2 & & 0 \\ \vdots & & & \ddots & \\ \hat{C}_m & 0 & 0 & & A_m \end{bmatrix}, \ B_a = \begin{bmatrix} B_0 \\ \hat{M}_1 \\ \hat{M}_2 \\ \vdots \\ \hat{M}_m \end{bmatrix}, \tag{21}$$
$$C_a = \begin{bmatrix} C_0 & 0 & 0 & \ldots & 0 \end{bmatrix}.$$

Next, define R_{ij} as an output matrix for this augmented system, designed to access the jth element of z_i so that $z_{ij} = R_{ij}x_a$. Then, the only nonzero element of R_{ij} is the $(\sum_{l=1}^{i-1} m_{l2} + j)^{\text{th}}$ term, which is 1.

Note that, although extra dynamics associated with the multipliers have been added to the system $\tilde{G}_a(s)$, since A_a is lower block triangular, it can easily be shown that

$$\tilde{G}_a(s) = C_a(sI - A_a)^{-1}B_a + M^{-1} = C_0(sI - A)^{-1}B_0 + M^{-1} = \tilde{G}(s). \tag{22}$$

Hence, by pole-zero cancellation in each input-output loop, the frequency domain representations of $\tilde{G}(s)$ and $\tilde{G}_a(s)$ are equivalent in terms of their input-output properties.

From Corollary 1, it is known that in the development of the storage functions and supply rates, we can consider each input-output pair of $G(s)$ independently. Hence, in the following, without loss of generality, we consider the development of supply rates and storage functions for the separate single-input single-output nonlinearities coupled with the appropriate multiplier. Since the goal is to demonstrate that the combination of the nonlinear function and the multiplier, as in Fig. 3, is passive, it follows from Section 3 that the appropriate supply rate is the product of the system inputs and outputs. However, a modification of this supply rate is required for the multipliers in Eq. 9 if $M \neq 0$ [19,38]. In this case, with s denoting the standard Laplace variable and $\hat{y}_i = W_i(s)\tilde{y}_i$, we consider signals of the form

$$\overline{y}_i = W_i(s)\tilde{y}_i + \beta_{i0}M_{ii}^{-1}su_i. \tag{23}$$

It will be seen that this additional term is equivalent to the quadratic term added to the Lur'e-Postnikov Lyapunov function to account for a direct transmission term in the plant dynamics [7]. As will be seen in the following development, the term su_i is used to cancel an equivalent term from the Popov multiplier in the expression $\hat{y}_i = W_i(s)\tilde{y}_i$. The assumed differentiability of the shifted nonlinearities guarantees that the expression in Eq. 23 exists.

4.2 Monotonic Nonlinear Functions

In this section we develop the supply rate and appropriate storage functions for monotonic nonlinearities ($m_{i2} = m_{i1}$), and present robust stability conditions for the full system via algebraic Riccati equations. Using the definitions of $W_i(s)$ in Eq. 9, $\tilde{y}$ in Eq. 10, and the filtered outputs in Eq. 16, the signal in Eq. 25 becomes

$$\begin{aligned}\overline{y}_i &= \sum_{j=1}^{m_{i1}} \alpha_{ij}[1 - \frac{\alpha_{ij}}{\beta_{ij}(s+\eta_{ij})}]\tilde{y}_i + (\alpha_{i0} + \beta_{i0}s)(y - M^{-1}u)_i + \beta_{i0}M_{ii}^{-1}su_i \\ &= \sum_{j=1}^{m_{i1}} \alpha_{ij}(\tilde{y}_i - z_{ij}) + (\alpha_{i0} + \beta_{i0}s)y_i - \alpha_{i0}M_{ii}^{-1}u_i \\ &= \sum_{j=1}^{m_{i1}} \alpha_{ij}(\tilde{y}_i - z_{ij}) + \beta_{i0}sy_i + \alpha_{i0}\tilde{y}_i. \end{aligned} \tag{24}$$

We then construct the supply rate $r_i(\hat{y}_i, u_i)$ in terms of the time domain representation of $\overline{y}_i$ and u_i, which yields

$$r_i(\hat{y}_i, u_i) = \left[\sum_{j=1}^{m_{i1}} \alpha_{ij}(\tilde{y}_i - z_{ij}) + \beta_{i0}\dot{y}_i + \alpha_{i0}\tilde{y}_i\right] u_i. \tag{25}$$

An appropriate storage function for this supply rate is given in the following lemma.

Lemma 2. *Consider a differentiable monotonic nonlinear function $f_i(\cdot)$ that satisfies the slope restrictions $0 \leq \frac{\mathrm{d}f_i(\sigma)}{\mathrm{d}\sigma} < M_{ii}$. Define the differentiable monotonic nonlinearity $\tilde{f}_i(\cdot)$ that satisfies Eqs. 14 and 15 as in Eq. 11. Consider the dynamical system $\mathcal{G}_i$ which is a combination of $\tilde{f}_i(\cdot)$ and the multiplier $W_i^{-1}(s)$ from Eq. 9 ($m_{i2} = m_{i1}$), with corresponding state space representation given by Eqs. 2 and 3. Then, using the definition of z_i, the function V_{si} defined as*

$$V_{si}(\tilde{y}_i, z_i) = \beta_{i0}\left(\int_0^{\tilde{y}_i} \tilde{f}_i(\sigma)\,d\sigma + \frac{1}{2}M_{ii}^{-1}u_i^2\right) + \sum_{j=1}^{m_{i1}} \beta_{ij}\int_0^{z_{ij}} \tilde{f}_i(\sigma)\,d\sigma, \tag{26}$$

is a storage function for the supply rate in Eq. 25.

Proof. See [39]. □

While it is convenient to independently consider the supply rates and storage functions for the nonlinear functions, these must be combined to form the supply rate for the linear system $\tilde{G}(s)$ and the full multiplier $W(s)$. Vector notation will simplify this development, but we must first define the following matrices. These definitions are complicated by the fact that each $W_i(s)$ can have a different number of expansion terms in Eq. 9. This difficulty can be handled by defining extended values of α_{ij}, β_{ij}, and η_{ij}. Let $m_1 = \max_i(m_{i1})$. Then, for each $i = 1, \ldots, m$, and $j = 1, \ldots, m_1$, let $\alpha_{ij} = 0$, $\beta_{ij} = 0$, $\eta_{ij} = 0$, and $R_{ij} = 0$ if $j > m_{i1}$. Furthermore, define $H_j = \mathrm{diag}(\alpha_{1j}, \ldots, \alpha_{mj})$, $N_j = \mathrm{diag}(\beta_{j1}, \ldots, \beta_{jm})$, $S_j = \mathrm{diag}(\eta_{j1}, \ldots, \eta_{jm})$, $j = 0, 1, \ldots, m_1$. Finally, let $R_j = \left[R_{1j}^T, R_{2j}^T, \ldots, R_{mj}^T\right]^T$. Then as in Corollary 1, using Eq. 25 to form the overall supply rate for the LTI system $G(s)$ yields

$$R(\hat{y}, u) = \sum_{i=1}^{m} r_i(\hat{y}_i, u_i) \tag{27}$$

$$= [\, u_1 \cdots u_m \,] \begin{bmatrix} \sum_{j=1}^{m_{11}} \alpha_{1j}(\tilde{y}_1 - z_{1j}) + \beta_{10}\dot{y}_1 + \alpha_{10}\tilde{y}_1 \\ \vdots \\ \sum_{j=1}^{m_{m1}} \alpha_{mj}(\tilde{y}_m - z_{mj}) + \beta_{m0}\dot{y}_m + \alpha_{m0}\tilde{y}_m \end{bmatrix} \tag{28}$$

$$= u^T \left[\sum_{j=1}^{m_1} H_j(\tilde{y} - \hat{z}_j) + N_0\dot{y} + H_0\tilde{y}\right], \tag{29}$$

where $\hat{z}_j = [\, z_{1j} \cdots z_{mj} \,]^T$. This representation of $R(\cdot,\cdot)$ can be simplified further by using the definition of R_j to note that for each j, $\hat{z}_j \triangleq R_j x_a$.

Furthermore, using Eq. 10 for $\tilde{y}$, and noting that $\dot{y} = C_a\dot{x}_a = C_a(A_a x_a - B_a u)$, the overall supply rate can then be written as

$$R(\hat{y}, u) = u^T\hat{C}x_a - u^T(N_0C_aB_a + \sum_{j=0}^{m_1} H_jM^{-1})u. \tag{30}$$

where

$$\hat{C} \triangleq H_0C_a + N_0C_aA_a + \sum_{j=1}^{m_1} H_j(C_a - R_j). \tag{31}$$

Having developed the overall supply rate, we now present the stability condition for the full system.

Theorem 1. *Consider an LTI system $G(s)$ independently coupled to m differentiable monotonic nonlinearities that satisfy the slope restrictions $0 \leq \frac{\mathrm{d}f_i(\sigma)}{\mathrm{d}\sigma} < M_{ii}$. If for each input-output pair (u_i, y_i), there exist multipliers $W_i(s)$ as in Eq. 9 and a matrix $R = R^T > 0 \in \mathcal{R}^{n_a \times n_a}$ such that with the preceding definitions of H_j, N_j, and R_j,*

(i) $\hat{R}_0 \triangleq [(N_0C_aB_a + \sum_{j=0}^{m_1} H_jM^{-1}) + (N_0C_aB_a + \sum_{j=0}^{m_1} H_jM^{-1})^T] > 0,$

(ii) *there exists a symmetric matrix $P > 0$, satisfying*

$$A_a^TP + PA_a + R + [\hat{C} - B_a^TP]^T\hat{R}_0^{-1}[\hat{C} - B_a^TP] = 0 \tag{32}$$

then the negative feedback interconnection of the system $G(s)$ and the nonlinearities is asymptotically stable. Furthermore, a Lyapunov function for the combined system is given by

$$V(x_a) = x_a^TPx_a + 2\sum_{i=1}^{m}\left[\beta_{i0}\left(\int_0^{\tilde{y}_i} \tilde{f}_i(\sigma)\,d\sigma + \frac{1}{2}M_{ii}^{-1}u_i^2\right) + \sum_{j=1}^{m_{i1}} \beta_{ij}\int_0^{z_{ij}} \tilde{f}_i(\sigma)\,d\sigma\right] \tag{33}$$

Proof. See [19,39]. □

Remark 1. $V(x_a)$ in Eq. 33 is an extended Lur'e-Postnikov function since it depends explicitly on the nonlinearity $\tilde{f}_i(\cdot)$. Similarly, in the linear case, where $\tilde{f}_i(\tilde{y}_i) = F_i\tilde{y}_i$, $V(x_a)$ is then a parameter-dependent Lyapunov function. The uncertain parameters are then not allowed to be arbitrarily time-varying, resulting in a refined framework for constant real parametric uncertainty. □

4.3 Odd Monotonic Nonlinear Functions

We now consider nonlinearities with odd monotonic restrictions. The procedure is identical to the one discussed in the previous section, the main difference now being that the transformed nonlinear function $\tilde{f}(\cdot)$ satisfies Eqs. 14, 15, and an additional constraint (see Ref. [7])

$$0 \leq \sigma_1\tilde{f}(\sigma_1) + \sigma_2\tilde{f}(\sigma_2) + \sigma_1\tilde{f}(\sigma_2) - \sigma_2\tilde{f}(\sigma_1), \quad \sigma_1, \sigma_2 \in \mathcal{R}. \tag{34}$$

The definition of the supply rate and multiplier augmentation process are as discussed in Sections 4.1 and 4.2. In this case, we consider $m_{i2} > m_{i1}$ in $W_i(s)$ of Eq. 9.

Using the simplification of $r_i(\hat{y}_i, u_i)$ in Eq. 25, the definition of z_{ij} in Eq. 16, and noting the form of the multiplier terms in $W_i(s)$ for $j = m_{i1}+1, \ldots, m_{i2}$, the supply rate can be rewritten as

$$r_i(\hat{y}_i, u_i) = \left[\sum_{j=1}^{m_{i1}} \alpha_{ij}(\tilde{y}_i - z_{ij}) + \sum_{j=m_1+1}^{m_2} \alpha_{ij}(\tilde{y}_i + z_{ij}) + \beta_{i0}\dot{y}_i + \alpha_{i0}\tilde{y}_i\right] u_i. \tag{35}$$

A storage function for this supply rate is given in the following lemma.

Lemma 3. *Consider a differentiable odd monotonic nonlinear function $f_i(\cdot)$ that satisfies the slope restrictions $0 \leq \frac{\mathrm{d}f_i(\sigma)}{\mathrm{d}\sigma} < M_{ii}$. Define the differentiable odd monotonic nonlinearity $\tilde{f}_i(\cdot)$ that satisfies Eqs. 14 and 34 as in Eq. 11. Consider the dynamical system $\mathcal{G}_i$ which is a combination of $\tilde{f}_i(\cdot)$ and the multiplier $W_i^{-1}(s)$ from Eq. 9 ($m_{i2} > m_{i1}$). The corresponding state space representation is given by Eqs. 2 and 3. Then the function V_{si} defined as*

$$V_{si}(\tilde{y}_i, z_i) = \beta_{i0}\left(\int_0^{\tilde{y}_i} \tilde{f}_i(\sigma)\,d\sigma + \frac{1}{2}M_{ii}^{-1}u_i^2\right) + \sum_{j=1}^{m_{i2}} \beta_{ij}\int_0^{z_{ij}} \tilde{f}_i(\sigma)\,d\sigma, \tag{36}$$

is a storage function for the supply rate in Eq. 35.

Proof. See Ref. [38,39]. □

To prove overall system stability, we again form augmented matrices using m_1 and $m_2 = \max_i(m_{i2})$. Then, from Eqs. 30 and 35, the overall supply rate can be written as

$$R(\hat{y}, u) = u^T \tilde{C} x_a - u^T (N_0 C_a B_a + \sum_{j=0}^{m_2} H_j M^{-1}) u. \tag{37}$$

where

$$\tilde{C} \triangleq H_0 C_a + N_0 C_a A_a + \sum_{j=1}^{m_1} H_j (C_a - R_j) + \sum_{j=m_1+1}^{m_2} H_j (C_a + R_j). \tag{38}$$

We can now state the following theorem governing the overall stability of the system.

Theorem 2. *Consider an LTI system $G(s)$ independently coupled to m differentiable odd monotonic nonlinearities that satisfy the slope restrictions $0 \leq \frac{df_i(\sigma)}{d\sigma} < M_{ii}$. If for each input-output pair (u_i, y_i), there exist multipliers $W_i(s)$ as in Eq. 9 and a matrix $R = R^T > 0 \in \mathcal{R}^{n_a \times n_a}$ such that, with the preceding definitions of H_j, N_j, and R_j,*

(i) $R_0 \triangleq [(N_0 C_a B_a + \sum_{j=0}^{m_2} H_j M^{-1}) + (N_0 C_a B_a + \sum_{j=0}^{m_2} H_j M^{-1})^T] > 0,$

(ii) *there exists a symmetric matrix $P > 0$, satisfying*

$$A_a^T P + P A_a + R + [\tilde{C} - B_a^T P]^T R_0^{-1} [\tilde{C} - B_a^T P] = 0 \tag{39}$$

then the negative feedback interconnection of system $G(s)$ and the nonlinearities is asymptotically stable. In this case, a Lyapunov function for the combined system is given by

$$V(x_a) = x_a^T P x_a + 2 \sum_{i=1}^{m} \left[\beta_{i0} \left(\int_0^{\tilde{y}_i} \tilde{f}_i(\sigma)\, d\sigma + \frac{1}{2} M_{ii}^{-1} u_i^2 \right) + \sum_{j=1}^{m_{i2}} \beta_{ij} \int_0^{z_{ij}} \tilde{f}_i(\sigma)\, d\sigma \right] \tag{40}$$

Proof. See Ref. [39]. □

Remark 2. To consider nonlinearities with both upper and lower slope constraints, we employ both transformations in Fig. 4. In particular, we define $\tilde{f}(\cdot)$ and $\tilde{G}(s)$ in Eqs. 11 and 13 in terms of $\hat{f}(\cdot)$ and $\hat{G}(s)$ where

$$\hat{f}(y) = f(y) - M_1 y, \quad y \in \mathcal{R}^m \tag{41}$$

$$\hat{G}(s) = (I + G(s) M_1)^{-1} G(s) \sim \left[\begin{array}{c|c} A - B_0 M_1 C_0 & B_0 \\ \hline C_0 & 0 \end{array} \right]. \tag{42}$$

The previous analysis can then be repeated, starting with a system $\hat{G}(s)$ and a differentiable (odd) monotonic nonlinearities $\hat{f}(\cdot)$ with upper slope bounds $M_2 - M_1$. The appropriate Riccati equations can then be obtained from Theorems 1 and 2 by redefining A_a in Eq. 21, replacing u with $\tilde{u} = u - M_1 y$, and then substituting $M_2 - M_1$ for M. □

In the next section, we make explicit connections between the time and frequency domain stability conditions. As a result, we provide sufficient conditions for the existence of positive definite solutions to Eqs. 32 and 39. These conditions also enable then us to make explicit connections between absolute stability theory and mixed-μ.

5 Frequency Domain Stability Conditions

The utility of absolute stability criteria such as the Popov criterion [40] is its simplicity via graphical interpretation in the frequency domain. The previous section developed state space conditions for the stability of an LTI system $G(s)$ coupled with two main classes of nonlinear functions. One advantage of the frequency domain criteria is the insight that they provide to the role of the frequency domain multipliers. The following provides a powerful tool for converting from state space to frequency domain stability conditions within the supply rate framework.

Lemma 4. (Trentelman and Willems [41]). *Consider an LTI system $y = -Gu$ with supply rate $r(y, u)$. Define a system $z = -Hu$, $z \in \mathcal{R}^{m+l}$ with state space representation $H(s) \overset{\min}{\sim} \left[\begin{array}{c|c} A_h & B_h \\ \hline C_h & D_h \end{array} \right]$, with supply rate $\hat{r}(z, u) = z^T L z$, where $L = L^T \in \mathcal{R}^{(m+l)\times(m+l)}$ and $\hat{r}(z, u) \triangleq r(y, u)$. If A_h has no poles on the $\jmath\omega$-axis, then the following statements are equivalent:*

(i) *$H(s)$ is dissipative with respect to the supply rate $\hat{r}(z, u)$;*

(ii) *$H^*(\jmath\omega) L H(\jmath\omega) \geq 0$, $\omega \in \mathcal{R}$.*

The system $H(s)$ represents a modification of the system $G(s)$ to provide functions of both the inputs $-u$ and outputs $y = -Gu$ in the vector z. Consider the simple example of the supply rate $\hat{r}(z, u) = -u^T y$, where

$$z = \begin{bmatrix} y \\ -u \end{bmatrix}, \; H(s) = \begin{bmatrix} G(s) \\ I_m \end{bmatrix}, \text{ and } L = \begin{bmatrix} 0 & I_m \\ I_m & 0 \end{bmatrix}. \tag{43}$$

The frequency domain test for dissipativity is then

$$H^*(\jmath\omega) L H(\jmath\omega) = G^*(\jmath\omega) + G(\jmath\omega) \geq 0, \quad \omega \in \mathcal{R}, \tag{44}$$

which, along with the condition that $G(s)$ is asymptotically stable, is the standard matrix positive real test for the system $G(s)$. Note that similar statements can also be developed with the strongly dissipative conditions.

Thus Lemma 4 provides a tool for transforming the time domain stability criteria given in the previous sections into equivalent frequency domain stability criteria. Furthermore, while these tests are interesting in terms of the extra interpretation that they provide for the multipliers, they also provide conditions for the existence of the positive definite matrices P in Theorems 1 and 2.

Following the approach of How and Hall [19], introduce a system $H(s)$ with outputs that form the components of the negative overall supply rate. Let $W(s) = \text{diag}(W_i(s))$, and define the output vector z

$$z = H(s)(-\tilde{u}) = \begin{bmatrix} W(s) & -N_0(M_2 - M_1)^{-1} s \\ 0 & I \end{bmatrix} \begin{bmatrix} \tilde{y} \\ -\tilde{u} \end{bmatrix},$$

$$= \begin{bmatrix} W(s) & -N_0(M_2 - M_1)^{-1} s \\ 0 & I \end{bmatrix} \begin{bmatrix} \tilde{G}_a \\ I \end{bmatrix} (-\tilde{u}). \tag{45}$$

With the matrix L as in Eq. 43, it follows that the supply rate is $-R(\hat{y}, u) = z^T L z$. From Lemma 4, the test for dissipativeness is then whether $H^*(\jmath\omega) L H(\jmath\omega) \geq 0$. Substituting the definition of $H(s)$ and noting that N_0, M_1, and M_2 are diagonal, we obtain

$$H^* L H = \begin{bmatrix} \tilde{G}_a^* & I \end{bmatrix} \begin{bmatrix} W^*(\jmath\omega) & 0 \\ N_0(M_2 - M_1)^{-1} \jmath\omega & I \end{bmatrix} \begin{bmatrix} 0 & I \\ W(\jmath\omega) & -N_0(M_2 - M_1)^{-1} \jmath\omega \end{bmatrix} \begin{bmatrix} \tilde{G}_a \\ I \end{bmatrix},$$

$$= \tilde{G}_a^* W^* + W \tilde{G}_a. \tag{46}$$

Consequently, an equivalent test for stability is that $T_1(s) \triangleq W(s)\tilde{G}_a(s)$ be positive real. Furthermore, since it follows from Eq. 22 that $\tilde{G}(s)$ and $\tilde{G}_a(s)$ are equivalent in terms of their input-output properties, we need only consider the positive realness of $T(s) \triangleq W(s)\tilde{G}(s)$. Hence, it follows that if A_a is asymptotically stable and $T(s)$ is strongly positive real, then there exists an $n_a \times n_a$ symmetric matrix $P > 0$ which satisfies Eq. 32 or 39, depending on the form of the multiplier $W(s)$, as discussed in Theorems 1 and 2. Conversely, for a given selection of $W(s)$, if there exists a $P > 0$ for all $R > 0$, then A_a is asymptotically stable, and $T(s)$ is strongly positive real.

In the following, we consider the case with finite upper and lower bounds on the slope of the nonlinearity. In particular, assume that the upper bound satisfies $M_2 > 0$, the lower bound satisfies $M_1 < M_2$. In this case, since the double shift illustrated in Fig. 4 must be utilized, we replace Eq. 13 with

$$\tilde{G}(s) = (I + G(s) M_1)^{-1} (I + G(s) M_2) (M_2 - M_1)^{-1}, \tag{47}$$

where it is assumed that M_1 is selected so that $I + GM_1$ is invertible at all frequencies. To clarify the physical interpretation of the stability criterion and develop connections with the upper bounds for mixed-μ, the symmetric bound $M_1 = -M_2$ will be used in the following development.

Theorem 3. *Consider the LTI system $G(s)$ with m independent nonlinearities $\tilde{f}_i(\cdot)$ with appropriate sector bounds given by M_1 and M_2. Assume $M_1 = -M_2 < 0$, and that $I - G(\jmath\omega) M_2$ is invertible for all $\omega \in \mathcal{R}$. For each $i = 1, \ldots, m$, select the multiplier $W_i(s)$ as in Eq. 9* *based on the characteristics of* $\tilde{f}_i(\cdot)$. *Furthermore, define* $W(s) = W_{\text{Re}}(s) + \jmath W_{\text{Im}}(s) = diag(W_i(s))$. *Then, if*

$$G^* W_{\text{Re}} M_2 G - \jmath (W_{\text{Im}} G - G^* W_{\text{Im}}) - W_{\text{Re}} M_2^{-1} \leq 0 \qquad (48)$$

for all $\omega \in \mathcal{R}$, *then the negative feedback interconnection of* $G(s)$ *and the* m *nonlinearities as illustrated in Fig. 2 is Lyapunov stable.*

Proof. With $M_1 = -M_2$, it can easily be demonstrated that the first two terms of $\tilde{G}(s)$ commute. Then from Eqs. 46 and 47, the condition for stability is that

$$0 \leq T(\jmath\omega) + T^*(\jmath\omega), \qquad \omega \in \mathcal{R}, \qquad (49)$$

where $T(s) = W(s)\,(I + G(s)M_2)\,(I - G(s)M_2)^{-1}\,(2M_2)^{-1}$. Since $I - G(s)M_2$ is assumed to be invertible and M_2 is positive, we can develop an equivalent test by pre- and post-multiplying Eq. 49 by $(I - G(\jmath\omega)M_2)^* M_2$ and $M_2(I - G(\jmath\omega)M_2)$ respectively. Performing this operation, and substituting for $T(s)$, the condition of Eq. 49 is equivalent to the requirement that, for all $\omega \in \mathcal{R}$,

$$\begin{aligned}
0 \leq & (I - GM_2)^*\, M_2 W\,(I + GM_2) + (I + GM_2)^*\, W^* M_2\,(I - GM_2) \\
= & M_2(W_{\text{Re}} + \jmath W_{\text{Im}}) + M_2(W_{\text{Re}} + \jmath W_{\text{Im}})GM_2 - M_2 G^* M_2(W_{\text{Re}} + \jmath W_{\text{Im}}) \\
& + M_2(W_{\text{Re}} - \jmath W_{\text{Im}}) - M_2(W_{\text{Re}} - \jmath W_{\text{Im}})GM_2 + M_2 G^* M_2(W_{\text{Re}} - \jmath W_{\text{Im}}) \\
& - M_2 G^* M_2(W_{\text{Re}} + \jmath W_{\text{Im}})GM_2 - M_2 G^* M_2(W_{\text{Re}} - \jmath W_{\text{Im}})GM_2.
\end{aligned}$$

Collecting terms and dividing through by $2M_2^2$ yields Eq. 48. □

Remark 3. The stability criterion in Eq. 48 of Theorem 3 is very general since it may involve a mixture of the time-invariant sector-bounded nonlinearities in a Popov test, the differentiable monotonic and odd monotonic nonlinearities discussed in Sections 4.2 and 4.3, and the time varying nonlinearities considered in a small gain test. □

Remark 4. As discussed by How and Hall [19], Eq. 48 has a graphical interpretation in the scalar case. Specifically, since $W_{\text{Re}} > 0$, Eq. 48 can be rewritten as

$$(G + \jmath \frac{W_{\text{Im}}}{W_{\text{Re}} M_2})^* (G + \jmath \frac{W_{\text{Im}}}{W_{\text{Re}} M_2}) - (\frac{W_{\text{Im}}}{W_{\text{Re}} M_2})^2 - \frac{1}{M_2^2} \leq 0. \qquad (50)$$

or, equivalently, letting $G = x + \jmath y$, Eq. 50 can be written as

$$x^2 + (y + \frac{W_{\text{Im}}}{W_{\text{Re}} M_2})^2 \leq \frac{1}{M_2^2} + (\frac{W_{\text{Im}}}{W_{\text{Re}} M_2})^2. \qquad (51)$$

This corresponds to a circle with a frequency dependent center at $-\frac{W_{\text{Im}}(\omega^2)}{W_{\text{Re}}(\omega^2) M_2}$ and constant real axis intercepts at $\pm M_2^{-1}$. This approach is reminiscent to the classical off-axis circle criteria of Cho and Narendra [25], where a single bounding circle is employed as opposed to a family of frequency dependent circles. Further discussions of the role of the multiplier phase and its relationship to the conservatism of the test are presented in Refs. [17,19,38]. □

Remark 5. For odd monotonic nonlinearities, Narendra and Taylor [7], and Thathachar and Srinath [24,34] discuss multiplier terms that contain complex poles and zeroes of the form

$$\tilde{W}_i(s) = (\text{Eq. 9}) + \sum_{j=m_{i2}+1}^{m_{i3}} \alpha_{ij} \frac{s^2 + a_{ij}s + b_{ij}}{s^2 + \lambda_{ij}s + \eta_{ij}}. \qquad (52)$$

These additional terms allow more rapid phase variations, the benefit of which is apparent in the frequency domain test of Theorem 3. □

5.1 Connections to Mixed-μ

In order to compare the upper bounds for real-μ and the frequency domain stability tests developed in the previous section, we present a brief summary of the notation in Fan *et al.* [18]. For the system matrix $G(s) \in \mathcal{C}^{m \times m}$, let m_r, m_c, and m_C ($m_t = m_r + m_c + m_C \leq m$) define the types and number of uncertainties expected in the system. The positive integers k_i ($\sum_{i=1}^{m_t} k_i = m$) then define the block structure and repetition of the uncertainties denoted by $\mathcal{K}(m_r, m_c, m_C) = (k_1, \ldots, k_{m_r}, \ldots, k_{m_r+m_c}, \ldots, k_{m_t})$. The set of allowable perturbations for the system G is then defined to be

$$\begin{aligned}
\mathcal{X}_{\mathcal{K}} = \{ \Delta = & \operatorname{blkdiag}(\delta_1^r I_{k_1}, \ldots, \delta_{m_r}^r I_{k_{m_r}}, \delta_1^c I_{k_{m_r+1}}, \ldots, \delta_{m_c}^c I_{k_{m_r+m_c}}, \ldots, \\
& \Delta_1^C, \ldots, \Delta_{m_C}^C) : \delta_i^r \in \mathcal{R}, \delta_i^c \in \mathcal{C}, \Delta_i^C \in \mathcal{C}^{k_l \times k_l}, {\scriptstyle l = m_r + m_c + i} \}. \qquad (53)
\end{aligned}$$

Definition 2. (Doyle [42]). *For* $G(s) \in \mathcal{C}^{m \times m}$, $\mu_{\mathcal{K}}(G)$ *is defined as*

$$\mu_{\mathcal{K}}(G) = [\min_{\Delta \in \mathcal{X}_{\mathcal{K}}} \{\sigma_{\max}(\Delta) : \det(I - \Delta G) = 0\}]^{-1}, \qquad (54)$$

where $\mu_{\mathcal{K}}(G) = 0$ *if no* $\Delta \in \mathcal{X}_{\mathcal{K}}$ *exists such that* $\det(I - \Delta G) = 0$.

The complexity inherent in the definition and computation of $\mu_{\mathcal{K}}(G)$ has led to the use of approximations by both upper and lower bounds. For purely complex uncertainties, the bounds $\rho(G) \leq \mu_{\mathcal{K}}(G) \leq \sigma_{\max}(G)$ involving the spectral radius and the maximum singular value are commonly employed. Note that these bounds can be refined with frequency dependent scaling matrices D.

As discussed by Doyle *et al.* [42,43], these scaled bounds are exact if $m_r = 0$ and $2m_c + m_C \leq 3$,but for a larger number of complex uncertainties, results from these papers demonstrate that the bounds are only correct to within approximately 15%, and with real parametric uncertainties, they can be quite poor. Recent developments have led to new upper and lower bounds for mixed-μ ($m_r \neq 0$), as discussed in Refs. [18,43]. For the upper bound of interest, define the Hermitian scaling matrices

$$\begin{aligned}
\mathcal{D}_{\mathcal{K}} = \{ & \operatorname{blkdiag}(D_1, \ldots, D_{m_r+m_c}, d_1 I_{k_{m_r+m_c+1}}, \ldots, d_{m_C} I_{k_{m_t}}) : \\
& 0 < D_i = D_i^* \in \mathcal{C}^{k_i \times k_i}, 0 < d_i \in \mathcal{R}\}, \qquad (55) \\
\mathcal{N}_{\mathcal{K}} = \{ & \operatorname{blkdiag}(N_1, \ldots, N_{m_r}, 0_{k_{m_r+1}}, \ldots, 0_{k_{m_t}}) : N_i = N_i^* \in \mathcal{C}^{k_i \times k_i}\} \qquad (56)
\end{aligned}$$

which are partitioned to be compatible with the uncertainty structure $\mathcal{X}_{\mathcal{K}}$. The set of matrices $\mathcal{D}_{\mathcal{K}}$ includes elements for all three types of uncertainties, but $\mathcal{N}_{\mathcal{K}}$ has nonzero terms only in those parts corresponding to the real uncertainties. Members of both $\mathcal{D}_{\mathcal{K}}$ and $\mathcal{N}_{\mathcal{K}}$ are frequency dependent weighting functions and are constrained to be Hermitian, but only the elements of $\mathcal{D}_{\mathcal{K}}$ must be positive. Note that, within the block definition of Eq. 56, the elements of the scaling matrix N are essentially arbitrary.

With these definitions, a bound was developed by Fan *et al.* in Ref. [18] by incorporating the particular constraints associated with the eigenvectors of the system ΔG where $\Delta \in \mathcal{X}_{\mathcal{K}}$ and $m_r \neq 0$. It is sufficient to note here the following definition of the upper bound for μ with real parametric uncertainties.

Definition 3. (Young *et al.* [43]). *For* $G(s) \in \mathcal{C}^{m \times m}$ *and compatible uncertainty block structure* $\mathcal{K}$, *if*

$$\alpha_* = \inf_{\substack{D \in \mathcal{D}_{\mathcal{K}} \\ N \in \mathcal{N}_{\mathcal{K}}}} [\min_{\alpha \in \mathcal{R}} \{\alpha : (G^* D G + \jmath(NG - G^* N) - \alpha D) \leq 0\}], \qquad (57)$$

then $\mu_{\mathcal{K}}(G) \leq \sqrt{\max(0, \alpha_*)}$.

Corollary 2. *Consider the diagonal case where* $k_i = 1$, $i = 1, \ldots, m_t$ *in* $\mathcal{K}$. *Then it follows from Theorem 3 with* $W_{\text{Re}} M_2$ *replaced by* D, W_{Im} *replaced by* N, *and* M_2^{-2} *bounded by* α, *that the conditions in Eq. 48 and Eq. 57 are identical. Furthermore, if we specialize to the case of* m_r *linear time-invariant and* $m_c + m_C$ *nonlinear time-varying functions, then the bound in Eq. 57 is recovered. Finally, for nonlinear time varying functions, take* $W_{\text{Re}} > 0$ *and* $W_{\text{Im}} = 0$

Proof. For the particular selection of $\mathcal{K}$, since $W_{\text{Re}} M_2 > 0$ and W_{Im} are real, they are members of the $\mathcal{D}_{\mathcal{K}}$ and $\mathcal{N}_{\mathcal{K}}$ respectively. The equivalence then follows by direct substitution. In the linear case, the only restriction on $W(s)$ is that it be a positive real function, and thus W_{Re} and W_{Im} can be any functions in the sets $\mathcal{D}_{\mathcal{K}}$ and $\mathcal{N}_{\mathcal{K}}$. □

The equivalence of these two stability criteria is even stronger if we recognize that the upper bound for mixed-μ is related to M_2^{-1}. Then, as in Eq. 57, minimizing over α for a particular selection of D and N functions, is equivalent to determining the largest sector (slope) bound M_2 that will destabilize the system for a given multiplier selection. This process is, of course, the foundation for the absolute stability theory, see Ref. [7,19]. Extensions of Theorem 3 to the block diagonal case have been addressed for the Popov criterion in Ref. [17].

(Continued in Part II)

Proceedings of the 31st Conference
on Decision and Control
Tucson, Arizona • December 1992

Extensions of Mixed-μ Bounds to Monotonic and Odd Monotonic Nonlinearities Using Absolute Stability Theory: Part II

Wassim M. Haddad
Dept. of Mechanical and
Aerospace Engineering
Florida Institute of Technology
Melbourne, FL 32901

Jonathan P. How
Space Eng. Research Center
Dept. of Aero. and Astro.
Mass. Inst. of Technology
Cambridge, MA 02139

Steven R. Hall
Space Eng. Research Center
Dept. of Aero. and Astro.
Mass. Inst. of Technology
Cambridge, MA 02139

Dennis S. Bernstein
Dept. of Aerospace Engineering
The University of Michigan
Ann Arbor, MI 48109-2140

6 Robust Stability and Performance Analysis

In this section, we specialize the results of Section 4 to linear uncertainty and introduce the Robust Stability and Performance Problems. As shown in Section 4.1, in order to account for the extra dynamics introduced by the frequency domain multiplier, the resulting state space model is of increased dimension. Hence, let $\mathcal{U} \subset \mathcal{R}^{n_a \times n_a}$ denote a set of perturbations ΔA_a of a given nominal augmented dynamics matrix $A_a \in \mathcal{R}^{n_a \times n_a}$. Within the context of robustness analysis, it is assumed that A_a is asymptotically stable and $0 \in \mathcal{U}$. We begin by considering the question of whether or not $A_a + \Delta A_a$ is asymptotically stable for all $\Delta A_a \in \mathcal{U}$. First, however, we note that since A_a in Eq. 21 is lower block triangular, it follows that if $A_a + \Delta A_a$ is asymptotically stable, then $A + \Delta A$ is asymptotically stable for all perturbations ΔA.

Robust Stability Problem. Determine whether the linear system

$$\dot{x}_a = (A_a + \Delta A_a)x_a, \tag{58}$$

is asymptotically stable for all $\Delta A_a \in \mathcal{U}$.

To consider the problem of robust performance, introduce an external disturbance model involving white noise signals as in standard LQG ($\mathcal{H}_2$) theory. The robust performance problem concerns the worst-case $\mathcal{H}_2$ norm, that is, the worst-case (over $\mathcal{U}$) of the expected value of a quadratic form involving outputs $z = Ex_a$, where $E \in \mathcal{R}^{q \times n_a}$, when the system is subjected to a standard white noise disturbance $w(t) \in \mathcal{R}^d$ with weighting $D \in \mathcal{R}^{n_a \times d}$.

Robust Performance Problem. For the disturbed linear system

$$\dot{x}_a = (A_a + \Delta A_a)x_a + Dw, \; z = Ex_a, \tag{59}$$

where $w(\cdot)$ is a zero-mean d-dimensional white noise signal with intensity I_d, determine a performance bound β satisfying

$$J(\mathcal{U}) \stackrel{\triangle}{=} \sup_{\Delta A_a \in \mathcal{U}} \limsup_{t\to\infty} \mathcal{E}\{\|z(t)\|_2^2\} \leq \beta. \tag{60}$$

The following result is immediate. For convenience define the $n_a \times n_a$ nonnegative-definite matrices $R \stackrel{\triangle}{=} E^TE$, $V \stackrel{\triangle}{=} DD^T$.

Lemma 5. *Suppose $A_a + \Delta A_a$ is asymptotically stable for all $\Delta A_a \in \mathcal{U}$. Then*

$$J(\mathcal{U}) = \sup_{\Delta A_a \in \mathcal{U}} \operatorname{tr} Q_{\Delta A_a} R, \tag{61}$$

where the $n_a \times n_a$ matrix $Q_{\Delta A_a} \stackrel{\triangle}{=} \lim_{t\to\infty} \mathcal{E}[x_a(t)x_a^T(t)]$ is given by

$$Q_{\Delta A_a} = \int_0^\infty e^{(A_a+\Delta A_a)t} V e^{(A_a+\Delta A_a)^T t}\, dt, \tag{62}$$

which is the unique, nonnegative-definite solution to

$$0 = (A_a + \Delta A_a)Q_{\Delta A_a} + Q_{\Delta A_a}(A_a + \Delta A_a)^T + V. \tag{63}$$

In order to draw connections with traditional Lyapunov function theory, we express the $\mathcal{H}_2$ performance measure in terms of a dual variable $P_{\Delta A_a}$ for which the roles of $A_a + \Delta A_a$ and $(A_a + \Delta A_a)^T$ are interchanged.

Proposition 1. (Haddad and Bernstein [17]). *Suppose $A_a + \Delta A_a$ is asymptotically stable for all $\Delta A_a \in \mathcal{U}$. Then*

$$J(\mathcal{U}) = \sup_{\Delta A_a \in \mathcal{U}} \operatorname{tr} P_{\Delta A_a} V, \tag{64}$$

where $P_{\Delta A_a} \in \mathcal{R}^{n_a \times n_a}$ is the unique, nonnegative-definite solution to

$$0 = (A_a + \Delta A_a)^T P_{\Delta A_a} + P_{\Delta A_a}(A_a + \Delta A_a) + R. \tag{65}$$

In the following sections our approach is to obtain robust stability as a consequence of sufficient conditions for robust performance.

6.1 Robust stability and performance

The key step in obtaining robust stability and performance is to bound the uncertain terms $\Delta A_a^T P_{\Delta A_a} + P_{\Delta A_a}\Delta A_a$ in the Lyapunov Eq. 65 by means of a parameter-dependent bounding function $\Omega(P, \Delta A_a)$ which guarantees robust stability by means of a family of Lyapunov functions. As shown in Ref. [17], this framework corresponds to the construction of a parameter-dependent Lyapunov function that guarantees robust stability. The following result is fundamental and forms the basis for all later developments. For notational convenience, let $\mathcal{S}^r$ and $\mathcal{N}^r$ denote the set of $r \times r$ symmetric and nonnegative definite matrices respectively.

Theorem 4. (Haddad and Bernstein [17]). *Let $\Omega_0 : \mathcal{N}^{n_a} \to \mathcal{S}^{n_a}$ and $P_0 : \mathcal{U} \to \mathcal{S}^{n_a}$ be such that*

$$\Delta A_a^T P + P\Delta A_a \leq \Omega_0(P) - [(A_a+\Delta A_a)^T P_0(\Delta A_a) + P_0(\Delta A_a)(A_a+\Delta A_a)], \tag{66}$$

for all $\Delta A_a \in \mathcal{U}$, and $P \in \mathcal{N}^{n_a}$, and suppose there exists $P \in \mathcal{N}^{n_a}$ satisfying

$$0 = A^T P + PA + \Omega_0(P) + R \tag{67}$$

and such that $P + P_0(\Delta A_a)$ is nonnegative-definite for all $\Delta A_a \in \mathcal{U}$. Then

$$(A_a + \Delta A_a,\ E) \text{is detectable}, \quad \Delta A_a \in \mathcal{U}, \tag{68}$$

if and only if $A_a + \Delta A_a$is asymptotically stable, $\Delta A_a \in \mathcal{U}$. In this case,

$$P_{\Delta A_a} \leq P + P_0(\Delta A_a), \quad \Delta A_a \in \mathcal{U}, \tag{69}$$

where $P_{\Delta A_a}$ is given by Eq. 65. Therefore,

$$J(\mathcal{U}) \leq \operatorname{tr} PV + \sup_{\Delta A_a \in \mathcal{U}} \operatorname{tr} P_0(\Delta A_a)V. \tag{70}$$

If, in addition, there exists $\bar{P}_0 \in \mathcal{S}^n$ such that

$$P_0(\Delta A_a) \leq \bar{P}_0, \quad \Delta A_a \in \mathcal{U}, \tag{71}$$

then $J(\mathcal{U}) \leq \beta$, where $\beta \stackrel{\triangle}{=} \operatorname{tr}\,[(P + \bar{P}_0)V]$.

Note that with $\Omega(P, \Delta A_a)$ denoting the right hand side of Eq. 66, this equation can be rewritten as

$$\Delta A_a^T P + P\Delta A_a \leq \Omega(P, \Delta A_a), \quad \Delta A_a \in \mathcal{U}, \quad P \in \mathcal{N}^{n_a}, \tag{72}$$

where $\Omega(P, \Delta A_a)$ is a function of the uncertain parameters ΔA_a. For convenience we shall say that $\Omega(\cdot,\cdot)$ is a parameter-dependent bounding function or, to be consistent with Ref. [17], a parameter-dependent Ω-bound. To apply Theorem 4, we first specify a function $\Omega_0(\cdot)$ and an uncertainty set $\mathcal{U}$ such that Eq. 72 holds. If the existence of a nonnegative-definite solution P to Eq. 67 can be determined analytically or numerically and the detectability condition Eq. 68 is satisfied, then robust stability is guaranteed and the performance bound Eq. 70 can be computed.

Finally, we establish connections between Theorem 4 and Lyapunov function theory. Specifically, we show that a parameter-dependent Ω-bound establishing robust stability is equivalent to the existence of a parameter-dependent Lyapunov function which also establishes robust stability. To show this, assume there exists a positive-definite solution

*Research was funded in part by NSF Grant ECS-9109558, NASA Grant NAGW-2014, NASA SERC Grant NAGW-1335, and AFOSR Grant F49620-92-J-0127.

CH3229-2/92/0000-2820$1.00 © 1992 IEEE

to Eq. 67, let $P_0 : \mathcal{U} \to \mathcal{N}^{n_a}$, and define the parameter-dependent Lyapunov function $V(x_a) \triangleq x_a^T(P + P_0(\Delta A_a))x_a$. Note that since P is positive definite and $P_0(\Delta A_a)$ is nonnegative definite, $V(x_a)$ is positive definite. Thus, the corresponding Lyapunov derivative is given by

$$\dot{V}(x_a)=x_a^T[A_a^T P + PA_a + \Delta A_a^T P + P\Delta A_a + A_a^T P_0(\Delta A_a) + P_0(\Delta A_a)A_a + \Delta A_a^T P_0(\Delta A_a) + P_0(\Delta A_a)\Delta A_a]x_a \quad (73)$$

or, equivalently, using Eq. 67

$$\dot{V}(x_a) = -x_a^T[\Omega(P, \Delta A_a) + R]x_a. \quad (74)$$

Thus, using Eq. 66 it follows that $\dot{V}(x_a) \leq 0$ so that $A_a + \Delta A_a$ is stable in the sense of Lyapunov. To show asymptotic stability using La Salle's Theorem [44] we need to show that $\dot{V}(x_a) = 0$ implies $x_a = 0$. Note that $\dot{V}(x_a) = 0$ implies $Rx_a = 0$, or, equivalently, $Ex_a = 0$. Thus, with $\dot{x}_a = (A_a + \Delta A_a)x_a$, $Ex_a = 0$ and the detectability assumption in Eq. 68, it follows from the PBH test that $x_a = 0$. Hence asymptotic stability is established.

6.2 Construction of parameter-dependent Lyapunov functions for monotonic and odd monotonic nonlinearities

Having established the theoretical basis for our approach, we now assign explicit structure to the set $\mathcal{U}$ and the parameter-dependent bounding function $\Omega(\cdot,\cdot)$. Specifically, the uncertainty set $\mathcal{U}$ is defined by

$$\mathcal{U} \triangleq \left\{\Delta A_a \in \mathcal{R}^{n_a \times n_a} : \Delta A_a = -B_a F \left(I + M^{-1}F\right)^{-1} C_a, F \in \mathcal{F}\right\}, \quad (75)$$

where $\mathcal{F}$ satisfies $\mathcal{F} \triangleq \{F \in \mathcal{R}^{m\times m} : F \geq 0\}$, and where $B_a \in \mathcal{R}^{n_a \times m}$ and $C_a \in \mathcal{R}^{m\times n_a}$ are fixed matrices denoting the structure of the uncertainty, $M \in \mathcal{R}^{m\times m}$ is a given diagonal positive-definite matrix, and $F \in \mathcal{R}^{m\times m}$ is a diagonal uncertain matrix.

Next we digress slightly to provide an alternative characterization of the uncertainty set $\mathcal{U}$. In order to state our next result, define the subset $\hat{\mathcal{F}}$ of $\mathcal{F}$ by

$$\hat{\mathcal{F}} \triangleq \left\{\hat{F} : \hat{F} = F\left(I + M^{-1}F\right)^{-1}, F \in \mathcal{F}\right\}, \quad (76)$$

where by Lemma 3.2 of Ref. [45], $\det(I + M^{-1}F) \neq 0$.

Proposition 2. *Let $M \in \mathcal{R}^{m\times m}$ be positive definite. Then*

$$\hat{\mathcal{F}} = \left\{\hat{F} \in \mathcal{R}^{m\times m} : \det(I - \hat{F}M^{-1}) \neq 0 \text{ and } \hat{F}M^{-1}\hat{F} \leq \hat{F}\right\}. \quad (77)$$

Proof. See Ref. [39]. □

Finally, we present a key Lemma that shows the equivalence of $0 \leq \hat{F} \leq M$ and the structure presented in Proposition 2.

Lemma 6. *Let $F \in \mathcal{R}^{m\times m}$ be a nonnegative definite diagonal matrix and $M \in \mathcal{R}^{m\times m}$ a positive-definite diagonal matrix. Then $\hat{F}M^{-1}\hat{F} \leq \hat{F}$ if and only if $0 \leq \hat{F} \leq M$.*

Proof. The proof is a direct consequence of Lemma 4.4 of Ref. [17]. □

Now, it follows from Proposition 2 and Lemma 6 that an equivalent representation for our uncertainty set $\mathcal{U}$ in Eq. 75 is

$$\mathcal{U} \triangleq \left\{\Delta A_a \in \mathcal{R}^{n_a\times n_a} : \Delta A_a = -B_a\hat{F}C_a, \hat{F} \in \hat{\mathcal{F}}\right\}. \quad (78)$$

For the structure of $\mathcal{U}$ satisfying Eq. 75, the parameter-dependent bound $\Omega(\cdot,\cdot)$ satisfying Eq. 66 can now be given a concrete form. Since the elements ΔA_a in $\mathcal{U}$ are parameterized by the elements F in $\mathcal{F}$, for convenience in the following results we shall write $P_0(F)$ in place of $P_0(\Delta A_a)$.

Proposition 3. *Let N_0, H_0, N_j, H_j, $S_j \in \mathcal{R}^{m\times m}$ be nonnegative-definite diagonal matrices such that, as in Theorem 2, $R_0 > 0$, and*

$$N_j S_j - H_j \geq 0, \qquad j = 1, \ldots, m_2. \quad (79)$$

Then, using Eq. 38, the functions

$$\Omega(P) = [\tilde{C} - B_a^T P]^T R_0^{-1}[\tilde{C} - B_a^T P], \quad (80)$$

$$P_0(F)=C_a^T(I + M^{-1}F)^{-1}\left[FN_0 + FM^{-1}N_0F\right](I + M^{-1}F)^{-1}C_a + \sum_{j=1}^{m_2} R_j^T FN_jR_j, \quad (81)$$

or, equivalently,

$$P_0(\hat{F}) = C_a^T\hat{F}N_0C_a + \sum_{j=1}^{m_2} R_j^T(I - \hat{F}M^{-1})^{-1}\hat{F}N_jR_j, \quad (82)$$

satisfy Eq. 66 with $\mathcal{U}$ given by Eq. 75.

Proof. The proof is a direct consequence of Theorem 2, with $\tilde{f}(\tilde{y}) = F\tilde{y} = F(I + M^{-1}F)^{-1}C_ax_a$. See Ref. [17]. □

Theorem 5. *Let N_0, H_0, N_j, H_j, $S_j \in \mathcal{R}^{m\times m}$ be nonnegative-definite diagonal matrices such that $R_0 > 0$ and Eq. 79 is satisfied. Furthermore, suppose that there exists a nonnegative-definite matrix P satisfying Eq. 39. Then*

$$(A_a + \Delta A_a, E) \text{ is detectable}, \qquad \Delta A_a \in \mathcal{U}, \quad (83)$$

if and only if

$$A_a + \Delta A_a \text{ is asymptotically stable}, \qquad \Delta A_a \in \mathcal{U}. \quad (84)$$

In this case,

$$J(\mathcal{U}) \leq \operatorname{tr}\ PV + \sup_{\hat{F}\in\hat{\mathcal{F}}} \operatorname{tr}\ [(C_a^T\hat{F}N_0C_a + \sum_{j=1}^{m_2} R_j^T(I - \hat{F}M^{-1})^{-1}\hat{F}N_jR_j)V]. \quad (85)$$

Proof. The result is a direct specialization of Theorem 4 using Proposition 3. We only note that $P_0(\Delta A_a)$ now has the form in Eq. 82. Since $\hat{F}N_j \geq 0$, $j = 0, \ldots, m_2$, for all $\hat{F} \in \hat{\mathcal{F}}$ it follows that $P + P_0(\hat{F})$ is nonnegative definite for all $\hat{F} \in \hat{\mathcal{F}}$ as required by Theorem 4. □

Theorem 5 is directly applicable to dynamic systems with m-mixed uncertainties. Specifically, it follows from Theorem 2 that if the nonlinearity m-vector $f(y)$ is composed of n_1 time invariant first and third quadrant functions, $n_2 - n_1$ monotone increasing functions, and $m - n_2$ odd monotone increasing functions, then the nominal system is robustly stable for all such mixed uncertainty. Furthermore, in the linear uncertainty case, $f(y) = \hat{F}y$ it was recently shown by Haddad and Bernstein [17] that under certain compatibility assumptions between N_0 and $\hat{\mathcal{F}}$ (for the Popov case), the set $\mathcal{U}$ allows a richer class of multivariable uncertainties in that F may represent a fully populated uncertainty matrix. Similar extensions for the monotonic and odd monotonic case are possible, however, for simplicity of exposition, we defer these results to a future paper. This of course, allows for nonscalar multiple uncertainty blocks within the analysis and synthesis framework.

7 Robust dynamic output feedback synthesis

In this section, we introduce the Robust Stability and Performance dynamic output-feedback control problem. Since the multiplier dynamics increase the plant order from n to n_a, to allow for greater design flexibility, the compensator dimension n_c is fixed to be less than the augmented plant order n_a. Hence, define $\tilde{n} = n_a + n_c$. Note that in this context, an nth-order controller can be regarded as a reduced order design. As in Ref. [46], this constraint leads to an oblique projection that introduces additional coupling in the design equations along with additional design equations. This coupling shows that regulator/estimator separation breaks down in the reduced-order controller case.

Dynamic Robust Stability and Performance Problem. Given the n_a^{th}-order stabilizable and detectable plant with constant structured real-valued plant parameter variations

$$\dot{x}_a = (A_a + \Delta A_a)x_a + Bu + D_1w, \; y = Cx_a + D_2w, \quad (86)$$

where $u(t) \in \mathcal{R}^{m_0}$, $w(t) \in \mathcal{R}^d$, and $y(t) \in \mathcal{R}^l$, determine an n_c^{th}-order dynamic compensator

$$\dot{x}_c = A_cx_c + B_cy, \; u = C_cx_c, \quad (87)$$

that satisfies the following design criteria:

(i) the closed-loop system in Eqs. 86-87 is asymptotically stable for all $\Delta A_a \in \mathcal{U}$; and
(ii) the performance functional in Eq. 88 is minimized.

$$J(A_c, B_c, C_c) \triangleq \sup_{\Delta A_a \in \mathcal{U}} \limsup_{t\to\infty} \frac{1}{t}\mathcal{E}\{\int_0^t [x_a^T R_{xx} x_a + u^T R_{uu} u]\, ds\} \quad (88)$$

Since we are only interested in controlling the actual system dynamics, i.e., the non-augmented dynamics, in accordance with the partitioning in Eq. 21, our control, measurement, disturbance, and state weighting matrices B, C,D, and R_{xx}, have the following structure

$$B = \begin{bmatrix} \hat{B} \\ 0 \end{bmatrix},\ C = \begin{bmatrix} \hat{C} & 0 \end{bmatrix},\ D = \begin{bmatrix} \hat{D} \\ 0 \end{bmatrix},\ R_{xx} = \begin{bmatrix} \hat{R}_{xx} & 0 \\ 0 & 0 \end{bmatrix}. \quad (89)$$

For each uncertain variation $\Delta A_a \in \mathcal{U}$, the closed-loop system in Eqs. 86-87 can be written as

$$\dot{\tilde{x}} = (\tilde{A} + \Delta\tilde{A})\tilde{x} + \tilde{D}w, \quad (90)$$

where

$$\tilde{x} \triangleq \begin{bmatrix} x_a \\ x_c \end{bmatrix},\ \tilde{A} \triangleq \begin{bmatrix} A_a & BC_c \\ B_c C & A_c \end{bmatrix},\ \Delta\tilde{A} \triangleq \begin{bmatrix} \Delta A_a & 0_{n_a\times n_c} \\ 0_{n_c\times n_a} & 0_{n_c\times n_c} \end{bmatrix}, \quad (91)$$

and where the closed-loop disturbance $\tilde{D}w(t)$ has intensity $\tilde{V} = \tilde{D}\tilde{D}^T$, where $\tilde{D} \triangleq \begin{bmatrix} D_1 \\ B_c D_2 \end{bmatrix}$, $\tilde{V} \triangleq \begin{bmatrix} V_1 & 0 \\ 0 & B_c V_2 B_c^T \end{bmatrix}$, $V_1 = D_1 D_1^T$, and $V_2 = D_2 D_2^T$. The closed-loop system uncertainty $\Delta\tilde{A}$ has the form

$$\Delta\tilde{A} = \tilde{B}_a \hat{F} \tilde{C}_a,\ \text{where},\ \tilde{B}_a \triangleq \begin{bmatrix} B_a \\ 0_{n_c\times m} \end{bmatrix},\ \tilde{C}_a \triangleq [\, C_a \ 0_{m\times n_c}\,]. \quad (92)$$

Finally, if $\tilde{A} + \Delta\tilde{A}$ is asymptotically stable for all $\Delta A_a \in \mathcal{U}$ for a given compensator (A_c, B_c, C_c), then it follows from Proposition 1 that the performance measure in Eq. 95 is given by

$$J(A_c, B_c, C_c) = \sup_{\Delta A_a \in \mathcal{U}} \operatorname{tr} \tilde{P}_{\Delta\tilde{A}} \tilde{V}, \quad (93)$$

where $P_{\Delta\tilde{A}}$ satisfies the $\tilde{n}\times\tilde{n}$ Lyapunov equation

$$0 = (\tilde{A} + \Delta\tilde{A})^T \tilde{P}_{\Delta\tilde{A}} + \tilde{P}_{\Delta\tilde{A}}(\tilde{A} + \Delta\tilde{A}) + \tilde{R}, \quad (94)$$

and $\tilde{E} = [\, E_1\ E_2 C_c\,]$, $\tilde{R} = \tilde{E}^T\tilde{E}$.

Next, we proceed as in the previous section where we replace the Lyapunov Eq. 94 for the dynamic problem with a Riccati equation that guarantees that the closed-loop system is robustly stable. Thus, for the dynamic output feedback problem, Theorem 5 holds with A_a, R, V replaced by $\tilde{A}$, $\tilde{R}$, $\tilde{V}$.

Dynamic Optimization Problem. Determine (A_c, B_c, C_c) that minimizes

$$\mathcal{J}(A_c, B_c, C_c) \triangleq \operatorname{tr} \tilde{P}\tilde{V} + \sup_{\hat{F}\in\hat{\mathcal{F}}} \operatorname{tr} [(\tilde{C}^T \hat{F} N_0 \tilde{C} + \sum_{j=1}^{m_2} \tilde{R}_j^T (I - \hat{F}M^{-1})^{-1} \hat{F} N_j \tilde{R}_j)\tilde{V}], \quad (95)$$

where $\tilde{R}_j \triangleq [\, R_j\ 0_{m\times n_c}\,]$, and $\tilde{P} \in \mathcal{N}^{\tilde{n}}$ satisfies

$$\begin{aligned}0 =& \tilde{A}^T\tilde{P} + \tilde{P}\tilde{A} + \tilde{R} \\ &+ [H_0\tilde{C} + N_0\tilde{C}\tilde{A} + \sum_{j=1}^{m_1} H_j(\tilde{C} - \tilde{R}_j) + \sum_{j=m_1+1}^{m_2} H_j(\tilde{C} + \tilde{R}_j) - \tilde{B}_a^T\tilde{P}]^T \tilde{R}_0^{-1} \\ & [H_0\tilde{C} + N_0\tilde{C}\tilde{A} + \sum_{j=1}^{m_1} H_j(\tilde{C} - \tilde{R}_j) + \sum_{j=m_1+1}^{m_2} H_j(\tilde{C} + \tilde{R}_j) - \tilde{B}_a^T\tilde{P}], \quad (96)\end{aligned}$$

$$\tilde{R}_0 \triangleq (N_0\tilde{C}\tilde{B}_a + \sum_{j=0}^{m_2} H_j M^{-1}) + (N_0\tilde{C}\tilde{B}_a + \sum_{j=0}^{m_2} H_j M^{-1})^T, \quad (97)$$

and such that (A_c, B_c, C_c) is controllable and observable, and Eq. 79 holds.

The relationship between the Dynamic Optimization Problem and the Robust Stability and Performance Problems is straightforward as shown by the following observation.

Proposition 4. *If $P \in \mathcal{N}^{n_a}$ and (A_c, B_c, C_c) satisfy Eq. 95 and the detectability condition Eq. 68 holds, then $\hat{A} + \Delta A_a$ is asymptotically stable for all $\Delta A_a \in \mathcal{U}$, and*

$$J(A_c, B_c, C_c) \le \mathcal{J}(A_c, B_c, C_c). \quad (98)$$

Proof. Since Eq. 95 has a solution $P \in \mathcal{N}^{n_a}$ and the detectability condition Eq. 68 holds, the hypotheses of Theorem 4 are satisfied so that robust stability with robust performance bound is guaranteed. The condition in Eq. 98 is merely a restatement of Eq. 70. □

By deriving necessary conditions for the Dynamic Optimization Problem as in Section 6.1 we obtain sufficient conditions for characterizing fixed-order dynamic output feedback controllers that guarantee robust stability and performance. The following lemma is required for the statement of the main theorem.

Lemma 7. (Bernstein and Haddad [47]). *Let $\hat{Q}, \hat{P}$ be $n_a \times n_a$ nonnegative definite matrices and suppose that rank $\hat{Q}\hat{P} = n_c$. Then there exist $n_c \times n_a$ G, Γ and $n_c \times n_c$ invertible M, unique except for a change of basis in $\mathcal{R}^{n_c}$, such that*

$$\hat{Q}\hat{P} = G^T M \Gamma,\ \Gamma G^T = I_{n_c}. \quad (99)$$

Furthermore, the $n_a \times n_a$ matrices

$$\tau \triangleq G^T\Gamma,\ \tau_\perp \triangleq I_{n_a} - \tau, \quad (100)$$

are idempotent and have rank n_c and $n_a - n_c$ respectively.

We now state the main results of this section concerning the reduced-order controllers. For convenience, recall the definitions of R_0 and $\tilde{C}$, and define $\overline{\Sigma} \triangleq C^T V_2^{-1} C$, and

$$R_{2a} \triangleq R_{uu} + B^T C_a^T N_0 R_0^{-1} N_0 C_a B, \quad (101)$$

$$P_a \triangleq B^T P + B^T C_a^T N_0 R_0^{-1} (\tilde{C} - B_a^T P), \quad (102)$$

$$A_P \triangleq A_a - B_a R_0^{-1}\tilde{C}, \quad (103)$$

$$A_{\hat{Q}} \triangleq A_P - Q\overline{\Sigma} + B_a R_0^{-1} B_a^T P, \quad (104)$$

$$A_{\hat{P}} \triangleq A_P + B_a R_0^{-1} B_a^T P + (B_a R_0^{-1} N_0 C_a - I) B R_{2a}^{-1} P_a \quad (105)$$

for arbitrary $Q, P \in \mathcal{R}^{n_a\times n_a}$.

Theorem 6. *Let $n_c \le n_a$, and assume $R_0 > 0$ and Eq. 79 holds. Furthermore, suppose there exist $n_a \times n_a$ nonnegative-definite matrices $P, Q, \hat{P}, \hat{Q}$ satisfying*

$$\begin{aligned}0 =& A_P^T P + P A_P + R_{xx} + \tilde{C}^T R_0^{-1}\tilde{C} + P B_a R_0^{-1} B_a^T P - P_a^T R_{2a}^{-1} P_a \\ &+ \tau_\perp^T P_a^T R_{2a}^{-1} P_a \tau_\perp, \quad (106)\end{aligned}$$

$$\begin{aligned}0 =& (A_P + B_a R_0^{-1} B_a [P + \hat{P}])Q + Q(A_P + B_a R_0^{-1} B_a [P + \hat{P}])^T \\ &+ V_1 - Q\overline{\Sigma}Q + \tau_\perp^T Q\overline{\Sigma}Q\tau_\perp, \quad (107)\end{aligned}$$

$$0 = A_{\hat{Q}}^T \hat{P} + \hat{P} A_{\hat{Q}} + \hat{P} B_a R_0^{-1} B_a^T \hat{P} + P_a^T R_{2a}^{-1} P_a - \tau_\perp^T P_a^T R_{2a}^{-1} P_a \tau_\perp, \quad (108)$$

$$0 = A_{\hat{P}}\hat{Q} + \hat{Q} A_{\hat{P}}^T + Q\overline{\Sigma}Q - \tau_\perp Q\overline{\Sigma}Q\tau_\perp^T \quad (109)$$

$$\operatorname{rank} \hat{Q} = \operatorname{rank} \hat{P} = \operatorname{rank} \hat{Q}\hat{P} = n_c \quad (110)$$

and let A_c, B_c, C_c be given by

$$A_c = \Gamma\left[A_{\hat{P}} - Q\overline{\Sigma}\right]G^T,\ B_c = \Gamma Q C^T V_2^{-1},\ C_c = -R_{2a}^{-1} P_a G^T. \quad (111)$$

Then $(\tilde{A} + \Delta\tilde{A}, \tilde{E})$ is detectable for all $\Delta A_a \in \mathcal{U}$ if and only if $\tilde{A} + \Delta\tilde{A}$ is asymptotically stable for all $\Delta A_a \in \mathcal{U}$. In this case the performance of the closed-loop system Eq. 90 satisfies the parameter-dependent $\mathcal{H}_2$ bound

$$J(A_c, B_c, C_c) \le \operatorname{tr} [(P + \hat{P})V_1 + \hat{P}Q\overline{\Sigma}Q] \quad (112)$$

$$+ \sup_{\hat{F}\in\hat{\mathcal{F}}} \operatorname{tr} [(C_a^T \hat{F} N_0 C_a + \sum_{j=1}^{m_2} R_j^T (I - \hat{F}M^{-1})^{-1} \hat{F} N_j R_j) V_1]. \quad (113)$$

Proof. The proof follows as in the proof of Theorem 5.1 of Ref. [17] with additional terms arising due to the reduced-order dynamic compensation structure and odd monotonic constraints. For details of a similar proof see Ref. [48]. □

Remark 6. Several special cases can immediately be discerned from Theorem 6. For example, in the full-order case, set $n_c = n_a$ so that $\tau = G = \Gamma = I_{n_a}$ and $\tau_\perp = 0$. In this case the last term in each of Eqs. 106-109 is zero and Eq. 109 is superfluous. Alternatively, letting $B_a = 0$, $C_a = 0$ and retaining the reduced-order constraint $n_c < n_a$ yields the result of Ref. [46]. Finally, setting $m_2 = 0$ yields the results of Ref. [17] for the case in which F is diagonal. □

Theorem 6 provides constructive sufficient conditions that yield reduced-order dynamic feedback gains A_c, B_c, C_c for robust stability and performance. Note that when solving Eqs. 106-109 numerically, the matrices M, N_j, H_j, and S_j, $j = 0, \ldots, m_2$, and the structure of B_a and C_a appearing in the design equations can be adjusted to examine tradeoffs between performance and robustness. Furthermore, to further reduce conservatism, one can view the multiplier matrices N_j, H_j, and S_j as a free parameters and optimize the worst case $\mathcal{H}_2$ performance bound with respect to them Ref. [17,39]. The basic approach is to employ a numerical algorithm to design the optimal compensator and the multipliers simultaneously, thus avoiding the need to iterate between controller design and optimal multiplier evaluation. [38,39]

8 Conclusions

The goal of this paper has been to make explicit connections between the classical absolute stability theory and modern mixed-μ analysis and synthesis. To this end, we extended previous results on absolute stability theory for monotonic and odd monotonic nonlinearities to provide a tight approximation for constant real parametric uncertainty. Specifically, using a parameter-dependent Lyapunov function framework in which the uncertain parameters appear explicitly in the Lyapunov function, the allowable time-variations of the parameter are restricted, thereby reducing conservatism with respect to constant real parametric uncertainty.

Connections to the μ-analysis and synthesis are made through frequency domain tests, demonstrating that the stability multipliers are a parameterizations of the D, N-scales in mixed-μ. Combining the parameter-dependent Lyapunov functions with fixed-order optimization techniques leads to a Riccati equation characterization for robust $\mathcal{H}_2$ controllers. An advantage of this approach is that reduced-order controllers with optimal frequency domain multipliers can be designed directly, while avoiding the standard $D, N - K$ iteration.

Acknowledgments

The authors would like to thank M. G. Safonov for his valuable discussions. The first two authors are grateful to the Department of Aerospace Eng. at the Univ. of Michigan for their hospitality during the summer of 1992.

References

[1] M. A. Aizerman and F. R. Gantmacher, *Absolute Stability of Regulator Systems.* San Francisco: Holden-Day, Inc., 1964.

[2] S. Lefschetz, *Stability of nonlinear control systems.* New York: Academic Press, 1965.

[3] V. M. Popov, *Hyperstability of control systems.* New York: Springer, 1973.

[4] M. G. Safonov, *Stability and Robustness of Multivariable Feedback Systems.* MIT Press, 1980.

[5] M. Vidyasagar, *Nonlinear systems analysis.* Englewood Cliffs, NJ: second edition, Prentice-Hall, 1992.

[6] G. Zames, "On the input-output stability of time-varying nonlinear feedback systems part I and II," *IEEE TAC*, vol. AC–11, no. 2,3, pp. 228–238,465–476, 1966.

[7] K. S. Narendra and J. H. Taylor, *Frequency Domain Criteria for Absolute Stability.* New York: Academic Press, 1973.

[8] D. D. Siljak, Personal Communication. 1987,1990.

[9] W. Haddad and D. Bernstein, "Explicit construction of quadratic Lyapunov functions for the small gain, positivity, circle, and Popov theorems and their application to robust stability part I and II:," *submitted to Int. J. Robust and Nonlinear Control.*

[10] B. A. Francis, *A Course in $\mathcal{H}_\infty$ Control Theory.* Springer-Verlag, 1987.

[11] G. Leitmann, "Guaranteed asymptotic stability for some linear systems with bounded uncertainties," *Trans. of the ASME*, vol. 101, pp. 212–216, Sept. 1979.

[12] P. P. Khargonekar, I. R. Petersen, and K. Zhou, "Robust stabilization of uncertain linear systems: quadratic stabilizability and $\mathcal{H}_\infty$ control theory," *IEEE TAC*, vol. AC–35, pp. 356–361, 1990.

[13] A. Packard and J. C. Doyle, "Quadratic stability with real and complex perturbations," *IEEE TAC*, vol. AC–35, pp. 198–201, 1990.

[14] B. D. O. Anderson and S. Vongpanitlerd, *Network analysis and synthesis: a modern systems theory approach.* Englewood Cliffs, NJ: Prentice-Hall, 1973.

[15] I. R. Petersen, "Disturbance attenuation and $\mathcal{H}_\infty$ optimization," *IEEE TAC*, vol. AC–32, pp. 427–429, 1987.

[16] J. C. Doyle, K. Glover, P. P. Khargonekar, and B. A. Francis, "State-space solutions to standard $\mathcal{H}_2$ and $\mathcal{H}_\infty$ control problems," *IEEE TAC*, vol. AC–34, pp. 831–847, Aug. 1989.

[17] W. M. Haddad and D. S. Bernstein, "Parameter-dependent Lyapunov functions, constant real parameter uncertainty, and the Popov criterion in robust analysis and synthesis, parts I and II," in *Proc. CDC*, pp. 2274–2279, 2632–2633, Dec. 1991, submitted to the *IEEE Trans. Autom. Contr.*

[18] M. K. H. Fan, A. L. Tits, and J. C. Doyle, "Robustness in the presence of mixed parametric uncertainty and unmodelled dynamics," *IEEE TAC*, vol. AC–36, pp. 25–38, Jan. 1991.

[19] J. P. How and S. R. Hall, "Connections between the Popov stability criterion and bounds for real parametric uncertainty," Submitted to the *IEEE TAC* and the *1993 ACC*, May 1992. MIT SERC Report 9-92-J.

[20] J. C. Willems, "Dissipative dynamical systems part I and II," *Archive Rational Mechanics Analysis*, vol. 45, pp. 321–351,352–393, 1972.

[21] M. G. Safonov and G. Wyetizner, "Computer-aided stability analysis renders popov criterion obsolete," *IEEE TAC*, vol. AC–32, pp. 1128–1131, 1987.

[22] K. S. Narendra and C. P. Neuman, "Stability of a class of differential equations with a single monotone nonlinearity," *SIAM J. of Control*, vol. 4, pp. 295–308, 1966.

[23] K. S. Narendra and Y.-S. Cho, "Stability of feedback systems containing a single odd monotonic nonlinearity," *IEEE TAC*, pp. 448–450, Aug. 1967.

[24] M. A. L. Thathachar, M. D. Srinath, and H. K. Ramapriyan, "On a modified Lur'e problem," *IEEE TAC*, vol. AC–12, pp. 731–739, Dec. 1967.

[25] Y.-S. Cho and K. S. Narendra, "An off-axis circle criterion for the stability of feedback systems with a monotonic nonlinearity," *IEEE TAC*, pp. 413–416, Aug. 1968.

[26] G. Zames and P. Falb, "Stability conditions for systems with monotone and slope-restricted nonlinearities," *SIAM J. of Control*, vol. 6, pp. 89–108, Jan. 1968.

[27] R. Lozano-Leal and S. Joshi, "Strictly positive real transfer functions revisited," *IEEE TAC*, vol. AC–35, pp. 1243–1245, 1990.

[28] J. T. Wen, "Time domain and frequency domain conditions for strict positive realness," *IEEE TAC*, vol. AC–33, pp. 988–992, 1988.

[29] B. D. O. Anderson, "A system theory criterion for positive real matrices," *SIAM J. of Control and Opt.*, vol. 5, pp. 171–182, 1967.

[30] S. Pinzoni and J. C. Willems, "The dissipation inequality for systems described by high-order differential equations," in *Preprint, to appear in the Proc. IEEE Conf. on Decision and Control*, Dec. 1992.

[31] D. J. Hill and P. J. Moylan, "Dissipative dynamical systems: basic input-output and state properties," *Journal of the Franklin Institute*, vol. 309, pp. 327–357, 1980.

[32] D. J. Hill, "Dissipativeness, stability theory and some remaining problems," in *Analysis and Control of Nonlinear Systems*, North Holland, 1988.

[33] R. W. Brockett and J. L. Willems, "Frequency domain stability criteria – part I and II," *IEEE TAC*, vol. AC–10, no. 3,4, pp. 255–261,407–412, 1965.

[34] M. A. L. Thathachar and M. D. Srinath, "Some aspects of the Lur'e problem," *IEEE TAC*, vol. AC–12, pp. 451–453, Aug. 1967.

[35] E. A. Guillemin, *Synthesis of Passive Networks.* Chapman and Hall Ltd., 1957.

[36] Z. V. Rekasius and J. E. Gibson, "Stability analysis of nonlinear control systems by the second method of Liapunov," *IRE Transactions on Automatic Control*, pp. 3–15, January 1962.

[37] C. A. Desoer and M. Vidyasagar, *Feedback Systems: Input-Output Properties.* New York: Academic Press, 1975.

[38] J. P. How, *Control Techniques for Structural Systems with Real Parametric Uncertainties.* PhD thesis, Department of Aeronautics and Astronautics, M.I.T., Cambridge, MA, manuscript in preparation 1993.

[39] W. M. Haddad, J. P. How, S. R. Hall, and D. S. Bernstein, "Extensions of mixed-μ bounds to monotonic and odd monotonic nonlinearities using absolute stability theory," Submitted to the *Int. J. Cont.*, Oct. 1992.

[40] V. M. Popov, "On absolute stability of non-linear automatic control systems," *Avtomatika I Telemekhanika*, vol. 22, no. 8, pp. 961–979, 1961.

[41] H. L. Trentelman and J. C. Willems, "The dissipation inequality and the algebraic Riccati equation," in *The Riccati Equation*, pp. 197–242, Springer-Verlag, 1991.

[42] J. C. Doyle, "Analysis of feedback systems with structured uncertainties," *IEE Proceedings*, vol. 129, Part D, pp. 242–250, Nov. 1982.

[43] P. M. Young, M. P. Newlin, and J. C. Doyle, "μ analysis with real parametric uncertainty," in *Proc. CDC*, pp. 1251–1256, Dec. 1991.

[44] J. P. LaSalle, "Some extensions of Liapunov's second method," *IRE Trans. Cir. Theory*, pp. 520–527, 1960.

[45] W. M. Haddad and D. S. Bernstein, "Robust stabilization with positive real uncertainty: Beyond the small gain theorem," *Sys. and Con. Let.*, vol. 17, pp. 191–208, 1991.

[46] D. C. Hyland and D. S. Bernstein, "The optimal projection equations for fixed-order dynamic compensation," *IEEE TAC*, vol. AC–29, pp. 1034–1037, Nov. 1985.

[47] D. S. Bernstein and W. M. Haddad, "Robust stability and performance via fixed-order dynamic compensation with guaranteed cost bounds," *Math. Contr. Sig. Sys.*, vol. 3, pp. 139–163, 1990.

[48] D. S. Bernstein and W. M. Haddad, "LQG control with an $\mathcal{H}_\infty$ performance bound: A Riccati equation approach," *IEEE TAC*, vol. AC–34, pp. 293–305, Mar. 1989.

Proceedings of the 31st Conference
on Decision and Control
Tucson, Arizona • December 1992

REDUCED-ORDER ROBUST COMPENSATOR DESIGN WITH REAL PARAMETER UNCERTAINTIES

Hsi-Han Yeh, Chin S. Hsu* and Siva S. Banda
Wright Laboratory, WL/FIGC, WPAFB, OH 45433-6553

Abstract

An approach based on observer theory to obtain reduced-order robust H_∞ compensator is presented in this paper. For a class of linear uncertain systems with real parameter uncertainties and output disturbances, this new approach offers a numerical algorithm for designing reduced-order observer-based robust controllers.

I. Introduction

Various approaches have been proposed for solving the standard H_∞ design problem such as the two-Riccati formulation, the conjugation method and the J-spectral factorization among others. Extension of H_∞ control theory includes the recently developed H_∞/LTR and H_2/H_∞ design methodologies [1], [2]. A remarkable feature common to these different approaches on H_∞ design is that the resultant H_∞ compensators are observed-based controllers. With few exceptions, the underlying observers are full-order observers; that is, the compensator is of the same dimension as that of the plant.

In this paper, we address the issue of obtaining reduced-order observer-based compensators which stabilize a given plant with real-parameter uncertainties and satisfies an H_∞-norm bound from the disturbance input to the controlled output. Unlike several recent results [3], [4] on the robust H_∞ design based on full-order observers, our proposed approach offers a design procedure which incorporates the notion of disturbance decoupling leading to reduced-order robust H_∞ compensators.

II. Problem Statement

The robust control design problem is considered. The plant is assumed to have bounded real parameter uncertainties in A and B matrices and disturbance signals are injected at both the input and the output. For clarity of presentation, the C matrix of the plant is assumed to have no parameter uncertainty. The proposed design method developed in this paper is applicable to the case $\Delta C \neq 0$ as well.

The plant is described by $\dot{x}(t) = (A + \Delta A)x(t) + (B + \Delta B)u(t) + G_1 w_1(t)$

$$y(t) = Cx(t) + G_2 w_1(t) \qquad (2.1)$$

$$z_1(t) = H_1 x(t) \quad , \quad z_2(t) = H_2 u(t)$$

*Visiting Scientist on leave from WSU,Pullman WA

where, x is an n-dimensional state vector, u is an m-dimensional control vector, w_1 is a q_1-dimensional disturbance input, y is the p-dimensional output, and z_1 and z_2 are controlled outputs of dimensions p_1 and p_2 respectively. It is assumed that {A, B, C} is a stabilizable and detectable triple. The matrices B and H_2 are of full column rank. {A, H_1} is a detectable pair.

The uncertainties in the plant parameters are assumed to be time invariant parameter variations which may be represented by

$$\Delta A = \sum_{j=1}^{r} D_j \Delta_j E_j \,, \quad \Delta B = \sum_{j=1}^{r} D_j \Delta_j F_j \qquad (2.2)$$

where Δ_j has the dimension of $k_j \times k_j$. The matrices D_j, E_j and F_j are fixed, and are scaled so that for all j the uncertainty Δ_j has the same singular value bound:

$$\overline{\sigma}(\Delta_j) \leq \rho \qquad \forall j \qquad (2.3)$$

The structure of the feedback controller K(s) is described by the following observer equations:

$$\dot{x}_o(t) = Fx_o(t) + Gu(t) + K_f y(t) \qquad (2.4)$$

$$u(t) = -Nx_o(t) - My(t)$$

where the dimension of the observer state x_o is n-p. The observer state is an estimate of a linear transformation of the plant state Tx. An error signal can be defined as:

$$e = Tx - x_o \qquad (2.5)$$

The closed-loop system is internally stable and the error signal e approaches zero asymptotically if the following Luenberger constraint equations are satisfied:

$$TA - FT = K_f C \qquad (2.6)$$

$$G = TB \qquad (2.7)$$

$$NT + MC = K_c \qquad (2.8)$$

for any K_c such that $A - BK_c$ is Hurwitz.

III. Theoretical Results

In this section, we follow our earlier work on the procedure of constructing Riccati equations for robust controller design [5]. The basic idea is to break down the robust performance design problem into smaller problems with each of the design goals taken individually, namely: H_∞ norm constraint for

CH3229-2/92/0000-2824$1.00 © 1992 IEEE

nominal system, stabilization with respect to ΔA, and stabilization with respect to ΔB. With state space representation obtained for each subproblem an overall Riccati constraint for the solution of the robust performance design can then be constructed.

To begin with, we impose (3.1) for the purpose of disturbance decoupling,

$$MG_2 = 0 \quad (3.1)$$

Without the above assumption, the ensuing robust compensator design will involve tightly-coupled Riccati and Sylvester equations which are not mathematically tractable. Justification of this assumption will become clear in the sequel. Equipped with (3.1), we obtain the following equations:

$$u = -K_c x + Ne \quad (3.2)$$

$$\dot{x} = (A-BK_c)x + BNe + \Delta Ax + \Delta Bu \quad (3.3)$$

$$\dot{e} = Fe + T\Delta Ax + T\,\Delta Bu + (\gamma^{-1}TG_1 - K_f\gamma^{-1}G_2)w_1 \quad (3.4)$$

$$z_2 = -H_2K_c x + H_2Ne \quad (3.5)$$

and the closed-loop transfer function matrix from the disturbance to the controlled output is obtained as $\{A_{cl}, B_{cl}, C_{cl}\}$, where

$$A_{cl} = \begin{bmatrix} A - BK_c & BN \\ 0 & F \end{bmatrix}, \quad x_{cl} = \begin{bmatrix} x \\ e \end{bmatrix}$$

$$B_{cl} = \begin{bmatrix} G_1 & \tilde{D} & \tilde{D} \\ \gamma^{-1}(TG_1 - K_fG_2) & T\tilde{D} & T\tilde{D} \end{bmatrix}$$

$$C_{cl} = \begin{bmatrix} H_1 & 0 \\ -H_2K_c & H_2N \\ \rho\tilde{E} & 0 \\ -\rho\tilde{F}K_c & \rho\tilde{F}N \end{bmatrix} \quad z = \begin{bmatrix} z_1 \\ z_2 \\ z_3 \\ z_4 \end{bmatrix} \quad (3.6)$$

$$w^T = \begin{bmatrix} w_1 & w_2 & w_3 \end{bmatrix}^T$$

From equations (3.3) and (3.4) or from (3.6) it is easily seen that the coefficient matrices of the disturbance rejection of the nominal system, and the stabilization with respect to DA and DB are $\{A_i, B_i, C_i\}$, i = 1, 2, 3, respectively, as given below.

$$A_1 = A_2 = A_3 = A_{cl}$$

$$B_1 = \begin{bmatrix} G_1 \\ \gamma^{-1}(TG_1 - K_fG_2) \end{bmatrix} \quad (3.7)$$

$$B_2 = B_3 = \begin{bmatrix} \tilde{D} \\ T\tilde{D} \end{bmatrix}, \quad C_1 = \begin{bmatrix} H_1 & 0 \\ -H_2K_c & H_2N \end{bmatrix}$$

$$C_2 = \begin{bmatrix} \rho\tilde{E} & 0 \end{bmatrix}, \quad C_3 = \begin{bmatrix} -\rho\tilde{F}K_c & \rho\tilde{F}N \end{bmatrix}$$

The overall Riccati constraint for robust performance can now be formulated as

$$R(P) \triangleq PA_{cl} + A_{cl}^T P + P\sum_{i=1}^{3} d_i^2 B_i B_i^T P + \sum_{i=1}^{3} d_i^{-2} C_i^T C_i \le 0 \quad (3.8)$$

We shall look for a solution to (3.8) in the block diagonal form:

$$P = \begin{bmatrix} P_1 & 0 \\ 0 & P_2 \end{bmatrix} \quad (3.9)$$

Substituting (3.9) into (3.8) and taking advantage of the inequality

$$X^TY + Y^TX \le aX^TQX + \frac{1}{a}Y^TQ^{-1}Y \quad (3.10)$$

for any a > 0 and Q > 0, we obtain an upper bound for R(P)

$$R(P) \le \begin{bmatrix} U_{11} & U_{12} \\ U_{12}^T & U_{22} \end{bmatrix} \quad (3.11)$$

Setting $U_{12} = 0$, and U_{11} and U_{22} to be negative semi-definite, we obtain

$$K_c = \phi^{-1} B^T P_1 \quad (3.12)$$

$$U_{11} = P_1 A + A^T P_1 + Q_1 + P_1 X P_1 \le 0 \quad (3.13)$$

$$U_{22} = P_2 F + F^T P_2 + Q_2 + P_2 Y P_2 \le 0 \quad (3.14)$$

$$\phi = d_1^{-2} H_2^T H_2 + d_3^{-2} \rho^2 \tilde{F}^T\tilde{F} \quad (3.15)$$

$$X = d_1^2 (1+a)\gamma^{-2} G_1 G_1^T + a\tilde{D}\tilde{D}^T - B\phi^{-1}B^T \quad (3.16)$$

$$a = d_2^2(1+b) + d_3^2(1+c) \quad (3.17)$$

$$Q_1 = d_1^{-2} H_1^T H_1 + d_2^{-2} \rho^2 \tilde{E}\tilde{E}^T \quad (3.18)$$

$$Q_2 = N^T \phi N \quad (3.19)$$

$$Y = d_1^2(1 + a^{-1}) SS^T + b\,T\tilde{D}\tilde{D}^T T^T \quad (3.20)$$

$$S = \gamma^{-1}(TG_1 - K_fG_2) \quad (3.21)$$

$$b = d_2^2(1 + b^{-1}) + d_3^2(1 + c^{-1}) \quad (3.22)$$

It is noticed here that all matrices in (3.15)-(3.18) are given data and all scalar tuning parameters are positive numbers.

IV A Numerical Algorithm

The robust H_∞ state feedback K_c can be computed via (3.12) after a Riccati equation (3.13) is solved to yield P_1, $P_1 \geq 0$. The matrix U_{22} needs to be negative semi-definite. Due to the fact that the observer system matrix F is a stability matrix, there exists a P_2, $P_2 \geq 0$ such that

$$P_2F + F^T P_2 + Q_2 \leq 0 \tag{4.1}$$

which is a special case of (3.14), i.e., Y= 0. Note that if Y $\neq$ 0, real solution to (3.14) may not exist. To satisfy the sufficient condition Y= 0 it suffices that

$$T\tilde{D} = 0\,,\; S \triangleq \gamma^{-1}(T\,G_1 - K_f\,G_2) = 0 \tag{4.2}$$

We are now in a position to propose a numerical algorithm for reduced-order robust H_∞ compensators using the following design equations:

$$\begin{aligned} &TA - FT = K_fC \\ &TG_1 = K_f\,G_2 \\ &T\tilde{D} = 0 \quad , \quad MG_2 = 0 \end{aligned} \tag{4.3}$$

Design Algorithm:

1. Perform QR decomposition of G_2, (assuming rank $(G_2) = q_1$ without loss of generality)

$$G_2 = Q\begin{bmatrix} R_1 \\ 0 \end{bmatrix}, \qquad Q^TQ = I, \;\; |R_1| \neq 0$$

$$E \triangleq Q^TC \triangleq \begin{bmatrix} E_1 \\ E_2 \end{bmatrix}$$

2. Solve a Sylvester equation for T,

$$T(A - G_1\,R_1^{-1}E_1) - FT = LE_2 \tag{4.4}$$

and $\quad T\tilde{D} = 0 \qquad (4.5)$

This step can be accomplished using the algorithm proposed by Monahemi et.al. [6].

3. $K_f = [T\,G_1\,R_1^{-1} \quad L]\,Q^T \qquad (4.6)$

4. $[M \quad N] = [K_c \quad 0]\begin{bmatrix} C & G_2 \\ T & 0 \end{bmatrix}^{-1} \qquad (4.7)$

This step requires the compensator order be chosen as $n_c = n - p + q_1$ and $p > q_1$ for the condition $MG_2 = 0$ to be satisfied. .The proposed approach requires solving only one Riccati equation in conjunction with a Sylvester equation. It should also be noted that Monahemi's algorithm requires that the rank of $\tilde{D}$ be less than p, $C\tilde{D}$ has full rank, and (A, $\tilde{D}$, C) has no transmission zeros. As long as the parameter uncertainties are not too densely distributed throughout A and B, there is flexibility in selecting $\tilde{D}$ to meet this requirement.

V. Conclusions

In this paper, a new algorithm to design reduced-order H_∞ compensators for a class of linear uncertain control systems with real-parameter uncertainties is presented. The effects of disturbances at the input and the output are incorporated in the controller synthesis. The order of Luenberger compensator has been shown to be affected by the number of disturbance variables and the number of uncertain real parameters. The challenging issue of designing reduced-order H_∞ compensators for a general class (4-block) of uncertain systems with structured uncertainties has yet to be fully explored.

References:

[1] J. Stoustrup and H. H. Niemann, "An H_∞/LTR Method for Robust Controller Design," AIAA Guidance, Navigation and Control Conference, New Orleans, Louisiana, 1991, pp 1160-1171.

[2] A. G. Sparks, H. H. Yeh and S. S. Banda, "Mixed H_2 and H_∞ Optimal Robust Control Design," Optimal Control Applications and Methods, Vol 11, No. 4, Oct-Dec 1990, pp 307-325.

[3] Y. J. Wang and L. S. Shieh, "Observer-Based Robust H_∞ Control Laws, "AIAA Guidance, Navigation and Control Conference, New Orleans, Louisiana, 1991, pp 741-751.

[4] F. Jabbari and W. E. Schmitendorf, "Effects of Using Observers on Stabilization of Uncertain Linear Systems," American Control Conference, Boston MA, 1991, pp 3131-3136.

[5] H. H. Yeh, J. L. Rawson and S. S. Banda, "Robust Control Design with Real-Parameter Uncertainties" American Control Conference, Chicago IL, 1992, pp.3249-3256.

[6] M. M. Monahemi, J. B. Barlow and D. P. O'Leary, "The Design of Reduced-Order Luenberger Observers with Precise LTR," AIAA Guidance Navigation and Control Conference, New Orleans, Louisiana, 1991, pp 1145-1159.

Proceedings of the 31st Conference
on Decision and Control
Tucson, Arizona • December 1992

DECOMPOSITION OF NEAR-OPTIMUM REGULATORS WITH NONSTANDARD MULTIPARAMETER SINGULAR PERTURBATIONS

N. Eva Wu† Yue-Y. Wang* Paul M. Frank‡ H. Wang*

†Department of Electrical Engineering
SUNY–Binghamton
Binghamton, NY 13902-6000

*Department of Electrical and Computer Engineering
Syracuse University
Syracuse, NY 13244

‡Meß und Regelungstechnik, FB9, Duisburg University
4100 Duisburg 1, Germany

Abstract

Near-optimal control of multiparameter nonstandard singularly perturbed systems is studied using the descriptor variable approach. The slow models of the considered systems may exhibit impulses. Under stabilizable and detectable assumptions on the slow models and block-D stabilizable assumption on the fast or boundary layer systems, the near-optimum regulators are completely decomposed. It is shown that the near optimal control is a cascade connection of a slow and fast subregulators that are independent of perturbed parameter vector ϵ, and that both the slow and the fast subregulators can be designed separately. In addition, the frequency responses of the near-optimum systems are derived, which give further insights into the behavior of near-optimal controls. Several numerical design examples are presented.

1.Introduction

Due to its wide applications in power system dynamics, control of large scale systems and electrical networks, etc. [5], the singular perturbation theory has been extended to systems with multiparameter singular perturbations. Most earlier results obtained in this area were centered in analyzing asymptotic stability of such systems (e.g. [7,9,2]). Some related control issues were studied in [8]. The result ,however, was available only for the systems with multiparameter standard singular perturbations. That means some nonsingularity condition is imposed on the subsystems of fast time scales. For a two time scale system, $\dot{x}_1 = A_1x_1 + A_2x_2, \epsilon\dot{x}_2 = A_3x_1 + A_4x_2$, standard singular perturbation requires that A_4 be nonsingular.

This paper studies the minimization problem

$$J = \frac{1}{2}\int_0^\infty (y'y + u'Ru)dt \tag{1}$$

subject to a singularly perturbed system, which may not be standard,

$$\begin{aligned} \dot{x} &= A_0x + \sum_{j=1}^{N} A_{0j}z_j + B_0u \quad x(0) = x_0 \\ \epsilon_i\dot{z}_i &= A_{i0}x + \sum_{j=1}^{N} A_{ij}z_j + B_iu \quad z_i(0) = z_{i0} \end{aligned} \tag{2}$$

where

$$y = C_0x + \sum_{j=1}^{N} C_jz_j$$

and the vectors x, z_i, u and y are respectively n_0, n_i, m and p dimensional with $n = \sum_{i=0}^{N} n_i$. $\epsilon_1, \epsilon_2 \dots \epsilon_N$ are small positive scalars, representing small unknown parasitic, such as small masses, stray capacitances and inductors. Denote the vector $\epsilon = [\epsilon_1, \cdots, \epsilon_N]'$. ϵ is limited to lie within a certain set. According to the ratios of ϵ_i/ϵ_j, the optimal control problem (1) and (2) is classified into the problem with multiple-time scale singular perturbations, or the one with multiparameter singular perturbations. The former is governed by $\epsilon_{i+1}/\epsilon_i \to 0$ as $\|\epsilon\| \to 0$, and the latter is governed by that the possible values of ϵ are restricted to a cone $H \in R^N$ [8]. The nonstandard singular perturbations mean that the reduced system of (2) by setting $\epsilon = 0$ may exhibit impulses.

Since ϵ, stimulating the higher order of the system, may not be known exactly, the difficulty arises when solving the Riccati equation related to the problem (1)(2). Hence, we turn to seeking for a state feedback control of the form

$$u(t) = K[x' \ z_1' \cdots z_N']' \tag{3}$$

where the $m \times n$ constant matrix K is independent of ϵ, and the control law (3) serves as a near-optimal solution for the problem (1)(2) as $\|\epsilon\|$ is sufficiently small.

In [6,4,13] and [12] etc., the researchers studied the optimal control problem with single parameter singular perturbation, and the output control problem as well. Among them were the factorization approach in the frequency domain [12] and the descriptor variable approach [13], which extended the result of composite control [4] to nonstandard singularly perturbed systems where A_4 may be singular.

Near-optimum control of multitime scale systems (2) has been discussed in the past, for example, in [18,16]. The pioneer work on the multiparameter standard optimal control problem was due to [8]. Under the assumption on block D-stability, [8] gave a near-optimal solution to the problem. The resulting near-optimal regulator, however, depended on the perturbed parameters ϵ. The solution was decoupled from ϵ only when the reduced part of control was applied and the fast subsystem matrix was block-D stable.

This paper presents a unified approach to the optimal control problem expressed in (1) and (2), with both standard and nonstandard multiparameter singular perturbations. A complete answer to how one can find the near-optimal control law in the form of (3) is given. The paper is organized as follows. In Section Two, stabilizability and detectability on the slow subsystem of (2) and the block-D stabilizability on its fast subsystem are defined. Then Section Three discusses how to decompose the near optimum regulators for the problem (1) and (2). The near-optimal regulator gains can be found by solving a generalized Riccati equation related to the reduced system of (2). The solution of the generalized Riccati equation is decomposed into the solutions of a slow and a fast subregulator problems, both of which are mutually independent and decoupled from ϵ. As a result, the near-optimal control can be realized as a cascade connection of the fast and the slow regulators. It is also found that the near-optimum regulator in the form of (3) is not unique, all of near-optimal regulators achieve $O(\|\epsilon\|)$ approximation of the optimal value of the performance index. Furthermore, from the frequency point of view, spectral factorization of the near-optimal systems is derived, Based on the theory of two-frequency-scale transfer functions, the asymptotic properties of the near-optimal systems are discussed from their frequency responses. In the final Section, several design examples are presented.

2. Preliminaries

CH3229-2/92/0000-2827$1.00 © 1992 IEEE

Before investigating the control problem, we describe some structural properties of the original system (2) for latter use.

The reduced or slow system of (2) is referred to the one there by setting $\epsilon = 0$.

$$\begin{aligned} \dot{\bar{x}} &= A_0\bar{x} + A_{0f}\bar{z} + B_0 u \quad x(0) = x_0 \\ 0 &= A_{f0}\bar{x} + A_f\bar{z} + B_f u \\ y &= C_0\bar{x} + C_f\bar{z} \end{aligned} \tag{4}$$

where the matrices A_{0f}, A_{f0}, A_f and B_f as well as C_f are formed of the block matrices A_{0i}, A_{i0}, A_{ij} and B_i, C_j $i, j = 1, \ldots, N$, respectively. If the perturbed parameters in (2) are of the same order, the ratios of $\epsilon_1, \epsilon_2 \ldots \epsilon_N$ are therefore assumed bounded by some positive constants $\bar{m}_{ij}$, $\bar{M}_{ij}$,

$$\bar{m}_{ij} \le \frac{\epsilon_i}{\epsilon_j} \le \bar{M}_{ij}, \; i, j = 1, \ldots, N \tag{5}$$

The system (2) can be rewritten in a form resembling a single parameter system as.

$$\begin{aligned} \dot{x} &= A_0 x + A_{0f} z + B_0 u \quad x(0) = x_0 \\ \mu\dot{z} &= DA_{f0}x + DA_f z + DB_f u \\ y &= C_0 x + C_f z \end{aligned} \tag{6}$$

where $\mu = \mu(\epsilon) = (\epsilon_1\epsilon_2\ldots\epsilon_N)^{1/N}$, $z' = [z_1', \cdots, z_N']$ and

$$D = D(\epsilon) = block - diag[\frac{\mu}{\epsilon_1}I_1, \ldots, \frac{\mu}{\epsilon_N}I_N] \tag{7}$$

The elements of D are bounded by

$$m_i \le \frac{\mu}{\epsilon_i} \le M_i, \; \forall\epsilon \in H \;, i = 1, \ldots, N \tag{8}$$

where H is the cone defined by (5). The following theorem shows that the asymptotic stability of the full system (2) is guaranteed by the stabilities of the reduced system and the boundary layer system below

$$\begin{aligned} \frac{d\eta(\tau)}{d\tau} &= D(\epsilon)A_f\eta(\tau) + D(\epsilon)B_f u, \tau = t/\mu \\ \xi &= C_f\eta(\tau) \end{aligned} \tag{9}$$

Theorem 1: If

$$Re\lambda(A_0 - A_{0f}A_f^{-1}A_{f0}) < 0 \tag{10}$$

$$Re\lambda(DA_f) < 0, \forall\epsilon \in H \tag{11}$$

the original system (2) is asymptotically stable for all $\epsilon \in H, 0 < \|\epsilon\| \le \nu$, for some positive scalar ν [8].

There is an important class of matrices satisfying (11), which are called block-D stable matrices. A matrix A_f is said to be block D-stable if $Re\lambda(DA_f) < 0$ for all $D(\alpha) = block - diag[\alpha_1 I_1, ..., \alpha_N I_N]$ with arbitrary positive scalars α_i [7]. If we extend the cone H to $\epsilon_i > 0$ and $m_i \to 0$ and $M_i \to \infty$, the set H covers the regions of multitime scale perturbations.

In the slow system (4) when the matrix A_f is nonsingular, the system (2) is referred to the one with standard singular perturbations. When A_f is singular, the system (2) is referred to a nonstandard singularly perturbed system. In the case of nonstandard singular perturbations, the reduced system (4) will exhibit impulses [10]. Besides, A_f will never be a block-D stable matrix, for

$$det(D(\epsilon)A_f) = 0 \;\; \forall\epsilon \in H \tag{12}$$

Therefore, the decomposition theory about the nonstandard singular perturbation problem (1)(2) must be further investigated.

As we know, controllability is the essential condition for stabilizing a control system by state feedback. In order to stabilize an unstable system (2), we define some necessary structural notions related to controllability and observability.

Definition 1: The finite modes of the slow system (4) is said to be stabilizable and detectable if

$$rank\begin{bmatrix} sI - A_0 & -A_{0f} & B_0 \\ -A_{f0} & -A_f & B_f \end{bmatrix} = n, \; rank\begin{bmatrix} sI - A_0' & -A_{f0}' & C_0' \\ -A_{0f}' & -A_f' & C_f' \end{bmatrix} = n, \tag{13}$$

Definition 2: The boundary layer system (9) is said to be block D-stabilizable by constant gains, if there exists a constant feedback P_f such that $(A_f - B_f P_f)$ is block-D stable.

Definition 3: The boundary layer system (9) is said to be block D-detectable if $(D(\alpha)A_f, C_f)$ is detectable for all $\alpha_i > 0$.

In modelling of large scale systems, A_f may be singular, Even if A_f is nonsingular, A_f is in general not block D-stable. What Definition 2 requires is that there exists a constant feedback gain to make the fast system (9) block D-stable.

3. Near-Optimal Control

In this section, near-optimal control of the system (2) with multi-parameter singular perturbations is studied. Since A_f is not required to be nonsingular, it may be difficult to decompose the near-optimum regulators when A_f is singular, for any preliminary feedback making A_f invertible will change the performance index (1) too. Thus couplings between the fast and slow subregulators may become complex. However, the problem can be solved using the descriptor variable approach. First, we make the following assumptions.

Assumption 1: The finite modes of the slow system (4) is stabilizable and detectable.

Assumption 2: a).The fast system (9) is block D-stabilizable by constant gains; and b).the system (9) is block D-detectable.

Remark 1 A necessary condition for Assumption 2a). to hold is that $(D(\alpha)A_f, D(\alpha)B_f)$ is stabilizable for all $\alpha_i > 0$.

Suppose $\mu(\epsilon)$ is sufficiently small and known exactly, say at ϵ_0. If Assumption 1 holds and the triple $(D(\epsilon_0)A_f, D(\epsilon_0)B_f, C_f)$ is stabilizable and detectable, it is shown that a near-optimum feedback gain for the original problem (1)(2) is identical to the optimal one of the descriptor LQ regulator problem below [13].

$$\Im(v) = \frac{1}{2}\int_0^\infty (y'y + v'Rv)dt \tag{14}$$

$$\begin{aligned} \dot{\bar{x}} &= A_0\bar{x} + A_{0f}\bar{z} + B_0 v \quad x(0) = x_0 \\ 0 &= D(\epsilon_0)A_{f0}\bar{x} + D(\epsilon_0)A_f\bar{z} + D(\epsilon_0)B_f v \\ y &= C_0\bar{x} + C_f\bar{z} \end{aligned} \tag{15}$$

where (15) is obtained from (6) by setting $\mu(\epsilon) = 0$. Let v^* denote the optimal control for (14)(15). It turns out that

$$\begin{aligned} v^* &= -R^{-1}[B_0' \;\; B_f'D']P(\epsilon_0)\begin{bmatrix} \bar{x} \\ \bar{z} \end{bmatrix} \\ &= -R^{-1}[B_0' \;\; B_f'D']\begin{bmatrix} P_0 & 0 \\ P_{f0}(\epsilon_0) & P_f(\epsilon_0) \end{bmatrix}\begin{bmatrix} \bar{x} \\ \bar{z} \end{bmatrix} \end{aligned} \tag{16}$$

and the corresponding near-optimum control, denoted by $\hat{u}$, is [13]

$$\hat{u} = -R^{-1}[B_0' \;\; B_f'D']P(\epsilon_0)\begin{bmatrix} x \\ z \end{bmatrix} \tag{17}$$

In (16) P_0 is the positive semidefinite solution of an algebraic Riccati equation independent of ϵ, $P_{f0}(\epsilon_0)$ is a linear combination of P_0 and $P_f(\epsilon_0)$, in particular, $P_f(\epsilon_0)$ is the positive semidefinite solution of the Riccati equation

$$\begin{aligned} &P_f(\epsilon_0)D(\epsilon_0)A_f + A_f'D(\epsilon_0)P_f(\epsilon_0) \\ &-P_f(\epsilon_0)D(\epsilon_0)B_f R^{-1}B_f'D(\epsilon_0)P_f(\epsilon_0) + C_f'C_f = 0 \end{aligned} \tag{18}$$

As a whole, $P(\epsilon_0)$ satisfies the following ϵ-dependent generalized Riccati equation evaluated at ϵ_0

$$\begin{cases} (i) \; E'P(\epsilon) = P(\epsilon)'E \\ (ii) \; P(\epsilon)'A_D + A_D'P(\epsilon) - P(\epsilon)'B_D R^{-1}B_D'P(\epsilon) + C'C = 0 \end{cases}$$

where $E = diag[I,\ 0]$ and A_D, B_D are defined in (55). (18) defines a fast subregulator problem which optimizes the boundary layer system (9). In view of (18), unless $\epsilon \to 0$ along the particular path that keeps $D(\epsilon)$ a constant matrix equal to $D(\epsilon_0)$, the values of the elements of D will vary with ϵ. Hence for arbitrary sufficiently small $\epsilon \in H$, the solution $P_f(\epsilon)$, thus $P(\epsilon)$ will continuously depend on ϵ.

Now let the optimal control for the original problem (1)(2) be

$$u^* = -R^{-1}[B_0' \ \frac{1}{\mu}B_f' D']K(\epsilon)\begin{bmatrix} x \\ z \end{bmatrix} \tag{19}$$

where

$$K(\epsilon) = \begin{bmatrix} K_1(\epsilon) & \mu(\epsilon)K_2(\epsilon) \\ \mu(\epsilon)K_2'(\epsilon) & \mu(\epsilon)K_3(\epsilon) \end{bmatrix} \tag{20}$$

is the stabilizing solution of an algebraic Riccati equation. The limiting behavior of $K(\epsilon)$ as $\epsilon \to 0$ along any arbitrary path is given in the following theorem:

Theorem 2: If Assumptions 1 and 2 hold, then for sufficiently small $\epsilon \in H$,

$$\begin{aligned} K_1(\epsilon) &= P_0 + O(\|\epsilon\|) & (21)\\ K_2(\epsilon) &= P_{f0}(\epsilon) + O(\|\epsilon\|) & (22)\\ K_3(\epsilon) &= P_f(\epsilon) + O(\|\epsilon\|) & (23) \end{aligned}$$

where P_0, $P_{f0}(\epsilon)$ and $P_f(\epsilon)$ are the blocks in $P(\epsilon)$, and

$$P(\epsilon) = \begin{bmatrix} P_0 & 0 \\ P_{f0}(\epsilon) & P_f(\epsilon) \end{bmatrix}$$

is the stabilizing solution of the above generalized Riccati equation.

The proof of Theorem 2 is omitted due to the limited space. It is seen from (21,22,23) that the near-optimum feedback gains well approximate the optimal ones, and in general the near-optimal control is of the form (17) where $D(\epsilon_0)$ and $P(\epsilon_0)$ are replaced by $D(\epsilon)$ and $P(\epsilon)$. However, this near optimal control is difficult to realize, for one can not implement the feedback (17) unless ϵ is known exactly.

Taking robustness to unmodelled dynamics into account, we shall find an ϵ -independent control. The control is near-optimal for the performance index (1) as $\epsilon \to 0$ along any arbitrary path in the cone H.

Near-Optimal Control Independent of ϵ

Directly setting $\epsilon = 0$ in the problem (1)(2) yields the reduced or the descriptor variable LQ regulator problem.

$$\Im(v) = \frac{1}{2}\int_0^\infty (y'y + v'Rv)dt \tag{24}$$

$$\begin{aligned} E\dot{w} &= Aw + Bv \quad x(0) = x_0 \\ y &= Cw \end{aligned} \tag{25}$$

where $w' = [\bar{x}' \ \bar{z}']$ and

$$E = \begin{bmatrix} I & 0 \\ 0 & 0 \end{bmatrix} \ A = \begin{bmatrix} A_0 & A_{0f} \\ A_{f0} & A_f \end{bmatrix} \ B = \begin{bmatrix} B_0 \\ B_f \end{bmatrix} \ C = \begin{bmatrix} C_0 & C_f \end{bmatrix} \tag{26}$$

This problem is equivalent to the problem (14)(15), but with no ϵ contained herein. The solution to the problem (24)(25) is obtained in [11,13]. The results are briefly summarized as follows.

Theorem 3: Under Assumption 1 and the triple (A_f, B_f, C_f) is stabilizable and detectable (the latter is assured by Asumption 2), there exists a constant matrix P satisfying the generalized Riccati equation

$$\begin{cases} (i)\ E'P = P'E \\ (ii)\ P'A + A'P - P'BR^{-1}B'P + C'C = 0 \end{cases} \tag{27}$$

and the stabilizing optimal control for (24)(25) is

$$v^* = -R^{-1}B'Pw \tag{28}$$

The generalized Riccati equation (27) was first derived by Clements [17]. The stabilizing solution P appears the lower-triangular block form

$$P = \begin{bmatrix} P_0 & 0 \\ P_{f0} & P_f \end{bmatrix} \tag{29}$$

Calculating the matrix P is divided into two major steps as shown below [11]. Denote

$$T_1 = \begin{bmatrix} A_0 & -B_0R^{-1}B_0' \\ -C_0'C_0 & -A_0' \end{bmatrix} T_2 = \begin{bmatrix} A_{0f} & -B_0R^{-1}B_f' \\ -C_0'C_f & -A_{f0}' \end{bmatrix}$$
$$T_3 = \begin{bmatrix} A_{f0} & -B_fR^{-1}B_0' \\ -C_f'C_0 & -A_{0f}' \end{bmatrix} T_4 = \begin{bmatrix} A_f & -B_fR^{-1}B_f' \\ -C_f'C_f & -A_f' \end{bmatrix} \tag{30}$$

- Step One: Compute

$$\begin{bmatrix} F_0 & -S_0 \\ -Q_0 & -F_0' \end{bmatrix} = T_1 - T_2T_4^{-1}T_3 \tag{31}$$

Then find the positive semidefinite solution P_0 from the algebraic Riccati equation

$$P_0F_0 + F_0'P_0 - P_0S_0P_0 + Q_0 = 0 \tag{32}$$

- Step Two: Find the positive semidefinite solution form the Riccati equation

$$P_fA_f + A_f'P_f - P_fB_fR^{-1}B_f'P_f + C_f'C_f = 0 \tag{33}$$

Calculating P_{f0} is trivial, which is the solution of the algebraic equation below.

$$P_{f0} = [P_f \ \ I]T_4^{-1}T_3\begin{bmatrix} I \\ P_0 \end{bmatrix} \tag{34}$$

Remark 2: In Step Two, instead of solving the Riccati equation (33), one can also choose P_f arbitrarily as long as $(A_f - B_fR^{-1}B_f'P_f)$ is nonsingular. Each of these different choices will also result in an optimal feedback gain to (28), hence the optimal feedback gains are parametrized [11]. However, the resulting matrix P in (29) will no more satisfy the generalized Riccati equation (27) except the one where its block P_f satisfies the Riccati equation (33). As an important property pointed out here, the computational steps one and two are mutually separated.

Next we shall prove that, a subset of the optimal regulators defined by (28) constitute near-optimum regulators for the original problem (1)(2). This subset, say Ω, is defined as follows. The elements in Ω are those optimal regulator gains P in (29) where the block P_f makes $(A_f - B_fR^{-1}B_f'P_f)$ not only nonsingular but also block D-stable. For this subset of optimal feedback gains, we have

Theorem 4: Under Assumptions 1 and 2, each gain matrix P in the subset Ω becomes the near-optimum regulator gain for the original problem (1)(2); and the corresponding control

$$u_{nopt} = -R^{-1}B'P\begin{bmatrix} x \\ z \end{bmatrix} \tag{35}$$

is near-optimal for the performance index (1), which results in $J(u_{nopt})$ satisfying

$$J(u_{nopt}) = J(u^*) + O(\|\epsilon\|),\ \forall\epsilon \in H \tag{36}$$

Proof: Based on (20), the optimal value of the performance index is

$$J(u^*) = \frac{1}{2}[x_0' \ z_0']\begin{bmatrix} K_1(\epsilon) & \mu(\epsilon)K_2(\epsilon) \\ \mu(\epsilon)K_2'(\epsilon) & \mu(\epsilon)K_3(\epsilon) \end{bmatrix}\begin{bmatrix} x_0 \\ z_0 \end{bmatrix} \tag{37}$$

note that $K_1(\epsilon), K_2(\epsilon), K_3(\epsilon)$ are bounded and $\mu(\epsilon) = k\|\epsilon\|$, where k is a scalar. Substituting (21,22,23) into (37) yields

$$J(u^*) = \frac{1}{2}x_0'P_0x_0 + O(\|\epsilon\|) \tag{38}$$

From [3], the first term $1/2x_0'P_0x_0$ is just the optimal value of (14), hence

$$J(u^*) = \Im(v^*) + O(\|\epsilon\|) \tag{39}$$

On the other hand, Assumptions 1 and 2 guarantee the existence of that subset of optimal regulators P for the problem (24)(25). Suppose the feedback in (35) is implemented by a gain matrix P in the subset Ω, we substitute (35) into (2), and rewrite the resulting closed-loop system into a compact form like (25)

$$E(\epsilon)\begin{bmatrix} \dot{x} \\ \dot{z} \end{bmatrix} = (A - BR^{-1}B'P)\begin{bmatrix} x \\ z \end{bmatrix} \tag{40}$$

where $E(\epsilon) = diag[I, \epsilon_1 I, \ldots, \epsilon_N I]$ and

$$(A - BR^{-1}B'P) = \begin{bmatrix} \bar{A}_0 & \bar{A}_{0f} \\ \bar{A}_{f0} & \bar{A}_f \end{bmatrix} . \tag{41}$$

Let $\epsilon = 0$ in (40), one gets the reduced system

$$\begin{aligned} \dot{\bar{x}} &= \bar{A}_0\bar{x} + \bar{A}_{0f}\bar{z} \;\; x(0) = x_0 \\ 0 &= \bar{A}_{f0}\bar{x} + \bar{A}_f\bar{z} \end{aligned} \tag{42}$$

where $\bar{A}_f = (A_f - B_fR^{-1}B_f'P_f)$ is invertible. Obviously, the solution of (42) corresponds to the resulting optimal trajectory of the descriptor system (25) controlled by (28), thus denote the solution by $w^* = [\bar{x}^{*\prime}, \bar{z}^{*\prime}]'$. Since $\bar{A}_f$ is block D- stable, and $(\bar{A}_0 - \bar{A}_{0f}\bar{A}_f^{-1}\bar{A}_{f0})$ is stable (the latter is true, for the resulting optimal control system (25)(28) is stable). From [8], it is known that for all $\epsilon \in H$, $0 < \|\epsilon\| \le \nu$, the trajectories (40) can be approximated by $\bar{x}^*, \bar{z}^*$

$$\begin{aligned} x(t) &= \bar{x}^*(t) + O(\|\epsilon\|) \\ z(t) &= \bar{z}^*(t) + \eta(\tau) + O(\|\epsilon\|) \end{aligned} \tag{43}$$

where $\eta(\tau)$ defines the responses of the boundary layer system

$$\frac{d\eta}{d\tau} = D(\epsilon)(A_f - B_fR^{-1}B_f'P_f)\eta, \;\; \eta(0) = z_0 - \bar{z}_0 \tag{44}$$

and $\tau = t/\mu$ is the 'stretched' time scale. Substituting (43) into (35), then both u_{nopt} and (43) into the performance index (1), after arrangement, we get

$$J(u_{nopt}) = \frac{1}{2}\int_0^\infty (w^{*\prime}C'Cw^* + v^{*\prime}Rv^*)dt + J_0 \tag{45}$$

The first term is just the optimal value of (24), and the second term is given by

$$\begin{aligned} J_0 = \frac{\mu}{2}\int_0^\infty &(\eta'(\tau)(C_f'C_f + P_f'B_fR^{-1}B_f'P_f)\eta(\tau) \\ &+2\eta'(\tau)(C_f'C_0x^* + C_f'C_fz^* - P_f'B_fv^*))d\tau \end{aligned} \tag{46}$$

Since the integration $\int_0^\infty *d\tau$ in (46) is bounded, we have from (45)(46) that

$$J(u_{nopt}) = \Im(v^*) + O(\|\epsilon\|) \tag{47}$$

The optimal performance values of (14)(24) must be equal, subtracting (39) from (47) yields

$$J(u_{nopt}) = J(u^*) + O(\|\epsilon\|) \;\; \square\square \tag{48}$$

Remark 3: In review of the structure of (29), the near-optimum regulators are completely decomposed regardless of standard or nonstandard singular perturbations. The near-optimum regulators consist of two subregulators, designs of which are mutually independent. As we see, design of the slow regulator is to solve the Riccati equation (32). Such a solution always exists under Assumptions 1 and 2. Design of the fast regulator is to find a constant P_f to make $\bar{A}_f = (A_f - B_fR^{-1}B_f'P_f)$ block-D stable. This D-stabilization problem is isolated from the decomposition result and thus can be studied individually. A designer can choose an arbitrary block D-stabilizer P_f as the fast regulator gain. Each of this choice, when placed in (29), contributes a near-optimum control (35) to the original problem (1)(2). This set of near-optimal controls are attractive in application for they can be implemented without the exact knowledge of ϵ. Therefore, they are robust with respect to unmodelled dynamics.

In the fast regulator design, even though $(D(\epsilon)A_f, D(\epsilon)B_f, C_f)$ is stabilizable and detectable for all $\epsilon_i > 0$, a constant P_f which makes $\bar{A}_f$ block D-stable may not be always obtainable. However, when the coupling between the fast state variables is 'limited' in the sense described in [15], one is able to make $(A_f - B_fR^{-1}B_f'P_f)$ block D-stable. The detailed steps about how to do this were presented in [15]. In addition, in most multiparameter problems of interest, the small parameters $\epsilon_1, \ldots, \epsilon_N$, are of the same order of magnitude, hence the bounds m_i, M_i may be close to one. Let

$$D(\epsilon) = I + \bar{D}(\epsilon) \tag{49}$$

then the elements of $\bar{D}$ are "small". In this case, one may be likely to find a block D-stabilizing P_f for all $\epsilon \in H$

In fact, when $\epsilon_1, \ldots, \epsilon_N$, are of different order or of the same order but with sufficiently different magnitudes, we suggest to treat the problem (1)(2) as multitime scale singular perturbations [18].

Frequency Responses of Near-Optimum Systems

Suppose P_f is the positive semidefinite solution of (33) such that $(A_f - B_fR^{-1}B_f'P_f)$ is block D-stable or $D(A_f - B_fR^{-1}B_f'P_f)$ is stable for all $\epsilon \in H$. We call this P_f the quadratic D-stabilizer or simply quadratic stabilizer. As we see before, the gain matrix P in (29) with the quadratic stabilizer satisfies the generalized Riccati equation (27). From (35), denote the corresponding near-optimum feedback gain $K = -R^{-1}B'P$, we can derive the following factorization result.

Lemma 1: With the quadratic stabilizer, the near-optimum feedback gain K satisfies the equality (or spectral factorization):

$$\begin{aligned} &[I - K(-j\omega E(\epsilon) - A)^{-1}B]'[I - K(j\omega E(\epsilon) - A)^{-1}B] = \\ &I + B'(-j\omega E(\epsilon) - A')^{-1}C'C(j\omega E(\epsilon) - A)^{-1}B + \Delta_1(j\omega, \epsilon) \end{aligned} \tag{50}$$

where without loss of generality, R is taken as an identity matrix and

$$\Delta_1(s, \epsilon) = sB'(-sE(\epsilon) - A')^{-1}L(\epsilon)(sE(\epsilon) - A)^{-1}B \tag{51}$$

and

$$L(\epsilon) = P'E(\epsilon) - E(\epsilon)P \tag{52}$$

Proof: Since from (27) $E'P = P'E$, it can be checked that

$$-sE(\epsilon)P + sP'E(\epsilon) = sL(\epsilon) \tag{53}$$

Subtracting the equation (27,(ii)) from the identity (53) yields

$$P'(sE(\epsilon) - A) + (-sE(\epsilon) - A')P + K'R^{-1}K = C'C + sL(\epsilon) \tag{54}$$

Multiplying on the left by $B'(-sE(\epsilon) - A')^{-1}$ and on the right by $(sE(\epsilon) - A)^{-1}B$, then adding I to each side obtains the factorization identity (50). The proof is accomplished.

From optimal control theory [1], if $\Delta_1(s, \epsilon)$ were identically zero in (50), $u_{nopt} = K[x'z']'$ would be the optimal control. Therefore, $\Delta_1(j\omega, \epsilon)$ is the error term or performance loss in the frequency domain introduced by the near-optimal control. Define

$$A_D = \begin{bmatrix} A_0 & A_{0f} \\ D(\epsilon)A_{f0} & D(\epsilon)A_f \end{bmatrix} E(\mu) = \begin{bmatrix} I & 0 \\ 0 & \mu(\epsilon)I \end{bmatrix} B_D = \begin{bmatrix} B_0 \\ D(\epsilon)B_f \end{bmatrix} \tag{55}$$

Then $\Delta_1(s, \epsilon)$ can be rewritten as

$$\begin{aligned} \Delta_1(s, \epsilon) &= \mu(\epsilon)\bar{\Delta}_1(s, \epsilon) \\ &= \mu sB_D'(-sE(\mu) - A_D')^{-1}L(sE(\mu) - A_D)^{-1}B_D \end{aligned} \tag{56}$$

and

$$L = P'\begin{bmatrix} 0 & 0 \\ 0 & D^{-1}(\epsilon) \end{bmatrix} - \begin{bmatrix} 0 & 0 \\ 0 & D^{-1}(\epsilon) \end{bmatrix}P \tag{57}$$

Therefore, for sufficiently small $\epsilon \in H$, one has

$$\|\Delta_1(j\omega, \epsilon)\| \le \mu\|\bar{\Delta}_1\| \to O(\|\epsilon\|) \;\; \forall 0 \le \omega \le \omega_1 \tag{58}$$

where ω_1 is close to ∞ as $\mu \to 0$. In other words, the frequency response error $\Delta_1(j\omega, \epsilon)$ tends to $O(\|\epsilon\|)$ on compact set of ω, covering the slow frequency of interest. Hence, for sufficiently small ϵ, $\Delta_1(s,\epsilon)$ presents merely some limited high frequency errors, which is due to not having optimized the fast subsystem through (18). Even so, near optimality is still achieved (see(36)) for we do have optimized the slow system (4), the dominant part.

Consider a special case[1] of a standard two-time-scale singular perturbation LQR problem, where $D(\epsilon) = I$, an identity matrix, and $\mu(\epsilon) = \epsilon$ is a scalar. In this case, the Riccati equation (18) reduces to the Riccati equation (33), Hence, the quadratic stabilizer P_f optimizes the fast subsystem. This in turn suggests that there may be no high frequency error larger than $O(\epsilon)$. To show this, we note that the error term $\Delta_1(s,\epsilon)$ becomes

$$\Delta_1(s,\epsilon) = \epsilon s B'(-sE(\epsilon) - A')^{-1} \begin{bmatrix} 0 & P'_{f0} \\ -P_{f0} & 0 \end{bmatrix} (sE(\epsilon) - A)^{-1} B \quad (59)$$

It is a two-frequency-scale transfer function [14]. Using the theory of [14], we find the slow and the fast transfer functions of $\Delta_1(s,\epsilon)$, denoted as $H_s(s)$ and $H_f(p)$,

$$\begin{aligned} H_s(s) &= \Delta_1(s,0) = 0 & (60)\\ H_f(p) &= \Delta_1(\frac{p}{\epsilon},\epsilon)|_{\epsilon=0} & (61)\\ &= p[0, B'_f(pI - A'_f)^{-1}] \begin{bmatrix} 0 & P'_{f0} \\ -P_{f0} & 0 \end{bmatrix} \begin{bmatrix} 0 \\ (pI - A_f)^{-1} B_f \end{bmatrix} = 0 \end{aligned}$$

One has from [14] that

$$\begin{aligned} \|\Delta_1(j\omega,\epsilon)\| &= \|\Delta_1 - H_s(s) - H_p(\epsilon s)\| & (62)\\ &= O(\epsilon) \ \forall 0 \le \omega \le \infty \end{aligned}$$

The error term Δ_1 in (50) converges to $O(\epsilon)$ uniformly in all frequency ω in $j\omega$-axis or

$$\begin{aligned} &[I - K(-j\omega E(\epsilon) - A)^{-1}B]'[I - K(j\omega E(\epsilon) - A)^{-1}B] = \\ &\quad I + B'(-j\omega E(\epsilon) - A')^{-1}C'C(j\omega E(\epsilon) - A)^{-1}B + O(\epsilon) \quad (63) \end{aligned}$$

This gives an alternative description of near-optimal controls in the frequency domain.

Remark 4: For a nonstandard two-time scale LQR problem, the parallel result does not hold. This can be seen by using a counter example given by [11]

Example 1: Given the performance index

$$J = \frac{1}{2}\int_0^\infty (x_1^2 + x_2^2 + u^2)dt \quad (64)$$

subject to the nonstandard singular perturbation system

$$\begin{aligned} \dot{x}_1 &= x_2 & (65)\\ \epsilon \dot{x}_2 &= x_1 + u \end{aligned}$$

After solving the near-optimal control that contains the quadratic stabilizer P_f, we calculate the frequency performance loss

$$\Delta_1(j\omega,\epsilon) = \frac{2(1+\sqrt{2})\epsilon\omega^2}{(\epsilon\omega^2+1)(\epsilon\omega^2+1)} \quad (66)$$

This error Δ_1 will converge to $O(\epsilon)$ on compact set of ω, but not uniformly in ω, since (66) has a peak value of $(1+\sqrt{2})/2$ independent of ϵ. Thus we should take care in the nonstandard case while $\Delta_1(s,\epsilon)$ no longer belongs to the category of two-frequency-scale transfer functions defined in [14]. However, that (66) converges on compact set of ω means that the system at slow frequency is optimized, and near optimality can be achieved too, see (36).

To compare other near-optimum LQ regulators with the one containing the quadratic stabilizer, we carry out the spectral factorization of the other near-optimum systems.

For distinction, let $\bar{K} = -R^{-1}B'\bar{P}$ be a near-optimal feedback gain but with an arbitrary stabilizer P_f which makes $\bar{A}_f$ block D-stable. It can be shown that $\bar{K}$ satisfies the spectral factorization

$$\begin{aligned} &[I - \bar{K}\phi(-s)B]'[I - \bar{K}\phi(s)B] = \\ &\quad I + B'\phi'(-s)C'C\phi(s)B + \Delta_1(s,\epsilon) + \Delta_2(s,\epsilon) \quad (67) \end{aligned}$$

[1]the authors are indebted to the anonymous professor for this analysis

where $\phi(s) = (sE(\epsilon) - A)^{-1}$, $\Delta_1(s,\epsilon)$ is the same as that defined by (51) and

$$\Delta_2(s,\epsilon) = B'\phi'(-s)(\Delta P'A_c + A'_c\Delta P + \Delta P'BR^{-1}B'\Delta P)\phi(s)B \quad (68)$$

where $A_c = A - BR^{-1}B'P$ and $\Delta P = P - \bar{P}$. Both gain matrices P and $\bar{P}$ are defined by (29) but P contains the quadratic stabilizer P_f in its blocks. It is seen that $\Delta_2(s,\epsilon)$ is in general not $O(\|\epsilon\|)$ order. $\Delta_2(s,\epsilon) = 0$ only when $\bar{P} = P$. To this point, it is seen that the near-optimal control, which stabilizes the fast subsystem by the quadratic stabilizer, is superior to others, while the other near-optimum controls may result in larger frequency response errors.

4.Design Examples

Example 2: Given a performance index

$$J = \frac{1}{2}\int_0^\infty (y'y + 10u'u)dt \quad (69)$$

for the system

$$\begin{aligned} \begin{bmatrix} \dot{x} \\ \epsilon_1 \dot{z}_1 \\ \epsilon_2 \dot{z}_2 \end{bmatrix} &= \begin{bmatrix} -0.6 & 1.2 & 3 \\ 3 & 1 & 2 \\ 0.9 & 1 & 2 \end{bmatrix} \begin{bmatrix} x \\ z_1 \\ z_2 \end{bmatrix} + \begin{bmatrix} 1 \\ 2 \\ 3 \end{bmatrix} u & (70)\\ y &= x + 31.62z_1 + 31.62z_2 \end{aligned}$$

and ϵ_1 and ϵ_2 are restricted to a cone $H \in R^2$ i.e.

$$m_{21} \le \frac{\epsilon_2}{\epsilon_1} \le M_{21}, \quad m_{12} \le \frac{\epsilon_1}{\epsilon_2} \le M_{12} \quad (71)$$

where $m_{12}, m_{21}, M_{12}, M_{21}$ are some positive scalars. This is a multiparameter singularly perturbed optimal control problem. The boundary layer system of (70) is defined as

$$\frac{d\eta}{d\tau} = D(\epsilon)A_f\eta \quad (72)$$

where

$$D(\epsilon) = diag(\sqrt{\frac{\epsilon_2}{\epsilon_1}}, \sqrt{\frac{\epsilon_1}{\epsilon_2}}), \quad A_f = \begin{bmatrix} 1 & 2 \\ 1 & 2 \end{bmatrix} \quad (73)$$

Here A_f is singular, hence it is a nonstandard singularly perturbed control problem.

We try to find the near-optimal control by first solving the fast regulator problem

$$P_fA_f + A'_fP_f - 0.1P_fB_fB'_fP_f + C'_fC_f = 0 \quad (74)$$

where $B_f = [2\ \ 3]'$ $C_f = [31.62\ \ 31.62]$. The positive definite solution of (74) leads to a quadratic stabilizer for the closed-loop boundary layer system

$$\frac{d\eta}{d\tau} = D(\epsilon)\bar{A}_f\eta, \ \forall \epsilon \in H \quad (75)$$

for it can be checked that the resulting matrix

$$\begin{aligned} \bar{A}_f &= A_f - B_fR^{-1}B'_fP_f & (76)\\ &= \begin{bmatrix} -19.8223 & -19.6446 \\ -30.2335 & -30.4668 \end{bmatrix} \end{aligned}$$

is D-stable. Next, according to (32) we have the Riccati equation for the slow regulator

$$-0.0004p_0^2 - 22.882p_0 + 44.1002 = 0 \quad (77)$$

Solving (77) yields $p_0 = 1.9272$, then we can combine the slow and the fast regulator gains to get the near-optimum control.

$$u_{nopt} = K\bar{x} = [-1.1649\ \ -10.4111\ \ -10.8223]\bar{x} \quad (78)$$

where $\bar{x} = [x'\ z'_1\ z'_2]'$. Here the near optimal feedback gain K satisfies the equality (50). For the values $\epsilon_1 = 0.1$ and $\epsilon_2 = 0.2$, the optimal control for this example is

$$u_{opt} = [-1.3753\ -10.4426\ -10.8865]\bar{x} \quad (79)$$

For the initial condition $\bar{x}_0 = [0,1,1]'$, the values of the performance index are

$$J_{nopt} = 6.1005 \tag{80}$$

$$J_{opt} = 6.1003 \tag{81}$$

The performance loss is almost negligible. In fact, (78) constitutes a near optimal control for the problem (69)(70) as $\epsilon \to 0$ along any arbitrary path, because the stabilized boundary layer system (75) is D-stable.

Example 3: Given the same LQR problem (69)(70), we can find the near-optimal control by solving the slow and fast regulator problems like Example 2. As we know, the slow regulator problem has a unique solution, but there are various solutions to the fast regulator problem. For instance, by pole placement algorithm we find an P_f

$$P_f = \begin{bmatrix} 23 & 14 \\ 3 & 17 \end{bmatrix} \tag{82}$$

With this P_f, it can be checked that

$$A_f - B_f R^{-1} B_f' P_f = \begin{bmatrix} -10 & -13.8 \\ -15.5 & -21.7 \end{bmatrix} \tag{83}$$

which is D-stable too. Then from (29), we combine the obtained slow regulator gain $p_0 = 1.9272$ with the fast regulator gain (82) to find another near-optimum control.

$$u_p = [-15.2744 \;\; -5.5000 \;\; -7.9000]\bar{x} \tag{84}$$

For the values $\epsilon_1 = 0.1$, $\epsilon_2 = 0.2$, and the initial condition $\bar{x}(0) = [0 \; 1 \; 1]$, the performance value of the index (69) with this near-optimum control is

$$J_p = 8.2071 \tag{85}$$

Comparing (85) with (80)(81), we see that the near-optimum control u_{nopt} is even better than u_p, since in Example 2, the quadratic stabilizer is used to D-stabilize the fast subsystem (75). Instead, u_p converges to u^* slower. For further testing, given the same initial condition, but $\epsilon_1 = 0.001$ and $\epsilon_2 = 0.002$, the performance values of the index (69) under the controls u_{opt}, u_{nopt}, and u_p are respectively

$$J_{opt} = 0.0612, \quad J_{nopt} = 0.0612, \quad J_p = 0.0737 \tag{86}$$

5. Conclusion

This paper presents the basic theorems and a design approach for near-optimal control of systems with multiparameter nonstandard singular perturbations. Assumptions 1 and 2 are sufficient to guarantee the existence of an near optimum solution for the performance index (1).

Our work generalizes the standard result obtained in [4] for the single parameter problem. In multiparameter case, the near-optimum regulators can also be decomposed into a slow and a fast regulators. The slow regulator design requires to solve the algebraic Riccati equation, and the fast regulator design requires to find a stabilizing feedback gain P_f for the fast subsystem $(D(\epsilon)A_f, D(\epsilon)B_f)$, for all $\epsilon \in H$. Among the set of feedback gains which D-stabilize the fastsystem, the quadratic stabilizer P_f gives a better near-optimal control in the sense that the resulting return difference matrix will converge to that of the optimal system ,with an estimated slow frequency error of $O(\|\epsilon\|)$ order. The topics of the further study include how to select the weighting matrices in (33) so as to obtain the quadratic stabilizer P_f, as well as how to select other block D-stabilizers from the robust theory of interval matrices.

It can be seen that the descriptor variable theory has clarified the structural properties of the near-optimum regulators, and has led to an efficient algorithm for decomposing the regulators. The descriptor variable approach is especially useful in solving the nonstandard singularly perturbed LQR problems. Since the near-optimum controls are applicable without exact knowledge of ϵ, they are robust in the face of unmodelled dynamics in controlled systems.

References

[1] Anderson, B.D.O. and J.B. Moore (1971). Linear Optimal Control. *Prentice-Hall*,

[2] Abed, E.H. (1986). Decomposition and stability of multiparameter singular perturbation problems. *IEEE Trans. Aut. Control*, **AC-31**, 925-934.

[3] Bender, D.J. and A.J. Laub (1987). The linear quadratic regulator problem for descriptor variable systems. *IEEE Trans. Aut. Control*, **AC-32**, 672-688.

[4] Chow, J.H. and P.V. Kokotovic (1976). A decomposition of near-optimum regulators for systems with slow and fast modes. *IEEE Trans. Aut. Control*, **AC-21**, 701-705.

[5] Kokotovic, P.V. (1984). Applications of singular perturbation techniques to control problem. *SIAM Rev.*, **26**, 501-550.

[6] Kokotovic, P.V. and R.A. Yackel (1972). Singular perturbation of linear regulators: Basic theorems *IEEE Trans. Aut. Control*, **AC-17**, 29-37.

[7] Khalil, H.K. and P.V: Kokotovic (1979a). D-stability and multiparameter singular perturbation. *SIAM J. Control and Optimization*, **17**, 56-65.

[8] Khalil, H.K. and P.V. Kokotovic (1979b). Control of linear systems with multiparameter singular perturbations. *Automatica*,**15**, 197-207.

[9] Ladde, G.S. and D.D. Siljak (1983). Multiparameter singular perturbations of linear systems with multiple time scales. *Automatica*, **19**, 385-394.

[10] Verghese, G.C., B.C. Levy and T. Kailath (1981). A generalized state-space for singular systems. *IEEE Trans. Aut. Control*, **26**, 811-831.

[11] Wang, Y.Y., P.M. Frank and D.J. Clements (1990). The robustness properties of the linear quadratic regulators for singular systems. to appear in *IEEE Trans. Aut. Control*, Sept. 1992.

[12] Khalil, H.K., (1989). Feedback control of nonstandard singularly perturbed systems. *IEEE Trans. Aut. Control*, **AC-34**, 1052-1062.

[13] Wang, Y.Y., S.J. Shi and Z.J.Zhang (1988). A descriptor variable approach to singular perturbation of linear regulators. *IEEE Trans. Aut. Control*, **AC-33**, 370-373.

[14] Luse, D.W. and H.K.Khalil (1986). Frequency domain results for sytems with slow and fast modes. *IEEE Trans. Aut. Control*, **AC-31**, 918-924.

[15] Khalil, H.K. (1979). Stabilization of multiparameter singularly perturbed systems.*IEEE Trans. Aut. Control*, **AC-24**, 790-791.

[16] Özgüner, Ü.(1979). Near-optimal control of composite systems: The Multi Time-Scale Approach. *IEEE Trans. Aut. Control*, **AC-24**, 653-656.

[17] Clements, D.J. (1989). Spectral factorization of linear regulators for singular systems. Technical Report, Dept. Systems and Control, UNSW, Australia.

[18] Wang,Y.Y. and Paul M. Frank (1991). Parallel design of near-optimum regulators with multitime scale singular perturbations. *30 CDC Conf.* Brighton, UK.

Proceedings of the 31st Conference
on Decision and Control
Tucson, Arizona • December 1992

ON MODEL REDUCTION BY OUTPUT COVARIANCE APPROXIMATION

P. Resende† and M.M.J. Martini‡

†Departamento de Engenharia Eletrônica, Universidade Federal de Minas Gerais,
UFMG, C.P. 1294, Belo Horizonte, MG - 30160, Brazil
‡Departamento de Eletricidade, Fundação Ensino Superior de São João Del Rei,
FUNREI, São João Del Rei, MG - 36300, Brazil

ABSTRACT

This paper presents a model order reduction method for multivariable linear systems by output covariance matrix approximation. The reduced model is obtained from parameters of the complete model that are more important in the output covariance, by using a block-Schwarz state-space representation with stationary white noise inputs. For a class of multivariable systems the stability of the reduced model can be predictable.

1. INTRODUCTION

For multivariable systems the white noise response is proved useful in model reduction problems [1]. The present paper utilizes a block-Schwarz state-space form [2] of the complete model driven by white noise inputs to obtain a model reduction method based on an output covariance approximation. Using a stability condition [3] associated to the block-Schwarz form the stability of the reduced model can be predictable from the complete model.

It is considered an m-input/p-output linear system represented in the frequency-domain by a $p \times m$ matrix fraction description

$$G(s) = N(s)D^{-1}(s) \tag{1}$$

$$D(s) = Is^n + D_{n-1}s^{n-1} + D_{n-2}s^{n-2} + \cdots + D_1 s + D_0 \tag{2}$$

$$N(s) = N_{n-1}s^{n-1} + N_{n-2}s^{n-2} + \cdots + N_1 s + N_0 \tag{3}$$

where $D(s)$ and $N(s)$ are, respectively, $m \times m$ and $p \times m$ matrix polynomials with real constant matrix coefficients. It is assumed that $N(s)$ and $D(s)$ are right coprime and D_{n-1} is non-singular.

2. BLOCK-SCHWARZ FORM

For the system description (1)-(3) it is intended to obtain the corresponding state equation in the block-Schwarz form given below

$$\begin{aligned} \dot{x}(t) &= Mx(t) + Eu(t) \\ y(t) &= Kx(t) \end{aligned} \tag{4}$$

where

$$M = \begin{bmatrix} -M_1 & -M_2 & 0 & \cdots & 0 & 0 \\ I & 0 & -M_3 & \cdots & 0 & 0 \\ 0 & I & 0 & \cdots & 0 & 0 \\ \vdots & \vdots & \vdots & \ddots & \vdots & \vdots \\ 0 & 0 & 0 & \cdots & 0 & -M_n \\ 0 & 0 & 0 & \cdots & I & 0 \end{bmatrix}; E = \begin{bmatrix} I \\ 0 \\ 0 \\ \vdots \\ 0 \\ 0 \end{bmatrix}$$

$$K = \begin{bmatrix} K_1 & K_2 & K_3 & \cdots & K_{n-1} & K_n \end{bmatrix}$$

the state vector is defined as a block state vector, $x = (x_1', x_2', \cdots, x_n')'$, consisting of n block $m-$vectors. The $m \times m$ block matrices $\{M_i, i = 1, 2, \cdots, n\}$ in (4) can be computed from the matrix coefficients $\{D_i, i = 0, 1, \cdots, n-1\}$ in (2) using the following equations [2]

$$M_1 = D_{n-1} \tag{5}$$

$$Q_{k,i} = Q_{k-1,i-1} + M_{n-k+2}Q_{k-2,i}$$

$$((k+i) \text{ even}; k = 2, 3, \cdots, n; i = 0, 1, \cdots, k-1) \tag{6}$$

$$Q_{k,i} = \begin{cases} I & \text{if } k = i \geq 0 \\ 0 & \text{if } k < i \text{ or } k < 0 \text{ or } i < 0 \end{cases} \tag{7}$$

$$Q_{n,i} = D_i, ((i+n) \text{ even}; i = 0, 1, \cdots, n-1) \tag{8}$$

$$Q_{n-1,i} = D_{n-1}^{-1} D_i, ((i+n) \text{ odd}; i = 0, 1, \cdots, n-2) \tag{9}$$

The matrix equations in (6) can be solved recursively since equations with only one unknown matrix is provided taking k in descending order from n to 2 and i from k-2 to 0 [2]. Next, the $p \times m$ matrices $\{K_i, i = 1, 2, \cdots, n\}$ in (4) can be computed recursively from the matrix coefficients $\{N_i, i = 0, 1, \cdots, n-1\}$ in (3), using the equations below [2]

$$K_i = N_{n-i} - \sum_{j=1}^{i-2} K_j Q_{n-j,n-i}$$

$$(i \text{ and } j \text{ even or } i \text{ and } j \text{ odd}; i = 1, 2, \cdots, n) \tag{10}$$

In a similar manner, which is shown by [3], it can be derived that the block-Schwarz matrix M in (4) is stable if, for some $m \times m$ positive-definite symmetric matrix W, there exists an $m \times m$ real matrix V, so that

$$V + V' = W \tag{11}$$

$$P_i = P_i' > 0, (i = 1, 2, \cdots, n) \tag{12}$$

with the $m \times m$ matrices $\{P_i, i = 1, 2, \cdots, n\}$ defined by

$$\begin{aligned} P_1 &= V^{-1}M_1 \\ P_i &= P_{i-1}M_i, (i = 2, 3, \cdots, n) \end{aligned} \tag{13}$$

3. OUTPUT COVARIANCE AND MODEL REDUCTION

Consider that the block-Schwarz equation (4) satisfies the stability condition in (11)-(13). The vector input $u(t)$ is presumed to be zero mean stationary white random-process with covariance matrix W. The steady-state output covariance matrix [4] is defined by

$$C = \lim_{t \to \infty} \mathcal{E}[y(t)y'(t)] \tag{14}$$

With null initial conditions in (4) it is easy to verify that

$$C = K \int_0^\infty \exp(M\tau)EWE'\exp(M'\tau)d\tau K' \tag{15}$$

Thus, the above equation is equivalent to

$$C = KSK' \tag{16}$$

where S is the solution of the matrix equation

$$MS + SM' = -EWE' \tag{17}$$

Taking the positive-definite symmetric matrix W in (11) as input covariance and since (4) satisfies (11)-(13) it is easy to show that

$$S = \text{bloc-diag}\,(P_1^{-1}, P_2^{-1}, \cdots, P_n^{-1}) \tag{18}$$

Hence, the output covariance can again be written as

$$C = \sum_{i=1}^{n} K_i P_i^{-1} K_i' \tag{19}$$

CH3229-2/92/0000-2833$1.00 © 1992 IEEE

that can be computed from matrices $\{M_i, K_i; i = 1,2,\cdots,n\}$ in the block-Schwarz form (4) using (13).

The following *cost function* associated to the output covariance is defined

$$J = \text{trace } C = \sum_{i=1}^{n} \text{trace}(K_i P_i^{-1} K_i') \quad (20)$$

An approximation by output covariance can be obtained by comparing the values of the terms $\{\text{trace } (K_i P_i^{-1} K_i'), i = 1,2,\cdots,n\}$ in (20), such that those terms which give more contribution for J are selected to form the reduced model. These terms are indexed as $\{j_i, i = 1,2,\cdots,r\}$, with $j_1 < j_2 < \cdots < j_r$. Thus, the selected terms form the following expression for the cost function associated to the output covariance approximation

$$J_r = \sum_{i=1}^{r} \text{trace}(\bar{K}_i \bar{P}_i^{-1} \bar{K}_i') \quad (21)$$

where

$$\bar{K}_i = K_{j_i} \quad (22)$$

$$\begin{aligned} \bar{P}_1 &= V^{-1}\bar{M}_1 \\ \bar{P}_i &= \bar{P}_{i-1}\bar{M}_i, (i = 2,3,\cdots,r) \end{aligned} \quad (23)$$

with

$$\bar{M}_i = M_{j_{i-1}+1} M_{j_{i-1}+2} \cdots M_{j_i} \quad (24)$$

and $j_0 = 0$. It is easy to note that a block-Schwarz form in the same pattern as (4), with r block states, can be associated to the selected terms with matrices $\{\bar{K}_i, \bar{M}_i; i = 1,2,\cdots,r\}$ given in (22) and (24):

$$\begin{aligned} \dot{\bar{x}}(t) &= \bar{M}\bar{x}(t) + \bar{E}u(t) \\ \bar{y}(t) &= \bar{K}\bar{x}(t) \end{aligned} \quad (25)$$

where

$$\bar{M} = \begin{bmatrix} -\bar{M}_1 & -\bar{M}_2 & 0 & \cdots & 0 & 0 \\ I & 0 & -\bar{M}_3 & \cdots & 0 & 0 \\ 0 & I & 0 & \cdots & 0 & 0 \\ \vdots & \vdots & \vdots & \ddots & \vdots & \vdots \\ 0 & 0 & 0 & \cdots & 0 & -\bar{M}_r \\ 0 & 0 & 0 & \cdots & I & 0 \end{bmatrix}; \bar{E} = \begin{bmatrix} I \\ 0 \\ 0 \\ \vdots \\ 0 \\ 0 \end{bmatrix}$$

$$\bar{K} = \begin{bmatrix} \bar{K}_1 & \bar{K}_2 & \bar{K}_3 & \cdots & \bar{K}_{r-1} & \bar{K}_r \end{bmatrix}$$

It should be stressed that if (4) satisfies the stability condition (11)-(13), then the block-Schwarz form (25) corresponding to the reduced model also satisfies (11) with

$$P_i = P_i' > 0, (i = 1,2,\cdots,r) \quad (26)$$

where $\{P_i, i = 1,2,\cdots,r\}$ are defined in (13).

4. ILLUSTRATIVE EXAMPLE

Consider the two-input/two-output system given by

$$\hat{G}(s) = (N_0 s^3 + N_1 s^2 + N_2 s + N_3)(D_0 s^4 + D_1 s^3 + D_2 s^2 + D_3 s + I)^{-1}$$

with

$$D_0 = \frac{1}{12}\begin{bmatrix} 17 & -4 \\ 4 & 12 \end{bmatrix}, D_1 = \frac{1}{132}\begin{bmatrix} 305 & -246 \\ 247 & 14 \end{bmatrix}$$

$$D_2 = \frac{1}{66}\begin{bmatrix} 237 & -25 \\ 37 & 394 \end{bmatrix}, D_3 = \frac{1}{2}\begin{bmatrix} 2 & -1 \\ 2 & 1 \end{bmatrix}$$

$$N_0 = \begin{bmatrix} 2 & 0 \\ 0 & 3 \end{bmatrix}, N_1 = \begin{bmatrix} 12 & 4 \\ 0 & 16 \end{bmatrix} N_2 = \frac{1}{2}\begin{bmatrix} 6 & 3 \\ 3 & 6 \end{bmatrix}, N_3 = \frac{1}{4}\begin{bmatrix} 2 & 2 \\ 1 & 3 \end{bmatrix}$$

To obtain a low frequency approximation the reciprocal [5] of the above matrix fraction description is considered, $G(s) = \hat{G}(1/s)/s$ (output covariance matrix and stability property are invariant under reciprocal transformation)

$$G(s) = (N_3 s^3 + N_2 s^2 + N_1 s + N_0)(I s^4 + D_3 s^3 + D_2 s^2 + D_1 s + D_0)^{-1}$$

From (8) and (9) we have $Q_{40} = D_0$, $Q_{42} = D_2$, $Q_{31} = D_3^{-1}D_1$. Using (5-7) and (10) the block-Schwarz form (4) is obtained with: $M_1 = D_3$, $K_1 = N_3$, $K_2 = N_2$ and

$$M_2 = \frac{1}{2}\begin{bmatrix} 3 & 1 \\ 2 & 8 \end{bmatrix}, M_3 = \frac{1}{11}\begin{bmatrix} 12 & -6 \\ -3 & 18 \end{bmatrix}, M_4 = \frac{1}{6}\begin{bmatrix} 6 & -2 \\ -1 & 2 \end{bmatrix},$$

$$K_3 = \frac{1}{264}\begin{bmatrix} 2950 & 912 \\ -51 & 3892 \end{bmatrix}, K_4 = \frac{1}{4}\begin{bmatrix} -3 & 2 \\ -4 & 10 \end{bmatrix}$$

Choosing W in (14) as the 2×2 identity matrix it can be verified that the stability condition in (11)-(13) is satisfied with

$$V = \frac{1}{2}\begin{bmatrix} 1 & -1 \\ 1 & 1 \end{bmatrix}$$

which implies that $G(s)$ is stable and its output covariance approximations are assured to be stable.

Using (13) the terms of the cost function (20) are computed from $\{M_i, K_i, i = 1,2,\cdots,4\}$:

i	1	2	3	4
$\text{trace}(K_i P_i^{-1} K_i')$	0.97	5.52	79.85	3.38

Considering a reduced model with $r = 2$, the third and second terms are those that give more contribution to the cost function (20). However, to obtain the same stationary responses for G and its reduced model for step inputs, the first term is held instead of the second. Thus, the selected terms are referred by the indices $j_1 = 1$ and $j_2 = 3$. Hence, using (22) and (24) it is obtained $\bar{M}_1 = M_1$, $\bar{M}_2 = M_2 M_3$, $\bar{K}_1 = K_1$ and $\bar{K}_2 = K_3$, that are used to form the block-Schwarz equation (25). Its corresponding right matrix-fraction description can be easily obtained from equations like (6-10) used in reverse order, as follows:

$$R(s) = (\bar{N}_1 s + \bar{N}_0)(I s^2 + \bar{D}_1 s + \bar{D}_0)^{-1}$$

with $\bar{D}_1 = \bar{M}_1$, $\bar{D}_0 = \bar{M}_2$, $\bar{N}_1 = \bar{K}_1$ and $\bar{N}_0 = \bar{K}_2$. Thus, the reduced model for $\hat{G}(s)$ is $\hat{R}(s) = R(1/s)/s$.

CONCLUSION

A model reduction method for a class of multivariable linear systems is proposed by using the terms of cost function associated to the output covariance matrix corresponding to zero-mean white noise inputs. The reduced model stability can be assured since the complete model satisfies certain stability condition.

ACKNOWLEDGMENT

This work has been supported in part by Brazilian Research Council-CNPq, under Grant 300818/88.

REFERENCES

[1] R.E. Skelton, "Cost decomposition of linear systems with application to model reduction", *Int. J. Control*, **32**, pp. 1031-1055, 1980

[2] P. Resende and V.V.R. Silva, "On the model order reduction of linear multivariable systems using a block-Schwarz realization", **2**, pp. 266-270, in *Pre-prints of the XI IFAC World Congress*, Tallinn, 1990

[3] P. Resende and E. Kaszkurewicz, "A sufficient condition for the stability of matrix polynomials", *IEEE Trans.*, **AC-34**, pp. 539-541, 1989

[4] R.E. Skelton, "*Dynamics systems control: linear systems analysis and synthesis*", New York:Wiley, 1988

[5] M.F. Hutton and B. Friedland, "Routh approximation for reducing order of linear time invariant systems", *IEEE Trans.*, **AC-20**, pp. 329-337, 1975

Proceedings of the 31st Conference
on Decision and Control
Tucson, Arizona • December 1992

OVERLAPPING BLOCK-BALANCED CANONICAL FORMS AND PARAMETRIZATIONS: THE STABLE SISO CASE

Bernard Hanzon [1]
Dept. Econometrics, Free University Amsterdam
De Boelelaan 1105
1081 HV Amsterdam
Holland

Raimund J. Ober
Center for Engineering Mathematics, University of Texas at Dallas
Programs in Mathematical Sciences, UTD
Richardson
Texas 75083-06688
USA

Abstract: The balanced canonical form and parametrization of Ober for the case of SISO stable systems are extended to block- balanced canonical forms and related input-normal forms and parametrizations. They form an overlapping atlas of parametrizations of the manifold of stable SISO systems of given order. This extends the usefullness of these parametrizations, e.g. in gradient algorithms for system identification.

1. Introduction

In [5],[6] a canonical state space form was presented for the set of asymptotically stable linear systems, with the property that it is balanced, i.e. for each system represented in canonical form the corresponding observability and controllability Grammians are equal and diagonal (and positive definite). One motivation for studying balanced realizations and balanced canonical forms is their close relation to model reduction (see [6] and the references given there), Another motivation mentioned in [6] is the potential usefullness of balanced realizations for system identification. In many cases, in system identification as well as in related areas, one can reduce the problem at hand to an optimization problem in which some criterion function is optimized over a set of systems. Very often one cannot solve the optimization problem analytically and one has to use search algorithms (e.g. gradient algorithms), in which an initial point in the set of systems is adapted iteratively to give a hopefully good approximation of the optimal system. In such search algorithms one often uses a parametrization of the set of relevant systems. The balanced parametrization of [6] has the advantage that by construction, problems of identifiability are to a large extend avoided in such a search algorithm. The parametrization has the property that it contains structural indices (i.e. discrete–valued parameters), and with each possible choice of values for these indices corresponds a particular submanifold of systems, for which a parametrization in terms of real-valued parameters is given. To each system corresponds a unique set of structural indices. As the structural indices can take a large number of values, even for rather low-order systems (the number of possibilities increases fast with increasing order of the system), this means that in a search algorithm one has to either identify the structural indices by other means or one has to apply the search algorithm to a large number of parametrized submanifolds of systems. This is due to the fact that the parametrizations are disjoint.

Several authors (see e.g. [7,1] and the references given there) have investigated the possibility of using socalled overlapping parametrizations (in differential geometric terms: an atlas of coordinate charts). If one uses overlapping parametrizations, one does not have to search through each and every of the submanifolds, but instead one can search through the manifold as a whole, using the parametrizations to describe the manifold locally and changing from one parametrization to another when required. In case the search algorithm is of the gradient type, one can make sure that the decision rule for changing from one parametrization to another has little effect on the search algorithm by using a Riemannian gradient, with respect to some suitable Riemannian metric on the manifold (cf. [1] and the references given there).

In view of this it would be very desirable if the balanced parametrization of [6] could be extended to give a set of overlapping parametrizations. In this paper such an extension will be presented for the case of SISO stable systems. In the extension balancedness of the realization no longer holds for all realizations. Instead (what we will call) block-balanced realizations are used and the corresponding input-normal realizations. With a block-balanced canonical form we mean a canonical form for which the observability and controllability Grammian are equal and block-diagonal (and of course positive definite).

2. Canonical forms, balanced realizations and block-balanced realizations

Let us consider continuous time SISO systems of the form $\dot{x}_t = Ax_t + bu_t, y_t = cx_t$ with $t \in \mathbf{R}, u_t \in \mathbf{R}, x_t in \mathbf{R}^n, y_t \in \mathbf{R}, A \in \mathbf{R}^{n\times n}, b \in \mathbf{R}^{n\times 1}, c \in \mathbf{R}^{n\times 1}, (A, b, c)$ a minimal triple. Let for each $n \in \{1, 2, 3, \cdots\}$ the set C_n be given by $C_n = \{(A, b, c) \in \mathbf{R}^{n\times n} \times \mathbf{R}^{n\times 1} \times \mathbf{R}^{1\times n} | (A, b, c)$ minimal and the spectrum of A is contained in the open left half plane$\}$. As is well-known two minimal system representations (A_1, b_1, c_1) and (A_2, b_2, c_2) have the same transfer function $g(s) = c_1(sI - A_1)^{-1}b_1 = c_2(sI - A_2)^{-1}b_2$, and therefore describe the same input-output behaviour, iff there exists an $n \times n$ matrix $T \in Gl_n(\mathbf{R})$ such that $A_1 = TA_2T^{-1}, b_1 = Tb_2, c_1 = c_2T^{-1}$. In that case we say that (A_1, b_1, c_1) and (A_2, b_2, c_2) are i/o-equivalent. This is clearly an equivalence relation; write $(A_1, b_1, c_1) \sim (A_2, b_2, c_2)$. A unique representation of a linear system can be obtained by deriving a canonical form:

[1] E-mail: bhnz@sara.nl; Fax +31-20-6461449. A first version of this paper was written while the first named author visited the Dept. Engineering, Cambridge University

CH3229-2/92/0000-2835$1.00 © 1992 IEEE

Definition 2.1 *A canonical form for an equivalence relation '$\sim$' on a set X is a map $\Gamma : X \to X$ which satisfies for all $x, y \in X$: (i) $\Gamma(x) \sim x$ and (ii) $x \sim y \iff \Gamma(x) = \Gamma(y)$*

Equivalently a canonical form can be given by the image set $\Gamma(X)$; a subset $B \subseteq X$ describes a canonical form if for each $x \in X$ there is precisely one element $b \in B$ such that $b \sim x$. The mapping $X \to B, x \mapsto b$ then describes a canonical form.

Let $(A, b, c) \in C_n$. As is well-known the controllability Grammian W_c can be obtained as the unique, positive definite symmetric solution of the Lyapunov equation $AW_c + W_cA^T = -bb^T$. In a dual fashion, the observability Grammian W_o is the unique, positive definite symmetric solution of the Lyapunov equation $A^TW_o + W_oA = -c^Tc$.

Definition 2.2 *Let $(A,b,c) \in C_n$, then (A,b,c) is called* balanced *if the corresponding observability and controllability Grammians are equal and diagonal, i.e. there exist positive numbers $\sigma_1, \sigma_2, \ldots, \sigma_n$ such that*

$$W_o = W_c = diag(\sigma_1, \ldots, \sigma_n) =: \Sigma \qquad (1)$$

The numbers $\sigma_1, \ldots, \sigma_n$ are called the (Hankel) singular values of the system.

The singular values are known to be uniquely determined by the input-output behaviour of the system.

Theorem 2.3 (Moore 1981) *Let $(A, b, c) \in C_n$ with*

$$\Sigma = diag(\sigma_1 I_{n(1)}, \ldots, \sigma_k I_{n(k)}), \sigma_1 > \sigma_2 > \ldots \sigma_k > 0 \text{ and } \sum_{j=1}^{k} n(j) = n.$$

Then (A, b, c) is unique up to an orthogonal state-space transformation of the form

$$Q = diag(Q_1, Q_2, \ldots, Q_k)$$

with orthogonal $Q_i \in \mathbf{R}^{n(i)\times n(i)}, i = 1, \ldots, k$.

Definition 2.4 *Let $(A, b, c) \in C_n$, then (A, b, c) is called* input-normal *if $W_c = I_n$ and will be called* σ-input-normal *if $W_c = \sigma I_n$.*

Similarly (A, b, c) is called output-normal *if $W_o = I_n$ and* σ-output-normal *if $W_o = \sigma I_n$.*

It is not difficult to show that an input-normal realization is unique up to an arbitrary orthogonal state-space transformation.

The following definition is new and basic to our considerations in this paper.

Definition 2.5 *Let $(A, b, c) \in C_n$, then (A, b, c) will be called* block-balanced, *with indices $n(i) \in \mathbf{N}, i = 1, \ldots, k$,adding up to n, if the observability Grammian and the controllability Grammian are equal and block-diagonal, i.e. there exist $n(i) \times n(i)$ positive definite matrices $\Sigma_i, i = 1, \ldots, k$, such that*

$$W_o = W_c = diag(\Sigma_1, \ldots, \Sigma_k)$$

It will be convenient to call an arbitrary system representation $(A, b, c) \in \mathbf{R}^{n\times n} \times \mathbf{R}^{n\times 1} \times \mathbf{R}^{1\times n}$ block-balanced if the pair of Lyapunov equations $A\Sigma + \Sigma A^T = -bb^T, A^T\Sigma + \Sigma A = -c^Tc$ has a positive definite solution of the form $\Sigma = diag(\Sigma_1, \ldots, \Sigma_k)$ (assuming neither asymptotic stability nor minimality).

Remark. The matrices $\Sigma_i, i = 1, \ldots, k$ are in general *not* uniquely determined by the input-output behaviour of the system. However the eigenvalues $\lambda_1(\Sigma_i) \geq \lambda_2(\Sigma_i) \geq \ldots \geq \lambda_{n(i)}(\Sigma_i)$ of the matrices $\Sigma_i, i = 1, \ldots, k$ together form the set of Hankel singular values of the system, which are uniquely determined by the input- output behaviour of the system, as remarked before.

Theorem 2.6 *Suppose $(A, b, c) \in C_n$ is block-balanced with indices $n(j) \in \mathbf{N}, j = 1, \ldots, k, \sum_{j=1}^{k} n(j) = n$ and with the additional property $\lambda_1(\Sigma_1) \geq \lambda_{n(1)}(\Sigma_1) > \lambda_1(\Sigma_2) \geq \lambda_{n(2)}(\Sigma_2) > \ldots > \lambda_1(\Sigma_k) \geq \lambda_{n(k)}(\Sigma_k) > 0$. This uniquely determines (A, b, c) up to an orthogonal state-space transformation of the form $Q = diag(Q_1, \ldots, Q_k)$ with orthogonal $Q_i \in \mathbf{R}^{n(i)\times n(i)}, i = 1, \ldots, k$*

Proof. See [2].

The following theorem will be fundamental for our results.

Theorem 2.7 (Pernebo and Silverman, [8], Kabamba, [3]) *Let $(A, b, c) \in \mathbf{R}^{n\times n} \times \mathbf{R}^{n\times 1} \times \mathbf{R}^{1\times n}$ be conformally partitioned as follows:*

$$A = \begin{pmatrix} A_{11} & A_{12} \\ A_{21} & A_{22} \end{pmatrix}, b = \begin{pmatrix} b_1 \\ b_2 \end{pmatrix}, c = \begin{pmatrix} c_1 & c_2 \end{pmatrix},$$

with $A_{ii} \in \mathbf{R}^{n(i)\times n(i)}, i = 1, 2$ and let (A, b, c) be block-balanced with indices $n(1), n(2)$ such that $\Sigma_1, \Sigma_2 > 0$ have no eigenvalues in common.

Then $(A, b, c) \in C_n \Leftrightarrow (A_{ii}, b_i, c_i) \in C_{n(i)}, i = 1, 2$.

3. The case k=1: a Schwarz-like canonical form for stable SISO systems in continuous time

Theorem 3.1 *Consider the set B_n of all $(A, b, c) \in C_n$ of the following form:*

$$A = \begin{pmatrix} a_{11} & \alpha_1 & & 0 \\ -\alpha_1 & 0 & \ddots & \\ & \ddots & \ddots & \alpha_{n-1} \\ 0 & & -\alpha_{n-1} & 0 \end{pmatrix}, a_{11} = -\frac{b_1^2}{2} < 0,$$

$$\alpha_i > 0, i = 1, \ldots, n-1,$$

$$b = \begin{pmatrix} b_1 \\ 0 \\ \vdots \\ 0 \end{pmatrix}, b_1 > 0,$$

$$c = \begin{pmatrix} c_1 & \gamma_1 & \ldots & \gamma_{n-1} \end{pmatrix}, c_1 \in \mathbf{R}, \gamma_j \in \mathbf{R}, j = 1, \ldots, n-1.$$

Each triple $(A, b, c) \in B_n$ is input-normal.

Let S_n be the set of values of the vector of parameters $(b_1, \alpha_1, \ldots, \alpha_{n-1}, c_1, \gamma_1, \ldots, \gamma_{n-1})$ such that the corresponding triple $(A, b, c) \in B_n$, i.e. such that $b_1 > 0, \alpha_i > 0, i = 1, \ldots, n$ and $c_1, \gamma_1, \ldots, \gamma_{n-1}$ such that the pair (c, A) is observable.

The set B_n describes a continuous canonical form and the parametrization mapping $S_n \longrightarrow B_n$, which maps each parameter vector to the corresponding triple (A, b, c), is a homeomorphism.

If $(\gamma_1, \ldots, \gamma_{n-1}) \neq 0 \in \mathbf{R}^{n-1}, n \geq 2$, then the system has several different singular values.

Proof See [2].

Remarks (i) If $c_1 \neq 0$ we define $\sigma := \left|\frac{c_1}{b_1}\right| > 0$, which we will call a pseudo-singular value. If the vector $\gamma = (\gamma_1, \ldots, \gamma_{n-1})$ is close enough to zero the pseudo-singular value will be close to the true singular values of the system, because of continuity of the singular values as a function of γ and the fact that if $\gamma = 0$, the system has only one singular value and its value is σ. If $c_1 \neq 0$ the system can be brought simply into σ-input-normal form by multiplying c by $\sigma^{-\frac{1}{2}}$ and b by $\sigma^{\frac{1}{2}}$. The resulting σ-input -normal form is a *canonical* form locally around $\gamma = 0$, but not globally because the systems which have $c_1 = 0$ in the previous canonical form cannot be represented in this way. (ii)

Clearly the canonical forms presented are controllable (because they are input-normal, resp. σ−input-normal), but observability will fail for certain choices of c; the observability Grammian will be singular for such a choice of c. If $\gamma = 0, c_1 \neq 0$, the system is observable, because the observability Grammian will be $\sigma^2 I$, resp. σI. Therefore also in some open neighbourhood around such a system, observability will still hold. (iii) This canonical form is closely related to the so-called Schwarz canonical form, cf. e.g. [4], [9].

4. An input-normal and a block-balanced canonical form

Let $n(1), \ldots, n(k) \in \{1, 2, \ldots, n\}, \sum_{j=1}^{k} n(j) = n$, denote a partition of n as before. Let $C_{n(1),n(2),\ldots,n(k)}$ denote the subset of all systems in C_n with the property that their n Hankel singular values (multiplicities *included*) $\sigma(1) \geq \sigma(2) \geq \ldots \geq \sigma(n) > 0$ can be partitioned into k disjoint sets of singular values (again with multiplicities included) in the following way:

$$\begin{aligned}\sigma(1) &\geq \ldots \geq \sigma(n(1)) > \sigma(n(1)+1) \geq \\ &\geq \ldots \geq \sigma(n(1)+n(2)) > \sigma(n(1)+n(2)+1) \geq \\ &\geq \ldots \geq \sigma(\sum_{j=1}^{l} n(j)) > \sigma((\sum_{j=1}^{l} n(j))+1) \geq \\ &\geq \ldots > 0 \end{aligned} \tag{2}$$

So we require that $\sigma(\sum_{j=1}^{l} n(j)) > \sigma((\sum_{j=1}^{l} n(j)) + 1)$ for $l = 1, 2, \ldots, k-1$ and $\sigma(n) > 0$ of course. Note that the notation is consistent with the fact that C_n denotes the set of stable systems which have as their only "restriction" that there are n positive singular values (multiplicities included), i.e. that the order of the system is n.

The other extreme is $C_{1,1,\ldots,1}$, which denotes the set of n−th order stable systems with n *distinct* singular values. For this set of systems a balanced canonical form was derived in [3].

Next we will present a canonical form on $C_{n(1),\ldots,n(k)}$.

Theorem 4.1 *Consider the set $B_{n(1),\ldots,n(k)}$ of triples (A, b, c) of the following form:*

$$\begin{aligned}
A &= (A(i,j))_{1\leq i,j\leq k}, \\
A(i,j) &\in \mathbf{R}^{n(i)\times n(j)}, i,j \in \{1,\ldots,k\} \\
b &= \begin{pmatrix} b(1) \\ b(2) \\ \vdots \\ b(k) \end{pmatrix}, b(i) \in \mathbf{R}^{n(i)}, i = 1,\ldots,k, \\
c &= (c(1),\ldots,c(k)), c(j)^T \in \mathbf{R}^{n(j)}, j = 1,\ldots,k, \\
A(i,i) &= \begin{pmatrix} a(i,i)_{11} & \alpha(i)_1 & 0 & \ldots & 0 \\ -\alpha(i)_1 & 0 & \alpha(i)_2 & \ddots & \vdots \\ 0 & -\alpha(i)_2 & & \ddots & 0 \\ \vdots & \ddots & \ddots & & \alpha(i)_{n(i)-1} \\ 0 & \ldots & 0 & -\alpha(i)_{n(i)-1} & 0 \end{pmatrix}, \\
a(i,i)_{11} &= -\frac{b_i^2}{2}, \\
\alpha(i)_j &> 0, j = 1,\ldots,n(i)-1, \\
b(i) &= \begin{pmatrix} b_i \\ 0 \\ \vdots \\ 0 \end{pmatrix}, b_i > 0, \\
c(i) &= \left(c_i, \gamma(i)_1, \ldots, \gamma(i)_{n(i)-1}\right), i = 1,\ldots,k,
\end{aligned}$$

where the parameters are to be taken such that the corresponding observability Grammians $\Sigma_i^2, i = 1,\ldots,k$, which satisfy the observability Lyapunov equations

$$\Sigma_i^2 A(i,i) + A(i,i)^T \Sigma_i^2 = -c(i)^T c(i) \tag{3}$$

are fulfilling the following matrix inequalities

$$\Sigma_1^2 > \Sigma_2^2 > \ldots > \Sigma_k^2 > 0; \tag{4}$$

for each pair $(i,j), i \neq j$, the matrices $A(i,j), A(j,i)$ are determined (uniquely!) from the following pair of linear matrix equations:

$$\begin{aligned} A(i,j) + A(j,i)^T &= -b(i)b(j)^T \\ \Sigma_i^2 A(i,j) + A(j,i)^T \Sigma_j^2 &= -c(i)^T c(j) \end{aligned} \tag{5}$$

The set $B_{n(1),\ldots,n(k)}$ describes a continuous canonical form on $C_{n(1),\ldots,n(k)}$. The $2n$ "free" parameters of the canonical form are

$$b_i, \alpha(i)_1, \ldots, \alpha(i)_{n(i)-1}, c_i, \gamma(i)_1, \ldots, \gamma(i)_{n(i)-1}, i = 1,\ldots,k.$$

Let $S_{n(1),\ldots,n(k)} \subset \mathbf{R}^{2n}$ be the set of all values of the parameter vector for which the corresponding triple $(A,b,c) \in B_{n(1),\ldots,n(k)}$, i.e. for all $i \in \{1,\ldots,k\}$: $b_i > 0, \alpha(i)_j > 0, j = 1,\ldots,n(i)-1$, and $c_i, \gamma(i)_1, \ldots, \gamma(i)_{n(i)-1}$ such that the matrices $\Sigma_i, i = 1,\ldots,k$, found in (3) satisfy the inequalities (4). The mapping $S_{n(1),\ldots,n(k)} \longrightarrow B_{n(1),\ldots,n(k)}$ which maps a parameter vector to the corresponding triple (A,b,c) is a homeomorphism.

The form is input-normal, i.e. $A + A^T = -bb^T$ and has block-diagonal observability Grammian $\Sigma^2 := diag(\Sigma_1^2, \ldots, \Sigma_k^2) > 0$.

Let $\sigma(1) \geq \sigma(2) \geq \ldots \geq \sigma(n) > 0$ denote the n positive Hankel singular values of the system (with their multiplicities).

If for some $i \in \{1,\ldots,k\}$ the vector $\gamma(i) = 0$, then Σ_i^2 is a scalar matrix $\Sigma_i^2 = \sigma^2\left(1 + \sum_{j=1}^{i-1} n(j)\right) . I_{n(i)}$, and

$$\sigma\left(\sum_{j=1}^{i-1} n(j)\right) > \sigma\left(1 + \sum_{j=1}^{i-1} n(j)\right) = \ldots = \sigma\left(\sum_{j=1}^{i} n(j)\right) > \sigma\left(1 + \sum_{j=1}^{i} n(j)\right)$$

If for all $i \in \{1,\ldots,k\}, \gamma(i) = 0$, then the observability Grammian is consequently diagonal.

Remark. A block-balanced realization can be obtained from the presented canonical form by applying a state-space transformation

$$T := \Sigma^{\frac{1}{2}} = \text{diag}\left(\Sigma_1^{\frac{1}{2}}, \ldots, \Sigma_k^{\frac{1}{2}}\right) > 0 \tag{6}$$

The corresponding controllability and observability Grammians will both be equal to

$$\Sigma = \text{diag}(\Sigma_1, \ldots, \Sigma_k) > 0$$

Proof. See [2].

5. An atlas of overlapping block-balanced canonical forms

Theorem 5.1 *Let the state space dimension n be fixed. The continuous canonical forms $C_{n(1),\ldots,n(k)} \longrightarrow B_{n(1),\ldots,n(k)}$, $n(j) \in \{1,\ldots,n\}; j = 1,\ldots,k$; $\sum_{j=1}^{k} n(j) = n$; $k \in \{1,\ldots,n\}$, form an overlapping set of continuous canonical forms covering C_n. Each of the sets $C_{n(1),\ldots,n(k)}$, $\sum_{j=1}^{k} n(j) = n$, is an open subset of C_n and together they cover C_n.*

Proof. See [2].

Corollary 5.2 *The set of mappings $C_{n(1),\dots,n(k)}/\sim \longrightarrow S_{n(1),\dots,n(k)} \subset \mathbf{R}^{2n}$, $(n(1),\dots,n(k)) \in P(n;k), k = 1,\dots,n$, which map each equivalence class of triples to the corresponding parameter vector in the canonical form, forms an atlas for the manifold of stable SISO i/o-systems of order n.*

Proof. See [2].

Remark. A motivation for using this atlas rather than e.g. just the Schwarz-like canonical form B_n is the following. Suppose one wants to use *balanced realizations.* Then one can use the balanced canonical form of [6]. However this form is discontinuous at all points of $C_{n(1),\dots,n(k)} \setminus C_{1,\dots,1}$, i.e. in all triples $(\tilde{A},\tilde{b},\tilde{c})$ which have two or more coinciding singular values. And the complement $C_{1,\dots,1}$, of the set of discontinuity points consists of 2^n topological components, one component for each sign pattern; this should be compared to C_n which has only $n+1$ topological components (the Brockett components). It appears that this is a serious disadvantage if one wants to use balanced realizations and canonical forms in e.g. search algorithms for system identification.

In order to overcome these difficulties one could use the overlapping block-balanced canonical forms as follows. If $(\tilde{A},\tilde{b},\tilde{c})$ has k distinct Hankel singular values $\sigma_1 > \sigma_2 > \dots > \sigma_k > 0$ with multiplicities resp. $n(1),\dots,n(k)$, then one can use the block-balanced continuous canonical form on $C_{n(1),\dots,n(k)}$ *locally around* $(\tilde{A},\tilde{b},\tilde{c})$. If one is moving away from $(\tilde{A},\tilde{b},\tilde{c})$ in a search algorithm for example, one has to decide whether the canonical form corresponding to a different partition should be used: if the largest $n(1)$ singular values differ sufficiently from each other one could use e.g. $C_{1,\dots,1,n(2),\dots,n(k)}$ (where there are $n(1)$ ones in the subindex before $n(2)$) etc. In this way one would use balanced realizations and "almost- balanced" realizations while moving around in the set of $n-$th order systems, without encountering discontinuity points.

References

[1] B. Hanzon, *Identifiability, Recursive Identification and Spaces of Linear Dynamical Systems*, CWI Tracts 63,64, CWI, Amsterdam, 1989.

[2] Bernard Hanzon and Raimund Ober, *Overlapping block-balanced canonical forms and parametrizations: the stable SISO case*, Research Memorandum 1992-29, Dept. Econometrics, Free University Amsterdam, Amsterdam, 1992.

[3] P.T. Kabamba, *Balanced forms: canonicity and parametrization*, IEEE Transactions on Automatic Control **AC-30** (1985), pp.1106-1109.

[4] R.E. Kalman, *On partial realizations, transfer functions, and canonical forms*, Acta Polyt. Scand. Ma. **31** (1979),pp.9-32.

[5] R. Ober, *Asymptotically stable allpass transfer functions: canonical form, parametrization and realization* in: Proceedings IFAC World Congress, Munich, 1987.

[6] R. Ober, *Balanced realizations : canonical form, parametrization, model reduction*, Int. Journal of Control **46**, No.2 (1987), pp.643-670.

[7] A. J. M. van Overbeek and L. Ljung, *On-line structure selection for multivariable state space models*, Automatica **18**, pp.529-543.

[8] L. Pernebo and L.M. Silverman,*Model reduction via balanced state space representations*. IEEE Transactions on Automatic Control **AC-37**(1982),pp.382-387.

[9] H.R. Schwarz, *A method for determining stability of matrix differential equations*(in German), Z.angew.Math.Phys. **7** (1956),pp.473-500.

Proceedings of the 31st Conference
on Decision and Control
Tucson, Arizona • December 1992

Robust Generation of Analytical Redundancy Equations and Applications to Diagnosis

Frédéric Kratz, Didier Maquin and José Ragot

Centre de Recherche en Automatique de Nancy - CNRS UA 821
BP 40 - Rue du doyen Marcel Roubault
54 501 Vandoeuvre Cedex - FRANCE

Abstract

This paper describes a procedure for generating analytical redundancy equations for process fault detection and isolation. The proposed procedure uses a direct approach by considering the relations between the inputs and the outputs of the process ; systems with known or partially known inputs are considered and it is pointed out that the same technique yields for these different situations.

1. Introduction

The design of fault tolerant control systems requires failures to be detected, identified and taken into account within acceptable time interval as not to affect excessively the system operation. Surveys on design methods for failure detection and isolation are given in the papers of Isermann (1984) and Gertler (1991). In the past decade, interest has been focused on the use of analytic redundancy equations rather than massive redundancy. In particular, analytical redundancy equations have been designed to form and process residuals ; these residuals are closed to zero if no failure occurs and differ significantly from zero when a failure occurs.

When certain inputs of the process cannot be measured, the state observation must be designed by eliminating the unknown inputs. This problem has received considerable attention in the literature. Several researchers have investigated state or input observers when some inputs are unknown ; among them, the reader is referred to Kurek (1983), El-Tohami (1983) or Miller (1984) for example. In the study of Park (1988) a closed-loop observer that can identify simultaneously states and inputs has been developed. A systematic investigation of unknown input observers was carried out recently by Wünnenberg (1990) using the Kronecker canonical form transformation. A more simple design procedure is given by Hou (1991) using only algebraic approach. In the present work, we introduce a generation scheme of analytical redundancy equations.

2. Decoupling properties

In practise, the process models are not perfectly known : parameter variations, unknown inputs and component faults significantly influence processes behavior. If these perturbations have not been included in the models, the model behavior differs from those of the actual process. Consequently, the residuals are as much sensitive to fault as to inherent uncertainties of the models. To separate the faults from the model inaccuracy, we must use robust redundancy relations which are insensitive to undesired perturbations. We consider a linear system described by the following state space equation :

$$\lambda\, x(k) = A\, x(k) + B\, u(k) + \bar{B}\, f(k) \qquad (1a)$$

$$y(k) = C\, x(k) + D\, u(k) + \bar{D}\, f(k) \qquad (1b)$$

where f(k) is a perturbation vector, λ states for the s or z operator depending on the representation of the model in continuous or discrete form and $\bar{B}$ and $\bar{D}$ represent the perturbation matrices of appropriate dimensions. It is assumed in the following that C is a full rank matrix. Robust redundancy equations can be generated by the elimination of the state x(k) and the perturbation vector f(k). Therefore, using (1), the equation to solve may be generally written :

$$\left(p^T(\lambda)\;\; q^T(\lambda) \right) \begin{pmatrix} A - \lambda I & \bar{B} \\ C & \bar{D} \end{pmatrix} = (0\;\; 0) \qquad (2)$$

where $p(\lambda)$ and $q(\lambda)$ are two vectors with compatible dimensions with the state matrices. If this condition holds, multiplying equation (1) with $(p^T(\lambda)\;\; q^T(\lambda))$, yields the redundancy equations :

$$(p^T(\lambda)\;\; q^T(\lambda)) \left(\begin{pmatrix} 0 \\ I \end{pmatrix} y(k) - \begin{pmatrix} B \\ D \end{pmatrix} u(k) \right) = 0 \qquad (3)$$

If the system (2) has no solution, due for example to an excessive number of perturbations, then one may look for an approximate solution. In order to find the optimal approximation, a certain performance index that contains a measure of the effects of the disturbances must be defined. If the perturbation matrices do not exist, the preceding formalism is still valid and allows to generate the redundancy relations.

3. General solution of the decoupling problem

To solve equation (2), rewrite it in the equivalent form :

$$(p^T(\lambda)\;\; q^T(\lambda)) \left(\begin{pmatrix} A & \bar{B} \\ C & \bar{D} \end{pmatrix} - \lambda \begin{pmatrix} I & 0 \\ 0 & 0 \end{pmatrix} \right) = (0\;\; 0) \qquad (4)$$

Solving (4) corresponds to the well-known problem of the determination of the left nullspace of a binomial

CH3229-2/92/0000-2839$1.00 © 1992 IEEE

matrix (Gantmacher, 1977). Taking the transpose of equation (4) yields :

$$\left(\begin{pmatrix} A_0 & C_0 \\ B_0 & D_0 \end{pmatrix} - \lambda \begin{pmatrix} I & 0 \\ 0 & 0 \end{pmatrix}\right)\begin{pmatrix} p(\lambda) \\ q(\lambda) \end{pmatrix} = \begin{pmatrix} 0 \\ 0 \end{pmatrix} \qquad (5)$$

The solution of equation (5) is obtained through a two-step algorithm. First, we solve (5) with $\bar{D} = 0$ (i.e. $D_0 = 0$). Second, we transform system (5) in order to satisfy this property. Then, the first step leads to solve the following system :

$$(\lambda I - A_0)\, p(\lambda) = C_0\, q(\lambda) \qquad (6a)$$
$$B_0\, p(\lambda) = 0 \qquad (6b)$$

Equation (6b) implies that $p(\lambda)$ belongs to the kernel of B_0, so that the general solution of (6b) is expressed in term of an appropriate vector of constants :

$$p(\lambda) = \bar{B_0}\; v(\lambda) \qquad (7)$$

Substituting $p(\lambda)$ in (6a) and pre-multiplying first by N the left annihilator of C_0 ($N\,C_0 = 0$) and second by $\bar{C_0}$ the left inverse of C_0 ($\bar{C_0} C_0 = I$) yields the two equations :

$$N\,(\lambda I - A_0)\, \bar{B_0}\, v(\lambda) = 0 \qquad (8a)$$
$$\bar{C_0}\,(\lambda I - A_0)\, \bar{B_0}\, v(\lambda) = q(\lambda) \qquad (8b)$$

Equation (8a) may be solved in respect to $v(\lambda)$ (Gantmacher, 1977, p. 29-30), then $p(\lambda)$ is deduced from (7) and $q(\lambda)$ from (8b). At the second step, consider the complete system (5) in which D_0 may always be expanded into :

$$D_0 = H \begin{pmatrix} R & 0 \\ 0 & 0 \end{pmatrix} K^T \qquad (9)$$

where H and K are two orthogonal matrices and R a regular matrix. With $K^T q(\lambda) = \bar{q}(\lambda)$ and $\bar{B_0} = H^T B_0$, we have from the second equation of system (5) :

$$\bar{B_0}\, p(\lambda) + \begin{pmatrix} R & 0 \\ 0 & 0 \end{pmatrix} \bar{q}(\lambda) = 0 \qquad (10)$$

which may be expanded into :

$$\bar{B}_{01}\, p(\lambda) + R\, \bar{q}_1(\lambda) = 0 \qquad (11a)$$

$$\bar{B}_{02}\, p(\lambda) = 0 \qquad (11b)$$

The solution $\bar{q}_1(\lambda)$ of equation (11a) is then substituted in (6a) which gives :

$$(\lambda I - \bar{A})\, p(\lambda) = C_{02}\, \bar{q}_2(\lambda) \qquad (12)$$

$$\text{with : } \bar{A} = A_0 - C_{01} R^{-1} \bar{B}_{01} \qquad (13)$$

Thus, the remaining equations (11b) and (12) have the same structure as the system (6) and therefore may be solved by using the same technique. Therefore, the residual generation consists in the extraction of the redundancy relations from the system state equations. In fact, this extraction is possible by eliminating the unknown variables (state variables or perturbations). If the disturbance directions are unknown, the elimination affects only the state variables ($\bar{B} = 0$, $\bar{D} = 0$), otherwise a priori knowledge can be used to design the matrices $\bar{B}$ and $\bar{D}$.

4. Unknown inputs systems

Consider the linear system, with a p dimensional unknown input vector d(k), described by :

$$\lambda\, x(k) = A\, x(k) + B\, u(k) + G\, d(k) \qquad (14a)$$
$$y(k) = C\, x(k) + D\, u(k) + F\, d(k) \qquad (14b)$$

where G and F are known constant matrices. Without loss of generality, we assume that :

$$\text{rank} \begin{pmatrix} G \\ F \end{pmatrix} = \dim(d) \qquad (15)$$

The classical conditions of observability (Kurek, 1983), are not necessary because we do not try to estimate system state. With obvious definition of the matrices M, N, $\bar{A}$, Q and $\bar{B}$, the equations (14) can be expressed as :

$$M\, \lambda\, x(k) + N\, y(k) = \bar{A}\, x(k) + Q\, d(k) + \bar{B}\, u(k) \qquad (16)$$

Assuming that equation (15) holds, there exist two orthogonal matrices H and K and a non-singular matrix R such that :

$$Q = H \begin{pmatrix} R \\ 0 \end{pmatrix} K^T \qquad (17)$$

Introducing a new vector $\bar{d}(k)$ defined by :

$$\bar{d}(k) = K^T d(k) \qquad (18)$$

and pre-multiplying equation (16) with H^T, we can partitioned equation (16) as :

$$M_1\, \lambda\, x(k) + N_1\, y(k) = \bar{A}_1\, x(k) + R\, \bar{d}(k) + \bar{B}_1\, u(k) \qquad (19a)$$
$$M_2\, \lambda\, x(k) + N_2\, y(k) = \bar{A}_2\, x(k) + \bar{B}_2\, u(k) \qquad (19b)$$

The first equation (19a) may be used to estimate the unknown input $\bar{d}(k)$ since R is regular. The second one (19b) may contain redundancies between the input u(k) and the output y(k) provide it is possible to eliminate the state x(k). For that purpose (19b) is rewritten :

$$(\lambda\, M_2 - \bar{A}_2)\, x(k) = \bar{B}_2\, u(k) - N_2\, y(k) \qquad (20)$$

Redundancy equations are then obtained by searching the left null space of the pencil $(\lambda\, M_2 - \bar{A}_2)$:

$$p^T(\lambda)\,(\lambda\, M_2 - \bar{A}_2) = 0 \qquad (21)$$

The solution of equation (21) is the same as those of the equation (4). The redundancy equations can be described by :

$$p^T(\lambda)\,(\bar{B}_2\, u(k) - N_2\, y(k)) = 0 \qquad (22)$$

The previously described method can be used to detect and localize faulty actuators. Consider a standard system where the ith input has been isolated :

$$\lambda\, x(k) = A\, x(k) + B_i\, u_i(k) + \bar{B}_i\, \bar{u}_i(k) \qquad (23a)$$

$$y(k) = C\, x(k) + D_i\, u_i(k) + \bar{D}_i\, \bar{u}_i(k) \qquad (23b)$$

where B_i is the ith column of B and $\bar{B}_i$ is the n(r-1) matrix obtained from B by deleting B_i (with the same definition for D_i and $\bar{D}_i$). Let $u_i(k)$ be the ith entry of u(k) and $\bar{u}_i(k)$ the (r-1) column vector obtained from u(k) by deleting $u_i(k)$. The structure of the system (23) is similar to the structure of the system (14). We can consider $u_i(k)$ as an unknown input and generate the redundancy relations independently of this input. If we exchange the roles of $u_i(k)$ and $\bar{u}_i(k)$ which is therefore considered as unknown inputs, the redundancy equations depend on all but one input, therefore it make faulty actuators isolation easier. This approach should be compared with the dedicated observer approach (Frank, 1989).

5. Numerical example

Consider a third order system described by equation (14), where :

$$A = \begin{pmatrix} 1 & 0 & 0 \\ 0 & 0.5 & 0 \\ 0 & 0 & 0.5 \end{pmatrix} \quad B = \begin{pmatrix} 0 \\ 1 \\ 1 \end{pmatrix} \quad G = \begin{pmatrix} 1 \\ 1 \\ 0 \end{pmatrix}$$

$$C = \begin{pmatrix} 1 & 1 & 0 \\ 0 & 1 & 1 \\ 0 & 0 & 1 \end{pmatrix} \quad D = \begin{pmatrix} 0 \\ 0 \\ 0 \end{pmatrix} \quad F = \begin{pmatrix} 0 \\ 0 \\ 0 \end{pmatrix}$$

Using the decomposition (17), we obtain :

$$H = \begin{pmatrix} H_0 & 0 \\ 0 & I \end{pmatrix} \text{ with } H_0 = \begin{pmatrix} a & -a & 0 \\ a & a & 0 \\ 0 & 0 & 1 \end{pmatrix}$$

where $a = \sqrt{2}/2$ and K = 1.

The decomposition (19) allows to define the pencil :

$$\lambda\, M_2 - \bar{A}_2 = \begin{pmatrix} -a(\lambda-1) & a(\lambda-0.5) & 0 \\ 0 & 0 & \lambda-0.5 \\ -1 & -1 & 0 \\ 0 & -1 & -1 \\ 0 & 0 & -1 \end{pmatrix}$$

from which we find the left orthogonal matrix :

$$p^T(\lambda) = \begin{pmatrix} 1 & 0 & a(1-\lambda) & a(2\lambda-1.5) & -a(2\lambda-1.5) \\ 0 & 1 & 0 & 0 & \lambda-0.5 \end{pmatrix}$$

Applying (22), the redundancy equations are then expressed :

$$u(k) - (\lambda-0.5)\, y_3(k) = 0$$
$$2u(k) - 2(1-\lambda)\, y_1(k) - (4\lambda-3)\, y_2(k) + (4\lambda-3)\, y_3(k) = 0$$

6. Conclusion

The problem of analytical redundancy equation design has been considered. We have pointed out a systematic procedure to design these equations. The technique can be used for systems with unknown inputs.

References

P.M. FRANK, I. WUNNENBERG. "Robust fault diagnosis using unknown input observer schemes". In Fault diagnosis in dynamic systems, R. Patton, P.M. Frank, R. Clark (eds), p. 47-97, Prentice Hall, 1989.

F.R. GANTMACHER. The theory of matrices. Chelsea publishing company, 1977.

J. GERTLER. "Analytical redundancy in fault detection and isolation". Proc. of the IFAC / IMACS symp. on fault detection supervision and safety for technical processes, Safeprocess, p. 9-22, Baden-Baden, 1991.

M. HOU, P.C. MULLER. "Design of robust observers for fault isolation". Proc. of the IFAC / IMACS symp. on fault detection supervision and safety for technical processes, Safeprocess, p. 295-300, Baden-Baden, 1991.

R. ISERMANN. "Process fault detection based on modelling and estimation methods". Automatica, vol. 20, n° 4, p. 387-404, 1984.

J.E. KUREK. "The state vector reconstruction for linear systems with unknown inputs". IEEE trans. on automatic control, vol. AC-28, n° 12, p. 1120-1122, 1983.

J.F. MAGNI, P. MOUYON. "A generalized approach to observers for fault diagnosis". Proc. of the 30th IEEE conf. on decision and control, vol. 3, p. 2236-2241, Brighton, 1991.

R.J. MILLER, R. MUKUNDAN. "On designing reduced-order observers for linear non-invariant systems subject to unknown inputs". Int. J. control, vol. 35, n° 1, p. 183-188, 1984.

J. PARK, J.L. STEIN. "Closed-loop, state and input observer for systems with unknown inputs". Int. J. control, vol. 48, n° 3, p. 1121-1136, 1988.

M. EL-TOHAMI, V. LOVASS-NAGY, M. MUKUNDAN. "On the design of observers for generalized state space systems using singular value decomposition". Int. J. control, vol. 38, n° 3, p. 673-683, 1983.

J. WÜNNENBERG. Observer-based fault detection in dynamic systems. PhD, Duisbourg, 1990.

FA9 - 9:30

Proceedings of the 31st Conference
on Decision and Control
Tucson, Arizona • December 1992

Optimal and Suboptimal Placement of Actuators and Sensors for Steady–State Stochastic Systems

Mifang Ruan and Ajit K. Choudhury
Dept. of Electrical Engineering
Howard University
Washington, DC 20059

Abstract

The main objective of this paper is to explore the inherent properties associated with the optimal actuator and sensor placement. Since the strict optimal method may require large computational time, suboptimal methods are also considered.

Introduction

Most of the works that appeared in the literature for actuator/sensor placement consider the transient response of deterministic system. To achieve desirable transient response, one generally places the actuators and sensors so that the degrees of controllability and observability are large. However large degree of controllability generally requires large influence coefficients of the actuators, and the large coefficients will make the system very sensitive to the actuator noises. Thus, the steady–state performance may be degraded. Refs. 1–2 are the typical works in the literature which consider the steady–state performance.

In this paper, we explored several properties associated with optimal placement of actuators and sensors using analytical method and a strictly optimal methods. Suboptimal methods are suggested for large systems.

Methodology

Let the system model be

$$\dot{x} = A\,x + B\,(u+w) + G\,\bar{w} \tag{1a}$$

$$y = H\,x \tag{1b}$$

$$z = M\,x + v \tag{1c}$$

where x = state, y = output, u = control force, z = measurement, w = actuator noise, v = sensor noise, and $\bar{w}$ = plant disturbance. w, v and $\bar{w}$ are all white Gaussian noises with zero mean and covariance matrices of W, V and $\bar{W}$ respectively. We will see that the separation of Actuator noises and plant disturbances is important in optimal actuator and sensor placement. Matrices A, H and G are given. We need to find matrices B and M to minimize the following cost function:

$$J = E_{\infty}\,[\,y^TQy + u^TRu\,] \tag{2}$$

where Q and R are given weighting matrices. LQG (linear quadratic Gaussian) method is assumed to find the feedback and Kalman gain matrices.

Examples

In order to explore the properties associated with optimal placement of actuators and sensors, we studied two simple systems. The first system is a second order mechanical system specified by Eq. (1) and following matrices:

$$A = \begin{bmatrix} 0 & 1 \\ 0 & 0 \end{bmatrix};\quad B = \begin{bmatrix} 0 \\ b \end{bmatrix}\quad G = \begin{bmatrix} 0 \\ 1 \end{bmatrix}$$

$$H = \mathrm{diag}\,[1, 1]\,;\qquad M = [c\ \ 0] \tag{3}$$

where b is a parameter depending on actuator location, and c is a parameter depending on sensor orientation. Because of physical limitation, b and c must be within 0 and 1. More information about the system can be found in Ref. (2).

When observation is complete, the total cost can be simplified to

$$J = \sqrt{rq}\,(\,W + \frac{\bar{W}}{b^2})\ \sqrt{2b\sqrt{r/q} + b^2} \tag{4}$$

To minimize the cost, the parameter b should satisfy:

$$b^3 + \sqrt{r/q}\,b^2 - \bar{W}/W\,b - 3\sqrt{r/q}\,\bar{W}/W = 0 \tag{5}$$

Eq. (5) shows that the optimal b (actuator location) depends on the noise ratio and weight ratio.

Because of the constraint $b \leq 1$, b should be set to 1 when the solution of Eq. (5) is greater than 1. It follows that b should be set to 1 if

$$\bar{W}/W \geq (\,1 + \sqrt{r/q}\,)/(\,1 + 3\sqrt{r/q}\,) \tag{6}$$

To optimize transient response, we usually select the most "efficient" actuator, i.e. largest b. However, according to Eq. (5), we may prefer to use a less "efficient" actuator in order to optimize the steady–state performance. The reason is that when the actuator is more efficient, the system is more sensitive to the actuator noise and thus degrade the system.

Fig. 1 shows the optimal b as a function of r/q for several values of $\bar{W}/W$. It can be seen that b tends to increase with r/q and $\bar{W}/W$. This phenomenon can easily be interpreted. When r/q increases, the energy becomes more important, and a more "efficient" actuator is desired to save control energy. When $\bar{W}/W$ increases, greater part of the noise is not related to the actuator location; therefore, increasing b will save control energy and will not significantly increase the introduced noise.

Fig. 2 shows the ratio of optimal cost (J) to the cost with b = 1 (J_0) as a function of r/q for several values of $\bar{W}/W$. Clearly, selecting optimal actuator location may greatly reduce the steady–state cost.

When the plant disturbance $\bar{w}$ is zero, Eq.(5) gives b = 0. However, in that case, the feedback will become infinite. To avoid singularity, we should select a

CH3229-2/92/0000-2842$1.00 © 1992 IEEE

very small b, and the cost will be nearly zero. On the other hand, when energy is not important, r is very small, and optimal b approaches sqrt($\bar{W}/W$).

When the observation is incomplete, the feedback control is based on the estimated state which contains errors. In that case, we can obtain another cost function which containing parameters b and c. By examining that cost function, we find that the cost monotonously decreases with c. To minimize the cost, we should select the sensor as sensitive as possible. Because of the constraint of $0 \leq c \leq 1$, c should be selected as 1. This selection is clearly consistent with the selection of optimizing the transient response.

The consistence between the optimization of transient response and the optimization of steady-state performance in sensor selection is expected to be true for general control systems, if the noise strengths of the sensors are independent of their locations. The reason is that unlike the introduced actuator noises which depend on the actuator locations, the introduced sensor noises are independent of the sensor locations.

We can also use this simple example to show that the optimal actuator locations may depend on the sensor locations. For a specific sensor location (c), we can find an optimal actuator location (b) and the optimal cost (J). Those data show that the optimal b and J are functions of c. More informations can be found from Ref. (2).

In this simple example, optimal sensor location does not depend on the actuator location because the optimal c equals to 1 regardless of actuator location. However, the dependence of the optimal sensor location on the actuator location can be seen from a fourth order example.

The fourth order system is a coupled mass-spring system, the plant disturbance acts on one of the mass, one actuator can be placed on either mass and one sensor can be placed on either mass. We find that when the actuator location is given, the sensor should be placed on the same mass as that of the actuator in order to minimize the cost. This fact shows that the optimal sensor location also depends on the actuator location. Therefore, the actuator placement and sensor placement are coupled.

Also, using this example we find that the cost with two actuators (one on each mass) could be greater than the cost with one actuator because two actuators will introduce an additional actuator noise.

When System is Large

Strictly optimal method, as used in previous section, generally requires large computational efforts. For large system, it is suggested to use suboptimal methods. Ref. 1 gives a good example of suboptimal methods, where cost is decomposed. However, we can use other algorithms to improve the optimality.

The main reason for the strictly optimal method to be computationally inefficient is that it requires to examine all the combinations of possible actuator locations and sensor locations. When the number of possible actuator locations and sensor locations is large, the number of the combinations is very large. To reduce the computational burden, we can use algorithms which deletes bad actuators and sensors one by one or changes for better actuators and sensors one by one. Also we may use suboptimal methods to reduce the possible number of actuator and sensor locations, and then use strictly optimal method to make final placement.

Summary

The optimal placement of actuators and sensors for stochastic systems is investigated. Conflict between the transient response and steady-state performance in optimal actuator placement has been found. However, in sensor placement, we expect that the objectives of transient response and steady-state performance are still consistent. It is also found that the actuator placement and sensor placement are coupled.

References

1. Skelton, R.E., and Delorenzo, M., "Space Structure Control Design by Variance Assignment," *J. Guidance,* Vol. 8, No. 4, 1985, pp. 454–462.
2. Ruan, M. and Choudhury, A.K. "On Optimal Selection of Actuators and Sensors for Stochastic Systems," *Proceedings of the Eighth VPI&SU Symposium on Dynamics and Control of Large Structures,* Blacksburg, VA, May 6–8, 1991.

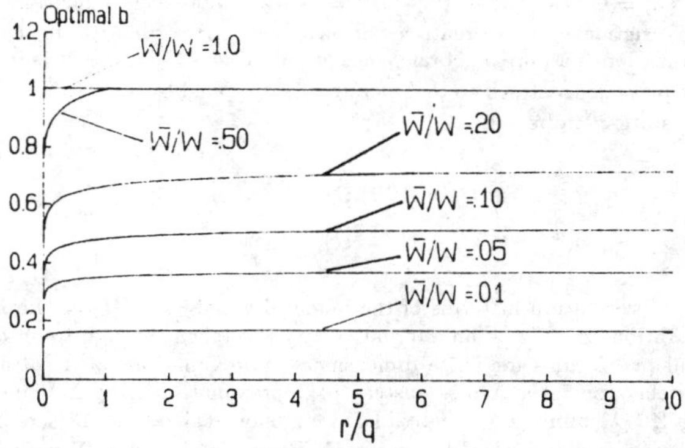

Figure 1. The optimal b as a function of r/q for several values of $\bar{W}/W$.

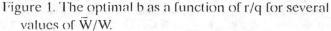

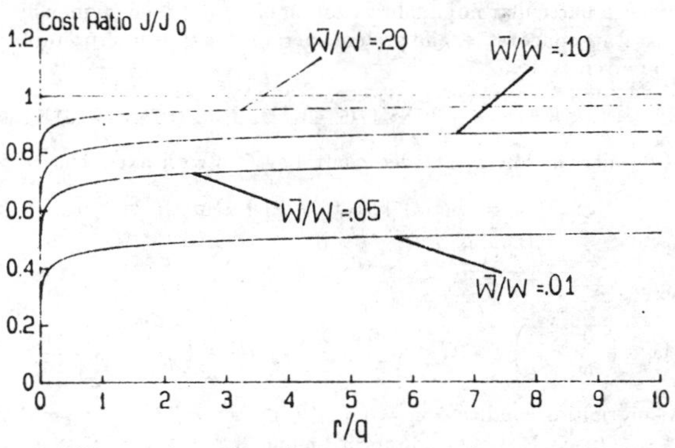

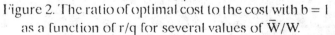

Figure 2. The ratio of optimal cost to the cost with b = 1 as a function of r/q for several values of $\bar{W}/W$.

Proceedings of the 31st Conference
on Decision and Control
Tucson, Arizona • December 1992

MINIMAL REALIZATION OF AVERAGING APPROXIMATIONS BY AVERAGING APPROXIMATIONS

W. Prager

Institut für Mathematik, Universität Graz
Heinrichstraße 36, A–8010 Graz, Austria

Abstract

For the averaging approximating systems Σ^N of the controlled linear retarded oscillator in the state space $\mathbf{R}^2 \times L^2$ minimal realizations $\tilde{\Sigma}^{\tilde{N}}$ are constructed, which can be interpreted as averaging approximations of another delay control system, again in a state space with structure $\mathbf{R}^2 \times L^2$. Furthermore a state transformation from Σ^N to $\tilde{\Sigma}^{\tilde{N}}$ is derived, which enables the application of $\tilde{\Sigma}^{\tilde{N}}$ to approximating optimal control problems posed subject to Σ^N, by this means allowing economization of time and storage without loss of accuracy.

1. Introduction

Suppose a dynamical process Σ is modeled by

$$\begin{aligned} \dot{z}^N(t) &= A^N z^N(t) + B^N u(t), \quad t \geq 0, \\ y^N(t) &= C^N z^N(t), \quad t \geq 0, \end{aligned} \qquad (\Sigma^N)$$

where $A^N \in \mathbf{R}^{\nu(N)\times\nu(N)}, B^N \in \mathbf{R}^{\nu(N)\times\ell}, C^N \in \mathbf{R}^{m\times\nu(N)}$ with $\ell, m, N, \nu(N) \in \mathbf{N}$. It is well known that to the proper rational transfer matrix $Z^N(\lambda) = C^N(\lambda I_{\nu(N)} - A^N)^{-1}B^N \in \mathbf{C}^{m\times\ell}[\lambda]$, there exists a unique (up to coordinate changes) minimal realization, i. e. a triple $\tilde{A}^{\tilde{N}} \in \mathbf{R}^{\tilde{\nu}(\tilde{N})\times\tilde{\nu}(\tilde{N})}$, $\tilde{B}^{\tilde{N}} \in \mathbf{R}^{\tilde{\nu}(\tilde{N})\times\ell}, \tilde{C}^{\tilde{N}} \in \mathbf{R}^{m\times\tilde{\nu}(\tilde{N})}$ with minimal $\tilde{\nu}(\tilde{N}) \leq \nu(N)$, such that $\tilde{Z}^{\tilde{N}}(\lambda) = \tilde{C}^{\tilde{N}}(\lambda I_{\tilde{\nu}(\tilde{N})} - \tilde{A}^{\tilde{N}})^{-1}\tilde{B}^{\tilde{N}} = Z^N(\lambda)$. Moreover, algorithms for the construction of these matrices are available. But algorithmical minimal realization is a pure algebraic concept, and if $\tilde{\nu}(\tilde{N}) < \nu(N)$, it is in general not possible to attach to the operator $\tilde{A}^{\tilde{N}}$ supplied by an algorithm, and to the state $\tilde{z}^{\tilde{N}}$, obeying

$$\begin{aligned} \dot{\tilde{z}}^{\tilde{N}}(t) &= \tilde{A}^{\tilde{N}}\tilde{z}^{\tilde{N}}(t) + \tilde{B}^{\tilde{N}}u(t), \quad t \geq 0, \\ y^N(t) &= \tilde{C}^{\tilde{N}}\tilde{z}^{\tilde{N}}(t), \quad t \geq 0, \end{aligned} \qquad (\tilde{\Sigma}^{\tilde{N}})$$

an interpretation in terms of the modeled variables $z^N(t)$ or the modeled operations A^N. In [4] an alternative way was taken through to obtain minimal realizations for finite dimensional approximations of a certain class of delay control systems. Instead of approximating Σ by Σ^N and realizing $Z^N(\lambda)$ minimally to obtain any minimal realization, Σ is replaced in the infinite dimensional by an i/o-equivalent delay system $\tilde{\Sigma}$ with smallest possible delay and appropriate controllability/ observability properties. $\tilde{\Sigma}$ is approximated by $\tilde{\Sigma}^{\tilde{N}}$ in essentially the same manner as Σ by Σ^N. The result is a particular minimal realization of $Z^N(\lambda)$ (synonymously: of Σ^N), whose state space, state and system operator are thus interpretable in terms of the given system.

2. Minimal realizations for the linear retarded oscillator

Consider the linear retarded control system with fixed delay $r > 0$

$$\begin{aligned} \dot{x}(t) &= A_0 x(t) + A_1 x(t-r) + B_0 u(t), \quad t \geq 0, \\ y(t) &= C_0 x(t), \quad t \geq 0, \end{aligned} \qquad (1)$$

where

$$A_0 = \begin{pmatrix} 0 & 1 \\ \gamma_0 & \delta_0 \end{pmatrix}, \quad A_1 = \begin{pmatrix} 0 & 0 \\ \gamma_1 & \delta_1 \end{pmatrix}, \quad B_0 = \begin{pmatrix} 0 \\ b \end{pmatrix}, \quad C_0 = (c \quad 0). \qquad (2)$$

For any initial condition $\phi = (\phi^0, \phi^1) \in M^2 = \mathbf{R}^2 \times L^2(-r, 0; \mathbf{R}^2)$ and $u \in L^2_{loc}(0,\infty;\mathbf{R}^2)$, (1) admits a unique solution $x \in W^{1,2}_{loc}(0,\infty;\mathbf{R}^2) \cap L^2_{loc}(-r,\infty;\mathbf{R}^2)$ satisfying (1) for $t \geq 0$ a.e. and $x(0) = \phi^0$, $x(s) = \phi^1(s), -r \leq s < 0$. Defining $x_t(s) = x(t+s), -r \leq s \leq 0$, and the state at time t by $z(t) = (x(t), x_t)$, (1) can be formulated abstractly in M^2 as

$$\begin{aligned} \dot{z}(t) &= Az(t) + Bu(t), \quad t \geq 0, \\ y(t) &= Cz(t), \quad t \geq 0, \end{aligned} \qquad (\Sigma)$$

where $A : \operatorname{dom} A \to M^2$ is given by

$$\operatorname{dom} A = \{\phi \in M^2 | \phi^1 \in W^{1,2}(-r, 0, \mathbf{R}^2), \phi^1(0) = \phi^0\},$$
$$A\phi = (A_0\phi^1(0) + A_1\phi^1(-r), \dot{\phi}^1),$$

and $B : \mathbf{R} \to M^2$, $Bu = (B_0 u, 0)$, $C : M^2 \to \mathbf{R}$, $C\phi = C_0\phi^0$. Given an initial condition $\phi \in M^2$, then $z(0) = \phi$. The matrix representation of the transfer function is given by $Z(\lambda) = C_0\Delta(\lambda)^{-1}B_0$, $\lambda \in \mathbf{C} \setminus \sigma(A)$, where $\Delta(\lambda) = \lambda I - A_0 - A_1 e^{-\lambda r}$. We deal with the matrix representation $\Sigma^N = (A^N, B^N, C^N)$ of the averaging approximations (see [1]) for $N = 1, 2, \ldots$, where

$$A^N = \begin{pmatrix} A_0 & 0 & \ldots & 0 & A_1 \\ \frac{N}{r}I_2 & -\frac{N}{r}I_2 & 0 & \ldots & 0 \\ 0 & \ddots & \ddots & \ddots & \vdots \\ \vdots & \ddots & \ddots & \ddots & 0 \\ 0 & \ldots & 0 & \frac{N}{r}I_2 & -\frac{N}{r}I_2 \end{pmatrix} \in \mathbf{R}^{2(N+1)\times 2(N+1)},$$

and $B^N = \operatorname{col}(B_0, 0, \ldots, 0) \in \mathbf{R}^{2(N+1)\times 1}$, $C^N = (C_0, 0, \ldots, 0) \in \mathbf{R}^{1\times 2(N+1)}$. The initial value for Σ^N is defined by $z^N(0) = p^N\phi$, where $p^N : M^2 \to \mathbf{R}^{2(N+1)}$ is given by

$$[p^N\phi]_0 = \phi^0, \quad [p^N\phi]_j = \frac{N}{r}\int_{-jr/N}^{-(j-1)r/N} \phi^1(\tau)d\tau, \quad j = 1, \ldots, N.$$

The transfer function of Σ^N is given by $Z^N(\lambda) = C_0\Delta^N(\lambda)^{-1}B_0$ with $\Delta^N(\lambda) = \lambda I_2 - A_0 - A_1(1 + \lambda r/N)^{-N} \in \mathbf{C}^{2\times 2}[\lambda]$. Since $\det A_1 = 0$, $(1 + \lambda r/N)^N$ can be removed in all entries of $\Delta^N(\lambda)^{-1}$, hence there exist systems $\tilde{\Sigma}^{\tilde{N}}$ with dimension $\leq N+2$ and transfer function $\tilde{Z}^{\tilde{N}}(\lambda) = Z^N(\lambda)$ for all $\lambda \in \mathbf{C} \setminus \sigma(A^N)$.

Consider the linear autonomous neutral delay control system with fixed $\tilde{r} > 0$

$$\begin{aligned} &\frac{d}{dt}\left(x(t) - \tilde{A}_{-1}x(t-\tilde{r})\right) = \tilde{A}_0 x(t) + \tilde{A}_1 x(t-\tilde{r}) + \tilde{B}_0 u(t), \quad t \geq 0, \\ &y(t) = \tilde{C}_0 x(t) + \tilde{C}_{00} x_t, \quad t \geq 0, \end{aligned} \qquad (3)$$

with

$$\tilde{A}_i = \begin{pmatrix} \tilde{\alpha}_i & \tilde{\beta}_i \\ \tilde{\gamma}_i & \tilde{\delta}_i \end{pmatrix}, \ i = -1, 0, 1, \quad \tilde{B}_0 \in \mathbf{R}^{2\times 1}, \quad \tilde{C}_0 \in \mathbf{R}^{1\times 2},$$

and $\tilde{C}_{00} : L^2(-\tilde{r}, 0; \mathbf{R}^2) \to \mathbf{R}$ bounded and linear. Defining $w(t) = x(t) - \tilde{A}_{-1}x(t-\tilde{r})$, $t \geq 0$, (3) admits for any initial condition $w(0) = \phi^0$, $x(s) = \phi^1(s)$, $-\tilde{r} \leq s < 0$, $\phi = (\phi^0, \phi^1) \in \tilde{M}^2 = \mathbf{R}^2 \times L^2(-\tilde{r}, 0; \mathbf{R}^2)$, and any $u \in L^2_{loc}(0,\infty;\mathbf{R}^2)$ a unique solution $x \in L^2_{loc}(-\tilde{r},\infty;\mathbf{R}^2)$ such that $w \in W^{1,2}_{loc}(0,\infty;\mathbf{R}^2)$. Defining the state at time t by $z(t) = (w(t), x_t)$, it was shown in [2] that (3) can be formulated as an abstract system in $\tilde{M}^2$,

$$\begin{aligned} \dot{z}(t) &= \tilde{A}z(t) + \tilde{B}u(t), \quad t \geq 0, \\ y(t) &= \tilde{C}z(t), \quad t \geq 0, \end{aligned} \qquad (\tilde{\Sigma})$$

where $\tilde{A} : \operatorname{dom} \tilde{A} \to \tilde{M}^2$ is given by

$$\operatorname{dom} \tilde{A} = \{\phi \in M^2 | \phi^1 \in W^{1,2}(-\tilde{r}, 0, \mathbf{R}^2), \phi^0 = \phi^1(0) - \tilde{A}_{-1}\phi^1(-\tilde{r})\},$$
$$\tilde{A}\phi = (\tilde{A}_0\phi^1(0) + \tilde{A}_1\phi^1(-\tilde{r}), \dot{\phi}^1),$$

CH3229-2/92/0000-2844$1.00 © 1992 IEEE

$\tilde{B}: \mathbf{R} \to \tilde{M}^2$, $\tilde{B}u = (\tilde{B}_0 u, 0)$, and $\tilde{C}: \tilde{M}^2 \to \mathbf{R}$, $\tilde{C}\phi = \tilde{C}_0(\phi^0 + \tilde{A}_{-1}\phi^1(-\tilde{r})) + \tilde{C}_{00}\phi^1$. The transfer function has the matrix representation $\tilde{Z}(\lambda) = (\tilde{C}_0 + \tilde{C}_{00}e^{\lambda\cdot})\tilde{\Delta}(\lambda)^{-1}\tilde{B}_0$, $\lambda \in \mathbf{C} \setminus \sigma(\tilde{A})$, with $\tilde{\Delta}(\lambda) = \lambda I - \tilde{A}_0 - (\tilde{A}_1 + \lambda\tilde{A}_{-1})e^{-\lambda\tilde{r}}$. The matrix representation $\tilde{\Sigma}^N = (\tilde{A}^N, \tilde{B}^N, \tilde{C}^N)$ of the averaging approximations for $N = 2, 3, \ldots$ is given by

$$\tilde{A}^N = \begin{pmatrix} \tilde{A}_0 & 0 & \cdots & \cdots & 0 & \tilde{A}_0\tilde{A}_{-1} + \tilde{A}_1 \\ \frac{N}{\tilde{r}}I_2 & -\frac{N}{\tilde{r}}I_2 & 0 & \cdots & 0 & \frac{N}{\tilde{r}}\tilde{A}_{-1} \\ 0 & \ddots & \ddots & \ddots & \vdots & 0 \\ \vdots & \ddots & \ddots & \ddots & 0 & \vdots \\ \vdots & & \ddots & \ddots & \ddots & 0 \\ 0 & \cdots & \cdots & 0 & \frac{N}{\tilde{r}}I_2 & -\frac{N}{\tilde{r}}I_2 \end{pmatrix} \in \mathbf{R}^{2(N+1)\times 2(N+1)},$$

$\tilde{B}^N = \mathrm{col}(\tilde{B}_0, 0, \ldots, 0) \in \mathbf{R}^{2(N+1)\times 1}$ (see [3]), and $\tilde{C}^N \in \mathbf{R}^{1\times 2(N+1)}$ to be specified below. For the initial value of $\tilde{\Sigma}^N$ we have $\tilde{z}^N(0) = \tilde{p}^N\phi$, where the action of $\tilde{p}^N: \tilde{M}^2 \to \mathbf{R}^{2(N+1)}$ is defined as that of p^N with r replaced by $\tilde{r}$. The characteristic matrix of $\tilde{\Sigma}^N$ is given by $\tilde{\Delta}^N(\lambda) = \lambda I_2 - \tilde{A}_0 - [\tilde{A}_1 + \lambda\tilde{A}_{-1}](1 + \lambda\tilde{r}/N)^{-N} \in \mathbf{C}^{2\times 2}[\lambda]$.

Theorem 1. *For the abstract system Σ corresponding to (1) with matrices (2) and transfer function $Z(\lambda)$ there exists a system $\tilde{\Sigma}$ corresponding to (3) with minimal delay $\tilde{r} = r/2$ and transfer function $\tilde{Z}(\lambda) = Z(\lambda)$ for $\lambda \in \mathbf{C} \setminus \sigma(A)$.*

Proof. Form $\tilde{A}$ with $\tilde{A}_0 = A_0$, $\tilde{\alpha}_{-1} = \tilde{\beta}_{-1} = \tilde{\delta}_{-1} = 0$ and $\tilde{\gamma}_{-1} \neq 0$, $|\tilde{\gamma}_{-1}|$ sufficiently large, such that

$$\tilde{\alpha}_1 = -\frac{1}{2}(\tilde{\gamma}_{-1} + \frac{\delta_0\delta_1}{\tilde{\gamma}_{-1}}) \pm \sqrt{\frac{1}{4}(\tilde{\gamma}_{-1} + \frac{\delta_0\delta_1}{\tilde{\gamma}_{-1}})^2 + (\frac{\delta_1}{\tilde{\gamma}_{-1}})^2\gamma_0 + \gamma_1} \in \mathbf{R},$$

$$\tilde{\beta}_1 = \delta_1/\tilde{\gamma}_{-1}, \quad \tilde{\gamma}_1 = \tilde{\alpha}_1\delta_0 - \tilde{\beta}_1\gamma_0, \quad \tilde{\delta}_1 = -(\tilde{\alpha}_1 + \tilde{\gamma}_{-1}).$$

Take $\tilde{B}_0 = B_0$, $\tilde{B} = B$, and replace the action of the output operator by

$$y(t) = (\,\tilde{c}_0 \;\; 0\,)x(t) + (\,\tfrac{\tilde{\alpha}_1}{\tilde{\beta}_1} \;\; 1\,)\int_{-r/2}^{0} \tilde{c}(s)x(t+s)ds, \quad t \geq 0,$$

where

$$\tilde{c}_0 = c\frac{e^{-(\tilde{\alpha}_1/\tilde{\beta}_1)r/2}}{\tilde{\beta}_1 + e^{-(\tilde{\alpha}_1/\tilde{\beta}_1)r/2}}, \quad \tilde{c}(s) = c\frac{\tilde{\beta}_1 e^{-(\tilde{\alpha}_1/\tilde{\beta}_1)s}}{\tilde{\beta}_1 + e^{-(\tilde{\alpha}_1/\tilde{\beta}_1)r/2}}, \quad -r/2 \leq s \leq 0.$$

Minimality of $\tilde{r}$ is clear by the definitions of $\Delta(\lambda)$, $\tilde{\Delta}(\lambda)$, and $Z(\lambda) = \tilde{Z}(\lambda)$ is shown by direct calculation. □

Beyond minimal dimension of the state space, minimal realizations are characterized equivalently by the property of being controllable and observable. Suitable generalizations of these finite dimensional system properties hold for $\tilde{\Sigma}$ constructed in the previous theorem.

Proposition 1. *(a) Σ with matrices (2) is neither approximately controllable nor strictly observable. (b) $\tilde{\Sigma}$ constructed according to Thm.1 is approximately controllable and strictly observable.* □

Define for $N = 1, 2, \ldots$ approximations to the output operator $\tilde{C}$ by

$$\tilde{C}^N = (\tilde{c}_0^N \;\; 0 \;|\; \frac{\tilde{\alpha}_1}{\tilde{\beta}_1}\tilde{c}_1^N \;\; \tilde{c}_1^N \;|\; \cdots \;|\; \frac{\tilde{\alpha}_1}{\tilde{\beta}_1}\tilde{c}_N^N \;\; \tilde{c}_N^N) \in \mathbf{R}^{1\times 2(N+1)},$$

where

$$\tilde{c}_0^N = c\frac{(1 - \frac{\tilde{\alpha}_1}{\tilde{\beta}_1}\frac{\tilde{r}}{N})^N}{\tilde{\beta}_1 + (1 - \frac{\tilde{\alpha}_1}{\tilde{\beta}_1}\frac{\tilde{r}}{N})^N}, \quad \tilde{c}_j^N = c\frac{\tilde{r}}{N}\frac{\tilde{\beta}_1(1 - \frac{\tilde{\alpha}_1}{\tilde{\beta}_1}\frac{\tilde{r}}{N})^{j-1}}{\tilde{\beta}_1 + (1 - \frac{\tilde{\alpha}_1}{\tilde{\beta}_1}\frac{\tilde{r}}{N})^N}, \quad j = 1, \ldots, N.$$

Proposition 2. *$\tilde{\Sigma}^{N/2} = (\tilde{A}^{N/2}, \tilde{B}^{N/2}, \tilde{C}^{N/2})$ is a minimal realization of $\Sigma^N = (A^N, B^N, C^N)$, $N = 2K$, for $K = 1, 2, \ldots$ in case $\delta_1 = 0$ and for $K = 2, 3, \ldots$ with at most one exception (if $\gamma_1 = \delta_1 N/r$) in case $\delta_1 \neq 0$.* □

In finite-dimensional systems theory two systems $\Sigma = (A, B, C)$ and $\tilde{\Sigma} = (\tilde{A}, \tilde{B}, \tilde{C})$ in the state space $\mathbf{R}^n$ are defined to be algebraically equivalent if there exists a regular $T \in \mathbf{R}^{n\times n}$ such that $\tilde{A} = TAT^{-1}$, $\tilde{B} = TB$ and $\tilde{C} = CT^{-1}$. In that case, the same input $u(t)$ fed to the dynamical systems associated with Σ and $\tilde{\Sigma}$ produces the same output $y(t) = \tilde{y}(t)$ if the initial values x_0 of Σ and $\tilde{x}_0$ of $\tilde{\Sigma}$ are related by $\tilde{x}_0 = Tx_0$. In our situation a generalization would be desirable in order to enable transformation of the approximating initial values $p^N\phi$ of Σ^N to the initial values of $\tilde{\Sigma}^{N/2}$ corresponding in the same sense.

Theorem 2. *Let be given Σ with matrices (2) and let be $\tilde{\Sigma}$ a corresponding replacement as in thm 1. If $\tilde{\Sigma}^{N/2}$ is the minimal realization of Σ^N, then there exist $U^N \in \mathbf{R}^{(N+2)\times 2(N+1)}$, such that $\tilde{A}^{N/2} = U^N A^N (U^N)^+$, $\tilde{B}^{N/2} = U^N B^N$, $\tilde{C}^{N/2} = C^N(U^N)^+$ for $N = 4K$, $K = 1, 2, \ldots$, where $(U^N)^+$ is the Moore-Penrose inverse of U^N.*

Proof. U^N is derived from $U^N A^N = \tilde{A}^{N/2}U^N$ under the constraint $U^N B^N = \tilde{B}^{N/2}$. It can be shown that $C^N = \tilde{C}^{N/2}U^N$ and $\mathrm{rank}\, U^N = N+2$ for $N = 4K$, $K = 1, 2, \ldots$. Hence $(U^N)^+ = (U^N)^T[U^N(U^N)^T]^{-1}$, from which it follows $U^N(U^N)^+ = I_{N+2}$ and therefore $U^N A^N (U^N)^+ = \tilde{A}^{N/2}$ and $C^N(U^N)^+ = \tilde{C}^{N/2}$. The general form of $U^N \in \mathbf{R}^{(N+2)\times 2(N+1)}$ is

$$U^N = \begin{pmatrix} U_{00}^N & U_{01}^N & \cdots & \cdots & \cdots & \cdots & \cdots & \cdots & U_{0N}^N \\ U_{10}^N & U_{11}^N & \cdots & \cdots & \cdots & U_{1,N/2+1}^N & 0 & \cdots & 0 \\ \vdots & \vdots & \ddots & & & & \ddots & \ddots & \vdots \\ \vdots & \vdots & & \ddots & & & & \ddots & 0 \\ U_{N/2,0}^N & U_{N/2,1}^N & \cdots & \cdots & U_{11}^N & \cdots & \cdots & \cdots & U_{1,N/2+1}^N \end{pmatrix}$$

where all occuring $U_{ij}^N \in \mathbf{R}^{2\times 2}$ can be computed explicitely. □

U^N approximates an operator $U: \mathcal{D} \to \tilde{M}^2$, where $\mathcal{D}$ is a dense subset of $\mathrm{dom}\, A$, which transforms a given state ϕ of Σ to the corresponding state $U\phi = ([U\phi]^0, [U\phi]^1)$ of $\tilde{\Sigma}$. One can compute U explicitely and show

Theorem 3. *Let $\phi \in \mathcal{D}$, then $U\phi \in \mathrm{dom}\, \tilde{A}$ and $\tilde{A}U\phi = UA\phi$.* □

3. Application to an optimal control problem

As an example for the applicability of the minimal realizations constructed in the previous section, consider the following infinite time horizon optimal control problem for the averaging approximating systems Σ^N:

Minimize $J^N(u, p^N\phi) = \int_0^\infty [y^N(t)^2 + u(t)^2]dt$ over $u \in L^2_{loc}(0, \infty; \mathbf{R})$

subject to Σ^N with coefficient-, input- and output-matrices (2) and

initial value $z^N(0) = p^N\phi$, $\phi \in M^2$. (P)

Since Σ^N is stabilizable and detectable there exists a unique positive semidefinite solution $\Pi^N \in \mathbf{R}^{2(N+1)\times 2(N+1)}$ of the corresponding algebraic Riccati equation and the optimal control $\bar{u}^N$ minimizing J^N is given by $\bar{u}^N(t) = -(B^N)^T\Pi^N\bar{z}^N(t)$, $t \geq 0$, where $\bar{z}^N$ is the solution of the corresponding closed loop system. Moreover, $J^N(\bar{u}^N, p^N\phi) = z^N(0)^T\Pi^N z^N(0)$.

Theorem 4. *Let be $\tilde{\Sigma}^{N/2} = (\tilde{A}^{N/2}, \tilde{B}^{N/2}, \tilde{C}^{N/2})$ minimal realizations of $\Sigma^N = (A^N, B^N, C^N)$, $N = 4K$, $K = 1, 2, \ldots$ constructed as in Thm.1. Then the solution of (P) is also given by $\bar{u}^N(t) = -(\tilde{B}^{N/2})^T\tilde{\Pi}^{N/2}\bar{\tilde{z}}^{N/2}(t)$, $t \geq 0$, where $\tilde{\Pi}^{N/2} \in \mathbf{R}^{(N+2)\times(N+2)}$ is the unique positive semidefinite solution of the algebraic Riccati equation corresponding to $\tilde{\Sigma}^{N/2}$, and $\bar{\tilde{z}}^{N/2}$ is the solution of*

$$\dot{\tilde{z}}^{N/2}(t) = (\tilde{A}^{N/2} - \tilde{B}^{N/2}(\tilde{B}^{N/2})^T\tilde{\Pi}^{N/2})\tilde{z}^{N/2}(t), \quad t \geq 0,$$
$$\tilde{z}^{N/2}(0) = U^N p^N\phi.$$

Moreover,

$$J^N(\bar{u}^N, p^N\phi) = \tilde{z}^{N/2}(0)^T\tilde{\Pi}^{N/2}\tilde{z}^{N/2}(0),$$

and

$$\tilde{\Pi}^{N/2} = ((U^N)^+)^T\Pi^N(U^N)^+. \quad \square$$

References

[1] H.T. Banks and J.A. Burns, *Hereditary control problems: Numerical methods based on averaging approximation*, SIAM J. Control Opt. **16** (1978), 169-208.

[2] J.A. Burns, T.L. Herdman and H.W. Stech, *Linear functional differential equations as semigroups on product spaces*, SIAM J. Math. Anal. **14** (1983), 98-116.

[3] F. Kappel, *Approximation of neutral functional differential equations in the state space $\mathbf{R}^n \times L^2$*, Colloquia mathematica societatis János Bolyai, 30. Qualitative Theory of Differential Equations **I** (1982), 463-506, János Bolyai Math. Soc. and North Holland Publ. Comp., New York.

[4] W. Prager, *Minimal realizations of averaging approximations by averaging approximations*, Institute for Mathematics, University of Graz, Preprint No. 180-1991.

Proceedings of the 31st Conference
on Decision and Control
Tucson, Arizona • December 1992

QUADRATICALLY DECENTRALIZED STABILIZATION FOR UNCERTAIN STRUCTURED INTERCONNECTED SYSTEMS

Cheng-Fa Cheng* Wen-June Wang† Yu-Ping Lin‡
*Institute of Electronics, National Chiao-Tung University, Hsin-Chu 30050, Taiwan, R.O.C.
†Department of Electrical Engineering, National Central University, Chung-Li 32054, Taiwan, R.O.C.
‡Department of Control Engineering, National Chiao-Tung University, Hsin-Chu 30050, Taiwan, R.O.C.

ABSTRACT

The quadratically decentralized stabilizability conditions for a general uncertain interconnected system is determined via Riccati equation approach. Willems' lemma [4] is applied to improve the stabilizability condition such that the local feedback control gain K_i can be selected easily. Moreover, the linear quadratic controller is synthesized by solving a modified Riccati-type equation. It is also shown that, the quadratically decentralized stabilization problem for the general uncertain interconnected system can be broken down as N decoupled H_∞ control problems.

1. Introduction

When the interconnected system are considered, the supposition of centrality of the above works fails to hold due to the lack of centralized information and computing capability. Consequently, interconnected systems are more susceptible to the uncertainties than other systems for their parameters cannot be calculated with sufficient accuracy for on-line controllers. Thus, the problem of determining local feedback controls to stabilize the interconnected systems with uncertainties has received some attention in recent literatures [1-3].

In this paper, a general model of uncertain interconnected system is considered. The Riccati equation approach is extended to attack the quadratically decentralized stabilization of the interconnected system. Associated algebraic stability conditions are evaluated and related H_∞ control problems will be investigated. The system matrix of each subsystem is not required to be stable and there is no a priori restriction on interconnections.

2. System Description and Problem Formulation

Let S be a interconnected system composed of N structured subsystems S_i. S_i is described by the equation

$$\begin{aligned}\dot{x}_i(t) &= [A_{io} + \Delta A_i(r_i(t))]x_i(t) + [B_{io} + \Delta B_i(s_i(t))]u_i(t) \\ &\quad + \sum_{j\neq i}^{N} D_i C_{ij}(t) E_j x_j \qquad (1)\end{aligned}$$

where (A_{io}, B_{io}) is the nominal system matrix, $x_i \in R^{n_i}$, $B_{io} \in R^{n_i \times p_i}$, $D_i \in R^{n_i \times d_i}$, $E_j \in R^{e_j \times n_j}$, and $C_{ij}(t) \in R^{d_i \times e_j}$. The interconnection is of the structured form $D_i C_{ij}(t) E_j$, where $C_{ij}(t)$ is Lebesgue measurable and only its upper bound is known

$$C_{ij}^T(t) C_{ij}(t) \leq \alpha_{ij} I \qquad (2)$$

with $\alpha_{ij} \geq 0$. The vectors $r_i(t) \in R^{k_i}$ and $s_i(t) \in R^{l_i}$ are Lebesgue measurable and range in the known compact sets Υ_i and Π_i

$$\begin{aligned}\Upsilon_i &= \{r_i \in R^{k_i} : |r_{im}| \leq \bar{r}_{im}, m = 1, 2, \cdots, k_i\} \qquad (3a) \\ \Pi_i &= \{s_i \in R^{l_i} : |s_{iq}| \leq \bar{s}_{iq}, q = 1, 2, \cdots, l_i\} \qquad (3b)\end{aligned}$$

The uncertainty matrices are assumed to be

$$\Delta A_i(r_i(t)) = \sum_{m=1}^{k_i} A_{im} r_{im}(t) \; ; \; \Delta B_i(s_i(t)) = \sum_{q=1}^{l_i} B_{iq} s_{iq}(t) \qquad (4)$$

where A_{im} and B_{iq} can be expressed as

$$A_{im} = A'_{im} A''_{im} \quad ; \quad B_{iq} = B'_{iq} B''_{iq} \qquad (5)$$

For simplicity, let

$$\begin{aligned}\mathcal{T}_i &= \sum_{m=1}^{k_i} \bar{r}_{im} A'_{im} {A'_{im}}^T \; ; \; \mathcal{U}_i = \sum_{m=1}^{k_i} \bar{r}_{im} {A''_{im}}^T A''_{im} \qquad (6a) \\ \mathcal{V}_i &= \sum_{q=1}^{l_i} \bar{s}_{iq} B'_{iq} {B'_{iq}}^T \; ; \; \mathcal{W}_i = \sum_{q=1}^{l_i} \bar{s}_{iq} {B''_{iq}}^T B''_{iq} \qquad (6b)\end{aligned}$$

By hypothesis all pairs (A_{io}, B_{io}) are controllable. Let u_i be the linear local state feedback control law for S_i,

$$u_i = K_i x_i \qquad (7)$$

Definition 1 : The uncertain interconnected system (1) is said to be quadratically decentralized stabilizable if there exist a linear local state feedback control (7), a positive definite symmetric matrix P_i and a constant $\beta > 0$ such that the following condition holds :

Given any admissible uncertainty, the Lyapunov derivative corresponding to the resulting closed-loop system and Lyapunov function $V(X) = X^T P X$ satisfies

$$\begin{aligned}\dot{V} &\triangleq \sum_{i=1}^{N} \{x_i^T [A_{io}^T P_i + P_i A_{io}] x_i + 2 x_i^T P_i \Delta A_i(r_i(t)) x_i \\ &\quad + 2 x_i^T P_i [B_{io} + \Delta B_i(s_i(t))] K_i x_i \\ &\quad + 2 x_i^T P_i \sum_{j \neq i}^{N} D_i C_{ij}(t) E_j x_j\} \leq -\beta \|X\|^2 \qquad (8)\end{aligned}$$

where $X^T = [x_1^T, \cdots, x_N^T]$ and $P = block - diag[P_1, \cdots, P_N]$.

3. Quadratically Decentralized Stabilizability

Let the Laypunov function V for the whole system be

$$V = \sum_{i=1}^{N} v_i = \sum_{i=1}^{N} x_i^T P_i x_i \qquad (9)$$

According to the Lyapunov direct method, we have the following results.

Theorem 1 : If there exists a solution $P_i = P_i^T > 0$ in

$$\hat{A}_{io}^T P_i + P_i \hat{A}_{io} + P_i \Phi_i P_i + (\Psi_i + Q_i) = 0 \qquad (10)$$

where $\hat{A}_{io} = A_{io} + B_{io} K_i$, $\Phi_i \doteq \mathcal{T}_i + \mathcal{V}_i + \delta_i D_i D_i^T$, $\Psi_i = \mathcal{U}_i + K_i^T \mathcal{W}_i K_i + (\sum_{j\neq i}^{N} \alpha_{ji}) E_i^T E_i$, $Q_i = Q_i^T > 0$, and $\delta_i \triangleq \sum_{j\neq i}^{N} \Delta_{ij}$ with

$$\Delta_{ij} = \begin{cases} 1 & , \text{if } C_{ij} \neq 0, \, i \neq j \\ 0 & , \text{otherwise} \end{cases} \qquad (11)$$

for some K_i, then the local feedback control (7) will ensure the whole uncertain structured interconnected system S to be quadratically decentralized stabilizable.

CH3229-2/92/0000-2846$1.00 © 1992 IEEE

With the aid of Lemma 5 of Willems [4], the following theorem gives us an explicit technique to determine the controller gain K_i.

Theorem 2 : The uncertain structured interconnected system S will be quadratically decentralized stabilizable via the linear local state feedback control (7) if the following relations hold for some K_i, $i = 1, 2, \cdots, N$.

(a) $Re\lambda(\hat{A}_{io}) < 0$ (12)

(b) There exists a constant $\gamma_i < 1$ such that

$$\|\Psi_i^{1/2}(sI - \hat{A}_{io})^{-1}\Phi_i^{1/2}\|_\infty \le \gamma_i \qquad (13)$$

4. Linear Quadratic Controller Synthesis

Consider a linear quadratic controller as

$$u_i(t) = K_i x_i = -\frac{1}{\epsilon_i} R_i^{-1} B_{io}^T P_i x_i \qquad (14)$$

where the constant $\epsilon_i > 0$, and both R_i and P_i are symmetric positive-definite matrices.

Theorem 3 : If there exists a solution $P_i = P_i^T > 0$ in

$$A_{io}^T P_i + P_i A_{io} - \frac{1}{\epsilon_i} P_i B_{io}(2R_i^{-1} - R_i^{-1}\mathcal{W}_i R_i^{-1}) B_{io}^T P_i + P_i(\mathcal{T}_i + \frac{1}{\epsilon_i}\mathcal{V}_i + \delta_i D_i D_i^T) P_i + [\mathcal{U}_i + (\sum_{j \ne i}^{N} \alpha_{ji}) E_i^T E_i + \epsilon_i Q_i] = 0 \qquad (15)$$

for some $\epsilon_i > 0$, $i = 1, 2, \cdots, N$, where both Q_i and R_i are symmetric positive-definite matrices, then the uncertain structured interconnected system S is quadratically decentralized stabilizable via the linear local feedback control (14).

Remark : If we select R_i such that $\hat{R}_i^{-1} \triangleq (2R_i^{-1} - R_i^{-1}\mathcal{W}_i R_i^{-1}) > 0$ and consider systems [5]

$$\dot{x}_i = A_{io} x_i + B_{io} u_i + G_i \xi_i, \quad z_i = \begin{pmatrix} C_i x_i \\ \hat{R}_i^{1/2} u_i \end{pmatrix} \qquad (16)$$

where $G_i G_i^T = \mathcal{T}_i + \frac{1}{\epsilon_i}\mathcal{V}_i + \delta_i D_i D_i^T$, $C_i^T C_i = \mathcal{U}_i + (\sum_{j \ne i}^{N} \alpha_{ji}) E_i^T E_i + \epsilon_i Q_i$, (A_{io}, B_{io}) is controllable, (A_{io}, C_i) is observable and z_i is the output to be regulated. Now, define

$$\beta_{i\infty} = \inf_{u_i \in L_2[0,\infty)} \sup_{\xi_i \in L_2[0,\infty)} \frac{\|z_i\|_2}{\|\xi_i\|_2} \qquad (17)$$

the problem becomes adjusting ϵ_i such that the restriction,

$$\beta_{i\infty} < 1 \qquad (18)$$

for $i = 1, 2, \cdots, N$, can be satisfied.

5. Illustrative Example

This example is the 6-plate gas-absorber system adopted from [6]. The system is decomposed into two subsystems as

$$\dot{x}_1(t) = \begin{bmatrix} -1.173 & 0.634 & 0.000 \\ 0.538 & -1.173 & 0.634 \\ 0.000 & 0.538 & -1.173 \end{bmatrix} x_1(t) + \begin{bmatrix} 0.000 & 0.000 & 0.000 \\ 0.000 & 0.000 & 0.000 \\ 0.634 & 0.000 & 0.000 \end{bmatrix} x_2(t) + \begin{bmatrix} 0.538 \\ 0.000 \\ 0.000 \end{bmatrix} u_1(t)$$

$$\dot{x}_2(t) = \begin{bmatrix} -1.173 & 0.634 & 0.000 \\ 0.538 & -1.173 & 0.634 \\ 0.000 & 0.538 & -1.173 \end{bmatrix} x_2(t) + \begin{bmatrix} 0.000 & 0.000 & 0.538 \\ 0.000 & 0.000 & 0.000 \\ 0.000 & 0.000 & 0.000 \end{bmatrix} x_1(t) + \begin{bmatrix} 0.000 \\ 0.000 \\ 0.880 \end{bmatrix} u_2(t)$$

If we assume that the parameters variation to be

$$\Delta A_1 = \begin{bmatrix} r_{11} & r_{12} & 0.0 \\ r_{13} & r_{11} & r_{12} \\ 0.0 & r_{13} & r_{11} \end{bmatrix}; \quad \Delta B_1 = \begin{bmatrix} s_{11} \\ 0.0 \\ 0.0 \end{bmatrix}$$

$$\Delta A_2 = \begin{bmatrix} r_{21} & r_{22} & 0.0 \\ r_{23} & r_{21} & r_{22} \\ 0.0 & r_{23} & r_{21} \end{bmatrix}; \quad \Delta B_2 = \begin{bmatrix} 0.0 \\ 0.0 \\ s_{21} \end{bmatrix}$$

$$D_1 C_{12} E_2 = \begin{bmatrix} 0.0 \\ 0.0 \\ 1.0 \end{bmatrix} (0.634 + r_{12}) [\ 1.0 \quad 0.0 \quad 0.0\ ]$$

$$D_2 C_{21} E_1 = \begin{bmatrix} 1.0 \\ 0.0 \\ 0.0 \end{bmatrix} (0.538 + r_{23}) [\ 0.0 \quad 0.0 \quad 1.0\ ]$$

with

$$\Upsilon_1 = \{r_1 \in R^3 : |r_{11}| \le 0.1, \quad |r_{12}| \le 0.1, \quad |r_{13}| \le 0.12\}$$
$$\Upsilon_2 = \{r_2 \in R^3 : |r_{21}| \le 0.2, \quad |r_{22}| \le 0.12, \quad |r_{23}| \le 0.15\}$$
$$\Pi_1 = \{s_1 : |s_{11}| \le 0.3\} \quad ; \quad \Pi_2 = \{s_2 : |s_{21}| \le 0.15\}$$

Moreover, we have $\alpha_{12} = 0.539$, $\alpha_{21} = 0.473$. Apply Theorem 3 to the system and choose $\delta_1 = \delta_2 = 1$, $\epsilon_1 = 0.001$, $\epsilon_2 = 0.01$, $R_1 = R_2 = 1$ and $Q_1 = Q_2 = I_3$. From (14), the stabilizing control laws are

$$u_1(t) = -[\ 16.2451 \quad 12.0468 \quad 11.0062\ ]\, x_1(t)$$
$$u_2(t) = -[\ 5.4306 \quad 6.5360 \quad 3.4020\ ]\, x_2(t)$$

6. Conclusion

Using the Riccati equation approach, we have derived the sufficient conditions for the quadratically decentralized stabilization of a class of uncertain structured interconnected systems. All the results are developed by Lyapunov theorem and the solutions of modified Riccati-type equations. It is noted that when the Riccati equation approach is used, the uncertainties can be time-varying and the so-called "matching conditions" are not needed. Furthermore, the associated algebraic stability conditions are evaluated and related H_∞ problems have been investigated.

Acknowledgment

This research is supported by the National Science Council of Taiwan, R.O.C., under contract NSC 81-0404-E009-523.

References

[1] M. J. Chen and C. A. Desoer, "Algebraic theory for robust stability of interconnected systems: necessary and sufficient conditions," *IEEE Trans. Automat. Contr.*, vol. AC-29, pp. 511-519, 1984.

[2] A. A. Bahnasawi, A. S. Al-Fuhaid, and M. S. Mahmoud, "Decentralized and hierarchical control of interconnected uncertain system," *Proc. IEE, Pt. D*, vol. 137, pp. 311-321, 1990.

[3] C. J. Mao and W. S. Lin, "Decentralized control of interconnected systems with unmodeled nonlinearity and interaction," *Automatica*, vol. 26, pp. 263-268, 1990.

[4] J. C. Willems, "Least squares stationary optimal control and the algebraic Riccati equation," *IEEE Trans. Automat. Contr.*, vol. AC-16, pp. 621-634, 1971.

[5] R. J. Veillette and J. V. Medanic, "H_∞-norm bounds for ARE-based designs," *Syst. Contr. Lett.*, vol. 13, pp. 193-204, 1989.

[6] M. Darwish and J. Fantin, "Stabilization and control of absorber tower chemical process," *IFAC/IFIP/IFORS Conf.*, Rabat, Morocco, pp. 153-158, November, 1980.

Proceedings of the 31st Conference
on Decision and Control
Tucson, Arizona • December 1992

Generalized Frequency Weighted Balanced Reduction

Pepijn M.R. Wortelboer
Philips Research Laboratories
P.O.Box 80.000, 5600 JA Eindhoven, The Netherlands

Okko H. Bosgra
Mech. Eng. Systems and Control Group
Delft University of Technology
Mekelweg 2, 2628 CD Delft, The Netherlands.

Abstract

A generalization of frequency weighted balanced reduction is worked out that comprises: [1] Enns' original frequency weighted balanced reduction, and an extension of [2] frequency interval balanced reduction. In this extension scalar quadratic frequency functions can be assigned in separate frequency intervals. Weighting in frequency points is also developed. This generalization provides a very direct and flexible way of specifying frequency weightings. In a MATLAB implementation the frequency weightings can be adjusted and refined until the reduction error after weighted balanced reduction is satisfactory. This is shown in an example.

Balanced truncation [1, 2, 3] hinges on the calculation of 1) the controllability and observability Gramian, 2) the balancing transformation, and 3) the balanced realization which can then be truncated. The result is balanced and stable. Frequency weighted balanced reduction leaves steps 2 and 3 unaltered and aims at the construction of relevant frequency weighted Gramians. Stability is not guaranteed in general.

In the work reported here the aim is to build frequency weighted Gramians in a step by step procedure and to use scalar frequency functions that can be constructed pointwise. In each step these frequency functions can be adjusted/refined based on an inspection of the previous reduction results. The approach taken is that of building the controllability and observability Gramians of different parts, each associated with a certain frequency range and a specific input or output.

The idea of defining Gramians over a fixed frequency interval is due to Gawronski and Juang [2]. We will call their method *frequency interval balanced reduction*. In this paper non-uniform frequency weights within these intervals are worked out. The limit case of the frequency interval going to a single frequency point is introduced. All methods fit into one framework. The construction of the observability Gramian is similar to the construction of the controllability Gramian and will not be discussed separately.

1 Enns' frequency weighted Gramians [1]

In normal balanced reduction the inputs are assumed to be unit intensity white noise processes. Colored noise processes can be generated by means of a weighting function (filter) $W_i(s)$ receiving white noise inputs u_i. Let the state space matrices of the system $G(s)$ be given by (A, B, C, D) and those of $W_i(s)$ by (A_i, B_i, C_i, D_i). The frequency response of the states of (A, B, C, D) to the inputs u_i on $W_i(s)$ is

$$x(s) = (sI-A)^{-1}BW_i(s)u_i(s) = \Pi_s(sI-A_s)^{-1}B_su_i(s) \quad (1)$$

with $\Pi_s = \begin{bmatrix} O & I \end{bmatrix}$, $A_s = \begin{bmatrix} A_i & O \\ BC_i & A \end{bmatrix}$ and $B_s = \begin{bmatrix} B_i \\ BD_i \end{bmatrix}$. Using (1) and the fact that $u_i(j\omega)u_i^H(j\omega) = I$ we get the weighted controllability Gramian as

$$\begin{aligned} P_{W_i} &= \tfrac{1}{2\pi}\int_{-\infty}^{\infty} x(j\omega)x^H(j\omega)\mathrm{d}\omega \\ &= \tfrac{1}{2\pi}\int_{-\infty}^{\infty}(j\omega I-A)^{-1}B\Omega(\omega)B^H(j\omega I-A)^{-H}\mathrm{d}\omega \quad (2) \\ &= \Pi_sP_s\Pi_s^H \end{aligned}$$

with $P_s = \frac{1}{2\pi}\int_{-\infty}^{\infty}(j\omega I-A_s)^{-1}B_sB_s^H(j\omega I-A_s)^{-H}\mathrm{d}\omega$, the solution to $A_sP_s + P_sA_s^H + B_sB_s^H = 0$, and $\Omega(\omega) = W_i(j\omega)W_i^H(j\omega)$.

2 Frequency weighted interval Gramians

[2] is extended in two ways: more intervals and quadratic weighting within each interval. Each input and each interval contributes to the final weighted controllability Gramian. The idea is worked out for one input ($B \in \mathbb{R}^{n\times 1}$) and one frequency interval $[\omega_a, \omega_b]$ with $\omega_b > \omega_a \geq 0$. The Gramian part of concern can be written as

$$p_{\omega_a,\omega_b} = \frac{1}{2\pi}\int_{\omega_a}^{\omega_b}(j\omega I-A)^{-1}B\Omega(\omega)B^H(j\omega I-A)^{-H}\mathrm{d}\omega \quad (3)$$

(ω_a=$-\infty$, ω_b=∞ and $\Omega(\omega) = W_i(j\omega)W_i^H(j\omega)$ gives Enns' integral for input weighting W_i (2)). In [2] the plain interval case with $\Omega(\omega) = 1$ is analysed. Here we introduce a generalization with $\Omega(\omega) = \sum_{k=0}^{2}\Gamma_k\omega^k$. Γ_0, Γ_1 and Γ_2 should be such that $\Omega(\omega)$ is non-negative for $\omega \in [\omega_a, \omega_b]$.

For the solution of p_{ω_a,ω_b}, $B(\Gamma_0 + \Gamma_1\omega + \Gamma_2\omega^2)B^H$ is written as

$$(j\omega I-A)X + X^H(j\omega I-A)^H + (j\omega I-A)B\Gamma_2B^H(j\omega I-A)^H$$

with X the solution to the following linear matrix equation

$$AX + XA^H + B\left[\Gamma_0B^H + j\Gamma_1B^HA^H - \Gamma_2B^HA^{2H}\right] = 0 \quad (4)$$

The integration in (3) can now be performed term by term:

$$\begin{aligned} p_{\omega_a,\omega_b} &= \tfrac{1}{2\pi}\int_{\omega_a}^{\omega_b}\left\{X(j\omega I-A)^{-H} + (j\omega I-A)^{-1}X^H + B\Gamma_2B^H\right\}\mathrm{d}\omega \\ &= \tfrac{1}{2\pi}\left\{XS^H(\omega_a,\omega_b) + S(\omega_a,\omega_b)X^H + B\Gamma_2B^H(\omega_b-\omega_a)\right\} \end{aligned}$$

with $S(\omega_a, \omega_b) = -j\ln[(j\omega_aI-A)^{-1}(j\omega_bI-A)]$. $S(\omega_a, \omega_b)$ can be solved by first transforming A to diagonal form.

3 Frequency pulse Gramians

Suppose we have $\Gamma_0 = (\omega_b-\omega_a)^{-1}$, $\Gamma_{1,2} = 0$ then $\int_{\omega_a}^{\omega_b}\Gamma_0\mathrm{d}\omega = 1$. For $\omega_a \to \omega_b$ the interval $[\omega_a, \omega_b]$ deforms into a frequency point and the weight Γ_0 goes to infinity. For such a frequency pulse function (3) simplifies to

$$p_{\omega_b} = \frac{1}{2\pi}(j\omega_bI-A)^{-1}BB^H(j\omega_bI-A)^{-H}$$

For $B \in \mathbb{R}^{n\times 1}$ the rank of this matrix is one. By choosing n_k pairs of frequency pulses (at positive and negative frequencies) for m inputs the order of the controllability Gramian will be $2n_km$. If we require a positive definite controllability Gramian for minimal systems we have to take $n_k \geq \frac{1}{2}n/m$.

CH3229-2/92/0000-2848$1.00 © 1992 IEEE

4 Implementation of generalized frequency weighted balanced reduction

A MATLAB model reduction tool has been developed that builds weighted controllability and observability Gramian in steps, showing the reduction result (frequency response, reduction error) after each step and allowing the user to add or change weightings by

- changing the scaling factor of the normal unweighted Gramians (this applies to all frequencies)
- specifying an interval with a quadratic frequency function by picking three points (and two for each neighboring interval) in a plot with the magnitude of the system transfer function
- specifying frequency pulses by picking points in the same plot that define the frequencies and scaling factors

Enns' frequency weighted balanced reduction is incorporated by applying the above procedure to $W_o(s)G(s)W_i(s)$ ($W_o(s)$ is the output weighting) and by extracting the system-state part from the controllability and observability Gramians of $W_o(s)G(s)W_i(s)$.

5 Example

Frequency weighted reduction of a 34th-order siso model (Fig.1) of the tracking mechanism in a Compact Disc player is analysed. The

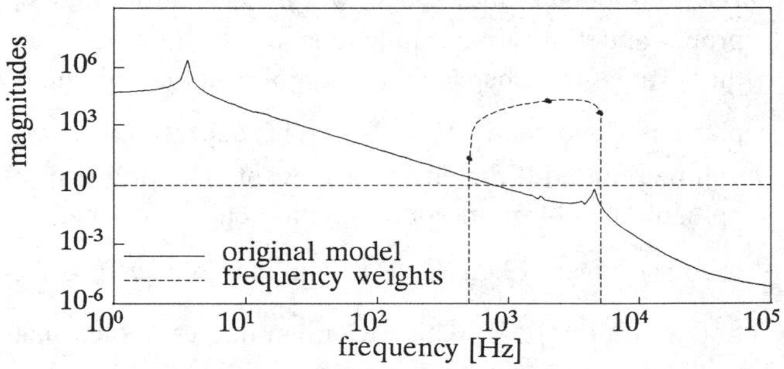

Fig. 1: Magnitude of model and frequency weights

reduced order model will be used for the design of a controller that achieves a bandwidth of about 1 kHz. The lightly damped system poles turn unstable if the model reduction error near 1 kHz is too large. Here it will be shown that our frequency weighting concept is very efficient for controller-relevant reduction. Let us first analyse standard balanced reduction (unweighted case). The Hankel

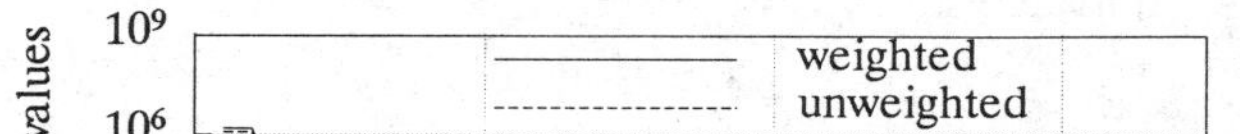

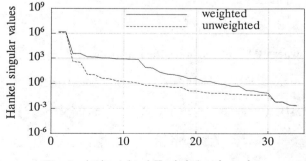

Fig. 2: (un)weighted Hankel singular values

singular value (HSV) plot (Fig.2) promises a good fit for relatively small reduced orders: $n_r = 2, 4, 6$. The dynamics in the bandwidth frequency range however are truncated even for $n_r = 12$. To force a better fit around 1 kHz a quadratic frequency function is created. Figure 1 shows the three points that were picked and the quadratic fit within the chosen interval. The frequency weighted HSV's (Fig.2) point to a 12th order reduced model. Normalized LQG control design on a 12th order frequency weighted reduced model gave a controller that performed well on the 34th order model whereas the same design on a balanced approximation of order 12 resulted in a controller that destabilized the original model. Figures 3 and 4 show reduction errors for $n_r = 4, 12, 20, 30$ in the weighted and unweighted case.

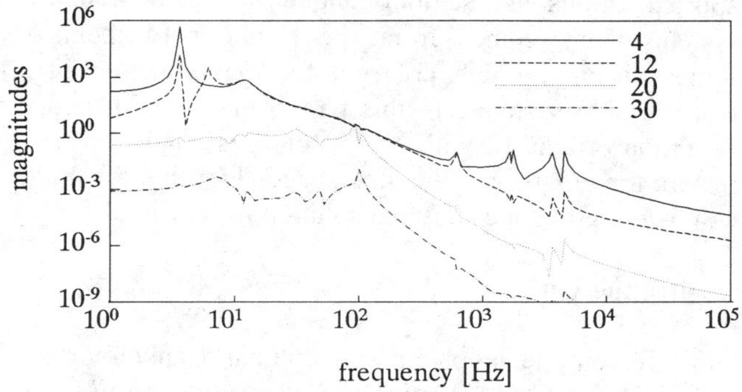

Fig. 3: Reduction errors with frequency weighting

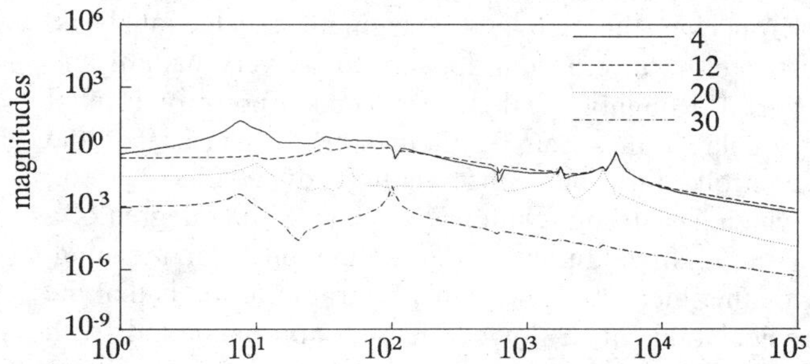

Fig. 4: Reduction errors without frequency weighting

Balanced reduction gives reduction errors distributed almost uniformly over all frequencies. In the weighted case the accuracy is clearly improved within the interval. At low frequencies the fit is worse than with normal balanced reduction, but this does not hamper successful control design.

Weighting only with the quadratic part the 3.5 Hz mode turns unstable in most reduced models.

The general experience is that over the whole frequency range a minimal weighting has to be applied to ensure that the reduced order models remain stable. The easiest way to do this is to incorporate the unweighted Gramians times a small scaling factor in the total weighted Gramians. Very small reduction errors at specific frequencies can be obtained with frequency pulse weightings.

References

[1] D. Enns, "Model reduction with balanced realization: an error bound and a frequency weighted generalization," *Proc. 23rd CDC, Las Vegas, USA*, pp.127–132, 1984.

[2] W. Gawronski, and J-N Juang, "Model Reduction for Flexible Structures," *Control and Dynamic Systems*, vol.36, pp.143–222, 1990.

[3] R.E. Skelton, *Dynamic Systems Control*, New York: John Wiley & Sons, 1988.

Proceedings of the 31st Conference
on Decision and Control
Tucson, Arizona • December 1992

SUPER-OPTIMAL HANKEL-NORM APPROXIMATIONS

Fang-Bo YEH and Lin-Fang WEI
Department of Mathematics, Tunghai University
Taichung, Taiwan, Republic of China

Abstract

It is well-known that optimal Hankel-norm approximations are seldom unique for multivariable systems. This comes from the Hankel-norm being somewhat of a crude criterion for the reduction of multivariable systems. In this paper, the strengthened condition originated with N. J. Young is employed to restore the uniqueness. A state-space algorithm for the computation of super-optimal solution is presented.

I. Introduction

Recent developments in optimal Hankel-norm approximations [1]-[3] have held great attention in the control society. As pointed out in [1], based on the Kronecker theorem and the singular value analysis, Hankel-norm criterion appears to be very natural and useful. Roughly speaking, the Hankel-norm or Hankel singular values can be thought as a measure of the controllability and observability of a LTI system, which has strong relations to the McMillan degree of a system. In addition to this sound physical meaning, another merit of using Hankel-norm criterion is that the calculation of a lower degree approximation with minimum error or a minimum degree approximation within a given tolerance can be easily computed. These features bail the design out of endless iterations in the face of large-scale multivariable systems. The celebrated paper [2] gives detailed state-space solutions and their L^∞ error bounds to these problems. For single input or single output systems, it is well-known that the optimal Hankel-norm approximation is unique and can be easily determined by the Schmidt pair of the associated Hankel operator. However, the optimal solutions are seldom unique for multivariable systems. The problem how to choose the best solution then naturally arises. A simple example reported in [2] is used to clarify the situation. Consider the system

$$\hat{G}(s) = \begin{bmatrix} \frac{2s+0.45}{s^2+1.25s+0.09} & 0 \\ 0 & \frac{1}{s+0.5} \end{bmatrix}$$

A bit of calculation shows that all optimal Hankel-norm approximations with McMillan degree two are given by

$$\hat{G}(s) = \begin{bmatrix} \frac{0.5s+0.675}{s+0.15} & 0 \\ 0 & \hat{g}(s) \end{bmatrix}$$

where is any stable function of McMillan degree one and $\left\| \frac{1}{s+0.5} - \hat{g} \right\|_H \le \frac{1}{2}$. Here $\|\cdot\|_H$ stands for the Hankel-norm of a system. It is trivial to see that $\hat{g}(s) = \frac{1}{s+0.5}$ would be the best choice in a natural sense. However, it is not at all clear how this could be generalized in [2]. The non-uniqueness comes from that the Hankel-norm is somewhat of a crude criterion for the reduction of multivariable systems. To make the solution unique, some finer measure should be imposed. In this paper, the strengthened condition in [9] is employed to restore the uniqueness.

To Formulate the problem more precisely, we begin with the nomenclature. The symbol $\mathbf{RL}^{\infty,p\times q}$ denotes the space of $p\times q$ matrix-valued, proper and real-rational functions in s with no poles on the $j\omega$ axis. Its subspace with no poles in the right half-plane is denoted as $\mathbf{RH}_+^{\infty,p\times q}$. The superscript $p\times q$ will be omitted if the size is irrelevant. The problem of optimal Hankel-norm approximation can be defined as follows: given $G\in\mathbf{RH}_+^{\infty,p\times q}$ and an integer $k\ge 0$, find a $\hat{G}\in\mathbf{RH}_+^{\infty,p\times q}$ with McMillan degree k such that the error's Hankel-norm is minimized, i.e.,

$$\min_{\hat{G}\in\mathbf{RH}_+^{\infty,p\cdot q}\text{and }\deg(\hat{G})=k} \left\|G-\hat{G}\right\|_H \qquad (1.1)$$

It is known that Hankel-norm is only related to the stable part of a system, and the addition of an antistable function does not affect the norm. Thus, the problem of optimal Hankel-norm approximations is equivalent to

$$\min_{\hat{G}\in\mathbf{RH}_{-,k}^{\infty,p\cdot q}} s_1^\infty\left(G-\hat{G}\right) \qquad (1.2)$$

where $\mathbf{RH}_{-,k}^{\infty,p\times q}$ denotes the subset $\mathbf{RL}^{\infty,p\times q}$ of with no more than k poles in the left half-plane, $\hat{G}\in\mathbf{RH}_{-,k}^{\infty,p\times q}$ $\mathbf{RH}_{-,0}^{\infty,p\times q}$ is abbreviated to $\mathbf{RH}_{-}^{\infty,p\times q}$, and

$$s_j^\infty(E) := \max_{\omega\in\mathbf{R}} s_j\left(E(j\omega)\right)$$

The symbol $s_j(\cdot)$ denotes the jth largest singular value of a constant matrix. Then, deleting the antistable part of $\hat{G}$ obtained from (1.2) will give the optimal solution to (1.1). In (1.2), it is seen that only the first frequency dependent singular value of the error system $G-\hat{G}$ is

CH3229-2/92/0000-2850$1.00 © 1992 IEEE

minimized. A generalization is that we seek to minimize all the singular values whenever possible. Therefore, the problem of super-optimal Hankel-norm approximation is defined as follows: given $G \in \mathbf{RL}^{\infty,p\times q}$ and an integer $k \geq 0$ find a $\hat{G} \in \mathbf{RH}^{\infty,p\times q}_{-,k}$ such that the sequence

$$s_1^{\infty}(G-\hat{G}),\ s_2^{\infty}(G-\hat{G}),\ \cdots$$

is minimized lexicographically. To our knowledge, this problem was first studied in [10], wherein the existence and uniqueness of the super-optimal solution had been proved using the conceptual operator-theoretic constructions. In this paper, a different approach that requires only simple state-space calculations will be studied. The approach is much more straightforward and comprehensible. Besides, the pole-zero cancellations in the algorithm will be analyzed in detail.

II. Mathematical Preliminaries

In the development of this work, the state-space approach is adopted. For a proper and real-rational matrix function $G(s)$, $G^{\sim}(s)$ is synonymous with $G^T(-s)$, and the data structure $[A,B,C,D]$ denotes a realization of $G(s)$, i.e.,

$$G(s)=C(sI-A)^{-1}B+D=[A,B,C,D]=\left[\begin{array}{c|c} A & B \\ \hline C & D \end{array}\right]$$

where A, B, C and D are constant matrices with compatible dimensions. A collection of state-space operations using this data structure can be found in [6]. For a stable system $G(s)$ with the realization $[A,B,C,D]$, the corresponding controllability and observability Gramian are defind as the uniquely non-negative definite solutions to the following Lyapunov equations, respectively,

$$AP + PA^T + BB^T = 0$$
$$A^TQ + QA + C^TC = 0$$

If the realization is minimal, then P and Q must be positive definite. When both P and Q are equal and diagonal, we say that the realization $[A,B,C,D]$ is balanced and the jth largest diagonal element, denoted $\sigma_j(G)$, is defined as the jth Hankel singular values of $G(s)$. It is always possible to get a balanced realization by use of the state similarity transformation. The problem of approximating a given Hankel matrix by a lower rank one had been studied in [4] and [5]. Their striking results stated that the restriction of the solution to be a Hankel matrix does not affect the achievable error. These remarkable results will be briefly reviewed here. We denote $\mathbf{RL}^{2,p}$ the space of p vector-valued, strictly proper and real-rational functions with no poles on the $j\omega$ axis, and is a Hilbert space under the inner product

$$\langle u,v\rangle := \frac{1}{2\pi}\int_{-\infty}^{\infty} u^T(-j\omega)\, v(j\omega)\, d\omega$$

The subspace of $\mathbf{RL}^{2,p}$ with no poles in the right-half plane is denoted $\mathbf{RH}^{2,p}$, and its orthogonal complement is denoted $\mathbf{RH}^{2,p}_{\perp}$. For a system G in $\mathbf{RH}^{\infty,p\times q}_{+}$, the Hankel operator with symbol G, denoted Γ_G maps $\mathbf{RH}^{2,p}_{\perp}$ to $\mathbf{RH}^{2,p}$. For $x \in \mathbf{RH}^{2,p}_{\perp}$, $\Gamma_G x$ is defined as

$$\Gamma_G x := \Pi(G\,x)$$

where Π is the orthogonal projection operator which maps $\mathbf{RL}^{2,p}$ onto $\mathbf{RH}^{2,p}$. An important result is that the Hankel operator Γ_G is of finite rank and its singular values are just the Hankel singular values of $G(s)$. The pair (v_j,w_j) satisfies

$$\Gamma_G\, v_j := \sigma_j(G)\, w_j$$
$$\Gamma_G^*\, w_j := \sigma_j(G)\, v_j$$

is called the jth Schmidt pair of Γ_G corresponding to $\sigma_j(G)$. The following Lemma relates the Schmidt pair of Γ_G to any optimal solution, and is central to this study.

Lemma 2.1: Given $G \in \mathbf{RH}^{\infty,p\times q}$ and an integer $k \geq 0$ then

(1) $\min_{\hat{G}\in \mathbf{RH}^{\infty,p\times q}_{-,k}} s_1^{\infty}(G-\hat{G}) = \sigma_{k+1}(G) := \sigma,$

(2) if $\sigma_k(G) > \sigma$ then for any optimal $\hat{G}$ we have $(G-\hat{G})v = \sigma w$ and $(G-\hat{G})^{\sim} w = \sigma v$. Here (v,w) denotes the $(k+1)$st Schmidt pair of the Hankel operator Γ_G. In case of $k=0$, $\sigma_0(G)$ is interpreted as $+\infty$.

A matrix G in $\mathbf{RL}^{\infty}$ is all-pass if $G^{\sim}(s)G(s)=I$, and is inner if G is restricted to $\mathbf{RH}^{\infty}_{+}$. An inner matrix that has a left-inverse in $\mathbf{RH}^{\infty}_{+}$ is called a minimum-phase inner matrix [8]. The following lemmas give some important properties of minimum-phase inner matrices.

Lemma 2.2: [8] Let $G(s)=[A,B,C,D]$ be inner together with the controllability Gramian P and observability Gramian Q. Then G is minimum-phase inner if and only if $\|PQ\|<1$, and

$$G^{-L}(s)=\left[A, BD^T+PC^T, D^TC(PQ-I)^{-1}, D^T\right]$$

is one of stable left-inverses of G.

Lemma 2.3: For a strictly tall inner matrix $G(s)=[A,B,C,D]$ having nonsingular controllability Gramian P, the right-coprime factorization of $G(-s)$ over $\mathbf{RH}_+^\infty$, i.e. $G(-s)=N_G(s)M_G^{-1}(s)$ can be written as

$$N_G(s)=\left[A^T,-P^{-1}B,CP+DB^T,D\right]$$
$$M_G(s)=\left[A^T,-P^{-1}B,B^T,I\right]$$

where M_G is inner and N_G is minimum-phase inner.

III. Diagonalizing Matrices

In this section, we shall study how to construct all-pass matrices which will diagonalize each possible error system. It is seen that in order to decide the second singular value, namely $s_2^\infty(G-\hat{G})$, and keep $s_1^\infty(G-\hat{G})$ unchanged at the same time, a natural way is to diagonalize the error system $G-\hat{G}$ by pre-multiplying and post-multiplying two suitable all-pass matrices. Not to sacrifice any generality, it is important to have this diagonalizing process hold for all optimal solutions. Recalling that any optimal solution should satisfy Lernrna 2.1 part (2), it is clear that the Schmidt pair (v,w) serve as a starting point in this diagonlization process. We begin by assuming that a minimal balanced realization $[A,B,C,D]$ of a stable system G with McMillan degree n is given. Then the associated controllability and observability Gramians are both equal and can be arranged as diag(σ,Σ), where $\sigma=\sigma_{k+1}(G)$ and Σ is also diagonal. To have a clear presentation, hereinafter, we shall assume that σ is distinct, i.e.,

(Al) $\sigma_k(G)>\sigma>\sigma_{k+2}(G)$

Relaxing this assumption is possible but only leads to a more messy indexing notation. Partition matrices A, B and C as

$$A=\begin{bmatrix} a_{11} & A_{12} \\ A_{21} & A_{22}\end{bmatrix},\quad B=\begin{bmatrix} B_1 \\ B_2\end{bmatrix},\quad C=\begin{bmatrix} C_1 & C_2\end{bmatrix}$$

where a_{11} is a scalar, B_1 is a row vector and C_1 is a column vector. Obviously, the following Lyapunov equations are hold

$$\begin{bmatrix} a_{11} & A_{12} \\ A_{21} & A_{22}\end{bmatrix}\begin{bmatrix} \sigma & 0 \\ 0 & \Sigma\end{bmatrix}+\begin{bmatrix} \sigma & 0 \\ 0 & \Sigma\end{bmatrix}\begin{bmatrix} a_{11} & A_{21}^T \\ A_{12}^T & A_{22}^T\end{bmatrix}+\begin{bmatrix} B_1B_1^T & B_1B_2^T \\ B_2B_1^T & B_2B_2^T\end{bmatrix}=0$$

$$\begin{bmatrix} a_{11} & A_{21}^T \\ A_{12}^T & A_{22}^T\end{bmatrix}\begin{bmatrix} \sigma & 0 \\ 0 & \Sigma\end{bmatrix}+\begin{bmatrix} \sigma & 0 \\ 0 & \Sigma\end{bmatrix}\begin{bmatrix} a_{11} & A_{12} \\ A_{21} & A_{22}\end{bmatrix}+\begin{bmatrix} C_1^TC_1 & C_1^TC_2 \\ C_2^TC_1 & C_2^TC_2\end{bmatrix}=0$$

To simplify the notation, we will assume that G has been scaled such that $B_1B_1^T=C_1^TC_1=1$, which is without loss of any generality. The $(k+1)$st Schmidt pair v and w for the above balanced realization can be written as

$$v(-s)=\left[A^T,e_1/\sqrt{\sigma},B^T,0\right],\quad w(s)=\left[A,e_1/\sqrt{\sigma},C,0\right]$$

where e_1 is the first column of an $n\times n$ identity matrix. By direct state-space calculation, it is easy to verify that $v^\sim v=w^\sim w$. Thus, it is possible to factorize $v(-s)$ and $w(s)$ such that

$$v(-s)=\hat{v}(s)\alpha(s),\quad w(s)=\hat{w}(s)\alpha(s)$$

where $\hat{v}$ and $\hat{w}$ are all-pass vectors, and α is a scalar function. The state-space realizations of $\hat{v}$ and $\hat{w}$ can be written as

$$\hat{v}(s)=\left[A_v,A_{12}^T,C_v,B_1^T\right],\quad \hat{w}(s)=\left[A_w,A_{21},C_w,C_1\right]$$

in which

$$A_v=A_{22}^T+\sigma A_{12}^T\left(A_{21}^T-A_{12}X_v\right),\quad C_v=B_2^T+\sigma B_1^T\left(A_{21}^T-A_{12}X_v\right)$$
$$A_w=A_{22}+\sigma A_{21}\left(A_{12}-A_{21}^TX_w\right),\quad C_w=C_2+\sigma C_1\left(A_{12}-A_{21}^TX_w\right)$$

where X_v and X_w satisfy the following algebraic Riccati equations,

$$\left(A_{22}+\sigma A_{21}A_{12}\right)X_v+X_v\left(A_{22}+\sigma A_{21}A_{12}\right)^T - \sigma X_vA_{12}^TA_{12}X_v-\sigma A_{21}A_{21}^T=0 \quad (3.1)$$

$$\left(A_{22}+\sigma A_{21}A_{12}\right)^TX_w+X_w\left(A_{22}+\sigma A_{21}A_{12}\right) - \sigma X_wA_{21}A_{21}^TX_w-\sigma A_{12}^TA_{12}=0 \quad (3.2)$$

A direct computation will show that any solutions to (3.1) and (3.2) will make $\hat{v}$ and $\hat{w}$ all-pass. Two special kinds of solutions are of interest, namely, the stabilizing and antistabilizing solutions. For our purpose, we choose X_v and X_w to be the antitabilizing solutions, which is inspired from [13]. It will be shown that the use of antistabilizing solutions can be of great help in analyzing the pole-zero cancellations. Hence to ensure the existence of antistabilizing solutions, it is natural to assume that

(A2) $\left(A_{22}^T,A_{12}^T\right)$ and (A_{22},A_{21}) are controllable.

Then, it can be verified that $Q_v=\sigma X_v+\Sigma$ and $Q_w=\sigma X_w+\Sigma$ satisfy the following Lyapunov equations, respectively

$$A_v^TQ_v+Q_vA_v+C_v^TC_v=0$$
$$A_w^TQ_w+Q_wA_w+C_w^TC_w=0$$

In case of $k=0$, it can be proved that Q_v and Q_w are nonsingular provided that σ is distinct. However, for general k, the situation is not clear and we will assume that

(A3) Q_v and Q_w are non-singular.

The all-pass completions [7] of $\hat{v}$ and $\hat{w}$ are given by

$$\hat{V}_\perp(s)=\left[A_v,-Q_v^{-1}B_2B_\perp^T,C_v,B_\perp^T\right]$$
$$\hat{W}_\perp(s)=\left[A_w,-Q_w^{-1}C_2^TC_\perp,C_w,C_\perp\right]$$

where $\begin{bmatrix} B_1^T & B_\perp^T \end{bmatrix}$ and $\begin{bmatrix} C_1 & C_\perp \end{bmatrix}$ are orthogonal. Denote $\hat{V} = \begin{bmatrix} \hat{v} & \hat{V}_\perp \end{bmatrix}$ and $\hat{W} = \begin{bmatrix} \hat{w} & \hat{W}_\perp \end{bmatrix}$. Then, the error $G - \hat{G}$ can be diagonalized as shown in the following lemma.

Lemma 3.1: Given that G satisfies Assumptions (Al), (A2) and (A3), then for any optimal solution $\hat{G}$ we have

$$\hat{W}^{\sim}(G - \hat{G})\hat{V}(-s) = \begin{bmatrix} \sigma g_1 & 0 \\ 0 & \hat{W}_\perp^{\sim}(G - \hat{G})\hat{V}_\perp(-s) \end{bmatrix}$$

where g_1 is all-pass and independent of $\hat{G}$.

IV. Super-Optimal Solutions

The super-optimal model matching problem, i.e. $k = 0$, was first studied in [9] wherein a high-level algorithm is released. The implementations of Young's algorithm have been reported in [11] using the polynomial approach, and in [12] and [13] using the state-space approach. The basic idea of Young's algorithm can be summarized as follows. First, the minimum achievable L^∞ norm of the error is calculated. Then, two all-pass matrices are constructed to diagonalize the error system, which results in the same model-matching problem but with one less dimension. Hence by induction, this dimension peeling process can be continued layer by layer until a unique solution is found. Finally, the super-optimal solution of the original problem is constructed from the solution of each layer. However, difficulty arises when apply this idea to the general case $k > 0$. The reason is that the addition and multiplication of two $\mathbf{RH}_-^\infty$ functions are still a $\mathbf{RH}_-^\infty$ function, but is generally not true for $\mathbf{RH}_{-,k}^\infty$ functions. This causes the minimum achievable norm of the subsequent layer hard to determine. In order to ensure that the final super-optimal solution of the original problem is in $\mathbf{RH}_{-,k}^{\infty,p\times q}$, the solution set of each subsequent layer should be precisely characterized, and will be studied in this section.

We now continue our work in the previous section. By [2], an optimal solution is given by

$$\hat{G}_1(s) = \left[\begin{array}{c|c} \Gamma^{-1}(\sigma^2 A_{22}^T + \Sigma A_{22}\Sigma + \sigma C_2^T C_1 B_1 B_2^T) & \Gamma^{-1}(\Sigma B_2 - \sigma C_2^T C_1 B_1) \\ \hline C_2\Sigma - \sigma C_1 B_1 B_2^T & D + \sigma C_1 B_1 \end{array}\right]$$

where $\Gamma = \Sigma^2 - \sigma^2 I$ is nonsingular by (Al). Recalling Lemma 3.1, we see that any optimal solution G should satisfy

$$\hat{W}^{\sim}(\hat{G} - \hat{G}_1)\hat{V}(-s) = \begin{bmatrix} 0 & 0 \\ 0 & \hat{W}_\perp^{\sim}(\hat{G} - \hat{G}_1)\hat{V}_\perp(-s) \end{bmatrix}$$

Thus, $\hat{G}\hat{V}(-s)$ and $\hat{G}_1\hat{V}(-s)$ have the same first column, i.e.,

$$\hat{G}\hat{V}(-s) = \begin{bmatrix} \hat{G}\hat{v}(-s) & \hat{G}\hat{V}_\perp(-s) \end{bmatrix} := \begin{bmatrix} h & H \end{bmatrix}$$
$$\hat{G}_1\hat{V}(-s) = \begin{bmatrix} \hat{G}_1\hat{v}(-s) & \hat{G}_1\hat{V}_\perp(-s) \end{bmatrix} := \begin{bmatrix} h & H_1 \end{bmatrix}$$

Besides, it is required that $\hat{w}^{\sim}(H - H_1) = 0$. Now observe that

$$\hat{W}^{\sim}(G - \hat{G})\hat{V}(-s) = \hat{W}^{\sim}(G - \hat{G}_1)\hat{V}(-s) - \hat{W}^{\sim}(\hat{G} - \hat{G}_1)\hat{V}(-s)$$
$$= \begin{bmatrix} \sigma g_1 & 0 \\ 0 & F_1 \end{bmatrix} - \begin{bmatrix} 0 & 0 \\ 0 & \hat{W}_\perp(H - H_1) \end{bmatrix}$$

where

$$F_1(s) = \hat{W}_\perp^{\sim}(G - \hat{G}_1)\hat{V}_\perp(-s)$$

Hence to compute the super-optimal solution, we require finding H such that

(Cl) $\begin{bmatrix} h & H \end{bmatrix} \in \mathbf{RH}_{-,k}^{\infty,p\times q}\hat{V}(-s)$,

(C2) $\hat{w}^{\sim}(H - H_1) = 0$,

(C3) $s_j^\infty(F_1 - \hat{W}_\perp^{\sim}(H - H_1))$, for j = 1,2,. . . is minimized lexicographically.

To solve the above problem, the first step is to parametrize all the functions H that satisfy both Conditions (Cl) and (C2) in terms of some free function. And then the minimizing process (C3) can be carried out. These parametrizations are studied in Theorems 4.1 and 4.2.

First of all, according to Lemma 2.3, we introduce the following factorizations which play an important role in the riddance of (Cl) and (C2)

$$\hat{v}(s) = n_v(-s)m_v^{-1}(-s)$$
$$\hat{w}(s) = n_w(-s)m_w^{-1}(-s)$$

with

$$n_v(s) = \left[A_v^T, -P_v^{-1}A_{12}^T, C_vP_v + B_1^TA_{12}, B_1^T\right]$$
$$n_w(s) = \left[A_w^T, -P_w^{-1}A_{21}, C_wP_w + C_1A_{21}^T, C_1\right]$$

where P_v and P_w are the uniquely negative definite solutions to the Lyapunov equations, respectively,

$$A_vP_v + P_vA_v^T + A_{12}^TA_{12} = 0$$
$$A_wP_w + P_wA_w^T + A_{21}A_{21}^T = 0$$

It is important to notice that $n_v(-s)$ and $n_w(-s)$ are minimum-phase inner and, hence, have stable left inverse. With these notations, we have the following theorem.

Theorem 4.1: $[h \quad H] \in \mathbf{RH}_{-,k}^{\infty,p\times q}\hat{V}(-s)$ if and only if

$$H \in \hat{G}_1 n_v l_{n_v}(-s)\hat{V}_\perp(-s) + \mathbf{RH}_{-,k}^{\infty,p\times q} := \mathscr{H}$$

where $l_{n_v}(-s)$ is any stable left inverse of $n_v(-s)$.
Now by Theorem 4.1, it follows that H and H_1 can be parametrized as

$$H = \hat{G}_1 n_v l_{n_v}\hat{V}_\perp(-s) + R$$
$$H_1 = \hat{G}_1 n_v l_{n_v}\hat{V}_\perp(-s) + R_1$$

for some R and R_1 in $\mathbf{RH}_{-,k}^{\infty,p\times(q-1)}$. In other words, R_1 can be computed as follows

$$R_1 = \hat{G}_1\left(I - n_v l_{n_v}\right)\hat{V}_\perp(-s)$$

wherein the function $V_2 := \left(I - n_v l_{n_v}\right)\hat{V}_\perp(-s)$ has at most n - 1 states and is antistable.This can be verified by letting

$$l_{n_v}(s) = \left[A_v^T, C_v^T, A_{12}\left(I - P_v^{-1}Q_v^{-1}\right), B_1\right]$$

according to lemma 2.2. Then a series of state-space calculations yields that

$$V_2(s) = \left[A_v^T, P_v^{-1}Q_v^{-1}B_2 B_\perp^T, C_v P_v + B_1^T A_{12}, B_\perp^T\right]$$

As R_1 is found, Condition (C2) is equivalent to $n_w^T(R - R_1) = 0$. The following theorem characterize all such function R.

Theorem 4.2: $R \in \mathbf{RH}_{-,k}^{\infty,(p-1)\times(q-1)}$ and $n_w^T(R - R_1) = 0$ if and only if

$$R \in l_{n_w}^T n_w^T R_1 + \hat{W}_\perp \mathbf{RH}_{-,k}^{\infty,(p-1)\times(q-1)}$$

where $l_{n_w}(-s)$ is any stable left inverse of $n_w(-s)$.

Hence by theorem 4.2, there exist functions $\hat{Q}$ and Q_1 in $\mathbf{RH}_{-,k}^{\infty,(p-1)\times(q-1)}$ such that

$$R = l_{n_w}^T n_w^T R_1 + \hat{W}_\perp \hat{Q}$$
$$R_1 = l_{n_w}^T n_w^T R_1 + \hat{W}_\perp \hat{Q}_1$$

and, in other words

$$Q_1 = \hat{W}_\perp^{\sim}\left(I - l_{n_w}^T n_w^T\right)\hat{G}_1 V_2$$

To compute the function $W_2 := \hat{W}_\perp^{\sim}\left(I - l_{n_w}^T n_w^T\right)$, we choose

$$l_{n_w} = \left[A_w^T, C_w^T, A_{21}^T\left(I - P_w^{-1}Q_w^{-1}\right), C_1^T\right]$$

according to Lemma 2.3. then a series of state-space calculations gives that

$$W_2 = \left[A_w, P_w C_w^T + A_{21}C_1^T, C_\perp^T C_2 Q_w^{-1}P_w^{-1}, C_\perp^T\right]$$

which is also antistable and has McMillan degree no more than n-1. Finally, the function which need to be minimized in (C3) becomes

$$F_1 - \hat{W}_\perp^{\sim}(H - H_1) = F_1 - \hat{W}_\perp^{\sim}(R - R_1) = F_1 + Q_1 - \hat{Q}$$

Define

$$Q := F_1 + Q_1 = \hat{W}_\perp^{\sim}\left(G - \hat{G}\right)\hat{V}_\perp(-s) + W_2\hat{G}_1 V_2 \qquad (4.1)$$

Then Condition (C3) is reduced to the problem that finds a $\hat{Q}$ in $\mathbf{RH}_{-,k}^{\infty,(p-1)\times(q-1)}$ such that the sequence

$$s_1^\infty\left(Q - \hat{Q}\right), s_2^\infty\left(Q - \hat{Q}\right), \cdots$$

is minimized lexicographically. This is just the same super-optimal problem but with one less dimension. This dimension peeling process can be recursively invoked until the row or column dimension is reduced to one, wherein the solution can be uniquely determined. By induction, the super-optimal solution is clearly unique. And once the super-optimal solution $\hat{Q}$ is found, $\hat{G}$ can be recovered as follows .

$$\hat{G}(s) = \hat{G}_1 + \hat{W}_\perp\left(\hat{Q} - Q_1\right)\hat{V}_\perp^T \qquad (4.2)$$

Although this algorithm is conceptually workable, the computation should not follow (4.1) and (4.2) directly. The reason is that the sizes of the A-matrices of Q and $\hat{G}$ will blow up very rapidly if pure state-space additions and multiplications are used, and a further computation is required in order to get the minimal realizations of Q and $\hat{G}$. Owing to this observation, the pole-zero cancellations occur in (4.1) and (4.2) should be analyzed in detail.

It can be proved that Q is stable and can be realized with n - 1 states. Then, the super-optimal solution $\hat{G}$ will require no more than deg($\hat{Q}$) + n - 1 states, where $\hat{Q}$ is the super-optimal Hankel-norm approximation of Q. As a result, the required computation time in each recursive step will gradually decrease rather than increase. Since our intension is to apply the theory to large-scale multivariable systems, it is clear that a feasible computer program can be setup only when the full analysis of the pole zero cancellations is carried out. Thus, the result of this section is valuable from the pratical consideration. The proof requires a series of tedious state-space calculations and is omitted.

Theorem 4.3: A realization of Q is given by

$$Q(s) = \left[\begin{array}{c|c} -A_v & Q_v^{-1}B_2 B_\perp^T \\ \hline \sigma^2 C_\perp^T C_2 Q_w^{-1}(X_w X_v - I) & 0 \end{array}\right]$$

which is stable and minimal. Moreover, Let $\hat{\underset{\sim}{Q}} = \left[\breve{A}, \breve{B}, \breve{C}, \breve{D}\right] \in \mathbf{RH}_{-,k}^{\infty,(p-1)\times(q-1)}$ be any optimal Hankel-norm approximation to Q with L^{∞} error $\sigma_{k+1}(\underset{\sim}{Q})$. Then a realization for $\hat{G}$ is

$$\hat{G}(s) = \left[\begin{array}{cc|c} A_w & -\underset{\sim}{Q}_w^{-1} C_2^T C_\perp \breve{C} & \underset{\sim}{Q}_w^{-1}\left(\Sigma B_2 - \sigma C_2^T C_1 B_1 - C_2^T C_\perp \breve{D} B_\perp\right) \\ 0 & \breve{A} & \breve{B} B_\perp + P_{12}^T C_v^T \\ \hline C_w & C_\perp \breve{C} & D + \sigma C_1 B_1 + C_\perp \breve{D} B_\perp \end{array}\right]$$

where P_{12} satisfies $A_v P_{12} - P_{12} \breve{A}^T = \underset{\sim}{Q}_v^{-1} B_2 B_\perp^T \breve{B}^T$.

V. Concluding Remarks

Throughout this paper, we have concentrated on the computation of super-optimal Hankel-norm approximations. The existence and uniqueness of the super-optimal solution are proved by use of simple state-space calculations. The approach is unlike the work in [10], which based on conceptual operator-theoretic constructions. In addition, we have given a detailed analysis of pole-zero cancellations in the algorithm and a bound on the McMillan degree of the super-optimal solution, which generalize the regults in [13].

References

[1] S. Y. Kung and D. W. Lin, "Optimal Hankel-norm model reductions: Multivariable systems," *IEEE Trans. Automat. Contr.*, vol. AC-26, pp. 832-852, 1981.

[2] K. Glover, "All optimal Hankel-norm approximations of linear multivariable system and their L^{∞} error bounds," *Int. J. Contr.*, vol. 39, pp. 1115-1193, 1984.

[3] J. A. Ball and A. C. M. Ran, "Optimal Hankel-norm model reductions and WienerHopf factorizations II: the noncanonical case," *Integral Eqn. Operator Theory*, vol. 10, pp. 416-436, 1987.

[4] V. M. Adamjan, D. Z. Arov, and M. G. Krein, "Analytic properties of Schmidt pairs for a Hankel Operator and the generalized Schur-Takagi problem," *Math. of the USSR: Sborink*, vol. 15, pp. 31-73, 1971.

[5] V. M. Adamjan, D. Z. Arov, and M. G. Krein, "Infinite block Hankel matrices and related extension problems," *AMS Transl.*, ser. 2, vol. 111, pp. 133-156, 1978.

[6] B. A. Francis, A *Course in H^{∞} Control Theory.* New York: Springer-Verlag, 1987.

[7] J. C. Doyle, "Lecture Notes in Advances in Multivariable Control," *ONR / Honeywell Workshop*, Minneapolis, MN, 1984.

[8] F. B. Yeh and L. F. Wei, "Inner-outer factorizations of right-invertible real-rational matrices," *Syst. Contr. Lett.*, vol. 14, pp. 31-36, 1990.

[9] N. J. Young, "The Nevanlinna-Pick problem for matrix-valued functions," *J. Operator Theory*, vol. 15, pp. 239-265, 1986.

[10] N. J. Young, "Super-optimal Hankel-norm approximations," in *Modeling Robustness and Sensitivity Reduction in Control Systems*, R. F. Curtain, Ed. New York: Springer-Verlag, 1987.

[11] F. B. Yeh and T. S. Hwang, "A computational algorithm for the super-optimal solution of the model matching problem," *Syst. Contr. Lett.*, vol. 11, pp. 203-211, 1988.

[12] M. C. Tsai, D. W. Gu, and I. Postlethwaite, "A state-space approach to superoptimal H^{∞} control problems," *IEEE Trans. Automat. Contr.*, vol. AC-33, pp. 833-843, 1988.

[13] D. J. N. Limebeer, G. D. Halikias, and K. Glover, "State-space algorithm for the computation of super-optimal matrix interpolating functions," *Int. J. Contr.*,vol.50, pp. 2431-2466, 1989.

Proceedings of the 31st Conference
on Decision and Control
Tucson, Arizona • December 1992

Globally Stabilizing Controllers for Flexible Multibody Systems

by

Atul G. Kelkar, Suresh M. Joshi, and Thomas E. Alberts

M/S 230, NASA Langley Research Center
Hampton, VA 23681

Abstract

In this paper global asymptotic stability of nonlinear multibody flexible space-structures under static dissipative compensation is established. Furthermore, the stability is shown to be robust to certain actuator and sensor nonlinearities, modeling error, and parametric uncertainty. The results are applicable to a wide class of systems, including large flexible space-structures with articulated flexible appendages, which are classified as "Class II/IV structures" by NASA's Controls-Structures Interaction (CSI) program. The stability proof uses the Lyapunov approach and exploits the inherent passivity of such systems.

Introduction

Many space missions envisioned for the future will require large multibody space systems. Examples of such structures include large space platforms with multiple articulated payloads and space-based manipulators for on-orbit assembly and satellite servicing. Such systems are expected to have significant flexibility in the structural members as well as joints. Control systems design for multibody flexible systems is a difficult problem because of the large number of significant elastic modes with low inherent damping and the inaccuracies and uncertainties in the mathematical model. Furthermore, the dynamics of such systems are highly nonlinear. The literature contains a number of important stability results for certain subclasses of this problem; e.g., linear flexible structures, nonlinear multibody rigid structures, and most recently multibody flexible structures. Under certain conditions the input-output maps for such systems can be shown to be "passive" [1]. A stability theorem based on Popov's hyperstability concepts [2] states that a passive linear system controlled by strictly passive compensator is closed-loop stable. The Lyapunov and passivity approaches are used in [3] to demonstrate global asymptotic stability of linear flexible space structures (with no articulated appendages) for a class of dissipative compensators. These include Collocated Attitude Controllers (CAC) and Collocated Damping Enhancement Controllers (CDEC). The stability properties were shown to be robust to first-order actuator dynamics and certain actuator/sensor nonlinearities. Multibody rigid structures comprise another class of systems for which stability results have been advanced. Ideally, subject to certain restrictions, these systems can be categorized as "natural systems" [4]. Such systems are known to exhibit global asymptotic stability under proportional-and-derivative (PD) control. Upon recognition that rigid manipulators belong to the class of natural systems, a number of researchers [5], [6], [7], etc., have established global asymptotic stability of rigid manipulators employing PD control with gravity compensation. Stability of tracking controllers was investigated in [8] and [9] for rigid manipulators. In [10] an extension of the results of [9] to the exponentially stable tracking control for flexible multilink manipulators, local to the desired trajectory, was done. Lyapunov stability of multilink flexible systems was addressed in [11]. However, the global asymptotic stability for nonlinear multilink flexible space-structures, in the presence of a wide range of actuator/sensor nonlinearities, has not been addressed in the literature, and that is the subject of this paper.

This paper considers the stability of flexible multibody systems controlled by the static dissipative controllers. Furthermore, the effect of realistic nonlinearities in the actuators and sensors are investigated. The proof given here uses Lyapunov's stability theorem along with Lasalle's theorem and is based on [3]. The Lyapunov function used is an energy-type quadratic function augmented with an appropriate positive definite function to prove global asymptotic stability. The stability proof by Lyapunov's method can take a simpler form if the Work-Energy Rate principle [11] is used. However, since the Work-Energy Rate principle is applicable only when the system is holonomic in nature, we have used a more direct approach, so that the results are applicable to nonholonomic as well as holonomic systems.

Stability with Static Dissipative Control Law

Typically, the mathematical model of flexible multibody space structures is given by the

CH3229-2/92/0000-2856$1.00 © 1992 IEEE

following dynamical equation of motion:

$$M(p)\ddot{p} + C(p,\dot{p})\dot{p} + D\dot{p} + Kp = B^T u \qquad (1)$$

where $\{p\} = \{\theta^T, q^T\}^T$, θ is the k-vector of measurable coordinates (joint angles in the case of manipulators) and q is the $(n-k)$ vector of the flexural coordinates. $M(p) = M^T(p) > 0$ is the configuration-dependent mass-inertia matrix, $C(p,\dot{p})$ corresponds to Coriolis and centrifugal forces, D is the symmetric, positive semidefinite damping matrix, K is the symmetric, positive semidefinite stiffness matrix, and u is the k vector of applied torques. B is the influence matrix of the control input u and has the form $B = [I_{k\times k} \quad 0_{k\times(n-k)}]$. It should be noted that such systems always have zero-frequency modes associated with rigid-body coordinates.

Consider, the static dissipative control law u, given by:

$$u = -G_p y_p - G_r y_r \qquad (2)$$

where,

$$y_p = Bp \qquad and \qquad y_r = B\dot{p} \qquad (3)$$

y_p and y_r are measured angular position and rate vectors.

Theorem 1. Suppose G_p and G_r are symmetric and positive definite. Then, the closed-loop system given by equations (1) and (2) is globally asymptotically stable.

Proof.

Consider the Lyapunov function

$$V = \frac{1}{2}\dot{p}^T M(p)\dot{p} + \frac{1}{2}p^T(K + B^T G_p B)p \qquad (4)$$

V is clearly positive definite since $M(p)$ and $(K + B^T G_p B)$ are positive definite symmetric matrices. Taking the time derivative and letting $\overline{K} = (K + B^T G_p B)$,

$$\dot{V} = \dot{p}^T M\ddot{p} + \frac{1}{2}\dot{p}^T \dot{M}\dot{p} + \dot{p}^T\overline{K}p \qquad (5)$$

Using (1) in (5), we get,

$$\dot{V} = \dot{p}^T[B^T u - C\dot{p} - D\dot{p} - Kp] + \frac{1}{2}\dot{p}^T\dot{M}\dot{p} + \dot{p}^T\overline{K}p \qquad (6)$$

Now substituting (2) and (3) in (6),

$$\dot{V} = \dot{p}^T B^T(-G_p\theta - G_r\dot{\theta}) + \dot{p}^T(\frac{1}{2}\dot{M} - C)$$

$$\dot{p} - \dot{p}^T D\dot{p} - \dot{p}^T Kp + \dot{p}^T\overline{K}p \qquad (7)$$

$$\dot{V} = \dot{p}^T(\frac{1}{2}\dot{M} - C)\dot{p} - \dot{p}^T\overline{K}p + \dot{p}^T\overline{K}p - \dot{p}^T(D + B^T G_r B)\dot{p} \qquad (8)$$

Now, using a very important property of the system, that $(\frac{1}{2}\dot{M} - C)$ is a *skew symmetric* matrix [9], which is the characteristic of the systems whose dynamical equations of motion have the same form as eq.(1), we get, $\dot{p}^T(\frac{1}{2}\dot{M} - C)\dot{p} = 0$ and, after some cancellations, we obtain

$$\dot{V} = -\dot{p}^T(D + B^T G_r B)\dot{p} \qquad (9)$$

Since $(D + B^T G_r B)$ is the positive definite symmetric matrix,

$$\dot{V} \le 0 \qquad (10)$$

i.e., $\dot{V}$ is negative semidefinite in p and $\dot{p}$ and

$$\dot{V} = 0 \Rightarrow \dot{p} = 0 \Rightarrow \ddot{p} = 0 \qquad (11)$$

Substituting in the closed-loop equation we get

$$(K + B^T G_p B)p = 0 \qquad \Rightarrow p = 0 \qquad (12)$$

Thus, $\dot{V}$ is not zero along any trajectories, then by Lassalle's theorem, system is globally asymptotically stable.

The significance of this result is that *any nonlinear multibody system in this class can be robustly stabilized with this control law.* In the case of manipulators, this means that one can accomplish any terminal position from any initial position with *guaranteed asymptotic stability.*

Robustness to Actuator/Sensor Nonlinearities

Although, as shown in the proof of theorem 1, the static dissipative controller (2) globally asymptotically stabilizes the nonlinear system (1) in the presence of perfect (i.e., linear, instantaneous) actuators and sensors, in practice, these devices have nonlinearities and phase lags. Therefore, for practical applications, the controller (2) should be robust to the nonlinearities and the phase shifts in the actuator/sensor. The following theorem extends the results of [3] to the case of nonlinear flexible multibody systems. That is, the robust stability property of the static dissipative controllers is proved in the presence of a wide class of actuator/sensor nonlinearities. In particular, it is proved that the static dissipative controller preserves global asymptotic stability when actuators have monotonically increasing nonlinearities and sensors have nonlinearities that belong to the $(0, \infty)$ sector. [A function $\psi(\nu)$ is said to belong to the $(0, \infty)$ sector if $\psi(0) = 0$ and $\nu\psi(\nu) > 0$ for $\nu \neq 0$: ψ is said to belong to the $[0, \infty)$ sector if $\nu\psi(\nu) \geq 0$].

In the presence of actuator/sensor nonlinearities, the actual input is given by:

$$u = \psi_a[-G_p\psi_p(y_p) - G_r\psi_r(y_r)] \quad (13)$$

where ψ_a, ψ_p, and ψ_r denote the actuator nonlinearity and the position and rate sensor nonlinearities, respectively. Assuming G_p and G_r are diagonal,

$$u_i = \psi_{ai}[-G_{pi}\psi_{pi}(y_{pi}) - G_{ri}\psi_{ri}(y_{ri})]$$

We assume that ψ_{ai}, ψ_{pi}, and ψ_{ri} $(i = 1, 2, ..., k)$ are continuous single-valued functions: $\mathbf{R} \rightarrow \mathbf{R}$. The following theorem gives the sufficient conditions for stability.

Theorem 2. Consider the closed-loop system given by (1), (2), (3), and (13), where G_p and G_r are diagonal with positive entries. Suppose ψ_{ai}, ψ_{pi}, and ψ_{ri} are single-valued continuous functions, and that, for $i = 1, 2, ..., k$,

(i)$\psi_{ai}(0) = 0$, ψ_{ai} are time invariant and monotonically nondecreasing.

(ii)ψ_{pi}, ψ_{ri} belong to the $(0, \infty)$ sector and ψ_{pi} are time invariant.

Then, the closed-loop system is globally asymptotically stable.

Proof.

(The proof closely follows [3].) Let $w = -y_p = -\theta$ (k-vector). Define

$$\overline{\psi}_{pi}(\nu) = -\psi_{pi}(-\nu) \quad (14)$$

$$\overline{\psi}_{ri}(\nu) = -\psi_{ri}(-\nu) \quad (15)$$

If $\psi_{pi}, \psi_{ri} \in (0, \infty)$ or $[0, \infty)$ sector then $\overline{\psi}_{pi}$, $\overline{\psi}_{ri}$ also belong to the same sector. Now, consider the following Luré-Postnikov Lyapunov function :

$$V = \frac{1}{2}\dot{p}^T M(p)\dot{p} + \frac{1}{2}q^T\overline{\overline{K}}q + \sum_{i=1}^{k}\int_0^{w_i}\psi_{ai}\{G_{pi}\overline{\psi}_{pi}(\nu)\}d\nu \quad (16)$$

where, $\overline{\overline{K}}$ is the symmetric positive definite part of K. Taking the time derivative and using (1),

$$\dot{V} = \dot{p}^T[B^T u - C\dot{p} - D\dot{p} - Kp] + \frac{1}{2}\dot{p}^T\dot{M}\dot{p}$$

$$+\sum_{i=1}^{k}\dot{w}_i\psi_{ai}\{G_{pi}\overline{\psi}_{pi}(w_i)\} + \dot{q}^T\overline{\overline{K}}q \quad (17)$$

Upon several cancellations and using the "*skew symmetric*" property of $(\frac{1}{2}\dot{M} - C)$,

$$\dot{V} = \sum_{i=1}^{k}u_i\dot{\theta}_i - \dot{q}^T\overline{D}\dot{q} + \sum_{i=1}^{k}\dot{w}_i\psi_{ai}\{G_{pi}\overline{\psi}_{pi}(w_i)\} \quad (18)$$

where, matrix $\overline{D}$ is the positive definite part of D.

$$\dot{V} = -\dot{q}^T\overline{D}\dot{q} - \sum_{i=1}^{k}\dot{w}_i(\psi_{ai}[G_{ri}\overline{\psi}_{ri}(\dot{w}_i) + G_{pi}\overline{\psi}_{pi}(w_i)]$$

$$-\psi_{ai}[G_{pi}\overline{\psi}_{pi}(w_i)]) \quad (19)$$

If ψ_{ai} are monotonic nondecreasing and ψ_{ri} belong to the $(0, \infty)$ sector, $\dot{V} \leq 0$, and it can be concluded that the system is at least Lyapunov-stable. Now we will prove that in fact the system is globally asymptoticaly stable. First, let us consider a special case when ψ_{ai} are monotonic increasing. Then $\dot{V} \leq -\dot{q}^T\overline{D}\dot{q}$, and $\dot{V} = 0$ only when $\dot{q} = 0$ and $\dot{w} = 0$, which implies $\dot{\theta} = 0 \Rightarrow \dot{p} = 0 \Rightarrow \ddot{p} = 0$. Substituting in the closed-loop equation,

$$Kp = B^T\psi_a[-G_p\psi_p(\theta)] \quad (20)$$

$$\begin{bmatrix} 0 \\ \overline{Kq} \end{bmatrix} = \begin{bmatrix} \psi_a\{-G_p\psi_p(y_p)\} \\ 0 \end{bmatrix} \quad (21)$$

$$\Rightarrow \psi_a[-G_p\psi_p(\theta)] = 0, \qquad \textit{and} \qquad q = 0$$

If ψ_{pi} belong to the $(0, \infty)$ sector, $\psi_{ai}(\nu) = \psi_{pi}(\nu) = 0$ only when $\nu = 0$. Therefore, $\theta = 0$. Thus, $\dot{V} = 0$ only at the origin, and the system is globally asymptotically stable.

In the case when actuator nonlinearities are of the monotonic nondecreasing type (such as saturation nonlinearity), $\dot{V}$ can be 0 even if $\dot{w} \neq 0$. However, we will show that every system trajectory along which $\dot{V} \equiv 0$, has to go to the origin asymptotically. When $\dot{w} \neq 0$, $\dot{V} \equiv 0$ only when all actuators are saturated. Then, from the equations of motion, it means that system trajectories will go unbounded which is not possible since we have already proved that the system is Lyapunov-stable. Hence, system trajectories have to approach the origin asymptotically and again the system is globally asymptotically stable.

Remarks

It is proved that, under static dissipative control, nonlinear, multibody, flexible space structures exhibit global asymptotic stability. The stability is not only robust to the modeling errors and parametric uncertainties, but also to a wide class of nonlinearities in the actuators and sensors. This has a significant practical value since the mathematical models of the system usually have substantial inaccuracies, and the actuation and sensing devices available are not perfect. Future research effort will address the extension of these results to the case when the actuator dynamics, as well as other types of nonlinearities, such as hysteresis, dead-zone, etc., are present. Future work will also address methods for controller *synthesis* and also dynamic dissipative compensators.

References

[1]. Desoer, C. A., and Vidyasagar, M.: *Feedback Systems: Input-Output Properties.* Academic Press, Inc., New York, 1975.

[2]. Popov, V. M.: *Hyperstability of Control Systems.* Springer-Velag, Berlin, 1973, pp. 118-239.

[3]. Joshi, S. M.:*Control of Large Flexible Space Structures.* Berlin Springer-Verlag, 1989 (Vol. 131, Lecture Notes in Control and Information Sciences).

[4]. Meirovitch, L.: *Methods of Analytical Dynamics.* McGraw-Hill, New York, 1970.

[5]. Takegaki, M., and Arimoto, S.: A New Feedback Method for Dynamic Control of Manipulators. ASME Journal of Dynamic Systems, Measurement and Control, Vol. 102, June 1981.

[6]. Koditschek, D. E.: Natural Control of Robot Arms. Proc. 1984, I.E.E.E Conference on Decision and Control, Las Vegas, Nevada., pp. 733-735.

[7]. Arimoto, S., and Miyazaki, F.: Stability and Robustness of PD Feedback Control With Gravity Compensation for Robot Manipulator. ASME Winter Meeting, Anaheim, California, December 1986, pp. 67-72.

[8]. Wen, J. T., and Bayard, D. S.: A New Class of Control Laws for Robotic Manipulator. Int. Journal of Control, 1988, Vol. 47, No. 5, pp. 1361-1385.

[9]. Paden, B., and Panja, R.: Globally Asymptotically Stable PD+ Controller for Robot Manipulators. Int. Journal of Control, 1988, Vol. 47, No. 6, pp 1697-1712.

[10]. Paden, B., Riedle, B., and Bayo, E.: Exponentially Stable Tracking Control for Multi-Joint Flexible-Link Manipulators. Proc. 1990, American Control Conference, San Diego, California, May 23-25, 1990, pp. 680-684.

[11]. Juang, J.-N., Wu, S.-C., Phan, M., and Longman, R. W.: Passive Dynamic Controllers for Non-linear Mechanical Systems. NASA Technical Memorandum, TM 104047, March 1991.

Proceedings of the 31st Conference
on Decision and Control
Tucson, Arizona • December 1992

CONTROL THEORETIC MODELS OF ENVIRONMENT-ECONOMY INTERACTIONS

D. A. Carlson
Dept. of Mathematics
University of Toledo
Toledo, Ohio

A. Haurie
Dept. of Business and Industrial Economics
University of Geneva
Geneva, Switzerland

1 INTRODUCTION.

The long term control of our environment is becoming increasingly important. The simplest way to eliminate pollution is to stop production. This is not economically practical. Thus, we need to consider economic growth models which take environmental effects into account. In our model we describe a class of infinite horizon optimal control models that link economic growth and pollution control models by coupling the standard investment-consumption equations to a pollution control equation. We focus on a distributed parameter model to take into account spatial distributions of capital and pollution. The performance of the system is measured by an integral cost functional which, for $T > 0$, measures the total net benefit or utility accumulated over the time interval $[0,T]$. Our goal is to maximize this utility over the time interval $[0,+\infty)$. These ideas have been pursued by several authors (see e.g., [6], [5], [4], and [1]) from a variety of viewpoints. In all cases, a discounted objective is considered. We consider an undiscounted utility to avoid weighting our decisions to the present at the expense of the future. This consideration leads to a divergent objective functional which requires us to consider the *overtaking optimality concept.* Using the results in [3] (see also, [2]) we provide conditions under which there exists an overtaking optimal solution and an associated steady state which attracts the optimal trajectory. Thus, we arrive at an "acceptable" level of pollution.

2 THE MODEL.

The model we consider describes the flow of goods and pollution throughout a fixed spatial region. We let $\Omega \subset I\!R^n$ $(n = 1,2,3)$ be a bounded spatial domain with smooth boundary, Γ, and we consider a capital accumulation process with two types of capital. The first, $k_1(\cdot,\cdot) : \Omega \times [0,+\infty) \to I\!R$, is used to produce goods that are consumed while the second, $k_2(\cdot,\cdot) : \Omega \times [0,+\infty) \to I\!R$, is used to control pollution. We let $P(\cdot,\cdot) : \Omega \times [0,+\infty) \to I\!R$ denote the concentration of the pollutant. The dynamics of this process is given by the following system of partial differential equations defined on $\Omega \times [0,+\infty)$

$$\begin{aligned} \frac{\partial}{\partial t}k_1 &= i_1 - \mu_1(x)k_1 - \mathrm{div}_x(\alpha_1(x)\nabla_x k_1) \\ \frac{\partial}{\partial t}k_2 &= i_2 - \mu_2(x)k_2 - \mathrm{div}_x(\alpha_2(x)\nabla_x k_2) \qquad (1) \\ \frac{\partial}{\partial t}P &= i_3 - \nu(x)P - \mathrm{div}_x(\beta(x)\nabla_x P) \end{aligned}$$

In the above for $s = 1,2$, $i_s(\cdot,\cdot) : \Omega \times [0,+\infty) \to I\!R$ and $\mu_s(\cdot) : \Omega \to I\!R$, denote the rates of investment in capital and depreciation of capital $k_s(\cdot,\cdot)$ while $\nu(\cdot) : \Omega \to I\!R$ denotes the natural abatement of the pollutant $P(\cdot,\cdot)$. The function $i_3(\cdot,\cdot) : \Omega \times [0,+\infty) \to I\!R$ is viewed as the "rate of investment in pollution." The coefficients $\alpha_s(\cdot)$ and $\beta(\cdot)$, defined on Ω, denote proportionality coefficients relating the flow of capital $k_s(\cdot,\cdot)$ and pollutant $P(\cdot,\cdot)$ out of a point $x \in \Omega$. We assume that the functions $\alpha_s(\cdot)$, $\mu_s(\cdot)$, $\nu(\cdot)$, and $\beta(\cdot)$ are smooth. As written (2) is a set of uncoupled equations. The coupling occurs in the form of point-wise of investment in capital satisfy a standard investment-consumption relation. That is,

$$c(x,t) + i_1(x,t) + i_2(x,t) \leq f(k_1(x,t)), \qquad (2)$$

in which $c(\cdot,\cdot) : \Omega \times [0,+\infty) \to I\!R$ denotes the rate of consumer consumption. The production function $f : I\!R \to I\!R$ is assumed to be smooth, increasing, concave and to satisfy the usual growth conditions

$$\lim_{k\to 0^+} f'(k) = +\infty \text{ and } \lim_{k\to+\infty} f'(k) = 0.$$

The investment in pollution is linked to the economic growth model through a pollution control function $G(\cdot,\cdot) : I\!R^2 \to I\!R$ which is a smooth convex function, increasing in k_1, decreasing in k_2. This constraint takes the form

$$i_3(x,t) \geq G(k_1(x,t), k_2(x,t)). \qquad (3)$$

Finally, we impose the following nonnegativity constaints

$$c(x,t),\ i_s(x,t),\ k_s(x,t), \text{ and } P(x,t) \geq 0 \qquad (4)$$

and assume there exists functions $\bar{k}_s : \Omega \to I\!R$, $s = 1,2$ such that

$$k_s(x,t) \leq \bar{k}_s(x) \text{ a.e. } t \geq 0 \qquad (5)$$

We view (2)–(5) as a control system with state variables $z(\cdot,\cdot) = (k_1(\cdot,\cdot), k_2(\cdot,\cdot), P(\cdot,\cdot))$ and control variables

$$u(\cdot,\cdot) = (i_1(\cdot,\cdot), i_2(\cdot,\cdot), i_3(\cdot,\cdot), c(\cdot,\cdot)).$$

To make this system well-posed we impose following initial conditions and boundary conditions.

$$k_s(x,0) = k_s^0(x) \text{ for } s = 1,2 \text{ and } P(x,0) = P^0(x) \qquad (6)$$

for $x \in \Omega$ and

$$\begin{aligned} \frac{\partial}{\partial \mathbf{n}}k_s(x,t) + \gamma_s(x)k_x(x,t) &= 0 \text{ for } s = 1,2 \\ \frac{\partial}{\partial \mathbf{n}}P(x,t) + \gamma_3(x)P(x,t) &= 0, \end{aligned} \qquad (7)$$

for $(x,t) \in \Omega \times [0,+\infty)$, where $\partial/\partial\mathbf{n}$ denotes the outward normal derivative on Γ.

We measure the performance of our economy by accumulating utility derived from consumption and pollution throughout the region Ω over the unbounded time interval $[0,+\infty)$. The utility flow is given by a function $\mathcal{U}(\cdot,\cdot) : I\!R^2 \to I\!R$ and our accumulated utility up to time $T > 0$ is given by

$$J_T(z(\cdot,\cdot), u(\cdot,\cdot)) \doteq \int_0^T \int_\Omega \mathcal{U}(c(x,t), P(x,t))\,dx\,dt. \qquad (8)$$

CH3229-2/92/0000-2860$1.00 © 1992 IEEE

We assume that the function $\mathcal{U}(\cdot,\cdot)$ is smooth, strictly concave, increasing in its first argument and decreasing in the second one. Our goal here is to determine (z^*, u^*) so that the total accumulated utility, (8) is maximized as $T \to \infty$. Generally letting $T \to +\infty$ leads to a divergent improper integral, and thus the usual definition of a maximum is inadequate. To circumvent this difficulty we consider the *overtaking optimality concept* as defined below.

Definition 2.1 We say a pair (z^*, u^*) is *overtaking optimal* if for any other pair (z, u) one has

$$\liminf_{T\to+\infty} [J_T(z(\cdot,\cdot), u(\cdot,\cdot) - J_T(z^*(\cdot,\cdot), u^*(\cdot,\cdot)] \le 0.$$

We will show it is possible to give conditions under which the above problem has an overtaking optimal solution.

3 SEMIGROUP SETTING

We now recast our problem to the semigroup setting described in Chapter 9 of [2] (see also [3]). We begin by defining the following Hilbert spaces

$$E \doteq L^2(\Omega; \mathbb{R}^3) \text{ and } F \doteq L^2(\Omega; \mathbb{R}^4)$$

with the usual norms, and let A be the linear differential operator with domain

$$\mathcal{D}(A) \doteq \left\{ z = (k_1, k_2, P) \in H^2(\Omega; \mathbb{R}^3) : z \text{ satisfies } (8) \right\},$$

(where $H^2(\Omega; \mathbb{R}^3)$ denotes the usual Sobolev space of twice differentiable functions on Ω (in the sense of distributions)), defined for $z = (k_1, k_2, P)$ by

$$Az \doteq - \begin{pmatrix} \operatorname{div}_x(\alpha_1(x)\nabla_x k_1(x)) + \mu_1(x)k_1(x) \\ \operatorname{div}_x(\alpha_2(x)\nabla_x k_2(x)) + \mu_2(x)k_2(x) \\ \operatorname{div}_x(\beta(x)\nabla_x P(x)) + \nu(x)P(x) \end{pmatrix}, \quad (9)$$

further define

$$X \doteq \{z = (k_1, k_2, P) \in E : P(x) \ge 0 \text{ and } k_s(x) \ge 0,\ s = 1, 2, \text{ a.e. } x \in \Omega\}, \quad (10)$$

the bounded linear operator $B : F \to E$ by, $u = (i_1, i_2, i_3, c)$

$$Bu = (i_1, i_2, i_3) \quad (11)$$

and let $U : X \to 2^F$ be the set-valued mapping from X into the nonempty closed subsets of F defined, for $z = (k_1, k_2, P)$, by

$$\begin{aligned} U(z) \doteq \{u = (i_1, i_2, i_3, c) \in F :& \\ c(x) + i_1(x) + i_2(x) \le f(k_1(x))& \quad (12) \\ i_3(x) \ge G(k_1(x), k_2(x))\}.& \end{aligned}$$

Finally define $F_0 : E \times F \to \mathbb{R}$ by the formula, for $z = (k_1, k_2, P)$ and $u = (i_1, i_2, i_3, c)$

$$F_0(z, u) \doteq \int_\Omega \mathcal{U}(c(x), P(x))\, dx. \quad (13)$$

With this notation the optimal control system described in the previous section can be briefly written as

$$\begin{aligned} \dot{z}(t) &= Az(t) + Bu(t) \text{ a.e. } t \ge 0, \\ z(0) &= z^0 = (k_1^0, k_2^0, P^0), \\ z(t) &\in X \text{ for } t \ge 0, \\ u(t) &\in U(z(t)) \text{ a.e. } t \ge 0. \end{aligned} \quad (14)$$

Also, the objective functional J_T is written as

$$J_T(z(\cdot), u(\cdot)) = \int_0^T F_0(z(t), u(t))\, dt. \quad (15)$$

The optimal control problem under consideration is now written in the form used in [3] and thus the results presented there may be applied to our problem. In fact, under the hypotheses given here, the operator A is the infinitesimal generator of a C_O-semigroup (see e.g., [2] and the references therein).

3.1 Assumptions

H1. There exists $K_1, K > 0$ such that

$$\|P\|^2 + \|c\|^2 > K_1 \Rightarrow F_0(z, u) \le -K\left(\|P\|^2 + \|c\|^2\right),$$

where $\|\cdot\|$ denotes the $L^2(\Omega, \mathbb{R})$ norm.

H2. (OSSP) The optimal steady state problem (OSSP)

$$\text{maximize } \{F_0(z, u)\colon 0 = Az + u,\ z \in X,\ u \in U(x)\}$$

has a solution $(\bar{z}, \bar{u})$, with $\bar{z}$ uniquely determined.

H3. There exists $\bar{p} \in D(A^*)$ so that for all (z, u), $z \in X$ and $u \in U(x)$

$$L_0(x, u) = F(z, u) - F_0(\bar{z}, \bar{u}) + \langle z, A^*\bar{p}\rangle + \langle Bu, \bar{p}\rangle \le 0$$

3.2 Results

We summarize our main results in the the following theorem.

Theorem 3.1 Suppose the set

$$\mathcal{G} \doteq \{z \in E :\ L_0(z, u) = 0 \text{ for some } u \in U(z)\} \quad (16)$$

is a singleton and that there exists a pair (z, u) such that

$$\int_0^{+\infty} L_0(z(t), u(t))\, dt > -\infty. \quad (17)$$

Then there exists a feasible pair (z^*, u^*) which satifies $z^*(t) \xrightarrow{w} \bar{z}$ as $t \to +\infty$ and which is overtaking optimal over all pairs (z, u) for which $t \to z(t)$ is bounded in E on $[0, +\infty)$.

The convergence of the optimal states $z^*(\cdot)$ to $\bar{z}$, the solution to OSSP, indicates that optimal capital stocks and pollution levels approach steady state levels. From this we see that the optimal steady state is indeed a sustainable development level for which the economic goods and environmental nuisances are maintained according to an efficient tradeoff.

4 ACKNOWLEDGEMENT

This research supported by FNRS and a visiting professor grant from the University of Toledo.

REFERENCES

[1] B. Beavis, " Optimal Pollution in the Presence of Adjustment Costs," J. Envir. Econ. and Mgt., vol. 6, pp. 1-10, 1979.

[2] D.A. Carlson, A. Haurie, and A. Leizarowitz, Infinite Horizon Optimal Control, 2nd ed., New York: Springer-Verlag, 1991.

[3] D.A. Carlson, A. Haurie, and A. Jabrane, "Existence of Overtaking Solutions to Infinite Dimensional Problems on Unbounded Time Intervals," SIAM J. Control and Opt., vol. 25, pp. 1517-1541, 1987.

[4] R.C. Griffin, "Environmental Policy for Spatial and Persistent Pollutants," J. Envir. Econ. and Mgt., vol. 14, pp. 41-53, 1987.

[5] C. Plourde and D. Yeung, "A Model of Industrial Pollution in a Stochastic Environment," J. Envir. Econ. and Mgt., vol. 16, pp. 97-105, 1989.

[6] F. Wirl, "Evaluation of Management Strategies under Environmental Constraints," European Journal of Oper. Res., vol. 55, pp. 191-200, 1991.

Proceedings of the 31st Conference
on Decision and Control
Tucson, Arizona • December 1992

Balanced truncation is Hankel-norm optimal when used to approximate a finite-difference model of a parabolic PDE

Denis Mustafa
Department of Engineering Science
University of Oxford, Parks Road, Oxford, OX1 3PJ, UK

Abstract

A standard finite-difference model is developed for a parabolic partial differential equation. The model is shown to belong to a class of state-space systems whose balanced approximants are optimal with respect to the Hankel-norm.

1 Introduction

In a recent paper [1] the following result was proved on the relations between balanced truncation [2] and optimal Hankel-norm approximation [3] of stable linear multivariable systems.

Proposition 1.1 *Let* $(\mathbf{A},\mathbf{B},\mathbf{C})$ *denote a system with transfer function* $\mathbf{C}(s\mathbf{I}-\mathbf{A})^{-1}\mathbf{B}$ *and define the class of systems*

$$\mathcal{S}^{p,m}_{n;stable} := \{(\mathbf{A},\mathbf{B},\mathbf{C}) \in \mathbf{R}^{n\times n}\times\mathbf{R}^{n\times m}\times\mathbf{R}^{p\times n} \\ : \mathbf{A}=\mathbf{A}^T<0, \mathbf{B}\mathbf{B}^T=\mathbf{C}^T\mathbf{C}=\mathbf{I}_n\}.$$

If $\mathbf{G}\in\mathcal{S}^{p,m}_{n;stable}$ *and has distinct poles then* $\mathbb{BT}(\mathbf{G},k)$*, the k-state balanced approximant of* $\mathbf{G}$*, is also a k-state optimal Hankel-norm approximant of* $\mathbf{G}$*.*

The primary interest in the above result is that it gives an affirmative answer to the question of whether balanced truncation is ever optimal with respect to any meaningful criterion. A question that was not addressed in [1] was whether the (at first glance very restricted) class $\mathcal{S}^{p,m}_{n;stable}$ contains any systems of engineering interest. It is the purpose of the present paper to point out that $\mathcal{S}^{p,m}_{n;stable}$ does indeed contain systems arising in practice: a specific but typical example is given from the realm of finite-difference models of partial differential equations.

2 Modelling of PDEs

Consider the parabolic partial differential equation

$$\frac{\partial\theta}{\partial t}=\frac{\partial^2\theta}{\partial x^2}+u$$

for $0\le x\le 1$ and $t\ge 0$, subject to boundary conditions $\theta(0,t)=\theta(1,t)=0$ and with given initial conditions. The equation represents the temperature $\theta(x,t)$ in a one-dimensional (normalized) heat conduction problem, with a heat input term $u(x,t)$. Given a desired temperature profile $y(x,t)$ the control problem is to choose $u(x,t)$ to make $\theta(x,t)$ track $y(x,t)$. As pointed out in [4] for a similar problem, such a goal is related to determining optimal policies for heating metals in a continuous furnace.

In the present paper we will not be dealing with the control issues but will focus on modelling. We use a standard finite-difference approach [5, Chapter 3], sometimes called the *method of lines* [6]. That is, replace the partial differential equation by a system of ordinary differential equations in the following way. Define a solution grid by $x=ih$ where $i=0,1,\dots,N,N+1$ and $(N+1)h=1$. Apply the finite-difference approximation

$$\frac{\partial^2\theta}{\partial x^2}=\frac{\theta(x-h,t)-2\theta(x,t)+\theta(x+h,t)}{h^2}+O(h^2)$$

to each of the N interior grid points. Then at each grid point $x=ih$ the exact solution $\theta(ih,t)$ is approximated by $v_i(t)$ solving

$$\dot{v}_i=\frac{v_{i-1}-2v_i+v_{i+1}}{h^2}+u_i(t), \quad i=1,\dots,N$$

where $v_0=v_{N+1}=0$ and $u_i(t):=u(ih,t)$.

To express the above system of N equations for the v_i in state-space form, define the input vector $\mathbf{u}(t)=(u_1(t),\dots,u_N(t))^T$, let the state vector be $\mathbf{x}(t)=(v_1(t),\dots,v_N(t))^T$, and let the output vector be $\mathbf{y}(t)=(v_1(t),\dots,v_N(t))^T$. The system of interest is hence

$$\dot{\mathbf{x}}=\mathbf{A}\mathbf{x}+\mathbf{B}\mathbf{u}, \quad \mathbf{y}=\mathbf{C}\mathbf{x},$$

where $\mathbf{A}$ is the $N\times N$ tridiagonal matrix

$$\mathbf{A}=\frac{1}{h^2}\begin{bmatrix} -2 & 1 & 0 & & \\ 1 & -2 & 1 & \ddots & \\ 0 & 1 & -2 & \ddots & 0 \\ & \ddots & \ddots & \ddots & 1 \\ & & 0 & 1 & -2 \end{bmatrix}$$

and $\mathbf{B}=\mathbf{C}=\mathbf{I}_N$. Note that $\mathbf{A}$ is stable as may be shown using Gershgorin's Theorem together with the fact that $\det\mathbf{A}=(-1)^N(1+N)^{2N+1}\neq 0$.

It is immediate that $(\mathbf{A},\mathbf{B},\mathbf{C})\in\mathcal{S}^{N,N}_{N;stable}$, and Proposition 1.1 applies.

3 Approximation

The fact that $(\mathbf{A},\mathbf{B},\mathbf{C})\in\mathcal{S}^{N,N}_{N;stable}$ means that all of the useful approximation properties proven in [1] will hold in the present case. This is fortunate, since accuracy may require N to be large (perhaps of order 100) and approximate lower order models may be desirable to reduce the computational burden. A particular attraction is that, as we will show, the approximation error is quantifiably modest.

Lack of space prevents us from stating the many consequences of applying the results of [1], which are in terms of the poles of the system. Here it is enough to show that the system poles

CH3229-2/92/0000-2862$1.00 © 1992 IEEE

can be written explicitly in terms of the problem data, as can the Hankel singular values. The Hankel singular values are of interest because—as has been convincingly argued in [3]—they contain fundamental information about approximation properties.

Proposition 3.1 *Let* $\mathbf{G} = (\mathbf{A}, \mathbf{B}, \mathbf{C})$ *be as defined in Section 2.*

1. *The poles of* $\mathbf{G}$ *are distinct and are given by*

$$\lambda_i = -4(N+1)^2 \sin^2\left(\frac{\pi i}{2(N+1)}\right), \qquad i = 1, \ldots, N.$$

2. *The Hankel singular values of* $\mathbf{G}$ *are distinct and are given by*

$$\sigma_i = \left(\frac{1}{8(N+1)^2}\right) \operatorname{cosec}^2\left(\frac{\pi i}{2(N+1)}\right), \qquad i = 1, \ldots, N.$$

Proof. The proof of 1 is straightforward: evaluate the eigenvalues of $\mathbf{A}$ using standard results, as on page 154 of [5]. The proof of 2 follows easily from part 1 applied to Corollary 2.4 of [1]. □

The expression for σ_i allows us to deduce that σ_i decreases rapidly with i—approximately as $1/i^2$—indicating that good low-order models are possible. The following results quantify that observation.

Proposition 3.2 *Let* $\mathbf{G} = (\mathbf{A}, \mathbf{B}, \mathbf{C})$ *be as defined in Section 2, and recall that* $\mathbb{BT}(\mathbf{G}, k)$ *denotes the* k*-state balanced approximant of* $\mathbf{G}$ *(where of course* $1 \le k \le N$*). Then*

$$\|\mathbf{G}\|_\infty = \|\mathbb{BT}(\mathbf{G}, k)\|_\infty = \left(\frac{1}{4(N+1)^2}\right) \operatorname{cosec}^2\left(\frac{\pi}{2(N+1)}\right)$$

and

$$\|\mathbf{G} - \mathbb{BT}(\mathbf{G}, k)\|_\infty = \left(\frac{1}{4(N+1)^2}\right) \operatorname{cosec}^2\left(\frac{\pi(k+1)}{2(N+1)}\right)$$

where $\|\mathbf{G}\|_\infty := \sup_\omega \lambda_{\max}^{1/2}\{\mathbf{G}^T(-j\omega)\mathbf{G}(j\omega)\}$ *is the usual* $\mathcal{H}_\infty$*-norm.*

Proof. Immediate from Proposition 3.1 and [1]. □

To illustrate the usefulness of the above result, we now indicate how it can be used to predict the order of a reduced-order model that will satisfy a prescribed relative error condition.

Suppose we seek a reduced-order model such that

$$\|\mathbf{G} - \mathbb{BT}(\mathbf{G}, k)\|_\infty \le \|\mathbf{G}\|_\infty / p^2$$

where $p > 1$ is given. From Proposition 3.2 we deduce that k must satisfy

$$p \sin\alpha < \sin(k+1)\alpha$$

where $\alpha := \pi/(2(N+1))$. Clearly it is necessary that $p \sin\alpha < 1$. By exploiting the bounds $2x/\pi \le \sin x \le x$ for $0 \le x \le \pi/2$ one can readily verify that k should be chosen to satisfy

$$(\pi p/2) - 1 \le k \le N.$$

For example, suppose $N = 100$ and we want a relative error of no more than 1%. Then $p = 10$ and so k must satisfy $14.71 \le k \le 100$. Thus a balanced approximant with as few as 15 of the 100 full-order states satisfies the 1% relative error condition.

4 Concluding Remark

Systems in the class $\mathcal{S}^{p,m}_{n;stable}$ have many desirable approximation properties as explained in [1]. Physical examples of systems in $\mathcal{S}^{p,m}_{n;stable}$ were not provided in [1] so the goal of the present paper has been to give a physically motivated example of a system in $\mathcal{S}^{p,m}_{n;stable}$. The method-of-lines technique used to obtain the example—the replacement of the spatial derivative in a PDE by a finite-difference approximation—is standard and widely applicable. There may well be other PDEs that would similarly lead to systems in $\mathcal{S}^{p,m}_{n;stable}$. Work is under way to investigate those other cases, which may have wider implications for the numerical solution of PDEs by finite-difference methods.

References

[1] D. Mustafa. A class of systems for which balanced truncation is Hankel-norm optimal. In *Proceedings of the IEEE Conference on Decision and Control*, Brighton, England, December 1991.

[2] B. C. Moore. Principal component analysis in linear systems: controllability, observability and model reduction. *IEEE Transactions on Automatic Control*, 26(1):17–32, 1981.

[3] K. Glover. All optimal Hankel–norm approximations of linear multivariable systems and their $\mathcal{L}_\infty$–error bounds. *International Journal of Control*, 39(6):1115–1193, 1984.

[4] A.G. Butkovskii. Some approximate methods for solving problems of optimal control of distributed parameter systems. *Automation Remote Control*, 22:1429–1438, 1962.

[5] G.D. Smith. *Numerical Solution of Partial Differential Equations: Finite Difference Methods.* Oxford University Press, third edition, 1985.

[6] W.E. Schiesser. *The Numerical Method of Lines: Integration of Partial Differential Equations.* Academic Press, 1991.

Proceedings of the 31st Conference
on Decision and Control
Tucson, Arizona • December 1992

CLOSED-LOOP MODEL REDUCTION METHOD

Grigore Braileanu

Electrical Engineering Department
Gonzaga University, Spokane, WA 99258

Abstract

The problem of specialized model reduction of plants destined for closed-loop operation is first given an algebraic form, by the means of a transformation based on Chebyshev series expansion, then solved through a least squares approach. The main feature of the method — the frequency-dependent design — is used to advantage to tune the model to closed-loop specifications.

I. Introduction

In this paper, we consider the problem of reducing the model of a linear time invariant plant, which may contain open-loop or closed-loop time lags. Many of the existing model reduction techniques involve some form of mode elimination [1-3], and do not lend themselves well to a closed-loop approach. To some extent, this is also true of the balancing technique of Moore [2], which has proved to be of great value for general linear systems. Model reduction involves a trade-off between model order and the degree to which the characteristics of the plant are reflected by the model. The relative importance of various plant characteristics is highly dependent upon the application, and so, while the high-frequency modes appear unimportant open-loopwise, they will be more important in the closed-loop behavior of the control system. The class of methods based on [2] reduces the model order by deleting state combinations that contribute less than a certain tolerance to the *Hankel singular values of the transfer function* of the plant. Yet, the Hankel singular values are measures of the contributions of the state components to the input-output relationship, which is essentially an open-loop entity. Since feedback can significantly alter the open-loop behavior of a plant, the singular values lose their significance when the plant is closed by a feedback loop. Finally, plants which contain a cluster of poles near the origin and many lightly damped structural poles at higher frequencies near the imaginary axis (e.g., linear models of flexible aircraft) generate numerical problems due to poorly conditioned Liapunov equations involved in *Hankel-norm reduction* methods.

In the following, we shall address these problems by using a modified least-squares model reduction method which is based on a new transformation with features similar to the Laplace transformation but numerically oriented [4]. The mathematical background is outlined in Section II. Section III introduces the main idea of the proposed method as an extension of a previous result [4] to plants which may contain time lags. The details of the method implementation are presented in Section IV, where the model order reduction is performed according to closed-loop plant behavior.

II. Preliminaries

As shown in [4], according to the theory of Chebyshev polynomial approximation in normed spaces of square integrable functions, signals may be represented as vectors of Chebyshev coefficients, differential operators as upper triangular matrices with integral entries, and linear subsystems as precomputed matrices. For convenience, the above mapping of functions into the space of vectors defined over the field of real numbers is referred to as the *A transformation*. Its main properties are summarized in [4]. For example, a signal $x(t)$ defined on the normalized time interval [-1,1] is associated with the vector $\mathbf{x} = A\,x(t)$, whose components are the coefficients of the Chebyshev series expansion of $x(t)$. The A-transform of the k-th derivative of $x(t)$ is $D^k\mathbf{x}$, where D, defined in [4], is an upper triangular matrix whose entries are integers. The *modified A-transform* of $x(t)$, which is the main tool of the proposed model reduction technique, was also given in [4]:

$$\hat{\mathbf{x}} = \hat{A}\,x(t) = T^{-1}\mathbf{x} \qquad (1)$$

where $\mathbf{x} = A\,x(t)$, and T is the optimal solution, in the sense of numerical conditioning, of the Diophantine matrix equation

$$TJT^{-1} = D \qquad (2)$$

The matrices J, T and T^{-1} are upper triangular and all their entries are integers; J has only the first upper diagonal nonzero.

The modified A transform, $\hat{\mathbf{x}}$, of a signal $x(t)$ has interesting properties which will be used to advantage in the following. First of all, the elements of $\hat{\mathbf{x}}$, are the coefficients of a polynomial expansion of $x(t)$, $t \in [-1,1]$, which, except for a multiplicative constant, tend to the values of the coefficients in the Taylor series expansion of $x(t)$ about $t=0$ when the size of the approximation space tends to infinity. Nevertheless, the truncated series corresponding to $\hat{\mathbf{x}}$ preserves properties of least squares (LS) approximations. It is this trade-off between the accuracy of Taylor series around $t=0$ and the accuracy of Chebyshev series expansions toward the ends of the approximation interval [-1,1] which makes the *modified A transform* behave better than other LS methods used in reduced-order modeling. Moreover, unlike the A transform, the implementation of the modified A transform results in *very sparse matrices* and *fast algorithms* (see, e.g., Algorithm 1 in [4]), especially in model reduction problems.

III. Basic Model Reduction Method

The procedure we shall outline here is analogous to well-known LS identification methods based on the series expansion of the pair $(u(t),y(t))$ of the input signal and system response. The proposed method, which uses the modified A transformation, was introduced in [4] for finite-dimensional linear time-invariant plants, and proven superior to two traditional model order reduction methods based on the Routh and Pade approximations. In [5], the method was extended to a large class of infinite-dimensional systems containing time lags, and compared to Hankel-norm model reduction methods. In this section, we shall present this approach for "open-loop plants," then, in Section IV, consider the *specialized model reduction* for plants destined for *closed-loop operation.*

The building blocks of the original mathematical model of the plant are subsystems of the form presented in the figure below.

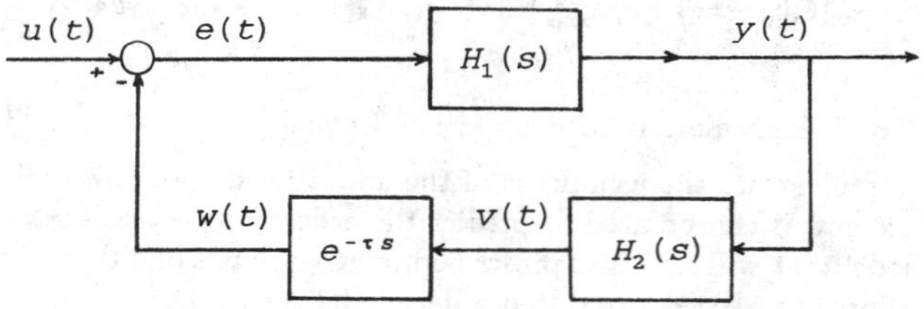

Figure 1. Block diagram of the plant

The main steps of the method, illustrated on a simple plant which consists of just one such subsystem, are as follows.

1) Choose a "persistent" input which excites the significant system modes.

2) Compute the modified A transforms $\hat{\mathbf{u}}_i$, i = 1, 2, ..., for consecutive Chebyshev approximation intervals of length τ, and round them off to reduced-word internal representations. Replace the former vectors $\hat{\mathbf{u}}_i$, and denote the new excitation by $u(t)$. The properties of the A transformation [4] let us compute the response of a feedback system with a time lag τ through a recursion of open-loop responses on time intervals of length τ.

3) Apply the fast A-transform method, defined by Algorithms 1 and 2 in [4], to the plant; compute the modified A transforms $\hat{\mathbf{y}}_i$ on the time intervals chosen above. The parameters of the original plant can also be rounded off to reduced-word representations. The errors introduced by the latter will be definitely smaller than

CH3229-2/92/0000-2864$1.00 © 1992 IEEE

those involved in the engineering process of plant modeling. At the same time, the reduced-word representations of both input and plant parameters, together with the fact that all the elements of the $\hat{A}$-operators D, T, J, and T^{-1} are integers, will allow for very good numerical properties of the model reduction algorithm.

4) Determine an n-th order model of the plant,

$$H_n(s) = \frac{Y(s)}{U(s)} = \frac{P_m(s)}{Q_n(s)}, \qquad m \le n, \tag{3}$$

where the coefficients of the polynomials $P_n(s)$ and $Q_m(s)$ are unknown; we group these coefficients, ordered for increasing powers of s, in the vectors **p** and **q**, respectively. The operator D, which plays the role of the Laplace differential operator s, can be applied to an alternative form of (3),

$$Q_n(s)\, Y_i(s) = P_m(s)\, U_i(s),$$

to give

$$Q_n(D)\, \mathbf{y}_i = P_m(D)\, \mathbf{u}_i, \tag{4}$$

for each one of the time intervals considered before. Now, we plug (2) in (4) and, based on (1) and a known property of the similarity transformation, we get

$$Q_n(J)\, \hat{\mathbf{y}}_i = P_m(J)\, \hat{\mathbf{u}}_i. \tag{5}$$

In these equations, the $\hat{A}$-transforms $\hat{\mathbf{u}}_i$ and $\hat{\mathbf{y}}_i$ are known, and so we reorder the terms as to evidence the unknowns **p** and **q**; moreover, we catenate all the vectors $\hat{\mathbf{u}}_i$ and $\hat{\mathbf{y}}_i$ into only two vectors, $\hat{\mathbf{u}}$ and $\hat{\mathbf{y}}$, obtaining:

$$[\hat{\mathbf{y}} \mid J\hat{\mathbf{y}} \mid \dots \mid J^n\hat{\mathbf{y}} \mid -\hat{\mathbf{u}} \mid -J\hat{\mathbf{u}} \mid \dots \mid J^m\hat{\mathbf{u}}]\begin{bmatrix}\mathbf{q}\\ \mathbf{p}\end{bmatrix} = 0. \tag{6}$$

This is a linear LS problem with equality constraints, as the latter are being added to (6) in order to satisfy conditions imposed on the new model (e.g., here we may have $q_{n+1} = 1$, by normalization, and $p_1 = y_{ss}\, q_1$, where y_{ss} is the imposed steady state gain).

5) Finally, we solve the linear LS problem with equality constraints for different pairs (n,m). A set of candidate models will be selected by inspecting the overall LS errors and validated through simulation.

IV. Closed-Loop Model Reduction Method

In this section, we discuss the main feature of the proposed model reduction method which allows for models which exhibit the behavior of the original plant within a given feedback control system. Specifically, since the method described in Section III above is based on the *test signal* $u(t)$ built analytically, the resulting reduced-order model will be sensitive to the specific excitation. Models of different properties will be generated with different signals $u(t)$. The criteria for choosing such test signals are similar to those used in the practice of system identification. Here, we shall restrict ourselves to a rule-of-thumb approach, based on the relationship between the settling time specification and the bandwidth of the closed-loop system. This will be shown an the following example. Let the plant of Fig. 1 be defined by

$$H_1(s) = \frac{(s+12)(s^2+4s+30)}{s(s+2)(s+5)(s^2+5s+10)},$$

$$H_2(s) = \frac{(s+10)(s+0.3)(s^2+30)}{20(s^2+2s+7)(s^2+6s+11)}, \qquad \tau = 2.$$

The loop gain of this plant is the product of a *9*-th order transfer function and e^{-2s}. A preliminary analysis of the closed-loop system specifications established the need for a bandwidth of approximately 7 rad/s. Accordingly, we chose a wideband input signal $u(t) = 20t^{10}e^{-5t}(1 + 5\sin 6.5t),\ t \ge 0$. An exact computation of the response of this plant to the input $u(t)$ was performed with the fast *A*-transform algorithm [4]. The procedure described in Section III provided a *4*-th order model defined by the input-output transfer function

$$\frac{Y(s)}{U(s)} = \frac{.0007s^4 - .0144s^3 + 1.2552s^2 + 4.8554s + 36.915}{s^4 + 4.0532s^3 + 13.25s^2 + 7.265s + 2.1574}.$$

In Fig. 2, relevant responses of this model are compared to those of a model obtained with the Hankel-norm method proposed in [2].

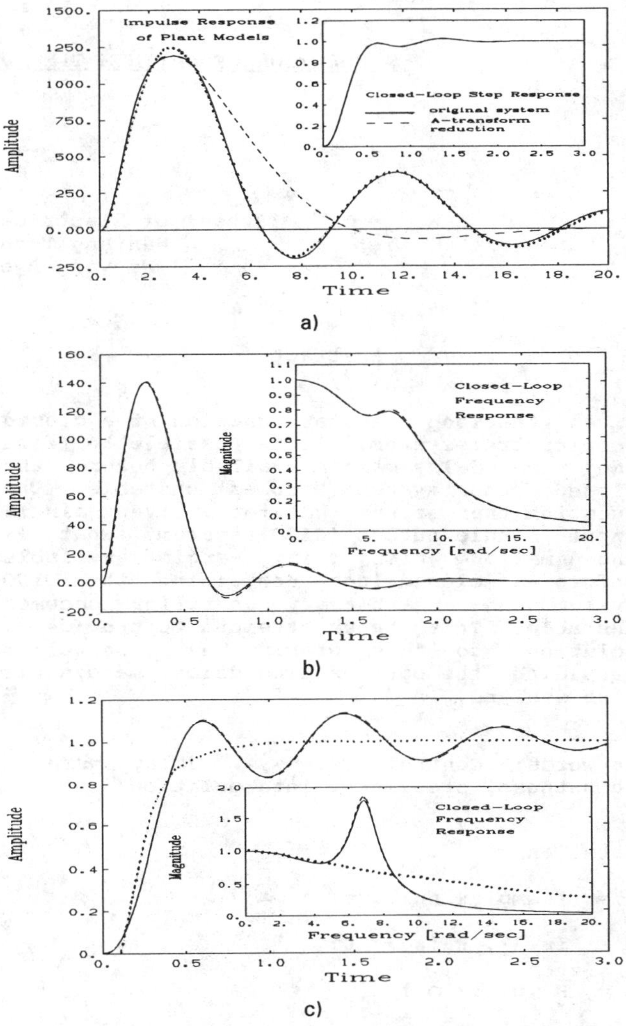

Figure 2. Step and impulse responses of open- and closed-loop systems: *solid lines* — the exact model; *dashed lines* — the *A*-transform model; *dotted line* — Hankel-norm reduced model of order four.

Fig. 2 a) shows that although the open-loop step response of the *A*-transform reduced model sensibly differs from the exact plant response, it is remarkably accurate in reproducing the closed-loop response. By contrast, the Hankel-norm reduced model of order four yields an accurate open-loop step response but unstable closed-loop response for the chosen controller. Accordingly, the impulse and frequency responses in Fig. 2 b) are shown only for the exact and *A*-transform models. Finally, in Fig. 2 c), a different controller was chosen in order to compare the three models.

References

[1] M. Aoki, "Control of large scale dynamic systems by aggregation," *IEEE Trans. Aut. Contr.*, vol. AC-13, pp. 246-253, 1968.

[2] B. C. Moore, "Principal component analysis in linear systems Controllability, observability, and model reduction," *IEEE Trans. on Aut. Contr.*, vol. AC-26, pp. 17-32, 1981.

[3] M.G., Safonov, R.Y.,Chiang, and D.J.N., Limebeer, "Hankel model reduction without balancing - A descriptor approach," *Proc. of the 26th CDC,* Los Angeles, CA, Dec. 1987, pp. 112-117.

[4] G. Braileanu, "Matrix operators for numerically stable representation of stiff, linear dynamic systems," *IEEE Trans. Automat. Contr.*, vol. AC-35, pp. 974-980, 1990.

[5] G. Braileanu, "Application of a new transformation to reduced order modeling," *Proc. of the 34th Midwest Symposium on Circuits and Systems,* Monterey, CA, May 1991, pp. 19-22.

Proceedings of the 31st Conference
on Decision and Control
Tucson, Arizona • December 1992

A SOLUTION OF OPTIMUM TIME DELAY MARGIN PROBLEM FOR SISO SYSTEMS

R.Devanathan

School of Electrical and Electronic Engineering
Nanyang Technological University
Nanyang Avenue, Singapore 2263

Abstract

Given the loop transfer function of a closed loop control system, it is possible to find the time delay margin available before the closed loop system becomes unstable. The question then arises that for a given single input single output (SISO) system, what is the maximum time delay margin available before a closed loop containing the SISO system and an arbitrary controller becomes unstable. This paper attempts to provide a solution to this problem.The controller providing the optimum time delay margin is also derived.

Keywords:- Control systems, delay systems, robustness, stability, interpolation.

Notation

$\mathcal{C}$ = {Complex numbers}

$\bar{H}$ = $\{s \in \mathcal{C}, \mathrm{Re}(s) >= 0\}$

$\tilde{H}$ = $\bar{H} \cup \{\infty\}$

$\bar{s}$ = (Complex conjugate of s)

1. INTRODUCTION

For time delay systems, Walton and Marshall [1] have given a complete solution for finding the maximum time delay before a given closed loop system with a known loop transfer function becomes unstable. Suppose we ask: Given a SISO system with time delay uncertainty, what is the maximum time delay margin available before the closed loop system containing the SISO system and an arbitrary controller becomes unstable ? This paper attempts to answer this question. The method of solution proposed is based on Khargonekar's [2] general framework for the robust stabilization of SISO systems with parameter uncertainty using the Nevanlinna - Pick interpolation theory [3] and the theory of conformal mapping [4].

2. PRELIMINARIES

Consider Fig.1 . Let p(s) be a SISO time delay system given by

$$p(s) = p_o(s) \exp(-hs) , \quad h >= 0 \qquad (2.1)$$

Let the nominal system $p_o(s)$ have n arbitrary poles $\{p_i , i= 1,2,..n\}$ in $\tilde{H}$ and m arbitrary zeros $\{z_l , l =1,2,...m\}$ in $\tilde{H}$, multiplicities included. Let $p_o(s)$ be strictly proper. Then the problem of stabilization of $p_o(s)$ is well posed [5]. The nominal system of Fig. 1 is stable if and only if [2]

(i) T(s) is analytic in $\tilde{H}$

$$\text{(ii) } T(p_i) = 1, \quad i =1,2,..n \qquad (2.2)$$

(iii) $T(z_l) = 0$, $l =1,2,..m$

where T(s) is the complimentary sensitivity function defined by

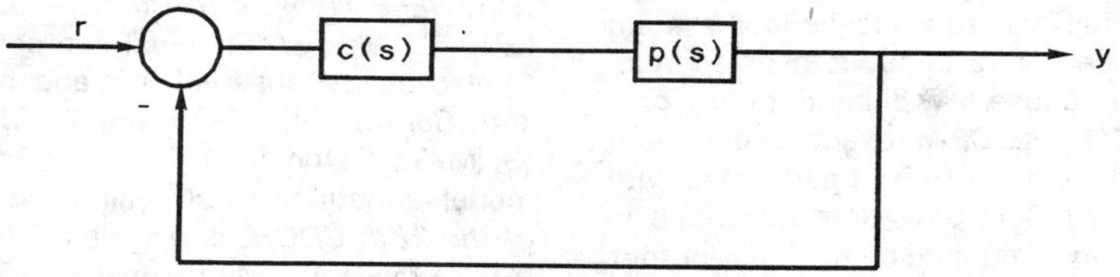

Fig. 1 Closed Loop System

CH3229-2/92/0000-2866$1.00 © 1992 IEEE

$$T(s) = p_o(s)\, c(s) / \{1 + p_o(s)\, c(s)\} \quad (2.3)$$

Let T(s) be such that

$$T(s) : \tilde{H} \rightarrow G \quad (2.4)$$

where G is a simply connected region containing points 0 and 1. If the mappings ϕ and θ are defined such that

$$\phi : G \rightarrow D \quad (2.5)$$

with

$$\phi(0) = 0 \quad (2.6)$$

and

$$\theta : \tilde{H} \rightarrow D \quad (2.7)$$

with

$$\theta(p_i) = \phi(1), \quad i=1,2,...n$$

$$\theta(z_l) = 0, \quad l=1,2,..m \quad (2.8)$$

where D is a open unit disk, then

$$T(s) = (\phi)^{-1}\, \theta \quad (2.9)$$

The necessary and sufficient condition for T(s) to exist [1] is that

$$|\phi(1)| < \alpha_{max} \quad (2.10)$$

where α_{max} is a parameter defined in [2] on the interpolation data of Eq.(2.2).

3. TIME DELAY UNCERTAINTY

Assume that the closed loop of Fig. 1 with the nominal system (h=0) is stable. Then for the robust stabilization of the closed loop of Fig. 1, it is required that no solution exists for

$$1 + p_o(s)\, c(s) \exp(-hs) = 0, \quad s \in \tilde{H} \quad (3.1)$$

Eq. (3.1) can be put in the form

$$Y(s)=\{1-T(s)\}/\, T(s)=-\exp(-hs), \quad s \in \tilde{H} \quad (3.2)$$

Since the nominal system is stable, the solution to Eq.(3.2) for s=jw is of interest. That is,

$$Y(jw) = -\exp(-jhw), \quad h > 0, \quad w >= 0 \quad (3.3)$$

The time delay h is given by

$$h=[(2k+1)\pi - \arg\{Y(jw)\}]/w, \quad k=0,1,2,.. \quad (3.4)$$

where Y(jw) is an arbitrary point on the unit circle centre origin as shown in Fig. 2. Following Walton and Marshall [1] , the smallest h in Eq.(3.4) is obtained when k=0 and when $w=w_o$ is the largest value of w for which Eq.(3.3) has a solution for some h> 0. That is,

$$h = [\pi - \arg\{Y(jw)\}]/w_o \quad (3.5)$$

If we now limit the feasible solution for Y(jw) on the unit circle to those outside the hatched portion as shown in Fig. 2, then the smallest h, for instability, should at least be such that

$$h = \gamma / w_o \quad (3.6)$$

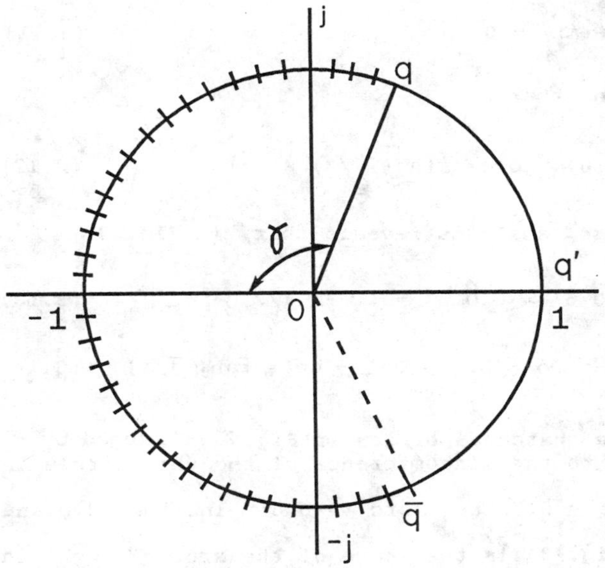

Fig. 2 Y- Plane

where γ is as shown in Fig. 2.

Denote the simply connected region, viz., the Y- plane minus the hatched portion as shown in Fig. 2 as G'. Then G' in the Y-plane is the image of G in the T-plane under the transformation shown on the L.H.S. of Eq.(3.2) denoted by ψ. Then

$$\psi : G \rightarrow G' \quad (3.7)$$

and

$$\psi(0) = \infty \quad (3.8)$$

$$\psi(1) = 0$$

Since G' is a simply connected region, consider the mapping

$$\eta : G' \subseteq Y \to D \subseteq U \tag{3.9}$$

given by

$$\eta(y) = \left\{ \frac{\sqrt{\{(y-q)/(y-\bar{q})\}\exp(j\gamma)} - \exp(j\gamma/2)}{\sqrt{\{(y-q)/(y-\bar{q})\}\exp(j\gamma)} + \exp(-j\gamma/2)} \right\} \cdot [\exp\{j(\pi-\gamma)/2\}] \tag{3.10}$$

The image of the point at infinity and that of the origin in the Y-plane (Fig.2) under the mapping η are given by the origin and the point v respectively as in Fig. 3. That is,

$$\eta(\infty) = 0 \tag{3.11}$$

and that

$$\eta(0) = Ov = \sin(\gamma/2) \tag{3.12}$$

Also, analysis reveals that, in Fig. 3,

$$\bar{u}\hat{O}o' = u\hat{O}o' = (\pi-\gamma)/2 \tag{3.13}$$

$$Ob = Oo' - bo' = Oo' - uo' = \tan(\gamma/4) \tag{3.14}$$

The hatched portion in Fig. 2 is mapped by η on to the circumference of the unit circle in Fig. 3. The arc '$\bar{u}$ b u' in the U-plane (Fig. 3) is the image of the arc 'q q' $\bar{q}$ ' in the Y-plane (Fig. 2) which corresponds to the set of feasible y-solutions (for instability) of Eq.(3.3). The region inside the unit circle in Fig. 2 is mapped into the region enclosed by the arc '$\bar{u}$ b u ' and the minor arc 'u $\bar{u}$ '.

Notice that

$$\phi = \eta \cdot \Psi \tag{3.15}$$

and

$$\phi(1) = \eta(0) = \sin(\gamma/2) \tag{3.16}$$

The necessary and sufficient condition , viz., Eq.(2.10) for T(s) to exist reduces to

$$\gamma < 2\sin^{-1}(\alpha_{max}) \tag{3.17}$$

Consider the mapping θ defined by Eq.(2.7) and (2.8). Since $p_o(s)$ is strictly proper , $\theta(s) \to 0$ as $s \to \infty$. If

$$|\theta(jw)| < Ob = \tan(\gamma/4), \quad w \geq w' > 0 \tag{3.18}$$

then, there are no common points between the open disk of radius Ov and the arc '$\bar{u}$ b u ' , and thus no solution to Eq.(3.3) exists for w >= w'. One can then take that

$$w_o = w' \tag{3.19}$$

and Eq.(3.6) becomes, for instability,

$$h = \gamma / w' \tag{3.20}$$

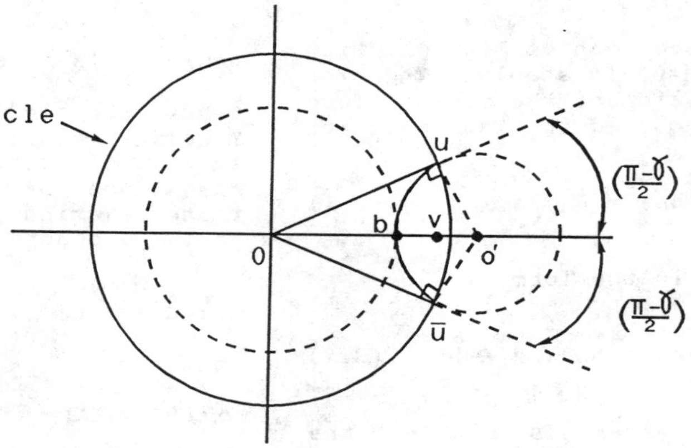

Fig. 3 U- Plane

Clearly as γ increases subject to the constraint of Eq.(3.17), Ob increases and w' in Eq.(3.18) decreases and h, given by Eq.(3.20), increases.

Combining Eq.(3.2), (3.9), (3.10) and (2.7), an expression for T(s) can be derived and c(s) can be solved for from Eq.(2.3).

Thus we have proved the following theorem.

Theorem 1

Consider the time delay SISO system p(s) connected in a closed loop with an arbitrary controller as shown in Fig. 1. The time delay h before the closed loop becomes unstable should be such that

$$h < \gamma / w' \tag{3.21}$$

where γ can be chosen subject to the constraint of Eq.(3.17) and w' satisfies the constraint of Eq.(3.18). The optimum (maximum) time delay margin with an arbitrary controller is obtained by choosing γ as close to the limit given by Eq.(3.17) as possible. The corresponding complimentary sensitivity function T(s) satisfying Eq.(2.2) is given by

$$T(s) = \frac{\{\theta \sin(\gamma/2) - 1\}(\theta)}{\sin(\gamma/2)(\theta^2 - 1)} \tag{3.22}$$

and the controller c(s) becomes

$$c(s) = \{1/p_o(s)\}\{\theta \sin(\gamma/2)-1\}(\theta)/\{\theta - \sin(\gamma/2)\} \tag{3.23}$$

4. EXAMPLE

Let the SISO system have a pole at s=2 and a zero at infinity.Since $\alpha_{max} = 1$ in this case, $0 < \gamma < \pi$. $\theta(s): \tilde{H} \text{--> } D$ is given by

$$\theta(s) = \{(2+t)(s-\beta)\}/\{(s+t)(s+\beta)\}, \quad t > 0$$

where

$$\beta = 2\{1 - \sin(\gamma/2)\}/\{1 + \sin(\gamma/2)\}$$

The controller is given by

$$c(s) = \frac{(2+t)[(s+t)(s+\beta)-(2+t)(s-\beta)\sin(\gamma/2)](s-\beta)}{K[\sin(\gamma/2)s-(2+t)+\sin(\gamma/2)(t+\beta+2)](s+t)(s+\beta)}$$

where K is the dc gain of the SISO system.

$$|\theta(jw)| = (2+t)/(w^2 + t^2)^{1/2} < \tan(\gamma/4)$$

Since t is arbitrarily close to 0, it is required that

$$w > w_o = w' = 2 / \{\tan(\gamma/4)\}$$

Hence for stability

$$h < \gamma \tan(\gamma/4) / 2, \quad 0 < \gamma < \pi$$

The limits on the time delay are then given as

$$0 < h < \pi/2$$

5. CONCLUSIONS

Given a SISO system with arbitrary poles and zeros in the closed right half plane and having time delay uncertainty , the question arises as to the maximum time delay margin available before the closed loop system becomes unstable when an arbitrary controller is connected to the SISO system in the closed loop.This paper has provided a solution to this problem and shown how to derive the controller corresponfing to the optimum condition.

REFERENCES

[1] Walton K and J.E.Marshall, "Direct Method for TDS Stability Analysis", IEE Proceedings, 134, Pt. D, 2, 1987, pp. 101-107.

[2] Khargonekar, P.P. and R.Tannenbaum , "Non-euclidean Metrics and the Robust Stabilization of Systems with Parameter Uncertainty", IEEE Trans. Autom. Contr., AC-30, 10 , 1985 , pp. 1005-1013.

[3] Nevanlinna, R., <u>Analytic Functions</u>, Springer - Verlag, New York, 1953.

[4] Churchill,R.V.,J.W. Brown and R.F.Verhey, <u>Complex Variables and Applications</u>, Tokyo: McGraw- Hill, 1974.

[5] Zames, G., and B.A.Francis, " Feedback , Minimax Sensitivity and Optimal Robustness", IEEE Trans. Automat. Contr., AC-28, 5, 1983, pp. 585-601.

Proceedings of the 31st Conference
on Decision and Control
Tucson, Arizona • December 1992

Deterministic Control of Uncertain State Delayed Systems

Suthee PHOOJARUENCHANACHAI and Katsuhisa FURUTA
Department of Control Engineering
Tokyo Institute of Technology
2-12-1, Oh-okayama, Meguro-ku, Tokyo 152 JAPAN

Abstract

In this paper, we consider a robust stabilization problem in a state-space setting for a class of uncertain time-delay systems. The system under consideration are described by linear state delayed equation whose matrices contain norm-bounded time-varying elements. Two novel designs of nonlinear controllers to cope with the uncertainties are presented. The first design is made possible by appropriately combining a state transformation technique with the second method of Lyapunov. Specifically, the transformation technique is employed to convert the stabilization problem into an equivalent one which is solvable via the finite dimensional Lyapunov min-max approach. The second design is mainly based on solving a certain infinite dimensional Riccati equation arising in the optimal control theory for hereditary systems. The relative merits of the two designs are compared. An application of the main results to a certain model following control problem is also demonstrated.

1 Introduction

Deterministic control of uncertain dynamical systems in presence of extraneous disturbances has been the subject of considerable study; e.g. see [1] and references therein. One approach to deterministic control design is by means of the so-called "second method" of Lyapunov. The design approach is based on a nominal linearization of a given system, with (possibly fast) time-varying, non-linear uncertain elements of the system and the extraneous disturbances grouped into an unknown but bounded function. In principle, one start with a positive definite function V corresponding to the nominal system. Then an appropriate control function must be designed so that V robustly decreases along the system trajectory against all uncertainties. In contrast with many stochastic control schemes, only knowledge of compact sets bounding the system uncertainties is required. Furthermore, if they satisfy certain matching conditions, complete insensitivity to the system variations can be achieved.

Sharing the direction of early research workers such as Leitmann, Corless, Ryan, Gutman and Palmor [2][3], this paper is mainly concerned with the stabilization problem of a class of linear time-delay systems containing uncertain parameters and additive disturbances. As in [4], we restrict our attention to systems with norm-bounded time-varying uncertainties. The main difference between our approach and that of [4] is allowing the nominal models to be linear time-invariant systems with state delay. This feature therefore provides feasibility of utilizing more information on the models than that of [4], and extends the applicability of the Lyapunov theory approach to robust control of time-delay systems.

In this paper, we propose two new methods of controller design, namely, indirect design and direct design. The indirect design basicly draws on some ideas in [5] and [6] to determine a suitable structure of stabilizing controllers. Determination of controller parameters can be divided into two phases. In the first phase, the transformation technique proposed by Fiagbedzi and Pearson [6] is utilized to transform the original stabilization problem to an equivalent one which is easier to solve. Next, by using the well-known Lyapunov min-max approach of Gutman [5], a suitable stabilizing control law is derived in the second phase.

On the other hand, the direct design is achieved by direct application of infinite dimensional version of the Lyapunov min-max approach without any transformation step. Specifically, we employ a solution to the infinite dimensional Riccati equation arising in the linear quadratic optimal control theory [8] to construct a Lyapunov functional candidate. Then we design a control law such that the resulting closed-loop system robustly admits the Lyapunov functional against all system uncertainties.

The present paper is organized as follows. First, we begin with statement of the problem and some assumptions. In Section 3 and 4, the indirect design and the direct design are presented respectively. Some comparisons between both two designs are given. As an application, the results are employed to solve a certain model following control problem in Section 5. The final Section 6 contains some remarks on controller implementation and possible extension.

2 Statement of the Problem

The following notation will be adopted throughout this paper. Let $\mathcal{R} = (-\infty, \infty), \mathcal{R}^+ = [0, \infty), \mathcal{R}^n$ by any real n-dimensional linear vector space over the reals with $\|\cdot\|$ referring to the standard Euclidean norm. The identity matrix is denoted by I. Let X and Y be $n \times n$ symmetric matrices. Then the notation $X < (\leq) Y$ denotes the fact that $X - Y$ is a negative definite(semidefinite) matrix. For given $h \in \mathcal{R}^+$, let $\mathcal{C}_d$ denote the Banach space of continuous vector-valued functions defined on an interval $[-h, 0]$ taking values in $\mathcal{R}^n$ with norm :

$$\|\psi\|_d := \sup_{-h \leq \eta \leq 0} \|\psi(\eta)\| , \quad \text{where } \psi \in \mathcal{C}_d([-h, 0]; \mathcal{R}^n)$$

Furthermore, if x is a continuous function of t defined on $-h \leq t \leq T$; $h > 0$ and $T > 0$. Then for fixed $t \in [0, T)$, x_t

denotes the restriction of x to the interval $[t-h, t]$ translated to $[-h, 0]$; i.e $x_t \in \mathcal{C}_d$ and $x_t(\eta) = x(t+\eta)$, $-h \le \eta \le 0$.

Consider a dynamical time-delay system of the form (S_d):

$$\begin{aligned} \dot{x}(t) &= [A + \Delta A(t)]x(t) \\ &\quad + [A_h + \Delta A_h(t)]x(t-h) \\ &\quad + [B + \Delta B(t)]u(t) + Bw(t) \end{aligned} \tag{2.1}$$

where

$x \in \mathcal{R}^n$ is the current value of the system state,

$u(t) \in \mathcal{R}^m$ is the control function,

$w(t) \in \mathcal{R}^l$ is the additive disturbance,

A, A_h, B are constant matrices of appropriate dimensions,

$\Delta A(t), \Delta A_h(t), \Delta B(t)$ are matrices whose elements are continuous, unknown but bounded functions,

$h \in \mathcal{R}^+$ is a known constant delay time.

Further, let the initial functions of the system (S_d) be specified as $x_0(\eta) \in \mathcal{C}_d([-h, 0]; \mathcal{R}^n)$.

The following assumptions are made throughout this paper.
Assumption 1: The nominal system of (S_d); i.e., the system (S_d) with $\Delta A(t) = \Delta A_h(t) = 0, \Delta B(t) = 0, w(t) = 0$ is spectrally stabilizable [9].
Assumption 2: For all $t \in \mathcal{R}^+$, there exist continuous matric functions $H(t), H_h(t)$ and $E(t)$ of appropriate dimensions such that

a) $\Delta A(t) = BH(t)$,
b) $\Delta A_h(t) = BH_h(t)$,
c) $\Delta B(t) = BE(t)$,
d) $I + \frac{1}{2}(E(t) + E^T(t)) \ge \delta I$ for some scalar $\delta > 0$.

Note here that if the matching conditions defined in Assumption 2 are satisfied, the system (S_d) takes the form

$$\dot{x}(t) = Ax(t) + A_h x(t-h) + B(u(t) + v(t)) \tag{2.2}$$

where

$$v(t) = H(t)x(t) + H_h(t)x(t-h) + E(t)u(t) + w(t) \tag{2.3}$$

The main objective is to determine a state feedback control law such that the resulting closed-loop system is asymptotically stable for all realizations of the uncertainties.

The following notations are also made throughout this paper. For given any square matrix M, let

$$\sigma(M) = \{s \in \mathcal{C}; det(sI - M) = 0\}$$

denotes the spectrum of M.

3 Indirect Design

In this section, a finite dimensional approach to derive a formula for constructing a suitable state feedback law is presented. We begin with the following linear transformation T_c defined by

$$\begin{aligned} z(t) &= (T_c(x))(t) \\ &= \Gamma x(t) + \int_{-h}^{0} e^{A_c\theta}\Gamma A_h x(t-h-\theta)d\theta \end{aligned} \tag{3.1}$$

where $A_c \in \mathcal{R}^{nN \times nN}$ is a matrix yet to be defined, and

$$\Gamma = \sum_{i=1}^{N}(e_i \otimes I_m) \tag{3.2}$$

where e_i denotes the N-dimensional unit column vector of which entries are zero except for unity in the ith entry.

Proposition 3.1 *Let the matrix A_c in (3.1) be defined by*

$$A_c = \otimes_{i=1}^{N} A_i \tag{3.3}$$

where $A_i, i = 1, \ldots, N$, satisfy

$$A_i = A + e^{-hA_i}A_h, \tag{3.4}$$

$$\sigma(A_i) \cap \sigma(A_j) = \phi, \tag{3.5}$$

whenever $i \ne j$, and

$$\sigma_u(S_d) \subset \cup_{i=1}^{N}\sigma(A_i) \subset \sigma(S_d), \tag{3.6}$$

where

$$\sigma(S_d) = \left\{s \in \mathcal{C}; det(sI - A - e^{-hs}A_h) = 0\right\},$$

and

$$\sigma_u(S_d) = \{s \in \sigma(S_d); Re(s) \ge 0\}.$$

Then, subject to (3.1), $x(t)$ satisfies (2.2), and hence (2.1), if and only if $z(t)$ satisfies the system of the form (S_0):

$$\dot{z}(t) = A_c z(t) + \Gamma B(u(t) + v(t)) \tag{3.7}$$

Furthermore, asymptotic stability of $z(t)$ implies asymptotic stability of $x(t)$.

Proof: By using the *Leibniz's formula* [10], it is straightforward to verify that (2.2) in conjunction with the transformation (3.1) yields (3.7). To prove the second part of the Proposition, Laplace transform (3.1) to obtain, after some rearrangement,

$$Z(s) = (sI - A_c)^{-1}\Gamma\Delta(s)X(s) \tag{3.8}$$

where $\Delta(s) = [sI - A - e^{-hs}A_h]$. Partitioning $z(t)$ as

$$z(t) = [z_1^T(t)|z_2^T(t)|\ldots|z_N^T(t)]^T;$$

$z_i(t) \in \mathcal{R}^n$, $i = 1, \ldots, N$, we then obtain

$$X(s) = \Delta^{-1}(s)(sI - A_i)Z_i(s) \tag{3.9}$$

where $Z_i(s), i = 1, \ldots, N$, is the Laplace transform of $z_i(t)$. By combining (3.9) with (3.4) $\sim$ (3.6), it can be verified that if there exists any unstable mode in $X(s)$, it must be appears in $Z_i(s)$ for some i. Consequently, asymptotic stability of $z(t)$ implies asymptotic stability of $x(t)$. □

Remark 3.1 *The transformation (3.1) satisfying hypothesis of Proposition 3.1 can be considered as a class of that proposed in Theorem 3.2 of [6]. Therefore, the computational method proposed in [6] for finding the transformation can be applied here as well.*

We shall show the following theorem that existence of the required transformation in Proposition 3.1 implies stabilizability of the system (S_d).

Theorem 3.1 *Suppose there exists a transformation (3.1) satisfying the hypothesis of Proposition 3.1. Then there exist a positive definite solution P to the Riccati equation*

$$A_c^T P + PA_c - P\Gamma BB^T\Gamma^T P + I = 0 \qquad (3.10)$$

Furthermore, a stabilizing control law is given by

$$u(t) = -\rho(x_t)\frac{B^T\Gamma^T Pz(t)}{\|B^T\Gamma^T Pz(t)\|}; \qquad \forall z(t) \notin \Psi_z \qquad (3.11)$$

$$u(t) = \{u \in \mathcal{R}^m; \|u\| \le \rho(x_t)\}; \qquad \forall z(t) \in \Psi_z \qquad (3.12)$$

where

$$\Psi_z = \left\{z \in \mathcal{R}^{nN}; B^T\Gamma^T Pz = 0\right\}, \qquad (3.13)$$

$$\rho(x_t) > \Big[\tfrac{1}{2}\|B^T\Gamma^T Pz(t)\| + \|H(t)\|\|x(t)\| + \|H_h(t)\|\|x(t-h)\| + \|w(t)\|\Big] / \delta \qquad (3.14)$$

and δ is the positive scalar defined in Assumption 2-d.

Proof: First note that the existence of transformation (3.1) satisfying the hypothesis of Proposition 3.1 implies that

$$\text{rank}\Big[\Gamma B|A_c\Gamma B|\ldots|A_c^{nN-1}\Gamma B\Big] = nN$$

(see Theorem 3.2 of [6] for proof). In other words, $(A_c, \Gamma B)$ is a controllable pair. Hence, (3.10) holds for some $P > 0$. Now we take the positive definite function

$$V_z(t) = z^T(t)Pz(t) \qquad (3.15)$$

as a Lyapunov function for the system (3.7) with the control (3.11) ~ (3.14). Applying the Riccati equation (3.10), the following is obtained for the bound of derivative of the Lyapunov function (3.15)

$$\begin{aligned}\dot{V}_z(t) &= -z^T(t)z(t) + 2z^T(t)P\Gamma B\Big[u(t) + v(t) + \tfrac{1}{2}B^T\Gamma^T Pz(t)\Big] \\ &\le -z^T(t)z(t) + 2z^T(t)P\Gamma B(I + E(t))u(t) \\ &\quad +2\|B^T\Gamma^T Pz(t)\|\Big[\tfrac{1}{2}\|B^T\Gamma^T Pz(t)\| + \|w(t)\| \\ &\quad +\|H_h(t)\|\|x(t-h)\| + \|H(t)\|\|x(t)\|\Big]\end{aligned}$$

Substitution of the control law (3.11) ~ (3.14) in the last inequality yields

$$\dot{V}_z(t) \le -\|z(t)\|^2 \qquad (3.16)$$

(3.15) and (3.16) imply that $z(t)$ is asymptotically stable. Consequently from Proposition 3.1, $x(t)$ is also asymptotically stable. This completes the proof. □

We end this section with an illustrative example, consider the system (S_d) with $\|w(t)\| \le w_{max}$ and

$$A = \begin{bmatrix} 0 & 1 \\ 0 & 0 \end{bmatrix}, \quad \Delta A(t) = \begin{bmatrix} 0 & 0 \\ \alpha(t) & 0 \end{bmatrix},$$

$$A_h = \begin{bmatrix} 1 & 0 \\ 0 & 0 \end{bmatrix}, \quad \Delta A_h(t) = \begin{bmatrix} 0 & 0 \\ 0 & \beta(t) \end{bmatrix},$$

$$\|\alpha(t)\| \le \alpha_{max}, \quad \|\beta(t)\| \le \beta_{max},$$

$$B = \begin{bmatrix} 0 \\ 1 \end{bmatrix}, \quad \Delta B(t) = \begin{bmatrix} 0 \\ \gamma(t) \end{bmatrix}, \quad \|\gamma(t)\| \le 0.5,$$

Note that the theoretical results in [4] cannot be directly applied to this example because the A_h is a non-zero matrix and cannot be considered as an uncertain part which satisfies the matching conditions [4]. Now following the procedure given in [6], one can verifies that the unstable poles of the nominal system is $\{0, s_0\}$ where $s_0 \approx 0.567$. Furthermore, by using the result in [9], it can be verified that both of these unstable modes are controllable. The required parameters of the transformation (3.1) is then determined to be

$$A_c = \begin{bmatrix} s_0 & 1 \\ 0 & 0 \end{bmatrix}, \qquad \Gamma = I_n,$$

so that (3.9) becomes

$$X(s) = \begin{bmatrix} \frac{s(s-s_0)}{p(s)} & 0 \\ 0 & 1 \end{bmatrix} Z(s)$$

where $p(s) = s(s - e^{-hs})$ is the characteristic equation of transfer function of the nominal system. Clearly, $x(t) \to 0$ provided that $z(t) \to 0$ as $t \to \infty$. To derive the control law, we solve the Riccati equation (3.10) to get

$$P = \begin{bmatrix} p_1 & p_2 \\ p_2 & p_3 \end{bmatrix},$$

where $p_1 \approx 4.853, p_2 \approx 2.55, p_3 \approx 2.47$. A suitable stabilizing control law is then given by (3.11) ~ (3.13) with

$$\rho(x_t) > 2\Big[\tfrac{1}{2}\|B^T\Gamma^T Pz(t)\| + \alpha_{max}\|x(t)\| + \beta_{max}\|x(t-h)\| + w_{max}\Big].$$

4 Direct Design

This section provides an alternative technique to solve the proposed stabilization problem. The technique involves solving a certain infinite dimensional Riccati equation which can be accomplished by an existing reliable computational algorithm. Before proceeding with the main result, first note that Assumption 1 implies the existence of matric solutions $P_0, P_1(\cdot), P_2(\cdot,\cdot)$ to the following Riccati equation

$$A^T P_0 + P_0 A + P_1(0) + P_1^T(0) - P_0 BB^T P_0 + Q = 0 \qquad (4.1)$$

$$\frac{d}{d\alpha}P_1(\alpha) = P_1(\alpha)\{A - BB^T P_0\} + P_2(\alpha, 0) = 0 \qquad (4.2)$$

$$\left\{\frac{\partial}{\partial\alpha}+\frac{\partial}{\partial\beta}\right\}P_2(\alpha,\beta)+P_1(\alpha)BB^TP_1^T(\beta)=0 \qquad (4.3)$$

$$P_0=P_0^T \qquad (4.4)$$

$$P_1(-h)=A_h^TP_0 \qquad (4.5)$$

$$P_2(\alpha,-h)=P_1(\alpha)A_h \qquad (4.6)$$

$$P_2(\alpha,\beta)=P_2^T(\beta,\alpha); \qquad \alpha,\beta\in[-h,0], \qquad (4.7)$$

for any $Q > 0$; see [8] for full derivation of these equations. Furthermore, these solutions also have a property such that for any non-zero $x \in \mathcal{R}^n$ and any

$$x_1(\beta)\in\mathcal{R}^n; \qquad \int_{-h}^{0}x_1^T(\beta)x_1(\beta)d\beta\neq 0,$$

$$\begin{aligned}&x^TP_0x+2x^T\int_{-h}^{0}P_1^T(\beta)x_1(\beta)d\beta\\&+\int_{-h}^{0}\int_{-h}^{0}x_1^T(\alpha)P_2(\alpha,\beta)x_1(\beta)d\alpha d\beta>0\end{aligned} \qquad (4.8)$$

The main result of this section is summarized in the following theorem.

Theorem 4.1 *The system (S_d) is stabilizable via the control law*

$$u(t)=-\mu(x_t)\frac{y(t)}{\|y(t)\|}; \qquad \forall x_t\notin\Psi_x \qquad (4.9)$$

$$u(t)=\{u\in\mathcal{R}^m;\|u\|\leq\mu(x_t)\}; \qquad \forall x_t\in\Psi_x \qquad (4.10)$$

where

$$y(t)=B^TP_0x(t)+\int_{-h}^{0}B^TP_1^T(\beta)x(t+\beta)d\beta \qquad (4.11)$$

$$\Psi_x=\{x_t\in\mathcal{C}_d([-h,0];\mathcal{R}^n); \quad y(t)=0\}, \qquad (4.12)$$

$$\begin{aligned}\mu(x_t) \;>\; &\Big[\tfrac{1}{2}\|y(t)\|+\|H(t)\|\|x(t)\|\\&+\|H_h(t)\|\|x(t-h)\|+\|w(t)\|\Big]\,/\,\delta\end{aligned} \qquad (4.13)$$

and δ is the positive scalar defined in Assumption 2-d, and $P_0, P_1(\cdot), P_2(\cdot,\cdot)$ are the solutions to (4.1) $\sim$ (4.8).

Proof: Using P_0, $P_1(\cdot)$, and $P_2(\cdot,\cdot)$ to take the following positive definite function

$$\begin{aligned}V_x(t)&=x^T(t)P_0x(t)\\&+2x^T(t)\int_{-h}^{0}P_1^T(\beta)x(t+\beta)d\beta\\&+\int_{-h}^{0}\int_{-h}^{0}x^T(t+\alpha)P_2(\alpha,\beta)x(t+\beta)d\alpha d\beta\end{aligned} \qquad (4.14)$$

as a Lyapunov functional to be admitted by the resulting closed-loop system. Differentiating (4.14) with respect to time and evaluate it along the solution of (S_d) with the control (4.9) $\sim$ (4.13) yields, after some lengthy rearrangement,

$$\begin{aligned}\dot{V}_x(t)&=-x^T(t)Qx(t)+y^T(t)y(t)\\&+2y^T(t)(u(t)+v(t))\end{aligned} \qquad (4.15)$$

where $v(t)$ is defined in (2.3), it now becomes routine to verify that the proposed control law yields

$$\dot{V}_x(t)\leq-\lambda_{min}(Q)\|x(t)\|^2 \qquad (4.16)$$

Consequently,

$$\lambda_{min}(Q)\int_{0}^{\tau}\|x(t)\|^2dt\leq V_x(0)-V_x(\tau).$$

Since $V_x(\tau)\geq 0$, we can conclude that

$$\int_{0}^{\tau}\|x(t)\|^2dt\leq\lambda_{min}^{-1}(Q)V_x(0).$$

The last inequality implies asymptotic stability of $x(t)$. This completes the proof. □

Remark 4.1 *The Riccati equation (4.1) $\sim$ (4.7) has the same form as that arises in linear quadratic control theory for state delayed systems [8][11]. An efficient computational procedure for solving the Riccati equation can be found, for example, in [11].*

Regarding the merits of both controller designs, one can observe that the indirect design is much computationally simpler than the direct design. This stems from the fact that the former does not involve solving an infinite dimensional Riccati equation while the latter does. Nevertheless, one merit of the direct design is that it can be easily modified to stabilize a larger class of state delayed systems. To show this, we modify the original system (S_d) to the following uncertain state delayed system of the form $(\tilde{S}_d)$

$$\begin{aligned}\dot{x}(t)\;=\;&[A+\Delta A(t)]x(t)+[B+\Delta B(t)]u(t)\\&+[A_h+\Delta A_h(t)]x(t-h)\\&+Bw(t)+Bf(x(t-r))\,,\end{aligned} \qquad (4.17)$$

where

$f(x(t-r))\in\mathcal{R}^l$ is the state-dependent additive disturbance satisfying
$\|f(x(t-r))\|\leq\zeta\|x(t-r)\|$ for some finite $\zeta>0$
$r\in\mathcal{R}^+$ is an *unknown* but constant delay time.

Note that the modified system $(\tilde{S}_d)$ cannot be converted back into the form (S_d) since the delay time r is allowed to be unknown. The problem of stabilizing this class of uncertain systems seems to be intractable via the the indirect approach used in the previous section. In the following corollary, it is shown that the problem is still tractable by the direct approach.

Corollary 4.1 *Suppose that the system $(\tilde{S}_d)$ satisfies Assumption 1 and 2. Then the control law proposed in Theorem 4.1 also stabilizes the system $(\tilde{S}_d)$ for all delay time $r<\infty$ if Q in (4.1) and μ in (4.13) are chosen so that*

$$Q>\zeta^2I \qquad (4.18)$$

and

$$\begin{aligned}\mu(x_t) \;>\; &\Big[\|y(t)\|+\|H(t)\|\|x(t)\|\\&+\|H_h(t)\|\|x(t-h)\|+\|w(t)\|\Big]\,/\,\delta\end{aligned} \qquad (4.19)$$

Proof: The result can be proved by taking the positive definite functional

$$\tilde{V}_x(t)=V_x(t)+\zeta^2\int_{t-r}^{t}x^T(\tau)x(\tau)d\tau, \qquad (4.20)$$

where $V_x(t)$ is defined in (4.15), as a Lyapunov functional to be admitted by the resulting closed-loop system. Note here that the idea of adding the second term of the right-hand side of (4.20) has already been proposed in many early researchs; e.g., see [13][15][16]. The derivative of (4.20) is then given by

$$\dot{\bar{V}}_x(t) = \dot{V}_x(t) + \zeta^2\|x(t)\|^2 - \zeta^2\|x(t-r)\|^2, \tag{4.21}$$

Combine the Riccati equation (4.1) $\sim$ (4.7) with (4.21) to get

$$\begin{aligned}\dot{\bar{V}}_x(t) &= -x^T(t)Qx(t) + y^T(t)y(t) \\ &\quad + 2y^T(t)(u(t)+v(t)) \\ &\quad + 2y^T(t)f(x(t-r)) + \zeta^2\|x(t)\|^2 \\ &\quad - \zeta^2\|x(t-r)\|^2 \\ &\leq -x^T(t)[Q-\zeta^2 I]x(t) + 2y^T(t)y(t) \\ &\quad + 2y^T(t)(u(t)+v(t)) \\ &\leq -x^T(t)[Q-\zeta^2 I]x(t)\end{aligned}$$

The last inequality is obtained using the control law (4.9) $\sim$ (4.12) with $\mu(\cdot)$ satisfying (4.19). Furthermore, the above inequality, in conjunction with (4.18), yields

$$\dot{\bar{V}}_x(t) \leq -\sigma\|x(t)\|^2 \tag{4.22}$$

for any positive scalar σ such that $\sigma < \lambda_{min}(Q-\zeta^2 I)$. The above analysis shows that the proposed controller renders the closed-loop system asymptotically stable. □

5 Application to Model-following Control

In this section, we apply the results obtained in the previous sections to model following control for the system (S_d). Let a reference model be given by

$$\dot{x}_m(t) = A_m x_m(t) + A_{mh}x_m(t-h) + B_m u_m(t) \tag{5.1}$$

where u_m is a command signal. Next, define the tracking error state

$$e(t) = x(t) - x_m(t). \tag{5.2}$$

Then

$$\begin{aligned}\dot{e}(t) &= A_m e(t) + A_{mh}e(t-h) \\ &\quad + [B+\Delta B(t)]u(t) - B_m u_m(t) \\ &\quad + [A - A_m + \Delta A(t)]x(t) \\ &\quad + [A_h - A_{mh} + \Delta A_h(t)]x(t-h) \\ &\quad + Bw(t).\end{aligned} \tag{5.3}$$

In addition to Assumption 1 and 2 , we make the following assumptions

Assumption 3: The reference model (5.1) is asymptotically stable; i.e., all poles of (5.1) in absence of u_m lie within the open left-half plane.
Assumption 4: There exist constant matrices H_m, H_{mh} and E_m of appropriate dimensions such that

a) $A - A_m = BH_m$,
b) $A_h - A_{mh} = BH_{mh}$,
c) $B - B_m = BE_m$.

Now let the control function $u(t)$ be given by

$$u(t) = u_e(t) + u_m(t)\ , \tag{5.4}$$

where $u_e(t)$ is yet to be defined. By using (5.4), in conjunction with Assumption 1, 2, 3 and 4, (5.3) can be rewritten into the following form

$$\dot{e}(t) = A_m e(t) + A_{mh}e(t-h) + B(u_e(t) + v_e(t))\ , \tag{5.5}$$

where

$$\begin{aligned}v_e(t) &= [H(t) + H_m]x(t) + [H_h(t) + H_{mh}]x(t-h) \\ &\quad + [E(t) + E_m]u_m(t) + E(t)u_e(t) + w(t)\end{aligned}$$

Note here that asymptotic stability of $e(t)$ implies that $x(t)$ is asymptotically tracks $x_m(t)$ for all bounded command signal $u_m(t)$. Consequently, the problem can be solved by finding $u_e(t)$ which robustly stabilizes (5.5) against $v_e(t)$. Clearly, both Theorem 3.1 and Theorem 4.1 can now be applied to solve this stabilization problem. For instance, applying the indirect design in Section 3, we first define the transformation

$$z_e(t) = e(t) + \int_{-h}^{0} e^{A_c\theta}A_{mh}e(t-h-\theta)d\theta\ , \tag{5.6}$$

where

$$A_c = A_m + e^{-hA_c}A_{mh}, \tag{5.7}$$

to get the system

$$\dot{z}_e(t) = A_c z_e(t) + B(u_e(t) + v_e(t)) \tag{5.8}$$

Remark 5.1 *Since we have assumed that no unstable poles in the reference model, the process of finding a suitable transformation of the form (5.6) becomes much simpler; i.e., any real $A_c \in \mathcal{R}^{n\times n}$ which satisfies (5.7) is usable.*

The final step is to apply the Lyapunov min-max control [4]. It can be shown that a suitable control law is given by (5.4) with

$$u_e(t) = -\rho(e_t)\frac{B^T P z_e(t)}{\|B^T P z_e(t)\|}; \qquad \forall z_e(t) \notin \Psi_{z_e} \tag{5.9}$$

$$u_e(t) = \{u \in \mathcal{R}^m; \|u\| \leq \rho(e_t)\}; \qquad \forall z_e(t) \in \Psi_{z_e} \tag{5.10}$$

where

$$\Psi_{z_e} = \left\{z \in \mathcal{R}^n; B^T P z = 0\right\}, \tag{5.11}$$

$$\begin{aligned}\rho(e_t) > \Big[&\|H(t) + H_m\|\|x(t)\| \\ &+ \|H_h(t) + H_{mh}\|\|x(t-h)\| \\ &+ \|E(t) + E_m\|\|u_m(t)\| + \|w(t)\| \Big] / \delta\end{aligned} \tag{5.12}$$

and δ is the positive scalar defined in Assumption 2-d, and $P > 0$ is a solution to the Riccati equation

$$A_c^T P + PA_c + I = 0. \tag{5.13}$$

6 Concluding Remarks

We have proposed two methods of robust controller design for achieving stability of uncertain state delayed systems. When the nominal system is spectrally stabilizable and its uncertain parts satisfy certain matching conditions, a formula for constructing the stabilizing state feedback law can be derived. The indirect design is achieved by first computing all unstable poles of the nominal system to find a suitable transformation of the form (3.1). Then a standard finite dimensional technique is employed to derive a stabilizing control law. In contrast with the indirect design, the direct design is made possible by solving an infinite dimensional Riccati equation which requires more computational effort than the indirect one. Nevertheless, the direct design can be easily modified to cope with a larger class of uncertainty.

The common shortcomings of both controller designs are that the resulting control laws are generally discontinuous; therefore, a solution to the closed-loop equation may not exist in the usual sense. Moreover, such discontinuous control laws cannot be directly implemented owing to the problem of unmodeled dynamics excitation. To overcome this difficulty, it is possible to apply various existing techniques proposed in early works such as [3],[4] and [12] to modify the control laws. For example, if the control law (4.9) $\sim$ (4.10) is replaced by the continuous control law

$$u(t) = -\frac{\mu^2(x_t)y(t)}{\mu(x_t)\|y(t)\| + \epsilon} \tag{6.1}$$

for some $\epsilon > 0$. Then it can be verified that the inequality (4.16) becomes

$$\dot{V}_x(t) \leq -\lambda_{min}(Q)\|x(t)\|^2 + \frac{\mu(x_t)\|y(t)\|\epsilon}{\mu(x_t)\|y(t)\| + \epsilon}, \tag{6.2}$$

which implies that

$$\dot{V}_x(t) \leq -\lambda_{min}(Q)\|x(t)\|^2 + \epsilon \tag{6.3}$$

Although asymptotically stability of $x(t)$ cannot be guaranteed by this continuous control law, uniform ultimate boundedness of $x(t)$ is still obtained with arbitrarily small ultimate bound provided that ϵ is sufficiently small; this can be verified by utilizing the results given in [14] and [16].

Another area for future research would involve the extension to allow for uncertain systems with distributed state delay. For the indirect design case, it is expected that this extension can be achieved via the modified transformation proposed in [7]. Also, one might combine the results given here with the state transformation of exponential type used in [17] to derive a stabilizing control law with specified degree of stability.

References

[1] A.S.I. Zinober, *Deterministic Control of Uncertain Systems*, IEE Control Engineering Series 40, Peter Peregrinus, 1990.

[2] M. Corless, G. Leitmann and E.P. Ryan, "Tracking in the presence of bounded uncertainties", *Fourth International Conference on Control Theory*, Cambridge, UK,Sept,1984.

[3] S. Gutman and Z.J. Palmor, "Properties of min max controller in uncertain dynamicals systems," *SIAM J. Contr. Optimiz.*, vol.20, pp.850-861, 1982.

[4] E. Cheres, S. Gutman and Z.J. Palmor, "Stabilization of uncertain dynamical systems including state delay," *IEEE Trans. Automat. Contr.*, vol.AC-34, no.11, pp.1199-1203, 1989.

[5] S. Gutman, "Uncertain dynamical systems- A Lyapunov min max approach," *IEEE Trans. Automat. Contr.*, vol.AC-24, no.6, pp.437-443, 1979.

[6] Y.A. Fiagbezi and A.E. Pearson, "Feedback stabilization of linear autonomous time lag systems," *IEEE Trans. Automat. Contr.*, vol.AC-31, no.9, pp.847-855, 1986.

[7] Y.A. Fiagbezi and A.E. Pearson, "A multistage reduction technique for feedback stabilizing distributed time-lag systems," *Automatica*, vol.23, no.3, pp.311-326, 1987.

[8] M.C. Delfour, C. MaCalla and S.K. Mitter, "Stability and the infinite-time quadratic cost problem for linear hereditary differential systems," *SIAM J. Contr. Optimiz.*, vol.13, no.1, pp.48-88, 1975.

[9] A.W. Olbrot, "Stabilizability, detectability and spectrum assignment for linear autonomous systems with general time delays," *IEEE Trans. Automat. Contr.*, vol.AC-23, pp.887-890, 1978.

[10] I.S. Sokolnikoff and R.M. Redheffer, *Mathematics of Physics and Modern Engineering*, McGraw-hill, 1966.

[11] J.S. Gibson, "Linear quadratic optimal control of hereditary differential systems: infinite dimansional Riccati equations and numerical approximation," *SIAM J. Contr. Optimiz.*, vol.21, no.1, pp.95-139, 1983.

[12] A. Thowsen, "Uniform ultimate boundedness of the solutions of uncertain dynamic delay systems with state-dependent and memoryless feedback control," *Int. J. Contr.*, vol.37, no.5, pp.1135-1143, 1983.

[13] T.A. Burton, *Stability and Periodic solutions of Ordinary and Functional Differential Equations.* Academic Press, New York, 1985.

[14] T.A. Burton and S. Zhang, "Uniform boundedness, periodicity, and stability in ordinary and functional differential equations," *Annali di Matematica Pura ed Applicata* , vol.145, pp.129-158, 1986.

[15] S. Phoojaruenchanachai and K. Furuta, "Memoryless stabilization of uncertain linear systems including time-varying state delays", *IEEE Trans. Automat. Contr.*, vol.AC-37, no.7, pp.1022-1026, 1992.

[16] J.E. Gayek, "Lyapunov functional approach to uncertain systems governed by functional differential equations with finite time-lag," *Control and Dynamic Systems* , vol.35, part 2, Academic Press, 1990.

[17] A. Thowsen, "Stabilization of a class of linear time delay systems ," *Int. J. Syst. Sci.*, vol.12, no.12, pp.1485-1492, 1981.

Proceedings of the 31st Conference
on Decision and Control
Tucson, Arizona • December 1992

OUTPUT FEEDBACK CONTROL OF SAMPLED–DATA SYSTEMS WITH PARAMETRIC UNCERTAINTIES

Peng Shi[1], Lihua Xie[2] and Carlos E. de Souza[1]

1. Department of Electrical and Computer Engineering
University of Newcastle, NSW 2308
Australia

2. School of Electrical and Electronic Engineering
Nanyang Technological University, 2263
Singapore

Abstract

This paper is concerned with the design of robust control for linear sampled–data systems with parametric uncertainty. Interest is focused on linear dynamic output feedback controllers and two problems are addressed. The first one is the robust stabilization whereas the other is H_∞ synthesis in which both robust stability and robust performance in an H_∞ sense are required to be achieved. A technique is proposed for designing stabilizing controllers for both problems by converting them into H_∞ synthesis for sampled–data systems without parameter uncertainties.

1. Introduction

Robust control of uncertain dynamical systems has been a subject of recurring interest in the past few years. Considerable attention has been paid to both the problems of robust stabilization and robust performance of uncertain continuous–time and discrete–time dynamical systems; see, e.g. ([2], [4], [10] and the references therein). Therefore, the topic of robust control of uncertain sampled–data systems has also attracted a lot of interest. Analysis and synthesis of linear sampled–data systems with parametric uncertainty has been tackled in ([1], [3], [5] and [6]).

In this paper we consider the robust control of a class of uncertain linear systems under sampled output measurements. The class of uncertain linear systems is described by an uncertain state space model with time–varying norm–bounded parameter uncertainties in both the state and the input matrices. Here attention is focused on the design of linear dynamic controllers using sampled output measurements and two problems are addressed. The first one is the robust stabilization by dynamic feedback of sampled output measurements and the other is the problem of robust H_∞ synthesis in which both robust stability and robust performance in the face of parameter uncertainty are required to be achieved. In the latter problem both the cases of finite and infinite horizon will be considered. We show that both the above problems can be recast into H_∞ synthesis ones for sampled–data systems without parameter uncertainties.

Notation. Throughout the paper $L_2[0,T]$ (respectively, $\ell_2(0,T)$) stands for the space of square integable functions (respectively, square summable sequences) over the interval $[0,T]$ (respectively, $(0,T)$). $\|\cdot\|$ will refer to the Euclidian vector norm whereas $\|\cdot\|_{[0,T]}$ denotes the usual $L_2[0,T]$ norm over $[0,T]$ and $\|\cdot\|_{(0,T)}$ is the usual $\ell_2[0,T]$ norm over $(0,T)$.

2. Problem Formulation And Preliminaries

Consider uncertain linear time–varying systems described by a state space model of the form

$$(\Sigma_1): \dot{x}(t) = [A + \Delta A]x(t) + B_1 w(t) + [B + \Delta B]u(t)\,, x(0) = x_0 \quad (2.1a)$$

$$z(t) = C_1 x(t) + D_1 u(t)\,, \quad t \in (0,T) \quad (2.1b)$$

$$y(ih) = C_2 x(ih) + D_2 v(ih)\,, ih \in (0,T) \quad (2.1c)$$

where $x(t) \in \mathbb{R}^n$ is the state, x_0 is an unknown initial state, $u(t) \in \mathbb{R}^m$ is the control input, $w(t) \in \mathbb{R}^q$ is the disturbance input which is from $L_2[0,T]$, $v(ih) \in \mathbb{R}^\ell$ is the measurement noise which belongs to $\ell_2(0,T)$, $y(ih) \in \mathbb{R}^r$ is the sampled output measurement, i is an integer, h is the sampling period, $z(t) \in \mathbb{R}^p$ is the controlled output, A, B, B_1, C_1, C_2, D_1 , and D_2 are known real matrices of appropriate dimensions that describe the nominal system and ΔA and ΔB are real–valued uncertain matrix functions of the form

$$[\Delta A \quad \Delta B] = HF[E_1 \quad E_2] \quad (2.2)$$

CH3229-2/92/0000-2876$1.00 © 1992 IEEE

where $H \in \mathbb{R}^{n\times i}$, $E_1 \in \mathbb{R}^{j\times n}$ and $E_2 \in \mathbb{R}^{j\times m}$ are known real matrices and $F \in \mathbb{R}^{i\times j}$ is an unknown matrix function satisfying

$$F^T(t)F(t) \le I, \quad \forall t \tag{2.3}$$

with the elements of F being Lebesgue measurable. Note that all the matrices in (2.1) are allowed to be time–varying. For the sake of notation simplification, we will omit the dependence on t or ih for all the matrices throughout the paper.

Remark 2.1. Note that the uncertainty matrix F is allowed to be state dependent as long as (2.3) is satisfied. Also, any possible parameter uncertaintyin B_1 and D_2 are assumed to absorbed in w and v. □

In relation to the system (2.1), in this paper we first consider the problem of robust stabilization via dynamic feedback of sampled output measurements. More specifically, setting $w(t) \equiv 0$ we are concerned with *designing a dynamic feedback control law based on the sampled output measurements of (2.1), i.e.* $u(t) = \mathcal{G}\{y(ih),\ ih \in [0,\infty)\}$, *such that the closed–loop system is exponentially stable for all admissible uncertainties.* In this situation, we say that the system (2.1) is *robustly stabilizable by sampled output feedback* and the closed–loop system is *robustly stable.*

We will also investigate the design of a linear causal sampled–data controller ($\mathcal{G}$) for (2.1) that reduces z uniformly for any w, v and x_0 in the sense that given a scalar $\gamma > 0$, the closed-loop system of (2.1) with the controller ($\mathcal{G}$) satisfies

$$\left[\| z \|^2_{[0,T]} + x^T(T)Sx(T)\right]^{\frac{1}{2}} < \gamma\left[\| w \|^2_{[0,T]} + \| v \|^2_{(0,T)} + x_0^T R x_0\right]^{\frac{1}{2}} \tag{2.4}$$

for any non-zero (w, v, x_0) and for all admissible parameter uncertainties, where $R = R^T > 0$, $S = S^T \ge 0$ are given weighting matrices for x_0 and $x(T)$, resp^ectively. In this situation, the closed-loop system of (2.1) with ($\mathcal{G}$) is said to have robust H_∞ performance γ.

In the case when the initial state of (2.1) is known to be zero and infinite horizon is considered, i.e. $T = \infty$, (2.4) will be replaced by

$$\| z \|_{[0,T]} < \gamma\left[\| w \|^2_{[0,T]} + \| v \|^2_{(0,T)}\right]^{\frac{1}{2}}. \tag{2.5}$$

It should be noted that (2.5) can be viewed as the limit of (2.4) as the smallest eigenvalue of R approaches infinity.

The robust performance problem we address in this paper is as follows: *Given a scalar $\gamma > 0$, design a causal controller $\mathcal{G}$ based on the sampled measurements, $y(ih)$, such that:*

- *In the finite horizon case, the closed–loop system of (2.1) with $\mathcal{G}$ has robust H_∞ performance γ;*
- *In the infinite horizon case, i.e. $T = \infty$, the closed loop system of (2.1) with $\mathcal{G}$ is exponentially stable and has robust H_∞ performance γ.*

Note that as (2.1a,b) is in the continuous–time, the controller $\mathcal{G}$ should generate a continuous–time control input u using the sampled output measurements.

In the reminder of this section we shall introduce a definition related to linear time–varying systems with discrete jumps of the form (see also [8]):

$$\dot{x}(t) = Ax(t) + Bu(t),\ t \ne ih \tag{2.6a}$$

$$x(ih) = A_d x(ih^-) + B_d u_d(ih) \tag{2.6b}$$

$$y(t) = Cx(t) \tag{2.6c}$$

$$y_d(ih) = C_d x(ih) \tag{2.6d}$$

where the time–varying matrices A, B, C, A_d, B_d and C_d are assumed to be bounded.

Note that both continuous–time and discrete-time linear systems are special cases of the linear system (2.6) with $A_d = I$, $B_d = 0$ and $C_d(ih) = C(ih)$ and $A = 0$, $B = 0$ and $C = 0$, respectively.

The stability of the unforced system of (2.6) (setting $u(t) \equiv 0$ and $u(ih) \equiv 0$) is equivalent to that of the following time–varying discrete–time system

$$x(ih) = \Phi_A^{A_d}(ih)x[(i-1)h] \tag{2.7}$$

where

$$\Phi_A^{A_d}(ih) = A_d(ih)\Phi(ih, ih-h) \tag{2.8}$$

with $\Phi(t,s)$ being the transition matrix of $A(t)$.

Definition 2.1([8]). The system (2.6) or $(A,\ C,\ A_d,\ C_d)$ is said to be detectable if there exist bounded matrix functions $L(t)$ and $L_d(ih)$ such that $\Phi_{A+LC}^{A_d+L_dC_d}(ih)$ is exponentially stable.

3. Bounded Real Lemma For Linear Systems With Finite Discrete Jumps

In this section we will develop the bounded real lemma for linear time-varying systems with finite discrete jumps which will be fundamental in the derivation of our main results.

Consider the linear time-varying system with finite discrete jumps:

$$(\Sigma_2): \dot{x}(t) = Ax(t) + Bw(t), \ t \in [ih, ih+h); \quad x(0) = x_0 \qquad (3.1a)$$

$$x(ih) = A_d x(ih^-) + B_d v(ih), ih \in (0,T) \qquad (3.1b)$$

$$z(t) = Cx(t) \qquad (3.1c)$$

where $x \in \mathbb{R}^n$ is the state, x_0 is an unknown initial state, $w(t) \in \mathbb{R}^p$ and $v(ih) \in \mathbb{R}^q$ are the disturbance inputs which belongs to $L_2[0,T)$ and $\ell_2(0,T)$, respectively, $z \in \mathbb{R}^m$ is the controlled output, A, B, C, A_d and B_d are real time-varying bounded matrices of appropriate dimensions with A, B and C being piecewise continuous.

Next, motivated by [8], we introduce the following H_∞ – like performance for the system (3.1):

$$J(\Sigma_2, R, S, T) = \sup_{0 \neq (w,v,x_0)} \left[\frac{\| z \|^2_{[0,T]} + x^T(T)Sx(T)}{\| w \|^2_{[0,T]} + \| v \|^2_{(0,T)} + x_0^T R x_0} \right]^{\frac{1}{2}} \qquad (3.2a)$$

where $R = R^T > 0$ and $S = S^T \geq 0$ are given weighting matrices for x_0 and $x(T)$, respectively.

When the initial state is known to be zero and the infinite horizon case is considered, (3.2a) is replaced by

$$J_0(\Sigma_2) = \sup_{0 \neq (w,v)} \left[\frac{\| z \|^2_{[0,\infty)}}{\| w \|^2_{[0,\infty)} + \| v \|^2_{(0,\infty)}} \right]^{\frac{1}{2}} \qquad (3.2b)$$

We now present a version of bounded real lemma in finite horizon for systems of the form (3.1). Due to the space limination, all the proofs of the results will be omitted here. They can be found in the full version of this paper[7].

Theorem 3.1 ([7]). Consider the system (3.1) and let $\gamma > 0$ be a given scalar. Then the following statements are equivalent:

(a) $J(\Sigma_2, R, S, T) < \gamma$;

(b) There exists a matrix function $P(t) = P^T(t) \geq 0$, $t \in [0,T]$ such that

$$-\dot{P} = A^TP + PA + \gamma^{-2}PBB^TP + C^TC, \quad \forall t \in (ih, ih+h] \qquad (3.3a)$$

$$P(T) = S \qquad (3.3b)$$

$$\gamma^2 I - B_d^T P(ih^+) B_d > 0 \qquad (3.4a)$$

$$P(ih) = A_d^T P(ih^+) A_d + A_d^T P(ih^+) B_d \cdot [\gamma^2 I - B_d^T P(ih^+) B_d]^{-1} B_d^T P(ih^+) A_d \qquad (3.4b)$$

$$P(0) < \gamma^2 R \qquad (3.5)$$

(c) There exists a matrix function $Q(t) = Q^T(t) > 0$, $t \in [0,T)$ such that

$$\dot{Q} + A^TQ + QA + \gamma^{-2}QBB^TQ + C^TC < 0, \quad \forall t \in (ih, ih+h] \qquad (3.6a)$$

$$Q(T) = S \qquad (3.6b)$$

$$\gamma^2 I - B_d^T Q(ih^+) B_d > 0 \qquad (3.7a)$$

$$Q(ih) > A_d^T Q(ih^+) A_d + A_d^T Q(ih^+) B_d \cdot [\gamma^2 I - B_d^T Q(ih^+) B_d]^{-1} B_d^T Q(ih^+) A_d \qquad (3.7b)$$

$$Q(0) < \gamma^2 R. \qquad (3.8)$$

In the following we extend the previous theorem to the case of infinite horizon.

Theorem 3.2 ([7]). Consider the system (3.1) and let $\gamma > 0$ be a given scalar. Then the following statements are equivalent:

(a) The system (3.1) is exponentially stable and $J(\Sigma_2, R, 0, \infty) < \gamma$;

(b) There exists a bounded matrix function $P(t) = P^T(t) \geq 0$, $t \in [0, \infty)$ such that

$$-\dot{P} = A^TP + PA + \gamma^{-2}PBB^TP + C^TC, \quad \forall t \in (ih, ih+h] \qquad (3.9)$$

$$\gamma^2 I - B_d^T P(ih^+) B_d > 0 \qquad (3.10)$$

$$P(ih) = A_d^T P(ih^+) A_d + A_d^T P(ih^+) B_d \cdot [\gamma^2 I - B_d^T P(ih^+) B_d]^{-1} B_d^T P(ih^+) A_d \qquad (3.11)$$

$$P(0) < \gamma^2 R \qquad (3.12)$$

and the system

$$\dot{\eta}(t) = [A + \gamma^{-2}BB^TP(t)]\eta(t), \ t \neq ih \qquad (3.13a)$$

$$\eta(ih^+) = \Big\{A_d + B_d\big[\gamma^2 I - B_d^T P(ih^+) B_d\big]^{-1} \cdot B_d^T P(ih^+) A_d \Big\} \eta(ih) \quad (3.13b)$$

is exponentially stable;

(c) There exists a bounded matrix function $Q(t) = Q^T(t) > 0$, $t \in [0, \infty)$ such that

$$\dot{Q} + A^T Q + QA + \gamma^{-2} QBB^T Q + C^T C < 0 \quad \forall t \in (ih, ih + h] \quad (3.14)$$

$$\gamma^2 I - B_d^T Q(ih^+) B_d > 0 \quad (3.15)$$

$$Q(ih) > A_d^T Q(ih^+) A_d + A_d^T Q(ih^+) B_d \cdot [\gamma^2 I - B_d^T Q(ih^+) B_d]^{-1} B_d^T Q(ih^+) A_d \quad (3.16)$$

$$Q(0) < \gamma^2 R. \quad (3.17)$$

Remark 3.1. Theorems 3.1 and 3.2 establish the Bounded Real Lemma for linear systems with finite discrete jumps over both finite and infinite horizon, respectively. Note that when the initial state of (3.1) is known to be zero, (3.12) and (3.17) will no longer be required. ❑

4. Robust Stabilization

This section investigates the robust stabilization of sampled–data systems with parameter uncertainty. We will show that this problem can be converted to an H_∞ synthesis for sampled–data systems which do not involve parameter uncertainties.

First, we consider the problem of robust stability analysis for a class of linear systems with finite discrete jumps. The motivation for this is that, as it will be shown latter, the robust stability of the closed–loop system of (2.1) with a linear sampled–data controller can be recast into the robust stability of an uncertain linear systems with finite discrete jumps of the form:

$$(\Sigma_3): \dot{x}(t) = [A + \Delta A]x(t), \; t \in [ih, ih + h) \quad (4.1a)$$

$$x(ih) = A_d x(ih^-), \; ih \in (0, \infty) \quad (4.1b)$$

where $x \in \mathbb{R}^n$ is the state, A and A_d are real bounded matrices with A being piecewise continuous and ΔA is as in (2.2).

In order to solve the stability problem of (4.1), we introduce the following system associated with (4.1):

$$(\Sigma_3^a): \dot{x}(t) = Ax(t) + Hw(t), \quad t \in [ih, ih + h); \; x(0) = 0 \quad (4.2a)$$

$$x(ih) = A_d x(ih^-), \; ih \in (0, \infty) \quad (4.2b)$$

$$z(t) = E_1 x(t) \quad (4.2c)$$

where $x \in \mathbb{R}^n$ is the state, $w \in \mathbb{R}^i$ is the disturbance input, $z \in \mathbb{R}^j$ is the controlled output and A, A_d, E_1 and H are the same as in (4.1).

Then, we have the following result.

Theorem 4.1. The system (Σ_3) is exponentially stable for all admissible uncertainties if the system (Σ_3^a) is exponentially stable and satisfies $J_0(\Sigma_3^a) < 1$.

Remark 4.1. Theorem 4.1 implies that the stability analysis of uncertain linear systems of the form (4.1) can be recast into an H_∞ analysis problem for systems without parameter uncertainty and thus, by Theorem 3.1, it can be performed in terms of Riccati equations. ❑

In connection with the robust stabilization problem of (2.1) (setting $w(t) \equiv 0$ and $v(ih) \equiv 0$), we introduce an auxiliary system as below:

$$\dot{x}(t) = Ax(t) + H\tilde{w}(t) + Bu(t), \quad t \in [ih, ih + h); \; x(0) = 0 \quad (4.3a)$$

$$\tilde{z}(t) = E_1 x(t) + E_2 u(t) \quad (4.3b)$$

$$y(ih) = C_2 x(ih), \quad ih \in (0, \infty) \quad (4.3c)$$

where $x \in \mathbb{R}^n$ is the state, $u \in \mathbb{R}^m$ is the control input, $\tilde{w} \in \mathbb{R}^i$ is the disturbance input, $\tilde{z} \in \mathbb{R}^{p+j}$ is the controlled output $y \in \mathbb{R}^r$ is the sampled measurement and all the matrices are the same as in (2.1).

Motivated by results in [8], we consider the following form of controller for (2.1) and (4.3)

$$\dot{x}_c(t) = F_c x_c(t), \; t \in [ih, ih + h); \; x_c(0) = 0 \quad (4.4a)$$

$$x_c(ih) = F_d x_c(ih^-) + G_d y(ih), \; ih \in (0, \infty) \quad (4.4b)$$

$$u(t) = K x_c(t) \quad (4.4c)$$

where the dimension of the controller and the time–varying matrices F_c, F_d, G_d and K are to be chosen. Note that (4.4) can be viewed as a sampled-data controller with a generalized hold function. Observe that a controller with a zero-order hold is a special case of (4.4) with $F_c = 0$ and K being a constant matrix.

Theorem 4.2. The system (2.1) with $w(t) \equiv 0$ is robustly stabilizable for all admissible uncertainties by a controller (4.4) if the closed-loop system (Σ_{cl}) of (4.3) with (4.4) is exponentially stable and satisfies $J_0(\Sigma_{cl}) < 1$.

Remark 4.2. Theorem 4.2 shows that the sampled-data robust stabilization of (2.1) can be converted into an H_∞ control problem for sampled-data systems without parameter uncertainty. Hence, existing results on sampled-data H_∞ control such as those in [8, 9] can be used to solve the robust stabilization of (2.1). More specifically, a solution can be obtained in terms of Riccati differential and difference equations. □

5. H∞ Synthesis

In this section, the robust stabilization technique discussed in Section 4 will be extended to deal with the H_∞ synthesis problem for the uncertain system (2.1) to achieve both robust stability and robust H_∞ performance in the face of parameter uncertainties.

In order to solve the robust H_∞ synthesis for (2.1), we first consider a related problem which is that of robust H_∞ analysis for uncertain linear systems with finite discrete jumps. To this end, consider the system (4.1) augmented with a controlled output z and bounded energy disturbance inputs w and v, i.e.

$$(\Sigma_4): \dot{x}(t) = [A + \Delta A(t)]x(t) + Bw(t), \quad t \in [ih,\ ih + h);\ x(0) = x_0 \quad (5.1a)$$

$$x(ih) = A_d x(ih^-) + B_d v(ih),\ ih \in (0, T) \quad (5.1b)$$

$$z(t) = Cx(t) \quad (5.1c)$$

where x_0 is an unknown initial state, $w \in \mathbb{R}^q$ and $v \in \mathbb{R}^a$ are the disturbance inputs, $z \in \mathbb{R}^p$ is the controlled output, B and B_d are known real bounded matrices with B being picewise continuous and all the other matrices are the same as in (4.1).

Next, introduce the following scaled system associated with (5.1):

$$(\Sigma_4^a): \dot{x}(t) = Ax(t) + [\gamma H/\varepsilon \quad B]\tilde{w}(t), \quad t \in [ih,\ ih + h);\ x(0) = \tilde{x}_0 \quad (5.2a)$$

$$x(ih) = A_d x(ih^-) + B_d v(ih),\ ih \in (0, T) \quad (5.2b)$$

$$\tilde{z}(t) = \begin{bmatrix} \varepsilon E_1^T & C^T \end{bmatrix}^T x(t) \quad (5.2c)$$

where $x \in \mathbb{R}^n$ is the state, $\tilde{x}_0$ is an unknown initial state, $\tilde{w} \in \mathbb{R}^{q+i}$ and $v \in \mathbb{R}^a$ are the disturbance inputs, $z \in \mathbb{R}^{p+j}$ is the controlled output, $\gamma > 0$ is the desired H_∞ performance, $\varepsilon > 0$ is a scaling parameter to be chosen and all the matrices are the same as in (5.1).

Our first result deals with the H_∞ performance analysis for (5.1) over finite horizon.

Theorem 5.1. Given a scalar $\gamma > 0$, the system (Σ_4) has robust H_∞ performance γ over $[0, T]$ if there exists an $\varepsilon > 0$ such that the system (Σ_4^a) satisfies $J(\Sigma_4^a, R, S, T) < \gamma$.

Theorem 5.1 can be extended to the case of infinite horizon ($T \to \infty$).

Theorem 5.2. Given a scalar $\gamma > 0$, the system (Σ_4) is exponentially stable and has robust H_∞ performance γ over $[0, \infty)$ if there exists an $\varepsilon > 0$ such that system (Σ_4^a) is exponentially stable and satisfies $J(\Sigma_4^a, R, 0, \infty) < \gamma$.

Remark 5.1. In view of Theorems 5.1 and 5.2 the H_∞ analysis problem of an uncertain linear system with finite discrete jumps of the form (5.1) can be converted into a scaled H_∞ analysis for system (5.2) which involves no parameter uncertainty and thus, by Theorem 3.1, can be carried out by Riccati equations. □

Similar to the robust stabilization problem of Section 4, in connection with the robust H_∞ control of system (2.1) we introduce the following scaled linear sampled-data system associated with (2.1):

$$\dot{x}(t) = Ax(t) + [\gamma H/\varepsilon \quad B_1]\hat{w}(t) + Bu(t), \quad x(0) = \tilde{x}_0 \quad (5.3a)$$

$$\hat{z}(t) = \begin{bmatrix} \varepsilon E_1 \\ C_1 \end{bmatrix} x(t) + \begin{bmatrix} \varepsilon E_2 \\ D_1 \end{bmatrix} u(t) \quad (5.3b)$$

$$y(ih) = C_2 x(ih) + D_2 \hat{v}(ih),\ ih \in (0,\ T) \quad (5.3c)$$

where $x \in \mathbb{R}^n$ is the state, $\hat{w} \in \mathbb{R}^{q+i}$ is the disturbance input, $\hat{z} \in \mathbb{R}^{p+j}$ is the controlled output, $y \in \mathbb{R}^r$ is the sampled measurement, $\hat{v} \in \mathbb{R}^\ell$ is the measurement noise, $\gamma > 0$ is the prescribed H_∞ performance we wish to achieve for the system (2.1), $\varepsilon > 0$ is a scaling parameter to be chosen and all the matrices are the same as in (2.1).

The next theorem provides an H_∞ synthesis result for (2.1) in the finite horizon case.

Theorem 5.3. Let $\gamma > 0$ be a given scalar and a controller $\mathcal{G}$ be given in the form of (4.4). Then, the closed-loop system of (2.1) with $\mathcal{G}$ has robust H_∞ performance γ over $[0, T]$ if there exists an $\varepsilon > 0$ such that the closed-loop system, (Σ_{cl}), of (5.3) with $\mathcal{G}$ satisfies $J(\Sigma_{cl}, R, S, T) < \gamma$.

The next Theorem deals with the robust H_∞ control of (2.1) over infinite horizon.

Theorem 5.4. Let $\gamma > 0$ be a given scalar and a controller $\mathcal{G}$ be given in the form of (4.4). Then, the closed-loop system of (2.1) with $\mathcal{G}$ is robustly stable and possesses robust H_∞ performance γ over $[0, \infty)$ if there exists an $\varepsilon > 0$ such that the closed-loop system, (Σ_{cl}), of (5.3) with $\mathcal{G}$ is exponentially stable and satisfies $J(\Sigma_{cl}, R, 0, \infty) < \gamma$.

Remark 5.2. Theorems 5.3 and 5.4 show that the robust H_∞ synthesis for (2.1) can be converted into a scaled H_∞ control problem for a sampled-data system without parameter uncertainty. Hence, the corresponding scaled H_∞ synthesis problem can be solved via existing techniques such as those in [8, 9], which involve solving Riccati differential and difference equations. ❑

6. Conclusion

This paper has considered the design of output feedback control for a class of sampled-data systems with parameter uncertainties. Both the problems of robust stabilization and robust H_∞ performance have been tackled. It has been shown that the above problems can be converted to H_∞ synthesis for sampled-data systems without parameter uncertainties.

References

[1] D.S. Berstein and C.V. Hollot, "Robust stability for sampled-data control systems," Systems & Control Letters, Vol. 13, No.3, pp. 217-226, May 1989.

[2] C.E. de Souza, M. Fu and L. Xie, "H_∞ analysis and synthesis of discrete-time systems with time-varying uncertainty," IEEE Trans. Automat. Control, Vol. AC-37, 1992, to appear.

[3] R.M. Dolphus and W.E. Schmitendorf, "Robust control design for uncertain sampled data systems," in Proc. 29th IEEE Conf. Decision & Control, Honolulu, Hawaii, Dec. 1990, pp.1918–1920.

[4] P.P. Khargonekar, I.R. Petersen and K. Zhou, "Robust stabilization of uncertain linear systems : quadratic stabilizability and H_∞ control theory," IEEE Trans. Automat. Control, Vol. AC-35, No.3, pp. 356–361, March 1991.

[5] A. Linnemann, "On robust stability of continuous-time systems under sampled-data control," in Proc. 29th IEEE Conf. Decision & Control, Honolulu, Hawaii, Dec. 1990, pp.1921–1922.

[6] N. Sivashankar and P.P. Khargonekar, "Robust stability and performance analysis of sampled-data systems," in Proc. 30th IEEE Conf. Decision & Control, Brighton, England, Dec. 1991, pp.881–885.

[7] P. Shi, L. Xie and C.E. de Souza, "Robust control and filtering of sampled-data systems with parametric uncertainties," Tech. Report EE9216, Dept. of Electrical and Computer Engineering, University of Newcastle, NSW 2308, Australia, 1992.

[8] W. Sun, K.M. Nagpal and P.P. Khargonekar, "H_∞ control and filtering for sampled-data systems," in Proc. 1991 American Control Conf., 1991, pp.1652–1657.

[9] G. Tadmor, "H_∞ optimal sampled-data control in continuous time systems," Int. J. Control, Vol. 56, No.1, pp.99–141, July 1992.

[10] L. Xie, M. Fu and C.E. de Souza, "H_∞ control and quadratic stabilization of systems with parameter uncertainty via output feedback," IEEE Trans. Automat. Control, Vol. AC-37, 1992, to appear.

FA10 - 9:40

Proceedings of the 31st Conference
on Decision and Control
Tucson, Arizona • December 1992

SYNTHESIS OF COMPENSATORS FOR A NON-OVERSHOOTING NOMINAL STEP RESPONSE IN LINEAR UNCERTAIN PLANTS

by
Jay-Wook Song and Suhada Jayasuriya
Department of Mechanical Engineering
Texas A&M University
College Station, TX 77843-3123

ABSTRACT

Synthesis of a compensator for assuring a non-overshooting step response is a fundamental problem that has received very little attention. In this paper we present a design methodology for synthesizing such compensators for SISO, stable and minimum phase plants with significant plant variations. The basic idea of the technique is to appropriately locate the closed loop poles with respect to fixed and added zeros as developed by the authors [1]. An example illustrating the design procedure is included.

I. INTRODUCTION

Overshoot during set point changes is undesirable in many practical problems. In process control for example, the optimum set point may be close to an economic or safety constraint. Consequently, overshoot of a set point could lead to a violation of a constraint and endanger process operation. Therefore, a non-overshooting step response is often desirable if not required. For example, in a pressurized water reactor(PWR) nuclear power plant, if the average core coolant temperature exceeds its set point by a certain prespecified tolerance, the plant protection system will initiate a shutdown. Such shutdowns could be minimized by proper control system design. Automatic focusing of an electron beam in a C.R.T is another example, where non-overshooting of the controlling voltage is desired to prevent the tube from burning. In contour machining, overshooting of the position of the cutter results in an incorrect shape of the work-piece.

Recently a characterization was given by Jayasuriya and Franchek [2] for a class of transfer functions to have a non-negative impulse response. This characterization was subsequently extended and enlarged by Jayasuriya and Song [1] to characterize non-overshooting step responses. These results are based on the identification of some primitive transfer functions with the desired properties which can then be convolved to get a wider class of systems with the same properties. A remarkable feature of these sufficiency conditions is that they can be easily checked by inspection, thus making them very useful as a design tool.

II. PROBLEM FORMULATION

The control objective is to synthesize a control system so that that the nominal system output reaches its steady state value y_{ss} without overshooting it (i.e., $y(t) \leq y_{ss} \forall t \in [0, \infty)$)while robust stability, disturbance rejection and sensitivity reduction are achieved. In particular we consider linear, SISO, stable, minimum phase systems with parametric uncertainty.

We consider the two degrees of freedom control system configuration shown in Fig. 1 where P denotes the uncertain plant, G a feedback compensator inside the loop and F a prefilter. The controller G will be used primarily for guaranteeing robust stability, disturbance rejection and sensitivity reduction. The prefilter F is for the purpose of assuring a non-overshooting step response. Design of the feedback compensator can be based on any robust control design technique such as Quantitative Feedback Theory (QFT), H_∞, l_1 or LQG/LTR. The exact methodology to be used will depend on the type of performance specifications imposed. The crucial part in the synthesis is the selection of an appropriate F. In order to select F we make use of the recent results of Jayasuriya and Song [1].

For the purpose of this paper we assume that the performance specifications are given in the QFT format.

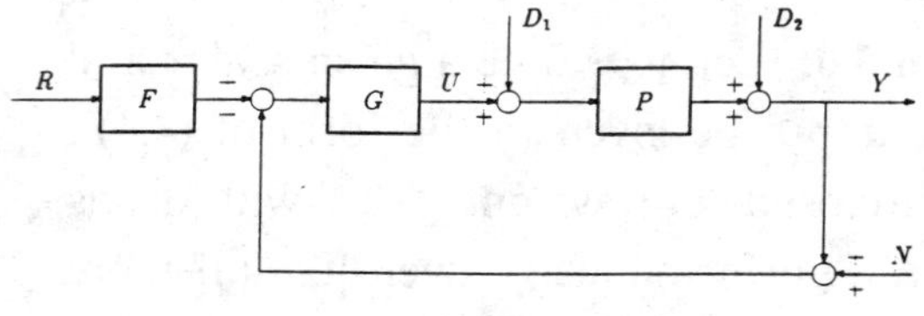

Figure 1: Feedback Structure

III. DESIGN METHODOLOGY

The QFT approach is adopted since it is a frequency domain method, and it allows direct manipulation of poles and zeros and can handle large plant uncertainties while guaranteeing robust performance as well as robust stability. In the QFT approach, the amount of plant uncertainty is quantitatively related to control system specifications which manifests itself as frequency domain bounds on a pre-defined nominal loop transfer function. The desired controller G is realized by loop shaping.

The proposed feedback structure has two degrees of freedom in the sense that the transmission function, $T_{yr}(s) = \frac{GPF}{1+GP}$, characterizing the overall dynamic response, and the loop transfer function, $L(s) = GP$, characterizing the sensitivity, can be independently specified.

In what follows we suppose that the feedback controller has been designed using the QFT technique [3] or any other method. So it remains only to design F to guarantee that the nominal step response from r to y does not overshoot.

Design of prefilter $F(s)$

The purpose of the prefilter is two fold when employing QFT type specifications: (1) ensure a non-overshooting nominal step response by appropriately locating the poles and zeros of the nominal loop transmission function T_{yr}^o, and (2) position $\| T_{yr}(j\omega) \|$ for all possible plants within the uncertain set inside the frequency domain target transfer functions (see any standard reference on QFT such as [3] for QFT specifications). It should be noted that the latter requirement does not arise when using other robust control methods for choosing G.

Now the goal is to choose F to satisfy,

$$\| F(j\omega) \| = \| T_{yr}(j\omega) \| - \| T(j\omega) \|_{\text{mean}} \tag{1}$$

where $T(s) = \frac{L(s)}{1+L(s)}$ and $T_{yr}(s) = T(s)F(s)$. The prefilter can now be designed by following the steps given below:

1. Get the Bode plot of $\| T(j\omega) \|$ for all plant parameter variations.
2. Find the mean value of $\| T(j\omega) \|$ for each frequency corresponding to the templates used in the design of $L_o(j\omega)$.
3. Plot $\| T_{yr}(j\omega) \| - \| T(j\omega) \|_{\text{mean}}$ vs. ω which defines a frequency domain region in which $\| F(j\omega) \|$ must lie.

CH3229-2/92/0000-2882$1.00 © 1992 IEEE

4. Use straight-line approximations to synthesize $F(s)$ so that $\| F(j\omega) \|$ falls inside the region defined in step 3.

In step 4, pole-zero cancellation and/or addition can be done by utilizing the results of Jayasuriya and Song [1]. In executing this step, it is important to locate any additional poles and zeros inside the loop transmission bandwidth in order to achieve a non-overshooting nominal step response.

IV. DESIGN EXAMPLE

To illustrate the design procedure, consider the following linear uncertain plant

$$P(s) = \frac{ka}{s(s+a)} \tag{2}$$

where the uncertainty in k and a are given by $1 \leq k \leq 5$ and $1 \leq a \leq 5$. Let us choose $a_o = 1$ and $k_o = 1$, which yield the minimum plant gain amplitude for all $\omega > 0$, as the nominal parameters. A rise time, t_r, of 2 seconds and a bandwidth, ω_b, of 2 rad/sec is the minimum design specifications to achieve. For zero steady state tracking error, $L(s)$ will be a Type 1 system since the input is a step input. The resulting $L_o(j\omega)$ is shown in Fig. 2 and is given below:

$$L_o(s) = \frac{1}{s(s+1)} \frac{(1+\frac{s}{.4})(1+\frac{s}{113})^2}{(1+\frac{s}{41})(1+\frac{2}{237}s+\frac{1}{237^2}s^2)^2} \tag{3}$$

Division of $L_o(s)$ by the nominal plant transfer function $1/s(s+1)$ determines the compensator which has three zeros at $-.4$, -113 and -113 and one pole at -41 and quadruple poles at -237.

Now the prefilter $F(s)$ needs to be synthesized to yield the desired step response. Using the sufficient conditions discussed in [1], poles and zeros of $T^o_{yr}(s)$ are modified to have a non-overshooting nominal step response while matching the desired tracking bounds. The synthesized prefilter is given by

$$\begin{aligned} F(s) &= F_1(s) \cdot F_2(s) \\ &= \frac{1+\frac{s}{.31}}{1+\frac{s}{.4}} \cdot \frac{1}{1+\frac{s}{2.21}} \cdot \end{aligned} \tag{4}$$

The time and frequency response of the final closed loop system for a step input are shown in Fig. 3 and Fig. 4. The time response shows no overshoot for the nominal plant but has a maximum of about 10 % overshoot for non-nominal plants. We notice that the rise time is well within 2 seconds but the settling time is too large to be acceptable. The slow settling time is due to the very low frequency zero of the controller which tend to attract the pole at the origin thus yielding a low frequency closed loop pole. This problem persists for all systems except

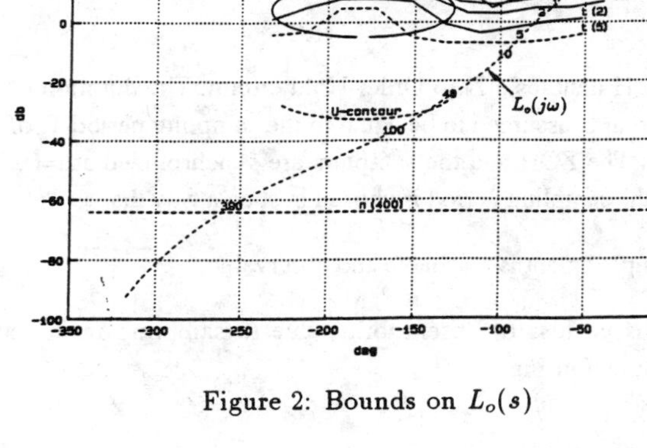

Figure 2: Bounds on $L_o(s)$

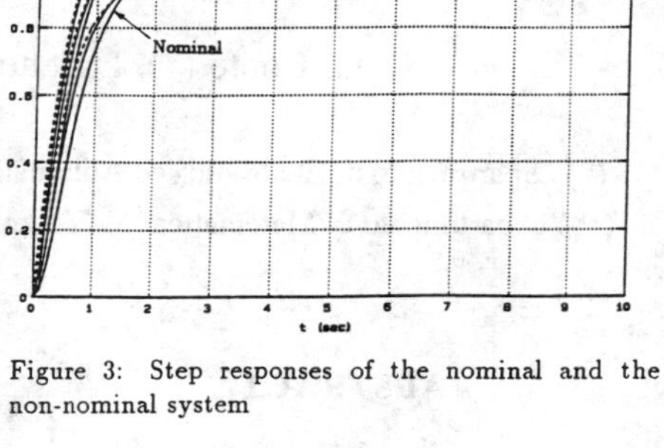

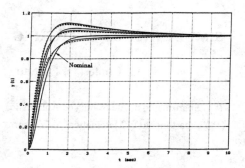

Figure 3: Step responses of the nominal and the non-nominal system

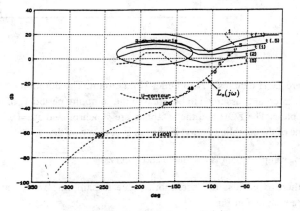

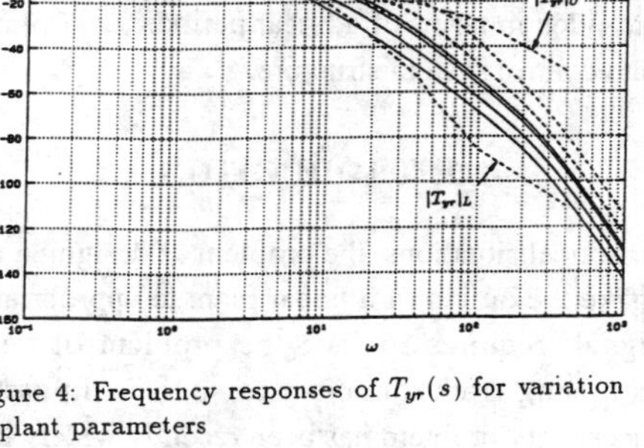

Figure 4: Frequency responses of $T_{yr}(s)$ for variation in plant parameters

for the nominal case where the low frequency pole is canceled by $F(s)$. In order to resolve this slow settling time problem without increasing the loop bandwidth, the variation of $|T_{yr}(jw)|$ in the low frequency range needs to be widened to produce lower tracking bounds.

V. SUMMARY

In this short paper we outlined a design methodology for synthesizing control systems for a class of SISO systems to guarantee a non-overshooting nominal step response in addition to satisfying standard feedback requirements such as robust stability, disturbance rejection and sensitivity reduction. While it is desirable to be able to assure non-overshooting step responses for any plant in the set our present technique can do so only for the nominal plant. We believe what is presented in this paper is the first synthesis technique to deal with the non-overshooting requirement. The robust performance with respect to the non-overshooting requirement is a much harder problem than many other robust performance problems. To date the robust performance (in time domain) problem even in more simple situations (i.e., without the non-overshooting requirement) remains an open problem.

REFERENCES

[1] Jayasuriya, S., and Song, J., "*On the Synthesis of Compensators for Non-Overshooting Step Response,*" Proc. of 1992 ACC, Chicago, IL, June 1992.

[2] Jayasuriya, S., and Francheck, M.A., "*A Class of Transfer Functions with Non-negative Impulse Response* ," Journal of Dynamic Systems, Measurement and Control, Vol. 113, pp. 313-315, June 1991.

[3] Horowitz, I.M. and Sidi, M., "*Synthesis of feedback systems with large plant ignorance for prescribed time-domain tolerances,*" Int. J. Control, vol. 16, no. 2, pp. 287-309, 1972

Proceedings of the 31st Conference
on Decision and Control
Tucson, Arizona • December 1992

SIMULTANEOUS TRACKING FOR A FAMILY OF LINEAR PLANTS

G. Conte (*), L. Jetto (*), S. Longhi (*) and A.M. Perdon (**)

(*) Dipartimento di Elettronica ed Automatica, Università di Ancona, via Brecce Bianche, 60131 Ancona, Italy.
(**) Dipartimento di Matematica "V. Volterra", Università di Ancona, via Brecce Bianche, 60131 Ancona, Italy.

ABSTRACT

For a given family of linear, continuous-time parameter dependent plants, we consider the problem of designing a controller which assures the exact tracking of an external reference signal. The solution is provided in terms of a linear periodic controller acting on a sampled-data control system structure.

I. INTRODUCTION

In many practical situations, the problem of designing a controller in order to make the output of a linear plant to appropriately track a reference signal, requires to face the problem of dealing with uncertainties arising from a number of sources. In particular, the parameter uncertainty problem has been recently widely investigated and various robust control techniques (see e.g.[1-6]) have been proposed.

A similar problem is considered here; namely we investigate the conditions that ensure the solvability of the exact output tracking problem for a family of continuous-time linear plants whose parameters may take values in a known, finite set. This problem refers e.g. to situations where different operating conditions of the same plant have to be taken into account [2,3].

The problem can be stated in the following terms. Let $\Sigma(\beta)$ be the linear, time-invariant continuous-time system described by

$$\dot{x}(t) = A(\beta)x(t) + B(\beta)u(t), \qquad t \in \mathbb{R}^+,\ x(t_0)=x_0 \tag{1.1}$$

$$y(t) = C(\beta)x(t) + D(\beta)u(t) \tag{1.2}$$

where:

- β is an r-vector of physical parameters which may take value in a known, finite set $\Theta = \{\beta_1, \beta_2, ..., \beta_N\}$;

- $x(t) \in \mathbb{R}^n =: \boldsymbol{X}$ is the state, $u(t) \in \mathbb{R}^p =: \boldsymbol{U}$ is the control input, $y(t) \in \mathbb{R}^q =: \boldsymbol{Y}$ is the output to be controlled (which is assumed to be measurable) and $A(\beta)$, $B(\beta)$, $C(\beta)$ and $D(\beta)$ are real matrices whose entries depend on β. The value of these matrices for a particular β_i (i=1,2,...,n) is denoted by $A(\beta_i)$, $B(\beta_i)$, $C(\beta_i)$ and $D(\beta_i)$.

Assume that $\Sigma(\beta)$ is controllable and detectable for all $\beta \in \Theta$.

Let r(t) be a reference signal generated as the free output response of the linear time-invariant continuous-time system Σ_G described by

$$\dot{\xi}(t) = A_G \xi(t), \qquad t \in \mathbb{R}^+,\ \xi(t_0)=\xi_0 \tag{1.3}$$

$$r(t) = C_G \xi(t) \tag{1.4}$$

where $\xi(t) \in \mathbb{R}^m$ is the state. Then, the problem we consider consists in finding a linear controller Σ_c such that, independently of the actual value of β, for every ξ_0, x_0 and for every internal initial state of Σ_c, the output y(t) of $\Sigma(\beta)$ track the reference signal r(t) with zero steady-state error. We call this problem as the Robust Output Tracking Problem (ROTP). The solution proposed is based on a sampled-data control system structure where the compensator is chosen as a discrete linear periodic system. It is shown that if the plant contains an internal model of the reference generator, then the periodic controller guarantees an exact tracking for each $\beta \in \Theta$.

II. CONTROL SYSTEM STRUCTURE

The stabilization problem for linear plants with parameter uncertainty like that admitted for the system (1.1),(1.2) has been considered in [7,8], where a discrete-time linear periodic compensator has been proposed. Following the line of [7], we search for a solution of the ROTP by means of a linear discrete-time periodic compensator Σ_c. Therefore we make resort to a hybrid control scheme like that described in Fig.1

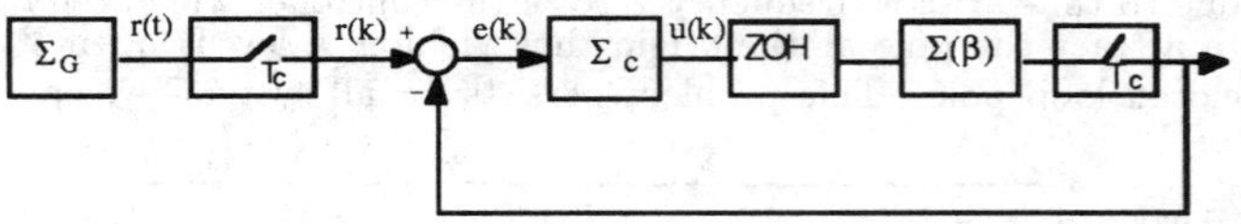

Figure 1

where ZOH denotes a Zero Order Hold circuit. The duration of the hold interval is assumed to be equal to the sampling period T_c of the samplers. The ZOH and the samplers are synchronized at t=0. The value of the sampling period is chosen in such a way that

a) the sampled plant is reachable and observable

b) there is no loss of observability due to sampling in the error system shown in Fig.2.

CH3229-2/92/0000-2884$1.00 © 1992 IEEE

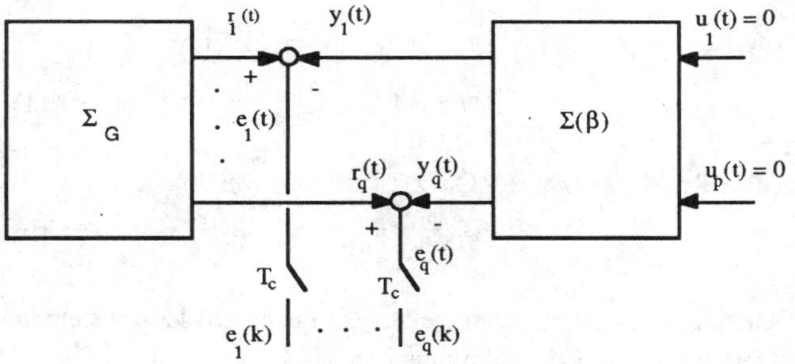

Figure 2

Explicitly, assumption b) means that $e(k) := [e_1(k),e_2(k),...,e_q(k)]^T= 0, (k \geq k_a)$ if and only if $e(t) =: [e_1(t),e_2(t),...,e_q(t)]^T = 0, (t \geq k_aT_c)$. Conditions ensuring the fulfilment of this requirement are well-known [9].

III. PROBLEM SOLUTION

In the light of the foregoing formulation of the ROTP, we look for a linear periodic compensator Σ_c described by

$$w(k+1) = A_C(k)w(k) + B_C(k)e(k), \qquad w(0)=w_0,\ k\in \mathbb{Z}^+ \tag{3.1}$$

$$u(k) = C_C(k)w(k) + D_C(k)e(k) \tag{3.2}$$

where:

- $w(k) \in \mathbb{R}^{\mu}$ is the state;

- and $A_C(\cdot)$, $B_C(\cdot)$, $C_C(\cdot)$ and $D_C(\cdot)$ are periodic matrices of period ω (or, more briefly, ω-periodic);

such that, for all $\beta \in \Theta$, for any initial state $x(0)$ of system $\Sigma(\beta)$, $\xi(0)$ of the reference generator Σ_G and $w(0)$ of the compensator Σ_C, the reference error $e(k) := r(k) - y(k)$ and the control input $u(k)$ satisfy

$$e(k) = 0, \quad u(k)= 0, \quad \forall k \geq k_a \tag{3.3}$$

where k_a is a finite positive integer.

Remark 1. Taking into account the assumption that there is no loss of observability in the error system, condition (3.3) guarantees an exact tracking for each time instant and not only in correspondence of the sampling instants.

The assumption that the parameter set Θ is finite, implies that the family of controllable and detectable model $\Sigma(\beta)$ is finite. By well-known results on linear time-invariant systems (see, for example [10,11]), there exists, for each β_i, i= 1, ..., N, a linear time-invariant compensator Σ_C^i ,with state $w_i(k) \in \mathbb{R}^{\mu}$, $\mu \leq n$, described by

$$w_i(k+1) = A_iw(k) + B_ie(k), \qquad w_i(0)=w_{i0} \tag{3.4}$$

$$u(k) = C_iw(k) + D_ie(k) \tag{3.5}$$

such that, for the zero initial state $\xi(0) = 0$ of generator Σ_G and arbitrary initial states $x(0)$ of $\Sigma(\beta)$ and $w_i(0)$ of Σ_C^i the state $[\,x'(k)\ \ w_i'(k)\,]'$ of the closed-loop system, described by

$$\begin{bmatrix} x(k+1) \\ w_i(k+1) \end{bmatrix} = \begin{bmatrix} A(\beta_i)-B(\beta_i)D_iC(\beta_i) & B(\beta_i)C_i \\ -B_iC(\beta_i) & C_i \end{bmatrix} \begin{bmatrix} x(k) \\ w_i(k) \end{bmatrix} \tag{3.6}$$

goes to zero in a finite time, lesser than or equal to 2n, i.e:

$[\,x'(k)\ \ w_i'(k)\,]' = 0$ for all $k \geq 2n$.

A possible compensator Σ_C^i with the above properties may be obtained as the series connection of a dead-beat observer Σ_O^i together with a dead-beat feedback F_i as reported in Fig.3.

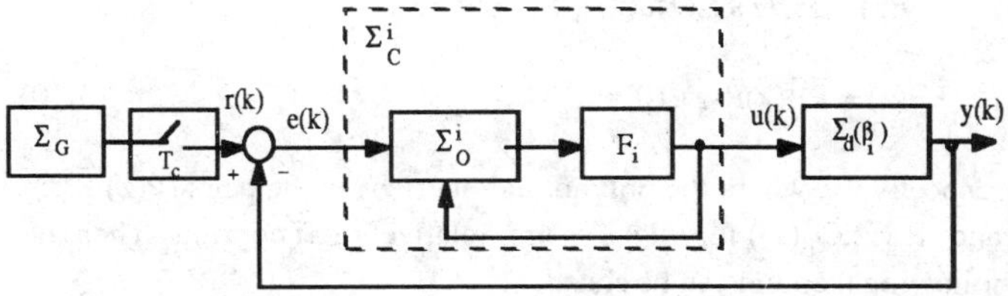

Figure 3

where $\Sigma_d(\beta_i)$ denoted the series connection of the ZOH, of $\Sigma(\beta_i)$ and of the sampler.

Equivalently, Σ_C^i may be found by using a transfer function approach and related polynomial equations [12,13].

Making use of these N linear time-invariant compensators, Σ_C^i define the ω-periodic matrices of Σ_C in the following way:

let $\qquad k_j := 4nj$, for $j = 0,...,N$; $\omega = 4nN$

then

$$A(k) := A_i,\ \ B(k) := B_i,\ \ C(k) := C_i,\ \ D(k) := D_i \qquad \forall k \in [\,k_{i-1}, k_i\,),\ i = 1, ..., N \tag{3.7}$$

$$A(k+\omega) := A(k),\ \ B(k+\omega) := B(k),\ C(k+\omega) := C(k). \tag{3.8}$$

In this way, in each interval $[\,k_{i-1}, k_i\,)$, the periodic compensator Σ_C coincides with the dead-beat controller Σ_C^i, which guarantees the state dead-beat control of system $\Sigma(\beta)$ with $\beta = \beta_i$.

With this choice of periodic compensator Σ_C, if the initial state $\xi(0)$ of generator Σ_G, is zero, the state [x'(k) w(k)']' of the closed-loop system described by (3.6) goes to zero, as well as y(k) and e(k), in a finite time interval less or equal to ω, for all $\beta \in \Theta$ and any arbitrary initial states x(0) of $\Sigma(\beta)$ and w(0) of Σ_C. In fact, if the parameter β of system (1.1), (1.2) is equal to $\beta_i \in \Theta$, in the time interval $[\,0, k_{i-1})$, the compensator Σ_C does not produce any significative effect on the state of the closed-loop system, while, in the interval $[\,k_{i-1}, k_i)$ it coincides with Σ_C^i and, therefore it reduces to zero the state [x'(k) w'(k)]' of the closed-loop system, as well as e(k) = – y(k) and u(k). After this time interval, the input e(k) and the state w(k) of the compensator Σ_C are zero, then the control action produced by the compensator Σ_C is zero and the state of the closed-loop system is kept equal to zero.

A periodic compensator Σ_C described by (3.4), (3.5), (3.7) and (3.8) is a solution of the ROTP for the special case of initial state $\xi(0) = 0$ of generator Σ_G, i.e.reference signal r(k) = 0 for all $k \in \mathbb{Z}^+$.

In order to derive the solvability conditions of ROTP for the general case $\xi(0) \neq 0$ (i.e. $r(t) \neq 0$) with the same linear periodic compensator Σ_C, it is convenient to make reference to polynomial matrix descriptions of the sampled plant and reference generator. Then the z-transform of y(k) and r(k) can be written as:

$$y(z) = P_\beta^{-1}(z)Q_\beta(z)u(z) + P_\beta^{-1}(z)R_\beta(z)x(0) \tag{3.9}$$

$$r(z) = \bar{P}^{-1}(z)\bar{R}(z)\xi(0) \tag{3.10}$$

respectively, where the polynomial matrices in the pairs $(\bar{P}(z),\bar{R}(z))$ and $(P_\beta(z),Q_\beta(z))$ for all $\beta \in \Theta$ are relatively left coprime. Then the following theorem can be stated.

Theorem. *The* ROTP *is solvable with the periodic compensator* Σ_C *described by* (3.4) (3.5),(3.7) and (3.8) *if the polynomial matrix* $\bar{P}(z)$ *is a right divisor of* $P_\beta(z)$ *for all* $\beta \in \Theta$, i.e. *there exist polynomial matrices* $G_{\beta_i}(z)$ *such that* $P_{\beta_i}(z) = G_{\beta_i}(z)\bar{P}(z)$ *for* $\beta_i \in \Theta$, (i=1,2,...,n).

Proof. By the structure of the periodic compensator Σ_C, if the parameter β is equal to $\beta_i \in \Theta$, in the time interval $[\,0, k_{i-1})$ the compensator Σ_C does not produce significative effects on the state [x'(k) w'(k)]' of the closed-loop system. While, in the time interval $[\,k_{i-1}, k_i)$ the compensator Σ_C coincides with Σ_C^i and for $\xi(k_{i-1}) = 0$ and for any arbitrary $\bar{x} \in \mathbb{R}^n$ and $\bar{w} \in \mathbb{R}^\mu$ such that $x(k_{i-1}) = \bar{x}$ and $w(k_{i-1}) = \bar{w}$, the state response of the closed-loop system starting at time k_{i-1} goes to zero before time k_i.

Then, in the interval of time $[\,k_{i-1}, k_i)$ it is sufficient to analyze the tracking error response starting at time k_{i-1} with initial state $x(k_{i-1}) = 0$, $w(k_{i-1}) = 0$ and arbitrary $\xi(k_{i-1}) \in \mathbb{R}^m$.

Now, denote by $W_i(z)$ the transfer matrix of the compensator Σ_C and let $(T_i(z), S_i(z))$ be a pair of right coprime matrices such that $W_i(z) = T_i(z), S_i^{-1}(z)$.

Under the latter initial conditions, it is easy to see that, if the condition of the Theorem holds, then,

$$e(k) = S_i(z)\,[\,P_{\beta_i}(z)S_i(z) + Q_{\beta_i}(z)T_i(z)\,]^{-1}\,G_{\beta_i}(z)\bar{R}(z)\xi(k_{i-1}) \quad \forall\, k \in [\,k_{i-1}, k_i\,) \tag{3.11}$$

$$u(k) = T_i(z)\,[\,P_{\beta_i}(z)S_i(z) + Q_{\beta_i}(z)T_i(z)\,]^{-1}\,G_{\beta_i}(z)\bar{R}(z)\xi(k_{i-1}) \quad \forall\, k \in [\,k_{i-1}, k_i\,) \tag{3.12}$$

Denoting with $\hat{A}_i$ the dynamic matrix of the closed-loop system in the interval $[\,k_{i-1}, k_i)$, the following holds [14]

$$\det[zI_{n+\mu} - \hat{A}_i] = \det[zI_n - A(\beta_i)]\,\det[zI_\mu - A_i]\,\det[I_q + P_{\beta_i}^{-1}(z)Q_{\beta_i}(z)T_i(z), S_i^{-1}(z)] \tag{3.13}$$

Taking into account the dead-beat property of the state response of the closed-loop system for $\xi(k_{i-1}) = 0$, it follows that

$$\det[zI_n - A(\beta_i)] = z^{r_i}\det Q_{\beta_i}(z)\ \text{cost} \tag{3.14}$$

$$\det[zI_\mu - A_i] = z^{\rho_i}\det S_i(z)\ \text{cost} \tag{3.15}$$

$$\det[zI_{n+\mu} - \hat{A}_i] = z^{n+\mu}. \tag{3.16}$$

Then, equations. (3.13)-(3.16) give

$$\det[\,P_{\beta_i}(z)S_i(z) + Q_{\beta_i}(z)T_i(z)\,] = z^{n+\mu-r_i+\rho_i} \tag{3.17}$$

which, together with (3.11) and (3.12) implies that e(k) = 0 and u(k) = 0 for all $k \geq k_{i-1} + n + \mu - r_i - \rho_i \in [\,k_{i-1}, k_{i-1} + h_i)$. After the time instant $k_{i-1} + n + \mu - r_i - \rho_i$ the input e(k) of the compensator Σ_C is zero, as well as the control action u(k) produced by Σ_C. The subcompensator Σ_C^i is composed by the series connection of a dead-beat observer Σ_O^i with a dead-beat feedback F_i as reported in Fig. 3, and before time k_i also the state of Σ_O^i goes to zero. Therefore, after time k_i the compensator does not produce control action u(·) and the tracking error e(·) is kept equal to zero. Taking into account Remark 1, it follows that an exact continuous time tracking is achieved after a finite number of sampling intervals ❑

Remark 2. The condition stated in the previous theorem is equivalent to require that the sampled plant contains an internal model of the sampled reference generator. [15]. If this condition is not " a priori" satisfied, it may be fulfilled by suitably precompensating the plant.

REFERENCES

[1] E.Emre, Simultaneous stabilization with fixed closed loop characteristic polynomial, *IEEE Trans. Autom. Contr.*, **28** (1983) 103-104.

[2] J. Ackermann, Multi-model approaches to robust control system design, *Technical report, DFVLR Institut fur Dynamic der Flugsysteme*, Oberpfaffenhofen, Germany,1985.

[3] J.Ackermann, Sampled-data control systems, *Springer Verlag*, Berlin, Heidelberg, New York,1985.

[4] Y.C. Soh, R.J. Evans, I.R. Petersen, and R.E. Betz, Robust pole assignment, *Automatica*, **23** (1987) 601-610.

[5] K. Wei and B. Barmish, An iterative design procedure for a simultaneous stabilization of MIMO systems, *Automatica*, **24** (1988) 643-652.

[6] Y.C.Soh, R.J. Evans, Characterization of robust controllers, *Automatica*, (1989) 115-117.

[7] A.W. Olbrot, Robust stabilization of uncertain systems by periodic feedback, *Int. J. Control*, **45**, (1987),747-758.

[8] P. P. Khargonekar, K. Poolla and A. Tannenbaum, Robust control of linear time-invariant plants using periodic compensation, *IEEE Trans. Autom. Control*, **AC-30**,(1985) 1088-1096.

[9] C.T. Chen, Linear system theory and design, *Holt, Rinehart and Winston*, New York,1984.

[10] R. Isermann, Digital control systems, *Springer Verlag*, Berlin,1981.

[11] J O' Reilly, The dead-beat control of linear multivariable systems with inaccessible state, *Int.J.Contr.*, **31** (1980) 645-654.

[12] V Kucera, Discrete linear control: the polynomial equation approach, *J Wiley*, Chichester, England,1987.

[13] J.O' Reilly, The discrete linear time invariant optimal control problem- an overview, *Automatica*, **17** (1981) 363-370.

[14] H.H. Rosembrock, Computer-aided control system design, *Academic press*, London, 1974.

[15] G. Bengtsson, Output regulation and internal models - A frequency domain approach, *Automatica*, **13**, (1977), 333-345.

FA10 - 10:10

Proceedings of the 31st Conference
on Decision and Control
Tucson, Arizona • December 1992

A Set Induced Norm Approach to the Robust Control of Constrained Systems

Mario Sznaier
Electrical Engineering, University of Central Florida, Orlando, FL, 32816–2450
email: msznaier@frodo.engr.ucf.edu

Abstract

Most realistic control problems involve both some type of time–domain constraints and model uncertainty. However, there presently exist few design methods capable of simultaneously addressing both issues. We recently proposed to address this class of problems by using a "constrained robustness measure", generated by a constraint–set induced operator norm, to assess the stability properties of a family of systems. In this paper we explore the properties of this constrained–robustness measure and we extend the theoretical framework to include control as well as state constraints. These results are applied to the problem of designing fixed–order stabilizing feedback controllers for systems subject to structured parametric model uncertainty and time–domain constraints.

1. Introduction

A substantial number of control problems can be summarized as the problem of designing a controller capable of achieving acceptable performance under system uncertainty and design constraints. However, this problem is far from solved, even in the simpler case where the system under consideration is linear. Several methods have been proposed recently to deal with constrained control problems under the assumption of exact knowledge of the model (see [1] and references therein). However, such an assumption can be too restrictive, preventing their application in realistic problems.

On the other hand, during the last decade a considerable amount of time has been spent analyzing the question of whether some relevant properties of a system (most notably asymptotic stability) are preserved under the presence of unknown perturbations. This research effort has led to procedures for designing "robust" controllers, capable of achieving desirable properties under various classes of plant perturbations while, at the same time, satisfying frequency–domain constraints. However, most of these design procedures cannot accommodate directly time domain constraints (which precludes their use in cases such as when there exist physically motivated "hard" bounds on the states or control effort), although some progress has been recently made in this direction [2–5].

In [6–7] we proposed to approach time–domain constrained systems using an operator norm–theoretic approach. We introduced a simple robustness measure that indicated how well the family of systems under consideration satisfied a given set of time–domain constraints and we proposed a design method yielding controllers that maximized this robustness measure. In this paper we extend our formalism to include control as well as a more general description of state constraints and we explore the properties of the resulting constrained robustness measure. These theoretical results are applied to the problem of designing stabilizing controllers for systems subject to structured parametric model uncertainty and time–domain constraints. We show that in cases of practical interest the synthesis problem can be reduced to a convex, albeit in general non–differentiable, optimization problem.

The paper is organized as follows: In section II we introduce the concepts of *constrained stability* and *robust constrained stability* and we use these concepts to give a formal definition of the *robust constrained stability analysis* and *robust constrained stability design* problems. The *analysis* problem is studied in section III where we give necessary and sufficient conditions for constrained stability. We use these results to define a constrained robustness measure and we show that, under mild dynamics of the system. In section IV we apply the results of section III to the *design* problem and we show that in cases of practical interest our approach yields a well behaved optimization problem. Finally, in section V, we summarize our results and we indicate directions for future research.

Supported in part by NSF under grant ECS–9211169 and by Florida Space Grant Consortium

2. Definitions and Background Results

2.1 Preliminary Definitions

• **Def. 1:** Consider the linear, time invariant, discrete time, autonomous system modeled by the difference equation:

$$\underline{x}_{k+1} = A\underline{x}_k,\ k = 0,1\dots \qquad (S^a) \text{ subject to the constraint:}$$

$$\underline{x} \in \mathcal{G} \subset R^n \tag{1}$$

where $A \in R^{n*n}$ and where $\underline{x}$ indicates x is a vector quantity. The system (S^a) is *Constrained Stable* if for *any* point $\underline{\tilde{x}} \in \mathcal{G}$, the trajectory $\underline{x}_k(\underline{\tilde{x}})$ originating in $\underline{\tilde{x}}$ remains in $\mathcal{G}$ for all k.

Remark 1: A nonempty subset $\mathcal{S} \subset R^n$ is a *positively invariant set* of the system (S^a) if for any initial state $\underline{x}_o \in \mathcal{S}$, the trajectory $\underline{x}_k(\underline{x}_o) \in \mathcal{S}\ \forall\ k$, or equivalently [8] if and only if $\underline{x} \in \mathcal{S}$ implies $A\underline{x} \in \mathcal{S}$. Therefore, it follows that the system (S^a) is constrained stable *iff* it has the set $\mathcal{G}$ as a positively invariant set.

• **Def. 2:** Consider the family of linear discrete–time systems modeled by the difference equation:

$$\underline{x}_{k+1} = (A+\Delta)\underline{x}_k \tag{S^a_Δ}$$

where Δ belongs to some perturbation set $\mathcal{D} \subseteq R^{n*n}$. The system (S^a) is *Robustly Constrained Stable* with respect to the set $\mathcal{D}$ if (S^a_Δ) is constrained stable for all perturbation matrices $\Delta \in \mathcal{D}$.

We proceed now to restrict the class of constraints allowed in our problem. The introduction of this restriction, while not affecting significantly the number of real-world problems that can be handled by our formalism, introduces more structure into the problem. This additional structure plays a key role in section III where we derive necessary and sufficient conditions for constrained stability.

2.2 Constraint Qualification Hypothesis

In this paper, we will limit ourselves to constraints of the form:

$$\underline{x} \in \mathcal{G} \subset R^n \tag{2}$$

where $\mathcal{G}$ is a convex, compact, balanced set (i.e a convex compact set such that $\underline{x} \in \mathcal{G} \Rightarrow \lambda\underline{x} \in \mathcal{G}$ for $|\lambda| \leq 1$ [9]) containing the origin in its interior.

• **Def. 3:** [9] The *Minkowsky Functional* (or gauge) p of a balanced convex set $\mathcal{G}$ containing the origin in its interior is defined by

$$p(\underline{x}) = \inf_{r>0}\left\{r: \frac{\underline{x}}{r} \in \mathcal{G}\right\} \tag{3}$$

A well known result in functional analysis (see for instance [9]) establishes that p defines a seminorm in R^n. Furthermore, when $\mathcal{G}$ is compact, this seminorm becomes a norm. In the sequel, we will denote this norm as $\|\underline{x}\|_{\mathcal{G}} \triangleq p(\underline{x})$

Remark 2: The set $\mathcal{G}$ can be characterized as the unity ball in $\|.\|_{\mathcal{G}}$ i.e. $\mathcal{G} = \{\underline{x}: \|\underline{x}\|_{\mathcal{G}} \leq 1\}$.

2.3 Statement of the Problem

Consider the LTI system represented by the following state–space realization:

$$\underline{x}_{k+1} = A\underline{x}_k + B\underline{u}_k \qquad (S)$$

subject to the constraint:

$$\underline{x}_k \in \mathcal{G} \subset R^n$$

where $\underline{x} \in R^n$ represents the state and $\underline{u} \in R^m$ represents the control input. Then, the basic problems that we address in this paper are the following:

• **Robust Constrained Stability Analysis Problem:** Given the nominal system (S) and a linear feedback control law $\underline{u}_k = F\underline{x}_k$, determine if the resulting closed–loop system is constrained–stable. If the nominal closed–loop system is constrained–stable, determine the maximum allowable level of model uncertainty (in the sense of some previously defined norm) such that the constraints are satisfied for any initial condition $\underline{x} \in \mathcal{G}$.

• **Linear Robust Constrained Control Synthesis Problem:** Given the system (S) find a *linear* controller such that the resulting closed–loop system is constrained stable and satisfies some additional specifications such as:

i) maximum robustness against structured model uncertainty of the form $A = A_o + \Delta,\ \Delta \in \mathcal{D}$

ii) bounds on the control effort of the form $\underline{u}_k \in \Omega \subset R^m$, where Ω is a compact, convex balanced set containing the origin in its interior.

3. Constrained Stability Analysis

Consider the system (S^a) and let $\|.\|_{\mathcal{G}}$ denote the operator norm induced in R^{n*n} by $\mathcal{G}$ (i.e. $\|A\|_{\mathcal{G}} \triangleq \sup_{\|\underline{x}\|_{\mathcal{G}}=1} \|A\underline{x}\|_{\mathcal{G}}$). From definition 1 it follows that (S^a) is constrained stable *iff* $\|A\|_{\mathcal{G}} \leq 1$. Moreover, (S^a) is robustly constrained stable with respect to a given set $\mathcal{D}$ *iff* $\|A + \Delta\|_{\mathcal{G}} \leq 1$ for all $\Delta \in \mathcal{D}$. This observation can be used to define a robustness measure as follows:

• **Def. 4:** Consider the system (S^a). The *constrained stability measure* $\varrho_{\mathcal{G}}^{\mathcal{N}}$ is defined as:

$$\varrho_{\mathcal{G}}^{\mathcal{N}} \triangleq \begin{cases} 0 & \text{if } \|A\|_{\mathcal{G}} > 1; \\ \max_{\Delta \in \mathcal{D}} \|\Delta\|_{\mathcal{N}} & \text{if } \|A + \Delta\|_{\mathcal{G}} < 1\ \forall\, \Delta \in \mathcal{D}; \\ \min_{\Delta \in \mathcal{D}} \{\|\Delta\|_{\mathcal{N}}: \|A + \Delta\|_{\mathcal{G}} = 1\} & \text{otherwise.} \end{cases}$$

where $\|.\|_{\mathcal{N}}$ denotes a suitable operator norm defined in $\mathcal{D}$. In the special case where the induced operator norm $\|.\|_{\mathcal{G}}$ is used in the set $\mathcal{D}$, we will denote the constrained stability measure as $\varrho_{\mathcal{G}}$.

Remark 3: Let the set $\mathcal{B}\Delta^{\mathcal{N}}$ be the intersection of $\mathcal{D}$ with the origin centered ball of radius $\varrho_{\mathcal{G}}^{\mathcal{N}}$, i.e:

$$\mathcal{B}\Delta^{N} = \left\{\Delta \in \mathcal{D}: \|\Delta\|_{\mathcal{N}} \leq \varrho_{\mathcal{G}}^{\mathcal{N}}\right\}$$

Then, from definition 4 it follows that the family (S^a_{Δ}) is constrained stable for all perturbations $\Delta \in \mathcal{B}\Delta^{\mathcal{N}}$.

Remark 4: In principle $\varrho_{\mathcal{G}}^{\mathcal{N}}$ can be a *non–continuous* function of A. In the sequel we will show that under some assumptions that are commonly verified in practice, $\varrho_{\mathcal{G}}^{\mathcal{N}}$ is a *continuous, concave* function of the dynamics matrix A.

• **Theorem 1:** Assume that the perturbation set $\mathcal{D}$ is is a *closed cone* with vertex at the origin [10], (i.e. $\Delta^o \in \mathcal{D} \iff \alpha\Delta^o \in \mathcal{D}\ \forall\, 0 \leq \alpha$). Then $\varrho_{\mathcal{G}}^{\mathcal{N}}$ is a *continuous, concave* function of A.

Proof: The proof of the theorem is given in Appendix A.

Remark 5: Note that the class of sets considered in this theorem includes as a particular case sets of the form:

$$\mathcal{D} = \left\{\Delta: \Delta = \sum_{1}^{m} \mu_i E_i;\ \mu_i \geq 0,\ E_i \text{ given}\right\} \qquad (4)$$

which has been the object of much interest lately ([11–13] and references therein).

In the next lemma we introduce a *lower bound* of the constrained stability measure and we show that for *unstructured* perturbations (i.e. the case where $\mathcal{D} \equiv R^{n*n}$) this lower bound is saturated.

• **Lemma 1:**

$$\varrho_{\mathcal{G}} \geq 1 - \|A\|_{\mathcal{G}} \qquad (5)$$

Furthermore, for the unstructured perturbation case, i.e. the case where $\mathcal{D} \equiv R^{n*n}$, condition (5) is saturated.

Proof: The first part of the lemma can be easily proved from definition 4 and the triangle inequality. The second part follows by noting that for $\Delta^o \triangleq \frac{(1-\|A\|_{\mathcal{G}})A}{\|A\|_{\mathcal{G}}}$ (5) is saturated $\diamond$.

Remark 6: Note that the results of Lemma 1 can be used to find a lower bound for the constrained robustness measure in the general case when an operator norm different from $\|.\|_{\mathcal{G}}$ is used in the set $\mathcal{D}$. Since all finite dimensional matrix norms are equivalent [14], it follows that, given any norm $\mathcal{N}$ in the set $\mathcal{D}$, there exist a constant c such that $\|.\|_{\mathcal{G}} \leq c\|.\|_{\mathcal{N}}$. Hence $\varrho_{\mathcal{G}}^{\mathcal{N}} \leq \frac{\varrho_{\mathcal{G}}}{c}$.

3.1 Quadratic Constraints Case:

In this section we particularize our theoretical results for the special case where the constraint region is an hyperellipsoid. In this case, without loss of generality, we have:

$$\mathcal{G} = \{\underline{x}: \underline{x}'P\underline{x} \leq 1,\ P \in R^{n*n} \text{ positive definite}\}$$

Hence $\|\underline{x}\|^2_{\mathcal{G}} = \underline{x}'P\underline{x}$ and:

$$\begin{aligned} \|A\|^2_{\mathcal{G}} &= \max_{\underline{x}} \left\{\frac{\underline{x}'A'PA\underline{x}}{\underline{x}'P\underline{x}}\right\} \\ &= \max_{\underline{x}} \left\{\frac{\underline{x}'L'L'^{-1}A'L'LAL^{-1}L\underline{x}}{\underline{x}'L'L\underline{x}}\right\} \\ &= \max_{\|\underline{y}\|_2=1} \|LAL^{-1}\underline{y}\|^2_2 = \|LAL^{-1}\|^2_2 = \|\tilde{A}\|^2_2 \end{aligned} \qquad (6)$$

where $L'L = P$ and $\tilde{A} = LAL^{-1}$. In this case our approach yields a generalization of the well known technique of estimating the robustness measure by using quadratic based Lyapunov functions, (see [15] and references therein).

• *Example 1:* (multilinearly correlated perturbations) In the case of quadratic constraints and multilinearly correlated uncertainty, the lower bound on ϱ given by (5) can be tightened as follows. Assume that the set $\mathcal{D}$ is given by:

$$\mathcal{D} = \left\{\Delta \in R^{n*n}: L\Delta L^{-1} = U\begin{pmatrix}\tilde{\Delta} \\ 0\end{pmatrix}\right\}$$
$$\tilde{\Delta} \in R^{m*n},\ U'U = I_n,\ L'L = P \qquad (7)$$

Since the euclidian norm is invariant under multiplications by a unitary matrix we have:

$$\begin{aligned}\|A+\Delta\|_{\mathcal{G}} &= \|L(A+\Delta)L^{-1}\|_2 \\ &= \|\tilde{A}+U\begin{pmatrix}\tilde{\Delta}\\0\end{pmatrix}\|_2 = \|\begin{pmatrix}A_1\\A_2\end{pmatrix}+\begin{pmatrix}\tilde{\Delta}\\0\end{pmatrix}\|_2 \\ &= \|\begin{pmatrix}A_1+\tilde{\Delta}\\A_2\end{pmatrix}\|_2\end{aligned} \tag{8}$$

A well known result on matrix dilations establishes [16] that:

$$\|\begin{pmatrix}X\\A_2\end{pmatrix}\|_2 \le 1 \iff \|A_2\|_2 \le 1$$

$$\text{and } X = Y(I-A_2'A_2)^{\frac{1}{2}},\ \|Y\|_2 \le 1$$

hence it follows that:

$$\|A+\Delta\|_{\mathcal{G}} = 1 \iff \|(A_1+\tilde{\Delta})N\|_2 = 1 \tag{9}$$

where $N \triangleq (I-A_2'A_2)^{\frac{-1}{2}}$. Finally, by defining $\|\Delta\|_{\mathcal{N}} \triangleq \|\tilde{\Delta}N\|_2$ and using the results of Lemma 1, we get:

$$\varrho_{\mathcal{G}}^{\mathcal{N}} = 1 - \|A_1N\|_2 \tag{10}$$

Remark 7: Note that when $A_2 = 0$ we recover the results of Lemma 1, since in this case $\varrho = 1-\|A_1\|_2 = 1-\|A\|_{L'L}$

• *Example 2:* (unstructured perturbation)

In this case, Theorem 2 yields $\varrho_{\mathcal{G}} = 1-\|A\|_{\mathcal{G}}$ where:

$$\|A\|_{\mathcal{G}}^2 = \|A\|_P^2 = \max_{\underline{x}}\left(\frac{\underline{x}'A'PA\underline{x}}{\underline{x}'P\underline{x}}\right) \tag{11}$$

Consider now the case where $\varrho_{\mathcal{G}} > 0$. Then, there exists Q positive definite such that:

$$A'PA - P = -Q \tag{12}$$

and:

$$\|A\|_{\mathcal{G}}^2 = \max_{\underline{x}}\left(1-\frac{\underline{x}'Q\underline{x}}{\underline{x}'P\underline{x}}\right) \le 1-\frac{\sigma_{min}(Q)}{\sigma_{Max}(P)} \tag{13}$$

Hence:

$$\varrho_{\mathcal{G}} = 1-\|A\|_{\mathcal{G}} \ge 1-\left(1-\frac{\sigma_{min}(Q)}{\sigma_{Max}(P)}\right)^{\frac{1}{2}} \tag{14}$$

A common technique in state space robust analysis is to obtain robustness bounds from equation (12) ([17–18]). This case can be accommodated by our formalism by recognizing the fact that once P is selected, the system becomes effectively constrained to remain within an hyperellipsoidal region. It has been suggested ([17–18]) that good robustness bounds can be obtained from (12) when P is selected such that $Q = I$. In this case our approach yields:

$$\varrho_{\mathcal{G}} = 1-\|A\|_{\mathcal{G}} = 1-\left(1-\frac{1}{\sigma_{Max}(P)}\right)^{\frac{1}{2}} \tag{15}$$

which coincides with the robustness bound found in [18].

• *Example 3:* (Unstructured perturbation, A semisimple) Consider the case where A is semisimple, i.e.

$$\begin{aligned}A &= L^{-1}\Lambda L \\ \Lambda &= diag\left\{\begin{pmatrix}\sigma_1 & \omega_1\\-\omega_1 & \sigma_1\end{pmatrix},\ldots,\begin{pmatrix}\sigma_p & \omega_p\\-\omega_p & \sigma_p\end{pmatrix},\sigma_{p+1},\ldots,\sigma_n\right\}\end{aligned} \tag{16}$$

Then, the maximum of the stability measure, $\varrho_{\mathcal{G}}$, over all possible positive definite matrices P, is achieved for $P = L'L$.

Proof: The proof follows by noting that $\|A\|_{L'L} = \rho(A)$ where $\rho(.)$ denotes the spectral radius, which is a lower bound for any matrix norm [14] ⋄.

3.2 Polyhedral Constraints

Consider now the case where the region $\mathcal{G}$ is polyhedral, i.e. the case where:

$$\mathcal{G} = \{\underline{x}: |G\underline{x}| \le \underline{\omega}\} \tag{17}$$

where $G \in R^{p*n}$, rank$(G) = n$, $\underline{\omega} \in R^p$, $\omega_i > 0$ and the $|.|$ should be interpreted on a component by component sense. Although this case is of practical importance, up to date a technique to estimate the robustness of such systems was unavailable, except perhaps to fit an hyperellipsoidal region within the admissible region and then use some of the bounds available for the quadratic case. Such a technique is clearly inappropriate since it guarantees robust stability *only* in a certain subregion of the region of interest. In this section we show that polyhedral regions fit naturally within our formalism and that in this case $\varrho_{\mathcal{G}}^{\mathcal{N}}$ can be efficiently computed as the minimum of the solution of p Linear Programming problems.

• **Theorem 2:** Let $\varrho_i^{\mathcal{N}}$ be the solution of the following optimization problem:

$$\varrho_i^{\mathcal{N}} = \min_{\Delta\in\mathcal{D}}\{\|\Delta\|_{\mathcal{N}}: \|H+\Delta H\|_1^{(i)} \ge 1\} \tag{18}$$

where:

$$W = diag\{w_i\},\ H \triangleq W^{-1}GA(G'G)^{-1}G'W$$
$$\Delta H \triangleq W^{-1}G\Delta(G'G)^{-1}G'W$$

and where $\|M\|_1^{(i)}$ indicates the l_1 norm of the ith row of the matrix M. Then:

$$\varrho_{\mathcal{G}}^{\mathcal{N}} = \min_{1\le i\le p}\{\varrho_i^{\mathcal{N}}\} \tag{19}$$

Proof: It is easily shown that:

$$\|\underline{x}\|_{\mathcal{G}} = \max_{1\le i\le p}\left\{\frac{|G\underline{x}|_i}{\omega_i}\right\} = \|W^{-1}G\underline{x}\|_\infty \tag{20}$$

From the definition of H we have that $W^{-1}GA = HW^{-1}G$. Hence:

$$\|A\underline{x}\|_{\mathcal{G}} = \|W^{-1}GA\underline{x}\|_\infty = \|HW^{-1}G\underline{x}\|_\infty$$

and $\|A\|_{\mathcal{G}} = \|H\|_\infty$ Assume that the lemma is false and that there exist $\tilde{\varrho}$ and $\tilde{\Delta}$ such that:

$$\|A+\tilde{\Delta}\|_{\mathcal{G}} = 1;\ \|\tilde{\Delta}\|_{\mathcal{N}} = \tilde{\varrho} < \varrho_{\mathcal{G}}^{\mathcal{N}} \tag{21}$$

Since $\|A+\tilde{\Delta}\|_{\mathcal{G}} = 1$ there exists i^o such that $\|H+\tilde{\Delta}H\|_1^{(i^o)} = 1$, $\|H+\tilde{\Delta}H)\|_1^{(j)} \le 1$, $j \ne i^o$, but this implies (eq. (18)) that $\varrho_{i^o}^{\mathcal{N}} \le \tilde{\varrho}$ which contradicts (21) ⋄.

• *Example 4:* (unstructured perturbation) Consider the following case:

$$A = \begin{pmatrix}0.8 & 0.5\\-0.0208 & 0.5083\end{pmatrix}\ G = \begin{pmatrix}1.0 & 2.0\\-1.5 & 2.0\end{pmatrix}\ \underline{\omega} = \begin{pmatrix}5.0\\10.0\end{pmatrix} \tag{22}$$

Then, from the definition of H, we have that:

$$H = \begin{pmatrix}0.7583 & 0.0\\-0.2083 & 0.55\end{pmatrix},\ \|A\|_{\mathcal{G}} = 0.7583 \tag{23}$$

and, from Theorem 2,

$$\varrho_i = \min_{\|\Delta\|_{\mathcal{G}}}\left\{\|\Delta\|_{\mathcal{G}}: \sum_{j=1}^{2}|H+\Delta|_{ij} = 1\right\}\ i = 1,2 \tag{24}$$

Casting the problems (24) into a linear programming form and solving we have that:

$$\varrho_1 = 0.2417,\ \varrho_2 = 0.2417 \text{ and } \varrho_{\mathcal{G}} = \min_{1\le i\le 2}\varrho_i = 0.2417$$

Note that $\varrho_{\mathcal{G}} = 1-\|A\|_{\mathcal{G}} = 0.2417$ as shown in Lemma 1.

4. Application to Robust Controllers Design

Consider the *Linear Robust Constrained Control Synthesis Problem* introduced in section 2.3. Let $p_\Omega(u)$ be the Minkowsky gauge for the set Ω and denote by $\|.\|_\Omega$ the corresponding norm induced in R^m. It follows that, given a feedback control law of the form $\underline{u}_k = F\underline{x}_k$, the control bounds are satisfied if and only if:

$$\|F\|_{\mathcal{G},\Omega} \triangleq \sup_{\|\underline{x}\|_{\mathcal{G}} \leq 1} \|F\underline{x}\|_\Omega \leq 1$$

Hence a full state feedback matrix F that solves the synthesis problem can be found solving the following optimization problem:

$$\max_F \{\varrho_{\mathcal{G}}^{\mathcal{N}}(F)\} \tag{25}$$

subject to:

$$\begin{aligned} &\varrho_{\mathcal{G}}^{\mathcal{N}}(F) \triangleq \min_{\Delta \in \mathcal{D}} \{\|\Delta\|_{\mathcal{N}}: \|A + BF + \Delta\|_{\mathcal{G}} = 1\} \\ &\|F\|_{\mathcal{G},\Omega} \leq 1 \end{aligned} \tag{26}$$

Since from Theorem 1, $\varrho_{\mathcal{G}}^{\mathcal{N}}(F)$ is a concave function, and since $\|F\|_{\mathcal{G},\Omega} \leq 1$ is a convex constraint, it follows that (25) has a global optimum. Hence, the problem of finding the *maximally* robust controller leads to convex, albeit non-differentiable, optimization problems, which can be solved using a number of techniques ([19]). In the remainder of this section, we give several design examples using the proposed technique.

- *Example 5:* Consider the following system:

$$\begin{aligned} &A = \begin{pmatrix} 0 & 1 \\ 0.505 & -0.51 \end{pmatrix} \; B = \begin{pmatrix} 1 \\ 0 \end{pmatrix} \\ &\mathcal{G} = \{\underline{x}: \|\underline{x}\|_2 \leq 1\} \end{aligned} \tag{27}$$

The open–loop system has poles at $s_1 = 0.5$ and $s_2 = -1.01$. Assume that the perturbation set is such that changes the position of the poles while maintaining constant their sum, i. e:

$$\mathcal{D} = \left\{\Delta: \Delta = \mu E,\; E \triangleq \begin{pmatrix} 0 & 0 \\ 0 & 1 \end{pmatrix},\; \mu \in \Re\right\} \tag{28}$$

Note that $\|E\|_2 = 1$ hence $\|\Delta\|_2 = |\mu|$.

In this case, the solution to the unconstrained maximally robust control problem can be computed by solving a matrix dilation problem [16]. Rewrite the dynamics matrix as:

$$A = \begin{pmatrix} x_1 & x_2 \\ a_1 & a_2 \end{pmatrix}$$

where x_i denote elements that can be modified using state–feedback. Since matrix dilations are norm–increasing we have that:

$$\begin{aligned} \|A + \mu E\|_2 &\geq \max\{\|(\, a_1 \quad a_2 + \mu\,)\|_2\} \\ &= \sqrt{a_1^2 + (a_2 + \mu)^2} \end{aligned} \tag{29}$$

Define now:

$$\begin{aligned} \mu^0 &= \text{argmin}\,\{|\mu|, \mu \in \Re: a_1^2 + (a_2 + \mu)^2 = 1\} \\ &= \sqrt{(1 - a_1^2)} - |a_2| \end{aligned} \tag{30}$$

From (29) and (30) it follows that $\|A + \mu^o E\|_2 \geq 1$ which implies that $\varrho_2(F) \leq \mu^o$ for all F. Furthermore, from the definition of μ^o it follows that if F is selected such that $x_1 = x_2 = 0$, then $\varrho_2(F) = \mu^o$. Hence, this choice of F yields the solution to the unconstrained problem. In this particular example we have:

$$F^o = (\,0 \quad 1\,),\; \varrho_2 = 0.3531 \tag{31}$$

Consider now a feedback matrix F and let A_{cl} be the corresponding *closed-loop* matrix, i.e:

$$A_{cl} = A + BF = \begin{pmatrix} a_{11} & a_{12} \\ a_{21} & a_{22} + \mu \end{pmatrix} \tag{32}$$

The corresponding value of the robustness measure can be computed using standard results on matrix dilations [16] as follows: The set Υ of numbers μ such that $\|A_{cl}\|_2 \leq 1$ can be parametrized as:

$$\Upsilon = \left\{\mu: \mu = -a_{22} - y a_{11} z + (1 - y^2)^{\frac{1}{2}} w (1 - z^2)^{\frac{1}{2}}\right\} \tag{33}$$

where:

$$\begin{aligned} y &= \frac{a_{21}}{(1 - a_{11}^2)^{\frac{1}{2}}} \\ z &= \frac{a_{12}}{(1 - a_{11}^2)^{\frac{1}{2}}} \\ w &\in \Re,\; |w| \leq 1 \end{aligned} \tag{34}$$

From (33) it follow that the constrained stability margin of A_{cl} is given by:

$$\varrho_2(F) = |a_{22} + y a_{11} z - (1 - y^2)^{\frac{1}{2}} (1 - z^2)^{\frac{1}{2}} \text{sign}(a_{22} + y a_{11} z)|$$

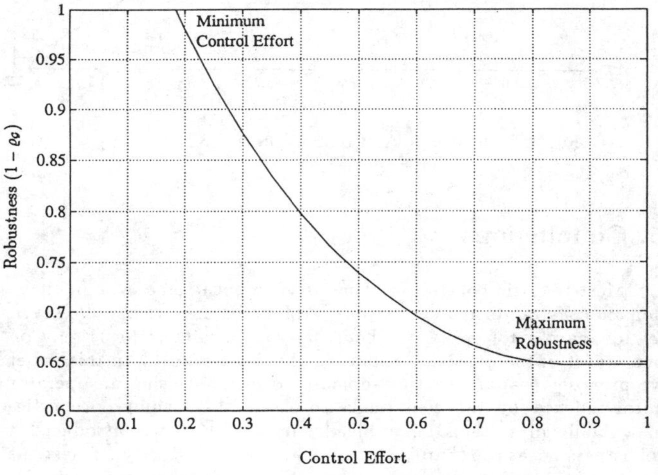

Fig 1. Robustness vs. Control Effort for Example 5

Figure 1 shows $\varrho_2(F)$ versus $\|F\|_2$, the norm of the solution to (25). For $\|F\|_2 = 1$, we recover the unconstrained solution, for $\|F\|_2 = 0.1850$, we get the minimum control effort capable of stabilizing (in the constrained sense) the nominal system. Note the trade–off between control effort and robustness. In particular, there exist a region where the curve is flat, i.e. the control effort can be reduced while essentially maintaining the same robustness obtained with a "maximum robustness" type design.

- *Example 6:* Polyhedral constraints, unstructured perturbation

Consider the following system:

$$\begin{aligned} &A = \begin{pmatrix} 0.8 & 0.5 \\ -0.4 & 1.2 \end{pmatrix} \; B = \begin{pmatrix} 0 \\ 1 \end{pmatrix} \\ &G = \begin{pmatrix} 1.0 & 2.0 \\ -1.5 & 2.0 \end{pmatrix} \; \underline{\omega} = \begin{pmatrix} 5.0 \\ 10.0 \end{pmatrix} \; \Omega = \{u: |u| \leq \gamma\} \end{aligned} \tag{35}$$

Since the constraint sets $\mathcal{G}$ and Ω are polyhedral, the synthesis problem can be cast in the following format:

$$\min_F \epsilon$$

subject to:

$$\begin{aligned} \|A + BF\|_{\mathcal{G}} &\leq \epsilon \\ \|F\|_{\mathcal{G},\Omega} &\leq 1 \end{aligned}$$

which can be transformed into an LP problem and solved using the simplex method. Note that a similar design algorithm was proposed by Vassilaki et. al. [20], although in their case the goal was to find admissible linear controllers for systems under polyhedral constraints, without taking into account robustness considerations. Figure 2 shows the constrained robustness measure versus γ, the bound on the control effort. Note that the minimum control effort required to stabilize the system is $\gamma = 2.6$.

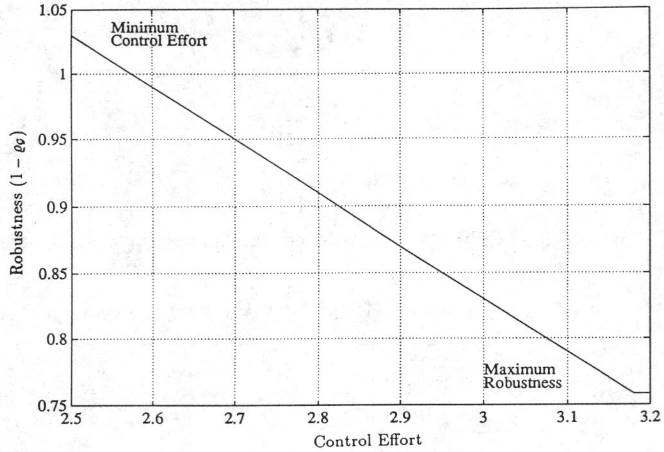

Fig 2. Robustness vs. Control Effort for Example 6

5. Conclusions

Most realistic control problems involve both some type of time–domain constraints and certain degree of model uncertainty. However, few of the control design methods currently available focus only on one aspect of the problem. Following the spirit of [6–7], in this paper we propose to approach time–domain constraints using an operator norm induced by the constraints to assess the stability properties of a family of systems. Specifically, in section II we introduced a robustness measure that indicates how well the family of systems under consideration satisfies a given set of constraints. In section III we explored the properties of this robustness measure for the case of additive parametric model uncertainty and we showed that our formalism provides a generalization of the well known technique of estimating robustness bounds from the solution of a Lyapunov equation. We then proposed, in section IV, a synthesis procedure for fixed order controllers, based upon maximization of the robustness measure subject to additional performance constraints such as bounds on the control effort. There we showed that the proposed design procedure leads to convex optimization problem. We believe that the results presented here will provide a valuable new approach to the problems of robust controllers analysis and design for linear systems. Further, since our approach is based purely upon time–domain analysis, we have reasons to believe the theory could be extended to encompass non–linear systems in a much more direct fashion than other currently used techniques.

Perhaps the more severe limitation of the theory in its present form arises from the fact that the incorporation of additional performance constraints of the form of a bound on the norm of a relevant transfer function results in non–convex optimization problems. We are currently looking into a solution to this problem by using an observer–based parametrization of all stabilizing controllers. It is expected that this formulation will be able to handle more general performance constraints as well as dynamic uncertainty, at the price of resulting in higher order controllers.

References

[1]. M. Sznaier, "Suboptimal Feedback Control of Constrained Linear Systems," *Ph.D. Dissertation,* University of Washington, 1989.

[2]. S. Boyd et. al., "A New CAD Method and Associated Architectures for Linear Controllers," *IEEE Trans. Automat. Contr.,* Vol 33, 3, pp 268–283, March 1988.

[3]. E. Polak and S. Salcudean, "On the Design of Linear Multivariable Feedback Systems Via Constrained Nondifferentiable Optimization in H_∞ Spaces," *IEEE Trans. Automat. Contr.,* Vol 34, 3, pp 268–276, March 1989.

[4]. J.W. Helton and A. Sideris, "Frequency Response Algorithms for H_∞ Optimization With Time Domain Constraints, *IEEE Trans. Automat. Contr.,* Vol 34, 4, pp. 427–434, April 1989.

[5]. A. Sideris and H. Rotstein, "H_∞ Optimization with Time Domain Constraints Over a Finite Horizon," *Proc. of the 29th IEEE CDC,* Hawaii, Dec. 5–7, 1990, pp. 1802–1807.

[6]. M. Sznaier, "Norm Based Robust Control of State– Constrained Discrete Time Linear Systems," *IEEE Trans. Automat. Contr.,*, Vol 37, 7, pp. 1057–1062, July 1992.

[7]. M. Sznaier and A. Sideris, "Norm Based Optimally Robust Control of Constrained Discrete Time Linear Systems," *Proc. of the 1991 ACC,* Boston, Massachusetts, June 23–25, 1991, pp 2710–2715.

[8]. J. P Lasalle, "The Stability and Control of Discrete Processes," *Vol 62 in Applied Mathematics Series,* Springer–Verlag, New–York, 1986.

[9]. J. B. Conway, "A Course in Functional Analysis," *Vol 96 in Graduate Texts in Mathematics,* Springer–Verlag, New–York, 1990.

[10]. D. G. Luenberger, "Optimization by Vector Space Methods," Wiley, New–York, 1969.

[11]. R. K. Yedavalli, "Improved Measures of Stability Robustness for Linear State Space Models," *IEEE Trans. Automat. Contr.,* Vol AC-30, pp. 557–579, June 1985.

[12]. K. Zhou and P. P. Khargonekar, "Stability Robustness Bounds for Linear State–Space Models with Structured Uncertainty," *IEEE Trans. Automat. Contr.,* Vol AC-32, 7, pp 621–623, July 1987.

[13]. L. H. Keel et. al., " Robust Control With Structured Perturbations," *IEEE Trans. Automat. Contr.,* Vol 33, 1, pp. 68-77, January 1988.

[14]. R. A. Horn and C. R. Johnson, "Matrix Analysis," Cambridge University Press, 1985.

[15]. D. D. Siljak, " Parameter Space Methods for Robust Control Design: A Guided Tour, " *IEEE Trans. Automat. Contr.,* Vol 34, 7, pp. 674–687, July 1989.

[16]. J. C. Doyle, "Lecture Notes in Advances in Multivariable Control," ONR /Honeywell Workshop, Minneapolis, 1984.

[17]. R. V. Patel and M. Toda, "Quantitative Measures of Robustness for Multivariable Systems," *Proc. Joint Automat. Contr. Conf.,* San Francisco, CA, 1980, paper TP-8A.

[18]. M. E. Sezer and D. D. Siljak, "Robust Stability of Discrete Systems," *Int. J. Control.,* Vol 48, 5, pp. 2055–2063, Nov. 1988.

[19]. F. H. Clarke, "Optimization and Nonsmooth Analysis," Canadian Mathematical Society Series of Monographs and Advanced Texts, Wiley, New–York, 1983.

[20]. M. Vassilaki et. al., "Feedback Control of Discrete–Time Systems Under State and Control Constraints," *Int. J. Control,* Vol 47, 6, pp. 1727–1735, 1988.

[21]. A. Naylor and G. R. Sell, "Linear Operator Theory in Engineering and Science," *Vol 40 in Applied Mathematical Sciences,* Springer–Verlag, New York, 1982.

Appendix A

Proof of Theorem 1 We begin by introducing two preliminary results:

• **Lemma 2:** Consider the system (S^a). Assume that the perturbation set D is *a closed cone with vertex at the origin* [10], i.e. $\Delta^o \in \mathcal{D} \iff \alpha\Delta^o \in \mathcal{D}\ \forall\ 0 \leq \alpha$ and that (S^a) is constraint stable (i.e. $\|A\|_{\mathcal{G}} < 1$). Let:

$$\Delta^o = \operatorname*{argmin}_{\Delta \in \mathcal{D}} \{\|\Delta\|_{\mathcal{N}} : \|A + \Delta\|_{\mathcal{G}} = 1\} \tag{A1}$$

and consider a sequence $A^i \to A$ such that $\|A^i\|_{\mathcal{G}} < 1$. Finally, define the sequence λ^i as:

$$\lambda^i = \min_{\lambda \in \Re^+} \left\{\lambda : \|A^i + \lambda\Delta^o\|_{\mathcal{G}} = 1\right\} \tag{A2}$$

Then the sequence λ^i has an accumulation point at 1.

Proof: Since $\|A^i\|_{\mathcal{G}} < 1$ and since $\mathcal{D}$ is a closed cone it follows that λ^i is well defined. Furthermore, from (A2) it follows that:

$$\lambda^i \leq \frac{1 + \|A^i\|_{\mathcal{G}}}{\|\Delta^o\|_{\mathcal{G}}} \leq \frac{2}{\|\Delta^o\|_{\mathcal{G}}} \tag{A3}$$

Hence from Bolzanno–Weierstrass' theorem [21] it follows that λ^i has an accumulation point $\tilde{\lambda}$ and that there exist a subsequence $\tilde{\lambda}^i \to \tilde{\lambda}$. Hence:

$$\|A^i + \tilde{\lambda}^i\Delta^o\|_{\mathcal{G}} = 1$$

and since $A^i \to A$ then:

$$\|A + \tilde{\lambda}\Delta^o\|_{\mathcal{G}} = 1 \tag{A4}$$

Assume that $\tilde{\lambda} < 1$ and let $\hat{\Delta} \triangleq \tilde{\lambda}\Delta^o$ Then $\|\hat{\Delta}\|_{\mathcal{N}} < \|\Delta^o\|_{\mathcal{N}}$, $\|A + \hat{\Delta}\|_{\mathcal{G}} = 1$ and $\hat{\Delta} \in \mathcal{D}$ (since $\mathcal{D}$ is a cone) which contradicts (A1). Assume now that $\tilde{\lambda} > 1$. Then, for i large enough, $\tilde{\lambda}^i > 1$, which together with (A2) implies that:

$$\|A^i + \Delta^o\|_{\mathcal{G}} < 1 \tag{A5}$$

and hence:

$$\|A + \Delta^o\|_{\mathcal{G}} < 1 \tag{A6}$$

which contradicts (A1). Therefore $\tilde{\lambda} = 1$ ⋄.

• **Lemma 3:** Let $\rho_1 > 0, \rho_2 > 0$ and $0 \leq \lambda \leq 1$ be given numbers and assume that $\mathcal{D}$ is a cone with vertex at the origin. Consider the following sets:

$$\begin{aligned}\rho_1 B\Delta &= \{\Delta \in \mathcal{D} : \|\Delta\|_{\mathcal{N}} \leq \rho_1\}\\ \rho_2 B\Delta &= \{\Delta \in \mathcal{D} : \|\Delta\|_{\mathcal{N}} \leq \rho_2\}\\ \rho B\Delta &= \{\Delta \in \mathcal{D} : \|\Delta\|_{\mathcal{N}} \leq \rho \triangleq \lambda\rho_1 + (1-\lambda)\rho_2\}\end{aligned} \tag{A7}$$

Then $\rho B\Delta \subseteq \lambda\rho_1 B\Delta + (1-\lambda)\rho_2 B\Delta$

Proof: Consider any $\Delta^o \in \rho B\Delta$. Then:

$$\begin{aligned}\Delta^o &= \frac{\|\Delta^o\|_{\mathcal{N}}}{\rho}\left[\frac{\rho\Delta^o}{\|\Delta^o\|_{\mathcal{N}}}\right]\\ &= \frac{\|\Delta^o\|_{\mathcal{N}}}{\rho}\left[\lambda\rho_1\frac{\Delta^o}{\|\Delta^o\|_{\mathcal{N}}} + (1-\lambda)\rho_2\frac{\Delta^o}{\|\Delta^o\|_{\mathcal{N}}}\right]\\ &= [\lambda\Delta_1 + (1-\lambda)\Delta_2]\end{aligned} \tag{A8}$$

where:

$$\begin{aligned}\Delta_1 &= \alpha\rho_1\frac{\Delta^o}{\|\Delta^o\|_{\mathcal{N}}}\\ \Delta_2 &= \alpha\rho_2\frac{\Delta^o}{\|\Delta^o\|_{\mathcal{N}}}\\ \alpha &= \frac{\|\Delta^o\|_{\mathcal{N}}}{\rho} \leq 1\end{aligned} \tag{A9}$$

The proof is completed by noting that from (A9) and the hypothesis it follows that $\Delta_1 \in \rho_1 B\Delta$ and $\Delta_2 \in \rho_2 B\Delta$ ⋄.

Proof of Theorem 1

Assume that $\varrho_{\mathcal{G}}^{\mathcal{N}}$ is *not* continuous. Then, given $\epsilon > 0$, for every $\delta > 0$ there exist A_δ such that $\|A_\delta - A\|_{\mathcal{G}} \leq \delta$ and $|\varrho_{\mathcal{G}}^{\mathcal{N}}(A_\delta) - \varrho_{\mathcal{G}}^{\mathcal{N}}| > \epsilon$. Hence there exist a sequence $A^i \to A$ such that $\varrho_{\mathcal{G}}^{\mathcal{N}^i} \not\to \varrho_{\mathcal{G}}^{\mathcal{N}}$. Furthermore, it is easily seen that the sequence $\varrho_{\mathcal{G}}^{\mathcal{N}^i}$ is bounded and therefore is contains a convergent subsequence. It follows that there exist a sequence $A^i \to A$ such that $\varrho_{\mathcal{G}}^{\mathcal{N}^i} \to \tilde{\varrho} \neq \varrho_{\mathcal{G}}^{\mathcal{N}}$. Let:

$$\Delta^i = \operatorname*{argmin}_{\Delta \in \mathcal{D}} \left\{\|\Delta\|_{\mathcal{N}} : \|A^i + \Delta\|_{\mathcal{G}} = 1\right\} \tag{A10}$$

From (A10) it follows that $\|\Delta^i\|_{\mathcal{G}} \leq 1 + \|A^i\|_{\mathcal{G}}$. It follows then that the sequence Δ^i is bounded and therefore, since $R^{n\times n}$ with a finite dimensional matrix norm is complete and since $\mathcal{D}$ is a closed set, it has an accumulation point $\tilde{\Delta}$ (Bolzano Weierstrass) and a convergent subsequence $\tilde{\Delta}^i \to \tilde{\Delta}$ such that $\|A + \tilde{\Delta}\|_{\mathcal{G}} = 1$. Furthermore, from the definition of Δ^o it follows that

$$\tilde{\varrho} = \|\tilde{\Delta}\|_{\mathcal{N}} > \|\Delta^o\|_{\mathcal{N}} = \varrho_{\mathcal{G}}^{\mathcal{N}} \tag{A11}$$

Hence, for i large enough,

$$\|\tilde{\Delta}^i\|_{\mathcal{N}} > \|\Delta^o\|_{\mathcal{N}} \tag{A12}$$

Applying Lemma 3, we have that there exist a sequence $\lambda^i \to 1$ such that:

$$\lambda^i = \min_{\lambda \in \Re^+} \left\{\lambda : \|A^i + \lambda\Delta^o\|_{\mathcal{G}} = 1\right\} \tag{A13}$$

From (A12) and since $\lambda^i \to 1$ it follows that for i large enough

$$\begin{aligned}\|\lambda^i\Delta^o\|_{\mathcal{N}} &< \|\tilde{\Delta}^i\|_{\mathcal{N}}\\ \|A^i + \lambda^i\Delta^o\|_{\mathcal{G}} &= 1\end{aligned} \tag{A14}$$

and, since $\mathcal{D}$ is a cone, $\lambda^i\Delta^o \in \mathcal{D}$, which contradicts (A10). The proof is completed by noting that since *all* finite dimensional matrix norms are equivalent [14] then continuity in the $\|.\|_{\mathcal{G}}$ norm implies continuity in any other norm defined over $R^{n\times n}$ ⋄.

To prove concavity, start by considering a convex linear combination $A = \lambda A_1 + (1-\lambda)A_2$, $\lambda \leq 1$ of given matrices A_1 and A_2. Then, from Lemma 4 it follows that:

$$\begin{aligned}\max_{\Delta \in \rho B\Delta} \|A + \Delta\|_G &\leq \max_{\substack{\Delta_1 \in \rho_1 B\Delta\\ \Delta_2 \in \rho_2 B\Delta}} \|\lambda(A_1 + \Delta_1) + (1-\lambda)(A_2 + \Delta_2\|_{\mathcal{G}}\\ &\leq \lambda \max_{\Delta_1 \in \rho_1 B\Delta} \|A_1 + \Delta_1\|_{\mathcal{G}} + (1-\lambda) \max_{\Delta_2 \in \rho_2 B\Delta} \|A_2 + \Delta_2\|_{\mathcal{G}}\end{aligned} \tag{A15}$$

Consider now the case where $\rho_1 = \varrho_{\mathcal{G}}^{\mathcal{N}}(A_1)$ and $\rho_2 = \varrho_{\mathcal{G}}^{\mathcal{N}}(A_2)$. Then it follows from the definition of $\varrho_{\mathcal{G}}^{\mathcal{N}}$ that both maximizations in the right hand side of (A15) yield 1 and therefore:

$$\max_{\Delta \in \rho B\Delta} \|A + \Delta\|_{\mathcal{G}} \leq 1 \tag{A16}$$

Hence, from the definition of $\varrho_{\mathcal{G}}^{\mathcal{N}}$:

$$\varrho_{\mathcal{G}}^{\mathcal{N}}[\lambda A_1 + (1-\lambda)A_2] \geq \varrho = \lambda\varrho_{\mathcal{G}}^{\mathcal{N}}(A_1) + (1-\lambda)\varrho_{\mathcal{G}}^{\mathcal{N}}(A_2) \diamond$$

Proceedings of the 31st Conference
on Decision and Control
Tucson, Arizona • December 1992

DUALITY IN DESIGNING STABILIZING CONTROL OF UNCERTAIN LINEAR SYSTEMS

Keqin Gu

School of Engineering
Southern Illinois University at Edwardsville
Edwardsville, IL 62026-1805

ABSTRACT

The design of stabilizing linear output feedback control of uncertain systems is a computationally demanding optimization problem as local minima may exsit which are distinct from the global minimum. However, in some special cases, quasi-convexity can be proven through a simple re-definition of variables and/or a re-formulation of the problem in the *dual* form. Such special cases include state feedback control for both discrete and continuous time systems.

1. INTRODUCTION

Linear systems with exact mathematical models can be analyzed by well established theory. Unfortunately, uncertainties often exist in practical systems. Therefore, there are strong interests among control researchers to study the control of uncertain systems. This paper is mainly concerned about uncertainties which may be time varying. Much work has been done for this subject. For example, Barmish [1] studied the necessary and sufficient conditions for an uncertain linear system to be stabilizable by state feedback control. The uncertainties of A and B are assumed independent, and the control he proposed was nonlinear. Petersen [12] showed that stabilizability by a nonlinear control does not necessarily mean stabilizability by a linear control. Hollot and Barmish [9] proposed a necessary and sufficient condition of linear state feedback stabilizability for systems without uncertainties in B matrix. Barmish [2] derived an interesting relation between stabilizability of systems with and without uncertainties in B. Gu, *et. al.*, [8] later formulated the condition into an convex optimization problem. Zhou, *et. al.*, [13] solved a linear state stabilization problem of a systems with uncertainties of a special structure through a Riccati equation approach, which was later found to have interesting connections with H^∞ problems [11].

Optimization approach can be used to a very large class of uncertain linear stabilization problem. For example, Horisberger and Belanger [10] proposed an algorithm to check stabilizability of a very general class of uncertainties. Unfortunately, the algorithm was extremely complicated. A computationally much less demanding algorithm was proposed later [6]. The algorithm consists of an optimization of a nonsmooth function. Yet, the algorithm is still rather costly. Especially, it is possible for local minima to be distinct from the global minimum. Other works along this line include Structured Lyapunov Function approach[5] and Linear Optimization approach [3].

This paper discusses a number of special cases of output stabilization problem. Through a transformation of variables and/or consideration of the dual formulation, these special cases can be formulated in a quasiconvex problem, therefore, any local minima will be identical to the global minimum. These special cases include such important problems as state feedback control for both continuous and discrete time systems.

2. DUALITY IN QUADRATIC STABILITY

We will mainly discuss the discrete time case. The continuous time case is more straightforward, and will be commented on later. Consider an uncertain system described by the following equations:

$$x(k+1) = (A + \Delta A(k))x(k) + (B + \Delta B(k))u(k), \tag{1}$$
$$y(k) = (C + \Delta C(k))x(k), \tag{2}$$

where $A \in R^{n\times n}$, $B \in R^{n\times p}$, $C \in R^{m\times n}$, $x \in R^n$, $y \in R^m$, $u \in R^p$. $\Delta A(k)$, $\Delta B(k)$ and $\Delta C(k)$ represent the uncertainties of the system, which can be arbitrarily time varying, but a bounding set is known:

$$(\Delta A(k), \Delta B(k), \Delta C(k)) \in \Omega \qquad \text{for all } k, \tag{3}$$

where Ω is a compact set in $R^{n\times n} \times R^{n\times p} \times R^{m\times n}$. This form of uncertainty expression is very general. We also assume that $0 \in \Omega$, which means that the nominal system is also included in the set of all the possible systems. It is desired to design an output feedback controller to stabilize the system. It is sufficient to discuss the static feedback stabilization

$$u(k) = Ky(k), \tag{4}$$

as the dynamic output feedback control problem can be transformed into a static one[6]. The closed loop feedback system can be written as

$$x(k+1) = (A_c + \Delta A_c(k))x(k), \tag{5}$$

where

$$A_c = A + BKC. \tag{6}$$

The closed loop uncertainties satisfy

$$\Delta A_c(k) \in \Omega_c = \{E + FKC + BKG + FKG \mid (E, F, G) \in \Omega\}, \quad \text{for all } k, \tag{7}$$

We will restrict to the case where the convex hull of the uncertainty set $\text{conv}(\Omega)$ is polytopic. This is a more general assumption than polytopic uncertainty set. For example, the uncertainty may be a sum of multilinear functions of l uncertain parameters q_i's. It should also be noted that if $\text{conv}(\Omega)$ is polytopic, then $\text{conv}(\Omega_c)$ is also polytopic. The following quadratic stability criterion has become standard for such a problem:

Definition 1. A system of the form (5) and (7) is said to be quadratically stable if there exists a positive definite matrix P and a scalar $\alpha > 0$ such that

$$x^T[(A_c + E_c)^T P(A_c + E_c) - P]x^T \leq -\alpha\|x\|^2 \tag{9}$$

holds for arbitrary $x \neq 0$ and $E_c \in \Omega_c$. □

In a number of important cases, it is more convenient to formulate the quadratic stability in a *dual form* as described in the following lemma. Although the continuous time counterpart has been widely used and a number of important results are derived through its application [2,3,8], there does not seem to exist any such applications for the discrete time systems previously.

Lemma 1. A closed loop system is quadratically stable if and only if there exists a matrix $P > 0$ or a $S > 0$ such that one of the following conditions is satisfied for arbitrary $E_c \in \Omega_c$:

$$A. \quad P - (A_c + E_c)^T P(A_c + E_c) > 0; \tag{10}$$
$$B. \quad \begin{pmatrix} P & P(A_c + E_c) \\ (A_c + E_c)^T P & P \end{pmatrix} > 0; \tag{11}$$
$$C. \quad S - (A_c + E_c)S(A_c + E_c)^T > 0. \tag{12}$$
$$D. \quad \begin{pmatrix} S & S(A_c + E_c)^T \\ (A_c + E_c)S & S \end{pmatrix} > 0; \tag{13}$$

□

Proof: A is necessary and sufficient according to the definition and the compactness of Ω_c. Since $P > 0$, A is equivalent to

$$\begin{pmatrix} P & 0 \\ 0 & P - (A_c + E_c)^T P(A_c + E_c) \end{pmatrix} > 0. \tag{14}$$

But LHS of (11) and (13) can be obtained by multiplying the LHS of (14) by a nonsingular matrix on the right, and by its transpose on the left (let $S = P^{-1}$), B and D are proven. C can be proven similarly. ■

(12) and (13) are the dual forms of (10) and (11), which is analogous to the duality between control problems and observer problems. In this paper, the dual forms will be used to derive a method to design state feedback control of uncertain systems through quasi-convex optimization.

As discussed in [6], the general stabilization problem leads to an optimization problem with possibly multiple local minima. Therefore, the following sections will discuss a number of special cases in which the optimization becomes quasiconvex.

3. MAIN RESULTS

We will first consider the state feedback stabilization problem of discrete time systems. A number of extensions to other cases will be commented on later. Consider system (1) with uncertainty expressed as

$$(\Delta A(k), \Delta B(k)) \in \Omega_\mu = \{(\mu E, \mu F) \mid (E, F) \in \Omega_1\}, \quad \text{for all } k. \tag{15}$$

Since $\text{conv}(\Omega_\mu)$ is polytopic according to the assumption, the convex hull of Ω_1 is also polytopic. We want to design a linear state feedback control

$$u(k) = Kx(k) \tag{17}$$

to stabilize the system. Denote the resulting closed loop system as $\mathbf{S}_\mu(K)$. This system is obtained by setting $C = I$ and $\Delta C = 0$ in the general system (6)–(8) and scaling the size of the uncertainty set. We want to compute

CH3229-2/92/0000-2894$1.00 © 1992 IEEE

$$\bar{\mu} = \sup\{\mu \mid \mathbf{S}_\mu(K) \text{ is quadratically stable for some } K\}. \quad (22)$$

Applying the dual form of quadratic stability criterion (13), then

$$\bar{\mu} = \sup_{\substack{S>0\\K}} \left\{\mu \,\middle|\, \begin{pmatrix} S & S[(A+BK)+\mu(E+FK)]^T \\ [(A+BK)+\mu(E+FK)]S & S \end{pmatrix} > 0, \quad \text{for all } (E,F)\in\Omega_1\right\}. \quad (23)$$

Let

$$\underline{\nu} = \frac{1}{\bar{\mu}}, \quad (24)$$

and define

$$\hat{K} = KS. \quad (25)$$

Then it is easily seen that

$$\underline{\nu} = \inf_{\substack{S>0\\\hat{K}}} \left\{\nu \,\middle|\, \nu \begin{pmatrix} S & (AS+B\hat{K})^T \\ (AS+B\hat{K}) & S \end{pmatrix} + \begin{pmatrix} 0 & (ES+F\hat{K})^T \\ (ES+F\hat{K}) & 0 \end{pmatrix} > 0, \quad \text{for all } (E,F)\in\Omega_1\right\}. \quad (26)$$

For a pair of given $S > 0$ and $\hat{K}$ define

$$\nu_2(S,\hat{K}) = \inf \left\{\nu \,\middle|\, \nu \begin{pmatrix} S & (AS+B\hat{K})^T \\ (AS+B\hat{K}) & S \end{pmatrix} + \begin{pmatrix} 0 & (ES+F\hat{K})^T \\ (ES+F\hat{K}) & 0 \end{pmatrix} > 0, \quad \text{for all } (E,F)\in\Omega_1\right\}. \quad (27)$$

Note that the infimum of an empty set is defined as $+\infty$ by convention. Then

$$\underline{\nu} = \inf_{\substack{S>0\\\hat{K}}} \nu_2(S,\hat{K}) = \inf_{\substack{S>0\\\mathrm{tr}(S)=1\\\hat{K}}} \nu_2(S,\hat{K}). \quad (29)$$

From the above discussion, $\underline{\nu}$ can be computed by a minimization process. It turns out that this minimization problem is strictly quasiconvex as defined in the following:

Definition 2. A function $\phi(z)$ on a convex domain Θ is said to be quasiconvex if any arbitrary $z_1, z_2 \in \Theta$ and $0 < \gamma < 1$ satisfy

$$\phi(\gamma z_1 + (1-\gamma)z_2) \le \max\{\phi(z_1), \phi(z_2)\}.$$

It is said to be strictly quasiconvex if the inequality is strict whenever $\phi(z_1) \neq \phi(z_2)$. □

Remark 1. An important fact about quasiconvax functions is that any local minimum is automatically the global minimum. To be exact, if there exists a neighborhood $N(z_0)$ of z_0 such that $\phi(z) > \phi(z_0)$ for all $z \in N(z_0), z \neq z_0$, then z_0 is also the global minimum. If $\phi(\cdot)$ is strictly quasi-convex, then $\phi(z) \ge \phi(z_0)$ for all $z \in N(z_0)$ will be sufficient to guarantee z_0 to be a global minimum point. Most algorithms designed for convex optimizations can be directly used for quasi-convex problem as is discussed in Boyd and Barratt [4]. □

Theorem 1. $\nu_2(S,\hat{K})$ is strictly quasi-convex in the set

$$\Theta = \{(S,\hat{K}) \mid \mathrm{tr}(S) = 1, \nu_2(S,\hat{K}) < +\infty\}. \quad (30)$$

□

Proof: Similar to [6], omitted. ■

In order to compute ν_2, define another function $\nu_1(S,\hat{K},E,F)$ as the greatest eigenvalue of the generalized eigenvalue problem:

$$-\begin{pmatrix} 0 & (ES+F\hat{K})^T \\ ES+F\hat{K} & 0 \end{pmatrix}\xi = \nu \begin{pmatrix} S & (AS+B\hat{K})^T \\ (AS+B\hat{K}) & S \end{pmatrix}\xi \quad (39)$$

Notice that the matrix on the right hand side of (39) is positive definite for $(S,\hat{K}) \in \Theta$. Well estabilished algorithms are available to solve such generalized eigenvalue problem. Recall that $\mathrm{conv}(\Omega_1)$ is polytopic. Let $(E_i,F_i), i = 1,2,\ldots,n_v$ be the vertices of the convex hull of Ω_1. Then it is not difficult to conclude that

$$\nu_2(S,\hat{K}) = \max_{(E,F)\in\Omega_1} \nu_1(S,\hat{K},E,F) = \max_{i=1,\ldots,n_v} \nu_1(S,\hat{K},E_i,F_i). \quad (41)$$

Therefore, $\nu_2(S,\hat{K})$ can be computed by comparing n_v values of ν_1. After the optimizing or "almost optimizing" parameter $(S^*,\hat{K}^*)$ is obtained, the feedback control with gain

$$K = \hat{K}^* S^{*-1} \quad (42)$$

stabilizes the uncertain system with uncertainty set Ω_μ for any

$$\mu < \frac{1}{\nu_2(S^*,\hat{K}^*)}, \quad (43)$$

and the corresponding Lyapunov function is

$$V(x) = x^T S^{*-1} x. \quad (44)$$

Therefore, the design process can be summarized as follows:

1. Use any method (such as LQR) to design a feedback gain K to stabilize the nominal system.
2. Choose an arbitrary $Q > 0$ and solve Lyapunov equation
$$A_c S A_c^T - S = -Q$$
to obtain a $S > 0$. Scale S to make $\mathrm{tr}(S) = 1$. Let $\hat{K} = KS$. Then $(S,\hat{K}) \in \Theta$.
3. Use the above S and $\hat{K}$ as initial condition to minimize ν_2. The feedback gain and the size of stabilizable uncertainty are obtained as (42) and (43).

Remark 2. A problem dual to the above state feedback stabilization problem is equations (1–4) with

$$B = I, \quad \text{and} \quad \Delta B \equiv 0. \quad (45)$$

Derivation parallel to the above discussion but using (11) instead of (13) and P instead of S leads to a similar algorithm. □

Remark 3. Continuous time case can also be solved in a similar manner. In which case

$$PA_c + A_c^T P < 0 \quad (46)$$

should be used instead of (13) in state feedback stabilization case. Compared to the approaches in [3] and [8], this approach offers the advantage of automatically computing the stabilizable uncertainty bound. Similar to Remark 3, a dual version can also be formulated. It is noted that a similar transform to (25) was used in [3] to solve the continuous stabilization problem. □

REFERENCES

[1] Barmish, B. R., "Necessary and Sufficient Conditions for Quadratic Stabilizability of an Uncertain System," *J. of Optimization Theory and Applications*, Vol. 46, No. 4, pp 399–408, 1985.

[2] Barmish, B. R., "Stabilization of Uncertain Systems via Linear Control," *IEEE Transac. Automatic Control*, Vol. 28, No. 8, pp 848, 1983.

[3] Bernusseu, J, Peres, P. L. D. and Geromel, J. C., "An Linear Programming Oriented Procedure for Quadratic Stabilization of Uncertain Systems," *Systems and Control Letters*, Vol. 13, No. 3, pp. 65–72, 1989.

[4] Boyd, S. and Barratt, C., *Linear Controller Design: Limits of Performance*, Prentice-Hall: Englewood Cliffs, N.J., 1990.

[5] Boyd, S. and Yang, Q., "Structured and Simultaneous Lyapunov Functions for System Stability Problems," *Int. J. Control*, Vol. 49, No. 6, pp. 2215–2240, 1989.

[6] Gu, K., "Fixed Order Output Feedback Stabilization of Uncertain Systems," *1991 ASME Winter Annual Meeting*, Atlanta, GA, DSC-Vol. 33, pp. 7–13, December 1991.

[7] Gu, K., Zohdy, M. A. and Loh, N. K., "Necessary and Sufficient Conditions of Quadratic Stability of Uncertain Linear Systems," *IEEE Trans. on Automatic Control*, Vol. 35, No. 5, pp. 601–604, 1990.

[8] Gu, K., Chen, Y. H., Zohdy, M.A. and Loh, N. K., "Quadratic Stabilizability of Uncertain Systems: a Two Level Optimization Setup", *Automatica*, Vol. 27, No. 1, pp. 161–165, 1991.

[9] Hollot, C. V. and Barmish, B. R., "Optimal Quadratic Stabilizability of Uncertain Linear Systems," *Proceedings of the 18th Allerton Conference on Communication, Control and Computing*, University of Illinois, Monticello, IL, pp. 697–706, 1983.

[10] Horisberger, H. P. and Bélanger, P. R., "Regulators for Linear, Time Invariant Plants with Uncertain Parameters," *IEEE Trans. on Automatic Control*, Vol. AC-21, pp. 705–708, 1976.

[11] Khargonekar, P. P., Petersen, I. R. and Zhou, K., "Robust Stabilization of Uncertain Linear Systems: Quadratic Stability and H^∞ Control Theory," *IEEE Transactions on Automatic Control*, Vol. 35, No. 3, pp. 356–361, 1990.

[12] Petersen, I. R., "Quadratic Stabilizability of Uncertain Linear Systems: Existence of a Nonlinear Stabilizing Control Does not Imply Existence of a Linear Stabilizing Control," *IEEE Transactions on Automatic Control*, Vol. 30, No. 3, pp. 291-293, 1985.

[13] Zhou, Kemin and Khargonekar, Pramod P., "Stability Robustness for Linear State-Space Models with Structured Uncertainty," *IEEE Transactions on Automatic Control*, Vol. AC-32, No. 7, pp. 621–623, 1987.

Proceedings of the 31st Conference
on Decision and Control
Tucson, Arizona • December 1992

THE COUPLED MODELING AND ROBUST CONTROL DESIGN PROBLEM

Gerald A. Brusher, Pierre T. Kabamba and A. Galip Ulsoy

Department of Mechanical Engineering and Applied Mechanics
2250 G. G. Brown Laboratory
The University of Michigan
Ann Arbor, MI 48109-2125

Abstract

We present a formulation of the coupled modeling and robust control design problem. The distinguishing feature is the treatment of the *control-design model* as a *degree of design freedom.* A simple example highlights the potential improvement in closed-loop performance under the coupled approach. Quantification of the *strength* of the coupling in terms of *performance boundaries* is also described.

1. Introduction

Mathematical modeling of a physical system traditionally precedes the design of a compensator [1]. A potential weakness of this sequential approach stems from the independent treatment of the modeling and control-design problems: the model is constructed without regard to the subsequent control-design task. However, control-design issues affect several modeling decisions, including the idealization of the physical system [2], the structure of the resulting mathematical model [3] and the number and type [4] of shape functions used to obtain a discrete model. Therefore, this paper proposes a framework within which the influence of the control-design problem upon the modeling problem may be quantified.

2. Problem Formulation

We limit the present research to finite-dimensional, linear, time-invariant (FDLTI) models of physical systems. We assume that the system may be represented by a family of models parametrized by uncertain physical parameters, such as damping coefficients, and by modeling parameters, such as the shape functions in a Ritz-type approximation. For a given control-design method, the compensator may also be parametrized (e.g., by the gains in a PID controller or the control weight in an LQ design). Thus, the goal of our combined modeling and control-design problem is to *choose simultaneously the model and controller parameters to guarantee the desired closed-loop performance over the entire family of models.*

We formulate the Coupled Modeling and Robust Control Design Problem (CMRCD) in terms of a constrained minimization problem. Given

- a vector of performance specifications J_i^*; $i = 1, ..., m$,
- a family of parametrized models $M(P)$, where P is a set of admissible model-parameter vectors $\boldsymbol{p}$,
- a control-design method which, given a vector of model parameters $\boldsymbol{p}$ and a vector of controller parameters $\boldsymbol{q}$ belonging to an admissible set Q, returns a controller $C(\boldsymbol{p},\boldsymbol{q})$,
- a vector of performance operators $J_i(M,C)$; $i = 1, ..., m$, which map a model and a controller into measures consistent with the performance specifications,
- a control-effort operator $J_u(M,C)$, which maps a model and controller into an appropriate measure of control effort,

we seek

- a set of evaluation-model parameter vectors

$$P_e^* = \left\{ \boldsymbol{p}_{e1}^*, \boldsymbol{p}_{e2}^*, \ldots, \boldsymbol{p}_{em}^*; \boldsymbol{p}_{ei}^* \in P,\ i = 1, \ldots, m \right\},$$

- a control-design model parameter vector $\hat{\boldsymbol{p}}^*$,
- a control-parameter vector $\hat{\boldsymbol{q}}^*$,

satisfying

$$\min_{\substack{\hat{\boldsymbol{q}} \in Q \\ \hat{\boldsymbol{p}} \in P}} \ \max_{\boldsymbol{p} \in P} \ J_u\big(M(\boldsymbol{p}),\ C(\hat{\boldsymbol{p}}, \hat{\boldsymbol{q}})\big) \tag{1}$$

subject to

$$\max_{\boldsymbol{p} \in P} \ J_i\big(M(\boldsymbol{p}),\ C(\hat{\boldsymbol{p}}, \hat{\boldsymbol{q}})\big) \le J_i^*; \quad i = 1, \ldots, m \tag{2}$$

CH3229-2/92/0000-2896$1.00 © 1992 IEEE

Note that the problem formulation features the *simultaneous* determination of the controller, $C(\hat{p}^*, \hat{q}^*)$, the control-design model, $M(\hat{p}^*)$, and the set of evaluation models, $M(p^*_{ei})$; $i = 1, ..., m$, (to each performance constraint is associated an evaluation model for which the *worst-case* performance with respect to that particular measure is realized). Moreover, by minimizing the maximum control effort expended subject to the constraint that the worst-case performance with respect to each objective is tolerable, the resulting controller not only exhibits *minimum intervention*, but also *guarantees robust performance over the family of admissible models M(P).*

The coupled problem also includes as special cases the traditional sequential approaches. Indeed, fixing the control-design model to be some nominal model $M(p_0) \in M(P)$ *prior to the control-design stage* yields the Sequential Modeling and Robust Control Design Problem (SMRCD). If, in addition, the evaluation models are restricted to be nominal, the Sequential Modeling and Non-Robust Control-Design Problem (SMCD) is obtained. The additional *degree of design freedom* (DODF) introduced in the coupled methodology by allowing the control-design model to vary rather than employing a prespecified nominal model for controller synthesis is the critical feature from which arises the potential for superior closed-loop performance.

3. Design Example

To illustrate the potential advantages of the coupled approach, we consider the design of a robust controller to regulate the response of a first-order system with a single uncertain parameter and subject to both process and sensor noise. The structure of the controller is that of an observer-based state feedback designed via the standard LQG method; therefore, the controller parameter $\hat{q}$ is simply the control weight appearing in the typical LQG objective functional. Our goal is to guarantee a prescribed level of regulation cost *over all possible models* with a minimum amount of control effort.

"Brute-force" solution of the corresponding SMRCD and CMRCD yields the loci of worst-case control effort vs. worst-case performance trade-off curves, or *performance boundaries* [5], shown in Figure 1. Since the minimization of the control-effort objective in the CMRCD problem formulation occurs over a superset of control-design model parameters in comparison to the SMRCD approach, the control effort expended by a CMRCD controller must be less than or equal to that required by its SMRCD counterpart to satisfy an identical performance goal. Indeed, this difference in control effort provides a convenient quantification of the strength of the coupling between the modeling and control-design problems.

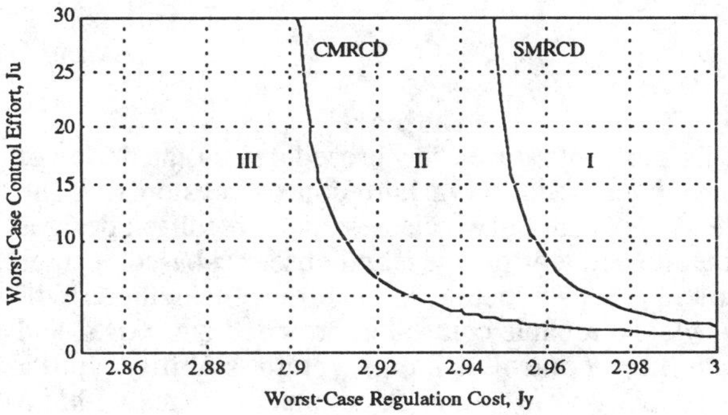

Figure 1: Performance Boundaries for CMRCD and SMRCD Controllers

Using such a measure, we note that, as the performance specifications are tightened, the coupling between the modeling and control-design problems increases. Although not demonstrated herein, we also have evidence which suggests that the strength of the coupling also depends upon the range of the parameter uncertainty as well as the characteristics of the exogenous inputs. Thus, the use of performance boundaries to measure coupling appears to exhibit great generality.

References

[1] Skelton, R. E., Dynamic Systems Control: Linear Systems Analysis and Synthesis, John Wiley & Sons,New York, 1988.

[2] Spector, V. A. and Flashner, H., "Sensitivity of Structural Models for Noncollocated Control Systems," ASME Journal of Dynamic Systems, Measurement and Control, Vol. 111, Dec. 1989, pp. 646-655.

[3] Forrest-Barlach, M. G., "On Modelling a Flexible Robot Arm as a Distributed Parameter System with Nonhomogeneous, Time-Varying Boundary Conditions," Control of Distributed Parameter Systems, Proceedings of the 4th IFAC Symposium, Los Angeles, CA, 1986, No. 3, pp. 75-81.

[4] Hu, A., Skelton, R. E. and Yang, T. Y., "Modeling and Control of Beam-Like Structures," Journal of Sound and Vibration, Vol. 117, No. 3, 1987, pp. 475-496.

[5] Boyd, S. P. and Barratt, C. H., Linear Controller Design: Limits of Performance, Prentice Hall, Englewood Cliffs, NJ, 1991.

Proceedings of the 31st Conference
on Decision and Control
Tucson, Arizona • December 1992

Reliable State Feedback and Reliable Observers

R. J. Veillette
The University of Akron
Department of Electrical Engineering
Akron, OH 44325-3904

Abstract

This paper introduces new procedures for the design of reliable linear, time-invariant control systems. The reliable designs are of two classes: state-feedback designs that tolerate actuator outages, and observer-based designs that tolerate sensor outages. The design methods will always yield a reliable controller, provided one exists, and the resulting control systems retain stability and a prescribed H_∞-norm bound despite any outages within a prespecified subset of actuators or sensors. The development of the reliable observers assumes that the sensor failures are detected and the observer dynamics accordingly adjusted. An alternative output-feedback design approach is also given for which failure detection is not required; however, the reliability of the resulting designs must be verified by checking an auxiliary sufficient condition, or by analysis. An example shows that this condition is not always necessary for reliability.

1. Introduction

It may be possible to stabilize a given MIMO plant using only a subset of the available actuators or sensors. In such a case, it is of interest to design a controller that uses all available actuators and sensors, but that will continue to provide stability and adequate performance despite the failure of certain actuators and sensors. Such a controller will be referred to as *reliable*. Standard H_∞- and H_2-optimal design techniques provide no guarantee of yielding reliable controllers. In the common case when the optimal controller is not reliable, it follows that any reliable controller will be suboptimal with respect to the given H_∞ or H_2 cost.

Some past results on reliable state feedback are given in [1], [2], and [3], while observer-based pole placement for reliable decentralized systems is developed in [4]. Observer-based centralized and decentralized controllers that guarantee reliable H_∞ performance are derived in [5], where an overview of past reliable control results also appears.

This paper extends and improves the results of [5] for the centralized case. It first develops a procedure for the design of state-feedback controls that will tolerate outages within a prespecified subset of actuators, while maintaining a prescribed H_∞-norm bound. This procedure will never fail to yield the desired reliable state-feedback control unless there exists no such control. Thus, the existence of the derived solution is a necessary and sufficient condition for the existence of any state-feedback control with the desired reliability properties.

The paper then develops an analogous procedure for the design of observer-based controls that will tolerate outages within a prespecified subset of sensors, while maintaining a prescribed H_∞-norm bound. Again, the existence of the derived solution is a necessary and sufficient condition for the existence of any output-feedback control with the desired reliability properties. The solution assumes that the observer dynamics may adapt to the sensor failures, which therefore must be detected. The problem of failure detection is not itself addressed. For the case where failure detection is not practical, an alternative design is developed, which uses the same observer gains, but assumes the controller dynamics fixed. In this case, the design might not be reliable in all contingencies; the determination of reliability depends on an auxiliary sufficient condition, or must be determined by analysis of the closed-loop system.

For both the state-feedback and the observer designs, if the design equations cannot be solved, a solution to the given reliability problem does not exist, and the prescribed H_∞-norm bound must be relaxed. As the specified bound approaches ∞, the design equations can always be solved, assuming stability can be achieved despite the possibility of all admissible failures occurring at once. The resulting designs then provide reliable stability, but do not guarantee an H_∞-norm bound *a priori*.

2. Background for Reliable H_∞ Control

2.1. H_∞-norm-bounding control

Consider a generalized plant of the form

$$\dot{x} = Ax + Bu + Gw_p, \quad z = \begin{pmatrix} Hx \\ u \end{pmatrix}, \tag{2.1}$$

where x is the state of the plant, u is the control input, w_p is a disturbance, and z is a system output to be regulated. Associated with the plant is the measurement vector

$$y = Cx + w_m, \tag{2.2}$$

where w_m is a measurement noise vector. A common criterion for disturbance rejection is that the H_∞ norm of the closed-loop transfer-function matrix $T(s)$ from w_p and w_m to z be smaller than a prescribed positive bound α. A complete solution to the problem of H_∞-norm-bounding control is given in [6], where the existence of solutions to two AREs is shown to be necessary and sufficient for the existence of a stabilizing controller that guarantees the desired bound. The following lemma, an extension of Lemma 1 of [7] that is proved in [8], can be used to derive these AREs as control design equations.

Lemma 2.1: Let $T(s) = H(sI-F)^{-1}G$, with (F,H) a detectable pair. If there exists a matrix $X \geq 0$ such that

$$F^TX + XF + \frac{1}{\alpha^2}XGG^TX + H^TH \leq 0, \tag{2.3}$$

then F has all its eigenvalues in the open left-half plane, and $\|T\|_\infty \leq \alpha$.

The simplest application of Lemma 2.1 is in the derivation of an H_∞-norm-bounding state-feedback control $u = Kx$, with K a constant.

CH3229-2/92/0000-2898$1.00 © 1992 IEEE

Theorem 2.1: Suppose (A,H) is detectable, and let the state-feedback gain matrix be given by $K = -B^TX$, where $X \geq 0$ satisfies

$$A^TX + XA + \frac{1}{\alpha^2}XGG^TX - XBB^TX + H^TH = 0. \quad (2.4)$$

Then the state-feedback system is stable, and the closed-loop transfer-function matrix

$$T(s) = \begin{pmatrix} H \\ K \end{pmatrix}(sI-A-BK)^{-1}G$$

satisfies $\|T\|_\infty \leq \alpha$.

The derivation of an H_∞-norm-bounding output-feedback control based on Lemma 2.1 can proceed in two steps: First, assume a suitable state-feedback control as given by Theorem 2.1. Then, based on this state-feedback solution, change variables in the output-feedback problem in such a way that an observer can be separately designed to preserve the H_∞-norm bound achieved by the state feedback. This change of variables is discussed next, following the development in [6].

Assume $K = -B^TX$, where $X \geq 0$ satisfies (2.4) for a given value of $\alpha > 0$. This K characterizes a state-feedback control that guarantees a closed-loop H_∞ norm no greater than α. Now define new variables

$$r_p = w_p - K_d x \quad (2.5)$$

and

$$v = u - Kx, \quad (2.6)$$

where K_d is to be determined. Substituting (2.5) into the dynamic equation of the plant (2.1), and assuming the regulated output v given by (2.6), we obtain a new plant

$$\dot{x} = (A+GK_d)x + Bu + Gr_p, \quad v = -Kx + u. \quad (2.7)$$

It is clear that, for a given control input, the states of the plants (2.1) and (2.7) are identical if (2.5) holds. It also happens that if we choose $K_d = \frac{1}{\alpha^2}G^TX$, and if we define $T(s)$ and $T_v(s)$ by

$$z(s) = T(s)\begin{pmatrix} w_p(s) \\ w_m(s) \end{pmatrix}, \quad v(s) = T_v(s)\begin{pmatrix} r_p(s) \\ w_m(s) \end{pmatrix},$$

then $\|T\|_\infty \leq \alpha$ if and only if $\|T_v\|_\infty \leq \alpha$. If a stabilizing feedback control for the plant (2.7) can be generated from the measurement (2.2) such that $\|T_v\| \leq \alpha$, then the same control law will stabilize the plant (2.1) and guarantee $\|T\|_\infty \leq \alpha$. This fact is derived in [6].

Now, an observer for the plant (2.7) is

$$\dot{\xi} = (A+GK_d)\xi + Bu + L(y-C\xi),$$

while the control can be given by $u = K\xi$. A candidate control law can thus be written as

$$\dot{\xi} = (A+GK_d+BK-LC)\xi + Ly, \quad u = K\xi. \quad (2.8)$$

Combining (2.7) and (2.8) while using (2.2) gives the "error dynamics"

$$\dot{e} = (A+GK_d-LC)e + (-G \;\; L)\begin{pmatrix} r_p \\ w_m \end{pmatrix}, \quad v = Ke, \quad (2.9)$$

where the error is defined as $e = \xi - x$. Note that only this "error subsystem" is observable through v; hence, provided the "state-feedback subsystem" is stable, the overall closed-loop system is adequately represented by (2.9). The design equations for computing an observer gain L so that the transfer-function matrix $T_v(s)$ representing (2.9) satisfies $\|T_v\|_\infty \leq \alpha$ can be derived using a dual form of Lemma 2.1. The result follows:

Theorem 2.2: Suppose (A,G) is a stabilizable pair, and set $L = WC^T$, where $W \geq 0$ satisfies

$$(A+GK_d)W + W(A+GK_d)^T + \frac{1}{\alpha^2}WK^TKW - WC^TCW + GG^T = 0. \quad (2.10)$$

Then (2.9) is stable and the transfer-function matrix $T_v(s)$ from r_p and w_m to v satisfies $\|T_v\|_\infty \leq \alpha$.

The following theorem summarizes the results of the foregoing discussion:

Theorem 2.3: Assume (A,H) is detectable and (A,G) is stabilizable. Let the gains K, K_d, and L be given by

$$K = -B^TX, \quad K_d = \frac{1}{\alpha^2}G^TX, \quad L = WC^T,$$

where $X \geq 0$ and $W \geq 0$ satisfy (2.4) and (2.10), respectively. Suppose that $A+GK_d+BK$ has all its eigenvalues in the open left-half plane. Then the control law (2.8) stabilizes the plant, and gives a closed-loop transfer-function matrix $T(s)$ from w_p and w_m to z satisfying $\|T\|_\infty \leq \alpha$.

The problem of selecting a particular controller that is reliable from among all controllers that provide a specified H_∞-norm bound is addressed in [5], where design methodologies for reliable centralized and decentralized control are derived. One of the main results derived in [5] is now summarized.

2.2. A solution to the reliability problem [5]

Consider first the design of a controller that can tolerate the outage of certain sensors. Let Ω denote the subset of all sensors that are considered susceptible to outage, and let Ω' denote the complementary subset of sensors. Without loss of generality, suppose that the measurement y is decomposed as

$$y = \begin{pmatrix} y_{\Omega'} \\ y_\Omega \end{pmatrix} = \begin{pmatrix} C_{\Omega'} \\ C_\Omega \end{pmatrix}x + \begin{pmatrix} w_{m\Omega'} \\ w_{m\Omega} \end{pmatrix},$$

where y_Ω denotes the measurement vector associated with the susceptible set Ω, and $y_{\Omega'}$ denotes the measurement vector associated with Ω'. Any admissible contingency can be described as the outage of all sensors in an arbitrary subset ω of Ω; that is, the susceptible measurements may be decomposed, without loss of generality, as

$$y_\Omega = \begin{pmatrix} y_{\Omega\backslash\omega} \\ y_\omega \end{pmatrix} = \begin{pmatrix} C_{\Omega\backslash\omega} \\ C_\omega \end{pmatrix}x + \begin{pmatrix} w_{m\Omega\backslash\omega} \\ w_{m\omega} \end{pmatrix},$$

and a contingency then modeled as $y_\omega = 0$. Note that a sensor failure is assumed to eliminate both the measurement and its associated noise. Then the following theorem holds:

Theorem 2.4 [5]: Assume (A,H) is detectable. Let the gains K, K_d, and L be given by

$$K = -B^TX, \quad K_d = \frac{1}{\alpha^2}G^TX, \quad L = WC^T,$$

where $W = (I-\alpha^{-2}YX)^{-1}Y > 0$, and $X \geq 0$ and $Y > 0$ satisfy the design equations

$$A^TX + XA + \frac{1}{\alpha^2}XGG^TX - XBB^TX + H^TH + \alpha^2 C_\Omega^T C_\Omega = 0, \quad (2.11)$$

$$A^TY + YA + \frac{1}{\alpha^2}YH^THY - YC_{\Omega'}^T C_{\Omega'}Y + GG^T = 0. \quad (2.12)$$

Suppose that $A+GK_d+BK$ has all its eigenvalues in the open left-half plane. Then the control law (2.8) stabilizes the plant (2.1) with measurement (2.2), and gives a nominal closed-loop transfer-function matrix $T(s)$ from w_p and w_m to z satisfying $\|T\|_\infty \le \alpha$. Furthermore, if measurements y_ω fail for any $\omega \subseteq \Omega$, then the closed-loop system is still stable, and the transfer-function matrix $T_{\omega'}(s)$ from w_p and $w_{m\omega'}$ to z also satisfies $\|T_{\omega'}\|_\infty \le \alpha$.

A similar design that can tolerate the outage of certain actuators is also given in [5].

The design procedure suggested by Theorem 2.4 has the limitation that, while the existence of solutions to the given design equations is sufficient to guarantee the existence of reliable H_∞-norm-bounding controllers, it is not necessary. The following sections derive new design methods for reliable H_∞ control that avoid this limitation.

3. Reliable State Feedback

3.1. A reliable state-feedback control

Introduce the decompositions

$$u = \begin{pmatrix} u_{\Omega'} \\ u_\Omega \end{pmatrix}, \quad B = \begin{pmatrix} B_{\Omega'} & B_\Omega \end{pmatrix},$$

where B_Ω corresponds to those actuators that are susceptible to outage, and $B_{\Omega'}$ to the complementary subset of actuators. The submatrices B_Ω and $B_{\Omega'}$ are fixed as part of the problem definition. Define also the submatrix B_ω, which corresponds to those actuators that actually experience an outage in a given situation. Since $\omega \subseteq \Omega$ is arbitrary, B_ω is not fixed, but consists of any subset of the columns of B_Ω. It will also be convenient to define the submatrices $B_{\omega'}$ and $B_{\Omega\backslash\omega}$, which naturally correspond to the indicated sets of actuators. Any actuator failure is assumed to eliminate not only the control input, but also its associated cost in the regulated ouptut z; that is, the controls associated with failed actuators are not considered in computing the H_∞ norm of the system in a contingency situation.

Theorem 3.1: Suppose (A,H) is detectable, and define the state-feedback gain matrix as

$$K = -B^TX \quad (3.1)$$

where $X \ge 0$ satisfies

$$A^TX + XA + \frac{1}{\alpha^2}XGG^TX - XB_{\Omega'}B_{\Omega'}^TX + H^TH = 0. \quad (3.2)$$

Then the nominal state-feedback system is stable, and the nominal closed-loop transfer-function matrix

$$T(s) = \begin{pmatrix} H \\ K \end{pmatrix}(sI-A-BK)^{-1}G$$

satisfies $\|T\|_\infty \le \alpha$. Furthermore, if control inputs u_ω fail for any $\omega \subseteq \Omega$, then the closed-loop system is still stable and the transfer-function matrix

$$T_{\omega'}(s) = \begin{pmatrix} H \\ K_{\omega'} \end{pmatrix}(sI-A-B_{\omega'}K_{\omega'})^{-1}G$$

satisfies $\|T_{\omega'}\|_\infty \le \alpha$.

Proof: Let $\omega \subseteq \Omega$, and decompose K as

$$K = \begin{pmatrix} K_{\omega'} \\ K_\omega \end{pmatrix},$$

consistent with the corresponding decomposition of B. Now, manipulate (3.2) to obtain

$$(A+B_{\omega'}K_{\omega'})^TX + X(A+B_{\omega'}K_{\omega'}) + \frac{1}{\alpha^2}XGG^TX + \begin{pmatrix} H^T & K_{\omega'}^T \end{pmatrix}\begin{pmatrix} H \\ K_{\omega'} \end{pmatrix} = -XB_{\Omega\backslash\omega}B_{\Omega\backslash\omega}^TX \le 0.$$

Since (A,H) is detectable, so is $\left(A+B_{\omega'}K_{\omega'}, \begin{pmatrix} H \\ K_{\omega'} \end{pmatrix}\right)$; thus, by Lemma 2.1, the control law $u = \begin{pmatrix} K_{\omega'} \\ 0 \end{pmatrix}x$ stabilizes (2.1) and provides the desired H_∞-norm bound regardless of any actuator failures within the susceptible set Ω. This includes the nominal case of no actuator failures, $\omega = \varnothing$ and $K_{\omega'} = K$. Q.E.D.

3.2. Necessity and sufficiency of the condition

Theorem 3.1 gives a sufficient condition for the existence of a reliable state-feedback control; however, the given condition is also necessary for the existence of any state-feedback control with the desired reliability property. The loss of *all* the susceptible actuators ($\omega = \Omega$) is an admissible contingency; therefore, a reliable H_∞ state-feedback control exists only if the bound $\|T\|_\infty \le \alpha$ can be achieved using the non-susceptible actuators $B_{\Omega'}$ alone. This condition is equivalent to the existence of a solution $X \ge 0$ to the ARE (3.2). Thus, if the design procedure given by Theorem 3.1 fails, then there exists no state-feedback solution to the given reliability problem.

3.3. The effect of including the susceptible actuators

The solution of (3.2) depends only on the non-susceptible control inputs Ω'; however, once the solution $X \ge 0$ is found, it is used for computing state-feedback gains for any additional actuators in the reliable design, whatever their input directions B_Ω may be. One may ask how the system is affected by including these additional actuators. For the case where $\alpha = \infty$ (reliable stability only), a relatively straightforward answer can be given in terms of the H_2 norm of the closed-loop system. The state-feedback system including the actuators in $\Omega\backslash\omega$ is

$$\dot{x} = (A+B_{\omega'}K_{\omega'})x + Gr_p, \quad z_{\omega'} = \begin{pmatrix} H \\ K_{\omega'} \end{pmatrix}x. \quad (3.3)$$

The H_2 norm of the closed-loop transfer function matrix $T_{\omega'}(s)$ from r_p to $z_{\omega'}$ can be found from the relation

$$\|T_{\omega'}\|_2^2 = \mathrm{Tr}(G^TQ_{\omega'}G), \quad (3.4)$$

where $Q_{\omega'}$, the observability Gramian of (3.3). The state-feedback system excluding all susceptible actuators Ω is

$$\dot{x} = (A+B_{\Omega'}K_{\Omega'})x + Gr_p, \quad z_{\Omega'} = \begin{pmatrix} H \\ K_{\Omega'} \end{pmatrix} \tag{3.5}$$

The H_2 norm of the corresponding transfer function $T_{\Omega'}(s)$ from r_p to $z_{\Omega'}$ is given by

$$\|T_{\Omega'}\|_2^2 = \mathrm{Tr}(G^T Q_{\Omega'} G), \tag{3.6}$$

where $Q_{\Omega'}$ is the observability Gramian of (3.5). Note that the actuators in Ω' are used optimally with respect to the cost (3.6), assuming they are the only actuators present. However, a comparison or the integrals defining the two Gramians, using the design equations (3.1) and (3.2), reveals that $Q_{\omega'} \le Q_{\Omega'}$. Thus, the cost (3.6) is greater than the cost (3.4). This is true even though the additional actuators also represent additional system outputs in $z_{\omega'}$; that is, the beneficial effect of the additional actuators more than makes up for the extra energy they exert. In fact, the total output energy of the system is reduced by the very amount of energy exerted by those additional actuators.

The following points summarize the discussion of the reliable state-feedback design:

- When the non-susceptible actuators Ω' are used alone, they are used optimally with respect to an H_2 criterion.
- When additional actuators Ω are included, the total H_2 norm decreases, even taking into account the control energy exerted by the additional actuators.
- The more control energy the additional actuators exert, the more they improve the overall H_2 norm of the system.
- The additional actuators Ω are used non-optimally.

4. Reliable Observers

4.1. An adapting reliable output-feedback control

The following theorem gives a control law that tolerates the admissible sensor outages.

Theorem 3.1: Suppose (A,H) is detectable and (A,G) is stabilizable. Define the gain matrices

$$K = -B^T X, \quad K_d = \frac{1}{\alpha^2} G^T X, \quad L = WC^T, \tag{4.1}$$

where $X \ge 0$ and $W \ge 0$ satisfy the design AREs

$$A^T X + XA + \frac{1}{\alpha^2} XGG^T X - XBB^T X + H^T H = 0, \tag{4.2}$$

$$(A+GK_d)W + W(A+GK_d)^T + \frac{1}{\alpha^2} WK^T KW - WC_{\Omega'}^T C_{\Omega'} W + GG^T = 0, \tag{4.3}$$

and $A+GK_d+BK$ is Hurwitz. Let the nominal controller be given by

$$\dot{\xi} = (A+GK_d+BK-LC)\xi + Ly, \quad u = K\xi. \tag{4.4}$$

Then the nominal control system is stable, and the closed-loop transfer-function matrix $T(s)$ from w_p and w_m to z satisfies $\|T\|_\infty \le \alpha$.

Now suppose that the measurements y_ω fail for some $\omega \subseteq \Omega$, and let the controller in this contingency be

$$\dot{\xi} = (A+GK_d+BK-L_{\omega'}C_{\omega'})\xi + L_{\omega'}y_{\omega'}, \quad u = K\xi. \tag{4.5}$$

Then the closed-loop system is still stable, and the closed-loop transfer-function matrix $T_{\omega'}(s)$ from w_p and w_m to z satisfies $\|T_{\omega'}\|_\infty \le \alpha$.

Proof: By the introduction of the variables (2.5) and (2.6), we pass to the equivalent problem of control of the plant (2.7). Using the controller (4.5) with the plant (2.7), and transforming to error coordinates, yields the closed-loop system defined by the triple

$$\left\{ \begin{pmatrix} A+GK_d+BK & BK \\ 0 & A+GK_d-L_{\omega'}C_{\omega'} \end{pmatrix}, \begin{pmatrix} G & 0 \\ -G & L_{\omega'} \end{pmatrix}, (0 \; K) \right\}.$$

Given that $A+GK_d+BK$ is Hurwitz, this system is characterized by its observable part, which is defined by the triple

$$\left\{ (A+GK_d-L_{\omega'}C_{\omega'}), \left(-G \; L_{\omega'}\right), K \right\}. \tag{4.6}$$

Manipulation of the design ARE (4.3), and using the relation $L_{\omega'} = WC_{\omega'}^T$, gives

$$(A+GK_d-L_{\omega'}C_{\omega'})W + W(A+GK_d-L_{\omega'}C_{\omega'})^T + \frac{1}{\alpha^2} WK^T KW + \left(-G \; L_{\omega'}\right)\begin{pmatrix} -G^T \\ L_{\omega'}^T \end{pmatrix}$$

$$= - WC_{\Omega\backslash\omega}^T C_{\Omega\backslash\omega} W \le 0. \tag{4.7}$$

It is trivial to show that $(A+GK_d-L_{\omega'}C_{\omega'}, \left(-G \; L_{\omega'}\right))$ is a stabilizable pair; thus, by the dual form of Lemma 2.1, (4.6) is a stable system, and the transfer function from r_p and w_m to v has H_∞ norm less than or equal to α. Q.E.D.

It is important to note that this approach to reliable control system design assumes the capability to detect sensor outages as they occur: The presence of the $L_{\omega'}C_{\omega'}$ term in the dynamics of (4.5) implies that the controller adapts to the failure of measurements y_ω, since it is not determined *a priori* what set ω' of sensors will be available. Independent of the problem of failure detection, which is not addressed here, it is interesting to note that the observer gains do not need to be redesigned in case of sensor failures; rather, all the observer gains are computed ahead of time, and are then simply included or excluded depending on which measurements are available.

4.2. Necessity and sufficiency of conditions

Similar to the actuator outage case, the loss of *all* the susceptible sensors ($\omega = \Omega$) is an admissible contingency; therefore, a reliable H_∞ output-feedback control exists only if the bound $\|T\|_\infty \le \alpha$ can be achieved using the non-susceptible sensors $C_{\Omega'}$ alone. This condition is equivalent to the existence of solutions $X \ge 0$ and $W \ge 0$ to the design AREs (4.2) and (4.3). Thus, if the design procedure given by Theorem 4.1 fails, then there exists no solution to the given output-feedback reliability problem. Hence, the conditions given are necessary and sufficient for the existence of a reliable control law.

4.3. Adding sensors to a control system

The closed-loop system omitting all the susceptible sensors is equivalent to the error system representation

$$\dot{e} = (A+GK_d-L_{\Omega'}C_{\Omega'})e + \begin{pmatrix}-G & L_{\Omega'}\end{pmatrix}\begin{pmatrix}r_p \\ w_{m\Omega'}\end{pmatrix},\quad v = Ke,$$

while including the sensors $\Omega\backslash\omega$ yields

$$\dot{e} = (A+GK_d-L_{\omega'}C_{\omega'})e + \begin{pmatrix}-G & L_{\omega'}\end{pmatrix}\begin{pmatrix}r_p \\ w_{m\omega'}\end{pmatrix},\quad v = Ke.$$

The H_2 norms of these two systems can be found as

$$\|T_{\Omega'}\|_2^2 = \mathrm{Tr}(KP_{\Omega'}K^T), \tag{4.8a}$$

$$\|T_{\omega'}\|_2^2 = \mathrm{Tr}(KP_{\omega'}K^T), \tag{4.8b}$$

where $P_{\Omega'}$ and $P_{\omega'}$ are the respective controllability Gramians. It is a simple matter to show that $P_{\omega'} \le P_{\Omega'}$; hence, from comparing the expressions in (4.8), that $\|T_{\omega'}\|_2 \le \|T_{\Omega'}\|_2$.

The following points summarize the results:

- When the non-susceptible measurements $y_{\Omega'}$ are used alone, they are used H_2-optimally.
- When additional measurements $y_{\Omega\backslash\omega}$ are included, the total H_2 norm of the system is decreased, even though the additional measurement noise inputs are included.
- The decrease in the total output energy of the system is equal to the output energy that results from the added noise inputs.
- The measurements $y_{\Omega\backslash\omega}$ are used non-optimally.

4.4. A non-adapting reliable output-feedback control

This section presents a design procedure for a non-adapting reliable output-feedback control law. Since the design is based the same "necessary and sufficient" design equations as the adapting control law given by Theorem 4.1, the controller can be computed as long as the reliability problem has a solution; however, the particular controller developed does not necessarily provide reliability for every admissible contingency. The following theorem presents the design along with an auxiliary condition that is sufficient to guarantee system reliability.

Theorem 4.2: Assume (A,H) is detectable and (A,G) is stabilizable. Define K, K_d, and L as

$$K = -B^TX,\quad K_d = \frac{1}{\alpha^2}G^TX,\quad L = WC^T, \tag{4.9}$$

where $X\ge0$ and $W\ge0$ satisfy the design AREs (4.2) and (4.3) and $A+GK_d+BK$ is Hurwitz. Suppose also that there exists a solution $Z \ge 0$ to the equation

$$(A+GK_d+BK+L_\Omega C_\Omega)Z + Z(A+GK_d+BK+L_\Omega C_\Omega)^T + ZC_\Omega^TC_\Omega Z + LL^T = 0. \tag{4.10}$$

Then the control law

$$\dot{\xi} = (A+GK_d+BK-L_{\Omega'}C_{\Omega'})\xi + Ly,\quad u = K\xi \tag{4.11}$$

stabilizes the plant (2.1) with measurement (2.2), and gives a closed-loop transfer-function matrix $T(s)$ from w_p and w_m to z satisfying $\|T\|_\infty \le \alpha$. Furthermore, if measurements y_ω fail for any $\omega \subseteq \Omega$, then the closed-loop system is still stable, and the transfer-function matrix from w_p and $w_{m\omega'}$ to z satisfies $\|T_{\omega'}\|_\infty \le \alpha$.

Remark: The auxiliary condition (4.10) essentially is a requirement that the transfer-function matrix $M(s) = C_\Omega[sI-(A+GK_d+BK+L_\Omega C_\Omega)]^{-1}L$ be stable and have H_∞ norm less than unity.

Proof: Let $\omega \subseteq \Omega$ denote the arbitrary set of sensors that experience an outage. In this contingency, the controller is effectively described by

$$\dot{\xi} = (A+GK_d+BK-L_{\Omega'}C_{\Omega'})\xi + L_{\omega'}y_{\omega'},\ u = K\xi. \tag{4.12}$$

We again pass to the equivalent problem of H_∞-norm-bounding control of the plant (2.7) with measurement (2.2). Applying (4.12) to (2.7) and changing to error coordinates yields the closed-loop system

$$\dot{x}_c = F_cx_c + G_cw_{\omega'},\quad v = H_cx_c,$$

where

$$F_c = \begin{pmatrix}A+GK_d+BK+L_{\Omega\backslash\omega}C_{\Omega\backslash\omega} & -L_{\omega'}C_{\omega'} \\ L_{\Omega\backslash\omega}C_{\Omega\backslash\omega} & A+GK_d-L_{\omega'}C_{\omega'}\end{pmatrix},$$

$$G_c = \begin{pmatrix}0 & L_{\omega'} \\ -G & L_{\omega'}\end{pmatrix},\quad H_c = (0\ \ K).$$

Define the matrix $X_c \ge 0$ as

$$X_c = \begin{pmatrix}Z & 0 \\ 0 & W\end{pmatrix};$$

then, routine algebra using the design equation (4.3) and the assumed condition (4.10) yields

$$X_cF_c^T + F_cX_c + \frac{1}{\alpha^2}X_cH_c^TH_cX_c + G_cG_c^T \le 0.$$

Routine arguments show that (F_c,G_c) is a stabilizable pair; thus, the proof is thus completed. Q.E.D.

4.5. Example of non-adapting reliable control

This section presents the results of the reliable control given by Theorem 4.2 applied to a simple two-output 2nd-order example. The highlights are as follows:

- Using a standard H_2-optimal design via Theorem 2.3, any sensor outage results in a loss of stability for the closed-loop system.
- Attempted reliable design via methods given in [5] is unsuccessful.
- Using a design via Theorem 4.2, system stability is preserved despite the outage of either sensor, even though the auxiliary sufficient condition is not satisfied.

The plant is described by the matrices

$$A = \begin{pmatrix}0 & 1 \\ -1 & 1\end{pmatrix},\quad B = \begin{pmatrix}0 \\ 1\end{pmatrix},\quad C = \begin{pmatrix}1 & 0 \\ 0 & 1\end{pmatrix},$$

$$G = \begin{pmatrix}-1 \\ -1\end{pmatrix},\quad H = (1\ \ 0).$$

This system is oscillatory and unstable. The goal of the design will be to guarantee reliable stability despite sensor outages.

Using Theorem 2.3 with $\alpha = \infty$, a stabilizing H_2-optimal controller is obtained, with the gains

$$K = (-0.4142 \quad -2.3522), \quad L = \begin{pmatrix} 1.3643 & 0.6276 \\ 0.6276 & 1.5923 \end{pmatrix}.$$

Performance results are given in Table 1, where the column $\|T\|_2$ represents the H_2 norm of the closed-loop transfer-function matrix including all measurement noise inputs; $\|T_1'\|_2$ and $\|T_2'\|_2$ represent the H_2 norms of the submatrices omitting the first and second measurement noises, respectively; and $\|T_{wp}\|_2$ represents the H_2 norm omitting both measurement noises. The rows represent the various sensor-failure contingencies. For this controller, the failure of either measurement destabilizes the closed-loop system. Clearly, a reliable control design methodology could be useful in this case.

A reliable design is first attempted via the methods given in [5], as in Theorem 2.4. The second measurement is assumed susceptible to outage, so that $C_{\Omega'} = (1 \quad 0)$ and $C_\Omega = (0 \quad 1)$. The design approach is unsuccessful, because the state-feedback design ARE (2.11) cannot be solved for any value of α. An alternative design method is required.

The second attempt at reliable design uses the non-adapting approach of Theorem 4.2, with $\alpha = \infty$. The second measurement is again assumed susceptible to outage. From the design AREs (4.2) and (4.3) and the gain formulas (4.9), the controller parameters are found as

$$K = (-0.4142 \quad -2.3522), \quad L = \begin{pmatrix} 1.3643 & 0.6276 \\ 0.6276 & 1.5923 \end{pmatrix}.$$

The auxiliary sufficient condition that there exist $Z \geq 0$ satisfying (4.10) is not satisfied; in fact, the matrix $A+GK_a+BK+L_\Omega C_\Omega$ is not Hurwitz. Nevertheless, the controller is in fact reliable, as the closed-loop system can tolerate the failure of either measurement without becoming unstable. The performance results are given in Table 2.

5. Conclusion

Methodologies have been developed for the design of reliable state feedback and reliable observers. The derived design equations improve upon those previously available, which sometimes fail to have solutions even though reliability is achievable. For all the designs presented here, the non-existence of solutions to the design equations indicates that the prescribed performance bound is too strict. A sufficient relaxation of the bound will then allow the design to proceed. Thus, the new reliable control laws represent in some cases an essential alternative to previously existing reliability methods.

Acknowledgment

The author gratefully acknowledges the helpful comments of Professor J. V. Medanić and the encouragement of Professor T. T. Hartley.

References

[1] J. Ackermann, *Sampled-Data Control Systems*, Springer-Verlag, Heidelberg, 1985.

[2] S. M. Joshi, "Failure-accommodating control of large flexible spacecraft," in *Proc. 1986 American Control Conference*, Seattle, WA, 1986, pp. 156-161.

[3] M. Mariton and P. Bertrand, "Improved multiplex control systems: dynamic reliability and stochastic optimality," *International Journal of Control*, vol 44, pp. 219-234, 1986.

[4] R. A. Date and J. H. Chow, "A reliable coordinated decentralized control system design," in *Proc. 28th Conference on Decision and Control*, Tampa, FL, 1989, pp. 1295-1300.

[5] R. J. Veillette, J. V. Medanić, and W. R. Perkins, "Design of Reliable Control Systems," *IEEE Transactions on Automatic Control*, vol. AC-37, no. 3, pp. 290-304, 1992.

[6] J. C. Doyle, K. Glover, P. P. Khargonekar, and B. A. Francis, "State-space solutions to standard H_2 and H_∞ control problems," *IEEE Transactions on Automatic Control*, vol. AC-34, no. 8, pp. 831-847, 1989.

[7] J. C. Willems, "Least squares stationary optimal control and the algebraic Riccati equation," *IEEE Transactions on Automatic Control*, vol. AC-16, no. 6, pp. 621-634, 1971.

[8] R. J. Veillette, J. V. Medanić, and W. R. Perkins, "Robust stabilization and disturbance rejection for systems with structured uncertainty," in *Proc. 28th Conference on Decision and Control*, Tampa, FL, 1989, pp. 936-941.

	$\|T\|_2$	$\|T_1'\|_2$	$\|T_2'\|_2$	$\|T_{wp}\|_2$
Nominal	4.045	3.728	3.243	2.839
y_1 fails	unstable	unstable	unstable	unstable
y_2 fails	unstable	unstable	unstable	unstable

Table 1: Performance of the standard H_2 controller.

	$\|T\|_2$	$\|T_1'\|_2$	$\|T_2'\|_2$	$\|T_{wp}\|_2$
Nominal	8.252	7.423	5.065	3.558
y_1 fails	7.480	6.717	4.475	3.031
y_2 fails	10.99	9.647	6.158	3.279

Table 2: Performance of the reliable controller via Theorem 4.2.

Proceedings of the 31st Conference
on Decision and Control
Tucson, Arizona • December 1992

DYNAMIC ANALYSIS OF VOLTAGE COLLAPSE IN POWER SYSTEMS

David J. Hill and Ian A. Hiskens
Department of Electrical & Computer Engineering, The University of Newcastle,
University Drive, Callaghan, NSW 2308, Australia

Abstract
An analytical framework is presented for analysis of voltage collapse as a dynamic phenomenon. The approach depends on linking static and dynamic aspects within differential-algebraic models which preserve network structure and facilitate use of a novel approach to modelling aggregate (dynamic) load. Directions for stability/bifurcation analysis are indicated. In particular, new Lyapunov functions for large-disturbance voltage stability analysis can be derived.

1 INTRODUCTION

Development of stability theory for power systems continues to be an interesting subject. The three major problem areas are transient (angle) stability, oscillations and voltage collapse. Particularly in the nonlinear aspects of these problems, there are many questions to answer. The features which motivate this paper are:

- the interplay between static and dynamic behaviour;
- the interplay between angle and voltage behaviour;
- the network structure (which constrains power exchange) and the consequent differential -algebraic structure of models;
- the implications of using aggregate rather than device oriented nonlinear load models.

Each of these has of course received at least some attention in numerous results by researchers including the authors. No attempt will be made here to identify the detailed contributions; relevant surveys and collected works which refer to these are available [1-6]. The intention here is to present a framework for stability analysis with emphasis on the problem of voltage collapse analysis. The ideas follow earlier work on angle stability with related voltage behaviour [6-10] and voltage stability including system collapse [11, 12]. The main consideration is towards integrating a nonlinear dynamic load modelling approach [11, 13] into the broader stability analysis picture.

2 LOAD MODEL STRUCTURES

2.1 General Form

Models for dynamical analysis of power systems typically have a consistency problem. While it is scientifically possible to give quite detailed models for generators, lines, transformers and control devices, load modelling can often only be treated on an ad hoc basis. In stability analysis for instance, we need a representation of effective power demand at high voltage buses. This may include the aggregate effect of numerous load devices such as lighting, heating and motors plus some levels of transformer tap-changing and other control devices. Building up the aggregate effect by combining device characteristics may not be possible. Thus, in many cases, quite simplified aggregate load representations like impedances are used alongside detailed generator models. This section summarises an approach for developing aggregate models of nonlinear dynamic form [11, 13].

Consider a high voltage bus as in Figure 1. The real and reactive power demands P_d and Q_d are considered to be dynamically related to the voltage V.

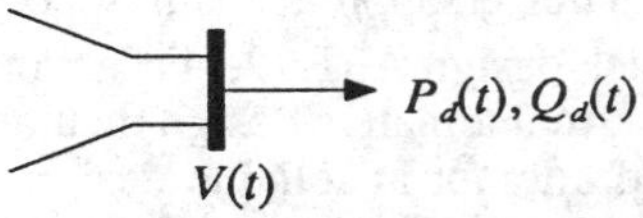

Figure 1 High Voltage Load Bus

Measurements in the laboratory and on power system buses [13, 14] show that the load response to a step in voltage V is of the general form shown in Figure 2. (The responses for real and reactive power are similar qualitatively; only the real power response is shown.) The significant features of the response are as follows: 1) a step in power immediately follows a step in voltage; 2) the power recovers to a new steady-state value; 3) the recovery appears to be of exponential (sometimes underdamped) form, at least approximately; 4) the size of the step and the steady-state value are nonlinearly related to voltage. These features are easily connected to physical aspects of specific loads [11]: for recovery time constants of the order of a second, the effect of motors (transient period) is captured; on a time-scale of minutes, network and load regulating devices are modelled.

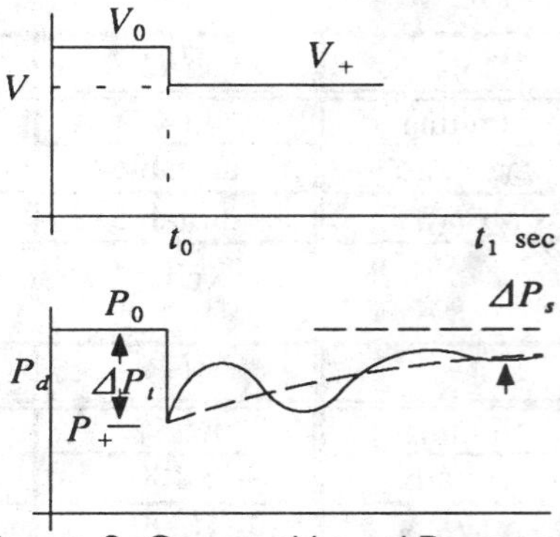

Figure 2 General Load Response

As in [11], we can propose a general load model as an implicit differential equation

CH3229-2/92/0000-2904$1.00 © 1992 IEEE

$$f(P_d^{(n)}, P_d^{(n-1)}, \ldots, \dot{P}_d, P_d, V^{(m)}, V^{(m-1)}, \ldots \dot{V}, V) = 0 \qquad (1)$$

where $P_d^{(j)}, V^{(i)}$ denote the higher order derivatives of P_d, V respectively. (A similar equation applies for reactive power Q_d.)

An input - output version of this model is illustrated simply in Figure 3 where V is chosen as the input to a nonlinear dynamical system with output P_d.

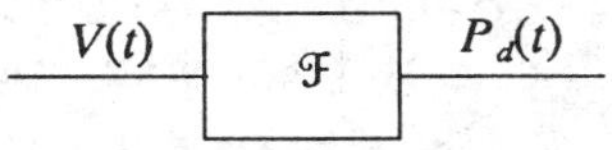

Figure 3 Input - Output Load Representation

2.2 Linear Recovery

The above discussion of Figure 2 suggests there are (at least) two nonlinearities in a reasonable model; one describing a steady-state relationship, i.e. the steady-state offset ΔP_s, and the other a transient one, i.e. the jump ΔP_t. Further, linear dynamics can approximately describe the transient recovery. Assuming first order dynamics, it was proposed in [11] that the load response in Figure 2 can be regarded as the solution of the scalar differential equation

$$T_p \dot{P}_d + P_d = P_s(V) + k_p(V)\dot{V} \qquad (2)$$

The motivation is easy to see.

Setting derivatives to zero gives the steady-state model

$$P_d = P_s(V) \qquad (3)$$

Rewriting (2) as

$$T_p \frac{dP_d}{dt} + P_d = P_s(V) + T_p \frac{d}{dt}(P_t(V)) \qquad (4)$$

where

$$P_t(V) := \frac{1}{T_p} \int_0^V k_p(\sigma) d\sigma + c_0 \qquad (5)$$

c_0 a constant, clearly shows that $P_t(.)$ defines the fast changes in load according to $P_d = P_t(V)$.

For solving equation (2) analytically or numerically, the fact that all solutions satisfy an equivalent normal form model is used. This form is expressed as

$$\dot{x}_p = -\frac{1}{T_p} x_p + N_p(V) \qquad (6)$$

$$P_d = \frac{1}{T_p} x_p + P_t(V) \qquad (7)$$

where

$$N_p(V) := P_s(V) - P_t(V) \qquad (8)$$

The solution of differential equation (2) to the voltage step is easily derived and

$$\Delta P_t := P_d(t_0-) - P_d(t_0+) = P_t(V_0) - P_t(V_+) \qquad (9)$$

$$\Delta P_s := P_d(t_0-) - P_d(\infty) = P_s(V_0) - P_s(V_+) \qquad (10)$$

So the nonlinear functions $P_s(\cdot)$ and $k_p(\cdot)$ (or $P_t(\cdot)$) independently determine the steady-state and transient power increments ΔP_s and ΔP_t respectively.

Karlsson [13] has identified models of the form (6)-(8) with the special load functions:

$$P_s(V) := P_0 \left(\frac{V}{V_0}\right)^{a_s} \qquad (11)$$

$$P_t(V) := P_0 \left(\frac{V}{V_0}\right)^{a_t} \qquad (12)$$

The steady-state model $P_d = P_0 \left(\frac{V}{V_0}\right)^{a_s}$ corresponds to the widely used static model.

Now, it is convenient to note that the normal form model is easily given the block diagram representation shown in Figure 4 where $G(s) = \frac{1}{T_p s + 1}$, $N_{p1} = N_p$ and $N_{p2} = P_t$.

In summary, the scalar nonlinear load model and its equivalent normal form (6)-(8) can be viewed as a block diagram interconnection of nonlinear functions and a linear transfer function.

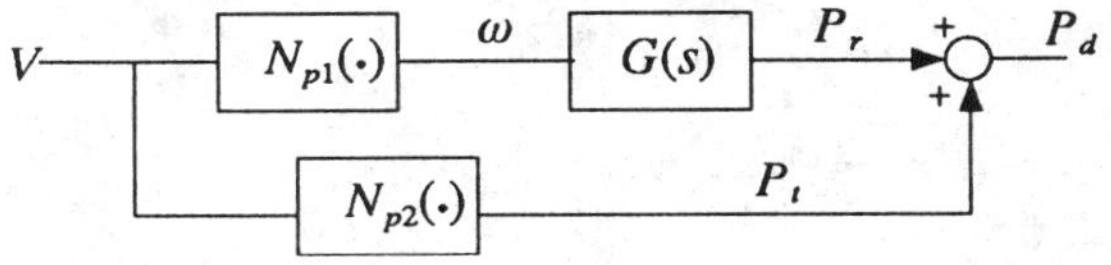

Figure 4 General Load Model with Linear Recovery Dynamics

For higher order dynamics

$$G(s) = \frac{b_m s^m + b_{m-1} s^{m-1} + \ldots + b_0}{s^n + a_{n-1} S^{n-1} + \ldots + a_0} \qquad (13)$$

the response of Figure 2 can be obtained with more exotic recovery, i.e. multiple time-constants and/or oscillatory behaviour.

Just using a second-order $G(s)$ can create the oscillatory response shown in Figure 2. and observed in field tests [14]. The only restriction needed in the general case is that $G(s)$ be strictly proper, i.e. $m < n$, to ensure the recovery response P_r is continuous.

Consider the second-order dynamic case

$$G(s) = \frac{b_1 s + b_0}{s^2 + a_1 s + a_0} \qquad (14)$$

In terms of the internal signal notation shown on Figure 4, we have

$$\ddot{P}_r + a_1 \dot{P}_r + a_0 P_r = b_1 \dot{\omega} + b_0 \omega \qquad (15)$$

The steady-state behaviour is obtained by setting derivatives to zero; this gives

$$P_d = \frac{b_0}{a_0} N_{p1}(V) + N_{p2}(V) \qquad (16)$$

The fast (jump) behaviour is obtained from the higher-order scalar form.

$$\ddot{P}_d + a_1 \dot{P}_d + a_0 P_d = f(\ddot{V}, \dot{V}, V) \qquad (17)$$

where

$$f(\ddot{V}, \dot{V}, V) := N_{p2}'(V).\ddot{V} + N_{p2}''(V).\dot{V}^2 + (a_1 N_{p2}'(V) + b_1 N_{p1}'(V)).\dot{V} + a_0 N_{p2}(V) + b_0 N_{p1}(V) \qquad (18)$$

Equality of the second-order terms gives

$$\frac{d}{dt}(\dot{P}_d) = \frac{d}{dt}(N_{p2}'(V).\dot{V})$$

This gives the transient load as $P_d = N_{p2}(V)$.

So just as in the simple first-order case, the steady-state and fast load behaviour can be directly connected to the nonlinearities $N_{p1}(\cdot)$ and $N_{p2}(\cdot)$.

The first-order normal form (6)-(8) is a special case of an alternative general representation to the scalar higher-order form. Translating the transfer function $G(s)$ to an equivalent state-space representation gives

$$\begin{aligned} \dot{x}_p &= F_p x_p + G_p \omega \\ P_r &= H_p^T x_p \end{aligned} \quad (19)$$

where x_p is an n-dimensional vector and F_p, G_p, H_p are appropriately dimensioned matrices. Combining (19) with the structure of Figure 4 gives

$$\begin{aligned} \dot{x} &= Fx + GN_1(V) \\ \begin{bmatrix} P_d \\ Q_d \end{bmatrix} &= H^T x + N_2(V) \end{aligned} \quad (20)$$

In the analysis of specific devices [11], it is easy to see that the P_d and Q_d models are related. To allow for this, the model (20) is presented in coupled form with $x = (x_p, x_q)$.

Simple MATLAB exercises show that a second order $G(s)$ gives responses close to those reported in Shackshaft et. al. [14] by appropriate choice of parameters $a_0, a_1, b_0, b_1, \alpha_s, \alpha_t$ (and reactive counterparts).

The model (20) can be easily incorporated into simulation programs for power system dynamics. The parameters which determine matrices F, G, H and nonlinear functions $N_1(\cdot)$ and $N_2(\cdot)$ must be obtained from measured data [13]. In [15] some comments are made about the need for second-order linear dynamics plus a nonlinearity response for load; this corresponds to $N_1 = I$ and a two pole $G(s)$ in Figure 4.

3 SYSTEM STABILITY ANALYSIS

3.1 System Model

Consider a power system subjected to a large disturbance. We model its post-disturbance behaviour with traditional assumptions on the generator and network.

Synchronous machines are represented by a constant voltage E_i in series with transient reactance. The network is assumed to be lossless, so all lines are modelled as series reactances.

Suppose the network consists of n buses connected by transmission lines. At m of these buses there are generators. The buses which have load but no generation are labelled $i = 1, \ldots n - m$. The network is augmented with m fictitious buses representing the generator internal buses in accordance with the classical machine model. They are labelled $i + m$ where i is the bus number of the corresponding generator bus.

Let the complex voltage at the ith bus be $V_i \angle \delta_i$ where δ_i is the bus phase angle with respect to a synchronously rotating reference frame. Let $V \in \mathbb{R}^{n+m}$ denote the vector of voltage magnitudes. Using the $(n+m)$th bus as the reference, we define internodal angles $\alpha_i := \delta_i - \delta_{n+m}$. The bus frequency deviation is given by $\omega_i = \dot{\delta}_i$. Let $\alpha \in \mathbb{R}^{n+m-1}$ and $\omega_g \in \mathbb{R}^m$ denote the vectors of internodal angles and generator frequencies respectively.

Let P_{b_i} and Q_{b_i} denote the total real and reactive power leaving the ith bus via transmission lines. Then

$$P_{b_i}(\alpha, V) := \sum_{j=1}^{n+m} V_i V_j B_{ij} \sin(\alpha_i - \alpha_j) \quad (21)$$

$$Q_{b_i}(\alpha, v) := -\sum_{j=1}^{n+m} V_i V_j B_{ij} \cos(\alpha_i - \alpha_j) \quad (22)$$

where jB_{ij} are the elements of the bus admittance matrix. For convenience, we have set $V_i = E_{i-n}$ and $\alpha_{n+m} = 0$. We denote by $-P_\ell, P_g$ the vectors of bus powers associated with load and generator buses respectively.

Combining generator swing equations and power flow equations gives [7]

$$\dot{\gamma}_g = -S(P_g(\alpha_g, \theta, V) - P_M) \quad (23)$$

$$\dot{\alpha}_g = \gamma_g \quad (24)$$

$$0 = P_\ell(\alpha_g, \theta, V) + P_d \quad (25)$$

$$0 = [V]^{-1}(Q_\ell(\alpha_g, \theta, V) + Q_d) \quad (26)$$

where damping has been ignored, S is an inverse inertia matrix, α has been written as (α_g, θ) to identify the generator and load bus angles separately and γ is the vector of internodal velocities. $[V]$ denotes the diagonal matrix with elements of $V \in \mathbb{R}^n$ as the diagonal elements.

Combining (23)-(26) with the load model (20) (or special case forms in Section 2) gives a differential-algebraic model with dynamic variables (α_g, γ_g, x) and algebraic variables (θ, V).

3.2 Static Loads Case

Assume $P_d = P_d^o$, constant; $Q_d = Q_s(V)$

Stability theory for this case is rather well-developed. Following an earlier result [6] showing that well-defined Lyapunov functions exist for the PV bus case, it was shown that the usual kinetic plus potential energy form can be used in this case [16]. It has been pointed out to the authors that such results were already known in Russia [17] (but certainly not accessible to Western readers).

A systemmatic derivation of the Lyapunov function (and various extended versions) is provided in [7] following first integral analysis on eqns. (23)-(26).

Fact 1 The first integral energy function for the model (23)-(26) is given by

$$\mathcal{V}_g(\alpha_g,\gamma_g) = \frac{1}{2}\gamma_g^T S^{-1}\gamma_g + \int_{\alpha_g^s}^{\alpha_g} (P_g(\mathcal{S},\theta,V) - P_M)^T d\mathcal{S} \quad (27)$$

$$= \frac{1}{2M_T}\sum_{i=n+1}^{n+m-1}\sum_{j=i+1}^{n+m} M_i M_j(\gamma_i - \gamma_j)^2$$

$$-\frac{1}{2}\sum_{i=1}^{n+m}\sum_{j=1}^{n+m} B_{ij}(|V_i\|V_j|\cos\alpha_{ij} - |V_i^s\|V_j^s|\cos\alpha_{ij}^s)$$

$$+\sum_{i=1}^{n} P_{d_i}(\alpha_i - \alpha_i^s) - \sum_{i=n+1}^{n+m-1} P_{M_i}(\alpha_i - \alpha_i^s) + \sum_{i=1}^{n}\int_{|V_i^s|}^{|V_i|}\frac{Q_{d_i}(\mu_i)}{\mu_i}\,d\mu_i \quad (28)$$

To establish this as a Lyapunov function required extending that stability theory to differential - algebraic systems [8]. Then it is possible to show that the function V at (27) is a Lyapunov function showing stability of an equilibrium $(\alpha_g^o,\gamma_g^o,V^o)$.

The equilibrium $(\alpha_g^o,\gamma_g^o,V^o)$ is given by

$$\gamma_g = 0 \quad (29)$$

$$P_g(\alpha_g,\theta,V) = P_M \quad (30)$$

and (25), (26), i.e. the usual power flow equations augmented by power balance at generator internal nodes. These can be written in the form

$$\tilde{f}_g(\alpha_g,\theta,V) = 0 \quad (31)$$

$$f_\ell(\alpha_g,\theta,V) = 0 \quad (32)$$

$$g(\alpha_g,\theta,V) = 0 \quad (33)$$

where $\tilde{f}_g, f_\ell, g$ are $\mathbb{R}^{m-1}, \mathbb{R}^n$ and $\mathbb{R}^n$ valued functions respectively. The power flow Jacobian has the form

$$J = \begin{bmatrix} F_{\alpha g} & F_{\theta g} \quad F_{vg} \\ F_{\alpha\ell} & \\ G_{\alpha\ell} & J_{\ell\ell} \end{bmatrix} \text{ where } J_{\ell\ell} = \begin{bmatrix} F_{\theta s} & F_{vs} \\ G_{\theta s} & G_{vs} \end{bmatrix}$$

The subscripts 's' refer to use of static load models. These power flow matrices are significant in establishing two important facts. We use the notation A^* to denote the Schur complement of A within a partitioned matrix.

Fact 2 The first-integral energy function (32) is a locally positive definite function if $F^*_{\alpha g}|_{(\alpha_g^o,\theta^o,V^o)} > 0$.

If generator damping is included this property implies small disturbance (asymptotic) stability. For large-disturbance stability, we have that the small-disturbance criterion ensures a valid Lyapunov function.

Fact 3 [7,8] Suppose that $\det J_{\ell\ell}|_{(\alpha_g^o,\theta^o,V^o)} \neq 0$. Then in some neighbourhood of $(\alpha_g^o,\theta^o,V^o)$ the DA system is equivalent to an ordinary differential equation (ODE) model of the form

$$\dot{\gamma}_g = -S(P_g(\alpha_g,\psi(\alpha_g)) - P_M) \quad (34)$$

$$\dot{\alpha}_g = \gamma_g \quad (35)$$

where continuous function ψ relates the algebraic variables to generator angles α_g.

These observations motivate a structural picture developed in [9, 10]. The impasse surfaces I on which $J_{\ell\ell}$ is singular separate open sets where the DA system model is equivalent to a local ODE model. The Lyapunov functions $\mathcal{V}$ must be seen globally as multivalued. Within a particular open set C_i, the Lyapunov stability of an equilibrium point is established using one "sheet" of this function. Discussion of regions of transient stability can proceed as for other ODE models (based on impedance loads) [18] if trajectories do not meet impasse surfaces. This depends on the load indices. In general, the critical energies must take account of the proximity of I to the stable equilibrium.

It is shown in [9] that trajectory impact on I corresponds to a short-term voltage collapse. In energy terms, this could correspond to a jump between energy "sheets". Related reasoning has led DeMarco and Overbye to a voltage security index [19].

In summary, we have static regularity conditions on the power flow Jacobian - see Facts 2, 3 - playing a key role in determining the large disturbance behaviour. Load indices are very important and voltage dip phenomena are associated with the angle behaviour.

3.3 Load System Stability

Assume $P_g = P_M^o$, constant.

We consider the case where the system is operating outside the transient period following a major disturbance, i.e. angle stability is not a concern. The stability analysis with first-order linear recovery load models has been considered in [12] to the extent of static regularity properties and their connections to small disturbance stability. Here we briefly review this and make some progress on large-disturbance behaviour.

System behaviour is determined by both static load and dynamic load characteristics. Let J_t be the Jacobian corresponding to the dynamic power flow

$$P_t(V) - P_\ell(\theta,V) \quad (36)$$

$$Q_t(V) - Q_\ell(\theta,V) \quad (37)$$

The angles α_g can be eliminated from (31) by the usual Implicit Function Theorem argument to give 2n power flow equations in (θ, V). The static power flow Jacobian is $J^*_{\ell\ell}$. (We drop the * for notational convenience through taking multiple Schur complements.)

For simplicity, we make a further restriction to assume slow real load recovery, i.e. $P_t \simeq P_s$. For this case, set $x_p = x_p^o$, constant.

Voltage regularity [12] requires that (for any PQ bus) $\Delta Q_i > 0$ gives all $\Delta V_j > 0$.

Fact 4 [12] Assume $\det F_{\theta s} \neq 0$. The load system is at a voltage regular operating point if $G^{\bullet}_{Vs}$ is an irreducible M-matrix.

The matrix $G^{\bullet}_{Vs}$ features in much recent research on static voltage collapse proximity indicators - see [12] for references.

Again, using the Implicit Function Theorem argument, if $F_{\theta s}$ is nonsingular, we can eliminate θ according to $\theta = \psi_p(V)$ for some differentiable function $\psi_p(\cdot)$ in a region of the operating point. (This assumption on $F_{\theta s}$ was used in deriving the QVsensitivity.) Then the DA model becomes

$$\dot{x}_q = Q_s(V) - Q_\ell(\psi_p(V), V) \quad (38)$$

$$0 = -Q_\ell(\psi_p(V), V) + T_q^{-1}x_q + Q_t(V) \quad (39)$$

Theorem 1 [12] Assume $\det F_{\theta s} \neq 0$ and $\det G^{\bullet}_{Vt}(V^o) \neq 0$. The load system with slow real power recovery and at operating point (x_q^o, V^o) is small disturbance stable if the matrix $G^{\bullet}_{Vs}(V^o)C_{qd}^{-1}(V^o)$ is stable where $C_{qd}(V^o) := -T^q(G^{\bullet}_{Vt}(V^o)$.

In [11, 12], various possibilities were discussed depending on relative values of $G^{\bullet}_{Vs}, T_q$ and $G^{\bullet}_{Vt}$ at the operating point. In particular, it is possible to operate at an irregular point with small disturbance stability.

The dynamics can be expressed in terms of the voltages. From equation (39), we have

$$x_q = T_q(Q_\ell(\psi_p(V), V) - Q_t(V)) \quad (40)$$

So $\dot{x}_q = -T_q G^{\bullet}_{Vt}(V)\dot{V}$ and (38)-(39) becomes

$$C_{qd}(V)\dot{V} = g^{\bullet}_s(V) \quad (41)$$

where

$$g^{\bullet}_s(V) := Q_s(V) - Q_\ell(\psi_p(V), V) \quad (42)$$

So the condition $C_{qd}(V^o) \neq 0$ is seen to correspond to avoidance of an impasse surface. The relevance to voltage collapse behaviour for the single load case is considered in [12].

The above results enable classification of operating points according to regularity and stability. According to Lyapunov theory, if the operating point is small disturbance (i.e. asymptotically) stable, then it is asymptotically stable for the nonlinear model. We are then interested in determining the region of attraction around the operating point. Again, use is made of Lyapunov stability theory for DA systems [8].

To illustrate briefly, consider the slow real recovery model (41) and assume $\det F_{\theta s} \neq 0$ and $\det G^{\bullet}_{Vt}(V^o) \neq 0$. Consider the Lyapunov function candidate

$$\mathcal{V}_\ell(V) = g^{\bullet}_s(V)^T \Lambda g^{\bullet}_s(V) \quad (43)$$

where Λ is some positive definite matrix. Differentiating $\mathcal{V}$ along the trajectories of (42) gives

$$\dot{\mathcal{V}}(V) = (\dot{g}^{\bullet}_s(V))^T \Lambda g^{\bullet}_s(V) + (g^{\bullet}_s(V))^T \Lambda \dot{g}^{\bullet}_s(V)$$
$$= (g^{p\bullet}_s)(V))^T M(V;\Lambda) g^{\bullet}_s(V) \quad (44)$$

where

$$M(V;\Lambda) := (G^{\bullet}_{Vs}C_{qd}^{-1}(V))^T \Lambda + \Lambda(G^{\bullet}_{Vs}(V)C_{qd}^{-1}(V)) \quad (45)$$

If Λ can be chosen to make $M(V;\Lambda)$ negative definite for $V \neq V^o$ in some region V^o, a stability result is established.

Theorem 2 Under the conditions of Theorem 1, the equilibrium V^o of load system (41) is asymptotically stable and $\mathcal{V}(\cdot)$ is a Lyapunov function.

In Section 3.2, Lyapunov functions were presented for voltage collapse studies in the transient period with generator dynamics; Theorem 2 now establishes one Lyapunov function $\mathcal{V}(\cdot)$ for load systems.

4 VOLTAGE COLLAPSE THEORY EXTENSIONS

Section 3 has summarised some results on generator (angle) and load stability which clearly say something about voltage collapse in a static or dynamic sense. The viewpoint established is to use the power flow equations as a basis and then include generator or aggregate load (including all static and dynamic components) as the case may be. Clearly in both cases, the load characteristics and network structure represented by appropriate Jacobian matrices play a key role. For large - disturbance angle and voltage stability analysis, we saw many similarities:

1. DA models based around the power flow equations, i.e. (23)-(26) and (38)-(39) respectively;
2. Connections between static regularity, small-disturbance stability and existence of valid Lyapunov functions;
3. The possibility of multiple stable equilibria related to different 'causal regions' separated by the impasse surfaces.

However, there are many questions to consider in more detail before the theory yields a firm basis for static and dynamic voltage collapse indicators. One direction concerns geometric analysis [15, 20] of the bifurcations associated with voltage collapse and the related issue of characterizing regions of stability. Lyapunov function techniques are useful in giving estimates of these regions of stability.

While the Lyapunov function $\mathcal{V}_g$ at (27) appears quite suitable for computation of stability regions, it is severely limited by relying on constant real loads for the potential energy component to be well-defined. (The usual transient energy function based on impedance loads [18] is on even shakier ground theoretically since it is only well-defined for no real load.) The Lyapunov function $\mathcal{V}_\ell$ at (43) for the load system can be easily generalised to real and reactive linear recovery loads [21], and does not involve any path dependence problems. The more general nth order linear recovery dynamics (20) can be accounted for by an additive term x^TKx, K positive definite. However, the ability of these Lyapunov functions to estimate regions of stability must be tested further. These issues and the question of well-defined potential energy type functions for load systems are investigated further in [21]. To illustrate for the single load case, we have the following result.

Theorem 3 For the single load version of system (41) under the conditions of Theorem 2, we have the potential energy function

$$W(x_q, x_q^o) = -\int_{x_q^o}^{x_q} g_s^*(V)dx_q$$

is a valid Lyapunov function.

W is positive definite if the small-disturbance stability condition of Theorem 1 is satisfied. In the multi-load case we again face the issue of path dependence. Fortunately, the nature of $Q - V$ relationships allows resolution of this issue under certain conditions [21].

5 CONCLUSION

This paper has presented an analytical framework for analysis of voltage collapse as a dynamic phenomenon. By giving equal emphasis to generator and load dynamics, short-term and longer-term behaviour can be studied by selecting appropriate parameter values.

It has been shown that there are many similarities between generator (angle) and load system stability analysis revolving around dynamic behaviour of differential - algebraic models. It has been argued [10] that in study of system collapse we should consider both angle and voltage dynamics (especially in the initial and final stages). Thus there is motivation for further unification. One issue on which more understanding is needed is existence of well-defined potential energy functions for use in computation of stability regions with both generator and load dynamics.

6 REFERENCES

[1] M.A. Pai, Energy Function Analysis for Power System Stability, Kluwer Academic Publ., Boston, 1989.

[2] M. Ribbens-Pavella and F.J. Evans, "Direct Methods for Studying Dynamics of Large-scale Electric Power Systems – A Survey", Automatica, Vol.21, No.1, January 1985, pp.1–21.

[3] Y. Mansour, (ed.), "Voltage Stability of Power Systems: Concepts, Analytical Tools, and Industry Experience", IEEE Task Force Report, Publication 90TH 0358-2-PWR.

[4] Proceedings: Bulk Power System Voltage Phenomena – Voltage Stability and Security, Potosi, Missouri, September 1988 (EPRI EL-6183, January 1989.)

[5] L.H. Fink (ed.), Proceedings of Bulk Power System Voltage Phenomena II Voltage Stability and Security, Deep Creek Lake, Maryland, August 1991.

[6] A.R. Bergen and D.J. Hill, "A Structure Preserving Model for Power System Stability Analysis", IEEE Trans. Power Apparatus and Systems, Vol. PAS-100, No.1, January 1981, pp.25–35.

[7] D.J. Hill and C.N. Chong, "Energy Functions for Power Systems Based on Structure Preserving Models", Proc. 25th Conf. on Decision and Control, Athens, Greece, December 1989, pp.1218–1223.

[8] D.J. Hill and I.M.Y. Mareels, "Stability Theory for Differential/Algebraic Systems with Application to Power Systems", IEEE Trans. Circuits and Systems, Vol.CAS-37, No.11, pp.1416–1423, November 1990.

[9] I.A. Hiskens and D.J. Hill, "Energy Functions, Transient Stability and Voltage Behaviour in Power Systems with Nonlinear Loads", IEEE Trans. on Power Systems, Vol.4, pp.1525–1533, November 1989.

[10] I.A. Hiskens and D.J. Hill, "Failure Modes of a Collapsing Power System", in [5].

[11] D.J. Hill, "Nonlinear Dynamic Load Models with Recovery for Voltage Stability Studies", IEEE Trans. Power Systems, to appear. (Publication 92 WM 102-4 PWRS, Power Engineering Society.)

[12] D.J. Hill, I.A. Hiskens and D. Popovic, "Stability Analysis of Load Systems with Recovery Dynamics", submitted to IEEE 1993 Winter Meeting.

[13] D. Karlsson and D.J. Hill, "Modelling and Identification of Nonlinear Dynamic Loads in Power Systems", submitted to IEEE 1993 Winter Meeting.

[14] G. Shackshaft, O.C. Symons and J.G. Hadwick, "General -purpose Model of Power-system Loads", Proc. IEE, Vol.124, No.8, August 1977.

[15] V. Venkatasubramanian, H. Schättler and J. Zaborszky, "A Taxonomy of the Dynamics of the Large Power System with Emphasis on its Voltage Stability", in [5].

[16] N. Narasimhamurthi and M.T. Musavi, "A Generalized Energy Function for Transient Stability Analysis of Power Systems", IEEE Trans. Circuits and Systems, Vol.CAS-31, No.7, July 1984, pp.637–645.

[17] V.P. Vasin, "Energy Integral for Equations of Electric Power System Transients with Loads Represented by Steady-state Characteristics", Izvestia AN SSR, Energetika i Transport, 6, 1974, pp.26–35. (In Russian).

[18] A.A. Fouad and V. Vittal, "The Transient Energy Function Method", International Journal of Electrical Power and Energy Systems, Vol.10, No.4, October 1988, pp.233–246.

[19] C.L. de Marco and T.J. Overbye, "An Energy Based Security Measure for Assessing Vulnerability to Voltage Collapse", IEEE Trans. Power Systems, Vol.5, No.2, May 1990, pp.419–427.

[20] I. Dobson and H.-D. Chiang, "Towards a Theory of Voltage Collapse in Electric Power Systems", Systems and Control Letters, Vol.13, 1989, pp.253–262.

[21] D.J. Hill and I.A. Hiskens, "Large Disturbance Stability Analysis of Load Systems", University of Newcastle Technical Report EE9260, October 1992.

Proceedings of the 31st Conference
on Decision and Control
Tucson, Arizona • December 1992

Efficient Computational Methods for a Practical Performance Index and the Exact Voltage Collapse Point in Electric Power Systems

René Jean-Jumeau, Hsiao-Dong Chiang, Robert J. Thomas
School of Electrical Engineering
Cornell University
Ithaca, NY 14853
U.S.A.

Abstract—**A number of performance indices intended to measure the severity of the voltage collapse problem have been proposed in the literature. These performance indices however can generally not answer questions such as "Can the system withstand another 100 MVar increase on bus 11?" This paper presents (i) a new performance index, developed in the load demand space, and providing a direct relationship between its value and the amount of load demands that the system can withstand before collapse and (ii) a new, more convenient method to determine the exact point of voltage collapse. The new index can thus answer questions such as "Can the system withstand a simultaneous increase of 70 MW on bus 2 and 50 Mvars on bus 6?". The exact point of collapse is computed using a novel (n +1)-dimensional characteristic equation instead of the generic (2n+1)-dimensional one generally offered in the literature. The computation involved in the proposed performance index and exact point determination are inexpensive in comparison with those required in the existing ones. Simulation results on the IEEE 39-bus system and the Tai-power 234-bus system are presented with promising results.**

I. INTRODUCTION

Voltage collapse is generally caused by one of the two types of system disturbances: load variations and contingencies. Several recent power system blackouts were related to voltage collapse. Voltage collapse has been especially experienced by heavily loaded power systems subject to an increase in load demands. Voltage collapse due to contingencies has been studied by several researchers, see for example [1–7], where the key issues are the feasibility of the stable equilibrium point after contingency and the estimate of its stability region (region of attraction). Voltage collapse has also been attributed to a lack of reactive power support, which can be equivalently regarded as due to increases in load demand [8]. In the present paper, we will thus concentrate on voltage collapse due to load variations.

A number of performance indices intended to measure the severity of the voltage collapse problem have been proposed in the literature [9–15]. These performance indices can be viewed as providing some measure relative to the "distance" between the current operating point and the bifurcation point. Note that all these performance indices are defined in the state space of power system models instead of in the parameter space. Thus, these performance indices can not directly answer questions such as "Can the system withstand a 100 MVar increase on bus 11?" or "Can the system withstand a simultaneous increase of 70 MW on bus 2 and 50 Mvars on bus 6?".

Determination of margins in the parameter space is still a new concept which may have been spurred by the concept presented in [16]. In [17], a method was proposed to compute the reactive power margin. However, the method utilizes a general optimization technique and only deals with reactive powers. Another method which solves an extended $(2n+1)$-dimensional system of equations characterizing the saddle-node bifurcation point was proposed in [18]. The method attempts to compute the saddle-node bifurcation point directly. The success of these two methods depends greatly on a good initial guess of the desired saddle-node bifurcation point. Otherwise, the methods may diverge or converge to an undesired saddle-node bifurcation point. These two methods compute the margins in the parameter space instead of giving a performance index of the margins. As such, these two methods demand a great deal of computational efforts.

In this paper we present a *new performance index* that provides a direct relationship between its value and the amount of load variations that the system can withstand before collapse. The new performance index is more practical than existing ones in terms of its ability to provide a direct relationship between its value and the amount of parameter variation that the system can withstand before collapse. More specifically, it can be readily interpreted by operators of power systems to answer questions such as "Can the system withstand a simultaneous increase of 70 MW, 40 Mvars on bus 2, 100 MW on bus 5 and 50 Mvars on bus 16?".

Additionally, we utilize these results and exploit the particular structure of parameterized power system equations to compute the *exact voltage collapse point.* This complements performance index information by yielding an accurate margin to collapse. The proposed index and collapse point computation present the distinct advantage of reduced computation and of using basic power system technology (load flows,

CH3229-2/92/0000-2910$1.00 © 1992 IEEE

etc.). Simulation results on the IEEE 39-bus system and the Tai-power 234-bus system are presented with promising results.

II. PRELIMINARY

In general, a power system can be described by

$$\dot{x} = f(x, \lambda) \tag{1}$$

where $x \in R^n$ is a state vector including bus voltage magnitudes and angles, $\lambda \in R^m$ is a parameter vector representing real and reactive power demands at each load bus. The parameter vector λ is subject to variation (due to load variations) causing a structure change in the load flow solutions of (2). If the parameter λ varies slowly or quasi-statically with respect to the dynamics of (1), then the power system can be appropriately modeled by

$$0 = f(x, \lambda) \tag{2}$$

Typically, the power system (2) operates at a stable load flow solution, say at x_1 with parameter λ_1, (the system Jacobian at x_1, $D_x f(x_1, \lambda_1)$, has eigenvalues only with negative real parts). For λ sufficiently close to λ_1, the stable load flow solution x_1 persists and its stability type remains unchanged. The stable operating point x_1 may disappear or become unstable depending on the system structure and the way in which the parameter is varied.

One typical way in which the stable operating point x_1 disappears is that x_1 and another load flow solution x_2 coalesce and disappear in a saddle-node bifurcation as parameter λ passes through a bifurcation value λ^* (Fig 1). When $\lambda = \lambda^*$, x_1 and x_2 coalesce to form an equilibrium point x^* whose corresponding system Jacobian has one zero eigenvalue and the real parts of other eigenvalues are negative. Furthermore, there are no load flow solutions nearby. In the terminology of nonlinear dynamical systems, λ^* is called the bifurcation value and (x^*, λ^*) called the bifurcation point. This center manifold voltage collapse model proposed in [19] was based on the saddle-node bifurcation and its consequent dynamics. Physical explanations of this center manifold voltage collapse model can be found in [20].

In what follows, we are interested in the situation when the system (2) is the so-called one-parameter dynamical system. In power system applications, a one-parameter dynamical system is a system together with one of the following conditions:

1. the reactive (or real) power demand at one load bus varies while the others remain fixed,
2. both the real and reactive power demand at a load bus vary and their variations can be parameterized. Again the others remain fixed,
3. the real and/or reactive power demand at some collection of load buses varies and their variations can be parameterized while the others are fixed.

In the next section, we will present a new performance index that provides a direct relationship between its value and the amount of load demands that the system can withstand before collapse.

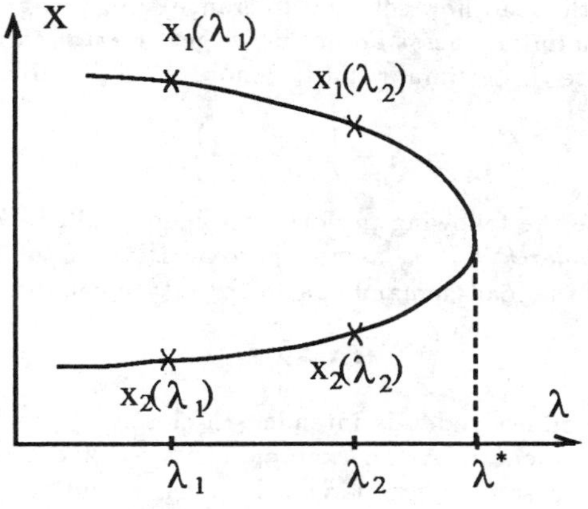

Figure 1: Coalescence of two equilibrium points x_1 and x_2 in a saddle-node bifurcation.

III. THE NEW PERFORMANCE INDEX

Given a power system (specifically, a set of data for load flow study), a stable operating point (say x_1, from a state estimator perhaps), its current load demand pattern (i.e. the load demand at each bus); and a change in load demand pattern, derived from a short-term load forcasting program in the energy management system perhaps (say λb, where $\lambda \in \mathcal{R}$ and $b \in \mathcal{R}^n$ define the variation in load demand pattern). The stable equilibrium point x_1 will lose stability and disappear when the load demand pattern increases to $\lambda^* b$, where λ^* is the bifurcation value. We investigate in this section the task of directly estimating the corresponding bifurcation value λ^*. Our estimate τ of the margin from the current value λ_1 to λ^* is a performance index (in the load demand space).

In order to explain the new performance index, we refer to equation (2). Consider the following test function [21],

$$t_s(x, \lambda) \triangleq e_l^T J(x, \lambda)\nu, \tag{3}$$

where $J = \frac{\partial f}{\partial x}$, It is shown in [21] that this test function has the following desired characteristics: (a) it is a function of x and λ and (b) it has a value of *zero* at the bifurcation value λ^* of (2). Seydel suggests in [21] that the above test function (3) is expected to be, but is not restricted to, a parabolic function symmetrical about the λ-axis. We have found that the test function can be approximated by a quadratic or quartic model when a change of both real and reactive demands on multiple buses of a power system occurs (Fig. 3, 5). This leads, after some algebraic manipulations, to the following expressions for the value λ^*:

$$\lambda^* \approx \lambda_1 - \frac{1}{c}\frac{t_s(x_1, \lambda_1)}{t_s'(x_1, \lambda_1)}, \tag{4}$$

where $c = 2$ or 4 for the quadratic or quartic models respectively and t_s' is the derivative of t_s with respect to time.

In practice, in order to avoid the large amount of computation needed to evaluate t_s' exactly, an approximation of t_s' can be found via finite difference. In order to apply the finite difference, one needs to run another load flow with $\lambda = \lambda_1 + \delta\lambda$ for small $\delta\lambda$ (in practice, we take $\delta\lambda$ equal to five times the load flow solution tolerance) and perhaps, using x_1 as an initial guess. From this we get an *estimate* of λ^*. Let $\bar{\lambda}$ denote this estimate and $\bar{t}'_s$ denote the finite-difference derivative:

$$\bar{\lambda} = \lambda_1 - \frac{1}{c} \frac{t_s(x_1, \lambda_1)}{\bar{t}'_s(x_1, \lambda_1)}. \qquad (5)$$

We propose the following performance index, called *Voltage Collapse Index* (VCI) τ, as an approximation of the exact margin (in the load-demand space) to voltage collapse:

$$\tau = \bar{\lambda} - \lambda_1. \qquad (6)$$

This performance index is intended to closely approximate the exact value $\Delta\lambda$. As an example of its effectiveness, see figure 4 for a sample comparison between τ and the *exact* margin $\Delta\lambda$.

The computation required in the above procedure basically involves calculations of two load flow solutions (the current one and a new one) and some matrix computations, making the proposed performance index easier and cheaper to compute than those proposed in the literature and which require calculations such as eigenvalues, eigenvectors, singular values, energy functions, conditional numbers. In addition, the proposed performance index provides a direct measure regarding the amount of load increases that the system can withstand before collapse. It should be pointed out that the proposed index does not require the computation of unstable load flow solutions (unstable equilibrium points); the two load flow solutions needed to compute the proposed index are stable load flow solutions corresponding to two (slightly) different loading conditions.

IV. A NEW METHOD FOR EXACT SADDLE-NODE BIFURCATION POINT COMPUTATIONS

We've suggested the following linearly parameterized formulation for the power system equations (2) in [22]:

$$f(x) - \lambda b = 0, \qquad (7)$$

where $\lambda \in \mathcal{R}$ and $b \in \mathcal{R}^n$. When attempting to compute the (exact) bifurcation point x^* associated with λ for a nonlinear system with equilibria defined by (2) or (7), it is well known (e.g. [23]) that the following extended system can be solved:

$$\begin{bmatrix} f(x, \lambda) \\ f_x(x, \lambda)\nu \\ \nu^T \nu - 1 \end{bmatrix} = 0, \qquad (8)$$

where $\nu \in \mathcal{R}^n$ is an auxiliary vector. However, this means attempting to solve a nonlinear system of equations twice as big as the original. In this section, we will use the particular form of the load flow equations (7) and a defining equation for a change of parameters (equation (9) below) to design a scalar characteristic equation for the saddle-node bifurcation point. In assocation with the original equilibrium equations, we will then only need to solve an $(n+1)$-dimensional system of equations.

In order to avoid ill-conditioning of the system Jacobian matrix involved in computations near the bifurcation point, we suggest in [22] a change of parameters from λ to κ defined by the equation:

$$\lambda - \kappa b^T x = 0. \qquad (9)$$

Also, by replacing λ from (9) in (7), we have seen that solutions of (7) can equivalently be obtained by solving:

$$g(x, \kappa) = f(x) - \kappa b^T x b = 0. \qquad (10)$$

Let $c \in \mathcal{R}^n$ be the solution of the linear equation

$$g_x(x, \kappa) c = b, \qquad (11)$$

so that

$$c^* = \left[g_x(x, \kappa)|_{(x^*, \kappa^*)} \right]^{-1} b$$

where $\kappa^* = \frac{\lambda^*}{b^T x^*}$ when evaluated at the saddle-node bifurcation point.

Now we present a new test-function which can characterize the SNBP. Consider the following scalar function:

$$t_r : \mathcal{R}^{n+1} \to \mathcal{R}; \quad t_r(x, \kappa) = \kappa b^T c + 1. \qquad (12)$$

We propose to use this function to compute the exact saddle-node bifurcation value λ^*. It can be proven that the new test-function t_r is only equal to zero at the SNBP for equation (7). Also it is a monotone function in the neighborhood of the SNBP by virtue of its relationship to the derivative of λ with respect to κ. Specifically, at the saddle-node bifurcation point (x^*, λ^*):

$$\kappa^* b^T c^* + 1 = 0$$

and

$$\kappa b^T c + 1 \neq 0 \text{ for } (x, \lambda) \neq (x^*, \lambda^*).$$

Thus, saddle-node bifurcations of equilibria for system 7 can be completely defined by the following $(n+1)$-dimensional system of equations:

$$H(x, \kappa) = \begin{bmatrix} g(x, \kappa) \\ t_r(x, \kappa) \end{bmatrix} = 0. \qquad (13)$$

We note that, although the linear system $g_x(x, \kappa) c = b$ must be solved at each iteration of the nonlinear solver being used, the matrix $g_x(x, \kappa)$ is already decomposed for the current step to compute $g(x, \kappa) = 0$. Consequently, obtaining $c = [g_x(x, \kappa)]^{-1} b$ is a simple matter.

Now, in order to solve (13), good initial values for the unknowns x and κ (or λ) are needed. We have already computed a good approximation of the bifurcation value, $\bar{\lambda}$ to obtain the VCI. Next, an initial guess of the bifurcation point is needed. Near the bifurcation point, the curve of equilibrium points $x(\lambda)$ defined by (7) as λ is varied is nearly parabolic [24]. Hence, assuming a quadratic form for $x_i(\lambda)$ of each component i of the state vector x (Fig. 2), we can approximate the bifucation value in the state space with $\bar{x}$ such that:

$$\bar{x}_i = x_{1i} + 2(\lambda_1 - \bar{\lambda}) \cdot x'_{1i}. \qquad (14)$$

Therefore, $\bar{\lambda}$ is a first guess for the parameter λ and $\bar{x}$ serves as initial guess in terms of x. The approximation $(\bar{x}, \bar{\lambda})$, allows us to compute the new parameter value

$$\bar{\kappa} = \frac{\bar{\lambda}}{b^T \bar{x}}. \qquad (15)$$

which is to be used as an initial guess in solving for the exact bifurcation value λ^*.

An actual solution of (7) can be obtained given $\bar{\kappa}$ and $g(x, \kappa) = 0$: let x_2 be that solution. Finally, using the approximation (x_2, κ), the bifurcation point (x^*, λ^*) can be obtained via (x^*, κ^*), by solving (13). Additionally, the *exact voltage collapse margin* (VCM) is simply:

$$\Delta\lambda = \lambda^* - \lambda_1. \qquad (16)$$

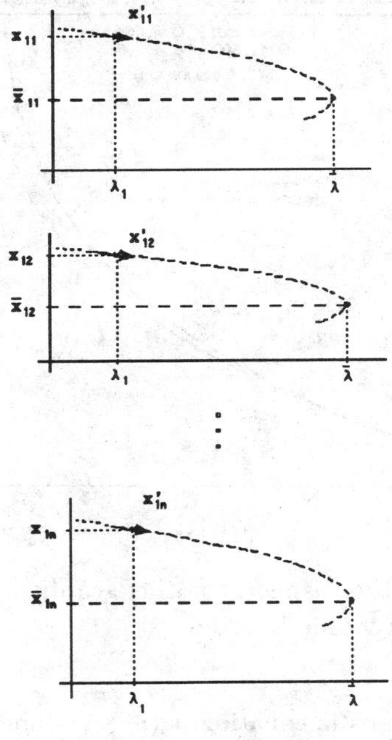

Figure 2: Determination of the approximation $\bar{x}$ of the bifurcation point x^*.

V. TEST RESULTS

The proposed performance index has been tested and evaluated on several power systems. In this section we present test results on the Ward-Hale system, the New England 39-bus system and the Tai-power 234-bus system. Both single-bus load variations and multiple-bus load variations are considered. The test function which the proposed index is based on is modeled as a polynomial function: second-order for load variation on a single bus, fourth-order for multiple-bus load variation. To evaluate its accuracy, we compare the values from the proposed index against the exact bifurcation values.

Single-bus load variations

We consider a power system under study with the reactive power demand at one of its load buses varying. Let k denote the index of the component corresponding to voltage magnitude on the chosen bus of state variable x. The load flow equations (2) have the form:

$$f(x, \lambda) = f(x) - \lambda b, \tag{17}$$

where $b = e_k$, the kth unit vector. λ represents the varying parameter.

In order to validate the quadratic model chosen to represent the test function t_s, we plot both t_s and its quadratic model versus the varying parameter λ. Both graphs start at $(x, \lambda) = (x_1, \lambda_1)$ and initially have the same derivative value $t_s'(x_1, \lambda_1)$. Figure 3 shows the simulation result when the reactive load demand on bus 5077 of a Tai-Power 234-bus system varies. In this case, as for a number of single-bus load variations simulated, the two curves are practically indistinguishable from each other; thus validating the chosen quadratic model for the test function.

Next, the accuracy of the proposed performance index for estimating the amount of load variation that the system can

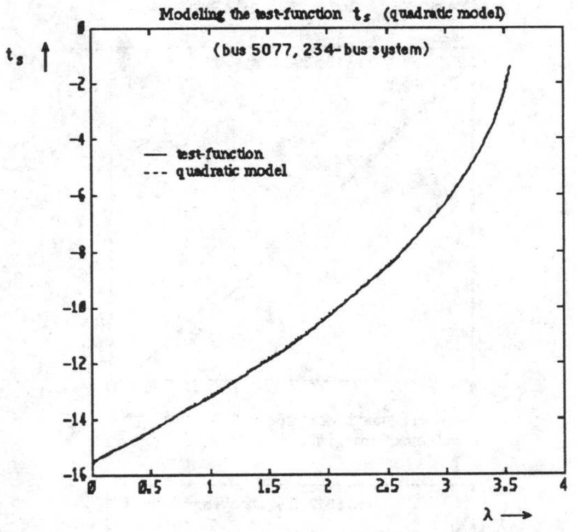

Figure 3: Comparison of test-function t_s with a quadratic model on a 234-bus test system.

withstand before collapse is assessed. To this end, both the proposed index (VCI), which is the estimated margin $\tau = \bar{\lambda} - \lambda_1$, and the *exact* margin are both plotted versus the varying parameter λ. Simulations on a 39-bus system and a 234-bus system yield the graphs in Fig. 4. These simulation results clearly favor that the proposed index can yield very accurate results.

Multiple-bus load variations

We consider in this case that the real and/or reactive power demand at some collection of load buses varies and their variations can be parameterized while the others are fixed. Here, the corresponding load flow equations have the form similar to that in (17) except that $b = \sum_{i \in J} \alpha_i e_i$, where $J \subset \{1, 2, \ldots, n\}$, $\alpha_i \in \mathcal{R}$ and the e_i are the ith standard unit vectors in $\mathcal{R}^n$.

In the multiple-bus case, a fourth degree polynomial better approximates the test-function. As an illustration, the test-function t_s and the model are plotted when real and reactive load are varied simultaneously on 4 different load buses of the 234-bus system. We have also plotted the test-function t_s and the model when both the real and reactive loads are varied simultaneously on 10 or more different load buses of the 234-bus system with similar results.

It has been observed that the new performance index performs quite well in all the simulations including both the single-bus and multiple-bus load variations. For instance, in the 234-bus system with multiple-bus load variations (see Fig. 6), when the load demand is with $\lambda = 4$, the proposed performance index predicts the margin is 8.21 p.u. while the exact margin is 8.215 p.u. When the load demand is with λ greater than 4, the proposed performance index predicts the margin quite accurately.

Exact Voltage collpase Point Computations

The appearance of a typical curve of the test-function t_r and its use in computing the exact bifurcation point (x^*, λ^*) are illustrated in Fig. 7. This figure is a plot of t_r vs. κ as both the real and reactive power are increased on bus 4 of a 6-bus test system. A net advantage of our new approach over the conventional $(2n+1)$-dimensional characteristic equation is that the solution can be bracketed with respect to $t_r(\kappa)$,

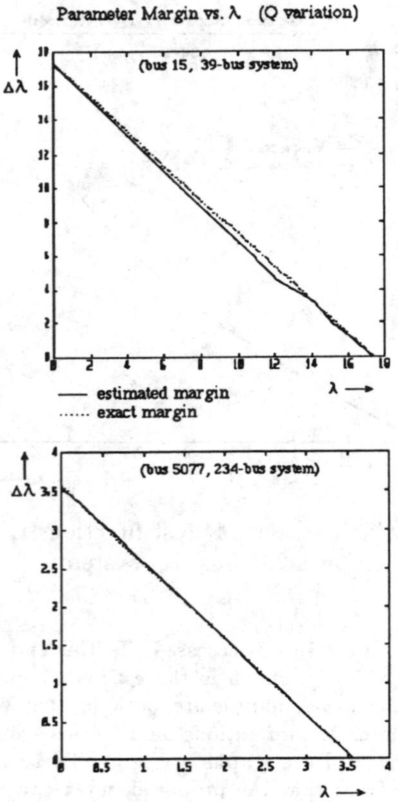

Figure 4: Comparison of the estimated margin τ estimated using the test-function t_s with the exact margin $\Delta\lambda$.

i.e. in the process of solution, we can find an interval $[\kappa_a, \kappa_b]$ such that $t_r(\kappa_a) \geq 0$ and $t_r(\kappa_b) < 0$, so that we are guaranteed that the solution $\kappa^* \in [\kappa_a, \kappa_b]$. As the solution process continues, the bracket shrinks considerably, allowing halt the process either (a) when t_r becomes smaller than some chosen tolerance or (b) when $|\kappa_a - \kappa_b|$ becomes less than a predetermined value. In the present case, we found $\kappa^* = 1.4437$ with $t_r = 0.0029$ and when $|\kappa_a - \kappa_b|$ had shrunk to $1.10e^{-4}$. The resulting bifurcation value in terms of λ was determined to be $\lambda^* = 0.8392$ p.u.

A practical case in which real and reactive power was varied on 15 buses of a Tai-Power 234-bus system resulted in $\kappa^* = 7.7757$ for a relatively large $t_r = 0.5955$, but within a bracket of width $|\kappa_a - \kappa_b| < 9.10e^{-5}$. The bifurcation value in this case is $\lambda^* = 0.0903$.

VI. CONCLUSIONS

In this paper, we have developed a new performance index based on the center manifold voltage collapse model as well as a new characteristic equation for computation of the exact collapse point. This index provides a direct relationship between its value and the amount of load demands that the system can further withstand before collapse. This relation makes the performance index readily interpretable to operators, making it more practical than existing ones to assess a potential voltage collapse. This new performance index will also allow us to identify the "weak areas" subject to voltage collapse and to derive preventive controls for the problem.

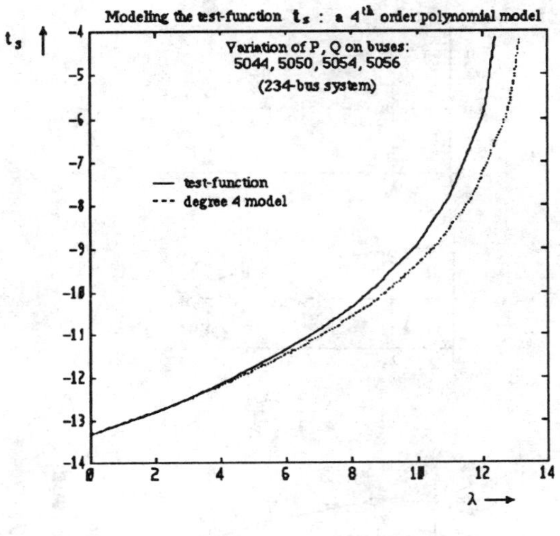

Figure 5: Relative aspects of the graphs of t_s and a fourth-order polynomial model.

The new characteristic equation is $(n+1)$-dimensional and therefore computationally more efficient that the standard $(2n+1)$-dimensional formulation generally utilized. Moreover, it allows to monitor the approach to the collapse point during the solution process, offering in this manner a net advantage over the conventional approach in terms of flexibility. The exact collapse point therewith obtained complements the index since the latter is essentially approximative in nature.

One of the features that distinguishes the performance index developed in this paper from existing ones is that this one is developed in the parameter space. It can answer questions such as "Can the system withstand a 100 MVar increase on bus 11?" or "Can the system withstand a simultaneous increase of 70 MW on bus 2 and 50 Mvars on bus 6?". We are encouraged by the promising simulation results on several power systems. We hope to extend the performance index to more detailed power system models. We reiterate that the proposed index and exact collapse point computation present the distinct advantage of reduced computation and of using basic power system technology (load flows, etc.).

References

[1] H.D. Chiang and F.F. Wu, "On Voltage Stability", *Proc. 1986 IEEE International Symposium on Circuits and Systems*, vol. 3, pp. 1039-1043, (May 1986).

[2] F. Mercede, J.C. Chow, H. Yan and R. Fischl, "A Framework to Predict Voltage Collapse in Power Systems", *IEEE Trans. on Power Systems*, Vol. 3, Nov. 1988, pp. 1807-1813.

[3] T.V. Cutsem, "Dynamic and static aspects of voltage collapse", *Proceedings: Bulk Power System Voltage Phenomena - Voltage Stability and Security* EPRI Report EL-6183, 1989, pp. 6-55 6-80.

[4] R. A. Schleuter, A. G. Costi, J. E. Sekerke, H. L. Forgey, "Voltage Stability and Security Assessment," EPRI Final Report EL-5967, August 1988.

[5] C.C. Liu and K.T. Vu, "Analysis of tap-changer dynamics and construction of voltage stabilityregions", *IEEE Trans. On Circuits and Systems*, Vol. 36, pp. 575–590, April 1989.

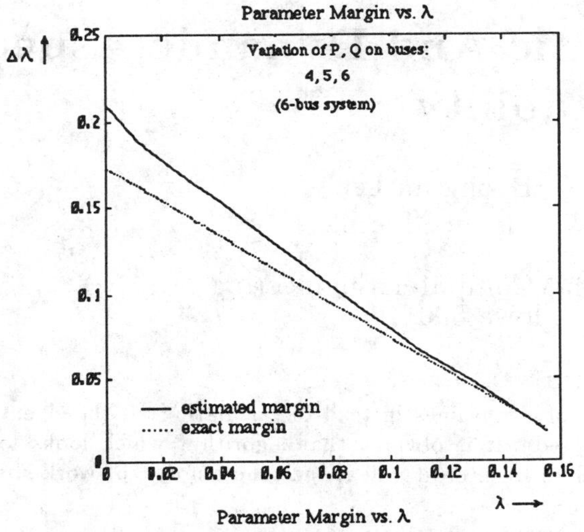

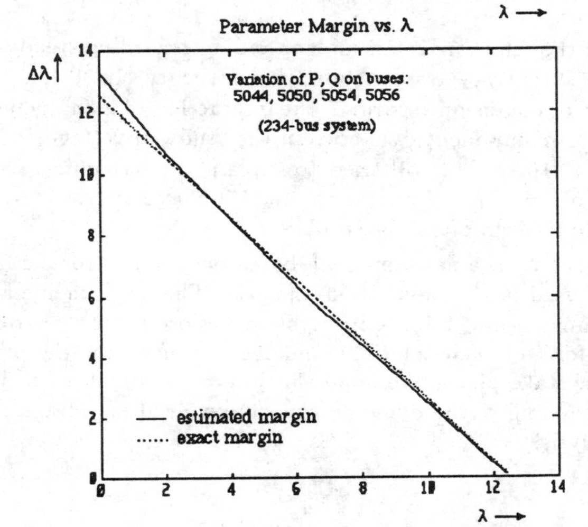

Figure 6: Comparison of the estimated margins τ by the proposed performance index with the exact margins $\Delta\lambda$ in the multiple-bus load variations case.

[6] R. A. Schleuter, J.C. Lo, T. Lie, T.Y. Guo and I. Hu, "A fast accurate method for midterm transient stability simulation of voltage collapse in power systems", *IEEE Proceedings of 28th Conference on Control and Decision,* Dec. 1989, Tampa, FL., pp. 340–344.

[7] B.H. Lee and K.Y. Lee, "A study on voltage collapse mechanism in electric power systems", *IEEE Trans. on Power Systems,* Vol. 6, August 1991, pp. 966–974.

[8] W.R. Lachs, "Insecure System Reactive Power Balance Analysis and Countermeasures", *IEEE Trans. Power Apparatus and Systems,* vol. PAS-104, no. 9, Sept. 1985, pp. 2413–2419.

[9] A. Tiranuchit and R. J. Thomas, "A Posturing Strategy Against Voltage Instabilities in Electric Power Systems," *IEEE Trans. on Power Systems,* Vol. 3, Feb. 1988, pp. 87-93. pp. 424-430.

[10] P.A. Löf, T. Smed, G. Andersson and D.J. Hill, "Fast calculation of a voltage stability index", IEEE/PES Winter Meeting, 1991, WM 203-0 PWRS.

[11] M.G. O'Grady and M.A. Pai, "Analysis of voltage collapse in power systems", *North American Power Symposium,* 1989.

[12] Y. Tamura, H. Mori, and S. Iwamoto, "Relationship Between Voltage Stability and Multiple Load Flow Solutions in Electric Power Systems," *IEEE Trans. Power Apparatus and Systems,* vol. PAS-102, no. 5, May 1983, pp. 1115-1125.

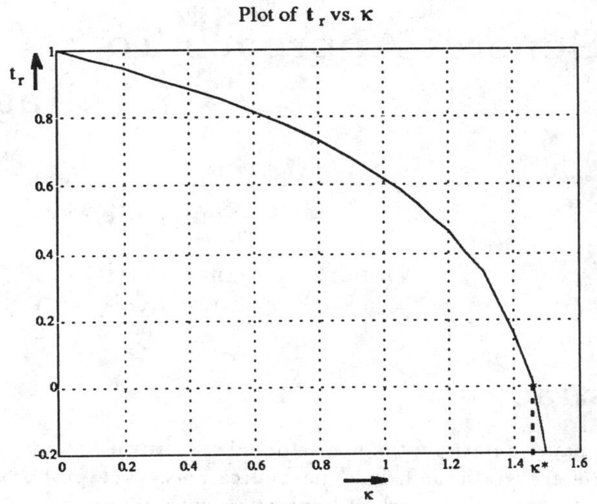

Figure 7: Sample test function t_r for the determination of the exact bifurcation point.

[13] Y. Tamura, "Voltage instability proximity index based on multiple load-flow solutions in ill-conditioned power systems", *IEEE Proceedings of 27th Conference on Control and Decision,* Dec. 1988, Austin, TX., pp. 2114-2119.

[14] C.L. DeMarco and T.J. Overbye, "An Energy Based Security Measure for Assessing Vulnerability to Voltage Collapse", IEEE/PES 1989 Summer Meeting, SM 712-1.

[15] T.J. Overbye and C.L. DeMarco , "Voltage Security Enhancement Using An Energy Based Sensitivity", *IEEE Trans. on Power Systems,* Vol. 6, August 1991, pp. 1196-1202.

[16] I. Dobson, H.D. Chiang, J.S. Thorp and F.-A. Lazhar, "A Model of Voltage Collapse in Electric Power Systems". IEEE Proceedings of 27th Conference on Control and Decision, Dec. 1988, Austin, TX.

[17] T. Van Cutsem, "A Method to Compute Reactive Power Margins with respect to Voltage Collapse," *IEEE Transactions on Power Systems,* Vol. 6, No. 1, Feb. 1991.

[18] F.L. Alvarado and T.H. June, "Direct Detection of Voltage Collapse Conditions," *Proceedings: Bulk Power System Voltage Phenomena - Voltage Stability and Security* EPRI Report EL-6183, 1989.

[19] I. Dobson and H.D. Chiang, "Towards a Theory of Voltage Collapse in Electric Power Systems", *System and Control Letter* Vol. 13, pp. 253-262, Sept. 1989.

[20] H.D. Chiang, I. Dobson, R.J. Thomas, J.S. Thorp and F.-A. Lazhar, "On the Voltage Collapse in Power Systems", *IEEE Trans. on Power Systems,* Vol. 5, No. 2, May 1990, pp. 601-611.

[21] R. Seydel, "Numerical Computation of branch points in Nonlinear Equations," *Numer. Math.* 33, pp. 339-352, 1979.

[22] R. Jean-Jumeau, H.D. Chiang, "Parameterizations of the Load-Flow Equations for Eliminating Ill-conditioning Load Flow Solutions," to appear in *Transactions on Power Systems,* Fall 1992.

[23] R. Seydel, *From Equilibrium to Chaos: Practical Bifurcation and Stability Analysis,* New York: Elsevier.

[24] J. Guckenheimer, P. Holmes, *Nonlinear Oscillations, Dynamical Systems, and Bifurcations of Vector Fields,* Springer-Verlag, NY 1983.

Proceedings of the 31st Conference
on Decision and Control
Tucson, Arizona • December 1992

A General Approach To Study Static And Dynamic Aspects Of Voltage Instability

Venkataramana Ajjarapu
Senior Member

Byongjun Lee
Student Member

Department of Electrical Engineering and Computer Engineering
Iowa State University, Ames, Iowa 50011

Abstract

In this paper an attempt is made to understand the voltage stability related static and dynamic bifurcations. The problem is analyzed in the framework of bifurcation with group symmetry. The observed power system phenomena where collision and splitting of eigenvalues take place is identified with a special type of bifurcation, called node-focus bifurcation.

1 Introduction

In any system (physical or natural), if some of the parameters of that system continue to vary, a critical stage may be reached at which point the system exhibits a sudden jump from one state to another. The other state may be qualitatively and sometimes quantitatively different from the original state. Consider a power system dynamic model which is represented by an autonomous differential equation of the form

$$\dot{x} = F(x,p), \quad F : X \times R \rightarrow X, \quad X \in R^n \tag{1}$$

where x is the n–dimensional state vector and p is a parameter. If the parameter is varied, the corresponding state vector x and the eigenvalues of the Jacobian $[\partial F/\partial x]$ evaluated on this path change accordingly. Near an equilibrium point the left hand side term $\dot{x}$ becomes zero :

$$F(x,p) = 0 \tag{2}$$

Equation (2) specifies the position of the equilibrium point x as a function of p. The power system state on the surface defined by (2) is asymptotically stable if the eigenvalues of the Jacobian have negative real parts at that point. One of the ways the system can reach a critical state is if a real eigenvalue becomes zero (static bifurcation) or a pair of complex conjugate eigenvalues cross the imaginary axis (Hopf bifurcation).

A numerical procedure for obtaining the bifurcation behavior would involve the following two phases:

- Solving and tracing the equilibrium path, and
- Identifying bifurcation points as well as the directions of new branches.

The first phase amounts to solving a system of nonlinear equations by e.g., a predictor-corrector type continuation method [1], whereas the second amounts to determining the bifurcation points. Usually, eigenvalue analysis is used, in which the potential critical eigenvalues are observed with respect to the parameter variation. In [2], an algorithm which calculates a set number of eigenvalues instead of calculating all the eigenvalues is proposed. It is obvious that algorithm which looks for the eigenvalues of interest will reduce the amount of work spent in the phase ii).

Even though there is fair understanding regarding steady-state voltage stability via static bifurcation theory [1, 3, 4, 5, 6, 7], there exist confusion regarding the interaction of load , generator and other component dynamics in the study of voltage collapse. Ref [8, 9, 10, 11, 12, 13] tried to explain the dynamical aspects of this phenomena. Ref [14, 15, 16, 17] contain both static and dynamic aspects of voltage problem.

In this paper, a new approach based on group theoretic framework is used in the bifurcation analysis. The basic theme is that the symmetries of bifurcating system impose strong restrictions on the form of their solutions and the way in which the bifurcation may take place. We hope this approach may lead to better understanding of the dynamic aspects involved in voltage stability analysis.

2 Background

Let us consider the dynamical system for a particular parameter value represented by Eq. 1, where F satisfies the following equivariance condition

$$F(\gamma x,p) = \gamma F(x,p), \qquad x \in X, p \in R, \gamma \in \Gamma. \tag{3}$$

which implies that γ represents the symmetry of Eq. 1, which, in turn, belongs to a symmetry group Γ. For detail, see the ref.[18]. Obviously, the Jacobian matrix $A = \partial F(x,p)/\partial x$ of Eq. 1 share the equivariance condition

$$\gamma A = A\gamma \quad \text{for all} \quad \gamma \in \Gamma. \tag{4}$$

Various bifurcation phenomena can be identified by observing the movement of the eigenvalues of A with respect to the variation of any parameter in the system. However, instead of observing the eigenvalue of entire A matrix, one can decompose A into irreducible representation. If it is possible to find an orthonormal basis which can be divided into a number of independent sets of basis vectors, the original representation Γ is a reducible representation. Conversely, an irreducible representation can not have such a basis. Irreducible representation of the matrix A is represented by block-diagonal matrix by the orthogonal trans-

CH3229-2/92/0000-2916$1.00 © 1992 IEEE

formation such as

$$H^*AH = \begin{bmatrix} A_1 & & & \\ & A_2 & & \\ & & \ddots & \\ & & & A_k \end{bmatrix},$$

where A_k is a square matrix. In block-diagonal matrix, the original vector space is divided into a number of invariant subspaces which correspond to each blocks. Symmetry property of matrix A can be represented by a direct sum of the symmetry properties of each block matrix, A_k, where the meaning of direct sum is equal to that of superposition. For example, eigenvalues of the original matrix consist of the eigenvalue of each block matrix. It is similar to Jordan canonical canonical form of a matrix.

Now we consider eigenvalues and the corresponding eigenspaces of Γ-symmetric A. Let ϑ be the symmetry type of the eigenvalue μ which is either real or complex. It is obvious that μ is an eigenvalue of A iff there is a diagonal block A_k such that μ is an eigenvalue of A_k. We call μ a ϑ_k-eigenvalue of A iff μ is an eigenvalue of the symmetry type ϑ_k which corresponds to the block A_k. In [19], Werner defined Γ-simple eigenvalue and derived the significant fact as follows,

> An eigenvalue μ of a Γ-symmetric matrix A_k is called Γ-simple iff i) There is one and only one ϑ_k such that μ belongs to ϑ_k, ii) μ is an algebraically simple eigenvalue of the ϑ_k-block. This Γ-simple eigenvalue is generic.

The above facts imply that if eigenvalues of irreducible block A_k are real, then the block A_k contain algebraically simple eigenvalues which are generic. Similarly, if eigenvalues of a particular block A_k are complex, then the block A_k contain algebraically simple complex eigenvalues.

Werner [19] defines the conventional static and Hopf bifurcations in terms of Γ-simple bifurcation point as follows,

> (x_0, p_0) is a Γ-simple bifurcation point of symmetric type ϑ_k if $\mu(p_0)$ is a Γ-simple eigenvalue of $A(p_0)$ with symmetric type ϑ_k and $\mu(p)$ crosses the imaginary axis for $p = p_0$ with non-zero speed. If $\mu(p_0) = 0$, it is a static bifurcation and a Hopf bifurcation otherwise.

Consider a smooth path $A(p)$. Since algebraically simple eigenvalues of a matrix A depend smoothly on A, there is a neighborhood U of p_0 such that for all $p \in U$ the matrices $A(p)$ have Γ-simple eigenvalues $\mu(p)$ of symmetry type ϑ_k of $A(p)$ depending smoothly on $p \in U$ and satisfying $\mu(p_0) = \mu_0$. Interesting question comes such as what if there exist critical parameter p_c for which $A(p_c)$ loses the generic property. Then $A(p_c)$ has no Γ-simple eigenvalues. In the generic case the path $A(p)$ crosses certain manifolds of codimension one, transversally. At this condition, a block A_k corresponding to an irreducible representation ϑ_k of real type has one real, algebraically double and geometrically simple eigenvalue μ_0. Then the answer to the above question comes from new bifurcation point which is non Γ-simple bifurcation. The transversal crossing of the manifold by the path $A(p)$ is caused by the collision of two Γ-simple eigenvalues of $A(p)$ for $p = p_c$. Figure 1 shows a non Γ-simple bifurcation, in which two Γ-simple real eigenvalues of symmetry type ϑ_k collide at p_c and split into two non-real (complex conjugate) Γ-simple eigenvalues after collision (or vice versa). The collision occurs with infinite speed controlled by $\sqrt{|p - p_c|}$.

A non-Γ-simple bifurcation point can be considered as a dynamical bifurcation point since the dynamic changes after bifurcation (a spiraling around the equilibrium vanishes or is borne).

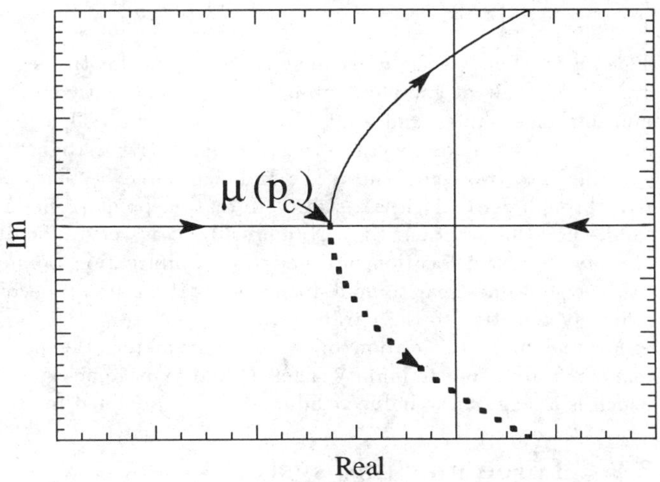

Figure 1: Eigenvalue movement near a node-focus bifurcation point

It is denoted as a node-focus bifurcation point [19]. In case $(\mu(p_c))$ is close to $\mu = 0$, there will be a Γ-simple static bifurcation point of symmetry type ϑ_k close to a node-focus bifurcation point since the speed of $\mu(p)$ is infinite at p_c. Moreover, a node-focus bifurcation point is very often accompanied by Hopf point (of the same symmetry type). In that case, Hopf points and static bifurcation points of the same symmetry type ϑ_k are close together.

The following power system examples are formulated in the framework of above group theory concepts. All the relevant bifurcation phenomena observed in ref [8, 9, 10] is reported.

3 Numerical example

3.1 Three bus system

The first example is a sample power system developed by Chiang et al. [12] based on Walve's work [20]. Many researchers investigated this model to study the behavior of the power system with respect to variation of system parameter. Original system is formed by four dimensional first order differential equation and reactive load is chosen as the system parameter. The bifurcation diagram is shown in Figure 2, in which two Hopf and one saddle-node bifurcation points appear [8].

If we neglect the time derivatives of load variables, then the system is represented by differential-algebraic(DA) form as follows,

$$\begin{bmatrix} \dot{\delta} \\ \dot{\omega} \\ 0 \\ 0 \end{bmatrix} = f(\delta, \omega, \delta_L, V, Q_1),$$

where δ_L and V become algebraic variables. It is linearized as follows,

$$\begin{bmatrix} \Delta\dot{\delta} \\ \Delta\dot{\omega} \\ 0 \\ 0 \end{bmatrix} = \begin{bmatrix} A & B \\ C & D \end{bmatrix} \begin{bmatrix} \Delta\delta \\ \Delta\omega \\ \Delta\delta_L \\ \Delta V \end{bmatrix}$$

Eliminating $\Delta\delta_L$ and ΔV gives the system matrix

$$\begin{bmatrix} \Delta\dot{\delta} \\ \Delta\dot{\omega} \end{bmatrix} = \begin{bmatrix} A - BD^{-1}C \end{bmatrix} \begin{bmatrix} \Delta\delta \\ \Delta\omega \end{bmatrix}$$

The Figure 3 shows the movement of eigenvalue for the reduced system. A pair of complex conjugate eigenvalues becomes two multiple eigenvalues and split into two different real eigenvalues. This is a node-focus bifurcation point (NF). Right after splitting into two eigenvalues, one real eigenvalue becomes positive through zero which is a static bifurcation point (generically saddle-node bifurcation, SD). Numerically, it is very difficult to identify the exact location of a node-focus bifurcation point because eigenvalue near a node-focus bifurcation point becomes infinitely sensitive to the parameter.

For the further variation of system parameter, the positive real eigenvalue goes to infinity when D matrix becomes singular, which is a singularity induced bifurcation (SI) defined in [10].

3.2 Three machine system

The second example is the three machine system [21] analyzed by Rajagopalan et al. [9]. In the paper, the authors used the linearized system equations represented by DA equation as follows,

$$\begin{bmatrix} \Delta\dot{x} \\ 0 \end{bmatrix} = \begin{bmatrix} A & B & \\ C & D_{11} & D_{12} \\ & D_{21} & D_{22} \end{bmatrix} \begin{bmatrix} \Delta x \\ \Delta y_1 \\ \Delta y_2 \end{bmatrix},$$

where Δx denotes the state variables involving machine, exciter and voltage regulator, Δy_1 consists of stator algebraic variables and other algebraic variables except the standard load flow variables and Δy_2 is the set of standard load flow variables.

Eliminating the algebraic variables $\Delta y_1, \Delta y_2$ gives the system matrix,

$$\Delta\dot{x} = [A - BD^{-1}C]\Delta x. \tag{5}$$

Figure 9 in [9] is redrawn in Figure 4 which shows the movement of critical eigenvalue of the system matrix. A pair of complex conjugate crosses the imaginary axis through A (Hopf bifurcation). They become two multiple real eigenvalue at B (node-focus bifurcation) and split into two real eigenvalues in opposite direction. The first of these two eigenvalues becomes negative by passing through infinity on the real axis. The system becomes

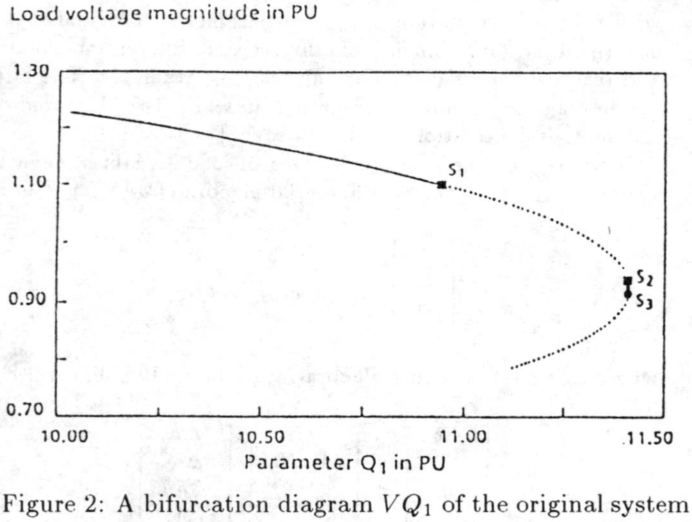

Figure 2: A bifurcation diagram VQ_1 of the original system

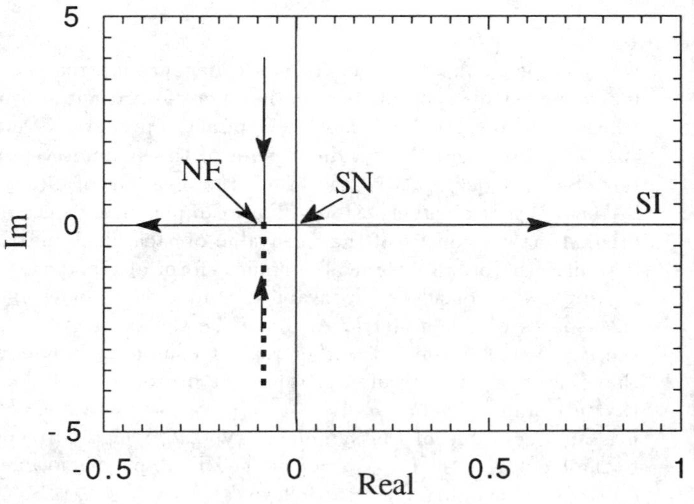

Figure 3: The movement of eigenvalues

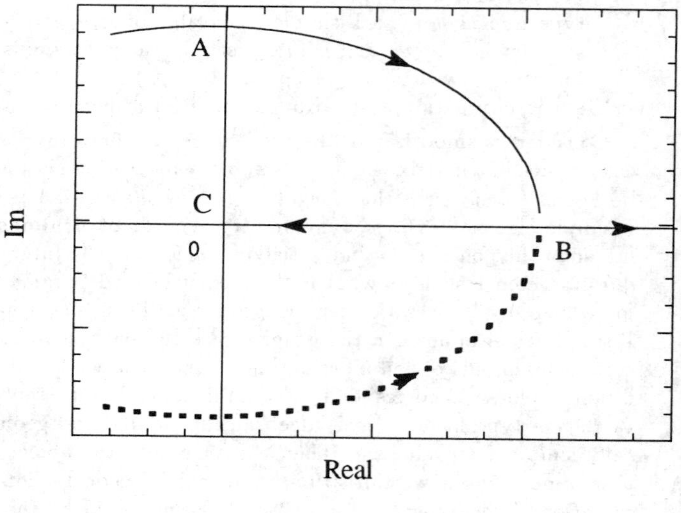

Figure 4: The movement of critical eigenvalue, ref [9]

stable when the later of two crosses the imaginary axis (point C, static bifurcation). The point C is the static bifurcation point because one eigenvalue is zero and the others have negative real part of eigenvalues. Generically, it is a saddle-node bifurcation point. The reason it does not locate at the peak point of PV curve is that the parameter chosen in their work is not independent. The PV curve used in the paper is not a bifurcation diagram.

4 conclusion

This paper tries to explain the nonlinear bifurcation phenomena observed in the literature in the framework of group theory. One of the contribution of the work is to identify node-focus bifurcation in power system network. We hope this approach may lead to better understanding of the dynamic aspects involved in voltage stability analysis.

5 Acknowledgement

This work was supported in part by National Science Foundation Grant ECS91-08356, and the Electrical Power Research Center of the Iowa State University.

References

[1] V. Ajjarapu and C. Christy "The Continuation Power Flow: A Tool For Steady-State Voltage Stability Analysis", *IEEE Trans. on Power Systems*,no.1, pp.416-423, Feb.1992.

[2] B. Gao, G. K. Morison and P. Kundur, "Voltage Stability Evalustion Using Modal Analysis", IEEE PES Summer Power Meeting, paper 91 SM 420-0 PWRS, 1991.

[3] A. Tiranuchit and R. J. Thomas, "A Posturing Strategy Against Voltage Instabilities In Electric Power Systems", IEEE Trans. on Power Systems, vol.3, pp.87-93, Feb. 1988.

[4] Y. Tamura, K. Iba and S. Iwamoto, "Relationship between voltage Instability and Multiple Load Flow Solutions In Electric Power Systems", IEEE Transactions on PAS, vol.PAS-102, no.5, pp.1115-1125, May 1983.

[5] F. L. Alvarado and T. H. Jung, "Direct Detection Of Voltage Collapse Conditions", Proceedings:Bulk Power System Voltage Phenomena-Voltage Stability and Security, Jan, 1989.

[6] P. W. Sauer and M. A. Pai, "Power System Steady State Stability and The Load Flow Jacobian", IEEE Trans. on Power Systems, vol.5, no.4, Nov. 1990.

[7] I. Dobson, "An Iterative Method To Compute A Closest Saddle Node Or Hopf Bifurcation Instability In Multidimensional Parameter Space", Proceedings of the International Symposium on Circuits and Systems, San Diego, May 1992, pp.2513-2516.

[8] V. Ajjarapu and B. Lee. "Bifurcation Theory And Its Application To Nonlinear Dynamical Phenomena In An Electrical Power System". *IEEE Trans. on Power Systems* , vol. 7, no.1, pp.424-431, Feb. 1992.

[9] C. Rajagopalan, B. Lesieutre, P. W. Sauer, M. A. Pai, "Dynamic Aspects of Voltage/Power Characteristics", IEEE Trans. on Power Systems, vol. 7, no.13, pp.990-1000, Aug. 1992.

[10] V. Venkatasubramaniam, H. Schattler and J. Zaborsky, "A Taxonomy of The Dynamics of A Large Power System With Emphasis On Its Voltage Stability", Proc. Bulk Voltage Stability and Security, Deep Creek Lake, Aug.4-7, 1991.

[11] V. Venkatasubramaniam, H. Schattler and J. Zaborsky, *A Stability Theory Of Large Differential Algebraic Systems - A Taxonomy*, Report SSM 9201 - Part I, Department of Systems Science And Mathematics, Washington University, Aug. 1992.

[12] H. D. Chiang, I. Dobson, R. J. Thomas, J. S. Thorp, and L. Fekih-Ahmed. "On Voltage Collapse in Electric Power Systems" *IEEE PICA Conference,* Seattle, Washington, U.S.A., May, 1989.

[13] Y. Tamura, "A Scenario Of Voltage Collapse In A Power System With Induction Motor Loads With A Cascaded Transition Of Bifurcation", Proceeding of the workshop on bulk power system voltage phenomena, Voltage Stability and Security, Deek Creek Lake, MD, U.S.A., Aug. 4-7, 1991.

[14] L. H. Fink, ed., Proceedings: Bulk Power System Voltage phenomena Voltage Stability and Security, EPRI Report, EL-6183, Potosi, MO, Jan. 1989.

[15] L.H. Fink, ed., Proceedings: Bulk Power System Voltage Phenomena, Voltage Stability and Security, ECC/NSF Workshop, Deep Creek Lake, MD, Aug. 1991, ECC Inc., 4400 Fair Lakes Court, Fairfax, VA 22033-3899.

[16] IEEE Working Group, "Voltage Stability Of Power Systems: Concepts, Analytical Tools, and Industry Experience", IEEE 90 TH 0358-2-PWR.

[17] NERC Report, "Survey Of The Voltage Collapse Phenomenon", Summary of the Interconnection Dynamics Task Force's Survey on the Voltage Collapse Phenomenon, North American Electric Reliability Council, 101 College Roak East. Princeton, New Jersey, Aug. 1991.

[18] M. Golubitsky, I. Stewart and D. G. Schaeffer, *Singularities and Groups in Bifurcation Theory*, Vol 2, New York: Springer-Verlag, 1985.

[19] B. Werner, "Eigenvalue Problems With The Symmetriy Of A Group And Bifurcations", *Continuation and Bifurcations: Numerical Techniques and Applications*, NATO ASI Seriec C, Vol.313, 1989.

[20] K. Walve, "Modeling Of Power System Components At Severe Disturbances", CIGRE paper 38-18, International Conference On Large High Voltage Electric Systems, 1986.

[21] P. M. Anderson and A. A. Fouad, *Power System Control and Stbility*, Iowa State University, 1977.

Proceedings of the 31st Conference
on Decision and Control
Tucson, Arizona • December 1992

ANALYSIS OF THE TAP CHANGER RELATED VOLTAGE COLLAPSE PHENOMENA FOR THE LARGE ELECTRIC POWER SYSTEM

Vaithianathan Venkatasubramanian[1], Heinz Schättler and John Zaborszky
Department of Systems Science and Mathematics
Campus Box 1040, Washington University
St. Louis, Missouri, 63130-4899

Abstract

It is shown in this paper that nonsmooth events such as tap changing can be integrated into the framework of a recently presented general theory [15, 17] on the dynamics of the large power system modelled by smooth equations. The boundary to the region of feasible operations starting at a stable operating point, the feasibility boundary, is shown to be composed of five types of segments including two types connected with tap changers. These segments are defined as zero sets of explicitely known functions and thus are relatively computable (at roughly the load flow level). These new results offer a powerful new approach to system security when tap changers are present. No reference needs to be made to $P-V$ curves of the large power system, an obstacle in more conventional approaches.

1 Introduction

Voltage collapse caused by locally controlled tap changers is investigated in this paper. The methods proposed can be extended to include other local discrete controls such as locally controlled load dropping. Locally controlled automatic tap changers have been the subject of much interest and intensive studies [1, 8, 9, 10, 15, 19, 20]. Tap changers are used in two different ways in the power transmission system:

1. Within the transmission system itself in a way where there are loops of transmission lines closed through them. The purpose is to control the distribution of reactive power flow within the transmission system (or active power flow with out of phase taps) and hence its voltage and stability. This application, normally under secondary control, has no conspicuous special role in voltage collapse.

2. At subtransmission or distribution substations to keep the voltage on the secondary side within tolerances in face of voltage variations on the primary side. Under uncoordinated local control, combined with load characteristics and system conditions, this type can cause a puzzling cascading operation of rising taps and falling secondary voltage into collapse as well as other irregularities.

Analysis of the cascading voltage descent problem associated with the latter type of tap changers is the main objective of this paper. Algebraic conditions have been developed recently to locate the cascading voltage descent problem associated with the on-line automatic tap changers (or ULTC's) under certain assumptions [15, 19]. It will be shown that the voltage descent is typically due to a saddle-node type bifurcation associated with the discrete tap changer control. Algebraic conditions are also developed here to detect possible 'hunting' phenomena among several tap-changers.

The cascading descent problem mentioned above is a parameter space event in the sense that the inception of this phenomenon is marked by an operating parameter such as the load and voltage set point etc. crossing a 'feasibility limit'. The notion of the 'feasibility region' is proposed to include all the parameters which correspond to stable system operation and the boundary of this region, the feasibility boundary, then defines the feasibility limits for the parameters. For a differential-algebraic system described by smooth functions, the concept of feasibility region was introduced in [23, 14, 15, 17]. Feasibility is a generic concept and distinct definitions exist in other problem areas, such as load flow [3] or optimization. In this paper, we extend the notion of the feasibility region as defined in [15, 17] to include local control devices such as the automatic tap changers. Even though we consider only the automatic tap changers, other local controls such as locally controlled load dropping, can also be treated similarly. The main feature of such control mechanisms is that the control is purely local. For instance, the tap changer will increase (reduce) the tap setting, when its terminal voltage is below (above) the voltage set point. The setpoint itself is an input, manual or coming from a secondary control. Exploiting this decoupled nature of the control, we will develop a simple algebraic condition to test the stability of the control dynamics [15]. (In [19], the same stability condition was proposed independently). This will then allow us to directly identify the new segments of the feasibility

[1] now at *School of Electrical Engineering and Computer Science, Washington State University, Pullman, WA 99164-2752.*

CH3229-2/92/0000-2920$1.00 © 1992 IEEE

boundary introduced by these local control devices. One type of the new boundary segments corresponds to the cascading voltage descent problem, which will be shown to be typically caused by a static saddle-node type bifurcation associated with the discrete tap changer control. We show that the entire feasibility boundary is composed of zero sets of explicit analytic functions and thus is accessible to computation. Hence, it fits directly into the security monitoring and operation planning process and can be very useful there. Questions of transient behavior involving tap changers seem to be less pressing, however, it has a sizeable literature.

2 Modeling Issues

Many problems in the dynamic analysis of the large power system like voltage stability analysis can be modeled (*within the quasi-stationary range* [15]) by a parameter dependent differential-algebraic system of the form

$$\dot{x} = f(x, y, p) \quad , \quad f : \mathbb{R}^{n+m+p} \to \mathbb{R}^n \tag{1}$$

$$0 = g(x, y, p) \quad , \quad g : \mathbb{R}^{n+m+p} \to \mathbb{R}^m \tag{2}$$

$$x \in X \subset \mathbb{R}^n, \quad y \in Y \subset \mathbb{R}^m, \quad p \in P \subset \mathbb{R}^p$$

In the state space $X \times Y$ dynamic state variables x and instantaneous state variables y are distinguished. Typical dynamic state variables are the time dependent values of generator voltages and rotor phases, instantaneous variables are bus voltages and other load flow variables. The parameter space P is composed of system parameters (the system topography, i.e. what is energized, and equipment constants e.g. inductances), and operating parameters (such as loads, generation, voltage setpoints etc.).

Modeling and problem formulation aspects of tap changers are treated in detail next.

(1) *Physical features:* Traditionally, to operate a tap changer under load, it will be necessary first to bridge two transformer taps through appropriate inductors or resistors with center taps, and then interrupt the circulating current at the old tap. This involves considerable mechanical operation of contacts and switches and the use of small interruptors (often vacuum type). Operating these mechanisms takes a considerable and fixed amount of time which is independent of the deviation of the bus voltage from the set point. The point is that the system experiences one (or two) fixed magnitude step jumps of the transformation ratio per tap change (usually less than 1%) and a minimum time interval, also of fixed magnitude of many seconds between consecutive changes (depending on type, 40-100 sec., 3-8 seconds, or, in new thyristor devices, under 1 sec. This paper applies mostly to the former two types which represent the vast majority here.) It is clear that the response to small step increases, placed many seconds apart, will break down to two decoupled phenomena.

1. A transient response to the small step jump is normally quite harmless, unless the system is operating with inadequate security that is much too close to its feasibility or stability boundary. This transient decays during a small initial part of the minimum delay preceding the next tap change.

2. Thus the success of a tap change is expressed by the steady state voltage E_L^s reached after the last tap change and before the next one. This voltage is independent of the transient which has died out by that time. So typically only the steady state value is significant.

The traditional tap changer operating control is a simple device which switches one step up or down when the deviation between the steady state load voltage E_{Li}^s and the reference setting E_{ri} reaches or exceeds the tolerance value, ΔE [9, 4]. Thus

$$n_{k+1} = \begin{cases} n_k + \Delta n & \text{if } E_{Li}^s - E_{ri} < -\Delta E \\ n_k & \text{if } |E_{Li}^s - E_{ri}| \leq \Delta E \\ n_k - \Delta n & \text{if } E_{Li}^s - E_{ri} > \Delta E \end{cases} \tag{3}$$

where Δn is the tap step size (usually equal or slightly less than ΔE) and E_{Li}^s is the steady state voltage at the load i and as such the solution of

$$0 = f(x, y, n, p), \quad 0 = g(x, y, n, p). \tag{4}$$

In equation (4), for convenience, the vector n of tap positions is displayed separately from other parameters. Its nature may vary depending on the problem. If we denote bus voltages by $z_1 = \{E_i^s\}$, the load voltages (secondaries of tap changers) by $z_2 = \{E_{Li}^s\}$ and the remaining (dynamic and instantaneous) state variables by z_3, and all the other parameters are grouped as p, then the system at steady state can be written comprehensively as

$$0 = h(z_1, z_2, z_3, n, p). \tag{5}$$

This model is analyzed in detail in section 3. Note that this analysis (See also subsection (2) below) applies to the traditional,i.e. slow tap changer.

There are a variety of questions connected with the tap changer that can be asked. From an operating, security monitoring and planning angle point of view, the most useful is the following one.

(2)*Effect of the tap changer on the feasibility boundary in the parameter space :* The actual postchange stationary voltage will depend both on the tap size and the system composition. As long as there is a rise in the secondary voltage above the set-point, no critical problem arises. However, because of load characteristics and other system features, the stationary voltage can actually go down in response to an upward tap change. When that happens, the simple automatic control device (3) set for local control (and even an inexperienced operator) will order another upward step initiating a chain reaction of descending and ultimately collapsing voltage. When a number of tap changers are involved, this scenario generalizes to the positive definiteness of the sensitivity matrix, i.e. the sensitivity of the load voltages to the tap changer positions. When this matrix is positive definite, for any load bus then, if the tap ratio is raised, the load bus voltage also increases, which is the desired effect. But, if the sensitivity matrix is not positive definite, then there is at least one direction in which the raise of the tap ratio may lead to a cascaded voltage collapse. It is later shown that the tap changer control becomes unstable when the sensitivity matrix is not positive definite. In the parameter space the boundary of after switch tap positions where the cascading phenomenon commences is where the sensitivity matrix becomes indefinite. Thus the boundary can be pinpointed by postulating an arbitrarily small tap size. For a finite tap size the actual boundary for the initiating step would then be lower by one step size. Over a region of system conditions, this condition defines a section of the feasibility boundary in parameter space where the cascading phenomenon sets in, i.e. local stability of the tap changer control is lost beyond this boundary. This leaves the small step boundary as the core of the answer which can be readily adjusted to the actual step size (See the details of analysis in section 3).

(3) *Transient phenomena connected with the tap changer in the state space :* The traditional slow tap changer has no real dynamic characteristics beyond (3) and (5). In other words, when the voltage is off the reference by the tap size, a tap change is implemented regardless of anything else. So on this type of device there is no $\dot{n}$ that is tap rate term–in fact the tap change is always strictly one tap at a time regardless of the *size* of the deviation. So this is a discrete element. Continuous models have been proposed [9] to approximate the discrete model by assuming that the tap size is small and that the tap changer has an option of not switching when the voltage is within the neutral band $|E_L^s - E_r| \leq \Delta E$. It is tempting to introduce a continuous approximation [1] of the form

$$\dot{E}_L^s = \frac{1}{T}(E_L^s - E_r) \tag{6}$$

or

$$\dot{n} = \frac{1}{T}(E_L^s - E_r). \tag{7}$$

No convincing justification of this approximation seems to exist for the traditional slow tap changers. In fact, originally it was simply stated as an unsupported assumption [1] and later users simply quote this reference. Since the normal dynamic elements (generator, rotor angle, excitation control, load dynamics etc.) are much too fast to couple into the tap changing cycle, the practical meaning of this assumption is dubious unless a much slower f, g system is substituted, like boiler dynamics or some AGC action. On the other hand, using very fast thyristor type tap changers, it would be quite easy to implement an actual controller with the control law of (6) or (7) in the usual transient stability range. In this latter case, if such a control dynamics exists, then the system itself would need to be modeled in its full dynamic form, that is $\dot{x} = f(x, y, p), 0 = g(x, y, p)$ where now $\dot{x}$ includes $\dot{E}_{Li}^s$ or $\dot{n}$ and f includes equation (6) or (7). Technology is still evolving for such fast switching of tap changers and this issue will not be discussed in this paper. These phenomena however are directly covered by the general dynamic theory presented in [15, 17].

The new introduction in subsection (2) above of studying the effect of the tap changer on the feasibility boundary seems to be providing the most pertinent information connected with the special type of collapse (the cascading one) which is not quite dynamic in the ordinary sense and which is an important practical problem in operation.

3 Mathematical Analysis

3.1 Feasibility boundary without tap changers

As a first step, it will be assumed that automatic tap changer controls are not present, i.e the transformer tap settings are taken as parameters. Let us define the equilibrium points or the steady state solutions as EQ,

$$\begin{aligned} EQ &= \{(x, y, p) \in X \times Y \times P : f(x, y, p) = 0, \\ &\qquad g(x, y, p) = 0\} \\ OP &= \{(x, y, p) \in EQ : \ D_y g \text{ is nonsingular and} \\ &\qquad J = D_x f - D_y f (D_y g)^{-1} D_x g \\ &\qquad \text{has eigenvalues with negative real part}\}; \end{aligned}$$

It can be seen that OP is the subset of stable equilibria, hence consists of possible candidates for system operating points.

Definition 1 *Given a stable equilibrium $z_s^0 = (z_0, y_0)$ (connected to a load flow solution) for parameter value p_0, the connected component F of OP which contains (z_0, y_0, p_0) is called the* feasibility region *of z_s^0. Its boundary (relative to EQ) is the* feasibility boundary.

Thus the feasibility region is defined as a subset of $X \times Y \times P$, which consists of all possible operating conditions which can be reached from (z_0, y_0, p_0) by continuous variations of the parameters while maintaining stability. The boundary of the feasibility region is composed of zero sets and thus is accessible to numerical computation.

3.2 Feasibility boundary without tap changers

Theorem 1 *[15] For a system defined in equations (1) and (2), the feasibility boundary of a feasibility region F consists of three zero sets*

$$\partial F = (\partial F \cap C_S) \cup (\partial F \cap C_Z) \cup (\partial F \cap C_H), \tag{8}$$

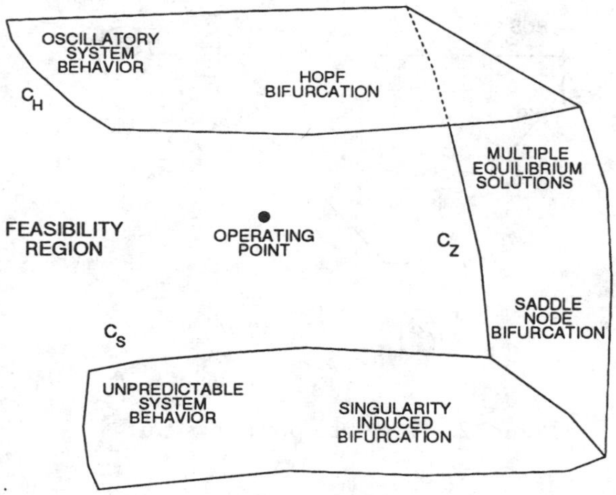

Figure 1: Conceptual sketch of the feasibiity boundary

where

$$\begin{aligned} C_S &= \{(x,y,p) \in EQ : \det D_y g = 0\}, \\ C_Z &= \{(x,y,p) \in EQ : \det D_y g \neq 0, \\ &\quad \det \begin{pmatrix} D_x f & D_y f \\ D_x g & D_y g \end{pmatrix} = 0\} \\ C_H &= \{(x,y,p) \in EQ : \det D_y g \neq 0, \\ &\quad \det \begin{pmatrix} D_x f & D_y f \\ D_x g & D_y g \end{pmatrix} \neq 0, \\ &\quad \det(H_{n-1}(J)) = 0\} \end{aligned}$$

with

$$J = D_x f - D_y f (D_y g)^{-1} D_x g \tag{9}$$

and $H_{n-1}(J)$ is the corresponding Hurwitz matrix of order $n-1$.

Thus the feasibility boundary can be computed as zero sets and generically most of the boundary points consist of either a saddle-node bifurcation (C_S, zero eigenvalue), a Hopf bifurcation (C_H, pure imaginary eigenvalue) or a singularity induced bifurcation (C_Z, eigenvalue at infinity) [15, 17]. For a conceptual sketch see Figure 1.

3.3 Feasibility boundary with tap changers

As pointed out in Section 2, the pertinent model for studying the difficulties resulting from locally controlled tap changers is the one describing the steady state reached after the tap change. This model (5) will now be analyzed for the restriction of the feasibility region caused by these types of tap changers. Let d denote the number of buses with tap changers. If bus i is equipped with tap changer, then $z_{2i} = z_{1i} n_i$, so that $d = dim(z_1) = dim(z_2) = dim(n)$. Substituting $z_{1i} = z_{2i}/n_i$ in (5) gives

$$0 = h_1(z_2, z_3, n, p) \tag{10}$$

where the number of equations is equal to the sum of the dimensions of z_2 and z_3. Suppose n and p are within the feasibility region and away from the feasibility boundary as introduced in this section. Then, since $(D_{z_2} h_1, D_{z_3} h_1)$ is nonsingular, locally near any such point the solutions to equation (10) can be represented as

$$z_2 = z_2(n,p), \quad z_3 = z_3(n,p). \tag{11}$$

The tap-changer continues switching until $z_{2i} = E_{ri}$ for all load buses. Hence, for the system with automatic tap changer control, the solution of (10) with $z_{2i} = E_{ri}$ defines the equilibrium state. The stability of the tap-changer control is determined by the Jacobian

$$J_d := -D_n z_2 = (I_d, 0)\left(D_{z_2} h_1, D_{z_3} h_1\right)^{-1} D_n h_1. \tag{12}$$

The equilibrium is stable if $-D_n z_2$ is stable (negative definite) and stability is lost when it has eigenvalues on the imaginary axis. This can be demonstrated by approximating the discrete dynamics of the tap-changer control with a continuous model as follows: The tap-changer changes the tap size by a small step if the voltage differs from the set point voltage by more than its allowed tolerance value. Usually, the tap-step size and the tolerance values are small, about 1% in p.u. Hence, we can approximate the discrete model by a continuous model of the form

$$\dot{n}_i = f_i(z_{2i}) = \begin{cases} c & \text{if } z_{2i} < E_{ri} \\ -c & \text{if } z_{2i} > E_{ri} \end{cases} \tag{13}$$

where c is a suitable positive constant representing the tap size as justified by any safety margins included in the actual switching law [9]). The system described by equation (13) is not smooth at $z_{2i} = E_{ri}$, but the 'slope' $D_{z_{2i}} f_i$ can be taken arbitrarily large negative. To analyze the local stability, choose a suitable smooth function $\tilde{f}_i$ which approximates f_i to the desired accuracy and, in addition is such that $D_{z_2} \tilde{f} = -k I_d$, where k is a positive constant. The Jacobian of the system $\dot{n}_i = \tilde{f}_i(z_{2i})$ is then given by $k J_d$. Hence, the approximated system is stable if all the eigenvalues of J_d are in $\mathbb{C}^-$, and is unstable if there exists at least one eigenvalue in $\mathbb{C}^+$. Considering the limiting case for $f_i(z_{2i})$, as the slope goes to $-\infty$, the result follows. Hence, the segment(s) of the feasibility boundary induced by the tap changer are pinpointed (at least after taking the closure, if necessary) by the parameters where the sensitivity matrix $(-J_d)$ has eigenvalues on the imaginary axis.

The new boundary is then where J_d loses stability, i.e. where the eigenvalues of J_d cross the imaginary axis (as we stay away from ∂F, eigenvalues do not blow up), i.e. for either det $(J_d) = 0$ (zero eigenvalue) or $\det(H_{n-1}(J_d)) = 0$ (purely imaginary eigenvalues) [24]. The Jacobian, J_d can be simplified as

$$J_d(z_2, z_3, n, p) = J_d(E_{ri}, z_3, \left(\frac{E_r}{z_1}\right)_i, p) = \tilde{J}_d(z_1, z_3, p).$$

Essentially, the tap-changer tries to track the solutions of

$$0 = h(z_1, z_2, z_3, n, p) \quad \text{where} \quad z_{2i} = E_{ri}, \quad n_i = \frac{E_{ri}}{z_{1i}}.$$

Rewriting these equations, the stationary solutions with the tap-changer control are the solutions of

$$0 = h_2(z_1, z_3, p) \tag{14}$$

and the new feasibility boundaries are defined by the conditions

$$0 = det(\tilde{J}_d) \text{ and } 0 = \det(H_{n-1}(\tilde{J}_d)).$$

This discussion can now be summarized as follows: The set of steady-state solutions with tap-changer control is defined by

$$\begin{aligned} EQ_{TC} &= \{(z_1, z_2, z_3, n, p) : h(z_1, z_2, z_3, n, p) = 0, \\ &\quad z_{2i} = E_{ri},\ n_i = \frac{E_{ri}}{z_{1i}}\} \end{aligned}$$

The set of stable solutions OP_{TC} within set EQ_{TC} is defined by those points where J_d is negative definite

Definition 2 *The connected component of the stable equilibria OP_{TC} which contains a specific solution (z_s, n_s, p_0) is defined as the* feasibility region with tap-changer control *and is denoted by F_{TC}.*

Theorem 2 (Feasibility Boundary Theorem with Tap Changer) *For a system containing locally controlled automatic tap changers, the feasibility boundary ∂F_{TC} of a stable equilibrium point (z_s, n_s, p_0) consists of five zero sets*

$$\partial F_{TC} = \partial F_{TC} \cap \{C_S \cup C_Z \cup C_H \cup C_{Z,TC} \cup C_{H,TC}\}.$$

The sets C_S, C_Z and C_H are as defined in Theorem 1; $C_{Z,TC}$ and $C_{H,TC}$ are defined as

$$\begin{aligned} C_{Z,TC} &= \{(z_1, z_3, p) : h_2(z_1, z_3, p) = 0, \\ &\quad \det(\tilde{J}_d)(z_1, z_3, p) = 0, \\ &\quad \det(D_{z_2} h_1(z_2, z_3, p), D_{z_3} h_1(z_2, z_3, p)) \neq 0\} \end{aligned}$$

and

$$\begin{aligned} C_{H,TC} &= \{(z_1, z_3, p) : h_2(z_1, z_3, p) = 0, \\ &\quad \det(\tilde{J}_d)(z_1, z_3, p) \neq 0, \\ &\quad \det(D_{z_2} h_1(z_2, z_3, p), D_{z_3} h_1(z_2, z_3, p)) \neq 0, \\ &\quad \det H_{n-1}(\tilde{J}_d)(z_1, z_3, p) = 0, \} \end{aligned}$$

where $H_{n-1}(\tilde{J}_d)$ is the Hurwitz determinant of the Jacobian $\tilde{J}_d$ and is defined exactly as in the definition of C_H.

4 Illustrative Examples

The results of this paper are applicable to any size system, however in this paper, we illustrate the results on a triangular three bus system shown in Figure 2, consisting of two generators and a load bus. For convenience, the figure shows both the generator and the ULTC together

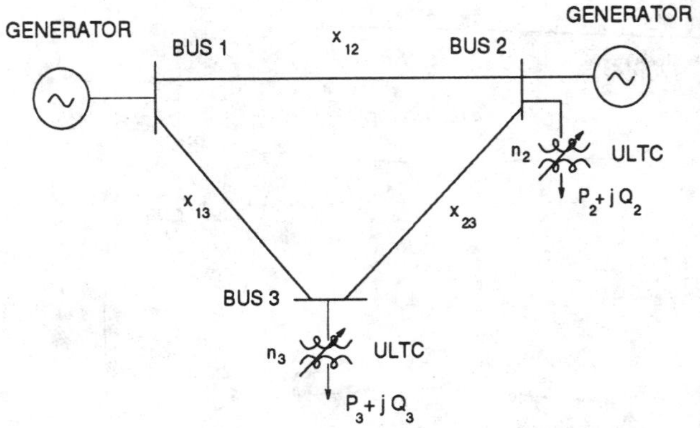

Figure 2: Three bus power system

at Bus 2, though only one of the two is assumed to be active. For both generators, the generator voltage dynamics is represented by the single axis model (E'_i). The excitation (E_{fd_i}) control which regulates the respective bus voltage E_i is modeled by a first order differential equation as in [14, 18].

$$T'_{do_i}\dot{E}'_i = -\frac{x_{d_i}}{x'_{d_i}}E'_i + \frac{x_{d_i} - x'_{d_i}}{x'_{d_i}}E_i\cos(\theta_i - \delta_i) + E_{fd_i} \tag{15}$$

$$T_i\dot{E}_{fd_i} = -K_i(E_i - E_{ref_i}) - (E_{fd_i} - E^0_{fd_i}) \tag{16}$$

for $i = 1, 2$. The dynamics of the generator angles (θ_i) are represented by the swing equations,

$$J_i\ddot{\theta}_i + b_i\dot{\theta}_i = P_{T_i} - \frac{1}{x'_{d_i}}E'_iE_i\sin(\theta_i - \delta_i) \tag{17}$$

for $i = 1, 2$. The loads P_2, Q_2, P_3 and Q_3 are asssumed to be static consisting of the three basic components

$$P_i = P^0_i + M_i(E_in_i) + G_i(E_in_i)^2 \tag{18}$$

$$Q_i = Q^0_i + H_i(E_in_i) + B_i(E_in_i)^2 \tag{19}$$

for $i = 2, 3$. Here n_i stands for the tap ratio of the ULTC's and hence E_in_i (transformer secondary voltage) corresponds to the load voltage. The reference voltage for the tap changer controls are taken to be 1 p.u. and the transmission lines are assumed to be lossless. Thus equations (15)-(17) together define the dynamic equations f in (1) and the power balance equations which can be easily written for buses 1, 2 and 3 define the algebraic constraints g in (2). As explained in Sections 3.2 and 3.3, we can now construct the feasibility regions for our system in the parameter space using Theorems 1 and 2. For practical parameter values, two cross sections of the feasibility boundary for the normal operating point are shown in Figures 3 and 4, for variation of the parameters K_1, the excitation control gain for Generator 1 and P_3, the real power load at Bus 3.

For the simulation shown in Figure 3, the generator at Bus 2 is not considered, i.e. the system consists of the

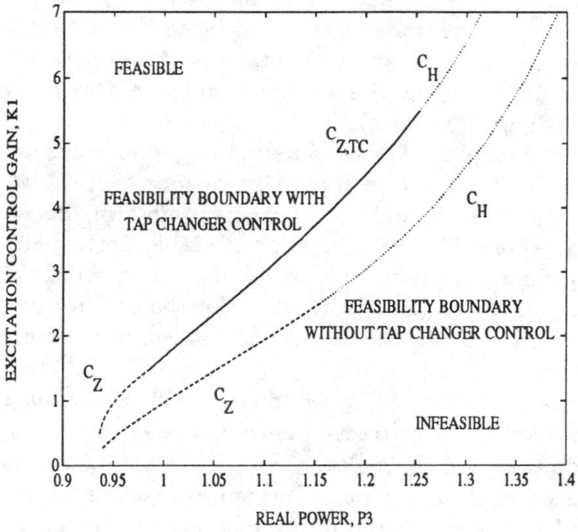

Figure 3: Feasibility boundaries with and without tap changer control with one generator and two load buses. The parameter values are $x_{12} = 0.2$, $x_{13} = 0.3$, $x_{23} = 0.5$, $x_{d_1} = 1.0$, $x'_{d_1} = 0.2$, $b_1 = 1.5$, $J_1 = 5$, $T_{d_{o_1}} = 8$, $T_1 = 1.0$, $E_{fd_{o_1}} = 1.9$, $E_{ref_1} = 1.1$, $Q_2 = 0.5P_2$, $P_2^0 = 0.1$, $M_2 = 0.05$, $G_2 = 0.05$, $Q_3 = 0.4P_3$ where P_3 consists of 80% constant power, 10% current and 10% impedance components.

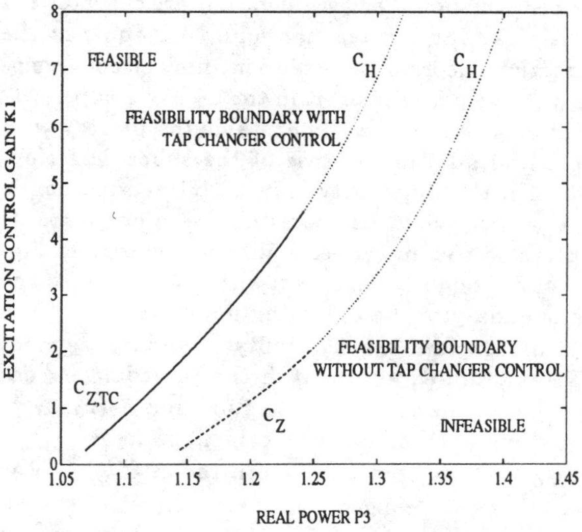

Figure 4: Feasibility boundaries with and without tap changer control with two generators and one load bus. The parameter values are $n_2 = 1$, $x_{d_2} = 1.0$, $x'_{d_2} = 0.25$, $b_2 = 1$, $J_2 = 3$, $T_{d_{o_2}} = 6$, $T_2 = 0.8$, $K_2 = 4$, $E_{fd_{o_1}} = 1.8$, $E_{ref_2} = 1.0$, $E_{ref_1} = 1.05$, $Q_2 = 0.4P_2$, $P_2^0 = 0.1$, $M_2 = 0.03$, $G_2 = 0.03$, $Q_3 = 0.5P_3$ where P_3 consists of 80% constant power, 10% current and 10% impedance components. Other parameter values are as shown in Figure 3.

generator at bus 1 and two load buses 2 and 3. When the tap changer controls are not present (i.e. with $n_2 = n_3 = 1$), the feasibility boundary consists of two zero sets, namely, C_Z (the saddle node bifurcation) at low control gain values of K_1, and C_H (the Hopf bifurcation) at higher gain values. When the tap changer controls are active, the load voltages are maintained at the reference value (1 p.u.), but it can be seen that the feasibility region is now bounded by three zero sets, C_Z and C_H along with $C_{Z,TC}$ (at medium control gains) which corresponds to the tap changer related saddle node bifurcation.

Both generators are considered for the simulation in Figure 4, but the tap ratio setting is $n_2 = 1$ for the ULTC at bus 2. When the tap changer control at Bus 3 is inactive (i.e. with $n_3 = 1$), the feasibility boundary again consists of the two zero sets C_Z and C_H as in Figure 3. When the tap changer control is present, the feasibility boundary now only consists of two zero sets C_H and $C_{Z,TC}$ which is connected with the cascading descent instability of the tap changer control.

For this three bus system, for the parameter values as shown, the feasibility boundary for the operating point consists of the three zero sets C_Z, C_H and $C_{Z,TC}$. However for the large system, by Theorem 2, the feasibility boundary may also consist of two other zero sets C_S and $C_{H,TC}$. Since the feasibility boundary can be solved as the (five) zero sets of analytic functions, *Theorem 2 provides a practical method for computing rigorous security criteria for the large power system*, when discrete control devices such as tap changers are included in the analysis.

5 Discussions

The power system with locally controlled tap changers, is a hybrid dynamic system with continuous dynamics defined in equations (1) and (2), along with the slower discrete dynamics defined in equations (3). We will next consider the physical implications of the five different segments in the feasibility boundary.

After a tap change (more generally after any parameter change), if the system finds itself on the other (unstable) side of the three boundary segments, C_2, C_S and C_H then the transient will diverge away from the operating point, since the equilibrium point for the dynamics (1) and (2) is unstable there. The exact nature of the transient may depend on the specific type of boundary encountered. For points near the saddle node bifurcations in the set C_Z, by the Center manifold theorem and by the saddle node bifurcation theorem, the transient will slide along the center manifold [12]. For points near the Hopf bifurcations in the set C_H, the transient will be oscillatory on the center manifold by the Hopf bifurcation theorem. Also generally near the feasibility boundary segment C_H, all the other $(n-2)$ eigenvalues are stable, hence it follows that locally the center manifold is attracting [5]. Hence the behavior of the transients near the boundary segment C_H are charaterized by the dynamics on the center manifold. Now applying the Hopf bifurcation theorem, the transient may either degenerate into a stable periodic orbit in the case of a supercritical Hopf bifurcation, or may be unstable (oscillatory with increasing amplitude) in the case of the subcritical Hopf bifurcation. For the singularity induced bifurcation points in the set C_S, the system behavior becomes unpredictable [15]. C_S plays no role in the feasibility boundaries in Figures 3 and 4, although it can be shown to be active for the feasibility boundary of the operating point [16].

Next let us consider the feasibility boundary segment $C_{Z,TC}$. The system equilibria with the tap changer control correspond to the zero set of the functions $h_2(z_2, z_3, p)$ in the equation (14) and the Jacobian of h_2 is $\tilde{J}_d = (D_{z_2} h_2, D_{z_3} h_2)$. Since $\det(\tilde{J}_d) = 0$ in the set $C_{Z,TC}$, the Jacobian $\tilde{J}_d$ has zero eigenvalues there. By the genericity of the saddle node bifurcation theorem, then most of the points in the set $C_{Z,TC}$ (this excludes the points of transition in Figures 3 and 4) correspond to the ocurrence of static saddle node bifurcations. For applying the static version of the saddle node bifurcation theorem, let us introduce the necessary transversality conditions first.

- (SN1TC) The Jacobian $\tilde{J}_d$ has a simple zero eigenvalue with right eigenvector v and left eigenvector w and there is no other eigenvalue on the imaginary axis.

- (SN2TC) $w(D_p h_2) \neq 0$

- (SN3TC) $w(D_{\tilde{x}}^2 h_2(v, v)) \neq 0$ where $\hat{x}$ corresponds to the x-coordinates $\tilde{x} = (x_2, x_3)$. □

Define the set $B_{SN,TC}$ as the points in the set $C_{Z,TC}$ where the transversality conditions (SN1TC), (SN2TC) and (SN3TC) are satisfied. Then by Sotomayor's saddle node bifurcation theorem [11], we conclude that a Saddle node bifurcation occurs for points in the set $B_{SN,TC}$ and moreover generically

$$\overline{B_{SN,TC}} = C_{Z,TC}. \qquad (20)$$

Therefore, near the bifurcation points in the set $B_{SN,TC}$, two solutions of the equation (14) meet and disappear. In other words, for most of the boundary points in the set $C_{Z,TC}$, the stable equilibrium of the tap changer control meets an unstable equilibrium and disappears at the boundary.

Suppose the system parameter crosses the feasibility boundary segment $B_{SN,TC}$. Then, since the tap changer control has no solution on the other side of the boundary (such as to the right of the curve $C_{Z,TC}$ in Figure 3), the tap changer control dynamics in equations (3) will diverge away from the previous operating point. The tap changer control will continuously order more tap changes, chasing a new equilibrium for the control dynamics in (3). This sequence may either take the system across the feasibility boundary (so that the continuous dynamics in (1) and (2)

itself becomes unstable, and the transient diverges as explained above) or the taps may run out (then the voltage is nonviable) (see [21]). In any case, stable system operation at the equilibrium stops before the combined feasibility boundary (Figures 3 and 4).

Since the Jacobian $\tilde{J}_d$ has purely imaginary eigenvalues in the set $C_{H,TC}$, from Hopf bifurcation theorem, we conjecture that the feasibility boundary points in the set $C_{H,TC}$ correspond to the emergence of stable limit cycles for the tap changers (possible hunting phenomenon) or the annihilation of the region of attraction for the equilibrium. The exact nature of the instability mechanism needs more investigation.

In summary, the feasibility boundary with tap changer control consists of five zero sets, hence it is relatively accessible to computation. It is clear that the system operation at the stable equilibrium point becomes unstable at the feasibility boundary. Therefore *the distance of the operating point to the feasibility boundary serves as a rigorous, computable and efficient, and so, a powerful tool for secure system operation.*

6 Conclusion

A principal objective of system operation is to maintain security. For system security, it is essential that the operating point has adequate safety margins of feasibility to survive normal operating activities and changes and also a specified set of potential disturbances such as the "first contingencies." In the parameter space the *feasibility boundary* is identified as being composed of five zero sets defined by algebraic equations and hence relatively easily computable. This then provides information not just on whether the (existing or proposed) operating point is feasible (this can be checked by direct computation) but also on its distance from the boundary that is its margin of security. Information is further gained on the security of a proposed operating change in the parameter space, specifically its security against any collapse event during the proposed change. This includes the consideration of the troublesome automatic tapchanger problem which can be directly computed as one (or two) additional types of (restrictive) segments to the feasibility boundary.

References

[1] S. Abe, Y. Fukunaga, A. Isono and B. Kondo, "Power system voltage stability", *IEEE Trans. PAS*, Vol. PAS-101, No. 10, October 1982, pp. 3830-3840

[2] J. H. Chow and A. Gebreselassie, "Dynamic voltage stability analysis of a single machine constant power load system", *Proceedings of the CDC*, Hawaii, December, 1990, pp. 3057-3062.

[3] J. Jargis and F. D. Galiana "Quantitative analysis of steady state stability in power networks", *IEEE*

Trans. PAS , Vol. PAS-100, No. 1, January 1981, pp. 318-326.

[4] M. Ilic and F. Mak, " Mid-range voltage dynamics modelling with the load controls present", *Proceedings of the CDC* Los Angeles, December 1987, pp. 45-52.

[5] A. Isidori, *Nonlinear control systems*, Springer-Verlag, Second Edition, 1989.

[6] H. G. Kwatny, A. K. Pasrija and L. Y. Bahar, "Static bifurcation in power networks: loss of steady state stability and voltage collapse", *IEEE Transactions on Circuits and Systems*, Vol. CAS-33, No. 10,

[7] W. R. Lachs, "Voltage Collapse in EHV Power Systems", Paper A78 057-2, 1978, New York, *IEEE Winter Power Conference.*

[8] C. C. Liu and K. T. Vu, "Analysis of tap-changer dynamics and construction of voltage stability regions", *IEEE Transactions on Circuits and Systems*, Vol. 36, No. 4, April, 1989, pp. 575-590.

[9] J. Medanic, M. Ilic-Spong, J. Christensen, "Discrete models of slow voltage dynamics under load tap-changing transformer coordination", *IEEE Trans. PAS*, Nov. 1987, pp. 873-882.

[10] H. Ohtsuki, A. Yokoyama and Y. Sekine, "Reverse action of on-load tap changes in association with voltage collapse", *IEEE Trans. PAS*, Vol. 6, No. 1, Feb. 1991, pp. 300-306.

[11] J. Sotomayor, "Generic bifurcations of dynamical systems", *Dynamical systems*, Edited by M. M. Peixoto, Academic Press, NY, 1973.

[12] I. Dobson, H. D. Chiang, "Towards a theory of volatge collapse in electric power systems", *Systems and Control letters*, Vol. 13, 1989, pp. 253–262.

[13] C.W. Taylor, F.R. Nassief and R.L. Cresap, "Northwest power pool transient stability and load shedding controls for generation-load imbalances", *IEEE Trans. PAS*, Vol. PAS-100, No.7, July 1981, pp. 3480-3495

[14] V. Venkatasubramanian, H. Schättler and J. Zaborszky, "Global voltage dynamics: study of a generator with voltage control, transmission and matched MW load", *Proceedings of the CDC* Hawaii, December 1990, pp. 3045-3056.

[15] V. Venkatasubramanian, H. Schättler and J. Zaborszky, "A taxonomy of the dynamics of the large electric power system", *Proceedings of the International Workshop on Bulk Power System Voltage Phenomena - II: Voltage Stability and Security*, Maryland, August 1991, pp. 9-52.

[16] V. Venkatasubramanian, H. Schättler and J. Zaborszky, "A Stability Theory of Differential Algebraic Systems such as the Power System", *Proceedins of the ISCAS* San Diego, May 1992, pp. 2517–2520.

[17] V. Venkatasubramanian, H. Schättler and J. Zaborszky, "A Stability Theory of Large Differential Algebraic Systems - a Taxonomy", Report SSM 9201 -Part I, Department of Systems Science and Mathematics, Washington University, School of Engineering and Applied Science, Saint Louis, Missouri, 63130, August 1992.

[18] V. Venkatasubramanian, H. Schättler and J. Zaborszky, "Voltage dynamics: study of a generator with voltage control, transmission and matched MW load", *IEEE Transactions on Automatic Control*, November 1992, to appear.

[19] N.Yorino, H.Sasaki, A.Funahashi, F.Galiana, M.Kitawaga "On the condition for inverse control action of tap changers", *Proceedings of the International Workshop on Bulk Power System Voltage Phenomena - II: Voltage Stability and Security*, Maryland, August 1991, pp. 193-199

[20] J. Zaborszky *Some basic issues in voltage stability and viability*, Proceedings of Bulk-Power Voltage Phenomena – Voltage Stability and Security, Potosi, MO September, 1988, pp. 1.17-1.60.

[21] J.Zaborszky, Comments on 'Dynamic Ststic Voltage Stability Criteria' by R.A.Schlueter et al.,*Proceedings of the International Workshop on Bulk Power System Voltage Phenomena - II: Voltage Stability and Security*, Maryland, August 1991, pp. 306-307

[22] J. Zaborszky and J.W. Rittenhouse, *Electric Power Transmission*, The Rensselaer Bookstore, New York, 1969.

[23] J. Zaborszky and B. Zheng, "Structure features of the dynamic state space for studying voltage-reactive control", *Proceedings of the PSCC* , Graz, Austria, August 1990, pp. 319-326.

[24] L.A. Zadeh and C.A. Desoer, Linear system theory: the state space approach, McGraw Hill, New York, 1963

Proceedings of the 31st Conference
on Decision and Control
Tucson, Arizona • December 1992

SENSITIVITY OF HOPF BIFURCATIONS TO POWER SYSTEM PARAMETERS

Ian Dobson **Fernando Alvarado** **Christopher L. DeMarco**
Electrical and Computer Engineering Dept.
University of Wisconsin, Madison, WI, 53706 USA
e-mail: dobson@engr.wisc.edu

Abstract: Hopf bifurcation of a power system leads to oscillatory instabilities and it is desirable to design system parameters to ensure a sufficiently large loading margin to Hopf bifurcation. We present formulas for the sensitivity of the Hopf loading margin with respect to any power system parameter. These first order sensitivities determine an optimum direction in parameter space to change parameters to increase the loading margin. We compute the Hopf bifurcation sensitivities of a simple power system with a voltage regulator and a dynamic load model. Parameter sensitivities of the Hopf and saddle node bifurcations are compared. An idea for eliminating some Hopf bifurcations is presented.

1 Introduction

Power systems require parameters or controls to be chosen so that oscillatory instabilities are avoided. This has previously been done by linearizing the power system model about an operating point and designing the linearized system to avoid instabilities [16]. More recently, starting with the work of Abed and Varaiya [1], the onset of oscillatory instability is studied in a nonlinear context as a Hopf bifurcation [20, 2, 3]. We formulate the design to avoid oscillatory instabilities in the nonlinear context as avoiding the Hopf bifurcation in the following manner [11]:

Suppose a stable operating equilibrium with a vector of nominal parameter values p_0 is given. If loads increase, then stability is lost in a Hopf bifurcation and the proximity of the base case to the Hopf bifurcation is measured by a loading margin M. M changes as the parameters p are varied from their nominal values p_0. We want to compute the first order sensitivity of M with respect to the power system parameters p in order to obtain an optimum direction of parameter change to increase M. Increasing M improves the system robustness to oscillatory instability caused by slow load increase. This paper derives and illustrates the computation of the sensitivity of M with respect to any power system parameters.

We review previous work [12] on avoiding of saddle node bifurcations since this is similar to the proposed method of avoiding Hopf bifurcations. Saddle node bifurcation is associated with voltage collapse of the power system [14] and always occurs for sufficently high loading. The dynamical consequences of saddle node bifurcation [9] seem consistent with some observed voltage collapses, in which voltage magnitudes decline monotonically. However, some simplified power system models become oscillatory unstable in a Hopf bifurcation before the saddle node bifurcation occurs [1,20,2,3]. The load power margin to a saddle node bifurcation is computed by continuation or direct methods by increasing the loading until saddle node bifurcation is first encountered [e.g. 21, 6]. (We assume throughout the paper that the distribution of load increase is specified.) The next step is to compute the normal vector to the saddle node bifurcation surface at the critical loading; the formula for the normal vector follows from one of the transversality conditions of bifurcation theory [10]. It turns out that the first order sensitivity of the load power margin to any power system parameters or controls is trivial to compute from the normal vector. This sensitivity determines (at least locally) the combination of parameters and controls to be varied in order to optimally increase the load power margin.

Since there are computations for the first Hopf bifurcation as loading increases [21, 5] and there is a formula for the normal vector to the Hopf bifurcation surface [13,7,15], the sensitivity to any power system parameters of the load power margin to Hopf instability can similarly be computed. In a Hopf bifurcation, a complex pair of eigenvalues of the linearized system crosses the imaginary axis and the normal vector essentially contains sensitivities with respect to parameters of the real parts of these eigen-

CH3229-2/92/0000-2928$1.00 © 1992 IEEE

values. (The Hopf bifurcation hypersurface is determined by the vanishing of the real parts of these eigenvalues.)

The sensitivity of the margin M also shows which parameters couple most strongly with the Hopf bifurcation. For example, one expects the voltage regulator parameters to strongly influence the margin to Hopf bifurcation. The sensitivity of the load power margin to both the Hopf and the saddle node bifurcation are compared to determine the extent to which different sets of parameters affect both margins.

2 Hopf parameter sensitivity

Consider a power system modeled by smooth parameterized differential equations

$$\dot{z} = f(z, \lambda), \qquad z \in \mathbf{R}^n, \ \ \lambda \in \mathbf{R}^{m+1} \qquad (1)$$

The parameter vector $\lambda = (\ell, p)$ consists of a real loading parameter ℓ and a vector p of m system design parameters. We write $x \in \mathbf{R}^n$ for a particular equilibrium of (1) and assume that x is asymptotically stable at the parameter vector $\lambda_0 = (\ell_0, p_0)$. p_0 is a nominal choice of design parameters. We assume that when the loading parameter ℓ slowly increases to some critical value ℓ_* and the design parameters are held fixed at p_0, the equilibrium x loses stability in a Hopf bifurcation. The loading margin to instability is then $M = \ell_* - \ell_0$. If we write $\lambda_* = (\ell_*, p_0)$, then the loading margin may also be expressed as $M = |\lambda_* - \lambda_0|$. The measure M of closeness to Hopf bifurcation takes full account of system nonlinearity.

The question we address is: What is the optimum direction for first order change in the design parameters p from p_0 in order to increase the load margin M? That is, we regard the load margin M as a function of the design parameters p and want to compute the gradient or sensitivity $M_p|_{p_0}$ so that the design may be incrementally improved by changing parameters in the direction $M_p|_{p_0}$.

The sensitivity $M_p|_{p_0}$ is essentially a scaled projection of a normal vector to the Hopf bifurcation hypersurface. The details follow: Write Σ^{hopf} for the set of $\lambda'_* \in \mathbf{R}^{m+1}$ for which equation (1) has a Hopf bifurcation at (x_*, λ'_*) with $f_x|_*$ having a simple pair of eigenvalues $\pm j\omega_*$, $\omega_* \neq 0$ and all other eigenvalues with nonzero real parts and satisfying the transversality condition (4) presented below. Since $f_x|_*$ is invertible, the implicit function theorem implies that there is a smooth function u defined in a neighborhood of λ_* with $u(\lambda_*) = x_*$, $u(\lambda) = x$ and $f(u(\lambda), \lambda) = 0$. $u(\lambda)$ specifies the position of the equilibrium of interest as a function of the parameters and its Jacobian u_λ is given by solving

$$f_x u_\lambda = -f_\lambda \qquad (2)$$

There is also a smooth function μ evaluating to an eigenvalue defined in a neighborhood of λ_* with $\mu(\lambda_*) = j\omega_*$ and $\mu(\lambda)$ an eigenvalue of $f_x|_{(u(\lambda),\lambda)}$. The real part of the eigenvalue μ is a function

$$\alpha(\lambda) = \text{Re}\{\mu(\lambda)\} \qquad (3)$$

which is smooth near λ_* and, if the transversality condition

$$\alpha_\lambda \neq 0 \qquad (4)$$

is satisfied, then there is a neighborhood $U \ni \lambda_*$ in which Σ^{hopf} is a smooth hypersurface specified by the zero set of α:

$$\Sigma^{\text{hopf}} \cap U = \{\lambda \in U \mid \alpha(\lambda) = 0\} \qquad (5)$$

That is, Σ^{hopf} is locally specified by the vanishing of the real part of the complex pair of eigenvalues associated with the Hopf bifurcation. It follows that a normal vector to Σ^{hopf} is given by the sensitivities of the real part with respect to the parameters:

$$N(\lambda_*) = D_\lambda(\text{Re}\{\mu(\lambda)\})|_{\lambda_*} = \alpha_\lambda|_{\lambda_*} \qquad (6)$$

Now we compute the gradient $\alpha_\lambda|_{\lambda_*}$ in (6) in terms of the equations f in (1). Write v_* and w_* for the right and left complex eigenvector of $f_x|_*$ corresponding to $j\omega_*$; these eigenvectors are normalized according to $|v| = 1$ and $wv = 1$ (It is convenient to regard w as a row vector). Using this normalization, it is easy to show that

$$\mu(\lambda) = w f_x v$$

Differentiate with respect to λ to obtain

$$\begin{aligned} \mu_\lambda &= w D_\lambda(f_x) v + w_\lambda f_x v + w f_x v_\lambda \\ &= w D_\lambda(f_x) v + \mu D_\lambda(wv) \\ &= w D_\lambda(f_x) v \\ &= w\,(f_{xx} u_\lambda + f_{x\lambda})\,v \end{aligned}$$

Take the real part and use $D_\lambda(\text{Re}\{\mu(\lambda)\})|_{\lambda_*} = \text{Re}\{\mu_\lambda\}|_{\lambda_*}$ and (6) to obtain

$$N(\lambda_*) = \alpha_\lambda|_{\lambda_*} = \text{Re}\{w(f_{xx} u_\lambda + f_{x\lambda})v\}\big|_* \qquad (7)$$

where u_λ is given by solving equation (2). (To exemplify the notation, note that $f_{xx} u_\lambda$ is an

$n \times n \times (m+1)$ tensor; contraction with w and v yields an $(m+1)$ vector.) A different scaling of formula (7) and similar computations appear in [15,13]. If the power system equations (1) are linearized *before* the sensitivity of the real parts of the critical eigenvalues are computed, then the term $\mathrm{Re}\{w(f_{xx}u_\lambda)v\}|_*$ does not appear. This omission arises because the linearization effectively fixes the equilibrium at the origin. Since this term has often been overlooked in the past, it would be interesting to know whether this term is significant in typical power system applications.

The loading increase from $\lambda_0 = (\ell_0, p_0)$ to $\lambda_* = (\ell_*, p_0)$ can be written as

$$\lambda(\ell) = \lambda_0 + (\ell - \ell_0)\, e_1 \tag{8}$$

where $e_1 = (1, 0, 0, ..., 0)$. We assume a transversality condition

$$D_\ell \mathrm{Re}\{\mu(\lambda(\ell))\}|_{\ell_*} = \alpha_\lambda|_{\lambda_*} e_1 \neq 0 \tag{9}$$

so that the critical eigenvalues pass the imaginary axis with nonzero speed as the Hopf bifurcation occurs. (This condition is generically satisfied.) We prove below that the gradient of M with respect to λ is

$$M_\lambda\big|_{\lambda_0} = -(N(\lambda_*)e_1)^{-1} N(\lambda_*) \tag{10}$$

Then it is clear from $M_\lambda = (M_\ell, M_p)$ that $M_p|_{p_0}$ is the projection π of $M_\lambda\big|_{\lambda_0}$ onto the m dimensional design parameter space:

$$M_p|_{p_0} = \pi M_\lambda\big|_{\lambda_0} \tag{11}$$

That is, if

$$N(\lambda_*) = (n_0^\ell, n_1^p, n_2^p, n_2^p, ..., n_m^p)$$

then (10) and (11) yield

$$M_p|_{p_0} = -(n_0^\ell)^{-1}(n_1^p, n_2^p, n_2^p, ..., n_m^p) \tag{12}$$

Our assumption (9) implies that $n_0^\ell \neq 0$.

The proof of (10) is adapted from the proof in [12] by substituting the Hopf bifurcation hypersurface Σ^{hopf} for a saddle node bifurcation hypersurface. $\lambda_*(\lambda)$ is a well defined smooth function of the parameters λ near λ_0 because Σ^{hopf} is a smooth hypersurface near λ_* and (9) implies that $N(\lambda_*)e_1 \neq 0$ so that the direction of load increase e_1 intersects Σ^{hopf} transversally at λ_*. It follows that $M(\lambda) = |\lambda_*(\lambda) - \lambda|$ is a well defined smooth function of the parameters λ near λ_0. Then

$$0 = \alpha(\lambda_*) = \alpha(\lambda + M(\lambda)e_1)$$

and differentiating with respect to λ and evaluating at λ_0 yields

$$\begin{aligned} 0 &= \alpha_\lambda|_{\lambda_*}(I + e_1 M_\lambda)\big|_{\lambda_0} \\ &= N(\lambda_*) + N(\lambda_*)e_1 M_\lambda\big|_{\lambda_0} \end{aligned}$$

and the result (10) follows by rearranging terms. The geometric content is clear: the optimum direction to increase the distance in a given direction e_1 of a point λ_0 to a hypersurface Σ^{hopf} is antiparallel to the outward normal to Σ^{hopf}.

3 Saddle node parameter sensitivity

We summarize formulas from [12] for the sensitivity of loading margin to saddle node bifurcation to any power system parameters to establish notation used in the example and to compare with the corresponding results for Hopf bifurcation.

The saddle node bifurcation occurs at loading ℓ_{SN} and the loading margin is $M^{SN} = \ell_{SN} - \ell_0$. At the saddle node bifurcation f_x is singular and has left eigenvector w_{sn} corresponding to the zero eigenvalue of $f_x|_{sn}$. Under suitable transversality assumptions the saddle node hypersurface has at λ_{sn} a normal vector

$$N(\lambda_{sn}) = w_{sn} f_\lambda\big|_{sn} \tag{13}$$

The normal vector formula (13) is simpler than the corresponding formula (7) for Hopf.

The sensitivity of the loading margin follows from the normal vector in the same way as developed for the Hopf bifurcation in section 2:

$$\begin{aligned} M_p^{SN}|_{p_0} &= \pi M_\lambda\big|_{\lambda_0} = -\pi(N(\lambda_{sn})e_1)^{-1} N(\lambda_{sn}) \\ &= (w_{sn} f_\ell|_{\ell_{SN}})^{-1} w_{sn} f_p|_{p_0} \end{aligned}$$

4 Illustrative Example

We illustrate the computation of the sensitivity of the Hopf load power margin in a simple power system example.

Chow and Gebreselassie [8] compute a Hopf bifurcation in a power system model consisting of single machine with a voltage regulator supplying a constant power load through a single line. Our example (see Fig. 1) is based on this model and we refer to [8] for most of the model equations and their description. We simplify the treatment of the voltage regulator set points in [8] by setting the reference voltage setpoint E_{ref} = 1.1 pu and computing the transformer high side voltage E_s in terms of other state variables.

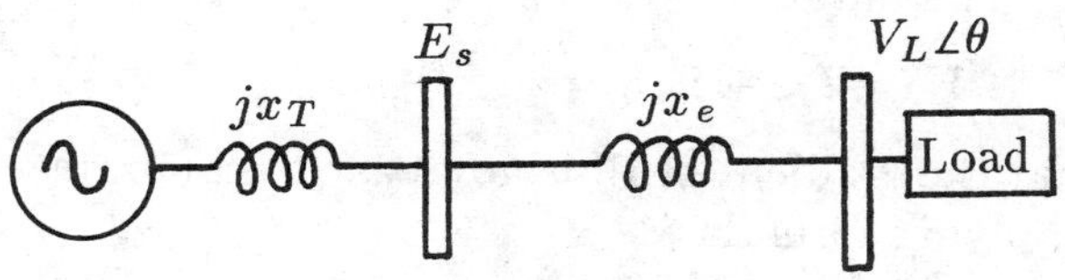

Fig. 1 One machine system with dynamic load

One problem in computing Hopf bifurcations is that the Hopf bifurcation depends on dynamical details of the models such as time constants and little reliable information is known about the dynamics of loads. (In contrast, saddle node bifurcations are somewhat independent of the details of load dynamics as argued in [Dobsonw].) We address the problem of poorly known but possibly significant load dynamics by assuming a crude form of dynamic load model and roughly estimating parameter values of an appropriate order of magnitude and then computing the sensitivity to the estimated parameters to assess the validity of the results.

The dynamic load model represents an aggregate load and replaces the constant real and reactive power loads of [8] by

$$\ell\,\mathrm{PF} + D\dot{\theta} + a\dot{V}_L$$

$$\ell\sqrt{1-\mathrm{PF}^2} + b\dot{\theta} + k\dot{V}_L$$

respectively, where ℓ parameterizes the increase of the constant power part of the load, PF stands for power factor and D, a, k, b are time constants of the load dynamics. Induction load models with similar terms are discussed in [19,9]. The nominal load parameters are $\mathrm{PF} = 0.95$, $D = 0.05$, $a = 0$, $b = 0$, $k = 0.1$. The order of magnitude of D and k is consistent with power system tests in [17,18,4].

The model of [8] could be written as 5 differential equations and 2 algebraic equations, together with a procedure for determining settings for E_{ref} at different loadings that yield a specified value of E_s. Our modifications to the model of [8] can be summarized as including a dynamic load model with a lower power factor and fixing $E_{ref} = 1.1$ pu for all loading levels. We use the nominal generator, machine and voltage regulator parameters of [8] except that the stabilizer gain $K_f = 0.1$. These modifications to the model of [8] allow us to write the model as 7 differential equations with loading parameter ℓ. The state vector is $(E'_d, E'_q, V_R, E_{FD}, R_f, \theta, V_L)$ where E'_d and E'_q are machine voltages, V_R is the voltage regulator output voltage, E_{FD} is the field voltage, R_f is the state of the stabilizer, and $V_L\angle\theta$ is the load voltage phasor. Specifying the load dynamics resolves the singularity of the load algebraic equations encountered in [8].

The first instability encountered by the stable equilibrium as the loading is increased from $\ell_0 = 0$ is a Hopf bifurcation at $\ell_H = 0.370$ so that the loading margin $M = 0.370$. The power system parameters are $p = (D, a, b, k, \mathrm{PF}, E_{ref}, K_A, T_A, T_E, K_f, T_f, x_T, x_e, x_d, x_q, x'_d, T'_{d0}, T'_{q0})$ where K_A and T_A are the gain and time constant of the voltage regulator, T_E is the exciter time constant, K_f and T_f are the gain and time constant of the stabilizer, x_T and x_e are the reactances of the step up transformer and the transmission line, x_d and x_q are the machine synchronous reactances, x'_d is the machine transient reactance, and T'_{d0} and T'_{q0} are the open circuit machine time constants. The sensitivities M_p are shown in the second row of Table 1. We verified the results by increasing K_f by 0.01 and recomputing M. M increased by 0.020 whereas the sensitivity predicts M increasing by 0.022. Increasing T_f by 0.1 caused M to decrease by 0.015 whereas the sensitivity predicts M decreasing by 0.015.

In our example, the dynamic load parameters b and k are moderately sensitive, but since their base values are small, the effect on the loading margin of, say, letting a, b, k, D tend to 0 is small. This is of interest since setting $a = b = k = D = 0$ effectively makes the load differential equations into algebraic equations. Further modeling and experiments along these lines are required to obtain a general conclusion about the relative importance of the dynamic load model.

The saddle node bifurcation occurs at the loading margin $M^{SN} = 1.03$. The sensitivities of M^{SN} to the power system parameters were computed according to the formulas of section 3 and are shown in the third row of Table 1. As expected, the voltage regulator parameters do not affect M^{SN} (The slight dependence of M^{SN} on K_A can be attributed to our simplified modeling of E_{ref} as a constant). This suggests that it might be desirable to design the voltage regulator system of this example to avoid Hopf bifurcations and system oscillations before addressing the avoidance of saddle node bifurcations and voltage collapse. Increasing the power factor PF or the reference voltage E_{ref} increases the margins to both the Hopf and saddle node bifurcations.

5 Eliminating Hopf bifurcations

In cases in which a further increase in loading past the Hopf bifurcation yields a "reverse" Hopf bifurcation which restores the stability of the equilibrium, we suggest that parameters be optimally changed to eliminate the Hopf bifurcation by making it coalesce with the "reverse" Hopf bifurcation.

Suppose the Hopf bifurcation occurs at a loading ℓ_H and the reverse Hopf bifurcation occurs at a higher loading $\ell_{RH} > \ell_H$. The inevitable saddle node bifurcation occurs at a loading $\ell_{SN} > \ell_{RH}$. Write x for the stable equilibrium at low loading. One of the possible situations as loading increases is that the Hopf bifurcation at ℓ_H makes x unstable and creates a stable periodic orbit γ which persists until it coalesces with the unstable equilibrium x at the reverse Hopf bifurcation at ℓ_{RH}. The reverse Hopf bifurcation restores the stability of x and the stability of x persists as the loading further increases until x disappears in the saddle node bifurcation at ℓ_{SN}. One possibility is that the stable periodic orbit γ can period double to chaos and reverse period double back to a stable periodic orbit in the interval (ℓ_H, ℓ_{RH}).

We measure the extent to which Hopf bifurcation is present in the system by the extent of the interval or "window" over which the Hopf bifurcation destabilizes the system. That is, we define the index $W = \ell_{RH} - \ell_H$ and suggest that decreasing W to zero will eliminate the Hopf bifurcation by causing the Hopf and "reverse" Hopf bifurcations to coalesce and disappear. We compute the sensitivity of W with respect to power system parameters. This sensitivity could be used to obtain the optimum direction in which to change the power system parameters so that W is decreased. Driving W to zero eliminates the Hopf bifurcations so that the stability of the equilibrium x is only limited by the saddle node bifurcation.

The sensitivity of W with respect to power system parameters is easy to obtain from the previous sensitivity results. Write $M^H = \ell_H - \ell_0$ and $M^{RH} = \ell_{RH} - \ell_0$ for the respective loading margins of the Hopf and reverse Hopf bifurcations. The index

$$W = \ell_H - \ell_{RH} = M^H - M^{RH}$$

so that the gradient of W is now easy to compute by applying formula (11):

$$W_p|_{p0} = \pi(M_\lambda^H|_{\lambda_0} - M_\lambda^{RH}|_{\lambda_0}) \qquad (14)$$

That is, using (12), the ith element of $W_p|_{p0}$ is

$$[W_p|_{p0}]_i = n_i^H/n_0^H - n_i^{RH}/n_0^{RH} \qquad i = 1, ..., m.$$

6 Conclusions

Exact formulas for the first order sensitivity of the loading margin to Hopf bifurcation to any power system parameters have been obtained. The formulas are illustrated using a small power system example and verified by numerically computing some of the sensitivities. These sensitivities could be used to optimally increase the loading margin to Hopf bifurcation. The loading margin and the sensitivity computation take full account of the system nonlinearities. The sensitivity results follow easily from computing a normal vector to a Hopf bifurcation hypersurface in parameter space. The normal vector contains the sensitivities of the real part of the critical pair of eigenvalues associated with the Hopf bifurcation. The formulas include a term associated with movement of the equilibrium which has been neglected in eigenvalue sensitivity studies of linearized power system models. Our results involve eigenvalue sensitivities but are exact first order sensitivities of loading margins. We also compare the sensitivities of the Hopf and saddle node bifurcations in our example.

The Hopf bifurcation depends on dynamic aspects of the load models and these are not well known. We approach this problem by choosing a crude dynamic load model with roughly estimated parameters and then computing the sensitivity of our margins to the estimated parameters. The dynamic load model allowed the model to be differential equations rather than differential-algebraic equations.

We have suggested a method of computing first order parameter changes which in some cases would tend to make two Hopf bifurcations coalesce and disappear.

Support in part by NSF grants ECS-9157192, ECS-8907391, ECS-8857019 and EPRI contract RP 8010-30 is gratefully acknowledged.

References

[1] E.H. Abed, P.P. Varaiya, Nonlinear oscillations in power systems, International Journal of Electric Energy and Power Systems, vol. 6 no. 1, Jan 1984, pp. 37-43.

[2] E.H. Abed, J.C. Alexander, H. Wang, A.M.A. Hamdan, H-C. Lee, Dynamic bifurcations in a power system model exhibiting voltage collapse, *Intl. Symp. on Circuits and Systems*, San Diego, CA, May 1992, pp. 2509-2512.

[3] V. Ajjarapu, B. Lee, Bifurcation theory and its application to nonlinear dynamical phenomena in an electrical power system, *IEEE Trans. on Power Systems*, vol. 7, no. 1, Feb 1992, pp. 424-431.

[4] A.M.Y.Akhatar, Frequency dependent dynamic representation of induction motor loads, Proceedings of the IEE, vol. 115, June 1968, pp. 802-812.

[5] F.L. Alvarado, Bifurcations in nonlinear systems: computational issues, IEEE ISCAS, New Orleans, LA, May 1990.

[6] C.A. Canizares, F.L. Alvarado, Computational experience with the point of collapse method on very large AC/DC power systems, in [14].

[7] S.N. Chow, J. Hale, *Methods of bifurcation theory*, Springer-Verlag, NY, 1982.

[8] J.H. Chow, A. Gebreselassie, Dynamic voltage stability of a single machine constant power load system, 29th IEEE CDC conference, Honolulu HI, Dec. 1990, pp. 3057-3062.

[9] I. Dobson, H.-D. Chiang, Towards a theory of voltage collapse in electric power systems, *Systems and Control Letters*, Vol. 13, 1989, pp. 253-262.

[10] I. Dobson, Observations on the geometry of saddle node bifurcation and voltage collapse in electric power systems, to appear in *IEEE Trans. on Circuits and Systems, Part 1* (scheduled vol. 39, March 1992).

[11] I. Dobson, Power system instabilities, bifurcations and parameter space geometry, Proceedings of *EPRI/NSF Workshop on application of advanced mathematics to power systems*, Redwood City, CA, September 1991.

[12] I. Dobson, L. Lu, Computing an optimum direction in control space to avoid saddle node bifurcation and voltage collapse in electric power systems, in [14] and *IEEE Trans. on Automatic Control* (scheduled Oct. 1992).

[13] I. Dobson, An iterative method to compute the closest saddle node or Hopf bifurcation instability in multidimensional parameter space, *International Symposium on Circuits and Systems*, San Diego, CA, May 1992, pp. 2513-2516.

[14] L.H. Fink, ed., Proceedings: Bulk power system voltage phenomena, voltage stability and security ECC/NSF workshop, Deep Creek Lake, MD, Aug. 1991, ECC Inc., 4400 Fair Lakes Court, Fairfax, VA 22033.

[15] M. Golubitsky, D.G. Schaeffer, *Singularities and groups in bifurcation theory, Vol.1*, pp. 352–355, Springer Verlag,NY,1985.

[16] *Eigenanalysis and frequency domain methods for system dynamic performance*, IEEE publication 90TH0292-3-PWR, 1990.

[17] IEEE\PES Committee, System load dynamics: simulation effects and determination of time constants, *IEEE Trans. on Power Apparatus and Systems*, vol. PAS-93, no.2, March/April 1973, pp. 600-609.

[18] F. Illecito, A. Ceyhan, G. Ruckstuhl, Behaviour of loads during voltage dips encountered in stability studies: Field and laboratory tests, *IEEE Trans. on Power Apparatus and Systems*, vol. PAS-91, no.6, Nov/Dec 1972, pp. 2470-2479.

[19] K. Jimma, A. Tomac, K. Vu, C.-C. Liu, A study of dynamic load models for voltage collapse analysis, in [14].

[20] C. Rajagopalan, B. Lesieutre, P.W. Sauer, M.A. Pai, 91 SM 419-2 PWRS, Dynamic aspects of voltage/power characteristics, IEEE PES Summer meeting, San Diego, CA, July 1991.

[21] R. Seydel, *From equilibrium to chaos: practical bifurcation and stability analysis*, Elsevier, NY 1988.

[22] S. Wolfram, *Mathematica: A system for doing mathematics by computer*, 2nd edition,Addison-Wesley, 1991.

	D	a	b	k	PF	E_{ref}	K_A	T_A	T_E	K_f	T_f
p	0.05	0.00	0.00	0.10	0.95	1.10	30.0	0.40	0.56	0.10	1.30
M_p	–0.05	–0.01	–0.33	–0.31	0.48	0.70	0.00	–0.27	–0.18	2.20	–0.15
M_p^{SN}	0.00	0.00	0.00	0.00	2.20	1.88	0.01	0.00	0.00	0.00	0.00

	x_T	x_e	x_d	x_q	x'_d	T'_{d0}	T'_{q0}
p	1.00	0.3406	1.00	1.00	0.18	5.00	1.50
M_p	–0.25	–0.31	0.00	0.00	–0.26	0.04	0.00
M_p^{SN}	–0.10	–2.45	–0.10	0.00	0.00	0.00	0.00

Table 1. Sensitivities of Hopf and saddle node loading margins to parameters

Proceedings of the 31st Conference
on Decision and Control
Tucson, Arizona • December 1992

A New Concept of an Aggregate Model for Tertiary Control Coordination of Regional Voltages

Xiaojun Liu Marija Ilić Michael Athans
Massachusetts Institute of Technology
Cambridge, MA 02139, USA

Christine Vialas Bertrand Heilbronn
Electricité De France
92141 Clamart Cedex, France

Abstract

In this paper a theoretical formulation of a discrete process representing regulation of steady-state voltages over mid-term horizons is introduced. The main interest is in regulating critical steady-state voltages in response to significant load and generation variations, under the assumption that the system is transiently stable. Practical motivation for this work comes from the need to study coordination of different electrically connected regions of the French power network.

Control theoretical problems that arise in solving this practical problem are new under the assumption that the system is transiently stable. Discrete process representing a sequence of steady state voltages over a mid-term horizon is only system input (load and control) driven, since the natural response is assumed stable and instantaneous. The fact that only controls drive the process leads to many interesting results in postulating an aggregate model for coordinating very large scale power systems. It is shown for the first time in a theoretically rigorous way that the natural coordinating variables are reactive power flows among the regions.

The approach proposed here for deriving an aggregate model for hierarchical voltage coordination could be generalized to many classes of other large-scale discrete processes.

1 Practical problem of interest

Theoretical results presented in this paper are motivated by the practical need to regulate load voltage changes over mid-term time horizons according to a specified performance by changing generator voltages. While the voltage control of an interconnected large-scale power system is widely recognized as a very important problem, its basic formulation and solutions are often utility specific. Most often the voltage control is viewed as an entirely static problem, whose solution is identical to a centralized open-loop optimization-based Var/voltage management. The most common tool for solving this optimization problem is an optimal power flow (OPF) type algorithm. This approach automatically assumes availability of the full information structure and is often referred to as a tertiary control. The OPF-based approach computes changes in generator voltages to regulate load voltages on the entire interconnected system. This approach requires a large amount of data which is sometimes missing when the system is experiencing unusual operating conditions (the data transmission network becomes saturated with alarm signals). Moreover, this approach does not lend itself to an easy balance between coordination and competition, which is of interest when trying to coordinate regions.

A second approach to voltage control coordination relies on decomposition of a large system into regions and an on-line decentralized close-loop reduced information structure for controlling regions. For instance, the French system has been committed to a full automation of system-wide voltage regulation while employing an intuitive reduced information structure at the regional level (the pilot load voltages are controlled within each region by regional generators, assuming that the neighboring regions have negligible effect.) In this case the responsibility for coordinated voltage regulation is shared among regional closed-loop controllers (called secondary voltage controllers) and the operators at the national control center (coordination level).

As the French power network has become increasingly meshed during the past decade and is also operated closer to the prespecified voltage limits, Electricité De France is considering the automation of this tertiary level in order to improve the security and economics of the entire system.

This paper provides first an overview of a system theoretic formulation of the presently employed regional voltage regulation in France. Next, new results are reported on defining a model for tertiary control coordination, as well as a model for improved secondary control strategy which accounts for the interactions with the neighboring regions. An important breakthrough is reported on theory required to introduce the role of physically measurable variables, such as reactive tie line flows, into the proposed two-level control concept. Based on better understanding of this model, a two-level control approach can be proposed with the well defined information flow for coordinating regional controllers, as well as for enhancing the existing secondary controllers.

CH3229-2/92/0000-2934$1.00 © 1992 IEEE

2 New theoretical problem formulation

A detailed literature survey of present state-of-the-art modeling and control in large scale interconnected systems [1] shows that one often uses as a starting point some sort of decomposition into subsystems. Two major general approaches are suggested in [1] for analyzing and controlling these subsystems: The first approach is referred to as a decomposition method based on relatively weak couplings among the subsystems, while the second proposes the idea of model aggregation into an aggregate model essential for analyzing and controlling subsystem interactions. Large scale power systems consisting of weakly interconnected regions in general possess time-scale separation property, due to the sparse interconnections among regions. Disturbances in one region are typically sensed by other regions in a slower fashion than they are by the region where they occur. This inherent time-scale separation property is particularly important for coordination of regions. It is also useful for controlling each particular region in a decentralized manner, while accounting for the effects of neighboring regions [2].

To relate these general approaches to their applications to the power system studies, we point out that the ideas of decomposition date back to the early work of Kron on diakoptics, and that they have been incorporated into many numerical methods, particularly for computing steady state solutions.

Possibly the most ingenious applied concept of system decomposition for automated small signal stabilization of very large scale systems in existence is the example of the Automatic Generation Control (AGC) [3], [4].

Here we explain a qualitatively different theoretical problem which arose in context of mid-term voltage regulation, and give its full formulation. This qualitative difference can be summarized by stating that since the natural dynamics are assumed instantaneous, the system matrix is zero, and the entire process can be viewed as a discrete quasi-static process which is entirely control driven. Consequences of this are very serious in terms of minimum complexity of information flow for decentralized controls relative to the full problem discussed in [1], [2]. For example, while it is well known that it is necessary to know the eigenstructure of the system matrix in order to define an aggregate model essential for coordination, this model can be arrived at in a fairly simple way for the class of problems studied here. Most importantly, a proposed aggregate model has an interesting interpretation in terms of actual physical variables, which is absolutely required for the aggregate model to be useful.

One last general observation prior to reporting specific theoretical results is of high importance. While many standard techniques based on system decomposition assume weak (or sparse) coupling among the regions, the general modeling approach adopted in this paper does not depend critically on this assumption. The rationale for this is described in the paper.

2.1 Modeling assumptions

As we are interested only in the mid-term behavior of the power system, the transient response of generators and their primary controls are assumed stable and instantaneous. Under this assumption, only steady state load flow equations are considered. Loads are modeled as constant power devices. The real and reactive power balances take in general the form

$$P_L = \mathcal{M}(V_L,\, V_G,\, \theta,\, X,\, R)$$

$$Q_L = \mathcal{N}(V_L,\, V_G,\, \theta,\, X,\, R)$$

where

$P_L = [P_{L1} \;\cdots\; P_{Ln}]^T$: real power
$Q_L = [Q_{L1} \;\cdots\; Q_{Ln}]^T$: reactive power
$V_L = [V_{L1} \;\cdots\; V_{Ln}]^T$: load voltage
$V_G = [V_{G1} \;\cdots\; V_{Gg}]^T$: generator voltage
$X,\, R =$ line parameters
$\theta =$ voltage angle differences

The number of loads and generators is n and g. For a typical power network, it is generally true that $n > g$, which we assume throughout this paper. It is shown here that this structural property gives rise to an important feature of large scale power systems, i.e. to the separation of variables into the ones which only change when interactions among the subsystems change, and the others.

Just as an illustration of the form of the nonlinear functions $\mathcal{M}$ and $\mathcal{N}$, let us take a simple example shown in Figure 1

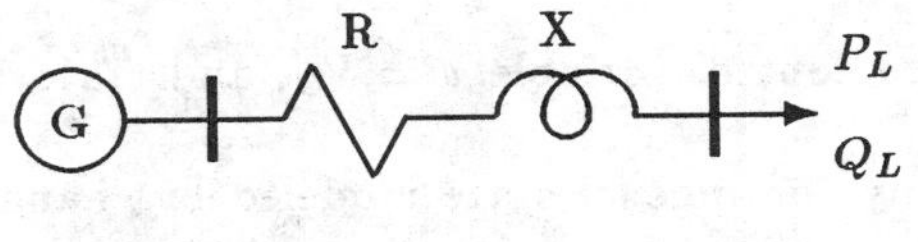

Figure 1: A simple example

The load flow equations are

$$P_L = G(V_L V_G \cos\theta - V_L^2) + B V_L V_G \sin\theta$$

$$Q_L = B(V_L V_G \cos\theta - V_L^2) - G V_L V_G \sin\theta$$

where $B = X/(X^2 + R^2)$, $G = R/(X^2 + R^2)$, and $\theta = \theta_G - \theta_L$.

As a preliminary practice, we first assume that angle differences are constant. Under this assumption, power balances are decoupled into a reactive power/voltage problem, neglecting the coupling with the real power. The reactive power balance can now be written as

$$Q_L = \mathcal{N}(V_L,\, V_G,\, X,\, R) \tag{1}$$

This is the model we focus on. It should be noticed that this is just the basic load flow equation for reactive power. There are no dynamics involved.

The control problem under consideration in this paper is not a conventional one. The basic posing of the problem could be quite confusing, since, in effect, no natural dynamics are modeled, i.e. they are assumed instantaneous. We are actually defining a sequence of steady state load voltages, which are changing in response to changes of steady state set points of generators. A typical scenario is to have the system operate at nominal, steady state conditions defined by equation (1). Changes are initiated by a load deviation Q_L from the nominal, or by topological changes of X and R. In order for the equation (1) to be satisfied for these new parameters, one needs to compute changes in set points of generator voltages ΔV_G which would bring the load voltage deviations ΔV_L back to zero. In this context, load voltages can be thought of state variables, which are regulated by changes in system controls ΔV_G. It is worth noticing that one should view the model as a discrete process in which time does not play any explicit role.

2.2 Discrete process of interest

Based on the above, generator voltages are recognized as feedback (control) signals occurring in response to load voltage deviations at selected nodes (pilot nodes) from their set values. Load voltages are denoted as variables x. For notational simplicity we use a parameter variable p to represent all possible contingencies, including changes of reactive power loading, line parameters, and system topologies.

Using this notation, the reactive power balance equation (1) can be rewritten as

$$f(x, v, p) = 0 \tag{2}$$

where control variable $v = V_G$, and state variable $x = V_L$.

Any contingencies are modeled by changing parameter variable p. Let us use p_k to denote one value of system parameters. The quasi-steady state reactive power balance (2) can be restated as

$$f(x_k, v_k, p_k) = 0 \quad \forall k \tag{3}$$

Note that the equation has to be true for all k, since power balance has to be maintained all the time (steady state assumption).

Let us investigate the linearized process. Then, linearizing (3) for $(k+1)$ around the same equation for event of k yields

$$f_x^k(x_{k+1} - x_k) + f_v^k(v_{k+1} - v_k) + d_k = 0$$

where f_x^k and f_v^k are the partial derivatives of f with respect to the indicated variables evaluated at event k. The term d_k reflects the effect of disturbances. If we consider the post-contingency process, then the term d_k is zero, because the post-contingency system does not experience any perturbation for all time later than the initial moment, until the next contingency comes along. This would always be true if the frequency of contingencies is much smaller than that of the control actions. In other words, the effect of d_k is to bring out nonzero initial conditions for the post-contingency system. With this understanding, the post-contingency process can be written as

$$x_{k+1} = x_k + C_v u_k \tag{4}$$

where $C_v = -(f_x^k)^{-1} f_v^k$ is the sensitivity matrix, and $u_k = v_{k+1} - v_k$ is the control signal to be designed. Note that this system (4) has nonzero initial conditions, resulted from the contingency.

Equation (4) defines load voltage changes due to changes of generator voltages. The equivalent continuous representation of (4) is

$$\dot{x} = C_v u \tag{5}$$

where $u = \dot{v}$ (To be more precise, $\dot{x} = \frac{dx}{d\tau}$, where τ is the duration between two control actions).

It should be noted that the sensitivity matrix C_v is dependent on operating conditions, and therefore is not constant. Since we only consider small perturbations in this paper, C_v is assumed constant.

We emphasize that the linearized models (4) and (5) are control-driven models in the sense that if no control action is taken, the variables x do not change. Clearly control actions are needed to bring the system (4) or (5), which has nonzero initial conditions due to perturbations, back to nominal operation.

A particularly important feature for the above models is that they are not fully controllable. The controllability matrix, for (5), for instance, is simply

$$[B \; AB \; \cdots \; A^{n-1}B] = [C_v \; 0 \cdots 0]$$

which does not have a full row rank, since the sensitivity matrix C_v $(n \times g)$ is not of full row rank for $n > g$.

Because the system is not fully controllable, and because it is not desirable to measure all voltages, a measurement (output) feedback to the control signal u_k or u is introduced instead of a feedback to all variables x. Since (4) and (5) are equivalent, the more convenient continuous model (5) is used for our further discussions. All results obtained for (5) hold for the discrete model (4).

Suppose the measurements (one possible particular case, pilot node load voltages) are

$$m = Cx \tag{6}$$

Assume the feedback signal is proportional to the deviation of measurements from their set values

$$u = K(m - m^s) = KC(x - x^s) \tag{7}$$

where m^s and x^s denote their set values.

With this output feedback control, the closed-loop process describing load voltage changes can be written as

$$\dot{x} = A(x - x^s) \tag{8}$$

where $A = C_v K C$. To simplify notations, equation (8) is still written as

$$\dot{x} = Ax \tag{9}$$

with the understanding that x is in fact the deviation from the set value. Models (4), (5) and (9) have been used by the French researchers in [5]. The follow-up material in this paper is new.

For the objective of coordination, one first needs to quantify the interactions among the regions. For this purpose one must ask the question: what does best represent the interactions among regions? In the next section, such variables are defined, termed coordination variables, which represent the effect of interactions among the regions. The proposed aggregate model for coordination is explicitly expressed in terms of these variables.

3 Separation of interactions

Without loss of generality, let us consider a two-region network shown in Figure 2.

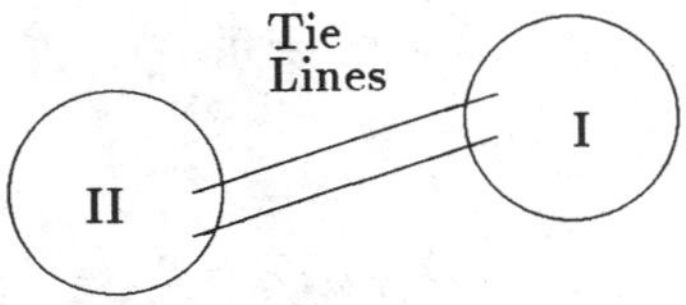

Figure 2: A two-region system

The sensitivity matrix for the interconnected system can be written as

$$C_v^* = C_v + \epsilon C_v^1 \tag{10}$$

where $0 \le \epsilon \le 1$. When $\epsilon = 0$, regions are assumed to be disconnected. Superscript * on any variable stands for the fully interconnected system. Variables without superscript are associated with the disconnected system. With the expression (10) for C_v^*, the system matrix of the closed-loop process in equation (9) becomes $A+\epsilon A_1$, where $A = C_v K C$ and $A_1 = C_v^1 K C$.

The equation defining the closed-loop process (9) now becomes

$$\dot{x} = (A + \epsilon A_1)x \tag{11}$$

As an illustrative example, let us consider a 9-bus example divided into two regions as shown in Figure 3.

For this system, the states are $x = [V_1 \cdots V_6]^T$, and the measurements (the pilot node voltages) are $m = [V_1 \cdots V_3]^T$. The sensitivity matrix when the regional interconnections are neglected can be found as

$$C_v = \begin{bmatrix} 1 & 0 & 0 \\ 0 & .6 & .4 \\ 0 & .4 & .6 \\ 0 & .6 & .4 \\ 0 & .4 & .6 \\ 0 & .4 & .6 \end{bmatrix}$$

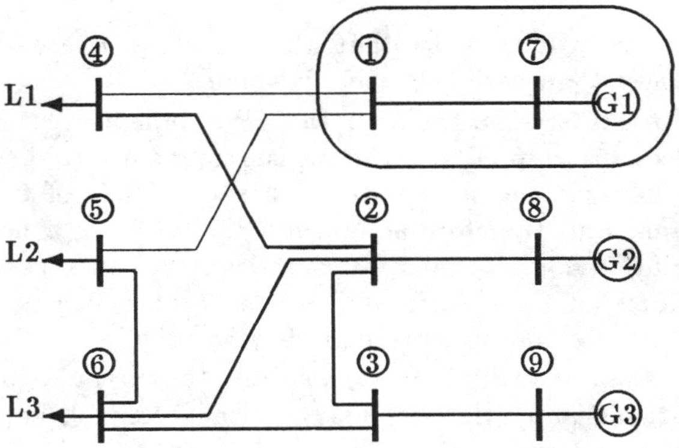

Figure 3: A 9-bus example

and the effect of the neighboring regions are seen through the sensitivity matrix

$$C_v^1 = \begin{bmatrix} -1.2 & .6 & .6 \\ .5 & -.3 & -.3 \\ .5 & -.2 & -.2 \\ .7 & -.4 & -.4 \\ 1.3 & -.6 & -.7 \\ .8 & -.4 & -.4 \end{bmatrix}$$

To extract only variables whose variations represent the effects of interconnections, let us define coordination variables y, as

$$\dot{y} = 0 \;\; at \;\; \epsilon = 0 \tag{12}$$

In other words, the coordination variables y do not change when interconnections are removed. Therefore for the interconnected system, any change in y is entirely due to the interactions among regions.

It should be noticed from the definition that the coordination variables are not unique. In fact, any linear combination of the coordination variables is still a coordination variable.

From this definition we can easily prove that the coordination variables can be expressed as a linear combination of the state variables as

$$y = Px \tag{13}$$

where P satisfies

$$PA = 0 \tag{14}$$

because

$$\dot{y} = P\dot{x} = PA\,x = 0 \;\; at \;\; \epsilon = 0$$

Since A is a rank-deficient matrix, equation (14) has a nonzero solution for P. For the example in Figure 3, we can find

$$P = \begin{bmatrix} 0 & 1 & 0 & -1 & 0 & 0 \\ 0 & 0 & 0 & 0 & 1 & -1 \\ 0 & -.4 & -.6 & 0 & 1 & 0 \end{bmatrix} \tag{15}$$

The corresponding coordination variables are

$$y = \begin{bmatrix} y_1 \\ y_2 \\ y_3 \end{bmatrix} = \begin{bmatrix} x_2 - x_4 \\ x_5 - x_6 \\ x_5 - .4x_2 - .6x_3 \end{bmatrix}$$

Simulations show that the variables y are indeed slower than variables x if ϵ is small.

An interesting point is that P is independent of the feedback gain K and the measurement matrix C. This is because the sensitivity matrix C_v is not of full row rank, and therefore equation $PC_v = 0$ has a nonzero solution for P, which automatically satisfies (14). Of course, if the matrix KC is also rank-deficient, there could be more independent slow variables.

The fast variables z are basically those regulated measurements by the secondary voltage control, e.q. the pilot node voltages, which can be written as

$$z = Rx \tag{16}$$

where R consists of independent rows in A.

In a more compact form let us write

$$\begin{bmatrix} y \\ z \end{bmatrix} = Tx \tag{17}$$

where the transformation matrix

$$T = \begin{bmatrix} P \\ R \end{bmatrix} \tag{18}$$

Under this transformation, we derive the model which also has a standard time-scale separation model interpretation

$$\dot{y} = \epsilon(B_{11}y + B_{12}z) \tag{19}$$

$$\dot{z} = \Lambda z + \epsilon(B_{21}y + B_{22}z) \tag{20}$$

For the 9-bus example, the above reads

$$\dot{y} = \epsilon \begin{bmatrix} -.2 & .2 & -.0 \\ .5 & -.1 & -.4 \\ .3 & -.0 & -.3 \end{bmatrix} z$$

$$\dot{z} = \begin{bmatrix} 1 & 0 & 0 \\ 0 & .6 & .4 \\ 0 & .4 & .6 \end{bmatrix} K \left(I + \epsilon \begin{bmatrix} -1.2 & .8 & .4 \\ .5 & -.4 & -.1 \\ .5 & -.3 & -.2 \end{bmatrix} \right) z$$

Model (19)-(20) can be used as a starting model for performing both tertiary coordination, and improved secondary voltage regulation, which takes into account effects of a neighboring region.

It is proposed here that model (19) forms the basis for the slower tertiary coordination, while (20) should be used for the improved secondary voltage control, at the regional level. In order to show the required information flow, and interactions among the regions, it is necessary to:

a) interpret model (19) in terms of physically meaningful, i.e. measurable, variables y, fundamental for coordination strategies;

b) interpret model (20) in terms of variables measurable on the regional level.

3.1 Analysis of the model to be used for slower tertiary coordination

In order to explain this in a systematic way it is most important to recognize that when regions are disconnected, i.e. $\epsilon = 0$ in (12) the variables y remain constant. They, therefore, only reflect changes due to interactions. The real question is what could be a way to define these variables, so they preserve physical meaning on the system. This is a very hard question, to which, we believe, one possible approach is as follows:

Consider a single region with sensitivity matrix

$$C_v = -D_{LL}^{-1} D_{LG} \tag{21}$$

where D_{LL} and D_{LG} are submatrices given by the following linearized relationship:

$$\begin{bmatrix} \Delta Q_L \\ \Delta Q_G \end{bmatrix} = \begin{bmatrix} D_{LL} & D_{LG} \\ D_{GL} & D_{GG} \end{bmatrix} \begin{bmatrix} \Delta V_L \\ \Delta V_G \end{bmatrix} \tag{22}$$

in which Δ represents a small change in the corresponding quantities.

The condition for slow variables:

$$PC_v = 0 \tag{23}$$

can be restated as:

$$PD_{LL}^{-1} D_{LG} = 0 \tag{24}$$

Next, define the matrix

$$S = PD_{LL}^{-1} \tag{25}$$

which obtains from (25)

$$SD_{LL} = P \tag{26}$$

or

$$[P\ 0] = S[D_{LL}\ D_{LG}] \tag{27}$$

Recall the relationship

$$\dot{Q}_L = [D_{LL}\ D_{LG}] \begin{bmatrix} \dot{V}_L \\ \dot{V}_G \end{bmatrix} \tag{28}$$

where Q_L is the vector of all load reactive demands, V_L, V_g are vectors representing load and generator voltages. Now, by pre-multiplying this relationship (28) by the matrix S, one obtains:

$$S\dot{Q}_L = S[D_{LL}\ D_{LG}] \begin{bmatrix} \dot{V}_L \\ \dot{V}_G \end{bmatrix} \tag{29}$$

$$= [P\ 0] \begin{bmatrix} \dot{V}_L \\ \dot{V}_G \end{bmatrix} \tag{30}$$

$$= P\dot{V}_L = P\dot{x} = \dot{y} \tag{31}$$

It follows from (31) that

Result 1:

$$\dot{y} = S\dot{Q}_L \tag{32}$$

i.e. the slow variable changes are just linear combinations of load demand changes, which are zero under the constant power load modeling assumption, and, therefore

$$\dot{y} = 0 \tag{33}$$

Result 2: Each load bus which is not connected to a generator yields a slow variable. Sufficient condition defining matrix S is

$$SD_{LG} = 0 \tag{34}$$

which cancels the effect of controls on load voltage changes.

To find slow variables y one does not need C_v matrix, all that is needed is a linear combination S of load demand changes satisfying (34).

Since D_{LG} has maximum rank g, it follows that any linear combination S such that (34) holds results in $(n-g)$ slow variables.

3.2 The dynamic equality of slow variables and tie line flows on an interconnected system

Let us consider the simpler case where generators are not boundary nodes between two areas (this restriction is eliminated in Section 4 using a fully decentralized formulation of the aggregate model).

Two types of "slow" variables exist on such a system: the variables y defined via (12), and "slow" variables y^* evolving on the fully connected system. Variables y^* satisfy a linear combination of x in the form of $y^* = P^* x$, where P^* is defined using (23) as $P^* C_v = 0$. Under the constant load power modeling assumption made in this paper, $\dot{y}^* = 0$, which directly follows from (33) when the system is treated as a single region.

Further, notice that on the system where generators are not boundary nodes

$$D_{LG} = D_{LG}^* \tag{35}$$

where superscript $*$ on any variable stands for the fully interconnected system. Variables without superscripts are associated with the disconnected system. Basically, load-to-generator connections are not affected by disconnecting tie-line flows. Therefore the matrix which defines slow variables is in this case the same for the disconnected system, as for the interconnected system. Using the same notation as above

$$S = S^* \tag{36}$$

The linear combination operation is the same for both types of slow variables, i.e.

$$SD_{LG} = 0 \tag{37}$$

and

$$S^* D_{LG}^* = 0 \tag{38}$$

Therefore

$$\dot{y} = S\dot{Q}_L \tag{39}$$

and

$$\dot{y}^* = S^* \dot{Q}_L^* \tag{40}$$

Note that Q_L^* corresponds to load demands on the interconnected network, and Q_L to the disconnected subsystems.

It is straightforward to show that

$$Q_L^* = Q_L + F \tag{41}$$

where F is the vector whose each element represents net tie line flow into a particular load bus, i.e. (see Fig. 4)

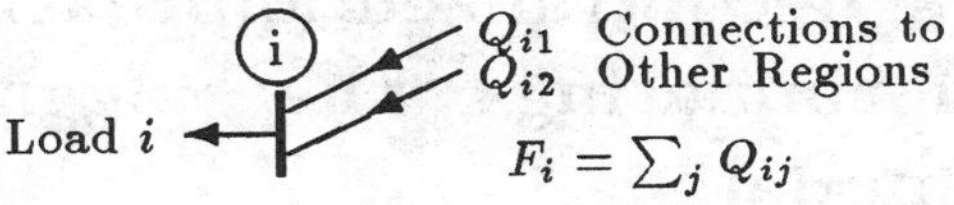

Figure 4: Net tie line flow into bus i: $F_i = Q_{i1} + \cdots$

$$F = \begin{bmatrix} F_1 \\ \cdot \\ \cdot \\ \cdot \\ F_n \end{bmatrix} \tag{42}$$

If i has only one interconnection to the other areas, then F_i = tie line flow. If bus i has more than one interconnection to the other areas, then F_i corresponds to the sum of all tie line flows into this bus. In other words, only the net flows into each bus can be distinguished. If bus i is not connected to other regions, obviously $F_i = 0$.

Now following an analogous process as in the case of a single area, multiply (41) by S (or S^*), which obtains

$$y^* = y + SF \tag{43}$$

or

$$y = y^* - SF \tag{44}$$

Since we are looking for a dynamic relationship (reflecting changes in variables), by differentiating (44) one obtains

$$\dot{y} = \dot{y}^* - S\dot{F} \tag{45}$$

However, $\dot{y}^* \equiv 0$ by definition, since these are slow variables on the interconnected system, resulting in

$$\dot{y} = -S\dot{F} \tag{46}$$

Relationship (46) is the main result stating that slow variable changes $\dot{y}$ defined on the disconnected system via (12) are linear combinations of the tie line flow changes. The interpretation for S is given by (32).

Note that result (46) should not be surprising at all, since this is just another form of the reactive power balance equation, which can be obtained by rewriting (46) as

$$\dot{y} + S\dot{F} = 0 \tag{47}$$

$$S\dot{Q}_L + S\dot{F} = 0 \tag{48}$$

or

$$S(\dot{Q}_L + \dot{F}) = 0 \tag{49}$$

or

$$S\dot{Q}_L^* = 0 \tag{50}$$

which is apparently true under the constant load power assumption $\dot{Q}_L^* = 0$.

This also helps us identify those slow variables which still do not move when regions are connected.

4 A decentralized approach to formulating the aggregate model

After deriving a relationship (46) between coordination variables and tie line flows, it is possible to have a fully decentralized interpretation of the results in the previous sections. This can be achieved by generalizing definition (12) of slow variables y on the disconnected system by relaxing the condition of disconnecting the system (i.e. $\epsilon = 0$) into the condition that variables at subsystem levels, $y = Px$, do not change when $\dot{F}_L = 0$, i.e. when the tie line flows into load buses are kept constant.

With this definition in mind, it is straightforward to show that the mathematical formulation of the model (11) can be re-written as

$$\dot{x} = C_v u - D_{LL}^{-1}\dot{F}_L \tag{51}$$

In this decentralized formulation flows F_L can be viewed as independent inputs as seen from each subsystem level.

Coordination variables y introduced in Section 4 are easily shown to be

$$\dot{y} = -S\dot{F}_L \tag{52}$$

where S is the same transformation as in (25).

5 Conclusions

The basic concept which has evolved throughout this paper is that the model to be used for coordination should be expressed in terms of variables which only depend on the degree of interactions between the subsystems.

To put this approach in perspective with the time scale separation based thinking, a first formulation was presented in Section 3. The coordinating variables were defined in such a way that when the subsystems are not connected, these variables do not change.

Next, a generalization based entirely on a structural decomposition was presented in Section 4. In this approach, the coordinating variables were defined such as if the reactive flows between the subsystems do not change, these variables remain constant.

The aggregate model (19)-(20) derived in this paper should play a fundamental role in coordinating set values of output variables at the subsystem levels. Its use for coordination design will be reported in a separate publication.

6 Acknowledgments

We greatly acknowledge the financial support provided by the Electricité De France in performing this work.

References

[1] N. Sandell Jr., P. Varaiya, and M. Athans, "A Survey of Decentralized Control Methods for Large Scale Systems", *Proc. of the Power Systems Conference*, Henniker, NH, pp. 334-352, 1975.

[2] J. Chow, *Time-Scale Modeling of Dynamic Networks with Applications to Power Systems*, Springer-Verlag, 1982.

[3] D. Siljak, *Large-Scale Dynamic Systems*, Elsevier North-Holland, New York, 1978.

[4] M. Calovic, "Linear Regulator Design for a Load and Frequency Control", *IEEE Transactions, PAS-91*, pp. 2271-2285, 1972.

[5] J. Paul and J. Leost, "Improvements of the Secondary Voltage Control in France", *IFAC Symposium on Power Systems and Power Plants Control*, Beijing, 1986.

Proceedings of the 31st Conference
on Decision and Control
Tucson, Arizona • December 1992

SOME ASPECTS OF THE ENERGY FUNCTION APPROACH TO ANGLE AND VOLTAGE STABILITY ANALYSIS IN POWER SYSTEMS

T. J. Overbye, M. A. Pai and P. W. Sauer

University of Illinois
Dept. of Electrical and Computer Engineering
Urbana, IL 61801

Abstract

Direct methods using Lyapunov/Energy functions for angle stability analysis have matured over the past two decades. For voltage stability the application of energy functions is somewhat recent. In this paper we discuss both angle and voltage stability in the context of the energy function approach. It is also shown that complete decoupling of the two phenomena is not possible in general, particularly in stressed systems. We demonstrate that the u.e.p concept is a valid one for both types of stability analysis.

1 Introduction

Energy function methods for transient stability analysis are now well documented in the literature [1,2] though interest is gaining in extending the technique to structure-preserving models [3-6]. The use of energy function methods for voltage stability is somewhat recent [7-9]. In this paper we seek to reconcile the two approaches and raise issues for further work towards developing a unified framework.

2 Mathematical Model

The dynamics of a multimachine power system are characterized by a set of differential-algebraic equations (DAE's) of the form

$$\dot{x} = f(x,z,v) \quad (1)$$
$$0 = g_1(x,z,v) \quad (2)$$
$$0 = g_2(z,v) \quad (3)$$

(1), (2) and (3) represent respectively the dynamic equations of the generating unit, the stator algebraic equations and the network power balance algebraic equations. The underlying dynamics in (2) and (3) are the 60Hz transients in the generator windings, transmission lines and the loads which are assumed infinitely fast. The use of singular perturbations [10] allows us to represent them as algebraic equations. While the DAE model is faster to simulate, analytically it poses a difficult problem in terms of constructing an energy function. For the special case where the loads are constant impedance type it is possible to solve for z and v in terms of x and get a differential equation model only [13]. In Refs. [3] and [4] energy functions are derived for the classical model assuming (2) and (3) to be differential equations which have fast dynamics compared to (1). The vectors x, z, v are defined as follows

x = state variables of the generating units.

z = stator currents I_{di}, I_{qi} $(i = 1, ..., m)$ where m = number of machines. The currents are in the machine reference frame.

v = bus (internal and external) network voltages and angles $|V_i|$, θ_i $i = 1,2,...,m+n$.

3 Energy Function For Transient Stability

For a general DAE model of the type (1)-(3) with 2 axis model along with exciter an energy function has been proposed in Ref. [6]. But it will be path dependent. For the special case where the machine is represented as a classical model with real power loads of constant power type and reactive loads which are voltage dependent, a path independent energy function is possible using the first integral. For a m machine n bus the system equations are [11,12].

$$\dot{\theta}_i = (\omega_i - \omega_s) \quad i = 1,2,...,m \quad (4)$$

$$M_i\dot{\omega}_i = P_{Mi} - \sum_{j=1}^{m+n} B_{ij}V_iV_j \sin(\theta_i - \theta_j)$$
$$i = 1,2,...,m \quad (5)$$

$$P_{load_i} = -\sum_{j=1}^{m+n} B_{ij}V_iV_j \sin(\theta_i - \theta_j)$$
$$i = m+1,...,m+n \quad (6)$$

$$Q_{load_i} = \sum_{j=1}^{m+n} B_{ij}V_iV_j \cos(\theta_i - \theta_j)$$
$$i = m+1,...,m+n \quad (7)$$

where B_{ij} are elements of the augmented Y_{bus} matrix [1].

In this formulation, the generator internal buses numbered $1 - m$ have constant voltage magnitude.

CH3229-2/92/0000-2941$1.00 © 1992 IEEE

State Equations in COI Reference Frame

We adopt the COI notation because it enables easier analytic calculations. The COI of a m-machine system is defined as

$$\theta_o = \frac{1}{M_T}\sum_{i=1}^{m} M_i\theta_i \text{ where } M_T = \sum_{i=1}^{m} M_i \quad (8)$$

It follows that

$$\omega_o = \frac{1}{M_T}\sum_{i=1}^{m} M_i\omega_i \,,\ \dot{\omega}_o = \frac{1}{M_T}\sum_{i=1}^{m} M_i\dot{\omega}_i \quad (9)$$

The machine rotor angles and bus voltage phase angles referred to the COI are

$$\tilde{\theta}_i \triangleq \theta_i - \theta_o \quad i = 1, ..., m+n \quad (10)$$

$$\tilde{\omega}_i \triangleq \omega_i - \omega_o \quad i = 1, ..., m \quad (11)$$

$$M_T\dot{\omega}_o = \sum_{i=1}^{m} P_{Mi} - \sum_{i=m+1}^{m+n} P_{load_i} \triangleq P_{COI} \quad (12)$$

We can now rewrite (after some algebra) the swing equations (4) and (5) in the COI reference frame as

$$\dot{\tilde{\theta}}_i = \tilde{\omega}_i \quad (13)$$

$$M_i\dot{\tilde{\omega}}_i = P_{Mi} - \sum_{j=1}^{m+n} B_{ij}V_iV_j\sin(\tilde{\theta}_i - \tilde{\theta}_j) - \frac{M_i}{M_T}P_{COI} \quad (14)$$

The network equations (6) and (7) remain the same except that θ_i and θ_j get replaced by $\tilde{\theta}_i$ and $\tilde{\theta}_j$. The energy function turns out to be [1,11]

$$V = \frac{1}{2}M_i\tilde{\omega}_i^2 + V_{p1}(\tilde{\theta}, V) + V_{p2}(\tilde{\theta}, V) \quad (15)$$

where

$$V_{p1}(\tilde{\theta}, V) = -\sum_{i=1}^{m} P_{Mi}(\tilde{\theta}_i - \tilde{\theta}_i^s) + \sum_{i=m+1}^{m+n}\int_{V_i^s}^{V_i}\frac{Q_i(V_i)}{V_i}dV_i$$

$$-\frac{1}{2}\sum_{i=1}^{m+n}\sum_{j=1}^{m+n} B_{ij}[V_iV_j\cos\tilde{\theta}_{ij} - V_i^sV_j^s\cos\tilde{\theta}_{ij}^s]$$

$$V_{p2}(\tilde{\theta}) = \sum_{i=m+1}^{m+n} P_i(\tilde{\theta}_i - \tilde{\theta}_i^s) \quad (16)$$

It is possible to express $V_{p1} + V_{p2}$ in a compact integral form as [12]

$$V_{PE} = \int_{\underline{W}^s}^{\underline{W}} < \underline{h}(\underline{\lambda}), d\underline{\lambda} > \quad (17)$$

where $\underline{W}^s$ denotes a stable e.p. and $\underline{h}$, $\underline{\lambda}$, $\underline{W}$ are appropriately defined vectors from (4)-(7). Thus

$$V = V_{K.E} + V_{P.E} \quad (18)$$

In transient stability studies we have the DAE model for the faulted and the post-fault state. The region of stability is defined with respect to the stable equilibrium point (SEP) of the post-fault state. It is defined as $V < V_{cr}$ where V_{cr} is the value of V at the controlling UEP. The faulted DAE model is integrated to find either (i) critical clearing time, i.e., when $V = V_{cr}$ or (ii) for a given clearing time t_{cl}, examine if the system is stable by checking at the time instant t_{cl} if $V_{cl} < V_{cr}$. In either case for a contingency a single value of V_{cr} is computed. The potential energy boundary surface (PEBS) method is generally used for computing V_{cr} in stability analysis [6] because of the difficulty in identifying the correct controlling UEP. Through a simple example we show in Sec. 5 that the (UEP) concept is valid for both transient and voltage stability analysis.

4 Energy Function For Voltage Stability

The development of analytic tools to aid in the quantification of power system voltage stability has proven challenging, partly because of the many different time frames of the underlying system dynamics. These dynamics range from the relatively fast dynamics of, for example, the generator excitation system and various load dynamics, to the midrange dynamics of LTC transformers and generator reactive power limiters, to the longer term dynamics associated with average load variation. Energy function methods have been applied to the longer term dynamic aspects of the problem, where they have proved useful in proving a measure of the maximum loadability of the system. In this context loss of system stability occurs as the result of a saddle-node bifurcation between the operable system solution and a type 1 UEP solution on the system stability boundary. As the system approaches this point the potential energy difference between this UEP and the SEP approaches zero. Voltage stability in a number of different regions of the system can be quantified by monitoring these energy differences for the various type 1 UEPs of the system. Thus unlike in transient stability we are not dealing with a single controlling UEP or one value of V_{cr} for a fixed potential energy well. Rather we are dealing with a difference in the potential energy between UEPs and the SEP (14) as the system parameters (e.g., average aggregate load at individual buses) slowly vary. This in turn results in a variation in the shape of the potential energy well, which can be represented pictorially in two dimensions as shown in Figure 1.

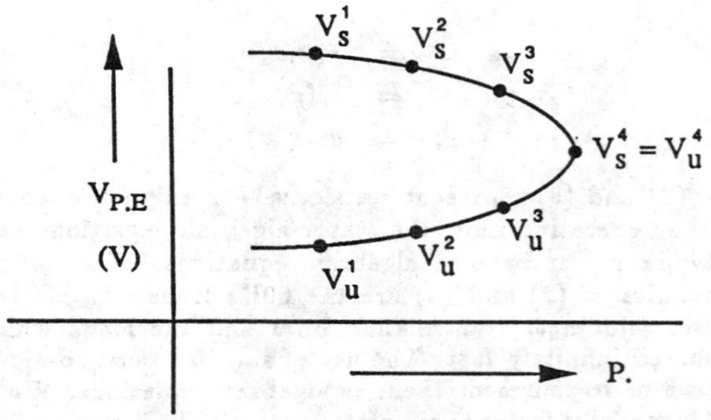

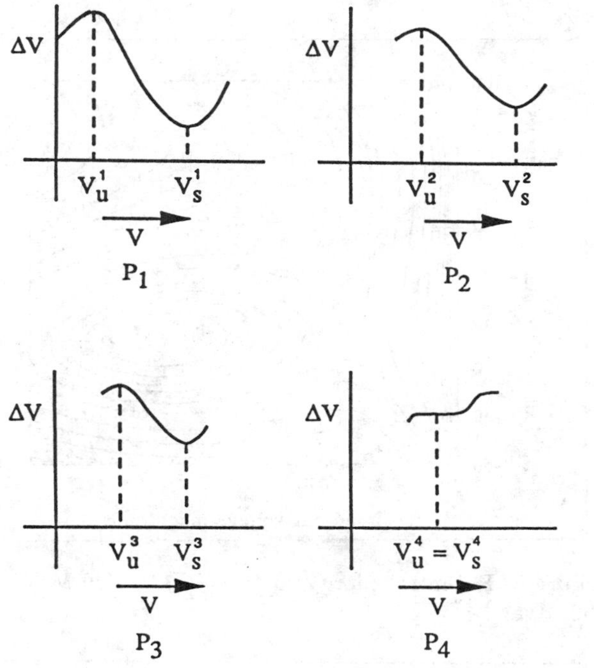

Figure 1:

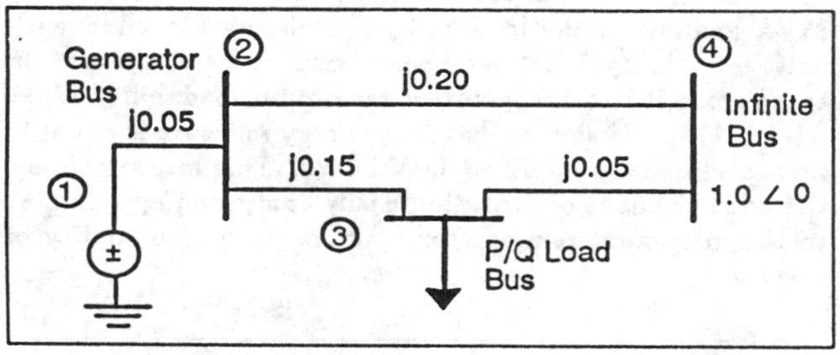

Figure 2: Three bus system diagram

The difference between the two energies can be calculated by the same potential energy function namely

$$\Delta V = \left[V_{p1}\left(\tilde{\theta}^u, V^u\right) + V_{p2}\left(\tilde{\theta}^u, V^u\right)\right] \quad (19)$$
$$- \left[V_{p1}\left(\tilde{\phi}^s, V^s\right) + V_{p2}\left(\tilde{\phi}^s, V^s\right)\right] \quad (20)$$

where the parameters pertain to post-fault state.

5 Unified Energy Function Framework

As was seen in the last section, two major similarities in the application of energy methods to the transient and voltage stability problems is that the same energy formulation can be used in both, and that both make use of the energy of the UEPs. In this section a new framework is explored to unify the two separate applications. In particular we focus on the nature of UEP's.

The development of this framework is aided by examining the application of energy methods on the small lossless system shown in Figure 2 (all per unit values are on 100 MVA base). The system consists of a generator, a load, and an infinite bus. The generator is modeled using the classical model of a constant internal voltage behind transient reactance. Let the generator internal voltage be designated as bus 1, its terminal as bus 2, the load as bus 3, and the infinite bus as bus 4.

Since there is an infinite bus there is no need to adopt the COI notation. The swing equations are of the form

$$\dot{\theta} = \omega - \omega_s$$
$$\dot{\omega} = \frac{1}{M}\left(P_M - D\omega - \sum_{j=1}^{4} B_{1j}V_1V_j\sin(\theta_1 - \theta_j)\right) \quad (21)$$

with P_M = 300 MW. The real/reactive power balance constraints at the generator terminal are

$$0 = -\sum_{j=1}^{4} B_{2j}V_2V_j\sin(\theta_2 - \theta_j) \quad (22)$$

$$0 = \sum_{j=1}^{4} B_{2j}V_2V_j\cos(\theta_2 - \theta_j) \quad (23)$$

The load model from [15], [16] is used to provide an approximation of the dynamics of a 3rd-order induction motor model with flux dynamics. The power balance constraints at the load bus are

$$P_L = -\sum_{j=1}^{4} B_{3j}V_3V_j\sin(\theta_3 - \theta_j)$$

$$Q_L + K\dot{V}_3 = \sum_{j=1}^{4} B_{3j}V_3V_j\cos(\theta_3 - \theta_j)$$

The use of this load model leads to the same energy function derived using singular perturbation [17]. The requirement on K is that it be strictly positive. A value of $K = 1$ was used for the simulations shown in this paper.

We consider four pre-fault loading conditions.

1. $P_L = 200$ MW $Q_L = 100$ MVar
2. $P_L = 400$ MW $Q_L = 200$ MVar
3. $P_L = 600$ MW $Q_L = 300$ MVar
4. $P_L = 800$ MW $Q_L = 400$ MVar

The assumption for each case is that the generator internal voltage is set to maintain a pre-fault terminal voltage magnitude of 1.0 per unit. Figures 3-6 show the potential energy surface for the various loadings in the $\theta_1 / \mid V_3 \mid$ plane. For case (1), shown in Figure 3, there are four distinct e.p's of which one is stable, two are type 1 and one is type 2. Let the SEP be designated as x^s, the type one UEP with a large generator angle as the transient stability UEP x^t, the type-one UEP with a low voltage magnitude as the voltage stability UEP x^v, and the type 2 UEP as x^2. As the load is increased in case (2), Figure 4 shows that x^t and x^2 are quite close to coalescing in a saddle-node bifurcation. For case (3) this bifurcation has occurred, leaving only x^s and x^v. Case (4) shows that these two points are coming together; further increase in loading would result in a final saddle-node bifurcation with subsequent loss of the system operating point. Figure 7 shows the smooth variation of the energy for each of the UEP solutions with respect to the MVA loading at the load bus. The energy associted with the voltage stability UEP has been shown to be a measure of system proximity to the point of maximum loadability [7],[8],[9]. Also in [18] it is shown that this energy measure is equal to the area enclosed by modified Q-V curve. This interpretation has the added benefit of providing easily computed approximations to the maximum real and/or reactive power loadability of the system.

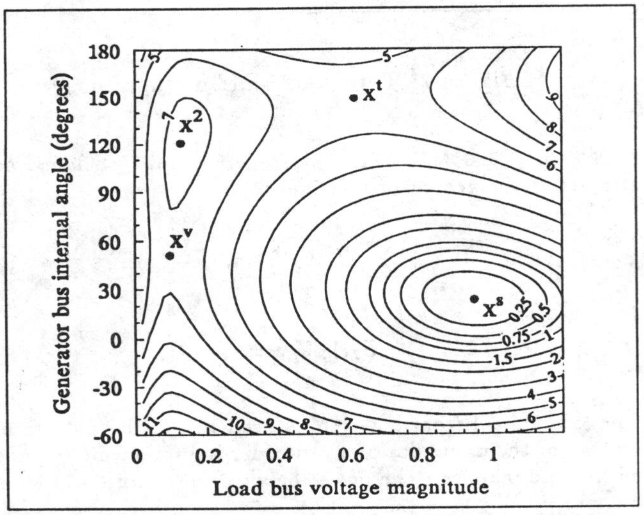

Figure 3: Potential energy surface - $P_L = 200$ MW, $Q_L = 100$ Mvar

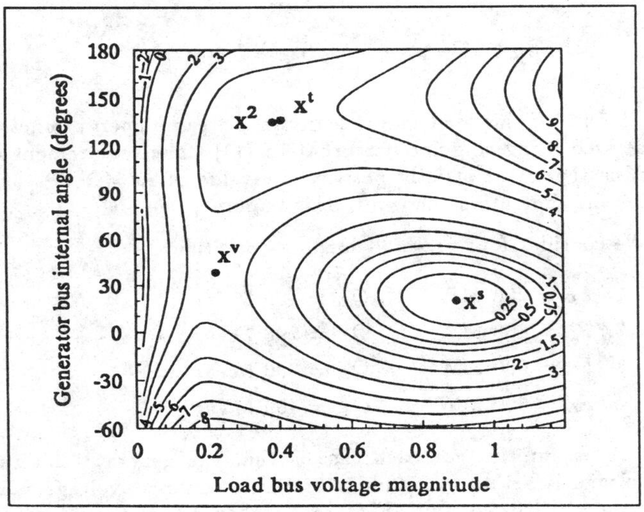

Figure 4: Potential energy surface - $P_L = 400$ MW, $Q_L = 200$ Mvar

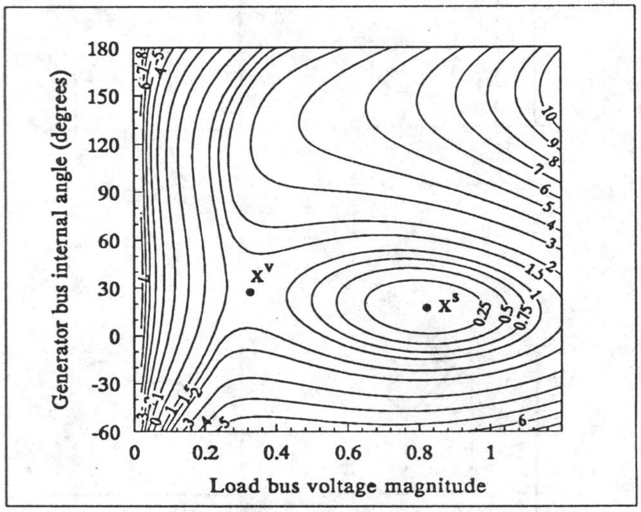

Figure 5: Potential energy surface - $P_L = 600$ MW, $Q_L = 300$ Mvar

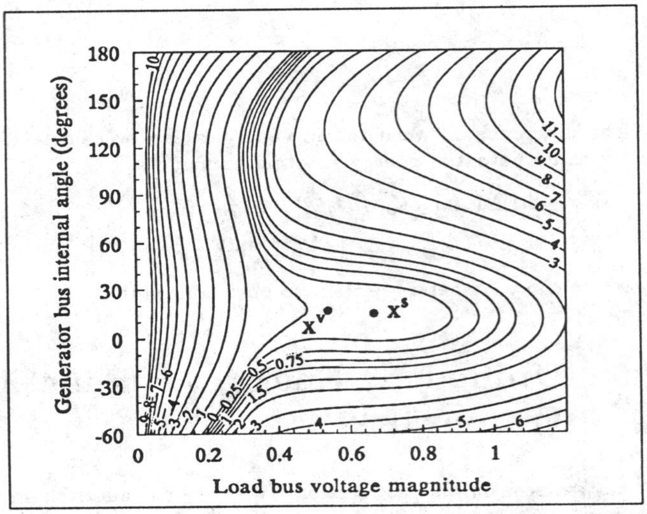

Figure 6: Potential energy surface - $P_L = 800$ MW, $Q_L = 400$ Mvar

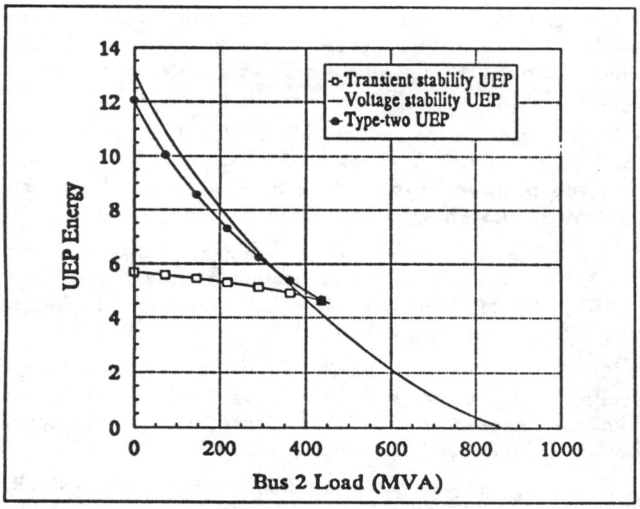

Figure 7: Variation in energy of UEPs

The potential for a unified energy function framework can be illustrated by applying a self-clearing fault at the terminal of the generator (thus the pre-fault and post-fault systems are identical). For the first scenario the system loading is 200 MW and 100 Mvar, with the fault being cleared after 0.235 seconds. Figure 8 shows the system trajectory superimposed on the potential energy surface from Figure 3 (note that the fault-on trajectory is shown only to illustrate the state space trajectory during the fault; the energy "well" is only valid for post-fault trajectory). For this marginally stable case the trajectories approaches the potential energy boundary surface (PEBS) in the vicinity of the transient stability UEP x^t with a high generator angle but with reasonably good load voltage magnitude. Thus the disturbance is clearly an angle instability phenomenon. However for the second scenario the system is assumed to be more stressed, with a loading of 600 MW and 300 Mvar; the fault is cleared after just 0.17 seconds. Figure 9 shows the trajectory superimposed on the potential energy surface for this loading from Figure 5. For this disturbance the system loses stability. However now the system crosses the PEBS in the vicinity of the voltage stability UEP x^v with a low voltage magnitude while the angle deviation is still relatively small. Note that the solution trajectory vanishes in the vicinity of the point ($\theta_1 = 84.8°$, $| V_3 | = 0.301$) when the power flow fails to converge. This is because the constant power load model is no longer valid at such a low voltage; a more suitable model should be used when modeling load response at low voltages. Thus this more heavily loaded system experiences voltage collapse before angle stability is lost. The key point is that the same energy measure (i.e., that associated with x^v) can be used both as the energy of the controlling UEP for the transient stability problem and as the energy measure for quasi-static voltage stability problem.

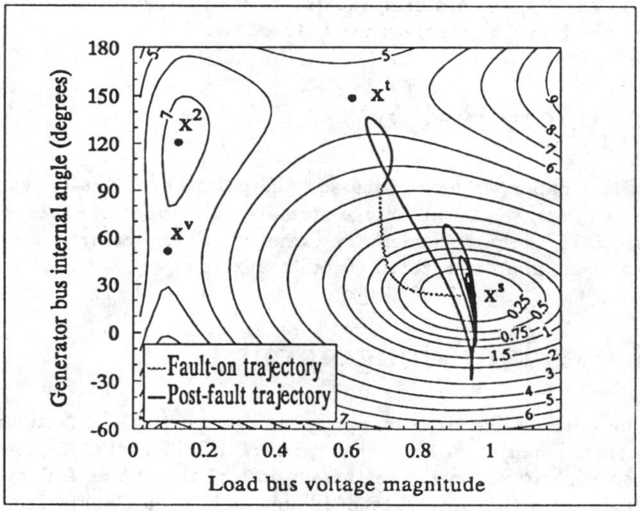

Figure 8: System trajectory for $P_L = 200$ MW, $Q_L = 100$ Mvar

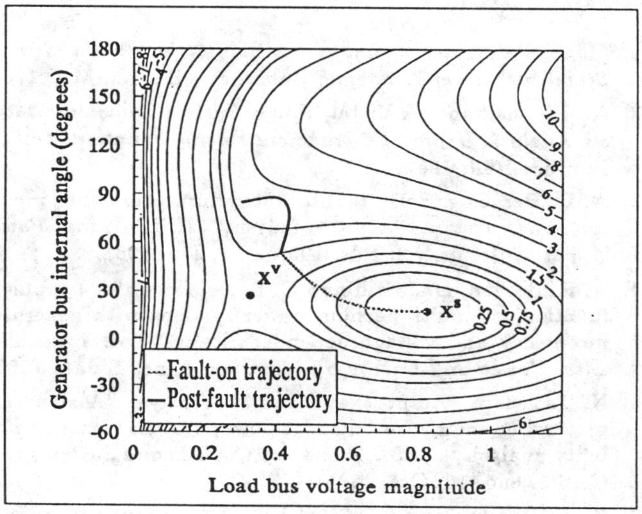

Figure 9: System trajectory for $P_L = 600$ MW, $Q_L = 300$ Mvar

The following interesting issues are posed:

1. The controlling UEP method is presently the most accurate technique for computing the stability region for the transient stability problem [14]. For multi-bus systems there can be a number of UEPs, including the low voltage power flow solutions used for voltage stability assessment. The above simple example has shown that for at least some system models the controlling UEP can be identical to one of the low voltage solutions. The interaction between these UEPs needs to be investigated for both more detailed system models and for larger systems.
2. An efficient technique to compute all the pertinent UEPs is needed. For transient stability one possibility is the gradient system proposed in [14]. For voltage stability a fast screening method has been proposed [8],[9]. An integrated approach needs to be investigated.

3. For stressed systems, the nature of the UEPs for transient and voltage stability needs to be clarified.

6 Conclusion

In this paper we have discussed the nature of UEPs relevant to both transient and voltage stability. A number of issues are raised through an elementary example. These research issues may lead to an integrated stability analysis.

7 Acknowledgement

The authors like to acknowledge the support of the National Science Foundation through it grant NSF ECS 87-19055. They also wish to acknowledge the support of the Power Affiliates Program of the University of Illinois at Urbana-Champaign.

References

[1] M. A. Pai, *Energy Function Analysis for Power System Stability*, Kluwer Academic Publishers, Boston, MA, 1989.

[2] A. A. Fouad and V. Vittal, *Power System Transient Stability Analysis Using The Transient Energy Function Method*, Prentice Hall, 1992.

[3] A. R. Bergen and D. J. Hill, "Structure preserving model for power system stability analysis," *IEEE Trans. Power Appar. Systems*, vol. PAS 100, no. 1, Jan. 1981.

[4] A. R. Bergen, D. J. Hill and C. L. DeMarco, "A Lyapunov function for multi-machine power systems with generator flux decay and voltage dependent loads," *Int. Journal of Elec. Power and Energy Systems*, vol. 8, no. 1, Jan. 1986.

[5] N. Tsolas, A. Arapostathis and P. Varaiya, "A structure preserving energy function for power system transient stability analysis," *IEEE Trans. on Circuits and Systems*, vol. CAS 32, no. 10, Oct. 1985.

[6] K. R. Padiyar and K. K. Ghosh, "Direct stability evaluation of power systems with detailed models and structure preserving energy functions," *Int. Journal of Elec. Power & Energy Systems*, vol. 11, no. 1, Jan. 1989.

[7] C. L. DeMarco and T. J. Overbye, "An energy based security measure for assessing vulnerability to voltage collapse," *IEEE Trans. on Power Systems*, vol. 5, no. 2, May 1990.

[8] T. J. Overbye and C. L. DeMarco, "Improved techniques for power system voltage stability assessment using energy methods," *IEEE Trans. on Power System*, vol. 6, no. 4, Nov. 1991.

[9] T. J. Overbye, "Use of energy methods for on-line assessment of power system voltage security," *IEEE PES 1992 Winter Power Meeting*, paper no. 92 WM 121-4 PWRS, January 1992.

[10] P. W. Sauer and M. A. Pai, "Modeling and simulation of power system dynamics," *Control and Dynamic Systems*, vol. 43, (Ed. C. T. Leondes), Academic Press, 1991.

[11] Th. Van Cutsem and M. Ribbens Pavella, "Structure preserving direct methods for transient stability analysis of power systems," Proc. 24th IEEE CDC Conference on Decision and Control, Ft. Lauderdale, FL, Dec. 1985.

[12] I. A. Hiskens and D. J. Hill, "Energy functions, transient stability and voltage behaviour in power systems with nonlinear loads," *IEEE Trans. on Power Systems*, vol. 4, pp. 1525-1533.

[13] P. W. Sauer, A. K. Behera, M. A. Pai, J. R. Winkelman and J. H. Chow, "A direct method for transient stability analysis of power systems with detailed models," *Electric Machines and Power Systems*, vol. 15, no. 1, 1988.

[14] H. D. Chiang, F. F. Wu and P. Varaiya, "A BCU method for direct analysis of power system transient stability," *IEEE/PES 1991 Summer Meeting*, paper no. 91 SM423-4 PWRS, San Diego, CA, July 28-August 1, 1991.

[15] K. Jimma, et al., "A study of dynamic load models for voltage collapse analysis," *Proceedings: Bulk Power System Voltage Phenomena II - Voltage Stability and Security*, pp. 423-429, McHenry, MD, Aug. 1991.

[16] K. T. Vu, "An analysis of mechanisms of voltage instability," *Proc. IEEE 1992 ISCAS*, pp. 2529-2532, San Diego, CA, May 1992.

[17] C. L. DeMarco and A. R. Bergen, "Application of singular perturbation techniques to power system transient analysis," *Proc. IEEE 1985 ISCAS*, pp. 597-601, Montreal, May 1985.

[18] T. J. Overbye, I. Dobson and C. L. DeMarco, "Q-V curve interpretations of energy measures for voltage security," submitted for presentation at the 1993 IEEE PES Winter Meeting, Columbus, OH, Feb. 1993.

Proceedings of the 31st Conference
on Decision and Control
Tucson, Arizona • December 1992

Modelling Hybrid Systems as Games

Anil Nerode and Alexander Yakhnis[1]
Mathematical Sciences Institute
Cornell University, Ithaca, New York 14853
anil@math.cornell.edu ayakh@msiadmin.cit.cornell.edu

Abstract

We propose a game framework for analyzing, extracting and verifying digital control programs for continuous plants by regarding such programs as finite state winning strategies in associated games. We call such interacting systems of digital control programs and continuous plants "hybrid systems" and model them as networks of interacting concurrent digital programs or automata, following [17] . This extends to hybrid systems the paradigm introduced by A. Nerode, A. Yakhnis, and V. Yakhnis [19] for analyzing concurrent digital programs meeting program specifications as winning finite state strategies in associated two person games. This formulation is intended to facilitate the transfer of recent tools from logic and concurrency and dynamical systems to extraction and verification of digital control programs for continuous systems.

INTRODUCTION

A typical example of a hybrid system is a closed loop system consisting of a continuous plant and a digital control program. The digital control program is event-driven by sensor data from the plant, which are the data supplied to the program. Occasionally, the digital control program outputs a new control law, which is then immediately imposed on the plant. The digital program and and the continuous plant should be modelled as a single hybrid system.

Digital technology allows us to program inexpensive digital control chips which change the plant control law used every few microseconds. This flexibility puts us beyond conventional control theory, linear or optimal or geometric or variational, so far as having tools for knowing what control programs can achieve. We must put as much emphasis on understanding how to exploit patterns of changes in continuous control laws to achieve our objectives as has been put on exploitation of single continuous control laws. We need to study problems of continuous plants and digital control programs at the same intellectual level of sophistication as has been applied in computer science for the analysis of purely digital programs and in control theory for the analysis of single continuous control laws for continuous plants.

Three Fundamental Problems

- Supply a clear mathematical and algorithmic model (semantics) for hybrid systems; that is, for networks of interacting continuous and digital devices connected by digital to analogue and analogue to digital transducers.
- Find general methods which, when presented with models of the continuous controllers and the continuous plants and the network configuration and the performance specification, extract digital control programs for the digital controllers.
- Either prove that the extraction process guarantees that the programs enforce the plant performance specifications, or give computer simulation procedures which are convincing that performance specifications are met.

At present, if we somehow manage to extract a digital control program from the performance requirements on a highly non-linear plant, we have little theory justifying its validity. What we have available is simulation. There we face directly the problem of computational infeasibility of simulation of large systems over all operating regimes.

The Digital Side We should take full advantage of the advances in programming languages of the last thirty years such as the denotational semantics of programs , the state semantics of Salwicki's Algorithmic Logic, Pratt and Harel's Dynamic Logics, and Manna and Pnueli's temporal logics [15]. Hybrid models should smoothly incorporate these ideas. There is very little such incorporation of these ideas yet in the control community. This is perhaps because traditional control engineers still tend to associate computer science with FORTRAN programs, without realizing that advances in theoretical computer science in the areas of concurrent and distributed computing may have something to contribute.

The Continuous Side We should take full advantage of the "traditional semantics" of continuous plants. We take this to include how to specify the mathematical spaces over which the disturbances, controls, plant states, and plant trajectories range. We include also the accompanying apparatus of ordinary differential and integral equations, vector fields on manifolds, Lie semigroups, variational calculus, generalized curves, relaxed solutions, Radon measures, linear and non-linear functional analysis, dynamic programming, linear and convex and non-smooth mathematical programming, etc., that have been applied

[1]Research supported by ORA Corp., DARPA-US ARMY AMCCOM (Picatinny Arsenal,NJ) DAAA21-92-C-0013, U.S. Army Research Office DAAL03-91-C-0027

CH3229-2/92/0000-2947$1.00 © 1992 IEEE

to understand and compute nonlinear controls for the past forty years. Hybrid models should be defined so as to be able to incorporate these ideas smoothly too. Our purpose here is to bring commonalities between the digital and continuous side to the surface.

SEQUENTIAL NON-DETERMINISTIC AUTOMATA

Definitions A non-deterministic input-output automaton is based on a non-empty set of states, an input alphabet, an output alphabet, a state transition relation relating each pair (i, s) consisting of an input symbol i and a state s to a set $M(i, s)$ of states, and an output function O assigning to each state s an output alphabet symbol $O(s)$. It is a sequential deterministic automaton if $M(i, s)$ always has one member. We think of $M(i, s)$ as the set of immediate successor states that could occur at time $n+1$ if i was the input and s was the state at time n. For any sequence $i_0, i_1 \ldots$ of input symbols there is a corresponding set of all possible execution sequences $s_0, s_1 \ldots$ of the automaton. These are the sequences such that for all j, $s_{j+1} \in M(i_j, s_j)$. The output function, applied to an execution sequence, produces an output sequence. So there is a set of output sequences corresponding to a given input sequence. To deal with a no input (respectively no output) automaton, leave out all reference to the input (respectively output) alphabet. To deal with a many input line automaton, use the Cartesian product of the input alphabets as the input alphabet. Treat multiple output lines similarly.

A standard automaton construction shows that if we are given a collection of non-deterministic input and output machines and we connect them into a network by connecting output lines to input lines with the same alphabet, we get naturally another non-deterministic sequential automaton. The set of states for this automaton is the Cartesian product of the states sets for the component automata and of the alphabets of ouput lines which are connected to input lines. The input alphabet is the Cartesian product of the input alphabets of input lines not connected to output lines. The output alphabet is defined similarly.

Why Non-Deterministic Automata? We have worded this carefully to allow non-determinism. Why do we insist on allowing sequential non-deterministic programs, which can have a continuum of execution sequences for a given input sequence, instead of sticking to sequential deterministic automata? Because real time digital control programs can best be modelled as denoting non-deterministic automata networks. A digital control program run on hardware, and started twice in the same state, may terminate differently for one of many reasons. This hardware is certainly then to be regarded as a non-deterministic automaton. Some of the reasons are listed below.

- There is a time limit on computation. The number of iterations achieved in that time limit may vary from run to run.
- Default values may sometimes have to be used because time has run out before termination.
- A multi-tasking system may be allocating computing time to many competing programs. These programs compete for execution time. The competitors may be different from run to run, so the portion of the run completed by time-out may be different.
- Fault tolerant systems deliberately introduce redundancy, and always use alternate runs when malfunctions destroy an intended run.
- Hybrid systems are inherently concurrent. Models for proving concurrent programs correct can utilize non-determinism. See Apt and Olderog [1] and Nerode, Yakhnis, and Yakhnis [19], [20].

DIGITAL PROGRAMS AS GAME STRATEGIES

Concurrent Programs and Two-Person Game Semantics In [19] we model concurrent programs with a shared memory model roughly as follows. We envisage a pool of instructions, in which the Programmer (Player 1) at her turn casts sets of instructions in the pool for possible execution by Computer. (The instructions cast into the pool by Programmer are envisaged as tagged with what kind of device is supposed to execute that instruction and also tagged with where the result is to be delivered.) We envisage the Computer (Player 2), at his turn, as taking some or no instructions out of the pool, after which they disappear from the pool. We think of Computer as delivering the instructions to the appropriate machine for execution, about which Programmer knows nothing except that they are gone from the pool. Programmer and Computer communicate only through the pool. This is formulated there as a two person game.

A play of the game is a finite sequence of alternate successive moves of both players. A program specification is a set of such plays, and this is the "winning set" of plays for the associated game. In this conception, a concurrent program is a finite state winning strategy for Programmer which, if followed, forces the plays to satisfy the program specification no matter what the behavior of the Computer.

Distributed Computing and n-Player Games Here is a generalization of this two player game for distributed and hybrid systems based on message-passing. It is a little different from that of [20]. For distributed computing among n communicating programs, we formulate a Game with n players. Each player takes its turn without knowing the strategy of the other players and communicates none or some symbols as input to all other players. We fix the order in which players take turns. If all players take turns once in order, we call this a Game Move. A Play is a sequence of Game Moves. A Winning Set is a set of

Plays. A Strategy is an n -tuple of strategies, one for each player in order of turns. A Strategy is Winning if whenever all players follow their respective strategies, the play is in the Winning Set. Distributed programs denote such Strategies, the peformance specification is the Winnning Set. This fits the hybrid systems network model below quite well.

CONTINUOUS PLANTS AND GAMES

Differential game theory dates at least back to Isaacs [10]. There have been many important contributors, including Pontryagin [21] and his school , Friedman [6], Warga [23]. We adapt the description of some differential games for purposes of comparison with the digital games above.

Continuous Plants Every engineer knows the material of this paragraph. But we have to say it a particular way to bring out common features of the digital and the continuous case. Let $x'(t) = f(x(t), t, c, d)$ be a vector ordinary differential equation with f , $x(t), t, c, d$, respectively interpreted as plant description, current plant state, time, control value, and disturbance value. We assume as given also a vector function $g(x)$ of plant state x intended as the value of sense data about the plant state $x(t)$ at time t available to the controller. A control law (disturbance) defined on a time interval is a function $c(t)$ (respectively $d(t)$), telling that the control value is $c = c(t)$ (the disturbance is value $d(t)$) at time t in that interval. If a control law $c(t)$ and a disturbance $d(t)$ are imposed, the plant state $x(t)$ is supposed to satisfy $x'(t) = f(x(t), t, c(t), d(t))$. We have to specify the function f and what the admissible function spaces are for the plant states $x(t)$, the sensor function $g(x)$, the control law $c(t)$,and the disturbances $c(t)$. Often the assumption is that f and g are continuous, c is sectionally continuous, and d is measurable. If there is no Lipschitz condition assumed we may get many trajectories through initial condition. There may be a whole range of trajectories $x(t)$ solving the equation and satisfying the initial conditions, but not capable of being extended over the whole time interval. Without ensuring the hypotheses of standard existence theorems, there may be no trajectories at all. There may be maximal trajectories, that is, extending over maximal intervals of time, but of short duration. These are a form of non-determinism for continuous plants. It is reflected in the theory of differential inclusions. We used differential equations as our formulation, but this phenomenon is unchanged for integral or differential-difference equation or variational inequalities.

If for each choice of initial conditions and function $c(t)$, there is a unique trajectory extending over the whole interval, we call the plant sequential deterministic.

Why Allow Non-Determinism for Continuous Plants? There are several reasons.

- Differential equations are not perfect descriptions of the physical processes they model. Coefficients, boundary conditions, initial conditions, constraints will be slightly off.
- There may be unspecified small disturbances, statistical or deterministic.
- Non-linear equations do not lend themselves easily to computing error bounds for solutions resulting from inaccurate coefficients and boundary conditions, etc.
- Striking changes of qualitative behavior may result from small changes of these values, which we cannot prevent, and may be best modelled by non-determinism.
- To be of any value, correctness proofs that trajectories meet performance specifications had better apply to trajectories resulting from small variations in the numerical descriptions.

Two-Person Differential Game for Plants
Suppose we are given plant equation $x'(t) = f(x(t), t, c, d)$, as above.

We assume that the performance specification is written as a set of acceptable trajectories. We wish to choose a control function such that we get a trajectory in the performance specification no matter what the disturbance, we have a two person game with Disturbance (player 2) playing against Control (Player 1).

There is a traditional two-person differential game in which a move takes place at each time t, with the choice at time t in plant state x of control c so as to make sure that $x'(t) = f(x(t), t, c(t), d(t))$. We wish to select a control function $c(t)$ from the space of admissible control functions, based on past sense data so that for all disturbances, all plant trajectories under this control are in the performance specification.

Analogies To decide how to model digital programs and continuous plants in common terms, homologous parts need to be identified. Plant differential equations correspond to automata describing computer behavior. Plant states correspond to automaton states. Trajectories solving a differential equation correspond to execution sequences of an automaton arising from a program execution. Values of plant control correspond to automaton input symbols. Plant measurement functions g correspond to automaton output functions O. Performance specifications for plants correspond to digital program specifications. In our games Control corresponds to Programmer and Disturbance to Computer.

A Hybrid System Model

Continuous devices work on a continuous time scale, digital devices work on a discrete time scale. For the last forty years, since the theory of linear servomechanisms expounded in the MIT Radiation Laboratory book series, it has been commonplace in control and systems theory books to write a common generalization of machines with discrete and continuous time [22]. But this does not of itself model what happens when devices on the two time scales interact, such as digital programs and continuous plants in hybrid systems. This is a problem that has to be faced by anyone modelling hybrid systems. Continuous and digital device have to exchange information at discrete times $t(i)$. The delicate part is what information is exchanged at these times, and what kind of plants are involved. We only look at hybrid systems consisting of one plant and one digital control program.

Envisage a digital controller, or digital chip with control program which has as its input functions of sampled plant state, and computes, and occasionally decides, on this basis, to intervene at discrete times $t(i)$ and alter the control law applied from the current control law c_{i-1} used in the time interval from $t(i-1)$ to $t(i)$ to another control law c_i to be used instead during the interval from $t(i)$ to $t(i+1)$, or forever if there is no further intervention and no such $t(i+1)$.

Autonomous Equations Suppose we are given an autonomous differential equation

$$x'(t) = f(x(t), c, d).$$

Autonomous means that time is not an independent variable of f. Suppose we are also given a sensor function g which gives a value $g(x(t(i)))$at time $t(i)$ determined by measurements on the plant .

The Δt-Plant Automaton Suppose now we choose a fixed interval between measurements, $\Delta t = (t(i+1) - t(i))$. The choice of Δt and the autonomous differential equation determine an infinite state sequential input-output automaton, which we call the Δt-plant automaton. The Δt-plant automaton has as set of states the set of plant states. It has two infinite input alphabets, the set C of all control functions of duration Δt, and the set D of all disturbance functions of duration Δt. The Δt-plant automaton has one infinite output alphabet, the set of possible values of $g(x)$. It changes state exactly at times $t(i), i = 0, 1, \ldots$.

Since the equation is autonomous, we get a well-defined transition function for a non-deterministic automaton. That is, knowing the plant state x at $t(i)$ and the control c and disturbance d on $[t(i), t(i+1)]$, we know the set of possible plant states at $t(i+1)$ corresponding to the end-points of trajectory solutions of the differential equation that extend from time $t(i)$ to time $t(i+1)$. We regard this non-deterministic sequential automaton as having output $g(x(t+1))$ at time $t(i+1)$. If the equation happens to be sequential deterministic, we get a sequential deterministic Δt-automaton.

A Simple Autonomous Hybrid System To form a simple autonomous hybrid system, assume as given also a digital control automaton with input alphabet the possible values of $g(x)$ and with output alphabet the set C of controls of duration Δt. At time $t(i)$, the Δt-plant automaton communicates its output symbol, $g(x)$, a sense data measurement, as input symbol to the digital control automaton. At time $t(i)$, the digital control automaton communicates its output symbol, a control c of duration Δt, to the plant as input symbol. The Δt-plant automaton takes in an additional second input symbol at time Δt, a disturbance of duration Δt . There are no interactions at other times. These are to form a "hybrid system".

The plant is regarded as a Δt-plant automaton, the digital control automaton is also regarded as a Δt-automaton. If we drop reference to Δt, this is simply an automaton network of two non-deterministic sequential automata, and hence, as described earlier, it is a non-deterministic automaton. Because we have the Δt present, this hybrid network is a Δt-automaton. Its state transitions take place at times $t(n) = t(0) + n(\Delta t)$, that is, in a coset of the discrete subgroup of the reals generated by Δt. In summary, the simple autonomous hybrid system can be regarded as a Δt-automaton, operating in the discrete time of the discrete subgroup generated by Δt.

But, on the other hand, we can alternately regard the digital control automaton as a continuous time device by having it maintain its state at time $t(i)$ for all t with $t(i) \leq t < t(i+1)$, changing state at $t(i+1)$. This makes the hybrid system a real time network of real time devices because the digital control automaton is now regarded as a continuous time dynamical system in which the state functions have jump discontinuities. This is the same mathematical artifice used to regard discrete probability distributions on countable discrete sets of reals as distributions on all of the reals, or for relating the δ-function to step functions. So, looking at the plant as a Δt-automata in discrete time, or replacing the control automaton by a real time dynamical systems, are dual discrete and continuous views of the hybrid system. It is because of these dual interpretations that one can use a combination of automata theoretic methods from the Δt-automata interpretation and analytic methods from the continuous dynamical systems interpretation to investigate the behavior of hybrid systems.

It is better for applications to work in phase space, rather than plant state space, but we have omitted that here[13]. For an earlier exposition, see [17]. The present exposition evolved from discussions with W. Kohn on modelling declarative control [11] and A. Yakhnis on modelling control with games. Modelling hybrid systems as Δt-automata networks or as continuous dynamical system networks is exactly the same for arbitrary networks of digital control automata and plants. Finally, there is no restriction in choosing times for sampling sense data and times for issuing control laws to be the same. At those times one wants to sample but not control, one can issue from the control automaton the instruction to continue the same control law. At those times one wants to issue a control law but not receive sampled information, one can design the control automaton to ignore new sampled information.

GAMES AND HYBRID SYSTEMS

Continuous Games with Discrete Time We discuss only simple hybrid systems, one plant and one controller, and in this section the equation is not assumed autonomous. We introduce a two person discrete time game between Control and Disturbance, each casting successively a control law and a disturbance for the next interval in sequence. No approximations are involved. We merely fix discrete times at which information exchange takes place. Fix initial conditions. A play in the game is a sequence of alternate moves of the two players that lasts as long as the control is supposed to last, possibly forever. A position is an initial segment of a play. The performance specification for the system is defined as a set of such plays. This specification has to be computed by analysis from endpoint conditions and any trajectory restraints imposed on the plant along the course of the play.

A deterministic input-output state automaton winning strategy for the game chooses, for each position in which Control is to move, a control law for the next time interval, based on automaton state and on supplied sensed data, functions of game position. A winning strategy for Control is one which, in the presence of any Disturbance (sequence of disturbance moves which constitute an admissible disturbance function), produces a play in the performance specification. To repeat, a winning strategy is a state strategy Control program which, in the face of a Disturbance, always produces a "control script" of successively applied control laws leading to a trajectory in the winning set.

Finite Approximation Games Replace the space of control functions (resp. disturbance functions) by a finite set of control (resp. disturbance) functions that is closely spaced, i.e., an ϵ-net for a small ϵ. Then we get a finite game in which the players each have only a finite number of moves. The object is to extract a finite automaton winning strategy for Control in that approximation game.

Note We use the real trajectories of the original game throughout the intervals, based on this small selection of controls and disturbances, to compute the results of moves. Functional analysis and solving differential equations are hidden exactly at this point, to compute the legal moves of the game.

The approximation games that we have examined have turned out to be naturally formulated as Büchi and Landweber games [3](pp.525-542), Gurevich and Harrington games [7], Yakhnis and Yakhnis games with restraints [26], or McNaughton games [16]. All these classes of games possess generally infeasible algorithms to determine which player wins and what that player's winning strategy is. These algorithms are part of the proofs of the "forgetful determinacy theorems". If Control wins, her winning strategy is the desired control program meeting performance specifications. If Disturbance wins, she can examine his winning strategy, and can see where her strategy failed by exhibiting a disturbance which cannot be controlled. She may then try again with a finer discretization of time and a finer ϵ-net. If she does not win in any such game, she may have to change the system structurally to get a game which she can win. Alternately, she may lower her expectations for performance by weakening the performance specification. The games discussed here were introduced at the Hybrid Systems Workshop (Mathematics and AI Conference, Ft. Lauderdale, Jan., 1991).

CLOSING REMARKS

Extensions Recently Nerode and Remmel and Yakhnis [18] (in prep.) have found a wide variety of cases in which winning strategies and control programs can be extracted practically. This happens when the performance specification is not too complicated. In a sequel to [19], similar games are introduced to extract concurrent digital programs from program specifications. The extraction of strategy procedure is a kind of game analogue to solving dynamics problems by the inverse method. This apparatus also applies to the non-unique solution case as well. The algorithms lend themselves to modular construction of strategies for complicated games from strategies for simpler subgames. This has the same "divide and conquer" effect as W. Kohn's [11] use of series-parallel decompositions of automata in his declarative control.

For hybrid systems stability requires new definitions. One was developed by Kohn and Nerode and announced at CACSD92 [12]. See also the Kohn and Nerode paper at this conference. Kohn and Nerode identify stability with continuity of winning strategies, that is, of control programs in not necessarily Hausdorff topologies on the continuous plant and the control automaton. Further stability requirements are then expressed by requiring continuity of these strategies with respect to parameters such as initial conditions and performance specifications.

Limitations Such discretized games are played on complex plant dynamical systems with possibly unfathomably diverse qualitative behavior.

- The more complex the plant dynamics, the more daunting the computation for extracting control from specification, or proving there is no control.
- An approximation game may have no winning strategy, while its original continuous counterpart has many.
- The winning strategies for approximation games may not even be close to having the same effect on performance as a winning strategy for the original game.
- The analytic problems in trying to prove that a fully discretized game winning strategy is close to a winning strategy for the original game are very formidable.

Nerode and Kohn are working on these problems for Kohn's Declarative Control based on relaxed control theory [23] and Lie semigroups of transformations [9]. In another direction, in our Darpa DSSA effort, we intend to simulate the effect of extracted control programs by use of DSTOOL, which has already been modified to allow digital control programs to interact with continuous systems.

See also the Guckenheimer and Nerode paper at this conference.

Historical Note Differential games have often been used for optimal control. Many of the theorems and computational procedures are restricted to the case of additive separation of control and disturbance. This is typical of the differential games approach to optimal control of Isaacs 1965 [10], Friedman 1971 [6], Warga 1972, 1989 [23] and [24], Hajek 1975 [8], Krasovskii-Subbotin 1988 [14], and Berkovitz 1989 [2]. Our method does not have this restriction. On the other hand, the question our method addresses is not the same. Our approach is not in the spirit of "games of degree", in which objective functions are to be minimized. It is more in the spirit of a two-person version of the "games of kind" introduced by Isaacs [10], only seeking "satisfactory" strategies forcing plays into a winning set. Also we do not look for solutions given by analytical formulas. This allows us to consider a much larger class of process control problems. This change is what allows substituting extraction of strategies in discrete games for calculus of variations methods.

References

[1] K.R. Apt and E-R. Olderog, *Verification of Sequential and Concurrent Programs*, Springer-Verlag, 1991.

[2] L.D. Berkovitz, Thirty Years of Differential Games, in Emilio O. Roxin (editor), *Modern Optimal Control*, Marcel Dekker, Inc., 1989.

[3] J. R. Büchi, *The Collected Works of J. Richard Büchi* (S. MacLane,. Siefkes, eds.), Springer-Verlag, 1990. 1990.

[4] K. M. Chandy and J. Misra, *An Introduction to Parallel Program Design*, Addison-Wesley, 1988.

[5] A. F. Filippov, Differential Equations with Discontinuous Right Hand Side, Kluwer Academic Publishers, 1988.

[6] A. Friedman, *Differential Games*, Wiley-Interscience, 1971.

[7] Y.Gurevich and L. Harrington, Trees, Automata and Games, Proc. of the 14th Ann. ACM Symp. on Theory of Comp., pp. 60-65, 1982.

[8] O. Hajek,*Pursuit Games*, Mathematics in Science and Engineering, vol. 120, Academic Press, New York, 1975.

[9] J. Hilgert, K. H. Hofmann, J. Lawson, Lie Groups, Convex Cones, and Semigroups, Oxford Clarendon Press, 1988.

[10] R. Isaacs,*Differential Games*, SIAM Series in Applied Mathematics, John Wiley and Sons, Inc., 1965.

[11] W. Kohn, Hierarchical Control Systems for Autonomous Space Robots, Proc. AIAA, 1988.

[12] W. Kohn and A. Nerode, An Autonomous Control Theory, Proc. CACSD92, IEEE, 1992

[13] W. Kohn and A. Nerode, Foundations of Hybrid Systems, Proc. Hybrid Systems Workshop, 19-21 Oct 1992, Technical University, Lyngby, Denmark, to appear.

[14] N.N. Krasovskii and A.I. Subbotin, *Game-Theoretical Control Problems*, Springer-Verlag, 1988.

[15] Z. Manna and A. Pnueli, *The Temporal Logic of Reactive and Concurrent Systems*, Springer-Verlag, 1992.

[16] R. McNaughton, Infinite Games Played on Finite Graphs, Tech. Report 92-14, Depart. of Comp. Sci., RPI, Troy, New York, May 1992.

[17] A. Nerode and J.B. Remmel, A Model for Hybrid Systems, Hybrid System Workshop Notes, MSI, Cornell University, Ithaca, NY, June 1990.

[18] A. Nerode, J.B. Remmel and A. Yakhnis, Playing Games on Graphs: Extracting Concurrent and Hybrid Control Programs, in prep.

[19] A. Nerode, A. Yakhnis, V. Yakhnis, Concurrent Programs as Strategies in Games, in *Logic from Computer Science*, (Y. Moschovakis, ed.), Springer-Verlag, 1992.

[20] A., Nerode, A. Yakhnis, V. Yakhnis, Distributed Programs as Strategies in Games, in prep.

[21] L.S. Pontryagin, On the Theory of Differential Games, Russian Mathematical Surveys 21 (No.4), pp. 193-246, 1966.

[22] E. D. Sontag, Mathematical Control Theory, Springer-Verlag, 1990.

[23] J. Warga, *Optimal Control of Differential and Functional Equations*, Academic Press, 1972.

[24] J. Warga, Some Selected Problems of Optimal Control, in Emilio O. Roxin (ed.), Modern Optimal Control, Marcel Dekker, Inc., 1989.

[25] A. Yakhnis, Game-Theoretic Semantics for Concurrent Programs and Their Specifications, Ph. D. Diss., Cornell University, 1990.

[26] A. Yakhnis, V. Yakhnis, Extension of Gurevich-Harrington's Restricted Memory Determinacy Theorem, Ann. Pure and App. Logic 48, 277-297, 1990.

Proceedings of the 31st Conference
on Decision and Control
Tucson, Arizona • December 1992

Viewing hybrid systems as products of control systems and automata

R. L. Grossman* and R. G. Larson†

Department of Mathematics, Statistics, & Computer Science
University of Illinois at Chicago
Box 4348, Chicago, IL 60680

1. Introduction

The purpose of this note is to show how hybrid systems may be modeled as products of nonlinear control systems and finite state automata. By a hybrid system, we mean a network of consisting of continuous, nonlinear control system connected to a discrete finite state automaton. Our point of view is that the automaton switches between the control systems, and that this switching is a function of the discrete input symbols or letters that it receives. There are several ways in which this may be modeled. The approach we take is a simple one: we show how a nonlinear control system may be viewed as a pair consisting of a bialgebra of operators coding the dynamics, and an algebra of observations coding the state space. We also show that a finite automaton has a similar representation. A hybrid system is then modeled by taking suitable products of the bialgebras coding the dynamics and the observation algebras coding the state spaces. An important advantage of viewing hybrid systems in an operator representation is that it is easy to specify algebraically how the various components of the hybrid system are "glued" together. One can then pass from the operator representation back to the state space representation when it is convenient. We simply outline the ideas in this note: for more details and examples, see [3].

In Section 2, we show how nonlinear control systems may be modeled in this way. In Section 3, we do the same for finite state automata. In Section 4, we define a suitable product of such systems, which we identify with a hybrid system. Section 5 contains some discussion and examples.

*This research was supported in part by NASA grant NAG2-513 and NSF grant DMS 910-1089.

†This research is supported in part by NSF grant DMS 910-1089.

2. Operator representations of nonlinear control systems

In this section, we describe how to pass from the state space representation of a nonlinear control system to the operator representation. Let X denote the state space, let E_1 and E_2 denote vector fields, and let $t \rightarrow u_j(t)$ denote controls. Then the dynamics are described by

$$\begin{aligned} \dot{x}(t) &= u_1(t)E_1(x(t)) + u_2(t)E_2(x(t)), \\ x(0) &= x^0 \in X. \end{aligned}$$

Let $k = \mathbf{R}$ denote the real numbers, and let X denote the state space of the system. Define the *observation algebra* R to be the commutative algebra of all functions

$$R = \{ f : X \longrightarrow k \}.$$

If X carries an additional structure (such as being smooth or algebraic), then we require the same of the maps f. We also define a map (the *augmentation*)

$$\epsilon : R \longrightarrow k, \quad f \mapsto f(x^0).$$

Let $H = k{<}E_1, E_2{>}$ denote the free associative algebra generated by the *symbols* E_1 and E_2. Here we are overloading E_j, so that it denotes either a vector field or a symbol depending upon the context. We call H the *dynamical algebra* associated with the control system. There

CH3229-2/92/0000-2953$1.00 © 1992 IEEE

is a natural action of H, given by the action of differential operators on observation functions. The algebraic structure of H is very rich: H is a bialgebra, so that the dual of H also carries an algebraic structure. This bialgebra structure interacts nicely with R, making R an H-module algebra. See [2] for further details.

To summarize, given a nonlinear control system in its state space representation, this process yields a pair (H, R), which we call the *operator representation* of the control system.

3. Operator representations of finite state automata

In this section, we describe how to pass from the state space representation of a finite state automaton to the operator representation. Let k be a field of characteristic 0. Let Ω denote the input symbols and $W = \Omega^*$ denote the free semigroup consisting of the words generated by Ω. Let

$$S = \{s^0, s^1, \ldots, s^n\}$$

denote the state space, and s^0 denote the intital state. The dynamics are described by specifying a transition for each input symbol α and each state s^j

$$S \times \Omega \longrightarrow S, \quad (s^j, \alpha) \mapsto s^j \cdot \alpha.$$

As usual, this action extends to an action of the words on the states

$$S \times W \longrightarrow S, \quad (s, w) \mapsto s \cdot w.$$

Define the observation algebra R and augmentation as before:

$$R = \{\, f : S \longrightarrow k \}$$

$$\epsilon : R \longrightarrow k, \quad f \mapsto f(s^0).$$

The dynamical algebra H is the semigroup algebra kW, defined to be the algebra over k consisting of finite formal sums of words. Again there is a natural action of the dynamical algebra H on the observation R,

$$L_w(f)(s) = f(s \cdot w), \quad w \in W, \quad s \in S, \quad f \in R.$$

Here L_w denotes a left action of W on R.

To summarize, given a finite state automaton, this process yields a pair (H, R), which we call the *operator representation* of the automaton.

4. Hybrid Systems

The basic idea is that a hybrid system is constructed out of components consisting of nonlinear control systems and discrete automata by taking suitable products of the components in their operator representations. Assume now that we have a finite state automaton, as described in the section above, with alphabet Ω and states $s \in S$, and that these states parametrize operator representations (H_s, R_s) of nonlinear control systems, one for each state s. For simplicity, assume that

$$H_s \cong H_0 = k\langle \xi_1, \ldots, \xi_M \rangle, \quad \text{all } s \in S,$$

where the ξ_i are formal symbols. This simply means that the dynamics of each control system are driven by the same number of vector fields—the control systems themselves may be different, since this is coded by the action of the H_i on the R_i.

The hybrid system has an operator representation given by

$$H = k\Omega^* \amalg H_0, \quad R = \bigoplus_{s \in S} R_s,$$

where $\amalg$ denotes the free product of the indicated associative algebras.

5. Remarks

In this section, we make some general remarks about this construction. First, observe that in the case that the hybrid system contains no automaton component ($\Omega = \emptyset$ and $S = \{s_0\}$), then the hybrid system reduces to a nonlinear control system, as described in [2]. Second, observe that in the case in which the hybrid system contains no nonlinear system components (H_s and R_s are both the trivial algebra k, for all $s \in S$) then the hybrid system reduces to a finite state automaton, as described in [3]. Finally, it is not hard to extend the realization theorem proved in [2] to hybrid systems of the form described here.

This theorem reduces in the two special cases just described to the Fliess nonlinear realization theorem [1] and the Myhill-Nerode theorem respectively [4].

References

[1] M. Fliess, "Nonlinear realization theory and abstract transitive Lie algebras," *Bulletin American Mathematical Society (NS)*, vol. 2, pp. 444–446, 1980.

[2] R. Grossman and R. G. Larson, "The realization of input–output maps using bialgebras," *Forum Math.*, vol. 4, pp. 109–121, 1992.

[3] R. L. Grossman and R. G. Larson, "A bialgebra approach to hybrid systems," submitted for publication.

[4] J. E. Hopcroft and J. D. Ullman, *Introduction to Automata Theory, Languages, and Computation.* Massachusetts: Addison-Wesley, 1979, pp. 65–67.

[5] R. G. Larson, "Cocommutative Hopf algebras," *Canad. J. Math.* vol. 19, pp. 350–360, 1967.

[6] M. E. Sweedler, *Hopf algebras,* NY: W. A. Benjamin, 1969.

FA12 - 9:10

Proceedings of the 31st Conference
on Decision and Control
Tucson, Arizona • December 1992

Multiple Agent Autonomous Hybrid Control Systems

Wolf Kohn
Intermetrics

Anil Nerode
MSI
Cornell University

Abstract

This paper overviews current research efforts in the development of a formal model for the control of dynamic, realtime autonomous systems in which multiple decision makers control the plant. The model provides a formal framework for expressing the interaction between evolution (continuous) and knowledge (discrete) components. This is the central characteristic of hybrid systems. The essence of the interaction between the evolution and discrete components is the continuity of the behavior of the system, as viewed by each of its agents, with respect to the topology defined by the knowledge [1]. The model is being developed via a computational architecture whose central functionality is on-line, distributed theorem proving and reactivity.. The central elements of this architecture are the controlling agents and the interagent communication network. Each agent is provided with an inferencer, a theorem planner and a communication protocol device.

1 Introduction

The theory of multiple agent hybrid control systems and the associated model provides a formal, computational framework for the design and implementation of reactive, intelligent distributed controllers [4].In general, intelligent controllers implement, with some formalism, a combination of planning and feedback control principles. Their chief objective is to tune the control law, distributed over the controlling agents, to respond to unpredicted events and uncertainty.on the basis of incomplete knowledge. The central functions of each agent are to monitor the performance on the basis of sensory data and to redesign or adjust its control law in realtime when desired behavior requirements are violated.

In general, autonomous systems are modelled by dynamic evolution operators that encode the evolution of trajectories, termed behavior trajectories, of the state of the system and a logic model representing discrete information about the system structure,multiple region of validity requirements, goals and agent coordination. In general, the state trajectories are functions of a continuous time while the logic model deals with discrete objects. This is why these models are termed hybrid models.

In many of the proposed models for hybrid systems, the continuous components are presented and differential equations, while the discrete components are presented by finite event automata. [2]. In our model, each agent models the evolution of the system with a set of coupled ordinary differential equations over a continuous manifold and the logic model with a set of Logical Horn Equational Clauses (termed the knowledge base) [3] together with a set of Inference Principles and mechanisms for deducing local control actions at each decision time interval as a function of goal, sensory data, and current instance of the knowledge base.

The central functions of each agent are: 1- to construct on-line an automaton, the Inference Automaton, 2- to simulate its behavior as a means to generate the control actions that satisfy encoded requirements and brings the system closer to its goal. and 3- to adapt the system behavior if failure of the inference occurs.

From an operational point of view, in our model, each agent carries out algebraic equational inference in the context of a variational representation of the dynamics of the systems under consideration [5]. This operation is accomplished in an architectural framework termed Multiple Agent Hybrid Control Architecture. In this paper we describe this architectural framework and use it to summarize the current status of our research.

The central mechanism of the architecture is on-line distributed mechanical theorem proving over a theory of rational forms [6]. The architecture is composed of a collection of agents and an inter-agent connectivity network. The architecture is especially tailored for applications in which the decision makers are not coordinated by a central umpire but rather by coordinating knowledge flowing through the network .

Autonomous system applications that can be represented by our model are characterized by a plant controlled by multiple decision makers not sharing a common Knowledge pool (see fig.1). This applications include, among many others, flexible manufacturing [6], Intelligent Avionics [7], and Distributed Computing [8]. We will illustrate some of our framework with a simple two-agent two-cell manufacturing example. For the purpose of fixing concepts and motivating the structure chosen for our architecture, a non inclusive list of required features for these applications is briefly discussed next.

CH3229-2/92/0000-2956$1.00 © 1992 IEEE

Real-Time: Constraints for real-time performance, which include both the agent's computational constraints and the plant dynamics should be explicit in its knowledge base. A valid inference therefore satisfies those constraints.

Reactive: The inference function of each agent on the architecture should be depend on plant and environmental events. This requires a feedback path between the agent and the part of the plant it controls.

Adaptive: The knowledge base of each agent should be open and modifiable by sensory and inter-agent data. Inference failure should at an agent should be localized and trigger tuning and corrective action throughout the network.

Distributive with Coordination: The inference should carried out distributively over the agents the coordination scheme is implicit without umpire; since for most applications, the presence of an umpire would require impractical amount of information flow.

Dynamic Hierarchization: The architecture should operate simultaneously at different levels of abstraction

Performance: The behavior of the closed loop distributive system must be stable and the system goals must be reachable as the least desirable performance requirements.

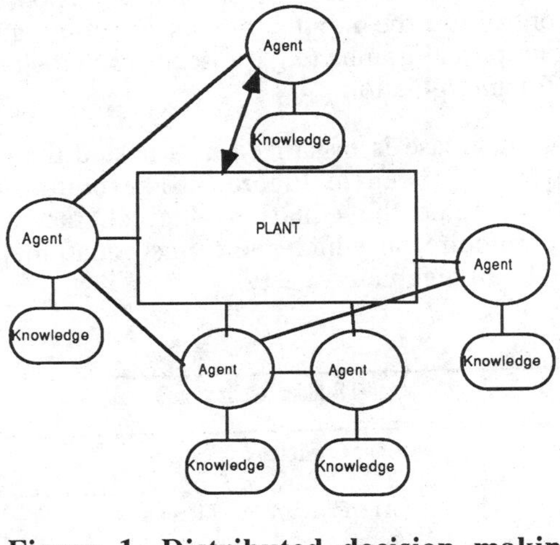

Figure 1. Distributed decision making

The proposed architecture is an outgrow of a model for intelligent control systems referred to as Declarative Control Architecture [4]. The model discussed in the next section, extends this architecture to the multiple-agent environment and refine its components.

2 Multiple Agent Hybrid Declarative Control Architecture

Each agent of the hybrid control architecture is provided with a real-time theorem prover whose domain is a theory, called relaxed variational theory. This theory is an encoding of the Relaxed variational theory developed by Young [14]. We will overview some of its main features in section 3. In this section, we proceed to discuss the main features of the architecture.

The architecture is composed of two items: The agent Controllers, and The Controller Network. These items are illustrated in Figures 2 and 3 respectively.

– *Architectural Elements of the a control agent*

A control agent is composed of five functional elements - a *Knowledge Base*, a *Theorem Planner* an *Inferencer*, a *Knowledge Decoder* and an *Adapter*. We discuss their functionality next.

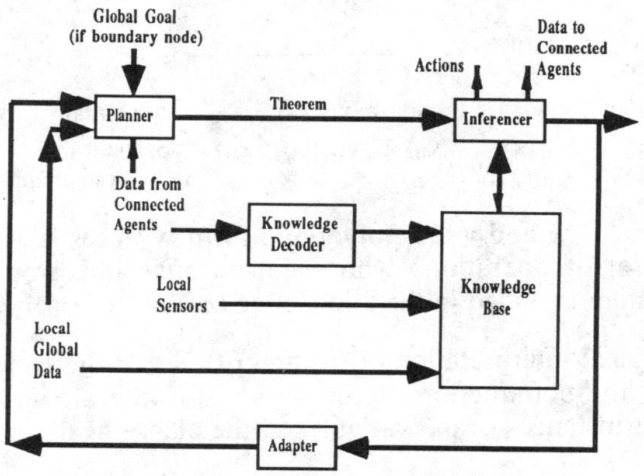

Figure 2 Agent Controller

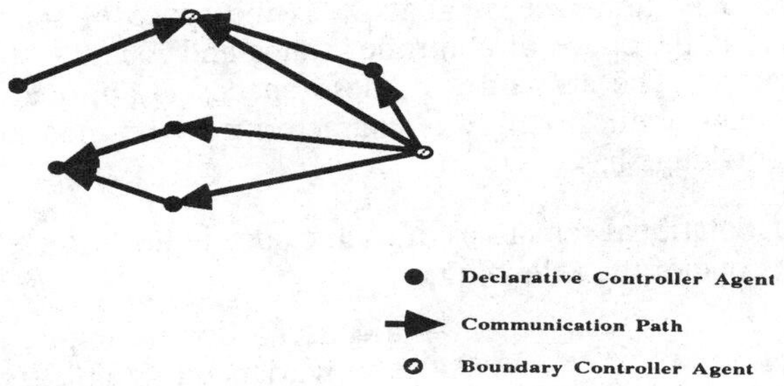

Figure 3 Declarative Hybrid Control Network

– *Knowledge Base*

The Knowledge Base consists of a set of equational first order logic clauses with second order extensions. The syntax of clauses is similar to the ones in the Prolog language. Each clause is of the form

$$\text{Head} \Leftarrow \text{Body} \quad (1)$$

where Head is a functional form, $p(x_1, \ldots x_n)$, where $x_1, x_2, \ldots x_n$ are variables or parameters in the domain of the controller. The symbol $\Leftarrow$ stands for logical implication.

The Body of a clause is a conjunction of one or more terms,

$$e_1 \wedge e_2 \wedge \ldots \wedge e_m \quad (2)$$

where $\wedge$ is the logical. Each term in (2) is an equational form, an inequational form, a covering form, or a clause head. The generic structure of these forms is illustrated in Table 1 below.

Table 1

Form	Structure	Meaning
equational	$w(x_1,\ldots,x_n) \approx v(x_1,\ldots,x_n)$	equal
inequational	$w(x_1,\ldots,x_n) \not\approx v(x_1,\ldots,x_n)$	not equal
covering	$w(x_1,\ldots,x_n) \leq v(x_1,\ldots,x_n)$	partial order
clause head	$q(x_1,\ldots,x_n)$	recursion, chaining

In Table 1, w and v are polynomic form with respect to a finite set of operations whose definitional and property axioms are included in the knowledge base.

The logical interpretation of (1) and (2) is that the Head is true if the conjunction of the terms of Body are jointly true for instances of the variables in the clause head.

The domain D in which the variables in a clause head take values is a Cartesian product of the form

$$D = G \times S \times X \times A \quad (3)$$

where G is the space of goals, S is the space of sensory data, X is the space of controller states and A is the space of actions. The structure of these spaces is application-dependent, their structure is characterized with clauses in the knowledge base.

The denotational semantics of each clause in the knowledge base is one of the following:

1 – a conservation principle,
2 – an invariance principle,
3 – a constraint principle.

Conservation principles are one or more clauses about balance of a particular process in the dynamics of the system, the goals or the computational resources. For instance, in the manufacturing example described later on, a conservation clause describes the material balance in the manufacturing process as viewed by each of the two agents.

As another example, consider the following clause representing conservation of computational resources,

comp(Load, Process Op_count, Limit) $\Leftarrow$

process(process_count) $\wedge$

process_count • Load1 -

Op_count Load

$\wedge$

Load1 $\leq$ Limit

$\wedge$

comp (Load1, Process, Op_count, Limit)

where Load corresponds to the current computational burden, measured in VIPS (Variable Instantiations Per Second), Process is a clause considered for execution, and Op_count is the current number of terms in process.
In general, conservation principles always involve recursion, not necessarily a single clause as in the example above, but with chaining throughout several clauses.

Invariance principles are one or more clauses establishing constant of motion in a general sense. These principles include stationarity which plays a pivotal role in the formulation of the theorems proved by the architecture, and geodesics.

The importance of invariant principles lies in the reference they provide for the detection of unexpected events. for example, in an adiabatic chemical process, the enthalpy is constant. Under normal operating conditions, An equational clause that states this invariance has a ground form that is constant any deviation from this value, represents deviation from normality.

Constraint principles are clauses representing engineering limits to actuators or sensors and, most importantly, behavioral policies. For instance in the chemical reactor example, the characteristics of the speed of response of the values controlling the input products or temperature are given by empirical graphs (e.g. pressure *vs* velocity) and a strategy for interpolation.

The clause database is organized in a nested hierarchical structure as illustrated in Figure 3. The bottom of this hierarchy contains the equations that characterize the algebraic structure on which the terms equational forms are defined: an algebraic variety.

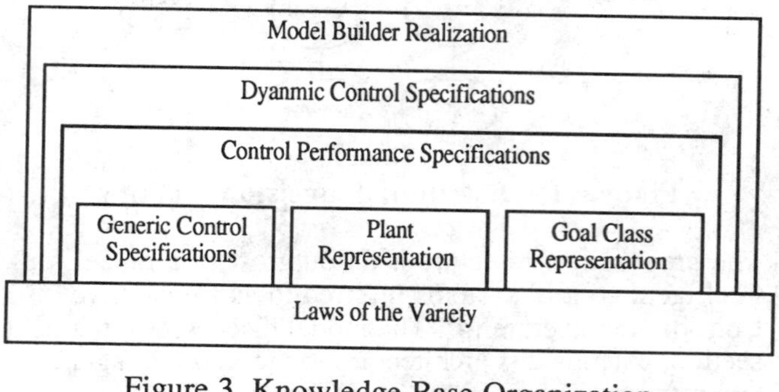

Figure 3 Knowledge Base Organization

At the next level of the hierarchy, three types of clauses are stored: Generic Control Specifications, Plant Representation and Goal Class Representation.

The generic control specifications are clauses expressing general desired behavior of the system. They include statements about stability, complexity and robustness that are generic to the class of declarative rational controllers. These specifications are written by constructing clauses that combine laws of the kind which use the Horn clause format described earlier.

The Plant Representation is given by clauses characterizing the dynamic behavior and structure of the plant, which includes sensors and actuators. These clauses are written as conservation principles for the dynamic behavior and as invariance principles for the structure. As for the generic control specifications, they are constructed by combining variety laws in the equational Horn clause format.

The next level of the hierarchy involves the Control Performance Specifications. These are typically problem-dependent criteria and constraints. They are written in equational Horn clause format. They include generic constraints such as speed and time of response, and qualitative properties of state trajectories [14]. Dynamic control specifications are equational Horn clauses whose bodies are modified as a function of the sensor and goal commands (see Figure 2).

Finally, model builder realization clauses constitute a recipe for building a procedural model for variable instantiation and theorem proving.

– *Theorem Planner*

The function of the theorem Planner, which is domain-specific, is to generate, for each update interval, a symbolic statement of the desired behavior of the system, as viewed by the agent (j, say) throughout the interval. The theorem statement that it generates has the following form:

"There exists action A_j in the control interval [t, t+D) such that A_j minimizes the functional

$$\int_t^{t+D} L_j(X,C)\,dA_j(X) \qquad (4)$$

subject to the following constraints:

$g_j\,(S, X\,[t+D)) \approx$ local goal for the interval,

$$\sum_m C^j_m(X,t) \approx V_{inter\,j}(X,t) \quad , \text{ and} \qquad (5)$$

$\int dA_j \approx 1.$ "

In (4), L_j is called the Local criterion of the system as viewed by agent j for the current interval of control [t,t+D). This function, which maps the cartesian product of the state and control spaces into the real line with the rational topology of intervals, captures the dynamics, constraints and requirements of the system as viewed by agent j. The relaxed Lagrangian function is a continuous projection in the topology defined by the knowledge base (see [27]), in the coordinates of the agent of the global Lagrangian function L that characterizes the system as a whole.

In (4), X represents the state of the system as viewed by the agent, C is a variable taking values in the countable set of possible actions (the control space) agent j can generate. These actions are constructed by a mixing process from a basic set of primitive actions. This mixing process is described next.

The term A_j in (1), is a Radon probability measure [6] on the set of primitive command or control actions the agent can execute for the interval [t,t+D). It measures, for the interval, the percentage of time C is to spend in each of the primitive actions. This is the nature of the mixing process. The central function of the control agent is to determine this mixture of actions for each control interval. This function is carried out by each agent by infering from the current status of the knowledge base whether a solution of the optimization problem stated by the current theorem exists, and, if so, to generate corresponding actions and state updates.

The expressions in (5) constitute the constraints imposed in the relaxed optimization problem solved by the agent. The first one is the local goal constraint expressing the general value of the state at the end of the current interval. The second represents the constraints imposed on the agent by the other agents in the network. Finally, the third one indicates that A_j is a probability measure.

Under relaxation, and the appropriate selection of the domain (see [1]), the optimization problem stated in (4) and (5) is a **convex** optimization problem. This is important because it guarantees that if a solution exists, it is unique up to probability, and also, it guarantees the computational effectiveness of the inference method that the agent uses for proving the theorem.
The construction of the theorem statement given by (4) and (5) is the central task carried out in the planner. It characterizes the desired behavior of the system as viewed by the agent in the current interval so that its requirements are satisfied and the system "moves" towards its goal in an optimal manner.

– *adapter*

The functional in (4) includes a term, referred to as the "catch-all" potential, which is not associated with any particular clauses in the Knowledge Base. Its function is to measure unmodelled dynamic events. This monitoring function is carried out by the Adapter which implements a generic commutator principle [8]. Under this principle, if the value of the catch-all potential is empty, the current theorem statement adequately models the status of the system. On the other hand, if the theorem fails, meaning that there is a mismatch between the current statement of the theorem and system status, the catch-all potential carries the equational terms of the theorem that caused the failure. These terms are negated and conjuncted together by the Inferencer according to the commutation principle (which is itself defined by equational clauses in the Knowledge Base) and stored in the Knowledge Base as an adaptation dynamic clause. The Adapter then generates a potential symbol, which is characterized by the adaptation clause and corresponding tuning constraints. This potential is added to criterion for the theorem characterizing the interval.

The new potential symbol and tuning constraints are sent to the Planner which generates a modified Local criterion and goal constraint. The new theorem, thus constructed, represents adapted behavior of the system. This is the essence of reactive structural adaptation in the our model

At this point, we pause in our description to address the issue of **robustness**. To a large extent, the adapter mechanism of Declarative Control provides the system with a generic and computationally effective means to recover from failures or unpredictable events. Theorem failures are symptoms of mismatches between what the agent thinks the system looks like and what it really looks like. The adaptation clause incorporates knowledge into the agent's Knowledge Base which represents a recovery strategy. The Inferencer, discussed next, effects this strategy as part of its normal operation.

– *Inferencer*

The Inferencer is an online equational theorem prover. The class of theorems it can prove are represented by statements of the form of (4) and (5). We will first present a brief overview of its functionality next.

The class of theorems proved by the inferencer are those expressed by an existentially quantified conjunction of equational terms of the form:

$$\exists Z \mid W_1(Z,p)\, rel_1\, V_1(Z,p) \wedge \bullet\bullet\bullet \wedge W_n(Z,p)\, rel_n\, V_n(Z,p) \qquad (6)$$

where Z is a tuple of variables each taking values in some domain D, p is a list of parameters in D, and $\{W_i, V_i\}$ are polynomial terms in the semiring polynomial algebra [40]

$$\tilde{D}\langle\Omega\rangle \qquad (7)$$

with

$$\tilde{D} = \{D, \langle +, \bullet, 1, 0\rangle\}$$

a semiring algebra with additive unit 0 and multiplicative unit 1.

In (6), rel_i, i = 1,..., n are binary relations on the polynomial algebra. Each rel_i can be either an equality relation ($\approx$), inequality relation ($\neq$), or a partial order relation. In a given theorem, more than one partial order relation may appear. In each theorem, at least one of the terms is a partial order relation that defines a complete lattice on the algebra; that is, it has a minimum element.

Given a theorem statement of the form of (6) and a knowledge base of equational clauses, the inferencer determines whether the statement logically follows from the clauses in the Knowledge Base, and if so, as a side effect of the proof, generates a non-empty subset of tuples with entries in D giving values to Z. These entries determine the agent's actions. Thus, a side effect is instantiation of the agent's decision variables.

In (7), Ω is a set of primitive unary operations, $\{ f_i \}$, called the infinitesimal operators. Each f_i maps the semiring algebra, whose members are power series involving the composition of operators, on Z to itself:

$$f_i : \tilde{D}\langle\langle Z\rangle\rangle \rightarrow \tilde{D}\langle\langle Z\rangle\rangle \qquad (8)$$

These operators are characterized by axioms in the Knowledge Base, and are problem dependent. In formal logic, the implemented inference principle can be stated as follows: Let Σ be the set of clauses in the Knowledge Base. Let $\vdash$ represent implication. Then, proving the theorem means to show that it logically follows from Σ, i.e.,

$$\Sigma \vdash \text{Theorem} \qquad (9)$$

The proof is accomplished by sequences of applications of the following inference axioms:

- equality axioms
- inequality axioms
- partial order axioms
- compatibility axioms
- convergence axioms
- knowledge base axioms
- limit axioms

The specifics of these inference axioms can be found in [7] where it is shown that each of the inference principles can be expressed as an operator on the cartesian product:

$$\tilde{D}\langle W\rangle \times \tilde{D}\langle W\rangle \tag{10}$$

Each inference operator transforms a relational term into another relational term. The inferencer applies sequences of inference operators on the equational terms of the theorem until these terms are reduced to either a set of ground equations of the form of (9), or determine that no such ground form exists.

$$z_i \approx \alpha_i, \quad \alpha_i \in \tilde{D} \tag{11}$$

We now provide an overview of the mechanism by which the inferencer carries out the procedure described above. The inferencer builds a procedure for variable goal instantiation: a locally finite automaton . We refer to this automaton as the Proof Automaton. This is a unique important feature of our approach. The proof procedure is customized to the particular theorem statement and knowledge base instance it is currently handling.

The structure of the proof automaton generated by the inferencer is illustrated in Figure 4.

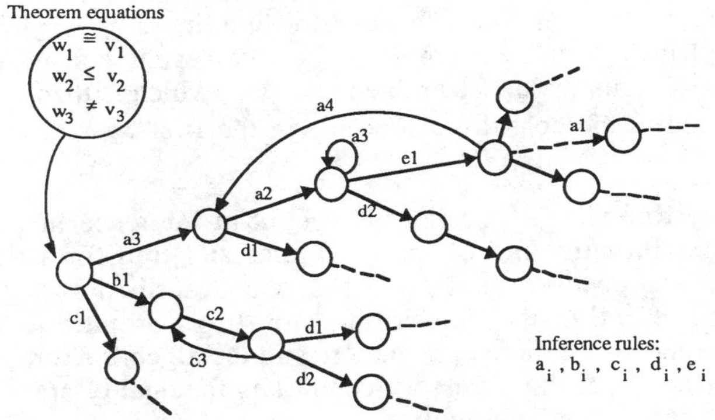

Figure 4: Conceptual Structure of the Proof Automaton

In Figure 4, the initial state represents the equations associated with the theorem. In general, each state corresponds to a derived equational form of the theorem through the application of a chain of inference operators to the initial state that is represented by the path,

$$s_0 \xrightarrow{\text{Inf}_1} s_1 \xrightarrow{\text{Inf}_2} s_2 \longrightarrow \cdots \xrightarrow{\text{Inf}_k} s_k$$

Each edge in the automaton corresponds to one of the possible inferences. A state is terminal if its equational form is a tautology, or it corresponds to a canonical form whose solution form is stored in the Knowledge Base.

In transversing the automaton state graph, values or expressions are assigned to the variables. In a terminal state, the equational terms are all ground states (see (11)). If the automaton contains at least one path starting in the initial state and ending in a terminal state, then the theorem is true with respect to the given Knowledge Base, and the resulting variable instantiation is a valid one. If this is not the case, the theorem is false.

The function of the complete partial order term, present in the conjunction of each theorem provable by ERI, is to provide a guide for constructing the proof automaton. This is done by transforming the equational terms of the theorem into a canonical fixed point equation, called the Kleene-Schutzenberger Equation (KSE) [2], which constitutes a blueprint for the construction of the proof automaton. The general form of KSE is :

$$Z \approx E(p)\cdot Z + T(p) \tag{12}$$

In (12), E is a square matrix, with each entry a rational form constructed from the basis of inference operators described above, and T is a vector of equational forms from the Knowledge Base. Each non-empty entry, E_{ij}, in E corresponds to the edge in the proof automaton connecting states i and j. The binary operator between E(p) and Z represents the "apply inference to" operator. Terminal states are determined by the non-empty terms of T. The p terms are custom parameter values in the inference operator terms in E(.).

A summary of the procedure executed by the inferencer is presented in Figure 4.

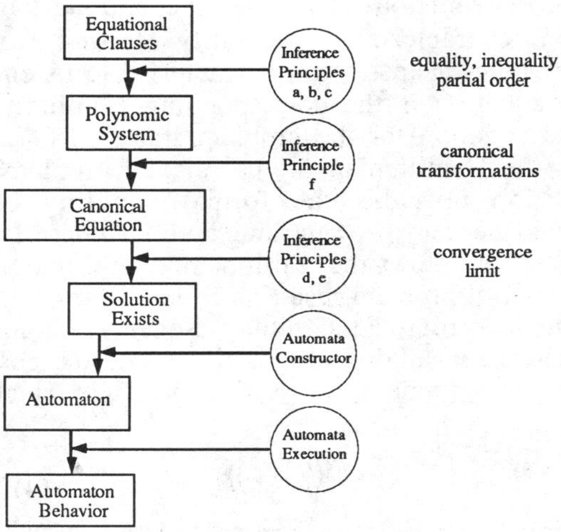

Figure 4: Summary of Inferencer Procedure

Note that the construction of the automaton is carried out from the canonical equation and not by non-deterministic application of the inference rules. This is done because the complexity of the computation of the canonical equation (low polynomic) is far better than applying the inference rules directly (exponential).

The automaton is simulated to generate instances of the state, action and evaluation variables using an automaton decomposition procedure [1] which requires $n \log_2 n$ time, where $n \approx$ # of states of the automaton. This procedure implements the recursive decomposition of the automaton into a cascade of parallel unitary (one initial and one terminal state) automata. Each of the resulting automaton on this decomposition is executed independently of the others. The behavior of the resulting network of automata is identical with the behavior obtained from the original automaton, but with feasible time complexity.

We proceed to explain how the inferencer is used by an agent. It fulfills two functions: to generate a proof for the system behavior theorem of each agent generated by the Planner ((4), (5)), and it is the central element in the Knowledge Decoder. We now describe its function for proving the behavior theorem. Later, we will overview its function as part of the Knowledge Decoder.

To show how the inferencer is used to prove the Planner theorem, (4),(5), first, we show how this theorem is transformed into a pattern of the form of (6). Since (4), (5) is a convex optimization problem, a necessary and sufficient condition for optimality is provided by the following dynamic programming formulation:

$$\begin{aligned} V_j(Y,\tau) &\approx \inf_{A_j} \int_{\tau} L_j(X,C)\,dA_j(X(\sigma)) \\ \partial V_j/\partial t &= \inf_{A_j} H(Y,\partial V_j/\partial Y, A_j) \\ X(t) &= Y \\ \tau &\in [t,t+D) \end{aligned} \qquad (13)$$

In (13), the function V_j, called the optimal cost-to-go function, characterizes minimality starting from any arbitrary point inside the current interval to its end. The second equation is the corresponding Hamilton Jacobi Bellman equation for the problem stated in (4) and (5) with H the Hamiltonian of the relaxed problem. This formulation provides the formal coupling between deductive theorem proving and optimal control theory. The inferencer allows the real-time optimal solution of the formal control problem resulting in intelligent distributed real-time control of the multiple-agent system. The central idea for solving (10) is to expand the cost-to-go function, in a rational power series, V(.,.), in the algebra:

$$\tilde{D}\langle\langle (Y,\tau) \rangle\rangle \qquad (14)$$

Replacing V for V_j in the second equation in (13), gives two items: a set of polynomic equations for the coefficients of V and a partial order expression for representing the optimality. Because of the convexity and rationality of V, the number of equations to characterize the coefficients of V is finite. The resulting string of conjunctions of coefficient equations and the optimality partial order expression are in the form of (6).

In summary, for each agent, the inferencer operates according to the following procedure.

Step 1: Load current theorem (4), (5).

Step 2: Transform theorem to equational form (6) via (13).

Step 3: Execute proof according to figure 5.

If the theorem logically follows from the Knowledge Base (i.e., it is true), the inferencer procedure outlined will terminate on step 3 with actions A(t+D). If the theorem does not logically follow from the Knowledge Base, the Adapter is activated, and the theorem is modified by the theorem Planner according to the strategy outlined above. This mechanism is the essence of reactivity in the agent. Because of relaxation and convexity, this mechanism ensures that the controllable set of the domain is strictly larger than the mechanism without this correction strategy.

– *Knowledge decoder*

The function of the Knowledge Decoder is to translate knowledge data from the network into the agent's Knowledge Base, by updating the inter-agent specification clauses. These clauses characterize the second constraint in (2). Specifically, they express the constraints imposed by the rest of the network on each agent. They also characterize the global-to-local transformations (see [3]). Finally, they provide the rules for building a generalized multiplier for incorporating the inter-agent constraints into a complete unconstrained criterion, which is then used to build the cost-to-go function in the first expression in (13).

The Knowledge Decoder has a built-in Inferencer used to infer the structure of the multiplier and transformations by a procedure similar to the one described for (13). Specifically, the multiplier and transformations are expanded in rational power series in the algebra defined in (14). Then, the necessary conditions for duality are used to determine conjunctions of equational forms and a partial order expression to construct a theorem of the form of (6) to generate the multiplier.

The conjunction of equational forms for each global-to-local transformation is constructed by applying the following invariant embedding principle:

"For each agent, the actions at given time t in the current interval, as computed according to (10), are the same actions computed at t when the formulation is expanded to include the previous, current, and next intervals."

By transitivity and convexity of the criterion, the principle can be analytically extended to the entire horizon. The invariant embedding equation has the same structure as the dynamic programming equation given in (13), but with the global criterion and global Hamiltonians instead of the corresponding local ones.

The local-to-global transformations are obtained by inverting the global-to-local transformations, obtained by expressing the invariant embedding equation, as an equational theorem of the form of (6). These inverses exist because of convexity of the relaxed Lagrangian and rationality of the power series.

It is important at this point to interpret the functionality of the Knowledge Decoder of each agent in terms of what it does. The multiplier described above has the effect of aggregating the rest of the system and the other agents into an equivalent companion system and companion agent, respectively. This is illustrated in Figure 5

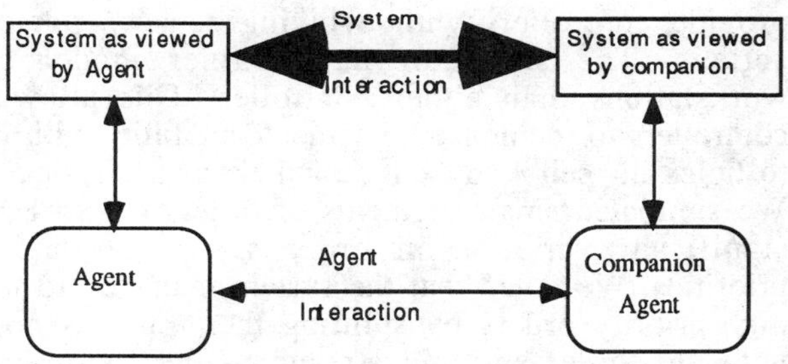

Figure 5: Dipole Network Equivalent

The aggregation model (figure 5) describes how each agent perceives the rest of the network. This unique feature allows us to characterize the scalability of the architecture in a unique manner: In order to determine computational complexity of an application, we have only to consider the agent with the highest complexity (i.e., the agent with a local with the most complex criterion) and its companion.

– *Inter-agent Network*

The inter-agent communication network's main function is to transfer inter-agent constraints among agents according to a protocol written in equational Horn clause language. These constraints include application dependent data and, most importantly, inter-agent synchronization. The inter-agent synchronization strategy is very simple. An agent is synchronous with respect to the network if its inter-agent constraint multiplier is continuous with respect to the current instance of its active knowledge. Since the equational Horn clause format allows for the effective test of continuity (which is implicitly carried out by the inferencer in the Knowledge Decoder), a failure in the Knowledge Decoder theorem results in a corrective action toward re-establishing continuity with respect to the topology defined by the current instance of the knowledge base(see [27])).

The specification of the geometry of the network, as a function of time, is dictated primarily by global observability. By global observability, we mean closure of the knowledge of the system as whole relative to the scope the system reactivity. One of the central research tasks of this study is to provide knowledge in the equational clause format to characterize global observability. Preliminary analysis suggests that stability and goal reachability are closely dependent on global observability among the agents.
this concludes the description of our architecture. In the next section we illustrate some of its characteristic with a two- cell autonomous manufacturing example.

3 Current Status, Conclusions and Future Research

Multiple Agent Hybrid Control Architecture is a computational frameworprk for the real-time implementation of reactive, intelligent distributed controllers. The central mechanism of the architecture is on-line restrictive mechanical theorem proving over a theory of rational forms. The architecture is composed of a collection of agents and an inter-agent connectivity network. The architecture is especially tailored for systems involving multiple decision makers without a prespecified inter decision protocol. A particular example of this type of system is manufacturing shop floor control systems. We list the main characteristics of the architecture next.

Real-Time: Constraints for real-time performance are explicit and part of the knowledge base

Reactive: The theorem proving function of each agent on the architecture operates according to a first principles feedback paradigm

Adaptive: The knowledge base of each agent is open and modifiable by sensory data. Theorem failure triggers tuning and corrective action

Distributive with Coordination: The theorem proving is carried out distributively over the agents the coordination scheme is implicit without umpire

Dynamic Hierarchization: The architecture can operate simultaneously at different levels of abstraction

Figure of Merit: The behavior of the closed loop distributive system is determined by proving that there exist a command trajectory that minimizes a goal functional

This architecture has been exercised with a simulation of a simple 2-agent 2-cell factory.

A number of theoretical results are summarized next.. We formulated the multiple-agent problem in terms of our Declarative Hybrid Control model and identified architectural elements and their functionality. In particular we have indentified the inference mechanisms required. for deductive, distributed inference without umpire. We determined that. communication actions, represented as constraints in the variational problem stating the desired behavior of each agent, fulfill these coordination functions.

We also developed a canonical way to represent networks of controllers. We showed that, given a connectivity graph with N nodes (controllers) and the corresponding agent's knowledge bases, a network of 2N agents can be constructed with the same input-output characteristics, so that each agent interacts only with another companion agent, whose knowledge base is an abstraction of the knowledge in the network. This, in general, the multiple-agent controller for any topology is reduced to a set of agent pairs.

We called that agent of the pair that represents network information the Thevenin agent, after the author of a similar theorem in electrical network theory. The proof carried out by the Thevenin agent generates, as a side effect, coordination rules that define what and how often to communicate to other agents. These rules also define what the controller needs from the network to maintain intelligent control of its physical plant.

We developed a canonical way to represent the state of an agent by representing the agent's theorem-proving process as a data structure called a "proof automaton." The inference process within a controller is represented by a KSE Equation, in which the Matrix elements represent rule applications to states in the automaton and states contain the true sentences. States having sentences about the network form a submatrix that pulls communication from those specific other agents. We derived the required communication protocol among agents this way. The protocol is pulled in the sense that what an agent sends to other agents is defined by what the other agent uses in its proof automaton.

The inference process is represented as a recursive variational problem in which the criterion is an integral of a function called the logic Lagrangian over time. The logic Lagrangian maps the cartesian product of equational rules and inference principles to the real line, thus effectively providing a hill-climbing heuristic for the inference strategy of the theorem prover. In declarative control, the inference steps play a role analogous to action signals in conventional control.

We showed that for any measure of performance of the system, one obtains performance better than, or at least equal to, that of the multiple-agent version by constructing a single super-controller whose knowledge base contains all the knowledge of the system.

Finally, we discuss some preliminary results on the basis of our manufacturing example. We characterized movement of material through a factory in terms of a production rate function. The model we developed provides one controller for each workstation in the workcell. That controller needs an evaluation function on its own set of possible actions to measure the effect of each possible action on system performance. We sought to define a flow function in terms of movement of material available in the holding area. If a workstation broke, no network effect was observed until the holding area for that workstation was saturated.

A simulator was built to generate workstation states and provide controllers with evolving sensor updates and actions. The simulator had a finer model of the workstations than either controller. This allowed the controllers to demonstrate some adaptability with respect to holes in their knowledge of their own physical plant. We simulated small segments of orders (10 to 80) with significant variation in order size, due dates, and priorities. We noted that the system managed to react to high-priority orders by splitting them (it was not built with any order splitting strategy – the figure-of-merit measurements forced this behavior).

In our implementation, the proof automata had several hundred states, and terminal states had 50 to 100 associated data items. This sizing resulted from a model with order priority, order size, due date, and holding area as the key operative constraints.

This model could be extended in essentially two directions. In the upstream direction, the model could contain means for anticipated order arrivals. The other direction would be to provide finer models of the physical plants involved, potentially down to the level of physical device control (e.g., numerically controlled machining). This study suggested that the model size (number of states of proof automaton) is cubic in the number of constraints; addition of a constraint such as cutter speed rate limits would increase automata size by about 50%.

Finally we have characterized the coupling between the evolution an logic elements of our model as a continuity condition. This aspect is central on our future research plans.

References

[1] Kohn, W. "Declarative Control Architecture" CACM Aug 1991, Vol 34, No 8.

[2] Nerode A. and W. Kohn " An Autonomous Systems Control Theory: An Overview" Proc. IEEE CACSD'92, March 17-19 Napa Ca.

[3] Mettala, E. "Domain Specific Architectures" Proceedings of the Workshop on Domain Specific Software Architectures, pp 193–231, July 9–12, 1990, Hidden Valley, PA.

[4] Coleman, N. "An Emulation/Simulation for Intelligent Controls" Proceedings of the Workshop on Software Tools for Distributed Intelligent Control Systems, pp 37– 52, July 17–19, 1990, Pacifica, CA.

[5] Mann, R.C. "A Hardware/Software Environment to Support R&D Intelligent Machines and Mobile Robotic Systems" Proceedings of the Workshop on Software Tools for Distributed Intelligent Control Systems, pp 203–207, July 17–19, 1990, Pacifica, CA.

[6] Iserman, R. "Digital Control Systems" Springer Verlag, NY, 1977.

[7] Rutherford, A. "The Mathematical Theory of Diffussion and Reaction in Permeable Catalysts" Vol. 2, Clarendon Press, Oxford, 1975.

[8] Singh, M.G. "Dynamical Hierarchical Control" North Holland, MY, 1977.

[9] Garcia, H.E. and A. Ray "Nonlinear Reinforcement Schemes for Learning Automata" Proceedings of the 29th IEEE CDC Conference, Vol. 4, pp 2204–2207, Honolulu, HA, Dec. 5–7, 1990.

[10] Kaplan, M. "Modern Spacecraft Dynamics & Control" John Wiley & Sons, NY, 1976.

[11] Kohn, W. "Declarative Hierarchical Controllers" Proceedings of the Workshop on Software Tools for Distributed Intelligent Control Systems, pp 141–163, Pacifica, CA, July 17–19, 1990.

[12] Kohn, W. and T. Skillman "Hierarchical Control Systems for Autonomous Space Robots" Proceedings of AIAA Conference in Guidance, Navigation and Control, Vol. 1, pp 382–390, Minneapolis, MN, Aug. 15–18, 1988.

[13] Kohn, W. "A Declarative Theory for Rational Controllers" Proceedings of the 27th IEEE CDC, Vol. 1, pp 131–136, Dec. 7–9, 1988, Austin, TX.

[14] Mesarovic, M. and Y. Tashahara "Theory of Hierarchical Multilevel Systems" Academic Press, NY, 1970.

[15] Young, L.C. "Optimal Control Theory" Chelsea Publishing Co., NY, 1980.

[16] Padawitz, P. "Computing in Horn Clause Theories" Springer Verlag, NY, 1988.

[17] Schoppers, M. "Automatic Synthesis of Perception Driven Discrete Event Control Laws" Proceedings of the 5th IEEE International Symposium on Intelligent Control, Sept. 1990, Philadelphia, PA.

[18] Kohn, W. "Declarative Multiplexed Rational Controllers" Proceedings of the 5th IEEE International Symposium on Intelligent Control, pp 794–803, Philadelphia, PA, Sept. 5, 1990.

[19] Kowalski, R. "Logic for Problem Solving" North Holland, NY, 1979.

[20] Lloyd, J.W. "Foundations of Logic Programming" second extended edition, Springer Verlag, NY, 1987.

[21] Robinson, J.A. "Logic: Form and Function" North Holland, NY, 1979.

[22] Kohn, W. "Rational Algebras; a Constructive Approach" IR&D BE-499, Technical Document D-905-10107-2, July 7, 1989.

[23] Kohn, W. "The Rational Tree Machine: Technical Description & Mathematical Foundations" IR&D BE-499, Technical Document D-905-10107-1, July 7, 1989.

[24] Skillman, T., W. Kohn, et.al. "Class of Hierarchical Controllers and their Blackboard Implementations" Journal of Guidance Control & Dynamics, Vol. 13, N1, pp 176–182, Jan.–Feb., 1990.

[24] Nii, P.H. "Blackboard Systems: The Blackboard Model of Problem Solving and the Evolution of Blackboard Architectures" the AI Magazine, Vol. 7, No. 2, Summer 1986, pp 38–53.

[25] Dodhiawala, R.T., V Jagoenathan and L.S. Baum "Erasmus System Design: Performance Issues" Proceedings of Workshop on Blackboard Systems Implementation Issues, AAAI, Seattle, WA., July 1987.

[26] Kohn, W. "Application of Declarative Hierarchical Methodology for the Flight Telerobotic Servicer" Boeing Document G-6630-061, Final Report of NASA-Ames research service request 2072, Job Order T1988, Jan. 15, 1988.

[27] Kohn, W. "Advanced Architectures and Methods for Knowledge-Based Planning and Declarative Control" IR&D BCS-021, will be presented as a paper in ISMIS'91, Oct. 1991.

[28] Kohn, W. and A. Murphy "Multiple Agent Reactive Shop Floor Control" will be presented as a paper in ISMIS'91, Oct. 1991.

[29] Liu, J.W.S. "Real-Time Responsiveness in Distributed Operating Systems and Databases" proceedings of the Workshop on Software Tools for Distributed Intelligent Control Systems, Pacifica, CA., July 17–19, 1990, pp 185–192.

[30] Warga, K. "Optimal Control of Differential and Functional Equations" Academic Press, NY., 1977.

[31] Kuich, W., A. Salomaa "Semirings, Automata, Languages" Springer Verlag, NY., 1985.

Proceedings of the 31st Conference
on Decision and Control
Tucson, Arizona • December 1992

On Markovian Fragments of COCOLOG for Logic Control Systems

Y.J.Wei* Peter E. Caines+

*Department of Electrical Engineering, McGill University, 3480 University Street, Montreal, P.Q, Canada H3A 2A7, Tel(514)-398-8207

+Department of Electrical Engineering, McGill University, 3480 University Street, Montreal, P.Q., H3A 2A7, and the Canadian Institute for Advanced Research, Canada
Tel.(514)-398-7129,
em:peterc@moe.mcrcim.mcgill.edu,

Abstract

The COCOLOG system is a partially ordered family of first order logical theories that describe the controlled evolution of the state of a given partially observered finite machine $\mathcal{M}$. The initial theory of the system, $Th_0 \triangleq Th(o_1^o)$, gives the general theory of $\mathcal{M}$ without any data being given on the initial state. Later theories, $Th(o_1^k), k \geq 1$, depend upon the (partially ordered lists of) observed input-output trajectories, where new data is accepted sequentially into the subsequent theories in the form of the *new axioms* $AXM^{obs}(L_k)$. The inputs u^k are determined via the solution of control problems posed in each theory in the form of *control axioms* $AXM^{cntl}(L_k)$. In order to create an efficient, recursive, control reasoning system, it is natural to seek a restricted version of COCOLOG in which a minimum amount of information is communicated from one theory to the next, and which is associated with the restricted set of candidate (present and future state dependent) control problems. Due to the overall dynamical setting of the problem, it is natural to formulate what we shall call Markovian fragment of the general COCOLOG system. This idea is introduced in this paper via the definition of a restricted set of axioms $MTh(o_1^k)$. Along with other results, we have proved that these fragments have the same state estimation and control theoretic power compared to full COCOLOG theories.

1 Introduction

The COCOLOG system, introduced by P.E. Caines and S. Wang [CW90,91,W91], is a partially ordered family of first order logical theories that describe the controlled evolution of the state of a given partially observered finite machine $\mathcal{M}$. The initial theory of the system, $Th_0 \triangleq Th(o_1^o)$, gives the general theory of $\mathcal{M}$ without any data given on the initial state. Later theories, $Th(o_1^k), k \geq 1$, through their axiom sets Σ_k, depend upon the (partially ordered lists of) observed input-output trajectories, where new data is accepted sequentially into the subsequent theories in the form of the *new axioms* $AXM^{obs}(L_k)$. The inputs $U(k)$ are determined by the solution to control problems posed in each theory in the form of *control axioms* $AXM^{cntl}(L_k)$. An important class of control problems involve the reachability predicate $Rbl(x, y, l)$, defined axiomatically in each theory, corresponding to the reachability of y from x in l steps. The solution to one problem that we may specify of this type would be the first control in a sequence of controls giving a minimal length path to y from the current state x.

The implementation of control reasoning in COCOLOG requires efficient automatic theorem proving methodologies, and the development of FE-resolution is one step in this direction which has recently been shown to be implementable. Another important development which is required is the definition of tractable fragments of the full COCOLOG theory with the following property: they would carry enough information to enable classes of significant control problems to be posed (through the axioms) and solved within the resulting framework of partially ordered first order theories. Due to the overall dynamical setting of the problem, it is natural to formulate a Markovian fragment of the general COCOLOG system. This idea is introduced in this paper via the definition of a restricted set of axioms for the new set of theories $MTh(o_1^k)$. The new set of axioms communicate only the basic dynamical properties of the machine under control plus the most recent new observations and the state estimate generated in the most recent COCOLOG theory; in addition, an updated version of a single control problem is carried along and this must be phrased only in terms of the other predicates and axioms available in the restricted theories $MTh(o_1^k), k \geq 1$. It is shown that this construction yields the desired family of Markovian fragments of COCOLOG theories .

2 COCOLOG

The reader is referred to [CW90,91,W91,CW92] for a full exposition of all terms and expressions which are not completely explained here.

2.1 Syntax and Semantics of COCOLOG

As stated above, COCOLOG theories describe finite machines. Formally these are defined as follows:

Definition 2.1 A *(partially observed) finite (input-state-output) machine is a quintuple* $\mathcal{M} = (\mathbf{X}, \mathbf{U}, \mathbf{Y}, \mathbf{\Phi}, \eta)$, where $\mathbf{X}$ is a (finite) set of *states*, $\mathbf{U}$ is a (finite) set of *inputs*, $\mathbf{Y}$ is a (finite) set of *outputs*, $\mathbf{\Phi} : \mathbf{X} \times \mathbf{U} \rightarrow \mathbf{X}$ is a *transition function*, $\eta : \mathbf{X} \rightarrow \mathbf{Y}$ is an *output function.* □

We always use bold face letter to distinguish semantic objects from the symbols in the first order language that describe them. For the purpose of describing such machines in first order language, we need the symbol set $S(L_0)$ that contains the *Constant Symbol Set* $Const(L_0)$; the *Variable Symbol Set* $Var(L_0)$; the *Function Symbols* $Func(L_0)$ that consist of the symbols $\{\Phi(\ ,\ ), \eta(), +_{K(N)}, -_{K(N)}\}$.; the *Atomic Predicate Symbols* $Pre(L_0) = \{Eq\ ,\ Rbl\}$ and *Logical Symbol Set* $\{\forall, \rightarrow, \perp\}$.

Any *well formed formula* wff of L is given by the standard *Backus-Naur* syntactic rules, The set of such formulas will be denoted $WFF(L_0)$.

CH3229-2/92/0000-2967$1.00 © 1992 IEEE

2.2 Axiomatic Theory of Th_0 and $Th(o_1^k)$

Th_0 has a set of *logical axioms*, a set of *equality axioms* for an equality predicate, a set of *arithmetic axioms* and a set of *special axioms* which specify true facts concerning the subject that the logic describes (in at least one of its interpretations). Correspondingly, $Th(o_1^k)$ is a logical theory that has the *observation axioms*, the *state estimation axioms* and the *control axioms* (see below) added to the logical theory Th_0.

Finite Machine Axioms

The special axiom set of Th_0 corresponds exactly to the state transitions and output map relations of the given machine $\mathcal{M}$:
State Transition Axioms:

$$AXM^{dyn}(L) \equiv \{Eq(\Phi(x^i,u^i),x^j);x^i,x^j \in X,u^i \in U\}$$

, where $\mathbf{\Phi(x^i,u^i)=x^j}$, and
Output Axioms:

$$AXM^{out}(L) \equiv \{Eq(\eta(x^i),y^j);x^i \in X,y^j \in Y\},\ \text{where } \eta(\mathbf{x^i})=\mathbf{y^j}.$$

Reachability Axioms: denoted by $AXM^{Rbl}(L)$, are recursively defined for the *reachability predicate Rbl* by the following:

0. $\forall x \forall x', Eq(x,x') \longleftrightarrow Rbl(x,x',0)$
1. $\forall x \forall x', (\exists u, Eq(\Phi(x,u),x')) \longleftrightarrow Rbl(x,x',1)$
2. $\forall x \forall x'' \forall l, Eq(l,K(N)+1) \vee [\{\exists x' \exists u, Rbl(x',x'',l) \wedge Eq(\Phi(x,u),x')\} \leftrightarrow Rbl(x,x'',l+_L 1)]$

The reachability axioms specify the l step reachability relation $Rbl(x,x',l)$ among any pair of states x,x'. We note that in these formulas the variables x,x',x'' range over X, the variable u ranges over U and l ranges over the integers $0,1,\cdots,K(N)+1$. We note that Axiom 2 excludes consideration of the infinity case in order to characterize reachability on the finite numbers in the arithmetic.
Size Axioms: Denoted by AXM^{size}, these specifies such a restriction that any model of this axiom set must have a domain that contains exactly $|\mathbf{X}|$ state objects, $|\mathbf{Y}|$ output objects, $|\mathbf{U}|$ input objects and $\mathbf{K(N)}$ integers. This restriction is natural since the controlled machine is fixed.
Finite Arithmetic Axioms:AXM^{arith}, these define the arithmetical operations($+_K(N)$ and $-_{K(N)}$) on the initial segment of the natural number $[0,K(N)]$.
Equality Axiom: Denoted by AXM^{Eq}, these consist of the basic definition for equality and the substitution axioms for every functional symbols and predicate symbols in the language.
Logic Axioms: AXM^{logic}, This is a set of axiom schematas for first order logic.
We write Σ_0 for the union of the above axiom sets of L_0, i.e.,

$$\Sigma_0 = \{AXM^{arth}(L), AXM^{dyn}(L), AXM^{out}(L), AXM^{Rbl}(L), AXM^{Eq}(L), AXM^{logic}\}.$$

Rules of Inference:
The rules of inference in Th_0(and all $Th(o)1^k$)) will be *Modus Ponens* and *Generalization* .

2.3 Axiomatic Theory of $Th(o_1^k)$

At each instant k, the system $\mathcal{M}$ generates the observed value of $Y(k)$ and the value for the new action $U(k)$ is needed to decide, we need to extend our language as well as the axiom set to reason about the system.

The language $L_k \triangleq L(o_1^k)$ is an extension of the language L_0 obtained by adding new constant symbols and predicates symbols in the following way:

$$L_k \triangleq L(o_1^k) = L_0 \bigcup_{j=1}^{k}\{U(j),Y(j)\} \bigcup_{j=1}^{k}\{CSE_j()\},$$

here $U(j),Y(j)$ are new constants, and $CSE_j()$ is new predicate which is the current state estimation predicate at time j.

Concerning the syntax of L_k, we shall only remark that the variables and constants are sorted and the well formed formulae parse according to the Bakus-Naur rules in each Σ_k(and hence in each Th_k). We also adopt the conventional set theoretic model theory for each of the axiomatic systems (see [CW90,91,W91]).

Observation Axioms: At each instant k the observer receives $\mathbf{U(k-1)}$ and $\mathbf{Y(k)}$; for the constants $\mathbf{u^i}$ and $\mathbf{y^i}$ such that $\mathbf{u^i=U(k-1)}$ and $\mathbf{y^i=Y(k)}$, the following set of formulas express the fact that these observations are added incrementally, as axioms, to Th_0 to form the theory $Th(o_1^k)$ for $k \geq 1$, we subject to the convention that the second axiom below holds only in case $k>1$):

$$AXM^{obs}(L^k) \equiv \{Eq(Y(k),y^i), Eq(U(k-1),u^i)\}.$$

State Estimation Axioms: $AXM^{est}(L_k)$. The following axioms express in axiomatic form the recursive formulas for the current state estimate sets. In case $k=1$:

$$Eq(\eta(x^i),Y(k)) \longleftrightarrow CSE_1(x^i). \qquad 1 \leq i \leq |X|$$

In case $k>1$:

$$\exists x, CSE_{k-1}(x) \wedge Eq(\Phi(x,U(k-_L 1)),x^i) \wedge Eq(\eta(x^i),Y(k)) \leftrightarrow CSE_k(x^i). \qquad 1 \leq < i \leq |X|$$

Control Axioms: $AXM^{cntl}(L_k)$) The following is the general form of a set of *Conditional Control Axioms*, where $C^j(\cdot)$ is a *conditional control formula* lying in $WFF(L(o_1^k))$:

$$\begin{aligned} C^1(WFF(L_k)) &\longrightarrow Eq(U(k),u^1) \\ \neg C^1(WFF(L_k)) \bigwedge C^2(WFF(L_k)) &\longrightarrow Eq(U(k),u^2) \\ \vdots \quad &\longrightarrow \quad \vdots \\ \bigwedge_{j=1}^{m-1}(\neg C^j(WFF(L_k))) \bigwedge C^m(WFF(L_k)) &\longrightarrow Eq(U(k),u^m) \\ \bigwedge_{j=1}^{m}(\neg C_j(WFF(L_k))) &\longrightarrow Eq(U(k),u^*). \end{aligned}$$

□

We shall write

$$\Sigma_k = \Sigma_0 \bigcup_{j=1}^{k}\{AXM^{obs}(L_j), AXM^{est}(L_j), AXM^{cntl}(L_j)\},$$

and let $Th(o_1^k)$ denote the theory generated from Σ_k.

$AXM^{cntl}(L_k)$ is central to the construction of COCOLOG. They have the following interpretation: If the condition $C^1(\cdot)$ is provable in the theory $Th(o_1^k)$, then, invoking the first axiom, we obtain the defined constant value u^1 as the value of the control constant $U(k)$; if not, but if $C^2(\cdot)$ can be proved, then the second axiom gives the defined value u^2 to the control constant $U(k)$; and so on. If none of the conditions $C^1, C^2, \cdots, C^m$ hold, then the last axiom sets the control function equal to the arbitrary constant u^*. This procedure uniquely determines the value of $U(k)$. When $k \rightarrow k+1$, we make the meta-logical step of passing to the theory $Th(o_1^{k+1})$ carrying along all the previous axioms including the constant value u^i chosen above. This is formally enforced by the above definition of the axiom set generating $Th(o_1^k)$. Hence, in the new $Th(o_1^{k+1})$, the observed control action $U(k)$ is precisely the constant value u^i determined in $Th(o_1^k)$.

Concerning the size of the axiom set at time k, we have:

Linear Bound Lemma For the axiom sets Σ_k defined above, it is the case that

$$|\Sigma_k| = |\Sigma_0| + k(|X| + |U| + 2). \quad (1)$$

□

The increment of the sizes of axiom sets is linear with respect to the time index.

Some important properties of COCOLOG families of theories are given in the following theorems.

Theorem 2.1 (Unique Model Property)[CW90,W91] *The logical theory generated by Σ_k has a unique model up to isomorphism.* □

Theorem 2.2 (Decidable Theoremhood)[CW90,W91] *The logical theory as generated by Σ_k for any given finite machine $\mathcal{M}$ is decidable.* □

Theorem 2.3 (The Nesting Theorem)[CW90,W91] $Th(o_1^k) \subset Th(o_1^{k+1})$ *for all* $k \geq 0$ □

Since the discussion in this paper is independent of the observation information o_1^k, we omit, from now on, the indication of the particular observation sequence when the context is clear. For example, we will write Th_k to replace $Th(o_1^k)$.

3 Language Fragment

The full COCOLOG language defined in the previous section has the power to express the whole observation history of the system and this gives rise to a monotonic evolution of the theories $\{Th_k, k \geq 0\}$(see the Nesting Theorem above) which, in particular, permits reference to the past. For example, one may write down a formula to express the following control law: *If the first control has not been invoked since the beginning of the process, then invoke it now.* The desired well-defined formula will involve the whole collection of languages. On the other hand, such expressive power is unnecessary for the purposes of control with respect to control criteria depending on present and future states and outputs, since the dynamical system $\mathcal{M}$, by definition, is current state dependent. In particular this avoids the unbounded increment in ATP complexity due to the increase in the number of eligible well-formed formulas in the successive axiom sets Σ_k. In this section, we introduce *Markovian language fragments* of the languages L_k, in order to deal only with the information expressed in a certain moving windows of specifical syntactic terms.

Definition 3.1 The *Markovian fragment* L_k^m of L_k is defined as

$L_k^m = L_0 \cup \{CSE_k, CSE_{k-1}\} \cup \{U(k-1), U(k), Y(k-1), Y(k)\}$.

□

From this definition, we immediately know that this language is a sublanguage of the full COCOLOG language L_k and we note that L_k^m has a fixed number of symbols which are fewer than those in L_k.

Syntax of L_k^m:

(1) $Const(L_k^m) = Const(L_0) \bigcup \{U(k-1), U(k), Y(k-1), Y(k)\}$,

which is a proper subset of $Const(L_k)$. Hence, L_k^m has fewer constant symbols.

(2) $Pre(M\Sigma_k) \equiv \{Rbl, Eq, CSE_{k-1}, CSE_k\}$,

(3) L_k^m has the same set of functional symbols as L_k, i.e. $Func(L_k^m) = Func(L_k)$.

□

Definition 3.2 The set of *well-formed-formulas(wffs)* $WFF(L_k^m)$ of L_k^m is defined in the usual way except that the only permitted atomic formulas are instances of Rbl, CSE_k, CSE_{k-1}, Eq. □

Any well-formed formula consisting of predicates beyond those above ones will be illegal in $WFF(L_k^m)$ by the above definition. Consequently, L_k^m can not, for instance , express the state estimation formulas concerning the state at time $k-2$. In general we have for $k \geq 1$:

$$WFF(L_k^m) \subset WFF(L_k).$$

Intuitively, this language fragment can only express information that relates to the most recent change and the current configuration of the controlled system.

4 Construction of $M\Sigma_k$

Here we give the axiom sets $M\Sigma_k$ for the Markovian fragment theories MTh_k, each of which is nested within the corresponding full COCOLOG theory Th_k. We shall make the restriction that the admissible control objectives within MTh_k shall only refer to the current and future state (estimate) behavior of the controlled system. Then we select axiom sets for the fragmentary theories $MTh_k \subset Th_k$ so that they should carry enough information to permit the deduction of control actions for this class of control objectives. This enables us to limit the information to be transferred into MTh_k at any instant k to that necessary to deduce the current state estimation at k, or more precisely, to deduce the set of states satisfying the predicate CSE_k in Th_k.

Consider that the temporal structure of the fragment sequences $\ldots, MTh_{k-1}, MTh_k, MTh_{k+1}, \ldots$, each containing the state estimate predicates $\ldots, CSE_{k-1}, CSE_k, CSE_{k+1}, \ldots$ and each of them nested respectively within $\ldots, Th_{k-1}, Th_k, Th_{k+1}, \ldots$ We observe in passing that (1) this is the logical analogue of the generation of the state estimate in a linear stochastic control problem , and (2), by handing on a critical subset of the theorems of MTh_{k-1} as a part of axiom set for MTh_k, a certain form of learning may be said to take place, since these theorems do not have to be deduced again from more elementary information in axiomatic form.

The following definition below specifies the axiom set $M\Sigma_k$ to be a certain combination of *(i)* the system dynamics, reachability and system size axioms , *(ii)*, a set of axioms carrying the most recent state estimate theorems, together with *(iii)* the most recent observation axioms expressed via the equality predicate.

Definition 4.1 The axiom set of a Markovian fragment theory is defined as follows:

$$M\Sigma_0 = \Sigma_0 \quad (2)$$

$$M\Sigma_1 = \Sigma_1 = \Sigma_0 \cup AXM^{special}(L_1), \quad (3)$$

$$M\Sigma_k = M\Sigma_0 \cup AXM^{special}(L_k) \cup K(\Sigma_{k-1}), \quad k > 1, \quad (4)$$

where $AXM^{special}(L_k)$ denotes the following union:

$$AXM^{cntl}(L_k) \cup AXM^{est}(L_k) \cup AXM^{obs}(L_k), \quad (5)$$

and where $K(\Sigma_{k-1},)$ is defined as follows:

$$K(\Sigma_0) = \emptyset; \quad (6)$$

$$K(\Sigma_{k-1}) =$$

$$\{?CSE_{k-1}(x^i); ?CSE_{k-1}(x^i) \in Th_{k-1}, \quad ? \in \{\neg, \emptyset\}, x^i \in X\}, \quad (7)$$

The notation ? indicates the positive or negative assertion of the predicate which follows it. □

Compared to Σ_k, $M\Sigma_k$ contains fewer axioms. But $M\Sigma_k \not\subset \Sigma_k$. Informally, $K(\Sigma_k)$ carries the state estimate information from MTh_k to MTh_{k+1}.

Lemma 4.1 (Bounded Size Lemma) With the definition of Markovian fragment given above,

$$|M\Sigma_k| = |\Sigma_0| + |X| + |U| + 2. \qquad k \geq 1. \quad (8)$$

□

5 Semantics of $M\Sigma_k$

The goal of this section is to investigate the semantics of the Markovian fragment theories. A comparison will be made between the fragments and their full version counterparts. The investigation will be carried out by using Herbrand structure (or $H-$structure for short) because of the following well known theorem.

The Herbrand Theorem A set of first order clauses is inconsistent if and only if it is unsatisfiable under all $H-$ structure.

□

Definition 5.1 The *Herbrand universe* of Σ_k (respectively $M\Sigma_k$), denoted by $H(\Sigma_k)$ (respectively $H(M\Sigma_k)$) is given by

$$\{x^i, u^j, \ldots U(k-2), \Phi(x^i, u^j), \Phi(\Phi(x^i, u^j), u^p), \eta(x^i), \eta(\eta(x^i)), \ldots\},$$

and, respectively,

$$\{x^{\cdot}u^j, \Phi(x^i, u^j), \Phi(x^l, U(k-1)), \Phi(x^l, U(k)), \ldots\},$$

where x^i, u^j range over X and U respectively. □

It is clear that

$$H(M\Sigma_k) \subset H(\Sigma_k).$$

Definition 5.2 Let $Pre(\Sigma_k)$ be the set of predicates of Σ_k, then the *Herbrand base* of Σ_k, denoted $HB(\Sigma_k)$, is the following set

$$\{Rbl(t_1, t_2, t_3), Eq(t, t), CSE_{k-1}(t_1), CSE_k(t_2) : t_1, t_2, t_3, t \in H(\Sigma_k)\}.$$

$HB(M\Sigma_k)$, which is a proper subset of $HB(\Sigma_k)$, can be defined in the same way. □

Definition 5.3 A *Herbrand Structure* of Σ_k is a pair $< I, \mathbf{D} >$, denoted by $\mathcal{H}$, where I is the *interpretation mapping* which maps terms into the domain $\mathbf{D}$, where $\mathbf{D} \neq \emptyset$. For this structure, each formula in Σ_k will be recursively assigned a truth value $V(F) \in \{0,1\}$ in the following way:
(1) for any $F = P(t) \in H(\Sigma_k)$, $V(F) = 1$ iff $I(F) \in I(P)$;
(2) for $F = F_1 \vee F_2$ $V(F) = 1$ iff either $V(F_1) = 1$ or $V(F_2) = 1$;
(3) for $F = \neg F_1$, $V(F) = 1$ iff $V(F_1) = 0$;
(4) for $F = F_1 \rightarrow F_2$, $V(F) = 1$ iff either $V(F_1) = 0$ or $V(F_2) = 1$;
(5) for $F = \forall x F_1$, $V(F) = 1$ iff $V(F_1(t/x) = 1$ for all $t \in H(\Sigma_k)$;
(6) for $F = \exists a F_1$, $V(F) = 1$ iff $V(F_1(t/x) = 1$ for some $t \in H(\Sigma_k)$.
If all of formulas in Σ_k have truth value 1 under this structure, then we called it $\mathcal{H}$ *model* for Σ_k and denoted it by $\mathcal{H} \models \Sigma_k$. □

Definition 5.4 Let $\mathcal{H}$ be an $H-$model of Σ_k, then we can define an $H-$ model for $M\Sigma_k$, denoted $\mathcal{H}_m$, any $H-$structure satisfying the following:

< i >	$\mathbf{D_m} = \mathbf{D}$;
< ii >	$I_m(\Phi) = I(\Phi)$
< iii >	$I_m(\eta) = I(\eta)$
< iv >	$I_m(+_{K(N)}) = I(+_{K(N)})$
< v >	$I_m(-_{K(N)}) = I(-_{K(N)})$
< vi >	$I_m(c) = I(c)$ for all $c \in L_k^m$
< vii >	$I_m(Eq) = I(Eq)\|_{L_k^m} = \{(\mathbf{d}, \mathbf{d}) : \mathbf{d} \in \mathbf{D}\}$
< viii >	$I_m(Rbl) = I(Rbl)\|_{L_k^m} \subset \mathbf{X} \times \mathbf{X} \times \mathbf{I_{K(N)}}$
< ix >	$I_m(CSE_{k-1}) = I(CSE_{k-1})\|_{L_k^m} \subset \mathbf{X}$
< x >	$I_m(CSE_k) = I(CSE_k)\|_{L_k^m} \subset \mathbf{X}$

□

As the matter of fact, $< I_m, \mathbf{D_m} >$ can be seen to be a restriction of $< I, \mathbf{D} >$ in the following sense:

$$\mathbf{D_m} = \mathbf{D}, \qquad \text{and} \qquad I_m = I|_{L_k^m}\ .$$

It is clear that $\mathcal{H}_m = \mathcal{H}|_{HB(M\Sigma_k)}$. We list some trivial results concerning the models of Σ_k and $M\Sigma_1$ below:

Theorem 5.1 Let $I_m, \mathbf{D_m}$ be defined as the above, then
(1) for any term $t \in L_k^m$, $I_m(t) = I(t)$;
(2) for any ground atomic formula $S \in WFF(L_k^m)$, $\mathcal{H}_m \models S$ if and only if $\mathcal{H} \models S$;
(3) for a propositional formula $P \in WFF(L_k^m)$, $\mathcal{H}_m \models P$ if and only if $\mathcal{H} \models P$;
(4) for a formula $F \in WFF(L_k^m)$ with free variable x, $\mathcal{H}_m \models \exists x F$, implies $\mathcal{H} \models \exists x F$.
(5) for a formula $F \in WFF(L_k^m)$ with free variable x, $\mathcal{H} \models \forall x F$ implies $\mathcal{H}_m \models \forall x F$.
Proof (1) and (2) are obvious since they follows directly from the definitions. (3) is obtained from the truth tables. (4) and (5) are also obvious since $HB(M\Sigma_k) \subset HB(\Sigma_k)$. □

One may ask such question as : *Is it the case that* $\mathcal{H} \models \exists x F \Rightarrow \mathcal{H}_m \models \exists x F$?. Generally speaking, the answer will be *no* without conditions on F . The reason is that the *witness* of this bound variable may lie outside $H(M\Sigma_k)$ and hence is not eligible as a witness with respect to $\mathcal{H}_m$. An analogous problem may arise for the universal quantifier, as in : $\mathcal{H}_m \models \forall x F \Rightarrow \mathcal{H} \models \forall x F$?

In the rest of this section, we are going to give positive answers for the above two questions. First we show

Theorem 5.2 For any formula $F \in WFF(L_k^m)$ with free variables x, if $\mathcal{H} \models \exists x F$, then $\mathcal{H}_m \models \exists x F$.

□

Theorem 5.3[CW90,W91]. $M\Sigma_k$ has a unique model up to isomorphism.

□

Theorem 5.6 For any $F \in WFF(L_k^m)$,

$$\mathcal{H} \models F \quad \text{iff} \quad \mathcal{H}_m \models F.$$

□

6 Proof Theory of $M\Sigma_k$

The theorems in the last section permit us to conclude that for any formula $S \in WFF(L_k^m)$ has the same truth value under both $\mathcal{H}$ and $\mathcal{H}_m$. This result, along with the completeness of each COCOLOG theory , yields the same provability power for $M\Sigma_k$ and Σ_k in terms of $WFF(L_k^m)$. In this section, we prove this result by use of proof theory.

Definition 6.1 A *proof sequence* for a formula A with respect to axiom set Σ_k (respectively $M\Sigma_k$) is an indexed list of formulas, in which A is the last one in the list, and any other formula in the list is either a logic axiom or a member of Σ_k (respectively $M\Sigma_k$), or a formula deduced from previous formulas in this list through modus pones or the generalization rule. This is written as $\Sigma_k \vdash A$ (respectively $M\Sigma_k \vdash A$).

□

Theorem 6.1 Let $F \in WFF(L_k^m)$, then

$$M\Sigma_k \vdash F \quad \Rightarrow \quad \Sigma_k \vdash F.$$

□

For the reverse implication, we will begin in Theorem 6.2 with the case where there are no logical connectives in F and treat the full case later on by using induction on the complexity of F.

Theorem 6.2 Let $S \in HB(M\Sigma_k)$ be an atomic formula with respect to Eq, Rbl, CSE_{k-1}. Assume $U(k)$ does not appear in S , then

$$\Sigma_k \vdash S \quad \text{implies} \quad M\Sigma_k \vdash S.$$

□

By induction on the length of an arbitrary propositional formula Q , we have

Theorem 6.3 For any propositional formula Q of L_k^m in which CSE_k and $U(k)$ do not appear,

$$\Sigma_k \vdash Q \quad \text{implies} \quad M\Sigma_k \vdash Q.$$

□

Theorems 6.2, 6.3 generalize to first order formulas:

Theorem 6.4 Let $F \in WFF(L_k^m)$,and assume CSE_k and $U(k)$ do not appear in F. Then

$$\Sigma_k \vdash F \quad \Rightarrow \quad M\Sigma_k \vdash F.$$

□

Theorem 6.5

$$\Sigma_k \vdash CSE_k(x^i) \quad \text{implies} \quad M\Sigma_k \vdash CSE_k(x^i) \quad \text{for any } x^i \in X.$$

Proof Suppose $\Sigma_k \vdash CSE_k(x^i)$, then $\Sigma_k \vdash E^i(L_k)$ by MP, $AXM^{est}(L_k)$. Now since CSE_k and $U(k)$ do not appear in $E^i(L_k)$, so $M\Sigma_k \vdash E^i(L_k)$ by Theorem 6.4. Then by MP, we have ,$AXM^{est}(L_k)$, $M\Sigma_k \vdash CSE_k(x^i)$.

□

Furthermore, we have

Theorem 6.6 Let $F \in WFF(L_k^m)$ and $U(k)$ does not appear in F. Then

$$\Sigma_k \vdash F \quad \text{implies} \quad M\Sigma_k \vdash F.$$

□

The following theorem says that the fragments defined in this paper have the same control theoretic power.

Theorem 6.7

$$\Sigma_k \vdash Eq(U(k), u^p) \quad \text{implies} \quad M\Sigma_k \vdash Eq(U(k), u^p).$$

□

7 Appendix: A Three State Machine Example

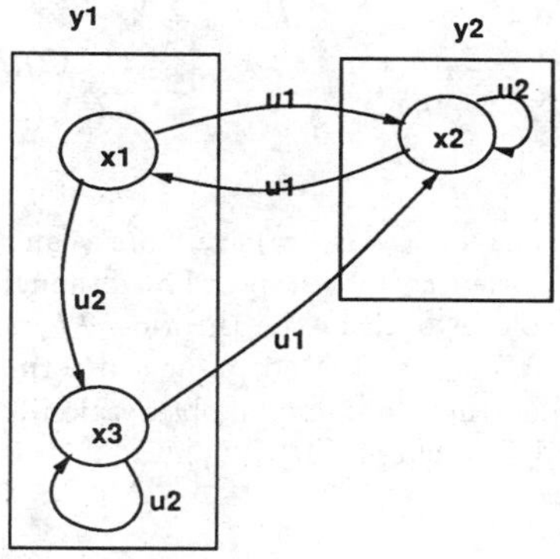

Figure 1: A Three State Machine

The axioms set AXM^{dyn} for this machine is the set of formulas:

$$Eq(\Phi(x^1, u^1), x^2) \quad Eq(\Phi(x^1, u^2), x^3) \quad Eq(\Phi(x^2, u^1), x^1)$$
$$Eq(\Phi(x^2, u^2), x^2) \quad Eq(\Phi(x^3, u^1), x^2) \quad Eq(\Phi(x^3, u^2), x^3)$$

and the set of axioms AXM^{out} are given as

$$Eq(\eta(x^1), y^1) \quad Eq(\eta(x^3), y^1) \quad Eq(\eta(x^2), y^2)$$

The following example will show how the proof in Σ_k can be shorten into a proof in $M\Sigma_k$.

Suppose the observation sequence is $(\emptyset, y^1), (u^1, y^2)$, the proof for $CSE_2(x^2)$ with respect to Σ_2 is:

$$
\begin{array}{lll}
(1) & Eq(Y(1), y^1) & AXM^{obs}(L_1) \\
(2) & Eq(Y(2), y^2) & AXM^{obs}(L_2) \\
(3) & Eq(U(1), u^1) & AXM^{obs}(L_2) \\
(4) & \forall x, Eq(\eta(x), Y(1)) \rightarrow CSE_1(x) & AXM^{est}(L_1) \\
(5) & Eq(\eta(x^1), y^1) & AXM^{out} \\
(6) & Eq(\eta(x^1), y^1) \rightarrow CSE_1(x^1) & (3), \text{Logic Axiom } MP \\
(7) & CSE_1(x^1) & (5), (4), MP \\
(8) & \exists x CSE_1(x) \wedge Eq(\Phi(x, U(1)), x^2) & \\
 & \wedge\, Eq(\eta(x^2), Y(2)) \rightarrow CSE_2(x^2) & AXM^{est}(L_2) \\
(9) & CSE_1(x^1) \wedge Eq(\Phi(x^1, U(1)), x^2) \wedge & \\
 & Eq(\eta(x^2), Y(2)) \rightarrow CSE_2(x^2) & AXM^{est}(L_2) \\
(10) & CSE_1(x^1) \wedge Eq(\Phi(x^1, U(1)), x^2) \wedge & \\
 & \wedge\, Eq(\eta(x^2), Y(2)) & \\
(11) & CSE_2(x^2) & (8), (9), MP
\end{array}
$$

However, a proof for $CSE_2(x^2)$ with respect to $M\Sigma_2$ is :

$$
\begin{array}{lll}
(1) & Eq(Y(2), y^2) & AXM^{obs}(L_2) \\
(2) & Eq(U(1), u^1) & AXM^{obs}(L_2) \\
(3) & CSE_1(x^1) & K(\Sigma_1) \\
(4) & \exists x CSE_1(x) \wedge Eq(\Phi(x, U(1)), x^2) & \\
 & \wedge\, Eq(\eta(x^2), Y(2)) \rightarrow CSE_2(x^2) & AXM^{est}(L_2) \\
(5) & CSE_1(x^1) \wedge Eq(\Phi(x^1, U(1)), x^2) & \\
 & \wedge\, Eq(\eta(x^2), Y(2)) \rightarrow CSE_2(x^2) & AXM^{est}(L_2) \\
(6) & CSE_1(x^1) \wedge Eq(\Phi(x^1, U(1)), x^2) & \\
 & \wedge\, Eq(\eta(x^2), Y(2)) & (1), (2), etc \\
(7) & CSE_2(x^2) & (5), (6), MP
\end{array}
$$

Here we see that having $CSE_1(x^1)$ available as an axiom in $M\Sigma_2$ has reduced the proof by four steps. Now although this formula would be available in both of the theories, Th_2 and MTh_2 for further proofs, MTh_2 has the advantage in terms of ATP efficiency that it does not contain the observation and estimation axioms $AXM^{obs}(L_1)$ and $AXM^{est}(L_1)$.

□

REFERENCES

[Cai88] P.E. Caines . Linear Stochastic Systems. *John Wiley ans Sons Inc.* New York, 1988

[CL73] Chin-Liang Chang and R. Char-Tung Lee. *Symbolic Logic and Mechanical Theorem Proving.* Academic Press, New York, 1973.

[CW90] P.E. Caines and S. Wang. "COCOLOG: A Conditional Controller and Observer Logic for Finite Machines",*Proc. of The 29th IEEE Conference on Decision and Control*, Hawaii, 1990). pp. 2845-2850. Complete version submitted for publication.

[CW91] Caines, P.E. and S.Wang," On a Conditional Observer and Controller Logic (COCOLOG) For Finite Machines and its Automatic Reasoning Methodology", *Proceeding of Recent Advances in Mathematical Theory of Systems, Control, Networks and Signal Processing, II*, Eds. H Kimura, S.Kodama , MTNS-91, KObe, Japan, June 1991, pp 49-54.

[W91] S. Wang. *Classical and Logic Based Control Theory for Finite State Machines*, Ph.D. Thesis, McGill University, Montreal, October, 1991.

[WC92] S.Wang and P.E. Caines. " Automated Reasoning with Function Evaluation for COCOLOG with Examples" *The 31 IEEE Conference on Decision and Control*, 1992. Complete version: Research Report N° 1713, INRIA-Sohhia-Antipolis, 1992,

Proceedings of the 31st Conference
on Decision and Control
Tucson, Arizona • December 1992

An Implementation for Hybrid Continuous Variable/Discrete Event Dynamic Systems

Pam Binns, Mike Jackson, and Steve Vestal*

binns@src.honeywell.com jackson@src.honeywell.com vestal@src.honeywell.com
Honeywell Systems & Research Center
3660 Technology Drive
Minneapolis, Minnesota 55418

1 Introduction

Honeywell and the University of Maryland are currently developing a software architecture for intelligent (adaptive) guidance, navigation and control (AGN&C) as part of the DARPA DSSA program. In addition to providing a classification of operations and data types appropriate for the AGN&C domain, our architecture will also provide standard interfaces (standard control and data flow mechanisms) that will facilitate component reuse.

There are several aspects of our program that we feel are interesting. Our architectures will be captured in a formal architecture representation language that can be used to drive analysis, architecture configuration, and component integration and/or code generation tools. Multiple views or representations will be provided for the multiple technologies and disciplines required to build an actual AGN&C system. We have developed preliminary specifications for two domain-specific representation languages: one to capture the control algorithms aspects, and one to capture the fault-tolerant, real-time, secure, multiprocessing aspects. We are also relying heavily on formal models, such as finite state machines, systems of differential and difference equations, rate monotonic scheduling theory, queueing theory, and Markov processes. Our approach is to derive the software architecture from formal models rather than attempt to construct approximate models for some convenient or pre-existing software architecture. Finally, we are building a layered architecture, where each layer can be described using a particular formal model. The challenge in this is to provide the right set of layers, and the right functionality in each layer to support the ones above it, in a way that is modular, automatically configurable, efficient and analyzable.

The focus of this paper, however, is on implementation techniques for hybrid control systems. In this paper we use the term "hybrid" to refer to a system that combines Continuous Variable Dynamic System (CVDS) and Discrete Event Dynamic System (DEDS) models, which is a broader problem than discrete-time fixed rate sampling. We will discuss several mechanisms for implementing hybrid systems. Switches allow efficient, limited control law changes in response to events. Scheduling modes provide more expensive but more flexible control law and implementation changes in response to events. Either type of change can be triggered by either external (hardware interrupt) or internal (software-raised) events. We will discuss support for fixed rate control processes, hard deadline aperiodic processes, and queueing processes to exchange data and share resources within the same hybrid AGN&C system.

To motivate the need for such techniques, we list a few guidance, navigation and control situations that require hybrid systems:

- changes in active actuators, such as switching from aerodynamic surfaces to reaction control jets when leaving the atmosphere.
- changes in active sensors, such as switching from star trackers to inertial measurement units on a satellite, or receiving navigation signals only while passing over a ground tracking station.
- sensors that provide aperiodic updates, such as a star tracker.
- expected changes in mission segment, such as switching from climb mode to cruise mode for an aircraft, or staging a rocket.
- turning onboard systems on or off, such as activation of a camera on an interplanetary probe, or calibration of an inertial measurement unit upon power up.
- change in mode of operation due to unexpected external influences, such as a fighter aircraft responding to a detected threat or switching means of communication between a spacecraft and ground controllers due to interference.
- change in mode of operation due to unexpected internal situation, such as detecting the failure of

*Supported by DARPA/ONR Contract No. N00014-91-C-0195.

CH3229-2/92/0000-2973$1.00 © 1992 IEEE

an onboard processor, sensor, actuator, or control surface.

- response to discrete pilot commands such as changing cockpit switches.

In this paper we will briefly describe seven implementation layers in bottom-up order. These layers together support both continuous variable and discrete event AGN&C systems. The seven layers we will discuss are:

1. At the lowest layer, the automatically configurable kernel we are developing supports a set of processes that switch from state to state. At this level, each process is modeled as a deterministic finite state machine (FSM).

2. A stream of fixed rate periodic events is introduced into the system, causing a series of fixed rate periodic process dispatches. The execution and communication of multiple processes must be scheduled in a way that meets a set of real-time deadlines, and a combination of static and rate monotonic scheduling is used to achieve this.

3. Using fixed rate real-time periodic process execution, continuous variable control theory is used as a basis for specifying and generating fixed rate discrete-time control process code. This code can either implement a set of difference equations or integrate a set of differential equations. At this level and above, processes are modeled as (computations of) mathematical control functions.

4. We introduce the concept of switches into control specifications in order to support logico-differential systems. Switches allow discrete if-then-else or case decisions to determine the exact set of continuous variable control laws in effect at any point in time.

5. Aperiodic server algorithms allow real-time aperiodic (event-driven) process execution to occur simultaneously with real-time periodic process execution. Events can either come from the external environment or from other periodic or aperiodic processes.

6. Queueing networks are used to support processing of events whose arrival times and/or service times are stochastic. Each queue server can be supported by an underlying aperiodic server, with server parameters set in a way that should allow us to analytically bound or estimate queueing network performance metrics.

7. A scheduling mode is a set of processes (periodic, aperiodic, queue server) together with all communication, scheduling, and resource allocation characteristics needed for fault-tolerant, real-time operation. Multiple scheduling modes are typically implemented in a system, and event-triggered switches between scheduling modes can implement fundamental changes in the characteristics of a system.

2 Process State Machines

The basic unit of computation in the system we are building is the process. Processes are entities that are scheduled, communicate with each other, are bound to processors, can provide fault-containment, etc. Associated with each process is a state, which consists of all the information needed to execute that process: memory and register values, memory map and address state information, execution time and scheduling data, error and fault data, etc. At the lowest level, however, we define a small set of process scheduling states for each type of process supported by the system, together with transitions between those states. At this level, the implementation is modeled as a set of concurrent, deterministic finite state machines, one per process.

Each process scheduling state transition is triggered by some event, internal or external. Each transition can be implemented by a specific block of nonpreemptable, sequential code (a critical section). These sections of code perform any changes to process state variables required at that transition. Figure 1 shows the scheduling state transition diagram and table for a simple periodic process, taken from the design document for the automatically configurable kernel that we are developing. Our current design supports three process types (periodic, aperiodic, period transformed) together with additional scheduling state transitions for error handling and semaphore locking/unlocking. In general, each type of process supported by a kernel can be described by a state transition diagram similar to that in Figure 1.

3 Periodic Scheduling

Almost all real-time systems incorporate a source of fixed rate periodic artificial events, for example a periodic interrupt from a hardware timer. This artificial event stream can be used to trigger periodic dispatches of some set of processes. However, this means that multiple processes may be ready for execution on the same processor at the same time. Such processes must also typically communicate with each other and must complete execution at some deadline prior to their next dispatch. A schedule that determines which specific action a processor is performing at which specific time is needed in order to insure

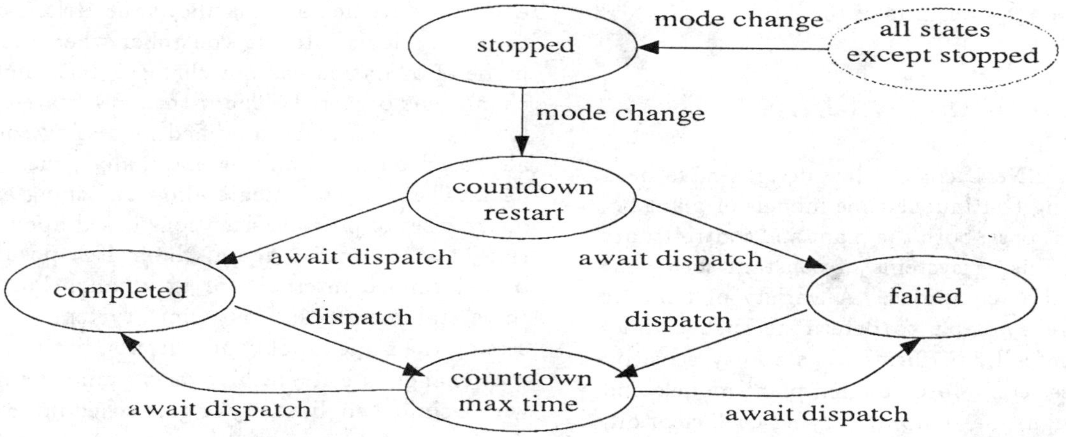

Transition From	Transition To	Event	Implemented by Kernel Component(s)
stopped	countdown restart	clock & mode change	scheduler → internal → processor
countdown restart	completed	await dispatch call	kernel → internal → processor
	failed	await dispatch call	kernel → internal → processor
	stopped	clock & mode change	scheduler → internal → processor
completed	countdown max time	clock & dispatch	scheduler → internal → processor
	stopped	clock & mode change	kernel → internal → processor
failed	countdown max time	clock & dispatch	scheduler → internal → processor
	stopped	clock & mode change	kernel → internal → processor
countdown max time	completed	await dispatch call	kernel → internal → processor
	failed	await dispatch call	kernel → internal → processor
	stopped	clock & mode change	scheduler → internal → processor

Figure 1: Example Periodic Process Scheduling State Transitions

real-time process communication and execution. In our system this is done using two sub-layers, static scheduling for communication and dispatching and fixed priority preemptive scheduling for process execution.

Our implementation assumes that a periodic hardware interrupt is provided almost simultaneously to all processors in a multi-processor system. At each interrupt, a predetermined set of assignments of values between process buffers followed by a predetermined set of process dispatches is performed. Periodic process execution at lower frequencies than the hardware interrupt is easily achieved by dispatching only at every 2nd, 3rd, 4th, etc. interrupt.

Among all ready processes on a particular processor, fixed priority preemptive scheduling is used to determine which process a processor is currently executing. A rate monotonic priority assignment is used, where periodic processes of higher rates are assigned higher priorities. This has been shown to be an optimal fixed priority assignment when a process need not complete any earlier than the next time it is to be dispatched ("single sample delay").

Periodic process execution times are generally assumed to be bounded. Analysis tools are available to determine whether a system of processes will be feasibly scheduled using these techniques before the system is actually implemented and executed. These tools can also perform various sensitivity analyses, such as determining allowed changes in process rates or execution times.

Our current design supports multiprocessing. Each process is to be statically assigned to a specific processor.[1] We have developed an algorithm to automatically generate fault-tolerant, hard real-time, multi-processor communication protocol code based on the scheduling, processor binding, and message passing patterns specified for an application. Communication actions are statically scheduled for execution at the appropriate time after periodic hardware interrupts. We will note that a hard real-time communication protocol between multiple processes executing at multiple rates on multiple processors must support "distributed control state update" in a way

[1] A form of process migration can be implemented using scheduling mode changes, which are described in a subsequent section.

that meets the requirements of the next layer.

4 Continuous Systems

Traditionally, GN&C control law designs have been developed using continuous-time models of plant behavior. At first pass, both the plant and controller are often modeled using systems of constant-coefficient linear differential equations. A variety of analytic techniques are available to determine such characteristics as controller performance, stability, sensitivity, robustness, etc. More detailed plant models and controller designs may employ varying-coefficient differential equations (e.g. gain scheduled and adaptive controllers) and they may also employ continuous nonlinear differential equations. Such designs can be analyzed using simulation, linearization, and describing functions.

In an actual implementation, a set of control computations is executed at some fixed sampling rate by the application processes. Transformations from continuous-time to fixed rate discrete-time models are known and can be automatically applied (e.g. z-transforms and systems of linear difference equations). Implementations are often produced directly from the continuous-time models using runtime integration rather than a discrete-time model. In either case, the actual control software can be automatically generated from a high-level specification of the continuous-time controller.

Some complexity arises when rates are not uniform over the full set of physical measurements, or over the full set of control laws, or over the full set of actuator commands. Slowly changing inputs need not be measured often, slowly changing outputs need not be updated often, and slow state dynamics within the GN&C laws need not be propagated often. Sometimes the element with the fastest sampling requirements is used to specify a uniform rate for the entire computation; every input, output, and subsystem is updated regardless of how wasteful this may be in terms of consumption of computational resources. When the waste of computational resources is unacceptable, the GN&C algorithms are partitioned into subsystems, and the various subsystem calculations are given various repetition rates. This is an implementation-dependent trade-off that involves both the GN&C engineer and the computer systems engineer.

5 Logico-Differential Systems

A logico-differential control specification translates naturally into software that implements traditional control computations embedded within if-then-else and case statements. This allows the GN&C designer to include logic into the controller when the overall mode of operation has not changed, but some minor change has occurred. The if-then-else statements allow the process to be modified by testing conditionals based on internally or externally generated signals. The case statements allow switching among a discrete set of possible situations based upon the setting of a "switch". Again, the switch position can be determined internally or externally. For both of these statements, the continuous system is still processing the same inputs and outputs, but the process may change in some fashion that cannot be modeled with a simple nonlinearity or gain schedule. Such systems have to be analyzed in a case by case fashion, using traditional analysis methods for each case, or by extensive simulation. Often, transient behavior between switch cases is important to analyze.

An interesting impact of including logico-differential systems is that the execution path through the software can vary depending on system state, which means the execution time can vary. This can be handled at analysis time by determining the worst-case execution time. Alternatively, for simple cases a timing analysis could be performed on a case-by-case bases for different switch settings. However, this will probably require automated support before it becomes practical.

6 Aperiodic Scheduling

An aperiodic process is one that is dispatched each time an event arrives, where event arrivals cannot be predicted in advance. If a minimum bound can be placed on the time between event arrivals, and a maximum bound can be placed on the process execution time in response to an event, then the demand for processing is no worse than a periodic process whose rate corresponds to the minimum event interarrival time. This is the intuition behind several server algorithms used to schedule aperiodic processes with hard deadlines using a fixed priority preemptive discipline. It can be shown that if each artificial periodic process created to service aperiodic events is feasibly scheduled in conjunction with all true periodic processes, then the actual mixture of aperiodic and periodic processes will be feasibly scheduled. These aperiodic server algorithms allow both periodic and aperiodic processes to be scheduled in the same system.

In many cases, aperiodic processes naturally meet both these assumptions. The amount of execution time required to service each event can be tightly bounded. Either the physical plant cannot produce events at higher than some rate (e.g. disk sector interrupt) or closely arriving events are so infrequent that they can be ignored or considered an error (e.g. near-coincident faults). Where these assumptions hold, a scheduling feasibility analysis can be used to determine whether an aperiodic process will always complete within some fixed deadline. This is called hard real-time aperiodic response.

In our compiled kernel, the minimum period and maximum execution time assumptions are enforced. That is, dispatches of aperiodic processes will not be performed any higher than some specified rate, and an aperiodic process that executes longer than some specified time after a dispatch will be suspended. Controlled time-slicing is implemented in order to insure that an aperiodic process meets the assumptions stated above. This provides security against timing faults and forms a basis for implementing higher layers of event-handling, such as queueing networks.

7 Queueing Networks

What happens when an event occurs and the aperiodic process intended to service it is still busy servicing a previous event? For hard real-time aperiodic processing, this is an error. However, in many cases such events are queued until they can be serviced. If event arrivals and/or aperiodic process service times can be characterized by probability distributions, then each aperiodic process and its queue may be approximated with models from queueing theory. Aperiodic processes may themselves raise events, and in fullest generality such a system can be modeled as a queueing network. Standard performance metrics in such soft real-time systems are average response times, mean throughput, and average queue lengths. Other performance metrics are also of interest, such as the probability that the response time falls within some bound. In these cases there are no hard deadlines, and such systems are often called soft real-time.

Figure 2 shows an example of a simple queueing network, where each queue is implemented on top of an aperiodic server, and the execution of each aperiodic process is provided by the underlying aperiodic server. Event interarrival rates λ and service rates μ are used to model the event production and service processes. Queues supported by aperiodic servers can be controlled by server parameters, namely the maximum rate at which the aperiodic process is dispatchable and the maximum time slice allowed for each dispatch. Together these parameters reserve a guaranteed amount of processor utilization, and when the ratio of service times to time slices is large, the results of traditional queueing analysis are likely to be reasonably accurate.

We are investigating a number of other questions in this area. For example, how reasonable is a continuous approximation for the analytic predictions when arrival rates and service times are small compared to the aperiodic server parameters? What happens when two or more aperiodic servers share time slices, so their associated service times are no longer independent in the queueing network model? How can computationally tractable predictions be made for more than average values, such as response time percentiles or transient behavior?

8 Scheduling Modes

A collection of periodic and aperiodic processes, together with all communication, scheduling, and resource allocation information, is sometimes called a mode within the real-time computing community.[2] Our automatically configurable kernel allows users to specify multiple scheduling modes, together with the set of transitions between modes that are allowed and supported at runtime. A transition between scheduling modes at runtime occurs in response to any one of a specified set of events, which may be either internal (raised by software) or external (e.g. an interrupt). The set of scheduling modes and allowed mode transitions can be modeled as a finite state machine and/or a Markov state space, depending on the type of analysis to be performed.

A scheduling mode change can be used to implement fundamental changes in the internal as well as the external operation of a system. Resource allocations can be changed, for example removing a failed processor by switching to a mode in which no processes are bound to that processor. A switch to completely new control law implementations can be performed, for example a control law whose implementation has a different rate structure. It is important to support submodes of operation, and to allow sets of processes to be shared among multiple scheduling modes (i.e. the execution of these processes is not disrupted during changes between scheduling modes that share those processes).

The basic actions performed at a scheduling mode change are to stop and start processes. However, care must be taken to synchronize mode change actions in a fault-tolerant way across multiple processors, to avoid transient processor overloads that might occur if processes in both modes execute concurrently for a time, and to maintain consistent shared state.

[2] The term "mode" is sometimes used within the control community to denote any discrete change between continuous control algorithms, which encompasses both our "switch" and "scheduling mode" mechanisms.

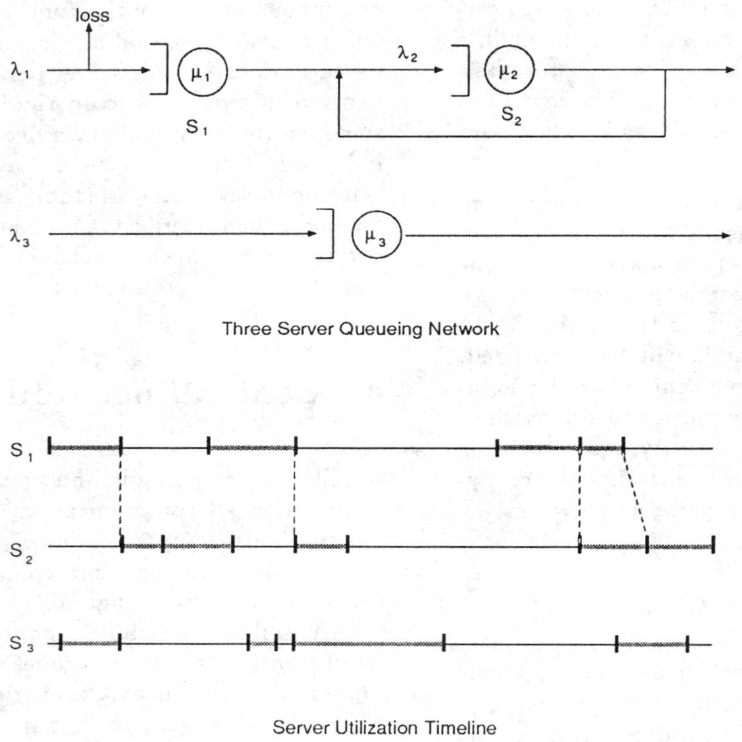

Figure 2: Queueing Network Supported by Aperiodic Servers

9 Summary

The techniques we have described provide two mechanisms to implement hybrid systems, switches and scheduling modes. Switches provide an efficient, fine-grained mechanism to make limited discrete changes in the control algorithm. Scheduling modes allow more dramatic changes in both the control algorithm and resource allocations, but with greater overhead and time lag. The techniques we describe also support the handling of hard deadline events described by bounds on interarrival and service times, and soft deadline events described by stochastic interarrival and service times.

The implementation and analysis of hybrid systems introduces a number of complex problems, some of which we have alluded to earlier. We will conclude by listing a few of the many areas in which further research is needed.

- Formal analysis techniques are needed to model the transient dynamic behavior of systems across switch and scheduling mode changes, and in response to aperiodic events.

- A number of difficult implementation problems arise when an attempt is made to integrate solutions to real-time, fault-tolerant, secure, and multi-processor requirements in the same system.

- Stochastic performance modeling techniques must be extended to capture important aspects of both the implementation and application (e.g. resource sharing between servers, solutions for higher moments than just the means).

- Further work is needed on techniques to verify that implementations are sufficiently accurately modeled by the various formalisms used during specification and analysis (i.e. the formal specifications, analysis results, and implementation are consistent with each other).

References

[1] Ashok Agrawala, James Krause, and Stephen Vestal. Domain-specific software architectures for intelligent guidance, navigation and control. *Conference on Computer Aided Control System Design*, 1992.

[2] Christos G. Cassandras and Peter J. Ramadge. Towards a control theory for discrete event systems. *IEEE Control Systems Magazine*, June 1990. special section on discrete event systems.

[3] Yu-Chi Ho. Scanning the issue. *Proceedings of the IEEE, special issue on Dynamics of Discrete Event Systems*, January 1989. special issue on dynamics of discrete event systems.

[4] James Krause and Stephen Vestal. Automatic scheduling challenges in guidance, navigation and control. *Conference on Computer Aided Control System Design*, 1992.

[5] Lui Sha and John B. Goodenough. Real-time scheduling theory and ada. *IEEE Computer*, April 1990.

[6] Brinkley Sprunt, Lui Sha, and John Lehoczky. Aperiodic task scheduling for hard-real-time systems. *Real-Time Systems*, January 1989.

[7] Stephen Vestal. On the accuracy of predicting rate monotonic scheduling performance. *Tri-Ada '90*, December 1990.

FA12 - 10:40

Proceedings of the 31st Conference
on Decision and Control
Tucson, Arizona • December 1992

SIMULATION FOR HYBRID SYSTEMS AND NONLINEAR CONTROL

John Guckenheimer and Anil Nerode[1]
Center for Applied Mathematics and Mathematical Sciences Institute
Cornell University
Ithaca, New York 14853
gucken@macomb.tn.cornell.edu anil@math.cornell.edu

Abstract

We discuss implementation issues for software tools to validate experimentally nonlinear and hybrid control by fully exploring qualitative behaviour. We discuss the effort of Guckenheimer and Back at Cornell under the DARPA DSSA program to extend the functionality of the dynamical System Simulator DSTOOL for this purpose by implementing atlases of manifolds and event-driven moves from chart to chart.

Introduction Will there ever be software tools which allow a line engineer of no special expertise to take a performance specification for an interacting system of non-linear plants not previously analyzed, use off the shelf components, connect them in block diagrams, and have a software tool which extracts a control program which then can be convincingly seen to enforce the performance specification?

What the control designer of today wants is to extract control system designs from performance specifications and equations of continuous non-linear plants. But even very simple appearing nonlinear equations can have very complex qualitative behaviour, and, further, nothing is actually linear anyway except in a limited range. System interactions often force supposedly linear systems out of that range, often putting these systems in modes not examined during the design phase. This affects mightly our ability to extract robust control laws. Linear and quasi-linear control theory mostly have avoided directly confronting these qualitative phenomena, even though they are common in high performance systems such as inherently unstable aircraft and the space station. Much the same can be said for control theory based on variational and vector fields on manifolds, which are less used in practice anyway. Further reasons that nonlinear control problems will probably always require experience and intelligence beyond the routine can be seen by examining the general methodology common to all control engineering.

Stages

- Develop the mathematical models for the plant, for its disturbances, for its available controllers, and for its plant state sensors. This is based on the the physics of the devices and can be done with more or less thouroughness as required.
- Develop the performance specifications based on the use intended. This involves detailed understandings with all clients.
- Extract a putative control program by some available methodology which is thought to apply to that kind of plant. This may involve conceptual analysis, and symbolic and numeric computing.
- Simulate and tune the closed loop system.
- Iterate the process.

At each iteration this process identifies the strong and weak points of the models and information about plant, controller, and sensors. Needs are identified for more detailed plant and controller and sensor characteristics, that is, a better understanding of the physics of plant, sensor, controller, and of their interactions. This is not a routine activity. It requires insight and experience, not likely to be turned into a recipe. It requires detailed knowledge of the plant and the performance specifications, not just of models handed to contol engineers.

Famous mathematicians spend generations on understanding a single non-linear equation. Many practical systems are much more complicated, and are therefore not likely to be amendable to recipes. Even computer testing of a proposed control law completing a closed loop non-linear system is hard to validate, estimating error propagation in algorithms intended to solve non-linear systems is notoriously difficult. Finally, under presently adopted methodologies, one has little conviction that all relevant qualitative behaviors have been adequately explored.

On the other hand, if all that is required is to achieve a small variation of a previous successful design, then of course a routine line engineer can be expected to modify the design adequately with software tools. But nevertheless, as we keep extending the frontiers of performance, we are constantly in a non-routine mode.

Hybrid Systems We use the term "hybrid system" when the controller is a digital program responding to sense data on the plant, and the plant is continuous responding to the contoller. These are non-linear dynamical systems with even less understood mathematical behavior

[1]Research supported by ORA Corp., DARPA-US ARMY AMC-COM (Picatinny Arsenal,NJ) DAAA21-92-C-0013, U.S. Army Research Office DAAL03-91-C-0027

CH3229-2/92/0000-2980$1.00 © 1992 IEEE

than using standard continuous nonlinear control for a continuous plant. Yet such programs are a powerful addition to the control arsenal, if we can figure out how to use their power and validate the results. So a vista of control still not well explored is the extraction and validation digital control programs for continuous devices. Fast, cheap, computer chips allow us to change the continuous plant control law used every few microseconds. These changes can be event-driven by decisions based on sensor measurements. Even if we somehow manage to extract a digital control program from performance requirements on the continuous plant, we have little theory justifying the validity of such control, and we then face directly the problem of reliable simulation in all operating regimes.

Dynamical Systems Simulators Dynamical systems simulators calculate by iteration solutions for a system of ordinary differential equations from chosen initial conditions, giving rise to pictures of qualitative behavior in phase space. If the resulting pictures are backed up by solid validated algorithms, such a simulator is a powerful tool for exploring what happens before and after a digital control regime has been imposed. Indeed, this is the basis of the work of Zhao and Bradley under Abelson and Sussman at MIT.

Beyond Existing Tools With further development, dynamical systems simulators are potentially equally useful for examining the effects of imposing continuous and hybrid control. To provide this enhanced functionality, there are several steps that must be taken beyond existing tools. Some of these steps present fundamental mathematical and algorithmic challenges, but existing theory is an adequate substrate for beginning this process. We sketch a scenario of for such a simulation environment, emphasizing those aspects that go beyond existing tools. There is an *ORA/MSI/Cornell* project under DARPA-DSSA funding being carried out by John Guckenheimer and Allen Back which is directed at creating the functionality described here by modifying and building extensions to the software package DSTOOL.

Encoding Hybrid Systems The first step required by a simulator of hybrid systems is that there must be a representation of mixed discrete and continuous times in the system. At certain continuous times within the evolution of a system, one would like to apply discrete mappngs to the state of the system. If one idealizes the invocation of a new control law as an instantaneous event that changes the state of the system and perhaps the law defining the vector field, then several new types of dynamical systems can be represented systematically, including systems with impacts or other discontinuities, systems with many regimes determined by digital controllers switching among control laws, and vector fields on manifolds. A consistent mathematical model for such hybrid systems is needed that allows for simulation and numerical analysis of the systems.

Charts and Atlas The hybrid dynamical systems themselves will typically have a much more complex structure than a single system of ordinary differential equations defined in a region of Euclidean space. The domains associated with different charts on which the system is given by a smooth vector field must be specified. The crossing of the boundary of a chart is an event that must be calculated accurately for some applications. Therefore the specification of the domains must include provision for the detection and calculation of these events. For each chart in the system, there will be a set of formulas defining the vector field in this region. Each part of the boundary of a chart will have an associated transformation that determines in which chart and where a new continuous time trajectory segment will start. For even simple systems, the amount of information that is required to specify the system will be large enough that tools for generating such descriptions are desired.

Consistency Checks In addition to tools to help in the process of encoding systems, one would also like to have the facilities that check the internal consistency of models. For example, one wants to be sure that transformations at the boundary of a coordinate chart that give changes of a control law map a system into the interior of the same or another coordinate chart, so that trajectories do not reach a "dead-end" from which no further evolution can be computed. For vector fields defined on manifolds, one wants to know that the representations of a vector field on the overlap of two different coordinate charts transform to one another in the appropriate manner. This can be determined by symbolic methods.

Existing Tools Existing tools for exploring dynamical systems rely upon visualization as a means of interpreting the large amounts of data contained in computed trajectories. With hybrid systems, the display of data presents new difficulties. One approach is to build simulations of devices that represent all of their degrees of freedom at any instant of time. Such simulations seldom retain data of past behavior, and achieving acceptable speeds for such animations is itself a difficult computational task. On the other hand, display of data from several charts in phase space will make large demands on screen "territory". At a higher level, one would like to represent visually the discrete time data of the events associated with changes of control laws.

The Logic Layer Our intuition is that the logic layer of the hybrid systems will be substantial. If one views organisms as machines, they combine robust nonlinear behaviors for automatically performing tasks, like walking or grasping, with complex reasoning about their environment. We expect that the design of intelligent control systems for many purposes will need to combine this same type of functionality. Whatever software environment is developed for control engineering should enhance the design and testing of these systems. The tools should not be a source of cumbersome overhead and confusion, separating the designer from the detailed understanding of the problem. We hope that this software will provide the designer with a clearer picture of the effects of nonlinear control. There are many mathematical, algorithmic, and implementation challenges ahead.

Proceedings of the 31st Conference
on Decision and Control
Tucson, Arizona • December 1992

Visual Observation for Hybrid Intelligent Control Implementation

Tarek M. Sobh*, Ruzena Bajcsy†, and John R. James‡

Abstract

We address the problem of design and implementation of a discrete event dynamic system (DEDS) observer for the execution of commands sent to a robotic arm during grasping and screwing tasks of an assembly operation. We discuss the resulting robot arm mechanism as a hybrid intelligent system. We argue that the non-intrusive observation mechanism offers reliability and robustness advantages over other sensor systems used to detect errors. Hybrid systems contain both continuous and discrete components. Hybrid systems analysis differs from conventional digital control systems analysis in that the discrete component is appropriately modeled using a finite-state machine (DEDS) to describe high-level machine dynamics and the continuous component contains both analog and digital models of the lower-level continuous variable dynamic system (CVDS) portion of the device. Another description of hybrid systems is that they are networks consisting of continuous physical devices controlled by discrete digital programs, requiring digital-analog transfer of information for their control (logical-linguistic techniques may be used to derive control laws for continuous systems and machines are structured as an interconnected network of continuous, nonlinear systems and finite-state automata). Our work has focused on the use of the DEDS linguistic approach for abstracting high-level knowledge concerning the current state of the machine in the presence of errors, mistakes and uncertainties in the manipulation system.

1 Introduction

A previous paper has discussed a new framework and representation for the general problem of observation [9]. That paper also asserted that the system being studied can be considered a "hybrid" one, due to the fact that we need to report on *distinct* and *discrete* visual states that occur in the *continuous*, *asynchronous* and three-dimensional world, from two-dimensional observations that are sampled periodically. In this work we expand that discussion to elaborate on the application of the new framework to detect error states and sequences.

The problem of observing a moving agent has been addressed in the literature extensively. It was discussed in the work addressing tracking of targets and determination of the optic flow [1], recovering 3-D parameters of different kinds of surfaces [7], and also in the context of other problems [2]. However, the need to recognize, understand, and report on different visual steps within a dynamic task was not sufficiently addressed. In particular, there is a need for high-level symbolic interpretations of the agents' actions. A previous work [9] closely examined the possibilities for errors, mistakes, and uncertainties in the visual manipulation system, observer construction process and event identification mechanisms, leading to a DEDS formulation with uncertainties.

2 DEDS

Discrete event dynamic systems (DEDS) are dynamic systems (typically asynchronous) in which state transitions are triggered by the occurrence of discrete events in the system. DEDS are usually modeled by finite state automata with partially observable events together with a mechanism for enabling and disabling a subset of state transitions [4,6]. We propose that this model is a suitable framework for many vision and robotics tasks, in particular, we use the model as a high-level structuring technique.

3 Observer Construction

Manipulation actions can be modeled efficiently within a discrete event dynamic system framework. We use the DEDS model to preserve and make use of information we know about the way in which each manipulation task should be performed. Each state in the automaton would represent a symbolic description of a stage in the manipulation process. In order to know the current state of the manipulation process we need to observe the sequence of events occurring in the system and make decisions regarding the state of the automaton. The goal will be to make the system a strongly output stabilizable one and/or construct an observer to satisfy specific task-oriented visual requirements.

*Robotics and Vision Laboratory, Department of Computer Science, University of Utah, Salt Lake City, Utah 84112.
†GRASP Laboratory, Computer and Information Science Department, University of Pennsylvania, Philadelphia, PA 19104.
‡United States Army Training and Doctrine Command, VA.
Acknowledgments : This research was supported in part by Air Force AFOSR Grants 88-0244, 88-0296; Army/DAAL Grant 03-89-C-0031PRI; NSF Grants CISE/CDA 88-22719, IRI 89-06770; DARPA Grant N0014-88-0630 and DuPont Corporation.

CH3229-2/92/0000-2982$1.00 © 1992 IEEE

4 Error States and Sequences

We utilize the observer framework to recognize error states and sequences. The idea behind this recognition task is to be able to report on *visually incorrect* sequences. In particular, if there is a predetermined observer model of a particular task under observation, then it would be useful to determine if something went wrong with the manipulation actions. The goal of this reporting procedure could be to alert operators or possibility to supply feedback to the manipulating robot so that it could correct its actions. Some examples of errors in manipulation include unexpected behaviour of the system, such as objects falling unexpectedly from the manipulating hand during a grasp and lift operation or some visual errors like unexpected occlusions between the observer camera and the manipulation environment.

There are a number of ways in which these problems could be reported. One such way can be to comply with the navigation strategy that was described in [8] in order to capture the current state (i.e. incrementally update belief in the current state and sequence as event matches occur while "navigating" the automata). If no match occurs, then the error would have to be reported. The correct sequences of automata state transitions can be formulated as the set of strings that are acceptable by the observer automaton. This set of strings represents precisely the language describing all possible visual task evolution steps.

5 Conclusions

The underlying mathematical representations of complex computer-controlled systems is still insufficient to create a set of models which accurately captures the dynamics of the system over the entire range of system operation. We remain in a situation where we must tradeoff the accuracy of our models with the manageability of the models. Closed-form solutions of mathematical models are almost exclusively limited to linear system models. Computer simulations of nonlinear and discrete-event models provide a means for off-line design of control systems through iterative search but such simulations cannot perform exhaustive search due to the complexity of the problem. Guarantees of system performance are limited to those regions where the robustness conditions apply. These conditions may not apply during startup and shutdown or during periods of anomalous operation. Excellent results are available for cases where adequate mathematical models are known and the system is operating "close enough" to a linear region. Also, effective tools are available to model high-level system changes as a finite state machine. Several attempts to improve our modeling capabilities are focused on mapping the continuous world into a discrete one However, repeated results are available which indicate that large interactive systems evolve into states where minor events can lead to a catastrophe. We are left with the result that there is a pressing need for a more adequate theory and mathematical basis for representing and predicting the performance of hybrid dynamical systems. Some current work has focused on providing a mathematical basis for the coupling of numerical and symbolic computing [3,5]. In the near term we will probably be able to mathematically prove (automatically verify) that the implementation of a subset of software for computer-controlled systems performs to specifications but will have to use conventional metrics for verification of the majority of the software being used. In this paper we have summarized a new framework and representation for the general problem of observation, emphasizing its' application to determining error states and sequences for the highest levels of computer-controlled systems. We assert that the framework provides a means for the explicit realization of transitions from low-level to high-level, goal-oriented knowledge in computer-controlled systems.

References

[1] P. Anandan, "A Unified Perspective on Computational Techniques for the Measurement of Visual Motion". In *Proceedings of the* 1st *International Conference on Computer Vision*, 1987.

[2] F. Chaumette and P. Rives, "Vision-Based-Control for Robotic Tasks", In *Proceedings of the IEEE International Workshop on Intelligent Motion Control*, Vol. 2, pp. 395-400, August 1990.

[3] Kohn, W and A. Nerode, Proceedings of the 1992 IEEE Symposium on Computer-Aided Control System Design, March 1992.

[4] C. M. Özveren, *Analysis and Control of Discrete Event Dynamic Systems : A State Space Approach*, Ph.D. Thesis, Massachusetts Institute of Technology, August 1989.

[5] Platek, R and A. Nerode, "Distributed Intelligent Control", Proceedings of the 1992 IEEE Symposium on Computer-Aided Control System Design, March 1992.

[6] P. J. Ramadge and W. M. Wonham, "Modular Feedback Logic for Discrete Event Systems", *SIAM Journal of Control and Optimization*, September 1987.

[7] T. M. Sobh and K. Wohn, "Recovery of 3-D Motion and Structure by Temporal Fusion". In *Proceedings of the* 2nd *SPIE Conference on Sensor Fusion*, November 1989.

[8] Sobh, T. M. "Active Observer: A Discrete Event Dynamic System Model for Controlling an Observer Under Uncertainty", Ph. D. Thesis, Department of Computer and Information Sciences, University of Pennsylvania, December 1991.

[9] Sobh, T. M. and R. Bajcsy, "A Model for Visual Observation Under Uncertainty", Proceedings of the 1992 IEEE Symposium on Computer-Aided Control System Design, March 1992.

Proceedings of the 31st Conference
on Decision and Control
Tucson, Arizona • December 1992

Real-Time Object Oriented Intelligent Control Environment

Roger D. Horn and J. Douglas Birdwell
Department of Electrical and Computer Engineering
University of Tennessee
Knoxville, TN 37996–2100 U.S.A.

Abstract

An approach to the design of intelligent control systems is proposed that provides for design optimization and parallelization of intelligent control structures in an object oriented environment. Embedded in the object oriented formulation is an inheritance mechanism that gives the design a multi-level hierarchical structure and yields an efficient design migration path from top level specifications to low level control functions. The object oriented modeling methodology, a colored Petri net extension, provides an optimization platform that is expected to mitigate the problem of combinatorial explosion. The Petri net structure allows easy parallelization, mapping to a heterogeneous computing environment based upon capabilities of the individual processing elements.

Introduction

Intelligent control is typically a mix of traditional control methods and either heuristic or deductive algorithms; the goal is a controller design that is better able to handle system complexity. A design environment that allows a hierarchical description is a natural choice for dealing with complex systems because the description can be resolved to varying degrees of abstraction at each level in the hierarchical structure. The specifications are decomposed to match the level of the system description, and controller design becomes a search for a structure that operates the system to the satisfaction of the specifications.

A design environment that supports a hierarchical description must insure that information in the lower levels reflects the system properties at the higher abstraction levels. An object oriented environment provides such an inter-level connection, and the Petri net paradigm, with an extension to allow object oriented net element descriptions, is a candidate for an intelligent control design environment.

Object Oriented Petri Net

An object oriented approach provides a multi-level, hierarchical structure for the Petri net system model. The model is composed of the standard Petri net objects: transition, place, token, and arc. One additional object type, the *branch*, has been added to the basic list. Branches are similar in function to *switches* described in [5]. The binary valued function, $\gamma(\cdot)$, in Figure 1 may be either a probablistic or deterministic function, and it controls along which arc tokens will flow following the branch.

Associated with each Petri net object is an object oriented information structure. A feature of the object oriented formulation is that, at each abstraction level, object structures inherit properties from structures in the higher levels. The information in the Petri net object structures is refined as the system description extends down to the next level. The top level is an abstract description that provides a comprehensive view of the aggregated system operation. The operation description perspective becomes more detailed in lower levels. At the lowest level, the object information elements are linked to basic, dynamical equations (discrete event or continuous value) describing the plant state trajectory or actions by the controller. In a distributed control environment, full dynamical information may not be available for neighboring agents, so the lowest abstraction level may be a simple input-output relationship of adjacent components.

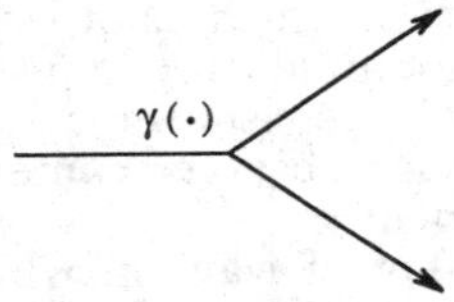

Figure 1: A Petri net branch type object.

Petri Net Coalgebraic Structure

There are parallels between Petri net and decision tree optimization, an area where significant advances have been made. It may be possible to define a coalgebraic structure for Petri nets that is similar to methods used in discrete decision theory [1] and stochastic decision theory [2]. Decision trees are defined as terms of a decision theory in the development of a coalgebraic structure for decision trees. Models of decision theory are decision coalgebras in category theory. Recent work [3, 4] has explored the algebraic formulation of Petri nets from the category theoretical perspective, and a coalgebraic view is a natural extension to be investigated.

The coalgebraic structure for decision trees uses properties of the trees to formulate a calculus of manipulations for decision expressions. This formulation allows the definition of an *irreducible form* which plays a central role in the optimization of decision trees. In a similar vein, a Petri net coalgebraic structure could provide manipulations for generating equivalent nets from which all irreducible forms, or all forms that satisfy a design specification, are extracted. For example, given one net satisfying a specification set, all irreducible forms in an equivalence class could be determined. The definition of coalgebraic manipulation operations and irreducible forms for Petri net models is an area of current research.

Optimization

Petri net optimization can exploit the hierarchical structure of the object oriented environment. An initial optimization would be carried out over a gross level description of hierarchical objects that define simple input-output relationships. Optimization of the top level Petri net structure is performed first to find a set of controller designs that satisfy the system specifications at that level. The problem then descends to the next level, and optimization proceeds with objects that have inherited properties. The optimization would exploit the structure that has been introduced by descending a level. This allows a natural decomposition of the problem for solution in a multicomputing environment. The object oriented structure permits optimization at the higher levels of the hierarchy to occur over a space of lower cardinality containing Petri nets modeling an aggregated system description. As the design optimization progresses to

CH3229-2/92/0000-2984$1.00 © 1992 IEEE

lower levels the cardinality increases, but large subsets of infeasible designs will have already been rejected. In this way, it is expected that the problem of combinatorial explosion can be controlled.

Petri nets have features that are desirable for the optimization operation. Petri nets employ a small set of modeling primitives so optimization can be accomplished using a small set of net modification operators. This further improves the computational growth factors. Our experience in the area of decision tree optimization indicates that this class of problems typically has many equivalent solutions, and this tends to ease the burden of finding a good solution. An analogy is subspaces in a linear space. Any element within a particular subspace qualifies as a solution, and the existence of the subspace reduces the problem size.

Optimization over Petri net models may require an intelligent branch-and-bound approach. A cost function is defined that is a weighted function of the specification satisfaction by a Petri net iterate. A net perturbation operator is applied to generate the next iteration. Intelligence can be added to the method by concentrating the perturbation operators on portions of the net identified as producing the worst cost function performance. For example, the build up of tokens in a place that violates a safety condition or processing implementation capability specification might cause a local refinement of the net's structure. A globally optimal solution with a non-zero cost function means the specification set is inconsistent. A zero cost solution could be used to generate the set of all irreducible forms that satisfy the specifications. One, or more "best cut" designs for the current level of abstraction taken from the set of all irreducible forms can then be examined.

Parallelization

Parallelization of a Petri net description provides a natural method to design net partitions mapping subnets to the available processors. Closed paths through the net arcs (Figure 2) define the processors' input/output boundaries while constraints on the paths can insure that a feasible partition conforms to the physical realities of the system. The interior region of a closed path defines the part of the net model to be implemented on a single processing element. Complexity measures, given by the size of the enclosed net or the number of tokens that must be processed within the region, can be used to quantify system limitations. Hardware-imposed constraints, such as throughput or interprocessor communications limitations, will determine the set of allowed net decompositions.

Inter-process communication constraints can be modeled by Petri net-based communication network models integrated into the net. Parallelization for a distributed computing system is an iterative process. Each partition requires that subnets modeling communication structure restrictions be inserted at intersections of closed paths and arcs, and the resulting Petri net model must be checked that it satisfies the aggregated system specification.

Discussion

Three application areas in electric power systems for a real-time object oriented design environment are smart transmission (FACTS) control, intelligent distributed alarm processing, and coordination of pricing and energy transfer among groups of utilities and non-utility generation systems (NUGS). These systems are composed of interacting agents, and a goal is to model and design control structures for electric energy distribution. For the producer, a controller is need that will set price and deliver energy to maximize revenue. On the energy consumption side, the customer desires a strategy for production scheduling and energy expenditure that will maximize profit. Integrating utility and customer is the transmission system with the goal of maintaining a stable energy highway. Each controller may have only a simple input-output relationship for neighboring systems, and controller design must be accomplished using this abstract description. In this case, the hierarchical structure provides a useful model and design environment.

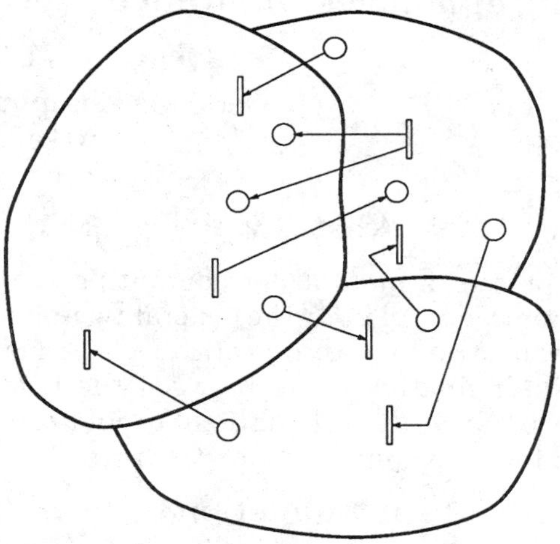

Figure 2: Parallelization of a Petri net system description.

The methodology has a "megaprogramming" flavor. System specifications can be supplied in a hierarchical framework, and the controller design can be delivered in a compact Petri net format that can be directly mapped to implementation code. This allows automatic design roll-out. Controller design verification is inherent in the approach because only controller net topologies that meet the system specifications are allowed as feasible solutions at each stage of the optimization and parallelization procedures.

Acknowledgements

This research was funded by the Oak Ridge National Laboratory under subcontract 99732-PAS21 from Martin Marietta Energy Systems, Inc.

References

[1] Cockett, J. R. B., "Optimizing Decision Expressions", *Fundamenta Informaticae X*, pp. 93–114, 1987.

[2] Cockett, J. R. B., J. Zrida and J. D. Birdwell, "Stochastic Decision Theory," *Probability in the Engineering and Informational Sciences*, vol. 3, no. 1, pp. 13–54, January, 1989.

[3] Marti-Oliet, N., and J. Meseguer, "From Petri Nets to Linear Logic", *Category Theory and Computer Science, LNCS 389*, Springer-Verlag, 1989.

[4] Meseguer, J., and U. Montanari, "Petri Nets are Monoids: A new algebraic foundation for net theory", *IEEE Conf. on Logic in Comp. Sci., LICS '88*, Edinburgh, 1988.

[5] Peterson, J. L., *Petri Net Theory and the Modeling of Systems*, Prentice-Hall, 1981.

Proceedings of the 31st Conference
on Decision and Control
Tucson, Arizona • December 1992

Eigenfrequencies of Non-collinearly Coupled Beams with Dissipative Joints

William H. Paulsen
Dept. of Computer Science, Math, and Physics
Arkansas State University
State University, AR 72401, USA

Abstract

We present a method for computing the in-plane eigenfrequencies for the general two-dimensional beam structure asymptotically. The method allows dissipative joints between the beams. We then perform the calculations to find the vibrations for an example beam structure.

1. Introduction

In the construction of large space structures, such as a large communication satellite or a space platform, different types of dampening devices, or dampers, are commonly installed at the joints of the beams to suppress the vibrations. Without these dampers, small vibrations in the structure would persist indefinitely, or even slowly build up, which could lead to disastrous results. Thus, the structural stability of such a configuration of beams depends on finding the natural vibrations of that structure. Determining how well these vibrations are suppressed involves careful analysis. For NASA's proposed space platform, as many as 40 beams are required in order to make construction possible. (Information obtained at a NASA workshop.)

In this paper, we will calculate eigenvalue asymptotics for the general n-beam non-collinear structure. That is, we will consider the case of n slender beams of lengths $l_1, l_2, \ldots, l_n$ with dissipative joints between the beams, as in figure 1.

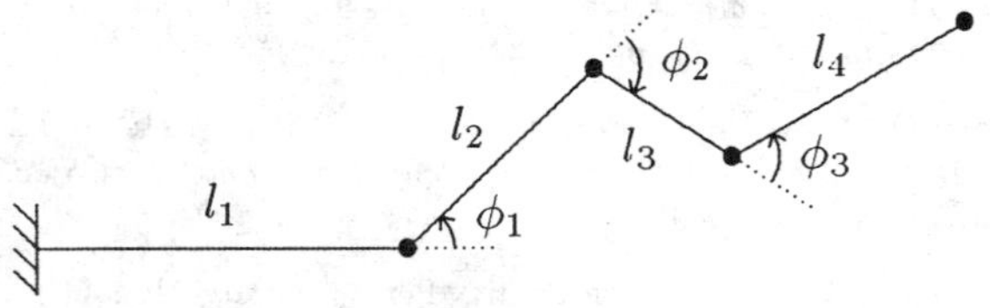

Figure 1

We will assume for simplicity that the mass per unit length, m, and the flexural rigidity of the beams, EI, are constant throughout the structure. We let $y^j(x,t)$, $0 \le x \le l_j$, $t \ge 0$ denote the transverse displacement function of the jth beam. Since these are Euler-Bernoulli beams, each of the y^j's satisfies the PDE

$$m\frac{\partial^2}{\partial t^2}y^j(x,t) + EI\frac{\partial^4}{\partial x^4}y^j(x,t) = 0. \qquad (1)$$

Here, we use superscripts instead of subscripts to ease the notation of partial derivatives.

We will make three simplifying assumptions for this model:

(H1) The frame can vibrate only in the plane of the frame.

(H2) The beams are essentially non-compressible, that is, the change of length of the beams due to the forces exerted at the ends are negligible.

(H3) Forces exerted on a beam in the direction parallel to the length of the beam are propagated in a negligible amount of time.

Because of assumption *(H2),* the longitudinal displacement is independent of the position within a given beam. Thus, we can denote the longitudinal displacement of the jth beam by $z^j(t)$. Also, because of assumption *(H3),* the longitudinal force of a given beam depends only on time, so we can let $H^j(t)$ denote the longitudinal force of the jth beam. Assumptions *(H2)* and *(H3)* together say that the time required for sound to travel the length of the structure is negligible compared to the period of the vibrations.

To simplify the calculations required for a given structure, we will consider the following new functions

$$\mathrm{Hya}(x) = \frac{\cosh(x) + \cos(x)}{2},$$
$$\mathrm{Hyb}(x) = \frac{\sinh(x) - \sin(x)}{2},$$
$$\mathrm{Hyc}(x) = \frac{\cosh(x) - \cos(x)}{2}, \qquad \text{and}$$
$$\mathrm{Hyd}(x) = \frac{\sinh(x) + \sin(x)}{2}.$$

One property of these "hybrid" functions is that the derivative of each one is the next, that is,

$$\frac{d}{dx}\mathrm{Hya}(x) = \mathrm{Hyb}(x); \quad \frac{d}{dx}\mathrm{Hyb}(x) = \mathrm{Hyc}(x);$$

$$\frac{d}{dx}\mathrm{Hyc}(x) = \mathrm{Hyd}(x); \quad \frac{d}{dx}\mathrm{Hyd}(x) = \mathrm{Hya}(x).$$

The general solution to (1) can be found using a separation of variables. The result can be

CH3229-2/92/0000-2986$1.00 © 1992 IEEE

expressed in terms as the sum of the eigenfunctions

$$y^j_\lambda(x,t) = e^{\lambda t\sqrt{EI/m}}(A_j\text{Hya}(\eta x) + B_j\text{Hyb}(\eta x) + C_j\text{Hyc}(\eta x) + D_j\text{Hyd}(\eta x)). \quad (2)$$

Here, $\eta = (1-i)\sqrt{\lambda/2}$ so that $i\eta^2 = \lambda$. We also can express $z^j(t)$ and $H^j(t)$ using similar notation:

$$z^j_\lambda(t) = E_j e^{\lambda t\sqrt{EI/m}}, \quad \text{and}$$
$$H^j_\lambda(t) = F_j EI\eta^3 e^{\lambda t\sqrt{EI/m}}. \quad (3)$$

Other relevant physical qualities are represented by:
Rotation $= y^j_x(x,t) \equiv \theta^j(x,t)$,
Shear Force $= -EIy^j_{xxx}(x,y) \equiv V^j(x,t)$, and
Bending Moment $= -EIy^j_{xx}(x,y) \equiv M^j(x,t)$.

2. The Transfer Matrices

Rather than letting each matrix represent all of the information about one beam, we will let each matrix, called a *transfer matrix,* represent one aspect of the structure, whether it be a length of a beam, an angle between two beams, or a damper. We begin by letting

$$\mathbf{v}_j = (A_j, B_j, C_j, D_j, E_j, F_j)$$

be the components of the wave in (2) and (3) after the jth piece of the structure. For example, the beam structure in figure 2 could be described as follows:

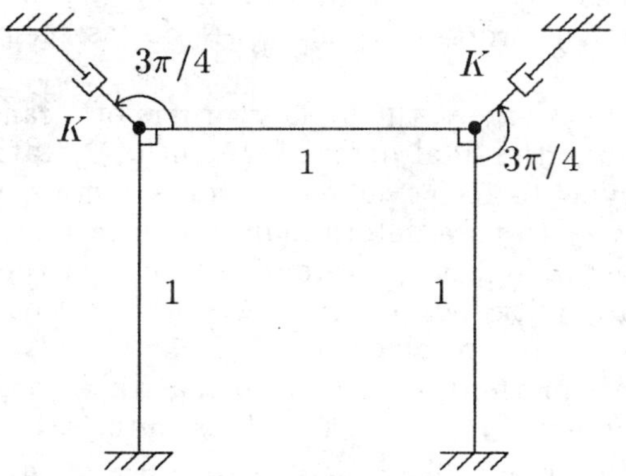

Figure 2

Begin with a clamped end.
Put in a beam of length 1.
Add a damper 45° to the left.
Turn 90° to the right.
Put in a beam of length 1.
Add a damper 45° to the left.
Turn 90° to the right.
Put in a beam of length 1.
Finish by clamping the end.

For each line in the description, there will be matrix which relates $\mathbf{v}_j$ to $\mathbf{v}_{j+1}$ by

$$\mathbf{v}_{j+1} = M_j \cdot \mathbf{v}_j.$$

The M_j is the transfer matrix for that line in the description. Thus, in this example there will be 9 transfer matrices.

The dampers in this example are called type III dampers, according to numbering system in [4]. That is, the displacement vector and beam coupling angle are rigid, while the bending moment and shear are discontinuous. An example of a type III damper in shown in figure 3.

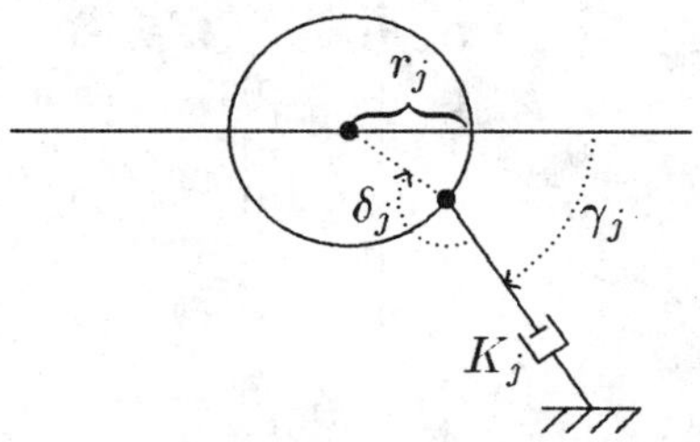

Figure 3

The boundary conditions are for this damper are as follows, considering only linear approximations to the angle displacement:

$$\begin{aligned}
V^{j+1} &= V^j + a\sqrt{mEI}y^{j+1}_t + b\sqrt{mEI}z^{j+1}_t \\
&\quad + c\sqrt{mEI}\theta^j_t, \\
H^{j+1} &= H^j + b\sqrt{mEI}y^{j+1}_t + d\sqrt{mEI}z^{j+1}_t \\
&\quad + e\sqrt{mEI}\theta^j_t, \\
M^{j+1} &= M^j - c\sqrt{mEI}y^j_t - e\sqrt{mEI}z^j_t \\
&\quad - f\sqrt{mEI}\theta^j_t, \\
y^j &= y^{j+1}, \\
z^j &= z^{j+1}, \qquad \text{and} \\
\theta^j &= \theta^{j+1}.
\end{aligned}$$

Here,

$$\begin{aligned}
a_j &= K_j \sin^2\gamma_j/\sqrt{mEI}, \\
b_j &= K_j \cos\gamma_j \sin\gamma_j/\sqrt{mEI}, \\
c_j &= K_j r_j \sin\delta_j \sin\gamma_j/\sqrt{mEI}, \\
d_j &= K_j \cos^2\gamma_j/\sqrt{mEI}, \\
e_j &= K_j r_j \sin\delta_j \cos\gamma_j/\sqrt{mEI}, \qquad \text{and} \\
f_j &= K_j r_j^2 \sin^2\delta_j/\sqrt{mEI}.
\end{aligned}$$

After multiplying by a factor of η, the transfer matrix is

$$M_j^{III} = \begin{pmatrix} \eta & -ia_j & ic_j\eta & 0 & 0 & ib_j \\ 0 & \eta & 0 & 0 & 0 & 0 \\ 0 & 0 & \eta & 0 & 0 & 0 \\ 0 & -ic_j\eta & if_j\eta^2 & \eta & 0 & ie_j\eta \\ 0 & -ib_j & ie_j\eta & 0 & \eta & id_j \\ 0 & 0 & 0 & 0 & 0 & \eta \end{pmatrix}.$$

Other types of dampers can also be considered.

We can write similar matrices which represent turning an angle, or placing a straight beam of a certain length in place. The matrix for turning an angle ϕ is given by $M_j^{\text{angle}} =$

$$\begin{pmatrix} \cos\phi & 0 & 0 & 0 & \sin\phi & 0 \\ 0 & \cos\phi & 0 & 0 & 0 & -\sin\phi \\ 0 & 0 & 1 & 0 & 0 & 0 \\ 0 & 0 & 0 & 1 & 0 & 0 \\ -\sin\phi & 0 & 0 & 0 & \cos\phi & 0 \\ 0 & \sin\phi & 0 & 0 & 0 & \cos\phi \end{pmatrix}.$$

Likewise, the matrix $M_j^{\text{length}} =$

$$\begin{pmatrix} \text{Hya}(l\eta) & \text{Hyd}(l\eta) & \text{Hyc}(l\eta) & \text{Hyb}(l\eta) & 0 & 0 \\ \text{Hyb}(l\eta) & \text{Hya}(l\eta) & \text{Hyd}(l\eta) & \text{Hyc}(l\eta) & 0 & 0 \\ \text{Hyc}(l\eta) & \text{Hyb}(l\eta) & \text{Hya}(l\eta) & \text{Hyd}(l\eta) & 0 & 0 \\ \text{Hyd}(l\eta) & \text{Hyc}(l\eta) & \text{Hyb}(l\eta) & \text{Hya}(l\eta) & 0 & 0 \\ 0 & 0 & 0 & 0 & 1 & 0 \\ 0 & 0 & 0 & 0 & 0 & 1 \end{pmatrix}$$

gives the transfer matrix for a single straight beam of length l.

Finally, we need to consider the transfer matrices for the two ends of the structure. The final transfer matrix will be given by either

$$M_n^{\text{clamp}} = \begin{pmatrix} 1 & 0 & 0 \\ 0 & 0 & 0 \\ 0 & 0 & 0 \\ 0 & 1 & 0 \\ 0 & 0 & 1 \\ 0 & 0 & 0 \end{pmatrix} \quad \text{or}$$

$$M_n^{\text{free}} = \begin{pmatrix} 0 & 0 & 0 \\ 1 & 0 & 0 \\ 0 & 1 & 0 \\ 0 & 0 & 0 \\ 0 & 0 & 0 \\ 0 & 0 & 1 \end{pmatrix},$$

depending on whether the end is clamped or free. Other types of end boundary conditions can be obtained by combining a damper with an end. For example, the boundary control in [8] can be obtained by considering a type III joint immediately followed by a free end.

By combining all of the matrices together, we get the equation

$$\mathbf{v}_1 \cdot (M_1 \cdot M_2 \cdot M_3 \cdots M_n) = \mathbf{0}.$$

This gives us three equations with six unknowns. The other three equations come from the fact that the beginning end is clamped, which can be put in the form $\mathbf{v}_1 \cdot M_n^{\text{clamp}} = \mathbf{0}$. This says that $A_1 = D_1 = E_1 = 0$, so only the second, third, and sixth columns of $(M_1 \cdot M_2 \cdot M_3 \cdots M_n)$ will be important. We can express this by letting

$$M_0^{\text{clamp}} = \begin{pmatrix} 0 & 1 & 0 & 0 & 0 & 0 \\ 0 & 0 & 1 & 0 & 0 & 0 \\ 0 & 0 & 0 & 0 & 0 & 1 \end{pmatrix}.$$

Then there will be a non-trivial solution to the wave equations if and only if

$$\det(M_0 \cdot M_1 \cdot M_2 \cdots M_n) = 0. \quad (4)$$

Let us denote $G = M_0 \cdot M_1 \cdot M_2 \cdots M_n$. Then whenever η is a solution to $|G| = 0$, $\lambda = i\eta^2$ will be an eigenfrequency of the structure.

3. Asymptotic Estimates

So far, we have a way to compute the *exact* eigenfrequencies of the linear equations. However, for any non-trivial structure, the roots of the determinant in (4) will be almost impossible to obtain. Thus, asymptotic approximations seem reasonable. We will need to find (4) to the highest two orders of η as $|\eta| \to \infty$, as seen in [4].

Unfortunately, the highest orders of η cancel as we take the final determinant in (4). Rather than trying to keep even more orders of η, we will find a way to take determinants in each matrix, causing the highest powers of η to cancel *before* they are multiplied together. We will do this by using some tensor algebra.

We can think of a 6 by 6 matrix as a linear function $F_* : V_6 \to V_6$. This induces a linear map on the set of all tensors $F^* : \mathcal{T}(V_6) \to \mathcal{T}(V_6)$. In particular, F^* maps all alternating covariant tensors of order 3 onto itself; thus $F^* : \Lambda^3(V_6) \to \Lambda^3(V_6)$. This induction map is an anti-homomorphism, that is, if $H_* = F_* \circ G_*$, then $H^* = G^* \circ F^*$. See [1] for details. By combining this transformation with a transpose, we can create a suitable homomorphism.

What this amounts to is this: There are 20 ways to pick 3 rows out of the six, and 20 ways of picking 3 columns in the matrix. We can consider all 400 ways of forming a 3 by 3 submatrix from the original, and take the determinants of all 400 matrices, forming a 20 by 20 matrix N_j. Then

$$\det(M_0 \cdot M_1 \cdots M_n) = N_0 \cdot N_1 \cdots N_n. \qquad (5)$$

The 3×6 and 6×3 matrices convert into 1×20 and 20×1 matrices, respectively.

We will use the following basis for $\Lambda^3(V_6)$.

$$\begin{array}{ll}
\lambda_1 = \omega_1 \wedge \omega_2 \wedge \omega_3 & \lambda_{11} = \omega_1 \wedge \omega_2 \wedge \omega_6 \\
\lambda_2 = \omega_1 \wedge \omega_2 \wedge \omega_4 & \lambda_{12} = \omega_1 \wedge \omega_3 \wedge \omega_6 \\
\lambda_3 = \omega_1 \wedge \omega_3 \wedge \omega_4 & \lambda_{13} = \omega_1 \wedge \omega_4 \wedge \omega_6 \\
\lambda_4 = \omega_2 \wedge \omega_3 \wedge \omega_4 & \lambda_{14} = \omega_2 \wedge \omega_3 \wedge \omega_6 \\
\lambda_5 = \omega_1 \wedge \omega_2 \wedge \omega_5 & \lambda_{15} = \omega_2 \wedge \omega_4 \wedge \omega_6 \\
\lambda_6 = \omega_1 \wedge \omega_3 \wedge \omega_5 & \lambda_{16} = \omega_3 \wedge \omega_4 \wedge \omega_6 \\
\lambda_7 = \omega_1 \wedge \omega_4 \wedge \omega_5 & \lambda_{17} = \omega_1 \wedge \omega_5 \wedge \omega_6 \\
\lambda_8 = \omega_2 \wedge \omega_3 \wedge \omega_5 & \lambda_{18} = \omega_2 \wedge \omega_5 \wedge \omega_6 \\
\lambda_9 = \omega_2 \wedge \omega_4 \wedge \omega_5 & \lambda_{19} = \omega_3 \wedge \omega_5 \wedge \omega_6 \\
\lambda_{10} = \omega_3 \wedge \omega_4 \wedge \omega_5 & \lambda_{20} = \omega_4 \wedge \omega_5 \wedge \omega_6
\end{array}$$

This basis determines the ordering of the elements of N_j. For example, the entry in the 16th row and 9th column of N_j is given by taking the determinant of the submatrix obtained from the 3rd, 4th and 6th columns, and 2nd, 4th and 5th rows of G_j. Notice that the transpose is included to make the mapping a homomorphism. Using this pattern, we can compute the matrices N_j. The end matrices N_0 and N_n are small enough to display:

$$N_0^{\text{clamp}} =$$

$$(0,0,0,0,0,0,0,0,0,0,0,0,0,1,0,0,0,0,0,0),$$

$$N_n^{\text{clamp}} =$$

$$(0,0,0,0,0,0,1,0,0,0,0,0,0,0,0,0,0,0,0,0)^T,$$

and $\quad N_n^{\text{free}} =$

$$(0,0,0,0,0,0,0,0,0,0,0,0,0,1,0,0,0,0,0,0)^T.$$

Here, A^T denotes the transpose of the matrix A.

Fortunately, the 20 by 20 matrices N_j are fairly sparse. We can describe N_j^{III} as follows. All elements along the main diagonal are η^2. The other non-zero entries are given by

$$\begin{aligned}
n_{3,4} &= n_{6,8} = n_{7,9} = n_{12,14} = n_{13,15} = n_{17,18} \\
&= -ia_j\eta, \\
n_{1,14} &= n_{2,15} = n_{3,16} = n_{5,18} = n_{6,19} = n_{7,20} \\
&= n_{6,1} = n_{7,2} = n_{19,14} = n_{20,15} = ib_j\eta, \\
n_{10,4} &= n_{17,11} = -ib_j\eta, \\
n_{3,1} &= n_{7,10} = n_{10,8} = n_{13,16} = n_{16,14} = n_{17,19} \\
&= ic_j\eta^2, \\
n_{2,4} &= n_{5,8} = n_{7,5} = n_{11,14} = n_{13,11} = n_{20,18} \\
&= -ic_j\eta^2, \\
n_{5,11} &= n_{6,12} = n_{7,13} = n_{8,14} = n_{9,15} = n_{10,16} \\
&= id_j\eta, \\
n_{2,11} &= n_{3,12} = n_{4,14} = n_{5,1} = n_{17,12} = n_{18,14} \\
&= ie_j\eta^2, \\
n_{7,3} &= n_{7,17} = n_{9,4} = n_{9,18} = n_{10,19} = n_{20,16} \\
&= -ie_j\eta^2, \\
n_{2,1} &= n_{7,6} = n_{9,8} = n_{13,12} = n_{15,14} = n_{20,19} \\
&= if_j\eta, \\
n_{7,16} &= n_{10,14} = (b_je_j - c_jd_j)\eta, \\
n_{5,14} &= n_{7,11} = -(b_je_j - c_jd_j)\eta, \\
n_{7,4} &= n_{7,18} = (b_jc_j - a_je_j)\eta, \\
n_{3,14} &= n_{17,14} = -(b_jc_j - a_je_j)\eta, \\
n_{2,14} &= n_{7,1} = n_{7,19} = n_{20,14} = (c_je_j - b_jf_j)\eta^2, \\
n_{6,14} &= n_{7,15} = (a_jd_j - b_j^2), \\
n_{7,8} &= n_{13,14} = (a_jf_j - c_j^2)\eta^2, \\
n_{7,12} &= n_{9,14} = -(d_jf_j - e_j^2)\eta^2, \qquad \text{and} \\
n_{7,14} &= \\
& i(a_jd_jf_j + 2b_je_jc_j - c_j^2d_j - e_j^2a_j - b_j^2f_j)\eta.
\end{aligned}$$

The matrix N_j^{angle} can also be described by giving the non-zero entries, as follows.

$$\begin{aligned}
n_{6,6} &= n_{7,7} = n_{14,14} = n_{15,15} = 1, \\
n_{3,3} &= n_{4,4} = n_{5,5} = n_{10,10} = n_{11,11} = n_{16,16} \\
&= n_{17,17} = n_{18,18} = \cos\phi, \\
n_{1,1} &= n_{2,2} = n_{8,8} = n_{9,9} = n_{12,12} = n_{13,13} \\
&= n_{19,19} = n_{20,20} = \cos^2\phi, \\
n_{3,10} &= n_{5,17} = n_{16,4} = n_{18,11} = \sin\phi, \\
n_{4,16} &= n_{10,3} = n_{11,18} = n_{17,5} = -\sin\phi, \\
n_{1,8} &= n_{1,12} = n_{2,9} = n_{2,13} = n_{19,8} = n_{19,12} \\
&= n_{20,9} = n_{20,13} = \cos\phi\sin\phi, \\
n_{8,1} &= n_{8,19} = n_{9,2} = n_{9,20} = n_{12,1} = n_{12,19} \\
&= n_{13,2} = n_{13,20} = -\cos\phi\sin\phi, \qquad \text{and} \\
n_{1,19} &= n_{2,20} = n_{8,12} = n_{9,13} = n_{12,8} = n_{13,9} \\
&= n_{19,1} = n_{20,2} = -\sin^2\phi.
\end{aligned}$$

Rather than describing N_j^{length} at this time, we will first make some asymptotic approximations. Since the complex conjugate of any eigenfrequency is another eigenfrequency, we can consider only eigenfrequencies with non-negative imaginary part. Also, the real part of λ must be non-positive, since the dampers are all dissipative. Ignoring the false solution $\lambda = 0$, we have that

$$\pi/2 \le \arg(\lambda) \le \pi.$$

Since $\lambda = i\eta^2$, $0 \le \arg(\eta) \le \pi/4$. We can use this to take estimates on the exponential powers of η. For example, $e^{l\eta} \gg e^{il\eta}$ as $|\eta| \to \infty$. If we let $x = e^{i\eta}$, then by removing the terms of N_j^{length} which are exponentially small in comparison, we obtain (after multiplying by $8e^{l(i\eta-\eta)}$)

$$\widetilde{N}_j^{\text{length}} = \left(\begin{array}{c|c|c|c} \widetilde{A} & 0 & 0 & 0 \\ \hline 0 & \widetilde{B} & 0 & 0 \\ \hline 0 & 0 & \widetilde{B} & 0 \\ \hline 0 & 0 & 0 & \widetilde{A} \end{array}\right),$$

where

$$\widetilde{A} = \begin{pmatrix} 2x^l & 2x^l & 2x^l & 2x^l \\ 2x^l & 2x^l & 2x^l & 2x^l \\ 2x^l & 2x^l & 2x^l & 2x^l \\ 2x^l & 2x^l & 2x^l & 2x^l \end{pmatrix}$$

and $\widetilde{B}$ is a 6×6 matrix described as follows:

$$\begin{aligned}
b_{1,1} &= b_{3,3} = b_{3,4} = b_{4,3} = b_{4,4} = b_{6,6} = x^{2l} + 1, \\
b_{1,6} &= b_{6,1} = -x^{2l} - 1, \\
b_{2,2} &= b_{5,5} = 2x^{2l} + 2, \\
b_{2,5} &= -2ix^{2l} + 2i, \\
b_{5,2} &= 2ix^{2l} - 2i, \\
b_{1,3} &= b_{1,4} = b_{3,6} = b_{4,6} = -ix^{2l} + i, \\
b_{3,1} &= b_{4,1} = b_{6,3} = b_{6,4} = ix^{2l} - i, \\
b_{1,2} &= b_{2,3} = b_{2,4} = b_{3,5} = b_{4,5} = b_{5,6} \\
&= (1-i)x^{2l} + 1 + i, \\
b_{2,1} &= b_{3,2} = b_{4,2} = b_{5,3} = b_{5,4} = b_{6,5} \\
&= (1+i)x^{2l} + 1 - i, \\
b_{1,5} &= b_{2,6} = (-1-i)x^{2l} - 1 + i, \qquad \text{and} \\
b_{5,1} &= b_{6,2} = (i-1)x^{2l} - 1 - i.
\end{aligned}$$

If we denote by $\widetilde{N}_j$ the approximate each N_j to the first two orders of η, then the eigenfrequencies can be approximated to two orders of η by the equation

$$\widetilde{N}_0 \cdot \widetilde{N}_1 \cdot \widetilde{N}_2 \cdots \widetilde{N}_n = 0. \tag{6}$$

To compute the positions of the streams of the eigenvalues we can use the results from [9]. If the lengths are all integers, we can divide (6) by factors of η to obtain an equation of the form

$$f(x) + \frac{g(x)}{\eta} = O(\eta^{-2})$$

for some polynomial functions $f(x)$ and $g(x)$. If we let r be a root of $f(x)$, then there will be two possibilities. If $|r| = 1$, then $\eta \sim 2k\pi + \arg(r)$, and

$$\lambda_k \sim -2\text{Re}\left\{\frac{g(r)}{f'(r)r}\right\} + (2k\pi + \arg(r))^2 i. \tag{7}$$

If $|r| \neq 1$ then

$$\begin{aligned}
\lambda_k \sim{} & 2\log|r|(2k\pi + \arg(r)) - 2\text{Re}\left\{\frac{g(r)}{f'(r)r}\right\} \\
& + \left((2k\pi + \arg(r))^2 - 2\log^2|r|\right) i.
\end{aligned}$$

4. An Example

We can now use the large matrices to find the approximate eigenfrequencies of the structure in figure 2. Let us suppose that the two dampers are both of type III, with $a = b = d = K/2$ and $c = e = f = 0$. After multiplying the 9 matrices, we find that

$$\begin{aligned}
f(x) = {} & 4(x+i)((1+i)x^2 + x + (1-i)) \\
& \cdot (3x^3 - (2+i)x^2 + (1+2i)x - 3i)
\end{aligned}$$

and

$$g(x) = 4K(2x^6 + (2-3i)x^4 + (-2-3i)x^2 - 2).$$

By plugging the 6 roots of $f(x)$ into (7), we have that the 6 streams of eigenfrequencies are given approximately as follows:

$$\begin{aligned}
\lambda_{1,k} &= -0.73654372K + (2\pi k - 1.15536425)^2 i, \\
\lambda_{2,k} &= -0.54770276K + (2\pi k + 2.24217507)^2 i, \\
\lambda_{3,k} &= -0.04908685K + (2\pi k + 0.48398550)^2 i, \\
\lambda_{4,k} &= 0 + (2\pi k - \pi/2)^2 i, \\
\lambda_{5,k} &= 0 + (2\pi k - 2.71756161)^2 i, \qquad \text{and} \\
\lambda_{6,k} &= 0 + (2\pi k + 1.14676529)^2 i.
\end{aligned}$$

Notice that three of the six streams lie on the imaginary axis, making it appear as though there were only 4 streams of eigenfrequencies. The following table gives the exact values of the first three eigenfrequencies in each stream for $K = 1$, comparing these with the values given above. One can see a rapid convergence in each stream to the vertical line, especially when the real part of the stream is small. The exact values were computed using Newton's method on equation (4).

Exact value		Approximate value	
$-.736389+$	$5.03000i$	$-.547703+$	$5.02735i$
	$12.64804i$		$12.71367i$
	$22.37329i$		$22.20661i$
$-.732192+$	$26.18663i$	$-.736544+$	$26.29455i$
$-.048892+$	$45.79855i$	$-.049087+$	$45.79460i$
	$55.19808i$		$55.20417i$
$-.547705+$	$72.66205i$	$-.547703+$	$72.68177i$
	$96.99872i$		$96.99904i$
	$120.90339i$		$120.90265i$
$-.736236+$	$130.18992i$	$-.736544+$	$130.21107i$
$-.049045+$	$170.31035i$	$-.049087+$	$170.31179i$
	$188.05008i$		$188.05010i$
$-.547563+$	$219.28154i$	$-.547703+$	$219.29302i$
	$260.24124i$		$260.24124i$
	$298.55554i$		$298.55553i$
$-.736402+$	$313.07072i$	$-.736544+$	$313.08442i$
$-.049068+$	$373.78483i$	$-.049087+$	$373.78582i$
	$399.85286i$		$399.85286i$

Much of the work in this paper required the use of the symbolic manipulator MATHEMATICA running on a SUN Microsystems workstation for the computation of the large matrices.

Although this paper analyses a linear model of a physical system, the experimental data cited in [5] and [6] indicate that this model is an accurate one. However, we still are only considering vibrations which occur within the plane of the structure. The analysis for the structures which do not lie in the plane can be done using a similar technique, and this will be treated in a future work.

References

[1] W. Boothby, *An Introduction to Differentiable Manifolds and Riemannian Geometry,* Orlando: Academic Press, 1986, ch. 5, p. 204.

[2] G. Chen, M. C. Delfour, A. M. Krall, and G. Payne, "Modeling, stabilization, and control of serially connected beams," *SIAM J. Control Optim.*, vol. 25, pp. 526-546, May 1987.

[3] G. Chen, S. G. Krantz, D. W. Ma, D. W., C. E. Wayne, and H. H. West, "The Euler-Bernoulli beam equation with boundary energy dissipation," *Operator Methods for Optimal Control Problems,* New York: Marcel Dekker, 1987, pp. 67-96.

[4] G. Chen, S. G. Krantz, D. L. Russell, C. E. Wayne, H. H. West, and M. P. Colman, "Analysis, designs and behavior of dissipative joints for coupled beams," *SIAM J. Appl. Math.*, vol. 49, pp. 1665-1693, Dec. 1989.

[5] G. Chen, S. G. Krantz, D. L. Russell, C. E. Wayne, H. H. West, and J. Zhou, "Modeling, Analysis and testing of dissipative beam joints — experiments and data smoothing," *Math. Comput. Modelling*, vol. 11, pp. 1011-1016, Nov. 1988.

[6] G. Chen and H. Wang, "Asymptotic locations of eigenfrequencies of Euler-Bernoulli beam with nonhomogeneous structural and viscous damping coefficients," *SIAM J. Control and Optim.,* vol. 29, pp. 347-367, March 1991.

[7] G. Chen and J. Zhou, "The wave propagation method for the analysis of boundary stabilization in vibration structures," *SIAM J. Appl. Math.,* vol. 50, pp. 1254-1283, Oct. 1990.

[8] A. M. Krall, "Asymptotic Stability of the Euler-Bernoulli Beam with Boundary Control," *J. Math. Anal. Appl.,* vol. 137, pp. 288-295, Jan. 1989.

[9] S. G. Krantz and W. Paulsen, "Asymptotic eigenfrequency distributions for the N-beam Euler-Bernoulli coupled beam equation with dissipative joints," *J. Symbolic Comp.,* vol. 11, pp. 369-418, Nov. 1991.

Proceedings of the 31st Conference
on Decision and Control
Tucson, Arizona • December 1992

STABILIZATION OF MULTI-LINK FLEXIBLE BEAMS

H. Unbehauen and A. Gnedin
Department of Electrical Engineering
Ruhr-University, 4630 Bochum,
Germany

Abstract

This contribution deals with the problem of stabilizing vibrations in a system of interconnected multi-link flexible beams by applying point actuators located at the joints. Boundary and in-span conditions describing linear vibrations are derived within the framework of an Euler-Bernoulli approach. A stability criterion and a linear feedback scheme will be proposed for the clamped-free cofiguration. Through this the eleastic energy will be damped and thus the system can be kept close to a nominal position.

1. Introduction

Structural flexibility is important to the dynamics of high performance mechanical systems such as robot manipulators, antennas, drilling machines etc. Structures with low natural damping are often modeled as Euler-Bernoulli beams. Disturbances acting on these systems result in vibrations which can significantly reduce the system performance. Various vibration suppression methods have been suggested in the literature [1,2].

In this contribution, a flexible multi-link Euler-Bernoulli beam with adjustable angles between the links is considered. Our study is related to problems of control of multi-link robot manipulators driven by actuators located at the joints. The primary purpose of these actuators is to control the trajectory or position of the tip (equipped with a working organ) by changing the angles between the links. The control inputs are calculated by a computer which processes the angle measurements.

We present a method to damp elastic energy of vibrations by the very same actuators used for trajectory control. This method is based on a linear feedback scheme which requires sensing of bending moments at the joints.

In Section 2 we discuss elementary geometrical properties of planar composite beams. In Section 3 we derive boundary and in-span conditions for a system of Euler-Bernoulli equations governing free transversal vibrations. In Section 4 we suggest two feedback schemes related to different stability criteria such as the elastic energy and elastic energy plus an additional term responsible for rigid-body deflection from a reference position. In Section 5 we discuss some problems related with stabilizability by suggested control schemes. The last section presents a short discussion of critical points of our approach and some engineering implications.

2. Geometry of composite beams

Consider a composite beam represented in its reference position by a chain of n identical line segments $[A_{i-1}, A_i]$ · i=1,...,n of unit length, as shown in Fig. 1. The geometrical configuration of the chain is dertermined uniquely by the angles α_i between the vectors $A_{i-1}A_i$ and A_iA_{i+1}. We shall be concerned with small deflections of the beam from the reference position resulting from elastic deformations and changing the angles between the links. We suppose that the root of the beam, A_0, is fixed and the tip of the beam, A_n, is free.

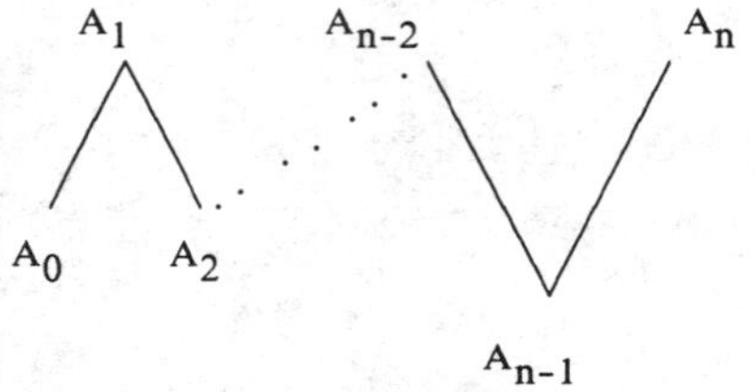

Figure 1. Geometrical configuration of a composite beam

Each point y on the i-th deflected link i displaced from the nominal position in two directions as shown in Fig. 2, where $A'_{i-1}A'_i$ represents the deflected position.

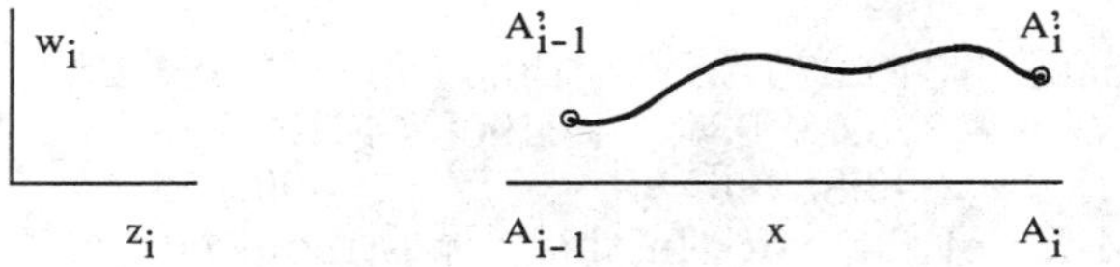

Figure 2. Beam deflection from nominal position

Denote by $w_i(x,t)$ and $z_i(x,t)$, $x \in [0,1]$, the transversal and longitudinal defelctions of a point x at time t. Assume further that the beam is made of inextensible material, then for small deflections, we obtain approximately

$$\frac{\partial z_i(x,t)}{\partial x} \approx 0 ,$$

and thus the longitudinal deflections are approximately the same for all points of the i-th link. An important implication of the above reasoning is that the deflection of each link can be represented as a superposition of a distributed transversal deflection $w_i(x,t)$ and a rigid-body longitudional deflection $z_i(t)$.

The deflection of each joint can be represented in two coordinate systems related to neighboring links (as in Fig. 2). The transfer from one of these systems to another is just a rotation by the angle α_i, therefore,

$$z_{i+1}(t) = z_1 \cos\alpha_i + w_i(1,t) \sin\alpha_i \tag{2.1}$$

$$w_{i+1}(0,t) = -z_i(t) \sin\alpha_i + w_i(1,t) \cos\alpha_i . \tag{2.2}$$

Equation (2.1) gives a recursive method to express longitudinal deflections through transversal deflections of the end-points. Iterating this formula we obtain

$$z_i(t) = \sum_{j=1}^{i-1} \cos \alpha_{i-1} \cos\alpha_{i-2} \dots \cos\alpha_{j+1} \sin\alpha_j \, w_j(1,t) . \tag{2.3}$$

Note that $z_1(t) = w_1(0,t) = 0$ since the root of the structure is fixed.
Denote by $\gamma_i(t)$ the angles between deflected links, i=1,...,n-1, i.e. between the curves $A_{i-1}A_i$ and A_iA_{i+1} at the joint A_i. Similarly, denote by α_0 the angle between the reference link A_0A_1 and a fixed direction on the plane, and by $\gamma_0(t)$ the anlge between the deflected link $A'_0A'_1$ and this direction. It is easy to see that if the angles γ_i are close to α_i, then in the first-order approximation we obtain

$$\left.\begin{aligned} w'_1(0,t) &= \gamma_0(t) - \alpha_0 \\ -w'_{i+1}(1,t) + w'_i(0,t) &= \gamma_i(t) - \alpha_i \end{aligned}\right\} \tag{2.4}$$

3. Dynamics of free vibrations

To simplify the presentation, consider only the case of homogeneous uniform links, assuming that the i-th link has a mass density m_i, a flexural rigidity E_iI_i (the product of Young modulus and the cross-sectional moment of inertia) and the length one. Each joint A_i is loaded by a concentrated mass M_i with no rotational inertia. In the following we shall derive the equations of free vibrations, when only inertial and elastic forces are acting on the beam, the root is clamped, the tip is free and the angles between the links are fixed.

The relevant physical quantities are represented by:

m_i mass of the i-th link	$-E_iI_iw_i''(x,t)$ bending moment
E_iI_i flexural rigidity	$-E_iI_iw_i'''(x,t)$ shear
$w_i(x,t)$ transversal deflection	$\dot{w}_i'(x,t)$ angular velocity
$w_i'(x,t)$ rotation .	

We accept the basic assumption of the Euler-Bernoulli theory: the potential of a flexible deformation of a small beam element is proportional to its bending [3], which results in the following formula for the potential of a flexible deformation of the i-th link:

$$U_i(t) = \frac{1}{2} E_i I_i \int_0^1 w_i''(x,t)dx. \tag{3.1}$$

The kinetic energy of the vibration is given by

$$T_i(t) = \frac{1}{2}(m_i+M_i)\left[\dot{z}_i(t)\right]^2 + \frac{1}{2} M_i \left[\dot{w}_i(1,t)\right]^2$$

$$+ \frac{1}{2} m_i \int_0^1 \left[\dot{w}_i(x,t)\right]^2 dx , \tag{3.2}$$

where the first term represents the kinetic energy of the longitudinal motion, the second term represents the kinetic energy of the transversal motion of the load and the third is the kinetic energy of the beam's transversal vibration. The potentional and the kinetic energy of the composite beam are given by

$$U(t) = \sum_{i=1}^{n} U_i(t) \quad , \ T(t) = \sum_{i=1}^{n} T_i(t), \tag{3.3}$$

respectively.

To derive the dynamic equations of vibration note that the elastic energy (U+T) is constant and one can apply Hamilton's principle:

$$\delta \int (T(t)-U(t))dt = 0 \tag{3.4}$$

for any virtual displacement $\{\delta w_i(x,t),\ i=1,\ldots,n\}$ satisfying

$$\delta w_i(x,t_0) = \delta w_i(x,t_1) = 0. \tag{3.5}$$

From (3.1) - (3.4) integrating by parts twice, varying the virtual displacements and applying Lagrange's lemma yields the familiar Euler-Bernoulli equation

$$E_iI_iw_i''''(x,t) + m_i\ddot{w}_i(x,t) = 0,\ i=1,\ldots,n, \tag{3.6}$$

which should be accomplished by 4n boundary and in-span conditions at the joints. The boundary conditions are easy to find. Indeed, since the root is clamped, we have

$$w_1(0,t) = 0 \ \text{ and } w_1'(0,t) = 0. \tag{3.7}$$

The tip is free, loaded by the point mass M_i, therefore

$$w_n''(1,t) = 0, \tag{3.8}$$

i.e. the bending moment is equal to zero, and

$$- E_iI_iw_n'''(1,t) + M_n\ddot{w}(1,t) = 0. \tag{3.9}$$

The derivation of the balance of moments acting through the i-th joint yields

$$E_iI_iw_i''(1,t) = E_{i+1}I_{i+1}w_{i+1}''(0,t),\ i=1,\ldots,n-1, \tag{3.10}$$

which can be derived also from mechanical considerations.

Eliminating longitudinal deflections in (2.2) by substituting (2.3) we obtain in-span conditions for end-point deflections

$$w_{i+1}(0,t) = -\sin\alpha_1 \sum_{j=1}^{i-1}\left[\prod_{k=j+1}^{i-1}\cos\alpha_k\right]\sin\alpha_j\ w_j(1,t) + \cos\alpha_i\ w_i(1,t), \tag{3.11}$$

which are essentially "non-local". The remaining (n-1) conditions, which tie shears at the joints, can be derived from mechanical considerations as a balance of shearing and inertial forces at joints, however we prefer the direct variational method.

The virtual longitudinal and transversal endpoint displacement are given by

$$\delta z_i(t) = \sum_{j=1}^{i-1}\left[\prod_{k=j+1}^{i-1}\cos\alpha_k\right]\sin\alpha_j\ \delta w_j(1,t) \tag{3.12}$$

$$\delta w_{i+1}(0,t) = -\sin\alpha_i\ \delta z_i(t) + \cos\alpha_i\ \delta w_i(1,t)\ , \tag{3.13}$$

whereas the neighboring links are related through

$$\delta z_{i+1}(t) = \cos\alpha_i\ \delta z_i(t) + \sin\alpha_i\ \delta w_i(1,t). \tag{3.14}$$

The last n-1 in-span conditions can be derived as

$$-E_iI_iw_i'''(1,t) + E_{i+1}I_{i+1}w_{i+1}'''(0,t)\cos\alpha_i + (m_{i+1}+M_{i+1})\sin\alpha_i\ \ddot{z}_{i+1}(t)$$

$$+ M_i\ddot{w}_i(1,t) + \sum_{p=i+1}^{n-1}\Big(-E_{p+1}I_{p+1}\sin\alpha_p\ w_{p+1}'''(0,t)$$

$$+ (m_{p+1}+M_{p+1})\ddot{z}_{p+1}(t)\cos\alpha_p\Big)\left[\prod_{k=i+1}^{p-1}\cos\alpha_k\right]\sin\alpha_i = 0 \tag{3.15}$$

Summarizing, free vibrations of the composite beam satisfy the Euler-Bernoulli equations (3.6) with boundary conditions (3.7) - (3.9), in-span conditions (3.10), (3.11), (3.15) and finally

$$w_i'(1,t) = w_{i+1}'(0,t) \quad i=1,\ldots,n-1. \tag{3.16}$$

4. Control and dissipation

Consider now the vibrations with varying angles $\gamma_0(t),\ldots,\gamma_{n-1}(t)$ at the root and between the links. From (2.4) we obtain

$$\dot{w}_{i+1}'(0,t) - \dot{w}_i'(1,t) = u_i(t) \quad i=1,\ldots,n-1 \tag{4.1}$$

$$\dot{w}'(0,t) = u_0(t)\ , \tag{4.2}$$

where we define

$$u_i(t) \equiv \dot{\gamma}_i(t) \quad i=1,\ldots,n-1 .$$

In what follows we regard the $u_i(t)$ as control variables.

Controlled vibrations of the composite beam satisfy, as in the case of free vibrations, the Euler-Bernoulli equations (3.6). At the root of the beam the boundary conditions (3.7) and (4.2) hold. The tip conditons (3.8) and (3.9) are still valid in the controlled case. In-span continuity of moments (3.10) holds along with (4.1). Finally, (3.11) means, as above, the continuity of end-points deflections, and (3.15) also hold because the virtual contribution of shearing forces is equal to zero.

To discuss the issues related with stability we need to introduce an appropriate stability criterion. A natural choice is the elastic energy $\mathcal{E}=T+U$, since the equality $\mathcal{E}=0$ implies that the beam rests and the links are undeformed. A free vibrating beam is a conservative mechanical system, i.e., the energy keeps a constant value, but the energy of controlled vibrations may be different at different instants of time. By integrating by parts and using (4.1), (4.2) one can see that the rate of change of energy is

$$\dot{\mathcal{E}}(t) = \sum_{i=1}^{n} E_i I_i (w_i''(1,t)\dot{w}_i'(1,t) - w_i''(0,t)\dot{w}_i'(o,t))$$
$$= \sum_{i=0}^{n-1} E_{i+1} I_{i+1} w_{i+1}''(0,t) u_i(t). \qquad (4.3)$$

Note that the control u_i is applied only at the i-th joint. Any linear feedback scheme with pointwise sensing at joints requires that the control is dependent only on pointwise "germs" of deflection, that is on deflection and/or its derivatives at joints. (4.3) suggests a feedback scheme of the form

$$u_i(t) = \sum_{j=1}^{n} k_{ij} \; w_j''(o0,t) \qquad i=0,\dots,n-1 \qquad (4.4)$$

with some positive matrix $K=(k_{ij})$. The closed-loop controlled beam with this feedback is a dissipative system, as it follows from (4.2), i.e.

$$\dot{\mathcal{E}}(t) \leq 0 \quad \text{for all } t \geq 0. \qquad (4.5)$$

In the case of a diagonal matrix K we shall speak of *local* feedback since in this case the control at each joint depends exclusively on sensing at this joint.

An energy criterion is not suitable in some practical problems, because in general the terminal position, stabilized by (4.1) and (4.2), differs from the reference one. It seems reasonable in these situations to include into the stability criterion some additional terms responsible for the rigid-body deflections of links from their reference positions.

For example, the criterion

elastic energy + distance between A_n and A_n'

would be suitable as stabilization goal, though it has a serious deficiency. A more refined criterion is

$$\mathcal{E}_1 \equiv \frac{1}{2} \sum_{i=1}^{n} a_i |w_i'(0,t) - w_{i-1}'(1,t)|^2 + \mathcal{E}, \qquad (4.6)$$

where $a_i>0$ and by definition $w_0'(0,t) \equiv 0$. The derivation similar to (4.3) shows that

$$\dot{\mathcal{E}}(t) = \sum_{i=0}^{n-1} \left[E_{i+1} I_{i+1} w_{i+1}''(0,t) + a_i(\gamma_i(t)-\alpha_i) \right] u_i(t) \qquad (4.7)$$

and the feedback

$$u_i(t) = -\sum_{i=0}^{n-1} k_{ij} \left[E_{i+1} I_{i+1} w_{i+1}''(0,t) - a_i(\gamma_i(t)-\alpha_i) \right] \qquad (4.8)$$

with positive $K=(k_{ij})$ satisfies the "dissipativity" in the form

$$\dot{\mathcal{E}}(t) \leq 0, \; t \geq 0 \qquad (4.9)$$

for any initial state of the beam. It is easy to see that $\mathcal{E}_1=0$ implies that the elastic energy is zero and the rotation is continuous at each joint, thus the beam rests at its reference position.

The dynamics of the compositve system can be described in state-space notation as a Cauchy problem

$$\dot{W}(t) = A \; W(t) \, , \quad W(0) = W_0 \qquad (4.10)$$

in an infinite-dimensional normed space of pairs W (deflection velocity), where the norm is associated with a selected stability criterion and A is an unbounded operator with a compact resolvent and discrete spectrum.

5. Stabilizability

Once a stability criterion has been selected (say $\mathcal{E}$), the problem is to design a control algorithm stabilizing the beam. However, stabilizability can be introduced only as an asymptotic decay property. The most desirable stabilizability form is the uniform exponential decay

$$\mathcal{E}(t) \leq c \; e^{-\omega t} \mathcal{E}(0), \; \mathcal{E}(0) > 0 \quad \text{for all } t \geq 0, \qquad (5.1)$$

where c and ω are some positive constants independent on initial conditions. A necessary condition for (5.1) is $\mathrm{Re}\,\lambda < 0$ for all eigenvalues λ of the operator A in (4.10). Another necessary condition is the uniform boundedness of the resolvent operator on the imaginary axis:

$$\sup_{\mathrm{Re}\,\lambda=0} \|R(\lambda,A)\| \leq \infty \, . \qquad (5.2)$$

It was proved recently in [4] that (5.2) along with the above spectral inequality constitute a necessary and sufficient condition for (5.1).

At present little is known about the uniform exponential stabilizability of composite Euler-Bernoulli beams with pointwise feedback control. In [5] it was proved by the energy multipliers method that this property holds for a cantilever beam with shear and bending moment feedback control applied at the free end. In [6] this result was extended to the case of moment control.

In the contrary to the uniform exponential stabilizability, strong stabilizability

$$\mathcal{E}(t) \to 0, \text{ as } t \to 0 \qquad (5.3)$$

for any initial state, does follow solely from the inequality $\mathrm{Re}\,\lambda < 0$ [4].

The general composite solution $\{w_j(x,t), j)1,\dots,n\}$ to the system (3.6) can be expressed as a linear combination of functions of the form

$$W_\lambda(x,t) = \{w_{\lambda j}(x) e^{\lambda t}\} \qquad (5.4)$$

with eigenmodes given by

$$w_{\lambda j}(x) = A_j e^{\beta_j \eta} + B_j e^{i\beta_j \eta} + C_j e^{-\beta_j \eta} + D_j e^{-i\beta_j \eta}. \qquad (5.5)$$

Here

$$i = \sqrt{-1}, \quad \beta_j = \left[\frac{E_j I_j}{m} \right], \quad \eta = \frac{1-i}{\sqrt{2}} \sqrt{\lambda} \; .$$

The 4 boundary and 4n-4 in-span conditions give us 4n equations with 4n unknowns $\{A_j, B_j, C_j, D_j\}$. Thus, λ will be an eigenfrequency if the determinant of the matrix formed by this equations is zero.

The problem of finding the eigenfrequencies requires monumental calculations. At present the results available concern only the asymptotics in the case of collinear links with dissipative end-point conditions similar to (4.4), in this case the spectrum has n vertical asymptotic streams [7]. The same is true also for a cantilever one-link beam controlled by (4.8).

To prove strong stabilizability we need to verify the inequality $\mathrm{Re}\,\lambda < 0$ for all eigenfrequencies. It is easy to see that for both types of feedback (4.4) or (4.8) the inequality $\mathrm{Re}\,\lambda \leq 0$ is a consequence of dissipativity (4.5) or (4.9), respectively. Hence, it is sufficient to find out whether there exists an imaginary eigenfrequency λ. For this eigenfrequency the function (5.4) with the eigenmode defined by (5.5) is a solution of (3.6) satisfying $\mathcal{E}=0$ (for the feedback (4.4)) or $\mathcal{E}_1=0$ (for the feedback (4.8)).

Consider the case of a local feedback according to (4.4), i.e.

$$u_j(t) = -k_j w_j''(0,t),\ k_j \geq 0,\ \text{for } j=0,...,n-1 \tag{5.6}$$

and similarly, a local feedback according to (4.8)

$$u_j(t) = -k_j \left[E_j I_j w_j''(0,t) - a_j(\gamma_j(t) - \alpha_j) \right] . \tag{5.7}$$

It can be shown that if a feedback coefficient k_j is zero then the j-th joint is uncontrolled and there is no any dissipation at this joint. In the contrary, if $k_j > 0$ then $\mathcal{E}=0$ (or $\mathcal{E}_1=0$) implies $w_{\lambda j}''(0)=0$. Consequently, the question whether there existis an imaginary eigenfrequency is reduced to the following problem: *Let $k_{j_1}, ..., k_{j_p}$ be positive and the others k's in (5.6) or (5.7) be zero. Does there exist an eigenmode related with this feedback, satisfying*

$$w_{\lambda j1}''(0) = ... = w_{\lambda jp}''(0) = 0\ ? \tag{5.8}$$

It seems a priori that the answer is always negative. However, consider a composite beam of two identical collinear links with hinged ends and ideal coupling at the mid-span (thus, it is actually one beam). In this case the boundary conditions are

$$w_1(0) = w_1''\ 0) = 0,$$

$$w_2(1) = w_2''\ 1) = 0,$$

and in-span conditions are

$$w_1^{(p)}(1) = w_2^{(p)}(0),\ p=0,1,2,3\ .$$

The function

$$w(x) = \sin \pi x \quad x \in [0,2]$$

is an eigenmode corresponding to the eigenvalue $\lambda=i$, satisfying $w''(1)=0$ in the mid-span. Of course, this "counter example" is related with hinged end-points and leaves the problem open for the fixed-free conditions of a composite beam.

Let us now prove that there is no eigenmode of a composite beam driven by (5.6) or (5.7) such that

$$w_{\lambda 1}''(0) = w_{\lambda 2}''(0) = 0. \tag{5.9}$$

Apparently, $u_1(t) \equiv 0$ implies

$$w_{\lambda 1}'(0) = 0. \tag{5.10}$$

Writing eigenmodes in a form slightly different from (5.5)

$$w_{\lambda 1}(x) = c_1 \sin \eta x + c_2 \cos \eta x + c_3 \sinh \eta x + c_4 \cosh \eta x \tag{5.11}$$

we obtain from (5.9), (5.10) and (3.7) that the coeffiecients satisfy

$$c_2 = c_4 = 0,\ c_1 = c_3.$$

We find that either

$$w_{1\lambda}(x) \equiv 0 \tag{5.12a}$$

or

$$w_{\lambda 1}(x) = c(\sin \eta x - \sinh \eta x),\ c \neq 0. \tag{5.12b}$$

In the first case we have

$$w_{2\lambda}'(0) = w_{2\lambda}'(0) = w_{2\lambda}''(0) = 0\ ,$$

and a similar alternative holds for $w_{2\lambda}$, and so on for $w_{j\lambda}$, $j>2$. Hence, it is sufficient to prove that (5.12b) is not compatible with the boundary condition $w_{1\lambda}''(1)=0$. Indeed, we have

$$\sin \eta + \sinh \eta = 0$$

for some real η, but this equation has no solutions.

A pratical implication of this result is that the composite beam is strongly stabilizable just by two dampers located at the end-points of the second link. We conjecture that even one damper located at the root is sufficient for strong stabilizability (at least to "almost all" composite beam configurations): it seems intuitively clear that clamped-free end-point conditions are incompatible with zero bending moment at the root.

6. Concluding remarks

To be more realistic, a model of a flexible manipulator should include also some other factors omitted in our discussion. Locality of a feedback scheme, introduced in Section 4, is not very important for active damping governed by a computer, but plays a crucial role in passive damping devices. Local feedback (5.6) corresponds to Type II dissipative in-span conditions, according to the classification in [8], where one can find also some hints regarding the possible design of passive dampers. The feedback coefficients in (5.6) should be taken as large as possible to have better stability properties [7,9].

The stability criterions $\mathcal{E}_1$, defined by (4.6), has a nice mechanical interpretation: it represents energy of a composite beam with rotational springs between the links. In this context, the reference position of the beam agrees with the unstrained state of the springs.

Realization of active damping by the scheme (4.8) requires measuring of bending moments at the joints. The moments can be measured directly by special sensors with output voltage proportional to the bending (direct measurements) or, indirectly, by measuring the displacement of a point close to the joint, with successive estimation of the curvature of the deflection. Experiments on a physical 4 link-beam are just under preparation.

Acknowledgement

This work was partially supported by the A.v. Humboldt Foundation for the second author.

References

[1] Balas, M.J.: Trends in large space structure control theory: fondest hopes, wildest dreams. IEEE Transactions Autom. Contr. AC-27(1982), pp. 15-33.

[2] Johnson, T.L.: Progress in modelling and control of flexible spacecraft. J. of Franklin Institute 315(1983), pp. 495-520.

[3] Russel, D.: Mathematical models for elastic beams and their control theoretic implications. In: Semigroups, Theory and Applications, volume II. Breis et al. (eds.), Harlow, U.K.: Longman, 1986.

[4] Huang, F.L.: Characteristic conditions for exponential stablity of linear dynamical systems in Hilbert spaces. Ann. Diff. Eqs.1 (1985), pp. 43-53.

[5] Chen, G., Delfour, M.C., Krall, A.M. and Payne, G.: Modelling, stabilization and control of serially connected beams. SIAM J. Control Optim., 25(1987), pp. 526-546.

[6] Chen, G., Krantz, S.G., Ma, D.W., Wayne, C.E., West, H.H.: The Euler-Bernoulli beam equation with boundary energy dissipation. In: Operator Methods for Optimal Control Problems. Sung J. Lee (ed.), Marcel Dekker, New York, 1988.

[7] Krantz, S.G. and Paulsen, W.H.: Asymptotic eigenfrequency distributions for the N-beam Euler-Bernoulli coupled beam with dissipative joints. Dept. Math. Washington University at St. Lois, preprint.

[8] Chen, G., Krantz, S.G., Russel, D.L., Wayne, C.E., West, H.H. and Coleman, M.P.: Analysis, design and behaviour of dissipative joints for coupled beams. SIAM J. Appl. Math. 49(1989), pp. 1665-1693.

[9] Conrad, F.: Stabilization of beams by pointwise feedback control, SIAM J. Control Optim. 28(1990), pp. 423-437.

Proceedings of the 31st Conference
on Decision and Control
Tucson, Arizona • December 1992

Modelling, Control and Asymptotics for a model of multi-dimensional flexible structure

Enrique Zuazua
Departamento de Matemática Aplicada
Universidad Complutense
28040 Madrid. Spain.

Abstract

We present some recent joint work with J. P. Puel on the modelling, exact controllability and stabilization of multi-dimensional flexible structures. We also describe some recent joint research with C. Conca on some asymptotic properties of the spectrum of the multi-dimensional elliptic systems involved in these models.

1. Modelling

Multi-structures are of very common ocurrence in practice. They are found, for example, in the study of aerials, plates, shells with stiffeners such as in solar panels, etc. However, despite their enormous practical importance, it is only recently that research in this area has begun from a rigorous point of view.

One of the first theoretical studies of mathematical models in multi-structures is the 1989 article by Ph. Ciarlet, H. Le Dret and R. Nzengwa [2]. Roughly speaking, the method that they propose for dealing with a junction of two bodies, say, one of dimension N and another of dimension N-1 consists of carrying out an asymptotic analysis of the structure. In the first place, the body of dimension N-1 is assimilated to another of dimension N, one of whose dimensions is very small compared to the N-1 others. Let us say that its size is ε while the (N-1)-dimensional volume of the original body is 1. Then a change is introduced to the scale of this body, so that the small dimension becomes of the same characteristic size as the rest, and one concludes letting ε go to zero. In the general case, the main idea is always the same, that is, of re-scaling the different parts of the multi-structure independently of each other and passing to the limit in all those dimensions which are small with respect to the others. It will be appreciated that in this approach the contact or junction condition is dealt with implicitly, since the multi-structure is approximated by a sequence of N-dimensional domains, and the junction of bodies thus comes about naturally. The ideas of the paper [2] have been extensively developed in the books by Ph. Ciarlet [2] and H. Le Dret [4].

Recently, J. P. Puel and E. Zuazua [7]-[10] have considered a number of multi-dimensional flexible structures in which the junction is dealt with explicitly. They have considered, in particular, a system made of a N-dimensional body and a one-dimensional string in which both bodies satisfy the wave equation.The junction is described by means of a special boundary condition that connects one of the extremes of the string with the boundary of the N-dimensional body. This a local (not a pointwise) condition since a neighborhood of the contact point intervenes in it; we will call it the contact region. This boundary condition introduces some rigidity of the contact region that is deformed following a prescribed profile. Let us describe briefly this model.

Let Ω be a bounded and smooth domain of $\mathbf{R}^N$ with $N\geq 2$ and $\omega=(A, B)$ a one-dimensional straight string attached to Ω at the extreme A which is assumed to belong to the boundary of Ω that we denote by Γ. Let Γ_0 be a subset Γ of the form

$$\Gamma_0 =\{ x \in \Gamma: (x-x_0)\cdot\nu(x) >0\}$$

for some x_0 of $\mathbf{R}^N$. We denote by ν the unit outward normal to Ω and by $\cdot$ the scalar product in $\mathbf{R}^N$. Let γ be the contact region, an open neighborhood of A contained in $\Gamma- \Gamma_0$. We set $\Gamma_1 = \Gamma- \Gamma_0$. Let $T >0$ and define $Q=\Omega x(0, T)$, $q=\omega x(0, T)$, $\Sigma_0= \Gamma_0 x(0, T)$ and $\Sigma_1= \Gamma_1 x(0, T)$. We denote by $x=(x_1, ..., x_N)$ a generic point in Ω, by s the abscissa of a point in the string so that A corresponds to s=0 and B to s=l and by t the time variable. Let θ be the contact profile, a smooth and nonnegative function with compact support on γ such that $\theta(A)=1$ and $0\leq\theta\leq 1$ everywhere. By $y=y(x, t)$ (resp. $z=z(s, t)$) we denote the displacement of a point x of the N-dimensional domain (resp. of a point of the string) at time t.

The system introduced by J. P. Puel and E. Zuazua is as follows:

$$\begin{array}{ll}
y''-\Delta y = 0 & \text{in } Q \\
z''-\partial^2 z = 0 & \text{in } q \\
y = 0 & \text{on } \Sigma_1 \\
\partial z(l, t) =0 & \text{for t in } (0, T) \\
y =v & \text{on } \Sigma_0 \\
y = z(0, t)\theta & \text{on } \gamma x(0, T) \\
\partial z(0, t) = \int_\gamma \partial_\nu y\, \theta\, d\Gamma & \text{for t in } (0, T) \\
y(x, 0) = y_0(x),\ y'(x, 0) = y_1(x) & \text{in } \Omega \\
z(s, 0) = z_0(s),\ z'(s, 0) = z_1(s) & \text{in } \omega.
\end{array} \quad (1)$$

In (1) ' denotes the derivative with respect to the time variable , ∂ the derivative with respect to s and ∂_ν the normal derivative.The function $v=v(x, t)$ represents a control function acting

CH3229-2/92/0000-2997$1.00 © 1992 IEEE

on a subset of the boundary of the N-dimensional domain (Γ_0) that excludes the control region.

This system is shown to be well posed in suitable spaces. When v=0, i.e. in the absence of control, the H^1 -energy of solutions is conserved along the time.

2. Controllability

The main controllability result that J. P. Puel and E. Zuazua prove in [7] and [8] is as follows:

Theorem. *There exists* $T_0>0$ *such that when* $T>T_0$ *for every initial data*

(y_0, z_0) *in* $L^2(\Omega)\mathrm{x}\, L^2(\omega)$

(y_1, z_1) *in* **V'**

there exists a control function

v *in* $L^2(\Sigma_0)$

such that the solution (y, z) *of* (1) *is at rest at time* t=T.

In this theorem **V'** denotes the dual space of

$$\mathbf{V}=\{(y, z) \text{ in } H^1(\Omega)\mathrm{x}\, H^1(\omega): y = 0 \text{ on } \Gamma_0 \text{ and } y = z(0)\theta \text{ on } \gamma\}$$

endowed with the norm of $H^1(\Omega)\mathrm{x}\, H^1(\omega)$.

When the initial data and boundary control belong to the spaces above system (1) admits a unique solution in the sense of transposition.

This theorem extends to the multi-dimensional system (1) well known results on the controllability of wave equations. It shows that the whole multi-structure can be controlled by acting only on the N-dimensional domain. In [7] and [8] an explicit estimate of the control time T_0 is given. It depends on Ω, ω, the point x_0 and the contact profile θ.

The method of proof combines HUM (cf. J. L. Lions [6]) and multiplier techniques.

The ideas of [7] and [8] can be adapted and applied in various situations. Let us mention the case where several strings are attached to the N-dimensional domain [10] or fourth order systems modelling plate-beam configurations (see also J. Lagnese [5] in this direction).

3. Asymptotics for the spectrum

In a recent joint work with C. Conca [4] we describe the qualitative behavior of the eigenfrequencies and eigenmotions of the multi-dimensional elastic system (1) in the absence of control when the contact region tends to disappear and converges to a set of (N-1)-dimensional zero measure. This goal is achieved in terms of the convergence of Green's operator and the spectral family associated with this problem.

The spectral problem corresponding to system (1) is the one associated with the Laplace operator in both regions of the multi-structure (Ω and ω) coupled with the special boundary condition above that models the junction between the two bodies. The resulting limit problem is the one associated with Laplace operator in both regions. In the N-dimensional domain Dirichlet boundary conditons are obtained and the boundary condition we get on the extreme A of the string depends , in particular, on N and the Hausdorff dimension of the limit set of the contact-region. The techniques we use are inspired by the theory of homogenization and the obtention of the limit system relies on the use of suitable test functions.

References

[1] Ph. Ciarlet, Plates and junctions in elastic multi-structures. An asymptotic Analysis, Paris, RMA 14, Masson, 1990.

[2] Ph. Ciarlet, H. Le Dret and R. Nzengwa, "Junctions between 3d and 2d linerly elastic structures", J. Math. Pures Appl., vol. 68, pp. 261-295, 1989.

[3] C. Conca and E. Zuazua, "Asymptotic Analysis of a multi-dimensional vibrating structure", preprint, 1992.

[4] H. Le Dret, Problèmes Variationnels dans les Multi-Domaines. Modélisation des Jonctions et Applications, Paris, RMA 19, Masson, 1991.

[5]J. Lagnese, "Modelling and controllability of plate-beam systems", preprint, 1992.

[6] J. L. Lions, Contrôlabilité exacte, perturbations et stabilisation de systèmes distribués. Tome 1. Contrôlabilité exacte, Paris, Masson, RMA 8, 1988.

[7] J. P. Puel and E. Zuazua, "Contrôlabilité exacte et stabilisation d'un modèle de structure vibrante multidimensionnelle", C. R. Acad. Sci. Paris, vol. 314, pp. 121-125, 1992.

[8] J. P. Puel and E. Zuazua, "Exact controllability for a model of multidimensional flexible structure", Proc. Roy. Soc. Edinburgh, to appear.

[9] J. P. Puel and E. Zuazua, "Controllability of a multi-dimensional system of Schrödinger equations: Application to a system of plate and beam equations", in Proceedings of the Coference State and frequency domain approaches for the control infinite-dimensional systems, Springer-Verlg, to appear.

[10] J. P.Puel and E. Zuazua, "Exact controllability for some models of multidimensional vibrating structures", in Prooceedings of the Conference Mathematics, Climate and Environment, J. I. Díaz and J.L. Lions eds., Masson, to appear.

Proceedings of the 31st Conference
on Decision and Control
Tucson, Arizona • December 1992

DECAY ESTIMATES FOR NONLINEAR STRUCTURES

Vilmos Komornik

Département de Mathématiques, Université Louis Pasteur
7, rue René Descartes, 67084 Strasbourg Cédex, France

Abstract

Using a special integral inequality instead of the usual Liapunov method we obtain sharp energy decay rate estimates for several nonlinear distributed systems. The method applies in particular for the internal or boundary stabilization of the wave equation and of some plate models.

1. INTRODUCTION

In many problems of nonlinear stabilization the Liapunov method provides a very useful tool not only for proving the convergence of the solutions but also for estimating the decay rate of the energy. This method was recently applied by various authors to obtain strong decay rate estimates for several wave or plate models, cf. e.g. [2], [3], [4], [7], [10], [11], [14], [15]. In most of these papers first the energy is shown to satisfy an integral inequality of a special type and then this integral inequality permits to construct a suitable Liapunov function.

Sometimes, mainly in linear problems, decay estimates were obtained directly from the integral inequalities, with or without explicit constants, cf. e.g. [1], [5], [6], [8], [12], [13]. This approach may lead to more precise estimates, cf. [8].

The purpose of this paper is to describe a general method based on the systematic application of integral inequalities instead of Liapunov functions and to illustrate this approach by improving several earlier results of Conrad, Haraux, Lasiecka, Leblond, Marmorat, Nakao, Rao and Zuazua obtained in [3], [4], [6], [7], [10], [12], [13], [14], [15]. As we shall see, this method also allows us to simplify the proofs by avoiding the introduction of a small parameter as in the Liapunov method.

We remark that other applications of this method will be given in [9].

2. AN INTEGRAL INEQUALITY

Let $E : \mathbb{R}_+ \to \mathbb{R}_+$ ($\mathbb{R}_+ := [0,+\infty)$) be a decreasing function, not identically zero, and assume that there exist two positive numbers α and A such that

$$\int_t^\infty E^{\alpha+1} ds \le AE(t) \qquad \text{for all} \qquad t \ge 0. \tag{2.1}$$

The purpose of this section is to give an optimal estimate on $E(t)$ under this condition.

Set

$$T := AE(0)^{-\alpha}$$

and introduce a decreasing function $I : \mathbb{R}_+ \to \mathbb{R}_+$ by the following formula :

$$I(\alpha, A, E(0); t) := \begin{cases} E(0), & \text{if } 0 \le t \le T; \\ E(0)\left(\frac{T+\alpha T}{T+\alpha t}\right)^{1/\alpha}, & \text{if } t \ge T. \end{cases}$$

Then we have the following result :

THEOREM 2.1. — *Under the above conditions we have*

$$E(t) \le I(\alpha, A, E(0); t) \qquad \text{for all} \qquad t \ge 0. \tag{2.2}$$

Conversely, for every $C > 0$ and $t' \ge 0$ there exists a decreasing function $E : \mathbb{R}_+ \to \mathbb{R}_+$ satisfying (2.1) and such that

$$E(0) = C \quad \text{and} \quad E(t') = I(\alpha, A, E(0), t').$$

REMARK 2.2. — In linear problems one often obtains the inequality (2.1) with $\alpha = 0$. In this case we have the estimate (2.2) with

$$I(0, A, E(0); t) := \begin{cases} E(0), & \text{if } 0 \le t \le T; \\ E(0)\exp(1 - \frac{t}{T}), & \text{if } t \ge T. \end{cases}$$

This case is well-known, cf. e.g. [4], [8], [11]. This estimate is also optimal.

REMARK 2.3. — If E is continuous (or at least rightcontinuous), then we have also the strict inequlities

$$E(t) < I(\alpha, A, E(0); t) \quad \forall t \ge T;$$

in particular we have $E(T) < E(0)$.

3. STABILIZATION BY DISTRIBUTED FEEDBACKS. A GENERAL RESULT

Let Ω be an open domain in $\mathbb{R}^n$, $H = L^2(\Omega)$ and let V be another real Hilbert space such that $V \subset H$ with dense and continuous imbedding. We denote by L the unique linear operator $L \in \mathcal{L}(V, V')$ such that $< Lu, v >= (u,v)_V$ for all $(u,v) \in V \times V$, and we set $D(L) := \{v \in V \ : \ Lv \in H\}$.

Let $g : \mathbb{R} \to \mathbb{R}$ be a non-decreasing, continuous function such that $g(0) = 0$ and consider the nonlinear evolution system

$$u'' + Lu + g(u') = 0 \quad \text{in} \quad \Omega \times \mathbb{R}_+, \tag{3.1}$$

$$u(0) = u_0 \quad \text{and} \quad u'(0) = u_1 \quad \text{in} \quad \Omega. \tag{3.2}$$

Many results are available on the well-posedness and on the weak, strong or uniform stabilization properties of this system, cf. e.g. [3], [6], [7]. The purpose of this section is to improve the earlier decay rate estimates by weakening the growth assumptions on g, and at the same time to provide simpler proofs which adapt more easily to the study of more complex systems.

CH3229-2/92/0000-2999$1.00 © 1992 IEEE

We recall from [6] that if

$$V \cap L^\infty(\Omega) \cap L^1(\Omega) \quad \text{is dense in} \quad V. \tag{3.3}$$

and

$$\{v \in V : \ g(v) \in H\} \quad \text{is dense in} \quad H, \tag{3.4}$$

then for every $(u_0, u_1) \in V \times H$ the system (3.1)-(3.2) has a unique solution in

$$C_b(\mathbb{R}_+; V) \cap C_b^1(\mathbb{R}_+; H),$$

and that the "energy" $E : \mathbb{R}_+ \to \mathbb{R}_+$ of the solution, defined by

$$E(t) := \frac{1}{2}\|u(t)\|_V^2 + \frac{1}{2}\|u'(t)\|_H^2, \ t \in \mathbb{R}$$

is non-increasing.

We have the following result.

THEOREM 3.1. — *Assume that Ω has finite measure and that the conditions (3.3), (3.4) are satisfied. Assume also that there exist four positive numbers $p \in [1, +\infty)$, $q \in [1, +\infty]$, c_1 and c_2 such that*

$$c_1|x|^p \le |g(x)| \le c_2|x|^{1/p} \quad \text{if} \quad |x| \le 1, \tag{3.5}$$

$$c_1|x| \le |g(x)| \le c_2|x|^q \quad \text{if} \quad |x| > 1 \tag{3.6}$$

and

$$V \subset L^{q+1}(\Omega). \tag{3.7}$$

(For $q = +\infty$ we write $q + 1 = +\infty$ in (3.7); in this case the last inequality of (3.6) is automatically satisfied, independently of the value of c_2.) Then for every $(u_0, u_1) \in V \times H$ there exists a constant C, depending only on $E(0)$, such that

$$E(t) \le Ct^{2/(1-p)} \ \forall t > 0 \quad \text{if} \quad p > 1 \tag{3.8}$$

and

$$E(t) \le C \exp(1 - t/C) \ \forall t > 0 \quad \text{if} \quad p = 1. \tag{3.9}$$

REMARK 3.2. — In the earlier works the estimates (3.8)-(3.9) were proved under stronger hypotheses on g than (3.5) and (3.6). We conjecture that the assumption (3.5) is optimal.

REMARK 3.3. — Let m be a positive integer, $\rho > -1$ a real number, $g(s) := |s|^\rho s$ and choose $V = H^m(\Omega)$ or $V = H_0^m(\Omega)$. Then the conditions (3.5) and (3.6) are satisfied. Furthermore, (3.7) is satisfied for all $q \le (n + 2m)/(n - 2m)$ if $n > 2m$, for all $q < +\infty$ if $n = 2m$, and for all $q \in [1, +\infty]$ if $n < 2m$. This follows at once from the Sobolev inequalities.

4. STABILIZATION BY DISTRIBUTED FEEDBACKS. A FURTHER RESULT FOR THE WAVE EQUATION

Let us now consider the special case $V = H_0^1(\Omega)$ i.e. the system

$$u'' - \Delta u + g(u') = 0 \quad \text{in} \quad \Omega \times \mathbb{R}_+ \tag{4.1}$$

$$u = 0 \quad \text{on} \quad \Gamma \times \mathbb{R}_+ \tag{4.2}$$

$$u(0) = u_0 \quad \text{and} \quad u'(0) = u_1 \quad \text{in} \quad \Omega. \tag{4.3}$$

Then E is given by

$$E = \frac{1}{2}\int_\Omega (u')^2 + |\nabla u|^2 \, dx.$$

Let $g : \mathbb{R} \to \mathbb{R}$ be a non-decreasing, locally absolutely continuous function such that $g(0) = 0$ and assume that there exist three numbers $p_0 \in [1, +\infty)$, $q \in [1, +\infty]$ and $r \in [0, 1]$ and two positive constants c_1 and c_2 such that

$$c_1|x|^{p_0} \le |g(x)| \le c_2|x|^{1/p_0} \quad \text{if} \quad |x| \le 1, \tag{4.4}$$

and

$$c_1|x|^r \le |g(x)| \le c_2|x|^q \quad \text{if} \quad |x| > 1 \tag{4.5}$$

It is easy to verify that then the conditions (3.3), (3.4) of the preceeding section are satisfied.

REMARK 4.1. — If $q = +\infty$, then the right hand side inequality of (4.5) is automatically satisfied. Similarly, if $r = 0$, then the left hand side inequality of (4.5) is automatically satisfied (observe that $|g(x)| \ge |g(\text{sgn } x)| \ge c_1$ if $|x| > 1$). Hence in case $q = +\infty$ and $r = 0$ the condition (3.31) is empty i.e. we have no growth assumption on g near $+\infty$.

THEOREM 4.2. — *Assume that*

$$H^2(\Omega) \subset L^{q+1}(\Omega) \tag{4.6}$$

and choose $p \in [p_0, +\infty)$ such that

$$2(p - 1) \ge (n - 2)(1 - r), \tag{4.7}$$

$$2(p + 1) \ge n(1 - 1/q) \tag{4.8}$$

and

$$p > 1 \quad \text{if} \quad n = 2 \text{ and } r < 1. \tag{4.9}$$

Then for every $(u_0, u_1) \in (H^2(\Omega) \cap H_0^1(\Omega)) \times H_0^1(\Omega)$ such that $g(u_1) \in L^2(\Omega)$, the solution of (4.1)-(4.3) satisfies the energy estimates

$$E(t) \le Ct^{2/(1-p)} \ \forall t > 0 \quad \text{if} \quad p > 1 \tag{4.10}$$

and

$$E(t) \le C \exp(1 - t/C) \ \forall t > 0 \quad \text{if} \quad p = 1. \tag{4.11}$$

REMARK 4.3. — We recall that by the Sobolev imbedding theorems (4.6) is satisfied for all $q \in [1, +\infty]$ if $n < 4$, for all $q \in [1, +\infty)$ if $n = 4$, and for all $q \in [1, (n + 4)/(n - 4)]$ if

$n > 4$.

REMARK 4.4. — Assume that $n < 4$ and choose $q = +\infty$ and $r = 0$. Then (4.6)-(4.9) are satisfied if we choose a sufficiently large $p : p > \max\{p_0, n/2\}$. Thus in these cases we obtain explicit decay rate estimates without any growth assumption on g near $+\infty$. Let us remark that for $n = 2$ and $n = 3$ the available existence and regularity results do not imply that $u' \in L^\infty(\mathbb{R}_+; L^\infty(\Omega))$; hence the absence of this growth assumption is not obvious *a priori*. This possibility of having decay rate estimates for not uniformly bounded functions without growth assumptions at infinity was conjectured by F. Conrad (private communication).

REMARK 4.5. — Theorem 4.2 improves an earlier result of Nakao [13] by weakening the growth assumptions on g. (He also applied integral inequalities to prove his result, but of a different kind.) Our result also answers a question raised by Zuazua in [14].

5. STABILIZATION OF THE WAVE EQUATION BY BOUNDARY FEEDBACKS.

In this section we assume that Ω is a **bounded** open set in $\mathbb{R}^n$ with a smooth boundary Γ. We shall denote by ν the outward unit normal vector to Γ. Fix a point $x_0 \in \mathbb{R}^n$, set

$$m(x) := x - x_0,\ x \in \mathbb{R}^n,$$

and fix an open subset Γ_- of Γ such that setting $\Gamma_+ = \Gamma \backslash \Gamma_-$ we have

$$m \cdot \nu \geq 0 \text{ on } \Gamma_+ \text{ and } m \cdot \nu \leq 0 \text{ on } \Gamma_-. \tag{5.1}$$

Let $g : \mathbb{R} \to \mathbb{R}$ be a non-decreasing, continuous function such that $g(0) \neq 0$, let α be a nonnegative number and consider the following seedback system :

$$u'' - \Delta u = 0 \quad \text{in} \quad \Omega \times \mathbb{R}_+ \tag{5.2}$$

$$u = 0 \quad \text{on} \quad \Gamma_- \times \mathbb{R}_+ \tag{5.3}$$

$$\frac{\partial u}{\partial \nu} + (m \cdot \nu)(\alpha u + g(u')) = 0 \quad \text{on} \quad \Gamma_+ \times \mathbb{R}_+ \tag{5.4}$$

$$u(0) = u_0 \quad \text{and} \quad u'(0) = u_1 \quad \text{in} \quad \Omega. \tag{5.5}$$

This system is well-posed in the following sense (cf. [15]) : introducing the Hilbert space V by

$$V = \{v \in H^1(\Omega) : v = 0 \text{ on } \Gamma_-\},$$

for every $(u_0, u_1) \in V \times L^2(\Omega)$ the system (5.2)-(5.5) has a unique solution u satisfying

$$C_b(\mathbb{R}_+; V) \cap C_b^1(\mathbb{R}_+; L^2(\Omega));$$

furthermore its energy $E : \mathbb{R}_+ \to \mathbb{R}_+$ defined by

$$E = \frac{1}{2}\int_\Omega (u')^2 + |\nabla u|^2\, dx + \frac{\alpha}{2}\int_\Gamma (m \cdot \nu) u^2\, d\Gamma$$

is non-increasing.

In order to obtain decay estimates for the energy let us assume that

$$\text{either } \Gamma_- \neq \emptyset \text{ or } \alpha > 0, \tag{5.6}$$

$$\inf_{\Gamma_+}(m \cdot \nu) > 0, \tag{5.7}$$

and assume that there exist $p \in [1, +\infty)$ and two positive constants c_1, c_2 such that

$$c_1|x|^p \leq |g(x)| \leq c_2|x|^{1/p} \quad \text{if} \quad |x| \leq 1 \tag{5.8}$$

and

$$c_1|x| \leq |g(x)| \leq c_2|x| \quad \text{if} \quad |x| > 1. \tag{5.9}$$

THEOREM 5.1. — *Assume* (5.1) *and* (5.6)-(5.9). *Then for every* $(u_0, u_1) \in V \times L^2(\Omega)$ *the solution of* (5.2)-(5.5) *satisfies the energy estimates*

$$E(t) \leq Ct^{2/(1-p)}\ \forall t > 0 \quad \text{if} \quad p > 1 \tag{5.10}$$

and

$$E(t) \leq C\exp(1 - t/C)\ \forall t > 0 \quad \text{if} \quad p = 1. \tag{5.11}$$

REMARK 5.2. — The first nonlinear result of this type was obtained by Zuazua in [15] under stronger growth assumptions on g and assuming that α is not too large. (See also [12].) His result was extended for large α by Conrad and Rao in [3] by using a new multiplier. Our condition (5.8) is weaker than their one and we conjecture that this condition is optimal.

REMARK 5.3. — The condition (5.7) is probably unnecessary : we refer e.g. to [8], [10], [15] for a discussion of this question.

Our next result shows that the condition (5.9) may be weakened if the solution is more regular.

THEOREM 5.4. — *Assume that g is also locally absolutely continuous and assume again* (5.1), (5.6) *and* (5.7). *Furthermore, assume that there exist three numbers* $p_0 \in [1, +\infty)$, $q \in [1, +\infty]$, $r \in [0, 1]$ *and two positive constants* c_1 *and* c_2 *such that*

$$c_1|x|^{p_0} \leq |g(x)| \leq c_2|x|^{1/p_0} \quad \text{if} \quad |x| \leq 1, \tag{5.12}$$

$$c_1|x|^r \leq |g(x)| \leq c_2|x|^q \quad \text{if} \quad |x| > 1 \tag{5.13}$$

and

$$q \leq n/(n-2) \text{ if } n > 2. \tag{5.14}$$

Let us choose $p \in [p_0, +\infty)$ *such that*

$$p - 1 \geq (n-2)(1-r) \text{ if } n > 2, \tag{5.15}$$

$$2p \geq n + (4-n)/q \text{ if } q > 1 \tag{5.16}$$

and

$$p > 1 \quad \text{if} \quad n = 2 \quad \text{and} \quad r < 1. \tag{5.17}$$

Then for every $(u_0, u_1) \in H^2(\Omega) \times H^1(\Omega)$ *such that*

$$u_0 = u_1 = 0 \quad \text{on} \quad \Gamma_-$$

and

$$\frac{\partial u_0}{\partial \nu} + (m \cdot \nu)(\alpha u_0 + g(u_1)) = 0 \quad \text{on} \quad \Gamma_+,$$

the solution of (5.2)-(5.5) *satisfies the estimates* (5.10)-(5.11).

REMARK 5.5. — If $r = 0$, then (5.15) is satisfied if we choose a sufficiently large p. Hence we may obtain explicit decay rate estimates for bounded feedback functions g.

References

[1] G. Chen, Energy decay estimates and exact boundary value controllability for the wave equation in a bounded domain, J. Math. Pures Appl. 58(1979), 249-274.

[2] G. Chen and H. K. Wang, Asymptotic behavior of solutions of the one-dimensional wave equation with a nonlinear elastic dissipative boundary stabilizer, SIAM J. Control Opt. 22(1989), 758-775.

[3] F. Conrad, J. Leblond and J. P. Marmorat, Stabilization of second order evolution equations by unbounded nonlinear feedback, to appear.

[3] F. Conrad and B. Rao, Decay of solutions of wave equation in a star-shaped domain with nonlinear boundary feedback, to appear.

[4] A. Haraux, Oscillations forcées pour certains systèmes dissipatifs non linéaires, preprint n° 78010, Laboratoire d'Analyse Numérique, Université Pierre et Marie Curie, Paris, 1978.

[5] A. Haraux, Semi-groupes linéaires et équations d'évolution linéaires périodiques, preprint n° 78011, Laboratoire d'Analyse Numérique, Université Pierre et Marie Curie, Paris, 1978.

[6] A. Haraux, Semilinear hyperbolic problems in bounded domains, Mathematical reports, J. Dieudonné editor, Harwood Academic Publishers, Gordon and Breach, New York, 1987.

[7] A. Haraux and E. Zuazua, Decay estimates for some semilinear damped hyperbolic problems, Arch. Rat. Mech. Anal. (1988), 191-206.

[8] V. Komornik, Rapid boundary stabilization of the wave equation, SIAM J. Control Opt. 29(1991), 197-208.

[9] V. Komornik, to appear.

[10] V. Komornik and E. Zuazua, A direct method for the boundary stabilization of the wave equation, J. Math. Pures Appl. 69(1990), 33-54.

[11] J. Lagnese, Boundary stabilization of thin plates, SIAM Studies in Applied Mathematics 10, SIAM, Philadelphia, PA, 1989.

[12] I. Lasiecka, Global uniform decay rates for the solutions to wave equation with nonlinear boundary conditions, to appear.

[13] M. Nakao, On the decay of soluions of some nonlinear dissipative wave equations in higher dimensions, Math. Z. 193(1986), 227-234.

[14] E. Zuazua, Stability and decay for a class of nonlinear hyperbolic problems, Asymptotic Anal. 1,2(1988), 161-185.

[15] E. Zuazua, Uniform stabilization of the wave equation by nonlinear boundary feedback, SIAM J. Control Opt. 28(1990), 466-477.

Proceedings of the 31st Conference
on Decision and Control
Tucson, Arizona • December 1992

On hyperbolic systems associated with the modelling and control of vibrating networks

J.E. Lagnese
Department of Mathematics
Georgetown University
Washington DC 20057
USA;

G.Leugering
Mathematisches Institut
Universitaet Bayreuth
Postfach 101 251
D-8580 Bayreuth
Germany.

E.J.P.G. Schmidt
Department of Mathematics
and Statistics
McGill University
805 Sherbrooke Street West
Montreal, Quebec
Canada, H3A2K6;

This work was partially supported by the following grants: Air Force Office of Scientific Research grant F49620-92-J-0031 (J.E.L. and G.L.), Deutsche Forschungsgemeinschaft, Heisenbergreferat L-595-3-1 (G.L), National Research Council of Canada grant A7271 and an FCAR grant from the Ministére del'Education du Quèbec (G.S.).

1 Introduction

In this paper we shall

- describe a general model for vibrating networks;
- show how the model applies to networks of elastic strings or Timoshenko beams;
- indicate how hyperbolic energy estimates for first order systems yield the a priori estimates needed to prove exact controllability and stabilizability;
- show how the model can be generalized to allow "mixed networks" of strings and beams;
- mention generalizations of the Timoshenko beam and their relation to our model.

In modelling networks of strings and beams located in space, it is essential that one allows for both longitudinal and transverse displacements of each element and that one adequately describes the interactions which occur where different elements are joined. An appropriate model for string networks was introduced and studied in (Schmidt, [10]). Subsequently string and Euler beam networks were discussed in (Leugering and Schmidt, [6]) and Timoshenko beam networks in (Lagnese, Leugering and Schmidt, [3]). Another paper [4] deals more specifically with a variety of beam models involving partial differential equations. Recent work of Schmidt and Wei Ming [12] allows point masses in the network, a situation which will not be considered here. One common feature of this work is that large displacements of the network must be described by non-linear equations and that linear systems are obtained by linearization about equilibria. These linearized equations are the subject of this talk.

A complete exposition of the subject matter of this talk is given in [5].

2 The model

In this section we introduce a rather general coupled system of linear second order hyperbolic equations associated with a network and show how this can be rewritten as a system of first order hyperbolic equations.

We begin by describing the network configuration and establishing notation. We suppose that the network consists of n curves in R^3 indexed by $i = 1, \ldots, n$ and parametrized by $x \in [0, l_i]$. The curves are connected at some of the endpoints in such a way that the resulting graph is connected. Every curve could, for example, describe the location of a curvilinear physical object such as a string or of the centerline of a three dimensional object such as a beam. Associated

CH3229-2/92/0000-3003$1.00 © 1992 IEEE

with each curve we have a configuration function $\mathbf{r}_i(x,t) : [0,l_i] \times [0,T] \mapsto R^p$. For string networks one has $p = 3$ and $\mathbf{r}_i(x,t)$ describes the displacement of the i-th string at time t and at the point parametrized by x. More generally some components of $\mathbf{r}_i$ may describe the spatial displacement while others may describe changes in "internal variables" such as shear angles. We refer to the points where one or more curves end as *nodes.* We need to distinguish between *multiple nodes* where several curves meet and *simple nodes* which lie at the endpoint of only one string. We suppose that there are m nodes with index set $\mathcal{I} = \{1 \le k \le m\}$ having subsets $\mathcal{I}_M$ and $\mathcal{I}_S$ corresponding respectively to multiple and simple nodes. We introduce also

$$\mathcal{E}_k = \{1 \le i \le n \ : \ \text{the } k\text{-th node is an endpoint of the } i\text{-th curve}\}.$$

For $i \in \mathcal{E}_k$ we let x_{ik} be the parameter value 0 or l_i corresponding to the endpoint of the ith string located at the kth node and set ϵ_{ik} equal to -1 if $x_{ik} = 0$ or 1 if $x_{ik} = l_i$. For $k \in \mathcal{I}_S$ $\mathcal{E}_k$ consists of a single index i_k and then we set $x_k = x_{i_k k}$ and $\epsilon_k = \epsilon_{i_k k}$.

At multiple nodes we impose the *geometric node condition*

$$\mathbf{C}_{ik}\mathbf{r}_i(x_{ik},t) \text{ is the same for all } i \in \mathcal{E}_k, \qquad (1)$$

where the $\mathbf{C}_{ik}$'s are given invertible $p \times p$ matrices and where for $k \in \mathcal{I}_S$ we set $\mathbf{C}_k = \mathbf{C}_{i_k k}$. This is a continuity condition on the configuration functions at the k-th node. We shall generalize it later to cover more complex situations than those we had in mind initially.

As far as simple nodes are concerned we distinguish between *clamped* nodes with indices in $\mathcal{I}_S^C$ and *geometrically controlled* nodes with indices in $\mathcal{I}_S^G$. These respectively correspond to the conditions

$$\mathbf{r}_{i_k}(x_k,t) = \mathbf{0} \text{ for } k \in \mathcal{I}_S^C \qquad (2)$$

and

$$\mathbf{r}_{i_k}(x_k,t) = \mathbf{u}_k(t) \text{ for } k \in \mathcal{I}_S^G, \qquad (3)$$

where $\mathbf{u}_k(t)$ is a preassigned control function.

We let $\mathbf{r}$ denote the n-tuple $(\mathbf{r}_i)$ of configuration functions. We suppose that the system has associated with it kinetic and potential energy functions of the form

$$\mathcal{K}(\mathbf{r}) = \sum_{i=1}^{n} \frac{1}{2} \int_0^{l_i} \mathbf{P}_i^2 \dot{\mathbf{r}}_i \cdot \dot{\mathbf{r}}_i(x)dx,$$

and

$$\mathcal{U}(\mathbf{r}) = \sum_{i=1}^{n} \frac{1}{2} \int_0^{l_i} \mathbf{Q}_i^2[\mathbf{r}_i' + \mathbf{R}_i\mathbf{r}_i] \cdot [\mathbf{r}_i' + \mathbf{R}_i\mathbf{r}_i](x)dx,$$

Here $\dot{\mathbf{r}}_i$ and $\mathbf{r}_i'$ denote the derivatives of $\mathbf{r}_i$ with respect to t and x respectively while $\mathbf{P}_i(x)$, $\mathbf{Q}_i(x)$ and $\mathbf{R}_i(x)$ are continuous $p \times p$ matrix valued functions the first two being assumed symmetric and positive definite. The matrices $\mathbf{P}_i^2$ and $\mathbf{Q}_i^2$ have been expressed as squares since this simplifies the notation at a later stage. One then derives a system of network equations by applying Hamilton's principle to the Lagrangian

$$\mathcal{L}(\mathbf{r}) = \int_0^T [\mathcal{K}(\mathbf{r}) - \mathcal{U}(\mathbf{r})]\, dt$$

allowing only variations in $\mathbf{r}$ which respect the geometric conditions (1), (2) and (3). In a standard way one obtains the system of equations

$$\mathbf{P}_i^2 \ddot{\mathbf{r}}_i = \left[\mathbf{Q}_i^2(\mathbf{r}_i' + \mathbf{R}_i\mathbf{r}_i)\right]' - \mathbf{R}_i^t\mathbf{Q}_i^2(\mathbf{r}_i' + \mathbf{R}_i\mathbf{r}_i), \quad (4)$$

(where $\mathbf{R}_i^t$ denotes the transpose of $\mathbf{R}_i$) along with the *dynamic node conditions*

$$\sum_{i \in \mathcal{E}_k} \epsilon_{ik}[\mathbf{C}_{ik}^{-1}]^t \mathbf{Q}_i^2 \, (\mathbf{r}_i' + \mathbf{R}_i\mathbf{r}_i)\,(x_{ik},t) = \mathbf{0}, \quad (5)$$

at the multiple nodes, i.e. for $k \in \mathcal{I}_M$, and, for for $k \in \mathcal{I}_S - \left(\mathcal{I}_S^C \cup \mathcal{I}_S^G\right)$.

$$[\mathbf{C}_k^{-1}]^t \mathbf{Q}_{i_k}^2 \left(\mathbf{r}_{i_k}' + \mathbf{R}_{i_k}\mathbf{r}_{i_k}\right)(x_k,t) = \mathbf{0}. \quad (6)$$

In fact we now impose a dynamic control on simple nodes indexed by $k \in \mathcal{I}_S^D = \mathcal{I}_S - \left(\mathcal{I}_S^C \cup \mathcal{I}_S^G\right)$:

$$[\mathbf{C}_k^{-1}]^t \mathbf{Q}_{i_k}^2 \left(\mathbf{r}_{i_k}' + \mathbf{R}_{i_k}\mathbf{r}_{i_k}\right)(x_k,t) = \mathbf{u}_k(t). \quad (7)$$

To summarize, *the dynamics of the network model are described by the equations (4) which are coupled by the geometric and dynamic node conditions (1) and (5) imposed at the multiple nodes, and subject to the simple node conditions*

(2), (3) or (7) as well as initial conditions on $\mathbf{r}(x,0)$ *and* $\dot{\mathbf{r}}(x,0)$.

One can also impose geometric or dynamic controls at multiple nodes, but here we choose not to pursue the issues raised by doing so. We also do not discuss the existence of solutions to the network system, which can be proved in appropriate spaces using variational and semigroup methods.

3 Networks of strings or Timoshenko beams

The string networks of [10] and the Timoshenko beam networks of [3] fit our scheme.

For the string network the equations were linearized about a configuration in which the i-th string is stretched by a factor $s_i > 1$ in the direction $\mathbf{v}_i$. For a network in R^3 one takes $p = 3$ and $\mathbf{r}_i(x,t)$ is just the displacement function of the i-th string. In the multiple node condition (1) each $\mathbf{C}_{ik}$ is set equal to the identity matrix so that the condition expresses the continuity of the displacements across multiple nodes. If ρ_i and h_i denote respectively the density and the Hooke's law constant of the i-th string one has $\mathbf{P}_i^2 = \rho_i\mathbf{I}$ and $\mathbf{R}_i = \mathbf{O}$, where $\mathbf{I}$ and $\mathbf{O}$ denote the identity and the zero matrix. Moreover $\mathbf{Q}_i^2$ is the matrix with eigenvalues h_i and $h_i(1 - s_i^{-1})$, whose corresponding eigenspaces are the space spanned by $\mathbf{v}_i$ and the orthogonal complement of that space. Each equation in (4) decomposes into transverse and longitudinal wave equations having different wave speeds. The components corresponding to different strings are "scrambled" at the multiple nodes by the node conditions (1) and (5).

In the above model it is natural for the matrices to have no spatial dependence. If the network is subject to gravity the equilibrium configuration will "sag" and the matrices in the linearized equations do depend on x (see [12]).

We turn now to Timoshenko beam networks. here one uses a planar beam model in which the configuration of the beam is given by the location of the centre line and the shear angle between the cross section and the normal to the centre line. We suppose that in the equilibrium configuration the centre line is straight and the shear angle vanishes. Denoting the direction of the centre line by the unit vector $\mathbf{v}_i$ and letting $\mathbf{v}_i^{\perp}$ denote an orthogonal unit vector in the plane of the network, we denote by $u_i\mathbf{v}_i + w_i\mathbf{v}_i^{\perp}$ the displacement of the centre line and by ψ_i the change in the shear angle of the i-th beam. Now we take $p = 3$ and set $\mathbf{r}_i = (u_i, w_i, \psi_i)$. One certainly needs the continuity of the displacements at multiple nodes, and for "rigid joints" one also demands continuity of the change in the shear angle. These conditions can be expressed in the form (1) if we set $\mathbf{C}_{ik}$ equal to the orthogonal matrix having as columns $\mathbf{v}_i$, $\mathbf{v}_i^{\perp}$ and $\mathbf{n}$, where $\mathbf{n}$ is the unit normal to the plane in which the network is constrained to lie. The physical constants entering the model are the density ρ_i, the cross-sectional length A_i, the Young's modulus E_i, the second moment of inertia I_i, the polar moment of inertia of a cross section I_{ρ_i} and the shear modulus K_i. Now, using the notation $\mathrm{diag}(\lambda_1, \lambda_2, \ldots, \lambda_p)$ for the diagonal $p \times p$ matrix with entries $\lambda_1, \lambda_2, \ldots, \lambda_p$, one sets

$$\mathbf{Q}_i^2 = \mathrm{diag}(\rho_i, \rho_i, I_{\rho_i}), \quad \mathbf{P}_i^2 = \mathrm{diag}(E_iA_i, K_i, E_iI_i),$$

and

$$\mathbf{R}_i = \begin{pmatrix} 0 & 0 & 0 \\ 0 & 0 & 1 \\ 0 & 0 & 0 \end{pmatrix}.$$

One then obtains the following system of equations for each beam:

$$\begin{aligned} \rho_i\ddot{u}_i &= E_iA_iu_i'', \\ \rho_i\ddot{w}_i &= K_i(\psi_i + w_i')', \\ I_{\rho_i}\ddot{\psi}_i &= E_iI_i\psi'' - K_i(\psi_i + w_i'). \end{aligned} \tag{8}$$

In this model the longitudinal displacements are governed by a wave equation, while the transverse displacements interact with the shear angle in the Timoshenko equations. One also obtains the same node condition as in [3].

4 Hyperbolic systems, energy estimates and consequences

In the papers [10] and [3] the a priori inequalities needed to establish exact controllability and stabilizability were obtained using multiplier methods. These did not yield transparent expressions for the minimal time in which exact controllability becomes possible. In fact it is now generally realized that for one dimensional problems the sharpest results can be expected from the theory of hyperbolic systems and their characteristics. Indeed control of such systems was treated in the seventies by David Russell and his beautiful results are described in his survey article [9]. In [11] the key estimates obtained previously by multiplier methods were derived for hyperbolic systems and an applications to a simple network was given. In fact similar estimates had been obtained by Rauch and Taylor in [8].

These methods can be applied to our general model. Each of the second order equations (4) can be rewritten as a first order hyperbolic system:

$$\dot{\mathbf{y}}_i(x,t) = \mathbf{A}_i(x)\mathbf{y}_i'(x,t) + \mathbf{B}_i(x)\mathbf{y}_i(x,t) \qquad (9)$$

where $\mathbf{y}_i$ and $\mathbf{A}_i$ respectively are given by

$$\begin{pmatrix} \mathbf{P}_i\dot{\mathbf{r}}_i \\ \mathbf{Q}_i(\mathbf{r}_i' + \mathbf{R}_i\mathbf{r}_i) \end{pmatrix} \text{ and } \begin{pmatrix} O & \mathbf{P}_i^{-1}\mathbf{Q}_i \\ \mathbf{Q}_i\mathbf{P}_i^{-1} & O \end{pmatrix},$$

while

$$\mathbf{B}_i = \begin{pmatrix} O & \mathbf{S}_i \\ \mathbf{T}_i & O \end{pmatrix},$$

with $\mathbf{S}_i = \mathbf{P}_i^{-1}\mathbf{Q}_i' - \mathbf{P}_i^{-1}\mathbf{R}_i^t\mathbf{Q}_i$ and $\mathbf{T}_i = \mathbf{Q}_i\mathbf{P}_i^{-1} - \mathbf{Q}_i\mathbf{P}_i^{-1}\mathbf{P}_i'\mathbf{P}_i^{-1}$.

By writing out the eigenvalue/eigenvector equation for the symmetric matrix $\mathbf{A}_i$ it is not difficult to check that the eigenvalues are $\{\pm\lambda_{ij}\}_{j=1}^p$ corresponding to orthogonal eigenvectors $\{(\mathbf{a}_{ij}, \pm\lambda_{ij}\mathbf{a}_{ij})\}_{j=1}^p$, where $\{\lambda_{ij}^2\}_{j=1}^p$ are the eigenvalues of the symmetric, positive definite matrix $\mathbf{P}_i^{-1}\mathbf{Q}_i^2\mathbf{P}_i^{-1}$ (with the convention that λ_{ij} is positive and non-decreasing with j) and $\{\mathbf{a}_{ij})\}_{j=1}^p$ are the corresponding eigenvectors.

The eigenvalues determine the characteristics of the system through a given point (ξ,τ), namely the solution curves $\mathcal{C}_j^{\pm}(\xi,\tau)$ of the equations

$$\frac{dx^j}{dt} = -\lambda_{ij}(x^j,t),\ x^j(\tau) = \xi.$$

These characteristics can be used to analyse the hyperbolic system (7) and to derive energy estimates (see [2], [8] or [11]). To compactly formulate the estimates we introduce some notation. Let $\mathbf{r}_i(x,t)$ be a solution of (4). We denote

$$e_i = \mathbf{Q}_i^2[\mathbf{r}_i' + \mathbf{R}_i\mathbf{r}_i]\cdot[\mathbf{r}_i' + \mathbf{R}_i\mathbf{r}_i] + \mathbf{P}_i^2\dot{\mathbf{r}}_i\cdot\dot{\mathbf{r}}_i;$$

We also introduce the following notation for *space-like* and *time-like energy integrals*:

$$E_i(t) = \int_0^{l_i} e_i(x,t)dx,\ E_i^\dagger(x;t_1,t_2) = \int_{t_1}^{t_2} e_i(x,t)dt.$$

Intuitively energy flows in a time like direction towards (ξ,τ) from the region between the "outermost characteristics" $\mathcal{C}_p^+(\xi,\tau)$ and $\mathcal{C}_p^-(\xi,\tau)$. Reversing the role of t and x one concludes also that energy flows in a space like direction towards (ξ,τ) from the region between $\mathcal{C}_1^+(\xi,\tau)$ and $\mathcal{C}_1^-(\xi,\tau)$. As a consequence of this one can prove the following lemma (which is most easily understood with the aid of a sketch of the characteristics in the (x,t)-plane).

Lemma 1 *Let $\mathbf{r}_i$ be a (classical) solution of (6) and suppose that*

$$T \geq T_i = 2\int_0^{l_i} \frac{1}{\lambda_{i1}(x)}dx.$$

Set $T_i = 2S_i$. Then for some constant C_i

$$\begin{aligned} E_i(\frac{T}{2}) \ &+\ E_i^\dagger(0;S+S_i,T-S-S_i) \\ &\leq\ C_iE_i^\dagger(l_i;S,T-S) \end{aligned}$$

for all nonnegative S satisfying $2S + T_i \leq T$. This also holds with 0 and l_i interchanged.

We also need the following lemma giving an estimate on the energy flow through multiple nodes.

Lemma 2 *Let $\mathbf{r}$ be a (classical) solution of the system (6) satisfying the multiple node conditions (1) and (7). For $k \in \mathcal{I}_M$ and $i_0 \in \mathcal{E}_k$ one has, with a suitable constant C,*

$$E_{i_0}^\dagger(x_{i_0k};t_1,t_2) \leq C \sum_{i\in\mathcal{E}_k-\{i_0\}} E_i^\dagger(x_{ik};t_1,t_2).$$

We can now state our main result on exact controllability and indicate its proof. Let $[i_1, i_2, \ldots, i_r]$ denote the path leading along the network elements labelled successively by $i_1, i_2, \ldots, i_r$.

Theorem 1 *Suppose that the network contains no closed circuit, that all multiple nodes and no simple nodes are free, that exactly one simple node is clamped and that the remaining simple* *nodes are controlled either geometrically or dynamically. Then the network is exactly controllable for each time* $T \geq \underline{T}$ *where*

$$\underline{T} = \min\{T_{i_1} + T_{i_1} + \cdots + T_{i_r} : [i_1, i_2, \ldots, i_r] \text{ goes from the clamped to a controlled node}\}$$

The statement of the theorem is of course incomplete since the state space and the space of controls needs to be specified. Various choices are possible, depending on one's approach to exact controllability. Whether one uses HUM (see [7]) or the equivalence of the surjectivity of a linear map $T : U \mapsto H$ between Hilbert spaces to an estimate $C \parallel T^*x \parallel_U \geq \parallel x \parallel_X$ on the hermitian adjoint mapping T^*, the proof depends on an a priori inequality on the system adjoint to the controlled system. This adjoint system is governed by the same equations, but the conditions at the simple nodes are homogeneous and initial data are replaced by data at time T. In fact exact controllability (and results on stabilizability) are a consequence of the a priori inequality

$$\sum_{i=1}^{n} E_i(\frac{T}{2}) \leq C \sum_{k \in \mathcal{I}_S^G \cup \mathcal{I}_S^D} E_{i_k}^{\dagger}(x_k; 0, T), \qquad (10)$$

where C is a suitable positive constant, which holds for solutions $\mathbf{r}$ of (4) subject only to (1) and (5). The proof of this inequality is by induction on n, and uses the lemmas to "follow back" the energy along paths leading from the clamped node to the various controlled nodes. Since for the adjoint system one has homogeneous conditions at the simple nodes, energy is conserved and one can replace $\frac{T}{2}$ by T in the latter estimate. The homogeneous conditions at the controlled simple nodes also simplify the expressions for the corresponding integrals $E_{i_k}^{\dagger}(x_k; 0, T)$.

5 Mixed systems and generalization of the model

We consider, for example, networks which contain both Timoshenko beams and strings. Then the configuration functions of different elements will have different dimensions and hence also the matrices $\mathbf{C}_{ik}$ entering into the continuity conditions (1) may not be square. We generalise our model in two ways:

a. let $\mathbf{r}_i(x,t) : [0, l_i] \times [0, T] \mapsto R^{p_i}$, and suppose that $\mathbf{P}_i$, $\mathbf{Q}_i$ and $\mathbf{R}_i$ are $p_i \times p_i$ matrices;

b. in condition (i) let $\mathbf{C}_{ik}$ be given $q_k \times p_i$ matrices of rank q_k, where q_k is a natural number associated with the kth (multiple) node, necessarily satisfying $q_k \leq p_i$ for $i \in \mathcal{E}_k$..

Before considering the consequences of these assumptions for the network system, we illustrate them at a multiple node (indexed by k) in a planar network where a string (indexed by i) and a Timoshenko beam (indexed by j) meet. Here $p_i = 2$ and $p_j = 3$. We set $q_k = 2$, let $\mathbf{C}_{ik}$ be the 2×2 identity matrix and $\mathbf{C}_{jk}$ be the 2×3 matrix whose columns are $\mathbf{v}_j$ and $\mathbf{v}_j^{\perp}$. Then

$$\mathbf{C}_{ik}\mathbf{r}_i(x_{ik}, t) = \mathbf{C}_{jk}\mathbf{r}_j(x_{jk}, t)$$

simply expresses the continuity of the string and beam displacements at the node. There is no meaningful continuity requirement on the shear angle. which is therefore free at the node.

We return to the generalized model where now the weakened geometric condition has to be respected. The equations (6) are unchanged but the dynamic node conditions at the multiple nodes have to be modified. In order to write down these conditions let $\mathbf{P}_{ik}$ denote orthogonal projection onto the the nullspace of the transformation defined by $\mathbf{C}_{ik}$ and $\mathbf{P}_{ik}^{\perp}$ be orthogonal projection onto the orthogonal complement of that nullspace. Also let $\mathbf{C}_{ik}^{+}$ denote the generalized inverse of $\mathbf{C}_{ik}$. This is the $p_i \times q_k$ matrix which satisfies

$$\mathbf{C}_{ik}\mathbf{C}_{ik}^{+} = \mathbf{I}_k \quad \text{and} \quad \mathbf{C}_{ik}^{+}\mathbf{C}_{ik} = \mathbf{P}_{ik}^{\perp}.$$

In place of the conditions (5) Hamilton's principle now gives two dynamic conditions at the multiple nodes:

$$\sum_{i \in \mathcal{E}_k} \epsilon_{ik} [\mathbf{C}_{ik}^{+}]^t \mathbf{Q}_i^2 \left(\mathbf{r}_i' + \mathbf{R}_i \mathbf{r}_i\right)(x_{ik}, t) = \mathbf{0} \qquad (11)$$

as well as the following condition applicable separately for each $i \in \mathcal{E}_k$

$$\mathbf{P}_{ik}\mathbf{Q}_i^2 \left(\mathbf{r}_i' + \mathbf{R}_i \mathbf{r}_i\right)(x_{ik}, t) = \mathbf{0}. \qquad (12)$$

It is now again possible to obtain estimates of the form (10) and hence results on exact controllability and stabilizability. The main problem lies with Lemma 2 which no longer holds in full generality. As an example consider two strings and a beam meeting at a single multiple node. If the end of one string is clamped and the ends of the beam and the second string are controlled the system is exactly controllable. If, on the other hand, the end of the beam is clamped and the ends of the strings are controlled one does not appear to have exact controllability. The reason is that, because of the weakened multiple node conditions, the "nodal energy" of the beam is not dominated by the sum of the nodal energies of the two strings.

6 Generalizations of the Timoshenko beam

The Timoshenko equations can be generalized in various ways. Ironically one of these generalizations leads to beam equations due to Bresse [1] which predate the Timoshenko equations by the better part of a century! These equations are derived also in [4] where it is suggested that the Timoshenko beam should more appropriately be called the *Bresse-Timoshenko* beam. In this model the equilibrium represents a curved planar beam. Excluding the terms whose order is higher than quadratic, the expressions for the kinetic and potential energies given in ([4] equations (66)) conform to our model, so that the results of section 4 apply to networks containing curved beams.

A second generalization described in [5] is to a threedimensional Timoshenko beam model and to spatial networks of such beams.

A third generalization of the Timoshenko beam involves describing the beam by the displacement of the centre line as well as higher order normal derivatives of the displacement function along the centre line. This model will be derived in detail elsewhere; it motivates a further generalization in the form the potential energy, but does not appear to change the nature of the results.

References

[1] J.A.C. Bresse, *Cours de Mechanique Applique*, Mallet Bachelier (1859).

[2] P.R. Garabedian,*Partial Differential Equations*, John Wiley and Sons (1964).

[3] J.E. Lagnese, G.Leugering and E.J.P.G. Schmidt, *Control of Planar Networks of Timoshenko Beams,* to appear in SIAM J. Cont. and Opt..

[4] J.E. Lagnese, G.Leugering and E.J.P.G. Schmidt, *Modelling of dynamic networks of thin thermoelastic beams,* to appear in Math. Meth. Applied Sc..

[5] J.E. Lagnese, G.Leugering and E.J.P.G. Schmidt, *On the analysis and control of hyperbolic systems associated with vibrating networks*, submitted manuscript.

[6] G.Leugering and E.J.P.G. Schmidt, *On the control of networks of vibrating strings and beams*, Proc. 28th IEEE CDC Conf., **3**(1989), pp. 2287-2290.

[7] J.L. Lions,*Contrôlabilité Exacte, Perturbations et Stabilisation de Systẽmes Distribués¡* Tome 1, Contrôlabilité Exacte, Collection RMA **8**, Masson, Paris (1988).

[8] J. Rauch and M. Taylor, *Exponential Decay of Solutions to Hyperbolic Equations in Bounded Domains*, Indiana Univ. Math. Jour, **24** (1974), pp.79-86.

[9] D.L. Russell, *Controllability and stabilizability theory for linear partial differential equations: recent progress and open questions*, SIAM Rev. **20**(1978), pp.639-739.

[10] E.J.P.G.Schmidt, *On the modelling and exact controllability of networks of vibrating strings,* SIAM J. Cont. and Opt., **30** (1992), pp.229-245.

[11] E.J.P.G.Schmidt, *On an energy estimate and exact boundary controllability for hyperbolic systems in one space variable*, Report #91-12 from the Department of Mathematics and Statistics, McGill University (1991).

[12] E.J.P.G.Schmidt and Wei Ming, *On the modelling and analysis of networks of vibrating strings and masses*, Report #91-13 from the Department of Mathematics and Statistics, McGill University (1991).

Proceedings of the 31st Conference
on Decision and Control
Tucson, Arizona • December 1992

BOUNDARY CONTROLLABILITY OF SYSTEMS OF CONNECTED STRINGS AND VECTOR EXPONENTIAL FAMILIES

Sergei A. Avdonin and Sergei A. Ivanov

Department of Applied Mathematics and Control
St.Petersburg State University
St.Petersburg, 198904 Russia

ABSTRACT

We consider boundary controllability problems for a class of systems described by hyperbolic equations for vector-functions of one spatial variable. With the help of Fourier method the study of controllability reduces to the investigation of families of vector exponentials $\{\eta_n \exp(i\omega_n t)\}$ with ω_n being eigenfriquencies of of the system and η_n being the traces of eigenfunction derivatives at the boundary points where control is applied. Since the strings are nonhomogeneous only the asymptotics of ω_n and η_n is known which is not always accurate enough to separate the exponentials. It is known [1,Ch.2] that in the case of vector exponentials the asymptotics is not by itself sufficient for the investigation of the minimality and the basis property. In fact there is practically one way to to examine vector exponential families, that is, to construct and study so called generating function (GF). Remarkably, the problem under consideration, which have physical foundation, naturally give rise to the GF, since it expressed via solution to ordinary differential equations of Helmholtz type. The known behavior of those solutons as functions of the spectral parameter enables one to make conclusion about the basis property of corresponding exponential family.

1. INTRODUCTION

The systems under consideration are treated within the framework of a general scheme, which represents for the case the method of moments for systems in Hilbert spaces. We describe it briefly now (see [1,Ch.3] for details).

Let V and H be Hilbert spaces, V being dense and continuously imbedded in H. Identify H with the dual to it space, and denote V' the space dual to V. Let $a[\varphi,\psi]$ be a continuous symmetric positive definite bilinear form on V. Then a selfadjoint positive definite operator A uniquely corresponds to the form a :

$$D(A)\subset V; \quad (A\varphi,\psi)_H = a[\varphi,\psi]; \quad \varphi\in D(A), \quad \psi\in V.$$

The norm generated by form a is equivalent to the norm of space V

$$D(A^{1/2}) = V; \quad (A^{1/2}\varphi, A^{1/2}\psi) = a[\varphi,\psi]; \; \varphi,\psi \in V. \quad (1)$$

We assume that operator A has a set of eigenvalues $\{\lambda_n\}$, $n\in\mathbb{N}$, and eigenfunctions $\{\varphi_n\}$, which form an orthnormal basis in space H. For string systems which will be investigated this assumption is valid.

With operator A one can associate spaces $V_r := D(A^{r/2})$, $r>0$, which are domains of powers of operator A; $V_0 := H$, $V_{-r} := (V_r)'$. Note that (1) implies $V_1 = V$, therefore $V_{-1} = V'$.

Boundary control systems under consideration can be represented in the form

$$\frac{d^2}{dt^2}y(t) + Ay(t) = Bu(t), \quad 0<t<T, \quad (2)$$

where control u belongs to space $U:= L^2(0,T;\mathcal{N})$, $\mathcal{N}$ is a Hilbert space and B is a linear bounded from $\mathcal{N}$ to V_{-r} for some $r\in\mathbb{R}$. (In our case $\mathcal{N}$ is finite dimensional. It does not matter for general scheme but is very important for the properties of exponential families and then for controllability of strings systems). Using B one can define a linear bounded operator B* from V_r to $\mathcal{N}$ by an equality

$$\langle B\eta,\varphi\rangle = (\eta, B^*\varphi)_{\mathcal{N}}, \quad \eta\in\mathcal{N}, \quad \varphi\in V_r,$$

where $\langle f,\varphi\rangle$ stands for the value of functional $f\in V_{-r}$ on element $\varphi\in V_r$ and $(\cdot,\cdot)_{\mathcal{N}}$ denotes scalar product in space $\mathcal{N}$.

By means of Fourier method, generalized solution to the equation (2) is constructed which is continuous function in time with the values in one of the spaces from the scale V_r, and the exponential family $E:=\{B^*\varphi_n \exp(\pm i\sqrt{\lambda_n}\, t)\}$ arises. Using the asymptotics of $B^*\varphi_n$ and λ_n which are known for string systems, we demonstrate in the following sections that equation (2) with zero initial conditions

$$y(0) = \dot{y}(0) = 0, \qquad \dot{y} := \frac{d}{dt}y, \quad (3)$$

has a unique solution such that

$$y\in C([0,T];H), \quad \dot{y}\in C([0,T];V'). \quad (4)$$

Note that this result is exact one.

Denote by R(T) a reachability set of the system (2),(3) in time T: $R(T)=\{(y(T),\dot{y}(T)) : u\in U\}$. It follows from (4) that $R(T)\subset W$, $W:= H\times V'$.

Definition 1. System (2) is called to be:

a) B-controllable relative to W in time T if $R(T) = W$;

b) M-controllable in time T, if for any n inclusions $(\varphi_n,0)\in R(T)$, $(0,\varphi_n)\in R(T)$ hold;

c) W-controllable in time T, if $cl_W R(T) = W$.

Along with the family E let us introduce a family $E' := \{(\sqrt{\lambda_n})^{-1} B^*\varphi_n \exp(\pm i\sqrt{\lambda_n}\, t)\}$, whose elements differ from elements of the first one by scalar factors. The next proposition establish a relation between the introduced controllability types and the properties of the family E'. It is a special case of a more general results proved in [1,Ch.3].

Theorem 2. The following assertions are true:

a) system (2) is B-controllable relative to W in time T if and only if the family E' is a Riesz basis in closure of its linear span in space U

CH3229-2/92/0000-3009$1.00 © 1992 IEEE

(shortly L-basis);

b) system (2) is M-controllable in time T if and only if the family E′ is minimal in U;

c) system (2) is W-controllable in time T if and only if the family E′ is w-linear independent in U.

(A family $\{u_n\} \subset U$ is called to be w-linear independent if conditions

$$\{a_n\} \in \ell^2, \quad \sum a_n u_n \to 0 \text{ weakly in } U$$

imply that $a_n = 0$ for all n).

Theorem 2 reduces controllability problems to investigation of the properties of the family E′. It is convinient to write E′ in the form $\{\eta_n \exp(i\omega_n t)\}$, $n \in \mathbb{K}$, $\mathbb{K} := \pm 1, \pm 2, \ldots$,

$$\omega_n = (\operatorname{sgn} n)\sqrt{\lambda_{|n|}}, \quad \eta_n = \omega_n^{-1} B^* \varphi_n .$$

To prove L-basis property of E′ in U we check that

i) spectrum $\{\omega_n\}$ allows a splitting into a finite number of separable sets; vectors η_n corresponding to any group of "close" points are uniformly in group linearly independent;

ii) there exist an entire matrix-function F(z) of exponential type (generating function) such that $F(\omega_n)\eta_n = 0$ and

$$\sup_{x \in \mathbb{R}} \{ \| F(x-i) \| , \| F^{-1}(x+i) \| \} < \infty .$$

2. SYSTEM OF STRINGS CONNECTED ELASTICALLY AT ONE POINT

Consider a system of strings linked at one point. The system is described by the following initial boundary problem (s = 1, 2, ..., N) :

$$\rho_s(x)\frac{\partial^2}{\partial t^2} y\ (x,t) = \frac{\partial^2}{\partial x^2} y\ (x,t), \quad 0<x<l, \quad 0<t<T; \tag{5}$$

$$y_s(0,t) = u_s(t), \quad u_s \in L^2(0,T), \quad 0<t<T; \tag{6}$$

$$y_1(l\ ,t) = y_2(l\ ,t) = \ldots = y_N(l\ ,t), \quad 0<t<T; \tag{7}$$

$$y_1'(l\ ,t) + y_2'(l\ ,t) + \ldots + y_N'(l\ ,t) = 0, \tag{8}$$

$$y_s(x,0) = \dot{y}_s(x,0) = 0, \quad 0<x<l \quad . \tag{9}$$

It is supposed that $\rho_s \in C\,[0,l\,]$, $\rho_s(x)>0$.

Functions u_s are the control actions; we consider them as components of vector-function $u \in U := L\ (0,T;\mathbb{C}^N)$. Conditions (7), (8) at the linking point are implied by the requirement of the system energy conservation.

Let us imbed classically formulated problem (5)-(9) into the scheme of Sec.1 and thus attach a strict meaning to it. Introduce spaces H and V:

$$H := \bigoplus_{s=1}^{N} L^2_{\rho_s}(0,l_s); \quad V := \{v=(v_1,\ldots,v_N) :$$

$$v_s \in H^1(0,l_s), \quad v_1(l_1) = v_2(l_2) = \ldots = v_N(l_N)\}$$

and specify a bilinear form on V

$$a[\varphi,\psi] = \sum_{s=1}^{N} \int_0^{l_s} \varphi_s'(x)\,\overline{\psi_s'(x)}\,dx \quad ; \ \varphi,\psi \in V.$$

Operator A corresponding to it acts according to the rule

$$\{\varphi_s(x)\}_{s=1}^{N} \longmapsto \left\{ \frac{1}{\rho_s(x)}\ \varphi_s''(x) \right\}_{s=1}^{N},$$

and its domain consists of sets $\{\varphi_s\}_{s=1}^{N}$ of functions such that

$$\varphi_1(l_1) = \varphi_2(l_2) = \ldots = \varphi_N(l_N)$$

and $\quad \varphi_1'(l_1) + \varphi_2'(l\ ) + \ldots + \varphi_N'(l\ ) = 0.$

Operator B from the scheme of Sec.1 here acts from space $\mathbb{C}^N$ to space V_{-2} by the rule

$$\langle B\eta,\varphi\rangle = \sum_{s=1}^{N} \eta_s\,\overline{\varphi_s'(0)}\ , \ \varphi \in V_2 := D(A)\ .$$

Lemma 3. Solution $y = \{y_s\}_{s=1}^{N}$ to the problem (5)-(9) exists, is unique, and

$$\{y,\dot{y}\} \in C([0,T];W), \ W := H \times [\ \bigoplus_{s=1}^{N} H^{-1}\ (0,l_s\ )].$$

The plan for the study of system (5)-(9) controllability is as follows. First, we relate with operator A an entire matrix-function G whose zeros coincide with the set $\{0\} \cup \{\pm\sqrt{\lambda_n}\}$. Then, after some manipulations with G, we produce a GF of the exponential family arising from the problem of moments. Demonstrating with the help of GF L-basis property of the family in space $L^2(0,T_0;\mathbb{C}^N)$ for some T_0, we thus prove system (5)-(9) controllability as well.

Let us introduce functions $\varphi_s\,(x,k)$ as solutions to the problems (s = 1,...,N) :

$$-\varphi_s''(x,k) = k^2 \rho_s(x)\varphi_s(x,k),$$

$$\varphi_s(0,k) = 0, \quad \varphi_s'(0,k) = k.$$

Set $\psi_s(k) := \varphi_s(l_s,k)$, $\zeta(k) := k^{-1}\ \varphi_s'(l_s,k)$,

and intriduce an entire matrix-function

$$G(k) = \begin{pmatrix} \psi_1(k) & -\psi_2(k) & 0 & \cdot & 0 \\ \psi_1(k) & 0 & -\psi_3(k) & \cdot & 0 \\ \cdot & \cdot & \cdot & \cdot & \cdot \\ \psi_1(k) & 0 & 0 & \cdot & -\psi_N(k) \\ \zeta_1(k) & \zeta_2(k) & \zeta_3(k) & \cdot & \zeta_N(k) \end{pmatrix}$$

Lemma 4. a) Any eigenfunction of operator A is of the form

$$\Phi_n = \operatorname{diag}[\ \varphi_s(x,\omega_n)]\ \eta_n, \ \eta_n \in \mathbb{C}^N, \text{ with } G(\omega_n)\eta_n = 0.$$

b) Conversely, any function of the form $\operatorname{diag}[\varphi_s(x,k_0)]\,\eta$ with $k_0 \neq 0$, $\eta \neq 0$ and $G(k_0)\,\eta = 0$ is an eigenfunction of operator A corresponding to the eigenvalue k_0 .

Basing on the structure of eigenfunctions Φ_n let us intriduce, according to Sec.1, exponential family E′:

$$E' = \{e_n\} \subset L^2(0,T;\mathbb{C}^N), \ e_n(t) = \eta_n \exp(i\omega_n t),$$

$\eta_n = |\omega_n^{-1}|\ \Phi_n'(0)$. B-contrtollability of system (5)-(9) follows from L-basis property of E′.

Theorem 5. System (5)-(9) is B-controllable in space W in time T_0,

$$T_0 := 2 \max_{s \in \{1,\dots,N\}} \int_0^{l_s} \sqrt{\rho_s(x)}\, dx.$$

This theorem is in good agreement with results of [2] obtained by Hilbert Uniqueness Method.

Remark 6. If one of control functions u is identically equal to zero then analog of the theorem is valid with T_0 replaced to $2T_0$.

3. CONTROL OF MULTICHANNEL ACOUSTIC SYSTEM

Consider following system of equations for vector-function $y(x,t) \in \mathbb{C}^N$:

$$P(x)y_{tt}(x,t) = y_{xx}(x,t),\ 0<x<l,\ 0<t<T, \tag{10}$$

$$y(0,t) = u(t),\ y(l,t) = 0,\ 0<t<T, \tag{11}$$

$$y(x,0) = y_t(x,0) = 0,\quad 0<x<l. \tag{12}$$

Here $P(x)$ is a C^2-diagonable positive definite matrix-function, i.e. $P(x) = Q(x)\operatorname{diag}[\rho_j(x)]Q^{-1}(x)$ with $Q(x)$ being unitary-valued matrix function, $\rho_j(x) > 0$, $j = 1,2,\dots,N$.

System (10)-(12) describes N wave channels interconnected by means of the transmission coefficients $p_{ij}(x)$. Boundary control function belongs to $L^2(0,T;\mathbb{C}^N)$.

To solve a control problem for the system we are going, just as in Sec.2, to construct a GF, to study its behavior in $\mathbb{C}$ and to demonstrate B-controllability for the time T_0,

$$T_0 := 2\int_0^l \|P(x)\|^{1/2} dx = 2\int_0^l \max_{j=1,\dots,N} \{\sqrt{\rho_j(x)}\}dx.$$

Considering system (10)-(12) by the plan of Sec.1, we set

$$H = L^2_P(0,l;\mathbb{C}^N),\quad \|z\|^2_H = \int_0^l (P(x)z(x),z(x))_{\mathbb{C}^N} dx,$$
$$V = H^1_0(0,l;\mathbb{C}^N).$$

Determine on space $V \times V$ a bilinear form

$$a[\varphi,\psi] = \int_0^l (\varphi'(x), \psi'(x))_{\mathbb{C}^N}\, dx$$

which generates operator A :

$$(A\varphi)(x) = -P^{-1}(x)\varphi''(x)$$

with the domain $H^2(0,l;\mathbb{C}^N) \cap H^1_0(0,l;\mathbb{C}^N)$. Operator $B: \mathbb{C}^N \mapsto V_{-2}$ we assign by the formula

$$\langle B\eta, \varphi\rangle = (\eta, \varphi'(0))_{\mathbb{C}^N},\quad \varphi \in V_2 := D(A).$$

As in Sec.2, the solution of system (10)-(12) satisfies the relation

$$\{y,y_t\} \in C([0,T];W),\ W := H \times H^{-1}(0,l;\mathbb{C}^N).$$

Consider matrix equation

$$-Y''(x,k) = k^2 P(x)Y(x,k),\ 0<x<l,$$

with conditions

$$Y(0,k) = 0,\ Y'(0,k) = kI.$$

Set $G(k) = Y(l,k)$, which is obviously entire function of parameter k of exponential type. As an analog of Lemma 4 serves

Lemma 7. a) Any eigenfunction of operator A is of the form $\Phi_n(x) = Y(x,\omega_n)\eta_n$ with $G(\omega_n)\eta_n = 0$.

b) Conversely, any function of the form $Y(x,k_0)\eta$ is the eigenfunction of operator A corresponding to its eigenvalue k_0 if only $k_0 \neq 0$,

$$\eta \in \mathbb{C}^N,\ \eta \neq 0,\ G(k_0)\eta = 0.$$

Let us introduce a vector exponential family

$$E' = \{\eta_n \exp(i\omega_n t)\},\ n \in \mathbb{K},\ \eta_n = |\omega_n|^{-1}\Phi'_{|n|}(0).$$

The properties of this family determine the controllability type of the system.

Theorem 8. System (10)-(12) is B-controllable in space W in time T_0.

Remark 9. If $Q(x)$ = const, then the system is B-controllable in the time T_1,

$$T_1 = 2\max_j \int_0^l \sqrt{\rho_j(x)}\, dx \le T_0.$$

References

[1] Avdonin, S.A., and S.A. Ivanov, Controllability theory for distributed parameter systems and families of exponentials, UMKVO Kiev, 1989 (in Russian).

[2] Schmidt, E.J.P.G., " On the Modelling and Exact Controllability of Networks of Vibrating Strings ", SIAM J. Control and Optimization, Vol. 30, No. 1, January 1992, pp. 229-245.

[3] Avdonin, S.A., and S.A. Ivanov, "A Generating Matrix-Valued Function in Problems of Control of Connected Strings", Soviet. Math. Dokl., Vol. 40, No. 1, 1990, pp. 179-184.